Maximum firepower. The P90 submachine gun.

Close up, the P90 is quick to aim and effortless to carry. The first submachine

gun with 50 instant stopping power rounds per magazine, on top of the barrel. And downward case ejection to suit lefthanders too. Made lighter, and less bulky, by the use of high tech composite materials. Comfortably rounded. Lethally effective. And deadly accurate. Complete with accessories and full range of 5.7 x 28 mm ammunition.

Full Force Partnership

DDB & Co.

Jane's

INFANTRY WEAPONS

Edited by Terry J Gander

Twenty-second Edition

1996-97

ISBN 0 7106 1354 7
"Jane's" is a registered trade mark

In the USA and its dependencies
Jane's Information Group Inc, 1340 Braddock Place, Suite 300, Alexandria, VA 22314-1651, USA

British Library Cataloguing-in-Publication Data.
A catalogue record for this book is available from the British Library.

Printed and bound in Great Britain by Biddles Ltd, Guildford and King's Lynn

CIS 40AGL

THE INFANTRY'S OWN COMBAT power multiplier.

A high velocity automatic grenade launcher for immediate support fires up to 2.2 km.

Comprehensive range of munitions:

- HEDP S411 for penetrating 50mm of RHA steel
- HE S412 for anti-personnel applications
- TPT S415 & TP S416A for realistic training.

The 'Lock & Fire Mount' for integration onto mobile platforms – Jeeps, Light Strike Vehicles, APCs and Naval Craft – for enhanced firepower.

A product from Chartered Industries of Singapore, the company with proven technological innovation, uncompromising quality standards and total customer satisfaction.

CIS 40AGL, the Infantry's combat power multiplier and CIS, the company that goes out of its way to meet customer needs.

The CIS 40AGL integrated with the 'Lock and Fire Mount'.

The CIS 40AGL comes with ammunition belted in 16, 32 and 48 rounds for immediate use.

Chartered Industries of Singapore

A member of **Singapore Technologies**

Marketed by Unicorn International, 3 Lim Teck Kim Road, #11-01/02 Singapore Technologies Building Singapore 088934. Tel: (65) 225 4788 Fax (65) 224 8862 Telex: RS 39882 UIPL

Contents

ADMINISTRATION

Publishing Director: Robert Hutchinson

Managing Editor: Peter Howard

Production Database Manager: Ruth Simmance

Editorial Services Manager: Sulann Staniford

Production Editor: Jane Stimson

EDITORIAL OFFICES

Jane's Information Group Limited, Sentinel House,
163 Brighton Road, Coulsdon, Surrey CR5 2NH, UK

Tel: +44 181 700 3700
Telex: 916907 Janes G
Fax: +44 181 700 3900
e-mail:yearbook@janes.co.uk

SALES OFFICES

Send enquiries to International Sales Manager:
Fabiana Angelini (Europe, CIS, Africa, Middle East)
David Eaton-Jones (Scandinavia, Far East, UK)
Jane's Information Group Limited, UK address as above

Tel Enquiries: +44 181 700 3759
Fax Enquiries: +44 181 763 1006
Fax Orders: +44 181 763 1005

Send USA enquiries to:
Joe McHale, Senior Vice-President Product Sales,
Jane's Information Group Inc, 1340 Braddock Place, Suite 300,
Alexandria, VA 22314-1651

Tel: +1 703 683 3700
Telex: 6819193
Fax: +1 703 836 0029

ADVERTISEMENT SALES OFFICES

Advertisement Sales Manager: Richard West

Australia: Brendan Gullifer, Havre & Gullifer (Pty) Ltd, Level 50, 101 Collins Street, Melbourne 3000

Tel: +61 3 9650 1100
Fax: +61 3 9650 6611

Benelux: Richard West, Jane's Information Group (see UK/Rest of World)

Brazil: L Bilyk, Brazmedia International S/C Ltda, Alameda Gabriel Monterio da Silva, 366 CEP, 01442 São Paulo

Tel: +55 11 853 4133
Telex: 32836 BMED BR
Fax: +55 11 852 6485

France: Patrice Février, Jane's Information Group – France, BP 418, 35 avenue MacMahon, F-75824 Paris Cedex

Tel: +33 1 45 72 33 11
Fax: +33 1 45 72 17 95

Germany and Austria: Janet Scott, Jane's Information Group (see UK/Rest of World)

Hong Kong: Jeremy Miller, Major Media Ltd, Room 1402, 14F Capitol Centre, 5-19 Jardine's Bazaar, Causeway Bay

Tel: +852 890 3110
Fax: +852 576 3397

Israel: Oreet Ben-Yaacov, Oreet International Media, 15 Kinneret Street, IL-51201 Bene-Berak

Tel: +972 3 570 6527
Fax: +972 3 570 6526

Italy and Switzerland: Ediconsult Internazionale Srl, Piazza Fontane Marose 3, I-16123 Genoa, Italy

Tel: +39 10 583684
Telex: 281197 EDINT I
Fax: +39 10 566578

Japan: Intermart/EAC Inc, 1-7 Akasaka 9-chome, Minato-Ku, Tokyo 107

Tel: +81 3 5474 7835
Fax: +81 3 5474 7837

Korea, South: Young Seoh Chinn, JES Media International, 6th Floor Donghye Building, 47-16 Myungil-Dong, Kangdong-Gu, Seoul 134-070

Tel: +82 2 481 3411
Fax: +82 2 481 3414

Scandinavia: Gillian Thompson, The Falsten Partnership, 11 Chardmore Road, Stamford Hill, London N16 6JA, UK

Tel: +44 181 806 2301
Fax: + 44 181 806 8137

Singapore, Indonesia, Malaysia, Philippines, Taiwan and Thailand: Hoo Siew Sai, Major Media (Singapore) Pte Ltd, 6th Floor, 52 Chin Swee Road, Singapore 0316

Tel: +65 738 0122
Telex: RS 43370 AMPLS
Fax: +65 738 2108

South Africa: Janet Scott, Jane's Information Group (see UK/Rest of World)

Spain: Jesus Moran Iglesias, Varex SA, Modesto Lafuente 4, E-28010 Madrid

Tel: +34 1 448 7622
Fax: +34 1 446 0198

UK/Rest of World: Janet Scott, Jane's Information Group, Sentinel House, 163 Brighton Road, Coulsdon, Surrey CR5 2NH

Tel: +44 181 700 3740
Telex: 916907 Janes G
Fax: +44 181 700 3744

USA
Advertising Production Manager/US & Canada – Maureen Nute
Jane's Information Group Inc, 1340 Braddock Place, Suite 300, Alexandria, VA 22314 USA

Tel: +1 703 683 3700
Fax: +1 703 836 0029

Eastern United States and Canada
Kimberley S Hanson
Global Media Services Inc, 299 Herndon Parkway, Suite 308, Herndon, VA 22070 USA

Tel: +1 703 318 5054
Fax: +1 703 318 9728

Southeastern USA
Kristin Schulze
Global Media Services Inc, PO Box 290706, Temple Terrace, FL 33617 USA

Tel: +1 813 987 2359
Fax: +1 813 980 0187

Western United States and Canada
Anne Marie St. John-Brooks
Global Media Services Inc, 25125 Santa Clara Street, Suite 290, Hayward, CA 94544 USA

Tel: +1 510 582 7447
Fax: +1 510 582 7448

Administration: UK: Fay Lenham
USA and Canada: Maureen Nute

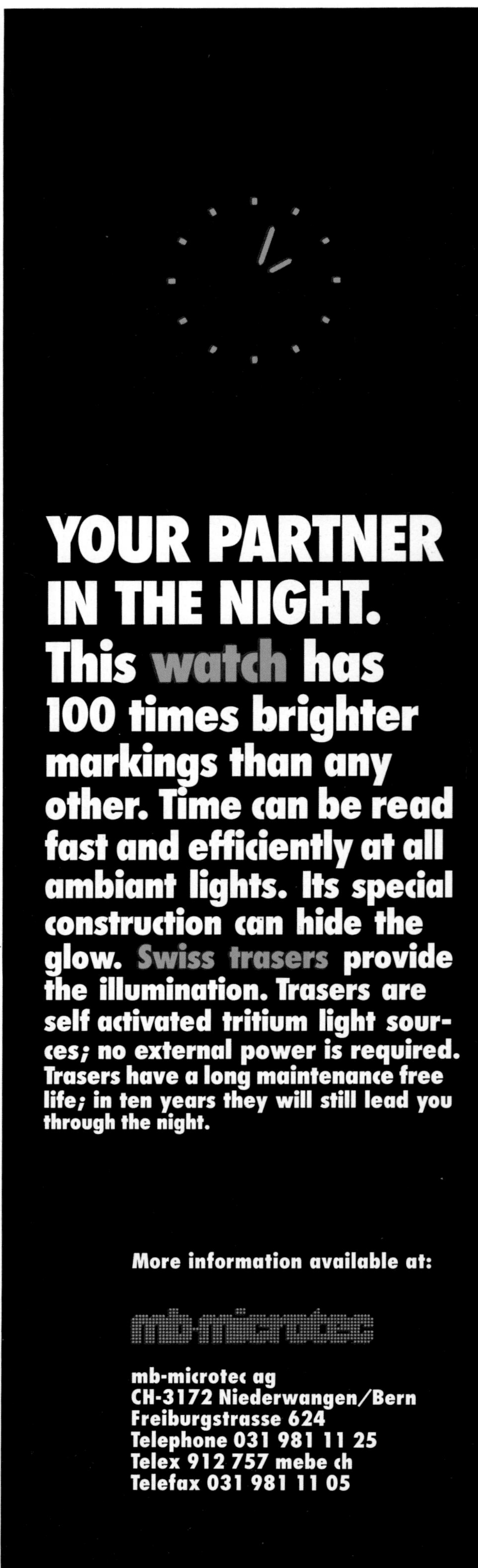

Alphabetical list of advertisers

How to use this book

The book is laid out in a series of main sections, each of which relates to a particular weapon type, for example, Pistols, Rifles, and so on. Within each section details are provided of Development, giving the outline history of the weapon, Description, providing the technical details, and Data, the hard facts relating to each weapon as far as can be determined. Then follows Status and Manufacturer. The details provided for the Manufacturer are the outline address only (where positively identified) — telephone and fax information is provided in the Manufacturers Index at the rear of the book.

Within each weapon type the section is laid out in country order. Within each national section the newest models are placed first. If only a weapon designation is available it may be found using the main Index.

A National Inventories section contains a nation-by-nation list of the known types of weapons in service. It cannot be fully comprehensive so additional or corrective information would be welcomed by the Editor.

To help users of this title evaluate the published data, Jane's Information Group has divided entries into three categories:-

● ***VERIFIED*** The editor has made a detailed examination of the entry's content and checked its relevancy and accuracy for publication in the new edition to the best of his ability.

● ***UPDATED*** During the verification process, significant changes to content have been made to reflect the latest position known to Jane's at the time of publication.

● ***NEW ENTRY*** Information on new equipment and/or systems appearing for the first time in the title.

In future all new pictures will be dated with the year of publication. New pictures this year are dated 1996.

Foreword

Too few people are paying attention to the current huge toll of death and destruction being inflicted by infantry weapons — which are killing more people every year than any other type of weapon or weapon system. They are not always in the hands of the soldiers for whom they were intended.

That this has been so for some considerable time takes on extra importance when considering the funds which have been spent on weapons and weapon systems of every other kind. Wealth on a lavish scale has been expended on missile systems, warships, all manner of aircraft platforms, and armoured vehicles by the thousand, yet relatively few of them have ever been used in anger for they were not designed and manufactured for the type of low-intensity conflict now prevalent. Infantry weapons outnumber all others. The United Nations Secretary General, Boutros Boutros Ghali, has stated that infantry weapons account for some 90 per cent of the deaths and injuries in armed conflicts.

As with all other weapons, rifles and hand guns are made with the expectation that they will be used. However, the main reason that infantry weapons are having such an effect on the human race at present is the scale of their proliferation.

All around the world, recent years have witnessed a massive increase in the quantities of infantry weapons in all manner of hands with irregular or unlawful purposes. The result is a rising tally of death and injury in locations where once such escalations were low scale or almost unknown. Locations such as Central and West Africa, the Indian subcontinent and parts of the former Soviet Union come readily to mind, while there are many others. In South African townships, where conflict is no stranger, the current most common cause of death among males aged between 20 and 30 years is gunshot wounds.

There appears to be one main reason why this proliferation has taken place. Given that the world has become a very violent place, and that the aftermath of major conflicts and political confrontations has resulted in infantry weapons being distributed to allies of one side or other of the major power blocs during the last 50 years, it has to be accepted that in parts of the world the abject failure of some states to govern has led to the current enormous demand for infantry weapons. As fragile governments and regimes crumble under the pressures of corruption, incompetence or old unsettled quarrels, populations tend to lose whatever affiliations they might have had to the state and revert to other loyalties such as religious, ethnic or tribal groupings. As states lose control, crime flourishes and adds further tensions. Weapons are obtained from whatever sources are convenient for protection against supposed or real enemies. Local disputes once settled with clubs, edged weapons or, at best, sporting firearms, are instead conducted using more lethal weapons. The demand for more and more weapons grows. Being relatively compact and light items, they can almost always be supplied at a profit by established smuggling and criminal routes. The ratchet of misery is thereby nudged a further notch.

There is no shortage of weapons to be obtained. A United Nations Institute for Disarmament Research paper (no 34) lists nearly 300 manufacturers in over 50 countries manufacturing small arms, associated equipment and accessories. This represents a 25 per cent increase in companies over a 10-year period when the end of the Cold War resulted in a glut of weapons anyway.

That glut has produced some bizarre results. The United Nations paper already quoted estimates that in Mozambique alone there are 1.5 million Kalashnikov series weapons unaccounted for. In the Northwest region of the Indian subcontinent stockpiles of rifles numbering some two to three million have been reported. The latter have been appearing as far away as Mumbai (Bombay) and Sri Lanka. Some of the Mozambique weapons changed hands in Rwanda for as little as a bag of maize.

Something needs to be done. After the understandable efforts to establish treaties relating to nuclear stockpiles, chemical weapons and land mines, the proliferation of infantry weapons must be accorded a higher priority than the subject has received to date. It may be impossible to control fully the movement of light weapons but the efforts of the agencies attempting to do so would be considerably aided if a central pool of information could be established relating to batch numbers, quantities, end users, stocks held, and so on. It would also be of considerable importance to the cause of peace if some nations and concerns currently manufacturing and marketing relatively low-tech weaponry would take a more responsible view of their activities.

At present there is no centralised method of supporting data-gathering organisations or ways of placing sanctions on rogue supplier nations. The United Nations has no effective means of dealing with the proliferation of infantry weapons. The UN Charter even restrains the body's authority to intervene in the internal affairs of member states. Perhaps a start could be made by UN member nations giving their organisation more 'teeth'.

Seen in this light the contents of this Yearbook take on an even greater importance.

Ammunition

Having stated the above, consideration of the past year's infantry weapon developments has to relate to the official aspects of small arms. There have been several recent highlights but most come under the headings of ammunition and systems rather than the usual categories of new pistols, rifles and so on. As a short perusal of the New Entries pages of this Yearbook will soon reveal, most of the completely new material relates to ammunition types.

The Ammunition section has been completely revised this year to provide a more in-depth study of what, after all, are the real weapons delivered by firearms, namely the bullets and other projectiles. It should be borne in mind that the coverage provided relates to military cartridges only as a Yearbook of this nature cannot attempt to cover the many and various commercial and sporting cartridges produced today.

Yet the term ammunition does not include just cartridges. For our purposes it also includes mortar bombs and it is here that the past year has witnessed some significant developments in the so-called smart munition field normally associated with artillery but now extending to infantry mortars. Although these have been around for a while, this past year has seen some programmes come to full technical fruition.

During 1995 the Bofors/Saab 120 mm Strix was the first smart mortar munition to reach full production status and is now being issued to the Swedish Army, thereby becoming the first of its kind to enter service anywhere in the world. Strix will not be the last. The Diehl Bussard guided 120 mm mortar bomb is involved in the Precision Guided Mortar Munition (PGMM) programme for possible US Army service in association with Lockheed Martin, while the 120 mm Gran (Facet) laser-guided mortar bomb has been offered to the world's export markets by the Russians. The UK self-homing 81 mm Merlin is also well advanced along its path to production, although a firm order has yet to emerge. In the USA development work on fibre optic guidance methods for mortar munitions are reported to be progressing well.

Other significant ammunition developments which have come closer to production during the last year relate to fuze systems. Advanced projects such as the US Objective Individual Combat Weapon (OICW) and Objective Crew Served Weapon (OCSW) have to involve highly accurate time fuzes which are set by the respective fire control systems only an instant before firing to ensure that their airburst projectiles function exactly when required. The airburst function is, after all, the main reason why the OICW and OCSW are being developed and the latest available information relating to these weapons will be found in their relevant sections. However, the technology involved has passed from the developmental to the hardware stage over the last few months and a new era will soon be upon us. The electronics and other associated techniques employed with the OICW and OCSW will eventually find many other applications in other areas of the infantry weapons field. Armourers will soon have to add oscilloscopes to their tool boxes.

To revert to less technically advanced topics, new ammunition types continue to appear on the scene. One, mentioned in these pages for the first time, is the 0.357 SIG, a combination of the 0.357 Magnum bullet and the necked case and loading from the 0.40 S & W. SIG have already produced pistols to fire this potent round so we can now await the course of ready acceptance or relegation as yet another byway of ammunition development, the latter being something which has happened to many other cartridges which seemed promising but failed to find sufficient takers.

One well established cartridge which will take some effort to topple from its position as the most widely used pistol and sub-machine gun cartridge is the 9 × 19 mm Parabellum. This venerable design has been around since the turn of the century, with various forecasts that it would fade away proving wrong.

There seem to be many sound reasons for the 9 mm Parabellum's replacement. Among others, in modern terms it is relatively low powered, has an indifferent armour penetration performance (an important consideration in these times of the increasing use of body armour by military, police and paramilitary forces), while retaining a high level of potentially lethal energy capable of harming non participants for some distance in excess of the ranges at which it is normally employed operationally. Yet the 9 mm Parabellum soldiers on apparently heedless of such considerations.

It now seems that at least one concern is determined to make inroads into the 9 mm Parabellum's predominance. The ever-innovative Fabrique Nationale Herstal (FN) has launched a major long-term programme relating to its 5.7 × 28 mm SS190 cartridge. Once again, this cartridge is nothing new as it has received coverage within this Yearbook for some years. Until now only one weapon was available to fire it, FN's futuristic P90 Personal Defence Weapon. That will change during 1996 with the introduction by FN of its Five-seveN pistol, also firing the SS190. (See the entry in the Pistols section for available details on the Five-seveN.)

FN is expending considerable resources to market what it already refers to as a weapon system firing the 5.7 × 28 mm SS190 which, it claims, is the true replacement for the 9 mm Parabellum. FN outlines numerous reasons, covering headings such as logistics and low recoil to ballistic and on-target performance. The seeming advantages of the new round over the 9 mm Parabellum veteran are indeed many and we wish FN well with its endeavours. The company is obviously aware of the long and difficult path it will have to traverse if it is to achieve even a partial replacement of the 9 mm Parabellum but the potential rewards are substantial and likely to be very long-lasting. We await results, although it may be some years before they are forthcoming. Apparently FN is willing to wait 10 to 15 years before assessing the degree of market penetration success. We look forward to charting the progress.

Innovations

While the past year cannot be said to have been a vintage one regarding infantry weapon designs (other than the first views of the OICW and OCSW), there have been some items of note. Included in this edition are some entirely new products on the market although several are already familiar in other guises.

For instance the GLOCK compact pistols are exactly as stated, compact versions of already well established and successful models. The Czech CZ Model 100 and Polish MAG 95 are indications that designers from the former Eastern Bloc have taken note of safety and other design trends in the West to ensure their products have at least a chance in the market place. Both pistols fire the 9 mm Parabellum cartridge. In this area the Russians are being somewhat subdued and no doubt have something in the pipeline more advanced than their PMM, an update of their PM Makarov pistol, which can now fire a more powerful 9 × 18 mm cartridge which merely places it in the run-of-the-mill 9 mm Parabellum category. What that something is we will have to wait and see. Their 9 mm Gyurza, fearsome in body armour-piercing performance as it might be, is definitely in the special forces category and is unlikely to find acceptance on any wide scale as a standard service pistol. We await developments from that quarter.

On the rifle front, there have been relatively few innovations this year, other than the little 5.56 mm Galil Micro, unless one contemplates the specialised sniper rifles. The French PGM series makes its appearance for the first time while mention of a missing link in the Russian armoury can be made at last. This is the 12.7 mm V-94 sniper rifle.

As with its counterparts in the West, the V-94 is not really a sniper rifle but an anti-materiel rifle, capable of taking out critical items of equipment and light vehicles up to ranges of about 1,500 m. The usual limitations of the 12.7 mm cartridges, from both East and West, preclude employment against anything smaller than communications shelters, trucks and aircraft at such ranges but the economic single shot from an anti-materiel rifle can wreak havoc and makes the development of such weapons worthy of increasing consideration.

However, it seems that designers are unwilling to place limits on the calibre of such weapons. The inclusion this year of the extraordinary 20 mm RT-20 produced in Croatia is an indication of what could well be an avenue for further development. The designers of the RT-20 seem to have deduced that if a 12.7 mm round can be usefully employed in the anti-materiel role, why not go one step further? That step has led to the RT-20 firing 20 mm cannon ammunition. No doubt the on-target effects are worthy of the designers' efforts but the introduction of such a cartridge creates all manner of user problems. Quite apart from the weights involved, the recoil produced on firing the RT-20 cannot be comfortable or bearable for any prolonged use. One is reminded of the First World War 12.7 mm Mauser anti-tank rifle which created damaged collar bones when fired but once by any particular individual. If the result was a damaged tank this was apparently considered acceptable but such deployments are unlikely to be considered economic in personnel management terms these days. Hence the recoilless arrangements and large muzzle brake of the RT-20. No doubt these palliatives do work to a certain extent but the sight of generous amounts of foam rubber padding on shoulder/butt interfaces does not inspire confidence. Neither does the prospect of the rearwards-facing venturi kicking up debris to indicate the firer's position. Yet for all this the RT-20 is a formidable addition to the soldier's array of potential weaponry.

The RT-20 already has a stable mate. As this Foreword was being prepared news arrived of the South African AEROTEK NTW 20 anti-materiel rifle firing 20 × 82 mm ammunition. This, as with the RT-20, appears to be a formidable piece of kit which, it is claimed, has a very low recoil for its calibre, no doubt because of the large muzzle brake and combined hydraulic/pneumatic damping and buffer systems. Available details regarding the NTW 20 appear in the *Addenda*.

Future anti-materiel rifles might well utilise the Eastern Bloc 14.5 × 114 mm round. One of the Hungarian anti-materiel rifles, the Gepard M3, already employs this cartridge. As we go to press (and unfortunately too late for inclusion elsewhere in this edition) news has arrived of a Czech design foray utilising the 14.5 × 114 mm cartridge. This is not a rifle but a heavy machine gun, a project employing the KPV 14.5 mm machine gun on a ground tripod as a heavy fire support weapon; the new combination has been named the Pirat. Until now KPVs have either been deployed on air defence carriages or armoured vehicles. Utilising the KPV on a ground tripod makes some sort of sense until one contemplates the total weight involved with the Pirat, namely 202 kg. In action the Pirat requires a crew of three (one gunner and two ammunition handlers) and a heavy softmount arrangement has to be included to reduce the considerable recoil and judder produced on firing. From the little information on hand regarding the Pirat it seems that the end result will be something of a handful for any infantry unit to contemplate. As yet the Pirat is in the prototype stage. An earlier Russian development along similar lines, the 14.5 mm PKP dating from the 1950s, found only limited acceptance and soon faded from the scene.

One other machine gun innovation worthy of note is the 12.7 mm technology demonstrator produced by CTA International, a joint concern formed by Giat Industries of France and Royal Ordnance of the UK. The prime purpose of CTA International is to develop and produce a 45 mm vehicle cannon firing Case Telescoped Ammunition (CTA) but their 12.7 mm Gatling-type machine gun is another possible application for such ammunition.

At this stage the CTA International 12.7 mm machine gun remains as stated, a technology demonstrator, but the advantages of CTA are such that it has numerous logistic, ammunition handling and economic attractions (among others). The CTA concept is no novelty as anyone with any knowledge of small arms history will already be well aware, but until now it has failed to achieve the acceptance level that mass production requires. That now seems certain to change and it could well be that this time the CTA International 12.7 mm machine gun and its experimental rounds mark a significant landmark in infantry weapon development. That is something for the longer term.

Sights

The section on Sighting Equipment has some interesting innovations this year. Once again all manner of remarkable but expensive vision and sighting devices are included, yet there are also some other significant innovations covered for the first time. One is the Bushnell HOLOsight, which introduces holography to the firearms user's selection of available technology, while on a more down to earth level there is the Trijicon self-luminous series which is used with both eyes open.

The latter is but one example of a now widespread 'red dot' type, although the Trijicon ACOG series has understandably gained widespread acceptance. It is one of the options available for another innovation mentioned in these pages for the first time, namely the Modular Weapon System (MWS) produced by the Knight Armament Company and intended for use by US Special Forces. The MWS is basically an M16A2 with its forestock and sight table configured to accept a wide range of sighting and other accessories to meet specific mission requirements. This latter point has to be stressed for some publicity shots issued to illustrate the MWS resemble gun enthusiasts' Christmas trees, with every possible accessory in place. Such cluttered combinations are most unlikely to see operational use. Instead the MWS will carry only whatever accessory or sight is needed for each particular mission. The rest of the MWS options will stay secure in their packing cases until needed.

Suppressors

This edition contains a new section on Suppressors. It has to be acknowledged that it is incomplete and very much a first attempt but it is hoped that inclusion of this section will prompt the information to make it more encyclopaedic. This new section is deemed necessary as the suppressor has already moved out of the category of an attachment for a special purposes weapon into more general use, although many potential users will not have noticed this as yet.

The suppressor, or silencer as it is commonly known, does two things to the firing signature of a weapon. It reduces the sound of a shot considerably, to a level where it is difficult for it to be identified as a weapon discharge, and it virtually eliminates muzzle flash. The tactical utility of these two properties is that a target can have no idea from where an unexpected burst of fire is originating, even in general terms. Taking cover from fire or some form of reaction such as return fire is thus made difficult. This alone makes the widespread use of suppressors an attractive proposition. Some armed forces in Northern Europe have already made provisions for the general issue of suppressors but have no intention of breaking into their stockpiles until they are really needed, thus enhancing tactical surprise.

Most current suppressors still come within the special weapons category. Even so, their utility is certain to expand into other areas. Once again, we await developments. In the meantime, any extra information that could expand this section will be much appreciated.

Acknowledgements

This edition marks the Editor's first solo foray, an exercise which has been greatly assisted by many. A top priority when acknowledging such assistance must be accorded to Ian Hogg who has continued to be a fount of advice, assistance and guidance during what has been a very busy period. Also from within the Jane's network, Christopher F Foss and Ian Kemp have been sources of information and help of all kinds. My thanks go to them all. Thanks are also due to many other individuals and organisations, too many to mention by name here although I hope they will recognise their input. Information supplied by many within the industry has, as always, been invaluable and the editor's gratitude is proffered to all concerned.

The production of a Yearbook of this nature and coverage cannot be accomplished without professional help. It is here that the considerable efforts of the Jane's production team comes to the fore and I can only repeat my admiration and thanks to all concerned for yet another job well done. The teams headed by Ruth Simmance and Sulann Staniford have been more than helpful while remaining cheerfully tolerant of the many hitches a book of this nature experiences during preparation. A special mention of thanks is more than due to Jane Stimson who has acted as the main contact point between the editor and the production teams. Her competence has been of the greatest assistance on many occasions. My repeated thanks to all.

Terry J Gander

The Editor would welcome any extra information, corrections or comments regarding the contents of this Yearbook for he is well aware of how isolated an individual can become in the defence market place without such inputs. Please direct all correspondence to the Editor at *Jane's Infantry Weapons*, Jane's Information Group, Sentinel House, 163 Brighton Road, Coulsdon, Surrey CR5 2NH, United Kingdom.

New entries in this Edition

Entry	Country
9 mm/0.380 GLOCK 25 pistol	Austria
9 mm GLOCK 26 subcompact pistol	Austria
0.40 GLOCK 27 subcompact pistol	Austria
5.7 mm FN Five-seveN personal defence weapon	Belgium
9 mm P-9 Gurza Pistol	CIS
9 mm PMM self-loading pistol	CIS
9 mm HS 95 pistol	Croatia
CZ Model 100 and 101 pistols	Czech Republic
9 mm Minimax 9 pistol	Hungary
Revolver 0.32 Mark 1	India
9 mm Beretta Model 92 Brigadier FS	Italy
Beretta Model 92, 96 and 98 Stock	Italy
Beretta Model 92, 96 and 98 Combat	Italy
9 mm P-93 pistol	Poland
9 mm MAG 95 pistol	Poland
Romtehnica 9 mm self-loading pistol	Romania
SIG-Sauer P239 pistols	Switzerland
Kel-Tec 9 mm P-11 pistol	USA
9 mm Smith & Wesson Model 410 pistol	USA
9 mm Smith & Wesson Model 909 and 910 pistols	USA
Ciener 0.22 LR Beretta 92/96 Conversion Kit	USA
9 mm 9A-91 sub-machine gun	CIS
Kiparis 9 mm sub-machine gun	CIS
9 mm sub-machine gun	Croatia
9 mm sub-machine gun	Romania
Ruger 9 mm MP-9 sub-machine gun	USA
12.7 mm V-94 sniper rifle	CIS
RT-20 20 mm heavy sniper rifle	Croatia
MACS 12.7 mm sniper rifle	Croatia
PGM UR Intervention 7.62 mm sniping rifle	France
PGM UR Commando 7.62 mm sniping rifles	France
PGM Hecate II 12.7 mm sniping rifle	France
5.56 mm Galil MAR Micro assault rifle	Israel
KAC 5.56 mm Modular Weapon System	USA
Ruger M77 Mark II rifle	USA
Ciener 0.22 LR Conversion Kits	USA
Ciener AR-15/M16 belt feed mechanism	USA
Maadi 40 mm grenade launcher	Egypt
Objective Crew Served Weapon (OCSW)	USA
MPIM/SRAW	USA
CTA International 12.7 mm rotary machine gun	International
Nag anti-tank missile system	India
Type 87 Chu-MAT anti-tank missile system	Japan
PG-7M 110 anti-tank projectile	Slovakia
MACAM anti-tank weapon system	Spain
Predator Short Range Assault Weapon	USA
PIMA 82 mm Mortar M82	Bulgaria
Lockheed Martin Lightweight 120 mm Mortar	USA
Leica SG12 Digital Goniometer	Switzerland
0.22 in Long Rifle	USA
7.62 × 42 mm SP-4	CIS
9 × 21 mm Gyurza	CIS
9 × 39 mm SP-5 and SP-6	CIS
0.357 SIG	USA
FM rifle grenades	Argentina
MECAR HE-AP-RFL-35, BTS M235 CLAW	Belgium
FAMAE GM 78-F7 offensive/defensive hand grenade	Chile
RGO-78 defensive hand grenade	CIS
RGN-86 hand grenade	CIS
RG-42 offensive hand grenade	CIS
ZDP incendiary smoke hand grenade	CIS
Kaha No 1 defensive hand grenade	Egypt
Kaha No 1 offensive hand grenade	Egypt
Giat 40 mm AP/AV rifle grenade	France
Ruggieri Type 0052 and Type 0052A hand grenades	France
ALSETEX SAE 210 offensive grenade	France
ALSETEX SAE 310 controlled fragmentation grenade	France
ALSETEX screening smoke grenades	France
ALSETEX coloured smoke grenades	France
Nico 40 mm training grenade cartridges	Germany
Elviemek EM 03 practice hand grenade	Greece
IOF 36M hand and rifle grenades	India
GT-5PE A2 hand grenade	Indonesia
POF plastic hand grenade	Pakistan
POF metal hand grenade	Pakistan

Entry	Country
POF 75 mm P-2 Mk 1 HEAT rifle grenade	Pakistan
POF WP P-1 smoke grenade	Pakistan
POF target indication grenades	Pakistan
Dezamet rifle grenades	Poland
Swartklip 40 mm grenades	South Africa
Swartklip 40 mm high-velocity grenades	South Africa
MFA HG 85 defensive hand grenade	Switzerland
MFA OHG 92 offensive hand grenade	Switzerland
MFA SIM HG 93 training hand grenade	Switzerland
MKEK hand grenades	Turkey
Haley and Weller N110 screening smoke grenade	UK
FAMAE 60 mm 60-M-61-A HE bomb	Chile
FAMAE 81 mm 81-M-57-DA HE bomb	Chile
FAMAE 120 mm 120-44/66 HE bomb	Chile
60 mm Type 63-1 fragmentation projectile	PRC
82 mm Type 53 smoke bomb	PRC
82 mm Type 53 illuminating mortar bomb	PRC
NORINCO ammunition for 100 mm Type 71 mortar	PRC
NORINCO ammunition for 120 mm Type 55 mortar	PRC
120 mm Gran laser-guided mortar projectile	CIS
Thomson-Brandt ammunition for 81 mm mortars	France
82 mm dual purpose mortar bomb	Hungary
IOF 81 mm Illuminating mortar bomb	India
IOF 120 mm Illuminating mortar bomb	India
120 mm Precision Guided Mortar Munition (PGMM)	International
TAAS 120 mm CL3144 ICM mortar bomb	Israel
EPD M797 electronic point detonating mortar fuze	Israel
CIS 60 mm mortar bombs	Singapore
CIS 81 mm mortar bombs	Singapore
CIS 81 mm Extended Range mortar bombs	Singapore
CIS 120 mm mortar bombs	Singapore
CIS 120 mm Extended Range mortar bombs	Singapore
60 mm Long Range mortar bombs	South Africa
81 mm Long Range mortar bombs	South Africa
MKEK 81 mm mortar bombs	Turkey
MKEK 107 mm mortar bombs	Turkey
MKEK 120 mm mortar bombs	Turkey
HE Ammunition for the 81 mm Mortar M252	USA
81 mm Smoke RP M819	USA
81 mm Illuminating M853A1	USA
NORINCO 11 mm pyrotechnic pistol	PRC
POF flare tripwire Mark 2/2	Pakistan
Singapore Technologies alarm flare	Singapore
M49A1 surface trip flare	USA
Zeiss NSV 80 II night sight attachment	Germany
Passive Night Sight	India
El-Op INTIM thermal binocular	Israel
Simrad IS2000 eye-safe laser gun sight	Norway
Leica VECTOR 1500 DAES range finding binocular	Switzerland
Maxi-Bino-Kite night vision binocular	UK
Contraves SACMFCS II	USA
Magnavox MAG-600 Individual Weapon Thermal Sight	USA
Magnavox AN/PAS-19 Multipurpose Thermal Sight	USA
BUSHNELL HOLOsight	USA
ITT Integrated Day/Night Scope F7111	USA
Trijicon Advanced Combat Optical Gunsights (ACOG)	USA
Trijicon ACOG Reflex Sight	USA
Trijicon Night Sights	USA
Insight M30 Boresighting Equipment	USA
PBS-1 Silencer for Assault Rifles	CIS
Valme Silencers	Finland
BRÜGGER + THOMET IMPULS-II Pistol Silencer	Switzerland
BRÜGGER + THOMET IMPULS-III Pistol Silencer	Switzerland
BRÜGGER + THOMET SUC Pistol Silencer	Switzerland
BRÜGGER + THOMET Silencers for Sub-machine Guns	Switzerland
BRÜGGER + THOMET Detachable Silencer for MP5	Switzerland
BRÜGGER + THOMET Silencers for Assault Rifles	Switzerland
BRÜGGER + THOMET Integrated Silencer Systems	Switzerland
Ciener Sound Suppressors	USA
Gemtech SOS Suppressor for Pistols	USA
Gemtech Vortex-9 Suppressor for Pistols	USA
Gemtech MK-9K Suppressor for MP5	USA
Gemtech MINI-TAC Suppressor for MP5	USA
Gemtech COMMANDO Suppressor for 5.56 mm Carbines	USA
Gemtech SPEC-OP 2 and 3 Suppressors	USA

Entries deleted from this Edition

(Page entries are from the 1995-96 edition)

Entry	Country
9 mm Mauser Models 80 and 90 pistols	Germany
9 mm Model P88 pistol	Germany
9 mm Model 92FC	Italy
9 mm Beretta Model 92FCM	Italy
9 mm Models 92FS Inox and 98FS Inox	Italy
9 mm Model 2000	USA
Colt Offensive Hand gun Weapon System	USA
10 mm Smith & Wesson Model 1000 series pistol	USA
Ruger KP89 Convertible pistol	USA
9 mm Calico M950 pistol	USA
Calico 9 mm M-900 carbine	USA
5.56 mm Colt M231 firing port weapon	USA
7.62 mm SAR-8 rifle	USA
M2A1-7 portable flame-thrower	USA
M9E1-7 portable flame-thrower	USA
66 mm M202A2 multishot portable flame weapon	USA
Marquardt MultiPurpose Individual Munition (MPIM)	USA
7.62 mm Uirapuru Mekanika machine gun	Brazil
SIG 7.62 mm 710-3 general purpose machine gun	Switzerland
20 mm M40 anti-aircraft cannon	Sweden
2K8 Falanga anti-tank guided missile system	CIS
82 mm M59 and M59A recoilless guns	Czech Republic and Slovakia
Aries anti-tank missile system	Spain
Saber multipurpose missile system	USA
81 mm fragmentation projectiles	PRC
82 mm M30 HE/fragmentation bomb	PRC

Entry	Country
82 mm HE/fragmentation bomb Type 20	PRC
RDG-1 smoke hand grenade	CIS
Losfeld fragmentation grenade	France
Losfeld multipurpose anti-personnel grenade	France
ETSQ electronic mortar fuze	Germany
Elviemek EM10 rifle grenade	Greece
Elviemek EM14 rifle grenade	Greece
Elviemek EM12 screening smoke rifle grenade	Greece
Borletti FB 282 mortar point detonating fuze	Italy
Round, 40 mm, HE M848A1	South Africa
M545 Universal fragmentation grenade	USA
M560 series anti-personnel fragmentation grenades	USA
Ring Airfoil Grenade (RAG)	USA
82 mm shrapnel bomb Model 81	Vietnam
Alsetex audio-visual chemical alarm	France
Model 56 hand smoke grenade	France
1 in Beretta signal pistol	Italy
71 mm Lyran illuminating system	Sweden
Federal Laboratories coloured smoke grenades	USA
Federal Laboratories Parachute and Meteor flares	USA
Contraves SACMFCS integrated sight	USA
Magnavox Short-Range Thermal Sight (SRTS)	USA
M921 submersible night vision sight	USA
M937/M938 individual weapon sights	USA
Litton AIM-1D/DLR laser aiming light	USA
ITT F4961 day/night combat rifle sight	USA
ITT F4960 Stinger Night Sight	USA
F4965 rifle night sight	USA

Glossary

AAT Arme Automatique Transformable. French equivalent of general purpose machine gun

Accelerator Component of a recoil-operated weapon intended to give extra velocity to the bolt or breech block during recoil. Best known application is the Browning machine gun.

ACLOS Automated Command to Line Of Sight. Advanced anti-tank missile guidance system in which the missile identifies the target and steers to it without operator intervention. Also called third generation and fire-and-forget.

ACP Automatic Colt Pistol. Differentiates certain pistol cartridges from others of similar calibre but different dimensions. Applied to cartridges originally developed for Colt automatic pistols.

ACR Advanced Combat Rifle

Advanced primer ignition System of operation common in sub-machine guns in which the primer is struck and the cartridge fired whilst the bolt is still moving forward. The explosion force must arrest the bolt before causing it to return, introducing a fractional delay between firing and breech opening, permitting the bullet to leave the barrel. Also absorbs some recoil energy, allowing the bolt to be lighter.

AGL Automatic Grenade Launcher

AIFV Armoured Infantry Fighting Vehicle

AK Avtomat Kalashnikova (CIS); Automatische Karabiner (Switzerland); Automat Karbin (Sweden)

AKM Avtomat Kalashnikova Modernizirovannyi

AKS Avtomat Kalashnikova Skladyvayushchimsaya

Annulus Recessed ring around the cartridge primer, formed when securing cap to case. Frequently coloured to provide identification of cartridge type.

AP Armour-Piercing. Signifies a solid projectile for kinetic energy attack of armour.

APC (1) Armoured Personnel Carrier. (2) Armour-Piercing, Capped. An AP projectile with a penetrating cap over its point.

APCBC Armour-Piercing, Capped, Ballistic Capped. An APC projectile with a light ballistically shaped nose cap to improve flight characteristics.

APDS Armour-Piercing, Discarding Sabot. A kinetic energy attack projectile with a heavy core, supported in the gun by a light metal or plastic sabot which is discarded at the muzzle.

APFSDS Armour-Piercing, Fin-Stabilised, Discarding Sabot. Kinetic energy projectile using a heavy, dart-like subprojectile supported in the gun bore by a light sabot. Used generally with smoothbore anti-tank weapons but also found with rifled weapons, in which case the driving band merely seals and is free to rotate, so as to impart minimum spin to the subprojectile.

APS Avtomaticheskiy Pistolet Stechkina

ATGW Anti-Tank Guided Weapon. A missile guided to its target and intended for the attack of armoured vehicles.

AUG Armee Universal Gewehr

Automatic pistol (Loosely) Any self-loading pistol

Ball ammunition Ammunition using a solid, inert bullet

Ballistic coefficient Measure of a projectile's carrying power. A function of the projectile mass, shape, diameter and coefficient of steadiness.

Ballistite Obsolete propellant principally used for launching rifle grenades. By extension, any cartridge used for launching rifle grenades, even though the propellant is no longer Ballistite.

Barrel That part of a gun through which the projectile is launched and given direction.

Barrel extension A frame attached to the barrel of a recoil-operated automatic weapon and carrying the breech block or bolt, and usually carrying some means of locking the bolt to the extension and so to the barrel.

Baton round Cartridge for large calibre riot control weapons using a low-velocity, non-lethal plastic or rubber projectile intended to stun or bruise.

Bayonet Knife-like weapon with a short blade attached to the rifle muzzle for close combat.

Belted A cartridge case with a raised belt near the base which positively locates the case in the gun chamber and reinforces the base for use with high pressure propellant charges.

Belt feed System of supplying ammunition to automatic weapons by means of a flexible canvas or metal belt into which cartridges are fixed. May retain its form when empty or be of the disintegrating link type with the belt falling into individual links when the cartridges are removed.

Bent A notch cut into the weapon's bolt or breech block, striker or hammer, into which the sear engages to hold the component ready to fire.

BFA Blank Firing Attachment. Device fitted to the muzzle of an automatic weapon to restrict the exit of gas from the barrel when firing blank cartridges and thus provide sufficient pressure to actuate the weapon's self-loading cycle.

Blank Small arms ammunition with a charge of powder and no bullet, so as to provide noise on firing. Used for training, to simulate weapons firing.

Blowback System of operation of self-loading and automatic weapons in which breech pressure overcomes inertia of breech closure. Used principally in low-powered weapons such as pistols and some sub-machine guns.

Blow-forward System of operation in self-loading weapons analogous to blowback, but in which the breech block is firmly anchored to the weapon frame and the barrel moves away from it, returning to chamber the next round. Rarely encountered.

BMS Battalion Mortar System (USA)

Boat-tailed A bullet in which the base diameter is less than the maximum diameter and the base is tapered, permitting airflow over the bullet to converge rapidly behind the base and thus reduce drag. Streamlined in British usage.

Bolt Part of a small arm which closes the breech. Generally indicates some rotary motion. Where no rotary motion is involved, the component is usually called a breech block.

Bolt action Breech closure by means of a hand-operated bolt moving in prolongation of the barrel axis. May be a turnbolt in which the bolt is pushed forward and given a partial turn to lock in; or a straight-pull bolt in which the manual action is a simple reciprocating movement and the bolt is rotated by cams.

Bore The interior of a weapon's barrel, from the rear face of the chamber to the muzzle.

Bottle-necked A cartridge case in which the diameter is reduced at the mouth to accommodate the bullet, which is smaller in diameter than the body of the case. Provides a greater volume in a given length of cartridge.

Box magazine Ammunition supply in the form of a metallic box, either integral with the weapon or detachable from it. May be mounted above, below or to one side of the weapon.

Breech Rear end of the weapon barrel, into which the ammunition is loaded.

Buffer A resilient unit at the rear of a machine gun body, against which the recoiling bolt impacts, absorbing recoil energy. Resilience may be caused by springs, rubber, oil or pneumatic force.

Bullet Projectile fired from a small arm.

Bulleted blank Blank ammunition using a light wooden, paper or plastic bullet which disintegrates at the muzzle. Frequently adopted in order to avoid feed mechanism troubles which can arise with unbulleted blank cartridges.

Bullpup A rifle in which the breech mechanism is set back in the stock so that the end of the receiver is against the firer's shoulder. It permits the use of a full-length barrel in a weapon which is, overall, shorter than a conventionally stocked weapon of the same barrel length.

Butt That part of a small arm which is held against the shoulder and which transfers recoil force to the firer's body.

Calibre The nominal diameter of a weapon's bore, measured across the lands (qv). Also used as a measure of length, example 26 calibres length and applied to barrels and bullets.

Cannelure A groove in a bullet enabling the metal of the cartridge case mouth to be pressed in to secure the bullet and case together or to contain lubricant to ease the passage of the bullet up the bore. Also a groove pressed into a cartridge case to facilitate location into a feed belt.

Carbine A short rifle. Traditionally the self-defence arm of cavalry, engineers and artillery whose primary task was not using a shoulder arm.

Cartridge (Small arms) A unit of ammunition sufficient to fire one shot, comprising cap, case, propellant and bullet; also called a round (qv).

Cartridge case Brass or steel metal container holding propellant and supporting bullet and primer cap.

Cartridge headspace The distance between the face of the bolt or breech block and the base of the cartridge case when the weapon is loaded. A critical dimension in a weapon's assembly; if insufficient, the bolt or breech block will not close or lock; if excessive, the case can move back on firing and burst or split.

Caseless A cartridge with no metal case and with the propellant formed into a block, into which the bullet and cap are embedded.

Centre fire That class of small arms ammunition which has the primer cap located centrally in the base of the cartridge case.

Chamber That portion of the barrel into which the cartridge is positioned for firing.

Change lever A lever or switch which allows the firer to select single shots or automatic fire in some weapons. Also called selector, and may be incorporated with the safety catch.

Charger A metal frame holding a number of cartridges. The loaded charger is placed at the mouth of the weapon magazine and the cartridges pressed down into the magazine with the thumb, after which the empty charger is discarded. Popular with bolt-action rifles and occasionally used with automatic pistols.

Clip A metal frame holding a number of cartridges. The loaded clip is inserted into the weapons' magazine and becomes an integral part of the feed system. As the last cartridge is loaded the clip is ejected from the magazine. Used with rifles, notably the bolt-action Mannlicher and the self-loading Garand. The principal defect is that the magazine can only be loaded with a full clip; it cannot be topped-up when partly emptied.

Closed bolt A weapon in which the breech is closed, though not necessarily locked, before the trigger is pulled.

CN Chloracetophenone. A chemical irritant agent used in riot control munitions.

Coaxial Machine gun on an armoured vehicle mounted with the main armament so as to traverse and elevate with it. In some designs the gun can be unlocked and elevated independently of the main gun, but not traversed independently.

Commence-ment of rifling The point in a gun bore at which the rifling grooves first reach maximum depth.

Compensator A device attached to a weapon barrel (usually an automatic weapon) to divert some of the muzzle blast upward and thus counteract the tendency for the muzzle to rise during automatic fire.

Composition B RDX/TNT mixture, commonly 60 per cent TNT and 40 per cent RDX, but proportions vary between manufacturers.

Compound bullet Small arms bullet made of different substances, for example a partly lead and partly steel core held within a steel jacket and coated with a gilding metal envelope. Such construction gives the mass advantage of lead in the core but allows the steel jacket to engrave into the rifling without leaving deposits.

Cook-off (Colloquial) Premature ignition of a cartridge caused by heat induced after loading into a gun chamber made hot by previous firing.

CS O-chlorobenzolmalonitrile. Chemical irritant agent used in riot control munitions. CS, adopted as a code designation, from the names of the inventors, Carson and Stoughton.

Cycle of operation The successive processes involved in firing a round: feeding, chambering, firing, extraction, ejection, cocking and storing energy in a return spring.

Cyclic The theoretical rate of fire of a weapon operated continuously and with an infinite supply of ammunition, that is, ignoring the need to change magazines or belts.

Cylinder Component of a revolver in which chambers are bored; held behind the barrel on an arbor or axis pin, so that it can be rotated by a mechanism coupled to the trigger or hammer so as to present each chamber to the barrel in succession.

DA (Of fuzes) Direct Action
(Of revolvers) Double-Action

Defensive grenade A hand grenade producing heavy lethal fragments over a radius greater than the distance it can be thrown. The thrower must therefore be able to take cover, and this type of grenade is usually classed as being for use in defensive positions.

Delayed blowback An automatic arm in which the bolt or breech block is not positively locked to the barrel for the entire period of the bullet's passage through the bore. Basically a blowback weapon in which the opening of the breech is artificially slowed.

Direct fire Fire in which the weapon has a direct line of sight to the target. Cf indirect fire.

Disconnector Part of the firing mechanism of a self-loading weapon which disconnects the trigger from the remainder of the firing mechanism as soon as the shot is fired and does not reconnect it until the firer positively releases the trigger and renews pressure for the next shot, thus preventing the weapon firing more than one shot for each trigger pressure. In a selective fire weapon the disconnector can be disengaged by the change lever or selector.

DM Diphenylaminochloroarsine. Irritant chemical agent used in some riot control munitions. More powerful than CN or CS. Called Adamsite in the USA after its discoverer.

Double-action Pistol firing mechanism allowing two methods of discharging a shot. The firer may either thumb back the hammer to full cock and then press the trigger, or pull the trigger through to raise and then drop the hammer. The latter option is generally used for rapid fire, the former for deliberate fire.

Double pull Trigger mechanism on certain selective fire weapons in which the first pressure on the trigger fires single shots, and further pressure, overcoming a check spring, provides automatic fire.

Double trigger A firing mechanism on some selective fire weapons in which one trigger provides single shot fire and another automatic fire.

Drag Resistance to motion of the projectile through the air, caused by a region of low pressure behind the base.

Drift Lateral deviation of the bullet in flight resulting from the spin imparted by the rifling.

Drum magazine A circular magazine in which rounds are loaded axially and driven to the feed lips by a spring.

Ejector Component which throws the empty cartridge case clear of the weapon after firing.

Extractor Component which pulls the empty cartridge case from the chamber after firing.

FAL Fusil Automatique Légère

FA-MAS Fusil Automatique de Manufacture d'Armes de St Etienne

FEBA Forward Edge of Battle Area

Feed That portion of the firing cycle during which a fresh cartridge is removed from the ammunition supply and aligned before loading into the chamber.

Feedway Area of weapon mechanism in which a cartridge is aligned with the chamber after being removed from the ammunition supply.

FIBUA Fighting In Built-Up Areas

Flash eliminator A device fitted to the muzzle to cool emergent gases, preventing the formation of flash or flame.

Flash hider Conical attachment to the muzzle for concealing muzzle flash from an observer. Also acts as a flash eliminator, though it is less efficient than a properly designed eliminator.

Flechette Thin, subcalibre, fin-stabilised projectile resembling a small arrow. Has been tried as a discrete projectile for rifles with indifferent success; more successful as a cluster weapon fired from major calibre arms.

Flip rearsight A double backsight consisting of two notches or apertures mounted at right-angles, either of which may be rotated into position, providing two alternative range settings.

Fluted chamber Chamber of a firearm with thin grooves cut longitudinally through most of its length and beyond the mouth of the cartridge case when loaded. On firing, some propellant gas flows down these flutes, equalising the pressure inside and outside the case, thus floating the case on a layer of gas and easing extraction in blowback weapons, particularly when using bottle-necked cases. The grooves do not extend to the mouth of the chamber, so obturation is not affected.

FM Code designation for titanium tetrachloride smoke mixture, used in mortar bombs.

Follower The spring-driven platform in a magazine, upon which the cartridges rest.

FS Code designation for a sulphur trioxide/chlorsulphonic acid mixture used as a screening smoke composition in mortar bombs.

Furniture Parts of a weapon solely for facilitating handling, example pistol grip, stock, fore-end, butt. Traditionally wood, but in modern weapons frequently of synthetic material or metal.

Fuze Initiating device for an explosive charge or projectile. May function on impact, after short delay, after a specified time of flight, or when it detects the proximity of a target.

GPMG General Purpose Machine Gun. A machine gun capable of operating as a squad light automatic, on a bipod and fired from the shoulder, or, tripod mounted, as a company support weapon.

Grip safety A lever or plunger let into the grip of a pistol or sub-machine gun so as to lock positively the firing mechanism or bolt unless the weapon is correctly held and the grip compressed. Acts as an automatic safety device to prevent discharge if the weapon is dropped.

Grooves Spiral cuts in the interior of the barrel which firm the rifling (qv) and thus impart spin to the projectile. Number, contour and dimensions vary according to the weapon and designer.

Hangfire Ammunition malfunction in which cartridge ignition takes place between a fraction of a second and several seconds after the cap has been struck. In British military specifications any delay greater than 0.1 second.

HB Heavy Barrel. Nomenclature used with a particular model of 0.50 Browning machine gun (the M2)

HC Hexachloroethane/zinc screening smoke mixture used in smoke grenades and mortar bombs. Also called Berger Composition after its inventor.

HE High Explosive. A chemical substance which, suitably initiated, undergoes molecular disruption accompanied by evolution of heat and gas.

Headstamp Markings on a cartridge case base, usually indicating manufacturer, calibre, date of manufacture or batch number.

HEAT High Explosive, Anti-Tank. An anti-armour munition using the hollow or shaped charge principle.

Hinged frame A revolver in which the barrel and cylinder form a movable unit hinged to the butt frame by a pivot pin. In firing condition, the two units are locked together; when the barrel is tipped down about the hinge, the rear of the cylinder is exposed for loading.

HITP High Ignition Temperature Propellant

HMG Heavy Machine Gun; an imprecise term but usually implying a calibre of 12.7 mm or more.

Hold-open device A mechanism actuated by the magazine platform and which holds the bolt or breech block to the rear after the last round in the magazine has been fired. Indicates that reloading is necessary, retains the mechanism in readiness for reloading, and when a fresh magazine has been inserted releases the bolt to run forward and chamber a fresh round ready for firing.

II Image Intensifier or Intensifying. An electro-optical sight incorporating electronic amplification of ambient light to present an enhanced view of the target area. Completely passive and undetectable, though some early units emitted a high-pitched noise which alarmed animals.

Indirect fire Fire in which the weapon and the target are not within sight of each other.

IR Infra-Red. System of night viewing relying on the infra-red portion of the spectrum. May be passive, relying on emission of heat by the target, or active in which an infra-red light source is used to illuminate the target. The passive version is now generally called thermal imaging.

IS Internal Security. Generally describes the use of troops to aid the civil power.

Lands Those portions of the barrel of a rifled weapon between the grooves, and between which the calibre is measured.

LAW Light Anti-armour Weapon. Any light, shoulder-fired, unguided rocket weapon for the attack of armoured targets.

Lead (Pronounced leed) The conical progression of the bore of a firearm from the chamber into the rifling.

LMG Light Machine Gun; a machine gun capable of being carried and operated by one man.

Lock time The interval between pressing the trigger and ejection of the bullet from the muzzle. Shortest in closed bolt weapons, where only the firing pin or hammer has to move, longest in open bolt weapons where the entire bolt unit has to run forward and chamber a round before firing.

Long recoil Operating system for automatic weapons in which barrel and locked breech recoil for a distance greater than the length of a cartridge. The breech is then unlocked and held, while the barrel runs back into the firing position, after which the breech is released to run forward and chamber the next round. Presently only encountered in some 40 mm grenade launchers.

LR Long Rifle; a particular type of 0.22 calibre cartridge

Machine gun An automatic firearm capable of delivering continuous fire so long as ammunition is provided and the trigger pressed.

Machine pistol Ambiguous term which can mean either a sub-machine gun or a pistol adapted so as to be capable of automatic fire.

Machine rifle An automatic rifle used as a light machine gun; generally a rifle with a heavy barrel and a bipod. Differs from a true LMG by usually not having a quick-change barrel and being directly derived from a rifle design rather than designed as a machine gun.

Magazine Ammunition feed system in which cartridges are held in a detachable or integral container. Detachable magazines may be of box or drum form, and are usually of metal, though plastic is becoming more common. Disposable magazines have yet to find military employment.

Magazine safety A safety device on pistols which prevents firing once the magazine has been removed. Prevents accidental discharge of the cartridge left in the chamber after removal of the magazine.

Magnum Commercial cartridge or weapon with greater power and velocity than is usual from that particular calibre.

Match ammunition Small arms ammunition carefully manufactured and graded for very consistent performance for competition and sniping purposes.

MCLOS Manual Command to Line Of Sight. Anti-tank missile guidance system in which the missile is steered to the target by the operator, using hand controls. Also called first generation guidance.

MFC Mortar Fire Controller

MICV Mechanised Infantry Combat Vehicle.

MMG Medium Machine Gun. A machine gun firing rifle-calibre ammunition but capable of sustained long-range fire for long periods. Generally water-cooled; example Vickers, Maxim, Browning M1917.

MOUT Military Operations in Urban Terrain

MPi Maschinen Pistole

MPI Mean Point of Impact. The centre point of a group of shots fired at the same line and elevation.

MRTF Mean Rounds To Failure

MTBF Mean Time Between Failures

MTBR Mean Time Between Repairs

Muzzle brake Attachment to the muzzle of a weapon designed to deflect some of the emergent gases and direct them against surfaces so as to generate a thrust on the muzzle countering the recoil force. Widely used on artillery but less popular in small arms since an efficient brake will divert too much gas to the sides and rear, to the discomfort of the firer and his companions.

NDI Non-Developmental Item (US)

NV Night Vision

Obturation Sealing against the unwanted escape of gas to the rear of a firearm. In small arms almost always performed by the expansion of the cartridge case against the chamber walls; in caseless ammunition it is a function of the breech block.

Offensive grenade A hand grenade producing small fragments over a limited radius and which can be thrown further than its danger radius, permitting it to be thrown by advancing troops who need not take cover, and thus considered suitable for use in offensive operations.

Open bolt A weapon in which the bolt or working parts are held to the rear when the weapon is cocked and there is no cartridge in the chamber. Used on automatic weapons so as to permit a flow of air through the barrel when not firing, allowing the barrel to cool. On pressing the trigger the bolt must run forward and chamber a round before actuating the firing mechanism. This movement tends to throw the weapon off the aim, and thus open bolt weapons are considered to be less accurate than closed bolt weapons for the first shot.

Parabellum Designs of cartridge and weapons originating with the Deutsche Waffen und Munitionsfabrik of Berlin; from their telegraphic address and trademark.

Patridge sight A pistol sight in which the foresight is a rectangular blade with vertical rear face and the rearsight is a plate with a rectangular notch. Named after E E Patridge, the inventor.

Pentolite A denatured form of PETN used as an explosive filling in cannon projectiles.

PETN Pentaeryhtrytoltetranitrate. A high explosive used in cannon projectiles.

Ph Hit probability

PIBD Point Initiating, Base Detonating. A type of fuze used with shaped charge projectiles and which is located at the tip of the projectile in order to strike the target first, but which initiates the shaped charge from the rear end.

Pk Kill probability

PM Pulemyot Kalashnikova

Primer Percussion or electrically initiated cap in the base of a cartridge case.

Propellant Low explosive used to propel a projectile from a firearm. Usually a nitro-cellulose compound, though nitro-glycerine and nitro-guanidine are often found as additives. Made in a wide variety of shapes and sizes designed to control the burning so as to provide the required propulsive force.

PSG Präzisions Schutzen Gewehr

Pump action A mechanism used in rifles and shotguns in which reciprocating movement is given to a grip beneath the barrel. Pulling it back and pushing it forward again causes the breech to be opened, a cartridge loaded, the breech closed and the action cocked. Also called slide or trombone action.

Pyrotechnic An explosive store which burns to provide flame, smoke, illumination or other effects.

RAW Rifleman's Assault Weapon

RCL Recoilless. A weapon which eliminates the recoil force by balancing the momentum of the projectile's forward movement by the momentum of a body of gas or solid material ejected rearwards from the breech. Adopted so as to save the weight of a conventional recoil system and carriage or mounting.

RDT&E Research, Development, Test and Evaluation

RDX Research Department formula X. British designation for the explosive variously known as Hexogen or Cyclonite. Too sensitive to be used alone, and generally denatured by admixture with TNT or wax.

Rebated rimless A type of cartridge case in which the extraction rim is smaller than the base of the case. Found in some 20 mm and small arms cases in military use.

Receiver The body of a weapon. The housing for the bolt or breech block.

Recoil intensifier Attachment to the muzzle of a recoil-operated machine gun to impede the muzzle blast and cause the barrel to recoil with greater force, making automatic action more positive and providing a reserve of energy to overcome friction caused by dust and dirt.

Regulator A device on gas-operated automatic weapons through which gas is channelled from the barrel to the operating piston. Generally with various sizes of port for selection of the amount of gas delivered to the piston, so that more gas can be admitted when the gun's action slows because of dirt or fouling and there is no opportunity to strip and clean it. Regulators often allow the gas system to be shut off completely so that the weapon can be used for launching grenades.

Return spring Spring in an automatic or self-loading weapon which is loaded by the recoiling bolt and which then forces the bolt forward again to recommence the cycle of operation.

Revolver (1) A pistol in which a rotating cylinder presents successive chambers to the rear of the barrel for discharge.
(2) A machine gun or cannon in which a revolving mechanism strips ammunition from a belt and presents it to the chamber.
(3) A machine gun or cannon in which a group of barrels rotate, each barrel being loaded and fired at specific points in the rotation.

RF Rimfire. A small arm cartridge in which the initiating composition is packed into the hollow rim of the cartridge case, instead of being held in a central cap. Cheap and effective, but not generally considered for high pressure ammunition used in military weapons.

Ricochet Rebound of a bullet from a struck surface. The angle of ricochet and the distance to which the bullet may fly are totally unpredictable.

Rifling Spiral grooves cut into a gun barrel in order to impart spin to the projectile so that it becomes gyroscopically stabilised and flies point-first.

Rimless A cartridge case with a deep groove in the rear end, so providing an extraction rim of the same diameter as the body.

Rimmed A cartridge case with a prominently raised rim at its rear end, which positively locates the case in the chamber and affords purchase for the extractor.

ROC Required Operational Characteristics

Round A complete item of ammunition, containing all those components - cap, propellant, case and bullet - required to fire one shot.

RPD Ruchnoi Pulemyot Degtyarova

RPK Ruchnoi Pulemyot Kalashnikova

SACLOS Semi-Automatic Command to Line Of Sight. Anti-tank missile guidance system in which the firer maintains his sight on the point of aim and the missile is automatically guided into the sightline by sensors and electronic circuits in the firing post. Also called second generation system.

SAWS Squad Automatic Weapon System

Sear Part of the firing mechanism, linked to the trigger, which engages with the hammer, firing pin, striker or bolt and when pulled clear by trigger action releases the component and thus allows the weapon to fire.

Selective fire A firearm which can be used as a single shot or automatic weapon.

Selector A device on a selective fire weapon which allows a choice of single shots or automatic fire. Also called change lever.

Self-cocking A firing mechanism on some automatic pistols in which pressure on the trigger first cocks the firing pin or striker and then releases it, preventing the pistol being fired if accidentally dropped. Also called double action only.

Set trigger Trigger, or stud or lever, on some target and sniping rifles in addition to the normal trigger which is operated first to cock the weapon and set the firing trigger so that very light pressure on it will fire the weapon.

SG (1) Schutzen Gewehr (Germany)
(2) Stankovyi Goryunova (CIS)

Shell A hollow projectile, filled with explosive or other agent, fired by cannon and artillery.

ShKAS Shpitalny, Komaritsky, Aviatsionnyi, Skorostrelnyi

Shot Solid projectile fired by cannon or artillery

Silencer A device attached to the muzzle or surrounding the barrel and intended to reduce the noise of discharge. Generally a Maxim silencer in which a number of internal baffles causes emergent gas to swirl round and lose velocity before escaping. If gas escapes elsewhere with sufficient velocity to cause noise, the effectiveness of the silencer is reduced; thus it is impractical to silence revolvers, since there is sufficient gas escaping from the cylinder/barrel junction to override the silencer's effect. Absolute silence is rarely attained, so it is now more common to refer to these devices as sound moderators or suppressors.

SK Selbstlader Karabiner (Switzerland)

SKS Samozaryadnaya Karabin Simonova

SLAP Saboted Light Armor Penetrator

SLAW Shoulder-Launched Assault Weapon

SLR Self-Loading Rifle

SLUFAE Surface-Launched Unit, Fuel-Air Explosive

SMAW Shoulder-launched Multipurpose Assault Weapon

SMG Sub-Machine Gun

SSKP Single Shot Kill Probability

STANAG (NATO) Standardisation Agreement

Stgw Sturmgewehr (Switzerland)

Stock Traditionally the wooden support into which the rifle's mechanism is fitted. Now more generally that part of the furniture which supports the weapon against the shoulder.

StuG Sturmgewehr (Germany)

Stun grenade A hand grenade producing a very loud report and bright flash, intended to disorient and unnerve terrorists and rioters without injuring them.

Sub-machine gun A lightweight, one-man weapon usually firing a pistol cartridge and with limited range and accuracy.

SVD Snayperskaya Vintovka Dragunova

Thermal imaging Night vision system relying upon the emanation of heat from the target; previously known as passive infra-red.

TNT Tri-Nitro-Toluene. High explosive used as a filling for various munitions.

Toggle lock A breech locking system for recoil-operated weapons, notably the Maxim and Vickers machine guns and the Parabellum (Luger) pistol.

TOP Total Obscuring Power. A measure of the effectiveness of a smoke screen.

TOW Tube-launched, Optically tracked, Wire-guided. US anti-tank missile system.

Tracer Ammunition carrying a small flare in the base of the bullet to mark out the trajectory in flight and so permit corrections to be made.

Trajectory The flight path of a projectile from gun to target

Twist of rifling The degree of rotation of the rifling grooves in a barrel. Can be constant or progressive, that is, increasing in pitch as it nears the muzzle. The degree of twist can be defined: by specifying the distance required for the groove to make one complete revolution - example, one turn in 250 mm; by specifying the number of calibres required for the groove to make one complete revolution - example, one turn in 30 calibres; or by specifying the angle between the line of the groove and the axis of the bore - example, 3.5°.
Sense of twist - right-hand or left-hand, as seen from the breech, should also be specified, though very few weapons use a left-hand twist.

Velocity Measure of the bullet's speed. V_0 or Muzzle Velocity (MV) refers to the speed as the bullet leaves the gun; Observed Velocity (OV) to the speed at any particular point in flight; and Remaining Velocity (RV) to the speed at the end of flight. Modern usage often specified the precise distance at which OV is measured, example, V_{25}, V_{50}, in which the figures indicate the distance in metres from the muzzle.

Vertex The highest point of the bullet's trajectory.

Vz Vzor - model (Czech)

WP White Phosphorus. An incendiary and smoke producing chemical agent used in grenades and mortar bombs.

Wz Wzor - model (Polish)

Yaw The angle between the axis of the projectile and the trajectory.

Zero The adjustment of sights so that the bullet will strike the point of aim at some specific range. From this zero adjustment, the sight's adjusting mechanism will provide the necessary compensation to the sightline so that bullet and sightline will coincide at other ranges.

PERSONAL WEAPONS

Pistols
Sub-Machine Guns
Rifles
Light Support Weapons

PISTOLS

ARGENTINA

9 mm FN High Power pistol

Development

The 9 mm High Power pistol is the standard pistol of the Argentinian armed services, produced under licence from FN Herstal SA at the Argentinian government factory at Rosario. The pistol is very similar to the original Belgian design, but there are dimensional differences outlined in the data table below.

Description

Argentine production covers three distinct variations; the 'Militar' model is the standard fixed sight, blued-finish pattern with plastic anatomical grip surfaces; it is generally the same as the original Belgian design. The Model M90 has a lengthened slide stop lever and a reshaped safety catch, making both easier to operate.

Pistol, Hi-Power, M90

The anatomical grip is fitted and there is a plastic addition to the base of the magazine to give a greater area for the hand. The 'Detective' model has the same changes as the M90 but is shorter and more convenient for concealed carrying.

All models are available in a variety of finishes, including matt or polished chrome, blued, phosphated, Parkerised green or 'military', a combination of phosphating and oven-baked finish.

Data

Cartridge: 9 × 19 mm Parabellum
Operation: recoil, semi-automatic, single action
Method of locking: projecting lugs
Feed: 14-round box magazine
Weight: Militar, 929 g; M90, 968 g; Detective, empty without magazine, 916 g
Length: Militar, M90, 197 mm; Detective, 172 mm
Barrel: Militar, M90, 118 mm; Detective, 93 mm
Rifling: 6 grooves, rh, 1 turn in 250 mm
Sights: fore, adjustable blade; rear, laterally adjustable notch

Manufacturer

Fabrica Militar de Armas Portatiles, 'Domingo Matheu', Avenida Ovidio Lagos 5250, 2000-Rosario.

Status

In production.

Service

Argentinian armed forces.

VERIFIED

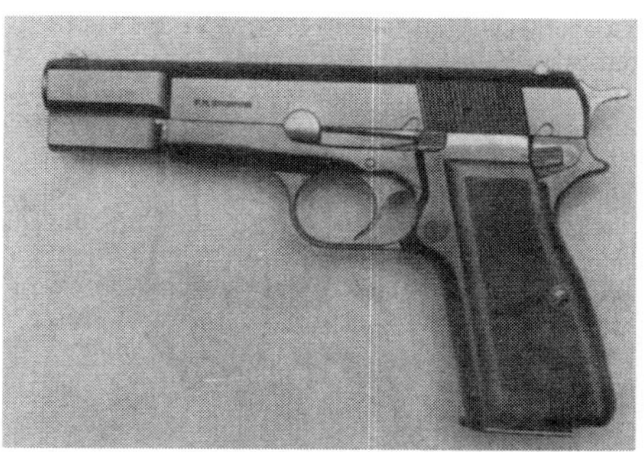

Argentine manufactured Browning GP35 'Militar'

The compact 'Detective' model of the M90 pistol

AUSTRIA

GLOCK semi-automatic Safe Action pistols

Development

The GLOCK pistol is a product of modern technology, incorporating many innovative design features which result in ease and safety of operation, extreme reliability, simple function, reduced maintenance, durability and light weight.

GLOCK pistols were initially adopted by the Austrian Army and Austrian Police. Since then they have been sold and are carried on duty in more than 45 countries worldwide by special forces, police and military, among them approximately 4,000 police departments and agencies in the USA at state, federal and local levels. Approximately 400,000 GLOCK 17 and GLOCK 19 pistols are carried daily in the United States as service hand guns.

The GLOCK pistol trigger mechanism differs from most conventional and traditional double-action/single-action semi-automatic pistols.

All models fire in a smooth self-cocking, constant double-action (Safe Action) mode, with an invariable trigger pull (adjustable between 2 and 5.5 kg) from the first to last shot. With each pull of the trigger the firing pin is fully cocked. When pulling the trigger three automatic safety devices disengage and automatically re-engage by simply removing the finger from the trigger without manipulating any manual safety devices or decocking levers. This enhances the user's ability to gain the highest first hit probability, instinctive reacting/aiming/shooting, and to concentrate fully on tactical tasks.

These unique features, together with others such as fewest number of components (only 33, including magazine components) and the use of high resistant polymer material, have been approved and recognised as the trendsetting pistol technology.

Manufacturer

GLOCK GmbH, Hausfeldstrasse 17, A-2232 Deutsch-Wagram.

UPDATED

9 mm GLOCK 17 and 17L

Description

The 9 mm GLOCK 17 is a locked-breech pistol using a modified Colt-Browning tilting barrel system; connection with the slide is by a squared section around the barrel breech which engages with the slide and ejection port. Firing is by striker, controlled by the operation of the trigger; the pressure on the trigger disengages the trigger safety, then (5 mm of travel) cocks the striker (which is in the half-cocked and secure position at all times when the weapon is ready to fire) and releases two further internal safety devices, the integral firing pin lock and the safety ramp; then pressure releases the striker to fire the pistol. The pressure required to operate the trigger can be adjusted between 2 and 5.5 kg. There is no conventional manual safety catch, since the trigger safety, firing pin safety and drop safety will not permit the firing of a cartridge unless the trigger is correctly pulled. All three safeties work independently and automatically. They disengage one by one when pulling the trigger, and re-engage automatically after each round has been fired and the trigger released.

For the requirements of special forces, the standard GLOCK 17 can be converted into an amphibious version with only slight modifications.

(Left, centre and right): 9 mm GLOCK 19, 17 and 17L pistols compared (not to same scale)

GLOCK 17L

This model is of the same design as the GLOCK 17 except for a longer barrel and slide. Most components are interchangeable with the GLOCK 17. The GLOCK 17L is particularly favoured by sports shooters because of its low standard trigger pull weight (2 kg) and the long sight radius.

Data

DATA FOR GLOCK 17; WHERE GLOCK 17L DIFFERS, SHOWN IN PARENTHESIS:
Cartridge: 9 × 19 mm Parabellum
Operation: short recoil, semi-automatic
Method of locking: tilting barrel
Feed: 17-round box magazine; 19-round optional
Weight: empty, without magazine, 625 g (670 g)
Weight of magazine: empty, 78 g; full, ca 280 g
Length, slide: 186 mm (225 mm)
Barrel: 114 mm (153 mm)
Rifling: hexagonal profile, rh, 1 turn in 250 mm
Sights: fixed or click-adjustable; fore, blade; rear, notch
Sight radius: 165 mm (205 mm)
Muzzle velocity: ca 385 m/s
Muzzle energy: ca 520 J

Manufacturer

GLOCK GmbH, Hausfeldstrasse 17, A-2232 Deutsch-Wagram.

Status

In production.

Service

AQAP, ÖNORM, DIN, CIP and SAAMI standards. Introduced into NATO countries, armed forces, police departments and special forces units in more than 45 countries.

UPDATED

9 mm GLOCK 19 pistol

Description

The 9 mm GLOCK 19 is a smaller and more compact version of the GLOCK 17. Except for different dimensions it is to the same basic design and most components are interchangeable.

Data

Cartridge: 9 × 19 mm Parabellum
Operation: short recoil, self-loading
Method of locking: tilting barrel
Feed: 15-round box magazine; 17-round optional
Weight: empty, without magazine, 595 g
Weight of magazine: empty, 70 g; full, ca 225 g
Length: 174 mm
Barrel: 102 mm
Rifling: hexagonal profile, rh, 1 turn in 250 mm
Sights: fixed or click-adjustable; fore, blade; rear, notch
Sight radius: 152 mm
Muzzle velocity: ca 350 m/s
Muzzle energy: ca 490 J

Manufacturer

GLOCK GmbH, Hausfeldstrasse 17, A-2232 Deutsch-Wagram.

Status

In production.

VERIFIED

9 mm GLOCK 26 subcompact pistol

Description

Introduced during the second half of 1995, the GLOCK 26 subcompact pistol is the smallest of all the GLOCK pistols but retains all the well-proven features of the other members of the GLOCK family.

Data

Cartridge: 9 × 19 mm Parabellum
Operation: short recoil, self-loading
Method of locking: tilting barrel
Feed: 10-round box magazine
Weight: empty, without magazine, 560 g
Weight of magazine: empty, 56 g; full, ca 180 g
Length, slide: 160 mm
Barrel: 88 mm
Rifling: hexagonal profile, rh, 1 turn in 250 mm
Sights: fixed or click-adjustable; fore, blade; rear, notch
Sight radius: 138 mm
Muzzle velocity: ca 340 m/s
Muzzle energy: ca 460 J

Manufacturer

GLOCK GmbH, Hausfeldstrasse 17, A-2232 Deutsch-Wagram.

Status

In production, starting second half of 1995.

NEW ENTRY

9 mm GLOCK 26 subcompact pistol

1996

9 mm GLOCK 18C selective fire sub-machine pistol

Description

The 9 mm GLOCK 18C was developed from the GLOCK 17 by adding a fire selector assembly and offering a larger magazine capacity, so turning it into a submachine pistol capable of automatic fire. It has a built-in compensator with four ports in the barrel and a special slot in the slide, enabling the firer to hold the pistol stable and which assists accuracy when firing in the auto mode.

For security reasons the main components of the GLOCK 18C are not interchangeable with the GLOCK 17.

The GLOCK 18C provides the user with the option of firing trigger-controlled bursts from a fully concealable lightweight pistol without the need for a shoulder stock. The optional 33-round box magazine provides a sufficient amount of immediate firepower.

By simply changing the position of the fire selector the GLOCK 18C may be used in the same semi-automatic mode as the other GLOCK models.

The manufacturer strongly and expressly recommends that only special law enforcement SWAT teams and specially trained military personnel should, as a general rule, be allowed to use the GLOCK 18C, since the effective and safe operation of a pistol in full-automatic mode requires special training and high personal discipline.

9 mm GLOCK 18C with 33-round magazine; note fire selector at rear end of slide

Data

Cartridge: 9 × 19 mm Parabellum
Operation: short recoil, selective fire
Locking system: tilting barrel
Feed: 17/19- or 33-round box magazines
Weight: empty, without magazine, 586 g
Length, slide: 186 mm
Barrel length: 114 mm
Rifling: hexagonal, rh, 1 turn in 250 mm
Sight radius: 165 mm
Rate of fire (theoretical): ca 1,200 rds/min
Muzzle velocity: ca 340 m/s
Muzzle energy: ca 460 J

Manufacturer

GLOCK GmbH, Hausfeldstrasse 17, A-2232 Deutsch-Wagram.

Status

In production.

UPDATED

9 mm/0.380 GLOCK 25 pistol

Description

In order to meet the demands of private firearm users in several parts of the world who value the advantages of the compact 9 mm Short/0.380 Auto cartridge, GLOCK introduced their GLOCK 25 pistol during the second half of 1995. The GLOCK 25 carries over all the well-proven features of other products in the GLOCK family.

Data

Cartridge: 9 mm Short/0.380 Auto
Operation: short recoil, self-loading
Method of locking: tilting barrel
Feed: 15-round box magazine
Weight: empty, without magazine, 570 g
Weight of magazine: empty, 68 g; full, ca 204 g
Length, slide: 174 mm
Barrel: 102 mm
Rifling: hexagonal profile, rh, 1 turn in 250 mm
Sights: fixed or click-adjustable; fore, blade; rear, notch
Sight radius: 152 mm
Muzzle velocity: ca 350 m/s
Muzzle energy: ca 350 J

Manufacturer

GLOCK GmbH, Hausfeldstrasse 17, A-2232 Deutsch-Wagram.

Status

In production, starting second half of 1995.

NEW ENTRY

0.40 GLOCK 22 and 23 pistols

Description

The GLOCK 22 is generally the same as the GLOCK 17, and the GLOCK 23 generally as the GLOCK 19, although both are chambered for the 0.40 S&W cartridge. The chambering is obviously different but they do correspond in frame size.

Data

DATA FOR GLOCK 22; WHERE GLOCK 23 DIFFERS, SHOWN IN PARENTHESIS:
Cartridge: 0.40 Smith & Wesson
Operation: short recoil, semi-automatic
Locking system: tilting barrel
Feed: 15-round (13-round) box magazine
Weight: empty, without magazine, 650 g (600 g)
Length, slide: 186 mm (174 mm)
Barrel length: 114 mm (102 mm)
Rifling: hexagonal, rh, 1 turn in 400 mm
Sight radius: 165 mm (152 mm)
Muzzle velocity: ca 300 m/s (ca 295 m/s)
Muzzle energy: ca 520 J (ca 510 J)

Manufacturer

GLOCK GmbH, Hausfeldstrasse 17, A-2232 Deutsch-Wagram.

Status

In production.

UPDATED

0.40 GLOCK 23 Compact pistol

0.40 GLOCK 27 subcompact pistol

Description

Introduced during the second half of 1995, the GLOCK 27 subcompact pistol bears the same relationship to the other GLOCK 0.40 S&W pistol models as its 9 mm equivalent, the GLOCK 26. The GLOCK 26 and 27 are visually identical (see separate entry for an illustration of the GLOCK 26) and differ only in calibre-related detail. As with the GLOCK 26, the GLOCK 27 retains all the well-proven features of the other members of the GLOCK family.

Data

Cartridge: 0.40 S&W
Operation: short recoil, self-loading
Method of locking: tilting barrel
Feed: 9-round box magazine
Weight: empty, without magazine, 560 g
Weight of magazine: empty, 60 g; loaded, ca 205 g
Length, slide: 160 mm
Barrel: 88 mm
Rifling: hexagonal profile, rh, 1 turn in 250 mm
Sights: fixed or click-adjustable; fore, blade; rear, notch
Sight radius: 138 mm
Muzzle velocity: ca 285 m/s
Muzzle energy: ca 475 J

Manufacturer

GLOCK GmbH, Hausfeldstrasse 17, A-2232 Deutsch-Wagram.

Status

In production, starting second half of 1995.

NEW ENTRY

GLOCK 20 and 21 pistols

Description

Inheriting the same main features of design, material, technology and 'Safe Action' operation, serial production of the GLOCK 20 (in 10 mm Auto calibre) and GLOCK 21 (in 0.45 ACP calibre) began in 1991.

Data

DATA FOR GLOCK 20; WHERE GLOCK 21 DIFFERS, SHOWN IN PARENTHESIS:
Cartridge: 10 mm Auto (0.45 ACP)
Operation: short recoil, semi-automatic
Locking system: tilting barrel
Feed: 15-round (13-round) box magazine
Weight: empty, without magazine, 785 g (745 g)
Length: 193 mm
Barrel length: 117 mm
Rifling: octagonal, rh
Sight radius: 172 mm
Muzzle velocity: ca 370 m/s (250 m/s)
Muzzle energy: 750 J (460 J)

Manufacturer

GLOCK GmbH, Hausfeldstrasse 17, A-2232 Deutsch-Wagram.

Status

In production.

UPDATED

0.45 ACP GLOCK 21, shown with fixed sights

0.40 GLOCK 24 and 24C

Description

The GLOCK 24 is generally the same as the GLOCK 17L, but chambered for the 0.40 S&W cartridge. It is available in two versions, with a solid barrel (GLOCK 24) or a factory-produced compensated barrel (GLOCK 24C).

Data:

DATA FOR GLOCK 24; WHERE GLOCK 24C DIFFERS, SHOWN IN PARENTHESIS:
Cartridge: 0.40 S&W
Operation: short recoil, semi-automatic
Locking system: tilting barrel
Feed: 15-round box magazine
Weight: empty, without magazine, 750 g
Length: 225 mm
Barrel length: 153 mm
Rifling: hexagonal, rh, 1 turn in 250 mm
Sight radius: 205 mm
Muzzle velocity: ca 310 m/s (ca 305 m/s)
Muzzle energy: ca 560 J (ca 545 m/s)

Manufacturer

GLOCK GmbH, Hausfeldstrasse 17, A-2232 Deutsch-Wagram.

Status

In production.

UPDATED

0.40 S&W GLOCK 24

Steyr 9 mm Special Purpose Pistol

Description

The Special Purpose Pistol (SPP) is a locked-breech weapon in 9 × 19 mm Parabellum calibre. There are only 41 component parts and the frame and top cover are made from a plastic material. The locking system relies upon a rotating barrel, using a single lug which moves in a groove in the frame to rotate and thus lock and unlock from the bolt. The cocking handle is at the rear of the weapon, beneath the rearsight, and is pulled back to cock.

Data

Cartridge: 9 × 19 mm Parabellum
Operation: recoil, semi-automatic
Locking system: rotating barrel
Feed: 15- or 30-round magazines
Weight: empty, 1.3 kg
Length: 282 mm
Barrel: 130 mm
Sights: fore, blade; rear, notch
Sight radius: 188 mm

Manufacturer

Steyr-Mannlicher AG, Postfach 1000, A-4400 Steyr.

Status

In production.

VERIFIED

9 mm Steyr Special Purpose Pistol (SPP)
1996

BELGIUM

9 mm BDA 380 double-action pistol

Description

The 9 mm BDA 380 double-action pistol is intended as a general purpose defensive pistol for military and police forces. It is a blowback-operated self-loading pistol. With a light alloy frame and steel slide, the overall weight is kept as low as possible.

The trigger is double action and the pistol can be carried with perfect safety when set on double action. However, there is another safety and setting the manual safety lever will disconnect the trigger from the hammer. To fire, all that is necessary is to move the lever with the thumb and pull the trigger. A particular feature of this safety lever is that it projects on both sides of the slide enabling it to be used by a left- or right-handed shooter without any alteration to the pistol.

The pistol was manufactured in both 7.65 mm and 9 mm Short calibres.

Data

Cartridge: 9 mm Short or 7.65 mm
Operation: blowback

9 mm BDA 380 double-action FN pistol

Feed: 9 mm Short, 13-round box magazine; 7.65 mm, 12-round box magazine
Weight: unloaded, 640 g
Length: 173 mm
Rifling: 6 grooves, rh
Muzzle velocity: 9 mm Short, 270-290 m/s; 7.65 mm, 290-300 m/s

Manufacturer

FN Nouvelle Herstal SA, Voie de Liège 33, B-4040 Herstal.

Status

Production complete.

Service

Belgian and some foreign police forces.

VERIFIED

9 mm Browning High Power pistol

Development

The Browning High Power pistol was designed in 1925 by John M Browning and granted a US Patent in February 1927, some three months after his death. It was produced in 1935 and in the USA is still often referred to as the 1935 Model. When first introduced in that year, two versions were available: the 'ordinary model' had fixed sights and the alternative, which had a tangent rearsight graduated to 500 m and a dovetail slot in the rear of the pistol grip to take a wooden shoulder stock attached to a leather holster, was known as the 'adjustable rearsight model'. In Belgium the pistol is known as the Grande Puissance or GP.

The High Power has been manufactured in Belgium by FN Herstal SA, near Liège, for sale worldwide. It was in service in Belgium, Denmark, Lithuania, Netherlands and Romania. During the Second World War the pistol was made in Liège for the use of the German SS troops and by the John Inglis Company in Toronto for use by Australian, British, Canadian and Chinese troops. It is still in production in India.

Description

The Browning High Power pistol consists of the frame, the slide and the barrel. The barrel fits into the slide and on its underside, beneath the chamber where the barrel is strongly reinforced, is a lug into which is cut a forward and upward sloping camway. A pin is riveted to the frame in such a way that when the barrel moves backwards the pin enters the upward sloping path and pulls the rear of the barrel down. The receiver also has a stop to arrest the slide, the firing mechanism and the magazine well with the magazine-retaining catch. The trigger mechanism consists of a trigger pivoted at its top front edge so that the rear end of the trigger assembly rises when the trigger is pulled. This forces a short straight trigger lever up under the front end of a long horizontal, centrally pivoted sear lever mounted in the slide. The lowering of the rear end of this sear lever pivots the sear out of engagement with the bent of the hammer. The safety catch, mounted on the left of the frame behind the pistol grip, is pivoted to lock the slide and a projection on its lower side fits into the rear of the sear to prevent the release of the hammer. There is no grip safety fitted.

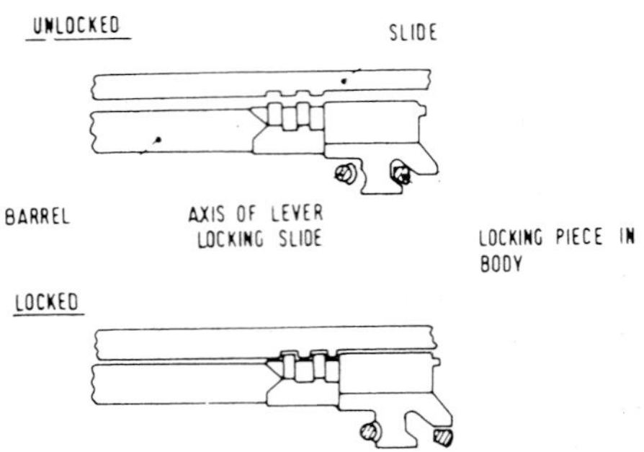

Breech locking and unlocking

The slide contains the integral breech block at the rear. The front end has a single recess for the muzzle and a solid face below in which the return spring is seated. The slide travels on the frame and has an ejection opening on the right side. Within the top of the slide are the recesses in which the barrel locking lugs engage.

On firing, the gas pressure forces the slide rearwards and the barrel is pulled back since its locking lugs are engaged in the recesses in the slide. After 5 mm of free travel the cam path below the barrel rides over the pin and the rear end of the barrel is pulled down to unlock it from the slide, and the barrel movement is terminated. The slide continues back to ride over the hammer and the extractor on the breech block pulls out the empty case which hits the ejector and is thrown out of the ejection port on the right of the slide. As the slide recoils the return spring is compressed around its guide rod. Rearward movement ceases when the slide hits the stop in the frame. The slide is then driven forward by the compressed return spring and the breech block feeds a round from the magazine into the chamber. The breech block forces the barrel forward and the cam forces the breech up and holds it up when the locking lugs on the barrel enter the recesses in the slide ceiling.

When the firer pulls the trigger again, the trigger lever is raised, the tail of the sear lever is forced up and the nose is forced down to contact the sear in the receiver and rotate it out of engagement with the notch of the hammer. The mainspring, lying in the rear of the pistol grip, drives the hammer forward to strike the firing pin driving it forward to hit the cap and fire the cartridge.

When the ammunition is expended, the magazine follower forces up the slide stop and the slide is held to the rear. When a fresh magazine has been inserted into the pistol grip, the slide stop can be pressed down to allow the return spring to force the slide forward. When the magazine is taken out a spring-loaded safety lever is forced out into the magazine well. This lever is linked to the trigger lever and forces it forward from beneath the tail of the sear lever. Thus, when the magazine is removed and a cartridge is left in the chamber, the pistol cannot be discharged inadvertently and the magazine must be replaced before firing is possible.

Data

Cartridge: 9 × 19 mm Parabellum
Operation: recoil, semi-automatic
Locking: projecting lug
Feed: 13-round box magazine
Weight: empty, 882 g
Length: 200 mm
Barrel: 118 mm
Rifling: 6 grooves, rh, 1 turn in 250 mm
Sights: fore, adjustable barleycorn; rear, square notch
Sight radius: 159 mm
Muzzle velocity: 350 m/s
Chamber pressure: 212 MP

9 mm Browning High Power pistol

1996

9 mm Browning High Power pistol

1996

Manufacturer
FN Novelle Herstal SA, Voie de Liège 33, B-4040 Herstal.

Status
In production.

Service
Belgian forces and most Western nations. In use in more than 55 countries.

Licence production
ARGENTINA
Manufacturer: Fabrica Militar de Armas Portatiles, 'Domingo Matheu'
Type: FN High Power
Remarks: See separate entry under *Argentina*.
INDIA
Manufacturer: Rifle Factory, Ishapore
Type: Pistol Auto 9 mm 1A
Remarks: Standard specifications.

UPDATED

9 mm Browning High Power Mark 2 pistol

Description
The Browning High Power Mark 2 pistol is a variant of the standard 9 mm High Power described previously; having the same technical characteristics. It is fitted with anatomical grip plates, an oblong ambidextrous safety catch, wider sights and has an external phosphated anti-glare finish.

Data
Cartridge: 9 × 19 mm Parabellum
Operation: recoil, semi-automatic
Locking: projecting lug
Feed: 13-round box magazine
Weight: empty, 882 g
Length: 200 mm
Barrel: 118 mm
Rifling: 6 grooves, rh, 1 turn in 250 mm
Sights: fore, adjustable barleycorn; rear, square notch
Sight radius: 159 mm
Muzzle velocity: 350 m/s

Manufacturer
FN Nouvelle Herstal SA, Voie de Liège 33, B-4040 Herstal.

Status
Production complete.

Service
Belgian police forces.

UPDATED

9 mm Browning High Power Mark 2 pistol

9 mm Browning High Power Mark 3 pistol

Description
The 9 mm Browning High Power Mark 3 pistol is more or less the Mark 2 with improvements, being manufactured on new machinery which has improved the quality. The frame and slide have been slightly redimensioned to make them stronger and obviate the danger of cracking. The ejection port has been enlarged and recontoured, the grips redesigned, and the sights are now mounted in dovetails for easier adjustment so that the rearsight can be removed and replaced with target sights should the user so wish.

A Mark 3S version is manufactured for police use; this is exactly the same but also incorporates an automatic firing pin safety system. This model is marketed by Browning SA.

Data
Cartridge: 9 × 19 mm Parabellum
Operation: recoil, semi-automatic
Locking: projecting lug
Feed: 13-round box magazine
Weight: empty, 930 g; loaded, 1.085 kg
Length: 200 mm
Barrel: 118 mm
Rifling: 6 grooves, rh, 1 turn in 250 mm
Sights: fore, adjustable barleycorn; rear, square notch
Sight radius: 159 mm
Muzzle velocity: 350 m/s

Manufacturer
FN Herstal Nouvelle SA, Voie de Liège 33, B-4040 Herstal.

Status
In production.

UPDATED

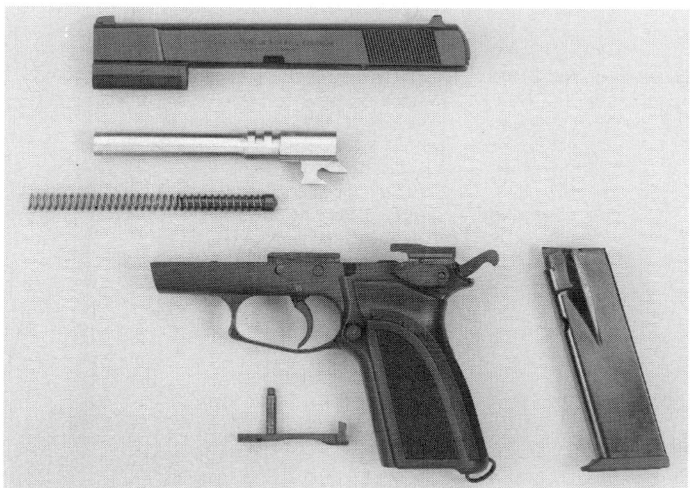

9 mm Browning Mark 3 dismantled

9 mm Browning Mark 3 pistol

9 mm BDA 9 pistol

Description
The 9 mm BDA 9 pistol is derived from its predecessor, the FN/Browning High Power and functions in the same way, using short recoil and a cam beneath the breech to disengage the barrel from the slide.

It differs in having a double-action trigger and a hammer decocking lever in place of the safety catch. This decocking lever is duplicated on both sides of the frame and can thus be used with either hand. The magazine release is normally fitted for right-handed users, but it can be removed easily and reassembled for left-handed use. The trigger guard is shaped to facilitate a two-handed grip.

The pistol is loaded in the usual way, by inserting a magazine, pulling back the slide and releasing it to chamber the first round, leaving the hammer cocked. The pistol can then be fired, or, by pressing down the decocking lever, the hammer can be safely lowered. Pushing down the decocking lever causes a decocking safety to interpose itself between the hammer and the firing pin. There is an automatic firing pin safety system which locks the firing pin except during the last movement of pressing the trigger to fire the pistol. Once the hammer has been lowered, the pistol can be safely carried, and it can be fired instantly by simply pulling through on the trigger for double-action firing. Once a shot has been fired, subsequent shots are in the single-action mode.

Data
Cartridge: 9 × 19 mm Parabellum
Operation: short recoil, semi-automatic
Locking: dropping barrel, cam actuated
Feed: 14-round box magazine
Weight: empty, 915 g; loaded, 1.07 kg
Length: 200 mm
Barrel: 118.5 mm
Rifling: 6 grooves, 1 turn in 250 mm
Sights: fore, adjustable blade; rear, fixed notch, adjustable laterally by an armourer
Sight radius: 160 mm
Muzzle velocity: 350 m/s

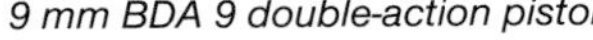

9 mm BDA 9 double-action pistol

9 mm BDA 9 pistol dismantled

Manufacturer
FN Nouvelle Herstal SA, Voie de Liège 33, B-4040 Herstal.

Status
In production.

UPDATED

---

9 mm Browning BDAO pistol

Description

The BDAO (Browning Double Action Only) version was developed especially to meet the new requirements of law enforcement agencies. This pistol contributes to reducing the risk of accidental or non-voluntary discharges so is suitable for non-specialised officers.

The main advantages of the BDAO are: it is always ready and this in total safety for the officer and the environment; no additional manipulations are asked of the shooter other than to draw the weapon, aim and shoot; it minimises the risk of accidental discharge under stress; it gives the same behaviour of the weapon with every shot, giving improved accuracy; and training is simplified, in that there needs to be less technical training leaving more time for behaviour analysis.

The BDAO is to the same general design as the BDA described previously, apart from the self-cocking firing mechanism. There is an automatic firing pin safety which ensures that the pistol cannot possibly be fired unless the trigger is correctly pulled through, and as the slide runs forward after each shot the hammer is positively stopped before reaching the firing pin. The only visible change is the absence of a cocking spur on the hammer.

Data

Cartridge: 9 × 19 mm Parabellum
Operation: short recoil, semi-automatic
Locking: dropping barrel, cam actuated
Feed: 14-round box magazine
Weight: empty, with empty magazine, 870 g
Length: 200 mm
Barrel: 118.5 mm
Rifling: 6 grooves, rh, 1 turn in 250 mm
Sights: fore, adjustable blade; rear, fixed notch, adjustable laterally by an armourer
Sight radius: 160 mm
Muzzle velocity: 350 m/s

Manufacturer

FN Nouvelle Herstal SA, Voie de Liège 33, B-4040 Herstal.

Status

Available.

UPDATED

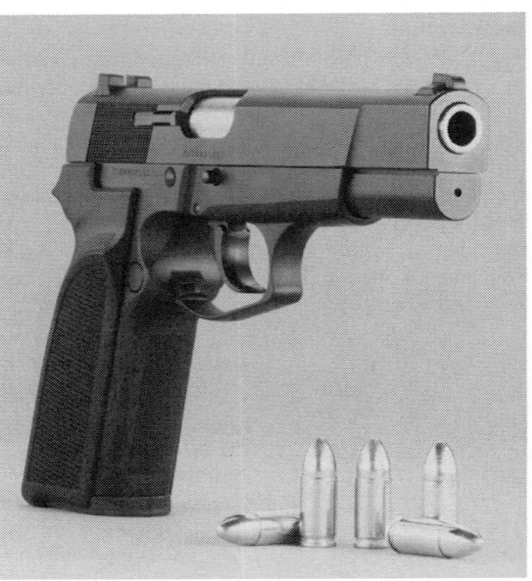

9 mm Browning BDAO pistol

5.7 mm FN Five-seveN personal defence weapon

Development

When it was first announced in August 1995 the 5.7 mm Five-seveN personal defence weapon had been under development for some time as a hand-held adjunct to the FN P90 personal defence weapon (see entry under *Sub-machine guns*) developed in response to the NATO D/296 Personal Defence Weapon (PDW) requirement. The Five-seveN fires the same 5.7 × 28 mm cartridge as the P90.

Many details regarding the Five-seveN have yet to be disclosed and even the limited contents of this description must be taken as provisional. Despite the dimensions of the cartridge involved the pistol is relatively compact for a weapon holding 20 rounds and the length overall is limited to 209 mm. Much use is made of synthetic materials throughout the construction. The method of locking has not been disclosed as it is understood to be entirely novel and many other details relating to the Five-seveN are still lacking.

Data

Cartridge: 5.7 × 28 mm
Operation: self-loading, double action
Method of locking: not disclosed
Feed: 20-round box magazine
Weight: empty, 644 g; loaded, 770 g
Length: 209 mm
Height: 138 mm
Width: 34 mm

Manufacturer

FN Nouvelle Herstal SA, Voie de Liège 33, B-4040 Herstal.

Status

Development.

NEW ENTRY

5.7 mm FN Five-seveN personal defence weapon
1996

BRAZIL

9 mm and 0.45 IMBEL pistols

Description

The IMBEL pistols are redesigned versions of the Colt 0.45 M1911A1. The pistol is made in both 0.45 ACP and 9 mm Parabellum calibres, the former for competition and export, the latter for use by the Brazilian armed forces.

The standard pistol is in all respects similar to the original Colt; the 'TP' competition model uses a special barrel bushing and barrel contour, available only to special order, to improve accuracy.

In 1986 IMBEL developed a 0.45 ACP pocket pistol, based upon the M973 design. Known as the 0.45 ACP IMBEL MD1, this model has a smaller slide and frame than standard and entered series production in 1989.

Data

Cartridge: 0.45 ACP or 9 × 19 mm Parabellum
Operation: recoil, semi-automatic, single action
Feed: 8-round (9 mm) or 7-round (0.45) box magazine
Weight: empty, without magazine; 9 mm, 1.01 kg; 0.45, 1.035 kg
Length: 216 mm
Barrel: 128 mm
Rifling: 9 mm, 6 grooves, rh, 1 turn in 254 mm; 0.45, 6 grooves, lh, 1 turn in 406 mm
Muzzle velocity: 9 mm, 349 m/s; 0.45, 250 m/s

Manufacturer

Indústria de Material Bélico do Brasil - IMBEL, Rua São Joaquim 329, Liberdade, CEP-01508-001, São Paulo SP.

Imbel pistols; from left: M973, M911A1, M911A1(TP)

Status

In production.

Service

Brazilian armed forces and export (see text).

VERIFIED

0.380 (9 mm Short) IMBEL pistol MD1

Description

The MD1 is a blowback pistol firing the 9 mm Short (0.380 Auto) cartridge and is based on the IMBEL M1911A1 0.45 weapon. It uses various components of the M973 and M1911A1 pistols described previously but dispenses with the breech locking system. The size is reduced, making it convenient to carry and handy in operation. An added feature is the positioning of the extractor at the right side of the breech in such a manner that it acts as a loaded chamber indicator.

The pistol is available in various finishes, shown in the picture: (1) black epoxy; (2) black epoxy slide with chromium-plated frame; or (3) stainless steel.

Data

Cartridge: 9 × 17 mm Short/0.380 Auto
Operation: blowback, semi-automatic, single action
Feed: 7-round box magazine
Weight: empty, without magazine, 943 g
Length: 192 mm
Barrel: 104 mm
Rifling: 6 grooves, rh, 1 turn in 254 mm
Muzzle velocity: 270 m/s

Manufacturer

Indústria de Material Bélico do Brasil - IMBEL, Rua São Joaquim 329, Liberdade, CEP-01508-001, São Paulo SP.

0.380 IMBEL MD1 pistol showing various finishes

Status

In production.

VERIFIED

0.380 (9 mm Short) IMBEL Pistol MD2

Description

The 0.380 MD2 is a blowback pistol firing the 0.380 Auto (9 mm Short) cartridge and is based upon the IMBEL M911A1 0.45 calibre weapon. It uses various components of the M973 and M911A1 pistols described previously, but dispenses with the breech locking system. The MD2 uses the same frame as that of the 0.45 M911A1; the slide is reduced, similar to that of the 0.380 MD1 pistol, so making the resulting weapon light and compact to carry and handy in operation. An added feature is the positioning of the extractor at the right side of the breech in such a manner that it acts as a loaded chamber indicator.

Data

Cartridge: 9 × 17 mm Short/0.380 Auto
Operation: blowback, semi-automatic, single action
Feed: 9- or 10-round box magazine
Weight: empty, 1.016 kg
Length: 192 mm
Barrel: 104 mm
Rifling: 6 grooves, rh, 1 turn in 254 mm
Muzzle velocity: 270 m/s

Comparison between IMBEL 0.380 pistols: 1 MD1, 2 MD2

IMBEL 0.380 MD2 pistol

Manufacturer
Indústria de Material Bélico do Brasil - IMBEL, Rua São Joaquim 329, Liberdade, CEP-01508-001, São Paulo SP.

Status
Advanced development.

VERIFIED

0.38 Super IMBEL pistol MD1 and MD2

Description
The design of the MD1 and MD2 0.38 Super pistols is based on the Colt model, and almost all pieces of the M973 9 mm pistol are used. The 0.38 Super model pistol is intended primarily for export and civilian markets. The MD1 model utilises a standard (not ramped) barrel, the MD2 has a fully supported ramp barrel. Other components are common to both models.

Data
Cartridge: 0.38 Super Auto +P
Operation: recoil; semi-automatic, single action
Locking: tilting barrel
Feed: 9-round box magazine
Weight: empty, 1.078 kg
Length: 219 mm
Barrel: MD1 ramped barrel, 128 mm; MD2 plain barrel, 129 mm
Rifling: 6 grooves, lh, 1 turn in 406.4 mm
Muzzle velocity: ca 390 m/s

Manufacturer
Indústria de Material Bélico do Brasil - IMBEL, Rua São Joaquim 329, Liberdade, CEP-01508-001, São Paulo SP.

Status
In production.

VERIFIED

0.38 Super IMBEL MD1 pistol

7.65 mm IMBEL pistol MD1

Description
This is a blowback pistol firing the 7.65 mm Browning (0.32 ACP) cartridge, and is based upon the IMBEL M911A1 weapon. It uses various components of the M911A1 pistol described previously, but dispenses with any form of breech locking. The size is reduced, making it convenient to carry and handy in operation, but it remains rather larger and heavier than the average 7.65 mm pistol, making it accurate and easy to shoot. As with other IMBEL designs, the extractor is positioned at the right side of the breech where it fills the additional role of loaded chamber indicator.

Data
Cartridge: 7.65 × 17 mm Browning/0.32 ACP
Operation: blowback, semi-automatic, single action
Feed: 9-round box magazine
Weight: empty, 943 g
Length: 192 mm
Barrel: 104 mm
Rifling: 6 grooves, rh, 1 turn in 305 mm
Muzzle velocity: 270 m/s

Manufacturer
Indústria de Material Bélico do Brasil - IMBEL, Rua São Joaquim 329, Liberdade, CEP-01508-001, São Paulo SP.

Status
Production on request.

7.65 mm IMBEL pistol MD1

VERIFIED

Pistol conversion kit 0.45/9 mm

This kit is designed to allow conversion of any Colt M1911A1 0.45 pistol to 9 mm Parabellum calibre. The conversion is reversible and requires no modification of the pistol. It can be performed by IMBEL or by the user, IMBEL supplying full instructions with the kit. The kit consists of a replacement slide, barrel, firing pin with spring and retainer, extractor, magazine, slide stop pin and sights, together with replacement pins and cleaning equipment.

Status
Available.

Manufacturer
Indústria de Material Bélico do Brasil - IMBEL, Rua São Joaquim 329, Liberdade, CEP-01508-001, São Paulo SP.

VERIFIED

CHILE

0.38 Special FAMAE revolver

Description
The 0.38 Special FAMAE revolver is a conventional design of solid-frame, double-action six-shot revolver. The cylinder is released by a catch on the left side of the frame and swings out for extraction and reloading. A slightly smaller version in 0.32 Long Colt chambering is also manufactured.

Data
Cartridge: 0.38 Special
Operation: double action revolver
Feed: 6-shot cylinder
Weight: empty, 696 g
Length: 181 mm
Barrel: 63.5 mm
Sights: fore, blade; rear, frame notch

Manufacturer
FAMAE Fabricas y Maestranzas del Ejercito, Avenida Pedro Montt 1568, PO Box 4100, Santiago.

Status
In production.

Service
Chilean armed forces and Gendarmerie.

VERIFIED

FAMAE 0.38 Special revolver

CHINA, PEOPLE'S REPUBLIC

7.62 mm Type 54 pistol

Description
The 7.62 mm Type 54 pistol is a direct copy of the Soviet 7.62 mm Tokarev TT-33 pistol and is the standard pistol in the People's Liberation Army. The Chinese version may be distinguished from the Soviet or Polish equivalents by the serrations on the slide. The Soviet and Polish pistols have a series of alternate wide and narrow vertical cuts and the Chinese pistols have uniform narrow vertical serrations. The Chinese pistol can be further distinguished from the Hungarian Model 48 and Serbian Model M57 (which also have uniform narrow slots) by the Chinese markings on the receiver or top of the slide. The Serbian pistol carries '7.62 mm M57' on the left side of the slide, and the Hungarian Model 48 has an emblem consisting of a star, wheatsheaf and hammer, in a wreath, on the grip.

Data
Cartridge: 7.62 × 25 mm 'P' cartridge
Operation: recoil, semi-automatic
Locking: projecting lugs

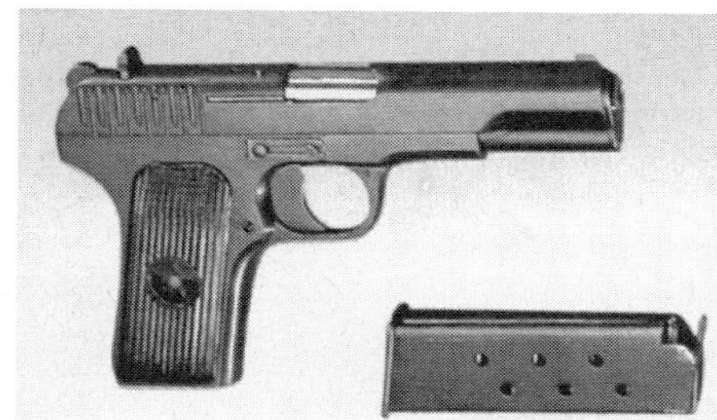

7.62 mm Type 54 pistol

Feed: 8-round box magazine
Weight: empty, 890 g
Length: 195 mm
Barrel: 115 mm
Rifling: 4 grooves, rh
Sights: fore, blade; rear, notch
Muzzle velocity: 420 m/s

Manufacturer
State arsenals.

Status
In production.

Service
Chinese armed forces.

VERIFIED

7.65 mm Type 64 and Type 67 silenced pistols

Description
The Type 64 is a pistol produced solely in silenced form. It may be used either as a manually operated single shot arm or as a self-loader.

When the maximum silencing effect is required the selector bar is pushed to the left and the lugs of the rotating bolt in the slide engage in recesses in the receiver and the weapon fires from a locked-breech. After the round has been fired the slide is hand-operated to unlock the bolt, retract the slide and extract the fired case. When the selector bar is pushed to the right, the locking lugs do not engage in the recesses in the receiver and the pistol functions as a blowback-operated semi-automatic. This results in a more noisy method of operation as the slide reciprocates and the empty case is ejected. The cartridge is 7.65 × 17 mm. It is rimless and unique. No other round can be used in this pistol.

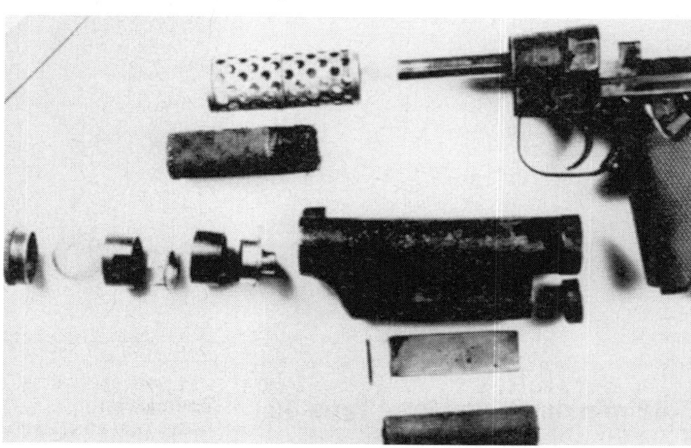

7.65 mm Type 64 pistol: baffles, gauze and mesh make up interior of silencer

7.65 mm Type 64 pistol stripped

The silencing effect is obtained by placing a large bulbous attachment on the front of the receiver extending well forward of the muzzle. The gases leave the muzzle and expand into a wire mesh cylinder surrounded by an expanded metal sleeve. The bullet passes through a series of rubber discs which trap the gases. Used as a single-shot manually operated pistol it is extremely quiet but its reduced muzzle velocity greatly affects its powers of penetration.

The Type 67 is an improved model; in essentials it is the same as the Type 64 above but the silencer has been reshaped into a plain cylindrical unit which makes the weapon easier to carry in a holster and gives it better balance. There are some small changes in the assembly of the silencer but the principle of operation remains the same. It is chambered for the 7.62 mm Type 64 cartridge, a rimless, necked round of reduced velocity.

7.65 mm Type 67 pistol with slide retracted

Data

DATA FOR TYPE 64; WHERE TYPE 67 DIFFERS, SHOWN IN PARENTHESIS

Cartridge: 7.65 × 17 mm rimless (7.62 Type 64 rimless)
Operation: (a) manual, single shot or (b) blowback, semi-automatic
Locking: (a) rotating bolt or (b) nil
Feed: 9-round box magazine
Weight: empty, 1.81 kg (1.05 kg)
Length: 222 mm (226.2 mm)
Barrel: 95 mm (89 mm)
Rifling: both 4 grooves, rh
Sights: fore, blade; rear, notch
Muzzle velocity: 205 m/s (240 m/s)

7.65 mm Type 64 pistol

Manufacturer

State arsenals.

Status

In production.

Service

Chinese armed forces.

VERIFIED

7.62 mm pistol Type 77

Development

The 7.62 mm Type 77 is a light pistol intended for equipping senior officers, military attachés and police. It fires the rimless 7.62 mm Type 64 cartridge, the round used with the Type 64 silenced pistol (see previous entry).

There have been several 'single-handed cocking' pistol designs. This version, one of the few to have any commercial success, was originated in 1913-15 and marketed by Bergmann and later by Lignose in Germany in the early 1920s.

Description

The design of the Type 77 is rather unusual, reviving a long-defunct German system of operation. The pistol is a simple blowback weapon which can be carried with a loaded magazine and empty chamber. When required for use, the trigger finger is hooked around the front edge of the trigger guard and the guard pulled to the rear. This moves the slide back against its return spring and if the finger is then released the slide will run back and chamber a cartridge. By simply transferring the finger to the trigger the weapon is ready to fire almost instantly. The trigger guard is not permanently linked to the slide and so does not move during normal firing. This method of operation also has the advantage that in the event of a misfire the cartridge can be ejected single handed.

The chamber is fluted in an endeavour to reduce the slide recoil velocity by reducing the gas pressure acting on the base of the cartridge case.

Data

Cartridge: 7.62 × 17 mm
Operation: blowback
Feed: 7-round box magazine
Weight: 500 g
Length: 148 mm
Muzzle velocity: 318 m/s

Manufacturer

China North Industries Corporation, PO Box 2137, Beijing.

VERIFIED

7.62 mm pistol Type 77

7.62 mm Type 77 pistol dismantled

7.62 mm machine pistol Type 80

Development

The 7.62 mm machine pistol Type 80 is based on the long-obsolete Mauser Model 712 (System Westinger) machine pistol. This was manufactured from 1932 to 1936, with an estimated 70,000 being supplied to China. Numerous copies were also locally manufactured in the late 1930s, but the design then fell into disuse.

Description

The Type 80 uses the same basic mechanism, an internal reciprocating bolt locked by a plate beneath it, with an external hammer, but has a canted magazine (probably for more reliable feed) and a lengthened barrel. There are also a few minor differences in the contours of the frame and the grip angle has also been improved. A clip-on folding stock is available, as is a bayonet, and with the stock attached the weapon can be used at ranges up to 150 m with good effect.

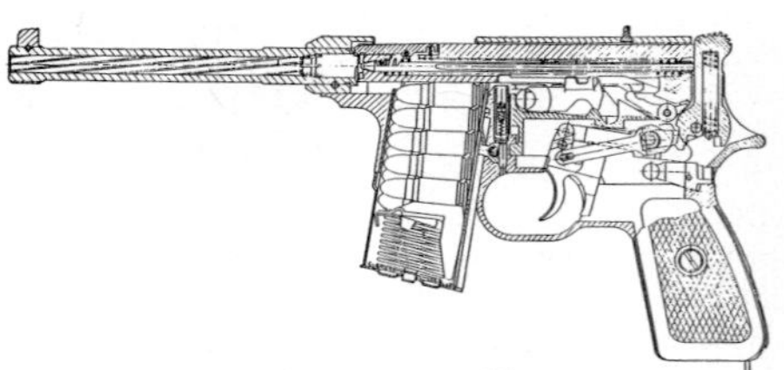

7.62 mm machine pistol Type 80

Data

Cartridge: 7.62 mm Tokarev (Type 51, 7.63 × 25 mm Mauser)
Operation: locked-breech, recoil-operated, selective fire
Feed: 10- or 20-round detachable magazines
Weight: with empty 10-round magazine, 1.1 kg
Length: 300 mm
Rate of fire: effective, 60 rds/min
Muzzle velocity: 470 m/s

Manufacturer

China North Industries Corporation, PO Box 2137, Beijing.

Status

In production.

UPDATED

7.62 mm Type 80 pistol with folding stock attached

9 mm pistol Type 59

Description

The 9 mm Type 59 pistol is a NORINCO manufactured copy of the CIS PM Makarov pistol so is a blowback weapon with a double-action trigger. For further information on the design, refer to the CIS Makarov entry.

Data

Cartridge: 9 × 18 mm Soviet (known as Type 59 in China)
Operation: blowback, double action
Feed: 8-round box magazine
Weight: 730 g
Length: 162 mm
Barrel: 93.5 mm
Rifling: 4 grooves, rh
Sights: fore, blade; rear, U-notch
Sight radius: 130 mm
Muzzle velocity: 314 m/s

Manufacturer

China North Industries Corporation, PO Box 2137, Beijing.

Status

In production.

VERIFIED

9 mm pistol Type 59

NORINCO 0.22 Type 85 Bayonet Pistol

Description

This unusual weapon appears to be a Chinese version of a Spetsnaz pistol/knife. Known as the 0.22 Type 85 Bayonet Pistol it contains four barrels in the grip, chambered for standard 0.22 rimfire cartridges. These barrels are aligned so that two lie each side of the blade with the muzzles at the base of the blade. By pressing in a spring-loaded catch at the rear of the handle, the rear cap is removed, exposing the four chambers. After loading, the cap is replaced and locked in place. There is a serrated rotary catch at the front end of the handle which acts as a safety catch. Rotated clockwise, this locks the trigger (which forms half the handguard of the knife) and makes the pistol inactive so that the knife can be used. When rotated anti-clockwise, the safety catch uncovers a red warning mark alongside the foresight and allows the trigger to function. Pulling the trigger then fires one barrel at a time. To reload, the rear cap is removed and the trigger pulled, whereupon the empty cases are ejected from two barrels; a further pull ejects the cases from the other two barrels. The safety catch is applied and the chambers can then be reloaded.

A notch rearsight is built into the rear end of the handle and a foresight forms part of the handguard. To fire, the pistol is held at arm's length so that aim can be taken.

It is claimed that the projectiles can penetrate pine boards 8.5 mm thick and a covered quilt with cotton wadding 40 mm thick 5 m from the muzzle. It is claimed that the projectile fired from the Type 85 cannot penetrate an aircraft's pressure bulkhead.

The bayonet blade can be folded for concealment.

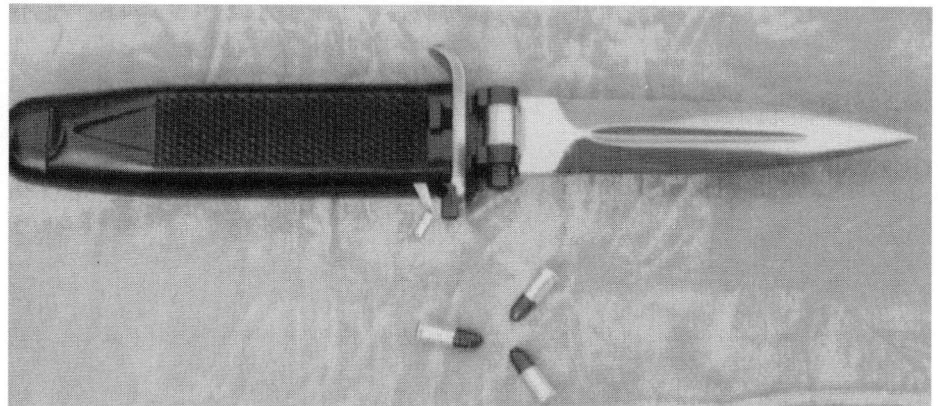

NORINCO 0.22 Type 85 Bayonet Pistol

1996

Data

Cartridge: 0.22 rimfire
Operation: single shot
Feed: 4 preloaded barrels
Weight: empty, 330 g
Length overall: extended, 265 mm; folded, 150 mm
Barrels: 86.3 mm
Rifling: 4 grooves, rh, 1 turn in 224 mm
Muzzle velocity: ca 140 m/s
Effective range: 5-8 m

Manufacturer

China North Industries Corporation, PO Box 2137, Beijing.

Status

In production.

UPDATED

COMMONWEALTH OF INDEPENDENT STATES

5.45 mm PSM pistol

Development

The 5.45 mm PSM pistol was introduced into service in 1975. It is the standard sidearm of all police and internal security forces and also the security element of the armed forces. PSM, stands for 'Pistolet Samozaryadniy Malogabaritniy' ('pistol, self-loading, small'). It fires the 5.45 × 18 mm PMTs bottle-necked cartridge designed by A D Denisova. The features of the pistol and cartridge suggest low performance and an excess of complication for the chosen role. However, tests in the UK have shown that it has a remarkable penetrative ability, defeating standard bulletproof vests with ease, and can penetrate up to 55 layers of Kevlar at a practical range.

5.45 mm PSM pistol, left side (J Lenaerts)

5.45 mm PSM pistol, right side (J Lenaerts)

Description

The PSM resembles the Walther PP (see under Germany) in size and appearance although the general mechanical arrangements are similar to those of the Makarov PM, being a fixed-barrel blowback weapon with double-action lock. If the safety lever is rotated forward to the 'safe' position the hammer will drop onto a block which prevents it from striking the firing pin; the slide is also locked when the lever is on 'safe'. The safety catch is fitted so as to protrude at the rear of the slide; this is in order to reduce the thickness of the pistol as much as possible for concealed carrying (overall width is only 17.5 mm). The barrel is chromed internally.

Examples produced for commercial rather than military sales have moulded plastic grips with a more comfortable outline than the standard flat alloy castings.

Data

Cartridge: 5.45 ×18 mm PMTs
Operation: blowback, double action
Feed: 8-shot detachable box magazine
Weight: empty, 460 g; loaded, 500 g
Length overall: 160 mm
Width: 17.5 mm
Barrel length: 85 mm
Rifling: 6 grooves, rh
Muzzle velocity: 293 m/s

Agency

Rosvoorouzhenie, 18/1 Ovchinnikovskaya Emb, 113324 Moscow.

Status

In production.

Service

CIS and Bulgarian armed forces, police and paramilitary forces. Offered for export sales.

Licence production

BULGARIA
Manufacturer: Arcus Company
Type: 5.45 mm PSM
Remarks: Standard specifications. Offered for export sales.

UPDATED

7.62 mm TT-33 Tokarev pistol

Development

The TT-33 Tokarev is now obsolete in the former Warsaw Pact countries. Production ceased in 1954 in its native country, although it continued for longer than that in Yugoslavia and China. There are therefore many of these pistols still to be found worldwide. The TT-33 was derived from the Browning design in the 1920s at the Tula Arsenal by Feodor Tokarev. He simplified parts of the design and modified others, but the basis was the Model 1911 Colt. The main differences are the lock mechanism, the magazine and the safety arrangements.

Description

The Tokarev was fully described in *Jane's Infantry Weapons 1978*, page 53.

Data

Cartridge: 7.62 × 25 mm Type P
Operation: recoil, single action
Feed: 8-round box magazine
Weight: empty, 0.85 kg

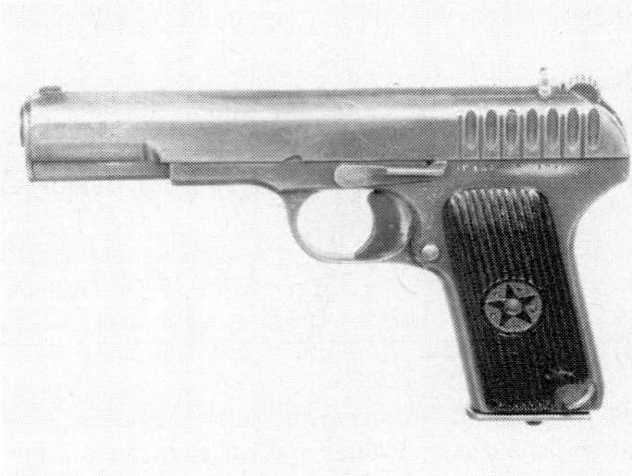

TT-33 Tokarev pistol

Length: 196 mm
Barrel: 116 mm
Rifling: 4 grooves, rh
Muzzle velocity: 420 m/s

Status

No longer in production.

Service

CIS and with many Third World and guerrilla forces.

Licence production

CHINA, PEOPLE'S REPUBLIC
Manufacturer: NORINCO
Type: 7.62 mm Pistol Type 54
Remarks: Production probably complete. Standard specifications.
HUNGARY
Manufacturer: State arsenals
Type: 7.62 mm Model 48
Remarks: Production complete. See separate entry under *Hungary*.
YUGOSLAVIA
Manufacturer: Zastava
Type: 7.62 mm Model M57
Remarks: Probably no longer in production. See separate entry under *Yugoslavia (Serbia and Montenegro)*.

UPDATED

9 mm Stechkin automatic pistol (APS)

Description

The Stechkin (APS) pistol is no longer in service with regular CIS forces, and probably many other forces, but it continues to turn up in many parts of the world. As a pistol it is unusual in having the facility for full automatic fire. Such firing is an almost certain waste of ammunition since the weapon is too light to be controllable, even when fitted with its optional carbine-style wooden butt, and it would seem that the Soviets came to this conclusion and withdrew the weapon.

Mechanically the Stechkin is similar to the Makarov in that it is a blowback pistol firing the standard Eastern Bloc 9 × 18 mm cartridge. The main difference lies in the fact that it has a selector lever which permits single shots or automatic fire, and it is larger and heavier. The design is very complex and expensive to manufacture. The standards of finish and fitting are of the highest and much above that usually found in Soviet weapons.

A variant, the Stechkin APS-B, was produced with an eccentrically bored sound suppressor. When not in use the suppressor could be clipped to a skeleton wire shoulder stock which could be bolted to the butt.

The Stechkin (APS) is fully described in *Jane's Infantry Weapons 1978*, page 54.

Data

Cartridge: 9 × 18 mm Makarov
Operation: blowback, single or double action, selective fire

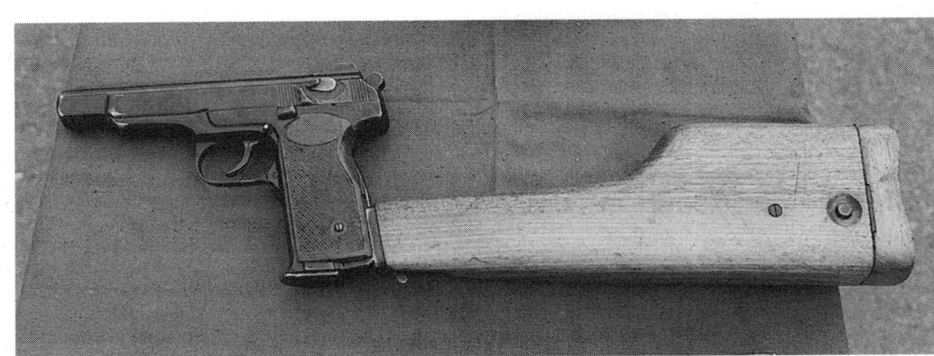

9 mm Stechkin automatic pistol with holster butt

1996

Locking: nil
Feed: 20-round box magazine
Weight: 1.03 kg; with holster stock, 1.58 kg
Length: 225 mm; with stock attached, 540 mm
Barrel: 127 mm
Rifling: 4 grooves, rh
Sights: fore, blade; rear, notch, adjustable by rotating drum to 25, 50, 100 or 150 m
Muzzle velocity: 340 m/s
Rate of fire: cyclic, 750 rds/min; practical, 80 rds/min

Manufacturer

State arsenals.

Status

No longer produced.

Service

May still be encountered, but the pistol was not exported in quantity.

UPDATED

9 mm Makarov pistol (PM)

Development

The 9 mm Makarov (PM) pistol dates from the early 1950s and is the standard pistol for the CIS forces and most ex-members of the Warsaw Pact. It also appeared in several countries which received Soviet military aid.

The Makarov (PM) pistol is a copy of the Walther PP in its general size, shape and handling. It is well made and uses good quality steel, but the handling is considered to be a little awkward by some observers because of the bulky grip.

The Makarov (PM) pistol remains in production both in the CIS and in at least two other countries.

Description

For operation the magazine is inserted into the butt and the pistol is loaded and cocked by drawing the slide back and releasing it. The safety catch is on the slide above the left-hand grip. When the selector is rotated upwards to the 'safe' position the red dot is covered. Setting the safety catch to 'safe' with the hammer

9 mm Makarov self-loading pistol (T J Gander)
1996

cocked will drop the hammer, but the firing pin is blocked. When it is necessary to use the pistol the safety is moved to the 'fire' position, and the trigger can then be operated either in the single-action mode, by cocking the external hammer with the thumb, or in double-action mode by a long pull-through on the trigger which cocks and releases the hammer. When the magazine is empty the slide remains to the rear. It can be released by depressing the slide stop latch on the left side of the frame or by releasing the magazine, then pulling back and releasing the slide.

The trigger and firing mechanism is unusual. Assuming the pistol is fired at double action, sustained pull on the trigger, pivoted about its mid-point, moves the trigger bar forward. The cocking lever is pivoted to enter a notch in the hammer and the further movement of the trigger bar forces the cocking lever round to rotate the hammer back. This movement continues until the cocking lever slips out of the notch in the hammer, taking the sear with it. The hammer is then free to go forward and drive the firing pin forward to fire the cap. This double-action pull is poor from the point of view of the shooter.

The recoiling slide forces the cocking lever sideways, clear of the sear. The sear is then impelled against the hammer by its spring. As the hammer is rocked back by the slide, the sear engages and holds the hammer notch. When the slide runs out the hammer remains held back. The slide completes its forward movement but the trigger is still pressed and must be released. When this occurs the cocking lever rotates forward and moves under the sear. When the trigger is pressed the cocking lever engages the sear and moves it out of engagement with the hammer which swings forward, under the influence of its own spring, to drive the firing pin into the cap.

The safety lever is moved up to the 'safe' position and interposes a block between the hammer and pin; shortly afterwards a projection meets a tooth on the sear and lifts the sear from the hammer. The hammer falls and is locked in its forward position by the safety. When the safety is applied the slide is locked.

It has been stated that the Makarov (PM) can fire 9 × 17 mm (0.380 ACP) ammunition but this is not a recommended practice.

A version of the Makarov with an integral twin-chamber suppressor has been reported and is understood to be designated the PB. This version is 369 mm long overall and weighs about 1 kg empty. The slide has to be actuated manually for each shot.

9 mm Makarov self-loading pistol with holster, magazine and cleaning tool
1996

Data

Cartridge: 9 × 18 mm Makarov
Operation: blowback, self-loading, double action
Locking: nil
Feed: 8-round detachable box magazine
Weight: without magazine, 684 g; empty, 730 g; loaded, 810 g
Length: 161 mm
Height: 127 mm
Barrel: 93.5 mm
Rifling: 4 or 6 grooves, rh
Sights: fore, blade; rear, fixed notch
Sight radius: 130 mm
Muzzle velocity: ca 310 m/s

Manufacturer

Arsenal, Rozova dolina str 100, 6100 Kazanluk.

Agency

Rosvoorouzhenie, 18/1 Ovchinnikovskaya Emb, 113324 Moscow.

Status

In production in Russia and other former Soviet Bloc countries. Offered for export sales.

Service

Former Soviet forces, including the armies of the former Warsaw Pact.

Licence production

BULGARIA
Manufacturer: Kintex
Type: 9 mm Makarov
Remarks: Standard specifications. Offered for export sales.
CHINA, PEOPLE'S REPUBLIC
Manufacturer: NORINCO
Type: 9 mm Type 59
Remarks: Standard specifications. See separate entry under *China, People's Republic*.

UPDATED

9 mm P-9 Gyurza pistol

Description

The 9 mm P-9 Gyurza (Snake, or Viper) pistol falls into the category of a special forces pistol for it fires a special armour-piercing 9 × 21 mm cartridge intended to penetrate body armour. It is also intended for use against materiel such as communications equipment and similar high value targets. The 9 × 21 mm cartridge has a muzzle velocity of about 240 m/s and contains a hard metal insert (possibly tungsten) which enables the projectile to penetrate one or two 1.4 mm titanium plates and 30 layers of Kevlar at ranges up to 100 m, or 4 mm steel plates up to 60 m.

The Gyurza pistol is of advanced design. The frame is a polymer material with steel rails for the slide. A partially concealed hammer indicates that the pistol has both a single and double action. Although it has yet to be confirmed, it is very likely that locking uses barrel locking lugs of the Colt/Browning type. The box magazine holds 18 rounds in a staggered row. Trigger and grip safeties ensure that a positive trigger pull is required to discharge the pistol.

The Gyurza is claimed to be able to produce an 80 mm group at 25 m and a 320 mm group at 100 m. The fixed sights can be used at ranges up to 100 m.

Data

Cartridge: 9 × 21 mm Gyurza
Operation: recoil, semi-automatic, double or single action
Feed: 18-round detachable box magazine
Weight: empty, 995 g
Length: 195 mm
Sights: fore, fixed blade; rear, fixed rectangular notch
Muzzle velocity: 415 m/s

Manufacturer

Tzniitochmash, 2 Zavodskaya Street, Klimovsk 142080, Moscow Region.

Agency

Rosvoorouzhenie, 18/1 Ovchinnikovskaya Emb, 113324 Moscow.

Status

In production. Offered for export sales.

9 mm P-9 Gyurza pistol (T J Gander)
1996

Service

Russian and other CIS special forces.

NEW ENTRY

12.3 mm UDAR revolver

Description

Quite apart from its unusual one-off calibre, the 12.3 mm UDAR revolver is an unusual weapon. It is described as a multipurpose revolver for use by security units, the multipurpose aspect being emphasised by the array of cartridges produced for possible operational use. These include a lethal ball, a plastic baton round, a 'pyroliquid' cartridge which is assumed to be some form of disabling fluid such as CS or a similar behaviour moderation substance, and blank.

This array of possible cartridges is loaded into five-round revolver drums which are loaded into the revolver frame as and when required. The drums can be changed rapidly using a release catch on the left-hand side of the frame. Operation is either single or double action with provision of a shaped trigger guard for two-handed aiming and firing.

The 12.3 mm ball cartridge has an effective range of up to 50 m and is claimed to have twice the lethal power of the 9 × 18 mm Makarov cartridge. Accuracy is such that most bullets will fall within a 40 mm radius at 25 m.

The plastic baton round can strike and partially disable a target without causing serous physical injury at ranges up to 15 m. At 25 m most baton projectiles will

12.3 mm UDAR revolver

1996

maintain sufficient accuracy to fall within a 200 mm radius.

The 'pyroliquid' cartridge will discharge an irritant fluid within a 150 mm radius cone at 4 m.

The blank cartridge is apparently intended to disorientate as well as simulate firing and signalling sounds as the cartridge will create a sound pressure level of not less than 150 dB at 1.5 m.

Data

Cartridge: 12.3 mm (see text)
Operation: double or single action revolver
Feed: 5-round removable drum
Weight: empty, 950 g
Length: 210 mm
Sights: fore, fixed blade; rear, notch

Manufacturer

Tzniitochmash, 2 Zavodskaya Street, Klimovsk 142080, Moscow Region.

Agency

Rosvoorouzhenie, 18/1 Ovchinnikovskaya Emb, 113324 Moscow.

Status

Offered for export sales.

Service

Russian and other CIS special forces.

NEW ENTRY

7.62 mm PSS silent pistol

Description

The 7.62 mm PSS, also known as the Vul, is a small blowback pistol firing the SP-4 7.62 × 42 mm special cartridge capable of generating a sound level comparable to that of an air pistol. The trigger mechanism is single or double action with an external hammer, and there is a slide-mounted safety catch which lowers the hammer safely on a loaded chamber.

There is no conventional silencer in the weapon. Instead, the SP-4 special cartridge contains a piston between the propelling charge and the blunt nosed slug-type bullet which weighs 10 g. On firing, the propelling charge ignites, driving the piston forward; this impels the bullet forward. The piston is then arrested by the cartridge shoulder, which is securely held by a shoulder in the pistol chamber. Thus the noise and flash signatures of the propellant detonation are retained inside the brass cartridge case, and the only noise is that of the air trapped between the piston and the bullet as it escapes from the muzzle.

The bullet has an effective range of 50 m. It can penetrate 2 mm of vertical steel plate or a standard steel helmet at 25 m range and still have sufficient energy to inflict a lethal wound.

7.62 mm PSS silent pistol (T J Gander)

1996

7.62 mm PSS silent pistol (T J Gander)

1996

The PSS can be employed over a temperature range of from −50 to +50°C.

Data

Cartridge: 7.62 × 42 mm SP-4
Operation: blowback, semi-automatic, single or double action
Feed: 6-round detachable box magazine
Weight: empty, 700 g, loaded, 840 g
Length: 165 mm
Height: 140 mm
Width: 30 mm
Sights: fore, fixed block; rear, wide notch
Muzzle velocity: ca 200 m/s

Manufacturer

Tzniitochmash, 2 Zavodskaya Street, Klimovsk 142080, Moscow Region.

Agency

Rosvoorouzhenie, 18/1 Ovchinnikovskaya Emb, 113324 Moscow.

Status

In production. Offered for export sales.

Service

Russian and other CIS special forces.

UPDATED

7.62 mm NRS Scouting Knife

Description

The NRS scouting knife is similar to the Chinese Type 85 described elsewhere in this section insofar as it incorporates a firearm in a utility knife. The handle of the knife carries a chamber and short barrel, into which is loaded one 7.62 × 42 mm SP-4 special cartridge of the type described under the 7.62 mm silent pistol in this section. This is a piston-type cartridge which ejects the bullet by means of a piston which seals in the explosion noise and gas, so that the firearm part of this weapon is virtually silent.

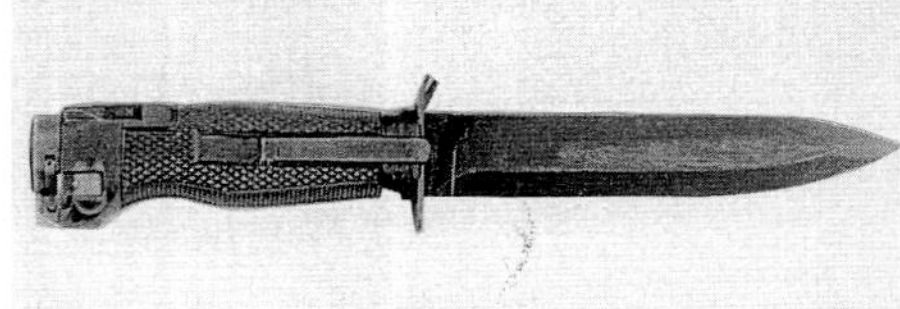

7.62 mm NRS Scouting Knife

The muzzle of the pistol is at the end of the knife handle. The knife is reversed in the hand and fired by pressure in a trigger bar set into the handle. A notch in the crosspiece acts as a rudimentary sight. There is a sliding safety catch which prevents accidental discharges.

The knife is a substantial tool which can cut steel rods up to 10 mm diameter, is insulated to permit cutting electrical cables, and also incorporates a screwdriver. The metal scabbard incorporates a wire cutter and can be used for other purposes.

Data

Cartridge: 7.62 × 42 mm SP-4
Operation: single shot
Weight: knife, 350 g; knife with scabbard, 620 g
Length: 284 mm
Blade: 162 × 28 × 3.5 mm
Effective firearm range: 25 m

Manufacturer

Tzniitochmash, 2 Zavodskaya Street, Klimovsk 142080, Moscow Region.

Agency

Rosvoorouzhenie, 18/1 Ovchinnikovskaya Emb, 113324 Moscow.

Status

In production. Offered for export sales.

Service

Russian and other CIS special forces.

UPDATED

4.5 mm SPP-1 Underwater pistol

Description

The 4.5 mm SPP-1 underwater pistol is a four-barrelled pistol firing a drag-stabilised dart with sufficient power to deliver a lethal effect at 50 m range in air or up to 17 m range underwater, the exact range being dependent upon the depth at which the pistol is fired; at 40 m depth the lethal range is reduced to 6 m. The lethal range involves the penetration of a padded underwater suit or a 5 mm thick glass facepiece.

The dart is part of the special SPS rounds, four of which are held in a special clip which is loaded into the barrels of the pistol from the rear and secured in place by the firing chamber and breech unit. A double-action firing mechanism then fires one cartridge for each pull of the trigger. During reloading, the clip with the spent cartridge cases is automatically ejected.

The 4.5 mm dart is 115 mm long, the complete SPS round being 145 mm long; each round weighs 21 g.

A complete set of SPP-1 equipment includes the pistol, 10 cartridge clips, a holster, a device for loading SPS cartridges into the clips, and a special sling to carry the pistol and three containers with loaded clips.

Data

Cartridge: 4.5 mm SPS
Operation: breech loading four-shot repeater, tip-down barrels
Feed: 4-round clip (see text)
Weight: empty, 950 g; loaded, 1.03 kg
Length: 244 mm

4.5 mm SPP-1 underwater pistol and cartridges (T J Gander)

Height: 138 mm
Width: 25 mm
Effective range: 50 m (in air); 6 m at 40 m depth, varying to 17 m at 5 m depth underwater
Muzzle velocity: in air, 250 m/s

Manufacturer

Tzniitochmash, 2 Zavodskaya Street, Klimovsk 142080, Moscow Region.

Agency

Rosvoorouzhenie, 18/1 Ovchinnikovskaya Emb, 113324 Moscow.

Status

In production. Offered for export sales.

Service

Russian and other CIS special forces.

UPDATED

CROATIA

9 mm HS 95 pistol

Development

The design of the 9 m HS 95 pistol would appear to owe much to that of the Swiss SIG-Sauer 9 mm P226, and some other pistols as well, thereby allowing the manufacturers to bypass a prolonged and expensive development programme. The HS 95 does differ in detail from the P226 in the grip outline, the shape of the trigger guard, the sights, and the machining on the slide, while the grip plates are wood instead of plastic. There are also slight dimensional differences. The double stack box magazine holds 15 rounds.

Although only one illustration has been seen to date, it is assumed that the HS 95 carries over the SIG P226 safety/decocking lever on the left-hand side of the frame. The ambidextrous magazine release catch is certainly present.

Data

Cartridge: 9 × 19 mm Parabellum
Operation: short recoil, self-loading, single or double action
Method of locking: projecting lug
Feed: 15-round box magazine
Weight: 995 g
Length: 185 mm
Barrel: 103 mm
Rifling: 6 grooves, rh
Sights: fore, fixed blade; rear, laterally adjustable notch
Muzzle velocity: 360 m/s

Agency

RH-ALAN d.o.o., Stančićeva 4, 41000 Zagreb.

Status

In production.

Service

Croatian armed forces.

NEW ENTRY

9 mm HS 95 pistol

1996

CZECH REPUBLIC

7.62 mm Model 52 pistol

Description

The 7.62 mm Model 52 (vzor 52) pistol was designed to fire the Model 48 cartridge which has the same physical dimensions as the Mauser 7.63 × 25 mm round, and the Soviet Type P pistol cartridge. The Czech round, however, has a larger propellant load than either the Soviet military round or most commercial cartridges of this type. This is probably the reason for the use of the unusual, but rigid, roller locking system employed.

The roller locking system used is unique to the Model 52 pistol. There is a rectangular barrel lug over the chamber and carried in the lower surface of this is a locking block and two rollers. Two slots in the barrel lug permit the rollers to move outwards. At the rear end of the locking block a tongue is formed and a projection, the unlocking lug, reaches down and rests against a projection in the frame. When barrel and slide are fully forward the locking block forces the two rollers outwards and they enter recesses in the slide, thus locking barrel and slide together. On firing, the barrel and slide begin moving rearwards. The unlocking lug remains in contact with the projection in the frame and so the

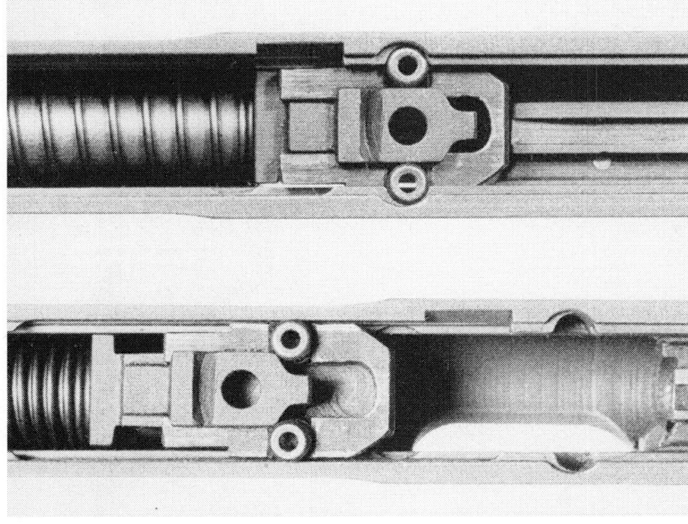

7.62 mm Model 52 pistol's roller locking system

7.62 mm Model 52 pistol (T J Gander)

1996

locking block remains still. After 5 mm of recoil the rollers are carried over the narrow tongue of the locking block and are forced into their housings in the barrel lug by the edges of the recesses in the slide, thus unlocking the barrel from the slide.

The slide continues to the rear, cocking the hammer, and the barrel comes up against the barrel stop in the frame and is halted. One advantage of this system is that the barrel movement is purely one of reciprocation and there is no tilting of the barrel to unlock. Thus there is no need for a loose fit at the front barrel bearing and very little wear apparent, even after a large number of rounds have been fired. This results in a long accuracy life for this weapon.

The backward movement of the slide terminates when the underside of the front barrel bearing in the slide comes up against the receiver and the return spring then forces the slide forward. The feed horn below the breech face picks up a round from the magazine and the breech face pushes it against the bullet guide and into the chamber.

When the recesses in the slide come up to the rollers the latter are forced outwards by the barrel lug, along the tongue of the locking block and in the final 5 mm of forward movement of both barrel and slide, the full width of the locking block holds the rollers so that they are engaging both the slide and the barrel lug. This provides a firm and rigid locking system.

Data

Cartridge: 7.62 mm bottleneck M48, 7.63 mm Mauser, 7.62 mm Type P
Operation: recoil, self-loading, single action
Locking: rollers
Feed: 8-round box magazine
Weight: empty, 960 g
Length: 209 mm
Barrel: 120 mm
Rifling: 4 grooves, rh
Sights: fore, blade; rear, square notch
Muzzle velocity: 396 m/s

Manufacturer

Česká Zbrojovka a.s., CZ-68827 Uherský Brod.

Status

Production ended mid-1950s.

VERIFIED

7.65 mm Model 50 pistol

Description

The 7.65 mm Model 50 pistol is generally similar to the German Walther PP and PPK. The safety catch has been located at the top of the left grip instead of the slide and the trigger guard is integral with the frame, so that dismantling is controlled by a stripping catch on the right side of the frame, ahead of the trigger guard. Together with the later Model 70 it is the main hand gun of the Czech security forces, but is not issued to the Army. Production of the Model 50 ceased in 1969 when it was replaced by the Model 70.

Variant

7.65 mm Model 70 pistol

The Model 70 was a somewhat improved version of the Model 50, which, in its early days, had a poor reputation for reliability. The Model 70 addressed these problems and was a more serviceable weapon. The changes are detail items and the appearance of the pistol is almost unchanged, except for the markings and the patterning on the butt-grips. Production of the Model 70 ceased in 1983.

Data

Cartridge: 7.65 mm/0.32 ACP
Operation: blowback, semi-automatic, double action only
Locking: nil
Feed: 8-round detachable box magazine
Weight: 681 g
Length: 173 mm
Barrel: 97 mm
Rifling: 6 grooves, rh
Sights: fore, blade; rear, notch
Muzzle velocity: 280 m/s

Manufacturer

Česká Zbrojovka a.s., CZ-68827 Uherský Brod.

7.65 mm Model 50 pistol

Status

Production complete (see text).

VERIFIED

9 mm Model 75 pistol

Development

The 9 mm Model 75 pistol is not a development of previous models so it represents a completely new design in which the best features of several other pistols, not necessarily Czech, were incorporated. Standards of workmanship are excellent and the 'balance' of the design is exactly right. The designers are the brothers Josef and František Koucký and they have not only aimed for a good product, but have also looked to the manufacturing processes and the final cost.

Description

The system of operation is by short recoil, the barrel being locked to the slide for sufficient time to allow the bullet to clear the muzzle and the chamber pressure to drop to a safe level. The slide and frame are diecastings, the barrel is forged steel and the grips can be either walnut or plastic. The angle and size of the grip is comfortable and convenient for shooting. The slide catch, safety lever and magazine catch are all placed on the left side of the frame. Sights are available with either a fixed back sight or one on a dovetail which allows for lateral adjustment.

A particular feature of the pistol is the large double-row magazine holding 15 rounds. Another round can be carried in the chamber so that the total load is 16. The firing pin is an inertia type and it is out of contact with the base of the cartridge until fired. Drop tests have shown that repeated falls of 2 m onto concrete, muzzle first, do not fire the pistol. The double action fires the first round without needing the slide to be cocked, hence the pistol can be carried perfectly safely with a round in the chamber and, on drawing out of the holster to shoot, the only action required is to trip the safety lever and pull the trigger.

The pistol is in production but has not been adopted by the Czech armed forces, probably because they hold sufficient stocks of the previous models. It is, however, sold in the West as a police or military arm.

Variants

CZ Model 75 Compact and SemiCompact

These two models are essentially similar to the standard CZ Model 75 but, as their designations imply, are more compact for ease of carrying and handling. Both are 186.3 mm long and the barrel length is reduced to 100 mm. The SemiCompact version retains the 15-round magazine capacity but on the Compact it is reduced to 13 rounds. Weight, empty, of the Compact is 960 g, and 920 g for the SemiCompact.

Data

Cartridge: 9 mm Parabellum
Operation: short recoil
Locking: projecting lugs (Browning)
Feed: 15-round box magazine
Weight: with empty magazine, 1 kg
Length: 206 mm
Height: 138 mm
Width: 33 mm
Barrel: 120 mm
Rifling: 6 grooves, rh, 250 mm pitch
Sights: front, blade integral with slide; rear, square notch
Sight radius: 160.5 mm
Muzzle velocity: 360 m/s

9 mm Model 75 pistol

9 mm Model 75 pistol showing thumb-operated levers

Manufacturer
Česká Zbrojovka a.s., CZ-68827 Uherský Brod.

Status
In production.

Service
Some units of Czech police force. Commercial sales and police use in various countries.

UPDATED

9 mm Model 75 full-automatic pistol

Description
The 9 mm Model 75 retains all the functions of the CZ 75 described previously and can be fired in the semi-automatic mode in either double or single action. In addition, a selector permits firing full automatic at a cyclic rate of about 1,000 rds/min. A shoe in the frame, in front of the trigger guard, permits fitting a laser spot projector or an inverted spare magazine which then acts as a forward handgrip. The additional weight of this magazine, when loaded, will also aid in keeping the muzzle under control during automatic fire.

On early versions of this model the barrel was lengthened, with three slots in the upper surface close to the muzzle, to counter muzzle rise. On later models this feature has been eliminated.

Data
Cartridge: 9 × 19 mm Parabellum
Operation: short recoil, selective fire
Locking: projecting lugs (Browning)
Feed: 15-round box magazine; a 25-round extended magazine is optional
Weight: with empty 15-round magazine, 1.02 kg
Height: 138 mm; with front grip, 160 mm
Width: 38 mm
Rifling: 6 grooves, rh, 250 mm pitch
Sight radius: 160 mm
Cyclic rate of fire: 1,000 rds/min
Muzzle velocity: ca 370 m/s

Manufacturer
Česká Zbrojovka a.s., CZ-68827 Uherský Brod.

Status
In production.

UPDATED

9 mm CZ 75 full-automatic pistol with laser sight
1996

CZ Model 82 and 83 pistols

Description
The CZ Model 82 and 83 pistols are basically identical apart from their calibres. The CZ Model 82 is chambered for the 9 × 18 mm Makarov cartridge. The CZ Model 83 may be chambered for the 9 mm Short/0.380 ACP or the 7.65 mm Browning cartridge. Whereas the CZ Model 82 is intended primarily for military service, the CZ Model 83 is regarded as a police or personal defence pistol.

The CZ Model 82 and 83 pistols are fixed-barrel blowbacks of conventional pattern. Of all-steel construction, they have an ambidextrous safety catch and a magazine catch behind the trigger guard which is operable from both sides of the pistol. The double-action lockwork includes an automatic safety feature which blocks the hammer until the trigger has been fully squeezed. The trigger guard is large enough to accommodate a gloved finger, and also controls dismantling. A sprung detent holds the guard either open or closed and it is interlinked so that the weapon cannot be stripped if the magazine is in place, and the magazine cannot be inserted unless the trigger guard is closed into the frame. The sides of the slide may be polished or matt but the top is always left matt to prevent sight glare.

7.65 mm CZ Model 83 pistol

Data
Cartridge: 7.65 mm ACP; 9 mm Short/.0380 ACP; 9 mm Makarov
Operation: blowback
Feed: 15-round (7.65 mm) or 12-round (9 mm) box magazine
Weight: empty, 7.65 mm, 750 g; 9 mm 800 g
Length overall: 172 mm
Barrel: 96 mm
Rifling: 6 grooves, rh, 1 twist in 250 mm, except in 9 mm Makarov calibre which has 4 grooves, rh.
Sights: fore, fixed blade with white insert; rear, adjustable square notch with 2 white spots
Sight radius: CZ 82, 128 mm; CZ 83 126 mm

Manufacturer
Česká Zbrojovka a.s., CZ-68827 Uherský Brod.

Status
In production.

UPDATED

9 mm CZ Model 85 pistol

Description
The CZ 9 mm Model 85 is an updated version of the Model CZ 75 pistol. In size, shape and general design it retains the best features of the Model CZ 75 although the manual safety and slide stop have been remodelled to permit ambidextrous operation. The top surface of the slide is ribbed, to reduce reflection, and minor internal mechanical changes have been introduced to improve the action and reliability.

The standard Model CZ 85 has fixed sights. A full-automatic version of the CZ 85 similar to the CZ 75 equivalent has been reported but no information is available.

The CZ 85 Combat has adjustable sights to suit it for the competitive combat shooting role. It is also possible to fit a laser sight and spare magazine forward grip to the Model CZ 85 Combat in the same manner as those used on the full-automatic version of the Model CZ 75. However, it is believed that in this configuration the Model CZ 85 Combat remains a semi-automatic weapon only.

The CZ 85 Champion is a competitive target shooting model with numerous modifications for the role. These include a single-action-only adjustable trigger with a revised contour, the ambidextrous slide stop manual safety is no longer ambidextrous, a revised magazine rapid-release is introduced, and a muzzle compensator is provided. There is a 0.40 S&W model of the CZ 85 Champion with a magazine capacity of 12 rounds.

Data
Cartridge: 9 × 19 mm Parabellum
Operation: short recoil, double action
Locking: projecting lugs (Browning)
Feed: 15-round box magazine
Weight: empty, 1 kg
Length: 206 mm
Height: 138 mm
Barrel: 120 mm
Rifling: 6 grooves, rh
Sights: fore, fixed blade; rear, notch
Muzzle velocity: ca 360 m/s

9 mm CZ Model 85 pistol
1996

Manufacturer
Česká Zbrojovka a.s., CZ-68827 Uherský Brod.

Status
In production.

UPDATED

CZ Model 100 and 101 pistols

Description
With the CZ Model 100 and 101 pistols the designers at Česká Zbrojovka have made the leap into the latest generation of pistols, complete with the introduction of ergonomic outlines, polymer materials, advanced safeties and the introduction of the 0.40 S&W cartridge for general use in a pistol from what was once the Eastern Bloc.

The main difference between the CZ Model 100 and 101 is the magazine capacity although both are available chambered for either the 9 × 19 mm Parabellum or 0.40 S&W cartridge. In both calibres the CZ Model 100 holds nearly twice the number of rounds; 13 9 mm rounds on the CZ 100 as opposed to 7 on the CZ 101 and 10 0.40 rounds on the CZ 100 against 6 for the CZ 101.

Both models have the same smooth but solid-looking ergonomic outline with no protruding levers or other impediments to snag clothing or hinder fast drawing from a holster. However, it is intended to be a standard procedure to load the pistol following magazine insertion using one hand only, by pressing the rearsight or a slide protrusion against a fixed edge object to operate the slide. The frame is high impact polymer plastic with the slide being steel. Noticeable features on the frame include the width of the trigger guard.

The CZ 100 and 101 are double-action-only 'load and forget' designs with a manual safety/decocking catch on the frame. Once a cartridge has been loaded into the chamber all springs are released and none remains under tension. The trigger pull weight is given as between 27 and 35 N, considered an optimum pressure by the designers. Other points regarding ease of handling include a low weight and the convenient overall dimensions intended to suit a wide range of hand sizes.

When firing, the grip contours and overall balance of the weapon make the pistol easy to grip and point, improving firing comfort and reducing recoil effects. Aiming is assisted by an illuminated white dot on the fixed foresight and a white outline around the rearsight notch. A further aid to aiming is an optional laser sight module on which the laser beam is switched on as the initial trigger pressure is applied. The laser sight is mounted forward of the trigger guard in slots on the frame.

Data

Cartridge: 9 × 19 mm Parabellum; 0.40 S&W
Operation: short recoil, double action only
Locking: projecting lugs
Feed: CZ 100, 9 mm, 13-round box magazine; CZ 100, 0.40, 10-round box magazine; CZ 101, 9 mm, 7-round box magazine; CZ 101, 0.40, 6-round box magazine
Weight: empty, CZ 100, 680 g; CZ 101, 670 g
Length overall: 177 mm
Barrel: 95 mm
Sights: fore, fixed blade with white insert; rear, adjustable square notch with white outline
Sight radius: 148 mm

Manufacturer

Česká Zbrojovka a.s., CZ-68827 Uherský Brod.

Status

In production from end of 1995 onwards.

NEW ENTRY

0.40 CZ 100 pistol

1996

FINLAND

9 mm Model 35 and Model 40 pistols

Development

In 1935 the Lahti Model 35 replaced the 9 mm Luger which had been the standard pistol in the Finnish Army since its adoption in 1923. The Model 35 was designed by Aimo Lahti and the weapon was manufactured by Valtion Kiväärithedas (VKT) Jyväskylä. It is a mixture of the Luger and the Bergmann-Bayard with a feature from Browning's designs.

This pistol was also used in a slightly modified form in Sweden where it was known as the Model 40. Some of the Swedish weapons, in turn, were taken to Denmark from Sweden by Danish troops after the Second World War and were there taken into service as the Model 40(S).

Description

In appearance the Lahti Model 35 resembles the Luger but its action is different. The barrel is screwed to the slide and the rear end of the slide is enlarged to hold the breech block. The breech block travels in the slide and cocks the hammer as it recoils. It has thumb pieces by which it can be pulled back to the rear to cock the action. When the breech block is fully forward the locking piece is engaged in mortises on each side of the slide and its inner surfaces are engaged in slots in the breech block. Cam faces on the sides of the locking

9 mm Model 35 pistol

piece raise and lower it during rearward and forward travel. One very unusual device is the accelerator which, although a common feature on a short recoil operated machine gun, is unique in a pistol. It is similar to that on the Browning 0.30 calibre pistols and when the barrel ceases its rearward travel the accelerator is rotated and the breech block thrown rearwards with increased velocity. The magazine, like the Luger, has a button which relieves the weight of the spring while loading is carried out. The button operates a pivoted lever so the breech block remains open when the ammunition is expended.

The Lahti Model 35 is extremely well sealed to keep out mud, snow and sand and is very difficult to strip without workshop facilities. It is also very reliable and this probably accounts for the retention in service of a relatively heavy pistol of rather old design.

Data

Cartridge: 9 × 19 mm Parabellum
Operation: short recoil, semi-automatic
Locking: dropping-block
Feed: 8-round detachable box magazine
Weight: 1.22 kg
Length: 246 mm
Barrel: 107 mm
Rifling: 6 grooves, rh
Sights: fore, blade; rear, notch
Muzzle velocity: ca 350 m/s

Manufacturer

Valtion Kiväärithedas Jyväskylä. Formerly made by Husqvarna in Sweden as the m/40.

Status

No longer manufactured.

Service

Finnish armed forces.

VERIFIED

FRANCE

9 mm Model 1950 MAS pistol

Description

The 9 mm Model 1950 MAS pistol was designed at the French Arsenal at St Étienne and manufactured there and at the Chatellerault factory.

The pistol is loaded by removing the magazine, inserting nine 9 × 19 mm Parabellum cartridges and replacing the magazine in the butt. The magazine release catch is on the left side of the receiver, behind the trigger. Pulling back on the rear end of the slide and then releasing it positions the cartridge in the chamber. There is an indicator to the rear of the ejection slot which projects above the slide to indicate that the chamber is loaded.

The safety catch is on the left rear of the slide. When the lever is horizontal the pistol is safe. When it is below the horizontal a red dot is exposed to indicate that this is the 'fire' position. The hammer can be lowered by pulling the trigger if the pistol is set to 'safe' because the safety blocks the hammer from hitting the firing pin.

When the pistol is fired, the gas pressure forces the slide rearwards. The locking ribs on top of the barrel are

9 mm Model 1950 MAS pistol

1996

fitted into the grooves on the inside of the slide and so the barrel goes back with it. The lower rear end of the barrel is attached to the receiver by a swinging link. The barrel and slide move back together for a short distance and then, since the lower portion of the link is attached to the non-recoiling frame, the link pulls the rear end of the barrel down. This separates the locking ribs of the barrel from the recesses inside the slide and when the locking system is disconnected the barrel is at rest and the slide continues to the rear under its own momentum. The empty case comes out on the breech face and stays there until it strikes the fixed ejector. It is then rotated about the extractor and flung out to the right. The hammer is cocked and the return spring compressed.

The return spring forces the slide forward and the face of the breech block carries the top cartridge in the magazine forward into the chamber. The breech block contacts the barrel and pushes it forward where the link raises the breech and the ribs on the top of the barrel enter the grooves on the slide and lock the two together. Forward movement ceases when the lug on the bottom of the barrel contacts the slide stop pin.

Data

Cartridge: 9 × 19 mm Parabellum
Operation: recoil, self-loading, single action
Locking: projecting lug
Feed: 9-round box magazine

Weight: empty, 860 g
Length: 195 mm
Barrel: 112 mm
Rifling: 4 grooves, lh, 1 turn in 254 mm
Sights: fore, blade; rear, square notch
Muzzle velocity: 354 m/s

Manufacturer
Manufacture Nationale d'Armes de Chatellerault; Manufacture Nationale d'Armes de Saint Étienne.

Status
No longer in production.

Service
French and French ex-colonial forces.

UPDATED

9 mm PAMAS G1 pistol

The PAMAS G1 pistol is the Beretta Model 92G manufactured under licence by the Manufacture Nationale d'Armes de Saint Etienne, now part of Giat Industries. It has been adopted by the Gendarmerie Nationale and the French Air Force, and adoption by the French Army is anticipated.

The PAMAS G1 generally resembles the Beretta Model 92F but is fitted with a decocking lever only, rather than the combined safety/decocking lever of the Model 92F. When the decocking lever has been released, after the hammer has been lowered, it springs back into the ready to fire position. There is no manual safety catch.

Data
Cartridge: 9 × 19 mm Parabellum
Operation: short recoil, semi-automatic with single or double action
Locking: falling block
Feed: 15-round detachable box magazine
Weight: empty, 960 g
Length: 217 mm
Barrel: 125 mm
Rifling: 6 grooves, rh, 1 turn in 250 mm
Sights: fore, barrel integral with slide; rear, notched bar dovetailed to slide
Sight radius: 155 mm
Muzzle velocity: nominal, 347 m/s

Manufacturer
Giat Industries, 13 route de la Minière, Satory, F-78034 Versailles Cedex.

Status
In production.

Service
Gendarmerie and French Air Force.

UPDATED

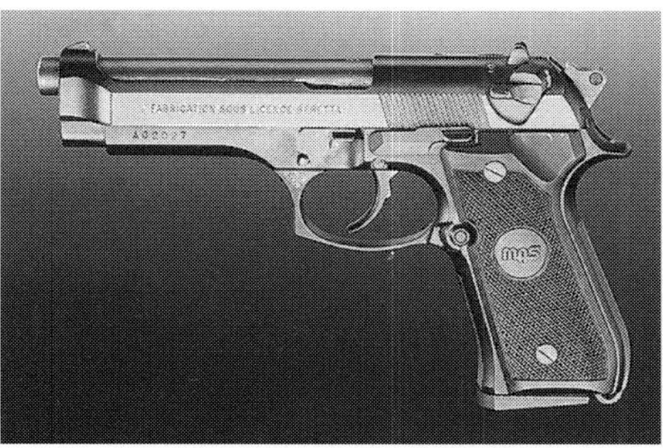

9 mm PAMAS G1 pistol

1996

9 mm PA15 MAB pistol

Development
Manufacture d'Armes Automatiques Bayonne (MAB) manufactured a number of commercial and target pistols. Among these was the 9 mm PA15 MAB used by the French Army. The pistol has the bulky grip required to accommodate 15 9 mm Parabellum cartridges. It has a very prominent spur at the rear of the receiver and a ring-type hammer.

In late 1991 it was announced that the PA15 was being manufactured by Zastava Arms of Serbia for export. See under *Yugoslavia (Serbia and Montenegro)* for further details.

Description
The PA15 has a delayed blowback action relying upon a rotating barrel. The barrel carries two lugs, above and below the chamber. The lower lug engages in a slot cut in the return spring guide, which is pinned to the frame, so that the lug can rotate but cannot move back or forward. The upper lug engages in a track cut in the inner surface of the slide; this track is shaped so that the initial opening movement of the slide will rotate the barrel through about 35°, after which the track is straight so that the slide can recoil while the barrel remains still. Rotation of the barrel is initially resisted by a combination of its inertia and the torque effect of the bullet passing through the rifling. When the chamber pressure has reached a safe level the slide is blown back, compressing the return spring, rotating the hammer and depressing the trigger bar. The empty case is extracted and ejected to the right of the gun.

The return spring forces the slide forward. The top round in the magazine is pushed forward up the bullet guide and enters the chamber. The cam groove on the slide then causes the barrel to rotate and the extractor grips the rim of the case. The trigger bar rises into its recess when the slide is fully forward and the pistol is ready to fire another round.

There is a magazine safety which prevents the hammer going forward when the magazine is out of the gun. There is also a hold-open device.

Stripping
The magazine catch is on the left of the grip immediately behind the trigger. When the magazine is out the slide should be retracted and the chamber and feedway checked.

With the slide held back about 6 mm, the slide stop pin can be pushed to the left and then removed.

The slide can be pulled forward off the receiver. With the slide upside down the return spring guide and barrel seat can be pressed towards the muzzle. They can then be disengaged and lifted out. The barrel will lift out of the slide if it is pushed forward about 8 mm.

9 mm PA15 MAB pistol

Data
Cartridge: 9 mm Parabellum
Operation: delayed blowback, semi-automatic
Locking: barrel lug rotates in cam path in slide
Feed: 15-round box magazine
Weight: 1.09 kg
Length: 203 mm
Barrel: 114 mm
Rifling: 6 grooves, rh
Sights: fore, blade; rear, notch
Muzzle velocity: 350 m/s

Manufacturer
Manufacture d'Armes Automatiques de Bayonne (MAB), Lotissement Industriel des Pontots, F-64100 Bayonne.

Status
No longer made in France, but may be in licence production in Serbia.

Service
French armed forces. May be in service with some former Yugoslav forces.

Licence production
YUGOSLAVIA
Manufacturer: Zastava
Type: P15S MAB
Remarks: Licence production reported to have commenced in 1991. Current status uncertain.

UPDATED

GERMANY

7.65 mm Models PP and PPK pistols

Development
The 7.65 mm Model PP was produced as the 'Police Pistol' in 1929 for carrying on the uniform belt and was widely adopted by police forces. Originally the pistol was made in 7.65 mm calibre but was later constructed in 0.22 Long Rifle (LR) and 9 × 17 mm Short; a small number of 6.35 mm models was also made. Pistols in 0.22 and 6.35 mm calibres have not been made for many years.

In 1931 a smaller pistol called the PPK appeared in the same calibres. This was intended for police use as a concealed weapon and the initials stood for 'Polizei Pistole Kriminal'. This differed in size and also in the construction of the handgrip. In the PP the butt was forged to shape and had simple plastic side plates but in the PPK it was rectangular with a plastic butt providing the contour. Both pistols were blowback-operated and well constructed and finished. They also used a pin, with centre fire cartridges, to indicate a loaded chamber which protruded from the rear of the slide.

After the 1939-45 war the pistols were copied and used by several countries, including Turkey and Hungary. The French firm of Manurhin produced the pistols under licence in 0.22 LR (PP), 7.65 mm (PPK) and 9 mm Short (both).

The South Korean Daewoo 0.22 LR DP52 and 0.380 ACP DH-380 display definite Walther PP design influences.

Description
The Walther PP and PPK are straight blowback pistols with external hammers, double-action triggers and very adequate safety arrangements. The hammer is

7.65 mm Model PPK pistol

9 mm Short/0.380 ACP Walther PPK/S pistol

7.65 mm Model PP pistol sectioned

7.65 mm Model PP pistol

prevented from reaching the firing pin until the sear movement, when the trigger is pulled, moves the block clear. The disconnector works into a recess in the slide and until the disconnector can rise, the sear cannot rotate. This only occurs when the slide is fully forward.

A variant model, the PPK/S, has been manufactured for Walther under licence in the USA for commercial sales. Designed to meet US import regulations, it has the frame of the PP model allied to the barrel and slide of the PPK, the whole weapon being made in stainless steel.

Data

Cartridge: 7.65 mm/0.32 ACP; 9 mm Short/0.38 ACP
Operation: blowback
Feed: 8-round (PP) or 7-round (PPK) box magazine
Weight: PP, 682 g; PPK, 568 g
Length: PP, 173 mm; PPK, 155 mm
Barrel: PP, 99 mm; PPK, 86 mm
Rifling: 6 grooves, rh
Sights: fore, blade; rear, notch
Muzzle velocity: 7.65 mm PP, 290 m/s; PPK, 280 m/s

Manufacturer

Carl Walther Waffenfabrik, PO Box 4325, D-89033 Ulm.

Status

Production in 9 mm and 7.65 mm calibres.

Service

Widespread use by armed forces, police and government agencies in Germany and other European countries.

Licence production

TURKEY
Manufacturer: MKEK
Type: 9 mm and 7.65 mm MKE pistols
Remarks: See separate entry under *Turkey* in this section.

UPDATED

---

9 mm Model P1 pistol

Development

The P1 is the current version of the Walther P38 used by the German Army in the Second World War. After the war the P38 was produced in a lightweight frame and this has been continued in the P1. The difference in markings on the two weapons is small. Both have 'Walther' and then 'Carl Walther Waffenfabrik Ulm/Do'. The Army pistol has 'P1 Cal 9 mm' and the commercial weapons 'P38 Cal 9 mm'.

Description

The main parts of the P1 pistol are the barrel, receiver, slide and the locking lug. There is a double-action trigger mechanism that enables the hammer to be cocked and released by a single trigger pull. The hammer can also be cocked manually to produce a single action with a reduced trigger pull. If the weapon is set to 'safe' with the hammer cocked, the hammer will go forward but the spindle of the change lever locks the firing pin which cannot go forward.

In addition to the standard 9 mm Parabellum calibre, the P1 is also available in 0.22 Long Rifle and 7.65 mm Parabellum calibres.

Data

Cartridge: 9 × 19 mm Parabellum
Operation: short recoil, self-loading, double action
Locking: hinged locking piece
Feed: 8-round box magazine
Weight: empty, 772 g
Length: 218 mm
Barrel: 124 mm
Rifling: 6 grooves, rh, 1 turn in 254 mm
Sights: fore, blade; rear, U-notch
Muzzle velocity: 350 m/s

9 mm Model P1 pistol

Sectional view of P38 pistol

Manufacturer
Carl Walther Waffenfabrik, PO Box 4325, D-89033 Ulm.

Status
Production as required.

Service
Chilean, German, Norwegian, Portuguese and other armed forces and government agencies.

VERIFIED

---

9 mm Model P5 pistol

Development
The Walther 9 mm Model P5 was developed in order to provide a reliable and safe pistol for police and military use. The basic specification was provided by the German police forces, who demanded a very high standard of safety in handling, together with the ability to fire double action without having to release cumbersome safety devices before operating the trigger.

Description
The P5 is a locked-breech recoil-operated pistol of conventional appearance, but with several unusual features incorporated in the lockwork. There are four built-in safety operations:

(a) the firing pin is held out of line with the hammer nose until the hammer is released by the trigger
(b) until the moment of firing the firing pin is held opposite a recess in the face of the hammer. Should the hammer be released by any means other than the trigger, it will not move the firing pin even though it strikes with full force
(c) the hammer has a safety notch
(d) the trigger bar is disconnected unless the slide is fully closed and the barrel locked to it.

The effect of these various safeties is that the hammer can strike the firing pin only when the slide is fully forward and locked to the barrel and when the trigger is pulled through to the limit of its movement. At any other time the pistol is completely safe from such actions as dropping, sharp blows or inadvertent tripping of the hammer when cocking by hand.

A particular feature of the pistol is the decocking lever on the left side of the body, just behind the trigger. The function of this lever, as its name implies, is to take the spring tension off the hammer and moving parts after loading. The pistol is loaded with a magazine in the normal way and the slide allowed to go forward and chamber the first round. The decocking lever is now pushed down with the thumb and allowed to come up. This action operates all the safeties and releases the hammer which moves forward to rest against the rear of the slide with its recess enclosing the projecting end of the firing pin. The firing pin is then completely protected from any external blows and is also held in place by a notch in the pin engaging a lug in the slide. The hammer is held by the safety sear.

When the trigger is pulled through, its first action is to release the safety sear and then to cock the hammer. At full cock a trip is released which moves up to engage with the firing pin at the moment that the hammer moves. As the hammer arrives the pin is waiting for it, ready to be driven forward to fire the cartridge. For single-action shooting the hammer has to be fully cocked by the thumb. This will allow the safety sear to be pushed clear of the hammer breast and free it to move forward. If the hammer is not fully cocked, should the thumb slip off it for instance, the safety sear will catch it and will prevent the firing pin trip from working.

The locking system works by a loose locking piece held between two lugs beneath the barrel. A projection on each side of this locking piece engages with a slot in the slide on each side. When the locking piece is pushed upwards, by a ramp on the body when the slide is fully forward, the barrel and slide are locked together by the projections. Barrel and slide remain locked during the initial recoil movement, but after a short distance a sliding pin in the rear barrel lug meets a vertical face on the body. This pin then pushes against the locking piece, camming it down out of engagement.

In addition to the standard 9 mm Parabellum chambering, the P5 is available in 7.65 mm Parabellum and 9 × 21 mm chambering.

Data
Cartridge: 9 × 19 mm Parabellum; 7.65 mm Parabellum; 9 × 21 mm
Operation: recoil, locked-breech
Locking: loose locking piece
Feed: 8-round detachable box magazine
Weight: empty, 795 g; loaded, 885 g
Length: 180 mm
Barrel: 90 mm
Rifling: 6 grooves, rh
Sights: fore, blade; rear, square notch with white contrast markings. Adjustable for elevation and line
Sight radius: 134 mm
Muzzle velocity: 350 m/s

Manufacturer
Carl Walther Waffenfabrik, PO Box 4325, D-89033 Ulm.

Status
In production.

Service
Adopted by the Netherlands police and the police forces of Baden-Württemberg and Rheinland-Pfalz. Export orders include Nigeria, Portugal, the United States and various South American countries.

VERIFIED

9 mm Model P5 pistol

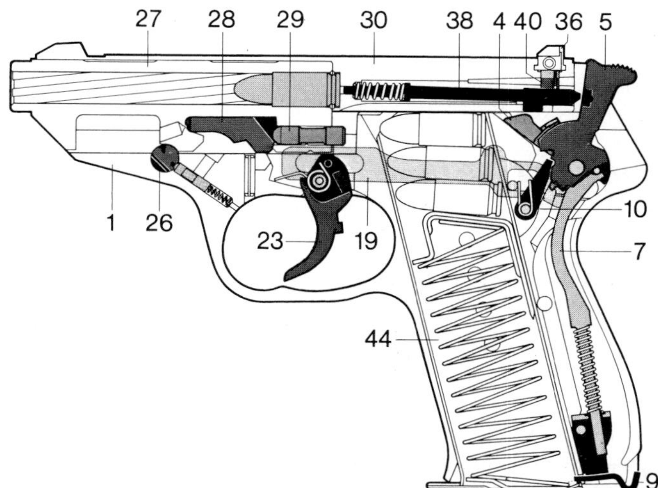

Cutaway view of 9 mm Model P5 pistol, loaded and decocked **(1)** *frame* **(4)** *trip lever* **(5)** *hammer* **(7)** *hammer strut* **(9)** *magazine catch* **(10)** *sear* **(19)** *trigger bar* **(23)** *trigger* **(26)** *barrel catch* **(27)** *barrel* **(28)** *locking piece* **(29)** *locking pin* **(30)** *slide assembly* **(35)** *detent* **(36)** *rearsight* **(37)** *sight adjustment screw* **(38)** *firing pin* **(40)** *insert* **(44)** *magazine*

9 mm Model P5 Compact pistol

Description

The Walther 9 mm Model P5 Compact pistol is essentially the same as the P5 described in the previous entry but is shorter and lighter, making it more convenient for concealed carrying. The size and shape make it well suited to the smaller hand. There is a lateral magazine release which falls conveniently under the thumb when required. The pistol has a light alloy frame, a polished finish and plastic grips, and the hammer has been rounded so as not to catch in clothing when drawn.

Data

Cartridge: 9 × 19 mm Parabellum
Operation: recoil, locked-breech
Locking: loose locking piece
Feed: 8-round detachable box magazine
Weight: empty, 822 g
Length: 181 mm
Barrel: 97 mm
Rifling: 6 grooves, rh
Sights: fore, blade; rear, square notch with white contrast markings. Adjustable for elevation and line
Sight radius: 141 mm
Muzzle velocity: 350 m/s

Manufacturer

Carl Walther Waffenfabrik, PO Box 4325, D-89033 Ulm.

Status

In production.

VERIFIED

9 mm P5 Compact pistol

6.35 mm Walther TPH pistol

Description

The Walther TPH (Taschen Pistole Hahn) is virtually a miniaturised PP pistol and is available in 6.35 mm and 0.22 LR calibres. It is widely used as a police back-up and staff officers' pistol. A simple blowback, it is handy in use and easily concealed. The trigger is double action; applying the safety catch when the hammer is cocked will lock the firing pin securely and then drop the hammer. To fire, all that is required is to push the safety catch up and pull through on the trigger to cock and release the hammer.

Data

Cartridge: 6.35 mm ACP or 0.22 LR
Operation: blowback
Feed: 6-round detachable box magazine
Weight: empty, 325 g
Length: 135 mm
Barrel: 71 mm
Rifling: 6 grooves, rh
Sights: fore, blade; rear, square notch with white contrast markings
Muzzle velocity: 6.35 mm, 232 m/s

Manufacturer

Carl Walther Waffenfabrik, PO Box 4325, D-89033 Ulm.

Status

Production as required.

VERIFIED

0.22 LR TPH pistol; 6.35 mm model identical

9 mm Model P88 Compact pistol

Development

The 9 mm Model P88 Compact pistol is a development of the Model P88 which is no longer in production. It employs a modified Colt-Browning method of breech locking: the squared-off chamber section of the barrel locks into the ejection recess and is released by a cam beneath the barrel striking an actuating lug. Normally supplied in 9 mm Parabellum calibre, it is also available chambered for the 9 × 21 mm cartridge.

Description

The 9 mm Model P88 Compact is a double-action, hammer-fired semi-automatic, with an ambidextrous decocking lever which also functions as the slide release. There is also an ambidextrous magazine catch located in the front edge of the butt, just below the trigger guard. Safety is ensured by a complex firing pin system: the firing pin normally rests at an angle and the corresponding face of the hammer is recessed, so that even if it should fall no pressure will be placed on the firing pin. When the trigger is operated, the rear end of the firing pin is moved up so as to align it with the solid face of the hammer, which then strikes the pin as it falls.

Releasing pressure on the trigger then returns the firing pin to its safe rest position until the next shot is required. There is no way in which the firing pin can be propelled forward by impact or accidental blows.

Data

Cartridge: 9 × 19 mm Parabellum; 9 × 21 mm IMI
Operation: recoil, semi-automatic, double action
Locking: dropping barrel
Feed: 14- or 16-round box magazine
Weight: with empty magazine, 822 g
Length: 181 mm
Barrel: 97 mm
Rifling: 6 grooves, rh
Sights: fore, blade; rear, adjustable square notch

Manufacturer

Carl Walther Waffenfabrik, PO Box 4325, D-89033 Ulm.

Status

In production.

UPDATED

9 mm Model P88 Compact pistol

9 mm Model P88 Competition pistol

Description

The 9 mm Model P88 Competition pistol is based upon the P88 Compact, but with a considerably modified mechanism. It has a tuned single-action trigger and an outside safety lever. With this pistol the hammer is locked in place and not decocked, so that it is possible to carry the weapon 'cocked and locked'.

Data

Cartridge: 9 × 19 mm Parabellum
Operation: recoil, semi-automatic, single action
Locking: dropping barrel
Feed: 14-round box magazine
Weight: 800 g; with long barrel, 845 g
Length: 188 mm; with long barrel, 211 mm
Barrel: 100 mm or 125 mm
Muzzle velocity: 341 m/s; with long barrel, 350 m/s

Manufacturer

Carl Walther Waffenfabrik, PO Box 4325, D-89033 Ulm.

Status

In production.

VERIFIED

9 mm Model P88 Competition pistol

Heckler and Koch HK4 pistol

Development

The HK4 Heckler and Koch pistol is a self-loading, double-action pocket pistol designed for easy conversion from centre fire to rimfire, or vice versa, through a variety of calibres: 9 mm Short (0.380), 7.65 mm (0.32), 6.35 mm (0.25) and 0.22 Long Rifle (LR).

Description

The safety catch, on the left side of the receiver, is down for 'safe' and up for 'fire'. A white spot indicates 'safe', a red spot 'fire'. The magazine is slotted to show the number of rounds loaded. To chamber a round, the slide must be retracted and then released. When a round is in the chamber, the extractor on the right of the slide is proud and this can be seen by day and felt with the forefinger at night. When the pistol is set to 'safe' the hammer can be uncocked by pulling the trigger. The adoption of the 'safe' position blocks the firing pin from the hammer. When the magazine is emptied the slide remains to the rear. Putting in a loaded magazine releases the slide stop and the slide goes forward to chamber a round.

To change from one centre fire barrel to another of a different calibre is merely a matter of changing barrels, springs and magazines. To change from centre to rim fire the firing pin must also be realigned. This is done by removing the barrel, holding the extractor clear of the breech face plate with a pin or a nail and removing the breech face plate. When using 0.22 Long Rifle rim fire the face plate is turned so that the side marked 'R' is showing. The firing pin protrudes from the top hole. When converting back to centre fire the plate marking 'Z' is to the front and the firing pin comes through the lower holes.

Data

COMMON TO ALL CALIBRES
Operation: blowback, self-loading, double action
Locking: nil
Feed: box magazine
Weight: empty, 520 g
Length: 157 mm
Barrel: 85 mm
Rifling: 6 grooves, rh, 1 turn in 254 mm
Sights: fore, blade; rear, U-notch

Data

Calibre	Magazine capacity	Muzzle velocity
0.22 LR	8	300 m/s
0.25 Automatic Colt Pistol (ACP) (6.35 mm)	8	257 m/s
0.32 ACP (7.65 mm)	8	302 m/s
0.380 ACP (9 mm)	7	299 m/s

Manufacturer

Heckler and Koch GmbH, D-78722 Oberndorf-Neckar.

Status

No longer in production.

Service

Wide commercial sales and some military use.

VERIFIED

Heckler and Koch HK4 pistol

Cutaway view of Heckler and Koch HK4 pistol

---

9 mm Heckler and Koch P9S pistol

Description

The 9 mm Heckler and Koch P9S self-loading pistol is a double-action weapon. The barrel is polygonally rifled, which permits a somewhat higher muzzle velocity than obtained with normal rifling, reduces barrel wear and bullet deformation and also reduces the accumulation of fouling.

The safety catch on the rear of the left side of the slide is pressed down for 'safe'. This uncovers a white spot. When rotated clockwise up to the horizontal position the weapon is set to 'fire' and a red spot is revealed.

To remove the magazine for loading, the magazine catch, below the pistol grip, is pushed to the rear. The magazine will then be eased out and can be withdrawn. The magazine is inserted into the pistol grip until the catch engages. To place a cartridge in the chamber the slide is pulled fully back and then released. This also cocks the hammer. When the hammer is cocked, a pin protrudes from the rear of the slide when the latter is fully forward. When a cartridge is chambered the extractor stands proud. Both these indicators can be felt at night and seen by day. When the ammunition is expended the pistol will cease firing with the slide to the rear. As soon as a loaded magazine is inserted the slide may be released by pressing down on the cocking lever on the left side of the receiver behind the trigger or by pulling the slide back and letting the return spring drive it forward. As the slide goes forward a round is chambered and the hammer is left cocked. The cocked hammer may be released by first setting the safety, pressing down the cocking lever, pulling the trigger and holding it back while the cocking lever is allowed to rise again and releasing the trigger. After disengaging the safety, the model P9S can be fired from this position by a long trigger pull.

Although the main production was in 9 mm Parabellum calibre, an export version for the USA was manufactured in 0.45 ACP. For a short time some were made in 7.65 mm Parabellum.

The pistol operates by delayed blowback. The method of connecting the slide to the barrel is to use a two-part breech block consisting of a bolt head containing the two rollers and a heavy bolt body which, by means of angled faces, forces the rollers out into barrel extensions. This method is derived from Heckler and Koch's G3 rifle and a more detailed account of the action will be found in the relevant entry. When the pistol fires, the gas pressure forces back the breech face but movement is severely limited because the projecting rollers are engaged in recesses in the barrel extension. The rollers must be free of these seatings before the breech face can move back significantly. The reaction of the recesses drives the rollers inwards but their inward movement is resisted by the angled faces of the bolt body and the strength of the return spring. The velocity ratio obtained by the angles of the recesses in the barrel extension and the angle of the faces of the bolt body results in the heavy bolt body having a rearward movement four times as great as the bolt face. Eventually the rollers are forced fully in, the bolt is then blown back by the residual pressure and the empty case comes out, held to the breech face by the extractor. The hammer is rotated back by the bolt body, the empty case is thrown out by the ejector and the rearward movement of the bolt ceases on contact with the plastic buffer. The return spring around the barrel expands and the slide goes forward, feeding a round into the chamber. To fire another round the trigger must be released to allow the disconnector to bear on the sear which is engaged in the hammer notch.

Heckler and Koch 9 mm P9S pistol

Data

Cartridge: 9 × 19 mm Parabellum or 0.45 ACP
Operation: delayed blowback, self-loading, double action
Locking: rollers
Feed: 9-round (9 mm) or 7-round (0.45 ACP) box magazine
Weight: 9 mm empty, 880 g; 0.45 ACP empty, 750 g
Length: 192 mm
Barrel: 102 mm
Rifling: polygonal, rh
Sights: fore, blade; rear, square notch
Sight radius: 147 mm
Muzzle velocity: 9 mm, 351 m/s; 0.45 ACP, 260 m/s

Manufacturer

Heckler and Koch GmbH, D-78722 Oberndorf-Neckar.

Status
No longer in production by Heckler and Koch. Still in licence production in Greece.

Service
German police forces. Military and police forces in many other countries. Wide commercial sales.

Licence production
GREECE
Manufacturer: Hellenic Arms Industry (EBO) SA
Type: 9 mm EP9S
Remarks: Standard specifications. 9 mm version only.

UPDATED

Heckler and Koch 9 mm P7 pistol

Development
The 9 mm P7 pistol was developed by Heckler and Koch with the requirements of police forces primarily in mind. It is blowback operated, with a recoil braking system which delays breech opening, reduces the felt recoil and aids steadier shooting.

Description
The action of the P7 is self-locked by the gas pressure developed when a round is fired. When the pistol is fired, part of the propellant gas is channelled through a small vent in the barrel ahead of the chamber and into a cylinder lying beneath the barrel. A piston, attached to the front end of the slide, enters the front end of this cylinder, and thus when the slide begins to move rearward under the recoil pressure, the movement of the piston in the cylinder is resisted by the gas pressure. This delays the movement of the slide, and hence delays the opening of the breech; it also tends to absorb some of the recoil shock. This system gives the advantage of a fixed barrel and does away with the need for a locking mechanism.

The pistol is loaded in the conventional way, by pulling back and releasing the slide. When the firer grasps the pistol, his fingers automatically depress the squeeze-cocking grip at the front of the pistol grip. This cocks the firing pin ready for the first shot, and it remains engaged in the cocked position so long as light pressure is maintained. As soon as the pistol is released, however, the cocking grip snaps forward, automatically decocking the firing pin. Should the weapon be dropped, it will be uncocked and safe before it hits the ground. The cocking grip also releases the slide stop after a fresh magazine has been inserted. Since there is no slide release lever, nor any safety catch to be manipulated, the P7 can be used with equal facility by right- or left-handed firers.

Heckler and Koch 9 mm P7 M8 pistol

Heckler and Koch 9 mm P7 M13 pistol

1996

Although the pistol grip is well angled at about 110° to the axis of the bore, the magazine enters almost vertically, providing optimum feed for the cartridges, even when using ammunition with unusual bullet configurations. This makes the P7 much less liable to feed malfunctions than other weapons. Should a misfire occur, the firer merely releases and recocks the grip, then pulls the trigger for a second attempt; there is no need to use two hands to recock. The pistol can be silently decocked by simply pulling back the slide about 10 mm, releasing the cocking grip, and then allowing the slide to go forward.

The rearsight is adjustable for windage; adjustment for elevation is done by interchanging the foresight blade. These blades are available in different heights, and have their height engraved on the base. Both rear notch and fore blade are fitted with durable contrasting white markers for shooting in poor light, and Betalight luminous markers are available if required.

There are two distinct models of the P7; the P7 M8 which takes an 8-round magazine and the P7 M13 which takes a 13-round magazine. This involves some minor differences, which are noted in the following data table in parenthesis for the P7 M13.

Data
Cartridge: 9 × 19 mm Parabellum
Operation: delayed blowback, self-loading
Feed: 8-round (13-round) box magazine
Weight: loaded, 950 g (1.135 kg)
Length: 171 mm (175 mm)
Barrel: 105 mm
Sights: fore, blade; rear, notch (see text)
Sight radius: 148 mm
Muzzle velocity: ca 351 m/s

Manufacturer
Heckler and Koch GmbH, D-78722 Oberndorf-Neckar.

Status
In production.

Service
German police, special forces and army; US police forces; military and police forces in many other countries, and wide commercial sales.

Licence production
GREECE
Manufacturer: Hellenic Arms Industry (EBO) SA
Type: 9 mm EP7
Remarks: Standard specifications. See also separate entry under *Greece* in this section.

VERIFIED

Heckler and Koch 0.40 S&W P7 M10 pistol

Description
The Heckler and Koch P7 M10 is the same basic design as the Heckler and Koch P7 pistols but is chambered for the 0.40 S&W pistol cartridge. The frame is identical to that of the 9 mm P7 M13 but the slide is heavier and larger in order to handle the more powerful cartridge. It is otherwise identical to the P7 M13 in external appearance and operation, delivers superior ballistics and carries 11 rounds (10 in the magazine and one in the chamber) in complete safety.

Manufacturer
Heckler and Koch GmbH, D-78722 Oberndorf/Neckar.

Status
No longer in production.

VERIFIED

Heckler and Koch 0.40 S&W P7 M10 pistols in nickel and standard blue finish

Heckler and Koch 9 mm Short/0.380 P7 K3 pistol

Description

The Heckler and Koch P7 K3 is the smallest model in the P7 series, and its design and operation are broadly similar to the earlier P7 M8 and P7 M13 models. The principal differences are that the P7 K3 is a blowback weapon without the gas delay system, using the less powerful 9 mm Short (0.380 ACP) cartridge, and conversion kits are available to allow firing 0.22 Long Rifle or 7.65 mm (0.32 ACP) cartridges.

The P7 K3 is simple to operate. The filled magazine is pushed into the closed pistol until the magazine catch engages. The slide is fully retracted and then allowed to snap forward, chambering the top round from the magazine.

Instead of a conventional double-action trigger the P7 K3 features a squeeze cocker. When the user grasps the pistol in the firing position, his fingers simultaneously depress the squeeze cocker, automatically cocking the firing pin for the first shot. This cocking grip remains engaged in the cocked position for subsequent shots, but as soon as the grip on the pistol is released it snaps forward and automatically uncocks the firing pin. If a round fails to fire, the user simply eases his grip, squeezes again to recock the firing pin and pulls the trigger. The pistol may be silently uncocked by pulling the slide approximately 10 mm to the rear, releasing the squeeze cocker, and manually letting the slide go forward.

This system gives light trigger pull from the first shot onward, contributing to accuracy. It eliminates the need for a conventional firing pin system incorporating a hammer, which is more expensive, complex and space-consuming. There is no need for a lateral slide catch and release lever; thus the P7 K3 is equally suited to right- and left-handed users. It is an important safety feature for, should the pistol be dropped, it is uncocked before it strikes the ground.

Although the pistol grip is at the ergonomically favourable angle of 110° relative to the barrel, it has been possible to position the magazine almost vertically to the barrel. This provides optimum cartridge feed from magazine to chamber, even when special ammunition with unconventional bullet configurations is used. The P7 K3 is claimed to be much less susceptible to malfunctions than conventional pistols.

After the last round has been fired, the slide remains open. To continue firing, the empty magazine is replaced with a filled one and then either the squeeze cocker is depressed thus snapping the slide forward and automatically cocking the firing pin, or the slide is pulled back and allowed to snap forward. The pistol is then ready to fire again. The P7 K3 may be unloaded by actuating the ambidextrous magazine release, removing the magazine and fully retracting the slide to eject the chambered cartridge. After ensuring that the chamber is clear, the slide is allowed to snap forward again. The catch lever at the rear of the trigger guard may be used to keep open the action without a magazine being inserted.

Data

Cartridge: 9 × 17 mm Short/0.380 ACP (also see text)
Operation: blowback, self-loading, single action
Feed: 8-round detachable box magazine
Weight: empty, 750 g
Length: 160 mm
Barrel: 96.5 mm
Sights: fore, blade; rear, notch
Sight radius: 139 mm

Manufacturer

Heckler and Koch GmbH, D-78722 Oberndorf-Neckar.

Status

In production.

Service

Military and police forces in various countries and wide commercial sales.

VERIFIED

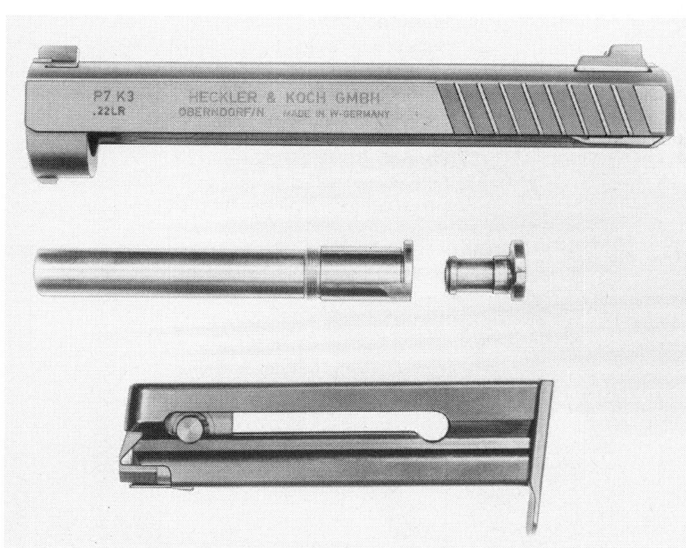

0.22 LR conversion kit for the P7 K3 pistol. A similar conversion kit for 0.32/7.65 mm is also available

Heckler and Koch 9 mm/0.380 P7 K3 pistol

Heckler and Koch USP pistols

Development

The Heckler and Koch USP (Universal Self-loading Pistol) was designed to incorporate features demanded by civil, law enforcement and military users and was introduced to the market in 1993. The USP can be safely carried 'cocked and locked' or, by use of the control lever, a combination safety and decocking lever can be used in single- or double-action modes. The frame-mounted control lever has a positive stop and returns to the 'fire' position after decocking.

The USP was designed from the outset for the 0.40 S&W cartridge but has also been produced in the widely used 9 mm Parabellum. As the USP was intended for the 0.40 S&W, 9 mm models can handle the latest and more powerful 9 mm rounds. In 1995 a slightly larger and heavier 0.45 ACP version of the USP was introduced for the American market and is in production.

Description

Using a modified Browning action with a patented Heckler and Koch recoil reduction system the USP was developed specifically for the 0.40 Smith & Wesson cartridge, thus giving an ample margin of strength and permitting the pistol to be chambered for the 9 mm Parabellum cartridge as an option. The polymer frame was designed with the benefit of experience gained with early synthetic designs such as the VP70 and P9S pistols. Metal components are corrosion resistant, with outer surfaces protected by an extremely hard nitro-gas carburised black oxide finish. Internal metal parts, including springs, are coated with a special Dow Corning anti-corrosion process which reduces friction and wear. From 1995 onwards all USP pistols have polygonal rifling for their chromium steel barrels.

Heckler and Koch USP pistols, from top, 0.45 ACP, 0.40 S&W, 9 mm Parabellum

1996

By using a modular approach to the internal components, the control lever of the USP can be switched from the left to the right side for left-handed firers. The pistol can also be converted from a conventional double-action lock to a self-cocking (double-action-only) lock. Potential users are thus provided with the choice of no fewer than nine variants with differing control lever configurations, as follows:

Variant 1. Double action/single action with 'safe' position and control lever (manual safety/decocking lever) on the left side of the frame.

Variant 2. Double action/single action with 'safe' position and control lever (manual safety/decocking lever) on the right side of the frame.

Variant 3. Double action/single action without 'safe' position and control lever (manual safety/decocking lever) on the left side of the frame.

Variant 4. Double action/single action without 'safe' position and control lever (manual safety/decocking lever) on the right side of the frame.

Variant 5. Double action only with 'safe' position and control lever (manual safety) on the left side of the frame.

Variant 6. Double action only with 'safe' position and control lever (manual safety) on the right side of the frame.

Variant 7. Double action only without control lever (no safety/decocking lever).

Variant 8. Designation not officially assigned although it was temporarily applied to a Variant 7 tested by a US Government agency.

Heckler and Koch 0.45 USP pistol **1996**

Heckler and Koch 0.45 USP pistol fitted with Universal Tactical Light (UTL) **1996**

Variant 9. Double action/single action with 'safe' position and control lever (manual safety/no decocking lever) on the left side of the frame.

Variant 10. Double action/single action with 'safe' position and control lever (manual safety/no decocking lever) on the left side of the frame.

All variants are available in 0.40 S&W, 9 mm Parabellum and 0.45 ACP.

The recoil reduction mechanism is incorporated into the recoil/buffer spring assembly located below the barrel. Designed primarily to buffer the slide and barrel and reduce recoil shock to the pistol components, the system also lowers the recoil forces felt by the firer and aids improved accuracy. The system is insensitive to ammunition variations and requires no adjustments or special maintenance.

The forward end of the frame is grooved for fitting a Heckler and Koch Universal Tactical Light (UTL) laser spot projector (weight 150 g). For competition shooting a Heckler and Koch 'HK Quik-Comp' muzzle brake/compensator is available, as is a scope mount.

Data

DATA GIVEN FOR 0.40; DATA FOR 9 mm IN PARENTHESIS

Cartridge: 0.40 S&W (9 × 19 mm Parabellum)
Operation: short recoil, semi-automatic
Locking: dropping barrel
Feed: 13- (15-) round box magazine
Weight: 780 g (750 g)
Length: 194 mm
Barrel: 108 mm
Rifling: polygonal, 1 twist in 380 mm
Sights: fore, blade; rear, square notch, adjustable for windage and elevation; 3-dot (tritium sights optional)
Sight radius: 158 mm
Muzzle velocity: ca 285 m/s

Cartridge: 0.45 ACP
Operation: short recoil, semi-automatic
Locking: dropping barrel
Feed: 10- or 12-round box magazine
Weight: 840 g
Length: 200 mm
Barrel: 112 mm
Rifling: polygonal, 1 twist in 406 mm
Sights: fore, blade; rear, square notch, adjustable for windage and elevation; 3-dot (tritium sights optional)
Sight radius: 158 mm
Muzzle velocity: ca 270 m/s

Manufacturer

Heckler and Koch GmbH, D-78722 Oberndorf/Neckar.

Status

In production.

Service

Commercial sales. Under consideration by several armed forces and police units.

UPDATED

0.45 Heckler and Koch SOF Offensive Handgun

Development

The Heckler and Koch 0.45 SOF Offensive Handgun was developed by Heckler and Koch GmbH of Oberndorf, as part of a Phase 1 contract awarded to Heckler and Koch Inc of Sterling, Virginia, on August 28, 1991, by the US Special Operations Command (USSOCOM). In August 1992 30 complete prototype systems were delivered to the Naval Surface Warfare Center for testing. In January 1994, Heckler and Koch was awarded a Phase II contract along with a letter contract for Phase III. A further 30 Phase II prototype systems were delivered for further testing in November 1994. It is anticipated that a total of 7,500 pistols will be required along with 1,950 laser aiming modules (LAM) developed under a separate contract. Production will extend until April 1997.

Phase II Heckler and Koch 0.45 SOF Offensive Handgun with suppressor and laser aiming module in position (Scott Gourley) **1996**

Description

The Heckler and Koch 0.45 SOF Offensive Handgun is a 0.45 ACP calibre semi-automatic single- or double-action weapon based on technology developed for the Heckler and Koch USP pistol (see previous entry); a flash and noise suppressor (developed by the Knight's Armament Company, Vero Beach, Florida) and a laser aiming module (LAM) can be readily attached and removed.

A mechanical recoil reduction system is incorporated in the design, reducing the recoil force felt by the pistol and the firer by some 30 per cent. The frame is of polymer plastic material, the slide is a one-piece machined component, and the barrel employs the Heckler and Koch polygonal bore. The polymer magazine has a capacity of 12 0.45 ACP rounds and the

Phase II Heckler and Koch 0.45 SOF Offensive Handgun with suppressor in position **1996**

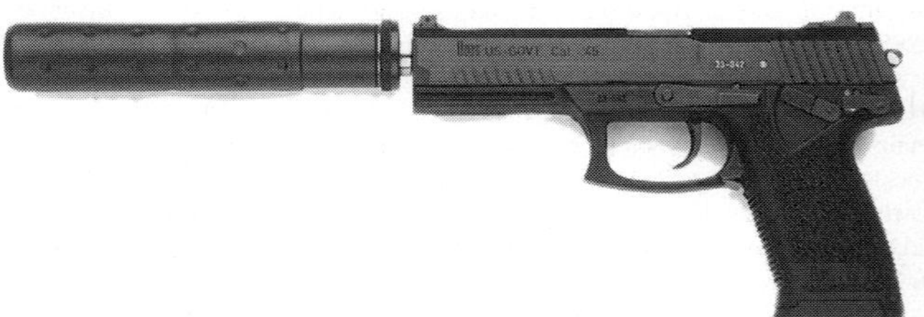
Phase II Heckler and Koch 0.45 SOF Offensive Handgun with suppressor in position **1996**

ammunition fired may be standard 0.45 ACP or Olin +P 185 gr JHP.

The pistol is both single and double action, with an ambidextrous manual safety lever. A separate decocking lever, which can be operated by gloved hands, is provided which will lower the hammer safely and silently. The slide release is extended, and both it and the ambidextrous magazine release are easily reached and operated by the firing hand without any alteration of grip. A slide lock is provided to lock the slide closed for single-shot operation in conjunction with the sound suppressor.

The weapon may be aimed using iron sights or by the laser aiming module. The iron sights provide a three-dot picture with white or tritium dots. The laser module, powered by two lithium batteries, provides visible or infra-red aiming marks alone or in conjunction with a visible or infra-red illuminator. The laser aiming mark is adjustable for windage and elevation, and the entire module may be removed and replaced on the pistol without affecting the zero.

Data

Cartridge: 0.45 ACP; Olin +P 185 grain JHP
Operation: recoil, semi-automatic
Locking: modified Browning drop-barrel
Feed: 12-round box magazine
Weight: with empty magazine, 1.21 kg; with full magazine (ACP) and suppressor, 1.92 kg
Length: without suppressor, 245 mm; with suppressor, 421 mm
Length, suppressor: 192 mm
Barrel: 149 mm
Rifling: polygonal, rh
Sights: iron 3-dot or laser aiming module; see text
Sight radius: 197 mm
Muzzle velocity: M1911 Ball, 270 m/s; Olin +P 185 gr JHP, 348 m/s

Manufacturer

Heckler and Koch GmbH, D-78722 Oberndorf/Neckar.

Status

Advanced Phase II development; under evaluation by USSOCOM.

UPDATED

GREECE

9 mm EP7 pistol

Description

The Hellenic Arms Industry (EBO) SA manufactures several weapons, one of which is the Heckler and Koch P7 pistol. Designated the EP7 it is exactly the same as the German pistol, to which reference should be made for specifications. It is marked Mod 'EP7' with the Hellenic Arms Industry monogram (in the form of a diamond) and has been adopted by the Greek armed forces as well as being exported to various markets.

The Hellenic Arms Industry (EBO) SA also licence manufactures the Heckler and Koch P9S pistol as the EP9S.

Manufacturer

Hellenic Arms Industry (EBO) SA, 160 Kifissias Avenue, GR-11525 Athens.

Status

In production.

Service

Greek military and security forces and for export.

UPDATED

9 mm Hellenic Arms Industry EP7 pistol

HUNGARY

7.62 mm Model 48 pistol

Description

The former Soviet 7.62 mm Tokarev (TT-33) was manufactured in Hungary as the Model 48. The Hungarian pistol is identified by the crest on the grip (a star, wheatsheaf and a hammer surrounded by a wreath) and the uniform narrow vertical cuts on the slide for the firer to grip while cocking the pistol. All other details are the same as the Tokarev (TT-33) (see entry under CIS). The pistol was produced by state arsenals and is still in service.

Data

Cartridge: 7.62 × 25 mm Type P cartridge
Operation: short recoil, semi-automatic
Locking: projecting lugs
Feed: 8-round box magazine
Weight: 846 g
Length: 196 mm
Barrel: 116 mm
Rifling: 4 grooves, rh
Sights: fore, blade; rear, notch
Muzzle velocity: 420 m/s

Manufacturer

State arsenals.

Status

Production complete.

Service

Hungarian armed forces.

VERIFIED

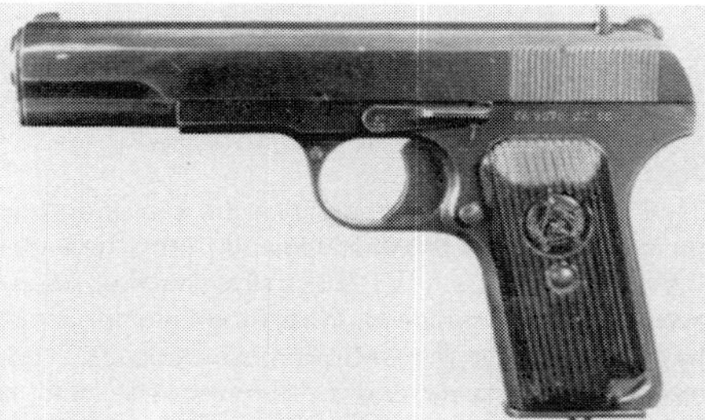

7.62 mm Model 48 pistol, Hungarian copy of Tokarev TT-33

9 mm (and 7.65 mm) PA-63 and AP pistols

Description

Based on the German Walther PP pistol, these designs are virtually the same weapon, the distinction being that the PA-63 is made only in 9 mm Makarov calibre and is the official sidearm of Hungarian military and police forces, while the Model AP is produced in both 7.65 ACP and 9 mm Short (0.380 ACP) calibres for commercial sales. Both weapons are made with an aluminium frame and steel slide, the PA-63 having the frame left bright while the AP has the frame anodised black.

Data

Cartridge: 9 mm Makarov or 7.65 mm ACP
Operation: blowback, double action, semi-automatic
Feed: (9 mm) 7- or (7.65 mm) 8-round detachable box magazine
Weight: empty, 595 g
Length: 175 mm
Barrel: 100 mm
Rifling: 6 grooves, rh
Sights: fore, blade; rear, notch
Muzzle velocity: 7.65 mm, ca 310 m/s

Manufacturer

Fegyver es Gazkeszuelekgyara NV, Soroksári út 158, H-1095 Budapest.

Status

In production.

Service

Hungarian forces and export.

VERIFIED

9 mm AP pistol

9 mm FEG Model P9 pistol

Description

The FEG Model P9 is a copy of the Browning High Power M1935GP, described under Belgium. The copy is so exact that many parts are completely interchangeable, although the Hungarian model adds a ventilated rib to the slide on some examples. It has not been adopted by Hungarian forces, because of the calibre, but is commercially available throughout Europe and is believed to have been adopted by some police forces.

Data

Cartridge: 9 × 19 mm Parabellum
Operation: short recoil, semi-automatic
Locking: projecting lug
Feed: 13-round detachable box magazine
Weight: empty, 900 g; loaded, 1.07 kg
Length: 198 mm
Barrel: 118 mm
Rifling: 6 grooves, rh, 1 turn in 250 mm
Sights: fore, fixed post; rear, notch
Muzzle velocity: 350 m/s

Manufacturer

Fegyver es Gazkeszuelekgyara NV, Soroksári út 158, H-1095 Budapest.

Agency

Technika Foreign Trading Company, Salgotarjani ut 20, H-1475 Budapest.

Status

In production.

Service

Believed to be in service with some European police forces. Commercial sales.

Licence production

BULGARIA
Manufacturer: Arcus Company
Type: 9 mm Belitza
Remarks: Apparently a direct copy of the P9, exhibiting all the signs of a direct Browning High Power GP35 clone.

9 mm FEG Model P9 pistol

1996

UPDATED

9 mm FEG Model P9R and P9RA pistols

Description

These pistols are derived from the FEG 9 mm Model P9 (see previous entry), but with locally designed modifications to give them a more modern specification. The principal change is the adoption of double-action lockwork with a slide-mounted safety catch which lowers the hammer when applied. The operation of the safety catch locks the firing pin and interposes a positive stop between hammer and firing pin. In other respects the mechanism is identical to that of the Browning GP35 pistol.

The Model P9R has a steel frame; the Model P9RA has a light alloy frame. An interesting and unusual variation is the manufacture of a completely left-handed version of this pistol which has the safety catch, slide release and magazine catch on the right side of the frame. There is also a Model P9RK with a short barrel of unspecified length.

Data

Cartridge: 9 × 19 mm Parabellum
Operation: short recoil, semi-automatic
Locking: projecting lug
Feed: 14-round detachable box magazine
Weight: empty, P9R, 1 kg; empty, P9RA, 820 g; loaded, P9R, 1.17 kg; loaded, P9RA, 990 g
Length: 203 mm
Barrel: 118.5 mm
Muzzle velocity: ca 380 m/s

Manufacturer

Fegyver es Gazkeszuelekgyara NV, Soroksári út 158, H-1095 Budapest.

Agency

Technika Foreign Trading Company, Salgotarjani ut 20, H-1475 Budapest.

Status

In production.

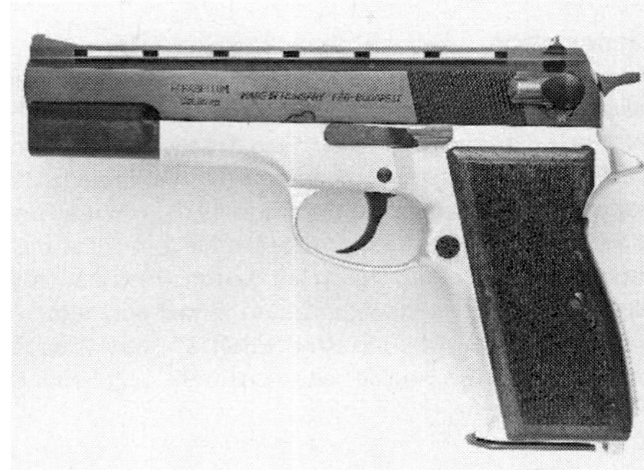

9 mm FEG Model P9RA pistol

Service

Commercial sales.

UPDATED

9 mm FEG Model B9R pistol

Description

The B9R pistol generally resembles the P9R in outline, but is a fixed-barrel blowback weapon firing the 9 mm Short cartridge. It is fitted with a slide-mounted safety lever which, when applied, lowers the hammer, locks the firing pin to the rear and interposes a positive stop between the firing pin and the hammer. The pistol is then brought to the firing condition by manually cocking the hammer or by releasing the safety lever and pulling through on the trigger in double-action mode.

Data

Cartridge: 9 × 17 mm Short (0.380 Auto)
Operation: blowback, self-loading
Feed: 14-round box magazine
Weight: empty, 700 g; loaded, 840 g
Length: 174 mm
Barrel: 101 mm

Manufacturer

Fegyver es Gazkeszuelekgyara NV, Soroksári út 158, H-1095 Budapest.

Agency

Technika Foreign Trading Company, Salgotarjani ut 20, H-1475 Budapest.

Status

In production.

Service

Commercial sales.

UPDATED

9 mm FEG Model B9R pistol

1996

9 mm Minimax 9 pistol

Description

Although the 9 mm Minimax 9 pistol hardly qualifies as an infantry weapon it nevertheless provides an example of just how compact pistols can be designed and considered for operational use, and it could conceivably have applications as a personal defence or survival weapon. It certainly qualifies as an undercover or assassination weapon.

The Minimax 9 has overall dimensions of 96 mm long and 68 mm high; the width is 24 mm. The magazine in the butt holds only four rounds which may be either 9 mm Short (0.380), 9 × 19 mm Parabellum, or 9 × 18 mm Makarov. The Minimax 9 can accommodate any of these cartridges without modification, for the rounds are simply fired from a magazine directly into the barrel. Rifling grooves are provided in the barrel but they are straight, without twist, apparently to ensure the bullet is immediately unstable and will tumble on striking a target to release its maximum energy.

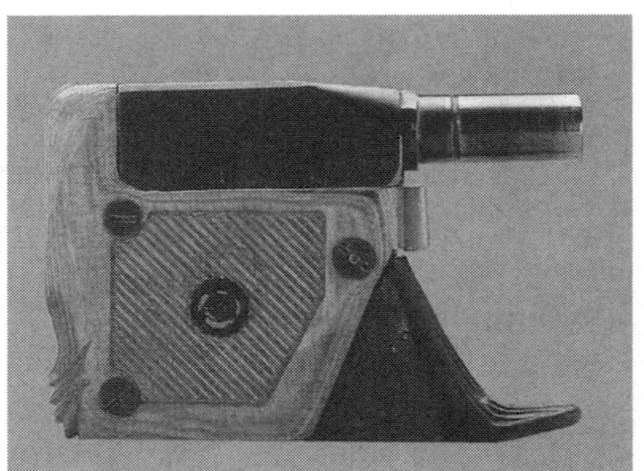

9 mm Minimax 9 pistol

1996

The pistol is loaded from the top using a solid magazine arrangement which holds the four rounds. The top of the magazine is solid and aligns with the barrel so that the pistol can be carried safely, either in a pocket or in its special wallet-shaped pouch. When the pistol is required for use the front of the pistol grip is squeezed to index the magazine upwards (and out of the receiver) and cock the hammer. The pistol is fired when a button trigger, just above the indexing lever, is pressed. There is no spent case ejection as the next round is raised and aligned with the barrel by a further squeeze on the indexing lever. There are no sights. When the pistol is being carried there are no springs or other components under tension.

The Minimax 9 can be used to fire ball cartridges but may also fire blanks or an unspecified disabling gas projectile. It is also possible to fit a miniature ball-shaped baton round over the muzzle to be projected by a blank cartridge.

Data
Cartridge: 9 × 17 mm Short; 9 × 19 mm Parabellum; 9 × 18 mm Makarov
Operation: single shot
Feed: 4-round manually indexed magazine
Weight: empty, 40 g
Length: 96 mm
Height: 68 mm
Width: 24 mm
Barrel: total, 81 mm
Rifling: 2 or 4 grooves, straight
Sights: none

Agency
Technika Foreign Trading Company, Salgotarjani ut 20, H-1475 Budapest.

Status
Available. Offered for export sales.

NEW ENTRY

INDIA

Revolver 0.32 Mark 1

Description
Although it has no doubt been on the local scene for a considerable time it was not until recently (1995) that the Revolver 0.32 Mark 1 appeared on the defence market scene. It is based on the old British Enfield revolver which was itself an enhanced version of the Webley series revolvers dating back to the last Century.

For the Revolver 0.32 Mark 1, the basic Enfield opening frame downwards hinge for reloading, the fixed cylinder and automatic spent case ejection mechanism have all been retained. Some alterations have been introduced to the butt outline compared to the 0.38 Enfield series, but otherwise the overall simple and rugged design is identical. Firing may be single or double action, with the prominent hammer with its long spur being thumb-actuated for the single-action mode usually employed by all revolver users requiring any semblance of accurate fire. Only when the hammer is cocked are any springs placed under tension.

The main feature in firearms terms regarding this revolver is the choice of cartridge, namely the 0.32 S&W Long, a low-powered cartridge usually associated with police and security firearms from a previous era and not normally utilised by military hand guns. It would therefore appear that the Revolver 0.32 Mark 1 is intended more for the police and security force market, where this revolver's simplicity and durability would make it an ideal side arm.

Data
Cartridge: 0.32 S&W Long (7.65 × 32 mm)
Operation: revolver, single or double action
Feed: 6-round cylinder
Weight: empty, 700 g
Length: 177.8 mm
Barrel: 76.2 mm
Rifling: 6 grooves, rh, 1 turn in 381 mm
Sights: fore, fixed blade; rear, fixed notch
Muzzle velocity: ca 215 m/s
Muzzle energy: ca 50 J

Manufacturer
Indian Ordnance Factories, Small Arms Factory, Kanpur.

Status
In production. Offered for export sales.

Revolver 0.32 Mark 1

1996

Service
Indian police and security forces.

NEW ENTRY

ISRAEL

9 mm Uzi pistol

Description
The 9 mm Uzi pistol is a shortened and lightened modification of the Uzi sub-machine gun, with a mechanism permitting semi-automatic fire only. Although appearing cumbersome by comparison with conventional pistols, it has the advantages of an exceptional magazine capacity for a pistol and a shape and size which allows a very firm two-handed grip. The bulk also helps to absorb recoil so that it is easy to control during the firing of a rapid succession of shots. It was designed for civilian use, but there are obvious applications to military and security forces.

Data
Cartridge: 9 × 19 mm Parabellum
Operation: blowback, closed breech
Feed: 20-round box magazine
Weight: empty, 1.65 kg; loaded, 2 kg
Length overall: 240 mm
Barrel: 115 mm
Rifling: 4 grooves, rh, 1 turn in 254 mm
Muzzle velocity: 115 gr bullet, 358 m/s

Manufacturer
TAAS - Israel Industries Limited, PO Box 1044, Ramat Hasharon 47100.

Status
In production.

VERIFIED

0.45 ACP calibre Uzi pistol

0.357 Desert Eagle pistol

Description
The Desert Eagle has been available for some time as a sporting pistol, but its undoubted utility has caused interest to be taken in other fields and it is now offered as a military weapon.

The design is unusual in that it is a gas-operated locked-breech pistol using a rotating bolt. The barrel is fixed to the frame and has a gas port just ahead of the chamber. This connects with a channel in the frame which runs forward to a point beneath the muzzle and there turns down to a gas cylinder inside which is a short-stroke piston. The slide is formed into a receiver at the rear end, which contains the bolt, and into two side arms which run forward below the barrel. On firing, a portion of the gas is diverted through the port and channel to drive the piston backwards. This drives the slide to the rear, against a spring. Movement of the slide first rotates the bolt by a cam, to unlock it from the chamber, and then withdraws the bolt to extract and eject the fired case. The return spring, under the barrel, then drives the slide forward again; the face of the bolt chambers a fresh cartridge and the final movement of the slide rotates the bolt and locks it into the chamber.

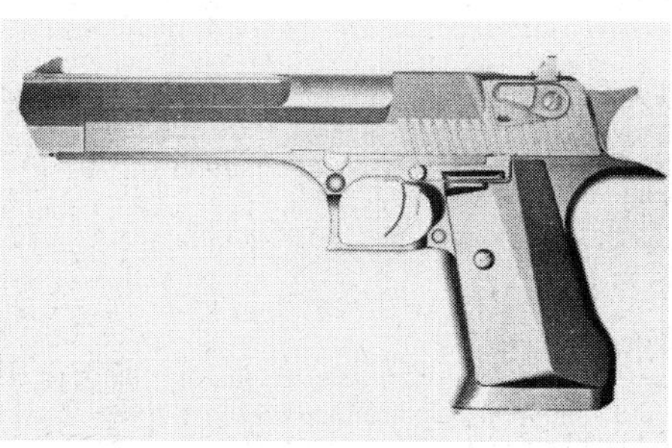

0.357 Desert Eagle pistol

Another unusual feature of this weapon is that it fires the 0.357 Magnum rimmed cartridge, widely available as revolver ammunition. It will feed lead, semi-jacketed or full-jacketed cartridges with equal facility. A larger version in 0.44 Magnum calibre, also firing rimmed revolver cartridges, and the standard model chambered for the 0.41 Action Express cartridge are now available. In 1992 a version chambered for the 0.50 Action Express cartridge was announced.

The pistol is equipped with an ambidextrous safety catch on the rear of the slide which locks the firing pin and also disconnects the trigger from the hammer mechanism. The standard sear assembly can be removed and replaced by a special assembly which allows adjustment of trigger pull for both length and weight. The pistol is normally fitted with combat sights, but an adjustable rearsight is available as an option and the barrel is grooved to accept mounts for a sighting telescope.

Data
Cartridge: 0.357 Magnum, 0.44 Magnum, 0.41 Action Express or 0.50 Action Express
Operation: gas, semi-automatic, single action
Locking: rotating bolt
Feed: 9-round box magazine

Weight: steel frame, 1.76 kg; aluminium frame, 1.466 kg
Length: 260 mm
Barrel: 152 mm (355 mm also available)
Rifling: 6 grooves, rh, 1 turn in 355 mm
Sights: fore, blade; rear, U-notch
Sight radius: 225 mm

Manufacturer
TAAS - Israel Industries Limited, PO Box 1044, Ramat Hasharon 47100.

Status
In production.

VERIFIED

9 mm/0.40 Jericho pistol

Description
Although resembling the Desert Eagle, the Jericho pistol is a more conventional type of recoil-operated pistol relying upon the Browning system of cam-dropped barrel to unlock the breech. The slide moves on internal rails, a feature generally conceded to improve accuracy and rigidity. The most novel feature of this weapon is its ability to change calibres; it is simply a matter of field-stripping the pistol and reassembling it with the proper components to make the conversion. There are two basic models:
Model R ; in this model the safety catch is on the slide and also acts as a decocking lever. In the 'Safe' position the firing pin is blocked and the trigger neutralised.
Model F: In this model the safety catch is mounted on the frame. The firing pin is blocked until the trigger is pulled, and with the safety catch in the 'Safe' position the trigger is blocked.

The Model F is offered in three different sizes; 'F' standard, 'FS' short and 'FB' baby.

Data
Cartridge: 9 × 19 mm Parabellum; 0.40 S&W
Operation: recoil, semi-automatic, double action
Breech lock: cam-operated dropping barrel
Feed: box magazine; 16-round (9 mm) or 12-round (0.40)
Weight: F, R, 1.1 kg; FS 900 g; FB, 860 g
Length: F, R, 207 mm; FS, FB, 184 mm
Barrel: F, R, 112 mm; FS, FB, 90 mm
Rifling: 9 mm, 6 grooves, rh, 1 turn in 254 mm; 0.40 S&W, 1 turn in 407 mm
Sights: fore, blade with luminous dot; rear, square notch with 2 dots; both sights adjustable laterally
Sight radius: F, R, 150 mm; FS, FB, 134 mm

Manufacturer
TAAS - Israel Industries Limited, PO Box 1044, Ramat Hasharon 47100.

Status
In production.

9 mm Jericho pistol

Service
Israeli police and police special anti-terrorist units.

VERIFIED

ITALY

9 mm Model 1951 pistol

Development
The Beretta Model 1951 was the standard pistol of the Italian armed forces and is now being gradually replaced by pistols from the Beretta Model 92 series. The Model 1951 is still used by both the Israeli and Egyptian armies (the Model 1951 is licence produced in Egypt as the Helwan) and in Nigeria.

Description
The Model 1951 has three main parts; the frame, barrel and slide. The frame holds the magazine, trigger and firing mechanism and has a forward extension to take the slide. The barrel carries a swinging locking piece pivoting from a lug on the underside, and the slide fits over the frame, sliding in grooves.

To load the pistol the magazine release in the lower left side of the grip is pressed and the magazine removed. When the loaded magazine is in place the pistol is cocked by pulling the slide fully back and then releasing it. The safety catch is a push-through type mounted at the top rear of the butt. 'Safe' comes from pushing from right to left.

When the pistol is fired, the breech block, integral with the slide, goes back and the barrel which is locked to the slide goes back with it. After a short period of free travel of about 13 mm the unlocking plunger on the rear barrel lug strikes the receiver and stops. As the barrel and slide continue back the locking piece strikes the stationary plunger and is forced down into recesses in the slide in a manner similar to that of the Walther P38. The barrel comes to rest but the slide continues rearward for a further 50 mm. The hammer is recocked and the empty cartridge case is extracted. The return spring then pushes the slide forward and the feed rib pushes the next cartridge into the chamber. The slide picks up the barrel and the locking piece on the barrel lug is lifted up by the receiver cam to lock the barrel to the slide. The forward motion of the slide and barrel stops when the barrel reaches the take-down lever spindle.

9 mm Beretta Model 1951 pistol

When the trigger is pulled to fire the first shot, the trigger bar moves back and rotates the sear on its pivot to release the hammer. The hammer spring rotates the hammer on to the firing pin and the cartridge is fired. If there is a misfire the hammer must be manually recocked for a second blow. There is no double action.

Data
Cartridge: 9 × 19 mm Parabellum
Operation: short recoil, self-loading, single action
Locking: swinging arm
Feed: 8-round box magazine
Weight: with steel slide, 870 g; with alloy slide, 780 g
Length: 203.2 mm
Barrel: 114.2 mm
Rifling: 6 grooves, rh, 1 turn in 254 mm
Sights: fore, blade; rear, square notch adjustable for windage
Sight radius: 140 mm
Muzzle velocity: 350 m/s

Manufacturer
Armi Beretta SpA, Via Beretta 18, I-25063 Gardone VT (Brescia).

Status
No longer in production.

Service
Italian armed forces. Also in use with Egyptian and Israeli forces, Nigerian police and in some other countries.

Licence production
EGYPT
Manufacturer: Maadi Company for Engineering Industries
Type: 9 mm Helwan
Remarks: Standard specifications, although length overall given as 187 mm and weight loaded as 1.025 kg. 'Presentation' finish models available.

UPDATED

7.65 mm Model 81 pistol

Development
Model 81 is one of three Beretta pistols which entered full-scale production in 1976. The other two were the 9 mm Short Model 84, which is identical to the Model 81 in most respects, and the larger Model 92 which fires the 9 mm Parabellum round and operates on the short recoil principle whereas the two smaller weapons are operated by blowback. All three pistols have a number of design features in common, so while the description relates primarily to Model 81 it covers points that are relevant to the other two weapons.

Description
To load the weapon a loaded magazine is inserted in the butt and the slide operated manually to chamber a round. The firing pin is spring-loaded and shorter than the breech block, requiring a sharp blow to cause it to overcome the spring resistance and fire the cartridge so the hammer can be safely lowered under thumb restraint without firing a round. A manual safety, operable from either side of the weapon, locks both the trigger mechanism and the slide in the closed position.

When the round is fired the pressure in the chamber drives the cartridge case, together with the slide assembly, back against the combined pressure of the recoil spring and hammer spring, the case being held on the face of the breech block by the extractor until it strikes the ejector. If the magazine is empty the slide is held open. If the magazine is not empty the slide will move forward to chamber a round and when it has done so the extractor will protrude laterally, showing red, and can thus be seen and felt. The hammer will remain cocked and the next round can be fired by single action.

In the event of a misfire, there will be no blowback action and the hammer will be forward. A second attempt to fire the round can then be made by releasing the trigger and pulling it again.

Other features of the weapon include a 12-round magazine with a staggered loading arrangement, a reversible magazine release button to suit right- or left-handed firers, a stripping catch arrangement, which makes stripping easy but guards against accidental disassembly and an optional magazine safety which prevents the weapon from being fired when the magazine is removed and a cartridge remains in the chamber.

Data

Cartridge: 7.65 mm ACP
Operation: blowback, single or double action
Feed: 12-round detachable box magazine
Weight: with empty magazine, 670 g
Length: 172 mm
Barrel: 97 mm
Rifling: 6 grooves, rh, 250 mm pitch
Sights: fore, blade, integral with slide; rear, notched bar dovetailed to slide
Sight radius: 124 mm
Muzzle velocity: nominal, 300 m/s

Manufacturer

Armi Beretta SpA, Via Beretta 18, I-25063 Gardone VT (Brescia).

Status

No longer in production.

Service

Many law enforcement agencies in and outside Europe. Also extensive commercial sales.

UPDATED

7.65 mm Beretta Model 81 pistol

9 mm Model 84 pistol

Description

The Beretta Model 84 resembles the 7.65 mm Model 81 described previously in all respects save those that are relevant to the change of calibre to 9 mm Short. Components affected are the barrel, magazine box, magazine follower and magazine spring and it should be noted that the magazine capacity of the Model 84 is 13 rounds instead of 12.

The Beretta Model 84 is one of the Beretta Medium Frame pistols marketed under the name Cheetah.

Details of the characteristics that distinguish the Model 84 from the Model 81 are given below: all other characteristics are the same.

Data

Cartridge: 9 mm Short
Feed: 13-round detachable box magazine
Weight: with empty magazine, 660 g
Muzzle velocity: nominal, 280 m/s
Muzzle energy: 235 J

Manufacturer

Armi Beretta SpA, Via Beretta 18, I-25063 Gardone VT (Brescia).

Status

No longer in production.

Service

Law enforcement agencies in many parts of the world and commercial sales.

UPDATED

9 mm Model 84 pistol

Beretta Model 82BB, 83F, 85BB, 85F, 87BB and 87BB/LB pistols

Description

These Beretta pistols are derivatives of the Models 81 and 84 and in most of the essential features they are the same. The Model 82BB is in 7.65 mm, the 83F, 85BB and 85F in 9 mm Short and the Models 87BB and 87BB/LB in 0.22 RF calibre. For a general description reference should be made to the entry for the Model 81.

These Beretta models are among the Beretta Medium Frame pistols marketed under the name Cheetah.

The differences between these models and the 81 are as follows:

BB Models

Chamber loading indicator: a pin projects laterally from the slide to indicate when a round is in the chamber. This pin is coloured red and it gives both a visual and a tactile indication.

Magazine capacity: the magazine is smaller so allowing a lighter and thinner grip. This favours the smaller hand and also makes the pistols easier to conceal.

New safety system: this is really divided into four sections: the manual safety breaks the connection between the trigger and the sear, the firing pin is permanently locked until the last stage of the trigger pull when it is released, the firing pin is operated by inertia and so there is no direct contact with the primer, there is a half-cock position. It is worth noting that these pistols can be fired from the half-cock position by simply pulling the trigger to raise and drop the hammer.

F Models

These have all the features of the BB models described before, with the addition of a hammer decocking facility built in to the applied safety system. To lower the hammer safely, the manual safety catch is applied; this drops the hammer against an interceptor, locks the slide in the closed position and interrupts the connection between trigger and sear. The half-cock hammer position is not available in these models. The F models also have the barrel and chamber chromium-plated. The 83F differs from the 85F in having a 101 mm barrel and a 7-round magazine.

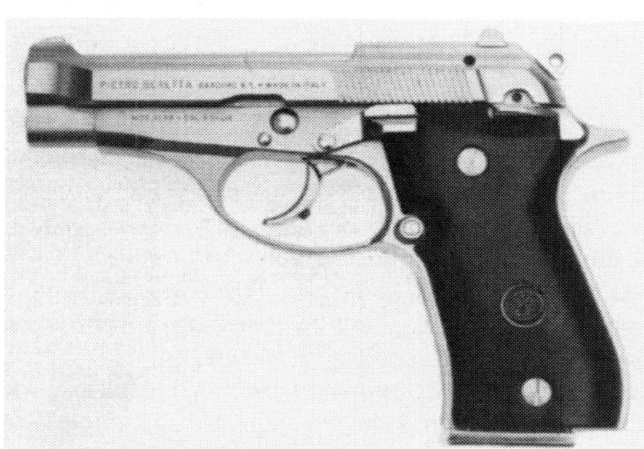

9 mm Short Model 85BB pistol, nickelled version

7.65 mm Beretta Model 82BB pistol. Chamber loaded indicator is directly below and just forward of rearsight

0.22 LR Model 87BB/LB pistol

9 mm Short Model 83F pistol

0.22 RF Model 87LR pistol

9 mm Short Model 85F pistol

Model 87BB/LB

This model differs from the remainder of the group in being single action only. The notation '/LB' indicates the use of a long (150 mm) barrel.

Data

AS FOR MODELS 81 AND 84 EXCEPT:

Cartridge: Model 82BB, 7.65 mm ACP; Models 83F, 85BB, 85F, 9 mm Short; Models 87BB, 87BB/LB, 0.22 Long Rifle RF
Feed: Model 82BB, 9-round magazine; Models 85BB, 85F, 87BB, 87BB/LB, 8-round magazine; Model 83F, 7-round magazine
Weight: Model 82BB, 630 g; Models 85BB, 85F, 620 g; Model 87BB, 570 g; Model 87BB/LB, 660 g

Manufacturer

Armi Beretta SpA, Via Beretta 18, I-25063 Gardone VT (Brescia).

Status

Current production models feature the characteristics of the F series.

Service

Law enforcement agencies in many parts of the world, and commercial sales.

UPDATED

Beretta Model 81BB, 84BB and 84F pistols

Description

These Beretta models are among the Beretta Medium Frame pistols marketed under the name Cheetah.

Although of greater magazine capacity than Models 82 and 85 described previously, Models 81BB and 84BB are, like the former, derivatives of Models 81 and 84 and resemble them in most essential features. Model 81BB is in 7.65 mm; Model 84BB in 9 mm Short. For the general method of operation and stripping, reference should be made to the entry for Model 81.

Beretta Model 84 Cheetah

1996

Models 81BB and 84BB are double-action pistols, incorporating staggered detachable box magazines (12-round capacity in the Model 81BB; 13-round capacity in the Model 84BB); chamber loading indicator; reversible magazine release button; stripping catch arrangement (as described for Model 81) and a manual safety operable from either side of the weapon. The safety system is as described for the Models 82BB and 85BB. A magazine safety, acting on the trigger mechanism when the magazine is removed, is optional.

The Model 84F, now known as the Model 84 Cheetah, is similar to the BB series but has the additional feature of a hammer decocking facility, operated by the safety catch. Pressing the safety catch will allow the hammer to fall safely on to an interceptor bar, lock the slide and disconnect trigger and sear.

The front and back straps of the grips are longitudinally grooved to ensure a firm hold in wet conditions or during rapid firing. Plastic grips are optional. It is also available in a limited deluxe series featuring gold-plated hammer and trigger, walnut grips and blued or gold-plated finish.

Data

MODEL 81BB AS FOR MODEL 81, EXCEPT:

Operation: blowback, semi-automatic, double action

MODEL 84BB AS FOR MODEL 84

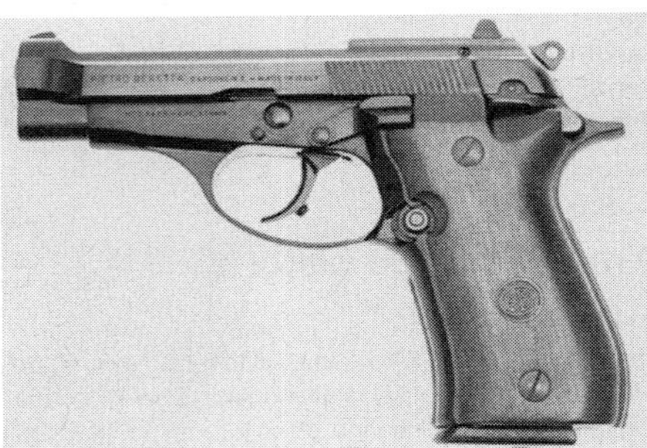

7.65 mm Beretta Model 81BB pistol

Manufacturer

Armi Beretta SpA, Via Beretta 18, I-25063 Gardone VT (Brescia).

Status

In production.

Service

Law enforcement agencies in many parts of the world and commercial sales.

UPDATED

9 mm Model 86 pistol

Description

Introduced in 1985, the Beretta Model 86 is a pocket blowback pistol which has applications in police and security roles. It is unusual in employing a tip-up barrel, a design feature which was once common but which has not been seen for many years.

The Beretta Model 86 is one of the Beretta Medium Frame pistols marketed under the name Cheetah.

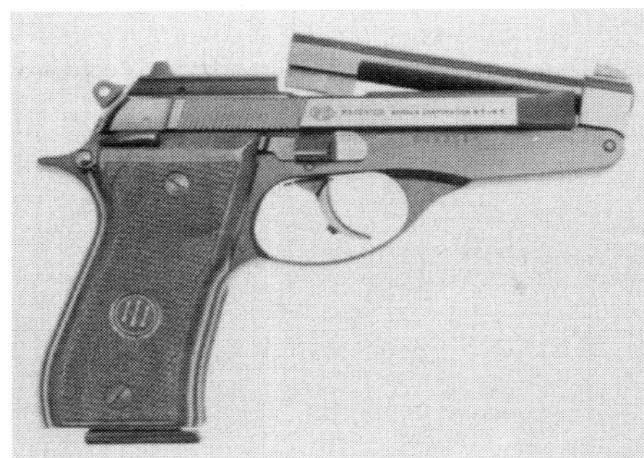

9 mm Model 86 pistol with barrel raised

The Model 86 is of modern appearance and has double-action lockwork with an external hammer. There is a manual ambidextrous safety catch which acts to block the firing pin, a hammer release lever which allows the hammer to be safely lowered, and a red chamber-loaded indicator. The barrel is pivoted beneath the muzzle, and by pressing a catch the breech end is released and allowed to rise, under spring pressure. This simplifies cleaning the barrel without having to dismantle the gun, it also permits an additional cartridge to be loaded into the chamber when a full magazine has been inserted into the butt. The pistol frame is of anodised light alloy, and walnut or plastic grips are available.

Data

Cartridge: 9 × 17 mm Short/0.380 ACP
Operation: blowback, semi-automatic
Feed: 8-round box magazine
Weight: empty, 660 g
Length: 185 mm
Barrel: 111 mm
Rifling: 6 grooves, rh
Sights: fore, blade; rear, U-notch
Sight radius: 127 mm
Muzzle velocity: ca 310 m/s

9 mm Model 86 pistol

Manufacturer

Armi Beretta SpA, Via Beretta 18, I-25063 Gardone VT (Brescia).

Status

Available.

VERIFIED

0.22 LR Model 89 pistol

Description

The Beretta Model 89 was designed as a target pistol which, because of its general configuration, can also be used as a training pistol for heavier service weapons. It is marketed under the name Model 89 Gold Standard.

The Model 89 is a simple blowback semi-automatic with a heavy fixed barrel mounted on a light alloy frame. There is an external hammer, ambidextrous safety catch, a rearsight fully adjustable for both elevation and windage and an adjustable trigger stop. The front sight is interchangeable, allowing infinite degrees of adjustment for zero. A magazine safety device is fitted.

Data

Cartridge: 0.22 Long Rifle rimfire
Operation: blowback, semi-automatic
Feed: 8-round box magazine
Weight: empty, 1.16 kg
Length: 240 mm
Barrel: 152 mm
Rifling: 6 grooves, rh, 1 turn in 350 mm
Sights: fore, interchangeable blade; rear, fully adjustable U-notch
Sight radius: 185 mm
Muzzle velocity: nominal, 345 m/s

Manufacturer

Armi Beretta SpA, Via Beretta 18, I-25063 Gardone VT (Brescia).

0.22 LR Model 89 pistol

Status

In production.

UPDATED

9 mm Model 92 pistol

Description

Firing the 9 × 19 mm Parabellum round, the Beretta Model 92 entered production in 1976, forming the basis of the Beretta Large Frame pistol series. It is both larger and more powerful than the Models 81 and 84. It also employs a short recoil operating system in place of the blowback system suited to the less powerful rounds of the smaller weapons.

In most general design respects the Model 92 resembles the other pistols. It has a double-action trigger system working on the same principles, a similar firing pin assembly and a similar arrangement for stripping. The short recoil system uses a falling locking block which is driven down to disengage the slide from the barrel and halt the rearward motion of the barrel but otherwise the extraction, cocking and loading operations are similar to those of the smaller weapons; the extractor provides the same loaded-chamber indication and the slide is held to the rear when the magazine is empty.

Data

Cartridge: 9 mm Parabellum
Operation: short recoil, semi-automatic with single or double action
Locking: falling block
Feed: 15-round detachable box magazine
Weight: with empty magazine, 950 g
Length: 217 mm
Barrel: 125 mm
Rifling: 6 grooves, rh, 250 mm pitch
Sights: fore, blade integral with slide; rear, notched bar dovetailed to slide
Sight radius: 155 mm
Muzzle velocity: nominal, 390 m/s

Manufacturer

Armi Beretta SpA, Via Beretta 18, I-25063 Gardone VT (Brescia).

Status

No longer in production.

Service

Italian forces and some foreign armies.

UPDATED

Beretta 9 mm Model 92 pistol

1996

Beretta 9 mm Model 92 pistol

9 mm Model 92S pistol

Description

The Beretta 9 mm Model 92S closely resembles the Model 92 described previously but has a modified safety mechanism. Whereas in the Model 92 the safety is mounted on the frame the Model 92S safety is mounted on the slide and provides a safe decocking facility. When applied it deflects the firing pin from the hammer head, releases the hammer and breaks the connection between the trigger bar and the sear.

If the weapon is cocked, application of the safety will allow the hammer to fall safely into the uncocked position; if the weapon is not cocked, operation of the trigger will not operate the hammer; and if the hammer should be inadvertently operated by some other means the weapon will not fire even though there may be a round in the chamber.

Data

GENERALLY AS MODEL 92 (SEE PREVIOUS ENTRY) EXCEPT:
Weight: with empty magazine, 980 g

Manufacturer

Armi Beretta SpA, Via Beretta 18, I-25063 Gardone VT (Brescia).

Status

No longer in production.

Service

Italian armed forces and police and some foreign armies.

VERIFIED

9 mm Model 92S pistol showing manual safety mounted on slide

9 mm Model 92 SB pistol

Description

The Beretta Model 92 SB is a direct development of the Model 92S to which reference should be made for the main details. The Model 92 SB differs in the following features: the safety lever is on both sides of the slide, allowing the pistol to be used by left-handers without alteration; the magazine release button has been moved to a position underneath the trigger guard where it can be pressed without moving the hand from the grip. The button can be switched from left to right side to allow for left-handed firing.

A new series of safeties comprises the following items: the manual safety disengages the trigger from

9 mm Model 92 SB pistol

9 mm Model 92 SB pistol

the sear; the firing pin is permanently locked until the last movement of the trigger on firing; the firing pin is inertia operated; there is a half-cock position.

The butt is grooved in the front and rear to improve the grip.

Data
All data is identical with the Model 92S pistol.

Manufacturer
Armi Beretta SpA, Via Beretta 18, I-25063 Gardone VT (Brescia).

Status
In production.

Service
Italian armed forces (special units) and police forces; some foreign armies.

VERIFIED

9 mm Model 92 SB Compact pistol

Description
The Beretta 9 mm Model 92 SB Compact is a smaller and handier version of the Model 92 SB. The main differences are in the size and the magazine capacity. All other features are the same as for the Model 92 SB as stated previously.

Data
Feed: 13-round box magazine
Length: 197 mm
Barrel: 109 mm
Height: 135 mm

Manufacturer
Armi Beretta SpA, Via Beretta 18, I-25063 Gardone VT (Brescia).

Status
No longer in production.

Service
Police forces in Italy and abroad.

VERIFIED

9 mm Model 92 SB Compact pistol

9 mm Model 92 SB Compact Type M pistol

Description
The Beretta Model 92 SB Compact Type M is almost the same as the SB-C but has an eight-round single-column magazine with a special base providing a small rest for the finger, and a slightly curved front face to the grip frame. Other features, and dimensions, are as for the SB-C.

Manufacturer
Armi Beretta SpA, Via Beretta 18, I-25063 Gardone VT (Brescia).

Status
No longer in production.

Service
Law enforcement agencies and commercial sales.

VERIFIED

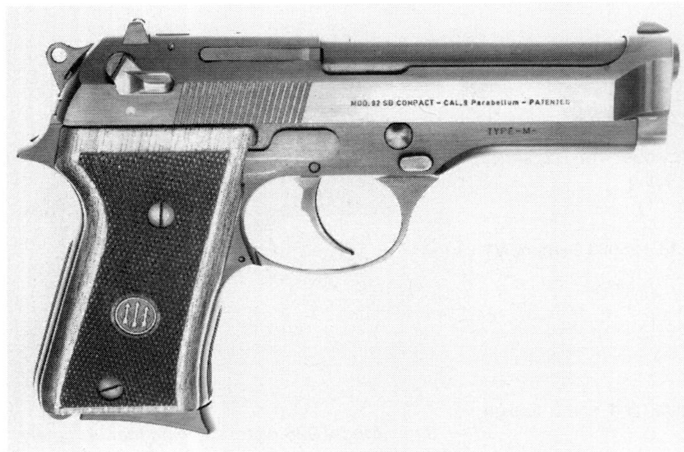

9 mm Model SB-C Type M pistol

9 mm Model SB Compact Type M pistol

9 mm Model 92FS pistol

Development
The Beretta Model 92FS is a component in the Beretta Large Frame Pistol series and has the distinction of being one of the most widely used military and paramilitary pistols of recent years. It gained this distinction primarily as a result of being the successful contestant following the prolonged series of trials held during the early 1980s which led to the selection of a new pistol for the US Army. The Beretta Model 92FS, slightly modified in detail to suit the contest requirements, thus became the M9 Pistol and was produced in both Italy and the United States.

Description
The Beretta 9 mm Model 92FS is dimensionally and mechanically the same as the Model 92 SB. It differs in having the trigger guard formed to suit a two-handed grip, an extended base to the magazine to improve the grip, a curved front edge to the grip frame, new grip plates and a new lanyard ring. The barrel is chromed

9 mm Model 92FS Inox stainless steel pistol

9 mm Model 92FS pistol

internally and the pistol is externally finished in 'Bruniton', a Teflon-type material.

As with all Model 92 pistols, the Model 92FS operates on the short recoil principle. It is based on an anodised lightweight alloy frame with a combat rigger guard. The open slide design involves a completely open ejection port which allows easy access to the chamber should a stoppage occur or for loading a cartridge directly into the chamber. The magazine release offers a rapid release for the 15-round box magazine while the magazine release button can be reversed for left-handed firers.

The pistol features a firing pin block which firers can check visually for confidence. There is also an external chamber loaded indicator to denote when a round is loaded in the chamber. The trigger action features a double action which helps to reduce the incidence of inadvertent discharges. To fire the initial shot, after the manual safety is turned off, the user must apply enough pressure to pull the trigger the full length of draw. After the double-action pull all subsequent shots require a single-action trigger pull for rapid firing, with far less pressure needed on the trigger. When the safety is returned to the 'safe' position the double-action safety feature is reset automatically.

When the safety is applied the trigger is disengaged and the hammer is lowered, with the firing pin rotated away from the hammer so that it cannot touch the firing pin. At the same time the safety automatically engages the trigger bar disconnect mechanism and disengages the connection of the trigger with the sear mechanism. These features prevent the pistol from firing caused by an inadvertent trigger pull.

For competition purposes, the pistol can be converted by replacing the existing barrel with a 185 mm barrel, adding a counterweight and fitting anatomical wooden grips and target sights. These items, together with a carrying case for the converted pistol, are available as a kit. Walnut and rubber grips are available as options.

Various models, such as the Model 98FS De Luxe, are available as 'presentation' weapons. There is also a Model 92FS Inox in stainless steel.

Manufacturer
Armi Beretta SpA, Via Beretta 18, I-25063 Gardone VT (Brescia).

Status
In production.

Service
US Army, Navy, Marine Corps, Air Force and Coast Guard. French Gendarmerie Nationale. Law enforcement agencies worldwide and commercial sales.

Licence production
UNITED STATES OF AMERICA
Manufacturer: Beretta USA Corporation
Type: 9 mm Pistol M9
Remarks: See text and entry under *United States of America*

UPDATED

9 mm Model 92FS Compact pistol

Description
The Beretta 9 mm Model 92FS Compact is the 92 SB-C modified in the same manner as the Model 92FS previously; that is, with improvements designed to facilitate handling and shooting comfort. It is dimensionally the same as the Model 92 SB-C.

Manufacturer
Armi Beretta SpA, Via Beretta 18, I-25063 Gardone VT (Brescia).

Status
In production.

Service
Law enforcement agencies and commercial sales.

VERIFIED

9 mm Model 92FS Compact pistol

9 mm Model 92FS Compact Type M

Description
The 'Type M' variation of the Beretta 9 mm Model 92FS Compact pistol merely reduces the magazine capacity to eight rounds, in order to provide a slimmer grip and better concealment. The weight is reduced to 875 g but other dimensions remain the same.

Manufacturer
Armi Beretta SpA, Via Beretta 18, I-25063 Gardone VT (Brescia).

Status
No longer in production.

VERIFIED

9 mm Model 92FS Compact Type M pistol

9 mm Model 92G

Description
For the Beretta 9 mm Model 92G, the 'G' indicates 'Gendarmerie', as the Gendarmerie Nationale de France adopted this design in 1989; it is licence-produced in France by Giat Industries as the 9 mm PAMAS G1 (qv). The Model 92G generally resembles the Model 92FS but is fitted with a decocking lever only, rather than the combined safety/decocking lever of the Model 92FS. When the decocking lever is released, after the hammer has been lowered, it springs back up to the ready to fire position. There is no manual safety catch.

This pistol is also available chambered for the 0.40 S&W cartridge, when it is known as the Model 96G.

Manufacturer
Armi Beretta SpA, Via Beretta 18, I-25063 Gardone VT (Brescia).

Status
In production.

Service
In service with police and gendarmerie forces in many parts of the world (including France) and French Air Force.

Licence production
FRANCE
Manufacturer: Giat Industries
Type: 9 mm PAMAS G1
Remarks: See separate entry in this section

UPDATED

9 mm Model 92G pistol

9 mm Model 92DS

Description
The Beretta 9 mm Model 92DS is a self-cocking (or 'double action only') version of the Model 92F in which the hammer always follows the slide forward to come to rest in the double-action position. Each pull of the trigger draws the hammer back and then releases it. The hammer spur has been removed and is flush with the rear of the slide, and the DA trigger pull has been reduced. The safety lever on the slide provides only safety functions and has no effect upon the hammer.

Manufacturer
Armi Beretta SpA, Via Beretta 18, I-25063 Gardone VT (Brescia).

Status
In production.

Service
Police forces and commercial sales.

UPDATED

9 mm Beretta Model 92DS

9 mm Model 92D

Description

The Beretta 9 mm Model 92D is similar to the Model 92DS in being a self-cocking design, the difference being that there is no manual safety device and there are no levers on the slide at all. This feature is termed by Beretta as a 'slick slide'.

Manufacturer

Armi Beretta SpA, Via Beretta 18, I-25063 Gardone VT (Brescia).

Status

In production.

Service

Police forces and commercial sales.

VERIFIED

9 mm Model 92D pistol

9 mm Beretta Model 92 Brigadier FS

Description

The Beretta 9 mm Model 92 Brigadier FS is a development of the Model 92FS series primarily intended for combat competition shooting and other applications where a large number of rounds will be fired over prolonged periods. To this end the Brigadier models feature a heavily reinforced slide with revised contours.

In all other respects the Model 92 Brigadier FS is essentially similar to other models in the Model 92FS series.

A double-action-only model, the Model 96 Brigadier D, is available. As well as the standard 9 × 19 mm Parabellum model, versions chambered for the 0.40 S&W (Model 96 Brigadier) and 9 × 21 mm IMI (Model 98 Brigadier FS) are available.

Manufacturer

Armi Beretta SpA, Via Beretta 18, I-25063 Gardone VT (Brescia).

Status

In production.

Service

US Border Patrol and Immigration and Naturalisation Service. Commercial sales.

NEW ENTRY

Beretta 9 mm Model 96 Brigadier D

1996

Beretta 9 mm Model 96 Brigadier D

1996

Beretta Model 92, 96 and 98 Stock

Description

The Beretta Stock series is produced for competition shooting. The reinforced slide introduced on the Brigadier models is employed and a replaceable wear adjustable bushing is mounted on the muzzle. A special competition safety on the frame prevents any trigger action, locks the sear and slide, and allows the pistol to be carried in a cocked and locked condition.

The standard sights are dovetailed front and rear and can be replaced; a three-dot sighting system is used. Fully adjustable and tritium sights are available on request, as are Ultrathin aluminium grips and an oversized magazine release button and safety levers.

The Model 92 Stock is chambered for the 9 × 19 mm Parabellum cartridge.

The Model 96 Stock is chambered for the 0.40 S&W cartridge.

The Model 98 Stock is chambered for the 9 × 21 mm IMI cartridge.

All models are dimensionally similar to the Model 92FS other than that the sight radius is 161 mm and the overall width is 44.5 mm; weight empty is 1 kg.

Manufacturer

Armi Beretta SpA, Via Beretta 18, I-25063 Gardone VT (Brescia).

Status

In production.

Service

Commercial sales.

NEW ENTRY

Beretta 9 mm Model 92 Stock

1996

Beretta 9 mm Model 92 Stock

1996

Beretta Model 92, 96 and 98 Combat

Description

The Beretta Combat series is produced for competition shooting and features most of the characteristics of the Beretta Stock series (see previous entry). To meet international combat shooting standards the action is single action only. A 150 mm barrel with an aluminium counterweight at the muzzle is provided. The trigger action is adjustable.

The standard sights are micrometer adjustable at the rear. Tritium sights are available on request, as are Ultrathin aluminium grips and over-sized magazine release button and safety levers.

The Model 92 Combat is chambered for the 9 × 19 mm Parabellum cartridge.

The Model 96 Combat is chambered for the 0.40 S&W cartridge.

The Model 98 Combat is chambered for the 9 × 21 mm IMI cartridge.

Data

Cartridge: 9 × 19 mm Parabellum; 0.40 S&W; 9 × 21 IMI
Operation: short recoil, semi-automatic with single action
Locking: falling block
Feed: 15-round detachable magazine (0.40 11-rounds)
Weight: with empty magazine, 1.135 kg
Length: 242 mm
Width overall: 44.5 mm
Barrel: 150 mm
Rifling: 6 grooves, rh, 250 mm pitch
Sights: fore, blade integral with slide; rear, micrometer adjustable
Sight radius: 173 mm

Manufacturer

Armi Beretta SpA, Via Beretta 18, I-25063 Gardone VT (Bresica).

Status

In production.

Service

Commercial sales.

NEW ENTRY

Beretta 9 mm Model 92 Combat

1996

Beretta 9 mm Model 92 Combat

1996

9 mm Beretta Model 92FS Conversion Kit

Description

This conversion kit converts a standard Model 92FS into a competition pistol. The kit includes a 185 mm barrel with counterweight and elevated front sight, a fully adjustable rear target sight, and ergonomic walnut grips. The kit comes, only with the standard 9 × 19 mm Parabellum pistol, in a special carrying case.

Data

CONVERTED PISTOL
Feed: 15-round magazine
Weight: empty, 1.12 kg
Length: 277 mm
Barrel: 185 mm
Sight radius: 230 mm

Manufacturer

Armi Beretta SpA, Via Beretta 18, I-25063 Gardone VT (Brescia).

UPDATED

9 mm Model 92FS Competition Conversion

Beretta Model 96 Series pistols

Description

The Beretta Model 96 series of pistols comprises the Models 96, 96 Compact, 96D, 96DS, 96G, 96 target, 96 Competition Conversion Kit, 96 De Luxe and 96 Centurion. They are the same in all respects as the equivalent Beretta Model 92 series, but are in 0.40 Smith & Wesson chambering. The only difference in data is the magazine capacity, which is 10 rounds for all models except the 96 Compact, which has a nine-round magazine.

Manufacturer

Armi Beretta SpA, Via Beretta 18, I-25063 Gardone VT (Brescia).

Status

In production.

VERIFIED

Beretta Centurion series pistols

Description

The standard Centurion pistols have the frame and magazine capacity of the basic 92FS or 96 models with the reduced length of barrel and slide of the Compact versions. Special G Centurion, D Centurion and DS Centurion models are available in some countries, which apply the same principle but with the addition of the special features of the D, DS and G models.

Data

MODEL 92FS CENTURION; MODEL 96 THE SAME EXCEPT FOR 10-ROUND MAGAZINE
Cartridge: Model 92FS, 9 × 19 mm Parabellum; Model 96, 0.40 S&W
Operation: recoil, semi-automatic, double action
Feed: 15-round (10-round) box magazine
Weight: empty, 940 g
Length: 197 mm
Barrel: 109 mm
Sight radius: 147 mm

Beretta 0.40 S&W Model 96 Centurion

Manufacturer

Armi Beretta SpA, Via Beretta 18, I-25063 Gardone VT (Brescia).

VERIFIED

Beretta 9 mm Model 92D Centurion; note absence of safety devices characteristic of the D variation

Beretta Model 98 Series pistols

Description
The Beretta Model 98 was hitherto a 7.65 mm Parabellum version of the Model 92 SBC pistol. This model is no longer in production, so the designation '98' now applies to a more recent product.

The new Model 98 Series comprises the 98FS, 98FS Inox, 98FS De Luxe and 98FS Target. These correspond to the equivalent 92 models but are chambered for the 9 × 21 mm IMI cartridge. Their data and appearance are exactly the same as the equivalent Model 92 models.

Manufacturer
Armi Beretta SpA, Via Beretta 18, I-25063 Gardone VT (Brescia).

Status
In production.

Service
Commercial sales.

VERIFIED

9 mm Model 98FS Target pistol

7.65 mm Beretta Model 99 pistol

Description
Intended for police use, the Beretta 7.65 mm Model 99 is the same as the Model 92 SB-C Type M but chambered for the 7.65 mm Parabellum cartridge. Data are as for the Type M except that the empty weight is approximately 900 g.

Manufacturer
Armi Beretta SpA, Via Beretta 18, I-25063 Gardone VT (Brescia).

Status
No longer in production.

Service
Commercial sales.

VERIFIED

7.65 mm Model 99 pistol

Beretta 8000/8040 Cougar Series pistols

Development
The Beretta 8000/8040 Cougar series, part of the Beretta Compact Frame Pistol family, represents a totally new design, using an all-enveloping slide and a rotating barrel lock for the breech. The barrel carries lugs around the chamber area, acting in slots in the slide so as to retain the breech closed for a short distance and then rotate so as to allow the slide to continue moving and thus open the breech.

The Cougar series has the same range of variants as the Model 92 series, that is to say F models with manual safety, decocking lever with rotating firing pin striker; G models without manual safety; D models with double action only and without manual safety. All models have an external hammer, but the D models have no hammer spur.

9 mm Beretta 8000 Cougar Series pistol

Description
The frame is of light alloy, sandblasted and anodised black. The Inox model has the frame sandblasted. Slide and barrel are of steel, the slide being phosphated and Bruniton coated. The Inox model is of stainless steel and sandblasted. All barrels are internally chromium-plated.

9 mm Beretta 8000 Cougar Series pistol with parts to convert it into 0.41 AE, showing rotating lugs on the barrel

All models have an automatic firing pin blocking device; F models have a manual safety and decocking lever with rotating firing pin striker and trigger-bar disconnect; G models have a decocking lever, rotating firing pin striker, and no manual safety; D models have no manual safety.

Data
Cartridge: 9 × 19 mm Parabellum; 9 × 21 mm IMI; 0.41 Action Express; 0.40 Smith & Wesson (Model 8040)
Operation: Short recoil, semi-automatic
Locking: rotating barrel
Feed: 9 mm, 15-round box magazine; 0.40 S&W, 11 rounds; 0.41AE, 10 rounds
Weight: 950 g; D Model, 940 g
Length: 180 mm
Barrel: 92 mm
Rifling: 9 mm calibres, 6 grooves, rh, 1 turn in 250 mm; 0.40 and 0.41 calibres, 6 grooves, rh, 1 turn in 400 mm
Sights: fore, blade; rear, square notch, both dovetailed to slide. Interchangeable 3-dot system
Sight radius: 132 mm

Manufacturer
Armi Beretta SpA, Via Beretta 18, I-25063 Gardone VT (Brescia).

Status
F and D models in production.

Service
Commercial sales.

UPDATED

9 mm Model 951R selective fire pistol

Description
The Beretta 9 mm Model 951R selective fire pistol was directly derived from the Beretta 9 mm Model 1951 so reference should be made to the entry for that weapon.

The major characteristics of the Model 951R are apparent from the accompanying photographs (note the forward handgrip and the fire selector switch) and data. The manufacturer states that data for rate of fire and muzzle velocity is dependent on ammunition quality and atmospheric conditions.

Data
Cartridge: 9 × 19 mm Parabellum
Operation: short recoil, semi- or full-automatic
Locking: hinged block
Feed: 10-round detachable box magazine
Weight: with empty magazine, 1.35 kg

9 mm Model 951R pistol

Length: 170 mm
Barrel: 125 mm

9 mm Model 951R pistol

Rifling: 6 grooves, rh, 1 turn in 250 mm
Sights: fore, blade; rear, notch

Muzzle velocity: 390 m/s
Rate of fire: practical, ca 750 rds/min
Selector: automatic or semi-automatic

Manufacturer
Armi Beretta SpA, Via Beretta 18, I-25063 Gardone VT (Brescia).

Status
No longer in production.

Service
Italian special forces.

UPDATED

9 mm Model 93R selective fire pistol

Development
The Beretta 9 mm Model 93R is an advanced self-loading pistol which can fire either single shots or three-round bursts and so falls more into the category of 'machine pistol' in the true sense of the word. However, it is meant to be carried and normally used in the same way as a single-handed pistol and as such it handles as a slightly large 9 mm self-loading pistol of conventional design. It can be used and fired just like any other pistol, but should the firer wish to engage a target beyond the normal pistol range, or even a difficult one close at hand, they can quickly fold down the front handgrip and hold the pistol with both hands. This hold is far steadier than the much-publicised modern method of clasping the butt with two hands for, with the forward handgrip, the two hands are a finite distance apart and able to exert a sensible control on the direction of the barrel.

For firing three-round bursts at any range it is essential that both hands are used and if there is time, the firer is recommended to fit the folding carbine stock and take proper aim with the weapon in the shoulder. With both hands holding the pistol the right hand holds the butt in the normal way; the left hand grasps the forehand grip and the thumb is looped through the enlarged trigger guard. To assist in holding the weapon on the target when firing bursts there is a small muzzle brake, which also acts as a flash hider for night shooting.

Description
The basic frame of the Model 93R is similar to that of the Model 92, but there is a burst-controlling mechanism in the right-hand butt grip. The fire selector lever is added above the left grip and the lower frame forward of the trigger guard is deepened to carry the hinges of the forehand grip. The fire selector lever can be moved with the right thumb to select either single shots (one white dot) or three-round bursts (three white dots) without disturbing the aim, and the safety can be similarly applied or released.

The metal folding stock quickly clips on to the bottom of the butt without interfering with the magazine and provides a reasonably steady hold for burst fire.

Although normal cleaning can be carried out by the firer, it is recommended that stripping of the burst control mechanism be carried out by an armourer.

Data
Cartridge: 9 × 19 mm Parabellum
Operation: short recoil, single shot or 3-round burst
Locking: hinged block
Feed: 15- or 20-round detachable box magazine
Weight: with 15-round magazine, 1.12 kg; with 20-round magazine, 1.17 kg
Length: 240 mm
Barrel: including muzzle brake, 156 mm
Rifling: 6 grooves, rh, 1 turn in 250 mm
Sights: fore, blade integral with slide; rear, notched bar dovetailed to slide
Sight radius: 160 mm
Rate of fire: cyclic, ca 1,100 rds/min
Selector: manual, single self-loading shot or 3-round bursts
Muzzle velocity: 375 m/s
Metal stock: length folded, 195 mm; extended, 368 mm
Weight of stock: 270 g

Manufacturer
Armi Beretta SpA, Via Beretta 18, I-25063 Gardone VT (Brescia).

Status
In production.

Service
Adopted by Italian and foreign special forces.

UPDATED

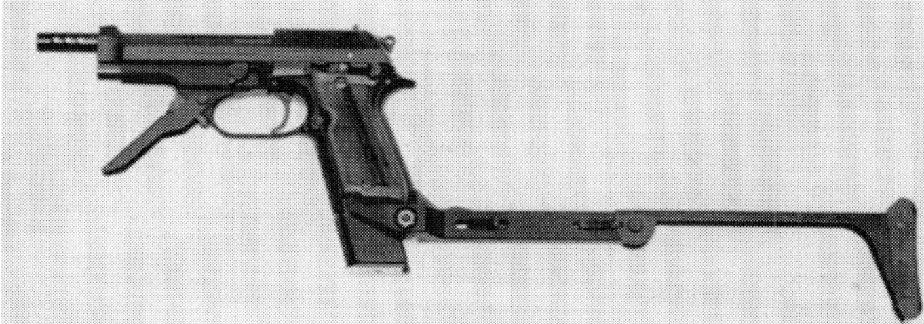

Beretta 9 mm Model 93R selective fire pistol with 20-round magazine, ready for single-handed firing

Beretta 9 mm Model 93R selective fire pistol with extended stock fitted and forehand grip folded down

Bernardelli 9 mm P-018 pistol

Description
The P-018 is a military automatic pistol which combines functionality with strong design and convenient features.

The most important characteristics are an all-steel frame, double-action lockwork and a magazine capacity of 15 rounds. The P-018-9 with 9 mm Parabellum chambering was specially produced for use by military and police forces.

The pistol is a semi-automatic double-action design, with the breech locked on discharge. The outline is squared-off but compact and the trigger guard is shaped for a two-handed grip. The magazine holds 15 rounds in staggered column, plus one round in the

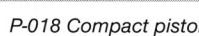

P-018 Compact pistol

P-018-9 pistol

chamber. The P-018 is said to handle well, be well-balanced and have a relatively soft recoil due to the block locking system employed.

P-018 Compact

This model is generally the same as the P-018 but smaller in all dimensions. The magazine capacity is reduced by only one round.

Data

VARIATIONS OF MODEL P-018 COMPACT SHOWN IN PARENTHESIS

Cartridge: 9 mm Parabellum
Operation: semi-automatic, double action
Locking: block locking-breech
Feed: 15-round box magazine (14)
Weight: empty, 998 g (950 g)
Length: 213 mm (190 mm)
Barrel: 122 mm (102 mm)
Rifling: 6 grooves, rh
Sights: fore, blade; rear, notch adjustable for windage
Sight base: 160 mm (137 mm)

Manufacturer

Vincenzo Bernardelli SpA, I-25063 Gardone Val Trompia, Brescia.

Status

In production.

VERIFIED

Bernardelli Mod. USA Pistol

Description

The Bernardelli Mod. USA pistol is available in three calibres: 9 mm Short, 7.65 mm ACP and 0.22 LR. It is a single-action semi-automatic employing a locked-breech and is made entirely of steel. There is a loaded chamber indicator and provision for lowering the hammer on a loaded chamber. The rearsights are adjustable for windage and elevation.

Data

Cartridge: 9 × 17 mm (9 mm Short/0.380 Auto)
Operation: semi-automatic, single action
Feed: 7-round box magazine
Weight: empty, 690 g
Length: 164 mm
Barrel: 90 mm
Rifling: 6 grooves, rh
Sights: fore, fixed blade; rear, adjustable
Sight base: 113 mm
Muzzle velocity: 300 m/s

Manufacturer

Vincenzo Bernardelli SpA, I-25063 Gardone Val Trompia, Brescia.

Status

In production.

VERIFIED

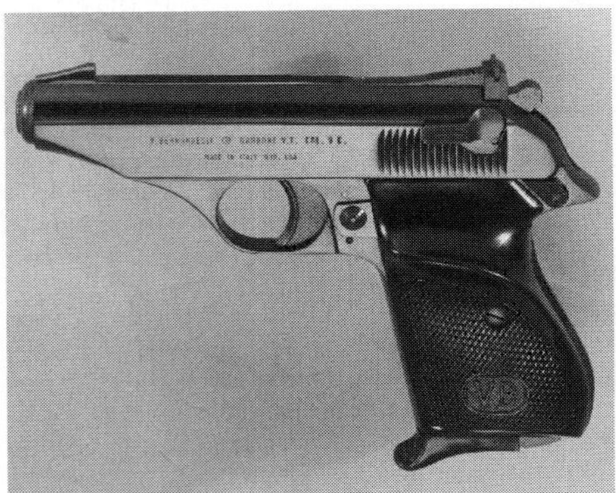

Bernardelli Mod. USA pistol

Tanfoglio double-action pistols

Description

All Tanfoglio double-action pistols are locked-breech weapons with a dropping barrel on the Colt/Browning system and a staggered magazine. The pistols are manufactured for firing the standard 9 mm Parabellum cartridge, the 0.40 S&W, the 0.41 AE, the 0.45 ACP, the 10 mm Auto and the 'Italian' calibres, 9 × 21 mm IMI and 0.45 HP.

The frame is of cast steel, while the slide and barrel are machined from forged steel; all steels are chrome-molybdenum alloy. All pistols are supplied in black finish or hard-chromed, and are also available with frame and slide in stainless steel.

On the 'Standard' models there is a manual safety catch on the slide which operates to lock the firing pin and releases the hammer and the trigger, ensuring safety at all times.

On the 'Combat' models there is a manual cocked-and-locked safety mounted on the frame, with automatic firing pin safety that operates to lock the firing pin until the trigger is pulled.

The following data refers to the Tanfoglio Standard Models TA90, TA40, TA41, TA10 and TA45.

9 mm Tanfoglio TA90 Standard model

Data

Cartridge: 9 × 19 mm Parabellum; 9 × 21 mm; 0.40 S&W; 0.41 AE; 10 mm Auto; 0.45 ACP; 0.45 HP
Operation: recoil, semi-automatic, double action
Locking: dropping barrel
Feed: box magazines: 9 mm, 15 rounds; 0.40, 12 rounds; 0.41, 11 rounds; 10 mm, 12 rounds; 0.45, 10 rounds
Weight: empty, ca 1.015 kg
Length: 202 mm
Barrel: 120 mm
Rifling: 9 mm, 6 grooves, rh, 1/250 mm; 10 mm, 0.40, 0.41, 8 grooves, rh, 1/400 mm; 0.45, 6 grooves, rh, 1/400 mm
Sights: fore, fixed blade; rear, adjustable U-notch

Manufacturer

Fratelli Tanfoglio SpA, Via Val Trompia 39/41, I-25063 Gardone Val Trompia, Brescia.

Status

In production.

VERIFIED

Tanfoglio Baby Standard Models TA90, TA40, TA41, TA10 and TA45

Description

The Tanfoglio Baby models are similar to the Standard models but of compact dimensions; they fit well in the hand, are easily concealable, deliver exceptional accuracy and are comfortable to shoot.

Data

Cartridge: 9 × 19 mm Parabellum ; 9 × 21 mm; 0.40 S&W; 0.41 AE; 10 mm Auto; 0.45 ACP; 0.45 HP
Operation: recoil, semi-automatic, double action
Locking: dropping barrel
Feed: box magazines: 9 mm, 12 rounds; 0.40, 9 rounds; 0.41, 8 rounds; 10 mm, 9 rounds; 0.45, 7 rounds
Weight: empty, ca 850 g
Length: 175 mm
Barrel: 90 mm
Rifling: 9 mm, 6 grooves, rh, 1/250 mm; 10 mm, 0.40, 0.41, 8 grooves, rh, 1/400 mm; 0.45, 6 grooves, rh, 1/400 mm
Sights: fore, fixed blade; rear, adjustable U-notch

Manufacturer

Fratelli Tanfoglio SpA, Via Val Trompia 39/41, I-25063 Gardone Val Trompia, Brescia.

Status

In production.

VERIFIED

Tanfoglio Baby Standard model pistol

Tanfoglia Ultra Model

Description

The Tanfoglia 'Ultra' is a sport version of the Combat models, with longer, ported, barrel and slide so as to reduce the muzzle rise on firing. The 'Ultra' is approved by the IPSC (International Practical Shooting Confederation).

Data is as for the Standard models except that the weight is 1.04 kg, the length 210 mm and the barrel length 130 mm. The front sight is fixed, the rear fully adjustable.

Manufacturer

Fratelli Tanfoglio SpA, Via Val Trompia 39/41, I-25063 Gardone Val Trompia, Brescia.

Status

In production.

VERIFIED

Tanfoglio Combat Models

Description

The Tanfoglio Combat models comprise the TA90, TA40, TA41, TA10 and TA45 Combat models which differ from the Standard models solely in their safety arrangements. The dimensions are the same in all respects.

Manufacturer

Fratelli Tanfoglio SpA, Via Val Trompia 39/41, I-25063 Gardone Val Trompia, Brescia.

Status

In production.

VERIFIED

Tanfoglio TA90 Combat model pistol

Tanfoglio TA90 Baby Combat; similar to the Baby Standard but with different safety

Tanfoglio Model S

Description

The Tanfoglio Model S is the compensated version of the Combat models, fitted with a muzzle compensator that reduces both recoil and muzzle rise on firing. All data is as for the Standard models except that the weight is 1.2 kg, the length 250 mm and the barrel length 135 mm.

Manufacturer

Fratelli Tanfoglio SpA, Via Val Trompia 39/41, I-25063 Gardone Val Trompia, Brescia.

VERIFIED

Tanfoglio Model S pistol

JAPAN

0.38 Model 60 New Nambu revolver

Description

This Smith & Wesson type pistol in 0.38 Special has been the Japanese police pistol since 1961 and 130,000 have been sold. The revolver is also issued to the Japanese Maritime Safety Guard.

Data

Cartridge: 0.38 Special
Operation: manual - revolver, single or double action
Feed: 5-chamber cylinder
Weight: 680 g
Length: 197 mm
Barrel: 77 mm
Sights: fore, fixed half-tapered serrated ramp; rear, square notch
Muzzle velocity: 220 m/s
Effective range: 40 m

Manufacturer

Omori Factory, Division of Minebea Co Limited (Tokyo), 18-18 Omori-nishi 4-chome Ohta-ku, Tokyo 143.

Status

In production.

Service

Japanese police forces and Maritime Safety Guard.

UPDATED

0.38 Model 60 New Nambu revolver

KOREA, NORTH

7.62 mm Type 68 pistol

Development

The 7.62 mm Type 68 pistol is a much modified Tokarev TT-33 (qv). It is shorter and bulkier than either the former Soviet TT-33 or the Chinese Type 51 or Type 54. It may further be distinguished from these pistols by the serrations on the rear of the slide intended to give a grip while the weapon is being cocked. The variations are: the Soviet TT-33: vertical, alternately wide and narrow; the Chinese Type 51 or 54: vertical, narrow; the North Korean Type 68: sloping forward, narrow.

Description

Internally the Tokarev TT-33 has been reworked considerably. The link system, used to lift and lower the barrel ribs into and out of the grooves in the slide, has been replaced by a cam cut into a lug under the chamber in a manner similar to that used in the Browning 9 mm High Power pistol. The magazine catch has been relocated and is now at the heel of the magazine. The magazine from the Tokarev TT-33 will work in the Type 68 pistol but the reverse is not true as the Type 68 magazine lacks the necessary cut-out for the magazine catch. The firing pin is retained by a plate instead of a cross pin and

North Korean 7.62 mm Type 68 pistol

the slide stop is a robust pin instead of the rather fragile clip in the TT-33.

A poor feature of the TT-33 which has had to be retained is the large-radius curve at the junction of butt and slide which presses into the web of the thumb when firing. It was not possible to machine this into a smaller radius without a complete redesign of the hammer mechanism.

Data

Cartridge: 7.62 × 25 mm Type P; 7.63 mm Mauser
Operation: short recoil, self-loading, single action
Locking: projecting lug
Feed: 8-round box magazine
Weight: empty, 795 g
Length: 185 mm
Barrel: 108 mm
Rifling: 4 grooves, rh, 1 turn in 305 mm
Sights: fore, blade; rear, notch
Sight radius: 160 mm
Muzzle velocity: 395 m/s

Manufacturer

State factories.

Status

In service with North Korean forces.

VERIFIED

7.65 mm Type 64 pistol

Description

In introducing the 7.65 mm Type 64 pistol the North Korean design authority responsible resurrected the Browning Model 1900. A photograph of Browning's pistol is shown.

North Korean 7.65 mm Type 64 pistols

The North Korean Type 64 has the stamping '1964 7.62' on the left side but in fact takes the 7.65 × 17 SR cartridge, the American 0.32 ACP.

There is also a silenced version of the Type 64. It has a shortened slide to allow the muzzle to protrude and the end of the barrel is threaded to take the silencer attachment.

The basic parameters of the pistol are the same as the Browning Model 1900.

Data

Cartridge: 7.65 mm (0.32 ACP)
Operation: blowback, semi-automatic
Locking: nil
Feed: 7-round detachable box magazine
Weight: 624 g
Length: 171 mm
Barrel: 102 mm
Rifling: 6 grooves, rh
Sights: fore, blade; rear, notch
Muzzle velocity: approx 290 m/s
Effective range: 30 m

Manufacturer

State factories.

7.65 mm Model 1900 Browning pistol

Status

Probably no longer manufactured.

Service

North Korean forces.

VERIFIED

KOREA, SOUTH

Daewoo 9 mm DP51 pistol

Description

The Daewoo 9 mm DP51 is a semi-automatic pistol in 9 mm Parabellum calibre operating on the delayed blowback system. Designed for military and police use, it has a double-action trigger mechanism and the sights are fitted with luminous dots for operation in poor light conditions. The frame is high tensile aluminium alloy.

A 9 mm DP51C compact version is available. This weighs 737 g, is 177 mm long with a 89 mm barrel, and has a 10-round magazine capacity.

Data

Cartridge: 9 × 19 mm Parabellum
Operation: delayed blowback, semi-automatic, single or double action
Feed: 10- or 13-round box magazine
Weight: with empty magazine, 800 g
Length: 190 mm
Barrel: 105 mm
Rifling: 6 grooves, rh, 1 turn in 254 mm
Sights: fore, blade with luminous dot; rear, square notch, 2 luminous dots
Sight radius: 145 mm
Muzzle velocity: 351 m/s

Manufacturer

Daewoo Precision Industries Limited, PO Box 25, Kumjung, Pusan.

Status

In production.

Service

South Korean Army.

UPDATED

Daewoo 9 mm DP51 pistol

Daewoo 0.40 DH40 pistol

Description

The Daewoo 0.40 DH40 is broadly the same pistol as the DP51 described above except for being chambered for the 0.40 S&W cartridge. A semi-automatic, double-action weapon using delayed blowback operation it is well-suited to military and police use.

Data

Cartridge: 0.40 S&W
Operation: delayed blowback, semi-automatic, double action
Feed: 10- or 12-round box magazine
Weight: with empty magazine, 907 g
Length: 190 mm
Barrel: 170 mm
Rifling: 6 grooves, lh
Sights: fore, blade with white dot; rear, square notch, adjustable for windage, 2 white dots
Muzzle velocity: 300 m/s

Manufacturer

Daewoo Precision Industries Limited, PO Box 25, Kumjung, Pusan.

Status

In production.

UPDATED

Daewoo 0.40 S&W DH40 pistol

Daewoo 0.22 LR DP52 pistol

The Daewoo 0.22 DP52 is a fixed-barrel semi-automatic blowback pistol in 0.22 LR calibre, intended for police and security duties. The design was obviously influenced by the Walther PP series (qv). The DP52 has a double-action trigger mechanism, and the sights have self-luminous markers inserted for firing at night or in poor visibility.

The Daewoo DH-380 is similar to the DP52 but is chambered for the 0.380 ACP cartridge. Magazine capacity for the DH-380 is limited to eight rounds.

Data

Cartridge: 0.22 LR rimfire
Operation: blowback, semi-automatic, double or single action
Feed: 10-round box magazine
Weight: 652 g
Length: 170 mm
Barrel: 97 mm
Rifling: 6 grooves, rh, 1 turn in 405 mm
Sights: fore, blade with self-luminous insert; rear, square notch with 2 self-luminous dots
Muzzle velocity: 290 m/s

Manufacturer

Daewoo Precision Industries Limited, PO Box 25, Kumjung, Pusan.

Status

In production.

Service

South Korean police.

UPDATED

Daewoo 0.22 LR DP52 pistol

POLAND

9 mm P-64 self-loading pistol

Development

The Polish armed forces were at one time equipped with the 7.62 mm Pistolet TT which was identical to the Soviet Tokarev TT-33 pistol except for the handgrips. This pistol is now obsolete and has been replaced by the blowback-operated P-64 which, although an original Polish design, looks rather like the CIS Makarov and has some design features which originated with the German Walther PP pistol. It has the inscription '9 mm P-64' on the left side of the slide.

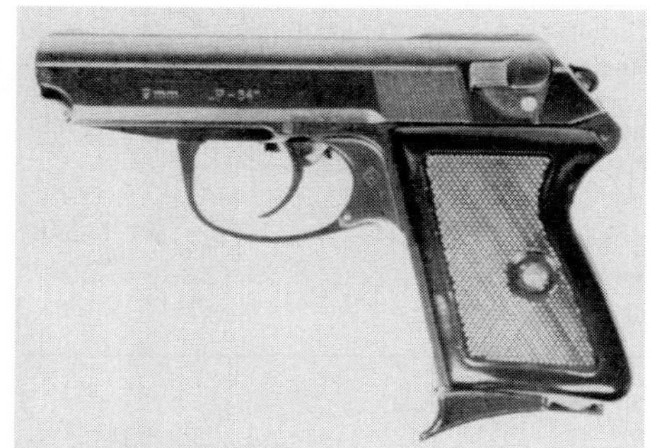

9 mm P-64 pistol

Description

The P-64 is a blowback-operated pistol of conventional form. It is prepared for firing by inserting a magazine, pulling back the slide and releasing it to chamber a round. The hammer is then cocked and the pistol may be fired. Alternatively, the safety catch may be pressed down; this rotates two lugs into place to protect the firing pin and allows the hammer to fall on to these lugs. The pistol can then be fired by pushing up the safety catch and pulling through on the trigger in double-action mode. When the magazine is emptied the slide is held by a slide stop.

Data

Cartridge: 9 × 18 mm
Operation: blowback, self-loading, single or double action
Locking: nil
Feed: 6-round box magazine
Weight: empty, 620 g
Length: 160 mm
Barrel: 84.6 mm
Rifling: 4 grooves, rh
Sights: fore, blade; rear, notch
Sight radius: 117 mm
Muzzle velocity: 310 m/s

Manufacturer

Zaklady Metalowe LUCZNIK, ul 1905 Roku 1/9, 26-600 Radom.

Status

No longer manufactured.

Service

Polish armed forces.

UPDATED

9 mm VANAD P-83 pistol

Description

Development of the VANAD P-83 pistol began in the late 1970s, the object being to replace the P-64 with a similar pistol but one which would be cheaper and easier to manufacture. The resulting weapon was introduced as the VANAD P-83, a simple fixed-barrel blowback chambered for the 9 mm Makarov or 9 mm Short cartridges and generally similar to the P-64 in its principles. The difference lies in manufacture, much use being made of pressings, forgings and welding which has reduced cost at the expense of a considerable increase in weight.

The trigger is double action, and an external safety lever drops the cocked hammer when applied; at the same time the rear of the firing pin is lowered so that it is opposite a hammer arm and thus cannot strike a cartridge in the chamber. There is also a loaded chamber indicator.

The standard finish is a black oxide coating, although the whole pistol or some of its components can be delivered with a bright or dull chrome finish. Other options include high contrast sights and adjustable rear sights.

9 mm VANAD P-83 pistol

1996

P-83G Gas Pistol

The P-83G gas pistol is basically a blank-firing VANAD P-83 pistol with a muzzle attachment capable of projecting disabling gas pellets. An alternative muzzle attachment can be used to launch up to four flares. If required the P-83G can fire blanks as a starting or training pistol.

Data

Cartridge: 9 × 18 mm Makarov; 9 × 17 mm Short
Operation: blowback, self-loading, single or double action
Locking: nil
Feed: 8-round detachable box magazine
Weight: empty, 730 g
Length: 165 mm
Barrel: 90 mm
Rifling: 4 grooves, rh
Sights: fore, blade; rear, notch
Sight radius: 120 mm
Muzzle velocity: 9 mm Makarov, 312 m/s; 9 mm Short, 284 m/s

Manufacturer

Zaklady Metalowe LUCZNIK, ul 1905 Roku 1/9, 26-600 Radom.

Status

In production.

Service

Polish armed forces.

UPDATED

9 mm P-93 pistol

Description

The 9 mm P-93 is a developed version of the earlier VANAD P-83 pistol (see previous entry) chambered for the 9 × 18 mm Makarov cartridge, although the design could be adapted for the 9 mm Short cartridge. The main changes are a revised outline to make the pistol suitable for two-handed firing, and a safety that allows safe carrying of the pistol with the external hammer lowered while a round is chambered. This operation is carried out using a lever below the slide on the left of the pistol. Pulling the trigger releases this safety, allowing the round to be fired. There is also a drop safety which blocks hammer movement.

The standard finish is a black oxide coating, although the whole pistol or some of its components can be delivered with a bright or dull chrome finish. Other options include adjustable rear sights; high contrast sights are standard.

Data

Cartridge: 9 × 18 mm Makarov
Operation: blowback, self-loading, double action
Locking: nil
Feed: 8-round detachable box magazine
Weight: empty, 750 g

Length: 178 mm
Barrel: 100 mm
Rifling: 4 grooves, rh
Sights: fore, blade; rear, notch adjusted to 25 m
Sight radius: 135 mm
Muzzle velocity: 316 m/s

Manufacturer
Zaklady Metalowe LUCZNIK, ul 1905 Roku 1/9, 26-600 Radom.

Status
In production. Offered for export sales.

Service
Polish armed forces.

NEW ENTRY

9 mm P-93 pistol
1996

9 mm MAG 95 pistol

Description
With the MAG 95 pistol the designers at the Zaklady Metalowe LUCZNIK have abandoned their previous design practices based on the use of the 9 × 18 mm Makarov cartridge and have instead utilised a Western-influenced approach involving the 9 × 19 mm Parabellum cartridge and some of the latest safety and other design techniques. The MAG 95 incorporates most of the user facilities to which firers in the West have become accustomed, but which have been absent from most Eastern Bloc pistols until now.

The MAG 95 relies to a large extent on design features from the later Browning designs, and others, combining features from several weapons into one pistol. The locking system is the widely used dropping barrel and cam action of the Browning series. This is combined with a double-action trigger mechanism combined with a hammer decocking lever which allows the external hammer to be lowered in safety when a round is chambered. Only a positive pull on the trigger will allow the hammer to recock for firing the round. The barrel is chrome plated to prolong its life and to permit easier cleaning.

The MAG 95 is suitable for use by both right- and left-handed users and is designed for two-handed firing. It is claimed that the pistol is well balanced to produce greater accuracy when rapid firing.

The box magazine contains 15 rounds; an optional 20-round magazine is available. Other options include a laser target indicator to be mounted in front of the trigger guard. The standard finish is a black oxide coating but a bright or matt chrome finish is optional.

9 mm MAG 95 pistol
1996

Data
Cartridge: 9 × 19 mm Parabellum
Operation: short recoil, double action with external hammer
Locking: dropping barrel, cam actuated
Feed: 15-round detachable box magazine
Weight: empty, 1.1 kg
Weight of magazine: empty, 80 g
Trigger pull: double action, max 50 N
Length: 200 mm
Barrel: 115 mm
Sights: fore, fixed blade; rear, notch adjusted to 25 m; high contrast with illumination
Sight radius: 143 mm
Muzzle velocity: 356 m/s
Muzzle energy: approx 506 J

Manufacturer
Zaklady Metalowe LUCZNIK, ul 1905 Roku 1/9, 26-600 Radom.

Status
In production. Offered for export sales.

NEW ENTRY

SOUTH AFRICA

Vektor 9 mm Z-88 pistol

Development
In 1985 the South African police expressed a requirement for a new pistol. Because of the then-current arms embargo, the acquisition of foreign pistols was very difficult in South Africa, so it was decided to investigate the possibility of local manufacture. Subsequent discussions between the police and Armscor led to the drawing up of a specification in April 1986, and LIW, a Division of Denel (Pty) Limited, was instructed to proceed with the project, the aim being to start production within two years. The project plan proceeded according to schedule and by August 1988 200 pistols had been manufactured to confirm the production capability and to allow a quantity of pistols to be subjected to field testing. The testing involved the firing of some thousands of rounds from 20 pistols, as well as the usual environmental tests.

The Z-88 pistol takes its name from the late Mr T D Zeederberg (former general manager of LIW, who was instrumental in the successful handling of the project by the company) and from its year of introduction.

Vektor 9 mm Z-88 pistols
1996

Description
The pistol is a double-action locked-breech weapon in 9 mm Parabellum calibre. The barrel is locked by a floating wedge between barrel and frame. Rotation of a manually applied safety lever on the slide separates the firing pin from the hammer, lowers the hammer and interrupts the connection between the trigger and sear.

The design of the Vektor Z-88 was based on that of the Beretta Model 92 although there are some locally introduced innovations such as titrium luminous night sights and a reversible magazine release button for the convenience of left-handed firers.

Data
Cartridge: 9 × 19 mm Parabellum
Operation: short recoil, semi-automatic with double action
Locking: falling block
Feed: 15-round detachable box magazine
Weight: with empty magazine, 995 g
Length: 217 mm
Barrel: 125 mm
Rifling: 6 grooves, rh, 1 turn in 254 mm
Sights: fore, blade integral with slide; rear, notched bar dovetailed into slide; titrium luminous night sights provided front and rear
Sight radius: 158 mm
Muzzle velocity: 330-360 m/s

Manufacturer
LIW Division of Denel (Pty) Limited, 68 Selborne Avenue, Lyttelton, Pretoria 0001.

Status
In production.

Service
South African police and security forces, as well as the commercial market.

UPDATED

Vektor 9 mm SP1 pistol

Development

The Vektor SP1 was developed by LIW, a Division of Denel (Pty) Limited, and adopted by the South African National Defence Force. It is a recoil-operated semi-automatic weapon with double-action mechanism and a 15-shot magazine. Design detail includes the use of an ambidextrous manual safety lever and a reversible magazine release button for left-handed firers.

A Vektor SP2 was developed to fire the 0.40 S&W cartridge but has yet to be produced in quantity. On the SP2 the magazine capacity is 11 rounds (plus one in the chamber), the rifling is six polygonal grooves, right-hand, one turn in 254 mm, and the muzzle velocity lies between 300 and 350 m/s. A conversion kit can be provided for the SP2 to permit firing 9 mm Parabellum cartridges. This consists of a barrel, recoil spring and magazine, the only components affected by the calibre change, and the conversion is simply a matter of field-stripping the pistol and substituting these parts for the corresponding 0.40 parts.

There is also a Vektor SP1 Sport for competition shooting. This model has modified sights and a muzzle compensator.

Vektor 9 mm SP1 pistol

1996

Description

On the Vektor SP1 the enveloping slide is machined from solid bar steel, eliminating possible material defects. The frame is of aircraft-quality aluminium alloy, and the principal components of the pistol are manufactured to close tolerances on modern CNC machinery.

An automatic firing pin blocking safety ensures that no discharge can take place unless the trigger is pressed. A manual safety catch which locks the slide and sear is fitted to both sides of the weapon, and the magazine release can be removed and repositioned on either side of the butt to suit the user. A slide stop ensures the slide remains open when the last round has been fired, thus indicating that the 15-round staggered box magazine is empty. Despite the magazine capacity the use of a wraparound grip ensures that the pistol is suitable for users with small hands. The hammer-forged barrel has polygonal rifling with four grooves, reducing bullet friction and wear.

Data

Cartridge: 9 × 19 mm Parabellum
Operation: recoil, semi-automatic
Locking: dropping wedge
Feed: 15-round box magazine
Weight: with empty magazine, 995 g
Length: 210 mm
Barrel: 118 mm
Sights: fore, blade; rear, square notch. Tritium inserts optional
Sight radius: 156 mm
Rifling: 4 grooves, rh, polygonal, 1 turn in 254 mm
Muzzle velocity: 330 to 360 m/s

Manufacturer

LIW Division of Denel (Pty) Limited, 68 Selborne Avenue, Lyttelton, Pretoria 0001.

Status

In production.

Service

South African National Defence Force.

UPDATED

Vektor 9 mm CP1 Compact pistol

Description

The Vektor 9 mm CP1 is an entirely new design using the latest materials technology. The frame is made from a sophisticated polymer, resulting in low weight, and the trigger, safety slide, magazine bottom covers and trigger safety are also of polymer. The cold-forged polygon barrel is made from stainless steel and the slide is made from a high tensile steel and surface treated to ensure a scratch-free and rust-resistant finish.

A gas buffer delayed blowback system is used, delaying the rearward movement of the slide until the chamber pressure has dropped to a safe level. There is an internal hammer and a slide hold-open device. Safety is catered for by means of a safety slide on the front of the trigger guard, a trigger safety on the trigger and a half-cock notch on the hammer. The rearsight is integrated in a polymer cap fitted to the rear of the slide; the front sight is adjustable for windage.

A unique feature of the Vektor CP1 is that no device, lever or catch is fitted on the pistol to increase the effective width. The envelope, together with the ergonomical design, contributes to the fact that the CP1 is principally intended to be carried as a concealed pistol.

9 mm Vektor CP1 Compact pistol

Data

Cartridge: 9 × 19 mm Parabellum
Operation: delayed blowback
Feed: 13-round (standard) or 12-round (compact) box magazine
Weight: standard magazine, 700 g; compact magazine, 690 g
Length: 176 mm
Barrel: 99 mm
Rifling: 4 grooves, rh, polygonal, 1 turn in 254 mm
Sight radius: 136 mm
Muzzle velocity: 330 to 360 m/s

Manufacturer

LIW, a Division of Denel (Pty) Limited, 68 Selborne Avenue, Lyttleton, Pretoria 0001

Status

Ready for production.

UPDATED

9 mm ADP Mk II pistol

Description

The ADP Mk II lightweight pistol takes its designation from the designer, Alex Du Plessis, who also developed the LDP sub-machine gun. The original ADP design was replaced by the ADP Mk II in 1994, introducing several enhancements mainly based around the use of a more ergonomic outline to enable the relatively small dimensioned pistol to fit into the hand with greater ease and user comfort. One innovation involves a fully ambidextrous safety lever while the magazine release catch configuration can be reversed for left-handed users. Other changes include a magazine extension for grip comfort, and the sights have been raised for easier aiming. An optional 15-round box magazine was also introduced.

The pistol frame is a synthetic moulding with embedded slide rails. The slide is of cast and machined steel and the entire pistol is covered by a black oxide surface finish for durability. The operation is by retarded blowback; a gas port in the barrel venting into a cylinder in the frame where the high-pressure gas acts against a piston attached to the front end of the slide. So long as the pressure in the barrel and cylinder remains high, the slide is prevented from moving rearward. As soon as the bullet has left the barrel the gas is able to exhaust from the cylinder and the slide can then move back to perform the reloading cycle.

The firing mechanism is a self-cocking striker. Pulling the trigger causes a sear to engage in the striker and force it back, further compressing the striker spring. As the spring becomes fully compressed so the operating arm is disengaged and the striker then flies forward to fire the round in the chamber. For full drop safety a manual safety catch interposes a block behind the trigger, preventing it from engaging the striker. There is also an open breech disconnector.

9 mm ADP Mk II pistol

1996

Data

Cartridge: 9 × 19 mm Parabellum
Operation: retarded blowback, semi-automatic
Locking: gas pressure delay
Feed: 10-round box magazine (15-round magazine optional)
Weight: empty, 572 g
Length: 160 mm
Barrel: 91 mm
Width: 28 mm
Rifling: 6 grooves, rh, 1 turn in 254 mm
Sights: fore, fixed blade; rear, square notch adjustable for windage

Manufacturer

Aserma Manufacturing, 115 Escom Road, New Germany 3620

Status

In production.

UPDATED

SPAIN

Llama 0.45 Model IX-C

Description

The Model IX-C is the latest version of a design which first appeared in 1936. It is a conventional semi-automatic based upon the well-known Colt M1911A1 pattern, and, except for some minor differences in the shape of the slide, is almost identical to the Colt. It uses the same components, strips and reassembles in the same way, and fires the standard 0.45 ACP cartridge. The major change lies in the magazine, which holds 13 rounds in a double column, leading to some changes in the frame structure around the magazine aperture. Though fractionally shorter than the M1911A1 it is slightly heavier and appears to be a robust and reliable design. The finish is to a high standard, the basic pistol being blued and polished, with hard rubber grips; engraving and gold damascene embellishments can be provided as required.

Data

Cartridge: 0.45 ACP
Operation: short recoil, semi-automatic
Locking: dropping barrel
Feed: 13-round box magazine
Weight: empty, 1.2 kg
Length: 216 mm
Sights: fore, blade; rear, square notch, adjustable for windage
Muzzle velocity: ca 255 m/s

Manufacturer

Llama-Gabilondo y Cia SA, Portal de Gamarra 50, Apartado 290, E-01080 Vitoria.

Status

In production.

VERIFIED

0.45 ACP Llama Model IX-C pistol

9 mm Llama M-82

Description

The 9 mm Llama M-82 is a locked-breech 9 mm Parabellum pistol using a double-action lock. Breech locking is performed by a dropping block, similar to that used in the Walther P38 pistol, a system which generally provides more reliable feed, firing and extraction since the barrel remains fixed in relation to the rest of the weapon. There is a slide-mounted safety catch which, when operated, conceals and locks the firing pin and disconnects the trigger bar.

To load, the safety is lowered to the 'safe' position, a magazine inserted, and the slide drawn back and released. This feeds a round into the chamber and as the slide closes so the hammer drops in a safe condition. To fire, all that is necessary is to move the safety up to the 'fire' position and pull the trigger; this will cock and release the hammer. Subsequent shots are fired in the single-action mode. After the last shot in the magazine has been fired the magazine follower rises and holds the slide in the rearward position.

Data

Cartridge: 9 × 19 mm Parabellum
Operation: recoil, semi-automatic
Locking: dropping block
Feed: 15-round box magazine
Weight: steel frame, 1.11 kg; alloy frame, 875 g
Length: 209 mm
Barrel: 114 mm
Rifling: 6 grooves, rh
Muzzle velocity: 345±3 m/s (V_{15})

Manufacturer

Llama-Gabilondo y Cia SA, Portal de Gamarra 50, Apartado 290, E-01080 Vitoria.

Status

In production.

Service

Spanish Army.

VERIFIED

9 mm Llama M-82 pistol

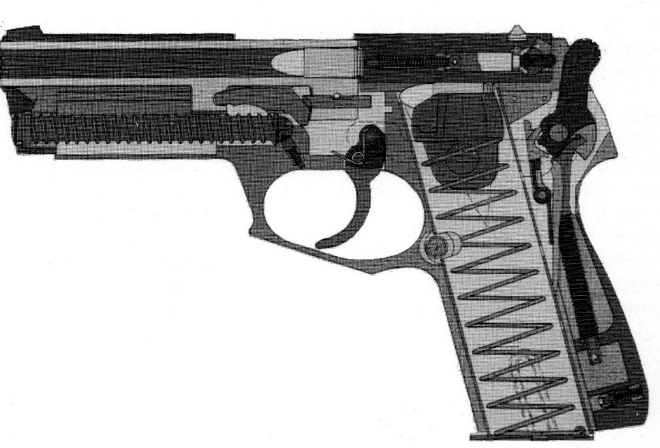

Llama M-82 pistol mechanism

9 mm Llama M-87 pistol

Description

The Llama M-87 is essentially an enlarged and improved version of the M-82; the basic mechanism remains unchanged, using the same dropping block system of breech locking. The barrel has been lengthened, and the slide has been extended to match by an attachment which acts as a balance weight and as a muzzle brake, being ported at the upper sides. The frame and magazine are in a new chemically treated nickel finish which is attractive as well as being highly resistant to corrosion. The magazine catch has been enlarged and the magazine opening is bevelled to ease insertion of the magazine. The trigger is micrometrically adjustable for length and sensitivity of pull and the single-action pull has been improved. A version without muzzle brake and with a fully adjustable sight is available for competition shooting.

Data

Cartridge: 9 × 19 mm Parabellum
Operation: recoil, semi-automatic, double action
Locking: dropping block
Feed: 15-round box magazine
Weight: empty, 1.235 kg
Length: 245 mm
Barrel: 133 mm
Rifling: 6 grooves, rh, 1 turn in 250 mm
Sights: fore, blade; rear, square notch, laterally adjustable
Sight radius: 190.5 mm
Muzzle velocity: ca 370 m/s

Manufacturer

Llama-Gabilondo y Cia SA, Portal de Gamarra 50, Apartado 290, E-01080 Vitoria.

Status

In production.

VERIFIED

9 mm Llama M-87 pistol

---

9 mm M-43 Firestar pistol

Description

The Firestar is among the smallest of 9 mm Parabellum pistols. It uses the Colt-Browning system of dropping barrel to lock the breech, with three locking lugs on the barrel and a shaped cam in the breech lump. The muzzle end of the barrel is conical and shaped so as to lock with the slide after each shot and render a barrel bushing redundant. The interaction of the shaped muzzle and the slide also improves the accuracy of the pistol. Operation is single action only, with an ambidextrous applied safety catch and an automatic firing pin safety.

Data

Cartridge: 9 mm Parabellum
Operation: short recoil, semi-automatic, single action
Locking: projecting lugs
Feed: 7-round detachable box magazine or 8 rounds with special extension magazine
Weight: empty, 798 g
Length: 163 mm
Barrel: 86 mm
Rifling: 6 grooves, rh
Sights: fore, blade; rear, notch
Sight base: 115 mm
Muzzle velocity: 348-380 m/s

Manufacturer

Star Bonifacio Echeverria SA, PO Box 10, E-20600 Eibar.

Status

In production.

VERIFIED

9 mm M-43 Firestar pistol

M-40 and M-45 Firestar pistols

Description

Just as the 9 mm M-43 Firestar lays claim to being the smallest of its class, so the M-40 and M-45 models are extremely compact pistols in 0.40 S&W and 0.45 ACP calibres respectively. They are slimmer than revolvers of similar calibre and can be carried easily and unobtrusively, yet deliver considerable power in spite of their short barrels.

The general design of both the M-40 and the M-45 is of a modern recoil-operated, double-action weapon with simple but effective safeties. The only manual safety device is an ambidextrous safety catch which locks slide and hammer if the hammer is down, hammer only if the hammer is cocked, thus permitting the slide to be withdrawn to check the chamber. A magazine safety, automatic firing pin safety and a half-cock safety notch on the hammer are also provided.

Data

DATA FOR M-40; WHERE M-45 DIFFERS, SHOWN IN PARENTHESIS:
Cartridge: 0.40 S&W (0.45 ACP)
Operation: recoil, semi-automatic, double action
Locking: Browning cam and tilting barrel
Feed: 6-round detachable box magazine
Weight: 855 g (1.025 kg)
Length: 165 mm (170 mm)
Barrel: 86 mm (97 mm)
Sights: fore, blade; rear, square notch
Sight radius: 130 mm
Muzzle velocity: 302 m/s (260 m/s)

Manufacturer

Star Bonifacio Echeverria SA, PO Box 10, E-20600 Eibar.

Status

In production.

VERIFIED

Star 0.40 M-40 Firestar pistol

Star 0.45 M-45 Firestar pistol

10 mm/0.45 Megastar pistols

Description

The Megastar is a full-sized, heavy calibre pistol available either in 10 mm Auto or 0.45 ACP calibre. It is a recoil-operated, double-action design with decocking lever and, for these calibres, large capacity magazines.

Star 0.45 Megastar pistol

The Megastar breech is locked by the traditional lugs on top of the barrel mating with recesses in the slide top, movement of the barrel being controlled by a shaped cam slot beneath the chamber. The manual safety catch on the slide is ambidextrous and can be moved easily between 'safe' and 'fire' positions; when set to 'safe' the firing pin is retracted into the slide and locked, so that no movement of the hammer can affect it. If the safety lever is pressed beyond the normal 'safe' position, it becomes the decocking lever, releasing the hammer to fall safely against the slide surface. When the decocking lever is released, it immediately springs back to the 'safe' position. There is also a magazine safety system: when the magazine is withdrawn, the trigger is disconnected from the firing mechanism.

Dimensions and data are identical for the two calibres except for weight and ballistic performance.

Data

Cartridge: 10 mm Auto; 0.45 ACP
Operation: short recoil, semi-automatic, double action
Locking: Browning cam and tilting barrel
Magazine capacity: 10 mm, 14 rounds; 0.45, 12 rounds
Weight: 10 mm, 1.4 kg; 0.45, 1.36 kg
Length: 212 mm
Barrel: 116 mm
Sights: fore, blade; rear, square notch, white dots
Sight radius: 170 mm
Muzzle velocity: 10 mm, ca 370 m/s; 0.45, ca 270 m/s

Manufacturer

Star Bonifacio Echeverria SA, PO Box 10, E-20600 Eibar.

Status

In production.

VERIFIED

9 mm Star Model 30M and Model 30PK pistols

Description

These two pistols are updated versions of the Models 28 and 28PK, descriptions of which will be found in the 1986/87 edition of *Jane's Infantry Weapons*. They are of the same construction and appearance, but incorporate an ambidextrous safety catch which locks the firing pin when applied. The Model 30M is made entirely of forged steel, while the Model 30PK has a light alloy frame. The pistol is somewhat unusual in having the slide running in internal frame rails; this gives excellent support throughout the slide movement, with a minimum bearing of over 110 mm. There is an ambidextrous safety catch on the slide which retracts the firing pin into its tunnel, out of reach of the hammer. The trigger and hammer action are quite unaffected by the safety catch action and it is possible to pull the trigger so as to drop the hammer after applying the safety and also to pull the trigger to rise and drop the hammer for 'dry firing' practice, without needing to unload the weapon. It is possible to draw the pistol, set at safe, and pull the trigger through, expecting it to fire. The normal safety system at least prevents the trigger or hammer moving and thus gives a quick reminder that the safety is set. There is also a magazine safety which prevents firing if the magazine is removed, but in recognition of the fact that some users do not like these, the manufacturers have built it so that it is possible to remove this feature at will.

The forward part of the trigger guard is shaped for the two-handed grip, and there is a loaded chamber indicator which stands proud of the slide when the weapon is loaded. The sights are clear, the rearsight being adjustable for windage, and the gun is very accurate.

Data

Cartridge: 9 ×19 mm Parabellum
Operation: short recoil, semi-automatic
Locking: Browning cam
Feed: 15-round box magazine
Weight: 30M, 1.14 kg; 30PK, 860 g
Length: 30M, 205 mm; 30PK, 193 mm
Barrel: 30M, 119 mm; 30PK, 98 mm
Rifling: 6 grooves, rh
Sights: fore, blade; rear, notch, adjustable for windage
Muzzle velocity: ca 380 m/s

Manufacturer

Star Bonifacio Echeverria SA, PO Box 10, E-20600 Eibar.

Status

In production.

Service

Spanish armed and police forces; Peruvian police and security forces.

VERIFIED

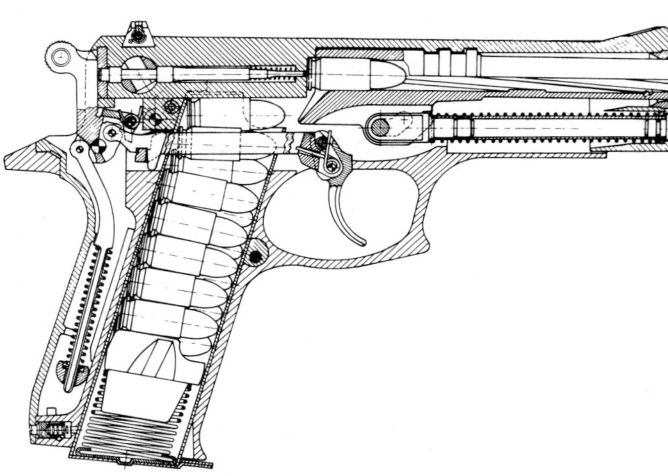

Sectioned view of Star 30M pistol

9 mm Star Model 30M double-action pistol

9 mm Model 105 Compact pistol

Description

As a development of their well-known 30 series and the newer Megastar models, Star designed a new and compact double-action pistol in 9 mm Parabellum chambering to fulfil the requirements of law enforcement agencies for special operations, as well as for the civilian user in search of a dependable and compact pistol.

The Model 105 is a recoil-operated, double-action pistol of modern appearance, breech locking being performed by the well-tried Browning link and tilting barrel. There is a manual ambidextrous safety catch which retracts the firing pin and locks it when set to safe; as with the Megastar, further movement of the safety catch actuates the decocking mechanism, dropping the hammer safely. There is also an automatic firing pin lock, released only by pulling the trigger to fire a

Right side of Star 9 mm Model 105 pistol showing ambidextrous safety and rearsight adjuster

9 mm Star Model 105 Compact pistol

shot, and the usual locking safety system which prevents firing unless the slide is fully forward and the breech locked.

Data

Cartridge: 9 × 19 mm Parabellum
Operation: recoil, semi-automatic, double action
Locking: Browning link and tilting barrel
Feed: 9-round box magazine
Weight: empty, 810 g
Length: 176 mm
Barrel: 89 mm
Sights: fore, blade; rear, fully adjustable square notch, white dots
Sight radius: 135.5 mm
Muzzle velocity: ca 355 m/s

Manufacturer

Star Bonifacio Echeverria SA, PO Box 10, E-20600 Eibar.

Status

In production.

VERIFIED

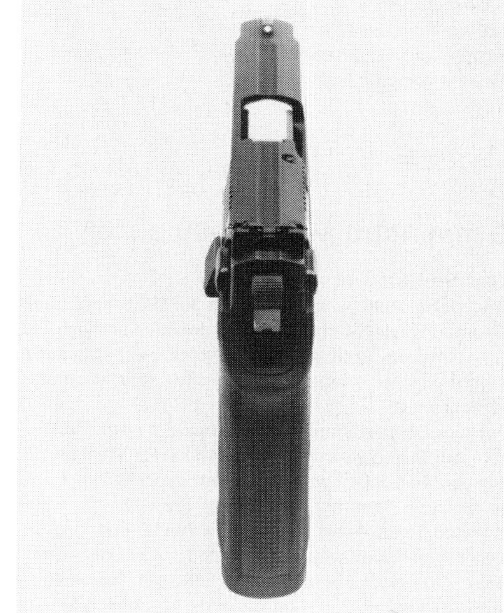

Top view of Star 9 mm Model 105 pistol, showing sights and loaded chamber indicator behind ejection port

9 mm Astra Model A-70 pistol

Description

The Astra A-70 is a locked-breech single-action semi-automatic pistol. The powerful cartridge and small size make it ideally suitable for self-defence, police or military use.

The pistol employs three independent safety devices: an automatic firing pin safety which ensures that the firing pin cannot move unless the trigger is deliberately pressed, a manual safety applied with the thumb which, with the hammer cocked, blocks the trigger and slide, and a half-cock notch on the hammer which prevents the hammer striking the firing pin should the hammer be allowed to slip during cocking or uncocking the weapon.

Data

Cartridge: 9 × 19 mm Parabellum or 0.40 S&W
Operation: short recoil, semi-automatic
Locking: barrel/slide insert
Feed: 8-round (9 mm) or 7-round (0.40) box magazine
Weight: with empty magazine, 830 g
Length: 166 mm
Barrel: 89 mm

Manufacturer

Astra-Unceta y Cía SA, Apartado 3, E-48300 Guernica, Vizcaya.

Status

In production.

VERIFIED

9 mm Astra Model A-70 pistol

9 mm/0.40 Astra A-75 pistol

Description

The Astra A-75 is a compact pistol designed to meet the needs of police and military forces. A recoil-operated weapon, it uses the usual cam-dropped barrel system, the squared-off chamber area locking into the ejection port of the slide. There are three independent safety systems: an automatic firing pin safety, which keeps the firing pin securely locked except during the last movement of the trigger when deliberately pulled; a hammer safety which, should the hammer slip during thumb-cocking, arrests the hammer in the rebounded position and prevents it striking the firing pin (which is, in any case, locked); and a decocking lever, whereby the hammer can be lowered safely on to a loaded chamber. The firing mechanism is double action, and the pistol can be fired from the decocked condition by simply pulling through the trigger, which releases both the firing pin and hammer safeties.

In 1994 it was announced that the A-75 was available with an aluminium frame, and that a version in 0.45 ACP chambering would be available from October 1994.

Data

Cartridge: 9 × 19 mm Parabellum or 0.40 S&W
Operation: recoil, self-loading, double action
Locking: barrel/slide insert
Feed: 8-round (9 mm) or 7-round (0.40) box magazine
Weight: empty, 880 g
Length: 166 mm
Barrel: 89 mm

Manufacturer

Astra-Unceta y Ciá, SA, Apartado 3, E-48300 Guernica, Vizcaya.

Status

In production.

VERIFIED

9 mm/0.40 Astra A-75 pistol

9 mm Astra Model A-80 pistol

Description

The A-80 is reminiscent of the SIG-Sauer P220 in its angular outlines and in the use of a decocking lever. When the gun has been loaded the decocking lever is pressed; this releases the hammer which falls into a first step on the decocking lever and goes down until arrested by a notch on the sear, preventing the hammer reaching the firing pin. The firing pin itself is restrained from movement by a spring-loaded plunger which is engaged at all times except when the trigger is pulled; when this is done the action of the sear in releasing the cocked hammer also forces the plunger out of engagement and permits the pin to move when struck by the hammer. Unless the trigger is consciously pulled, therefore, even an accidental fall of the hammer cannot drive the firing pin forward. In view of this, there is no manual safety catch on the Model A-80.

The decocking lever is on the left side of the frame where it can be conveniently operated by the (right-handed) firer's thumb; for left-handed firers it is possible to remove the lever and install a replacement lever on the right side of the frame.

The box magazine holds 15 rounds and has holes which show when there are 5, 10 or 15 rounds in the magazine. Front and rearsights are of the Stavenhagen pattern with white inlays to provide an aiming mark in poor visibility.

Data

Cartridge: 7.65 mm Parabellum, 9 mm Steyr, 9 mm Parabellum, 0.38 Super Auto and 0.45 ACP
Operation: short recoil, semi-automatic
Locking: Browning swinging link (cam)
Feed: 15-round box magazine
Weight: empty, 985 g
Length: 180 mm

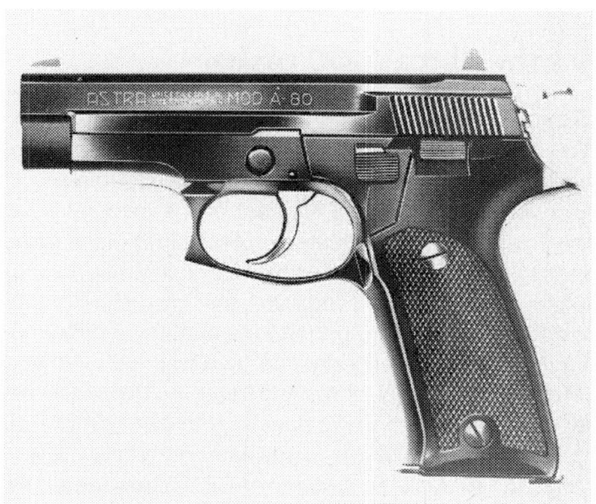

9 mm Astra Model A-80 pistol

Barrel: 96.5 mm
Rifling: 6 grooves, rh
Sights: fore, blade; rear, notch, adjustable for windage
Muzzle velocity: 350 m/s

Manufacturer
Astra-Unceta y Ciá SA, Apartado 3, E-48300 Guernica, Vizcaya.

Status
Production discontinued.

Service
In service in several unspecified countries.

VERIFIED

---

9 mm Astra Model A-90 pistol

Description
The Astra A-90 was introduced in 1985 and is an updated version of the A-80 (see previous entry). It has an improved double-action mechanism, adjustable sights, large magazine capacity and compact dimensions.

Safety in operation has been made a particular feature of the design and the facilities on the A-90 permit the user to adopt a variety of safety procedures. There is firstly a hammer decocking lever which when pressed, releases the sear to allow the hammer to drop, but the hammer stroke is arrested by the decocking lever and it does not contact the firing pin. There is also a manual safety catch on the slide which operates to rotate a portion of the two-piece firing pin out of the hammer path; this safety can be applied whatever the position of the hammer. Finally the major (front) portion of the firing pin is locked by an automatic block which is only released when the trigger is pulled to its full extent, as when firing. At any other time the forward part of the firing pin cannot move, irrespective of what other safety is or is not in operation.

9 mm Astra Model A-90 pistol

Data
Cartridge: 9 × 19 mm Parabellum
Operation: recoil, semi-automatic, double action
Locking: barrel/slide insert
Feed: 17-round box magazine
Weight: empty, 985 g
Length: 180 mm
Barrel: 96.5 mm

Manufacturer
Astra-Unceta y Ciá SA, Apartado 3, E-48300 Guernica, Vizcaya.

Status
In production.

VERIFIED

---

9 mm Astra Model A-100 pistol

Description
The Astra A-100 is a further development in the A-70/80/90 series; it is more or less the same basic pistol as the A-90 but the manual safety has been removed and operation now relies entirely upon automatic safeties and a decocking lever. After loading the pistol in the usual manner, the decocking lever can be pressed; this lowers the hammer under control until it engages in a safety notch in the sear and is thus held out of contact with the firing pin. A spring-loaded block also locks the firing pin so that it cannot move forward far enough to contact the cartridge cap. To fire, all that is required is to either pull the trigger or cock the hammer by using the decocking lever. In either case, pressure on the trigger will rotate the sear and allow the hammer to fall. The sear also has an upper arm which presses on the firing pin lock and, at the instant the hammer is about to be released, lifts it so as to allow the firing pin free movement. As soon as the shot has been fired and the trigger released, the firing pin lock automatically moves back into place.

Data
Cartridge: 9 × 19 mm Parabellum, 0.40 S&W or 0.45 ACP
Operation: short recoil, semi-automatic
Locking: barrel/slide insert
Feed: 17-round (9 mm), 13-round (0.40), or 9-round (0.45) box magazine
Weight: 985 g (0.45, 955 g)
Length: 180 mm
Barrel: 96.5 mm
Sights: fore, blade; rear, notch, adjustable for windage
Muzzle velocity: 340-380 m/s

Manufacturer
Astra-Unceta y Cía SA, Apartado 3, E-48300 Guernica, Vizcaya.

9 mm Astra Model A-100 pistol

Status
In production.

VERIFIED

---

9 mm Astra A-50 pistol

Description
The Astra A-50 is a compact pistol designed for personal defence and police use. It is of the conventional blowback type, with the barrel fixed to the frame, and in spite of its appearance it is a single-action weapon. It is also available in 7.65 mm ACP chambering.

Data
Cartridge: 9 mm Short (7.65 mm ACP)
Operation: blowback
Feed: 7-round (9 mm) or 8-round (7.65 mm) box magazine
Weight: 650 g (660 g)
Length: 168 mm
Barrel: 89 mm
Rifling: 6 grooves, rh

Manufacturer
Astra-Unceta y Ciá SA, Apartado 3, E-48300 Guernica, Vizcaya.

Status
Production discontinued.

VERIFIED

9 mm Astra A-50 pistol

---

9 mm Astra A-60 pistol

Description
The Astra 9 mm A-60 pistol is a fixed-barrel blowback weapon of conventional pattern, using double-action lockwork. An innovative element of the design is that the safety catch is duplicated on both sides of the slide, so that it can be operated with equal facility by a right- or left-handed firer. In addition, the magazine catch located in the forward edge of the butt, behind the trigger guard on the left side, can be easily removed and refitted into the right side of the frame, so making it easier for left-handed firers to operate with their firing hand.

The Astra A-60 is available in either 9 mm Short/0.380 Auto or 7.65 mm ACP chambering. The dimensions are the same in both calibres, the only change being in the magazine capacity.

Data
Cartridge: 9 mm Short; 7.65 mm ACP
Operation: blowback, semi-automatic
Feed: 12-round (7.65 mm) or 13-round (9 mm) box magazine
Weight: empty, 700 g
Length: 168 mm
Barrel: 89 mm
Rifling: 6 grooves, rh
Sights: fore, fixed blade; rear, square notch, adjustable for windage

Manufacturer
Astra-Unceta y Ciá SA, Apartado 3, E-48300 Guernica, Vizcaya.

Status
In production.

VERIFIED

9 mm Astra A-60 pistol

9 mm Astra Falcon pistol

Description

The Astra Falcon 9 mm pistol is the sole remaining example of the style which made Astra famous, a design which has its roots in the Campo-Giro pistol of 1913. Astra pistols in this distinctive 'round-barrel' pattern formed the Spanish service sidearm for many years and were widely distributed throughout Europe. The Falcon is available in 7.65 mm or 9 mm Short calibres for police or military use. The mechanism is simple and robust, with the return spring coiled around the removable barrel and a visible hammer. The safety catch operates on the trigger, and there is a magazine safety incorporated.

Data

Cartridge: 9 mm Short (7.65 mm ACP)
Operation: blowback
Feed: 7 round (9 mm) or 8 round (7.65 mm) box magazine
Weight: 646 g (668 g)
Length: 164 mm
Barrel: 98.5 mm

Manufacturer

Astra-Unceta y Ciá SA, Apartado 3, E-48300 Guernica, Vizcaya.

Status

In production.

VERIFIED

9 mm Astra Falcon pistol

0.38 Special Astra Model 960 revolver

Description

Astra produces a number of service pattern revolvers, of which the Model 960 is a typical example. It is a conventional solid-frame weapon with swing-out cylinder, simultaneous ejection and double-action lockwork. The rearsight is fully adjustable for elevation and windage and there is a regulator which permits adjustment of the pressure of the mainspring. Various barrel lengths are available, and while the standard finish is blue with checkered wooden grips, there are alternative finishes.

Data

Cartridge: 0.38 Special
Operation: double action revolver
Feed: 6-shot cylinder
Weight: empty, with 102 mm barrel, 1.15 kg
Length: with 102 mm barrel, 241 mm
Barrel: 102 or 152 mm
Sights: fore, blade; rear, notch, fully adjustable
Muzzle velocity: 265 m/s

Manufacturer

Astra-Unceta y Ciá SA, Apartado 3, E-48300 Guernica, Vizcaya.

Status

In production.

Service

In service in several unspecified countries.

VERIFIED

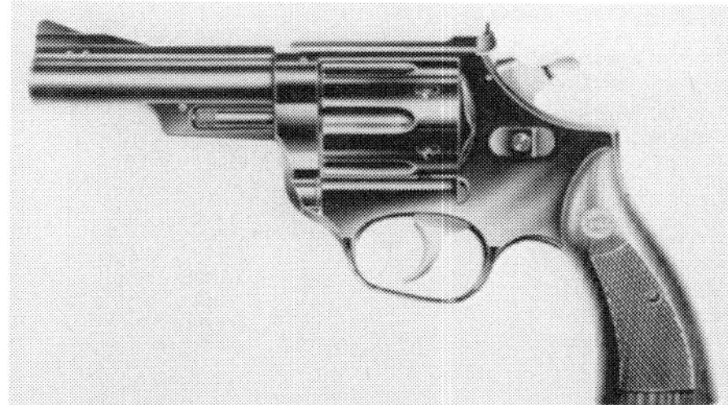

0.38 Special Astra Model 960 revolver with 102 mm barrel

Astra Model 680 revolver

Description

The Astra Model 680 is a lightweight and compact revolver, a solid-frame weapon with a swing-out cylinder and simultaneous ejection. The sights are fixed, and there is an adjustment which allows the pressure of the mainspring to be varied. The cylinder is released by the usual thumb-catch on the left side of the frame, and the ejector rod is carried in a shrouded housing under the barrel.

The standard steel model is the 680A; variants are the 680 Inox, in stainless steel, and the 680AL which has a light alloy frame. The 680A is also available in 0.32 S&W Long chambering, and in 0.22 LR or 0.22 Magnum chambering with an eight-chambered cylinder.

Data

Cartridge: 0.38 Special (see also text)
Operation: double action revolver
Cylinder capacity: 6 shots
Weight: empty, 630 g
Length: 167 mm
Barrel: 51 mm
Muzzle velocity: ca 250 m/s

Manufacturer

Astra-Unceta y Ciá SA, Apartado 3, E-48300 Guernica, Vizcaya.

Status

No longer in production.

VERIFIED

Astra Model 680A 0.38 Special revolver

Astra Police revolver

Description

The Astra Police revolver is a conventional solid-frame, swing-out cylinder, double-action revolver chambered for the 0.357 Magnum cartridge and intended for use by police and security forces. The sights are fixed (the rearsight being merely a groove in the top strap) and the weapon has been designed with the rigours of service well in mind. The pistol is available in the three calibres enumerated below, and in addition it can be supplied with a spare cylinder in either 9 mm Parabellum or 9 mm Steyr calibre. Cylinders for rimless cartridges are provided with loading clips.

Data

Cartridge: 0.357 Magnum, 0.38 Special, 9 mm Parabellum
Cylinder capacity: 6 rounds
Weight: 1.04 kg
Length: 212 mm
Barrel: 77 mm

Manufacturer

Astra-Unceta y Ciá SA, Apartado 3, E-48300 Guernica, Vizcaya.

Status

In production.

VERIFIED

0.357 Astra Police revolver showing 9 mm cylinder and loading clip

SWITZERLAND

9 mm SIG P210 and Model 49 pistols

Development

Schweizerische Industrie-Gesellschaft (SIG), based at Neuhausen Rheinfalls, took up Charles Petter's patents from Société Alsacienne de Constructions Mécaniques (SACM). Development resulting in important technical improvements continued over the years 1938-46 and a series of weapons was produced. The 9 mm Model 44/16 held 16 rounds and the 44/8 held 8 rounds but these were produced in small numbers only. The 9 mm Model 49 is the Swiss Army pistol (designated 9 mm Pistole 49) but is being replaced by the SIG-Sauer P220 (see separate entry). It is identical to the P210-2. The SIG P210 pistol has been produced in several versions: the P210-1, polished finish and wooden grip plates; P210-2, sandblast finish and plastic grip plates; P210-4, a special production model for the German Border Police; P210-5, a target version with 150 mm barrel; and the P210-6, also a target model but with a 120 mm barrel.

The P210-1, -2 and -6 are produced in either 9 mm or 7.65 mm Parabellum. The calibre can be changed by substituting the other barrel, of the alternative calibre, with its own return spring. The pistols can also be converted to 0.22 Long Rifle by changing the barrel, return spring, slide and magazine, reducing training costs.

SIG Pistole 49

Description

The magazine is removed for loading by pressing the magazine catch at the heel of the pistol grip rearward. The magazine takes eight cartridges. With the magazine in place, pulling back the slide and then releasing it drives a cartridge into the chamber and pulling the trigger fires one round.

On firing the slide and barrel, locked together by lugs on the barrel mating with recesses in the slide, recoil together until the shaped cam beneath the chamber, acting against the slide stop pin, pulls the lugs out of engagement. The barrel then stops and the slide recoils, cocking the hammer at the end of its stroke. On returning, the slide collects a fresh round from the magazine and chambers it, then lifts the barrel back into engagement with the slide before coming to rest. When the ammunition is expended the slide is held to the rear by the slide stop which is forced into the notch on the slide by a lip on the magazine follower.

9 mm P210-5 pistol

9 mm P210-6 pistol

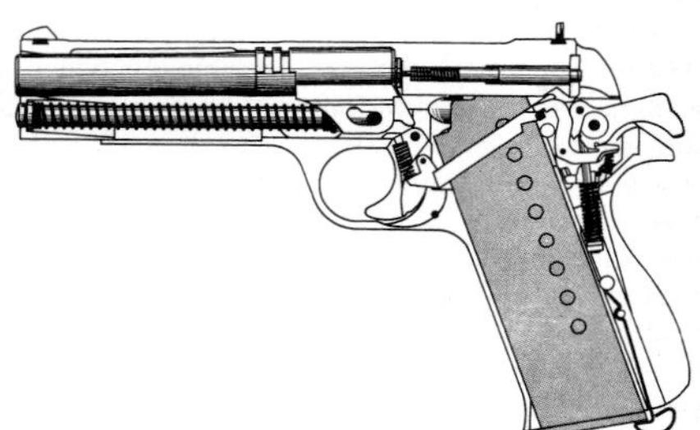

P210-1 or P210-2 pistol

Data

DATA BELOW IS COMMON TO 9 mm, 7.65 mm OR 0.22 VERSIONS OF P210-1 OR P210-2 UNLESS OTHERWISE SPECIFIED

Cartridge: 9 × 19 mm Parabellum (or 7.65 mm Parabellum or 0.22 Long Rifle)
Operation: short recoil, self-loading
Locking: projecting lug
Feed: 8-round box magazine
Weight: empty, 900 g (845 g for 0.22)
Length: 215 mm
Barrel: 120 mm
Rifling: 6 grooves, rh (4 for 7.65 mm); twist - 1 turn in 250 mm (450 mm for 0.22)
Sights: fore, blade; rear, notch
Sight radius: 165 mm
Muzzle velocity: 9 mm, 335 m/s; 7.65 mm, 385 m/s; 0.22, 330 m/s
Muzzle energy: 9 mm, 454 J; 7.65 mm, 453 J; 0.22, 141 J

Manufacturer

SIG: Swiss Industrial Company, CH-8212 Neuhausen-Rheinfalls.

Status

In production.

Service

Swiss Army has complete stocks and takes no more deliveries. Also in use by Danish Army. Commercial sales of target versions continue.

UPDATED

---

9 mm SIG-Sauer P220 pistol

Description

The SIG-Sauer P220 is a short recoil-operated, self-loading, single- or double-action pistol with a light alloy frame. The magazine catch is below the heel of the butt (except for the P220-1 0.45 ACP model, where it is a button in the front edge of the butt) and the magazine is loaded in the usual way by being pushed up into the butt until the catch clicks into place.

The slide is pulled to the rear and released to feed a round into the chamber. If it is not intended to fire the weapon immediately the cocked hammer is lowered by pressing down on a hammer decocking lever above and slightly behind the trigger on the left side of the receiver. The effect of depressing this lever is to lift the sear out of engagement with the hammer which is rotated by its spring until the safety notch is caught by the sear and it comes to rest held clear of the firing pin. The firing pin itself is locked by a pin which is forced through it by a spring and cannot move even if the pistol is dropped.

The pistol can be fired double action by a long pull on the trigger or it can be used as a single-action weapon by cocking the hammer by hand and then using a shorter, lighter trigger pull.

When the round is fired the pressure in the chamber forces the cartridge case back against the breech face. The slide and barrel, locked together, recoil for about 3 mm, after which the barrel is unlocked. The locking

SIG-Sauer 9 mm P220 pistol

P220-1 pistol in 0.45 ACP calibre, showing magazine release button

system is a modification of the Browning dropping barrel method. The barrel has a large shaped lug above the chamber which, in the locked position, engages in a recess formed around the ejection slot in the slide. A shaped lug beneath the chamber strikes a transom in the frame and this lowers the rear of the barrel, disengaging it from the slide.

The slide continues rearward and extracts and ejects the empty case. The return spring is compressed and the hammer is cocked. The slide comes to rest when it reaches a stop in the frame above the butt and is then thrown forward. The cam formed beneath the chamber lifts the rear of the barrel into the recess in the slide. The cam slides forward another 3 mm along the flat top of a supporting ramp and this keeps the rear end of the barrel firmly locked into the slide.

When the ammunition is expended the magazine follower rises and lifts the slide stop, located on the left of the receiver above the butt, into a recess in the slide, thus holding the slide in an open position.

When a loaded magazine is inserted the slide can be released either by pressing down on the slide catch with the thumb of the right hand or by gripping the slide, pulling back slightly and then releasing it. In both cases the slide goes forward and chambers a cartridge and the hammer remains cocked for immediate action.

In addition to the standard Stavenhagen high contrast sights, tritium front and rear luminous night sights are available as an option, as is a hard rubber floorplate pad for the magazine.

Data

DATA FOR 9 mm VERSION IS COMMON TO ALL VERSIONS EXCEPT WHERE INDICATED

Cartridge: 9 × 19 mm Parabellum, 7.65 mm Parabellum, 0.45 ACP, 0.38 Super
Operation: short recoil, self-loading, single or double action
Locking: projecting lug
Feed: 9-round (9mm, 7.65 mm and 0.38 Super) or 7-round (0.45 APC) detachable box magazine
Weights: pistol without magazine; 9 mm, 750 g; 7.65 mm, 765 g; 0.45 ACP, 730 g; 0.38 Super, 750 g
Empty magazine: 9 mm, 7.65 mm, 80 g; 0.45 ACP, 0.38 Super, 70 g
Length: 198 mm
Barrel: 112 mm
Rifling: 9 mm, 6 grooves, rh, 1 turn in 250 mm; 7.65 mm, 4 grooves, 1 turn in 250 mm; 0.45 ACP, 6 grooves, 1 turn in 400 mm; 0.38 Super, 6 grooves, 1 turn in 250 mm
Sights: fore, blade 3 mm wide, white spot on surface; rear, square notch 3 mm wide, white spot below notch; (graduation) nil; (zeroing) lateral, rearsight moves in dovetail. Elevation. Change rearsight. 8 sizes available in steps of 0.27 mm, corresponding to 4.2 mm at 25 m or 8.4 mm at 50 m. Tritium luminous night sights available
Sight radius: 160 mm (Stavenhagen contrast sights)
Muzzle velocity: 9 mm, 345 m/s; 7.65 mm, 365 m/s; 0.45 ACP, 245 m/s; 0.38 Super, 355 m/s

Manufacturer

SIG: Swiss Industrial Company, CH-8212 Neuhausen-Rheinfalls.
J P Sauer and Sohn GmbH, Eckernförde, Germany.

Status

In production.

Service

Orders for 35,000 of the 9 mm version were placed with SIG by the Swiss Government and the weapon is in service under the Swiss Army designation 9 mm Pistole 75. Also in service with some foreign police and special forces. Well over 150,000 are in service.

UPDATED

9 mm/0.380 SIG-Sauer P230 pistol

Description

The SIG -Sauer P230 is a blowback pocket pistol which, although designed for police work, has numerous military applications.

The P230 can be used as a single- or double-action pistol and has the same facility for lowering the hammer, by depressing the hammer decocking lever, as on the P220 (see previous entry). Similarly the firing pin is permanently locked except when released by the trigger bar immediately before the hammer falls.

The pistol is available chambered for the 9 mm Short/0.380 ACP whilst a version firing the 7.65 mm ACP cartridge is no longer offered. Both blue-black light metal alloy (standard) and stainless steel models are available.

SIG-Sauer 9 mm/0.380 P230 pistol

1996

The magazine platform forces up the slide stop on the left of the receiver when the ammunition is expended. When a loaded magazine has been inserted, the slide is pulled back to the stop and then allowed to snap forward. This cocks the hammer and feeds a cartridge into the chamber.

In addition to the standard Stavenhagen high contrast sights, tritium front and rear luminous night sights are available as an option.

Data

Cartridge: 9 mm Short (0.380 Short)
Operation: blowback, semi-automatic, double action
Feed: 7-round detachable box magazine
Weight: standard, empty, 460 g; stainless steel, empty, 590 g
Trigger pull: single action, 17 N; double action, 45 N
Length: 168 mm
Barrel: 92 mm
Rifling: 6 grooves, rh, 1 turn in 250 mm
Sights: fore, blade; rear, notch; Stavenhagen pattern
Sight radius: 120 mm
Muzzle velocity: 275 m/s
Muzzle energy: approx 226 J

Manufacturer

SIG: Swiss Industrial Company, CH-8212 Neuhausen-Rheinfalls.
J P Sauer and Sohn GmbH, Eckernförde, Germany.

SIG-Sauer 9 mm/0.380 P230 pistol

1996

Status

In production; light alloy and stainless steel versions.

Service

Several Swiss state police corps and smaller Swiss police units. Numerous American police forces.

UPDATED

9 mm SIG-Sauer P225 pistol

Description

The 9 mm SIG-Sauer P225 pistol is slightly smaller and lighter than the P220 and carries one round fewer in the magazine. It is similar in operation in that it is a mechanically locked recoil-operated weapon with an automatic firing-pin lock, double-action trigger, decocking lever and external slide catch lever. An additional safety has been built in which provides an absolute lock even if the pistol is accidentally dropped with the hammer cocked, decocked, or halfway to being cocked.

SIG-Sauer 9 mm P225 pistol

1996

As the P225 was designed specifically for employment by police forces, much design emphasis went into ensuring that a shot can be fired only by actually pulling the trigger. Since there is no safety catch to be pushed off, the weapon is remarkably quick to bring into action.

SIG-Sauer 9 mm P225 pistol

1996

The design of the grips and the positioning of the centre of balance ensure a good hold and positive control of the weapon while firing. The finish is excellent and all parts are interchangeable between pistols of the same designation.

The sights are tool adjustable. Adjustment for windage is available on either the fore or rearsight. Elevation is altered by inserting a different sight element. Six rear notches and five foresight posts are available.

For training, the P225 PT has been designed to fire the 9 × 19 mm PT (Plastic Training) ammunition.

In Germany the P225 is known as the P6.

In addition to the standard Stavenhagen high contrast sights, tritium front and rear luminous night sights are available as an option, as are wooden grip plates.

Data

Cartridge: 9 × 19 mm Parabellum
Operation: recoil, single or double action trigger
Locking: projecting lug
Feed: 8-round detachable box magazine
Weight: without magazine, 740 g
Weight of magazine: 80 g
Trigger pull: single action, 20 N; double action, 55 N
Length: 180 mm

Height: 131 mm
Barrel: 98 mm
Rifling: 6 grooves, rh, 1 turn in 250 mm
Sights: see text
Sight base: 145 mm
Muzzle energy: approx 500 J

Manufacturer
SIG: Swiss Industrial Company, CH-8212 Neuhausen-Rheinfalls.
J P Sauer and Sohn GmbH, Eckernförde, Germany.

Status
In production.

Service
Swiss and German police forces.

UPDATED

9 mm SIG-Sauer P226 pistol

Development
The SIG-Sauer P226 pistol was conceived in late 1980 as SIG's candidate in the competition for a new automatic pistol in 9 mm Parabellum for the US armed forces, in which it featured as a 'technically acceptable finalist'. About 80 per cent of its parts come from current production P220 and P225 pistols and, like those pistols, it is a positively locked short recoil weapon with automatic firing pin lock, double-action trigger, decocking lever and external slide latch lever.

In spite of losing out in the US Army pistol contest to the Beretta M9, numbers of P226 pistols have been acquired by the US Coast Guard and more have been procured by the United Kingdom Ministry of Defence. Numerous other armed forces around the world, including the New Zealand Army and Navy, have either acquired the P226 for standard issue or for special forces.

The design of the Croatian 9 mm HS 95 pistol would appear to owe much to the P226.

SIG-Sauer 9 mm P226 pistol

1996

Description
The dimensions of the P226 are similar to those of the P220, from which it differs in magazine capacity and in the provision of an ambidextrous magazine catch. The detailed technical description given for the P220 applies equally to the P226.

In addition to the standard Stavenhagen high contrast sights, tritium front and rear luminous night sights are available as an option, as is a hard rubber floorplate pad for the magazine.

Data
Cartridge: 9 × 19 mm Parabellum
Operation: short recoil, self-loading, single or double action
Locking: projecting lug
Feed: 15-round detachable box magazine
Weight: without magazine, 790 g
Weight of magazine: empty, 95 g
Trigger pull: single action, 20 N; double action, 55 N
Length: 196 mm
Barrel: 112 mm
Rifling: 6 grooves, rh, 1 turn in 250 mm
Sights: fore, blade; rear, square notch
Sight radius: 160 mm
Muzzle velocity: 350 m/s
Muzzle energy: approx 500 J

Manufacturer
SIG: Swiss Industrial Company, CH-8212 Neuhausen-Rheinfalls.
J P Sauer and Sohn GmbH, Eckernförde, Germany.

Status
In production. Well over 400,000 units of the P220 family have been produced to date.

Service
US Coast Guard, UK Ministry of Defence, New Zealand Army and Navy, and numerous other armed forces around the world.

UPDATED

9 mm SIG-Sauer P228 pistol

Development
The P228 has been designed to provide a compact pistol with large magazine capacity. It is a mechanically locked recoil-operated autoloader in 9 mm Parabellum calibre which is particularly suitable for concealed carrying and for individuals with smaller hands.

The P228 was selected by the US Army as its 9 mm Compact Pistol M11 and is also in US service with the Federal Bureau of Investigation, the Drug Enforcement Administration, the Bureau of Alchohol, Tobacco and Firearms, the Internal Revenue Service, the Federal Aviation Administration, and numerous other Federal, State and law enforcement departments. The P228 has also been procured by the United Kingdom Ministry of Defence.

Description
A double-action pistol, the P228 has a magazine capacity of 13 rounds. It has the automatic firing pin safety system used on previous SIG-Sauer pistols and the magazine catch can be mounted on the left or right side to suit the firer's preference. The magazine floorplate, made of high-impact synthetics, gives the magazine a considerable degree of protection against damage caused by dropping.

An essentially closed design concept renders the P228 particularly resistant to dust and dirt. The majority of wearing parts are identical to those of the P225 and P226 pistols and most of the accessories for these weapons are suitable for use with the P228.

In addition to the standard Stavenhagen high contrast sights, tritium front and rear luminous night sights are available as an option.

Data
Cartridge: 9 × 19 mm Parabellum
Operation: short recoil, self-loading, single or double action
Locking: projecting lug
Feed: 13-round detachable box magazine
Weight: empty, 830 g
Trigger pull: single action, 20 N; double action, 55 N
Length: 180 mm
Barrel: 98 mm
Rifling: 6 grooves, rh, 1 turn in 250 mm
Sights: fore, blade; rear, square notch
Sight radius: 145 mm
Muzzle energy: approx 500 J

SIG-Sauer 9 mm P228 pistol

1996

SIG-Sauer 9 mm P228 pistol, showing the conventient location of all control catches and levers close to the firer's thumb

1996

Manufacturer
SIG: Swiss Industrial Company, CH-8212 Neuhausen-Rheinfalls.
J P Sauer and Sohn GmbH, Eckernförde, Germany.

Status
In production.

Service
US Army, UK Ministry of Defence, the Federal Bureau of Investigation, the Drug Enforcement Administration, the Bureau of Alchohol, Tobacco and Firearms, the Internal Revenue Service, the Federal Aviation Administration, and numerous other Federal, State and law enforcement departments, both in the USA and elsewhere.

UPDATED

0.40 SIG P229 pistol

Description
Except for some minor changes in the slide contours, this is virtually the same pistol as the P228 but chambered for the 0.40 Smith & Wesson cartridge. It was developed primarily for law enforcement officers, using high-contrast sights and having the usual SIG double-action, automatic firing pin safety and decocking lever.

There are three variant models: the P229 is the basic model in 0.40 S&W calibre and with a steel slide and aluminium alloy frame. The P229 SL is the same but with the slide in stainless steel and may also be obtained in 9 × 19 mm Parabellum calibre. Production is concentrated on the 0.40 S&W model. All that is required for a 0.40 version to fire 0.357 SIG is a change of barrel.

In addition to the standard Stavenhagen high contrast sights, tritium front and rear luminous night sights are available as an option, as are wooden butt plates.

The SIG-Sauer P239 is a variant of this pistol. See following entry for details.

Data
Cartridge: 0.40 S&W, 9 × 19 mm Parabellum or 0.357 SIG
Operation: short recoil, self-loading, double action
Locking: dropping barrel
Feed: 12-round box magazine (9 mm, 13 rounds)
Weight: with empty magazine, 865 g
Trigger pull: single action, 20 N; double action, 55 N
Length: 180 mm
Barrel: 98 mm
Rifling: 6 grooves, rh, 1 turn in 380 mm (0.357 SIG, 1 turn in 406 mm)
Sights: fore, blade; rear, square notch; both adjustable for windage by lateral movement and adjustable for elevation by substitution
Sight radius: 145 mm
Muzzle energy: 0.40, approx 600 J; 9 mm, approx 500 J; 0.357 SIG, approx 685 J

Manufacturer
SIG: Swiss Industrial Company, CH-8212 Neuhausen-Rheinfalls.

Status
In production.

UPDATED

SIG 0.40 P229SL pistol

1996

SIG-Sauer P239 pistols

Description
The SIG-Sauer P239 pistols are variants of the P229 pistols (see previous entry), the only change being that their box magazine capacity is reduced to eight (9 mm) or seven (0.40 and 0.357) rounds to reduce the width of the butt and make them easier to carry as concealed weapons by police and security personnel.

The P230 is available chambered for either 9 mm Parabellum, 0.40 S&W or 0.357 SIG. The handling, decocking feature and other such details remain identical to their equivalents on the P229.

A 10-round magazine is available as an option, as are SIGLITE sights.

Data

Cartridge	9 mm Para	0.40 S&W	0.357 SIG
Operation	short recoil, self-loading, double-action		
Locking	dropping barrel		

	9 mm Para	0.40 S&W	0.357 SIG
Magazine capacity	8	7	7
Weight, empty	780 g	820 g	820 g
Length	132 mm	132 mm	132 mm
Barrel	92 mm	92 mm	92 mm
Rifling grooves	6, rh	6, rh	6, rh
Twist	250 mm	380 mm	406 mm
Sights	fore, blade; rear, square notch; both adjustable for windage by lateral movement and adjustable for elevation by substitution		
Sight radius	132 mm	132 mm	132 mm
Muzzle energy	500 J	600 J	685 J

Manufacturer
SIG: Swiss Industrial Company, CH-8212 Neuhausen-Rheinfalls.

Status
In production.

NEW ENTRY

SIG-Sauer P239 pistol

1996

9 mm Sphinx AT 2000 S pistol

Description
The AT 2000 S pistol was originally the Czech CZ 75 manufactured under licence but now, after many improvements developed by the Swiss manufacturers, it can be considered a new and independent design. The assembly tolerances and quality of finish are much improved and the barrel has been slightly changed in dimensions so that it is not interchangeable with that of the CZ 75. The safety catch can now be applied whether the pistol is cocked or uncocked. In 1987 an automatic firing pin safety was introduced, which prevents movement of the firing pin except when the trigger is approaching the final movement of firing, and the most recent innovation is the fitting of an ambidextrous safety catch and, if desired, an ambidextrous slide stop pin.

The pistol is available chambered for the 9 mm Parabellum, 9 × 21 mm or 0.40 S&W cartridges. A conversion kit allows changing a 9 mm pistol to other calibres by simply changing the barrel and magazine.

9 mm Sphinx AT 2000 S pistol

Data
Cartridge: 9 mm Parabellum, 9 × 21 mm or 0.40 S&W
Operation: recoil, semi-automatic, double action
Locking: Browning cam
Feed: 15-round (9 mm) or 11-round (0.40 S&W) box magazine
Weight: empty, 1.03 kg
Length: 204 mm
Barrel: 115 mm
Rifling: 6 grooves, rh, 1 turn in 250 mm
Sights: fore, blade; rear, square notch, adjustable
Sight radius: 161 mm
Muzzle velocity: 352 m/s

Manufacturer
Sphinx Engineering SA, chemin des Grandes-Vies 2, CH-2900 Porrentruy.

Status
In production.

Service
Australia, Hong Kong, USA, Norway police forces.

VERIFIED

9 mm Sphinx AT 2000 P pistol

Description
As with the AT 2000 S (see previous entry) the AT 2000 P is based on the design of the Czech CZ 75, but it is a shortened and lightened version which has been designed by Sphinx. As with the AT 2000 S, an automatic safety firing pin, ambidextrous safety catch and ambidextrous slide stop pin have been added to this model, and there is also a similar conversion kit to allow changing a 9 mm Parabellum pistol to 9 × 21 mm or 0.40 S&W calibre.

Data
Cartridge: 9 mm Parabellum, 9 × 21 mm or 0.40 S&W
Operation: recoil, semi-automatic, double action
Locking: Browning cam
Feed: 13-round box magazine
Weight: empty, 940 g
Length: 184 mm
Barrel: 93 mm
Rifling: 6 grooves, rh, 1 turn in 250 mm
Sights: fore, blade; rear, square notch, adjustable
Sight radius: 150 mm

Manufacturer
Sphinx Engineering SA, chemin des Grandes-Vies 2, CH-2900 Porrentruy.

Status
In production.

VERIFIED

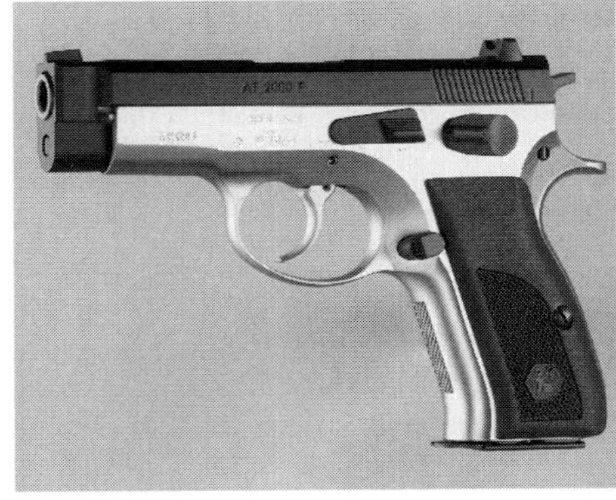

Sphinx AT 2000 P pistol

9 mm Sphinx AT 2000 H pistol

Description
The AT 2000 H is the 'Hideaway' version of the AT 2000 family. It is mechanically similar to the other AT 2000 models and exhibits the same safety features, but is smaller in all dimensions and has had the slide and barrel redesigned. It is normally supplied in 9 mm Parabellum chambering but can be converted to 9 × 21 mm or 0.40 S&W calibres. A patented design of magazine has been developed which will accept any of these calibres without requiring modification.

Data
Cartridge: 9 × 19 mm; 9 × 21 mm; 0.40 Smith & Wesson
Operation: recoil, semi-automatic, double action
Locking: Browning cam
Feed: 10-round box magazine
Weight: empty, 915 g
Length: 178 mm
Barrel: 90 mm
Rifling: 6 grooves, rh
Sight base: 140 mm

Manufacturer
Sphinx Engineering SA, chemin des Grandes-Vies 2, CH-2900 Porrentruy.

Status
In production.

VERIFIED

9 mm Sphinx AT 2000 H pistol

9 mm Sphinx AT 2000 SDA, PDA and HDA pistols

Description
This family of pistols developed by Sphinx is to the same basic design as the AT 2000 series described previously, and the S P and H variations have the same dimensions and dimensional differences. The fundamental change is the adoption of 'double-action-only' firing mechanisms together with the Sphinx Safe System inherent safety. The Safe System ensures that the pistol is in permanent readiness, there being no requirement to cock the hammer or operate a safety device before opening fire. The pistol is automatically decocked and made safe after every shot and is not brought into the firing condition until the user consciously pulls the trigger with the intention of firing. These pistols have the same ambidextrous features - slide catch lever and reversible magazine release - as the AT 2000 series and have the same ability to interchange barrels between 9 mm and other calibres.

Manufacturer
Sphinx Engineering SA, chemin des Grandes-Vies 2, CH-2900 Porrentruy.

Status
In production.

VERIFIED

9 mm Sphinx AT 2000 SDA double-action pistol

9 mm Sphinx AT-2000S/PS Decocking pistols

Description
The AT-2000S and PS models are available with the firing mechanism modified by the addition of a decocking lever. When the decocking lever is pressed down the sear is forced out of engagement with the hammer notch. By releasing the decocking lever the main spring pressure pushes the hammer and the sear into the safety intercept notch. While decocking the pistol it is impossible for the hammer to have any contact with the firing pin. During and after the decocking function the automatic firing pin safety is constantly effective.

The decocking lever has three positions: 'S' for the manual safety of the cocked hammer; 'F' released, to fire; and 'D' to decock the hammer. The manual safety is activated 'on' when the hammer is in the cocked position.

Manufacturer
Sphinx Engineering SA, chemin des Grandes-Vies 2, CH-2900 Porrentruy.

Status
In production.

UPDATED

9 mm Sphinx AT2000S/PS pistol with decocking lever

9 mm Sphinx AT 2000 PS (Police Special) pistol

Description
The AT 2000 PS has been designed with the needs of law enforcement, military and security departments firmly in mind. It uses the AT 2000 full-sized frame but with a compact slide and barrel, producing a compact weapon but with a 16-round capacity. It can be supplied in standard single/double-action form or in self-cocking or 'double-action only' form. It is available chambered for the 9 × 19 mm Parabellum or the 0.40 S&W cartridge.

Manufacturer
Sphinx Engineering SA, chemin des Grandes-Vies 2, CH-2900 Porrentruy.

Status
In production.

Service
Belgium (police); Switzerland (police and security services); Venezuela (police).

VERIFIED

9 mm AT 2000 PS Police Special pistol

9 mm Sphinx AT 0.380 pistol

Description
This pistol was developed primarily for police use, but it has undoubted applications for second-line military employment. The action is self-cocking (or double action only) with a light and very smooth trigger action. It is fitted with new and patented automatic decocking, automatic firing pin safety, ambidextrous magazine release and slide locking catch systems. The frame is of stainless steel and the remaining parts are of MnCrV or CrMo steels. In addition to the standard 10-round magazine, an extended 15-round magazine is available when extra ammunition capacity is desirable.

The standard model is equipped with fixed rearsights and rubber grips. Optionally, adjustable sights, tritium luminous sights and wooden grips may be fitted.

A variant model, the Model 380M, is fitted with a manual safety catch in order to meet requirements of the US Bureau of Alcohol, Tobacco and Firearms regulations.

Data
Cartridge: 9 × 17 mm Short/0.380
Operation: blowback
Feed: 10- or 15-round box magazine
Weight: empty, 710 g
Length: 155 mm
Barrel: 85 mm
Rifling: 6 grooves, rh
Muzzle velocity: 270 m/s

Manufacturer
Sphinx Engineering SA, chemin des Grandes-Vies 2, CH-2900 Porrentruy.

Status
In production.

VERIFIED

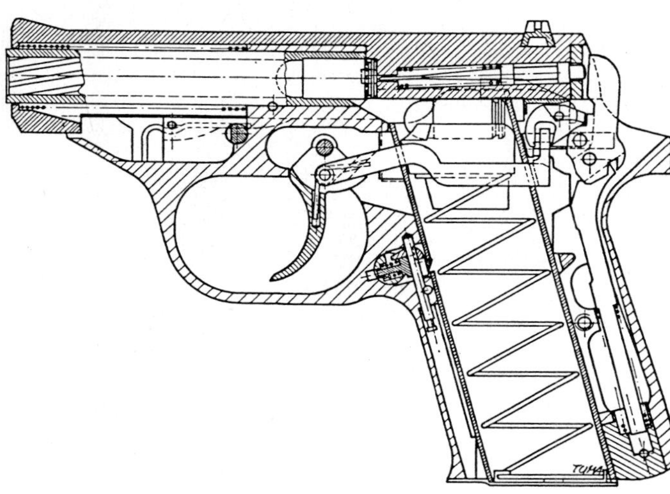
Sectioned drawing of the 9 mm Sphinx AT 0.380 pistol

9 mm Sphinx AT 0.380 pistol

TURKEY

MKE 9 mm and 7.65 mm pistols

Description
These pistols are produced by Makina ve Kimya Endüstrisi Kurumu (MKEK) at Kirikkale, Ankara. They are based on the Walther PP. The pistols are marked 'MKE' on the grip and MKEK, together with the calibre, on the slide. A furher distinguishing feature is the 'bobble' surface finish on the plastic grip plates.

The method of operation is exactly the same as the Walther PP, including the double-action trigger mechanism, chamber status indicating pin and safety system. There are only minor external modifications such as the shape of the magazine finger rest.

Data
Cartridge: 9 mm Short or 7.65 mm
Operation: blowback, double action trigger
Locking: nil
Feed: 7-round detachable box magazine
Weight: empty, 680 g
Length: 170 mm
Barrel: 98 mm
Rifling: 6 grooves, rh
Sights: fore, blade; rear, notch
Muzzle velocity: 260-280 m/s

Manufacturer
Makina ve Kimya Endüstrisi Kurumu (MKEK), Tandogan Medyani, 06330 Ankara.

Status
In production.

Service
Turkish Army.

UPDATED

MKE 9 mm (top) and 7.65 mm (bottom) pistols
1996

UNITED KINGDOM

9 mm Spitfire pistol

Description

The Spitfire is a conventional recoil-operated self-loading pistol which uses the well-tried Browning dropping barrel method of locking the breech. A cam beneath the chamber acts as the controlling device to lower and raise the barrel so that lugs on the upper surface lock into recesses in the slide. The entire pistol, except for the grip surfaces and springs, is made from cast and machined stainless steel, and the machining is performed on computer-controlled tools to a tolerance of 5 μm. As a result the pistol is 'tight' throughout, with no slack in any of the components and with a very crisp trigger action.

The sights are adjustable for windage and zeroing and are extremely well-defined and clear. The action allows for double-action firing or for single-action operation, carrying the pistol cocked and locked. There is an ambidextrous safety catch on the frame where it can be operated easily by the thumb of either hand. The G1 Spitfire magazine holds 15 rounds and one further round can be loaded into the chamber. There is a filler piece on the bottom of the magazine which gives additional support to the little finger for firers with large hands.

The G1 Spitfire incorporates several minor improvements, including an increase in magazine capacity from 13 to 15 rounds. Variant models include the 'Stirling Spitfire G2' with fully adjustable rearsight; The Stirling Spitfire G2/LS' with 120 mm barrel and long slide; the 'Spitfire Pilgrim G3/LS' with 150 mm barrel and dual-port compensator; and the 'Spitfire Carry-Comp G7' with ported 110 mm barrel.

9 mm Stirling Spitfire G2/LS long slide pistol

To allow owners of standard CZ75 pistols to bring their weapons up to date, John Slough of London make the 'Pistol Unit' series of pistol conversion kits, each comprising a barrel/slide/compensator combination which can be fitted to the frame of a CZ75 to bring it up to G2, G2/LS, G3/LS or Carry-Comp standards. Conversions are available in 9 × 19 mm, 9 × 21 mm or 0.40 S&W calibres and are made entirely from stainless steel.

Data

STANDARD SPITFIRE G1
Cartridge: 9 × 19 mm Parabellum, 9 × 21 mm or 0.40 S&W
Operation: short recoil
Locking: Browning cam
Feed: 15-round magazine
Weight: with empty magazine, 1 kg; loaded, 1.17 kg
Length: 180 mm
Barrel: 94 mm
Rifling: 6 groove, rh, 1 turn in 254 mm
Sights: fore, blade; rear, adjustable square notch
Sight base: 154 mm
Muzzle velocity: 352 m/s

Manufacturer

John Slough of London, Old Forge, Peterchurch, Hereford HR2 0SD.

Status

Available for production.

Service

Commercial sales; undergoing military evaluation and has been adopted by a number of European police forces.

UPDATED

9 mm Stirling Spitfire G2 pistol

9 mm Spitfire pistol field-stripped

UNITED STATES OF AMERICA

0.45 Model 1911A1 automatic pistol

Development

Development of this pistol, by John Browning, began in 1896 and culminated in a comprehensive trial by the US Army in 1908. This generally approved of the Colt/Browning design but required some modifications and the perfected design was approved as the M1911, entering US service in that year.

The pistol was used during the First World War, after which some minor defects were remedied; the rear edge of the butt was given a more curved shape and chequered, a shorter grooved trigger was fitted and the frame chamfered off behind the trigger, the grip safety was slightly lengthened and the hammer spur was shortened. With these changes, which took effect from 1926, the pistol became the M1911A1 and has remained unchanged ever since.

The M1911 pistols have been replaced by the 9 mm Pistol M9 (qv) in US Army service but large numbers remain in use. A virtual cottage industry has grown up devoted to supplying M1911 series spares and accessories.

0.45 Model 1911A1 automatic pistol

Description

The M1911 introduced the locking system since known as the Colt/Browning 'swinging link' in which the barrel has raised lugs on its upper surface which fit into recesses on the inner surface of the top of the slide. Beneath the chamber of the barrel is a hinged link, pinned to the barrel at its upper end and secured by the slide stop lever pin into the frame at its lower end. With the slide forward, held there by the force of the return spring beneath the barrel pressing against the front of the slide, the link is vertical and holds the barrel up so that the lugs engage the slide. On firing, the slide recoils and carries the barrel with it, keeping the breech closed. As the barrel moves back, so the link pivots about the slide lock pin and thus the top end of the link describes an arc, drawing the rear end of the barrel downwards until the lugs come free from the slide. The barrel then stops moving while the slide is free to continue rearward, extract and eject the spent case, cock the hammer and compress the return spring.

On the return stroke the slide collects a fresh round from the magazine and loads it into the chamber. Continued pressure against the base of the cartridge now forces the barrel forward and, because of the link, the rear end is raised until the lugs re-engage with the slide. The pistol is now ready to fire the next shot. A manual safety catch is fitted to the left rear of the frame, and there is a grip safety in the rear of the grip which locks the firing mechanism except when pressed in by being grasped.

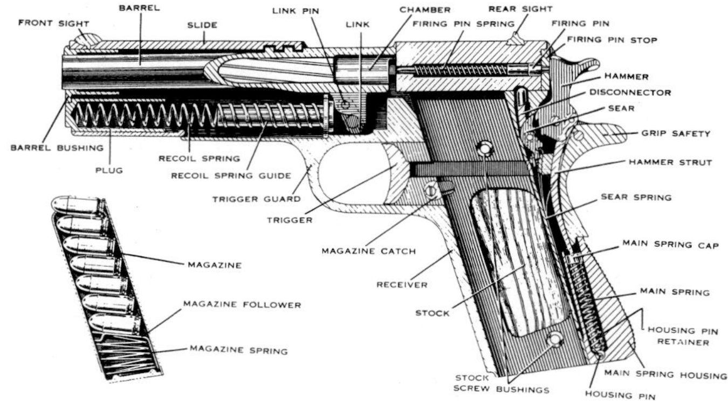

Section view of 0.45 Model 1911A1 automatic pistol

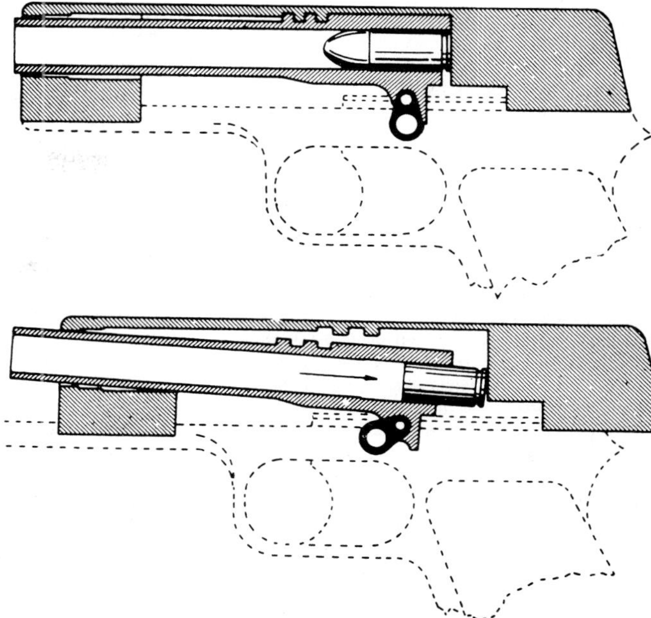

Locking system of 0.45 Model 1911A1 automatic pistol

Data

Cartridge: 0.45 ACP
Operation: short recoil, self-loading
Locking: projecting lug
Feed: 7-round box magazine
Weight: empty, 1.13 kg
Length: 219 mm
Barrel: 127 mm
Rifling: 6 grooves, lh, 1 turn in 406 mm
Sights: fore, blade; rear, U-notch, adjustable for windage
Sight radius: 164.6 mm
Muzzle velocity: 253 m/s

Manufacturers

Colt's Patent Firearms Manufacturing Co, Hartford, Connecticut. Firearms Division, Colt Industries, Hartford, Connecticut. Ithaca Gun Co, Ithaca, New York. Remington Rand Inc, Syracuse, New York. Remington Arms-Union Metallic Cartridge Co, Bridgeport, Connecticut. Springfield Armoury, Springfield, Massachusetts. Union Switch & Signal Co, Swissvale, Pennsylvania. Also made under licence at various times in Argentina, Canada and Norway.

Status

Current for commercial production. Obsolescent for US military service.

Service

US armed forces and many others. One of the most widely used pistols in the world today.

UPDATED

Pistol, 9 mm M9

Description

This pistol is the Beretta Model 92F or FS, first manufactured in Italy and then by a Beretta subsidiary in the USA. Reference should be made to the entry under Italy for full details and dimensions.

The initial US Army order was for 315,930 pistols at a cost of $53 million.

Manufacturer

Beretta USA Corporation, 17601 Beretta Drive, Accokeek, Maryland 20607.

Status

In service with the US Army.

UPDATED

The 9 mm Pistol M9 (right) with the pistol it replaced, the 0.45 M1911A1, on the left
1996

0.45 Model 15 General Officers' pistol

Development

During the Second World War the US Army issued general officers with Colt Pocket Model automatic pistols in 0.380 ACP calibre. When Colt took the pistol out of production in 1946 the Army still had sufficient stocks to last for several years, but in the early 1970s these stocks had dwindled and it became necessary to consider a new pistol for issue to senior ranks. The requirement was that it should be more compact than the issue Model 1911A1 but should have sufficient power to make it a practical combat weapon. A design developed by Rock Island Arsenal as their XM70 was selected and was standardised as the Model 15 in 1972. Production ceased in the mid 1980s.

Description

The Model 15 is a 0.45 Model 1911A1 cut down in size and rebuilt. It operates in precisely the same way as the Model 1911A1 but is shorter in all dimensions. Because of the shorter barrel it tends to develop rather more flash and muzzle blast than the standard pistol, but this is felt to be acceptable since it is anticipated that the weapon will only be used in emergencies. It can be recognised by the dark blue finish and the inscription 'General Officer Model RIA' on the slide. The left grip is inlet with a brass plate on which the individual officer's name is engraved, and the right grip has an inlaid Rock Island Arsenal medallion. The sights are raised rather more prominently from the slide than is the case with the Model 1911A1.

0.45 Model 15 General Officers' pistol

Data
Cartridge: 0.45 ACP
Operation: short recoil, semi-automatic
Locking: Browning swinging link
Feed: 7-round box magazine
Weight: empty, 1.02 kg
Length: 200 mm
Barrel: 106 mm
Rifling: 6 grooves, lh
Sights: fore, blade on ramp; rear, notch
Muzzle velocity: 245 m/s

Manufacturer
Rock Island Arsenal, Rock Island, Illinois 61299-5000.

Status
Production complete.

Service
US Army general officers.

UPDATED

La France silenced Colt 0.45 pistol

Description
The La France silenced Colt 0.45 pistol is a suppressor-equipped Colt 0.45 ACP Government Model pistol. The design uses the straight blowback system of operation by attaching the silencer directly to the slide, with a special fixed barrel. The mass of the suppressor/slide assembly has been engineered to produce sufficient dwell time so as to delay the opening of the breech, thereby gaining maximum bullet velocity while eliminating breech flash and inherent mechanical noise. Sights are incorporated directly on to the 800 g silencer, which is fabricated of aluminium and heat-resistant synthetic materials.

Since the design does not depend upon bullet wipes to achieve noise reduction, accuracy is excellent and noise level remains constant regardless of the number of rounds fired. The system has also been developed for the Beretta 92 pistol series and the Heckler and Koch P7 series.

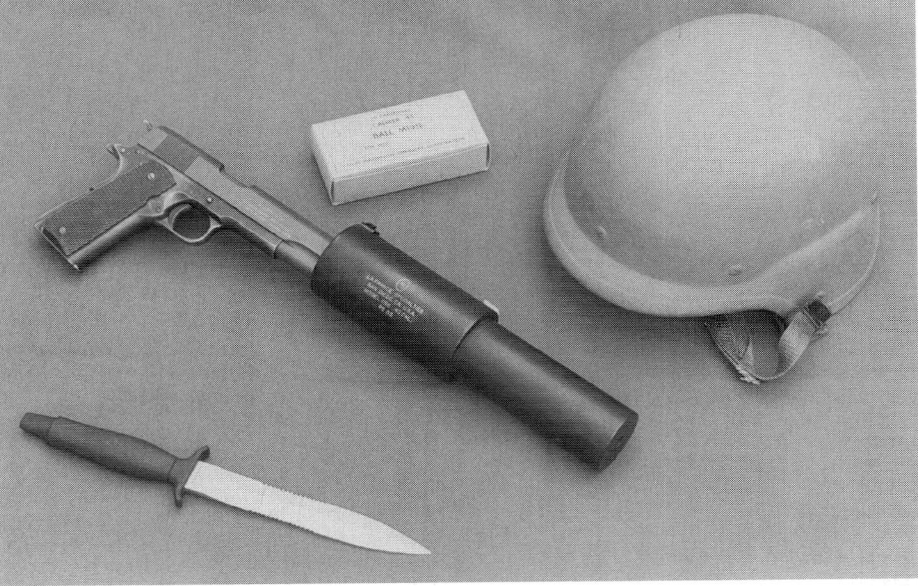
La France silenced 0.45 pistol

Data
Cartridge: 0.45 ACP
Operation: blowback
Feed: 7-round box magazine
Weight: empty, 1.9 kg
Length: 485 mm
Barrel: 127 mm
Muzzle velocity: 259 m/s
Effective range: 25 m

Manufacturer
La France Specialties, PO Box 178211, San Diego, California 92177-8211.

Status
In production.

VERIFIED

Kel-Tec 9 mm P-11 pistol

Description
The Kel-Tec 9 mm P-11 pistol entered production in early 1995 and is claimed to be the smallest and lightest production pistol ever made capable of firing a 9 × 19 mm Parabellum cartridge. It is intended to be a back-up weapon for police and military personnel.

The P-11 box magazine holds 10 rounds in a staggered pattern. The pistol can thus be carried with a full magazine load of 10 rounds plus another in the chamber. If required, the P-11 will also accept standard Smith & Wesson large capacity magazines.

The P-11 is fired double action only with the breech locked by the usual Colt/Browning swinging barrel and locking lug system. SAE 4140 steel is used for the barrel and slide, while the rectangular frame is machined from solid aluminium. The grip is high impact polymer and forms the magazine well and trigger guard. The hammer is driven by a novel free-floating extension spring and connects via a floating bar to the hammer where a lightweight firing pin transmits the hammer energy to a primer when required.

For aiming, high contrast fixed sights are provided. The double-action trigger pull is 45 N.

Data
Cartridge: 9 × 19 mm Parabellum
Operation: short recoil, semi-automatic, double action
Locking: dropping barrel
Feed: 10-round box magazine
Weight: empty, 400 g; loaded, 600 g
Length: 142 mm
Barrel: 78 mm
Sights: fore, fixed blade; rear, fixed notch; both high contrast
Sight radius: 116 mm
Muzzle energy: 400 J

Manufacturer
Kel-Tec CNC Industries Inc, PO Box 3427, Cocoa, Florida 32924-3427.

NEW ENTRY

Kel-Tec 9 mm P-11 pistol

1996

9 mm P-12 pistol

Description
The P-12 is a blowback semi-automatic pistol chambered for the 9 mm Short/0.380 ACP cartridge. The firing mechanism is self-cocking (double action only), with a novel low-inertia safety hammer system.

The pistol has six main component groups: barrel, slide, frame, grip, firing mechanism and magazine. The barrel is turned and milled from SAE 4140 ordnance steel with a tensile strength of over 140,000 lbs/in^2. The bore is to standard SAAMI dimensions.

The slide is milled from a solid block of LaSalle Stressproof steel and contains the firing pin and extractor. The frame is an all-steel construction which holds the barrel and firing mechanism, and also forms the slide rails and ejector. The frame is stamped with the serial number.

The grip is made of the ultra-high impact strength polymer Dupont ST-800. The grip also forms the magazine well and trigger guard; it attaches to the frame with three screws. The magazine is made from glass-reinforced Zytel.

The trigger, with a lever, connects via a bar to the hammer. The hammer is driven by a novel power spring. Virtually the whole mass of the hammer is centred around its axis. The lightweight firing pin transmits the energy of the hammer to ignite the primer.

Data
Cartridge: 9 mm Short/0.380 ACP
Operation: blowback, semi-automatic, self-cocking
Feed: 11-round box magazine
Weight: empty, 370 g
Length: 135 mm
Barrel: 76 mm
Sight radius: 114 mm

9 mm Grendel P-12 pistol

Manufacturer
Grendel Inc, PO Box 560909, Rockledge, Florida 32956-0909.

VERIFIED

0.45 Springfield 1911A1 pistol

Description

This is a standard M1911A1 pistol, the parts of which are completely interchangeable with any other model. In addition to being made in the usual 0.45 calibre, some models are available in 9 mm Parabellum or 0.38 Super chambering. Although of standard design, this version is entirely made of hardened and heat-treated steel machined forgings, and all parts have been surface treated for longer life and better wearing qualities.

The MIL-SPEC model, with Parkerised finish and plastic grips, is supplemented by a number of others. The Standard is in polished blue finish with wood grips, a ring hammer and bevelled magazine aperture. The Trophy Match has an improved trigger pull, micro-adjustable rearsight and match-grade barrel and bushing. It may be had in blued finish or made of stainless steel. The Stainless 1911A1 is similar to the Standard but of stainless steel, while the Stainless Champion has a shorter, Commander-length, slide and barrel.

The Champion MIL-SPEC model is to the same general design as the normal MIL-SPEC model but has a 102 mm barrel and is about 8 mm shorter than the normal MIL-SPEC. The grip frame is also shorter and holds a seven-round magazine. The Champion 1911A1 is similar but with polished blue finish and walnut grips and a somewhat larger grip to accommodate an eight-round magazine. The Compact MIL-SPEC resembles the Champion MIL-SPEC but with the grip frame reduced even more and taking a 6-round magazine. The Compact 1911A1 has a seven-round magazine, and is in polished blue finish with walnut grips, while the Lightweight Compact is similar but with an alloy frame.

0.45 Springfield PDP series Factory Comp 1911-A1 pistol

0.45 Springfield Standard 1911A1 pistol

PDP Series Pistols

The PDP (Personal Defense Pistol) Series is based upon the M1911A1 pattern but represents the incorporation of features developed for competition pistols into weapons meant for service use. The principal feature is the adoption of various forms of muzzle compensator to reduce the recoil and muzzle rise on firing, thus assuring a rapid return to the aim after each shot.

The Factory Comp 1911A1 is the basic model and is generally as a standard 1911A1 but with a 143 mm barrel fitted with a triple-port muzzle compensator. It also has a fully adjustable rearsight, speed trigger, bevelled magazine housing and an extended thumb safety catch. It is available in 0.45 ACP (eight-round magazine) or 0.38 Super (nine-round magazine) chambering.

The Champion Comp 1911-A1 is similar but with a single-port muzzle compensator and expansion chamber which considerably reduces recoil. With fixed sights and an eight-round magazine, this is an excellent combat pistol. The Compact Comp and Lightweight Compact Comp have single-port compensators and smaller frames to take seven-round magazines and 114 mm barrels. The Defender Comp is a full-sized model with a two-port compensator and other features similar to the Factory Comp. Finally, the High-Capacity Factory Comp model is similar to the Factory Comp but with a special grip frame which accepts a 13-round 0.45 ACP magazine, or a 17-round 0.38 Super magazine.

0.45 Springfield compact Champion 1911A1 pistol

Data

STANDARD MODEL

Cartridge: 0.45 ACP, 9 ×19 mm Parabellum or 0.38 Super
Operation: recoil, semi-automatic
Locking: dropping barrel
Feed: 8-round box magazine (9 rounds in 9 mm and 0.38 Super)
Weight: empty, 1.01 kg
Length overall: 219 mm
Barrel: 127 mm
Rifling: 6 grooves, lh (0.45) or rh (9 mm/0.38 Super), 1 turn in 406 mm
Sights: fore, blade; rear, square notch, 3 aiming dots
Sight radius: 158 mm

Manufacturer

Springfield Inc, 420 West Main Street, Geneseo, Illinois 61254.

Status

In production.

VERIFIED

9 mm Ruger P89 pistol

Description

Announced early in 1987, the 9 mm P89 pistol was the first military automatic pistol to be developed by Sturm, Ruger & Company. Originally called the P85, an improved version is known as the P89. It is a conventional double-action pistol, using the familiar Browning type of swinging link to unlock the breech during recoil. Instead of using lugs on the top surface of the barrel, the chamber section is squared off and locks into the ejection opening in the slide. The barrel is stainless steel, as are the hammer, trigger and most internal components. The frame is of lightweight aluminium alloy, is hardened to resist wear and is finished in matt black. The slide is of chrome-molybdenum steel and is similarly matt black. There is an external hammer and the safety catch is on the rear of the slide. This catch is ambidextrous and locks the firing pin, blocks the hammer, and disconnects the trigger when applied.

The firing mechanism is double action, the trigger guard being proportioned so that the pistol can be fired by a gloved hand, and the forward edge of the trigger guard is shaped to form a grip for the non-firing hand. The magazine release is in the forward edge of the butt and can be operated from either side of the pistol without adjustment. The sights are provided with white dot inserts to assist night firing.

Two versions of this basic model are available; the P89 with chrome-molybdenum steel slide and blue finish, and the KP89 with stainless steel slide.

De-Cocker version P89DC

This is the same as the basic P89 but with ambidextrous decocking levers on the slide in place of the normal safety lever. By pressing this lever the hammer may

9 mm Ruger P89DC De-Cocker pistol

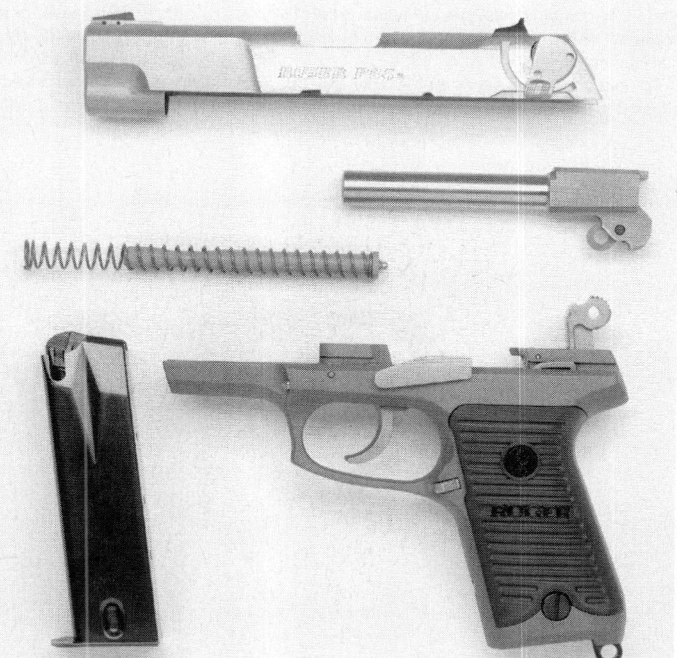

9 mm Ruger P89 stainless pistol dismantled

be safely lowered, since the lever simultaneously blocks the firing pin movement. Once the hammer is down the lever is released. It springs back to its original position and thereafter the pistol can be fired by a double-action pull or by thumb-cocking the hammer and a single-action pull.

Double-Action-Only P89

This is a further variant of the P89, allowing self-cocking (or 'double-action-only') operation. The hammer follows the slide after each shot and cannot be manually cocked.

Data

Cartridge: 9 × 19 mm Parabellum
Operation: short recoil, semi-automatic, double action
Locking: Browning link and tipping barrel
Feed: 15-round box magazine
Weight: empty, 907 g
Length: 200 mm
Barrel: 114 mm
Sights: fore, blade; rear, square notch, adjustable for windage
Sight radius: 155 mm

Manufacturer

Sturm, Ruger & Co Inc, Lacey Place, Southport, Connecticut 06490.

Status

In production.

VERIFIED

0.45 in Ruger KP90DC De-Cocker pistol

Description

The 0.45 in Ruger KP90DC De-Cocker pistol is essentially the same as the 9 mm Ruger De-Cocker pistol P89DC (see previous entry) but chambered for the 0.45 ACP cartridge. The ability to decock a loaded pistol by a thumb-lever is becoming more and more popular and the KP90DC caters for those who require this facility in a heavy calibre. The pistol is approximately the same size (200 mm long) and weight (964 g) as many 9 mm pistols and is made entirely from stainless steel except for the grip frame which is of investment cast aluminium alloy. An oversized trigger guard permits use with a gloved hand and a lanyard loop stud is standard. High visibility sights have white dot inserts and the rearsight can be laterally adjusted for windage.

Manufacturer

Sturm, Ruger & Co Inc, Lacey Place, Southport, Connecticut 06490.

Status

In production.

VERIFIED

9 mm Ruger KP93 Compact pistol

Description

The Ruger 9 mm KP93 Compact pistol is slightly shorter and some 14 per cent lighter than the KP89 pistol, and there are slight changes in the contours which result in a sleeker appearance. The frame is of aluminium alloy with a matt silver hard finish. The shorter length places the centre of gravity closer to the centre of mass, giving the pistol excellent balance.

The barrel, slide, trigger, hammer and most internal metal parts are of stainless steel, requiring minimal maintenance. The magazine holds 15 rounds in a double column, and is interchangeable with the magazine of the KP89. The KP93 will fire all commercially manufactured 9 mm Parabellum ammunition, including +P loads.

Two models are offered: a double-action-only version and a decocker version. Sights are provided with white dots for aiming in poor light, and the rearsight is drift-adjustable for windage.

Data

Cartridge: 9 × 19 mm Parabellum
Operation: recoil, semi-automatic, double action
Locking: dropping barrel
Feed: 15-round box magazine
Weight: empty, 878 g
Length: 185 mm
Barrel: 99 mm
Rifling: 6 grooves, rh, 1 turn in 254 mm
Sights: fore, blade; rear, square notch, adjustable for windage. White dots
Sight radius: 127 mm
Muzzle velocity: ca 355 m/s

Manufacturer

Sturm, Ruger & Co Inc, Lacey Place, Southport, Connecticut 06490.

Status

In production.

VERIFIED

9 mm Ruger KP93 Compact pistol

9 mm/0.40 Ruger KP94 pistol

Description

Sized midway between the full-sized P series and the compact KP93, the Ruger KP94 uses the link actuated short recoil action. The frame is of hard-coated aircraft aluminium and the slide is of stainless steel, the operating rails being of particularly robust dimensions so as to manage any commercially manufactured ammunition. The pistol is available in manual safety (KP94), decocker (KP94D) or double-action-only (KP94DAO) configurations.

Three-dot Patridge sights assist rapid and accurate alignment. The windage-adjustable rearsight locks into place with a precision setscrew and the frontsight blade is replaceable. Ergonomically designed unbreakable Xenoy grip panels are particularly suitable for users with smaller hands.

Index marks on the slide and frame, together with enlarged take-down slots, facilitate disassembly and re-assembly. The index mark identifys the point at which the frame-captive slide stop pin can be withdrawn or inserted.

All models have a fully ambidextrous magazine release. Ambidextrous slide-mounted decocking or safety levers on respective models feature reduced rotation to drop a cocked hammer safely. The patented Ruger 'push-forward' safety system positively engages and restrains the firing pin. moving it forward and away from all possible contact with the hammer when the decocking lever or safety lever is engaged. This independent, redundant safety feature functions in addition to the passive firing pin block, the operation of which arrests the firing pin until the trigger is pulled fully rearward.

Double-action-only models cannot be thumb-cocked; their slide-flush hammers transferring no energy to the firing pin even if dropped on the rear of the slide. When loaded, the double-action pistol's hammer rests in a position away from all contact witt the firing pin.

9 mm Ruger KP94 pistol

9 mm Ruger KP94DAO double-action pistol

Data
Cartridge: 9 × 19 mm Parabellum or 0.40 S&W
Operation: short recoil, semi-automatic
Locking: tilting barrel
Feed: 15-round (9 mm) or 11-round (0.40) box magazine
Weight: empty, 935.5 g
Length: 193 mm
Barrel: 107 mm
Rifling: 6 grooves, rh, 1 turn in 254 mm (9 mm) or 1 turn in 406 mm (0.40)
Sights: fore, blade with white dot; rear, notch with 2 white dots. Rear sight driftable for windage.
Sight radius: 134.1 mm

Manufacturer
Sturm, Ruger & Co Inc, Lacey Place, Southport, Connecticut 06490.

Status
In production.

UPDATED

9 mm Ruger KP94D decocker pistol

9 mm Ruger KP94L laser pistol

Description
The Ruger KP94L is exactly the same pistol as the 9 mm KP94 described previously but with the addition of an integral laser aiming marker. The pistol frame is deepened ahead of the trigger guard to form a housing for the laser and its associated circuitry and power supply. The projected spot is visible to the naked eye and projects a positive aiming mark to the limits of effective hand gun range. This model is only available in 9 × 19 mm Parabellum calibre, and is dimensionally the same as the KP94, apart from the slight change in frame contour.

Manufacturer
Sturm, Ruger & Co Inc, Lacey Place, Southport, Connecticut 06490.

Status
In production.

VERIFIED

9 mm Ruger KP94L laser pistol

Ruger SP101 revolver

Description
Following the introduction of the large frame 0.357 Magnum GP100 and the 0.44 Magnum Super Redhawk, Ruger completed their basic double-action family with the SP101, a small-frame, five-shot 0.38 calibre revolver. The all stainless steel SP101 incorporates the engineering refinements of earlier Ruger models into a new compact revolver.

Frame width has been increased in the critical areas which support the barrel, and both frame sidewalls are solid to provide great strength and rigidity. The design of an offset ejector rod has allowed the building of a thicker and stronger frame in the forcing cone area, that which undergoes the most severe pressure.

The lock mechanism is contained within the trigger guard, which is inserted into the frame as a single subassembly without the need for frame-weakening sideplates.

The SP101 cylinder provides all the strength necessary to withstand the pressures of modern high-velocity 0.38 cartridges. The cylinder locking notches are offset, and are cut into the thick part of the cylinder walls, between the chamber centres. The crane and cylinder assembly swings out of the frame in the usual manner, but when the cylinder is in the firing position it is securely locked to the frame in two places: the traditional cylinder pin at the rear and at the front of the crane by a large spring-loaded latch. Invented by Ruger, this forward lock ensures correct barrel and chamber alignment and also allows a larger thread diameter on the barrel and a thicker frame.

Barrels, cylinders and frames for the SP101 are made from ordnance quality 400-series stainless steel, as are the hammer, trigger and other internal parts.

The SP101 is available in two barrel lengths; 57 mm, the pistol weighing 709 g; and 76 mm, the pistol weighing 765 g. It is now available chambered for the 0.22

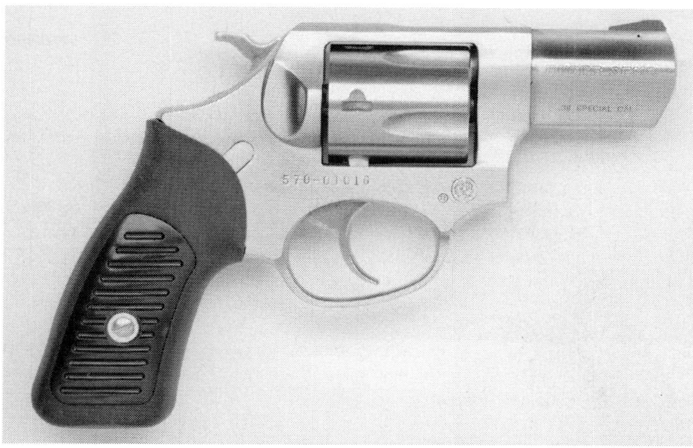

0.38 Ruger SP101 double-action revolver

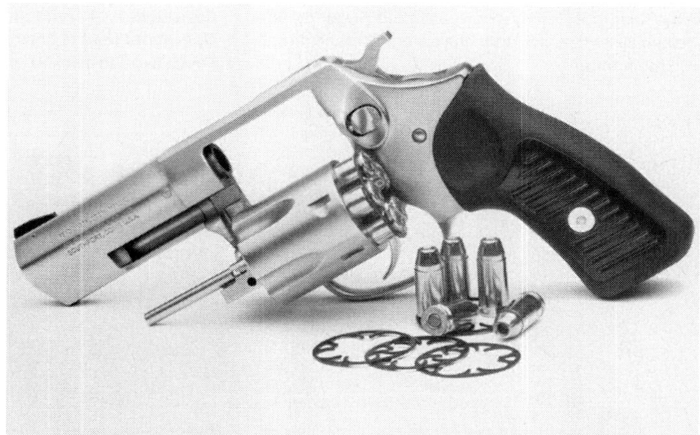
Ruger SP101 revolver in 9 mm calibre, showing five-shot full-moon clips

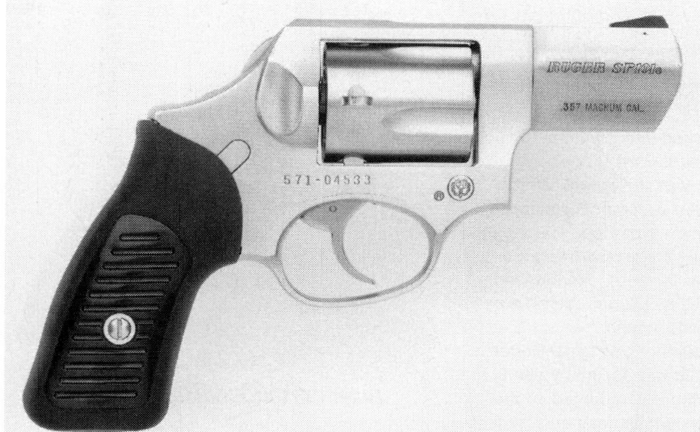

0.38 Special Ruger Spurless SP101 revolver

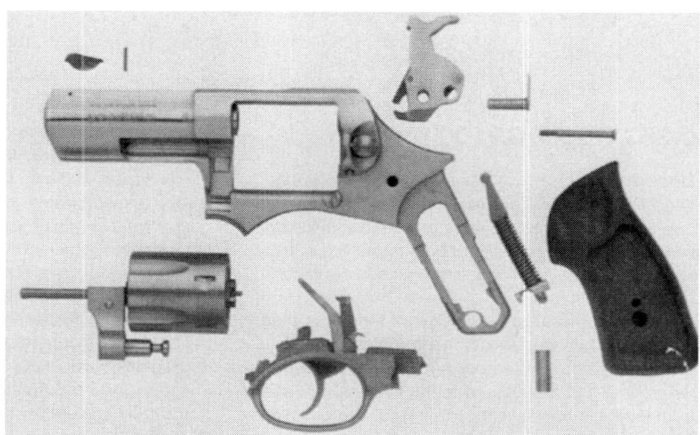
0.38 Ruger SP101 revolver dismantled

Long Rifle, 0.32 H&R Magnum, 0.357 S&W Magnum and 9 mm Parabellum cartridges. The frame, cylinder, crane and ejector have all been lengthened approximately 1.6 mm, making it suitable for all Magnum loads; thus the 0.357 calibre version is no longer restricted to use with the 125 gr (8.1 g) jacketed bullet round.

A double-action-only version, having a spurless hammer, was introduced in January 1993. The hammer has no single-action notch on it, so that it cannot be cocked. It is available in 0.38 Special (Model KSP821L) and 0.357 Magnum (Model KSP321XL) chambering.

The 9 mm Parabellum version uses full-diameter spring steel clips to retain the rimless cartridges in the cylinder and to provide a bearing surface for the ejector. These clips are provided with the pistol.

Manufacturer
Sturm, Ruger & Co Inc, Lacey Place, Southport, Connecticut 06490.

Status
In production.

VERIFIED

0.44 Magnum Ruger Super Redhawk revolver

Description
The Ruger Super Redhawk revolver incorporates the mechanical design features and patented improvements of the GP100 model, with a number of important additional features, the most significant of which is the massive extended frame and use of the exclusive Ruger Integral Scope Mounting System on the wide top strap, which provides a solid scope mounting surface. The extended frame also provides lengthened bearing surfaces and relocated barrel threads for greater strength and rigidity in barrel mounting.

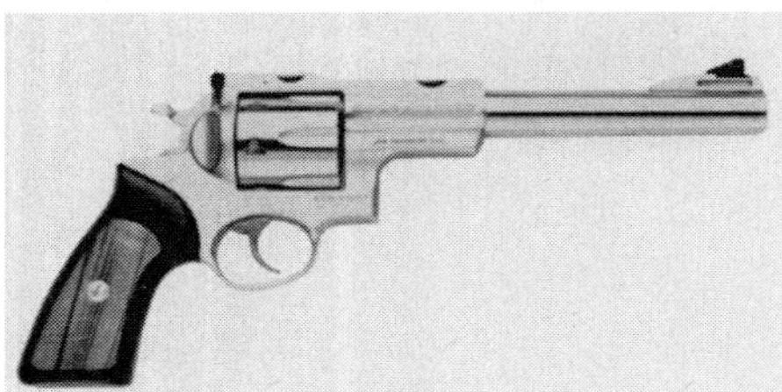

0.44 Magnum Ruger Super Redhawk revolver

The pistol is built of corrosion-resistant stainless steel in a brushed satin finish. It is available with 190 or 241 mm barrels. Both barrels are equipped with ramp front sight base with interchangeable insert blades. A steel adjustable rearsight with white outline square notch is standard.

Ruger Cushioned Grip panels are live rubber grip panels with Goncalo Alves wood inserts. The grip frame has been designed to allow installation of custom grips of a variety of shapes and sizes.

The pistol also incorporates a number of other Ruger features, including the floating firing pin mounted in the frame, transfer-bar safety, hammer and cylinder interlock, and the exclusive use of stainless steel springs throughout.

Data
Cartridge: 0.44 Magnum
Operation: double action revolver
Feed: 6-round cylinder
Weight: empty, 1.502 kg
Length: with 190 mm barrel, 330 mm
Barrel: 190 or 241 mm
Rifling: 6 grooves, rh, 1 turn in 508 mm
Sights: fore, ramp with insert blade; rear, square notch with white outline, adjustable

Manufacturer
Sturm, Ruger & Co Inc, Lacey Place, Southport, Connecticut 06490.

Status
In production.

VERIFIED

0.44 Ruger Redhawk revolver

Description
The Ruger Redhawk revolver is based on an entirely new mechanism and design philosophy. It encompasses a series of unique improvements and exclusive new features, making it outstanding in its field.

The frame has extra metal in the top strap and in critical areas below and surrounding the barrel threads. The frame has no side plate, so preserving both sidewalls intact as integral sections of the frame, resulting in increased strength and rigidity. The cylinder is locked in the firing position by a new, patented Ruger cylinder locking system, which bolts the swinging crane directly into the frame.

With the new Ruger patented single-spring mechanism, the hammer and trigger are powered by opposite ends of the same coil spring, and the components which link the trigger and hammer to this spring function smoothly with minimum friction loss.

The Redhawk is available in blued or stainless steel finish. Two kits are available as accessories; the first includes four interchangeable foresights of glassfibre-reinforced nylon, coloured light blue, fluorescent orange, ivory and yellow. The second kit has a steel gold bead foresight with matching V-notch rearsight.

Data
Cartridge: 0.44 Magnum
Operation: double or single action
Feed: 6-round cylinder, side-loading
Weight: with 140 mm barrel, 1.474 kg
Length: with 140 mm barrel, 280 mm
Barrel: 140 or 190 mm
Rifling: 6 grooves, rh, 1 turn in 580 mm
Sights: interchangeable red insert front; interchangeable rear

Manufacturer
Sturm, Ruger & Co Inc, Lacey Place, Southport, Connecticut 06490.

Status
In production.

VERIFIED

0.44 Magnum Ruger Redhawk revolver with 140 mm barrel

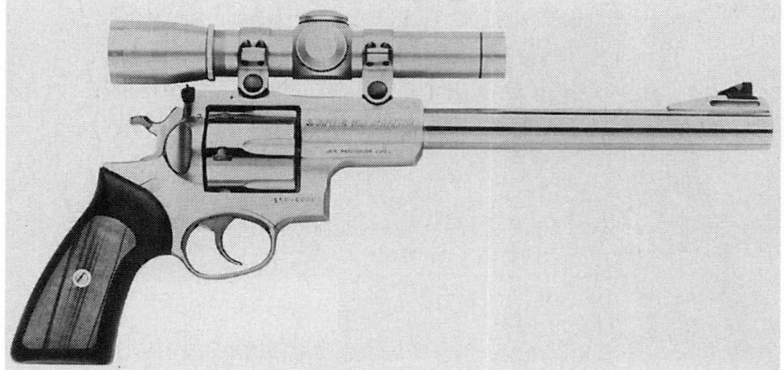

0.44 Magnum Ruger Redhawk revolver with 190 mm barrel and Ruger Integral Scope Mounting System

1996

0.357 Ruger GP100 revolver

Description
The GP100 is a new revolver in 0.357 Magnum calibre. The frame width has been increased in critical areas which support the barrel, and both frame sidewalls are solid and integral to provide strength and rigidity to the whole weapon.

The lock mechanism is contained within the trigger guard which is inserted into the frame as a single subassembly. The cylinder locking notches are substantially offset and are located in the thickest part of the cylinder walls between the centres of the chambers. The crane and cylinder assembly swings out of the frame in the normal manner. When the cylinder is in the firing position it is securely locked into the frame by a unique new Ruger-invented mechanism.

The heavy 102 mm barrel with full-length ejector shroud is made from a hot-rolled section of ordnance-quality 4140 chrome-molybdenum alloy steel. The long shroud helps to achieve the slightly muzzle-heavy balance generally considered desirable by experienced users. The Ruger Cushioned Grips (Ruger patent) are of live rubber, with polished wood inserts.

The design incorporates a number of original Ruger innovations which have been in use for many years. These include the floating firing pin mounted in the frame, the transfer-bar safety system hammer and cylinder interlock, and the exclusive use of coil springs throughout the mechanism. The hammer, trigger, and

Ruger GP100, Fixed Sight Model

most small internal parts are of durable, corrosion-proof stainless steel. The frame and cylinder are of ordnance quality chrome-molybdenum steel alloys. The revolver is available in blued finish or stainless steel. The 102 mm barrel is available with a full shroud, the 152 mm barrel with full or short shroud and adjustable sights.

GP100 Fixed Sight Model

The 76 mm and 102 mm barrel models are available with full or short ejector shroud in 0.357 Magnum and 0.38 Special, with fixed sights, in blued or stainless steel finish. An optional red insert foresight is available from the factory. The Fixed Sight Model also uses a patented smaller round butt with one-piece cushioned grip.

Data

Cartridge: 0.357 Magnum or 0.38 Special
Operation: double action revolver
Feed: 6-round cylinder
Weight: empty, 1.247 kg
Length: 238 mm
Barrel: 102 mm
Rifling: 5 grooves, rh, 1 turn in 52 calibres
Sights: fore, interchangeable blade; rear, adjustable square notch with white outline

Manufacturer

Sturm, Ruger & Co Inc, Lacey Place, Southport, Connecticut 06490.

Status

In production.

VERIFIED

0.22 LR Ruger Government Target Model pistol

Description

The Ruger Government Target Model is a blowback automatic pistol firing the 0.22 rimfire cartridge for the training and target practice roles. The Mark II version of the Ruger design retains the accuracy and handling characteristics of the earlier model.

Data

Cartridge: 0.22 LR
Operation: blowback, semi-automatic
Feed: 10-round box magazine
Weight: empty, 1.247 kg
Length: 283 mm
Barrel: 175 mm
Rifling: 6 grooves, rh, 1 turn in 381 mm
Sights: fore, blade; rear, notch, adjustable for windage
Sight radius: 235 mm

Manufacturer

Sturm, Ruger & Co Inc, Lacey Place, Southport, Connecticut 06490.

Status

In production.

VERIFIED

Ruger 0.22 LR Government Target Model

Smith & Wesson Sigma Series pistols

Development

Introduced in March 1994, the Sigma series was developed over a period of some 10 years. Using high-strength polymer material for the frame, and incorporating a self-cocking ('double-action-only') striker firing system, the pistol is extremely robust, simple to maintain and operate, and has a smooth and consistent trigger pull. The basic Model SW40F is chambered for the 0.40 S&W cartridge, but a variant Model SW9F is provided for those who prefer the 9 × 19 mm Parabellum round. A subcompact Sigma variant chambered for the 9 mm Short/0.380 Auto cartridge has been developed, as have subcompact variants of the 0.40 S&W and 9 mm pistols.

A total of 12 design patents has been issued or allowed for the Sigma series pistols.

Description

The Sigma series pistols employ a double-action-only operating principle allied with dropping barrel locking. A high degree of safety has been engineered into the series, including a trigger safety intended to prevent the pistol from firing in the event of it being dropped. Only when the trigger is fully depressed, and an internal striker safety plunger is lifted, can the pistol be fired. Even so, the internal safety was designed for immediate firing access.

0.40 Smith & Wesson Sigma Series Model SW40F

Other features of the Sigma series include a Melonite finished slide and a teflon coated carbon steel box magazine. The frame was designed using human factor analysis to produce an ultimate fit involving web angle, grip girth, angle of grasp, and trigger reach to meet human shooting characteristics for a very wide range of shooters, including those with small hands. These factors, and others, support easy target acquisition and promote shooter comfort and control.

Data

DATA FOR SW40F; THAT FOR SW9F IN PARENTHESIS
Cartridge: 0.40 S&W (9 × 19 mm Parabellum)
Operation: recoil, semi-automatic, double action only
Locking: dropping barrel
Feed: 15-round magazine (17-round magazine)
Weight: empty, 737 g
Length: 188 mm
Barrel: 114 mm
Sights: 3-dot (patent pending); Tritium optional
Sight radius: 162 mm

Manufacturer

Smith & Wesson Inc, 2100 Roosevelt Avenue, PO Box 2208, Springfield, Massachusetts 01102-2208.

Status

In production.

UPDATED

0.40 Smith & Wesson Model 410 pistol

Description

The 0.40 Smith & Wesson Model 410 double-action pistol is chambered for the 0.40 S&W cartridge and features a single-column 11-round capacity (10 rounds in the magazine and one in the chamber) and a one-piece Xenoy grip. The frame is aluminium alloy with a carbon steel slide, with a matt bead-blasted blue external finish. The Model 410 also has a slide-mounted manual safety/decocking lever on the left-hand side, enabling the pistol to be carried in safety with a round chambered but ready for immediate firing when required.

Data

Cartridge: 0.40 S&W
Operation: recoil, semi-automatic, double action only
Locking: dropping barrel
Feed: 10-round box magazine
Weight: empty, 810 g
Length: 190.5 mm
Barrel: 101 mm
Sights: 3-dot; fixed

Manufacturer

Smith & Wesson Inc, 2100 Roosevelt Avenue, PO Box 2208, Springfield, Massachusetts 01102-2208.

Status

In production.

NEW ENTRY

0.40 Smith & Wesson Model 410 pistol

1996

9 mm Smith & Wesson Model 909 and 910 pistols

Description

The 9 mm Smith & Wesson Model 909 and 910 double-action pistols are chambered for the 9 × 19 mm Parabellum cartridge. The main difference between the two models is that the Model 909 features a single column 9-round magazine and a curved backstrap grip while the Model 910 has a 10-round double stack magazine and a straight backstrap grip.

Both models have aluminium alloy frames with carbon steel slides, all with a matt bead-blasted blue external finish. The grips are lightweight and durable Xenoy. Both models have a slide-mounted manual safety/decocking lever on the left-hand side, enabling the pistol to be carried in safety with a round chambered but ready for immediate firing when required.

Data

DATA FOR MODEL 909; THAT FOR MODEL 910 IN PARENTHESIS
Cartridge: 9 × 19 mm Parabellum
Operation: recoil, semi-automatic, double action only
Locking: dropping barrel
Feed: 9-round box magazine (10-round box magazine)
Weight: empty, 765 g (794 g)
Length: 187 mm
Barrel: 101 mm
Sights: 3-dot; fixed

Manufacturer

Smith & Wesson Inc, 2100 Roosevelt Avenue, PO Box 2208, Springfield, Massachusetts 01102-2208.

Status

In production.

NEW ENTRY

9 mm Smith & Wesson Model 910 pistol
1996

Smith & Wesson Third-Generation semi-automatic pistols

Description

The Third Generation series of semi-automatic pistols replaced earlier models in production and was designed with the assistance of many US law enforcement agencies who were encouraged to make suggestions as to their requirements. Features incorporated in these new pistols include fixed barrel bushings for better accuracy and simpler dismantling, a greatly improved trigger pull, three-dot sights which allow a quicker sight picture in all conditions, improved wraparound grips, bevelled magazine aperture for quicker reloading, and a triple safety system comprising an ambidextrous manual safety catch, automatic firing pin safety system and magazine safety. Several other areas have been redesigned to improve service life.

Numbering System

Smith & Wesson Third-Generation pistols adopt a fresh numbering system which reflects the physical characteristics of the weapon. The first two digits are the basic model designation and indicate the calibre; the third digit indicates the type of model; and the fourth digit indicates the material from which the pistol is made:

Model
39 = 9 mm
59 = 9 mm
69 = 9 mm
10 = 10 mm
40 = 0.40 S&W
45 = 0.45 ACP

Type
0 = Standard
1 = Compact
2 = Standard, with decocking lever
3 = Compact, with decocking lever
4 = Standard, double action only
5 = Compact, double action only
6 = Non-standard barrel length
7 = Non-standard barrel length and decocking lever
8 = Non-standard barrel length, double action only

Material
3 = Aluminium alloy frame, stainless steel slide
4 = Aluminium alloy frame, carbon steel frame
5 = Carbon steel frame and slide
6 = Stainless steel frame and slide
7 = Stainless steel frame, carbon slide

Thus Model 4506 is a standard 0.45 ACP pistol with stainless steel frame and slide.

VERIFIED

9 mm 5900 series Smith & Wesson pistols

Description

The 5900 series consists of four models, the 5903, 5904, 5906 and 5946. The 5903 has an alloy frame and stainless steel slide, fixed sights; 5904 has an aluminium alloy frame, carbon steel slide and stainless steel barrel and is finished in blue; 5906 is entirely of stainless steel, is satin finished and has adjustable sights; and 5946 is self-cocking or 'double action only', all stainless steel and with fixed sights. All are fitted with one-piece wraparound 'Delrin' grips with curved backstrap.

Data

Cartridge: 9 × 19 mm Parabellum
Operation: recoil, semi-automatic
Locking: dropping barrel
Feed: 15-round magazine
Weight: 5903, 808 g; 5904, 752 g; 5906, 1.063 kg (all with fixed sights; adjustable sight models 15 g more)
Length: 190.5 mm
Barrel: 101.6 mm
Sights: fore, post with white dot; rear, U-notch fixed with 2 white dots, or adjustable for windage and elevation. Novak night sight optional on 5906.

Manufacturer

Smith & Wesson Inc, 2100 Roosevelt Avenue, PO Box 2208, Springfield, Massachusetts 01102-2208.

Status

In production.

VERIFIED

9 mm Smith & Wesson Model 5904, fixed sight

9 mm 3900 series Smith & Wesson pistols

Description

The 3900 series generally resembles the 5900 series but is slimmer, slightly lighter in weight, and has a reduced magazine capacity. Three basic models are manufactured: the 3913 with aluminium alloy frame and stainless steel slide, the 3914 which has an alloy frame and carbon steel slide, and the 3953 which is self-cocking (double action only), with fixed sights, alloy frame and stainless steel slide. In addition, a variant model of the 3913 known as the 3913LS is the 'Ladysmith' model, specially finished and provided with a purse-like soft carrying case and intended for the female hand.

Data

Cartridge: 9 × 19 mm Parabellum
Operation: recoil, semi-automatic
Locking: dropping barrel
Feed: 8-round box magazine
Weight: 709 g
Length: 173 mm
Barrel: 89 mm
Sights: fore, post with white dot; rear, U-notch fixed with 2 white dots, or adjustable for windage and elevation

Manufacturer

Smith & Wesson Inc, 2100 Roosevelt Avenue, PO Box 2208, Springfield, Massachusetts 01102-2208.

Status

In production.

VERIFIED

9 mm Smith & Wesson Model 3914, fixed sight

9 mm 6900 series Smith & Wesson pistols

Description

The 6900 series are compact semi-automatics, using the same general design features as the rest of the Third Generation pistols but of smaller dimensions. The Model 6904 has an aluminium alloy frame, carbon steel slide and stainless steel barrel, Model 6906 has an alloy frame and stainless steel slide, and Model 6946 is 'double action only' with alloy frame and stainless steel slide.

Data

Cartridge: 9 × 19 mm Parabellum
Operation: recoil, semi-automatic, double action
Locking: dropping barrel
Feed: 12-round box magazine
Weight: 751 g
Length: 174.6 mm
Barrel: 89 mm
Sights: fore, post with white dot; rear, fixed U-notch with 2 white dots. Novak night sight optional on 6906.

Manufacturer

Smith & Wesson Inc, 2100 Roosevelt Avenue, PO Box 2208, Springfield, Massachusetts 01102-2208.

Status

In production.

VERIFIED

9 mm Smith & Wesson Model 6906 pistol with alloy frame and stainless steel slide

1996

0.40 Smith & Wesson Model 4000 series pistol

The Model 4000 introduced the 0.40 S&W cartridge into the Third Generation range; the 0.40 S&W might be described as the 'Imperial Measurement' alternative to the 10 mm cartridge, being some 2.5 mm shorter but with similar power. The size reduction makes it possible to adapt a 9 mm frame without having to make serious changes to the magazine well dimensions. Apart from the change in calibre and magazine capacity, the 4006 is generally the same as the other standard models of this class. The Model 4043 is double action only with alloy frame and stainless steel slide, while the Model 4046 is also double action only but entirely of stainless steel. The Model 4053 is a compact model with an alloy frame and a stainless steel slide.

Data

Cartridge: 0.40 S&W
Operation: recoil, semi-automatic, double action
Locking: dropping barrel
Feed: 11-round box magazine
Weight: fixed sight, 1.091 kg; adjustable sight, 1.106 kg
Length: 190.5 mm
Barrel: 101.6 mm
Sights: fore, post with white dot; rear, U-notch fixed with 2 white dots, or adjustable for windage and elevation

Manufacturer

Smith & Wesson Inc, 2100 Roosevelt Avenue, PO Box 2208, Springfield, Massachusetts 01102-2208.

Status

In production.

UPDATED

0.40 Smith & Wesson Model 4053 pistol, fixed sight, with alloy frame and a stainless steel slide

1996

0.45 Smith & Wesson Model 4500 series pistol

Description

The 4500 series pistols are chambered for the 0.45 ACP cartridge and provide the various modern features of the Third Generation with the power of the well known heavy calibre bullet. The 4506 is of stainless steel throughout, in satin stainless finish, and is available with fixed or adjustable sights.The 4566 is similar to the 4506 but with a slightly shorter barrel. The 4586 is similar in size to the 4566 but is double action only.

Data

Cartridge: 0.45 ACP
Operation: recoil, semi-automatic, double action
Locking: dropping barrel
Feed: 8-round box magazine
Weight: 4506, 1.162 kg; 4566/4586, 1.091 kg
Length: 4506, 219 mm; 4566/4586, 200 mm
Barrel: 4506, 127 mm; 4566/4586, 108 mm
Sights: fore, post with white dot; rear, U-notch fixed with 2 white dots, or adjustable for windage and elevation

Manufacturer

Smith & Wesson Inc, 2100 Roosevelt Avenue, PO Box 2208, Springfield, Massachusetts 01102-2208.

Status

In production.

VERIFIED

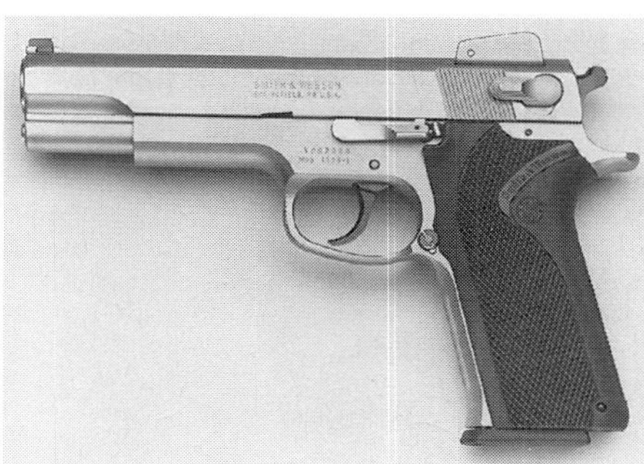

0.45 Smith & Wesson Model 4506-1, adjustable sight

1996

0.38 Model 64 Military and Police Stainless revolver

Description

The Smith & Wesson Model 10 Military & Police revolver has been well known and popular for many years. The Model 64 is the stainless steel version which has obvious attractions for military service under arduous conditions. The metal is satin finished and the butt grips, formerly of walnut, are now supplied with combat grips. The standard 4 in (101 mm) barrel is designated as a 'heavy' barrel, being heavier in section than is usual in this calibre; a 2 in (50 mm) barrel of normal section is also available.

From May 1994 all Smith & Wesson service-type revolvers were supplied as standard with synthetic grips.

Data

Cartridge: 0.38 Special
Operation: double action revolver
Feed: 6-round cylinder
Weight: empty, 865 g
Length: 235 mm
Barrel: 101.6 mm
Sights: fore, serrated ramp; rear, square notch

Manufacturer

Smith & Wesson Inc, 2100 Roosevelt Avenue, PO Box 2208, Springfield, Massachusetts 01102-2208.

Status

In production.

Service

Widely used by police and security forces.

UPDATED

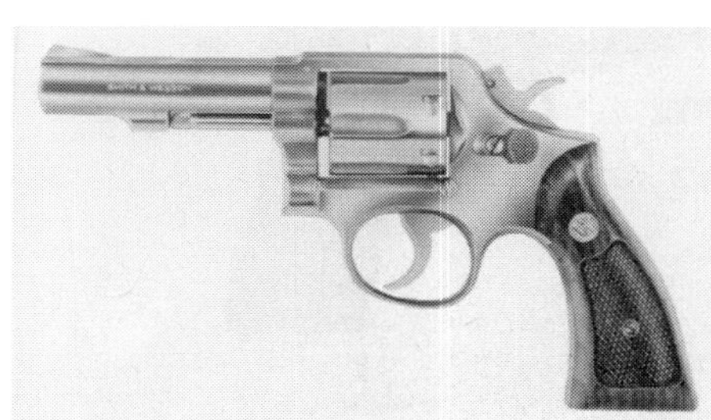

0.38 Smith & Wesson Model 64 Military and Police Stainless revolver

0.44 Magnum Smith & Wesson Model 29 revolver

Description

When the utmost power is required from a hand gun, the Model 29 is one solution, with its 0.44 Magnum cartridge delivering upwards of 1,600 J of muzzle energy with the longer barrels. The Model 29 is of the conventional Smith & Wesson design, a solid-frame revolver with swing-out cylinder and double-action lock, but the proportions are substantial in order to accommodate the powerful cartridge.

As from May 1994 all Smith & Wesson service-type revolvers are supplied as standard with synthetic grips.

Data

Cartridge: 0.44 Magnum
Operation: double action revolver
Feed: 6-round cylinder
Weight: with 165 mm barrel, 1.332 kg
Length: with 165 mm barrel, 302 mm
Barrel: 101 mm, 165 mm or 212 mm
Sights: fore, S&W Red Ramp; rear, micrometer notch

Manufacturer

Smith & Wesson Inc, 2100 Roosevelt Avenue, PO Box 2208, Springfield, Massachusetts 01102-2208.

Status

In production.

UPDATED

Smith & Wesson Model 29 0.44 Magnum revolver with 212 mm barrel

1996

Vector 22 Shooting System

Description

The Vector 22 Shooting System is a 10-shot enclosed cylinder revolver which has the capability to eject and reload either 0.22 RF Long Rifle or 0.22 Magnum chambered cylinders quickly. These patented cylinders are defined under the trade name 'Ammo Cassette'.

Ammo Cassette ejection and reloading is performed when the barrel assembly is unlocked and tilted forward. The cylinder can be removed and replaced, and when exchange is completed the assembly is swung back and locked closed. The gun is then ready to fire by compressing the firing handle against the grip. The firer holds the Vector like a hand saw, with the index finger extended parallel to the gun's magazine and barrel housing in a natural pointing position. The remaining fingers are compressed against the grip in order to fire. The barrel is aligned with the anatomical centreline of the firer's hand, capitalising on the natural pointing instinct and giving quick target acquisition and a high hit probability as a result of the recoil energy passing in a straight line to the shoulder and not promoting muzzle rise.

The operation of the Ammo Cassette cylinder and the firing mechanism is controlled by a patented action rod, the Autobolt. This allows the Vector to function as a mechanical double-action weapon, or as a gas-operated weapon capable of firing in the semi-automatic mode.

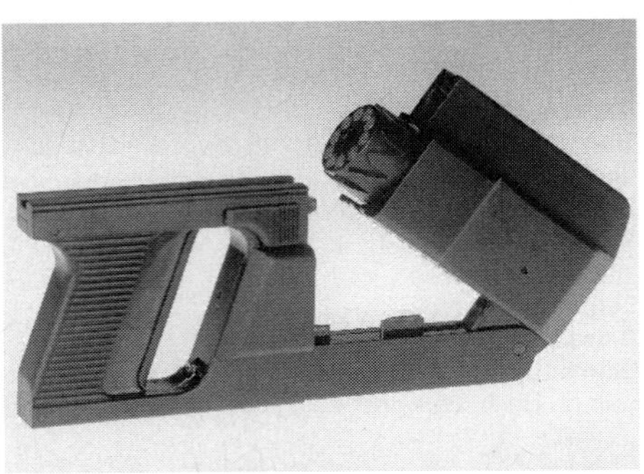

Vector with Ammo Cassette and barrel in released position

The system is modular, and by changing components the weapon can be configured as a carbine or rifle. The basic kit includes a rifle barrel which can rapidly be changed for the pistol barrel, and the carrying case converts into a butt-stock. An optical sight is also part of the basic kit.

The system can also be produced as a selective fire mechanism, allowing semi- or full-automatic fire, for special service. The Vector's barrel and cassette housing design allows it to be sound-suppressed internally. This integral feature, with the system in its assembled rifle configuration and using 0.22 Magnum ammunition, converts the weapon into a silent sniper system capable of highly accurate fire out to 200 m. In conjunction with the system's optical sight fire control system, this offers a competitive and economical sniping system for many security applications.

Data

Cartridge: 0.22 RF Long Rifle or Magnum
Operation: revolver; mechanical repeater, double action; or gas-operated semi-automatic; or gas-operated selective fire
Locking: Autobolt (see text)
Feed system: 10-round Ammo Cassette cylinder
Weight: loaded, 680 g
Length: 229 mm
Barrel: 114 mm, 6 grooves, 1 turn in 355 mm
Sights: fore, post; rear, channel
Construction: principal components are injection-moulded thermoplastic resins and 4340 steel

Manufacturer

Mark Three, 1410 Central Avenue SW, Unit No 23, Albuquerque, New Mexico 87104.

Status

Development complete.

VERIFIED

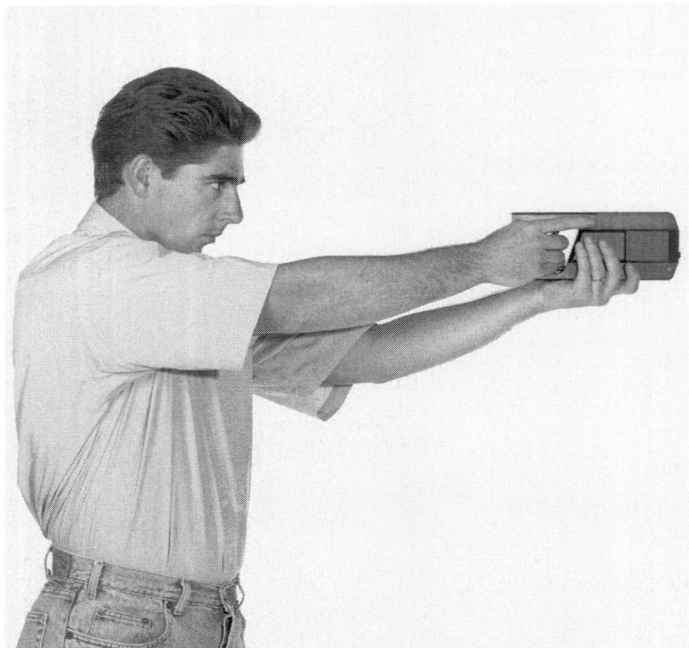

Using the Vector as a hand gun

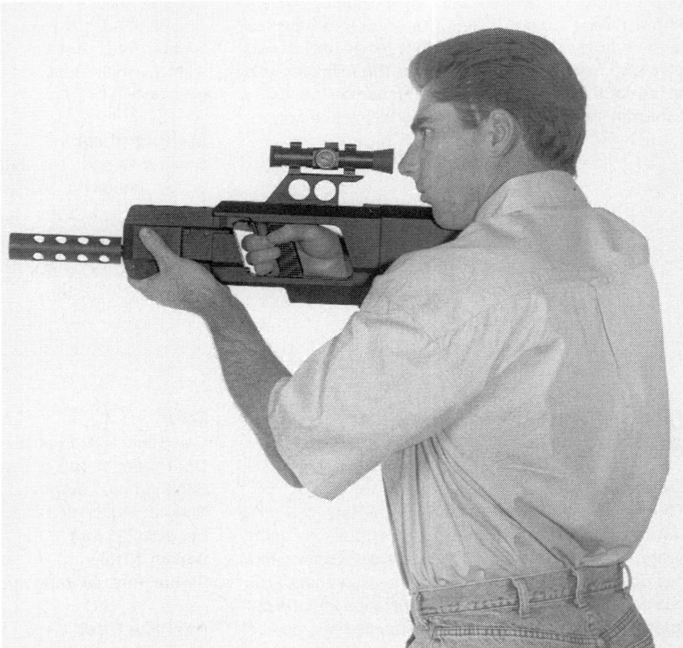

Using the Vector as a special operations weapon system

Ciener 0.22 LR M1911A1 Conversion Kit

Description

The Ciener 0.22 Conversion Kit permits conversion of M1911A1 0.45 automatic pistols to fire 0.22 LR rimfire cartridges. The conversion kit fits all full size M1911A1 pistols whose dimensional tolerances meet those of the US Government drawings, both Series 70 and Series 80. The unit consists of a 0.22 LR slide, barrel, return spring and guide rod, recoil buffer and 10-round 0.22 LR magazine in a fitted plastic case. These items are assembled on to the dismantled weapon replacing the standard 0.45 ACP parts. Change over time is 10 seconds or less.

Manufacturer

Jonathan Arthur Ciener Inc, 8700 Commerce Street, Cape Canaveral, Florida 32920.

Status

In production.

VERIFIED

The assembled Ciener 0.22 LR weapon is externally indistinguishable from the standard 0.45 ACP weapon

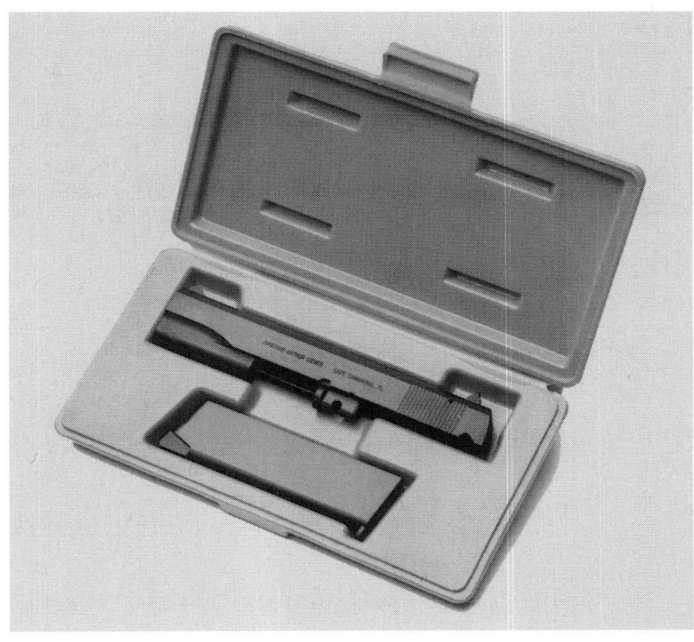

The Ciener 0.22 LR Conversion Kit in plastic case

Ciener 0.22LR Beretta 92/96 Conversion Kit

Description

The Ciener 0.22 Conversion Kit permits conversion of Beretta 92/96 automatic pistols to fire 0.22 LR rimfire cartridges. There are versions for standard and Compact models. A limited production unit is available for the Beretta 93R machine pistol, allowing semi- and automatic fire.

The unit consists of a 0.22 LR slide, barrel, return spring and guide rod, recoil buffer and 10-round 0.22 LR magazine in a fitted plastic case. These items are assembled on to the dismantled weapon replacing the standard 0.45 ACP parts. Changeover time is 10 seconds or less.

Essentially similar 0.22 LR conversion kits are available for the Taurus PT92/99.

Manufacturer

Jonathan Arthur Ciener Inc, 8700 Commerce Street, Cape Canaveral, Florida 32920.

Status

In production.

NEW ENTRY

Beretta 93R fitted with a Ciener 0.22 LR Conversion Kit

1996

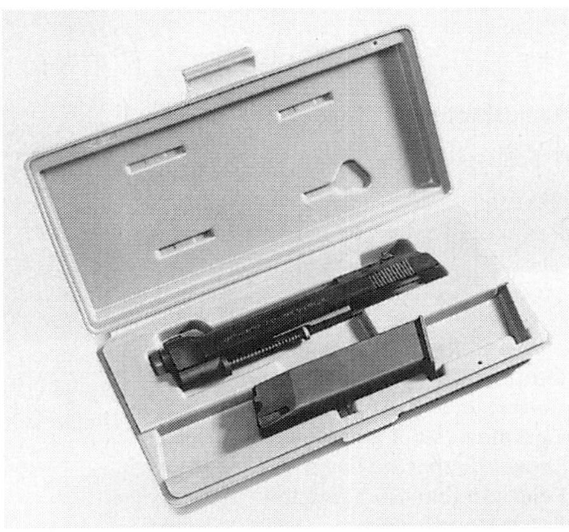

The Ciener 0.22 LR Conversion Kit for the Beretta 92/96 in its plastic case

1996

As with other Ciener conversions the assembled Ciener 0.22 LR weapon is externally indistinguishable from the standard Beretta 92/96

1996

YUGOSLAVIA (SERBIA AND MONTENEGRO)

7.62 mm Model M57 and 9 mm Model M70 and M70A pistols

Description

The Model M57 is the former Yugoslav model of the Tokarev TT-33 (qv) and can be distinguished from it by the maker's name on the slide and/or the emblem on the grip). The principle of operation and general mechanical details are similar to those of the TT-33.

The Model M70 is the M57 built to accept the 9 × 19 mm Parabellum cartridge. Apart from the change of calibre and the use of six-groove rifling instead of four grooves it is mechanically similar to the M57. The Model M70A is an improved M70 which has had a slide-mounted safety catch added which locks the firing pin when applied.

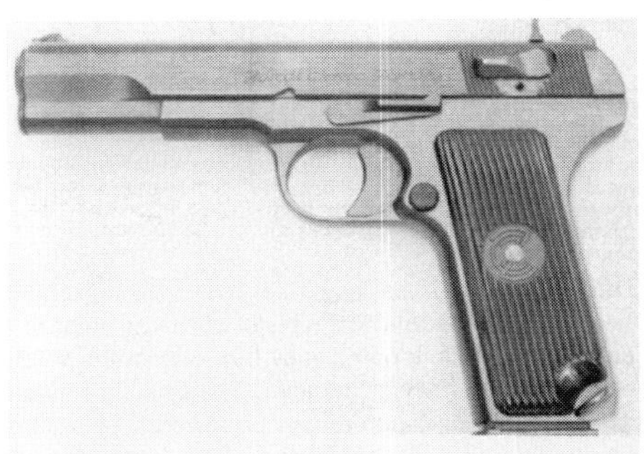

9 mm Model 70A with added safety catch

Data

	M57	M70 and M70A
Cartridge	7.62 × 25 mm	9 × 19 mm Parabellum
Operation	short recoil	
Feed	9-round box magazine	
Weight	approx 900 g	
Length	200 mm	
Barrel length	116 mm	
Rifling	4 grooves	6 grooves
Muzzle velocity	450 m/s	330 m/s
Penetration at 25 m: sand	350 mm	250 mm
Fir plank	60 mm	50 mm

Manufacturer
Zastava Arms, 29 Novembra 12, YU-11000 Beograd.

Status
Production status uncertain.

Service
Former Yugoslav forces.

VERIFIED

9 mm Model M70 pistol

7.62 mm Model M57 pistol

7.65 mm Model M70 and 9 mm Model M70(k) pistols

Description
These two pistols, in 7.65 mm and 9 mm (Short) calibre are further developments of the Models M57 and M70 pistols (see previous entry) with improvements in operation, handling and accuracy in addition to the changes in calibre. The distribution of masses has been improved to permit better control and consequent firing accuracy.

The side-mounted manual safety secures both the firing mechanism and the slide when applied and there is an automatic safety which blocks the sear when the magazine is removed.

The two weapons are chambered for 7.65 mm and 9 mm Browning cartridges and comparative data follow.

Data

	M70	M70(k)
Cartridge	7.65 mm	9 mm Short
Operation	short recoil	
Feed	8-round box magazine	
Weight	740 g	720 g
Length	165 mm	
Barrel	94 mm	
Rifling	6 grooves, 240 mm twist	
Muzzle velocity	300 m/s	260 m/s
Penetration at 25 m: sand	250 mm	350 mm
Fir plank	100 mm	70 mm

Manufacturer
Zastava Arms, 29 Novembra 12, YU-11000 Belgrade.

Status
Production status uncertain.

7.65 mm Model M70 pistol

Service
Former Yugoslav forces.

UPDATED

9 mm M88 and M88A pistols

Description
The M88 was developed as a reduced size and modernised version of the 9 mm M70 (see previous entry). The general principle of operation, the Browning link and dropping barrel, is retained, but the entire appearance was altered so that the resemblance to the original TT-33 Tokarev is no longer apparent. The size of the pistol was reduced, and the shape and material of the grips entirely changed.

As with the M70, the M88 has a frame-mounted safety catch, while the M88A has the safety catch mounted at the rear of the slide, where it blocks the firing pin as well as interrupting the link between trigger and hammer.

Data
Cartridge: 9 × 19 mm Parabellum
Operation: short recoil, semi-automatic
Locking: dropping barrel
Feed: 8-round box magazine
Weight: empty, 780 g
Length: 175 mm
Barrel: 96 mm
Rifling: 6 grooves, rh twist
Muzzle velocity: 325 m/s

Manufacturer
Zastava Arms, 29 Novembra 12, YU-11000 Belgrade.

Status
Production status uncertain.

VERIFIED

9 mm Zastava M88A pistol

9 mm CZ-99 pistol

Description
The CZ-99 is a double-action pistol with large magazine capacity and with a completely new concept of safety elements and manual controls. There is no manual safety catch; the hammer can be lowered by a combined slide stop lever and decocking lever which is duplicated on both sides of the weapon. The firing pin is fitted with an automatic safety system which prevents the pin moving except during the final movement of the trigger; and the firing pin and hammer are automatically separated by 20° except, again, for a brief period during the final release of the trigger. The magazine release button is also duplicated on both sides of the butt, so that the pistol can be used with equal facility in either hand.

The pistol frame is of aluminium alloy, the slide of forged steel, and other parts of steel. The grips are of plastic, with an ergonomic, hand-filling design giving a firm hold. The surface finish can be performed to suit the customer, either blued or painted in various shades.

Data

Cartridge: 9 × 19 mm Parabellum
Operation: short recoil, semi-automatic
Locking: dropping barrel
Feed: 15-round box magazine
Weight: with empty magazine, 965 g
Length: 190 mm

Manufacturer

Zastava Arms, 29 Novembra 12, YU-11000 Belgrade.

Status

Production status uncertain.

VERIFIED

9 mm Zastava CZ-99 pistol

9 mm MAB P15S pistol

Description

This pistol, originally designed and manufactured by the Manufacture d'Armes, Bayonne, France, and in service with the French Army, has been manufactured in Serbia. The dimensions and data for the Serbian model are the same as those for the original French pattern, and there does not appear to be any significant difference between the two. For details of the pistol consult the entry in the French section on 9 mm PA15 MAB pistol.

Manufacturer

Zastava Arms, 29 Novembra 12, YU-11000 Belgrade.

Status

Production status uncertain.

UPDATED

0.357 Magnum Model 1983 revolver

Description

In order to complete their production programme of weapons, Zastava developed the Model 1983 revolver, designed for the 0.357 Magnum cartridge. It can also chamber and fire the 0.38 Special cartridge, and with a change of cylinder and the use of a special adaptor can fire rimless 9 mm Parabellum cartridges as well.

The revolver is a solid-frame, side-opening pattern with double-action lockwork. The barrel carries a ventilated rib and the extractor rod is fully shrouded. Cylinder release is by the usual thumb-operated latch on the left side of the frame.

In addition to the standard version, the revolver can be made with special handgrips or with adjustable sights, and sporting versions are available with 64, 102 and 152 mm barrel lengths. On special request, the revolver can be made in a deluxe version, with engraving and chrome plating.

Data

Cartridge: 0.357 Magnum (and see text)
Operation: double action revolver
Feed: 6-round cylinder
Weight: empty, 900 g
Length: 188 mm
Barrel: 64 mm (see text)

Manufacturer

Zastava Arms, 29 Novembra 12, YU-11000 Belgrade.

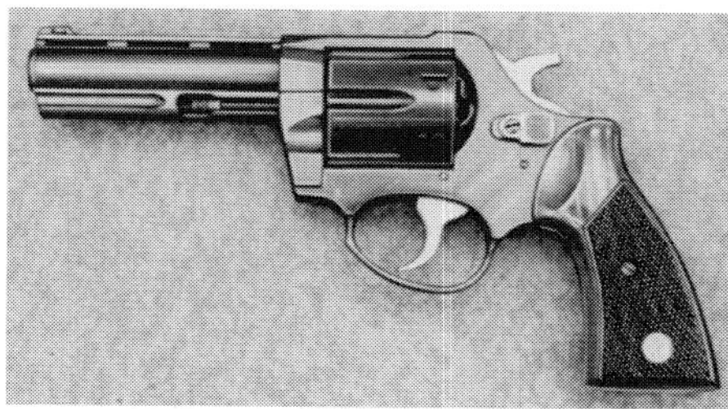

Zastava 0.357 Magnum Model 1983 revolver

Status

Production status uncertain.

VERIFIED

SUB-MACHINE GUNS

ARGENTINA

9 mm FMK-3 Mod 2 sub-machine gun

Description

The 9 mm FMK-3 Mod 2 is the current sub-machine gun manufactured by Fabrica Militar de Armas Portatiles 'Domingo Matheu', Rosario. It is a blowback-operated weapon of modern design.

It was formerly produced in two models, one with a fixed plastic butt and the other with a sliding butt modelled on the US M3 (qv). The current 'Modification 2' model is produced only in sliding butt form. The body of the gun is a metal pressing and there is a screw-threaded cap at the front end to allow easy release of the barrel. A plastic fore-end grip is under the receiver. The 25- or 40-round box magazine fits into the pistol grip which has a grip safety at the back. There is also a safety position on the selector.

The cocking handle is on the left side of the receiver, well forward, and there is a slide which covers the cocking slot to keep out dirt.

The FMK-3 is designed with a wraparound bolt. The bolt encloses 180 mm of the barrel which itself is 290 mm long. This, the manufacturer claims, leads to good control and stability in firing, as well as reducing the length.

9 mm FMK-3 Mod 2 sub-machine gun

Variant

9 mm FMK-5 carbine

The FMK-5 carbine is the same basic weapon as the FMK-3 Mod 2 except that it is fitted with a different safety/selector lever which permits only semi-automatic firing. The markings are changed accordingly and the weapon is prominently marked 'SEMI-AUTO'. The FMK-5 is intended for use by security services, prison staffs and for commercial sale.

Data

Cartridge: 9 × 19 mm Parabellum
Operation: blowback, selective fire
Feed: 25- or 40-round box magazine
Weight: empty, 3.6 kg
Length: butt retracted, 520 mm; butt extended, 690 mm
Barrel: 290 mm
Rifling: 6 grooves, rh, 1 turn in 250 mm
Sights: fore, pillar; rear, flip aperture, 50 and 100 m
Sight radius: 320 mm
Muzzle velocity: 400 m/s
Rate of fire: cyclic, 600 rds/min

Manufacturer

Fabrica Militar de Armas Portatiles 'Domingo Matheu', Avenida Ovidio Lagos 5250, 2000 Rosario.

Status

In production.

Service

Argentinian forces.

VERIFIED

AUSTRIA

9 mm MPi 69 and MPi 81 Steyr sub-machine guns

Description

The Steyr MPi 69 sub-machine gun is a simple and robust weapon. The receiver pressing is of light-gauge steel and is welded into a hollow box with two gaps on the right-hand side: one in the middle for ejection of the spent case and one approximately 76 mm from the front to accept an insert to take the barrel seating, barrel release catch and barrel securing nut. The cocking slide, a simple pressing, is on the front left-hand side of the receiver.

The ejector is a simple bent strip riveted in position in the middle of the base of the receiver to run in a groove in the bottom of the breech block. Spot-welded under the receiver at the back is a small bracket which provides guides for the spring steel telescoping butt and spring-loaded release plungers. The strip-down catch is also at the rear. The moulded nylon receiver cover fits under the receiver and carries the trigger mechanism, the pistol grip and magazine housing.

The barrel is 250 mm long and is cold-forged on a rifling mandrel. The breech block has a fixed firing pin on the bolt face which is half-way along the bolt length. Thus the bolt wraps round the barrel and has a slot cut along the right-hand side for ejection. The long barrel maintains its pressure longer and so the bolt has to be somewhat heavier than it would be with a normal barrel length.

The weapon fires either single shot or full automatic. The choice is controlled by the safety bar which has three positions, safe, single fire and automatic.

The applied safety is a cross bolt which is pressed through the receiver. One end is marked 'S' in white and projects when the gun is safe. The other end is marked 'F' in red and this projects from the receiver when the gun is set to fire. When the pin is in the middle position, the safe button is only half-way through and the gun will fire single shot only.

In order to prevent accidental discharge caused by the bolt being jarred or mishandled during cocking, the breech block is provided with three safety bents. The first bent is the front edge of the block and this engages the sear before the bolt has moved back enough to pass over the base of the round in the magazine. This is about 40 mm of backward travel. The second bent allows the bolt a further 10 mm or so of travel and the same distance further back is the third bent. It is impossible to bounce the bent off the sear in any of the three positions.

The third bent is the normal working bent, and the second bent is provided to prevent a runaway gun when using 9 mm ammunition of lower impulse than usual. The weapon cannot be accidentally discharged and is at least as safe as others with more complicated safety arrangements.

The cocking lever is a pressing running in a groove at the top of the left side of the receiver at the front. The method of cocking is unusual. The sling is attached to the cocking lever and the soldier cocks the gun by pulling back on the sling. To prevent this happening unintentionally a bracket is welded to the top of the receiver which prevents movement of the cocking lever backwards unless the sling is held out at right angles to the gun.

9 mm MPi 69 Steyr sub-machine gun

Cocking arrangements: (top) *MPi 69 is cocked by pulling on sling* (bottom) *MPi 81 uses conventional cocking handle*

MPi 81 Firing Port version with stock folded

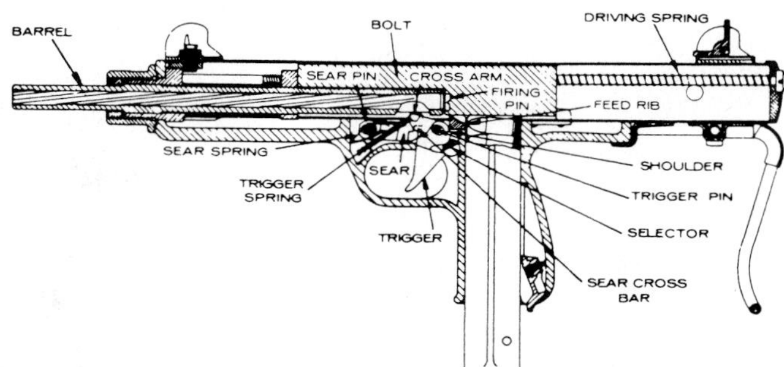

MPi 69 sub-machine gun

Variants

MPi 81

On the MPi 81 the sling is attached to a conventional swivel on the right side of the receiver and cocking is accomplished by a normal cocking handle which protrudes at the top left side of the receiver. In addition the rate of fire has been increased to about 700 rds/min by slight internal improvements.

Firing port model

A special version of the MPi 81 was developed for firing from APCs, IFVs and similar vehicles with firing ports. The receiver is fitted with the optical sight of the AUG rifle carried in special brackets, and the barrel is extended and fitted with a locking collar. The rear positioning of the optical sight allows it to be used with a vision block, if the block position is suitable.

Data

Cartridge: 9 × 19 mm Parabellum
Operation: blowback, selective fire
Feed: 25- or 32-round box magazine
Weight: empty, 3.13 kg
Length: butt retracted, 465 mm; butt extended, 670 mm
Barrel: 260 mm
Rifling: 6 grooves, rh, 1 turn in 254 mm
Sights: fore, blade; rear, flip aperture, 100 and 200 m
Sight radius: 326 mm
Muzzle velocity: 381 m/s
Rate of fire: cyclic, 550 rds/min; practical, 100 rds/min

Manufacturer

Steyr-Mannlicher AG, Postfach 1000, A-4400 Steyr.

Status

Production completed; superseded by TMP and AUG Para.

Service

Various military and police forces.

VERIFIED

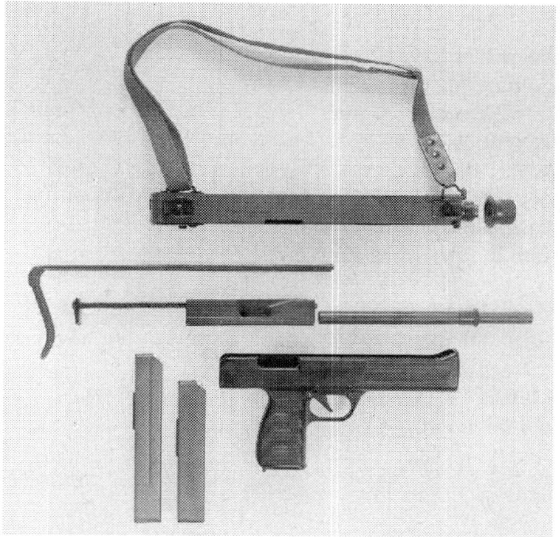
9 mm MPi 69 sub-machine gun field-stripped

Steyr AUG 9 mm Para

Description

The Steyr AUG 9 mm Para is a sub-machine gun version of the standard AUG assault rifle. It uses the existing stock and receiver units but is fitted with a new barrel of 9 mm calibre, a special bolt group, a magazine adaptor and a magazine.

Data

Cartridge: 9 × 19 mm Parabellum
Operation: blowback, selective fire, closed bolt
Feed: 25- or 32-round box magazine
Weight: empty, 3.5 kg
Length: 665 mm
Barrel: 420 mm
Rifling: 6 grooves, rh, 1 turn in 250 mm
Sights: integral ×1.5 telescope
Muzzle velocity: ca 400 m/s
Rate of fire: cyclic, 670-770 rds/min

Manufacturer

Steyr-Mannlicher AG, Postfach 1000, A-4400 Steyr.

Status

In production.

UPDATED

Steyr AUG 9 mm Para sub-machine gun with silencer

Steyr AUG 9 mm Para

Steyr 9 mm Tactical Machine Pistol

Description

The Steyr 9 mm Tactical Machine Pistol (TMP) could equally be considered as a pistol, but its full-automatic ability and general configuration places it into the 'personal defence weapon' category and gives it equal credence as a sub-machine gun.

The Tactical Machine Pistol (TMP) is a locked-breech weapon in 9 × 19 mm Parabellum calibre, although other calibres are planned. There are only 41 component parts and the frame and top cover are made from a plastic material. An integrated mounting rail is provided over the receiver for optical and optronic sighting devices.

The locking system uses a rotary barrel, controlled by a single lug which engages in a groove in the frame to turn the barrel and thus unlock it from the bolt. The cocking handle is at the rear of the weapon, beneath the rearsight, and is pulled back to cock. Selection of single shot or semi-automatic fire is performed by a three-position safety bar. Box magazines hold 15 or 30 rounds.

An indication of the compact nature of the TMP is that its size allows it to fit inside the dimensions of a sheet of A4 paper.

The TMP has facilities for attaching a sound suppressor or other muzzle attachments.

Variant

The TMP is also produced in semi-automatic only form, whereupon it becomes the 'Special Purpose Pistol' (SPP). See entry in *Pistols* section for details of the SPP.

Data

Cartridge: 9 × 19 mm Parabellum
Operation: recoil, semi-automatic (TMP selective fire)
Locking: rotating barrel
Feed: 15- or 30-round box magazine

Steyr 9 mm Tactical Machine Pistol (TMP)

Weight: empty, 1.3 kg
Length: 282 mm
Barrel length: 130 mm
Height: 162 mm
Width: 45 mm
Sights: fore, blade; rear, notch
Sight radius: 188 mm
Rate of fire: cyclic, ca 900 rds/min

Manufacturer
Steyr-Mannlicher AG, Postfach 1000, A-4400 Steyr.

Status
In production.

VERIFIED

Steyr 9 mm Tactical Machine Pistol (TMP) fitted with sound suppressor

BELGIUM

FN 5.7 × 28 mm P90 personal defence weapon system

Development
This weapon system, comprising the FN P90 Personal Weapon and the 5.7 × 28 mm SS190 cartridge, was developed by FN Herstal to equip military personnel whose prime activity is not that of operating small arms - artillery, signals, transport and similar troops whose duties require that they should be effectively armed for self-protection but who do not wish to be burdened by a heavy weapon whilst performing their normal tasks. It is also a suitable weapon for special forces who require very compact firepower.

The 5.7 × 28 mm SS190 was developed as the optimum cartridge for the P90 and is capable of imparting 90 per cent of its kinetic energy upon impacting with a target, a wounding capability enhanced by a distinct propensity to tumble inside the target. It is understood that the bulk of the filling for the lead-free ball projectile is a dense plastic material. The ball projectile can penetrate more than 48 layers of Kevlar body armour at 150 m. Recoil forces are stated to be one-third of those produced by a 5.56 mm cartridge.

Apart from the ball round, tracer, blank and subsonic rounds are under development.

A pistol firing the 5.7 × 28 mm cartridge is known as the Five-seveN; (see separate entry under *Pistols* for details).

5.7 mm FN P90 personal weapon

1996

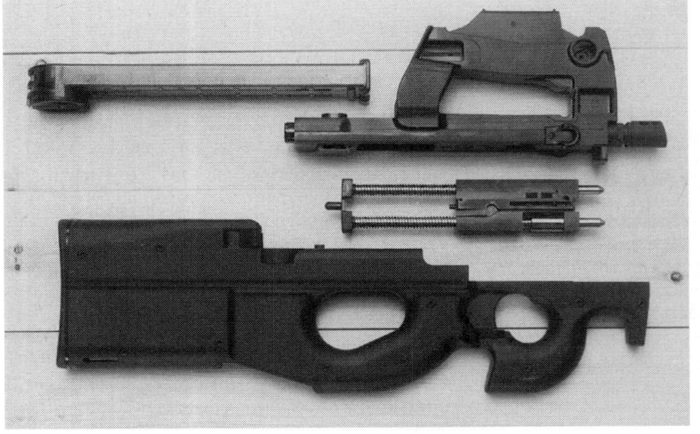

5.7 mm FN P90 personal weapon field-stripped

5.7 mm FN P90 personal weapon with tactical light projector mounted on sight assembly

Forward end of 5.7 mm FN P90 personal weapon showing laser aiming pointer under muzzle and system push-button switches

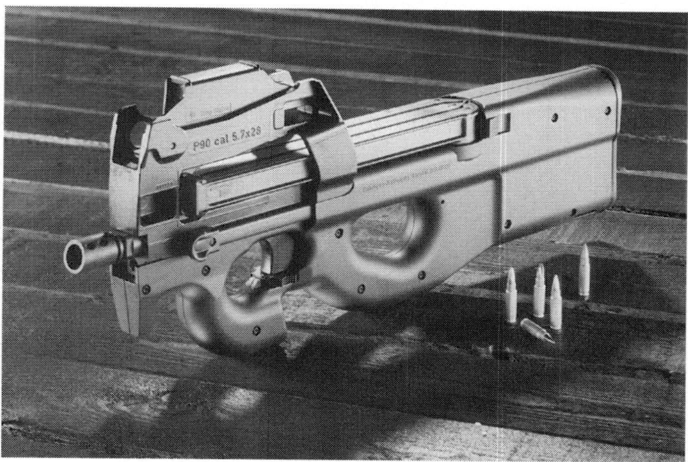

5.7 mm FN P90 personal weapon

Description

The P90 is a blowback weapon firing from a closed bolt. The overall design places great reliance on ergonomics to the extent that the pistol grip, with a thumb-hole stock, is well forward on the receiver so that when gripped, the bulk of the receiver lies along the firer's forearm. The controls are fully ambidextrous; a cocking handle is provided on each side, and the selector/ safety switch is a rotary component located under the trigger. Even the forward sling swivel can be located on either side of the weapon as required.

The magazine lies along the top of the receiver, above the barrel, and the cartridges are aligned at 90° to the weapon axis. The 50 rounds lie in double-row configuration, and as they reach the mouth of the magazine they are ordered into a single row by a fixed ramp. A spiral ramp then turns the round through 90° as it is being guided down into the feedway so that it arrives in front of the bolt correctly oriented for chambering. The magazine is of translucent plastic so its contents can be visually checked at any time.

The sight unit is a reflex collimating sight, the graticule being a tritium illuminated circle and dot; it can be used with both eyes open and allows very rapid acquisition of the target and accurate aiming. Should the sight be damaged two sets of iron sights are machined into the sight base, one on each side; this duplication is another aspect of the ambidextrous design. Detail design relating to ergonomics is such that the P90 can be carried about the user's body in such a manner that the smooth outlines impose a minimum of hinderance to movement and user comfort, no matter how it is slung.

The weapon strips easily into three basic groups for field maintenance. Much of the body and internal mechanism is of high-impact plastic material, only the bolt and barrel being of steel. Moving parts require little or no lubrication. Empty cartridge cases are ejected downwards through the pistol grip, which is shaped so as to offer a grip for the disengaged hand when firing from the shoulder.

5.7 mm FN P90 personal weapon showing ease of carriage

Optional accessories for the P90 include a laser aiming pointer located in front of the foregrip and just under the muzzle. Externally mounted tactical lights and laser aiming devices can be located either side of the sight assembly. Also available is a combat sling, a blank firing attachment, a cleaning kit and a magazine filler. A sound suppressor is under development.

Data

Cartridge: 5.7 × 28 mm
Operation: blowback, selective fire, closed bolt firing
Feed: 50-round magazine
Weight: empty, 2.54 kg; with full magazine, 3 kg
Length: 500 mm
Barrel: 263 mm
Height: 210 mm
Width: 55 mm
Muzzle velocity: 715 m/s
Rate of fire: 900 rds/min
Combat range: 150 m

Manufacturer

FN Nouvelle Herstal SA, Voie de Liège 33, B-4040 Herstal.

Status

In production.

Service

Not confirmed but understood to include Saudi Arabia and some Far Eastern nations.

UPDATED

BRAZIL

9 mm Mtr M 9 M1 - CEV

Description

The 9 mm Mtr M 9 M1 - CEV sub-machine gun is a modified Bergom BSM/9 M3 (see *Jane's Infantry Weapons, 1983-84*, page 77), which was never placed in production. The project was taken over in mid-1982 by Companhia de Explosivos Valparaiba, an explosives manufacturer. The company has stated it is tooling up for series manufacture of the gun.

The Mtr M (Metralhadora de Mão, or 'Hand machine gun') 9 (millimetre) M (Model) 1 - CEV (the manufacturer's initials) is a conventional selective fire blowback-operated sub-machine gun firing from the open-bolt position. The receiver is a tubular structure with barrel-cooling perforations in the forward end, which also has a plastic handguard. The foresight is a protected post and the rearsight is a protected aperture, factory-set for 100 m. The cocking handle, which remains stationary when the gun is firing, is located at about 45° to the left at the forward end of the receiver, Beretta M12 style, though the handle proper is much smaller than that of the Italian weapon. The ejection port is located to the right, and is covered by a spring-loaded cover that snaps open when the weapon is cocked or fired.

Mtr M 9 M1 - CEV

Mtr M 9 M1 - CEV

Mtr M 9 M1 - CEV

The firing mechanism, together with the plastic pistol grip and trigger assembly, is an integral unit that fits under the receiver, just aft of the magazine housing. The magazine catch is a small blade at the rear of the housing, having to be pulled back to release the empty magazine. The fire selector lever is on the left side, above the trigger, with the positions 'S' (Safety) forward, '30' (Automatic) middle, and '1' (Semi-automatic) rear. A grip safety is provided against accidental firing from drops or hits on the butt, blocking the bolt forward unless firmly squeezed by the firer.

The weapon has a wire-type telescoping stock, with the catch located under the receiver, making extension/retraction a two-hand operation. The stock will come out of the gun if pulled all the way.

Data

Cartridge: 9 × 19 mm Parabellum
Operation: blowback, selective fire
Feed: 30-round box magazine
Weight: empty, 3 kg; loaded, 3.7 kg
Barrel: 228 mm
Sights: fore, protected post; rear, aperture, fixed for 100 m
Sight radius: 393 mm
Muzzle velocity: 400 m/s
Rate of fire: cyclic, 600 rds/min

Manufacturer

Companhia de Explosivos Valparaiba, Praia do Flamengo, 200-20° andar 22210 Rio de Janeiro.

Status

Final development.

VERIFIED

9 mm Uru Model II sub-machine gun

Development

The original Uru sub-machine gun was designed by Olympio Vieira de Mello in mid-1974. Design work took him one month and he was able to transform drawings into a working prototype in three months. When this was ready, in early 1975, the Mekanika company was formed to undertake manufacture and sales of the new gun.

The Uru was handed over to the Brazilian Army's Marambaia proving grounds, in Rio de Janeiro, for test and evaluation. Initial results were promising, though, as might be expected, some minor modifications were made: the shape of the removable stock was changed, the capacity of the box magazine was reduced from 32 to 30 rounds, the magazine guide was enlarged to act as a forward grip and better protect the magazine, the fire selector was enlarged and the wooden pistol grip was replaced by a plastic one.

Meanwhile, Mekanika had tooled up for series manufacture of the gun, and the official Brazilian Army Ministry production certification was granted in July 1977. Production of the Uru has since grown to meet local (Brazilian Army and state police forces) and foreign orders (Africa, Latin America and Middle Eastern countries). In 1988 Mekanika handed over all rights in the design to Bilbao SA Indústria e Comércio de Máquinas e Peças, Division FAU Guns, who developed the Uru Model II design and now produce the weapon.

Description

The Uru Model II sub-machine gun operates on the blowback principle, being largely built from tubular elements and stampings. The main parts are assembled by welding and spot welding. Heat-treated alloy steels are used in parts subject to wear or shocks. The body comprises a plastic rear pistol grip, trigger guard (large enough to allow gloved use) and a metal forward grip, into which the 30-round box magazine is inserted. The magazine catch, at the rear of the forward grip, is pulled back to release the magazine.

The receiver is a tubular structure closed at the end, where there is a spring-loaded button on top for the stock attachment. The forward end acts as a thermal jacket for barrel cooling and hand protection. The 175 mm barrel is fitted into place by its own mounting nut, which screws to the forward end of the receiver. A screw-on compensator/muzzle brake, adjustable by the firer for right- or left-hand use, is available.

The weapon is sighted for 50 m; the rearsight is of the aperture type, and the foresight is a flat blade, with a radius of 235 mm. The cocking handle is 45° to the right, and the rearward travel before firing is 60 mm. On firing, the bolt's rearward travel is 80 mm. The bolt remains closed after the last round is fired.

The fire selector is on the left side, above the plastic pistol grip, and has the three usual positions: 'S' (Safe) forward, 'SA' (Semi-Automatic) upward, and 'A' (Automatic) backward. In the 'S' position the trigger action is solidly blocked to prevent accidental firing when the gun is carried cocked. Accidental discharge caused by a fall or a hard blow, a common problem in free-bolt weapons, is guarded against by a new grip safety fitted behind the pistol grip. This positively locks the bolt at all times except when pressed in by the firer's hand.

As the Uru Model II has only 33 parts, including the magazine and the stock, it can be totally disassembled without tools in about 30 seconds. All parts of the Uru Model II are interchangeable, and only the body of the weapon bears the manufacturer's number.

A silencer is offered to transform the Uru Model II into a fully silenced gun. The original barrel is retained and the silencer is merely screwed on to the muzzle. The gun can then be used with standard ammunition and at full-automatic fire. This new silencer is smaller than the original combined barrel and silencer which had to be replaced as a complete unit.

In addition to the Uru Model II in 9 mm Parabellum calibre, the design is also available as the FAU Model 1 in 9 mm Short (0.380 ACP) calibre. The dimensions are the same but the weight is about 100 g less. A carbine version, the FAU Carbine Model 1, firing only in the semi-automatic mode and with 406 mm or 482 mm barrels, is available in 9 mm Short (0.380 ACP). Conversion kits are also available to allow Uru Model 1 or

Uru Model II sub-machine gun

FAU Model 1 carbine

Model 2 9 mm Parabellum sub-machine guns to be converted into 9 mm Short (0.380 ACP) calibre or into semi-automatic carbines in either calibre.

Data

9 mm PARABELLUM VERSION

Cartridge: 9 × 19 mm Parabellum
Operation: blowback, selective fire
Feed: 30-round box magazine
Weight: without stock and magazine, 2.58 kg; with stock, without magazine, 3 kg; complete weapon, loaded, 3.69 kg
Magazine: empty, 350 g; loaded, 690 g
Length: without stock, 433 mm; overall, 671 mm
Barrel: 175 mm
Rifling: 6 grooves, rh
Sights: fore, blade; rear, aperture, sighted for 50 m
Sight radius: 235 mm
Muzzle velocity: 389 m/s
Rate of fire: cyclic, 750 rds/min

Manufacturer

Bilbao SA Indústria e Comércio de Máquinas e Peças, Division FAU Guns, Rua Bom Pastor, 2190 - 04203 São Paulo SP.

Status

In production.

Service

Brazilian Army, Navy and some state police forces. Exported to unspecified countries.

UPDATED

0.45 INA MB 50 and Model 953 sub-machine gun

Description

The MB 50 is a copy of the Danish Madsen Model 1946 made by the Indústria Nacional de Armas SA, São Paulo, Brazil which acquired all the manufacturing rights. It has the Brazilian crest on the left side of the receiver and is marked 'Exercito Brasileiro'. The models used by the police are marked 'Policia Civil'. The weapon remains a true copy of the Madsen.

The Model 953 has an enlarged magazine housing which has a wire clip around the mouth to keep the two halves together. The cocking handle was moved from the top of the receiver to the right side.

A 9 mm version of this weapon has also been produced in Brazil. For details see the following entry.

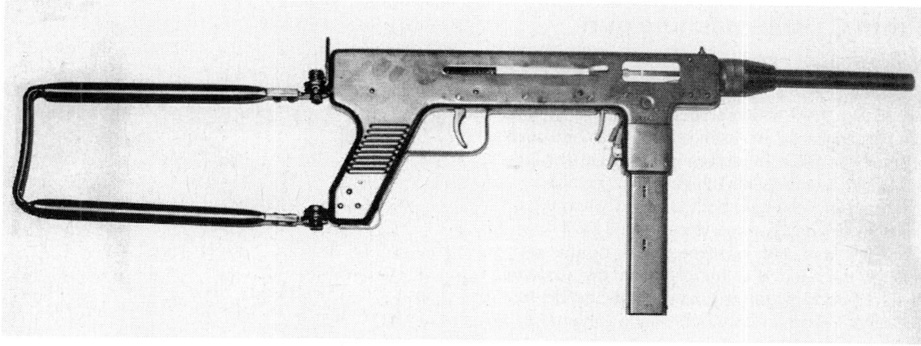

0.45 INA MB 50 sub-machine gun

Data

Cartridge: 0.45 ACP
Operation: blowback, automatic
Feed: 30-round box magazine
Weight: empty, 3.4 kg; loaded, 4.32 kg
Length: stock retracted, 546 mm; stock extended, 794 mm
Barrel: 213 mm
Rifling: 4 grooves, rh
Sights: fore, blade; rear, aperture 100 m
Muzzle velocity: 280 m/s
Rate of fire: cyclic, 650 rds/min

Manufacturer

Indústria Nacional de Armas SA, São Paulo.

Status

No longer in production.

Service

Police and paramilitary forces. Also military reserve units.

UPDATED

---

9 mm Madsen sub-machine gun

Description

As noted in the previous entry, the Madsen sub-machine gun was manufactured in Brazil, in 0.45 calibre, by the Indústria Nacional de Armas (INA).

IMBEL, at their Fábrica de Itajubá factory, developed and manufactured a conversion from 0.45 ACP calibre to 9 mm calibre to permit the use of the 9 × 19 mm Parabellum cartridge.

To improve the aiming and holding of the weapon during automatic fire, the IMBEL designers added a highly efficient muzzle compensator. It was found in practical testing that apart from the vibration of the weapon caused by the reciprocating bolt there is no undue motion and the gun remains stable and under full control at all times.

The first lots of this 'new' weapon were delivered to Brazilian Army and police forces after approval of the prototypes at the Marambaia Military Proof Establishment; all proof and testing was in accordance with NATO standards.

Data

Cartridge: 9 × 19 mm Parabellum
Operation: blowback, selective fire
Feed: 30-round box magazine
Weight: 3.74 kg
Rifling: 6 groove, rh
Muzzle velocity: 392 m/s
Rate of fire: cyclic, 600 rds/min

Manufacturer

Indústria de Material Bélico do Brasil - IMBEL, Rua São Joaquim 329, Liberdade, CEP-01508-001 São Paulo.

Status

In production.

Service

Brazilian armed forces and police forces.

VERIFIED

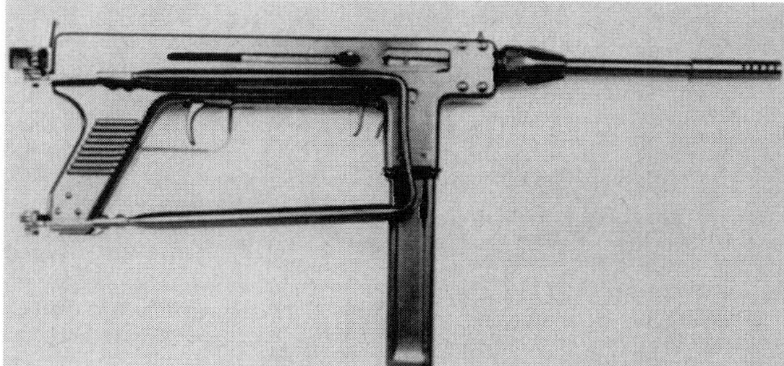

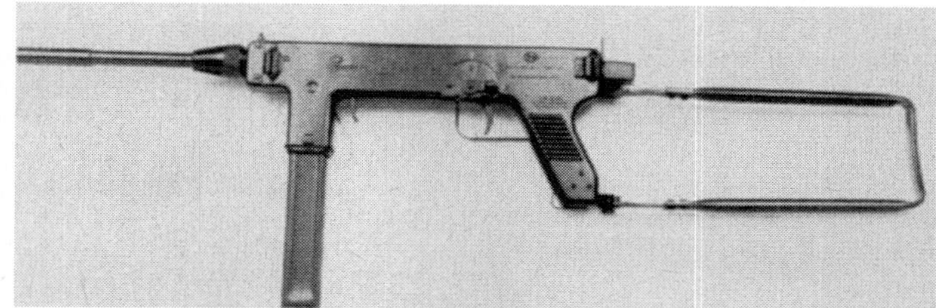

9 mm Madsen with butt extended

9 mm Madsen with butt folded

BULGARIA

5.56 mm AK-74U sub-machine gun

Development

Despite its AK-74U designation, the 5.56 mm AK-74U sub-machine gun is actually a variant of the 5.45 mm AKS-74U Kalashnikov sub-machine gun as it has a side-folding metal skeleton butt. The main change from the original is the change of calibre to 5.56 × 45 mm NATO to attract possible export sales outside the old Warsaw Pact bloc. As far as is known no sales have yet been made but the 5.56 mm AK-74U is on offer by the government-owned Kintex manufacturing and marketing organisation.

Kintex also supplies a standard 5.45 × 39.5 mm version of their AK-74U for issue to the Bulgarian armed forces. This variant is also offered for export sales.

Description

The Bulgarian 5.56 mm AK-74U sub-machine gun is virtually identical to the standard AKS-74U other than the changes made necessary to suit the NATO cartridge. The magazine continues to hold 30 rounds and maximum sighting range is 500 m. It is claimed that skilled firers can effectively engage targets by firing single shot at ranges up to 350 m. Maximum effective aiming range against multiple targets is 250 m when firing bursts.

Each 5.56 mm AK-74U is issued with four magazines in a pouch, a sling, a cleaning rod, an oiler, and various spares and other accessories.

Data

Cartridge: 5.56 × 45 mm NATO
Operation: gas, selective fire
Feed: 30-round box magazine
Weight: empty, 2.7 kg; loaded, 3.1 kg
Weight of magazine: loaded, 600 g
Length: butt folded, 490 mm; butt extended, 730 mm
Barrel: 207 mm
Muzzle velocity: ca 800 m/s
Rate of fire: cyclic, 700-750 rds/min

Marketing Agency

Kintex, 66 James Boucher Street, Sofia 1407.

Status

Offered for export sales.

UPDATED

CANADA

9 mm C1 sub-machine gun

Description

When the British L2A1 Sterling sub-machine gun was developed the Canadian Army made some modifications and Canadian Arsenals Limited started producing the C1 in 1958. The differences between the C1 and the British Sterling and L2 series are not great. The magazine was changed both in capacity and design. It holds 30 rounds, having previously held 34. A 10-round magazine was also produced. The design change involved the removal of the rollers and the substitution of an orthodox magazine follower. The bayonet for the C1 is that from the Fusil Automatique Léger (FAL) rifle whereas the L2A3 takes the No 5 bayonet. The components are largely interchangeable between the two guns, although the C1 is said to be cheaper to manufacture. The performance of the two weapons is much the same.

9 mm C1 sub-machine gun

Data

Cartridge: 9 × 19 mm Parabellum
Operation: blowback, selective fire
Feed: 30-round curved box magazine; 10-round box optional
Weight: empty, 2.95 kg; loaded, 3.46 kg
Length: butt folded, 493 mm; butt extended, 686 mm
Barrel: 198 mm
Rifling: 6 grooves, rh
Sights: fore, blade; rear, flip aperture 300 and 600 ft (90 and 180 m)
Muzzle velocity: 366 m/s
Rate of fire: cyclic, 550 rds/min

Manufacturer

Canadian Arsenals Limited, Long Branch, Ontario.

Status

No longer in production.

Service

Reserve use with Canadian Forces. To be phased out as C7/C8 rifles become available.

UPDATED

CHILE

9 mm S.A.F. sub-machine gun

Description

The 9 mm S.A.F. sub-machine gun fires from a closed bolt and the design is substantially based upon the SIG 540 rifle manufactured in Chile under licence by FAMAE. Certain modifications to the SIG mechanism were made, notably in a cocking-handle release catch, in the firing mechanism and in a new three-round burst mechanism, but the basic hammer and floating firing pin system of operation is retained. Operation is by blowback. Three forms are manufactured: the standard model with fixed butt, standard with side-folding skeleton butt, and a silenced model, also with side-folding butt.

There are 20- or 30-round magazines available. The 30-round magazines are of translucent synthetic material which permits visual checking of the magazine contents at any time. These synthetic magazines also have studs and slots in their sides which permit two or more magazines to be clipped together; in this configuration one magazine can be inserted into the gun's magazine housing and fired, and when empty it can be quickly slipped out and the connected, loaded, magazine slipped into the housing.

Variant

9 mm Mini-S.A.F. sub-machine gun
See following entry.

Data

Cartridge: 9 × 19 mm Parabellum
Operation: blowback, selective fire with 3-round burst
Feed: 20- or 30-round plastic box magazine
Weight: fixed butt standard, 2.7 kg without magazine; folding butt standard, 2.9 kg without magazine; silenced model, 3 kg without magazine
Length: standard, butt folded 410 mm; standard, butt extended, 640 mm; silenced, butt folded, 570 mm; silenced, butt extended, 810 mm;
Barrel: standard, 200 mm; silenced, 220 m
Rifling: 6 grooves, rh, 1 turn in 250 mm
Sights: fore, post, adjustable for elevation; rear, aperture, adjustable for windage
Sight radius: standard, 310 mm; silenced, 440 mm
Muzzle velocity: standard, 390 m/s; silenced, 300 m/s
Rate of fire: cylic, standard, 1,200 rds/min; cyclic, silenced, 980 rds/min

Comparison of standard 9 mm S.A.F. sub-machine gun with fixed stock (top) and sliding stock (bottom); note translucent magazines

9 mm S.A.F. sub-machine gun with twin 30-round magazines

Manufacturer
FAMAE Fabricas y Maestranzas del Ejercito, Av Pedro Montt 1568, PO Box 4100, Santiago.

Status
In production.

Service
Chilean Army and police.

VERIFIED

9 mm S.A.F. sub-machine gun models; from top, standard with fixed stock, standard with folding stock, silenced, and Mini-S.A.F. (bottom)

Silenced version of 9 mm S.A.F. sub-machine gun

9 mm S.A.F. sub-machine gun field-stripped

9 mm Mini-S.A.F. sub-machine gun

Description
The Mini-S.A.F. uses the same basic mechanism as the standard S.A.F. described above, but has a much shorter barrel, no shoulder stock, and a forward handgrip which also has a guard to prevent fingers slipping in front of the muzzle. The standard 30-round magazine can be used, although a special 20-round short magazine is made for this weapon to provide maximum compactness. The Mini-S.A.F. is particularly suited to covert and bodyguard operations.

The Mini-S.A.F. can be carried concealed and rapidly brought into action

Data
Cartridge: 9 × 19 mm Parabellum
Operation: blowback, selective fire, with 3-round burst
Feed: 20- or 30-round box magazine
Weight: without magazine, 2.3 kg
Length: 310 mm
Barrel: 115 mm
Rifling: 6 grooves, rh, 1 turn in 250 mm
Sights: fore, post, adjustable for elevation; rear, aperture, adjustable for windage
Sight radius: 250 mm
Muzzle velocity: 370 m/s
Rate of fire: cyclic, 1,200 rds/min

Manufacturer
FAMAE Fabricas y Maestranzas del Ejercito, Av Pedro Montt 1568, PO Box 4100, Santiago.

Status
In production.

Service
Chilean Army and police.

VERIFIED

9 mm Mini-S.A.F. sub-machine gun with 20-round magazine

CHINA, PEOPLE'S REPUBLIC

7.62 mm Type 79 light sub-machine gun

Description

The Type 79 sub-machine gun is an extremely lightweight weapon, firing the 7.62 × 25 mm pistol cartridge. The receiver is rectangular and made from steel stampings, and it has a safety lever and fire selector on the right side, above the pistol grip, which appears to have been modelled upon that of the Kalashnikov AK series rifles. A man trained in using the AK system will find no difference if he picks up one of these weapons.

Operation is by gas, using a short-stroke tappet above the barrel. This drives a short piston attached to a bolt carrier, operating a rotating bolt. This is a rather complex way of operating a sub-machine gun but, as with the external controls, it has some similarity to the rifle design and thus facilitates training, and, more importantly, does away with the need for a heavy bolt and long bolt travel, making the weapon lighter and more controllable.

Data

Cartridge: 7.62 × 25 mm Pistol
Operation: gas, selective fire
Locking: rotating bolt
Feed: 20-round box magazine
Weight: with empty magazine, 1.9 kg
Length: stock folded, 470 mm; stock extended, 740 mm
Muzzle velocity: 500 m/s
Rate of fire: cyclic, ca 650 rds/min

Manufacturer

China North Industries Corporation, 7A Yue Tan Nan Jie, PO Box 2137, Beijing.

Status

In production.

Service

Not known; available for export.

VERIFIED

7.62 mm Chinese Type 79 light sub-machine gun

Type 79 light sub-machine gun; mechanism details

7.62 mm Type 85 light sub-machine gun

Description

The 7.62 mm Type 85 is a modified and simplified version of the Type 79. It is a plain blowback weapon, with a cylindrical receiver into which the barrel is fitted and which carries the bolt and return spring. There is a folding butt and the weapon uses the same magazine as the Type 79. The manufacturers claim the ability to fire reduced-velocity Type 64 pistol ammunition as well as the standard Type 51 cartridge.

Data

Cartridge: 7.62 × 25 mm Pistol
Operation: blowback, selective fire
Feed: 30-round box magazine
Weight: empty, 1.9 kg
Length: butt folded, 444 mm; butt extended, 628 mm
Sights: fore, blade; rear, flip aperture
Muzzle velocity: ca 500 m/s
Rate of fire: cyclic, 780 rds/min

Manufacturer

China North Industries Corporation, 7A Yue Tan Nan Jie, PO Box 2137, Beijing.

7.62 mm Type 85 sub-machine gun

Status

In production.

Service

Not known; available for export.

VERIFIED

7.62 mm Type 64 silenced sub-machine gun

Development

The 7.62 mm Type 64 silenced sub-machine gun is a Chinese-designed and constructed weapon gun which combines a number of features taken from various European weapons. The bolt action is the same as that of the Type 43 sub-machine gun which was copied from the Soviet PPS-43. The trigger mechanism, giving selective fire, may have been taken from the Bren gun, numbers of which fell into Chinese hands during the Korean War, although the Bren mechanism was derived from that of the series of light machine guns purchased by pre-war Chinese government forces from the former Czechoslovakia.

Description

The Type 64 is blowback-actuated using the 7.62 × 25 mm pistol cartridge held in a curved magazine under the receiver. The chamber contains three flutes each 0.1 mm wide and 0.075 mm deep, extending from the commencement of the small cone to just beyond the mouth of the chamber: a total length of 10 mm. The suppressor is of the Maxim type. The barrel is 200 mm long and is plain for the first 131 mm after which it is perforated by four rows of holes for a distance of 57 mm, each following the rifling groove and each having nine holes of 3 mm diameter, making 36 holes in all. The tube surrounding the barrel continues forward for a further 165 mm and then there is a muzzle cap. Between the end of the barrel and the cap is a stack of baffles each of which is dished with a central hole of 9 mm diameter; two rods passing down through the baffles keep the stack together and properly lined up. The rods can be rotated and then are free to come out through the baffles, allowing ready disassembly.

7.62 mm Type 64 silenced sub-machine gun

7.62 mm Type 64 sub-machine gun with silencer removed

The suppressor is reasonably effective and also has the virtue of preventing any flash from either muzzle or breech.

This unusual sub-machine gun was specifically designed and made as a silenced weapon. In most other cases the silencer is fitted to an existing gun, which at least reduces the cost and manufacturing effort.

Data

Cartridge: 7.62 × 25 mm Type P Ball
Operation: blowback, selective fire
Feed: 30-round curved box magazine
Weight: empty, 3.4 kg
Length: stock closed, 635 mm; stock open, 843 mm
Barrel: 244 mm
Rifling: 4 grooves, rh
Muzzle velocity: 513 m/s
Rate of fire: cyclic, 1,315 rds/min
Effective range: 135 m

Manufacturer

State factories.

Status

Not in production.

Service

Chinese People's Army.

VERIFIED

7.62 mm Type 85 silenced sub-machine gun

Description

The 7.62 mm Type 85 is a simplified and lightened version of the silenced 7.62 mm Type 64 sub-machine gun, produced principally for export. It appears to be based on the simple mechanism of the Type 85 light sub-machine gun, but is of about the same size as the Type 64 and uses similar silencing arrangements.

The Type 85 is regulated for the Type 64 silenced cartridge but it is also possible to fire the standard Type 51 pistol cartridge, although the silencing effect is much reduced and there will also be the problem of bullet noise. With the Type 64 cartridge, the sound of discharge is reduced to less than 80 dB.

Data

Cartridge: 7.62 × 25 mm Type 64 (and see text)
Operation: blowback, selective fire
Feed: 30-round box magazine
Weight: empty, 2.5 kg
Length: butt folded, 631 mm; butt extended, 869 mm
Sights: fore, blade; rear, flip aperture
Muzzle velocity: ca 300 m/s
Rate of fire: cyclic, 800 rds/min
Effective range: 200 m

Manufacturer

China North Industries Corporation, 7A Yue Tan Nan Jie, PO Box 2137, Beijing.

Status

In production.

VERIFIED

7.62 mm Type 85 silenced sub-machine gun

7.62 mm Type 85 silenced sub-machine gun (left) compared to Type 64 (right)

COMMONWEALTH OF INDEPENDENT STATES

Bison 9 mm sub-machine gun

Development

The Bison 9 mm sub-machine gun was first seen publicly during 1993 and is a product of the Izhmash factory at Izhevsk; the development team leader being Victor Kalashnikov. The Bison makes considerable use of components and assemblies from existing Kalashnikov designs but is mainly noticeable for the introduction of a new 64-round helical-feed magazine located under the forestock. The cartridges fired may be the standard 9 × 18 mm Makarov or the more potent 9 × 18 mm 'Special' with the designation 57-N-181SM.

It is intended that users of the Bison will include Russian Interior Ministry personnel and special armed forces units.

Description

The Bison 9 mm sub-machine gun is a compact, blowback-operated weapon with selective fire modes. The blowback action is buffered to provide stability during firing. About 60 per cent of the parts used in the Bison are taken from other Kalashnikov models, mainly from the AK-74 series, including the folding butt stock, the trigger mechanism and the receiver cover.

The helical-feed magazine holds 64 rounds although this may be increased on production models. The prototypes use steel although full production versions will be made of high strength plastic. To fit the magazine two lugs at the forward end protrude over two pins under the muzzle attachment. The feed end of the magazine is then lifted to engage with the existing Kalashnikov magazine catch. When fitted the magazine can be used as a foregrip.

The Bison can fire both the standard 9 × 18 mm Makarov or the newer 9 × 18 mm 57-N-181SM 'Special' cartridges. When firing the latter the effective range of the Bison is 150 m compared to 100 m for the standard 9 mm Makarov round.

Drawings of Bison 9 mm sub-machine gun with stock folded (top) and extended (below)
(L Haywood)

Data

Cartridge: 9 × 18 mm Makarov; 9 × 18 mm Makarov 57-N-181SM
Operation: blowback, selective fire
Feed: 64-round helical feed magazine
Weight: without magazine, 2.1 kg; with empty magazine, 2.47 kg
Length: butt folded, 425 mm; butt extended, 660 mm
Rate of fire: cyclic, 700 rds/min

Manufacturer

IZHMASH Factory, Kalashnikov Joint Stock Company, 3 Derjabin Street, 426006 Izhevsk, Russia.

Status

Prototypes.

VERIFIED

MA 9 mm sub-machine gun

Development

The MA 9 mm sub-machine gun is referred to in descriptive brochures as a small automatic rifle (MA - malogabaritnyi avtomat) and was for some time after its original public appearance in 1992 thought to be the A-91. The A-91 has emerged as a separate design for the MA forms the basis for the BA and BV silenced assault and sniper rifles (see entries under *Rifles*). The A-91 is described in the following entry. The design bureau for both weapons was the Institute of Precise Mechanical Engineering at Klimovsk.

The SP-5 round fired by the MA is the 9 × 26 mm high penetration cartridge capable of penetrating 1.2 mm titanium plates and 30 layers of Kevlar or 6 mm of steel plate, both at a range of 200 m, much of the penetration capability being provided by a steel or tungsten insert which protrudes from the bullet nose.

The MA was developed as an armour-piercing weapon which could be fairly easily concealed and may thus be regarded as a special forces weapon rather than a front-line weapon.

Description

The MA 9 mm sub-machine gun is a gas-operated weapon with selective fire. The combination of short barrel and an automatic fire capability place the MA more in the sub-machine gun than in the assault rifle category. With the short overall length being only 380 mm when the folding steel stock is folded up and over the receiver, the weapon can be readily concealed. Length overall with the stock extended is 620 mm.

At least two models of the MA have been produced, the most recent being recognisable by the rearsight assembly which is more prominent than on the earlier model. Other changes from the earlier model include a ribbed forestock to improve handling and a larger muzzle attachment.

The MA utilises the same 10- or 20-round box magazines as those used with the BA and BV silenced rifles.

Drawing of early model of MA 9 mm sub-machine gun (L Haywood)

Data

Cartridge: 9 × 26 mm SP-5
Operation: gas, selective fire
Feed: 10- or 20-round box magazine
Weight: empty, 2 kg
Length: butt folded, 380 mm; butt extended, 620 mm
Sights: fore, fixed post; rear flip aperture
Muzzle velocity: ca 240 m/s

Manufacturer

Institute of Precise Mechanical Engineering, 2 Zavodskaya Street, 142080 Klimovsk, Russia.

Status

Ready for production. Offered for export sales.

Service

Unknown.

VERIFIED

A-91 sub-machine gun

Development

The A-91 sub-machine gun, described in brochures as a small sized assault rifle, is a product of the Institute of Precise Mechanical Engineering at Klimovsk. Designed from the outset as a special forces weapon the A-91 is available in a variety of calibres including 5.45 × 39.5 mm, 5.56 × 45 mm NATO, and 7.62 × 39 mm. The A-91 is also available in 9 mm so that, fitted with a sound suppressor, it can fire the 9 × 39 mm SP-5 and SP-6 low-velocity cartridges; (see following entry).

Description

The A-91 sub-machine gun is gas operated with gas being tapped from a position close to the muzzle. Firing modes are single shot and fully automatic at a cyclic rate of 700 to 900 rds/min, depending on the ammunition. The dimensions of the A-91 place it in the compact category for once the simple steel butt is folded up and over the receiver the overall length is only 384.5 mm and overall width is 44 mm.

The size of the magazine housing will vary according to the calibre involved but all magazines carry 20 rounds.

Variant

9A-91
See following entry.

A-91 sub-machine gun (T J Gander)

Data

Cartridge: 5.45 × 39.5 mm; 5.56 × 45 mm NATO; 7.62 × 39 mm
Operation: gas, selective fire
Feed: 20-round box magazine
Weight: without magazine, 1.75 kg
Weight, magazine with 20 rounds: 5.45 × 39.5 mm, 416 g; 5.56 × 45 mm, 436 g; 7.62 × 39 mm, 534g
Length: butt folded, 384.5 mm; butt extended, 604 mm
Height: 187 mm
Width: 44 mm
Sights: fore, fixed post; rear flip aperture
Muzzle velocity: 5.45 × 39.5 mm, 670 m/s; 5.56 × 45 mm, 680 m/s; 7.62 × 39 mm, 570 m/s
Rate of fire: cyclic, 700-900 rds/min
Effective range: 250 m

Manufacturer

Institute of Precise Mechanical Engineering, 2 Zavodskaya Street, 142080 Klimovsk, Russia.

Status

Ready for production. Offered for export sales.

Service

Unknown.

VERIFIED

9 mm 9A-91 sub-machine gun

Development

The 9 mm 9A-91 sub-machine gun is a variant of the A-91 sub-machine gun (see previous entry) developed to fire the 9 × 39 mm SP-5 and SP-6 low-velocity cartridges. As it is intended for use by special purpose forces the 9A-91 can accommodate tactical accessories such as a laser target designator, a suppressor and an optical sight assembly.

For some reason the manufacturers of the 9A-91 are the KBP Instrument Design Bureau at Tula rather than their colleagues at Klimovsk.

Description

The 9A-91 sub-machine gun follows exactly the same lines as the standard A-91 so reference to the previous entry should be made for details. Where the 9A-91 differs is the ammunition and the accessories.

Accessories include a laser designator which can be mounted alongside the forestock and a long cylindrical sound suppressor; the muzzle has an extension to accommodate the latter and lacks the muzzle attachment of the A-91. The optical sight assembly appears to be a suitably modified PSO-1 used with the Dragunov SVD rifle. Night sights may also be employed.

The SP-5 and SP-6 cartridges have a muzzle velocity of 270 m/s, both having a bullet weight of 16.2 g. The SP-6 is a general purpose Ball round while the SP-5 has a steel or tungsten tip capable of penetrating a 6 mm high density steel plate at 100 m.

The 9 × 39 mm SP-5 and SP-6 low-velocity cartridges limit the maximum effective range of the 9A-91 to 200 m.

Data

Cartridge: 9 × 39 mm SP-5 and SP-6
Operation: gas, selective fire
Feed: 20-round box magazine
Weight: without magazine, 1.75 kg
Weight, magazine with 20 rounds: 670 g
Length: butt folded, 384.5 mm; butt extended, 604 mm
Height: 187 mm
Width: 44 mm
Sights: fore, fixed post; rear flip aperture (see also text)
Muzzle velocity: 270 m/s
Rate of fire: cyclic, 700-900 rds/min
Effective range: 200 m

Manufacturer

KBP Instrument Design Bureau, Tula 300001, Russia.

Agency

Rosvoorouzhenie, 18/1 Ovchinnikovskaya Emb, 113324 Moscow, Russia.

Status

Ready for production. Offered for export sales.

9A-91 sub-machine gun with tactical accessories in place

1996

Service

Unknown.

NEW ENTRY

Kiparis 9 mm sub-machine gun

Description

Relatively few details regarding the Kiparis (Cypress) 9 mm sub-machine gun have yet been released. It seems certain that it fires the 9 × 18 mm Makarov cartridge and is unlikely to be able to accommodate the more potent 9 × 18 mm 57-N-181SM round. The weapon appears to be a straightforward blowback design with a conventional 'machine pistol' appearance. Construction is pressed steel with a plastics-based pistol grip. Pressed steel is also used for the vertical box magazines which may hold either 10 or 30 rounds. A rudimentary steel folding butt is provided with the butt located over the receiver, when folded, in such a manner that the simple base plate outline extends downwards either side of the muzzle.

It would appear that the Kiparis was developed as a weapon for security forces but this conclusion is partly denied by the availability of a suppressed version on which a long suppressor is secured over the muzzle, though it is uncertain if this is a permanent or removable accessory. The suppressed version can also carry a fairly sizeable laser aiming device which clips on forward of the magazine housing in such a way that the bottom of the laser aiming device can act as a forward grip during aiming.

Data

Cartridge: 9 × 18 mm Makarov
Operation: blowback
Feed: 10- or 30-round box magazine
Weight: empty, 10-round magazine, 1.58 kg; empty, 30-round magazine, 1.6 kg
Length: butt folded, 317 mm; butt extended, 600 mm
Sights: fore, fixed post; rear, notch sighted to 100 m
Muzzle velocity: 320 m/s
Rate of fire: cyclic, 750-900 rds/min

Manufacturer

Enterprise Metallist, 1 Urdinskaya Street, Uralsk, Kazakhstan 417024.

Status

Apparently in production. Offered for export sales.

Service

Unknown.

NEW ENTRY

Kiparis 9 mm sub-machine gun fitted with suppressor and laser aiming device

1996

Kiparis 9 mm sub-machine gun fitted with suppressor and laser aiming device

1996

---

9 mm PP-90 sub-machine gun

Development

The 9 mm PP-90 sub-machine gun appears to have been based upon an American Ares design dating from the 1980s for it is a folding sub-machine gun intended to be used as a concealed self-defence weapon by special duties personnel such as plain clothes police or bodyguards. However it is conceivable that it could be carried as a survival weapon by aircrew or others.

The PP-90 is a product of the Design Bureau for Instrument Engineering at Tula.

Description

When carried in the folded configuration the 9 mm PP-90 sub-machine gun could be mistaken for an assault rifle magazine for only a securing hook at one end provides any indication of its true purpose. To prepare the weapon for action all that is required is to grasp the two halves and pull them apart. The weapon then hinges open and the loaded magazine snaps down into place as one half becomes a rudimentary butt stock. The entire sequence takes only 2 seconds.

The 9 mm PP-90 operates on the blowback principle.

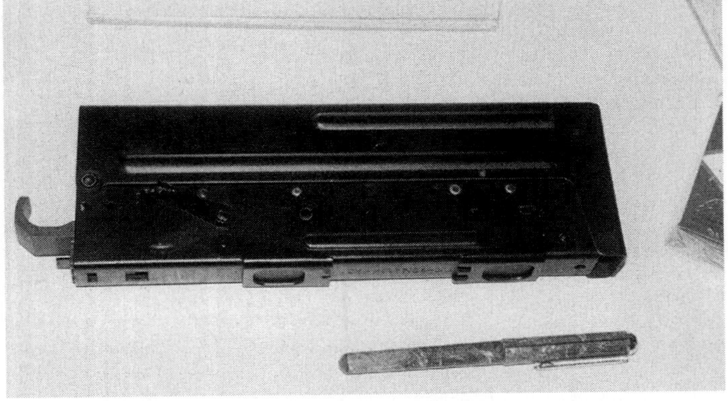

9 mm PP-90 sub-machine gun in folded configuration; note fountain pen for size comparison (T J Gander)

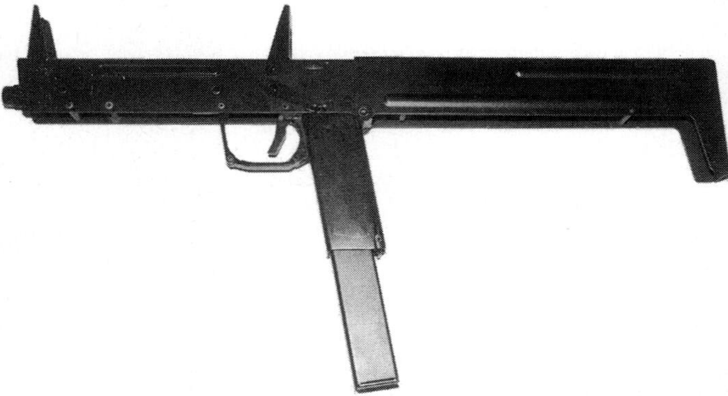

9 mm PP-90 sub-machine gun in firing configuration (T J Gander)

Construction overall is steel with most parts being simple steel pressings, so most handling surfaces are unsympathetic to the touch and will no doubt cause discomfort during firing. However this matters little as the PP-90 is intended for emergency use only.

The round fired from the PP-90 is the 9 × 18 mm Makarov. Firing is fully automatic only and there does not appear to be any form of safety catch. The pressed steel box magazine holds 30 rounds.

The PP-90 is primarily a close quarter weapon for firing from the hip. If aimed fire is required the rear and fore sights can be raised; the pressed steel sights being fixed for 100 m. The claimed accuracy of the PP-90 is such that, firing at a range of 25 m from the sitting position, 2- to 5-round bursts will be grouped within a 100 mm circle. Firing from the standing position without support, also at a range of 25 m, a full 30-round burst can be grouped within a 450 mm circle.

The PP-90 has been observed fitted with a suppressor.

Data

Cartridge: 9 × 18 mm Makarov
Operation: blowback, automatic only
Feed: 30-round folding box magazine
Weight: empty, 1.83 kg; loaded; 2.23 kg
Length: unfolded, 490 mm
Dimensions: folded, 270 × 90 × 32 mm
Sights: fore, folding frame; rear, folding slot
Muzzle velocity: 320 m/s
Rate of fire: cyclic, 600-700 rds/min

Manufacturers

Design Bureau for Instrument Engineering, Tula, Russia.
Also: Enterprise Metallist, 1 Urdinskaya Street, Uralsk, Kazakhstan 417024.

Status

In production. Offered for export sales.

Service

Unknown.

UPDATED

9 mm KEDR and Klin Machine Pistols

Development

During the early 1970s the Soviet Army issued a requirement for a machine pistol to replace the 9 mm Stechkin automatic pistol. Two competing designs emerged, one being by Evgeni Dragunov and the other by Nikolai Afanasyev, but the contest was terminated at that time because of ammunition performance limitations at ranges over 50 m. During the early 1990s the Dragunov design was resurrected and modified for mass production as the KEDR (Konstruktsiya Evgeniya Dragunova - designed by Evgeni Dragunov).

The KEDR fires the standard 9 × 18 mm Makarov cartridge. Further development carried out at the Izhevsk facility resulted in a number of variants including the KEDR model PP-91-01 with a flash-hiding silencer, a single-shot assault pistol without a folding butt, and the Klin machine pistol firing the modernised 9 × 18 mm 57-N-181SM cartridge. A trial batch of Klin machine pistols was manufactured at the Zlatoust Engineering Plant during 1994.

The Byelorussian Opto-Mechanical Association has developed a laser target designator which can be installed on both the KEDR and Klin machine pistols.

Description

The KEDR and Klin machine pistols are identical in overall construction and operation, the main difference being that the Klin can fire the 9 × 18 mm 57-N-181SM cartridge which has a 25 per cent higher chamber pressure than the standard Makarov cartridge. The Klin can also fire the standard cartridge.

Both machine pistols operate on the blowback principle. Safeties include a lock which engages both the trigger and the bolt. A stamped steel butt can be folded up-and-over the receiver with, when folded, the rudimentary butt plate located close to the muzzle. Ammunition feed is from 20- or 30-round box magazines inserted through a magazine housing in front of the trigger assembly.

The high impulse cartridge fired by the Klin machine pistol has a muzzle velocity of 430 m/s; firing a 5.6 g bullet the muzzle energy is 510 J, sufficient to penetrate standard Zh-81 armoured jackets at up to 10 m. Aimed fire is possible at ranges of up to 150 m using battle sights.

The laser target designator which can be fitted under the muzzle of both the KEDR and the Klin provides a laser spot diameter of not less than 30 mm at 50 m. Output radiation power is 5 mW, with a 'Blik' battery providing power for up to 20 hours of constant use. Overall dimensions are 80 × 38 × 20 mm.

Data

DATA GIVEN FOR KEDR; DATA FOR KLIN IN PARENTHESIS
Cartridge: 9 × 18 mm Makarov; 9 × 18 mm 57-N-181SM
Operation: blowback, selective fire
Feed: 20- or 30-round box magazine
Weight: empty, 1.4 kg (1.41 kg); loaded, 1.82 kg (1.83 kg)
Length: stock folded, 300 mm (305 mm); stock extended, 540 mm
Barrel: 120 mm
Sights: fore, blade; rear, notch; laser target designator optional
Muzzle velocity: V_{10} 310 m/s (430 m/s)
Muzzle energy: 285 J (510 J)
Rate of fire: cyclic, 800-850 rds/min (1,030-1,200 rds/min)

Manufacturer

Klin - Zlatoust Engineering Plant.

Status

KEDR in production. Klin in preproduction stage.

Service

KEDR apparently in service with some Russian and other CIS internal security forces.

UPDATED

9 mm KEDR PP-91-01 machine pistol with butt extended (L Haywood)

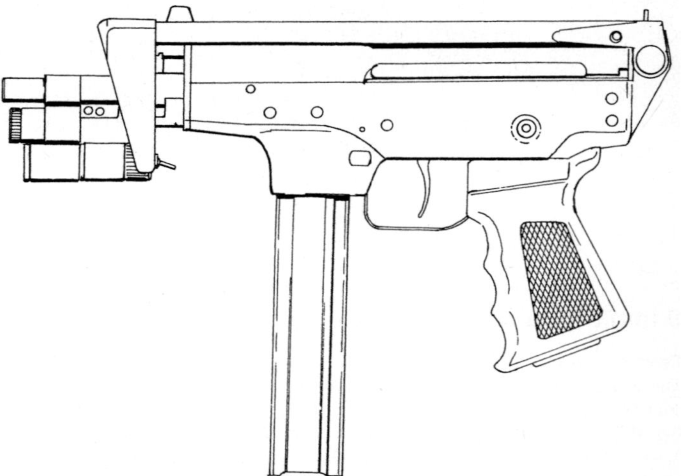

9 mm Klin machine pistol fitted with laser target designator (L Haywood)

9 mm BAKSAN personal defence weapon

Development

The precise status of the 9 mm BAKSAN personal defence weapon is uncertain as the few details available to date were taken from a text display at a defence exhibition. The weapon can fire either the 9 × 19 mm Parabellum or the new 9 × 18 mm 'Special' cartridge, the 57-N-181SM. However, it is possible that the BAKSAN could be chambered to fire the 9 × 26 mm cartridge also used by the P-9 Gyurza automatic pistol (qv).

It is probable that the BAKSAN was designed to fulfil the requirement for a handy personal defence weapon once filled by the Czech Model 61 machine pistol (qv) and its ilk.

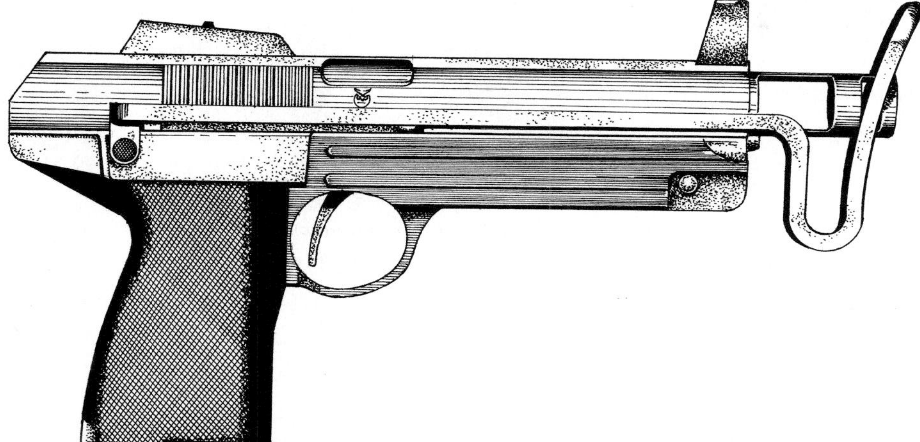

Drawing of 9 mm BAKSAN personal defence weapon (L Haywood)

Description

From what little information is available, the 9 mm BAKSAN selective fire personal defence weapon is a recoil-operated weapon with a positive breech locking system of an, as yet, unspecified type. There is a prominent muzzle brake to reduce recoil forces during bursts which would be limited by the capacity of the 20-round double-stack magazine in the pistol grip. For shoulder-aimed fire to the maximum effective range of 200 m there is a folding steel butt stock, which can be folded to lay either side of the receiver, while the extended body under the receiver could act as a foregrip. It is presumed that the BAKSAN could be carried in a holster.

Data

Cartridge: 9 × 18 mm 57-N-181SM; 9 × 19 mm Parabellum; see text
Operation: recoil, selective fire
Feed: 20-round magazine
Weight: 1.8 kg
Length: stock folded, 480 mm; stock extended, 650 mm
Muzzle velocity: 9 × 18 mm 57-N-181M, ca 450 m/s

Manufacturer

Institute of Precise Mechanical Engineering, 2 Zavodskaya Street, 142080 Klimovsk, Russia.

Status

Uncertain; probably still under development.

VERIFIED

5.45 mm AKS-74U sub-machine gun

Description

The 5.45 mm AKS-74U sub-machine gun, also known as the AKSU-74, is a variant of the AKS-74 rifle, but exhibits a number of new features. Its construction varies from that of previous AK series models by having the receiver top hinged to the gas tube retainer block so that it hinges forward on opening rather than lifting off.

The short barrel is fitted with a cylindrical muzzle attachment which incorporates a bell-mouthed flash hider. It is understood that the body of the attachment forms an expansion chamber which will perform two functions. Firstly, it will give a sudden pressure drop before the bullet exits the muzzle, so reducing the otherwise high pressure in the gas tube and on the gas piston, caused by tapping off propellant gas very close to the chamber. Secondly, it will also permit a degree of flame damping which will reduce flash and blast which would otherwise be inseparable from firing a full-charge rifle cartridge from such a short barrel.

The plastic magazine has been strengthened by moulded-in ribs on the front rebate. The rearsight is a simple flipover pattern with two U-notches, marked for 200 and 400 m. The skeleton metal stock folds to the left and locks into a spring-loaded lug on the side of the receiver. Apart from the shortness of the gas piston, return spring and spring guide rod, the internal mechanism is virtually identical to that of the AK-74 rifle.

The AKS-74U can be fitted with the GP-25 grenade launcher but a barrel extension adaptor is required.

Variants

AKS-74U variants chambered for the 5.56 × 45 mm NATO cartridge have been produced in Bulgaria, Poland and the former Yugoslavia (qv). All have been offered for export sales.

Data

Cartridge: 5.45 × 39.5 mm
Operation: gas, selective fire
Feed: 30-round curved box magazine
Weight: empty, 2.7 kg; loaded, 3.2 kg
Length: butt folded, 490 mm; butt extended, 730 mm
Barrel: 206.5 mm
Rifling: 4 grooves, rh
Muzzle velocity: 735 m/s
Rate of fire: cyclic, 650-735 rds/min

Manufacturer

Kalashnikov Joint Stock Company, 373 Pushkinskaya Street, 426000 Izhevsk, Russia, and others.

Status

In production.

AKS-74U sub-machine gun, left side, stock folded

5.45 mm AKS-74U sub-machine gun, right side

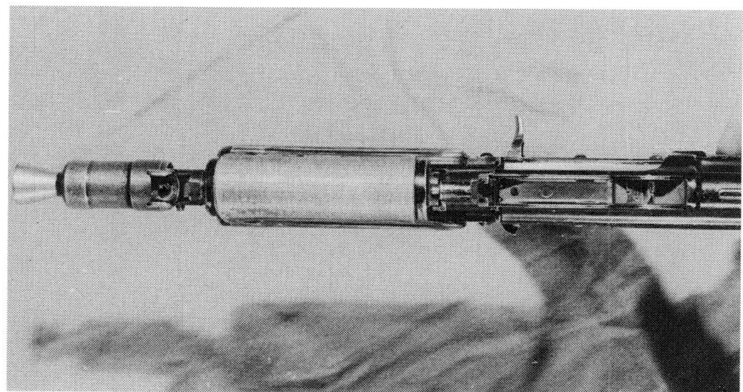

AKS-74U sub-machine gun, top view; note flip-over sight

5.45 mm AKSU-74 sub-machine gun, left side

Service

CIS and some former Warsaw Pact armed forces. The AKS-74U is known to have fallen into the hands of irregular forces, such as the dissident militias within some of the former CIS republics, and Afghanistan.

Licence production

BULGARIA
Marketing Agency: Kintex
Type: AK-74U
Remarks: Although described as an AK-74U the Bulgarian variant is actually an AKS-74U with a folding butt and is available for both the 5.45 × 39.5 mm and 5.56 × 45 mm NATO calibres. (See separate entry under *Bulgaria* for details of the 5.56 mm variant).
POLAND
Manufacturer: Zaklady Metalowe LUCZINK, Radom
Type: ONYX
Remarks: Basically a standard AKS-74U fitted with a side-folding butt stock similar to that used on the former East German MPiKMS-72 and with a three-round burst limiter. Available for both the 5.45 × 39.5 and 5.56 × 45 NATO calibres. (See separate entry under *Poland* for details).
YUGOSLAVIA (SERBIA AND MONTENEGRO)
Manufacturer: Zastava Arms, Belgrade
Type: 5.56 mm Zastava M85
Remarks: May no longer be in production. (See separate entry for details).

AKS-74U sub-machine gun fitted with GP-25 grenade launcher under the muzzle (T J Gander)

UPDATED

Kalashnikov 'Hundred Series' sub-machine guns

Development

There are five weapons in the Kalashnikov 'Hundred Series' of weapons, two of them being assault rifles while the other three fall into the sub-machine category (although they are still labelled as assault rifles). All five weapons incorporate the very latest modifications introduced to the Kalashnikov family and are manufactured using modern materials and production techniques. In order to make these 'export model' weapons attractive to a wide range of potential purchasers, the models are produced in a variety of calibres. An outline of the complete series is provided here:

AK101 - Assault rifle chambered for 5.56 × 45 mm NATO cartridge
AK102 - Sub-machine gun chambered for 5.56 × 45 mm NATO cartridge
AK103 - Assault rifle chambered for 7.62 × 39 mm cartridge
AK104 - Sub-machine gun chambered for 7.62 × 39 mm cartridge
AK105 - Sub-machine gun chambered for 5.45 × 39 mm cartridge

The AK105 firing the 5.45 × 39 mm cartridge is destined to replace the AKS-74U (see previous entry) in service with the CIS armed forces.

Description

The three sub-machine guns in the Kalashnikov 'Hundred Series' are the AK102, AK104, and the AK105. The AK102 fires 5.56 × 45 mm NATO standard ammunition while the AK104 fires the standard Eastern Bloc 7.62 × 39 mm cartridge. The 'domestic use' AK105 fires the 5.45 × 39 mm cartridge and is destined to replace the AKS-74U in service with the CIS armed forces. All three models are visually identical and are constructed with provision to accommodate a wide range of optical and night sighting equipment.

All models are provided with black plastic furniture and the solid butts can be folded forward to lay on the left-hand side of the receiver; the butts can also accommodate an accessories case. The plastic fore-end is ribbed to improve the forward grip. On the AK104 the standard AK-74 pattern muzzle attachment can be replaced by a PBS flash silencer which reduces the firing signature and muzzle flash. All models can be fitted with 40 mm GP-25 grenade launchers.

The overall standard of manufacture of these sub-machine guns is very high and all the many attributes of the Kalashnikov series have been carried over. These attributes include a high degree of reliability, a complete interchangeability of components between models and general ease of maintenance.

5.56 mm AK102 sub-machine gun

Data

Cartridge: AK102, 5.56 × 45 mm; AK104, 7.62 × 39 mm; AK105, 5.45 × 39 mm
Operation: gas, selective fire
Feed: 30-round plastic box magazine
Weight: (less magazine) AK102, 3 kg; AK104, 2.9 kg; AK105, 3 kg
Weight: (loaded) AK102, 3.23 kg; AK104, 3.15 kg; AK105, 3.25 kg
Length: butt folded, 586 mm; butt extended, 824 mm
Barrel: 314 mm
Sights: fore, post; rear, U-notch; provision for fitting optical or night sights
Muzzle velocity: AK102, 850 m/s; AK104, 670 m/s; AK105, 840 m/s
Rate of fire: cyclic, 600 rds/min

Manufacturer

IZHMASH, 3 Derjabin Street, 426006 Izhevsk, Russia.

Status

In production.

Service

AK105 to replace AKS-74U in CIS service. Other models offered for export sales.

VERIFIED

CROATIA

9 mm ERO sub-machine gun

Description
The designation of this 9 mm sub-machine gun is ERO MP. It is a direct copy of the metal stocked version of the Israeli Uzi 9 mm sub-machine gun, correct in every detail apart from the markings. Reference should be made to the entry relating to the Uzi for further details.

Marketing Agency
RH-ALAN d.o.o., Stančićeva 4, 41000 Zagreb.

Status
In production.

Service
Croatian armed forces.

NEW ENTRY

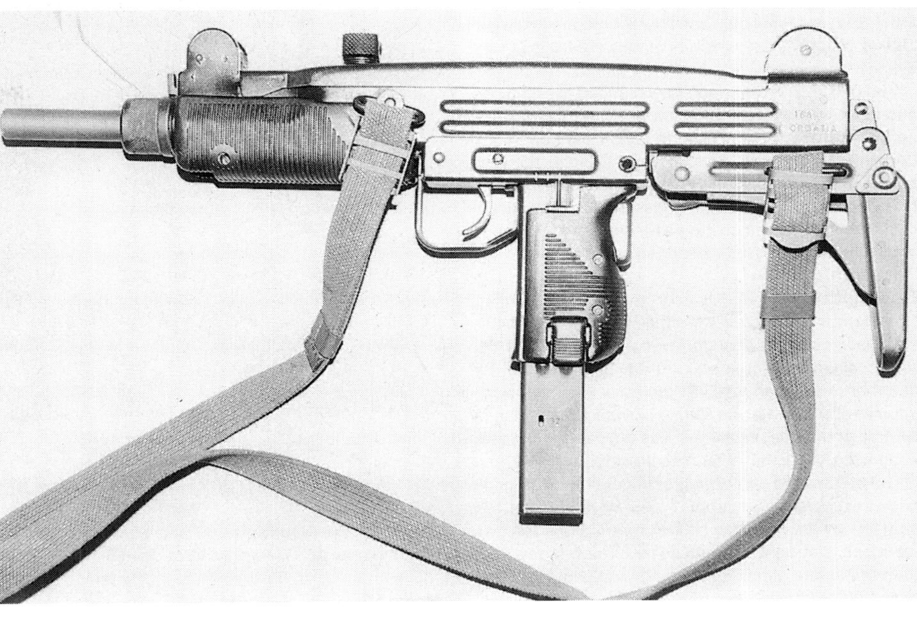

Croatian-produced ERO, a copy of 9 mm Uzi sub-machine gun ((T J Gander)
1996

CZECH REPUBLIC

Models 23, 24, 25 and 26 sub-machine gun

Description
These sub-machine guns were all designed by Vaclav Holek and were produced from 1949 onwards. They differ only in cartridge and butt-stock.

The Model 23 fires 9 × 19 mm Parabellum cartridges and has a wooden stock.

The Model 25 fires 9 × 19 mm Parabellum cartridges and has a folding metal stock.

The Model 24 fires the 7.62 × 25 mm Pistol cartridge and has a wooden stock.

The Model 26 fires the 7.62 × 25 mm Pistol cartridge and has a folding metal stock.

All the guns are blowback-operated, selective fire weapons and use box magazines.

The 9 mm magazine capacity is 24 or 40 rounds while the 7.62 mm magazines hold 32 rounds.

The weapons offer selective fire with control exercised by trigger pull. A short pull produces single shots and a long pull produces automatic fire. The bolt is of the wraparound variety with the firing pin and breech face well back from the front of the bolt which envelops the breech at the moment the firing pin reaches the chambered round.

All the weapons carry a magazine filler on the right of the foregrip. The Models 24 and 26 have magazines that slope forward whereas those of the Models 23 and 25 are vertical.

The Models 23 and 25, firing the 9 mm Parabellum cartridge, were in service with the Czechoslovak Army in 1951 and 1952 and were then replaced by the 7.62 × 25 mm Models 24 and 26, which remained in service until 1962. This was but one example of many of the then USSR forcing other former Warsaw Pact countries to conform and use the standard Soviet cartridge.

Data
MODELS 23 AND 25
Cartridge: 9 × 19 mm Parabellum
Operation: blowback, selective fire
Feed: 24- or 40-round box magazine
Weight: Model 23, 3.27 kg; Model 25, 3.5 kg
Length: stock folded, 445 mm; stock extended, 686 mm
Barrel: 284 mm
Rifling: 6 grooves, rh
Muzzle velocity: 381 m/s
Rate of fire: cyclic, 650 rds/min

MODELS 24 AND 26
As Models 23 and 25 except
Cartridge: 7.62 × 25 mm Pistol Type P
Feed: 32-round box magazine
Weight: 3.41 kg
Rifling: 4 grooves, rh
Muzzle velocity: 550 m/s

Manufacturer
Ceská Zbrojovka as, Brno.

Status
No longer manufactured.

Service
No longer in Czech military service. Supplied to Cuba and Syria where some are probably still in use (Models 23 and 25). Cambodia, Cuba, Czech Republic, Guinea, Mozambique, Nicaragua, Nigeria, Romania, Slovakia, Somalia, Syria, Tanzania. (Models 24 and 26).

VERIFIED

Czech 9 mm Model 23 sub-machine gun

Czech 7.62 mm Model 26 sub-machine gun

7.65 mm Model 61 Skorpion machine pistol

Development
During the latter years of the 1950s Omnipol, which then controlled Czechoslovak arms production, hastened the pace of weapon development and the first post-war weapons were replaced. The Mark 52 pistol was superseded by a dual-purpose pistol/sub-machine gun called the vz 61 Skorpion (vzor - model) of unusual, but effective, design. Like the great majority of dual-purpose weapons it carries out neither role to perfection and is really neither a pistol nor a sub-machine gun but is best described as a 'machine pistol'.

The Skorpion fires the American 0.32 ACP cartridge and this feature distinguishes it from other weapons of the former Soviet Bloc. The 0.32 ACP (7.65 mm) cartridge with its light bullet and low muzzle velocity produces a muzzle energy which is less than that of a high velocity 0.22 Long Rifle cartridge and it seems likely that commercial considerations dictated the choice of a cartridge which is readily available almost everywhere.

The Skorpion is a weapon ideally suited to being carried by vehicle crews and others who have to work in a confined space, or who have to carry heavy and cumbersome loads, such as signallers. Despite the limitations of the ammunition the Skorpion enables its user to engage targets at greater ranges than a conventional pistol would allow, since it is easily held with two hands and, with shoulder stock extended, it shoots well.

Description

The Skorpion can produce either single shots or full-automatic fire. The change lever is marked '1' at the rear position for single shot, '0' for safe and '20' for automatic. The cartridge produces a very low recoil impulse and this enables simple blowback operation to be employed. Gas pressure drives the case back in the chamber against the resistance provided by the inertia of the bolt and its two driving springs. The block goes back, extracting the empty case which is ejected straight upwards and can hit the firer full in the face. As the block is so light it is necessary to have a device to control the rate of fire which, otherwise, would be well over 1,000 rds/min.

The rate reducer operates as follows: when the bolt reaches the end of its rearward stroke it strikes, and is caught by a spring-loaded hook mounted on the back plate. At the same time it drives a light, spring-loaded plunger down into the pistol grip. The light plunger is easily accelerated and passes through a heavy weight which due to its inertia is left behind. The plunger, having compressed its spring, is driven up again and then meets the descending inertia pellet. This slows down the rising plunger which, when it reaches the top of its travel, rotates the hook, releasing the bolt which is driven forward by the compressed driving springs. This reduces the rate of fire to 840 rds/min. The bolt feeds a round from the 20-shot magazine, chambers it and supports the case in the chamber. Since it is a simple blowback-operated system there is no bolting or delay device and the sole support to the cartridge is the inertia of the bolt and the strength of the driving springs. The magazine platform operates a plunger which retains the bolt in the rear position when the ammunition supply is exhausted.

Variants

Export models

There were three main export models based on the Model 61, the Models 64, 65 and 68. These differed from the original Model 61 only in detail.

7.65 mm Model 61 Skorpion machine pistol

Model 82

Chambered for the 9 mm Makarov cartridge, with 113 mm barrel. Utilised straight box magazines. Apparently not produced in quantity, but is still offered for possible export sales.

Model 83

Chambered for the 9 mm Browning Short cartridge. Utilised straight box magazines. Apparently not produced in quantity.

CZ 91S

Although production of the original Model 61 Skorpion ended some years ago a virtually identical semi-automatic variant known as the CZ 91S is still produced. CZ 91S models are available to fire the original 0.32 ACP (7.65 mm), 9 mm Browning Short, or 9 × 18 mm Makarov cartridges. Overall finish is to a very high 'collectors' standard with a stainless steel receiver and a black enamel coating for other external parts.

Data

Cartridge: 0.32 ACP (7.65 mm); 0.380 ACP; 9 × 18 mm Makarov
Operation: blowback with selective fire
Feed: 10- or 20-round box magazine
Weight: empty, 1.59 kg
Length: butt retracted, 269 mm; butt extended, 513 mm
Barrel: 115 mm
Rifling: 6 grooves, rh, 1 turn in 305 mm
Sights: fore, post; rear, flip aperture, 75 and 150 m
Sight radius: 171 mm
Muzzle velocity: 317 m/s; with silencer, 274 m/s
Rate of fire: normal, 840 rds/min; silenced, 950 rds/min

Manufacturer

Ceska Zbrojovka as, Brno.

Status

Model 61, production complete.
CZ 91S in production.

Service

Some Czech and Slovak units, Afghanistan, Angola, Egypt, Libya, Mozambique and Uganda.

Licence production

YUGOSLAVIA (SERBIA AND MONTENEGRO)
Manufacturer: Zastava Arms, Belgrade
Type: 7.65 mm Model 84
Remarks: Probably no longer in production.

UPDATED

DENMARK

9 mm Model 49 Hovea sub-machine gun

Description

The 9 mm Model 49 Hovea sub-machine gun was designed in Sweden at the Husqvarna arms factory and competed with the Carl Gustaf weapon in arms trials at which the latter weapon (the m/45) was selected for Swedish use. The Hovea sub-machine gun was preferred by the Danish authorities, so they purchased a quantity together with a manufacturing licence; Danish-built weapons are still in service with Danish forces.

The weapon was originally designed to take the Finnish 50-round magazine, but this was subsequently replaced by the 36-round Carl Gustaf box magazine. Fitted with a rectangular folding stock, the Model 49 Hovea bears a superficial resemblance to both the Carl Gustaf m/45 and the Madsen weapons (see entries following).

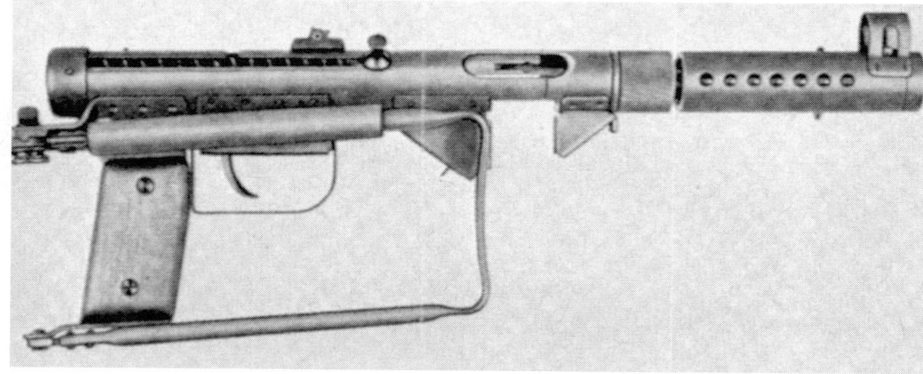

9 mm Model 49 Hovea sub-machine gun with butt folded and without magazine

Data

Calibre: 9 × 19 mm Parabellum
Operation: blowback, automatic
Feed: 36-round detachable box (see text)
Weight: empty, 3.4 kg; loaded, 4 kg
Length: stock retracted, 550 mm; stock extended, 810 mm
Barrel: 215 mm
Sights: fore, blade; rear, V-notch flips 100/200 m
Rate of fire: cyclic, 600 rds/min

Manufacturer

Vabenarsenalet, Copenhagen. (Closed 1970).

Status

Not in production.

Service

Danish forces.

VERIFIED

9 mm Madsen sub-machine guns

Development

Madsen has produced a series of sub-machine guns of the same basic simple design, known as the Model 1946, Model 1950, Model 1953 and the Mark II. All have the same rectangular section folding stock. The Model 1946 has a shaped cocking handle which fits over the receiver and down on each side; the Model 1950 has a small cylindrical knob on top of the receiver; the Model 1953 has a knurled cylindrical barrel nut that screws on to the barrel whereas the 1946 and 1950 models have a grooved nut fitting on to the receiver. The magazines of the Model 1946 and Model 1950 are 32-round, single-position feed, straight flat-sided types. Those of the Model 1953 and Mark II are two-position feed types also holding 32 rounds. The Mark II is the only one giving selective fire: it may have a perforated barrel jacket.

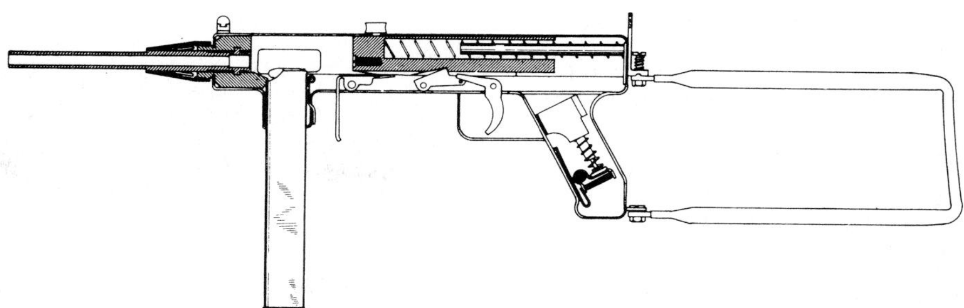

9 mm Model 1950 Madsen sub-machine gun

Description

A feature of all the Madsen sub-machine guns is a grip safety behind the magazine wall. This has to be pressed forward before the weapon can be cocked or fired. It engages the bent in the bolt to hold it when the bolt is cocked and blocks the sear when the bolt is forward.

The 32-round magazine is most easily loaded by employing a filler which is carried inside the pistol grip. To get at it means opening up the two halves of the receiver shell and so putting the gun out of action for a short while. As a result the filler is not always used and hand filling is adopted which can be difficult as the magazine spring is progressively compressed. The filler fits over the top of the magazine and has a plunger to depress the magazine follower. The base of the cartridge slides under the lips while the plunger takes the weight of the spring, and is pushed fully home when the plunger is released.

When the grip safety is pushed forward, and the trigger operated, the sear will release the bolt which will go forward under the action of the compressed return spring. The feed rib on the bottom of the bolt picks up the top cartridge in the magazine and feeds it up over the bullet guide into the chamber. The cartridge is chambered, the extractor snaps into the groove and the fixed firing pin on the bolt-face fires the cap while the bolt is still moving forward. The gas pressure forces the cartridge case rearward and the bolt movement is halted and reversed. The extractor holds the case to the bolt-face until it hits the fixed ejector and is ejected to the right of the gun. The return spring is compressed. At the end of rearward travel the bolt is impelled forward again. If the trigger or grip safety has been released the bolt will be held back; otherwise the cycle is repeated until the magazine is empty.

Data

Cartridge: 9 × 19 mm Parabellum
Operation: blowback, automatic
Feed: 32-round detachable box magazine
Weight: empty, 3.2 kg
Length: stock folded, 528 mm; stock extended, 794 mm
Barrel: 198 mm
Rifling: 4 grooves, rh
Muzzle velocity: 390 m/s
Rate of fire: cyclic, 550 rds/min

Manufacturer

Dansk Industri Syndikat, Madsen, Copenhagen.

Status

No longer in production.

Service

Some South American and South-east Asian countries.

Licence production

BRAZIL
Manufacturer: IMBEL, Sao Paulo
Type: 9 mm Madsen
Remarks: See separate entry for details

Manufacturer: Industrias Nacional de Armas SA, Sao Paulo
Type: 0.45 INA MB 50 and Model 953
Remarks: See separate entry for details

VERIFIED

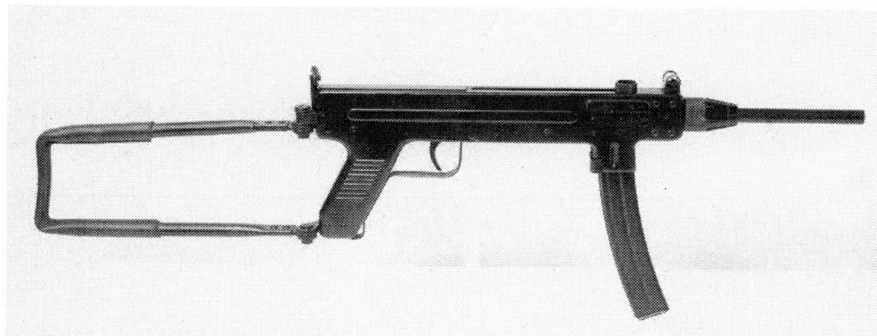

9 mm Model 1953 Madsen sub-machine gun

9 mm Model 1950 Madsen sub-machine gun

EGYPT

9 mm Port Said sub-machine gun

Description

The 9 mm Port Said sub-machine gun is a licence-produced version of the Swedish 9 mm m/45 Carl Gustaf and is still in production for local use and possible export sales by the government-owned Maadi Company for Engineering Industries. The Port Said is visually identical to the m/45 although Egyptian data differs in detail and is provided.

9 mm Port Said sub-machine gun

Data

Cartridge: 9 × 19 mm Parabellum
Operation: blowback, automatic
Feed: 36-round box magazine
Weight: empty, 3.65 kg; loaded, 4.33 kg
Length: butt folded, 550 mm; butt extended, 808 mm
Barrel: 212.5 mm
Rifling: 6 grooves, rh
Sights: fore, blade; rear, flip
Muzzle velocity: 400 m/s
Rate of fire: cyclic, 600 rds/min

Manufacturer

Maadi Company for Engineering Industries, Maadi, Cairo.

Status

In production.

Service

In service with the Egyptian armed forces.

VERIFIED

FINLAND

9 mm Jati-Matic sub-machine gun

Description

The Jati-Matic is a design with some unusual features. The designers have aimed at producing a light weapon which is easy to use and control, and the weapon uses a patented system in which the bolt recoils up an inclined plane at an angle to the barrel. At the same time the bolt presses against the bottom of the receiver. This upward movement of the bolt allows the pistol grip to be set higher than normal, so that the firer's hand lies on the axis of the bore. This, it is claimed, reduces the torque effects, easing the strain on the firer's wrist and also making the gun particularly steady when firing automatic.

The Jati-Matic consists of a pressed steel receiver with a hinged top cover. Into this fits the bolt and barrel, retained by the top cover. A folding forward grip handle beneath the barrel acts as a steadying grip and is also the cocking handle. When closed, this handle acts as a positive bolt stop in either the cocked or uncocked positions and prevents inadvertent firing even if the trigger is pressed. The trigger is a two-stage type in which the first pressure fires single shots, and further pressure, against a spring stop, fires automatic.

The box magazine is inserted into the receiver ahead of the trigger; Smith & Wesson or Carl Gustaf magazines will also fit the Jati-Matic.

A range of accessories, including magazines, silencer and a laser aiming spot is available. The weapon is optimised for use with 9 mm Parabellum ammunition having a bullet weight of 8 g and a muzzle velocity of 360-400 m/s. With other ammunition it may be necessary to modify the strength of the return spring.

This weapon is now produced by Oy Golden Gun Ltd as the GG-95 PDW.

Data

Calibre: 9 × 19 mm Parabellum
Operation: blowback, selective fire
Feed: detachable 20- or 40-round box
Weights: empty, 1.65 kg; 20-round magazine, 300 g; 40-round magazine, 600 g; gun and 20-round magazine, 1.95 kg
Length: 375 mm
Barrel: 203 mm
Rifling: 6 grooves, rh, 1 turn in 250 mm
Sights: fore, fixed blade; rear, fixed U-notch; (zero) 100 m
Sight radius: 290 mm
Muzzle velocity: 360-400 m/s
Rate of fire: cyclic, 600-650 rds/min

Manufacturer

Oy Golden Gun Ltd, Humalistonkatu 9, FIN-20100 Turku.

Status

Available.

UPDATED

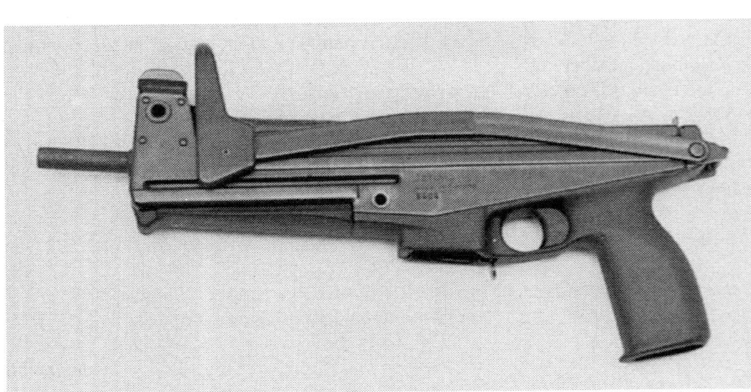

9 mm Jati-Matic with front grip folded; note angle between axis of bore and axis of receiver

9 mm Jati-Matic with front grip folded down and 40-round magazine

---

FRANCE

9 mm MAT 49 sub-machine gun

Description

The MAT 49 is a very solid, strongly made weapon making considerable use of stampings from heavy gauge steel sheet. It is of conventional blowback design, firing the 9 × 19 mm Parabellum cartridge, but has a number of features which, although not original, are seldom met elsewhere.

The magazine housing can be pivoted forward in front of the trigger to allow the magazine to lie along the underside of the barrel jacket. A housing release is forward of the trigger guard; when this is operated the housing, holding the 32-round magazine, can be rotated forward. When under the barrel jacket it is retained by the housing catch. The magazine housing must be pulled manually into the firing position. It is shaped to afford a firing grip for the left hand. When the housing is under the barrel it seals off the feed opening into the receiver and this, together with the spring-loaded ejection port cover, makes the weapon proof against the ingress of sand, dirt and so on into the working parts. When the magazine is housed under the barrel the weapon is completely safe since forward movement of the bolt, whether accidental or otherwise, cannot feed, chamber or fire a round. When the magazine is swung into the firing position, safety is provided by the grip safety which has two functions. It locks the bolt in the forward position and locks the trigger when the bolt is cocked. This ensures that if the weapon is inadvertently dropped or jarred, the bolt cannot move back from its forward position under its own inertia to chamber and fire a round, nor can the cocked bolt go forward. When the grip safety is operated by the act of grasping the pistol grip, the trigger restraint is released and the lock preventing bolt movement is lowered. Since bolt movement is possible only when the grip safety is operated, no manual safety of the usual 'fire' and 'safe' type is required or provided.

The bolt is of interesting shape. The modern conception of design is to shorten the overall length of the sub-machine gun by using a wraparound bolt, enveloping the chamber, with a fixed firing pin placed towards the rear of the bolt. The head of the bolt of the MAT 49 enters an extension of the chamber thus producing a

9 mm MAT 49 sub-machine gun

9 mm MAT 49 sub-machine gun stripped

wraparound barrel. The MAT 49 has a fixed firing pin and the weapon fires while the bolt is still moving forward; the design of the bolt face and breech ensures that shielding is provided to minimise the effects of both early firing and a hang-fire.

The length of this French sub-machine gun can be adjusted by a sliding wire stock, similar to that of the US M3 sub-machine gun. There are two indentations in which the stock latch can engage: the stock may be locked in an intermediate position as well as in the fully extended position required for firing from the shoulder.

Data

Cartridge: 9 × 19 mm Parabellum
Operation: blowback, automatic
Feed: 32- or 20-round box magazine
Weight: empty, 3.5 kg
Length: butt retracted, 460 mm; butt extended, 720 mm
Barrel: 228 mm
Rifling: 4 grooves, lh
Sights: fore, hooded blade; rear, flip aperture 100 and 200 m
Muzzle velocity: 390 m/s
Rate of fire: cyclic, 600 rds/min

Manufacturer

Manufacture d'Armes de Tulle, Tulle.

Status

No longer in production.

Service

French police, and many former French colonies.

VERIFIED

GERMANY

Heckler and Koch 9 mm MP5 sub-machine gun

Development

The Heckler and Koch MP5 sub-machine gun was developed from the G3 rifle (qv) and has the same method of operation. Some components are interchangeable with those used on the rifle. The sub-machine gun was adopted in late 1966 by the police forces of the former West Germany and by the border police. It has since been purchased by many military and police forces worldwide.

Description

The Heckler and Koch 9 mm MP5 sub-machine gun usually offers only a choice of single shot or full automatic but a burst-fire device, allowing three-round bursts each time the trigger is operated, is offered by the manufacturers to those requiring this facility. This trigger arrangement can be fitted to all automatic weapons in the Heckler and Koch series. The top position of the selector lever is 'safe' and the selector spindle lies over the trigger lug and prevents sufficient movement of the trigger to disengage the sear from the hammer notch.

9 mm Heckler and Koch MP5A3 sub-machine gun

The breech mechanism is of the same design as that used in the G3 rifle. It may be described, in brief, as a two-part bolt with rollers projecting from the bolt head. The more massive bolt body lies up against the bolt head when the weapon is ready to fire and inclined planes on the front lie between the rollers and force them out into recesses in the barrel extension. The gas pressure forces back on the bolt head which is unable to go back since the rollers are in the recesses in the barrel extension and must move in against the inclined planes of the heavy bolt body. The selected angles of the recesses and the incline on the bolt body produce a velocity ratio of about 4:1 between the bolt body and the bolt head. Thus the bolt head moves back only about 1 mm while the bolt body moves some 4 mm. As soon as the rollers are fully in, the two parts of the breech are driven back together. The empty case is held to the breech face by the extractor until it strikes the ejector arm and is thrown out of the ejection port to the right of the gun. The return spring is compressed during the backward movement of the bolt and drives the bolt forward. A round is fed into the chamber and the bolt face come to rest. The bolt body continues to move forward and the inclined planes drive the rollers out again into the barrel extension recesses. The bolt body closes up to the bolt head and the weapon is ready to fire another round. The MP5 thus fires from a

9 mm Heckler and Koch MP5A2 sub-machine gun with passive night sight

9 mm Heckler and Koch MP5A5 sub-machine gun with automatic and burst-fire facilities

9 mm Heckler and Koch MP5A4 sub-machine gun with three-round burst facility

9 mm Heckler and Koch MP5A2 sub-machine gun

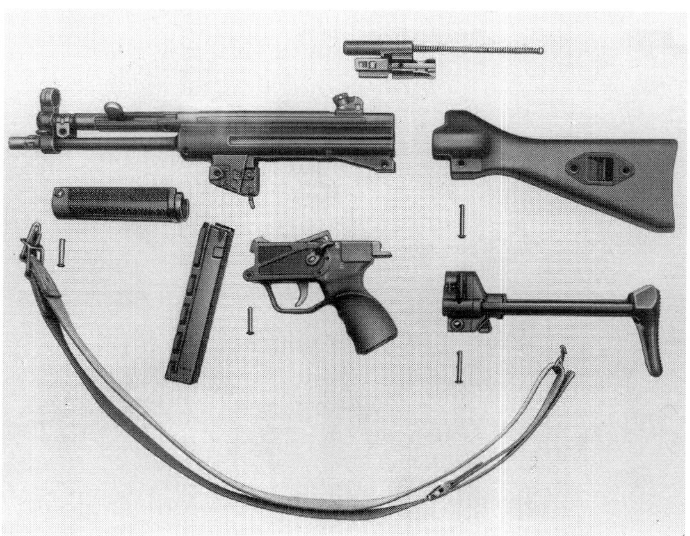

9 mm Heckler and Koch MP5A2 sub-machine gun stripped

closed bolt, making it more accurate than the conventional blowback sub-machine gun.

If the selector is set to full automatic the trigger has moved up sufficiently for the nose of the sear to be depressed so far that it does not re-engage the hammer and the next round is fired as soon as the bolt is fully closed, and the safety sear is moved out of engagement with the hammer. If the selector is for a single shot the trigger is unable to rise fully and the spring-loaded sear holds the hammer until the trigger is released and repressed. Provided the bolt is fully closed the hammer will be released.

If a burst-fire control is fitted a ratchet counting device in the trigger mechanism holds the sear off the hammer until the allotted number of rounds have been fired. This device ensures the correct number of cartridges are discharged in a single burst and any interruption, for example an empty magazine, starts a fresh count. After the burst the trigger must be released to set the counter back to zero and another sustained pressure on the trigger fires another burst of three rounds.

The first models were fitted with a straight magazine, but in 1978 this was changed to one with a slight curve. It was found that the curved shape improved the feeding characteristics with the many different types of bullet and nose shapes in 9 × 19 mm Parabellum ammunition.

The standard telescopic sight has ×4 magnification, has range settings for 15, 25, 50, 75 and 100 m and is adjustable for windage and elevation. A passive night sight may also be fitted.

A blank firing attachment is available. It is a conventional device with a restricted gas flow. It is placed on the muzzle using the locking lugs and is retained by a catch. It is marked with a red ring to enable the firer to identify it as a training aid.

Variants

The basic MP5 mechanism is employed in several different sub-machine guns in the MP5 series. Set out below is data relating to the MP5A2 and MP5A3 versions which differ only in that the former has a fixed butt-stock whereas the MP5A3 has a single metal strut stock which may be slid forward to give a considerable reduction to the overall length of the weapon. Three silenced variants with the general designation MP5 SD and the sixth weapon, the short-barrelled MP5K are described in the following entries.

9 mm MP5 SD sub-machine gun

This is a silenced version of the standard MP5. (See separate entry for details).

10 mm MP5/10 sub-machine gun

A product improved version of the MP5 chambered for the 10 mm Auto cartridge. (See separate entry for details).

0.40 MP5/40 sub-machine gun

This is a product-improved version of the MP5 chambered for the 0.40 S&W cartridge.

9 mm MP5-N sub-machine gun

A special variant (N - Navy) produced in the United States. (See entry under *United States of America* for details).

Data

MP5A2, A3, A4 and A5

Cartridge: 9 × 19 mm Parabellum
Operation: delayed blowback, selective fire
Locking: rollers
Feed: 15- or 30-round curved box magazine
Weight: empty, 2.55 kg
Length: butt retracted, 490 mm; butt extended, 660 mm; fixed butt, 680 mm
Barrel: 225 mm
Rifling: 6 grooves, rh
Sights: fore, fixed post; rear, apertures for different eye relief adjustable for windage and elevation. A telescopic or night sight or an aiming projector may be fitted
Sight radius: 340 mm
Muzzle velocity: 400 m/s
Rate of fire: cyclic, 800 rds/min

Manufacturer

Heckler and Koch GmbH, D-78722 Oberndorf/Neckar.

Status

In production.

Service

In service with Afghanistan, Argentina, Australia, Bahrain, Belgium (police), Brazil (police), Cameroon, Chile, Columbia, Denmark (police), El Salvador, France, Germany (police and border police), Ghana, Greece (licence production), Honduras, Iceland (police), India (special forces), Iran, Ireland, Italy (police), Japan (police), Jordan, Kenya, Kuwait, Luxembourg, Malta, Mauritius, Mexico (licence production), Morocco, Netherlands, New Zealand, Niger, Nigeria, Norway, Pakistan (licence production), Qatar, Saudi Arabia, Singapore, Spain, Sri Lanka, Sudan, Switzerland (police), Thailand, Turkey (licence production), UAE, UK (police and special forces), USA (special forces), Uruguay, and Venezuela.

Licence production

GREECE
Manufacturer: Hellenic Arms Industry
Type: 9 mm EMP5
Remarks: See entry under *Greece*.
IRAN
Manufacturer: Defence Industries Organisation
Type: 9 mm Tondar
Remarks: Produced in MP5A2 AND MP5A3 forms.
MEXICO
Manufacturer: Not known
Type: MP5
Remarks: Licence production reported but current status uncertain.
PAKISTAN
Manufacturer: Pakistan Ordnance Factories
Type: MP5A2 and MP5A3
Remarks: Manufactured for Pakistan armed forces and offered for export.
TURKEY
Manufacturer: Makina ve Kimya Endustrisi Kurumu (MKEK)
Type: MP5A3
Remarks: See entry under *Turkey*.
UNITED KINGDOM
Manufacturer: Heckler and Koch (UK)
Type: Various MP5 models
Remarks: Originally Royal Ordnance plc, Small Arms Factory, Enfield Lock, and later Royal Ordnance plc, Small Arms Division, Nottingham. Production mainly for export to nations such as Nigeria.

UPDATED

Heckler and Koch 9 mm MP5 SD sub-machine gun

Description

The Heckler and Koch 9 mm MP5 SD is a silenced version of the MP5 (see previous entry). Its mechanism is the same as that of the MP5 but the weapon differs in having a barrel in which 30 holes are drilled. The silencer on the barrel features two separate chambers, one of which is connected to the holes in the barrel and serves as an expansion chamber for the propulsive gases, thus reducing the gas pressure to slow down the acceleration of the projectile. The second chamber diverts the gases as they exit the muzzle, so muffling the exit report. The bullet leaves the muzzle at subsonic velocity, so it does not generate a sonic shock wave in flight. The silencer requires no maintenance; only rinsing in an oil-free cleaning agent is prescribed.

There are six versions of the weapon. The MP5 SD1 has a receiver end cap and no butt-stock; the SD2 has a fixed butt-stock and the SD3 a retractable butt-stock, the components being identical to those of the MP5A2 and MP5A3 respectively. The MP5 SD4 resembles the SD1 but has a three-round burst facility in addition to single and automatic fire: the SD5 is the SD2 with three-round burst facility: and the SD6 is the SD3 with the three-round burst facility. It will be noted that the three

9 mm MP5 SD3 sub-machine gun

9 mm Heckler and Koch MP5 SD4 sub-machine gun with 15-round magazine and 3-round burst facility

9 mm MP5 SD2 sub-machine gun

9 mm MP5 SD1 sub-machine gun

latest models have a slightly changed contour of the pistol grip. Each may be used with iron sights, a telescopic sight, the Hensoldt Aiming-Point-Projector, or an image intensifier sight.

Data

Cartridge: 9 × 19 mm Parabellum
Operation: delayed blowback, selective fire
Delay: rollers
Feed: 15- or 30-round curved box magazine
Weight: SD1, 2.9 kg; SD2, 3.2 kg; SD3, 3.5 kg
Length: SD1, 550 mm; SD2, 780 mm; SD3, 610 or 780 mm
Barrel: 146 mm
Rifling: 6 grooves, rh
Sights: fore, fixed post; rear, apertures for different eye relief adjustable for windage and elevation. A telescopic or night sight or an aiming projector may be fitted
Sight radius: 340 mm
Muzzle velocity: 285 m/s
Rate of fire: cyclic, 800 rds/min

Manufacturer

Heckler and Koch GmbH, D-78722 Oberndorf/Neckar.

Status

In production.

Service

German special forces and numerous military and police forces worldwide.

UPDATED

Heckler and Koch 9 mm MP5K series sub-machine guns

Description

Heckler and Koch introduced the 9 mm MP5K series sub-machine guns for use by special police and anti-terrorist squads. They are extra-short versions of the standard MP5 and are meant for carriage inside clothing, in the glove-pocket of a car, or any other limited space. They offer all the fire options of the MP5: the MP5K is fitted with adjustable iron sights, or a telescope if need be; the MP5KA1 has a smooth upper surface with a very small fore and rearsight so that there is little to catch in clothing or a holster as the gun is withdrawn. The MP5KA4 is similar to the MP5K but has a three-round burst facility as well as single and automatic fire. The MP5KA5 is similar to the A1, with the addition of the three-round burst facility. The prominent forehand grip gives the firer the best possible control for all types of fire. The butt-stock is not fitted and both versions are meant to be fired from a hand hold at all times, with no other support.

Variant

9 mm MP5K-PDW

This variant is assembled in the United States as a personal defence weapon (PDW). (See entry under *United States of America* for details).

Data

Calibre: 9 × 19 mm Parabellum
Operation: delayed blowback, selective fire
Delay: rollers
Feed: 15- or 30-round detachable box magazine
Weight: without magazine, 2 kg
Weight of magazines: empty, 15-round, 100 g; empty, 30-round, 160 g; loaded, 30-round, 520 g
Length: 325 mm
Barrel: 115 mm
Rifling: 6 grooves, rh
Sights: (MP5K, MP5KA4) fore, post; rear, open rotary adjustable for windage and elevation, or ×4 telescopic (MP5KA1, MP5KA5) fore, fixed blade; rear, fixed notch
Sight radius: (MP5K, MP5KA4) 260 mm; (MP5KA1, MP5KA5) 190 mm
Muzzle velocity: 375 m/s
Muzzle energy: 570 J
Rate of fire: cyclic, 900 rds/min

Manufacturer

Heckler and Koch GmbH, D-78722 Oberndorf/Neckar.

Status

In production.

Service

German special forces and numerous military and police forces worldwide.

Licence production

TURKEY
Manufacturer: Makina ve Kimya Endustrisi Kurumu (MKEK)
Type: 9 mm MP5K
Remarks: See entry under *Turkey* for illustration

UPDATED

9 mm MP5KA5 sub-machine gun

9 mm Heckler and Koch MP5K sub-machine gun

Heckler and Koch 10 mm MP5/10 sub-machine gun

Description

The Heckler and Koch 10 mm MP5/10 is a product-improved version of the Heckler and Koch MP5 sub-machine gun which has been designed particularly for use by US law enforcement agencies, though it obviously has applications elsewhere. The operation and functioning of the MP5/10 is exactly the same as that of the standard MP5, the difference being that this weapon is chambered for the 10 mm Auto cartridge.

A newly designed carbon fibre reinforced magazine has been developed with a straight shape in place of the usual curve. A metal dual magazine clamp allows two magazines to be held together on the weapon, allowing instantaneous change.

The standard model will have the usual HK trigger assembly allowing selective fire with a three-round burst, but trigger mechanisms allowing any combination of firing modes can be supplied. This model also marks the introduction of a new two-round burst feature, which will become available on all MP5 models. This simply permits the firing of two rounds for a single trigger pressure, producing an automatic 'double-tap' and doing away with the third round which is frequently a 'flyer' high and right of the target.

As an option, the sights can be fitted with self-luminous Beta modules for firing in poor light conditions. By removing the fore-end and replacing it with the HK Tactical Forearm Light, the user is provided with an integral 15,000 candela spotlight.

10 mm MP5/10 sub-machine gun with fixed stock and three-round burst limiter

Variant

0.40 in HK MP5/40 sub-machine gun

This weapon is identical in all respects to the MP5/10 differing only in that it is chambered and rifled for the 0.40 S&W cartridge.

Data

Cartridge: 10 mm Auto - see also text
Operation: delayed blowback, selective fire with 2- or 3-round burst
Locking system: rollers
Feed: 30-round box magazine
Weight: fixed stock, 2.67 kg; retractable stock, 2.85 kg
Length: stock retracted, 490 mm; stock extended, 660 mm; with fixed stock, 680 mm

Barrel: 225 mm
Rifling: 6 grooves, rh, 1 turn in 380 mm
Sights: fore, hooded post; rear, rotary aperture, fully adjustable
Sight radius: 340 mm
Muzzle velocity: ca 442 m/s
Muzzle energy: ca 650 J

Manufacturer
Heckler and Koch GmbH, D-78722 Oberndorf/Neckar.

Status
In production.

UPDATED

Heckler and Koch 5.56 mm HK 53 sub-machine gun

Description
The Heckler and Koch HK 53 sub-machine gun is chambered for the 5.56 × 45 mm cartridge, so can be deployed either as a sub-machine gun or as an assault rifle. The method of operation is similar to the other rifles and sub-machine guns made by Heckler and Koch and uses the system of rollers which delays the rearward movement of the bolt head until the pressure has dropped sufficiently to allow the complete bolt unit to be blown back in safety.

In appearance the HK 53 sub-machine gun is very similar to the HK 33K rifle and it can be fitted with either a conventional plastic butt or a double strut telescoping butt-stock. The HK 53 is the shortest of the 5.56 mm weapons. Length, with butt retracted, is only 563 mm whereas that of the HK 33K is 670 mm and the HK 33A3 is 730 mm.

The HK 53 is capable of either single shots or full automatic, with a selector lever on the left of the receiver above the pistol grip. Optionally, it may be provided with a different grip with burst-fire control and an ambidextrous selector/safety lever. It is now produced with a type of flash suppressor which totally eliminates the muzzle flash which was previously experienced with this weapon.

The Heckler and Koch HK 53 may be regarded as a scaled-down version of the G3 rifle, and many of the parts of the two weapons are interchangeable with each other and the HK 33. The 9 mm MP5 sub-machine gun also has parts interchangeable with the G3 and HK 33 and it is important that the bolts, (particularly the locking pieces), are not interchanged between weapons of different calibre and different recoil impulse.

5.56 mm Heckler and Koch HK 53A3 sub-machine gun with current style flash hider and three-round burst limiter

Accessories for the HK 53 include a luminous front sight, various optical and night sight mountings, a dual magazine clip, a blank attachment, a cleaning kit, and magazine fillers and emptiers.

Data
Cartridge: 5.56 × 45 mm
Operation: delayed blowback, selective fire
Method of delay: rollers
Feed: 25-round box magazine
Weight: empty, 3.05 kg
Length: butt retracted, 563 mm; butt extended, 755 mm
Width: 52 mm
Barrel: 211 mm
Rifling: 6 grooves, rh, 1 turn in 178 mm or 305 mm as required
Sights: fore, post; rear, apertures for 200, 300, 400 m; V 100 m battle sight
Sight radius: 390 mm
Muzzle velocity: 750 m/s
Muzzle energy: 1,300 J
Rate of fire: cyclic, 700 rds/min

Manufacturer
Heckler and Koch GmbH, D-78722 Oberndorf/Neckar.

Status
In production.

Service
Special military and police units in several countries.

VERIFIED

GREECE

9 mm EMP5 sub-machine gun

Description
The Hellenic Arms Industry (EBO) SA manufactures the Heckler and Koch MP5 as the EMP5, which has been issued to the Greek security forces and police. The weapon is also offered for export. Details will be found in the MP5 entry under *Germany*.

Manufacturer
Hellenic Arms Industry (EBO) SA, 160 Kifissias Avenue, GR-11525 Athens.

Status
In production.

Service
Greek security forces and police; offered for export.

VERIFIED

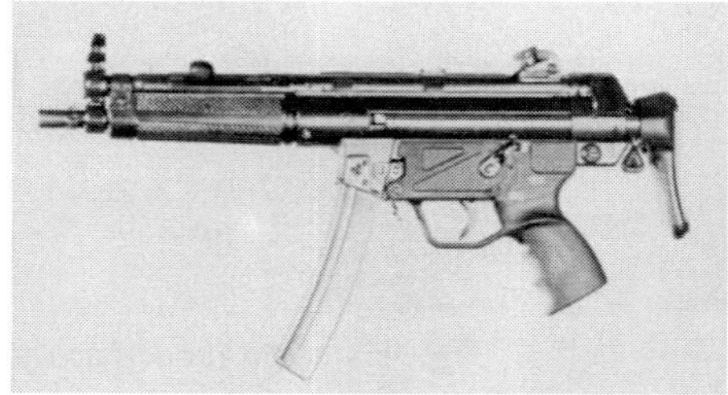

9 mm Greek EMP5 sub-machine gun

HUNGARY

9 mm KGP-9 sub-machine gun

Description
The KGP-9 sub-machine gun is a blowback weapon with a side-folding steel butt. The receiver is of steel plate stiffened by precision castings. The bolt is fitted with a floating firing pin and the firing mechanism is of the hammer type. The safety catch and fire selector are in front of the trigger. The magazine is of a two-column pattern and is of steel plate. The rearsight is a two-range flip aperture. The weapon is easily dismantled. One feature is that the barrel can be removed and replaced by a longer (250 mm) barrel to enable the weapon to be deployed as a carbine.

The magazine housing is in front of the trigger, and there is a substantial synthetic fore-end offering a firm grip. The cocking handle is on top of the receiver, offset from the sight line, and does not reciprocate during firing.

Data
Cartridge: 9 × 19 mm Parabellum
Operation: blowback, selective fire
Feed: 25-round box magazine
Weight: empty, 2.75 kg
Length: butt folded, 355 mm; butt extended, 615 mm
Barrel: sub-machine gun, 190 mm; carbine, 250 mm
Sights: fore, post; rear, two-position flip aperture
Sight radius: 252 mm
Muzzle velocity: 390 m/s
Rate of fire: cyclic, 900 rds/min

Manufacturer
Fegyver-és Gázkészülékgyár, Soroksári út 158, H-1095 Budapest.

Status
In production.

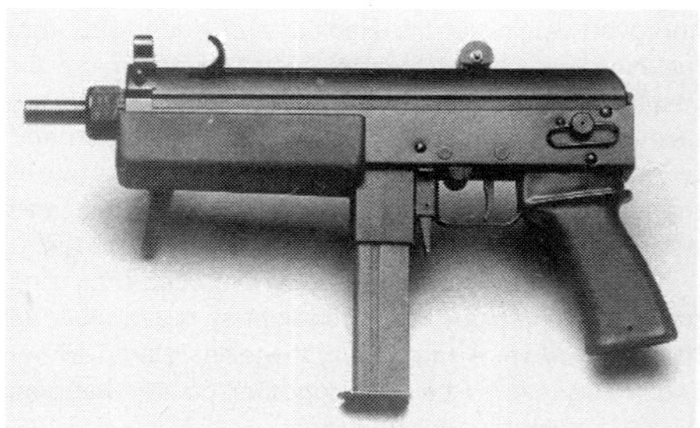

9 mm KGP-9 sub-machine gun

Service
Hungarian security forces and police; offered for export.

VERIFIED

INDIA

Sub-Machine Gun Carbine 9 mm 1A1

Description

The Sub-Machine Gun Carbine 9 mm 1A1 is virtually identical to the British 9 mm L2A3 Sterling sub-machine gun in almost every respect. However some of the data provided by Indian sources differs somewhat from the British original so that for the 9 mm 1A1 is provided below. The 9 mm 1A1 remains in production at Kanpur and is currently being offered for export sales.

Data

Cartridge: 9 × 19 mm Parabellum
Operation: blowback, selective fire
Feed: 34-round curved box magazine
Weight: empty, 2.835 kg; loaded with bayonet, 3.798 kg
Length: butt retracted, 482.6 mm; butt extended, 685.8 mm
Barrel: 198.12 mm
Rifling: 6 grooves, rh, 1 turn in 250 mm
Sights: fore, blade; rear, flip aperture
Muzzle velocity: ca 390 m/s
Rate of fire: cyclic, 550 rds/min

Manufacturer

Indian Ordnance Factories, Small Arms Factory, Kanpur.

Status

In production.

Service

In service with the Indian armed forces.

VERIFIED

ISRAEL

9 mm Uzi sub-machine gun

Description

The 9 mm Uzi is a blowback-operated sub-machine gun using advanced primer ignition in which the round is fired while the bolt is still travelling forward. This produces a reduced impulse to the bolt and as a result this component can be designed to weigh less than half the amount that would be required for a static firing breech block.

The 9 mm Uzi sub-machine gun is 445 mm long from the muzzle to the rear of the breech casing and has a very pronounced advantage in that in its short length it achieves a 260 mm barrel. This is managed by wrapping the bolt around the chamber and putting the breech face not on the front face of the bolt but 95 mm further back. Thus at the moment of firing the bolt completely surrounds the rear end of the barrel except for a cutout section on the right-hand side which allows ejection of the fired case.

The magazine is inserted into the pistol grip; 25- and 32-round vertical box magazines are available. This has the advantage of making magazine changing very easy in the dark and also giving positive support to the magazine over a greater length than is usual. The gun stops firing with the bolt to the rear. When the trigger is operated the bolt goes forward, collects a round from the top of the magazine and feeds it over the bullet guide into the chamber. The cartridge is held in the magazine at an angle with the nose slightly elevated so that it does not line up with the fixed firing pin on the breech face until the cartridge case enters the chamber.

9 mm Uzi sub-machine gun with folding stock extended

The change lever provides the three positions of Automatic (A), Single Shot (R) and Safe (S). There is a grip safety which must be fully depressed before the gun can be either cocked or fired. In addition, extra safety is provided by fitting a ratchet in the cocking handle slide to prevent accidental discharge if the hand cocking the gun should slip off the handle, after the breech block has come back far enough to pass behind a round in the magazine. After the cocking handle has come back 475 mm, it cannot go forward again until it has been withdrawn fully to the rear, a distance of 80 mm.

Data

Cartridge: 9 × 19 mm Parabellum
Operation: blowback, open bolt, selective fire
Feed: 25- and 32-round box magazines
Weight: metal stock, 3.5 kg; wood stock, 3.8 kg
Length: metal stock, retracted, 470 mm; metal stock,

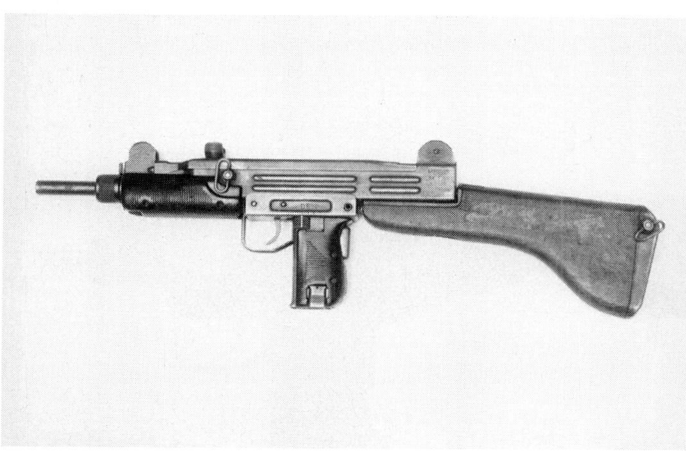

9 mm Uzi with fixed wooden stock

9 mm Uzi with stock retracted

Folding stock version of Uzi in use by German soldier of ACE Mobile Force (T J Gander)

extended, or wood, 650 mm
Barrel: 260 mm
Rifling: 4 grooves, rh, 1 turn in 254 mm
Sights: fore, post; rear, flip aperture 100 and 200 m
Sight radius: 309 mm
Muzzle velocity: 376 m/s
Rate of fire: cyclic, 600 rds/min

Manufacturer
TAAS - Israel Industries Limited, PO Box 1044, Ramat Hasharon 47100.

Status
In production.

Service
Israeli armed forces; also Algeria, Angola, Belgium, Bolivia, Central African Republic, Chad, Chile, Columbia, Dominican Republic, Ecuador, Ethiopia, Gabon, Germany, Guatemala, Haiti, Honduras, Ireland, Kenya, Liberia, Luxembourg, Netherlands, Nicaragua, Niger, Nigeria, Panama, Paraguay, Peru, Philippines, Portugal (known as MP2), Rwanda, Somalia, South Africa (known as S-1), Sudan, Surinam, Swaziland, Thailand, Togo, Uganda, USA (special forces) Uruguay, and Venezuela.

Licence production
CHINA, PEOPLE'S REPUBLIC
Manufacturer: China North Industries (NORINCO)
Type: 9 mm sub-machine gun
Remarks: Production not licenced. May no longer be in production. Was offered for export sales.
CROATIA
Agency: RH-ALAN
Type: 9 mm ERO sub-machine gun
Remarks: Production almost certainly unlicenced. See separate entry under *Croatia* for illustration.

UPDATED

9 mm Mini-Uzi sub-machine gun

Description
In 1987 Israel Military Industries (now TAAS - Israel Industries Limited) produced a new smaller version, designated the Mini-Uzi, of the 9 mm Uzi sub-machine gun described in the previous entry. In operation it exactly resembles its larger 'parent', differing only in size, weight and firing characteristics as set out below. It will accept a 20-round magazine for its 9 × 19 mm Parabellum pistol ammunition, as well as 25- and 32-round magazines.

However, there is one significant difference in operation. The Mini-Uzi does not utilise the advanced primer ignition system of operation and fires from a closed bolt, using a floating firing pin.

Since it can be easily concealed under ordinary clothing, and carried in the minimum of space in vehicles, the Mini-Uzi is particularly intended for security and law enforcement personnel and for use in commando operations. It can be fired full or semi-automatic from the hip or, with stock extended, from the shoulder, and is said to maintain the high standards of reliability and accuracy set by the Uzi. The stock lies alongside the receiver when folded.

Data
Cartridge: 9 × 19 mm Parabellum
Operation: blowback, selective fire
Feed: 20-, 25- and 32-round box magazines
Weight: empty, 2.7 kg
Length: butt retracted, 360 mm; butt extended, 600 mm
Barrel: 197 mm
Rifling: 4 grooves, rh, 1 turn in 254 mm
Sights: fore, post; rear, flip aperture 50 and 150 m
Sight radius: 235 mm
Muzzle velocity: 366 m/s
Rate of fire: cyclic, 1,500-1,900 rds/min

Manufacturer
TAAS - Israel Industries Limited, PO Box 1044, Ramat Hasharon 47100.

Status
In production.

Service
Known to be in service with Israel, Columbia, Guatemala, Haiti, Panama, USA (special forces), and Uruguay.

VERIFIED

9 mm Mini-Uzi sub-machine gun with stock extended

9 mm Mini-Uzi sub-machine gun with stock folded

9 mm Micro-Uzi sub-machine gun

Description
The Micro-Uzi sub-machine gun is the third member of the Uzi family. In effect, it is the Uzi sub-machine gun reduced to its smallest practical size, and is marginally larger than the Uzi pistol (qv). The operation is the same as that of the larger weapons, but the bolt has a tungsten insert which increases the mass and thus keeps the rate of fire down to a reasonably practical figure. The stock lies alongside the left of the receiver when folded.

The Micro-Uzi has been produced in 0.45 ACP calibre, using a 16-round magazine.

Data
Cartridge: 9 × 19 mm Parabellum
Operation: blowback, closed bolt, selective fire
Feed: 20-round box magazine
Weight: empty, 2 kg
Length: stock retracted, 282 mm; stock extended, 486 mm
Barrel length: 134 mm
Rifling: 4 grooves rh, 1 turn in 254 mm
Sights: fore, post; rear aperture
Sight radius: 180 mm
Rate of fire: cyclic, ca 1,250 rds/min
Muzzle velocity: 350 m/s

Manufacturer
TAAS - Israel Industries Limited, PO Box 1044, Ramat Hasharon 47100.

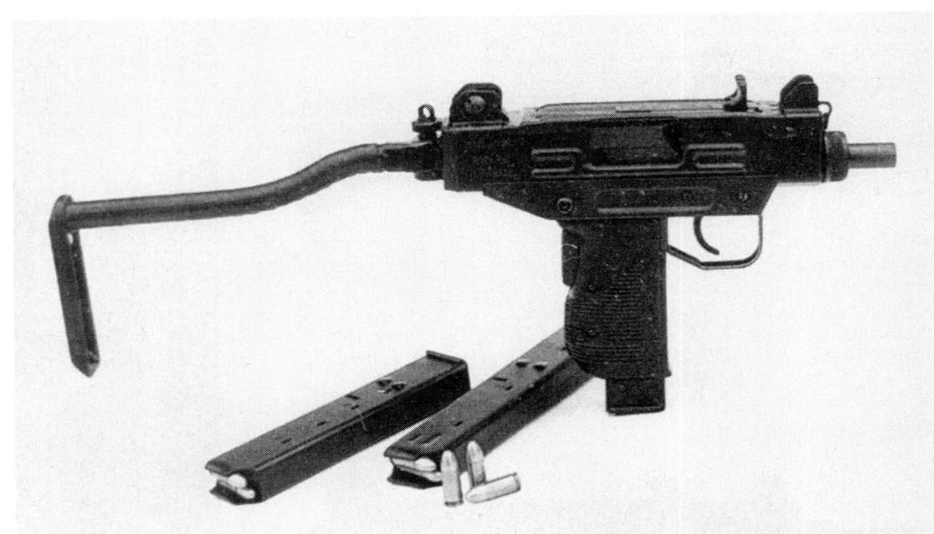

9 mm Micro-Uzi sub-machine gun with stock extended

Status
In production.

Service
Uncertain. Understood to have been sold to some police forces.

VERIFIED

ITALY

9 mm Beretta Model 12 sub-machine gun

Development

A series of Beretta sub-machine gun developments in the mid-1950s culminated in the 9 mm Model 12 which was first produced in 1958 and went into series production in 1959. It was adopted by the Italian Army and was later sold to Brazil, Gabon, Libya, Nigeria, Saudi Arabia and Venezuela.

Description

On the Beretta Model 12 sub-machine gun heavy sheet metal stampings spot-welded together form the receiver and magazine housing. There are longitudinal grooves extending the full length of the receiver to ensure efficient functioning in conditions of dust, sand or snow. The receiver, forward pistol grip, magazine housing, trigger housing and pistol grip are all one unit. The breech block is of the wraparound type and envelops the barrel before firing. The fixed firing pin is well back and, of the total barrel length of 200 mm, 150 mm lies inside the breech block at the moment of firing.

The weapon has two safety systems. There is a grip safety in the front of the pistol grip, below the trigger, which locks the bolt when released in either the forward or cocked position. It must be held in before the action can be cocked. There is also a push-button safety just above the pistol grip and this locks the grip safety until pushed to the right. The selector lever is a push-through type. The weapon normally has a metal stock which folds laterally to the right; it can be fitted with a quickly detachable wooden butt.

Data

Cartridge: 9 × 19 mm Parabellum
Operation: blowback, selective fire
Feed: 20-, 32- or 40-round detachable box magazine
Weight unloaded: folding metal stock, 3 kg; wooden stock, 3.4 kg
Length: metal stock folded, 418 mm; metal stock extended, 645 mm; wooden stock, 660 mm
Barrel: 200 mm
Rifling: 6 grooves, rh
Sights: fore, blade; rear, flip aperture 150 and 250 m
Muzzle velocity: 381 m/s
Rate of fire: cyclic, 550 rds/min

Manufacturer

Armi Beretta SpA, Via Beretta 18, I-25063 Gardone VT (Brescia).

Status

No longer manufactured in Italy.

Service

Italian Army, also Brazil, Gabon, Libya, Nigeria, Saudi Arabia and Venezuela. Manufactured under licence in Indonesia and Brazil.

VERIFIED

9 mm Model 12 sub-machine gun with stock folded

9 mm Model 12 sub-machine gun

9 mm Beretta Model 12S sub-machine gun

Description

Generally similar to the Model 12, the Model 12S is the latest in the sequence of Beretta sub-machine guns. The principal differences from the earlier model are a new design of the manual safety and fire selector and modifications to the sights. In addition the rear-cap catch has been modified, there is a new butt-plate and an epoxy resin finish, resistant to corrosion and wear, has been introduced. The manual safety and fire selector have been incorporated in a single-lever mechanism which can be operated by the right thumb without removing the hand from the grip. The three positions are marked (clockwise) 'S' (safe), 'I' (semi-automatic) and 'R' (automatic fire). When the manual safety is applied, with the lever at 'S', both the trigger mechanism and the grip safety are blocked.

The front sight of the Model 12S is adjustable for elevation and windage and both fore and rearsight supports have been strengthened.

9 mm Model 12S sub-machine gun with stock extended

The rear-cap catch has been moved to the top of the receiver, both to facilitate fastening and unfastening and to enable the user to see at a glance whether or not the catch is correctly fastened.

The butt-plate of the metal stock has been modified by the addition of a catch which gives positive locking in the open and closed position. As with the Model 12, a detachable wooden stock is available.

Data

Cartridge: 9 × 19 mm Parabellum
Operation: blowback, selective fire

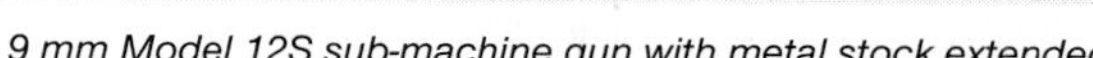

9 mm Model 12S sub-machine gun with metal stock extended

9 mm Beretta Model 12S sub-machine gun with metal stock folded

Feed: 32-round detachable box magazine with options of 20- or 40-round box magazines
Weight: folding metal stock, 3.2 kg without magazine; wooden stock, 3.6 kg without magazine
Length: metal stock folded, 418 mm; metal stock extended, 660 mm; wooden stock, 660 mm
Barrel: 200 mm
Rifling: 6 grooves, rh, 1 turn in 250 mm
Sights: fore, post, adjustable; rear, flip aperture
Muzzle velocity: 380-430 m/s
Rate of fire: cyclic, 500-550 rds/min

Manufacturer

Armi Beretta SpA, Via Beretta 18, I-25063 Gardone VT (Brescia).

Status

In production.

Service

Italian and Tunisian armed forces and some other forces.

VERIFIED

9 mm Spectre M-4 sub-machine gun

Description

This weapon was announced in 1983 and has several features which support its claim to being a revolutionary design. In appearance it is a conventional enough weapon of its type, with the receiver and barrel shroud of pressed steel and with a folding steel butt-stock which folds up and forward to lie along the top of the receiver when not in use. The first unusual feature is the patented four-column magazine which allows a capacity of 50 rounds in the length usually associated with 30 rounds. There is also a four-column 30-round magazine which is only 160 mm long.

The mechanism is blowback, but fires from a closed bolt and uses an independent hammer unit combined with a unique double-action trigger system. To fire, the magazine is inserted in the usual way and the cocking handle pulled back and released; this allows the bolt to close, chambering a cartridge. The hammer unit remains at the rear of the receiver; but depressing a decocking lever allows the hammer to move forward under control to stop a short distance behind the bolt. The weapon is now safe to carry, but when necessary to fire a pressure on the trigger will retract the hammer from its rest position and allow it to go forward, impelled by its own spring to strike the firing pin in the bolt and thus fire the chambered round. Thereafter the action is normal, the hammer being automatically released to go forward and fire each round after the bolt has closed.

A closed bolt system predisposes the weapon to heating of the barrel, so to counter this a forced draught system is employed; this is controlled by the movement of the bolt and ensures that air passes through and around the barrel during firing, thus keeping the barrel cool. It is claimed that the 'sinusoidal' rifling employed, which reduces bullet friction, also assists in keeping the barrel temperature down.

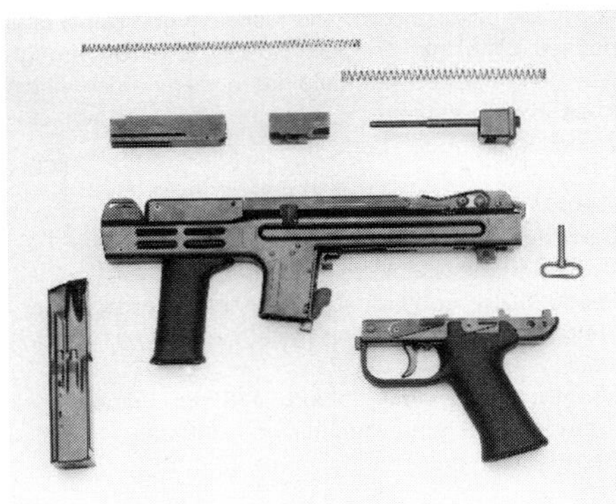

Spectre M-4 sub-machine gun field-stripped

There is no applied safety: accidental firing is impossible because of a safety lock preventing the hammer striking the firing pin unless the trigger is pressed. The double-action trigger mechanism allows the user to open fire instantly without having to perform any operation on the weapon.

It can be appreciated from this description that the makers' claim that the Spectre initiates a totally new concept in sub-machine guns, has considerable validity. The weapon is based on an absolutely instinctive and automatic usage: the user can open fire instantly, without having to consider the safety condition or perform any manual action other than pulling the trigger. This is the most important feature of the weapon and makes it ideal for unconventional warfare, counter-terrorism and similar operations.

Variants

The SITES company manufactures two variant models for the civil market, both derived from the Spectre sub-machine gun.

Spectre C carbine

This is a semi-automatic carbine; it has all the innovative features of the sub-machine gun but fires only single shots and has a barrel 220 mm long. It is only available in 9 mm calibre.

Spectre P pistol

This is a semi-automatic pistol, also with all the sub-machine gun features. As with the carbine the mechanism permits only single shots to be fired, and there is no stock and no foregrip. It accepts the standard sub-machine gun magazines. In addition to the standard 9 mm model it is also available in 0.40 S&W calibre (with 22- and 35-round four-column magazines) and in 0.45 ACP calibre (with a 32-round magazine.)

Spectre PCC (Police Compact Carbine)

The Spectre PCC was developed from the Spectre M-4 and is a semi-automatic carbine which has been specially developed for police and security use. It retains the double-action characteristics of the sub-machine gun and is of the same dimensions. It is available in both 9 mm Parabellum and 0.40 S&W calibres, and a

Firing the Spectre with butt folded

9 mm Spectre M-4 sub-machine gun with standard magazine

9 mm Spectre P pistol

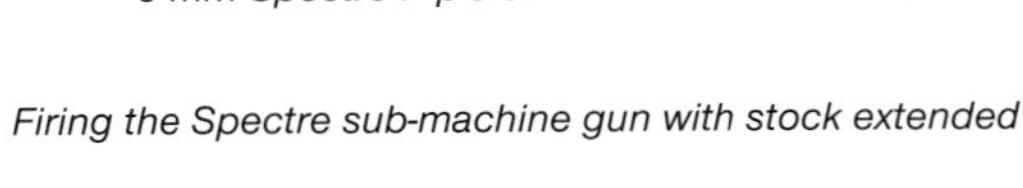

Firing the Spectre sub-machine gun with stock extended

flash suppressor and sound moderator are also available.

Data

Cartridge: 9 × 19 mm Parabellum
Operation: blowback, closed breech, selective fire, double-action trigger mechanism, decocking lever
Feed: 30- or 50-round four-row box magazines
Weight: empty, 2.9 kg
Length: butt folded, 350 mm; butt extended, 580 mm
Barrel: 130 mm
Sights: fore, post adjustable; rear, U-notch
Muzzle velocity: ca 400 m/s
Rate of fire: cyclic, 850 rds/min

Manufacturer

SITES SpA, Via Magenta 36, I-10128 Torino.

Status

In production.

Service

No sales disclosed but offered for export.

VERIFIED

---

9 mm AGM-1 semi-automatic carbine

Description

The AGM-1 is a semi-automatic weapon of bullpup design. It is configured for the 9 × 19 mm Parabellum cartridge as standard, but there are conversion kits which, by changing barrel, bolt, recoil springs and magazine, allow it to be changed to 0.45 ACP, 0.40 S&W or 0.22 LR chambering. The changeover takes about 10 minutes and can be accomplished without special tools or training.

The weapon fires from a closed bolt, and has a set trigger and a heavy target-grade barrel, giving the weapon a degree of accuracy above average for its class. The bullpup layout makes it extremely compact and well balanced.

Data

Cartridge: 9 × 19 mm Parabellum, or 0.45 ACP, 0.40 S&W, or 0.22 LR
Operation: blowback, semi-automatic
Feed: box magazines; 9 mm, 13 or 20 rounds; 0.45 ACP, 6 or 10 rounds; 0.40 S&W, 10 rounds; 0.22 LR, 15 or 20 rounds
Weight: 9 mm version, empty, 3 kg
Length: 670 mm
Barrel: 410 mm
Sights: fore, post; adjustable for elevation; rear, two position flip aperture, adjustable for windage and elevation

Manufacturer

ALGIMEC Srl, Via Monte San Genesio 31, I-20158 Milan.

Status

In production.

VERIFIED

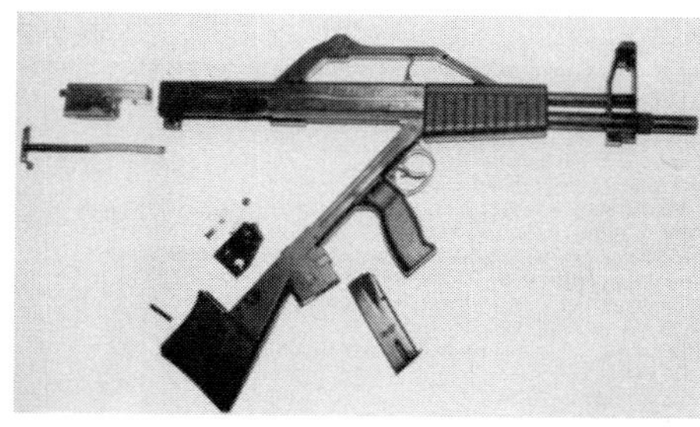

9 mm AGM-1 carbine field-stripped

9 mm AGM-1 semi-automatic carbine

KOREA, SOUTH

5.56 mm K1A sub-machine gun

Description

The 5.56 mm K1A sub-machine gun is based on the mechanism and components of the 5.56 mm K2 assault rifle (qv). It is provided with a highly efficient muzzle compensator/flash suppressor which reduces muzzle climb and minimises flash. A telescoping metal stock is fitted, and ejection is forward so as to ensure the ejected cases are well clear of the firer's face. Tritium inserts in the sights permit accurate shooting in poor light conditions.

Data

Cartridge: 5.56 × 45 mm M193
Operation: blowback, selective fire with 3-round burst
Feed: 20- or 30-round box magazine
Weight: without magazine, 2.87 kg
Length: stock retracted, 653 mm; stock extended, 838 mm
Barrel: 263 mm
Rifling: 6 grooves, rh, 1 turn in 305 mm
Sights: fore, post; rear, aperture
Sight radius: 403 mm
Muzzle velocity: 820 m/s
Rate of fire: cyclic, 700-900 rds/min
Effective range: 250 m

Daewoo 5.56 mm K1A sub-machine gun

Manufacturer

Daewoo Precision Industries Limited, PO Box 25, Kumjeong, Pusan.

Status

In production.

Service

South Korean armed forces.

VERIFIED

PERU

9 mm MGP-79A sub-machine gun

Description

The 9 mm MGP-79A sub-machine gun was designed and manufactured in Peru and is a simple and robust blowback design. The butt is hinged at the rear of the receiver and folds sideways to lie along the right side of the weapon, allowing the shoulder pad to be gripped, together with the magazine, by the non-firing hand. The applied safety and fire selector switches are separated so that they can be operated by both hands. There is a perforated barrel jacket, allowing a different grip to be taken when deliberate aim is required and the barrel and jacket can be replaced with a silencer assembly for special operations.

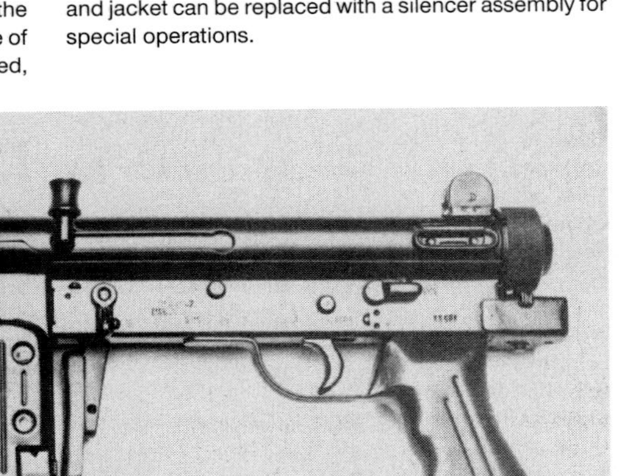

9 mm MGP-79A sub-machine gun

Variant

9 mm MGP-87 sub-machine gun
(See following entry).

Data

Cartridge: 9 × 19 mm Parabellum
Operation: blowback, selective fire
Feed: 20- or 32-round magazines
Weight: empty, 3.085 kg
Length: butt folded, 544 mm; butt extended, 809 mm
Barrel: 237 mm
Rifling: microgroove, 12 grooves, rh
Sight radius: 258.5 mm
Sights: fore, adjustable blade; rear, two-position notch, 100 m and 200 m
Muzzle velocity: 410 m/s
Rate of fire: cyclic, 600-850 rds/min

Manufacturer

SIMA-CEFAR, Av Contralmirante Mora 1102, Base Naval, Callao.

Status

No longer in production.

Service

Peruvian armed forces.

VERIFIED

9 mm MGP-87 sub-machine gun

Description

The 9 mm MGP-87 sub-machine gun is built to the same basic design as the MGP-79A (see previous entry) but is shorter by virtue of a shorter barrel and shorter folding butt. There is no barrel jacket, and the cocking handle is turned up into the vertical position and made more prominent so that it can be grasped and operated more rapidly. The design is particularly intended for use in special operations and for counter-insurgency troops where rapid action is essential.

A sound suppressor is available. This is a complete replacement unit which is installed by unscrewing the barrel retaining nut and removing the nut and barrel, then offering up the suppressor unit and screwing it, and its contained barrel, into the receiver.

9 mm MGP-87 sub-machine gun with suppressor and accessories

Data

Cartridge: 9 × 19 mm Parabellum
Operation: blowback, selective fire
Feed: 20- or 32-round box magazines
Weight: empty, 2.9 kg; with loaded 32-round magazine, 3.485 kg
Length: butt folded, 500 mm; butt extended, 766 mm
Barrel: 194 mm
Rifling: microgroove, 12 grooves, rh, 1 turn in 250 mm
Sight radius: 258.5 mm
Sights: fore, adjustable blade; rear, fixed aperture 100 m
Muzzle velocity: 362 m/s
Rate of fire: 600-850 rds/min

Manufacturer

SIMA-CEFAR, Av Contralmirante Mora 1102, Base Naval, Callao.

Status

In production.

Service

Peruvian armed forces.

VERIFIED

9 mm MGP-84 sub-machine gun

Description

The 9 mm MGP-84, at one time known as the MGP-15, is a very small sub-machine gun designed for use by special forces and security guards who require compact firepower. It is a blowback weapon, the mechanism being the same as the larger weapons described previously, but there is no exposed barrel and the magazine housing is incorporated into the pistol grip. There is a folding butt, hinged at the rear of the receiver and folding to the right side so that when folded the shoulder-piece can be utilised as a front grip. The safety and fire selector are incorporated into a single switch, located at the front end of the trigger unit in a position to be operated by the non-firing hand. The weapon is

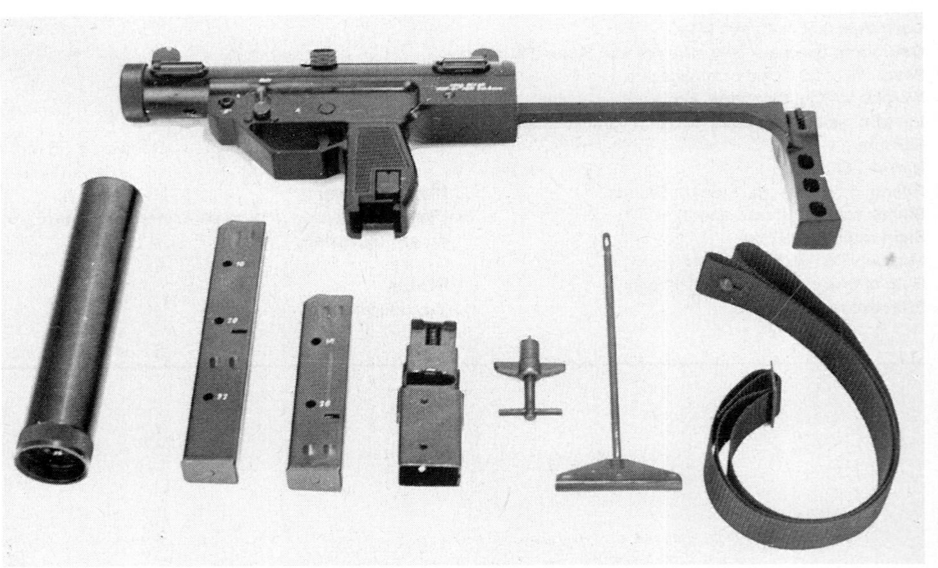

9 mm MGP-84 sub-machine gun with suppressor and accessories

well-balanced and can be fired single-handed if necessary. The bolt is of the wraparound type, allowing the maximum length of barrel, and, like the other SIMA-CEFAR designs, the magazine is compatible with the Uzi design so that Uzi magazines can be used.

A sound suppressor is available; this can be fitted to the weapon very easily by unscrewing the barrel retaining cap and screwing the suppressor on to the receiver in its place.

The 9 mm MGP-84, and its variants, are manufactured by SIMO-CEFAR (Servicios Industrials de la Marino-Centro de Fabrication de Armas), the Peruvian Navy's small-arms manufacturing organisation.

Variants

9 mm MGP-14 pistol carbine

This is the same basic weapon as the MGP-84, except that it will operate only in the semi-automatic mode, firing single shots. The dimensions are exactly the same except that, because of internal changes and perhaps a slightly heavier bolt, the empty weight is 2.45 kg. The same sound suppressor can be fitted, in the same manner.

9 mm MGP-14 Micro assault pistol

The 9 mm MGP-14 Micro assault pistol is based on the MGP-84 but lacks the folding butt. One addition is a folding foregrip and another an additional applied safety on the forward left-hand side of the weapon, acting on the firing mechanism. The MGP-14 Micro can use either 20- or 32-round magazines. Weight with a loaded 20-round magazine is 2.84 kg.

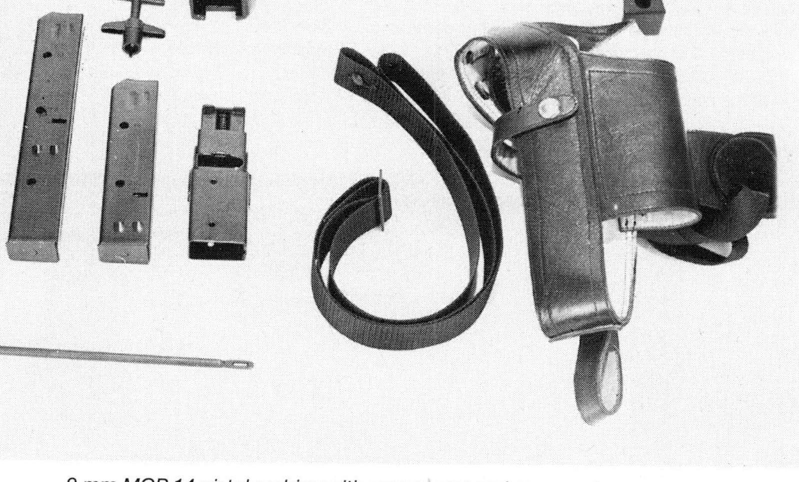

9 mm MGP-14 pistol carbine with suppressor and accessories

Data

Cartridge: 9 × 19 mm Parabellum
Operation: blowback, selective fire
Feed: 20- or 32-round box magazines
Weight: empty, 2.31 kg; with loaded 32-round magazine, 2.895 kg
Length: butt folded, 271 mm; butt extended, 490 mm
Barrel: 152 mm
Rifling: microgroove, 12 grooves, rh, 1 turn in 250 mm
Sight radius: 232.5 mm
Sights: fore, adjustable blade; rear, two-position notch, 100 m and 200 m
Muzzle velocity: 342 m/s
Rate of fire: 650-700 rds/min

Manufacturer

SIMA-CEFAR, Av Contralmirante Mora 1102, Base Naval, Callao.

Status

In production.

Service

Peruvian armed forces.

UPDATED

POLAND

GLAUBERYT PM-84 and PM-84P machine pistols

Development

The GLAUBERYT machine pistols are available in two models, the PM-84 chambered for the 9 × 18 mm Makarov cartridge and the PM-84P chambered for the 9 × 19 mm Parabellum cartridge for potential export sales. The maximum effective range for both models is stated to be 150 m.

Description

The GLAUBERYT machine pistols may be regarded as updated versions of the 9 mm Wz 63 (PM-63) machine pistol (see following entry). One of the main changes introduced on the GLAUBERYT is that the receiver no longer reciprocates during firing as all motions are now carried out by an internal bolt. It may otherwise be assumed that the blowback and trigger operations remain the same as for the earlier model. However, field stripping will differ.

The main changes on the GLAUBERYT are to the muzzle, which now lacks the 'trough' of the Wz 63, and the outline of the folding foregrip has been altered. The retracting butt-stock has a different configuration and the rearsight housing has been moved to the rear of the receiver to increase the sight radius. An ambidextrous cocking system has been introduced. The safety catch, when engaged, locks the hammer, breech and trigger.

Using either available model it is possible to fire the GLAUBERYT using only one hand. For accurate fire and when firing fully automatic it is advisable to fire from the shoulder using the stock and the foregrip. It is possible to remove the foregrip and replace it with a laser aiming device with a fixed foregrip mounted underneath.

Each GLAUBERYT is issued with three 25-round and one 15-round box magazines to be inserted through the pistol grip.

A semi-automatic version of the GLAUBERYT is available.

GLAUBERYT machine pistol

Data

Cartridge: PM-84, 9 × 18 mm Makarov; PM-84P, 9 × 19 mm Parabellum
Operation: blowback, selective fire
Feed: 25-round (long) and 15-round (short) box magazine
Weight: (empty) PM-84, 2.07 kg; PM 84P, 2.17 kg; (loaded) PM-84, 2.45 kg; PM-84P, 2.6 kg
Length: butt folded, 375 mm; butt extended, 575 mm
Barrel: 185 mm
Sights: fore, blade; rear, 75 and 150 m aperture settings
Sight radius: 280 mm
Muzzle velocity: PM-84, 330 m/s; PM-84P, 360 m/s
Muzzle energy: PM-84, 327 J; PM-84P, 360 J
Rate of fire: (cyclic) PM-84, 600 rds/min; PM-84P, 640 rds/min

Manufacturer

Zaklady Metalowe LUCZINK, Radom.

Status

In production. Offered for export sales.

Service

PM-84 understood to be in service with the Polish armed forces.

VERIFIED

9 mm Wz 63 (PM-63) machine pistol

Description

Popularly known as the RAK (Reczny Automat Komandosa), the 9 mm Wz 63 (PM-63) machine pistol combines the characteristics of the self-loading pistol and the fully automatic sub-machine gun. It has the advantage that single shots can be fired using only one hand. If fully automatic fire is needed the shoulder stock can be pulled out and the fore-end dropped to provide a steady hold, which is necessary with the 9 × 18 mm Makarov cartridge.

The magazine catch is at the heel of the pistol grip; 15 or 25 round magazines are available. Since the gun fires from the open-breech position, the slide must be pulled back to cock the action. Alternatively, the compensator can be placed against a vertical surface and the weapon pushed forward.

When the trigger is pulled the slide is released and driven forward, feeding a round from the magazine into the chamber. As soon as the cartridge is lined up with the chamber the extractor grips the cannelure and the gun fires while the slide is still going forward. The firing impulse halts the slide and drives it back against the return spring. The extractor grips the empty case until the ejector pushes it through the ejection port in the right of the slide. The slide continues to the rear and the return spring, under the barrel, is fully compressed. The slide rides over a retarder lever which snaps up and holds the slide to the rear. The retarder, an inertia pellet in the rear of the slide, continues rearward under its own momentum and compresses its own spring. When the spring is fully compressed it throws the retarder forward and this pushes the retarder lever down out of engagement with the slide and, provided the trigger is still depressed and ammunition remains in the magazine, the slide goes forward to repeat the cycle. The retarder keeps the cyclic rate of fire down to 600 rds/min from a natural frequency of about 840 rds/min.

In some ways the Wz 63 resembles the Czech Skorpion (qv) but it has a few drawbacks which make it rather less attractive. It is difficult and expensive to manufacture the complex one-piece slide and the sights move during firing, so making it almost impossible to correct the fall of shot during a burst.

The GLAUBERYT machine pistol is an updated version of the Wz 63 (see previous entry).

Data

Cartridge: 9 × 18 mm Makarov
Operation: blowback, with selective fire
Feed: 15- or 25-round box magazine
Weight: empty, 1.8 kg
Length: butt retracted, 333 mm; butt extended, 583 mm
Barrel: 152 mm
Rifling: 6 grooves, rh, 1 turn in 203 mm
Sights: fore, blade; rear, flip aperture 75 and 150 m
Sight radius: 112 mm
Muzzle velocity: 323 m/s
Rate of fire: cyclic, 600 rds/min

Manufacturer

State factories.

Status

No longer in production.

Service

Polish armed forces.

UPDATED

9 mm Wz 63 machine pistol (T J Gander)

9 mm Wz 63 machine pistol field-stripped, showing slide, barrel, return spring, frame and magazine

ONYX sub-machine guns

Development

ONYX sub-machine guns are available in two models, the Model 89 and the Model 91. Both are Polish modifications of the basic Kalashnikov AKS-74U sub-machine gun but several changes have been introduced on the ONYX to suit local requirements.

The Model 89 is chambered for the 5.45 × 39.5 mm cartridge while the Model 91 is produced for potential export sales chambered for the 5.56 × 45 mm NATO cartridge.

Description

The ONYX sub-machine gun is very similar to the basic AKS-74U but differs in having a three-round burst limiter option on the fire selector mechanism and the side-folding steel butt has a profile very similar to that provided on the MPiKMS-72, a variant of the 7.62 mm AKM assault rifle produced in the former East Germany.

Another change introduced on the ONYX sub-machine gun is a lengthened rearsight with settings for 100, 200 and 400 m. Apart from the usual aperture and post iron sights it is possible to use the extended rear sight as an attachment point for a variety of optical and other sights, including a RED-POINT collimator sight. The same attachment is also used with a more complex sight utilised when the muzzle attachment is used to fire rifle grenades to a range between 200 and 300 m. It is also possible to use the ONYX in conjunction with a laser aiming device.

All other operating details are virtually identical to those for the AKS-74U.

ONYX sub-machine gun

Data

Cartridge: Model 89, 5.45 × 39.5 mm; Model 91, 5.56 × 45 mm NATO
Operation: gas, selective fire
Feed: 30-round box magazine
Weight: empty, 2.9 kg; loaded, 3.7 kg
Length: butt folded, 519 mm; butt extended, 720 mm
Barrel length: 207 mm
Sights: fore, post; rear, 100, 200 and 400 m U-notch settings; see also text
Sight radius: 223 mm
Muzzle velocity: Model 89, 700 m/s; Model 91, 710 m/s
Muzzle energy: Model 89, 857 J; Model 91, 1,010 J
Rate of fire: cyclic, 700-750 rds/min

Manufacturer

Zaklady Metalowe LUCZINK, Radom.

Status

In production. Offered for export sales.

Service

Model 89 understood to be in service with the Polish armed forces.

VERIFIED

PORTUGAL

9 mm INDEP Lusa A2 sub-machine gun

Description

The 9 mm Lusa A2 sub-machine gun replaced the earlier Lusa A1 and is a blowback-operated automatic weapon with a detachable barrel and retractable butt-stock. It employs an overhung bolt moving in a double-cylinder configured receiver. The magazine housing doubles as a front grip and there is an extending steel rod stock which slides into the recess between the two cylinders on the receiver. Firing modes include single-shot, three-round controlled burst and fully automatic. Rounds are fed vertically from a detachable 30-round steel box magazine.

Optional attachments include a laser aiming device and a suppressor.

Data

Cartridge: 9 × 19 mm Parabellum
Operation: blowback, selective fire with 3-round burst limiter
Feed: 30-round box magazine
Weight: empty, 2.85 kg; loaded, 3.4 kg
Length: stock folded, 458 mm; stock extended, 585 mm
Barrel: 160 mm
Rifling: 6 grooves, rh, 1 turn in 250 mm
Sights: fore, protected post; rear, flip aperture
Muzzle velocity: 390 m/s
Rate of fire: cyclic, 900 rds/min

Manufacturer

INDEP SA, Fabrico de Braco de Prata, Rua Fernando Palha, P-1802 Lisboa-Codex.

Status

In production.

Service

Portuguese Army. Available for export.

UPDATED

9 mm INDEP 'Lusa A2' sub-machine gun

1996

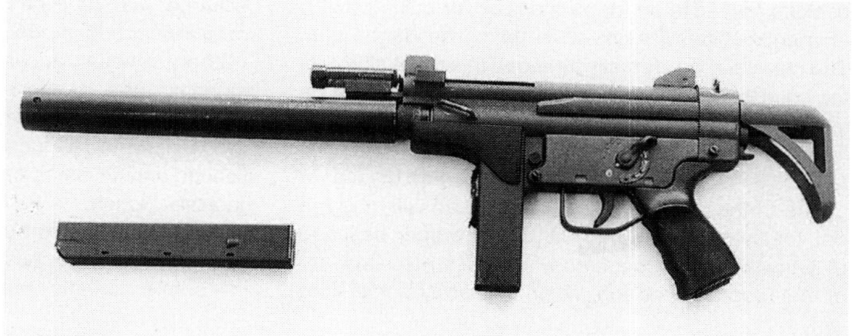

9 mm INDEP 'Lusa A2' sub-machine gun with suppressor and laser sighting device

1996

ROMANIA

9 mm sub-machine gun

Description

The designation of this 9 mm sub-machine gun is not known although it is described in a sales brochure as an 'automatic pistol'. Chambered for the 9 × 19 mm Parabellum cartridge, it is a blowback-operated weapon with a cyclic rate of fire of 1,600 rounds per minute. To render the weapon manageable at such a high rate of fire the three-position safety lever on the right-hand side of the receiver has a bottom position for three-round bursts. The central position is for semi-automatic single-shot fire and the top position for safe; there is no fully automatic fire mode. The forward part of the wraparound breech block surrounds the barrel at the instant of firing, leading sales personnel to refer to this weapon as an 'Uzi'.

The exact capacity of the vertical box magazine is not known but appears to be about 30 rounds. The cocking lever is on top of the receiver and is thus slotted to retain the sightline. Construction is pressed steel over-all secured by rivets, while the butt, which folds to the left of the receiver, is a shaped single strut thick wire rod. An adjustable shoulder sling is connected to a single swivel on the rear face of the receiver.

One unusual design feature is that the exterior of the barrel appears to be ribbed to assist cooling, leading to the introduction of a perforated shroud around the exposed barrel to protect the firer's hands.

Data

Cartridge: 9 × 19 mm Parabellum
Operation: blowback, selective fire with 3-round limiter
Feed: box magazine
Weight: empty, 1.88 kg; loaded, 2.43 kg
Length: stock folded, 370 mm; stock extended, 600 mm
Sights: fore, post; rear, notch, fixed for 100 m
Muzzle velocity: ca 370 m/s
Rate of fire: cyclic, 1,600 rds/min

Manufacturer

Romtehnica, 9-11 Drumul Taberei Street, Bucharest.

Status

In production. Offered for export sales.

NEW ENTRY

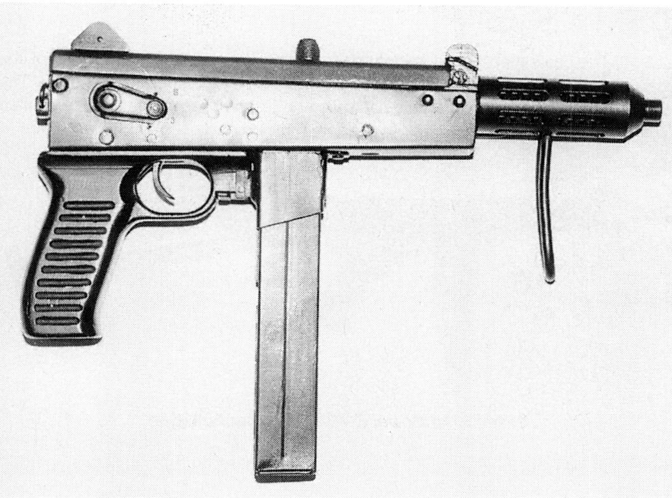

Romanian 9 mm sub-machine gun

1996

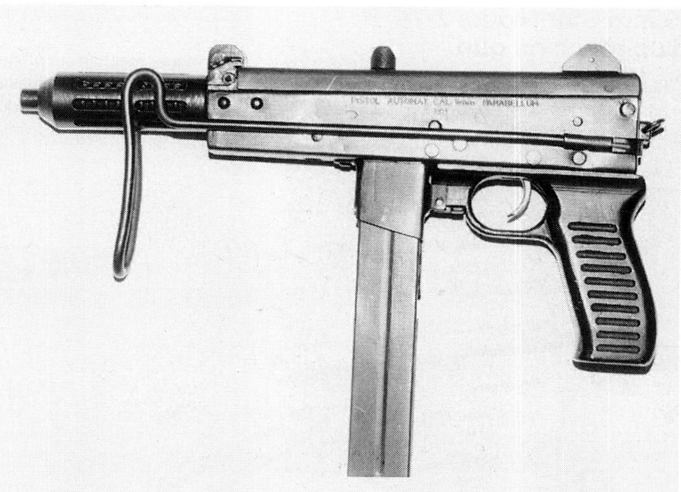

Romanian 9 mm sub-machine gun

1996

SOUTH AFRICA

9 mm BXP sub-machine gun

Description

The 9 mm BXP sub-machine gun is a conventional blowback weapon, constructed from stainless steel pressings and precision castings. It is simple to operate and fires from an open bolt. It is extremely compact and can be fired single-handed with the butt folded, like a pistol.

The bolt is of the 'telescoping' type which envelops the rear end of the barrel, thus allowing the overall length to be short whilst accommodating the maximum length of barrel. It also helps to keep the centre of gravity over the pistol grip and thus keep down the oscillations caused by the movement of the bolt during automatic fire.

With the bolt in the forward position all apertures are closed so dirt and dust cannot enter the receiver. The cocking handle is on top of the receiver but allows an unimpeded line of sight when taking deliberate aim. The barrel nut is externally threaded to accept a protector/compensator or a sound suppressor which can be used with standard or subsonic ammunition.

An ambidextrous change lever/safety catch is located on both sides of the receiver behind the trigger guard. In the horizontal position, marked with a green dot, the catch is on 'Safe', locking the trigger, sear and bolt, the latter being secured whether it is in the forward or rear (cocked) position. When the safety is moved to the red dot position the weapon is free to fire at automatic or single shot; the first pull on the trigger gives single shots, a further pull against a felt resistance, gives automatic fire. There is an extra sear recess on the bolt to prevent accidental firing should the weapon be dropped on its butt, the hand slip during cocking, or the weapon be fired with a weak cartridge.

The exterior surfaces are coated with a rust-resisting finish which also acts as a lifelong dry-film lubricant. The shoulder stock folds under the receiver when not required for aimed fire, and in this position the butt plate acts as a front grip and deflects heat from the barrel away from the hand.

A 32-round magazine fits into the pistol grip; though an optional 22-round magazine was produced at one time. The rearsight is a combination of aperture and notch; the aperture is used for 100 m range, while the notch above it is used when firing at 200 m range. An occluded eye (collimating) sight may be fitted to improve shooting in poor light conditions.

Other optional accessories include: a grenade-launching tube, by means of which various grenades such as CS anti-riot grenades, may be fired; a muzzle compensator to reduce climb during automatic fire; a silencer; and a holster, carrying bag, carrying sling and magazine pouch.

Stripping and assembly of the BXP is simple and takes only a few seconds.

A semi-automatic version is available for civilian sales. This model fires from a closed breech and is available with a short barrel.

Data

Cartridge: 9 × 19 mm Parabellum
Operation: blowback, selective fire
Feed: 32-round detachable box magazine
Weight: empty, 2.6 kg
Length: stock folded, 387 mm; stock extended, 607 mm
Barrel: 208 mm
Rifling: 6 grooves, rh, 1 turn in 245 mm
Sights: fore, adjustable cone; rear, aperture and V-notch or OEG (Occluded Eye Gunsight)
Muzzle velocity: ca 380 m/s
Rate of fire: cyclic, 800 rds/min

Enquiries to

MECHEM, PO Box 912454, Silverton 0127.

Status

Available for production.

Service

South African National Defence Force. South African police.

UPDATED

9 mm BXP sub-machine gun with Occluded Eye Gunsight

1996

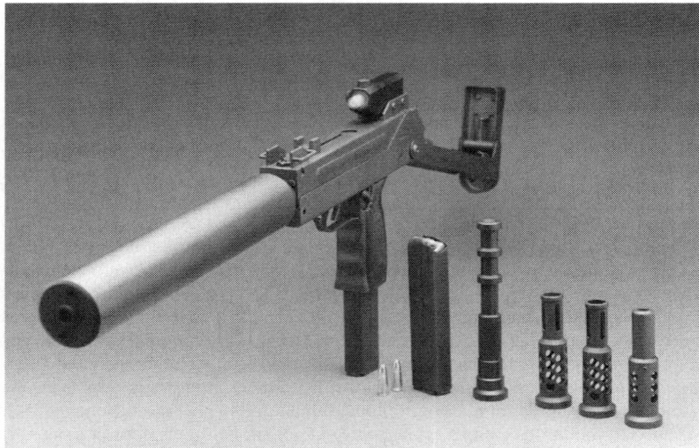

9 mm BXP sub-machine gun with silencer fitted, showing grenade-launching tube and alternative jacket/compensator units

SPAIN

9 mm Star Model Z-70/B sub-machine gun

Description

The Star 9 mm Model Z-70/B was introduced into the Spanish Army in 1971, as a result of difficulties with the trigger mechanism of the earlier Z-62. The basic mechanism of the Z-62 remains unchanged but a new trigger mechanism was introduced. Instead of firing single shot and automatic from the two halves of the trigger, a conventional system with a selector on the left of the receiver above the pistol grip has been installed. The safety catch lies under the trigger guard and is pulled back to the safe position and pushed forward to fire.

9 mm Star Model Z-70/B sub-machine gun

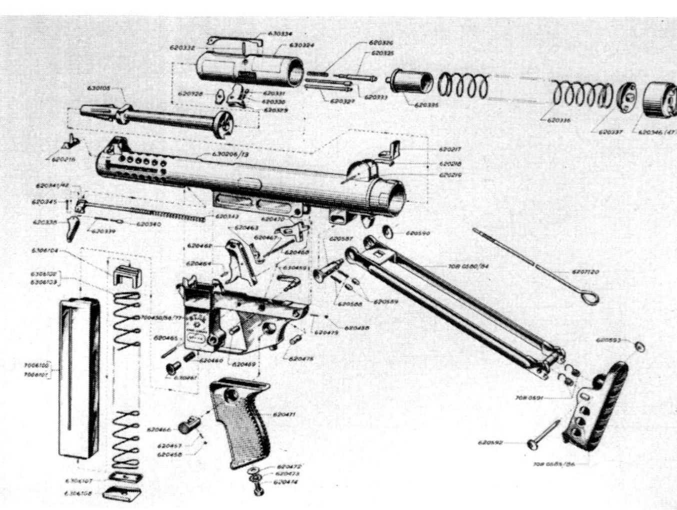

9 mm Model Z-70/B components

Data

Cartridge: 9 mm Bergmann Bayard; 9 × 19 mm Parabellum
Operation: blowback, selective fire
Feed: 20-, 30-, or 40-round box magazine
Weight: empty, 2.87 kg; loaded, 30 rounds, 3.55 kg
Length: stock folded, 480 mm; stock extended, 701 mm
Barrel: 201 mm
Muzzle velocity: 9 mm Parabellum, 380 m/s
Rate of fire: cyclic, 550 rds/min

Manufacturer

Star Bonifacio Echeverria SA, PO Box 10, E-20600 Eibar.

Status

No longer in production.

Service

Spanish armed forces and some Spanish police forces.

UPDATED

9 mm Star Model Z-84 sub-machine gun

Description

The Star 9 mm Model Z-84 sub-machine gun was developed to meet a requirement for a light, simple, reliable and effective personal weapon. Special attention was paid during design to permit manufacture with an extensive use of steel stampings and investment castings. Because of its specially designed feed system it will fire soft-point and semi-jacketed bullets equally as well as full-jacketed military ammunition. Its underwater resistance and performance capabilities make it suitable for marine and commando units as well as for special services.

The Z-84 fires from an open bolt, using the advanced primer ignition principle. The breech block is recessed so that it wraps around the barrel when closed, allowing the maximum barrel length within a compact overall length. The centre of gravity is above the pistol grip, allowing the weapon to be fired single-handed with remarkable stability. The magazine housing is located in the pistol grip, allowing magazines to be replaced, even in darkness, with the utmost speed.

There are no external moving parts thus, after the breech block has been cocked, the cocking handle returns to its forward location by spring power and remains still during firing. There is a dust cover in the cocking handle slot so the interior of the weapon is open only when an empty case is being ejected. The breech block is suspended on two rails, on four small contact points, and there is a minimum of 1 mm clearance on the other surfaces, so that any dust which finds its way into the weapon is unlikely to derange it; moreover there is ample space in the receiver for dirt to settle out of the way of any moving parts.

There is a sliding fire selector on the left side of the receiver which pushed forward, exposing one white dot, gives single shots; pushed rearwards, exposing two white dots, gives automatic fire. The safety catch is a lateral sliding button inside the trigger guard. In addition to the manual safety, the bolt is provided with three bents which allow it to be securely held in any position. Should the hand slip while cocking, for example, the central safety bent will arrest the breech block before it can fire a round. There is also an automatic inertia safety unit which locks the breech block when it is in the closed position, so that a sudden shock will not cause the block to move. This mechanism is overridden by the cocking handle and is automatically put out of action when the weapon is being fired.

Data

Cartridge: 9 × 19 mm Parabellum
Operation: blowback, selective fire
Feed: 25- or 30-round box magazine
Weight: without magazine, 3 kg
Length: (with standard barrel) stock folded, 410 mm; stock extended, 615 mm
Barrel: standard, 215 mm; optional, 270 mm
Sights: fore, post, adjustable for elevation; rear, flip aperture for 100 m or 200 m
Sight radius: 325 mm
Muzzle velocity: 400 m/s
Rate of fire: cyclic, 600 rds/min

Manufacturer

Star Bonifacio Echeverria SA, PO Box 10, E-20600 Eibar.

Status

In production.

Service

Spanish armed forces (special units) and security forces. Military and security services in various countries.

VERIFIED

9 mm Star Z-84 sub-machine gun

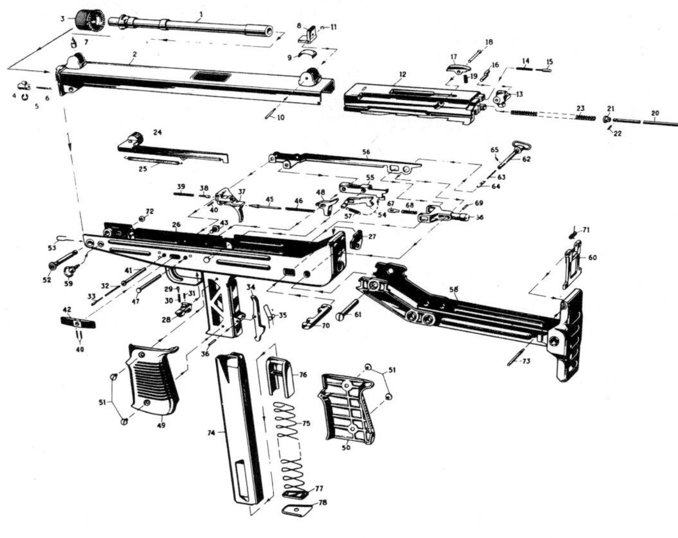

Component parts of Star Z-84 sub-machine gun

9 mm C2 sub-machine gun

Description

Originally developed by CETME, the 9 mm C2 sub-machine gun is a compact and handy weapon which can be used with either the 9 mm Largo or 9 mm Parabellum cartridge. Particular attention has been paid to the balance of the weapon which can be controlled with one hand.

Blowback-operated, the C2 differs from many weapons of similar size in not having a fixed firing pin, and incorporates a bolt-holding device as one of its three safeties. The bolt can also be locked forward for transport; and the third safety is a position on the selector lever.

Other characteristics of the weapon include a horizontal magazine holding 32 rounds in a double stack; a grooved bolt surface to reduce susceptibility to jamming by dust; a folding metal stock and a generally robust but economical construction.

9 mm C2 sub-machine gun

Data
Cartridge: 9 × 19 mm Parabellum; 9 mm Largo
Operation: blowback, selective fire
Feed: 32-round detachable magazine
Weight: empty, 2.87 kg
Length: stock retracted, 500 mm; stock extended, 720 mm
Barrel: 212 mm
Rifling: 4 grooves, rh
Sight radius: 400 mm
Muzzle velocity: 9 mm Parabellum, 325 m/s; 9 mm Largo, 340 m/s
Rate of fire: cyclic, 600 rds/min

Manufacturer
SANTA BARBARA SA, Julian Camarillo 32, E-28037 Madrid.

Status
Available.

VERIFIED

SWEDEN

9 mm m/45 sub-machine gun

Description
Generally known as the Carl Gustaf, the 9 mm m/45 sub-machine gun (*Kulsprutepistol*) has been largely relegated to reserve status. Originally the gun used the Suomi 50-round box magazine from the Model 37-39 but in 1948 an excellent two-column magazine holding 36 rounds was introduced. This was a wedge-shaped magazine which has been distinguished by its reliability and has been widely copied and used, or a similiar type developed, in the former Czechoslovakia, other Scandinavian countries and in Germany.

The change of magazine led to the introduction of a detachable housing to allow both types of magazine to be used, the gun incorporating this feature being known as the m/45. The m/45B has an improved backplate lock, smaller holes in the barrel jacket, uses only the 36-round magazine and has a baked green enamel finish. The m/45C is as the B model but with a bayonet lug, while the m/45E has the option of selective fire.

Derivatives of the m/45 included a silenced version, used by US special forces in South-east Asia, and the Model 49 Hovea used by the Danish Army and resulting from further development of the m/45 by Husqvarna. The m/45 was also copied and manufactured in Indonesia.

Accessories for the m/45 include a magazine filler, and a catcher device which clips over the ejection port to collect spent cases during training.

9 mm m/45 sub-machine gun

Data
Cartridge: 9 mm m/39B
Operation: blowback, automatic
Feed: 36-round box magazine
Weight: empty, 3.9 kg; loaded, 4.5 kg
Length: stock folded, 552 mm; stock extended, 808 mm
Barrel: 213 mm
Rifling: 6 grooves, rh
Sights: fore, post; rear, flip
Muzzle velocity: 410 m/s
Rate of fire: cyclic, 600 rds/min

Manufacturer
Bofors Carl Gustaf, Eskilstuna.

Status
Not in production in Sweden. Licence production in Egypt.

Service
Swedish forces (reserve). Also sold to Egypt, Indonesia and Ireland.

Licence production
EGYPT
Manufacturer: Maadi Company for Engineering Industries, Maadi, Cairo
Type: 9 mm Port Said
Remarks: See separate entry under *Egypt*

UPDATED

TURKEY

9 mm MKEK MP5A3 and MP5K sub-machine guns

Description
The MKEK MP5A3 sub-machine gun is, as the name implies, the Heckler and Koch MP5A3 manufactured under licence by Makina ve Kimya Endüstrisi Kurumu (MKEK). The same organisation also licence produces the 9 mm MP5K. Both MKEK weapons are identical to the German-made models, to which reference should be made for operating information and data.

Manufacturer
Makina ve Kimya Endüstrisi Kurumu (MKEK), 06330 Tandogan, Ankara.

Status
In production.

Service
Turkish forces; offered for export.

UPDATED

9 mm MP5A3 (top) and MP5K sub-machine guns as licence produced by MKEK

1996

UNITED KINGDOM

9 mm L2A3 Sterling sub-machine gun

Development

The L2A3 sub-machine gun is the military version of the Mark 4 Sterling sub-machine gun manufactured by the Sterling Armament Company at Dagenham. The gun was named the Sterling Mark 2 in 1953 and adopted by the British Army as the L2A1. The L2A2 Mark 3 Sterling, which featured some small modifications, came into service in 1955 and the L2A3 in 1956. The L2A3 is no longer in service with the UK armed forces but remains in production in India (see separate entry); and was also licence produced in Canada (again, see separate entry).

The weapon was also available with single shot (open bolt) capability only and in this version was known as the Sterling Police Carbine Mark 4. Also available was a single shot, closed bolt, floating firing pin version known as the Sterling CBS Mark 8.

Description

The weapon consists of a tubular, perforated receiver, a cylindrical bolt and the stock.

The magazine fits horizontally into the left-hand side of the receiver. It holds 34 9 mm Parabellum cartridges. The magazines of the Australian F1 and Canadian C1 will fit the L2A3. The butt-stock may be extended and the weapon used from the shoulder. Alternatively it may be folded under the barrel and the gun fired from the hip. Fixed butts for both the Mark 4 and Mark 5 (see entry for 9 mm L34A1 sub-machine gun) were introduced in 1981. The combination of a variable-length butt and a high cheek comb increases the efficiency of these weapons. The selector lever is directly above the trigger and the three positions are '34', '1' and 'SAFE'.

When the weapon is in the 'ready to fire' position, the bolt is held to the rear by the sear and the ammunition is in the magazine. When the trigger is pulled, the bolt goes forward under the influence of the return spring and the first round in the magazine is fed forward into the chamber. Since the magazine is on the side of the gun the round is angled inwards towards the chamber and the cap is not aligned with the fixed firing pin on the breech face. As the round enters the chamber it becomes lined up with the axis of the bore and as the extractor grips into the groove of the cartridge, the firing pin touches the cap of the cartridge case. The friction force between the cartridge and the chamber wall slows the cartridge and the fixed firing pin fires the cartridge while the bolt is still moving forward. The rapid development of the pressure quickly brings the bolt to a standstill and then reverses its direction and blows it back against the return spring. The weight and forward velocity of the bolt at the moment of cap initiation ensure that the rate of return of the bolt is not enough to allow the residual pressure to blow out the case. The cartridge case is forced back by the gas pressure and drives the bolt back. The extractor holds the case to the bolt face until the cartridge hits the fixed ejector which throws it out of the ejection port to the right. The compressed return spring drives the bolt forward and if the selector is at '34' the automatic cycle continues; if the selector is at 'single shot', the bolt is held to the rear.

The trigger mechanism operates as follows. The sear is fitted into a spring-loaded sear carrier. The disconnector is ¬ shaped and is pivoted at the junction of the two arms to the sear carrier. The nose of the disconnector is spring-loaded to bear against the step on the rear of the sear and the lower arm reaches down towards the spindle of the selector lever. The selector lever spindle has an inner arm which limits the movement of the lower end of the disconnector. The trigger is pivoted at the top and has a shoulder fitting under the sear carrier, which lifts the front of the carrier when the trigger is pulled.

When the selector is set for automatic, the selector spindle is rotated forward so that the inner arm is not in contact with the lower end of the disconnector. When the trigger is pulled the sear carrier is rotated up at the front and the nose of the sear is lowered off the bent of the bolt which goes forward. The bolt reciprocation continues as long as the trigger is pulled and ammunition remains in the magazine.

When the selector is set to '1' the inner arm is rotated upwards toward the lower end of the disconnector. When the trigger is pulled the sear comes down as before and the bolt is released. The lower arm of the ¬ shaped disconnector hits the inner arm of the selector lever and rotates clockwise. This causes the nose to slip off the spring-loaded sear which promptly rises to hold up the bolt. When the trigger is released the sear carrier rises and the top arm of the disconnector springs back over the shoulder of the sear.

When the selector is set to 'SAFE' the inner arm moves rearward under the lower arm of the disconnector. The disconnector is firmly held because it is pivoted on the sear carrier that also is held. Since the sear carrier does not descend the sear holds the sear notch of the bolt or, if the bolt is forward, the safety notch at the rear of the bolt.

Variant

9 mm L34A1 Sterling sub-machine gun
See following entry.

Data

Cartridge: 9 × 19 mm Parabellum
Operation: blowback, selective fire
Feed: 34-round curved box magazine; 10- and 15-round magazines and twin stacked 10 × 2, 15 × 2 and 34 × 2 magazines are also available
Weight: empty, 2.72 kg
Length: butt retracted, 483 mm; butt extended, 690 mm
Barrel: 198 mm
Rifling: 6 grooves, rh, 1 turn in 250 mm
Sights: fore, blade; rear, flip aperture
Sight radius: 410 mm
Muzzle velocity: 390 m/s
Rate of fire: cyclic, 550 rds/min

Manufacturer

The L2A3 was made by the Sterling Armament Company at Dagenham. In addition the L2A3 was manufactured at the Royal Ordnance Factory at Fazackerley. The Mark 4 Sterling and the Patchett/Sterling Mark 5 were manufactured at Dagenham. All rights in Sterling products and designs were purchased by Royal Ordnance plc in 1988.

Status

No longer in production in the United Kingdom. Still produced in India (qv).

Service

The L2A3 is no longer in service with the British armed forces. The Mark 4 Sterling was sold abroad on a considerable scale, some 90 countries having purchased the gun in varying quantities. The principal purchasers were Canada (licence production), Ghana, India (which still manufactures under licence), Libya, Malaysia, Nigeria and Tunisia. Many are also in service in the Arabian Gulf States.

Licence production

CANADA
Manufacturer: Canadian Arsenals Limited
Type: 9 mm C1 sub-machine gun
Remarks: No longer in production. See separate entry under *Canada*.
INDIA
Manufacturer: Indian Ordnance Factories, Small Arms Factory, Kanpur.
Type: Sub Machine Gun Carbine 9 mm 1A1
Remarks: See separate entry under *India*.

9 mm Sterling L2A3 sub-machine gun

VERIFIED

9 mm L34A1 Sterling sub-machine gun

Description

The L34A1 is the silenced version of the L2A3 sub-machine gun; the commercial version was the Patchett/Sterling Mark 5 sub-machine gun. It was also available with single shot (repetition) capability only; this weapon was known as the Sterling Police Carbine Mark 5.

The bore of the barrel has 72 radial holes drilled through it, allowing some of the propellant gas to escape and thus reduce the muzzle velocity of the bullet.

The gases which pass through the radial holes in the barrel enter a diffuser tube. This has a series of holes bored through it, and the gases pass into an expanded metal wrap. They emerge from this and are contained in the barrel casing. Eventually they seep forward through the barrel-supporting plate or return to the barrel. The silencer casing is extended forward of the barrel and contains a spiral diffuser. The bullet passes down the centre of this but the gases following behind are given a swirling action by the diffuser. The gases which follow closely behind the bullet are deflected back by the end cap and mingle with the gases coming forward through the diffuser. The result of this action is to ensure that the gas velocity on leaving the weapon is low.

9 mm Patchett/Sterling Mark 5 silenced sub-machine gun

As the effective pressure is reduced it is necessary to have a lightened bolt and only a single return spring to enable the blowback action to function efficiently.

The Sterling L34A1 uses standard 9 × 19 mm Parabellum ammunition, unlike other silent weapons which require special subsonic ammunition.

Data

Cartridge: 9 × 19 mm Parabellum
Operation: blowback, selective fire
Feed: 34-round curved box magazine
Weight: empty, 3.6 kg
Length: butt retracted, 660 mm; butt extended, 864 mm
Barrel: 198 mm
Rifling: 6 grooves, rh, 1 turn in 250 mm
Sights: fore, blade; rear, flip aperture
Sight radius: 502 mm
Muzzle velocity: 293-310 m/s
Rate of fire: cyclic, 515-565 rds/min

Manufacturer
Sterling Armament Company, Dagenham.
All rights in Sterling products and designs were purchased by Royal Ordnance plc in 1988.

Status
No longer in production in United Kingdom.

Service
Known to be in service with Argentina.

VERIFIED

---

9 mm Bushman IDW

Description
The Bushman IDW (Individual Defence Weapon) is a very compact sub-machine gun which incorporates a patented rate regulator. This is an electrically actuated cam which, rotating, catches the bolt and then releases it according to the speed at which it revolves. Driven by a lithium battery, the mechanism is factory adjusted to a standard rate of 450 rds/min, though this can be altered to suit the user's specification to any rate between 1 and 1,400 rds/min, the latter being the natural cadence of the weapon.

The selected rate of 450 rds/min has been chosen as being the optimum rate which balances the various forces in the weapon to give absolutely steady operation; at this rate of fire there is no muzzle rise whatever, the gun merely oscillating gently about its centre of gravity and firing on the forward oscillation so that every shot is delivered from approximately the same muzzle location. As a result the weapon fires close groups whether held in one hand or two. Should the rate regulator fail for any reason, the gun reverts to its natural rate of 1,400 rds/min, but even at this rate it is still quite controllable as there is minimal climb, so it can still be fired single-handed with quite acceptable accuracy.

A further result of using the rate regulator is that the rate of fire, once set, is entirely unaffected by ammunition variations, and the gun will fire service ball, plastic practice or frangible ammunition with equal facility and at the same rate.

The Bushman IDW consists of an alloy frame and machined stainless steel receiver and barrel. The light bolt moves inside the receiver, and there is a cocking handle mounted at the rear of the receiver which can be gripped and pulled back, a similar action to cocking a common automatic pistol, after which the cocking handle is pushed back in to the receiver and remains stationary during firing. The box magazine is inserted into a housing in front of the pistol grip. The weapon fires from the open bolt, but the bolt is so light that there is no discernible shift of balance when the trigger is pressed, and single shots can be fired with accuracy comparable to a single-action pistol.

There are four independent safety devices. A thumb lever acts as a manual safety catch and fire selector, with positions for 'safe', 'fire' and 'auto'. With the selector on 'auto' the trigger becomes a double-pull type, the first 7 mm pull giving single shots, and a further 7 mm pull, against extra pressure, giving automatic fire. When set to 'safe' the lever directly locks the sear.

The grip safety positively locks the sear and totally inhibits the weapon from firing unless it is firmly pushed in by the act of gripping the pistol grip. It also acts as an indicator of the state of the weapon: when the weapon is cocked, the grip safety protrudes, when the weapon is uncocked the grip safety lies flush with the rear surface of the grip. The direction of action of the grip safety and the manual safety/selector switch are at 90° to each other, thus if the weapon is dropped, any impact which might tend to force one safety out will force the other into deeper engagement.

The magazine safety relies upon the position of the magazine in its housing; when inserted to a first stop the top round in the magazine is not in the feedway, and if the trigger is pressed and the bolt goes forward, it will not load a round. By giving the magazine a tap to seat it into the second, higher stop the rounds are brought into the feedway and the gun will load and function normally. Operation of the magazine does not need a violent blow and can, with the short magazines, be conveniently performed with the little finger of the forward hand which is gripping the weapon. The weapon can thus be carried cocked, with a loaded magazine inserted to the first stop, in perfect safety and be ready for instant operation with the minimum of manipulation.

The bolt lock safety locks the bolt until the trigger is pulled and totally prevents the weapon from self-feeding and firing should it be dropped directly on to the end of the receiver. Even if the grip safety, manual safety and sear should all fail at once, the bolt lock will still prevent the weapon firing.

The sights are set into an 'Instafit' rib which can be inserted into and removed from the top of the receiver very easily. With the stock sight rib removed, alternative ribs carrying iron sights with tritium night markers, laser spot markers, telescopes, or collimating sights can be fitted. Each alternative can be zeroed to the weapon and thereafter requires only about 20 seconds work, without special tools, to change.

Disassembly for cleaning and routine maintenance is simple and requires no tools. The mechanism can be exposed by removing cover plates and the mechanism can be operated and the gun fired with these cover plates removed. The weapon is entirely made from milled steels and alloys, precision machined; no plastics or stampings are used. The production weapons are currently in 9 mm Parabellum calibre, but by simply changing the barrel it becomes possible to fire 0.41 Action Express ammunition. A version in 10 mm calibre is under development.

Data
Cartridge: 9 × 19 mm Parabellum; 0.41 AE and 10 mm optional
Operation: blowback
Feed: 20-, 28- or 32-round box
Weight: empty, 2.92 kg
Length: with standard barrel, 276 mm
Barrel: standard, 82.5 mm; optional, 152 mm and 254 mm
Rifling: 6 grooves, rh, 1 turn in 406 mm
Sights: fore, blade; rear, V-notch, fully adjustable
Sight radius: 171.5 mm
Muzzle velocity: 352 m/s, varying with ammunition
Rate of fire: regulated, 450 rds/min; unregulated, 1,400 rds/min
Effective range: 150 m

Manufacturer
Bushman Limited, 10 Park Industrial Estate, Frogmore, St Albans, Hertfordshire AL2 2DR.

Status
Available.

Service
Undergoing military evaluation.

VERIFIED

9 mm Bushman IDW showing one-hand use

9 mm Bushman Individual Defence Weapon

7.62 mm MC51 sub-machine gun

Description

The 7.62 mm MC51 sub-machine gun is probably the only weapon in its class to fire the 7.62 × 51 mm NATO rifle cartridge. It is based upon Heckler and Koch practice, is slightly longer than the Heckler and Koch MP5A3 and weighs about 1.2 kg more than that weapon. The combination of short weapon and powerful cartridge produces ballistics sufficient to defeat any type of body armour, and even 13 mm steel plate can be penetrated using AP bullets. The recoil mechanism of the weapon has been considerably modified so that the weapon is remarkably controllable, even when fired in the full automatic mode.

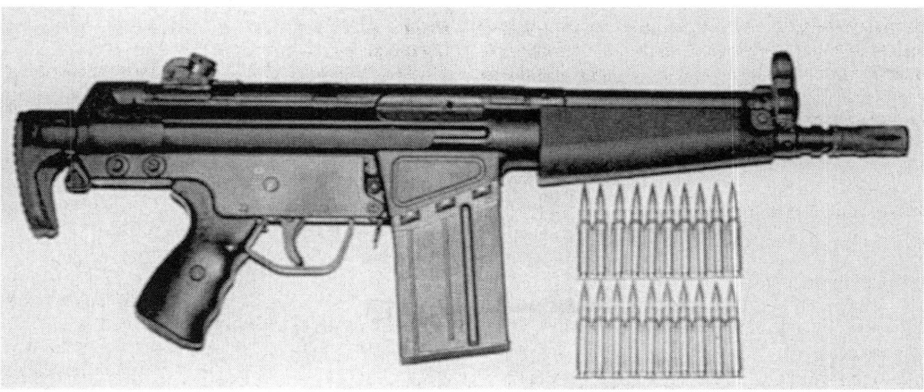

7.62 mm F R Ordnance MC51 sub-machine gun

Data

Cartridge: 7.62 × 51 mm NATO
Operation: delayed blowback, selective fire
Feed: 20-round box magazine
Weight: 4.4 kg
Length: stock extended, 625 mm
Barrel: 230 mm
Muzzle velocity: 690 m/s
Effective range: >300 m

Manufacturer

F R Ordnance International Limited, Frog's Hall, Hertfordshire CM22 6PE.

VERIFIED

UNITED STATES OF AMERICA

0.45 M3 sub-machine gun

Development/Description

When the USA entered the Second World War it had only one sub-machine gun in production, the Thompson, which was heavy, expensive to produce and demanding in labour and machine tool resources.

In October 1942 work was authorised on a new sub-machine gun. The requirement called for an all-metal weapon, easily disassembled and easily converted to accept the 9 × 19 mm Parabellum as well as the 0.45 ACP cartridge. The cost and performance were to be at least comparable to the British Sten gun. In December 1942 it was adopted as the Sub-machine Gun 0.45 calibre M3.

The M3 was manufactured by the Guide Lamp Division of General Motors Corporation at Anderson, Indiana. The bolts were made by the Buffalo Arms Company. Many difficulties were met in construction and the schedule fell back but during 1944 production was running at about 8,000 a week.

A silencer was made for the M3 but only about 1,000 silenced guns were made, all for the OSS. A flash hider was also developed and could be added to the weapon.

Over 600,000 M3s were made, but it was reported in early 1944 that the bolt-retracting handle gave trouble and the opportunity was taken to improve the design still further to reduce manufacturing effort.

The modified gun became the M3A1 and the chief difference lay in the fact that the bolt was pulled back by inserting the right forefinger into a hole in the bolt. To allow for this the ejection port and its spring-loaded cover were both made larger. A flash suppressor was fitted to some of the production run. There was a further order for M3A1s during the Korean War, but it was never completed and the total number made did not exceed 50,000.

Thus there was a grand total of roughly 650,000 of both types actually made in the USA, and although copies have appeared in the Far East from time to time there cannot have been many of these. Of the 650,000, 25,000 were produced in 9 mm with a Sten magazine and a replacement barrel. These were all supplied to various underground movements and it is possible that there are still some in use somewhere.

The M3 was unusual in having a very slow rate of fire which allowed the firer to control the movement of the gun while firing bursts and it was also sufficiently slow to enable a single shot to be snatched off without too much disturbance of the aim. Despite its advantages the M3 was never popular with the US Army so in 1957 it was declared Substitute Standard and relegated to the reserve. It was little used after 1960, many being passed to some friendly countries under offshore aid programmes. Portugal bought components, and perhaps complete guns, to make the M/948 FMBP.

A full description and history appears in *Jane's Infantry Weapons 1978*, page 97.

Data

DATA FOR M3, WHERE M3A1 DIFFERS, SHOWN IN PARENTHESIS

Cartridge: 0.45 ACP; also 9 × 19 mm Parabellum
Operation: blowback, automatic
Feed: 30-round box magazine
Weight: 3.63 kg; (3.47 kg)
Length: stock retracted, 579 mm; stock extended, 757 mm
Barrel: 203 mm
Rifling: 4 grooves, rh
Sights: fore, fixed blade; rear, notch
Muzzle velocity: 280 m/s
Rate of fire: cyclic, 450 rds/min

Manufacturers

Guide Lamp Division, General Motors Corporation, Anderson, Indiana. Ithaca Gun Company, Ithaca, New York. Buffalo Arms Company, Buffalo (bolts only).

Status

Obsolescent. No longer in production.

Service

Still in limited use by armoured vehicle crews in US forces; may also be encountered in South America and South-east Asia.

VERIFIED

0.45 M3 sub-machine gun showing cocking handle

0.45 M3A1 sub-machine gun on which the cocking handle was removed and a hole in the bolt took the firer's finger; the flash hider was an optional screw-on feature (T J Gander)

9 mm Colt sub-machine guns

Development

The Colt 9 mm sub-machine gun is a light and compact blowback weapon which embodies the same straight line construction and design as the Colt 5.56 mm M16A1 rifle. This straight line construction, coupled with the lower recoil impulse of the 9 × 19 mm Parabellum cartridge, provides highly accurate fire with reduced muzzle climb.

The US Drug Enforcement Agency (DEA) adopted the weapon, after conducting extensive tests which included other 9 mm sub-machine guns. Other US armed forces and law enforcement agencies also use the weapon and it has been evaluated by various governmental agencies throughout the world.

Description

As with the M16 rifle, the Colt sub-machine gun fires from a closed bolt, with the bolt remaining open after the last round has been fired. This feature allows the user to replace magazines and reopen fire more rapidly. It is equipped with a rigid sliding butt-stock and is readily field stripped without the need for special tools. Operation and training is similar to M16 series rifles and

9 mm Colt sub-machine gun, right side; note spent case deflector behind ejection port

9 mm Colt sub-machine gun, left side

carbines, thus eliminating the need for extensive cross-training.

The Colt 9 mm sub-machine gun fires all standard 9 × 19 mm Parabellum ammunition, fed from 20- or 32-round box magazines.

Variants

The Colt Model 635 has safe, semi-automatic and full automatic fire modes.

The Colt 634 is essentially the same as the Model 635 but is capable of semi-automatic fire only.

The Colt Model 639 is again similar to the Model 635 but has safe, semi-automatic and three-round burst fire modes.

There were two further variants of the Colt 9 mm sub-machine gun family. The Model 633HB has a barrel only 178 mm long, compared with the 260 mm barrel of the Model 635, and is provided with a hydraulic buffer to reduce action forces during firing. The basic Model 633 has a mechanical buffer.

Data

COLT MODEL 635

Cartridge: 9 × 19 mm Parabellum
Operation: blowback, selective fire
Feed: 20- or 32-round box magazine
Weight: without magazine, 2.59 kg
Length: stock retracted, 650 mm; stock extended, 730 mm
Barrel: 267 mm
Sights: fore, post adjustable for elevation; rear, flip aperture (50 m and 50-100 m), adjustable for windage and elevation
Muzzle velocity: ca 396 m/s
Muzzle energy: 584 J
Rate of fire: cyclic, 800-1,000 rds/min

Manufacturer

Colt's Manufacturing Company Inc, PO Box 1868, Hartford, Connecticut 06144-1868.

Status

Available.

Service

US Marine Corps, the US Drug Enforcement Agency (DEA), and other countries.

UPDATED

Ingram sub-machine gun

Description

Gordon B Ingram designed a series of sub-machine guns after he returned to the USA following the Second World War. All his weapons have been simple designs firing automatic only.

The Models 10 and 11 are of the same basic design and differ in the weight and length dictated by the cartridge used. The Model 10 is chambered for the 0.45 ACP or the 9 × 19 mm Parabellum. The Model 11 takes the 9 mm Short (0.380 ACP) cartridge. The weapons are very short, very compact, solidly built and made from steel pressings. The bolts are of the wraparound type with the bolt face and fixed firing pin placed well back to allow the greater part of the bolt to envelop the breech. With the bolt forward all openings are closed and no dirt can get into the action.

Both Models 10 and 11 are externally threaded at the muzzle to take the MAC suppressor. The suppressor differs from the conventional silencer in that the bullet is allowed to reach its full velocity and therefore becomes supersonic. The suppressor is added to the muzzle and is intended to reduce the emergent gas velocity to a subsonic level. The suppressor consists of helical channels going forward from the muzzle of the gun which meet similar channels coming back from the front of the screwed-on tube. The meeting of the two gas streams results in a dissipation of their energy inside the suppressor. The suppressor tube is covered with Nomex-A heat-resistant material.

The cocking handle is on top of the receiver and, like the early Thompson sub-machine guns, has a U-notch to allow an unimpeded line of sight. When the bolt is in the closed position, rotating the cocking handle through 90° locks it. The firer is, of course, warned that his bolt is locked because he can no longer use his sights. There is a second safety catch on the right of the trigger guard, forward of the trigger. When it is pulled to the rear the bolt is locked in either the forward or cocked positions.

The shoulder stock pulls out for firing from the shoulder and pushes in when the gun is to be fired from the hip.

Accessories: long barrel with special receiver to increase the effective range and improve the power of the bullet. Full carrying equipment in leather.

Manufacturers

Various.

Status

No longer in production as the Ingram. See entry following.

Service

Bolivia, Colombia, Greece, Guatemala, Honduras, Israel, Portugal, US Navy and Venezuela.

VERIFIED

Ingram Model 10 sub-machine gun

Data

	Model 10	Model 10	Model 11
Cartridge	0.45 ACP	9 mm Parabellum	9 mm Short (0.380 ACP) or 9 mm Parabellum
Operation	blowback	blowback	blowback
Feed, box magazine	30 rounds	32 rounds	16 or 32 rounds
Firing mode	selective	selective	selective
WEIGHTS			
Gun			
empty	2.84 kg	2.84 kg	1.59 kg
loaded, 16-round magazine	–	–	1.872 kg
loaded, 30-round magazine	3.818 kg	–	–
loaded, 32-round magazine	–	3.46 kg	2.1 kg
Suppressor	0.545 kg	0.545 kg	0.455 kg
LENGTHS			
Gun			
no stock	267 mm	267 mm	222 mm
stock telescoped	269 mm	269 mm	248 mm
stock extended	548 mm	548 mm	460 mm
Barrel	146 mm	146 mm	129 mm
Suppressor	291 mm	291 mm	224 mm
MECHANICAL FEATURES			
Barrel	6 grooves, rh 1 turn in 508 mm	6 grooves, rh 1 turn in 305 mm	6 grooves, rh 1 turn in 305 mm
FIRING CHARACTERISTICS			
Muzzle velocity	280 m/s	366 m/s	293 m/s
Rate of fire: cyclic,	1,145 rds/min	1,090 rds/min	1,200 rds/min

Cobray M11 sub-machine guns

The Ingram M11 design, described previously, is currently being manufactured as the 'Cobray M11'. Two types are produced, the M11 for the 9 mm Short (0.380 Auto) cartridge and the M11/9 for the 9 × 19 mm Parabellum cartridge.

Data

As for Ingram M11 above

Manufacturer

FMJ, PO Box 759, Copperhill, Tennessee 37317.

Status

In production.

UPDATED

Heckler and Koch 9 mm MP5K-PDW Personal Defence Weapon

Description

The Heckler and Koch 9 mm MP5K-PDW Personal Defence Weapon was developed as a compact weapon for vehicle or aircraft crewmen, or for other roles in which a rifle or full-sized sub-machine gun is inappropriate and a pistol a poor compromise. It can be easily carried in a holster or concealed under a coat or inside a document case or soft bag.

9 mm Heckler and Koch MP5-N sub-machine gun

The MP5K-PDW is essentially the familiar Heckler and Koch MP5K with the addition of a side-folding stock and with three lugs on the barrel to permit the fitting of either a flash or sound suppressor, blank-firing device or grenade launcher. When the stock is not required it can be quickly removed and replaced with a butt cap. This folding stock can also be purchased separately and fitted to existing MP5K weapons of any type. The sights are fitted with a front tritium aiming light; rear tritium lights are optional. The standard rearsight can be replaced by a dioptric sight for long-range firing, and mounts on the receiver will accept all types of optional sighting and aiming equipment. The 9 mm ammunition fired may be standard NATO Ball or Olin +P 115 g JHP.

All controls are ambidextrous; the standard firing mechanism permits selective fire, and an optional three-round burst mechanism is available.

Variant

9 mm MP5-N sub-machine gun

The MP5-N (N for Navy) sub-machine gun is the standard MP5 but with a threaded muzzle to permit the attachment of a stainless steel sound suppressor, an ambidextrous safety/fire selector switch, and a tritium insert in the foresight to assist aiming in poor light conditions.

9 mm MP5K-PDW with stock folded, in comparison with US M9 pistol

9 mm MP5K-PDW with stock extended and suppressor

Data

Cartridge: 9 × 19 mm Parabellum; Olin +P
Operation: delayed blowback, selective fire, optional 3-round burst
Locking: rollers
Feed: 15- or 30-round box magazines
Weight: with stock, 2.79 kg; with butt cap, 2.09 kg; with stock and suppressor, 3.36 kg
Length: stock folded, 368 mm; stock extended and suppressor fitted, 800 mm
Barrel: 127 mm
Rifling: 6 grooves, rh, 1 turn in 380 mm
Sights: fore, hooded post; rear, rotary aperture, fully adjustable
Sight radius: 260 mm
Muzzle velocity: Ball, ca 375 m/s; +P, 398 m/s
Muzzle energy: Ball, ca 570 J; +P, 590 J

Manufacturer

Heckler and Koch Inc, 21480 Pacific Boulevard, Sterling, Virginia 20166.

Status

In production.

UPDATED

La France 0.45 M16K sub-machine gun

Description

The La France 0.45 M16K sub-machine gun is described as an assault carbine and is a variant of the La France 5.56 mm M16K carbine (see entry in *Rifles* section). The 0.45 M16K was designed for the unique requirements of special operations missions, providing more stopping power than provided by 9 mm sub-machine guns. The 0.45 M16K functions reliably with M1911 ball ammunition and most brands of jacketed hollow point 0.45 ACP ammunition.

As with M16-series rifles, the M16K fires from a closed bolt to provide good accuracy in the semi-automatic mode but operates using the blowback principle. The straight-line configuration combines with optimised weight distribution to provide good control when firing fully automatic.

The four-pronged Vortex flash suppressor eliminates muzzle flash even when firing high performance ammunition. A screw-on silencer is in the development stage.

Data

Cartridge: 0.45 ACP
Operation: blowback
Feed: 30-round box magazine
Weight: empty, 3.86 kg

La France 0.45 M16K sub-machine gun

Length: 676 mm
Barrel: 184 mm
Sights: fore, aperture; rear, flip aperture, adjustable for windage
Sight radius: 317 mm
Muzzle velocity: ca 260 m/s
Rate of fire: cyclic, 625 rds/min

Manufacturer
La France Specialties, PO Box 178211, San Diego, California 92177-8211.

Status
Ready for production.

VERIFIED

Ruger 9 mm MP-9 sub-machine gun

Development
The Ruger 9 mm MP-9 sub-machine gun was the result of a joint design venture between Bill Ruger and the Israeli Uzi Gal. The MP-9 first entered production in late 1994, with some minor changes incorporated during 1995. Some MP-9 components are manufactured at the Ruger facility in Prescott, Arizona, with final assembly taking place in Newport, New Hampshire.

The MP-9 has been tested by US special operations forces and has drawn interest from several unidentified South American military forces.

Description
The Ruger 9 mm MP-9 sub-machine gun may be regarded as an updated Uzi sub-machine gun modified internally to fire from a closed bolt. Another visual change from the Uzi is the provision of a new design of telescopic metal butt which, when telescoped to its minimum length, can be folded downwards to form a close union with the frame extending under the receiver and behind the grip. The entire lower receiver is fibreglass reinforced moulded Zytel, a hard wearing plastic-based material. The upper receiver is polished and blued steel alloy while the barrel is heat-treated chrome molybdenum steel. If required the barrel can be easily removed in the field by unscrewing a locking nut.

As with the Uzi, the 32-round double column box magazine is inserted upwards through the centrally-placed pistol grip. To the left of the grip is a sliding safety catch with three positions; safe (rear), semi-automatic, and full automatic. There is also a separate trigger actuated disconnector and a firing pin block.

Standard accessories include a tactical sling and two 32-round magazines. It is possible to fit optical sights, night vision devices, and a suppressor to the MP-9.

Ruger 9 mm MP-9 sub-machine gun with stock extended

1996

Data
Cartridge: 9 × 19 mm Parabellum
Operation: blowback, closed bolt, selective fire
Feed: 32-round box magazine
Weight: empty, 3 kg
Length: stock folded, 375.9 mm; stock extended, 556.3 mm
Barrel: 172.7 mm
Rifling: 6 grooves, rh, 1 turn in 255 mm
Sights: fore, protected post, adjustable for elevation; rear, dual aperture, adjustable for windage; optical sights optional
Muzzle velocity: ca 350 m/s
Rate of fire: cyclic, 600 rds/min

Manufacturer
Sturm, Ruger & Co Inc, Lacey Place, Southport, Connecticut 06490.

Status
In production.

NEW ENTRY

---

9 mm C-MAG 100-round magazine

Description
The C-MAG is a twin-drum high-capacity ammunition magazine for machine and sub-machine guns and other specialised weapons in 9 mm calibre. Covered by US and other patents, the main components are constructed of high-impact plastic. A full description of the operating principles, construction and appearance of these magazines can be found in the Beta Company entry in the *Rifles* section.

Because of the geometry of its construction, the C-MAG can be used on almost any modern 9 mm weapon without modification. Initial production models include magazine interfaces for the Heckler and Koch MP5 and Colt 9 mm sub-machine guns. Other models may be adapted by substituting a feed clip designed for the particular weapon.

Mechanical and automatic loading devices holding 5, 10 and 100 cartridges for speed loading are available, as well as carrying pouches.

Data
Capacity: 100 9 × 19 mm Parabellum cartridges
Ammunition types: ball, hollow-point
Weight: empty, 1 kg
Dimensions: 265 × 50 × 120 mm
Rate of fire: >1,000 rds/min

Manufacturer
The Beta Company, 2137B Flintstone Drive, Tucker, Georgia 30084.

Status
In production.

UPDATED

YUGOSLAVIA (SERBIA AND MONTENEGRO)

7.62 mm M56 sub-machine gun

Description
The M56 sub-machine gun is similar in appearance to the German MP 40. The barrel looks very long and rather fragile and it takes a knife pattern bayonet. The cocking handle is on the right of the receiver and there are locking slots at each end of the cocking way allowing the bolt to be secured in either the forward or cocked positions. The folding stock is of the same pattern as the MP 40.

The internal mechanism is much simpler than the MP 40 and there appears to be less emphasis on keeping dirt away from the bolt. A feature of the body is that there are several large cutouts in the lower half of it, and the return spring is actually exposed for a part of its length. The bolt slot has no cover so it can be assumed that the entire gun is vulnerable to the effects of dirt and snow clogging the mechanism. The body is longer than with most contemporary sub-machine guns, with the magazine at the forward end. This makes it difficult to find a proper hold with the left hand, and the designer has incorporated the forward handgrip in the lower part of the body, just behind the magazine. The firing position is thereby distinctive and unusual.

Data

Cartridge: 7.62 × 25 mm Type P Pistol or 7.63 mm Mauser
Operation: blowback, selective fire
Feed: 35-round curved box magazine
Weight: 3 kg
Length: stock folded, 591 mm; stock extended, 870 mm
Barrel: 250 mm
Muzzle velocity: 500 m/s
Rate of fire: cyclic, 600 rds/min

Manufacturer

Zastava Arms, 29 Novembra 12, YU-11000 Belgrade.

Status

Not in production.

Service

Former Yugoslav forces.

VERIFIED

Yugoslav troops with 7.62 mm M56 sub-machine gun

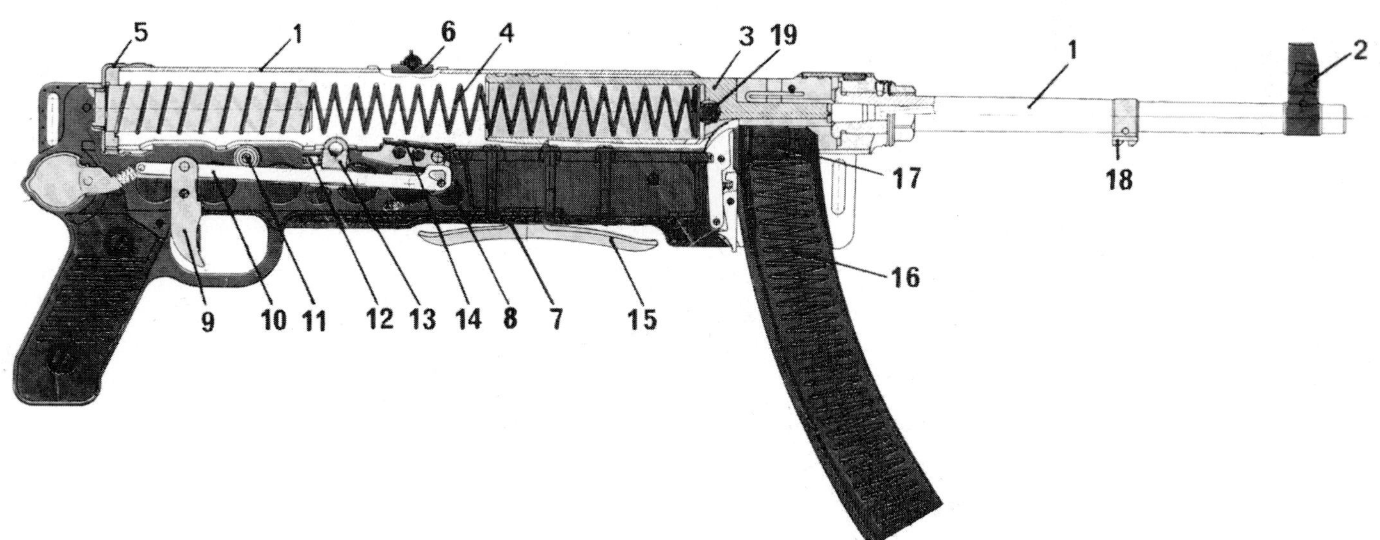

7.62 mm M56 sub-machine gun
(1) *receiver and barrel* **(2)** *foresight* **(3)** *bolt* **(4)** *return spring* **(5)** *rear cap* **(6)** *rearsight* **(7)** *left-hand grip* **(8)** *lower frame* **(9)** *trigger* **(10)** *trigger bar* **(11)** *safety button* **(12)** *selector catch* **(13)** *selector lever* **(14)** *sear* **(15)** *folded metal stock* **(16)** *30-round magazine* **(17)** *7.62 mm cartridges on the magazine platform* **(18)** *bayonet lug* **(19)** *cocking handle*

7.65 mm Model 84 machine pistol

Description

Zastava manufactured the Czech 7.65 mm Model 61 pistol (Skorpion) under licence. The pistol is identical with the Czech model (qv), but is probably no longer in production.

Manufacturer

Zastava Arms, 29 Novembra 12, YU-11000 Belgrade.

Status

Probably no longer in production.

Service

Former Yugoslav forces.

VERIFIED

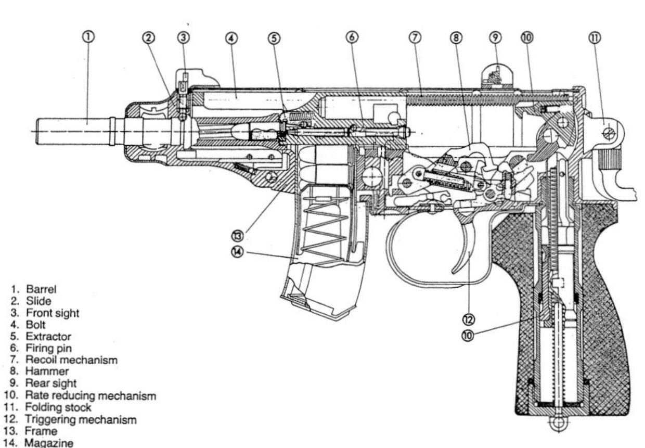

Sectioned drawing of Model 84 machine pistol

7.65 mm Model 84 machine pistol

5.56 mm Zastava M85 sub-machine gun

Description

The 5.56 mm Zastava M85 sub-machine gun appears to be a variant of the M80 assault rifle, resembling the Kalashnikov AKS-74U in general appearance and intention. It uses the standard Kalashnikov-pattern rotating bolt gas action with a short barrel which is fitted with a tubular muzzle attachment which serves as a flash hider and also as a ballistic regulator to compensate for firing the 5.56 × 45 mm cartridge from a short barrel. The folding stock is a two-strut type which folds beneath the receiver to rest with its butt under the fore-end.

According to the manufacturers, the M85 was intended for use by personnel in combat vehicles, helicopters and aircraft, and by special forces and internal security forces.

5.56 mm Zastava M85 sub-machine gun

Data

Cartridge: 5.56 × 45 mm
Operation: gas, selective fire
Lock system: rotating bolt
Feed: 20- or 30-round box magazine
Weight: without magazine, 3.2 kg
Length: butt folded, 570 mm; butt extended, 790 mm
Barrel: with flash hider, 315 mm (estimated)
Sights: fore, hooded post; rear, flip U-notch 200 and 400 m; an optical sight may be fitted
Muzzle velocity: ca 790 m/s
Rate of fire: cyclic, 650-750 rds/min

Manufacturer

Zastava Arms, 29 Novembra 12, YU-11000 Belgrade.

Status

May no longer be in production.

VERIFIED

RIFLES

ARGENTINA

7.62 mm FN pattern rifles

Description

Formerly equipped with Mauser-type 7.65 mm rifles, the Argentinian Army has for some time used the Fabrique Nationale Fusil Automatique Léger (FN-FAL) NATO-pattern rifle and the FN heavy-barrelled rifle made under FN licence at the ordnance factory at Rosario.

Four models have been manufactured. The Fusil Automatico Liviano Modelo IV is the standard FN-FAL Model 50-00 and is identical in all respects other than being slightly heavier than the Belgian original. The Fusil Automatico Liviano Modelo Para III is the FN-FAL Model 50-64 with side-folding butt and standard barrel. This differs slightly from the Belgian model, being 1,092 mm long with extended butt, 897 mm long with butt folded, and 4.25 kg in weight when empty.

The third model is the Fusil Automatico Pesado Modelo II, which is the heavy-barrelled FN-FAL Model 50-41 with bipod. This also differs slightly in dimensions from the Belgian version, weighing 6.45 kg and being 1,125 mm long.

The fourth model is the Fusil Semiautomatico Liviano, which is simply the FAL Modelo IV without the selective fire option. This is produced for police and security use and for commercial sale.

Argentine 7.62 mm Fusil Automatico Liviano Modelo IV

Data

Cartridge: 7.62 × 51 mm NATO
Operation: blowback, selective fire
Feed: 20-round box magazine
Weight: empty, 4.6 kg
Length: 1.09 m
Barrel: 533 mm
Rifling: 6 grooves, rh, 1 turn in 400 mm
Sights: fore, protected post; rear, two-range flip, 50 and 100 m
Rate of fire: cyclic, 800 rds/min
Muzzle velocity: 330 m/s

Manufacturer

Fabrica Militar de Armas Portatiles 'Domingo Matheu', Avenida Ovidio Lagos 5250, 2000 Rosario.

Status

In production.

Service

Argentinian forces and offered for export.

VERIFIED

5.56 mm FAL

Description

In order to produce an inexpensive 5.56 mm rifle, the Fabrica 'Domingo Matheu' developed a version of the FAL in 5.56 mm calibre. Many of the existing components of the 7.62 mm FAL are used in this design, the only special manufacture being that of the barrel, bolt, magazine and similar cartridge-sensitive items. It has also been necessary to 'tune' the gas system to the different propelling charge. Two versions have been developed, the 'Modelo Infanteria' and the 'Modelo Paracaidista' (infantry and parachutist models).

Data

DATA FOR PARACHUTIST MODEL IN PARENTHESIS
Cartridge: 5.56 × 45 mm SS109/NATO
Operation: gas, selective fire
Locking system: tipping bolt
Feed: 30-round box magazine
Weight: without magazine, 4.325 kg (4.2 kg)
Length: 1.09 m (1.02 m); Parachute model with butt folded, 770 mm
Barrel: 533 mm (458 mm)
Sights: fore, blade; rear, two-range flip aperture; 150 and 250 m, both adjustable for zero
Sight radius: 553 mm
Muzzle velocity: 965 m/s (950 m/s)

Manufacturer

Fabrica Militar de Armas Portatiles 'Domingo Matheu', Avenida Ovidio Lagos 5250, 2000 Rosario.

Status

Advanced development.

VERIFIED

5.56 mm FARA 83 assault rifle

Description

The 5.56 mm FARA 83 assault rifle is a gas-operated selective-fire rifle using a conventional gas piston arrangement to actuate a bolt carrier and two-lug rotating bolt. Manufacture is from steel forgings and metal pressings, the receiver being a welded sheet metal construction to which subassemblies are attached by rivets. The butt is of glass-reinforced plastic and is hinged so as to fold alongside the receiver. The fore-end is of plastic material.

The rearsight is a three-position drum, adjustable for windage by means of two screws. Two of the positions are for daylight firing, giving ranges of 200 m and 400 m, while the third position is a notch with tritium inserts for firing at 100 m in poor light. There is also an auxiliary foresight fitted with a tritium source for use in poor light, which can be folded down when not in use.

A bipod is available as an optional fitment; this can be attached below the gas block, and a special fore-end is fitted which has a recess to take the bipod legs when folded.

Issue of the FARA 83 rifle began in early 1984 but was halted after about 1,000 rifles were made as a result of budgetary restrictions. Small numbers have been made since that time, but production has now ceased.

5.56 mm FARA 83 rifle field-stripped

5.56 mm FARA 83 assault rifle

Data
Cartridge: 5.56 × 45 mm M193 or SS109
Operation: gas, selective fire
Locking: rotating bolt
Feed: 30-round box magazine
Weight: 3.95 kg
Length: butt folded, 745 mm; butt extended, 1 m
Barrel: 452 mm
Rifling: 6 grooves, rh, 1 turn in 229 mm
Sights: see text
Muzzle velocity: 965-1,005 m/s, depending upon ammunition
Rate of fire: cyclic, 750-800 rds/min

Manufacturer
Fabrica Militar de Armas Portatiles 'Domingo Matheu', Avenida Ovidio Lagos 5250, 2000 Rosario.

Status
Production complete.

Service
Argentinian forces.

VERIFIED

AUSTRALIA

5.56 mm F88 Weapon System

Description

The 5.56 mm F88 System is the Australian manufactured version of the Steyr AUG. At one time there was an Australian armed forces requirement for at least 84,000 weapons.

The F88 is an extremely simple weapon to operate. The large safety catch requires a simple but definite lateral movement to lock or unlock the trigger. The gun fires in semi-automatic mode when the trigger is pressed to a clearly felt point, then fully automatic when it is fully depressed. This gives the soldier total control over weapon function without the need to operate a change lever. Stripping for field cleaning and maintenance is performed without the need for any tools.

In all F88 models the barrel is connected to the receiver by eight locking lugs, an arrangement which gives maximum security of barrel location whilst retaining ease of barrel removal. Removal requires only a 22.5° rotation of the barrel in the housing assembly. This allows the weapon to be stripped easily and stowed in a knapsack. Its longest component is no more than 550 mm in length. All barrel styles are fully interchangeable with either receiver type. This allows a weapon to be instantly converted between the standard assault rifle, the machine carbine, the commando weapon or the heavy-barrelled light support weapon.

Australian troops armed with 5.56 mm F88 rifles

The F88 barrel is rifled 1 twist in 178 mm to accommodate SS109 5.56 mm ammunition.

The F88 features an optical sight integrated into the weapon's carrying handle. This sight offers a magnification of ×1.5, which allows a wide field of view (45 m at a range of 300 m) and allows the soldier to fire with both eyes open.

In adverse light conditions the optical sight will perform effectively long after open sights have ceased to be useful. The standard circular graticule covers a height of 1.8 m at a range of 300 m. This system allows for target acquisition in typically 1.5 seconds compared with 3 seconds for open sights. The circular graticule also provides an effective means for the soldier to estimate and compensate for target range without having to adjust the sight. A bayonet, the KCB 77, was supplied direct from Austria.

The 'bullpup' design makes the F88 an extremely short weapon. The overall length with the standard 548 mm barrel (with flash hider) is only 800 mm. With a 407 mm barrel fitted the overall length is reduced to 690 mm which eliminates the need for any consideration of a folding stock version of the weapon.

Full technical data on the functioning of the F88 can be found in the Steyr AUG entry in the *Austrian* section.

Manufacturer
Australian Defence Industries Limited, Bondi Junction, New South Wales 2022.

Status
In production.

Service
Australian and New Zealand defence forces (18,000 units).

UPDATED

7.62 mm Model 82 sniper rifle

Description

Introduced into service in 1978, the Australian 'Rifle 7.62 mm Sniper System' comprises a Parker-Hale Model 82 rifle with a 1200TX barrel which has been matched to a Kahles Helia ZF69 telescopic sight.

The scope is fitted with two bridge clamps dovetailed in the base to slide into the rear and front blocks of the receiver. It features a reticule pattern No 7, fixed magnification of 6 × 42 and visual and click adjustments for a range of 100 to 800 m in 50 m increments. Further click adjustments up to a range of 900 m are available although visual markings extend only to 800 m.

The Model 82 is also fitted with a Parker-Hale 5E aperture rearsight and tunnel foresight. The 5E iron sights enable the weapon to be used as a sniping rifle in the event of the telescopic sight being damaged. The rear iron sight is mounted to the rearsight block on the receiver and contains a vernier adjustment for elevation and windage, which allows for zeroing at ranges of 100 to 1,000 m. A carrying container is provided for the rearsight which cannot be attached to the rifle until the telescopic sight is removed.

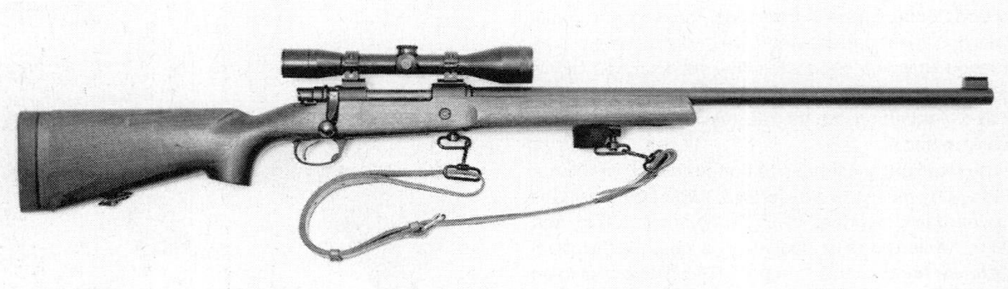

7.62 mm Model 82 sniper rifle

The remaining details of this system are given in the *United Kingdom* section.

Manufacturer
Parker-Hale Limited, Bisley Works, Golden Hillock Road, Birmingham B11 2PZ, UK.

Status
No longer manufactured.

Service
Australian armed forces.

UPDATED

AUSTRIA

Steyr 15.2 mm IWS 2000 Anti-Material Rifle (AMR)

Development

In order to fill a perceived gap in the number of support weapon types available to the infantry, Steyr-Mannlicher began the development of what they termed an anti-materiel rifle (AMR) during the mid 1980s. At that time they envisaged a two-man single shot rifle capable of accurate fire up to 1,000 m, the intended targets being armoured personnel carriers, soft-skin vehicles, electronics equipment and helicopters. The ammunition was to be APDS or APFSDS with early development concentrating on APDS.

Early trials with ammunition based on 12.7 mm cartridges demonstrated that an APFSDS approach would be more beneficial, resulting in the construction of a small number of 14.5 mm AMR 5075 semi-automatic rifles with smooth-bore barrels. Trials with these weapons resulted in a change of calibre to 15.2 mm and a change of programme name to Infantry Weapon System 2000 (IWS 2000).

Development work is continuing with a view to perfecting the ammunition and increasing the muzzle velocity to 1,500 m/s. Work on the weapon is concentrating on reducing the overall weight and dimensions with a view to producing a prototype of a five-shot semi-automatic rifle. Other possible future options include a low-rate automatic fire version and the use of a rifled barrel so as to explore other ammunition design possibilities.

Steyr 15.2 mm IWS 2000 Anti-Materiel Rifle (AMR)

1996

Description

The 15.2 mm IWS 2000 is a heavyweight precision rifle designed as a relatively inexpensive system for the long-range attack of equipment such as light armoured vehicles, aircraft on the ground, fuel and supply dumps, radar installations and similar targets. It can be dismantled into two units for man carriage.

The rifle is a semi-automatic bullpup, using plastics and light metal to reduce the weight as far as is consistent with the strength demanded by its role. The mechanism employs the long recoil principle of operation, the barrel and bolt recoiling for about 200 mm, after which the bolt is unlocked and held while the barrel is returned to battery. The bolt is then released, collects a cartridge from the magazine and chambers it, locking into the barrel by a rotary motion.

Recoil of the barrel is reduced by a cylindrical multi-port muzzle brake of considerable efficiency and is controlled by a hydropneumatic annular system carried in a ring cradle forming the front portion of the tubular receiver. The weapon is supported by a bipod, attached above the recoil cradle, and by an adjustable firing pedestal beneath the butt. A ×10 optical sight is fitted as standard.

The five-round box magazine is inserted from the right side, at an angle of about 45° below the horizontal.

The complete 15.2 mm APFSDS round weighs 150 g and is 207 mm long. The cartridge case (maximum diameter at the base is 26 mm) is of part-synthetic construction, conventional bottle-necked in form and carries a 20 g fin-stabilised tungsten dart projectile which has a muzzle velocity of 1,450 m/s; the complete projectile assembly with its four sabot segments weighs 35 g. The projectile has a practical range of 1,000 m and a probable range of 1,500/2,000 m depending upon the type of target. At 1,000 m range the current projectile has been demonstrated to pierce 40 mm of RHA plate and the secondary fragmentation behind the plate is considerable. The high velocity bestows an exceptionally flat trajectory; the vertex at 1,000 m range is no more than 800 mm above the line of sight.

Steyr 15.2 mm IWS 2000 Anti-Materiel Rifle (AMR)

1996

Data (PROVISIONAL)

Cartridge: 15.2 mm Special
Operation: long recoil, semi-automatic
Locking: rotating bolt
Feed: 5-round box magazine
Weight: ca 18 kg
Length: 1.8 m
Barrel: 1.2 m
Rifling: smoothbore
Sight: ×10 telescope
Muzzle velocity: ca 1,450 m/s

Manufacturer

Steyr-Mannlicher AG, Postfach 1000, A-4400 Steyr.

Status

Development.

UPDATED

7.62 mm SSG 69 rifle

Description

The 7.62 mm SSG 69 is the standard sniper rifle for the Austrian Army under the designation Scharfschützengewehr 69 (SSG 69).

The barrel is made by the cold forging process, originally developed by Steyr and now used by many barrel makers, which results in work-hardening both the bore and the outside surface of the barrel. It also allows, as in the case of the SSG 69, a slightly tapered bore.

The bolt is manually operated and moves through 60°. It has six rotating locking lugs, symmetrically arranged in pairs, set at the rear. The trigger mechanism gives a double pull. The length and weight of the pull are both adjustable externally. The safety catch on the top right of the rear of the receiver is of the sliding type, locking the bolt and the firing pin.

7.62 mm SSG-P Special

7.62 mm Steyr SSG-P Suppressed

7.62 mm Steyr SSG-P Police model

The standard magazine, holding five rounds, is the rotating spool type which has been used on Mannlicher sporting and military rifles for many years. The rifle can, however, be used with a 10-round box magazine, though this is no longer in production. The 10-round magazine was principally for competition use and is now no longer permitted because of changes in UIT regulations.

The stock is made of a synthetic material. It can be adjusted for length by the addition or removal of a butt pad.

The receiver has a longitudinal rib machined on the top. Mounts, which are available for all types of sight, can be fixed to this rib. The standard sight is the ZFM 6 × 42, which is graduated up to 800 m and internally adjusted. The sight is attached by lever clamped rings on to the rib. There are also iron sights brazed on to the barrel for emergency use. All optronic sighting devices may be mounted.

Variant models of the SSG 69 have been produced to meet the requirements of police and security forces. The SSG-P is basically the same rifle but the stock is in matt black finish, the heavy barrel of the Steyr Match-UIT rifle is fitted, the match rifle bolt-handle with extra-large grip is used, a detachable shooting sling is mounted on a fore-end rail and there is a sensitive adjustable set trigger. The rifle has no iron sights, a 6 × 42 telescope being provided as standard, though other telescopes can be fitted as preferred.

For specialised tasks the SSG-P (Suppressed) is fitted with a sound suppressor which, when used in conjunction with subsonic ammunition, is highly effective. In situations where a more compact rifle is desired, the SSG-P Special has a short barrel with a muzzle flash hider.

7.62 mm SSG 69 sniper rifle

Data

Cartridge: 7.62 × 51 mm or 0.243 Winchester
Operation: bolt action, single shot
Locking: rotating bolt
Feed: 5-round rotary magazine
Weight: empty, 3.9 kg; with telescope, 4.6 kg
Length: 1.14 m
Barrel: 650 mm
Rifling: 4 grooves, rh
Sights: ×6 telescope, internally adjustable to 800 m; emergency iron sights, fore, blade; rear, V-notch
Muzzle velocity: 860 m/s
Effective range: 800 m

Manufacturer

Steyr-Mannlicher AG, Postfach 1000, A-4400 Steyr.

Status

In production.

Service

Austrian Army and several foreign police and military forces.

VERIFIED

5.56 mm Steyr AUG rifle

Development

This modern assault rifle was developed by Steyr in conjunction with the Austrian Army and the first production models came off the line early in 1978. Since then the rifle has been extensively adopted by armies in various parts of the world.

The *Armee Universal Gewehr* (AUG) is designed to be converted to different modes; as a parachutist's rifle or sub-machine gun, with a 350 mm barrel; as a carbine, with a 407 mm barrel; as a standard assault rifle, with a 508 mm barrel; and as a heavy-barrel automatic rifle in the light machine gun role, using a 621 mm barrel. The rifle versions all use a 30-round box magazine, while the heavy-barrelled version uses a 42-round magazine and has a light bipod. All types have an optical sight integral with the carrying handle, but customers may specify a sight-mounting bracket instead; this can be used to mount a variety of telescope or night vision sights.

Description

The AUG is a bullpup design of novel appearance. The intention behind the weapon is to have a light, handy gun with particular emphasis on use in and from vehicles and commonality of parts for the different modes of use. It can be altered for use by a left-handed person by changing the bolt and moving a blanking cap from the left ejection opening to the right.

The barrel is made from high quality steel and the internal profile is obtained by the cold-hammering process. Both chamber and barrel are chromed to reduce wear and fouling. An external sleeve is shrunk on to the barrel and carries the gas port and cylinder, gas regulator and forward handgrip hinge jaw. There is a short cylinder which contains a piston and its associated return spring. The gas regulator has two positions, normal and large, the latter to be used when and if the gun is slowed by fouling or dirt. A flash suppressor is screwed to the muzzle and it is internally threaded for taking a blank firing attachment. The barrel locks into the body by a system of eight lugs arranged around its chamber end which engage with the locking sleeve in the body. The bolt locks into the rear half of the same sleeve.

The receiver is made from an aluminium pressure die casting which holds the bearings for the barrel lugs and the guide rods. The carrying handle and sight are integral with the receiver casting and the receiver itself does not carry or guide the bolt. The cocking handle works in a slot on the left side of the receiver, and is operated by the firer's left hand.

The hammer mechanism is contained in the rear of the butt, covered by the synthetic rubber butt-plate. Except for the springs and pins it is entirely made of plastic and is contained in an open-topped plastic box which lies between the magazine and the butt-plate, and the bolt group recoils over the top of it. Since the trigger is some distance away it transmits its pressure through a sear lever which passes by the side of the magazine. Operation of the firing pin is by a plastic hammer driven by a coil spring. There is no change lever for different modes of fire: single shots are fired by pulling the trigger a short distance to the first sear action. Further movement operates the sear in the automatic mode.

The mode of operation of the weapon can be changed by altering the hammer mechanism unit; thus for police use a mechanism which permits only

5.56 mm Steyr AUG standard rifle with 508 mm barrel

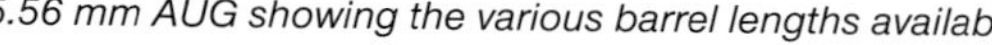

5.56 mm AUG showing the various barrel lengths available

1996

5.56 mm AUG rifle

1996

semi-automatic fire is available. For military use the standard mechanism offers the usual options of single shot or full automatic fire, but a third mechanism has a selector switch which permits the user to withdraw the mechanism and select either full automatic fire or three-round burst fire as the alternative to single shots. The option can only be changed by withdrawing the mechanism to make the choice.

The bolt is a light multi-lug rotating component contained in a carrier and operated by a camway in that carrier. The carrier has two guide rods brazed to it and these rods run in steel bearings in the receiver. All movement is therefore isolated from the receiver itself and the guide rods are hollow and contain the return springs. The left-hand rod takes the pressure from the cocking handle and it can be locked to that handle by a catch so that the firer can push the bolt home, though in normal firing the cocking handle is free of the rod and does not reciprocate. The right-hand rod is the piston which operates the bolt. The left-hand rod can also be used as a clearing rod if the gas cylinder becomes fouled. The bolt locks by seven lugs into the locking sleeve in the body and carries an extractor in place of an eighth lug.

The magazine is entirely plastic apart from the spring. The body is translucent so that the ammunition stock can be seen. The stock group is almost entirely plastic. At the forward end is the pistol grip with its large forward guard completely enclosing the firing hand. The trigger is hung permanently on this pistol grip, together with its two operating rods which run in guides past the magazine housing. Immediately above the grip is a horizontal button which is the safety lever. This lever acts on the trigger rods. Behind that is the locking catch for the stock group. Pressing this to the right allows the receiver and stock to be separated. The magazine catch is behind the housing, on the underside of the stock. Above the housing are the two ejector openings, one of which is always covered by a movable piece of plastic. The rear of the stock is the actual shoulder butt and it contains the hammer group and the rear end of the bolt-way. The butt is closed by the butt-plate which is held in place by the rear sling swivel. This swivel is attached to a pin which pushes in across the butt and locks the plate.

The AUG operates as a completely conventional gas-operated weapon and is unusual only in the gas cylinders being offset to the right and working on one of the

5.56 mm AUG rifle with special receiver group to accept alternative sighting systems

Irish soldier armed with a 5.56 mm AUG rifle

5.56 mm AUG rifle

5.56 mm AUG rifle with accessories

5.56 mm AUG rifle with Signaal day and night sight

two guide rods. For single-shot operation the gas piston drives the bolt unit back, cocking the hammer which is held by the disconnector lever. The return springs pull the bolt forward, loading a fresh cartridge, and the bolt is rotated to lock. For semi-automatic fire the trigger must be released between shots. For automatic fire the operation is the same but the hammer is held by the automatic fire lever which is released after the bolt has locked.

The ×1.5 optical sight has been optimised for battle ranges and the graticule is a black ring in the field of view. This can be placed quickly and easily around a man-sized target at ranges up to 300 m, though after that it may take some care to get the target in the centre. However, for all normal infantry engagements it offers an easily taught and rapidly used sight with very good accuracy and performance in poor light.

An alternative graticule with a fine dot in the centre of the aiming circle is also available, allowing a more precise aim. This graticule is supplied as standard on police models.

A special receiver group, with a flat mounting platform in place of the optical sight, permits the use of all types of optical, electro-optical or collimating sights.

Data

Cartridge: 5.56 × 45 mm
Operation: gas, selective fire
Locking: rotating bolt

DIMENSIONS

	Sub-machine gun	Carbine	Rifle	Heavy-barrelled
Length				
overall	626 mm	714 mm	805 mm	915 mm
barrel	350 mm	417 mm	508 mm	621 mm
Weights				
unloaded	3.52 kg	3.7 kg	3.5 kg	4.9 kg
loaded magazine	0.49 kg	0.49 kg	0.49 kg	0.66 kg

Magazine: 30- and 42-round detachable translucent plastic box
Rate of fire: cyclic, 650 rds/min

SIGHTS
Optic sight: set in carrying handle, ×1.5 power

ACCESSORIES
Bayonet, blank-firing attachment, special receiver for sights, grenade launcher M203 (AUG-8), muzzle cap, carrying sling, cleaning kit, bipod with wire-cutter, fixed and adjustable bipod (HBAR only), rifle grenades.

Manufacturer

Steyr-Mannlicher AG, Postfach 1000, A-4400 Steyr.

Status

In production.

Service

Austrian Army as the Stg 77. Adopted by the Australian, New Zealand, Omani, Malaysian, Saudi Arabian, Irish and other armed forces, including the Falkland Islands Defence Force.

Licence production

AUSTRALIA
Manufacturer: Australian Defence Industries
Type: 5.56 mm F88 Weapon System
Remarks: See entry under *Australia* for details.
MALAYSIA
Manufacturer: SME Technologies
Type: 5.56 mm AUG
Remarks: Standard specifications

UPDATED

BELGIUM

7.62 mm FN-FAL rifle

Development

The 7.62 mm FN-FAL rifle has been adopted by more than 90 countries all over the world. It is, or has been, manufactured under licence in several countries and some of these have incorporated their own minor modifications to suit their own particular needs. In some instances these modifications have resulted in deviations from the original specification to the extent that some licensed weapons are not interchangeable with others. There are several versions of the basic FAL and they are listed at the end of this entry.

7.62 mm FN-FAL Para rifle, butt extended (50-63)

Description

To load the FAL, the front end of a loaded magazine is placed in the housing and the rear end rotated back and up until the magazine catch clicks in to lock it in position.

The selector lever is above the pistol grip on the left of the receiver and is operated by the thumb of the right hand. The 'safe' position is at the top, 'repetition' lies below it and 'automatic' is further forward. In the UK, Canadian, Indian and Netherlands versions, there is no automatic fire. The cocking handle is on the left side of the receiver. On some models it folds flat and must be pulled back to extend it at a right angle before it can be used. When the cocking handle is retracted fully and then released, the bolt picks up the top round from the magazine and chambers it. The weapon always fires from the closed bolt position.

7.62 mm FN-FAL rifle heavy-barrelled version, with bipod and optical sight

When the trigger is pulled the hammer strikes the firing pin and fires the cartridge. When the bullet passes the gas port in the top of the barrel, some of the propellant gas is diverted into the gas cylinder where it expands and drives back a light, spring-loaded piston. This piston strikes the top of the front face of the bolt carrier, transfers its energy and, after a short travel, is returned by its spring to its forward position. The carrier is driven rearwards and has about 6 mm of free travel during which the chamber pressure is dropping to an acceptably low level. After this idle motion, known as 'mechanical safety after firing', the unlocking cam of the carrier moves under the bolt lug and lifts the rear end of the bolt out of the locking recess in the floor of the receiver. The bolt and carrier travel back together, compressing the return spring. The extractor withdraws the empty case from the chamber and holds it on the face of the bolt until it strikes the fixed ejector and is thrown out of the rifle to the right, through the ejection port.

The return spring drives the carrier and bolt forward; the feed rib on the bolt pushes the top cartridge out of the magazine and over the bullet guide into the chamber. The extractor snaps into the extraction groove of

7.62 mm FN-FAL Para and standard versions

the cartridge case and the bolt can travel forward no further. The carrier continues on and the locking cam rides over the bolt, forcing the rear of the bolt down into a recess in the receiver floor and holding it there. The final forward movement of the carrier trips the automatic sear when the selector is on 'automatic'. If the trigger is still pressed, the hammer will drive the firing pin into the cartridge cap.

When the last round is fired a rib on the magazine platform lifts the bolt stop into the path of the bolt. On the United Kingdom version the last 6 mm of the bolt stop is ground off and thus the bolt is in the forward position when the ammunition is expended. By pushing up the bolt stop by hand when the bolt is to the rear, the bolt can be held open whenever required.

The gas regulator works on the 'exhaust to atmosphere' principle. When the gun is clean and firing under ideal conditions, a large proportion of the gas is passed through the regulator and out into the atmosphere. When the need comes to increase the gas pressure on the piston-head, the regulator is screwed up and more gas is diverted into the cylinder.

The gas plug must be rotated when firing grenades to ensure all the gas pressure goes to drive the grenade out and none goes to the piston. To do this the plunger is depressed and the plug rotated until the marking is on the bottom. After each grenade the bolt must be retracted by hand and a fresh cartridge loaded into the propelling chamber.

Zeroing the sights is achieved by screwing the foresight up or down to adjust for elevation and the rearsight aperture can be moved across by loosening the screw underneath on the side to which movement is required and tightening the screw on the other side.

Variants

General characteristics and special characteristics of four versions of the weapon are given below. The four versions, with their FN reference numbers are:

FN 50-00: Standard model with fixed butt-stock and standard barrel

FN 50-63: Model with side-folding butt-stock and short barrel

FN 50-64: Model with side-folding butt-stock and standard barrel

FN 50-41: Model with fixed butt-stock, bipod and heavy barrel

Data

Cartridge: 7.62 × 51 mm NATO
Operation: gas, selective or semi-automatic fire
Locking: dropping bolt
Feed: 20-round steel or light box magazine
Weight: empty, without magazine, 4.455 kg
Barrel: (regulator) exhaust to atmosphere system
Rifling: 4 grooves, rh, 1 turn in 305 mm
Sights: fore, cylindrical post; rear, aperture, sliding on bed, adjustable 200-600 m × 100 m; or flip aperture 150/250 m; or fixed battle sight 300 m (see table)
Rate of fire: cyclic, 650-700 rds/min
Effective range: 650 m

Manufacturer

FN Nouvelle Herstal SA, Voie de Liège 33, B-4040 Herstal.

Status

In production (except 50-41 - production complete).

SPECIAL CHARACTERISTICS

Type	**50-00**	**50-64**	**50-63**	**50-41**
Length				
without bayonet	1.09 m	—	—	1.15 m
stock extended	—	1.095 m	1.02 m	—
stock folded	—	845 mm	770 mm	—
Weight: (without bayonet or magazine)	4.455 kg	4.305 kg	4.36 kg	6 kg
Barrel length	533 mm	533 mm	436 mm	533 m
Rearsight type	Sliding	Flip	Fixed	Sliding
Sight radius	553 mm	549 mm	549 mm	553 mm
Muzzle velocity	840 m/s	840 m/s	810 m/s	840 m/s
Muzzle energy	3,281 J	3,281 J	3,051 J	3,281 J

Service

Argentina, Australia, Austria, Barbados, Belgium, Brazil, Burundi, Cambodia, Canada, Chile, Cuba, Dominican Republic, Ecuador, Gambia, Germany, Ghana, Guyana, India, Indonesia, Ireland, Israel, Kuwait, Liberia, Libya, Luxembourg, Malawi, Malaysia, Mexico, Morocco, Mozambique, New Zealand, Nigeria, Norway, Oman, Paraguay, Peru, Portugal, Singapore, South Africa, United Arab Emirates, and Venezuela.

Licence production

ARGENTINA
Manufacturer: Fabrica Militar de Armas Portatiles 'Domingo Matheu'
Type: 7.62 mm FN-FAL
Remarks: See entry under *Argentina* for details.
AUSTRALIA
Manufacturer: Small Arms Factory, Lithgow
Type: 7.62 mm Rifle L1A1
Remarks: No longer in production or first-line service with Australian armed forces.
AUSTRIA
Manufacturer: Steyr Mannlicher
Type: 7.62 mm Stg 58
Remarks: No longer in production or first-line service with Austrian armed forces.
BRAZIL
Manufacturer: IMBEL
Type: 7.62 mm IMBEL
Remarks: See entry under *Brazil* for details.
CANADA
Manufacturer: Canadian Arsenals Limited
Type: Rifle, 7.62 mm C1A1
Remarks: Production ceased 1968. No longer in front-line service with the Canadian Forces.
INDIA
Manufacturer: Rifle Factory, Ishapore
Type: Rifle 7.62 mm 1A1 and Rifle 7.62 mm 1C
Remarks: Based on UK L1A1. Still in production and offered for export sales. Rifle 7.62 mm 1C is a special version for use when firing from Sarath (BMP-1) IFV firing ports. See entry under *India* for details.
ISRAEL
Manufacturer: Israel Military Industries (now TAAS - Israel Industries)
Type: 7.62 mm FN-FAL
Remarks: No longer in production. Many sold as surplus.

MEXICO
Manufacturer: Government facilities
Type: 7.62 mm FN-FAL
Remarks: Production complete.
SOUTH AFRICA
Manufacturer: Lyttleton Engineering Works
Type: 7.62 mm R1
Remarks: No longer in production. Retained as reserve weapon.
UNITED KINGDOM
Manufacturer: Royal Small Arms Factory, Enfield Lock
Type: 7.62 mm Rifle L1A1
Remarks: Production complete. No longer in service in the United Kingdom but may be encountered in service with many other nations. See entry under *United Kingdom* for details.
VENEZUELA
Manufacturer: C A Venezolana de Industrias Militares (CAVIM)
Type: Fusil Automatico Livano
Remarks: Production probably complete.

VERIFIED

5.56 mm FN-FNC assault rifle

Development

The 5.56 mm FN-FNC is a light assault rifle intended for use by infantry operating without continuous logistical support, or who are in jungle, mountain or other difficult country. The construction is from steel, aluminium alloy and, for non-working parts, plastic, great use being made of stampings and pressings.

There are two versions of the rifle; the first, known as the Standard, has a standard length barrel together with a folding tubular light alloy butt-stock encased in a plastic coating and braced by a plastic strut; the second model, the Para, is similar but has a shorter barrel. Optional on the Standard is a fixed polyamid stock. A special bracket to accept the US M7 bayonet is provided on the standard barrel.

The layout of both models follows the general pattern of FN rifles and the body opens on a front pivot pin to allow the working parts to be taken out in much the same way as the Fusil Automatique Léger (FAL).

The 5.56 mm FN-FNC is licence produced in Indonesia and, in a much modified form, in Sweden. See separate entries for details of these two programmes.

Description

Operation of the 5.56 mm FN-FNC is by gas, using a conventional piston and cylinder mounted above the barrel. The breech is locked by a rotating bolt, with a two-lug head locking into the barrel extension. On the FNC the bolt and carrier are among the very few items which need precision machining in manufacture. The gas regulator acts directly on the gas passage at the port opening and the escape hole is open or shut by the gas cylinder. The open hole is the standard setting and is used for all normal firing. The closed hole allows full gas to flow and is meant for use only in adverse conditions.

The body is made from pressed steel, with the bearing surfaces for the bolt and carrier inserted. The trigger frame is of light alloy. The top and bottom of the body are held together at the front by a pin on the under side and they are locked by a pin which pushes in from the right side just above the pistol grip. When this pin is pushed out the lower body can be dropped away and the working parts are exposed and can then be lifted out. No tools are needed and in the field no further stripping is necessary.

The cocking handle is on the right side, working in a slot which is normally protected by a cover held shut by a spring. This cover remains closed at all times and ensures that the working parts are protected from mud, dirt and snow. The movement of the cocking handle opens the cover by a simple cam action, but it closes as soon as the handle reaches the forward position.

The sights are conventional in that they are an adjustable post for the foresight and an aperture for the rearsight. There is lateral adjustment for the rearsight and flip settings for range. For grenade firing a gas-tap is folded up over the foresight and this cuts off the flow of gas to the cylinder, thereby allowing the maximum pressure behind the grenade. The skeleton tubular butt folds to the right, fitting below the cocking handle. This reduces the length of the Standard version to 776 mm and the Para version to 680 mm so allowing it to be used like a sub-machine gun. There is also a fixed butt version of the Standard model with a polyamid butt.

The receiver is provided with a telescope support capable of accommodating all sighting devices conforming to the NATO standard mounting plan. When firing with these sights it is recommended that an optional light bipod be clipped on to steady the weapon.

The steel magazine is interchangeable with the US M16 series rifle and both types of magazines can be used for the FNC, the Minimi light machine gun and the M16.

FN anticipated that possible users may have different ammunition requirements and the rifle is offered with two possible barrel riflings. The first is optimised for the NATO standard FN SS109 cartridge and has a twist of 1 in 178 mm (1:7 in). The second rifling, available to special order, is the same as the M16A1 with a twist of 1 in 305 mm (1:12 in) and will fire the M193 or FN SS92 cartridge.

Data

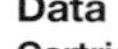

Cartridge: 5.56 × 45 mm (M193 or SS109)
Operation: gas, selective fire with 3-round burst controller
Locking: rotating bolt
Feed: 30-round box magazine
Weight: (empty magazine) Standard, 4.01 kg; Paratroop, 3.81 kg; Fixed butt, 4.06 kg: (loaded, 30-round magazine) Standard, 4.36 kg; Paratroop, 4.16 kg; Fixed butt, 4.41 kg
Length: Standard, butt extended, 997 mm; Standard, butt folded, 776 mm; Paratroop, butt extended, 911 mm; Paratroop, butt folded, 680 mm; Fixed butt, 1,012 mm
Barrel: standard, 449 mm; short, 363 mm
Rifling: 6 grooves, rh, 1 turn in 305 mm or 178 mm
Sights: fore, cylindrical post; rear, flip aperture 250 and 400 m
Sight radius: 513 mm
Rate of fire: cyclic, 625-750 rds/min
Muzzle velocity: M193, 965 m/s; SS109, 915 m/s

Manufacturer

FN Nouvelle Herstal SA, Voie de Liège 33, B-4040 Herstal.

Status

In production.

Service

Adopted by Belgium, Indonesia, Nigeria, Sweden and others.

Licence production

INDONESIA
Manufacturer: PT Pindad
Type: 5.56 mm SS1-V1, SS1-V2, SS1-V3
Remarks: See entry under *Indonesia* for details.
SWEDEN
Manufacturer: FFV (now Bofors Carl Gustaf AB)
Type: 5.56 mm CGA5/Ak5
Remarks: Modified to meet Swedish Armed Forces requirements. See entry under *Sweden* for details.

UPDATED

5.56 mm FN-FNC assault rifle undergoing harsh conditions trials

5.56 mm FN-FNC assault rifle, Para model

5.56 mm FN-FNC assault rifle, Standard model

5.56 mm FN-FNC assault rifle, Standard model

BRAZIL

7.62 mm IMBEL semi-automatic rifle SAR

Description
IMBEL produces the standard FN-FAL rifle in semi-automatic standard and paratroop form, for supply to the armed forces of Brazil and other countries. The dimensions and data are exactly as for the Belgian original (qv).

Manufacturer
Indústria de Material Bélico do Brasil - IMBEL, Rua São Joaquim 329, Liberdade, CEP-01508-001 São Paulo SP.

Status
In production.

Service
Brazilian and other military forces.

VERIFIED

IMBEL 7.62 mm semi-automatic rifles, SAR and Para-SAR

7.62 mm IMBEL light automatic rifle

Description
The IMBEL light automatic rifle in 7.62 × 51 mm calibre has been proved in use under the most severe conditions. It is the FN-FAL manufactured under licence and over 150,000 have been produced, 40,000 of which have been exported.

The rifle is gas-operated, with an adjustable regulator to ensure smooth and certain action without excessive recoil. The bolt is mechanically locked before firing and is not unlocked until the bullet has left the barrel.

Since the bolt is in the closed position when the trigger is pressed, there is no disturbance of aim caused by movement of a heavy mass, a drawback with many automatic weapons.

Data
Cartridge: 7.62 × 51 mm NATO
Operation: gas, selective fire
Locking: tilting bolt
Feed: 20-round box magazine
Weight: empty, 4.5 kg; Paratroop, 4.37 kg
Length: 1.1 m; Paratroop, 990 mm
Barrel: 533 mm; Paratroop, 437 mm
Rifling: 4 grooves, rh
Rate of fire: cyclic, 650-750 rds/min

Manufacturer
Indústria de Material Bélico do Brasil - IMBEL, Rua São Joaquim 329, Liberdade, CEP-01508-001 São Paulo SP.

Status
In production.

Service
Brazilian and other military forces.

VERIFIED

IMBEL light automatic rifles with bayonets

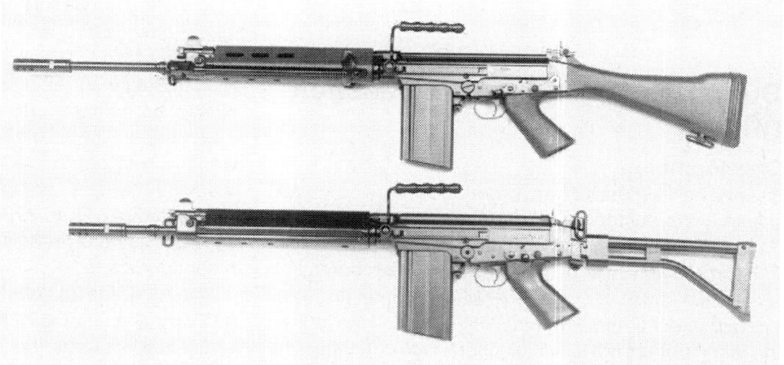

IMBEL standard and paratroop 7.62 mm light automatic rifles

5.56 mm IMBEL MD2 and MD3 rifles

Description
Using the 7.62 mm light automatic rifle (see previous entry) as a starting point, the Fábrica de Itajubá developed two 5.56 mm versions for use by the Brazilian Army and for export. The MD2 has a folding butt, is gas-operated, offers selective single shot or automatic fire, and will fire either M193 or SS109 ammunition with equal facility. The MD3 is to the same general specification but uses a fixed plastic butt. Both rifles are normally supplied in selective-fire form but if required can be supplied with irreversible semi-automatic operation.

The rifles have been designed and developed in conformity with various international and NATO standards and have the advantage of a high proportion of parts interchangeable with the 7.62 mm light automatic rifle LAR.

5.56 mm IMBEL MD2 rifle

Data

Cartridge: 5.56 × 45 mm M193 or SS109
Operation: gas, selective fire
Feed: 20-round box magazine or 30-round US M16 magazine
Weight: 4.404 kg; MD3, 4.565 kg
Length: MD3 and MD2 with stock extended, 1.01 m; MD2 with stock folded, 764 mm
Barrel length: without flash hider, 453 mm
Rifling: 6 grooves, rh, 1 turn in 305 mm or 1 turn in 178 mm, according to selected ammunition
Sights: fore, post; rear, flip aperture 150 m and 250 m (Paratroop version) or 200 m and 600 m (Standard version); adjustable for windage
Rate of fire: cyclic, 700-750 rds/min

Manufacturer

Indústria de Material Bélico do Brasil - IMBEL, Rua São Joaquim 329, Liberdade, CEP-01508-001 São Paulo SP.

Status

In production.

Service

Brazilian and other armies.

VERIFIED

0.22 IMBEL MD2A1 training rifle

Description

This is virtually a replica of the light 7.62 mm calibre automatic rifle (Fabrique Nationale's FN-FAL) in terms of size and weight but it fires the cheaper 0.22 rimfire cartridge. Rather than offering a conversion kit for the service rifle, the MD2A1 is factory-made in the smaller calibre. The 10-round 0.22 calibre non-removable magazine fits inside a false FAL magazine; the cocking handle is on the left of the receiver and the rifle is used in exactly the same manner as the 7.62 mm version.

Data

Cartridge: 0.22 LR rimfire
Operation: blowback, selective fire
Feed: 10-round box magazine
Weight: 4.14 kg

Manufacturer

Indústria de Material Bélico do Brasil - IMBEL, Rua São Joaquim 329, Liberdade, CEP-01508-001 São Paulo SP.

Status

In production. Prototypes have been evaluated by the Brazilian Army and the rifle is available for export.

VERIFIED

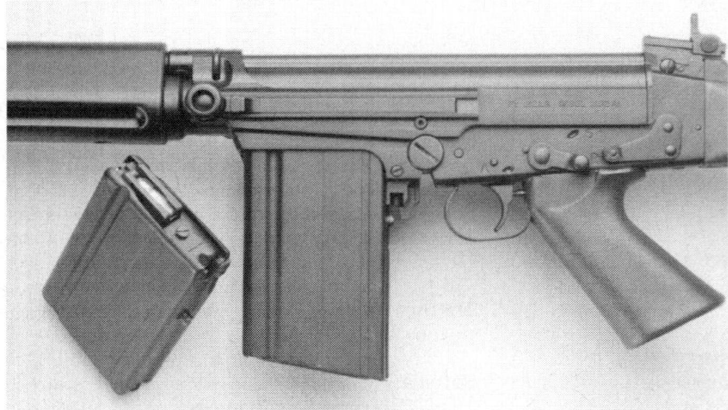

0.22 in IMBEL MD2A1 training rifle

0.22 IMBEL MD2A1 training rifle, showing cocking handle and a top view of the special magazine

BULGARIA

Bulgarian 7.62 mm AK-47 assault rifles

Development

Despite having passed from production elsewhere, Bulgarian state factories continue to manufacture the 7.62 mm AK-47 assault rifle and offer it for export sales. For some reason the Bulgarian weapons procurement authorities chose to skip the AKM generation and proceeded direct to the 5.45 mm AK-74 series (see following entry) but did not lose the capability to continue with the AK-47.

Four AK-47 models are on offer, plus a 5.6 mm training rifle, all of them produced to a high standard of finish and including some features also seen on the AKM series.

Description

The Bulgarian 7.62 mm AK-47s are identical in operation to the original Soviet CIS AK-47 so reference should be made to the AK-47 entry under *Commonwealth of Independent States* for full details.

The Bulgarian AK-47s make considerable use of plastic or compressed resin fibre furniture but are otherwise similar to other AK-47s. The basic AK-47 is virtually identical to late production CIS models, as is the AKS-47 with a folding butt.

The AKN-47 is an AK-47 fitted with an NSPU night sight providing an effective range of up to 600 m under night conditions.

The AK-47M1 has black all-plastic furniture and is fitted with a 40 mm GP-25 (Kastyor) grenade launcher under the forestock.

All four models are provided with a bayonet, four magazines in a pouch, a sling, spares, an oiler and some other accessories.

Data - AK-47

Cartridge: 7.62 × 39 mm
Operation: gas, selective fire
Locking: rotating bolt
Feed: 30-round detachable box magazine
Weight: without magazine, 3.63 kg; loaded, 4.45 kg
Length: 870 mm
Barrel: 414 mm
Rifling: 4 grooves, rh, 1 turn in 235 mm
Sights: fore, post, adjustable; rear, U-notch tangent, adjustable to 800 m with battle sight for 200 m
Sight radius: 376 mm
Muzzle velocity: 715 m/s
Rate of fire: cyclic, 600 rds/min

Manufacturer

Arsenal, 6100 Kazanluk, Rozova dolina str 100.
Marketed by: Kintex, 66 James Boucher Street, PO Box 209, 1407 Sofia.

Status

In production. Offered for export sales.

Service

Bulgarian armed forces.

UPDATED

Bulgarian 7.62 mm AK-47M1 assault rifle fitted with 40 mm GP-25 (Kastyor) grenade launcher

Bulgarian 7.62 mm AKN-47 assault rifle fitted with NSPU night sight

Bulgarian AK-74 assault rifles

Development

Bulgarian AK-74 assault rifles are manufactured in three basic models (see below) chambered for either the standard eastern bloc 5.45 × 39 mm round or 5.56 × 45 mm NATO standard ammunition, both versions being intended for export sales.

These AK-74 models are marketed by Kintex of Sofia.

Description

Bulgarian AK-74s are identical in operation to the CIS originals so reference should be made to the AK-74 entry under *Commonwealth of Independent States* for full details.

The Bulgarian AK-74s utilise plastic furniture and are generally built to late production AK-74 standards. The basic Bulgarian AK-74, as with the other Bulgarian AK-74 models, can be fitted with a 40 mm GP-15 or GP-25 (Kastyor) grenade launcher under the forestock. The AKS-74 has a folding butt.

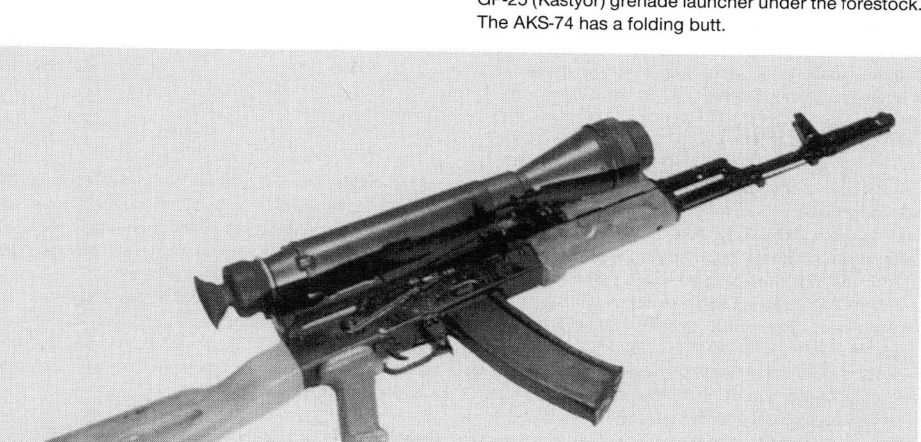

Bulgarian 5.56 mm AKN-74 assault rifle fitted with NSPU night sight

The AKN-74 is an AK-74 fitted with a NSPU night sight providing an effective range of up to 600 m under night conditions.

All three models are provided with a bayonet, four magazines in a pouch, a sling, spares, an oiler and some other accessories.

Data

Cartridge: 5.45 × 39 mm or 5.56 × 45 mm NATO
Operation: gas, selective fire
Locking: rotating bolt
Feed: 30-round detachable box magazine
Weight: AKS-74, without magazine, 3.2 kg; AKS-74, loaded, 3.736 kg; night sight, 2 kg
Length: AKS-74, butt folded, 690 mm; AKS-74, butt extended, 942 mm
Barrel: 415 mm
Rifling: 4 grooves, rh, 1 turn in 196 mm
Sights: fore, post, adjustable; rear, U-notch
Muzzle velocity: 5.45 mm, 900 m/s; 5.56 mm, 1,005 m/s
Rate of fire: cyclic, 5.45 mm, 600 rds/min; 5.56 mm, 650 rds/min

Manufacturer

Arsenal, 6100 Kazanluk, Rozova dolina str 100.
Marketed by: Kintex, 66 James Boucher Street, PO Box 209, 1407 Sofia.

Status

In production. Offered for export.

Service

5.45 mm, Bulgarian armed forces.

UPDATED

CANADA

7.62 mm C1 and C1A1 rifles

Description

The Canadian C1 is derived from the FN-FAL and differs from the British Army's L1A1 in having an opening in the feed cover through which the magazine can be recharged using chargers and guides. It also has a rearsight which is a disc offering five different ranges, from 200 to 600 m, in 100 m intervals, by rotating the outside of the disc. The range appears in an aperture at the bottom of the front face of the disc.

In 1960 the British L1A1 was found to be firing rounds before the breech was closed. This was caused by the long firing pin bending on firing and failing to withdraw into the breech block. When the next round was chambered the projecting firing pin could fire the cap. The long firing pin was replaced by a two-part firing pin. In Canada this modification produced the C1A1 rifle which also has a plastic carrying handle.

7.62 mm C1 rifle showing charger guides and disc rearsight

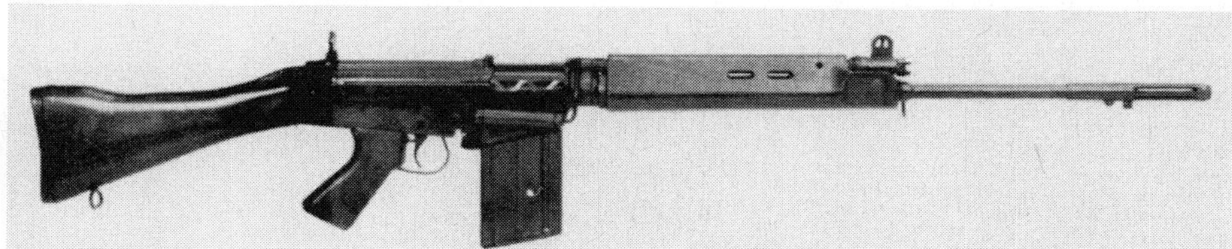

Canadian 7.62 mm C1A1 rifle

Data - C1A1

Cartridge: 7.62 × 51 mm
Operation: gas, semi-automatic
Feed: 20-round box magazine
Weight: 4.25 kg
Length: 1.136 m
Barrel: 533 mm
Sights: fore, post; rear, aperture, disc
Muzzle velocity: 840 m/s

Manufacturer

Canadian Arsenals Limited.

Status

Production complete.

Service

Reserve use with Canadian Forces.

UPDATED

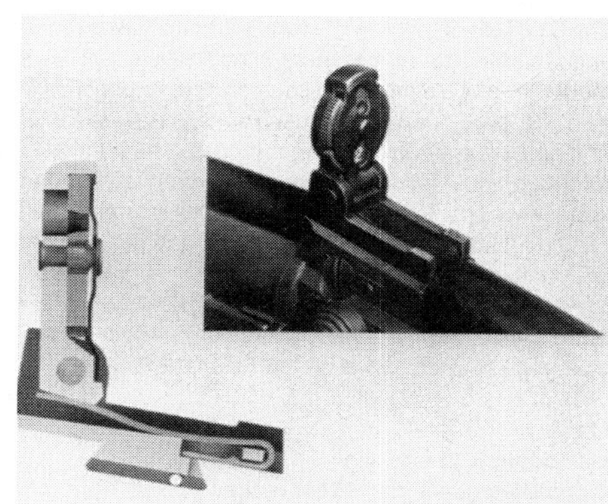

Canadian disc rearsight

7.62 mm C2 and C2A1 automatic rifles

Description

The heavy-barrelled Fabrique Nationale (FN) rifle was taken by the Canadian Army as its Squad Light Automatic. It has selective fire allowing either single shots or full automatic. There are box magazines holding 20 or 30 rounds. The rearsight is the same type of rotating disc as used on the C1 rifle, which can be adjusted from 200 to 900 m.

The C2 rifle may readily be recognised by the uncovered gas cylinder over the forward grip. The gas regulator takes a light bipod.

The C2 rifle was modified by the substitution of a two-part firing pin and a plastic carrying handle, to become the C2A1.

Data

Cartridge: 7.62 × 51 mm
Operation: gas, selective fire
Locking: tilting block
Feed: 20- or 30-round box magazine
Weight: 6.93 kg
Length: 1.136 m
Barrel: 533 mm
Sights: fore, post; rear, tangent, aperture 200-1,000 yds (182.8-914.4 m)
Muzzle velocity: 854 m/s
Rate of fire: cyclic, 710 rds/min

Manufacturer

Canadian Arsenals Limited.

Status

No longer in production.

Service

Canadian Forces.

VERIFIED

7.62 mm C3 and C3A1 sniper rifle

Description

The Canadian sniper rifle, the C3, is the British Parker-Hale Model 82 modified slightly to meet Canadian requirements. The stock is fitted with four 12.7 mm spacers to permit length adjustment. The rifle has two male dovetail blocks on the receiver, to accept either the Parker-Hale 5E vernier rearsight, or the Austrian Kahles ×6 telescope. All exposed metal parts of the rifle are non-reflective.

The Canadian Department of Defence later adopted a revised version of the C3 rifle which was redesignated C3A1. The rifle incorporates a number of important changes from the earlier C3 model which entered Canadian service in the mid-1970s. The stock configuration remains unchanged but major alterations were made to the action body which was strengthened and equipped with a six-round capacity steel box magazine with release catch located in the leading edge of the trigger guard. The trigger mechanism has a redesigned safety catch and is of all-steel construction. Rifling specifications were changed to provide optimum performance from both standard NATO 7.62 × 51 mm ball and heavier bullets which have been developed for sniping and competition use. A Parker-Hale bipod is provided which attaches to a handstop assembly. The bipod has extendable legs, allows approximately 14° of swivel and cant, and can be folded either forward or back in addition to being hand detachable.

The telescope sight, which was specifically designed for use in conjunction with the C3A1 rifle, is a ×10 magnification Unertl model similar to that adopted by the US Marine Corps. The body and controls are of all-steel construction and the consequent mass requires specially reinforced mounts which have been designed to withstand a very high degree of stress.

The bolt handle was modified by the addition of an aluminium extension knob which provides easier purchase particularly with a gloved hand, and greater clearance from the telescope sight than was the case with the C3 model.

7.62 mm C3A1 sniper rifle

Data - C3A1

Cartridge: 7.62 × 51 mm NATO
Operation: bolt-action, single shot
Feed: 6-shot box magazine
Weight: empty, with Unertl sight, 6.3 kg
Length: 1.14-1.21 m
Barrel: 660 mm
Rifling: 4 grooves, rh, 1 turn in 305 mm

Manufacturer

Parker-Hale Limited, Bisley Works, Golden Hillock Road, Birmingham B11 2PZ, UK.

Status

Production complete.

Service

Entered service with Canadian Forces 1989.

UPDATED

5.56 mm C7 assault rifle

Description

In 1984 the Canadian Forces adopted a new generation of 5.56 mm NATO calibre weapons for all services. The chosen weapons, designated C7, were based upon a M16A2 model licence produced by Diemarco; the Colt model number is 715.

In 1994 the C7 family of weapons was adopted by the Netherlands armed forces. The order included 50,680 C7 rifles, some being the C7A1 with ×3.5 optical sights, and 1,400 C8 carbines (for the Gendarmerie).

Description

The 5.56 mm C7 fires in the fully automatic firing mode and a single-shot mode; there is no three-round burst mode. It uses a simple two-position aperture rearsight with one large aperture for short-range firing and a second smaller aperture for longer ranges. The rifle has a cartridge case deflector present to permit left-handed firing and the barrel is a cold-forged chrome-lined design rifled to accept both SS109 and M193 type ammunition.

A full range of accessories for the C7 has been developed, including bayonet, adjustable length stock, cleaning kits, blank firing attachment, and a sling.

Data

Cartridge: 5.56 × 45 mm
Operation: direct gas, selective fire
Locking: rotating bolt
Feed: 30-round box magazine
Weight: empty, 3.3 kg; loaded, 3.9 kg
Length: normal butt, 1.02 m
Barrel: 510 mm
Rifling: 6 grooves, rh, 1 turn in 178 mm
Sights: fore, post; rear, 2-position aperture
Muzzle velocity: SS109 at 24 m, 920 m/s
Rate of fire: cyclic, 800 rds/min

5.56 mm C7 assault rifle

Manufacturer

Diemaco, 1036 Wilson Avenue, Kitchener, Ontario N2C 1J3.

Status

In production.

Service

Canadian and Netherlands armed forces.

UPDATED

5.56 mm C7A1 assault rifle

Description

The 5.56 mm C7A1 assault rifle is an improved version of the basic C7 combat rifle, incorporating a low-mounted optical sight. The C7A1 eliminates the carrying handle of the C7 and substitutes an optical sight mounting rail which positions the selected optic in a natural position for rapid target acquisition. A fully capable back-up iron sight fits into the butt trap of the C7A1 for use should the optic not be fitted. The C7A1 requires only five new components, plus the optic module, and was introduced into service alongside the C7.

The C7A1 is available with either a Weaver type rail configuration or a RARDE type rail.

In 1995 Denmark purchased 2,500 C7A1 rifles for use by Danish armed forces personnel serving with the United Nations in the former Yugoslavia.

Data

AS FOR C7 RIFLE EXCEPT:

Weight: with Canadian sight, empty, 3.9 kg; loaded, 4.5 kg

Sights: optical or fore, post; rear, 2-position aperture

Manufacturer

Diemaco, 1036 Wilson Avenue, Kitchener, Ontario N2C 1J3.

5.56 mm C7A1 rifle with back-up iron sight fitted

Status

In production.

Service

Canadian, Danish and Netherlands armed forces.

UPDATED

5.56 mm C8 assault carbine

Description

The 5.56 mm C8 assault carbine is a compact version of the Canadian forces standard C7 rifle. The C8 features a telescoping butt-stock and a shortened barrel, while retaining all normal field replaced parts common with the C7 rifle. The carbine is being issued to armoured vehicle crews and may subsequently be issued to special forces and other users requiring a more compact personal weapon than the C7.

In spite of the shortened barrel the C8 has demonstrated accuracy nearly equivalent to the C7 in service, and it shares most of the firing characteristics of the rifle. The carbine uses the same hammer-forged barrel as the rifle, for superior life and accuracy, and it also accepts all C7 accessories.

Data

AS FOR C7 RIFLE EXCEPT:

Weight: empty, 2.7 kg; loaded, 3.2 kg

Length: butt closed, 760 mm; butt extended, 840 mm

Barrel: 370 mm

Rate of fire: cyclic, 900 rds/min

Manufacturer

Diemaco, 1036 Wilson Avenue, Kitchener, Ontario N2C 1J3.

5.56 mm C8 assault carbine and accessories

Status

In production.

Service

Canadian Armed Forces and Netherlands Gendarmerie.

UPDATED

0.22 LR Small Bore Cadet rifle

Description

The Small Bore Cadet rifle is an exact copy of the C7 assault rifle, manufactured to fire the 0.22 LR rimfire cartridge in the semi-automatic mode only. The Small Bore rifle is a primary marksmanship trainer for use by cadets, militia units and regular force units on indoor range facilities. The rifle features a barrel designed for the 0.22 LR projectile and a 10- or 15-round capacity magazine which fits within a standard 5.56 mm magazine body for total weapon handling practice. Many components of the C7 have been redesigned in plastic, including the complete lower receiver and butt-stock assembly, to make the Small Bore rifle less expensive to acquire.

A kit is also available to convert existing C7/M16 type rifles to Small Bore rifle standard.

Data

AS C7 RIFLE, EXCEPT:

Feed: 10- or 15-round box magazine

Rifling: rh twist, 1 turn in 406 mm

Manufacturer

Diemaco, 1036 Wilson Avenue, Kitchener, Ontario N2C 1J3.

0.22 LR Small Bore Cadet rifle and conversion kit

Status

In production.

Service

Canadian Armed Forces.

UPDATED

CHILE

5.56 mm SG540-1 assault rifle
7.62 mm SG542-1 assault rifle

Description

These two rifles are the SIG SG540 and SG542, described more fully under *Switzerland*. The Chilean versions are built and tested by FAMAE in accordance with a licence from SIG and form part of a 'weapons family', with a high percentage of common components and identical handling, training and maintenance characteristics.

Both weapons are manufactured to weight-saving design concepts, with a heavy barrel hammer-forged to form the chamber and bore in one operation.

Other features include a three-round burst facility; rotating gas regulator; two-stage trigger pull with a distinctive let-off point; folded trigger arc; a bipod attached to the fore-end; sights adjustable for both windage and elevation; and, in the SG542-1 model, a hard-chromed chamber. Both rifles are produced in fixed or folding butt models.

Data

Model	**540-1**	**542-1**
Cartridge	5.56 × 45 mm	7.62 × 51 mm
Operation	gas, selective fire	
Locking	rotating bolt	
Feed	20- or 30-round box magazines	
Weight		
fixed butt	3.54 kg	3.83 kg
folding butt	3.59 kg	3.83 kg
Length		
fixed butt	950 mm	1.02 m
butt folded	720 mm	754 mm
Barrel	460 mm	465 mm
Rifling length	412 mm	412 mm
Rifling twist	6 grooves, rh 1/305 mm	4 grooves, rh 1/305 mm
Sights	rear aperture, front blade	
Sight radius	495 mm	528 mm
Muzzle velocity	980 m/s	820 m/s
Muzzle energy	1,716 J	3,187 J
Rate of fire	cyclic, 650-800 rds/min	

Manufacturer

FAMAE Fabricas y Maestanzas del Ejercito, Avenida Pedro Montt 1568, PO Box 4100, Santiago.

Status

In production.

Service

Chilean armed forces.

VERIFIED

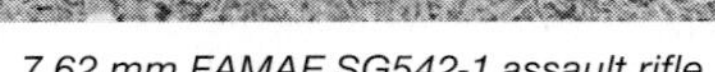

7.62 mm FAMAE SG542-1 assault rifle

5.56 mm FAMAE SG540-1 assault rifle

CHINA, PEOPLE'S REPUBLIC

7.62 mm Type 56 carbine

Description

The 7.62 mm Type 56 carbine is a Chinese copy of the Soviet Simonov self-loading carbine (SKS) and may be identified by the Chinese symbols on the left front of the receiver. The weapon is fashioned and functions in exactly the same way as the SKS (qv). Later versions of the Type 56 have a spike bayonet replacing the folding blade of conventional shape used on all other SKS variants.

The Type 56 is manufactured in various forms for sale as a sporting rifle.

Data

Cartridge: 7.62 × 39 mm
Operation: gas, semi-automatic
Locking: tilting block
Feed: 10-round internal magazine
Weight: unloaded, 3.85 kg
Length: bayonet folded, 1.025 m; bayonet extended, 1.332 m
Barrel: 521 mm
Sights: fore, cylindrical post; rear, tangent, notch
Muzzle velocity: 735 m/s
Effective range: 400 m

Manufacturer

China North Industries Corporation (NORINCO), 7A Yue Tan Nan Jie, PO Box 2137, Beijing.

Status

Manufactured for commercial sale.

Service

Reserve units only within China. In limited service in Cambodia.

UPDATED

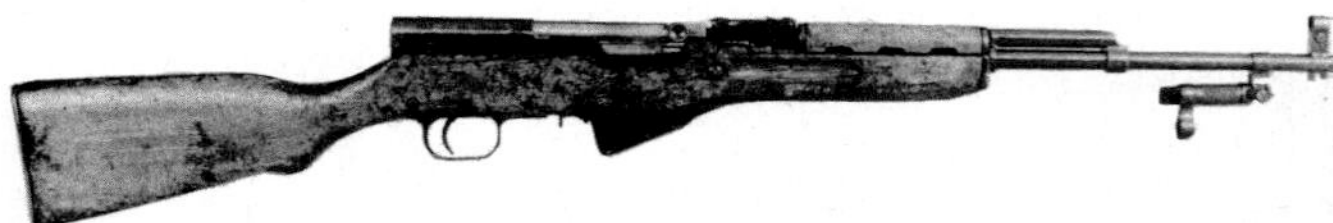

Chinese SKS 7.62 mm Type 56 carbine

Chinese militia receiving instruction on 7.62 mm Type 56 carbine

7.62 mm Types 56, 56-1, 56-2 and 56C assault rifles

Development

The original 7.62 mm Type 56 assault rifle was a copy of the later model of the Soviet AK-47 in which the rear end of the top surface of the receiver is straight. Earlier models had a down slope on the top rear surface of the receiver. As production progressed the Type 56 incorporated many of the changes introduced on the AKM but the Type 56 designation remained unaltered. Production and development of the Type 56 series continues, the latest model being the Type 56C.

Description

The Type 56 models can be identified by the Chinese characters on the right-hand side of the receiver showing the positions of the selector lever, the upper and lower positions being for full automatic and single shots respectively. Later production models show 'L' for the full automatic marking and 'D' for single shot. The basic fixed-butt Type 56 is fitted with a folding bayonet which acts as a recognition feature to differentiate the Type 56 from other AK-47 models. The bayonet hinges down and back to lie beneath the fore-end.

There is also a Type 56-1 assault rifle which has a folding metal stock. This may be distinguished from the

Chinese 7.62 mm Type 56-2 assault rifle with side-folding stock

Chinese 7.62 mm Type 56-1 assault rifle with folding stock

AK-47 not only by the Chinese markings or 'L' and 'D' for automatic and single shot respectively, but also by the prominent rivets in the arms of the butt-stock, which are not present on the original Soviet version. The Type 56-1 lacks the fixed bayonet of the Type 56.

The Type 56-2 generally resembles the 56-1 but has a butt-stock which folds sideways to lie along the right side of the receiver. This version also lacks the fixed bayonet of the Type 56.

All three of the above models are commercially available in semi-automatic form as the Type 56S sporting rifle series.

Type 56C

The latest model in the Type 56 series is the Type 56C. This may be regarded as a compact Type 56, incorporating many of the various design enhancements bestowed upon the CIS AK-74. However, the Type 56C retains the use of the 7.62 × 39 mm cartridge. Numerous changes have been made to the furniture, which is now all plastic-based, the muzzle has a revised flash hider, there is a new pattern of side-folding butt stock with an integral cheek pad, revised sights are provided, a cleaning kit is stowed inside the revised pistol grip, and various modifications have been introduced to the bolt assembly. There are numerous other modifications compared to earlier models. The Type 56C is 765 mm long with the butt extended, 563 mm with the butt folded. Barrel length is 345 mm. Weight empty is 3.5 kg.

Chinese 7.62 mm Type 56 rifle

Data

Cartridge: 7.62 × 39 mm
Operation: gas, selective fire
Locking: rotating bolt
Feed: 30-round detachable box magazine
Weight: 3.8 kg
Length: 874 mm; Type 56-1, stock folded, 645 mm; Type 56-2, stock folded, 654 mm
Barrel: 414 mm
Sights: fore, cylindrical post; rear, tangent, notch
Muzzle velocity: 710 m/s
Rate of fire: cyclic, 600 rds/min
Effective range: 300 m

Manufacturer

China North Industries Corporation (NORINCO), 7A Yue Tan Nan Jie, PO Box 2137, Beijing.

Status

In production.

Service

Chinese forces. Supplied to some South-east Asian countries, including Cambodia.

UPDATED

7.62 mm Type 68 rifle

Development

The 7.62 mm Type 68 rifle is of Chinese design and manufacture. In general appearance it resembles the Type 56 (SKS) carbine but the barrel is longer, the bolt action is based on that of the AK-47 and the rifle provides selective fire. It has a two-position gas regulator. It normally uses a 15-round box magazine but if the bolt stop is removed, or ground down, the 30-round magazines of the AK-47 and AKM can be used.

There are two versions of the Type 68. The earlier version has a receiver machined out of the solid whereas the later version has a pressed steel receiver. It can be recognised by the large rivets in the side of the receiver.

Description

The Type 68 can be loaded in one of four ways. First, using an empty 15-round magazine, cocking the action holds the bolt to the rear; a 10-round SKS-type charger can then be put into the feed guides and the rounds forced down. Five further rounds can then be pressed in from the next charger. Secondly, if chargers are not available, 15 rounds can be forced down, one after the other, into the magazine. Thirdly, the 15-round magazine can be pre-loaded off the gun and then placed in position. Lastly if the bolt stop has been removed or ground down the 30-round magazine can be used but it must be pre-filled off the gun as the bolt will automatically close on the empty chamber and close off the top feed opening.

When the bolt is retracted over the loaded magazine and released, the top round is fed into the chamber as the bolt flies forward.

The selector lever is positioned directly in front of the trigger on the right. When the selector is pulled to the rear, reading '0', the trigger is locked. The bolt, however, can be pulled back if required. The middle position, marked '1', with the lever vertical, provides semi-automatic fire and when the lever is fully forward, the gun will fire at full automatic. This position is marked '2'.

The sights are adjusted by pressing in on the side locks and moving the sight bar along the leaf to the required range. The position marked III is a battle sight position for all ranges up to 300 m.

Normally the gas regulator will be set with the smaller of the two holes nearer the barrel. If it is necessary to use the large hole to increase the gas flow the gas regulator retainer is pressed in towards the cylinder. When it disengages from the handguard, the retainer is pulled out of the gas cylinder and then the regulator is rotated to select the other hole. The retainer is replaced and rotated and the gun should then be ready to fire.

The bayonet is similar to that of the SKS. The handle is forced back against the spring and the bayonet then rotated forward and locked at the muzzle.

To fire the weapon, the selector lever is placed on '1' or '2'. When the trigger is pulled the hammer strikes the firing pin and the cartridge is fired. Some of the propellant gas passes through the vent and into the gas cylinder above the barrel. The piston is forced back, and the bolt carrier starts to move to the rear and rotate the hammer back. After a period of free travel of about 6 mm, the cam cut in the bolt carrier reaches the operating lug of the bolt. The bolt is rotated, providing initial extraction, and then withdrawn. The fired case is held to the bolt-face by the extractor until the ejector pivots it out to the right of the gun. The return spring is compressed. The bolt carrier hits the rear wall of the receiver and rebounds. As it goes forward the next round is fed and chambered. The extractor clips over the cartridge rim and the bolt comes to rest. The carrier continues on and the cam forces the operating lug across and rotates the bolt into the locked position. The carrier, in its final forward motion, after the bolt is fully locked, releases the safety, or automatic, sear and then stops as it reaches the front of the receiver.

Chinese 7.62 mm Type 68 rifle (T J Gander)

Data

Cartridge: 7.62 × 39 mm
Operation: gas, selective fire
Locking: rotating bolt
Feed: 15-round detachable box magazine (30-round box magazine from the AK-47 and AKM can be used only if the holding open stop is removed)
Weight: 3.49 kg
Length: 1.029 m
Barrel: 521 mm
Rifling: 4 grooves, rh
Sights: fore, cylindrical pillar; rear, tangent, notch
Muzzle velocity: 730 m/s
Rate of fire: cyclic, 750 rds/min

Manufacturer

China North Industries Corporation (NORINCO), 7A Yue Tan Nan Jie, PO Box 2137, Beijing.

Status

Available for export.

Service

Chinese armed forces.

VERIFIED

7.62 mm Type 79 sniper rifle

Description

The 7.62 mm Type 79 is a precise copy of the Soviet SVD Dragunov sniper rifle, except that the butt is slightly shorter. It is equipped with a ×4 magnification optical sight which is a copy of the Soviet PSO-1 and has the same ability to detect infra-red emissions.

Data

Cartridge: 7.62 mm × 54R Type 53
Operation: gas, short-stroke piston, self-loading
Feed: 10-round box magazine
Weight: rifle only, 3.8 kg; with optical sight, 4.4 kg
Length: 1.22 m
Barrel: 620 mm
Rifling: 4 grooves, rh, 1 turn in 254 mm
Sights: fore, adjustable post; rear, U-notch, tangent
Sight radius: 585 mm
Sight range: iron, 1,200 m; optical, 1,300 m
Muzzle velocity: 830 m/s

Manufacturer

China North Industries Corporation (NORINCO), 7A Yue Tan Nan Jie, PO Box 2137, Beijing.

Status

In production. Offered for export sales.

Service

Chinese armed forces.

VERIFIED

7.62 mm Type 81 assault rifle

Development

The 7.62 mm Type 81 assault rifle is one half of a family of Type 81 weapons developed for export sales, the other half being a light machine gun (see entry under *Machine guns* for details). The Type 81 assault rifle appears to be an amalgam of the Kalashnikov-based Type 56 assault rifle and the Type 68 rifle of purely Chinese origins, but overall must be considered as a modernised Kalashnikov series weapon produced to a high manufacturing standard.

Description

The 7.62 mm Type 81 assault rifle utilises a gas-operated mechanism based on that of the Type 68 rifle, all the major components of which are interchangeable with those on the Type 81 light machine gun. The gas regulator, a carry-over from the Type 68 rifle, can be altered without having to remove the upper handguard. Field stripping and assembly can be accomplished without recourse to special tools.

Ammunition is fed from a 30-round magazine although it is possible to utilise the 75-round drum magazine normally fitted to the Type 81 light machine gun.

The maximum effective range of the Type 81 is given as 400 m and the sights are graduated in five steps, numbered from 1 to 5. The required trigger pull can vary from 1.8 to 2.8 kg.

The muzzle can be used to fire rifle grenades and there is provision to fit a detachable bayonet which doubles as a field knife.

The standard Type 81 rifle has a solid butt which may be constructed using a plastic-based material. The Type 81-1 has a side folding frame butt.

7.62 mm Type 81 assault rifle

Data

Cartridge: 7.62 × 39 mm
Operation: gas, selective fire
Locking: rotating bolt
Feed: 30-round detachable box magazine or 75-round drum magazine
Weight: Type 81, 3.4 kg; Type 81-1, 3.5 kg with empty magazine
Length: Type 81, 955 mm; Type 81, with bayonet, 1.105 m; Type 81-1, stock folded, 730 mm
Rifling: 4 grooves, rh, 1 turn in 240 mm
Sights: fore, cylindrical pillar; rear, tangent, notch
Muzzle velocity: 720 m/s
Rate of fire: cyclic, ca 650 rds/min
Effective range: 400 m

Manufacturer

China North Industries Corporation (NORINCO), 7A Yue Tan Nan Jie, PO Box 2137, Beijing.

Status

Available for export.

Service

None known.

VERIFIED

5.56 mm Type CQ automatic rifle

Description

The 5.56 mm Type CQ automatic rifle appears to have been based upon the American M16A1, from which it differs only in minor details. It is rifled to suit the 5.56 mm cartridge Type CJ, the Chinese equivalent of the M193. The construction of the rifle differs very slightly, in that during dismantling the butt and pistol grip group is removed entirely, rather than hinged to the receiver, but the rotating bolt and trigger mechanism appear to be the same.

Data

Cartridge: 5.56 mm Type CJ (M193)
Operation: gas, selective fire
Locking: rotating bolt
Feed: 20-round box magazine
Weight: empty, 3.2 kg
Length: 987 mm
Barrel: 505 mm
Rifling: 1 turn in 305 mm
Muzzle velocity: 990 m/s

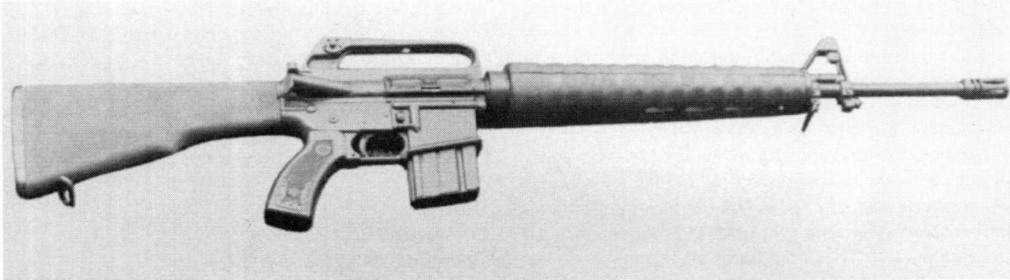

5.56 mm Type CQ automatic rifle

Manufacturer

China North Industries Corporation (NORINCO), 7A Yue Tan Nan Jie, PO Box 2137, Beijing.

Status

Production for export.

Service

Has appeared in Mojahedin hands in Afghanistan but in small numbers only.

VERIFIED

COMMONWEALTH OF INDEPENDENT STATES

9 mm AS silent assault rifle

Development

The 9 mm AS silent assault rifle (also referred to as the Val) can trace its origins back to the time when Gary Powers was shot down with his U-2 reconnaissance aircraft over the former USSR in May 1960. Among his captured equipment was a silenced pistol which so impressed those examining it that a requirement for a similar Soviet weapon was issued. The captured pistol was examined by the Central Institute for Precision Machinery Construction (TsNIITOChMASh) at Klimovsk, leading to the development of a series of silenced weapons by a team of designers including I Kas'yanov, A Deryagin, P Serdyukov and V Petrov.

Included in this series are two silenced rifles, the AS (or Val - Rampart or Shaft) and the VSS (or Vintorez - Thread Cutter - see following entry). Both these weapons are based on the receiver of the MA submachine gun (qv) and differ in detail, especially relating to the ammunition. The AS fires the 9 × 39 mm SP-6 while the SP-5 is intended to be used with the VSS.

The designers of the AS regard it not as a weapon but as a 'complex', a term encompassing not just the rifle but the special cartridge as well.

Description

The 9 mm AS silenced assault rifle uses a modified Kalashnikov action - a gas-operated rotating bolt - allied to an integral silencer assembly to provide a very low noise level allied to far greater range and penetrating power than is usual with silent weapons. The cartridge is the special 9 × 39 mm SP-6 round with a rimless necked case, firing a heavy 250 gr bullet at a muzzle velocity below the speed of sound. The weapon is capable of selective single shot or automatic fire without sustaining damage to the sound suppressor.

The suppressor used with both the AS and the VSS relies on a dual-chamber principle. As the propellant gases produced after firing a cartridge pass down the barrel they can escape through specially designed barrel perforations to enter the first chamber. Inside the chamber the expanding gases lose pressure and heat prior to passing through a series of mesh screens which break up the gas stream still further before passing them out into the outer portions of the suppressor. The resultant sound signature is far less than that from an unsuppressed rifle and from even a short distance away cannot be recognised as the discharge of a rifle.

The rifle has a short forestock and a side-folding tubular steel skeleton butt. A built-in sight bracket on the left side of the receiver will mount any CIS optical or electro-optical sight, including the PSO-1 used on the Dragunov SVD 7.62 mm sniper rifle or the 1PN52-1 night sight. Iron sights are also provided.

The 20-round magazine carries a unique series of indentations to provide tactile identification and prevent confusing this magazine with other, similar, Kalashnikov pattern magazines.

The special cartridge was developed with a view to defeating body armour at long range. It is capable of 100 per cent penetration of Levels I, II and III protection garments at ranges up to 400 m.

9 mm AS silent assault rifle with stock folded (T J Gander)

Data

Cartridge: 9 × 39 mm SP-6
Operation: gas, selective fire
Locking: rotating bolt
Feed: 20-round box magazine
Weight: with empty magazine, 2.5 kg
Length: butt folded, 615 mm; butt extended, 875 mm
Muzzle velocity: 295 m/s
Effective range: day, 400 m; night, 300 m

Manufacturer

Institute of Precise Mechanical Engineering, Izhevsk, Russia.
Marketed by: Rosvoorouzhenie, 18/1 Ovchinnikovskaya Emb, 113324, Moscow, Russia.

Status

In production. Offered for export sales.

Service

CIS special forces and Internal Ministry personnel.

9 mm AS silent assault rifle with stock extended (T J Gander)

UPDATED

9 mm VSS silent sniper rifle

Development

The development of the 9 mm VSS silent sniper rifle was carried out in parallel with that of the AS silent assault rifle (see previous entry) but whereas the AS is intended for use as a silenced rifle for special forces the VSS (also known as the Vintorez, or Thread Cutter) was developed for use as a silenced sniper rifle by Spetsnaz undercover or clandestine units, a role made evident by its ability to be stripped down for carriage in a specially fitted briefcase. The VSS forms part of the BSK silenced sniper system.

The VSS fires the armour-penetrating 9 × 39 mm SP-5 cartridge. Sniper fire is normally single shot only, using a 10-round magazine but it is possible to fit the 20-round AS magazine for fully automatic fire.

As with the AS silent assault rifle, the designers regard the combination of rifle and ammunition as a 'complex'. For the VSS the complex is enlarged to include sighting systems.

9 mm VSS silent sniper rifle equipped with PSO-1 optical sight (T J Gander)

1996

Description

The overall operating features and sound suppression system used on the 9 mm VSS silent sniper rifle are exactly the same as those for the Val assault rifle (see previous entry). The main change involved with the VSS is that it is optimised to fire the 9 × 39 mm SP-5 subsonic armour-penetrating cartridge which is provided with a hardened steel or tungsten tip to penetrate a 6 mm 'high-density steel plate' at 100 m; a 2 mm steel plate can be fully penetrated at 500 m. In both cases there will still be sufficient energy available to disable targets after the plates have been penetrated.

The VSS is normally provided with a 10-round magazine and is fired single shot. Should the need arise to employ the weapon's fully automatic fire capability the AS 20-round magazine can be utilised.

The butt-stock used on the VSS is a more rounded derivative of that provided on the Dragunov 7.62 mm SVD sniper rifle and can be removed when the rifle is dismantled for carriage and/or concealment in a special briefcase measuring 450 × 370 × 140 mm. The briefcase contains all components of the VSS along with a PSO-1 (1P43) optical sight, an NSPU-3 (1PN75) night sight and two magazines. If the optical and night sights are not required the usual iron sights remain available.

Drawing of 9 mm VSS silent sniper rifle equipped with NSPU-3 night sight (L Haywood)

Data

Cartridge: 9 × 39 mm SP-5 AP
Operation: gas, selective fire
Locking: rotating bolt
Feed: 10- or 20-round box magazine
Weight: with empty magazine, 2.6 kg
Length: 894 mm
Range: day, >400 m; night, >300 m
Optical sight weight: 580 g
Optical sight length: 375 mm
Night sight weight: 2.1 kg
Night sight length: 340 mm
Carrying case dimensions: 450 × 370 × 140 mm

Manufacturer

Institute of Precise Mechanical Engineering, Izhevsk, Russia.
Marketed by: Rosvoorouzhenie, 18/1 Ovchinnikovskaya Emb, 113324, Moscow, Russia.

Status

In production. Offered for export sales.

Service

CIS special forces and Internal Ministry personnel.

UPDATED

7.62 mm Simonov self-loading rifle (SKS)

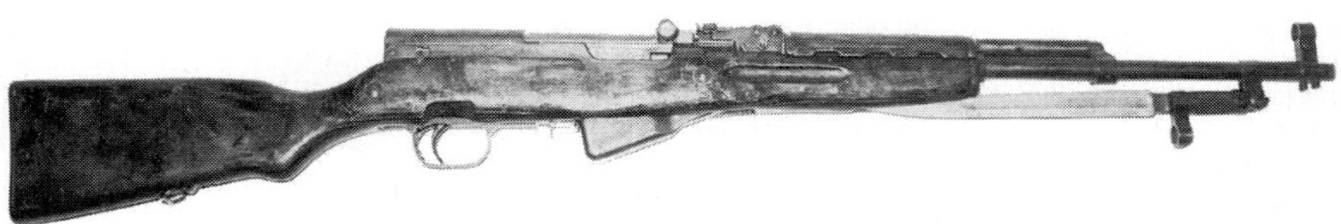

7.62 mm Simonov self-loading rifle (SKS) (T J Gander)

Description

The SKS is a gas-operated rifle of conventional design with a charger-loaded 10-round box magazine enclosed inside the receiver. There is a catch below the receiver, behind the magazine, which when pressed releases the bottom plate of the magazine to allow quick emptying. It has a permanently attached, folding blade bayonet. It is now regarded as obsolete in the former Warsaw Pact armed forces but is widely retained as a 'parade' weapon for honour guards.

To load the rifle the cocking handle on the right of the bolt is retracted. This is permanently attached to the bolt. If the magazine is empty the bolt will be held to the rear. The ammunition comes in 10-round chargers and is placed in the charger guides in the front of the bolt carrier and pressed fully down. When the 10 rounds are in the magazine the charger is removed; the bolt is pulled slightly back, released and goes forward to chamber the top round. The magazine can, if necessary, be loaded or topped up as required, by using individual rounds.

The safety catch is along the rear of the trigger guard and is pushed forward and up for 'safe' where it indicates its presence by obstructing the trigger finger as well as blocking the trigger.

Pulling the trigger releases the hammer which drives the firing pin into the cartridge cap. Some of the gases following the bullet up the bore are diverted through the gas port and impinge on the head of the piston. The piston is forced rearwards and the tappet strikes the bolt carrier. It is a short stroke action and the spring returns the tappet and piston to their forward position. The carrier is driven back and after about 8 mm of free travel, during which the gas pressure drops, it lifts up the rear end of the bolt out of engagement with the floor of the receiver. The bolt assembly then moves rearward as an entity. The hammer is rocked back to its cocked position and the return spring is progressively compressed. The extractor holds the empty case to the bolt face until it contacts the ejector and is expelled through the port on the right of the receiver.

The return spring drives the bolt assembly forward; the bolt picks up the top round in the magazine and chambers it. The extractor enters the cannelure of the cartridge case and bolt motion ceases. The carrier continues forward for about 8 mm and forces the rear of the bolt down into its recess in the receiver.

The rifle continues to fire single shots until the ammunition is expended. The magazine platform then rises and a small stud pushes up a bolt-retaining catch which holds the bolt to the rear. When the magazine platform is depressed by the insertion of ammunition, the catch continues to hold the bolt to the rear until the bolt is pulled slightly back, when the catch drops, and is released to feed the top round.

Data

Cartridge: 7.62 × 39 mm
Operation: gas, self-loading
Locking: tilting block
Feed: 10-round internal box magazine
Weight: empty, 3.85 kg
Length: bayonet folded, 1.021 m
Rifling: 4 grooves, rh
Sights: fore, post; rear, tangent, notch
Muzzle velocity: 735 m/s
Effective range: 400 m

Manufacturer

State factories.

Status

Production complete.

Service

No longer used in the CIS, except for ceremonial purposes, having been replaced in first-line service by the AK-47 (qv). Still encountered in some Asian countries.

Licence production

CHINA, PEOPLE'S REPUBLIC
Manufacturer: China North Industries Corporation (NORINCO)
Type: 7.62 mm Type 56
Remarks: Still available for export sales. See separate entry under *China, People's Republic* for details.
KOREA, NORTH
Manufacturer: State factories
Type: 7.62 mm Type 63
Remarks: Apparently no longer in production. Still in service with North Korean armed forces.
YUGOSLAVIA (SERBIA AND MONTENEGRO)
Manufacturer: Zastava Arms, Belgrade
Type: 7.62 mm M59/66A1
Remarks: No longer in production. See separate entry under *Yugoslavia (Serbia and Montenegro)* for details.

VERIFIED

7.62 mm Dragunov sniper rifle (SVD)

Description

The 7.62 mm Dragunov sniper rifle, or SVD, is a semi-automatic arm with a 10-round magazine and is chambered for the rimmed 7.62 × 54R cartridge. It is gas-operated with a cylinder above the barrel. There is a two-position gas regulator which may be adjusted using the rim of a cartridge case as a tool. The first position is employed in the usual operation of the rifle and the second is for extended use at a rapid rate or when conditions are adverse.

The bolt system is, in principle, exactly the same as that used in the AK-47, AKM and RPK, but the Dragunov bolt cannot be interchanged with that of the other weapons which fire the 7.62 × 39 mm rimless round. However, the assault rifle and light machine gun are operated on a long-stroke piston principle which is inappropriate for this rifle since the movement of the fairly heavy mass with the consequent change in the centre of balance militates against extreme accuracy. Therefore, in the Dragunov the designer utilised a short-stroke piston system. The piston, of light weight, is driven back by the impulsive blow delivered by the gas force and transfers energy to the bolt carrier which moves back and a lug on the bolt, running in a cam path on the carrier, rotates the bolt to unlock it. The carrier and the bolt go back together; the return spring is compressed and the carrier comes forward and locks the bolt before firing can take place. Mechanical safety is produced by the continued movement of the carrier after bolting is completed. When the carrier is fully home a safety sear is released and this frees the hammer which, when the trigger is operated, can come forward to drive the firing pin into the cap.

Since the trigger mechanism has to provide only for single-shot fire, it is a simple design using the hammer, the safety sear controlled by the carrier and a disconnector. The disconnector ensures that the trigger must be released after each shot to reconnect the trigger bar with the sear.

The PSO-1 sight is a ×4 telescopic sight with power for graticule illumination supplied by a small battery. It is rather longer than most telescopic sights at 375 mm but a rubber eyepiece is included in this length. The firer's eye is in contact with this rubber which automatically gives the correct eye relief of 68 mm. The true field of view is 6° which is comparable with that obtained in most military telescopes of modern design. The coating used on the lenses to reduce light loss on the interchange surfaces is extremely effective and the depth and uniformity of the deposit compares favourably with any other similar sight in service. Weight of the PSO-1 is 580 g.

The sight incorporates a metascope, meaning it is capable of detecting an infra-red source. It is not sufficiently sensitive to be used as a night vision sight, and since the development of thermal imaging sights the metascope is a somewhat redundant device.

When fitted with the PSO-1 optical sight the SVD has the designation 6V1. When fitted with the NSPU-3 night sight the designation changes to 6V1-N3. The NSPU-3 sight weighs 2.1 kg and allows accurate sighting to be carried out to a range of 1,000 m at night or under poor visibility conditions.

An unusual feature for a sniper rifle is that the muzzle of the SVD can accommodate a bayonet, the 6H4, which also doubles as a field knife.

Bullpup version of 7.62 mm SVD sniper rifle

1996

Variant

Bullpup SVD

A bullpup variant of the 7.62 mm SVD rifle was first observed in newsreels and other illustrations of MVD and Spetznaz troops operating in Chechnya during late 1994. No unclassified information is yet forthcoming regarding this weapon and details are limited to the one clear illustration seen to date (provided here). From this it appears that the basic SVD has been considerably shortened for ease of handling and carriage by simply rearranging the location of the trigger group and adding the necessary internal linkages. It appears that the barrel length has also been reduced. A prominent suppressor has been added to the muzzle. The PSO-1 optical sight has been retained and so has the 10-round magazine. A small cheek rest has been added to the left-hand side of the receiver for user comfort when using the optical sight and the butt stock has been reduced to a rudimentary plate assembly.

Data

Cartridge: 7.62 × 54R
Operation: gas, short-stroke piston, self-loading
Locking: rotating bolt
Feed: 10-round box magazine

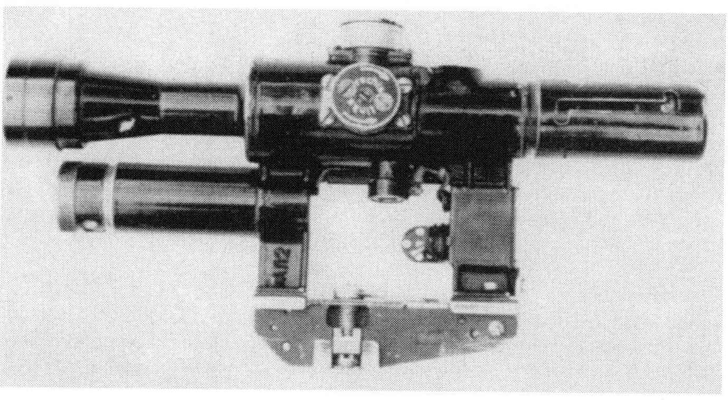

German version of PSO-1 telescope. Note enlarged battery case (lower left) for graticule illumination and omission of long rubber eyepiece

Left side of Dragunov rifle, showing PSO-1 telescope sight and mount

Weight: with PSO-1, 4.3 kg; with NSPU-3, 6.4 kg
Length: 1.225 m; with bayonet-knife, 1.37 m
Barrel: 622 mm
Rifling: 4 grooves, rh, 1 turn in 254 mm
Sights: fore, adjustable post; rear, U-notch, tangent.
PSO-1 telescope: 4 × 24, 68 mm eye relief, 6° field of view
Muzzle velocity: 830 m/s
Effective range: 1,000 m

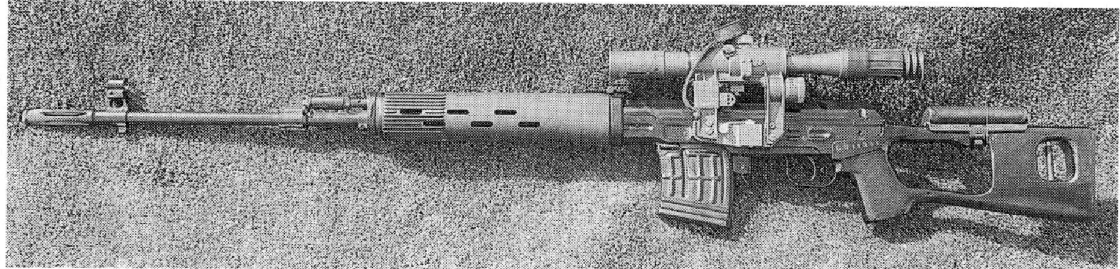

7.62 mm Dragunov sniper rifle (SVD) (T J Gander)

Manufacturer
IZHMASH, 3 Derjabin Street, 426006 Izhevsk, Russia. Marketed by: Rosvoorouzhenie, 18/1 Ovchinnikovskaya Emb, 113324, Moscow, Russia.

Status
In production. Offered for export sales.

Service
Former Warsaw Pact armies.

Licence production
CHINA, PEOPLE'S REPUBLIC
Manufacturer: China North Industries Corporation (NORINCO)
Type: 7.62 mm Type 79
Remarks: Available for export sales. See separate entry under *China, People's Republic* for details.
IRAN
Manufacturer: State factories
Type: 7.62 mm Al-Kadisa
Remarks: See entry under *Iran* for details.
ROMANIA
Manufacturer: Romtehnica
Type: Sniper Rifle 7.62 mm
Remarks: Identical to standard SVD. Offered for export sales.

UPDATED

5.66 mm APS underwater assault rifle

Description
Although underwater weapons have been in supply to various special forces for some years, this is the first of the class which has been formally acknowledged and of which some information has been released. It is intended for arming combat swimmers and it is also possible to use the rifle on land.

The underwater assault rifle is a selective-fire, gas-operated weapon, the mechanism of which appears to be broadly based upon the Kalashnikov rotating bolt system. It is of fairly conventional appearance, and has a folding butt. The most distinctive feature is the magazine, necessarily large because of the ammunition used.

The projectile is a drag-stabilised dart, the rear end of which is held in a conventional cartridge case apparently based on that for the 5.56 × 45 mm NATO. The dart is 120 mm long and the complete round 150 mm long, which is the reason for the abnormal front-to-back depth of the magazine. Each MPS cartridge weighs 26 g and the projectile is 120 mm long.

The muzzle velocity is 365 m/s, but the effective range in air is only 100 m. The range underwater varies according to the depth, from 30 m at 5 m depth to 11 m at 40 m depth. At these ranges the dart will penetrate the usual types of underwater suit and also 5 mm of glass (face masks) to inflict fatal wounds.

5.66 mm APS underwater assault rifle (T J Gander)

Data
Cartridge: 5.66 mm Model MPS
Operation: gas, selective fire
Feed: 26-round box magazine
Weight: empty, 2.4 kg; with empty magazine, 2.7 kg; loaded, 3.4 kg
Length: butt folded, 614 mm; butt extended, 840 mm
Height: 187 mm
Width: 65 mm
Muzzle velocity: in air, 365 m/s
Lethal range: in air, 100 m; in water 5 m deep, 30 m; in water 20 m deep, 20 m; in water 40 m deep, 11 m

Manufacturer
Institute of Precise Mechanical Engineering, Izhevsk, Russia.
Marketed by: Rosvoorouzhenie, 18/1 Ovchinnikovskaya Emb, 113324, Moscow, Russia.

Status
In production. Offered for export sales.

Service
CIS special forces.

UPDATED

7.62 mm AK-47 assault rifle

Development
The Kalashnikov assault rifles have become the most important and widespread weapons in the world. Starting with the AK-47, Kalashnikovs have been the standard rifles of the Soviet, CIS and most former Warsaw Pact armed forces since the early 1950s. To these can be added most communist inspired or sponsored countries elsewhere. Practically all communist-influenced guerrilla and nationalist movements of recent decades seem to have obtained, or been given, stocks of AK-47s.

The origins of the AK-47 have passed almost into legend. The designer was Mikhail Timofeyevich Kalashnikov who devised the basics of various weapon designs while recovering from wounds obtained during the Great Patriotic War. Following several attempts at producing infantry weapons such as sub-machine guns, plus numerous difficulties in attracting the sponsorship of the necessary executives, the AK-47

CIS 7.62 mm AK-47S with folding butt (T J Gander)

CIS 7.62 mm AK-47 rifle (T J Gander)

(Avtomat Kalashnikova obrazets 1947g) was finally accepted for Soviet service in 1947, firing the 7.62 × 39 mm M1943 cartridge. At least three basic Soviet-produced AK-47 models have been detected, differences mainly as a result of the production methods used. CIS production had ceased by the late 1960s but was continued in other countries.

Production of the AK-47 was undertaken in several of the former Warsaw Pact arsenals of Eastern Europe, China and elsewhere. AK-47 rifles are still being produced for export sales in Bulgaria, the only nation where AK-47s are apparently still in production.

It is impossible to estimate how many AK-47 rifles have been made, nor how many are in use, but even at a very rough estimate it must be well in excess of 20 million. Rifles manufactured outside the former Soviet Union frequently exhibit slight detail changes and reference should be made to individual country entries for details of these.

The basic Kalashnikov action, first used for the AK-47, has been taken for many weapons other than the Soviet Kalashnikov derivatives such as the AKM and AK-74 series (see appropriate entries) and the RPK light machine gun. Designs originating outside the CIS using the basic Kalashnikov action include the Israeli Galil series (and from that the South African R4 and R5) and the Finnish Valmet series of assault rifles. The Indian 5.56 mm INSAS assault rifle also has Kalashnikov affiliations.

Description

There are some features of the AK series which are a departure from the conventional but in general it is a simple, robust and handy weapon of sound design and generally excellent workmanship.

The AK-47 assault rifle is a compact weapon, capable of selective fire, and is robust and reliable. The former Warsaw Pact countries have now largely replaced it with the AKM, a modernised and improved version (see following entry), but it is still held in reserve by many former Eastern Bloc countries.

The AK-47 may be encountered in two basic configurations, one with a rigid butt (AK-47) and one with a double-strut folding metal butt-stock (AK-47S) controlled by a simple press-button release above the pistol grip.

There are also early and late production models. The early model has a built-up receiver which has an angular shape to the rear end, sloping noticeably down to the butt. The later type has a straight receiver.

The various countries in the former Warsaw Pact have produced a variety of materials for butt-stocks and forearms ranging from laminated sheets of plywood to various types of plastic and all-metal construction.

The magazine is loaded and positioned in the rifle by placing the front end into the magazine-well until the lug engages and then rotating the rear end back and up until the magazine catch engages. The cocking handle is permanently fixed to the right of the bolt carrier and reciprocates with it. This allows manual bolt closure if for some reason the bolt carrier does not go fully forward. If the cocking handle is retracted and released, the top round is fed into the chamber. When the trigger is operated the hammer hits the floating firing pin and drives it into the cap of the cartridge. Some of the propellant gases are diverted into the gas cylinder on top of the barrel. There is no gas regulator. The piston is driven back and the bolt carrier, built into the piston extension, has about 8.5 mm of free play while the gas pressure drops to a safe level. A cam slot in the bolt carrier engages the cam stud on the bolt and the bolt is rotated through 35° to unlock it from the receiver. There is no primary extraction during bolt rotation to unseat the case and so a large extractor claw is fitted which grips the empty case and holds it to the bolt face until it contacts the fixed ejector formed in the guide rail and is thrown out of the right-hand side of the gun. As the bolt travels back it rocks the hammer over and also compresses the return spring. The bolt is brought to a halt by hitting the solid rear end of the receiver. The return spring drives the bolt forward, another round is chambered and the bolt comes to rest. The carrier continues on for about 5.5 mm after locking is completed. During this last forward movement of the carrier the safety sear is released and control of the hammer is returned to the trigger sear.

When the trigger is pressed, the hammer is released from the trigger sear and goes forward to fire the round. The recoiling carrier rotates the hammer back and it is held by the safety sear. As soon as the carrier is fully forward the safety sear is disengaged, the hammer is freed, and another round is fired. So long as the trigger is pressed, ammunition is available in the magazine, and the bolt carrier goes fully forward, the gun will continue to fire at its cyclic rate of about 600 rds/min.

The change lever is mounted on the right-hand side of the receiver and is unique in its design. It is a long pressed-out bar pivoted at the rear and applied by the firer's thumb. The top position is 'safe'. Here it locks the trigger and physically prevents the bolt from coming back sufficiently to pass beyond the rear of a cartridge in the magazine, but does allow sufficient movement for checking that the chamber is clear.

The lever produces automatic fire at the centre position and single shots when fully depressed. It is stiff to operate, noisy in functioning and difficult to manipulate when wearing arctic mittens.

The foresight is a post, screw-threaded for adjustment when zeroing. The spanner in the combination tool kit is used for this. Lateral adjustment for zeroing is achieved by moving the foresight block in a dovetail.

The rearsight is an open U-shaped notch which will allow ranges up to 800 m by means of a slide and ramp. There is a battle sight setting for all ranges up to 200 m.

Variant

5.6 mm AKT-47 Training Rifle

The 5.6 mm AKT-47 is a single shot rifle intended for marksmanship training. Chambered for the 0.22 LR cartridge, this rifle has to be re-cocked manually after each shot. In outward appearance, dimensions and weight the AKT-47 is identical to the standard AK-47 but there have been some changes to the bolt block and, inevitably, to the barrel. Muzzle velocity is 400 m/s and the weapon is sighted to 400 m. Rounds are fed from a 30-round magazine outwardly identical to the standard curved box magazine.

The AKT-47 is apparently now manufactured only in Bulgaria. It is still offered by Kintex for export sales.

Data

Cartridge: 7.62 × 39 mm
Operation: gas, selective fire
Locking: rotating bolt
Feed: 30-round detachable box magazine
Weight: 4.3 kg
Length: butt folded, 699 mm; butt extended, 869 mm
Barrel: 414 mm
Rifling: 4 grooves, rh, 1 turn in 235 mm
Sights: fore, post, adjustable; rear, U-notch tangent, adjustable to 800 m with battle sight for 200 m
Sight radius: 376 mm
Muzzle velocity: 710 m/s
Rate of fire: cyclic, 600 rds/min

Manufacturer

Various state factories.

Status

Production complete other than in Bulgaria.

Service

Replaced in first-line Warsaw Pact service by the AKM (see following entry). Widely used in Asian countries and in Egypt and Syria. Still used in reserve units in former Eastern Bloc countries.

Licence production

BULGARIA
Manufacturer: Kintex
Type: 7.62 mm AK-47, AKS-47, AKN-47, AK-47M1, AKT-47
Remarks: Offered for export sales. See separate entry under *Bulgaria* for details.

UPDATED

7.62 mm AKM assault rifle

Development

The 7.62 mm AKM assault rifle is a modernised version of the AK-47 first produced in 1959. The forged and machined receiver of the AK-47 was replaced with a body of pressed steel construction with riveting employed extensively to join the 1 mm thick U section to the inserts which house the locking recess, the barrel bearing, and the rearsight block at the front and the butt at the back. The bolt locks into a sleeve and not directly into the barrel as in the AK-47. The slides on which the breech block reciprocates are pressed out and spot welded inside the receiver walls. The results of these changes are reduced manufacturing costs and a reduction in weight from 4.3 to 3.14 kg.

The AKM is produced with a wooden stock. It is also made with a folding butt-stock and is then known as the AKMS.

Production of the AKM and AKMS has ceased in the CIS nations in favour of the 5.45 mm AK-74 series but production continues elsewhere (see under *Licence production*.)

Description

There are several features which distinguish the 7.62 mm AKM assault rifle from the AK-47 and allow visual recognition. These are:

(a) There is a small recess in each side of the receiver centrally over the magazine; this is a magazine guide.
(b) The lower handguard has a groove for the firer's fingers.
(c) The receiver cover has transverse ribs.
(d) There is an additional bayonet lug under the gas tap-off point.
(e) The four gas escape holes on each side of the gas cylinder are omitted.
(f) The sight is graduated to 1,000 instead of 800 m.
(g) A small compensator is fitted to the muzzle on all but the earliest production models.

There is no essential difference between the functioning of the AKM and that of the AK-47. There is an additional assembly in the trigger mechanism which is a device to delay the fall of the hammer during automatic fire until the breech is closed and locked.

The AKM and AK-47 are supplied with the same accessories which are: bayonet, blank-firing device, combination tool kit, magazine carrier, night firing sight, oil bottle and cleaning fluid, and sling.

The bayonet of the AKM is different in shape from that of the AK-47. Whereas the AK-47 bayonet has an even taper to the point on each side, that of the AKM has an undercut reverse edge. The AKM bayonet has a slot in the blade into which a lug on the scabbard fits. The scabbard is electrically insulated and has a shearing edge so that, when linked with the bayonet, an effective wire cutter is produced. The blank-firing

7.62 mm AKMS assault rifle with butt folded (T J Gander)

device goes on to the muzzle thread which lies under the muzzle nut. It produces sufficient pressure in the bore to operate the piston.

A combination tool kit fits under the butt-plate. It is contained in a case which also acts as a handle for the cleaning rod. The combined drift, screwdriver and spanner is used, among other things, to adjust the foresight.

Data

Cartridge: 7.62 × 39 mm
Operation: gas, selective fire
Locking: rotating bolt
Feed: 30-round detachable box magazine
Weight: 3.14 kg
Length: 876 mm
Barrel: 414 mm
Rifling: 4 grooves, rh
Sights: fore, pillar; rear, U-notch
Muzzle velocity: 715 m/s
Rate of fire: cyclic, 600 rds/min
Effective range: 300 m

Manufacturer

State factories.

Status

No longer in production in CIS but still produced elsewhere (see below).

Service

Replaced in first-line service with CIS states by the AK-74. Used by all former Warsaw Pact countries since 1959. In service with numerous armed forces, both regular and irregular.

Licence production

CHINA, PEOPLE'S REPUBLIC
Manufacturer: China North Industries Corporation (NORINCO)
Type: 7.62 mm Type 56, 56-1, 56-2
Remarks: The original 7.62 mm Type 56 assault rifle was an AK-47 copy but when production switched to the AKM the Type 56 designation was not changed. Still in production and offered for export sales. See separate entry under *China, Peoples Republic* for details.
EGYPT
Manufacturer: Maadi Company for Engineering Industries
Type: 7.62 mm Misr
Remarks: In production and offered for export sales. See entry under *Egypt* for details.
FINLAND
Manufacturer: Sako Limited
Type: 7.62 mm Valmet M60 series
Remarks: Originally a Finnish modification of AK-47 later updated to include AKM standard changes. See entry under *Finland* for details.
GERMANY
Manufacturer: State factories
Type: 7.62 mm MPiKM and MPiKMS-72
Remarks: The MPiKM was a direct copy of the AKM identifiable by distinctive 'pebble-dash' plastic buttstock. The MPiKMS-72 had a side-folding stock. Both are no longer in production in Germany but are still encountered elsewhere, typically in Cambodia and Vietnam.
HUNGARY
Manufacturer: State factories
Type: 7.62 mm AMD-65
Remarks: Originally a local updated version of the basic AK-47 but later altered to include many AKM standard modifications. No longer in production. See entry under *Hungary* for details.
IRAQ
Manufacturer: State factories
Type: 7.62 mm Tabuk
Remarks: Apparently still in production. A 5.56 mm version has also been produced. See entries under *Iraq* for details.
KOREA, NORTH
Manufacturer: State factories
Type: 7.62 mm Type 68
Remarks: See entry under *Korea, North* for details.
POLAND
Manufacturer: Zaklady Metalowe LUCZNIK
Type: 7.62 mm AKM and AKMS
Remarks: Offered for export sales.
ROMANIA
Manufacturer: Romtehnica
Type: 7.62 mm AKM
Remarks: In production and ordered by India. See entry under *Romania* for details.
YUGOSLAVIA (SERBIA AND MONTENEGRO)
Manufacturer: 7.62 mm M70B1 and M70B2
Type: Zastava Arms, Beograd
Remarks: Probably no longer in production. See entry under *Yugoslavia (Serbia and Montenegro)* for details.

7.62 mm AKM assault rifle (T J Gander)

UPDATED

5.45 mm AK-74 and AKS-74 assault rifles

Development

Introduced in 1974, the 5.45 mm AK-74 assault rifle is a small calibre version of the AKM. It is similar in size though slightly heavier. The effective range has decreased, but not below what is considered to be normal infantry fighting ranges. The magazine has the same capacity (30 rounds) as that of the AKM; the 40-round plastic magazine of the RPK-74 light machine gun is however interchangeable with that of the AK-74.

Production of the AK-74 series is in progress in several nations outside the former Soviet Union. Some of these nations, including Bulgaria, Hungary and Poland, also manufacture the AK-74 to fire 5.56 × 45 mm NATO ammunition to make their products more attractive to prospective purchasers in the West. Similar models are also produced within Russia by IZHMASH (see following entry).

Description

The AK-74 is built around the receiver of the AKM and there appears to be no difference between the receivers, leading to the belief that they come from the same dies. The bolt carrier remains virtually the same, so that it runs in the same guideways while carrying a smaller bolt. This bolt is lighter than the AKM version and so gives a better ratio of bolt to carrier mass, leading to more efficient working.

There is a slight modification to the AKM bolt and carrier design: the AK-74 bolt has a small flat and the carrier a nib, which interacts to prevent the bolt falling free of the carrier during disassembly of the weapon. The AKM extractor is prone to breakage so for the AK-74 it was substantially enlarged and strengthened. The 30-round magazine is of laminated construction, a brownish plastic surface concealing an inner shell of steel. It has thick walls and is extremely hard-wearing and strong.

The plywood furniture closely resembles that of the AKM, the only obvious difference being that the AK-74 has a horizontal finger groove along each side of the butt. This is probably a recognition feature rather than an aid to holding and together with the muzzle brake it forms the quickest way of picking out an AK-74 from other AK-series rifles. Late production AK-74s have all-plastic furniture.

On the early production folding butt version, the AKS-74, the folding butt is a tubular skeleton assembly which folds by swinging to the left and lying alongside the receiver.

There is also a folding butt version with a solid plastic butt. This version, the AK-74M, has the butt folding to the right of the receiver, reducing the overall length (when folded) to 700 mm.

One further model, the AK-74M N3, has provision for mounting a NSPU-3 night sight weighing 2.1 kg.

All versions of the AK-74 can be used in conjunction with 40 mm GP-15 or GP-25 Kastyor grenade launchers located beneath the barrel.

The AK-74 is one of the very few rifles to have been fitted with a successful muzzle brake. The reason seems to be to allow the firer to fire bursts without the muzzle moving away from the line of sight. The brake works by allowing the emerging gases to strike a flat plate at the front of the assembly. This deflects the gas and produces a forward thrust. To counter the upward movement during automatic fire, gases escaping through three small ports on the upper part of the muzzle brake force the muzzle down, compensating for muzzle jump. The difficulty with all muzzle brakes is that the deflected

5.45 mm AK-74 assault rifle

Bullpup variant of 5.45 mm AK-74

1996

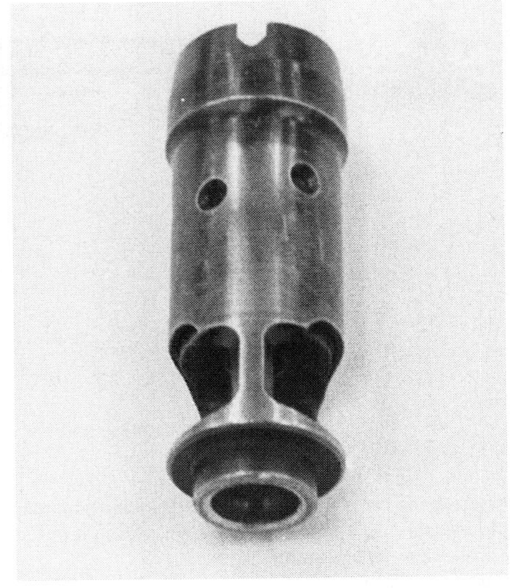

Front view of muzzle brake from AK-74 rifle. Note upward vents which are intended to counter muzzle climb and vent slots pointing forward so as to deflect side blast from muzzle brake ports

gas comes back towards the firer; the AK-74 overcomes this, to a great extent, by means of narrow slits on the forward end of the muzzle attachment. These deflect gases forward from the muzzle brake instead of allowing it to flow back to the firer. A plan view of the muzzle would show a pattern of gas flowing out from the muzzle in a fan on either side, rather like a pair of butterfly wings. While this protects the firer from the effect of his own shots, it does nothing for anyone on either side of him, and an article in the former Soviet military medical press expressed concern at the probable aural damage likely to occur on training ranges where firers are lined up on conventional firing points within a metre or two of each other.

Despite the chance of deafness, however, the muzzle brake is highly effective and it allows the rifle to have less recoil than is usually the case with a weapon of this calibre. There are defects however, one being that the muzzle brake does not reduce flash and the AK-74 has a substantial muzzle flash, of the order of three times the normal.

Variants

5.45 mm RPK light machine gun

This variant is described in the *Machine guns* section.

AK-74 Bullpup

No information is yet forthcoming regarding a bullpup variant of the 5.45 mm AK-74. The type was first observed in newsreels and other illustrations relating to MVD and Spetznaz troops operating in Chechnya during early 1995. From the one clear illustration seen to date (provided here) it appears that this weapon is not a 'one-off' or prototype venture but a full production variant of the AK-74. The weapon features a carrying handle which also carries the iron sights and it appears very likely that this assembly could also be used to mount optical or night sight systems.

6 mm Special assault rifle

During late 1993 a variant of the AK-74, believed to be experimental, was shown in 6 mm calibre, apparently firing a cartridge (not yet examined) based on the dimensions of the 5.45 × 39 mm round. The 6 mm cartridge was developed by the Institute of Precise Mechanical Engineering at Klimovsk.

The barrel of the revised AK-74 is new, as is the muzzle brake, and there is an enlarged gas return cylinder over the barrel. This return cylinder apparently conceals an action-smoothing buffer spring which contributes a major part to the objective of the design which is to produce an assault rifle with 'increased efficiency and balanced automatic functioning', resulting in improved weapon balance and handling with significantly improved firing results. One claim is that automatic fire accuracy when fired from the standing position is improved by a factor of 5.1. When fired hand-held from the prone position the improvement factor is claimed to be 1.5 (2.2 if a rest is used).

Data

Cartridge: 5.45 × 39 mm
Operation: gas, selective fire
Locking: rotating bolt
Feed: 30-round plastic box magazine
Weight: unloaded AK-74 and AKS-74, 3.3 kg; loaded AK-74 and AKS-74, 3.9 kg
Length: AK-74 overall, 943 mm; AKS-74 butt folded, 690 mm
Barrel: 415 mm
Rifling: 4 grooves, rh, 1 turn in 196 mm
Sights: fore, post; rear, U-notch
Muzzle velocity: 900 m/s
Rate of fire: cyclic, 600-650 rds/min

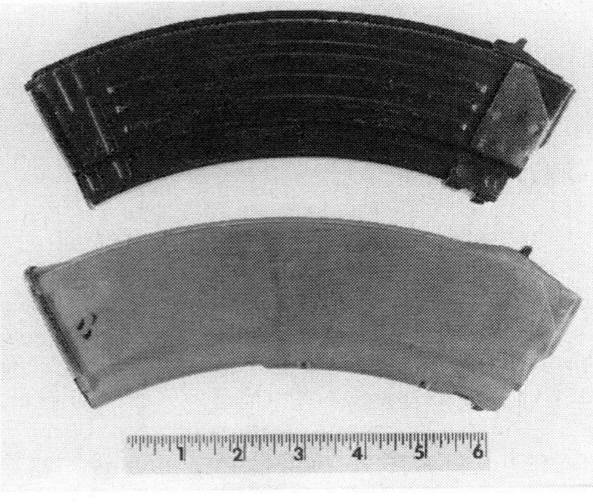

30-round plastic magazine of AK-74 (lower) compared with AKM 30-round magazine

Manufacturer

IZHMASH, 3 Derjabin Street, 426006 Izhevsk, Russia.
Marketed by: Rosvoorouzhenie, 18/1 Ovchinnikovskaya Emb, 113324 Moscow, Russia.

Status

In production.

Service

Former Soviet and some former Warsaw Pact forces.

Licence production

BULGARIA
Manufacturer: Kintex
Type: 5.45 or 5.56 mm AK-74, AKS-74 and AKN-74 assault rifles
Remarks: In production and offered for export sales. See entry under *Bulgaria* for details.
HUNGARY
Manufacturer: Technika
Type: 5.56 mm NGM assault rifle
Remarks: Was produced for possible export sales but none known to date. Production has now ceased. See entry under *Hungary* for available details.
POLAND
Manufacturer: Zaklady Metalowe LUCZNIK
Type: 5.45 or 5.56 mm wzor 1988 TANTAL automatic rifle
Remarks: See entry under *Poland* for details.
ROMANIA
Manufacturer: Romtehnica
Type: 5.45 mm sub-machine gun
Remarks: Produced with frame stock and pistol foregrip for AK-74 equivalent and simple single-strut folding butt for AKS-74 equivalent. A training rifle chambered for 0.22 cartridges is also produced. See entry under *Romania* for details.

UPDATED

Kalashnikov 'Hundred Series' assault rifles

Development

There are five weapons in the Kalashnikov 'Hundred Series' of weapons, two of them being assault rifles while the other three fall into the *sub-machine* category (see entry in Sub-machine section for full details). All five weapons incorporate the very latest modifications introduced to the Kalashnikov rifle family and are manufactured using modern materials and production techniques. In order to make these 'export model' weapons attractive to a wide range of potential purchasers the models are produced in variety of calibres. An outline of the complete line is provided here:

AK101 - Assault rifle chambered for 5.56 × 45 mm NATO cartridge
AK102 - Sub-machine gun chambered for 5.56 × 45 mm NATO cartridge
AK103 - Assault rifle chambered for 7.62 × 39 mm cartridge
AK104 - Sub-machine gun chambered for 7.62 × 39 mm cartridge
AK105 - Sub-machine gun chambered for 5.45 × 39 mm cartridge

Description

The two assault rifles in the Kalashnikov 'Hundred Series' are the AK101 and AK103, the AK101 firing 5.56 × 45 mm NATO standard ammunition while the AK103 fires the standard Eastern Bloc 7.62 × 39 mm cartridge. Both rifles are visually identical and are constructed with provision to accommodate a wide range of optical and night sighting equipment.

Both rifles are provided with black plastic furniture and have a black phosphate finish, leading to the collective name of 'Black Kalashnikovs'. On both rifles the solid butts can be folded forward to lay on the left-hand side of the receiver; the butts contain a compartment to carry an accessories case. On both rifles the plastic fore-end is ribbed to improve the forward grip. The muzzle brake previously fitted to the AK-74 is carried over and applied to the 7.62 mm AK103. On the AK103 the standard AK-74 pattern muzzle brake can be replaced by a PBS flash silencer which reduces the firing signature and muzzle flash. Both weapons can be fitted with 40 mm GP-25 grenade launchers.

The overall standard of manufacture of these rifles is very high and all the many attributes of the Kalashnikov series have been carried over. These attributes include a high degree of reliability, complete interchangeability of components between models and general ease of maintenance. Among claims made for the overall reliability are mentions of firing trials involving from 10,000 to 15,000 rounds without a single malfunction, trials which were terminated only when the bore was worn beyond further utility.

Data
Cartridge: AK101, 5.56 × 45 mm; AK103, 7.62 × 39 mm
Operation: gas, selective fire
Locking: rotating bolt
Feed: 30-round plastic box magazine
Weight: (less magazine) AK101, 3.4 kg; AK103, 3.3 kg; (loaded) AK101, 3.63 kg; AK103, 3.55 kg
Length: overall, 943 mm; butt folded, 700 mm
Barrel: 415 mm
Rifling: AK101, 4-groove, rh, 1 twist in 180 mm; AK103, 4-groove, rh, 1 twist in 240 mm
Sights: fore, post; rear, U-notch; provision for fitting optical or night sights
Muzzle velocity: AK101, 910 m/s; AK103, 715 m/s
Rate of fire: cyclic, 600 rds/min

Manufacturer
IZHMASH, 3 Derjabin Street, 426006 Izhevsk, Russia.
Marketed by: Rosvoorouzhenie, 18/1 Ovchinnikovskaya Emb, 113324 Moscow, Russia.

Status
In production.

Service
Not yet determined.

UPDATED

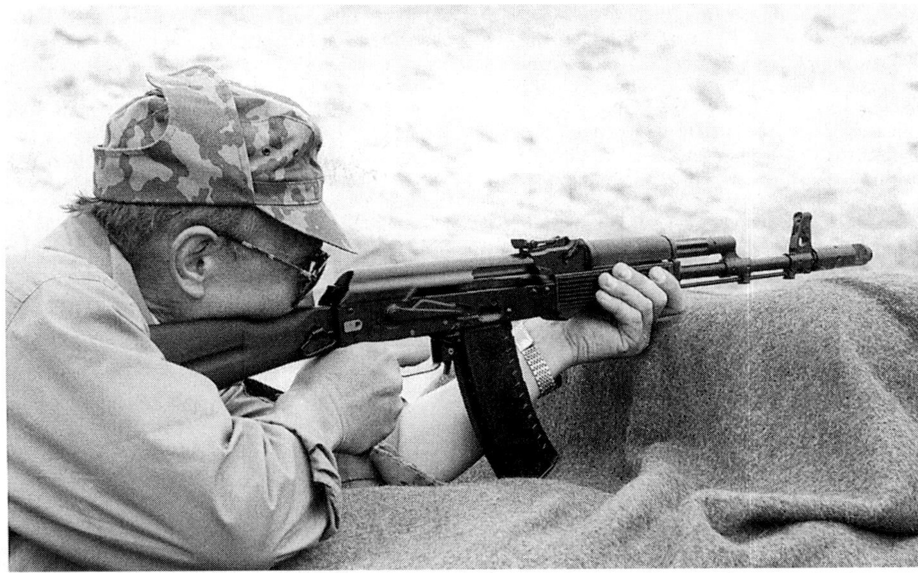

5.56 mm AK101 assault rifle (T J Gander)

1996

5.45 mm ASN assault rifle

Development
First shown publicly in May 1993, the 5.45 mm ASN assault rifle was designed by Gennadiy Nikonov at the IZHMASH facility at Izhevsk. Despite early reports that this rifle had been adopted as the next service rifle for the Russian armed forces it now appears that the ASN is in direct competition with a Kalashnikov design, probably based on the 'Hundred Series' rifles (see previous entry). Few details regarding the ASN have yet been made public so the following details should be regarded as provisional.

Description
From the few illustrations of the 5.45 mm ASN assault rifle seen to date it would appear that the ASN is a gas-operated weapon with the gases being tapped off from a point under the prominent muzzle attachment which appears to act as a combined twin-chamber compensator and muzzle brake. Standard 5.45 × 39 mm rounds are fed upwards into the weapon from a 30-round stamped steel box magazine and can be fired to an effective range of 1,000 m when the optical sight is used. This sight is mounted on a plate located on the left of the receiver. Standard iron sights are also provided.

The fire selector switch is on the left side of the pistol grip and can select semi-automatic, two-round burst and variable rate automatic fire. When the variable rate automatic fire mode is selected the first two rounds are fired at a cyclic rate of 1,800 rds/min, that is, they will be clear of the muzzle before the recoil forces generated can be felt by the firer. After those first two rounds the cyclic fire rate is reduced to a constant 600 rds/min.

5.45 mm ASN assault rifle

The intention appears to be that accuracy is improved by the reduced dispersion produced by the first two rounds fired at the high rate of fire. The mechanism involved in the variable fire rate system has not been disclosed but according to a Russian report it is based on a blowback shifted pulse principle.

The butt used on the ASN is a simple L-shaped design and appears to be fixed. Furniture is plastic. A new type of multipurpose field knife/bayonet is provided.

Data - (PROVISIONAL)
Cartridge: 5.45 × 39 mm
Operation: gas, selective fire, variable fire rate
Feed: 30-round steel box magazine
Weight: without magazine, 3.85 kg
Length: overall, 943 mm
Sights: fore, post; rear, U-notch; provision for fitting optical sights
Rate of fire: cyclic, first 2 rounds 1,800 rds/min; thereafter 600 rds/min
Effective range: 1,000 m with optical sight

Manufacturer
IZHMASH, 3 Derjabin Street, 426006 Izhevsk, Russia.

Status
Development.

VERIFIED

12.7 mm V-94 sniper rifle

Development
Information regarding the 12.7 mm V-94 sniper rifle first became publicly available during 1995. At first sight the overall design resembles that of the 14.5 mm PTRS 1941 anti-tank rifle of the Great Patriotic War years (1941-1945) but the resemblances are superficial for the V-94 is quite different in many details, not least the cartridge employed, the 12.7 × 107 mm. The V-94 is a product of the Instrument Design Bureau (KBP) at Tula, Russia.

According to promotional literature, potential targets for the V-94 include parked aircraft, radars, missile launchers and artillery, lightly armoured and soft-skin vehicles, small coastal craft and sea mines, all with emphasis on the fact that it is far more economical to utilise a large calibre sniper rifle against such targets than more costly ordnance.

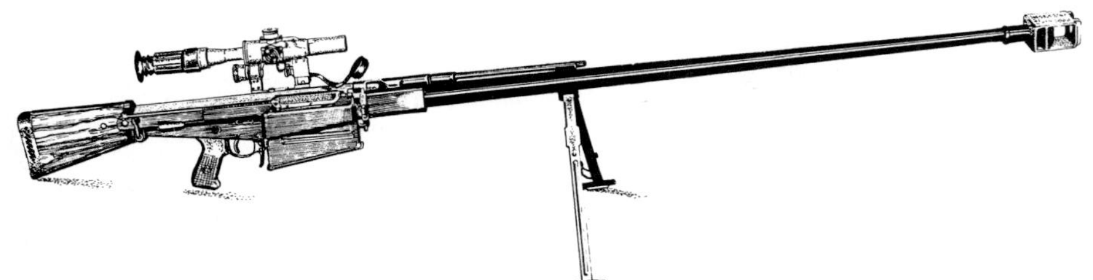

V-94 sniper rifle (L Haywood)

1996

Description
The 12.7 mm V-94 sniper rifle is a gas-operated single-shot rifle, the most prominent feature of which is the long slender barrel with its large muzzle brake device. The latter is a partial counter to the considerable recoil forces produced when firing the 12.7 × 107 mm cartridge, more usually fired from heavy machine guns. The firer is further protected against recoil by a rubber butt pad. Gas for the self-loading mechanism is tapped off from a point about one third along the length of the barrel. No firm information is available regarding the locking mechanism employed but it is assumed that it

is some form of projecting lug bolt operating with a rotary motion.

Ammunition is fed from a five-round box magazine under the receiver. The round quoted in the limited literature available mentions the B-32, an API with a muzzle velocity of 840 to 850 ms and a muzzle energy rating of 15,570 J. This provides the V-94 with an effective range of up to 2,000 m and an armour penetration of 20 mm against 20 mm plate at 100 m.

The V-94 has no iron sights so is provided with a suitably modified PSO-1 telescopic sight, the same unit fitted to the 7.62 mm SVD rifle. This sight is removed during moves.

To reduce the overall length of 1.7 m when travelling the V-94 can be folded about a point where the barrel meets the receiver assembly. This is accomplished by raising a large locking lever on the right of the receiver, just forward of the ejection port. This allows the receiver and butt to be swung to the right-hand side of the barrel; the bipod folds to the rear for carrying. In the folded state the overall length of the V-94 is reduced to 1.1 m, less than the overall length of the SVD sniper rifle.

Data

Cartridge: 12.7 × 107 mm
Operation: gas, single shot
Feed: 5-round box
Weight: without sight, 11.7 kg
Length: firing, 1.7 m; carry, 1.1 m
Barrel: estimated, 1.02 m
Sights: PSO-1 optical
Muzzle velocity: 840-850 m/s
Effective range: max, 2,000 m

Manufacturer

Instrument Design Bureau (KBP), Tula 300001, Russia

Status

In production. Offered for export sales.

NEW ENTRY

CROATIA

MACS 12.7 mm sniper rifle

Description

The MACS sniper rifle fires the 12.7 mm × 99 mm (0.50 Browning) cartridge. It is a single shot, bolt-action rifle with an optical sight and a multibaffle muzzle brake to reduce recoil forces. As far as can be determined there is no magazine, rounds being fed into the weapon individually by hand. Some individual user adjustments can be made on the bipod, cheek rest and butt plate.

Data

Cartridge: 12.7 × 99 mm (0.50 Browning)
Operation: bolt action, single shot
Feed: manual, single shot
Weight: 11.5 kg
Length: 1.47 m
Barrel: 780 mm
Sights: optical
Muzzle velocity: 855 m/s

Provisional side view drawing of MACS 12.7 mm sniper rifle (L Haywood)

1996

Marketing Agency

RH - ALAN d.o.o., Stančićva 4, 41000 Zagreb.

Status

In production.

Service

Croatian Army.

NEW ENTRY

RT-20 20 mm heavy sniper rifle

Description

The Croatian RT-20 20 mm sniper rifle falls into the category of the anti-material rifle as it has a maximum operational range of 2,000 m. At that distance the ballistic limitations of the round fired, the 20 × 110 mm Hispano Suiza HS 404, are such that hitting precise point targets such as an individual would be problematic. However, hitting a soft-skin vehicle or an aircraft on the ground would be much more viable and rewarding.

The RT-20 is a bullpup weapon. This configuration was doubtless selected to avoid the extreme overall length a conventional layout would have dictated, but it has also introduced some unusual design features.

One is that the RT-20 is a single-shot bolt-action rifle with the manually operated bolt handle on the left-hand side of the receiver. The bullpup layout has meant that the bolt is at the extreme rear of the weapon, above and to the rear of the butt plate. In view of the power of the cartridge employed the bolt has three sets of locking lugs opened by raising the bolt handle through 60°. Once the bolt is moved forward and locked the rifle is fired normally and the second set of unusual RT-20 features comes into play.

Firing 20 mm rifles from the shoulder has always introduced recoil problems, for any 20 mm cartridge carries a sizeable propellant load producing considerable recoil forces. Although no definite information is yet forthcoming is seems very likely that the round fired is API and very similar to the rounds fired from the Yugoslav M55 series of anti-aircraft cannon and it may also be possible to fire HE or HEI projectiles. The projectile will thus weigh about 130 to 140 g and have a muzzle velocity of 840 to 850 m/s. Firing such a round is likely to be a lively and painful procedure should a conventional rifle design be involved.

For this reason the RT-20 does not have a conventional design. About half way down the barrel is a block with an upwards-facing gas port. As the projectile passes this point some of the propellant gases are tapped off and directed along a recoil compensation tube to be vented into the atmosphere behind and above the bolt area. As on recoilless weapons a cone shaped area covering a rearwards arc of 60° is thus hazardous as the rifle is fired and has to be kept clear of obstacles. It follows that the RT-20 cannot be fired with structures, such as walls, in close proximity to the rear and that under some circumstances the firing position might be disclosed.

The rearward escaping gases also entail that the firer cannot assume an orthodox firing position. Instead he has to lay at an angle to the left of the receiver with his right shoulder supporting the weapon and assuming some of the recoil forces. To attenuate the firing and supporting loads on the shoulder recourse has to be made to pads of foam rubber which are simply taped into position according the few illustrations seen to date. Recoil forces are further reduced by a high efficiency multibaffle muzzle brake.

Aiming the RT-20 involves a telescopic sight with a long eye relief to ensure the firer's eye is not so close to the eyepiece that recoil might cause problems. To assist the firer a bipod is provided which can be adjusted in height from 215 to 295 mm.

Many RT-20 details, such as barrel or overall length, are not yet available but it would appear to be a very efficient if rather specialised weapon. The 20 mm API M60 round fired from M55 anti-aircraft guns retains sufficient energy to penetrate most soft-skinned targets at 2,000 m and may even be able to penetrate lightly armoured vehicles at reduced ranges. The HE M57 round contains a small charge of TNT or RDX/Aluminium.

Despite its weight of 26 kg, the standard RT-20 can be carried by one man using a special backpack. Mention has been made of a RT-20 variant with the weight reduced to 18 kg, a revised bipod arrangement, and with a small monopod added just forward of the butt plate. The drawing illustrates this model. No further details are yet available.

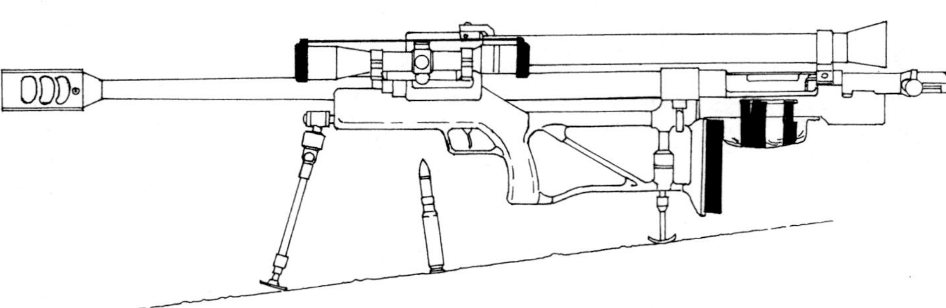

Provisional side view drawing of RT-20 20 mm heavy sniper rifle with bolt open (L Haywood)

1996

Data

Cartridge: 20 × 110 mm (20 mm Hispano Suiza HS 404)
Operation: bolt action, single shot
Feed: manual, single shot
Weight: standard model 26 kg; lightweight model, 18 kg
Sights: optical
Muzzle velocity: 840-850 m/s
Max effective range: ca 2,000 m

Marketing Agency

RH - ALAN d.o.o., Stančićva 4, 41000 Zagreb.

Status

In production.

Service

Croatian Army.

NEW ENTRY

CZECH REPUBLIC AND SLOVAKIA

LADA assault rifles

Development

The LADA assault rifle was first seen publicly during 1993. It is being promoted chambered for one of two possible rounds, the Eastern Bloc 5.45 × 39 mm or 5.56 × 45 mm NATO. The choice is made more significant by an announcement of intent made by the Czech armed forces to switch from Eastern Bloc to NATO standards for all munitions, from small arms to artillery. However the funding for such a drastic move has not yet been forthcoming so production of the LADA rifles has to date been limited pending a final decision. Indications would appear to favour the move to produce LADA rifles in 5.56 mm for the Czech armed forces but this may be overshadowed by procurement interest displayed by nations such as Poland and North Korea who are more likely to favour the retention of 5.45 mm.

There are three components in the LADA range of weapons; the standard assault rifle, a carbine and a light machine gun. Many components are interchangeable between all three weapons. For details of the light machine gun refer to the entry in the *Machine guns* section.

5.56 mm LADA carbine (top) and assault rifle (bottom) (T J Gander)

1996

Description

The LADA rifles are gas operated and utilise the widely used rotating breech block for locking to the barrel at the instant of firing. Three modes of fire are possible, semi-automatic single shot, automatic and three-round burst, with the selector lever on the right of the receiver. The receiver is constructed from steel pressings and there are obvious resemblances to Kalashnikov series rifles. The pistol grip and forward handguard are plastic.

Rounds are fed into the weapon from a transparent plastic curved box magazine holding 30 rounds although it is possible to utilise the 75-round drum magazine normally used on the LADA light machine gun. M16-type magazines can also be used.

On the LADA assault rifle there is an aperture rearsight graduated from 100 to 800 m in 100 m steps. The front and rearsight are provided with luminescent dots for shooting in poor light conditions. It is possible to utilise the optical sight mounting bar normally fitted to the LADA light machine gun on the assault rifle model (as a production modification) and it is also possible to attach the light machine gun bipod. Neither of these options apply to the LADA carbine which has a barrel length of 185 mm compared to the assault rifle's 382 mm, a factor which would prompt some to place the carbine in the sub-machine gun category. Muzzles are fitted with a flash suppressor. Other attachments for the assault rifle include a grenade launcher under the barrel and a bayonet.

The butt-stock fitted to both the assault rifle and carbine is of the side folding type, using tubular steel struts and a single-piece butt plate. If required the complete weapon can be broken down into four basic components.

Data

Cartridge: 5.45 × 39 mm or 5.56 × 45 mm NATO
Operation: gas, selective fire
Locking: rotary bolt, forward locking
Feed: 30-round plastic box magazine
Weight: (empty) assault rifle, 3 kg; carbine, 2.6 kg
Length: (overall, butt extended) assault rifle, 850 mm; carbine, 675 mm
Barrel: assault rifle, 382 mm; carbine 185 mm
Sights: fore, post; rear, aperture, graduated in 100 m steps from 100 to 800 m
Muzzle velocity: assault rifle, 910 m/s; carbine, 735 m/s
Rate of fire: cyclic, 750-850 rds/min

Manufacturer

Ceska Zbrojovka as, 688 27 Uhersky Brod.

Status

Limited production. Offered for export sales.

UPDATED

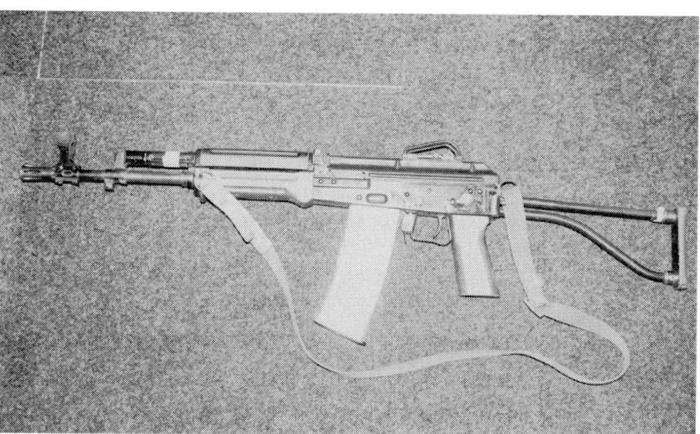

LADA assault rifle with butt extended (T J Gander)

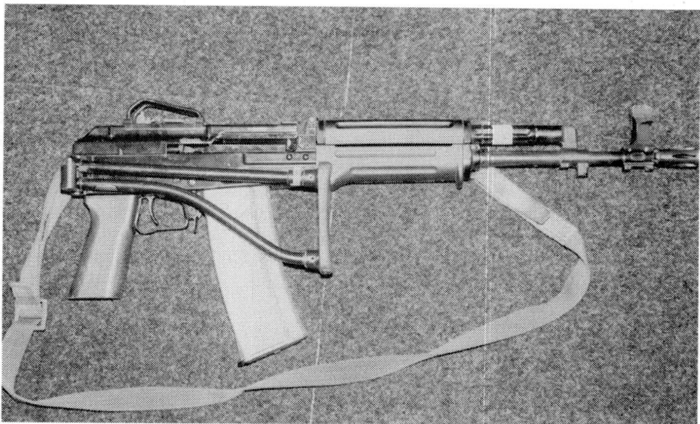

LADA assault rifle with butt folded (T J Gander)

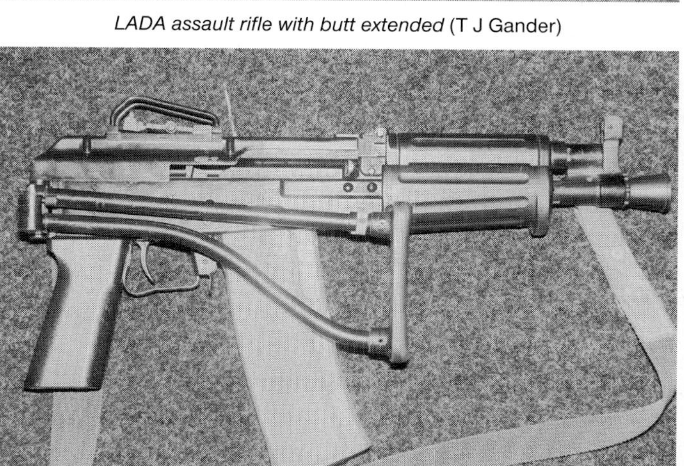

LADA carbine with butt folded (T J Gander)

LADA carbine with butt extended (T J Gander)

7.62 mm Model 58 assault rifle

Development

The Czech Army is equipped with the Model 58 rifle, an indigenous product of original Czech design. The earliest known versions had wooden butts, pistol grips and fore-ends, but most production weapons use a wood fibre-filled plastic for those parts. There are three standard versions:

Model 58 P which has a solid butt and is the normal infantry rifle; the Model 58 V which has a folding butt; and the Model 58 Pi which is a P version with a long dovetail bracket on the left side to accept a night sight. This version is usually seen with a light bipod and an enlarged conical flash hider.

The Model 58 bears a superficial resemblance to the Soviet AK-47 but there are considerable differences.

Description

The Model 58 is gas operated with a vent, 215 mm from the breech face, opening into a cylinder placed above the barrel. There is no gas regulator and the full gas force is exerted on the piston-head. The entire piston is chromium plated so as to remain free of fouling. The gas pressure can drive the piston back only 19 mm and the shoulder on the shank then butts up against the seating and no further movement is possible. There is a light return spring held between the piston shoulder and the seating which returns the piston to its forward position. The cylinder is vented on the underside and the gas pressure gives the piston an impulsive blow before exhausting to atmosphere after the piston has gone back 16 mm. There is a protective metal heat shield over the piston which has a covering of the wood-filled plastic used throughout for the furniture. It is exactly the same shape as that in the AK-47 and is removed by the withdrawal of a pin passing through the rearsight block in which the piston sits.

The short tappet-like stroke of the piston strikes the breech block carrier and drives it rearwards. After 22 mm of free travel an inclined plane on the carrier moves under the locking piece and lifts it out of engagement with the locking shoulders in the steel body. The locking piece swings and this movement provides the leverage required for primary extraction. The breech block is then carried rearwards extracting the empty case from the chamber. A fixed ejector in the receiver passes through a groove cut in the underside of the bolt and the case is flung upwards clear of the gun. The continued rearward movement of the carrier and bolt compresses two double-coiled helical springs. The larger of these fits into the top hole of three drilled in the carrier and the smaller rests in the hollow steel tube which acts as a hammer. The carrier is driven forward and the feed horns on the underside of the bolt face force a round out of the magazine and into the chamber. When the round is fully chambered the carrier still has 16 mm of travel and, as it advances, a transverse cam face forces the locking piece down and the two lugs enter the locking shoulders in the body. It should be noted that the arm can be assembled and fired without the locking piece, which could lead to a serious accident.

Unlike the majority of self-loading and automatic rifles the Model 58 does not have a rotating hammer which strikes a firing pin; the hammer is a steel bar hollowed from one end almost throughout its full length to take the hammer spring. At the open end is welded a bent and there is a groove cut in each side of this to slide on the receiver guideways. This hammer enters the hollow bolt and drives forward a fully floating firing pin.

The selector is on the right-hand side of the receiver; single shot is indicated by '1' and full automatic by '30'. At the 'safe' position with the selector pointing vertically downwards, the trigger bar and the disconnector are lowered and so there is no connection between the trigger and the semi-automatic sear which holds the hammer.

Data

Cartridge: 7.62 × 39 mm
Operation: gas, selective fire
Locking: pivoting locking piece
Feed: 30-round box magazine
Weight: 3.14 kg
Length: butt extended, 820 mm; butt retracted, 635 mm
Barrel: 401 mm
Rifling: 4 grooves, rh, 1 turn in 240 mm
Sights: fore, post; rear, tangent leaf V-notch
Sight radius: 356 mm
Muzzle velocity: 710 m/s
Rate of fire: cyclic, 800 rds/min
Effective range: 400 m

Manufacturer

Uherský Brod ordnance factory.

Status

Production complete. Large numbers held in reserve.

Service

Czech Army. Sold on the commercial market.

VERIFIED

7.62 mm Model 58 P assault rifle

EGYPT

7.62 mm Misr assault rifle

Development

The 7.62 mm Misr assault rifle is a direct copy of the Soviet AKM manufactured at 'Factory 54' in Egypt for the Egyptian armed forces. It is also offered for export sales and has reportedly been sold in significant numbers. Also produced is a semi-automatic version known as the ARM which has been sold on the sporting arms market but which would be suitable for many police and paramilitary applications.

Description

The 7.62 mm Misr assault rifle is virtually identical to the Soviet AKM in all respects so reference can be made to the AKM entry under *Commonwealth of Independent States* for operating and other details. The standard Misr rifle has a wood forward handguard and butt-stock while the pistol grip and upper handguard are plastic. A version with a side-folding butt has a rudimentary but light single-strut metal stock. Night sights can be fitted to the Misr.

Variant

The ARM is a semi-automatic version of the Misr and is available with fixed or side-folding butts which render the ARM visually identical to the Misr, especially when the standard 30-round magazine is employed. The only changes involved are to the trigger mechanism which allows single-shot only. One version of the ARM is available with a wooden 'sporting' butt-stock having an integral thumb grip. All versions of the ARM may be supplied with a 5-round box magazine.

Data

Cartridge: 7.62 × 39 mm
Operation: gas, selective fire
Locking: rotating bolt
Feed: 30-round detachable box magazine
Weight: with bayonet and loaded magazine, 3.86 kg
Length: 880 mm
Barrel: 415 mm
Rifling: 4 grooves, rh
Sights: fore, adjustable pillar; rear, U-notch
Sight radius: 378 mm
Muzzle velocity: 715 m/s
Rate of fire: cyclic, 600 rds/min
Effective range: 300 m

Manufacturer

Maadi Company for Engineering Industries, Maadi, Cairo.

Status

In production. Offered for export sales.

Service

Egyptian armed forces and some other countries in Africa and the Middle East.

VERIFIED

7.62 mm Misr assault rifle with side-folding butt

7.62 mm ARM, the semi-automatic rifle version of the Misr, seen here with a thumbhole butt-stock and 5-round magazine

FINLAND

Sako M90 automatic rifle

Description

The Sako M90 is an improved version of the M62/M76 series of assault rifles (see following entry). In short, the M90 is based on the Kalashnikov action but has been lightened and refined to a new high standard. A new side-folding butt has been developed, new rearsights with an adjustable tangent aperture and a fixed combat sight, and a new flash eliminator which also functions as a grenade launcher. Tritium auxiliary night sights are fitted as standard.

The M90 is available in two calibres, 7.62 × 39 mm or 5.56 × 45 mm. Optional equipment, in either calibre, includes a removable bipod, optical day or night sights, bayonet and a blank-firing device.

To accompany the M90, Sako have also developed a new hard core K413 bullet. Both the bullet and its cartridge case are made from special alloy steel.

Data

DATA FOR 7.62 × 39 mm VERSION
Cartridge: 7.62 × 39 mm
Operation: gas, selective fire
Locking: rotating bolt
Feed: 30-round magazine
Weight: 3.85 kg
Length: butt folded, 675 mm; butt extended, 930 mm
Barrel: 416 mm
Rifling: 4 grooves, rh, 1 turn in 240 mm
Sights: fore, hooded blade with tritium spot; rear, adjustable tangent aperture, 150-300-400 m, with tritium night spots
Sight radius: 485 mm
Rate of fire: cyclic, 600-750 rds/min
Muzzle velocity: ca 800 m/s

Manufacturer

Sako Limited, PO Box 149, FIN-11101 Riihimäki.

Status

Available.

VERIFIED

Top view of Sako M90 rifle showing improved sights

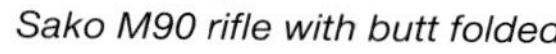

Sako M90 rifle with butt folded

7.62 mm Sako M90 automatic rifle

7.62 mm Valmet M60, M62 and M76 automatic rifles

Description

The Finnish assault rifle is essentially a locally manufactured version of the Kalashnikov AK-47. Finland adapted the design in the late 1950s and the first small pilot batch required for troop trials bore the designation M60. After some minor changes the Finnish Defence Forces adopted the weapon under the designation M62. The latest version is called the M76.

The M60 differed from the AK-47 in several respects. There was no wood on the M60. The metal fore-end and the pistol grip were both plastic covered. The butt-stock was a cylindrical tube with a cross bar riveted on for the shoulder piece. The cleaning rod and brush were carried in the tube of the stock. The forearm of the Model 60 had 10 large and 10 small holes to allow dissipation of heat. The receiver was made from milled steel bar.

Finnish 7.62 mm M76T automatic rifle

M76W automatic rifle

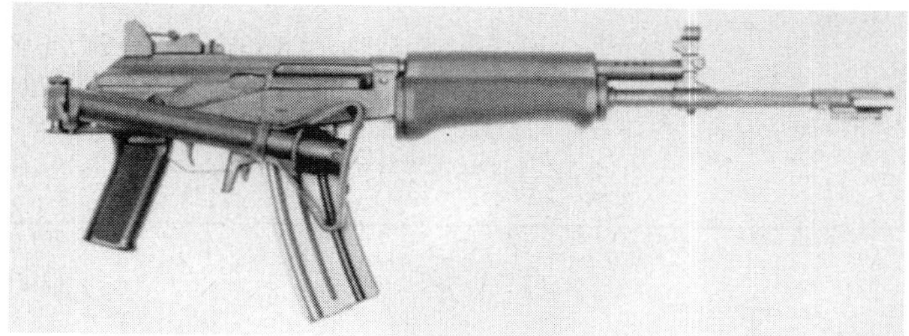

M76F automatic rifle

The barrel had a three-prong flash eliminator. The foresight was a hooded post over the front end of the gas cylinder. The rearsight was a tangent aperture protected by two wings. Both foresight and rearsight had night sight option. Since, unlike the Soviet practice, an aperture rearsight was used, it was placed at the back of the receiver cover. There was no trigger guard, but a protective bar dropped down in front of the trigger to prevent accidental release of the magazine. The safety catch and selector lever on the right of the receiver was exactly the same as that of the AK-47. The semi-automatic marking was a single dot, and next came full automatic with three dots. The M60 did not go into production and after the trials the remaining examples were recalled.

The 1962 modifications were not of great significance: the fore-end was made of plastic, ribbed, with 24 large cooling holes, a trigger guard was fitted, the pistol grip was ribbed plastic and swept back further at the rear upper end and the fitting of the shoulder piece to the tubular stock was amended somewhat. A recent innovation is the addition of tritium light sources to the sights to facilitate aiming in poor light.

The following versions of the M76 have been made both in 7.62 mm and in 5.56 mm:

- M76T tubular butt-stock
- M76F folding butt-stock
- M76P plastic butt-stock
- M76W wooden butt-stock.

Data

M76W

Cartridge: 7.62 × 39 mm or 5.56 × 45 mm
Operation: gas, selective fire
Locking: rotating bolt
Feed: 15-, 20-, or 30-round detachable box magazine
Weight: empty, 3.6 kg; loaded with 30 rounds 7.62 mm, 4.51 kg
Length: butt folded, 710 mm; butt extended, 950 mm
Barrel: 418 mm
Rifling: 7.62 mm, 4 grooves, rh; 5.56 mm, 6 grooves, rh
Sights: fore, hooded blade; rear, tangent aperture
Muzzle velocity: 7.62 mm, 719 m/s; 5.56 mm, 960 m/s
Rate of fire: cyclic, 600-750 rds/min

Manufacturer

Sako Limited, PO Box 149, FIN-11101 Riihimäki.

Status

In production.

Service

Finnish and Qatar forces, and Indonesian security forces.

VERIFIED

7.62 mm Valmet M78 long barrel automatic rifle

Description

The Valmet M78 is a variant of the M76 assault rifle. The two main differences are a longer and heavier barrel with a carrying handle, and a bipod fitted to the front of the barrel. Normal rifle magazines are used. The weapon is light and portable and good reliability is claimed, stemming from the considerable experience gained with the M76. The low weight makes it particularly suitable for areas in which infantry have to operate on their feet rather than from vehicles.

7.62 mm Valmet M78 long barrel rifle

Data

Cartridge: 7.62 × 39 mm; 5.56 × 45 mm; 7.62 × 51 mm in semi-automatic form only
Operation: gas, selective fire
Feed: 15- and 30-round box magazines
Weight: empty, 4.7 kg; loaded with 30 rounds, 5.9 kg
Length: 1.06 m
Muzzle velocity: 7.62 × 39 mm, 719 m/s
Rate of fire: cyclic, 650 rds/min

Manufacturer

Sako Limited, PO Box 149, FIN-11101 Riihimäki.

Status

No longer in production.

Service

Military trials.

VERIFIED

Sako-TRG sharpshooting system

Description

The Sako-TRG is a conventional bolt-action repeating rifle, Sako being of the opinion that for first-round effectiveness and accuracy the bolt action cannot be bettered. Two versions are manufactured, the TRG-21, firing 0.308 Winchester (7.62 × 51 mm NATO) cartridges, and the TRG-41, firing 0.338 Lapua Magnum cartridges.

The barrel is manufactured by the cold hammering method. The muzzle can be fitted with a detachable muzzle brake which also acts as an efficient flash hider. Silencers can be mounted on to the muzzle brake, where it forms part of the silencer and is not to be removed.

The receiver is constructed of solid steel and also formed by cold hammering. It incorporates three locking lugs for maximum strength, while the bolt lift angle of 60° provides minimum bolt movement. The receiver is stabilised with three fastening screws and utilises an oversized bedding surface for additional rigidity. A 17 mm integral dovetail is used for optical day or night sight mounting. Auxiliary folding steel peep sights are provided for emergency use.

The cartridges are fed from a detachable centre-fed magazine.

The base of the stock is of aluminium profile to which the polyurethane forestock is attached. The butt-stock is also made of polyurethane and is reinforced through the use of an aluminium skeleton. Since the receiver is bedded directly to the aluminium profile, no bedding material is required. Spacers allow the cheekpiece to be fully adjustable in height and infinitely adjustable in windage and pitch. The butt-plate is adjustable both for distance and angle through the use of spacers and is also infinitely adjustable in height and pitch. The fully adjustable and versatile stock design allows the rifle to be used as a sniper rifle, a UIT standard rifle, or for CISM military rapid fire. In addition, the stock is designed for both right- and left-handed users.

The double-stage trigger pull is adjustable from 1 to 2.5 kg. It is also adjustable in length and horizontal or vertical pitch. The entire trigger assembly including the trigger guard can be removed from the rifle without disassembling any other parts of the rifle.

The safety catch, which is silent in operation, is located inside the trigger guard. The safety locks the trigger mechanism and locks the bolt in the closed position with the firing pin blocked from the primer cap.

Sako-TRG sharpshooting system

Data

DATA FOR TRG-21, WHERE TRG-41 DIFFERS IN PARENTHESIS

Cartridge: 0.308 Winchester (7.62 × 51 mm); (0.338 Lapua Magnum)
Operation: bolt action, single shot
Feed : 10-round box magazine; (5-round)
Weight: empty, without spacers, 4.7 kg; (5.1 kg)
Length: without butt-plate spacers, 1.15 m; (1.2 m)
Barrel: 660 mm; (690 mm)
Rifling: 4 grooves, rh, 1 turn in 280 mm; (1 turn in 305 mm)
Sights: telescope mount and iron emergency sights

Manufacturer

Sako Limited, PO Box 149, FIN-11101 Riihimäki.

Status

In production.

VERIFIED

FRANCE

5.56 mm FAMAS F1 assault rifle

Description

A bullpup design, the 5.56 mm FAMAS F1 assault rifle can be fired from either shoulder without the difficulty inherent in most bullpup rifles of ejection of the spent case. This has been achieved by providing two extractor positions on the bolt face. The rifle is issued with the extractor on the right and with this arrangement the spent cases are ejected to the right. By moving the extractor to the other position the direction of ejection is reversed, the cheek rest being removed and positioned on the other side of the butt-stock where it closes off the ejection slot provided for operation on the other side. The cocking handle is placed centrally above the receiver to permit operation by either hand and the centrally mounted sights can be used regardless of the shoulder the soldier uses.

The barrel is of plain steel and has a fluted chamber. The muzzle is formed as a grenade launcher and an adjustable collar controls the position of the grenade and so varies the velocity and hence the range. A flash hider is fitted to the muzzle. The receiver is of light alloy and the other assemblies are pinned to it.

The breech block fits into the carrier, is drilled to take the firing pin and the ejector rod and spring, and has a detachable front section carrying the extractor and a dummy extractor plug.

The carrier has grooves at the bottom of the rear end which enable it to slide in the receiver. It is drilled at the top of the front face to take the cylinder containing the return spring.

The delay lever has two parallel, angled arms joined by a cross-piece. The arms connect the breech block to the carrier and the cross-piece controls the position of the firing pin. The lower ends of the arms bear against a hardened steel pin across the receiver and the upper ends rest against the back face of the breech block carrier. It is the means of holding up the breech face, while the chamber pressure is high, and transferring energy to the carrier. It also controls the trigger mechanism by operating the safety sear.

The trigger mechanism is self-contained in a plastic box which is pinned to the receiver. It provides single shots, automatic fire and three-round bursts by using two selector controls: the usual 'safe', 'semi-automatic' and 'automatic' control which is near the trigger, and the burst-fire controller under the trigger mechanism. This arrangement has been adopted because the burst-fire escapement is kept completely separate from the basic mechanism and should the burst-fire device be inoperable it will not affect the primary means of fire.

The raised foresight is mounted on a column pinned to the barrel. The foresight blade is mounted on a leaf spring and can be moved across the pillar, by using a notched screw. Each notch on the screw head moves the mpi 40 mm at 200 m (0.2 mil). There is a foresight cover with a luminous bead for night firing and a detachable open sight for firing grenades.

The rearsight is also on a column, above the return spring cylinder. It is an aperture with the choice of two diameters selected by the use of two hinged plates, one on the front of the sight column with a small aperture for good light, and one on the rear with a large aperture for poor light. At night both shutters are lowered and the top of the pillar makes a large aperture which is used in conjunction with the luminous bead on the foresight support.

5.56 mm FAMAS F1 assault rifle

1996

5.56 mm FAMAS F1 rifle fired from shoulder showing small size, short sight radius and general compactness

The rearsight is adjustable for elevation. There is a milled screw which compresses a spring to raise and lower the sight cradle. One turn of the screw moves the mpi 70 mm at 200 m (3.35 mils).

The plastic butt-stock carries a spring buffer in the top half to cushion the blow of the working parts as they recoil. There is a rubber shoulder pad to reduce the impact on the firer's shoulder. The carrying handle is plastic. It provides protection for the sights against accidental damage, for the return spring cylinder and against barrel heat. It carries a sight for the indirect fire of grenades at 45° or 75°.

The vertical box magazine holds 25 rounds, and is placed in the weapon by locating the front end in the well and rotating it rearwards to engage the magazine catch. The magazine has holes drilled in the sides to indicate that it contains (from top to bottom) 5, 10, 15, 20 or 25 rounds.

If the weapon is not to be fired immediately, the safety catch should be set to 'S'. The cocking handle is pulled to the rear as far as possible and then released, whereupon it flies forward and feeds a round into the chamber. As the bolt comes to rest the bolt carrier bears against the top of the delay lever, causing it to rotate forward into a vertical position. In this position the bottom end of the left arm of the lever presses down on a spring-loaded rod which releases the safety sear and at the same time the centre cross bar, connecting the two arms, rotates out of a notch in the firing pin, which is then free and can be driven forward when struck by the hammer. When the trigger is operated the hammer flies forward and hits the firing pin which goes forward and fires the cartridge.

The pressure developed by the expanding gases forces the cartridge case back against the breech face and the bolt starts to move back. The lower arms of the delay lever lie below the bearing pin in the receiver and the force exerted by the breech block causes the delay lever to rotate backwards. The top ends of the arms bear against the inside of the bolt carrier and since the top lever arms are longer than the bottom arms, a leverage differential exists and the carrier is accelerated back relative to the bolt which moves only a very short distance.

During the first 45° of backward rotation of the delay lever, the following occur: the bolt carrier is accelerated backwards; the cross bar of the delay lever engages in the notch in the firing pin and withdraws it into the bolt; the bottom of the left arm of the delay lever moves off the rod leading to the safety sear which is then released so that the hammer is held back regardless of the position of the trigger or the firing sear; the movement of the breech block is retarded and this ensures that the cartridge case does not emerge, unsupported, from the chamber whilst the gas pressure is high.

When the delay lever has rotated some 45°, it clears the pin in the receiver and the residual pressure in the chamber forces the cartridge case and bolt rapidly back. The bolt and its carrier then travel back together with the piston compressing the return spring in the cylinder. The empty case is extracted and ejected. The back of the bolt carrier hits the buffer and compresses it, and the bolt carrier and bolt are then reversed in direction and, impelled by the return spring, go forward to repeat the cycle.

Variants

FAMAS G1 and G2

See following entry.

FAMAS Export

This the standard 5.56 mm rifle but is modified so as to permit only single-shot firing. There is no grenade launching capability. The FAMAS Export is intended for the overseas commercial market.

FAMAS Civil

This model is intended for the French commercial market, and in order to comply with French firearms laws has had its barrel lengthened to 570 mm and the calibre changed to 0.222 Remington chambering. It can only fire single shots and has the grenade launching rings removed.

FAMAS Commando

The FAMAS Commando was a short version intended for use by Commando and other special forces. The barrel was shortened to 405 mm and there was no grenade launching facility, but in other respects it was the same as the service rifle and offered the full range of selective fire options. It was not produced in quantity.

FAMAS à Plombs

The FAMAS à Plombs is a sub-calibre training weapon firing 4.5 mm air gun pellets by means of a 12 g CO_2 gas cartridge. The weapon fires single shot only from a modified magazine but is otherwise a close replica of the FAMAS F1 service rifle. Muzzle velocity is 130 m/s and accuracy is such that 10 rounds can be placed in a 50 mm circle at 12 m.

Data

Cartridge: 5.56 × 45 mm
Operation: delayed blowback, selective fire and 3-round burst facility
Method of delay: vertical delay lever
Feed: 25-round box magazine
Weight: empty, 3.61 kg
Length: 757 mm
Barrel: 488 mm
Rifling: 3 grooves, rh, 1 turn in 305 mm
Sights: fore, blade; rear, aperture 0-300 m
Sight radius: 330 mm
Muzzle velocity: 960 m/s
Rate of fire: cyclic, 900-1,000 rds/min
Effective range: 300 m

Manufacturer

Giat Industries, Division Giat Vecteur, 13 route de la Minière, F-78034 Versailles-Satory.

Status

In production.

Service

In service with the French Army. Also exported to Djibouti, Gabon, Senegal and the United Arab Emirates.

VERIFIED

5.56 mm FAMAS G2 assault rifle

Development

The 5.56 mm FAMAS G2 assault rifle was developed by Giat Industries as a private venture, and is primarily intended for the export market. It is an updated version of the interim FAMAS G1 which could accommodate only the standard FAMAS F1 25-round box magazine; the FAMAS G1 is no longer offered.

Description

Mechanically the 5.56 mm FAMAS G2 is the same as the in-service FAMAS F1 and is chambered for the standard 5.56 × 45 mm cartridge. However the standard rifling is one turn in 228 mm, allowing it to be used with M193 or SS109 types of ammunition. Optionally, the barrel can be rifled one turn in 178 mm or one turn in 305 mm to suit specific requirements.

The outline has changed by the adoption of a full-hand trigger guard, allowing the rifle to be fired when wearing NBC or arctic mittens or heavy gloves. The breech block buffer has been reinforced in order to withstand better the firing of rifle grenades and the magazine housing now conforms to NATO STANAG 4179 and will accept all M16-type magazines, straight or curved, metal or plastic.

The selector lever/safety catch is now placed inside the trigger guard, and the front end of the handguard has been given a downward lip to prevent the hand sliding forward onto a hot barrel.

Other FAMAS G2 features include the provision of a NATO sight base to allow the fitting of optical day or night sights. A bipod can be fitted and there is provision for mounting a bayonet over or under the muzzle. The ambidextrous features of the FAMAS F1 are carried over to the FAMAS G2.

Data

Cartridge: 5.56 × 45 mm
Operation: delayed blowback, selective fire and 3-round burst facility
Locking: differential leverage
Feed: 20- or 30-round M16 type magazines
Weight: with empty magazine, 3.8 kg; with 30-round magazine, 4.17 kg
Length: 760 mm
Barrel: 488 mm
Rifling: 3 grooves, rh, available 1 turn in 178, 228 or 305 mm
Sights: fore, blade; rear, aperture to 450 m; provision for optical sights
Muzzle velocity: 925 m/s
Rate of fire: cyclic, 1,100 rds/min

Manufacturer

Giat Industries, Division Giat Vecteur, 13 route de la Minière, F-78034 Versailles Satory.

Status

In production. Available.

VERIFIED

5.56 mm FAMAS G2 assault rifles

1996

5.56 mm FAMAS G2 assault rifle

FR-F1 sniper rifle

Development

The French sniper rifle, the Fusil à Répétition Modèle F1, is a 10-shot, manually operated bolt-action rifle based on the action of the now obsolete 7.5 mm Model 1936. It has a detachable 10-round box magazine. The rifle was specifically designed for sniping but two other versions are in existence. The three models are known as: the Tireur d'Elite (sniper rifle); Tir Sportif (target rifle with target sights and a 1.5 to 1.9 kg trigger pull); Grande Chasse (which is described as being designed for big game hunting: it has an APX model 804 telescopic sight and 2 to 2.5 kg trigger pull).

The FR-F1 uses match quality 7.5 × 54 mm ball ammunition and the use of tracer or armour-piercing ammunition, which will damage the bore, is discouraged. The FR-F1 was manufactured in 7.62 × 51 mm NATO and the marking 7.5 or 7.62 mm is inscribed on the left-hand side of the receiver.

Description

The bolt is manually operated with a turn-down knob on the right-hand side. The locking lugs are rear-mounted. As the bolt is forced forward it feeds the top round into the chamber and as it closes the extractor grips the rim of the cartridge. At the same time the sear holds up the firing pin lug to retain the firing pin in the cocked position.

When the trigger is operated it rotates back around the trigger pin joining the trigger and sear until the trigger stud reaches the bottom of the receiver. Continued pressure pivots the sear and compresses its spring. When the top of the trigger adjusting screw reaches the receiver the first pressure is felt. When the second pressure is taken the sear releases the lug of the firing pin which goes forward under its own spring to hit the cap. The lock time is very short. It should be noted that if the trigger is held back and the bolt operated, the bolt stop which pivots on the trigger pin is depressed against its spring and the bolt comes freely out of the receiver.

The magazine catch on the right of the receiver is pressed to release the magazine. The magazine holds 10 rounds. If it is to be replaced in the rifle the rubber cap is left on the bottom but if the magazine is being carried as a loaded replacement the rubber cap is placed over the top to keep out dust and put back on the bottom, if time permits, when the magazine is placed in the rifle. It is possible to load single rounds through the bolt way to replenish a partially expended magazine but this procedure is not recommended in action.

The bipod is permanently fixed at the rear of the fore-end and the rifle is carried with the bipod legs swung forward and fitting into recesses on each side of the forearm. The bipod is always used unless firing over cover such as a low wall. The bipod legs can be lengthened by twisting the knurled collar above the spade and rotating it back when the spring-loaded lower leg has extended sufficiently. The legs should be fully retracted for stowing the bipod along the forearm.

The FR-F1 sniper rifle uses the Modèle 53 bis telescopic sight which is carried, with its adjusting tool, in a transit case. The mount is placed with the locking lever to the rear and when in place it is secured by rotating the lever forward. To zero the telescope the screws at the bottom of the mount ring are loosened, the

FR-F1 sniper rifle on bipod with telescopic sight fitted and iron sights folded down

elevation knob set to '2' and the rifle is boresighted on a distant object. The plastic rings between the telescope and the mount rings are rotated until the telescope reticle is on the aiming mark with zero windage set. When the bore and graticule are aligned on the same aiming mark the screws on the mount rings are tightened. A three-round check firing at 200 m should then be carried out to ensure that the graticule is on the mean point of impact. Once the sight is correctly zeroed it may be dismounted, cased, and remounted without change of zero.

There may be occasions when the telescope cannot be used. The fore and rearsights are normally laid flat and must be raised. The rearsight is a square shouldered notch. The foresight is a flat-topped pyramid. The shoulders of the rearsight and the centre of the foresight have three luminous green dots which at night are evenly spaced in a horizontal line and placed on the target. To use the iron sights the telescope must first be dismounted.

Data

Cartridge: 7.5 × 54 mm or 7.62 × 51 mm
Operation: bolt action, single shot
Locking: rotating bolt
Feed: 10-round box magazine
Weight: empty, 5.2 kg
Length: without stock spacers, 1.138 m
Barrel: 552 mm
Rifling: 4 grooves, rh, 1 turn in 305 mm
Sights: optical, ×4; iron fore, flat-topped pyramid with luminous spot; rear, square notch with luminous spots
Muzzle velocity: 852 m/s
Effective range: 800 m

Manufacturer

Giat Industries, 13 route de la Minière, F-78034 Versailles-Satory.

Status

Production complete.

Service

French Army and Mauretania.

VERIFIED

7.62 mm FR-F2 sniper rifle

Description

Introduced in late 1984, the 7.62 mm FR-F2 is an improved model of the FR-F1. The basic characteristics, action and dimensions are the same as the FR-F1 model, the changes being in the nature of functional improvements. The fore-end is now metal covered in matt black plastic material; the bipod is more robust and has been moved from its location at the front end of the fore-end to a position just ahead of the receiver. This allows it to be adjusted more easily by the firer, and it is now suspended from a yoke around the rear of the barrel where it is less likely to affect the stability of the rifle when firing. The most innovative change is the enclosure of the barrel in a thick plastic thermal sleeve; this, it is claimed, reduces the possibility of heat haze interfering with the line of sight and it is also likely that it will reduce the infra-red signature of the weapon.

The FR-F2 is chambered only for the 7.62 × 51 mm NATO cartridge producing a muzzle velocity of 820 m/s. There is no other change in data from the FR-F1, to which reference should be made.

7.62 mm FR-F2 sniper rifle

Variants

7.62 mm FR-G1 and FR-G2

The 7.62 mm FR-G1 and FR-G2 are further revisions of the FR-F1 and FR-F2 sniper rifles. On the FR-G1 and FR-G2 the forestock is replaced by a wood assembly and the thermal sleeve of the FR-F2 is removed. The main difference between the FR-G1 and FR-G2 is that the FR-G1 has a fixed bipod while that for the FR-G2 is articulated. Various optical sights may be fitted. Without sights the FR-G1 weighs 4.5 kg, with the FR-G2 weighing 4.6 kg.

Manufacturer

Giat Industries, Division Giat Vecteur, 13 route de la Minière, F-78034 Versailles-Satory.

Status

Available.

Service

French Army.

VERIFIED

7.62 mm FR-G1 sniper rifle

7.62 mm FR-G1 sniper rifle

PGM UR Intervention 7.62 mm sniper rifle

Description

The PGM UR Intervention 7.62 mm sniper rifle is the base model of a series of specialised sniper rifles grouped under the name of Ultima Ratio ®, or UR. All rifles within this series are manually operated bolt-action rifles with all components mounted separately on a central rigid aircraft grade aluminium alloy frame.

The special steel bolt has three front locking lugs and has overpressure vent holes. An instant-release five-round clip magazine is carried inside the receiver (which is also manufactured using aluminium alloy), together with an ambidextrous pistol grip and fully adjustable shoulder stock with an adjustable black rubber shoulder pad. The two-stage trigger mechanism is stated to be 'of military design but of match quality'. A sear safety is provided on the left of the receiver. Located at the front end of the forestock is a folding adjustable bipod provided with an axial locking system to prevent accidental tilt and for operations on any type of terrain.

The fully floating barrel is finned for cooling and is manufactured to match grade standards. An integral muzzle brake is provided. One feature of the PGM UR series is that, using a quick change system, the barrel can be rapidly exchanged for any other barrel with a similar calibre in the UR series, with a guaranteed return to zero. Standard barrel lengths for the UR Intervention are 470 and 600 mm long.

Using the barrel rapid change system, it is also possible to exchange a standard PGM UR barrel for a high performance modular suppressed barrel which permits the use of subsonic or supersonic ammunition.

The top of the receiver is provided with a rail which can accommodate Universal, Weaver or STANAG scope bases without any need for further fitting.

The PGM UR base model is produced chambered for 7.62 × 51 mm (0.308 Winchester) although other calibres can be produced on request.

Also produced are PGM UR models intended for civilian sales. One model is the UR Europa, which can be chambered for several calibres, and another the UR Magnum which is chambered for 0.300 Winchester Magnum, although once again, other Magnum calibres can be provided. The UR Magnum has been acquired by some French administration agencies.

Data

Cartridge: 7.62 × 51 mm (0.308 Winchester); see also text
Operation: bolt action, single shot
Locking: rotating bolt

Feed: 5-round box magazine
Weight: ca 5.5 kg, according to configuration
Length: with 470 mm barrel, 1.03 m
Barrel: 470 or 600 mm
Sights: optical, to choice

Manufacturer
PGM Precision SARL, 'Usine Rouge', F-73660 Les Chavannes/Maurienne.

Status
In production.

Service
French Government security agencies.

NEW ENTRY

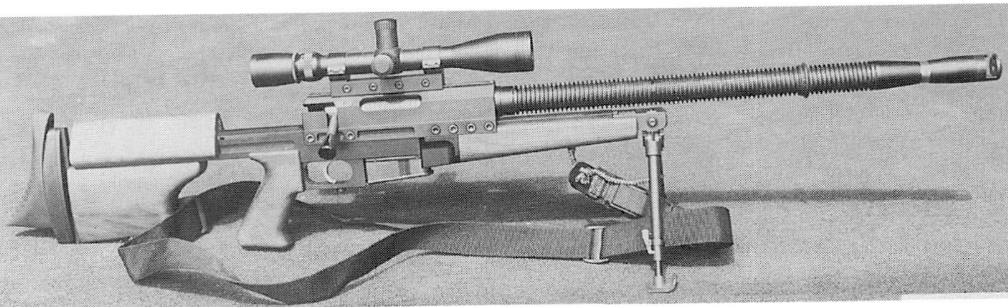

PGM UR Intervention 7.62 mm sniper rifle (T J Gander)

1996

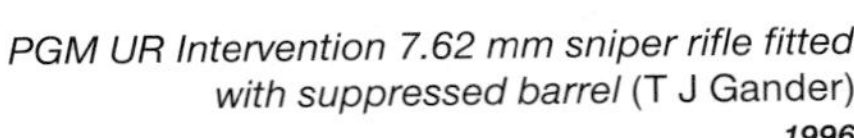

PGM UR Intervention 7.62 mm sniper rifle fitted with suppressed barrel (T J Gander)

1996

PGM UR Commando 7.62 mm sniper rifles

Description
The PGM UR Commando 7.62 mm sniper rifles are based on the PGM Ultima Ratio® Intervention model (see previous entry) but are shorter and lighter in construction to suit the requirements of special forces. All the construction features of the UR Intervention model are carried over, although on the Commando models the barrels are fluted rather than finned.

There are two UR Commando models. The UR Commando I has a fixed shoulder stock and may be fitted with barrels 470 or 550 mm long. On the Commando II the shoulder stock assembly can be folded to the left of the receiver to reduce the overall length for transport and stowage. The Commando II barrel length is normally limited to a length of 470 mm. It is also possible for both models to be fitted with the PGM suppressed barrel.

The PGM UR Commando models are produced chambered for 7.62 × 51 mm (0.308 Winchester) although other calibres can be produced on request.

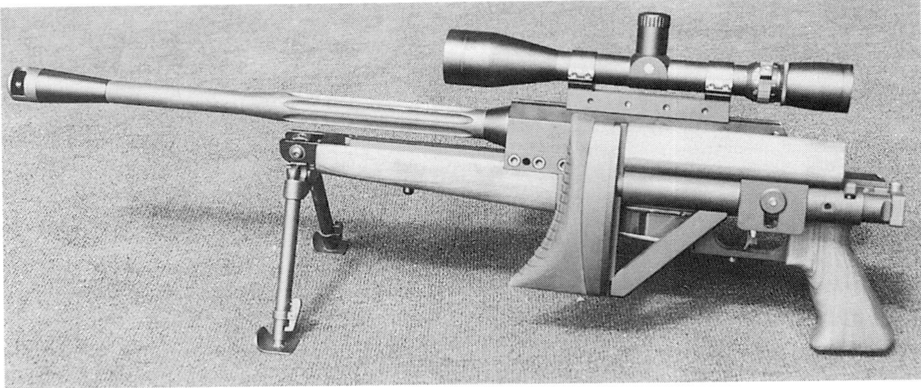

PGM UR Commando II 7.62 mm sniper rifle with shoulder stock folded (T J Gander)

1996

Data
Cartridge: 7.62 × 51 mm (0.308 Winchester); see also text
Operation: bolt action, single shot
Locking: rotating bolt
Feed: 5-round box magazine
Weight: ca 5.5 kg, according to configuration
Length: with 470 mm barrel, stock folded, 740 mm
Barrel: 470 or 550 mm
Sights: optical, to choice

Manufacturer
PGM Precision SARL, 'Usine Rouge', F-73660 Les Chavannes/Maurienne.

Status
In production.

Service
French Government security agencies.

NEW ENTRY

PGM Hecate II 12.7 mm sniper rifle

Description
The PGM Hecate II 12.7 mm sniper rifle is another PGM design based on the UR Intervention® (see separate entry) but with the Hecate II the overall scale is enlarged to accommodate the 12.7 × 99 mm (0.50 Browning) cartridge. Using this cartridge the Hecate II is claimed to be highly effective at ranges up to 1,500 m, although performance will depend on the type of ammunition involved.

The modular frame construction of the UR Intervention and Commando models is retained for the Hecate II although the overall dimensions are much enlarged. The magazine capacity is increased to seven rounds and the stock assembly can be removed for transport purposes. The heavy barrel, 700 mm long, is fluted and is fitted with a muzzle brake assembly which is claimed to be so efficient that the recoil forces imposed on the firer are no greater than those produced by a 7.62 mm rifle.

To assist in carrying the Hecate II a collapsible carrying handle is provided. Once the weapon is in the firing position an optional and adjustable stock support can be added under the stock assembly. Early Hecate II models employed an M60 pattern folding bipod, although this might be changed on later models. Other changes for the firer are that the bolt handle can be removed instantly for security or other purposes, and the sear safety is moved to the right of the receiver.

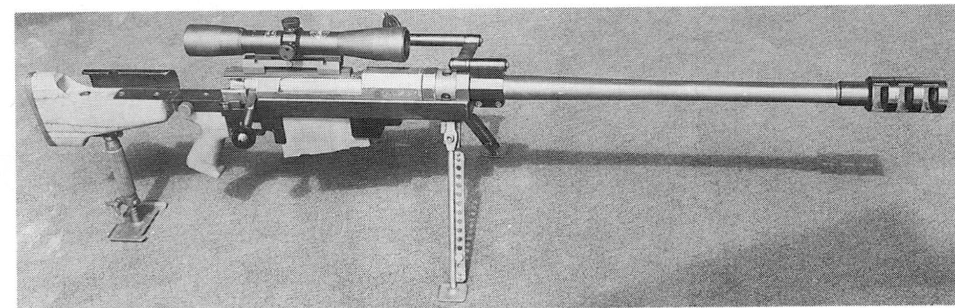

PGM Hecate II 12.7 mm sniper rifle (T J Gander)

1996

Data
Cartridge: 12.7 × 99 mm (0.50 Browning)
Operation: bolt action, single shot
Locking: rotating bolt
Feed: 7-round box magazine
Weight: ca 13.8 kg, according to configuration
Length: 1.38 m
Barrel: 700 mm
Sights: optical, to choice

Manufacturer
PGM Precision SARL, 'Usine Rouge', F-73660 Les Chavannes/Maurienne.

Status
Available.

NEW ENTRY

GERMANY

7.62 mm Mauser Model SP 66 sniper rifle

Description

The Model SP 66 is a heavy-barrelled bolt-action rifle of considerable accuracy which has been specifically designed and manufactured for use by law enforcement agencies and military snipers.

The Model SP 66 uses the well-known Mauser short-action bolt system in which the bolt handle is placed at the forward end of the bolt, just behind the locking lugs. By so doing the bolt movement is reduced by 90 mm and this length can then be added to the barrel. Since the body is correspondingly shorter, it is also lighter. Another advantage is that on opening the bolt there is less overhang behind the body and the firer does not have to move his head away from the sight.

Special emphasis has been placed on lock time and the operation of the sear/trigger combination. By using a very high rate on the firing pin spring and reducing the travel of that pin it has been possible to cut lock time to 50 per cent of that with the conventional bolt on the Model 98. The trigger has a 'shoe' 10 mm wide so that the firer can feel it even when wearing gloves.

The stock is fully adjustable to suit all builds of firer and features the thumb hole which is becoming common on many good quality target rifles. All surfaces which are touched with the hands are roughened to offer the best grip and the broad fore-end is designed to allow the left hand to have a comfortable and steady grasp whatever the length of the firer's arm.

The muzzle is fitted with a combined flash hider and muzzle brake. The requirement to reduce flash is critical with this and similar rifles since they are intended to be fired with high-magnification optical sights or image intensifiers and the presence of muzzle flash inhibits the firer with such magnified devices. Mauser claims that the chance of successive hits is high because of the lack of blinding from the muzzle.

Mauser offers the Model SP 66 rifle in standard form complete with telescopic mount, telescopic sight to customer's choice, a second mount for night vision devices and a case. It can be supplied with a night vision sight on request.

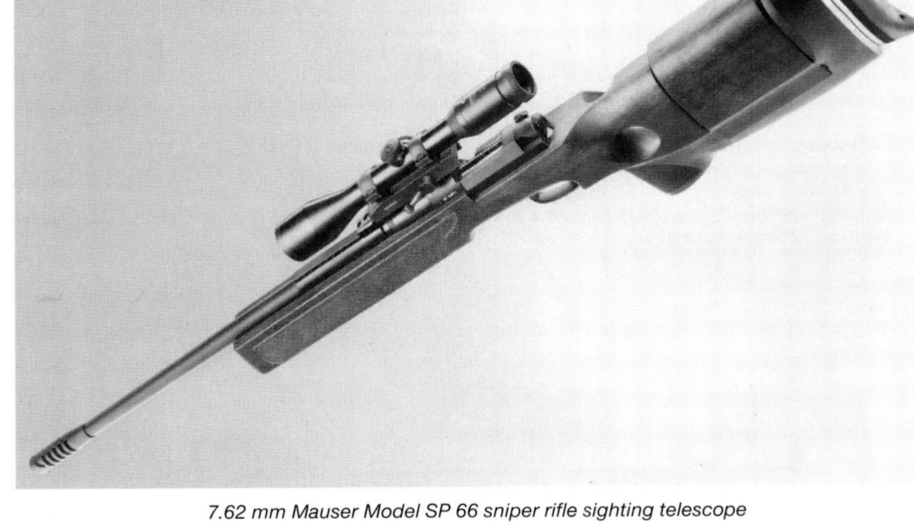

7.62 mm Mauser Model SP 66 sniper rifle sighting telescope

Data

Cartridge: 7.62 × 51 mm (selected sniper batches) and 0.300 Winchester Magnum
Operation: Mauser short-action bolt
Locking: forward lugs
Feed: 3-round integral magazine
Barrel: without muzzle brake, 650 mm; with muzzle brake, 730 mm

Manufacturer

Mauser-Werke Oberndorf Waffensysteme GmbH, Postfach 1349, D-78722 Oberndorf.

Status

In production.

Service

German forces and the armed forces of at least 12 other countries.

UPDATED

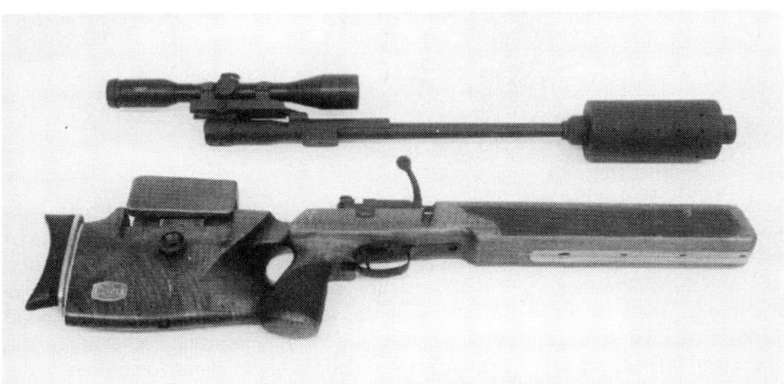

Silenced rifle dismantled for carriage

7.62 mm Mauser Model SP 66 sniper rifle with silencer

7.62 mm Mauser Model 86 sniper rifle

Description

The Model 86 is offered by Mauser as an alternative to the SP 66 (previously) and is the result of systematic development and years of experience. It uses a newly developed bolt action, locking by forward lugs, has a ventilated stock to provide additional heat-dissipating area and is fitted with an adjustable match trigger. The cold-forged free-floating barrel is fitted with a multibaffle muzzle brake to reduce recoil and damp down muzzle flash. The laminated wood match stock has an adjustable recoil pad and a rail in the forearm allows use of a firing sling or bipod. The rifle is not fitted with iron sights but a special telescope mount allows the use of optical telescope sights or night sights.

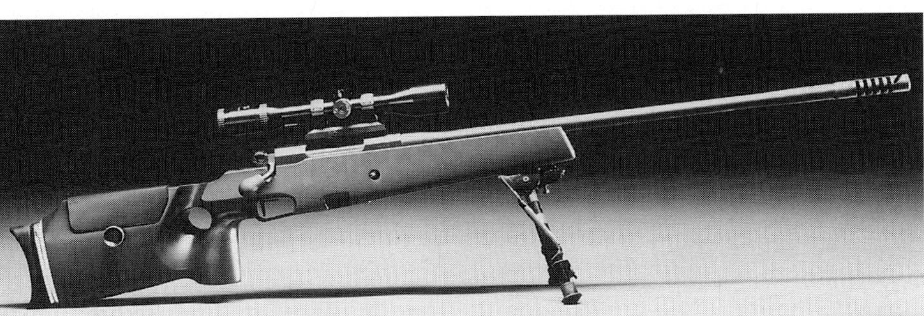

7.62 mm Mauser Model 86 sniper rifle

1996

Data

Cartridge: 7.62 × 51 mm NATO
Operation: bolt action, single shot
Feed: 9-shot double-row box magazine
Weight: without sight, 4.9 kg
Length: 1.21 m
Barrel: with muzzle brake, 730 mm
Rifling: 4 grooves, rh, 1 turn in 305 mm
Trigger pull: adjustable between 0.8 and 1.4 kg
Stock: Match stock with thumb hole or glassfibre stock optional

Manufacturer

Mauser-Werke Oberndorf Waffensysteme GmbH, Postfach 1349, D-78722 Oberndorf.

Status

In production.

UPDATED

Mauser Model SR 93 sniper rifle

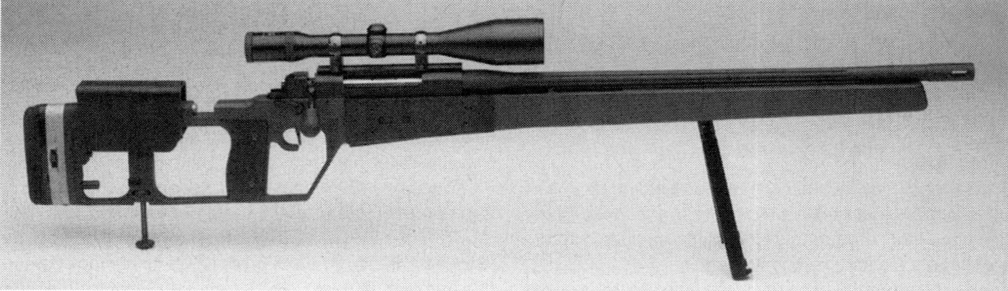

Mauser Model SR 93 sniper rifle

Description

The Mauser Model SR 93 sniper rifle was first shown publicly in 1993. It was developed in response to a German Army requirement which required a sniper rifle with the ability to penetrate armour at a range of 600 m. The Mauser-Werke adopted a 'no compromise' approach and consider that their solution, the SR 93, more than meets all the German Army's specifications.

The SR 93 can fire either 0.300 Winchester Magnum or 0.338 Lapua Magnum ammunition. For low-cost training purposes it is possible to use a changeover kit firing 7.62 × 51 mm NATO rounds.

Description

The Mauser Model SR 93 is a bolt-action sniper rifle with the front locking action using two lugs to create a locking surface of approximately 76 mm². Firing pin travel is 6.5 mm. The bolt action is so arranged that it can be rapidly changed over without tools to suit either left- or right-handed firers, a feature carried over to all other parts of the SR 93 which was designed from the outset for ambidextrous use. For instance, a safety catch is provided on both sides of the receiver.

The SR 93 uses a frame construction on which the cast magnesium/aluminium open frame stock is symmetrical in cross-section and has a synthetic cheek-piece adjustable on both sides; the butt plate is also adjustable in height and length to suit individual firers. The free-floating selected match barrel is manufactured using cold forged chrome-molybdenum steel and is provided with a muzzle brake to reduce recoil forces. The firer is further assisted by a synthetic forward hand-guard and a pistol grip made from a similar material; both are positioned to provide the optimum ergonomic firing posture. Trigger weight is adjustable between 1.5 and 2.5 kg.

A bipod, infinitely variable to suit circumstances and containing no sharp edges is integrated into the fore-stock. Overall the SR 93 has smooth-surfaced contours to prevent any possible snagging on clothing or other surfaces. A small monopod is provided under the butt-stock to allow the weapon to be held in the prone position for extended periods.

On the 0.300 Winchester Magnum model the box magazine contains six rounds; on the 0.338 Lapua Magnum model the capacity is five rounds.

Various types of optical and night sights can be fitted as well as a laser rangefinder. Accessories include a carrying case and a sling.

Firing match ammunition the SR 93 can place five shots within a 25 mm circle at 100 m.

Data

Cartridge: 0.300 Winchester Magnum; 0.338 Lapua Magnum (7.62 × 51 mm NATO changeover kit available for training)
Operation: bolt action, single shot
Feed: 0.300 Winchester Magnum, box magazine, 6 rounds; 0.338 Lapua Magnum, box magazine 5 rounds
Weight: without accessories, 5.9 kg
Length: 1.23 mm
Barrel: with muzzle brake, 690 mm; without muzzle brake, 650 mm
Rifling: 4 grooves, rh, 1 turn in 305 mm (1 turn in 254 mm optional)
Trigger pull: adjustable between 1.5 and 2.5 kg

Manufacturer

Mauser-Werke Oberndorf Waffensysteme GmbH, Postfach 1349, D-78722 Oberndorf.

Status

In production.

UPDATED

7.62 mm Heckler and Koch G3 rifle

7.62 mm G3A3 rifle

Development

The 7.62 mm Heckler and Koch G3 has been the service rifle of the German Army since 1959 and will probably remain so for some time to come. It remains in production, not only in Germany but in other countries where licenced production has been negotiated. Many of these licence-producer nations also have marketing rights so G3 rifles may be encountered almost anywhere throughout the world.

Description

The G3 receiver is a steel pressing, grooved on each side to guide the bolt and seat the back plate, and carrying the barrel. Above the barrel a tubular extension, welded on to the receiver, houses the cocking lever and the forward extension of the bolt. The cocking lever runs in a slot cut in the left side of this tubular housing and can be held in the open position by a transverse recess. The barrel is screw-threaded at the muzzle with a serrated collar to engage the retaining spring of the flash eliminator or blank-firing attachment. The rifling is orthodox and the chamber carries 12 longitudinal flutes extending back from the lead to 6 mm from the chamber face.

The bolt is shaped with a long forward overhang which is hollow and takes the return spring. It extends into the tube above the barrel. The bolt head carrier has long bearing surfaces on each side which slide in the grooves in the sides of the receiver. The bolt head carries two rollers which project on each side and are forced out by the inclined front faces of a 'locking piece'. When out, the rollers engage in recesses in the barrel extension. The bolt head and locking piece seat in the bolt head carrier and are held by the bolt head locking lever to prevent bounce on chambering the cartridge. The trigger mechanism fits into the grip assembly, secured to the receiver by locking pin.

When the trigger is squeezed the hammer flies forward and drives the firing pin through the hollow locking piece to fire the cap. The pressure generated in the chamber forces the cartridge case rearwards and exerts a force on the breech face which drives the bolt head back. Before the bolt head can go back, the rollers must be driven inwards. The rollers, which are carried in the bolt head, are pushed back and the angle of the recesses in the barrel extension is such that the rollers are forced inwards against the inclined planes on the front of the locking piece. In turn this inward force drives the locking piece back and with it the bolt head carrier. The angle of the locking piece face is such that the velocity ratio between bolt head carrier and bolt head is 4:1. Thus while the rollers are moving backwards and inwards the bolt head carrier is travelling back four times as far as the breech face.

As the carrier moves back the bolt head locking lever is disengaged. After the bolt face has moved back a little over a millimetre the rollers are clear of the recesses in the barrel extension, the entire bolt is blown back by the residual pressure with the bolt head and bolt carrier maintaining their relative displacement of about 5 mm. The bolt head carrier cocks the hammer and compresses the return spring. The cartridge case, held by the extractor, hits the ejector arm and is thrown out to the right. The rear end of the carrier hits the buffer. The buffer action, assisted by the return spring, drives the bolt carrier forward. The front face of the bolt head strips a cartridge from the magazine and chambers it. The extractor springs over the extraction groove of the cartridge and the bolt head comes to rest. The locking piece and bolt carrier then close up the gap of 5 mm and the rollers are pushed out into the recesses of the barrel extension. The bolt head locking lever engages the bolt head shoulder, thus preventing bounce. The weapon is then ready to be fired again.

A blank-firing attachment may be screwed on the muzzle instead of the flash eliminator. The attachment consists of an open-ended cylinder with a cross-bolt which closes the opening entirely. A groove is cut across the bolt so that by rotating the bolt the amount of gas escaping can be regulated to produce correct functioning of the rifle. This device has a dull chromium-plated finish to prevent confusion with the flash hider.

A special bolt marked 'PT' replaces the normal bolt and is used for firing plastic training ammunition. This works on pure blowback and has no delay rollers. It is not suitable for use with service ammunition. The normal magazine is used for plastic ammunition.

To conserve ammunition and allow troops to train in confined and built-up areas, a subcalibre training device is available. It consists of a subcalibre tube with a special bolt and magazine. The ammunition used is 5.6 × 16 mm (0.22 Long Rifle). The subcalibre tube is inserted from the breech and locked with a spring ring. In addition the tube is secured by a bolt on the magazine engaging in the end of the chamber face of the tube. The extractor enters a groove in the chamber face. The bolt assembly carries out the normal functions of chambering, supporting, firing, extracting and cocking the hammer. The magazine for 20 5.6 × 16 mm cartridges fits into the normal G3 magazine.

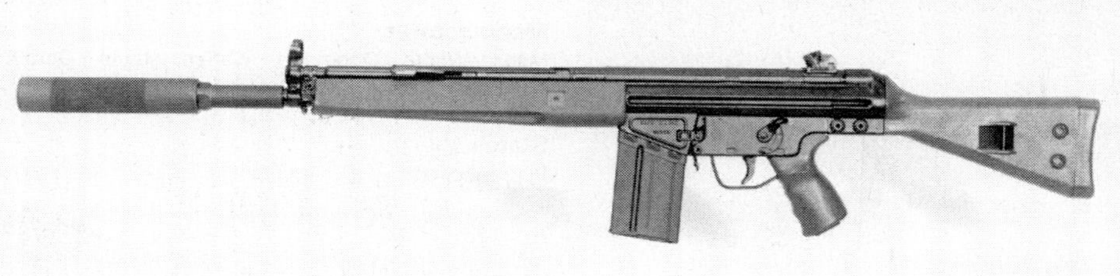

7.62 mm G3A3 rifle with fitted silencer

7.62 mm G3A4 rifle with butt retracted

A silencer may be screwed to the barrel in place of the flash hider and serves to reduce the muzzle report. It will fit all Heckler and Koch rifles with rifle grenade guides.

The use of subsonic ammunition is not recommended, in view of its reduced performance and the need for a special sight. The principal purpose of this silencer is for it to be used on firing ranges in order to reduce noise disturbance both for firers and local residents.

Variants

The standard rifle, in production since 1964, is known as the G3A3 and has a plastic butt-stock and plastic handguard.

When a telescopic sight is fitted to this rifle it is called the G3A3ZF.

When the plastic butt-stock is replaced by a retractable butt it is called the G3A4.

The G3A5 was a version produced for sale to Denmark.

The G3A6 is the Heckler and Koch designation given to a version developed for licence production in Iran. This version is apparently still in production in Iran at the Mosalsalsasi Weapons Factory.

The G3A7 is the Heckler and Koch designation given to a version developed for licence production in Turkey by MKEK.

Data

Cartridge: 7.62 × 51 mm
Operation: delayed blowback, selective fire
Method of delay: rollers
Feed: 20-round box magazine
Weight: empty, fixed butt, 4.4 kg; retractable butt, 4.7 kg
Length: fixed butt, 1.025 m; retracted butt, 840 mm
Barrel: 450 mm
Rifling: 4 grooves, rh, 1 turn in 305 mm
Sights: fore, post; rear, V battle sight at 100 m. Apertures for 200, 300 and 400 m
Sight radius: 572 mm
Muzzle velocity: 780-800 m/s
Rate of fire: cyclic, 500-600 rds/min
Effective range: 400 m

Manufacturer

Heckler and Koch GmbH, D-78722 Oberndorf-Neckar.

Status

In production.

Service

German Army. The armed forces and/or police forces of the following countries have adopted Heckler and Koch G3 rifles and are completely or partially equipped with them:

Europe: Cyprus, Denmark (G3A5), France*, Germany, Greece*, Italy, Netherlands, Norway*, Portugal*, Sweden*, Turkey* (G3A7).

Africa: Angola, Burkina Faso, Burundi, Chad, Ethiopia, Gabon, Ghana, Ivory Coast, Kenya, Libya, Malawi, Mauritania, Morocco, Niger, Nigeria, Senegal, Somalia, Sudan, Tanzania, Togo, Uganda, Zaire, Zambia, Zimbabwe.

Central and South America: Bolivia, Brazil, Chile, Columbia, Dominican Republic, El Salvador, Guyana, Haiti, Mexico*, Paraguay, Peru.

Middle East: Bahrain, Lebanon, Iran* (G3A6), Jordan, Qatar, Saudi Arabia*, United Arab Emirates, Yemen (North).

Far East: Bangladesh, Brunei, Burma (Myanmar)*, Indonesia, Pakistan*, Philippines.

* Countries marked with an asterisk are producing (or have produced) Heckler and Koch G3 series rifles under licence.

Licence production

BURMA (MYANMAR)
Manufacturer: Government facilities
Type: 7.62 mm G3
Remarks: Little information available. A locally modified model has a heavy barrel and a bipod.

FRANCE
Manufacturer: Manufacture Nationale d'Armes de St Etienne (MAS)
Type: 7.62 mm G3
Remarks: Manufactured for export to nations such as Lebanon. No longer in production.

GREECE
Manufacturer: Hellenic Arms Industry (EBO)
Type: 7.62 mm G3A3 and G3A4
Remarks: Most production for domestic use but some export sales made to nations such as Libya. See entry under *Greece* for details.

IRAN
Manufacturer: Mosalsalsasi Weapons Factory
Type: 7.62 mm G3A6
Remarks: Production still continuing despite facility damage suffered during Iran-Iraq war.

MEXICO
Manufacturer: State factories
Type: 7.62 mm G3A3 and G3A4
Remarks: No details available.

NORWAY
Manufacturer: Norsk Forsvarsteknologi A/S
Type: Gevaer, Automatisk AG3
Remarks: No longer in production. See entry under *Norway* for details.

PAKISTAN
Manufacturer: Pakistan Ordnance Factory, Wah Cantt
Type: 7.62 mm G3A3 and G3A4
Remarks: Production for domestic use and offered for export. Some examples in service in Bangladesh.

PORTUGAL
Manufacturer: INDEP
Type: 7.62 mm G3A2, G3A3 and G3A4
Remarks: Produced for domestic use and for export. See entry under *Portugal* for details.

SAUDI ARABIA
Manufacturer: Al-Khardi Arsenal
Type: 7.62 mm G3
Remarks: Production for domestic use and export to nations such as North Yemen.

SWEDEN
Manufacturer: FFV (now Bofors Carl Gustaf AB)
Type: 7.62 mm Ak4
Remarks: No longer in production. Some undergoing a refurbishment programme prior to issue to reserve units.

TURKEY
Manufacturer: Makina ve Kimya Endüstrisi Kurumu (MKEK)
Type: 7.62 mm G3A7
Remarks: Produced for domestic use and export to nations such as Cyprus. See entry under *Turkey* for details.

UNITED KINGDOM
Manufacturer: Heckler and Koch (UK) Limited
Type: Various models
Remarks: Production of various models to order.

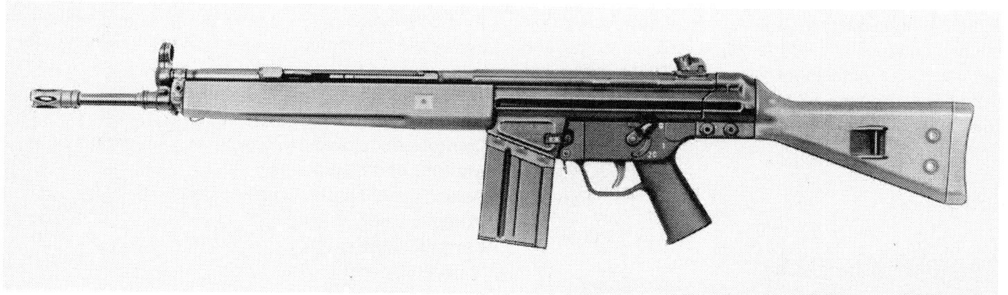

7.62 mm G3A3 rifle with plastic pistol grip

UPDATED

G3 SG/1 sniper rifle

Description

The G3 SG/1 sniper rifle is used by the German police. It is basically the G3A3 but has some differences. During proof and testing of the standard rifles, those that demonstrate their ability to produce minimum groups are set aside for modification to sniper rifles. The sniper rifle is fitted with a special trigger unit. This incorporates a set trigger which has a pull-off that can be varied. Setting the trigger can be done only when the selector lever is set to 'E'. After firing one round the trigger can be reset or the rifle can be fired again using the ordinary trigger setting which has been reduced to 2.6 kg. If the trigger has been set, it is automatically released as soon as the selector lever is changed from 'single shot' to 'safe' or 'automatic'.

The sight is not that normally used by the G3 but is a special Zeiss or Schmidt & Bender telescope with variable power from ×1.5 to ×6 and with windage and range adjustment from 100 to 600 m. The mount employed fits over the receiver and allows use of the iron sights without dismounting the telescope. The graticule pattern is based on the use of the mil, for example, the angle subtended by one unit of length at a distance of 1,000 units, or 1 m at 1,000 m subtends one mil. This system enables the firer to set the correct range provided he can align the telescope on an object of known size. For example a car of approximately 5 m length subtends 25 mils on the telescope. The range therefore is $5/25 \times 1{,}000 = 200$ m.

Data

Cartridge: 7.62 × 51 mm
Weight: unloaded, with sight, 5.54 kg
Trigger pull: normal, 2.6 kg; with set trigger, 0.9-1.5 kg
Sight: ×1.5 to ×6 zoom telescope optical; 6 positions, 100-600 m settings

Manufacturer

Heckler and Koch GmbH, D-78722 Oberndorf-Neckar.

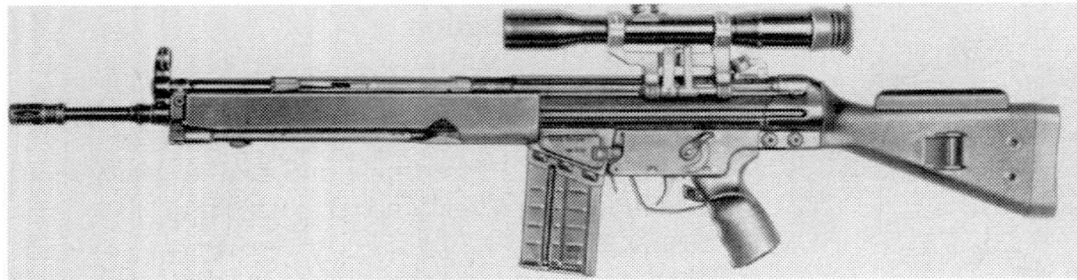

G3 SG/1 police sniper rifle

Status
Production suspended.

Service
German police and special forces, Malaysia, Italian Carabinieri and other police forces.

VERIFIED

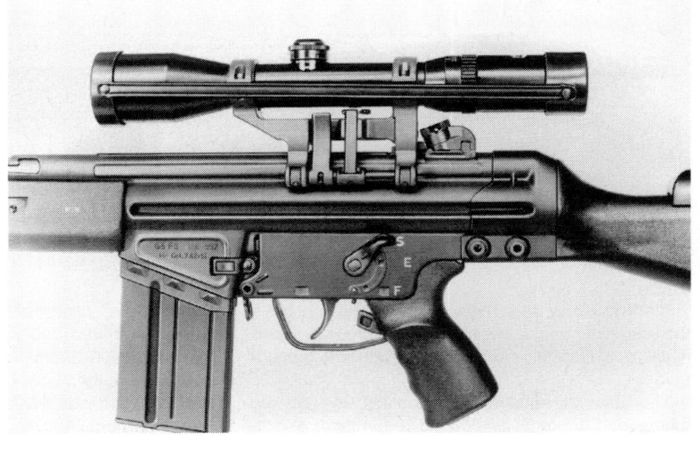

Sights and trigger of G3 SG/1 police sniper rifle

7.62 mm PSG1 high-precision marksman's rifle

Description
The Präzisionsschützengewehr (High-precision Marksman's Rifle) PSG1 is manufactured for police and service use. It is a semi-automatic, single-shot weapon (with 5- or 20-round magazine option) incorporating the company's roller-locked bolt system. Superb accuracy is claimed: an average dispersion diameter of ≤ 80 mm is quoted at a range of 300 m for test grouping (5 × 10 shots with Lapua 0.308 Winchester Match ammunition). A 6 × 42 telescopic sight with illuminated reticule is an integral part of the weapon. Windage and elevation adjustment is by moving lens, six settings from 100 to 600 m.

The manufacturer states that a special system provides for silent and positive bolt closing. A vertically adjustable trigger shoe enlarges the trigger width. The trigger pull is approximately 1.5 kg. Length of shoulder stock, vertical adjustment of cheekpiece and angular adjustment of butt to shoulder are all variable according to the firer's requirements. A precision tripod is available.

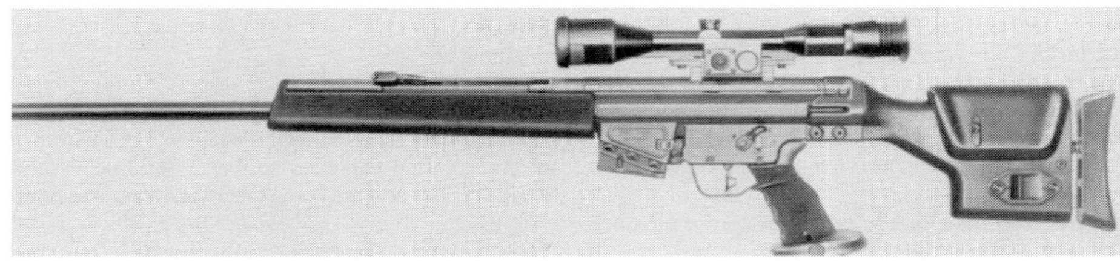

7.62 mm PSG1 high-precision marksman's rifle

Data
Cartridge: 7.62 × 51 mm
Operation: delayed blowback, single shot
Method of delay: rollers
Feed: option for 5- or 20-round box magazine
Weight: empty, 8.1 kg
Length: 1.208 m
Barrel: 650 mm
Rifling: polygonal, 4 grooves, rh
Sight: 6 × 42 telescope optical; 6 positions, 100-600 m settings

Manufacturer
Heckler and Koch GmbH, D-78722 Oberndorf-Neckar.

Status
In production.

Service
German and numerous other police and special forces worldwide.

VERIFIED

7.62 mm MSG90 sniper rifle

Description
The 7.62 mm MSG90 sniper rifle, introduced in 1987, uses the same roller-locked delayed blowback system which is common to all Heckler and Koch rifles. Designed to military specifications, the MSG90 uses a special cold-forged and tempered barrel and has a pre-adjusted trigger with a shoe which enlarges the trigger width for better control. The trigger pull is approximately 1.5 kg. The stock is adjustable for length and has a vertically adjustable cheek rest. There are no iron sights, the standard sight being a 10-power telescope with range settings from 100 to 1,200 m. The receiver is fitted with a STANAG mount which will accept any NATO standard night vision or other optical sight.

The fore-end is fitted with an internal T-rail which allows the attachment of a shooting sling or a bipod.

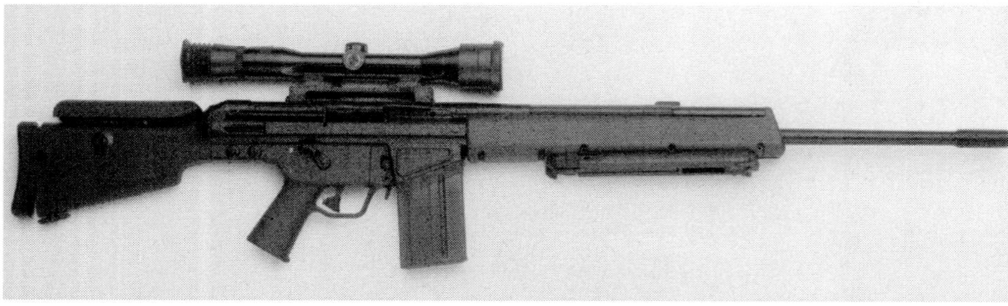

7.62 mm MSG90 sniper rifle showing latest form of bipod and Hensoldt ×10 telescope sight

Data
Cartridge: 7.62 × 51 mm
Operation: delayed blowback, semi-automatic only
Method of delay: rollers
Feed: 5- or 20-round box magazine
Weight: empty, 6.4 kg
Length: 1.165 m
Barrel: 600 mm

Manufacturer
Heckler and Koch GmbH, D-78722 Oberndorf-Neckar.

Status
In production.

VERIFIED

5.56 mm HK33 rifle

Description
The basic operating principle of the HK33 rifle is that of delayed blowback, firing from a closed breech, using the Heckler and Koch system which employs a two-part bolt, the action of which has been described in the entry on the G3 rifle (qv). The HK33 is a scaled-down version of the 7.62 mm G3 rifle and uses the same trigger and firing mechanism as the larger rifle. The mechanism provides automatic and single shot operation and a 'safe' position.

The sights fitted to the rifle have a V rearsight for battle setting and apertures for 200, 300 and 400 m. Zeroing for elevation and line is carried out on the rearsight. Optical sights can be fitted in a standard mount producing a four-power magnification with adjustment by hundreds of metres from 100 to 600. Infra-red and other special sights can also be accepted by the same mount.

There are five HK33 variants: the standard rifle with fixed butt (HK33A2), rifle with retractable butt (HK33A3), rifle with bipod, sniper rifle with telescopic sight, and HK33K carbine version. All have the same operating system.

Following a period of production at the Oberndorf-Neckar factory, production of the HK33 was switched to Heckler and Koch (UK) at Nottingham, UK. It was there that a production batch for Ecuador was completed.

Data
Cartridge: 5.56 × 45 mm
Operation: delayed blowback, selective fire
Method of delay: rollers
Feed: 25-round box magazine
Weight: (empty) fixed butt, 3.65 kg; retracting butt, 3.98 kg; carbine, 3.89 kg
Length: fixed butt, 920 mm; extending butt model, 940 mm and 735 mm; carbine model, 865 mm and 675 mm
Barrel: rifle, 390 mm; carbine, 322 mm
Rifling: 6 grooves, rh, 1 turn in 178 mm or 305 mm as required

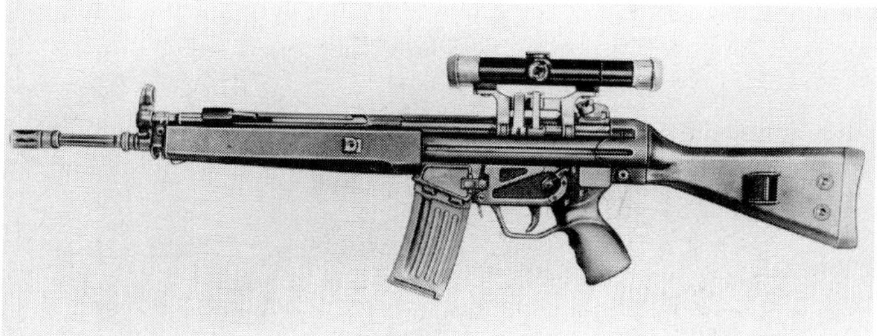

5.56 mm HK33 SG1 sniper rifle with fixed butt and telescope sight

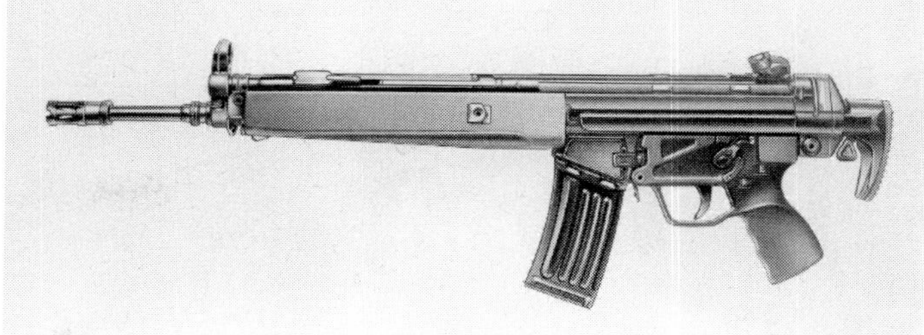

5.56 mm HK33A3 rifle with retracting butt

Sights: fore, post; rear, V battlesight 100 m. Apertures 200, 300 and 400 m; optical, ×4 telescope. 6 range increments from 100-600 m. Adjustable for windage and elevation
Sight radius: 480 mm
Muzzle velocity: rifles, ca 920 m/s; carbine, ca 880 m/s
Rate of fire: rifles, 750 rds/min; carbine, 700 rds/min

Manufacturer
Heckler and Koch GmbH, D-78722 Oberndorf-Neckar. Production line at Heckler and Koch UK, Kings Meadow Road, Nottingham NG2 1EQ, UK.

Status
Production in United Kingdom and Thailand.

Service
Malaysia, Chile, Thailand, and Brazilian Air Force. Also sold in varying quantities in South-east Asia, Africa and South America (including Ecuador).

UPDATED

5.56 mm HK33K carbine

5.56 mm G41 rifle

Description
The 5.56 mm G41 rifle was specifically developed to fire the NATO standard 5.56 × 45 mm cartridge and was based on the 7.62 mm G3. As well as the characteristics shared with that arm, including the 'roller-locked' bolt system (see entry for G3), the G41 is stated to incorporate several notable features: low-noise, positive-action bolt closing device; bolt catch, to keep open the bolt when the magazine has been emptied; dustproof cover for cartridge case ejection port; facilities for employment of 30-round magazines standardised according to STANAG 4179 (M16); burst control trigger for firing three-round bursts; fitting for bipod; new telescopic sight mount according to STANAG 2324 for all daylight and night sighting devices; twilight sighting device; acceptance of G3 bayonets; robust design for a service life of more than 20,000 rounds.

Data
Cartridge: 5.56 × 45 mm
Operation: delayed blowback, selective fire with 3-round burst facility
Method of delay: rollers
Feed: 30-round box magazine

5.56 mm G41A2 rifle with fixed butt and telescope sight

Weight: (empty) fixed butt, 4.1 kg; retracting butt, 4.35 kg; G41K, 4.25 kg
Lengths: fixed butt, 997 mm; retracting butt, long barrel, 996 and 806 mm; retracting butt, short barrel, 930 and 740 mm
Barrel: 380 (G41K) or 450 mm
Rifling: 6 grooves, rh, 1 turn in 178 mm
Sights: fore, post; rear, rotary. 4 settings; 200, 300, 400 m aperture V-notch for 100 m. Adjustable for windage and elevation
Rate of fire: cyclic, ca 850 rds/min

Manufacturer
Heckler and Koch GmbH, D-78722 Oberndorf-Neckar.

Status
Production complete.

VERIFIED

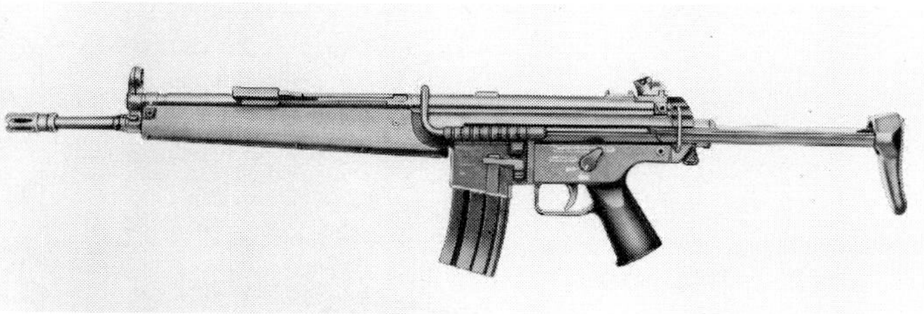

5.56 mm G41A3 rifle

G41K short rifle with butt retracted

7.62 mm G8 rifle

Description
This is a revised model of the pattern previously known as the HK11E; it was specially designed for use by anti-terrorist police and security forces.

The mechanism is the familiar Heckler and Koch roller-locked retarded blowback system and the rifle permits semi- or fully automatic fire and also has a three-round burst facility. Feed is by means of a standard box magazine, but a special 50-round drum magazine is available, and by means of an accessory feed mechanism outfit the weapon can be adapted rapidly to belt feed. There is a G8A1 variant which will not accept the belt feed mechanism and is restricted to use with magazines. The barrel is heavier than usual, precisely rifled, and capable of being changed rapidly when the weapon is being used in the automatic fire mode.

Iron sights are fitted, but the weapon is normally supplied with a four-power sighting telescope. The telescope mount is to STANAG specifications and will therefore accept all standard night vision devices.

7.62 mm G8 rifle

7.62 mm G8 rifle with accessories

Data

Cartridge: 7.62 × 51 mm NATO
Operation: delayed blowback, selective fire and 3-round burst facility
Method of delay: rollers
Weight: empty with bipod, 8.15 kg
Length: 1.03 m
Barrel: without muzzle brake, 450 mm
Sights: mechanical: adjustable 100-1,100 m in 100 m steps; optical: ×4 telescope, graticules graduated 100-600 m, fully adjustable for elevation and windage
Sight radius: 685 m
Muzzle velocity: 800 m/s
Muzzle energy: 3,000 J
Rate of fire: cyclic, ca 800 rds/min

Manufacturer

Heckler and Koch GmbH, D-78722 Oberndorf-Neckar.

Status

In production.

Service

German border police (Bundesgrenzschutz) and several other police forces.

VERIFIED

4.7 mm G11K3 caseless rifle

Development

In 1969 the former West German Government started a feasibility study into a future assault rifle and let contracts to three different firms, Diehl, IWKA and Heckler and Koch. The terms of reference were very general, calling for an improved infantry weapon with a better hit probability (Ph) than any then in existence, yet fulfilling the FINABEL range and rate of fire characteristics. Designers were given a free hand as to the methods used, but it quickly became apparent that for any significant improvement a radical approach was necessary.

Statistics showed that for a range of 300 m a circular dispersion of three shots would hit the target allowing for normal aiming errors. However, the recoil load of the 5.56 mm round was too great to produce the desired close grouping if a conventional three-round burst system was employed. With conventional systems the recoil of the first shot causes the muzzle to rise and thus the second and third shots are usually high and right. The only way to achieve the desired grouping was to have fired all three shots before the recoil impulse took effect on the weapon. Experiments showed that to beat the firer's reactions required a rate of fire of at least 2,000 rds/min, a virtual impossibility with conventional cased ammunition but not totally impossible if caseless was used. It was this overwhelming need for a better Ph which drove the research into caseless ammunition and a suitable weapon to fire it.

The G11 completed its development and in March 1988 was issued for technical trials. Troop trials with the weapon started in June 1988. The rifle passed its 'cook-off test' and received Safety Certification in May 1988, and type classification was granted towards the end of 1989. The first production weapons were issued to the former West German Special Forces in 1990. By this time sufficient data had been received from troop tests to show that the G11 gave average soldiers a considerable increase in their ability to hit targets when under simulated combat stress. A slightly modified version of the G11 was then evaluated in the USA as part of the now defunct Advanced Combat Rifle Programme.

In 1990, however, with the end of the Cold War and the reunification of Germany, the German government made severe cuts in its defence budget, resulting in the postponement *sine die* of the re-equipping of the bulk

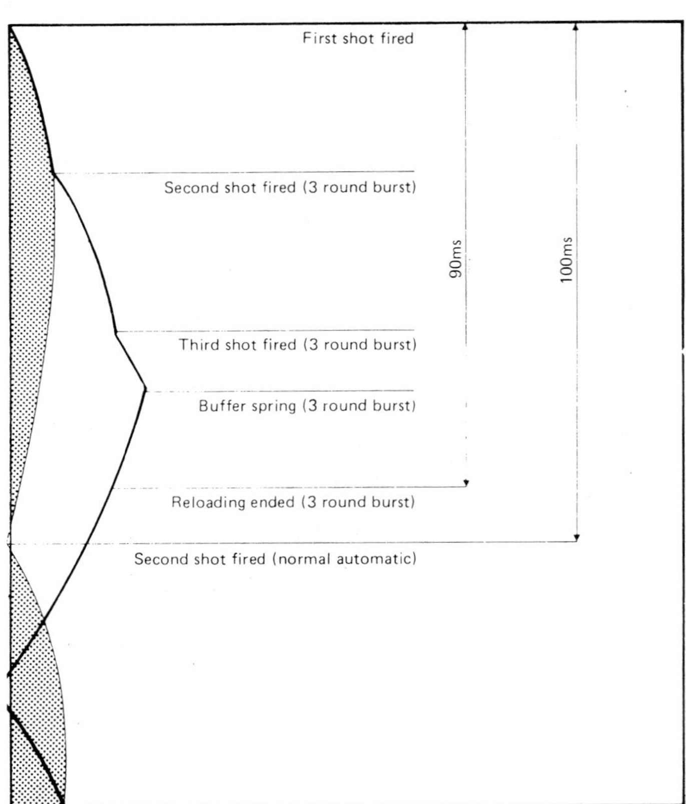

Shaded area is the graph of normal automatic fire at 600 rds/min, each shot being separated by an interval of 100 ms. The thick line graph above it shows the build up of movement of the recoiling mass when firing a three-round burst. The first shot accelerates the mass at the same rate as for automatic, the second shot then fires and the motion speeds up, the third shot increases this still more, and when the bullet is clear of the barrel the mass reaches the buffer and is returned to battery, reloading on the way forward

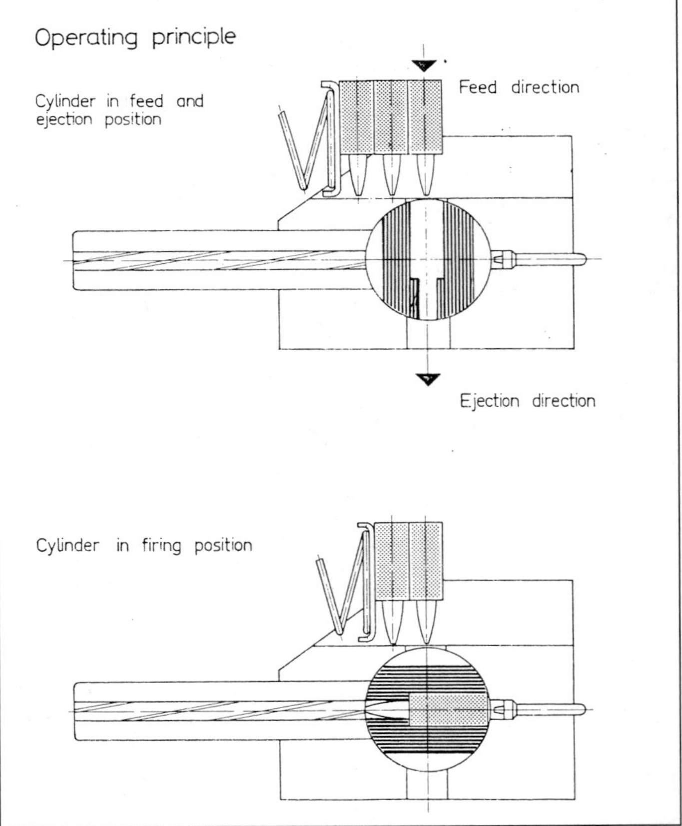

Operating principle of the cylinder breech. In the upper drawing the cylinder is aligned vertically, ready to receive a fresh round which will be pushed down into it. In the second drawing the cylinder is aligned with the barrel and is ready to fire. Everything in the drawings moves together so that loading and firing can take place while the mechanism is recoiling

4.7 mm G11K3 rifle with bayonet and Litton night sight (Robert Bruce)

of the German Army with the G11 rifle. Shortly afterwards the US Army completed their ACR trials and reached the conclusion that no entrant showed sufficient advantage over the M16A2 to warrant further development. As a result, the future of the G11 and its caseless ammunition concept is now in some doubt.

Description

It can be seen from the illustrations that the G11K3 is considerably different from other existing designs. There is a single smooth outer casing, with virtually no protruberances or holes. The pistol grip is at the point of balance and the trigger pivots inside a rubber bellows which seals the slot. The selector lever is above the trigger where it can be reached by the thumb of the firing hand. Just behind the trigger is a large circular knob which is the equivalent of a cocking handle. The optical sight is in the carrying handle above the receiver.

These are the only protruberances and there are only two holes, one being the muzzle, the other the ejection opening for clearing misfired rounds. While the muzzle is open at all times the ejection opening is sealed until it is necessary to clear a round. The casing provides complete protection for the mechanism against shocks, rough handling, immersion, frost, dust and dirt.

The bullet is set rigidly into a block of propellant which is square in section and longer than the bullet itself, so that the overall length of the complete round is just greater than the bullet length and less than half the length of an equivalent round with a conventional case. The criteria for such caseless propellants are demanding and Dynamit Nobel, who developed the cartridge, discovered that the normal nitro-cellulose propellants were not adequate. It therefore developed a special propellant by taking a high explosive of the Hexogen class and moderating it so as to burn as a propellant. Early designs using nitro-cellulose-based propellant suffered from cook-off in the chamber, as the result of the absence of a brass cartridge case to act as a 'heat sink', but the 'High Temperature Propellant' raises the cook-off threshold temperature by some 100°C and has thus solved the problem which has defeated every other caseless experiment. The primer is a small pellet of some other explosive with a different figure of sensitivity, pushed into a cavity in the base. Ahead of the primer is a small 'booster' charge of initiatory explosive; this, when ignited by the cap, generates sufficient pressure to drive the bullet forward and seat it in the rifling of the barrel before the ignition of the propellant takes place. While the bullet is thus being seated, the gas pressure from the booster causes the solid propellant to shatter and be ignited. The object of seating the bullet by means of the booster is to ensure that it is engaged in the rifling before the pressure of the main propellant explosive takes effect, so avoiding any gas leaks around an unseated bullet which would cause erosive wear and damage the barrel. It also ensures that the gas volume and pressure remain constant from round to round.

The magazine is positioned parallel to the bore axis, above the handguard, and slides into place from the front. It is a spring-powered, single-row magazine which holds 45 cartridges. The approximate number of cartridges remaining is shown on the outside by an indicator, even when the weapon is ready to fire.

The upper surface of the rifle above the barrel is shaped so as to hold two spare loaded magazines in 'parked' positions alongside the feeding magazine, from where they can be quickly moved to replace an empty magazine. Charging and reloading of magazines is done by prepacked reloading units, each containing 15 rounds. Three of these units are pressed, in succession, over the magazine mouth, and 45 rounds are loaded in under 5 seconds.

The essential part of the breech is a cylinder rotating about an axis at right angles to the line of the barrel. The chamber is a hole bored in this cylinder, across the line of the axis. By rotating the cylinder the chamber comes into line with the barrel and, by a turn of 90°, the chamber is turned upside down and is ready to receive a fresh round of ammunition. The first difference, therefore, between this weapon and any previous one is that the breech motion has been reduced to partial rotation of a fixed cylinder. There is no axial movement of a bolt or a breech block nor any complications of locking: the turning of the cylinder does it all. The breech cylinder is entirely contained within the barrel extension, and behind the cylinder is a further short extension containing the firing pin and the hammer. When a round is fired the entire mechanism recoils with the barrel and does not separate from it at any part of the cycle.

The cocking knob engages with the cylinder when in the forward position, but it disengages when the cylinder recoils and so there is no open slot as with most conventional systems. The trigger also stays still and the breech mechanism moves to and fro above it. When the mechanism comes into battery the trigger sear can then engage with whatever bent has been selected for the next shot. There is one exception to this rule: when the three-round burst is selected the trigger only fires the first round and the other two are controlled by a counter which is in the recoiling mass.

The magazine also moves with the recoiling mass so when a shot is fired the barrel recoils, taking with it the breech mechanism, the firing pin mechanism, the three-round counter and magazine. One immediate effect of this is that the recoiling mass is maximised and so the impulse is considerably reduced. The return spring lies alongside the barrel so as to reduce overall length.

From this description it can be seen that the Heckler and Koch G11 is formed of two separate parts. There is the outside casing which is completely smooth and seals from all dust and dirt. The second part is the internal mechanism which slides back and forth in this casing and which comprises most of the components that make up a complete conventional gun.

The first and most important aspect of the firing sequence is that the mechanism only has to load and fire the round. There is no empty case to be pulled out of the breech and thrown clear of the gun and this fact significantly reduces the time and movement required for the firing cycle. The first round is loaded by turning the cocking knob by hand. This cocks the firing pin and presents the chamber to the first waiting round at the end of the magazine. The round is fed downwards into the chamber by a feeding element under positive control and the cylinder rotates back into line with the barrel. The weapon is now loaded and cocked and if the selector lever is set at single shot the trigger will fire this round. On firing, the recoiling mass goes back and returns to battery. During this journey the cylinder rotates, a fresh round is loaded and the firing pin is cocked. A second shot can be taken at the instant that the recoiling mass is stationary again. The impulse to the firer is both low and long in duration, so that kick is very slight.

When the selector is at automatic the procedure varies only slightly. The first round is obviously fed and fired in the same way. The hammer remains cocked after the first rotation of the cylinder until the rearward motion of the recoiling mass against the pressure of the return spring is completed and this spring has driven the recoiling mass back into its initial forward position, where the second round is automatically fired. The rate of fire is set at approximately 600 rds/min, which is considered to be optimum for control of burst length and correction of aim.

With the selector lever set for three-round bursts a very different sequence is employed. The first shot is the same as for the others, but now the timing is not controlled by the trigger sear but by a timing device in the recoiling parts. The mechanism begins to recoil from the first shot, and the second round is loaded and fired; the recoil stroke is intensified, and the third round is loaded and fired. Only after this third shot is the recoil stroke completed and the firer actually feels the recoil

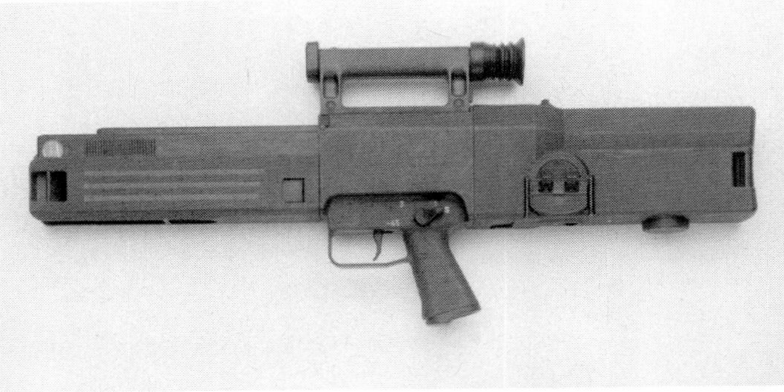

4.7 mm G11K3 rifle with carrying handle and integral optical sight

4.7 mm G11K3 rifle with bayonet, night sight, and spare magazines in the 'park' position. The carrying handle has been removed

impulse at his shoulder, so that by the time the inevitable upward movement of the muzzle begins the third bullet is well downrange.

It is this action which demonstrates the true value of caseless ammunition for, with the short mechanical distance and low inertia loadings involved, the rate of fire is about 2,200 rds/min. This is largely because of the absence of the extraction and ejection phases of the firing cycle, as well as the rotating breech and novel feed system. The diagram shows what happens but it may take some study to distinguish each item exactly. However, the main lesson it does show is that all three shots of a three-round burst are fired within 60 ms whereas the time between individual shots at the automatic rate of 450 rds/min is longer at 140 ms.

The final action requiring explanation is the ejection of a round which has failed to fire. For this operation the cocking knob is rotated so that the cylinder tilts into the vertical position and the round is allowed to fall downwards through a hole which is only ever open during this operation.

Heckler and Koch made many experiments but finally decided on 4.73 mm as providing the best compromise. The rifling is polygonal, which is known to reduce mechanical wear, and while no figures are yet quoted Heckler and Koch predicts a barrel life equivalent to the 5.56 mm M16A1.

It should be noted that some American sources refer to this rifle as being in 4.92 mm calibre; this figure is derived from measurement from the bottom of one rifling groove to the bottom of the opposite groove and flies in the face of all previous standards. The parentage of this unconventional method of determining calibre is unknown.

The optical sight is simple and designed with a view to mass production at the lowest cost. The lenses are toughened glass, protected by rubber at each end. Graticule adjustment is internal.

Data

Cartridge: 4.73 × 33 mm DM11 caseless
Operation: gas, closed breech
Breech system: cylindrical drum
Feed: 45-round box magazine
Type of fire: selective and 3-round burst
Weight: unloaded, 3.65 kg; loaded with 90 rounds, 4.3 kg
Length: 750 mm
Barrel: excluding chamber, 540 mm
Rifling: increasing twist rh, 1 turn in 155 mm at the muzzle
Rifling profile: polygonal
Muzzle velocity: 930 m/s
Rate of fire: 600 rds/min automatic; 3-round burst, >2,000 rds/min
Practical battle range: 300 m+
Penetration of steel helmet: 600 m+
Sights: optical 1:1

Manufacturer

Heckler and Koch GmbH, D-78722 Oberndorf-Neckar.

Status

Development suspended.

UPDATED

GREECE

Greek 7.62 mm G3A3 and G3A4 rifles

Description

The Hellenic Arms Industry (EBO) SA has been manufacturing the Heckler and Koch G3A3 and G3A4 rifles for over 10 years. Production is mainly for supply to the Greek armed forces for their operational needs and for export. In addition several accessories for the G3 rifle, such as bayonets and bipods, are also produced. The technical data for these weapons is exactly the same as that for the German-manufactured weapons, to which reference should be made.

Manufacturer

Hellenic Arms Industry (EBO) SA, 160 Kifissias Avenue, GR-11525 Athens.

Status

In production.

Service

Greek armed forces and offered for export. Sold to Libya.

VERIFIED

Greek-manufactured 7.62 mm G3A3 and G3A4 rifles

HUNGARY

7.62 mm AKM-63 assault rifle

Description

In accordance with what was then standard Warsaw Pact policy the Hungarian Army was equipped with Soviet designs of weapons and for some years the AK-47 was produced locally under licence, though with minor local modifications. In the 1960s the idea of replacing the wooden furniture with plastic was considered and this was incorporated in the production of the AKM. The resulting rifle was the AKM-63 in which the stock and forehand grip are made from polypropylene. The forehand grip is the distinctive feature of this weapon as it is a second pistol grip. It is attached to a perforated sheet steel forearm under the barrel. The resulting rifle is said to be 0.25 kg lighter than the standard AKM and cheaper to manufacture. All data, except for the reduced weight, is identical to the AKM (qv).

Manufacturer

State arsenals.

Status

Production complete.

Service

Hungarian armed forces.

VERIFIED

Hungarian 7.62 mm AK-47 assault rifle (T J Gander)

7.62 mm AKM-63 assault rifle

7.62 mm AMD-65 assault rifle

Description

The Hungarians took the unusual step of modifying their AKM-63 assault rifle to permit it to be used more easily inside vehicles and similar confined spaces. The modifications included a shorter barrel and a folding stock. The shorter barrel produces more muzzle flash which is counteracted by a prominent flash hider with two large holes in each side. A magazine article referred to this flash hider as a muzzle brake and claimed that it permits the rifle to be fired without using the stock at all.

There is also a version with a grenade launcher on the muzzle, and the length of this launcher is such that it almost takes the barrel back to the standard length.

The folding stock of the AMD-65 is supported with a single tubular strut. Grenade launching rifles have an extra shock-absorbing device on the stock and it may be that this is a spring-loaded plunger to allow some recoil movement. This same rifle has an optical sight, understood to be specifically for firing grenades, and a folding fore-pistol grip.

Data

Cartridge: 7.62 × 39 mm
Operation: gas
Locking: rotating bolt
Feed: 30-round detachable box magazine
Weight: 3.27 kg
Length: butt folded, 648 mm; butt extended, 851 mm
Barrel: with muzzle brake, 378 mm; without muzzle brake, 318 mm
Rifling: 4 grooves, rh
Muzzle velocity: 700 m/s
Rate of fire: cyclic, 600 rds/min

Manufacturer

State arsenals.

Status

Production complete.

Service

Hungarian forces.

7.62 mm AMD-65 assault rifle (T J Gander)

UPDATED

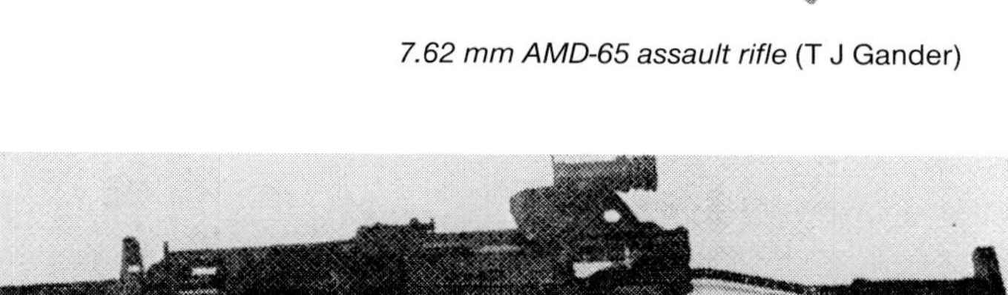

Magazine photograph of 7.62 mm AMD-65 grenade launching version. The long grenade muzzle attachment can be seen, together with the optical sight on its large mounting plate. The shock absorber on the stock is contained within the thicker section by the shoulder rest. The folding fore-pistol grip is not easily discerned in this photo. There is an unexplained small projection on the gas regulator: this could be a cut off for grenade firing

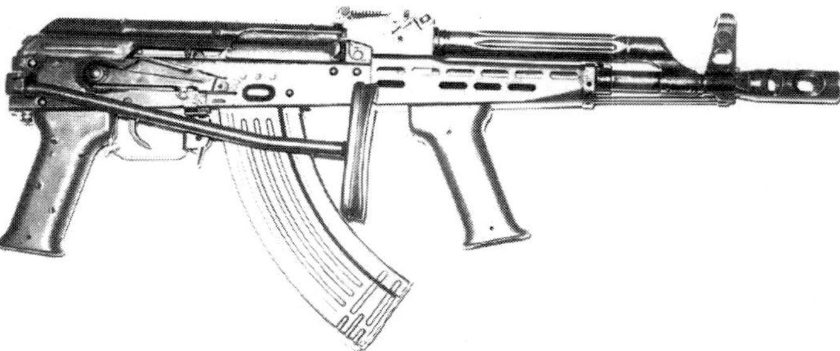

7.62 mm AMD-65 assault rifle with stock folded (T J Gander)

5.56 mm NGM assault rifle

Description

This weapon is officially known by the Hungarians as a sub-machine gun, but its calibre and proportions belie this description. It is basically the Hungarian version of the CIS AK-74, chambered for the 5.56 × 45 mm cartridge and manufactured for export. Production has now ceased.

The quality of manufacture is high. The barrel is cold forged from alloy steel and has 40 μm thick chromium plating in the bore. All other mechanical components are of high-strength alloy steel. The barrel is rifled one turn in 200 mm (one turn in 36 calibres), a compromise value which produces good results with both M193 and SS109 types of bullet, and the makers claim that after 10,000 rounds the dispersion is no more than twice the proof figure.

The NGM was manufactured only in standard form, with wooden furniture and selective-fire capability.

Data

Cartridge: 5.56 × 45 mm M193 or SS109
Operation: gas, selective fire
Locking: rotating bolt
Feed: 30-round box magazine

5.56 mm NGM assault rifle

Weight: empty, without magazine, 3.18 kg
Length: 935 mm
Barrel: 412 mm
Rifling: 4 grooves, rh, 1 turn in 200 mm
Sights: fore, hooded post; rear, tangent U-notch
Muzzle velocity: 900 m/s
Rate of fire: cyclic, 900 rds/min

Manufacturer

Technika, PO Box 125, Salgótarjáni u. 20, H-1475 Budapest.

Status

Production complete.

Service

Offered for export sales. None known to date.

UPDATED

12.7 mm Gepard M1 and M1A1 sniper rifles

Description

The 12.7 mm Gepard M1 and M1A1 sniper rifles are single-shot weapons with an unconventional breech mechanism. The pistol grip acts as the bolt handle and is attached to a multiple-lug breech block. The unit also contains a simple manually cocked hammer mechanism. To load, the pistol grip is partly rotated to unlock the breech block and then withdrawn, exposing the chamber. The cartridge is inserted and the breech block replaced, the hammer is cocked and then the trigger pressed. There is a grip for the disengaged hand on the butt and a substantial cheek pad. The rifle is provided with an adjustable bipod but a standard PKM machine gun tripod can also be used. A high-efficiency muzzle brake is fitted. The usual aiming system is a 12 × 60 telescopic sight.

The Gepard M1 and M1A1 rifles are chambered for the CIS 12.7 × 107 mm B32 or MDZ-3 AP-T cartridge. This is sufficiently accurate to permit a trained firer to achieve a 300 mm group with five shots at 600 m range, and the kinetic energy of the bullet makes it ideal for engaging targets behind the types of cover commonly found in urban areas. At 600 m the AP-T round can penetrate 15 mm of homogeneous armour or up to 30 mm at 100 m. Maximum effective aimed range at a vehicle-sized target is 2,000 m.

The Gepard M1A1 is the basically the same rifle as the Gepard M1 but the weapon is mounted on a backpack frame which can double as an aiming mount on soft ground or snow.

It has been proposed that both these rifles could be manufactured to fire Western Bloc 12.7 × 99 mm ammunition.

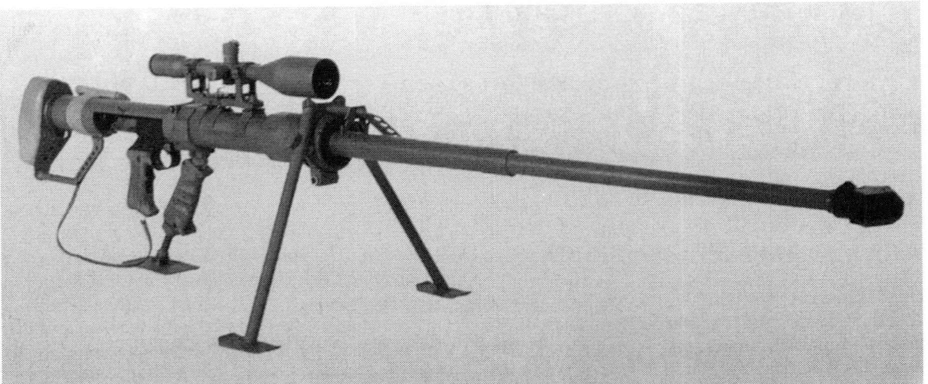

12.7 mm Gepard M1 sniper rifle

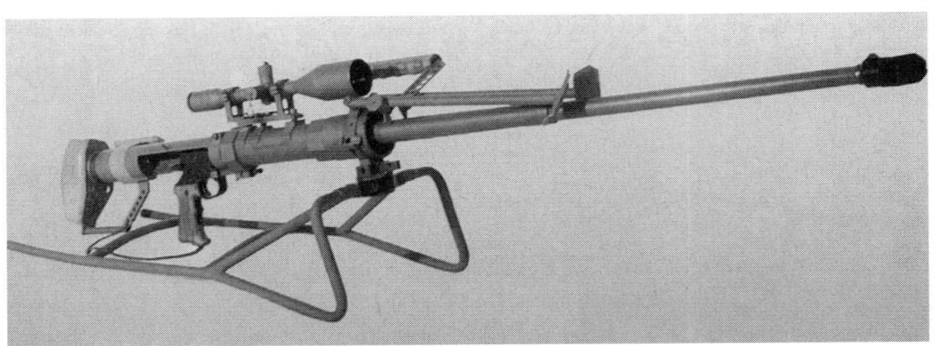

12.7 mm Gepard M1A1 sniper rifle

Data

Cartridge: 12.7 × 107 mm
Operation: bolt action, single shot
Weight: M1, 19 kg; M1A1, 22 kg
Length: 1.57 m
Barrel: 1.1 m
Muzzle velocity: 840 m/s
Muzzle energy: 17,181 J
Striking energy: >5,800 J at 1,000 m
Penetration: 30 mm RHA at 100 m; 15 mm at 600 m
Effective range: 2,000 m LAV; 1,200 m personnel

Manufacturer

Technika, PO Box 125, Salgótarjáni u. 20, H-1475 Budapest.

Status

In production.

Service

Offered for export.

VERIFIED

12.7 mm Gepard M2 and M2A1 rifles

Description

Although based on the Gepard M1 sniper rifle the Gepard M2 is intended to be used as more of an anti-material rifle and differs mainly in having a semi-automatic action in place of the single-shot mechanism of the Gepard M1 and M1A1. Using a positively locked long recoil system, the Gepard M2 and M2A1 can feed 12.7 × 107 mm rounds from a 5- or 10-round vertical box magazine located next to the pistol grip and trigger assembly.

All other aspects of the Gepard M2 are otherwise similar to the Gepard M1 and they fire the same ammunition, although the effective aimed range is reduced to 1,000 m.

The Gepard M2A1 is a shortened version of the Gepard M2 intended for use by airborne and special forces.

It has been stated that both of these rifles could be manufactured to fire 12.7 × 99 mm NATO rounds.

Data

Cartridge: 12.7 × 107 mm
Operation: long recoil, semi-automatic
Weight: M2, 12 kg; M2A1, 10 kg
Length: M2, 1.53 m; M2A1, 1.26 m
Barrel: M2, 1.1 m; M2A1, 830 mm
Muzzle velocity: M2, 840 m/s; M2A1, 790 m/s
Muzzle energy: M2, 16,700 J; M2A1, 15,100 J
Penetration: 25 mm RHA at 100 m; 15 mm at 600 m
Effective range: 1,000 m

Manufacturer
Technika, PO Box 125, Salgótarjáni u. 20, H-1475 Budapest.

Status
In production.

Service
Offered for export.

VERIFIED

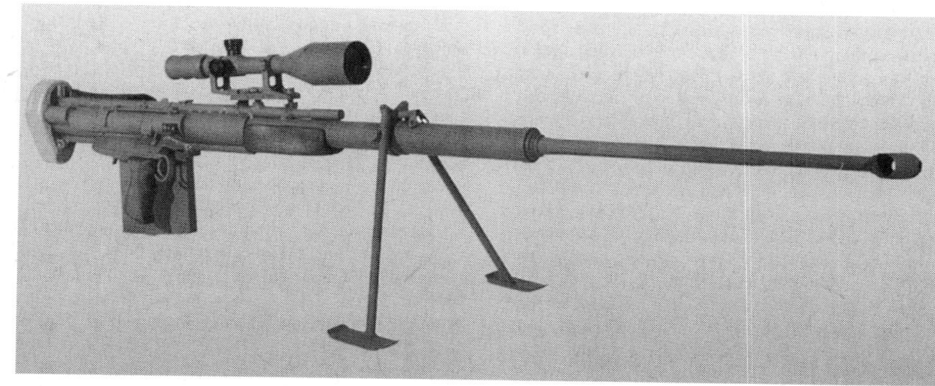

12.7 mm Gepard M2 rifle

12.7 mm Gepard M2A1 rifle

14.5 mm Gepard M3 heavy rifle

Description
The 14.5 mm Gepard M3 heavy rifle, also known as the Destroyer, is a self-loading, bipod-mounted weapon designed to provide accurate heavy-calibre fire for the engagement of lightly armoured vehicles, helicopters, field defences and targets at longer ranges.

The rifle is chambered for the Soviet 14.5 × 114 mm cartridge. It is recoil operated and the design incorporates a hydraulic buffer and a highly efficient muzzle brake, so that the felt recoil is of about the same level as a heavy game rifle.

Using CIS 14.5 × 114 mm B32 AP-T ammunition, the Gepard M3 can penetrate 30 mm of homogeneous armour at 100 m and 25 mm at 600 m. Maximum effective range is 1,000 m.

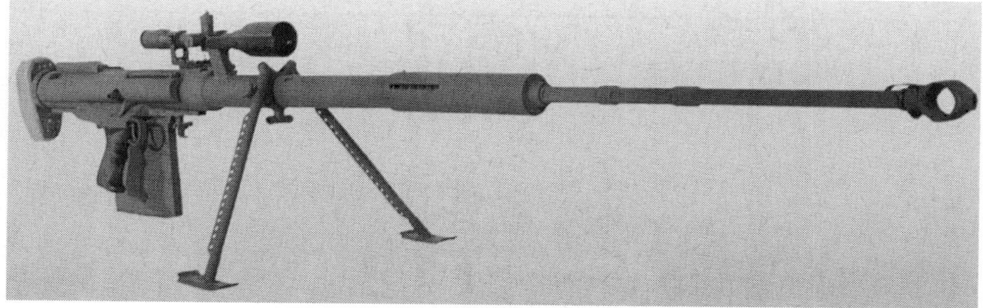

14.5 mm Gepard M3 heavy rifle

Data
Cartridge: 14.5 × 114 mm
Operation: long recoil, semi-automatic
Feed: 5- or 10-round box magazine
Weight: 20 kg
Length: 1.88 m
Barrel: 1.48 m
Muzzle velocity: 950-1,000 m/s
Muzzle energy: 32,000 J
Penetration: steel plate: 30 mm at 100 m; 25 mm at 600 m
Practical range: 1,000 m

Manufacturer
Technika, PO Box 125, Salgótarjáni u. 20, H-1475 Budapest.

Status
In production.

Service
Offered for export.

VERIFIED

INDIA

5.56 mm INSAS assault rifle

Development
The INSAS (INdian Small Arms System) rifle is one component in a pair of weapons with a similar design base, the other component being a light machine gun (see entry in *Machine guns* section).

Developed since the mid-1980s under the auspices of the Ordnance Factory Board and the Armaments Research and Development Establishment at Pune, the INSAS rifle is intended to replace all existing rifles in service with the Indian armed forces. A first batch of 7,000 was due for delivery by mid-1994 but it is understood that the service debut was delayed by the lack of an indigenous facility to manufacture the 5.56 × 45 mm SS109-based ammunition the INSAS was designed to utilise. It is further understood that although this round is based on the SS109 it is not NATO standard. General defence funding difficulties are also likely to delay the introduction of the INSAS into service to a significant degree.

By 1995 it was estimated that the INSAS programme was running almost four years behind schedule. As an interim measure the Indian government purchased a batch of 100,000 7.62 mm AKM rifles from Romania to replace at least some of the FN FAL rifles currently in front-line service with the Indian armed forces. See entry under *Romania* for details of this purchase.

5.56 mm INSAS assault rifle (T J Gander)

Description
The INSAS (INdian Small Arms System) 5.56 mm rifle is a gas-operated selective fire weapon which shows an interesting blend of features culled from a variety of sources. The receiver and pistol grip show Kalashnikov influence, the butt, gas regulator and flash hider show FN-FAL influence, the fore-end appears to rely upon the AR-15, and the cocking handle is based on Heckler and Koch practice.

The gas system operates the usual front locking

rotating bolt. Sheet metal pressings are used for the receiver and the barrel bore is chrome plated. All furniture is made from a plastic-based material. The magazine housing has been made to accommodate standard M16 magazines, although the standard magazine is made of semi-transparent plastic and holds 22 rounds; the 30-round magazine used with the INSAS light machine gun may also be used. The selector mechanism allows single shots and three-round bursts.

Both fixed and folding butt versions have been produced. Accessories include a blank firing attachment, a multipurpose bayonet and a sling.

One design detail of note is that the butt-stock has been configured to allow the continued use of the old Lee-Enfield No 1 Mark III butt plate, complete with trap for oil bottle and cleaning pull-through.

A carbine version is understood to be under development.

Data

Cartridge: 5.56 × 45 mm SS109-based Special
Operation: gas, selective fire (3-round bursts)
Locking: rotating bolt
Feed: 22-round plastic box magazine
Weight: empty, without magazine, 3.2 kg; loaded, 4.1 kg
Length: fixed butt, 945 mm; butt folded, 750 mm; butt extended, 960 mm
Barrel: 464 mm
Rifling: 6 grooves, rh, 1 turn in 200 mm
Sights: fore, blade; rear, flip aperture, 200 and 400 m
Muzzle velocity: 915 m/s
Rate of fire: cyclic, 650 rds/min
Max effective range: 400 m

Manufacturer

Rifle Factory, Ishapore.

Status

Not yet in full production.

Service

Destined for the Indian armed forces. Offered for export sales.

UPDATED

Rifle 7.62 mm 1A1

Description

The Rifle 7.62 mm 1A1 is a licence-produced copy of the British 7.62 mm L1A1 and reference should be made to the entry under *United Kingdom* for full details. The Indian 1A1 does differ from the British original in some details but also fires semi-automatic only. One design detail is that the wooden butt-stock is configured to allow the continued use of the butt plate from the old Lee-Enfield 0.303 No 1 Mark III, complete with trap for oil bottle and cleaning pull-through.

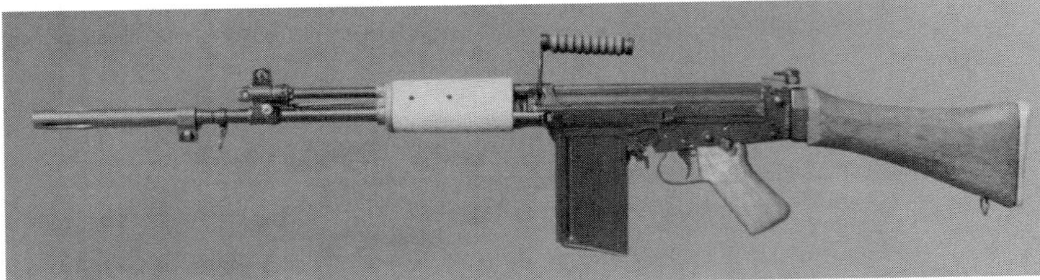

Rifle 7.62 mm 1C intended for firing through firing ports on Sarath (BMP-1) IFV

Variant

Rifle 7.62 mm 1C

The Rifle 7.62 mm 1C is a special version of the 1A1 designed to allow the weapon to be fired from the six rifle ports of the Sarath (Indian-produced BMP-2) IFV. To this end the muzzle is provided with an extension to allow the barrel to be secured in the firing ports and there is a fully automatic fire feature in addition to the normal semi-automatic mode.

The following data relate to this variant.

Data

RIFLE 7.62 mm 1C

Cartridge: 7.62 × 51 mm
Operation: gas, selective fire
Locking: tilting block
Feed: 20-round box magazine
Weight: empty without magazine, 4.5 kg
Length: 1.063 m
Barrel: 458 mm
Rifling: 6 grooves, rh, 1 turn in 304.8 mm
Sights: fore, blade; rear, flip aperture
Muzzle velocity: 825 m/s
Rate of fire: cyclic, 650-700 rds/min
Max effective range: 400 m

Manufacturer

Rifle Factory, Ishapore.

Status

In production.

Service

Indian Army. Offered for export sales.

VERIFIED

INDONESIA

5.56 mm SS1-V1, SS1-V2 and SS1-V3 assault rifles

Description

The SS1 series assault rifles are licensed copies of the Belgian FN-FNC rifle (qv) which are manufactured in Indonesia. The SS1-V1 is the Standard assault rifle with a 449 mm barrel, while the SS1-V2 is the Para carbine model with a 363 mm barrel; both weapons having folding tubular metal butt-stocks. The SS1-V3 is the Standard model with a fixed solid polyamid butt.

The dimensions and data are identical to the information given in the Belgian FN-FNC section except that the SS1-V2 is 910 mm long with butt extended and 666 mm long with butt folded. The muzzle velocities differ from the Belgian values; the SS1-V1 and SS1-V3 deliver 939 m/s and the SS1-V2 901 m/s, which probably reflects differences in the Indonesian propellant and environmental standards. The Indonesian ammunition (MU-5TJ, MU-5N, MU-5H, Blank) is based upon the SS109 round and all three weapons are rifled with six grooves, right-handed, one turn in 177.8 mm.

5.56 mm SS1-V1 assault rifle, showing close affinity to the Belgian FN-FNC Standard model

Manufacturer

PT Pindad (Persero), Jl Gatot Subroto Kiaracondong, PO Box 8, Bandung.

Status

In production.

Service

Indonesian armed forces.

VERIFIED

IRAQ

7.62 mm Tabuk assault rifle

Description

The 7.62 mm Tabuk rifle is a copy of the standard Kalashnikov AKM with some minor differences. The fore-end is differently shaped from the Russian original, and there is a folding anti-aircraft sight attached to the gas regulator block. The butt is contoured differently and is longer than that of other AK versions.

There is also a short-barrelled folding stock variant model, also called the 'Tabuk'. Though still based upon the AKM series, it is quite distinctive, with a different

7.62 mm Tabuk assault rifle

pistol grip, hooded foresight mounted on the gas regulator block and a reinforced muzzle which may be intended for grenade launching.

Data

DATA FOR SHORT VERSION IN PARENTHESIS
Cartridge: 7.62 × 39 mm M1943
Operation: gas, selective fire
Locking: rotating bolt
Feed: 30-round box magazine; (20- or 30-round magazine)
Weight: without magazine, 3.75 kg; (3.21 kg)
Length: 900 mm; (butt folded, 555 mm; butt extended, 800 mm)
Muzzle velocity: 700 m/s; (670 m/s)
Rate of fire: cyclic, 600 rds/min; (580 rds/min)

Manufacturer

State arsenal.

Status

In production.

VERIFIED

7.62 mm Tabuk short assault rifle

5.56 mm Tabuk assault rifle

Description

The Tabuk rifle is a copy of the standard Kalashnikov AKM or AKMS chambered to fire the 5.56 × 45 mm cartridge. Details are scant and there is no information available as to the rifling, but it is presumed that it is rifled to suit the widely distributed American M193 cartridge. Versions are available with fixed wooden butt or with a light metal folding butt. Appearance is identical to the standard AKM rifle.

The data given below is that provided by the manufacturers; we are at a loss to account for the unusually low muzzle velocity quoted.

Data

Cartridge: 5.56 × 45 mm
Operation: gas, selective fire
Locking: rotating bolt
Feed: 30-round box magazine
Weight: wooden butt, 3.2 kg; folding butt, 3.28 kg
Rate of fire: cyclic, 600 ±60 rds/min
Muzzle velocity: 700-710 m/s

Manufacturer

State arsenals.

Status

In production.

VERIFIED

7.62 mm Tabuk sniper rifle

Description

The 7.62 mm Tabuk sniper rifle is based on the standard 7.62 × 39 mm Kalashnikov action but fitted with a long barrel with a muzzle brake, a skeleton butt and an optical sight. The general effect is similar to the former Yugoslav M76 rifle, but firing the standard intermediate rimless cartridge rather than one of the more powerful cartridges more usually found in sniper weapons. The effective range is claimed to be 800 m.

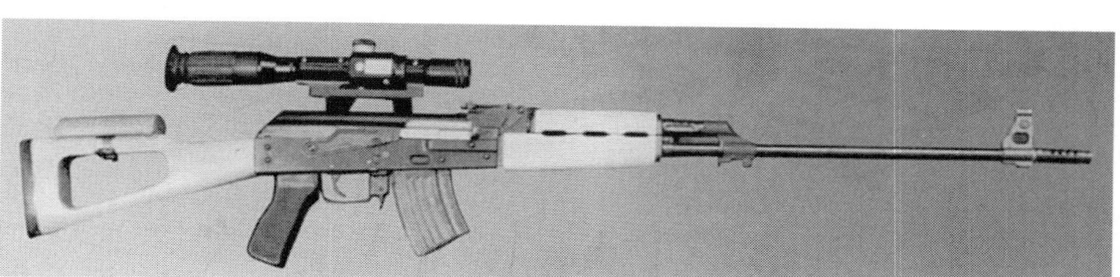

7.62 × 39 mm Tabuk sniper rifle

Data

Cartridge: 7.62 × 39 mm M1943
Operation: gas, semi-automatic
Locking: rotating bolt
Weight: with sight and empty magazine, 4.5 kg
Length: 1.11 m
Barrel: 600 mm
Muzzle velocity: 740 m/s
Effective range: 800 m

Manufacturer

State arsenals.

Status

In production.

VERIFIED

7.62 mm Al-Kadisa sniper rifle

Description

The 7.62 mm Al-Kadisa sniper rifle is the standard CIS Dragunov 7.62 mm SVD rifle manufactured in Iraq. The only points of difference are in the fore-end, pierced with four long slots on each side instead of six short, and the magazine, which has an ornamental relief pattern showing a stylised palm tree. There are some very small dimensional differences, though it is difficult to see precisely how these arise. It is probable that the butt may be a fraction longer in this version.

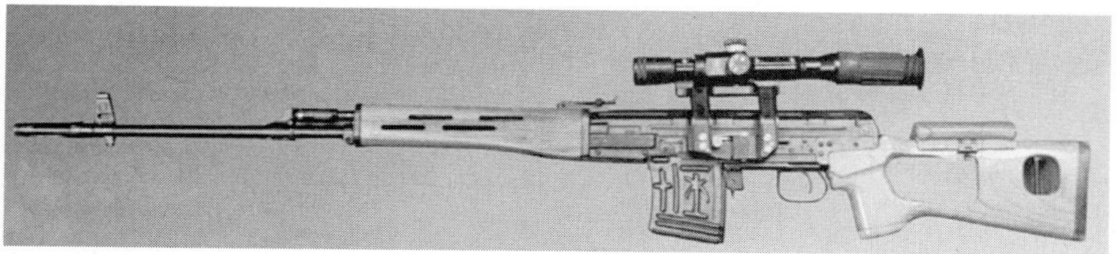

7.62 mm Al-Kadisa sniper rifle

Data

Cartridge: 7.62 × 54 R M1891
Operation: gas, semi-automatic
Locking: rotating bolt
Feed: 10-round box magazine
Weight: with telescope and empty magazine, 4.3 kg
Length: 1.23 m
Barrel: 620 mm
Muzzle velocity: 830 m/s

Manufacturer

State arsenals.

Status

In production.

VERIFIED

5.56 mm and 7.62 mm Galil assault rifles

Description

The Galil ARM was designed to fill the roles of the submachine gun, assault rifle and light machine gun. It can also be used to project anti-tank and anti-personnel grenades from the shoulder, and illuminating and signal grenades with the butt on the ground. The folding metal stock is of rugged and lightweight construction.

In addition to the ARM there is an assault rifle equipped with a folding metal stock without bipod and carrying handle, known as the AR (assault rifle), and a rifle with a shortened barrel and folding stock without bipod and carrying handle, known as the SAR (short assault rifle).

There are three magazine capacities: 12, 35 and 50 rounds. The 12-round magazine is only used with the ballistite cartridges required for grenade launching, the 35-round magazine is generally employed in the assault rifle role. When the weapon is employed as a light machine gun the 50-round magazine is used. On 5.56 mm models it is possible to fit an M16 pattern magazine by utilising an adaptor.

5.56 mm Galil AR assault rifle with full range of accessories

The bipod is carried lying under the barrel and fitting into a groove under the fore-grip. The powerful leverage which can be obtained when the bipod legs are swung down to the vertical position, is used to provide a wire cutter. Any barbed wire placed under the hook at the bipod fulcrum can be sheared, usually in one movement of the bipod legs.

The sights enable the weapon to produce aimed fire out to 600 m. The foresight is a cylindrical post which can be screwed up or down for zeroing and the rearsight is a flip aperture with the choice of 0 to 300 m and 300 to 500 m. There are also night sights set for use at 100 m, which are normally folded, but when lifted produce a pattern of three luminous spots. The centre spot is that of the foresight, and three luminous aiming marks are lined up horizontally and placed on the target.

The Galil in many respects is modelled on the AK-47 series and nearly approximates to the Finnish M62 assault rifle. It is a gas-operated rifle with no regulator. The gas block is pinned to the barrel and the gas track is drilled back at 30° into the gas cylinder. The piston-head and shank are chrome-plated. The bolt carrier is an extension of the piston end and is hollowed out over the bolt to take the return spring. The bolt has a cam pin operating in a slot in the carrier and has two locking lugs which are rotated into locking recesses in the chamber. The cocking handle is attached to the bolt carrier to give positive bolt closure. It emerges from the right side of the receiver and is bent upwards to allow cocking with either hand. The change lever on the right of the receiver, like that of the AK-47, is a long pressing which when placed on 'safe' both closes up the cocking way against the entry of dirt and restricts the carrier movement. It is possible to retract the bolt sufficiently to inspect the chamber but not enough to feed a round.

The Galil fires from a closed bolt position. The magazine, fitting under the gun, is placed in position and held by the magazine catch in front of the trigger guard. The change lever is taken off 'safe' and the cocking handle pulled to the rear. When it is released the carrier is driven forward and the top round pushed out of the magazine into the chamber. The bolt comes to a halt and the further forward travel of the carrier causes the cam pin, engaged in a cam slot in the carrier, to rotate the bolt.

When the trigger is pulled, the hammer drives the firing pin forward and the round is fired. Some of the propellant gas escapes through the gas vent into the cylinder and drives the piston rearwards. There is a brief period of free travel, while the gas pressure drops, as the width of the cam slot passes the cam pin. The slot then engages the pin which is rotated as the carrier proceeds back. The rotation of the bolt provides primary extraction and then the case is withdrawn as the bolt retracts. The return spring is compressed, the empty case is ejected and then the return spring energy drives the carrier forward and the cycle starts again.

A special Blank firing attachment can be fitted over the muzzle during training. This attachment permits firing in the fully automatic and semi-automatic firing modes and is used in conjunction with a special 35-round magazine which can accommodate Blank ammunition only as an added safety feature.

Variants

Galil 5.56 mm Marksman's Assault Rifle Mark 1

This is a special 'high accuracy' version of the 5.56 mm AR with accuracy enhancements such as an adjustable bipod, a dual position cheek rest, and a 3 × 21 Eyal telecopic sight on a mounting bar secured to the left-hand side of the receiver. The weapon fires NATO SS109 ammunition using a 400 mm long chromed barrel with 1 turn in 177.8 mm rifling. Weight of the 5.56 mm Marksman's Assault Rifle Mark 1 without a magazine or sling is 4.95 kg.

5.56 mm Galil MAR Micro assault rifle

See following entry.

Data

DATA REFERS TO 5.56 mm ARM/AR EXCEPT AS NOTED

Cartridge: 5.56 × 45 mm
Operation: gas, selective fire
Locking: rotating bolt
Feed: 35- or 50-round box magazine
Weight: ARM, with bipod and carrying handle, 4.35 kg; AR, without bipod or handle, 3.95 kg; SAR, without bipod or handle, 3.75 kg
Length: ARM/AR overall, 979 mm; with stock folded, 742 mm; SAR overall, 840 mm; with stock folded, 614 mm

Galil 5.56 mm Marksman's assault rifle Mark 1

5.56 mm Galil ARM rifle

5.56 mm Galil SAR short assault rifle

Barrel: ARM/AR, 460 mm; SAR, 332 mm
Rifling: 6 grooves, rh, 1 turn in 305 mm
Sights: fore, post, with protector; rear, flip aperture, 300 and 500 m. Tritium night sights
Sight radius: ARM/AR, 475 mm; SAR, 445 mm
Muzzle velocity: ARM/AR, 950 m/s; SAR, 900 m/s
Rate of fire: cyclic, ca 650 rpm

Data

DATA FOR 7.62 mm RIFLES
Cartridge: 7.62 × 51 mm
Operation: gas, selective fire
Locking: rotating bolt
Feed: 25-round box magazine
Weight: all without bipod or carrying handle: ARM, 4 kg; AR, 3.95 kg; SAR, 3.75 kg
Length: ARM/AR overall, 1.05 m; stock folded, 810 mm; SAR overall, 915 mm; stock folded, 675 mm
Barrel: ARM/AR, 535 mm; SAR, 400 mm
Rifling: 4 grooves, rh, 1 turn in 305 mm
Sights: fore, post with protector; rear, flip aperture, 300 and 500 m. Folding tritium night sights
Sight radius: ARM/AR, 475 mm; SAR, 445 mm
Muzzle velocity: ARM/AR, 850 m/s; SAR, 800 m/s
Rate of fire: ARM/AR, 650 rds/min; SAR, 750 rds/min

Manufacturer

TAAS - Israel Industries Limited, PO Box 1044, Ramat Hasharon 47100.

Status

In production.

Service

Israeli forces and other armies including Bolivia, Botswana, Chile, Colombia, Costa Rica, Guatemala, Haiti, Honduras, Nicaragua, Philippines, Swaziland, Trinidad and Tobago, Zaire.

Licence production

SOUTH AFRICA
Manufacturer: Lyttleton Engineering Works
Type: 5.56 mm R4 and R5
Remarks: See entry under *South Africa* for details.

UPDATED

5.56 mm Galil MAR Micro assault rifle

Description

The 5.56 mm Galil MAR Micro assault rifle is claimed to be the smallest and lightest assault rifle in the world. It is a shortened version of the 5.56 mm Galil assault rifle (see previous entry) intended for special forces and special applications such as commanders and tank crews. The basic user controls and mechanism of the Galil rifle are retained virtually unaltered, as are many components. The barrel has been shortened to 195 mm and only the 35-round curved box magazine is employed. The maximum effective range is stated to be 300 m.

It is noticeable that the rifling for this rifle has one twist in 178 mm and is thus intended for firing NATO SS109 ammunition, although it is assumed that the weapon could be produced with one twist in 305 mm for M193 series ammunition.

The fore-grip of the Galil MAR Micro is made from a shatterproof nylon-based material and is so contoured that the forward part acts as a partial guard against the firer's hand straying too far towards the muzzle. A folding aluminium alloy stock is provided to make the weapon even more compact when required.

Data

Cartridge: 5.56 × 45 mm
Operation: gas, selective fire
Locking: rotating bolt
Feed: 35-round box magazine
Weight: rifle only, 2.95 kg; loaded, 3.67 kg
Length: stock folded, 445 mm; stock extended, 690 mm
Barrel: 195 mm
Rifling: 6 grooves, rh, 1 turn in 178 mm
Sights: fore, post, with protector; rear, flip aperture, 300 and 500 m.
Sight radius: 300 mm
Muzzle velocity: 710 m/s
Rate of fire: cyclic, 600-750 rpm

Manufacturer

TAAS - Israel Industries Limited, PO Box 1044, Ramat Hasharon 47100.

Status

In production.

NEW ENTRY

5.56 mm Galil MAR Micro assault rifle with stock folded (T J Gander)
1996

5.56 mm Galil MAR Micro assault rifle with stock folded (T J Gander)
1996

7.62 mm Galil sniper rifle

Description

Developed in close co-operation with the Israeli Defence Forces, the Galil sniper rifle is a semi-automatic gas-operated rifle specially designed to meet the particular demands of sniping. It is chambered for the 7.62 × 51 mm NATO cartridge but will perform best with selected lots or with match-quality ammunition. The accuracy requirement is considered to be a 120 to 150 mm circle at 300 m range and a 300 mm circle at 600 m range and the Galil sniper consistently exceeds these requirements.

The mechanism and general configuration is that of the standard Galil rifle, but there are a number of features peculiar to this model. A bipod is mounted behind the fore-end and attached to the receiver, so it can be adjusted by the firer and also relieve the barrel of supportive stress. The barrel is heavier than standard, which contributes to the accuracy. The telescope sight mount is on the side of the receiver and is a precision cast, long-base unit giving particularly good support to the Nimrod 6 × 40 telescopic sight supplied as standard. The sight mount and telescope can be mounted and dismounted quickly and easily without disturbing the rifle's zero and any type of night sight can be fitted.

The barrel is fitted with a muzzle brake and compensator which reduces jump and permits rapid realignment of the rifle after firing. A silencer can be substituted for the muzzle brake if desired; when in use, it is recommended that subsonic ammunition be used.

There is a two-stage trigger. The safety catch is above the pistol grip and there is no provision for automatic fire in this weapon. The wooden stock can be folded, to reduce the overall length of the rifle for carriage, and when unfolded and locked is perfectly rigid. The butt-stock is fitted with a cheekpiece and a rubber recoil pad, both of which are fully adjustable.

Each rifle is supplied in a specially designed case, which also contains the sight, two optical filters for use with the sight, a carrying and firing sling, two magazines and the cleaning kit.

7.62 mm Galil sniper rifle

Data

Cartridge: 7.62 × 51 mm NATO
Operation: gas-operated, semi-automatic only
Locking: rotating bolt
Feed: 20-round box magazine
Weight: empty, including bipod and sling, 6.4 kg
Length: butt folded, 840 mm; butt extended, 1.115 m
Barrel: without muzzle brake, 508 mm
Rifling: 4 grooves, rh, 1 turn in 305 mm
Sights: iron sights, fore, blade; rear, aperture; optical, Nimrod 6 × 40 telescope
Sight radius: 475 mm
Muzzle velocity: FN match ammunition, 10.9 g bullet, 815 m/s; M118 match ammunition, 11.2 g bullet, 780 m/s

Manufacturer

TAAS - Israel Industries Limited, PO Box 1044, Ramat Hasharon 47100.

Status

In production.

Service

Israeli Defence Forces.

VERIFIED

ITALY

7.62 mm BM 59 rifle

Development/Description

At the end of the Second World War the firm of Pietro Beretta at Gardone Val Trompia began to manufacture the US M1 0.30 rifle, the Garand, for the Italian Army. These rifles were later supplied to Indonesia and Denmark. Some 100,000 rifles were produced by 1961.

The need to compete with weapons of more modern design (the M1 rifle came into US service in 1936) led to the production of a modified M1 rifle firing the NATO 7.62 × 51 mm cartridge. Studies to convert the Garand into a modern selective fire weapon were commenced in 1959 under the direction of Domenico Salza and completed by Vittorio Valle.

The earliest versions were given letter suffixes. The first BM 59 was a Garand, lightened and shortened, modified to take a 20-round detachable box magazine. Next came the BM 59 R which had a device to slow down the rate of fire. This was followed by the BM 59 D with a pistol grip and bipod and the BM 59 GL with a grenade launcher. The BM 60 CB had a three-round burst fire controller.

Of the BM 59 series the Mark E was the cheapest since it used the Garand barrel and gas cylinder. In all the others the barrel and gas cylinder was a Beretta design. The final production versions were as follows:

The BM 59 Mark I was a fixed wooden-stocked rifle with the gas cylinder under the barrel. It had a 20-round detachable box magazine which could be loaded and replenished using chargers and guides built into the receiver. The corresponding Italian Army version was known as the BM 59 Mark Ital and had a light bipod fitted to the gas cylinder.

7.62 mm BM 59 Ital Para airborne version with folding skeleton butt, bipod and grenade launcher

The BM 59 Mark II was similar to the Mark I but had a pistol grip, a winter trigger and a bipod.

The BM 59 Mark III had a folding metal stock and two pistol grips.

The Italian Army BM 59 Mark Ital TA and the BM 59 Mark Ital Para were derived from the Mark III. The Mark Ital TA was for Alpine troops and the Mark Ital Para was for airborne use.

The BM 59 Mark IV was a heavy-barrelled version intended for the squad light automatic role. It had a plastic butt-stock, a hinged butt-plate and a pistol grip. There was no forward pistol grip.

The BM 59 can fire any anti-personnel or anti-tank grenade with a tail tube having an inside diameter of 21 mm. Where the grenade launcher is fitted it is not intended that it should be removed, except for the Mark Ital Para, the parachutists' rifle. The erection of the grenade sight automatically cuts off the gas supply to the gas piston.

7.62 mm BM 59 Mark I rifle

Data

BM 59 MARK ITAL
Cartridge: 7.62 × 51 mm
Operation: gas
Locking: rotating bolt
Feed: 20-round detachable box magazine
Weight: 4.6 kg
Length: 1.095 m
Barrel: 490 mm
Rifling: 4 grooves, rh
Sights: fore, blade; rear, aperture, adjustable for range and windage
Muzzle velocity: 823 m/s
Rate of fire: cyclic, 750 rds/min
Effective range: 600 m

Manufacturer

Armi Beretta SpA, Via Beretta 18, I-25063 Gardone VT (Brescia).

Status

Production complete; spares available.

Service

Italian Army. Manufactured under licence in Indonesia and Morocco.

VERIFIED

Beretta 5.56 mm AR70/223 rifle

Development

In early 1968 Pietro Beretta started an initial survey for a new assault rifle. Work included a detailed evaluation of existing 5.56 mm systems and eventually a gas-operated system was adopted. It was decided to have a conventional piston and cylinder system with the gas port as far forward as possible. The forward locking, two-lug bolt system of the Garand 0.30 calibre M1 rifle and the Kalashnikov AK-47 was adopted. One of the advantages of forward locking, in modern design, is that the bolt can lock directly into the barrel extension. Thus only the barrel and bolt head are stressed and must be of high tensile material but the rest of the bolt and the receiver need not be produced from such expensive steel. Beretta, however, chose to weld a sleeve into the receiver and the bolt lugs close into recesses cut into this.

5.56 mm AR70 assault rifle

Description

On firing, the gas pressure drives the bullet forward up the bore and a proportion of the gas passes through the vent into the cylinder and drives the piston rearward. There is a free travel of the piston for about 9 mm before the carrier starts to rotate the bolt and produce unlocking. The bolt is cammed through 30° and is fully unlocked; it then travels back with the carrier, extracts the spent case, compresses the operating spring around the piston and, as the latter reasserts itself, the working parts go forward, feeding a new round into the chamber. When chambering is complete the bolt comes to a stop but the carrier continues forward and the cam rotates the bolt into the locked position.

The trigger and firing mechanism are orthodox in their action. Single shot is obtained using a disconnector between the trigger and the sear; the latter operating in a bent near the hammer axis. When firing at full automatic the hammer is controlled by a safety sear which is operated by the bolt carrier in its last forward motion. This ensures locking is completed before the hammer is released and it operates also in single shot. If the bolt is not locked the hammer cannot move.

Data

Cartridge: 5.56 × 45 mm (M193 and SS109)
Operation: gas, selective fire

	AR70	SC70	SC70 Short
Length			
overall	955 mm	960 mm	820 mm
butt folded	–	736 mm	596 mm
barrel	450 mm	450 mm	320 mm
Sight radius	507 mm	507 mm	455 mm
Rifling	4 grooves, rh, 1 turn in 304 mm		
Weight			
empty magazine	3.8 kg	3.85 kg	3.7 kg
full magazine	4.15 kg	4.2 kg	4.05 kg
Rate of fire (cyclic)	650 rds/min	650 rds/min	600 rds/min
Muzzle velocity	950 m/s	950 m/s	885 m/s
Muzzle energy	167 kgm	167 kgm	142 kgm
Grenade	40 mm	40 mm	–

Locking: rotating bolt
Feed: 30-round box magazine

Manufacturer

Armi Beretta SpA, Via Beretta 18, I-25063 Gardone VT (Brescia).

Status

No longer in production.

Service

Italian special forces, Jordan, Malaysia and some other countries.

VERIFIED

Beretta 5.56 mm 70/90 assault rifle

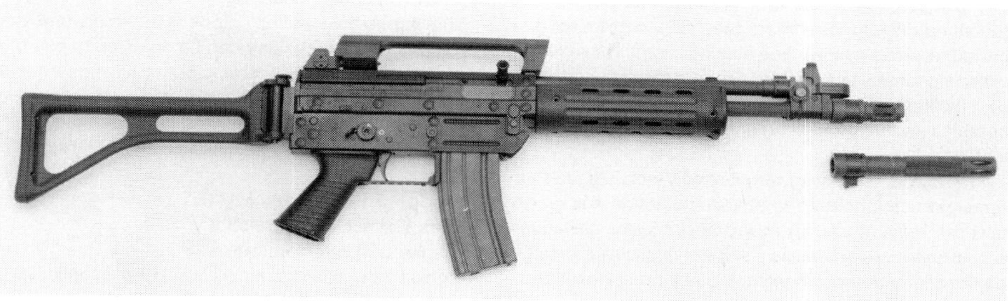

5.56 mm SCP70/90 carbine with grenade launcher

Development

The Beretta 70/90 system was developed primarily as a potential system for the Italian Army, the rifle being approved for service in July 1990. It consists of five weapons: the assault rifle AR70/90 for infantry; the carbine SC70/90 for special forces; the special carbine (short) SCS70/90 for armoured troops; the SCP70/90 with grenade launching attachment; and the light machine gun AS70/90, for use as the squad automatic weapon and which is described in the *Machine guns* section.

Description

The AR70/90 is, broadly, an improved version of the 70/223 described in the previous entry, and its design has been influenced by service experience gained with the earlier weapon. In the 70/223, for example, the receiver was a pressed-steel rectangle in which the bolt moved on pressed-in rails. Experience showed that this could distort under severe conditions, and thus the AR70/90 receiver is of trapezoidal section and has steel bolt guide rails welded in place. The method of operation is the same, using a gas piston mounted over the barrel which actuates a bolt carrier and a two-lug rotating bolt. A noticeable new feature is the carrying handle above the receiver. This clips into place by means of a spring-loaded catch and carries a luminescent source for use as an aiming aid in poor light conditions. Removing the handle reveals a dovetailed receiver cover which meets STANAG 2324 requirements for a sight mounting which will take any optical or electro-optical sight.

The trigger mechanism in the standard version permits selection of single shots, three-round bursts or full-automatic fire. An optional mechanism is available which restricts the available modes to single shots and three-round bursts. However, the three-round burst mechanism can be removed and replaced by components which will permit automatic fire or semi-automatic action only.

The rifle feeds from a 30-round box magazine, with the magazine housing interface to STANAG 4179, allowing the use of any M16-type magazine. The magazine is released by an ambidextrous button which is operated by the hand holding the weapon.

The gas cylinder is fitted with a regulator having three positions: open for normal use, further open for use in adverse conditions, or closed for grenade firing. The regulator is fitted with a lever which is raised for grenade firing; in this position it also impinges on the sight line, so reminding the firer that it is set. When lowered for normal firing, the lever will obstruct any attempt to load a grenade.

The normal iron sights consist of a blade foresight and a two-position aperture backsight. The supports of the carrying handle are open, so that the sight line passes through without obstruction, and the foresight can be adjusted for zeroing.

Accessories provided for the rifle include bayonet, bipod, slings, blank firing attachment and cleaning kit. The blank firing attachment is of thin metal and will rupture if a ball round is inadvertently fired. The trigger guard can be hinged down to give adequate clearance when wearing arctic gloves.

The carbine SC70/90 differs from the rifle only in that it has a folding tubular metal butt. It is worth noting that when the butt is folded, the patented ambidextrous magazine release can still be operated.

The special carbine SCS70/90 has a shorter barrel but no facility for launching grenades and hence no gas regulator lever or bayonet attachment; it has a folding tubular metal butt.

The SCP70/90 is similar to the SCS70/90 but is provided with a grenade-launching attachment which can be fitted in place of the standard flash hider when required.

Particular features of the 70/90 system which are stressed by the manufacturers include: a minimum number (105) of components; exceptionally easy field-stripping, essentially the same as that of the 70/223; accessibility of the gas cylinder for maintenance - it can be field-stripped without tools and is believed to be the only 5.56 mm rifle for which this claim can be made; it has no screws in its assembly; and it has a unique method of attaching the barrel to the receiver. Normal practice is simply to screw the barrel into the receiver, but this means a certain amount of hand-fitting to obtain the correct cartridge head clearance, and also introduces problems when it is necessary to fit a new barrel. The system used with the 70/90 is to make a plain hole in the receiver, and a plain rear end to the barrel. There is a collar on the barrel, and a screw cap which fits over the barrel which, when screwed on to the front end of the receiver, clamps the barrel into place. During manufacture the headspace gauging datum is carefully located and the locating collar and chamber face are carefully ground, by computer-controlled machines, to give the correct headspace when the barrel retaining nut is tightened to a specified torque. Subsequent changes of barrel will thus always have the correct headspace.

Data

Cartridge: 5.56 × 45 mm
Operation: gas, selective fire
Locking: rotating bolt
Feed: 30-round box magazine

WEIGHTS
AR70/90: 3.99 kg
SC70/90: 3.99 kg
SCS70/90: 3.79 kg
SCP70/90: 4.05 kg; with grenade launcher, 4.2 kg

5.56 mm SCS70/90 short carbine

5.56 mm SC70/90 carbine with butt folded

LENGTHS
AR70/90: 998 mm
SC70/90: 986 mm
SCS70/90: butt extended, 876 mm; butt folded, 647 mm
SCP70/90: butt extended, 910 mm; butt folded, 668 mm; with grenade launcher fitted, 798 mm and 1.04 m

BARREL
Lengths: (AR70/90, SC70/90) 450 mm; (SCS70/90) 352 mm; (SCP70/90) 369 mm
Rifling: 6 grooves, rh, 1 turn in 178 mm. Chrome-plated

Manufacturer
Armi Beretta SpA, Via Beretta 18, I-25063 Gardone VT (Brescia).

Status
In production.

5.56 mm AR70/90 infantry rifle

Service
Italian Army.

VERIFIED

Beretta 7.62 mm sniper rifle

Description
The Beretta sniper rifle is a conventional bolt-action repeating rifle chambered for the 7.62 × 51 mm NATO cartridge. It has a heavy free-floating barrel, with a flash eliminator on the muzzle, and a tube in the fore-end acts as a locating point for a bipod. It is possible that it also contains a harmonic balancer to reduce barrel vibration.

The rifle is fitted with target-quality iron sights, the foresight being hooded to obviate reflection and glare and the rearsight being fully adjustable for elevation and windage. However, as a sniper rifle, it will normally be fitted with an optical sight; the sight recommended by the makers being the Zeiss Diavari-Z, a ×1.5 to ×6 zoom telescope. The sight mount is to NATO STANAG 2324, and thus virtually any optical or electro-optical sight can be accommodated.

The thumb-hole pattern stock is of wood and the recoil pad and cheekpiece can be adjusted to suit the firer. Similarly, there is a hand-stop under the fore-end which is adjustable and which also serves to locate the firing sling, should one be desired.

Data
Cartridge: 7.62 × 51 mm NATO
Action: bolt action, single shot
Feed: 5-round box magazine
Weight: empty, 5.55 kg; bipod, 950 g; telescope, 700 g
Length: 1.165 m
Barrel: 586 mm
Rifling: 4 grooves, rh, 1 turn in 305 mm
Sights: fore, hooded blade; rear, fully adjustable V-notch; optical, Zeiss Diavari-Z ×1.5-×6 42T

Manufacturer
Armi Beretta SpA, Via Beretta 18, I-25063 Gardone VT (Brescia).

Status
No longer in production.

VERIFIED

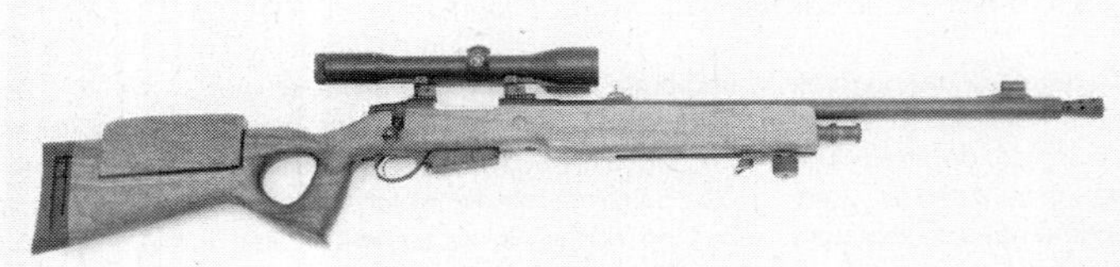
Beretta 7.62 mm sniper rifle

Beretta 7.62 mm sniper rifle with bipod

JAPAN

7.62 mm Type 64 rifle

Development
The formation of the Japanese Self-Defence Force led to the requirement for a new series of weapons and the Howa Machinery Company of Nagoya received a contract to develop a rifle. The project leader and designer was General K Iwashita. Several designs and modifications to those designs were produced until eventually the Type 64 emerged. This was the weapon accepted for the Japanese Self-Defence Force.

Description
The cartridge used with the 7.62 mm Type 64 rifle is a reduced load 7.62 × 51 mm, the propellant charge having been reduced by 10 per cent below NATO normal. This leads to a reduced muzzle impulse and as no effort has been made to reduce the weight significantly, the recoil energy that the soldier has to absorb is less than that generated by the conventional 7.62 mm cartridge. Should it ever be necessary to employ the full charge NATO round then the gas regulator must be set to reduce the amount and pressure of the gases reaching the piston head. The rifle was designed to launch rifle grenades and the gas regulator therefore includes a facility to cut off the gas supply to the piston entirely. The rifle is orthodox in appearance and action. The gas cylinder and piston are positioned above the barrel. The locking system incorporates a tilting block, lifted into engagement and subsequently lowered and carried back by the bolt carrier.

The butt is a straight-through design and this, with the bipod under the front of the fore-end, and the muzzle brake, produces extremely accurate single shot fire. The gun fires at 500 rds/min on full automatic and this relatively slow rate assists in maintaining the burst on the target. Three gas ports adjustable for load act as a regulator.

7.62 mm Type 64 rifles (K Nogi)

Data

Cartridge: 7.62 × 51 mm (reduced load)
Operation: gas, selective fire
Locking: tilting block
Feed: 20-round detachable box magazine
Weight: empty, 4.4 kg
Length: 990 mm
Barrel: 450 mm
Rifling: 4 grooves, rh, 1 turn in 250 mm
Sights: fore, blade; rear, aperture, adjustable for elevation and windage
Muzzle velocity: reduced charge, 700 m/s; full charge, 800 m/s
Rate of fire: cyclic, 500 rds/min
Effective range: 400 m

Manufacturer

Howa Machinery Limited, Shinkawacho, Nishikasugai-Gun, Aichi.

Status

Production ceased in 1990.

Service

Japanese Self-Defence Forces.

VERIFIED

5.56 mm Type 89 rifle

Development

The 5.56 mm Type 89 is a light assault rifle developed by the Technical Research and Development Institute, Japan Defense Agency. Howa Machinery Limited co-operated in the development based on a contract with the Japan Defense Agency.

The 5.56 mm Type 89 rifle comes in two versions; one has a folding tubular light alloy stock with a plastic butt-plate, and the other has a fixed plastic stock.

Description

The 5.56 mm Type 89 rifle operates by gas and uses a unique piston and cylinder system which has a long gas expansion chamber. The piston is stepped - the diameter of the front is smaller than that of the rear - and positioned in the centre of the cylinder some distance from the gas port. The gas is ported into the cylinder and there expands, as a result of which the piston starts the bolt carrier moving with a light kick so increasing the functional reliability and the life of the parts.

The breech is locked by a rotating bolt, a seven-lug head locking into the barrel extension. The receiver and the trigger housing are of steel, formed by pressing and welding.

A removable three-round burst device is installed in the rear portion of the trigger housing and is completely separated from the basic trigger mechanism for semi- and fully automatic firing.

The rifle is fitted with a high-efficiency muzzle brake to reduce the recoil energy. The rearsight is an aperture and incorporates a ballistic cam. The foresight is an adjustable square post. The bipod is a removable and folding type.

Accessories include a bayonet, a blank-firing attachment and a sling.

Data

Cartridge: 5.56 × 45 mm
Operation: gas, selective fire with 3-round burst controller
Locking: rotating bolt
Feed: 20- or 30-round box magazine
Weight empty, 3.5 kg
Length: stock folded, 670 mm; stock extended, 916 mm
Barrel: 420 mm
Rifling: 6 grooves, rh, 1 turn in 178 mm
Sights: fore, square post, adjustable for zero; rear, aperture, adjustable for elevation and windage
Sight radius: 440 mm
Muzzle velocity: 920 m/s
Rate of fire: cyclic, 650-850 rds/min

Manufacturer

Howa Machinery Limited, Shinkawacho, Nishikasugai-Gun, Aichi.

Status

In production.

Service

Japanese Self-Defence Force.

VERIFIED

5.56 mm Type 89 assault rifle, folding butt version

5.56 mm Type 89 assault rifle, fixed butt version

KOREA, NORTH

7.62 mm Type 68 assault rifle

Description

The North Korea armed forces were originally equipped with the 7.62 mm Type 58 rifle, a copy of the AK-47, which differed from the original only in not having the two finger grooves in the fore-end. This was followed by the 7.62 mm Type 68 assault rifle, based on the AKM. This also had the plain fore-end but was otherwise identical to the CIS model. Finally, the Type 68 was modified by the substitution of a folding metal stock for the normal fixed wooden stock. This folding stock is easily identified as it consists of two perforated rails joined at the shoulder piece, making the North Korean Type 68 among the lightest of the AKM-derived rifles.

North Korean 7.62 mm Type 58 assault rifle

Manufacturer

State arsenals.

Status

Production probably complete.

Service

North Korean armed forces and exported to some other nations, including Malta.

VERIFIED

7.62 mm Type 68 assault rifle showing perforated butt-stock

7.62 mm Type 68 assault rifle

KOREA, SOUTH

Daewoo 5.56 mm K2 assault rifle

Description

The Daewoo 5.56 mm K2 assault rifle is a gas-operated selective fire rifle with folding plastic stock. The operating system uses a long-stroke piston, similar to that of the Kalashnikov AK series rifles, operating a rotating bolt. The upper and lower receivers are machined from aluminium alloy forgings, and although the lower receiver may appear similar to that of the M16 there are considerable detail differences; the two are not interchangeable. The selector lever is rotatable in either direction and has four positions for safe, automatic fire, three-round burst fire and single shot. The burst control mechanism does not reset when the trigger is released: on pulling the trigger for a second time the burst picks up from where it stopped in the previous cycle.

The barrel is fitted with a muzzle brake/compensator which has no slot underneath so as to prevent dust being blown up. The folding stock is hinged to fold to the right of the receiver and uses a locking system similar to that of the FN-FNC. The rearsight has a two-position flip aperture with additional cam-actuated elevation control. The normal position of the flip gives a small aperture for precise shooting. The other position of the flip provides two small spots for night shooting. Fore and rearsights are provided with luminous spots for firing in poor light. In either position, rotation of the elevation control raises the selected aperture, the maximum marked range being 600 m.

Daewoo 5.56 mm K2 assault rifle

Data

Cartridge: 5.56 × 45 mm M193 or SS109
Operation: gas, selective fire with 3-round burst
Locking: rotating bolt
Feed: 30-round box magazine
Weight: empty, 3.26 kg
Length: butt folded, 730 mm; butt extended, 990 mm
Barrel: without compensator, 465 mm
Rifling: 6 grooves, rh
Muzzle velocity: M193, 960 m/s; SS109, 920 m/s

Manufacturer

Daewoo Precision Industries Limited, PO Box 25, Kumjeong, Pusan.

Status

In production.

Service

South Korean Army.

VERIFIED

NORWAY

Norwegian rifles

Description

The Norwegian Army was equipped with the original 6.5 mm Krag-Jørgensen rifle with a rimless cartridge. These rifles were developed and made at Kongsberg, Norway. The rifles were largely lost during the German Occupation in the Second World War.

After the war the Norwegian Army decided to replace German, UK, and US rifles with the G3A3. Norwegian G3A3 rifles were licence-produced at Kongsberg. The Norwegian model differs from the standard Heckler and Koch design by allowing silent cocking. When required, the bolt is allowed forward under control and then finally closed by finger pressure on grooves cut into the side of the bolt.

Manufacturer

Norsk Forsvarsteknologi A/S, PO Box 1003, N-3601 Kongsberg.

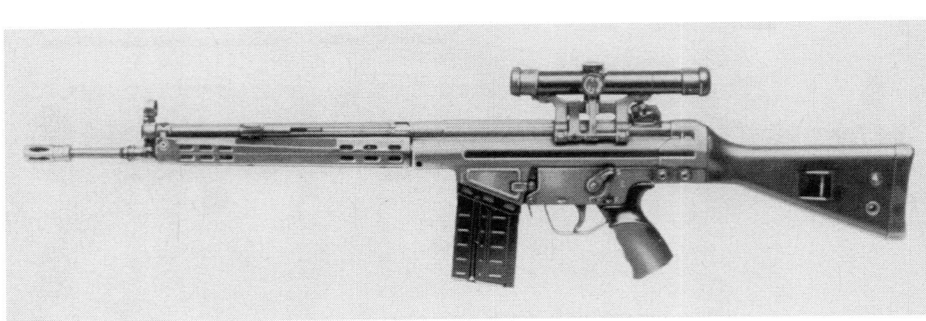

G3 sniper rifle

Status

No longer in production.

Service

Norwegian armed forces.

VERIFIED

G3A4 rifle manufactured by Norsk Forsvarsteknologi

Norwegian G3 rifle bolt closure

7.62 mm NM149S sniper rifle

Description

The 7.62 mm NM149S sniper rifle uses a Mauser M98 bolt action and was developed in co-operation with the Norwegian Army and police forces. Intended for use out to a range of 800 m, it is fitted with a Schmidt & Bender 6 × 42 telescope sight which may be mounted and removed without altering the weapon's zero. Iron sights are also fitted, and the telescope sight may be replaced by a Simrad KN250 image intensifying sight.

The stock is of impregnated and laminated beech veneer and may be adjusted for length by the use of butt-plate spacers; the police model also has an

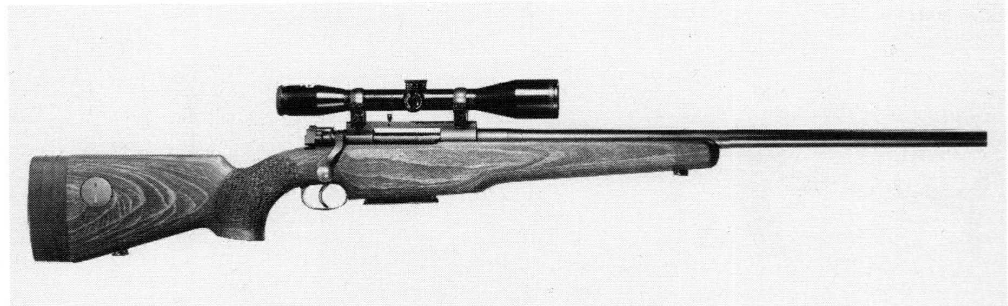

7.62 mm NM149S sniper rifle

adjustable cheekpiece. The rifle has a match trigger, adjusted to a 1.5 kg pull, and the weapon can be equipped with bipod and sound suppressor if required.

Data

Cartridge: 7.62 × 51 mm NATO
Operation: bolt action, single shot
Feed: 5-round box magazine
Weight: with telescope, 5.6 kg
Length: 1.12 m
Barrel: 600 mm
Rifling: 4 grooves, rh, 1 turn in 305 mm

Manufacturer

Våpensmia A/S, Box 86, N-2871 Dokka.

Status

Production complete.

Service

Norwegian Army and police forces.

UPDATED

PAKISTAN

POF 7.62 mm G3 rifles

Description

Pakistan Ordnance Factories (POF) manufacture Heckler and Koch G3 assault rifles under licence. Both the G3A3 (fixed butt) and G3A4 (telescopic butt) versions are manufactured for the Pakistan armed forces and potential export sales. These rifles are virtually identical to the German originals but POF-supplied data is provided below.

7.62 mm G3A3 assault rifle produced by Pakistan Ordnance Factories (POF)

Data

Cartridge: 7.62 × 51 mm NATO
Operation: delayed blowback, selective fire
Delay: rollers
Feed: 20-round box magazine
Weight: G3A3, 4.4 kg; G3A4, 4.7 kg
Length: G3A3, overall, 1.025 m; G3A4, butt retracted, 840 mm; butt extended, 1.02 m
Barrel: 450 mm
Rifling: 4 grooves, 1 turn in 305 mm
Sights: front, hooded fixed post; rear, rotary with 4 adjustments (100 m 'V' notch, 200, 300 and 400 m) and aperture adjustable for windage and elevation
Muzzle velocity: 780-800 m/s
Rate of fire: cyclic, 500-600 rds/min

Manufacturer

Pakistan Ordnance Factories, Wah Cantt.

Status

In production.

Service

Pakistan armed forces.

VERIFIED

POLAND

7.62 mm AK assault rifle

Description

After the Second World War the Polish forces were equipped with Mosin-Nagant rifles but these were displaced by the Kalashnikov assault rifles as soon as these appeared.

A Polish-built copy of the AK-47, originally known as the PMK but later redesignated as the AK, does not differ significantly from the original.

Numbers of AK rifles were modified for grenade launching. The muzzle on these rifles is cone-shaped and threaded to take the LON-1 grenade launcher which is externally 20 mm in diameter. The gas cylinder was modified to take a gas cut-off valve which prevents gas from the grenade cartridge reaching the operating piston. A special grenade-launching sight, the CN 70, is attached to the standard rearsight. A 10-round magazine, which will hold only grenade cartridges, is used for grenade launching. To absorb recoil the butt was modified in shape and a boot fitted which goes into the firer's shoulder. The only mechanical change is the fitting to the rear of the breech cover of a securing catch to the return spring rod. This must be released before the rod can be pushed in to allow removal of the breech cover.

Polish-made F1/N60 anti-personnel and PGN-60 anti-tank grenades have 20 mm internal diameters and will fit no other launcher. Conventional grenades with an internal diameter of 22 mm should never be used from the Polish AK rifle.

Polish 7.62 mm AK assault rifle, folding butt version, a Polish version of the AK-47S (T J Gander)

Status

Production complete.

Service

Polish armed forces.

UPDATED

7.62 mm PMKM assault rifle

Description

The 7.62 mm PMKM assault rifle is the Polish copy of the CIS AKM or AKMS. Like the later CIS AKM and AKMS models it is fitted with a compensator at the muzzle. It cannot launch grenades and the muzzle is not coned.

Both AKM and AKMS style variants are produced along with accessories such as a multipurpose field knife/bayonet. A folding bipod can be attached, as can a Polish-produced 40 mm PALLAD grenade launcher. Sighting options include a passive sniperscope, an optical sight, a collimated RED-POINT sight and a laser target marker.

Models have been produced with the pistol grip being plastic and the forward handguard made of hardwood. Also available are versions with all furniture being manufactured from high impact black plastic-based materials.

Also available is the RADOM-HUNTER, a semi-automatic version of the AKM with wooden furniture including a combined pistol grip and thumb-hole stock. It uses a five-round magazine.

Polish folding stock model of 7.62 mm PMKM assault rifle with butt extended and bayonet fitted

Data

AKMS MODEL
Cartridge: 7.62 × 39 mm
Operation: gas, selective fire
Locking: rotating bolt
Feed: 30-round detachable box magazine
Weight: empty, 3.165 kg; loaded, with bayonet, 4.226 kg
Length: butt folded, 757 mm; butt extended, bayonet fixed, 1.027 m
Barrel: 415 mm
Rifling: 4 grooves, rh
Sights: fore, pillar; rear, U-notch, calibrated from 100 to 1,000 m
Sight radius: 384 mm
Muzzle velocity: 710 m/s
Muzzle energy: 1,990 J
Rate of fire: cyclic, 600 rds/min
Effective range: 400 m

Manufacturer

Zaklady Metalowe LUCZNIK, ul 1905 Roku 1/9, PL-26-600 Radom.

Status

In production.

Service

Polish armed forces.

UPDATED

TANTAL assault rifles

Development

The TANTAL assault rifles may be regarded as the Polish versions of the CIS AK-74S since only folding-stock models are produced. Two models of the TANTAL are available, the wzor (wz - model) 88 firing the 5.45 × 39 mm cartridge and the wzor 90 for 5.56 × 45 mm NATO cartridges. Both models can be provided with a family of Polish-produced accessories and have been offered for export sales.

Description

Both models of the TANTAL assault rifle are based on the CIS AK-74S so for operation and similar details reference can be made to the entry relating to the AK-74 in this section. The main operating difference compared to the AK-74 is that the TANTAL has three fire modes; single shot, three-round burst, and fully automatic.

Only one version of the TANTAL is apparently made, having a simple single-strut butt-stock folding to the right-hand side of the receiver. All furniture is high impact black plastic, although one version does have a plastic pistol grip and hardwood handguard.

The muzzle attachment can accommodate a multipurpose field knife/bayonet and can also be used to launch rifle grenades using standard ball ammunition. Grenades weighing from 350 to 700 g can be launched to ranges of 250 and 150 m respectively with aiming assisted by a special sight which is mounted over the usual rearsight assembly when the grenade launching mode is required. Anti-armour HEAT rifle grenades launched from the muzzle attachment are effective at ranges up to 150 m. It is also possible for the TANTAL to mount the Polish-produced PALLAD 40 mm grenade launcher under the barrel and handguard.

To assist aim the TANTAL is provided with a light clip-on bipod. The normal rear and foresights are provided with luminescent dots to assist aiming in poor light conditions. Other sighting systems available include a passive night sight, an optical telescope, a collimated RED-POINT sight, and a laser target indicator.

TANTAL assault rifle with bipod and bayonet attached

Data

Cartridge: wz 88, 5.45 × 39 mm; wz 90, 5.56 × 45 mm NATO
Operation: gas, selective fire, including 3-round burst
Locking: rotating bolt
Feed: 30-round detachable box magazine
Weight: rifle empty, 3.4 kg; loaded, with bipod, 4.3 kg
Length: butt folded, 742 mm; butt extended, 943 mm
Barrel: 423 mm
Sights: fore, pillar; rear, U-notch, calibrated from 100 to 1,000 m
Sight radius: 370 mm
Muzzle velocity: wz 88, 880 m/s; wz 90, 900 m/s
Muzzle energy: wz 88, 1,316 J; wz 90, 1,620 J
Rate of fire: cyclic, 600-650 rds/min
Effective range: 600 m

Manufacturer

Zaklady Metalowe LUCZNIK, ul 1905 Roku 1/9, PL-26-600 Radom.

Status

In production.

VERIFIED

PORTUGAL

Portuguese 7.62 mm Heckler and Koch G3A2, A3 and A4 rifles

Description

Portugal manufactures the Heckler and Koch G3 rifles under licence. The rifles are identical to the German models.

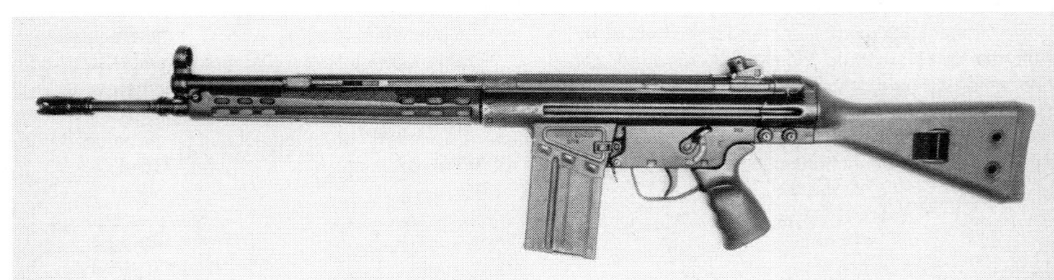

Portuguese Heckler and Koch G3A2 rifle

Data

Cartridge: 7.62 × 51 mm
Operation: delayed blowback, selective fire
Method of delay: rollers
Feed: 20-round box magazine
Weight: fixed butt, empty, 4.4 kg; retractable butt, empty, 4.7 kg
Length: retracted butt, 840 mm; fixed butt, 1.025 m
Barrel: 450 mm
Rifling: 4 grooves, rh, 1 turn in 305 mm
Sights: fore, post; rear, V battle sight at 100 m. Apertures for 200, 300 and 400 m
Sight radius: 572 mm
Muzzle velocity: 780-800 m/s
Rate of fire: cyclic, 500-600 rds/min
Effective range: 400 m

Manufacturer

INDEP SA, Fabrica de Braco de Prata, Rua Fernando Palha, P-1802 Lisboa-Codex.

Status

In production.

Service

Portuguese and other armed forces.

UPDATED

ROMANIA

Romanian 7.62 mm AKM assault rifles

'Reduced' model of Romanian 7.62 mm AKM assault rifle

Development

The Romanian Army was at one time equipped with a locally produced AK-47 copy, but this has now been relegated to militia and reserve use.

The standard Romanian armed forces assault rifle (now being supplemented by numbers of local variations of the 5.45 mm AK-74 - see following entry) is a copy of the CIS AKM. The Romanian version may be readily recognised by the forward pistol grip made from laminated wood. In addition the usual Romanian markings appear on the selector - 'FA' for full automatic and 'FF' on the lower position for single shot. Both 30- and 40-round box magazines are available. It is possible to attach a grenade launcher.

In August 1995 it was announced that India had contracted to purchase 100,000 AKM assault rifles from Romania at a cost estimated to be $85 to $90 per unit. This order was associated with an ammunition purchase for around 30 million rounds, reportedly from North Korea. These purchases were made necessary by delays in the Indian INSAS system (see entry under *India*).

Description

Romanian AKM models closely follow those from other nations and include the fixed stock AKM and folding butt AKMS. However, there is one 'Reduced' model which appears to be unique to Romania. This model has a 20-round magazine and a barrel only 305 mm long. The muzzle lacks any form of compensator and there is no provision for a bayonet, while the side-folding butt-stock is reduced to a simple single strut and one-piece butt plate. The data provided in the table below relates to this 'Reduced' model.

It should be noted that Romanian literature refers to all these rifles as 'sub-machine guns'.

Data

'REDUCED' MODEL
Cartridge: 7.62 × 39 mm
Operation: gas, selective fire
Locking: rotating bolt
Feed: 20-round detachable box magazine
Weight: rifle, less magazine, 3.1 kg
Weight: magazine, loaded, 640 g
Length: butt folded, 550 mm; butt extended, 750 mm
Barrel: 305 mm
Rifling: 4 grooves, rh
Sights: fore, pillar; rear, U-notch, calibrated from 100-500 m
Sight radius: 370 mm
Muzzle velocity: 670 m/s
Rate of fire: cyclic, 600 rds/min
Effective range: 400 m

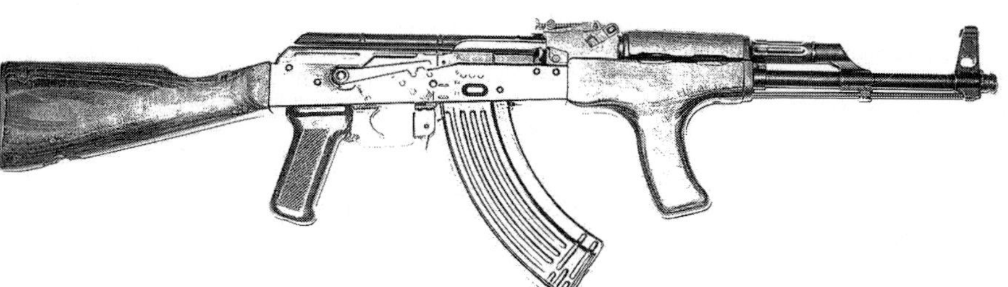

Romanian 7.62 mm AKM assault rifle (T J Gander)

Manufacturer

Romtehnica, 9-11 Drumul Taberei Street, Bucharest.

Status

In production.

Service

Romanian armed forces. Offered for export sales and purchased by India (see text).

UPDATED

Romanian 5.45 mm AK-74 assault rifles

Development

When the Romanian concern RATMIL began production of 5.45 mm assault rifles the initial result was not a direct copy of the CIS AK-74 but a straightforward conversion of the 7.62 mm AKM to accommodate 5.45 × 39 mm ammunition. This initial model lacked the muzzle brake of the AK-74 and featured a solid wooden butt. This model retained the usual Romanian trademark of a forward pistol grip and was placed into production but eventually it was replaced by a more direct copy of the standard AK-74 with the upper handguard extended forward to the gas tap-off port. This version has the muzzle brake and frame butt outline of the AK-74S.

Other models followed, one with a rudimentary single-strut butt-stock and conventional handguard, and another a 'Reduced' model apparently developed as the Romanian counterpart to the CIS AKSU-74 sub-machine gun. Although the 'Reduced' model is provided with a muzzle attachment very similar to that on the AKSU-74, the Romanian model is a cut-down variant of their AK-74 model with a 305 mm long barrel.

It should be noted that Romanian literature refers to all these rifles as 'sub-machine guns'.

Description

The Romanian 5.45 mm AK-74 models are identical in operation to their CIS counterparts so reference should be made to the appropriate entry for such details.

The base model of the Romanian AK-74 family is actually closer to the AK-74S in appearance but the open frame butt-stock does not normally fold, although the means to make it do so could be easily introduced; there is also a model with a solid wooden stock. The folding stock model has a side-folding butt-stock formed from a single length of steel strip. 30- or 40-round box magazines can be used on both these models.

The 'Reduced' model has a 305 mm long barrel, a side-folding frame butt-stock and is apparently provided with the 30-round magazine only.

There is one further member of the Romanian AK-74 family. This is a training rifle chambered for low-cost 0.22/5.6 mm ammunition but which is otherwise identical to the 5.45 mm AK-74, complete with bayonet; the butt-stock being solid hardwood. The full-size magazine holds only 20 rounds and the rear sights are calibrated for 50, 75 and 100 m only. Weight is 3.57 kg.

Data

Cartridge: 5.45 × 39 mm
Operation: gas, selective fire
Locking: rotating bolt
Feed: 30- or 40-round detachable box magazine; Reduced model, 30-round only
Weight: standard rifle, without magazine, 3.4 kg; Reduced model, without magazine, 3.15 kg; 30-round mazagine, loaded, 650 g
Length: standard and folding models, butt extended, 940 mm; folding model, butt folded, 715 mm; Reduced model, butt extended, 800 mm; butt folded, 600 mm
Barrel: standard, 415 mm; Reduced, 305 mm
Rifling: 4 grooves, rh
Sights: fore, pillar; rear, U-notch
Muzzle velocity: standard, 880 m/s; Reduced, 790 m/s
Rate of fire: cyclic, 700 rds/min

Folding stock model of Romanian 5.45 mm AK-74 assault rifle

Manufacturer
Romtehnica, 9-11 Drumul Taberei Street, Bucharest.

Status
In production.

Service
Romanian armed forces. Offered for export sales.

VERIFIED

Fixed stock model of Romanian 5.45 mm AK-74 assault rifle

SINGAPORE

5.56 mm SAR 80 assault rifle

Development
The SAR 80 was designed by Frank Waters of Sterling Armament in early 1976. Prototypes were made by Chartered Industries of Singapore in 1978 and evaluated by the School of Infantry Weapons of the Singapore Armed Forces Training Institute. Modified models were then subjected to extensive troop trials and design improvements were incorporated before commencement of full production in 1980. To minimise production costs, manufacturing concepts used in the production of the SAR 80 are based on the more economical production technologies. Approximately 45 per cent of components are sheet metal pressings. Another 40 per cent are standard parts readily available on the commercial market.

Description
The SAR 80 is a 5.56 mm magazine-fed, gas piston-operated, air-cooled assault rifle. It is capable of both semi-automatic and automatic modes of fire from the closed bolt position. Both the US M16-type 20- and 30-round magazines can be used. The barrel is provided with a flash suppressor which serves as a grenade launcher and as a front support for the bayonet.

The folding butt-stock variant of the SAR 80 is easy to handle, especially for combat at close quarters and in confined areas. Combined with its low recoil and capability of automatic fire, the folded weapon is ideal for paratroopers and troops of armoured units.

The barrel, breech block and butt-stock are assembled in a straight line, this construction improving controllability and reducing vertical dispersion in full-automatic fire by minimising barrel climb. The SAR 80 is fitted with a spring-loaded firing pin to prevent firing if the rifle is accidentally jolted or dropped. The gas regulator can be adjusted manually and enables the SAR 80 to be operated under adverse conditions by releasing more return gas. Firing rifle grenades is also possible by sealing off the return gas completely.

The gas system uses a piston operating a breech block carrier which carries the rotating breech block. Rotation is achieved by means of a cam pin projecting out of the cam path in the breech block carrier. The breech block is prevented from rotating as the carrier is pushed forward to feed a round into the barrel chamber. However, when the feed is complete, the continued movement of the carrier rotates the breech block and thus locks it to the barrel.

With the rifle armed, the hammer is controlled by a fire control selector and an additional safety lever, the automatic sear. The automatic sear is timed so as to ensure the breech block is fully closed before the hammer can be released. In the unlikely event of a mechanical failure, and supposing the breech block is not fully closed, the hammer will strike the rear of the carrier thereby dissipating its energy in pushing the carrier forward. No firing can therefore occur.

When firing a round, the carrier is forced backwards by the gas piston/push-rod. After a short delay, the cam slot on the carrier engages the cam pin in the breech block. The breech block unlocks, providing both primary extraction and loosening of the empty cartridge case in the chamber.

Once the carrier is forced back fully, the empty case is extracted and ejected and the return spring is compressed. The breech block carrier runs out, the next round is fed and, if the trigger is still depressed, fired. When the ammunition supply is exhausted, the breech block catch is automatically pushed up by the magazine follower to hold the breech block carrier open at the rear.

A full range of accessories and options is available for the SAR 80: US M16 type 20- and 30-round magazines; folding butt-stock; telescopic sight; blank-firing attachment; bayonet and scabbard; sling and a cleaning kit.

5.56 mm SAR 80 assault rifle on bipod with 30-round box magazine

Data
Cartridge: 5.56 × 45 mm (US M193)
Operation: gas (piston, short stroke)
Locking: rotating bolt, multilug head
Feed: 20- or 30-round magazine
Weight: unloaded, 3.7 kg
Length: butt folded, 738 mm; fixed butt, 970 mm
Barrel: 459 mm
Sight radius: 541 mm
Muzzle velocity: 970 m/s
Rate of fire: cyclic, 600-800 rds/min
Effective range: 400 m

Manufacturer
Chartered Firearms Industries Pte Limited, 249 Jalan Boon Lay, Singapore 2261.

Status
No longer in production.

Service
Singapore armed forces.

VERIFIED

5.56 mm SR88 assault rifle

Description
The SR88 is a lightweight gas-operated assault rifle which was developed to meet all current military requirements. It has a four-position adjustable gas regulator which also serves as an anti-fouling device and which can be set to shut off gas action for firing grenades. Field maintenance is greatly simplified by the chrome-plated gas piston system which virtually eliminates carbon fouling of the bolt carrier. The straight-line layout allows excellent control and accuracy.

The receiver, which is an aluminium forging, houses the operating mechanism. The stock is of glass-reinforced nylon, and its length is adjustable by adding or removing the butt-pads. The rifle will function with the stock removed.

The bolt group assembly is in modular form and consists of the bolt, bolt carrier, buffer springs and rods. The bolt carrier and bolt are made from special high grade steel. The seven-lug bolt locks into the barrel, following a rotation of 22.5°, the rotation being controlled by a cam in the carrier. The bolt is always kept forward by the firing pin spring, a concept which permits rapid assembly after cleaning.

The upper receiver assembly comprises the upper receiver, barrel and carrying handle. The receiver is a high grade steel stamping and houses the bolt group assembly. The barrel is forged of high grade steel and features a chrome-plated chamber. The barrel is fitted with a flash suppressor which vents sideways and upwards and is internally threaded to accept a blank-firing attachment. The barrel is connected to the receiver by a locknut and detent, a system which simplifies barrel replacement.

The gas piston assembly consists of a chrome-plated steel regulator and piston-cum-pushrod. The pushrod and spring are shrouded by the handguards, leaving the venting of gas to the atmosphere. This keeps the handguard relatively cool and comfortable under sustained fire. The handguards are interchangeable - left and right are identical.

An effective safety feature is a spring-loaded firing pin which prevents inadvertent firing should the rifle be dropped or jarred. Other features include the provision for night aiming with a luminous front sight, the option of a folding butt-stock where compactness is desired, a carrying handle, an arctic trigger and provision for mounting a sighting telescope. The cocking handle has an automatic lock to provide forward assist and silent bolt closure. A selector for full automatic or three-round burst fire can be fitted according to requirements. There is provision to mount a grenade launcher without the need to remove handguards.

The SR88 is produced in fixed butt standard, folding butt standard, and fixed and folding butt carbine versions.

Variant

5.56 mm SR88A assault rifle
See following entry.

Data

Cartridge: 5.56 × 45 mm M193 or SS109
Operation: gas, selective fire and 3-round burst
Locking: rotating bolt
Feed: 20- or 30-round magazine
Weight: unloaded, 3.66 kg
Length: butt folded, 746 mm; fixed butt, 970 mm
Barrel: 459 mm
Rifling: 6 grooves, rh, 1 turn in 32 or 55 calibres
Sight radius: 526 mm
Muzzle velocity: 970 m/s
Rate of fire: cyclic, 650-850 rds/min
Effective range: 400 m

Manufacturer

Chartered Firearms Industries Pte Limited, 249 Jalan Boon Lay, Singapore 2261.

Status

No longer in production.

Service

Singapore armed forces.

VERIFIED

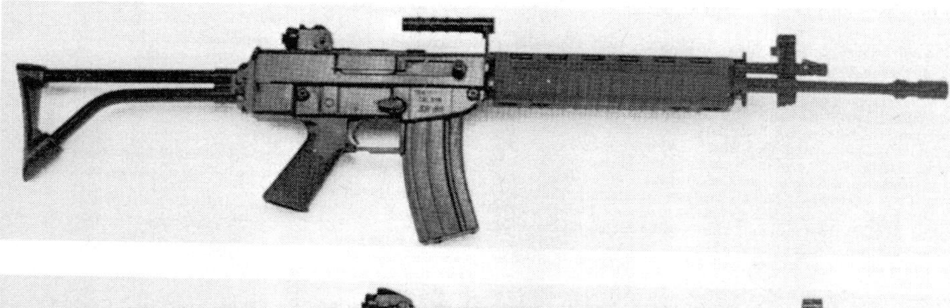

Folding butt versions of the SR88; (top) standard assault rifle, (lower) carbine

5.56 mm SR88A assault rifle

Description

The SR88A assault rifle was developed to supersede the SR88, described previously. It is basically similar, being a gas-operated weapon using a rotating bolt, but there are numerous detail changes in construction. The lower receiver is an aluminium forging, the upper receiver a steel pressing and the stock is of glassfibre-reinforced nylon. The bolt group is in modular form, as with the SR88. The barrel is hammer forged from high grade steel and has a chromium-plated chamber; optionally, the barrel bore can also be chromed. There is a flash suppressor which vents sideways and upwards, and a blank-firing attachment can be fitted directly to the suppressor. The barrel is connected to the receiver by a locknut and detent system, which is also oriented by locating flats on the barrel and a special slot profile on the extension, simplifying barrel replacement.

The gas system consists of a chromium-plated regulator and a long stroke piston assembly connected to the bolt carrier. The regulator has three positions, two for gas tap-off and the third to shut off the gas for grenade launching. The piston assembly is housed in a guide tube above the barrel and is shrouded by the handguard, allowing the gas to vent to atmosphere.

The standard rifle has a fixed butt-stock; optionally, a retracting butt-stock can be fitted where compactness is important. A carbine version, with shorter barrel and retracting butt-stock is available for airborne forces and other troops requiring a more easily carried weapon.

5.56 mm SR88A assault rifle and carbine

Data

SR88A RIFLE
Cartridge: 5.56 × 45 mm
Operation: gas, selective fire
Locking: rotating bolt
Feed: 30-round box magazine
Weight: with empty magazine, 3.68 kg
Length: butt retracted, 810 mm; fixed butt, 960 mm
Barrel: 460 mm
Rifling: 6 grooves, rh, 1 turn in 178 or 305 mm, as required
Sights: fore, hooded blade; rear, adjustable aperture; tritium night sights optional
Sight radius: 500 mm
Muzzle velocity: M193, 970 m/s; SS109, 940 m/s
Rate of fire: cyclic, 700-900 rds/min

SR88A CARBINE: AS FOR RIFLE EXCEPT
Weight: with empty magazine, 3.81 kg
Length: butt retracted, 660 mm; butt extended, 810 mm
Barrel: 292 mm
Sight radius: 425 mm

Manufacturer

Chartered Firearms Industries Pte Limited, 249 Jalan Boon Lay, Singapore 2261.

Status

In production.

VERIFIED

SOUTH AFRICA

Vektor 5.56 mm R4 and R5 assault rifles

Development

The 5.56 mm R4 assault rifle, which replaced both the Fabrique Nationale 7.62 mm Fusil Automatique Léger (R1) and the G3 as the South African Defence Force's (now the South African National Defence Force - SANDF) standard infantry weapons, is a modified version of the Israeli Galil (qv). The modifications consist, in the main, of improvements in the strength of material and construction to withstand better the severe conditions of bush warfare. The butt-stock was lengthened, since South African soldiers are generally of larger stature than Israelis, and made from a reinforced synthetic material; the handguard is similarly strengthened.

The Vektor 5.56 mm R4 assault rifle is also available in a short version, known as the Vektor R5 assault rifle. It is identical to the R4 except that it has a barrel shortened by 128 mm, no bipod and a shorter handguard.

The R5 was adopted by the South African Air Force and Marines. Semi-automatic versions of both the R4 and R5 (the LM4 and LM5 respectively) are in production by Vektor for use by police and paramilitary forces.

Description

Compared to the Israeli original, the Vektor 5.56 mm R4 assault rifle has small changes in the receiver, gas piston and bolt assemblies in order to facilitate manufacture. The rifle is supplied with a 35-round magazine manufactured from reinforced plastic material (nylon). A 50-round magazine made from steel is no longer available.

The sights incorporate tritium spots for night firing, set for 200 m engagement range.

Since its introduction the R4 has been improved in various details, these improvements being a result of experience gained operationally. The original butt was of aluminium covered with nylon; this was replaced with a glassfibre-filled nylon butt with a strengthening web halfway along its length, the result being stronger and more robust than the original. The metal magazine was replaced by a nylon/glassfibre design; this has little mechanical advantage but is much lighter than the metal version.

Other changes include a redesigned gas tube lock to prevent the tube shaking loose under the firing shock; the removal of the click adjustment from the foresight, thus allowing the mounting dovetail to be made stronger; and the introduction of a wider and stronger sear.

The most important improvement was a redesign of the bolt. The old pattern was capable of firing a cartridge with a sensitive primer accidentally, by inertia of the firing pin as the bolt closed. A resilient retracting spring, of oil-proof polyurethene, now keeps the firing pin inside the bolt except when positively driven forward. It has the additional advantage of spreading the rotational torque on bolt closure, thus relieving the firing pin locating pin of excessive stress.

The R4 has a removable bipod which incorporates a wire cutting facility. The flash hider on the muzzle is fitted with a spring-loaded retainer which allows the unit to be removed so that rifle grenades can be fired from the muzzle.

A cleaning kit is available.

Variant

5.56 mm R6 compact assault rifle
See following entry.

SANDF soldiers with 5.56 mm R4 assault rifles (T J Gander)

Data

DATA FOR R5 IN PARENTHESIS
Cartridge: 5.56 × 45 mm
Operation: gas, selective fire
Locking: rotating bolt
Feed: 35-round box magazine
Weight: empty, 4.3 kg (3.7 kg); with 35-round magazine, 4.44 kg (3.84 kg)
Length: butt folded, 740 mm (615 mm); butt extended, 1.005 m (877 mm)
Barrel: 460 mm (332 mm)
Rifling: 6 grooves, 1 turn in 305 mm
Sights: front, hooded adjustable post with tritium insert; rear, flip-over aperture graduated for 300 and 500 m
Sight radius: 475 mm (445 mm)
Muzzle velocity: 980 m/s (920 m/s)
Rate of fire: cyclic, 600-750 rds/min

Manufacturer

LIW Division of Denel (Pty) Limited, 368 Selborne Avenue, Lyttelton, Pretoria 0001.

Status

In production.

Service

South African National Defence Force.

UPDATED

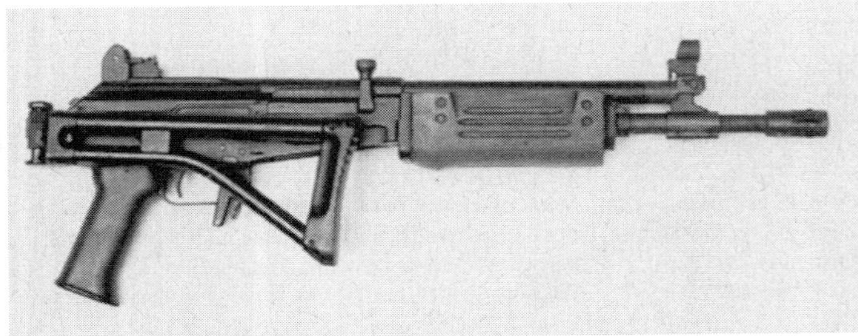

Vektor 5.56 mm R5 assault rifle with stock folded

Vektor 5.56 mm R4 assault rifle

Vektor 5.56 mm R6 compact assault rifle

Description

The Vektor 5.56 mm R6 compact assault rifle is a shortened version of the basic R4/R5 design (see previous entry), intended for use by vehicle-mounted, airborne and similar forces requring the most compact weapon available. The operation and general characteristics are exactly as the R4 and R5, only the dimensions differ.

Data

Cartridge: 5.56 × 45 mm
Operation: gas, selective fire
Locking: rotating bolt
Feed: 35-round box magazine
Weight: empty, 3.675 kg
Length: butt folded, 565 mm; butt extended, 805 mm
Barrel: 280 mm
Rifling: 6 grooves, rh, 1 turn in 305 mm
Sights: fore, adjustable post, tritium insert; rear, 2-position flip aperture, 300 m and 500 m
Muzzle velocity: 825 m/s
Rate of fire: cyclic, 585 rds/min

Vektor R6 5.56 mm compact assault rifle

1996

Manufacturer

LIW Division of Denel (Pty) Limited, 368 Selborne Avenue, Lyttelton, Pretoria 0001.

Status

In production.

Service

South African National Defence Force.

UPDATED

SPAIN

7.62 mm Model R firing port weapon

Description

The 7.62 mm Model R firing port weapon is a variant of the Model E 7.62 mm rifle (described in earlier editions of *Jane's Infantry Weapons*) using the same basic roller-locked mechanism but without a butt and with a shorter barrel. A flash-hider is fitted to the muzzle, and there is a substantial collar forward of the receiver which locks into a firing port mounting (Rotula CETME RH 762/BMR) installed in the vehicle sidewall. The mounting allows a traverse of 50° and elevation from −20 to +30°. The position of the locking collar and mount means that the usual forward cocking handle of the Model E cannot be used, and a modified cocking device, using the MG3 machine gun cocking handle, is fitted. Fitting the weapon to the mount takes only four or five seconds.

The Model R fires automatic only, from a 20-round box magazine.

Data

Cartridge: 7.62 × 51 mm NATO
Operation: delayed blowback, automatic only
Delay: rollers and locking lever
Feed: 20-round box magazine
Weight: 6.4 kg
Length: 665 mm
Barrel: 305 mm
Muzzle velocity: 690 m/s
Rate of fire: cyclic, 500-600 rds/min

Manufacturer

SANTA BARBARA, Julian Camarillo 32, E-28037 Madrid.

Status

Development.

VERIFIED

7.62 mm Model R firing port weapon mounted in Rotula CETME RH 762/BMR

7.62 mm Model R firing port weapon (T J Gander)

5.56 mm Models L and LC assault rifle

Development

This selective fire weapon in 5.56 mm calibre embodies Compañia de Estudios Técnicos de Materiales Especiales' (CETME's) extensive experience in developing its earlier range of similar weapons in 7.92 mm and 7.62 mm calibres.

Two versions of the weapon were designed: a standard model with a fixed butt-stock (Model L) and a short-barrelled version with a telescopic stock (Model LC). A 20-round magazine was first used, but the design was then modified to accept the standard 30-round M16-pattern magazine. Early rifles had a four-position sight graduated for 100, 200, 300 and 400 m ranges. Current models have a simple two-position flip-over aperture sight graduated for 200 and 400 m.

Description

Operation of the automatic mechanism is by a delayed blowback system essentially the same as that used on earlier CETME weapons and on the Heckler and Koch series of rifles, delay being achieved by the use of rollers.

In general, the operation of the trigger mechanism for single shots (T) or normal automatic fire (R) is similar to that of the German Heckler and Koch G3 rifle: similarly the safety setting (S) of the selector lever provides a conventional impediment to the operation of the trigger. Early models had a fourth (R) setting, giving a regulated three-round burst. This is no longer standard, since it has been found that the firer can regulate the length of the burst by using the trigger. However, if the three-round burst facility is desired, it can be fitted at the factory as an optional extra.

5.56 mm Model L assault rifle

1996

Data

MODEL LC, WHERE DIFFERENT, SHOWN IN PARENTHESIS
Cartridge: 5.56 × 45 mm NATO
Operation: delayed blowback, selective fire
Delay: rollers and locking lever
Feed: 10- or 30-round detachable box magazine
Weight: unloaded, 3.4 kg

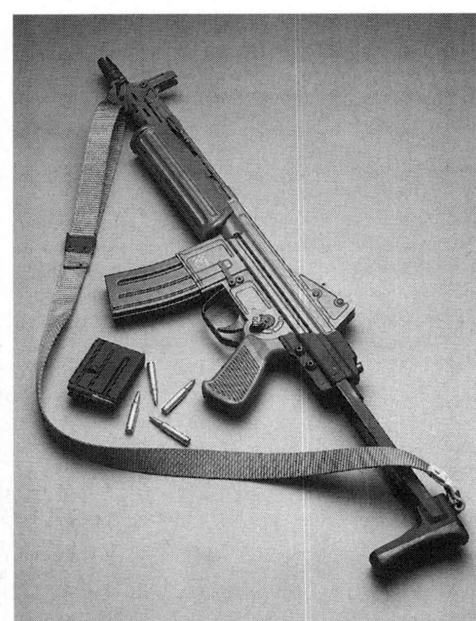

5.56 mm Model LC assault rifle with stock extended

1996

Weight: 30-round magazine empty, 210 g
Length: 925 mm (665 or 860 mm)
Barrel: 400 mm (320 mm)
Rifling: 6 grooves, rh, 1 turn in 178 mm
Sights: fore, protected conical post; rear, flip-over aperture for 200 and 400 m
Sight radius: 440 mm
Muzzle velocity: 875 m/s (832 m/s)
Muzzle energy: 1,531 J (1,384 J)
Rate of fire: cyclic, 600-750 (650-800) rds/min

Manufacturer
SANTA BARBARA, Julian Camarillo 32, E-28037 Madrid.

Status
In production.

UPDATED

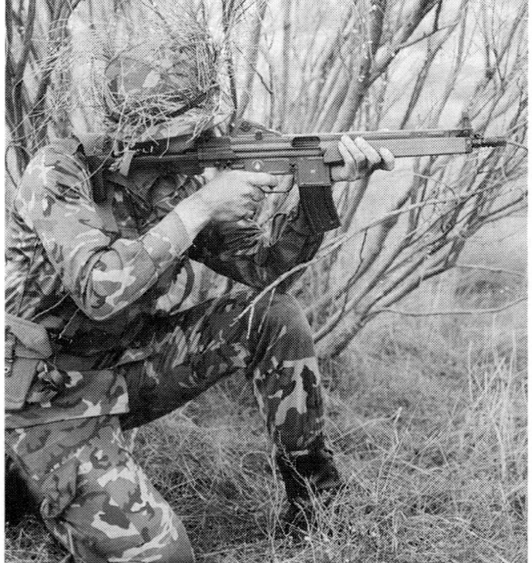

5.56 mm Model LC assault rifle
1996

7.62 mm C-75 special forces rifle

Description
The 7.62 mm C-75 special forces rifle is a conventional bolt-action rifle, using a Mauser action, intended for sniping, police and special forces tasks. Fitted with iron sights as standard, it also has telescope mounts machined into the receiver and can thus be equipped with virtually any optical or electro-optical sight. With a telescope, match ammunition and a skilled firer, it is capable of effective fire out to 1,500 m range. It is also provided with a muzzle cup launcher for discharging various riot control devices, such as rubber balls, smoke canisters or CS gas grenades, by use of a standard grenade-discharging cartridge. It is not suited to firing combat types of grenade.

No data other than the weight, 3.7 kg, is available.

7.62 mm C-75 special forces rifle

Manufacturer
SANTA BARBARA, Julian Camarillo 32, E-28037 Madrid.

Status
Available.

Service
Spanish police and paramilitary organisations.

VERIFIED

SWEDEN

5.56 mm CGA5 assault and infantry combat rifles

Development
The Swedish Army began looking for a new, light, intermediate calibre rifle in the mid-1970s, with a view to the eventual replacement of the 7.62 mm Ak4, the Heckler and Koch G3. Most existing 5.56 mm weapons were tested for reliability, endurance, accuracy, handling and other capabilities. The results of these studies led to the elimination of most designs until only two remained, the licence-built FFV-890C (a slightly modified Israeli Galil) and the Belgian FN-FNC.

After comprehensive troop and technical trials in 1979-80 the FFV-890C was rejected in favour of the FN-FNC, which, it was considered, had 'developable characteristics'. Further tests were carried out in the ensuing years and the prototype FNC rifles were continuously modified to meet Swedish requirements.

A production order for the resultant design, the CGA5, was placed with FFV (now Bofors Carl Gustaf AB) in 1986, with deliveries commencing during 1988. Production is expected to continue until at least the end of the 1990s. The total Swedish armed forces requirement is for 250,000 weapons.

5.56 mm CGA5D rifle
1996

5.56 mm CGA5 rifle, known to the Swedish Armed Forces as the Ak5
1996

Description
During development of the 5.56 mm CGA5 (known by the military designation Automatkarbin 5, or Ak5) changes were introduced to meet Swedish requirements which included low-life cycle costs, high performance and reliability, and being able to remain functional under extreme climatic conditions. Modifications included a new butt stock and butt stock lock, bolt, extractor, handguard, gas block, sights, cocking handle, magazine and magazine catch, selector switch, trigger guard and sling swivels; the three-round burst facility was removed. Accessories available include a bipod, a 40 mm grenade launcher, butt extensions, a cleaning kit, a bayonet, and a blank firing adaptor.

The CGA5B is a special version produced to accept the Sight Unit Small Arms Trilux (SUSAT) optical sight on a rail mount. This version has a butt-stock with an ergonomic cheek support and lacks open sights. It is used by the Swedish Armed Forces as the Ak5B.

The CGA5C was a bullpup development which was not pursued.

The CGA5D was developed to mount virtually any type of sighting equipment and is equipped with a MIL-STD Picatinny Rail, an ergonomic cheek support and open sights. Models produced to date include a version with an integral sight of the red dot type and a ×1.5 or ×3 optical unit. CGA5Ds can be produced from new or by upgrading CGA5Bs.

Data

CGA5/Ak5

Cartridge: 5.56 × 45 mm NATO
Operation: gas, selective fire
Locking: rotating bolt
Feed: 30-round box magazine
Weight: empty, 3.9 kg; loaded with sling, 4.7 kg
Length: stock folded, 750 mm; stock extended, 1.005 m
Barrel: 450 mm
Rifling: 6 grooves, rh, 1 turn in 178 mm
Sights: fore, protected post; rear, flip aperture, 250 m and 400 m
Muzzle velocity: 930 m/s
Rate of fire: cyclic, 650-700 rds/min

Manufacturer

Bofors Carl Gustaf AB, S-631 87 Eskilstuna.

Status

In production.

Service

Swedish Armed Forces.

UPDATED

SWITZERLAND

SIG 7.62 mm SG510-4 rifle

SIG 7.62 mm SG510-4 rifle

Development

In the early 1950s SIG's director Rudolf Amsler produced the AM55 which was a blowback-operated rifle using a roller-action delay and was based to some extent on the German Sturmgewehr 45 assault rifle. This rifle was adopted by the Swiss Army as the Stgw 57, replacing the Schmidt Rubin 7.5 mm bolt-operated rifle.

The SIG SG510 series was developed from the AM55 and consisted of:

the SG510-1 firing the 7.62 × 51 mm NATO cartridge;

the SG510-2 firing the same cartridge but lighter in construction with a correspondingly higher recoil;

the SG510-3 firing the Soviet 7.62 × 39 mm cartridge and

the SG510-4 using the 7.62 × 51 mm NATO cartridge.

Only small quantities of the SG510-3 were produced and this version is not discussed further here. A brief note on the differences between the SG510-4 and the Stgw 57 service rifle follows this entry.

Description

The SG510-4 is a delayed blowback-operated rifle utilising a two-part block and delay rollers to hold up the breech face.

The body of the gun is made from pressings wrapped round and hard-soldered. The rear end of the receiver has been strengthened to take the butt, with the return spring, and the front end has a strengthened section to house the breech and the roller seatings. The roller seatings may be taken out and replaced when worn.

The barrel is made by the cold swaging or 'hammering' process. It has integral gas rings for grenade launching off the muzzle and a muzzle brake, which also acts as a flash suppressor, is incorporated. The rear part of the barrel is surrounded by a perforated casing. A bipod is suspended behind the foresight and can be stowed by folding the legs and rotating the bipod until the legs lie along the top of the barrel.

The foresight is a pillar and the rearsight an aperture. The rearsight has press locks which allow the frame to be moved up or down the sight bed to the selected range. The front sight mount can be moved across in a dovetail for lateral zeroing. For elevation zeroing an Allen key is used to screw the pillar in or out.

A long Arctic trigger is permanently fitted. It is also used for grenade firing. When not in use it hinges up flat under the receiver. When pushed down it has a short projecting arm which operates the normal trigger.

A blank-firing device can be fitted over the muzzle for training. A bayonet can be fitted to the lug under the collar below the foresight. A ring fits over the muzzle brake. The rifle will fire grenades using a blank cartridge. These are usually contained in a special magazine which takes the place of the normal one. This helps to ensure a ball round is not fired into the grenade.

A loaded magazine is placed in the magazine-well and pushed home until the catch clicks. The gun is loaded by pulling back and then releasing the T-shaped cocking handle; as the bolt comes back, the hammer is cocked and held by the sear. On its forward travel the bolt chambers the cartridge and the gun is ready for action. The bolt is a two-part type with a light bolt head and a heavier body. The bolt head carries two light rollers. These are of unusual shape since they are not simply cylindrical in section but have small pivoting pieces. These are engaged in the bolt head. When the round is chambered the bolt head movement ceases and the bolt body, moving forward, is able to pass between the rollers. The front face of the bolt body is wedge-shaped and the rollers are forced out into recesses in the receiver.

When the cartridge is fired the bullet is forced up the bore and the cartridge case is driven back. The chamber is fluted and the case floats on a film of gas. A proportion of these flutes extend the full length of the chamber and appear at the chamber face, so that upon firing a small amount of high pressure gas is allowed to leak and strike the bolt face. There are two small holes in the bolt face through which this gas passes, through the bolt head, to impinge on the bolt body. The pressure of the cartridge case and the gas on the breech face forces the rollers to the rear face of their recesses and the angle of the recesses forces the rollers inwards against the inclined faces of the bolt body. Because of the angle of the wedge and the pressure of the leaked gas, the bolt body is accelerated back and the two parts of the bolt are separated. When the rollers are clear of the recesses the entire bolt is drawn back with a displacement between the two parts. The empty case is held to the breech face by the extractor. The ejector is of an unusual type. It is a rocking system attached to the top of the bolt head and as the bolt recoils the ejector contacts a ramp on the left wall of the receiver and the case is pushed through the ejection slot on the right of the receiver. This method is less violent than the usual fixed ejector system.

The return spring is compressed. The bolt comes up against the rear plate and the compressed return spring forces it forward. The bolt picks up the next round and chambers it.

Accessories include a telescopic sight, a blank-firing attachment, a cleaning kit, night sights, and an anti-tank grenade sight.

Data

SG510-4

Cartridge: 7.62 × 51 mm NATO
Operation: delayed blowback, selective fire
Feed: 20-round magazine
Weight: empty, 4.25 kg
Length: 1.016 m
Barrel: 505 mm
Rifling: 4 grooves, rh, 1 turn in 305 mm
Sights: fore, post; rear, aperture
Sight radius: 540 mm
Muzzle velocity: 790 m/s
Rate of fire: cyclic, 600 rds/min

Manufacturer

SIG Swiss Industrial Company, CH-8212 Neuhausen-Rheinfalls.

Status

No longer in production.

Service

In service as Stgw 57 with Swiss Army (see following entry), Chile and Bolivia.

VERIFIED

7.5 mm Stgw 57 service rifle

Description

The 7.5 mm Stgw 57 is the Swiss Army version of the SIG SG510-4 from which it differs in calibre and in several minor respects. The rifle is used with the Swiss 7.5 × 55 mm cartridge (Swiss Army designation GP 11) and has a rubber butt-stock, folding sights and a special bipod which can be used to support the weapon at either the rear or the front of the jacket.

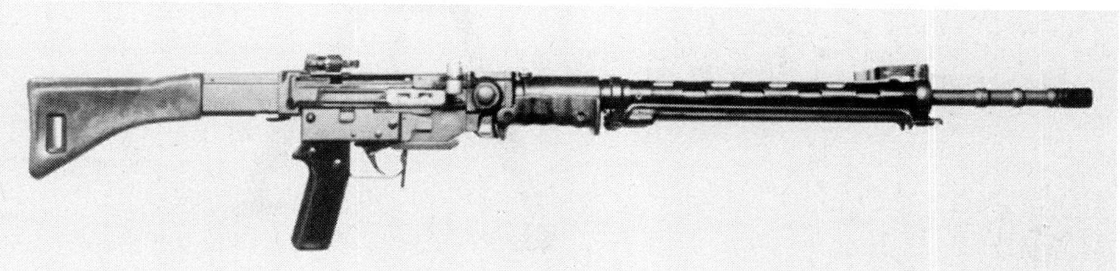

7.5 mm Stgw 57 Swiss service rifle

For a general description of the weapon mechanism reference should be made to the preceding entry for the SIG SG510-4. The Stgw 57 proved to be a very reliable weapon even under the extreme conditions that can be encountered in Switzerland.

Manufacturer
SIG Swiss Industrial Company, CH-8212 Neuhausen-Rheinfalls.

Status
Production complete. Being replaced by the Stgw 90.

Service
Swiss armed forces.

VERIFIED

SIG-Sauer SSG 2000 sniper rifle

Description
The SIG-Sauer SSG 2000 rifle was purpose built for military, law enforcement and target marksmanship. The weapon is based on the bolt action of the Sauer 80/90 repeating rifle. This bolt-action uses hinged lugs at the rear end of the bolt which are driven outwards, to lock into the receiver, by the action of cams driven by the rotation of the bolt handle. The bolt body is non-rotating, giving very positive case extraction. This design results in a reduction in the angular travel of the bolt to only 65° and gives a fast and smooth loading action.

The heavy hammer-forged barrel is equipped with a combination flash hider and muzzle brake, providing excellent weapon recovery and allowing a fast, controlled, follow-up shot.

The ergonomic thumb-hole stock can be optimally adjusted to suit the firer; right- and left-hand stocks are available.

The trigger is of the double set pattern. The sliding safety catch has a triple function: blocking of the sear, the sear pivot and the bolt itself. The bolt can be opened when the safety catch is applied; the set trigger can be decocked by pulling the trigger when the weapon is in the safe condition, and it is automatically decocked when the bolt is opened. There is a signal pin to indicate when a round is in the chamber.

No iron sights are fitted; standard equipment includes a Schmidt & Bender ×1.5–×6 × 42 or Zeiss Diatal ZA 8 × 56T telescope sight.

The SIG-Sauer SSG 2000 is available in 7.62 × 51 mm, 0.300 Weatherby Magnum, 5.56 × 45 mm and 7.5 × 55 mm Swiss calibres; the data which follows is for the 7.62 × 51 mm version.

SIG-Sauer SSG 2000 sniper rifle

Data
Cartridge: 7.62 × 51 mm (also see text)
Operation: bolt action, single shot
Feed: 4-round box magazine
Weight: with magazine and sight, 6.6 kg
Length: 1.21 m
Barrel: excluding flash hider, 610 mm
Rifling: 4 grooves, rh, 1 turn in 305 mm
Trigger pull: standard, 18 N; set, 3 N
Muzzle velocity: 750 m/s, depending upon ammunition
Muzzle energy: 3,500 J, depending upon ammunition

Manufacturer
SIG Swiss Industrial Company, CH-8212 Neuhausen-Rheinfalls.

Status
In production.

Service
Swiss police forces, Argentine Gendarmerie National, Police Force of Kingdom of Jordan, Royal Hong Kong Police, Malaysian Police Kuala Lumpur, Combined Service Forces of Taiwan, some UK police forces, and others.

VERIFIED

SIG-Sauer 7.62 mm SSG 3000 sniper rifle

Description
The SIG-Sauer 7.62 mm SSG 3000 sniper rifle is a precision bolt-action repeating rifle incorporating the latest firearms technology developed from the base of the Sauer 200 STR target rifle. The rifle is on a modular system: the barrel and receiver are joined by screw clamps, the trigger system and magazine form a single unit which fits into the receiver, and the stock is of non-warping wood laminate. The receiver housing is machined from a single billet; lock-up, with six locking lugs, is between bolt and barrel, thus the stress on the bolt upon firing is not transmitted to the receiver. A light firing pin and short striking distance endow the rifle with an exceptionally short lock time. The barrel is heavy, cold-swaged, and fitted with a combined muzzle brake and flash suppressor.

Two precision trigger actions, adjustable for both length and take-up weight, are available, these being single-stage or double-stage patterns. The sliding safety catch, above the trigger, locks the trigger, firing pin and bolt. A signal pin at the end of the striker head indicates whether the rifle is cocked.

The stock was ergonomically designed, allowing the weapon to be held in position for extended periods without fatigue. The butt-plate can be adjusted for height, length, offset and rake, and the entire system - stock, bolt and receiver - is available as a completely left-handed version. A rail integrated into the fore-end accepts the bipod and the carrying or firing sling.

There are no iron sights, the rifle being intended solely for use with a telescope; the normal mount is designed to operate with the recommended sight, the Hensoldt ×1.5–×6 × 24BL, which is manufactured specifically for the SSG 3000 rifle, but a NATO STANAG sight base is available if preferred.

For practice purposes a 0.22 LR conversion kit is available.

SIG-Sauer 7.62 mm SSG 3000 sniper rifle

Data
Cartridge: 7.62 × 51 mm NATO
Operation: bolt action, single shot
Feed: 5-round box magazine
Weight: with empty magazine and without sight, 5.4 kg
Length: ca 1.18 m (depending upon stock adjustment)
Barrel: excluding flash suppressor, 610 mm
Rifling: 4 grooves, rh, 1 turn in 305 mm
Muzzle velocity: 750 m/s, depending on ammunition
Muzzle energy: 3,500 J, depending on ammunition

Manufacturer
SIG Swiss Industrial Company, CH-8212 Neuhausen-Rheinfalls.

Status
In production.

VERIFIED

SIG 540 Series assault rifles

Development

The French firm Manurhin was licensed to manufacture three types of SIG rifles, the SG540, 542 and 543. The SG540 and 543 are chambered for the 5.56 × 45 mm cartridge and the SG542 for the 7.62 × 51 mm NATO round; but all three were made to the same basic design thus giving them a high parts commonality. In 1988 the licence operated by Manurhin was transferred to INDEP of Portugal, but at present the only licensed production is in Chile, where FAMAE manufacture the SG542.

Description

The SIG 540 Series assault rifles are gas operated with a rotating bolt. The cocking handle reciprocates with the bolt and provides a means of closing the bolt if for some reason the return spring is unable to force it fully home. The gas regulator is mounted at the front of the cylinder and has a milled cap which can be rotated to one of the positions marked respectively 0, 1 and 2. At '0' the valve is fully closed, no gas passes to the cylinder, and the entire gas force is used to project a grenade from the muzzle. The normal firing position is with the regulator set to '1', and '2' is an oversized port reserved for those occasions when the weapon is lacking in energy caused by fouling or the entry of sand or snow, for example.

The weapon can be fired at single shot, full automatic or with a three-round burst controller which provides for the release of three rounds at a cyclic rate of 725 rds/min from a single trigger operation. When firing at single shot, accuracy is improved by the provision of a double-pull pressure point trigger.

When the weapon is used in extremely cold conditions demanding the use of Arctic mittens, the trigger guard can be rotated so that the hand has direct access to the trigger. It is similarly rotated when rifle grenades are fired.

When the ammunition is expended the magazine platform operates a hold-open device. When a new, loaded, magazine is inserted the action can be released by a slight backward movement of the cocking handle and when this is released the bolt will chamber the next round. Alternatively the device may be released by pressing up a catch on the left of the receiver above and behind the magazine.

Data

	SG540	SG542	SG543
Cartridge	5.56 × 45 mm	7.62 × 51 mm	5.56 × 45 mm
Operation		gas, selective fire	
Locking		rotating bolt	
Feed		magazine	
WEIGHTS			
Rifle			
with fixed butt	3.26 kg	3.55 kg	2.95 kg
with folding butt	3.31 kg	3.55 kg	3 kg
Additional for bipod	0.28 kg	0.28 kg	–
Empty magazine			
20 rounds	0.2 kg	0.24 kg	0.2 kg
30 rounds	0.24 kg	0.35 kg	0.24 kg
Full magazine			
20 rounds	0.43 kg	0.73 kg	0.43 kg
30 rounds	0.585 kg	1.085 kg	0.585 kg
LENGTHS			
Overall			
with fixed butt	950 mm	1 m	805 mm
with folding butt	720 mm	754 mm	569 mm
Barrel			
without suppressor	460 mm	465 mm	300 mm
MECHANICAL FEATURES			
Barrel			
rifling	6 grooves	4 grooves	6 grooves
		rh, 1 twist in 305 mm. 2-position regulator	
Sights		fore, cylindrical post; rear, aperture	
Sight radius	495 mm	528 mm	425 mm
FIRING CHARACTERISTICS			
Muzzle velocity	980 m/s	820 m/s	875 m/s
Rate of fire cyclic,		650/800 rds/min	
Effective range		600 m	

The foresight is a pillar which may be rotated up or down to allow for elevation zeroing. The rearsight is an aperture in a tilted drum which may be rotated to give ranges of 100 to 500 m for the 5.56 mm rifle and 100 to 600 m on the SG542. Each rifle has range alteration in increments of 100 m.

A bipod is mounted under the front of the barrel casing. This is normally stowed under the barrel and is pulled down when needed.

The flash suppressor is the closed prong-type and when grenades are used these are slid straight down on to the barrel. The spike bayonet has a tubular handle which fits over the flash suppressor.

A telescopic sight, an infra-red sight or an image intensifier can be fitted.

The use of plastics for the butt, trigger guard, pistol grip and fore-end has reduced the weight and cost of the weapon without detracting from its efficiency in any way.

A conventional butt or a tubular one folding flat alongside the right side of the receiver can be fitted.

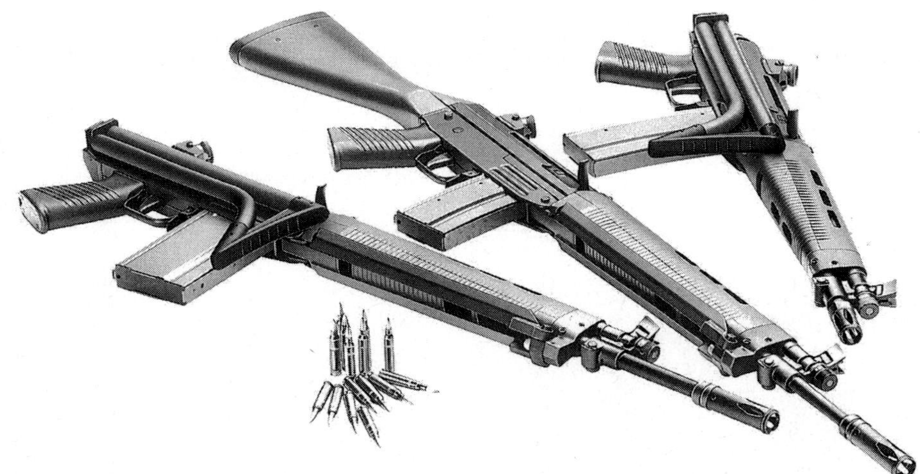

From left, 5.56 mm SG540 with butt folded, 7.62 mm SG542, and 5.56 mm SG543 (right)

Manufacturer

SIG Swiss Industrial Company, CH-8212 Neuhausen-Rheinfalls.

Status

Production in Chile only (see text).

Service

Bolivia, Burkina Faso, Chad, Chile, Djibouti, Ecuador, France, Gabon, Indonesia, Ivory Coast, Lebanon, Mauritius, Nicaragua, Nigeria, Oman, Paraguay, Senegal, Seychelles, Swaziland.

Licence production

CHILE
Manufacturer: FAMAE
Type: 5.56 mm SG540-1 and 7.62 mm SG542-1
Remarks: See entry under *Chile* for details.

VERIFIED

SIG 5.56 mm SG550 and SG551 (Stgw 90) assault rifles

Description

The SIG 5.56 mm SG550 and SG551 assault rifles were developed to meet a Swiss Federal Army specification and were accepted as the Swiss Army assault rifle in early 1984 under the type designation Stgw 90.

There are two main models. The SG550 is the standard model with a folding butt and bipod while the SG551 is shorter and lacks the bipod. There is also a SG551 SWAT (see under *Variants*). The SG550/551 SP is a semi-automatic sporting rifle produced for the Swiss civilian market - over 20,000 have been sold.

Description

In designing the 5.56 mm SG550 and SG551 SIG paid particular attention to weight saving, making extensive use of plastics for the butt, handguard and magazine. The magazine is made of transparent plastic, allowing a check to be kept on the ammunition supply even when firing. It is provided with studs and lugs on the side which allow a number of magazines to be clipped together so that changing from an empty to a full magazine is greatly facilitated. The butt folds to one side and it is claimed that, even in this configuration, the weapon's optimum weight distribution and ergonomically convenient shape allow accurate firing in all positions. SIG states that accuracy at 300 m is equal to that of the Stgw 57.

The SG550 is provided with a three-round burst controller and a trigger guard which can be rotated when rifle grenades are fired or so that a user wearing Arctic mittens may have direct access to the trigger. When the last round in the magazine is fired a hold-open device operates; a new magazine is inserted, the lock catch is actuated and the weapon is ready to fire once more.

The combined dioptre and alignment sight mounted on the breech housing is adjustable for traverse and elevation. The alignment sight has luminous spots for aiming during night firing, and when the daylight sight is adjusted the night firing sight is adjusted simultaneously. The foresight, with its tunnel and swing-up night firing sight with luminous spot, is permanently mounted. Telescopic and infra-red sights can be fitted to an integral telescope mount; this is dimensioned to Swiss Army standards, although a mount to NATO STANAG 2324 can be fitted if required.

Variants

SG551 SWAT

This is a special forces or law enforcement model virtually identical to the service SG551 apart from a few details such as a revised butt and provision for mounting various models of optical sight. Sights available include the Hensoldt 6 × 42 BL scope (specially developed for this rifle) or the Trijicon ACOG 3.5 × 35 combat scope. The latter scope may be fully integrated with the rifle. Tactical accessories such as a flashlight or a 40 mm M203 or similar grenade launcher are available; as are lightweight transparent magazines holding 5, 20 or 30 rounds.

SG550 sniper rifle

See following entry.

Data

Cartridge: 5.56 × 45 mm
Operation: gas, selective fire
Locking: rotating bolt
Feed: 20- or 30-round detachable box magazine
Weight: SG550, with empty magazine and bipod, 4.1 kg; SG551 with empty magazine, 3.4 kg
Weight of loaded magazine: 20-round, 340 g; 30-round, 475 g
Length: overall, butt extended; SG550, 998 mm; SG551, 827 mm; SG550, 772 mm; SG551, 601 mm
Barrel: SG550, 528 mm; SG551, 372 mm
Rifling: 6 grooves, rh
Sight radius: SG550, 540 mm; SG551, 466 mm
Muzzle energy: SG550, 1,700 J; SG551, 1,460 J
Rate of fire: cyclic, ca 700 rds/min

Manufacturer

SIG Swiss Industrial Company, CH-8212 Neuhausen-Rheinfalls.

Status

In production.

Service

Swiss Army.

UPDATED

SIG 5.56 mm SG551 SWAT rifle with Henso1dt scope and flashlight

1996

SIG 5.56 mm SG550 assault rifle, the Swiss Army's Stgw 90

SIG 5.56 mm SG551 assault rifle

SIG 5.56 mm SG550 assault rifle

SIG 5.56 mm SG550 sniper rifle

Description

The SIG 5.56 mm SG550 sniper rifle was developed from the standard SG550 assault rifle and is a semi-automatic version developed in close co-operation with special police units. The accuracy is the result of the heavy hammer-forged barrel and low recoil, and there is a sensitive double-pull trigger. The semi-automatic action gives the sniper the ability to fire a very fast second shot if necessary.

The rifle carries a fully adjustable bipod and folding butt, the pistol grip is adjustable for rake and has an adjustable hand-rest. The sighting telescope can be altered alongside and at right-angles to the sight base in such a way that the head leans naturally against the cheek rest and takes up the correct position relative to the line of sight. There are no iron sights fitted.

Accessories include a spare magazine, a sling, a mirage band, a bipod, carrying case, a cleaning kit, and sight cleaning equipment.

5.56 mm SG550 sniper, with telescope sight

Data

Cartridge: 5.56 × 45 mm
Operation: gas, semi-automatic
Locking: rotating bolt
Feed: 20- or 30-round box magazine
Weight: empty, without magazine, 7.02 kg
Weight of magazine: empty, 95 g; with 20 rounds, 340 g; with 30 rounds, 480 g
Length: butt folded, 905 mm; butt extended, 1.13 m
Barrel: 650 mm
Rifling: 6 grooves, rh, 1 turn in 254 mm
Trigger pull: first pressure, 800 g; second pressure, 1.5 kg
Muzzle energy: ca 1,820 J, depending on ammunition

Manufacturer

SIG Swiss Industrial Company, CH-8212 Neuhausen-Rheinfalls.

Status

In production.

Service

Swiss police; Kingdom of Jordan police.

VERIFIED

TAIWAN

5.56 mm Type 65 assault rifle

Description

The 5.56 mm Type 65 assault rifle generally resembles the American M16 and the firing characteristics are markedly similar, as are many of the dimensions. The rifle is made entirely in Taiwan, probably on tools and machinery intended for the M16.

The lower receiver was taken directly from the M16A1, while the bolt, bolt carrier and piston-type gas system come from the AR-18. Initial Type 65 prototypes were made with stamped steel receivers, but later weapons were machined from blocks of aluminium.

The barrel is the same length as the M16 and gives a similar muzzle velocity with M193 ammunition. The receiver has the cocking handle of the original M16 without the bolt-closure device incorporated in the M16A1. The general shape of the receiver pressings is all but identical to the M16, even to the Arctic trigger guard. The missing feature is the carrying handle, and the rearsight is mounted on a substantial bracket in place of the rear part of the carrying handle. The foresight appears to be the same as the M16, as does the plastic butt. 20- or 30-round M16 magazines are used.

A major recognition feature is the long plastic handguard which is somewhat larger than that on the M16. A light bipod can be fitted to the barrel under the foresight as required, though this is not a normal fixture.

Taiwan 5.56 mm Type 65 assault rifle (foreground) compared with Types 74 and 75 machine guns

A modified version of the original Type 65 design incorporates the M16A1-type tube gas system and a more reliable, single-piece bolt assembly. It also has a higher rate of fire and can use transparent plastic magazines. This later model was previously (and incorrectly) referred to as the Type 68.

Although the Type 65 initially proved to be neither reliable nor easy to manufacture, much effort was put into eradicating the various problems. The Type 65K1 incorporated modifications, which notably improved the hardness of the aluminium buffer and the adequacy of the heat insulation beneath the handguard. This was followed by the Type 65K2, which is rifled one turn in 178 mm to accept the SS109 bullet and also has a three-round burst facility in addition to full-automatic fire. The gas system was also redesigned in order to slow up the primary ejection of the spent case and thus overcome problems of case ejection which had occurred with the earlier models. The Type 65K2 underwent extensive testing and is in production.

The bayonet used on the Type 65 rifle is different from the standard US M7. It is 209 mm long overall, 38 mm longer than the M7. Fully bladed on the lower edge and 120 mm bladed on the upper edge, the Taiwanese bayonet has a lined handle for a secure grip. It is designated the Type 65 bayonet.

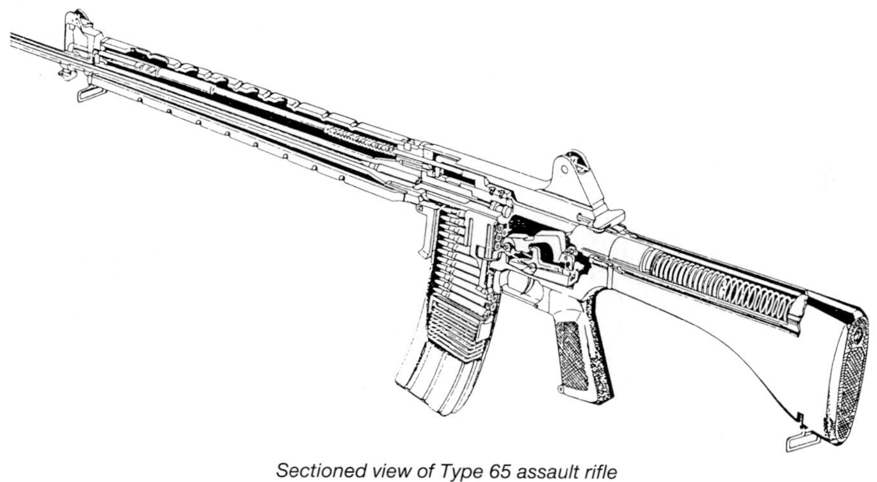

Sectioned view of Type 65 assault rifle

Data

Cartridge: 5.56 × 45 mm M193 or SS109 (see text)
Feed: 20- or 30-round M16 detachable box magazine
Weight: 3.17 kg
Length: 990 mm
Barrel: 508 mm
Rifling: 4 grooves, rh, 1 turn in 178 mm or 304 mm (see text)
Muzzle velocity: 990 m/s
Muzzle energy: 1,745 J
Rate of fire: cyclic, 700-800 rds/min
Effective battle range: 400 m

Manufacturer

Combined Service Forces, Hsing-Ho Arsenal, Kaohsiung.

Status

In production.

Service

In service with Taiwan Army Airborne/Special Forces, Military Police, and Taiwan Marine Corps.

VERIFIED

TURKEY

7.62 mm G3 rifle

Description
This is the Heckler and Koch G3 rifle (described under *Germany*) manufactured under licence in Turkey. From information supplied by the makers it appears that different manufacturing methods have resulted in slight changes in dimensions resulting in the Heckler and Koch designation of G3A7. The Turkish G3 is issued with a locally designed bayonet which fits above the muzzle. Currently, both G3A3 (fixed butt) and G3A4 (telescopic butt) versions are manufactured.

Data
Cartridge: 7.62 × 51 mm NATO
Operation: delayed blowback, selective fire
Feed: 20-round box magazine
Weight: empty, 4.25 kg; with full magazine, 5 kg
Length: 1.02 m
Barrel: 450 mm
Muzzle velocity: ca 780-800 m/s
Rate of fire: 480-620 rds/min
Effective range: 400 m

Manufacturer
Makina ve Kimya Endüstrisi Kurumu (MKEK), 06330 Tandogan, Ankara.

Licence-built G3A3 and G3A4 rifles by MKEK

Status
In production.

Service
Turkish armed forces.

VERIFIED

UNITED KINGDOM

7.62 mm L1A1 rifle

Development
The 7.62 mm Rifle L1A1 is a gas-operated self-loading rifle adapted from the very successful Belgian Fusil Automatique Léger (FAL) rifle, with modifications to suit the special requirements of the British forces. Most of the modifications were carried out at the former Royal Small Arms Factory at Enfield Lock. Production has ceased in the UK although an essentially similar weapon is still in production in India (qv). The L1A1 is no longer in service with the UK armed forces but is still held in reserve and is issued on occasion to the Royal Ulster Constabulary. It remains in widespread use by many other nations to whom it was either sold or provided as military aid.

Description
While the gas-operated 7.62 mm L1A1 rifle was normally fitted for the self-loading single-shot mode, minor component changes in the trigger mechanism enabled the rifle to be fully automatic. The bolt carrier was provided with oblique cuts on its outer surface which were designed to scour loose dirt from the inside of the receiver and eject it through the ejection port during firing. The cocking handle was fitted on the left-hand side of the weapon, allowing the right hand to remain on the trigger when cocking for firing. Both the cocking handle and the carrying handle fold down when not in use.

The L1A1 remained the standard British infantry rifle until 1985. Most rifles were fitted with plastic furniture although some early production examples used hardwood: the butt could be varied in length by using one of four different butt-plates, thus enabling the weapon to be adjusted to suit the stature of the individual firer.

Data
Cartridge: 7.62 × 51 mm
Operation: gas, single shots
Locking: tilting block
Feed: 20-round box magazine
Weight: empty, 4.3 kg; with full magazine, 5 kg
Length: 1.143 m
Barrel: 554 mm
Rifling: 6 grooves, rh, 1 turn in 305 mm
Sights: fore, blade; rear, aperture; Sight Unit Infantry Trilux (SUIT) may be fitted
Sight radius: 554 mm
Muzzle velocity: 838 m/s
Effective range: 600 m with SUIT

Manufacturer
Heckler and Koch (UK), British Aerospace Defence Limited, Kings Meadow Road, Nottingham NG2 1EQ.

Status
No longer manufactured in United Kingdom.

Service
Australia, Barbados, Canada, Gambia, Guyana, India, Malaysia, New Zealand, Oman, Royal Ulster Constabulary, and Singapore. Made under licence in Australia, Canada and India.

UPDATED

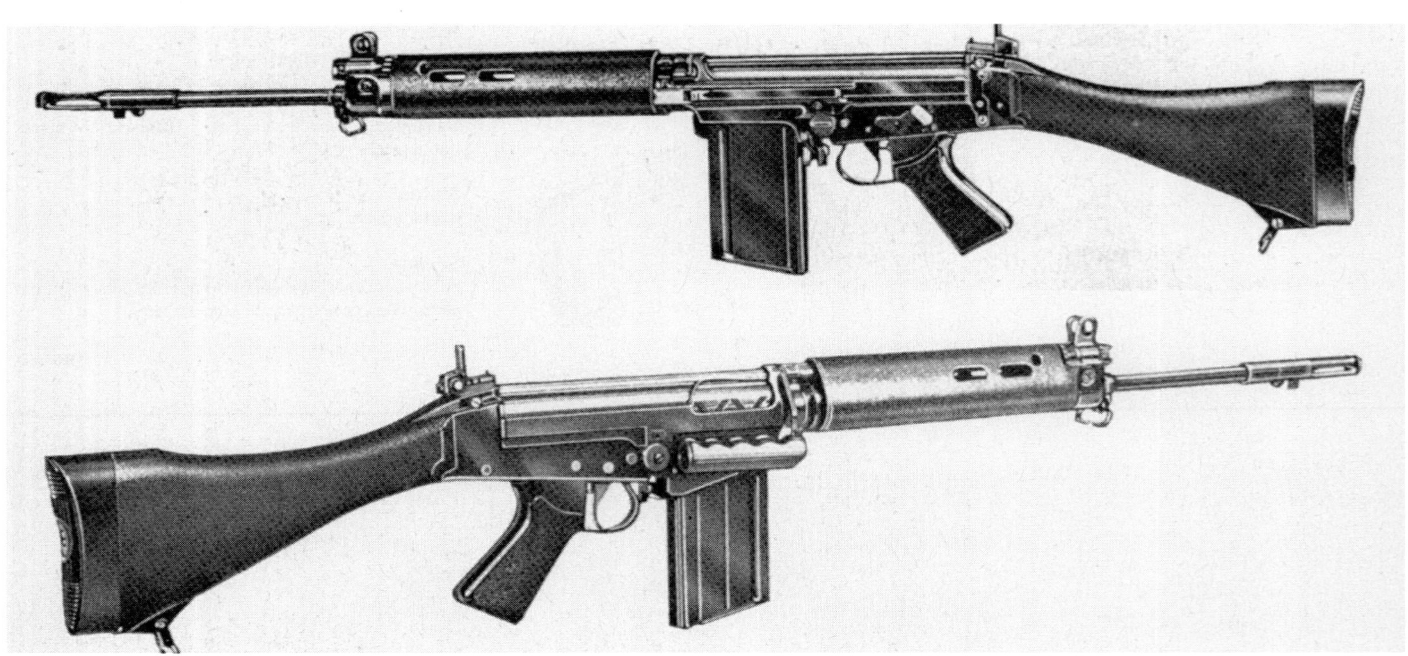

7.62 mm L1A1 rifle

7.62 mm Enfield L39A1 rifle

Development

The 7.62 mm L39A1 rifle was introduced to provide a satisfactory target rifle for competitive shooting for units of the British forces which normally would be equipped with the 7.62 mm L1A1 rifle. The L1A1 rifle was unsuitable for serious target shooting and the decision was made to produce target rifles from 0.303 No 4 rifles held in ordnance depots.

Description

The 7.62 mm L39A1 is a manually operated bolt-action single-shot rifle with a heavy 7.62 mm barrel, and with the necessary modifications carried out on the extractor and receiver to permit the use of 7.62 mm ammunition.

The barrel is produced using the cold-forging process. A foresight block is soldered to the barrel but no foresight is fitted. Commercial target sights are fitted by the using unit.

The barrel projects from the wooden fore-end for 381 mm. The butt is unchanged from the No 4 rifle except that a recess has been machined under the knuckle to take a container holding spare foresight blades. The handguard is the same pattern as that on the No 8 0.22 rifle.

The rifle is proved for 7.62 mm ammunition which produces a higher chamber pressure than the 0.303 Mark VII and, in addition to the usual military proof marks of crown, ER and crossed flags with the letter P, the receiver, bolt and bolt head are stamped 19T. There are four sizes of bolt head available, marked 0, 1, 2 and 3 in ascending size, for ensuring correct cartridge head space.

The prime purpose of the weapon is competition shooting, and since rounds are usually hand fed into the chamber the only service required from the magazine is the provision of a platform to support the round while this is done. For this purpose the original 0.303 magazines are retained. There is no positive ejector when this magazine is used. A groove in the left of the receiver shallows towards the rear and the friction between the side of the case and the groove tips the mouth of the case to the right and the extractor loses its grip. As an alternative a 7.62 mm magazine holding 10 rounds can be provided. It has an ejector plate spot-welded to the lip at the left rear.

Data

Cartridge: 7.62 × 51 mm
Operation: manual, single shot
Locking: rotating bolt
Feed: 10-round box magazine
Weight: 4.42 kg
Trigger pull: first pull, 1.13-1.59 kg; second pull, 1.81-2.04 kg
Length: 1.18 m
Barrel: 700 mm
Rifling: 4 grooves, rh, 1 turn in 305 mm
Sights: competition sights attached as required by user
Muzzle velocity: 841 m/s

Manufacturer

Heckler and Koch (UK), British Aerospace Defence Limited, Kings Meadow Road, Nottingham NG2 1EQ.

Status

No longer manufactured.

Service

UK armed forces.

VERIFIED

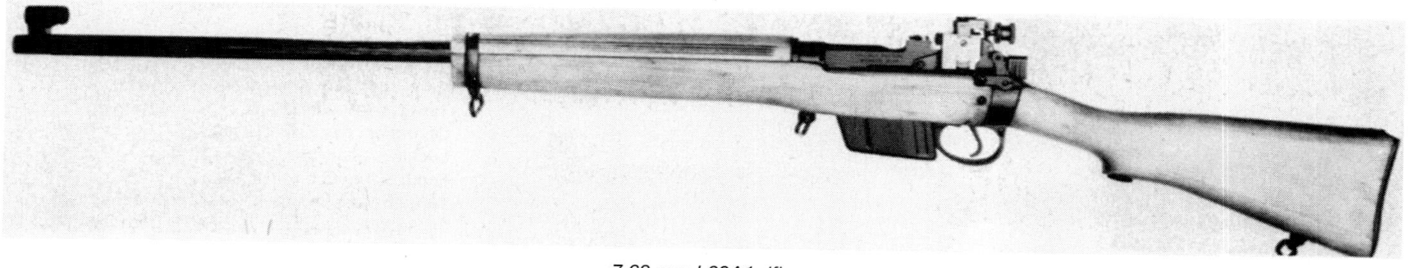

7.62 mm L39A1 rifle

5.56 mm L85A1 Individual Weapon

Development

The two 5.56 mm automatic weapons in the Enfield Weapon System (also known as SA-80) are the L85A1 IW (Individual Weapon - assault rifle) and the L86A1 light support weapon (LSW) which is a light machine gun (refer to entry in *Machine guns*). Both weapons fire the same 5.56 × 45 mm NATO ammunition and also use a majority of common components which gives increased flexibility, reduces spares requirements and simplifies maintenance in service.

The British Army search for small calibre cartridges and a rifle to fire them can be traced back to before 1914. During the 1970s much time was spent investigating a 6.25 × 43 mm cartridge before the decision was taken to adopt a 4.85 × 49 mm round along with a new weapon, known as the Individual Weapon or IW, the XL65E5. The 4.85 mm ammunition/IW combination performed well but ran into the problem that for the rest of NATO 4.85 mm was non-standard. Most NATO nations favoured a move to the American 5.56 × 45 mm round to take advantage of mass production availability, especially as the USA had already adopted the 5.56 mm M193 cartridge and the M16/AR-15 rifle to go with it. Most European nations other than the UK were already developing their own designs of 5.56 mm rifles.

The result was a series of weapon and ammunition trials to determine the future NATO standard round. As far as the UK was concerned there were two results. One was that their 4.85 × 49 mm cartridge fell by the wayside. The second was the adoption of a Belgian FN cartridge known as the 5.56 × 45 mm SS 109.

5.56 mm SA-80 Carbine

There was no agreement on the associated weapon - every NATO nation decided to go ahead with its own local design or preference. The IW XL65E5 was thus revamped to accommodate the 5.56 × 45 mm NATO, and became the XL70E3. The XL70E3 rifle was visually much different from the XL65E5 as at the same time many design changes were incorporated to allow the new model to be produced on up-to-date automated assembly lines. Troop trials followed and further design changes were incorporated before the decision was taken to adopt the fully fledged Rifle, 5.56 mm L85A1.

Production was initially undertaken at the old Royal Small Arms Factory at Enfield Lock before the production line was switched during 1988 to Royal Ordnance's Nottingham Small Arms Facility (NSAF), now Heckler and Koch (UK), a division of Royal Ordnance.

Production for the British Army ceased during 1994 after a total of 323,920 IWs had been produced. Development of the basic design continues. Recent innovations have included a three-position gas port (trialled environmentally in Alaska and Australia), a detachable 40 mm grenade launcher (see entry in *Light support weapons* section, and a revised ×3 or ×1.5 magnification optical sight assembly (see entry in *Sighting equipment* section).

5.56 mm L85A1 individual weapon, infantry version with SUSAT sight

5.56 mm L85A1 individual weapon with secondary iron sight system, comprising rearsight/carrying handle (adjustable in azimuth) and foresight (adjustable for elevation)

Description

The L85A1 Individual Weapon (IW) is made from steel, using the modern processes of pressing and welding, CNC and conventional machining, and is fully proofed and tested using computer-aided systems. The furniture is plastic, using high-impact nylon. Stripping and assembly is straightforward and can be done without using special tools.

The L85A1 is a conventional gas-operated rifle locked by a rotating bolt engaging in lugs behind the breech and carried in a machined carrier running on two guide rods; a third rod controls the return spring. The cocking handle is on the right side and has a cover which is spring-loaded to open when the carrier moves to the rear. The gas regulator has three positions, a normal opening for most firing, a large opening for use in adverse conditions and a closed position for grenade firing. The trigger is in front of the magazine and there is a long connecting rod running to the mechanism. On the left side is a selector lever which is set either to semi-automatic or automatic fire.

The L85A1 IW can be fitted with a robust, high-performance optical sight (SUSAT) of ×4 magnification which enables the weapons to be used operationally under poor light conditions and is also useful for surveillance. The sight is mounted on a bracket which incorporates range adjustment and zeroing. This bracket slides on to a dovetail sight base permanently fitted to the weapon body and which allows sight position adjustment. An emergency open sighting system is also included, being permanently fitted to the body of the primary optical sight. In the British Army the optical sight is issued only to front-line combat troops, other users employing 'iron' sights consisting of a double aperture rearsight housed in a carrying handle mounted on the dovetail sight base, with a foresight blade on the gas block.

An alternative optical sight of ×1.5 power, using a black ring reticle optimised for battle ranges, is also available.

The L85A1 IW can be fitted with image intensifying sights by dismounting the SUSAT from its dovetail base.

The L85A1 IW was designed to be simple to dismantle, without special tools, into the main subassemblies for cleaning and maintenance. The trigger mechanism is a self-contained assembly in a pressed steel housing which is located to the main weapon body by two pins and a small butt-plate. The main body is a steel pressing which houses the bolt and carrier assembly and guide rods which locate in the barrel extension welded into the body and into which the barrel is screwed.

A number of accessories are available including a sling, bayonet, blank-firing attachment, cleaning kit and a multipurpose tool. An adaptor kit can be fitted to allow the firing of 0.22 LR rimfire ammunition.

5.56 mm L85A1 individual weapon, showing major subassemblies and component parts, bayonet and scabbard and a 30-round magazine

Variants

SA-80 Carbine

A SA-80 Carbine was developed to be a short and handy counterpart to the IW. Most details are the same as for the L85A1 IW but the overall length is reduced to 709 mm, with the barrel length being 442 mm. Weight without magazine and iron sights is 3.71 kg; weight loaded with iron sights is 4.42 kg.

Cadet GP Rifle L98A1

This training rifle uses major components of the L85A1 but without the gas actuating system, thus converting the weapon into a manually operated single-shot rifle. It uses the secondary iron sight system as standard, has no provision for launching grenades and can be fitted with an adaptor to allow firing of 0.22 rimfire ammunition.

Data

INDIVIDUAL WEAPON

Cartridge: 5.56 × 45 mm NATO standard
Types: ball, tracer, blank, low power training
Operation: gas, selective fire
Locking: rotary bolt, forward locking
Feed: 30-round box magazine
Weight: weapon without magazine and optical sight, 3.8 kg; with loaded magazine and optical sight, 4.98 kg
Length: 785 mm
Barrel: 518 mm
Rifling: 6 grooves, rh, 1 turn in 177.8 mm
Trigger pull: 3.12-4.5 kg
Muzzle velocity: 940 m/s
Rate of fire: cyclic, 610-775 rds/min

Manufacturer

Heckler and Koch (UK), British Aerospace Defence Limited, Kings Meadow Road, Nottingham NG2 1EQ.

Status

Production complete for British Army but development continuing.

Service

In service with the British armed forces and some other nations, including Jamaica.

UPDATED

7.62 mm Parker-Hale Model 82 sniper rifle

Description

The Model 82 is a bolt-action rifle with a four-round magazine and the option of a variety of sighting systems. It is intended for use by military and security forces as a precision weapon capable of engaging a point target with a 99 per cent chance of a first-round hit at all ranges out to 400 m in good light or to the range limits of the sights employed when fitted with a passive night vision infantry weapon sight. It is also designed for use as a marksman training and competition rifle when fitted with aperture sights.

The action is a Mauser 98-type with an internal magazine and a positive non-rotating extractor: twin locking lugs engage the front receiver ring and a third safety lug engages the rear receiver ring. The heavy, free-floating barrel, weighing almost 2 kg, is of chrome-molybdenum steel, with rifling cold forged to increase tensile strength by 5 to 10 per cent and improve wear characteristics.

The trigger mechanism is a separate self-contained assembly with full adjustment for alteration of trigger pull and wear. The safety locks the trigger, bolt and sear in a unique triple action. The butt length is adjustable by detachable spacers. A detachable, folding bipod with spring-loaded extending legs is available.

As with all sniper rifles the ultimate performance from the weapon depends to a great extent on using the correct ammunition.

Data

Cartridge: 7.62 × 51 mm, selected batches
Operation: bolt action, single shot
Weight: unloaded, 4.8 kg; barrel, 1.93 kg
Length: 1.162 m
Barrel: 660 mm
Rifling: 4 grooves, rh
Sights: fitted with iron target sights at the factory; optical units may be fitted as desired to the machined dovetail on the receiver

Manufacturer

Parker-Hale Limited, Bisley Works, Golden Hillock Road, Birmingham B11 2PZ.

Status

No longer in production.

Service

In service with Australia, Canada and New Zealand.

VERIFIED

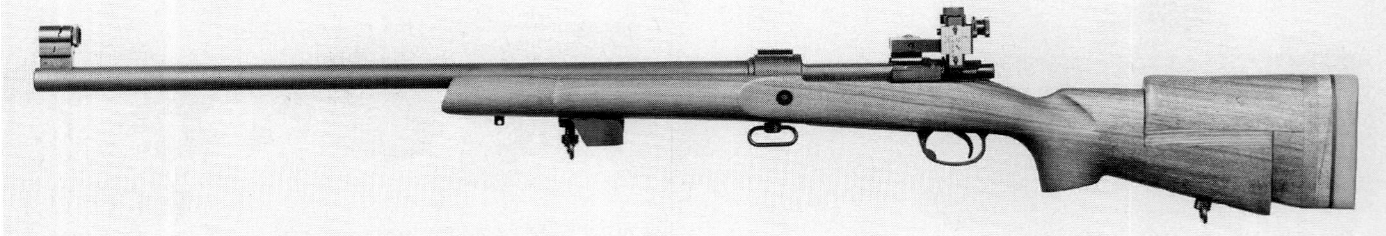

7.62 mm Parker-Hale Model 82 sniper rifle

7.62 mm Parker-Hale Model 83 target rifle

Description

This rifle was developed by Parker-Hale from their PH1200TX target rifle. It is a single-shot rifle, the absence of a magazine and feedway allowing the action to be stiffened and thus improving accuracy.

The rifling has been specially matched to the standard 144 grain 7.62 mm NATO bullet and is handbedded with Devcon F metal compound for the entire length of the action body. Accuracy is within half-minute of arc for 10 rounds, depending upon the quality of the ammunition.

The target trigger is single stage and fully adjustable for weight, creep and backlash. Lock time is relatively fast because of a striker travel of only 7 mm. The length of stock is adjustable to 317.5, 330 or 343 mm. The sights are fully adjustable and a variety of foresight elements are provided to suit individual users.

The Model 83 rifle was adopted by the UK Ministry of Defence as the Cadet Training Rifle L81A1 and entered service in early 1983.

Data

Cartridge: 7.62 × 51 mm NATO
Operation: bolt action, single shot
Weight: 4.98 kg; barrel, 2.04 kg
Length: 1.187 m
Rifling: 1 turn in 14 calibres
Sights: fore, tunnel with removable ring or blade; rear, fully adjustable, aperture, 6 holes

Manufacturer

Parker-Hale Limited, Bisley Works, Golden Hillock Road, Birmingham B11 2PZ.

(NOTE: Parker-Hale no longer manufacture these rifles; manufacturing rights were sold to Navy Arms USA and some Parker-Hale rifles are now manufactured by the Gibbs Rifle Company; see entry under USA.)

Status

No longer in production.

Service

British cadet forces.

VERIFIED

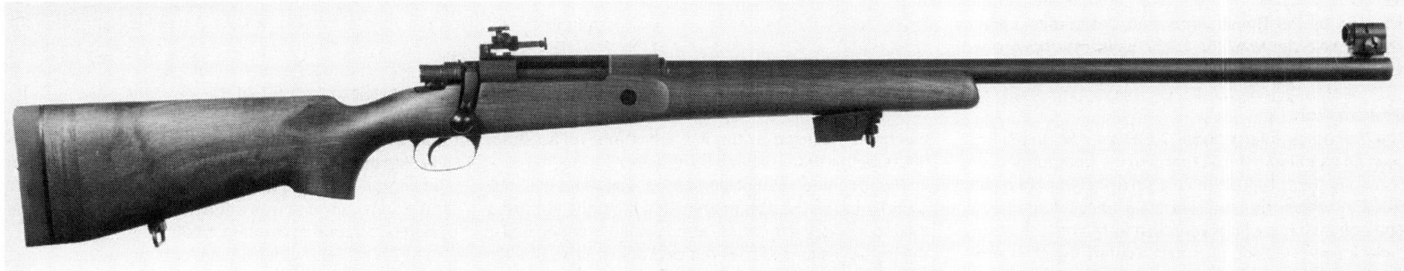

7.62 mm Parker-Hale Model 83 target rifle

5.56 mm Milcam rifles

Description

The Milcam is a bolt-action rifle, being the only military-standard bolt-action rifle available in the 5.56 × 45 mm calibre. Extremely robust, simple to operate and clean, it is an option for situations where an automatic rifle is neither necessary nor practical. The magazine interface is to NATO standard and accepts M16 type magazines, and the flash hider at the muzzle doubles as a standard diameter grenade-launching spigot.

The trigger has adjustment for pressure which is carried out by use of an Allen key and is thus unlikely to be accidentally deranged in the field. The safety catch is based upon that of the Garand rifle, being located inside the large trigger guard; a forward movement of the finger sets the catch to 'fire', pulling it back locks the trigger and also locks the bolt closed. The trigger guard is large enough to accept an Arctic-gloved finger.

The bolt action is strong, giving positive extraction, and the locking movement is through only 22°, giving rapid manipulation. The rifle cannot be fired unless the bolt is completely locked.

The butt and stock are of hardwood and there is a compartment inside the butt for the storage of a cleaning kit.

Variants

5.56 mm Milcam HB rifle

This is the sniper version of the Milcam rifle. It differs in having an adjustable final trigger pressure and in having a 19 mm dovetail sight mount as an integral part of the receiver in addition to the standard iron sights. Normally supplied with a 1/305 mm twist barrel, a 1/178 mm barrel for SS109 ammunition can be supplied if required. Other options include heavy barrels and long barrels.

5.56 mm Comcan rifle

The Comcan is a lighter and shorter carbine version of the Milcam rifle. Except for the barrel length there is no significant difference, the bolt action and sights being the same.

5.56 mm Snicam rifle

The Snicam is an advanced sniper rifle, a cam bolt-action repeater with a heavy barrel. It can be used with any 5.56 × 45 mm cartridge, though the manufacturers recommend the Federal 69 grain bullet for ultimate accuracy. There are no iron sights, a telescope mount and telescope being fitted as standard, and the fore-end carries a steel bipod with adjustable legs. The stock has adjustable shoulder and cheekpieces to suit the individual user.

5.56 mm Snicam advanced sniper rifle

5.56 mm Milcam bolt-action rifle

5.56 mm Comcan compact bolt-action rifle

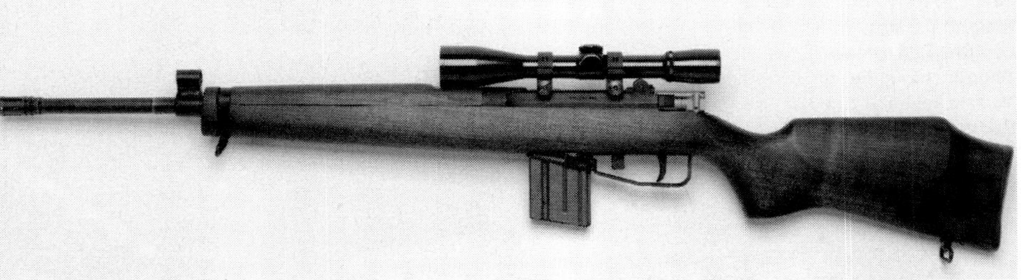

5.56 mm Milcam HB sniper rifle

The Snicam weighs 5 kg with bipod and sight, and has a 620 mm long barrel. A 20-round box magazine is provided.

Data

MILCAM RIFLE

Cartridge: 5.56 × 45 mm M193
Operation: bolt action, single shot
Feed: 30-round box magazine
Weight: empty, 3.6 kg
Barrel: 508 mm
Rifling: 1 turn in 305 mm
Sights: fore, hooded post; rear, two-position flip aperture

Manufacturer

BMS Trading Limited, BCM BMS Arms, London WC1N 3XX.

Status

Available.

VERIFIED

Accuracy International 7.62 mm PM sniper rifle system

Description

The 7.62 mm PM sniper rifle was designed from the start as a military sniper rifle, intended to put its first shot on target, in any conditions, from a clean or fouled barrel. Developed by Accuracy International, it was adopted by the British Army and designated L96A1. The initial order was for 1,212 units - all have been delivered.

Accuracy International 7.62 mm suppressed sniper rifle with PM 6 × 42 sight

Description

The Accuracy International 7.62 mm PM sniper rifle uses an aluminium frame to which the components are firmly attached. This is then clad in a high-impact plastic stock in which the stainless steel barrel floats freely. The bolt action is of conventional form, the bolt having three forward locking lugs and a safety lug at the handle. Bolt lift on opening is 60° and bolt throw is 107 mm allowing the firer to keep his head on the cheek rest while operating the bolt and thus keep observation on his target while reloading. The rifle is equipped with a light alloy bipod which is fully adjustable.

The infantry version has fully adjustable iron sights for use out to 700 m, but is routinely fitted with a special design Schmidt & Bender 6 × 42 telescopic sight designated the L1A1. The accuracy requirement stated by the British Army was for a first round hit at 600 m range and accurate harassing fire out to 1,000 m range. This has been achieved by the PM, which has an accuracy figure better than 0.75 minutes of arc.

The counter-terrorist version is fitted with a ×10, ×2.5-10 as well as the infantry 6 × 42 sight. A spring-loaded monopod is usually fitted, which is concealed in the butt. This can be lowered and adjusted so that the rifle can be laid on the target and supported while the firer observes, without having to support the weight of the rifle for long periods. A spiral flash hider is fitted to the muzzle. Iron sights are not normally fitted to this version.

A suppressed rifle is also produced. Using special subsonic ammunition, this rifle is accurate out to 300 m without unreasonable trajectory or wind deflection. This increases the accepted state of the art by at least 100 m.

The PM was produced in single-shot Magnum calibres for long-range anti-terrorist work; calibres available included 0.300 Winchester Magnum and 7 mm Remington Magnum for use up to 1,000 m range, and an 8.6 mm cartridge was under development for use at ranges up to and exceeding 1,000 m.

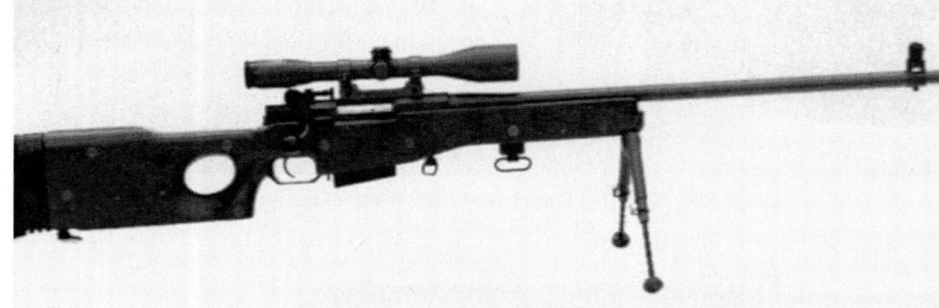

Accuracy International 7.62 mm 'PM Infantry' sniper rifle

Variant

'Covert' sniper rifle system
See following entry.

Data

INFANTRY VERSION

Cartridge: 7.62 × 51 mm NATO
Operation: bolt action, single shot
Feed: 10-round box magazine
Weight: 6.5 kg
Length: 1.124-1.194 m
Barrel: 655 mm
Trigger: 2-stage adjustable, 1-2 kg
Sights: Schmidt & Bender PM 6 × 42

Manufacturer

Accuracy International, PO Box 81, Portsmouth PO3 5SJ.

Status

Available to order.

Service

British Army (1,212), and several African, Middle East and Far East armies.

VERIFIED

Accuracy International 'Covert' sniper rifle system

Description

One variant of the Accuracy International PM system is the 'Covert PM'. The system consists of the suppressed PM rifle in a take-down version which, together with all its ancillaries, packs into an airline suitcase with fitted wheels and retractable handle.

The rifle is the 'Covert' folding PM bolt-action repeating rifle fitted with either the PM 6 × 42, 10 × 42 or 12 × 42 Schmidt & Bender military sights, two 10-shot magazines, bipod, and one box containing 20 rounds of subsonic ammunition.

Data

Cartridge: 7.62 × 51 mm subsonic
Operation: bolt action, single shot
Feed: 10-round box magazine
Weight: 6.5 kg
Length: 1.25 m
Trigger: 1-2 kg pull-off, two-stage detachable
Muzzle velocity: 314-330 m/s
Range: 0-300 m
Suppression: 85 dBA over 125 ms timebase with subsonic ammunition. With full power ammunition, 109 dBA over 125 ms timebase.

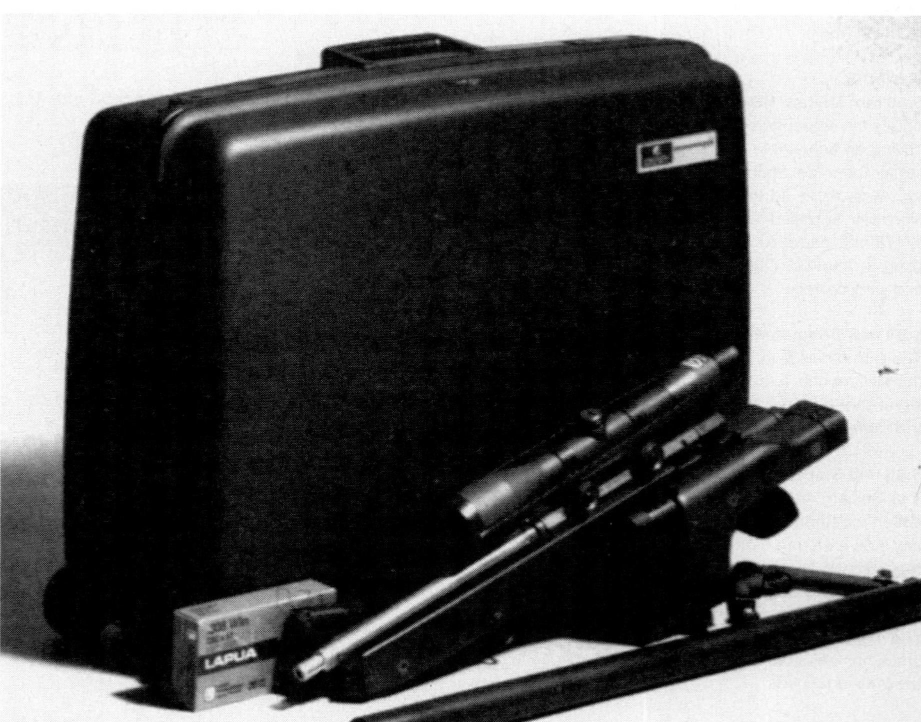

Accuracy International 'Covert' sniper rifle dismantled, with carrying case

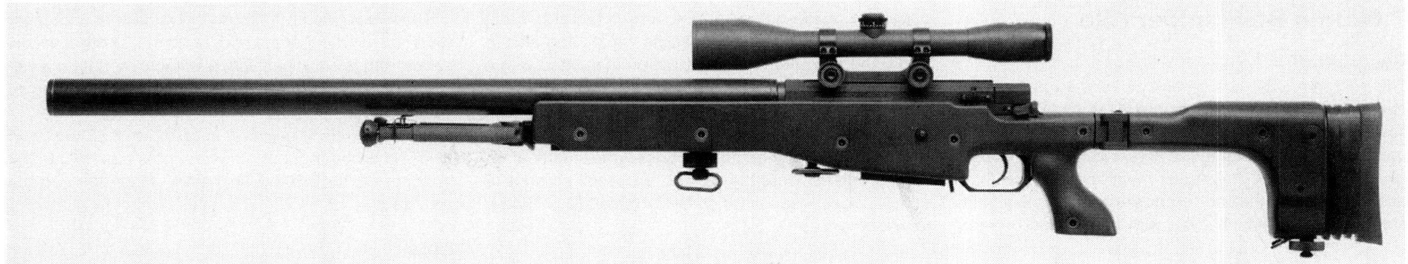

Accuracy International 7.62 mm 'Covert' sniper rifle

Manufacturer
Accuracy International, PO Box 81, Portsmouth PO3 5SJ.

Status
Available to order.

VERIFIED

Accuracy International 0.338 Super Magnum sniper rifle

Description
The Accuracy International 0.338 Super Magnum sniper rifle was designed as a dedicated sniper rifle giving guaranteed accuracy, ease of maintenance, reliability and military robustness. All the lessons learned during the development of the L96A1 sniper rifle, together with many new and innovative ideas, were combined in this weapon.

Using the new ×10 sight and 0.338 Lapua Magnum ammunition, the weapon is able to meet requirements for equipment destruction and light armour penetration as well as the normal anti-personnel capability, to ranges well beyond 1,000 m. The 16.2 g bullet is still supersonic at 1,400 m, at which range there are still over 1,000 J of energy remaining.

The rifle is supplied in three calibres: 0.338 Lapua Magnum, 0.300 Winchester Magnum and 7 mm Remington Magnum. Special accuracy ammunition is under development so as to provide a multiprojectile capability.

Data
Cartridge: 0.338 Lapua Magnum (and see text)
Operation: bolt action, single shot
Feed: 4- (0.338) or 5- (0.300, 7 mm) round box magazine
Weight: 6.8 kg
Length: 1.268 m
Barrel: 0.338, 686 mm; 0.300, 7 mm, 660 mm
Muzzle velocity: 914 m/s

Manufacturer
Accuracy International, PO Box 81, Portsmouth PO3 5SJ.

Status
Available to order.

VERIFIED

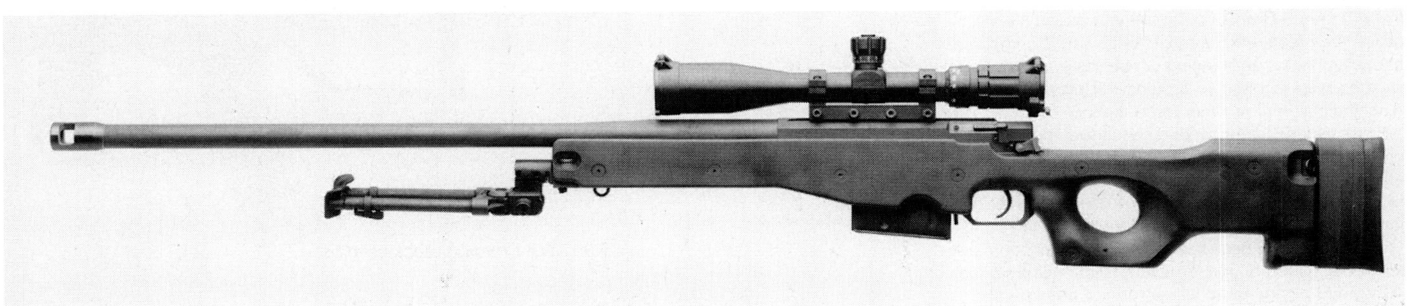

Accuracy International 0.338 Super Magnum sniper rifle

Accuracy International 7.62 mm Model AW sniper rifle

Development
The Accuracy International 7.62 mm Model AW sniper rifle is the result of further development and enhancement of the L96A1 sniper rifle (qv) to suit requirements stipulated during recent trials for a new army sniper rifle. This rifle is thus the 'second-generation' sniper rifle from Accuracy International and took over from the L96A1 when outstanding orders were fulfilled.

Designated the *Prickskyttegevär 90*, or *Psg 90*, by the Swedish armed forces, who have adopted it, the Model AW is to the same basic design concept as the L96A1. The Swedish Army had not hitherto had a dedicated sniper wing and so were able to take a completely fresh approach to their requirements. The rifle successfully underwent four years of trials and development in temperatures below −30°C.

Description
The design of the 7.62 mm Model AW sniper rifle incorporates anti-freeze mechanisms, a different shroud, three-way safety, smooth bolt manipulation, a muzzle brake, simpler and more robust detachable iron sights, an improved bipod, multipoint sling attachments for sling or carrying harness, and a nine-round magazine. For the Swedish Army the rifle is fitted with a Hensoldt 10 × 42 Mil Dot Reticle with tritium lighting for night operation.

Part of the sniping package provided by the Model AW is subcalibre saboted ammunition with a muzzle velocity of about 1,300 m/s. This subcalibre ammunition is in service with the Swedish armed forces, along with 7.62 × 51 mm Ball. Tracer and Blank.

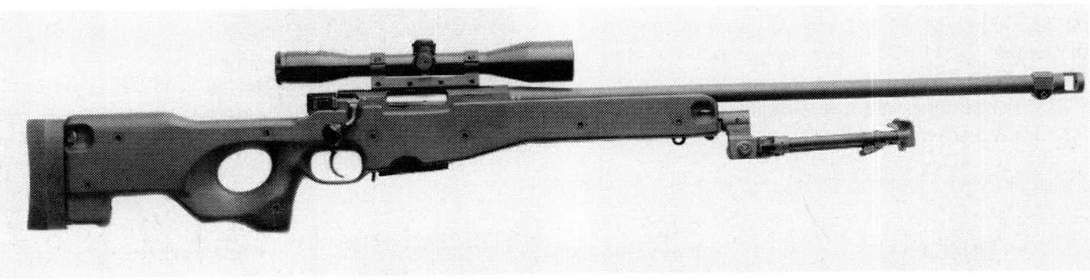

Accuracy International Model AW sniper rifle

Screw-on muzzle suppressors are available to remove flash and reduce the sound signature. Normal full power ammunition can be used with these suppressors.

Variants

Model AWP
The Model AWP is a further refinement of the Model AW. Enhancements introduced include a 610 mm long stainless steel barrel, a multi-adjustable butt pad and a corresponding handstop/bipod mounting which offers flexibility of firing height and better stability when firing over a parapet. The preferred sight for this model is a Schmidt & Bender military pattern ×3 to ×12 variable telescopic sight.

Model AWS
The Model AWS is a fully suppressed version of the Model AW on which a fully integrated barrel and suppressor is interchanged with the standard barrel; the operation takes about 3 minutes. In this form the rifle is effective up to 300 m using full power ammunition but recalibration of the sights is necessary.

Data
Cartridge: 7.62 × 51 mm NATO (see also text)
Operation: bolt action, single shot
Feed: 9-round box magazine
Weight: empty, 6.5 kg
Length: 1.2 m
Barrel: 650 mm
Rifling: rh, 1 turn in 250 mm
Stock: chassis with plastic furniture
Trigger: 2-stage, 2-3 kg
Muzzle velocity: Ball, ca 850 m/s; subcalibre, ca 1,300 m/s

Manufacturer
Accuracy International, PO Box 81, Portsmouth PO3 5SJ.

Status
Available to order.

Service
Swedish armed forces.

UPDATED

7.62 mm BGR sniper rifle

Description

The 7.62 mm BGR is a carefully designed bolt-action rifle manufactured from components from a number of specialist makers. The bolt action is a strengthened bench-rest action using three lugs and with a long and smooth bolt travel. The striker mechanism is carefully designed to have the shortest possible lock time and to ensure regular ignition even with the hardest primer caps. The match grade hammer-forged barrel is fluted to achieve stiffness without excess weight and also to afford the maximum heat dissipation. It is fitted with a combination muzzle brake and flash suppressor. The action and barrel are bedded into the stock by means of a specially developed compound, and stocks are of wood, carbon fibre-reinforced co-polymer material or, for the lightest possible weight, a special composite carbon fibre, Kevlar and GRP material is under development.

The rifle is fitted with an Anschutz pattern accessory rail which will accept a bipod, hand-rest or other accessories, and is equipped with a suitable mount for a variety of telescope sights. Iron sights may be fitted to special order.

The BGR rifle is normally supplied in 7.62 × 51 mm NATO chambering, but other chamberings from 0.243 Winchester to 0.300 Winchester Magnum can be provided.

Data

Cartridge: 7.62 × 51 mm NATO and others
Operation: bolt action, single shot
Feed: 5- to 20-round detachable box magazines
Weight: with bipod and telescope sight, 6.6 kg
Length: 1.2 m
Barrel: 700 mm
Rifling: 4 grooves, rh, 1 turn in 280 mm

Manufacturer

Armalon Limited, 44 Harrowby Street, London W1H 5HX.

Status

In production.

7.62 mm BGR sniper rifle

VERIFIED

UNITED STATES OF AMERICA

0.30 M1 carbine

Development

In 1940 the US Ordnance Department issued a specification for a light rifle not to exceed 5½ lb (2.5 kg) and capable of either self-loading or automatic action. This weapon was required to replace the pistol and submachine gun in arms other than infantry.

The cartridge for this weapon was developed by Winchester Repeating Arms Co from its 0.32 self-loading rifle cartridge of 1905 from which it deviated only very slightly in external measurements. It was known as Calibre 0.30 SR M1, where SR stood for 'Short Rifle'.

The Winchester design was based on its experimental model rifle produced for test in 1940. This weighed 4.2 kg and incorporated a new form of gas operation, developed by David M Williams, and now universally referred to as 'short-stroke piston operation'.

It was accepted on 30 September 1941 as the US Carbine Calibre 0.30 M1 and eventually more of these carbines were produced than any other single US model. According to the best available figures, a total of 6,232,100 M1, M1A1, M2 and M3 carbines were made, of which 6,117,827 were accepted by the US Government.

The M1 carbine was a self-loading model, of which 5,510,000 were made.

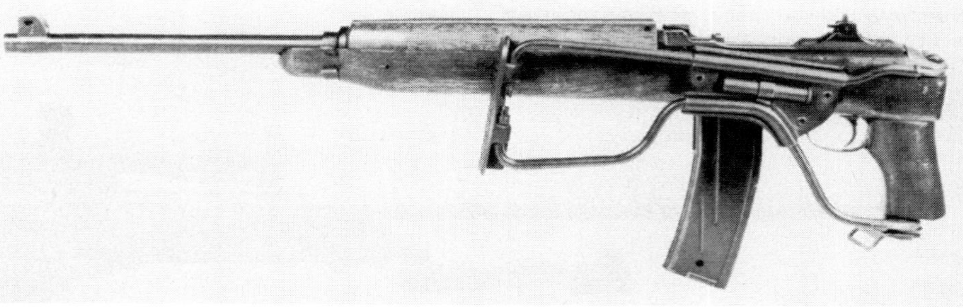

0.30 M1A1 folding stock carbine

The M1A1 carbine, standardised in May 1942, had a side-folding stock. It was intended primarily for use by airborne troops and 150,000 were made.

The M2 carbine was standardised in November 1944 as a selective fire carbine and 570,000 were made.

The M3 was an M2 with no sights and a flash hider. It was intended specifically to carry the infra-red Sniperscope and only 2,100 were made.

The carbine has been adopted, officially and unofficially, by many armies and is still in service use although it was relegated to the reserve in the USA in 1957. There are still small quantities being produced by private manufacturers in the USA for sporting use. The carbine is popular with police forces because of its small size and relatively low power, making it a safer weapon to use in crowded urban areas.

Description

To load the 0.30 M1 carbine the magazine is loaded and pushed up into the magazine-well. The cocking handle is on the right and is pulled back and released to chamber the top round in the magazine.

When the trigger is squeezed the hammer is released and the firing pin is driven into the cap of the cartridge. The gas pressure forces the bullet up the bore and some of the gas following it is diverted through a gas vent about 114 mm from the chamber. The pressure is high and it impinges on a very small piston which is driven back only 3.6 mm before its movement is stopped. This strikes the operating slide which acquires the momentum of the piston and moves back. There is a delay to allow the chamber pressure to fall while the slide travels back a little over 7.6 mm and then a cam recess engages the operating lug on the bolt. The bolt is rotated, providing primary extraction of the near parallel-sided case, and then unlocked. This rotation starts the cocking action on the hammer and also retracts the firing pin. The empty case is pulled out by the extractor and then the spring-loaded ejector in the bolt throws it forward to the right of the carbine. The hammer is now cocked and the return spring is compressed to drive the bolt forward.

The bolt chambers the round and rotates to the locked position, the operating slide pushes the piston forward inside its cylinder and the weapon is ready to fire again. Should the self-loading action fail, the cocking handle can be used to operate the cycle by hand.

The safety is a push-through type just forward of the trigger guard. When it is pushed to the right the solid diameter of the plunger moves under the forward end of the trigger and prevents it from moving down.

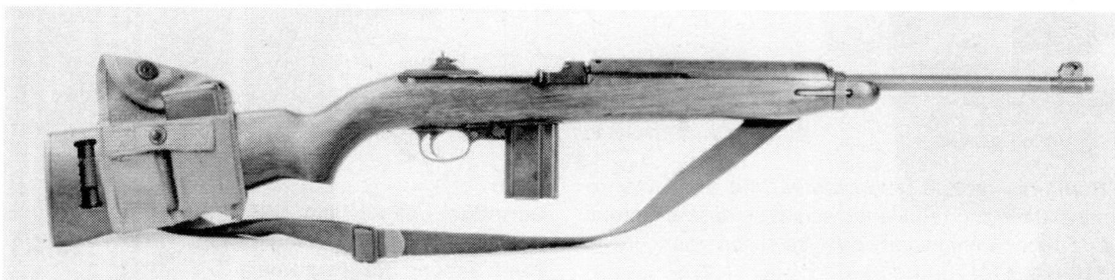

0.30 M1 US Army carbine with 15-round magazine

0.30 M2 carbine, showing selector lever alongside chamber

Data

Cartridge: 0.30 M1 Carbine
Operation: M1, gas, self-loading; M2 and M3, selective fire
Locking: rotating bolt
Feed: 15- or 30-round box magazine
Weight: M1 and M2 carbines, with unloaded magazine, 2.36 kg; with loaded magazine and sling, 2.63 kg
Weight: M1A1 carbine, with unloaded magazine, 2.53 kg; with loaded magazine and sling, 2.77 kg
Length: M1 carbine, 904 mm
Length: M1A1 carbine, stock folded, 648 mm; stock extended, 905 mm
Barrel: 458 mm
Rifling: 4 grooves, rh, 1 turn in 508 mm
Sights: fore, blade; rear, flip aperture 0-150, 150-300; leaf slide on M2 models
Sight radius: 546 mm
Muzzle velocity: 607 m/s
Rate of fire: cyclic, M2 and M3 only, 750 rds/min
Effective range: 300 m

Manufacturer

No longer in production.

Status

Obsolescent but still in widespread use.

Service

US National Guard, Chile, Ethiopia, Honduras, Japan, South Korea, Mexico, Norway (in reserve), Philippines, Taiwan and Tunisia. Made (M1 and M2) under licence by Beretta for the Italian Army and sold to Morocco and the Dominican Republic. Also adopted by many police forces throughout the world.

VERIFIED

7.62 mm NATO M14 rifle

Development

The 7.62 mm M14 rifle was adopted in 1957 as the successor to the M1 Garand and was the first weapon in the US Army to use the standardised NATO round. The M14 is basically an evolution of the M1 and has, for example, the trigger mechanism of that rifle with virtually no changes. However, it has incorporated some improvements and it could be fairly said that the M14 represents the ultimate development of the old M1. One major change is in the magazine: in the M14 the detachable 20-round box allows full magazines to be loaded on to the rifle without needing to use the unsatisfactory clip of the M1. Another feature is the gas cylinder which in the M1 ran right up to the muzzle. This interfered with the jump of the barrel and affected accuracy and consistency. On the M14 the gas port has been moved back to a more normal place about two thirds of the way up the barrel.

Another change is in the gas system. The M1 had a direct-action piston which gave the operating rod a sharp and heavy blow. This was necessary to start the mass of the rod and its cam moving backwards. Although the M14 uses the same sort of rod and cam the gas port allows a more gentle and progressive push on the piston-head by using a different gas cut-off and expansion system. The result is that it is a more pleasant and steady rifle to fire and successive shots are made more quickly since the sights can be brought back on to the target a little faster.

Numbers of variants on the basic M14 were proposed, but only one was ever put into production. This was the M14A1, a heavy barrel version to be used as the squad light automatic. It was an unsuccessful weapon and was never produced in any quantity.

US Government production of the M14 rifle ceased in 1964, total production being 1,380,346 weapons. Production was resumed on a commercial basis in 1974 by Springfield Armory Inc (see following entry).

Description

To load the 7.62 mm M14 the 20-round box magazine is placed in the rifle by inserting the front end first and then rotating the rear until the magazine catch snaps into engagement.

When the cocking handle is pulled back and released, the bolt picks up the top round and chambers it. The safety catch is in the front of the trigger guard and is pulled back for 'safe'.

The sights consist of a blade foresight and an aperture rearsight graduated from 200 to 1,000 m. The elevation control is on the left of the backsight and is rotated clockwise to elevate. The windage control is on the right of the rearsight. The foresight can be moved across on its block in a dovetail for lateral zeroing. The range scale can be adjusted for elevation zero.

The standard M14 will fire semi-automatic only. If a selector is fitted it must be rotated according to the type of fire required. When it is positioned with the face marked 'A' to the rear the rifle is set for automatic fire.

When the safety is pushed forward and the trigger squeezed, the hammer rotates forward and strikes the firing pin. The bullet passes up the bore and some of the gas passes through a port and into the hollow interior of the piston-head. This is filled with gas and the piston is forced rearwards, driving the operating rod and bolt with it. After the piston has travelled back slightly less than 4 mm, the gas ports are no longer aligned and no further gas can enter. This system requires no gas regulator as, by design, the pressure available to move the actuating rod back will automatically increase until it is sufficient to overcome the resistance to motion. The piston moves back 38 mm and the exhaust port in the bottom of the gas cylinder is then exposed to allow the expanded gases to escape into the atmosphere.

The operating rod has a free travel of about 9.5 mm to allow the chamber pressure to fall and then the camming surface inside the hump forces the bolt roller upward to rotate and unlock the bolt. The bolt is pushed back and the empty case is extracted and ejected. The compressed return spring forces the operating rod forward and the bolt chambers another round and is rotated into the locked position.

When the ammunition is expended the magazine follower operates the bolt lock and the bolt is held to the rear ready for a fresh magazine to be inserted.

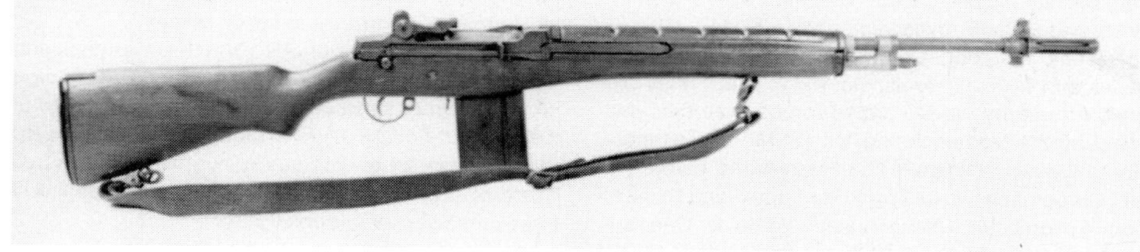

7.62 mm M14 rifle

Data

Cartridge: 7.62 × 51 mm
Operation: gas
Locking: rotating bolt
Feed: 20-round box magazine
Weight loaded: M14, 5.1 kg; M14A1, 6.6 kg
Length: 1.12 m
Barrel: 559 mm
Rifling: 4 grooves, rh, 1 turn in 305 mm
Sights: fore, fixed post; rear, tangent aperture
Sight radius: 678 mm
Muzzle velocity: 853 m/s
Rate of fire: cyclic, 700-750 rds/min

Manufacturer

Harrington and Richardson Arms Company, Worcester, Maryland; Thompson-Ramo-Wooldridge, Port Clinton, Ohio; Winchester-Western Arms Division of Olin Mathieson Corporation, New Haven, Connecticut; Springfield Armory (Government) Inc, Springfield, Maryland; Taiwan.

Status

Manufactured commercially by Springfield Armory Inc.

Service

US Army, Israel (reserve), Taiwan, South Korea.

VERIFIED

M14 (M1A, M1A-A1) Springfield Armory rifles

Description

Springfield Armory Inc is the only manufacturer of the M14 rifle and its variants, designated the M1A and M1A-A1. When the US Government Springfield Armory in Massachusetts was closed in the late 1960s the tooling was sold. It was then set up again in Illinois where the rifles remain in production. Sales are largely to commercial users and the majority of these rifles are semi-automatic.

7.62 mm Springfield Armory M1A Standard rifle

The standard M1A model is a duplicate of the original general issue M14 rifle. It is also available with National Match and the even heavier Super Match barrels. All three of these rifles, especially the two heavier barrel models, are consistent winners in civilian shooting competitions.

Springfield Armory also offers the M1A-A1 assault-paratrooper rifle.

Springfield Armory makes and supplies a comprehensive range of parts and accessories for all models, military or commercial, and additionally it offers factory service for all Government models of the M14.

Manufacturer

Springfield Armory Inc, 420 West Main Street, Geneseo, Illinois 61254.

Status

In production.

VERIFIED

La France 7.62 mm M14K assault rifle

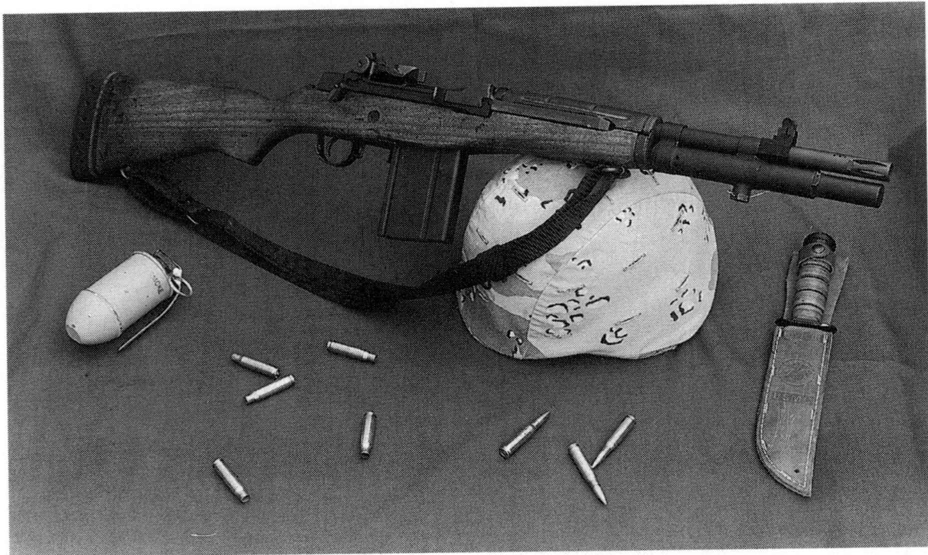

La France 7.62 mm M14K assault rifle

Description

The La France 7.62 mm M14K assault rifle is an attempt to improve the handling qualities and full-automatic controllability of the 7.62 mm M14 service rifle. The reduced overall length considerably improves the handling, while the shortened barrel causes a decrease in muzzle velocity and, in consequence, muzzle energy, thereby reducing muzzle climb. The high-efficiency muzzle brake helps to keep the sights on target, as does the lower cyclic rate produced by the large-volume modified M60 gas system. Perceived recoil is little more than that of most 5.56 mm rifles.

Data

Cartridge: 7.62 × 51 mm NATO
Operation: gas, selective fire
Locking: rotating bolt
Feed: 5- or 20-round magazine
Weight: empty, 3.74 kg
Length: 902 mm
Barrel: full automatic, 338 mm; semi-automatic, 406 mm
Rifling: 4 grooves, rh, 1 turn in 40 calibres
Sights: fore, fixed post; rear, tangent aperture
Sight radius: 445 mm
Muzzle velocity: 762 m/s
Rate of fire: cyclic, 650 rds/min

Manufacturer

La France Specialties, PO Box 178211, San Diego, California 92177-8211.

Status

In production.

VERIFIED

5.56 mm AR-15/M16 series rifles and carbines

5.56 mm M16A2 rifle fitted with FIRM, BOSS, 40 mm M203PI and a laser aiming device, demonstrating some of the various sight systems that can be accommodated by the FIRM system

1996

Development

The 0.223 (5.56 mm) AR-15 was designed by Eugene Stoner, then an employee of Armalite Inc. Upon being type classified as a military weapon in 1962 the AR-15 was designated the M16 and issued initially to the US Air Force. For the US Army the AR-15 rifle was modified as a result of combat experience in Vietnam and in 1967 became the M16A1.

The differences between the M16 and M16A1 are chiefly that the M16A1 has a bolt with serrations on the right-hand side and a forward assist assembly which protrudes from the upper receiver and can be used to force the bolt home if the return spring for some reason is unable to do so. This device, the forward assist assembly, allows the firer to close the bolt when a dirty cartridge or chamber fouling produces a high friction force. The M16A1 became the main US Army version of the M16 series and is still the most numerous variant in the AR-15/M16 series.

With the adoption of a revised specification for NATO 5.56 × 45 mm ammunition the barrel rifling was altered to one turn in 177.8 mm to accommodate all possible types of NATO cartridges. This variant became the M16A2. Other changes introduced with the M16A2 included a three-round burst capability, a heavier barrel, rear sights adjustable for windage, and new impact resistant materials.

The latest model of the AR-15/M16 rifle series is the M16A3 which incorporates various 'human engineering' and performance changes, the main one being the provision of a 'Picatinny' flat top rail in place of the usual carrying handle and rear sight assembly; the rail permits the fitting of various optical and night sights.

Co-incident with the development of the M16 rifle series has been the development of a series of carbines, usually virtually identical to the rifles in engineering and operation terms but with shorter barrels and telescopic butts. Following a number of tentative attempts to introduce carbine models into US service, this series resulted in the 5.56 mm M4 and M4A1 carbines, the first examples of which were handed over by Colt's to the Special Forces during August 1994 as part of an $11 million order for 24,000 units. (The Colt 9 mm sub-machine gun based on the AR-15/M16 is described in the *Sub-machine gun* section.)

Despite the long association of the AR-15/M16 series with Colt's Firearms, many M16 series rifles destined for use by the US armed forces were manufactured by FN Manufacturing Inc (FNMI) of Columbia, South Carolina (M16A2), the Hydra-Matic Division of GM Corporation (M16A1), and Harrington and Richardson of Worcester, Massachusetts (M16A1). One contract placed in April 1995 was awarded to FN Manufacturing Inc. This firm fixed price contract was for 16,000 M16A2 rifles and was worth $6,955,520. Contract completion was scheduled for March 1996.

Licence production, no longer in progress, has been undertaken in South Korea, Singapore and the Philippines. Licence production for the Canadian and other governments is currently by Diemarco of Kitchener, Ontario (qv). NORINCO of China produce an M16 clone known as the 'CQ' (qv).

Numerous designs based on the AR-15/M16 and its derivatives have been produced by numerous manufacturers and may be detected throughout the pages of this Yearbook. Single-shot sporting models for commercial sales have been produced (including one model chambered for the CIS 7.62 × 39 mm cartridge). Also apparent in the general defence market place is a thriving market in 'matched' accessories ranging from high capacity magazines to special slings.

Well over seven million AR-15/M16 series rifles and carbines have been produced to date and production continues. In order to provide a general impression of the range of Colt AR-15/M16 models produced to date the following list of Colt model numbers is provided. Some of these numbers relate to models mentioned elsewhere in this Yearbook.

Model number	Description
601	0400AR-15 rifle.
602	AR-15 rifle. Early US Government purchase - no forward assist.
603	M16A1 rifle. Originally the XM16E1. Standard US Government model with forward assist.
603K	M16A1 rifle. Produced for South Korea.
604	M16 rifle. Produced for US Air Force without forward assist.
605A	M16 carbine. Carbine version of M16A1 rifle with forward assist and fixed butt-stock.
605B	M16 carbine. Same as Model 605A but with additional three-round burst limiter; no forward assist.
606	Heavy Barrel Assault Rifle (HBAR) without forward assist. Also known as the Heavy Assault Rifle M1.
606A	Heavy Barrel Assault Rifle (HBAR) with forward assist.
606B	Heavy Barrel Assault Rifle (HBAR) with forward assist and three-round burst limiter.
607	Sub-machine gun with sliding butt-stock and 254 mm barrel.

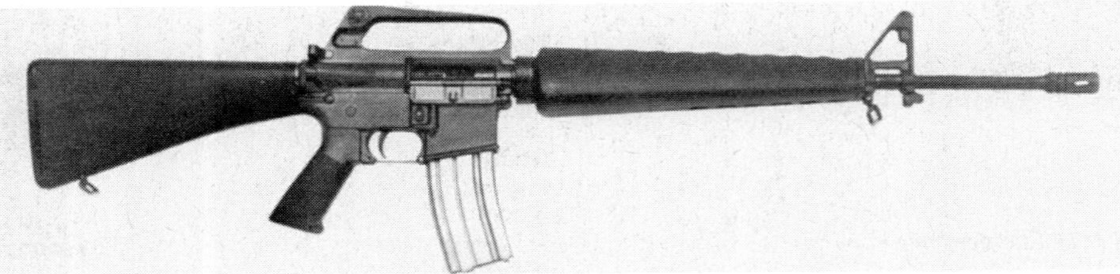

5.56 mm M16A1 rifle

Model number	Description
608	CAR-15 survival rifle with 254 mm barrel and fixed tubular butt-stock. About 10 produced.
609	XM177E1 sub-machine gun for US Army with 254 mm barrel and 89 mm long sound/flash suppressor.
610	XM177E1 sub-machine gun for US Air Force with 254 mm barrel.
610B	XM177E1 sub-machine gun for US Air Force with addition of three-round burst limiter.
611	Heavy Barrel Assault Rifle. Export model of Model 606.
611P	Heavy Barrel Assault Rifle. Export version of Model 611 for the Philippines.
613	M16A1 rifle. Export version of Model 603.
613P	M16A1 rifle. Export version of Model 613 for the Philippines.
614	M16 rifle. Export version of Model 604 without forward assist.
614S	M16 rifle. Export version of Model 614 licence produced by Chartered Industries of Singapore.

Model number	Description
616	Heavy Barrel Assault Rifle. Export model of Model 606.
619	Sub-machine gun. Export version of Model 609.
620	Sub-machine gun. Export version of Model 610.
621	Heavy Barrel Assault Rifle. Heavy barrel version of Model 603.
629	XM177E2 sub-machine gun for US Army with 292 mm barrel and 114 mm long sound/flash suppressor.
630	Sub-machine gun for US Air Force without forward assist and 292 mm barrel.
633	9 mm sub-machine gun with 178 mm barrel and mechanical buffer.
633HB	9 mm sub-machine gun with 178 mm barrel and hydraulic buffer.
634	9 mm sub-machine gun with 267 mm barrel, without forward assist.
635	9 mm sub-machine gun with 267 mm barrel, without forward assist.

5.56 mm M16A2 rifle (Colt Model 711)

5.56 mm Colt M4 carbine, right side

5.56 mm Colt Model 733 Commando, right side

5.56 mm Colt M4 carbine, with M203 grenade launcher attached

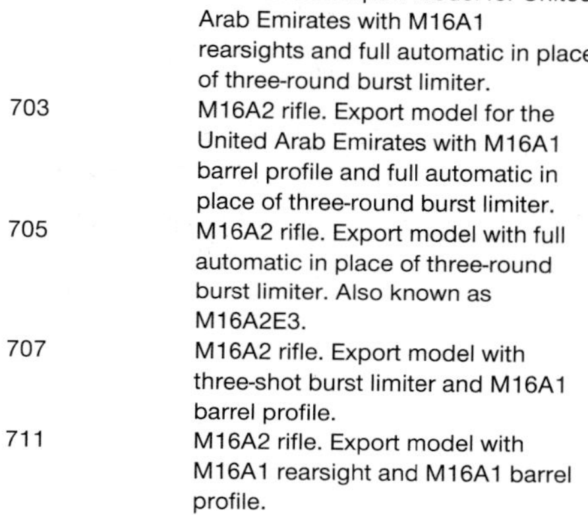

Model number	Description
639	Sub-machine gun. Export version of Model 629.
640	Sub-machine gun. Export version of Model 629 without forward assist.
645	M16A2 rifle. Standard US Government model with revised 1 in 177.8 mm rifling, 508 mm barrel and new rearsight.
645E	M16A2 rifle 'Enhanced'. Modified M16A2 with removable carrying handle for optical sight mounting and flip-up front sight.
649	Sub-machine gun. For US Air Force, with 356 mm barrel and 1 in 305 mm rifling.
651	M16A1 carbine. Export model with 368 mm barrel and full size stock.
652	M16 carbine. Export model with 368 mm barrel and full size stock, but without forward assist.
653	M16A1 carbine. Export model with 368 mm barrel and sliding stock.
653P	M16A1 carbine. Export model with 368 mm barrel and sliding stock, licence-produced in Philippines.
654	M16 carbine. Export model without forward assist. Fitted with 368 mm barrel and sliding stock.
655	M16A1 sniper rifle. Experimental model with high profile upper receiver.
656	M16A1 sniper rifle. Experimental model with low profile upper receiver and Sionics suppressor.
701	M16A2 rifle. Export model with full automatic in place of three-round burst limiter.
702	M16A2 rifle. Export model for United Arab Emirates with M16A1 rearsights and full automatic in place of three-round burst limiter.
703	M16A2 rifle. Export model for the United Arab Emirates with M16A1 barrel profile and full automatic in place of three-round burst limiter.
705	M16A2 rifle. Export model with full automatic in place of three-round burst limiter. Also known as M16A2E3.
707	M16A2 rifle. Export model with three-shot burst limiter and M16A1 barrel profile.
711	M16A2 rifle. Export model with M16A1 rearsight and M16A1 barrel profile.
715	M16A2 rifle. Canadian version of M16A2 rifle known as C7 with M16A1 rearsight, three-round burst limiter and M16A2 barrel. Licence produced by Diemarco.
719	M16A2 rifle. Version of Model 715 produced by Colt's with three-round burst limiter.
720	M4 carbine. Originally known as M16A2 carbine, with 370 mm barrel.
723	M16A2 carbine. Produced for United Arab Emirates and US Army Delta Force. With M16A1 rearsight, 370 mm barrel and full-automatic fire capability.
725	M16A2 carbine. Canadian version of M16A2 carbine known as C8 with M16A1 rearsight, three-round burst limiter and M16A2 barrel. Licence produced by Diemarco.
725A	M16A2 carbine. Export version of the M16A2 carbine produced by Colt's for United Arab Emirates with M16A1 profile barrel.
725A	M16A2 carbine. Export version of M16A2 carbine with M16A2 barrel.
727	M16A2 carbine. Carbine with full-automatic fire feature produced for (among others) the US Navy. With 370 mm barrel.
733	M16A2 Commando. Short-barrelled model with full-automatic fire, 290 mm barrel and M16A1 rearsight.

Model number	Description
733A	M16A2 Commando. Short-barrelled model with three-round burst limiter in place of fully automatic fire, 290 mm barrel and M16A1 rearsight.
735	M16A2 Commando. Similar to Model 733 but with three-round burst limiter.
737	M16A2 Heavy Barrel Assault Rifle (HBAR). With M16A1 rearsight.
741	M16A2 Heavy Barrel Assault Rifle (HBAR). As Model 737 but with M16A2 rearsight.
	M16A2 Squad Automatic Weapon (SAW). With 508 mm barrel, bipod and front handguard assembly. Fully automatic.
901	M16A3 rifle. Similar to Model 701 but with flat top rail under carrying handle. Fully automatic.
905	M16A3 rifle. Similar to Model 705 but with flat top rail under carrying handle. Three-round burst limiter.
925	M16A3 rifle. Similar to Model 725 but with flat top rail under carrying handle. Three-round burst limiter.
927	M4A1 carbine. Originally known as M16A3 carbine. Similar to Model 727 but with flat top rail under carrying handle. Fully automatic. In production for Joint DoD Special Forces.
941	M16A3 Heavy Barrel Assault Rifle (HBAR). Similar to Model 741 but with flat top rail under carrying handle. Semi- and fully automatic.
942	M16A3 Squad Automatic Weapon (SAW). Similar to Model 741 but with flat top rail under carrying handle. Fully automatic only.
950	M16A3 Squad Automatic Weapon (SAW). Similar to Model 741 but with flat top rail under carrying handle. Fully automatic.

5.56 mm Colt Model 733 Commando, left side

Description

On all AR-15/M16 weapons the operation is as follows: the charging handle positioned behind the carrying handle is withdrawn to cock the weapon. If the magazine is empty the magazine follower will rise under the force of the magazine spring and hold the bolt carrier to the rear. When a loaded magazine is in place the carrier will be driven forward by the return spring and the bolt will pick up a round from the magazine and feed it into the chamber.

The bolt motion stops when the cartridge is fully chambered but the carrier continues forward and the cam slot cut in the carrier rotates the bolt anti-clockwise (viewed from the rear) and the eight locking lugs move behind abutments in the barrel extension. The rifle is now ready and if the fire-control selection on the left side of the receiver is set either to automatic (or optional three-round burst) or semi-automatic, operation of the trigger will fire the round. When the trigger is pulled, the sear, extending forward of the trigger, is rotated down and moves out of the hammer notch; the hammer is then rotated forward by its spring and hits the firing pin which in turn strikes the cartridge cap and so fires the round. As the bullet passes the gas port some of the gas passes back along a stainless steel tube and through the bolt carrier key into the hollow interior of the carrier. The expanding gas forces the carrier back, the movement of the cam slot moves the cam pin, and the bolt rotates and unlocks. The momentum acquired by the carrier enables it to carry the bolt to the rear at a slightly reduced velocity. The extractor withdraws the cartridge from the chamber; the spring-loaded ejector rod emerges from the left of the bolt face and rotates the case around the extractor as soon as the case clears the chamber, the case passing through the ejection port on the right side of the receiver. The carrier continues rearward compressing the return spring and cocking the hammer. The action of the buffer and the return spring force the carrier forward and the cycle starts again.

One widely used accessory for the AR-15/M16 rifle and carbine series is the 40 mm Grenade Launcher M203. For details refer to the appropriate entry under *Light support weapons*.

On the M16A2 an integral mount is provided on the carrying handle for a ×4 telescopic sight. On the M16A3 rifle and M4A1 carbine models the entire carrying handle and rearsight assembly can be removed to reveal an integral 'Picatinny' mounting rail for low profile optical and other sights.

A form of reflex sight known as Close Combat Optics is scheduled to enter service in late 1996.

Other standard accessories for the AR-15/M16 rifles and carbines include 20- and 30-round magazines, multimagazine holders, a snap-on lightweight bipod and carrying case (the HBAR models normally have a permanently attached bipod), a bayonet (the US armed forces model is the M7), a blank firing attachment, a sling, and a cleaning kit designed to fit inside a butt-stock compartment on rifles. An armourer's tool kit is also available, as is the M30 boresighting device.

Variants

A full listing of all AR-15/M16 variants is provided under Development but an expanded differentiation between the main production versions is provided here to give a more complete account of the differences between the various models.

M16

The 5.56 mm M16 was the militarised version of the Armalite AR-15 rifle (produced by Colt's) as taken into service by the US Air Force in late 1961.

M16A1

When the US Army adopted a number of militarised AR-15 rifles for service in Vietnam they initially encountered problems with 'sticking' or ruptured empty cartridge cases, a problem eventually traced to a change of propellant which caused excessive fouling and an increase in cyclic fire rates. This was compounded by an unfortunate 'soldier's myth' that their new rifles did not require cleaning. By the time the overall problem had been addressed a forward assist device had been added to the right-hand side of the receiver to allow the user to manually force home a non-seated cartridge. Other changes introduced on the M16A1 included a revised buffer and some other changes. Type classification was in 1967.

M16A2

Originally the M16A1E1, the M16A2 was type classified in 1982, although adoption by the US Marine Corps was delayed for about a year; deliveries to the US Army began in 1985. The main innovation on the M16A2 was a revised barrel with 1 in 177.8 mm rifling to allow it to fire the complete range of 5.56 × 45 mm ammunition and a revised end profile to prevent bending damage to the section protruding from the forward handguard. Other changes included an optional three-round burst limiter, a revised rearsight arrangement with provision for windage (the foresight has provision for elevation adjustment). New high impact polymer composite materials with aluminium inserts were introduced for the furniture and the handguard profile is circular. The muzzle has a revised compensator which reduces muzzle climb when firing.

M16A3

The M16A3 is essentially a M16A2 with the provision of a rearsight option. The carrying handle and rearsight assembly can be removed to expose a low profile mounting rail for optical day and night sights. Various fire mode options are available according to model.

M4 Carbine

Following a series of trials and experimental models intended to provided a 'shorty' M16 the end result was the Colt Model 720 with a 370 mm long barrel and 1 turn in 177.8 mm rifling. Basically a short M16A2 rifle, the M4 Carbine is in production for the US Army. It has a telescopic butt-stock and can be fitted with the M203 grenade launcher.

M4A1 Carbine

The M4A1 is essentially the same as the M4 but with the removable carrying handle and rearsight assembly of the M16A3 rifle to allow optical sights to be fitted on the exposed sight rail.

Commando

Having a 290 mm long barrel, the Commando models are the shortest of all the M16A2-based variants and are intended for use by special forces or agencies where concealment of weapons is an advantage. Overall length with the telescopic butt retracted is 680 mm and empty weight 2.44 kg.

HBAR

There have been several models of Heavy Barrel Assault Rifle (HBAR) to provide sustained fire rather than short bursts, all involving the use of an open bolt. Typically, the Model 741 HBAR weighs 4.58 kg. All have provision for permanently attached bipods while some have light carrying handles and foregrips.

FIRM

FIRM stands for Floating Integrated Rail Mount, a patent pending design enhancement for the M16 rifle and M4 carbine series developed by FN Manufacturing Inc of Columbia, South Carolina. FIRM is intended to overcome inherent aiming variations produced by mounting various optical and other sighting systems onto the receiver of the M16 series weapons. After a period of firing the barrel will tend to heat and move, causing a shift in the zeroing of the sights. With FIRM the sight is mounted on a light alloy forward handguard with an extended Picatinny rail and other mounting points arranged above and around the barrel in such a manner that the mounting rails are able to float and follow the barrel as it moves. The point of impact compared to the

5.56 mm Colt Model 741 Heavy Barrel Assault Rifle (HBAR)

point of aim therefore remains much closer to the intended values during periods of extended firing.

FIRM has passed engineering tests and has been trialled in conjunction with the Browning Ballistic Optimizing Shooting System, or BOSS, a muzzle-mounted compensator device which allows the host weapon to maintain accuracy performance. FIRM can also be used together with the 40 mm M203PI quick disconnect system (see separate entry under *Light support weapons*). Channels are incorporated in the FIRM design to accommodate wiring and other connections for laser, optical and night sight systems. Numerous types of sighting system can be utilised with FIRM.

KAC 5.56 mm Modular Weapon System

See following entry.

Conversions

Throughout the life of the AR-15/M16 there have been various commercial projects involving conversions, accessories and modifications to cover just about every aspect of the M16 rifle series. An Atchisson conversion kit produced by Jonathan Arthur Ciener Inc which allows the M16 series to fire 0.22 LR ammunition is described in a separate entry in this section. Another Ciener conversion which provides M16 series rifles with a belt feed mechanism is also described in a separate entry in this section.

Data

M16 RIFLES

Cartridge: 5.56 × 45 mm
Operation: gas, direct action, selective fire
Locking: rotating bolt
Feed: 20- and 30-round box magazine
Weight: (rifle without magazine) M16, 3.1 kg; M16A1, 3.18 kg; M16A2 and M16A3, 3.4 kg
Weight, magazine, empty: 20-round, 91 g; 30-round standard, 117 g; 30-round nylon, 113 g
Weight, magazine, loaded: 20-round, 318 g; 30-round standard, 455 g
Length: with flash suppressor, 990 mm; with bayonet knife M7, 1.12 m
Barrel: 508 mm; with flash suppressor, 533 mm
Rifling: M16A1, 6 grooves, rh, 1 turn in 305 mm; M16A2 and M16A3, 6 grooves, rh, 1 turn in 177.8 mm
Sights: fore, cylinder on threaded base; rear, (M16A1) flip aperture, (M16A2) adjustable for elevation, (M16A3) as M16A2 but can be removed to expose integral optical sight rail on receiver
Muzzle velocity: 990-1,000 m/s
Rate of fire: cyclic, 700-950 rds/min
Effective range: 400 m

Data

M4 CARBINE

Cartridge: 5.56 × 45 mm
Operation: gas, direct action, selective fire
Locking: rotating bolt
Feed: 20- and 30-round box magazine
Weight: carbine without magazine, 2.52 kg
Weight, magazine, empty: 20-round, 91 g; 30-round standard, 117 g; 30-round nylon, 113 g
Weight, magazine, loaded: 20-round, 318 g; 30-round standard, 455 g
Length: stock retracted, 757 mm; stock extended, 838 mm
Barrel: 370 mm
Rifling: 6 grooves, rh, 1 turn in 177.8 mm
Sights: fore, cylinder on threaded base; rear, (M4) adjustable for windage, (M4A1) as M4 but can be removed to expose integral optical sight rail on receiver
Muzzle velocity: 921 m/s
Rate of fire: cyclic, 700-950 rds/min
Effective range: 360 m

Manufacturer

Colt's Manufacturing Company Inc, PO Box 1868, Hartford, Connecticut 06144-1868. (See also text.)

Status

In production.

Service

In service with US Forces. Also Australia, Barbados, Belize, Bolivia, Brazil, Brunei, Cameroon, Chile, Costa Rica, Denmark, Dominican Republic, Ecuador, El Salvador, Fiji, Gabon, Ghana, Greece, Grenada, Guatemala, Haiti, Indonesia, Israel, Jamaica, Jordan, Kampuchea, South Korea, Lebanon, Lesotho, Liberia, Malaysia, Mexico, Morocco, New Zealand, Nicaragua, Nigeria, Oman, Panama, Peru, Philippines, Qatar, Somalia, South Africa, Sri Lanka, Thailand, Tunisia, Turkey, Uganda, UAE, Uruguay, Vietnam, and Zaire.

Licence production

Licence production was undertaken in the Philippines by the Elisco Tool Company of Manila, in South Korea by Pusan Arsenal (now Daewoo Precision Industries), and in Singapore by Chartered Industries of Singapore. None of these concerns are still producing M16s. The only known licensees are the following:

CANADA
Manufacturer: Diemarco
Type: 5.56 mm C7 and C8
Remarks: See entry under *Canada* for details.

UPDATED

KAC 5.56 mm Modular Weapon System

Development

In August 1995 the Knight's Armament Company (KAC) announced a US Government contract award to produce a Rail Interface System (RIS) for the US Special Operation Command's (USSOCOM) Modular Weapon System requirement. The contract initially called for applications with USSOCOM 5.56 mm M4A1 Carbines but the system is equally applicable to other M16 series rifles.

The heart of the 5.56 mm Modular Weapon System (MWS) is KAC's aluminium RIS forestock which replaces the standard components and mounts four sections of Picatinny Rail around the barrel to facilitate the attachment of various tactical accessories. The forestock also acts as thermal protection for the user.

Description

The KAC Modular Weapon System (MWS) is based on M16 series rifles and M4 Carbines. The base modular weapon combines a standard M16 rifle or M4 Carbine with a non-permanent Rail Interface System (RIS) which replaces the factory standard forestock components of the host weapon. The RIS is composed of four parallel accessory mounting rails configured to MIL-STD-1913, the Picatinny Rails. These rails are mounted at 12, 3, 6 and 9 o'clock around the barrel and thus provide precise indexing points for the mounting of a wide range of tactical accessories such as sights and aiming devices.

The individual rails of the RIS contain several so-called precision 'recoil grooves' along their lengths. Odd numbered recoil grooves of each quadrant rail are sequentially numbered. Numbers on the top rail have a T prefix while those on the bottom rail have a B prefix; the rails to the shooter's right and left have R and L prefixes respectively. These numbers and prefixes are provided to assist the operator when remounting an accessory in the same position and to provide an 'address' for every position on the system. This explains to an operator exactly where to mount an accessory and can denote which addresses are incompatible with some accessories.

As accessories are added or repositioned, the battle sight zero of reflex sights and aim lights may be confirmed without firing. The zero confirmation of reflex sights is achieved by adjusting the point of aim of the optic to that of the prezeroed flip-up rearsight while simultaneously sighting through both. Zero confirmation of aim lights is achieved by sighting through the prezeroed flip-up rearsight and adjusting the point of aim.

Tactical accessories that can be used with the RIS include both standard and flip-up iron sights, telescopic sights, reflex sights, starlight telescopic sights, small night vision devices, visible and infra-red laser aim lights, a vertical foregrip, a quick-detach mounting for a 40 mm M203 grenade launcher, a quick-detach sound suppressor, police flashlights, and a KAC Masterkey breaching weapon, the latter being an adaptation of the Remington 870 shotgun. KAC can provide suitable mounts for many types of sighting system. Each RIS rail contains three holes provided for the attachment of camera or video accessories, typically a tripod.

Various lengths of RIS are available to suit a wide array of requirements and weapons. The RIS is claimed to provide excellent thermal protection for the firer and excellent barrel cooling.

M4A1 Carbine fitted with the KAC Rail Interface System (RIS) the heart of the KAC Modular Weapon System (MWS); the RIS can be seen mounting several types of combat accessory, including a vertical foregrip

1996

Manufacturer

Knight's Armament Company, 7750 9th Street SW, Vero Beach, Florida 32968.

Status

In production for USSOCOM (see text).

NEW ENTRY

7.62 mm Stoner SR-25 rifle

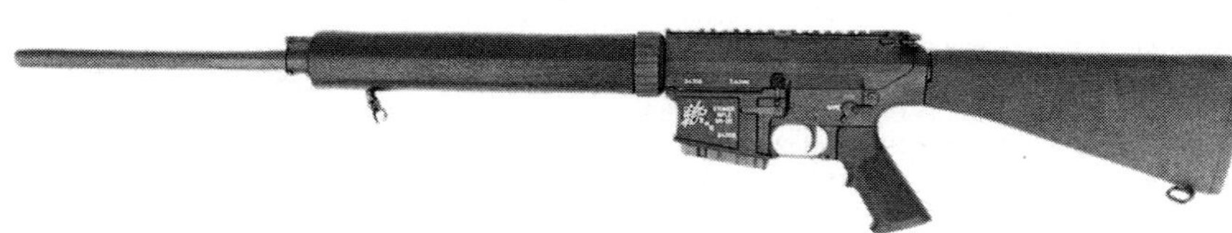

7.62 mm SR-25 Match Rifle (without sights)

1996

7.62 mm SR-25 Stoner Carbine (without sights)

1996

Development

Knight's Armament Company has designed and produced a 7.62 mm version of the M16 rifle, thereby marking a reversion to the Stoner 7.62 mm AR-10 of the mid-1950s. The design employs over 60 per cent of existing M16 parts and, with its similarity to the M16, costly retraining of personnel can be avoided. The SR-25 rifle has a possible US Army application as a support weapon in sniper teams where the 'second man' can have a local defence function in addition to using the SR-25 as a replacement rifle should the main sniper rifle become disabled for any reason.

Description

On the 7.62 mm SR-25 the upper receiver and gas block have a flat top Picatinny Rail design, allowing the user to adopt a variety of options, including a carrying handle and iron sights. The handguard is attached only at the receiver end, providing a fully floating barrel and avoiding any change in zero when used with a bipod. Initial testing has shown the ability to achieve groups of less than 19 mm at 100 m range.

Variants

7.62 mm SR-25 Match Rifle

Intended for target shooting, the SR-25 Match Rifle has a Remington 5R carbon steel free floating barrel with a twist of one turn in 286 mm, optimised for M118 Military Match ammunition with a 173 grain bullet. The barrel is 610 mm long and overall length is 1.105 m; weight is 4.87 kg.

7.62 mm SR-25 Lightweight Match Rifle

The SR-25 Lightweight Match Rifle has a 508 mm free floating barrel and the weight is reduced to 4.3 kg.

7.62 mm SR-25 Stoner Carbine

The SR-25 Stoner Carbine follows the same general lines as the other SR-25 series but the barrel length is reduced to 406 mm and the overall length is 914 mm. A special shortened fluted non-slip handguard is provided. Weight is 3.515 kg.

7.62 mm SR-25 Sporter

In many ways the SR-25 Sporter marks a return to the overall concept of the Stoner 7.62 mm AR-10 but there are many detail design changes. The overall appearance is that of an enlarged AR-15/M16 rifle, with the carrying handle and sight assembly removable to reveal a Picatinny Rail capable of mounting numerous optical sights and other sighting systems. The handguards are M16A2 pattern components. Overall length is 990 mm and the barrel length is 508 mm; weight is 3.97 kg. The barrel is not free floating on this model.

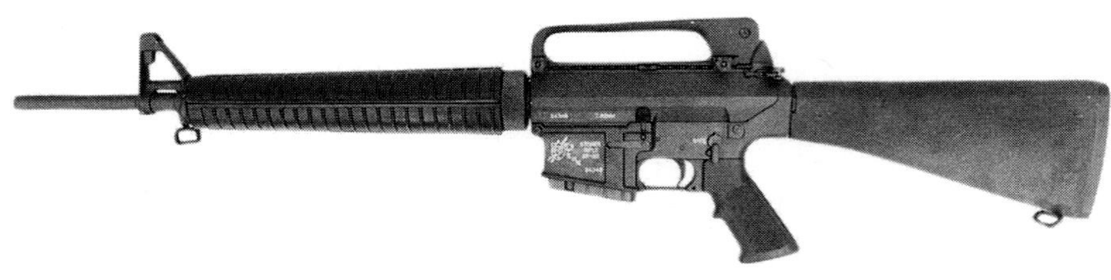

7.62 mm SR-25 Sporter

1996

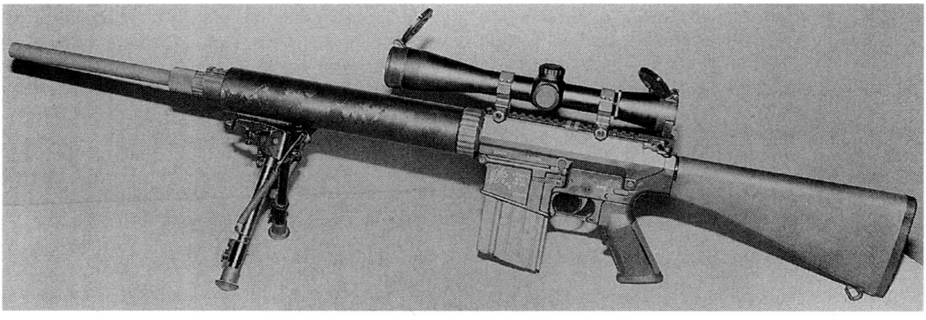

7.62 mm Stoner SR-25 rifle (T J Gander)

Data - Base rifle

Cartridge: 7.62 × 51 mm NATO
Operation: gas, semi-automatic
Locking system: rotating bolt
Feed: 10- or 20-round box magazine
Weight: with magazine, 4.88 kg
Length: 1.1175 m

Manufacturer

Knight's Armament Company, 7750 9th Street SW, Vero Beach, Florida 32968.

Status

Available.

UPDATED

La France 5.56 mm M16K assault carbine

Description

Among the smallest and lightest of the M16 variants, the La France 5.56 mm M16K assault carbine can be carried in the confined spaces of aircraft and motor vehicles. It was designed to give special operations units controllable firepower and high lethality in a weapon of minimum size and weight.

The specially engineered gas system results in a cyclic rate of fire under 600 rds/min, unusually low for weapons of this calibre. A vortex flash suppressor completely eliminates any trace of muzzle flash. The M16K utilises an aperture foresight in combination with the standard rearsight in an attempt to duplicate the effectiveness of the ring-type graticule in the optical sight of the Steyr AUG.

A molybdenum-disulphide dry-film lubricant is applied to all metal components. This maintenance-free finish prevents rust and corrosion and provides lubrication to moving parts in all extremes of climate and temperature.

Variant

La France 0.45 M16K sub-machine gun
See entry in *Sub-machine guns* section.

Data

Cartridge: 5.56 × 45 mm
Operation: gas, selective fire
Locking: rotating bolt
Feed: 20-, 30- or 90-round magazine
Weight: empty, 2.5 kg
Length: butt retracted, 610 mm; butt extended, 686 mm
Barrel: 213 mm; with flash suppressor, 254 mm
Rifling; 6 grooves, rh, 1 turn in 177.8 mm
Sights: fore, aperture; rear, flip aperture, adjustable for windage
Sight radius: 127 mm
Muzzle velocity: M193, 732 m/s
Rate of fire: cyclic, 550-600 rds/min

Manufacturer

La France Specialties, PO Box 178211, San Diego, California 92177-8211.

Status

In production.

VERIFIED

La France 5.56 mm M16K assault carbine

7.62 mm Model 85 Parker-Hale sniper rifle

Development

The Model 85 was originally made in the United Kingdom by Parker-Hale and was entered in the 'shoot-off' competition which led to the adoption of the Accuracy International L96A1 as the British Army's new sniper rifle. In 1990 Parker-Hale sold the manufacturing rights, and the Gibbs Rifle Company Inc was established to manufacture the entire Parker-Hale rifle line including the Model 85.

The Model 85 Sniper is an adaptable and robust high-precision rifle designed to provide a 100 per cent first round hit capability at all ranges up to 600 m, and is sighted up to 900 m.

7.62 mm Model 85 Parker-Hale sniper rifle with British Army camouflage pattern stock

Description

The specially designed action of the Model 85 has a built-in aperture rearsight adjustable up to 900 m for emergency use, or when the use of optical sights would be impractical. The integral dovetail mounting is designed for rapid attachment and removal of telescope or passive night sights and features positive return-to-zero and recoil stop.

A suppressor can be fitted for use with supersonic or reduced velocity ammunition. The suppressor eliminates all muzzle flash and firing signature and significantly reduces the recoil energy. The muzzle is threaded to accept the suppressor and the front sight assembly, which is clamped to the barrel and may be removed and replaced with the aid of an Allen key.

Bracketing is available for fitting the Simrad KN250 night vision sight to most daytime telescope sights, thus eliminating the need for an adjustable height butt-stock.

The Model 85 has an adjustable length butt-stock to suit all sizes of user, and is equipped with a quickly detachable militarised bipod with provision for both swivel and cant adjustment by the sniper when in the firing position.

A complete kit is offered, including high-duty welded aluminium foam-lined transit and storage case containing all necessary accessories for the sniper. These optional ancillary items may include passive night vision individual weapon sight, soft padded cover, spotting telescope and bipod, cleaning and maintenance tools, spare magazine and sling. Final specifications are matched to the user's requirements.

7.62 mm Model 85 Parker-Hale sniper rifle with black stock and night vision sight

Data

Cartridge: 7.62 × 51 mm NATO
Operation: bolt action, single shot
Feed: 10-round box magazine
Weight: with telescope sight, 5.7 kg
Length: min, 1.15 m; max, 1.21 m
Barrel: 700 mm
Rifling: 4 grooves, rh, 1 turn in 305 mm
Feed: 10-round box magazine

Manufacturer

Gibbs Rifle Company Inc, Cannon Hill Industrial Park, Hoffman Road, Martinsburg, West Virginia 25401.

Status

Available.

VERIFIED

7.62 mm M21 rifle

Description

The 7.62 mm M21 rifle was originally called the US Rifle 7.62 mm M14 National Match (Accurised). It was the standard US Army sniper rifle but is being phased out and replaced by the M24 Sniper Weapon System (qv).

The differences between this rifle and the M14 are that the barrels are gauged and selected to ensure correct specification tolerances. Barrels are not chromium plated; the stock is of walnut and is impregnated with an epoxy resin; the receiver is individually fitted to the stock using a glassfibre compound; the firing mechanism is hand fitted and polished to provide a crisp hammer release. Trigger pull is between 2 and 2.15 kg; the gas cylinder and piston are hand fitted and polished to improve operation and reduce carbon build up; and the gas cylinder and lower band are permanently attached to each other. The rifle must group consistently with an average extreme spread for a 10-round group not exceeding 150 mm at 300 m. A sound suppressor may be fitted to the muzzle. This does not affect the bullet velocity but reduces the velocity of the emerging gases to below that of sound. The suppressor is hand fitted, and reamed to improve accuracy and eliminate misalignment.

The telescope uses two stadia on the horizontal graticule which subtend 1.52 m at 300 m when viewed with the telescope variable magnification ring set to ×3 power. On the vertical graticule are two stadia which subtend 760 mm at 300 m at the same setting. As an illustration of this, the distance from the soldier's waist belt to the top of his helmet can be taken to be 760 mm. At 300 m range the two stadia on the vertical graticule will, with a ×3 setting, rest on the waist belt and top of the steel helmet. If the range is greater than 300 m the power ring is used to increase the size of the picture until once again the two stadia lines rest on the waist belt and steel helmet. Clearly the range is proportional to the power used, for example since the ×3 magnification just places the stadia on belt and helmet at 300 m then the ×9 magnification will put the stadia on belt and helmet at 900 m.

This system can be used to give the sniper a range readout which he can use in reporting enemy positions, for example. In addition, however, a ballistic cam is attached to the telescope power ring. This cam is cut for the cartridge in use and so the act of placing the

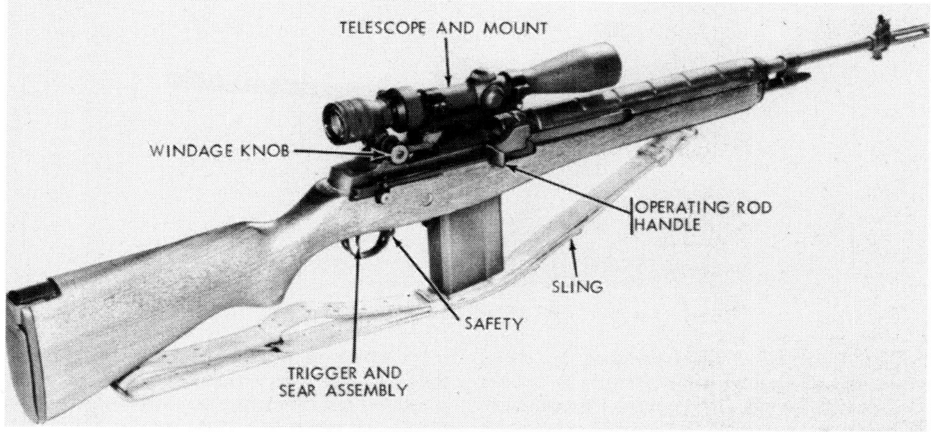

7.62 mm M21 rifle

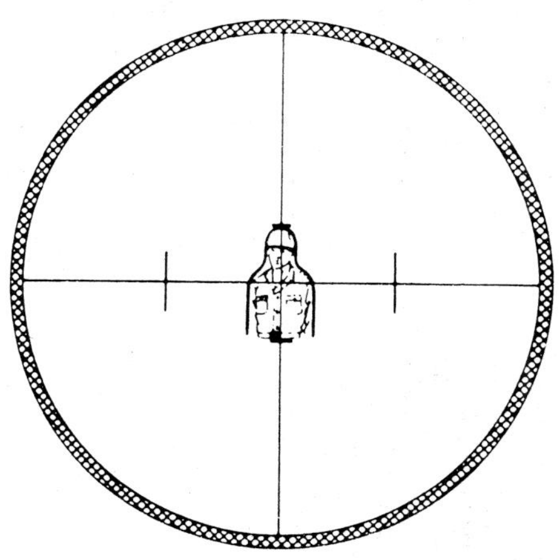

Sniper's sight picture-stadia correctly adjusted

stadia in the correct position not only records the range but displaces the telescope axis and so automatically applies the correct tangent elevation to the rifle.

Data

TELESCOPE
Weight: with cam, 455 g
Length: 324 mm
Magnification: variable ×3 to ×9
Eye reliefs: 76.2-95.3 mm
Adjustments: internal; 30 s graduations for elevation and windage
Graticule: cross-hairs with stadia marks
Ballistic cam: for M118 match ammunition
Objective lens diameter: 46 mm
Eyepiece diameter: 34 mm
Finish: black matt anodised
Mount: (weight) 170 g; (material) aluminium; (operation) hand fixed, spring-loaded base; (finish) black matt anodised

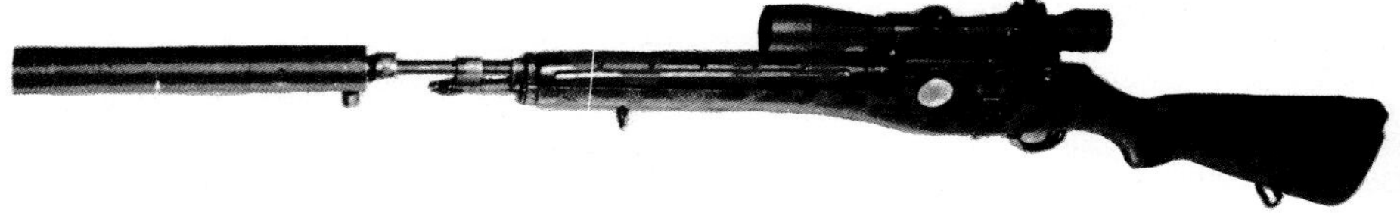

Suppressor on 7.62 mm M21 rifle

Status

Production complete.

Service

US Army. Being phased out in favour of M24 Sniper Weapon System.

VERIFIED

Springfield Armory M21 sniper rifle

Description

The Springfield Armory M21 sniper rifle is based on the M21 rifle (see previous entry) but with some improvements. The stock is of a new pattern, with adjustable cheekpiece and rubber recoil pad, and has the rifle action and barrel bedded in glassfibre resin. The bipod has been redesigned for ease of use and better stability. The heavy barrel is specially made by Douglas, air-gauged and has a twist of one turn in 254 mm; the operating rod guide has also been redesigned. The telescope mount is of Springfield's own design, and whilst the choice of sight is left to the purchaser, Springfield recommends the Leupold Stevens 3.5 × 10 variable-power sight.

Manufacturer

Springfield Armory Inc, 420 West Main Street, Geneseo, Illinois 61254.

Status

Available to special order only.

VERIFIED

7.62 mm Springfield Armory M21 sniper rifle

7.62 mm M40A1 sniper rifle

Description

The M40A1 is a conventional bolt-action magazine rifle with heavy barrel and wooden stock. There are no iron sights, a telescope sight being maintained on the rifle at all times. A catch inset in the forward edge of the trigger guard allows the bolt to be removed, and a catch in front of the magazine floor plate enables the magazine spring and follower to be removed for cleaning.

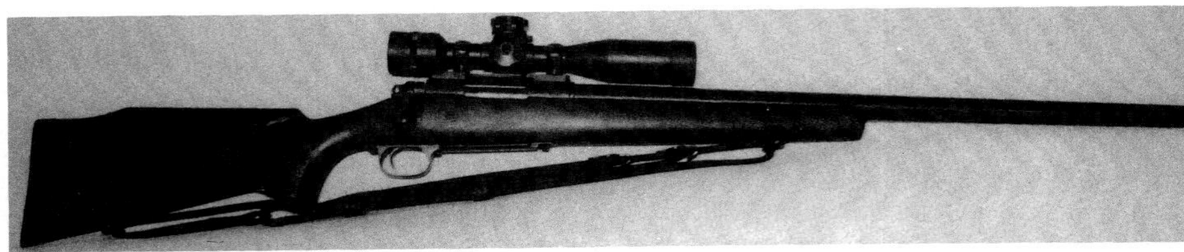

7.62 mm M40A1 sniper rifle

Data

Cartridge: 7.62 × 51 mm NATO
Operation: bolt action, single shot
Feed: 5-round integral magazine
Weight: 6.57 kg
Length: 1.117 m
Barrel: 610 mm
Rifling: rh, 1 turn in 305 mm
Muzzle velocity: 777 m/s
Sight: telescope, USMC sniper, ×10

Manufacturer

Remington Arms Company Inc, PO Box 179, Ilion, New York 13357-0179.

Status

In production.

Service

US Marine Corps.

VERIFIED

7.62 mm M24 Sniper Weapon System

Description

The M24 is the US Army's first complete sniping system and will eventually replace all other service sniper rifles. First issues, as a non-developmental item, took place in November 1988 and the system is being issued to all infantry battalions, special forces and Ranger units. It has been stated that the procurement objective is 2,510 systems.

The rifle M24 was developed by Remington, based upon their commercial M700 long bolt action and an M/40X custom trigger mechanism. The design was developed around the M118 special sniper ball cartridge, with the possibility of adapting to the 0.300 Winchester Magnum cartridge as a retrospective modification if required. The stock is synthetic, made of Kevlar-graphite, with an aluminium bedding block and adjustable butt-plate.

7.62 mm M24 sniper rifle

The complete system consists of the six-shot bolt-action rifle, bipod, a laser-hardened day optical sight, iron sights, deployment kit, cleaning kit, soft rifle carrying case, telescope carrying case and total system carrying case. The complete system in its carrying case weighs 25.4 kg.

The M24 Sniper Night Sight was due to be added to the M24 Sniper Weapon System by late 1996. Also scheduled for addition was a flash and blast attenuation muzzle attachment (late 1995), and the XM144 telescope (August 1996), and a straight tripod-mounted telescope for surveillance and target acquisition. Some form of anti-reflection device is under investigation for the M24's day optical sight.

Data

Cartridge: 7.62 × 51 mm M118 Special Ball
Operation: bolt action, single shot
Feed: 6-round integral magazine
Weight: rifle with sling, 5.49 kg; sight, 794 g; bipod, 318 g
Length: 1.092 m
Rifling: 5 grooves, rh, 1 turn in 285 mm
Sight: Leupold Ultra M3 ×10 telescopic
Muzzle velocity: ca 792 m/s
Max effective range: 800 m

Manufacturer

Remington Arms Company Inc, PO Box 179, Ilion, New York 13357-0179.

Status

In production.

Service

US Army.

UPDATED

7.62 × 36 mm Grendel S-16 rifle

Description

The Grendel Silent-16 (S-16) is an intermediate range (to 300 m) silenced sniper rifle firing a special 7.62 × 36 mm round. The S-16 uses standard M16 magazines and other components. By concept, the S-16 is primarily intended for accurate semi-automatic shooting but can also deliver full-automatic fire. Everything except the upper receiver is standard M16, and the S-16 is the same size and only slightly heavier than the M16.

The intent of the S-16 rifle is to increase performance compared to present systems in range and lethality, and also in simplified training and logistics.

The M16 was selected as the basis for this design because of the simplified training and logistic load and also the inherent accuracy. Within the subsonic velocity envelope an extensive study was conducted, as well as ballistic tests, to determine the optimum projectile.

A 7.83 mm streamlined pointed projectile with a weight of 14.3 g proved to give the best performance in both accuracy and velocity retention. Bullets with even higher sectional density were tried, but showed excessive dispersion even with very fast rifling twists. The extremely high ballistic coefficient of this bullet allows it to retain more energy at 300 m than a 9 mm ball has at the muzzle.

Data

Cartridge: 7.62 × 36 mm
Operation: gas, selective fire
Locking: rotating bolt
Weight: empty, 4.3 kg; with loaded magazine, 4.8 kg
Trigger pull: 2.3-3.8 kp
Length: 995 mm
Barrel: 409 mm
Rifling: 6 grooves, rh, 1 turn in 203 mm

Manufacturer

Grendel Inc, PO Box 560908, Rockledge, Florida 32956-0908.

Status

Available.

VERIFIED

7.62 × 36 mm Grendel silenced sniper rifle

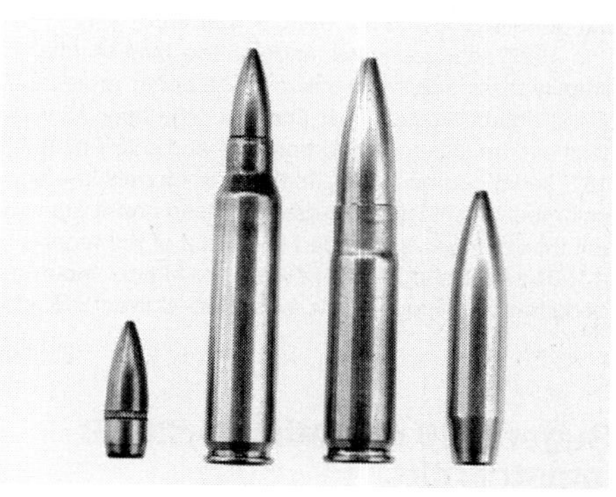

Comparison of 5.56 × 45 mm and Grendel 7.62 × 36 mm rounds

Ruger 7.62 mm Mini Thirty rifle

Description

The Ruger 7.62 mm Mini Thirty rifle is a modified version of the Ruger Mini-14 Ranch Rifle chambered for the CIS 7.62 × 39 mm M43 cartridge. The barrel, receiver and bolt have been engineered to accommodate the larger cartridge. The Mini Thirty is designed for use with telescope sights and features a low, compact telescope mounting which provides greater potential accuracy and carrying facility than found in other rifles of this calibre.

Ruger 7.62 × 39 mm Mini Thirty rifle

Data

Cartridge: 7.62 × 39 mm M43
Operation: gas, semi-automatic
Locking: rotating bolt
Feed: 5-round box magazine
Weight: 3.26 kg
Length: 948 mm
Barrel: 470 mm
Rifling: 6 grooves, rh, 1 turn in 254 mm
Sights: fore, bead on post; rear, flip aperture; integral telescope mounts
Muzzle velocity: 713 m/s

Manufacturer

Sturm, Ruger & Company Inc, Lacey Place, Southport, Connecticut 06490.

Status

In production.

VERIFIED

Ruger 5.56 mm Mini-14 rifle

Development

The Ruger 5.56 mm Mini-14 rifle was introduced commercially in 1973, and has since been adopted by various military and security forces. The mechanism is similar to that used on the M1 and M14 rifles in many respects, but the reductions made possible by the use of high tensile alloy steels have resulted in a light handy rifle with all the advantages conferred by the flat trajectory of the 5.56 × 45 mm cartridge.

Description

To load the first round the cocking handle is pulled back and released. The compressed return spring drives it forward and the bottom face of the bolt picks up the top cartridge in the magazine and pushes it into the chamber. The operating slide carries a cam path in which the bolt roller engages. When the round is chambered the bolt comes to rest but the slide continues on and the bolt is rotated clockwise and locked. The slide continues on for approximately 6 mm over the bolt roller and when the end of the cam path contacts the roller all forward movement ceases and the gun is ready to fire.

When the trigger is pulled the hammer drives the firing pin against the cap and the round is fired. The gas pressure forces the bullet up the bore and the gas follows it up towards the muzzle. Some of the gas passes down through a vent drilled radially through the barrel wall and through a stationary piston to impinge on the hollow interior face of the slide. The slide is driven back and the straight action of the cam path moves over the bolt roller. This free travel of approximately 6 mm allows the chamber pressure to drop to atmospheric before the roller is lifted by the cam path and the bolt is rotated. The firing pin is cammed back. The rotation of the bolt dislodges the cartridge case in the chamber and provides primary extraction. The bolt is then carried to the rear and the empty case is extracted and ejected to the right and forward. The bolt continues back over the rounds in the magazine and the return

Ruger 5.56 mm Mini-14 rifle with 20-round magazine

spring is fully compressed. The spring reasserts itself and drives the slide forward to repeat the cycle.

If the ammunition is expended the magazine follower operates the bolt hold open device. This can be released either by operating the bolt release on the left of the receiver, or by withdrawing the magazine, replacing with a full magazine and pulling back slightly on the cocking handle. When it is released the bolt will go forward and chamber the top cartridge.

The applied safety lies in the front of the trigger guard and can be operated by the movement of the forefinger of the trigger hand. It is pushed forward to 'fire' and pulled back to 'safe'. When at 'safe' it blocks the hammer and pushes it down off the sear which is thus disconnected. When the weapon is set to 'safe' the slide can be retracted and the weapon loaded.

In 1977 changes were made to the Mini-14 rifle to simplify the mechanism and provide better protection of the action against dust and dirt. The later Mini-14 rifles can be distinguished from earlier models by their '181' serial number prefix. In the later models, the bolt lock mechanism is completely enclosed and a manual bolt lock plunger is provided in the top of the receiver. In 1981 a flash suppressor of the old M14 type, incorporating radio-luminous sights, was fitted. Current models also feature a glassfibre ventilated handguard which protects the firer's hand from barrel heat and from contact with the moving slide.

A stainless steel version of the Mini-14 rifle was introduced in 1978 and is in production. Data and specifications are substantially the same as for the standard model.

A conversion kit to allow this weapon to fire 0.22 LR ammunition is available from Jonathan Arthur Ciener Inc. See separate entry in this section for details.

Data

Cartridge: 5.56 × 45 mm M193, SS109 or any commercial counterpart
Operation: gas, self-loading
Locking: rotating bolt
Feed: 5-, 20- and 30-round box magazine. 30-round magazine available only to government and law enforcement agencies
Weight: 2.9 kg
Length: 946 mm
Barrel: 470 mm
Rifling: 6 grooves, rh, 1 turn in 178 mm
Sights: fore, bead on post; rear, aperture. Adjustable for elevation and windage. Clicks of 1 min of angle (1 min is approx 30 mm at 100 m); (zeroing) elevation, rearsight. Line, rearsight
Sight radius: 561 mm
Muzzle velocity: 1,005 m/s
Effective range: 300 m

Manufacturer

Sturm, Ruger & Company Inc, Lacey Place, Southport, Connecticut 06490.

Status

In production.

UPDATED

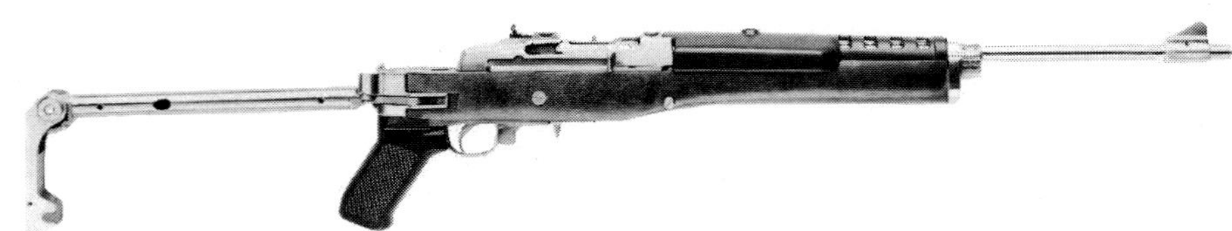

Ruger 5.56 mm Mini-14 Ranch Rifle, folding stock model in stainless steel

Ruger 5.56 mm Mini-14/20GB infantry rifle

Description

The Ruger Mini-14/20GB infantry rifle is an adaptation of the commercial Ruger Mini-14 rifle intended for military applications. Changes from the basic Mini-14 configuration include the addition of a protected front sight which incorporates a bayonet lug, a heat-resistant glassfibre handguard, and a flash suppressor. Technical data and specifications for the Mini-14/20GB rifle are substantially the same as for those of the standard Mini-14 rifle. The Mini-14/20GB rifle is available in both blued and stainless steel versions.

Manufacturer

Sturm, Ruger & Company Inc, Lacey Place, Southport, Connecticut 06490.

Status

In production.

UPDATED

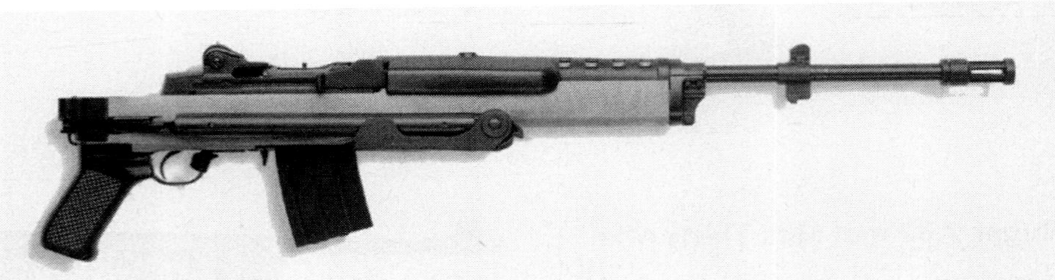

Ruger 5.56 mm Mini-14/20GBF, blued, folding stock, infantry rifle

Ruger 5.56 mm K-Mini-14/20GB stainless steel infantry rifle

Ruger 5.56 mm AC-556 selective fire weapon

Description

Although similar in appearance to the standard Mini-14/20GB infantry rifle, the Ruger AC-556 is a selective fire weapon specifically designed for military and law enforcement applications using standard 5.56 mm military or commercial ammunition. The AC-556 is equipped with a heat-resistant, ventilated glassfibre handguard, protected foresight which incorporates a bayonet lug, and a flash suppressor. The fire control mechanism consists of a positive three-position selector lever which provides semi-automatic, three-shot burst or fully automatic fire. The selector lever is at the right rear of the receiver and is readily accessible to right- or left-handed shooters.

Variations on the basic model include the KAC-556 in stainless steel with fixed stock; AC-556GF, folding stock and blued finish; and KAC-556GF, folding stock and stainless finish.

A conversion kit to allow this weapon to fire 0.22 LR ammunition is available from Jonathan Arthur Ciener Inc. See separate entry in this section for details.

Ruger 5.56 mm AC-556 selective fire weapon

Data

Cartridge: 5.56 × 45 mm M193, SS109 or any commercial counterpart
Operation: gas, self-loading
Locking: rotating bolt
Feed: 5-, 20- or 30-round box magazines
Weight: 2.89 kg
Length: 984 mm
Barrel: 470 mm
Rifling: 6 grooves, rh, 1 turn in 305 mm
Sights: fore, protected military type, with bayonet lug; rear, aperture adjustable for windage and elevation; clicks of 1 min of angle (1 min is approx 30 mm at 100 m)
Muzzle velocity: 1,058 m/s
Rate of fire: cyclic, ca 750 rds/min

Manufacturer

Sturm, Ruger & Company Inc, Lacey Place, Southport, Connecticut 06490.

Status

In production.

UPDATED

Ruger 5.56 mm AC-556F selective fire weapon

Description

The Ruger AC-556F is a short barrel, selective fire weapon optionally equipped with a steel folding stock. Employing the same basic mechanism as the Ruger AC-556 model, the AC-556F is adapted for aircraft and armour crews, paratroops, patrol vehicles, dignitary protection, law enforcement and other applications where a high rate of fire combined with compact configuration and short overall length are required. Except as listed below, technical data and specifications are the same as those for the Ruger AC-556 model. Both the Ruger AC-556 and AC-556F selective fire weapons are available in blued or stainless steel versions. In stainless steel they are designated the KAC-556 and KAC-556F respectively.

A conversion kit to allow this weapon to fire 0.22 LR ammunition is available from Jonathan Arthur Ciener Inc. See separate entry in this section for details.

Ruger 5.56 mm KAC-556GF selective fire weapon

Data

Cartridge: 5.56 × 45 mm M193, SS109 or any commercial counterpart
Operation: gas, self-loading
Locking: rotating bolt
Feed: 5-, 20- or 30-round box magazines
Weight: 3.15 kg; loaded, 20 rounds, 3.6 kg
Length: stock folded, with flash suppressor, 603 mm; stock opened, with flash suppressor, 851 mm
Barrel: 330 mm
Rifling: 6 grooves, rh, 1 turn in 254 mm

Manufacturer

Sturm, Ruger & Company Inc, Lacey Place, Southport, Connecticut 06490.

Status

In production.

UPDATED

Ruger M77 Mark II rifle

Description

The Ruger M77 Mark II rifle is described as a 'countersniper' rifle and is intended primarily for use by police and paramilitary units. It is an extremely well-made and well-finished weapon of 'classic' design, being a bolt action repeater.

The bolt has dual opposed front locking lugs and a non-rotating Mauser type extractor which cocks on opening. The entire action is heat treated investment cast chrome-molybdenum steel with a matt blue finish. A two-stage target type trigger mechanism is provided, fully adjustable for let-off and sear engagement. The match grade free floating barrel is hammer forged stainless steel, while the one-piece stock is high quality laminated black American hardwood.

The standard calibre is 0.308 Winchester (7.62 × 51 mm), although other calibres can be produced on request. The weapon has a four-round magazine and is intended to be carried with one further round in the chamber; the bolt has a three-position wing safety which permits unloading with the safety engaged. Accuracy using 0.308 Winchester ammunition is such that five shots can be grouped within a 25.4 mm mean radius at 100 yards/91.4 m.

The Ruger M77 Mark II is supplied without sights, although a pair of 25.4 mm scope rings is provided to accommodate an optical sight of the user's choice. A Harris detachable pivoting bipod is also provided as standard, as are a sling, sling swivels, and a cleaning kit.

A 5.56 mm version of this rifle is available.

Ruger M77 Mark II rifle

1996

Data

Cartridge: 0.308 Winchester
Operation: bolt action, single shot
Locking: rotating bolt
Feed: 4-round magazine
Weight: empty, approx 4.42 kg
Length: 1.184 m
Barrel: 660 mm
Rifling: 6 grooves, rh, 1 turn in 254 mm
Sights: optical, attached as required by user

Manufacturer

Sturm, Ruger & Company Inc, Lacey Place, Southport, Connecticut 06490.

Status

In production.

NEW ENTRY

McMillan 0.50 M-87R sniper rifle

Development

The McMillan M-87R is a heavy bolt-action magazine rifle chambered for the 12.7 × 99 mm (0.50 Browning) cartridge and which is capable of accurate shooting to ranges in excess of 1,500 m. The design is entirely conventional, requiring little maintenance and relatively brief familiarisation and training.

The M-87R is the magazine-fed repeater; the M-87 is a single shot weapon without magazine, while the M-88 is similar to the M-87R but designed for rapid takedown.

Description

The McMillan 0.50 M-87R sniper rifle is fitted with a highly efficient muzzle brake and the recoil is sufficiently reduced to permit firing from the shoulder in emergencies when there is no time for, or where the terrain denies, adoption of the prone position. The rifle can be fitted with a bipod and with a variety of optical and electro-optical sights. Additionally, the rifle can be supplied with a glassfibre stock which provides adjustments in length, drop and pull to suit the individual user.

Variants

McMillan M-93 and M92
See following entries.

Data

Cartridge: 12.7 × 99 mm (0.50 Browning)
Operation: bolt-action repeater
Feed: 5-shot magazine
Weight: without optical sight, 9.52 kg
Length: 1.346 m; when dismantled for carrying, 991 mm
Barrel: 736.5 mm

Manufacturer

McMillan Gunworks Inc, 302 W Melinda Lane, Phoenix, Arizona 85027.

Status

In production.

Service

French Army (48).

VERIFIED

McMillan 0.50 M-87R sniper rifle with bipod

McMillan 0.50 M-93 sniper rifle

Description

The McMillan M-93 sniper rifle is a revised version of the M-87R (see previous entry) and is thus a heavy bolt-action magazine rifle chambered for the 12.7 × 99 mm (0.50 Browning) cartridge capable of accurate shooting to ranges in excess of 1,500 m. The main change to the earlier weapon is the inclusion of a hinged butt-stock for ease of storage and transport. Another innovation is the provision of a 10- or 20-round detachable box magazine. It is claimed that with these magazines fitted it is possible for a trained user to fire accurately between 8 and 10 rds/min.

Data

Cartridge: 12.7 × 99 mm (0.50 Browning)
Operation: bolt-action repeater
Feed: 10- or 20-round magazine
Weight: without optical sight, 9.52 kg
Length: 1.346 m; with butt folded, 991 mm
Barrel: 736.5 mm

Manufacturer

McMillan Gunworks Inc, 302 W Melinda Lane, Phoenix, Arizona 85027.

Status

In production.

Service

French Army (18).

VERIFIED

McMillan 0.50 M-93 sniper rifle

McMillan 0.50 M-93 sniper rifle showing hinged butt-stock

McMillan 0.50 M-92 Bull Pup sniper rifle

Description

The McMillan M-92 Bull Pup sniper rifle is a shortened version of the M-87R (qv) and is thus a heavy bolt-action magazine rifle chambered for the 12.7 × 99 mm (0.50 Browning) cartridge capable of accurate shooting to ranges in excess of 1,500 m. The main change is to the butt-stock which has been reconfigured to provide a bullpup configuration, making the rifle much handier in confined areas and more convenient to carry and stow.

The butt-stock is fibreglass and has an adjustable cheekpiece. Recoil forces are reduced by the McMillan M-87R pattern muzzle brake.

The M-92 Bull Pup is provided with a bipod and can accept a variety of optical and night sights.

Weight of the McMillan M-92, with an optical sight, is approximately 10.9 kg.

A semi-automatic version of this model is now available.

Manufacturer

McMillan Gunworks Inc, 302 W Melinda Lane, Phoenix, Arizona 85027.

Status

Available.

UPDATED

McMillan 0.50 M-92 Bull Pup sniper rifle

Barrett 0.50 'Light Fifty' Model 82A1 rifle

Development

The Barrett 0.50 'Light Fifty' Model 82A1 is a semi-automatic rifle firing the 0.50 in (12.7 × 99 mm) Browning heavy machine gun cartridge. It is intended as an explosive ordnance disposal (EOD) and long-range interdiction weapon and is suggested as having uses in military and police employment and also as a suitable defensive weapon for ocean-going light craft.

Description

The Barrett 0.50 'Light Fifty' Model 82A1 rifle operates on short recoil principles. Upon firing, the bullet is forced through the barrel and considerable case-head thrust is delivered to the bolt face. This thrust transfers through the bolt lugs to the barrel extension and through the bolt body to the rear of the bolt carrier. The transfer to the carrier is unique to this design and dissipates much of the firing shock which might otherwise damage the locking assembly.

The barrel has recoiled approximately 13 mm when the bullet leaves the barrel. The barrel, bolt and bolt carrier continue to the rear under recoil, during which movement the cocking lever withdraws the firing pin and recocks it. At 25 mm of recoil travel an accelerator arm strikes an abutment in the receiver and swings forward, forcing the accelerator rod to separate the bolt carrier and the barrel extension. This action serves to rotate the bolt out of engagement with the barrel extension and to transfer energy from the barrel to the bolt carrier. The barrel continues to the rear for a total of 53 mm, stopping when a key, locked into a recess in the barrel, strikes the barrel stop. The bolt carrier continues to move under its own momentum and begins to separate from the barrel. A cam pin in the carrier, engaged in a helical cut in the bolt, causes the bolt to rotate 30° during the separation of carrier and barrel, thus moving the bolt locking lugs out of engagement with the barrel extension. At this point the bolt is fully extended from the carrier and the locking recess, on the bolt body, has been rotated so that a bolt latch, mounted in the carrier, engages so as to lock the bolt in the forward position.

As the carrier pulls the bolt to the rear, the fired cartridge case is withdrawn from the chamber, and as it clears the barrel extension so the case is expelled from the weapon by the ejector. As soon as the bolt is free of the barrel extension the barrel return springs restore the barrel to the forward battery position. The bolt carrier, having compressed the main return spring housed in the butt-stock, bottoms the buffer against the rear of the housing. The return spring then forces the bolt carrier forward, stripping a new cartridge from the magazine and chambering it. As the carrier moves forward the bolt latch is pivoted out of engagement with the bolt body by the pressure of the bolt latch trip, allowing the bolt to retract and rotate when it meets the breech face, locking the lugs into engagement with the barrel extension.

The barrel is fitted with a new high-efficiency muzzle brake, which reduces the recoil by 65 per cent. There is an adjustable bipod, and the rifle can also be mounted on small and large socket mounts with the use of the pintle adaptor, called the 'soft mount'.

The Model 82A1 is fitted as standard with a ×10 telescope, employing a ballistic reticle calibrated to the recommended ammunition types. The reticle allows adjustment of the point of impact from 500 to 1,800 m, considered to be the maximum effective range when using standard Browning ammunition.

The manufacturers recommend using the APEI multipurpose cartridge, but the rifle is capable of firing any standard 12.7 × 99 mm Browning machine gun ammunition. The preferred standard types are APHCI and AP M8.

Data

Cartridge: 12.7 × 99 mm (0.50 Browning)
Operation: short recoil, semi-automatic
Locking: rotating bolt
Feed: 10-round detachable box magazine
Weight: 12.9 kg

Length: 1.4478 m
Barrel: 736.7 mm
Sights: ×10 telescope
Muzzle velocity: M33 Ball, 853 m/s

Manufacturer
Barrett Firearms Manufacturing Inc, PO Box 1077, Murfreesboro, Tennessee 37133-1077.

Status
In production.

Service
US Marine Corps, Air Force, Army - EOD, Special Forces. Also France and UK armed forces.

UPDATED

Barrett 0.50 'Light Fifty' Model 82A1 rifle

Barrett 0.50 Model 82A2 semi-automatic rifle

Description
The Barrett 0.50 Model 82A2 semi-automatic rifle is the latest version of the Barrett M82 and is slightly smaller and lighter, somewhat simplified, but essentially the same mechanism. By relocating the pistol grip and trigger unit it has been converted, more or less, to a bull-pup configuration, with the shoulder rest now behind the magazine, allowing much of the receiver to pass across the firer's shoulder. This makes the weapon more handy and also reduces concealment problems.

Barrett 0.50 Model 82A2 semi-automatic rifle

Data
Cartridge: 12.7 × 99 mm (0.50 Browning)
Operation: short recoil, semi-automatic
Feed: 10-round box magazine
Weight: 12.24 kg
Length: 1.409 m
Barrel: 736 mm
Sights: optical

Manufacturer
Barrett Firearms Manufacturing Inc, PO Box 1077, Murfreesboro, Tennessee 37133-1077.

Status
In production.

VERIFIED

Barrett 0.50 Model 90A1 bolt-action rifle

Description
The Barrett 0.50 Model 90A1 bolt-action rifle was developed for potential users who prefer a bolt action weapon. The alteration of the mechanism has been accompanied by other modifications resulting in a shorter and lighter weapon than the semi-automatics. It is a bullpup design, with the action well back in the stock and with the pistol grip ahead of the magazine. A new high-efficiency muzzle brake has been fitted, and this, together with a 'Sorbothane' butt-pad, reduces the felt recoil to a quite manageable level. As with the other Barrett rifles there are no iron sights, but a mount is provided upon which virtually any preferred optical sight can be fitted.

Barrett 0.50 Model 90 bolt-action rifle as produced for US Navy

Data
Cartridge: 12.7 × 99 mm (0.50 Browning)
Operation: bolt action, single shot
Feed: 5-round box magazine
Weight: 9.98 kg
Length: 1.143 m
Barrel: 736 mm
Muzzle velocity: M33 Ball, 853 m/s

Manufacturer
Barrett Firearms Manufacturing Inc, PO Box 1077, Murfreesboro, Tennessee 37133-1077.

Status
In production.

Service
US Navy.

UPDATED

Ciener 0.22 LR conversion kits

Description
Jonathan Arthur Ciener Inc produce and market several conversion kits which allow standard service-calibre weapons to fire 0.22 LR rimfire ammunition for training or other purposes.

One kit is the Atchisson AR-15/M16 0.22 LR conversion kit which permits conversion of AR-15 and M16 series rifles. The unit replaces the standard bolt/bolt carrier assembly and allows the user to employ inexpensive 0.22 LR ammunition. Changeover time is 10 seconds or less. The unit is made from chrome-molybdenum heat-treated steel. With the addition of an automatic trip and anti-bounce weight it will permit semi- and full-automatic fire from weapons with a selective fire capability.

Ciener also markets the Hohrein Ruger Mini-14/AC-556 0.22 LR conversion kit which is applicable for both the 5.56 mm and 0.222 versions of the Ruger Mini-14 and AC-556 rifles. This unit consists of a

Atchisson AR-15/M16 0.22 LR conversion kit fitted to an M16 series rifle

1996

Hohrein Mini-14/AC-556 0.22 LR conversion kit fitted to a Mini-14 series rifle

1996

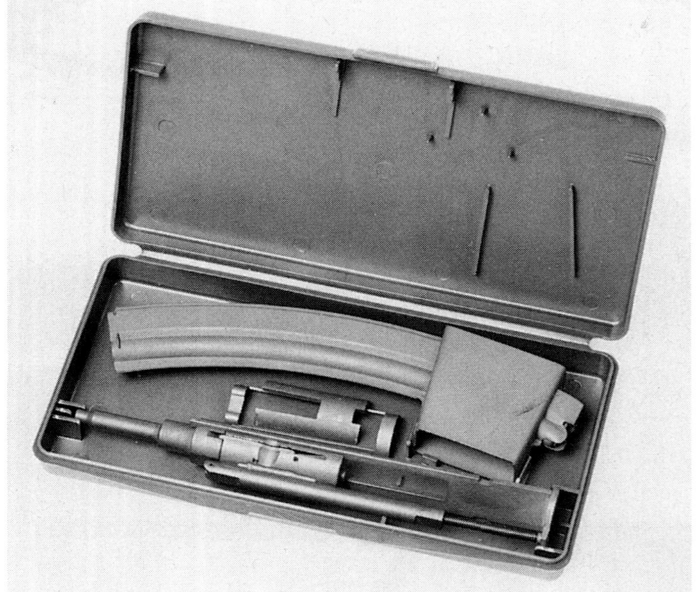

Ciener AK47/84 0.22 LR conversion kit fitted to a Type 56 rifle
1996

An Atchisson AR-15/M16 0.22 LR conversion kit in its fitted plastic case
1996

subcalibre chamber adaptor, a subcalibre bolt, a subcalibre operating rod, return spring and guide rod, guide rod locating adaptor and a 30-round 0.22 LR magazine, all in a fitted plastic case. These items are assembled into the dismantled firearm, replacing the standard parts, and the firearm is reassembled; changeover takes about one minute. The unit is made from chrome-molybdenum heat-treated steel. The standard kit fits Mini-14/AC-556 rifles with serial numbers prefixed by 188 through 187 and 190, 191. Another kit is available for 188 to current production firearms. The standard kit allows, semi- and full-automatic fire and three-round bursts on the AC-556 model.

A third Ciener-produced 0.22 LR kit, the AK47/84, permits the conversion of 5.56 and 7.62 mm AK series Kalashnikov rifles to fire 0.22 LR ammunition. This unit consists of a subcalibre chamber adaptor with a 228 mm barrel liner (no liner is needed for 5.56 mm weapons), a subcalibre bolt, return spring and return spring guide, a rod/locating block assembly and a 30-round 0.22 LR magazine. These items are assembled into the dismantled firearm, replacing the standard parts, and the firearm is reassembled; changeover takes about one minute. With the substitution of an automatic trip and anti-bounce weight the kit will permit semi- and full-automatic fire from weapons with a selective fire capability.

Manufacturer
Jonathan Arthur Ciener Inc, 8700 Commerce Street, Cape Canaveral, Florida 32920.

Status
Available.

NEW ENTRY

Ciener AR-15/M16 belt feed mechanism

Description
The Ciener AR-15/M16 belt feed mechanism is a factory installed conversion which allows AR-15 and M16 series rifles, and M4 series carbines, to be belt-fed in either semi- or fully automatic fire modes. The feed mechanism is driven by the reciprocating bolt carrier, and it can be readily removed from the firearm by opening the action and lifting it out. With the unit removed the firearm can then be used with the normal magazine with no adjustment. When the magazine is in place the holes in the receiver necessary for installation of the belt feed are closed off so that dirt or debris cannot enter.

The belt is of the disintegrating link pattern. There is an accessory belt carrier which clips on below the firearm and carries a 100-round belt. In addition it has a chute which deflects the empty links back into a partitioned section of the carrier, saving them for future use. The manufacture supplies the unit ready assembled to new AR-15 and M16 series rifles and carbines, or will convert existing weapons at its factory.

Manufacturer
Jonathan Arthur Ciener Inc, 8700 Commerce Street, Cape Canaveral, Florida 32920.

Status
Available.

NEW ENTRY

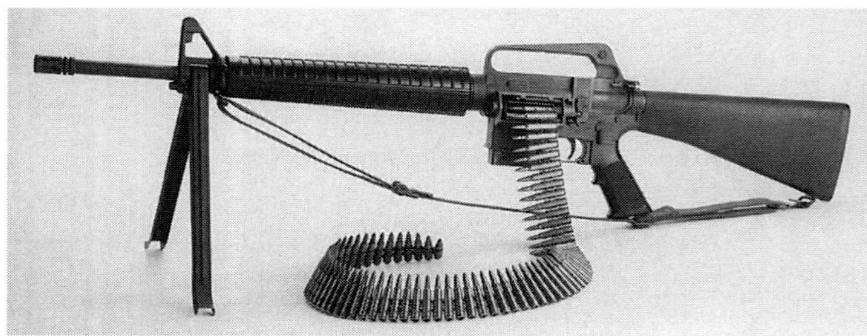

Ciener AR-15/M16 belt feed mechanism fitted to an M16 series rifle
1996

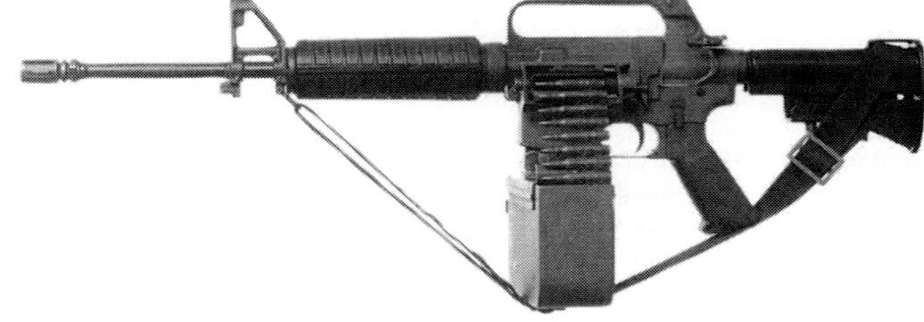

Ciener AR-15/M16 belt feed mechanism with belt box fitted to an M4 carbine
1996

C-MAG 5.56 mm 100-round magazine

Description
The C-MAG is a twin-drum high-capacity ammunition magazine for rifles, light support weapons, firing port and other specialised weapons in 5.56 mm calibre. Covered by US and other patents, its main components are constructed of high impact plastic. Avoiding trouble-prone compression springs, the C-MAG utilises a unique rotation device and low torsion stress springs which ensure positive control of each round and proper feeding. This design makes the C-MAG the world's first magazine capable of providing a full 100-round capacity and yet which can be preloaded and stored indefinitely. The design provides for equal distribution of the ammunition in each of the twin drums located on either side of the weapon. This results in a fixed centre of gravity during firing, an important attribute during full-automatic performance. The compact size allows for better ground clearance and light weight. The C-MAG loads and functions just like standard magazines. It may be reloaded and reused repeatedly.

Loading does not demand disassembly or special tools. The cartridges are loaded by pressing them into the feed aperture in the same way as a normal straight magazine. The cartridges will automatically feed alternately into each drum. Loading can be done by hand or by an appropriate mechanical loader. The springs are tensioned during the loading operation, without the need for any supplementary operations. The magazine does not need to be empty to be loaded and remains completely operational at all times. Loading may be stopped at any point, the magazine inserted into the weapon and the C-MAG is ready for immediate operation.

The key feature of the C-MAG is the formation of the double rows of cartridges in each of the drums. The diagram illustrates how these double rows are formed in the interior of the magazine. Also shown are 15 solid spacer rounds, two pusher arm assemblies and the linked spacer round assembly. The solid spacer rounds occupy the feed clip when the magazine is empty. The number required depends on the length and style of the feed clip, which is specific to an individual weapon system. The pusher arm assemblies transfer the spring energy to the cartridges while being loaded and unloaded. In addition, they serve to maintain the proper orientation of the leading solid spacer rounds. The linked spacer round assembly is very important; this assembly is positioned at the open end of the feed clip

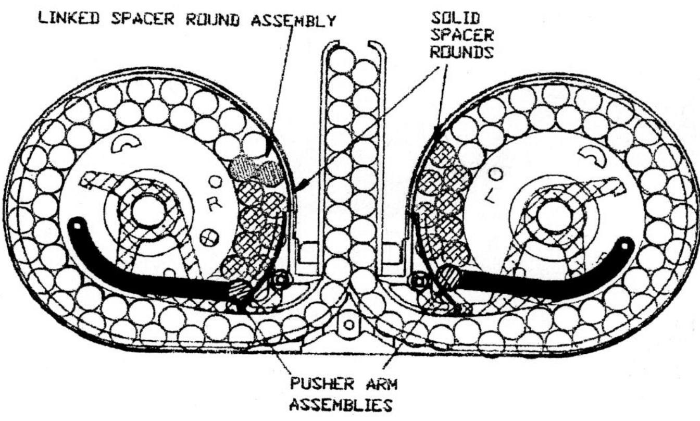

Diagram showing the feed system of the C-MAG magazine

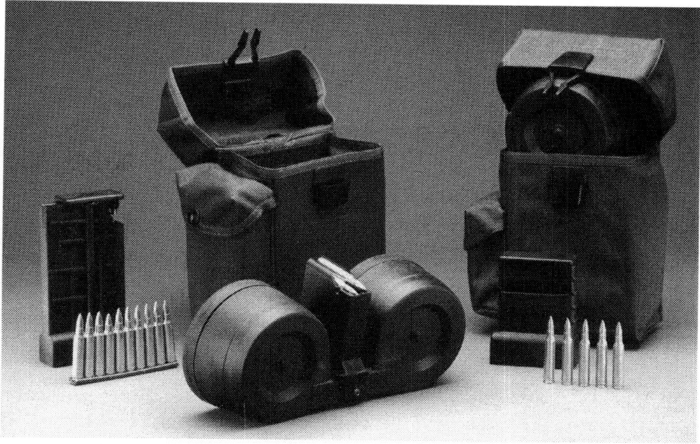

C-MAG 5.56 mm 100-round magazine with accessories

when the magazine is empty and is linked so that it cannot fall out of the feed clip. The upper half of the assembly is tapered so that the weapon bolt will pass over it (when the last round has been fired) and stop in the closed position.

During loading the live cartridges are pressed into the feed clip. The column of spacer rounds and live cartridges are, as loading progresses, pushed into the drums. The cartridge column splits at the point where the feed clip joins the storage housing; this allows for the concurrent division of the stored ammunition on each side of the magazine. Further into each drum the cartridges are allowed to divide into two concentric rows for storage. Loading stops when the maximum number of cartridges has been inserted.

Unloading during firing is the reverse of the above process. As cartridges are stripped from the feed clip, so the column of cartridges advances to fill the space. The springs in the C-MAG are designed to provide virtually a constant amount of tension, regardless of the number of cartridges remaining in the magazine.

Because of the geometry of its construction, the C-MAG can be used on almost any modern 5.56 mm combat rifle or other weapon without modification to the weapon being required. While the current production model employs the NATO STANAG 4179 (M16) magazine interface, it is easily adaptable by the manufacturer to any other 5.56 mm weapon by substituting a feed clip designed for the particular magazine housing. The feed clip is the centrepiece of the C-MAG, detachable from the storage housing for maintenance or weapon compatibility purposes. The main storage housing is standard for all types of weapon; magazines can be ordered with compatible feed clips in place, or the changing can be done by the user. With the correct feed clip installed the C-MAG snaps into the standard magazine housing in the normal manner and is released by the weapon's magazine catch in the usual way.

Accessories available include mechanical and automatic loading devices holding 5, 10 and 100 cartridges for speed loading, as well as carrying pouches.

Data

Capacity: 100 5.56 mm cartridges
Ammunition: ball, tracer, blank rounds
Weight: empty, 1 kg
Dimensions: 251 × 81 × 120 mm
Rate of fire: up to 1,300 rds/min
Materials: main components: filled thermoplastic polyester; minor components: non-corrosive steel, alloy materials
Shelf life, loaded: indefinite

Manufacturer

The Beta Company, 2137B Flintstone Drive, Tucker, Georgia 30084.

Status

In production.

Service

US armed forces and 10 countries worldwide

UPDATED

YUGOSLAVIA (SERBIA AND MONTENEGRO)

7.62 mm M59/66A1 rifle

Description

The 7.62 mm M59/66A1 gas-operated self-loading rifle closely resembles the CIS Simonov SKS rifle (qv) from which it may be said to have been derived by way of the earlier M59 weapon which it replaced.

Most noticeable of the differences between the M59/66A1 and the SKS is a spigot-type grenade launcher attached permanently to the muzzle. Associated with this is a folding grenade-launching sight which is normally folded flat behind the front sight. To launch grenades the gas supply must be cut off from the gas piston by pressing in the gas cut off valve catch and rotating it to the top of the gas cylinder. The grenade sight is then erected.

As with the SKS, a bayonet is permanently attached to a pivot positioned below the foresight; when not in use it is folded back under the barrel, the point being shielded by a recess in the stock. The presence of the grenade launcher slightly reduces the length of free blade when the bayonet is in the forward position.

7.62 mm M59/66A1 self-loading rifle

Data

Cartridge: 7.62 × 39 mm
Operation: gas, self-loading
Locking: tilting block
Feed: 10-round internal box magazine
Weight: ca 4.1 kg
Length: bayonet folded, 1.12 m; bayonet forward, 1.32 m
Sights: fore, post; rear, tangent notch, 100-1,000 m
Sight radius: 444 mm
Muzzle velocity: 735 m/s
Rate of fire: 30-40 rds/min
Effective range: 500 m
Grenade launcher diameter: 22 mm

Manufacturer

Zastava Arms, 29 Novembra 12, YU-11000 Belgrade.

Status

No longer in production.

Service

Former Yugoslav armed forces.

VERIFIED

7.62 mm M70B1 and M70AB2 assault rifles

Description

These two weapons are the rifle elements of the FAZ family of automatic arms based on the 7.62 × 39 mm cartridge, the other members of the family being the M72B1 and M72AB1 light machine guns (qv). The M70B1 and M70AB2 differ only in that the former has a fixed wooden stock whereas the M70AB2 has a folding metal stock. Both are generally similar in design to the CIS AK-47 and AKM (qv).

7.62 mm M70B1 assault rifle showing grenade sight pivoted at gas port

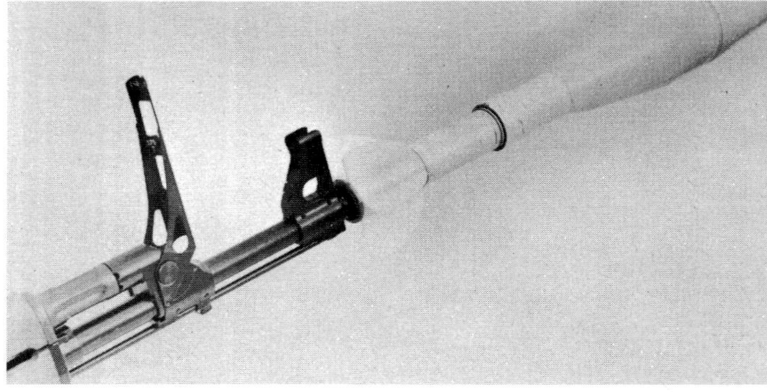

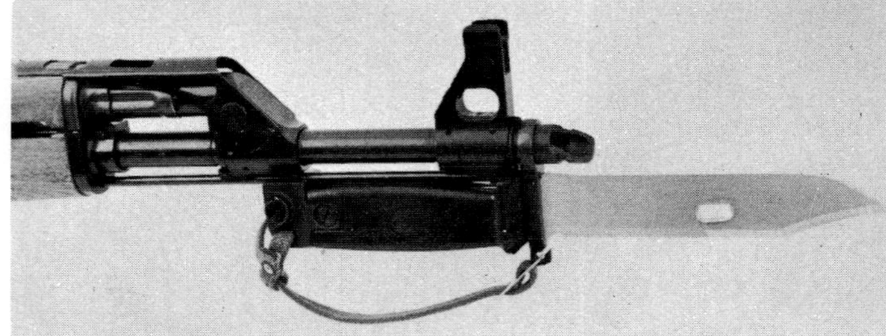

AKM-type bayonet on M70B1 rifle

Grenade launcher with AT grenade on M70B1, with grenade sight raised

As with the AK-47 and AKM, the weapons are gas operated and capable of selective fire. One difference is the provision of a grenade-launching sight which is permanently attached to the weapon at the gas port and which, when raised for use, cuts off the gas supply to the piston. At other times it lies flat on top of the gas cylinder. The launcher itself is not permanently attached but is fitted to the muzzle when required.

Equipment provided includes a bayonet similar to that of the AKM, with a slot on the blade into which a lug on the scabbard fits to form an insulated wire cutter.

Data

Cartridge: 7.62 × 39 mm
Operation: gas, selective fire
Locking: rotating bolt
Feed: 30-round detachable curved box magazine
Weight: M70B1, 3.7 kg; empty magazine, 360 g; loaded magazine, 870 g; grenade launcher, 210 g
Length: M70B1, 900 mm; M70AB2, stock retracted, 640 mm
Barrel: 415 mm
Rifling: 4 grooves, rh
Sights: fore, pillar; rear, tangent notch 100-1,000 m; (night sights) tritium spots on iron sights
Sight radius: 395 mm
Muzzle velocity: 720 m/s
Rate of fire: cyclic, 620-660 rds/min
Effective range: 400 m

Manufacturer

Zastava Arms, 29 Novembra 12, YU-11000 Belgrade.

Status

In production.

Service

Former Yugoslav armed forces.

VERIFIED

7.62 mm M70AB2 assault rifle with stock folded

7.92 mm M76 semi-automatic sniper rifle

Description

Information received on the M76 semi-automatic sniper rifle is scanty, but it was said to be in production for the former Yugoslav armed forces and to be based on the FAZ family of automatic arms. In the standard version the receiver is adapted to handle a 7.92 mm cartridge; variants have also been manufactured to accept the CIS 7.62 × 54 mm and 7.62 × 51 mm NATO cartridges. A 10-round detachable box magazine is fitted.

The M76 is fitted with a telescopic sight of ×4 magnification, apparently having much in common with the PSO-1 sight of the CIS 7.62 mm Dragunov SVD (qv). It is stated to have a field of view (true) of 5° 10′ and a sweeping range of 320 m for head silhouette (target height, 300 mm); 400 m for chest silhouette (500 mm); and 620 m for a running target (1.5 m). The most effective range is given as 800 m. The optical sight bracket is also designed to accept a passive optical night sight.

Data

Cartridge: 7.92 mm (see text)
Operation: gas, self-loading
Locking: rotating bolt
Feed: 10-round detachable box magazine
Weight: 4.2 kg
Length: 1.135 m
Barrel: 550 mm
Rifling: 4 grooves, rh, 1 turn in 240 mm
Sights: fore, blade; rear, U-notch tangent, ×4 telescope

7.92 mm M76 sniper rifle with passive optical night sight

Muzzle velocity: 720 m/s
Effective range: 1,000 m

Manufacturer

Zastava Arms, 29 Novembra 12, YU-11000 Belgrade.

Status

In production.

Service

Former Yugoslav armed forces.

VERIFIED

7.92 mm M76 sniper rifle with telescopic sight

7.62 mm M77B1 assault rifle

Description

The 7.62 mm M77B1 assault rifle, once stated to be in production for export along with the M77B1 light machine gun (qv), closely resembles the M70B1 assault rifle. It is based, however, on the 7.62 × 51 mm NATO cartridge and feeds from a 20-round detachable straight box magazine. It is provided with a grenade-launching sight which is not permanently attached (as is that of the M70B1) but is fitted to the muzzle, along with the launcher, when required. The AKM-type bayonet of the M70B1 is also used.

Data

Cartridge: 7.62 × 51 mm NATO
Operation: gas, selective fire
Locking: rotating bolt
Feed: 20-round detachable straight box magazine
Weight: with empty magazine, 4.8 kg
Length: 990 mm
Barrel: 500 mm

7.62 mm M77B1 assault rifle

Rifling: 6 grooves, rh, 1 turn in 240 mm
Sights: fore, pillar; rear, tangent notch
Sight radius: 485 mm
Muzzle velocity: 840 m/s
Rate of fire: cyclic, 600 rds/min
Effective range: 600 m

Manufacturer
Zastava Arms, 29 Novembra 12, YU-11000 Belgrade.

Status
Intended for export sales. Probably no longer in production.

VERIFIED

5.56 mm M80 assault rifle

Description
The development of a Yugoslav family of weapons for the 5.56 mm M193 and SS109 cartridges was completed. There are two rifles, the M80 with wooden stock and the M80A with folding metal stock, and two machine guns. All weapons are gas operated and are capable of selective semi- or fully automatic fire. A special design of gas regulator ensures reliability of operation even with ammunition of varying energy levels. Both rifles are capable of firing rifle grenades; a launcher attachment and accessory grenade-launching sight is included in the equipment of each rifle.

Data
Cartridge: 5.56 × 45 mm (M193)
Operation: gas, selective fire
Locking: rotating bolt
Feed: 30-round detachable curved box magazine
Weight: without magazine, 3.5 kg
Length: 990 mm
Barrel: 460 mm
Rifling: 6 grooves, rh, 1 turn in 178 mm
Sights: fore, pillar; rear, tangent notch
Sight radius: 439 mm
Muzzle velocity: 920 m/s
Effective range: 400 m

Manufacturer
Zastava Arms, 29 Novembra 12, YU-11000 Belgrade.

5.56 mm M80 rifle with wooden stock and M80A with folding metal stock

Status
Probably no longer available for export.

VERIFIED

LIGHT SUPPORT WEAPONS

BRAZIL

LC T1 M1 flame-thrower

Description

The LC T1 M1 flame-thrower was produced as a replacement for the American models formerly used by the Brazilian Army and Marine Corps. It uses the principle of a single pressure tank containing air or nitrogen and two fuel tanks. The normal fuel is a mixture of gasoline and diesel oil, but other mixtures such as fish or vegetable oil with gasoline or thickened mixtures may also be used. The pressure in the air tank is sufficient for the entire fuel load without loss of range and power at the end.

The specially designed ignition system is energised by an electronic unit which is supplied by eight 1.5 V alkaline batteries. The electronic circuit converts this into about 20,000 V which provides an ignition arc across two electrodes mounted at the muzzle of the projector.

The high pressure cylinder of the portable flame-thrower requires a charging unit that must be available during every operation, otherwise an excessive number of charged cylinders would be required. There is a mobile charging unit which brings logistic support to the front and permits weapon recharging in a very short time. The unit is designed to fit any standard military trailer and consists of a diesel- or gasoline-driven compressor with two 8 m³ air storage cylinders, a flame-thrower fuel tank of 600 litres capacity, fuel and oil tanks for the motor and a kit of tools and spares.

The UR1 is a portable recharging unit consisting of one plastic container with 18 litres fuel capacity (the lid is used as a filling funnel) and one complete high pressure cylinder, attached to a carrier frame of highly resistant lightweight tubular construction.

Where a UM T1 M1 mobile charging unit is not available to charge the high pressure flame-thrower cylinder, commercial compressed air or nitrogen cylinders can be used. Hydroar can furnish a separate DC1 charging device to be connected to the commercial cylinders and to the high pressure flame-throwing cylinders.

Data

Weight: unloaded, 21 kg; with 15 l fuel, 32 kg; with 18 l fuel, 34 kg
Max range: 70 m
Working pressure, fuel tank: 25 kg/cm^2
Working pressure, air tank: 200 kg/cm^2
Fuel hose length: 1.115 m
Weapon tube length: 635 mm

Manufacturer

Hydroar S/A, Rua do Rócio 196, CEP 04552 São Paulo, SP.

Status

In production.

Service

Brazilian Army and Marines.

VERIFIED

LC T1 M1 flame-throwers in operation

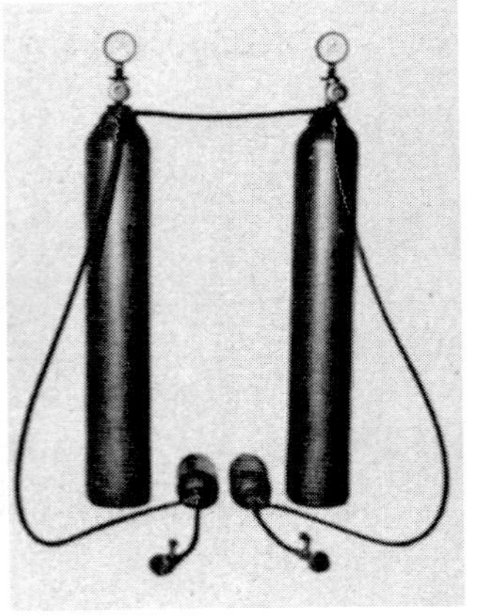

DC1 charging device

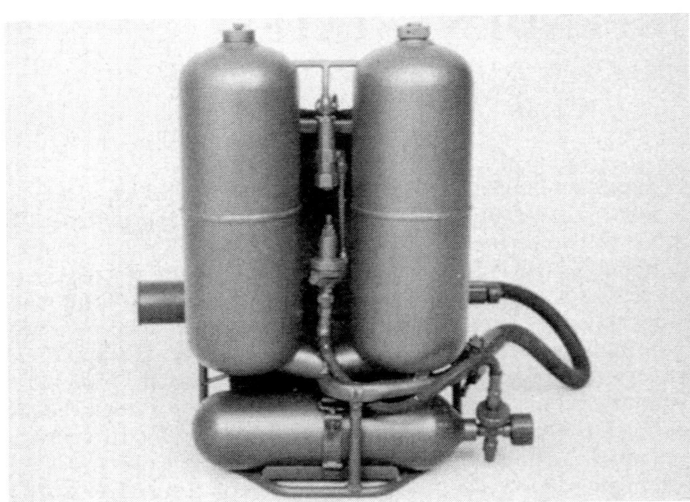

Flame gun of LC T1 M1 flame-thrower, equipped with electronic ignition

LC T1 M1 flame-thrower

CHINA, PEOPLE'S REPUBLIC

30 mm grenade launcher

Description

So far as can be determined, this 30 mm grenade launcher is the CIS AGS-17 grenade launcher manufactured in China. There appear to be some very minor dimensional differences, but in all major respects the weapon appears to be identical. The ammunition used is the standard CIS 30 mm type; HE and HEAT rounds are available.

Data

Calibre: 30 mm
Operation: blowback, selective fire
Feed: belt
Length: 1.28 m
Barrel: 300 mm
Height, zero elevation: 423 mm
Muzzle velocity: ca 183 m/s
Rate of fire: 40-65 rds/min
Max range: 1,700 m

Manufacturer
China North Industries Corporation (NORINCO), 7A Yue Tan Nan Jie, Beijing.

Status
Available.

Service
Chinese Army.

VERIFIED

Chinese troops with locally made 30 mm AGS-17 grenade launcher

35 mm Type W87 grenade launcher

Description
The 35 mm Type W87 grenade launcher is of Chinese design and manufacture, firing a unique 35 mm round. The weapon is smaller and lighter than the 30 mm AGS-17, even though it fires a larger projectile. The mechanism would appear to be simpler, and is probably a plain blowback. It is noticeable that the feed system is either a box magazine or a helical drum, so that there are no feed pawl complications and the operating system can be kept correspondingly simple. Also noteworthy is the prominent muzzle compensator, intended to keep the barrel climb under control during automatic fire, and the bipod folded beneath the barrel, suggesting a one-man light support role as well as the more common tripod-mounted role.

The ammunition available for this weapon has not been seen, but information from the manufacturers is that two rounds are provided, a high explosive/fragmentation projectile containing some 400 3 mm steel balls with a lethal radius of 10 m; and a HEAT projectile capable of penetrating 80 mm of steel armour at 0° incidence. Ammunition is fed to the weapon from either six- or nine-round vertical box magazines or from a 12-round helical drum magazine.

Chinese 35 mm Type W87 grenade launcher with helical drum magazine

Data
Calibre: 35 mm
Operation: blowback, selective fire
Feed: 6- or 9-round box, 12-round helical drum
Weight launcher: 12 kg
Weight, tripod: 8 kg
Weight projectile: 270 g
Muzzle velocity: 170 m/s
Rate of fire: 400 rds/min
Max range: 1,500 m
Effective range: 600 m

Manufacturer
China North Industries Corporation (NORINCO), 7A Yue Tan Nan Jie, Beijing.

Status
Available.

Service
Chinese Army.

VERIFIED

Portable Flame-thrower Type 74

Description
The Portable Flame-thrower Type 74 is a direct copy of the CIS LPO-50 flame-thrower so reference should be made to that entry for full details. Details provided by NORINCO differ in some respects from the Russian data so these details are provided below. The Type 74 has been offered for export sales.

Data
Calibre: 14.5 mm
Weight: loaded, 20 kg
Length of flame gun: 850 mm
Range: 40-45 m
Number of shots: 3
Time for firing 3 shots: 2-3 s
Time for reloading: 4-6 min
Recoil: ca 65 kg

Manufacturer
China North Industries Corporation (NORINCO), 7A Yue Tan Nan Jie, Beijing.

Status
Available. Offered for export sales.

Service
Chinese Army.

VERIFIED

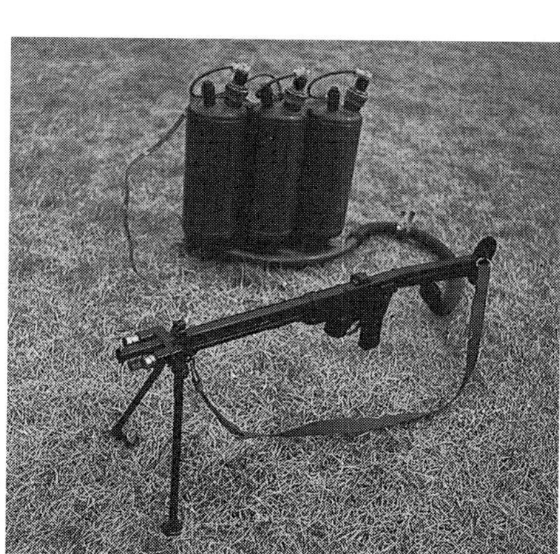

Portable Flame-thrower Type 74

COMMONWEALTH OF INDEPENDENT STATES

30 mm TKB-722K automatic grenade launcher

Description

First seen in 1994, the 30 mm TKB-722K automatic grenade launcher is scheduled to replace the 30 mm AGS-17 automatic grenade launcher (see following entry). Few details are yet available but the main change from the AGS-17, apart from the new weapon having 40 per cent fewer components, is that the TKB-722K is much lighter while retaining a similar range and rate of fire to the earlier weapon. Recoil forces are claimed to be low to the extent that a butt-stock to assist aiming is available.

Few details regarding the operation of the TKB-722K are known but it appears to utilise the blowback principle with extensive buffering. Rounds are fed into the weapon from a 30-round ammunition box. The system uses the same 30 mm VOP-17M fragmentation grenade as the AGS-17, fired at a cyclic rate of fire of 395 to 425 rds/min to a maximum range of 1,700 m.

The TKB-722K is mounted on a light tripod with large spade feet to allow the weapon to be fired from soft and unprepared surfaces; a large azimuth arc is possible. The optical sight mounted on a post to the left of the receiver is the same PAG-17 unit as employed on the AGS-17.

Data

Calibre: 30 mm
Operation: blowback, selective fire
Feed: 30-round belt
Weights: weapon mounted on tripod, less ammunition, 16 kg; magazine box with 30 rounds, 13.7 kg; sight, 1 kg
Length: 1.1 m
Height: 500 mm
Width: 360 mm
Muzzle velocity: 185 m/s
Max range: 1,730 m
Rate of fire: cyclic, 395-425 rds/min

VOP-17M GRENADE
Weight: round, 348 g; grenade, 275 g
Filling: 34 g RDX/Wax 94/6
Fuze: impact, self-destruct after 25 s
Cartridge case: belted, rimless, steel, 28.2 mm long

Manufacturer

Instrument Design Bureau (KBP), Tula, 300001 Russia.

Status

In production. Scheduled to replace the AGS-17 (see following entry).

UPDATED

30 mm TKB-722K automatic grenade launcher (Christopher F Foss)

30 mm TKB-722K automatic grenade launcher (Christopher F Foss)

30 mm TKB-722K automatic grenade launcher (Christopher F Foss)

---

30 mm AGS-17 automatic grenade launcher

Development

The AGS-17 'Plamya' ('Flame') 30 mm automatic grenade launcher was introduced into Eastern Bloc service in 1975 but did not achieve prominence until its employment during the Afghanistan campaign when it prompted a series of similar developments in the West. The AGS-17 is issued down to infantry company level, a section of two AGS-17 launchers being added to each infantry company. They are employed as the company's base of fire during the advance. Other AGS-17 variants have been mounted in helicopters and on various forms of vehicle mounting.

Description

The AGS-17 (Avtomaticheskyi Granatmyot Stankovyi-17) 30 mm automatic grenade launcher operates on the blowback system. Ammunition is belt-fed into the right side of the receiver and the bolt is cocked by pulling back on an operating handle at the rear of the weapon, which is attached by a steel cable to the bolt. On releasing this handle the bolt runs forward, propelled by two springs, and collects a round from the belt, loading it into the chamber. The firing pin is then released by means of a trigger mounted between the spade grips at the rear of the receiver. On firing, the breech pressure blows the bolt to the rear; its travel is damped by a hydraulic recoil buffer mounted on the left side of the receiver. The bolt then returns and chambers the next round. The empty case is extracted from the chamber by the rearward movement of the bolt and ejected through the bottom of the receiver. Belt feed is done by pawls actuated by a feed arm driven by the movement of the bolt.

A PAG-17 optical sight with a ×2.7 magnification is normally fitted to the left rear of the receiver and a plate on top of the receiver gives ranges and time of flight for low and high angle firing. Iron sights are also provided.

The standard infantry mounting is a SAG-17 tripod with elevation and traverse facilities; the tripod can be folded to form a back-pack for man-carriage. The tripod can also be mounted on most types of combat vehicle.

The 30 mm grenades fired from the AGS-17 consist of two main types, the VOP-17 and the VOP-17M, the latter being slightly the lighter of the two. Both are fragmentation grenades with their performance enhanced by the provision of a number of small steel spheres packed around the 36 g explosive payload (34 g on the VOP-17M). When the nose-mounted impact fuze functions the resultant fragments are spread over a 7 m lethal radius.

Variant

AG-17A

The AG-17A is a helicopter-mounted version of the AGS-17 intended for installation on the chin gimbal mounting of 'Hind' series attack helicopters. The AG-17A has the same receiver and mechanism as the AGS-17 with a revised barrel. Grenades are fed from a 300-round magazine and the cyclic rate of fire is increased to 420 to 500 rds/min.

A further helicopter version of the AGS-17 is intended for firing from the load compartments of helicopters. The special mounting employed for this role has the

AGS-17 launcher; belt feed and details of feed mechanism

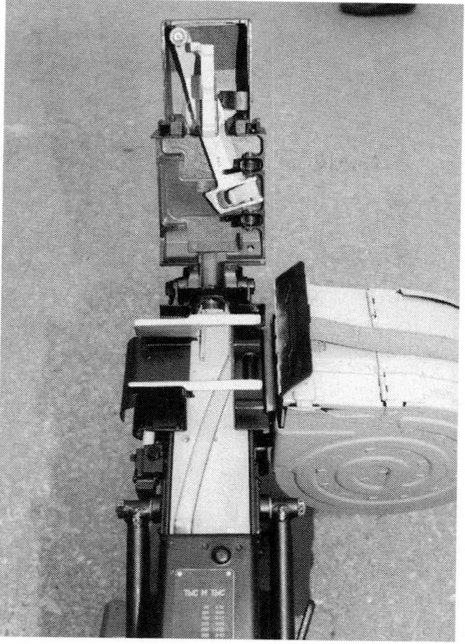

Top view of AGS-17 showing feed mechanism

30 mm training grenade, with fired cartridge case

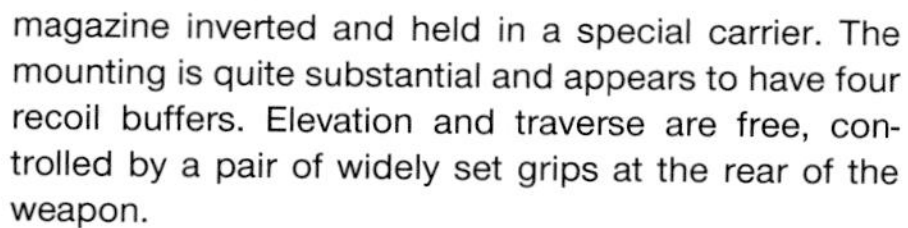

magazine inverted and held in a special carrier. The mounting is quite substantial and appears to have four recoil buffers. Elevation and traverse are free, controlled by a pair of widely set grips at the rear of the weapon.

Data

Calibre: 30 mm
Operation: blowback, selective fire
Feed: 29-round belt
Weights: weapon, ca 18 kg; mounted on tripod, ca 31 kg; tripod, 12 kg; magazine loaded, ca 14.5 kg; optical sight, 1 kg
Length: 840 mm
Barrel: 290 mm
Rifling: 16 grooves, rh
Muzzle velocity: 185 m/s
Range: max, 1,730 m; max effective, iron sight, 800 m; max effective, optical sight, 1,700 m
Rate of fire: cyclic, 350-450 rds/min; practical, 50-100 rds/min

VOP-17M GRENADE
Calibre: 30 mm
Weight: round, 348 g; grenade, 275 g
Filling: 34 g RDX/Wax 94/6
Length: round, 132 mm; grenade, 112 mm
Fuze: impact, self-destruct after 25 s
Cartridge case: belted, rimless, steel, 28.2 mm long

Manufacturers

Vyatsko-Polyanski Machine-Building Plant 612900, Vyatskiye Polyany, Kirov Region, Russia.
Instrument Design Bureau (KBP), Tula, 300001 Russia.
Marketed by: Rosvoorouzhenie, 18/1 Ovchinnikovskaya Emb, 113324, Moscow, Russia.

Status

In production. Scheduled to be replaced by the TKB-722K (see previous entry).

Service

Afghanistan, Angola, Chad, China, CIS, Cuba, Iran, Mozambique, Nicaragua, Poland and South Africa.

Licence production

CHINA, PEOPLE'S REPUBLIC
Manufacturer: NORINCO
Type: 30 mm automatic grenade launcher
Remarks: Virtually identical to original AGS-17. See entry under *China, People's Republic* for details.

UPDATED

Left side of AGS-17 launcher, showing details of PAG-17 optical sight

Firing the 30 mm AGS-17 automatic grenade launcher

Cocking and loading the 30 mm AGS-17 automatic grenade launcher

40 mm GP-25 rifle-mounted grenade launcher

Development

The first examples of what was to become the 40 mm GP-25 grenade launcher were first encountered in Afghanistan during 1984, mounted beneath an AK-74 rifle in the same manner as the American 40 mm M203 launcher. On these early models, initially known to the West as the BG-15, the sights had a three-position sight on the left of the launcher, with the three positions being marked '2', '3' and '42', understood to indicate indirect ranges of 200 m, 300 m and the maximum range of 420 m. The main production model, with the designation changed to GP-25, often known as the Kastyor or Kastjor, has a sight graduated to a maximum of 400 m with intermediate range markings between 200 and 400 m. Direct fire is possible at ranges of 50 to 100 m.

Description

The 40 mm GP-25 Kastyor grenade launcher can be mounted under the handguard of virtually any AKM or AK-74 series assault rifle, including the AKS-74U submachine gun models. A more involved conversion kit including a muzzle extension assembly is required for the latter.

The GP-25 launcher is short, being 323 mm long overall, having a trigger unit with a thumb grip, and with the grenade launching sight on the left side of the launcher barrel. On the GP-25 the sight is graduated to a maximum of 400 m, with intermediate range markings between 100 and 400 m.

To load the GP-25 the grenade is muzzle-loaded into the barrel base first and, after aiming, the self-cocking trigger is pulled. This detonates the KVM-3 primer, the nitrocellulose propellant inside the grenade body ignites, and the flow of gas from vents in the grenade base propels the grenade from the muzzle. As it travels along the barrel, gas pressure forces the grenade's driving bands outwards to engage in the 12-groove rifling of the barrel to provide spin stabilisation.

Accuracy at the maximum range of 400 m is 3 m in azimuth and 6 m in range. A practical rate of fire is quoted as 4.5 rds/min.

Two types of grenade are provided for the GP-25, the VOG-25 and the VOG-25P. Both are steel fragmentation grenades with the VOG-25 having a VMGK nose impact fuze capable of operating in snow or soft ground. By contrast the VOG-25P is a 'bounding' fragmentation grenade which strikes the ground before being propelled upwards to a height of 0.5 to 1.5 m, where it detonates to spread anti-personnel fragments over a lethal radius of at least 6 m. If either type of grenade does not operate on impact they will self-destruct after 14 to 19 seconds.

Data

GP-25
Calibre: 40 mm
Weight: unloaded, 1.5 kg
Length: 323 mm
Barrel: 120 mm
Muzzle velocity: 76.5 m/s
Max range: 400 m

Close-up of 40 mm GP-25 Kastyor grenade launcher mounted on 5.45 mm AK-74 assault rifle

VOP-25/VOP-25P GRENADES
Calibre: 40 mm/40 mm
Weight: 255 g/278 g
Explosive payload: VOP-25, 48 g TNS; VOP-25P, 37 g
Muzzle velocity: 76.5 m/s/75 m/s
Max range: 400 m/400 m
Min range: 50 m/50 m

Agency

VO GED, General Export for Defence, 113324 Moscow.

Status

In production. Offered for export sales.

Service

Former Warsaw Pact nations and others.

Licence production

BULGARIA
Manufacturer: Arsenal, marketed by Kintex
Type: 40 mm GP-25 Kastjor
Remarks: Direct copy of CIS original. VOP-25 grenades also produced.

UPDATED

40 mm GP-25 Kastyor grenade launcher mounted on 5.45 mm AK-74 assault rifle; grenade on left is a VOP-25P, with a VOP-25 on right (T J Gander)

RPO-A Schmel Rocket Infantry flame-thrower

Development

Although usually referred to in descriptive literature as a flamethrower, the RPO-A Schmel (Bumblebee) is a rocket-propelled incendiary/blast projectile launcher. Development of this weapon began in 1984 and it was accepted for service in 1988. The projectile fired is described by the Russians as 'thermobaric' as it appears to utilise advanced fuel-air explosive techniques which, on detonation, create deflagratation as the warhead cloud expands, thereby producing a relatively long-lasting blast effect in addition to the high temperature effects. When utilised against structures the blast effect of the projectile is stated to be equivalent to a 122 mm howitzer projectile.

There have been several versions of the RPO-A. The initial version was termed a capsule flame-thrower and relied upon its incendiary and blast capabilities for on-target effects. A later version combines the fuel-air warhead with a small hollow charge which penetrates light armour or structures to allow the main warhead to detonate inside a target, thereby considerably enhancing the projectile's destructive effects.

The RPO-A is issued to airborne and marine assault troops to destroy strongpoints (for example bunkers) or lightly armoured vehicles, and, using a special smoke projectile, to produce dense smoke screens from 55 to 90 m long. In Afghanistan the weapon was used to clear Mujahadeen from caves. It is also claimed that the RPO-A could have peaceful applications as it can be

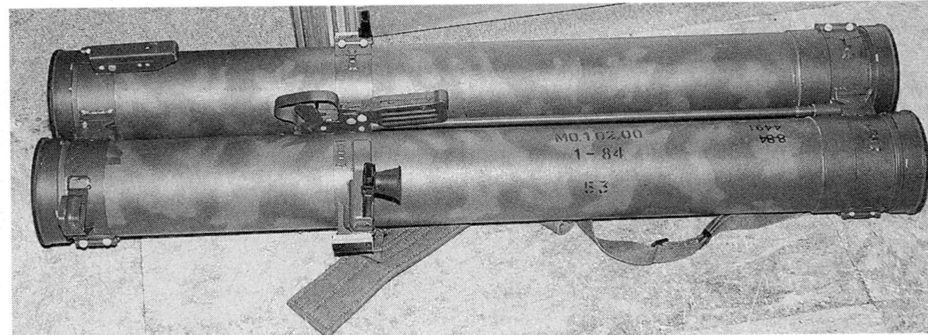

Pack of two RPO-A Schmel Rocket Infantry flame-throwers, as issued (T J Gander)

used to destroy ice jams in rivers, break up potential snow avalanches or extinguish fires by the blast effect.

Description

The RPO-A is a single-tube launcher which is normally issued in pairs joined together to be carried as a backpack using the two slings. To operate the launcher the two tubes are separated and a simple optical sight calibrated to 600 m is flipped up. Launching is from over the shoulder using a folding pistol grip and trigger assembly together with a forward grip under the muzzle. Once fired the launcher tube is discarded.

The 93 mm projectile warhead contains 2.1 kg of 'thermobaric' flammable mixture, 2.1 kg of an incendiary substance or 2.3 kg of an incendiary-smoke mixture. Initial velocity of the rocket-propelled projectile is 125 m/s (±5 m/s). Length overall of the blunt-nosed projectile is estimated to be around 700 mm.

The effect of the RPO-A projectile is such that when detonated inside a structure its lethal and destructive effects will cover an area of 80 m^3. When utilised against personnel in the open the lethal area will cover 50 m^2. Maximum range is 1,000 m but the sights are calibrated to only 600 m. Point blank range against a target 3 m high is 200 m.

Shelf life of the RPO-A is 10 years and it can be operated over a temperature range of from −50 to +50°C.

Data

Calibre: 93 mm
Weight: single launcher, 11 kg; pack of two launchers, 22 kg

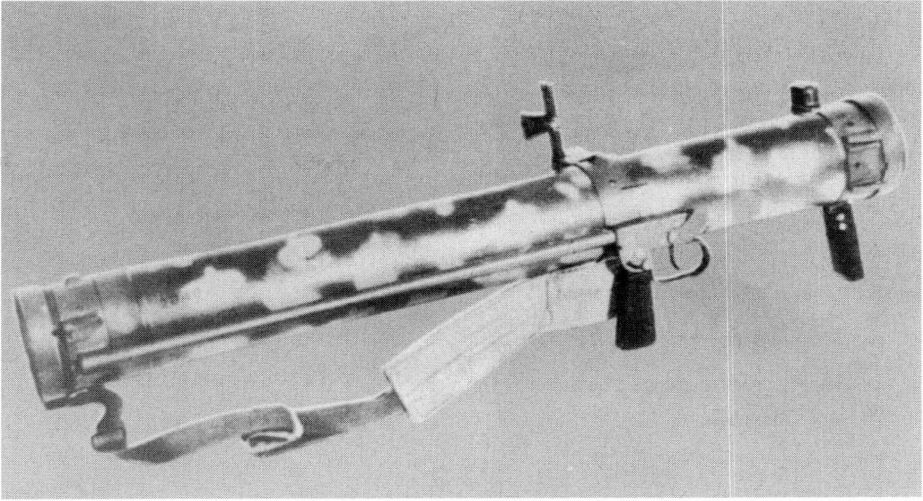

RPO-A Schmel Rocket Infantry flame-thrower

Weight of warhead: ca 6.5 kg
Length: 920 mm
Max range: 600 m
Effective range: 350-400 m
Initial velocity: 125 ±5 m/s

Agency

VO GED, General Export for Defence, 113324 Moscow.

Status

In production. Offered for export sales.

Service

Former Warsaw Pact nations and others.

UPDATED

LPO-50 flame-thrower

Development

The LPO-50 was, until recently, the standard flame-thrower in former Warsaw Pact armies, having replaced the earlier ROKS-2 and ROKS-3 flame-throwers. This flame-thrower is produced in China as the Portable Flame-thrower Type 74 and has been offered for export sales. See entry under *China, People's Republic* for details.

Description

The LPO-50 flame-thrower consists of a tank group, hose and gun group. There are three tanks, on the top of each of which are a pressure relief valve and a cap for the filling aperture which also contains the chamber for the pressurising cartridge.

Wires from the three containers are combined in a harness which is fastened to the hose and attached to the gun group. Outputs from the three tanks are connected to a manifold through one-way valves which prevent fuel flowing from one tank into another manifold, being connected to the hose.

Ignition is by means of a slow-burning pyrotechnic cartridge, three of which are grouped below the muzzle of the flame gun. A selector lever is mounted forward of the trigger guard on the gun and, when the trigger is pressed, energy is supplied from a power pack of four 1.5 V cells to one of the ignition cartridges and simultaneously to one of the tank pressurising cartridges. Pressure from the latter drives fuel from the tank through the appropriate non-return valve into the manifold and then by way of the hose to the flame gun where it is ignited by the pyrotechnic cartridge. The firer can thus fire three shots, changing the selector lever position between shots. The tank capacity is 3.3 litres which is sufficient for a flame burst of 2 to 3 seconds. A trigger safety is fitted.

Data

Calibre: 14.5 mm
Length of flame gun: 850 mm
Fuel capacity: 3 × 3.3 l
Weight: empty, 15 kg; loaded, 23 kg
Max range: with unthickened fuel, 20 m; with thickened fuel, 70 m

Manufacturer

State factories.

Status

In current use.

Service

Former Warsaw Pact armies.

Licence production

CHINA, PEOPLE'S REPUBLIC
Manufacturer: China North Industries Corporation (NORINCO)
Type: Portable Flame-thrower Type 74
Remarks: Virtually indentical to LPO-50. See entry under *China, People's Republic* for details.

VERIFIED

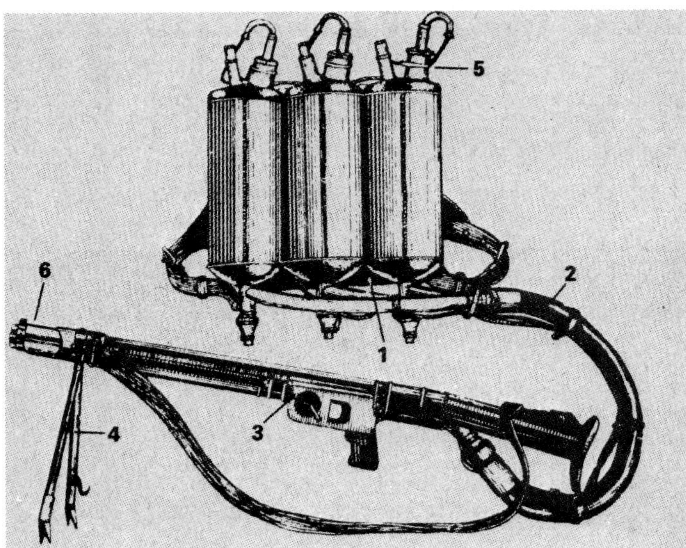

Components of LPO-50 flame-thrower
(1) *fuel manifold* **(2)** *hose* **(3)** *selector lever* **(4)** *bipod* **(5)** *container for pressure cartridge* **(6)** *ignition cartridges*

Firing LPO-50 flame-thrower

EGYPT

Home Guard anti-personnel weapon

Description

The Home Guard anti-personnel weapon is actually the Egyptian PG-7 anti-tank rocket launcher employing a specially developed anti-personnel warhead. It was designed to fill various operational requirements over ranges which fall between the hand grenade and larger weapons; it could be used to replace an infantry section mortar. The rocket and warhead can be fired from the standard PG-7 launcher thereby easing the logistic problem.

The warhead is cylindrical and consists of 210 g of Hexogen 5A-IX-1 surrounded by controlled fragmentation steel rings. An O-4M impact fuze is fitted. When the warhead detonates it creates a lethal radius of more than 25 m.

Data

Calibre: 40 mm
Length: 612 mm
Weight: 1.75 kg
Explosive content: 210 g Hexogen 5A-IX-1
Lethal radius: more than 25 m
Muzzle velocity: 154 m/s
Max range: 1,000 m
Min range: 300 m

Manufacturer

Sakr Factory for Developed Industries, PO Box 33, Heliopolis, Cairo.

Status

In production.

Service

Egyptian Army.

UPDATED

Sakr Industries Home Guard light support weapon

1996

Sakr Industries Home Guard light support weapon

1996

Maadi 40 mm grenade launcher

Description

The Maadi 40 mm grenade launcher appears to have been based on the US 40 mm M203 but was produced to fit under the forward handguard of AK series Kalashnikov assault rifles. With minimal modification it could be installed under other standard rifles such as the AUG, FN-FNC or M16 series.

The 40 mm grenade launcher is a single shot lightweight weapon constructed using Duraluminium 7075. The launcher may be installed on or removed from a host rifle within seconds and opens forward with a pump action once a barrel latch has been released. Grenades are then inserted into the open breech manually. The launcher can fire all standard NATO low-velocity 40 mm grenades to a maximum range of 400 m.

The sights for the launcher may be mounted on the rifle receiver or on the launcher. They have a post foresight and a rear 2 mm aperture with graduations from 50 to 400 m in 25 m increments. There is a safety which, when applied, not only blocks any trigger movement but also inserts an obstruction which prevents the user's finger from actually touching the trigger.

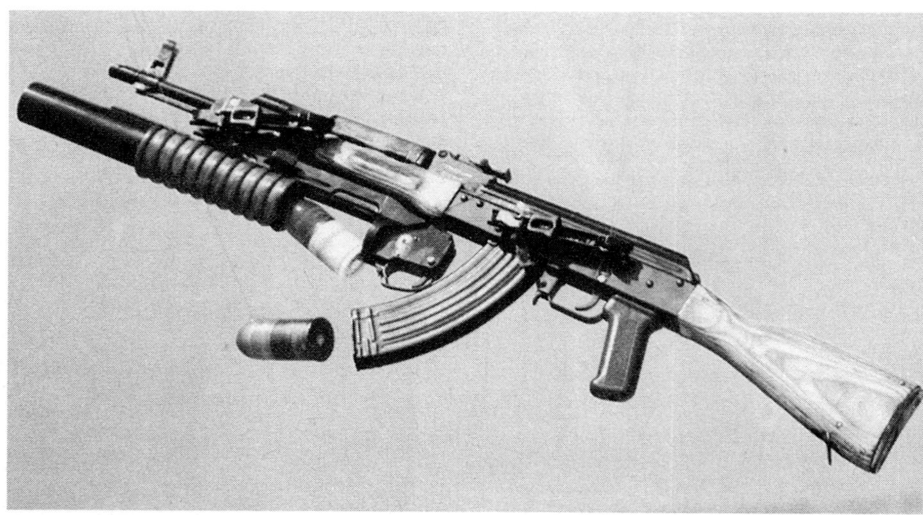

Maadi 40 mm grenade launcher on 7.62 mm AKM (Misr) assault rifle; note the two alternative sight locations

1996

Data

Calibre: 40 mm
Operation: single shot
Feed: breech loading, sliding barrel
Weight: empty, 1 kg; loaded, 1.3 kg
Length: 360 mm
Barrel: 305 mm
Rifling: 6 grooves, rh, 1 twist in 1.2 m
Sights: fore, post; rear, 2 mm aperture, graduated 50 to 400 m in 25 m increments
Sight radius: 142 mm
Muzzle velocity: 76 m/s
Max range: 400 m

Manufacturer

Maadi Company for Engineering Industries, PO Box 414, Maadi, Cairo.

Status

In production.

Service

Egyptian Army.

NEW ENTRY

FRANCE

Ruggieri SPIDER local defence system

Description

The Ruggieri SPIDER local defence system covers a wide range of applications and is produced in three main forms. The DREC is intended for the close-in defence of armoured vehicles. DIMP is intended for installation around sensitive or valuable static sites or positions. The third version of Spider, DIPS, is intended for the defence of infantry positions.

DIPS is a man-portable system consisting a control box and a five-barrelled launcher assembly in a tubular carrying frame, with each barrel launching one two-part munition. Once the launcher assembly has been installed and loaded at a selected position, and connected by cable to its control box, it is ready for use. Numerous DIPS units may be connected into an area defence system under the control of a single operator. It is also possible to control DIPS by various forms of sensor. Installation and removal take less than two minutes.

The munition most likely to be used by the infantry is known as CAPIRO. All five of these munitions are discharged instantaneously over a 120° quadrant under the control of a delay fuze which ensures the munitions are at least 50 m away when they detonate. As they detonate each munition releases 900 steel balls at a velocity of 1,100 m/s and produces a loud report to add to the shock effect. A DIPS salvo can cover an area of 2,400 m^2 within which there is a 0.5 probability of an enemy being incapacitated. None of the preformed fragments will be directed towards the launch point.

A CAPIRO munition weighs 1.7 kg and measures 190 × 100 × 40 mm. Other munitions available for use with DIPS include: MANTA, for underwater targets; SOUND, producing a deafening report but remaining non-lethal; LARM, producing disabling tear and choking vapours at least 100 m from the DIPS position; SEPFI, visible smoke; and SEPIA, visible and infra-red screening smoke. The weight of these munitions varies between 1.6 and 2.2 kg.

Manufacturer

Ruggieri Technologie, route de Gaudiès, F-09270 Mazères.

Status

In production.

UPDATED

Ruggieri DIPS, the infantry component of the Ruggieri SPIDER local area defence system
1996

ABB multipurpose weapon system

Description

Developed by Matra Manurhin in conjunction with SERAT and Giat Industries, the ABB (Arme Individuelle Antibunker Antiblinde) is a man-portable launcher which can be used against any type of target from tanks to landing craft and bunkers. At one time this weapon was known as the AB92.

The ABB system consists of a disposable launcher loaded with a shaped charge projectile and a countermass, thus giving recoilless launch. This 'Davis Gun' principle enables the weapon to be fired from a position only 1 m in front of an obstacle such as a wall. The warhead can be adjusted before firing to provide instantaneous or delayed action, thus allowing it to be suited to any particular target.

The ABB projectile warhead can blast a hole more than 250 mm in diameter in a 200 mm thick concrete wall with significant behind-protection effects. When used against rolled homogeneous armour the ABB can penetrate up to 400 mm.

For training several types of simulator are available, including laser-based and sub-calibre trainers. A night sight can be fitted.

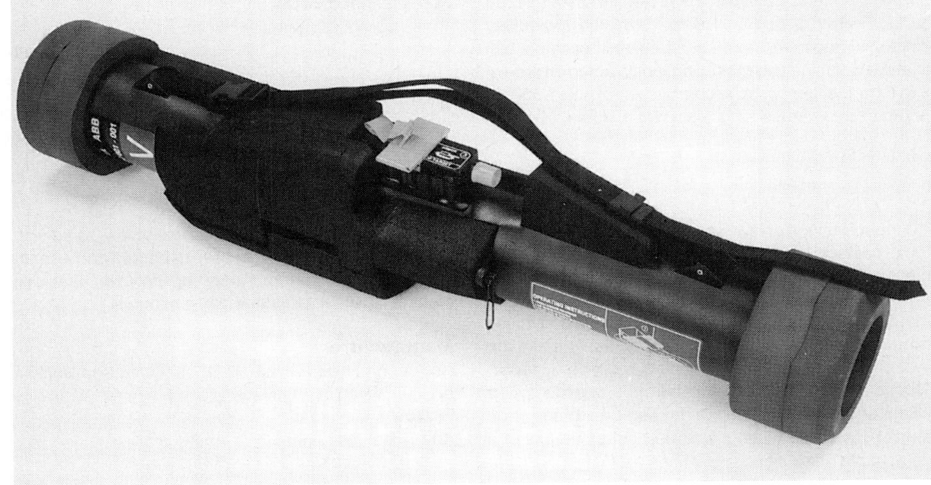

ABB multipurpose weapon system
1996

Data

Calibre of warhead: 92 mm
Launcher weight: loaded, 7.3 kg
Launcher length: 1 m
Muzzle velocity: 270 m/s
Effective range: 300 m
Penetration: >400 mm RHA

Manufacturer

Giat Industries, 13 route de la Minière, F-78034 Versailles-Satory.

Status

Advanced development. Undergoing qualification.

UPDATED

Lacroix Sphinx Perimeter Defence System

Description

Sphinx is used to increase the protection provided for fixed assets or for troop concentrations or command centres in temporary locations. Designed as a man-portable unit it may be set up on any type of surface (snow, sand, rocks, concrete) within 10 minutes, day or night and under all weather conditions, with minimal site preparation and in urban situations or any area without cover.

Sphinx is designed to operate as a self-contained, reusable unit assembled from the following elements: a firing frame ensuring ease of set-up and safety of operation; a control and firing pane with a power supply and 135 m of cable; operational rounds include armour-piercing, smoke, CS and warning rounds, the latter providing an audible signal. The operational rounds were derived from and are interchangeable with GALIX munition intended for the close-in defence of armoured vehicles.

Lacroix Sphinx Perimeter Defence System and key point protection

The Sphinx provides a high degree of operator safety during storage, transport and firing; the system is unarmed during storage and transport. Deployment is simple and rapid and, when deployed but not used, can be disarmed and reused. Detection is rendered difficult due to the low physical profile. Personnel wearing NBC protective clothing can handle the system and it can be decontaminated rapidly.

A series of modular Sphinx devices may be combined into a comprehensive system for site protection in conjunction with a centralised, interactive control panel. Devices may be positioned to provide interlocking coverage.

Data

Position	'Transportation'	'Firing'
Height	240 mm	220 mm
Length	550 mm	1.1 m
Width	455 mm	455 mm
Weight	25 kg	
Operating temperature range		
Firing	−40 to +60°C	
Storage	−40 to +70°C	

Manufacturer
Lacroix Etienne Tous Artifices SA, 6 boulevard de Joffrery, F-31607 Muret Cedex.

Status
In production.

Service
Undisclosed.

VERIFIED

GERMANY

HAFLA DM 34 hand flame cartridge launcher

Description
The HAFLA DM 34 hand flame cartridge launcher is a single shot, one-man operated, throwaway weapon designed to impel an incendiary smoke charge to a range of 70 to 80 m.

The weapon comprises an aluminium launch tube and three compressed sections of incendiary smoke composition contained in a projectile. A pivotal handgrip fuze and firing assembly is fitted at one end of the tube and a plastic cover at the other end. The launcher is watertight. Three units are individually packed and moisture-sealed in a plastic carrying case.

The HAFLA DM 34 is delivered with the handgrip in the 'safe' position. By depressing the safety button the grip can be moved outwards exposing the trigger. The trigger cannot be activated until the grip is moved completely to the rear and is locked in position. The weapon is then ready to fire.

The trigger is depressed and the firing pin strikes the primer which ignites the first propellant. Simultaneously the projectile is set in motion, the second propellant is ignited and the aluminium capsule is blown out of the tube.

HAFLA DM 34 and DM 38 cartridge launchers

As the second propellant burns, the delay fuze begins to smoulder. This action concurrently activates the incendiary smoke payload. After some two seconds, the burning mixture in the delay fuze reaches the dispersion charge. An immediate burst then demolishes the projectile, spreading burning fragments of incendiary smoke composition.

The projectile bursts after travelling 70 to 80 m and burning particles will be spread over an area 10 m wide and 15 m long. If, however, the projectile strikes a hard surface within a range of 8 to 70 m, the incendiary smoke charge immediately scatters on impact in a brilliant flash of blinding smoke and fire covering an area of 5 to 8 m. In either case the combustion of hot (1,300°C) fragments continues for at least two minutes.

Variant

Practice grenade DM 38
This has the same dimensions and weight as the HAFLA DM 34 but has an inert filling of lime and a smoke marker to indicate the point of impact.

Data
Calibre: 35 mm
Weight: 625 g
Length: 445 mm
Weight and type of filling: 240 g red phosphorous

Manufacturer
Buck Werke GmbH and Company, Postfach 2405, D-8230 Bad Reichenall.

Status
In production.

Service
German armed forces.

VERIFIED

40 mm salvo grenade launcher

Description
The salvo grenade launcher fires 10 40 mm grenades simultaneously; all standard 40 × 46 mm grenades may be fired. It can be employed for protective tasks such as the covering and defence of points. The grenades have a good fragmentation performance at their foreseen range of employment between 200 and 300 m.

Launching of the grenade is effected by a control device connected to the launcher by an ignition cable up to 200 m long. The control device can be linked to several launchers simultaneously. The Dynamit Nobel AG inductive ignition system does not require any electrical contacts, providing safe handling and employment without the danger of accidental ignition.

The launcher consists of the grenade container and the firing system. During storage and transport these two units are separated by a safety plate. To make the system ready to fire this safety plate must be removed. The detachable grenade launcher mount (two support pillars and one spur with spade) can be stored inside the firing system. Aiming is carried out by means of a detachable reflex sight.

The launcher is compact, measuring 330 × 440 × 144 mm in the transport mode. Weight, including the mount and ten grenades, is approximately 15 kg without the control device. It is easy to handle and operate and provides optimum safety for the user. Emplacement or removal takes only a few minutes.

Manufacturer
Heckler and Koch GmbH, D-78722 Oberndorf-Neckar.

Status
Advanced development.

VERIFIED

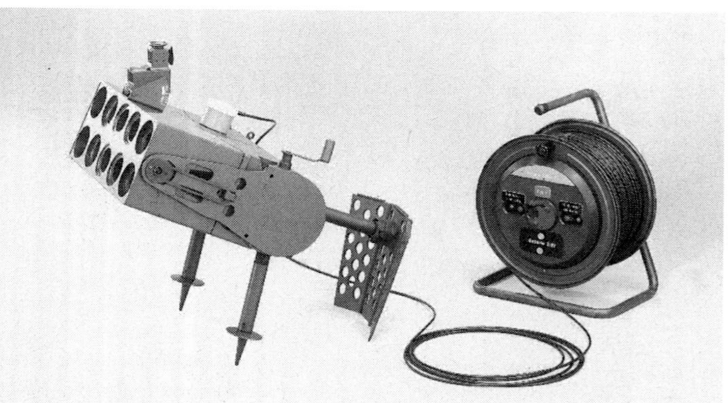

40 mm salvo grenade launcher emplaced; the reflex sight is removed before firing. The front is normally closed by aluminium foil to protect the grenades

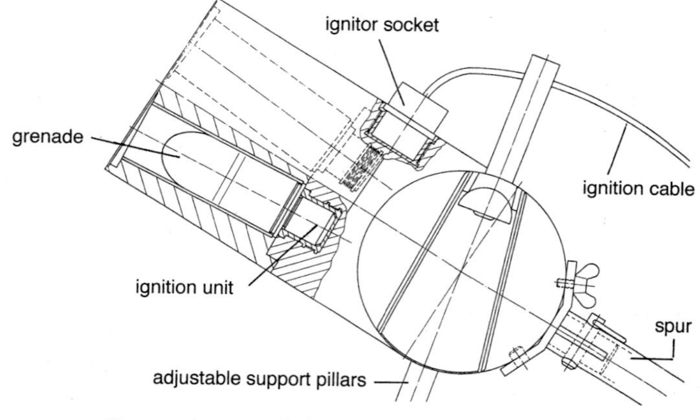

40 mm salvo grenade launcher, part-sectioned drawing

Heckler and Koch 40 mm grenade machine gun (GMG)

Development

The Heckler and Koch 40 mm grenade machine gun (GMG) was developed as a private venture. As of early 1995 four prototypes had been manufactured. One was retained by Heckler and Koch for their own use while the other three were handed over to the Germany Army in March 1995 for technical trials at Meppen. User trials were scheduled to commence during the Summer of 1995.

It is anticipated that the German Army has a requirement for 'several thousand' units of this type of weapon. Norway and the Netherlands have also expressed interests.

Description

The Heckler and Koch 40 mm grenade machine gun (GMG) was developed to be a two-man portable load and can deliver all types of 40 × 53 mm grenade to a range of 2,200 m.

Operation is by blowback using an inertia bolt fired from an open breech position. The feed system is derived from that of the MG3 machine gun and is optimised to fire standard linked 40 × 53 mm ammunition fed from a NATO PA 120 ammunition box holding 32 rounds. The feeding mechanism is driven by the bolt and positively transports and guides the linked cartridges in two steps; the first during the forward and the second during the rearward travel of the bolt. The feeding direction to the weapon can be changed from right to left without the use of tools.

To load the GMG the operator pulls back the breech to the cocked position, opens the feed cover and inserts the first round in the ammunition belt into the feed pawls of the feed tray, from where the rounds are directly chambered by the breech. After closing the feed cover the weapon is ready to fire once the gunner has selected the fire mode by switching the safety/fire lever from Safe to Single Fire or Automatic Fire.

Safeties include setting the safety/fire selector lever at Safe which not only blocks the trigger but also maintains the breech in the cocked position. Two independent firing pin safety mechanisms ensure that during the advanced ignition the firing pin can only reach the cartridge primer after the breech has chambered the round so that the case is fully supported by the chamber. A loading safety ensures that as the loader opens the feed cover the breech will automatically be arrested and prevented from moving forward. If, during the cocking operation, the gunner fails to fully pull back and engage the breech in its cocked position, an integral mechanism prevents the breech from travelling forward and accidentally firing. This safety is released once the breech has been fully moved to the rear and engaged there.

Stripping for field maintenance does not require the use of tools and reassembly can only be carried out in a correct sequence.

The 40 mm GMG can be mounted on a standard M3 tripod and may be installed on vehicles; a softmount is optional. A reflex sight is available although other sighting systems can be utilised.

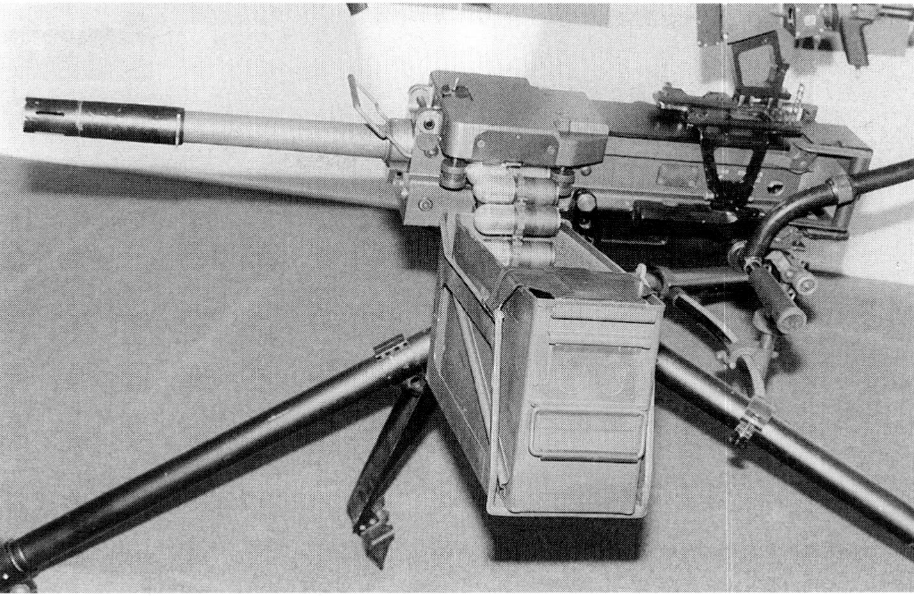

Heckler and Koch 40 mm grenade machine gun (GMG) (T J Gander)

1996

Heckler and Koch 40 mm grenade machine gun (GMG) (T J Gander)

1996

Data

Calibre: 40 × 53 mm grenades
Operation: blowback; selective fire
Feed: 32-round linked belt from right or left
Weight: gun, 29 kg; tripod, 15 kg; cradle, 10.9 kg; 32-round loaded ammunition box, 20 kg
Length: 1.18 m
Width: 140 mm
Height: 208 mm
Barrel: 415 mm
Sight: reflex or mechanical
Muzzle velocity: 241 m/s
Rate of fire: cyclic, ca 330 rpm
Range: max, 2,200 m

Status

Prototypes.

Manufacturer

Heckler and Koch GmbH, D-78722 Oberndorf-Neckar.

UPDATED

HK69A1 40 mm Granatpistole grenade launcher

Description

The HK69A1 40 mm Granatpistole grenade launcher is a light and handy weapon designed to fire 40 mm low velocity grenades at either low or high angle; it can project a 40 mm grenade to a range of 350 m. It thus fills the gap between maximum hand throwing and minimum mortar range.

The HK69A1 40 mm grenade launcher is a single-shot, break-action weapon with a retractable butt-stock, a fixed foresight and a folding ladder-pattern rearsight. The barrel is rifled.

To load the weapon, the locking lever is pulled to the rear until the barrel pivots to its loading position; the striker does not cock during this movement. If there is a spent case in the breech it may be extracted manually, a recess cut in each side of the breech-end enabling the firer to grip and withdraw the case. When the next round has been inserted into the chamber, the barrel is pivoted back to the receiver, where it is secured by the locking lever which engages in the locking slots on the barrel. The striker is then cocked by hand, and may be secured by applying the safety lever positioned on either side of the pistol grip.

Grenades may be launched with the butt-stock either extended or retracted. To extend the butt-stock, it is rotated 90° to the left or right, extracted all the way and rotated again 90°. Retraction of the butt-stock is carried out in the reverse manner. The ladder rearsight is raised for aiming if the range exceeds 100 m.

A 37 mm version of this weapon, for firing riot control projectiles, is also available.

Data

Calibre: 40 mm
Operation: single shot, manually loaded
Length: stock extended, 683 mm; stock retracted, 463 mm
Barrel: 356 mm
Height: 205 mm
Sights: fore, fixed barleycorn; rear, folding ladder
Sight radius: 342 mm
Muzzle velocity: ca 75 m/s

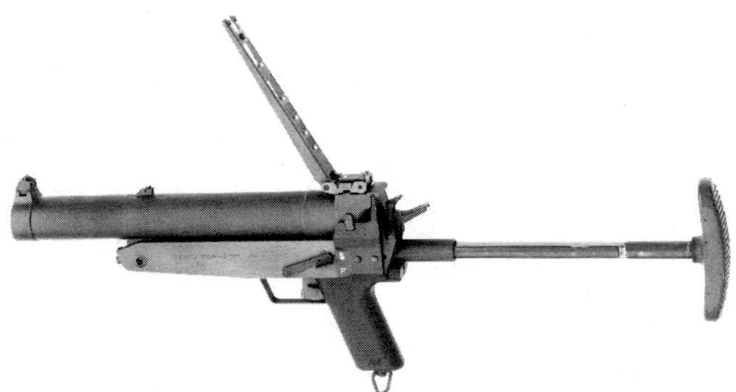

40 mm Granatpistole grenade launcher with butt-stock extended and rear ladder sight raised

40 mm Granatpistole grenade launcher

Manufacturer
Heckler and Koch GmbH, D-78722 Oberndorf-Neckar.

Status
In production.

Service
German Army. Several military and police forces worldwide, including Saudi Arabia and Sri Lanka.

VERIFIED

40 mm HK79 add-on grenade launcher

Description
The HK79 add-on 40 mm grenade launcher fits on all G3 and G41 rifles except the short K models, being assembled to the weapon in place of the handguard. Once attached the HK79 does not affect the accuracy of the rifle, since the barrel is still free to oscillate. Handling and operating functions of the rifle are not affected by the presence of the launcher.

A versatile range of ammunition, the low velocity 40 mm grenades based on those for the American M79 launcher, is available, enabling targets from 50 to 350 m distant to be engaged.

The HK79 launcher is a single-shot weapon with a dropping steel barrel. To load, the barrel is unlatched for it to drop down under its own weight; the round is then inserted into the chamber and the barrel closed. The tilting barrel will accept grenade rounds of any length. The firing pin is cocked by hand and the launcher is fired by pressing the trigger on the left-hand side of the handguard. At the rear of the launcher is a sliding safety catch, the status of which is shown by coloured rings - white for 'safe' and red for 'fire'; the weapon can be loaded and cocked with the safety set at either position. A mechanical tilting sight is fitted to the right side of the grenade launcher and permits range setting from 50 to 350 m.

The 40 mm SA-80 under slung grenade launcher was developed from the HK79. See separate entry under *United Kingdom* in this section for details.

Data
Calibre: 40 mm
Operation: single shot
Feed: breech loading, tilting barrel
Weight: launcher, 1.5 kg; G3 rifle and launcher, 5.6 kg; G41 rifle and launcher, 5.4 kg
Sights: 50 to 350 m in 50 m steps
Muzzle velocity: 76 m/s

Manufacturer
Heckler and Koch GmbH, D-78722 Oberndorf-Neckar.

Status
In production.

Service
Norway (ca 3,250).

UPDATED

G3 rifle with 40 mm HK79 add-on launcher opened for loading

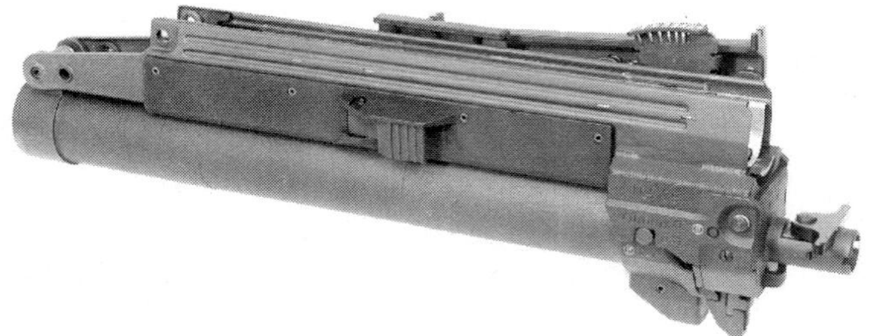

40 mm HK79 add-on launcher

ITALY

GLF-90 multiple grenade launcher

Description
The GLF-90 multiple grenade launcher permits the use of rifle grenades to their maximum possible range and fully exploits the energy of the propelling cartridge, producing performance superior to that achieved by rifle launching.

Two types of baseplate are available, optimised for use on hard or soft ground. Each has four stabilising arms which can be folded for transport. Attached to the baseplate is the launcher mechanism which has a traverse of 10° right and left of zero and is capable of elevation from +15 to +75°. There is a cross-levelling control which can compensate for up to 10° platform tilt.

The GLF-90 can be fitted with one or two launcher tubes; these are in two forms, suitable for using either 5.56 or 7.62 mm rifle cartridges as the propelling agent. The launcher tubes are not rifle barrels, but are bored in a tapering manner, narrower at the base than at the muzzle though the exteriors are parallel. A complete rifle round of the appropriate calibre is simply dropped into the bore and the grenade slipped over the exterior of the launcher tube. On firing, a firing pin passes through a hole in the bottom of the launcher tube and fires the cartridge primer. The bullet and propellant gas go forward to propel the grenade, while the gas pressure forces the cartridge case back. However, the loose fitting of the case in the tube does not accomplish obturation and a proportion of the gas passes the case to exit through the firing pin hole. This gas passes into channels in the platform, rebounds, passes back into the tube and adds to the propulsive force. The initial setback of the gas through the firing pin hole and into the platform acts to stabilise the platform and prevent it bouncing and thus upsetting the alignment of the tubes at the moment of grenade separation.

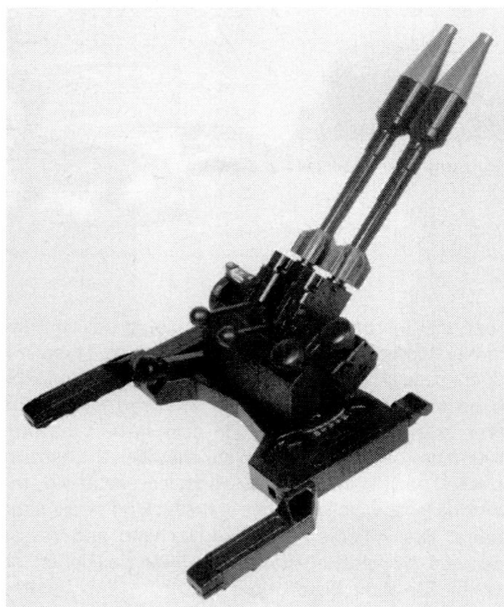

Franchi GLF-90 multiple grenade launcher

The minimum range is 100 m, governed by the minimum elevation and also by ground conditions. Maximum range depends upon the calibre of the projecting cartridge and the weight of the grenade. With 5.56 mm rounds a 400 g grenade will carry to 600 m, a 500 g grenade to 520 m. With a 7.62 mm round these figures improve to 900 m and 630 m.

The launcher unit weighs 8 kg with two tubes and the soft ground platform. Dimensions are 460 × 400 × 200 mm in the carrying mode.

Manufacturer
Luigi Franchi SpA, Via del Serpente 12, I-25131 Brescia, Fornaci.

Status
Development.

VERIFIED

12-gauge Franchi SPAS 12 shotgun

Development

The firm of Luigi Franchi is well known in the field of sporting shotguns. Franchi decided that it was time that a specific riot shotgun was produced since practically all the law enforcement shotguns in use are sporting guns with or without some modification. Since the requirements for a riot gun are different from those needed by a sports shooter it follows that there are deficiencies in current police and military weapons. Franchi set out to correct this by designing a gun meeting the requirements of the riot shotgun user. The result is the Franchi Special Purpose Automatic Shotgun (SPAS), first produced in October 1979.

Description

The SPAS is a short-barrelled semi-automatic shotgun with a skeleton butt and a special device enabling the firer to carry and fire the gun with one hand if necessary. The receiver is made from light alloy and the barrel and gas cylinder are chromed to resist corrosion. All external metal parts are sandblasted and phosphated black.

The characteristic scattering effect of a smoothbored barrel spreads the pellets to about 900 mm diameter at 40 m and progressively greater at longer ranges so that all that is needed to hit the target is a quick, approximate aim.

The automatic action will fire at about four shots a second. Using standard buckshot rounds it is possible to put 48 pellets a second onto a 1 m^2 target at 40 m. At that range the pellets each have a residual energy approximately 50 per cent greater than that of a 7.65 mm pistol bullet at the same range.

The wide range of ammunition available varies from buckshot and solid slug (with the same potential energy as a 7.62 mm NATO rifle round) to small pellets and teargas rounds which shoot a plastic container filled with CS gas up to a maximum range of 150 m.

When using the lighter ammunition the firer selects manual reloading, using the pump-action, but for all other types of round the gun will fire automatically.

With the grenade launcher fitted to the muzzle, a grenade can be fired to a maximum range of 150 m. There is also a special scattering device which fits to the muzzle and produces an instantaneous spread of pellets.

Data

Model	**11**	**12**
Calibre	12 bore	12 bore
Length	900 mm	930 mm
(with folded stock)	–	710 mm
Barrel	500 mm	460 mm
Weight	3.2 kg	4.2 kg

Operation: gas, semi-automatic or hand pump
Sights: fore, ramp; rear, notch
Rate of fire: 250 rds/min theoretical; 24-30 rds/min practical

Manufacturer

Luigi Franchi SpA, Via del Serpente 12, I-25131 Brescia, Fornaci.

Status

Available.

VERIFIED

Franchi SPAS 12 shotgun

12-gauge Franchi SPAS 15 MIL shotgun

Development

Following the success of the SPAS 12 shotgun, the Franchi company, after discussions with US and Italian authorities, decided to transform the SPAS 12 by replacing the tubular magazine with a box magazine. The resulting weapon was the SPAS 14, which, although technically effective, had some ergonomic shortcomings and went no further than the prototype stage. In 1983 Franchi decided to approach the problem from a completely different angle and develop a new system. Prototypes of the SPAS 15 were built early in 1984 and these matched both US JSSAP and Italian Ministry of the Interior requirements.

Description

The SPAS 15 uses both semi-automatic and manual pump action. Pump action, while not required by military authorities, is nevertheless considered necessary for police employment where the ability to generate a large volume of fire is not so important and where there is a requirement to fire non-lethal cartridges of varying ballistic performance. The semi-automatic functioning is performed by a gas piston which drives a bolt carrier and rotating bolt. The same components are manually operated by pulling back the sliding fore-end in the conventional pump-action manner. The bolt remains open after the last shot has been fired, and replacing the empty magazine by a loaded one will automatically release the bolt, which then loads a fresh round into the chamber. Pump or semi-automatic action is selected by a button in the fore-end.

The receiver is of die-formed and drawn nickel chromium steel; the external surface is sandblasted and phosphated. The barrel is of tempered alloy steel, is chromed internally and phosphated externally, and has a screwed section at the muzzle for the attachment of either a shot concentrator or one of three available grenade launching attachments.

The stock is of tubular steel and can be folded. There is a carrying handle above the receiver, which protects the cocking handle (used when operating in the semi-automatic mode) and can also be used to mount a variety of sights.

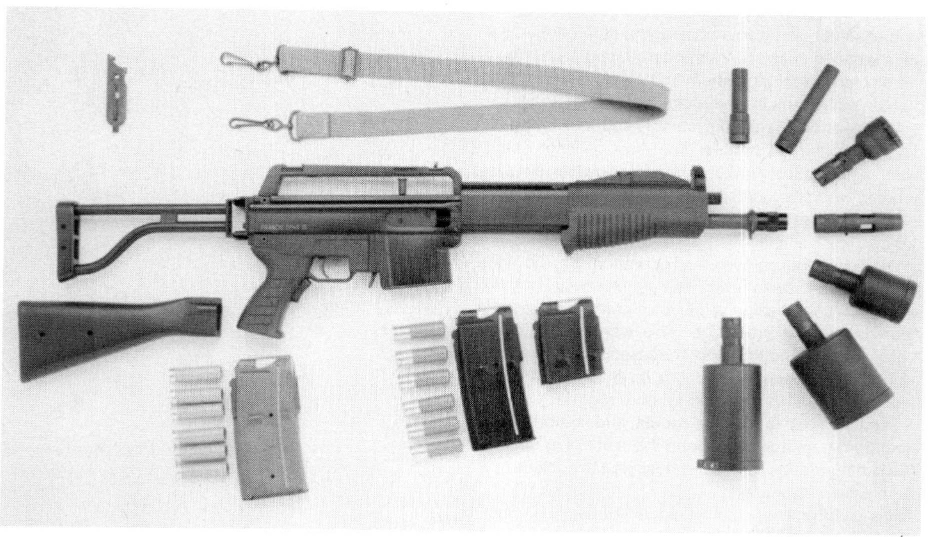

Franchi SPAS 15 MIL shotgun with accessories

Data

Cartridge: 12-gauge × 70 mm
Operation: gas, semi-automatic, or manual pump-action
Locking: rotating bolt
Feed: 6-round box magazine
Weight: without magazine, 3.9 kg
Length: 915 mm
Barrel: 406 mm

Manufacturer

Luigi Franchi SpA, Via del Serpente 12, I-25131 Brescia, Fornaci.

Status

In production.

VERIFIED

12-gauge Bernardelli B4 shotguns

Description

The Bernardelli B4 shotgun family was designed specifically for use by military and security forces. The frame is a single piece of aluminium alloy, giving high strength, the barrel is high grade steel and all external surfaces have an anti-corrosion and anti-glare finish. The breech mechanism employs a rotating bolt for additional security, and it is possible to launch grenades from the muzzle. The weapon feeds from a removable box magazine. There is a grip safety device which prevents the weapon being fired unless properly handled.

The B4 shotgun can be operated as a semi-automatic weapon or manually as a pump-action gun, the choice being selected by a lever. It was developed to meet a US JSSAP specification for a combat shotgun. The action is gas-operated in the semi-automatic mode. It can fire all normal plastic and brass-cased shotgun ammunition.

The B4/B model is derived from the B4 but dispenses with the semi-automatic action and is pump-action only. Its dimensions are the same as those of the B4 except that it weighs 450 g less.

Data

Cartridge: 12-gauge × 70 mm
Operation: gas, semi-automatic or manual pump-action
Feed: 3-, 5-, or 8-round box magazines

Weight: 3.45 kg
Length: stock folded, 730 mm; stock extended, 950 mm
Barrel: 460 mm, cylinder bored

Manufacturer
Bernardelli SpA, I-25063 Gardone Val Trompia, Brescia.

Status
In production.

VERIFIED

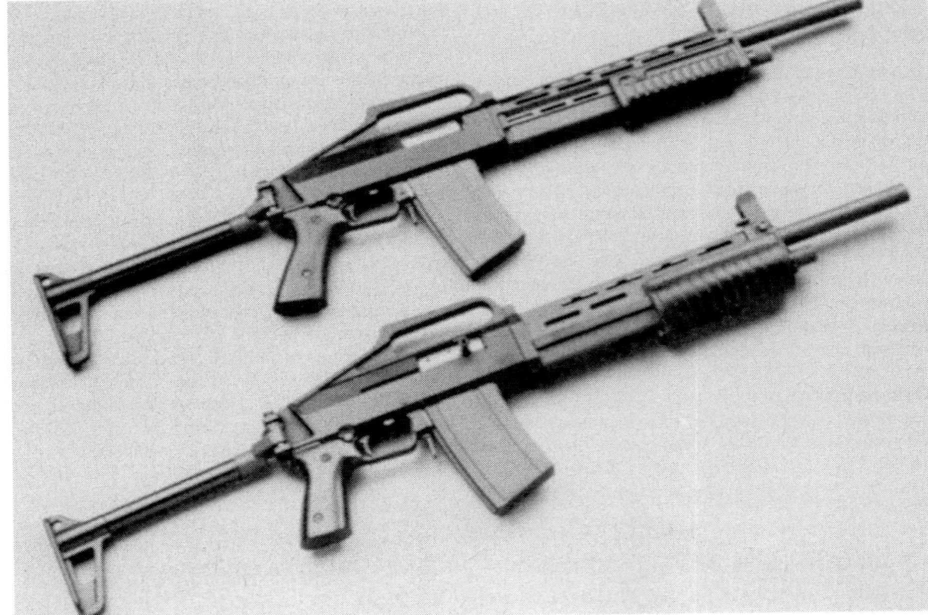

Bernardelli B4 (lower) and B4/B (upper) shotguns

Mod T-148 portable flame-thrower

Description
The Mod T-148 portable flame-thrower is a light, efficient and easy to operate weapon. It consists of a tank assembly, a flexible hose and a flame gun. Operation is based on two fundamental concepts: an innovative electronic ignition system and the elimination of the high pressure tank from the tank assembly.

The electronic ignition system is superior to the traditional ignition cartridges. The electronic ignition is silent and almost invisible, thus allowing surprise actions, whereas ignition cartridges may reveal to the enemy the presence of the operator, exposing him to danger.

The concept behind the operation of the tank assembly consists in pressurising the tank full of thickened fuel to the design value, and having the pressure decrease during projection. This way the traditional high pressure bottle, and the associated pressure reducer, have been eliminated from the tank assembly, making the design simpler and more rugged.

The flame gun is light and robust, and comprises the ejection nozzle and valves and the electronic ignition. The two handgrips have been designed to achieve optimum ease of operation.

The electronic ignition system is housed in a shock-resistant casing attached to the front handgrip. Its components are completely encapsulated in the casing, so that the ignition system will continue to function even if the flame-thrower has been completely immersed in water. The system is powered either by commercial alkaline batteries or by Ni/Cd rechargeable cells.

The tank assembly consists of two cylindrical containers, made of weldless aluminium alloy, surface treated with thick anodic coating. It is equipped with a safety valve with a built-in pressure indicator that signals the pressurised condition.

Before operation the tank assembly is filled with thickened fuel for two-thirds of its capacity and, after closing the plugs, the tank is pressurised with air or nitrogen through the filling valve by simply connecting it to a suitable compressed gas supply.

The flexible hose connecting the tank assembly to the flame gun is provided, at both ends, with ball-bearing couplings. This allows easy handling of the weapon, even when it is pressurised, by allowing free relative rotation between the flame gun and the tank.

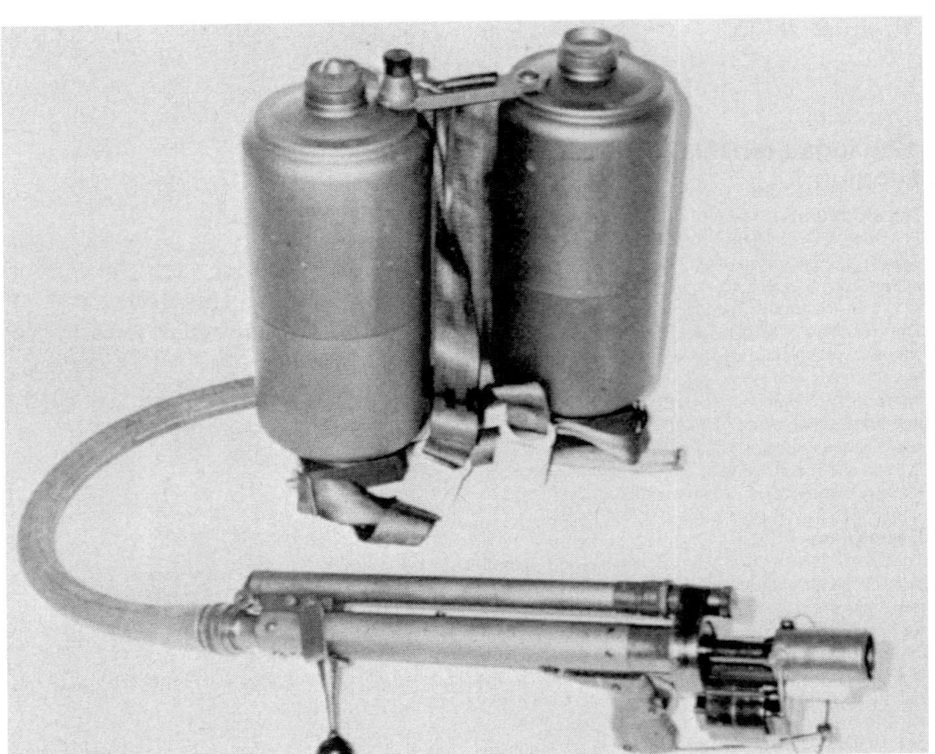

Mod T-148 portable flame-thrower

Data
Weight: total, 25.5 kg; empty tanks, 8.5 kg; flame gun, 4.3 kg; hose, 1 kg; 15 l of fuel, 11.2 kg; batteries, 0.5 kg
Operating range: in excess of 60 m
Working pressure: 28 kg/cm^2
Test pressure: 60 kg/cm^2
Operational temperature limits: −20 to +55°C

ACCESSORIES
Field Unit: steel container with accessories to prepare and transfer the thickened fuel. Capacity 45 l
Pressure Reducer: pressurises the flame-thrower directly from commercial cylinders up to 28 bar
Double Scale Tester: to check battery and circuit efficiency
Haversack: contains one complete refill of fuel and nitrogen

Manufacturer
Tirrena SIPA, via Monte d'Oro, I-00040 Pomezia, Rome.

Status
In production.

Service
In service with unspecified armies.

VERIFIED

KOREA, SOUTH

40 mm K201 grenade launcher

Description

The 40 mm K201 grenade launcher is based on the American M203 design, a sliding barrel breech loader which fits beneath the fore-end of the K2 assault rifle. It is accompanied by a special sight assembly which provides for day or night firing.

Data

Calibre: 40 mm
Operation: single shot
Weight: empty, 1.62 kg; with rifle, 4.88 kg
Length: 382 mm
Barrel: 305 mm
Rifling: 6 grooves, rh, 1 turn in 1.22 m
Muzzle velocity: 74.7 m/s
Effective range: area, 350 m; point, 150 m

Manufacturer

Daewoo Precision Industries Limited, PO Box 25, Kumjeong, Pusan.

Daewoo 40 mm K201 grenade launcher attached to K2 rifle

Status

In production.

Service

South Korean Army.

VERIFIED

POLAND

40 mm PALLAD and PALLAD-D grenade launchers

Description

The 40 mm PALLAD grenade launcher is a single-shot, add-on launcher intended for use with AKM pattern assault rifles. Converting an AKM rifle to carry the PALLAD takes only 60 seconds and removal 45 seconds. The PALLAD-D utilises the same launcher but allied to a twin-strut folding butt-stock assembly with a padded butt-plate and pistol grip to convert the launcher into an independent weapon.

Both types of PALLAD can be used to fire special low velocity (78 m/s) grenades in flat or high trajectories; maximum effective range is 430 m. It is possible to fire up to 8 rds/min.

The 40 mm low velocity grenade involved is the NGO, a FRAG-HE grenade capable of spreading fragments over a claimed effective radius of 180 to 200 m. Round weight is 250 g and projectile weight 188 g. The NGC is a 'flash-smoke' practice grenade while the NGB is another practice grenade with an inert projectile. All grenades are aimed using a quadrant range scale and the rifle or launcher sights.

Outlines of 40 mm PALLAD (top) and PALLAD-D (bottom) grenade launchers

Data

Calibre: 40 mm
Operation: single shot
Weight: empty, PALLAD on AKM, 5.55 kg; PALLAD-D, 2.3 kg
Length: PALLAD, 324 mm; PALLAD-D, butt folded, 390 mm, butt extended, 650 mm
Sights: quadrant
Sight radius: 85 mm
Muzzle velocity: 78 m/s
Effective range: up to 430 m

Manufacturer

Zaklady Mechaniczne Tarnow, 33-100 Tarnow, ul Kochanowskiego 30.

Status

In production.

Service

Polish armed forces. Offered for export sales.

UPDATED

ROMANIA

Snake 99 mm portable grenade launcher

Description

There are several aspects of the Romanian portable 99 mm grenade launcher known as the Snake which require a level of detailed explanation which is not yet forthcoming. It is claimed that there is no firing signature such as flash, smoke or noise and there is no firing recoil. Exactly how this can be made to happen is uncertain but it would appear that launching is achieved by detonating a fuel-air mixture within a sealed piston assembly, the rapid expansion of which drives the grenade from its position over the muzzle while still containing the detonated propellant gases. However, some form of exhaust-propelled

Snake 99 mm portable grenade launcher (T J Gander)

counter-weight is ejected from the rear of the launch tube for the minimum 'room size' from which the Snake can be fired is 3 × 3 m.

The 99 mm Snake is a shoulder-fired weapon some 1.4 m long when loaded and with a maximum effective range of 600 m; maximum possible range is 1,100 m. Each grenade weighs 4 kg and can be of the high explosive, fuel-air explosive or smoke type; there is also a 'penetrating' grenade which is probably HEAT. It is apparent that the grenades are fired with a high trajectory, allowing them to fall onto their target at a steep angle. Sighting is carried out using a small optical sight unit on the launch tube. A foregrip is provided and there is a carrying sling.

Data

Calibre: 99 mm
Weight: launcher, loaded, 9.5 kg; grenade, 4 kg
Length: launcher, loaded, 1.4 m; launcher tube, 1.068 m
Max range: theoretical, 1,100 m; effective, 600 m

Manufacturer

TURBOMECANICA, Boulevard Păcii no.244, sector 6, 77286 Bucharest.

Status

Available.

VERIFIED

40 mm automatic grenade launcher

Description

The Romanian 40 mm automatic grenade launcher is broadly based upon the CIS AGS-17 but is of 40 mm calibre as opposed to 30 mm and is chambered for a special high velocity cartridge which is longer than the universal US pattern 40 mm cartridge and appears to be slightly bottlenecked. The launcher is normally mounted on a tripod, with elevation and traverse locks. Feed is from a drum containing 10 rounds, mounted high on the left side of the receiver. The launcher can be fired in the single shot or full automatic modes, though descriptive literature suggests restricting burst fire to no more than five rounds.

Romanian 40 mm automatic grenade launcher

Data

Cartridge: 40 mm high velocity, special to weapon
Operation: blowback
Feed: 10-round drum magazine
Weight: gun, 23 kg; with tripod, without magazine, 32.5 kg
Weight, magazine: empty, 4.5 kg; loaded, 9.3 kg
Length: gun, 875 mm; on tripod, overall, 1.36 m
Barrel: ca 300 mm
Height of bore, on tripod: 600 mm
Muzzle velocity: 220 m/s
Rate of fire: cyclic, 300-400 rds/min
Elevation: −5 to +20°
Traverse: (total) 30°
Max range: 1,300 m
Min range: 100 m

Manufacturer

Romtehnica, 9-11 Drumul Taberei Street, Bucharest.

Status

In production.

Service

Romanian armed forces.

VERIFIED

40 mm grenade launcher

Description

This Romanian 40 mm grenade launcher was designed to be attached under the forward handguards of AKM and AK-74 pattern assault rifles but is apparently designed to fire Western pattern low velocity 40 mm grenades. Few details are available so there are no details of how the barrel is actuated to load the grenade; it is assumed that the barrel moves forward for loading. Firing is by a trigger on the left-hand side of the launcher.

Maximum range is 450 m and minimum range 50 m. Accuracy at 100 m is claimed to be 0.2 m in azimuth and 0.12 m in range. At maximum range (450 m) these figures are 3.5 m and 14 m.

Romanian 40 mm grenade launcher mounted on AK-74 rifle (T J Gander)

1996

Data

Calibre: 40 mm
Operation: single shot
Feed: manual loading
Weight: empty, 1.3 kg; loaded, 1.64 kg
Length: 450 mm
Barrel: 300 mm
Muzzle velocity: 80 m/s
Max range: 450 m
Min range: 50 m

Manufacturer

Romtehnica, 9-11 Drumul Taberei Street, Bucharest.

Status

Available.

UPDATED

Romanian 40 mm grenade launcher mounted on AK-74 rifle (T J Gander)

1996

SINGAPORE

40 mm CIS-40GL grenade launcher

Description

The 40 mm CIS-40GL is a modular-construction, single-shot grenade launcher capable of firing all types of standard 40 mm grenades. The weapon is built up from four basic units: receiver, barrel, leaf sight and butt-stock or rifle adaptor. In normal form it is fitted with a butt-stock as an individual weapon, but by removing the stock and replacing it with the rifle adaptor it can be fitted beneath the barrel of a wide range of assault rifles, in the manner of the US M203 system.

To load, the charging lever at the left side of the receiver is fully depressed; this simultaneously cocks the weapon, applies the safety catch and unlocks the barrel. The barrel can then be swung to the side and the grenade loaded, after which the barrel is swung back, when it automatically locks to the receiver. The safety catch must be manually released before the weapon can be fired.

The sight is a folding, graduated, leaf sight which can be adjusted for both elevation and azimuth. It is graduated to 350 m in steps of 50 m.

40 mm CIS-40GL grenade launcher

Data

Calibre: 40 mm
Operation: break open, single shot
Weights: empty, with stock, 2.35 kg; empty, without stock, 1.95 kg; loaded, 2.62 kg
Length: 655 mm
Barrel: 305 mm
Muzzle velocity: 76 m/s
Max range: 400 m

Manufacturer

Chartered Firearms Industries (Pte) Limited, 249 Jalan Boon Lay, Singapore 2261.

Status

In production.

VERIFIED

CIS 40 mm 40-AGL automatic grenade launcher

Development

Introduced in 1990, the CIS 40-AGL is an air-cooled, direct blowback operated automatic grenade launcher, firing standard US and locally produced 40 mm high velocity grenades. With an effective range exceeding 1,500 m (maximum range is 2,200 m), and the flexibility of being either tripod, turret/cupola or pedestal mounted on mobile platforms such as light vehicles or coastal patrol craft, the 40-AGL is an effective weapon against both personnel and light armoured vehicles.

During 1993 a licence production agreement was undertaken with PT Pindad to produce the 40-AGL in Indonesia for the Indonesian armed forces.

Description

The 40-AGL automatic grenade launcher fires high velocity grenades, belted by metal links, from an open bolt position. Modular in construction for ease of operation and maintenance, the 40-AGL comprises four main assemblies: bolt and backplate, trigger and sight, feed cover, and receiver.

The grenades are fed from the left by pawls actuated by movement of the bolt. As the bolt recoils, it causes a round to move half a cartridge distance. When the bolt is released by the trigger mechanism and moves forward, it causes the round to move another half cartridge distance and align with the barrel chamber. Just before the round is fully chambered, the firing pin is released to strike the primer and ignite the round. As the bolt is driven back, it both extracts and ejects the spent cartridge case to the right and at the same time causes the next round to be moved half a cartridge distance, and the whole operation is repeated. The launcher can be fired either manually or by remote control, using an electrical solenoid, in fully automatic mode at 350 rds/min.

CIS 40 mm 40-AGL automatic grenade launcher

Data

Calibre: 40 mm
Operation: direct blowback with advanced primer ignition, selective fire
Feed: disintegrating metal link belt
Weight: 33 kg
Length: 966 mm
Barrel: 350 mm
Muzzle velocity: 241 m/s
Max range: 2,200 m
Rate of fire: min 325-375 rds/min

Manufacturer

Chartered Firearms Industries (Pte) Limited, 249 Jalan Boon Lay, Singapore 2261.

Status

In production.

Service

Singapore armed forces and Indonesia.

Licence production

INDONESIA
Manufacturer: PT Pindad (Persero)
Type: Pelontar Granat Otomatis Kal. 40 mm
Remarks: Manufactured for the Indonesian armed forces with some locally introduced modifications to suit local mounting requirements.

VERIFIED

SOUTH AFRICA

Vektor 40 mm AGL automatic grenade launcher

Development

Originally developed as the AS88 by ARAM (Pty) Limited of Pretoria, and first shown publicly in 1992, the Vektor 40 mm AGL automatic grenade launcher is designed to fire a variety of high velocity 40 mm grenades. The patent rights for the original AS88 were purchased from ARAM by the LIW division of Denel.

The AGL is still under development, with full scale production scheduled to commence in April 1996. The AGL has been demonstrated firing from ground tripods and from various types of vehicle mounting.

Description

The Vektor 40 mm AGL automatic grenade launcher operates on the long recoil principle and firing is from an open breech to reduce the possibility of cook-offs; the weapon actually fires when in counter-recoil. All moving assemblies are buffered to reduce vibrations and recoil forces. Grenades are fed into the weapon in

Vektor 40 mm AGL automatic grenade launcher

Vektor 40 mm AGL automatic grenade launcher

M16A2 linked belts, with a pitch of 55.5 mm, from an ammunition box mounted on either the left- or right-hand side of the receiver. The feed direction can be altered in the field without requiring additional components. There is no conventional feed lever mechanism and the rate of fire can be altered by altering the position of the muzzle booster cup; maximum cyclic rate of fire is 350 rds/min.

When the AGL is ground-mounted on a tripod it is normally fired using spade grips and a manual trigger. However these components can be removed and replaced by a firing solenoid for vehicle or aircraft/helicopter mountings.

Among grenades specified for use with the AGL are the M383 and M648 HE, M430 HEDP and M385 Practice. Maximum range is 2,200 m.

Data

Calibre: 40 mm
Operation: long recoil
Feed: disintegrating metal link belt
Weight: 27 kg
Length: 861 mm
Height: 207 mm
Width: 238 mm
Barrel: 300 mm
Rifling: 24 grooves, rh, 1 turn in 1.219 m
Muzzle velocity: 244 m/s
Muzzle impulse: 6.26 kg/s
Rate of fire: max, cyclic, 350 rds/min
Max range: 2,200 m

Vektor 40 mm AGL automatic grenade launcher

Manufacturer

LIW, a division of Denel (Pty) Limited, PO Box 7710, Pretoria.

Status

Development. Production scheduled for April 1996.

UPDATED

Armscor 40 mm grenade launcher

Description

The 40 mm low velocity grenade launcher is a single-shot, break-open, breech loading, shoulder-fired weapon, fitted with a ranging Armson Occluded Eye Gunsight (OEG) system. It is an infantry small arm for firing low velocity grenades, whose tactical range lies between the maximum hand grenade throwing range and the minimum range of the mortar.

The high explosive grenades are effective against soft-skinned vehicles and personnel up to 375 m. The grenade has an effective casualty radius of 5 m and it may normally be expected that 50 per cent of any exposed personnel will become casualties.

The launcher is capable of firing smoke, illuminating, flash, inert and high explosive fragmentation rounds. It was produced using sheet-metal stampings, pressings and precision castings. The range quadrant is graduated in 25 m increments from 25 to 375 m and adaptable in poor light conditions.

Variant

37 mm Stopper grenade launcher

The 37 mm Stopper grenade launcher is a 37 mm riot grenade launcher version of the 40 mm launcher. Apart from the calibre difference the main change is that the 37 mm Stopper is provided with simple iron sights in place of the Armson OEG. Grenades launched include baton, stun, illuminating and irritant types. Maximum range is 300 m although the sights are set for 50 m. Weight is 3.7 kg complete with a sling.

Armscor 40 mm grenade launcher

Data

Calibre: 40 mm
Weight: empty, 3.7 kg
Length: butt retracted, 475 mm; butt extended, 665 mm
Barrel: 340 mm

Muzzle velocity: 76 m/s
Max effective range: point targets, 150 m; area targets, 375 m
Min range: training, 100 m; combat, 30 m

Enquiries to
Armscor, Private Bag X337, Pretoria 0001.

Status
Production complete but limited quantity available.

Service
South African National Defence Force.

UPDATED

Stopper 37 mm grenade launcher (T J Gander)

Mechem 40 mm MGL 6-shot grenade launcher

Development
The Mechem MGL (Multiple Grenade Launcher) 6-shot grenade launcher is a light, semi-automatic, low-velocity, shoulder-fired 40 mm grenade launcher with a six-round capacity and a progressively rifled steel barrel. It uses the well-proven revolver principle to achieve a high rate of accurate fire which can be rapidly brought to bear on a target. A variety of rounds such as HE, HEAT, anti-riot baton, irritant or pyrotechnic can be loaded and fired at a rate of one per second; the cylinder can be loaded or unloaded rapidly to maintain a high rate of fire. Although intended primarily for offensive/defensive use with high explosive rounds, with suitable ammunition the launcher is suitable for anti-riot and other security operations.

Description
The MGL 40 mm grenade launcher consists of a light rifled barrel, a sight assembly, a frame with firing mechanism, a spring-actuated revolving cylinder and a folding butt. The butt is adjustable to suit the eye relief and stance of the user, and the safety catch can be operated from either side. The launcher cannot be accidentally discharged if dropped.

The launcher is loaded by releasing the cylinder axis pin and swinging the steel frame away from the cylinder. By inserting the fingers into the empty chambers and rotating the aluminium cylinder it is then wound against its driving spring. The grenades are then inserted into the chambers, the frame closed and the axis pin re-engaged to lock. When the trigger is pressed, a double action takes place and the firing pin is cocked and released to fire the grenade. Gas pressure on a piston unlocks the cylinder and allows the spring to rotate it until the next chamber is aligned with the firing pin, whereupon the next round can be fired. If a misfire occurs the trigger can be pulled repeatedly.

The Armson Occluded Eye Gunsight (OEG) is a collimating sight which provides a single aiming spot; the firer aims with both eyes open and the effect is to see the aiming spot superimposed on the target, both target and spot being in sharp focus. The launcher is also fitted with an artificial boresight which can be used to zero the collimating sight. The collimating sight includes a radioluminous lamp which provides the spot contrast and which has a life of approximately 10 years.

The Armson sight was designed so that it can be used to determine the range to the target and be instantly adjusted accordingly. It enables the user to increase the hit probability at ranges up to 375 m. The range quadrant is graduated in 25 m increments and the aim is automatically compensated for drift.

Data
Calibre: 40 mm
Operation: semi-automatic, single shot
Feed: 6 chamber revolving cylinder
Weight: empty, 5.3 kg
Length: butt folded, 566 mm; butt extended, 778 mm
Barrel: 310 mm
Rifling: 6 grooves, rh, increasing twist to 1 turn in 1.2 m at muzzle
Rate of fire: practical, 18 rds/min
Muzzle velocity: 76 m/s
Rate of fire: practical, 18 rds/min
Max effective range: 400 m
Min range: training, 80 m; combat, 30 m

Manufacturer
MILKOR, PO Box 15148, Lynn East 0039.
Marketed by: Mechem, PO Box 912454, Silverton 0127.

Status
In production.

Service
South African National Defence Force and Croatian armed forces.

Licence production
CROATIA
Manufacturer: RH-Alan d.o.o.
Type: Ručni Bacač Granata 40 mm
Remarks: Standard specifications. 40 mm HE grenades also produced.

UPDATED

Mechem 40 mm MGL 6-shot grenade launcher

1996

Mechem 40 mm MGL 6-shot grenade launcher

1996

Mechem 40 mm MK 40 grenade launcher

Description
The Mechem 40 mm MK 40 is an add-on, single-shot grenade launcher which can be attached to virtually any standard assault rifle to fire standard low velocity 40 mm grenades. Once attached the MK 40 will not interfere with normal rifle operations in any way. It is loaded by swinging the barrel breech to one side and inserting a grenade directly into the breech. Target range is then set using a distance scale on the MK 40 and aiming is then carried out using the standard rifle sights. Maximum range is 400 m.

Although the MK 40 was designed to be attached to a rifle, an optional butt-stock is available to convert the MK 40 into an independent weapon; various sighting systems are available if this option is selected.

All standard 40 mm low velocity grenades can be fired, including anti-riot types.

Data
Calibre: 40 mm
Weight: 2.2 kg
Barrel: 300 mm
Rifling: rh, progressive
Muzzle velocity: from 76 m/s
Max range: 400 m

Manufacturer
Mechem, PO Box 912454, Silverton 0127.

Status
Available.

UPDATED

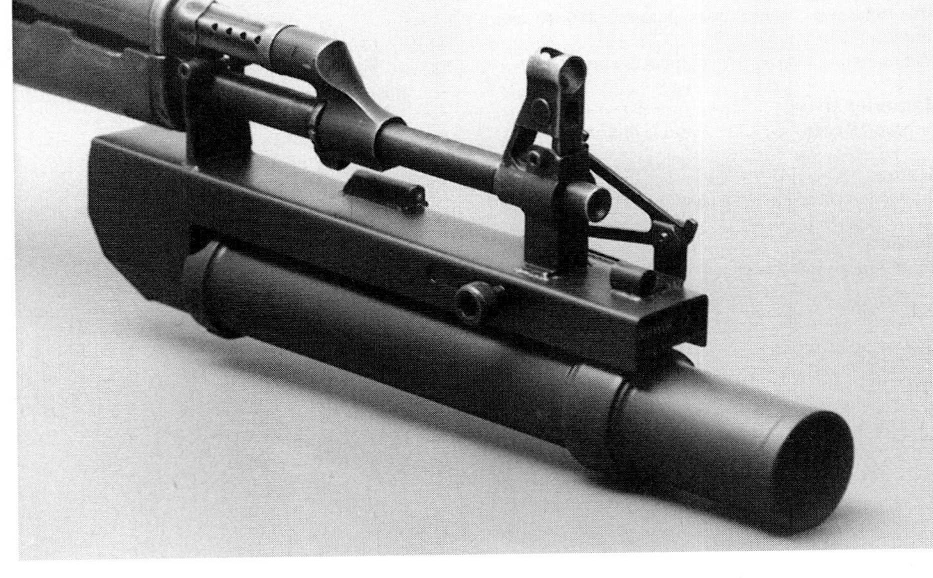

Mechem 40 mm MK 40 grenade launcher mounted on a 7.62 mm AK-47 assault rifle
1996

Mechem 40 mm M79 grenade launcher update

Description
Having appreciated that there are large numbers of US M79 40 mm grenade launchers still in service, Mechem have proposed an update programme to provide the M79 with an increased service life in a form more suited to modern warfare. Virtually the only component retained from the original M79 is the barrel. The original butt-stock is removed and replaced by a swing-around frame unit taken from the R4/R5 assault rifle. A pistol grip is added. These changes enable the M79 to be fired comfortably from the shoulder or hip. Originally a new type of flip-up ladder sight which will not snag on undergrowth or other obstacles was fitted but experience indicated the superior performance of the Armson Occluded Eye Gunsight (OEG - a red dot type reflex sight which permits aiming with both eyes open) so this sight is now offered as standard. All metal surfaces are re-coated with Gun Kote corrosion-resistant finish and permanent lubrication is provided for all moving parts.

The converted M79 is issued in a steel carrying case which also contains a sling, cleaning brushes and some other accessories.

Once converted the refurbished M79 can continue to fire all standard 40 mm low velocity grenades, including anti-riot types.

Updated 40 mm M79 grenade launcher from Mechem (T J Gander)
1996

Data
Calibre: 40 mm
Operation: break-open, single shot
Weight: 2.9 kg
Length: butt folded, 510 mm; butt extended, 720 mm
Barrel: 356 mm
Sight: Occluded Eye Gunsight (OEG)
Muzzle velocity: 76 m/s
Max range: ca 400 m
Max effective range: area targets, 350 m; point targets, 150 m
Min safe firing range: training, 80 m; combat, 31 m

Manufacturer
Mechem, PO Box 912454, Silverton 0127.

Status
Available.

UPDATED

Aserma Protecta and Bulldog 12-gauge compact shotguns

Description
The Aserma Protecta 12-gauge compact shotgun was designed for security, police and military applications where a combination of firepower and rapid deployment is required. The Protecta operates on the revolver principle, having a 12-round rotary magazine rotated to a fresh chamber by every pull of the double-action trigger. Cocking is effected by a cocking handle on the left-hand side of the barrel and spent cartridges are automatically ejected. Safety features include a conventional safety catch, a drop-safe trigger lock and a hammer lock. The weapon cannot be fired unless the double-action trigger is deliberately and fully pulled.

The Protecta is constructed using corrosion-resistant high tensile aluminium alloys, investment castings and glassfibre reinforced polycarbonates. The weapon field strips into three modules; magazine assembly, pistol grip and stock assembly, and the frame/barrel and trigger assembly. Cleaning is not necessary for extended periods unless non-lethal rounds are fired.

The folding twin strut butt-stock can be folded up and over the barrel to reduce the overall length to only 500 mm; recoil forces are stated to be low enough to allow the Protecta to be fired single-handed from the hip or shoulder. Despite the short barrel length (300 mm unchoked) the shot pattern up to 50 m is effectively the same as conventional riot guns with barrels twice as long.

The Protecta fires any 12-gauge cartridge up to 2.75 in/70 mm long. All natures of ammunition from non-lethal to solid slug can be employed.

Optional extras include an Armson Occluded Eye Gunsight (OEG), a sling and a carrying bag.

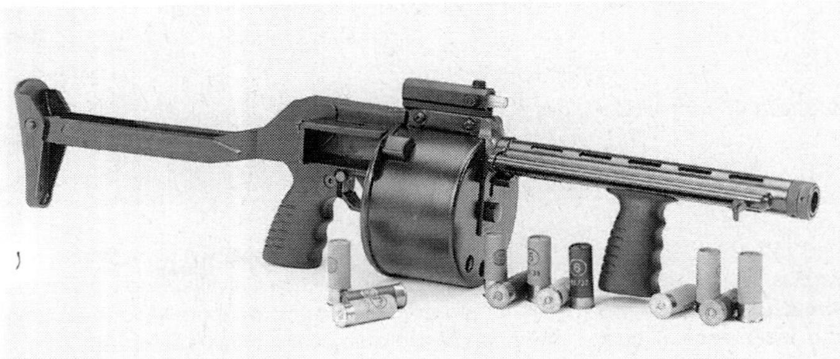

Aserma Protecta 12-gauge compact shotgun with folding butt extended
1996

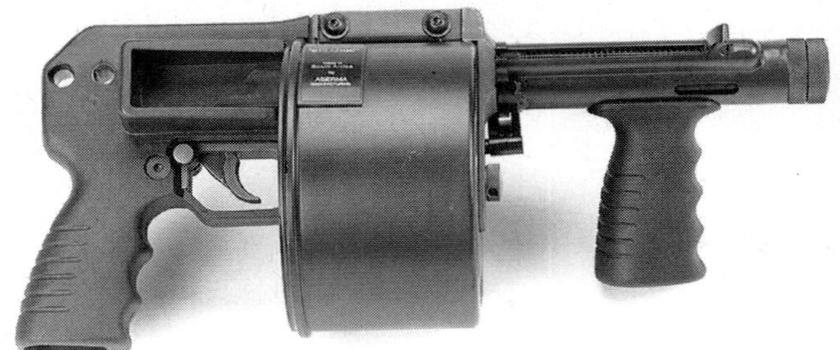

Aserma Bulldog 12-gauge compact shotgun
1996

Bulldog

The Bulldog is a more compact version of the Protecta shotgun. The length overall is reduced to 400 mm, although to achieve this length reduction there is no butt and the barrel length is reduced to 171 mm. The number of cartridges contained in the rotary magazine is reduced to 11. The Bulldog's operation and general principles remain the same as those for the Protecta but the Bulldog is a far more handy and potent weapon at close quarters, especially as the short barrel produces a greater spread of shot and a slightly higher muzzle velocity. For instance, firing No 8 24 g shot the spread produced by a Protecta at 10 m is 450 mm. The result achieved by a Bulldog at the same range is 510 mm.

The weight of the Bulldog, empty, is 2.167 kg.

Data - Protecta

Calibre: 12-gauge × 70 mm
Operation: double action, single shot
Feed: 12-round rotary magazine
Weight: loaded, 4.9 kg; empty, 4.2 kg
Length: butt folded, 500 mm; butt extended, 780 mm
Barrel: 300 mm unchoked

Manufacturer

Aserma Manufacturing, 115 Escom Road, New Germany 3620.

Status

Available.

UPDATED

Neostead 12-gauge pump action shotgun

Description

The Neostead 12-gauge pump action shotgun is a drastic departure from conventional shotgun designs and is an uncompromising military or police combat weapon. Developed from the outset as a combat shotgun, the Neostead has already attracted a great deal of attention from numerous potential users, including some in the USA.

The Neostead utilises an 'all-in-line' bullpup layout with the receiver located within the butt-stock. This not only greatly reduces the overall length of the weapon but provides the firer with a central point of balance above the pistol grip, enabling the weapon to be fired one-handed if necessary. The raised sight assembly above the weapon doubles as a carrying handle which may also be used to mount optical or night vision sights.

One of the main innovations on the Neostead is the ammunition feed. Two tubular alloy magazines are located side by side over the barrel, each holding up to six 12-gauge cartridges of up to 2.75 in/70 mm in length; a further cartridge may be carried in the chamber making it possible to carry 13 rounds ready for use. A selection switch lever associated with the feed can be used to select feed from either the left or right magazine. Leaving the selector switch in a central position causes the gun to feed alternately from either magazine until both are empty. To reload the magazines a release lever is actuated and the magazines are tipped upwards at the rear. Fresh rounds can then be fed directly into the two magazine tubes. Clear polymer slots over the magazines permit easy and rapid monitoring of each magazine tube's contents. Detachable twin tube polymer magazines are under development.

The Neostead's pump action operates in reverse to the usual. To eject a spent round the slide is pushed forward (the spent cartridges are ejected downwards) and to reload the slide is pulled back. A fixed breech is used so pushing the slide forward unlocks the barrel from the receiver and moves the entire barrel assembly forward while ejecting the spent case downwards. As the slide is moved to the rear a fresh cartridge is stripped from the appropriate magazine tube and carried backwards down a feed ramp to be placed against the fixed breech, ready for the barrel to be telescoped over the seated cartridge. As the barrel moves into battery a single heavy locking lug on the bottom of the receiver is forced upwards into a reinforced recess under the barrel to lock it with the receiver. Safety catches on both sides of the receiver are pushed aside by the slide as it comes back into battery to prevent the gun from firing unless the barrel and receiver are fully locked.

Virtually all the external surfaces of the Neostead shotgun are shrouded by a smooth-contoured high strength textured polymer stock and fore-guard with a non-slip surface. Only the barrel, breech block and receiver are steel; the magazine tubes and other unstressed metal components utilise high strength alloys. All controls (safety catch, slide action lock, magazine catch and selector) are centrally positioned to allow operation by right- and left-handed firers.

Provision is made for sling swivels and a bayonet could be accommodated, if required.

An 'ultra-short' version of the Neostead is planned. This will be only 550 mm long overall, with a nine-round capacity.

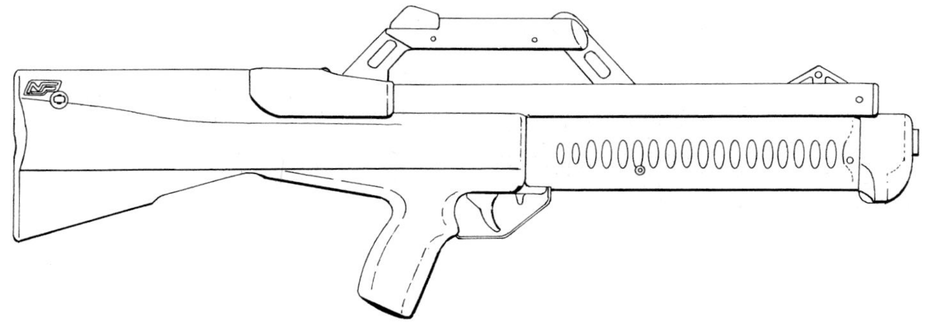

Outline drawing of Neostead 12-gauge pump action shotgun (L Haywood)

Data

Calibre: 12-gauge × 70 mm
Operation: pump action, single shot
Feed: 12-round twin tube magazines plus 1 round in chamber
Weight: 3.9 kg
Length: 690 mm
Barrel: 570 mm
Sights: fore, fixed blade; rear, fixed aperture; optical sights possible
Sight radius: ca 200 mm

Manufacturer

Neostead, PO Box 11410, Brooklyn 0011.

Status

Prototypes; pre-production development.

VERIFIED

SPAIN

40 mm LAG 40 SB-M1 automatic grenade launcher

Description

Originally known as the SB40 LAG, the LAG 40 SB-M1 is a belt-fed automatic grenade launcher with an effective range of about 1,500 m. The design is based on the long recoil principle, this being considered the most appropriate to achieve the lightest possible weight for the weapon by reducing the recoil loadings.

The long recoil system lends itself, in this calibre, to delivering the relatively low rate of fire of 215 rds/min which is tactically desirable and which allows the weapon to be aimed and controlled easily.

The LAG 40 SB-M1 was designed to be turret-, tripod- or pedestal-mounted on wheeled or tracked vehicles, helicopters or boats and as an infantry weapon.

An important feature is the ability to shift the feed to the right or left side as desired, a change which can be effected very rapidly without recourse to special tools. It is a simple weapon to operate and maintain. Field

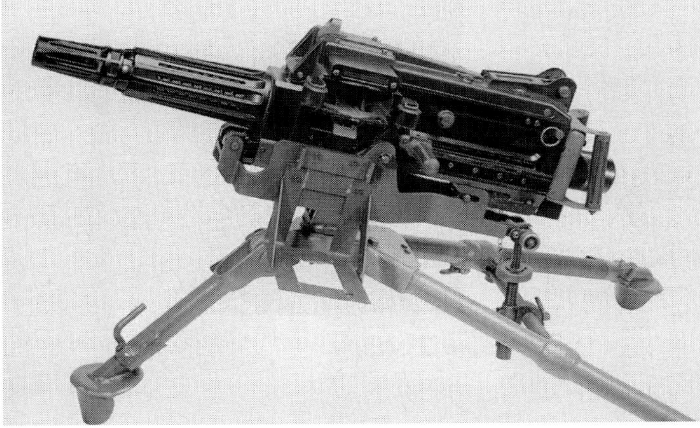

40 mm LAG 40 SB-M1 automatic grenade launcher

40 mm LAG 40 SB-M1 grenade launcher showing feed and sights

stripping can be carried out without tools and takes no more than 30 seconds.

The LAG 40 SB-M1 was designed to fire 40 mm high velocity grenades SB990 (M385), SB 991 (M384), SB 992 (M918), SB 993 (M430) and similar types. Maximum range is 2,200 m.

Data

Calibre: 40 mm
Operation: long recoil, automatic
Feed: 24- or 32-round linked belt
Weight: gun, 34 kg; tripod, 22 kg; cradle mount, 9.6 kg; feed box with 24 grenades, 12 kg
Length: 996 mm
Barrel: 415 mm
Rifling: 18 grooves, rh, 1 turn in 1,219 mm
Sight: mechanical and optical panoramic
Rate of fire: cyclic, 215 rds/min
Muzzle velocity: 240 m/s
Max range: 2,200 m
Effective range: 1,500 m

Manufacturer

SANTA BARBARA, Calle Julian Camarillo 32, E-28037 Madrid.

Status

In production.

Service

Spanish armed forces (reported to be between 200 to 300 units) and Portugal.

UPDATED

TAIWAN

Type 67 flame-thrower

Description

The Type 67 portable flame-thrower consists of a fuel/pressure unit and a hose/gun unit. The fuel and pressure group comprises a tubular frame with two interconnected fuel tanks, a spherical pressure unit mounted beneath the smaller fuel tank and other accessories. The hose supplies fuel from the tanks to the gun group. Then the fuel is ignited by a replaceable ignition cylinder in the nozzle end of the gun.

Data

Weight: basic, 12.58 kg; total with fuel, 23.51 kg
Height: 700 mm
Width: 530 mm
Length: 300 mm
Fuel: gasoline and napalm
Fuel capacity: 17 l
Range: 40-55 m

Manufacturer

CSF, Hsing Hua Company Limited, PO Box 8746, Taipei.

Status

In production and available.

Service

Taiwan Army and Marine Corps.

VERIFIED

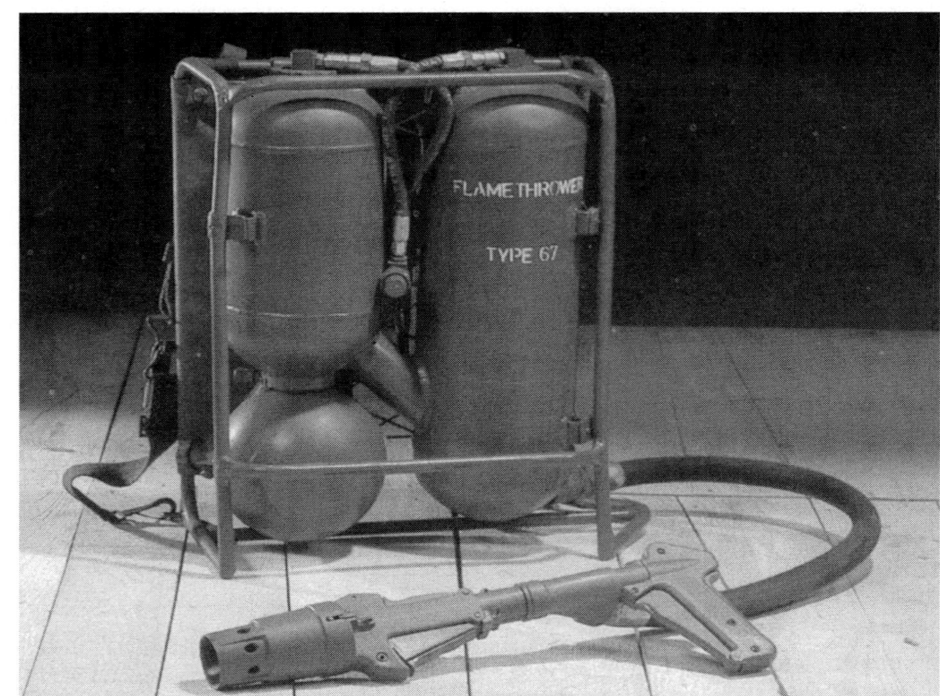

Type 67 flame-thrower

UNITED KINGDOM

40 mm SA-80 Under Slung Grenade Launcher

Description

The 40 mm SA-80 under slung grenade launcher was developed from the Heckler and Koch HK79 grenade launcher. The launcher is attached to the 5.56 mm Rifle L85A1 in place of the handguard. The handling and operation are not affected by the addition of the launcher and the operation of the HK79 remains unchanged.

Data

Calibre: 40 mm
Weight: grenade launcher, 1.89 kg; grenade launcher with L85A1 rifle and SUSAT sight, 6.14 kg
Sights: 50 to 350 m in 50 m steps
Muzzle velocity: 76 m/s

Status

Available.

Manufacturer

Heckler and Koch (UK), British Aerospace Defence Limited, Kings Meadow Road, Nottingham NG2 1EQ.

UPDATED

40 mm SA-80 under slung grenade launcher on L85A1 rifle

1996

Loading the 40 mm SA-80 under slung grenade launcher

1996

UNITED STATES OF AMERICA

Objective Individual Combat Weapon Program (OICW)

Development

The Objective Individual Combat Weapon Program (OICW) is supervised by the Joint Services Small Arms Program (JSSAP), under the Armament, Research Development and Engineering Center (ARDEC), directed towards providing the individual soldier with a weapon with the potential to replace the M16 rifle, the M4 Carbine, the M203 40 mm grenade launcher and the M249 5.56 mm Squad Automatic Weapon (SAW).

While the eventual OICW selected will continue to fire kinetic energy projectiles it will also be capable of firing high explosive bursting munitions producing fragments capable of penetrating body armour. The calibre of the latter munitions has been selected as 20 mm.

One of the programme goals is to produce a weapon capable of providing a 0.9 probability of a hit at 500 m and 0.5 probability of a hit at 1,000 m. The OICW will be capable of both day and night use, leading to the provision of electro-optical sights embodying uncooled infra-red sensor technology and a laser rangefinder, making the resultant sighting system more akin to a fire control system than most current small arms optical systems. Composite materials and plastics are utilised to reduce weight.

The OICW programme involves three phases. Phase 1, the design study phase, began in June 1994. In December 1994 JSSAP selected the designs deemed suitable for Phase 2, the system design and critical subsystem technology demonstration stage. Phase 2, including system design, critical systems, and technology demonstration (ie firing models), will last 12 months before Phase 3 commences. Phase 3 will involve prototype system design, fabrication and demonstration. During Phase 3, prototypes will be demonstrated in a non-firing Dismounted Battlespace Battle Lab experiment to be held during 1998. The following year, 1999, another experiment will be held at Fort Benning, Georgia, which will include safety-certified weapons and live firing demonstrations. It is intended that the first units will be fielded from 2002 onwards.

Three teams were originally formed to compete for Phase 1 of the OICW programme. AAI headed a group which originally included the Hughes Aircraft Company, Dyna East, Dynamit Nobel of Germany, and Mason and Hanger.

A second team was led by Alliant Techsystems, with Contraves Inc of Pittsburgh, Pennsylvania, Heckler and Koch GmbH, and Dynamit Nobel.

The third team was headed by the Olin Corporation who allied with FN Herstal of Liege, Belgium.

At the end of Phase 1 the three teams were whittled down to two, those led by AAI and Alliant Techsystems.

In October 1995 it was announced by AAI that their team had been joined by FN Herstal and Olin Ordnance, the losers following the Phase 1 stage of the programme. AAI and Olin will jointly develop and produce the OICW ammunition while FN Herstal will actually produce the weapon, probably involving its US subsidiary, FN Manufacturing Inc (FNMI). The AAI/Olin team thus consists of AAI (weapon and ammunition system development, integration, production and training), Dyna East (warhead development), Olin (ammunition production), Hughes (fire control and training), Omega (integration and training), and FN (weapon production.

In the other team, Alliant Techsystems is the prime system contractor, responsible for total weapon integration and development. Contraves Inc of Pittsburgh, Pennsylvania, is responsible for the development of the fire control system. Heckler and Koch GmbH of Germany are designing and manufacturing the weapon, while Dynamit Nobel AG, also of Germany, are responsible for the kinetic energy ammunition. Both German companies are to establish or licence a US base of supply to ensure a US manufacturing capability.

It has been estimated that the OICW programme could be worth $500 million plus from 2002 onwards.

Description

AAI and Alliant Techsystems both demonstrated OICW prototypes in October 1995. Few details of either were released, although both combined a 20 mm grenade launcher with a 5.56 mm rifle. Both are expected to meet the OICW requirements which include: a kinetic energy and high explosive/fragmentation capability; semi- and fully automatic fire with programmable bursts; a range of 500 m for direct-fire and 1,000 m for high explosive suppressive fire; a tactical weight of no more than 5.45 kg; a lethality greater than that of the MK 19 automatic grenade launcher due to accuracy of placing projectile bursts and the associated fuzing; and a high hit probability (Ph) due to laser rangefinding and accurate fuze timing.

The AAI mock-up depicted a partial bullpup all-in-line layout with the 20 mm barrel over the 5.56 mm barrel, the latter being associated with a virtually unchanged M16 series rifle mechanism and feed. The NATO standard 5.56 mm 30-round magazine is thus forward of the trigger group with the 20 mm box magazine being behind it in the butt. The fire control system, with an integral laser rangefinder, will enable the OICW to measure the distance to the target and then programme an electronic, air-bursting fuze prior to the projectile leaving the muzzle. The system also incorporates digital and direct view optics. A plug-in modular infra-red imager will be included to provide night and adverse weather capability. The AAI OICW weighs approximately 5 kg and is 810 mm long. Provision has been made in the design to provide fully ambidextrous controls.

Although no hard information is yet available, it is assumed that the fire control system for the Alliant Techsystems OICW will have similar capabilities to those of the AAI submission, featuring an uncooled infra-red imager utilising technology provided by Alliant Techsystems. However, on the Alliant Techsystems OICW, designed by Heckler and Koch, the 20 mm and 5.56 mm barrels are located side by side (the 20 mm barrel on the firer's left), the two magazines also being side by side in a location just forward of the trigger group. The 20 mm magazine holds six rounds while the 5.56 mm is NATO standard and holds 30 rounds. The butt stock houses the fire control system battery, some electronics and a recoil mitigation system. Provision has been made in the design to provide fully ambidextrous controls.

It would appear that the main design problem to be addressed with both OICW submissions is recoil attenuation when firing 20 mm ammunition.

Manufacturers

See text.

Status

Development.

UPDATED

Mock-up of the Alliant Techsystems Objective Individual Combat Weapon (OICW) proposal (T J Gander)

1996

Mock-up of the AAI Objective Individual Combat Weapon (OICW) proposal (T J Gander)

1996

Objective Crew Served Weapon (OCSW)

Development

The Objective Crew Served Weapon (OCSW) is under development for the US Army as part of Phase II of the Joint Services Small Arms Master Plan (JSSAMP). It is a two-man portable weapon system intended to provide a high probability of suppression and incapacitation against light vehicles, slow moving aircraft and watercraft to a range of 1,000 m, and protected personnel up to a range of 2,000 m. It is intended to replace the 40 mm MK 19 automatic grenade launcher and 0.50 M2 Browning heavy machine gun.

It is anticipated that prototype systems will be demonstrated in FY 1997-1999. Fielding is not expected much before 2015.

Following a period of examination of proposals, in early 1995 Olin Ordnance were awarded a 36-month contract to develop the OCSW concept. Olin are prime contractor, including ammunition development, and have combined with Dayron Inc of Florida who are responsible for fuzing. McDonnell Douglas Helicopters will produce the gun.

Description

Following an operational research programme involving various possible OCSW calibres, including a 19 mm cartridge, it was decided that the eventual calibre will be 25 mm. Target weight for the gun is 10.9 kg. It will be gas operated, firing from an open bolt, with ammunition being fed from either the left or right. Linked percussion-fired ammunition will be fed from modules, one containing 22 rounds weighing 10.9 kg, and the other containing 74 rounds and weighing 15.88 kg.

The gun will be mounted in a softmount cradle on a lightweight tripod weighing about 4.5 kg. The gun fire control system will be housed in a single module over the receiver and will weigh just over 2 kg. The rate of fire will be variable on the gun, with a maximum cyclic rate of 400 rds/min. For transport the barrel can be removed by a quick release mechanism.

Mock-up of the 25 mm Objective Crew Served Weapon (OCSW) (T J Gander)

1996

Mock-up of the 25 mm Objective Crew Served Weapon (OCSW) (T J Gander)

1996

Operational 25 mm ammunition for the OCSW will be of two natures, HE and dual purpose HEDP; there will also be a TP. Cartridge weight will be 173.5 g with the projectile weight 140.3 g. Muzzle velocity is 406 m/s. With this ammunition the apogee for a range of 2,000 m is 24.4 m, compared with 122 m for 40 mm rounds fired by the MK 19 to the same range.

Many details of the OCSW have still to be decided and only mock-ups for ergonomic trials and demonstrations have been produced to date. The illustrations provided here should not be regarded as anything more than a general impression of what the final form of the OCSW might take.

Main contractor

Olin Ordnance, 10101 9th Street North, St Petersburg, Florida 33716.

Status

Development for US Army - see text.

NEW ENTRY

Multipurpose Individual Munition/ Short-Range Assault Weapon (MPIM/SRAW)

Development

The Short Range Assault Weapon (SRAW) has been under development by Loral Aeronutronic for the US Marine Corps since 1990. The SRAW programme commenced as a competitive development concept for a lightweight shoulder-fired weapon that could defeat MBTs equipped with reactive armour and which would replace the AT4 and the M72 LAW series. The resultant product, later renamed Predator (see entry under *Anti-tank weapons* for further details of Predator), underwent further development to qualify for production. By then SRAW had demonstrated that it could meet its weight and guidance requirements.

The SRAW was designed to have a modular warhead so when the US Army produced a requirement for a shoulder-launched man-portable weapon to defeat personnel protected by buildings, bunkers or light armour it was decided that the successfully demonstrated SRAW, fitted with an alternative warhead, could be configured for this role. This resulted in the Multipurpose Individual Munition (MPIM).

In December 1994 an 18-month technology demonstration contract was awarded by the US Army's Missile Command to Loral Aeronutronic for the integrated MPIM/SRAW. This is now a joint US Army/US Marine Corps programme to develop a lightweight shoulder-fired multipurpose weapon. The US Army and the Marine Corps will share the same flight module and launcher, but the US Army missiles will have the MPIM warhead while the US Marine Corps will be primarily concerned with the anti-armour SRAW, or Predator (qv).

In July 1996, MPIM/SRAW (hopefully by then a less cumbersome acronym will be in use) will enter the engineering and development phase. Production could commence during FY 1999. The first field unit to be equipped with MPIM/SRAW is planned for FY 2001.

The principal MPIM/SRAW contractor is Loral Aeronutronic. Subcontractors include Aerojet (MPIM warhead), Hercules (propulsion system), and Systron Donner (inertial guidance).

Description

The MPIM/SRAW weighs less than 9 kg and is about 890 mm long. The launch tube contains a fire-and-forget missile, inertially guided to the gunner's line of sight at launch with a very high probability of hitting bunkers, buildings or light armour over a range of 17 to 200 m (400 m has been demonstrated), regardless of crosswinds or obscurants such as smoke.

MPIM/SRAW has a two-stage propulsion system. The first stage is soft launch which propels the missile from the launch tube at a velocity of about 25 m/s. The second stage is initiated about 5 m from the launch point and increases the velocity to approximately 300 m/s at a distance of 125 m down range; time to 300 m is 1.6 seconds. This two-stage launch system allows the missile to be fired from within structures or other enclosures. Aiming the missile, using a folding ×2 magnification telescopic sight, is stated to be exactly the same as aiming a rifle. If the target is moving the gunners needs to track it for less than 2 seconds. After firing, the gunner has no further need to track the missile, which proceeds towards the target under inertial guidance control, so the launcher can be immediately discarded.

The MPIM warhead is a two-stage system with an explosively formed penetrator (EFP) and a follow-through fragmentation grenade. On impact with the target, and following initiation by a crush switch, the EFP forms a long rod which passes through the centre of an annular follow-through fragmentation grenade and makes a hole through the structure or bunker wall. Following a pyrotechnic delay, the grenade then passes

The effects of a MPIM warhead on a concrete bunker protected by a layer of sandbags and earth

1996

Aiming a MPIM/SRAW from within a structure

1996

through the newly created hole, which is far larger than its own diameter, and detonates within the structure, scattering lethal fragments over the interior.

The MPIM warhead fuze has a built-in capability for immediate or delayed initiation, depending on the type of target involved. For instance, the delayed function could be employed when sandbag bunkers are attacked to allow the warhead to partially penetrate the sand prior to detonation.

The total MPIM warhead sub-system weight is 2.34 kg; total warhead weight is 3.116 kg. Overall diameter of the warhead is 141.5 mm and length (of the MPIM warhead assembly) is 325.6 mm. The warhead can penetrate double reinforced concrete walls, a triple brick wall, and earth and timber bunkers.

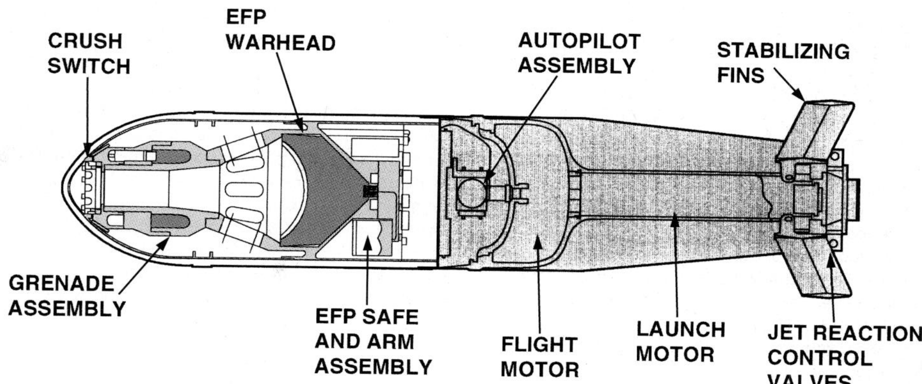

The main components of the Multipurpose Individual Munition (MPIM)

1996

Main contractor
Loral Aeronutronic, 22982 Arroyo Vista, PO Box 7004, Rancho Santa Margarita, California 92688-7004.

Status
Development for US Army - see text.

NEW ENTRY

140 mm Brunswick RAW rocket

Development
RAW (Rifleman's Assault Weapon), developed by the Brunswick Corporation in conjunction with the US Army Missile Command and the US Marine Corps. RAW, although rifle-launched, is comparable with firepower normally associated with light artillery and heavier tube-launched assault weapons.

The origin of RAW lies in the resurgence of interest in urban warfare and Fighting In Built-Up Areas (FIBUA) (or Military Operations in Urban Terrain - MOUT). In street warfare the enemy is protected by strong walls and sandbagged firing positions, and the same is true when infantry has to engage bunkers at close quarters. In these conditions it is rarely possible to bring tanks or large calibre guns to where they can bring fire to bear on the target and the infantry have to take unnecessary casualties in any subsequent attack. In these conditions there is a need for a weapon which will blow sizeable holes in masonry and concrete. In the past this has been overcome by firing anti-tank rounds or by emplacing a charge by hand. The latter is usually suicidal and the former results in a small hole only a few centimetres in diameter. To make a big hole demands a payload of explosive beyond that which a normal grenade carries and it has to be accurately placed and exploded. This is what RAW achieves.

Brunswick completed a company-funded Non-Developmental Item qualification of their HESH RAW in late 1988. During late 1993, US Marine Corps Phase 1 qualification tests with the HESH RAW were successfully completed in connection with the Rifleman's Breaching Munition (RBM) programme intended to provide a stand-off replacement for a satchel charge. A low risk operational Phase 2 test has still to be completed before the RBM is type classified.

A prefragmented anti-personnel/anti-materiel warhead, known as the Flying Claymore by some, is designated the MultiPurpose Munition - RAW (MPM-RAW).

RAW mounted on 5.56 mm M16A2 rifle

1996

The designation MultiPurpose was used as the round retains 90 per cent of the breaching/hard target capability in the direct-fire mode, while the gunner can choose to set a range for indirect-fire in the Flying Claymore mode.

Description
The RAW is a rocket-propelled projectile launched from the muzzle of an M16 series rifle, or can be easily modified to fit other rifles. The HESH RAW warhead is a sphere 140 mm in diameter containing 1 kg of high explosive which is detonated on striking the target giving a squash-head effect powerful enough to blow a hole 360 mm in diameter in 200 mm of double reinforced concrete. It will also defeat light armour, soft-skinned vehicles and similar battlefield targets.

The RAW projectile is a light metal sphere to which is attached a small rocket nozzle. The sphere contains three elements, the warhead, its associated safe/arm device and fuze, and the rocket motor. It is spun prior to launch and flies at a constant angle to the line of sight, using the thrust from its motor both for propulsion and overcoming gravity. The balance of thrust ensures that it flies on a flat and undeviating path for at least 300 m, after which it begins to assume a ballistic trajectory as long as 2,000 m for indirect- or interdiction fire. Within 200 m the firer can confidently aim directly at a target without having to make allowances for range. 200 m will encompass the majority of urban targets (the average range in street fighting is about 45 m).

The launcher was designed to be fired from M16 series rifles, but there is no reason why launchers could not be modified to fit any type of rifle, regardless of calibre. The launcher is a metal bracket which slips over the muzzle and engages the bayonet lug. When fitted it does not affect the use of the rifle in any way, and the rifleman can continue to fire normal ball ammunition, the only restriction being the weight of the RAW on the end of his rifle. The bracket of the launcher holds beneath it a turbine which is free to rotate on two simple bearings in the bracket. The turbine is activated by escaping gases emerging from it at right angles to the

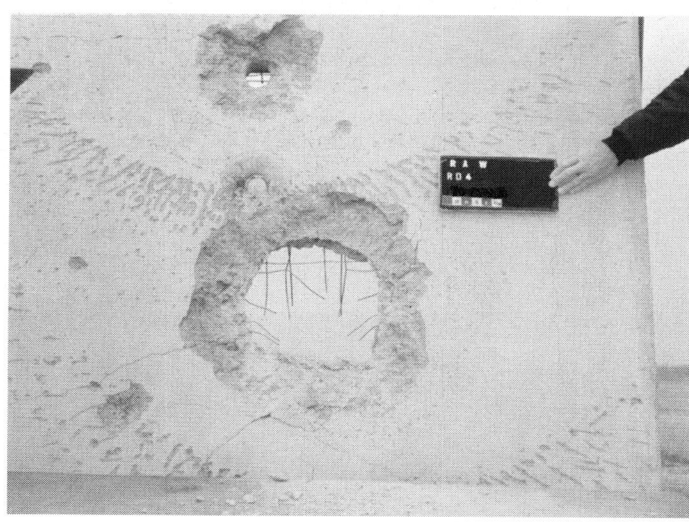

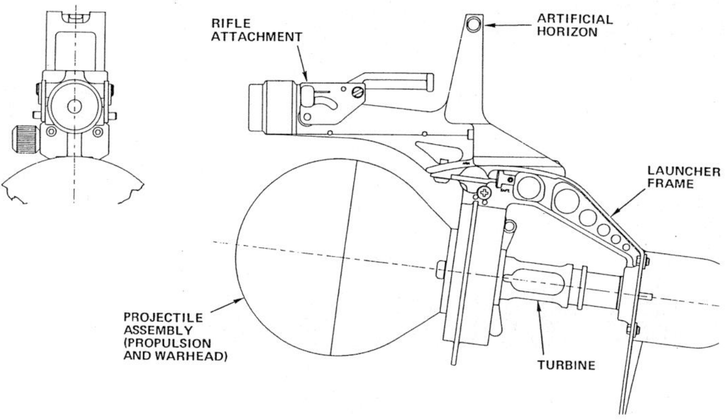

Diagram of RAW projectile and launcher assembly

Result of firing HESH RAW against a reinforced concrete wall

line of launch. Muzzle gases from the rifle flash suppressor will strike a firing pin to initiate the launch sequence.

Although the rifle can continue to be fired after RAW is mounted on it, the launch sequence can begin only after the rifleman has set an arming valve within the launcher allowing gases to reach the firing pin. Within the projectile is a matching ignition cap, aligned with the firing pin. The rotating tube is closed at its free end and there are gas escape slots cut into the walls.

The sequence of events on firing the projectile is as follows. The firer's first action is to turn the launcher arming valve. Having taken aim the firer then fires a normal ball round. The emergent gases at the muzzle are funnelled through the collar around the flash hider. Some gas runs down a tube and drives the firing pin forward to strike the ignition cap in the wall of the projectile. This cap fires the rocket motor and the efflux blows into the launch tube, emerging through the slots in the side walls of the tube but also emerging through several curved tubes. These tubes impart a spin to the launch tube and so to the projectile and before it has left the grip of the launch tube the spherical projectile is spin-stabilised for its flight. The projectile flies with a constant acceleration, reaching a velocity at 200 m of 173 m/s with a flight time of 1.9 seconds. An advantage of the spun projectile is that the absence of aerodynamic fins eliminates the tendency to weathercock into a crosswind. There is no back-blast on launch and the only effect on the gunner is a slight recoil and a sudden change of centre of gravity in his rifle.

A pre-fragmented anti-personnel/anti-materiel warhead utilising small diameter tungsten pellets and termed the MultiPurpose Munition - RAW (MPM-RAW) or Flying Claymore, has five to ten times the effective range and lethality of 40 mm grenades and is fitted with a variable-range fuze for direct fire and a height-of-burst fuze for indirect fire to 2,000 m. Practically every personnel or materiel target will be defeated by the thousands of tungsten pellets projected at high velocity from the 140 mm explosive charge.

Anti-armour multipurpose warheads offering shaped charge or explosively formed penetrative technologies, have been developed and successfully demonstrated on US Army contracts. In both cases stand-off for the anti-armour warhead is stabilised by a Motorola-designed optical proximity fuze. By proper selection of fuze mode the warhead can be detonated in the stand-off mode, impact mode, or impact mode with delay.

The large projectile payload capacity has permitted the development of designs for a number of solid and liquid payloads for smoke, illuminating, incendiary or any other aerosol or pyrophoric materials.

Data

Diameter: 140 mm
Weight: projectile and launcher, 4.73 kg
Length: overall, 305 mm
Weight of filling: 1 kg HE
Propulsion: rocket motor
Range: effective, 300 mm; max, ca 1,500 m
Velocity at 200 m: 173 m/s
Time to 200 m: 1.9 s
Accuracy: 5 mils

Manufacturer

Brunswick Corporation, Defense Division, 3333 Harbor Boulevard, Costa Mesa, California 92628-2009.

Status

Completed company funded qualification of HESH RAW in late 1988. Type classification for USMC in 1992 - see also under Development.

Service

Limited production for US Marine Corps.

VERIFIED

CMS 83 mm Shoulder-launched Multipurpose Assault Weapon (SMAW)

Development

The 83 mm Shoulder-launched Multipurpose Assault Weapon (SMAW) was originally a McDonnell Douglas Astronautics development for the US Marine Corps. Manufacture of the SMAW commenced during 1984. McDonnell Douglas' interests in the SMAW were later acquired by CMS of Shalimar, Florida, and development continued to MK 153 Mod 1 launcher standard involving a simplified design overall with fewer components, reduced cost and simplified maintenance. Operations in the Gulf region during 1990 and 1991 led to the US Army borrowing about 150 SMAW launchers and a quantity of rockets from the Marine Corps as an interim measure pending the adoption of a disposable Bunker Defeat Munition (BDM - see following entry).

Description

The Shoulder-launched Multipurpose Assault Weapon (SMAW) is a man-portable shoulder-launched 83 mm system consisting of a reusable launcher and individually encased High-Explosive Dual-Purpose (HEDP) or High-Explosive Anti-Armour (HEAA) rockets which are attached to the rear of the launcher for firing.

The reusable SMAW launcher features a ×3.8 optical sight and a 9 mm spotting rifle which uses tracer cartridges ballistically matched to the tactical rounds; the spotting rifle is an adaptation of the British LAW 80 light anti-tank weapon 9 mm spotting rifle modified for a longer service life. Cartridges for the spotter rifle are packaged in the end cap of each HEDP and HEAA rocket and enable the gunner to achieve a high probability when firing the main rocket. The reusable launcher is used for both tactical and training purposes.

The HEDP rocket was designed to defeat earth and timber bunkers, concrete and brick walls and light armoured vehicles. Its warhead and fuzing allows point detonation for hard targets and delayed detonation for soft targets. The HEDP warhead contains 1.1 kg of aluminised high explosive.

With the addition of the HEAA rocket, which completed development testing and certification in 1988, SMAW became a true multipurpose weapon suitable for close-in anti-armour and urban fighting. The HEAA warhead was designed to defeat 600 mm of rolled homogeneous armour (RHA) at angles of obliquity up to 70°. SMAW can engage tank-sized targets out to an effective range of 500 m using the HEAA rocket.

A Common Practice Round (CPR), with an inert warhead but ballistically and visually identical to the HEDP rocket, is available for training.

Changes introduced on the SMAW MK153 Mod 1 included an extended life launcher tube, improved spotting rifle mountings at both ends, a simplified spotting rifle bolt design, a field-removeable firing mechanism, and a sealed magnet generator.

83 mm SMAW in firing position

Reusable launcher for 83 mm Shoulder-launched Multipurpose Assault Weapon (SMAW) showing spotting rifle under launcher tube

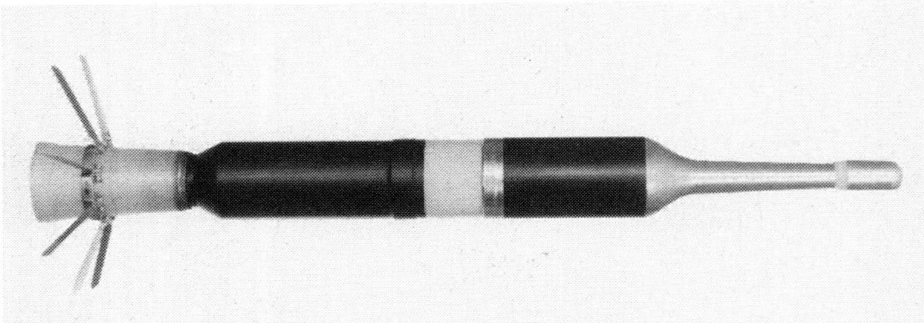

SMAW 83 mm High-Explosive Anti-Armour (HEAA) rocket

SMAW 83 mm High-Explosive Dual-Purpose (HEDP) rocket

Data

LAUNCHER

Calibre: 83 mm
Weight: 7.5 kg
Weight: (ready to fire) with HEDP, 11.8 kg; with HEAA, 12.62 kg
Length: 825 mm
Length: ready to fire, with HEDP or HEAA, 1.378 m
Optical sight: ×3.8, 6° field of view

ENCASED ROCKETS

HEDP
Length: 749.3 mm
Weight: 5.95 kg
Velocity: nominal, 220 m/s
Effective range: 500 m

HEAA
Length: 843.3 mm
Weight: 6.4 kg
Velocity: nominal, 220 m/s
Effective range: 500 m

Manufacturer

CMS, PO Box 896, Shalimar, Florida.

Status

Qualified for US military; more than 180,000 rockets produced to date.

Service

US Marine Corps.

UPDATED

Disposable SMAW (SMAW-D) Bunker Defeat Munition

Development

Following experience gained from operations in Grenada, Panama and South-east Asia, in September 1991 the US Army issued an Urgent Operational Requirement for a Bunker Defeat Munition (BDM). The BDM requirement followed an earlier but similar call for a Multi-Purpose Individual Munition (MPIM) which led to a development programme lasting six years. However, when the 1990-1991 operations in the Gulf region occurred the US Army was forced to borrow 150 SMAW launchers (see previous entry) from the US Marine Corps to meet their immediate bunker-busting needs. US Army experience with the SMAW led to their call for a disposable launcher with a similar performance.

Three companies submitted BDM proposals. Alliant Techsystems offered their AT8, a derivation of the AT4 anti-tank weapon, McDonnell Douglas and Talley Defense Systems offered their SMAW-D (Talley later acquired the McDonnell Douglas interests in the SMAW-D), while Giat Industries proposed their ABB system, only to withdraw in August 1993. Following a side by side test programme the US Army type classified the SMAW-D for limited procurement in September 1996 as their BDM.

Pending the release of $3.6 million in FY94, initial BDM production and first article testing was scheduled for January 1995. Additional FY95 procurement funding will allow production of the first 1,500 units to proceed to permit the first unit to be equipped at the beginning of 1996. At the end of August 1995 a further contract worth $5,321,438 was awarded for a further 1,012 BDM/SWAM-Ds.

The total of 6,000 BDM units (sufficient to equip the 82nd Airborne Division) was set by the US Congress as a cost-effective solution pending the resumption of a combined US Army/US Marine Corps programme for a renewed MultiPurpose Individual Munition and Short-Range Anti-Armor Weapon (MPIM/SRAW). For details of the MPIM/SMAW refer to the separate entry in this section.

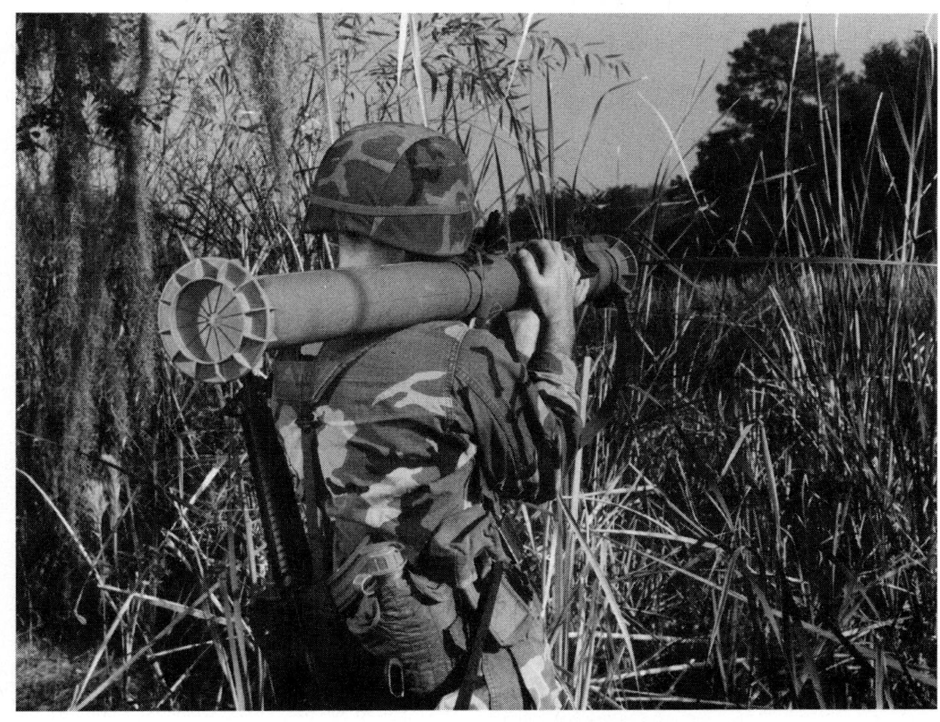

SMAW-D, the US Army's disposable Bunker Defeat Munition (BDM)

Description

The SMAW-D is a disposable combination of the launch tube originally developed for the MPIM project with the 83 mm HEDP rocket used with the US Marine Corps' SMAW. The 83 mm HEDP rocket was designed to defeat earth and timber bunkers, concrete and brick walls and light armoured vehicles. Its warhead and fuzing allows point detonation for hard targets and delayed detonation for soft targets. The HEDP warhead, containing 1.1 kg of aluminised high explosive, can penetrate 203 mm reinforced concrete walls and triple brick walls 305 mm thick.

The SMAW-D launcher accommodates the AN/PVS-4 night vision device and was designed to be compatible with other vision systems.

Data

Calibre: 83 mm
Weight: 7.26 kg
Carry length: 813 mm
Effective range: 250 m
Max range: 500 m

Manufacturers

Talley Defense Systems, 3500 North Greenfield Road, PO Box 849, Mesa, Arizona 85201.

Status

In production for US Army.

UPDATED

AT8 Bunker Buster

Description

The AT8 is a development of the AT4 anti-tank launcher described in the anti-tank weapons section (qv). Development was carried out by Alliant Techsystems, the object being to provide a light, disposable, multipurpose direct-fire weapon for bunker defeat and urban warfare. It is capable of breaching walls, destroying fortified targets and defending against attack by light armour.

The weapon utilises the existing AT4 shoulder-fired launcher loaded with a 83 mm SMAW HEDP projectile developed for the US Marine Corps. This is an explosive warhead device with a dual mode fuze which distinguishes between hard and soft targets. In use against hard targets the fuze detonates the 1.1 kg aluminised high explosive warhead on impact, allowing the explosive to develop its full effect on the target surface. In use against a soft target the fuze will delay the detonation so as to allow the warhead to penetrate some distance into the target before it takes effect. This will have devastating effects should the warhead still be within the target walls, or, if it has passed through the protection, it will have an anti-materiel and anti-personnel effect within the protected structure.

Since the AT8 uses precisely the same external launch tube and controls as the existing AT4, any soldier familiar with the AT4 can operate the AT8 without requiring further training. The AT8 incorporates an integral mounting bracket to use existing night vision and optical aids to improve accuracy under difficult conditions.

The M287 Trainer, originally developed for the AT4, can be used as a training simulator for the AT8. The M287 fires a 9 mm tracer bullet.

Data

Calibre: 84 mm
Weight: 7.3 kg

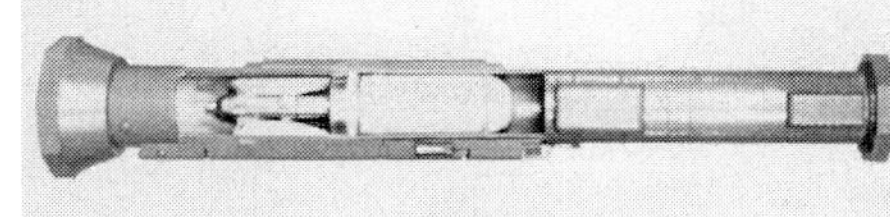

AT8 multipurpose weapon, sectioned

AT8 multipurpose weapon in firing position

Length: 1 m
Muzzle velocity: 219 m/s
Time of flight to 200 m: 0.99 s
Range: effective, 250 m
Accuracy at 250 m: <2 mils dispersion

Manufacturer

Alliant Techsystems Inc, Defense Systems, 600 Second Street NE, Hopkins, Minnesota 55343.

Status

Development completed.

VERIFIED

40 mm M203 grenade launcher

Development

The M203 40 mm grenade launcher was developed by the AAI Corporation of Cockeysville, Maryland, under the direction of the US Army Weapons Command, as a tactical accessory for the M16 rifle series. Development, as the XM203, started in May 1967 and it was type classified Standard 'A' in August 1969. The M203 replaced the M79 grenade launcher since it fulfilled the US Army requirement for a rifle/grenade launcher package, whereas the earlier M79 was a grenade launcher only.

By the end of 1986 well over 250,000 M203 grenade launchers had been produced by Colt's Manufacturing Company Inc, well over 100,000 of them being in US Government inventories.

A document dated April 1993 stated that the unit price for each of a batch of 1825 M203s supplied during FY90 to the US Marine Corps was $635.

Description

The 40 mm M203 grenade launcher is a lightweight, single-shot, breech-loaded, sliding barrel, shoulder-fired weapon designed especially for attachment to M16 series assault rifles. It allows the grenade launcher to fire a wide range of 40 mm low velocity high explosive and special purpose grenades, in addition to permitting normal use of the 5.56 mm M16 rifle.

The M203 consists of a receiver assembly made from high strength forged aluminium alloy, a barrel assembly again constructed using high strength aluminium alloy, a quadrant sight assembly, and a handguard and sight assembly group which replaces the normal M16 handguard. A quadrant sight assembly is mounted on the left side of the carrying handle of the M16 rifle or M4 carbine. This permits the sight to pivot on the range quadrant to the desired range setting (to 400 m) in 25 m increments. A leaf sight on the receiver is adjustable in 50 m increments to 250 m.

To load the M203 the barrel slides forward in the receiver and a grenade is inserted manually. The barrel then slides back to automatically lock in the closed position, ready to fire. The complete self-cocking firing mechanism is included in the receiver, thus allowing the M203 to be operated as an independent weapon even though it is mounted on the M16 rifle or (eventualy) on the M4 Carbine. A forward movement of the barrel ejects the spent case after firing.

The M203 attaches directly to any standard Colt M16 series rifle, utilising only two screws and without modification to the parent weapon. The rifle or carbine handguard is removed, and the M203 can then be easily installed in 5 minutes with the aid of a standard screwdriver. Pliers are required to install the lockwire that prevents the mounting screws from loosening. The M203 may be installed on an M4 Carbine in a manner similar to the rifle once a special kit has been incorporated; a kit to make the M203 compatible with the M4 Carbine was scheduled to enter service during late 1996.

40 mm M203 grenade launcher attached to M16A2 rifle

Although the maximum range of the M203 is 400 m the maximum effective range against area targets is quoted as 350 m; against point targets the maximum effective range is 150 m.

5.56 mm M16A1 assault rifle fitted with a 40 mm M203 grenade launcher (T J Gander)

1996

Colt Launcher System stand-alone 40 mm M203 grenade launcher (Scott Gourley)

1996

During peacetime training the minimum safe range for firing the M203 is 130 m. During combat this minimum safe range is reduced to 31 m.

Colt Launcher System

Colt's have proposed a variant of the M203 known as the Colt Launcher System. This allies the basic M203 launcher unit with an M16 series rifle stock, handguard and pistol grip to allow the M203 to be used as a stand alone weapon. If required the launcher unit may be allied with a M4 Carbine pattern telescopic stock, or the stock may be left off altogether, leaving the M203 launcher unit with just the M16 pistol grip and the handguard. A new interface receiver combines all the components.

A complete launcher with the rifle stock is 705 mm long and weighs 2.95 kg. If a carbine stock is involved the length extended is 711 mm, 628 mm with the stock retracted; weight is 2.77 kg. Wihout a stock the length overall is reduced to 457 mm and the weight is 2.5 kg.

Data - M203

Calibre: 40 mm
Operation: single shot
Feed: breech loading, sliding barrel
Weight: unloaded M16A1 and M203, 5.484 kg; loaded M4 and M203, 4.624 kg
Weight of launcher unit: empty, 1.36 kg; loaded, 1.63 kg
Length: 380 mm
Barrel: 305 mm
Muzzle velocity: with M406 grenade, 74.7 m/s
Max range: 400 m
Max range: (effective) area target, 350 m; point target, 150 m

Manufacturer

Colt's Manufacturing Company Inc, Hartford, Connecticut 06144-1868.

Status

In production.

Service

US Army since 1969 and US Marine Corps (1825), Australia (100), Brazil (200), Brunei, Burma (Myanmar) (500), Cameroons, Ecuador (40-50), El Salvador (224), Gabon (25), Greece (100), Grenada, Honduras (ca 1000), Indonesia (554 plus), Israel (10,280), Jordan (900), South Korea, Lebanon, Liberia (ca 275), Malaysia (over 8,500), New Zealand, Oman (250), Panama (ca 200), Philippines (ca 7,200), Qatar (ca 200), Singapore (300), Sri Lanka, Thailand (ca 5,700), Turkey, and United Arab Emirates.

Licence production

EGYPT
Manufacturer: Maadi Company for Engineering Industries
Type: 40 mm grenade launcher
Remarks: Essentially similar to M203 but intended for installation on Kalashnikov AK series assault rifles. See entry under *Egypt* for details.
KOREA, SOUTH
Manufacturer: Daewoo Precision Industries
Type: 40 mm K201
Remarks: Virtually identical to original M203. See entry under *Korea, South* for details.

UPDATED

40 mm M203PI grenade launcher

Description

The M203PI is a product-improved version of the M203 grenade launcher which can be rapidly mounted or removed from the rifle. Formerly the M203 has been available only on a limited range of rifles. With the production of interbars for any type of rifle, the M203PI launcher can be mounted on any assault rifle in use today. This is achieved by a unique locking mechanism on the launcher receiver which mates with an interbar secured to the barrel of the rifle. The interbar is attached to the rifle within the handguard and does not affect the operation of the weapon in any way.

For situations where only the launcher is required, the M203PI can be snapped on to a pistol grip assembly with a folding stock. This results in a highly portable weapon weighing less than 2 kg.

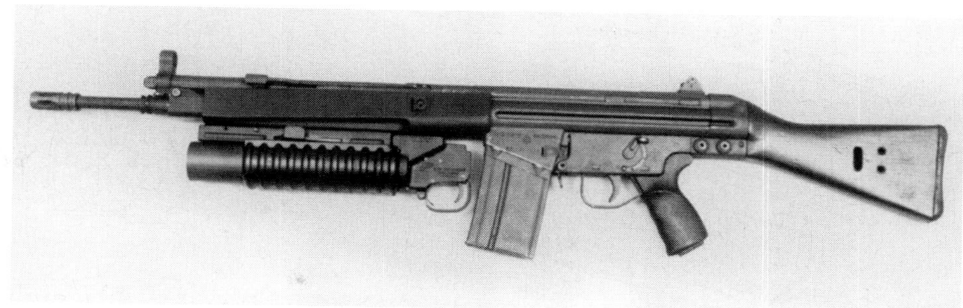

M203PI attached to G3 rifle

Data

Calibre: 40 mm
Weight: empty, 1 kg; loaded, 1.3 kg
Length: 305 mm
Barrel: 229 or 305 mm
Depth: 84 mm below rifle bore axis
Width: max, 84 mm
Muzzle velocity: 74.7 m/s
Max range: 400 m

Manufacturer

RM Equipment Inc, 6975 NW 43rd Street, Miami, Florida 33166.

Status

In production.

Service

Under test for certification and acceptance by US Marine Corps, Navy and Army, and British, French, Pakistan, Swedish and Venezuelan armies. In service with various US law enforcement agencies.

UPDATED

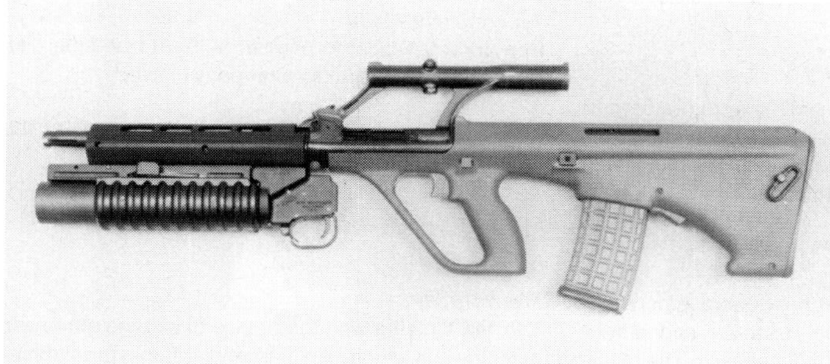

M203PI attached to Steyr AUG rifle

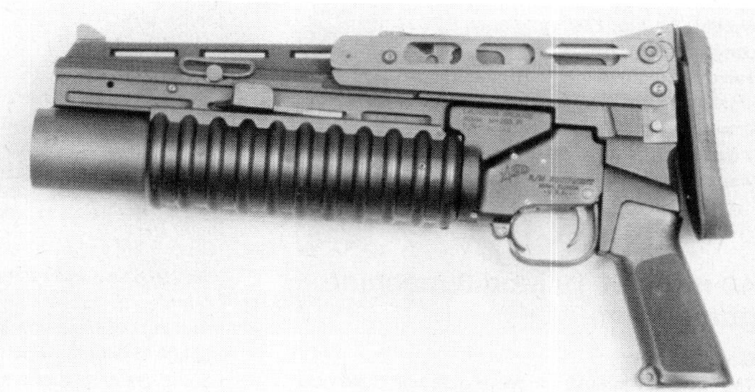

M203PI fitted to pistol grip unit, butt folded

M203PI attached to AKM rifle

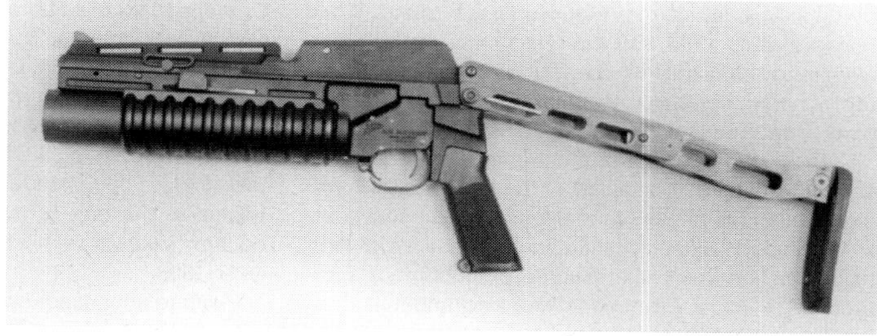

M203PI fitted to pistol grip unit, butt extended

40 mm M79 grenade launcher

Development

The M79 was the first weapon to come into service which was specifically designed to fire spin-stabilised grenades. It was first issued to the US Army in 1961 and some 350,000 were made in the next 10 years before production ceased in 1971. It was replaced by the M203 (see previous entry) in the early 1970s and is now obsolete in US service, although still widely used elsewhere. It is a light, handy weapon with an acceptable recoil and an adequate range. A trained man can put grenades through nominated windows in a house at 150 m range, although at greater ranges it is necessary to know the exact distance to the target since the grenade has a very high trajectory. This ability to place rounds on the target compensates to a great extent for the limited effective lethal radius of the grenade.

Description

The 40 mm grenade launcher, M79, is a single-shot, break-open, breech loading, shoulder-fired weapon. It consists of a receiver group, fore-end assembly, barrel group, sight assembly and stock assembly. A rubber recoil pad is attached to the butt of the stock to absorb part of the recoil. A sling is provided to carry the weapon.

To fire the launcher the safety must be in the forward position. In this position the letter 'F' is visible near the rear end of the safety. When the letter 'S' is visible just forward of the safety, the launcher will not fire. The safety is automatically engaged when the barrel locking latch is operated to open the breech.

The adjustable rearsight assembly consists of a rearsight lock, a windage screw and windage scale, an elevation scale and lock screw, a sight carrier and retainer lock-nut, an elevating screw wheel and elevating screw, and a rearsight frame and fixed sight.

The rearsight lock is spring-loaded and permits the rearsight frame assembly to be locked in either the up or down position. To unlock the sight frame push down on the flat surface of the rearsight lock. By releasing the pressure the frame is locked in the desired position. A knob at its right end turns the windage screw to adjust the rearsight for deflection.

The elevation scale is graduated from 75 to 375 m in 25 m increments and numbered at 100, 200, 300 and 375 m. As the rearsight carrier is moved up the scale, the rearsight is cammed to the left compensating for the normal right-hand drift of the projectile. The lock screw holds the elevation scale in position. When the rearsight frame is in the down position, the fixed sight

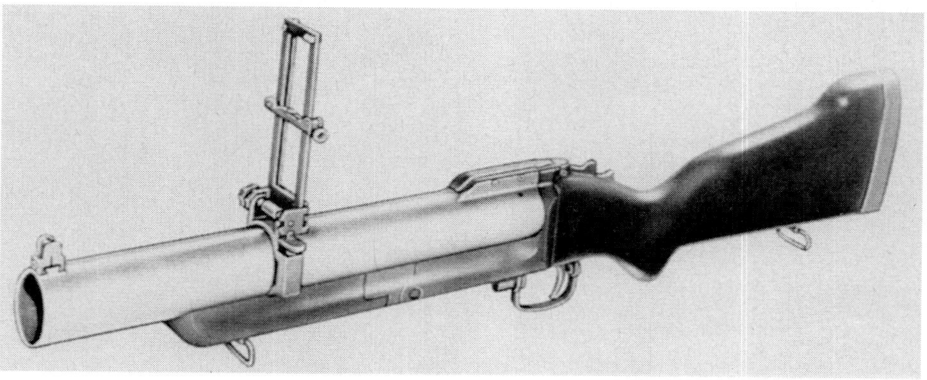

40 mm M79 grenade launcher

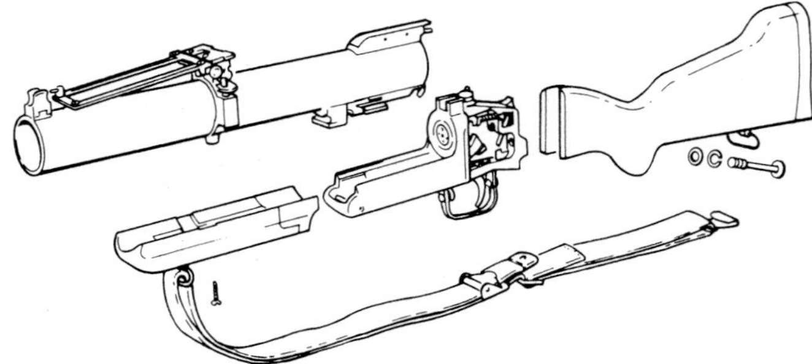

Major components of 40 mm M79 grenade launcher

40 mm M79 HE grenade

may be used to engage targets at ranges up to 100 m. The foresight consists of a tapered foresight blade and two foresight blade guards.

Variant

Mechem 40 mm M79 grenade launcher update.
See entry under *South Africa*.

Data

Calibre: 40 mm
Operation: break-open, single shot
Weight: loaded, 2.95 kg; empty, 2.72 kg
Length: 737 mm
Barrel: 356 mm
Sights: fore, blade; rear, folding leaf, adjustable
Chamber pressure: 210 kg/cm^2
Muzzle velocity: 76 m/s
Max range: ca 400 m
Max effective range: area targets, 350 m; point targets, 150 m
Min safe firing range: training, 80 m; combat, 31 m

Manufacturer

Originally manufactured by Colt; now made under licence by Daewoo Precision Industries Limited, South Korea.

Status

No longer in production in the USA. Superseded in US service by M203.

Service

US armed forces (reserve only). M79s were provided to the following nations but not all may remain in service: Australia, Burma (Myanmar) (3,000), Chad (100), Colombia (75), Costa Rica, Dominican Republic (111), Ethiopia (1,008), Fiji (103), Greece (2,666), Guatemala, Honduras (over 800), Iran (ca 2,900), Israel (1,240), Jordan (500), Kenya, South Korea, Lebanon, Malaysia, New Zealand, Nicaragua (64), Oman, Paraguay (24), Philippines (1,609), Portugal (404), Saudi Arabia (80), Somalia (150), Spain (80), Thailand (ca 2,800), Turkey (1,275), and North Yemen (200).

Licence production

KOREA, SOUTH
Manufacturer: Daewoo Precision Industries
Type: 40 mm KM79
Remarks: Identical to original M79. At one time also made by Tong Il Industry Company Limited.

VERIFIED

40 mm MK 19 Mod 3 machine gun system

Development

The 40 mm Mark 19 machine gun was developed by the US Naval Ordnance Station, Louisville, Kentucky, in order to provide the US Navy with a suitable weapon for riverine patrol work in Vietnam. It was designed around the high velocity 40 mm M384 grenade round. Work began in July 1966 and the first functioning models were ready early in 1967. The MK 19 Mod 0 was effectively used in combat in Vietnam. A total of 810 Mod 0 guns were produced.

A Product Improvement programme initiated by the US Navy in 1970-71 resulted in the MK 19 Mod 1 version; 583 of the original Mod 0 version were converted to this improved specification. In 1974 The Naval Ordnance Station Louisville resumed production of a further 761 Mod 1s. Production was also undertaken in Israel where a further 605 Mod 1 guns were produced by what was then Israel Military Industries. These were subsequently used by the Israeli armed forces on ground, vehicle and boat mountings. The MK 19 is licence-produced in South Korea.

Attempts to develop a streamlined MK 19 Mod 2 were deemed unsuccessful and no weapon production resulted.

In 1976 Louisville began a further improvement programme, aimed at improving reliability and safety, reducing cost and simplifying maintenance. The resulting MK 19 Mod 3, which appeared in 1980, has 47 per cent fewer parts and can be stripped without the need for special tools. A contract for production of the MK 19 Mod 3 weapon and the MK 64 Mod 4 Mount was awarded by the US Government to Saco Defense Inc in October 1983; this and previous orders amounted to a recorded 3,248 Mod 3 weapons. The 1983 contract and a subsequent contract awarded in December 1988 provided for deliveries of the weapon and mounts to the US Army, Air Force, Marine Corps and Navy through the early and mid-1990s. By 1995 well over 15,000 MK 19s had been produced and production was continuing at the rate of over 200 per month. This production rate is expected to be maintained until 1997-1998.

Planned enhancements for the MK 19 include a dual mount capable of accommodating both the MK 19 and 12.7 mm/0.50 Browning heavy machine gun, a soft-mount, and an ammunition box bracket.

It is understood that development work is being carried out on a MK 19 Mod 4 gun but no details are yet available.

40 mm MK 19 Mod 3 machine gun system

Description

The MK 19 is an air-cooled, blowback type automatic machine gun which fires a variety of 40 mm grenades at a muzzle velocity of 241 m/s. Grenades include the M430 High Explosive, Dual Purpose (HEDP) anti-personnel and armour-piercing round which can penetrate 50 mm of rolled homogeneous armour at a maximum range of 2,200 m; the M918 Flash-Bang practice round, and the M385 Target Practice inert round. Recent grenade rounds developed by Nico Pyrotechnic include an Impact Marker round, a Pepper round and a CS round. Maximum effective range exceeds 1,500 m.

The system can be ground-, vehicle- or turret-mounted. Ground mountings usually utilise the M3 tripod although a lightweight tripod weighing only 9.1 kg is available.

Ammunition is belted by a unique link which stays with the cartridge case and is ejected with the case after firing. The gun is fired from the open bolt position with the ammunition feed occurring during recoil similar to the 0.50 M2 machine gun. It can be fired manually or remotely by use of an electrical solenoid, a single round at a time, or fully automatic at 325 to 385 rds/min. The use of open bolt firing eliminates the chance of cook-offs while enhancing cooling between bursts, and allows sustained firing of three- to five-round bursts.

The round, which is belt-fed from ammunition containers holding either 32 or 50 rounds, is fed into the gun by pawls actuated by the movement of the bolt. The bolt recoils, withdrawing a round from the belt and, as the two travel rearward, a curved vertical cam forces the round down into a T-slot on the face of the bolt until it is aligned with the barrel chamber. When the bolt is released by the trigger mechanism, it is propelled forward by drive springs, chambering the round. Shortly before the cartridge case is fully closed in the chamber, the firing pin is released to impact the primer and ignite the round. The recoil force thus has to arrest the forward movement of the bolt and reverse it resulting in lower reaction forces and lower overall weight. The bolt is driven back by normal blowback action, extracting the empty case from the chamber. As it moves back and withdraws the next round from the belt, the vertical cam forces this round down into the T-slot and displaces the fired case, ejecting it through the bottom of the receiver.

The MK 19 comprises five major subassemblies: the bolt and back-plate, sear, top cover, feed slide and tray, and receiver. It is fitted with a sight, spade grips, and charger assembly similar to those on the M2 0.50 machine gun. The barrel can be removed and replaced without any requirement for headspace or timing adjustments.

A variety of fire control systems and mounts is available to enhance the capabilities of the MK 19. Recent fire control system enhancements to the MK 19 system have been the inclusion of the Ring Sight WC-30 Daylight Optic and the Saco Defense Adjustable Sight/Bracket. The Daylight Optic system replaces the existing 1,500 m iron sight with a ×1 graduated (200 to 2,000 m) day sight and a mount which contains an emergency iron sight. The Adjustable Sight/Bracket also replaces the existing iron sight and consists of a new iron sight and two mounting points for accessories (optics, laser aiming lights, and so on), all adjustable from 0 to 2,000 m. Both system decrease target acquisition time and improve accuracy, and are easily attached to, and removed from, the MK 19 by the gunner for storage or maintenance.

A MK 19 variant side-mounted on light helicopters such as the MD-500/OH-6 series is known as the MK 19 Thunder.

Data
Calibre: 40 mm
Operation: blowback
Feed: linked belt
Weight: gun, 34.3 kg; M3 tripod, 20 kg; lightweight tripod, 9.1 kg
Length: 1.095 m
Height: 224 mm
Width: 340 mm
Barrel: 412 mm
Rifling: 24 grooves, rh, 1 twist in 1.22 m
Muzzle velocity: 241 m/s
Effective range: point target, 1,500 m
Max range: 2,200 m
Rate of fire: cyclic, 325-385 rds/min

Manufacturer
Saco Defense Inc, 291 North Street, Saco, Maine 04072.

Status
Adopted by all US military forces and over 12 other nations, including Israel and Sweden (coastal defence forces).

Licence production
KOREA, SOUTH
Manufacturer: Daewoo Precision Industries Limited
Type: 40 mm Grenade Machine Gun K4
Remarks: Standard specifications. Can accommodate the KAN/TVS-5 night vision sight. Fires locally produced KM383 HE, KM212 HEDP, and KM395 TP.

UPDATED

40 mm EX-41 grenade launcher

Development
The US Naval Ordnance Station, Louisville, Kentucky, has developed a shoulder-fired 40 mm grenade launcher known as the EX-41 or Shoulder-Fired Weapon (SFW). The EX-41 was developed as part of a programme to determine the replacement for the US Marine Corps' M203 rifle-mounted grenade launchers but the development is also part of a programme officially designated the Bursting Munitions Technology Assessment, managed by the US Army's Picatinny Arsenal Research and Development Center. This programme is investigating all aspects of grenade launcher effectiveness, including laser ranging and airburst fuzing.

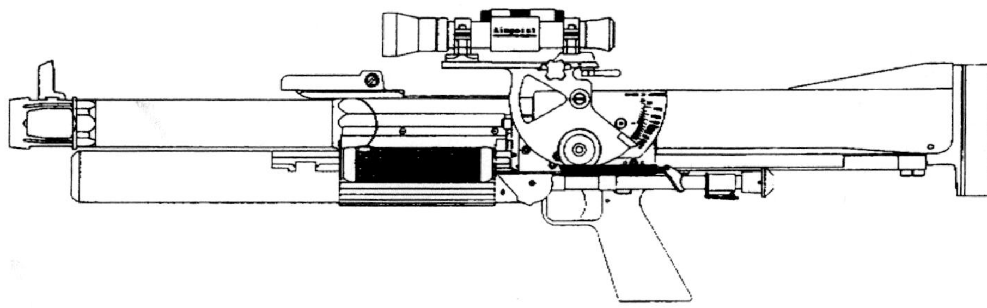

Outline drawing of engineering concept model of 40 mm EX-41 grenade launcher

Description
The 40 mm EX-41 is still at the engineering concept model stage although live firings have taken place. Early models are based around a four-shot pump-action launcher using a patented recoil buffering system; recoil forces are however estimated to be 'not insignificant'. Weight of the EX-41 concept model is almost 10 kg but further development is under way to bring this down to the projected 6.8 to 8.16 kg.

Demonstration rounds fired by the EX-41 include high velocity M430 HEDP and M918 Practice grenades coupled to a cut-down M433 cartridge case as used in the M203 launcher; the revised case was developed by Indiana Ordnance Inc. The resultant muzzle velocity is around 152 m/s, providing a maximum range of 1,500 m; the initial intention was to improve this to 3,000 m although this may not be achieved because of the excessive recoil forces anticipated. Optical sights are provided to take full advantage of the increased ranges possible.

Ammunition types envisaged for the EX-41 include flechette, incendiary, smoke and illumination; also planned is an improved HEDP round with increased armour penetration capabilities.

Data
Calibre: 40 mm
Operation: pump action
Feed: 4-round clips
Weight: engineering concept model, 9.98 kg; planned, 6.8 to 8.16 kg
Length: 914 mm
Height: 279 mm
Muzzle velocity: 152 m/s
Max range: 1,500 m

Development agency
US Naval Ordnance Station, Louisville, Kentucky.

Status
Development.

VERIFIED

40 mm MM-1 multiple grenade launcher

Description
The 40 mm MM-1 multiple grenade launcher is a revolver-type launcher capable of being operated by one man at a high rate of fire. It can be deployed in a variety of tactical situations and forms a convenient and highly effective source of emergency firepower for boat, helicopter and tank crews.

The MM-1 is a 12-shot revolver-type weapon which can be easily and quickly loaded, using any US or other 40 mm grenades up to 101 mm in length. The weapon is manufactured from aluminium, steel and high strength plastic materials and is extremely resistant to field conditions. It is easily maintained, requiring no more maintenance than a service revolver.

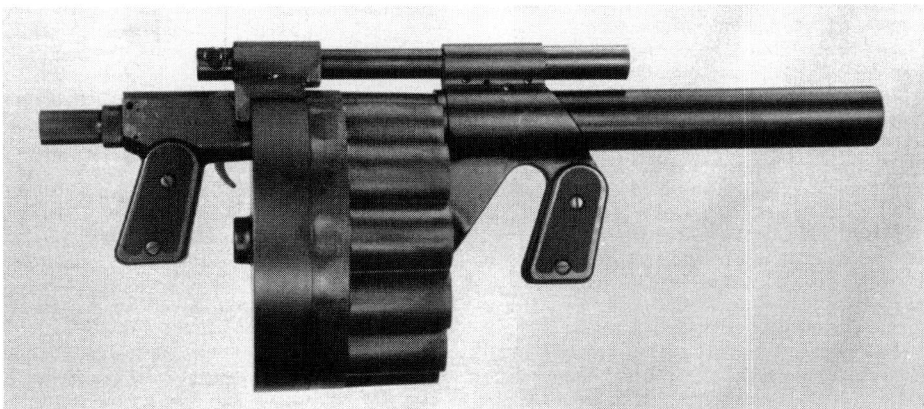

40 mm MM-1 grenade launcher

Data
Calibre: 40 mm
Operation: self-cocking revolver
Feed: 12-shot cylinder, spring-assisted
Length: 635 mm
Weight: empty, 5.7 kg
Rate of fire: cyclic, 144 rds/min; practical, 30 rds/min
Range: approx 350 m

Manufacturer
Hawk Engineering Inc, 42 Sherwood Terrace, Suite 101, Lake Bluff, Illinois 60044.

Status
In production.

Service
With special warfare units in the USA, Africa and Central America.

VERIFIED

30 mm Individual Grenade Launcher System (IGLS)

Description
The 30 mm Individual Grenade Launcher System (IGLS) includes a semi-automatic, shoulder-fired, magazine-fed launcher, and a family of three rounds of ammunition. The launcher is recoil-operated and uses a 10-round reloadable magazine. Ammunition includes High Explosive Dual Purpose (HEDP), marker and inert rounds. The programme was in Design Verification Testing, completed by the end of 1994. Future phases could include development of other types of ammunition such as buckshot, flechettes, smoke and illuminating rounds.

The 30 mm ammunition is lighter and smaller than its 40 mm equivalent, with no loss in performance. Armed with the IGLS the soldier can carry more rounds

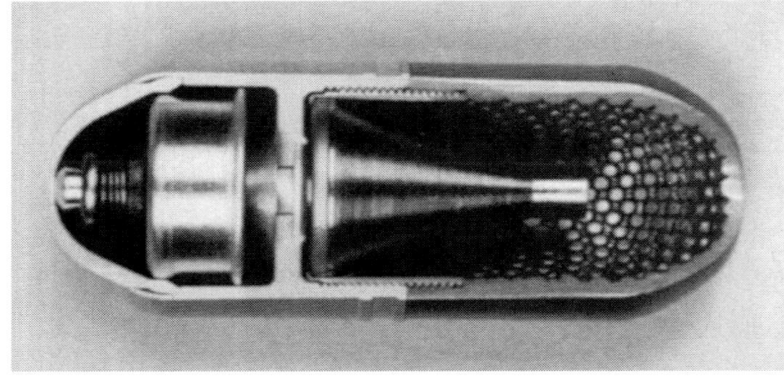

30 mm IGLS HEDP grenade

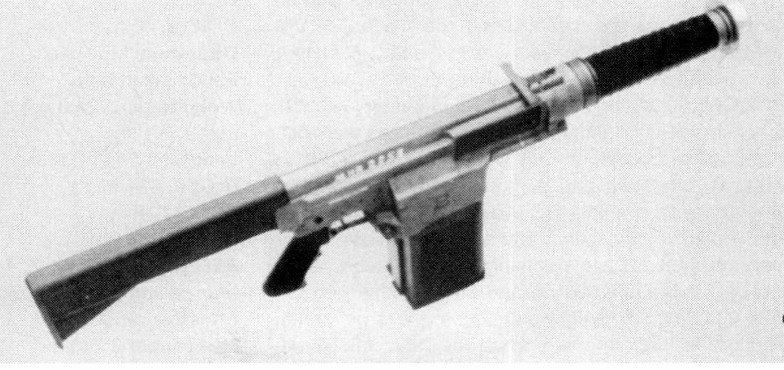

30 mm Individual Grenade Launcher System (IGLS)

pre-loaded in magazines without adding weight. With the weapon's semi-automatic capability the soldier can fire 10 rounds, replace the empty magazine with a full one and be ready to fire again within seconds.

Development of the IGLS system is a co-operative effort by Alliant Techsystems (systems contractor and ammunition development), Knox Engineering (launcher developer) and ARDEC (system manager).

Data

Calibre: 30 mm
Operation: recoil, semi-automatic
Feed: 10-round box magazine
Weight: with loaded 10-round magazine, 5.9 kg
Length: 813 mm
Muzzle velocity: 84 m/s
Noise level: <155 dB
Arming distance: 13-26 m
Penetration: 54 mm RHA
Lethal radius: equal to 40 mm HEDP M433
Range: >500 m

Manufacturer

Alliant Techsystems Inc, Defense Systems, 600 2nd Street, Hopkins, Minnesota 55343.

Status

Advanced development.

UPDATED

Pancor Jackhammer Mark 3-A2

Description

The Pancor Jackhammer is an automatic, gas-operated, 12-gauge shotgun which uses a preloaded rotating cylinder as its magazine. The cylinder has grooves cut into its outer surface which are engaged by a stud on an operating rod, so that as the rod oscillates it rotates the cylinder. The Jackhammer barrel floats, and is driven forward by gas pressure after the shot is fired. It is then returned by a spring giving movement to the cylinder operating rod. The significant point about this barrel movement is that it disconnects the barrel from a gas-tight connection with the cylinder, allowing the cylinder to be revolved to index the next round, and then, on the return stroke, reseals barrel and cylinder together.

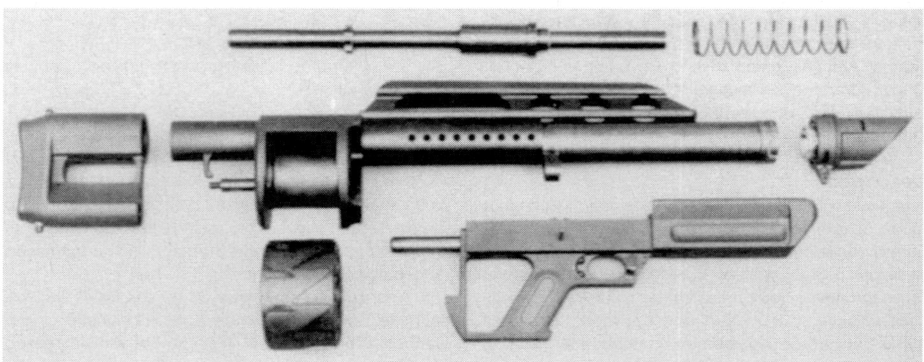

Jackhammer Mark 3-A2 dismantled

The barrel, flash eliminator, return spring, and 'Auto-bolt' (the Pancor patented name for the actuating rod which revolves the cylinder) are all of high quality steel; the remainder of the weapon is almost entirely plastic from Du Pont called Rynite SST™ which is a strong and durable polyethylene terephthalate (PET), plastic toughened by the admixture of glass fibre.

The cylinder, called by Pancor the 'Ammo Cassette', is also of Rynite SST™ plastic, contains 10 shots, and is preloaded and sealed with a shrink-film plastic, colour-coded to indicate the type of ammunition loaded. The sealing is removed by a pull-strip and the cassette simply clips into the weapon and immediately engages with its operating system. The weapon is then cocked by sliding the fore-end and it is ready to fire. It can fire at a rate of 4 rds/s. Once the cassette is emptied a simple movement of the fore-end and a hold-open latch allows it to be dropped out and a fresh cassette loaded. It is not possible to load single rounds into the gun and empty cases are not ejected whilst firing.

The Jackhammer Mark 3-A2 is the most recent step in the continuing development of this combat shotgun and is referred to by the company as its preproduction model. Whilst the Mark 3-A2 operates on the same principles as earlier models, described above, design modifications have been made which facilitate production and which have improved the various elements and subassemblies. One of the principal concerns was to keep the gun weight at an acceptable figure.

The Mark 3-A2 differs in outward appearance from previous models of the Jackhammer. The butt-stock assembly has been recontoured to improve control and absorption of recoil; the receiver has been redimensioned to allow smoother action; the sighting bridge and carrying handle have been reshaped; cooling ports have been added to the receiver, together with improved bearing surfaces for the floating barrel; the advancing track on the 'Ammo Cassette' has been recontoured to reduce firing stresses and improve cassette ejection and replacement time; and the pistol grip and front grip have been redesigned, together with the general exterior contour, to give a smoother and more compact outline.

There is a slight change in the method of operation. A decocking lever has been added, inside the butt-stock, which permits the hammer to be safely decocked when the weapon has been loaded. It can then be transported safely in the loaded condition. When required for action the lever is again operated, this time to cock the hammer, which is performed silently.

A particular feature is the rotating front sling swivel, mounted behind the flash guard. This allows the sling to take up the most convenient position according to the needs of the man carrying the weapon; it can be suspended in a firing position or slung across the back, in which position it automatically conforms to the body in the most comfortable manner.

In December 1987 Pancor Inc announced that the Jackhammer's receiver had been modified to accept and fire a special round that allows the payload or shot capacity of the standard shotgun cartridge (2.75 in/70 mm) to be increased by 100 per cent and perform with elevated pressures in the 20,000 psi (1,406 kg/cm²) region.

This cartridge, known as the 'Jack Shot', is a moulded cylindrical container of Du Pont Rynite SST™ which holds a sized — cut back to 50 mm length — standard 12-gauge shotgun cartridge pressed in from the rear end, with the powder, carrier and payload inserted from the front end. This allows unusual payloads to be

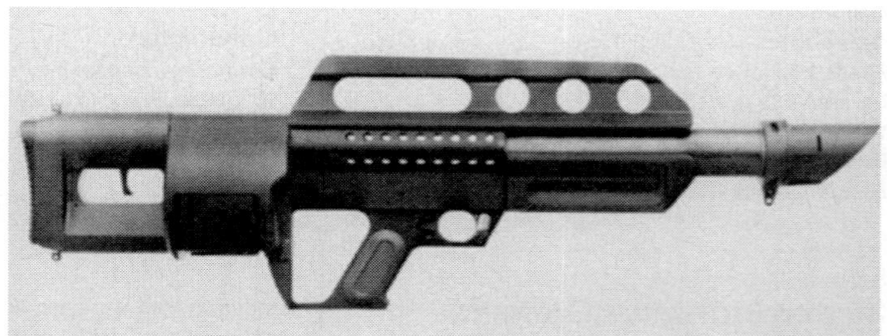

Pancor Jackhammer Mark 3-A2 combat shotgun

Rear view of Jackhammer Mark 3-A2, showing new butt-stock and improved form of 'Ammo Cassette'

loaded, such as armour penetrators, flechettes, fragmentation/shrapnel loadings, canister loadings, liquid or solid chemicals, rocket-assisted projectiles or simply larger loads of conventional lead shot. Moreover, the construction of the 'Jack Shot' permits ready reloading in the field.

The operation of the Jackhammer weapon using the 'Jack Shot' cartridge is similar to the normal use of the Ammo Cassette, except that the floating barrel is locked in its firing position (to prevent damage to the normal cyclic gas system from the extra pressure) and the auxiliary silent cocking lever is used to cock the hammer.

The company is studying a number of other ammunition concepts, with a view to developing a range of ammunition specifically suited to the Jackhammer shotgun. It is also involved in the development of a sound suppressor which will be relatively inexpensive and which can be discarded after a limited life.

Data

Cartridge: 12-gauge × 70 mm
Feed: 10-shot rotating cylinder
Weight: 4.57 kg
Length: 787 mm
Barrel: 525 mm
Height: 230 mm
Width: 125 mm
Rate of fire: cyclic, 240 rds/min

Manufacturer

Mark Three, 1410 Central Ave SW, Apartment 23, Albuquerque, New Mexico 87104.

Status

Development completed.

UPDATED

Ciener Ultimate over-under combination

Description

Rifles are particularly suited to long-range use; shotguns provide close-in defence when reaction time is at a minimum. The Ciener combination offers the opportunity to have both available as and when necessary. The Ultimate combination consists of a Remington 870 pump-action shotgun which attaches beneath the Colt AR-15/M16 rifle, utilising the bayonet lug as the basic fixing point.

A bayonet lug adaptor is attached to the shotgun barrel and a yoke adaptor to the receiver. This adaptor engages the ends of a special upper receiver to a lower receiver hinge pin. The bayonet lug adaptor has two spring pressure clips which fix the shotgun in place, allowing its removal by simply squeezing the clips.

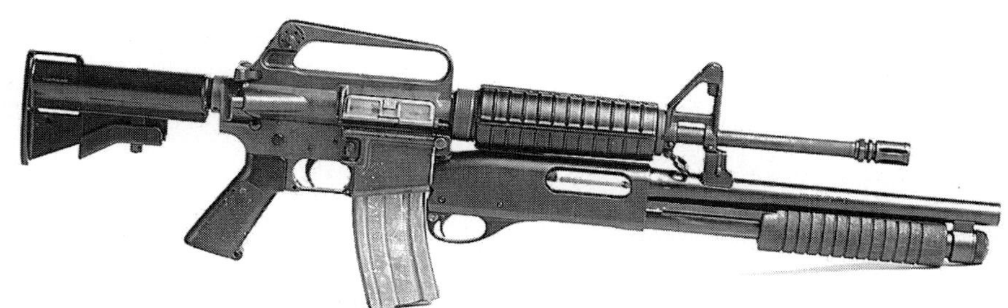

Ciener Ultimate over-under combination ***1996***

Manufacturer

Jonathan Arthur Ciener Inc, 8700 Commerce Street, Cape Canaveral, Florida 32920.

Status

In production.

VERIFIED

Master Key 'S'

Description

Knight's Armament Company produce a modified 12-gauge shotgun which can be attached underneath an M16 carbine. This unit is a modified Remington Model 870 12-gauge shotgun. The mounting brackets welded to the shotgun are made of hardened chrome-molybdenum steel. The system is designed so that the M16 magazine functions comfortably as a handgrip. The unit safely distributes the recoil loads through the barrel, and uses the bayonet lug to lock the weapons together. This unit also comes with a custom pistol grip adaptor which uses the standard M16 handgrip.

Manufacturer

Knight's Armament Company, 7750 9th Street Southwest, Vero Beach, Florida 32968.

Status

In production.

VERIFIED

Master Key 'S' shotgun attachment

Remington M870 Mark 1 US Marine Corps shotgun

Description

In 1966 the US Marine Corps conducted comparative trials of a number of shotguns and selected the Remington Model 870, slightly modified to suit the preferences of the Corps. The M870 Mark 1 is a rifle-sighted slide-action gun based on the commercial Remington 870 action, with a modified choke barrel. All metal parts are either Parkerised or have a black oxide finish, and there is a plain plastic butt-plate. The smooth extended forearm has prominent finger grooves. The magazine is extended to hold seven cartridges and the front end of the magazine tube is stepped to form a seating for the ring of a standard M7 bayonet; there is a bayonet locking lug forming part of the front sling swivel.

Data

Cartridge: 12-gauge × 70 mm
Operation: slide-action
Feed: 7-shot tubular magazine
Weight: 3.6 kg
Length: 1.06 m
Barrel: 533 mm, modified choke
Sights: fore, blade; rear, adjustable notch

Remington M870 Mark 1 US Marine Corps shotgun

Manufacturer

Remington Arms Company Inc, PO Box 179, Ilion, New York 13357-0179.

Status

Available.

Service

US Marine Corps.

VERIFIED

CREW-SERVED WEAPONS

Machine Guns
Cannon
Anti-Tank Weapons
Mortars
Mortar Fire Control

MACHINE GUNS

ARGENTINA

Argentine machine guns

Description

The Argentine armed forces have used a variety of machine guns in the past, including the French AA-52, the Browning 0.30 M3 converted to 7.62 × 51 mm and the 0.50 Browning M2 HB. Numbers of these are still in use, particularly the 0.50 M2 HB, but the standard company machine gun is now the locally manufactured version of the FN MAG. This is almost identical to the Belgian-made version, to which reference should be made for details of operation, differing only in the length (1.255 m) and empty weight (10.8 kg). The locally manufactured tripod appears to be based on the FN Model 04-41.

The squad automatic weapon is the Fusil Automatico Pesado (FAP) Model 2, a heavy-barrelled version of the FN-FAL rifle fitted with a bipod.

Data

FAP MODEL 2

Cartridge: 7.62 × 51 mm NATO
Feed: 20-round box magazine
Weight: unloaded, 6.45 kg
Length: 1.125 m
Barrel: 621 mm
Rifling: 4 grooves, rh, 1 turn in 305 mm
Sights: fore, blade; rear, aperture, adjustable to 600 m
Muzzle velocity: 840 m/s
Rate of fire: cyclic, 650-700 rds/min

Manufacturer

Fabrica Militar de Armas Portatiles 'Domingo Matheu', Avenida Ovidio Lagos 5250, 2000 Rosario.

Status

In production.

VERIFIED

Argentine manufactured 7.62 mm FN MAG machine gun

7.62 mm Fusil Automatico Pesado (FAP) Model 2

AUSTRALIA

5.56 mm F88 heavy barrel light support weapon

Description

The 5.56 mm F88 heavy barrel light support weapon is a 621 mm barrel length version of the F88 individual weapon. The individual modules of the weapon are all fully interchangeable with all other F88 models. It differs from the rifle model in having a slightly heavier barrel to cope with the increased thermal stresses of a higher rate of fire, an adjustable fold-up bipod attached at the muzzle end of the barrel, and a 42-round magazine in place of the usual 30-round magazine.

Data and specifications for the F88 light support weapon are as given for the Steyr AUG light support weapon under *Austria*, to which reference should be made.

Manufacturer

Australian Defence Industries Limited, Lithgow Facility, New South Wales (under licence from Steyr-Mannlicher AG).

Status

In production.

VERIFIED

5.56 mm F89 Minimi light support weapon

Description

The 5.56 mm F89 Minimi light machine gun was adopted by the Australian Army as its light support weapon, with production being undertaken from 1991 onwards at the Australian Defence Industries Lithgow Facility.

The F89 is a multipurpose machine gun capable of sustaining high rates of fire. It can be tripod-mounted and has the same capabilities as heavier calibre weapons. It can also be fired from the bipod or from the hip.

A full technical description of the F89 Minimi, together with all data, can be found in the entry under *Belgium*.

Manufacturer

Australian Defence Industries Limited, Lithgow Facility, New South Wales (under licence from FN Herstal SA).

Status

In production.

Service

Australian Defence Force.

UPDATED

Australian manufactured F89 Minimi machine gun, left side

Right side of Australian F89 Minimi machine gun

AUSTRIA

Steyr 5.56 mm AUG Light Support Weapon (LSW)

Description

One of the three versions of the Steyr AUG is a Light Support Weapon (LSW). All the details relevant to its action and operation are given in the section dealing with Austrian rifles and the data here refers to the LSW version only.

There are variant models of the LSW: the type firing only from a closed bolt (see below) is known as the 'Heavy Barrelled Automatic Rifle' (HBAR). The model firing from an open bolt is known as the 'Light Machine Gun' (LMG). Both open and closed bolt types are generically known as the 'Light Support Weapon' (LSW). All types use a barrel group with 178 mm pitch of rifling. A muzzle attachment acts as a flash hider and reduces recoil and muzzle climb during automatic fire.

There are two basic versions of this weapon; the LMG has the usual carrying handle and built-in optical sight of the AUG family. The LMG-T has a different receiver assembly which has, instead of the carrying handle and sight, a flat bar upon which any telescope sight or night-vision device can be mounted. In addition both types can be furnished with modifications which change the operation of the weapon so that it fires from an open bolt. This modification involves the substitution of a new hammer mechanism block for the standard type and changing the cocking piece in the bolt assembly; these components can also be supplied to permit modification to existing weapons; the conversion can be done in less than one minute. In the open bolt mode there is no change in the firing characteristics of the weapon.

Steyr 5.56 mm AUG LMG, with bipod extended

Data

Cartridge: 5.56 × 45 mm
Feed: 30- or 42-round box magazine
Weight: unloaded, 4.9 kg
Length: 915 mm
Barrel: 621 mm
Sights: optical, integral with carrying handle
Muzzle velocity: ca 1,000 m/s
Rate of fire: cyclic, ca 680 rds/min

Manufacturer

Steyr-Mannlicher AG, Postfach 1000, A-4400 Steyr.

Status

In production.

Service

Australia, Tunisia.

Licence production

AUSTRALIA
Manufacturer: Australian Defence Industries Limited
Type: 5.56 mm F88
Remarks: See entry under *Australia*.

UPDATED

BELGIUM

FN 0.50 Browning M2 heavy barrel machine gun

Description

The FN 0.50 M2 HB Browning machine gun is similar in design and operation to other types of Browning M2 machine gun, being a recoil-operated, air-cooled, belt-fed weapon. The mechanism operates in exactly the same manner as other Browning M2HBs so reference should be made to the appropriate entry under *United States of America*.

FN also produce various mountings for the M2 HB, including the universal M3 tripod and the anti-aircraft M63 mounting.

Variants

FN 0.50 M2 HB/QCB machine gun
See following entry.

FN M3M, M3P and M3 L/S
These are all aircraft or helicopter versions of the FN 0.50 M2 HB, the M3M for aircraft door mount applications, the M3P for pods (automatic only), and the M3 L/S for pintle mountings, with a light barrel and automatic firing only.

Data

Cartridge: 0.50 Browning (12.7 × 99 mm)
Operation: short recoil, selective fire
Feed: disintegrating link belt, M2 or M9 links
Weight: empty, 38.15 kg
Length: 1.656 m
Barrel: 1.143 m
Rifling: 8 grooves, rh; length 1.064 m
Muzzle velocity: 916 m/s with M33 ball
Rate of fire: cyclic, 450-550 rds/min
Max range: 6,765 m
Effective range: 1,500 m plus

Manufacturer

FN Herstal SA, B-4040 Herstal.

Status

In production.

Service

Belgium and other nations within Europe, plus countries in South America, and the Middle and Far East.

UPDATED

FN 0.50 Browning M2 HB heavy machine gun

FN 0.50 Browning M3M for airborne applications

FN 0.50 M2 HB/QCB machine gun

Description

Since the late 1970s the demand for the 0.50 M2 HB machine gun has been steadily growing, and this led FN to develop a quick-change barrel (QCB) version. This feature eliminates tedious headspace adjustment, ending the chance of dangerous errors, saves time in training and in operations, and reduces the risk taken in action whilst adjusting headspace. To accompany this machine gun FN introduced new APEI ammunition. There is also a 0.50 M2 HB/QCB variant for use as a coaxial machine gun on armoured vehicles, firing automatic only.

All the M2 HB and QCB machine guns made by FN are provided, as standard series production items, with a trigger safety, a height-adjustable frontsight and a timing requiring no adjustment on the user's part.

FN also supplies a kit allowing any M2 HB machine gun to be converted into the QCB version. This kit comprises: a barrel; a barrel extension; a set of breech locks; a set of shims for the barrel support sleeve; and an accelerator. A new accelerator has been developed to improve the performance of this machine gun.

With the exception of the parts specific to the QCB kit, all other parts are interchangeable with the M2 HB model.

A combined recoil booster accessory enables the 0.50 machine gun to be fired in the automatic mode with blank cartridges having brass or plastic cases (star or US crimping) by effectively reproducing live round operating conditions. It also allows the use of plastic bullet (PT) training ammunition by removal of the barrel muzzle plug.

Manufacturer

FN Herstal SA, B-4040 Herstal.

Status

In production.

Service

Europe, North America, Middle East, Far East, Africa and Latin America.

UPDATED

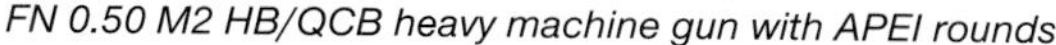

FN 0.50 M2 HB/QCB heavy machine gun with APEI rounds

FN 0.50 M2 HB/QCB heavy machine gun

FN 7.62 mm MAG general purpose machine gun

Development

In production since 1958, the FN MAG was designed by a team led by Ernest Vervier, hence the sometimes-used MAG 58 designation. The MAG (*Mitrailleuse d'Appui Generale*, also thought by some to refer to *Mitrailleuse a Gaz*) has become one of the western world's most successful and widespread of all post Second World War machine gun designs and it is still in production both in Belgium and by licenced manufacturers (see below) - well over 150,000 units have been manufactured. Although the vast majority of MAGs have been manufactured in 7.62 × 51 mm NATO, other calibres have been produced, notably for Sweden in a calibre of 6.5 mm as the Ksp 58 general purpose machine gun. Sweden was the first customer for the MAG; after 1962 these early models were converted to 7.62 × 51 mm NATO and licence production was carried out by FFV (now completed).

In addition to the infantry version of the MAG (FN 60-20) there is also a coaxial version intended for use on armoured vehicles (FN 60-40). The FN 60-30 is an aircraft version.

FN 7.62 mm MAG general purpose machine gun

Description

The FN MAG is gas operated, belt fed and has a quick-change barrel. It is light enough to be carried by infantry and is capable of producing sustained fire over considerable periods when mounted on a tripod. It fires the standard 7.62 × 51 mm NATO cartridge from a disintegrating link belt of the American M13 type; alternatively the 50-round continuous articulated belt can be used, but the two types of belt are not interchangeable.

The receiver is made of steel plates riveted together to make a rectangular section of considerable strength. It is reinforced at the front to take the barrel nut and gas cylinder and at the rear end for the butt and buffer.

Along the inside of the receiver are ribs which support and guide the breech block and piston extension in their reciprocating movement. The breech guides are shaped to force down the locking lever when the breech block is fully forward, and in the floor of the receiver is a substantial locking shoulder against which the locking lever rests when locking is completed.

There is a cutout section on the right of the receiver in which the cocking slide operates and the bottom of the receiver has a slot for ejection of the empty case.

The barrel is threaded externally at the rear end to fit into the barrel locking nut. This has an external interrupted buttress thread to engage into the receiver and is prevented from rotating by the barrel locking catch attached to the left side of the receiver. The carrying handle sleeve is engaged on flanges in the barrel locking nut and rotating the carrying handle locks the nut into the receiver. The handle is then clear of the line of sight.

A gas vent is drilled through the barrel leading down into the regulator. Gas comes into the regulator, which has a surrounding sleeve inside which is a gas plug with three gas escape holes. When the gun is clean and cold most of the gas passes out through these three holes and only the minimum required to operate the gun passes back to the piston head. When the need arises to increase the gas pressure to overcome frictional resistance the gas regulator knob is rotated, the gas regulator sleeve slides along the gas block and the three holes are progressively closed until eventually all the gas is diverted to the piston head. This same arrangement can be used to vary the rate of fire within the limits of 600 to 1,000 rds/min.

At the muzzle is the foresight block and the screw-threaded flash eliminator is of the closed-prong type. The bore is chromium plated and the barrel assembly can be changed without unloading.

The piston is attached to the piston extension which has a cutout section at the front to allow passage of the ejecting case from above. At the rear is a massive piston post, attached to which is the locking lever link, which in turn is connected to the locking lever. Finally the locking lever is pinned half way along the breech block. When the block is travelling forward the locking lever lies in recesses in both sides of the block. The block is hollowed to take the firing pin which is mounted in the piston post. On the front face is the extractor. The return spring fits into the hollow interior of the piston extension.

The trigger is centrally pivoted at its top edge so when it is pulled the rear end rises. Resting on the rear of the trigger is a long, centrally pivoted sear. When the trigger is pulled the front end of the sear (the tail) is forced up and the rear end (the nose) drops down out of engagement with a bent on the bottom of the piston extension. Also on the front top edge of the trigger is a pivoting tripping lever which projects up into the piston way. This tripping lever has a spring which pivots it forward. On its front face is a step.

When the trigger is pulled the central pivot causes the front, carrying the tripper, to go down and the back to rise, lifting the tail of the sear. As the tail of the sear rises, the spring-loaded tripping lever moves forward underneath it and holds it up.

When the trigger is released the tripping lever rises with it and holds up the tail of the sear so the nose cannot rise to engage the piston and hold it back. The tripping lever, as it rises, re-enters the bolt way. As the piston comes back it hits the top of the tripping lever and rotates it back against the spring. The tail of the sear flies down and its nose rises. The piston going back passes over the sear but when it comes forward again it is caught by the full width of the sear. The area of contact is large enough to prevent any chipping or bending.

The safety catch is a push-through plunger. When at safe it rests under the sear nose and prevents it falling. At fire a cutout section allows full sear movement.

The feed mechanism is a two-stage system with the belt moving half way across on the forward motion of the bolt and moving the other half pitch during bolt recoil. The top of the breech block carries a spring-loaded roller which engages in a curved feed channel in the feed cover over the receiver. This channel is pivoted near its rear end and its front end is attached to the end of the short feed link. This link swings about its centre so that as one end goes in towards the centre line of the gun the other end moves out. It carries an inner feed pawl at one end and two outer feed pawls at the other. Thus, as the breech block travels forward the roller will first travel down a straight section of the feed channel, while the cartridge is forced out of the stationary belt, and then enter the curved portion. This forces the feed channel to the right. This swings the feed link to the left and rotates it about its centre so that the inner feed pawl moves out over the waiting cartridge and the outer pawls force it in half the distance to the gun centre line.

As the breech block moves back, the roller swings the front of the channel to the left. The feed link swings to the right so that the inner pawl comes in, bringing the belt across half a pitch, and the first round comes up against the cartridge stop and is positioned for chambering.

The outer pawls move out over the next cartridge. Thus each set of pawls acts alternately as feed and stop pawls and the cartridge moves half way across for each forward and backward movement of the breech block.

The buffer assembly consists of a bush which receives the impact of the piston extension and moves back into a cone which it expands outwards. This grips the walls of the buffer cylinder and also moves back slightly. In moving back it flattens a series of 11 Belleville washers. These store the kinetic energy of the piston and in returning to shape they drive the cone and bush forward and force the piston forward again.

The foresight is a blade mounted on a screw-threaded base which fits into a block mounted on a transverse dovetail. The rearsight is a leaf which can be used folded down for ranges marked from 200 to 800 m by 100 m increments. A slide with two spring catches allows the aperture rearsight to be set to the correct range. When the leaf is raised ranges are marked from 800 to 1,800 m at intervals of 100 m. There is a V rearsight on the slider.

The gun is cocked by pulling the cocking handle fully to the rear and then returning it to the forward position. When the gun is cocked, and only then, the safety catch can be set to safe by pushing it from left to right through the gun body. The letter 'S' can then be read on the right side of the plunger facing the firer.

To load, the top cover is opened and lifted to the vertical position. The loaded belt is then inserted, open side down, across the feed tray so that the leading cartridge rests against the cartridge stop on the right. The top cover is lowered. The safety is pushed through from right to left. When the trigger is pressed the nose of the sear drops and the piston extension is forced forwards by the compressed return spring, carrying the bolt with it. The feed horn on the top edge of the breech block pushes the first round in the belt straight through the belt link and the bullet is directed down into the chamber by the bullet guide. The top surfaces of the locking lever are forced down when they hit the underside of the top breech block guide. The cartridge is fully chambered, the extractor slips into the groove and the base of the cartridge seats into the recessed face of the breech block. The ejector is forced back and the ejection spring is compressed. The breech block stops.

The piston extension continues forward. The locking lever continues to move down and drops in front of the locking shoulder in the bottom of the receiver. Further forward movement of the piston extension causes the locking lever link to rotate forward past the vertical position. The final forward movement of the piston post pushes the long firing pin through the front face of the breech block and the cartridge is fired. The forward motion of the piston extension ends when its shoulder hits the stop face of the gas cylinder. The gas pressure forces the bullet up the bore and some of the gas flows through the vent into the gas regulator and then to the piston head which is driven back.

The piston post withdraws the firing pin into the breech block. The locking lever link is rotated back to the vertical position. The continued movement of the piston post rotates the link further and this lifts the locking lever out of contact with the locking shoulder. The locking lever then pulls the breech block rearward, the empty case is withdrawn and as soon as it is clear of the chamber the compressed ejection spring pushes forward the ejector which rotates the case down, around the extractor, through the slot in the piston extension and out through the bottom of the gun. The rearward motion of the piston extension compresses the return spring. The piston extension hits the buffer and is thrown forward. If the trigger is still pressed the cycle starts again.

A bipod is mounted on the forward end of the gas cylinder and can rotate from side to side to allow firing across a slope, one leg higher than the other with the sights vertical. The legs are not adjustable for height.

The bipod can be folded back for carrying. A hook on each leg engages in a slot in the side of the receiver and is held securely in place by a sliding retainer catch.

The tripod mount is a spring-buffered assembly to which the gun is attached by a push-through pin which enters a hole above the trigger guard.

Variants

M240 series

The M240 is the US Military armed forces designation for the MAG in both coaxial and bipod-mounted forms. See separate entry under *United States of America* for details.

Data

Cartridge: 7.62 × 51 mm NATO
Operation: gas, automatic
Locking: dropping locking lever
Feed: belt
Weight: with butt and bipod, 11.65 kg
Length: 1.26 mm
Barrel: 548 mm; with flash hider, 630 mm
Rifling: 4 grooves, rh, 1 turn in 305 mm
Sights: fore, blade; rear, aperture when leaf is lowered, U-notch when leaf is raised
Sight radius: sight folded down, 848 mm; sight raised, 785 mm
Muzzle velocity: 840 m/s
Muzzle energy: 335 kg/m
Rate of fire: cyclic, 650-1,000 rds/min
Effective range: 1,500 m

Manufacturer

FN Herstal SA, B-4040 Herstal.

Status

In production in Argentina, Belgium, Egypt, India, UK and USA.

Service

Known to be in service with the armed forces of Argentina, Australia, Bahrain, Barbados, Belgium, Belize, Bolivia, Botswana, Brazil, Brunei, Burma (Myanmar), Burundi, Cameroons, Canada, Chad, Colombia, Cuba, Cyprus, Djibouti, Dominican Republic, Ecuador, Egypt, Gabon, Gambia, Ghana, Greece, Guatemala, Haiti, Honduras, India, Indonesia, Iraq, Ireland, Israel, Jamaica, Jordan, Kenya, Korea (South), Kuwait, Lebanon, Lesotho, Libya, Luxembourg, Malaysia, Mexico, Morocco, Netherlands, New Zealand, Nicaragua, Nigeria, Oman, Pakistan, Panama, Peru, Philippines, Portugal, Qatar, Rwanda, Saudi Arabia, Seychelles, Sierra Leone, Singapore, South Africa, Sri Lanka, Sudan, Surinam, Swaziland, Sweden, Switzerland, Tanzania, Thailand, Tunisia, Uganda, United Arab Emirates, United Kingdom, United States of America, Upper Volta, Uruguay, Venezuela, Zaire, and Zimbabwe. This list is probably not complete.

Licence production

ARGENTINA
Manufacturer: Direccion General de Fabricaciones Militares
Type: 7.62 mm Ametralladora Tipo 60-20
Remarks: In production. In service with Argentinian armed forces. Also supplied to Bolivia.

FN 7.62 mm MAG bipod-mounted as light machine gun

FN 7.62 mm MAG bipod-mounted as light machine gun

EGYPT
Manufacturer: Maadi Company for Engineering Industries
Type: 7.62 mm MAG GPMG
Remarks: Produced in bipod- and tripod-mounted forms. One bipod-mounted variant has no butt and is fitted with spade grips as well as the usual pistol grip.
INDIA
Manufacturer: Small Arms Factory, Kanpur
Type: 7.62 mm Gun Machine (MAG)
Remarks: In production and offered for export sales. Produced in infantry, coaxial, turret-mounted and aircraft versions.
SINGAPORE
Manufacturer: Ordnance Development and Engineering Company of Singapore (Pte) Limited
Type: 7.62 mm machine gun
Remarks: Apparently not a licenced producer and it appears that manufacture has ceased. Was available in infantry and coaxial forms.
UNITED KINGDOM
Manufacturer: Originally Royal Small Arms Factory, Enfield Lock, now Manroy Engineering
Type: 7.62 mm L7A2 GPMG
Remarks: See entry under *United Kingdom* for details.
UNITED STATES OF AMERICA
Manufacturer: FN Manufacturing Inc
Type: 7.62 mm M240 series
Remarks: See entry under *United States of America* for details.

UPDATED

FN 5.56 mm Minimi light machine gun

Development

Development of the 5.56 mm light machine gun which was to become the Minimi began in the early 1960s. Although some of the early development work centred around the 5.56 × 45 mm cartridge which was to become the M193, Ernest Vervier and his successor Maurice Bourlet carried out work on a different 5.56 mm cartridge which became the SS109. In the event the Minimi was developed to fire both cartridges. The first Minimi prototypes appeared during 1974 although production did not commence until 1982. Since then the Minimi has been adopted by many armed forces, including the US Army and Marine Corps (as the M249 Squad Automatic Weapon - SAW).

5.56 mm Minimi machine gun mounted on tripod

Description

The Minimi is gas operated, using gas tapped from the forward part of the barrel in conventional fashion. The rotary gas regulator is of a simple design, based on the earlier MAG machine gun type, and has two basic settings (normal and adverse conditions). Adjustment is by hand, even with a hot barrel.

The breech locking mechanism is an FN design, where the bolt is locked into the barrel extension by a rotational action. This action is initiated by a cam in the bolt carrier.

Normal gas operation, in which the gas piston is forced to the rear, moves the bolt carrier back, leaving the bolt still locked to the barrel extension. The residual chamber pressure is virtually zero by the time the cam action unlocks the bolt. Primary extraction of the spent case begins with the rotation of the bolt before it unlocks.

In the ammunition feed system on the Minimi the disintegrating link belt is held in a 200-round plastic box, which, apart from acting as an ammunition carrier when not on the gun, locks firmly to the gun and becomes virtually integral with it when in action.

The Minimi can accept either a magazine or a belt feed without any modification. The gun is normally regulated to fire using the belt and in this mode the mechanism has to lift the weight of the belt. When firing from a magazine the load is absent and thus the gun tends to have a higher rate of fire.

The gun is normally bipod mounted but can, if required, be mounted on a tripod. It can also be used with either a fixed or a sliding butt-stock.

Variants

Minimi Para

For those users who need a light machine gun shorter than the standard version there is a Para model with a sliding butt-stock and a shorter barrel. The chief advantage of this version is that it is much easier to take in and out of vehicles, helicopters and similar confined spaces. The latest Para models have a pepper-pot muzzle assembly in place of the usual slotted cone. Details of this version can be found in the Data table.

The Para butt-stock has been type classified by the US Army as the M5 collapsible butt-stock for their M249 Squad Automatic Weapon (SAW - see entry under *United States of America* for details).

Coaxial Minimi

Intended for use on armoured vehicles. Weight is 5.3 kg and length overall 793 mm.

M249 SAW

See entry under *United States of America* for details.

Data

Cartridge: 5.56 × 45 mm (FN SS109 NATO or M193)
Operation: gas, firing fully automatic
Locking: rotating bolt head
Feed: 200-round belts or 30-round M16A1 magazine
Weight: standard, 6.85 kg; Para, 7.1 kg
Length: standard, 1.04 m; Para, stock folded, 736 mm; Para, stock extended, 893 mm
Barrel: standard, 466 mm; Para, 347 mm
Rifling: 6 grooves, rh, 1 turn in 304 mm (M193) or 1 turn in 178 mm (SS109)
Sights: fore, semi-fixed hooded post, adjustable for windage and elevation; rear, aperture, adjustable for windage and elevation
Sight radius: 495 mm

FN 5.56 mm Minimi light machine gun

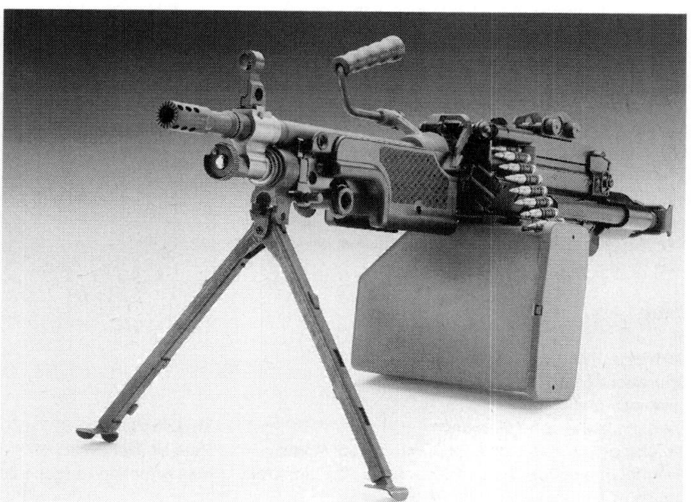

5.56 mm Minimi Para model

Muzzle velocity: M193, 965 m/s; SS109, 915 m/s
Rate of fire: cyclic, 700-1,000 rds/min
Effective range: up to 1,000 m

Manufacturer
Fabrique Nationale Nouvelle Herstal, B-4040 Herstal.

Status
In production.

Service
Adopted by Australia, Belgium, Canada (C9), Indonesia, Italy, New Zealand, Sri Lanka, Sweden, United Arab Emirates, United States of America, Zaire and other countries.

UPDATED

5.56 mm Minimi Para model with butt collapsed

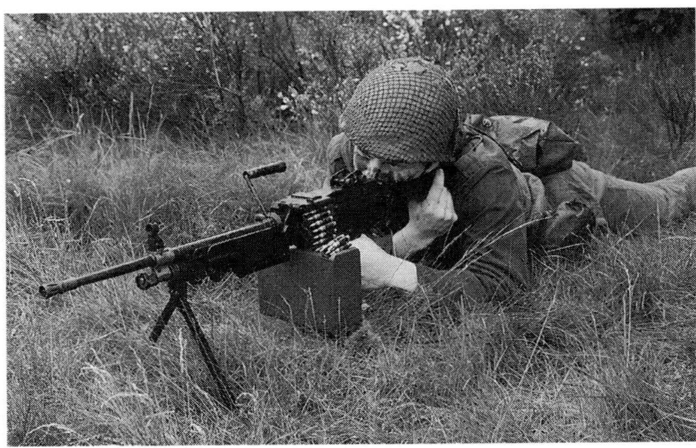

FN 5.56 mm Minimi light machine gun showing belt feed

BULGARIA

5.45 mm RPK-74 light machine gun

Description
The Bulgarian 5.45 mm RPK-74 light machine gun is produced under licence in Bulgaria and differs from the CIS original in few, if any, details. Production is for local armed forces requirements and also for potential export sales. A 45-round plastic box magazine is utilised as standard.

Manufacturer
Arsenal, 6100 Kazanluk, Rozova dolina str 100.
Marketed by: Kintex, 66 James Boucher Street, PO Box 209, 1407 Sofia.

Status
In production.

Service
Bulgarian armed forces.

UPDATED

Bulgarian 5.45 mm RPK-74 light machine gun

7.62 mm PK Kalashnikov machine gun

Description
The Bulgarian 7.62 mm PK Kalashnikov machine gun is the original PK model with a fluted barrel, produced under licence in Bulgaria. It differs from the CIS original in few, if any, respects but the details of this model provided below are from Bulgarian literature. Production is for local armed forces requirements and also for potential export sales.

A complete outfit for a single PK machine gun supplied by Kintex includes four magazine boxes, a spares and accessories kit, a cleaning rod, an oiler, a cover for the complete weapon, and a spare barrel. A tripod is available.

Also available is a 7.62 mm PKT machine gun suitable for mounting on armoured vehicles. Weight is 10.5 kg and length overall 1.098 m.

Bulgarian 7.62 mm PK Kalashnikov machine gun

Data
PK
Cartridge: 7.62 × 54R
Operation: gas, automatic
Locking: rotating bolt
Feed: belt; 100 and 250 rounds
Weight: empty less bipod, 9 kg; with bipod, 9.4 kg
Weight, magazine box: 100 rounds, 3.9 kg; 250 rounds, 8 kg
Length: 1.192 m
Barrel: 658 mm
Sights: fore, cylindrical post; rear, vertical leaf and windage scale adjustable to 1,500 m
Muzzle velocity: 825 m/s
Rate of fire: cyclic, 650 rds/min
Max effective range: 1,000 m

Manufacturer
Arsenal, 6100 Kazanluk, Rozova dolina str 100.
Marketed by: Kintex, 66 James Boucher Street, PO Box 209, 1407 Sofia.

Status
In production.

Service
Bulgarian armed forces.

UPDATED

CANADA

5.56 mm C7 light support weapon

Description

The C7 light support weapon (LSW) is a co-development between Diemaco and Colt Firearms (where the weapon is known as the Colt Model 715 or the M16 light machine gun). The C7 LSW features a select fire mechanism, an hydraulic recoil buffer and an adjustable bipod. The weapon also features a heavy contour Diemaco hammer-forged barrel. The improved life and durability of the barrel during testing has eliminated the need for a quick-change barrel system, greatly simplifying the design and making the LSW accurate to the maximum effective range of the ammunition.

The C7 LSW is fed from any M16 compatible box magazine, including high capacity types, and it uses the same family of accessories as the C7 rifle.

The C7 LSW uses the M16A2 type rearsight as standard but is also available mounting an optical sight as the C7A1 light support weapon - this version has been procured by the Netherlands Marine Corps.

5.56 mm C7 light support weapon and accessories

Data

AS FOR C7 RIFLE EXCEPT:
Weight: empty, 5.8 kg; loaded, 6.2 kg
Sights: fore, post; rear, adjustable aperture, 300 to 800 m; optical sight as C7A1
Rate of fire: cyclic, 625 rds/min

Manufacturer

Diemaco, 1036 Wilson Avenue, Kitchener, Ontario N2C 1J3.

Status

In production.

Service

Netherlands Marine Corps (C7A1). Under evaluation by several countries.

VERIFIED

CHINA, PEOPLE'S REPUBLIC

Chinese machine guns

Description

The armed forces of China have been equipped mainly with copies of weapons originally produced elsewhere. The weapons held are described individually in the sections devoted to the country of origin. They are:

Chinese Name	Origin
Type 53	Soviet 7.62 mm DPM light machine gun
Type 54	Soviet 12.7 mm Model 38/46 heavy machine gun
Type 54	Soviet 7.62 mm RPD light machine gun
Type 57	Soviet 7.62 mm SG43 medium machine gun
Type 58	Soviet 7.62 mm RP-46 Company machine gun
Type 63	Soviet 7.62 mm SGM medium machine gun

Status

Types 53, 54, 56 and 58 manufactured in China. All appear to remain in service although some are probably used only by reserve units or the People's Militia.

VERIFIED

12.7 mm Type 54 heavy machine gun mounted as air defence weapon on Type 59 MBT

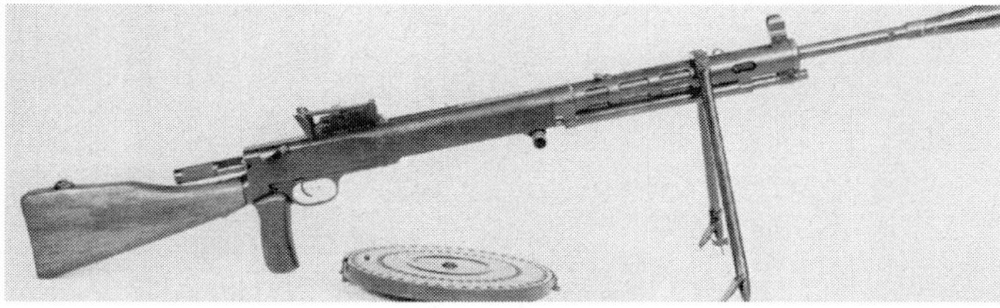

Chinese 7.62 mm Type 53 light machine gun corresponding to CIS DPM

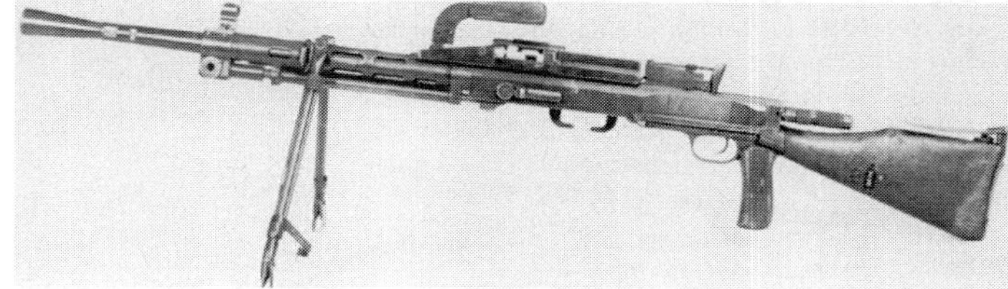

Chinese 7.62 mm Type 58 machine gun

7.62 mm Type 67 light machine gun

Description

The Type 67 is an indigenous Chinese design and has replaced the Types 53 and 58 in frontline units. It has been in production since the early 1970s and some of the early issues were given to North Vietnam. The gun is a sound design and bears all the hallmarks of being strong and reliable.

The gun is gas operated and belt fed and may be used with either a bipod or a tripod. Its design is a mixture of features from other guns, as follows:

Type 24 (Maxim)	Feed mechanism
Type 26 (ZB26)	Bolt and piston
Type 53 (DPM)	Trigger mechanism
Type 56 (RPD)	Gas regulator
Type 57 (SG43)	Barrel-change system

The Type 67 was originally issued as a light squad automatic, but in recent years the Type 67-2C appears to have largely replaced the original version. The 2C model fills the general purpose machine gun role, and can be used on its bipod or on a tripod in the support role. It can be quickly converted into an anti-aircraft mount. There do not appear to be any significant changes in the 2C model other than a slight reduction in weight by using an alloy steel barrel and the necessary attachment points for the tripod.

Description

The Type 67 uses the 7.62 × 54R cartridge, and has an open pocket metal belt with the pockets joined by spring metal coils. Each pocket carries a nib, extending to the rear, with a bent-over tab. The round is pressed into the link with the tab against the rear face of the cartridge. This type of belt allows the bolt to push the round out of the link and straight into the chamber.

To load the belt the feed cover catch at the rear of the receiver is pressed and the feed cover lifted about its front hinge. The cocking handle on the right of the receiver is normally folded vertically downwards and must be lifted through 45° before it can be pulled back to cock the action. Before cocking it is essential to check that the safety catch is back to fire. If forward to safe the sear will jam on the underside of the piston. The belt is placed in the feedway with the open side of the link facing down and the first round in the feedway slot. The cover is then closed. The gun fires from the open bolt position and when the ammunition is expended the bolt closes on an empty chamber.

When the bolt goes forward a cartridge is pushed into the chamber. Bolt movement ceases but the piston continues forward and the rear of the bolt is lifted up to lock into the ceiling of the receiver. The piston continues forward after locking is completed and the flat face of the piston post drives the firing pin into the cartridge cap.

After firing, the bullet travels up the bore and some of the gas following is diverted into the gas cylinder below the barrel, where it drives the piston rearward. There is a period of free travel allowing the gas pressure to drop, before the ramp on the rear face of the piston post pulls the rear end of the bolt down out of its recess in the top of the receiver. The piston then carries the bolt to the rear; the empty case is extracted and then ejected downwards out of the gun. The return spring is compressed along its guide rod. When the piston comes to rest the return spring drives it forward and the cycle is repeated.

Belts are fed from the right and the feed mechanism is operated by having a cam track on the top side of the piston extension at the front. This forces a roller on the bottom of the lower feed arm to move out and the rotation of the lower feed arm is transmitted by a vertical shaft to the upper feed arm. As the upper feed arm moves, a slot in it engages a roller on the feed slide and the feed slide moves outward to engage a round in the belt.

As the piston extension moves forward the bolt forces a cartridge through the belt and into the chamber. As the piston extension continues forward the lower feed arm is moved back to its original position and this in turn pulls in the upper feed arm and the next cartridge is pulled across from right to left by the feed slide and moved into the slot in the feed tray where it is pressed down by a pair of cartridge guides in the cover. A spring-loaded stop pawl on the feed tray prevents the belt from slipping out as the feed slide oscillates.

Chinese 7.62 mm Type 67-2C machine gun

The barrel will normally be changed after two minutes of firing at the rapid rate. The procedure is to lift the top cover plate, removing the belt if any cartridges remain in it, and then pressing the barrel retaining catch to the left. The carrying handle is then grasped and the barrel pushed forward off the gun. The new barrel is placed in position, with the gas cylinder in the gas cylinder tube, and the barrel pulled back. The retaining catch is pushed across to the right, a new belt put in and the top cover plate closed to resume firing.

The gas regulator works in the same way as that on the RPD. The nut on the left side of the regulator is loosened and the regulator is pushed through to the right. This disengages it from the index pin and the regulator can be rotated to the selected position. There are three settings, 1, 2 and 3, of which 1 is usually used. The regulator is then pushed back and the nut tightened.

The air defence sights consist of a pillar permanently attached to the top of the stirrup-like rearsight and a speed ring foresight, which fits into dovetails in the top of the receiver and is held by a spring catch. The tripod is up-ended and the gun attached by fitting the notches on the underside of the front of the receiver on to the pins in the mount and rotating the gun down until the catch locks on the front of the trigger guard.

Data

TYPE 67-2C

Cartridge: 7.62 × 54R
Operation: gas, automatic
Locking: tilting block
Feed: 250-round metal belt in magazine box or 50-round drum
Weight: total, 15.6 kg; gun only, 10 kg
Length: 1.25 m
Barrel: 606 mm
Rifling: 4 grooves, rh, 1 turn in 240 mm
Sights: fore, pillar; rear, leaf notch, adjustable for windage
Muzzle velocity: 840 m/s
Rate of fire: cyclic, 650 rds/min; practical, 300 rds/min
Effective range: light machine gun, 800 m; GPMG, 1,000 m

Manufacturer

China North Industries Corporation (NORINCO), 7A Yue Tan Nan Jie, PO Box 2137, Beijing.

Status

In production.

Service

China and Vietnam.

VERIFIED

7.62 mm Type 74 light machine gun

Description

The 7.62 mm Type 74 is a squad light machine gun, gas operated and fed from a drum magazine, although it is also possible to use the Type 56 rifle magazine in place of the drum. The mechanism is gas operated, using a piston to actuate a laterally locking bolt in a similar manner to the SG43 machine gun. The gas system has a four-position regulator mounted at the end of the gas cylinder, above the barrel. Barrel and chamber are chromium-plated and the muzzle carries a conical flash hider.

Data

Cartridge: 7.62 × 39 mm
Operation: gas
Locking: laterally moving bolt
Feed: 101-round drum magazine
Weight: empty, 6.4 kg
Length: 1.108 m
Sights: fore, hooded post, adjustable; rear, tangent leaf with square notch, adjustable to 800 m
Muzzle velocity: 735 m/s
Rate of fire: cyclic, 750 rds/min
Effective range: 600 m

Manufacturer

China North Industries Corporation (NORINCO) 7A Yue Tan Nan Jie, PO Box 2137, Beijing.

7.62 mm Type 74 light machine gun

Status

In production.

Service

Chinese armed forces.

VERIFIED

7.62 mm Type 80 machine gun

Description

The Type 80 is described by the Chinese as a multipurpose machine gun, their equivalent of the Western general purpose weapon. It is a copy of the CIS PK, a gas operated, rotating-bolt, belt-fed weapon, fully described in the *CIS* section. As is usual with Chinese designs, the tripod mount can be quickly converted to the air defence role, in which it is claimed to have an effective altitude of 500 m.

Also produced is a coaxial version of this machine gun for armoured vehicles.

Data

Cartridge: 7.62 × 54 mm
Operation: gas, automatic
Locking: rotating bolt
Feed: 100- or 200-round belt from magazine box or 50-round drum
Weight: total, 12.6 kg; gun, 7.9 kg; tripod, 4.7 kg
Length: 1.192 m
Barrel: 675 mm
Muzzle velocity: 825 m/s
Rate of fire: cyclic, 650 rds/min; practical, 350 rds/min
Effective range: 1,000 m

7.62 mm Type 80 machine gun

Manufacturer

China North Industries Corporation (NORINCO), 7A Yue Tan Nan Jie, PO Box 2137, Beijing.

Status

In production.

Service

Chinese armed forces.

VERIFIED

7.62 mm Type 81 light machine gun

Description

The 7.62 mm Type 81 light machine gun is intended as a squad automatic, though it seems likely that it has been produced principally for export as the squad support weapon counterpart to the 7.62 mm Type 81 assault rifle (qv). It is gas operated and uses a mechanism similar to that of the Type 68 rifle, a rotating bolt moving in a bolt carrier. Feed is normally from a drum magazine, though not the same pattern as that used with the Type 74 machine gun; the magazine of the Type 81 automatic rifle can be used. In addition, most of the working parts of the Type 81 light machine gun and rifle are interchangeable.

Data

Cartridge: 7.62 × 39 mm
Operation: gas, automatic
Locking: rotating bolt
Feed: 75-round drum or 30-round box magazines
Weight: empty, 5.15 kg
Length: 1.024 m
Rifling: 4 grooves, rh, 1 turn in 240 mm
Sights: fore, adjustable post; rear, tangent leaf with square notch, adjustable to 600 m
Muzzle velocity: 735 m/s
Rate of fire: cyclic, practical, 120 rds/min
Effective range: 600 m

7.62 mm Type 81 light machine gun

Manufacturer

China North Industries Corporation (NORINCO), 7A Yue Tan Nan Jie, PO Box 2137, Beijing.

Status

In production. Offered for export sales.

VERIFIED

12.7 mm Type 77 heavy machine gun

Description

The 12.7 mm Type 77 is an automatic weapon primarily intended for use for air defence, though it can also be used against ground targets. The operating mechanism is unusual for a heavy machine gun, using a direct-delivery gas tube similar to that used in the M16 rifle. The tube runs from a three-position regulator, beneath the barrel to the front of the receiver, where the gas is directed against the lower edge of the bolt carrier. Breech locking is performed by two hinged flaps on the bolt which, impelled by the bolt carrier, are moved outward to lock into recesses in the receiver.

Ammunition feed is from a belt, carried in a belt box on the left-hand side of the gun, and a special optical sight is provided for air defence firing. The tripod can be set at various heights and is provided with geared traverse and elevation controls which can presumably be unlocked to permit free movement against aerial targets.

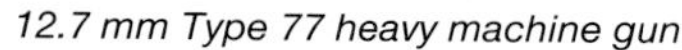

12.7 mm Type 77 heavy machine gun

Data

Cartridge: 12.7 × 107 mm Type 54
Feed: 60-round belt
Weight: with tripod, 56.1 kg
Length: 2.184 m
Barrel: 1.016 m
Rifling: 8 grooves, rh, 1 turn in 390 mm
Sights: fore, hooded adjustable post; rear, tangent U-notch adjustable to 2,400 m; optical AA sight
Muzzle velocity: 800 m/s
Rate of fire: cyclic, 750-800 rds/min
Range: (effective) air, 1,600 m; ground, 1,500 m

Manufacturer

China North Industries Corporation (NORINCO), 7A Yue Tan Nan Jie, PO Box 2137, Beijing.

Status

In production.

Service

Chinese Army.

VERIFIED

12.7 mm Type W-85 anti-aircraft machine gun

Description

The 12.7 mm Type W-85 anti-aircraft machine gun is a battalion automatic gun primarily intended for use against aerial targets but which can also be used as a heavy support machine gun against ground targets. The Type W-85 gun is gas operated and belt fed, and is exceptionally light, some 58 per cent lighter than the obsolescent Type 54, the Chinese version of the CIS DShK. The Type W-85 is provided with an adjustable tripod, for ground or aerial firing, and a telescope sight. Ammunition provided for this weapon covers AP, AP-T and AP-I with the addition of a bullet described as tungsten alloy cored which may be an APDS pattern.

12.7 mm Type W-85 machine gun on low tripod

Data

Cartridge: 12.7 × 107 mm Type 54
Feed: 60-round link belt
Length: 1.995 m
Weight: total, combat ready, 39 kg; gun, 18.5 kg; tripod, 15.5 kg
Muzzle velocity: AP, AP-T, AP-I, 800 m/s; tungsten alloy cored, 1,150 m/s
Rate of fire: practical, 80-100 rds/min
Range: (effective) air, 1,600 m; ground, 1,500 m

Manufacturer

China North Industries Corporation (NORINCO), 7A Yue Tan Nan Jie, PO Box 2137, Beijing.

Status

In production.

VERIFIED

14.5 mm Type 75-1 anti-aircraft machine gun

Description

The 14.5 mm Type 75-1 anti-aircraft machine gun is a Chinese version of the Soviet KPV machine gun mounting involving the 14.5 mm Type 56 heavy machine gun. It differs from the original in certain details such as the arrangement of the belt feed and the provision of cooling fins on the barrel. The Type 75-1 involves a tripod with two small wheels, the tripod folding to become a lightweight trailer. Mechanical elevation and traversing gears are fitted, there is a layer's seat and an optical course and speed sight mounted on a parallelogram arm so as to place the sightline in a convenient position. It will be noted that the picture shows an image intensifying sight fitted.

The Chinese 14.5 mm Type 56 heavy machine gun is also utilised on the Type 56 quadruple machine gun air defence mounting, the Chinese equivalent of the Soviet ZPU-4.

Data

Cartridge: 14.5 × 114 mm Type 56
Feed: 80-round link belt
Length: 2.39 m
Weight: with mount, 165 kg
Elevation: −10 to +85°
Traverse: 360°
Muzzle velocity: 995 m/s
Rate of fire: cyclic, 550 rds/min
Range: (effective) air, 2,000 m; ground, 1,000 m

Manufacturer

China North Industries Corporation (NORINCO), 7A Yue Tan Nan Jie, PO Box 2137, Beijing.

Status

In production.

Service

Chinese Army.

VERIFIED

14.5 mm Type 75-1 machine gun

COMMONWEALTH OF INDEPENDENT STATES

7.62 mm RP-46 company machine gun

Development

During the Great Patriotic War the USSR felt the need for a machine gun held at company level to produce a greater volume of fire than that supplied by the existing DPM, yet light enough to be carried during the assault. During 1944 designers developed the DPM and eventually produced a version which could be belt fed for sustained fire and by changing the top cover could make use of the flat drum of the DPM during the assault. Although the design was finalised during the war it did not come into service until 1946 and the gun was known as the 7.62 mm RP-46 company machine gun.

The RP-46 was officially adopted by the Soviet Army but saw relatively limited use. The Chinese manufactured the gun and called it the Type 58. The North Korean version is known as the Type 64. Apart from the markings the guns are the same.

7.62 mm RP-46 company machine gun

Description

On the 7.62 mm RP-46 company machine gun the belt feeder is operated by a somewhat crude use of the cocking handle which is fixed to the right-hand side of the slide and reciprocates with it. The handle fits into the claw-like feed operating slide and causes it to move backwards and forwards. The internal diameter of the claw is three times that of the cocking handle so that the bolt is able to accelerate to its full velocity in either direction before the feed slide is operated. This enables the full power of the gun to be used for unlocking the breech and accelerating the bolt back in one direction and similarly the full spring power is available to accelerate the bolt and push a round into the chamber.

As the feed slide moves back, a diagonal feed groove cut in the top surface moves the belt feed pawl inward, and the first round in the belt is moved up to the cartridge stop. The round is held by a retaining pawl in the top feed cover when the feed slide moving forward forces the feed pawl out to the right to slip over the next round to be fed.

The feed slide carries a pair of jaws which grip the rim of the case and extract it from the belt as the feed slide travels back. The cartridge is pulled back under a spring-loaded depressor arm mounted under the top feed cover plate and this arm pushes the cartridge down into the feedway where the bolt feed piece can drive it forward into the chamber.

The RP-46 takes a 250-round continuous closed pocket metallic belt. The cartridge is simply placed in the conical pocket as far as it will go. The belt is usually carried in a metal box.

An ammunition belt is loaded into the RP-46 by inserting the feed tab into the feedway on the right of the gun and pulling the tab through until the first round reaches the stops. To load the gun the cocking handle must be pulled twice. The first time the bolt is retracted the jaws cannot grip the cartridge rim. The second time the leading round is pulled back and forced down into the feedway. The gun is then cocked and ready to fire.

If for any reason the belt feed device is not required it can readily be removed by pressing the magazine catch behind the rearsight block. The cocking handle must be pulled back until it lies in the centre of the claw. Tilting the feeder forward frees it from the cocking handle. The front end is held by a T lug and can be moved back to free the feeder completely from the gun. If required a DP or DPM 47-round drum magazine can then be used on the gun.

The RP-46 differs from the DPM in some other minor details. There is a barrel-release lever on the left of the receiver to depress the barrel lock. The gas regulator has a catch that engages one of the three grooves in the bottom of the gas block and can be moved from position 1 to 2 and then 3 as required.

The Chinese Type 58 originally had this type of regulator but it was later changed to a rotary type similar to that used in the RPD light machine gun.

Data

Cartridge: 7.62 × 54R
Operation: gas, automatic
Locking: projecting lugs
Feed: 250-round belt
Weight: empty, 13 kg
Length: 1.283 m
Barrel: 607 mm
Sights: fore, cylindrical post; rear, leaf tangent with a V backsight
Sight radius: 635 mm
Muzzle velocity: 840 m/s
Rate of fire: cyclic, 600 rds/min
Effective range: 800 m

Manufacturer

State arsenals.

Status

No longer produced.

Service

Still likely to be found in parts of the world where there is (or was) a Soviet/CIS influence.

VERIFIED

7.62 mm RPD light machine gun

Development

Design work was started on the RPD by Degtyarev in 1943 to make use of the 7.62 × 39 mm cartridge then under development. The priority requirements of the armed services, all of whom used Degtyarev machine guns firing the 7.62 × 54R cartridge, caused slow progress to be made but, shortly after the war ended, production of the new short cartridge was accelerated and the RPD was the first machine gun to use this round.

The RPD was manufactured in large numbers and it formed the standard squad automatic for the Soviet Army and its allies. It was manufactured in China as the Type 56 and Type 56-1 and in North Korea as the Type 62 light machine gun.

The RPD is rated as obsolescent in the former Warsaw Pact countries but is still used in South-east Asia and by irregular forces in Africa.

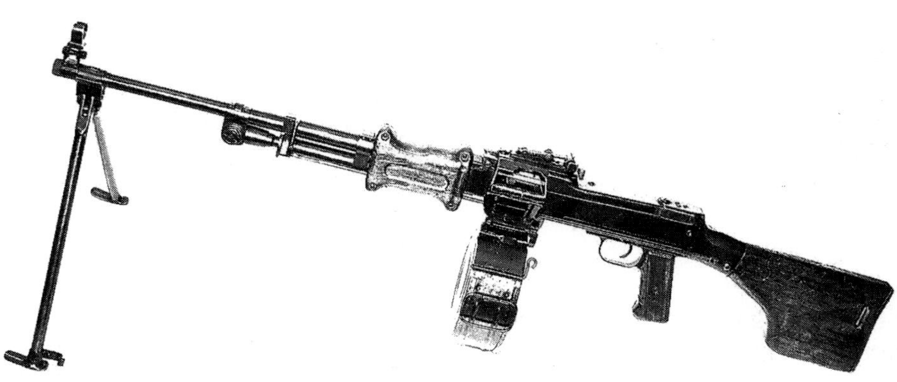

7.62 mm RPD light machine gun (T J Gander)

Description

The RPD is a belt-fed automatic weapon. It is gas operated and has a locking system which uses locking lugs pushed out to lock, not by the firing pin but by a wedge on top of the slide. The belt is a continuous, metal, open-pocket design. The links are connected by a short metal spiral spring link. The cartridge goes into the pocket and the nib at the rear of the belt fits into the extraction groove at the rear of the case. Some belts have no nib but a right-angled projection fits against the base of the cartridge. The 50-round belt can be joined to another by putting the end link of one belt against the joining link of the other. A cartridge is used to keep the two belts together. The cover on the belt carrier is opened and the belt inserted from the left, with the loading tab outside. The belt carrier is slid onto the mounting bracket under the gun from the rear and the mounting bracket lock is swung down to secure the belt carrier in place. On the Chinese Type 56-1 the dust cover over the feedway must be swung down and locked into place to act as a belt carrier bracket.

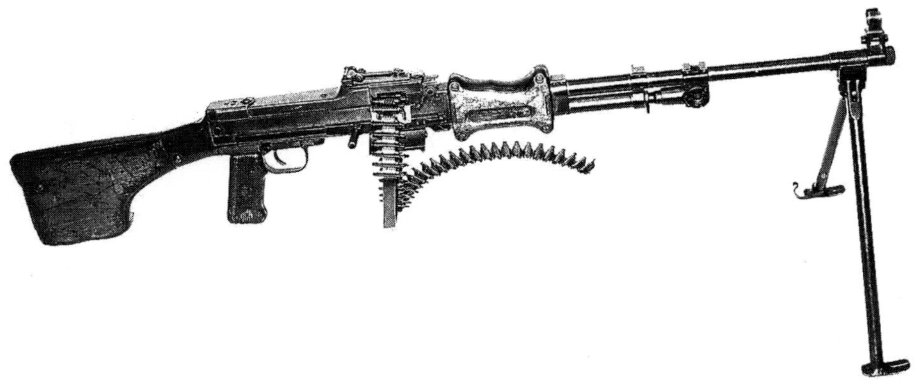

7.62 mm RPD light machine gun without belt drum (T J Gander)

The gun is fed by a feed arm oscillated across the receiver by a roller carried on the slide.

There is a regulator mounted in a block under the gas vent. It is adjusted by loosening the nut on the left side of the regulator and pressing it to the right. The regulator has three settings marked 1, 2 and 3 and is rotated to align the required number with the index pin. The regulator is then pushed back to the left and the nut tightened.

The barrel is fixed and cannot be changed.

The foresight is a post. The rearsight is a tangent U. The rearsight leaf registers from 100 to 900 m in 100 m steps.

For zeroing, the foresight is screwed in or out for elevation adjustment. For line, the lock nut must be loosened and the slide moved across its dovetail. The rearsight has a separate windage knob.

To load the RPD the belt tab is pushed through the feedway from left to right and pulled through to seat the first cartridge under the retaining pawl. If no loading tab is fitted the feed cover must be lifted and the belt put, open side down, on the feedway. The belt has a nib, already mentioned, fitting into the extraction cannelure, and there is also another nib at the front end level with the shoulder of the case. This nib must fit over a guide which pushes the nose of the bullet down for locating, ready for feeding. If the belt is not so positioned a stoppage is inevitable.

When the belt is in place and the top cover replaced, the cocking handle can be retracted; the cocking handle stays to the rear on the first two versions of the gun but is pushed forward and folded upward on the later models. The gun is then ready to fire. When the trigger is pulled the slide is driven forward carrying the bolt which pushes the first round out of the belt and into the chamber. The extractor slips into the extraction groove and the bolt comes to rest. The slide goes on and the solid wedge at the rear drives between the locking lugs and forces them out into the locking recesses in the receiver walls. The bolt is now firmly secured to the receiver. The slide goes on and the wedge hits the rear of the firing pin. The cap is struck and the bullet driven up the bore. A proportion of the gas passes through the gas port and after giving the piston-head a sharp blow, disperses to atmosphere from the open cylinder. The slide is attached to the piston and when it goes back there is a period of idle movement in which the breech block remains firmly locked while the chamber pressure drops to a safe level. Then the locking lugs, which rest in cam slots cut in the top face of the slide, are forced inwards against the sides of the bolt and unlocking is completed. The slide carries the breech block back. The empty case is held to the breech face and ejected downwards out of the gun when the ejector strikes the top of the base of the case and revolves it round the extractor. The return spring is fully compressed and the slide is driven forward again.

On the feed mechanism the slide carries a feed roller which, mounted on the left-hand side, fits into the feed arm under the top cover plate.

Pivoted at the centre of the feed arm is a feed slide which carries a spring-loaded feed pawl. When the feed arm is moved to the right, the feed pawl is moved to the left.

Starting with the gun cocked and ready for action, the cartridge is positioned by the feed pawl against the cartridge guide. It is lined up with the bullet guide and is ready for chambering.

When the trigger is pulled the bolt goes forward and pushes the round out of the belt into the chamber. During this action the cartridge belt must be stationary so the roller runs down a straight portion of the feed arm. As soon as the cartridge is clear of the belt, the roller pushes the feed arm to the right and the feed slide moves out of the gun to the left. The stop pawl on the feed tray prevents the belt being pushed to the left out of the gun as the spring-loaded feed pawl rides over the round in the belt. At the end of the forward stroke of the bolt the round is fully chambered and the feed pawl is gripping the next round in the belt.

As a round is fired the piston is driven to the rear. The roller moves back and there is an initial free travel before it bears on the feed arm. The feed arm is then pulled to the left, the feed slide is pivoted to the right. The feed pawl pulls the round it is gripping to the centre line of the gun where it is held by the cartridge guide and the feed pawl.

In brief the gun brings the belt across during the recoil stroke and positions the feed pawl over the next round, ready for feeding, during the firing stroke.

The RPD fires automatic only. The trigger has a hook at the front end which enters a window in the sear. When the trigger is pulled the hook pulls the sear down. The safety catch is on the right of the receiver, directly above the trigger. When it is applied (put forward), it locks the sear in the up position. If an attempt is made to cock the gun with the safety forward (at safe) the underside of the slide will jam on the immovable sear and remain locked. The gun will then be useless until the slide can be driven forward again. This presents no problem with the early version guns with fixed cocking handles but is not easy with non-reciprocating cocking handles.

Variants

During its service life several small variations were made to the RPD and as a result there were five recognised versions of the gun:

First version: Female gas piston fitting over male gas spigot. No dust cover. Rigid, reciprocating cocking handle. Windage knob on right of rearsight.

Second version: Male gas piston fitting into female cylinder. No dust cover. Rigid, reciprocating cocking handle. Windage knob on left of rearsight.

Third version: Male gas piston, female cylinder. Dust cover on feed mechanism. Folding non-reciprocating cocking handle. This is the version copied by the Chinese as their Type 56.

Fourth version: Male gas piston, female cylinder. Longer gas cylinder. Extra friction roller on piston slide. Dust cover on feed mechanism. Buffer in butt. (Sometimes called RPDM.)

Fifth version: As fourth version but with folding feed cover belt carrier. Multipiece cleaning rod carried in butt trap. (Also Chinese Type 56-1.)

Data

Cartridge: 7.62 × 39 mm
Operation: gas, automatic
Locking: projecting lugs
Feed: 100-round disintegrating link belt, carried in drum
Weight: empty, 7.1 kg
Length: 1.036 m
Barrel: 521 mm
Sights: fore, cylindrical post; rear, leaf tangent with V backsight
Sight radius: 600 mm
Muzzle velocity: 700 m/s
Rate of fire: cyclic, 700 rds/min
Effective range: 800 m

Manufacturer

State factories and similar establishments in most European satellites. Made in China and North Korea.

Status

Obsolescent in former Warsaw Pact countries. Current in many others.

Service

Known to be in service with China, Egypt, North Korea, Pakistan and Vietnam. Also likely to be found in other parts of the world.

Licence production

CHINA, PEOPLE'S REPUBLIC
Manufacturer: NORINCO
Type: Type 56 and Type 56-1
Remarks: Type 56-1 still available for production. In service with the Chinese armed forces and several other Asian armed forces and irregular organisations.

EGYPT
Manufacturer: Maadi Company for Engineering Industries
Type: 7.62 mm Suez
Remarks: Still in production and offered for export sales. In service with Egyptian armed forces.

KOREA, NORTH
Manufacturer: State factories
Type: 7.62 mm Type 62
Remarks: No longer in production but still in service.

UPDATED

7.62 mm SG43 and SGM (Goryunov) medium machine gun

Development

The designer Goryunov produced his SG43 to supplement the DP and DPM. The gun was modified and improved until there were eventually six versions, all with comparatively minor differences. These versions are:

SG43: This has a smooth barrel with no fins at all. It has the cocking handle lying horizontally between the two vertical spade grips of the firing gear. The sear is attached to the return spring guide. The barrel lock is a simple wedge which comes out to the side. There are no dust covers over the feed and ejection openings.

SG43B: This has a micrometer barrel lock and dust covers over feed and ejection openings.

SGM: This has longitudinal barrel fins and a separate sear housing. Dust covers are on late production models. The cocking handle is on the right-hand side of the receiver.

SGMT: This is the tank version of SGM with a solenoid on the back-plate of the receiver.

SGMB: Similar to SGM but has dust covers over the feed and ejection ports.

Hungarian SG general purpose machine gun: This gun has an RPD butt-stock with a pistol grip. It has an external resemblance to the PK general purpose machine gun but there is no hole in the butt-stock and the ejection slot is lozenge-shaped rather than rectangular.

The Chinese version of the SG43 is known as the Type 53 and the SGMB is called the Type 57 heavy machine gun.

7.62 mm SGM medium machine gun showing fluted barrel and dust cover over ejection port

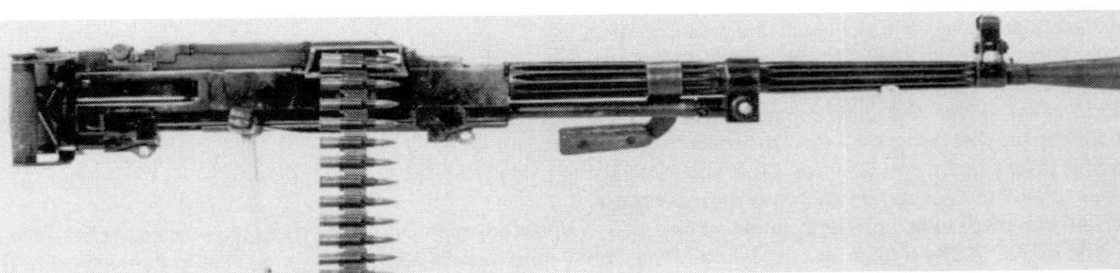

7.62 mm SGM medium machine gun showing cocking handle on right-hand side and fluted barrel

Description

The Goryunov is gas operated, belt fed and fires from a wheeled tripod which can be man-handled or pulled by a draught animal or vehicle; there are also a series of conventional tripods. It has a variable track gas regulator.

The barrel is changed readily. The SG43 has a simple wedge which drives into a dovetail slot on top of the barrel; all other models have a milled catch on the barrel lock which must be depressed before the lock can be withdrawn.

The belt is the standard CIS closed pocket type used in the RP-46, holding 250 7.62 × 54R cartridges. It is loaded by forcing the rounds into the metal pockets as far as they will go.

The gun fires from the open breech position and since it has the closed pocket belt, the cartridge is withdrawn from the belt, lowered to the level of the barrel and then rammed by the bolt.

All SG43 series fire at automatic only although the heavy barrel and the easy change make it possible to sustain a good rate of fire. The wheeled mounts incorporate coarse traverse and elevating gears and a fine elevating knob, but the traverse lock lever can be released so that free traverse through 360° is possible.

The foresight is a cylinder mounted between two protectors. It is zeroed for elevation by screwing in and out, and for line by undoing the nut at the base and pushing the foresight on its dovetail into the required position. The rearsight is a U-notch mounted on a tangent leaf.

For loading the feed cover is lifted to the vertical position and the belt is placed in the feedway from right to left with the first round positioned so that its rim lies inside the claws of the cartridge gripper. The cover is lowered and the gun cocked. The cocking handle on the SG43 and SG43M lies horizontally between the grips and, on the others, projects from the right of the receiver.

To fire the gun the trigger-locking lever must first be raised with the left thumb and then the trigger can be operated with the right thumb.

The return spring drives the slide forward carrying the bolt. The bolt picks up the cartridge which is lying above the feedway and pushes it into the chamber. As it enters the chamber the extractor on the bolt face grips the rim. The bolt comes to rest but the slide continues on. The raised piston post moves up the hollow interior of the bolt and forces the rear end to move sideways into a locking recess in the right-hand wall of the receiver. The piston post then continues on and strikes the firing pin to drive it into the cap of the cartridge.

As the slide travels forward the diagonal grooves cut in its top surface, engage ribs in the feed slide, which is pushed out to the right, and the feed pawl slips over the next cartridge in the belt.

Lying under the feed cover is the lower feed cover. Resting on top of this, with a lug which reaches down to the bolt, is the cartridge gripper. The lug engages in the bolt so the cartridge gripper reciprocates above it. The cartridge gripper has a pair of spring-loaded claws which grip over the rim of the cartridge in the belt when the bolt goes fully forward.

Gas is diverted into the gas cylinder and forces the piston and slide rearwards. The piston post on the slide moves back along the hollow interior of the bolt. There is a period of free travel while the chamber pressure drops to a safe level and then the piston post forces the rear end of the bolt out of the locking recess and carries it back. The empty case is withdrawn by the extractor and is ejected through the left of the gun.

As the bolt goes back the cartridge grippers pull the round out of the belt and withdraw it. The spring-loaded depressor arm projecting down from the underside of the feed cover forces the round down into the lower feed cover where it rests, nose slightly down, in the feedway where it can be fed by the next forward movement of the bolt.

As the slide goes back the grooves on the top surface pull the feed pawl inwards and the next round is moved over into position for the cartridge grippers to seize it at the end of the next feed stroke.

The SG43 series two-wheeled carriage can be adapted to form an anti-aircraft mount. An alternative, more conventional ground tripod constructed using heavy pressed steel components is known as the Sidorenko-Malinovski tripod.

Data

Cartridge: 7.62 × 54R
Operation: gas, automatic
Locking: tilting block
Feed: 250-round pocketed belt
Weight: empty, 13.6 kg
Length: 1.12 mm
Barrel: 722 mm
Rifling: 4 grooves, rh
Sights: fore, cylinder on dovetail; rear, tangent leaf with U backsight; anti-aircraft sights also available
Sight radius: 855.5 mm
Muzzle velocity: 800 m/s
Rate of fire: cyclic, 650 rds/min
Max range: effective, 1,000 m

Manufacturer

State factories, former Czechoslovakia, Egypt and Poland.

Status

No longer manufactured in the CIS. Still in production in Egypt.

Service

In service with Egypt and likely to be found in the Middle East, Africa, China and South-east Asia.

Licence production

CHINA, PEOPLE'S REPUBLIC
Manufacturer: NORINCO
Type: Type 57
Remarks: Copy of SGM. No longer in production but still in service with Chinese armed forces and others. Also known as M18. The earlier SG43 was known as the Type 53. The SGMT tank machine gun was produced as the Type 59T.
EGYPT
Manufacturer: Maadi Company for Engineering Industries
Type: 7.62 mm Asswan medium machine gun
Remarks: Copy of SGM with extra dust covers. Still in production and offered for export sales.

VERIFIED

7.62 mm RPK light machine gun

Description

The RPK is basically an AKM rifle with a longer and sturdier barrel. It will take magazines from the AK-47 or the AKM. It also has a 40-round box magazine of its own and a 75-round drum magazine. The drum magazine has a loading lever on the front which is used to compress the magazine spring before each round is inserted into the feed lips. It is released after each round is loaded. The drum has an extension piece which is inserted in the magazine-well under the gun, in the same way as a box magazine.

The RPK is gas operated with a vent opening into a cylinder placed above the barrel. There is no gas regulator, resulting in a slightly harsh action and a vigorous ejection to the right rear when the gun is clean, and a gradual slowing down of the rate of fire and force of ejection as firing progresses and fouling accumulates. The piston is forced back by gas pressure in the cylinder and there is a short period of free travel while the width of the cam slot cut in the piston extension travels across the unlocking lug of the bolt head and then the two locking lugs are caused to rotate anti-clockwise out of engagement with the locking shoulders. After unlocking is completed the piston carries the bolt to the rear.

The loading stroke of the bolt chambers the round; the bolt movement then ceases and further forward movement of the piston rotates the bolt to enable the locking lugs to engage in front of the locking shoulders.

The firing pin floats freely in the bolt. It cannot be reached by the hammer which comes up against a stop until such time as locking is completed. If the hammer is released early the bolt is forcibly closed and the reduced blow on the firing pin causes a misfire.

The cocking handle is integral with the bolt carrier and reciprocates as the gun fires. This allows positive closure if adverse conditions of sand or mud prevent completion of locking.

There is a single extractor located on the right-hand side of the bolt face which itself is recessed so that the base of the round is always fully supported in the chamber. A fixed ejector in the body runs along a groove in the bolt producing a strong ejection to the right rear of the gun. The barrel and chamber are both chromium plated.

There is no hold-open device so the mechanism always stops in the forward position.

The selector is on the right side of the machined receiver and is a long lever of similar shape to that in the rifle. There are three positions: fully automatic at the middle position of travel, single shot at the bottom and safe at the top.

The selector arm in the safe position carries out two functions. Its spindle carries an arm which, when rotated to safe, covers the trigger extension and prevents trigger movement. The selector itself comes up to block the cocking lever and thus prevents the weapon being cocked while set at safe. The limited bolt movement does not pass over the base of the round in the magazine but is sufficient for the firer to check that the chamber is empty, either by visual inspection by day or by touch at night.

7.62 mm RPK light machine gun with 40-round box magazine (T J Gander)

The rearsight is a leaf sliding on a ramp; it has a U rearsight. There is also a night sight with an enlarged U with a fluorescent white spot. The rearsight is graduated in hundreds of metres marked 1 to 10. There is a windage scale.

The foresight is a cylindrical post which can be screwed up or down for elevation adjustment when zeroing.

There is a battle sight which is a coarse U covering 100 to 300 m.

Accessories carried are the same as those for the AK-47 and AKM but there is no bayonet. A bipod, mounted well forward, can be folded under the barrel and clipped in place.

Some RPKs mount an infra-red night sight.

Data

Cartridge: 7.62 × 39 mm
Operation: gas, selective fire
Feed: 40-round box or 75-round drum magazine; can also use 30-round rifle magazine
Weight: empty, 5 kg
30-round magazine: empty, 340 g; loaded, 850 g
40-round magazine: empty, 368 g; loaded, 1.13 kg
75-round drum magazine: empty, 900 g; loaded, 2.1 kg
Length: 1.035 m
Barrel: 591 mm
Sights: fore, cylindrical post; rear, leaf notch
Sight radius: 570 mm
Muzzle velocity: 732 m/s
Rate of fire: cyclic, 660 rds/min
Range: effective, 800 m

Manufacturer

State factories.

Status

No longer in production in CIS. Still in production elsewhere.

Service

CIS and former Warsaw Pact countries.

Licence production

IRAQ
Manufacturer: State factories
Type: 7.62 mm Al-Quds
Remarks: Not so much a true RPK as a locally derived AKM heavy-barrelled machine rifle. See entry under *Iraq* for details.
ROMANIA
Manufacturer: Romtehnica
Type: 7.62 mm light machine gun
Remarks: Produced with fixed or AKMS-type folding butt. Offered for export sales.

UPDATED

75-round drum magazine of 7.62 mm RPK light machine gun

5.45 mm RPK-74 light machine gun

Development

The RPK-74 is essentially a heavy-barrelled version of the AK-74 rifle, bearing the same relationship to the rifle as the RPK does to the AK series of rifles.

There are four main models in the RPK-74 range, as produced at the IZHMASH complex in the CIS. The RPK-74 is the base model with a fixed stock. The RPKS-74 has a folding butt-stock which folds to the left. The RPKN3 has provision for mounting a night sight, usually the 1LH51; this model has also been displayed as the RPK-74N3. The RPKSN3 is the folding butt-stock equivalent of the RPKN3.

Description

The 5.45 mm RPK-74 receiver is of stamped steel, with bolt rails welded to the interior. Operation is by gas, a piston above the barrel driving a bolt carrier which contains a two-lug rotating bolt. A hammer mechanism is cocked during the rearward stroke and released either by trigger pressure or by a delay mechanism which allows the bolt to close and lock before releasing the hammer during automatic fire. A three-position selector/safety lever is on the right sight of the receiver.

The butt is of laminated wood, and the bipod of the RPK is attached to the barrel just below the foresight. A slotted flash hider/compensator is fitted to the muzzle. The standard magazine is a plastic 45-round box, though the 30-round magazine of the AK-74 rifle can also be used; 40-round box magazines also exist.

RPK-74 series light machine guns are usually issued together with eight magazines in a special carrying bag, and a shoulder sling.

Data

Cartridge: 5.45 × 39 mm
Operation: gas, selective fire
Locking: rotating bolt
Feed: 45-, 40- or 30-round box magazine
Weight, empty: RPK-74, 4.6 kg; RPKS-74, 4.7 kg
Length: 1.06 mm
Barrel: 616 mm
Rifling: 4 grooves, rh, 1 turn in 196 mm
Sights: fore, cylindrical post; rear, tangent leaf with U-notch; adjustable to 1,000 m; RPKN3 and RPKSN3, 1LH51 night sight
Sight radius: 556 mm
Muzzle velocity: ca 960 m/s
Rate of fire: cyclic, 600-650 rds/min; practical, 150 rpm
Max range: effective, 460 m

Manufacturer

Kalashnikov Joint Stock Company, 373 Pushkinskaya Street, 426000 Izhevsk, Russia.

Marketed by: Rosvoorouzhenie, 18/1 Ovchinnikovskaya Emb, 113324, Moscow, Russia.

Status

In production.

5.45 mm RPK-74 light machine gun with bipod folded (bottom) shown with 5.45 mm AK-74 assault rifle (top) for comparison

5.45 mm RPKN3 light machine gun with 1LH51 night sight in position (T J Gander)

Service

CIS and former Warsaw Pact countries.

Licence production

BULGARIA
Manufacturer: Kintex
Type: 5.45 mm RPK-74
Remarks: Identical to standard RPK-74. Offered for export sales. See entry under *Bulgaria* for illustration.

ROMANIA
Manufacturer: Romtehnica
Type: Light machine gun 5.45 mm
Remarks: Standard version is similar to RPK-74 but slightly longer at 1.082 m. Also produced in folding butt-stock form with frame butt-stock and handguard. See entry under *Romania* for details. Training version also produced.

UPDATED

6 mm Unified machine gun

Description

First shown publicly during 1993, the so called Unified machine gun fires a developmental 6 mm high-power cartridge. As far as can be determined this cartridge has the same overall dimensions as the standard CIS 5.45 × 39 mm round and probably uses the same case filled with a new and more powerful propellant producing a flatter trajectory out to extended ranges. This development work is also associated with a 6 mm variant of the AK-74 assault rifle (qv).

The 6 mm Unified machine gun is based on the 7.62 mm PK series but with a longer barrel. As far as can be determined the Unified machine gun continues to utilise the standard Kalashnikov rotating lock but within a revised receiver profile, resulting in what appears to be a longer barrel than that on the standard PK, even though the overall length of the weapon is 1.15 m, similar to that of a standard PK.

The combination of the developmental 6 mm cartridge and the new machine gun results in an effective range of 1,500 m and a reduction in overall weight to 6.5 kg with bipod. Combined with the enhanced range

6 mm Unified machine gun (T J Gander)

performance is a claimed improved fire effectiveness, by a factor of 2 to 2.5, largely due to a reduction in aiming errors - accuracy is claimed to increase by a factor of 2.5. Sighting is enhanced by the provision of a fixed optical sight mounted over the receiver to the left-hand side. This position leaves room on the right for a mounting rail to accept night sights such as the 1LH51.

This machine gun can be fired from standard tripods.

Data

Cartridge: 6 mm developmental, possibly 6 × 39 mm
Operation: gas, automatic
Locking: rotating bolt
Feed: belt, 100 rounds
Weight, empty: with bipod, 6.5 kg; on tripod, 11 kg
Length: 1.15 m
Sights: optical; provision for night sight
Max range: effective, 1,500 m

Manufacturer

Institute of Precise Mechanical Engineering, Tzniitochmash, 2 Zavodskaya Street, 142080 Klimovsk, Russia.

Status

Development - prototypes available and ammunition batches produced.

VERIFIED

7.62 mm PK machine gun family

Development

As the family of Kalashnikov weapons replaced those of Degtyarev the time came for the replacement of the RP-46 and the SGM series by a Kalashnikov design. The weapon selected for this was the PK which is a mixture of components and ideas from other weapons. The rotating bolt comes from the other Kalashnikov weapons, the AK-47 and RPK, the cartridge gripper comes from the Goryunov SGM and so does the barrel change. The system of feed operation using the piston to drive the feed pawls comes from the Czech Model 52 light machine gun. The trigger comes from the Degtyarev RPD.

The CIS 7.62 mm PK machine gun was first seen in 1964. Since then it has been modified and improved so that it has become a true general purpose machine gun and has replaced the RP-46 and SGM in frontline service.

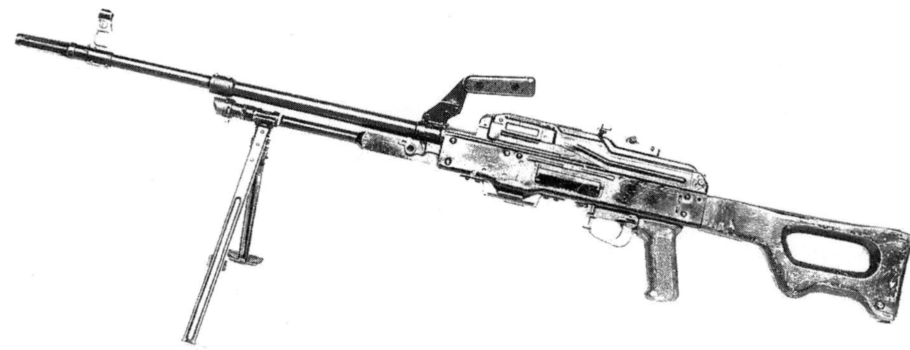

7.62 mm PKM light machine gun (T J Gander)

Description

The PK family of machine guns are gas operated, rotary bolt locked (Kalashnikov system), open bolt fired, fully automatic, belt-fed machine guns firing the 7.62 × 54R cartridge. The ammunition is fed by non-disintegrating metallic belts; current belts are composed of joined 25-round sections but earlier feed belts were made of one 250-round length. The belts are held either in 250-round ammunition boxes, in special large capacity boxes on tanks (for the PKT) or in a 100-round assault magazine attached to the bottom of the gun's receiver.

The receiver is constructed of riveted stampings. It carries the very simple, automatic only, trigger and the belt feed mechanism.

The barrel is chrome plated and has a change system which is not as quick as that employed in most modern guns. To change a barrel the gun must be unloaded; the feed cover comes up when the return spring guide is pressed and the feedway pivots up on the same pin. The barrel lock comes out to the side in the same way as the SGM and the barrel can be pulled forward out of the receiver with an asbestos glove. A new barrel is inserted, the lock is replaced, the feedway is lowered, the belt replaced and feed cover lowered and the gun is ready to fire.

A gas regulator is fitted. It is a variable track-type with three numbered positions, similar to that in the Goryunov SG-43, and the head of the regulator is disengaged from its detent and moved to the selected number.

The gas piston is attached to the slide. Running along the length of the slide is a cam path on which bears the roller of the belt feed lever. The roller follows the cam as the slide goes back and the feed lever pivots outward causing the feed pawl attached to it to move in with the next cartridge in the belt. This is placed in position ready for the cartridge gripper. As the bolt goes forward a second cam path forces the feed lever in at the bottom, out at the top and the spring-loaded feed pawl moves out over the next cartridge.

On the top of the slide is the bolt carrier. The relationship between this carrier and the bolt is the same as that in the AK-47. As soon as the bolt stops moving forward, having chambered the round, the cam path on the carrier engages the cam pin on the bolt and rotates it to lock into recesses.

At the rear of the slide, projecting forward over the bolt, is the cartridge gripper. When the bolt is fully forward the two spring-loaded claws of the gripper slip over the rim of the cartridge in the belt. When the bolt recoils the cartridge is pulled out of the belt and back. The spring-loaded cartridge depressor arm pushes the cartridge down out of the claws and it is held in the feed lips waiting for the bolt to drive it forward into the chamber.

The 250-round belt is loaded in the same way as the RP-46 or the Goryunov SG-43 series. The feed cover of the gun can be lifted when the return spring guide is pressed. The belt is placed in the feedway, from the right, with the first round gripped in the claws of the cartridge gripper. The feed cover is closed, the safety catch set to fire, and the cocking handle on the right of the receiver is retracted and returned to the forward position. If the safety catch is left in the rear position at safe the gun cannot be cocked.

With the safety at fire, pulling the trigger releases the sear and the compressed return spring drives the slide forward. The bolt picks up the round in the feed lips and chambers it. The extractor grips the rim and bolt movement ceases. As the slide goes forward the cam path along the side forces the feed pawl outwards over the next cartridge in the belt. The bolt carrier moves forward with the slide after the round is chambered and the bolt is rotated 30° to the locked position. The firing pin is carried on the bolt carrier and it comes through the bolt to fire the cap.

The gases force the bullet up the bore. Some of them pass through the gas vent and drive the piston and slide back. The bolt carrier moves back and the firing pin is retracted. There is a period of free travel while the pressure drops and then the cam slot on the bolt carrier engages the cam pin on the bolt and the bolt rotates to become unlocked. The bolt recoils. The empty case is held on the bolt face and the slide compresses the return spring. The cartridge gripper pulls the cartridge out of the belt and carries it back. The cam path on the side of the bolt carrier forces the feed roller outwards and the upper end of the pivoted feed arm moves inwards so the feed pawl pushes the belt into the gun and the next round reaches the cartridge stop where it remains for the gripper to come forward again.

The empty case is ejected by a fixed ejector. The depressor arm forces the live round down into the feed lips, waiting for the bolt to go forward again. This cycle is repeated as long as the trigger is pulled and ammunition remains in the gun.

The largest accessory is the general purpose tripod which can be quickly adapted for anti-aircraft fire.

Belt boxes for 100-, 200- and 250-round belts are available. The 100-round box can be attached to the underside of the PK for the assault role.

Accessories include spare parts, a combination tool and an oil-solvent container.

7.62 mm PKMS machine gun (T J Gander)

Variants

PK

The basic gun with a heavy fluted barrel, feed cover constructed from both machined and stamped components and a plain butt-plate. The PK weighs about 9 kg. Normally employed as a portable infantry fire support weapon and fired from the bipod.

PKS

The basic PK mounted on a tripod for the heavy machine gun fire support role. The lightweight tripod (4.75 kg) not only provides a stable mount for long-range ground fire, it also can be quickly opened up to elevate the gun for anti-aircraft fire.

PKT
The PK as altered for coaxial installation in an armoured vehicle. The sights, stock, tripod, and trigger mechanism have been removed, a longer heavy barrel is installed, and a solenoid is fitted to the receiver backplate for remote triggering. An emergency manual trigger and safety is fitted.
PKM
A product improved PK, with a lighter, unfluted barrel, a feed cover constructed wholly from stampings, and a hinged butt-rest fitted to the butt-plate. Excess metal is machined away wherever possible to reduce weight to about 8.4 kg.
PKMS
PKM mounted on a tripod (similar to the PKS).
PKB
The PKM with the tripod, butt-stock, and trigger mechanism removed and replaced by twin spade grips and a butterfly trigger similar to those on the SGMB. This gun may also be known as the PKMB. Late production models of this weapon, known as the Modernised Kalashnikov, may be fitted with night sight units with a night firing range of 300 m; weight of the sight unit is 2.95 kg.
6 mm Unified machine gun
See previous entry.

Data
PKS
Cartridge: 7.62 × 54R
Operation: gas, automatic
Locking: rotating bolt

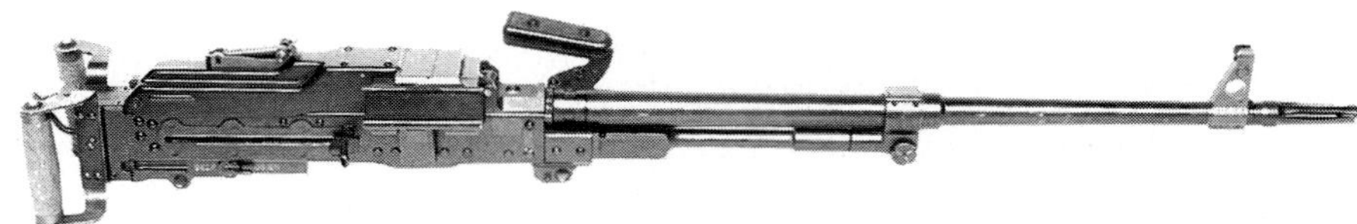

7.62 mm PKB light machine gun

Feed: belt; 100, 200 and 250 rounds
Weight: empty, 9 kg; tripod, 7.5 kg; 100-round belt, 2.44 kg
Length: 1.16 m; on tripod, 1.267 m
Barrel: 658 mm
Sights: fore, cylindrical post; rear, vertical leaf and windage scale adjustable to 1,500 m
Muzzle velocity: 825 m/s
Rate of fire: cyclic, 690-720 rds/min
Max range: effective, 1,000 m

Manufacturer
State factories.
Marketed by: Rosvoorouzhenie, 18/1 Ovchinnikovskaya Emb, 113324, Moscow, Russia.

Status
In production.

Service
CIS and allied armies.

Licence production
BULGARIA
Manufacturer: Kintex
Type: 7.62 mm Kalashnikov
Remarks: Identical to CIS PK. PKT also produced. See entry under *Bulgaria* for illustration.
CHINA, PEOPLE'S REPUBLIC
Manufacturer: NORINCO
Type: 7.62 mm Type 80
Remarks: See entry under *China, People's Republic* for details. A coaxial machine gun for armoured vehicles is also produced.
ROMANIA
Manufacturer: Romtehnica
Type: 7.62 mm machine gun
Remarks: Identical to CIS PKM. PKT also produced.

UPDATED

12.7 mm Degtyarev (DShK-38 and Model 38/46) heavy machine gun

Development
The DShK-38 was designed by Degtyarev and was based on an earlier model of his, the DK, which was produced in very limited numbers in 1934. The feed was designed by G S Shpagin and was a rotary type in which the rounds in the belt were successively removed from the links and then fed through a feed plate and collected by the bolt in its forward travel. This was a complicated process and required skill and training if stoppages were to be avoided.

In 1946 the rotary feed was replaced by the conventional shuttle-type feed used on the Degtyarev Model RP-46. This produces a ready recognition feature as the DShK-38 has the large circular drum-like feed mechanism and the Degtyarev Model 38/46 has a flat rectangular feed cover. There is no interchangeability between the parts of the two guns.

The 12.7 mm machine guns have been used extensively by CIS-influenced European and Asian countries. The European countries who originally used the gun on a tripod now fit it onto a number of armoured vehicles for anti-aircraft and anti-vehicle defence. It is still used on the ground mount by several Asian countries.

The Chinese version is a copy of the Model 38/46 known as the Type 54 heavy machine gun. It can be identified by the Chinese characters on the receiver behind the feeder. The Czechs made a towed four-gun air defence unit, the M53, which remains in service with a few nations.

Description
The DShK-38 feeds from the left and has a fixed barrel. The Model 38/46 can readily be adapted for feed from either side, by changing some parts in the feed mechanism, and has a quick-change barrel.

The belt is a continuous metallic-link type holding 50 rounds. Each link takes a cartridge which is located by the nib on the belt entering the groove of the cartridge.

The cover latch of the DShK-38 is at the rear of the cover and it permits the cover to be rotated about the hinge at the front.

An ammunition belt is placed in the gun from the left with the first link under the link stripper and the first cartridge in the top compartment of the rotating feed drum. The weight of the ammunition is taken off the feed drum and the ammunition and drum pushed round through some 120°. When the drum will rotate no further the feed cover is closed. The cocking handle between the spade grips is retracted and the slide held to the rear on the sear. The compartments of the rotating feed drum are then full and the first round is pressed through the feed plate into the path of the bolt.

CIS 12.7 mm DShK-38 heavy machine gun on Sokolov mount

The Model 38/46 should be regarded as a large version of the RP-46. The belt tab is inserted into the feed guide and pulled through until the first round passes over the stop pawl which will prevent it from falling out. When the cocking handle is fully retracted the gun is ready to fire.

When the trigger is pressed the compressed return spring drives the slide and bolt forward. The feed rib on the bolt passes under the feed drum of the DShK-38 and picks up the round which is projecting through the feed plate. In the Model 38/46 the round is held up to the catch stop. The round is pushed forward and into the chamber. The extractor grips into the cannelure at the base of the cartridge and the bolt comes to rest. On each side of the bolt is a long locking lug which is pivoted at the front end into a recess in the side of the bolt and opens out from the bolt at the rear end, like a flap, as the firing pin, which is attached to the slide, drives forward. The projecting lugs are cammed out by shoulders on the firing pin and engage in recesses in the side walls of the receiver. The firing pin goes in a further 16 mm after locking is completed and the cap is fired.

Some propellant gases are diverted into the cylinder and drive the piston back. The slide carrying the firing pin is retracted and there is about 16 mm of free play while the chamber pressure drops. The recesses cut into the top surface of the slide force the locking lugs into the sides of the bolt, and the slide, bolt and firing pin go back as one unit. The empty case comes out on the bolt face and is ejected through the bottom of the receiver. The return spring is compressed to provide energy for the next forward stroke.

As the operating stud on the slide comes back it enters the open stirrup of the feed lever and rotates it back. On the Model 38/46 this operates the feed slide, as in the RP-46. A cam path moves the belt feed slide inward and the feed pawl brings the next round up to the cartridge guide and over the stop pawl. As the feed slide goes forward with the operating stud, the feed pawl is moved out to engage the next round in the belt.

On the DShK-38 the backward movement of the operating stud rotates the feed lever and a pawl bears on a ratchet in the feed drum, causing it to revolve. At the same time another pawl prevents counter rotation. The drum rotation draws the belt into the gun. The first cartridge is pressed along the lips in the feed plate

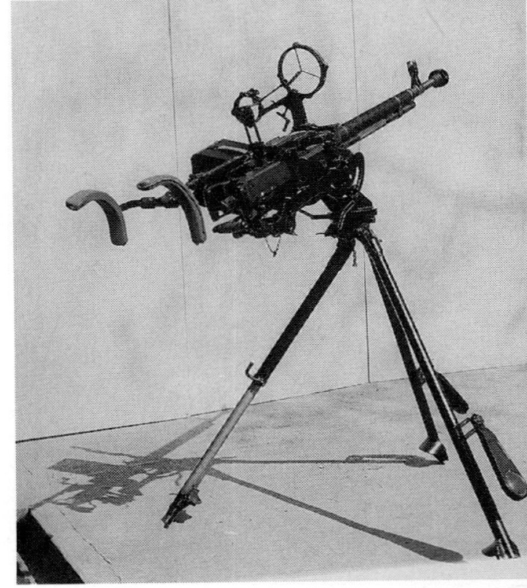

12.7 mm Type 54 heavy machine gun on air defence mounting as produced by Pakistan Ordnance Factories

and is forced forward on the next feed stroke of the bolt.

The ground sights are orthodox. The foresight is a pillar which screws up and down for zeroing. The bolt through the base of the foresight can be loosened and the sight moved totally for zeroing.

The rearsight has vertical twin pillars with a U backsight between them. When the sight is upright the elevating screw is at the top. There is a windage knob at the base of the sight.

The basic mount for both the DShK-38 and the Model 38/46 is the Model 1938. This is a two-wheeled mount moved by man, mule or vehicle; a shield is sometimes provided for ground use. The traverse and elevation are coarse with a fine elevation adjustment; the gun is locked in position before firing. The mount can be converted for anti-aircraft use by removing the gun and shield and then taking off the wheels and axle. The three legs are spread, the gun mounted above them on the saddle and the sight inserted in the dovetail on the left side of the receiver.

Accessories issued with the 12.7 mm heavy machine guns include: cleaning rod; chamber cleaning brush; oil and solvent container; combination tool; punch and separated case extractor.

Data

Cartridge: 12.7 × 108 mm
Operation: gas, automatic
Locking: projecting lugs
Feed: belt
Weight: empty, 35.7 kg
Length: 1.588 m
Barrel: 1.07 m
Sights: fore, cylindrical post; rear, vertical leaf with U backsight, adjustable to 3,300 m
Sight radius: 1.111 m
Muzzle velocity: 860 m/s
Rate of fire: cyclic, 575-600 rds/min
Range: effective, 1,500 m

Manufacturer

State factories.

Status

No longer in production in the CIS.

Service

CIS and former Warsaw Pact armed forces, Chinese armed forces, Pakistan, and various guerrilla forces.

Licence production

CHINA, PEOPLE'S REPUBLIC
Manufacturer: NORINCO
Type: 12.7 mm Type 54
Remarks: In production and offered for export sales. Usually mounted as armoured vehicle turret gun on Type 59 series tanks.
PAKISTAN
Manufacturer: Pakistan Ordnance Factories
Type: 12.7 mm Anti-aircraft Gun Type 54
Remarks: In production. Licence-produced Chinese Type 54. Offered for export sales.
ROMANIA
Manufacturer: Romtehnica
Type: 12.7 mm machine gun for armoured vehicles
Remarks: In production. Offered for export sales.

VERIFIED

12.7 mm NSV heavy machine gun

Development

Development of the 12.7 mm NSV heavy machine gun, known as the Utyos, began in 1969 and was carried out by a team of three designers, G I Nikitin, J M Sokolov and V I Volkov. It may be deployed in the ground role as a heavy support weapon or as an air defence weapon, and as an armoured vehicle machine gun. Apparently this machine gun is known as the NSV (or NSV-12.7) only when it is pintle-mounted on the turrets of late generation CIS MBTs such as the T-72 or T-80. When mounted on a standard ground tripod the NSV is known as the NSVS-12.7. Fitted with a night sight it becomes the NSV-12.7 N4 while as an armoured vehicle coaxial gun it becomes the NSVT.

In a document dated October 1992 the price of a single NSVT machine gun was quoted as $12,000.

12.7 mm NSV heavy machine gun on 6U6 multipurpose/air defence mounting (T J Gander)

1996

12.7 mm NSVS heavy machine gun (NSV-12.7)

Description

The 12.7 mm NSV heavy machine gun is a gas-operated weapon with an air-cooled barrel, fed by 50-round linked belts carried in a magazine box. Belts are formed from easily managed 10-round lengths using 6L19 metal links and joined together only when placed in the magazine box. (Polish sources refer to 70-round belts.) Feed direction may be from the left or right, the user's requirement being introduced at the factory; the normal feed for most mountings is from the right.

For operation the NSV series utilises gas operation with gas being tapped from the barrel via a three-position gas regulator located about half-way along the barrel. Pressure on a piston head forces back a relatively heavy breech block carrier on which is mounted the breech block proper. The breech block carrier operates within a receiver formed from pressed components held together by welds and rivets.

The NSV breech block is of a novel type on which there are three horizontally side-folding linked sub-blocks. For feeding a fresh round when the breech block is withdrawn the three sub-blocks are out of line with each other. As the breech block assembly moves forward and down a slight ramp one of the three sub-blocks guides the round towards the chamber as all three sub-blocks are moved by a system of lugs, ramps and levers to form an all-in-line formation for firing and locking. Only when the breech block carrier commences its rearwards travel can the three sub-blocks fall out of line for one to assist in ejecting the spent case as the feed sequence is repeated. The action of the breech block carrier also provides the drive motion for the feed system.

When mounted as the NSVS-12.7 on the standard ground tripod the weapon is fitted with a skeleton butt-stock on a strut. It is possible to utilise an SPP optical sight weighing 1.7 kg in this configuration although iron sights are provided. The SPP optical sight can be adjusted to provide either a ×3 or ×6 magnification. It is possible to utilise graduations on the SPP optics for ranges up to 2,000 m and the graticule also incorporates a ranging system against man-sized targets similar to that employed on the Dragunov SVD sniper rifle PSO-1 sight. The iron sights are graduated from 2 up to 20, each set of numerals referring to hundreds of metres.

An alternative mounting is the multipurpose machine gun mount 6U6, intended primarily as an air defence mounting against targets up to an altitude of 1,500 m. On the 6U6 the gunner is seated on the mounting and uses two control wheels to elevate and traverse the gun, using a 10P81 ×3.5 optical sight. Weight of the complete 6U6 mounting, with gun, is 92.5 kg. For transport the mounting can be broken down into five man-pack loads, with two men also carrying the gun between them.

The NSV is intended to fire two types of 12.7 × 107 mm round, the AP-I B-32 and the API-T BZT-74, the loading ratio being 3:1.

Variants

NSV-12.7 N4

This is the standard tripod-mounted NSVS machine gun fitted with a 1LN52-1 night sight.

12.7 mm NSVT

The NSVT is a NSV machine gun converted for coaxial use on tanks and other armoured fighting vehicles.

Firing is electrical using a 24 V solenoid. Weight of the NSVT alone is 26.8 kg.

Data

Cartridge: 12.7 × 107 mm
Operation: gas, automatic only
Locking: three-part horizontal side-folding breech block
Feed: 50-round linked belt from magazine box
Weight: gun only, 25 kg; barrel, 9.2 kg; tripod, 16 kg; 50-round belt, 7.7 kg
Length: 1.56 m
Rifling: 8 grooves, rh
Sights: fore, post; rear, folding tangent leaf with graduations from 2 (200 m) to 20 (2,000 m); sighted to 1,500 m for aerial targets, to 2,000 m for ground targets; SPP optical sight also available
Rate of fire: cyclic, 700-800 rds/min; practical, up to 270 rds/min
Muzzle velocity: 845 m/s
Range: effective, 2,000 m

Agency

GUSK, 18/1 Ovchinnikovskaya nab, 113324 Moscow, Russia.
Also marketed by: Rosvoorouzhenie, 18/1 Ovchinnikovskaya Emb, 113324, Moscow, Russia.

Status

In production.

Service

Former Warsaw Pact armies; also former Yugoslavia.

Licence production

BULGARIA
Manufacturer: Kintex
Type: 12.7 mm NSV Utyos
Remarks: In production in NSV, NSVS and NSVT forms. 6U6 multipurpose mounting also produced.
POLAND
Manufacturer: Zaklady Mechaniczne Tarnow
Type: 12.7 mm NSW Utios
Remarks: In production and offered for export sales on various mountings. See entry under *Poland* for details.
YUGOSLAVIA
Manufacturer: Zastava Arms, Belgrade
Type: NSV-12,7
Remarks: May no longer be in production.

UPDATED

14.5 mm KPV heavy machine gun

Development

The 14.5 mm KPV heavy machine gun was designed immediately after the Second World War by Vladimirov expressly to fire the high-velocity round produced for the PTRD-41 (Degtyarev) anti-tank rifle.

The gun was designed for simplicity of manufacture so the body is a simple metal cylinder to which the various attachments are riveted. There is some welding and considerable use of stampings, for example the feed tray.

While the gun was designed initially as an anti-aircraft gun it is suitable as an armoured fighting vehicle weapon. It has a short inboard length (600 mm), less than the 0.50 M2 Browning. The belt feed system, which allows breakdown in 10-round lengths, is applicable to vehicle use and so is the alternative left- or right-hand feed, but the forward barrel change is not an advantage in this role.

14.5 mm KPV heavy machine guns in one of their most widespread applications, on a North Korean-supplied ZPU-4 air defence mounting provided to Malta (T J Gander)

Description

The gun is short-recoil operated with gas assistance supplied by a muzzle booster. Before firing, the bolt assembly is securely locked to the barrel. The locking lugs consist of eight rows of castellated form, each 1.65 mm deep, which are rotated to lock into corresponding recesses on the barrel extension. The pressure built up forces the breech block rearwards and this in turn draws the barrel back with it against the resistance offered by the inertia of the assembly and the force of the return spring. Since the barrel has its own run out spring this also opposes the rearward movement. When the projectile reaches the hole in the muzzle cap the gases are briefly sealed and the resulting pressure build-up inside the booster drives the barrel backwards to give increased momentum to the recoiling parts. The breech block and barrel have a free recoil of slightly less than 5 mm which provides the mechanical safety after firing and allows the bore pressure to fall to a safe level. The bolt head carries a roller on each side running in a curved cam path in the bolt body and these rollers come up against curved cam paths in the body of the gun. The rotation imposed on the bolt head unlocks it from the barrel. There is a 6 mm pitch on the locking lugs and while the bolt head is being rotated it is drawn back this distance, providing the slow powerful leverage required for the initial extraction of the cartridge case. At the same time the firing pin is withdrawn. The unlocking is completed when the barrel and bolt have recoiled together for 18 mm. As the rollers are rotated by the cams in the body they are running in the slots in the bolt body and thus accelerate the bolt body to the rear. The relative displacement between the bolt head and bolt body is slightly less than 18 mm.

The fired case is held in a T-slot in the bolt head which grips the rim and extractor groove; rearward movement of the unlocked bolt withdraws the expended case from the chamber.

The next round to be fed is held by two claws which move back with the bolt and extract this round from the closed-pocket metal belt. When the bolt head is rotated to unlock, the T-slot is then lined up to receive the next round which is forced down by a cam-operated arm and displaces the case of the previous fired round which falls out of the bottom of the gun. However, this system cannot operate when the gun fires single shot or the last round of the belt is fired. To cater for these circumstances a spring-loaded arm is fitted behind the cam to depress the cartridge feed arm down far enough to eject the empty case.

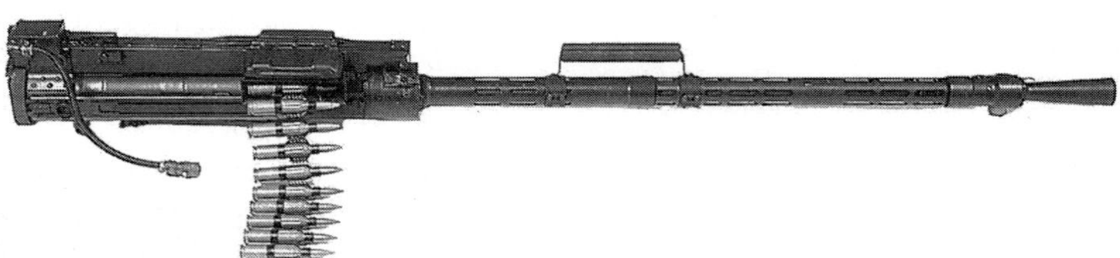

14.5 mm KPVT heavy machine gun for armoured vehicles

During bolt recoil the belt is fed across and a live round positioned ready to be gripped by the claws above the bolt, as the latter comes forward.

When the bolt has reached its rearmost position, energy is stored in the return spring and the buffer. This carries the bolt forward, driving a round into the chamber and slipping the feed claws over the waiting round in the belt. The bolt head must be correctly positioned to allow the locking lugs to enter the recesses in the barrel extension. This is accomplished in the KPV by locking the bolt head and releasing it to be free for rotation when the locking latch hits a cam in the body. The two bolt head rollers run on to locking cams in the side walls of the body and are rotated to lock the bolt head to the barrel. The bolt body then closes up to the bolt head and the firing pin is driven through into the cap and the round is fired. As the bolt goes forward the feed pawl is moved outwards to slip over the next round in the belt ready to draw it towards the barrel on recoil.

To load the KPV it is necessary to allow the claws that withdraw the round from the belt to grip the rim. Since the claws are mounted 51 mm behind the bolt face and the round is held in the belt at the shoulder it is necessary to withdraw the bolt about 63 mm before inserting the belt and pulling it through fully. The bolt is then released and the return spring will force it forward and allow the extractor claws to ride over the rim of the positioned round. The bolt is then withdrawn fully to the rear until held by the sear. This cocking movement must be continuous. If the bolt is checked half way back the live round will fall off the T-slot and out of the gun.

Variants

KPVT

This is the armoured vehicle variant of the KPV. Weight is 52.2 kg and length overall 1.98 m. Firing is electrical.

Towed-carriage air defence mountings exist for one, two or four guns: they are the ZPU-1, ZPU-2 and ZPU-4 respectively. Full details of these mountings are provided in *Jane's Land-based Air Defence 1995-96*.

Data

Cartridge: 14.5 mm
Operation: gas assisted, short recoil
Feed: belt, left- or right-hand
Weight: 49.1 kg
Length: 2.006 m
Barrel: 1.346 m
Sights: fore, cylindrical post; rear, tangent leaf U, 200-2,000 m by 100 m intervals
Sight radius: 736 mm
Muzzle velocity: 1,000 m/s
Rate of fire: cyclic, 600 rds/min

Manufacturer
State factories.

Status
In production.

Service
CIS, former Warsaw Pact armed forces, China and numerous other nations, including Malta.

Licence production
CHINA, PEOPLE'S REPUBLIC
Manufacturer: NORINCO
Type: 14.5 mm Anti-aircraft Machine Gun Type 56
Remarks: In production for air defence mountings such as Type 75-1 (ZPU-1) and Type 56 (ZPU-4).
KOREA, NORTH
Manufacturer: State Factory 67
Type: 14.5 mm KPV heavy machine gun
Remarks: Still in production for air defence mounting applications. Supplied to Malta.
ROMANIA
Manufacturer: Romtehnica
Type: 14.5 mm heavy machine gun
Remarks: Available for ZPU-2 and ZPU-4 mountings and as KPVT for armoured vehicles.

VERIFIED

CZECH REPUBLIC

LADA light machine guns

Development
The LADA range of weapons were first seen publicly during 1993. They are being promoted chambered for one of two possible rounds, the Eastern Bloc 5.45 × 39 mm or 5.56 × 45 mm NATO. The choice is made more significant by an announcement of intent made by the Czech armed forces to switch from Eastern Bloc to NATO standards for all munitions, from small arms to artillery. However the funding for such a drastic move has not yet been forthcoming so production of the LADA light machine guns and rifles has to date been limited pending a final decision. Indications would appear to favour the move to produce the LADA light machine gun, assault rifle and carbine in 5.56 mm for the Czech armed forces but this may be overshadowed by procurement interest displayed by nations such as Poland and North Korea who are more likely to favour the retention of 5.45 mm.

There are three components in the LADA range of weapons; a standard assault rifle, a carbine and a light machine gun. Many components are interchangeable between all three weapons. For details of the assault rifle and carbine refer to the LADA entry in the *Rifles* section.

Description
The LADA light machine gun is essentially similar to the LADA assault rifle but has a longer and heavier barrel and a bipod. It is gas operated and utilises the widely used rotating breech block for locking to the barrel at the instant of firing. Three modes of fire are available, semi-automatic single shot, automatic and three-round burst, with the selector lever on the right of the receiver. The receiver is constructed using steel pressings and there are obvious resemblances to Kalashnikov series rifles. The pistol grip and forward handguard are plastic.

Rounds are fed into the weapon from a 75-round drum magazine although it is possible to utilise the transparent plastic curved box magazine holding 30 rounds normally used on the LADA assault rifle and carbine. It is also possible to utilise M16-type magazines.

There is an aperture rearsight graduated from 300 to 1,000 m in 100 m steps. The front and rearsight are provided with luminescent dots for shooting in poor light conditions. There is also an optical or night sight mounting bar.

The butt-stock fitted to the LADA light machine gun is of the side folding type, using tubular steel struts and a single-piece butt plate. If required the complete weapon can be broken down into four basic components.

LADA light machine gun without magazine (T J Gander)

Data
Cartridge: 5.45 × 39 mm or 5.56 × 45 mm NATO
Operation: gas, selective fire
Locking: rotary bolt, forward locking
Feed: 75-round drum or 30-round plastic box magazine
Weight: empty, 4.1 kg
Length: butt extended, 1.05 m
Barrel: 577 mm
Sights: fore, post; rear, aperture, graduated in 100 m steps from 300 to 1,000 m
Muzzle velocity: 960 m/s
Rate of fire: cyclic, 750-850 rds/min

Manufacturer
Ceska Zbrojovka as, 688 27 Uhersky Brod.

Status
Limited production. Offered for export sales.

VERIFIED

7.62 mm Model 59 general purpose machine gun

Development
The 7.62 mm Model 59 general purpose machine gun fires the CIS 7.62 × 54R cartridge. As a squad automatic weapon with a light barrel and bipod or as a light machine gun with heavy barrel and bipod, it is known as the Model 59L.

As a medium machine gun with a heavy barrel and a light tripod it is called the Model 59. This tripod also enables the gun to fire in the air defence role. As a tank coaxial machine gun, fitted with a solenoid, it is referred to as the Model 59T.

The gun was also manufactured (for export) to use the NATO 7.62 × 51 mm cartridge and designated the Model 59N. The different cartridge contour required a different chamber and bolt face. As far as can be determined there were no sales of this model.

Heavy-barrelled version on tripod with 250-round container

Description
The 7.62 mm Model 59 general purpose machine gun uses the open pocket metal non-disintegrating Czech belt only. The open pocket allows the CIS rimmed cartridge to be pushed straight through the belt. The continuous metal belt can be reloaded as required; a belt filler which can be clamped on an ammunition box is available.

In the light machine gun role a 50-round metal box can be hung from the right-hand side of the gun. This is used for firing on the move during the assault. In the medium machine gun role a box holding 250 rounds in five belts is available.

The rearsight for the early model light machine gun is a horseshoe-shaped bar with odd ranges on the left-hand arm and even ranges on the right. The V-notch is adjusted up and down by means of knobs on each side. The sight reads from 1 to 20 in hundreds of metres. Later models, and particularly the Model 59N, have a light folding frame rearsight of conventional design with spring thumb catches for setting the range.

The foresight is a cylinder mounted eccentrically on a screw-threaded base and has a hooded protector. All zeroing is carried out on the foresight.

The adjustable bipod clamps into its seating below the foresight. It will pivot back along the barrel by closing the legs when they are hanging vertically below the gun and laying them under the barrel.

Belts are loaded by pressing the cartridges into the open links. Each link has a tab at its rear which fits up against the base of the cartridge - not into the cannelure if rimless NATO ammunition is being used.

The 50-round box has two grooves into which fit two lugs at the end of the feed tray on the right of the gun. When the box is fitted into place the cover will spring open. The loop fitted at the leading end of the belt is pulled out. The cover catch is pressed forward and the cover lifted. The belt exit cover is pressed back causing it to fly open. The belt is placed in position with the open side of the link downwards. It will be retained in the gun by the stop pawl when the cover is closed.

To cock the gun the trigger group latch, the cone-shaped projection mounted on the left side of the receiver above the pistol grip, is pulled down. Then, whilst squeezing the trigger, the pistol grip is pushed fully forward, releasing the trigger and pulling the pistol grip back, thus cocking the bolt. The pistol grip automatically reconnects with the receiver.

The safety catch also lies on the left of the receiver behind the trigger group latch. When pressed up it locks the sear. When the safety catch is moved down and the trigger is pressed the sear is depressed and the piston goes forward carrying the breech block. The feed ribs on top of the block force a cartridge through its link and into the chamber. The extractor snaps over the rim of the cartridge and the rim of the round hits the breech face and the breech block can advance no further. The piston continues on and the two ramps force the locking piece to rotate upwards and into two recesses in the receiver side walls. They ride under the locking piece and hold the lock in position. When locking is complete there is a short delay providing mechanical safety before firing and then the piston post hits the rear of the breech block, driving the firing pin into the primer.

The gun fires and some propellant gases are diverted into the gas port so the piston is driven to the rear. The piston post moves away from the breech block and the spring-loaded firing pin is retracted. The central unlocking cam has an inclined plane which pulls the lock down out of its recesses in the receiver and the piston carries the bolt back with it. The empty case is held to the bolt face and ejected downwards through the cut-away section in the piston. The return spring is compressed by the piston and rearward motion ceases when the back of the bolt hits the soft buffer. If the trigger remains pressed the cycle will be repeated.

In the feed system the flat sides of the piston are raised to form cam paths on which runs a roller attached to a feed arm. The cam forces the roller outwards. The roller in turn pushes out the bottom end of a feed arm, mounted vertically on the receiver, which is pivoted at its centre point. As the bottom moves out, the top of the lever must go in and the feed pawl attached to it forces the belt in towards the feeding position on the centre line of the gun, until the leading cartridge comes up against the cartridge stop of the feed tray.

When the piston goes forward another cam path controls the feed lever and the top is pivoted out. The feed pawl, which is spring loaded, slips over the next round in the belt. The belt cannot slip back because it is held up to the gun by a stop pawl on each side of the feed pawl. There is also another pair of stop pawls on the feed tray.

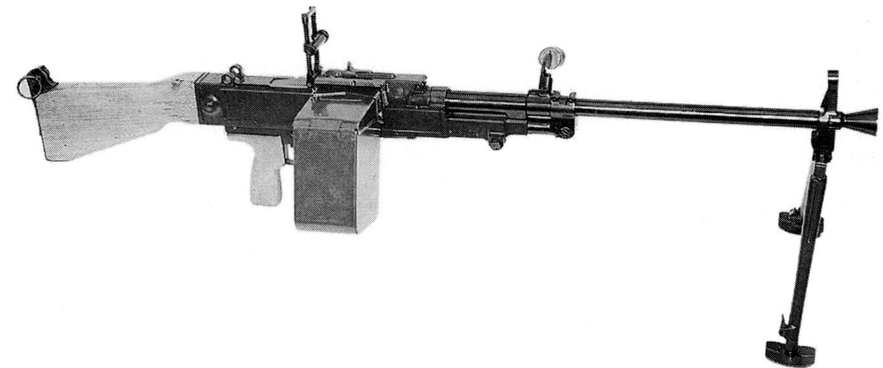

7.62 mm Model 59N general purpose machine gun (T J Gander)

In the trigger mechanism the trigger is L-shaped. When pressed the top arm rises and lifts the front of the sear, causing the rear end to fall and allowing the piston to go forward.

The safety catch, on the left of the receiver, is pushed up to engage and in pivoting it blocks the sear which cannot fall; at the same time a flange at its upper edge slides in a slot in the receiver and locks the trigger group. This prevents the gun from being cocked.

The gas regulator on all the Model 59 guns, except that chambered for NATO ammunition, has two positions. The normal position is marked '1' and this is changed to '2' using the combination tool.

The NATO guns have a four-position regulator. This is of unusual design in that movement is controlled by the carrying handle. To unlock the regulator the carrying handle lock, which lies parallel to the barrel at the foot of the handle is pressed; the carrying handle twisted and then rotated until the pointed lug at the front of the handle can be lined up with the notch at the top rear of the gas cylinder. Position '1' is used when the gun is functioning normally. The pointer on the right of the regulator is rotated to '1' and the regulator pressed fully in to the left. To get to position '2' the regulator is pressed to the right and rotated until the pointer is aligned with '2'. To get to '3' the regulator is rotated backwards still being held to the right and the pointer is at '3'. The position '4' is reached by pressing the regulator to the left and rotating it forward. To lock the regulator in the selected position, the carrying handle lock is pressed, then the carrying handle is twisted and rotated. It should be noted that position '4' produces a high rate of fire intended for anti-aircraft use.

A telescopic sight is available and can be used with the Model 59 mounted either on its bipod or a tripod. The telescope is attached to the receiver by a clamp and can be adjusted for windage and elevation. The graticule can be illuminated at night and the telescope has a dovetail to accept the lamp housing. An active infra-red source and receiver can also be used on the gun.

The bipod is standard with the light guns. To mount the gun on the tripod, the rearsight is lifted and the gun passed under the traversing arc and the mounting lugs on top of the rear of the receiver are then pinned to corresponding lugs under the main frame of the mounting. There is provision for free traverse and traversing stops are also provided. The tripod height can be adjusted. If required the tripod can be used for air defence by using extension pieces.

Data

Cartridge: 7.62 × 54R or 7.62 × 51 mm NATO
Operation: gas, automatic
Locking: swinging lock
Weight: empty with bipod and light barrel, 8.67 kg; with tripod and heavy barrel, 19.24 kg
Length: with heavy barrel, 1.215 m; with light barrel, 1.116 m
Barrel: heavy, 693 mm; light, 593 mm
Rifling: 4 grooves, rh, 1 turn in 240 mm
Sights: fore, pillar, adjustable for zeroing both line and elevation; rear, V-notch, adjustable from 100 to 2,000 m by 100 m increments
Sight radius: heavy barrel, 744 mm
Muzzle velocity: (light bullet) heavy barrel, 830 m/s; light barrel, 810 m/s; (heavy bullet) heavy barrel, 790 m/s; light barrel, 760 m/s
Rate of fire: cyclic, 700-800 rds/min
Max range: 4,800 m; effective, tripod, 1,500 m; bipod, 1,000 m

Manufacturer

Zbrojovka Brno, Narodni Podnik.

Status

No longer in production.

Service

Czech and Slovak armed forces.

UPDATED

FINLAND

7.62 mm M62 Valmet machine gun

Description

The 7.62 mm M62 Valmet machine gun is a gas-operated light machine gun which, although it bears little superficial resemblance, is based on the Czech ZB26 series and has a tilting block which locks into the roof of the receiver of the gun. It came into service with the Finnish Forces in 1966. It is belt fed and uses the Soviet 7.62 × 39 mm cartridge.

The M62 fires full-automatic only at a high rate of fire and is unusual in modern machine guns in feeding from the right. Its empty weight of 8.3 kg suggests that, with the cartridge it uses and a quick-change barrel, it should be capable of a large volume of fire. It has been supplied to the armed forces of Qatar.

Data

Cartridge: 7.62 × 39 mm
Operation: gas, automatic
Locking: tilting block
Feed: 100-round continuous link belt

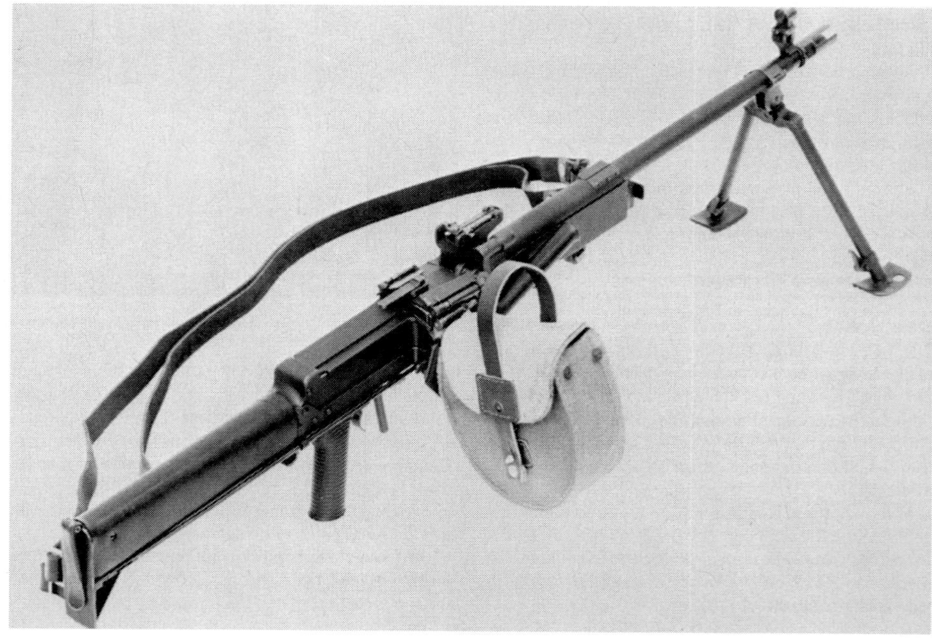

7.62 mm M62 Valmet machine gun

Weight: empty 8.3 kg; with full belt, 10.6 kg
Length: 1.085 m
Barrel: 470 mm
Rifling: 4 grooves, rh
Sights: fore, pillar; rear, aperture, setting 100 to 600 m in 100 m steps
Muzzle velocity: 730 m/s
Rate of fire: cyclic, 1,000-1,100 rds/min; practical, 300 rds/min
Range: effective, 350-450 m

Manufacturer
Sako Ltd, PO Box 149, FIN-11101 Riihimäki.

Status
No longer in production.

Service
Finnish and Qatar armed forces.

VERIFIED

FRANCE

7.62 mm N AAT mle F1 general purpose machine gun

Development
The Arme Automatique Transformable Model 52 was designed for ease of production, using stampings wherever possible. The receiver of the gun is made of fabricated semi-cylindrical shells welded together. As originally designed it fired the French 7.5 × 54 mm cartridge of 1929, and in this form it was known as the AAT 52. The majority of present guns have been converted to fire the NATO 7.62 mm round and the 7.5 mm version has been phased out. For both versions there is a choice of light and heavy barrels. The following description refers to the 7.62 × 51 mm version.

7.62 mm N AAT mle F1 general purpose machine gun

Description
The principle of operation is delayed blowback, but, since rifle calibre ammunition is used, special arrangements have to be made to ensure safe and smooth extraction of the spent case. There is a number of longitudinal grooves in the neck of the chamber running out half way towards the mouth. Gas enters these grooves and the case floats with equal pressure on each side of the brass wall and can move comparatively freely.

To prevent premature breech opening while pressure is still high a two-part block is employed, with the front part carrying a lever. The short end of the lever rests in a recess in the side of the receiver and the long arm bears against the massive rear part of the bolt. Gas pressure exerts a force on the bolt face and the lever is forced to rotate, thus accelerating the rear part of the block but restraining the front part until the lever is clear of the recess. The entire bolt is then blown back by the residual pressure. The empty cartridge case is held by the extractor to the bolt face until it strikes the double ejectors on the bottom rear of the feed tray and is deflected down through the bottom of the gun.

Cartridge headspace is critical and when wear occurs the bearing surface in the receiver can be quickly replaced. In spite of this ejected cases are deformed where they expand into the bullet guide. Any relaxation of tolerances in case manufacture or an increase in headspace will result in a blowout.

The 7.62 mm N AAT mle F1 fires from an open bolt. The firing pin floats freely in the front part of the bolt and is forced forward only when the rear part of the bolt closes up to the front part. This can occur only when the lever, which holds the two parts separated, can rotate into the recess in the receiver at the end of forward travel. Thus the round cannot be fired until the delay device is properly positioned and the bolt closed.

The 7.62 mm N AAT mle F1 gun uses a French disintegrating link belt which is based on the US M13 pattern but is somewhat more flexible and has a smaller extension for a given load. It can use either this type of belt or the M13 NATO belt. A cam groove on top of the bolt accommodates a lever which is swung from side to side to operate the feed pawl. Unlike the Belgian MAG or the British L7A2 GPMG, the mle F1 must be carried with the gun cocked when a loaded belt is in place.

The barrel change arrangements are awkward if the gun is being fired off its bipod. A barrel-release catch must be pressed in and then the wooden barrel-carrying handle rotated clockwise and pushed forward. Since the bipod is permanently attached to the barrel, once the barrel is drawn off there is no forward support and the gunner is left holding a hot gun.

The foresight is a hinged block with a slot in the top. For firing by day the slot is used and at night the entire foresight block is aligned with the notch of the rearsight, the sights being luminescent.

In addition to the dismounted infantry versions described above, guns can also be supplied in versions without a butt-stock for vehicle mounting; arrangements can be made for them to be fired electrically.

7.62 mm N AAT mle F1 general purpose machine gun on vehicle mounting (T J Gander)

	Light Barrel	Heavy Barrel
Weight of gun		
without bipod, empty	9.15 kg	10.55 kg
with bipod, empty	9.97 kg	11.37 kg
Weight of barrel	2.85 kg	4.25 kg
Weight of bipod	0.82 kg	0.82 kg
Weight of monopod	0.685 kg	0.685 kg
Weight of modified US M2 tripod	–	10.6 kg
Length of gun		
butt extended	1.145 m	1.245 m
butt retracted	980 mm	1.08 m
Length of barrel		
without flash hider	500 mm	600 mm
	AAT 52	**7.62 mm N AAT mle F1**
Muzzle velocity	840 m/s	830 m/s
Rate of fire cyclic	700 rds/min	900 rds/min
Practical range	800 m	1,200 m
Number of grooves	4	4

Data
COMMON TO BOTH VERSIONS UNLESS OTHERWISE STATED
Cartridge: 7.62 mm NATO
Operation: delayed blowback, automatic
Feed: disintegrating link belt (see text)
Sights: fore, slit blade; rear, leaf graduated 200-2,000 m

Manufacturer
Originally developed and produced by Manufacture Nationale d'Armes de Chatellerault. Production subsequently transferred to Manufacture Nationale d'Armes de Tulle. Enquiries to Giat, 13 route de la Minière, BP 1342, F-78034 Versailles-Satory Cedex.

Status
7.62 mm N AAT mle F1 in current production.

Service
French armed forces and many other countries.

VERIFIED

GERMANY

Heckler and Koch 7.62 mm HK 21 general purpose machine gun

Heckler and Koch 7.62 mm HK 21 general purpose machine gun with belt feed system (T J Gander)

Description

The HK 21 is no longer produced and has been replaced by the HK 21A1 and, more recently, the HK 21E (see following entries), but it was in production for some years and there are still numbers in service in some parts of the world. A full description was given in *Jane's Infantry Weapons 1978*, page 267.

The HK 21 is a belt-fed general purpose machine gun using the 7.62 × 51 mm NATO cartridge. This will normally be carried in the disintegrating link belt and the gun will function using either the German DM 60 belt, the US M13 belt or the French belt. However, the continuous link belt DM 1, can also be used if required. Furthermore, by changing the barrel, the belt feed plate and the bolt, the gun can be converted to firing the 5.56 × 45 mm or the 7.62 × 39 mm cartridge. The utility of the weapon is further increased by inserting a magazine adaptor in place of the feed mechanism. This takes any of the Heckler and Koch 7.62 mm magazines intended for the G3 rifle or the HK 11 light machine gun.

The gun has a practical and effective quick-change barrel that enables it to produce sustained fire when required.

The method of operation of the HK 21 is the same as that used in the G3 rifle (qv) employing a two-part breech block and delay rollers. The delayed blowback system, with a fluted chamber for easy cartridge movement, operates from the closed bolt position with a round in the chamber when the gun is ready to fire.

The feed system of the belt-fed gun functions as follows:

If the disintegrating belt has a feed tag this is pushed through the feed tray from left to right and pulled until the first round reaches the cartridge stop. Since the bolt will pass over the belt the open side of the belt links must be placed uppermost before the tag is inserted.

Where the belt has no tag the gun is first cocked and the cocking handle held to the rear in the recess of the cocking slideway. The feed mechanism catch is depressed and the mechanism moved over to the left. The first round is placed in the feed sprocket which is rotated to the right until it locks. The first cartridge is then in place and the feed unit can be pushed in. Releasing the cocking handle will feed the first round from the belt into the chamber.

If required the belt feed unit can be withdrawn and replaced with a magazine unit which fits into the receiver and is held by two locking pins. The unit allows the use of the 20-round box magazine or the 80-round double-drum plastic magazine.

The gun may be fired at single shot or full automatic. The selector lever is above the pistol grip and the trigger mechanism, disconnector and automatic sear are the same as those used in the G3 rifle.

Data

Cartridge: 7.62 × 51 mm, 5.56 × 45 mm or 7.62 × 39 mm
Operation: delayed blowback, automatic
Feed: disintegrating link belt; option of 20-round magazine or 80-round drum
Weight: with bipod, 7.92 kg
Length: 1.021 m; without butt for vehicle mounting, 820 mm
Barrel: 450 mm
Rifling: 4 grooves, rh, 1 turn in 305 mm
Sights: fore, hooded post; rear, drum with click adjustments from 200 to 1,200 m in 100 m steps, each adjustable for windage and elevation; optical sight available
Sight radius: 589 mm
Muzzle velocity: 800 m/s
Rate of fire: cyclic, 900 rds/min
Range: effective, 1,200 m

Manufacturer

Heckler and Koch GmbH, D-78722 Oberndorf-Neckar.

Status

No longer produced by Heckler and Koch, but manufactured in 7.62 × 51 mm calibre under licence in Portugal by Industrias Nacionais de Defesa EP.

Service

Portugal and some African and South-east Asian countries.

Licence production

PORTUGAL
Manufacturer: INDEP SA
Type: 7.62 mm HK 21
Remarks: See entry under *Portugal* for details.

UPDATED

Heckler and Koch 7.62 mm HK 21A1 general purpose machine gun

Heckler and Koch 7.62 mm HK 21A1 general purpose machine gun with loaded belt

Development

The HK 21A1 is a development of the HK 21 (see previous entry), and the principal change is the omission of the option of using a magazine feed; the HK 21A1 is belt-fed only. The HK 21A1 uses the under-fed mechanism of its predecessor, but the mechanism can be hinged down to allow the belt to be inserted which makes the operation of loading much quicker and easier.

The HK 21A1 is intended as a squad light gun, but it is also adaptable as a general purpose machine gun and there is a range of Heckler and Koch accessories which convert it to this role.

Description

The HK 21A1 is a closed bolt automatic weapon operated by delayed blowback utilising the same system of roller-locking of the bolt as the G3 rifle (qv). The feed system is a left-hand belt feed mechanism which will accommodate a jointed or disintegrating link belt, and the gun will accept either 7.62 × 51 mm or 5.56 × 45 mm ammunition. The latter requires that the barrel, bolt and feed mechanism be changed. As with other Heckler and Koch guns, when the weapon is cocked the bolt is closed and there is a round in the chamber.

The belt feed unit is detachable from the gun for cleaning. It is in two parts, the cartridge guide and the belt feed housing. The belt feed housing is hinged to the receiver by a socket pin and is held in place by a spring catch. Actuating the catch lever releases the feed unit and allows it to swing down. At the same time two cams cut in the control bolt move the follower and cartridge feed lever downwards below the level of the feed plate, so making it easier to insert a new belt, or to remove the existing one if necessary. When the feed unit is swung back into place the follower and feed lever are forced back into position.

Cartridges are fed into position for the bolt to engage by the combined actions of the feed sprocket, feed lever and follower. These are all driven by a camway cut in the bolt carrier which engages with a slide-bolt in the feed housing. As the bolt travels backward and forward it moves the slide-bolt from side to side and this motion rotates the sprocket. The belt is housed in a box attached to the gun by a catch, and it hangs under the centreline of the gun.

The usual mounting is the bipod, which has two positions on the gun; forward at the end of the receiver where it gives the maximum stability, or directly in front of the feed unit where it is near to the point of balance. The alternative mounting is the 1102 tripod.

A comprehensive range of accessories is offered for the HK 21A1, enabling it to be adapted for several different purposes. The range includes a belt box, a carrying sling, anti-aircraft sights, a telescopic sight, a passive night sight, and a tripod mount.

Data

Cartridge: 7.62 × 51 mm NATO
Operation: delayed blowback, selective fire
Feed: metal link belt
Weight: empty with bipod, 8.3 kg

Length: 1.03 m
Barrel: 450 mm
Sights: fore, hooded post; rear, drum with click adjustments from 200 to 1,200 m in 100 m steps, each adjustable for windage and elevation; optical sight available
Sight radius: 590 mm
Muzzle velocity: 800 m/s
Rate of fire: cyclic, 900 rds/min

Manufacturer
Heckler and Koch GmbH, D-78722 Oberndorf-Neckar.

Status
No longer in production by Heckler and Koch. In production in Greece.

Licence production
GREECE
Manufacturer: Hellenic Arms Industry SA
Type: 7.62 mm EHK21A1
Remarks: In production for local use and offered for export.

UPDATED

Heckler and Koch 7.62 mm HK 21A1 general purpose machine gun on tripod mount

Heckler and Koch 7.62 mm HK 21E and 5.56 mm HK 23E general purpose machine gun

Description
The 7.62 mm HK 21E and 5.56 mm HK 23E machine guns are the latest developments of the HK 21A1 design, based on the results of stringent testing of that weapon. Technical modifications have been made, resulting, it is claimed, in higher efficiency and greater robustness and durability.

The HK 21E and HK 23E have the following significant modifications in common: a 94 mm extension of the receiver, giving a longer sight radius; reduced recoil, contributing to greater accuracy in all modes of fire; standard fitting of a burst-control trigger for three-round bursts and provision of an attachable winter trigger; improved quick-change barrel grip; assault grip as a standard fitting; rear drum sight - 100 to 1,200 m (HK 21E); 100 to 1,000 m (HK 23E) - with windage and elevation adjustment and side-wind correction. Other features include: a special device for quiet closing of the bolt; carrying handle; cleaning kit housed in the grip; and replacement barrel for automatic firing of blank cartridges.

On the HK 21E the barrel is lengthened to 560 mm. The belt transport system in the feed is modified so that the belt is transported in two operations, ensuring that the feed unit and the belt itself are subjected to lower stress. As the bolt travels forward, and following the ejection of a cartridge from the belt, one operation is effected for the following cartridge. When the bolt has opened and has moved to the rear, a second operation brings the cartridge fully into the feed position.

Both weapons are provided with a bipod with three elevation settings and the ability traverse to 30° to either side. Both can also be mounted on the 1102 tripod and other Heckler and Koch mounts. The range of accessories is the same as for the HK 21A1.

Manufacturer
Heckler and Koch GmbH, D-78722 Oberndorf-Neckar.

Status
In production.

Service
Mexican Army and trials elsewhere.

UPDATED

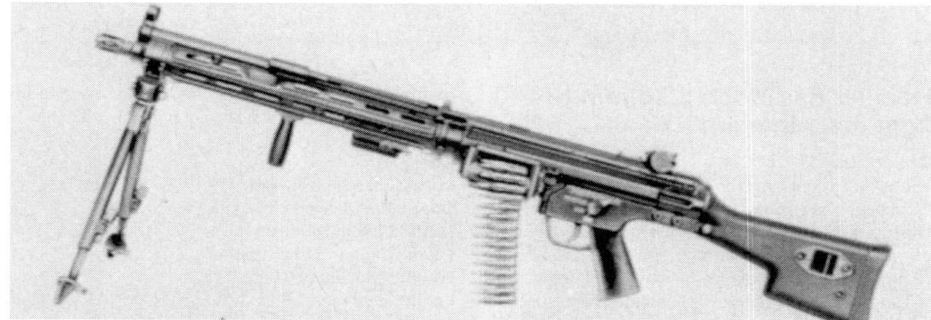

Heckler and Koch 5.56 mm HK 23E general purpose machine gun with belt feed system

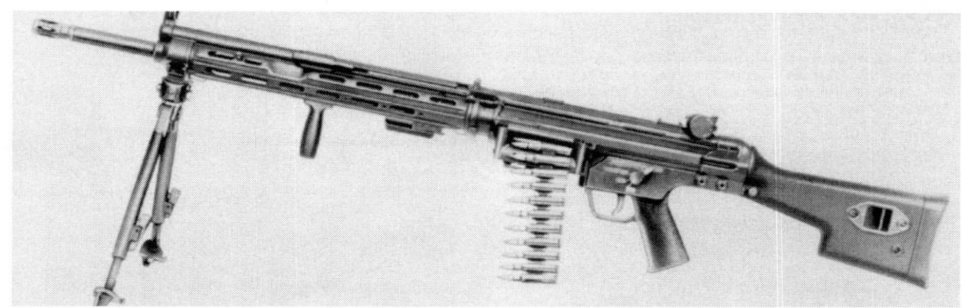

Heckler and Koch 7.62 mm HK 21E general purpose machine gun with belt feed system

Data	HK 21E	HK 23E
Cartridge	7.62 × 51 mm	5.56 × 45 mm (M193 or SS109)
Operation	delayed blowback, automatic	delayed blowback, automatic
Feed	metal link belt	metal link belt
Modes of fire	automatic, 3-round burst, single shot	automatic, 3-round burst, single shot
WEIGHTS		
Gun (unloaded with bipod)	9.3 kg	8.75 kg
Barrel	2.2 kg	1.6 kg
Bipod	0.55 kg	0.55 kg
LENGTHS		
Length overall	1.14 m	1.03 m
Barrel	560 mm	450 mm
Rifling	1 turn in 305 mm	1 turn in 178 mm
Sight radius	685 mm	685 mm
FIRING CHARACTERISTICS		
Rate of fire cyclic,	ca 800 rds/min	ca 750 rds/min
Muzzle velocity	ca 840 m/s	ca 950 m/s

Heckler and Koch 7.62 mm HK 11A1 light machine gun

Description
The HK 11A1 is a magazine-fed version of the HK 21A1 and employs all the same internal mechanisms. For the general system of operation reference should be made to the HK 21A1.

The major difference lies in the feed. Instead of the belt feed unit a magazine unit is pinned to the receiver, using the same method of attachment, and 20-round box magazines from the G3 rifle are clipped into it using a conventional small catch.

Data
Cartridge: 7.62 × 51 mm
Operation: delayed blowback, selective fire
Feed: 20-round box magazine
Weight: 7.7 kg; barrel, 1.7 kg; bipod, 600 g
Length: 1.03 mm
Barrel: 450 mm
Sights: fore, blade; rear, aperture; setting, 200-1,200 m, click adjustment per 100 m
Muzzle velocity: 800 m/s
Rate of fire: cyclic, 800 rds/min

Manufacturer
Heckler and Koch GmbH, D-78722 Oberndorf-Neckar.

Status
No longer in production.

Service
Africa, Greece, South-east Asia.

Licence production
GREECE
Manufacturer: Hellenic Arms Industry SA
Type: 7.62 mm EHK11A1
Remarks: Production complete. Was manufactured for local use and offered for export.

UPDATED

Heckler and Koch 7.62 mm HK 11A1 light machine gun

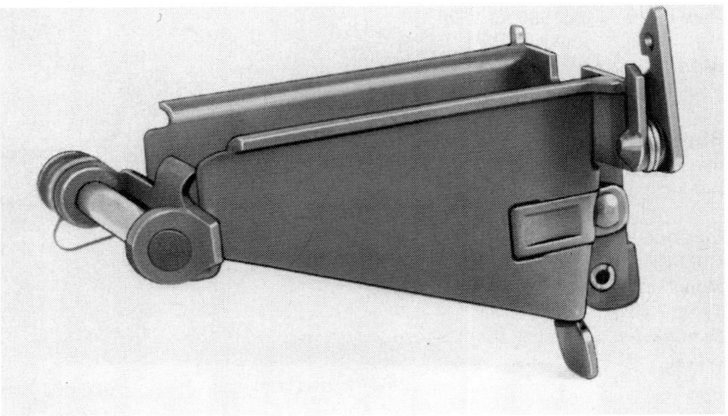

Magazine feed unit of Heckler and Koch 7.62 mm HK 11A1 light machine gun

Heckler and Koch 5.56 mm HK 13 light machine gun

Description
The HK 13, a light machine gun using the 5.56 × 45 mm cartridge, is essentially the same weapon as the HK 33 rifle (qv) and operates in exactly the same way. Its physical dimensions are similar but it has the heavier quick-change barrel necessary for its light machine gun role.

The HK 13 will take the 25-round magazines used in the HK 33 rifle.

A bipod may be fitted either at the front of the barrel casing or at a centre point just in front of the magazine.

The HK 13 light machine gun will produce selective fire. The change lever is above the pistol grip on the left of the receiver. The trigger and firing mechanism functions in the same way as that on the HK 33 rifle.

Data
Cartridge: 5.56 × 45 mm
Operation: delayed blowback, selective fire
Method of delay: rollers
Feed: 25-round box magazine
Weight: with bipod, 6 kg
Length: 980 mm
Barrel: 450 mm
Rifling: 4 grooves, rh, 1 turn in 305 mm
Sights: fore, blade; rear, 100 m V rearsight. Apertures for 200, 300 or 400 m. Adjustable for windage and elevation. Provision for fitting telescope sight
Sight radius: 541 mm
Muzzle velocity: 950 m/s
Rate of fire: cyclic, 750 rds/min
Range: effective, 400 m

Manufacturer
Heckler and Koch GmbH, D-78722 Oberndorf-Neckar.

Status
No longer in production.

Service
Some South-east Asian forces.

UPDATED

Heckler and Koch 5.56 mm HK 13 light machine gun with telescopic sight

Heckler and Koch 5.56 mm HK 13 light machine gun field-stripped (25-round magazine)

Heckler and Koch 7.62 mm HK 11E and 5.56 mm HK 13E light machine guns

The HK 11E and HK 13E light machine guns are developments of the HK 11A1 and HK 13 light machine guns described above and incorporate the technical modifications described in the entry for the Heckler and Koch HK 21E and HK 23E general purpose machine guns, including a burst-control trigger, rear drum sight and assault grip. Both are primarily magazine-fed weapons, but they can be converted to belt feed by exchanging the bolt assembly with the magazine adaptor for bolt assembly with belt feed unit.

As with the HK 11A1, the HK 11E will accept the 20-round magazine of the G3 rifle and the Heckler and Koch 50-round drum magazine. The HK 13E will accept the 20-round and the 30-round magazine of the G41 rifle, standardised according to STANAG 4179.

7.62 mm Heckler and Koch HK 13E light machine gun with 20-round box magazine

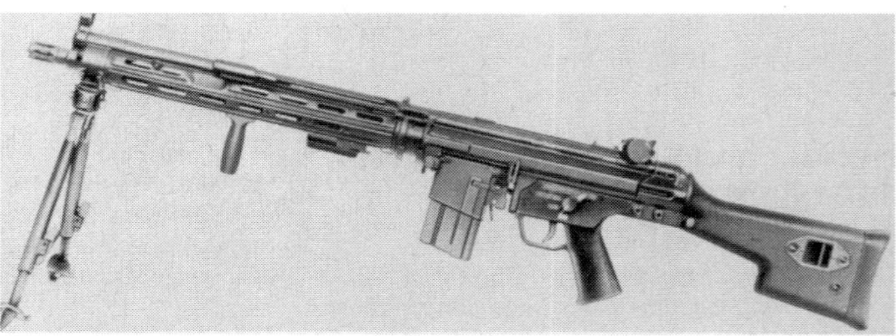

5.56 mm Heckler and Koch HK 11E light machine gun with 20-round box magazine

Data

	HK 11E	HK 13E
Cartridge	7.62 × 51 mm	5.56 × 45 mm
Operation	delayed blowback, automatic	delayed blowback, automatic
Feed	20-round box magazine and 50-round drum magazine	30-round box magazine and 20-round box magazine (M16 type)
Modes of fire	automatic, 3-round burst, single shot	automatic, 3-round burst, single shot
WEIGHTS		
Gun (unloaded with bipod)	8.15 kg	8 kg
Barrel	1.7 kg	1.6 kg
Bipod	0.55 kg	0.55 kg
LENGTHS		
Length (overall)	1.03 m	1.03 m
Barrel	450 mm	450 mm
Rifling	1 turn in 305 mm	1 turn in 178 mm (SS109)
Sight radius	685 mm	685 mm
FIRING CHARACTERISTICS		
Rate of fire cyclic,	ca 800 rds/min	ca 750 rds/min
Muzzle velocity	ca 800 m/s	ca 950 m/s

Manufacturer

Heckler and Koch GmbH, D-78722 Oberndorf-Neckar.

Status

In production.

Service

Military trials.

UPDATED

7.92 mm MG42 and 7.62 mm MG1, MG2 and MG3 machine guns

Development

The German Army entered the Second World War with the MG34 as its principal ground machine gun, both in the infantry squad and on armoured vehicles. It was a complex and expensive gun to produce, with forgings machined to close tolerances, so it became evident that a less costly successor would be needed. Prototypes of a new gun, which subsequently became the highly successful MG42, were made in 1938 with the maximum use of stampings and the very minimum of forgings.

When the West German forces entered NATO they decided to modify the MG42 from 7.92 mm calibre to 7.62 × 51 mm and adopt it as their standard general purpose machine gun. It was first manufactured by Rheinmetall in 1959 who called it the MG42/59; the Bundeswehr called it the MG1. All MG1 versions (MG1, MG1A1, MG1A2 and MG1A3) were chambered only for the 7.62 × 51 mm cartridge in a German 50-round continuous belt known as the DM 1. The MG1A3 had some small changes intended to speed production including the rounded-muzzle booster.

In parallel with this development process, some of the original MG42 weapons were converted from 7.92 to 7.62 mm calibre and were redesignated MG2. The current weapon is a further development known as the MG3 which came into service in 1968; it has the external shape of the MG1A3 and can be fed from the German DM 1 continuous belt, or the German DM 6 or US M13 disintegrating link belts. It has an AA sight and a belt retaining pawl to hold the belt up to the gun when the top cover plate is lifted.

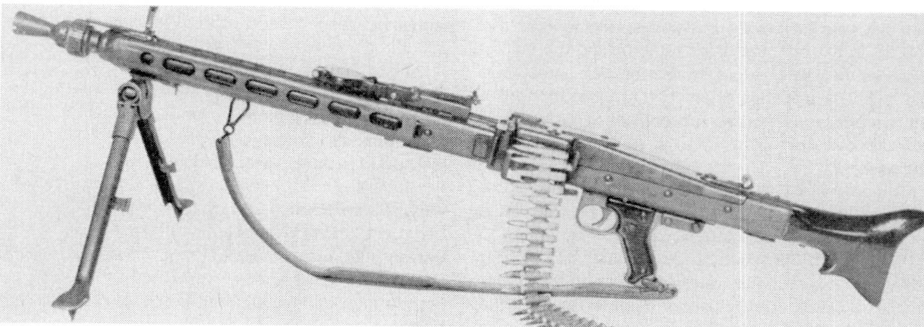

7.62 mm MG3 machine gun

Description

To operate the MG3 the T-shaped cocking handle is grasped and pulled straight back and the bolt cocked. When the bolt is to the rear, and only then, the safety catch above the pistol grip can be pushed through to the left to lock the sear.

Ammunition must be placed in the gun with the open side of the links downward so that the belt is above the cartridges. If a feed tab is fitted this can be pushed straight through the gun. If not, the top cover plate must be lifted and the belt laid across the feed tray, cartridges downwards. When the rear of the cover plate is pressed down, the gun is ready for action. Pushing the safety catch from left to right with the side of the right thumb places the gun in the ready-to-fire position. When the trigger is pressed the bolt goes forward, forcing a round out of the belt into the chamber.

The bolt has a slot in each side, in which are mounted two rollers, these having a shaft which protrudes above and below the roller portion. The upper and lower portions of the shaft ride in cam tracks in the bolt, while the rollers move freely in a wider part of the slot. The ends of the roller shafts protrude above and below the bolt and, as the bolt head enters the barrel extension, these shafts meet cam tracks in the barrel extension which force the rollers outwards. This movement is accelerated by the shaped nose of the firing pin carrier portion of the bolt moving forward under the pressure of the return spring, so that the two rollers are driven sideways

7.62mm MG3 machine guns (T J Gander)

Barrel change on 7.62 mm MG42/59

and, as the cartridge is seated in the chamber, the roller shafts lock into recesses in the barrel extension. Unless the rollers are in the locked position the firing pin cannot pass between them and strike the cartridge cap.

On firing the barrel and barrel extension, together with the locked bolt, all move rearwards because of recoil forces. After about 8 mm of movement, the rollers come into contact with cam surfaces in the walls of the receiver; these force the rollers inwards, so driving the roller shafts out of their locking recesses in the barrel extension. The inward movement of the rollers, acting on the shaped surface of the firing pin carrier, cause the latter to be accelerated to the rear.

As the bolt is freed from the barrel extension, the barrel is halted and then driven back into battery by a spring. The bolt continues to move back, extracting and ejecting the spent case, until it strikes the buffer spring which then returns the bolt at high speed.

Belt movement is produced by a stud on top of the bolt riding in a curved feed channel in a feed arm. This feed arm rests under the top cover plate and is pivoted at its rear end. As the bolt reciprocates, the front of the feed arm moves across the receiver and operates a lever attached to the belt feed slide. This slide has two sets of spring-loaded pawls mounted one on each side of the centrally positioned slide pivot. Thus when one set of pawls is moving out and springing over the rounds in the belt, the other set is pulling the belt in. Each set of pawls in turn moves the belt across one half pitch. This sharing of the load reduces the forces on the belt and the feed mechanism. It produces a smooth belt flow rather than a series of jerky belt movements.

The original MG42 fired at about 1,200 rds/min. The standard bolt used in the MG1A1, MG1A3 and MG3 weighs 550 g, producing a rate of fire of 1,100 ±150 rds/min. The MG1A2 uses a heavy bolt weighing 950 g and this produces rates of fire of about 900 rds/min. The German Army uses the lighter bolt but the Italian MG42/59 uses the heavier bolt.

The foresight is mounted on the front end of the barrel casing and hinges flat. The rearsight is a U-notch that is mounted on a slide moving on a ramp. Graduation is from 200 to 1,200 m on the MG3. Since the foresight is not on the barrel, zeroing must be carried out on the rearsight.

The barrel must be changed at frequent intervals during sustained firing. The gun is cocked, and the barrel catch on the right of the barrel casing is swung forward. The breech end of the hot barrel swings out and can be removed by elevating the gun. A fresh barrel is pushed through the barrel catch and the muzzle bearing. When the catch is rotated back the barrel is locked.

A buffered tripod to allow sustained fire is available. A dial sight allowing engagement of indirect targets and recording of previously registered targets can be fitted to the tripod. A blank-firing attachment can be fitted in lieu of the normal recoil booster at the muzzle.

One specialised accessory is a belt drum developed by Heckler and Koch for the German Army. It holds the normally loose belt of a length of approximately 800 mm. The belt drum holds 100 rounds and is latched on the left of the weapon feed unit. The rear side of the drum is provided with a transparent cover serving as a visual indicator for the amount of ammunition available.

For transport a number of drums may be linked together, forming convenient stacks. The drum is made of highly durable synthetic material and consists of only two parts. It may be safely employed in temperature ranges between −35 and +63°C. When empty it weighs 200 g.

Data

Cartridge: 7.62 × 51 mm
Operation: short recoil, automatic
Method of locking: roller locking
Feed: belt
Weight: with bipod, 11.05 kg
Length: 1.225 m; without butt, 1.097 m
Barrel: with extension, 565 mm; without extension, 531 mm
Sights: fore, barleycorn; rear, notch; also anti-aircraft sight
Sight radius: 430 mm
Muzzle velocity: 820 m/s
Rate of fire: cyclic, 700-1,300 rds/min
Range: (effective) bipod, 800 m; tripod, 2,200 m

Manufacturer

Rheinmetall Industrie GmbH, Pempelfurtstrasse 1, D-40880 Ratingen.

Status

Available for production.

Service

German armed forces. Also with the forces of Austria, Chile, Denmark, Greece, Iran, Italy, Norway, Pakistan, Portugal, Spain, Sudan and Turkey.

Licence production

GREECE
Manufacturer: Hellenic Arms Industry SA
Type: 7.62 mm MG3
Remarks: In production for local use and offered for export sales.
IRAN
Manufacturer: Defence Industries Organisation
Type: 7.62 mm MG3
Remarks: Produced in bipod- tripod- and pintle-mounted forms.
ITALY
Manufacturer: Beretta, Luigi Franchi and Whitehead Moto-Fides
Type: 7.62 mm MG42/59
Remarks: See separate entry under *Italy* for details.
PAKISTAN
Manufacturer: Pakistan Ordnance Factories
Type: 7.62 mm MG3
Remarks: In production for use by Pakistan armed forces and offered for export sales.
SPAIN
Manufacturer: SANTA BARBARA, Fabrica de Armas de Oviedo
Type: MG42/59, MG1A3 and MG3S
Remarks: Probably no longer in production.
TURKEY
Manufacturer: Makina ve Endüstrisi Kurumu (MKEK)
Type: 7.62 mm MG3
Remarks: In production. In service with Turkish armed forces. 4,220 sold to Norway. See separate entry under *Turkey*.

UPDATED

7.62 mm MG3 machine gun on anti-aircraft twin pedestal mount

German soldier carrying 7.62 mm MG3 machine gun fitted with blank adaptor (T J Gander)

GREECE

7.62 mm MG3, EHK21A1 and EHK11A1 machine guns

Description

Hellenic Arms Industry (EBO) SA manufactures the 7.62 mm MG3 general purpose machine gun as the principal machine gun for the Greek Army. In addition the company is also manufacturing the Heckler and Koch HK 21A1 (EHK21A1) machine gun and manufactured the HK 11A1 (EHK11A1) light machine gun; the latter was adopted by the Greek Army as its light support weapon.

The relevant technical specifications can be found by referring to the entries under *Germany* in this section.

Manufacturer

Hellenic Arms Industry (EBO) SA, 160 Kifissias Avenue, GR-11525 Athens.

Status

EHK21A1 in production.

Service

Greek Army.

UPDATED

7.62 mm EHK11A1 machine gun manufactured by Hellenic Arms Industry

7.62 mm MG3 machine gun manufactured by Hellenic Arms Industry

INDIA

5.56 mm INSAS light machine gun

Description

The 5.56 mm INSAS light machine gun is the light machine gun counterpart to the INSAS assault rifle. For full development, construction and other details of the INSAS family of weapons refer to the appropriate entry under *India* in *Rifles*. It is understood that the INSAS programme is running almost four years behind schedule.

The INSAS light machine gun differs from the rifle model primarily in having a longer and heavier barrel with revised rifling, and a bipod. The bipod is an adaptation of the same component fitted to Indian-produced Bren guns. The magazine capacity is enlarged to 30 rounds, although the 22-round magazine used with the INSAS rifle may also be used. Other changes are that there is no three-round burst selection feature on the light machine gun, and the iron sights are calibrated for ranges of 200 to 1,000 m in recognition of the higher ballistic performance potential of the longer chrome-plated barrel; the barrel cannot be rapid-changed. A grenade launcher adaptor is fitted to the muzzle but there is no provision for a bayonet.

As with the INSAS assault rifle, a model with a folding butt is available for use by parachutists or other special troops. For this model it is possible to install a shorter barrel to make the weapon more compact.

Mounting points are provided to allow the weapon to be placed on various vehicle or ground mountings.

5.56 mm INSAS light machine gun (T J Gander)

Data

Cartridge: 5.56 × 45 mm SS109-based Special
Operation: gas, selective fire
Locking: rotating bolt
Feed: 30-round plastic box magazine
Weight: loaded, 6.7 kg
Length: fixed butt, 1.05 m; butt folded, 800 mm; butt extended, 1.05 m
Barrel: standard, 535 mm; short, 510 mm
Rifling: 4 grooves, rh, 1 turn in 200 mm
Sights: fore, blade; rear, flip aperture, 200-1,000 m
Muzzle velocity: standard barrel, 954 m/s
Rate of fire: cyclic, 650 rds/min
Max range: effective, 1,000 m

Manufacturer

Rifle Factory, Ishapore.

Status

Not yet in full production.

Service

Destined for the Indian armed forces. Offered for export sales.

UPDATED

Indian machine guns

Description

Apart from the planned production of the 5.56 mm MNSAS light machine gun, two other types of machine gun are in production in India. The Indian Ordnance Factories at Kanpur continue to manufacture the 7.62 mm MAG general purpose machine gun under licence from FN; the model is the FN 60-20. For details of this machine gun refer to the appropriate entry under *Belgium* in this section. This weapon is available for export sales.

Also still in production at Kanpur and available for export sales is the Gun, Machine 7.62 mm 1B. This is a licence-manufactured version of the British L4A4, the 7.62 × 51 mm NATO version of the Bren Gun (qv). Data for this model is provided below. It is anticipated that production of this weapon will cease when the INSAS light machine gun enters service.

The Indian armed forces also have numbers of Browning 0.50 M2 HB heavy machine guns. At one time there were plans to manufacture these weapons in India but no information is forthcoming as to the current status of these intentions.

Data

GUN, MACHINE 7.62 mm 1B
Cartridge: 7.62 × 51 mm NATO
Operation: gas, selective fire
Locking: tilting block
Feed: 30-round box magazine
Weight: 9.185 kg
Length: 1.13 m
Barrel: 621.25 mm
Rifling: 6 grooves, rh, 1 turn in 305 mm
Sights: fore, adjustable blade; rear, flip aperture, 200-1,000 m
Muzzle velocity: ca 835 m/s
Rate of fire: cyclic, 500 rds/min
Max range: 1,830 m

Manufacturer

Small Arms Factory, Kanpur.

Status

In production (both types).

Service

Indian armed forces (both types). Offered for export sales.

VERIFIED

INTERNATIONAL

CTA International 12.7 mm rotary machine gun

Development

CTA International is a 50/50 joint concern formed by Royal Ordnance of the UK and Giat International of France to develop, promote and market the Cased Telescoped Ammunition (CTA) concept and associated weapons, initially with a view to a 45 mm CTA gun for light combat vehicles. They have also developed a 12.7 mm CTA rotary machine gun based on previous experimental work carried out in France between 1987 and 1994. This gun is intended to be a technology demonstrator involving the use of composite materials for the chambers, receiver and some other components. In the longer term it is proposed that this, or a similar gun, could have numerous infantry, vehicle and other applications because of its combination of relatively light weight and firepower.

Firing trials of the single prototype produced to date commenced during November 1995.

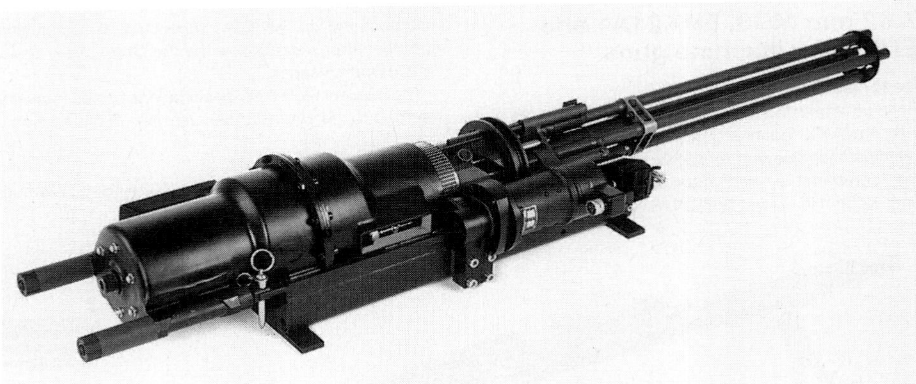

CTA International 12.7 mm rotary machine gun

1996

Description

The CTA International 12.7 mm rotary machine gun has four barrels and a cyclic rate of fire of 4,000 rds/min. Composite materials can be used for the chambers since the propellant is contained within a plastic cylinder (shorter than a conventional 12.7 mm round) inside which the projectile is completely enclosed. As the propellant ignites, the cylindrical plastic case acts as a thermal barrier to heat and thus metals are not required to contain the usual thermal stresses. The cylindrical cases need only to be aligned with the barrel as no ramming is required, greatly simplifying the feed and ammunition handling mechanisms.

Weight of the CTA International 12.7 mm rotary machine gun is 26 kg. No further details are yet available.

Manufacturer

CTA International, 7 route de Guerry, F-18023 Bourges Cedex.

Status

Technology demonstrator.

NEW ENTRY

IRAN

7.62 mm MG3 machine gun

Description

The German Rheinmetall 7.62 mm MG3 machine gun is manufactured under licence in Iran at the Mosalsalsasi weapons factory run by the Iranian Defence Industries Organisation, Armament Industries Group. This factory was originally established following an intergovernmental agreement between the former West German Government and the Shah of Iran and remained in production after the fall of the Shah. Heckler and Koch rifles are also made at this establishment.

MG3s produced in Iran lack some refinements, such as an anti-aircraft sight, but are otherwise identical to the German original so for details reference should be made to the appropriate entry under *Germany* in this section.

Iranian MG3s are produced in both bipod and tripod-mounted form. There is also a fixed deck pintle mounting for installation on vehicles or light vessels. On this mounting the gunner uses twin spade grips for aiming, with a mechanical link to the normal trigger mechanism. Weight of this mounting is 30.5 kg.

Manufacturer

Defence Industries Organisation, Armament Industries Group, Jangafzar Sazi, PO Box 17185, 618 Tehran.

Status

In production.

Service

Iranian armed forces.

VERIFIED

IRAQ

7.62 mm Al-Quds machine rifle

Description

This weapon, employed as a squad automatic by the Iraqi armed forces, is a heavy-barrelled version of the 7.62 mm AKM rifle. The barrel is longer than that of the rifle and has cooling fins beneath the gas cylinder. It is fitted with an easily removable bipod and uses the standard 30-round AKM rifle magazine.

Data

Cartridge: 7.62 × 39 mm M1943
Operation: gas, selective fire
Locking: rotating bolt
Feed: 30-round box magazine
Weight: empty, 5 kg
Length: 1.025 m
Barrel: 542 mm
Rate of fire: cyclic, 560-680 rds/min
Muzzle velocity: 745 m/s

Manufacturer

State arsenals.

Status

In production.

Service

Iraqi armed forces and for export.

VERIFIED

7.62 mm Al-Quds machine rifle

ISRAEL

5.56 mm and 7.62 mm Model ARM Galil light machine gun

Description

When the Israelis adopted the 5.56 mm Galil assault rifle family as standard weapons they did so with the intention that the new weapons should be used in the sub-machine gun, rifle and light machine gun roles. The ARM version of the weapon (described in the *Rifles* section) is fitted with a bipod and offers a choice of magazine sizes up to 50 rounds and is thus suitable for use in the light machine gun role.

Details of the weapon will be found in the corresponding rifle entry.

Manufacturer

TAAS - Israel Industries Limited, PO Box 1044, Ramat Hasharon 47100.

Status

In production.

Service

Israeli forces and other armies.

VERIFIED

7.62 mm Model ARM Galil light machine gun

5.56 mm Negev light machine gun

Description

The Negev light machine gun is a multipurpose weapon that can be fed from standard belts, drums or magazines and fired from the hip, bipod, tripod or ground/vehicle mounts. It is designed to fire standard 5.56 × 45 mm NATO SS109 ammunition, and with a replacement barrel will fire standard M193 rounds.

The weapon is gas operated with a rotating bolt which locks into a barrel extension. It fires from an open bolt. It can be operated either as a light machine gun or as an assault rifle. As an assault rifle, manoeuvrability can be increased and weight reduced by removal of the bipod and installation of a short barrel and magazine. It can be fired in semi-automatic and automatic modes from standard magazines fitting beneath the weapon; Galil 35-round magazines or M16 magazines with adaptors can be used. It may also be adapted to fire 200-round belts carried in a pouch or ammunition box.

The gas regulator has three positions: position 1 closes the gas vent for firing rifle grenades; position 2 provides a rate of fire of 700-850 rds/min; and position 3 provides a rate of fire of 850 to 1,000 rds/min.

The Negev can be easily and quickly stripped into six subassemblies (including the bipod). All parts, including the quick-change barrels, are fully interchangeable. Various types of telescope and sight mounts can be installed on the receiver.

Data

Cartridge: 5.56 × 45 mm SS109 or M193
Operation: gas, open bolt, selective fire
Locking: rotating bolt
Feed: 35- or 30-round box magazine, or belts, or drums
Weight: empty, 7.5 kg
Length: (long barrel) stock folded, 780 mm; stock extended, 1.02 m; (short barrel) stock folded, 650 mm; stock extended, 890 mm
Barrel: long, 460 mm; short, 330 mm
Rifling: 6 grooves, rh, 1 turn in 178 mm; optional, 1 turn in 305 mm
Sights: fore, post, adjustable for elevation and windage for zeroing; rear, aperture, 300-1,000 m; folding night sight with tritium illumination
Rate of fire: cyclic, 700-850 or 850-1,000 rds/min (see text)

Manufacturer

TAAS - Israel Industries Limited, PO Box 1044, Ramat Hasharon 47100.

Status

In production. Offered for export sales.

Service

Israeli Defence Forces.

UPDATED

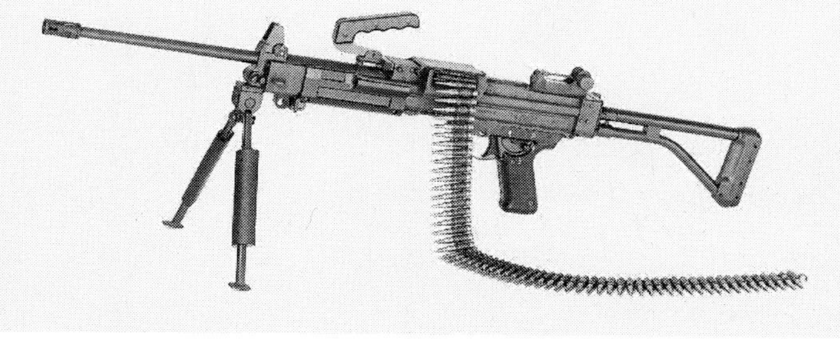

5.56 mm Negev light machine gun with belt feed unit

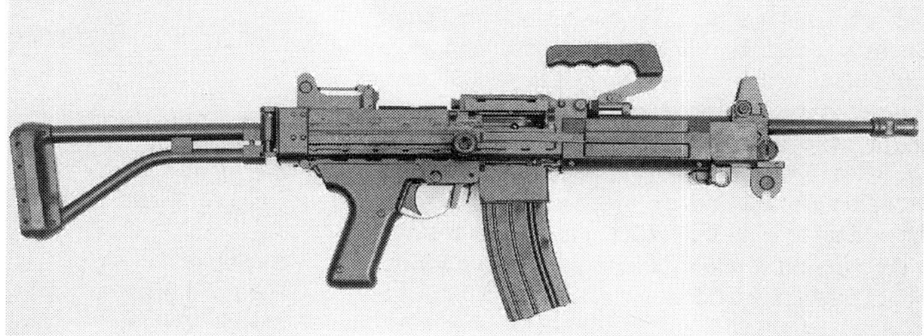

5.56 mm Negev machine gun using M16 magazine with adaptor

ITALY

7.62 mm MG42/59 general purpose machine gun

Description

The 7.62 mm MG42/59 is the standard Italian general purpose machine gun and was manufactured under licence from Rheinmetall.

Information on the method of operation is given under MG3 in *Germany*. However, there are some differences in dimensions and data, as noted below.

Data

Cartridge: 7.62 × 51 mm NATO
Operation: short recoil, automatic
Locking: rollers

7.62 mm MG42/59 general purpose machine gun

Feed: belt
Weight: with bipod, 12 kg
Length: 1.22 m
Barrel: with extension, 567 mm
Rifling: 4 grooves, rh, twist
Sights: fore, post; rear, notch, graduated 200-2,200 m
Sight radius: 432 mm
Rate of fire: cyclic, ca 800 rds/min
Muzzle velocity: 820 m/s
Range: bipod, 400-500 m; tripod, 800-1,000 m
Max range: 3,650 m at 33° elevation

Manufacturers
Pietro Beretta SpA, Brescia.
Luigi Franchi SpA, Brescia.
Whitehead Moto-Fides SpA, Livorno.

Status
No longer in production.

Service
Italian Army. Exported to Chile, Denmark, Mozambique, Nigeria and Portugal.

Licence production
AUSTRIA
Manufacturer: Steyr-Daimler-Puch
Type: 7.62 mm MG74
Remarks: Production now complete. MG42/59 licenced from Beretta.

VERIFIED

Beretta 5.56 mm AR70/84 light machine gun

Description
The AR70/84 is a light machine gun firing 5.56 × 45 mm cartridges and is based on the AR70/.223 Beretta rifle, with which it has a number of parts in common. It superseded the earlier Model 70/78, which also formed part of this rifle family. It operates on the same gas system and the main alterations are a heavier barrel with a perforated steel handguard, fixing points for vehicle or other types of mounting, and a new butt with shoulder rest and facilities for gripping it more firmly.

The AR70/84 fires from an open bolt; otherwise the gas system and rotating bolt are the same as those employed with the rifle. The quick-change barrel of the earlier 70/78 model has been abandoned. Iron sights are fitted as standard, and optical or electro-optical sights can be accommodated, provided that the receiver has been suitably modified during manufacture.

The bipod legs have a hinge in the middle of each leg to allow for coarse adjustment for height.

The muzzle is machined to act as a grenade launcher, and the grenade sight folds down in front of the foresight when not in use.

Beretta 5.56 mm AR70/84 light machine gun

Data
Cartridge: 5.56 × 45 mm
Operation: gas, selective fire
Locking: rotating bolt
Feed: 30-round box magazine
Weight: 5.3 kg
Length: 955 mm
Barrel: 450 mm
Rifling: 4 grooves, rh, 1 turn in 304 mm
Sight radius: 507 mm
Muzzle velocity: 970 m/s
Rate of fire: cyclic, 670 rds/min

Manufacturer
Armi Beretta SpA, Via Beretta 18, I-25063 Gardone VT (Brescia).

Status
Available.

UPDATED

Beretta 5.56 mm AS70/90 light machine gun

Description
The Beretta 5.56 mm AS70/90 light machine gun is a squad automatic weapon designed to accompany the AR70/90 rifle, with which it forms the 70/90 system.

The AS70/90 uses the same gas-operated, rotating bolt system as the rifle, but fires from an open bolt and has a heavier barrel which cannot be rapid-changed. There is a prominent metal handguard, and an articulated bipod which can be adjusted to several positions. Grenades may be launched from the muzzle, though the launcher is of a different pattern from that used with the rifles. The same carrying handle is fitted and, like those on the rifles, can be removed to expose a mount for optical or electro-optical sights. The butt-stock provides support at the shoulder and a good grip for the firer's disengaged hand.

Beretta 5.56 mm AS70/90 light machine gun

Data
Cartridge: 5.56 × 45 mm NATO
Operation: gas, selective fire
Locking: rotating bolt
Feed: 30-round box to STANAG 4179 (M16 standard)
Weight: without magazine and bipod, 5.34 kg
Length: 1 m
Barrel: 465 mm
Rifling: 6 grooves, rh, 1 turn in 178 mm
Sights: fore, post; rear, 2-position aperture, micrometer adjustment, for 300 and 800 m
Sight radius: 555 mm
Rate of fire: cyclic, ca 800 rds/min

Manufacturer
Armi Beretta SpA, Via Beretta 18, I-25063 Gardone VT (Brescia).

Status
Available.

Service
Italian Army.

UPDATED

JAPAN

7.62 mm Model 62 general purpose machine gun

Description
The 7.62 mm Model 62 general purpose machine gun was originally the Model 9M. It was accepted into service with the Japanese Self-Defence Forces in 1962.

The gas cylinder and piston are below the barrel. The locking system is unusual in that it is a tilting block with the front of the bolt forced up by cams on the piston extension (slide) to lock. Two wings level with the centreline move into recesses in the receiver and the bolt is held in position by the piston extension under it. The final movement of the piston, after locking is completed, carries the firing pin, fixed to the piston post, through the block, into the cartridge cap. Until the front of the block has risen the firing pin is not aligned with the cartridge cap.

After the round has fired, some of the propellant gases pass through the gas port and drive the piston to the rear. The firing pin is withdrawn and the front of the bolt is first cammed down and then pulled back.

The extraction is unusual. There is no spring-loaded extractor hook but while the round is in the chamber, before firing, there is a spring-loaded plunger forced up into the cannelure from below. When the front end of the bolt is carried down, a fixed hook on the bolt face above the firing-pin hole, drops down and grips the cannelure of the cartridge head. When the bolt starts back the case is withdrawn. There appears to be no primary extraction.

The feed system uses a bolt operated feed arm which oscillates as the bolt reciprocates and the belt is drawn across in two stages corresponding to the extraction and feed strokes.

The barrel-retaining catch is depressed by lifting the top cover plate. The carrying handle is rotated and then the barrel can be pushed forward. While the top cover plate is lifted, it prevents the front of the bolt from rising into the locked position and so the firing-pin hole is out of alignment. This prevents the feed and firing of a round when this is unlocked or there is no barrel in position.

Variant

Model 74

The 7.62 mm Model 74 is a vehicle-mounted variant of the Model 62. See following entry for details.

Data

Cartridge: 7.62 × 51 mm
Operation: gas, automatic
Locking: front tilting block
Feed: disintegrating link belt
Weight: 10.7 kg; barrel, 2 kg
Length: 1.2 m
Barrel: 524 mm
Sight radius: 590 mm
Muzzle velocity: 855 m/s
Rate of fire: cyclic, 600 rds/min

Manufacturer

Sumitomo Heavy Industries Limited, 2-1-1 Yato-Cho, Tanashi-Shi, Tokyo.

Status

In current use.

Service

Japanese Self-Defence Forces.

VERIFIED

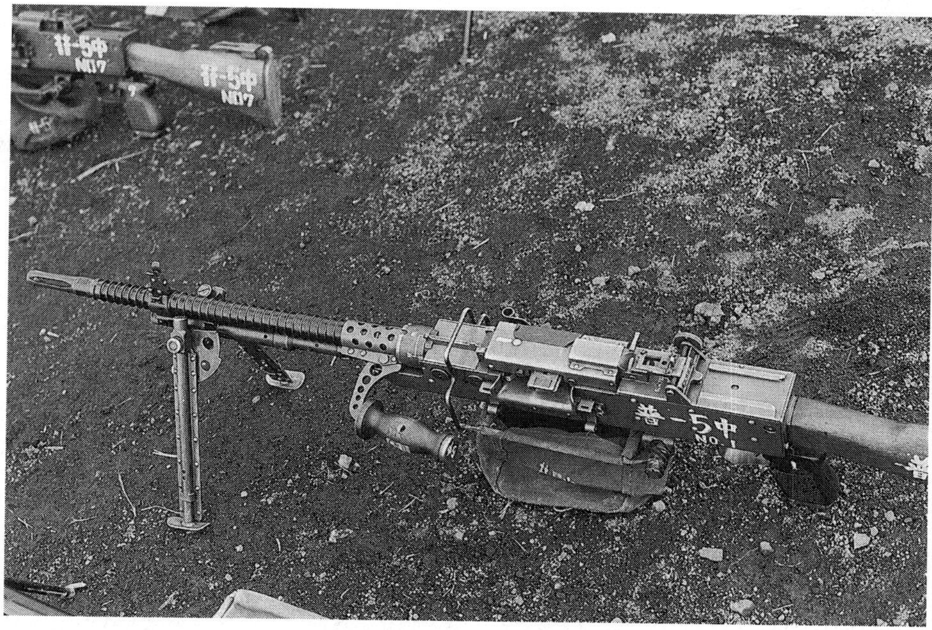

Bipod-mounted 7.62 mm Model 62 machine gun (K Nogi)

7.62 mm Model 74 machine gun

Description

The 7.62 mm Model 74 is a variant of the standard Model 62 and was adopted by the Japanese Defence Agency in 1974 as a coaxial gun for armoured vehicles. It is mounted in the tanks and armoured personnel carriers of the Japanese Self-Defence Forces, the guns being in a special mount in the front glacis plate.

The gun is very similar to the Model 62, but uses more robust components to maintain reliability. At an average rate of fire of 300 rds/min the gun can fire without difficulty for three minutes without any need to change the barrel. The trigger is operated either by a solenoid or by the manual triggers at the rear of the body and the firing rate is adjustable by means of a selector lever on the right lower side of the body.

All guns can be dismounted from their vehicle and mounted on the infantry tripod and used as support machine guns if necessary.

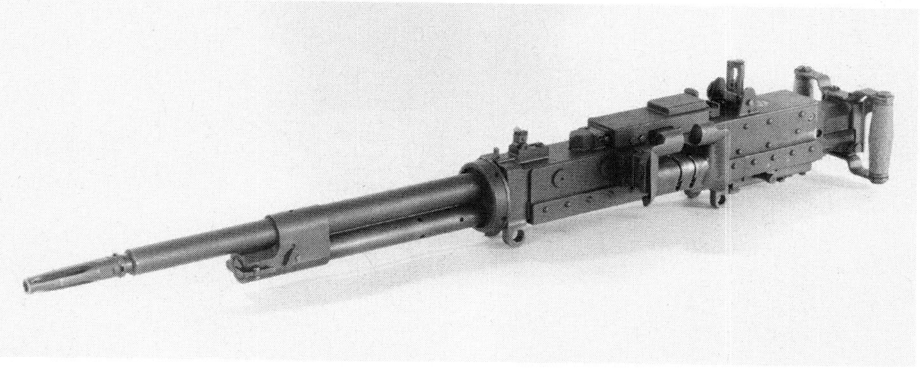

7.62 mm Model 74 vehicle machine gun

Data

Cartridge: 7.62 × 51 mm
Operation: gas
Locking: front tilting block
Feed: M14 link belt, left-hand side
Weight: 20.4 kg
Length: 1.085 m
Barrel: 625 mm
Sight radius: 306 mm
Muzzle velocity: 855 m/s
Rate of fire: cyclic, 700-1,000 rds/min

Manufacturer

Sumitomo Heavy Industries Limited, 2-1-1 Yato-Cho, Tanashi-Shi, Tokyo.

Status

In production.

Service

Japanese Self-Defence Forces.

VERIFIED

KOREA, SOUTH

5.56 mm K3 light machine gun

Description

The 5.56 mm K3 is a gas-operated, air-cooled, full-automatic light machine gun designed for one-man operation. The gas system uses a piston, bolt carrier and rotating five-lug bolt, together with a three-position adjustable gas regulator. The weapon will feed from a 200-round linked belt or from a 30-round box magazine.

The protected foresight is adjustable in elevation. The rearsight is adjustable in elevation and windage and is calibrated from 400 to 1,000 m.

The usual mounting is the integral bipod, which folds into the handguard. Alternatively, the gun can be mounted on the M122 tripod by means of an adaptor.

Data

Calibre: 5.56 × 45 mm M193 or K100 (SS109)
Operation: gas, fully automatic fire
Locking: rotating bolt
Feed: 200-round link belt in box or 30-round box magazine
Weight: with bipod, 6.85 kg
Length: 1.03 m
Barrel: 533 mm

K3 light machine gun showing magazine feed

K3 light machine gun showing belt feed and tripod

Rifling: 6 grooves, rh, 1 turn in 185 mm
Sights: fore, protected post; rear, graduated from 400 to 1,000 m, adjustable in elevation and windage
Muzzle velocity: M193, 960 m/s; SS109, 915 m/s
Rate of fire: cyclic, 700-1,000 rds/min
Max range: 3,600 m; effective, 800 m

Manufacturer
Daewoo Precision Industries Limited, PO Box 25, Kumjeong, Pusan.

Status
In production.

Service
South Korean Army.

VERIFIED

NORWAY

Vinghøgs NM152 Softmount

Description
The Vinghøgs NM152 Softmount is a recoil-absorbing cradle developed for the 0.50 Browning M2 HB heavy machine gun. The Softmount supports the gun receiver and interfaces with the normal ground, vehicle or naval mounting. It can be fitted with shoulder supports and can be adapted to carry accessory sights. It has been designed with operation in extreme climatic conditions in mind and has been successfully tested by the Norwegian Navy Materiel Command.

Data
Weight: 20 kg
Length: 850 mm; with shoulder rest, 1.2 m
Width: shoulder rest, 350-470 mm; hand grips, 500 mm

Manufacturer
Vinghøgs Mek Verksted A/S, Lindholmvn 14, N-3133 Duken.

Status
In production.

Service
Norwegian armed forces.

VERIFIED

Softmount in use, with shoulder support

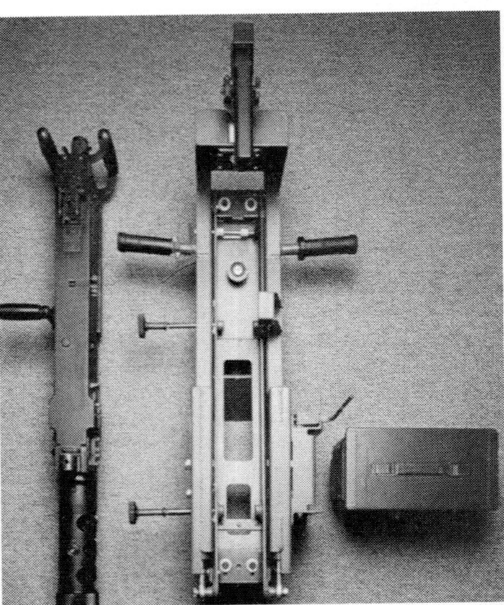

Top view of Softmount, with gun and ammunition box

PAKISTAN

7.62 mm MG3 machine gun

Description
The German Rheinmetall 7.62 mm MG3 machine gun is manufactured under licence in Pakistan by the Pakistan Ordnance Factories (POF) at Wah Cantt. Data relating to this weapon, originating from POF sources, is provided below.

Data
Cartridge: 7.62 × 51 mm
Operation: short recoil, automatic
Locking: rollers
Feed: belt
Weight: with bipod and sling, 11.05 kg; gun only, 10.5 kg; barrel, 1.7 kg
Length: 1.225 m
Barrel: with extension, 565 mm
Rifling: 4 grooves, rh
Sights: fore, barleycorn; rear, notch from 200 to 1,200 m; also anti-aircraft sight
Sight radius: 430 mm
Muzzle velocity: 820 m/s
Rate of fire: cyclic, 1,100-1,300 rds/min
Max range: 5,000 m

Pakistan Ordnance Factory 7.62 mm MG3 machine gun

Manufacturer
Pakistan Ordnance Factories, Wah Cantt.

Status
In production. Offered for export sales.

Service
Pakistan armed forces.

VERIFIED

POLAND

12.7 mm NSW Utios heavy machine gun

Description

The 12.7 mm NSW Utios heavy machine gun is the Polish licence-produced version of the CIS NSV machine gun. The Polish model is identical to the CIS original so for details reference should be made to the appropriate entry under *Commonwealth of Independent States* in this section.

Mountings produced in Poland for the NSW Utios include the 6U6 multipurpose air defence mounting and two naval mountings, the OHAR and DROP fixed pintle mountings, both of the latter involving the K10T collimated optical sight.

Manufacturer

Zaklady Mechaniczne Tarnow, 33-110 Tarnow, ul Kochanowski 30.

Status

In production.

Service

Polish armed forces.

VERIFIED

PORTUGAL

Heckler and Koch 7.62 mm HK 21 machine gun

Description

The Portuguese company INDEP SA manufactures the Heckler and Koch 7.62 × 51 mm HK 21 machine gun under licence. Full details of the weapon can be found under *Germany*.

Manufacturer

INDEP SA, Fabrica de Braco de Prata, Rua Fernando Palha, P-1802 Lisboa-Codex.

Status

In production.

Service

Portuguese Army.

UPDATED

7.62 mm HK 21 machine gun manufactured in Portugal

ROMANIA

Romanian machine guns

Description

The Romanian defence industry manufactures the following machine guns under licence from the CIS:

- 14.5 mm KPV heavy machine gun
- 12.7 mm Degtyarev Model 38/46 heavy machine gun
- 7.62 mm PK machine gun
- 7.62 mm RPK light machine gun
- 5.45 mm RPK-74 light machine gun.

All the above machine guns are offered for export sales and appear to be identical to their CIS counterparts, although slight variations can be observed. For instance there is a Romanian version of the 7.62 mm RPK fitted with the twin-strut folding butt-stock from the AKMS assault rifle. Another Romanian development is a training version of the RPK/RPK-74 firing low-cost 5.6 mm ammunition.

A more involved indigenous Romanian development is a revised version of the 5.45 mm RPK-74. The designation of this model is not known but it is an RPK-74 with a side-folding frame butt-stock and a revised perforated handguard made from what appears to be a black plastic-based material. The revised profile of the handguard includes the repositioning of the foresight assembly to a position at the front of the guard, instead of the more usual position near the muzzle; the folding bipod remains in its usual position. To maintain the length of the sight base the rearsight assembly (graduated to 1,000 m) is moved to the butt end of the receiver, directly above the pistol grip. The data in the table refers to this model.

Romanian version of 5.45 mm RPK-74 light machine gun with folding butt-stock and revised sight base

Data

Cartridge: 5.45 × 39 mm
Operation: gas, selective fire
Locking: rotating bolt
Feed: 40-round box magazine
Weight: without magazine, 5.5 kg
Length: butt folded, 895 mm; butt extended, 1.091 m
Barrel: 590 mm
Rifling: 4 grooves, rh, 1 turn in 196 mm
Muzzle velocity: 910 m/s
Rate of fire: cyclic, 600-700 rds/min; practical, 50-180 rds/min

Manufacturer

Romtehnica, 9-11 Drumul Taberei Street, Bucharest.

Status

In production. Offered for export sales.

Service

Romanian armed forces.

VERIFIED

SINGAPORE

5.56 mm Ultimax 100 light machine gun

Description

The Ultimax 100 is a magazine-fed, gas piston-operated, rotating bolt, air-cooled light machine gun. It is a one-man weapon and can be fired from the shoulder, hip or a bipod. It fires only in full-automatic mode, from the open bolt position, at 520 rds/min and feeds from 100-round drum magazines or standard 20- and 30-round box magazines.

The heavy, air-cooled, quick-change (Mark III only) barrel can sustain 500 rounds of full-automatic fire without heat damage. The foldable handle on the barrel is carefully situated for balance and is also used in the quick twist-action barrel change.

The rearsight is marked in 100 m increments to 1,000 m and is fully adjustable by moving the sight slide. The foresight is also fully adjustable for elevation and windage; thus each quick-change barrel can be pre-zeroed with respect to the receiver.

The trigger system's lockout mechanism prevents accidental firing if the weapon is improperly cocked or is dropped. The fire selector has two positions: fire and safe. On safe the trigger is blocked and the connection between trigger and sear moved out of position so that, if any part of the trigger mechanism fails, the gun will not fire.

Low recoil enables the gun to be fired without the easily detachable butt-stock. This, in conjunction with the short para-barrel, makes it a handy weapon in confined spaces or for paratroopers.

The bipod legs are positively locked, either in the down position or up flush with the sides of the receiver and are quickly adjustable for length. The bipod allows a 15° sweep and roll to either side and the entire assembly is easily detachable without tools.

The cocking handle does not cycle with the bolt and is locked in the forward position during firing. It is on the left side of the receiver to enable the gun to be cocked quickly while leaving the right hand ready for firing.

The gas regulator has three apertures. Use of a high pressure gas system prevents fouling and reduces the need to clean the gas piston and regulator. Both piston and regulator can be removed for cleaning if desired.

The Ultimax 100 fires from the open bolt position: the bolt group is held back behind the feed area and the bolt carrier is engaged by a sear. The front end of the sear makes contact with the sear actuator in the fire selector. Pulling the trigger releases the bolt carrier by means of the sear actuator and strips the top cartridge from the magazine. When chambering is complete, the continued forward movement of the bolt carrier rotates the bolt via a cam and locks it to the barrel. The forward momentum of the bolt carrier assembly drives the firing pin to ignite the cartridge. On firing, return gas is tapped off relatively close to the chamber, when barrel pressure is still high and the bolt carrier assembly is forced backwards by the gas piston, rotating the bolt via cam action. The bolt unlocks, providing primary extraction and loosening the empty case in the chamber, which is extracted, ejected and the return spring compressed. The bolt group has a long recoil travel and is designed to avoid impact with the receiver back-plate. The bolt group then moves forward and if pressure on the trigger is maintained, chambers the next round and fires it. If the trigger is released, the bolt carrier assembly will be engaged by the sear. When the ammunition supply is exhausted, the bolt carrier assembly will remain in the forward closed position.

Accessories and options available with the Ultimax 100 include: 100-round magazines, either in single pouches or in packs of four; standard 20- and 30-round box magazines; bipod; blank-fire attachment; sling; cleaning kit; quick-change barrels; and a twin-gun mount for vehicles.

Data

Cartridge: 5.56 mm × 45 mm (US M193 or SS109)
Operation: gas, automatic
Locking: rotating bolt, multilug head
Feed: 100-round drum magazine; 20- or 30-round box magazine
Weight: with bipod, 4.9 kg
Length: without butt, 810 mm; with butt, 1.024 m
Barrel: 508 mm
Rifling: 6 grooves, rh, 1 turn in 305 mm or 178 mm
Sights: fore, adjustable for elevation and windage; rear, adjustable from 100 to 1,000 m by 100 m increments; rapid shift; fine click adjustment for range; windage screw
Muzzle velocity: 970 m/s
Rate of fire: cyclic, 400-600 rds/min
Effective range: M193, 460 m; SS109, 1,300 m

Manufacturer

Chartered Industries of Singapore, 249 Jalan Boon Lay, Singapore 2261.

Status

Available for production.

Service

Singapore armed forces.

VERIFIED

5.56 mm Ultimax 100 Mark III light machine gun with quick-change barrel and 100-round drum magazine; bipod legs extended

5.56 mm Ultimax 100 Mark III light machine gun field-stripped

CIS 0.50 machine gun

Development

The Browning 0.50 M2 HB is a popular machine gun but its drawbacks are well known, namely adjustment of headspace, a closed bolt firing system, plus the complex design and subsequent high manufacturing cost. Knowing that there is still a need for a weapon to fill the gap between 7.62 mm machine guns and 20 mm cannon, Chartered Industries of Singapore (CIS) began designing a new 0.50 machine gun in late 1983, with a design team led by A C Cormack. The objective was a simpler and lighter weapon than the Browning M2 HB in order to improve the system portability and ease the problems of field maintenance. The ability to fire the developmental Saboted Light Armor Penetrator (SLAP) ammunition was another objective.

The design of the resultant machine gun is simple and modular in construction, consisting of five basic assemblies. The modular construction allows ease of assembly and maintenance. Simplicity of design means reliability and inherently allows for relative ease of manufacture and thus lower cost. There are only 180 components in the entire gun.

Description

The CIS .50 machine gun is gas-operated and fires from the open bolt position to prevent cook-off. The bolt carrier group is held back behind the feed area and the bolt carrier is engaged by a sear. Activating the trigger releases the bolt carrier, and as the bolt moves forward it strips the round centred on the feed tray from the ammunition belt feeding in from either the left- or the right-hand side of the gun.

When chambering is completed, the combined forward movement of the bolt carrier rotates the bolt by means of a cam and locks it to the barrel. The forward momentum of the bolt carrier assembly drives the firing pin forward to fire the cartridge.

On firing, propellant gas is tapped off into a gas cylinder. The bolt carrier is driven back by the gas pistons, causing the bolt to rotate, unlock and extract the empty case from the chamber.

As the bolt carrier continues to move rearward, the empty case is ejected via an ejection port at the bottom of the receiver. When the return springs are almost fully compressed, the bolt carrier begins to move forward and, if the pressure on the trigger is maintained, the bolt will chamber the next round and fire it. If the trigger is released the bolt carrier assembly will be engaged by the sear. When the ammunition belt is exhausted the bolt carrier group will remain in the forward closed position.

The design is simple and construction is modular, consisting of five basic groups:

Receiver body: made of pressed steel, including two tubes at the front end to house the pistons and recoil rods.

Feed mechanism: located on top of the receiver body, using a single sprocket. It is designed to facilitate either left- or right-hand feed of ammunition belts.

Trigger module: houses the trigger and sear mechanisms, with provision for safety lock. The selector has two modes, Automatic and Safe.

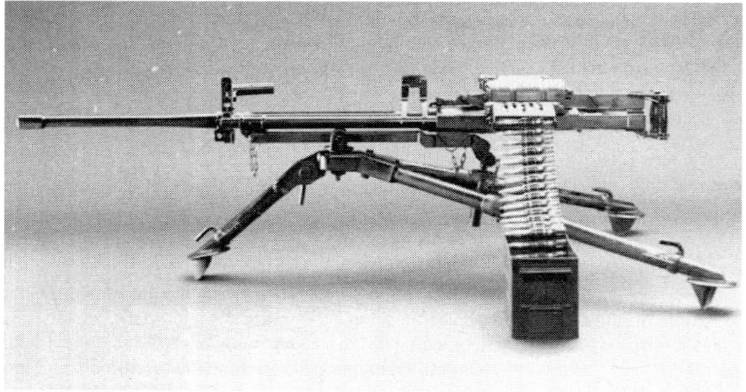

CIS 0.50 machine gun

CIS 0.50 machine gun mounted on armoured vehicle

Barrel group: a quick barrel change can be accomplished within seconds, without any headspace problem. The CIS gun has a fixed headspace, unlike the Browning M2 HB. The gas regulator has two positions to allow setting for normal and adverse environmental conditions.

Bolt carrier group: this group consists of a pair of pistons and recoil rods attached to the bolt carrier body by a pair of quick-release catches. The bolt is prevented from accidental firing out of battery by a sleeve lock device between the bolt and bolt carrier body.

Two types of mount are available; a pintle mount for the M113 APC and an adaptor for the M3 tripod.

Data

Cartridge: 12.7 × 99 mm (0.50 Browning)
Operation: gas, automatic
locking: rotating bolt
Feed: dual disintegrating M15A2 link belt
Weight: 30 kg; barrel, 10 kg; tripod, 23 kg
Length: 1.778 m
Barrel: 1.143 m
Rifling: 8 grooves, rh, 1 turn in 30 cal
Rate of fire: cyclic, 400-600 rds/min
Muzzle velocity: M8, 890 m/s; APDS, 1,200 m/s
Max range: 6,800 m; effective, 1,830 m

Manufacturer

Chartered Industries of Singapore, 249 Jalan Boon Lay, Singapore 2261.

Status

In production.

VERIFIED

SOUTH AFRICA

Vektor 7.62 mm SS-77 general purpose machine gun

Description

Development of the Vektor 7.62 mm SS-77 light machine gun began in 1977 and it entered service in 1986. It is of conventional type, gas operated, with quick-change barrel, bipod and folding butt.

Bolt locking is performed by swinging the rear end of the bolt sideways into a recess in the receiver wall, similar to the system used in the Goryunov machine gun. A gas piston beneath the barrel drives the bolt carrier back after firing. A post on the carrier, engaging in a cam path on the rectangular bolt, swings the bolt out of engagement with its locking recess and then withdraws it, extracting and ejecting the spent case downwards and forwards. A post on the top rear of the carrier engages in a belt feed lever in the top cover of the receiver. Because of the shape of this lever it is turned around a pivot so that the forward end, carrying the feed pawls, moves the incoming round half a step towards the feed position. On the return stroke the feed lever moves the round the remaining distance, the bolt loads it into the chamber, and as the bolt comes to rest the carrier continues and the post, engaged in the bolt cam path, swings the rear of the bolt into the locking recess. The post then strikes the firing pin and the round is fired. Ejection is downward and forward.

The barrel can be removed by depressing a locking lever and rotating the barrel to disengage its interrupted lugs from the receiver.

The SS-77 has no gas regulator, allowing easy maintenance.

If required, the quick-release foldable butt can be removed and replaced by spade grips or a remote control firing device.

Variant

Vektor Mini SS 5.56 mm light machine gun
See following entry.

Vektor 7.62 mm SS-77 general purpose machine gun (T J Gander)

Data

Cartridge: 7.62 × 51 mm
Operation: gas, automatic
Locking: transverse tilting block
Feed: disintegrating link belt R1M1 or M13; or non-disintegrating, DM 1 type links
Weight: gun, 9.6 kg; complete barrel, 2.5 kg
Length: butt folded, 940 mm; butt extended, 1.155 m
Barrel: without flash hider, 550 mm
Rifling: 4 grooves, rh, 1 turn in 305 mm
Sights: fore, conical post, with tritium spot; rear, tangent leaf with tritium spot; 200-800 m, aperture, 800-1,800 m, U-notch
Sight radius: short range, 747 mm; long range, 816 mm
Muzzle velocity: ca 840 m/s
Rate of fire: cyclic, 600-900 rds/min

Manufacturer

LIW (a division of Denel), PO Box 7710, Pretoria 0001.

Status

In production.

Service

South African National Defence Force.

UPDATED

Vektor 5.56 mm Mini SS light machine gun

Description

First shown in 1994 the Vektor 5.56 mm Mini SS light machine gun is a conversion of the 7.62 mm SS-77 general purpose machine gun to fire 5.56 × 45 mm ammunition. The conversion is made using a kit, weighing 3.76 kg, which includes a chromed-bore barrel assembly, a feed cover assembly, a breech assembly, a locking shoulder and a gas piston. When converted the Mini SS has an integral folding bipod and is normally provided with a fixed butt. A folding butt is an option, as is a 100-round capacity magazine pouch. Firing is from an open breech and can be from the bipod or a tripod.

Vektor 5.56 mm Mini SS light machine gun (T J Gander)

Data

Cartridge: 5.56 × 45 mm
Operation: gas, automatic
Locking: transverse tilting block
Feed: belt
Weight: gun, 7.96 kg; barrel, 1.5 kg
Length: standard butt, 1 m
Barrel: 513 mm
Rifling: 6 grooves, rh
Sights: fore, conical post, with tritium spot; rear, tangent leaf with tritium spot; 200-800 m, aperture, 800-1,800 m, U-notch
Rate of fire: cyclic, 700-850 rds/min
Muzzle velocity: 980 m/s
Max range: (effective) bipod, 500 m; tripod, 800 m

Manufacturer

LIW (a division of Denel), PO Box 7710, Pretoria 0001.

Status

Ready for production.

VERIFIED

Vektor 7.62 mm MG4 machine gun

Description

The Vektor MG4 is based upon the Browning M1919A4, updated and converted to 7.62 × 51 mm NATO calibre. In addition to the calibre change the feed mechanism was redesigned to improve reliability and facilitate the use of a disintegrating metal link belt while the trigger mechanism was altered to permit firing from an open bolt. A safety catch was added, and several minor changes, to improve reliability and maintainability, were incorporated.

The MG4 was designed to be mounted on a tripod or pintle mount on vehicles, or used from a firing port or otherwise installed on a vehicle. It is well suited to vehicle use as the barrel can be changed and all maintenance performed from inside the vehicle. A double mount is available for vehicle or light patrol boat mounting, as well as Combi mounts which permit the MG4 to be mounted alongside a 0.50 Browning M2 HB machine gun or a 20 mm cannon.

For use in the air defence role a special mount with a mechanical tachymetric sight is available. There is also a conversion kit to convert older 0.300 or 0.303 in Browning M1919A4 guns to the MG4 specification.

Vektor 7.62 mm MG4 machine guns (T J Gander)

Data

Cartridge: 7.62 × 51 mm
Operation: recoil, automatic
Locking: projecting lug
Feed: disintegrating metal link belt
Weight: ca 15 kg
Length: air defence, 1.015 m; secondary armament, 940 mm
Barrel: 595 mm
Rifling: 4 grooves, rh, 1 turn in 305 mm
Sight radius: 350 mm
Muzzle velocity: ca 840 m/s
Muzzle energy: ca 320 kpm
Rate of fire: cyclic, 700 rds/min
Max range: effective, 1,200 m

Manufacturer

LIW (a division of Denel), PO Box 7710, Pretoria 0001.

Status

In production.

Service

South African National Defence Force.

VERIFIED

SPAIN

5.56 mm Ameli machine gun

Description

The 5.56 mm Ameli machine gun operates on the delayed blowback system with the use of rollers, the system used in all CETME light weapons since 1956. Similarities with the CETME 7.62 mm Model C and 5.56 mm Model L rifles extend to the interchangeability of certain parts.

The Ameli is fed by a disintegrating link belt, hanging free or from a 100- or 200-round magazine box. The manufacturer states that the air-cooled barrel can be changed within five seconds.

The roller-locking device operates in the same way and gives the same advantages as described in the entry for the German G3 rifle.

The gun is loaded and cocked with the safety catch off, the sear holding the bolt immobile. When the trigger is pressed, the sear disengages and the bolt begins its forward travel, forcing the first round from the belt into the breech. Under the pressure of the bolt head, the rollers move outward to the sides of the breech. The extractor engages the groove of the cartridge case. When the bolt completes its forward travel, the rollers completely lock the assembly; the bolt head forces forward the firing pin and ignites the charge. As gas pressure falls, the recycled gases loosen the case and residual pressure in the barrel forces it back, to be ejected downward and forward.

Accessories include an integral ammunition state indicator, a tool set, a night vision equipment and a tripod, as well as the integral bipod.

Data

Cartridge: 5.56 × 45 mm (M193 or SS109)
Operation: gas, automatic
Locking: delayed blowback using rollers
Feed: disintegrating link belt or 100- or 200-round box
Weight: 5.7 kg; bipod, 0.65 kg; barrel 1 kg; 200-round box magazine, 3.1 kg; 100-round box magazine, 1.66 kg
Length: 980 mm
Barrel: 470 mm
Rifling: 6 grooves, rh, 1 turn in 178 mm
Sights: 4-position rear aperture graduated 200, 400, 600 and 800 m
Sight radius: 340 mm
Muzzle velocity: 875 m/s
Muzzle energy: 1,530 J
Rate of fire: cyclic, 900-1,250 rds/min
Max range: 1,650 m

5.56 mm Ameli machine gun in the light role, on bipod

5.56 mm Ameli machine gun on tripod, in the sustained fire role

Manufacturer
SANTA BARBARA, Calle Julian Camarillo 32, E-28037 Madrid.

Status
In production.

Service
Adopted by the Spanish Army.

VERIFIED

5.56 mm Ameli machine gun

SWITZERLAND

7.5 mm M51 machine gun

Description
The 7.5 mm M51 machine gun was designed by the government establishment at Bern. The design was based on the German MG42 but, largely because the design calls for machined parts in many places where the MG42 uses stampings, it is better made but also more expensive, and more than 4 kg heavier.

Features of the design include a quick-change barrel arrangement and a belt feed system, both of which are similar to those of the MG42. The locking system is also based on that of the German weapon but the Swiss design uses flaps instead of the rollers used on the MG42. Accessories include a tripod mount which can be fitted with an optical sight and a drum attachment which enables a 50-round belt to be carried conveniently.

Variant
7.5 mm MG83 tank machine gun
The 7.5 mm MG83 tank machine gun is based on the M51 and was developed to provide the coaxial armament for the Swiss Army's Pz 87 Leo (Leopard 2) MBTs. The main change is that the weapon can be folded at the centre for barrel removal. Feed is from belts up to 4,000 rounds long and weight is 32 kg; length overall is 1.41 m. Cyclic rate of fire is 500 or 1,000 rds/min.

7.5 mm M51 machine gun

Data
Cartridge: 7.5 mm M11
Operation: gas-assisted recoil, automatic
Locking: flaps
Feed: 50- or 250-round link belt
Weight: with bipod, 16 kg; with tripod, 27 kg
Length: 1.27 m
Barrel: 564 mm
Sights: fore, folding blade; rear, tangent
Rate of fire: cyclic, 1,000 rds/min
Range: (effective) bipod, 800 m; tripod, 1,000 m

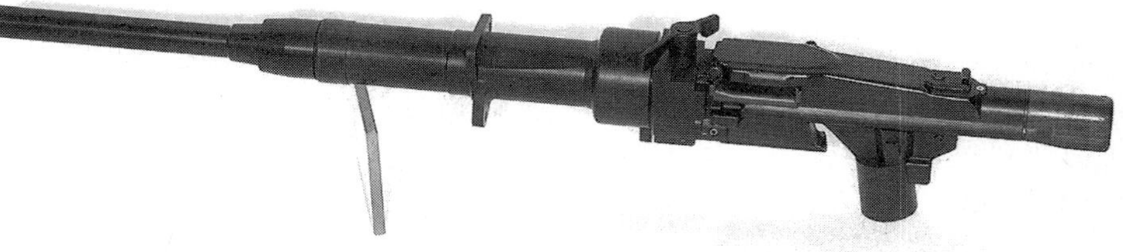

7.5 mm MG83 tank machine gun

Manufacturer
w + f Bern (formerly Swiss Federal Arms Factory), Stauffacherstrasse 65, CH-3000 Bern 22.

Status
In current use.

Service
Swiss Army.

UPDATED

TAIWAN

Machine guns in Taiwan

Description

The Chinese Nationalists carried to Taiwan a collection of machine guns, all of foreign origin. They included:

Type 24	—German Maxim MG'08 made in China
Type 26	—Czechoslovak ZB26 made in China
Type 30	—Czechoslovak ZB30 made in China
Madsen 7.92 mm	—Danish Madsen
Bren guns 7.92 mm	—Canadian manufacture
Browning 0.30 Model 1919A4	—US manufacture

Subsequently the Bren was rechambered for 0.30 calibre and manufactured in some quantity in Taiwan. Further quantities of Browning 0.30 Model 1919A4 and 0.50 M2 HB machine guns were supplied.

In 1968, Taiwan was provided with the necessary machinery to manufacture the M60 general purpose machine gun and since then tens of thousands of these have been produced, known as the Type 57 machine gun. Around 1978, the Combined Service Forces (CSF) developed a manufacturing capability for the 7.62 mm M134 Minigun. Whether this weapon was produced in any quantity is unknown.

More recently, an official statement suggested that the 0.50 M2 HB machine gun may be in production in Taiwan for use as anti-aircraft armament on Taiwan-built M41 tanks. A 7.62 mm coaxial machine gun (of unspecified type) is also made.

5.56 mm Type 75 squad automatic weapon

Type 74 light machine gun

The Type 74 was developed from the FN MAG general purpose machine gun and will replace the Type 57 machine gun as the platoon support weapon. The Type 74 differs from the FN MAG only in minor details, for example a slightly different shape of butt, different sights and a barrel with cooling ribs.

Type 75 squad automatic weapon

This closely resembles the FN Minimi (M249) and will replace the M14A1 as the squad automatic weapon so that both rifles and support weapon will use 5.56 mm ammunition. The Type 75 is rifled one turn in 178 mm, suggesting that the ROC army has standardised on the SS109 bullet.

Manufacturer

Combined Service Forces arsenals.

Status

The M60 is in production and service. Other older weapons are probably restricted to reserve units.

VERIFIED

TURKEY

7.62 mm MG3 machine gun

Description

The German MG3 is manufactured in Turkey for supply to the Turkish Army and also for export. It is identical in all respects with the German original, to which reference should be made for details.

In 1992 a contract was obtained to supply the Norwegian forces with a total of 4,220 MG3 machine guns over four years. Deliveries commenced in late 1992.

Manufacturer

Makina ve Kimya Endüstrisi Kurumu, TR-06330 Tandogan, Ankara.

Status

In production.

Service

Turkish forces, Norwegian Army, and for export.

VERIFIED

Turkish 7.62 mm MG3 machine gun on bipod mount

UNITED KINGDOM

7.62 mm L4A4 (Bren) machine gun

Development

The original Bren light machine gun was designed in Czechoslovakia being the Czech ZB26 modified to fire 0.303 ammunition. It went into production at Enfield Lock in September 1937.

The original gun was designated the Mark I. The Mark II had the same length barrel, a simplified rearsight, and the flash hider/gas regulator/foresight was fabricated rather than in one piece. The bipod was made with non-telescoping legs and the handle below the butt was omitted. These changes to assist production increased the weight from 10.04 to 10.52 kg. The Mark III simplified production, reduced barrel length, and reduced gun weight to 8.76 kg. The Mark IV had the shorter barrel and weight was reduced to the minimum compatible with the stresses imposed by the 0.303 Mark VII cartridge.

When the decision was made to adopt the 7.62 mm NATO cartridge, various conversions of the 0.303 Bren gun were made to adapt it for 7.62 mm; a brief summary of the L4 series follows.

L4A1: Converted Mark III. Two steel barrels. Obsolete.

L4A2: Converted Mark III. Two steel barrels. Obsolescent.

L4A3: Converted Mark II. One chromium-plated barrel. Obsolescent.

L4A4: Converted Mark III. One chromium-plated barrel. Current.

L4A5: Converted Mark II. Two steel barrels. Obsolescent for land and air service. Still in naval service.

L4A6: Converted L4A1. One chromium-plated barrel. Land service. Obsolescent.

Description

The 7.62 mm L4A4 light machine gun, still generally known as the Bren, is a magazine-fed, gas-operated gun using a tilting-block locking system lifting the rear end of the breech block into a locking recess in the top of the body.

During firing the body, barrel, breech block gas cylinder and bipod recoil on the butt slide approximately 6 mm. The movement is buffered by the piston buffer and spring. The piston buffer spring then reasserts itself and returns the body, barrel, cylinder and bipod to their normal positions on the butt slide. This movement reduces the shock experienced by the firer and makes for fewer breakages in the affected components.

When the gun is fired a proportion of the propellant gas is diverted through a tapping in the barrel, passes through the regulator and strikes the piston head, driving the piston back. Attached to the piston by a flexible joint is the piston extension on which is supported the breech block.

A piston post on the extension fits into the hollow interior of the breech block and two ramps hold the rear of the block up into the locked position engaged in the locking recess at the top of the body.

When the piston extension moves back there is a movement of about 32 mm during which the bolt remains fully locked. Further movement removes the ramp support under the block and then an inclined surface on the rear of the piston post forces the back end of the bolt down and unlocking is completed.

The tilting motion of the breech block provides primary extraction and the cartridge case is first unseated in the chamber and then withdrawn by the extractor claw as the breech block moves back. A fixed ejector rides in a groove on top of the block and it is chisel-shaped so that as it strikes the brass of the cartridge case above the primer cap, brass is burred over the cap to prevent the latter falling out and causing a stoppage. The empty case is pushed through a cutaway section in the piston extension and thrown downwards out of the gun. As the piston goes back the return spring is compressed, storing energy and this, plus the action of the soft buffer, throws the piston forward again. The soft buffer has a low coefficient of restitution and so the piston speed forward is not excessive and this keeps the cyclic rate to about 500 rds/min. The feed horns on top of the front of the block push a round out of the 30-round box magazine mounted vertically above the gun and the bullet is guided downwards into the chamber. As the cartridge goes forward the extractor claw clips over the rim of the round. When the round is fully chambered bolt movement ceases. The piston continues forward under its own momentum, and the remaining force in the return spring, and the two ramps at the rear end lift the rear of the breech block so that the locking surface on top of the rear of the block rises into the locking recess in the body. The ramps remain under the block and hold it locked. The forward movement of the piston continues for another 32 mm and the front face of the piston post acts as a hammer to drive the spring-retracted firing pin into the cap at the base of the cartridge.

The gas regulator has four tracks and a larger diameter gas track can be rotated into position as required. The gas impulse is applied only for a very short distance and then the gas escapes to atmosphere through vents bored in the cylinder walls. If excessive fouling occurs the bipod can be twisted and this cuts away any build-up of carbon which is then dispersed by the next blast of gas.

7.62 mm L4A4 machine gun

7.62 mm L4A4 machine gun on vehicle pintle mounting (T J Gander)

The barrel can be changed in a matter of seconds by raising the barrel latch and pulling the barrel forward using the carrying handle. With the gun fired at 120 rds/min (four magazines) the barrel requires changing every 2½ minutes. The hot barrel can be cooled by air after removal from the gun or by laying it in wet grass or even in a stream. In practice the latest chrome-plated barrels do not require changing other than at extreme rates of fire.

The weapon can be fired either at full automatic or at single shot, a change lever being fitted on the left side above the trigger. The applied safety disconnects the trigger from the sear.

Data

L4A4

Cartridge: 7.62 × 51 mm
Operation: gas, selective fire
Locking: tilting block
Feed: 30-round box magazine
Weight: 8.68 kg
Length: 1.156 m
Barrel: 635 mm
Sights: fore, blade; rear, aperture, tangent leaf
Sight radius: 788 mm
Muzzle velocity: 838 m/s
Rate of fire: cyclic, 520 rds/min

Manufacturer

Royal Small Arms Factory, Enfield; John Inglis Limited, Toronto, Canada; Small Arms Factory, Lithgow, New South Wales, Australia; Small Arms Factory, Kanpur, India.

Status

No longer manufactured in United Kingdom.

Service

British forces and those of many Commonwealth and former Commonwealth countries.

Licence production

INDIA
Manufacturer: Small Arms Factory, Kanpur
Type: Gun, Machine 7.62 mm 1B
Remarks: Still in production. Equivalent to L4A4. See entry under *India* for details.

VERIFIED

7.62 mm L7A2 general purpose machine gun

Development

The L7 General Purpose Machine Gun (GPMG) is based on the Fabrique Nationale MAG (see entry under *Belgium)* and was used by United Kingdom forces for many years both as a light machine gun (LMG) and a sustained fire machine gun (SFMG) - it is now retained only for the SFMG role. In the LMG role it is fired from the integral bipod mount, while in the SFMG role it is fired from a tripod. The essential dimensions, features and functioning of the gun are identical to those of the FN MAG, but parts are not interchangeable.

United Kingdom production was originally carried out at the Royal Small Arms Factory at Enfield Lock. When that establishment closed, L7A2 production and support was transferred to Manroy Engineering of Beckley, East Sussex.

Description

The 7.62 mm L7A2 general purpose machine gun is a fully automatic, belt-fed, gas-operated, air-cooled weapon capable of maintaining a high rate of fire at a cyclic rate of 750 rds/min. Control over the firing rate is available by means of the gas regulator which varies the quantity of gas escape. With the regulator set at maximum gas, 900 rds/min can be achieved. A safety feature of the gun is that the breech is positively locked before the round is fired and the gun fires from the open bolt. Risk of detection by the enemy is minimised by the design of the gas regulator and the attachment of a flash hider to the muzzle.

Its accuracy, together with the sustained fire kit, enable the gun to be fired effectively through smoke or fog and during the hours of darkness.

The 7.62 mm L4A1 machine gun mount was developed by the Royal Small Arms Factory at Enfield Lock to provide a stable light mounting for the L7A2. The mounting incorporates a recoil buffer unit, permits all-round traverse and has a quick-release mechanism allowing free traverse, elevation and depression.

Three tubular legs support the tripod head. At the foot of each leg is a shoe to give an improved ground grip. Each leg is locked by a cam lever which when pushed inwards secures the leg in any desired position.

The tripod can be set to high or low; these settings are notched on the legs and on the tripod head. A dial ring is held in position by friction and can be rotated by hand. It has a traverse scale of 0-3,200 mils in each direction with graduations every 250 mils. An indicator moves with the head and indicates the traverse on the scale.

The cradle and recoil unit is mounted on the tripod head. The cradle is attached to the tripod head by a ball joint which can be locked; when free it allows 600 mils elevation and all round traverse. The gun is secured to the recoil unit by front and rear mounting pins. The cradle has an enclosed buffer system of tubes, buffers and springs. When the first round is fired, the whole gun recoils and then the buffer returns it to battery. Before run out is completed the next round is fired and thereafter the gun movement is extremely small as it rests in equilibrium imposed between the recoil force and the buffer unit.

The elevation mechanism provides a fine adjustment using the handwheel on the left. It is controlled and locked by an eccentric cam tightened by a thumb lever. The fine traversing mechanism is controlled by the handwheel on the right. The range of controlled traverse is 200 mils and one turn of the handwheel traverses the gun through 6.6 mils. A clicker control operated by a sliding sleeve gives 2.2 mils to each click; three clicks per turn. If desired the sleeve may be pushed in to give a smooth, silent control.

The dial sight is the Sight Unit C2, Trilux, also used with the 81 mm mortar. It has a right-angled telescope with a magnification of ×1.7 and a field of view of 180 mils. The eyepiece may be rotated 1,600 mils (90°) right or left from the vertical. The sight unit is fitted with perspex bearing and elevation scales which, with the telescope graticule and the levelling bubbles, are Trilux illuminated.

When the L7A2 machine gun is used on the tripod the butt is removed and the recoil buffer is used in its place.

7.62 mm L7A2 general purpose machine gun shown with its full array of accessories including the L4A1, C2 dial sight in carrying case, spare barrels and carrying cases

Variants

L8A1

A tank version of the L7A2 with the trigger group replaced by a firing solenoid and modifications to the feed system. Some parts, such as the cocking handle, are reduced in size to save space. It is possible to use a kit to convert the L8A1 for the ground role but this is seldom used. The L8A1 was employed as the coaxial weapon for the Chieftain tank series.

L8A2

An improved version of the L8A1 which can also be used on the Challenger 1 tank.

L19A1

An L7 series model with a heavier barrel to remove the need for frequent barrel changing when carrying out sustained firing. Seldom issued.

L20A1

Intended for use in pods or other external mountings on helicopters and light aircraft, this version can be fed from the left- or right-hand side and is controlled electrically. No sights are fitted and the pronged flash hider at the muzzle is unique to this version.

L20A2

Only slight differences from the L20A1.

L37A1

This is an armoured combat vehicle version and is basically a combination of L7 and L8 components. A special barrel is used to permit a greater proportion of tracer rounds than normal to be fired for aiming purposes and the weapon can be dismounted and fitted with a bipod and butt to enable it to be employed in the ground role for local defence. This version was used in turreted FV432 armoured personnel carriers and as the commander's cupola weapon on the Chieftain tank.

L37A2

A generally improved L37A1 which can also be used on the Challenger 1 tank.

L41A1

A drill and training version of the L8A1 which cannot be fired.

L43A1

This version was specifically modified for use as the ranging machine gun for the 76 mm main gun on the Scorpion tracked reconnaissance vehicle. It also doubled as a coaxial weapon and is generally similar to the L8A1, although it lacks the latter's muzzle attachment. Now withdrawn.

L45A1

A drill and training version of the L37A1 which cannot be fired.

L46A1

A skeletonised training aid version of the L7A1/A2 which cannot be fired.

Data

L7A2

Cartridge: 7.62 × 51 mm
Operation: gas, automatic
Locking: dropping locking lever
Feed: belt
Weight: LMG role, 10.9 kg
Length: LMG role, 1.232 m; sustained fire role, 1.048 m
Barrel: including 50 mm overhang of carrying handle, with flash hider, 679 mm; without flash hider, 597 mm
Sights: fore, blade; rear, aperture
Sight radius: LMG role, sight down, 851 mm; sustained fire role, sight up, 787 mm
Muzzle velocity: 838 m/s
Rate of fire: cyclic, 750-1,000 rds/min

Data

Tripod L4A1

Weight: tripod, 13.61 kg; sight unit, cased, 2.58 kg; conversion kit, complete, 32.66 kg
Length, legs spread: across short legs, 1.118 m; front to rear, 1.118 m
Height of sightline above ground: 330-635 mm
Traverse: free, 360° (6,400 mils); mechanical traverse, max, 11° (200 mils)
Elevation: max, 22° (400 mils); mechanical elevation, max, 2°48′ (50 mils)
Depression: max, 11° (200 mils); mechanical depression, max, 2°48′ (50 mils)

Manufacturer

United Kingdom production was originally carried out at the Royal Small Arms Factory at Enfield Lock. When that establishment closed, L7A2 production and support was transferred to Manroy Engineering of Beckley, East Sussex.

Status

Available for production.

Service

British and some Commonwealth forces.

UPDATED

7.62 mm L7A2 general purpose machine gun in sustained fire role
(T J Gander)

7.62 mm L7A2 general purpose machine guns on a vehicle dual mounting
(T J Gander)

5.56 mm L86A1 Light Support Weapon - SA80 - Enfield Weapon System

Description

Most of the information relating to this weapon is already contained in the entry dealing with the Individual Weapon (IW) in the *Rifles* section. The light support weapon (LSW) is essentially a heavy-barrelled version of the IW and could more properly be termed a machine rifle; it shares the same design concept. Although a change-barrel version was investigated it was not followed up. Production ceased after 22,391 LSWs had been produced for the UK armed forces.

The gain from the light support weapon is in the longer and heavier barrel and the greater muzzle velocity and accuracy that this provides. When fired from the bipod it becomes an impressively accurate and consistent weapon which is able to take full advantage of the SUSAT sight which is a standard fitting.

Variant

SA80 LSW Cadet Rifle

It has been proposed that a single-shot rifle variant of the LSW could be produced using a manual cocking mechanism similar to that used on the SA80 Cadet L98A1 rifle (see under L85A1 IW in the *Rifles* section). The LSW Cadet rifle could be used for competition shooting while retaining the usual SUSAT sights (iron sights are an option), heavy barrel and bipod of the L86A1. The manual cocking mechanism is actuated between shots using a bolt handle on the left of the receiver. Ammunition feed is from a 10- or 30-round box magazine. Weight of the rifle is 6.3 kg.

Data

Cartridge: 5.56 × 45 mm NATO
Operation: gas
Locking: rotary bolt, forward locking
Feed: 30-round box magazine
Weight: without magazine and optical sight, 5.4 kg; with loaded magazine, 6.58 kg; optical sight, 0.7 kg; empty magazine, 0.12 kg; loaded magazine, 30 rounds, 0.48 kg;
Length: 900 mm
Barrel: including flash hider, 646 mm
Rifling: 6 grooves, rh, 1 turn in 180 mm
Muzzle velocity: 970 m/s
Rate of fire: cyclic, 610-775 rds/min

Manufacturer

Heckler and Koch (UK), British Aerospace Defence Limited, Kings Meadow Road, Nottingham NG2 1EQ.

Status

Production complete.

Service

In service with the British Army.

UPDATED

5.56 mm L86A1 light support weapon on air defence mount (T J Gander)

1996

5.56 mm SA80 LSW Cadet single-shot rifle

1996

7.62 mm Machine Gun, Chain, L94A1

Description

The Machine Gun, Chain, L94A1 is the British nomenclature for the long-barrelled Hughes EX-34 7.62 mm Chain Gun®, described more fully under *United States of America.* The L94A1 has a Stellite-lined barrel and a barrel shroud which provides the means of mounting the gun to the vehicle.

With the British Army the L94A1 is installed on Warrior IFVs, Challenger 2 MBTs and Sabre reconnaissance vehicles.

Manufacturer

Heckler and Koch (UK), British Aerospace Defence Limited, Kings Meadow Road, Nottingham NG2 1EQ.

Status

In production.

Service

British Army.

UPDATED

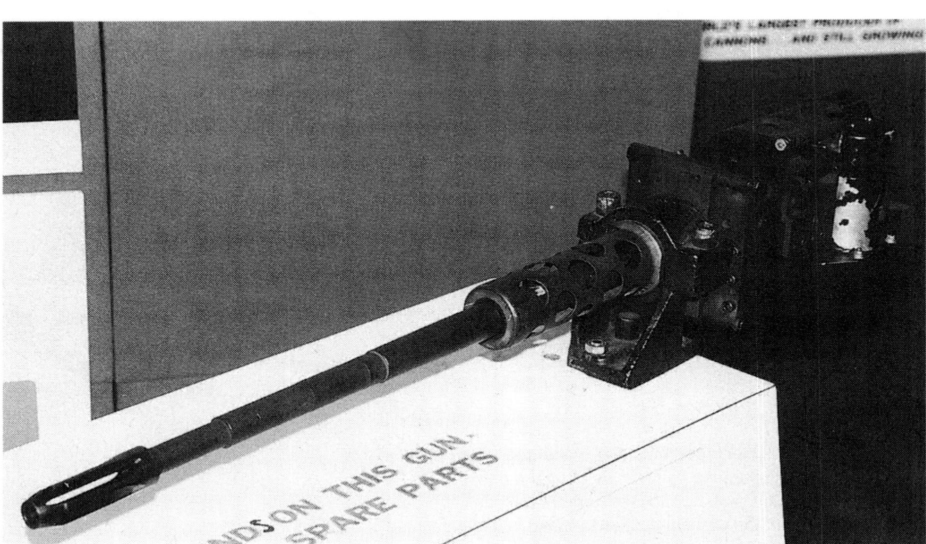

7.62 mm Machine Gun, Chain, L94A1 (T J Gander)

1996

Manroy 0.50 M2 HB heavy machine gun

Description

The Manroy 0.50 M2 HB heavy machine gun is the standard Browning 0.50 M2 HB heavy machine gun, manufactured in the United Kingdom by Manroy Engineering. It matches the standard US specification in every respect.

In addition to manufacturing the standard 0.50 M2 HB heavy machine gun, Manroy Engineering also produces a quick-change barrel version. As with other QCB systems the barrel can be changed by one man in under 10 seconds and there is no need to adjust headspace at any time. The change is made by rotating the cocking handle rearward and holding it, rotating the barrel, withdrawing it, inserting the new barrel, rotating it to stop and releasing the cocking handle.

Should it be necessary, the standard barrel can still be used with the gun, though in this case the normal headspace adjustment will have to be made.

In addition, a Manroy QCB conversion kit provides an inexpensive conversion for existing standard M2 HB guns. After fitting, the gun no longer requires the headspace gauging operation of standard M2 HB guns when changing barrels. Headspace is set to the correct size during assembly of the kit to the gun by suitable selection of the locking block.

A Manroy QCB kit utilises the standard barrel thread, so allowing standard M2 barrels to be fitted to the modified QCB gun if required, though normal headspacing must then be carried out and the barrel support locating stud removed. Existing standard barrels can, therefore, be used for training or in cases of emergency. A rigid insulated barrel carrying handle is supplied with the kit, but the standard carrying handle will fit the quick-change barrel and may be used if preferred. Barrels are supplied either standard unlined, chrome or Stellite-lined according to customer request.

Manroy also produce the M3 tripod mount, the M63 anti-aircraft mount, and a lightweight anti-aircraft mount adaptor. One further Manroy product is a Softmount system specifically intended for the 0.50 M2 HB.

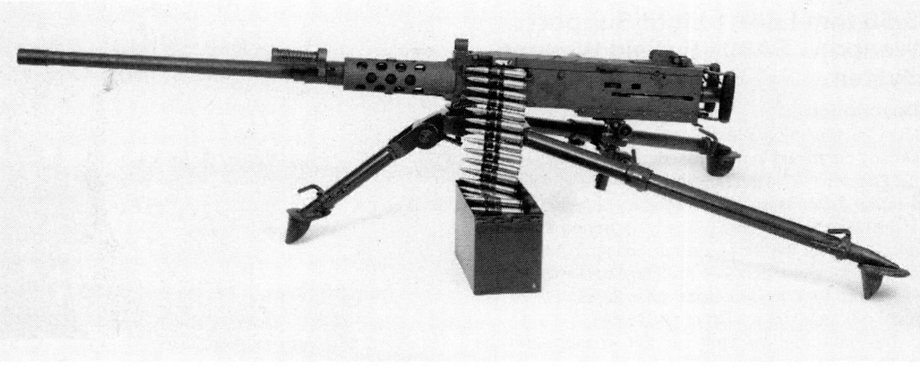

Manroy 0.50 M2 HB machine gun

Manroy 0.50 Softmount

Data

M2 HB

Cartridge: 12.7 × 99 mm (0.50 Browning)
Operation: short recoil, automatic
Locking: projecting lug
Feed: disintegrating link belt
Weight: 38.5 kg
Length: 1.651 m
Barrel: 1.143 m
Rifling: 8 grooves, rh, 1 turn in 381 mm
Muzzle velocity: 893 m/s
Rate of fire: cyclic, 450-500 rds/min
Max range: M2 ball, 6,766 m; effective, 1,850 m

Manroy M63 anti-aircraft mount

Manufacturer

Manroy Engineering Limited, Beckley, East Sussex TN31 6TS.

Status

In production.

Service

NATO and other forces worldwide.

VERIFIED

Manroy lightweight AA adaptor attached to M3 tripod

UNITED STATES OF AMERICA

Browning 0.30 calibre M1919A4 machine gun

Development

The M1919A4 was used as a fixed gun in tanks in the Second World War and on many post-war armoured fighting vehicles. It was also used by infantry as a company level weapon, mounted on the M2 tripod, with the flash hider M6 and a detachable carrying handle. Its evolution can be traced back to the 0.30 Browning tank machine gun which itself was derived from Browning's air-cooled aircraft machine gun designated the M1918.

Description

The mechanism of the M1919A4 is that of the original M1917 Browning; operating by short recoil of the barrel, barrel extension and bolt locked together. After 8 mm of movement the bolt is unlocked by cams in the receiver and is then accelerated to the rear while the barrel and extension move back into battery. During the rearward movement of the bolt a fresh round is extracted from the belt and fed into the T-slot in the bolt face, displacing the empty case just extracted and

0.30 calibre M1919A4 Browning machine gun on tripod M2 (T J Gander)

ejecting it downwards. On the return movement of the bolt the new cartridge is aligned and fed into the breech and the bolt is once more locked to the barrel extension.

For a more complete description of the basic Browning machine gun mechanism, readers are referred to earlier editions of *Jane's Infantry Weapons*.

Data

Cartridge: 0.30 M1 or M2 (also used in 7.62 mm form in Canada and South Africa)
Operation: short recoil, automatic
Locking: projecting lug
Feed: 250-round belt
Weight: 14.06 kg
Length: 1.044 m
Barrel: 610 mm
Rifling: 4 grooves, rh, 1 twist in 254 mm
Sights: fore, blade; rear, leaf aperture; windage scale
Sight radius: 354 mm
Muzzle velocity: 860 m/s
Rate of fire: cyclic, 400-500 rds/min
Range: effective, 1,000 m

Status

No longer in production in the United States.

Service

Used by many countries as a vehicle gun. Some have rechambered it to accept 7.62 × 51 mm NATO. In service with Canada, Denmark, Dominican Republic, Greece, Guatemala, Haiti, Iran, Israel, Italy, South Korea, Liberia, Mexico, South Africa, Spain, Taiwan, and Vietnam. In reserve in the USA.

Licence production

SOUTH AFRICA
Manufacturer: LIW
Type: 7.62 mm MG4
Remarks: In production. Fires 7.62 × 51 mm NATO ammunition and fires from open bolt. See separate entry under *South Africa* for details.

UPDATED

Browning 0.50 calibre heavy machine gun

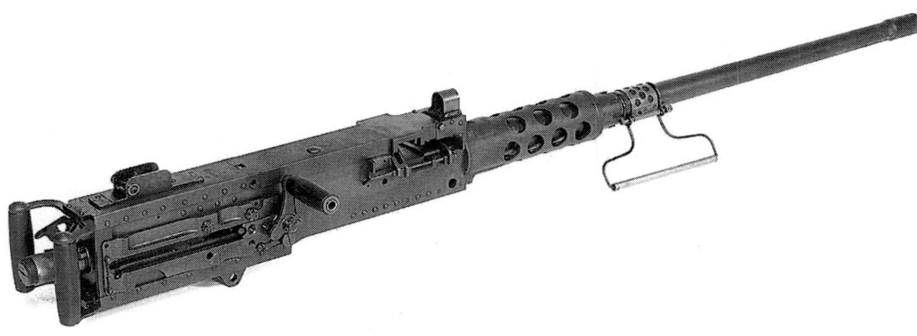

0.50 M2 HB heavy machine gun

Development

The air-cooled 0.50 Browning was developed for aircraft use and was adopted in 1923 as the Model 1921. In 1933 it was renamed the M2 and subsequently it was found necessary to increase the mass of the barrel and replace the oil buffer with a simpler design that eliminated the oil and improved service life and maintenance. It was then renamed the M2 HB (HB - heavy barrel) and in this form has been used and is still used extensively in the ground role and on vehicles. It is one of the most widespread and successful heavy machine guns in service and looks like continuing as such for many years.

Production is still undertaken in Belgium, the United Kingdom and the USA; in all three countries the output is primarily for export. The demand does not appear to be slackening.

For the US Army the eventual replacement for the M2 HB is intended to be the Objective Crew Served Weapon (OCSW). See entry in *Light support weapons* section for details.

Description

On the Browning 0.50 M2 HB the operating cycle begins when the firing pin strikes the cap and fires the cartridge. At this point the bolt is locked to the barrel/barrel extension assembly by the breech lock, a lug which is held up into a slot in the bolt by the locking cam fitted at the bottom of the receiver.

After the cartridge fires, the reaction to the explosion drives the moving parts to the rear. After approximately 20 mm of recoil the breech lock pin strikes the breech lock depressors, which push the pin and breech lock lug downward, out of engagement with the bolt. The bolt is thus unlocked, separates from the barrel/barrel extension, and travels to the rear. The barrel and barrel extension are completely halted by the barrel buffer body.

During rearward travel the barrel and extension assembly comes into contact with the accelerator, which pivots on the accelerator pin. The contact point between the barrel extension and the accelerator moves downwards progressively while the barrel extension slides backwards. This downward movement changes the ratio of leverage exerted by the tip of the accelerator against the bolt, so that the bolt is thrown backwards at an increased speed relative to the other recoiling parts. In addition, the gradual transfer of energy from the barrel extension to the bolt, via the accelerator, also damps the shock given by the barrel extension against the barrel buffer. A further advantage accruing from the accelerator is that the increase in bolt speed assists in extracting the spent case from the chamber in a progressive manner.

0.50 M2 HB heavy machine gun on M63 air defence tripod mounting

The rearward travel of the bolt compresses the return spring and is finally halted when the bolt strikes the buffer plate. The buffer plate discs absorb the shock and, rebounding, start the bolt forward once more, this movement being continued by the energy stored in the return spring. The lower edge of the bolt strikes the tip of the accelerator, turns it forward, and unlocks the barrel and barrel extension assembly from the barrel buffer, whereupon the barrel buffer spring drives the barrel and barrel extension assembly forward. This forward motion is augmented by that of the bolt and its driving spring as the bolt reaches its foremost position in the barrel extension.

At the end of the forward movement the breech lock lug, driven by the barrel extension, rides up the inclined plane of the breech lock cam and enters the recess in the bottom of the bolt, thereby locking the bolt to the barrel/barrel extension assembly.

During the recoil of the bolt the spent case in the breech is withdrawn by the T-slot in the face of the bolt. At the same time, a fresh cartridge is pulled from the belt by the extractor-ejector. The extractor is gradually forced down by the cover extractor cam until the fresh cartridge is located on the barrel axis. This downward movement also forces out the spent case which is then ejected downwards. As the bolt returns, so the fresh cartridge is inserted into the chamber.

Also during the recoil movement of the bolt, the belt feed lever engages in a diagonal groove in the top of the bolt and is pivoted as a result of the bolt's movement. This forces the belt feed slide to move to the left in the slot in the feed cover. The belt feed pawl engages behind the next cartridge. As the bolt moves forward the belt feed lever is pivoted in the reverse direction and the belt feed slide moves to the right, ready to begin the next feed.

The cartridge headspace is critical on the standard M2 HB. Ideally, it should be adjusted by means of go/no go gauges as described in the various field manuals relating to the M2 HB, but in default of the correct gauges an approximation may be achieved by screwing the barrel in fully, then unscrewing it two clicks. Should the gun fire sluggishly, the barrel should be unscrewed by one further click.

Special mountings and additional devices for the M2 HB abound. Typical of the latter is a trigger safety modification produced by w + f of Bern Switzerland, while RAMO, Saco and FN Herstal produce quick-change barrel kits, firing solenoids and special mountings.

Data

Cartridge: 12.7 × 99 mm (0.50 M2 Ball)
Operation: short recoil
Locking: projecting lug
Feed: disintegrating link belt
Weight: 38 kg
Length: 1.651 m
Barrel: 1.143 m
Rifling: 8 grooves, rh, 1 turn in 381 mm
Sights: fore, blade; rear, leaf aperture
Muzzle velocity: 930 m/s
Rate of fire: cyclic, 450-600 rds/min
Max range: 6,800 m; effective, 1,500 m

Manufacturers

Ramo Manufacturing Inc, 412 Space Park South, Nashville, Tennessee 37211.
Saco Defense Inc, 291 North Street, Saco, Maine 04072.
(See also Licence production).

Status

In production.

Service

US forces and at least 30 other countries.

Licence production

BELGIUM
Manufacturer: FN Herstal SA
Type: 0.50 M2 HB and M2 HB/QCB
Remarks: Produced in standard and quick-change barrel forms. See separate entries under *Belgium* for details.
UNITED KINGDOM
Manufacturer: Manroy Engineering Limited
Type: 0.50 M2 HB and M2 HB-QCB
Remarks: Produced in standard and quick-change barrel forms. See separate entry under *UK* for details.

UPDATED

Saco Defense 0.50 M2 HB QCB machine gun system

Description

In 1978 Saco introduced a modification to the 0.50 M2 HB machine gun to provide fixed headspace and a quick-change barrel. All the other features of the weapon have been retained, and ammunition may be fed to the gun from either side with a simple adjustment. The barrel is Stellite-lined and chromium-plated for long service life.

Almost all the standard M2 HB components are unchanged; the only new parts are the barrel, barrel extension assembly and barrel support. It is possible to convert a standard M2 HB gun to the fixed headspace configuration by incorporating QCB kit components. Fitting by a technician, followed by infrequent inspection, is sufficient to ensure continuous safe and reliable operation.

A variety of fire control systems and mounts is available. Day and night sights, laser aiming devices and laser rangefinders are attached to and detached from the weapon's bracket without loss of zero, using the US Army dovetail mounting configuration and advanced throw lever attachment device.

Data

Cartridge: 0.50 Browning (12.7 × 99 mm)
Operation: short recoil, selective fire
Locking: projecting lug
Feed: disintegrating link belt
Weight: 38.2 kg
Length: 1.651 m
Barrel: 1.143 m
Rifling: 8 grooves, rh, 1 twist in 381 mm
Muzzle velocity: 853 m/s
Rate of fire: cyclic, 450-600 rds/min
Max range: 6,800 m; effective, 1,500 m+

Manufacturer

Saco Defense Inc, 291 North Street, Saco, Maine 04072.

Status

In production.

Service

Several overseas armed forces.

UPDATED

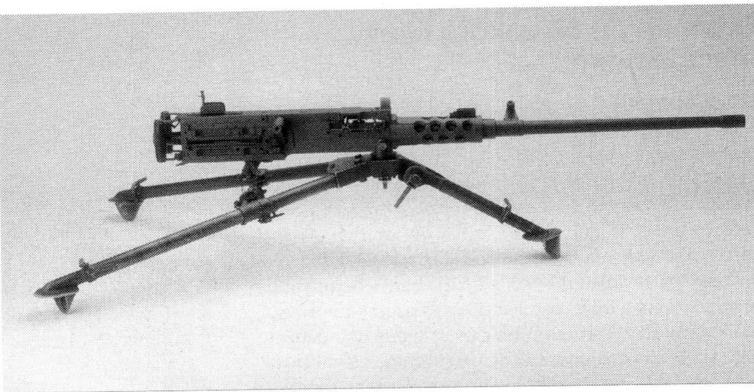

0.50 Saco M2 HB QCB machine gun system

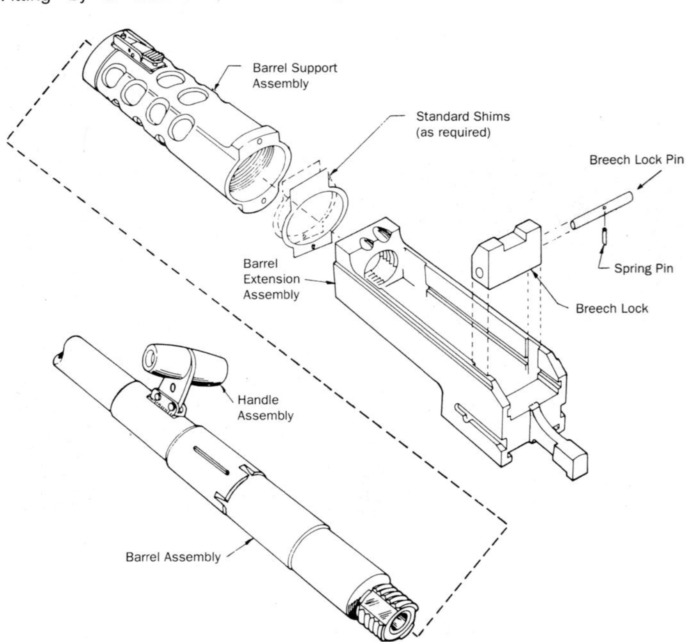

Drawing showing the components of the Saco QCB conversion kit

RAMO 0.50 M2 lightweight machine gun

Description

The RAMO 0.50 M2 lightweight was designed as a response to both ground and air requirements for a reduced weight 0.50 calibre weapon. Utilising the Browning M2 principles, the weapon was upgraded to include an adjustable buffer to vary the rate of fire, a patented quick-change Stellite-lined chrome bore barrel with flash suppressor, 'Max Safe' charging system, and a trigger safety switch. The weapon is constructed of the same basic material as the standard M2 but incorporates a specialised finish to enhance the maintainability of the weapon further, especially in marine environments.

The gun will fire all existing types of 0.50 (12.7 × 99 mm) ammunition, including multipurpose explosive projectiles and the SLAP Saboted Light Armour Penetrator. The rate of fire is adjustable from 550 to 750 rds/min so as to adapt the weapon to a ground support or an air defence role. The lined, chrome bore barrel maximises wear life and barrel accuracy.

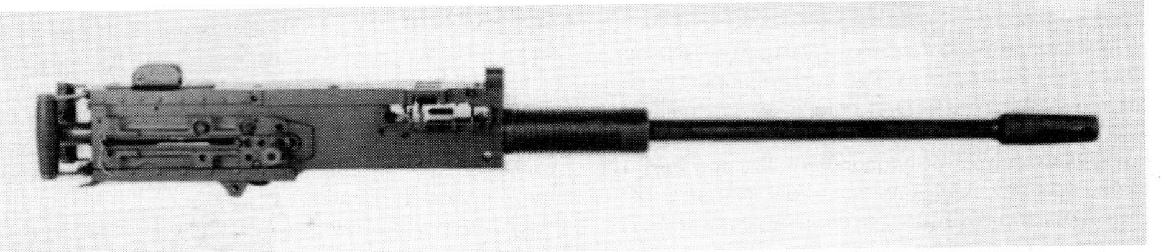

0.50 RAMO M2 lightweight machine gun

The lightweight M2 retains 75 per cent commonality of parts with the standard 0.50 M2 HB weapon. Consequently all these parts are logistically supportable through routine channels. The remaining items peculiar to the M2 lightweight are manufactured by RAMO and are available for worldwide distribution.

Data

Cartridge: 0.50 Browning (12.7 × 99 mm)
Operation: recoil, automatic
Locking: projecting lug
Feed: belt
Weight: complete, 26.72 kg; barrel, 7.48 kg
Length: 1.524 m; barrel, 914.4 mm
Muzzle velocity: 866 m/s
Rate of fire: cyclic, 550-750 rds/min

Manufacturer

RAMO Manufacturing Inc, 412 Space Park South, Nashville, Tennessee 37211.

Status

In production.

VERIFIED

RAMO 0.50 M2 HB-QCB machine gun and kit

Description

The RAMO 0.50 calibre M2 HB-QCB is a standard 0.50 M2 HB Browning machine gun fitted with the RAMO patented Quick-Change/Fixed Headspace Kit. This feature allows the gunner to change the barrel in a fraction of the time required by the standard model and without the need to adjust the cartridge headspace clearance. A feature of the RAMO system is that the gun can still be fitted with the standard M2 heavy barrel should the need arise. The advantages of the QCB are, of course, sacrificed, but the availability of the weapon is maintained.

The QCB can be supplied in a new weapon or the QCB feature can be retrofitted to an existing weapon. Only the barrel, barrel extension and barrel support assembly differ from the standard design while still maintaining 100 per cent interchangeability in the remaining weapon parts.

The RAMO quick-change barrel kit allows a quick-change, fixed headspace barrel to be retrofitted to any Browning 0.50 M2 HB machine gun, regardless of manufacturer. The kit may be installed without special tools.

RAMO is the only manufacturer fielding a Quick-Change Barrel Kit design that allows for the use of a standard 0.50 barrel should the operator not have access to a QCB barrel. The system allows for the modification of new conventional M2 HB barrels to the quick-change design. The barrel supplied in the kit can be either standard (unlined, without chrome bore), de luxe (unlined, with chrome bore), or Stellite-lined with chrome bore.

Data

Cartridge: 0.50 Browning (12.7 × 99 mm)
Operation: recoil, selective fire
Locking: projecting lug
Feed: belt
Weight: complete, 38.2 kg; barrel, 10.91 kg
Length: 1.651 m
Barrel: 1.143 m
Muzzle velocity: 929 m/s

Rate of fire: cyclic, 450-600 rds/min
Max range: 6,800 m; effective, 1,800 m

Manufacturer
RAMO Manufacturing Inc, 412 Space Park South, Nashville, Tennessee 37211.

Status
In production.

VERIFIED

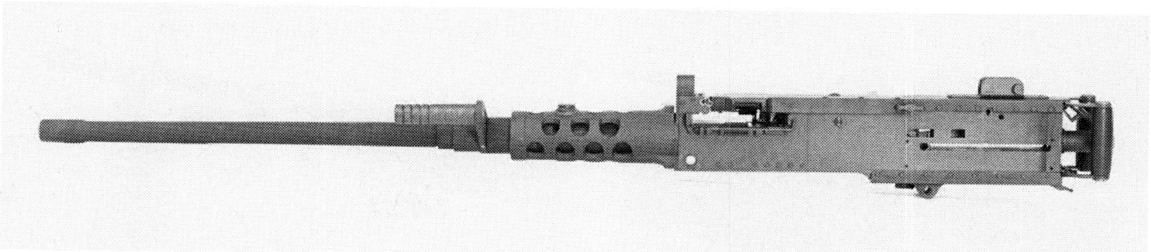

RAMO 0.50 M2 HB-QCB machine gun

7.62 mm M60 general purpose machine gun system

Development
The M60 is the US Army's general purpose machine gun and entered service in the late 1950s. The prime producer has been Saco Defense Inc, a subsidiary of the Chamberlain Manufacturing Corporation.

The original design had some interesting features. The straight line layout allows the operating rod and buffer to run right back into the butt and reduce some of the overall length. The large forehand grip is most convenient for carrying at the hip and the folded bipod legs continue the hand protection almost up to the muzzle. The gun can be stripped using a live round as a tool.

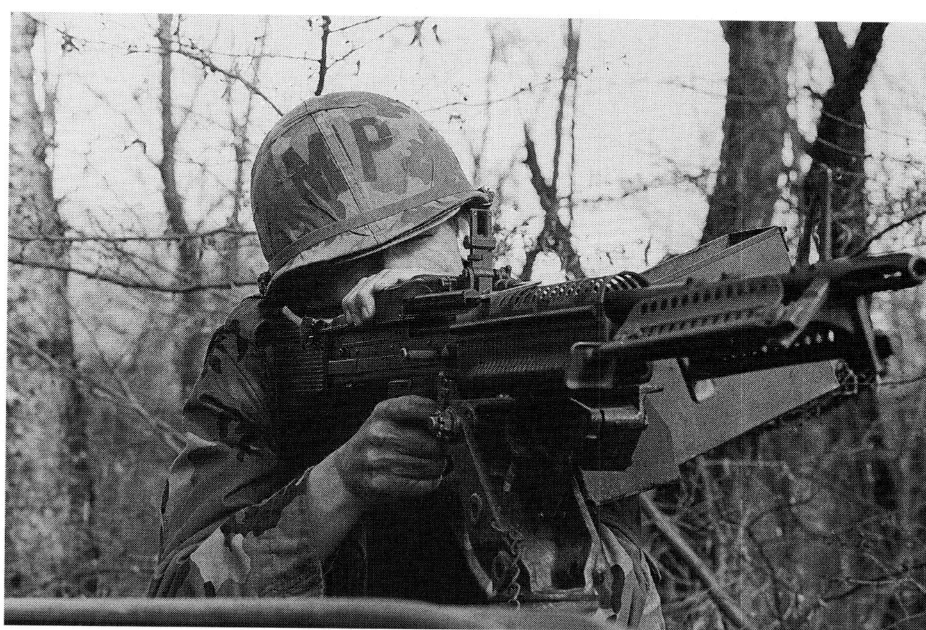

7.62 mm M60 general purpose machine gun

Description
The M60 is a gas-operated weapon. The barrel is drilled radially downwards some 200 mm from the muzzle so after the bullet has passed this point a small proportion of the propellant gas passes through the vent. The gas enters the gas cylinder, passes through a drilling in the side wall of the piston and expands to fill the interior of the piston and the forward section of the gas cylinder. When sufficient pressure has built up the piston is forced rearwards. As soon as it moves, the radial drillings through the piston are moved out of alignment with the gas vent and no further gas enters the cylinder.

The piston moves back and drives the operating rod, which carries the bolt rearwards. The action is that of a short-stroke piston since the piston travel is limited to 60 mm, and a sharp impulsive blow is given to the operating rod which imparts enough energy to carry out the complete cycle of operations.

The M60 has no method of gas adjustment; there is no regulator of any sort adjustable by the firer. The theory of operation is that the piston will move back when enough energy has been supplied to overcome fouling and external friction. When the piston moves, it automatically cuts off its own gas supply and can be said to be self-regulating. It is often referred to as a constant volume system.

The operating rod assembly has a post which rides in the hollow interior of the bolt and this post carries an anti-friction roller which is the bearing surface against the camway cut in the bolt. The post rides initially in the curved portion of the cam slot and tries to rotate the bolt which is restrained by the forward locking lugs riding in longitudinal grooves in the gun body. When the bolt is fully forward it can rotate clockwise and lock in the curved cam path cut in the barrel socket. When the rotation is completed the post is able to enter the last inch of the cam slot, which is parallel to the longitudinal axis, and travel straight forward. The firing pin, which is carried on the post, then strikes the cap.

After the bullet has passed the gas port the pressure starts to build up in the constant volume cylinder. Before there is sufficient pressure to move the piston the bullet has left the muzzle. The piston impinges on the operating rod and this moves back 21 mm before unlocking commences.

The M60 fires full automatic only, at a cyclic rate of 500-650 rds/min. This is slow enough for an accomplished firer to get off a single round. The tactical rate rapid is 200 rds/min.

The M60 originally used the German MG42 feed system employing a system of inner and outer feed pawls driven by a bolt operated feed lever, arranged to move in opposite directions and move the belt in two stages as the bolt moved back and forward again.

As the M60 was progressively developed the feed system was changed. The friction roller at the back of the bolt was retained as was the feed arm on the top of the feed cover in which it operated. The inner and outer pawl system was abandoned and a single pawl system substituted. In this the roller carried by the bolt moving forward swings the feed arm to the right. The feed arm is pivoted so the front end carrying the feed pawl slips to the left over the next round to be fed. As the bolt goes back the pawl moves to the right and lifts the entire weight of the belt through one complete pitch.

On the M60 the gas cylinder and the bipod are both permanently attached to the barrel. The M60E3 has a barrel with gas cylinder only, as the bipod is attached to the front of the receiver assembly. The barrel is chromium-plated and also has a Stellite liner in the 6 in/152 mm of barrel forward of the chamber.

The M60, having a fixed foresight, makes adjustment for zeroing on the rearsight. This means that the number one on the gun must first recognise which barrel is in the gun and secondly know the correct zero setting for both elevation and line for that barrel. In practice this means all barrels are fired from a common zero and the consequent loss of accuracy is accepted.

A variety of fire control systems and mounts is available. Day and night sights, laser aiming devices and laser rangefinders are attached and detached from the weapon brackets without loss of zero, with the US Army dovetail mounting configuration and advanced throw-lever attachment device.

Variants
M60C
Remotely fired, for external mounting on helicopters. No longer in production.
M60D
Helicopter, boat and vehicle weapon. Pintle-mounted in helicopter doorways and vehicle platforms.
M60E2
Solenoid-operated weapon for mounting on AFVs. Has barrel extension tube and gas evacuator.
M60E3
See separate entry.
M60E4
See separate entry.

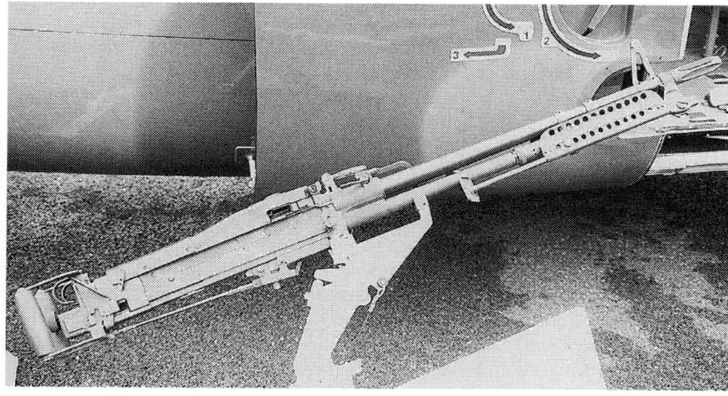

7.62 mm M60D machine gun (T J Gander)

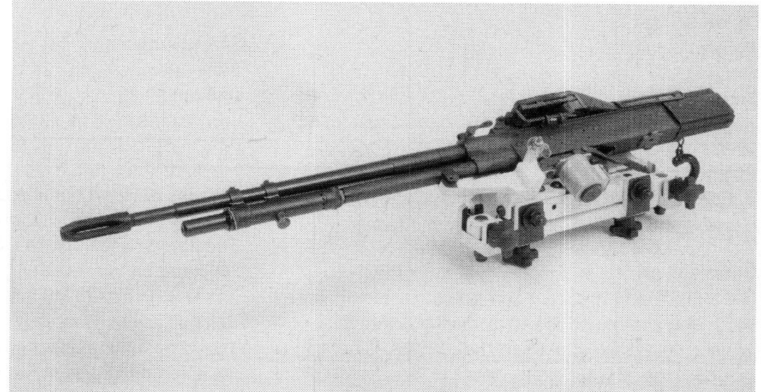

7.62 mm M60E2 machine gun system

7.62 mm M60 general purpose machine gun on bipod

Data
Cartridge: 7.62 × 51 mm
Operation: gas, automatic
Locking: rotating bolt
Feed: disintegrating link belt
Weight: 11.1 kg
Length: overall, 1.105 m
Barrel: excluding flash hider, 560 mm
Rifling: 4 grooves, rh, 1 twist in 305 mm
Sights: fore, fixed blade; rear, U-notch
Sight radius: 540 mm
Muzzle velocity: 853 m/s
Rate of fire: cyclic, 500-650 rds/min; practical, 200 rds/min
Max range: 3,750 m; (effective) bipod, 1,100 m; tripod 1,800 m

Manufacturer
Saco Defense Inc, 291 North Street, Saco, Maine 04072.

Status
In production.

Service
US forces and those of Australia, South Korea, Taiwan and many other countries.

Licence production
TAIWAN
Manufacturer: Combined Service Forces arsenals
Type: 7.62 mm Type 57
Remarks: In production for local service since 1968.

UPDATED

7.62 mm M60E3 light machine gun

Description

The M60E3 light machine gun was developed by Saco Defense Inc to provide a lighter, more versatile 7.62 mm machine gun maintaining all the capabilities of the earlier M60, with several additional features of its own.

The standard barrel is a lightweight assault barrel. Two optional barrels are available, lightweight/short length for assault and increased manoeuvrability and heavy barrel for missions demanding sustained fire. A lightweight bipod is mounted on the receiver eliminating the bipod on the spare barrel. The carrying handle is attached to the barrel for ease of changing and handling when hot without the need for a heat mitten. The forearm is replaced by a forward pistol grip with heat shield mounted forward under the bipod for better control of the weapon when firing. The feed system is modified to permit closing the cover with the bolt forward or in the seared position. A winter trigger guard allows the user to fire while wearing heavy gloves or mittens. The foresight on the standard and lightweight/short length barrels is adjustable for both windage and elevation allowing these barrels to be assembled to the weapon and zeroed without adjustment of the rearsight. The gas system is simplified and provided with interlocking cylinder nuts and a reversible piston. Major assemblies are interchangeable.

A variety of fire control systems and mounts is available that allow the soldier to respond to changing tactical situations. Day and night sights, laser aiming devices and laser rangefinders are attached and detached from the weapon brackets without loss of zero, with the US Army dovetail mounting configuration and advanced throw-lever attachment device.

An M60E3 conversion kit can be used to convert any serviceable, unwelded, M60 receiver to the lightweight M60E3 version. The kit was procured by the US Marine Corps prior to their acquisition of complete M60E3 machine guns. Conversions can be completed at the Saco Defense factory or technical personnel can be trained to carry out the conversion at a qualified facility.

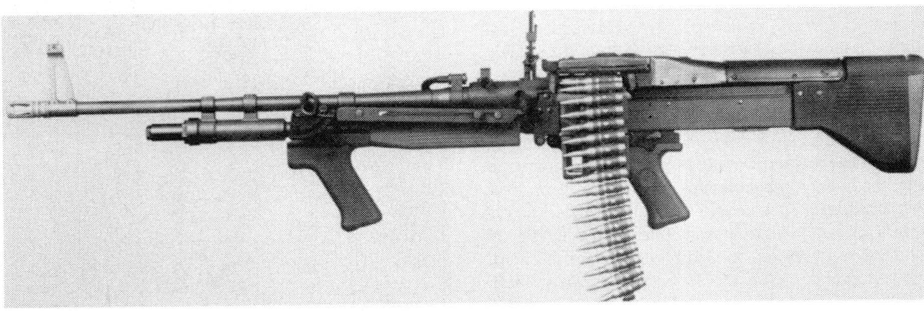

Saco Defense 7.62 mm M60E3 light machine gun

US Navy SEAL with M60E3 machine gun (Kevin Dockery)

Data
Cartridge: 7.62 × 51 mm
Operation: gas, automatic
Locking: rotating bolt
Feed: link belt
Weight: 8.8 kg
Length: 1.077 m
Barrel: 560 mm
Muzzle velocity: 853 m/s
Rate of fire: cyclic, 500-650 rds/min
Max range: 3,725 m; effective, 1,100 m

Manufacturer
Saco Defense Inc, 291 North Street, Saco, Maine 04072.

Status
In production.

Service
In service with US Marine Corps, Navy and Air Force and numerous foreign armed forces.

VERIFIED

7.62 mm M60E4 light machine gun series

Description

The M60E4 light machine gun series was designed by Saco Defense Inc to improve the M60 machine gun family's reliability and user friendliness. The series includes the M60E4 light machine gun (LMG), the M60E4 Mounted, and the M60E4 Coax (coaxial) machine guns. The M60E4 series retains the best features of its predecessors and introduces the following improvements:

The bipod has been strengthened and remains mounted on the receiver

There is an approximately 35 per cent improvement in belt pull power over previous M60 series machine guns. This reduces the effect of dirt or other obstructions on the strength of the pull during the feed process. The increased belt pull also introduces greater flexibility in how the belt is fed

A Picatinny Rail is integrated onto the feed cover, eliminating the bolt on fire control bracket of the early M60 machine guns

The forearm and pistol grip remains a one-piece assembly to provide protection from barrel heat, with added strength designed into the forearm

Three chromed-bore barrels are available for M60E4 series machine guns. A short length sustained fire barrel is standard for the M60E4 light machine gun; the Mounted and Coax M60E4s have the long length sustained fire barrel. The M60E3's short assault barrel remains available for use with any of the M60E4 machine guns. All barrels have a reversible gas piston, a redesigned gas cylinder extension to improve ease of maintenance, and a new firing suppressor that significantly reduces the flash signature

The reliability of the flat spring attaching the trigger assembly to the receiver has been improved. The M60E4 design prevents accidental detachment of the trigger assembly and a possible runaway gun.

Where the configuration of the Mounted and Coax versions allow, the above improvements are incorporated.

Upgrade kits to the M60E4 configuration are available to convert any of the early M60 machine guns. Various fire control systems and mounts are also available.

7.62 mm M60E4 light machine gun

Data

Cartridge: 7.62 × 51 mm
Operation: gas, automatic
Locking: rotating bolt
Feed: link belt from 50- or 100-round box
Weight: (LMG) short barrel, 10.2 kg; long barrel, 10.5 kg; assault barrel, 9.9 kg
Length: (LMG) short barrel, 958 mm; long barrel, 1.077 m; assault barrel, 940 m
Muzzle velocity: 853 m/s
Rate of fire: cyclic, 500-650 rds/min
Max range: 3,725 m; effective, 1,100 m

Manufacturer

Saco Defense Inc, 291 North Street, Saco, Maine 04072.

Status

In production.

Service

Scheduled to enter US Navy service in late 1995.

UPDATED

5.56 mm M249 Squad Automatic Weapon (SAW)

Development

The 5.56 mm M249 Squad Automatic Weapon (SAW) was type classified in 1982. Original plans called for 49,979 for the US Army and 9,974 for the US Marine Corps but these totals were revised downwards during 1985, pending a number of modifications deemed necessary. This lead to the hurried purchase of 1,000 units at the time of the 1991-92 Gulf War. Initial production forecasts were for 70,000 M249s.

In March 1995 FN Manufacturing Inc (FNMI) was awarded a $8,995,110 modification to a firm fixed price (with options) contract for 4,089 M249 SAWs.

5.56 mm M249 squad automatic weapon

Description

The licence-produced M249 is the FN 'Minimi' light machine gun (see entry under *Belgium* in this section) with modifications to meet US military specifications and manufacturing techniques. The main visual difference is that the fixed butt-stock has a revised profile to suit US requirements, as has the forward handguard. It is employed as the squad light machine gun in US Army and Marine Corps service.

During 1995 the US Army type classified the M5 collapsible butt-stock for the M249. Several thousands of these are being procured direct from FN Herstal. US Army Rangers also use an M249 with a barrel only 381 mm long; this model is also used with the M5 collapsible butt-stock.

Planned modifications for the M249 include a blank firing attachment, optical sights, and a mounting suitable for installation on the High Mobility Multipurpose Wheeled Vehicle (HMMWV) series.

Data

Cartridge: 5.56 × 45 mm
Operation: gas, firing fully automatic
Locking: rotating bolt head
Feed: 200-round belts or 30-round magazine
Weight: 6.85 kg; barrel, 1.7 kg
Length: 1.04 m
Barrel: overall, 523 mm or 381 mm (see text)
Rifling: 6 grooves, rh, 1 turn in 178 mm
Sights: fore, semi-fixed hooded post, adjustable for windage and elevation; rear, aperture, adjustable for windage and elevation
Sight radius: 490 mm
Muzzle velocity: 915 m/s
Muzzle energy: ca 175 kgm
Rate of fire: cyclic, ca 750 rds/min
Max range: ca 2,000 m; effective, up to 1,100 m

Manufacturer

FN Manufacturing Inc, Columbia, South Carolina.

Status

In production.

Service

US Army and Marine Corps.

UPDATED

7.62 mm M240G machine gun

Development

The M240 machine gun, also known as the M240D, was originally procured from FN of Belgium in 1976, being a variant of the FN MAG Model 60-40 (qv), subsequently licence-produced as a coaxial machine gun for mounting in M60 tanks - the type was later installed in the M1 Abrams. The original M240 was then followed by the M240C, the main change being that ammunition is fed from the right to suit installations in M2 and M3 Bradley infantry fighting vehicles. The attachment of spade grips in 1987 then resulted in the M240E1 for use on pintle mountings, such as those on the LAV series; the M240E1 is also suitable for mounting on light vessels or as a helicopter door weapon.

Several factors, including the high-reliability rates of the M240 series (demonstrated as 26,000 mean rounds between failure), resulted in a call for the M240 to be converted for ground-mounted use, both on an integral bipod or the M122E1 tripod, thereby bringing the M240 back to its general purpose machine gun origins as the FN MAG. The ground-mounted version is known as the M240G.

7.62 mm M240G machine gun

Initial issues were for Special Forces units and the type has been adopted by the US Marine Corps and US Army Rangers. It is understood that no new weapon production is involved as kits are issued to convert existing M240 machine guns.

Description

In virtually all aspects the M240G and MAG machine guns are identical so reference to the MAG entry under *Belgium* will reveal full details. There are some changes involved in the evolution from M240 to M240G, chiefly in the gas port area and bipod mounting arrangements. The M240G has provision for mounting electro-optical night sights.

All components in the M240 series are interchangeable.

Variants

M240E4

Following the adoption of the M240G by the US Marine Corps the US Army is considering a derivative, the M240E4, as a possible medium machine gun in direct competition with the M60E4. The main visual difference from the M240D is the provision of a forward hand-guard. Other changes include a Picatinny rail for optical sights on the feed cover, a self-levelling non-adjustable bipod, a combined barrel change and carry handle, a composite butt-stock and a new design of ammunition container bracket. A recoil-absorbing mount is available.

On 1st December 1995 the M240E4 was selected by the US Army as their medium machine gun following a shoot off between the SACO M60E4 and the M240E4.

M240T

The M240T is an export model with a weapon-mounted firing solenoid. It is intended primarily as a coaxial machine gun for armoured vehicles.

7.62 mm M240E4 machine gun

1996

Data

M240G

Cartridge: 7.62 × 51 mm
Operation: gas, automatic
Locking: dropping locking lever
Feed: M13 disintegrating link belt
Weight: 11.7 kg; barrel assembly, 3 kg
Length: 1.22 m
Barrel: 627 mm
Sights: fore, blade; rear, aperture and U-notch
Muzzle velocity: 853 m/s
Rate of fire: cyclic, 750 or 950 rds/min
Max range: 3,725 m

Manufacturer

FN Manufacturing Inc, Columbia, South Carolina.

Status

In production.

Service

US Marine Corps and US Army Rangers. M240E4 selected by US Army.

UPDATED

7.62 mm M134 Minigun machine gun

Description

The 7.62 mm M134 Minigun machine gun is based on the Gatling gun principle in which a high rate of fire is achieved by having a number of rotating barrels which fire in turn when the 12 o'clock position is reached.

The Minigun is driven by a 28 V DC or 115 V AC electric motor and produces a steady rate of fire which varies according to type from between 2,000 and 6,000 rds/min as a top rate, down to 300 rds/min as the slowest rate of fire. At a steady 6,000 rds/min the drive motor draws 130 A. The gun itself consists of four groups: the barrel group, gun housing, rotor assembly and bolt assembly.

Six 7.62 mm barrels fit into the rotor assembly and each is locked by a 180° turn. They pass through the barrel clamp which grips the barrel using a special tool to secure the bolt. The barrels are normally parallel but there are other clamps available which are designed to produce varying dispersion patterns by ensuring that the barrels converge at some selected range.

The gun housing is a one-piece casting holding the rotor assembly and providing a mounting for the safety sector and the guide bar. There is an elliptical cam path on the inner surface in which the bolt bearing roller runs.

The rotor assembly is the main structural component of the gun and is supported in the gun housing by ball bearings. The front part of the rotor assembly holds the six barrels. Six bolt tracks are cut into the rotor and six removable tracks are bolted on to the ribs on the rotor. Each bolt track carries an S-shaped triggering cam which cocks and fires the firing pin of one bolt.

There are six bolts each of which is mated to a barrel and locks into the barrel by means of a rotating head: when each is unlocked and withdrawn from the barrel the empty case is pulled out by a fixed single extractor. The firing pin is spring-operated and carries a projection or 'tang' at the rear end. This tang engages the S-shaped triggering cam on the rotor. The side slots on the bolt engage in the bolt tracks of the rotor and this causes the bolt to move round with the rotor.

In operation the red master switch cover must be elevated to expose the switch which controls all the electrics. The firing button on the left-hand trigger grip starts the rotor assembly and barrel assembly rotating in an anti-clockwise direction viewed from the rear. As the rotor assembly turns, the roller on each bolt follows the elliptical cam path on the inner surface of the gun housing. Each bolt in turn picks up a live round from the guide fingers of the guide bar.

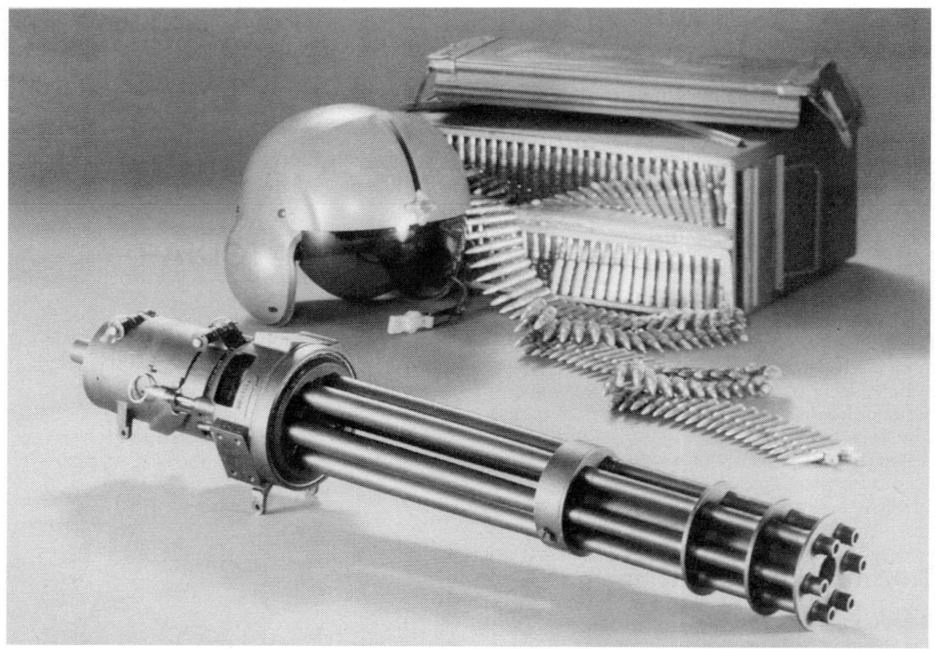

7.62 mm Minigun machine gun

As the bolt is carried round, the roller, working in the groove, moves the bolt forward to chamber the round. The bolt head is now rotated by the cam path in the bolt body and locks into the barrel. The firing pin has been cocked by the tang action in the triggering cam in the rotor and at the top position of the bolt the S-shaped groove releases the firing pin and the round is fired.

The elliptical cam path in the gun housing has a flat profile, or dwell profile, which holds the bolt locked until the bullet has gone and the pressure has dropped to a safe level. The bolt assembly roller enters the reverse segment of the path and the bolt head is unlocked. The cam path carries the bolt to the rear, extracting the empty case which is pushed away by the guide bar and ejected. The bolt has now completed a 360° cycle and is ready to pick up another live round. All six bolts repeat the same procedure in sequence.

Ammunition boxes carry a normal load of 4,000 linked rounds. This is pulled up a plastic chute to the gun. If the length of the chuting exceeds 1.5 m, or the radius of the curve is small, it is advisable to have another motor on the top of the ammunition boxes driving a sprocket which helps by pushing the belted ammunition up to the gun. It is possible to use unlinked ammunition.

Since the entire gun is driven by an external source, a misfired round is simply ejected and thrown down the large section hose with the empties either into a box or, more usually, out over the side.

When the gun ceases firing the bolts cease to pick up ammunition from the guide bar and the bolts close on an empty chamber. Thus there is no danger of a cook-off during the second it takes the barrels to come to rest.

Data

Cartridge: 7.62 × 51 mm
Operation: Gatling action, automatic, 6 revolving barrels
Locking: rotating bolt
Feed: linked belt or linkless
Weight: 16.3 kg; drive motor, 3.4 kg; recoil adaptors (2), 1.36 kg; barrels (6), 1.09 kg; feeder, 4.8 kg
Length: overall, 801.6 mm
Barrel: 559 mm
Sights: vary with employment
Power required: at 6,000 rds/min, steady rate, 130 A
Starting load: 260 A for 100 m/s
Reliability: gun only, not less than 200,000 MRBF
Gun life: 1.5 million rds
Muzzle velocity: 869 m/s
Rate of fire: up to 6,000 rds/min
Dispersion: 6 mrad max
Average recoil force: 0.5 kN

Manufacturer

Lockheed Martin, Armament Systems, Lakeside Avenue, Burlington, Vermont 05401.

Status

In production.

Service

US Army, Air Force, Marine Corps and Navy, and other countries armed forces, including Australia.

UPDATED

5.56 mm XM214 Microgun machine gun

Description

The 5.56 mm XM214 Microgun, also known as the Minigun or Six Pack, is similar to the 7.62 mm M134 described in the previous entry but has several design innovations. These include an access cover lever/safety lever which allows the gun to be made incapable of firing, bolts that are removable from the gun without tools, a clutch mechanism that stops the feeder drive gear upon trigger release and a case ejection sprocket.

The Microgun has been produced as a complete lightweight system which can be aircraft-, pod- or vehicle-mounted or fired from an M122 ground tripod. With 1,000 rounds of ammunition the system weighs 38.6 kg and can be quickly broken down into two equal loads for carrying.

Particular attention has been paid to ease of maintenance. The access cover, gun bolts and feeder can be removed without tools, and the feeder drive gear is a self-timing design which provides correct feeder-gun timing.

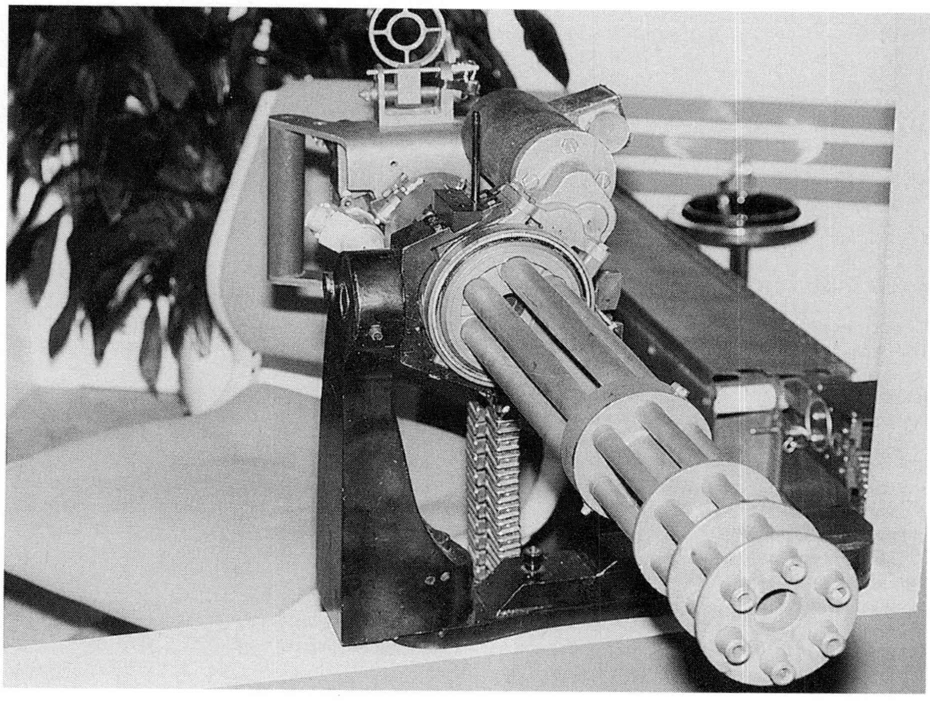

5.56 mm XM214 Microgun machine gun (T J Gander)

Data

Cartridge: 5.56 × 45 mm
Operation: Gatling action, 6 revolving barrels
Locking: rotating bolt
Feed: linked belt
Weight: 10.2 kg
Length: 731.5 mm
Rate of fire: 400 to 6,000 rds/min, selectable
Reliability: 240,000 MRBF
Dispersion: 8 mrad
Gun life: 500,000 rounds
Average recoil: 0.5 kN

Manufacturer

Lockheed Martin, Armament Systems, Lakeside Avenue, Burlington, Vermont 05401.

Status

Available for production.

VERIFIED

0.50 GAU-19/A machine gun

Development

First produced in 1983 and originally marketed as the GECAL 50, the 0.50 GAU-19/A provides lightweight lethal firepower for a large variety of air, land and sea platforms. It has special features for low recoil forces, extended and controlled bursts, and large ready-to-fire ammunition storage. It is a compact weapon that weighs only slightly more than conventional single-barrel 0.50 calibre guns but offers significantly lower recoil forces and two rates of fire. The GAU-19/A can fire 60 rounds in less than two seconds to provide ground suppressive fire out to 1,500 m or more.

The 0.50 GAU-19/A is currently in production for a wide range of applications, including helicopters and ground vehicles; a special mounting kit is available to install the GAU-19/A on the High Mobility Medium Wheeled Vehicle (HMMWV).

Description

Although the GAU-19/A has three barrels, some early models had six barrels. Operating on the Gatling principle and powered from rechargeable vehicle or other batteries, the operating mechanism and construction are virtually identical to that of the 7.62 mm M134 Minigun (apart from the reduced number of barrels) so reference can be made to that entry for details.

The GAU-19/A fires standard NATO 12.7 × 99 mm (0.50 Browning) ammunition and the more recent 0.50 multipurpose and SLAP rounds. These rounds provide improved capability for air-to-air, air-to-ground and air defence missions. A tracer round with visibility to about 1,600 m is also available.

A delinking feeder is used to feed the gun from standard linked belt ammunition containers.

Data

Cartridge: 12.7 × 99 mm (0.50 Browning)
Operation: Gatling action, 3 revolving barrels
Locking: rotating bolt
Feed: linked (M9) or linkless belts

0.50 GAU-19/A rotary machine gun (T J Gander)

0.50 GAU-19/A rotary machine gun (T J Gander)

Weight: 33.6 kg
Length: 1.181 m
Barrel: 914 mm
Rate of fire: dual rate, selectable between 1,000 and 2,000 rds/min
Time to rate: 0.3 s
Muzzle velocity: 884 m/s
Dispersion: 80 per cent of rounds fired, 3 mrad

Manufacturer
Lockheed Martin, Armament Systems, Lakeside Avenue, Burlington, Vermont 05402.

Status
Available.

VERIFIED

0.50 GAU-19/A rotary machine gun (T J Gander)

Ares 5.56 mm light machine gun

Description
The Ares 5.56 mm light machine gun was designed using a minimum number of parts so as to provide a reliable and lightweight weapon. The design uses components from the Stoner 63 system, as well as accessories from weapons such as the M60, M249 and M16. The modular components are made of corrosion-resistant material, stainless steel and aluminium with protective finishes. Modular design allows quick field-stripping without tools.

The 5.56 mm light machine gun is adaptable to both belt and magazine feeds. No variation in firing rates occurs because of environmental conditions, feed type or belt length. The use of a long bolt stroke provides low recoil forces, eliminating the need for a buffer and gas adjustments. The interchangeability of components provides versatility and enables rapid conversion to lightweight and assault gun versions. Barrels with rifling to suit either M193 or M855 (SS109) ammunition are available.

An assault gun variant (weighing approximately 4.2 kg) of the light machine gun is created by removing the bipod and butt-stock and replacing the standard barrel with a lighter/shorter barrel.

Both versions have quick-change barrels and both have an integral sight rail for optical or electro-optical sights.

Data
Cartridge: 5.56 × 45 mm
Operation: gas, automatic
Locking: rotating bolt
Weight: standard with 200 rounds, 8.51 kg; assault version with 200 rounds, 7.02 kg
Length: with butt-stock, 810 mm; assault version, 670 mm
Barrel: standard, 550 mm; assault version, 400 mm
Rate of fire: cyclic, 600 rds/min
Muzzle velocity: with SS109 ammunition, 945 m/s

Manufacturer
Ares Inc, Building 818, Front Street, Erie Industrial Park, Port Clinton, Ohio 43452-9399.

Status
Low rate initial production.

VERIFIED

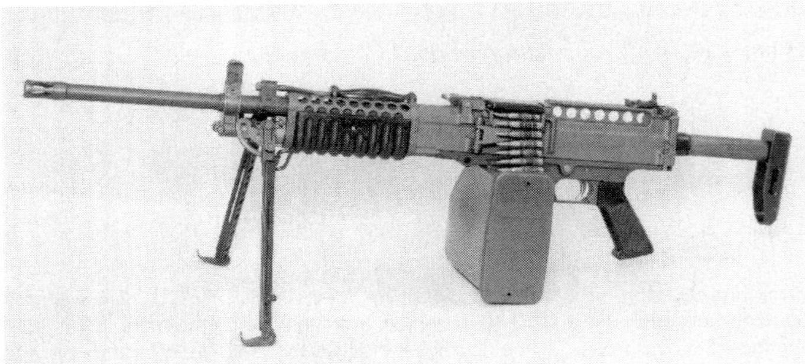

Ares 5.56 mm light machine gun

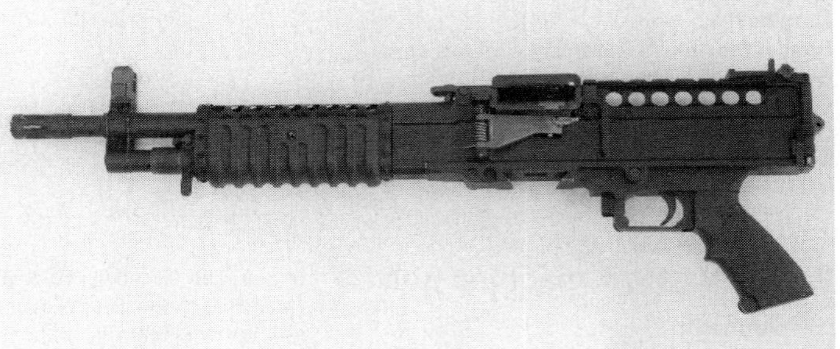

Ares 5.56 mm light machine gun, assault gun version

Colt 5.56 mm M16A2 light machine gun

Description
The Colt M16A2 light machine gun, the Colt Model 750, is a gas-operated, air-cooled, magazine-fed, fully automatic weapon which fires from the open bolt position. The heavier barrel increases stability, accuracy and heat dissipation. The M16A2 light machine gun features a rugged bipod design which deploys and retracts rapidly and easily with a single hand.

The M16A2 light machine gun shares common features with the M16A2 rifle. The light machine gun operating controls and handling characteristics are familiar to soldiers trained in operating the M16 family of weapons while the light machine gun uses the same wide variety of ammunition as the rifle.

The M16A2 light machine gun has wide and square handguards, and a forward grip on the bottom part of the handguard. It utilises the M16's standard magazine and other large-capacity magazines that can hold 90 to 100 rounds. The magazine-fed design reduces the number of parts and the ammunition weight carried by the soldier when the weapon is operated, while it has the ability to provide ammunition for sustained fire faster than most belt-fed weapons.

Data
Cartridge: 5.56 × 45 mm NATO
Operation: gas, fully automatic

Colt 5.56 mm M16A2 light machine gun, right side

Colt 5.56 mm M16A2 light machine gun, left side

Method of locking: rotating bolt
Feed: 30-round magazine; will accept any M16 interfacing magazine
Weight: empty, without magazine, 5.78 kg; with loaded magazine, 6.234 kg
Length: 1 m
Barrel: 510 mm
Rifling: 6 grooves, rh, 1 turn in 178 mm
Sights: fore, adjustable post; rear, aperture, adjustable for windage and elevation
Muzzle velocity: M193, 991 m/s; M855 NATO, 945 m/s
Muzzle energy: M193, 1,722 J; SS109, 1,765 J
Rate of fire: cyclic, 600-750 rds/min
Max range: effective, 800 m

Manufacturer
Colt's Manufacturing Company Inc, PO Box 1868, Hartford, Connecticutt 06144-1868.

Status
In production.

Service
US Marine Corps, US Drug Enforcement Administration, El Salvador, Brazil and various other countries.

VERIFIED

7.62 mm EX-34 Chain Gun® automatic weapon

Description
The 7.62 mm EX-34 Chain Gun® automatic weapon, specifically developed for coaxial use and the smallest member of the Chain Gun® family, is readily adaptable to a variety of armoured vehicle and aircraft gun roles. For coaxial installation, the weapon is configured to permit barrel changing in less than 10 seconds without removing the belt feed. It incorporates the basic features of the proven 30 and 25 mm Chain Guns® and addresses the critical problems of gas, brass and feed normally associated with all enclosed gun installations. As with the the M242, this weapon also ejects fired cases forward (overboard) and capitalises on long dwell time to eliminate gas build-up in cupola or turreted installations.

Compact and lightweight, the 7.62 mm machine gun is physically interchangeable with the M240 coaxial machine gun in M60 tanks and no mounting modification is required to replace the machine guns presently installed.

The coaxial version for armoured fighting vehicles has a long barrel jacket running right to the muzzle. A venturi system draws air forward through the jacket to cool the barrel and also acts as a fume extractor. Empty cases are thrown forward clear of the vehicle.

The weapon was tested by both the US Army and the US Naval Surface Weapons Center, Dahlgren, Virginia. The Navy tests were to determine reliability, durability, performance capabilities, safety and ease of handling. The Navy report stated that the performance of the EX-34 gun during all phases of testing was outstanding.

As a result of extensive tests conducted by the Royal Small Arms Factory, Enfield lock in 1978 and 1979, the British Army purchased a small number of EX-34 Chain Guns®. Following tests, the UK Ministry of Defence adopted the EX-34, initially for coaxial installation in the Warrior IFV. The gun is in serial production by Heckler and Koch (UK), under licence from McDonnell Douglas Helicopter Systems.

Although initially developed as an armoured vehicle weapon, the EX-34 has also been configured to a lightweight helicopter armament subsystem, designated HGS-55, for installation on the McDonnell Douglas 500MD 'Defender' light attack helicopter.

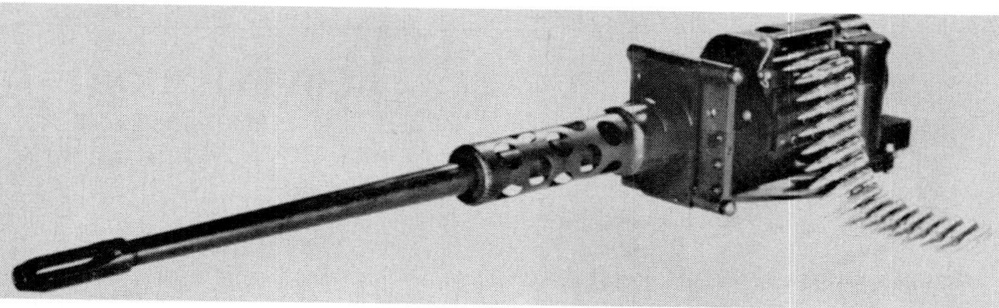

7.62 mm EX-34 Chain Gun® automatic weapon

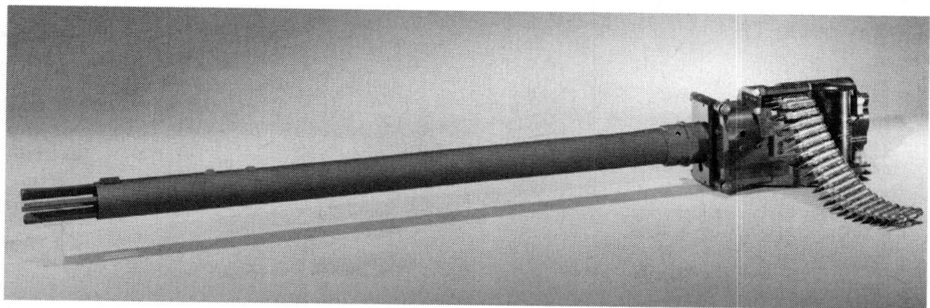

Coaxial version of 7.62 mm EX-34 Chain Gun® automatic weapon

Data
Cartridge: 7.62 × 51 mm NATO (MB linked)
Operation: externally powered from open bolt
Locking: rotating bolt
Feed: disintegrating link belt
Weight: (empty) long, 17.86 kg; short, 13.7 kg
Length: long, 1.25 m; short, 660.4 mm
Barrel: long, 703 mm; short, 580 mm
Rifling: 4 grooves, rh, 1 turn in 305 mm
Muzzle velocity: 862 m/s
Rate of fire: cyclic, 520 rds/min
Time to full rate: 0.15 s
Power required: 0.3 hp (0.22 kW)

Manufacturers
McDonnell Douglas Helicopter Systems, Mesa, Arizona 85205-9797.
Heckler and Koch (UK), British Aerospace Defence Limited, Kings Meadow Road, Nottingham NG2 1EQ, UK.

Status
In production in United Kingdom.

Service
British Army; trials elsewhere.

Licence production
UNITED KINGDOM
Manufacturer: Heckler and Koch (UK)
Type: 7.62 mm L94A1
Remarks: In production. This is the long-barrelled version of the EX-34. See also entry under *United Kingdom*.

VERIFIED

YUGOSLAVIA (SERBIA AND MONTENEGRO)

12.7 mm machine gun NSV-12,7

Description
This is the CIS NSV heavy machine gun (see separate entry) manufactured under licence for service with the former Yugoslav armed forces. It is precisely the same as the CIS weapon, to which reference should be made for dimensions and illustrations.

Manufacturer
Zastava Arms, 29 Novembra 12, YU-11000 Belgrade.

Status
May no longer be in production.

Service
Former Yugoslav armed forces.

VERIFIED

7.62 mm M72B1 and M72AB1 light machine guns

Description
These two light machine guns, together with the M70B1 and M70AB2 assault rifles, belong to the FAZ family of automatic weapons. They differ only in that the M72B1 has a fixed wooden stock whereas the M72AB1 has an easily detachable folding metal stock and is intended to be used by armoured or parachute units. All the weapons of this family have interchangeable magazines, are gas operated and have many parts in common. The main points of difference lie in the fins along the short stretch of the barrel under the gas cylinder and in the shape of the butt. The fins probably help slightly in cooling the hottest part of the bore.

The light machine gun can fire either single shots or automatically, the change lever being on the right of the weapon above the trigger guard and incorporating a safety position. Structurally, although there are many points of difference, the M72B1 and M72AB1 resemble the Kalashnikov RPK (qv) and have a similar performance.

Tactically the weapon is designed for use against ground targets up to 800 m and air targets up to 500 m. The rearsight is graduated from 100 to 1,000 m in 100 m steps and incorporates an engraved windage scale with two mil divisions. Both sights are marked with tritium spots for firing in conditions of poor visibility.

The M72B1 may be equipped with a drum magazine which fits into the normal magazine housing and holds 75 rounds. It weighs 2.175 kg when loaded.

When fitted with a bracket to carry the PN 5 × 80 night vision sight, the M72B1 machine gun takes the designation M72B1 N-PN.

Data
Cartridge: 7.62 × 39 mm
Operation: gas, selective fire
Locking: rotating bolt

Feed: 30-round curved box magazine or 75-round drum
Weight: empty, 5 kg
Length: 1.025 m
Barrel: 540 mm
Sights: fore, cylindrical post; rear, leaf, notch, 100-1,000 × 100 m; night sight, tritium marks
Sight radius: 525 mm
Muzzle velocity: 745 m/s
Rate of fire: cyclic, 600 rds/min
Range: effective, 800 m; anti-aircraft, 500 m

Manufacturer
Zastava Arms, 29 Novembra 12, YU-11000 Belgrade.

Status
May no longer be in production.

Service
Former Yugoslav armed forces.

VERIFIED

7.62 mm M72AB1 light machine gun with metal stock folded and bipod open

7.62 mm M72B1 light machine gun with 75-round drum magazine

7.62 mm M77B1 light machine gun

Description
Originally produced for export, the M77B1 light machine gun, like the M77B1 assault rifle (qv), is based on the 7.62 × 51 mm NATO cartridge. Otherwise it closely resembles the M72B1 light machine gun, although it appears to be supplied only with a 20-round detachable straight box magazine.

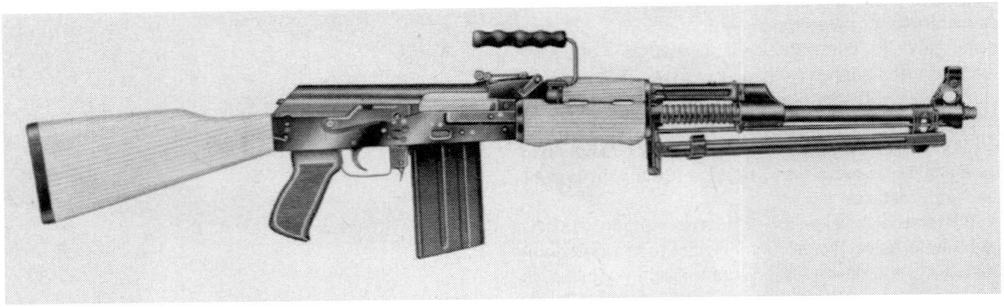

7.62 mm NATO M77B1 light machine gun with bipod folded

Data
Cartridge: 7.62 × 51 mm NATO
Operation: gas, selective fire
Locking: rotating bolt
Feed: 20-round detachable straight box magazine
Weight: gun with empty magazine, 5.1 kg
Length: gun, 1.025 m; barrel, 535 mm
Rifling: 6 grooves, rh, 1 turn in 240 mm
Sights: fore, pillar; rear, tangent notch; sight radius, 525 mm
Muzzle velocity: 840 m/s
Rate of fire: cyclic, 600 rds/min
Range: effective, 800 m

Manufacturer
Zastava Arms, 29 Novembra 12, YU-11000 Belgrade.

Status
May no longer be in production.

VERIFIED

5.56 mm M82 and M82A light machine guns

Description
These machine guns were developed as part of a 5.56 mm family, the other members of which are the M80 and M80A assault rifles. The design is obviously derived from the Kalashnikov-inspired M72 gun and is gas operated. The manufacturer claimed that a specially designed gas regulator ensures faultless operation even with ammunition of varying energy levels loaded into the magazine.

Data
Cartridge: 5.56 × 45 mm (M193 or SS109)
Operation: gas, selective fire
Locking: rotating bolt
Feed: 30-round detachable box magazine
Weight: empty, 4 kg
Length: 1.02 m
Barrel: 542 mm
Rifling: 6 grooves, rh, 1 turn in 178 mm
Sight radius: 498 mm
Muzzle velocity: 1,000 m/s
Range: effective, 400 m

Manufacturer
Zastava Arms, 29 Novembra 12, YU-11000 Belgrade.

Status
May no longer be in production.

VERIFIED

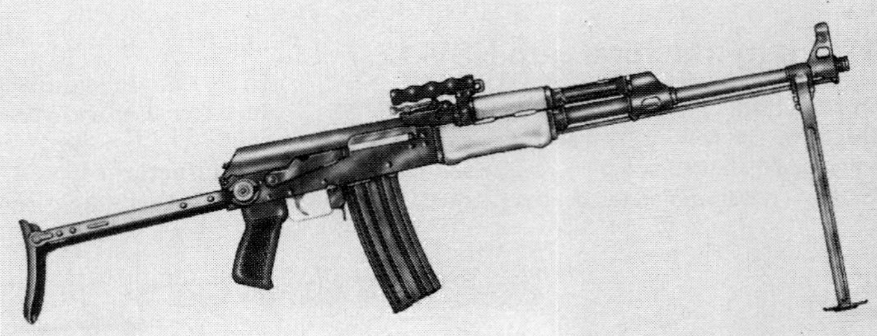

5.56 mm M82 machine gun

5.56 mm M82A machine gun

7.62 mm M84 general purpose machine gun

Description
Produced for the former Yugoslav armed forces, the M84 general purpose machine gun was designed to use the CIS 7.62 × 54 mm cartridge. It is a rotary bolt locked, belt-fed gun and all available data suggests that it closely resembles, both in appearance and mechanically, the PKM version of the CIS 7.62 mm PK machine gun family (qv).

The M84 fires either from a bipod, as shown, or from a tripod and is claimed to have an effective range of up to 1,000 m against both ground and aerial targets. The air-cooled barrel is capable of firing 500 rounds continuously before changing. It is fed by 100- or 250-round belts for which belt boxes are issued.

Also in production at one time was the 7.62 mm PKT coaxial machine gun for armoured vehicles.

Data
Cartridge: 7.62 × 54 mm
Operation: gas, automatic
Locking: rotating bolt
Feed: 100- or 250-round belt

Weight: 10 kg; tripod, 5 kg
Length: 1.175 m
Barrel: with flash hider, 658 mm
Rifling: 4 grooves, rh, 1 turn in 240 mm
Sights: fore, cylindrical post; rear, vertical leaf
Sight radius: 663 mm
Muzzle velocity: 825 m/s
Rate of fire: cyclic, ca 700 rds/min; practical, up to 250 rds/min
Range: effective, 1,000 m

Manufacturer
Zastava Arms, 29 Novembra 12, YU-11000 Belgrade.

Status
May no longer be in production.

Service
Former Yugoslav armed forces.

VERIFIED

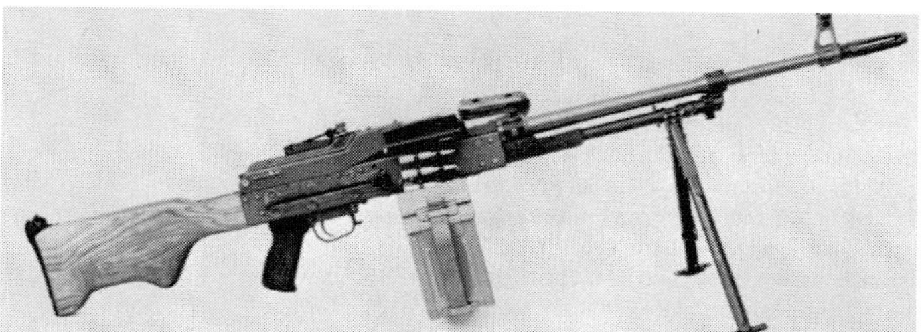

7.62 mm M84 general purpose machine gun with bipod open and belt box fitted. Compare with Soviet PKM

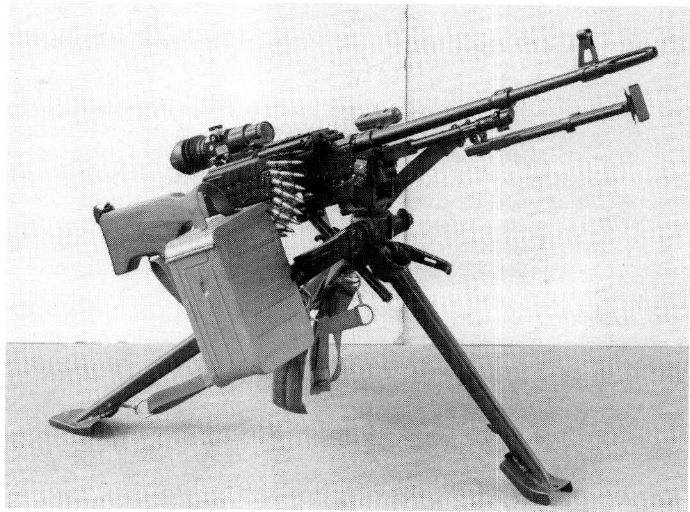

7.62 mm M84 machine gun on tripod, with optical sight ON-M80

7.92 mm M53 general purpose machine gun

Description
Intended for use either as a light machine gun, fired from its bipod, or as a sustained fire machine gun using a tripod, the M53 general purpose machine gun is generally similar to the German MG42 machine gun.

As with the the MG42, the M53 operates on the short-recoil principle with bolt acceleration and recoil intensification by a muzzle booster. A full description will be found in the appropriate entry under *Germany* in this section.

As a light machine gun the M53 is intended for use against ground targets up to 800 m with maximum effect up to 500 m. Corresponding ranges for the weapon on its tripod are 1,000 and 600 m, but two or more such weapons can provide effective fire against multiple targets at ranges up to 1,500 m.

The cyclic rate of fire is 800 to 1,050 rds/min with a practical combat rate of 300 to 400 rds/min. Unless special cooling arrangements are made the barrel must be changed after 150 rounds of sustained fire. The gun may be used either with a 50-round drum magazine or with 50-round belts.

A buffered tripod is fitted with an automatic recoil-operated search attachment. Another attachment enables it to be used for anti-aircraft fire.

Data
Cartridge: 7.92 × 57 mm
Operation: short recoil, gas assisted, automatic, air-cooled, quick-change barrel
Locking: projecting lugs
Feed: 50-round drum magazine or 50-round continuous metal link belt
Weight: gun only, 11.5 kg; bipod, 1 kg; tripod, ca 22 kg; anti-aircraft attachment, ca 1.7 kg; drum magazine, 50 rounds, ca 2.3 kg
Length: ca 1.21 m
Barrel: 560 mm
Sights: fore, barleycorn; rear tangent, notch, 200-2,000 m
Sight radius: 430 mm
Muzzle velocity: 715 m/s
Rate of fire: cyclic, 800-1,050 rds/min; practical, 300-400 rds/min
Range: (effective) bipod, 500-800 m; tripod, 600-1,000 m

Manufacturer
Zastava Arms, 29 Novembra 12, YU-11000 Belgrade.

Status
Probably no longer in production.

Service
Former Yugoslav armed forces.

VERIFIED

M53 in light machine gun role on bipod

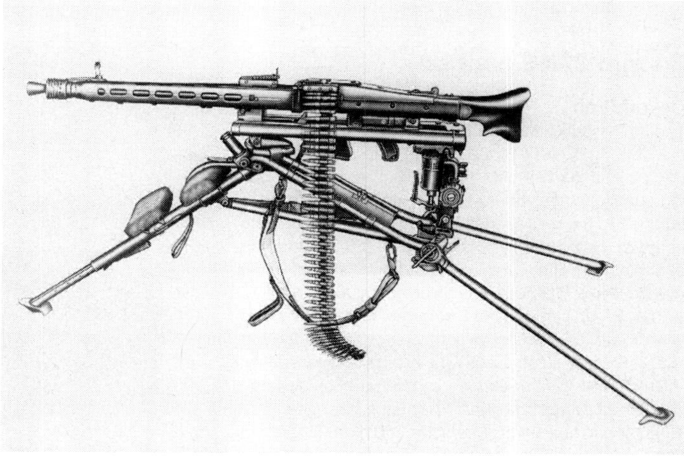

7.92 mm M53 general purpose machine gun on buffered tripod

CANNON

COMMONWEALTH OF INDEPENDENT STATES

23 mm ZU-23 cannon

Development

The ZU-23 is one of the most frequently used cannon in the armoury of the CIS ground forces and has gained an enviable reputation as an anti-aircraft weapon. In general concept it dates from a gun that appeared on Soviet aircraft just after the Second World War. The first specimens to be examined in the West came from fighters shot down during the Korean War. Since then the ZU-23 has remained in service with what appear to be very few alterations to either gun or ammunition, but some advances in mountings and tactical use. In various Middle East conflicts it demonstrated useful air defence when mounted in pairs and fours (ZSU-23-2 and ZSU-23-4), its most formidable features being its rate of fire and range.

Description

The 23 mm ZU-23 cannon is a gas-operated gun with a vertically moving breech block locking system which drops to unlock. The breech block is raised and lowered in guideways in the body of the gun and its position is controlled by cams on the piston extension.

The gun stops firing with the working parts held to the rear on the sear. When the trigger is operated and the sear is lowered, the return spring drives the working parts forward. A round is rammed by the separate rammer and extractor through the link and into the chamber. As the piston extension moves forward, three diagonal grooves cut across the top surface move a cross-feed slide outwards and the feed pawls are retracted. The extractor on the rammer engages the rim of the round in the chamber and the breech block is raised and pushed slightly forward by the cams on the piston extension raising it in the inclined guideways in the gun body. The firing pin is released during the last 20 mm of breech block movement. The piston extension holds the block in position and is prevented from bouncing.

The gun fires and some of the propellant gases are diverted through the gas regulator at a point 368 mm from the commencement of rifling. The regulator is chromium plated and has two ports of 3.2 and 3.4 mm diameter. The gases enter the gas cylinder which is also chromium plated. The gas piston is driven back, and after 10 mm of free piston movement the breech block is lowered by cams on the piston extension. The block moves back as it goes down and moves the extractor back to provide primary extraction. As the piston continues back, the extractor withdraws the empty case which is forced down through guides and out of the bottom of the gun. Further movement of the piston draws the feed slide across and the one-piece spring-loaded feed pawl pulls the belt across. The piston extension is checked by the buffer and thrown forward. If the ammunition is depleted a last round lever, operated by the linked round and connected to the firing mechanism, ensures that the last round is left on the feed tray ready for ramming so that the mechanism need not be recocked before a new belt is placed in the feed tray.

The gun can be fed from either side. To effect the change, only the feed pawl, the three-tracked cam plate on the piston extension and the retaining pawls have to be changed.

Data

Cartridge: 23 × 152 B
Operation: gas, automatic
Locking: vertically rising block
Feed: disintegrating link belt
Weight: 75 kg; barrel with muzzle attachment, 27.2 kg
Length: overall, 2.555 m; receiver, 953 mm
Barrel: with muzzle attachment, 2.01 m; barrel only, 1.88 m
Rifling: rh, 10 grooves increasing over 920 mm from 1 turn in 1.15 m to 1 turn in 575 mm; then constant to muzzle
Muzzle velocity: 970 m/s
Chamber pressure: 3,100 kg/cm^2
Rate of fire: cyclic, 800-1,000 rds/min
Max range: horizontal, 2,500 m; vertical, 1,500 m

Manufacturer

State factories.
Marketed by: Rosvoorouzhenie, 18/1 Ovchinnikovskaya Emb, 113324, Moscow, Russia.

Status

In production.

Service

CIS and former Warsaw Pact armies and other nations in Africa and the Middle East, including Egypt.

Licence production

BULGARIA
Manufacturer: Kintex
Type: ZU-23
Remarks: Produced for ZU-23-2 air defence mountings and various armoured vehicle installations.
EGYPT
Manufacturer: Abou Zaabal Company for Engineering Industries (F.100), 21487 West Cairo.
Type: 23 mm ZU
Remarks: Manufactured primarily for installation on air defence mountings, including ZSU-23-2, Ramadan 23 and Sinai 23.
POLAND
Manufacturer: Zaklady Mechaniczne Tarnow
Type: ZU-23
Remarks: Produced on ZU-23-2, ZUR-23-2S JOD land-based air defence mountings and ZU-23-2MT naval mounting.

UPDATED

Receiver and feed of 23 mm ZU-23 cannon

23 mm ZU-23 cannon

30 mm 2A42 cannon

Description

The 30 mm 2A42 cannon forms the main armament of the BMP-2 MICV, BTR-80A and BTR-90 APCs, and the BMD-2 and BMD-3 airborne combat vehicles. It is a long-barrelled, gas-operated gun with a dual feed. In addition to the dual feed the 2A42 also has a dual rate of fire which may be selected at either 200-300 rds/min or a minimum of 550 rds/min. The barrel is fitted with a muzzle brake. Service life is given as 6,000 rounds.

The 2A42 can be field stripped into eight basic assemblies. The number of components used in the gun is 578. Firing is electrical using a 27 V DC power supply.

The 2A42 is intended as a multipurpose weapon capable of defeating light APC armour at ranges up to 1,500 m and soft-skinned vehicles and material at up to 4,000 m. It also has a capability against slow-flying aircraft at altitudes up to 2,000 m and slant ranges of up to 2,500 m.

Ammunition fired includes HEI, HE-T, AP-T and APDS-T.

Data

Cartridge: 30 × 165 mm
Operation: gas, automatic
Feed: dual, disintegrating link belt
Weight: gun, 115 kg; barrel, 38.5 kg
Length: gun overall, 3.027 m
Rifling: 16 grooves, rh, 1 turn in 715.5 mm
Muzzle velocity: 960 m/s
Recoil force: 40-50 kN
Rate of fire: cyclic, 200-300 rds/min or a min of 550 rds/min

30 mm 2A42 cannon

Manufacturer

PA Tulamashzavod, Mosin Street 2, 300002, Tula, Russia.
Marketed by: Rosvoorouzhenie, 18/1 Ovchinnikovskaya Emb, 113324, Moscow, Russia.

Status

In production.

Service

CIS and former Warsaw Pact armies.

Licence production
BULGARIA
Manufacturer: Kintex
Type: 2A42
Remarks: Produced for various armoured vehicle installations.

CZECH REPUBLIC
Manufacturer: ZTS
Type: 2A42
Remarks: Produced for locally built BMP-2 MICV.

INDIA
Manufacturer: Indian Ordnance Factories
Type: 2A42
Remarks: Produced for Sarath, locally built BMP-2 MICV.

UPDATED

FRANCE

20 mm Type 20 M 621 cannon

Description

The 20 mm Type 20 M 621 cannon is an electrically controlled cannon designed for use on light vehicles but which also has applications in any field where its two major characteristics of light weight and low recoil are at a premium. It is therefore found on helicopters, light aircraft and rivercraft.

The locking system of the gun makes use of two swinging locking pieces, one in each side of the bolt, projecting outwards into the bolt way and engaging in recesses in the body. After the round is fired these locking pieces are retracted by gas action. There is a gas vent on each side of the barrel; gas from these drives back two pistons which force back the supports that secure the locking pieces, these are then driven inwards and the residual gas pressure enables the bolt to move back. When the locks are fully in, the block is blown to the rear.

The 20 M 621 can fire at single shot from a closed breech or at 750 rds/min. The rate of fire is controlled electrically. In the event of a misfire it is possible to recock the gun by hand using a wire cable provided with a T-handle or by an electrical or hydraulic recocking unit.

The 20 M 621 uses mainly the US 20 mm M56 cartridge, or the M55 or M53, all of which are electrically fired, allow ready variation of the rate of fire and also permit both open and closed breech firing. The US M12 link is used.

The first of the different feed modes is the usual flat feed with the belt coming in from one side and the empty links ejected on the opposite side. The direction of feed can be reversed by the reversal of the sprocket feed and by changing a few parts. The second method is referred to as the enveloping feed mechanism. This uses the same arrangements except that the axis of the sprocket drive has been raised so that the rounds enter the gun from a slightly elevated position and the links are ejected on the same side as the feed and flung out under the incoming belt. With both methods of feed, the electrical control box enables the gunner to select single-shot fire with a closed breech operation, or automatic fire from an open breech. At automatic the length of the burst can be fixed and a record maintained of the rounds fired.

20 mm Type 20 M 621 cannon mounted on HMMWV vehicle

Using the various available mounts, the 20 M 621 can be fitted to helicopters (Pod Mount 20-621; Puma Mount 19A; Gazelle Mount 22A or Ecureuil Mount 23A); trucks (Mount 15A); boats (Mount 15A), or in a number of turrets.

Data

Cartridge: 20 × 102 mm
Operation: gas, automatic
Locking: delayed blowback with locked breech
Weight: 49 kg; cradle, 12.5 kg
Length: 2.207 m
Width of gun cradle: 202 mm
Height of gun in cradle: 245 mm
Muzzle velocity: 990-1,026 m/s, according to ammunition
Rate of fire: cyclic, 750-800 rds/min
Recoil force: average, 250 daN

Manufacturer

Giat Industries, 13 route de la Minière, F-78034 Versailles-Satory Cedex.

Status

In production.

Service

French armed forces and several other armed forces.

VERIFIED

20 mm Type 20 M 693 cannon

Development

The 20 M 693 cannon embodies many features of the 20 M 621 from which it is derived but differs principally in that it fires cartridges of the more powerful HS820 family instead of the electrically primed US M56 and similar cartridges fired by the 20 M 621. As the HS820 cartridges are percussion primed an electrical power supply is not required for the firing operation. Although provision has been made in the 20 M 693 design for remote electrical operation, it can be fired under local control without electricity.

In French Army use the 20 M 693 forms the main armament of the 76 T2, 53 T4 and 53 T2 air defence mountings.

Description

The principle of operation is delayed blowback with positive breech locking. Immediately before firing, with the breech block fully forward and a round in the chamber, the breech block is locked in position by two swinging locking pieces which act as struts between the block and the main body of the gun. When the round is fired, some propellant gas is diverted through two vents a short distance from the breech and impinges on the faces of two pistons disposed symmetrically on either side of the chamber causing them to move rearwards (see below). As they do so they rotate the breech block locking pieces about their pivots; and when the locking pieces have been disengaged from the breech block the residual gas pressure drives it rearwards to carry out the mechanical cycle of operations.

While the breech remains locked, the gas pressure inside the bore causes the whole barrel and body assembly to recoil, moving rearwards in its mounting cradle. When the breech opens, the gas pressure ceases to accelerate the gun body directly and accelerates the breech block instead; this produces a rearward force on the gun body which rises to a peak when the breech block is at its rearmost point. Countering these rearward pressures, gases expanding through the muzzle brake after the bullet has left the barrel exert a forward pull on the barrel. The net recoil energy to which these factors contribute is absorbed by the recuperator assembly in the gun cradle: moreover the actual period of recuperation is arranged to be somewhat longer than the cycle time of the firing mechanism, so that the barrel and body are moving forwards in the counter-recoil motion when the next round is fired. With this floating firing arrangement much of the rearwards force produced by the propellant gas while the breech is locked is dissipated in extinguishing the counter-recoil momentum of the gun assembly, and the resultant pull of the gun on the mounting trunnions is very small. Furthermore the symmetrical arrangement of the moving parts about the centreline of the barrel ensures that lateral stresses which might deflect the gunner's aim are kept to a minimum.

Two ammunition belts, one on each side, can be coupled to the feed mechanism through flexible feed chutes, between which there is sufficient room for the

Dual 20 mm Type 20 M 693 cannon in 76 T2 air defence mounting

ejection of empty links. Two side sprocket shafts are driven independently through ratchet and pawl assemblies, and these are driven by a rack and pinion arrangement operated by the breech block as it reciprocates. A simple changeover lever, remotely actuated through a flexible cable, engages the rack with the left- or right-hand pinion according to the gunner's ammunition requirements. The feed mechanism is well to the rear of the gun, thus reducing vehicle installation problems.

In addition to the choice of ammunition type from the two belts, the gunner can choose single shot or burst fire, by using a selector switch which also incorporates a safety position. The gun will normally be fired remotely, the trigger being solenoid operated; in this case the firing mode is selected by a control on a remote-control box. With the gun selector switch set for automatic fire, the control provides for single shot limited burst or full-automatic fire and incorporates a safety position.

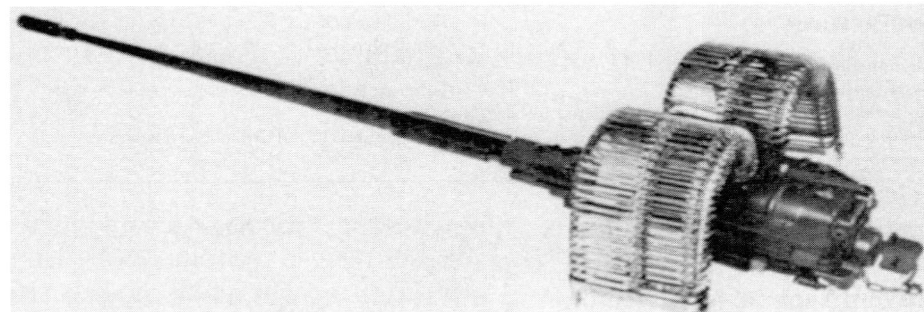

20 mm 20 M 693 cannon with flexible ammunition feed chutes

Variant

20 mm G12 cannon

Numbers of 20 M 693 cannon were supplied to South Africa where they were modified to the local G12 standard for local production. See separate entry under *South Africa* in this section for details.

Data

Cartridge: 20 × 139 mm
Operation: delayed blowback with locked breech, selective fire
Feed: dual-selectable belt
Weight: gun only, 72 kg; cradle, 10.5 kg; barrel with muzzle brake, 25 kg
Length: gun with cradle, 2.695 m
Barrel: with muzzle brake, 2.065 m
Width: gun with cradle, 205 mm
Height: gun only, 236 mm; gun with cradle, 266 mm
Muzzle velocity: standard rounds, ca 1,050 m/s; APDS, ca 1,300 m/s
Rate of fire: cyclic, 900 rds/min
Recoil force: max, 450 daN
Recoil distance: ca 40 mm
Counter-recoil run-out: ca 20 mm after last round
Power drain: 24 V DC, 6-7 A for remote operation
Belt traction: 2 m vertical

Manufacturer

Giat Industries, 13 route de la Minière, F-78034 Versailles-Satory Cedex.

Status

In production.

Service

French armed forces and South Africa.

Licence production

SOUTH AFRICA
Manufacturer: LIW
Type: 20 mm G12
Remarks: See separate entry under *South Africa* for details.

UPDATED

25 mm Type 25 M 811 cannon

Description

The 25 mm Type 25 M 811 is an automatic cannon with single or dual feed, intended for mounting in armoured or soft vehicles and on air defence mountings.

Operation is by an electric motor which drives a camshaft lying at the side of the receiver. This shaft has a spiral cam groove which engages with a lug on the bolt, so that as the shaft revolves the bolt is moved back and forth. The shaft is also geared to the feed mechanism, so that feed is in strict synchronisation with the bolt's movements; this is of importance since the design allows for various rates of fire as well as for single shots.

Data

Cartridge: 25 × 137 mm
Operation: mechanical, selective fire
Feed: dual belt
Length: with cradle, 2.64 m
Weight: with cradle, ca 105 kg, including control electronics
Rate of fire: single shot, 150 and 650 rds/min, or 150 and 400 rds/min
Recoil force: 1,300 daN

Manufacturer

Giat Industries, 13 route de la Minière, F-78034 Versailles-Satory Cedex.

Status

In production.

UPDATED

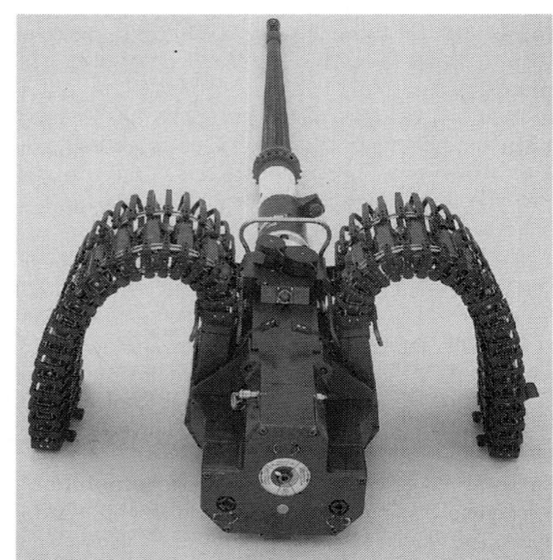

25 mm Type 25 M 811 cannon

30 mm Type 30 M 781 cannon

Description

The 30 mm Type 30 M 781 cannon uses a similar system of operation to the 25 mm 25 M 811 (see previous entry), an external electrically driven camshaft which drives the bolt back and forth. It was designed principally for helicopter applications, including the Tigre, but could be employed in a variety of ground and vehicle roles.

The 30 M 781 uses single feed and has a rate of fire which is variable up to a maximum of 750 rds/min. Single shots can also be fired, and the length of burst can be precisely limited.

The mechanical design makes it possible to actuate the gun for training or maintenance without the need to fire any ammunition.

The 30 M 781 cannon has been proposed as part of an add-on armament kit for the American HMMWV vehicle series.

Data

Cartridge: 30 × 150 B
Operation: externally powered, selective fire
Feed: belt
Length: 1.92 m
Weight: 65 kg, including control electronics
Rate of fire: variable to 750 rds/min
Recoil force: 600 daN

30 mm Type 30 M 781 cannon mounted on HMMWV

Manufacturer

Giat Industries, 13 route de la Minière, F-78034 Versailles-Satory Cedex.

Status

Preproduction.

UPDATED

GERMANY

Mauser MK 30 × 173 mm cannon

Description

The Mauser MK 30 × 173 mm cannon was developed to meet the emerging threats for light air defence and combat vehicle armament systems. Features such as high firepower, lightweight but rigid design, reliability and ease of operation make the MK 30 suitable for all applications in the infantry and air defence fields.

The Mauser MK 30 is fully gas operated by means of gas pistons for bolt movement and ammunition feed. All movable parts, such as return springs, buffers and gas pistons, are installed along the gun axis to eliminate transverse loads and improve dispersion. The breech system consists of double supporting flaps with a long rigid locking time.

The receiver assembly is forged and its special configuration ensures reliability under the most adverse environmental conditions. The gun itself is installed in a simple lightweight cradle in which it can recoil.

Mechanical simplicity ensures easy maintenance and the gun can be dismantled and reassembled without tools.

The Hellenic Arms Industry (EBO) holds a licence to manufacture MK 30 barrels.

Data

Cartridge: 30 × 173 mm
Operation: gas, automatic
Locking: double supporting flaps; modified Friberg-Kjellman system
Feed: single belt, double belt, each from sides or top
Weight: gun, complete, single belt, 148.5 kg; gun, complete, double belt, 155.7 kg; single belt, 21 kg; double belt, 28.2 kg; barrel, 63 kg
Length: with muzzle brake, 3.35 m
Muzzle velocity: HEI/SD-T, 1,035 m/s: APDS-T, 1,220 m/s
Rate of fire: cyclic, 800 rds/min
Recoil force: peak, 18 kN: during bursts, 16 kN
Recoil length 45 mm

Manufacturer

Mauser-Werke Oberndorf Waffensysteme GmbH, Postfach 1349, D-78722 Oberndorf-Neckar.

Status

In production.

Service

Installed in German Arrow twin field mount in service with Royal Thai Army; Greek Artemis 30 twin field mount; Breda twin Sentinel field and single and twin naval mounts (Italy - Guarda di Finanza). Also in the private venture ASCOD weapon station.

MK 30 × 173 mm Mauser Model F cannon, single feed version

UPDATED

Mauser MK 25 × 137 mm Model E cannon

Description

The Mauser MK 25 × 137 mm Model E was developed as a replacement weapon for the German Marder IFV, currently mounting a 20 mm RH 202.

The MK 25 Model E has the same design and operating principles as the Mauser MK 30.

The MK 25 fires standard 25 × 137 mm ammunition and uses a double belt feeder. The gun can be stripped down into its five main components (barrel, receiver, feeder, recoil system and bolt) without requiring any special tools, in about three minutes.

MK 25 × 137 mm Mauser Model E cannon with flexible feed chute

Data

Cartridge: 25 × 137 mm
Operation: gas, automatic
Locking: double supporting flaps; modified Friberg-Kjellman system
Feed: dual-selectable, disintegrating link belts
Weight: gun, complete, 112 kg; barrel, 38 kg; dual feed mechanism, 25 kg
Length: gun, 2.862 m
Barrel: without muzzle brake, 2.1 m
Muzzle velocity: HEI-T, 1,100 m/s; APDS-T, 1,360 m/s
Rate of fire: cyclic, 900 rds/min
Recoil force: peak, 8.5 kN; during bursts, 7.5 kN
Recoil length: 35 mm

Manufacturer

Mauser-Werke Oberndorf Waffensysteme GmbH, Postfach 1349, D-78722 Oberndorf-Neckar.
SACO Defense Inc holds a licence for the North American market.

Status

Development complete; final tests were carried out in 1988. It can be fitted in KUKA one- and two-man turrets, Norwegian Norsk Forsvarsteknologi towed AA gun FK 20-2 and their APC turrets (for example M113 and Pbv 302 vehicles).

UPDATED

Rheinmetall 20 mm MK20 Rh 202 automatic cannon

Description

The Rheinmetall MK20 Rh 202 was designed and developed to a broad spectrum of applications. It has a high rate of fire yet low recoil forces so that it can be adapted to numerous mounts hitherto deemed unsuitable for 20 mm high-performance weapons.

The gun is gas operated and has a rigid bolt: one feature being that the bolt is locked symmetrically by two locking pieces so that the forces are absorbed centrally. Recoil forces are reduced by the use of a muzzle brake and by firing each round before the recoil travel from the previous round is complete. The ammunition feed is gas operated and does not depend on the movement of the belt and weapon.

The MK20 Rh 202 uses the NATO standard 20 × 139 mm cartridge with disintegrating belt. Two different belt feed mechanisms are available. In one (Type 2) two

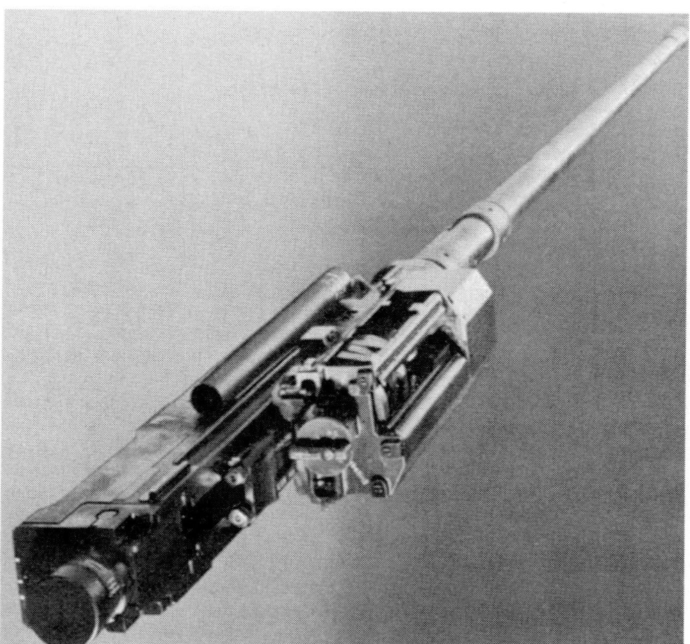

20 mm MK20 Rh 202 automatic cannon with Type 3 belt feeder

Type 3 belt feeder

standard cartridge belts can be introduced in parallel simultaneously from above and the operational belt can be selected by a simple lever control. In the other (Type 3) a single standard cartridge belt is used but can be introduced from the left, from the right or from above without making any mechanical changes.

The gun was designed to operate under the most arduous physical conditions, including temperatures below −54°C and exposure to water and heavy contamination. It can be dismantled and reassembled without the use of any tools.

The Hellenic Arms Industry (EBO) holds a licence to manufacture MK 20 Rh 202 barrels.

Data

Cartridge: 20 × 139 mm NATO
Operation: gas, automatic
Feed: Type 2, 2 belts attached parallel, selector; Type 3, feeding of one belt only from left, right or top, depending on mounting characteristics
Weight: ca 75 kg
Length: 2.612 m
Muzzle velocity: HEI-T, TP-T and disintegrating projectile, 1,050 m/s; API-T, 1,100 m/s; APDS-T, 1,250 m/s
Rate of fire: cyclic, 800-1,000 rds/min
Range: max, 7,000 m; effective, 2,000 m
Recoil force: 550-750 kgf
Recoil length: 26 mm

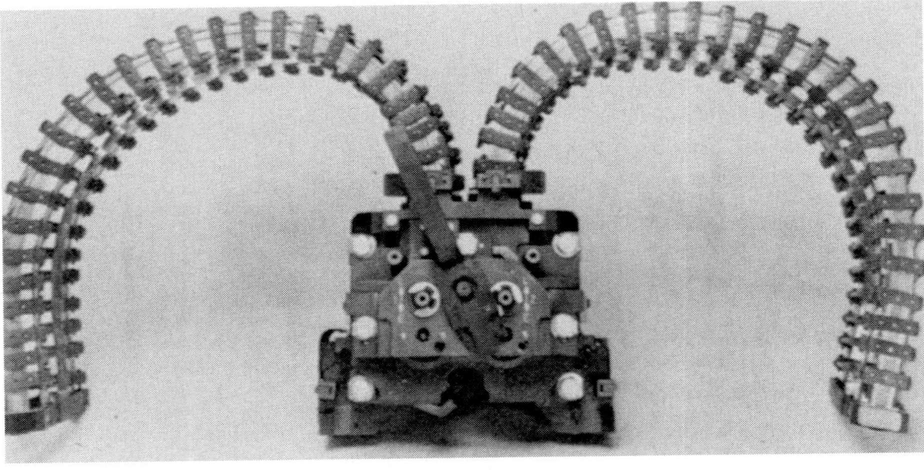

Type 2 belt feeder

Manufacturer

Rheinmetall Industrie GmbH, Postfach 1663, D-40836 Ratingen.

Status

In production.

Service

Armed forces of Argentina, Germany, Greece, Indonesia, Italy, Nigeria, Norway, Pakistan, Portugal, Saudi Arabia, Spain, Thailand and others.

UPDATED

SOUTH AFRICA

Vektor 20 mm G12 cannon

Description

The Vektor 20 mm G12 cannon is a development from the French 30 mm Type 30 M 693 (qv), used in South Africa for some years. In comparison with other cannon it is of reduced overall dimensions and weight and is therefore well suited for infantry applications, as well as for vehicle and helicopter armament.

The G12 is a gas-assisted blowback weapon. The bolt is locked in place by two side flaps; after firing, gas pressure drives two pistons back to rotate these flaps and unlock the bolt, after which residual chamber pressure forces the bolt backwards to complete the functioning cycle. There is a dual feed mechanism, with two ammunition belts permanently in place. Firing from one or other belt can be selected by a locally or remotely controlled selector.

Vektor 20 mm G12 cannon

The complete gun slides in a cradle, along the receiver guideways. The receiver is machined from a solid forging for maximum strength and stability, and the bore is nitrided to resist wear and prolong accuracy life.

An electrical fire control device is connected to the weapon by means of a cable; this allows the selection of single shots, limited bursts or sustained automatic fire.

The 20 mm G12 has a symmetrical arrangement of moving parts around its axis, which is instrumental in giving the weapon a very low dispersion figure. It can be installed upright, on its side or upside down without affecting performance.

Data

Cartridge: 20 × 139 mm
Operation: gas-assisted blowback
Locking: flaps
Feed: dual, disintegrating link belts; belt traction, 2 m vertical
Weight: 73.5 kg; barrel, 24.5 kg; cradle assembly, 10 kg
Length: 2.695 m
Rifling: 15 grooves, rh
Muzzle velocity: 1,050 m/s
Rate of fire: cyclic, 740 rds/min
Recoil force: 4,500 N
Power requirement: 24 V DC

Manufacturer

LIW (a division of Denel), PO Box 7710, Pretoria 0001.

Status

Available.

Service

South African National Defence Force.

UPDATED

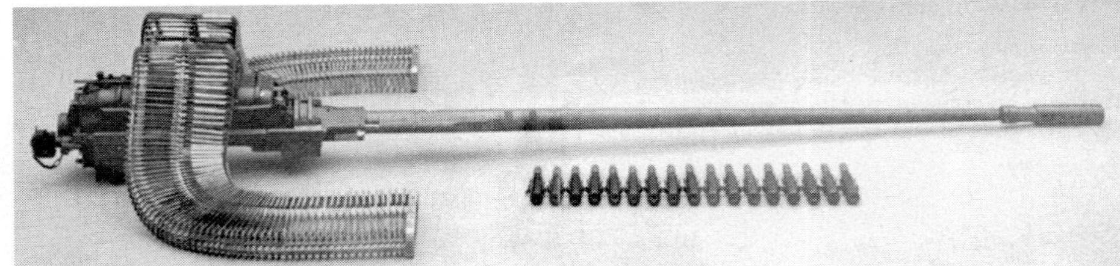

Vektor G12 cannon showing dual feed

SWITZERLAND

20 mm KAA (Type 204 GK) belt-fed automatic cannon

Description

The 20 mm KAA belt-fed automatic cannon is a gas-operated gun firing from a locked breech, also known by its Hispano-Suiza designation of Type 204 GK.

The gun fires from the open breech position. When the trigger is pressed the sear holding the breech block assembly is disengaged and the return spring drives the breech block forward. The head of the breech block forces a round through the link of the belt and into the chamber, stopping its movement when the front face of the breech block head strikes the stop insert. The breech block tail continues forward under force from the return spring and pushes out two pivoting locking pieces which engage with locking inserts in the receiver body. The breech block tail is prevented from rebounding by a spring-loaded detent holding it to the breech block head. The firing pin, attached to the tail, strikes the cap and the round is fired. In automatic firing, after the first round is fired, this occurs when the receiver and barrel are still moving forward in the cradle in counter-recoil from firing the previous round.

The gas pressure forces the projectile up the bore and as it passes the gas port some of the gas forces back a piston, which in turn drives the return spring housing back. A lug on the return spring housing is engaged with the breech block tail, which is pulled to the rear withdrawing the firing pin and releasing the pivoting locks. The breech block assembly is then blown back by the residual pressure in the bore. The chamber is fluted to reduce extraction friction. Each empty case is held by the extractor on the breech block face and ejected down out of the gun as the breech block passes under the fixed ejector. The breech block assembly moves back and is cushioned by a pneumatic/oil buffer. If the change lever is at single shot the sear retains the breech block assembly to the rear, the sear being spring loaded to cushion the shock of arresting the breech block. If the change lever is at automatic the breech block, under the influence of the return spring and breech block buffer, goes forward to fire again.

The feed mechanism is operated by the return spring housing. A lug on the housing engages in a spiral slot in the feed cylinder, rotating the cylinder in one direction during recoil and the other direction during counter-recoil. At the front end of the cylinder teeth are cut which engage in the feed slide and move the slide backwards and forwards across the gun. As the slide moves in, the spring-loaded feed pawls carried on it bring the next round in front of the breech block. When the feed slide moves out to collect the next round the ammunition belt is prevented from dropping back by spring-loaded retaining pawls.

Data

Cartridge: 20 × 128 mm
Operation: gas, positively locked, selective fire
Locking: pivoting locks
Feed: disintegrating link belt
Weight: complete gun, 87 kg; barrel with muzzle brake, 26.6 kg
Barrel: 1.7 m
Rifling: 12 grooves, increasing twist 0-6° 30′
Muzzle velocity: 1,050 m/s
Rate of fire: cyclic, 1,000 rds/min

Oerlikon-Contraves 20 mm KAA belt-fed cannon

Manufacturer

Oerlikon-Contraves AG, Birchstrasse 155, CH-8050 Zurich.

Status

Further production as required.

Service

Various countries.

UPDATED

20 mm KAB (Type 5TG) automatic cannon

Description

The 20 mm KAB (Type 5TG) automatic cannon is a positively locked gas-operated gun suitable for anti-aircraft or ground target applications. It is mechanically fired and uses a mechanical trigger, for either single-shot or automatic operation, with a trigger locking device which prevents the bolt from moving forward without feeding a round into the chamber. Ammunition feed is by either drum or box magazine. The barrel has Oerlikon-Contraves progressive rifling and is fitted with a four-stage muzzle brake. The gun may be field stripped without tools.

Data

Cartridge: 20 × 128 mm
Operation: gas, automatic
Feed: 50- or 20-round drum, 8-round box
Weight: gun, including barrel, 109 kg; barrel, 51.6 kg; magazine, 20-round drum, empty, 17 kg; 50-round drum, empty, 24.5 kg; 8-round box, empty, 4.5 kg
Barrel: 2.4 m (120 cal)
Rifling: 12 grooves, rh, progressive 0-6°
Mean recoil force: 800 kg
Recoil, barrel: 10 mm
Muzzle velocity: 1,100-1,200 m/s
Rate of fire: cyclic, ca 1,000 rds/min

Manufacturer

Oerlikon-Contraves AG, Birchstrasse 155, CH-8050 Zurich.

Status

Further production as required.

Service

Various countries.

UPDATED

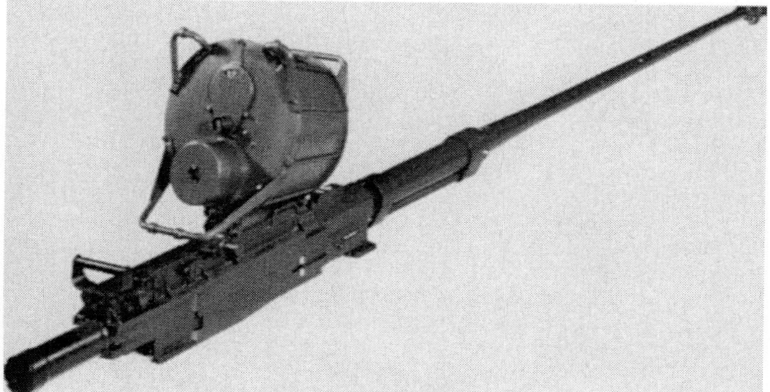

20 mm KAB-001 automatic cannon with drum magazine

20 mm KAB-001 automatic cannon with box magazine

25 mm KBA automatic cannon

Description

The Oerlikon-Contraves 25 mm KBA cannon is a fully automatic, positively locked, gas-operated weapon with a rotating bolt head and double belt feed.

The KBA offers a wide range of firing modes as single shot, programmable rapid single shot with a rate of fire of up to 200 rds/min, and full-automatic fire with 600 rds/min. The cannon functions, such as cocking and firing, are electrically actuated via remote control from the gunner's control box and in auxiliary mode mechanically by hand-crank and a trigger pedal.

Small size and low weight offer various integration possibilities such as one- and two-man gun turrets on APCs and AA tanks, field, naval and helicopter mounts.

For maintenance purposes the cannon can be field stripped to its main assemblies without tools.

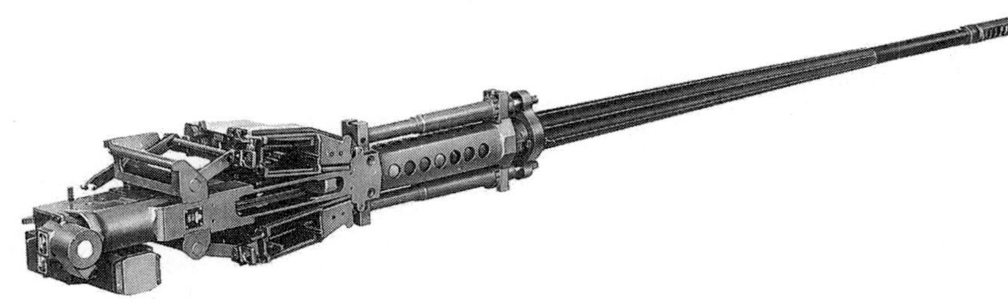

25 mm KBA automatic cannon

Data

Cartridge: 25 × 137 mm NATO standardised
Operation: gas, selective fire
Locking: rotating bolt
Feed: dual, selective belt, instantaneous change
Weight: gun, 112 kg; barrel, 37 kg
Length: overall, 2.888 m
Barrel: 2.173 m
Rifling: 18 grooves, increasing twist from zero at commencement of rifling to 7° (1 turn in 25.5 cal) at muzzle
Muzzle velocity: HEI/T, SAP/HEI/T, TP/T, 1,100 m/s; APDS/T, 1,335 m/s
Rate of fire: cyclic, 600 rds/min
Recoil length: max, 38 mm

Manufacturer

Oerlikon-Contraves AG, Birchstrasse 155, CH-8050 Zurich.

Status

In production.

Service

In service with NATO and other armies. Installed in Oerlikon-Contraves GBD-AOA turret on M113 C+R; in Oerlikon-Contraves GBD-COA on Mowag Piranha 6 × 6, 8 × 8 and M113; in FMC EWS on AIFV; four-barrel OTO-BREDA gun turret; OTOBREDA T25; Creusot-Loire 25 mm turret; Hägglunds 25 mm turret; Helio Mirror 25 mm turret.

UPDATED

UNITED KINGDOM

30 mm L21A2 RARDEN cannon

Development

The concept of the 30 mm RARDEN cannon dates from the early 1960s. The initial study concentrated on the defeat of armoured personnel carriers at a range of 1,000 m and a deterrent effect against low- and slow-flying aircraft. The then current Soviet armoured personnel carrier, the BTR-50P, had 14 mm thick armourplate sloping back at 45° from the vertical. It was considered that by the time the project came to fruition, the armour plate of comparable vehicles could have been increased to 40 mm and it was on this basis that the RARDEN calibre was decided; examination of current weapons and their penetration led to the view that 30 mm was the minimum acceptable calibre. The principal requirements influencing the design of the weapon were accuracy, low trunnion pull, short inboard length, low weight and low toxicity. The decision was made to design a completely new 30 mm gun to meet these requirements and the design work was done at the Royal Small Arms Factory at Enfield Lock (now closed). The name RARDEN derives from the initials of the Royal Armament Research and Development Establishment and ENfield.

Since the primary requirement was accuracy, the weapon was designed as an automatic, aimed, single-shot weapon which had a resulting cyclic fire rate of some 90 rds/min and a correspondingly low trunnion pull. No attempt was made to degrade accuracy to achieve a higher rate of fire. The requirement for low toxicity was met by arranging for forward external ejection of the empty case and by enclosing the entire gun mechanism.

The final design was the 30 mm L21A1 RARDEN operating on a long recoil system with a sliding block to reduce the length behind the breech and to cut down the inboard length further it was decided to feed from the side after initial rear-loading with three-round clips.

Various design improvements were incorporated into the L21A1 gun to upgrade it to L21A2. These were introduced in the mid-1980s and included the redesign of the barrel trough to allow easier barrel removal with the flash hider remaining attached; enlarging the rammer claw to provide better support to the round during loading, and other minor changes.

RARDEN applications in British Army service include the FV510 Warrior IFV, and the FV107 Scimitar and Sabre reconnaissance vehicles. On all these vehicles the RARDEN is compatible with day- and night-sighting systems, including image intensifiers and infra-red sights.

At one time a ground mount for the RARDEN was proposed but only a prototype was produced.

30 mm RARDEN cannon: handle at rear cocks action; clips are inserted from back and empty cases are ejected from chute on left. Gun casing is sealed and no fumes enter crew compartment

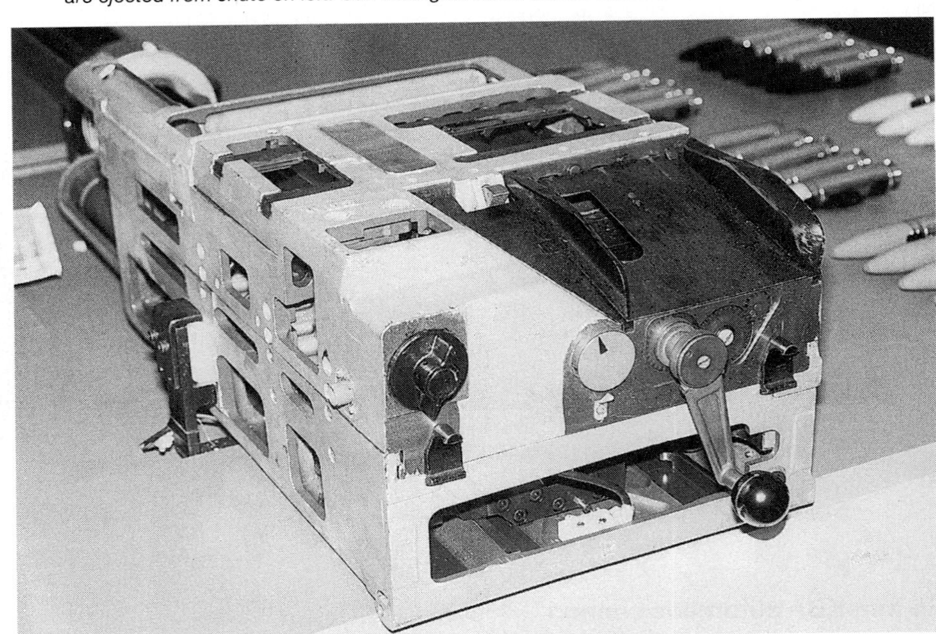

Cutaway example of 30 mm L21A2 RARDEN receiver (T J Gander)

Description

The 30 mm L21A2 RARDEN operates on the long-recoil system. When the gun fires, the breech block, breech ring, rammer, barrel and recuperator cylinder all move back some 330 mm inside the non-recoiling casing, against the buffer, compressing the air in the recuperator.

The breech block remains locked for the first 229 mm of recoil and is then opened by a cam attached to the block rolling round a radial wall of twice its own diameter, in the breech ring. This produces a movement of the breech block at right angles to the barrel axis, and at the same time the rammer is drawn down to grip the empty case.

When the recuperator drives the barrel and ring forward, the empty case is left on the rammer at the rear. The breech block is held open. After 203 mm of run-out, the empty case is clear of the breech and the rammer arm is rotated so that it carries the case upwards until it comes to rest in line with the feed slide, above the mechanism. The next round is now fed across and it pushes the empty case out of the rammer jaws and into the ejection chute, itself being gripped by the rammer.

The final forward movement of the breech ring ejects the empty case from the previous round forward out of the chute and out of the gun. The rammer is rotated to carry the live round down into line with the bore and then forward to feed the round into the chamber. The rammer releases the sliding block which pushes it into a recess in the breech ring. The gun is now ready to fire and the firing hammer is released from the sear by the gunner operating either the firing button or a solenoid.

Mechanical safety before firing is controlled by the breech block which pushes out a safety plunger when it is fully home. This releases the safety sear and leaves the hammer held by the sear. After firing, the free recoil of 230 mm ensures that the pressure is down to atmospheric before the breech opens.

The feed mechanism takes two three-round clips. The clip releases the rounds as soon as they are pushed in from the rear. For lightness the gun body is made of cast aluminium alloy in two sections. The bottom casing contains the recoiling masses and the top case houses the feed, firing and cocking mechanisms. The loading action is unusual in that the clips are pushed into the feedway from the rear, but the rounds travel sideways to be fed into the rammer claw. There is insufficient room for ammunition to come into the body from the side, so there is an opening in the top of the body which just accepts a three-round clip. Once the clip is inserted, it is taken by feed pawls across the top of the mechanism towards the left. Six rounds can be accommodated in this way. The rammer arm rotates up at the end of each cycle and collects a fresh round, taking it down into the body interior to present it to the breech. The length of time required for this activity is the main reason for the low rate of fire.

The barrel is of monobloc construction using high-yield steel, threaded at the breech end for attachment to the breech ring and at the muzzle for a flash suppressor. A barrel catch engages serrations at the breech end to prevent barrel rotation. The use of high quality steel produces a light barrel weighing 24.5 kg, with a length of 2.44 m. As with all light, long barrels, the main problem in obtaining accuracy and consistency is to control the vibration pattern on loading and firing. This is done on the RARDEN with a front bearing positioned about one quarter of the barrel length from the breech. This slides in the barrel sleeve. The vibrations are damped by four damping pads at the end of the barrel sleeve.

The range of ammunition for the RARDEN gun system includes HE, APDS-T and TP-T. There is a 'Reduced Range Training Round', while work is also underway to develop an 'Enhanced Armour-Piercing Round' that will probably be based on APFSDS-T concepts.

Data

Cartridge: 30 × 170 mm
Operation: long recoil
Feed: 3-round clips
Trunnion pull: 1,360 kg

Weight: gun, 113 kg; barrel, 24.5 kg; barrel with flash hider, ca 26.8 kg
Length: 2.959 m
Barrel: 2.44 mm
Muzzle velocity: HE, 1,070 m/s; APDS-T, 1,175 m/s
Rate of fire: cyclic, 90 rds/min

Manufacturer
Royal Ordnance Division, British Aerospace Defence Limited, Kings Meadow Road, Nottingham NG2 1EQ.

Status
Available.

Service
UK armed forces, also those of Belgium, Honduras, Malawi and Nigeria.

UPDATED

UNITED STATES OF AMERICA

30 mm ASP-30® Combat Support Weapon

Description
The McDonnell Douglas Helicopter Systems ASP-30® (Automatic, Self-Powered) cannon was developed as a combat support weapon readily interchangeable with 0.50/12.7 mm Browning M2 HB heavy machine guns. The ASP-30® has built-in dual recoil adaptors and can be used on virtually any existing vehicle mount or dismounted on a standard M3 heavy machine gun tripod for use in the direct infantry support role.

The gun is gas operated, using a rotating bolt to lock the breech securely during the firing phase. The feed is single, from the left side, and the short length of the receiver behind the feed allows simple and easily controlled spade grip operation from any cupola mounting structures.

Ammunition is the standard 30 × 113 B ADEN/DEFA M789 pattern, widely used throughout the world, carried in a disintegrating link belt.

Data
Cartridge: 30 × 113 B
Operation: gas, selective
Locking: rotating bolt
Feed: disintegrating link belt
Weight: 52 kg
Length: 2.06 m
Length behind feed: 283 mm
Barrel: 1.321 m; with blast suppressor, 1.473 m
Width: 286 mm
Height: 250 mm
Muzzle velocity: 820 m/s
Rate of fire: cyclic, 400-450 rds/min

Manufacturers
McDonnell Douglas Helicopter Systems, 5000 East McDowell Road, Mesa, Arizona 85215-9797.
Royal Ordnance Division, British Aerospace Defence Limited, Kings Meadow Road, Nottingham NG2 1EQ.

Status
Preproduction.

Licence production
UNITED KINGDOM
Manufacturer: Royal Ordnance
Type: ASP-30
Remarks: Licence agreement signed Spring 1989.

UPDATED

30 mm ASP-30® cannon on M3 tripod

1996

30 mm ASP-30® cannon mounted on Land Rover Defender 90 Multi Role Combat Vehicle

1996

25 mm M242 Chain Gun® weapon

Description
The McDonnell Douglas Helicopter Systems 25 mm M242 dual feed Chain Gun® weapon is in production for the US Army as the primary weapon for installation in the M2 and M3 Bradley Fighting Vehicles, the Light Armoured Vehicle (LAV-25), and on deck mountings on US Navy patrol boats. It has also been selected by the Spanish Government for installation in armoured personnel carriers. Over 9,400 M242 guns have been delivered to the US military since production began in 1981.

McDonnell Douglas has also delivered 124 M242 guns, plus spares, for installation in Swiss MOWAG vehicles sold to the Saudi Arabian Ministry of Defence and Aviation.

The entire gun, with integral dual feed, remote feed select and internal recoil mechanism, weighs 110 kg and provides for semi-automatic and automatic fire rates of 100, 200 and 500 rds/min.

The M242 fires both European and US 25 mm ammunition. Spent cases are ejected forward (overboard) while long dwell after firing eliminates gas build-up in turreted installations.

Data
Cartridge: 25 × 137 mm
Operation: external electric motor, selective fire

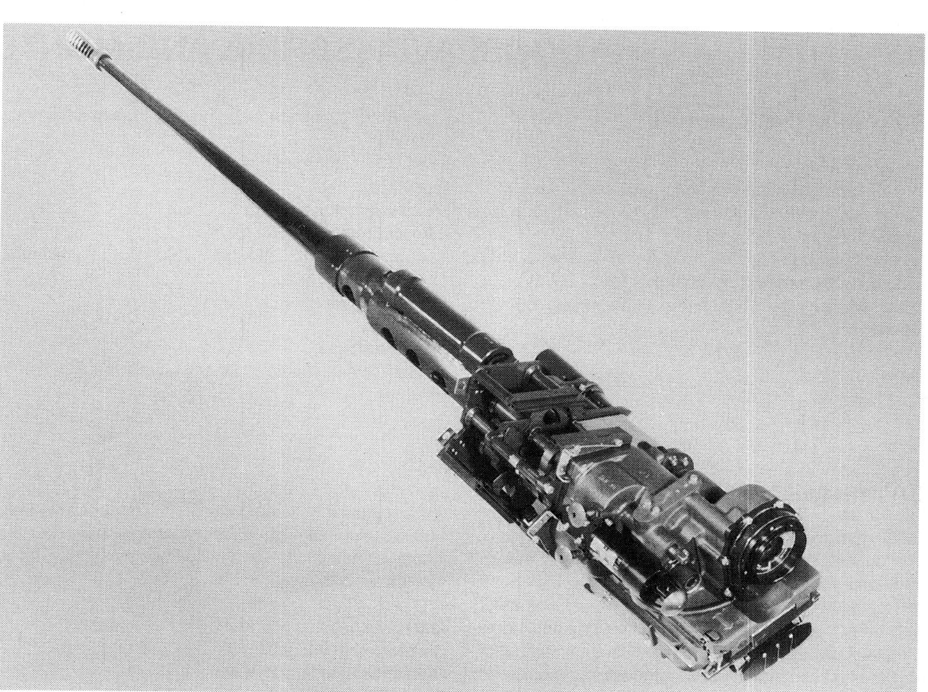

25 mm M242 cannon

Feed: belt
Weight: gun, 110.5 kg; barrel, 43 kg
Length: overall, 2.76 m
Width: 330 mm
Height: 380 mm
Barrel: 2 m
Rate of fire: single shot, 100, 200 or 500 rds/min
Time to rate: 0.15 s
Time to stop: 0.12 s
Power required: 1.5 hp for 200 rds/min; 8 hp for 500 rds/min
Clearing method: cook-off safe, open bolt
Safety: absolute hang-fire protection
Case ejection: forward
Peak recoil force: 3,175 kg
Dispersion: 0.5 mil
Demonstrated reliability: 22,000 mean rounds between stoppages

Manufacturer
McDonnell Douglas Helicopter Systems, 5000 East McDowell Road, Mesa, Arizona 85215-9797.

Status
In production.

Service
Fitted in US Army M2 and M3 Bradley Fighting Vehicles, in US Marine Corps and US Army Light Armoured Vehicle (LAV-25), and deck mountings on US Navy patrol boats. Also in service in Australia, Canada, Kuwait and Spain.

VERIFIED

30 mm Bushmaster II cannon

Description
The 30 mm Bushmaster II cannon was developed from the M242, to the extent that some 70 per cent of the components are common to both guns. The Bushmaster II fires standard 30 × 173 mm GAU-8 ammunition, using a side-stripping link developed by McDonnell Douglas. It can also fire RARDEN and Oerlikon KCB (30 × 170 mm) ammunition by changing the barrel, bolt and aft feed plate. As with the M242, the Bushmaster II has a dual feed capability.

The Bushmaster II is a candidate weapon for the Austrian, Spanish and Swiss combat vehicle programmes. In the USA it is a contender for the US Navy Advanced Minor Caliber Gun System, the US Marine Corps Advanced Attack Amphibious Vehicle and the US Army's Enhanced Bradley Fighting Vehicle programmes.

The Bushmaster II has been fitted into the Bradley fighting vehicle, the Swedish CV-30 and the AV Technology Multi-Gun Turret System (MGTS).

Data
Cartridge: 30 mm (see text)
Operation: mechanical (Chain Gun®)
Feed: dual, disintegrating link belt
Weight: 147.5 kg
Length: 3.499 m
Width: 335 mm
Height: 396 mm
Muzzle velocity: HEI, API and TP, 1,036 m/s; APDS-T, 1,220 m/s
Rate of fire: single shot, 200 or 400 rds/min
Peak recoil: 4,990 kg
Power required: to 200 rds/min, 1.5 hp; at 400 rds/min, 8 hp
Dispersion: <0.5 mil

Manufacturer
McDonnell Douglas Helicopter Systems, 5000 East McDowell Road, Mesa, Arizona 85215-9797.

Status
Undergoing evaluation in Austria, Spain and Switzerland. Selected by the Norwegian Army for their CV-9030 IFV.

UPDATED

20 mm M168 six-barrelled cannon

Description
The 20 mm M168 six-barrelled cannon is the M61A1 Vulcan aircraft gun modified by the omission of the automatic clearing function and the addition of a fixed casing. As with the M61A1 it is externally powered with a cluster of six barrels but the rate of fire has been reduced: whereas the airborne weapon fires 6,000 rds/min the M168 is capable of firing at 1,000 or 3,000 rds/min. The gun is basically a Gatling-type mechanism in which each of the six barrels fires only once during each revolution of the barrel cluster. Barrels are attached to the gun rotor by interrupted threads and no headspace adjustment is required. The muzzles are held in a muzzle clamp which allows the dispersion of shot to be spread into a flattened ellipse.

The mechanical arrangement and method of firing are similar to those of the 7.62 mm M134 Minigun (qv). The gun rotor rests on bearings inside the stationary outer housing and contains the six gun bolts. As the bolts rotate around the rotor, a cam follower on each bolt follows a stationary cam path fixed to the housing and this causes the bolt to reciprocate and carry out the functions of feeding, chambering, locking, firing, unlocking and extraction. Since each barrel fires only once for each revolution, a maximum of 500 rds/min/barrel, there is no chance of cook-off. Any misfires are thrown out of the gun and so are hang-fires. A declutching feeder is used to permit firing of all ammunition actually in the gun at the end of a burst.

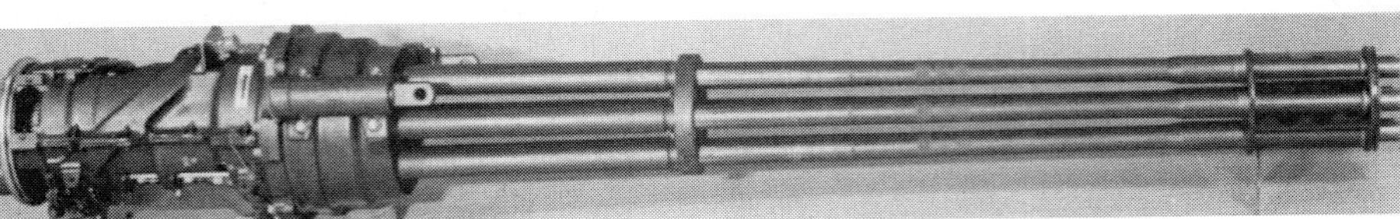

20 mm M168 six-barrelled cannon, modified M61A1 Vulcan gun

Data
Cartridge: 20 × 102 mm
Operation: externally powered. Gatling-type automatic
Locking: rotating bolt
Feed: boxed linked rounds
Weight: 136 kg
Barrel: 1.524 m
Rate of fire: selectable, 1,000 or 3,000 rds/min

Manufacturer
Lockheed Martin, Armament Systems, Lakeside Avenue, Burlington, Vermont 05401.

Status
In production.

Service
US forces. Also Belgium, Japan, South Korea, Saudi Arabia, and others.

UPDATED

YUGOSLAVIA (SERBIA AND MONTENEGRO)

20 mm M1955 cannon

Designed and manufactured in Serbia, this weapon closely resembles the Hispano-Suiza (later Oerlikon-Contraves) HS804 cannon and is believed to be substantially similar in design.

Long familiar in the M55 triple air defence mounting used by both land and naval forces, the gun has been observed on a single-barrel, two-wheel, towed mounting.

Data
Cartridge: 20 × 110 mm
Operation: blowback with positive breech locking, automatic
Feed: 60-round drum magazine
Barrel: ca 1.4 m (70 cal)
Muzzle velocity: 835 m/s
Rate of fire: cyclic, 800 rds/min
Range: practical, ground, 2,000 m; anti-aircraft, 1,200 m
Armour penetration: 25 mm at 1,000 m

Manufacturer
Zavodi Crvena Zastava, 29 Novembra 12, YU-11000 Belgrade.

Status
Probably no longer in production.

Service
Former Yugoslav armed forces.

VERIFIED

30 mm M86 automatic cannon

Description
The 30 mm M86 automatic cannon is of local design and is intended for mounting on armoured vehicles and for use with anti-aircraft systems. It is a gas operated, belt-fed gun which can feed from either side and ejects the spent case forwards. There is an integral gas buffer and spring recuperator which reduces the recoil load on the mounting. The barrel is easily removed and replaced. The feed system is powered by recoil energy and is capable of lifting a 40-round length of belt. Bolt cocking and sear release are both performed by hydraulic mechanisms and firing is controlled by a 24 V electric circuit.

Data
Cartridge: 30 × 210 B
Operation: gas, automatic
Feed: belt
Weight: gun, 200 kg; barrel, 60 kg
Length: 3 m
Height: 264 mm
Width: 242 mm
Barrel: 2.1 m
Rifling: 12 grooves, rh, increasing twist to 6° 24′
Muzzle velocity: 1,050-1,100 m/s
Rate of fire: cyclic, 650-750 rds/min
Gas pressure: 300 MPa
Recoil force: 25 kN
Recoil distance: 45 mm max

Manufacturer
Zavodi Crvena Zastava, 29 Novembra 12, YU-11000 Belgrade.

Status
Probably no longer in production.

Service
Former Yugoslav armed forces.

VERIFIED

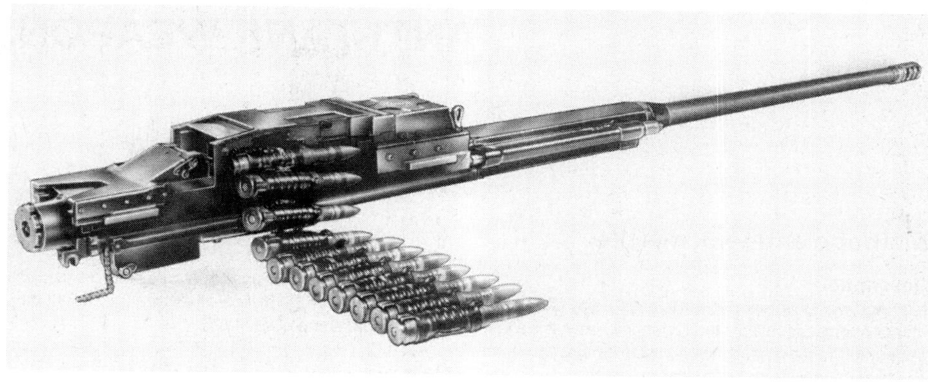

30 mm M86 automatic cannon showing right-hand belt feed

30 mm M89 cannon

Description
The 30 mm M89 cannon operates on the same principle and has the same component parts as the 30 mm M86 cannon (see previous entry). The two differ only in the feed system. The M89 is equipped with a dual belt feeder with direct ammunition changeover having second-round response. The feed system is powered by recoil energy.

Data
AS FOR M86 EXCEPT:
Feed: dual belt
Weight: 215 kg
Height: 300 mm
Width: 280 mm

Manufacturer
Zavodi Crvena Zastava, 29 Novembra 12, YU-11000 Belgrade.

Status
Probably no longer in production.

Service
Former Yugoslav armed forces.

VERIFIED

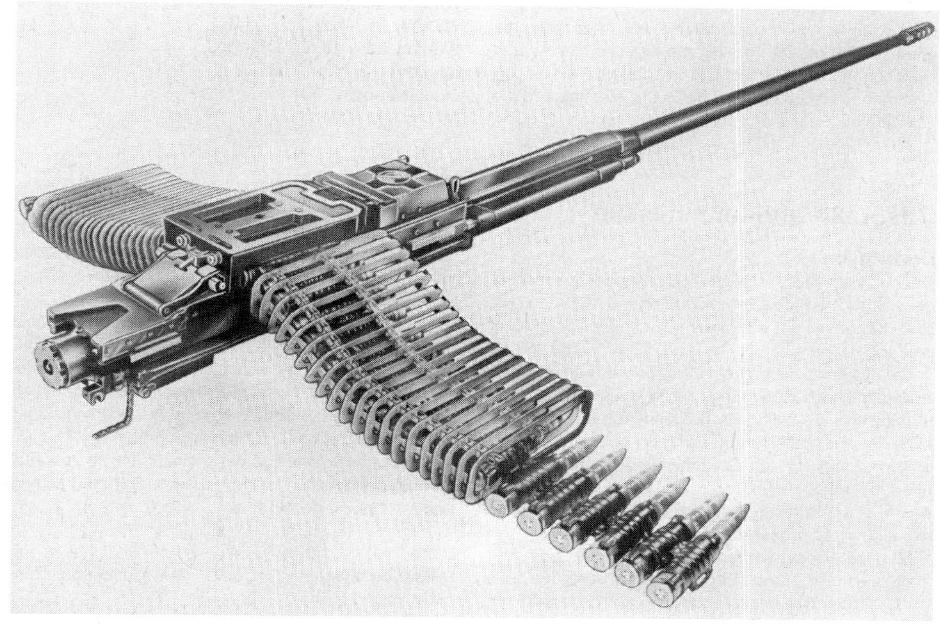

30 mm M89 cannon with dual feeder

ANTI-TANK WEAPONS

ARGENTINA

Mathogo anti-tank missile

Description

Mathogo was developed in the late 1970s by CITEFA to meet a specification from the Argentine Army. It is a first generation, wire-guided, man-portable missile which can also be vehicle or helicopter mounted.

The missile has a HEAT warhead and is carried in and launched from a watertight container. A control unit with telescope sight can be set up and connected with up to four launchers up to 50 m away. After firing, the operator gathers the missile into his field of view and then commands it in flight to impact with the target. Commands are transmitted down the wire link and are effected by means of spoilers in the wings. A boost rocket launches the missile, after which a sustainer motor takes over.

Two versions of this missile are known to exist; one has a range of 2,000 m, the other a range of 3,000 m. Both use the same control unit.

Data

Diameter: 102 mm
Length: 998 mm
Wingspan: 470 mm
Weight: 11.3 kg
Warhead: 2.8 kg
Launcher: 8.2 kg
Cruise velocity: 90 m/s
Min range: 400 m
Max range: 2,000 or 3,000 m (see text)
Armour penetration: 400 mm

Manufacturer

CITEFA, Instituto de Investigaciones Cientificas y Tecnicas de las Fuerzas Armadas, Zufriategui y Varela, 1603 Villa Martelli, Buenos Aires.

Status

Production probably complete.

Service

Argentine armed forces.

VERIFIED

CIBEL-2K anti-tank missile

Description

During the 1980s, CITEFA developed a second-generation missile system based on the Mathogo missile (previously) but with a semi-automatic line of sight guidance system.

The system utilises a near infra-red video area sensor to detect the location of the missile, and features countermeasures disability. For guidance the gunner maintains his aim at the target, using an optical telescope built into the firing post. Guidance commands are produced by means of a microprocessor-aided system which, at the same time, disables countermeasures by real-time video processing.

Missiles are not launched from the firing post, but from their own carrying box, which is designed as a launching platform, at distances up to 50 m from the firing post. This system allows connection of 12 missiles at a time, with a firing rate of two missiles per minute. Positioning the gunner at up to 50 m from the launch point means that firing smoke will not interfere with his sightline, and his position is unlikely to be discovered.

Trials have demonstrated that the hit probability on a medium-sized tank at 2,000 m range is over 95 per cent. The design has been aimed at keeping production costs low. First estimates of serial production prices were $4,000 for a missile and its carrying box/launcher, and $25,000 for each firing post.

The sighting telescope design includes the option of an image intensifier for night use; a thermal imaging system is being developed.

Data

Guidance system: SACLOS: Semi-automatic command to line of sight
Location sensor: near IR TV area sensor
Firing post weight: 12 kg with tripod
Missile connection capacity: 12 missiles simultaneously
Power supply: 6 × D-cell batteries, rechargeable
Battery life: 30 shots per charge
Telescope magnification: ×20
Elevation: −30 to +30°
Traverse: −180 to +180°

Manufacturer

CITEFA, Instituto de Investigaciones Cientificas y Tecnicas de las Fuerzas Armadas, Zufriategui y Varela, 1603 Villa Martelli, Buenos Aires.

Status

Under development.

VERIFIED

105 mm Mod 1974 FMK1 Mod 1 Czekalski recoilless gun

Description

The 105 mm Mod 1974 FMK1 Mod 1 Czekalski recoilless gun has a cylindrical perforated combustion chamber and a hinged breech block with a venturi mounted on it in such a way as to be easily replaced. There are two spiral vanes in the body of the venturi to compensate for the torque force resulting from the interaction of the barrel rifling and the shell. An oversized chamber provides a low loading density and gives optimum combustion of the powder and efficient use of the resultant gases. The rifled barrel allows firing of fin- or spin-stabilised ammunition.

The weapon is mounted upon a wheeled carriage which permits towing by any type of vehicle. The design of the carriage allows positioning of the gun at three different heights according to the operational requirement, with a high, medium or low axle.

Aiming is by means of a telescope which is calibrated up to 1,800 m and fitted with a stadiametric ladder anti-tank sight. An auxiliary spotting rifle (7.62 mm FAP - heavy automatic rifle) is fitted.

105 mm Mod 1974 FMK1 recoilless gun

Data

Calibre: 105 mm
Length: 4.02 m
Barrel: 3 m
Height in towing mode: 1.07 m
Min height of barrel: low angle position, 620 mm
Weight: in firing order, 397 kg; spotting rifle, 6.4 kg
Elevation: −7 to +45°
Max range: 9,200 m
Range of spotting rifle: effective, 1,200 m
Sight field of view: 12°
Magnification: ×4
Rear danger zone: 40 m (90°)
Crew: 4
Rate of fire: 3-5 rds/min

Ammunition	HE shell	Hollow charge (CHEA) shell
Weight	16.6 kg	14.7 kg
Muzzle velocity	400 m/s	514 m/s
Armour penetration	—	>400 mm

Manufacturer

Fábrica Militar Rio Tercero, Rio Tercero, Córdoba.

Status

Production complete.

Service

Argentine armed forces.

VERIFIED

AUSTRIA

106 mm M40A1 recoilless rifle

Description

The US M40A1 recoilless rifle is in service with the Austrian Army as a towed weapon, the 10.6 cm rPAK M40A1, on an Austrian-designed two-wheeled carriage developed and produced by Lohner GmbH. The carriage is notable for having two stable firing positions, the barrel being 260 mm higher in one position than in the other.

Data

Calibre: 106 mm
Barrel: 3.4 m
Weight: rifle, combat order, 113.9 kg; carriage, 170 kg; projectile, 7.71 kg
Muzzle velocity: 503 m/s
Max range: 6,900 m
Rate of fire: 5 rds/min
Elevation: −17 to +65°
Traverse: 360°

Status

Production complete.

Austrian Army 106 mm M40A1 recoilless rifle on Austrian carriage with gun fully raised

Service

Austrian Army.

VERIFIED

BELGIUM

MECAR 90 mm light gun

Description

The MECAR 90 mm light gun was designed primarily for anti-tank defence but fires a variety of other ammunition types in addition to a HEAT-T anti-armour round. It was supplied with or without a muzzle brake to meet customer requirements. Because of its low recoil force and light construction, the gun is particularly suited to mounting on light armoured vehicles and on small gun carriages for employment in mountainous and/or difficult terrain. Two versions of the gun exist, the heavier CAN-90H having a lower recoil velocity than the lighter CAN-90L.

Data

Calibre: 90 mm
Weight: 90L, 285 kg; 90H, 416 kg
Height: (overall) 90L, 350 mm; 90H, 400 mm
Width: (overall) 90L, 290 mm; 90H, 450 mm
Barrel: 2.9 m
Length of chamber: 368 mm
Chamber volume: 2,560 ml
Max pressure: 1,200 kg/cm^2
Max recoil distance: 420 mm
Max recoil velocity: (90L) 10.2 m/s; (90H) 7.8 m/s
Barrel life: up to 10,000 rounds

HE-T
Weight: projectile, 4.1 kg; round, 5.21 kg
Length: 520 mm
Muzzle velocity: 338 m/s
Max range: 4,200 m
Arming distance: 16 m
Lethal radius: 15 m
Effective radius: 25 m

SMOKE WP OR FM (TTC)
Weight: projectile, 4.1 kg; round, 5.21 kg
Length: 520 mm
Muzzle velocity: 338 m/s
Max range: 4,200 m
Arming distance: 16 m

CANISTER ROUND
Weight: projectile, 4.1 kg; round, 5.95 kg
Length: 373 mm
Contents: 1,120 balls
Muzzle velocity: 338 m/s
Range: 200 m
Cone angle forms at: ca 7 m

Manufacturer

MECAR SA, B-7181 Petit-Roeulx-lez-Nivelles.

Status

Production ceased 1991. Ammunition production continues.

Service

Sold to 15 countries.

VERIFIED

BRAZIL

57 mm M18A1 recoilless rifle

The Brazilian 57 mm M18A1 recoilless rifle is a copy of the US M18A1, manufactured in Brazil by Hydroar. Because of its portability and ease of operation, the weapon is still a viable defence against light armour. HE, HEAP and HEAT ammunition is readily available. Hydroar claims a barrel life of 2,500 rounds and a life for the breech block and throat of 500 rounds.

A 7.62 mm subcalibre training adaptor is available. Ammunition is made by IMBEL.

Manufacturer

Hydroar SA, Rua do Rócio 196, CEP 04552, São Paulo.

Status

Available.

Service

Brazilian Army.

VERIFIED

57 mm Hydroar RCL rifle

3.5 in Hydroar M20A1B1 rocket launcher

Description

Hydroar manufactures a copy of the US 3.5 in rocket launcher fitted with an electro-electronic grip of their own design. This replaces the original battery-powered M9 grip and the magneto-type M20 grip with a more modern circuit using two AA batteries and micro-electronics to produce the required power impulse. In other respects the launcher remains the same as the original US model, details of which will be found under *United States of America* in this section.

In addition, the grip can be supplied as a replacement kit for fitting to the 2.36 in M9A1 and 3.5 in M20A1B1 launchers. It will fit either launcher and replacement is very simple.

The 3.5 in rocket launcher fires a rocket weighing 4 kg. The rocket has a warhead containing 860 g of Composition B.

Manufacturer

Hydroar SA, Rua do Rócio 196, CEP 04552, São Paulo.

Status

In production.

VERIFIED

3.5 in Hydroar M20A1B1 launcher with PE-1 electro-electronic grip

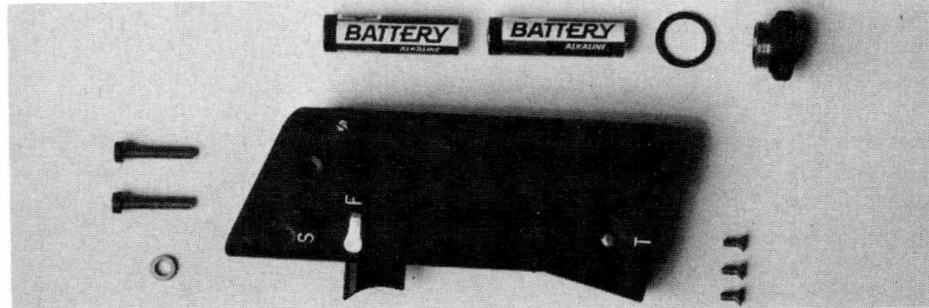

Grip kit PE-1 for modernisation of rocket launchers

CHINA, PEOPLE'S REPUBLIC

Red Arrow 8 guided weapon system

Description

Red Arrow 8 is a second-generation guided missile system intended for use by infantry against tanks and other armoured targets with a range of 100 to 3,000 m. It is a crew-portable weapon, fired from a ground tripod mount. It can also be configured for mounting in or on a variety of wheeled and tracked vehicles.

The system uses a tube-launched, optically tracked, wire-guided missile controlled by a SACLOS system based on an infra-red flare in the tail of the missile. The position of the flare is detected by the sight unit and corrections are automatically generated and signalled down the wire so as to fly the missile into the axis of the line of sight.

In general appearance the system is similar to MILAN, having a sight unit on to which the missile transport and launch tube is attached before firing. The tripod resembles that used with the TOW system. A full range of first and second echelon maintenance equipment and a training simulator is also manufactured to accompany the system in service.

Data

Operation: SACLOS, wire-guided
Warhead diameter: 120 mm
Missile length: 875 mm
Missile weight: at launch, 11.2 kg
Wingspan: 320 mm
Launch tube diameter: 255 mm
Launch tube length: 1.566 m
Launch tube weight, with missile: carry, 24.5 kg; firing, 22.5 kg
Sight unit weight: 24 kg
Tripod weight: 23 kg
Effective range: 100-3,000 m
Armour penetration: >800-900 mm
Velocity: initial, 70 m/s; max, 20-200 m/s

Manufacturer

China North Industries Corporation (NORINCO), 7A Yue Tan Nan Jie, PO Box 2137, Beijing.

Status

In production.

Service

Chinese Army.

VERIFIED

Red Arrow 8 launcher and missile

Red Arrow 8 launch unit showing sights

40 mm Type 56 anti-tank grenade launcher

Description

The 40 mm Type 56 anti-tank grenade launcher is a copy of the Soviet RPG-2 launcher and has the same weight and dimensions. It fires the Chinese Type 50 grenade, a better round as far as penetration is concerned than the Soviet PG-2 HEAT but its superiority is apparent only at normal impact; at 45° it may be less efficient.

Data

Calibre: launcher, 40 mm; warhead, 80 mm
Length: launcher, 1.194 m

RPG-2 launcher manufactured in China as a copy of the Soviet weapon. It is shown with practice grenade of Chinese origin

Weight: empty, 2.83 kg; projectile, 1.84 kg
Max range: effective, 150 m
Rate of fire: 4-6 rds/min
Armour penetration: Type 50 HEAT, 265 mm/0°

Manufacturer
China North Industries Corporation (NORINCO), 7A Yue Tan Nan Jie, PO Box 2137, Beijing.

Status
Production complete.

Service
Known to be used by militia and presumably a reserve weapon.

VERIFIED

40 mm Type 69 and Type 69-1 anti-tank grenade launchers

Description
The original 40 mm Type 69 anti-tank grenade launcher was a copy of the Soviet RPG-7 launcher first seen in 1972. It displayed some small improvements over the Soviet version, notably a heavier tube, a folding bipod, better insulation of the handgrip area on the tube, and a rearsight adjustable for windage. The performance was about the same as that of the Soviet version, though the Type 69 grenade has no self-destruct capability.

The Type 69 launcher was later supplemented by the slightly shorter Type 69-1 which incorporated some further slight improvements although the main change was to the ammunition. It is also possible to fit infra-red and electro-optical sights to the Type 69-1.

In place of the original Type 69 HEAT grenades the Type 69-1 is able to fire a family of nine NORINCO grenade types which considerably extends the launcher's tactical utility into the light support weapon field. These grenades may also be fired from RPG-7 grenade launchers.

Details of the NORINCO rocket-propelled grenades are as follows:

NORINCO Type 69-1 anti-tank grenade launcher

40 mm Anti-tank Rocket-assisted Hollow Charge Grenade Type 1
This is the 'standard' HEAT anti-tank grenade
Warhead diameter: 85 mm
Length: 926 mm
Weight: 2.1 kg
Range: effective, 300 m
Velocity: initial, 120 m/s
Armour penetration: 150 mm/65°

40 mm Anti-tank HEAT Grenade Type II
A revised grenade with a larger and more powerful warhead for use against heavily armoured vehicles and strongpoints
Warhead diameter: 94 mm
Length: 1.063 m
Weight: 2.8 kg
Range: effective, 200 m
Velocity: initial, 120 m/s
Armour penetration: 180 mm/65°

40 mm Anti-tank Grenade Type III
A rocket grenade with a revised warhead design and a self-destruct capability
Warhead diameter: 80 mm
Length: 970 mm
Weight: 2.26 kg
Range: effective, 290 m
Velocity: initial, 121 m/s; max, 295 m/s
Armour penetration: 180 mm/65°

40 mm Anti-tank HEAT Type 84
One of the latest types of grenade, the Type 84 is claimed to be relatively unaffected by side winds, no doubt because of its shorter overall length (not given). This long-range grenade can be used with both the Type 69 and Type 69-1 launchers.
Warhead diameter: 85 mm
Weight: 1.85 kg
Range: effective up to 400 m
Velocity: initial, 140 m/s
Armour penetration: 180 mm/65°

40 mm Airburst Anti-personnel Grenade
On this grenade the warhead strikes the ground only to be propelled upwards by a jump mechanism to a height of approximately 2 m where the explosive warhead detonates to scatter anti-personnel steel spheres over a lethal radius of 15 m.
Warhead diameter: 75 mm
Weight: 2.62 kg
Length: 805 mm
Range: effective up to 1,600 m
Velocity: initial, 102 m/s

40 mm HE/HEAT Grenade
On this multipurpose grenade the usual nose-mounted piezoelectric fuze has a graze component allowing the warhead to function as a long range HE/FRAG payload when used against troops in the open because of the provision of a preformed metal fragment sleeve which breaks up into 1,600 anti-personnel fragments scattering over a 20 m radius. The warhead retains the anti-armour HEAT capability.
Warhead diameter: 92 mm
Length: 919 mm
Weight: 2.67 kg
Range: effective up to 1,800 m
Velocity: initial, 102 m/s
Armour penetration: 150 mm/60° at 300 m

40 mm Incendiary/HE Grenade
Designed to have the same ballistic performance as the 40 mm Airburst Anti-personnel Grenade, this Incendiary/HE grenade was developed for jungle and mountain operations. Its anti-personnel effects are enhanced by a sleeve containing 900 steel balls around the warhead, together with 2,000 to 3,000 incendiary pellets, to be scattered over a radius of up to 15 m.
Warhead diameter: 76 mm
Length: 805 mm
Weight: 2.67 kg
Range: effective up to 1,500 m
Velocity: initial, 102 m/s

40 mm Illuminating Grenade
The latest of the Type 69-1 grenade launcher munitions to be developed, this grenade has a braking ring around the warhead to allow a parachute-suspended flare to be ejected at a range of 600 m (1,500 m with the braking ring removed) under the influence of a time fuze. The flare burns for 35 seconds at an intensity of 500,000 candles.
Warhead diameter: 75 mm; over braking ring, 101.3 mm
Length: 844 mm
Weight: 2.2 kg
Range: effective, 600 or 1,500 m
Velocity: initial, 115 m/s

40 mm HEAT Grenade Type 69
This is the original HEAT anti-armour grenade developed for the Type 69 launcher and provided with a Dian-2 piezoelectric fuze. It has no self-destruct capability.
Warhead diameter: 85 mm
Length: 919 mm
Weight: 2.25 kg
Range: effective, 300 m
Velocity: initial, 120 m/s; max, 295 m/s
Armour penetration: 110 mm/65°

Data
TYPE 69-1 LAUNCHER
Calibre: launcher, 40 mm; projectile, 85 mm
Length of launcher: 910 mm
Weight: without sight, 5.6 kg
Weight of grenade: see text
Muzzle velocity: see text

Manufacturer
China North Industries Corporation (NORINCO), 7A Yue Tan Nan Jie, PO Box 2137, Beijing.

Status
In production.

Service
Chinese armed forces. Offered for export sales.

UPDATED

82 mm Type 65 recoilless gun

Description
The 82 mm Type 65 is a copy of the Soviet B-10 recoilless gun. The performance of the Chinese version is the same as that of the Soviet weapon.

Data
Calibre: 82 mm
Operation: recoilless, with multi-vented breech block with enlarged chamber section
Length: 1.677 m
Barrel: 1.659 m
Weight: in travelling mode, 87.6 kg; HE round, 4.5 kg; HEAT round, 3.45 kg; HEAT projectile, 2.825 kg
Muzzle velocity: HEAT, 247 m/s
Max range: HE round, 4,470 m; effective anti-tank range, 390 m
Armour penetration: 120 mm/65°
Rate of fire: 6-7 rds/min

Status
No longer in production.

Service
Likely to be encountered in South-east Asia and Africa.

UPDATED

82 mm Type 65 recoilless gun

Type 70-1 62 mm portable anti-tank rocket launcher

Description
The Type 70-1 62 mm portable anti-tank rocket launcher is a light anti-armour weapon intended for one-man use against light and medium armour and strong-points. The launcher has a two-part launching tube made of glassfibre-reinforced plastic materials with a basic trigger mechanism and a simple post pistol grip; the rear part of the tube is removed for loading. Simple flip-up leaf sights are provided. Maximum direct-fire range is 150 m.

The Type 70-1 is fired from over the shoulder and is used with only one type of rocket, the 62 mm Type 70 HEAT weighing 1.18 kg and capable of penetrating 100 mm of armour set at an angle of 65°.

Data
Calibre: 62 mm
Weight: loaded, 3.21 kg; projectile, 1.18 kg
Length: overall, 1.2 m
Range: effective, 150 m
Armour penetration: 100 mm/65°

Manufacturer
China North Industries Corporation (NORINCO), 7A Yue Tan Nan Jie, PO Box 2137, Beijing.

Status
In production.

Type 70-1 62 mm portable anti-tank rocket launcher

Service
Chinese armed forces.

VERIFIED

Individual anti-tank rocket launcher Type PF89

Description
The individual anti-tank rocket launcher Type PF89 is a single-shot disposal anti-armour weapon intended for one-man use. Each Type PF89 is issued with the rocket ready loaded in the short launch tube.

The rocket has a warhead diameter of 80 mm and consists of three parts; the shaped charge warhead, the rocket motor, and the folding fin stabiliser assembly. The glass-reinforced plastics launch tube is provided with an optical sight and a pistol grip and trigger group. The tube is provided with an integral carrying handle and a sling.

The Type PF89 is intended for use against armoured vehicles at ranges up to 300 m; static targets can be engaged up to 400 m. Armour penetration is 200 mm against plates set at an angle of 65°. Dispersion at 200 m is 250 × 250 mm. After firing the launcher is discarded.

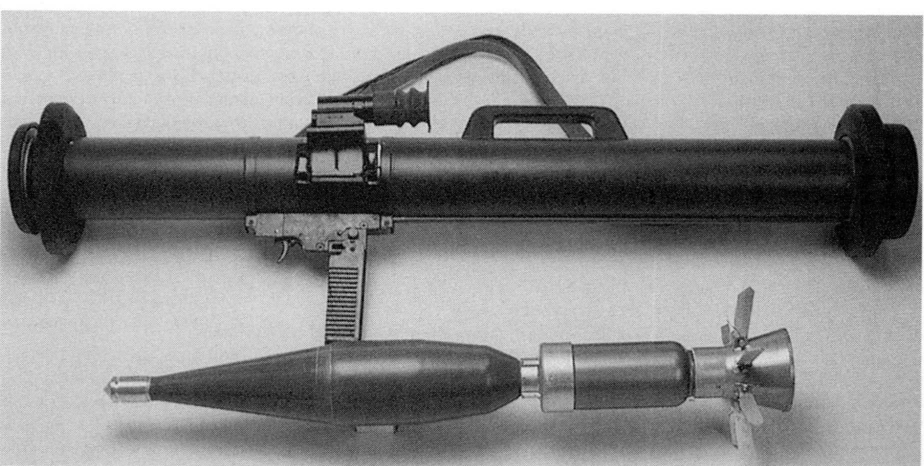

Individual anti-tank rocket launcher Type PF89 with rocket

Data
Calibre: 80 mm
Weight: 3.7 kg
Length: 900 mm
Range: effective, AFVs, 300 m; static targets, 400 m
Armour penetration: 200 mm/65°

Manufacturer
China North Industries Corporation (NORINCO), 7A Yue Tan Nan Jie, PO Box 2137, Beijing.

Status
In production. Offered for export sales.

Service
Chinese armed forces.

VERIFIED

COMMONWEALTH OF INDEPENDENT STATES

RPG-2 portable rocket launcher

Description
The RPG-2 was developed from the German Panzerfaust of the Second World War. It fires a fin-stabilised PG-2 HEAT projectile which is loaded into the muzzle of the launcher. The fins are made of a stiff flexible sheet and they can be rolled around the cylindrical container of the motor and retained in place with a ring. As the fins are pushed down into the barrel the ring slides off. The fins spring out and are held against the inside of the launcher tube. As soon as the missile leaves the muzzle the fins extend fully.

The metal tube of the launcher has a wood-encased section in the middle to protect the firer against heat and also to make the weapon easier to hold when it is used in cold climates. The firing mechanism is a spring-loaded hammer and this means that the round has to be indexed into position to ensure that the hammer and the cap are in line.

The rear end of the tube is usually a plain cylinder but some have been seen with a flange-like blast deflector at the rear of the launcher.

The Chinese manufactured the RPG-2 as the Anti-Tank Grenade Launcher (ATGL) Type 56 (qv).

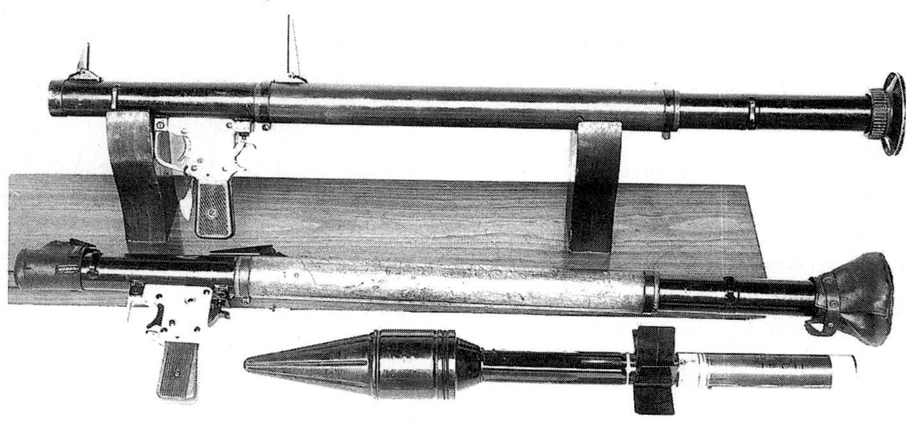

RPG-2 rocket launchers with PG-2 grenade (T J Gander)

Data
Calibre: launcher, 40 mm; warhead, 82 mm
Length: launcher, 1.494 m
Weight: empty, 2.83 kg; projectile, 1.84 kg
Max range: effective, 150 m
Rate of fire: 4-6 rds/min
Armour penetration: 150-180 mm

Manufacturer
State arsenals. Chinese state factories.

Status
Production complete.

Service
Still to be found with some militia and irregular forces in South-east Asia and the Middle East.

VERIFIED

RPG-7 portable rocket launchers

Description
The RPG-7 remains the standard man-portable short-range anti-tank weapon of the former Warsaw Pact countries and their allies. It has had wide operational use since it was introduced in 1962 and it remains an effective and efficient weapon.

The RPG-7 is similar to the RPG-2 in having a tube calibre of 40 mm but the maximum diameter of the basic PG-7 grenade is 85 mm. The RPG-7 has a stadia line, subtension type, rangefinding optical sight for day use and can accommodate an image intensifier sight for night firing. It is percussion fired with a positively indexed round. The gases emerge from the convergent/divergent nozzle at high velocity and the weapon rests on the firer's shoulder.

The grenade has large knife-like fins which spring out when the projectile emerges from the tube. At the rear end of the missile are small offset fins which give a slow rate of roll to improve stability. After a travel of 10 m the rocket motor fires. This gives a small increase to the velocity then sustains it out to 500 m. The point at which the rocket assistance cuts in is consistent and is a major factor in obtaining round-to-round matching of trajectory.

RPG-7V rocket launcher in action

The PG-7 grenade fuze is a VP-7M Point-Initiating, Base-Detonating (PIBD) type using a piezoelectric head connected to the actual fuze in the base of the shaped charge. Early versions of this fuze were prone to malfunction if the fuze was short-circuited by external contact, so screening a target with wire mesh gave a high probability of shorting-out the piezoelectric unit and thus rendering the fuze and warhead inert. Later versions of the fuze corrected this defect.

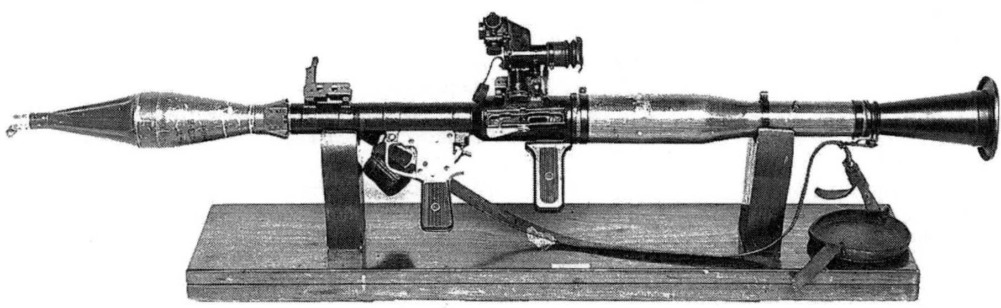

RPG-7V rocket launcher with rangefinding optical sight (T J Gander)

For firing the user screws the cardboard cylinder containing the propellant to the missile. The grenade is then inserted into the launcher muzzle with the small projection mating with a notch in the muzzle to line up the ignition cap with the percussion hammer. The nose-cap is then removed and the safety pin extracted. The cocked hammer is released when the trigger is fired and the missile is launched.

In 1968 a folding version of the RPG-7V was observed. This was initially taken to be a new launcher and was tentatively named the RPG-8. It was later realised that the revised design was a variant of the RPG-7V and was named the RPG-7D1. Originally used by airborne troops, the RPG-7D1 was gradually generally distributed throughout Warsaw Pact forces. Length of the RPG-7D1 is 945 mm; weight is 7.4 kg with the optical sight.

The PGO-7 and PGO-7V optical sights are marked with ranges from 200 to 500 m at intervals of 100 m. They have a 13° field of view with a ×2.5 magnification and a rangefinding stadia type sight. The NSP-2 infra-red night sight can also be used.

Details of the grenades fired by the RPG-7 series are as follows:

PG-7 and PG-7M

The PG-7 grenade, the original grenade used with the RPG-7, has a warhead cone diameter of 73 mm and a fluted nose cone cover. A more recent warhead, designated PG-7M, has a cone diameter of 70 mm, but because of better design of the fuzing and firing arrangements it has a better anti-armour performance. The updated warhead has a slightly longer nose cone so it appears to be slimmer. The muzzle velocity of the PG-7M round is 140 m/s, length 950 mm and weight 1.98 kg.

The complete round with the PG-7 grenade is the PG-7V. That for the PG-7M is the PG-7VM (produced in Poland as the PG-7WM).

PG-7N

Basically an updated PG-7M with a revised warhead containing 340 g of OKFOL-3.5 initiated by a VP-22 piezoelectric fuze. Also revised is the 4RN42 propelling charge which provides an initial velocity of 140 m/s and a maximum range of 500 m. Length overall is 965 mm. Armour penetration is 400 mm. The complete round is the PG-7VN.

PG-7L

The PG-7L grenade has a 93 mm diameter shaped charge warhead capable of penetrating 600 mm of armour. The grenade weighs 2.6 kg and is effective up to 300 m; muzzle velocity is 112 m/s. It can also penetrate over 1.1 m of reinforced concrete, 1.5 m of brick, and over 2.5 m thicknesses of logs and earth. The complete round is the PG-7VL.

PG-7VR

The PG-7VR is fitted with a tandem warhead for the attack of explosive reactive armour, having an extension boom ahead of the main warhead carrying a small shaped charge to ignite explosive reactive armour before the main 105 mm diameter shaped charge operates to penetrate the main armour. The complete PG-7VR round weighs 4.5 kg and has a maximum effective range of 200 m. It is claimed that the PG-7VR grenade can defeat the passive and active armour of all current MBTs. It can also penetrate over 1.5 m of reinforced concrete or brick, and over 2.7 m thicknesses of logs and earth.

PG-7LT

The PG-7LT is another tandem warhead grenade for the attack of explosive reactive armour. The complete round, the PG-7VLT, weighs 2.9 kg and is 1.13 m long. Maximum warhead diameter is 93 mm. Muzzle velocity is 100 m/s and direct fire range is 200 m. The PG-7LT grenade can penetrate 700 mm of RHA, with behind armour effects, or 550 mm of RHA after one layer of explosive reactive armour. It can also penetrate 1.4 m of reinforced concrete or brick and 2 m of logs and earth.

PG-7M 110

See separate entry under *Czech Republic and Slovakia.*

OG-7 and OG-7M

The OG-7 and OG-7M are high explosive anti-personnel grenades intended to expand the tactical utility of the RPG-7 launcher series. The main difference between the two is that the OG-7M uses a heavier propellant charge (4BN17) than the OG-7 (which uses the 4BN42) which results in a slightly reduced range (950 m for the OG-7M, 1,000 m for the OG-7). Both grenades use the O-4M impact fuze to detonate 210 g of A-IX-1 (RDX/Wax) explosive. Both grenades weigh 1.76 kg and are 595 mm long. The initial velocity for the OG-7 is 152 m/s while that for the OG-7M is 145 m/s. The complete round with the OG-7 is the OG-7V. That for the OG-7M is the OG-7VM.

OG-7G

The OG-7G grenade features an enlarged FRAG-HE warhead with an O-4M nose fuze, and a maximum range of 900 m; direct fire range is 150 m. Muzzle velocity is 66 m/s. The complete round for this grenade is the OG-7BG. It is produced in Bulgaria.

OG-7E

By contrast with other RPG-7 grenades, the OG-7E is a pure blast weapon initated by a GO-2H nose fuze. The complete round, the OG-7VE, has a direct fire range of 165 m and a maximum possible range of 1,000 m. Muzzle velocity is 70 m/s. It is produced in Bulgaria.

OFG-7V

The OFG-7V is another FRAG-HE warhead grenade (the complete round is also known as the OFG-7V) with a maximum possible range of 2,000 m; direct fire range is 270 m. Muzzle velocity is 99 m/s. It is produced in Bulgaria.

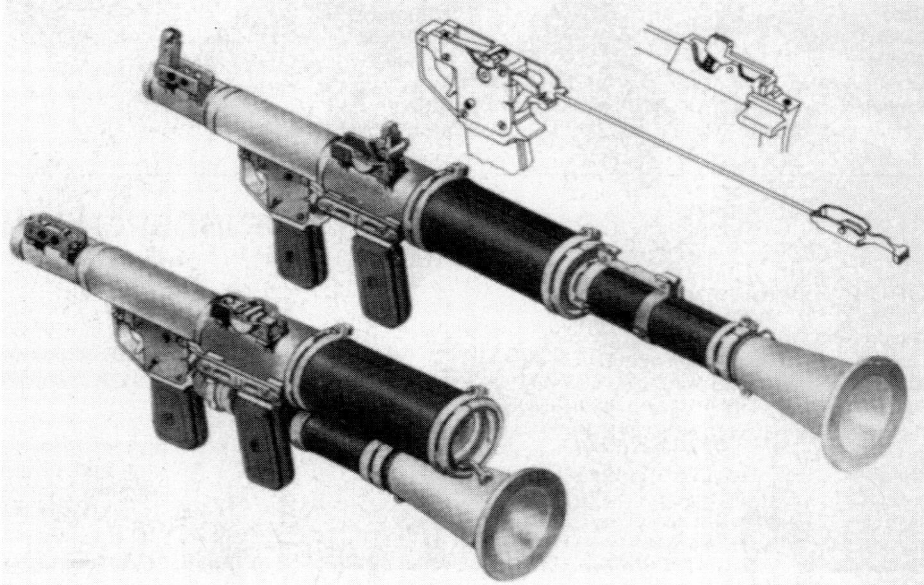

RPG-7D1 in folded and firing modes and showing firing mechanism

Upper warhead is the PG-7VR tandem warhead while that at the bottom is the 93 mm diameter PG-7VL (T J Gander)

KO-7V

The KO-7V grenade has a multipurpose warhead combining a shaped charge with a FRAG-HE collar. The HEAT charge is claimed to be able to penetrate more than 260 mm of armour. Direct fire range is 270 m and maximum possible range 2,000 m. Muzzle velocity is 98 m/s. It is possible that the KO-7V has its design origins in Bulgaria.

TBG-7

A fuel-air thermobaric warhead known as the TBG-7 has been developed for the RPG-7 launcher series. The blunt-nosed warhead is 105 mm in diameter and weighs 4.5 kg. Effective range is 200 m, with maximum range 700 m. The warhead is stated to be able to 'disable personnel in entrenchments or bunkers (with the grenade bursting at a distance of 2 m from the trench or embrasure) and for striking lightly armoured and soft-skinned vehicles'.

In addition to the above grenades, the RPG-7 series can also utilise Chinese NORINCO 40 mm grenades intended for the Type 69-1 launcher. For other RPG-7 launcher and ammunition suppliers refer to the list of licenced production manufacturers following, although it should be borne in mind that RPG-7s and their associated rocket grenades were produced in virtually every former Warsaw Pact country at one time or another.

Data

RPG-7V

Calibre: launcher, 40 mm
Length: launcher, 950 mm
Weight: launcher with sight, 6.9 kg
Muzzle velocity: 120 m/s
Max flight velocity: 300 m/s
Range: moving targets, 300 m; stationary targets, 500 m
Armour penetration: see text

Manufacturer

Kovrov Mechanical Plant, Kovrov, 601909 Vladimir Region, Russia.

Status

In production.

Service

All former Warsaw Pact countries and general use in Africa (including South Africa), Asia and the Middle East.

Licence production

BULGARIA
Manufacturer: Kintex
Type: RPG-7
Remarks: In production and offered for export sales. Also produced are PG-7, PG-7M, PG-7N, OG-7, OG-7M and other grenades (see text).

CHINA, PEOPLE'S REPUBLIC
Manufacturer: China North Industries Corporation (NORINCO)
Type: Type 69 and Type 69-1
Remarks: See separate entry under *China, People's Republic* for details of launchers and associated grenades.

EGYPT
Manufacturer: SAKR Factory for Develped Industries
Type: PG-7 Weapon System and Cobra
Remarks: See separate entries under *Egypt* for details.

IRAN
Manufacturer: Defence Industries Organisation, Armament Industries Group
Type: RPG-7
Remarks: As standard RPG-7 but with revised non-periscopic optical sight. Also produced are the Nader 40 mm HEAT rocket (PG-7), and the Saegheh 40 mm anti-personnel rocket (OG-7).

IRAQ
Manufacturer: State arsenals
Type: Al-Nassira
Remarks: See entry under *Iraq* for details.
PAKISTAN
Manufacturer: Pakistan Machine Tool Factory Limited
Type: RPG-7
Remarks: See entry under *Pakistan* for details. Rocket 40 mm HEAT P1 Mark 1 (PG-7 series) grenades are produced by Pakistan Ordnance Factories; the latter are actually Type 69-1 grenades produced under licence from NORINCO.
ROMANIA
Manufacturer: Romtehnica
Type: Portable Anti-tank Rocket Launcher
Remarks: Actually an RPG-7 with the addition of a bipod. An airportable version on which the barrel can be divided into two halves for transport is also available. Also produced are a 70 mm HEAT warhead grenade and two HE grenades, one similar to the OG-7 with the other having a 60 mm diameter warhead and a maximum range of 1,600 m. An 85 mm HEAT grenade was also produced but is apparently no longer available.

UPDATED

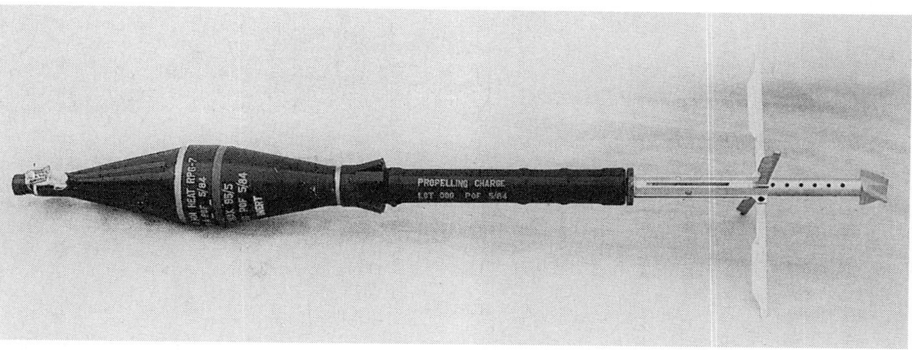

PG-7 HEAT rocket grenade as produced by Pakistan Ordnance Factories

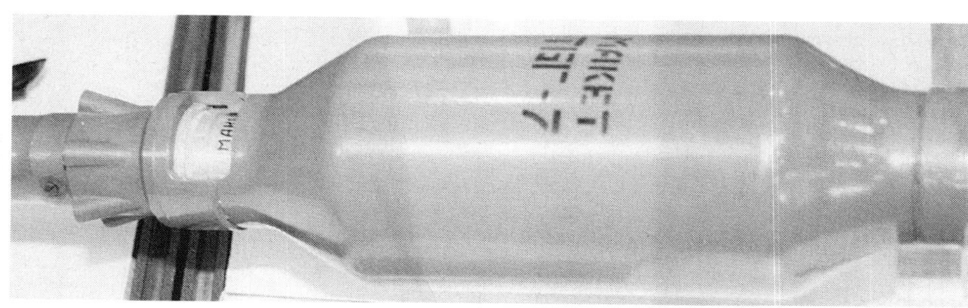

TBG-7 fuel-air grenade warhead (T J Gander)
1996

RPG-16 portable rocket launcher

Description

The RPG-16 appears to be a product-improved RPG-7 and retains most of the characteristics of that weapon.

The launcher seems to be the same in most details as the RPG-7, but the warhead is of unknown calibre though still of the over-calibre type. Some sources have suggested that it may be a tandem warhead, utilising two shaped charges, one to make the initial penetration and the second to enlarge the hole and do more damage behind the armour. A bipod appears to be fitted to the front of the tube.

Data

Calibre: 58.3 mm
Length: launch tube, 1.1 m
Weight: launcher, 10.3 kg; grenade, 3 kg
Muzzle velocity: 130 m/s
Flight velocity: max, 350 m/s
Range: practical, 500-800 m
Armour penetration: 375 mm
Sights: optical; IR night sight available

Status

Production probably complete.

Service

Limited service with Eastern Bloc armed forces.

VERIFIED

RPG-16 rocket launcher being fired. Picture suggests that general principles of the launcher are unchanged

RPG-16 rocket launcher being carried on exercise; bipod on muzzle can be identified

RPG-18 light anti-armour weapon

Description

The RPG-18 is a small rocket with a motor which is all-burnt on launch. The warhead is a shaped charge with a calibre of 64 mm. The launcher is disposable after firing.

The launcher is made of an extruded light alloy tube. It telescopes and the two halves are held together with a bayonet catch which has to be rotated to release the halves. The launcher is extended before firing. The firing mechanism is mechanical and not unlike that on the American M72 series. The sights are simple pop-up frames, made of plastic and graduated for ranges of 200, 250 and 300 m. There is also a stepped cutout slot which acts as a rudimentary rangefinder; an average length tank will fit inside one step of the slot at the three ranges noted above.

Each launcher carries a series of drawings showing the firer how to use it. There is no shoulder stop so the firer has to judge where to place his head when firing. One of the training instructions warns that the venturi should not be within 2 m of a wall or similar

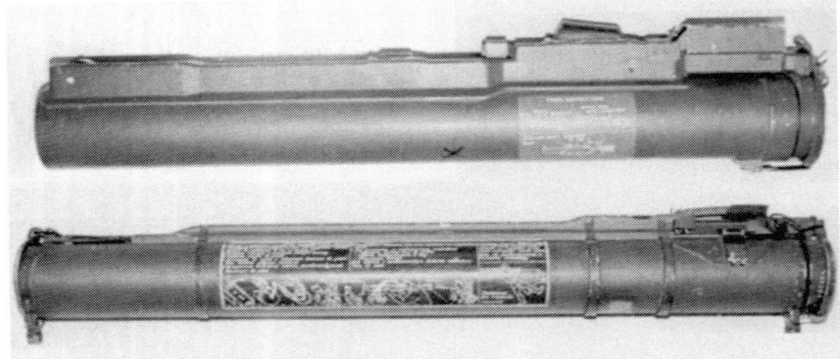

RPG-18, folded, compared with M72 LAW

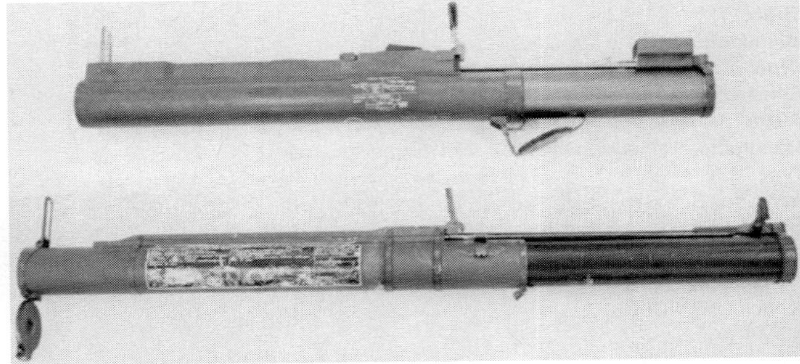

RPG-18 and M72 LAW extended to firing condition

solid obstruction. The simple drawings indicate that the weapon can be used by anyone in a defensive position.

Data
Calibre: 64 mm
Length: folded, 705 mm
Weight: overall, 2.7 kg; projectile, 1.4 kg
Range: combat, 200 m
Muzzle velocity: 115 m/s
Armour penetration: 375 mm

Status
Production probably complete.

Service
CIS and some former Warsaw Pact armies.

VERIFIED

RPG-22 light anti-tank weapon

Description
The RPG-22, first issued in 1985, is of the same general construction and appearance as the RPG-18 but fires a fin-stablised HEAT rocket of 72.5 mm calibre with a VP-16 or VP-22 fuze. The weapon has simple pop-up sights graduated for ranges of 50, 150 and 250 m, together with a temperature-compensating rearsight. The launcher consists of an outer glassfibre tube with an inner aluminium tube which is extended some 100 mm before firing; the launcher is 785 mm long when carried. The time necessary to prepare the weapon for firing is about 10 seconds.

For training the reusable RPG-22 UBS is available. This fires a 30 mm subcalibre projectile, the RPG-22 NETO weighing 350 g, to operational ranges. All handling and performance data are identical to those of the operational RPG-22.

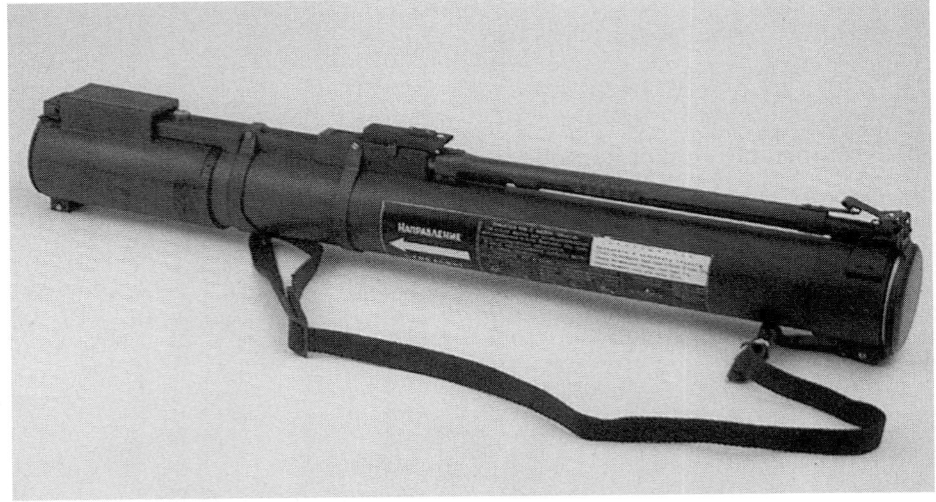

RPG-22 light anti-tank weapon

Data
Calibre: 72.5 mm
Length: carried, 785 mm; extended, 850 mm
Weight: 2.7 kg
Range: combat, 250 m
Muzzle velocity: initial, 133 m/s; max, 300 m/s
Armour penetration: 400 mm

Manufacturer
State factories.

Status
In production.

Service
Understood to be in service with Russian and some other armed forces.

Licence production
BULGARIA
Manufacturer: Vazov Engineering Plants
Type: RPG-22
Remarks: In production and offered for export.

UPDATED

RPG-26 light anti-tank weapon

Description
The RPG-26 appears to be a product-improved version of the RPG-22 (see previous entry). As such it is a general issue last ditch anti-armour weapon which can double as a bunker buster.

The telescoping launcher carries a shaped charge rocket with jack-knife fins which unfold after launch. The unitary HEAT warhead calibre is 72.5 mm and the complete round weighs 2.9 kg. It is claimed to defeat 500 mm of armour at the mean combat range of 250 m, and will also defeat 1 m of reinforced concrete, 1.5 m of brick, and 2.4 m of logs and earth.

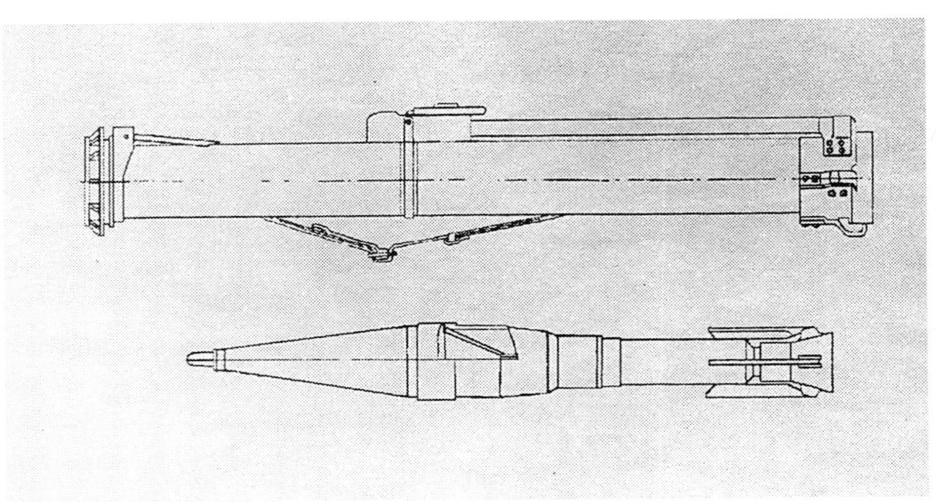

RPG-26 light anti-tank weapon

Data
Calibre: 72.5 mm
Weight: 2.9 kg
Range: combat, 250 m
Armour penetration: 500 mm
Operational temperature range: −50 to +50°C

Manufacturer
State factories.

Status
In production.

Service
Understood to be in service with Russian armed forces.

UPDATED

RPG-27 anti-tank rocket launcher

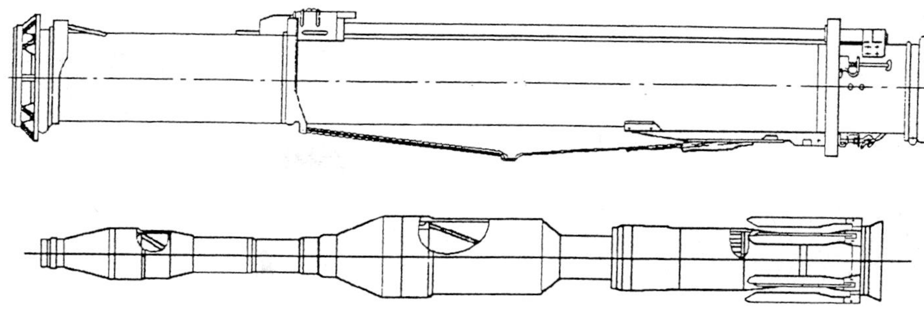

RPG-27 anti-tank rocket launcher

Description

The RPG-27 fires a 105 mm folding fin-stabilised rocket which carries a tandem shaped charge warhead. The main warhead is of 105 mm calibre, in front of which a tubular nose extension carries a smaller shaped charge warhead which is intended to detonate explosive reactive armour. It is claimed that the combination is sufficient to defeat the armour of 'all types of current tanks'. In the general support role the round will defeat over 1.5 m of reinforced concrete, 2 m of brick, and 2.7 m of logs and earth.

The complete weapon, in firing order, weighs 8 kg. The effective anti-armour range is 200 m.

Variant

105 mm multipurpose rocket launcher

This undesignated 105 mm multipurpose rocket launcher is believed to be a variant of the RPG-27 firing a revised rocket with a fuel-air thermobaric warhead. It appears that the intention is to provide a general purpose munition capable of defeating a wide array of battlefield targets from light armour to field fortifications. The warhead has an effective range against armoured targets of 300 m; against personnel the range is 700 m. Reinforced concrete up to 400 mm thick can be penetrated. Weight complete is 10 kg. After firing the launcher is discarded.

It is understood that this 105 mm multipurpose rocket launcher is still under development.

Data

RPG-27
Calibre: 105 mm
Weight: 8 kg
Range: combat, 200 m
Armour penetration: 1 m
Operational temperature range: −50 to +50°C

Manufacturer

GNPP Bazalt, 105058 Moscow, Russia.

Status

In production. Offered for export sales.

Service

Understood to be in service with Russian armed forces.

UPDATED

RPG-29 anti-tank rocket launcher

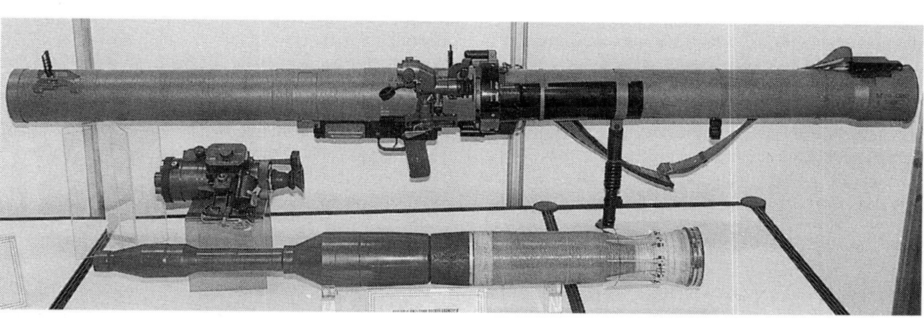

RPG-29 portable anti-tank rocket launcher and PG-29V round (T J Gander)

Description

Relatively few details have been released regarding the RPG-29 anti-tank rocket launcher, described in sales literature as a Portable Anti-tank Rocket Launcher. It resembles several other types of shoulder-fired rocket launcher in that it is a long tube folding into two halves for carrying. It is provided with an optical sight and a monopod to assume some of the rocket weight and steady the launcher before launch.

The main innovation on this weapon is the round launched, the PG-29V. This has tandem shaped charge warheads, the smaller being on a boom extension to defeat explosive reactive armour before the larger 105 mm diameter shaped charge can defeat the main target armour. It is claimed that the rocket warheads will defeat the armour of 'all current and other armoured vehicles at any target angles at a distance of up to 500 m'. It has been suggestd that this warhead is essentially similar to that used on the PG-7VR grenade fired by the RPG-7 portable rocket launcher series.

In the general support role the round will defeat over 1.5 m of reinforced concrete or brick, and 3.7 m of logs and earth.

There is a ground mounting for the RPG-29. Provided with laser-based sights and a fire control unit weighing 3 kg, this launcher/mounting combination can be used to engage armoured targets out to a range of 800 m. The complete launcher weight, with the mounting and fire control unit, is 20 kg.

Data

Calibre: 105 mm
Length: launcher, carrying, 1 m
Weight: launcher, 11.5 kg; round, 6.7 kg
Range: combat, 500 m
Operational temperature range: −50 to +50°C

Manufacturer

State factories.

Status

In production. Offered for export sales.

Service

Understood to be in service with Russian armed forces.

UPDATED

73 mm SPG-9 recoilless gun

Description

The 73 mm SPG-9 is a lightweight anti-tank gun, normally carried by two men, crewed by four men and mounted on a tripod for firing. It can be towed using a small two-wheel carriage. The PGO-9M optical sight is provided for direct fire. A PGO-K9 optical sight is also available.

The SPG-9 fires a fin-stabilised round with either a HEAT or HE warhead. The propellant charge is wrapped around a boom which extends behind the folded fins and terminates in a perforated disc of bore diameter. This construction bestows some ballistic advantages and also allows the extension boom to break free from the rest of the projectile but be ejected from the gun muzzle. The projectile consists of the warhead and a tubular rocket motor which is ignited by a delay element shortly after ejection from the muzzle of the gun.

The PG-9 grenade carries a HEAT warhead filled with 322 g of A-IX-1 (Hexogen); the generally similar PG-9N grenade has a warhead containing 340 g of OKFOL-3.5. The complete PG-9V round is 1.11 m long and weighs 4.4 kg as loaded; it develops a muzzle velocity of 300 m/s and a maximum range of 1,300 m. The PG-9N grenade used with the PG-9VN round can

73 mm SPG-9 recoilless gun

penetrate up to 400 mm of armour and has a muzzle velocity of 400 m/s.

A recent development is the PG-9VNT round with the PG-9NT grenade. This is a tandem warhead round intended for the defeat of explosive reactive armour and is claimed to be able to penetrate 550 mm of RHA with behind armour effects, or 400 mm after one layer of ERA. The warhead can also penetrate 1 m of reinforced concrete or brick, or 1.8 m of logs and earth.

The HE OG-9 grenade carries an HE warhead filled with 753 g of TNT. The complete round, the OG-9V, is 1.062 m long, weighs 5.4 kg, has a muzzle velocity of 316 m/s and a maximum range of 4,500 m. There is also an essentially similar round, the OG-9VM with an OG-9M grenade.

There are two further grenades, this time with FRAG-HE warheads. The OG-9BG1 round carries the OG-9G1 grenade which is fired at a muzzle velocity of 250 m/s to a range of 4,000 m. The OG-9BG round with the OG-9G grenade has rocket assistance to reach a range of 6,500 m.

Similar rounds to those mentioned above, with a different propelling charge system, are used with the 73 mm high/low pressure gun 2A28 mounted on BMP-1 IFVs.

Data

Calibre: 73 mm
Weight: launcher, 45 kg; tripod, 12 kg; trailer, 16 kg
Length: launcher, 2.11 m
Height: launcher, 820 mm
Width: max, over tripod, 1.075 m
Max barrel elevation: +25°
Muzzle velocity: PG-9, 300 m/s; OG-9, 316 m/s
Velocity: with rocket assistance, 700 m/s
Max range: direct, 1,300 m; indirect, 4,500 m; with rocket assistance, 6,500 m
Armour penetration: 400 mm

Manufacturer

State factories.

Status

In production.

Service

Former Warsaw Pact armies.

Licence production

BULGARIA
Manufacturer: Kintex
Type: SPG-9DMN
Remarks: In production. Offered for export sales.
ROMANIA
Manufacturer: Romtehnica
Type: 73 mm anti-tank rocket launcher
Remarks: Offered for export sales.

UPDATED

82 mm B-10 recoilless gun

Description

The B-10 is a smoothbore recoilless gun. The HEAT projectile fired is fin stabilised and resembles a mortar bomb. The weapon is towed on two wheels which are removed for firing. When in action the gun is fired off a tripod which is folded under the barrel in the travelling position and lowered when the gun is emplaced. If necessary, the weapon can be fired from the wheels. A bar attached to the muzzle is used to enable the weapon to be dragged into position. The gun has an optical sight without a rangefinding capability.

The B-10 was standard equipment in the former Soviet Army until the early 1960s, after when it was gradually phased out. It is still in service with the Egyptian and Syrian armies. It was copied by the Chinese as the Type 65 recoilless gun and may be found in the armies of several Asian countries.

Data

Calibre: 82 mm
Length: 1.677 m
Barrel: 1.659 m
Weight: in travelling order, 87.6 kg; HE round, 4.5 kg; HEAT round, 3.6 kg
Muzzle velocity: HE round, 320 m/s
Armour penetration: 240 mm
Max range: HE round, 4,500 m; effective anti-tank, 400 m
Rate of fire: 6-7 rds/min

Status

Production complete.

Service

In limited service within Europe but still in service in Egypt, Syria and the Far East.

Licence production

CHINA, PEOPLE'S REPUBLIC
Manufacturer: China North Industries Corporation (NORINCO)
Type: 82 mm Type 65
Remarks: No longer in production. See entry under *China, People's Republic* for details.

VERIFIED

107 mm B-11 recoilless gun

Description

The 107 mm B-11 recoilless gun fires a fin-stabilised mortar-like round. It is towed by the muzzle by a vehicle. The tripod legs are folded under the barrel in the travelling position and it is brought into action and normally fired off the tripod but, in an emergency, it can be fired off the wheels with reduced accuracy. The B-11 can be employed as an anti-tank weapon but the sights also allow it to be used in the indirect-fire role with an HE round.

Data

Calibre: 107 mm
Length: 3.314 m
Barrel: 2.718 m
Weight: 305 kg; HE round, 13.6 kg; HEAT round, 9 kg
Max range: HE round, 6,650 m; effective HEAT anti-tank range, 450 m
Armour penetration: 380 mm
Rate of fire: 6 rds/min

107 mm B-11 recoilless gun

Manufacturer

State arsenals.

Status

Obsolescent.

Service

No longer in use in former Warsaw Pact armies. May be found in some Middle East and South-east Asian forces.

VERIFIED

9K11 Malyutka anti-tank guided missile system

Description

The 9K11 Malyutka anti-tank guided missile system may be regarded as a follow-on from the earlier 2K15 Shmel system (Bumblebee - now unlikely to be encountered in combat) as it was developed at the Nepobidimy Design Bureau. Its NATO codename is 'Sagger' and the US alphanumeric designation AT-3. The CIS missile designation is 9M14.

First seen in 1963 using Manual Command to Line Of Sight (MCLOS), the Malyutka was extensively deployed in Europe and elsewhere in at least three ways: as a man-portable missile, mounted on the BRDM, and mounted on the BMP-1 IFV. As with the Falanga, the Malyutka was retrospectively converted to SACLOS operation some time in the late 1960s and, also like the Falanga, could be manually switched between MCLOS and SACLOS modes.

Malutka missiles have been produced in Bulgaria, Iran, Romania and Yugoslavia. The Taiwanese Kuen Wu 1 (qv) is a close copy.

Malyutka anti-tank missile mounted on armoured vehicle

Description

The 9M14M Malyutka-M missile, the most widely produced of the early Malyutka models, is carried into action in a waterproof glassfibre case (which can float), in which the warhead is separated from the motor. The lid of the case, the 9P111, forms a base for launching the missile which has a launch rail on the underside of the motor section. The two parts are taken out of the case and the front legs of the rail on the motor section are slotted into the lid so that there is a small angle of elevation. This is to ensure that the missile enters the operator's field of view when launched. The warhead is then clipped to the body. The missile is aligned on the target or the centre of the primary arc and the launcher is strapped to stakes driven into the ground. The leads from four missiles are taken to the 9S415 control box and connected up. At the back of the control box are the batteries. The 9Sh16 periscopic sight emerges from the centre and the control stick is in front of it. A trained crew could launch two missiles a minute.

The NCO in charge of the three-man Malyutka team is the system operator. If the target is under 1,000 m distant he does not use the sight but guides the missile to the target by eye. If the target is further away he gathers the missile to a line above the target and then controls it through the ×10 magnification periscopic sight.

The method of guidance is line of sight command and the corrections are transmitted through a multi-core cable fed out from the missile. There is a flare between the two upper fins. The missile has two solid propellant motors. The front annular motor is a booster. The rear motor controls the missile through two jetavator nozzles at the rear which swivel according to the commands transmitted through the wire. The four plastic fins are folded when the missile is stowed in the case. The HEAT warhead can penetrate about 400 mm of armour.

A late updated model of missile introduced in 1969, the 9M14P Malyutka-P utilised a SACLOS guidance system and featured a revised warhead with improved armour penetration (to 460 mm). This was followed in the early 1970s by the 9M14MP1, a further update of the Malyutka family with a stand-off probe for the warhead which increased its armour penetration to 520 mm. In 1992 the 9M14-2 Malyutka-2 appeared with a 3.5 kg shaped charge warhead capable of penetrating 800 mm of armour. A revised rocket motor increases flight velocity to 130 m/s.

The 9P110 BRDM-1 and 9P122 BRDM-2 armoured vehicle were modified to take a rectangular hatch lifted by a centrally mounted hydraulically operated column. Six launch rails are attached to the underside of the hatch and 9M14M missiles are clipped to these rails. Elevation, depression and swivelling are possible. The sight and control unit is between the two forward hatches. Control of the missiles is carried out in the same way as with the man-portable system. Eight missiles are carried as a reserve in the vehicles.

On the BMP-1 IFV the guide rail of the missile is fitted into a bracket above the 73 mm gun. The guide rail can be elevated and traversed at will. The missile is fired from inside the vehicle and controlled in the same way as the man-portable version. 9M14 Malyutka missiles are also carried by the BMD-1 airborne fighting vehicle.

Malyutka anti-tank missile on carrying case

Data

9M14M MALYUTKA-M
Type: wire-guided MCLOS system
Weight: at launch, 10.9 kg
Length: 860 mm
Diameter: 125 mm
Propulsion: 2 solid rocket motors
Guidance system: MCLOS wire link
Control: jetavator nozzles; roll rate stabilised
Warhead: 2.6 kg; HEAT, piezoelectric fuze
Range: 500-3,000 m
Armour penetration: 400 mm+
Flight velocity: 115-120 m/s
Operational temperature range: −50 to +50°C

Manufacturer

State factories.

Status

Production complete.

Service

CIS and former Warsaw Pact armies, Egypt, Iran, North Korea, Syria and the former Yugoslavia.

Licence production

BULGARIA
Manufacturer: Kintex - Vazov Engineering Plant
Type: Portable anti-tank guided missile 9M14M Malyutka
Remarks: Offered for export sales.
YUGOSLAVIA
Holding Company: YUGOIMPORT - SDPR
Type: Maljutka
Remarks: Probably no longer in production. See entry under *Yugoslavia (Serbia and Montenegro)* for details.

UPDATED

9K111 Fagot anti-tank guided missile system

Development

First disclosed in 1980, following development at the KBP Instrument Design Bureau, Tula, dating from the late 1960s (it entered service around 1973), the 9K111 Fagot (Bassoon) anti-tank guided missile system has the NATO codename of 'Spigot'; the US designation is AT-4. To complete the list of designations, the CIS missile designation is 9M111; there is also a 9M111M Faktorlya (Trading Post) missile and a 9M111-2.

The 9K111 Fagot SACLOS system is similar to MILAN (see entry under International) and uses the same general principle of operation. The chief outward difference is that the sight is smaller than for MILAN and the computing mechanism and goniometer are contained in a box below the launch rail, whereas in MILAN

9M111-2 guided missile with its container; this missile is part of the 9K111 Fagot anti-tank guided missile system

Iraqi Army ground-based 9K111 Fagot anti-tank guided missile system with missile containers (Christopher F Foss)

the sight container is larger and has the goniometer inside it.

The introduction of the 9K111 Fagot to the former Warsaw Pact armies greatly improved the capability of the Pact infantry to defend itself against armoured attacks, and it rapidly replaced the older manual command to line of sight systems.

Description

Fagot was first used on a 9P135 ground launcher, after which it appeared on the BMP-1 and BMD-1 infantry vehicles. The 9P135 launcher, when tripod mounted, is capable of being moved in fine control azimuth and elevation by gears. The missile is attached by means of a slide, with built-in electrical connections.

There are two current types of missile, the 9M111-2 and the 9M111M Faktorlya (Trading Post), the latter having an improved sustainer motor which increases the maximum range from 2,000 to 2,500 m and an improved warhead which increases armour penetration from 400 to 460 mm. The forward section of the missile is tapered in order to allow the flight control canards to be extended permanently.

The 9S451 control system uses three optical paths; one is visible light and is used for aiming, the other two are infra-red and track the flare at the rear of the missile during flight, deriving the deviation from the line of sight and generating corrections. The two infra-red paths have different beamwidths, one being for ranges to 1,000 m, the other for ranges above 1,000 m and change from one beam to the other is performed automatically. This ensures that the beam is no wider than necessary for control, reducing the possibility of infra-red jamming.

The 9M135 series of launchers can also fire the Konkurs missile (see following entry) without requiring any modification. The 9M135M1 and 9M135M2 are upgraded versions of the basic launcher while the 9M135M3 has a thermal imaging sight which clips over the usual 9Sh119 sight. The 9M135M3 launcher weighs 23.5 kg complete.

One detail has emerged regarding the packing of the missiles: two missiles are packed into a single bag within a container. The bags are sealed to the extent that Bulgarian sales descriptions maintain that the packed bags can be used as flotation devices when crossing water obstacles.

Data

Operation: SACLOS, wire-guided
Length: missile in canister, 1.1 m; missile only, 875 mm
Weight: launcher, 22.5 kg; complete round 9M111-2, 13 kg; complete round 9M111M, 13.4 kg; missile 9M111M, 7.3 kg; warhead, 2.5 kg
Warhead diameter: 120 mm
Range: 9M111-2, 75-2,000 m; 9M111M, 75-2,500 m
Velocity: max, 240 m/s; average, 180 m/s
Armour penetration: (armour at 90°) 9M111-2, 400 mm; 9M111M, 460 mm
Operational temperature range: −50 to +50°C

Manufacturer

Tzniitochmash, 2 Zavodskaya Street, 142080 Klimovsk, Russia.

Status

In production. Offered for export sales.

Service

Russian, other CIS and former Warsaw Pact armies.

Licence production

BULGARIA
Manufacturer: Kintex - Vazov Engineering Plant.
Type: 9K111 Fagot and 9M111-M Factoria
Remarks: In production. Launcher, 9M111-2 and 9M111M Factoria missiles produced and offered for export sales.

UPDATED

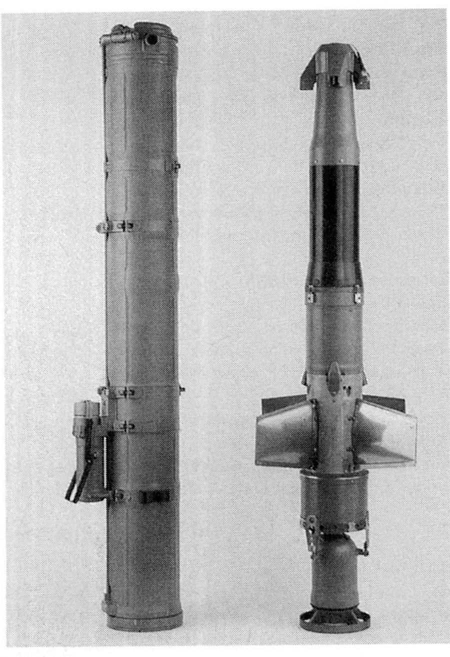

9M111M guided missile with its container; this missile is part of the 9K111 Fagot anti-tank guided missile system

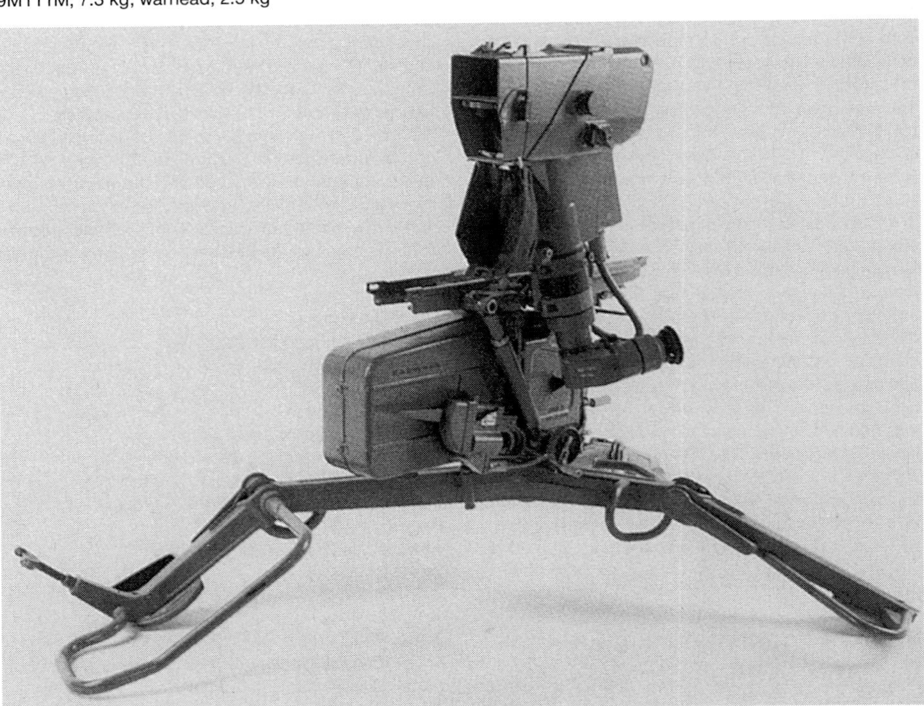

Launcher for 9K111 Fagot anti-tank guided missile system

9K113 Konkurs and Konkurs-M anti-tank guided missile system

Development

Developed by the KBP Instrument Design Bureau at Tula, the 9K113 Konkurs (Contest) anti-tank guided missile system entered service in 1974 and was given the NATO codename of 'Spandrel'; the US designation is AT-5. The CIS missile designation is 9K113.

The 9K113 Konkurs system has been licence-produced in various countries (see below). A document dated October 1992 gave a unit price for a launcher as $115,000, with $13,000 for one missile.

The Konkurs-M involves a missile with a revised warhead with a tandem configuration and an extending probe, capable of attacking reactive armour.

Although production of the 9K113 Konkurs, and Konkurs-M, is still continuing it is intended that the replacement for the system will be the Kornet laser-guided missile system (qv).

Description

In its ground-launched form the 9K113 Konkurs missile is launched from the same 9P135M-1 launcher as the 9K111 Fagot missile (qv). It may also be launched from various vehicle mountings for which there is a special system using a 9P148 launcher. The operational range of the missile is from 75 to 4,000 m using SACLOS wire guidance.

9K113 Konkurs missiles are supplied in sealed containers which are placed directly on the launcher in the same manner as Fagot and MILAN. After launch initial propulsion is from a booster unit which falls away to leave the main sustainer motor to provide propulsion to the target.

The Fagot/Konkurs 9P135M1 launcher can be provided with a thermal imager for night or poor visibility conditions. When fitted the thermal imager, which weighs 13 kg, allows engagements at ranges up to 2,500 m.

On target, the 9K113 Konkurs missile 135 mm diameter 9N131 HEAT warhead (weight 2.7 kg) can penetrate 750 to 800 mm of armour set at 90°. Against armour set at 60° the figure is 400 mm.

9K113 Konkurs anti-tank guided missile

A revised version of the Konkurs is known as the 9K113M Konkurs-M. It has a tandem warhead with an extending nose probe to defeat explosive reactive armour before the main warhead initiates to penetrate 750 to 800 mm of main armour.

Variant

Flame

See entry under *India* in this section for details.

Data

Operation: SACLOS, wire-guided
Weight: launcher, 22.5 kg; thermal sight, 13 kg; complete round, 25.9 kg; missile, 14.58 kg; warhead, 2.7 kg; (Konkurs-M, complete round, 26.5 kg)
Warhead diameter: 135 mm
Length: missile, with gas generator, 1.15 m; container, 1.263 m
Range: day, 75-4,000 m; night, 75-2,500 m
Velocity: 270 m/s
Armour penetration: armour at 90°, 750-800 mm; armour at 60°, 400 mm
Operational temperature range: −50 to +50°C

Manufacturers

Tzniitochmash, 2 Zavodskaya Street, 142080 Klimovsk, Russia.
KBP Instrument Design Bureau, Tula 300001, Russia.

Status

In production. Offered for export sales.

Service

Russian and other CIS armies, India, Czech Republic and Slovakia, and former Warsaw Pact forces.

Licence production

BULGARIA
Manufacturer: Kintex - Vazod Engineering Plant
Type: 9M113 Concurs missile and launcher
Remarks: In production. Offered for export sales.
INDIA
Manufacturer: Bharat Dynamics Limited
Type: Konkurs and Flame
Remarks: 9K113 Konkurs system in production since June 1989 at a facility in Bahnoor, Medak District, Andhra Pradesh. See entry under *India* for details of Flame.
SLOVAKIA
Manufacturer: Konštrukta Defence
Type: Konkurs
Remarks: In production. Offered for export sales.
Manufacturer: ZVS sp, Dubnica nad Váhom
Type: Konkurs
Remarks: In production. Offered for export sales.

UPDATED

Thermal imager sight installed on a Konkurs 9P135M1 launcher (T J Gander)

1996

9K115 Metis and 9M131 Metis-M anti-tank guided missile system

Development

Compared to other CIS anti-tank missile systems such as the 9K111 Fagot and 9K113 Konkurs, the 9K115 Metis (Mongrel) system is much smaller and has a reduced operational range (40 to 1,000 m). Development at the KBP Instrument Design Bureau at Tula is understood to have commenced during the early 1970s but it was the early 1980s before the first examples were observed, leading to the NATO codename 'Saxhorn'; the US designation is AT-7.

This missile did not replace any other system already in service but appears to have been developed in order to fill a perceived gap in Eastern Bloc anti-armour defensive measures.

The 9M131 Metis-M missile has an improved range and a choice of larger diameter warheads.

A document dated October 1992 gives the unit price for a 9P151 launcher as $70,000, with $13,000 for a 9M115 missile.

Description

In principle, Metis is similar to Fagot and Konkurs but smaller and with revised SACLOS electronics and a simpler design. It is normally fired from a 9P151 tripod ground mounting but there are also reports of a 9P152 wheeled carriage. The guidance system is the 9S816, powered by a thermal battery attached to the front of the launch tube prior to launch.

The 9M115 Metis missile is roll-stabilised in flight using three wraparound tail fins and the tracking flare is mounted eccentrically so that it traces a circle as the missile rolls. The infra-red tracking system derives deviation from the centre of this circle and steering sense from the movement of the flare. This is necessary because there are only two steering surfaces and thus their position must be accurately known for the correct instructions to be determined and transmitted.

The 9M115 missile is ejected from its launch container by a booster motor stage, after which the main rocket motor ignites. This allows the weapon to be fired from a confined space, though a clear area of at least 2 m behind the launch tube is necessary.

There is an updated missile known as the 9M131 Metis-M (AT-7B, 'Saxhorn'). This missile has a larger diameter warhead (130 mm as compared to 94 mm on the original Metis missile) and may have either a tandem

Metis-M anti-tank missile (T J Gander)

1996

9P151 launcher for 9K115 Metis anti-tank guided missile system with thermal imager fitted (T J Gander)

warhead weighing 4.6 kg capable of penetrating 800 to 900 mm of main armour once explosive reactive armour has been defeated by a precursor charge, or a 4.95 kg fuel-air explosive warhead for attacking bunkers and similar targets.

The 9M151 launcher can be elevated ±5° and traversed 20° either side of a fixed arc, although full traverse through 360° is possible. A ×6 optical sight is used for target tracking. A thermal imager can be mounted on the launcher.

For transport the Metis-M is carried in two backpacks. One, weighing 24.5 kg, contains the launcher and one missile. The other pack contains two missiles and weighs 28.6 kg.

Maximum range of the Metis-M is 1,500 m although the missile guidance is effective only to 1,000 m; time of flight to 1,500 m is 9 seconds. On target the 9M115 warhead can defeat up to 460 mm of armour.

Data

Operation: SACLOS, wire-guided
Warhead diameter: Metis, 94 mm; Metis-M, 130 mm
Length, complete round: Metis, 740 mm; Metis-M, 910 mm
Weight, launcher: 10 kg
Weight, complete round: Metis, 6 kg, Metis-M, 13.8 kg
Range: Metis, 40-1,000 m; Metis-M, 80-1,500 m
Velocity: Metis, 223 m/s; Metis-M, 200 m/s
Armour penetration: Metis, 460 mm; Metis-M, up to 900 mm
Operational temperature range: −50 to +50°C

Manufacturers

Tzniitochmash, 2 Zavodskaya Street, 142080 Klimovsk, Russia.
KBP Instrument Design Bureau, Tula 300001, Russia.

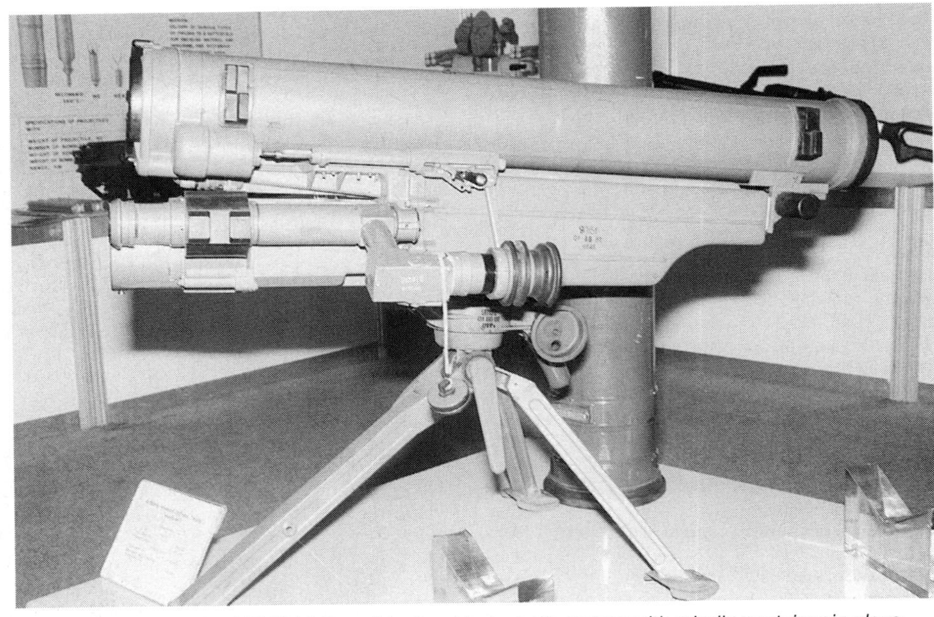

9P151 launcher for 9K115 Metis anti-tank guided missile system with missile container in place (T J Gander)

1996

Status

In production. Offered for export sales.

Service

CIS armed forces.

Licence production

BULGARIA
Manufacturer: Kintex - Vazov Engineering Plant
Type: 9K115 Metis
Remarks: In production.

UPDATED

Kornet anti-tank guided missile system

Development

First shown in the West in October 1994, the Kornet anti-tank guided missile system is intended to be the replacement for the 9K113 Konkurs system. When it was first shown it was stated that development of the Kornet (Cornet) was complete but production had yet to commence. Kornet was developed by the KBP Instrument Design Bureau at Tula. It has been allocated the US designation of AT-IX-14.

Description

The Kornet anti-tank guided missile system uses a laser beam-riding missile with SACLOS guidance; the missile appears to be a scaled-up 9M115 Metis derivative although Kornet has four tail fins in place of the three on the Metis. In operation all the operator has to do is keep his optical sight on the target and the system ensures the missile stays on track. Missile velocity has not been released but is understood to be of the order of 240 m/s; maximum range is 5,500 m.

Two types of warhead are available for the Kornet missile. For anti-armour use there is a 152 mm HEAT warhead which is claimed to be able to penetrate up to 1.2 m of armour protected by explosive reactive armour. The second warhead is a fuel-air explosive

Kornet anti-tank missile in its launch container on the Kornet launch unit (T J Gander)

Kornet anti-tank missile with tail fins deployed (T J Gander)

Kornet anti-tank missile showing laser receiver in tail (Christopher F Foss)

blast warhead for use against structures and troops in the open; it can also be effective against some types of armoured vehicle.

The Kornet missile weighs 27 kg in its sealed launch container. At launch it is boosted from the container before the main motor cuts in a short distance in front of the launcher. Four wraparound fins spring out as the missile leaves the container.

The Kornet launcher, an adjustable tripod, weighs 19 kg and is arranged with the launch tube/container over the main sight unit. The basic system has a day-only sight but a thermal night sight has been developed for use against targets at ranges up to 3,500 m.

Data

Operation: SACLOS, laser-guided
Warhead diameter: 152 mm
Length: complete round, 1.2 m
Weight: launcher, 19 kg; complete round, 27 kg
Max range: 5,500 m
Velocity: ca 250 m/s (not confirmed)
Armour penetration: up to 1.2 m

Development agency

KBP Instrument Design Bureau, Tula, Russia.

Status

Development complete. Not yet in production.

UPDATED

CZECH REPUBLIC AND SLOVAKIA

RPG-75 light anti-armour weapon

Description

The RPG-75 is a shoulder-fired rocket launcher which appears to be a local variant of the CIS RPG-18, and is one of several designs developed from the American M72 66 mm launcher.

The RPG-75 is made of an extruded light alloy tube. It telescopes and is 633 mm long when in the carrying position. The two halves are held together with a bayonet catch which has to be rotated to release the halves. The rear portion has a bell-shaped venturi not unlike the RPG-7. There are rubber end-caps to protect the tube and the projectile but these caps are not attached to the sling, as they are for the M72. The firing mechanism is mechanical and not unlike that on the M72. The sights are simple pop-up frames made of plastic; the foresight is marked for ranges of 200, 250 and 300 m and there is a simple rangefinder in the form of a stepped slot, indicating the apparent length of an average tank at the same ranges. The rearsight has three apertures on a revolving disc, marked 'N', '+' and '–' which adjust the sightline to compensate for temperature effects on the rocket motor.

Each launcher has a series of simple drawings showing the firer how it is used. Each drawing has a few instructions. The series translates as follows:

(1) Remove rubber cap from muzzle
(2) Tear off seal
(3) Turn chamber as far as possible
(4) Slide chamber as far as possible
(5) Turn chamber as far as possible
(6) Release catch of rearsight
(7) Erect foresight by pressure of button
(8) Erect rearsight and adjust correction of temperature
(9) Pull out carrying safety
(10) Squeeze the firing safety with forefinger as far as possible and hold it
(11) Fire with pressure of thumb.

Actions 3, 4 and 5 relate to the extending of the launch tube. Action 8 is again a similar one to the preparations involved in M72 and shows that the motor gives a different performance in cold weather. There is no shoulder stop so the firer has to judge where to put his head and eye when firing. One training instruction warns that the venturi should not be within 2 m of a wall or similar solid obstruction.

The projectile is a shaped charge design and is unusual in having no fins or other apparent stabilisation, relying entirely upon drag stabilisation over its short range. The complete projectile weighs 850 g and contains 323 g of explosive.

Data

Calibre: 68 mm
Length: folded, 633 mm; extended, 890 mm
Weight: firing, 3.1 kg

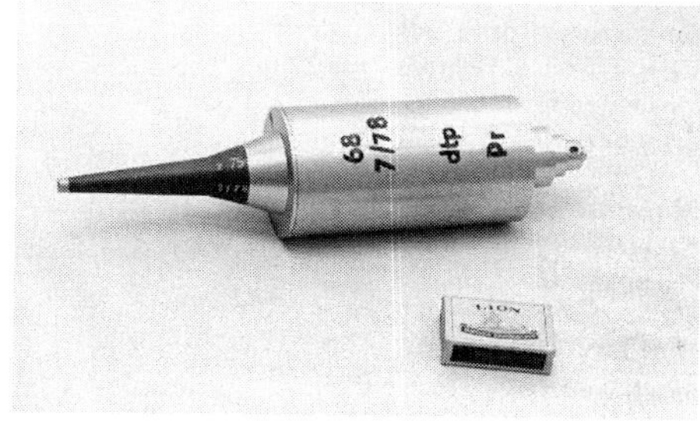

Shaped charge projectile of RPG-75

Launch velocity: 189 m/s
Range: (effective) moving targets, 200 m; stationary targets, 300 m
Time to self-destruct: 5 s
Armour penetration: 300 mm
Warhead: HEAT, 320 g

Status

Apparently in production.

Service

Former Warsaw Pact armies.

VERIFIED

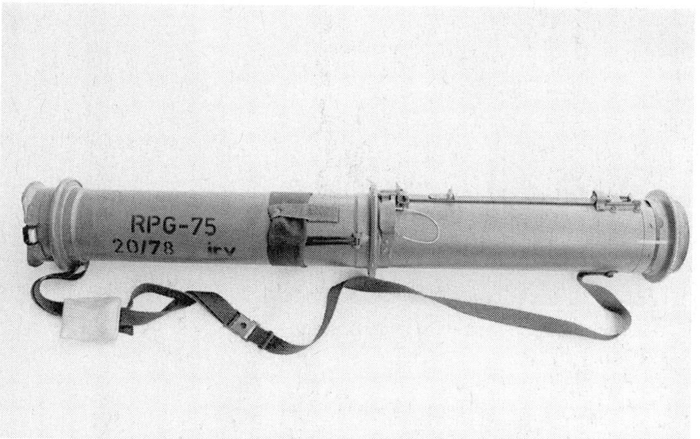

RPG-75 in carrying mode

RPG-75 in firing position

PG-7M 110 anti-tank projectile

Description

The PG-7M 110 anti-tank projectile was developed as a retrofit warhead option for use with the RPG-7 portable rocket launcher series grenades. The main change is the replacement of existing anti-armour warheads by a 110 mm warhead capable of defeating 600 to 700 mm of homogenous armour.

Modifying existing PG-7 series grenades consists of screwing the existing warhead from the rocket motor and removing the head and bottom components of the VP-7 piezoelectric fuze. The fuze components are then installed on the PG-7M 110 warhead and the new warhead is screwed onto the existing rocket motor. No modifications are required for the RPG-7 launcher sights.

Data

Calibre: warhead, 110 mm
Weight: warhead, 3.15 kg

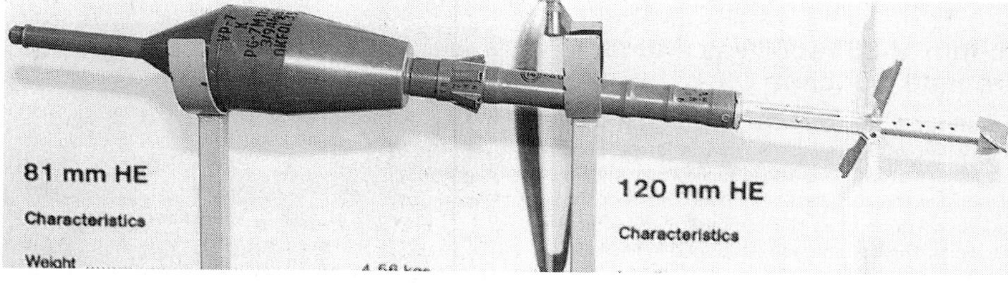

PG-7M 110 anti-tank projectile (T J Gander)

1996

Velocity: initial, 80 m/s; max, 180 m/s
Range: effective, 250 m
Armour penetration: 600-700 mm

Manufacturer

Konštrukta Defence, Kvystavisku 15, PO Box 62, 912 50 Trenčin, Slovakia.

Marketed by: Technopol, Kutlikova 17, 852 50 Bratislava, Slovakia.

Status

Available.

NEW ENTRY

EGYPT

PG-7 weapon system

Description

This is essentially the Soviet RPG-7 copied and manufactured in Egypt, with some modifications to suit Egyptian manufacturing methods.

Data

Calibre: launcher, 40 mm; warhead, 85 mm
Length: launcher, 950 mm; rocket, 930 mm
Weight: launcher, 6.8 kg; rocket, 2.25 kg; warhead, 900 g
Range: effective, AFVs, 300 m; other, 500 m
Range to self-destruct: 900 m
Velocity: initial, 120 m/s; max, 300 m/s
Max velocity: 300 m/s
Armour penetration: >260 mm

Manufacturer

Sakr Factory for Developed Industries, PO Box 33, Heliopolis, Cairo.

Status

In production.

Service

Egyptian Army.

UPDATED

Egyptian PG-7 anti-tank grenade launcher

1996

Cobra anti-tank weapon system

Description

This is an improved warhead for the standard PG-7 weapon system described previously. The Octal shaped charge has a new pattern liner manufactured by a specially developed process, giving an improvement in performance of approximately 20 per cent compared to standard copper cones. The fuzing system is electromagnetic and is initiated by contact between the double ogives of the warhead. It is capable of functioning at angles of 80° from normal and has a response time less that 10 μs.

The Cobra warhead has been demonstrated penetrating over 500 mm of armour steel at 3.5 calibres standoff distance in static tests.

Data

Length: grenade, 925 mm
Weight: grenade, 2.2 kg; warhead, 1 kg
Velocity: initial, 120 m/s; max, 300 m/s
Range: min, 200 m; effective, 350 m; max, 500 m
Armour penetration: >500 mm armoured steel

Manufacturer

Sakr Factory for Developed Industries, PO Box 33, Heliopolis, Cairo.

Status

In production.

Cobra weapon system in firing position

Service

Egyptian Army.

VERIFIED

FINLAND

55 mm M-55 recoilless anti-tank grenade launcher

Description

The 55 mm M-55 recoilless anti-tank grenade launcher is of rather more elaborate construction than many similar man-portable recoilless launchers. The weapons are issued on a scale of six per company in the motorised battalions. The weapon was developed in Finland.

Data

Calibre: 55 mm
Length: launcher, unloaded, 940 mm; loaded, 1.24 m
Weight: weapon, 8.5 kg; grenade, 2.5 kg
Range: effective anti-tank, 200 m
Rate of fire: 3-5 rds/min
Armour penetration: 200 mm

Status

Production complete.

Finnish 55 mm M-55 recoilless anti-tank grenade launcher

Service

Finnish armed forces.

VERIFIED

Raikka recoilless guns

Description

The firm of Raikka developed a series of anti-tank guns using the original recoilless principle of the countermass. In this arrangement the barrel is a plain tube and the cartridge is inserted into the centre of it. On firing, the shell goes forward and an equivalent weight of another substance is blown backwards, thus balancing the recoil.

The principle has not been used in this manner for many years, largely because of the extra weight of the countermass and the danger to troops behind. However, Raikka found that guns of this type are substantially cheaper to produce than normal recoilless models and by using water, sand or iron filings as the countermass the back blast danger is not excessive. Raikka offers a range of such guns in the following calibres: 41, 55, 81, 120 and 150 mm.

Barrels are either smoothbored or rifled and an unusual feature of the ammunition is that the 120 mm gun has a fin-stabilised APDS round (HVAPDS(FS)). It is claimed that this projectile can achieve 1,500 m/s and so is effective beyond 1,000 m against main battle tanks.

Manufacturer

Raikka Oy, PO Box 30, FIN-00251 Helsinki.

Status

Development completed.

VERIFIED

81 mm double-barrelled recoilless gun. Probably prototype as no sights fitted

41 mm launcher with sights extended. Tube beneath barrel may have some connection with trigger mechanism

Data

Weapon	Calibre	Length	Weight	Ammunition weight	Muzzle velocity	Effective range
VEHICLE-MOUNTED						
Raikka 150	150 mm	4.5 m	1,200 kg	42 kg	400 m/s	over 10 km
Raikka 120	120 mm	6 m	1,500 kg	6 kg	1,500 m/s	1.5 km
	120 mm	6 m	1,500 kg	15 kg	900 m/s	over 15,000 m
	120 mm	3.5 m	500 kg	5.3 kg	1,000 m/s	over 10,000 m
Raikka 81	81 mm	3 m	250 kg	3.5 kg	1,000 m/s	over 10,000 m
MAN-PORTABLE						
Raikka 81	81 mm	1.15 m	15 kg	2 kg	350 m/s	400 m
	81 mm	1.15 m	15 kg	3.3 kg	250 m/s	3,000 m
Raikka 55	55 mm	900 mm	4.5 kg	2 kg	170 m/s	200 m
Raikka 41	41 mm	760 mm	3 kg	1 kg	170 m/s	200 m

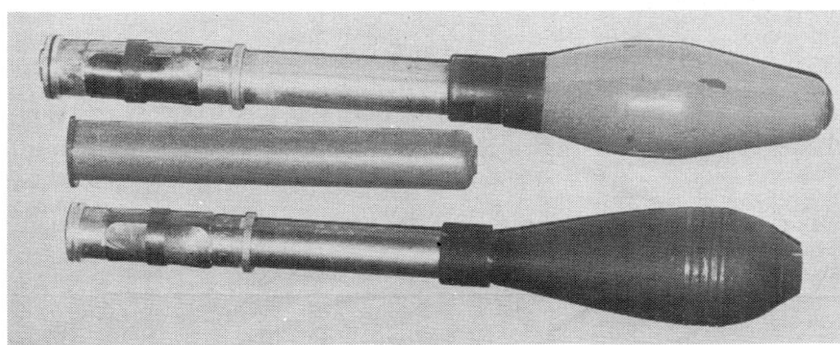

Two types of projectile with propellant cartridge between them. Note that with this size of launcher, warhead is outside barrel and stabilising fins are wrapped around propellant tube. Upper projectile appears to have warhead made from 81 mm mortar bomb

Raikka 41 mm countermass light anti-armour weapon. Note no shoulder stop to hold launcher in correct position for aiming

120 mm recoilless gun with 6 m barrel

120 mm recoilless gun with 3.5 m barrel and round

FRANCE

Wasp 58 light anti-armour weapon system

Description

Designed for the direct attack of light armoured vehicles at short range (250 m), the Wasp 58 is a light and disposable weapon system based on the Davis countershot principle using a plastic fibre countermass. Compact and easily handled, it may be fired with minimal signature, including firing from a confined space.

The Wasp 58 is composed of a fully equipped launching tube and a 58 mm hollow-charge warhead with significant piercing capabilities. Care has been taken in the design to facilitate bringing the weapon into use instantly, without lengthy preparation.

Data

Calibre: 58 mm
Length: 800 mm
Weight: complete, 3 kg; projectile, 615 g; countermass, 650 g
Muzzle safety: 8 m
Muzzle velocity: 250 m/s
Range: effective, 350 m; combat, 250 m; min 20 m
Armour penetration: 300 mm
Hit probability Ph 20-250 m: stationary target, 0.95; moving target, 0.88
Accuracy Standard Deviation: 0.3 m
Operating temperature range: −40 to +51°C

Manufacturer

Giat Industries, 13 route de la Minière, F-78034 Versailles-Satory Cedex.

Status

In production.

VERIFIED

Wasp 58 rocket launcher

Eryx short-range anti-tank missile

Development

Eryx was designed for use by forward infantry, providing them with a weapon effective against new types of armour to ranges up to 600 m. The objectives during development were to use the most simple and compact solutions to obtain high penetration against all armour likely to be in service during the remainder of this century and beyond, to have extreme precision, to be able to fire from enclosed spaces, and at the same time to keep the weight down to 12 kg and the unit cost compatible with broad distribution at section level. Mass production began in 1991.

In 1989 it was announced that a Memorandum of Understanding had been signed by France and Canada with the intention of producing Eryx as a co-operative venture.

To date, the French Army has ordered 400 Eryx firing posts and 4,700 missiles. Canada has ordered 425 firing posts and 4,500 missiles. Orders for Norway stand at 424 firing posts and 7,200 missiles. The Brazilian Army has ordered Eryx, as has one further unspecified customer.

Description

The Eryx system consists of a missile prepacked in a container tube in which it is transported and stored and from which it is launched, plus a compact firing unit containing the ignition, detection and timing systems, together with a remote control. The missile can be emplaced and made ready to fire in less than 5 seconds. During flight (4.2 seconds to 600 m) the firer is required only to keep the sight on the target. The missile carries an infra-red beacon which is detected by the sight unit, inside which corrections are derived, producing steering commands sent to the missile via a wire link which is unspooled as the missile proceeds.

The application of a new concept, the direct thrust flight control, efficient even at low speed, allows launching to be achieved using a small propulsion unit. The missile can thus be used in confined spaces. After launch, the rocket motor accelerates to the flight speed of 300 m/s.

A MIRABEL thermal imager can be added to the Eryx launcher for firing at night or under poor visibility conditions.

Data

Diameter: 160 mm
Length: 925 mm
Weight: 12 kg; firing post, 4.5 kg
Range: 50-600 m
Time of flight: 4.2 s to 600 m
Max velocity: 245 m/s
Armour penetration: 900 mm

Manufacturer

Aerospatiale-Missiles, 2 rue Béranger, BP 89, F-92322 Chatillon-Cedex.

Status

In production.

Service

In service with the French, Canadian and Norwegian armies.

UPDATED

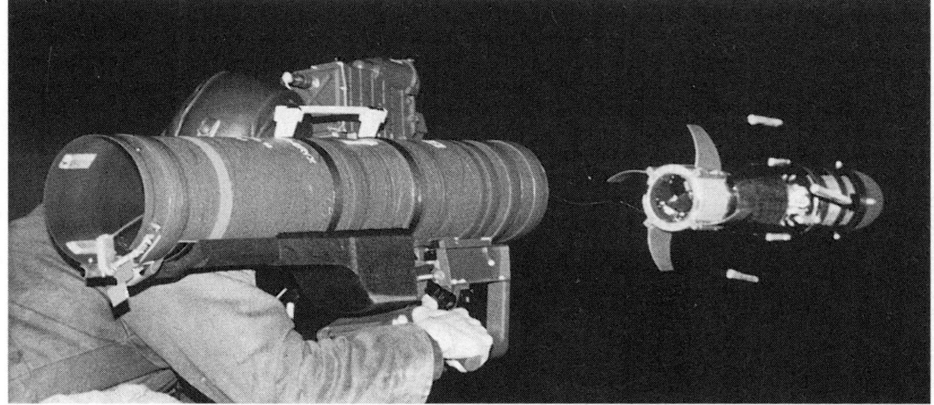

Eryx short-range anti-tank missile being launched using MIRABEL night sight

1996

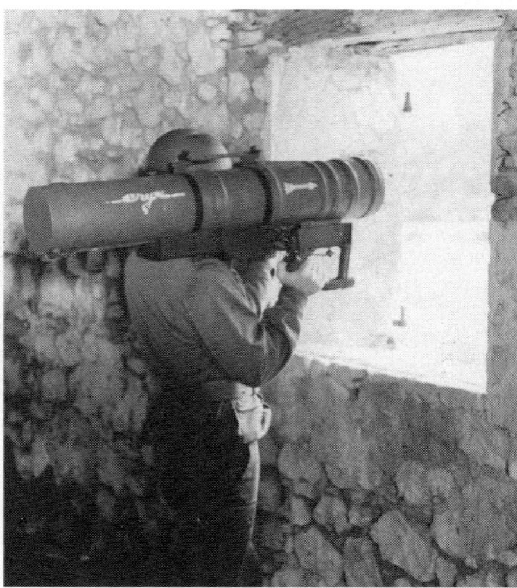

Eryx anti-tank missile being fired inside building

APILAS light anti-tank weapon

Description

APILAS was originally a product of Manurhin, now a subsidiary of Giat Industries. It is a disposable, man-portable, shoulder-fired launcher containing a projectile carrying a HEAT warhead propelled by a rapid-burning rocket motor.

The rocket is supplied prepacked in an aramid-fibre launch tube which has a retractable sight. The rocket motor nozzle is of wound aramid-fibre construction, with an aluminium venturi. The shaped charge warhead is initiated by an electrical fuze system powered and armed by gas pressure released when the rocket is launched. There is a mechanical delayed arming system which guarantees the warhead to be unarmed at 10 m from the launcher and fully armed from 25 m. The warhead has a long tapered nose cone which permits normal functioning up to incidence angles of 80°.

A training system comprises a subcalibre firing weapon using special 7.5 mm tracer bullet cartridges and a drill weapon.

Data

Calibre: 112 mm
Length: launcher, 1.29 m; projectile, 925 mm
Weight: overall, 9 kg; launcher, 4.7 kg; projectile, 4.3 kg; explosive charge, 1.5 kg
Range: effective, 330 m
Time of flight: 1.2 s to 330 m; 1.9 s to 500 m

Muzzle velocity: 293 m/s
Penetration: >RHA, 720 mm; reinforced concrete, >2 m

Manufacturer
Giat Industries, 13 route de la Minière, F-78034 Versailles-Satory Cedex.

Status
In production.

Service
France, Finland, Italy, Jordan, South Korea, Saudi Arabia and many others.

VERIFIED

APILAS fired from the shoulder

GERMANY

Panzerfaust 3 anti-armour weapon

Development
The Panzerfaust 3 (Pzf 3) is a compact, lightweight, man-portable, shoulder-fired, unguided weapon system providing effective engagement capabilities against current and future Main Battle Tanks (MBTs). The Pzf 3 system, which was developed under contractual agreement with the former West German Army, consists of a disposable cartridge with 110 mm warhead, and a reusable firing and sighting device.

The Pzf 3 concept was selected as it fulfils the following basic tactical requirements: high kill probability against current and future MBTs, even when frontally engaged; capability of being fired from enclosed spaces; and low cost, so that it can be widely distributed to all types of military formation.

Fire Salamander autonomous firing mount with TV sensor

Description
The Panzerfaust 3 is based on the Davis countermass principle. The adoption of this principle permits firing the weapon in enclosed spaces while allowing potential design growth, since it is possible to modify the warhead shape and calibre to meet different tactical requirements without affecting the launch tube. To emphasise the versatility of the system, the development of HEAT (DM 12), multipurpose fragmentation, smoke and illuminating warheads have formed part of the programme.

A version with a 110 mm diameter tandem warhead capable of defeating reactive armour is under development, with production scheduled for the end of 1996.

A night combat capability is achieved by use of an infra-red target marker fitted to the firing and sighting device and used in conjunction with night vision goggles.

An autonomous firing mount known as Fire Salamander has been developed. This carries four Pzf 3 systems around a central command module which can contain a variety of sensors for automatic firing. It can use a remote-controlled TV camera as an aiming device.

Data
Calibre: launcher, 60 mm; projectile, 110 mm
Length: 1.2 m
Weight: 12.9 kg; projectile, 3.9 kg; firing device, 2.3 kg
Range: moving targets, 300 m; stationary targets, 500 m
Velocity: initial, 163 m/s; max, 243 m/s
Vertex: 300 m range, 2.1 m
Time of flight: 1.36 s to 300 m
Armour penetration: >700 mm

Manufacturer
Dynamit Nobel AG, Postfach 1261, D-53839 Troisdorf.

Status
Adopted by German Army June 1987. Volume production commenced 1990.

Service
German Army and five other armies.

UPDATED

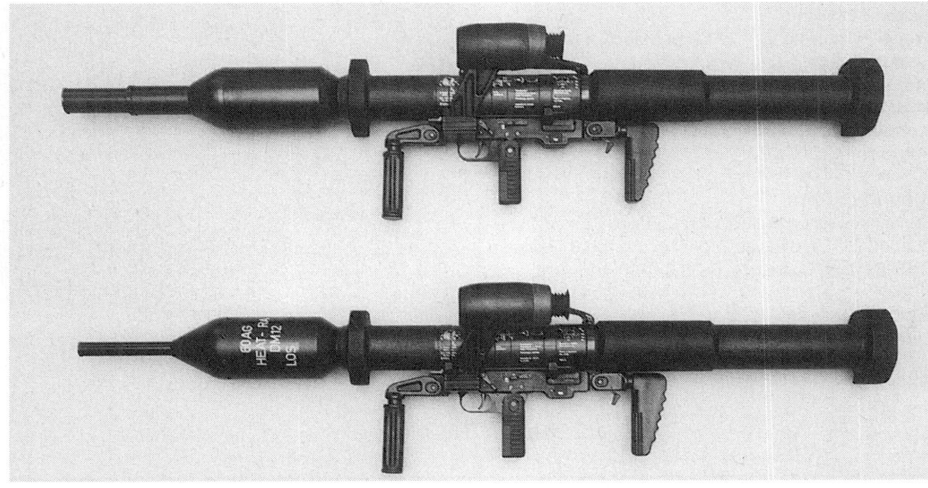

Top is a Panzerfaust 3 with a 110 mm tandem warhead for the attack of reactive armour; bottom is a standard Panzerfaust 3 with a standard HEAT warhead

1996

INDIA

Nag anti-tank missile system

Development/Description
Very little has been released regarding the Nag (Serpent) anti-tank missile system as it is still under development by the Defence Research and Development Laboratory at Hyderabad, and that development is understood to be behind schedule. Current forecasts are that Nag development will be complete by the end of 1996.

The Nag will be a third generation, top attack, fire-and-forget missile system with a range of 4,000 m. It is intended for ground, armoured vehicle and helicopter applications. It is claimed that it will be the first missile of its type to have a complete glassfibre composite structure. Initial guidance is provided from the launcher's target acquisition system, although for terminal guidance the Nag will have either a millimetric wave active seeker or an infra-red imaging passive unit, both of which are still under development. The infra-red unit incorporates a focal plane array sensor with a combination of charge-coupled devices and mercury-cadmium telluride arrays. Both types of seeker are stated to be resistant to countermeasures.

A tandem charge warhead will be fitted, enabling the Nag to destroy any known tank, including those protected by composite or active armours. The missile is propelled by booster and sustainer motors, both of which are ignited simultaneously, using low smoke composite modified double-base propellant. The sustainer motor exhausts through four canted, side-mounted venturi.

Development Agency
Defence Research and Development Laboratory, Hyderabad.

Status
Development

NEW ENTRY

Flame anti-tank missile system

Description

The Flame anti-tank missile system is a combination of two missile systems, in effect being the installation of a MILAN M2 missile launch capability on launch units normally used by the 9K113 Konkurs system. It is available in two forms, Flame V for vehicle mountings and Flame G for ground operations using a 9P135 launch unit. Both the MILAN and Konkurs systems are licence-produced in India. Development of Flame took place during 1992 and 1993.

For Flame G the original electronics units intended for use with the 9K113 Konkurs is replaced by a new Flame unit along with a launcher ramp unit adaptor to allow the MILAN M2 launch container to be fitted. The ramp adaptor also houses a battery pack and an electronic power supply module. Once the Flame conversion has been made Flame G can launch only the MILAN M2 missile. Performance of the MILAN M2 missile remains unchanged from the original (see MILAN entry under *International* in this section).

Flame V is an add-on adaptation to suit the BMP IFV. Once installed Flame V can launch both Konkurs and MILAN M2 missiles.

Manufacturer

Bharat Dynamics Limited, Kanchanbagh, Hyderabad.

Status

Available for production.

VERIFIED

The Flame anti-tank missile system, a combination of MILAN and Konkurs (T J Gander)

106 mm recoilless gun

Description

This 106 mm recoilless gun is a close variant of the US 106 mm M40A1C, available either on a field carriage formed from a Tripod T26 and 106 mm Rifle Mount T173, or mounted on a light vehicle such as a Mahindra jeep. The Indian weapon differs little from the American original and fires HEAT ammunition. It incorporates a 12.7 mm M8C spotting rifle manufactured at the Small Arms Factory, Kanpur.

Data

Calibre: 105 mm
Weight: complete, 217.72 kg; ordnance, 113.85 kg
Length: 3.4 m
Max range: direct-fire, 1,372 m
Muzzle velocity: 503 m/s
Armour penetration: 0°, 620 mm; 30°, 515 mm; 60°, 400 mm

Manufacturers

Ordnance Factory Board, Gun Carriage Factory, Jabalpur 482 011, and Indian Ordnance Factories, Gun and Shell Factory, Cossipore, Calcutta 700002.

Status

In production. Offered for export sales.

Service

Indian Army.

VERIFIED

Indian 106 mm recoilless gun mounted on Mahindra jeep (T J Gander)

INTERNATIONAL

MILAN portable anti-tank weapon system

Development

The Missile d'Infanterie Léger Anti-char (MILAN) was introduced by Euromissile, an international consortium of the French Aerospatiale Group and the German Deutsche Aerospace (DASA - now Daimler-Benz Aerospace). It is anticipated that the German sector of Euromissile will be assumed by the new Daimler-Benz missile company, LFK. MILAN has been in production since 1972.

MILAN was designed to be used by infantry and is a SACLOS wire-controlled missile. A flare on the missile emits an infra-red signature which enables the system computer to measure the error between its position and the line of sight. Missile velocity is twice that of most early portable missiles, allowing the MILAN missile to reach 1,500 m in 10 seconds and 2,000 m in 12.5 seconds.

Description

The complete MILAN weapon system is made up of two units: a round of ammunition, consisting of a missile, factory-loaded into a sealed launcher/container tube; a combined launching and guidance unit, consisting of a launcher combined with a periscopic optical sight and an infra-red tracking and guidance system, the whole being mounted on a tripod.

The round of ammunition comprises an assembled missile factory-loaded with wings folded into a sealed tube which serves the dual purpose of storage/transport container and launching tube.

The container/launcher tube is fitted with mechanical and electrical quick-connection fittings and a self-activating battery is mounted on the outside to provide electrical power for the firing installation.

MILAN has a night firing capability through the addition of the MIRA thermal imaging device adopted by the French, German, UK and other armies. This consists of a case weighing 9 kg which can be mounted on the standard firing post. Target detection is possible at a range of over 4,000 m and firing at 2,000 m.

The missile is an assembly of the following main components: an ogival head containing a shaped charge and fuze, a two-stage solid propellant motor discharging through an exhaust tube to a central nozzle located at the rear of the missile, a rear part containing the jet spoiler control system and guidance components. The guidance components include: a gas-driven, turbine-operated gyro; an infra-red flare; a spool carrying the

MILAN being fired from light vehicle

MILAN 3 anti-tank missile at the instant of launch

two guidance wires in one cable; a decoder unit; and a self-activating battery for internal power supply.

The missile is launched from its tube by a booster charge gas generator which is contained in the tube and burns for 45 ms. Initial velocity is 75 m/s.

The recoil effect is compensated but part of it is used to eject the tube to the rear of the gunner to a distance of 2 to 3 m.

The two-stage propulsion motor burns for 12.5 seconds and increases the velocity of the missile, at first rapidly, then more slowly to 210 m/s. The operator must keep his sight cross-hairs on the target throughout the engagement.

Guidance is achieved by means of a single jet spoiler operating in the sustainer motor exhaust jet. The jet spoiler operates on guidance command signals generated automatically by the launcher/sight unit (by measurement of the angular departure of the missile from the reference directions of the infra-red tracker in the sight unit) and transmitted to the missile via the guidance wires which unwind from the missile.

The guidance commands are decoded by a transistorised decoder unit within the missile. The self-activating battery which provides internal power is designed for long-term storage and use in worldwide temperature conditions.

For safety the missile is locked inside its tube and the solid propellant gas generator cannot be ignited until the missile is unlocked by the gunner. In addition the sustainer motor ignites when the missile is released from the tube and the wings have unfolded. The fuze cannot arm until the sustainer motor is ignited and an electrical safety device functions when the missile has flown approximately 20 m.

Variants

MILAN 2 Introduced in 1984, MILAN 2 has a warhead assembly intended to improve performance against the frontal arc of main battle tanks. The warhead was increased in diameter from 103 mm to 115 mm, the explosive weight was increased from 1.36 kg to 1.79 kg and the fuze initiating switch was relocated at the tip of a probe to add about 145 mm to the length of the warhead. These modifications considerably improved the piercing performance, largely by improving the stand-off distance and the explosive effectiveness of the shaped charge. The improvement was calculated to be some 65 per cent, and penetrations into steel armour of 1.06 m have been recorded.

The MILAN 2 missile does not affect the interface with the firing post, the overall missile weight remains the same and the tubular container is unchanged. Existing firing posts can therefore fire both MILAN 1 and MILAN 2 missiles.

MILAN 2T Introduced in 1993, MILAN 2T is an improved missile using a tandem warhead and an extended nose probe. The probe contains the primary shaped charge which is designed to detonate reactive armour protection and thus clear the line of attack for the main shaped charge. No modification is required to existing launchers.

MILAN 3 MILAN 3 is a new development, using the 2T warhead, and which gives the MILAN weapon system increased resistance to jamming. The missile is equipped with a flashing Xenon lamp. The localisation system, based on CCD sensor discriminations in both time and space, is synchronised before launch with the missile's lamp. This modification provides immunity against any natural or artificial infra-red jamming likely to be encountered on a battlefield.

Flame

See entry under *India* in this section for details.

Data

Type	**M1**	**M2**	**M2T**	**M3**
MUNITION IN TACTICAL PACK				
Weight in carrying mode	12.23 kg	12.23 kg	12.62 kg	12.62 kg
Weight in firing mode	11.52 kg	11.52 kg	11.91 kg	11.91 kg
Length in carrying mode	1.26 m	1.26 m	1.26 m	1.26 m
Length in firing mode	1.2 m	1.2 m	1.2 m	1.2 m
Diameter	133 mm	133 mm	133 mm	133 mm
MISSILE				
Weight	6.73 kg	6.73 kg	7.12 kg	7.12 kg
Length	769 mm	918 mm	918 mm non-extended nose 1,138 mm extended nose	918 mm non-extended nose 1,138 mm extended nose
Diameter, wings folded	125 mm	125 mm	125 mm	125 mm
Wing span	267 mm	267 mm	267 mm	267 mm
WARHEAD				
Weight	2.67 kg	2.7 kg	3.12 kg	3.12 kg
Diameter	103 mm	115 mm	117 mm	117 mm
Explosive filling	1.36 kg	1.79 kg	1.83 kg	1.83 kg
Cone diameter	101 mm	112.9 mm	112.9 mm	112.9 mm
TYPE OF FIRING POST	**MILAN (Munitions M1, M2)**			**MILAN 3 (All Munitions)**
Weight	16.4 kg			16.9 kg
Length	900 mm			900 mm
Height	650 mm			650 mm
Width	420 mm			420 mm

PERFORMANCE

Velocity: at launch, 75 m/s; at 2,000 m, 210 m/s

Time of flight: to max range, 12.5 s

Range: (effective) 25-2,000 m

Chance of a hit: 0-250 m, average 75%; 250-2,000 m, greater than 98% (manufacturer's figures)

Warhead: shaped charge. Detonated by electrical connection produced by crush-up of ogive when missile hits. Minimum performance is the triple penetration of NATO heavy tank plate

Manufacturers

Aerospatiale-Missiles, 2 rue Béranger, BP 89, F-92322 Chatillon-Cedex, France.

Daimler-Benz Aerospace, D-81663 München, Germany.

Marketed by: Euromissile, 12 rue de la Redoute, F-92260 Fontenay-aux-Roses, France.

Status

In production.

Service

In service in Croatia, France, Germany, India and the United Kingdom. Also supplied to other NATO and non-NATO countries.

Licence production

INDIA

Manufacturer: Bharat Dynamics Limited

Type: MILAN

Remarks: Licence production since January 1985. Missile type is MILAN M2. In production and offered for export sales.

UPDATED

Medium Range TRIGAT anti-tank weapon system

Development

Medium Range TRIGAT, or MR TRIGAT, is under development as the third-generation replacement for MILAN and similar anti-tank missiles by 2005. Few hard details have emerged but it is known that MR TRIGAT will follow the same general lines as MILAN regarding the launch post appearance and sealed missile container, and will have an operational range of from 50 to 2,000 m, with a secondary capability against attack helicopters.

Full development of MR TRIGAT began in September 1988 with France, Germany and the United Kingdom as the pilot nations who had conducted the earlier feasibility and definition phases. Belgium and the Netherlands are actively participating as Associate Nations.

Description

MR TRIGAT will consist of two main components; the missile and the firing post.

The missile is the AC3G, a laser-guided missile with dual shaped charge warheads to defeat active armours; weight of the missile is intended to be under 17 kg. The missile uses thrust vector control which imparts high manoeuvrability and control at low velocities while providing a soft launch and short-range accuracy. The missile is issued in a factory-sealed container which also acts as the launch tube. The missile is 950 mm long and has a body diameter of 152 mm. Flight velocity is 230 m/s.

The NBC-hardened, man-portable firing post uses modular construction and incorporates a thermal imager feature to allow firing at night or in poor visibility; weight of the firing post is intended to be under 17 kg. If required the thermal imager can be removed, leaving a basic optical sighting system. The MILAN MIRA night sight is compatible with MR TRIGAT. The firing post has its own go/no go test procedures prior to firing as well as built-in test equipment to isolate subassemblies which become defective in the field.

Using the MR TRIGAT firing post it is possible to fire 3 rds/min out to a maximum range of 2,000 m. Time of flight to 2,000 m is less than 12 seconds. Minimum range is 50 m.

In addition to the ground-mounted firing post MR TRIGAT will also be capable of being launched from vehicle roof, pintle or turret mountings; the British Army will deploy MR TRIGAT from specially modified Warrior MCVs. It may be launched from within buildings.

Following target tracking using either the firing post optical sight or thermal imager, the MR TRIGAT missile is guided by a scanning laser beam system. The firing post generates a guidance tunnel with a circular cross-section with the axis harmonised with the line of sight to the target. The missile, with its rearward-facing laser receiver, localises itself within the guidance tunnel and aligns itself on the guidance axis to impact with the target. The laser beam system thus provides high accuracy throughout the missile's flight while retaining high resistance to countermeasures.

Mock-up of a Medium Range TRIGAT (MR TRIGAT) firing post

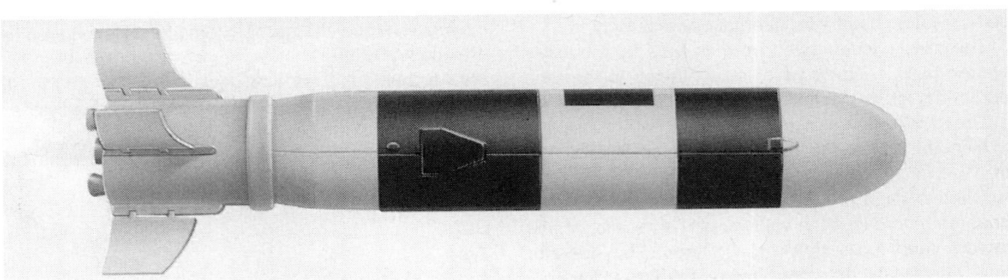

Artist's impression of a Medium Range TRIGAT (MR TRIGAT)

1996

Manufacturer

Euromissile Dynamics Group, 12 rue de la Redoute, F-92260 Fontenay-aux-Roses, France.

Status

Development.

UPDATED

IRAQ

Al-Nassira light anti-tank weapon

Description

The Al-Nassira weapon is a locally produced copy of the Soviet RPG-7 launcher and rocket system. From the little information released it would appear to be identical to the Soviet system except that the Al-Nassira has a simple tangent sight rather than an optical sight and is therefore slightly lighter.

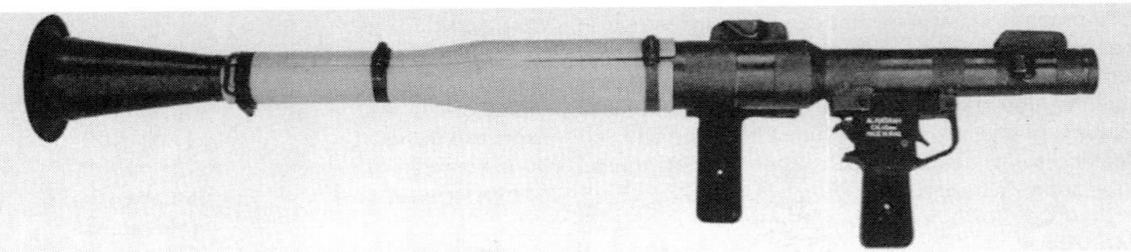

Al-Nassira light anti-tank weapon

Data

Weight: launcher, with sights, 7.4 kg
Muzzle velocity: 120 m/s
Flight velocity: 300 m/s

Manufacturer

State arsenals.

Status

In production.

VERIFIED

ISRAEL

B-300 light anti-armour weapon

Description

The B-300 is a man-portable, shoulder-fired, semi-disposable anti-armour system with an effective range of 400 m. It is carried, loaded and fired by one man. The launcher and three rounds weigh 19 kg.

The B-300 system has a calibre of 82 mm and is similar to the French LRAC 89. It breaks down into two major sections: the launcher, with a variety of sight options, which is reusable, and a projectile in a sealed container that is disposed of after firing. The GRP container acts as an extension to the launcher during firing and is coupled quickly to the launcher making all the necessary mechanical and electrical connections.

In addition to the basic Mk 1 HEAT round there is a Mk 2 HEAT round, slightly heavier than the Mk 1 but with significantly better penetration (ca 550 mm); a High Explosive Follow-Through (HEFT) round which

Assembling the round to the launcher prior to firing

Firing B-300 from the prone position

blows a hole in the target by means of a shaped charge, and then launches a secondary charge through the hole and into the interior of the target; an illuminating round of 600,000 candle-power and a range of 1,700 m; practice rounds with an impact marker for day or night use; and dummy rounds with a noise cartridge for loading and firing practice. The projectile has eight penknife fins folded forward around the motor nozzle. No details have been given of the motor which accelerates the projectile after an initial launch velocity of about 270 m/s.

The launcher is provided with an integral battle sight although a stadia telescopic sight is also available. The stadia sight is equipped with a Betalight for use in poor visibility. This can be switched on as required. For night operation, where white light is not available, a starlight telescopic sight can be mounted using an adaptor provided with each launcher.

The launcher is equipped with a folding bipod and shoulder rest and a broad sling. The B-300 is claimed to have a good hit probability.

Low cost training can be carried out using a subcalibre device firing a special 9 mm cartridge with a projectile that follows the trajectory of a live round.

Data

Calibre: 82 mm
Weight: unloaded, 3.5 kg; loaded, 8 kg; round, 4.5 kg; projected mass, 3.1 kg
Length: unloaded, 775 mm; loaded, 1.35 m
Muzzle velocity: initial, ca 270 m/s
Range: 400 m
Operational temperature range: −10 to +60°C
Armour penetration: >400 mm at a graze angle of 65°

B-300 launcher unit with sight

Manufacturer

TAAS - Israel Industries Limited, PO Box 1044, Ramat Hasharon 47100.

Status

Development.

VERIFIED

MAPATS anti-armour missile system

Description

The MAPATS anti-armour missile system consists of a laser beam-riding missile and a launcher. The missile is packed in a glassfibre launch and storage tube which is assembled to the launching unit. After launch, the missile follows a laser beam which is directed at the target and maintained there by the operator. An optical sight, with the laser slaved to its axis, is used for target tracking. In flight the missile detects the presence of the beam by using a rear-mounted sensor, detects any deviation from the beam axis, and generates correction manoeuvres so as to stay on the line of sight. The system is immune to jamming and entirely autonomous.

The components of the MAPATS system are the launch tube and missile, tripod, traversing unit, and the night vision system. The missile is housed in a factory-sealed container which is loaded into the launch tube. After firing, the empty container is discarded. Setting up takes less than one minute, including an automatic self-test routine. Loading subsequent rounds takes only a few seconds. On firing, the missile leaves the launch tube, the ejector motor falls away, and the laser beam modulator starts to transmit signals to the sensor in the missile. The laser beam cross-section is kept constant by a zoom system, creating a constant corridor which coincides with the line of sight. Beam intensity on the target is initially very low, increasing as the missile approaches.

Data

Calibre: 148 mm
Length: missile, 1.56 m
Weight: missile, 18.5 kg; tripod and battery, 14.5 kg; traversing unit, 25 kg; guidance unit, 21 kg; launch tube, 5.5 kg; night vision system, 6.5 kg; missile in container, 29.5 kg
Velocity: launch, 70 m/s; max, 305 m/s
Time of flight: 19.5 s to 4,000 m
Max range: 5,000 m
Firing angles: traverse, 360°; elevation, −20 to +30°
Warhead: shaped charge, 3.6 kg
Armour penetration: >800 mm

Manufacturer

TAAS - Israel Industries Limited, PO Box 1044, Ramat Hasharon 47100.

Status

In production.

MAPATS missile on tripod mount

Service

Israeli Army and some other armed forces.

UPDATED

ITALY

80 mm Folgore unguided anti-tank weapon

Description

Folgore is a lightweight (18 kg) one-man portable, reusable (not less than 600 rounds), recoilless multipurpose weapon system. It is intended primarily as an anti-tank weapon but may also be used to attack bunkers or sandbag emplacements.

The weapon is loaded by releasing a nozzle clamp, opening the nozzle and inserting a round. Three rounds can be fired in the first 30 seconds. Normally the weapon would redeploy before firing this number of rounds.

The optical sight has a magnification of ×5 and a weight of 350 g. It can be adjusted in azimuth and in elevation for boresighting, and in elevation to compensate for ambient temperature.

The nozzle design has been optimised to reduce the danger of blast on personnel, and improvements in propellant burning efficiency have also contributed to blast reduction. Low luminosity propellant has diminished the luminosity of hot nozzle gases.

Folgore ammunition is manufactured by BPD-Difesa e Spazio. The propellant charge is multi-perforated cylindrical grain containing 1 kg of US M10 single base explosive which gives the projectile a launch velocity of 385 m/s.

Folgore 80 mm unguided anti-tank weapon

The rocket-assisted projectile has a HEAT warhead. A pyrotechnic igniter ignites the rocket motor at a safe distance from the muzzle and increases the projectile velocity to about 500 m/s. Aerodynamic stabilisation is achieved by an assembly of six fins, canted to maximise accuracy. The result is a very flat trajectory and short time of flight, for example 1,000 m in 2.5 seconds, and a high accuracy and first round hit probability.

Target practice ammunition, having the same weight and ballistic performance as the HEAT round, is also available to minimise training costs.

According to official figures, Folgore is capable of penetrating about 450 mm of rolled homogeneous armour.

An optional night sight using a low-light intensifier is available.

Data

Calibre: 80 mm
Length: launcher, 1.8 m; round, 740 mm
Weight: 18 kg; round, 5.2 kg; rocket, 3 kg
Max range: 4,500 m; effective anti-tank, 1,000 m
Min range: 50 m
Muzzle velocity: 385 m/s
Max velocity: 500 m/s
Armour penetration: 450 mm

Manufacturers

Launcher: Breda Meccanica Bresciana, Via Lunga 2, I-25126 Brescia.
Ammunition: Breda Meccanica Bresciana in co-operation with BPD Difesa e Spazio.

Status

In production.

Service

Supplied to the Italian Army.

UPDATED

MAF man-portable missile system

Description

MAF is the name given to a long-range anti-tank weapon system of advanced SACLOS type under final development by OTOBREDA in co-operation with ORBITA of Brazil, in order to meet a requirement of the Brazilian Army.

The system consists of a missile in its container launcher and a firing post. Guidance is by laser beam-riding, the missile being subsonic. The firing post consists of a tripod, projector unit and sights.

The missile is provided with a two-stage solid fuel rocket motor. The first stage is all-burnt inside the launch tube; the second stage ignites at a safe distance from the launcher and boosts the missile to maximum speed.

The guidance beam is obtained from a laser modulating emitter operating in the near infra-red band. The beam is generated in the launcher unit and the receiver is in the rear of the missile body, giving a very high degree of protection against jamming. Missile guidance is ensured by a servo loop which corrects the missile trajectory as a function of the deviations from the line of sight. The sighting system in fair weather is by means of an optical sight and by night through a thermal imaging camera.

The missile configuration is of the canard type (roll free) with conical nose and tail. Folding stabilisers are mounted at the rear to provide high manoeuvring capabilities and aerodynamic damping. The warhead is a shaped charge with very high penetrative ability.

The maximum range is in excess of 3,000 m. The guidance system ensures a processable signal at more than 2,000 m in the worst conditions of optical visibility, humidity, temperature and turbulence.

The original system is man-portable; vehicle and helicopter-borne applications are being studied.

Data

Diameter: 130 mm
Length: launcher container, 1.38 m
Weight: missile, 14.5 kg; warhead, 4.1 kg; launcher post, 23 kg
Max range: 3,000 m
Min range: 70 m
Max speed: 290 m/s
Time of flight: 16 s to 3 km

Manufacturer

OTOBREDA SpA, Via Valdilocchi 15, Casella Postale 337, I-19100 La Spezia.

Status

Advanced development.

VERIFIED

MAF missile and firing post

MAF missile with operating crew

JAPAN

Type 64 MAT anti-tank system

Description

Work started on the Type 64 MAT (also known as the KAM-3D) manual command to line of sight (MCLOS) anti-tank guided missile in 1957. The prime contractor was Kawasaki Heavy Industries, responsible to the Technical Research and Development Institute of the Japanese Defence Agency. After prolonged trials it was adopted as standard equipment in the Japanese Ground Self-Defence Forces. It is now being phased from service.

The missile has a cylindrical body with cruciform wings incorporating full width spoilers for control. It carries a flare which enables the operator to track it by day using an optical system with a thumb-button control box. By night the sustainer rocket exhaust provides the required illumination. The missile is placed on a metal frame launcher, which gives an angle of 15°, by the two-man crew. When launched, a booster motor accelerates the missile to cruising velocity (85 m/s) in 0.8 second, after which a sustainer motor maintains this speed.

Missiles can be fired singly or in multiple units by infantry or from jeeps or helicopters. Some are mounted on Type 60 APCs acting as tank destroyers.

A field test set (KAM-3TE) and a simulator (KAM-3TP) were produced by the main contractor.

KAM-3D missiles on jeep mounting

Data

Diameter: 120 mm
Length: 1.015 m
Weight: at launch, 15.7 kg; warhead, 3 kg HEAT
Wing span: 600 mm
Cruising velocity: 85 m/s
Range: 350-1,800 m
Crew: 2

Manufacturer

Prime contractor: Kawasaki Heavy Industries Limited, World Trade Center Building, 4-1 Hamamatsu-cho 2-chome, Minato-ku, Tokyo.

Status

No longer in production.

Service

Being phased out of service with the Japanese Ground Self-Defence Forces.

UPDATED

Type 79 Jyu-MAT anti-tank missile system

Development

The Type 79 Jyu-MAT, sometimes known as the Type 79 Tan-SSM, commenced low-rate production in 1980 and entered service in 1984. The initial users were five coastal defence anti-landing battalions. Since then over 2,000 missiles have been produced. It is deployed on ground mountings, Type 73 Jeeps and the Type 89 MICV.

Description

The Type 79 Jyu-MAT is similar to the US TOW in configuration, size and performance. The missile, sometimes known as the KAM-9, is launched from a tubular container also used for transport and storage. A solid-propellant launch motor ejects the missile from the container to a safe distance from the operator. The flight motor, produced by Daicel Chemical Industries Limited, then ignites and accelerates the missile to its cruising velocity of approximately 200 m/s in a few seconds. Maximum range is approximately 4,000 m.

Prior to firing, the container is placed on the launcher which comprises firing control device, missile check-out device, tracking mechanism and sight unit. Guidance is semi-automatic command to line of sight (SACLOS). The optical sighting device, produced by the NEC Corporation of Tokyo, is designed to be operated by one man who, during the missile's flight, keeps his sight trained on the target. Sensors in the sight unit translate any course deviation by the missile (marked by a rear-facing Xenon lamp in the missile tail), into electrical signals fed into a computer which calculates and generates course correction signals communicated to the missile by way of the guidance cable.

Two launch modes are available, direct and remote launch; the latter being from up to 50 m away.

The Type 79 can be fitted with two types of warhead, both weighing 4.2 kg. One is HEAT and the other, intended for use against amphibious warfare targets, is semi-armour-piercing enhanced fragmentation with a variable delay fuze. No information is available regarding armour penetration.

Data

Guidance: SACLOS; optical sighting; wire command
Calibre: 152 mm
Length: missile, 1.5 m; container, 1.7 m
Span: 330 mm
Weight: complete system, 278 kg; loaded container, 19.9 kg; missile, 15.7 kg; warhead, 4.2 kg
Velocity: max, approx 200 m/s
Range: 320-4,000 m
Propulsion: solid propellant booster and sustainer

Manufacturer

Kawasaki Heavy Industries Limited, World Trade Center Building, 4-1 Hamamatsu-cho 2-chome, Minato-ku, Tokyo.

Status

In production. Not offered for export.

Service

Japanese Ground Self-Defence Force.

UPDATED

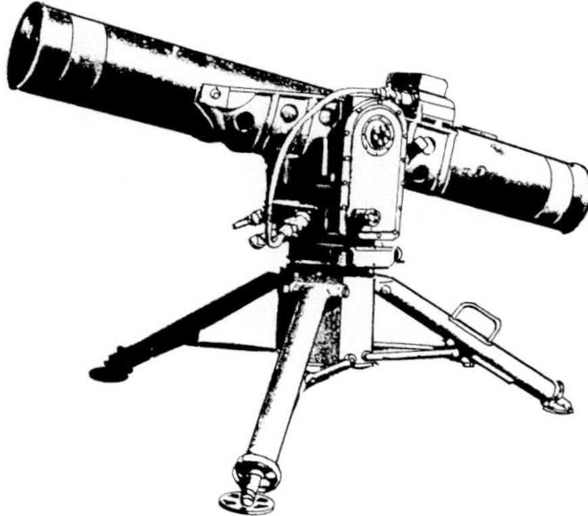

Ground mounting for Type 79 Jyu-MAT anti-tank missile system (Steve Zaloga/Teal Group)

1996

KAM-9 missile

Type 87 Chu-MAT anti-tank missile system

Development

Development of the Type 87 Chu-MAT was begun by the Japanese Defense Agency's Technical Research and Development Institute in 1976. Development was handed over to Kawasaki Heavy Industries in 1980, with a small batch of prototype missiles commencing trials during 1982. A complete prototype system was delivered in May 1985 and a second prototype in February 1986. Subsequent procurement was delayed until FY 1989. Production has continued at a slow rate since then (about 24 launchers each year).

In 1990 development of a new fibre-optic heavy anti-tank missile known as the XATM-4 commenced. An XATM-5 light man-portable anti-tank missile is also under development. Little information is yet forthcoming regarding either of these projects.

Description

The Type 87 Chu-MAT anti-tank missile system employs a semi-active laser guidance system. The missile, powered by a single-stage solid propellant motor, is contained in a container/launcher tube which is usually mounted on a low tripod served by a two-man crew from the prone position. A further crew member mans a laser target designator which may be established up to 200 m from the launcher, although it is possible to place the launcher on the laser designator mounting. The latter procedure can encounter problems caused by the ionised gas exhaust from the missile interfering with the laser beam. The system may be carried on a Type 73 Jeep.

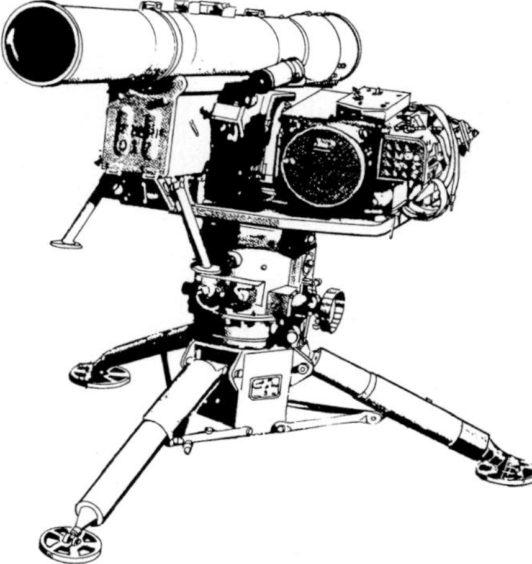

Type 87 Chu-MAT anti-tank missile system (Steve Zaloga/Teal Group)

1996

The laser target designator employs a Nd/YAG laser to provide target illumination for the missile seeker head. The seeker head is manufactured by the NEC Corporation of Tokyo.

No data relating to system performance is yet available. Range performance is classed as 'medium'.

Data

PROVISIONAL
Calibre: 120 mm
Length: missile, 1 m
Weight: at launch, 12 kg
Propulsion: solid propellant sustainer

Manufacturer

Kawasaki Heavy Industries Limited, World Trade Center Building, 4-1 Hamamatsu-cho 2-chome, Minato-ku, Tokyo.

Status

In production. Not offered for export.

Service

Japanese Ground Self-Defence Force.

NEW ENTRY

PAKISTAN

RPG-7 weapon system

Description

The Soviet RPG-7 anti-tank rocket launcher is manufactured in Pakistan. The design, operation and basic dimensions are precisely the same as the Soviet model, to which further reference should be made. The performance figures given below are those quoted by the makers and differ somewhat from those associated with the Soviet weapon.

Loading the RPG-7 rocket launcher

PG-7 rockets fired from this weapon are manufactured by Pakistan Ordnance Factories at Wah Cantt.

Data

Calibre: 40 mm
Length: launcher, 950 mm
Weight: with sight, 6.3 kg
Max range: aimed, 500 m
Rate of fire: 4-6 rds/min
Armour penetration: 260 mm

Manufacturer

Pakistan Machine Tool Factory Limited, Landhi, Karachi 34.

Status

In production.

Service

Pakistan armed forces.

VERIFIED

106 mm recoilless gun M40A1

Description

This is a locally manufactured version of the standard US design, and the entire system consists of the Gun M40A1, Mount M79, Spotting Rifle M8C, Elbow Telescope M92F, Telescope Mount M90 and Instrument Light M42. The Spotting Rifle is bought in from Empresa Nacional Santa Barbara of Spain and the sight assembly is manufactured in Pakistan by Mogul Industries.

The gun may be mounted on and fired from a light vehicle or it can be easily dismounted and brought into action on its own carriage.

106 mm HEAT ammunition for this gun, to the American M344A1 standard, is produced in Pakistan by Pakistan Ordnance Factories at Wah Cantt.

106 mm recoilless gun in use by Pakistan forces

Data

Calibre: 105 mm
Weight: gun, 115 kg; M79 mount, 82 kg
Max range: direct fire, 2,012 m; indirect fire, 7,680 m
Muzzle velocity: 507 m/s

Manufacturer

Pakistan Machine Tool Factory Limited, Landhi, Karachi 34.

Status

In production.

Service

Pakistan armed forces.

VERIFIED

POLAND

RPG-76 KOMAR disposable anti-tank weapon

Description

The RPG-76 KOMAR disposable anti-tank weapon is used to launch a PG-76 anti-tank grenade. The PG-76 has design affiliations with the anti-armour grenades used with the Soviet RPG-7 but the RPG-76 KOMAR launcher is intended for one-shot use only, along the lines of the German Second World War Panzerfaust series.

The RPG-76 KOMAR is normally carried over the shoulder on a sling. To prepare the unit for use takes about 10 seconds and entails raising the simple leaf sights and unfolding the rudimentary alumnium alloy shoulder stock; the sights normally holding the stock in the folded position. Using a point on the rocket grenade as a foresight the user can select ranges of 50, 150 or 250 m. Firing is then carried out using a mechanical trigger button mechanism (normally covered by a safety plate), after which the launcher is discarded.

At the instant of firing the firer is protected from the rocket exhaust by having its four venturi canted outwards to impart projectile roll. The rocket is armed by gas pressure during launch, although a delay element ensures that it is not fully armed until the rocket is some 10 m from the launcher.

RPG-76 KOMAR disposable anti-tank weapon (T J Gander)

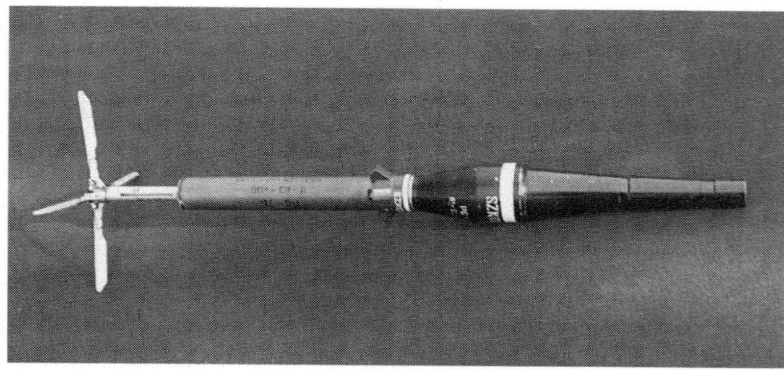

RPG-76 grenade used with the KOMAR disposable anti-tank weapon

1996

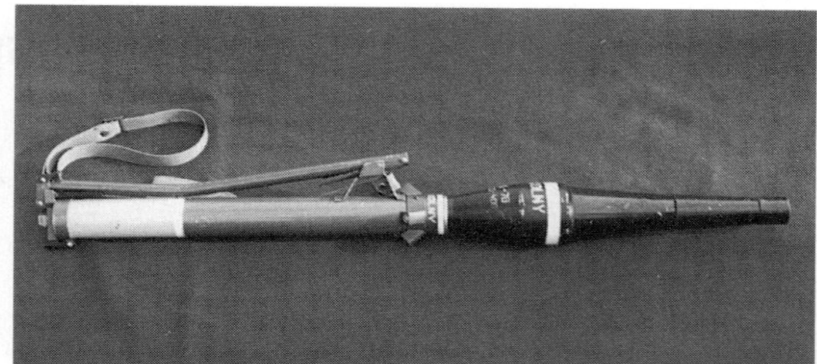

KOMAR disposable anti-tank weapon in carry configuration

1996

If the weapon is not fired it can be folded back to the carrying configuration for future use.

The HEAT warhead on the PG-76 grenade contains 320 g of A-IX-1 (RDW/Wax) activated by a DCR mechanical base impact fuze; the shaped charge cone is copper. The warhead can penetrate up to 260 mm of armour plate at 0° or 130 mm at 60°.

Inert units are available for training.

Data

Calibre: launcher, 40 mm; projectile, 68 mm
Length: transport, 805 mm; ready to fire, 1.19 m
Weight: complete, 2.1 kg; PG-76 grenade, 1.78 kg
Warhead: HEAT, 320 g A-IX-1
Sight radius: 340 mm
Muzzle velocity: 145 m/s
Max range: effective, 250 m
Armour penetration: 0°, 260 mm; 60°, 130 mm

Manufacturer

Zaklady Sprzetu Precyzyjnego, Ujazdm 97-170 Niewiadów.
Marketed by: Cenzin, Foreign Trade Enterprise, ul Frascati 2, PL-00489 Warsaw.

Status

In production.

Service

Polish Army.

UPDATED

SINGAPORE

Armbrust short-range anti-armour weapon

Development

The Armbrust short-range anti-armour weapon was originally designed and manufactured by Messerschmitt-Bölkow-Blohm GmbH of Germany who ceased production in 1988, at which time the designs and manufacturing rights were sold to Chartered Industries of Singapore.

Description

The Armbrust short-range anti-armour weapon has unique features which set it apart from most other similar weapons. It has no firing signature, emits neither smoke nor blast from the muzzle nor flash from the rear, is quieter than a pistol shot, can be fired from small enclosures or roofed foxholes without danger or discomfort to the firer, has no recoil, requires no maintenance and weighs only 6.3 kg.

The Armbrust's 67 mm diameter HEAT warhead can penetrate 300 mm of rolled homogeneous armour at 0°. In addition it will penetrate materials such as masonry, reinforced concrete and so on.

The Armbrust is a man-portable, shoulder-fired weapon with a maximum range of about 1,500 m and an operational range against armoured vehicles of up to 300 m. Time of flight to 300 m is 1.6 seconds; muzzle velocity is 210 m/s. The fuze will function at an impact angle of up to 78°. Once the weapon has been fired, the launcher is discarded.

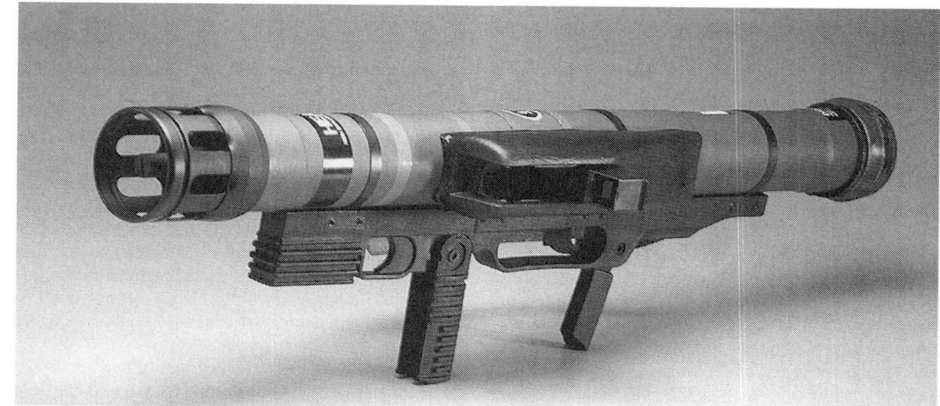

Armbrust short-range anti-armour weapon

1996

Data

Calibre: launcher, 75 mm; missile, 67 mm
Length: weapon, 850 mm; missile, 405 mm
Weight: weapon, 6.3 kg; missile, 1 kg
Muzzle velocity: 210 m/s
Time of flight: 1.5 s to 300 m
Range: operational, 300 m; max, 1,500 m
Armour penetration: 300 mm at 0°
Fuze impact angle: up to 78°
Operational temperature range: −40 to +52°C

Manufacturer

Chartered Industries of Singapore, a member of Singapore Technologies.
Marketed by: Unicorn International, 3 Lim Teck Road, 11-01/02 Singapore Technologies Building, Singapore 0208.

Status

Available.

UPDATED

SOUTH AFRICA

FT5 light anti-tank weapon system

Description

The FT5 is a recoilless shoulder-launched anti-armour weapon which can be handled and fired by one man. It consists of a reusable launcher and a high-performance subsonic rocket encapsulated inside a sealed disposable launch canister. The system can withstand rough handling and is air-transportable.

There is a choice of three warheads: the standard HEAT anti-armour warhead; High Explosive MultiPurpose (HEMP) for use against bunkers, structures, APCs and AFVs; and High Explosive Anti-tank Reactive Armour (HEAT/RA), a tandem hollow charge warhead for the attack of explosive reactive armour.

The standard sight is an optical unit of ×4 magnification with a rangefinding, lead-compensating reticle pattern. There is also an occluded eye type battle sight permanently mounted on the launcher. A passive infrared night sight is available.

Operational users have demonstrated that a trained soldier can consistently hit a 1.5 m × 1.5 m target at 400 m range.

Data

Calibre: rocket, 99 mm; warhead, 92 mm
Length: unloaded, 1.05 m; total, 1.618 m
Weight: launcher, unloaded, 5.9 kg; loaded, 11.3 kg
Range: operational, 40-400 m; max, 600 m
Warhead arming distance: 20-40 m
Predicted shelf life: 10 years
Armour penetration: HEAT, >650 mm; HEAT/RA, >630 mm RHA protected by ERA; HEMP, 1.5 m sandbags, 300 mm of lightly reinforced concrete, or 20 mm armoured steel at 90°
Muzzle velocity: HEAT, 275 m/s
Dispersion: <0.75 mil at 400 m

FT5 anti-tank launcher

Enquiries to

Somchem, a division of Denel (Pty) Limited, PO Box 187, Somerset West 7129.

Status

In production.

Service

South African National Defence Force.

UPDATED

SPAIN

MACAM anti-tank weapon system

Development

MACAM stands for Misil Avanzado Contra-Carro de Alcance Medio, often with the number 3 added to denote its third-generation status. It is being developed with Gyconsa as prime contractor under contract with the Spanish Ministry of Defence, in association with Hughes Aircraft Company of the USA and Grupo Indra of Spain.

Feasibility studies commenced during 1992, based on a previous demonstration and validation test of the fibre optic missile concept conducted by Hughes in 1988. Definition studies extended until mid 1994, with the design and development phase commencing at the beginning of 1995. This phase will extend until the end of 1999, with production commencing during 2000.

Planned in-service date for the Spanish Army and Marines is 2002. The requirement for the Spanish Government has been stated to be 10,000 to 14,000 missiles, although a proportion of this total would be intended for export sales.

Description

The MACAM, also known as the MACAM 3, is a man-portable shoulder-fired anti-armour system consisting of a sealed launch tube containing the missile, and a launcher unit. There is also an integral day/night sight. If required, the MACAM can be fired from TOW missile launchers, including the Light Weight Launcher (LWL) also being developed by Gyconsa.

The launcher unit, which clips onto the launch tube contains the aiming and control unit and the launcher electronics unit. The aiming and control unit provides all the operator's controls and displays and integrates a day/night sight. The interface with the missile is implemented by a quick-operating clamp providing mechanical, electrical and fibre optic connections.

The operator's controls and displays consist of two handgrips and a single eyepiece. Each handgrip has a trigger switch and a thumb control.

The eyepiece provides a visible light image from the day sight, infra-red imagery from the night sight, or missile seeker imagery, all as selected by the operator. The day/night sight provides two selectable fields of view. The night sight and the missile seeker operate in the medium wave infra-red band (3 to 5µm).

The launcher electronics unit is contained in a back-pack which also contains the replaceable launcher battery; the launcher battery life in action is four hours. The electronics receive the missile seeker video information through the fibre optic data link and operator commands from the aiming and control unit. Functions performed by the launcher electronics include system executive and system modes control, target tracking, and the generation of missile guidance commands.

The MACAM missile is issued as a sealed round containing the missile, plus a replaceable cryogenic gas bottle to cool down and operate the missile's infra-red detector prior to launch. If the missile is activated but not fired the external gas bottle can be changed in the field. The launch container is compatible with the TOW launch tube.

Mock-up of the main units of the MACAM anti-tank weapon system

1996

The missile itself contains a seeker module, the missile electronics, an armament module, a flight motor, a control module, a launch motor and a fibre optic dispenser.

The seeker module is a sealed unit containing the gimbal mounted infra-red sensor based on a staring focal plane array detector. Cooling is provided by the external gas bottle prior to launch and from an internal bottle during flight. The detector assembly is mounted on a two-axis rate-stabilised platform.

Video processing, target tracking and missile guidance are performed by the launcher electronics via the fibre optic data link so only onboard tasks, such as the internal mode control and timing, seeker video electrical-to-optical conversion, and auto pilot loop closure are carried out by the missile electronics.

The armament module consists of the warhead, the safety and arm device and the airframe supporting these units. The warhead is a tandem configuration with a small precursor charge to initiate reactive armour prior to the detonation of the main charge. The charges are canted downwards so that they do not need to defeat the nose-mounted seeker module prior to encountering the main target.

The flight motor uses a solid propellant and is ignited after the missile has left the launch tube and is a safe distance from the operator. The motor has twin venturi on the sides of the missile to allow the fibre optic cable to be dispensed from the back of the missile body.

The solid propellant launch motor provides the initial thrust to eject the missile from the launch tube and accelerate it to the minimum flying velocity. The motor burns out before the missile leaves the launch tube, thus providing a soft launch to allow the missile to be fired from within enclosures.

The control module includes the actuators and control surfaces to control the missile in flight. Also included in this module are the batteries to power the missile during flight, and the missile roll gyro.

The fibre optic dispenser pays out the optical fibre during flight and will support missile ranges up to 5,000 m, although the maximum effective combat range is of the order of 2,500 m.

There are two launch modes: Lock On Before Launch (LOBL), and Lock On After Launch (LOAL). LOBL is the primary mode.

For LOBL the operator uses the day/night sight to carry out surveillance and target search. Using the selectable fields of view the operator can detect and then recognise a target. The operator activates the missile once a target has been selected and the imagery in the sight automatically switches to missile seeker video in approximately 10 seconds. Using a two-dimensional thumb switch the operator positions a tracker gate over the desired target and commands track.

At this time the operator can select the trajectory required. The elevated trajectory is the normal choice against armoured vehicles as it provides a top attack terminal trajectory. Also available is a direct trajectory for use against hovering helicopters or bunkers.

Once lock on is achieved, the operator can fire the missile which will be automatically guided towards the target on a fire-and-forget basis. The sight will continue to display the live video to the operator so the operator can intervene to either improve the aimpoint, select an alternative target not previously seen, re-establish control if a lock is broken for various reasons, or abort an attack entirely.

The MACAM anti-tank weapon system in operation

1996

During the elevated trajectory the missile climbs to a mid-course elevation of about 2,000 m and levels off. The missile follows a proportional navigation law in the horizontal plane and a fixed altitude in the vertical plane. As the missile approaches the target, the look-down angle continues to increase. When this angle reaches a defined limit the terminal trajectory commences with proportional navigation utilised in both planes. The system automatically stores the last video images to allow the operator to conduct a real time battle damage assessment.

The secondary LOAL mode is used when the operator knows where a target is but for various reasons cannot, or chooses not to, initiate automatic tracking prior to launch. Typical LOAL situations thus include the operator being able to see the target in the launcher sight but not in the seeker video, the operator may not wish to risk exposure, or the operator knows where the target is but cannot establish a line of sight because of obstacles. In these cases the operator selects the LOAL mode and fires the missile in an elevated trajectory. Once at the mid-course altitude the seeker head is depressed to provide an optimum search footprint on the ground. The operator watches the live seeker video ready for when a target or target area is recognised. The track gate is then positioned over the desired point and track is commanded. From this point onward the operation proceeds as with the LOBL mode.

An alternative launch mode is remote launch using a kit including a tripod to support the launch tube and a 75 m electrical and fibre optic cable to allow the operator to be remotely located from the launch position. There will also be adaptor kits to permit MACAM to be mounted on various types of vehicle. These will probably be similar to existing kits which adapt TOW for M113 APCs, Jeeps and other combat vehicles.

Data

Provisional

Calibre: missile, <147 mm; launch tube, 174 mm
Length: missile, 1.05 m; launch tube, 1.2 m
Weight: system, <25 kg; missile, 13.8 kg; launcher, <9 kg
Max range: 5,000 m
Effective combat range: 150-2,500 m
Operational temperature range: −31.5 to +51°C

Manufacturer

Gyconsa, Joaquín, 11, E-28300 Aranjuez, Madrid.

Status

Advanced development.

NEW ENTRY

Alcotan weapon system

Description

The Alcotan anti-tank launcher is a two-unit system: the front one-third of the weapon is a reusable launch tube carrying an electronic fire director, while the rear two-thirds is a prepacked tube carrying the projectile. This is bayonet-jointed to the forward section prior to firing and discarded after the firing.

The fire director contains a day/night vision unit, a laser rangefinder and a ballistic computer which assesses the target speed, measures the range and takes into account the projectile type and temperature. A bright spot in the sight indicates the point of aim, which is displaced according to the lead angle determined by the computer. The result of all computations is then displayed in the sight by indicating the aimpoint to the operator.

The projectile is launched using the countermass (or Davis Gun) principle. Warheads developed include tandem, high-penetration single stage, and anti-bunker types.

Data

Calibre: 100 mm
Length: firing, 1.355 m; transport, 513 mm; ammunition, 1.01 m
Weight: firing, 15 kg
Launch velocity: 280 m/s
Max range: anti-tank, 600 m
Armour penetration: 600 mm

Manufacturer

Instalaza SA, Monreal 27, E-50002 Zaragoza.

Status

Undergoing Spanish Army acceptance trials.

UPDATED

88.9 mm M65 anti-tank rocket launcher

Description

The 88.9 mm M65 anti-tank rocket launcher is a Spanish-developed weapon, with an electromagnetic firing mechanism and a sight unit fitted with an adjustable light source which lights up the graticule for night sighting. The electrical connection between the round and launcher is established automatically during loading.

There are three types of ammunition available, CHM65 anti-tank, MB66 anti-personnel/anti-armour and FIM66 smoke/incendiary. The launcher and its ammunition are produced by the same manufacturer.

Data

Calibre: 88.9 mm
Length: firing, 1.64 m; folded, 850 mm
Weight: 6 kg

Manufacturer

Instalaza SA, Monreal 27, E-50002 Zaragoza.

Status

In production.

Service

Spanish and other armed forces.

UPDATED

88.9 mm rocket launcher M65

1996

AMMUNITION

	CHM65	MB66	FIM66
Rocket weight	2 kg	2.9 kg	2.7 kg
Initial velocity	215 m/s	145 m/s	155 m/s
Effective anti-tank range	450 m	300 m	–
Max range	–	1,000 m	1,300 m
Armour penetration	330 mm	250 mm	–

C90 (M3 Series) light anti-tank weapon

Description

The C90 (M3 Series) is the latest addition to the C90 family of anti-tank weapon systems developed and manufactured by Instalaza SA, providing the infantryman with simple and effective anti-tank weapons at ranges of up to 300 m for moving targets or 400 m for stationary targets.

An aramide resin container/launcher holds the projectile and also supports the firing mechanism, optical sight and carrying strap, and, optionally, a pistol-grip and a shoulder rest. The container/launcher is discarded after firing.

The round consists of a shaped charge warhead, an instantaneous fuze, a rocket motor and a fin stabilising unit. The firing mechanism is pyrotechnic, thus dispensing with the need for batteries or any other electrical device. The system is maintenance-free for its entire life.

The optical sight has a ×2 magnification and a reticle with the appropriate markings for distance and lateral predictions. An optional permanent light source fitted inside the sight permits the gunner to see the markings even at night. A night vision device is offered as an option.

There are a number of variants of the C90 (M3) system which differ in the type of warhead involved. Details are given in the data tables.

A specific training system, the TR90, has been developed. Based on the principle of the recoilless gun, it uses a cartridge to launch an aluminium alloy arrow and to compensate for the recoil by the backwards ejection of a jet of gas. The aluminium arrow can be reused many times provided it is fired against a soft target.

Manufacturer

Instalaza SA, Monreal 27, E-50002 Zaragoza.

Status

In production.

Service

Spanish and other armed forces.

UPDATED

C90 (M3 Series) light anti-tank weapon in firing position

1996

Data

Version	**C90-C**	**C90-C-AM**	**C90-CR** (M3)	**C90-CR-RB** (M3)	**C90-CR-AM** (M3)	**C90-CR-FIM** (M3)	**C90-CR-BK** (M3)
Function	Anti-tank	Anti-tank/ Anti-personnel	Anti-tank	Anti-tank	Anti-tank/ Anti-personnel	Smoke/ Incendiary	Anti-bunker
Calibre	90 mm	90 mm	90 mm	90 mm	90 mm	90 mm	90 mm
Length	840 mm	840 mm	984 mm	984 mm	984 mm	984 mm	984 mm
Weight	4.2 kg	4.2 kg	5 kg	5 kg	5 kg	5.6 kg	5.4 kg
Penetration, steel	400 mm	220 mm	400 mm	480 mm	220 mm	–	70 mm
Penetration, concrete	1 m	0.65 m	1 m	1.2 m	0.65 mm	–	–
Penetration, concrete, with follow-trough pass	–	–	–	–	–	–	250 mm
Anti-tank range	200 m	200 m	300 m	300 m	300 m	–	300 m
Anti-personnel range	–	600 m	–	–	800 m	750 m	350 m
Anti-personnel fragments	–	>1000	–	–	>1000	–	>400
Lethal radius	–	21 m	–	–	21 m	–	8 m

106 mm recoilless rifle

Description

This is a Spanish made version of the American M40A1 RCL rifle and is identical in all respects. It is fitted with an aiming rifle and optical sight and is provided with the usual range of HEAT, TP and anti-personnel ammunition. The standard mounting is the tripod ground mount, but it can also be mounted on any suitably modified light vehicle.

Data

Length: overall, 3.4 m
Height: on mount, 1.13 m
Weight: 219 kg
Rifling: 36 grooves, rh
Max elevation: over tripod, 27°; between legs, 65°
Max depression: −17°
Traverse on mount: 360°
Muzzle velocity: 503 m/s
Max range: 7,640 m

Manufacturer
SANTA BARBARA, Julian Camarillo 32, E-28037 Madrid.

Status
In production.

Service
Spanish Army.

UPDATED

106 mm recoilless rifle mounted on a UMM (4 × 4) light vehicle
1996

SWEDEN

84 mm AT4 light anti-armour weapon

Description

The AT4 is a preloaded, disposable and recoilless anti-armour weapon. It is light in weight and easy to handle.

The main parts of the weapon are: the barrel, made of glassfibre-reinforced plastic and fitted with an aluminium venturi to the rear; shock absorbers and muzzle cover; the firing mechanism, placed over the barrel; the sights; and the HEAT round.

On the barrel are labels with instructions for handling, safety and so on.

The HEAT round consists of a cartridge case assembly, similar to that used with Carl Gustaf ammunition, and a HEAT shell. The main parts of the HEAT shell are the fin assembly, base fuze, stand-off cap and HMX/TNT hollow charge with a special liner. The fins pop out after the shell has left the muzzle and stabilise the shell in flight. The fuze has an out-of-line detonator safety device to prevent accidental initiation. On impact, the fuze detonates the hollow charge, even at angles of impact as shallow as 80° to the normal. The special design of the hollow charge causes behind-armour damage by overpressure, extensive spalling and intense heat.

The sights are protected by sliding covers during transport. The rearsight is adjustable and factory-set to 200 m. The foresight has a centre post and two lead posts to be used depending upon target speeds.

The standard package is a handy and rugged plywood box with carrying handles, containing four weapons sealed in moisture-proof plastic bags.

There is a complete range of training devices for the AT4: for marksmanship training there is a 9 mm subcalibre AT4, a device with an integral subcalibre barrel firing 9 mm tracer ammunition matched to the ballistics of the HEAT round. For realistic training the subcalibre weapon can also use a back blast charge, separately or together with the 9 mm device, to simulate the bang and flash of the live HEAT weapon. For field exercises there is the AT4 TPT weapon, firing a full calibre inert projectile.

There is also a completely inert handling weapon for drill purposes.

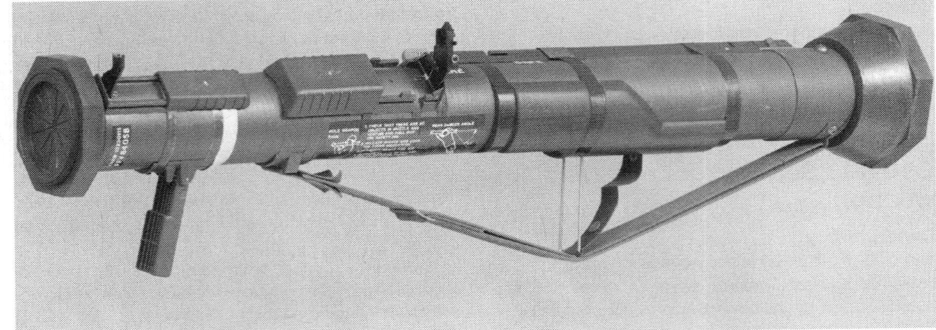

AT4 light anti-armour weapon

AT4 light anti-armour weapon in firing position

Variants

84 mm LMAW AT4

Bofors has developed the Light Multipurpose Assault Weapon (LMAW) AT4. This is a recoilless one-man portable weapon intended for fighting in built-up areas. The LMAW uses a slightly modified 84 mm HEDP round from the Carl Gustaf system.

The launcher is similar to that of the normal AT4 but is fitted with a fuze mode selector switch with two positions, 'D' for Delay burst and 'I' for Instantaneous burst on impact. The 'I' mode is used when engaging armoured vehicles, the 'D' mode being used when engaging troops protected by field fortifications or light cover. In the anti-armour 'I' mode the warhead causes special behind-armour damage. The warhead has a fragmentation steel shell which enables it to be used as an HE projectile against soft targets. The warhead can penetrate up to 150 mm of armour. Muzzle velocity is 235 m/s.

84 mm AT4 CS

A further development of the internal ballistics for AT4 has been completed and is in production. This gives the gunner the ability to fire AT4 from confined spaces as small as 22.5 m³. The employment of the counter-mass principle makes this possible. The AT4 CS can utilise either the HEAT or LMAW warheads.

AT4 HP

In this case HP stands for high penetration, being capable of defeating 600 mm of armour with the same behind-armour effects as AT4 HEAT. Weight is less than 7 kg and muzzle velocity is 290 m/s.

AT4 HP-T

This variant is under development and employs a tandem warhead capable of penetrating 600 mm of armour, even when protected by explosive reactive armour. To provide extra thrust the projectile incorporates a rocket booster motor which initiates after the projectile has left the launcher muzzle, raising the initial 210 m/s muzzle velocity to 325 m/s. Weight is 8.2 kg.

AT4 I

Also under development, AT4 I features an incendiary warhead.

Data

AT4 HEAT
Calibre: 84 mm
Length: 1.01 m
Weight: 6.7 kg; HEAT round, 3 kg
Range: effective, 300 m
Muzzle velocity: 290 m/s
Armour penetration: >420 mm

Manufacturer

Bofors AB, S-691 80 Karlskoga.

Status

In production.

Service

Swedish Army, US Army and Navy, Denmark, Netherlands, Brazilian Army and Navy, Venezuelan Army.

Licence production

UNITED STATES OF AMERICA
Manufacturer: Alliant Techsystems
Type: M136
Remarks: See entry under *United States of America* for details.

VERIFIED

120 mm AT12-T anti-tank weapon

Description

The 120 mm AT12-T is a shoulder-fired rocket anti-tank weapon developed for the Swedish Army. Feasibility studies were completed in 1991 with series production scheduled to commence during 1995. It is similar in general form to the AT4 although a folding mount allows it to be fired from the prone, kneeling or standing positions.

The AT12-T projectile uses a tandem warhead designed to defeat any reactive armour and penetrate over 950 mm of conventional armour after penetration of the reactive armour. The company claims that it will defeat all tanks likely to be fielded in the coming decade over their frontal arcs.

AT12-T is a preloaded, disposable, back blast weapon completely constructed of lightweight materials. The projectile is rocket assisted. The weapon is compatible with various day or night sights.

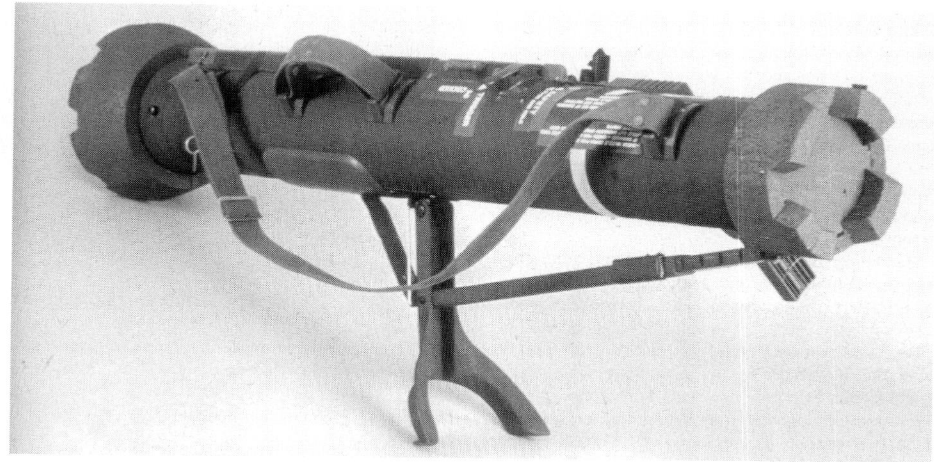

120 mm AT12-T anti-tank weapon

Data

Calibre: 120 mm
Length: 1.2 m
Weight: <14 kg
Range: effective, >300 m
Armour penetration: >950 mm, after reactive armours

Manufacturer

Bofors AB, S-691 80 Karlskoga.

Status

Initial production pending.

UPDATED

Firing the AT12-T from kneeling position, using steadying folding leg

84 mm Carl Gustaf M2 and M3 recoilless rifles

Development

The 84 mm Carl Gustaf is a one-man portable, recoilless gun, originally conceived for the anti-tank role but later upgraded by ammunition innovations to possess a true multipurpose capability, producing the 84 mm Carl Gustaf System.

Development of the Carl Gustaf System dates back to experimental work carried out during the early and mid-1940s. Following the development of a 20 mm recoilless anti-tank rifle the first prototypes of the 84 mm Carl Gustaf appeared in 1946 and the first service model was ready in 1948. Initial production was carried out by the state-owned Gevärsfaktoriet but production later switched to FFV Ordnance who also produced the ammunition. FFV Ordnance are now part of Bofors AB.

The original Carl Gustaf M2 gun and the later and lighter M3 are reliable and rugged designs, tailored for long operational life under adverse conditions. Recoilless functioning is obtained by allowing a proportion of

84 mm Carl Gustaf gun in foreground with, for comparison, M2 gun at the rear

the propellant gases to escape to the rear of the weapon under control. Barrels are rifled and fitted with a cone-shaped venturi funnel at the rear.

The Mark 3 gun is essentially a lightweight version of the Carl Gustaf M2 with the barrel consisting of a rifled steel liner, around which is wound a laminate of carbon fibre and epoxy. The venturi is made of steel, all other external parts being of aluminium or plastic. The M3 is fitted with a carrying handle to facilitate rapid moving of the firing position.

The M3 will accept and fire all existing types of 84 mm Carl Gustaf ammunition.

Both the M2 and M3 are usually served by a two-man crew; one fires the gun while the other loads and carries the ammunition.

Description

84 mm Carl Gustaf guns are breech-loaded and opened by releasing the venturi fastening strap and rotating the venturi sideways. Any empty case is removed and the next round can be loaded directly. Until the breech is rotated to the fully closed position the firing mechanism is inoperative. Since the round is percussion fired by a mechanism on the side of the chamber, it is necessary to index it into the chamber by a tongue and notch.

The firing mechanism is contained in a tube on the right side of the barrel and is cocked when the cocking lever, behind the pistol-grip, is pushed forward to compress the main spring. The safety catch is on the right side of the pistol-grip. The gun can be fired from the shoulder or from the prone position or it can be rested on the edge of a trench or fired from a mount on an armoured personnel carrier. When fired off a flat surface the weapon is supported on a flexible bipod immediately in front of the shoulder piece.

There is an open sight but the usual sighting system is a ×3 telescope with a 12° field of view. This is fitted with a temperature correction device and luminous front and rear adaptors are available for night work. A flare sight can be rapidly attached to the gun to provide the correct elevation when firing illuminating projectiles. Various types of electronic tracking fire control systems, some including laser rangefinders, have been produced for use with Carl Gustaf guns.

84 mm Carl Gustaf M3 gun with full range of available ammunition

For details of the ammunition for the Carl Gustav gun system refer to the following entry.

Data

Calibre: 84 mm
Length: M2, 1.13 m; M3, 1.07 m
Weight: M2, 14.2 kg; M3, 8.5 kg; (packed with accessories) M2, 29.5 kg; M3 21.5 kg; (mount) M2, 0.8 kg; M3, 0.5 kg

Manufacturer

Bofors AB, Carl Gustaf, S-691 80 Karlskoga.

Status

In production.

Service

In service with Swedish forces and Australia, Austria, Canada, Denmark, Germany, Ghana, India, Ireland, Japan, Malaysia, Netherlands, New Zealand, Nigeria, Norway, Portugal, Singapore, USA and Venezuela.

Licence production

INDIA
Manufacturer: Ordnance Factory Board, Gun and Shell Factory, Cossipore, Calcutta.
Type: 84 mm RL (Carl Gustaf M2)
Remarks: In production and offered for export.

VERIFIED

84 mm Carl Gustaf ammunition

The multipurpose capability of the 84 mm Carl Gustaf system is provided by light weight, compact dimensions, easy handling of the gun and a complete range of ammunition.

84 mm HEAT 751

The 84 mm HEAT 751 uses a rocket-assisted tandem warhead projectile intended to defeat reactive armour and penetrate up to 500 mm of conventional armour. This round is in production.

Data

Weight: complete round, 4 kg; projectile, 3.2 kg
Muzzle velocity: initial, ca 200 m/s; max, ca 320 m/s
Armour penetration: >500 mm

84 mm HEAT 551

The 84 mm HEAT 551 is effective against all types of armoured vehicles, including those fitted with standoff plates. The projectile is fin stabilised and has a rocket motor sustainer to flatten the trajectory and extend the range. It is fitted with a piezoelectric fuze system which will reliably detonate the hollow charge even at oblique impact.

84 mm HEAT 551 consists of a HEAT projectile and a cartridge assembly. The latter is a common feature of the entire Carl Gustaf ammunition range, having a light alloy case and a propelling charge of double-base strip propellant ignited by a lateral percussion primer. The rear end of the case is closed by a plastic blowout disc, one of the precision components instrumental in making the Carl Gustaf system recoilless.

The HEAT 551 projectile consists of: a nose, with ogive and standoff tube; the shell body with an Octol bursting charge, a copper liner, tetryl booster and an electric fuze system; the rocket motor of double-base smokeless propellant; a teflon slipping driving band to reduce shell spin; and a stabilising unit with six pop-out folding fins.

The rocket motor has a delay unit ignited by the propellant gases on firing. When the delay composition has burned for 45 ms it ignites the rocket propellant, which delivers 325 N of thrust for 1.5 seconds.

Training System 553B with 84 mm Carl Gusaf M3 gun at rear

The HEAT 551 is packed in rugged and durable high-density polyethylene twin containers. Three containers - six rounds - are packed together in a plywood transport box which is pressure-impregnated to resist rot and termite attack.

Data

Weight: complete round, 3.2 kg; projectile, 2.4 kg
Muzzle velocity: 255 m/s
Max velocity: ca 330 m/s
Armour penetration: >400 mm
Arming range: 5-15 m
Range: effective, 700 m

84 mm HE 441B

The HE 441B is intended for use against unprotected troops, entrenched troops, troops in machine gun posts, soft-skinned vehicles and similar targets.

The main parts of the projectile are the mechanical time and impact fuze 447 and the shell body with ball inserts and an RDX/TNT bursting charge.

Fuze 447 complies with MIL-STD-331. It is set by hand without tools; setting is between 40 and 1,250 m and is stepless. The impact mechanism functions at angles of impact as shallow as 85° to normal.

Data

Weight: complete round, 3.1 kg; projectile, 2.3 kg
Muzzle velocity: 240 m/s

Arming range: 20-70 m
Range: practical, to 1,100 m

84 mm Smoke 469B
Smoke 469B is intended for rapid laying of screening smoke. It is also suitable for marking targets for artillery and close support aircraft.

The projectile consists of a direct action and graze fuze, and a body with smoke composition and a central burster tube. The fuze impact mechanism functions at angles of impact as shallow as 85° to normal. The smoke composition consists of titanium tetrachloride adsorbed by powdered calcium silicate. It is non-toxic and leaves no harmful residues.

Data
Weight: complete round, 3.1 kg; projectile, 2.2 kg
Weight of smoke composition: 800 g
Muzzle velocity: 240 m/s
Range: practical, up to 1,300 m

84 mm Illuminating 545
The 84 mm Illuminating 545 is intended for battlefield illumination in support of direct-fire anti-tank weapons and missiles. It allows sub-units to supply their own illumination of targets as required.

The projectile consists of a pyrotechnic time fuze, a light alloy body, a canister with illuminating composition, and a nylon parachute.

The fuze setting ring has graduations from 200 to 2,300 m, with 50 m subdivisions. The canister is ejected in mid-air and falls slowly, braked by the parachute. It produces a sodium light of some 650,000 candela and has a burning time of about 30 seconds.

Data
Weight: complete round, 3.1 kg; projectile, 2.2 kg
Weight of illuminating composition: 500 g
Muzzle velocity: 260 m/s
Range: practical, from 300-2,100 m
Illuminated area, diameter: 400-500 m
Burning time: 30 s

84 mm HEDP 502
The High Explosive Dual Purpose (HEDP) 502 is tailored for Fighting In Built-Up Areas (FIBUA), being designed so the crew can shift quickly between HE and HEAT fire. The HEDP has two modes, selectable when loading. The impact mode is used when engaging armoured vehicles (that is HEAT) while the delay mode is for engaging troops protected by light cover and field fortifications. In the HEAT role the projectile causes special behind-armour damage.

The projectile consists of: a nose-cap providing standoff; a hollow charge designed to produce behind-armour damage in the HEAT role and optimum fragmentation in the FIBUA role; a base fuze; and a fin assembly with six pop-out fins.

The cartridge case contains essentially the same components as the other Carl Gustaf rounds. It is, however, fitted with two lateral percussion primers. The two primers are required since the fuze mode is set by selective orientation of the round on loading.

This round is also used in the AT4 LMAW (qv).

Data
Weight: complete round, 3.3 kg; projectile, 2.3 kg
Muzzle velocity: 230 m/s
Arming range: 15-40 m
Armour penetration: >150 mm
Range: (effective), moving targets, 300 m; bunkers, 500 m
Practical range, troops in open: up to 1,000 m

Training System 553B
Bofors produces a complete training system for the Carl Gustaf. The Training System 553B makes it possible to conduct marksmanship training and drill under realistic, near-live conditions, including battle noise and stress.

The core of the 553B system is the 7.62 mm 553B subcalibre adaptor, externally similar in shape to the HEAT 551. It is fitted with a subcalibre barrel firing 7.62 mm tracer ammunition matching the ballistics of the HEAT 551.

The subcalibre adaptor is loaded into the gun and the gun is fired in the normal manner, the firing pin striking a percussion cap which in turn operates a hammer mechanism to fire the 7.62 mm cartridge. The adaptor weighs 3.7 kg and is 600 mm long. There is also an optional back blast charge to simulate the bang and flash of full calibre Carl Gustaf ammunition. There are three ways of combining the ammunition components:

tracer round plus back blast charge and percussion cap for live fire exercises,

tracer round plus percussion cap for marksmanship training,

back blast charge plus percussion cap for combat exercises.

84 mm TP 552
84 mm TP 552 is used during training. The projectile contains no fuze, booster or bursting charge; it is fitted with a rocket motor to provide the same ballistic performance as the HEAT 551. The cartridge case contains a propelling charge.

Manufacturer
Bofors AB, Carl Gustaf, S-691 80 Karlskoga.

Status
In production.

Service
In service with Swedish forces and Australia, Austria, Canada, Denmark, Germany, Ghana, India, Ireland, Japan, Malaysia, Netherlands, New Zealand, Nigeria, Norway, Portugal, Singapore, USA and Venezuela.

Licence production
INDIA
Manufacturer: Indian Ordnance Factory, Khamaria, Jabalpur
Type: HE, HEAT, Illuminating, TP-T
Remarks: In production. Offered for export sales.

VERIFIED

3A-HEAT-T anti-armour round

Description
The 3A-HEAT-T anti-armour round was developed by Bofors in order to defeat reactive armour and ensure penetration of main armour protection. It is a fin-stabilised shaped charge projectile with an electric contact fuze and has a considerable behind-armour effect. It has been demonstrated to defeat a reactive armour protected sloping MBT glacis at 60° angle of impact and will defeat more than 700 mm of plate after detonating the reactive protection.

The round is manufactured in 106 mm and 90 mm calibres and could be produced in any required calibre. In general, the performance of the 106 mm 3A-HEAT-T projectile is double that of a conventional 106 mm HEAT round under any conditions.

A 106 mm AT-TP-T round is also produced for training purposes.

Data
106 mm 3A-HEAT-T
Calibre: 105 mm
Weight: projectile, 5.5 kg; complete round, 14.5 kg
Explosive filling: 1 kg Octol
Muzzle velocity: 570 m/s
Tracer duration: 5 s
Operating temperature: −40 to +55°C

Manufacturer
Bofors AB, S-691 80, Karlskoga.

Status
In production.

Service
Swedish Army for the 90 mm RCL gun and under evaluation elsewhere.

VERIFIED

106 mm 3A-HEAT-T projectile (front) with practice round

90 mm PV-1110 recoilless rifle

Description
The 90 mm PV-1110 recoilless rifle is employed mounted on a light wheeled vehicle, on the Bv 206 tracked vehicle, or on a two-wheel trailer. In the latter version the rifle is mounted on a turntable which forms part of the carriage and serves as an armrest for the gunner who fires from a kneeling position. There is a 7.62 mm spotting rifle mounted above the main barrel.

Data
Calibre: 90 mm
Barrel: 3.7 m
Muzzle velocity: 653 m/s
Elevation: −10 to +15°

PV-1110 recoilless gun on trailer mount

PV-1110 recoilless gun mounted on Bv 206 tracked carrier

Traverse: 360°
Weight: as a trailer, combat order, 260 kg; projectile, 3.8 kg
Range: effective, against moving target, 700 m
Armour penetration: 550 mm
Rate of fire: 6 rds/min
Crew: 3

Manufacturer
Bofors AB, S-691 80, Karlskoga.

Status
Production complete.

Service
Swedish armed forces and Ireland.

VERIFIED

RBS 56 BILL medium-range anti-tank system

Description
The RBS 56 BILL medium-range anti-tank system was developed by Bofors under a contract from FMV (the Swedish Defence Materiel Administration) in July 1979, the production contract being awarded in 1985.

BILL (Bofors, Infantry, Light and Lethal) is a man-portable, top-attack, medium-range, anti-tank missile system designed to overcome the technological advances made in special armour protection and to have the ability to combat any known or projected armour. The system consists of a day sight, an optional clip-on thermal imaging night sight, a tripod and the missile in its launch tube. The BILL system uses an advanced SACLOS technique with the laser beam-rider missile, using coded signals between missile and sight unit for guidance, making the system immune to jamming. The system was designed for the missile to fly 0.75 m above the line of sight. The BILL guidance system brings the missile under control immediately after launch, giving an extremely high hit probability, both at short ranges and at fast-moving targets.

The missile, which is kept in its sealed launch tube until the instant of firing, is 900 mm long and 150 mm in diameter. The warhead is canted 30° to the horizontal, incorporating a sophisticated warhead ignition system to detect the target and initiate the shaped charge at just the right moment to give maximum effect. There is also an impact fuze for igniting the warhead in the event of a direct hit.

Located at the rear of the launch tube is a gas generator to propel the missile at a velocity of 72 m/s before the sustainer motor accelerates the missile to 250 m/s. The sustainer motor burns for approximately two seconds, or 400 m down-range. The missile then continues in unpropelled flight, but throughout that time it is gyro-stabilised in roll, keeping the warhead pointing downwards. The point of aim when tracking the target is 0.75 m below the point where the missile ought to hit. On tanks, the ideal point of aim is the hull/turret junction (the turret ring), which means that the missile will be in position for the most effective mode of attack. Low parts of the target vertical section can also be engaged by aiming low and utilising the impact fuze. In some engagement situations, where it would be an advantage to operate without the sensor system, the gunner can switch it off before missile launch. The elevated flight path adds an important element of ground clearance to that of the line of sight, thus avoiding many terrain obstacles which might otherwise be limiting factors.

The dual proximity fuze has the effect of extending the target upwards, to the limit of the fuze sensitivity, so that it now becomes possible to inflict damage on targets displaying a small vertical section such as a tank in the hull-down position. This is still enough to give the gunner a good aiming mark and the dual proximity fuze will ensure an effective hit as the missile passes over.

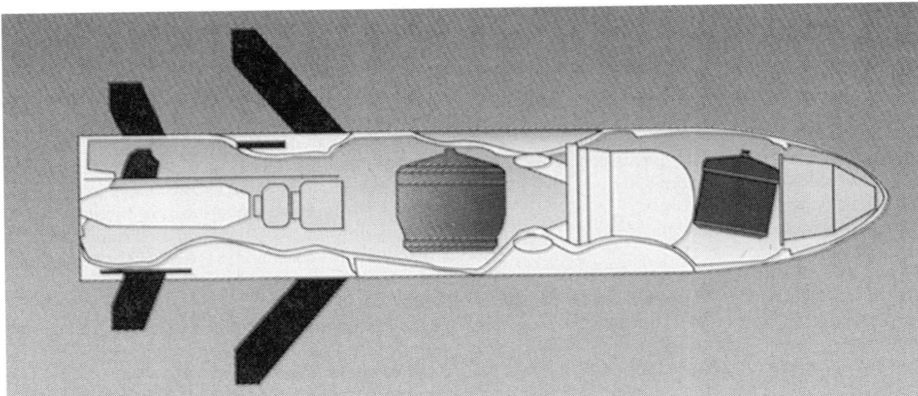
Outline drawing of BILL 2 missile showing position of the two warheads involved

Variant
BILL 2
Bofors Missile Division is developing an advanced form of BILL known as BILL 2 for use against explosive reactive armour. Externally, BILL 2 is identical to existing BILL missiles but internal changes involve more compact electronics, the sustainer rocket motor being moved slightly towards the rear, and two HEAT warheads. The two warheads are fired vertically downwards and are interacting.

The intention is that as the BILL 2 missile flies over a target protected by explosive reactive armour, at a height of about 1 m (the existing BILL height is 0.75 m), the two shaped warheads will strike downwards at almost exactly the same spot on the target front hull, centre or turret top. Using advanced magnetic and optical sensor electronics and algorithms for initiation, the 80 mm diameter first warhead in the missile nose, which is canted slightly to the rear, will remove the reactive armour ready for the second, main, 102 mm diameter warhead to penetrate the unprotected top armour after a suitable time delay and an effective angle of compensation to strike the same spot as the first warhead.

Various guidance changes will be introduced for BILL 2, including improved guidance accuracy as a result of rate gyros in the sight. BILL 2 will have five firing modes to increase operational flexibility. The main anti-armour mode has the sightline 1 m above the line of sight. An optional anti-armour sightline mode is available to suit the customer's perceived threat. A soft mode 1 utilises only an impact sensor and direct line of sight, while a soft mode 2 utilises the optical sensor and optimised ranging. A delayed initiation and impact sensor forms a further alternative anti-armour mode.

BILL 2 will be compatible with existing BILL launchers. For fire control a BILL 2 combined day/night sight will be introduced. Minimum range will be about 150 m and maximum range 2,200 m.

BILL 2 missile weight is planned as 18 kg, with the day/night sight weighing 9 kg; tripod weight is 11 kg. Time of flight to 1,000 m is 5.2 seconds; to 2,200 m it is 13 seconds.

Final verification firings of BILL 2 are scheduled to take place during 1997, with the first production missiles delivered during 1998.

Data
BILL 1
Length: 900 mm
Diameter: 150 mm
Weight: missile, 20 kg; system, 38 kg; emplaced, 37 kg; emplaced with night sight, 46 kg
Range: stationary targets, 150-2,200 m; moving targets, 300-2,200 m
Time of flight: 5.2 s to 1,000 m; 13 s to 2,000 m
Deployment time: 10-15 s
Loading/reloading time: 5 s

Manufacturer
Bofors Missile Division, S-691 80 Karlskoga.

Status
Swedish and Austrian armies (Panzerabwehrlenkwaffe 2000).

UPDATED

RBS 56 in firing position

RBS 56 in firing position

SWITZERLAND

90 mm PAK 50 and 57 anti-tank guns

These two light towed anti-tank guns are no longer in production but both are still in service with infantry formations of the Swiss Army. The more recently developed of the two (Model 57) is lighter and has a somewhat higher performance; but the two are functionally similar and both fire a HEAT round. The PAK 57 carries a 12.7 mm spotting rifle. Comparative data follows.

Manufacturer
Eidgenössische Konstruktionswerkstätte, CH-3602 Thun.

Status
Production complete.

Service
Swiss Army.

Data

	PAK 50	PAK 57
Calibre	90 mm	90 mm
Length of barrel	2.9 m	2.92 m
Length of rifling	2.53 m	2.55 m
Weight, emplaced	550 kg	530 kg
Weight, travelling	600 kg	580 kg
Elevation	−10 to +32°	−15 to +23°
Traverse	34-66°*	70°
Weight of HEAT shell	1.95 kg	1.95 kg
Weight of propellant	250 g	400 g
Muzzle velocity	600 m/s	650 m/s
Max range	4,000 m	4,000 m
Effective range	700 m	900 m
Max rate of fire	8-10 rds/min	8-10 rds/min
Crew	6	6

Sights (both models): sighting telescope for daylight. Infra-red night sight
*depending upon elevation

VERIFIED

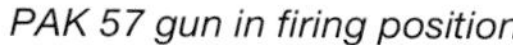
PAK 57 gun in firing position

PAK 57 anti-tank gun in towing configuration

TAIWAN

Kuen Wu 1 anti-tank guided missile

Description
The Kuen Wu 1 is a Manual Command to Line Of Sight (MCLOS) missile developed in 1974-78 by the Sun Yat-sen Scientific Research Institute of Taiwan. It resembles the Soviet 9K111 Malyutka (AT-3 Sagger) in general appearance and in many aspects of dimensions, weight and performance. It differs from the Malyutka in having a somewhat larger and differently shaped warhead.

The Kuen Wu 1 is powered by two solid rocket motors and control is by thrust vector control jetavator nozzles. Guidance is by a joystick, correcting to the line of sight via twin wires. The warhead is of the shaped charge type.

The missile is stored and carried in the same glassfibre case as Malyutka and launch preparations are similar. Normally the weapon system comprises a quadruple launcher on an M151 jeep or a similar vehicle; a maximum of eight missiles can be carried by the launch vehicle. The missile could also be fired by infantry and, presumably, from helicopters.

A computer-based missile simulator and field test set are made by the Sun Yat-sen Scientific Research Institute.

Kuen Wu 1 ATGW on M-151 jeep mount

Data
Type: surface-to-surface, anti-tank guided missile
Guidance: optical tracking, wire guidance, MCLOS
Control: TVC; roll rate stabilised
Propulsion: 2 solid rocket motors
Diameter: ca 119 mm
Length: ca 880 mm
Weight: ca 11.3 kg; missile and container, 15 kg
Weight and type of warhead: 2.72 kg HEAT
Range: effective, 500-3,000 m
Velocity: average, 120 m/s
Armour penetration: 400-500 mm

Manufacturer
Sun Yat-sen Scientific Research Institute, PO Box 2, Lung Tan, Taiwan 325.

Status
Probably no longer in production.

Service
Taiwan Army.

UPDATED

UNITED KINGDOM

LAW 80 light anti-tank weapon system

Description

LAW 80 is a one-shot low-cost disposable short-range anti-tank weapon developed to replace existing weapons which are either light and ineffective or of a size which demands crew operation to achieve lethality. It provides a capability for the individual soldier to engage current and future main battle tanks at ranges out to 500 m and achieve a high probability of killing the target. The performance of the HEAT warhead (armour penetration is in excess of 700 mm) ensures that main battle tanks, both current and future, can be defeated from any aspect, including frontal attack. LAW 80 is stored and transported in Ammunition Container Assemblies (ACA) holding 24 launchers or alternatively in single or twin weapon packs. LAW 80 is man-portable, each launcher being provided with a carrying handle and shoulder sling.

The tactical use of LAW 80 in the British forces is to provide defence against an armoured threat at ranges out to 500 m. Much attention has been placed on the need for the firer to obtain a first-time hit on the target. At the short ranges under consideration it is dangerous and time consuming for the firer to fire a second projectile from the same location if he misses with the first. To counter this LAW 80 utilises an integral semi-automatic spotting rifle containing five preloaded rounds, any number of which may be fired without revealing the position of the firer. The 9 mm spotting rifle ammunition, ballistically matched to the main projectile, is provided with a tracer and a flash head to indicate a hit on the target. At any time the operator can select and fire the main projectile.

The weapon is issued as a round of ammunition, complete with its projectile and the integral spotting rifle preloaded and precocked. Safety is assured by the provision of detents and links which require two failures before any hazard would be caused. In deploying and operating the weapon the safety links are progressively removed up to the point of firing the rocket projectile.

Shock absorbent end caps protect the weapon for carriage and storage. These also provide some side protection, in conjunction with the resilient carrying handle. The end caps provide sealing for the tubes to protect the projectile against the effects of water immersion, even though the projectile itself is sealed. The end caps also support the projectile against the effects of mishandling such as end drops.

After removal of the end caps, the tube containing the HEAT projectile is extended rearwards from the outer tube. The launch tube is automatically locked into position. This moves the centre of gravity of the weapon from the carrying handle to the shoulder rest to provide an ideal balance suitable for any firing position.

The spotting rifle is an integral part of the outer tube. Tests have shown that the five rounds of spotting ammunition sealed in the rifle are sufficient for two engagements. Also included in the external furniture is the firing handgrip/carrying handle and a folding shoulder rest. The sight is separately bonded to the front tube and a forward sliding protective cover allows the sighting prism to erect. The sight is of unit magnification with the graticule projected through the sighting prism into the firer's line of sight, like a head-up display. The sight can be used with one or both eyes open and for low-light use the graticule can be illuminated with a tritium light source selected by the firer, making the weapon effective down to starlight levels of illumination. A mounting shoe is provided for night sight attachment.

With the weapon extended and cocked and with the sight erected, the firer only has to select 'Arm' on the Safe/Arm lever and use the trigger to fire either the spotting rifle or the rocket projectile. When the rocket projectile is to be fired the change lever is moved forward by the thumb of the firing hand. The preparation drills are completely reversible if no target is engaged with the rocket projectile. As part of the safety design, the main round cocking lever covers the catch to unlock and close the weapon so that the weapon must be uncocked and in its spotting rifle mode before it can be closed down.

The projectile is initiated with a non-electric (and therefore radiation hazard-proof) system comprising a percussion cap in the launcher connected by a flash tube to the rocket igniter. The HTPB rocket motor is all burnt on launch, thus ensuring that no blast or debris strikes the firer.

The forward part of the projectile consists of the HEAT warhead and its fuzing unit, and a double ogive nose switch which also provides the optimum standoff distance. The fuzing unit generates electrical energy to fire the warhead by means of piezo crystals and contains various safety devices to ensure that the warhead does not arm until safe separation from the firer is achieved, that is, at a distance of between 10 and 20 m.

At the rear of the projectile the composite aluminium and filament wound motor case has an extruded vane HTPB propellant. Four wraparound fins are mounted on the rear of the motor. These are spring loaded to erect at muzzle exit to provide stability and to spin the projectile as it coasts to the target.

LAW 80 in the firing position

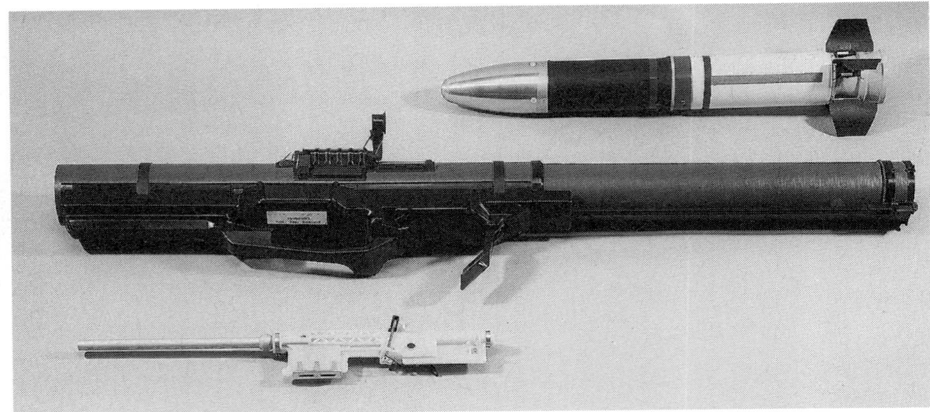

The main components of LAW 80, from top, the rocket, the launcher and the integral spotting rifle

A complete and simple training package is available. This includes: a drill weapon, made from a fired launcher; an indoor trainer; and an outdoor trainer.

The British Army night sight is fully compatible with LAW 80. After the projectile has been fired the sight is removed and the weapon discarded.

Data

Calibre: 94 mm
Length: folded, 1 m; extended, 1.5 m
Weight: carry, 10 kg; ready to fire, 9 kg; projectile, 4.6 kg
Range: effective, 20-500 m
Armour penetration: in excess of 700 mm
Operational temperature range: −46 to +65°C
Shelf life: 10 years

Prime contractor

Hunting Engineering Limited, Reddings Wood, Ampthill, Bedfordshire MK45 2HD.

Status

In production.

Service

British Army, Royal Marines, RAF Regiment, Jordan, Oman and other overseas customers.

UPDATED

NLAW: Next Light Anti-armour Weapon

Development

The Next Light Anti-armour Weapon (NLAW) was designed by the Defence Research Agency as one of a family of man-portable light anti-armour and assault weapons which incorporated the features essential for use in an urban or close-quarter battle. The NLAW Technical Demonstration Programme began in 1987, a significant factor being the concurrent adoption of explosive reactive armour. Since no single weapon appeared capable of meeting all the requirements, two were investigated, NLAW 1 and NLAW 4. Trials included successful firings of both models.

The NLAW test and development programme has been completed. Future NLAW developments await a Staff Requirement, possibly included with a European Staff Target, which will outline the necessary specifications for the future NLAW.

Draft requirements include the need for a man-portable system which may be reloadable or disposable. It will have an effective range of from 20 to 600 m and be capable of firing from enclosed spaces. There will be a high probability of kill against MBTs and LAVs from all aspects, with a secondary capability against bunkers and defensive positions.

Status

Awaiting Staff Requirement definitions.

UPDATED

UNITED STATES OF AMERICA

Olin RAAM Rifle-launched Anti-Armour Munition

Description

The Olin RAAM Rifle-launched Anti-Armour Munition is a rocket-boosted bullet-trap rifle grenade which can be launched from any unmodified M16A2 rifle by means of an expendable plastic launch adaptor. It has a direct-fire engagement range of 250 m, can defeat 400 mm of armour, and is safe during transport and handling, arming only after travelling 10 m from the rifle. The combination of rifle launch and rocket boost means that the firer is in no danger from back blast and can fire the weapon from inside a confined space in safety.

Olin RAAM launched from M16A2 rifle

Data

Length: 564 mm
Weight: 1.65 kg
Range: 250 m
Accuracy: 4 mrad
Armour penetration: 400 mm

Manufacturer

Olin Ordnance, 10101 Ninth Street North, St Petersburg, Florida 33716.

Status

Advanced development.

VERIFIED

66 mm Improved M72 Series LAW (Lightweight Anti-Armour Weapon)

Development

The M72 LAW was developed in the 1960s by the Hesse Eastern Company. It was a revolutionary design for its day, a pair of telescoped tubes with a preloaded HEAT rocket in place. The unit was light and compact and could be easily carried by an individual soldier. When required, the user extended the telescoped tube, which automatically cocked the launcher and erected the simple sight. The user placed the tube on his shoulder, took aim and squeezed the trigger to discharge the rocket to a range of up to 1,000 m, though for effective anti-tank use the maximum range was about 150 m against moving targets and 300 m against stationary tanks. The warhead could penetrate just over 300 mm of mild steel and the M72 was also effective against light field fortifications. The M72 was widely adopted by NATO and other armies and the design has since been copied in various countries.

In subsequent years the design has been improved and a manufacturing consortium set up. The current version is powered by a much-improved rocket motor and with a variety of more powerful warheads. Although primarily designed for the defeat of light armour, the weapon retains a moderate capability against main battle tanks when engaging them at the top, sides or rear.

The series of new variants, designated M72A4, M72A5 and M72A6, are man-portable, lightweight, direct-fire weapons. The tactical round consists of an in-tube burning free-flight rocket that is factory packed in a disposable telescoped launcher.

Firing position with fingers on trigger

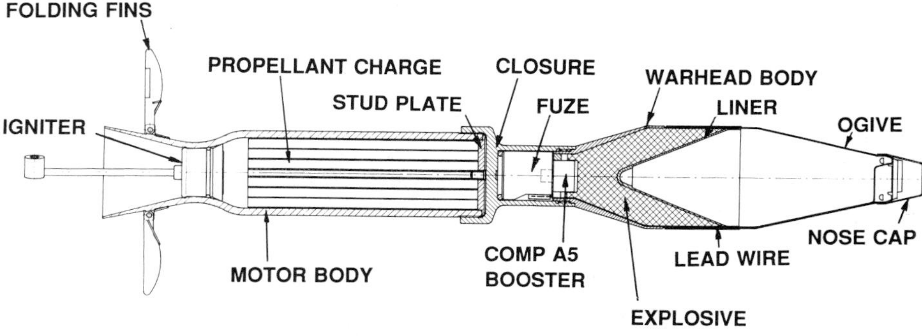

66 mm M72A5 HEAT rocket

Description

Each rocket consists of three major assemblies: a 66 mm high-explosive warhead; a Point-Initiating, Base-Detonating (PIBD) fuze and an improved rocket motor. Attached to the motor case are eight spring-loaded fins which are folded forward to lie alongside the rocket motor whilst in the launcher. Upon ignition of the rocket, gas pressure propels the rocket from the launcher and the fins spring out to stabilise flight.

The launcher, common to all three of the new series configurations, is a lightweight expendable assembly consisting of a high-strength aluminium inner tube, a glassfibre composite outer tube, sights, firing mechanism, safety interlocks, and a carrying sling. In the carry position the launcher serves as the field handling and storage container, allowing the weapon to be issued as a sealed single round of ammunition requiring no maintenance or field support.

Opening the launcher to the firing position releases the system safety interlocks, cocks the weapon and automatically deploys the pop-up rifle-type sights. In the firing position the launcher acts as the firing platform and initially guides the rocket on its relatively flat trajectory toward the target. After firing, the launcher is discarded.

Two important modifications made to the launcher are in the sights and the firing system. These improvements contribute to a substantial reduction in round-to-round dispersion at 250 m compared to earlier models of the M72. To enhance the gunner's ability to acquire and maintain an accurate sight picture, the launcher employs rifle-type sights. The redesigned firing mechanism provides a more uniform trigger action.

The improved rocket motor gives a better hit probability at extended ranges and decreases the effects of misestimation of range. Rocket motor velocity is increased from 150 to 200 m/s, producing a two-fold increase in hit probability at 250 m and a substantial increase in the system's operational range.

The M72A4 warhead is optimised to provide a minimum armour penetration of at least 355 mm, together with limited behind-armour effects. The M72A5 warhead retains the proven M72A3 warhead which provides moderate behind-armour effects and not less than 350 mm penetration. The M72A6 warhead provides

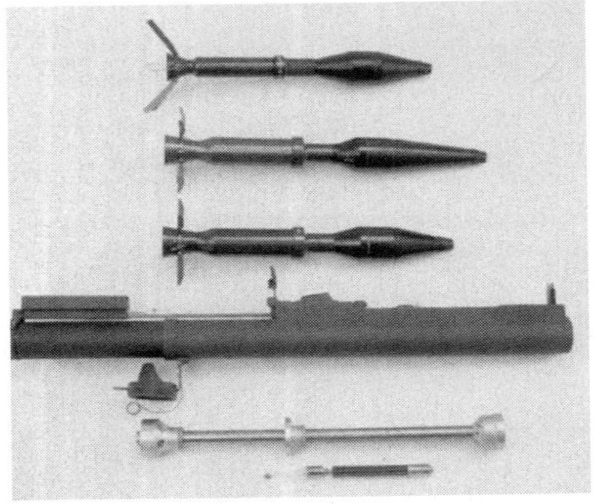

M72 launcher with (top to bottom) M72A3, M72A4 and M72A5 rockets, with subcalibre training insert below

further increases in effectiveness against single-, double- and triple-spaced armour on advanced infantry fighting vehicles, providing lethal behind-armour effects through reduced penetration and increased hole volume. The M72A6 also significantly increases fragmentation when fired against reinforced concrete and field fortifications.

Data

Model	M72A3	M72A4	M72A5	M72A6
Carry weight	2.5 kg	3.45 kg	3.45 kg	3.45 kg
Carry length	665 mm	775 mm	775 mm	775 mm
Firing length	899 mm	980 mm	980 mm	980 mm
Calibre	66 mm	66 mm	66 mm	66 mm
Muzzle velocity	150 m/s	200 m/s	200 m/s	200 m/s
Effective range	170 m	220 m	220 m	220 m
Operational range	250 m	350 m	350 m	350 m
Dispersion at 250 m	2.3 mil	1.5 mil	1.5 mil	1.5 mil
Time of flight to 250 m	1.9 s	1.4 s	1.4 s	1.4 s
Penetration mild steel	300 mm			
Penetration RHA		350 mm	300 mm	150 mm

Although the M72 warheads are optimised for use against a wide variety of armoured vehicles, they also possess the capability to deal with threats within urban buildings and hasty defences.

Manufacturers

Prime contractor USA: Talley Defense Systems, 3500 North Greenfield Road, PO Box 849, Mesa, Arizona 85211.
Prime contractor Europe: Raufoss A/S, PO Box 849, N-2831 Raufoss, Norway.
NI Industries Inc, 5215 South Boyle Avenue, Los Angeles, California 90058 (design and production of launchers).
Tracor Aerospace, 6500 Tracor Lane, Austin, Texas 78725-2070 (design and production warhead metal parts assemblies).

Status

In production.

Service

US forces and most NATO armies. Widely used in many countries all over the world.

UPDATED

3.5 in M20 rocket launcher

Description

The M20 rocket launcher (sometimes called the Super Bazooka) followed the 2.36 in (60 mm) M9A1 rocket launcher. The 3.5 in (89 mm) calibre launcher is a two-piece tube, open at both ends and with a smooth bore. Its function is to ignite the rocket which it launches and to give initial direction to its flight. Ignition is electrical.

To reduce the weight, the tube is made of aluminium and breaks into two sections for ease of carrying. The barrels of the M20 and M20A1 models are made from aluminium tube and the component parts are fastened by means of screws. The barrels of the M20A1B1 and M20B1 are aluminium castings and many component parts are cast integrally with the barrel. The sight is attached to the left side of the rear tube. It consists of a single lens and a graticule fitted into a housing. The graticule consists of a single broken vertical line up the centre of the lens and five broken horizontal lines. Each section of the vertical line and each space between sections represents 50 yards (45.7 m). The horizontal lines are marked in hundreds of yards and the appropriate range is laid on the target. Each section of the horizontal line and each space between the sections, represents 5 mph (2.2 m/s). The appropriate aim-off is applied when firing at a moving target.

The current to fire the rocket is provided by a magneto, the movement of the armature being produced when the trigger is pulled and also when it is released. Thus a firing impulse is produced for both movements of the trigger and a rocket that fails to fire when the trigger is pulled, may fire when the trigger is released.

The rocket HEAT 3.5 in is issued assembled and is loaded into the launcher as a unit. It consists of a rocket head, a fuze and a motor to which is attached a tail assembly. The head is cylindrical with a ballistic head made of steel. The fuze is a base percussion type. The motor is a steel tube with propellant held in spacer tubes. The tail assembly is attached to the rear.

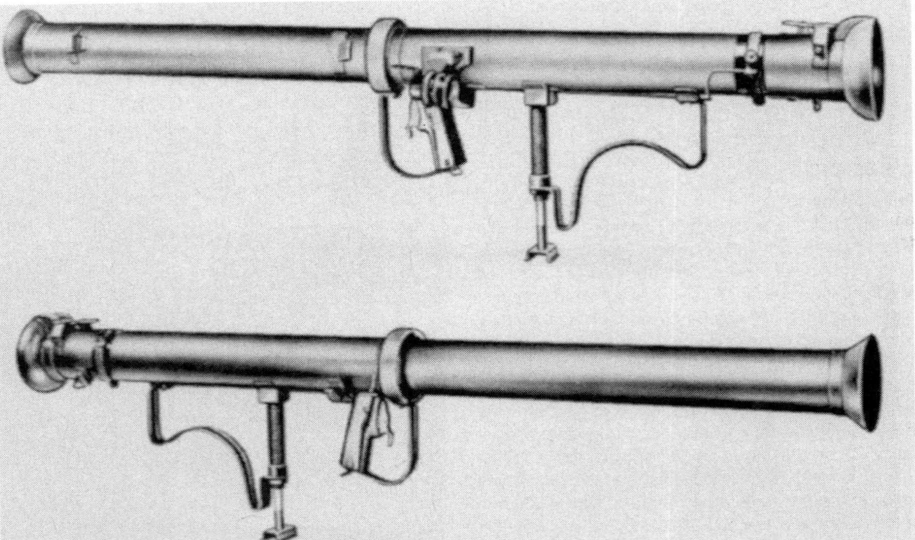

3.5 in M20 rocket launcher used during Korean War

The igniter is at the front of the propellant with two leads passing out through the nozzle at the rear. These convey the firing current from the magneto-trigger when connected up. When the rocket is fired, there is a flow of high-velocity high-temperature gas from the venturi at the rear and there is a triangular danger area with a base and height of 25 yards (23 m). The firer should wear a face mask to protect his eyes from particles of unburnt propellant.

Data

Calibre: 89 mm
Length: assembled launcher, 1.549 m; front tube, 768 mm; rear tube, 803 mm
Weight: launcher, 5.5 kg; rocket, 4.04 kg
Weight and type of explosive filling: 0.87 kg Comp B
Rocket motor propellant: M7
Weight of propellant: 163 g
Max range: against armour, 110 m; practical, 1,200 m

Status

No longer manufactured within the USA.

Service

Still in service in many countries.

Licence production

BRAZIL
Manufacturer: Hydroar SA
Type: 3.5 in M20A1B1
Remarks: See entry under *Brazil* for details.

VERIFIED

Rocket and ignition system

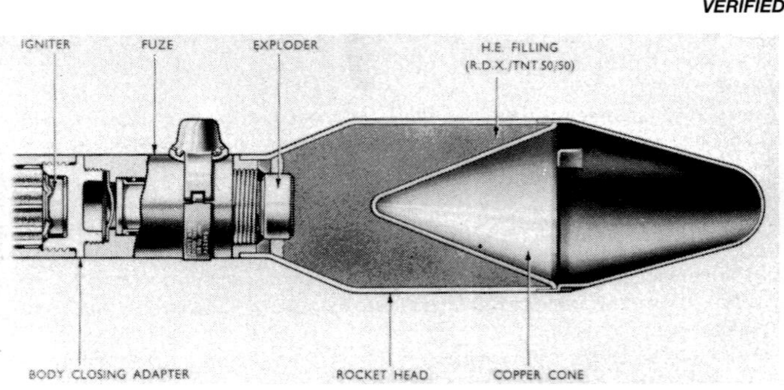

Warhead, filling and igniter

90 mm M67 recoilless rifle

Description

The 90 mm M67 recoilless rifle is a lightweight, portable, crew-served weapon intended primarily as an anti-tank weapon although it can also be employed against pill-boxes or field fortifications. It was designed to be fired from the ground off its bipod and monopod but may be fired from the shoulder.

The M67 is a breech-loaded single-shot weapon firing the HEAT round M371A1. The rifle is equipped with a manually operated breech mechanism and a centrally located percussion-type firing mechanism. It was designed for direct-fire only. The M103 sight has a fixed focus, ×3 telescope with a field of view of 10°. Its lead lines and stadia lines are provided for ranging on targets having a 10 ft (3 m) width and a 20 ft (6 m) length.

The HEAT round M371A1 is a 3.06 kg fin-stabilised projectile with a shaped charge head. It uses the PIBD M530, a point-initiating, base-detonating fuze with an inertially operated graze system. There is also a TP M371 practice round with a small charge. The complete HEAT round weighs 4.2 kg and the projectile has a muzzle velocity of 213 m/s.

Data

Length: 1.35 m
Weight: empty, 16.4 kg
Height: ground mounted, 432 mm
Max range: effective, 450 m; practical, 2,100 m

90 mm M67 recoilless rifle of Kia manufacture

Manufacturer

Now produced by Kia Machine Tool Company Limited, Kia Building, 4th Floor, 15 Yoido, Yeungdeungpo-gu, Seoul, South Korea.

Status

In production.

Service

South Korean Army.

VERIFIED

106 mm M40, M40A2 and M40A4 recoilless rifle

Description

The 106 mm M40 recoilless rifle is a lightweight weapon designed for both anti-tank and anti-personnel roles. It is air cooled, breech loaded and fires fixed ammunition. The marks of the gun are made up as follows and it will be seen that there is a substantial commonality between them:

the piece (cannon) M206
mount M79 (M40A2)
mount M92 with tripod M27 (M40A4)
spotting rifle M8C
elbow telescope M92F
telescope mount M90
cover, telescope and mount and
instrument light M42.

The 106 mm recoilless rifle is one of the most successful examples of its type. It came into service in the mid-1950s and is still used in substantial numbers in many parts of the world. It is lighter than its contemporaries, yet it is little reduced in performance. It was the first recoilless gun to have a spotting rifle for the gunner.

One of the lesser known details of the 106 mm is that it is actually 105 mm and is an improved version of the M27 105 mm. To avoid confusion at a time when both types were in service it was found convenient to measure the bore of the newer weapon from the bottom of the grooves and so give it a new calibre.

The barrel assembly consists of the tube, an enlarged reaction chamber and a mounting bracket. The tube is made of alloy steel, screw-threaded at the back to take the reaction chamber. The breech block is a short cylinder with an interrupted screw thread to mate with the segmental thread in the interior of the recoil chamber. The breech block is hinged on the left side of the breech and is drilled to take the percussion firing mechanism. The mount M79 is basically a tripod. The two rear legs have carrying handles and clamps and the front wheel has a hard rubber tyre. The mount has the traversing and elevating gear and part of the firing mechanism. The two rear legs of the mount are adjustable for lateral expansion. The mount can be lifted at the rear by two men and pushed forward on the single wheel, like a wheelbarrow. The mount can also be clamped in position on a vehicle. The traversing mechanism allows 360° of controlled or free traverse using the traverse knob. The elevating mechanism on the left of the mount provides slow or rapid rates of elevation. The handwheel gives rapid elevation (16 mils a turn) while the firing knob gives slow elevation (3.8 mils a turn). With the breech between the legs the gun can be elevated to +65° and depressed to −17° but over the legs the maximum elevation is 27°. The 0.50 spotting rifle (Cal 0.50 Spotting Gun M8C) is a gas-operated, magazine-fed, self-loading rifle, firing a bullet whose trajectory matches that of the recoilless rifle. The sight allows engagement up to 2,200 m and has stadia lines based on a target 20 ft (6 m) long and 10 ft (3 m) wide.

106 mm M40 recoilless rifle

Ammunition for the 106 mm rifle is issued in complete fixed cartridges. Propellant is contained in a plastic bag within the cartridge case. The cartridge case is perforated to permit expanding propellant gases to escape into the enlarged reaction chamber. The cartridge case has a well around the primer to permit the breech to be closed. Proper headspace is automatically provided for by the seating of the flange of the cartridge case against the vent bushing.

Depending on the type of projectile, ammunition for the 106 mm rifle is classified as High-Explosive Anti-Tank (HEAT), or High-Explosive Plastic with Tracer (HEP-T). The HEAT round has a maximum range of 3,000 m.

There was also an Anti-PERSonnel round with Tracer (APERS-T), which had a maximum effective range of 300 m.

The spotting rifle employs a spotter-tracer cartridge. The projectile contains a tracer element and an incendiary filler. On impact, the incendiary filler produces a puff of white smoke which aids in adjusting fire. The tracer burnout point is between 1,500 and 1,600 m. The cartridge can be identified by the red and yellow marking on the nose of the projectile.

To extend the utility of the 106 mm M40 series, Bofors AB of Sweden have introduced their 3A-HEAT-T round. See entry under *Sweden* for details.

Data

Calibre: 105 mm
Length: 3.404 m
Weight: empty, 209.5 kg
Height on M79 mount: 1.118 m
Width of M79 mount: legs spread, 1.524 m; legs closed, 800 mm
Length of bore: 2.692 m
Type of breech block: interrupted thread
Range: HEAT, max, 3,000 m; effective, 1,350 m
Rate of fire: sustained, 1 rd/min
Weight of HEAT round: 16.88 kg; projectile, 7.96 kg
Weight of HEP-T round: 16.95 kg; projectile, 7.96 kg

Manufacturer

Watervliet Arsenal, Watervliet, New York, New York 12189.

Status

Not in production in USA but still produced elesewhere (see under Licence production).

Service

In reserve in USA. Has been supplied to 32 countries.

Licence production

AUSTRIA
Manufacturer: Watervliet Arsenal; carriage Lohner GmbH (gun)
Type: 10.6 cm rPAK M40A1
Remarks: 106 mm on locally produced carriage. No longer in production. See entry under *Austria* for details.
INDIA
Manufacturer: Ordnance Factory Board
Type: 106 mm RL
Remarks: Produced in both vehicle-mounted and towed forms. See entry under *India* for details.
PAKISTAN
Manufacturer: Pakistan Machine Tool Factory Limited
Type: 106 mm M40A1
Remarks: See entry under *Pakistan* for details
SPAIN
Manufacturer: SANTA BARBARA
Type: 106 mm M40A1
Remarks: See entry under *Spain* for details.

UPDATED

M136 light anti-armour weapon

Description
M136 is the US Army designation for the Alliant Techsystems AT4, produced under licence from Bofors AB (formerly FFV) for American production, treatied and allied nations. It is a disposable recoilless weapon providing close-in armour defeat with better than 97 per cent reliability demonstrated after 500,000 units had been produced.

The M136 is the Bofors AT4 HEAT. For full details of this weapon refer to the appropriate entry under *Sweden*.

THe M287 trainer is a low-cost reusable system consisting of a 9 mm gun mounted in the M136 launch tube to fire a 9 mm tracer bullet.

Variant
AT8

See entry under *Light Support Weapons*.

Manufacturer
Alliant Techsystems Inc, Defense Systems, 7225 Northland Drive, Brooklyn Park, Minnesota 55428.

Status
In production.

Service
US Army, Navy, Air Force and Marine Corps.

VERIFIED

M136 light anti-armour weapon

Dragon Medium Anti-Armour Missile System

Description
Initial production of the Dragon Weapon System was carried out by the McDonnell Douglas Corporation. In December 1993 CMS Inc purchased all rights and assets associated with the Dragon system from McDonnell Douglas.

The Dragon Weapon System, in service with the US Army and Marine Corps since 1973, is a one-man portable anti-armour missile system capable of defeating main battle tanks, even when equipped with explosive reactive armour. Command to Line Of Sight (CLOS) wire guidance insures great accuracy against both stationary and moving targets out to maximum range. Its compact size and low weight make Dragon an excellent weapon for airborne and infantry use.

Dragon IIT anti-armour missile (T J Gander)

1996

Dragon II
The first Dragon product improvement includes an improved warhead which provides an 85 per cent increase in armour penetration performance when compared to the original Dragon I warhead. Dragon II is in production and is deployed with the US Army, Marine Corps and several foreign armies. Existing inventories of the Dragon I missile can be upgraded with the new Dragon II warhead.

SUPERDRAGON
SUPERDRAGON (originally known as Dragon II+) involves additional product improvements to maintain Dragon's effectiveness well into the next century, including a range increase to 2,000 m. An interim

Dragon improved Day Tracker

Dragon improved Night Tracker

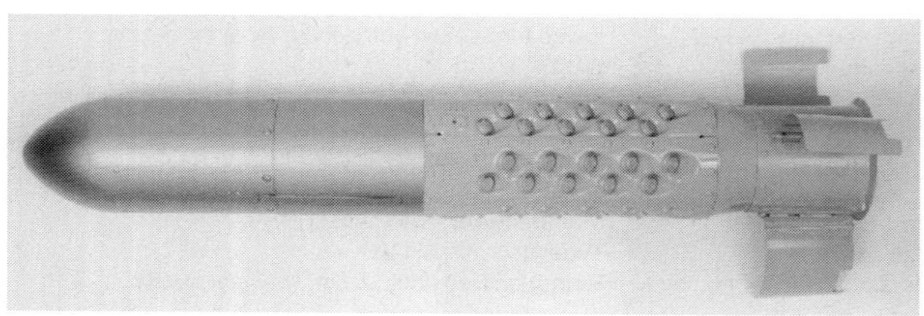

Dragon II anti-armour missile

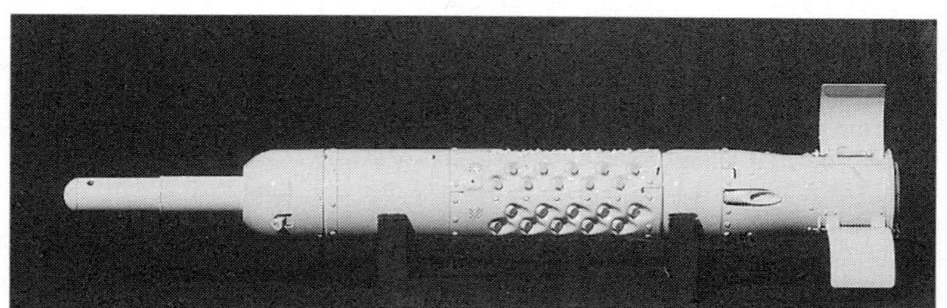

SUPERDRAGON anti-armour missile

Data

	Dragon	**Dragon II**	**SUPERDRAGON**
Range	65-1,000 m	65-1,000 m	65-2,000 m
Time of flight	11.2 s to 1,000 m	11.5 s to 1,000 m	<11 s to 2,000 m
Length	1.15 m	1.15 m	1.15 m
Weight:			
With day tracker	14 kg	15.4 kg	17.9 kg
With night tracker	20.7 kg	22.1 kg	24.6 kg

increase in the missile velocity was achieved by adding a sustainer rocket motor, greatly reducing the time of flight to the interim maximum range of 1,500 m. The range increase to 2,000 m was then completed by introducing aerodynamic refinements to the missile body to reduce drag and weight, and some reorientation of the rocket thrusters. SUPERDRAGON average velocity is more than 174 m/s.

A further improvement to the warhead is provided which includes an extendable probe containing a precursor warhead enabling Dragon to defeat explosive reactive armour and provides a 98 per cent increase in armour penetration over the original Dragon I. This represents the capability to defeat the Soviet T72 MBT equipped with reactive armour. New digital electronics in the day and night infra-red night trackers reduce the requirement for specialised test equipment. All existing Dragon missiles can be upgraded to the SUPERDRAGON configuration.

Dragon IIT

Dragon IIT was introduced during 1995 and is a short-range missile with a tandem warhead developed in response to a requirement from Turkey. Becasue of the weight of the round and its extended standoff probe carrying the initial charge (13.1 kg ready to fire), the range is limited to 750 m. Round length is 1.15 m.

Manufacturer

CMS Inc, 4904 Eisenhower Boulevard, Suite 310, Tampa, Florida 33634.

Status

Dragon I no longer in production. Dragon II and SUPERDRAGON in production. Dragon IIT available.

Service

US Army and Marine Corps and Iran, Israel, Jordan, Morocco, Netherlands, Saudi Arabia, Spain, Switzerland, Thailand, Yemen and the former Yugoslavia.

UPDATED

TOW heavy anti-tank weapon system

Development

TOW is a crew-portable, vehicle-mounted or helicopter-carried, heavy anti-tank weapon system. The name is derived from its description as a Tube-launched, Optically tracked, Wire command link guided missile.

Design work started in 1962, the first firings were carried out in 1963 and the weapon entered service with the US Army in 1970. Since then it has been proved in action and has been sold to more than 40 countries.

TOW, produced by the Hughes Missile Systems Company, is recognised as the most successful and widely used anti-tank missile system in the world. According to US Army statistics, the missile has a cumulative reliability record of over 93 per cent in over 12,000 test and training firings conducted since 1970. One contributory factor to its success is the fact that the training of the operator is far less complicated than with many previous types and various simulators enable training shots to be carried out under realistic conditions. Another TOW asset is that the same missile can be fired from ground, vehicle and helicopter mountings which simplifies logistics.

Although TOW is still in production its successor is under active consideration under the designation Advanced Missile System - Heavy (AMS-H).

TOW 2 missile launcher

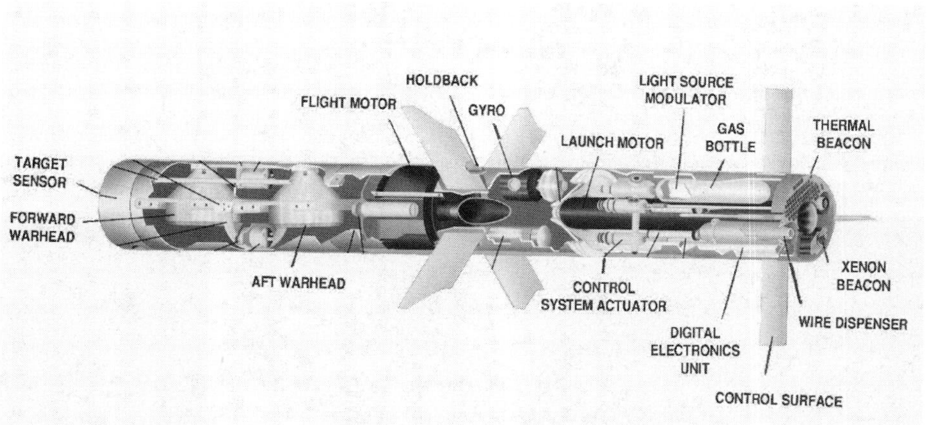

TOW 2B missile

Description

The TOW weapon system consists of six major units. These are: tripod; traversing unit; launch tube; optical sight; missile guidance set and battery assembly, housed in the missile guidance set.

The tripod provides a stable mounting base for the traversing unit and allows levelling on ground sloping up to 30°. The traversing unit is an electromechanical assembly attached to the tripod and is the mounting base for the optical sight and the launch tube. The launch tube is constructed of a lightweight honeycomb material, covered with laminated glassfibre. It provides initial guidance and stability to the missile and protects the gun crew from the missile launch blast on firing. The optical sight is used to track the target and to detect the infra-red signal from the missile in flight. The sight contains a ×13 telescope, boresighted to an infra-red tracker, with a field of view of 4°. The cross-hairs in the eyepiece may be illuminated and the task of the gunner is to keep these cross-hairs on the target. The missile guidance set consists of two rechargeable 50 V batteries and one rechargeable 24 V battery.

The basic TOW missile is the BGM-71A. It consists of the launch motor, the flight motor, four control surfaces, the infra-red source, two wire dispensers, a battery, a digital electronics unit, a gyro, safety and arming devices and the warhead. The launch motor is at the rear of the missile and consists of M7 propellant which is completely expended before the missile leaves the tube. The flight motor uses a solid propellant and is near the centre of gravity of the missile. The burning gases emerge from a pair of nozzles at 30° to the horizontal axis of the missile. This configuration eliminates interference with the wire link and minimises changes in the centre of gravity of the missile as the motor burns. The flight motor ignites about 12 m in front of the launch tube to protect the gunner. The wire command link is a two-wire system dispensed from two spools at the back of the missile. These wires carry steering commands from the launcher to the missile, and they are applied, together with signals from the gyro, to the four control surfaces. The infra-red source provides a beacon which is detected by the infra-red tracker in the optical sight to determine the missile's position. The IR lamp current is modulated to allow discrimination against a background of other strong infra-red emitters, such as the sun.

The M220B TOW 2 launcher is a modified version of the original launcher, with improvements aimed at permitting guidance through obscurants such as smoke or dust and at night. Changes include adding an AN/TAS-4 sight (the upper of three mounted sights) to track targets at night and function as a totally independent fire control sensor. Other new or modified components include the post amplifier electronics in the small rectangular box above the AN/TAS-4 sight, and the digital missile guidance set (shown beneath the tripod in the accompanying picture). All types of TOW missile - basic TOW, ITOW, TOW 2, TOW 2A, TOW 2B or any future TOW missile contemplated by the US Army - can be fired from this modified launcher.

When the missile is installed in the launcher, the gunner may use the self-test automatic sequencer, which checks the circuitry associated with weapon functioning. When this is completed the operator connects the weapon system to the encased missile by raising the arming lever. When a target appears the gunner moves the sight to place the cross-hairs on the target. When the target is in range the trigger is pressed. This activates the missile batteries and gas is released to spin the gyro up to speed. About 1.5 seconds later the launch motor is fired. This burns out completely within the tube but the missile acquires sufficient momentum to coast until the flight motor ignites. The missile wings and control surfaces are extended from the missile body. The infra-red sources on the missile start to operate and the two command-link wires are dispensed from the internal spools. The first stage of arming the warhead occurs. The flight motor is activated at the end of the 12 m coasting period and the warhead is fully armed after about 60 m.

The gunner operates the traversing unit and keeps the cross-hairs of the optical sight on the target. The infra-red sensor tracks the signal from the modulated lamp in the missile and detects any deviations from the line of sight path to the target. It provides continuous information over the wire link and to the missile guidance set which produces signals which are delivered, with those from the gyro, to the control surface to correct the flight path and bring the missile back to the line of sight.

Well over 314,000 basic TOWs (BGM-71A) have been delivered.

The US Army has conducted a number of programmes to improve the performance of the TOW warhead against advanced enemy armour. The first was an improved 127 mm warhead (ITOW, BGM-71C), roughly the same size and weight as the basic TOW warhead but with better performance against armour. The warhead has a telescopic nose probe which is retracted into the missile while in the launch tube but extends out when the missile has cleared the launcher and is in flight. The probe gives a longer standoff distance and enhances the jet so as to be better formed before striking the target plate. Over 60,000 ITOWs have been delivered.

The second upgrade, the TOW 2 (BGM-71D), includes a heavier 152 mm warhead which occupies the entire diameter of the missile body. With this warhead, the guidance system has been improved by introducing a digital guidance set based on a microprocessor which provides greater flexibility in programming and higher precision. The warhead increases the weight and, to compensate for this, the flight motor is loaded with an improved propellant. Over 77,000 TOW 2 missiles have been delivered.

TOW 2A (BGM-71E) adds a small warhead to the missile probe. This detonates explosive reactive armour and so clears a path for the primary warhead. More than 34,000 TOW 2As have been delivered.

TOW 2B (BGM-71F) is designed for top attack. It features a dual mode sensor and a new armament section with two warheads substantially different from those used in other TOW versions to form two explosively formed projectiles with substantial pyrophoric effects after penetration. The missile is programmed to fly over the target and the dual mode laser ranger/magnetic sensor system triggers the two warheads to shoot downwards at the proper instant.

In late 1987 THORN EMI was awarded a major contract by the United Kingdom Ministry of Defence for Further Improvements to the TOW system (FITOW). As prime contractor, THORN EMI led a team comprising Royal Ordnance, Hughes Aircraft and Westland Helicopters. The heart of the FITOW project is a new proximity fuze which, when combined with a new type of warhead, will provide TOW with a top-attack capability. The work is based on development work carried out over several years by THORN EMI Electronics and Royal Ordnance in conjunction with the Royal Armament Research and Development Establishment (RARDE), and it is anticipated that the enhanced performance of FITOW will make it capable of defeating the next generation of armour and maintaining TOW as a viable defence system well into the next century.

All versions of the TOW missile can be launched from any TOW launching system, but full system capability is attained when launched from launcher systems which have been upgraded through a software change, to the later versions. TOW 2B requires some further changes to the launch software to produce its overfly trajectory.

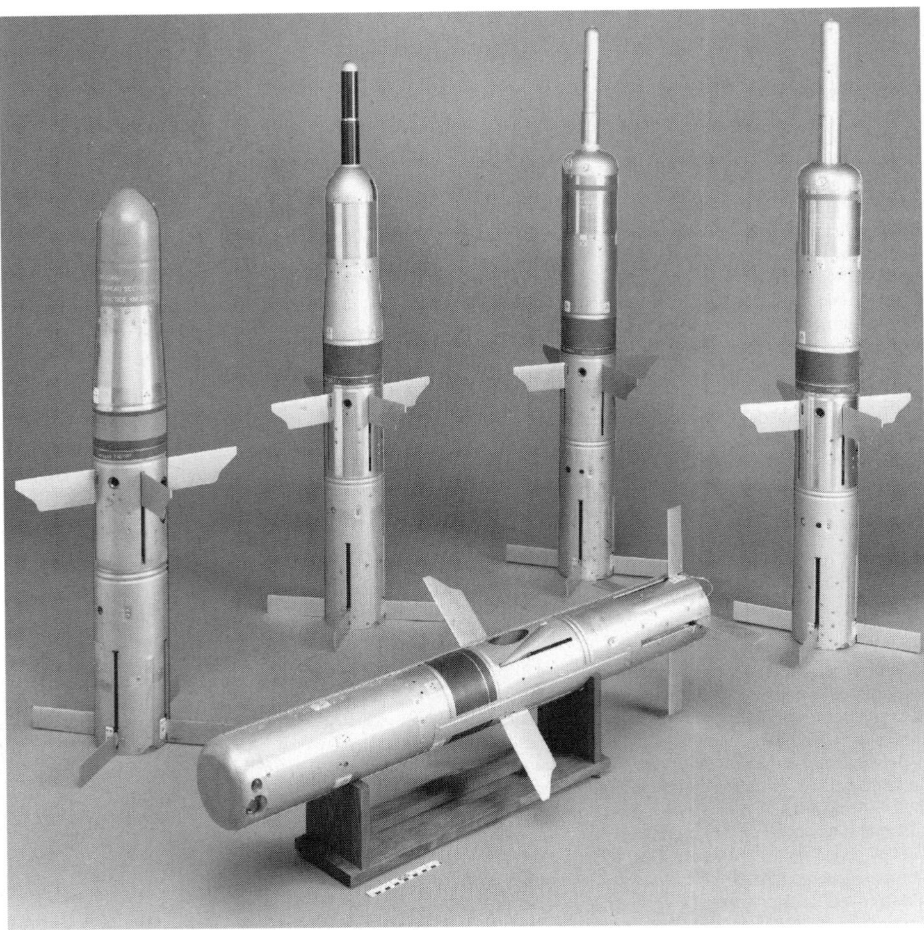

TOW family
(Left to right) *basic TOW: ITOW: TOW 2: TOW 2A.* (Foreground) *TOW 2B*

Data

Designation: BGM-71A (basic TOW); BGM-71C (ITOW); BGM-71D (TOW 2); BGM-71E (TOW 2A); BGM-71F (TOW 2B)
Guidance principle: automatic missile tracking and command to line of sight guidance from optical target tracker (SACLOS)
Guidance method: wire command link controlling aerodynamic surfaces
Propulsion: 2-stage, solid propellant motor
Warhead: HEAT
Missile velocity: 200 m/s
Max range: 3,750 m
Min range: 65 m
Crew: 4

TOW 2B missile approaching a target

1996

The effect of a TOW 2 missile warhead on a target tank

1996

	Length	Width	Height	Weight
Launcher, tubular, guided missile, M151E2	2.21 m	1.143 m	1.118 m	78.5 kg
Launcher, with AN/TAS-4				87.5 kg
Launcher, with TOW 2 Mods				93 kg
Tripod, retracted	1.064 m	645 mm	569 mm	9.5 kg
Traversing unit	297 mm	518 mm	511 mm	24.5 kg
Optical sight	544 mm	295 mm	315 mm	14.5 kg
Launch tube	1.675 m	191 mm	191 mm	5.9 kg
Missile guidance set, with battery	406 mm	406 mm	254 mm	24 kg
Battery assembly	394 mm	117 mm	178 mm	10.9 kg
Guided missile, basic TOW, BGM-71A	1.174 m	221 mm	221 mm	22.5 kg
Guided missile ITOW, BGM-71C	1.174 m	221 mm	221 mm	25.7 kg
Guided missile TOW 2, BGM-71D	1.174 m	221 mm	221 mm	28.1 kg
Guided missile TOW 2A, BGM-71E	1.174 m	221 mm	221 mm	28.1 kg
Guided missile TOW 2B, BGM-71F	1.168 m	221 mm	221 mm	22.6 kg
Guided missile practice (inert warhead, live motor) BTM-71A	1.285 m	221 mm	221 mm	24.5 kg

Manufacturer
Hughes Missile Systems, PO Box 11337, Building 807 MS A8, Tucson, Arizona 85734-1337.

Status
In production.

Service
US Army, Marine Corps. Purchasers include Bahrain, Belgium, Botswana, Cameroon, Canada, Chad, China, Colombia, Denmark, Egypt, Finland, Germany, Greece, Israel, Italy, Japan, Jordan, Kenya, South Korea, Kuwait, Lebanon, Luxembourg, Morocco, Netherlands, Norway, Oman, Pakistan, Portugal, Saudi Arabia, Singapore, Somalia, Spain, Sweden, Switzerland (licence production), Thailand, Tunisia, Turkey, UK and South Yemen.

UPDATED

JAVELIN Advanced Anti-tank Weapon System - Medium

Development
Now in production, JAVELIN is a lightweight man-portable fire-and-forget anti-armour weapon system. It was developed for the US Army and Marine Corps as a joint venture of Texas Instruments and Lockheed Martin (formerly Martin Marietta).

Texas Instruments began working on a fire-and-forget anti-armour weapon in the early 1980s, using, as a basis, a design first devised in DARPA's Tank Breaker programme. The Texas Instruments design uses infra-red homing and an advanced, top attack warhead with a high-performance airframe. In February 1988 Texas Instruments formed a joint venture with Martin Marietta (now Lockheed Martin) to continue the development of this missile system, and since the 1989 award both companies have continued working on the project. During the Engineering and Manufacturing Development (EMD) activity, the JAVELIN weapon system successfully completed all safety, qualification and performance evaluations required to obtain approval for initial production release.

During the EMD phase, 175 JAVELIN missiles were fired with a hit rate greater than 90 per cent. JAVELIN completed a successful initial operational test in December 1993. Production approval was granted by the US Department of Defense in June 1994. JAVELIN is currently in initial production to provide hardware for the first US Army fielding during 1996. Full rate production is scheduled to begin in May 1997.

The US Department of Defense has approved the JAVELIN system for foreign military sales.

Launch of a JAVELIN anti-tank weapon system

JAVELIN Command Launch Unit (CLU), left and Round (right)

Description
The JAVELIN weapon system consists of two major elements. The Command Launch Unit (CLU) is a compact, lightweight target acquisition device incorporating an integrated day/second-generation thermal sight, launch controls and gunner's eyepiece display. The second element is the Round, consisting of the Missile, 127 mm in diameter with a staring, imaging infra-red seeker, feature-based tracker, lethal warhead, dual in-line eject and flight motor, and gunner-selected direct or top attack engagement guidance options; and the Launch Tube Assembly, an expendable handling/launch tube to house the missile, with carrying straps and handles, power pack attachment and CLU interface.

The operating controls for the JAVELIN system are built into the CLU, so that the gunner's controls are at his thumb and fingertips. Quick connect and disconnect mating of the CLU and the round reduces the time required for firing and reloading.

The CLU is used for field surveillance, locating targets, and missile fire control. The dual field of view (×4 and ×9) thermal sight allows the user to engage targets by day or night and in degraded battlefield conditions. The CLU weighs 6.4 kg. Power for 4 hours of operation is provided by a BA5590 lithium battery.

The gunner can engage a target after mating the CLU and the round. By switching from CLU to missile seeker the gunner can lock the automatic target tracker to the target. The missile locks on to the target before launch, so the gunner does not have to undertake missile guidance. Launching can be carried out from within buildings or covered firing position. After launch, with the gunner free to reload or take cover, the missile flies itself in an arched trajectory towards the target and attacks the tank from the top, at its weakest point. JAVELIN can engage targets at a range of 2,000 m.

The JAVELIN system incorporates a classroom Basic Skills Trainer and a Field Tactical Trainer for training under tactical conditions.

Data
Calibre: missile, 127 mm; launch tube assembly, 142 mm
Guidance: passive imaging infra-red, lock on before launch, automatic self-guiding
Propulsion: 2-stage solid propellant
Warhead: tandem shaped charge
Weights: complete, transport, 22.3 kg; missile, 11.8 kg; launch tube assembly, 4.1 kg; command launch unit, 6.4 kg
Length: missile, 1.081 m; launch tube assembly, 1.198 m
Range: 2,000 m

Manufacturer
TI/Martin JAVELIN Joint Venture, 2501 South Highway 121, Lewisville, Texas 75067.

Status
Initial production.

UPDATED

JAVELIN in the firing position

Predator Short Range Assault Weapon

Development

The Predator Short Range Assault Weapon (SRAW) has been under development by Loral Aeronutronic for the US Marine Corps since 1990. The first missile was fired in 1991. The SRAW programme commenced as a competitive development concept for a lightweight shoulder-fired weapon that could defeat MBTs equipped with reactive armour and which would replace the AT4 and the M72 LAW series. The resultant product, later renamed Predator, underwent further development to qualify for production. By then SRAW had demonstrated that it could meet its weight and guidance requirements.

The SRAW was designed to have a modular warhead so when the US Army produced a requirement for a shoulder-launched man-portable weapon to defeat personnel protected by buildings, bunkers or light armour it was decided that the successfully demonstrated SRAW, fitted with an alternative warhead, could be configured for this role. This resulted in the Multipurpose Individual Munition (MPIM).

In December 1994 an 18-month technology demonstration contract was awarded by the US Army's Missile Command to Loral Aeronutronic for the integrated MPIM/SRAW. This is now a joint US Army/US Marine Corps programme to develop a lightweight shoulder-fired multipurpose weapon. The US Army and the Marine Corps will share the same flight module and launcher, but the US Army missiles will have the MPIM warhead while the US Marine Corps will be primarily concerned with the anti-armour SRAW, or Predator.

In July 1996, MPIM/SRAW (hopefully by then a less cumbersome acronym will be in use) will enter the engineering and development phase. Production could commence during FY 1999. The first field unit to be equipped with MPIM/SRAW is planned for FY 2001.

The principal MPIM/SRAW contractor is Loral Aeronutronic. Sub-contractors include Aerojet (warhead), Alliant Technology (formerly Hercules, propulsion system), and Systron Donner (inertial guidance).

TARGET DETECTION DEVICE
WARHEAD
FLIGHT
DETECTOR
LASER
WARHEAD
INERTIAL SENSOR
GUIDANCE PROCESSOR
ROCKET MOTOR
SOLENOID VALVES

Main component assemblies of a Predator missile
1996

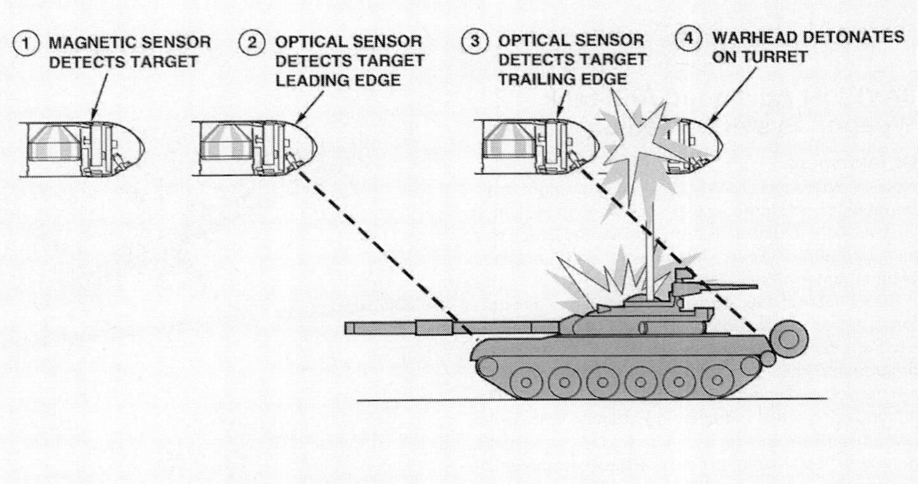

(1) MAGNETIC SENSOR DETECTS TARGET
(2) OPTICAL SENSOR DETECTS TARGET LEADING EDGE
(3) OPTICAL SENSOR DETECTS TARGET TRAILING EDGE
(4) WARHEAD DETONATES ON TURRET

Predator anti-armour target detection device (TDD) functioning sequence
1996

Description

Predator, the anti-armour component of the MPIM/SRAW, is issued ready assembled in its launcher. The launcher includes end caps to seal the launcher against moisture and other contaminants, with the end caps being released automatically during the firing sequence. There is also a carrying sling and a handle for hand carrying. The weapon is made ready for use by the operator erecting the eyepiece of the ×2.5 magnification scope and placing the weapon on the shoulder.

The firing controls are accessed by lifting the safety cover, exposing the arm button and the firing bar. Once the target has been acquired the arming button is pushed, causing a thermal battery in the missile to be actuated. This allows the operator to track moving targets or to target alternative vehicles. To fire the missile the fire bar is depressed while holding in the arm button. When the missile has completed a system self-check, the ordnance battery is initiated and the launch sequence commences. The missile leaves the launcher approximately 0.31 second after the fire bar is actuated.

Aiming a Predator short range assault weapon
1996

Predator is aimed in the same manner as a rifle. The missile uses an inertial autopilot to guide the missile above the line of sight to the target without the operator introducing super-elevation or ranging. After firing the operator has no further influence on the missile and can discard the disposable launcher.

Once positioned along its flight path the missile has an elevated trajectory 2.75 m above the line of sight. If the target is moving, maintaining the aim point on the target for less than 2 seconds will provide data sufficient for the autopilot to compute an average angular rate of the target relative to the gunner. The missile flight path is adjusted in flight to cause the missile to fly continuously along the predetermined operator to target line of sight. The operator makes no correction for range, velocity or the effects of firing up or down sloping terrain. Predator can accept current fielded night sights such as the AN/PVS-4.

To assist accuracy a two-stage solid propellant propulsion system combines high velocity with the ability to fire from within enclosed spaces. The initial launch velocity is approximately 25 m/s, with the flight motor accelerating the missile to a peak velocity of approximately 300 m/s at a distance 125 m down range. Time to 300 m is 1.6 seconds and to 600 m is 2.8 seconds.

Predator is intended to operate over a range of from 17 to 600 m. The warhead (the total payload of which weighs 3.26 kg) is installed in an injected moulded structure that is part of the missile structure, and incorporates the warhead mounting interface in the piece. The structure provides structural integrity for handling and flight loads and helps to mitigate violent reactions from bullet impact and fast and slow cook off.

Fuzing for the warhead module is provided by a Target Detection Device (TDD) mounted under the nose. The TDD uses a dual axis sensor including a laser ranger and a three axis magnetometer. As the missile flies over a target the cross beam laser ranger profiles the leading and trailing edges of the tank then, with magnetic confirmation, initiates the warhead. Both conditions, optical and magnetic, have to be met in order to initiate the warhead.

The warhead is understood to fire an explosively formed fragment which can defeat MBTs with explosive reactive armour (ERA). As the design is modular other types of warhead may be fitted in the future.

Data

PROVISIONAL

Calibre: 140 mm (not confirmed)
Weights: complete, <9 kg; missile, approx 6.4 kg; warhead, 3.26 kg
Length: 890 mm
Velocity: initial, approx 25 m/s; max, approx 300 m/s
Time to max range: 2.8 s
Operational range: 17-600 m

Status

Development for US Marine Corps - see text.

Main contractor

Loral Aeronutronic, 22982 Arroyo Vista, PO Box 7004, Rancho Santa Margarita, California 92688-7004.

NEW ENTRY

YUGOSLAVIA (SERBIA AND MONTENEGRO)

44 mm RB M57 anti-tank launcher

Description

The 44 mm RB M57 anti-tank launcher was derived from the German Panzerfaust and is generally similar to the Czech P-27. The M57 has a permanently attached bipod and is served by a two-man crew. The firer lies behind the weapon and the loader lies to his right.

Data

Calibre: launcher, 44 mm; projectile, 90 mm
Length: launcher, 1 m
Weights: launcher, 8.1 kg; projectile, 2.44 kg
Type of projectile: HEAT
Max range: effective, 200 m
Armour penetration: 380 mm
Muzzle velocity: 146 m/s

Holding Company

YUGOIMPORT - SDPR, PO Box 89, Bulevar umetnosti 2, YU-11070 Novi.

Status

Production complete.

Service

Former Yugoslav armed forces.

VERIFIED

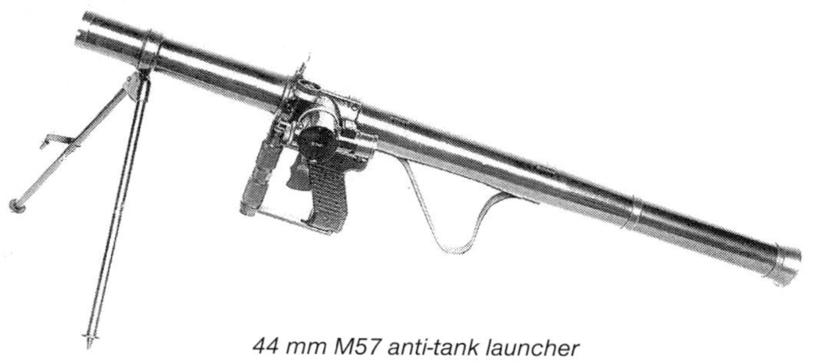

44 mm M57 anti-tank launcher

44 mm M57 anti-tank launcher in firing position

44 mm M80 anti-tank launcher

Description

The 44 mm M80 anti-tank launcher is a modernised version of the M57 (previously), operating in the same manner. The projectile uses the tail unit of the M57 round allied with a new and improved HEAT warhead fitted with a piezoelectric nose fuze. An optical sight ON-M80 is used for aiming and also for detecting sources of infra-red radiation. There is tritium illumination of the graticule for night firing.

Projectiles for the M57 launcher may be used in the M80 launcher.

Data

Calibre: launcher, 44 mm; projectile, 85 mm
Length: 1.04 m
Weight: launcher, 6.6 kg; sight, 830 g; projectile, 2.45 kg
Range: effective, 200 m
Armour penetration: 400 mm

Holding Company

YUGOIMPORT - SDPR, PO Box 89, Bulevar umetnosti 2, YU-11070 Novi.

Status

Production complete.

Service

Former Yugoslav armed forces.

VERIFIED

44 mm M80 anti-tank launcher

82 mm M60A recoilless gun

Description

The 82 mm M60A is a recoilless weapon of light and simple construction, intended for the destruction of tanks and other armoured vehicles, demolition of bunkers and field fortifications and anti-personnel firing. It may be towed by vehicle or animal and can be dismantled for pack-carriage.

Using the HEAT M60 projectile the gun will defeat 250 mm of armour steel at any impact angle over 25° to maximum range. This projectile is extremely accurate; at 500 m the lateral dispersion is 150 mm, the vertical dispersion 190 mm.

There is also a HEAT M72 with rocket propulsion; this has a flat trajectory and can be used successfully to a

Training with 82 mm M60A recoilless gun

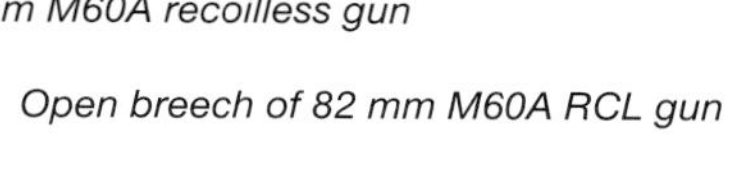

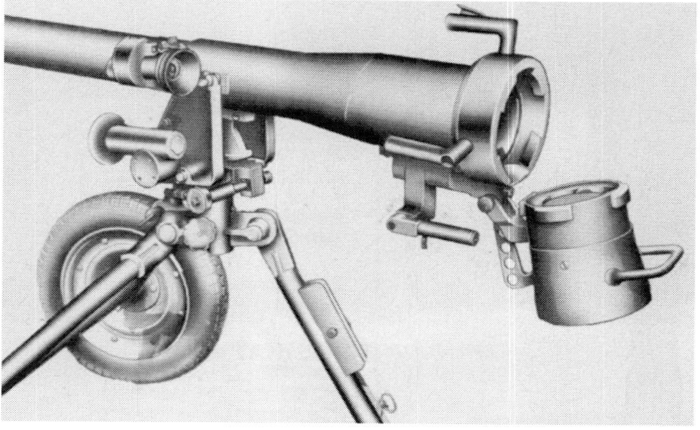

Open breech of 82 mm M60A RCL gun

range of 1,000 m. Penetration is the same as that of the M60, and at 1,000 m the lateral and vertical dispersions are 500 mm.

The firing safety zone behind the breech efflux is 45 m deep and 25 m wide at the rear boundary.

Data

Calibre: 82 mm
Length: 2.2 m
Weight: 122 kg; round, 7.2 kg
Traverse: 360°
Elevation: −20 to +35°
Muzzle velocity: 388 m/s
Max range: (effective, HEAT) against moving targets, 1,000 m; against stationary targets, 1,500 m
Armour penetration: 250 mm
Crew: 5

Holding Company

YUGOIMPORT - SDPR, PO Box 89, Bulevar umetriosti 2, YU-11070 Novi.

Status

Production complete.

Service

Former Yugoslav armed forces.

VERIFIED

82 mm M79 recoilless gun

Description

The 82 mm M79 recoilless gun may have had its initial inspiration from the Soviet B-10, but the weapon eventually developed displayed considerable differences. The gun is made from high-tensile steel and special alloys and uses the characteristic perforated case and annular chamber system. There is a single venturi at the rear of the breech block. A telescopic sight is mounted on the left side of the gun, and there are also open sights for emergency use.

For day firing the ON M72B optical sight is used, and for night firing the PN 5 × 80(J) night sight.

The gun is fired from a tripod, and can be easily and quickly moved to a new firing position. It can be moved either as a complete unit or with the barrel assembly separated from the tripod.

The mounting is a tripod with a flat top to which the gun is coupled; there is a rear pivot and a front support which screws up and down to alter elevation and can move laterally in a slot to give limited traverse. The gun-layer can thus control both movements of the gun with one hand while keeping the other on the firing lever. The firing mechanism is electromechanical and is interlocked so that firing is impossible unless the breech is closed and locked.

The ammunition is fixed, with a perforated cartridge case. The projectile is fin stabilised by six jack-knifed fins which flip out after shot ejection. There are two projectiles; the anti-tank rocket-assisted M79 HEAT and the M81 HE.

The gun is served by four men, and gun and crew are transported by TAM 110 T7 BV 4 × 4 or TAM 4500 DV vehicles. The gun, accessories and ammunition can be carried by pack animals, using special harnesses. For short distances an assembled or dismounted gun can be man-carried.

Data

Calibre: 82 mm
Length: 1.785 m
Weight: in action, loaded, with tripod and sight, 41.35 kg
Rifling: 1.2 m, 90 grooves, rh, 1°30′
Traverse: 10° right and left
Elevation: −3 to +10°
Max chamber pressure: 550 bar
Muzzle velocity: HEAT, 340 m/s; HE, 320 m/s
Max flight velocity: HEAT, 460 m/s
Max range: HE, 2,700 m
Armour penetration: HEAT, 380 mm
Rate of fire: 5-6 rds/min

Holding Company

YUGOIMPORT - SDPR, PO Box 89, Bulevar umetnosti 2, YU-11070 Novi.

Status

Probably no longer in production.

Service

Former Yugoslav armed forces.

VERIFIED

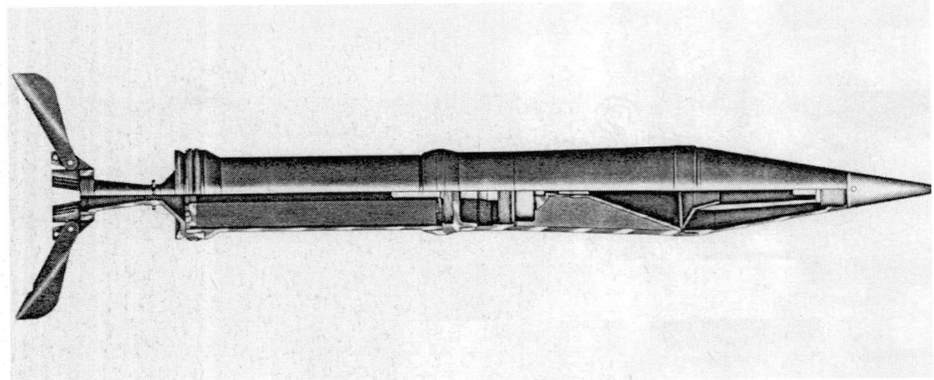

Sectioned M79 HEAT projectile for M79 RCL gun

82 mm M79 recoilless gun

90 mm M79 rocket launcher

Description

The 90 mm M79 is a rocket launcher loaded by coupling a prepacked rocket and combustion chamber to the rear end of the launch tube. After firing, the empty combustion tube is removed and discarded and another preloaded unit is attached. Connection of the combustion tube automatically connects the electrical firing circuits.

The projectile is a rocket carrying a shaped charge warhead, initiated by a piezoelectric impact fuze capable of acting at angles up to 70° from normal to the target.

Data

Calibre: 90 mm
Length: launcher, 1.432 m; launcher, loaded, 1.91 m; rocket, 672 mm
Weight: launcher, 6.2 kg; launcher, loaded, 11.2 kg; rocket, 3.5 kg

90 mm M79 rocket launcher with rocket

90 mm M79 rocket launcher

Launch velocity: 250 m/s
Max range: 1,960 m
Operational range: AFVs, 350 m; other, 600 m
Armour penetration: 400 mm

Holding Company
YUGOIMPORT - SDPR, PO Box 89, Bulevar umetnosti 2, YU-11070 Novi.

Status
May no longer be in production.

Service
Former Yugoslav armed forces.

Licence production
MACEDONIA
Manufacturer: EUROINVEST
Type: RBR 90 mm M79
Remarks: Production probably not sanctioned. In production and offered for export sales.

VERIFIED

64 mm RBR-M80 light anti-armour weapon

Description
The 64 mm RBR-M80 is another near-copy of the M72 Light Anti-armour Weapon (LAW). The launch tube is of telescopic construction, with the HEAT rocket packed inside and sealed. The six-finned rocket consists of a motor using tubular propellant, a percussion igniter, and a shaped charge warhead initiated by a piezo-electric fuze. The shaped charge uses phlegmatised Octol (HMX), and the fuze is bore safe, arming between 6 and 20 m from launch, with self-destruct after six seconds of flight should it miss the target.

Data
Calibre: 64 mm
Length: travelling, 860 mm; firing, 1.2 m; rocket, 644 mm
Weight: complete, 3 kg; launcher, 1.42 kg; rocket, 1.58 kg
Max range: 1,280 m
Operational range: 250 m
Muzzle velocity: 190 m/s
Armour penetration: 300 mm

Holding Company
YUGOIMPORT - SDPR, PO Box 89, Bulevar umetnosti 2, YU-11070 Novi.

Status
May no longer be in production in Serbia and Montenegro.

Service
Former Yugoslav armed forces.

Licence production
MACEDONIA
Manufacturer: EUROINVEST
Type: 64 mm RBR M80
Remarks: Production probably not sanctioned. In production and offered for export sales.

VERIFIED

64 mm RBR-M80 LAW in firing position

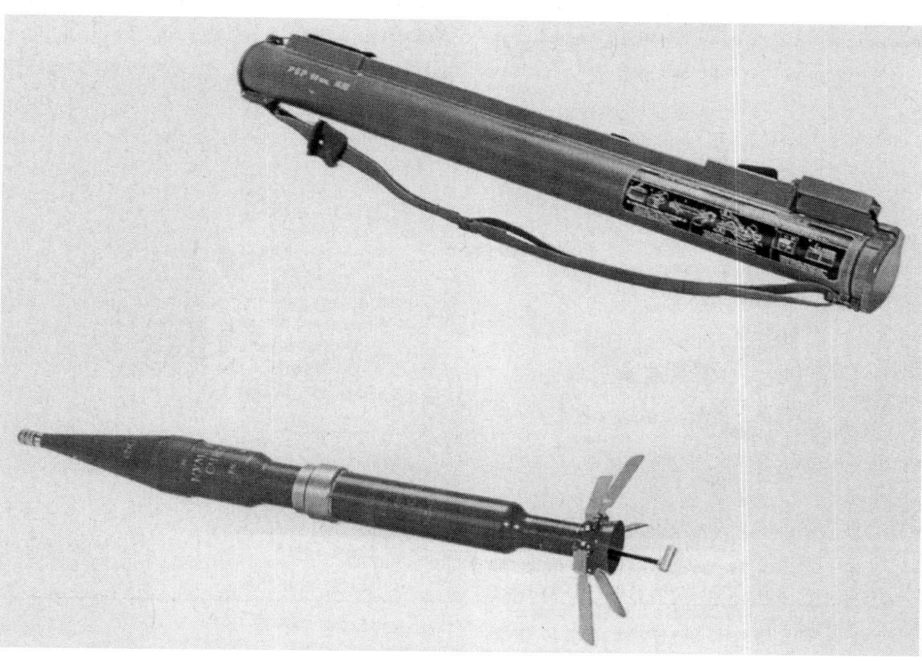

64 mm RBR-M80 LAW in carrying mode, with rocket

120 mm RBR M90 rocket launcher

Description
The 120 mm RBR M90 rocket launcher is a shoulder-fired weapon, designed to counter modern battle tanks from all angles, including head-on. It consists of a disposable man-portable launcher and a HEAT projectile propelled by an impulse-type sustainer.

The rocket is supplied prepacked in a glassfibre launching tube. The firing mechanism and an emergency open sight are already mounted on the launcher; a night sight can be fitted. The metal parts of the sustainer are of aluminium alloy.

The warhead, with HMX shaped charge, is initiated by a piezoelectric fuze. The pyrotechnic delayed arming safety device provides a muzzle safety distance of 8 to 25 m. It also ensures rocket self-destruction in case of target miss after four to six seconds of flight. The warhead fuze functions up to an incidence angle of 75°.

Training aids include a subcalibre launcher used with a special 7.62 mm tracer bullet, and a demonstration launcher and rocket set.

Data
Calibre: 120 mm
Length: launcher, 1.3 m; rocket, 950 mm
Weight: complete, 13 kg; launcher, 5.7 kg; rocket, 7.3 kg; explosive charge, 1.8 kg
Muzzle velocity: 205 m/s
Range: effective, 250 m
Time to 250 m: 1.3 s

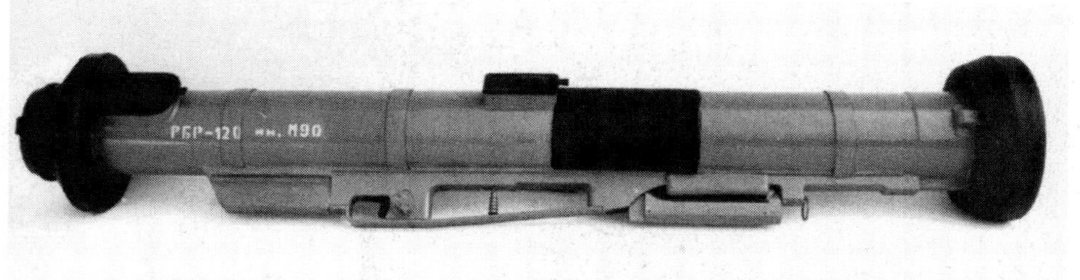

120 mm RBR M90 rocket launcher

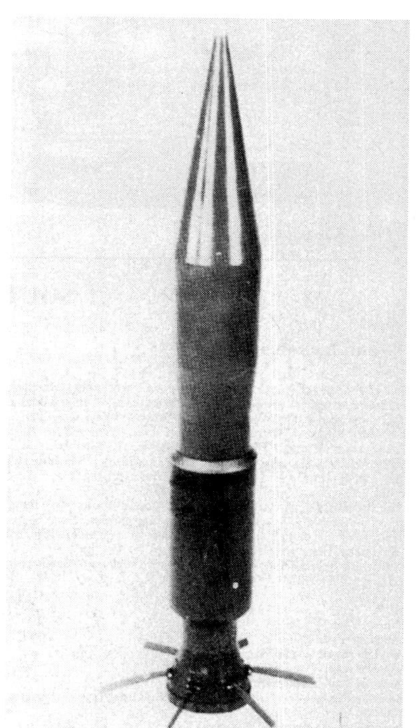

120 mm projectile for RBR M90 rocket launcher

Penetration: >800 mm RHA; >2 m reinforced concrete
Fuze graze angle: ≥15°
Dispersion: H+L ≤1.5 m
Operational temperature range: −31 to +51°C

Holding Company
YUGOIMPORT - SDPR, PO Box 89, Bulevar umetnosti 2, YU-11070 Novi.

Status
Probably not in production in Serbia or Montenegro.

Service
Former Yugoslav armed forces.

Licence production
MACEDONIA
Manufacturer: EUROINVEST
Type: RBR 120 mm M90
Remarks: Production probably not sanctioned. In production and offered for export sales.

VERIFIED

Maljutka anti-tank guided missile

Description
Known in Serbian as the Maljutka, this missile is generally similar to the Soviet 9M14 Malyutka (AT-3 Sagger) missile but with an enhanced performance warhead. There are two versions. One is manually guided and is used in conjunction with the Guidance Unit 9S415M when the two units constitute an MCLOS wire-guided missile system controlled by a joystick. The missile can be used at ranges from 500 to 3,000 m.

With the manually guided version the operator places the missile on its launch platform and connects the Guidance Unit. He visually observes the target sector, selects the target and then launches the missile by pressing the 'Launch' button. While continuously observing the target he guides it by operating the joystick, generating guidance corrections which are transmitted down the trailing wire. Relay and commutation logic enable selection of the launcher and preparation for launching; it is possible to connect a maximum of four launchers to one Guidance Unit. Maximum armour penetration of this version is 520 mm.

The late production of the Maljutka uses a SACLOS guidance system. This version features a standoff probe for the impact fuze to take full advantage of an enhanced shaped charge warhead providing armour penetration against homogeneous armour plate of over 600 mm.

Data
Weight: missile, 11 kg; launch platform, 7 kg; guidance unit, 12.5 kg
Max range: 3,000 m
Average flight velocity: 120 m/s
Armour penetration: manual, 520 mm; SACLOS, 600 mm
Preparation time: 2 min
Rate of fire: 2 missiles/min
Hit probability: 80%

Holding Company
YUGOIMPORT - SDPR, PO Box 89, Bulevar umetnosti 2, YU-11070 Novi.

Status
May no longer be in production.

Service
Former Yugoslav armed forces.

VERIFIED

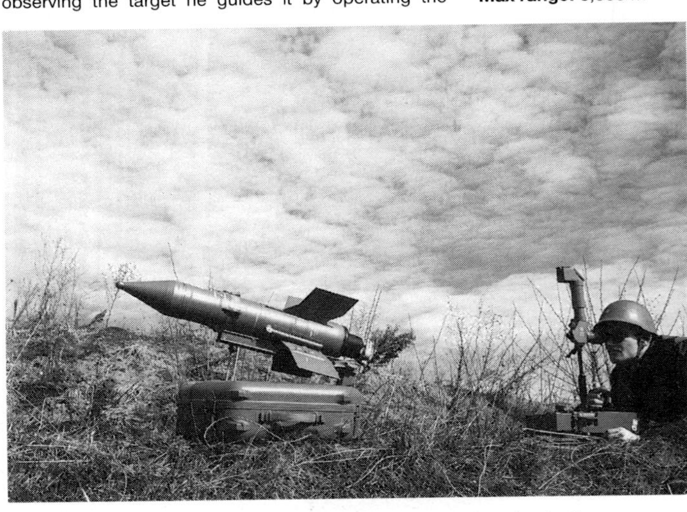

Manual guidance version of Maljutka anti-tank missile

SACLOS guidance version of Maljutka anti-tank missile

MORTARS

ARGENTINA

60 mm MC 1-60 FMK2 Mod 0 Commando mortar

Description

The 60 mm MC 1-60 FMK2 Mod 0 mortar was developed for commando groups, airborne troops and marines, and for any operation requiring fast movement and short-range engagements. The characteristics of this weapon are minimal weight, fast emplacement, high firepower, easy handling and reduced maintenance demands.

The mortar is of the standard drop-fired pattern and uses the sling, marked with ranges, as a fast method of laying. The firer places the baseplate on the ground, drops the sling ahead of the mortar, then places his foot on the desired range tag. Pulling the mortar up until the sling is tight sets the elevation accurately enough for the first shot. Alignment is done by a line painted on the barrel.

Data

Calibre: 60.75 mm
Length of barrel: 650 mm
Length, total: 685 mm
Weight: 5.68 kg
Rate of fire: 20 rds/min
Range: min, 100 m; max, 1,356 m

Manufacturer

Fabrica Militar Rio Tercero, Rio Tercero, Cordoba.

Status

Production complete.

Service

Argentine armed forces.

VERIFIED

60 mm MC 1-60 FMK2 Mod 0 Commando mortar

60 mm MA 1-60 FMK1 Mod 0 Assault mortar

Description

The 60 mm MA 1-60 FMK1 Mod 0 is an assault mortar for operations against armoured vehicles, field fortifications or other targets. The reduced size and weight allow this mortar to be easily carried by assault troops and rapidly brought into action.

The mortar is of an unusual design in which the baseplate and supporting bipod are attached permanently to the tube and fold up for transit, being quickly fitted to a special manpack carrier. The design also allows the mortar to be laid with the tube horizontal to permit firing a 'buckshot' cartridge for close-range anti-personnel defence; this cartridge projects 138 pellets, each weighing 3.67 g, to an effective range of 70 m.

The sight is an elbow telescope for elevations to 1,600 mils; it has two scales, one for quick approximation of range, the other for precision laying. The firing mechanism can be locked for drop-firing, set for trigger firing, or set to safe.

Data

Calibre: 60.75 mm
Length of barrel: 650 mm
Length, total: 770 mm
Weight: in firing position, 8 kg
Rate of fire: 20 rds/min
Range: normal mode, 2,200 m; horizontal mode with standard HE bomb, 250 m; with 'buckshot' cartridge, 70 m

Manufacturer

Fabrica Militar Rio Tercero, Rio Tercero, Cordoba.

Status

Production complete.

Service

Argentine armed forces.

VERIFIED

60 mm Assault mortar in carrying mode

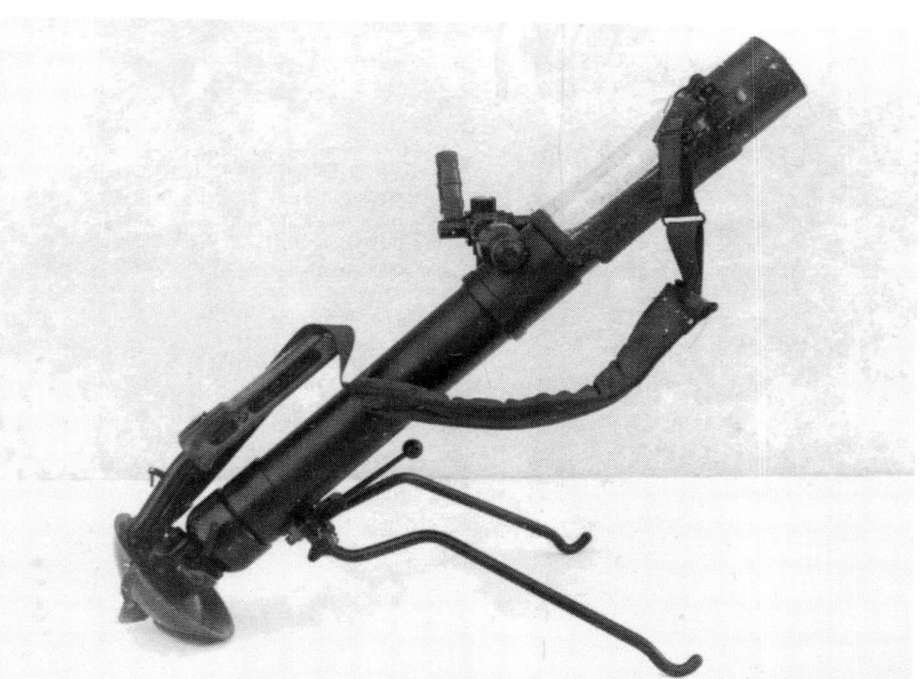

60 mm MA 1-60 FMK1 Mod 0 Assault mortar

60 mm MS 1-60 FMK3 Mod 0 Standard mortar

Description

The 60 mm MS 1-60 FMK3 Mod 0 Standard mortar is of Argentine design and manufacture and is intended for use as an infantry direct support mortar. It uses a tripod support which reduces recoil forces and thus prevents the mortar subsiding into soft ground. Accuracy is good, and combined with ease of use, simplicity, quick emplacement, low maintenance load and high rate of fire, provides a valuable support weapon for the infantry platoon. It can be carried complete by one man, but it can be dismantled into three loads each of which can be carried by one man using specially designed carrying packs.

Data

Calibre: 60.75 mm
Length of barrel: 650 mm
Length, total: 746 mm
Weight: in firing position, 15.57 kg
Weight of barrel: 3.58 kg
Weight of baseplate: 4.47 kg
Baseplate diameter: 342 mm
Tripod length: folded, 640 mm
Tripod weight: 5.15 kg
Elevation: 711-1,547 mils
Traverse: total, 300 mils
Rate of fire: 20 rds/min
Range: min, 100 m; max, 3,000 m

Manufacturer

Fabrica Militar Rio Tercero, Rio Tercero, Cordoba.

Status

Production complete.

Service

Argentine armed forces.

VERIFIED

60 mm MS 1-60 FMK3 Mod 0 Standard mortar

81 mm LR FMK2 Mod 0 mortar

Description

The 81 mm LR FMK2 Mod 0 mortar is a simple, mobile, light weapon able to engage targets with rapid precise fire. It can be used in any type of terrain, under all weather conditions and employed in any infantry role.

The design is conventional, using a baseplate and bipod; there appear to be two baseplates in use, one triangular, the other circular. The mortar is carried by three men, using specially designed carriers. It can also be mounted in various types of vehicle.

Data

Calibre: 81.4 mm
Length of barrel: 1.155 m
Length of barrel and breech: 1.28 m
Weight: in firing position, 42 kg
Weight of barrel: 13.75 kg
Weight of bipod: 13.65 kg
Weight of baseplate: 13.9 kg
Baseplate diameter: 550 mm
Elevation: 711-1,547 mils
Traverse: total, 167 mils
Rate of fire: 20 rds/min
Range: min, 100 m; max, 4,000 m

Manufacturer

Fabrica Militar Rio Tercero, Rio Tercero, Cordoba.

Status

Production complete.

Service

Argentine armed forces.

UPDATED

81 mm LR FMK2 Mod 0 mortar in firing position

81 mm mortar in carrying mode, with circular baseplate

120 mm LR FMK2 Mod 0 mortar

Description

The 120 mm LR FMK2 Mod 0 is a heavy support mortar, although, because of the design, it is light in relation to its power. It can be employed in any infantry role and in any terrain or weather conditions.

Of conventional tube/baseplate/bipod form, the mortar is normally transported on a two-wheeled trailer which attaches to the baseplate and allows towing by any motor vehicle. In conditions where this is not possible, the weapon can be dismantled and the three basic components man-carried for short distances. There are two baseplates in use, one triangular, the other circular. This mortar can also be animal-packed, air-dropped, or mounted in vehicles using special platforms. The firing mechanism permits drop or trigger firing.

Data

Calibre: 120.18 mm
Length of barrel: 1.5 m

120 mm LR FMK2 Mod 0 mortar

120 mm mortar in travelling mode

Length of barrel and breech: 1.663 m
Weight: in firing position, 116 kg
Weight of barrel and breech: 44.2 kg
Weight of baseplate: 45 kg
Weight of bipod: 25.5 kg
Elevation: 800-1,512 mils
Traverse: total, 177 mils
Rate of fire: 12-15 rds/min
Range: min, 500 m; max, 6,150 m

Manufacturer
Fabrica Militar Rio Tercero, Rio Tercero, Cordoba.

Status
Production complete.

Service
Argentine armed forces.

UPDATED

120 mm mortar in manpack mode

AUSTRIA

Hirtenberger 60 mm C6 Commando mortar

Description
The Hirtenberger 60 mm C6 mortar is a light mortar designed for easy carriage, especially in difficult terrain such as mountains or jungle. There are two models: the C6-110 has a fixed firing pin, while the C6-210 has a self-cocking firing pin controlled by a trigger built into a special handgrip.

The C6 mortar consists of the following parts: barrel with breech piece and firing pin assembly; baseplate; handgrip and trigger mechanism; and a set of auxiliaries and spare parts.

The barrel, of high-strength forged special steel, is provided with a collar at the muzzle, and is externally threaded at the rear end to receive the breech piece. The muzzle is bell-mouthed to facilitate loading. There is a protective sleeve of heat-resistant plastic, about 150 mm in length, shrunk on to the barrel behind the muzzle.

The breech piece is steel, internally threaded to mate with the barrel, and has a tapered sealing face in order to ensure a gastight seal. There is a central hole into which the firing pin assembly is mounted, and four threaded holes for attachment of the trigger mechanism. During factory assembly of the mortar the breech piece is screwed tightly to the barrel, the threads being treated with a special adhesive - the two cannot be separated in the field.

The C6-110 firing pin is a solid assembly fitted into the hole in the breech piece and retained by a cover. The C6-210 firing pin is a spring assembly which is retained in the breech piece by the trigger mechanism housing. The trigger assembly consists of an aluminium housing into which the stainless steel (and hence maintenance-free) mechanism is fitted. Operation of the firing pin is controlled by a thumb-actuated trigger set into the carrying handle part of the mechanism housing. There is a cross-bolt safety catch which, when applied, locks the mechanism. An advantage of this trigger firing system is that the mortar can be fired at flat trajectories.

The baseplate is circular and made of high tensile aluminium alloy. It carries a central stainless steel spike which serves to retain the baseplate to the breech piece as well as anchoring the mortar securely on hard ground.

The carrying strap is attached to the barrel and breech piece. It serves as a carrying sling and is marked with ranges; it is used by allowing the sling to drop on the ground when the mortar is supported in the firing position. The firer then kneels with his boot on the desired range mark and elevates the mortar until the vertical portion of the strap is taut. There is also an optional elevation indicator which can be clamped to the barrel.

The auxiliaries and spares consist of cleaning rod and brush, screwdrivers, spanners, oil can and a spare firing pin assembly of the appropriate type.

Data
Calibre: 60.7 mm
Length, total: 815 mm
Length of bore: 640 mm
Outer diameter: 72 mm
Weight of mortar: C6-210, 5.1 kg; C6-110, 4.3 kg
Weight of auxiliaries: 1.2 kg
Rate of fire: max, 30 rds/min
Range: max, 1,600 m

Manufacturer
Hirtenberger AG, A-2552 Hirtenberg.

Status
In production.

Service
Overseas armies; entered Austrian Army service in 1990.

VERIFIED

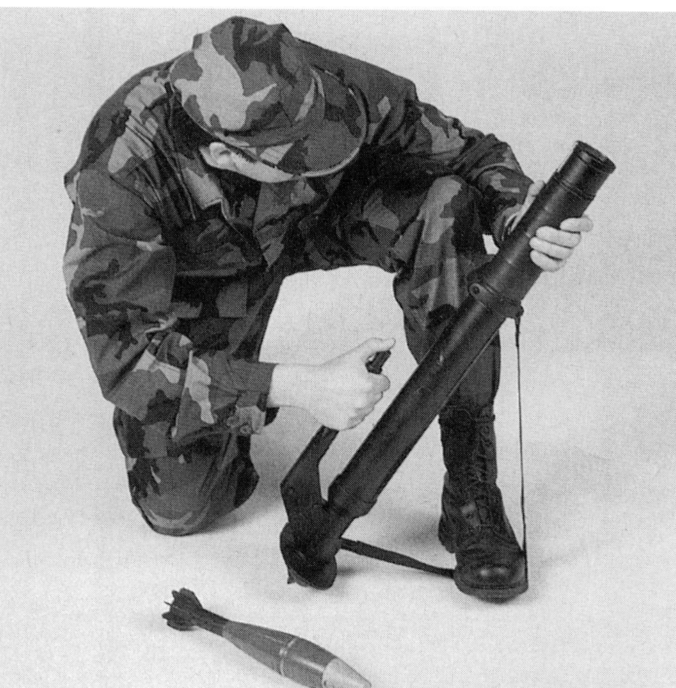

Hirtenberger 60 mm C6-110 mortar in firing position

Hirtenberger 60 mm C6-210 mortar in firing position

Hirtenberger 81 mm M8-111 and M8-211 medium mortars

Description

The Hirtenberger 81 mm M8-111 medium mortar was developed by Böhler Special Products for the Austrian Army, with whom it is in service as the 8.1 cm Granatwerfer 82, and with other armies. It is available in two models: the M8-111 standard version; and the M8-211 long-range. Two further models, the 82 mm M8-311 and M8-411 are apparently no longer offered.

The M8 barrel is forged from electro-slag refined steel, is smoothbored with a bottleneck at the breech end, and has a muzzle bell-mouthed to facilitate loading. There are radial fins on the rear section which add strength and aid cooling. The bipod barrel clamp is mounted on the cylindrical part of the barrel between the cooling ribs and an annular collar. The stainless steel breech piece seals the rear of the barrel and supports the barrel on the baseplate. A copper washer between barrel and breech piece provides a gastight seal. To facilitate assembly and disassembly for cleaning the barrel has two flats formed on it and the breech piece has a transverse hole. The breech piece is bored with an oblique threaded hole with tapered sealing face, to hold the firing pin. Because of this inclined position the firing pin can be removed and replaced when the mortar is in the firing position or can be completely retracted in the event of a misfire.

The bipod assembly consists of a barrel clamp, twin shock absorbers, traversing mechanism, sight bracket, cross-levelling mechanism, elevating mechanism and bipod legs. One leg of the asymmetrical bipod can be adjusted and locked in any desired position. The bipod is surmounted by the elevation mechanism, which is a screw device operated by a handwheel. By adjustment of the position of the barrel clamp on the barrel, in conjunction with the elevating mechanism, any desired elevation can be easily achieved.

The baseplate is a die-forged circular aluminium alloy unit, with ribs and spades on its undersurface. A stepped recess in the upper side holds the pivoted ball socket, which is secured by a retaining ring. Beneath the ball are damping rings and discs which absorb some of the recoil blow. A lateral opening in the ball socket allows insertion of the barrel and also permits lowering the barrel to an elevation of about 700 mils (39°). When the barrel is moved in azimuth the ball socket moves with it, and thus all-round firing is possible without repositioning the baseplate.

The auxiliaries and spare parts kit contains cleaning and maintenance equipment and tools, spare firing pin, breech piece washer and baseplate components, two aiming posts and two aiming post lamps.

All standard types of 81 mm bombs can be fired from these mortars, though optimum results will be obtained by using the HE 70 range of bombs produced by Hirtenberger.

Preparing the Hirtenberger 81 mm M8-111 mortar for action

Data

Calibre: 81 mm
Length of barrel: M8-111, 1.28 m; M8-211, 1.48 m
Weight in firing position: M8-111, 35.6 kg; M8-211, 36.6 kg
Weight of barrel: M8-111, 12 kg; M8-211, 13 kg
Weight of bipod: 11.6 kg
Weight of baseplate: 12 kg
Weight of sight unit: 1.35 kg
Baseplate diameter: 546 mm
Elevation: +39 to +84° (700-1,500 mils)
Range: min 100 m; max with HE 70 LD, standard, 6,300 m, long range, 6,600 m
Rate of fire: max, 25 rds/min; practical, 20 rds/min; sustained, 6 rds/min

Manufacturer

Hirtenberger AG, A-2552 Hirtenberg.

Status

In production.

Service

Austrian and other armies.

UPDATED

Hirtenberger 81 mm M8-111 mortar

Hirtenberger 81 mm M8-522 medium mortar

Description

The Hirtenberger 81 mm M8-522 medium mortar offers the same performance as the 81 mm M8-111 described previously but is of more conventional and somewhat heavier construction.

The barrel is externally smooth and without cooling fins and is made of special electro-slag refined steel. The breech piece is screwed and glued to the barrel but can be removed in the field if necessary. The bipod is of symmetrical pattern with the elevating gear mounted in a central tube and the traversing gear attached to the top, working on the barrel clamp. Two shock absorbers prevent the transmission of recoil force to the bipod during firing. The baseplate is similar to that of the M8-111 series. Auxiliaries and spares to the same specification as the M8-111 are provided (see previous entry).

Data

Calibre: 81 mm
Length of barrel: 1.28 m
Weight: in firing position, 42.1 kg
Weight of barrel: 13.5 kg
Weight of bipod: 15.6 kg
Weight of baseplate: 13 kg
Baseplate diameter: 546 mm
Range: with HE 70 LD, max, 6,400 m
Rate of fire: max, 30 rds/min

Manufacturer

Hirtenberger AG, A-2552 Hirtenberg.

Status

Available.

VERIFIED

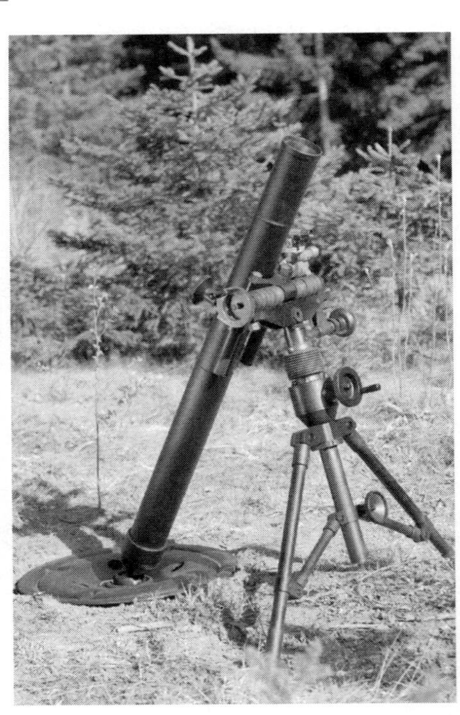

Hirtenberger 81 mm M8-522 medium mortar

Hirtenberger 120 mm M12-1111 heavy mortar

Description

The Hirtenberger 120 mm M12-1111 heavy mortar was jointly developed by Böhler Pneumatik and the Austrian Army, with whom it entered service in February 1985 as the 12 cm Granatwerfer 86. It is normally mounted on a towed carriage but can also be fitted into armoured vehicles of the M113/M125 type, when it is known as the M12-2330. Almost all types of 120 mm bombs can be fired from this mortar.

The barrel is of electro-slag refined steel. The muzzle is bell-mouthed to facilitate loading and the barrel clamp mounts between two collars. There is a carrying handle mounted towards the rear end. The breech piece is screwed into the barrel and prevented from rotating by a safety lock; a patented sealing washer between barrel and breech piece ensures a gastight seal. The firing mechanism, which is contained in the breech piece, can be adjusted to fire in fixed or trigger-actuated modes. For safety reasons the change to the fixed firing pin mode requires the use of two hands. There is an external trigger lever to which a firing lanyard can be attached, and there is also an external safety lever which withdraws and locks the firing pin when applied.

The bipod is of symmetrical type with a central elevating screw and a screw traversing mechanism attached to the barrel clamp. There are two shock absorbers, using hollow rubber springs, which attenuate the recoil stresses to both bipod and baseplate.

The baseplate is of sectional welded pattern. Handles on the upper side serve also as mounting devices for the carriage lifting gear. The bottom side has six radial spade ribs which provide stability on any type of ground. A rapid lock permits quick removal of the barrel from the baseplate.

The carriage is a suspended single-axle carriage of solid-frame construction. By use of a cable winch and tilting frame, the complete mortar can be either lowered into the firing position or pulled on to the carriage for transport. There is a locking mechanism which fastens the barrel and bipod in the travelling position. The drawbar is adjustable for height and there is a front castor wheel which aids manoeuvring and can be retracted when towing. The carriage frame is provided with hooks to facilitate helicopter lifting and handgrips for manhandling.

The auxiliaries and spare parts kit contains all the necessary cleaning and maintenance equipment and tools, spare firing pin and sealing washer, and two aiming posts and lamps.

Hirtenberger 120 mm M12-1111 mortar in firing position

Variants

M12-3222 and M12-2222

The 120 mm M12-3222 and M12-2222 are updated versions of the original M12-1111 capable of firing the later HP-HE 89 LD bombs but involving an overall decrease in weight. The M12-3222 in the firing position weighs 260 kg and uses a 1.9 m long barrel to achieve a maximum range of 9,500 m. The M12-2222 in the firing position weighs 255 kg and uses a 1.75 m long barrel to achieve a maximum range of 9,000 m.

M12-2330 APC version

The M12-2330 APC version uses a shorter barrel and instead of a bipod has a special support which consists of two tubular legs which carry the normal elevating gear in the centre but which end in rollers travelling on an arcuate racer set into the floor of the vehicle. Clamps are provided for locking the support at any desired point in the traverse. The rear end of the barrel fits into a ball socket of the usual type which is assembled into a heavy tubular member bolted to the vehicle sidewalls. There is sufficient freedom of movement in the ball and socket to accommodate the full range of elevation and traverse. Sighting, elevation and traversing mechanisms are the same as those for the towed mortar, though traverse is restricted by the length of the racer to about 45°.

The mortar is supplied with a bipod and standard baseplate, and can be removed from the vehicle and brought into action on the ground if desired. The bipod differs from standard in having no elevating gear; that from the APC mounting is removed with the barrel and fitted into a prepared fork in the bipod.

Hirtenberger 120 mm M12-1111 mortar on travelling carriage

Data

M12-1111
Calibre: 120.2 mm
Length of barrel: 1.9 m
Length of bore: 1.56 m
Weight: in firing position, 280 kg; travelling, 670 kg
Weight of barrel: 100 kg
Weight of bipod: 60 kg
Weight of baseplate: 117 kg
Weight of sight unit: 1.4 kg
Baseplate diameter: 1.11 m
Length of carriage: 3.7 m
Width of carriage: 1.71 m
Range: with HP-HE 78, max, 9,000 m
Rate of fire: max, 12 rds/min; sustained, 7 rds/min

Manufacturer

Hirtenberger AG, A-2552 Hirtenberg.

Status

In production.

Service

Austrian Army.

UPDATED

BELGIUM

PRB 60 mm NR 493 mortar

Description

The PRB 60 mm NR 493 mortar is a muzzle-loaded smoothbore mortar of conventional design with a fixed firing pin. It can fire all standard 60 mm ammunition but is particularly effective when used with the special NR 431 round.

Made of special steel, the barrel is a smoothbore tube of 60.7 mm diameter closed at one end by the breech with its fixed firing pin. The breech has a spherical projection which mates with the baseplate socket and there is a carrying belt attached at one end to this projection and at the other to the barrel near the muzzle. The belt carries a protective cap which is placed on the muzzle when the mortar is not in use.

The bipod carries the elevation and traverse mechanisms and the sight mounting and is fitted with two mechanical shock absorbers and a strap fastener. It is robustly made and the sliding parts are enclosed in sealed protective sheaths to prevent corrosion and provide permanent lubrication.

The baseplate is circular and dish-shaped. A socket on the upper face accepts the breech projection and is locked by a lever. On the underside are three ribs, one of which serves as an anchoring spade. The rim of the plate is perforated to ease extraction from the ground after firing.

The sight unit comprises a collimator sight, and an open sight for use in emergency, cross-levelling gauges and sighting controls. The deflection scale has 64 intervals of 100 mils each and the deflection knob is graduated in 50 intervals of 2 mils with the 10 mil intervals numbered. The elevation scale has nine 100 mil graduations (numbered every 200 mils) and the elevation knob is graduated in 50 intervals of 2 mils with the 10 mil intervals marked.

In normal service the mortar will be carried and operated by three men. In emergency, however, it is possible for one man to carry and fire the weapon.

PRB 60 mm NR 493 mortar

Data

Calibre: 60 mm
Length of barrel: 780 mm
Weight: in firing position, 22.1 kg
Weight of barrel: 6 kg
Weight of bipod: 10.4 kg
Weight of baseplate: 5.1 kg
Baseplate diameter: 320 mm
Traverse: on bipod, 135 mils
Range: max, with NR 431 bomb, 1,800 m

Manufacturer

Société Anonyme PRB, Département Défense, avenue de Tervueren 168, Bte 7, B-1150 Brussels.
This company is no longer trading.

Status

Production complete.

Service

Belgian Army.

VERIFIED

PRB 81 mm NR 475 A1 medium mortar

Description

The PRB 81 mm NR 475 A1 medium mortar was designed for use by infantry with either US or PRB bombs and especially with the NR 436 round. It is suitable for transport by land vehicle or helicopter or by pack animal. It may also be carried by three soldiers and is robust enough to be dropped by parachute.

The barrel is a smoothbore tube with radial fins on the lower half and is terminated by a breech piece in which is inserted a fixed firing pin. The exterior of the breech piece is terminated by a ball which engages in a socket in the baseplate.

The baseplate is circular and has four welded ribs and five spikes on the underside. Holes are provided to ease extraction from the ground. A socket in the centre of the plate accepts the ball projection from the breech piece and has a clamping device with a securing screw. Carrying handles are attached to the sides.

The bipod legs have spiked feet and are distanced by an adjustable spring-loaded chain. They are connected to the elevating mechanism which consists of an elevating screw rotating in a nut and driven through a gear mechanism. The elevating rod is terminated by the traversing mechanism which comprises the yoke, the traversing drive and two shock absorbers. Cross-levelling is effected by roughly adjusting the cross-levelling clamp on the left bipod leg and then completing the adjustment using the screw mechanism between the clamp and the elevating mechanism housing.

The BR1 sight unit is mounted at the left end of the yoke by a dovetail bracket and comprises a collimator, elevating and deflecting mechanisms and longitudinal and cross-levels. The elevation scale is graduated in seven intervals of 100 mils, numbered at the even numbers from 8(00) to 16(00) and the elevation knob is graduated in 2 mil intervals and numbered every 10 mils from 0 to 90. The deflection scale is graduated in 100 mil intervals, and numbered from 0 to 64(00) and the deflection knob is graduated in the same way as the elevation knob.

Standard accessories for the mortar include two aiming posts and aiming-post lamps.

Data

Calibre: 81 mm
Length of barrel: 1.35 m
Weight: in firing position, 43 kg
Weight of barrel: 15.3 kg
Weight of bipod: 12.5 kg
Weight of baseplate: 14.6 kg
Weight of sight unit: 600 g
Baseplate diameter: 500 mm
Elevation angle: 700-1,515 mils
Traverse: on bipod at 60° elevation, 140 mils
Range: min, 300 m; max, using NR 436 bomb and max charge, 5,500 m
Rate of fire: 15-20 rds/min

Manufacturer

Société Anonyme PRB, Département Défense, avenue de Tervueren 168, Bte 7, B-1150 Brussels.
This company is no longer trading.

Status

Production complete.

PRB 81 mm NR 475 A1 medium mortar

Service

Belgian Army.

VERIFIED

BULGARIA

PIMA 82 mm Mortar M82

Description

As well as producing several types of mortar with Soviet/CIS design origins, Bulgaria has introduced an indigenous 82 mm mortar known as the PIMA 82 mm Mortar M82. Relatively few details have been released regarding this mortar but it appears to incorporate conventional design features from several similar weapons.

The M82, with a crew of four, has a maximum range of 4,100 m firing a 3.1 kg HE fragmentation bomb with a lethal radius of 30 m against standing personnel. Barrel recoil is absorbed by a single recoil cylinder coupled with a circular baseplate.

The MUM telescopic sight employed is provided with a lighting system for use at night.

Data

Calibre: 82 mm
Elevation: +45 to +85°
Traverse: on bipod, 6°; by moving bipod, 30°
Range: min, 80 m; max, 4,100 m
Rate of fire: up to 30 rds/min
Bomb weight: FRAG-HE, 3.1 kg

Manufacturer

PIMA AEG, 3400 Montana.

Status

In production. Offered for export sales.

NEW ENTRY

PIMA 82 mm Mortar M82
1996

CHILE

60 mm Commando mortar

Description

This 60 mm Commando mortar is a conventional smoothbore drop-fired 60 mm mortar which appears to be of the Brandt Commando Type V pattern. The firing pin is fixed. Laying for direction is done visually, by means of an axis line painted on the mortar barrel, and laying for range is done either by a simple level sight or by using the carrying sling. By laying the sling on the ground and trapping it there with one foot on a range marker attached to the sling, the mortar can be given a reasonably accurate elevation sufficient for an opening round.

A variant model of this mortar has a trip-operated firing pin which can be withdrawn and set to safe if required.

Ammunition for this mortar are licence-manufactured copies of the Brandt 60 mm Mark 61 series.

Data

Calibre: 60 mm
Length of barrel: 650 mm
Length, total: 680 mm
Weight: 7.7 kg
Range: max, 1,050 m

Manufacturer

FAMAE Fabricaciones Militares, av Pedro Montt 1568/1606, Santiago.

Status

In production.

FAMAE 60 mm Commando mortar (T J Gander)

Service

Chilean Army.

VERIFIED

FAMAE 81 mm mortar

Description

Although referred to by the manufacturer as 'Brandt type' the FAMAE 81 mm mortar has some features which appear to be of Chilean design, though there are distinct leanings towards the Esperanza form of tripod. This is a conventional smoothbore mortar with fixed firing pin. The baseplate is a distinctive hexagonal shape with prominent spades and the whole equipment is rather heavier than is usual for this class of weapon.

The ammunition fired with this mortar is a licence-manufactured copy of the Brandt 81 mm M57 series.

Data

Calibre: 81 mm
Weight of barrel: 20.6 kg
Weight of tripod: 17.2 kg
Weight of baseplate: 28 kg
Range: min, 100 m; max, 4,200 m

Manufacturer

FAMAE Fabricaciones Militares, av Pedro Montt 1568/1606, Santiago.

Status

In production.

Service

Chilean Army.

VERIFIED

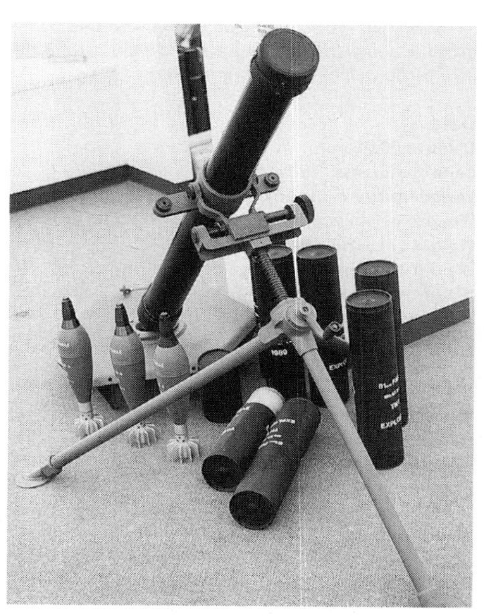

FAMAE 81 mm mortar (T J Gander)

FAMAE 120 mm mortar

The FAMAE 120 mm mortar has some features reminiscent of the Spanish Esperanza Model L, though it is considerably lighter. The mortar is a conventional drop-fired smoothbore supported by a circular baseplate and a tripod. It is moved by dismantling and fitting the three basic components on to a light two-wheeled trailer which can be vehicle-drawn or pulled by the mortar crew.

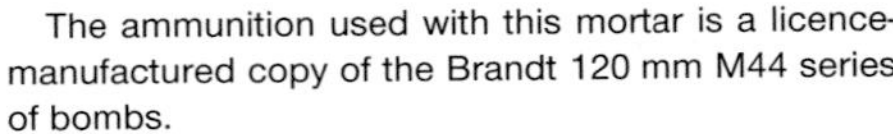

The ammunition used with this mortar is a licence-manufactured copy of the Brandt 120 mm M44 series of bombs.

Data

Calibre: 120 mm
Weight: travelling, 189 kg
Weight of barrel: 40 kg
Weight of tripod: 25 kg
Weight of baseplate: 44 kg
Weight of trailer: 80 kg
Range: min, 500 m; max, 6,650 m

Manufacturer

FAMAE Fabricaciones Militares, av Pedro Montt 1568/1606, Santiago.

Status

In production.

Service

Chilean Army.

VERIFIED

FAMAE 120 mm mortar

FAMAE 120 mm mortar on transport trailer

CHINA, PEOPLE'S REPUBLIC

NORINCO 60 mm Type M-83A

Description

The NORINCO 60 mm Type M-83A is a modern, lightweight weapon with good performance. It is intended for use by infantry companies, airborne forces and special operations groups where a combination of low weight and firepower is desirable.

The design is conventional, using a high-tensile steel barrel, a bipod and a baseplate. The barrel provides high strength with light weight and is externally ribbed to provide a heat-dissipating surface which also strengthens the tube. The barrel is screwed into the breech piece and there is a ground and lapped internal gastight joint. The breech piece contains a fixed firing pin.

A Type 63 (MS2) optical sight is mounted on the left end of the traversing slide. There is a carrying handle on the barrel and also a small pneumatic recoil buffer which takes some of the firing shock off the bipod and sight.

The 60 mm Type M-83A fires a Type M-83A HE bomb made of spheroidal graphite cast iron with a steel tail unit; bomb weight complete is 1.33 kg.

Data

Calibre: 60.75 mm
Length of barrel: 790 mm
Length of barrel assembly: 840 mm
Weight: in firing position, 14.7 kg
Weight of barrel assembly: 5.1 kg
Weight of mounting: 5.1 kg
Weight of baseplate: 4.1 kg
Baseplate diameter: 330 mm
Max chamber pressure: 350 kg/cm^2
Elevation: +45 to +85°
Traverse: on mount, 7°
Muzzle velocity: min, 80 m/s; max, 204 m/s
Range: min, 72 m; max, 2,655 m
Rate of fire: max, 30 rds/min
Bomb weight: M-83A HE, 1.33 kg

NORINCO 60 mm Type M-83A mortar

Manufacturer

China North Industries Corporation (NORINCO), 7A Yue Tan Nan Jie, PO Box 2137, Beijing.

Status

In production.

Service

Chinese Army.

VERIFIED

NORINCO 60 mm Type 63-1 mortar

Description

The NORINCO Type 63-1 light mortar is essentially an updated model of the earlier Type 31 mortar. It generally resembles the Model 83 described previously but is lighter, its baseplate is smaller and rectangular in shape and the bipod and mortar barrel are shorter. One design feature of the Type 63-1 is the provision of a carrying handle to enable the entire mortar, including the baseplate and bipod, to be carried over short distances by one person.

Data

Calibre: 60.75 mm
Length of barrel: 610 mm
Weight: in firing position, 11.5 kg
Muzzle velocity: 141 m/s
Range: min, 95 m; max, 1,550 m
Rate of fire: 15-20 rds/min

Manufacturer

China North Industries Corporation (NORINCO), 7A Yue Tan Nan Jie, PO Box 2137, Beijing.

Status

Available.

Service

Chinese armed forces and others.

Licence production

EGYPT
Manufacturer: Helwan Machine Tools Company, Factory 999
Type: 60 mm light mortar
Remarks: In production. See entry under *Egypt* for details.
PAKISTAN
Manufacturer: Pakistan Machine Tool Factory Limited
Type: 60 mm light mortar
Remarks: Some local modifications introduced. In production. See entry under *Pakistan* for details.

VERIFIED

NORINCO 60 mm Type 63-1 mortar

NORINCO 60 mm Type WX90 long-range mortar

Description

The NORINCO Type WX90 long-range mortar was originally known as the Type W89 and was designed as a lightweight company mortar with a relatively long high strength alloy steel barrel. Aluminium alloy is used for some parts of the mounting so that no part of the disassembled weapon weighs more than 9 kg for shoulder or manpack transport when the weapon is used by airborne or special troops. The baseplate is circular.

The 60 mm WX90 can fire a special Type WX90 long range bomb weighing 2.3 kg to a maximum range of 5,500 m; muzzle velocity is 314 m/s. Also available are a conventional HE bomb, Smoke and Illuminating bombs.

Data

Calibre: 60.75 mm
Length of barrel: 1.2 m
Weight: in firing position, 23 kg
Elevation: +45 to +85°
Traverse: ±4°
Max chamber pressure: 689 kg/cm^2
Muzzle velocity: long-range bomb, 314 m/s
Range: max, 5,500 m
Rate of fire: max, 30-35 rds/min

Manufacturer

China North Industries Corporation (NORINCO), 7A Yue Tan Nan Jie, PO Box 2137, Beijing.

Status

Available.

Service

Chinese armed forces and others.

VERIFIED

NORINCO 60 mm Type WX90 long-range mortar

NORINCO 60 mm Type WW90 long-range mortar system

Description

There are two NORINCO 60 mm Type WW90 mortars with high strength alloy steel barrels of differing lengths to suit various tactical applications. The mortars are the Type WW90-60L with a barrel 1.3 m long, and the Type WW90-60M, with a barrel length of 1.08 m. Performances vary with the barrel length.

On both types a new design of cylindrical buffer is employed to keep the weapon stable when firing. A self-locking mechanism is provided in the elevating mechanism to keep elevation angles constant during prolonged firing while the baseplate/barrel assembly junction utilises a special linkage for easier operation. The baseplate is constructed using a heat-treated steel alloy of high strength.

The Type WW90 mortars can fire the same special long-range ammunition as the 60 mm Type WX90 mortar (see previous entry) but to ensure maximum range using the length of the Type WW90-60L barrel a special extra charge can be employed to obtain a maximum range of 6,000 m - maximum full charge range using the same barrel is 5,775 m. Apart from the special 60 mm ammunition all other types of 60 mm bombs can be utilised with the Type WW-90 mortars.

Data

Calibre: 60.75 mm
Length of barrel: Type WW90-60L, 1.3 m; Type WW90-60M, 1.08 m
Weight in firing position: Type WW90-60L, 21.5 kg; Type WW90-60M, less than 20 kg
Weight of barrel: ca 8 kg
Weight of mounting: 7.3 kg
Weight of baseplate: 6.2 kg
Elevation: +45 to +85°
Traverse: on carriage, ±3.5°; full, 360°
Range: max, Type WW90-60L, special charge, 6,000 m, full charge, 5,775 m; max, Type WW90-60M, special charge, 5,000 m, full charge, 4,400 m
Dispersion: full charge, WW90-60L, 0.5 × 0.25%; full charge, WW90-60M, 0.4 × 0.2%
Rate of fire: max, 30-35 rds/min

Manufacturer

China North Industries Corporation (NORINCO), 7A Yue Tan Nan Jie, PO Box 2137, Beijing.

Status

Available.

Service

Chinese armed forces and others.

VERIFIED

NORINCO 60 mm Type WW90 long-range mortar

NORINCO 82 mm Type 67 mortar

Description

The NORINCO 82 mm Type 67 mortar replaced the earlier Type 53, itself a copy of the Soviet 82 mm M-37 design. The Type 67 is of conventional pattern, a high-strength tube carried on a triangular baseplate and a bipod. The tube screws into a breech piece in which is the firing pin; firing can be done by the usual drop method or by a lanyard, the pin being set to a self-cocking action.

Data

Calibre: 82 mm
Weight: in firing position, 35 kg
Muzzle velocity: 70-211 m/s
Range: min, 85 m; max, 3,040 m

Manufacturer

China North Industries Corporation (NORINCO), 7A Yue Tan Nan Jie, PO Box 2137, Beijing.

Status

Available.

Service

Chinese armed forces.

VERIFIED

NORINCO 82 mm Type 67 mortar

NORINCO 81 mm Type W87 mortar

Description

The NORINCO 81 mm Type W87 mortar may be regarded as an updated version of the earlier 82 mm Type 67 (see previous entry) with a revised sleeve-type buffer arrangement and a change of calibre to render the weapon more attractive for export sales. A new 81 mm ammunition family was introduced to be used with the Type W87. If required, the Type W87 can fire all existing types of 81 mm ammunition.

Apart from the sleeve-type buffer arrangement, the 81 mm Type W87 is entirely conventional. The barrel is high strength alloy steel and the elevating mechanism has a one-way locking catch. A sleeve-type buffer is provided. The baseplate has a trapezoidal configuration. The mortar is equipped with a Type ZMJ-7 light-emitting diode illuminating device for night operations.

Maximum range of the Type W87 using a Type 87 HE bomb is 5,700 m.

Data

Calibre: 81.4 mm
Weight: in firing position, 39.7 kg
Weight of barrel: 15.5 kg
Weight of mounting: 8.7 kg
Weight of baseplate: 15.5 kg
Weight of sight unit: 0.9 kg
Elevation: +45 to +85°
Traverse: on mounting, 45°, ±3.5°
Range: Type 87 HE, min, 120 m; max, 5,700 m
Rate of fire: max, 20 rds/min

Manufacturer

China North Industries Corporation (NORINCO), 7A Yue Tan Nan Jie, PO Box 2137, Beijing.

Status

Available.

UPDATED

NORINCO 81 mm Type W87 mortar

NORINCO 81 mm Type W91 long-range mortar

Description

The NORINCO 81 mm Type W91 mortar may be regarded as a long-range version of the Type W87 (see previous entry) retaining the sleeve-type buffer arrangement but changing the baseplate shape to circular. The barrel length is 1.65 m and it is claimed that, using special ammunition, the maximum range possible is 8,000 m.

As well as the retention of the sleeve buffer, the Type W91 has a self-locking device in the elevation mechanism to maintain elevation during prolonged firing.

The Type W91 can fire all existing types of 81 mm ammunition but a special HE bomb weighing 5.72 kg is available. Also available are Smoke, Illuminating and fragmentation bombs containing preformed anti-personnel fragments.

Data

Calibre: 81.4 mm
Length of barrel: 1.65 m
Weight: in firing position, 65 kg
Weight of barrel: 26.7 kg
Weight of mounting: 12.6 kg
Weight of baseplate: 25.7 kg
Weight of sight unit: 0.9 kg
Elevation: +45 to +85°
Traverse: on mounting, 45°, ±3.5°; full, 360°
Range: max, 8,000 m
Rate of fire: 25 rds/min

Manufacturer

China North Industries Corporation (NORINCO), 7A Yue Tan Nan Jie, PO Box 2137, Beijing.

Status

Available.

VERIFIED

NORINCO 81 mm Type W91 mortar

NORINCO 100 mm Type 71 mortar

Description

The NORINCO 100 mm Type 71 mortar introduced a new calibre, one not previously used by China or, indeed, anywhere else. The Type 71 is of conventional pattern with tube, bipod and triangular baseplate. The size is such that the three principal components can be man-carried or animal-packed, or the complete weapon can be carried on any light vehicle.

The Type 71 fires a Type 71 HE bomb weighing 8 kg and containing 961 g of TNT/Dinal. Also available are Smoke and Illuminating bombs.

Data

Calibre: 100 mm
Weight: in firing position, 74.5 kg
Muzzle velocity: 250 m/s
Range: min, 170 m; max, 4,750 m
Rate of fire: 15-20 rds/min

Manufacturer

China North Industries Corporation (NORINCO), 7A Yue Tan Nan Jie, PO Box 2137, Beijing.

Status

Available.

VERIFIED

NORINCO 100 mm Type 71 mortar

NORINCO 120 mm Type 55 mortar

Description

The NORINCO 120 mm Type 55 mortar appears to be an improved version of the earlier Type 53, which itself was a Chinese copy of the Soviet M1943. The transporting trailer is simplified and strengthened, though at the expense of an increase in weight. The general dimensions and performance remain the same as the M1943.

Data

Calibre: 120 mm
Length of barrel: 1.85 m
Weight: in firing position, 275 kg; travelling, 550 kg
Elevation: +45 to +80°
Traverse: on carriage, ±4°; total, 360°
Muzzle velocity: max, 272 m/s
Range: min, 450 m; max, 5,520 m
Rate of fire: 12-15 rds/min

Manufacturer

China North Industries Corporation (NORINCO), 7A Yue Tan Nan Jie, PO Box 2137, Beijing.

Status

Available.

VERIFIED

NORINCO 120 mm Type 55 mortar on travelling carriage

COMMONWEALTH OF INDEPENDENT STATES

82 mm M-36 mortar

Description

The 82 mm M-36 mortar (82-PM 36) was the first of the Soviet 82 mm mortars, being issued to the Red Army during 1936 and serving throughout the Great Patriotic War of 1941-1945. Despite its age the M-36 is still likely to be encountered in service in some parts of Africa and Asia.

The ammunition is conventional with a primary cartridge and six increments. Three different types of sight were used, the MP-1 and MP-82 optical sights and a sheet metal aiming circle-clinometer.

Data

Calibre: 82 mm
Length of barrel: 1.288 m
Weight: in firing position, 57.3 kg
Elevation: +45 to +85°
Traverse: top, 6°
Range: max, 3,100 m

Manufacturer

State factories.

Status

Not in production.

Service

Not in CIS or former Warsaw Pact armies, but still likely to be encountered in Asia and Africa.

UPDATED

82 mm M-36 mortar

82 mm M-37 mortar

Description

The 82 mm M-37 mortar (82-PM 37) is a modification of the M-36 with a circular baseplate. It may be distinguished by the cross-levelling and connecting rod on the right leg rather than the left. However, when using this as an identifying feature the position of the elevating handle should be noted. This should emerge from the gear case at the rear. If not, the bipod gear will be on the left. The circular baseplate was a great success and although a rectangular plate was introduced for use by mountain divisions, it is now rarely seen. The sight is the MPM-44.

A post-war addition to some M-37 mortars, especially those serving with the Polish Army, was a muzzle device to prevent double loading.

Data

Calibre: 82 mm
Length of barrel: 1.22 m
Weight: in firing position, 56 kg
Weight of barrel: 19 kg
Elevation: +45 to +85°
Traverse: top, 6°
Muzzle velocity: 211 m/s
Range: min, 85 m; max, 3,040 m
Rate of fire: 15-25 rds/min
Bomb weight: 3.05 kg

Manufacturer

State factories.

Status

Still offered for export sales.

Service

Most former Warsaw Pact armies. Was produced in China as the Type 53.

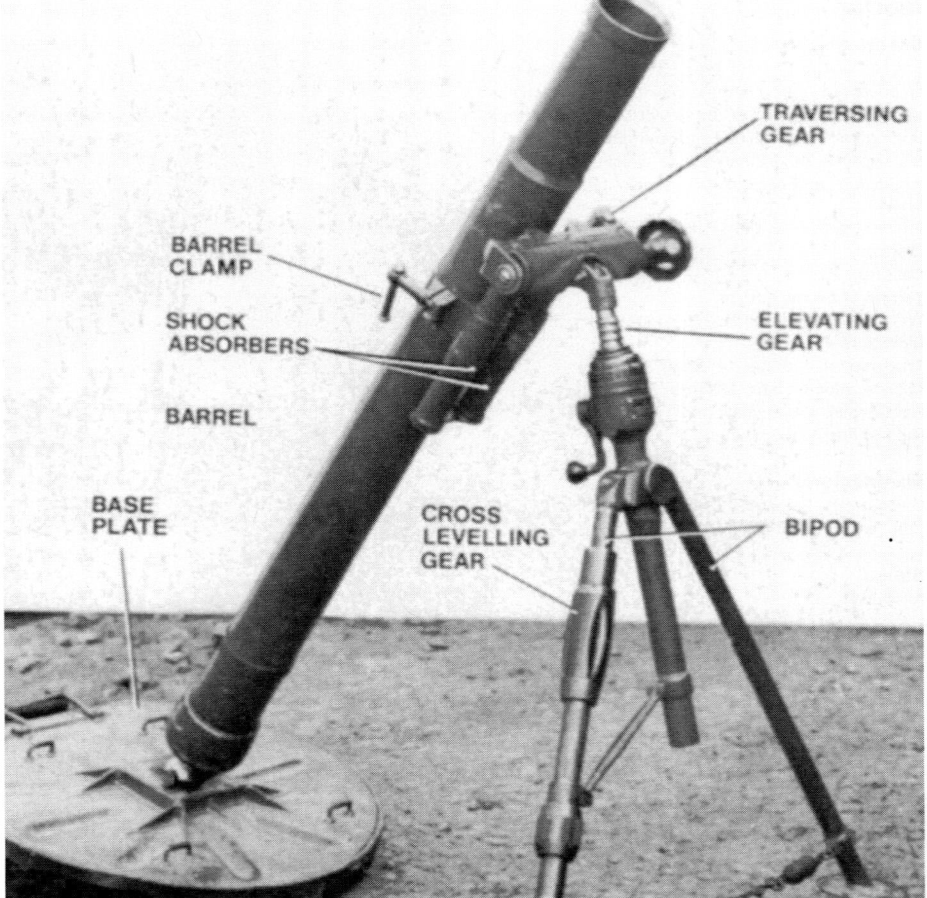

82 mm M-37 mortar

Licence production

BULGARIA
Manufacturer: Kintex
Type: 82 mm MOD 37 mortar
Remarks: In production. Offered for export sales.

EGYPT
Manufacturer: Helwan Machine Tools Company, Factory 999.
Type: 82 mm Model M 69 Model LMB
Remarks: See entry under *Egypt* for details.

UPDATED

82 mm M-41 mortar

Description

The 82 mm M-41 mortar (82-PM 41) was introduced to reduce the production demands of the M-37 mortar under difficult wartime conditions while at the same time improving tactical mobility. Instead of the conventional bipod and yoke, two short legs support the long elevation column and the traversing gear. Cross-levelling is accomplished using a linkage between the elevating shaft and the traversing screw. At the foot of each bipod leg is a stub axle to which a wheel can be fitted. When the mortar comes out of action the bipod is folded back and clamped to the circular baseplate. The wheels are attached and the mortar towed from the muzzle end, once a muzzle cap with a towing attachment is in place, by whatever type of transport is available - man, mule or motor vehicle.

The M-41 was found to be less successful and less stable than the M-37 when firing. It is no longer in first-line service, but has been supplied to many CIS-influenced countries and may still be encountered almost anywhere.

There was also an 82 mm M-43 (82-PM 43) mortar on which the travelling wheels became a permanent attachment, increasing the total weight to 58 kg. It is no longer likely to be encountered.

Data

Calibre: 82 mm
Length of barrel: 1.22 m
Weight: in firing position, 52 kg; travelling, 58 kg
Weight of barrel: 19.5 kg
Elevation: +45 to +85°
Traverse: top, 10°
Muzzle velocity: 211 m/s
Range: min, 85 m max, 3,040 m
Rate of fire: 15-25 rds/min
Bomb weight: 3.05 kg

Manufacturer

State factories.

Status

No longer in production.

Service

Some CIS-influenced and Third World countries.

UPDATED

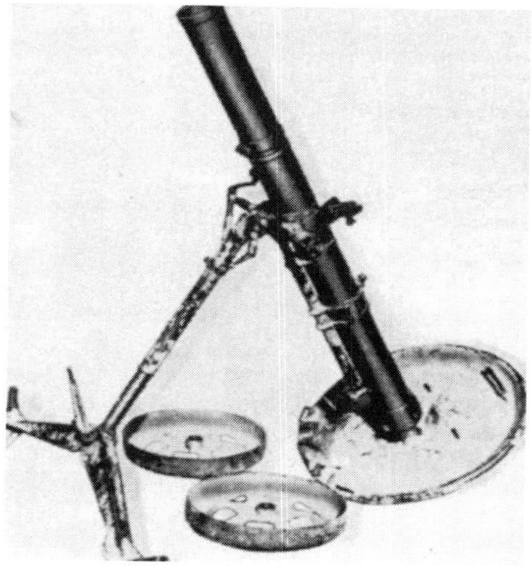

The 82 mm M-41 mortar has an unconventional mount with short legs and long elevation column; the wheels are removed before firing

82 mm 2B14 Podnos light mortar

Description

The 82 mm 2B14 Podnos light mortar may be regarded as an entirely new development owing little other than an overall similarity to previous CIS 82 mm designs such as the M-37. It is in production for use by airborne and other special force units.

The 2B14 Podnos utilises modern materials and manufacturing methods throughout but remains entirely conventional in overall design, being a bipod-mounted, drop-fired smoothbore mortar with a high strength steel alloy barrel. The muzzle is fitted with a device to prevent double loading. If required the entire mortar can be broken down into man-pack loads for carrying by the four-man crew.

Sighting equipment includes a MUM panoramic mortar sight and a K1 collimator, both provided with lighting equipment for use at night.

The 2B14 Podnos can fire all existing 82 mm mortar ammunition over a range of 80 to 3,200 m. However a new family of 82 mm ammunition is available to cover a range band of from 125 to 4,100 m.

In Bulgaria a self-propelled version of the 2B14 Podnos is carried in a variant of the MT-LB tracked carrier.

82 mm 2B14 Podnos light mortar

A document dated October 1992 gives the unit price for a single 2B14 Podnos light mortar as $15,000.

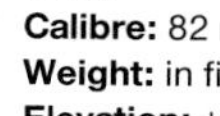

Data

Calibre: 82 mm
Weight: in firing position, 41.88 kg
Elevation: +45 to +85°
Traverse: 6°
Muzzle velocity: 220 m/s
Range: new ammunition, min, 125 m, max, 4,100 m
Rate of fire: max, 22-30 rds/min

Manufacturer

State factories.

Status

In production.

Service

CIS armed forces. Offered for export sales.

Licence production

BULGARIA
Manufacturer: Arcus Company
Type: 82 mm M-82 Podnos mortar
Remarks: In production. Offered for export sales.

UPDATED

82 mm AM 2B9 Vasilek automatic mortar

Description

The 82 mm AM 2B9 Vasilek automatic mortar was introduced into service some time in 1971 but its existence was not revealed until the late 1970s. It is officially designated AM for 'Avtomaticheskiy Minomet' or automatic mortar, but it also bears the name Vasilek (Cornflower).

The 2B9 is mounted on a light two-wheeled split-trail carriage resembling that of the 76 mm mountain gun M1969 and capable of being towed by almost any vehicle. Alternatively it can be mounted in and fired from suitable armoured vehicles. The field carriage has light spades on the trail legs and a screw-type firing jack at the front of the saddle, relieving the suspension of shock when firing.

A traversing top carriage carries the mortar assembly, balanced by spring-and-cable equilibrators running round the oversized trunnions. A hydrospring recoil system is fitted around the barrel, and a magazine-type rapid loading device is fitted behind the breech. The weapon can be muzzle loaded and fired at high angles, or breech loaded and fired at low angles. In the high-angle role up to three propelling charges can be used, giving the weapon a versatile trajectory for indirect-fire. In the low-angle role a separate-loading fixed propelling charge is used. The maximum rate of fire at automatic is 170 rds/min, though the sustained practical rate is approximately 120 rds/min.

In the mortar (high-angle) role standard 82 mm mortar bombs are used, with a special three-equal-increments propelling charge system. In the low-angle role a special anti-tank projectile is fired, using a fixed 75 g charge; this projectile weighs 3.2 kg and will defeat 100 mm of steel armour. In both firing modes fire control is carried out using a PAM-1 ×3 or ×2.5 optical sight and a K-1 collimator.

For transport the 2B9 may be either towed or carried on the cargo area of a GAZ-66 (4 × 4) 2,000 kg light truck.

82 mm Vasilek automatic mortar

Data

Calibre: 82 mm
Weight: in firing position, 632 kg; travelling, 645 kg
Elevation: −1 to +85°
Traverse: on carriage, 30° right and left
Muzzle velocity: low angle, automatic fire, 270 m/s; high angle, manual loading, Charge 3, 210 m/s
Range: 85°, min, 800 m; max, 4,720 m

Range 50% zone, max: 100 × 200 m
Rate of fire: max, practical, 120 rds/min

Manufacturer
State factories.

Status
Production probably complete in CIS.

Service
CIS and former Warsaw Pact armed forces.

Licence production
HUNGARY
Manufacturer: Army-COOP Kft.
Type: 82 mm automatic mortar Type DE 82
Remarks: Available for export sales in 81 mm calibre if required. A self-propelled variant with automatic fire control and mounted on a MT-LBu tracked carrier has been developed.

VERIFIED

107 mm M-38 mortar

Description
The 107 mm M-38 mortar (107-PM 38) was designed for mountain warfare and animal transport. The M-38 is carried complete on a two-wheeled trolley. It has now been replaced by an improved version, the M-107.

Data
Calibre: 107 mm
Length of barrel: 1.67 m
Weight: in firing position, 170 kg; travelling, 340 kg
Elevation: +45 to +80°
Traverse: top, 3°
Muzzle velocity: HE heavy, 302 m/s; HE light, 263 m/s
Range: min, HE heavy, 800 m; max, HE heavy, 5,150 m; HE light, 6,300 m
Rate of fire: 15 rds/min
Bomb weight: HE heavy, 9 kg; HE light, 7.9 kg

Manufacturer
State factories.

Status
Production complete.

Service
CIS reserves and in China, North Korea and possibly Vietnam.

VERIFIED

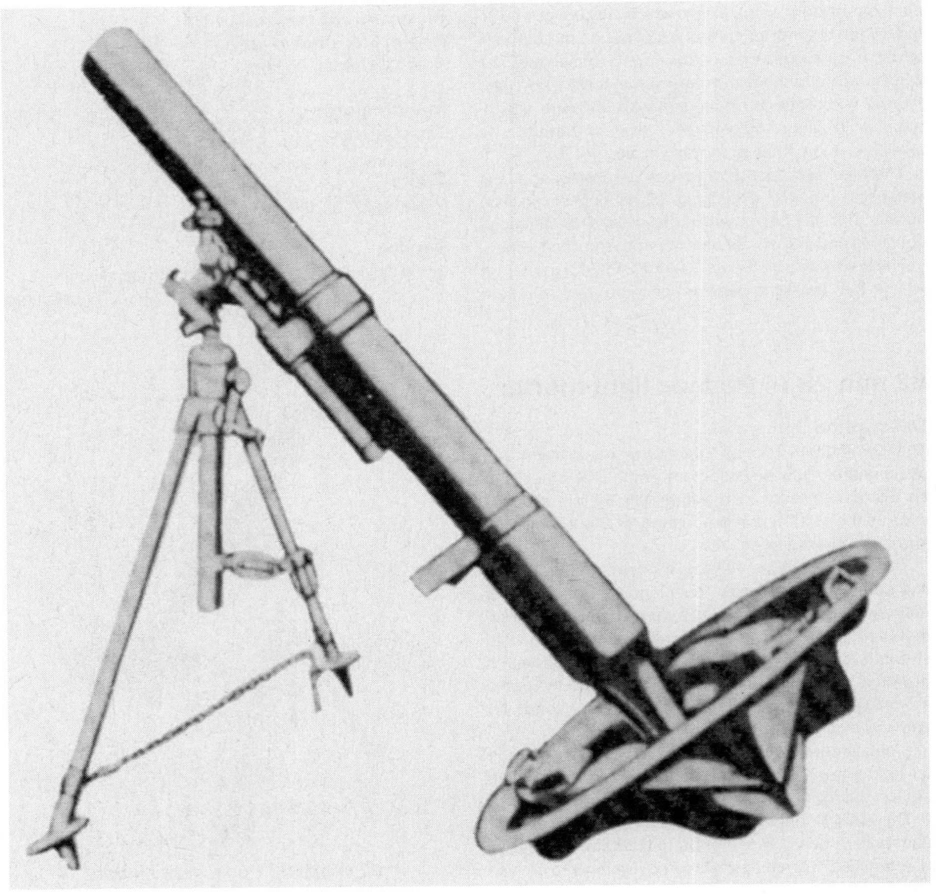

107 mm M-38 mortar

120 mm M-43 mortar

Description
The 120 mm M-43 (120-PM 43) was developed as a wartime expedient based on the earlier 120 mm M-38 which was more demanding on manufacturing resources. The main changes were that the M-43 has longer shock absorber cylinders while retaining the M-38's stamped circular baseplate and two-wheeled carriage.

The M-43 was produced in China as the Type 53; the similar Type 55 is a later development.

Data
Calibre: 120 mm
Length of barrel: overall, 1.854 m
Weight: in firing position, 280 kg; travelling 563-600 kg, depending upon carriage model
Width: travelling, 1.62 m
Height: travelling, 1.206 m
Track: 1.21 m
Elevation: +45 to +80°
Traverse: top, 8°
Muzzle velocity: 272 m/s
Range: min, 460 m; max, 5,700 m
Rate of fire: 12-15 rds/min
Bomb weight: 15.4 kg

Manufacturer
State factories.

120 mm M-43 mortar

120 mm M-43 mortar

Status
Not in production in CIS.

Service
CIS armies, also: Albania, China, Czech Republic and Slovakia, Egypt, Germany, Iraq, North Korea, Romania, Syria, Vietnam, South Yemen and Yugoslavia.

Licence production
EGYPT
Manufacturer: Helwan Machine Tools Company, Factory 999
Type: 120 mm heavy mortar Model UK 2
Remarks: See entry under *Egypt* for details.

UPDATED

120 mm M-43 mortar in travelling order

120 mm 2S11 Sani mortar

Description
The 120 mm 2S11 Sani mortar may be considered as being a product-improved version of the M-43 (see previous entry). The use of better-grade materials permitted a considerable weight reduction, aided by a complete redesign of the transporting carriage. The basic design remains unchanged, a smoothbore barrel on a conical baseplate and supported by a bipod, to which it is attached by the medium of a simple recoil buffer mechanism. The muzzle is fitted with a double-loading prevention system.

The 2S11 is usually carried in the assembled state on the two-wheeled transporting carriage 2L81, the tripod being carried uppermost and the baseplate remaining attached to the barrel. This carriage is very light (87 kg) so for this reason is restricted to short moves at reduced speed. It is therefore more usual to carry the mortar in a GAZ-66 truck provided with special fittings for carrying and off-loading the mortar. If required the mortar may be towed by the same vehicle. The vehicle also carries the five-man crew and ready-use ammunition. However, the three basic assemblies can be separated rapidly and loaded into any convenient transport, or animal packed. In such cases, additional transport has to be provided for the sights and accessories and ready-use ammunition. The sight is the MPM-44M.

In Bulgaria a self-propelled version of the 2S11 Sani is carried in a variant of the MT-LB tracked carrier.

Data
Calibre: 120 mm
Weight: in firing position, 210 kg; travelling, 297 kg
Track, carriage: 900 mm
Elevation: +45 to +80°
Traverse: on bipod, 5° right and left; by moving bipod, 26° right and left
Muzzle velocity: 325 m/s
Range: min, 480 m, max, 7,100 m
Rate of fire: 12-15 rds/min
Bomb weight: HE OF-843B, 16 kg

Manufacturer
State arsenals.

Status
In production.

Service
CIS forces and some other former Warsaw Pact armed forces.

Licence production
BULGARIA
Manufacturer: Kintex
Type: 120 mm 2S11 Sani
Remarks: In production. Offered for export sales.
Manufacturer: PIMA AEG
Type: 120 mm Wheeled Mortar 2S11
Remarks: In production. Offered for export sales.
HUNGARY
Manufacturer: Army-COOP Kft
Type: 120 mm Mortar
Remarks: Available.

UPDATED

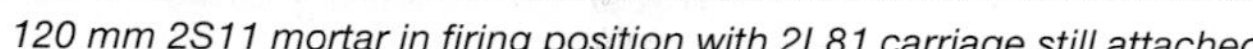

120 mm 2S11 mortar in firing position with 2L81 carriage still attached

120 mm 2S11 mortar in travelling mode

160 mm mortars

Description

The former Red Army introduced the 160 mm M1943 to provide infantry divisions with a weapon producing a heavy weight of high explosive without making undue demands on hard-pressed manufacturing resources. It was the heaviest mortar used by the Red Army during the Great Patriotic War. Because of the long barrel the M1943 had to be a breech-loading weapon with the barrel pivoted for loading about trunnions placed not far from the centre point. It was towed by the muzzle, using either an armoured personnel carrier or a heavy lorry. After the war it was used by Poland, Romania and Bulgaria and deployed in troops of four mortars. It has now been replaced throughout the former Warsaw Pact countries by the 160 mm M-160 mortar.

The M-160 is very similar in design and the method of breech loading has been preserved but it has a longer barrel and a greater range. It was originally used as a divisional mortar in all types of division but is now employed with the mountain divisions where its range, explosive projectile capacity and high angle of fire are very useful.

Data

	M-43	M-160
Calibre	160 mm	160 mm
Length of barrel	3.03 m	4.55 m
Length		
in travelling order	3.985 m	4.86 m
Width		
in travelling order	1.77 m	2.03 m
Height		
in travelling order	1.414 m	1.69 m
Weight		
in firing position	1,170 kg	1,300 kg
travelling	1,270 kg	1,470 kg
Track	1.75 m	1.75 m
Elevation		
min	45°	50°
max	80°	80°
Traverse	25°	24°
Rate of fire	3 rds/min	2-3 rds/min
Muzzle velocity	245 m/s	343 m/s
Max range	5,150 m	8,040 m
Min range	630 m	750 m
Bomb weight	40.8 kg	41.5 kg

Status

In current use, but no longer manufactured.

Service

M-160, CIS armies, also used by China, Egypt and India.

M-43, Albania, China, Czech Republic and Slovakia, Egypt, Germany, North Korea, and Vietnam.

VERIFIED

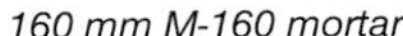

160 mm M-160 mortar

160 mm M-43 mortar

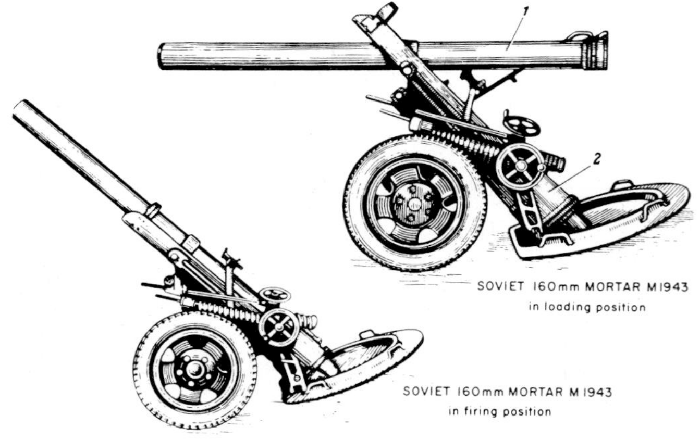

160 mm M-43 mortar in loading and firing positions

Detail of trunnions about which barrel is lowered for loading

CZECH REPUBLIC AND SLOVAKIA

Konštrukta 81 mm mortar

Description

The Konštrukta 81 mm mortar is of modern design and is intended to replace existing 82 mm mortars in service with the Czech and Slovak armed forces as well as being offered for possible export sales. The main design feature of this mortar is the mounting which, probably for ease of manufacture, uses an unorthodox inverted T-shaped monopod/bipod design formed from tubular steel. A single angled side arm is provided for on-mounting traverse. The elevating arm telescopes in the vertical part of the inverted T and is connected to the high strength steel barrel by a yoke engaging in twin recoil cylinders. A sight unit is mounted on the collar connecting the recoil assembly to the barrel. The weapon can be rapidly broken down by its three-man crew for transport and a light undercarriage can be provided for handling if required.

This mortar fires a special bomb known as the 81 HE weighing 4.56 kg and containing 750 g of RDX/TNT (another type of HE bomb contains 950 g of TNT). Using a seven charge system, plus a supercharge, the maximum range with a muzzle velocity of 352 m/s is 7,000 m.

Data

Calibre: 81.4 mm
Weight: in firing position, 90 kg
Elevation: +40 to +85°
Traverse: on mounting, ±5°; full, 360°
Muzzle velocity: 352 m/s
Range: min, 300 m; max, 7,000 m
Rate of fire: 15-20 rds/min

Manufacturer

Konštrukta Trenčin, Pod Sokolice 107, 911 90, Trenčin.

Status

Available.

VERIFIED

Konštrukta 81 mm mortar

EGYPT

Helwan 60 mm light mortar

Description

The Helwan 60 mm light mortar is a direct copy of the Chinese NORINCO 60 mm Type 63-1 and differs from the original in few details. Operated by a two-man crew the mortar is provided with a carrying handle on the barrel which allows the weapon to be carried as one load, including the mounting and attached baseplate, over short distances.

Locally produced ammunition includes HE, Smoke and Illuminating.

Data

Calibre: 60 mm
Length of barrel: 610 mm
Weight: in firing position, 12.3 kg
Elevation: max, +85°
Traverse: top, 8°
Muzzle velocity: 153 m/s
Range: max, 1,530 m
Rate of fire: 15-20 rds/min

Manufacturer

Helwan Machine Tools Company, Factory 999, 23 Talaat Harb Street, PO Box 1582, Cairo.

Status

In production.

Service

Egyptian Army.

UPDATED

Helwan 60 mm light mortar, a copy of the Chinese Type 63-1

Helwan 82 mm Model M 69 Model LMB mortar

Description

The Helwan 82 mm Model M 69 Model LMB mortar features a copy of the CIS 82 mm M-37 mortar barrel and bipod allied to a baseplate derived from a Thomson-Brandt design. The weapon can fire all former Warsaw Pact 82 mm HE, Smoke and Illumination rounds to a range of 3,045 m. A special long-range M 74 bomb can reach 5,500 m using a six-increment propellant charge system.

Data

Calibre: 82 mm
Length of barrel: 1.22 m
Weight: in firing position, 44.5 kg
Weight, barrel and breech piece: 14.5 kg
Weight of bipod: 14 kg
Weight of baseplate: 15 kg
Elevation: +45 to +85°
Traverse: 48 mils
Muzzle velocity: 205 m/s
Range: min, 85 m; max, 3,045 m (5,500 m with M74 bomb)
Rate of fire: 20 rds/min

Manufacturer

Helwan Machine Tools Company, Factory 999, 23 Talaat Harb Street, PO Box 1582, Cairo.

Status

In production.

Service

Egyptian Army.

UPDATED

Helwan 82 mm Model M 69 Model LMB mortar

Helwan 120 mm Model UK 2 heavy mortar

Description

The Helwan 120 mm Model UK 2 heavy mortar appears to be a direct copy of the CIS 120 mm M-43 with a few local modifications such as a permanently fitted muzzle device to prevent double loading. The Model UK 2 utilises the same two-wheel travelling carriage as the M-43. Some of the details provided for the Model UK 2 differ from those for the M-43.

Data

Calibre: 120 mm
Length of barrel: 1.85 m
Weight: in firing position, 282 kg; travelling, 620 kg
Elevation: +45 to +85°
Traverse: 4°
Muzzle velocity: 272 m/s
Range: min, 460 m; max, 5,520 m
Rate of fire: 8-12 rds/min

Manufacturer

Helwan Machine Tools Company, Factory 999, 23 Talaat Harb Street, PO Box 1582, Cairo.

Status

In production.

Service

Egyptian Army.

VERIFIED

Helwan 120 mm Model UK 2 heavy mortar in travelling configuration

FINLAND

Vammas 60 mm mortars

Description

The Vammas 60 mm mortars were originally designed by Tampella Defence Division which was merged to become Vammas in 1991. Produced with two barrel lengths, standard and long range, they are intended to serve as a support weapon for troops and guerrillas in conditions where lightness and mobility are essential. The mortar and 20 rounds of ammunition can be carried in special rucksacks by three men and bringing the weapon into action is rapid and easy.

Data

Calibre: 60 mm
Length of barrel: standard, 720 mm; long range, 870 mm
Weight in firing position: standard, 16 kg; long range, 18 kg
Range: min, 150 m; max, standard, 2,600 m; max, long range, 4,000 m
Rate of fire: 25 rds/min
Bomb weight: standard TAM 1.6, 1.6 kg; long range TAM 1.8, 1.8 kg

Manufacturer

Vammas Oy, PO Box 18, FIN-38201 Vammala

Status

Available.

Service

Finnish Defence Forces.

UPDATED

Standard version of Vammas 60 mm mortar

Vammas 81 mm mortars

Description

The Vammas 81 mm mortars were originally developed by Tampella Defence Division (merged to become Vammas in 1991) to meet the requirements of the Finnish Defence Forces. There are two basic models, standard and long range, both of which can have normal or long barrels. All models have a smooth, uniform barrel, a circular dished baseplate, and a bipod mounting with precision elevation and azimuth gears sealed against dust, sand and water.

Data

STANDARD VERSION
Calibre: 81 mm
Length of barrel: normal barrel, 1.305 mm; long barrel, 1.6 m
Weight in firing position: normal barrel, 40 kg; long barrel, 43 kg

Vammas 81 mm long-range mortar with normal barrel

Range: min, 150 m; max, normal barrel, 5,900 m; long barrel, 6,100 m
Rate of fire: 20 rds/min
Bomb weight: 4.2 kg

Data

LONG RANGE VERSION
Calibre: 81 mm
Length of barrel: normal barrel, 1.311 m; long barrel, 1.561 m
Weight in firing position: normal barrel, 56 kg; long barrel, 61 kg
Range: min, 150 m; max, normal barrel, 6,300 m; long barrel, 6,700 m
Rate of fire: 20 rds/min
Bomb weight: 4.2 kg

Manufacturer

Vammas Oy, PO Box 18, FIN-38201 Vammala.

Status

In production.

Service

Finnish Defence Forces.

VERIFIED

Vammas 81 mm M71 mortar

Vammas 120 mm Light mortar

Description

The Vammas 120 mm Light mortar was originally developed by Tampella Defence Division (merged to become Vammas in 1991) and later upgraded by Vammas. The weapon combines firepower and accuracy with mobility and relatively light weight. It can be transported by helicopter or dropped by parachute and rapidly prepared in its firing position.

The construction of the Vammas 120 mm Light mortar barrel, bipod and baseplate is similar to those of the Vammas 120 mm Long Range mortar (see following entry). The firing mechanism is assembled inside the breech piece and can be supplied either for drop firing or trigger firing.

Data

Calibre: 120.35 mm
Length of barrel: 1.729 m
Weight: in firing position, 146 kg; travelling, 266 kg
Weight of barrel: 50 kg
Weight of bipod and sight unit: 28 kg
Weight of baseplate: 66 kg
Weight of carriage: 120 kg
Max operating pressure: 100 MPa
Elevation: +45 to +80°
Traverse: on bipod, 4.5° right and left; by moving bipod, 360°
Range: min, 300 m; max, 7,300 m
Rate of fire: 15 rds/min

Manufacturer

Vammas Oy, PO Box 18, FIN-38201 Vammala

Status

In production.

Service

Finnish Defence Forces.

UPDATED

Vammas 120 mm Light mortar

Vammas 120 mm Long-Range mortar

Description

The Vammas 120 mm Long-Range mortar was originally developed by Tampella Defence Division (merged to become Vammas in 1991) and upgraded by Vammas to meet the requirements of the Finnish Defence Forces. The weapon combines firepower and accuracy and is adaptable to many types of tactical operation. The sturdy construction allows a rapid start of accurate firing because of the ability of the baseplate to 'set' itself into the ground during the first couple of shots.

The barrel is constructed using high tensile strength alloy steel with a close-tolerance bore. The design of the breech piece ensures tightness of the conical sealing surface and allows interchangeability of barrels and breech pieces. The firing mechanism is assembled inside the breech piece and can be supplied either for drop firing or for trigger firing.

The bipod frame is made of seamless high strength alloy steel tube. Elevation and azimuth gears are precision made, run smoothly, and are sealed against dust, sand and water.

The baseplate material is high strength steel, and the welded construction results in a sturdy unit giving exceptional stability in all types of terrain. The baseplate has been designed to accommodate 360° of traverse without having to be moved.

The transport carriage is designed for rapid deployment and redeployment by means of rear lifting forks. The mortar can be emplaced in less than one minute and removed in less than 20 seconds. The frame and rear forks are of welded steel, and the axle has torsion bar suspension which provides smooth riding over any surface. The carriage may be used as a service platform for the barrel.

Vammas 120 mm Long-Range mortar

Variant

Long Range trailer mortar
See following entry.

Data

Calibre: 120.25 mm
Length of barrel: 2.154 m
Weight: in firing position, 291 kg; travelling, 500 kg
Weight of barrel: 95 kg
Weight of bipod and sight: 63 kg
Weight of baseplate: 126 kg
Weight of carriage: 208 kg
Max operating pressure: 155 MPa
Elevation: +45 to +80°
Traverse: on bipod, 4.5° right and left; by moving bipod, 360°
Range: min, 300 m; max, with 12.8 kg bomb, 8,600 m
Rate of fire: 15 rds/min

Manufacturer

Vammas Oy, PO Box 18, FIN-38201 Vammala

Status

In production.

Service

Finnish Defence Forces. Swedish armed forces (early version).

UPDATED

Vammas 120 mm Long-Range mortar on transport carriage

Vammas 120 mm Long-Range trailer mortar

Description

The Vammas 120 mm Long-Range trailer mortar was developed by Vammas and Sisu to meet the requirements of the Finnish Defence Forces and is installed in the trailer unit of the Sisu NA-112 GT rubber-tracked all-terrain vehicle. The weapon involved is either the Vammas 120 mm Long-Range or Standard model, both of which can be either fired from the trailer or dismounted for firing on the ground. Maximum speed of the Sisu NA-112 GT on roads is 55 km/h and it can swim in water at a speed of 3.6 km/h using its tracks for propulsion.

The characteristics of the mortar are the same as those given in the previous two entries.

Manufacturers

Vammas Oy, PO Box 18, FIN-38201 Vammala.
Sisu Defence, PO Box 189, FIN-13101 Hämeenlinna.

Status

Under evaluation by the Finnish Defence Forces.

VERIFIED

Vammas 120 mm Long-Range trailer mortar installed on Sisu NA-112 GT rubber-tracked all-terrain vehicle

Vammas 160 mm M58 mortar

Description

160 mm mortars were developed only in the former Soviet Union and Finland (plus, from the latter, Israel). Soviet 160 mm mortars are both large and heavy and have to be loaded from the breech. This is both expensive and complicated since it requires balancing springs to allow the barrel to lower. In the Vammas mortar the traditional muzzle-loading feature has been retained. This does not affect the rate of fire since the time required to lift a bomb to the muzzle is much the same as that needed to open a breech and close it again.

The Vammas 160 mm M58 mortar makes use of a new series of HE bombs, which considerably extend the range and the terminal effectiveness. The use of a proximity fuze improves this effectiveness by up to three times.

Data

Calibre: 160.4 mm
Weight: in firing position, 1,450 kg
Range: max, 10,000 m
Rate of fire: 4-8 rds/min
Bomb weight: HE, 40 kg

Manufacturer

Vammas Oy, PO Box 18, FIN-38201 Vammala

Status

Available.

VERIFIED

Vammas 160 mm M58 mortar in firing position; the two-wheeled carriage assists in traversing weapon

FRANCE

Thomson-Brandt 60 mm Commando mortars

Description

The 60 mm Commando mortar is the lightest and most portable of the Thomson-Brandt mortars, produced in two variants, types V and A. Both have small spades and do not use a mounting. They are controlled for both line and elevation directly by the firer.

The type V has a short barrel and a fixed firing pin. As soon as the muzzle-loaded bomb reaches the firing pin it is automatically fired. The sight is a simple clamp-on type with a bubble fitted on the right-hand side. The firer sets the range on a drum and when the bubble is levelled the sight is elevated for the correct tangent elevation for the selected range. A white line is scribed along the front of the barrel and the azimuth is applied by lining up this mark with the target. A broad webbing strap is around the barrel to provide a grip for the firer's left hand which would otherwise find it difficult to control the hot barrel. A webbing sling is provided for carriage of the mortar and has a number of brass bars riveted to the fabric. These are graduated in ranges according to the firing charge involved - from left to right, Charge 0, Charge 1, Charge 2. If the gunner places his foot on the sling on the selected range bar and then pulls the sling tight by pulling on the barrel, the mortar will be at the correct elevation, provided the ground is horizontal. A muzzle cover is supplied to keep out intruding debris. The spade is small and button-shaped.

The mortar may be fired at a rate of 12 to 20 rds/min. There is a two charge system providing a maximum range of 1,050 m and a minimum range of 100 m. Length of the barrel is 680 mm; weight is 8.1 kg.

The type A is longer with a breech piece below the barrel holding the firing mechanism and connected to a small trough-shaped spade. The mortar is controlled by a firing lever which has a trip-over action. The bomb is dropped in from the muzzle and rests on the firing pin plug. The user pulls back on the lanyard attached to the firing lever and the firing pin is retracted, the spring is compressed and then the pin trips over to be driven forward to fire the round. There is no sight and the soldier lying behind the mortar lines up the barrel to control his fire for azimuth and estimates the correct quadrant elevation for his first round.

The type A mortar has an overall length of 861 mm but the barrel length is 650 m; weight is 8.9 kg.

The Commando mortars have a range of 60 mm ammunition. Primary cartridges for all bombs are 24 mm in diameter and 65 mm long, containing 4.2 g of ballistite. Secondary charges are generally of the horseshoe type, although earlier bombs used secondary charges fitted between the fins. The supercharge is horseshoe-shaped and contains 5 g of ballistite.

Data

Calibre: 60 mm
Length of barrel: type A, 650 mm; type V, 680 mm
Weight, complete: type A, 8.9 kg; type V, 8.1 kg
Range: max, both types, 1,050 m

Manufacturer

Thomson-Brandt Armements, Tour Chenonceaux, 204 Rond-Point du Pont de Sevres, F-92516 Boulogne-Billancourt.

Status

In production.

Service

The armed forces of 20 countries. Most use the type A.

VERIFIED

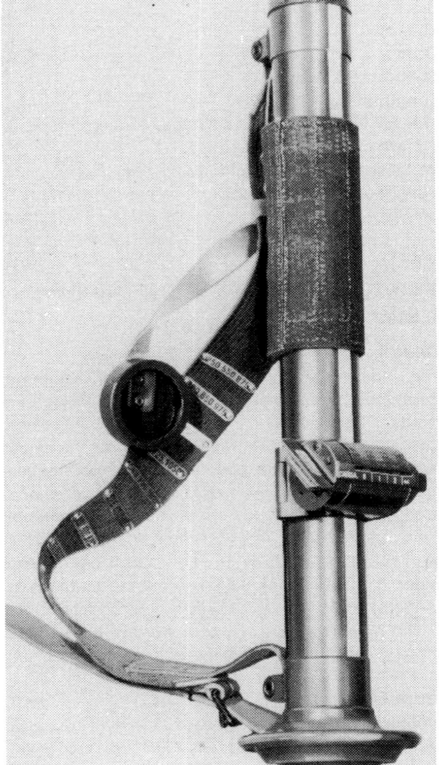

60 mm type V Commando mortar. This has a fixed firing pin and fires the bomb as soon as it drops. Note button-shaped spade

Thomson-Brandt 60 mm proximity mortar

Description

The 60 mm proximity mortar weapon derives its name from its intended use - in close proximity to the enemy by forward troops. Because of its lightness and handiness it can be carried easily and thus gives the advanced infantry squad accurate and efficient fire support. It can be carried in a shoulder bag by one man and is adjusted for elevation by sliding the bipod sleeve up and down the barrel. The curved ends of the bipod permit this movement to be made easily on any type of ground. The baseplate permits all-round firing.

All standard 60 mm bombs can be fired from this mortar.

Data

Weight, total: 6 kg
Length of barrel: with breech, 860 mm
Elevation: +45 to +82°
Range: HE, smoke, practice, 110-650 m; illuminating, 350-900 m

Manufacturer
Thomson-Brandt Armements, Tour Chenonceaux, 204 Rond-Point du Pont de Sevres, F-92516 Boulogne-Billancourt.

Status
In production.

Service
Swiss Army.

Licence production
SWITZERLAND
Manufacturer: w + f Bern
Type: 60 mm Model 87
Remarks: See entry under *Switzerland* for details.

VERIFIED

Thomson-Brandt 60 mm proximity mortar

Thomson-Brandt 60 mm light mortar

Description

This mortar, the MO-60-L, was originally produced as a private venture by Hotchkiss-Brandt during 1963. It has proved to be a very successful light mortar and is still in production.

The weapon is of orthodox configuration with a baseplate, barrel and bipod mounting. The barrel is smoothbored and made of nickel chrome steel. The base of the barrel is threaded to take the breech piece. The breech piece is made of steel and is screwed into the barrel. It is pinned in place and can be removed only by an armourer. The striker is dome headed and screws into the breech piece from the rear. It protrudes into the base of the chamber and projects about 1.5 mm. The bombs are dropped on the striker.

The baseplate is an equilateral triangle reinforced by three webs forming a star round the central cone. On the top surface there is a socket in which the breech piece fits.

The barrel is locked into position by rotating it in the socket. The baseplate allows all-round traverse and the webs give good stability in very wet soil. A lifting handle is fitted. The bipod supports the barrel. It has two legs made of aluminium alloy hinged to a cross-piece at the top. The legs can be moved out or in and the width between the feet is controlled by an adjustable chain. Passing upwards from the centre of the cross-piece, in a tube connecting the two bipod legs, is the elevating screw. This is attached to the yoke. The tube containing the elevating screw is connected to the left bipod leg by a rod. This rod can be slid up and down the tube and in so doing the tube is tilted. This provides the cross-levelling.

The yoke carries the traversing gear and at the end of it is the traversing handle. On the top of the yoke is the cradle into which the barrel fits, and at the other end from the traversing handle is the sight bracket. There is a shock-absorber which connects the yoke to the cradle and provides some isolation from the vibrations of firing.

The sight in use on the 60 mm light mortar is the F9. This has an elevating scale graduated from 40 to 60° at 1° intervals. The deflection micrometer is graduated from 0 to 8° 30′ on the right and from 360 to 351° 30′ on the left. Other graduation systems are available.

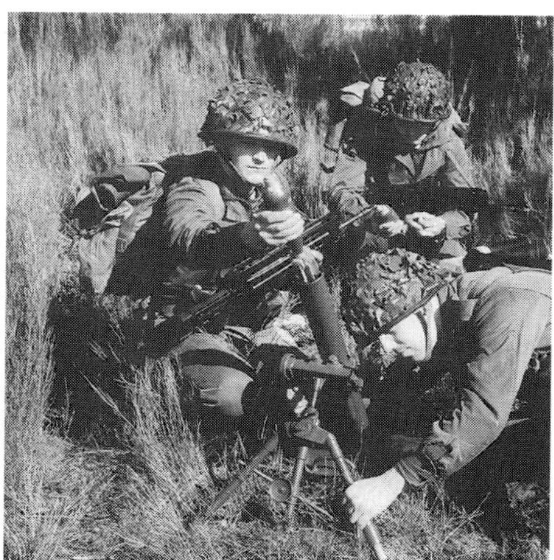

Thomson-Brandt 60 mm light mortar in action
1996

The mortar can be brought into action very quickly even on an unprepared platform. The baseplate should be level. If possible the first round should be fired at an angle greater than 60° and serves to bed the plate in.

If the ground is soft the baseplate will sink in but using the barrel as a lever it can be withdrawn speedily. The detachment can move on foot carrying the mortar over considerable distances.

The 60 mm light mortar was designed for the Brandt 60 mm Mark 61 bomb (qv). The mortar will also fire the US 60 mm bomb (old type), the Brandt Mark 35/47 (no longer manufactured by Thomson-Brandt), the M61 coloured marker, smoke and practice bombs and the M63 illuminating bomb, all of which are described in the *Mortar Ammunition* section.

Data

Calibre: 60 mm
Length of barrel: with breech piece, 724 mm
Weight: in firing position, 14.8 kg
Weight of barrel: 3.8 kg
Weight of baseplate: 6 kg
Weight of bipod: 5 kg

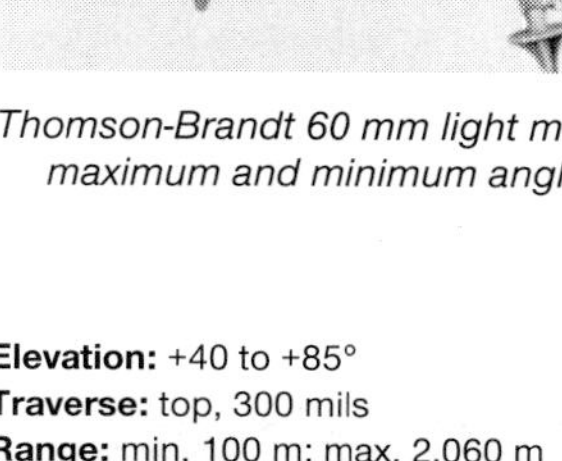

Thomson-Brandt 60 mm light mortar showing the maximum and minimum angles of elevation

Elevation: +40 to +85°
Traverse: top, 300 mils
Range: min, 100 m; max, 2,060 m
Rate of fire: 20 rds/min

Manufacturer

Thomson-Brandt Armements, Tour Chenonceaux, 204 Rond-Point du Pont de Sevres, F-92516 Boulogne-Billancourt.

Status

In production.

Service

French and several other armies.

VERIFIED

Thomson-Brandt 60 mm long-range mortar

Description

The Thomson-Brandt 60 mm long-range mortar, the MO-60-LP, was developed to combine the flexibility of the 60 mm light mortar and the firepower of the 81 mm mortar. For ease of transport it can be broken down into three components: barrel, baseplate and bipod.

It can fire the standard Thomson-Brandt M61 and M72 ranges of HE, smoke, practice and illuminating bombs. In addition a new long-range projectile called the LR has been developed for this mortar. The LR is of malleable pearlitic cast iron and has a fuze which detonates under any angle of impact, including grazing fire. The LR bomb weighs 2.2 kg and has a total length of 381 mm. It has a TNT charge which gives the bomb comparable efficiency with the 81 mm mortar bomb.

Data

Calibre: 60 mm
Length of barrel: including breech piece, 1.41 m
Weight: total, 33.6 kg
Weight of barrel: 10.1 kg
Weight of baseplate: 14 kg
Weight of bipod: 9.5 kg
Elevation: +40 to +85°
Traverse: 300 mils
Range: min, 100 m; max, 4,800 m
Rate of fire: max, 20 rds/min

Manufacturer

Thomson-Brandt Armements, Tour Chenonceaux, 204 Rond-Point du Pont de Sevres, F-92516 Boulogne-Billancourt.

Status

In production.

UPDATED

Thomson-Brandt 60 mm long-range light mortar

Thomson-Brandt 81 mm light mortars

Description

The Thomson-Brandt 81 mm light mortar was designed in 1961. It currently exists in three versions, the MO-81-LC, MO-81-LL and MO-81-LLR. The first two versions have barrels of 1.15 and 1.55 m respectively; for details of the third, MO-81-LLR, model see under Variants in this entry.

The mortar is typical of its type and consists of barrel, mounting and baseplate. The barrel is made of steel, reinforced at the rear end and is screwed on the breech piece which holds the firing lock. This allows the user to retract the firing pin when required in the safe position or allow it to protrude in the firing position. The end of the breech piece contains the ball which fits into the socket on the baseplate. It has two flats which allow for the insertion of the knob and subsequently its rotation.

The mounting consists of the bipod, elevating gear, traversing gear, cross-levelling gear and the barrel clamp. The bipod legs can be splayed at will and the distance between them is controlled by the chain connecting the feet. The elevating screw thread is enclosed in a tube. The tube is connected by a rod to the left bipod leg. By sliding the tube along the rod the elevating screw column is tilted and so the mortar is cross-levelled. The traversing gear is enclosed in a tube. The traversing handle is at one end of the tube and the sight mounting at the other.

The baseplate is made of chrome-molybdenum steel. It has, on the top surface, a socket in which the knob of the breech piece fits. This socket is movable to allow the equipment to rotate around the baseplate. The underside of the baseplate has three ribs which radiate outwards from under the socket and provide the strength to prevent the baseplate from buckling as the impulsive loading is applied.

The mortar may be carried in any suitable vehicle. It can also be broken down into mule loads or even carried by the detachment in framed rucksacks.

The sustained rate of fire for the mortar is 12 to 15 rds/min.

In addition to the standard HE bombs, M57D and M61, the following variants are available in both series: coloured, HE, target marker bomb (green, yellow or red) filled with TNT and colouring material; smoke bomb, filled with liquid titanium tetrachloride, also white phosphorus; practice bomb, filled with sulphur/naphthalene mixture and with dummy or live fuze.

Variant

81 mm Mortar Long Light Reinforced

The 81 mm Mortar Long Light Reinforced (MO-81-LLR) is the most recent of the Thomson Brandt 81 mm mortars and is intended for use by rapid deployment forces. It was designed from the outset to be compatible with automated fire control systems and can be provided equipped for either drop or trigger firing.

The 1.44 m long barrel is made of a new generation high strength steel alloy to allow it to fire any existing 81 mm mortar ammunition, including the new generation of smart munitions; design pressure is 127 MPa. Maximum possible range is 6,000 m although 5,600 m is more typical; minimum range is 250 m. Total weight is 45.2 kg (trigger firing) or 42.6 kg (drop firing).

Data

	Short barrel	Long barrel
Total weight	39.2 kg	42.5 kg
Length of barrel	1.15 m	1.55 m
Weight of barrel	12.7 kg	15 kg
Weight of baseplate	14 kg	15 kg
Weight of mount	12.5 kg	12.5 kg
Max range	4,140 m	5,000 m

Manufacturer

Thomson-Brandt Armements, Tour Chenonceaux, 204 Rond-Point du Pont de Sevres, F-92516 Boulogne-Billancourt.

Status

In production.

Service

French Army (short barrel). Also several other armies.

UPDATED

Thomson-Brandt 81 mm light mortar with short barrel

Thomson-Brandt 81 mm light mortar with long barrel

Thomson-Brandt 81 mm long-range mortar

Description

The Thomson-Brandt 81 mm long-range mortar (MO-81-LP) was designed in order to combine the flexibility of employment of the light 81 mm mortar with the range and efficiency of the 81 mm mortar-cannon CL 81 fitted to armoured vehicles, and to approach the capabilities of the 120 mm mortar.

It is of conventional design and can be broken down into three loads - barrel, baseplate and bipod - for man-carriage; it can also be carried by any vehicle, dropped by parachute, or transported by helicopter. It fires the 81 LP bomb to a range of 7,800 m or the 81 mm M82 bomb to a range of 5,800 m.

Data

Calibre: 81 mm
Length of barrel: 1.895 m
Weight: in firing position, 93.7 kg
Weight of barrel: 42.2 kg
Weight of baseplate: 35.8 kg
Weight of bipod: 15.7 kg
Weight of sight: 700 g
Elevation: +30 to +85° (533-1,511 mils)
Traverse: top, 520 mils
Rate of fire: 20 rds/min
Range: min, 100 m; max, with 81 LP bomb, 7,800 m

Manufacturer

Thomson-Brandt Armements, Tour Chenonceaux, 204 Rond-Point du Pont de Sevres, F-92516 Boulogne-Billancourt.

Status

In production.

UPDATED

Loading a Thomson-Brandt 81 mm long-range mortar (MO-81-LP)

Thomson-Brandt 120 mm MO-120-L light mortar

Description

The Thomson-Brandt 120 mm light mortar (MO-120-L) is a simple design in which mobility and firepower were allied. The mortar is made up of the barrel, mounting and baseplate and it fires a variety of ammunition providing all-round fire from a minimum range of 600 m out to 4,750 m.

The barrel is smoothbore and has a breech piece screwed over the end. The firing mechanism is enclosed in the breech piece and the firing pin can be set to protrude, thus firing a bomb dropped on it under gravity, or alternatively the pin can be retracted into a safe position. The bottom of the breech piece has a spherical ball, with two flats, which enables it to be secured in the baseplate. The mounting consists of the bipod legs, the elevating gear, the traversing gear, cross-levelling gear and the barrel clamp, which contains a shock-absorbing buffer. The bipod consists of two legs, each terminating in a shoe with a spike, and the spread of the legs is controlled by the length of an adjustable chain. The elevating gear in its tube is connected to the top of the bipod and reaches up to connect with the traversing gear. The traversing mechanism consists of a yoke which holds the traversing screw and also connects the elevating gear to the barrel clamp. On the right of the yoke is the traversing handwheel and on the left is the sight mounting.

To enable the sight to be upright at all times, there is a cross-levelling mechanism made up of the rod connecting the elevating screw tube to the left bipod leg. This can be moved across by a control on the bipod leg and so the elevating screw column can be moved to the vertical position regardless of the slope of the ground on which the mortar is standing.

Thomson-Brandt 120 mm light mortar (MO-120-L)

The baseplate is triangular in shape with a socket in the centre to take the breech piece ball and underneath are three ribs which provide the complete rigidity required and also prevent the baseplate slipping.

The mortar was designed to be operated in the simplest way possible and can be brought into action and maintained by a three-man detachment. The baseplate must be dug in if it is not possible to fire the initial round at an angle of elevation greater than 60°. If this can be done there is no need for a bedding-in round but there will be a five to eight per cent reduction in range for this first round.

In the event of a misfire, the firing mechanism can be set to safe which withdraws the firing pin and then the bomb extractor can be used to remove the bomb from the tube.

The 120 mm light mortar can be moved by vehicle, mule, manpack or by air.

Data

Calibre: 120 mm
Length of barrel: with breech piece, 1.632 m
Weight: total, 84 kg
Weight of barrel: 28 kg
Weight of bipod: 24 kg
Weight of baseplate: 32 kg
Elevation: +40 to +85° (711-1,511 mils)
Traverse: 360°
Range: min, 600; max, 4,750 m
Rate of fire: normal, 8 rds/min for 3 min; 15 rds/min for 1 min

Manufacturer

Thomson-Brandt Armements, Tour Chenonceaux, 204 Rond-Point du Pont de Sevres, F-92516 Boulogne-Billancourt.

Status

In production.

Service

French Army and several others.

UPDATED

Thomson-Brandt 120 mm MO-120-M65 light mortar

Description

This 120 mm mortar uses the baseplate and mount of the 120 mm light mortar but the barrel is reinforced and weighs some 10 kg more than that of the light mortar. There is also a difference in the method of moving this mortar. A trolley with pneumatic tyres and elastic suspension is used, on which the entire mortar rests. A muzzle cap with a towing eye is used to enable the mortar to be pulled by a light vehicle of the jeep type. Alternatively, if the necessity arises, the mortar can be dragged along by the detachment, using the towing handle which is also part of the muzzle cap. When the mortar is brought into action, the baseplate is lowered on to the required position and the bipod is used to connect the barrel to the baseplate. The mortar cannot be fired from the wheels.

The mechanism of the mortar allows conventional gravity firing but there is a spring-controlled firing pin which allows the bomb to be loaded into position and fired by means of a lanyard attached to the firing lever. The firing mechanism also allows the firing pin to be withdrawn as a safety measure. The mortar has an all round field of fire without needing to move the baseplate and with the barrel clamp giving an angle of elevation of 60°, there is a top traverse of 300 mils.

Data

Calibre: 120 mm
Length of barrel: with breech piece, 1.64 m
Weight: in firing position, 104 kg; travelling, 144 kg
Weight of barrel and breech piece: 44 kg
Weight of mount: 24 kg
Weight of baseplate: 36 kg
Weight of trolley and muzzle cap: 40 kg
Range: min, 500 m; max using rocket assistance, 9,000 m
Rate of fire: 8 rds/min for 3 min; 12 rds/min for 1 min

Manufacturer

Thomson-Brandt Armements, Tour Chenonceaux, 204 Rond-Point du Pont de Sevres, F-92516 Boulogne-Billancourt.

Status

No longer manufactured. Replaced in production by MO-120-LT (see following entry).

Service

Unspecified armies.

VERIFIED

120 mm MO-120-M65 strengthened light mortar

Thomson-Brandt 120 mm MO-120-LT mortar

Description

The Thomson-Brandt 120 mm MO-120-LT mortar was introduced to replace the MO-120-AM-50. It has a massive baseplate and is transported on a wheeled carriage. The carriage is substantial, with pneumatic tyres and torsion bar suspension, but the mortar is fired off the bipod mount. It is claimed that the mortar on its carriage can be pulled by two men but this can only be over short distances on favourable terrain.

Data

Calibre: 120 mm
Length of barrel: with breech piece, 1.702 m
Weight: in firing position, 168 kg; travelling, 247 kg
Weight of barrel: 62 kg
Weight of mount: 25 kg
Weight of baseplate: 80 kg
Weight of travelling carriage: including towing attachment, 80 kg
Elevation: +45 to +85° (711-1,511 mils)
Traverse: at 60° elevation, 17° (300 mils)
Range: max, M85 HE, 7,000 m
Rate of fire: normal, 8 rds/min; max, 20 rds/min for 1 min

Manufacturer

Thomson-Brandt Armements, Tour Chenonceaux, 204 Rond-Point du Pont de Sevres, F-92516 Boulogne-Billancourt.

Status

In production.

Service

In service with several countries.

VERIFIED

Thomson-Brandt 120 mm MO-120-LT mortar in travelling mode

1996

Thomson-Brandt 120 mm MO-120-LT mortar in action

Thomson-Brandt 120 mm MO-120-RT rifled mortar

Description

The Thomson-Brandt 120 mm MO-120-RT rifled mortar is probably the most complex of current mortars and in some aspects approaches very closely to the gun. It can be fired only off its wheels and can be deployed only in areas to which the towing vehicle has access. It is a massive piece of equipment which fires a heavy bomb out to 13,000 m. It has a rifled barrel and is muzzle loading. To overcome the windage problem the ammunition uses a pre-engraved driving band.

Main components are the barrel, cradle and undercarriage, and the baseplate. The barrel is a substantial forging with the towing eye at the muzzle. The outside is radially finned to increase the surface area for heat dissipation. The interior is rifled and with the pre-engraved driving band imparts rotation to the shell.

The cradle consists of a steel tube connecting the two wheels and carrying the torsion-bar suspension. The traversing gears are totally enclosed to exclude foreign matter. The elevating handwheel rotates a worm and gear assembly which transmits motion, through a multiple disc clutch, to a pinion meshing with the rack formed by the barrel finning. The collar sliding along the barrel produces the necessary change of elevation. The cross-levelling shaft, actuated by the upper left handwheel, tilts the traversing assembly together with the collar to which the sight is attached. The baseplate is very heavy with massive webs on the underside. After a prolonged period of firing the baseplate can be extricated by using the barrel as a lever and employing the towing vehicle to pull the baseplate up.

Although the MO-120-RT is a rifled mortar, it will fire smoothbore bombs except those types having spring-loaded tail fin assemblies with straight fins. Smoothbore bombs are frequently used for bedding in the baseplate (1 round charge 3, 1 charge 5, 1 charge 7) and also for cheaper training. Bombs produced specifically for the MO-120-RT are equipped with a tail tube carrying the primary and secondary cartridges. This tube is ejected just after the bomb has left the mortar and falls about 100 m from the muzzle. An anti-armour bomb has also been developed and is described in the *Mortar ammunition* section.

Variant

120 2R 2M

The 120 2R 2M is an adaptation of the MO-120-RT for mounting in light armoured vehicles such as MOWAG Piranha APCs. In this application, which weighs about one tonne, the rifled barrel is placed inside a recoil collar and allied with a hydraulic aiming system and a semi-automatic loading device. It is still under development.

Data

Calibre: 120 mm
Length of barrel: 2.06 m
Weight: total, 582 kg
Weight of barrel: 131 kg
Weight of baseplate: 194 kg
Weight of carriage: 257 kg
Elevation: +30 to +85°
Traverse: at 60° elevation, 250 mils
Range: min, 1,100 m; max, 13,000 m
Rate of fire: normal, 10-12 rds/min; max, 18 rds/min for a very limited period
Wheelbase: 1.73 m
Overall width: 1.93 m
Overall length: 3.01 m
Height in travelling order: 1.33 m
Ground clearance: 0.32 m
Wheel diameter: 0.7 m
Time into/out of action: 1.5 min/2 min

Manufacturer
Thomson-Brandt Armements, Tour Chenonceaux, 204 Rond-Point du Pont de Sevres, F-92516 Boulogne-Billancourt.

Status
In production.

Service
French Army. In service with 16 countries, including Canada.

Licence production
TURKEY
Manufacturer: MKEK
Type: 120 mm HY 12
Remarks: See entry under *Turkey* for details.

UPDATED

Thomson-Brandt 120 mm MO-120-RT rifled mortar in travelling mode

Thomson-Brandt 120 mm MO-120-RT rifled mortar in action

FLY-K Individual Weapon System TN 8111

Description

The basis of the FLY-K individual weapon system TN 8111 is the employment of a new propulsion unit known as the 'FLY-K' integrated inside the projectile stabiliser, that is the grenade tail unit. The concept provides noiseless, flashless, smokeless and heatless firing, thereby rendering the system undetectable to infra-red surveillance systems because of the absence of any thermal radiation. Tactical applications for the FLY-K system are therefore numerous.

The FLY-K propulsion unit is a thin-walled cylinder of high mechanical strength. The unit contains the propellant and is closed at its upper extremity. A piston is stopped at the end of its downwards travel by a conical stop bushing so propellant combustion forces the piston against the spigot head to propel the grenade. The FLY-K propulsion unit is fully sealed and retains propulsion gases during and after firing. The acoustic signature is 52 dbA at 100 m.

The firing rate is limited only by the skill of the firer (firing is manual) and there is no limit to the number of grenades which can be fired in any one firing sequence. Lateral dispersion is a maximum of 10 m at the maximum range of 675 m.

FLY-K ammunition is fired from a hand launcher which has five components: the propulsion unit, comprising the spigot and its guard tube; the firing unit, including the weapon body and the firing mechanism; a baseplate; a clinometer sight; and a carrying sling and protective cover.

The FLY-K individual weapon system is a reliable, light simple and rugged weapon which can fire at high angles (over 45°) as well as low angles (0-45°). Maximum range is 675 m at 45° elevation on compact terrain.

Ammunition types are the HE TN 208, Illuminating TN 209, Smoke-Incendiary TN 210, and Practice TN 315; all grenades have a 51 mm warhead apart from Illuminating which has a 47 mm calibre warhead. The HE TN 208 grenade weighs 765 g, of which 135 g is Composition B explosive surrounded by a fragmentation sleeve producing 580 fragments scattered over a minimum lethal radius of 16 m. Initial velocity is 88.5 m/s.

Variant

TN 8464 Multilauncher
An alternative launcher is the TN 8464 FLY-K Multilauncher which can launch up to 12 FLY-K grenades in any combination of types. Located on a ground- or vehicle-based traversing baseplate, the FLY-K Multilauncher weighs 100 kg, without ammunition, and is operated using a hand-held remote control unit.

Data

Calibre: 51 mm or 47 mm
Length: overall, 605 mm
Weight: with sight, 4.8 kg
Elevation: 0 to +85°
Range: max, 675 m

Manufacturer

TITANITE SA, F-21270 Pontailler-sur-Saone.

Status

In production.

Service

French armed forces.

VERIFIED

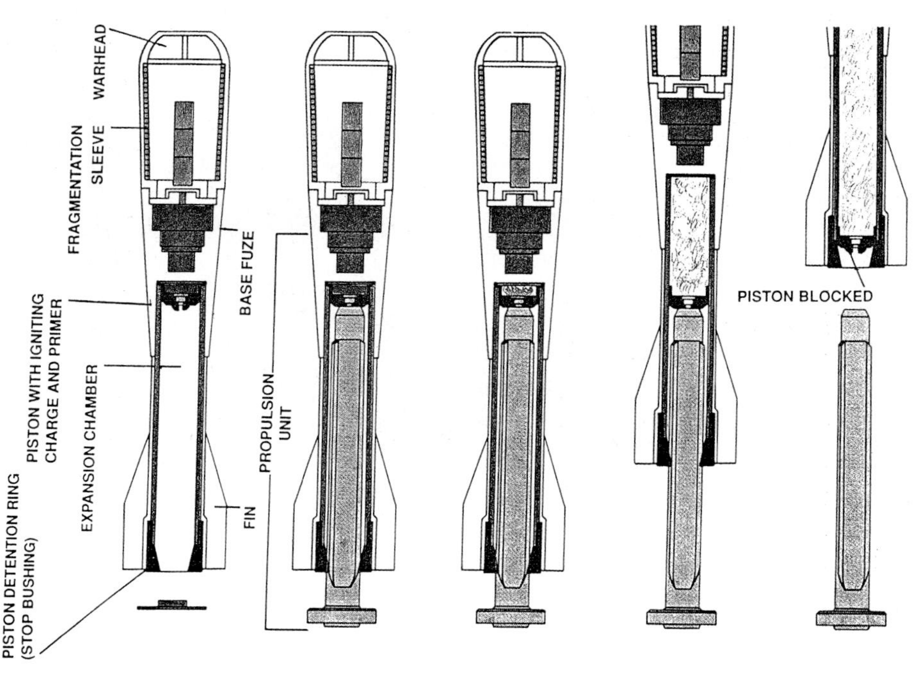

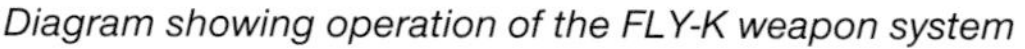
Diagram showing operation of the FLY-K weapon system

FLY-K individual weapon system TN 8111

GREECE

81 mm mortar Type E44

Description

The 81 mm mortar Type E44 was designed by Hellenic Arms Industry (EBO) SA to provide quick, accurate and effective firepower at battalion or company level. The E44 is a conventional drop-fire mortar using a K-type bipod mounting on which all levelling can be carried out on one leg.

The E44 mortar can fire most types of 81 mm projectile.

Data

Calibre: 81 mm
Length of barrel: 1.34 m
Weight: total, 40.9 kg
Weight of barrel: 15.1 kg
Weight of bipod: 14.1 kg
Weight of baseplate: 11.7 kg
Weight of sight: 2.9 kg
Elevation: +45 to +85°
Traverse: at 45° elevation, without moving bipod, ±5°; total, 360°
Max operating gas pressure: 850 bar
Design gas pressure: 1,250 bar
Range: min, 100 m; max, 5,900 m
Rate of fire: max, 30 rds/min; sustained, 16 rds/min

Manufacturer

Hellenic Arms Industry (EBO) SA, 160 Kifissias Avenue, GR-11525 Athens.

Status

In production.

Service

Greek Army.

VERIFIED

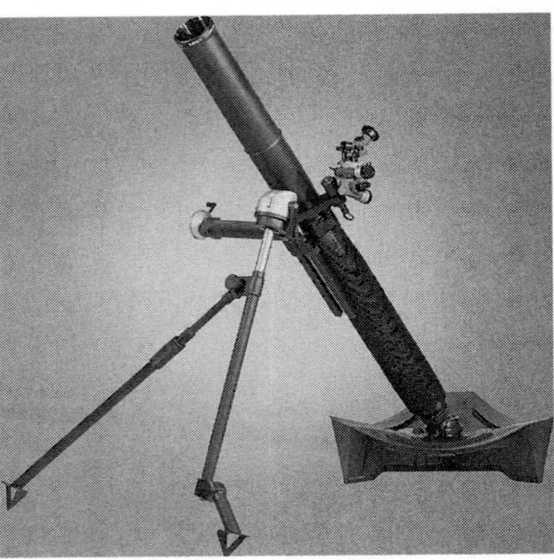

81 mm mortar Type E44

120 mm mortar Type E56

Description

The E56 mortar incorporates features of modern requirements which make it an important weapon in the conduct of land operations. It is mounted on a two-wheel carriage which can be towed by any suitable vehicle. The mortar supported on its special mounting can fire from any type of ground or from an armoured personnel carrier. High accuracy, long range, high rate of fire and robust construction are some of the main features of this mortar.

Data

Calibre: 120 mm
Weight: total, without carriage, 260 kg
Elevation: +40 to +85°
Traverse: at 45° elevation, ±8°; total, 360°
Max operating gas pressure: 1,800 bar
Design gas pressure: 2,350 bar
Range: min, 600 m; max, 9,000 m
Rate of fire: max, 18 rds/min; sustained, 8 rds/min

Manufacturer

Hellenic Arms Industry (EBO) SA, 160 Kifissias Avenue, GR-11525 Athens.

Status

Under development.

VERIFIED

INDIA

51 mm mortar E1

Description

The 51 mm mortar E1 is a direct copy of the British 2 in Mortar adopted by the British Army before 1939. It is still in production in India in the one-man assault form with a simple spade baseplate and a trigger-firing mechanism actuated by a short lanyard. There is no form of mounting as the barrel is held in the correct position by hand; a webbing gaiter is provided around the barrel to protect the firer's hand when the barrel becomes hot after prolonged firing. There are no sights other than an aiming line painted along the barrel. All aiming is carried out using the firer's judgement for matters such as elevation for range and correction for side winds. A rifle sling can be used for carriage.

Three types of 51 mm bomb are produced in India: HE, Smoke and Illuminating.

Data

Calibre: 51.28 mm
Length of barrel: 540 mm
Length overall: 665-670 mm
Weight: complete, 6.56 kg
Weight of barrel: 2.15 kg
Range: min, 200 m; max, 850 m
Rate of fire: normal, 8 rds/min; rapid, 12 rds/min

Manufacturers

Indian Ordnance Factory Board, 10A Auckland Road, Calcutta 700 001.
Rifle Factory, Ishapore, West Bengal.
Small Arms Factory, Kanpur.

Status

In production.

Service

Indian Army.

VERIFIED

81 mm mortar E1

Description

The 81 mm mortar E1 was designed to provide rapid, accurate and heavy sustained fire together with operational and logistic economy. The mortar E1 consists of the usual three basic components, barrel with breech piece, baseplate and bipod. Elevation, traverse and cross-level controls are carried on the bipod, which attaches to the barrel by means of a split collar. The triangular baseplate is formed with spade surfaces on the underside. Three men are required to man-pack the complete weapon which may be carried as a pack load by a single mule.

Ammunition available includes three types of HE bomb, a Smoke WP, and Illuminating. Using a nine-charge system the bombs can be fired to a maximum range of 5,000 m using a maximum muzzle velocity of 295 m/s. Bomb weight is 4.217 kg.

81 mm mortar E1 in action

Data

Calibre: 81.4 mm
Length of barrel: 1.55 m
Weight: total, including cases, sight and accessories, 135 kg
Weight of barrel: 14.5 kg
Weight of bipod: 11.7 kg
Weight of baseplate: 14.4 kg
Elevation: +30 to +85°
Traverse: 7° 30′ to 15° 3′
Range: max, 5,000 m
Rate of fire: with relaying, 6-8 rds/min; without relaying, 20 rds/min
Bomb weight: 4.217 kg

Manufacturer

Indian Ordnance Factory Board, 10A Auckland Road, Calcutta 700 001.
Gun Carriage Factory, Jabalpur.

Status

In production.

Service

Indian Army.

VERIFIED

120 mm mortar E1

Description

The 120 mm mortar E1 was designed to provide heavy and sustained firepower in all phases of battle in various types of terrain. It can provide support to amphibious forces immediately upon landing, and can be effectively deployed in mountainous country.

The mortar is a smoothbore, muzzle-loading weapon which can be easily transported by men, mule, boat, vehicle or aircraft. It can also be parachute-dropped. It consists of the usual three principal components, barrel, bipod and baseplate, and is also provided with a light two-wheeled carriage with pneumatic tyres. The bipod carries the elevating, traverse and cross-level controls and connects to a twin-cylinder recoil buffer mechanism which attaches to the barrel by means of two split collars.

Data

Calibre: 120 mm
Length of barrel: 1.75 m
Weight: total, with sight and accessories, 421 kg
Weight of barrel: with breech piece, 68 kg
Weight of bipod: 86 kg
Weight of baseplate: 80 kg
Weight of carriage: with wheels, 137 kg
Elevation: +45 to +85°
Traverse: 16-31°
Range: max, 6,650 m
Rate of fire: drop fired, 5-10 rds/min; trigger fired, 3-6 rds/min
Pack load: 5 mules; 3 mules without carriage

Manufacturer

Indian Ordnance Factory Board, 10A Auckland Road, Calcutta 700 001.

Status

In production.

120 mm mortar E1 in travelling mode

Service

Indian Army.

VERIFIED

IRAN

37 mm Marsh Mortar

Description

During their war with Iraq, Iranian forces were often deployed in swampy regions where conventional mortars could not be used to great effect as their recoil caused their baseplates to sink into the soft terrain. To overcome this a light mortar specifically designed to operate on marshy terrain was developed.

A series of trials demonstrated that 37 mm was the maximum calibre capable of firing an HE bomb with a viable explosive content while producing minimal launch recoil forces. The 37 mm HE bomb weighs less than 1 kg and maximum possible range is about 600 m. The mortar itself is hand held for drop firing and is provided with a spade baseplate; there are no sights. A webbing sling is provided for carrying along with a webbing gaiter to protect the user's hand against the heat produced during firing.

No other details are available. The term Marsh Mortar is apocryphal.

Manufacturer

Defence Industries Organisation, PO Box 13185-1153, Tehran.

Status

Available.

Service

Iranian armed forces.

VERIFIED

A 37 mm Marsh Mortar can be seen in the centre with an Iranian-produced RPG-7 anti-tank rocket launcher on the left and, on the right, a flare projector produced by converting an obsolete Mauser rifle (T J Gander)

60 mm T1 and T2 Commando mortars

Description

These two 60 mm Commando mortars, the T1 and T2, differ in several ways but both are of the one-man, hand-held assault type with no mounting or substantial baseplate.

The T1 Commando mortar is the longest of the two and provides the longest maximum range (1,050 m), having a spade-type baseplate on which elevation angles can be adjusted and maintained by using a locking screw. A carrying handle which doubles as a steadying handle during firing is provided and a webbing gaiter is fitted around the barrel to protect the user's hand against a hot barrel. Bombs fired from the T1 utilise a three-charge propellant system. It is possible that a manual trigger-firing system is used with the T1.

By contrast the T2 Commando mortar is a far simpler affair, being little more than a drop-firing mortar tube with a rudimentary button-shaped baseplate. Carried on a sling, the T2 has no sighting arrangements other than an aiming line painted along the barrel. A webbing gaiter is located around the barrel to protect the user's hand against the heat produced during firing. Bombs fired from the T2 utilise a two-charge system; maximum range is 800 m.

Data
Calibre: 60 mm
Length: overall, T1, 670 mm; T2, 570 mm
Weight: 6 kg
Elevation: T1, +45 to +85°; T2, max, +80°
Range: T1, 100-1,050 m; T2, 100-800 m
Rate of fire: 15-20 rds/min

Manufacturer
Defence Industries Organisation, PO Box 13185-1153, Tehran.

Status
In production.

Service
Iranian armed forces.

VERIFIED

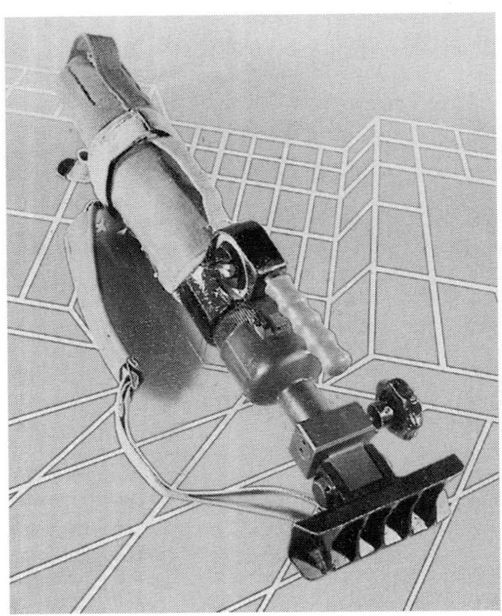

60 mm T1 Commando mortar

60 mm T2 Commando mortar

Hadid 60 mm mortar

Description
The Hadid 60 mm mortar appears to be a direct copy of the Israeli Soltam 60 mm Standard mortar, complete with the simple drum sight close to the muzzle for use when the mortar is deployed in the hand-held commando role. When mounted on a bipod and standard (380 mm maximum diameter) baseplate the Hadid 60 mm mortar utilises a rather complex sight unit clamped onto the bipod assembly. The mounting is separated from the barrel by two recoil cylinders.

Data
Calibre: 60 mm
Length of barrel: with breech piece, 740 mm
Weight: in firing position, 17.5 kg
Weight of barrel and breech piece: 6 kg
Weight of baseplate: 6.5 kg
Weight of sight unit: 1 kg
Elevation: +43 to +85°
Range: max, 2,500 m
Rate of fire: 30 rds/min

Manufacturer
Defence Industries Organisation, PO Box 13185-1153, Tehran.

Status
In production.

Service
Iranian armed forces.

VERIFIED

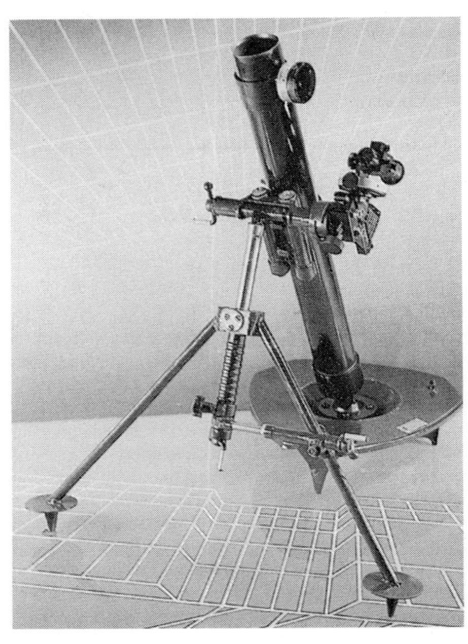

Hadid 60 mm mortar

Hadid 81 mm mortar

Description
The Hadid 81 mm mortar is another copy of an Israeli Soltam design, in this case the Soltam 81 mm Standard with a one-piece 1.56 m barrel. As far as can be determined the Iranian Hadid 81 mm mortar is identical to the Israeli original (qv).

Data
Calibre: 81 mm
Length of barrel: 1.56 m
Weight: in firing position, 50.5 kg
Weight, barrel and breech piece: 18 kg
Weight of baseplate: 17.5 kg
Weight of sight unit: 1.5 kg
Elevation: +43 to +85°
Range: max, 5,200 m
Rate of fire: 20 rds/min

Manufacturer
Defence Industries Organisation, PO Box 13185-1153, Tehran.

Status
In production.

Service
Iranian armed forces.

VERIFIED

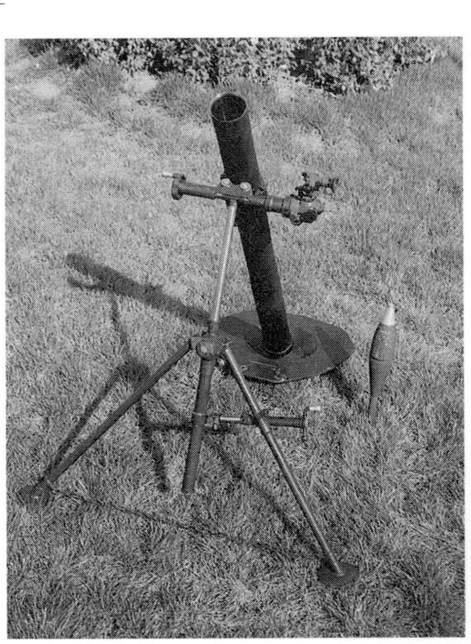

Hadid 81 mm mortar

Hadid 120 mm mortar

Description

The Hadid 120 mm mortar is another copy of an Israeli Soltam design, in this case the Soltam 120 mm Standard K6 with a 1.73 m barrel and complete with two-wheel travelling carriage and accessories. As far as can be determined the Iranian Hadid 120 mm mortar is identical to the Israeli K6 (qv).

Data

Calibre: 120 mm
Length of barrel: 1.73 m
Weight: in firing position, 138.5 kg
Weight, barrel and breech piece: 43 kg
Weight of sight unit: 1.5 kg
Elevation: +43 to +85°
Range: max, 6,200 m
Rate of fire: max, 10 rds/min

Manufacturer

Defence Industries Organisation, PO Box 13185-1153, Tehran.

Status

In production.

Service

Iranian Army.

VERIFIED

IRAQ

60 mm 'Al-Jaleel' light mortar

Description

The 60 mm 'Al-Jaleel' light mortar is a conventional drop-fired mortar. The bipod carries the cross-level, elevation and traversing apparatus and is connected to a twin recoil buffer unit which holds the barrel in a quick-release clamp. The barrel has a removable breech piece which is fitted with a ball unit engaging in the rectangular baseplate.

The mortar is fitted with an NSB-3 sight unit which is of the usual collimating pattern with Betalight illumination.

Data

Calibre: 60 mm
Weight: in firing position, 22 kg
Elevation: +45 to +85°
Range: max, 2,500 m
Rate of fire: 25-30 rds/min

Manufacturer

State arsenals.

Status

In production.

Service

Iraq Army and offered for export sales.

VERIFIED

60 mm 'Al-Jaleel' mortar

82 mm 'Al-Jaleel' mortar

Description

The 82 mm 'Al-Jaleel' mortar is a conventional drop-fired mortar and is generally similar in its arrangements to the 60 mm model described previously. The NSB-4A sight unit is fitted with an optical elbow telescope.

Data

Calibre: 82 mm
Weight: in firing position, 63 kg
Elevation: +45 to +85°
Range: max, 4,900 m
Rate of fire: 20-25 rds/min

Manufacturer

State arsenals.

Status

In production.

Service

Iraq Army and offered for export sales.

VERIFIED

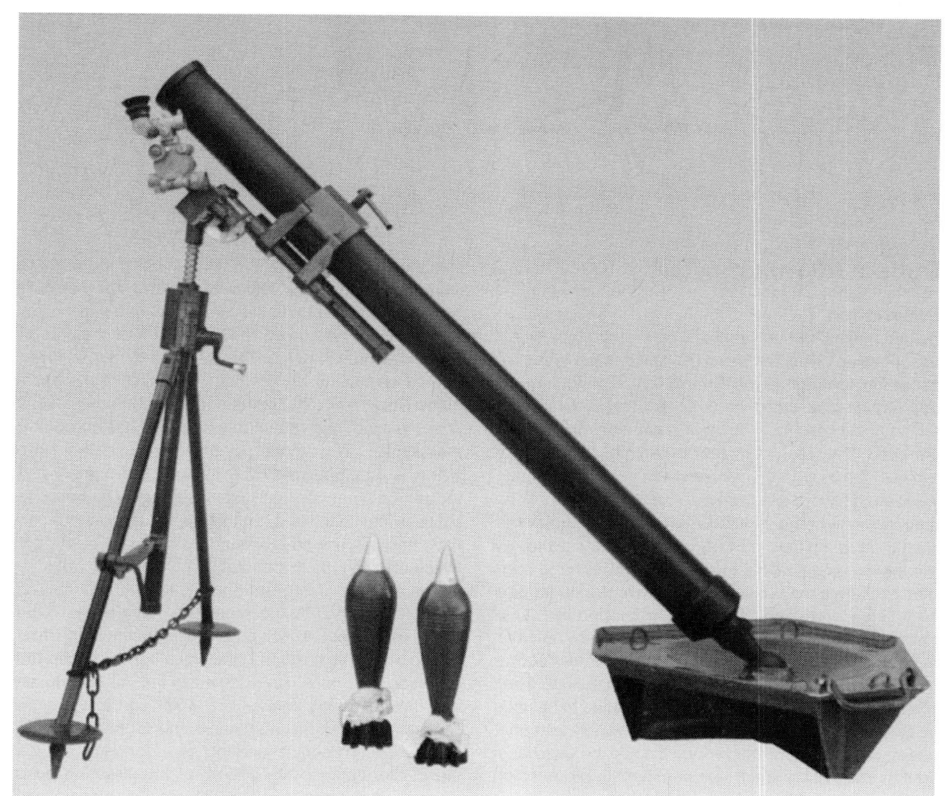

82 mm 'Al-Jaleel' mortar

120 mm 'Al-Jaleel' mortar

Description
The 120 mm 'Al-Jaleel' mortar is a conventional pattern of drop-fired heavy mortar. The barrel has a removable breech piece which fits into the rectangular baseplate. The bipod carries the cross-level, elevation and traverse mechanisms and attaches to a twin recoil buffer unit which holds the barrel in a quick-release clamp. The NSB-4A sight unit employs an optical elbow telescope.

Data
Calibre: 120 mm
Weight: in firing position, 148 kg
Elevation: +45 to +85°
Range: max, 5,400 m
Rate of fire: 5-8 rds/min

Manufacturer
State arsenals.

Status
In production.

Service
Iraq Army and offered for export sales.

VERIFIED

120 mm 'Al-Jaleel' mortar

ISRAEL

TAAS 52 mm mortar

Description
The TAAS 52 mm mortar is a conventional smoothbore muzzle-loaded mortar. It is of very simple design, consisting of a tube and a trough-like baseplate mounted across the base of the barrel. Connecting the barrel to the baseplate is the breech piece which contains the firing mechanism. The firer operates the firing mechanism by rotating the small wheel on the right of the barrel. There are no sights as such and aiming is carried out by using the white line on the barrel to align the bore. The mortar can be used for low- or high-angle fire, is easily managed by one man, and has a carrying handle and a muzzle cover. High explosive, illuminating and smoke bombs are all fired.

Data
Calibre: 52 mm
Length of barrel: 490 mm
Length: overall, with baseplate, 673 mm
Weight: overall, with baseplate, 7.9 kg
Weight of baseplate: 1.3 kg
Baseplate dimensions: 150 × 85 × 35 mm
Elevation: +20 to +85°
Traverse: total, 360°
Range: min, 130 m; max, 420 m
Rate of fire: 20-35 rds/min

Manufacturer
TAAS - Israel Industries Limited, PO Box 1044, Ramat Hasharon 47100.

Status
In production.

Service
Israel Defence Forces.

VERIFIED

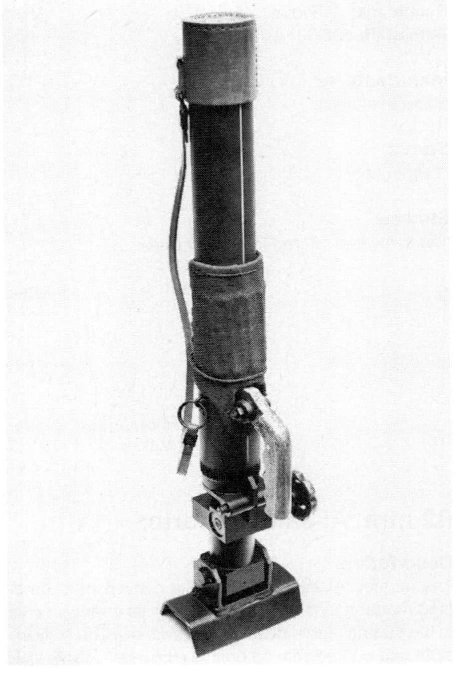

TAAS 52 mm mortar

Soltam 60 mm mortars

Description
Soltam makes four versions of a 60 mm mortar which it describes as 'Type Tampella' mortars because they are based on designs by the Finnish firm. The four weapons, which are designated Long Range, Standard, Commando and Under-Armour have many features in common. The Standard mortar can be used with or without its bipod; the Commando is a short-range weapon, is always used without a bipod and has a firing mechanism instead of a fixed pin. The Under-Armour mortar is a unique muzzle-loaded mortar which is mounted inside a tank's turret or an APC fighting compartment. This mortar is operated while the operator is completely protected. The main description which follows relates to the Standard and is followed by notes on the other weapons and a table of comparative data.

The Standard mortar was designed to fulfil three roles. Equipped with a bipod it was intended to be used in the normal way as a company support weapon; alternatively, without the bipod, it could be used as a light mortar in the assault role, or externally mounted on a tank turret or APC roof.

In the standard role, the mortar consists of three elements, barrel, bipod and baseplate. The barrel is made of alloy steel and the bore is given a particularly smooth finish. Screwed to the bottom of the mortar tube is the breech piece. This is rounded off at its lower end into a ball which fits into the socket of the baseplate. The striker pin is contained in the breech piece. It is of the fixed, non-retractable type and has no control mechanism. The impact of the bomb on the fixed striker is enough to set off the cap of the primary cartridge. A canvas sleeve is placed around the barrel to act as a heat shield and to allow the attachment of a carrying handle. Around the muzzle is a simple drum sight which is used only in the attack or Commando role.

The bipod is of conventional design with the spread of the legs limited by the construction of flanges in the plates where they are joined at the top of the legs. Passing up through the bipod is the elevating screw and the elevating column which is joined to the yoke. This carries the traversing screw. At the left end is the sight bracket, while at the right-hand end is the traversing handwheel. The cross-levelling gear connects the left leg of the bipod to the bottom of the elevating gear. Thus when the mortar is brought into action on uneven ground, the elevating screw and column can be swung to allow the sight to be brought to the upright position. Two recoil cylinders are interposed between the barrel collar and the yoke. The sight has a clamp which will normally be placed around the tubular yoke but can also be placed around the barrel when so required. There is a bearing scale which runs from 0 to 6,400 mils. The target or aiming post is viewed over a collimator. The elevation scale is set out in five columns corresponding to the ranges achieved by the primer and the four charges and the range reader is scribed on a perspex plate attached to the body of the sight. When the range is applied, the body of the sight is thrown out of level and to centre the elevation bubble the barrel must be elevated or depressed.

The baseplate is of welded construction with a flat, dished top plate and three webbed ribs which enable the baseplate to take the recoil force without slipping. In the centre of the baseplate is a socket to take the ball at the bottom of the breech piece. The socket has a clamping ring with a securing screw that can be used to clamp the barrel rigidly to the baseplate. The construction of the baseplate is such that a 360° traverse can be obtained. The second way of using the mortar is in the assault role. Here the bipod is not used. The normal sight can be secured to the muzzle or a simple drum sight can be used. The barrel is controlled by hand until it is required to go to fire for effect, the clamp on the baseplate is then used to lock the barrel at the required elevation in the desired line.

The Long Range mortar was previously described as the Type A. It differs from the Standard version by having a longer and heavier barrel designed to throw a bomb to 4,000 m with a rate of fire of 20 rds/min. For this a stronger breech is also required.

The Commando is the lightweight version of the Standard mortar and it is a much simplified design. The barrel is considerably shorter than the Standard and has no bipod. There is a simple trigger mechanism incorporated in the breech, and this is rigidly attached to a small baseplate. The whole mortar can be conveniently carried and fired by one man. A sight is clipped to the barrel, and is capable of indicating elevation and azimuth, but it is probable that the majority of firing would be undertaken by rough alignment and judgement.

The 60 mm Under-Armour mortar enables the user to conduct a complete mortar firing mission while being completely protected inside a tank turret or other fighting compartment.

The system covers ranges up to 3,500 m and provides screening, illumination and the advantage of hitting soft targets and threatened personnel with an effective, accurate, low-cost ammunition. The smoothbore muzzle-loading mortar fires all types of existing 60 mm mortar rounds, HE, smoke and illuminating.

The system, mounted on the turret or vehicle roof, pivots on trunnions to a bolted base which transmits the firing loads to the vehicle body.

For laying the mortar mount is parallel to the main armament. Deflection is controlled through the gunner's control handle, while the elevation is set by the mortar operator. The mortar is equipped with a quadrant elevation indicator which indicates the ranges according to the charge selected.

For loading, the barrel must be unlocked and the round can then be inserted into the muzzle. The barrel is then pushed forward and locked. For firing the mortar is fitted with a trip-action firing mechanism, and is also equipped with safety devices to prevent accidental double loading or firing when the barrel is in the loading position.

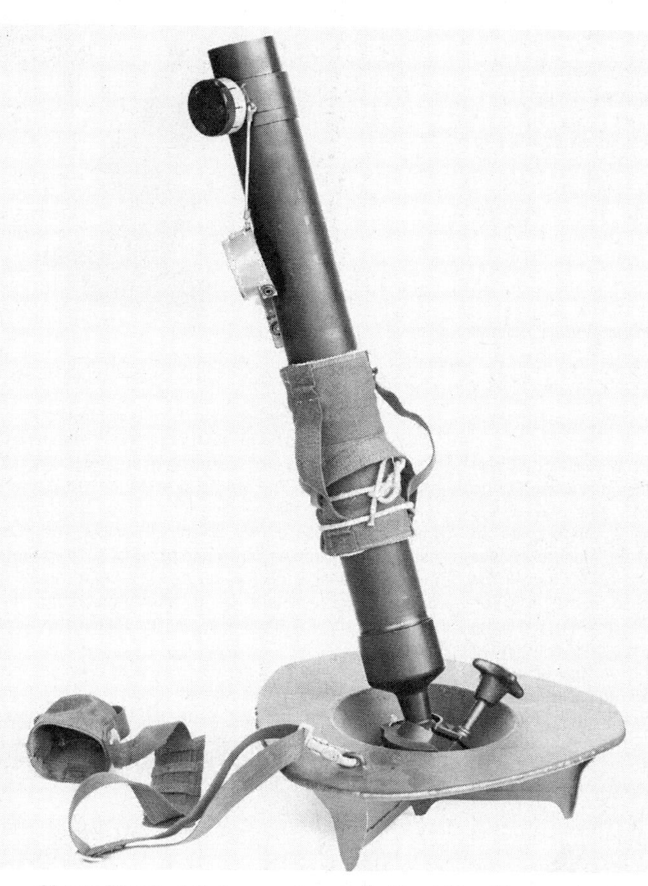

60 mm Standard Soltam mortar used without bipod in assault role

60 mm Standard Soltam mortar set up as conventional company support weapon showing sight on bipod and drum sight at muzzle

60 mm Long-Range Soltam mortar

60 mm Under-Armour mortar

Data

	Under-Armour	Long Range	Standard	Commando Regular	Commando Light
Type	**C-04**	**C-06**	**C-08**	**C-03**	**C-576**
Calibre	60 mm	60 mm	60 mm	60 mm	60 mm
Max range	3,500 m	4,000 m	2,550 m	1,000 m	1,600 m
Min range	150 m	150 m	150 m	100 m	100 m
Total weight in firing position	62 kg	18 kg	16.3 kg	6.7 kg	6.3 kg
Barrel	10 kg	6.7 kg	5.7 kg	6 kg	4.6 kg
Bipod	–	4.7 kg	4.5 kg	–	–
Baseplate	–	5.5 kg	5.1 kg	1.1 kg	1.7 kg
Sight	–	1.1 kg	1 kg	0.5 kg	0.5 kg
Barrel length	800 mm	940 mm	740 mm	725 mm	675 mm
Bipod length, folded	–	640 mm	540 mm	–	–
Baseplate diameter	–	350 mm	350 mm	–	200 mm

Manufacturer
Soltam Limited, PO Box 371, Haifa.

Status
All are in current use and available.

Service
Standard mortar in service with Israeli forces. All types supplied to many other countries.

Licence production
IRAN
Manufacturer: Defence Industries Organisation
Type: 60 mm Standard
Remarks: Production may not be authorised. See entry under *Iran* for details.

VERIFIED

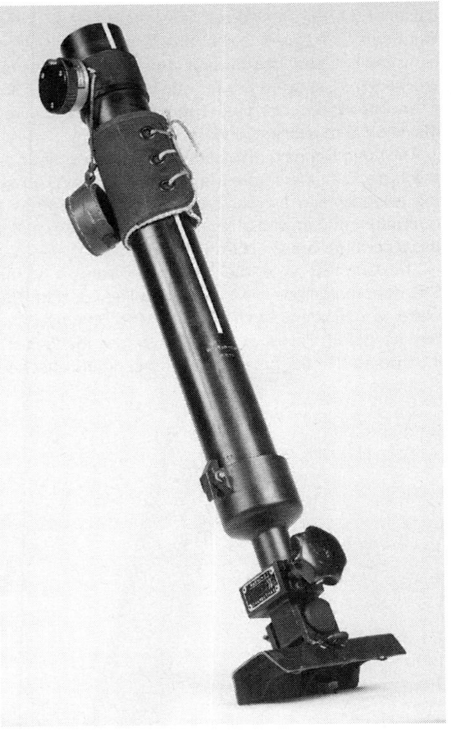

60 mm Commando Soltam mortar

Soltam 81 mm mortars

Description
As with the 60 mm versions, the intention with the 81 mm mortars made by Soltam is to provide variations in the design to suit all requirements and to do this with minimal alterations to the basic characteristics and manufacturing processes. The mortars are of a conventional pattern, based on the Tampella designs and using as many common items as possible.

There are four versions of the 81 mm, namely the Standard type with either a one-piece or split barrel, and the Long-Range type with a reinforced barrel similarly having a one-piece or a split barrel.

In each case the barrel is made from a high grade steel with a honed finish to the bore and a breech piece which fits into a socket in the baseplate. The breech piece carries a fixed firing pin although a controllable firing pin can be fitted.

The variety of barrels provides a flexibility of use which is unusual in modern mortars. The Long Range version is obviously less mobile than the others because of its extra weight, but it can provide an impressive range of 6,500 m and for many purposes the range justifies the slight extra load of the barrel. The Standard mortar could be looked on as the basic version and its performance is similar to others of the same size and weight. The split-barrelled mortar is intended for greater mobility when carrying on foot or in light vehicles, though the penalty for the reduced bulk is a small increase in weight because of the screw collar in the middle of the barrel.

The bipod is of orthodox pattern but has some interesting features. The two legs of the bipod have spiked feet and the distance between them is controlled by an adjustable chain. The elevating column is attached to the yoke and has a cross-levelling gear connecting it to the left bipod leg. The elevating screw thread is totally enclosed within the elevating tube and so is protected from the ingress of dirt, sand and snow. The yoke carries the traversing screw and protects it. The traversing handwheel is on the left and the sight mounting on the right. There is a shock-absorbing unit consisting of two cylinders joining the yoke to the barrel clamp.

The baseplate is of circular shape with welded ribs on the underside. The socket takes the ball of the breech piece and allows a full 360° traverse without need to move the baseplate. A carrying handle is attached and is needed to extract the baseplate when coming out of action.

Soltam 81 mm mortar with split barrel

Soltam 81 mm mortar

The sight is graduated in mils and covers the full 6,400 traverse. There is also a slipping scale. This enables the sight to be used on an aiming post and the target recorded in terms of a bearing rather than a switch from the point of aim. This permits registration of the target and subsequent engagement of the target without further ranging. If meteorological information is available, then the target data can be corrected and stored in a true form. The elevation scale and the cross-levelling are controlled by bubbles and a collimator is used for sighting. A quick-release is available for the traversing scale.

Soltam has developed an 81 mm mortar mounting suitable for installation in M113 and similar APCs.

Manufacturer

Soltam Limited, PO Box 371, Haifa.

Status

Available.

Service

Israeli forces. Exported to other countries and can be found in East Africa.

Three manpack loads of split barrel Soltam 81 mm mortar

Data

Mortar

	Standard		Long Range	
	Solid Barrel	**Split Barrel**	**Solid Barrel**	**Split Barrel**
Max Range	4,900 m	4,900 m	6,500 m	6,500 m
Min Range	150 m	150 m	150 m	150 m
Elevation	43-85°	43-85°	43-85°	43-85°
Traverse	360°	360°	360°	360°
Total weight	42 kg	44 kg	49 kg	51 kg
Barrel	15.5 kg	17.5 kg	22.5 kg	24.5 kg
Bipod	14 kg	14 kg	14 kg	14 kg
Baseplate	12.5 kg	12.5 kg	12.5 kg	12.5 kg
Length of barrel	1.56 m	1.56 m	1.583 m	1.583 m
Bipod length, folded	960 mm	960 mm	960 mm	960 mm
Baseplate diameter	518 mm	518 mm	518 mm	518 mm
Weight of sight	1.55 kg	1.55 kg	1.55 kg	1.55 kg
Traversing scale	6,400 mils	6,400 mils	6,400 mils	6,400 mils
Elevating scale	700-1,600 mils	700-1,600 mils	700-1,600 mils	700-1,600 mils

Licence production

IRAN

Manufacturer: Defence Industries Organisation

Type: 81 mm Standard

Remarks: Production may not be authorised. See entry under *Iran* for details.

VERIFIED

120 mm Soltam light mortar K5 and K6

Description

This mortar was designed to provide infantry with a mortar capable of firing the very lethal 120 mm ammunition but sufficiently light to enable it to be moved by a jeep-type vehicle, transported by a mule, carried by a detachment of three soldiers, moved in a helicopter or dropped by parachute. The light mortar can fire all the 120 mm bombs available including the rocket-assisted bomb. The K6 is an updated version of the older light mortar and has been adopted by the US Army as the M120/121 (see entry under *United States* for details).

The barrel is an alloy steel tube with a honed interior. The lower end is externally screw threaded to take the breech piece. The breech piece holds the striker. This is a fixed stud on which the bomb falls under gravity.

Soltam 120 mm light mortar K6 in travelling mode

South African National Defence Force Soltam 120 mm light mortar K6 in firing position (T J Gander)

However, to ensure safety while dealing with a misfire, there is a safety catch which, when rotated, draws the pin back into the interior of the breech piece. The lower end of the breech-piece is shaped into a ball which enters a socket in the baseplate. The barrel can be carried by a single man.

The bipod legs have spikes at the bottom and are joined at the elevation gear housing. The distance between the legs is controlled by an adjustable length chain. The lower end of the elevation column, which contains the elevation screw thread is attached to the left leg of the bipod by a cylinder containing a screw thread. The rotation of the handle at the end of this thread moves the elevation column out of the vertical and so allows the mortar to be cross-levelled to allow for irregularities in the ground. At the top end of the elevating column is the yoke which holds the sight at the left and the traversing handle at the right. The thread of the traversing gear is, like all the threads in this mortar, completely enclosed. The yoke is connected to the barrel clamp by a pair of shock-absorbing units which ensure that the barrel-yoke position does not change as the barrel recoils and then returns to its original firing position.

The baseplate is a heavy welded steel dish which has a socket at the centre to take the breech piece. The mortar can traverse through 360° without shifting the baseplate. The ribs welded on to the underside of the plate prevent the baseplate from slipping sideways.

The sight incorporates a 360° (6,400 mils) traverse scale and a slipping scale. Thus when the zero line has been recorded, the slipping scale can be adjusted to zero. The mortar can therefore engage targets, register at the conclusion of the action and re-engage without recourse to further ranging. Large switches are made easier by the incorporation of a quick-release lever which enables the sight head to be rotated rapidly on to a new bearing without laborious rotation of low geared handwheels. Elevation and cross-level are controlled by bubbles and a collimator is used to lay on the aiming post.

Tools supplied include one which can be inserted into the muzzle to engage around the fuze of a bomb and is used to withdraw it in the event of a misfire.

The carriage is a lightweight two-wheeled vehicle with a torsion-bar suspension. It enables the mortar to be pulled by troops. It is parachutable and can be towed behind any vehicle with a towing hook of the right height and size. In some roles the carriage is not used and the mortar is brought into action without it. In addition to supporting the mortar, the carriage carries the spare parts and tools for the weapon and also has six metal containers set across the axle, allowing the carriage of that number ready-to-use bombs.

The mortar takes the M48 and M57 series of rounds.

The K6 is an updated version of the light mortar and incorporates a new reinforced barrel. The firm quotes this barrel as being the one for use with the long-range M57 ammunition and it is this barrel which gives a maximum range of 7,200 m. All other dimensions and performance parameters are unaltered.

Soltam has developed a K6 mounting suitable for installation in M113 or similar APCs.

Data

	Standard	Long Range
Type	**K5**	**K6**
Weight travelling	266 kg	322 kg
Max range	6,250 m with M48 round	7,200 m with M57 round
Min range	250 m	200 m
Elevation	43-85°	40-85°
Traverse	6,400 mils	6,400 mils
Total weight, firing position	140 kg	144 kg
Barrel	46.5 kg	50 kg
Bipod	31.5 kg	32 kg
Baseplate	62 kg	62 kg
Length of barrel	1.758 m	1.79 m
Bipod length, folded	1.04 m	1.23 m
Baseplate diameter	880 mm	880 mm

Manufacturer
Soltam Limited, PO Box 371, Haifa.

Status
Available.

Service
Israeli Defence Forces and some other armies, including the South African National Defence Force.

Licence production
IRAN
Manufacturer: Defence Industries Organisation
Type: 120 mm K6
Remarks: Production may not be authorised. See entry under *Iran* for details.

UPDATED

Soltam 120 mm A7 long-range mortar

Description
The Soltam 120 mm A7 long-range mortar was designed for rapid deployment and for operation by fewer men than previous models. The A7 has a maximum range of 8,500 m using the M59 bomb and 9,500 m with the M100 bomb.

The Soltam A7 can be handled by only two or three men. The carriage has a lightweight torsion bar suspension from which the weapon is not dismounted for use, although when being fired the mortar does not rest on its wheels.

If necessary, the carriage can be removed if the weapon is to be carried by sleigh or pack animals.

Manufacturer
Soltam Limited, PO Box 371, Haifa.

Status
In production.

VERIFIED

Soltam 120 mm heavy mortar in travelling mode

Soltam 120 mm heavy mortar in firing position

Soltam 120 mm M-65 standard mortar

Description
The Soltam 120 mm M-65 standard mortar is a substantial mortar which can be used in the traditional way and can also be mounted in an armoured personnel carrier or similar vehicle.

The barrel is made of high tensile alloy steel and is given a very good internal finish. The breech piece screws into the bottom of the barrel and holds the firing mechanism. The spring-loaded striker is fired by pulling a firing lever which cocks the spring before allowing it to

120 mm M-65 Soltam standard mortar in travelling position

trip over to fire the bomb. The firing pin can be retracted to a safe position and this is of particular value when clearing a misfire. The ball-shaped end of the breech piece fits into the socket of the baseplate. At the front end of the barrel is a bayonet-type catch into which the towing eye is secured. This is sprung to reduce the shocks of travelling.

The bipod is, in principle, similar to that of the light 120 mm mortar. All exposed and sliding components are either chromium-plated or made of stainless steel to resist wear or corrosion and all the gears and screw-threaded columns are totally enclosed.

The baseplate is a circular dish-shaped plate of steel with strengthening ribs welded on. There are also carrying handles and eyes for securing the baseplate to the carriage welded onto the upper side of the baseplate.

The sight employed on the standard mortar is the same as that used on the 120 mm light mortar.

The carriage is more substantial than that used with the light mortar. It is of box construction made up of welded steel sheet. The top portion carries the barrel clamping collar and baseplate clamping hooks. The wheels are of standard jeep size and type. Drag rings are mounted on the wheel hubs for manhandling the mortar across country. The frame of the cross member is boxed in to take tools and has a lid. There are welded attachments to which can be secured the cleaning rods, carriage stay and baseplate levering rods during transit.

Data

Type	**Standard M-65**	**Long range A7**
Weight, travelling	351 kg	387 kg
Max range	6,500 m with M48 bomb	8,500 m with M59 bomb
Min range	400 m	400 m
Elevation	43-85°	43-85°
Traverse	360°	360°
Crew	6	4
Total weight, firing position	231 kg	387 kg
Barrel	82 kg	102 kg
Bipod	69 kg	70 kg
Baseplate	80 kg	85 kg
Length of barrel	1.95 m	2.154 m
Bipod length folded	1.675 m	1.52 m
Baseplate diameter	900 mm	900 mm
Sight weight	1.5 kg	1.5 kg
Traversing scale	6,400 mils	6,400 mils
Elevating scale	700-1,600 mils	700-1,600 mils

Manufacturer
Soltam Limited, PO Box 371, Haifa.

Status
Production complete.

Service
Israeli Defence Forces and some other armies.

VERIFIED

Soltam 160 mm M-66 mortar

Description
The Soltam 160 mm M-66 mortar mortar, an adaptation of the Finnish Vammas 160 mm M58 mortar, fires a 40 kg bomb out to 9,600 m. To do this requires a heavy weapon (1,700 kg in the firing position) and a detachment of six to eight men.

The barrel is a high tensile strength, alloy steel tube. At the bottom end is the breech piece which contains the firing mechanism. This is operated by a firing lever which allows the spring-loaded striker to trip off the sear and go forward to fire. The striker can be withdrawn to provide safety, particularly in the case of a misfire.

There is no bipod mounting for this weapon, instead the barrel is elevated and depressed by a single column which is part of the carriage. The weight of the bomb makes conventional loading difficult so it is necessary to lower the barrel to a loading position. This is done by folding back the elevating strut from a hinge at its mid-point. There is a counterbalance mechanism which makes it easy to elevate the barrel each time it is loaded. This consists of steel cables attached to the barrel some 2 m from the muzzle and led over sheaves into the lower pair of tubes that make up the chassis of the carriage.

The axles allow the road wheels to be turned in and locked by a cam plate attached to the axle tube. The off-side road wheel incorporates a clutch and handle to allow slow traverse. The carriage rolls on its wheels through a complete 360° circle without moving the baseplate.

The baseplate is a heavy flat disc welded with a number of webs to give stiffness and to prevent the tendency to slide sideways. Four handles are welded to the top-plate to allow carriage and baseplate extraction. There is a central socket into which the tail of the breech piece fits. This allows a full 360° traverse without movement of the baseplate. There is a spring-loaded locking arrangement to ensure that the breech piece cannot leave the baseplate.

Data
Calibre: 160 mm
Length of barrel: 2.85 m
Weight: in firing position, 1,700 kg; travelling, excluding baseplate, 1,450 kg
Weight of baseplate: 250 kg
Weight of sight: 1.57 kg
Elevation: +43 to +70° (764-1,244 mils)
Traverse: total, 360°
Range: min, 600 m; max, 9,600 m

Manufacturer
Soltam Limited, PO Box 371, Haifa.

Status
Available.

Service
Israeli Army.

UPDATED

160 mm M-66 mortar in travelling position

160 mm M-66 Soltam mortar in firing position

ITALY

Breda 81 mm light mortar

Description

The Breda 81 mm light mortar follows most other types in having a circular baseplate to permit all-round traverse without having to re-bed the plate, and a sturdy bipod. A special feature is the long barrel, specifically intended to give long range without resort to special bombs or unconventional means of propulsion.

Another feature is a means of withdrawing the striker when there is a misfire. The striker can be pulled out from the breech, leaving the barrel entirely safe for the bomb to be loosened and withdrawn without the danger of it slipping back on to the striker and exploding.

Two small recoil cylinders have been interspersed between the bipod clamp and the legs to reduce the shock load on the legs and the sight. This arrangement improves the consistency and reduces the number of times that the mortar has to be relaid. The cradle is made of aluminium alloy and it is held to the barrel by a half-ring locked by an over-centre clamp. The sight fits on to a bayonet fitting.

Traversing is by a horizontal screw on the lower side of the cradle and a nut on the top of the bipod legs travels along this screw, turned by a small wheel on the right-hand side.

Cross-levelling is by means of a screw adjuster running from one leg to the central column.

The mortar breaks down for carriage into three basic loads, each about 15 kg, which allows the weapon to be carried by three men, though more would be needed for the ammunition.

81 mm Breda light mortar with ammunition; the sight is not fitted

Data

Calibre: 81 mm
Length of barrel: 1.455 m
Weight: complete, 43 kg
Weight of barrel: 14.5 kg
Weight of baseplate: 13 kg
Weight of bipod: without sight, 15.5 kg
Elevation: +35 to +85°
Range: min, 75 m; max, 5,000 m
Rate of fire: max, 20 rds/min

Manufacturer

Breda Meccanica Bresciana SpA, Via Lunga 2, I-25126 Brescia.

Status

In production.

Service

Italian Army.

VERIFIED

KOREA, SOUTH

60 mm KM181 mortar

Description

The 60 mm KM181 mortar is a light mortar, similar in design to the US M224. It is capable of firing any current 60 mm ammunition, and has a fixed firing pin for drop-firing only. The CN81 barrel is finned for cooling, permitting a sustained rate of fire of 20 rds/min. The bipod is symmetrical, with a traversing screw at the top and a cross-levelling strut at the bottom of the central elevating tube. Cross-levelling is performed by screwing a sleeve on the left tripod leg up or down. The sight unit is provided with a self-powered tritium light source to permit operation at night.

The mortar can be carried by the two-man crew; one man carries the entire mortar, a load weighing 21.2 kg, while the second man carries eight bombs weighing 17.4 kg.

Data

Calibre: 60 mm
Length of barrel: 1 m
Weight: in firing position, 19.5 kg
Weight of barrel: 5.5 kg

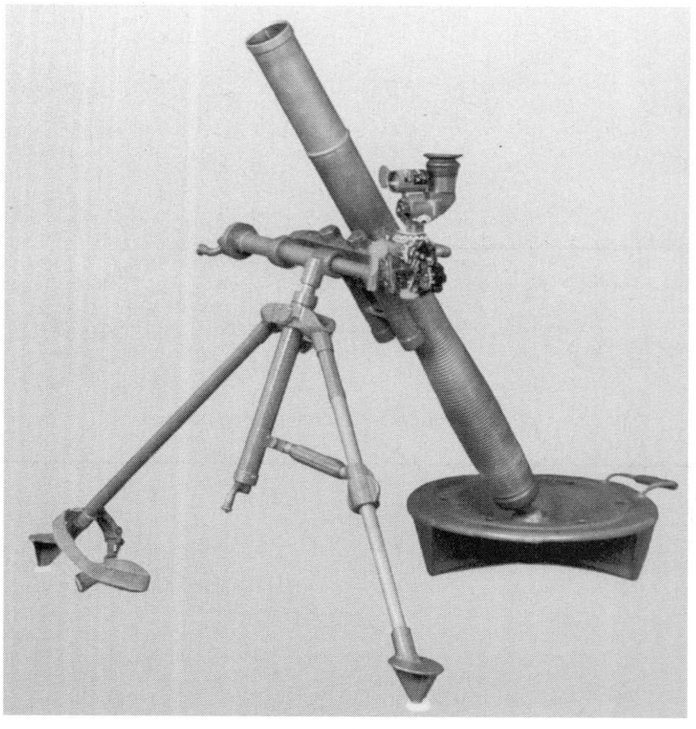

60 mm KM181 light mortar

Man-carrying the 60 mm KM181 light mortar

Weight of bipod: 6 kg
Weight of baseplate: 6.5 kg
Weight of sight: 1.5 kg
Elevation: 800-1,511 mils
Traverse: 250 mils
Range: max, 3,590 m

Manufacturer
Kia Machine Tool Company Ltd, Kia Building 4th Floor, 15 Yoido, Youngdeungpo-gu, Seoul.

Status
In production.

Service
Republic of Korea Army.

VERIFIED

PAKISTAN

60 mm light mortar

Description
This 60 mm light mortar is based on the Chinese NORINCO 60 mm Type 63 but has been improved in various details, notably by an increase in the weight and strength of the barrel. This permits a more powerful propelling charge to be used and thus provides greater range.

The light mortar is of conventional drop-fired pattern, using a rectangular baseplate and a symmetrical bipod. The bipod collar fits around the barrel and is connected to it by means of a recoil buffer, behind which is a carrying handle. The bipod has a simple screw traverse mechanism and cross-levelling is catered for by one telescopic leg on the bipod.

Data
Calibre: 60 mm
Length of barrel: 623 mm
Weight: in firing position, 14.8 kg
Weight of barrel: 4.5 kg
Weight of baseplate: 5.7 kg
Weight of bipod: 4.6 kg
Elevation: +45 to +80°
Range: min, 50 m; max, 2,000 m
Rate of fire: 8 rds/min

Manufacturer
Pakistan Machine Tool Factory Limited, Landhi, Karachi 34.

Status
In production.

Service
Pakistan armed forces.

VERIFIED

81 mm light mortar

Description
The design of this 81 mm mortar appears to be based upon that of the Thomson-Brandt 81 mm MO-81-LC. It is a conventional drop-fired weapon using a triangular baseplate and a symmetrical bipod. Elevation is provided by a central screw mechanism in the bipod, traverse by a cross-screw at the top of the bipod, and cross-level by a telescoping leg connected to the central, elevating screw casing. An optical panoramic sight is provided on the left side.

The mortar can be broken down into its three basic components rapidly and can be pack-loaded, man-carried or parachute-dropped.

Data
Calibre: 81 mm
Length of barrel: 1.45 m
Weight: in firing position, 41.5 kg
Weight of barrel: 14.5 kg
Weight of baseplate: 14.8 kg
Weight of bipod: 12.2 kg
Elevation: +45 to +85°
Range: min, 75 m; max, 5,000 m
Rate of fire: 15 rds/min

Manufacturer
Pakistan Machine Tool Factory Limited, Landhi, Karachi 34.

Status
In production.

Service
Pakistan armed forces.

UPDATED

81 mm Pakistan Machine Tool light mortar

120 mm heavy mortar

Description
This 120 mm mortar appears to be based upon a Thomson-Brandt design, though with small changes in detail to suit local military preferences. It is a conventional type of smoothbore mortar which can be drop- or trigger-fired and is transported by means of a light two-wheeled trailer. The mortar can be brought into action without disconnecting it from the trailer, by simply lowering the baseplate to the ground and then running the wheels forward until the bipod makes contact with the ground. Alternatively the entire mortar can be removed from the trailer and placed in action on its bipod in the usual manner. The former method is preferred when manpower is restricted, but the latter method can be accomplished easily by only four men.

The bipod is connected to the barrel through a two-cylinder recoil buffer and the usual traverse, elevation and cross-level facilities are provided. A towing eye is attached to the muzzle by means of an interrupted collar.

Data
Calibre: 120 mm
Length of barrel: with breech cap, 1.746 m
Weight: travelling, 402 kg
Weight of barrel: with muzzle towing attachment, 76 kg
Weight of baseplate: 80 kg
Weight of bipod: 86 kg
Weight of carriage: 137 kg
Elevation: +45 to +80°
Traverse: 17°
Range: max, 8,950 m
Rate of fire: 12 rds/min

Manufacturer
Pakistan Machine Tool Factory Limited, Landhi, Karachi 34.

120 mm Pakistan Machine Tool heavy mortar

Status
In production.

Service
Pakistan armed forces.

VERIFIED

PHILIPPINES

Philippine Army mortars

Description

The Philippine Army is equipped with two mortars of local design and manufacture. The 60 mm M75 uses a tripod and a thick-gauge round, stamped and welded baseplate. It is fitted with a locally manufactured M4 sight. The 60 mm M53 bomb uses a locally designed aluminium alloy tail fin unit, and the percussion fuze is also locally made. The 81 mm M2 mortar is another Philippine design; the lower portion of the barrel is radially finned and the circular baseplate is stamped from heavy gauge steel.

Manufacturers

Mortars, Philippine United Machinery & Foundry Company.
Ammunition, Dayton Metals Corporation, PO Box 435, Araneta Center Post Office, Fiesta Carnibal Building, Cubao, Quezon City.

VERIFIED

PORTUGAL

INDEP 60 mm M/965 light mortar

Description

The INDEP 60 mm M/965 mortar is a standard configuration weapon, designed for the close support of infantry and airborne troops. It is composed of three main groups: barrel, bipod mounting and baseplate.

The barrel, of high-quality alloy steel, carries a breech piece with a fixed firing pin, easily replaceable through the back of the breech piece. The barrel and baseplate are connected by a ball and socket joint, which provides 360° traverse without moving the baseplate. The barrel may, however, be locked on any desired azimuth to give the best accuracy. The baseplate, made of welded steel plate, is designed to provide good stability even in soft soil. The bipod mounting carries the elevation and traverse mechanism, the sight and the shock absorbers.

Data

Calibre: 60 mm
Length of barrel: 650 mm
Weight: in firing position, 15.5 kg
Range: with M49A2 bomb, min, 50 m; max, 1,820 m

Manufacturer

INDEP SA, Fabrica de Braco de Prata, Rua Fernando Palha, P-1802 Lisbon-Codex.

Status

Production complete.

Service

Portuguese armed forces and those of other countries.

VERIFIED

INDEP 60 mm M/965 light mortar

INDEP 60 mm M/968 Commando mortar

Description

The INDEP 60 mm M/968 is a light and simple mortar, with emphasis placed on extreme portability. It consists of a tube and a button-shaped baseplate which carries a fixed firing pin. Sighting is effected by a white line painted on the barrel and a series of engraved plates fixed to a sling which doubles as a carrying sling. For firing, the gunner aligns the white line with the target, steps on the plate corresponding to the desired range and pulls the sling taut for elevation.

An alternative model is available with a small rectangular baseplate, trip firing mechanism, and a simple clamp-on sight.

Although the barrel can withstand higher pressures (being manufactured to the same specifications as the one used for the 60 mm light mortar) the manufacturer recommends that, for the tactical situations for which this mortar is designed, a maximum of only two increments (of a possible four currently supplied with the ammunition) should be used. This will provide range coverage between 50 and 1,050 m.

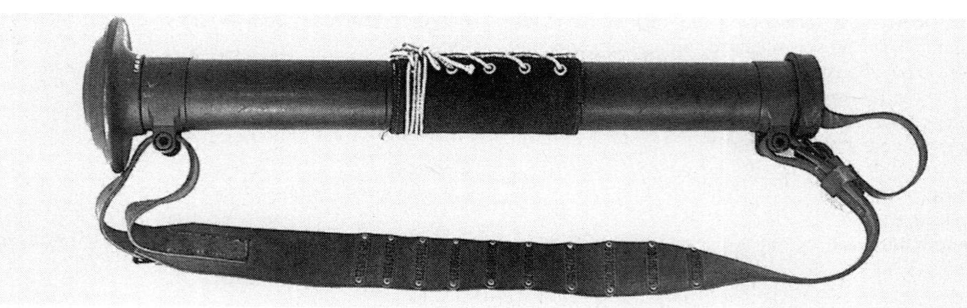

INDEP 60 mm M/968 Commando mortar

The mortar is supplied with a muzzle cover and a protecting sleeve to provide a grip for the hand supporting the barrel.

Data

Calibre: 60 mm
Length of barrel: 650 mm
Weight: in firing position, 6.6 kg
Range: min, 50 m; max, 1,050 m
Rate of fire: up to 20 rds/min

Manufacturer

INDEP SA, Fabrica de Braco de Prata, Rua Fernando Palha, P-1802 Lisbon-Codex.

Status

In production.

Service

Portuguese armed forces and other countries.

VERIFIED

INDEP 81 mm HP (FBP) mortar

Description

The INDEP 81 mm HP (FBP) mortar is a medium mortar designed for use by infantry units. The letters HP stand for High Pressure meaning that this mortar is prepared to accept high performance ammunition of the latest types.

The development of this mortar was completed in 1979. The main goals were to obtain a better performance than that of the mortar then in production (M/937), and to reduce weight as much as possible. These goals were achieved and this model replaced the M/937 in production.

The general configuration is orthodox. The barrel, of high grade alloy steel, may be supplied in two different lengths, the longer one giving a slightly better performance at the cost of a little more bulk. There is a reinforcement in the lower third of the barrel, near the breech. Ignition is by a fixed firing pin, which was designed so as to be easily replaceable through the back of the breech piece.

As with the 60 mm mortar, the design allows 360° traverse and, although the weapon has been lightened as much as possible, all the usual controls are included. The elevating, traversing and levelling mechanisms, the double shock absorbers and the sight are mounted on the bipod assembly. The weapon is supplied with two baseplates, a standard one of embossed, welded and reinforced steel plate and another smaller one for use over hard ground.

An adaptation of this mortar for firing from light vehicles was studied and produced in prototype form, but is not in production.

81 mm HP (FBP) mortar

Data

Calibre: 81 mm
Length of barrel: short, 1.155 m; long, 1.455 m
Weight in firing position: short, 42.5 kg; long, 46 kg
Weight of barrel assembly: short, 15 kg; long, 18.5 kg
Weight of bipod assembly: 13.5 kg
Weight of baseplate: normal, 13 kg; small, 2.8 kg
Weight of sight unit: 0.85 kg
Weight of accessories and tools: 13.4 kg
Elevation: +40 to +87°
Rate of fire: 25 rds/min
Range: short barrel, min 87 m; long barrel, min 100 m; short barrel, max 3,517 m; long barrel, max 3,837 m

Manufacturer

INDEP SA, Fabrica de Braco de Prata, Rua Fernando Palha, P-1802 Lisbon-Codex.

Status

In production.

VERIFIED

ROMANIA

60 mm Commando mortar

Description

In common with other Romanian mortar designs, the 60 mm Commando mortar marketed by the state-owned Romtehnica organisation owes virtually nothing to other Eastern Bloc designs and is an amalgamation of design features from Western designs. The design is a 60 mm mortar tube with spade-type baseplate and a simple drum sight close to the muzzle to determine elevation angles. Firing appears to be either drop-loading or with a mechanical trigger. A sling is provided along with a webbing gaiter intended to protect the firer's hand from a hot barrel.

The ammunition fired includes a 60 mm HE bomb weighing 1.6 kg.

Data

Calibre: 60 mm
Weight: in firing position, 7.6 kg
Elevation: +45 to +85°
Range: min, 100 m; max, 1,500 m
Rate of fire: 20 rds/min

Manufacturer

Romtehnica, 9-11 Drumul Taberei Street, Bucharest.

Status

Available.

VERIFIED

Romanian 60 mm Commando mortar

60 mm Standard mortar

Description

The 60 mm Standard mortar is an entirely orthodox drop-fired, bipod-mounted mortar with a rectangular baseplate. A recoil cylinder isolates the bipod carriage and sight unit from the main effects of the recoil forces.

The ammunition fired includes a 60 mm HE bomb weighing 1.6 kg.

Data

Calibre: 60 mm
Weight: in firing position, 7.6 kg
Elevation: +45 to +85°
Range: min, 100 m; max, 3,000 m
Rate of fire: 20 rds/min

Manufacturer

Romtehnica, 9-11 Drumul Taberei Street, Bucharest.

Status

Available.

VERIFIED

Romanian 60 mm Standard mortar

82 mm mortar

Description

The 82 mm Standard mortar is an entirely orthodox drop-fired, bipod-mounted mortar with a recoil cylinder isolating the bipod carriage and sight unit from the main effects of the recoil forces. An attachment at the muzzle prevents double loading.

The ammunition fired includes an 82 mm HE bomb weighing 3.1 kg capable of reaching 4,500 m, and an Illuminating bomb weighing 3.49 kg, and with a range of 2,670 m.

Data

Calibre: 82 mm
Weight: in firing position, 43 kg
Elevation: +45 to +85°
Traverse: on carriage, ±3°
Muzzle velocity: HE, 295 m/s
Range: min, 100 m; max, 4,500 m
Rate of fire: 20 rds/min

Manufacturer

Romtehnica, 9-11 Drumul Taberei Street, Bucharest.

Status

Available.

VERIFIED

Romanian 82 mm mortar

120 mm mortar

Description

The 120 mm mortar marketed by Romtehnica appears to owe little to other similar calibre mortars other than the general configuration and is apparently an all-Romanian design. It is a thoroughly state-of-the-art design being relatively light in weight and having up-to-date features such as a muzzle attachment to prevent double-loading of the drop-fired barrel. For transport the mortar is carried as one load on a two-wheeled carriage which also carries chests for spares, accessories and ready-use ammunition.

Ammunition fired by this mortar includes a HE bomb weighing 16.39 kg and having a maximum range of 5,775 m. Other available bombs include Smoke, Illuminating and a propaganda dispensing bomb.

Data

Calibre: 120 mm
Weight: in firing position, 181 kg; travelling, 341 kg
Elevation: +45 to +80°
Traverse: on carriage, 10°
Muzzle velocity: HE, 272 m/s
Range: min, 420 m; max, 5,775 m
Rate of fire: 8-10 rds/min

Manufacturer

Romtehnica, 9-11 Drumul Taberei Street, Bucharest.

Status

Available.

VERIFIED

Romanian 120 mm mortar in firing position

SINGAPORE

60 mm Commando mortar

Description

The 60 mm Commando mortar markete by Chartered Industries of Singapore is a light and simple one-man weapon intended for rapid deployment and use by infantry squads. There is no conventional mounting for it is a high tensile steel alloy barrel with an integrated steel breech piece and a small hemispherical (button-shaped) baseplate on which the mortar is hand-held for firing. A webbing sling is provided for carrying and incorporates a muzzle cover.

This mortar can be fired manually via a button on the end of a handle at the base of the barrel or the ammunition may be dropped onto a fixed firing pin. An extra hand-hold is provided further up the barrel and a webbing gaiter provides protection against heat when firing. The sighting system is a collapsible indexed leaf sight which is self-illuminated for use at night. Some models have a levelling sight and lack the manual firing mechanism.

The Commando mortar fires 60 mm HE or Smoke grenades weighing 1.68 kg by using a two-charge propellant system. Range is from 150 to 1,080 m.

Data

Calibre: 60 mm
Length overall: 746 mm
Weight: complete, 6 kg
Weight of barrel and breech: 5 kg
Weight of baseplate: 0.65 kg
Weight of sight: 0.35 kg
Elevation: +45 to +82°
Range: min, 150 m; max, 1,080 m

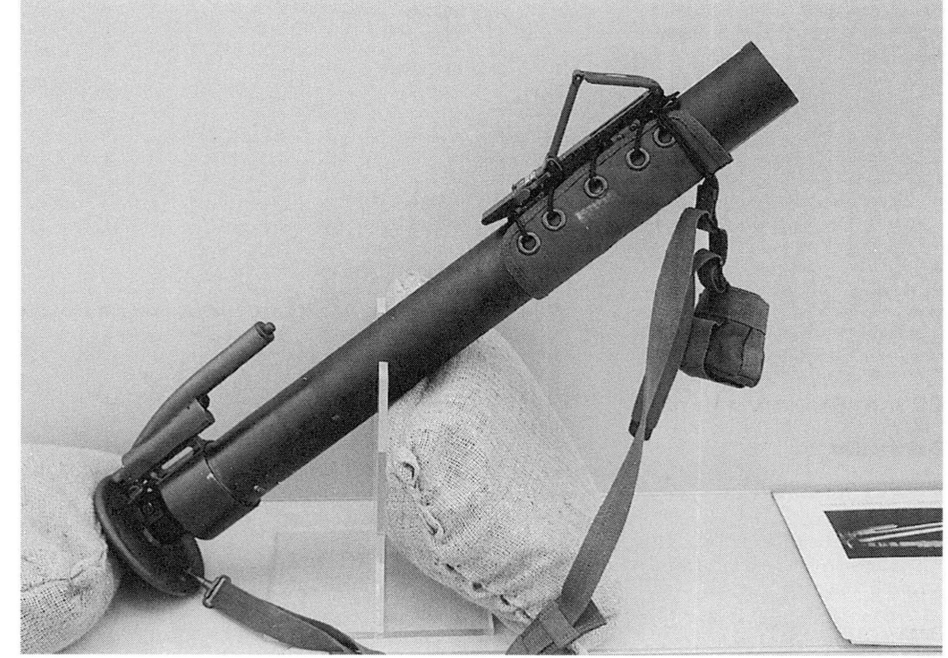

60 mm Commando mortar (T J Gander)

Manufacturer

Chartered Industries of Singapore, 249 Jalan Boon Lay, Singapore 2261.

Status

Available.

UPDATED

60 mm mortar

Description

The 60 mm mortar marketed by Chartered Industries of Singapore is a conventional drop-loaded mortar which appears to be produced to meet high specifications. It is notable that there are spring shock buffers on the bipod collar, a collimating dial sight with elevation scales, and a wheel-protected elevating screw. The baseplate is well ribbed for soft ground and of a triangular shape.

Data

Calibre: 60 mm
Length of barrel: 740 mm
Weight: in firing position, 15.5 kg
Weight of barrel: 5.4 kg
Weight of bipod: 4.8 kg
Weight of baseplate: 5.4 kg
Weight of sight: 1.2 kg
Elevation: 800-1,420 mils
Traverse: top, 116 mils
Range: min, 150 m; max, 2,555 m
Bomb weight: HE, 1.68 kg; FM Smoke, 1.68 kg; Illuminating 1.5 kg

Manufacturer

Chartered Industries of Singapore, 249 Jalan Boon Lay, Singapore 2261.

Status

Available.

UPDATED

CIS 60 mm mortar in firing position

81 mm mortar

Description

The 81 mm mortar marketed by Chartered Industries of Singapore is of the conventional drop-fired type. The barrel is of high-strength forged steel, the rear section being reinforced. The screwed-on breech piece contains the retractable firing pin and terminates in a ball unit which engages in the baseplate.

The baseplate is circular, of heavy duty steel alloy and is designed to facilitate extraction from the ground even in wet conditions. The bipod was ergonomically designed, with all the controls at the top for ease of operation.

The mortar can be carried in any convenient vehicle or can be broken down for man-carriage; manpack carriers are available.

Data

Calibre: 81 mm
Length of barrel: 1.319 m
Weight: in firing position, 43.6 kg
Weight of barrel: 14.8 kg
Weight of baseplate: 13.6 kg
Weight of bipod: 13.5 kg
Weight of sight: 1.7 kg
Baseplate diameter: 530 mm
Elevation: 800-1,550 mils
Traverse: top, 200 mils
Range: max, 6,200 m

Manufacturer

Chartered Industries of Singapore, 249 Jalan Boon Lay, Singapore 2261

Status

In production.

UPDATED

CIS 81 mm mortar with ammunition

120 mm Standard mortar

Description

The 120 mm Standard mortar marketed by Chartered Industries of Singapore is of conventional drop-fired design and is carried into action on a light two-wheeled trailer which has provision for a small supply of ammunition. The mortar consists of the usual barrel, baseplate and bipod and is clamped to the trailer without being dismantled.

Data

Calibre: 120 mm
Length of barrel: 1.94 m
Weight: in firing position, 236.7 kg; travelling, with trailer and accessories, 512 kg
Weight of barrel: 80 kg
Weight of bipod: 70 kg
Weight of baseplate: 85 kg
Weight of sight: 1.7 kg
Elevation: 800-1,422 mils
Traverse: top, 240 mils
Range: min, 400 m; max, 6,500 m
Bomb weight: HE, 13.2 kg; WP Smoke, 12.8 kg; Illuminating, 12.6 kg

Manufacturer

Chartered Industries of Singapore, 249 Jalan Boon Lay, Singapore 2261

Status

In production.

UPDATED

CIS 120 mm Standard mortar

SOUTH AFRICA

Vektor 60 mm M4 series Commando mortars

Description

The majority of portable and lightweight mortars made in South Africa are the 60 mm M4 series of Commando-type mortars of which the M4 and the M4 Mark 1 proved to be most popular designs. These mortars are of completely South African design to meet the requirements of infantry in bush war conditions.

The two models are identical except for the breech pieces in that the M4 model's breech-piece is equipped with a trigger mechanism which is pulled by a lanyard, with a toggle for better grip. As soon as the lanyard is released the trigger mechanism automatically recocks ready for the next firing. With this configuration foot patrols may walk with a bomb in the barrel to enable quick firing on contact. The M4 model can also be equipped with a fixed firing pin which is issued with the mortar as part of the accessories.

The M4 Mark 1 model is only issued with a fixed firing pin. Both models are muzzle loaded.

Both models are issued with a dome-type baseplate with an elevating action limiting elevation to one plane only. Sighting is by a simple clamp-on type sight unit. All particulars, for example distance in metres, elevation in degrees or mils, as well as the charge to be used, are engraved on the curved face plate. The face plate also houses a curved bubble for elevation and a small bubble for setting the weapon horizontal. The handgrip of the sight is used to elevate the barrel until the bubble corresponds with the required distance on the face plate; the charge to be used as well as the elevation can now be read off the face plate. This handle type grip is very convenient in handling or carrying the weapon.

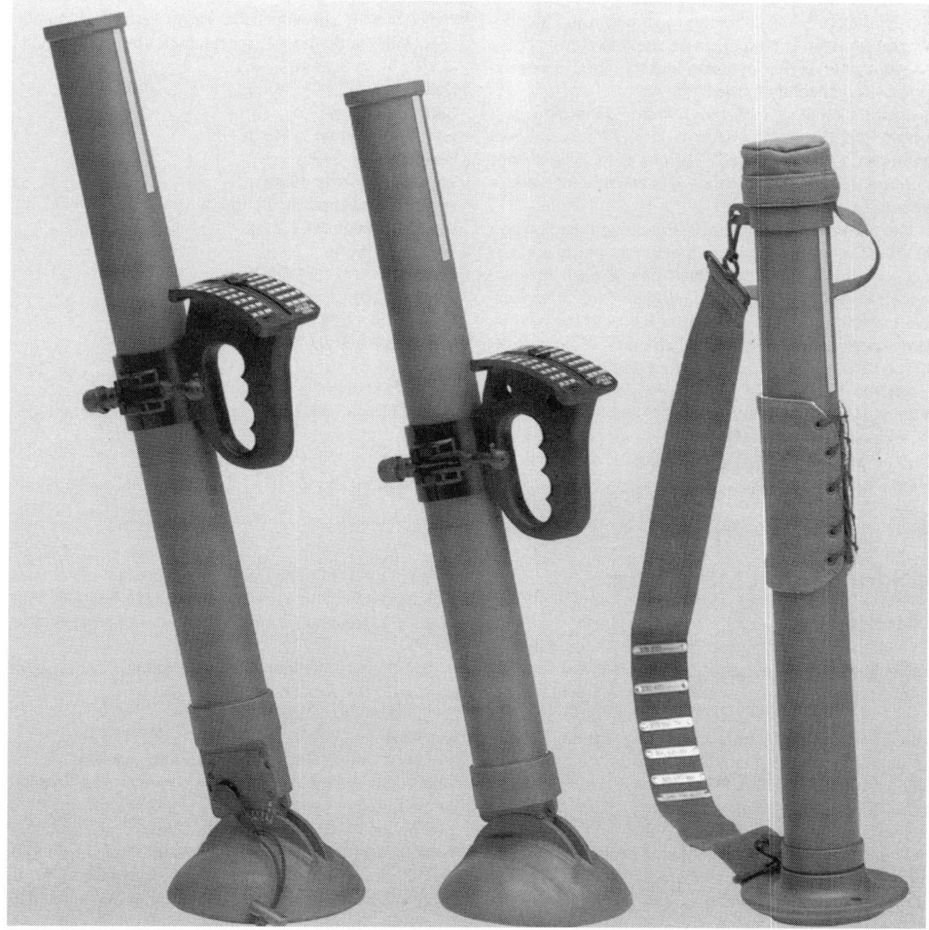

Vektor 60 mm Commando mortars; left to right: M4, M4 Mk 1, M4L3

Among a range of accessories and equipment, some of them optional, is a webbing sling with a muzzle cover to prevent dirt and sand from entering the barrel when not in use.

A third variant is the M4L3, fitted with a simplified baseplate and without the handgrip sight unit. Instead, the axis of the bore is marked with a white painted line, and the permanently fitted sling is marked at intervals with brass tags bearing the most usual ranges. The firer places the baseplate on the ground, drops the sling and places his foot against the tag bearing the range he requires. He then draws the barrel up until the sling is tight, thereby setting the correct elevation. Alignment is by visually lining up the white line on the barrel with the target. After the first round has been fired, subsequent adjustment is made by eye, moving the foot on the sling as necessary.

Data

	M4	M4 Mk 1	M4L3
Calibre	60 mm	60 mm	60 mm
Total mass of weapon as carried	7.6 kg	7 kg	7.5 kg
Length of barrel	650 mm	650 mm	650 mm

Manufacturer
LIW (A division of Denel), PO Box 7710, Pretoria 0001.

Status
In production.

Service
South African National Defence Force and some other armies.

UPDATED

Vektor 60 mm M1 mortar

Description
The Vektor 60 mm M1 mortar originated as a Hotchkiss-Brandt (now Thomson-Brandt) design manufactured in South Africa under licence. It remains in production, after the expiry of the licence agreement, with some small changes, largely intended to facilitate manufacture. The sight used is the H-019 which weighs 1.5 kg.

The bombs fired are of the M61 pattern and include HE and Smoke. Practice bombs are also available.

Data
Calibre: 60 mm
Length of barrel: 650 mm
Weight of barrel: 4 kg
Weight of baseplate: 6.7 kg
Weight of bipod: 5.6 kg
Weight of sight: 1.5 kg
Elevation: 710-1,600 mils
Traverse: top, 530 mils
Range: min, 100 m; max, 2,108 m
Muzzle velocity: 171 m/s

Manufacturer
LIW (A division of Denel), PO Box 7710, Pretoria 0001.

Status
Available.

Service
South African National Defence Force.

UPDATED

Vektor 60 mm M1 mortar

Vektor 60 mm LR long-range mortar system

Description
The Vektor 60 mm LR long-range mortar system (60 mm LR, also known as the 60 mm M6) originated as a conversion of the 81 mm M3 mortar (see following entry) to accommodate a long 60 mm barrel. The conversion was later produced as a complete system but is also available as a conversion kit requiring no special tools or modification facilities.

It is anticipated that the 60 mm LR system will replace the 81 mm M3 mortar in South African National Defence Force service, offering a significant increase in range performance with no great increase in weight.

The 60 mm LR is essentially the same basic design as the 81 mm M3 other than the 60 mm barrel and the associated mounting modifications based on two inserts around the barrel/mounting junction. It is possible to refit the 81 mm M3 mortar barrel in the field by removing the inserts using basic tools.

The 60 mm LR can continue to utilise the 81 mm M3 baseplate but to take full weight advantage of the 60 mm system a lighter 'patrol' baseplate has been developed. This weighs 6.7 kg as opposed to the 14.9 kg of the standard component. The sight is the H-019, weighing 1.5 kg.

The 60 mm LR was optimised to fire a new long-range 60 mm bomb over a range band from 175 to 6,180 m. The bomb, filled with TNT 80.2, weighs 2.4 kg and is fired at a maximum muzzle velocity of 325 m/s using a nine-charge propellant system in addition to the primary cartridge. At one time a Smoke bomb was under development. A Practice bomb is also available. The 60 mm LR can continue to fire the 60 mm M61 bombs used with the 60 mm M1 mortar (see previous entry).

Data
Calibre: 60 mm
Length of barrel: 1.445 m
Weight: total, 24 kg
Weight of barrel: 10 kg
Weight of baseplate: standard, 14.9 kg; patrol, 6.7 kg
Weight of bipod: 12.2 kg
Weight of sight: 1.5 kg
Elevation: 400-1,511 mils
Traverse: on carriage, 530 mils
Range: min, 175 m; max, 6,180 m
Muzzle velocity: 67-330 m/s

Manufacturer
LIW (A division of Denel), PO Box 7710, Pretoria 0001.

Status
In production.

Vektor 60 mm LR long-range mortar
1996

Service
South African National Defence Force.

UPDATED

Vektor 81 mm M3 mortar

Description
As with the 60 mm M1 (qv), the Vektor 81 mm M3 mortar is based on a Hotchkiss-Brandt (now Thomson-Brandt) design, in this case the MO-81-61 light mortar; the 81 mm M3 is almost identical to the French original. The 81 mm M3 can be converted to the 60 mm M6 long-range mortar system by changing the barrel and using inserts to hold the new barrel. It is anticipated that most 81 mm M3 mortars will be converted to the new system.

The 81 mm M3 is a conventional drop-fired smoothbore mortar, supported by a bipod and baseplate. The sight is the H-019, weighing 1,5 kg

The 81 mm M3 fires forged or cast steel HE bombs weighing 4.5 kg and containing RDX/TNT 60/40. Two types of Smoke bomb are available as well as a Practice bomb.

The 81 mm M3 mortar can be carried in a special mortar-carrier version of the Ratel IFV.

Variant
Vektor 81 mm LR long-range mortar system
The Vektor 81 mm LR long-range mortar system (also known as the 81 mm M8) is essentially the same overall design as the 81 mm M3 but utilising a heavier barrel (weight 16.3 kg) to withstand the stresses of firing a new family of long range 81 mm ammunition; barrel length is 1.445 kmg. The long-range TNT-filled HE bomb weighs 4.9 kg and has a range band of from 100 to 7,263 m. Muzzle velocities using a nine-charge system (plus the primary cartridge) range from 63 to 363 m/s. Accuracy at maximum range is such that all bombs will fall within an area measuring 270 × 68 m. Also available are two types of Smoke bomb and a Practice bomb, all with performances matching the long range HE.

Data
81 mm LR
Calibre: 81 mm
Length of barrel: 1.445 m
Weight of barrel: 16.3 kg
Weight of bipod: 12.2 kg

Weight of baseplate: 14.4 kg
Weight of sight: 1.5 kg
Elevation: 800-1,500 mils
Traverse: 530 mils
Range: min, 100 m; max, 7,263 m
Muzzle velocity: 63-363 m/s

Manufacturer
LIW (A division of Denel), PO Box 7710, Pretoria 0001.

Status
Available. Being converted to 60 mm M6 long-range mortar standard.

Service
South African National Defence Force.

UPDATED

Vektor 81 mm LR mortar
1996

SPAIN

60 mm Models L and LL M-86 mortars

Description
All the mortars used by the Spanish Army are manufactured by Esperanza y Cia SA, under the general trade designation ECIA. These mortars, and the ammunition manufactured by the same company, are in current use by the armies of many countries.

The Models L and M-86 are easily carried by one soldier and capable of firing high explosive, smoke or illumination bombs. The Model M-86 is heavier and can fire to a greater range, but otherwise the two are similar and the description which follows is appropriate to both models.

The barrel of the mortar is of steel with a thread turned on the outside of the lower end, over which the breech piece fits. At the bottom of the breech-piece is a ball-shaped extension which fits into the socket of the baseplate. The baseplate is circular; webs are welded on to the bottom to stiffen the body and prevent any slipping when the mortar is fired at a low angle of elevation. In the centre is the socket which takes the ball of the breech piece. This socket can revolve so the mortar has complete 360° movement.

In the L Model the breech piece contains the trigger mechanism. This allows the firer the choice of gravity firing or trigger operation. There is a safety device which withdraws the pin when required.

In the M-86 Model the firing pin is fixed for drop-firing and is externally removable. The sight has a telescope for viewing the aiming point and elevating and traverse scales.

The Model L mortar is provided with a tripod; the M-86 uses an asymmetrical bipod.

Data
MORTAR MODEL L
Calibre: 60 mm
Length of barrel: 650 mm
Weight: in firing position, 12 kg
Range: max, bomb Model N, 1,975 m; max, bomb Model AE, 3,800 m
Rate of fire: 30 rds/min
Number of charges: 5

MORTAR LL M-86
Calibre: 60 mm
Length of barrel: 1 m
Weight: in firing position, 17.75 kg
Range: max, bomb Model N, 2,350 m; max, bomb Model AE, 4,600 m
Rate of fire: 30 rds/min
Number of charges: 5

Manufacturer
Esperanza y Cia SA, Marquina (Vizcaya).

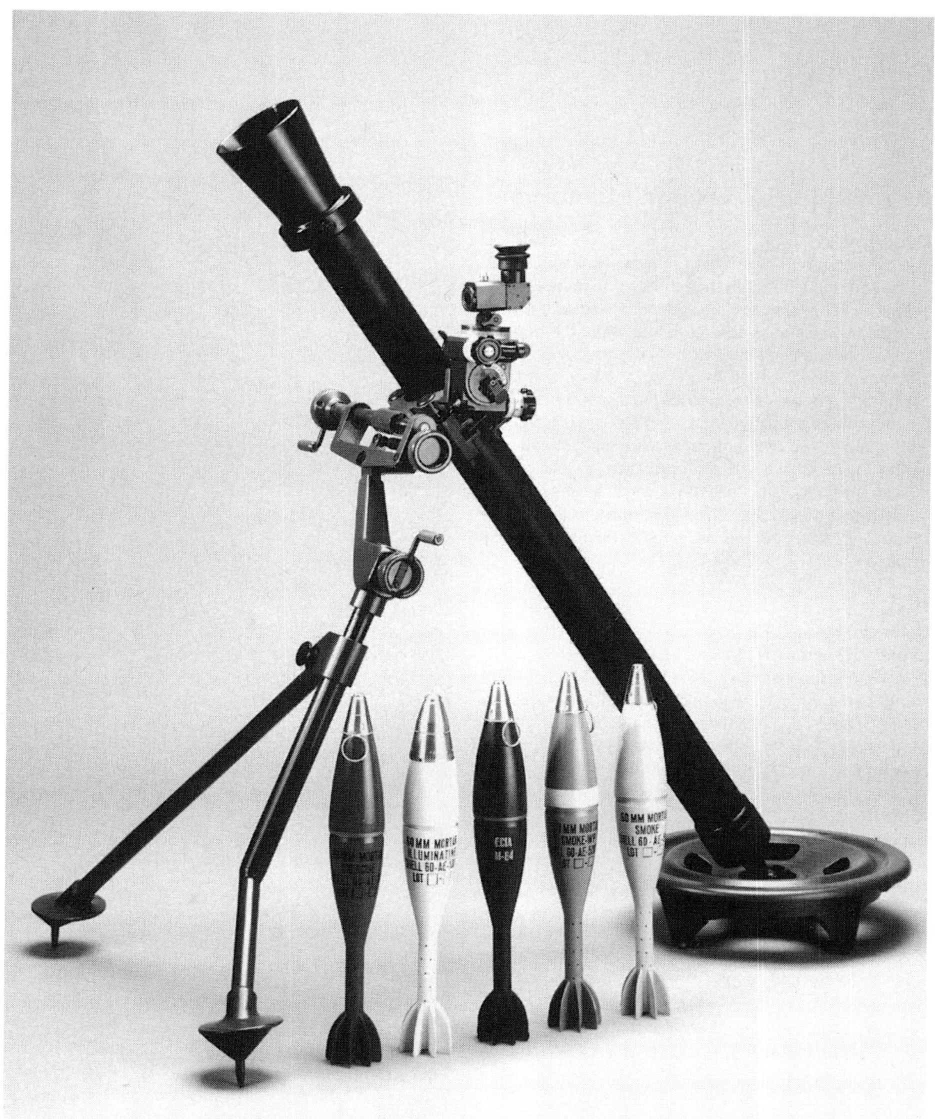

60 mm LL M-86 mortar in firing position, with the 60 mm AE family of ammunition

Status
In production.

Service
Spanish forces and several others.

VERIFIED

60 mm Commando mortar

Description

The 60 mm Commando mortar is a light mortar operated and carried by one man. The barrel is a plain tube with a small circular button-type baseplate at the bottom. There is no mount. The soldier supports the barrel with his left hand, grasping the canvas sleeve placed around it for protection. A very simple sight is held on the right of the tube by a 'hose clip' fastener.

Data

Calibre: 60 mm
Length of barrel: 650 mm
Weight: in firing position, 6.5 kg
Range: max, bomb Model N, 1,060 m; max, bomb Model AE, 1,290 m
Number of charges: bomb Model N, 2; bomb Model AE, 1

Manufacturer

Esperanza y Cia SA, Marquina (Vizcaya).

Status

In production.

Service

Spanish forces and several others.

VERIFIED

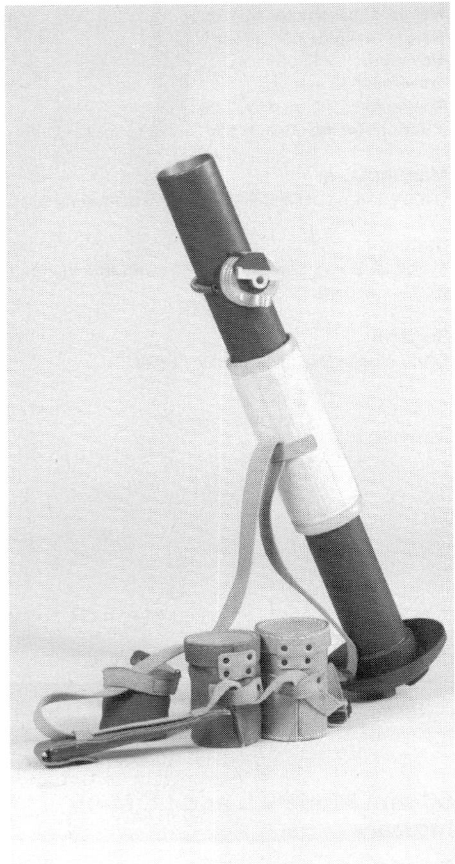

60 mm Commando mortar with accessories

81 mm Models LN and LL mortar

Description

The differences between these models are very slight and amount to little more than the length of the tube.

The barrel is a steel tube, strengthened at the rear end, and screw-threaded on the outside of the bottom to take the breech piece. This contains the firing mechanism and allows either drop-firing or firing by trigger operation. The ball extension of the breech piece fits into the socket of the circular baseplate and is held by a spring latch. Two different barrel lengths are available.

The tripod follows the same pattern as that of the 60 mm mortar, with the three legs arranged one forward and two to the rear, fitting into the fabricated tripod head. The mortar is designed to be man-portable in three loads.

Data

Calibre: 81 mm
Length of barrel: LN, 1.15 m; LL, 1.45 m
Weight in firing position: LN, 43 kg; LL, 47 kg
Range: (max, Model LN) bomb Model NA, 4,125 m; bomb Model AE, 6,200 m; (max, Model LL) bomb Model NA, 4,680 m; bomb Model AE, 6,900 m
Rate of fire: max, 15 rds/min

Manufacturer

Esperanza y Cia SA, Marquina (Vizcaya).

Status

In production.

Service

Spanish forces and several others.

VERIFIED

81 mm Model LL mortar in firing position

81 mm Model LN M-86 and LL M-86 mortars

Description

This is the latest version of the ECIA 81 mm mortar and introduces an asymmetrical bipod. It is manufactured in two different barrel lengths, 1.15 and 1.45 m. The barrel is made of special high-alloy steel and manufactured from a solid bar strengthened at the breech end. The ball extension fits into the socket of the baseplate, allowing a 6,400 mil movement. It has a drop-firing system, the firing pin being externally removable.

The mortar is equipped with a new sight with night vision and, like the earlier model, is man-portable in three loads.

Data

Calibre: 81 mm
Length of barrel: LN, 1.15 m; LL, 1.45 m
Weight in firing position: LN, 41 kg; LL, 43.5 kg
Range: (max, Model LN) bomb Model NA, 4,125 m; bomb Model AE, 6,200 m; (max, Model LL) bomb Model NA, 4,680 m; bomb Model AE, 6,900 m
Rate of fire: max, 15 rds/min

Manufacturer

Esperanza y Cia SA, Marquina (Vizcaya).

Status

In production.

Service

Spanish forces and several others.

VERIFIED

81 mm Model LL M-86 mortar in firing position

81 mm Model LN M-86 mortar in firing position

120 mm Model L mortar

Description

The Model L has all the characteristics of the other ECIA mortars with the tripod mount, barrel and circular baseplate. It can be fired either by gravity drop or by trigger control. The mortar is transported on a carriage towed by a light vehicle and can be mounted on armoured personnel carriers.

The baseplate lies over the axle and is held in position by a number of clips. The barrel and mount rest on a frame above the baseplate and are held in position by straps.

Data

Calibre: 120 mm
Length of barrel: 1.6 m
Weight: in firing position, 203 kg
Weight of barrel and breech piece: 61 kg
Weight of baseplate: 100 kg
Weight of mount: 40 kg
Weight of sight: 2 kg
Weight of carriage: with tools and accessories, 165 kg
Range: max, bomb Model L, 6,725 m; max, bomb Model AE, 8,000 m
Rate of fire: max, 20 rds/min

120 mm Model L mortar in firing position showing large baseplate

Manufacturer
Esperanza y Cia SA, Marquina (Vizcaya).

Status
Available.

Service
Spanish forces and several others.

VERIFIED

120 mm Model L mortar in travelling position

120 mm Model M-86 mortar

Description
The M-86 is lighter than the Model L and with two lengths of barrel. The design incorporates a light alloy bipod and an alloy steel barrel which is chromium-plated internally. The firing pin is locked for drop-firing and is externally removable. The mortar, when dismantled, fits on to a very light but strong two-wheeled trailer, the frame of which is of tubular steel with brackets for the retention of the mortar unit. It can also be mounted in several types of armoured personnel carrier.

Data
Calibre: 120 mm
Length of barrel: M120-13, 1.6 m; M120-15, 1.8 m
Weight in firing position: M120-13, 155 kg; M120-15, 160 kg
Rate of fire: max, 20 rds/min
Range: (max, M120-13) bomb Model L, 6,725 m; bomb Model AE, 8,000 m; (max, M120-15) bomb Model L, 7,000 m; bomb Model AE, 8,250 m

Manufacturer
Esperanza y Cia SA, Marquina (Vizcaya).

Status
In production.

Service
Spanish Army; South American, Middle Eastern and African armies.

VERIFIED

120 mm Model M-86 (M120-13) in firing position

120 mm M-86 (M120-13) barrel adapted for use in APCs

SWEDEN

81 mm m/29 mortar

Description

The 81 mm m/29 (1929) mortar is the standard light mortar of the Swedish Army and has been in service for over 60 years. It is basically the French 81 mm Stokes/Brandt Model 1917 mortar produced under licence in Sweden. In 1934 the French sights were replaced by those of Swedish design. In appearance the mortar is similar to most other mortars of that period: a rectangular baseplate with carrying handle, a bipod with chains connecting the legs and an anti-cant device on the left leg.

Data

Calibre: 81.4 mm
Length of barrel: 1 m
Weight: in firing position, 60 kg
Elevation: +45 to +80°
Traverse: 90°
Muzzle velocity: min, 70 m/s; max, 190 m/s
Range: max, 2,600 m
Rate of fire: 15-18 rds/min
Bomb weight: HE, 3.5 kg

Status

No longer in production.

Service

Swedish armed forces.

VERIFIED

81 mm m/29 mortar on trailer mounting

120 mm m/41D mortar

Description

The 120 mm m/41D (1941) mortar is the standard heavy mortar of the Swedish Army. It is basically the Finnish Tampella 120 mm Model 1940 mortar produced in Sweden. In 1956 Sweden built a number of Hotchkiss-Brandt M-56 baseplates under licence to replace the earlier Tampella baseplates. More recently the sights have been replaced. The mortar is issued on the scale of eight per mortar company of the infantry brigades and six per mortar company of the Norrland brigades.

Data

Calibre: 120.25 mm
Length of barrel: 2 m
Weight: in firing position, 285 kg; travelling, 600 kg
Elevation: +45 to +80°
Traverse: total, 360°
Muzzle velocity: min, 125 m/s; max, 317 m/s
Range: max, 6,400 m
Rate of fire: 12-15 rds/min
Bomb weight: HE, 13.3 kg

Status

No longer in production.

Service

Swedish Army and Ireland.

VERIFIED

120 mm m/41D mortar

SWITZERLAND

60 mm Model 87 mortar

Description

The 60 mm Model 87 mortar was accepted for service with the Swiss Army and entered production in 1988. Some elements of the design and ammunition are licensed from Thomson-Brandt Armements of France.

The mortar is a conventional drop-loaded smoothbore with fixed firing pin. A simple bipod is attached to the barrel by means of a collar, and the rear of the barrel rests on a small circular baseplate. Elevation of the barrel is manual. By slipping the collar up or down the barrel, the amount of elevation can be determined by a bubble-type unit mounted in the bipod collar and visible from behind the mortar. This unit can be adjusted to provide elevation data for different types of ammunition. Direction is assessed by the firer by using a white line marked on the axis of the barrel, aligning this with his target.

The ammunition is packed and carried in a special 'Duopack' of plastic, containing two bombs. The Duopack is designed so that any number can be clipped together for transport or stacking as a single unit.

60 mm Model 87 mortar with ammunition and 'Duopacks'

Data

Calibre: 60 mm
Length of barrel: 845 mm
Weight: in firing position, 7.6 kg
Elevation: Illuminating bombs, 800-1,100 mils; HE and smoke bombs, 800-1,540 mils
Traverse: 6,400 mils
Range: max, 1,000 m
Rate of fire: intense, 15 rds/min; normal, 3 rds/min

Manufacturer

w + f Bern (formerly Swiss Federal Arms Factory), Postfach 3000, Stauffacherstrasse 65, CH-3000 Berne 22.

Status

In production.

Service

Swiss Army.

VERIFIED

81 mm Model 1972 mortar

Description

The 81 mm Model 1972 mortar is an improvement on an earlier model introduced in 1933. The baseplate is a flat disc with webs welded on to the underside. The bipod has been cleaned up by having the screw threads fully enclosed, and the cross-levelling gear, although working in exactly the same way, is much easier to operate. The mortar breaks down into three manportable loads. Its military designation is 8.1 cm Mw 72.

Data

Calibre: 81 mm
Length of barrel: 1.28 m
Weight: complete, 45.5 kg
Weight of barrel: 12 kg
Weight of mount: 15 kg
Weight of baseplate: 16.5 kg
Weight of sight mount and sight: 2 kg
Baseplate diameter: 550 mm
Elevation: +45 to +85°
Traverse: top, 10°; by moving bipod, 360°
Range: max, 4,100 m
Rate of fire: 15 rds/min

Manufacturer

w + f Bern (formerly Swiss Federal Arms Factory), Postfach 3000, Stauffacherstrasse 65, CH-3000 Berne 22.

Status

Production complete.

Service

Swiss Army.

UPDATED

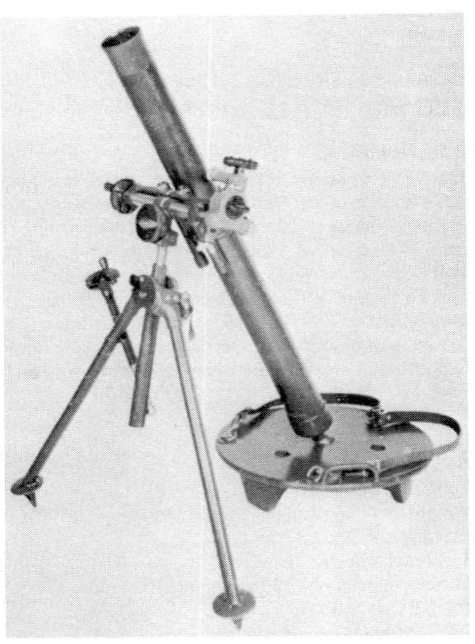

81 mm Model 1972 mortar

120 mm Model 87 mortar on two-wheel carriage (12 cm Mw 87)

Description

The 120 mm mortar Model 87 (12 cm Mw 87) is an improved version of the 120 mm Model 74 mortar which was produced from 1974 to 1978. The mortar consists of the smoothbore barrel with breech piece and trigger control, bipod, baseplate and carriage with equipment. The improved Model 87 has a reinforced baseplate to withstand maximum load on any type of ground. The mortar itself is the same as that used as the Model 64, installed in the M106 mortar carrier.

The Model 87 is designed to be towed by any vehicle, pulled by troops or carried as an underslung load by helicopters. The carriage is a two-wheeled vehicle with suspension. In addition to carrying the mortar, the carriage has an ammunition rack to stow six ready-use rounds, as well as equipment for firing and maintenance.

Data

Calibre: 120 mm
Length of barrel: with breech piece, 1.77 m
Weight of barrel: 85 kg
Weight of carriage: 250 kg
Weight of baseplate: 94 kg
Weight of bipod: with sight, 56 kg
Weight of equipment: 85 kg
Overall width: 1.53 m
Height loaded: 1.2 m
Wheelbase: 1.285 m
Ground clearance: 240 mm
Towing speeds on roads: max, 80 km/h
Elevation: +45 to +85°
Range: max, 7,500 m

Manufacturer
w + f Bern (formerly Swiss Federal Arms Factory), Postfach 3000, Stauffacherstrasse 65, CH-3000 Berne 22.

Status
In production.

Service
Swiss Army.

VERIFIED

120 mm Model 87 mortar in action position

120 mm Model 87 mortar on two-wheel carriage

TURKEY

MKEK 60 mm Commando mortar

Description
The MKEK 60 mm Commando mortar is an easily carried light infantry weapon. The special DIN 42 CrMo4 steel barrel screws into a breech piece which incorporates a novel bipod. The mortar can be fired from the ground or most types of vehicle. All NATO-standard 60 mm bombs (for example the HE M49 series) can be fired.

Data
Calibre: 60.4 mm
Length of barrel: 650 mm
Weight of barrel: 3 kg
Weight of breech piece: with bipod, 4.25 kg
Weight of sight: M4, 1.146 kg
Max pressure: 250 bar
Range: min, 360 m; max, 1,700 m

Manufacturer
Makina ve Kimya Endüstrisi Kurumu (MKEK), TR-06330 Tandogan, Ankara.

Status
In production.

Service
Turkish Army.

UPDATED

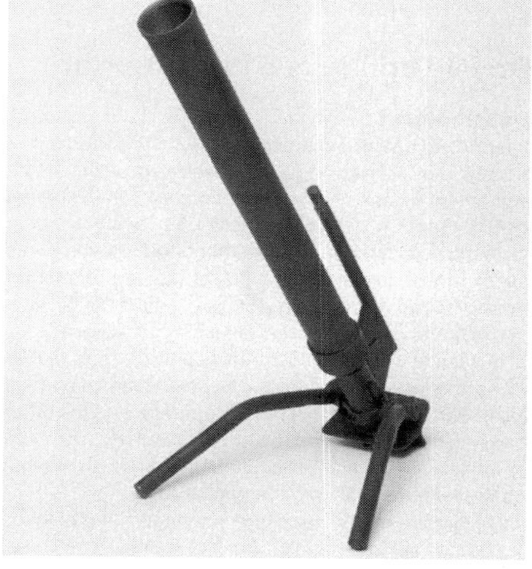

MKEK 60 mm Commando mortar

MKEK 81 mm UTI and NTI mortars

Description
The MKEK 81 mm UTI mortar is a purely Turkish design, being an entirely conventional drop-fired mortar developed to fire a wide array of existing 81 mm mortar ammunition. The UTI baseplate is circular while the bipod mounting is linked to the barrel via a single recoil cylinder. The UTI fires locally developed mortar bombs, including the MKE MOD 214 weighing 4.68 kg.

The MKEK 81 mm NTI mortar is essentially similar to the UTI but has a shorter barrel and was developed primarily to utilise the range of US 81 mm mortar ammunition produced in Turkey by MKEK.

Data
Calibre: 81 mm
Length of barrel: UTI, 1.453 m; NTI, 1.15 m
Weight of barrel: UTI, 28.1 kg; NTI, 21.6 kg
Weight of baseplate: UTI, 15 kg; NTI, 14.6 kg
Weight of bipod: UTI and NTI, 23.2 kg
Elevation: +40 to +85°
Traverse: 90 mils left and right
Range: UTI, min, 400 m; max, 6,000 m; NTI, min, 200 m, max, 3,120 m
Rate of fire: normal, 16 rds/min

MKEK 81 mm UT1 mortar

MKEK 81 mm NT1 mortar

Manufacturer
Makina ve Kimya Endüstrisi Kurumu (MKEK), TR-06330 Tandogan, Ankara.

Status
In production.

Service
Turkish Army.

UPDATED

MKEK 120 mm HY 12 mortar

Description
The MKEK 120 mm HY 12 mortar, also known as the Tosam, is a licence-produced version of the French Thomson-Brandt 120 mm MO-120-RT rifled mortar and differs from the original in few respects. In Turkish Army service this mortar is towed by a Unimog light truck carrying the six-man crew and 60 rounds. 120 mm ammunition for this mortar is produced in Turkey by MKEK.

Data
Calibre: 120 mm
Length of barrel: 1.9 m
Weight: in firing position, 570 kg
Weight of barrel: 143 kg
Weight of baseplate: 152 kg
Weight of bipod: 275 kg
Rifling: 40 grooves, 1 turn in 16.95 calibres
Elevation: +35 to +85°
Traverse: 250 mils left and right
Range: with MKEK Mod 209 HE, min, 1,500 m; max, 8,000 m
Rate of fire: normal, 5 rds/min
Weight of bomb: MKEK Mod 209 HE, 16 kg

MKEK HY 12 mortar

Manufacturer
Makina ve Kimya Endüstrisi Kurumu (MKEK), TR-06330 Tandogan, Ankara.

Status
In production.

Service
Turkish Army.

UPDATED

UNITED KINGDOM

Royal Ordnance 51 mm mortar

Development
Intended to be a replacement for the old 2 in mortar, the 51 mm Mortar was originally developed at the Royal Armament Research and Development Establishment (RARDE) at Fort Halstead in Kent. Early designs dating from the mid-1970s featured a monopod leg which was eventually abandoned in favour of hand-held support during firing. Production for the British Army was carried out between 1982 and 1988.

In service with the British Army as the 51 mm Mortar L9A1, the weapon is used by a two-man team at platoon level, with one man acting as the operator and the other carrying the ammunition in webbing satchels. The main emphasis is on firing Illuminating bombs to enable other weapons to engage targets at night.

Development of ammunition continues (see below).

Description
The barrel is a steel tube, bell-mouthed at the muzzle to assist in loading bombs and also to add strength to the vulnerable end of the barrel. The barrel is held to the breech by four spring-loaded plungers which lock into four holes in the barrel. The breech assembly incorporates a trigger-operated mechanism with a short firing lever tripping the firing pin when pulled over centre. A spade-type baseplate is provided and a webbing gaiter around the barrel protects the firer's hand from the heat produced when firing.

Aiming is carried out using a line painted along the barrel. Elevation adjustments are made with reference to a drum-type Trilux sight located just above the breech assembly.

A maximum range of 800 m can be achieved for all ammunition types. Originally the minimum range of 50 m was achieved by using a Short-Range Insert (SRI) dropped into the barrel from the muzzle. This was a rod with a long firing pin along its longitudinal axis. The lower end of the firing pin rested on the firing pin in the breech piece and was driven forward when the mortar was fired. Thus the single-charge bomb could not reach the bottom of the tube and the short range was achieved by increasing the chamber volume, reducing the working chamber pressure; at the same time there was a reduction in shot travel. When not in use the SRI was held in a clip attached to the muzzle cover.

Service experience demonstrated that the SRI concept was prone to difficulties such as inserting the SRI upside down or losing it during moves. This led to the development of a new range of 51 mm ammunition with which the bombs are provided with a two charge system matched to the existing sights. When short ranges are required the charge around the tail is removed; the SRI is no longer necessary.

A new family of improved 51 mm ammunition is under development following the award of a £50 million contract for 200,000 Illuminating bombs, 100,000 Smoke bombs and 100,000 HE bombs with cast iron fragmentation bodies. All three bomb types are ballistically identical. First deliveries will be made during 1997.

Data
Calibre: 51.25 mm
Length of barrel: 543 mm
Length overall: 750 mm
Weight: total, 6.3 kg
Weight of barrel and sight: 2.6 kg
Weight of spares and tool wallet: 1.4 kg
Barrel, outside diameter: 55 mm
Accuracy: 2% probable error in range; 3 mils in line
Range: min, 50 m; max, 800 m
Rate of fire: 3 rds/min for 5 min; 8 rds/min for 2 min

Firing a 51 mm mortar

1996

Manufacturer
British Aerospace Defence Limited, Royal Ordnance Division, Kings Meadow Road, Nottingham.

Status
Available.

Service
British Army.

UPDATED

81 mm L16A1 mortar

Development
The original 81 mm L16 mortar was developed during the late 1950s, with Canada being responsible for the baseplate and sight unit and the UK for the barrel and bipod. It entered service during the 1960s. A gradual process of improvement introduced the L16A1 while the L16A2 is a further development with the total weight being 38.6 kg, with the heaviest component weighing only 14 kg. The L16A2 was used as the basis for the US Army and Marine Corps model, the M252 (qv) type classified in 1984.

The L16A1 is licence produced in Japan by Howa, with Daikin manufacturing the ammunition. The licence was agreed in 1992 and was worth £28 million to Royal Ordnance. Following deliveries of complete mortars and ammunition from the UK, the first Japanese example was produced in early 1995.

Over 39 countries have purchased L16 series mortars and over 5,000 units have been produced.

Description
The 81 mm L16A1 mortar is normally deployed as part of a section used conventionally in a prepared position

on the ground. It was also developed to be fired from the FV432 armoured personnel carrier. The 81 mm L16A1 can be carried in a light vehicle and when necessary it can be broken down into three loads of 11.3, 11.8 and 12.25 kg which can be carried by the mortar section.

To enable the L16A1 mortar to fire bombs with the hot propellants used by some NATO countries, a barrel slightly heavier than that originally specified was adopted. Firing Royal Ordnance ammunition, the mortar can fire 15 rds/min indefinitely, the barrel temperature reaching an equilibrium value of 540°C.

The barrel is made from a forged steel tube which has been reduced in diameter at the bottom, and to save weight it has been screw-threaded internally for the insertion of a breech plug. The lower half of the barrel has been finned to increase the surface area available for heat dissipation, and the top half of the barrel has been left plain. There is a collar at the mouth of the barrel to strengthen the section there and a small internal taper is provided to assist in loading the bomb. Barrels produced for the US Army have a muzzle attachment to reduce firing blast - see M252 entry under *United States of America* for details.

A breech plug fits into the barrel at one end and has a ball shape to fit into the socket of the baseplate. It carries a longitudinal hole screw-threaded to take the firing pin.

The L5A5 mounting is of unusual shape and has been referred to as a 'K' mount. The shape was adopted because with the elevating screw incorporated in one of the legs there is a significant weight saving and no loss of function. All the screw threads associated with elevation and traverse of the mortar are enclosed to reduce wear and increase life.

The baseplate is of Canadian design and is of forged aluminium. It produces an adequate flotation area and the design of the web prevents any tendency for the plate to slip sideways. The circular baseplate allows all-round traverse without needing to disturb the plate.

The sight is the Canadian C2. It fits not only the mortar but also the L7A2 general-purpose machine gun for use in the sustained fire role. It allows either direct laying or indirect laying using a 45° angled telescope of some ×1.7 magnification. The sight is illuminated for night use with a tritium source.

The 81 mm L16A1 mortar can fire all NATO standard 81 mm mortar ammunition but is optimised for the Royal Ordnance range of bombs using a seven-charge propellant system. These include the HE L41A1, the Smoke WP L42A1 and the Practice L27.

A Mid Life Upgrade (MLU) package for the 81 mm L15/M252 mortar is under study by the UK, Canada and the USA. It is anticipated that this will include enhancements to the mortar, its ammunition (bombs and fuzes), and to the fire control elements.

Data

Calibre: 81.4 mm
Length of barrel: 1.28 m overall
Weight of barrel: 12.7 kg
Weight, mounting: 12.3 kg
Overall mounting length: folded, 1.143 m
Baseplate diameter: overall, 546 mm
Outside barrel diameter: muzzle, 86 mm; breech, 94 mm
Socket size: 50.8 mm
Elevation: over 125 mils left and right at elevations of greater than 950 mils; min, +45° (800 mils); max, +80° (1,422 mils)
Traverse: 100 mils left and right at 800 mils elevation
Range: min, 100 m; max, 5,800 m
Rate of fire: 15 rds/min

Manufacturer

British Aerospace Defence Limited, Royal Ordnance Division, Kings Meadow Road, Nottingham.

Status

In production.

Service

British Army, also Austria (designation 8 cm Granatwerfer 70), Bahrain, Brazil, Canada, Guyana, India, Kenya, Malawi, Malaysia, Netherlands (40), New Zealand, Nigeria, Norway (110), Oman, Portugal (22), Qatar, United Arab Emirates and North Yemen. Under the designation M252, the L16 is in service with the US Army.

UPDATED

81 mm L16 mortar in firing position

1996

81 mm L16 mortar (T J Gander)

1996

UNITED STATES OF AMERICA

60 mm M2 mortar

Description

Originally designed and made by Edgar Brandt in France, the 60 mm M2 mortar was licence-built in the USA from 1938 onwards. The designation M1 was applied to the samples purchased from France while US production mortars were designated M2 and used as the standard light infantry mortar for US forces in the Second World War. It was used mainly with HE ammunition (particularly the M49A2) but also fired the M83 illuminating round. It had a rectangular baseplate and a bipod of characteristic Brandt design.

Long withdrawn from US combat units, the M2 mortar can still be found in many US-supplied countries.

Data

Calibre: 60 mm
Length of barrel: 726 mm
Weight: in firing position, 19.07 kg
Elevation: +40 to +85°
Traverse: top, 14°
Muzzle velocity: 158 m/s
Range: min, 91 m; max, 1,816 m
Bomb weight: HE, 1.36 kg

Manufacturer

US arsenals.

Status

Production complete.

Service

No longer in US service. May be found in many countries supplied by the USA and ammunition is still manufactured by FMBP in Portugal. Probably still in service in Austria, Denmark, Greece, Guatemala, Haiti, Indonesia, South Korea, Morocco, Taiwan and several South and Central American countries.

UPDATED

60 mm M19 mortar

Description

The 60 mm M19 mortar is a conventional smoothbore, muzzle-loading weapon designed to produce high angle fire. The equipment consists of the barrel M19, bipod M2, sight M4 and baseplate M1 or M5. The bipod and baseplate together make up the M5 mount.

The M19 is a development of the M2 mortar, the main difference lies in the longer and heavier barrel. The weight is increased by roughly 2 kg, but range and accuracy are improved.

The M19 is fired by dropping a bomb down the muzzle. The setting of the selector lever determines the method by which the firing takes place. If the firing pin is locked out, the primer drops onto it and the bomb is fired. If the firing pin is retracted, the firing lever must be pulled back to compress the firing spring so that when the firing pin trips off the lever, it is driven forward to fire the bomb. At high angles of elevation either method may be used. At low angles only lever fire will provide the necessary impulse to fire the primer.

The M19 barrel is a steel tube with an external thread at the lower end to take the base cap. The firing mechanism is contained in a housing attached to the base cap by a threaded adaptor. The barrel is fitted to the baseplate by a spherical projection which fits into a socket and is locked in place. The firing mechanism consists of the firing pin, a spring, a trigger and pawl and a firing lever. A firing selector acts as a cam on the rear end of the firing pin and allows the mortar to be fired with or without the firing lever.

The M2 bipod consists of two legs, the elevating mechanism assembly and the traversing gear. The bipod legs are connected by a clevis joint which limits the spread of the legs. The elevating screw tube passes through the junction of the legs and connects up with the yoke. The left leg of the bipod carries the sleeve of the cross-levelling gear. When the sleeve is moved up or down the leg the connecting link which is attached to the elevating tube forces the elevating tube out of the vertical. This in turn moves the yoke out of the horizontal. Thus, when the mortar comes into action on uneven ground and the two spiked feet on the bipod legs are on different levels, the cross-levelling gear enables the yoke to be made level. The sight fits into a dovetail slot in the yoke.

The elevating gear consists of the screw thread in the tube with a handle at the bottom. This enables the yoke to be pushed up or down and the barrel clamp, mounted on the yoke, raises or lowers the muzzle of the barrel.

The traversing gear consists of a nut which is moved across inside a tube by the rotation of the traversing handwheel. This displaces the yoke to one side or the other. The lower half of the barrel clamp has two shock absorber cylinders which permit movement between the yoke and clamp assembly during firing.

The M5 baseplate is a rectangular plate with ribs on the underside and a centrally mounted socket to take the spherical end piece of the barrel. The barrel is held in place by a locking lever.

There is a small M1 baseplate which may be attached to the barrel and the weapon can then be hand held and fired without the bipod.

The M4 sight has an elevating scale covering the arc from 40 to 90°, in 10° divisions. The micrometer scale is divided up into 40 divisions each one representing ¼°. The traversing scale has 60 graduations each of 5 mils. The graduations are numbered every 10 mils from 0 to 150 on each side of the zero position.

The M4 sight weighs 0.57 kg. Alternative sights M34A2 (1.8 kg) or M53 (2.3 kg) may also be used with the mortar.

The high explosive round is the M49A4, having a steel body containing 154 g of TNT. A fin assembly is screwed on the rear of the body and a fuze is inserted in the nose. The fins envelop four secondary cartridges. The fuze is the M525 or M525A1.

The Smoke M302A2 is used as a screening, signalling or incendiary round. It is filled with White Phosphorus (WP) and has a burster charge. The fuze is the PD M527B1.

The M50A2 is a practice round. It differs from the HE round in colour and filling. It has the same ballistic characteristics and has a small black powder charge to indicate the point of fall.

In the Illumination M83A3 the illuminant assembly is expelled after 14.5 seconds of flight and drifts down on its parachute to illuminate the ground below. The illumination lasts for a minimum of 25 seconds with a minimum of 330,000 candela.

The M69 training bomb has a solid cast-iron body and the normal fin assembly. Only the primary cartridge is used and the maximum range is 225 m.

Data

Calibre: 60 mm
Length of barrel: 819 mm
Weight: mortar, complete with M5 baseplate, 21.03 kg; with M1 baseplate and no bipod, 9.3 kg

60 mm M19 mortar

Weight of barrel: 7.25 kg
Weight of bipod: 7.42 kg
Weight of baseplate: M5, 5.79 kg; M1, 2.03 kg
Elevation: with M5 mount, +40 to +85° (710-1,510 mils); with M1 baseplate, 0 to +85° (0-1,510 mils)
Traverse: 129 mils left or right
Rate of fire: max, 30 rds/min; sustained, 18 rds/min for 4 min or 8 rds/min indefinitely

Manufacturer

Watervliet Arsenal, Watervliet, New York 12189.

Status

Production complete.

Service

In USA, reserve stocks only. Supplied to many countries in South-east Asia and South and Central America which may well still have it in first-line service. It is also used in Belgium, Canada, Chile, Iran and Japan.

VERIFIED

60 mm M224 lightweight company mortar

Development

During the Vietnam campaign, the US Army found the 81 mm mortar to be excessively heavy and unwieldy for infantry use outside the firebases. As a result patrols had to rely for their own support on the 60 mm M19 mortar which was then obsolescent. In December 1970 the XM224 mortar was approved for development to succeed the 81 mm mortar in the Infantry, Airmobile Infantry and Airborne Infantry. It was intended to be a light portable weapon capable of providing the volume of fire required by the infantry and of operation at either high or low angle fire.

The mortar was developed at Watervliet and Picatinny arsenals: the advanced development objective was approved in 1971, engineering tests were completed at Aberdeen Proving Ground in 1972 and the engineering design was approved in 1973. The weapon was tested at Fort Benning and it was type classified standard in July 1977 and placed in production.

Description

To achieve the required range a barrel 254 mm longer than the M19's and with double the working pressure was developed. The lower part of the tube has been radially finned to increase the surface for heat dissipation and the wall thickness has been kept to much the same value as that of its predecessor by the use of a higher tensile steel. The aluminium forged baseplate is circular, with a pronounced web on the underside. There is also a light rectangular auxiliary baseplate, the M8, which is used when the weapon is hand-held. In the latter method of employment the weight of the mortar is only 7.8 kg. The firer grips the barrel towards the muzzle and at the rear there is a handle which holds the trigger mechanism. The mortar can be fired by either gravity or a spring-loaded firing pin controlled by a firing

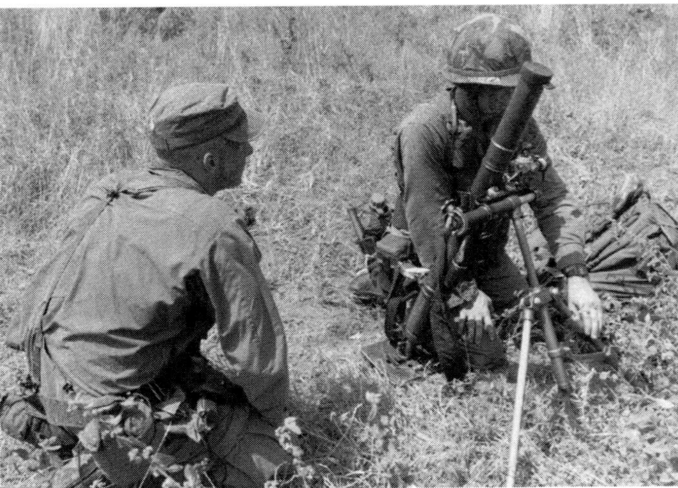

60 mm M224 lightweight company mortar

Comparison of M53 sight used on 81 mm mortar with M64 used on 60 mm M224 mortar

lever which cocks and fires the pin in one movement. The bipod has alloy legs and the elevating screw thread is contained in an alloy elevating tube. Attached to this tube is the connecting rod of the cross-levelling gear which is moved across by a sleeve on the left leg of the bipod. The M64 sight developed for the M224 mortar weighs only 1.1 kg and is self-illuminated for night operations. For easy transport the mortar breaks down into two man loads, the heavier being 11.8 kg.

Ammunition for the M224 comprises the M720 HE round with M734 fuze and M702 ignition cartridge, the M721 Illuminating, M722 Smoke, M723 Smoke (WP), M816 and M840 Practice rounds.

To reduce the number of different types of fuze required, running at 15 different types with the 81 mm mortar, a multi-option fuze, designated M734, was developed by the Harry Diamond Laboratories. This allows the user to select one of four options: high airburst, low airburst, point detonating and delay. Should the fuze fail to function on the chosen option it will automatically select the next one and function on that. If low airburst is selected and the fuze does not fire, it will fire on point detonating; should this not function then it will fire on delay. The fuze uses micro-circuitry and generates the required electrical power from a small turbine in the nose. The turbine unlocks with the setback of firing and air is fed through a hole in the extreme nose of the fuze. The spinning turbine operates a miniature generator. It takes a finite time before the current has built up to the required level, providing a safety feature in that the bomb is well clear of the baseplate position before the fuze is active. A plastic cover over the fuze must be removed before the round is fired but if it is inadvertently left in place, air pressure tears it off. The nose ring of the fuze provides the method of selection.

Manufacturer
Watervliet Arsenal, Watervliet, New York 12189.

Status
Production complete.

Service
US Army.

VERIFIED

81 mm M29 and M29A1 mortar

Development
This 81 mm mortar was originally given the service designation M29, its development designation being T106. The baseplate M3 was developed by the Canadian Armaments Research and Development Establishment (now DREV). A product-improved version of the mortar, with a barrel (designated M29E1 instead of M29) capable of sustaining a higher rate of fire, was standardised in 1970 with the designation M29A1. The M125A1 mortar carrier mounts the M29 mortar in a modified M113 carrier, originally designated T257E1.

The M29 mortars have now been almost entirely replaced in the US armed forces by the 81 mm M252 mortar (see following entry).

Description
The 81 mm M29 mortar is a smoothbore, muzzle-loaded weapon consisting of the barrel, the mount and the baseplate. The standard M29 mortar comprises the M29A1 barrel assembly, the M23A3 mount, the M53 sight and the M3 baseplate. It may be used in the conventional manner as a ground weapon or mounted into the Carrier, Mortar, M125A1.

The barrel is made up of the tube, externally threaded at the rear to take a base plug with a ball-shaped projection on its lower end to fit into the socket of the baseplate. There are two white lines painted 432 and 533 mm from the muzzle for the location of the mount attachment ring. The exterior of the barrel is radially finned to increase the cooling area. The mount comprises the bipod legs, elevating mechanism and the traversing mechanism. The bipod has two tubular steel legs hinged at the sides of the elevating mechanism: they have spiked feet and their spread is limited by an adjustable chain with a spring to relieve the shock on the legs during firing. The left leg carries the cross-level mechanism on a sliding bracket mounted on the leg with a locking sleeve and an adjusting nut. The sliding bracket is connected to the elevating housing by a connecting rod. When the mount is on uneven or sloping ground, the sliding bracket is moved across by rotation of the sleeve on the bipod leg and this in turn moves the elevating mechanism assembly across taking the barrel with it. This enables the sight, at the left end of the yoke, to be moved into an upright position. The elevating mechanism assembly includes a vertical elevating screw moving inside the elevating housing assembly. There is a handle projecting back from the gear case at the junction of the bipod legs, and rotation of this handle elevates or depresses the mortar barrel.

The traversing assembly consists of the yoke assembly, traversing mechanism and shock absorber.

81 mm M29A1 mortar, ground-emplaced with M53 sight and M3 aluminium baseplate

The yoke body supports the upper end of the barrel when the mortar is assembled. Older models of the mortar had a yoke with a levelling bubble. The sight unit is mounted in the dovetail sight slot on the left side of the yoke. The traversing mechanism is an internal screw shaft operating within a nut and tube. The handwheel turns a screw which forces the nut to traverse the yoke and take the barrel with it. The tube over the nut is connected to the elevating shaft which protrudes from the gear case of the bipod. The shock absorber is a compression spring, mounted in the yoke. When the barrel is located to the yoke by the barrel clamp the shock absorber connects the barrel to the bipod.

The M3 standard baseplate is a one-piece aluminium alloy forging. In the centre is the barrel socket which rotates through 360°.

Sight units M53 and M34A2 can be used with the mortar. The M53 is the standard sight, largely replacing the M34A2. Each incorporates a telescope mount and an elbow telescope fastened together into one unit. Both have fixed and slipping scales.

Data
Calibre: 81 mm
Length of barrel: 1.295 m
Weight of barrel: 12.68 kg
Weight of mount: 18.12 kg
Weight of baseplate M3: 11.3 kg
Weight of sight unit: M53, 2.84 kg; M34A2, 1.81 kg
Elevation: 800-1,500 mils
Traverse: top, right or left from centre, 95 mils

Manufacturer
Watervliet Arsenal, Watervliet, New York 12189.

Status
Production complete.

Service
Being replaced in US forces by the M252 (see following entry). Also many other countries including Australia, Austria (8.1 cm GrW M29/65) and Italy.

UPDATED

81 mm M252 mortar

Description
US Army interest in the British 81 mm mortar L16 began in 1975. In 1978 18 mortars were provided for test, together with a supply of ammunition. A new bomb, based on the British design, was developed as the M821, followed by the M889. A Smoke WP bomb and an Illuminating bomb were also developed. One change to the mortar barrel, introduced to meet muzzle blast reduction requirements, is a conical attachment to the muzzle to direct muzzle blast upwards. The M252 was type classified in 1984.

An initial order for the supply of mortars from Royal Ordnance plc was placed in December 1984, and a further order for 400 tubes and 350,000 bombs was placed in September 1985. Since then over 1,800 M252 mortars and some 3 million rounds of ammunition have been procured.

For a general description of the 81 mm system, reference should be made to the entry under *United Kingdom*. The data given here refers specifically to the US M252 version and shows some small differences in dimensions and performance.

The complete 81 mm M252 mortar consists of the M253 Cannon Assembly, the M177 Mount, the M3A1 Baseplate, and the M64A1 Sight Unit. The types of bomb fired include the M821/M989 HE, M819 Smoke RP, M853A1 Illuminating, M879 Practice (full range) and M880 Practice (short range). It is planned that rocket-assisted HE, Smoke and Illuminating bombs will be introduced, along with an 81 mm cargo bomb carrying dual purpose bomblets.

Data
Calibre: 81 mm
Length of barrel: 1.277 m
Weight: in firing position, 42.32 kg
Weight of barrel: 12.25 kg

Weight of bipod: 11.79 kg
Weight of baseplate: 11.34 kg
Elevation: +45 to +85°
Traverse: 5.62° (100 mils)
Range: M889 HE, min, 80 m; max, 5,700 m
Rate of fire: rapid, 30 rds/min for 2 min; sustained, 15 rds/min

Manufacturer
Watervliet Arsenal, Watervliet, New York 12189.

Service
US Army and Marine Corps.

UPDATED

US Army personnel in NBC gear test-firing an 81 mm M252 mortar

107 mm M30 rifled mortar

Description
The 107 mm mortar, M30 (T104), is a rifled, muzzle-loaded, drop-fired weapon which can be hand-carried for short distances when disassembled into five loads. The complete weapon comprises a barrel, a standard, a bridge assembly, a rotator assembly, a baseplate and a sight unit. The design is such that the recoil forces are absorbed by the main mechanical assemblies as a whole.

The barrel is a rifled tube 1.524 m long with an inside diameter of 107 mm. The rifling consists of 24 lands and grooves of which the first 228.6 mm, as measured from the base inside the barrel, are straight. The twist increases to the right from zero at this point to one turn in 19.98 calibres at the muzzle.

The standard assembly, consisting of the elevating, traversing and recoil mechanism, connects the barrel and the bridge assembly.

The traversing mechanism consists of an enclosed screw and bearing located at the top of the standard assembly. It is operated by turning the traversing crank on the left side of the mechanism.

The recoil mechanism consists of a series of springs mounted in the lower section of the standard assembly. These are designed to ease the downward shock of firing and return the mechanism to its pre-firing position.

The bridge assembly consists of two pieces joined by a swivel point. One end receives the cap at the base of the barrel: the other is fitted with a spade which facilitates the digging action of the bridge assembly during firing.

The rotator assembly is approximately 508 mm in diameter. It carries the barrel end of the bridge piece on its upper side and pivots on the baseplate.

The baseplate assembly is approximately 965 mm in diameter. A recess in the centre receives the bottom insert of the rotator assembly. The lower surface carries six ribs to increase the area in contact with the ground. Each rib has a depth of 165.1 mm. Two carrying handles are provided.

The M53 sight unit is the standard sighting device. It consists of an M128 telescope mount and an M109 elbow telescope fastened together into one unit for operation.

The M109 telescope is a lightweight, four-power, fixed-focus instrument, with a 10° field of view which provides the optical line-of-sight for aiming the weapon for line and elevation. There is a coarse elevation scale on the left side of the sight body with 18 graduations, each of 100 mils, numbered every 200 mils. The elevating knob has an adjustable micrometer with 100 graduations, each of 1 mil and numbered every 10 mils from 0 to 90. The deflection micrometer scale has 100 graduations numbered from 0 to 90. It is fastened to the deflection knob. There is a coarse deflection slip scale and an adjustable micrometer deflection slipping scale. There are two levelling bubbles at 90° to each other on the main sight housing.

Ammunition for the 107 mm M30 mortar is issued as complete rounds. The propelling charge consists of an ignition cartridge and 41 propellant increments assembled in a bag and sheets. To adjust the charge, individual increments are removed. Information on the standard rounds authorised for the 107 mm mortar M30 will be found in the *Mortar Ammunition* section.

Data
Calibre: 107 mm (4.2 in)
Length of barrel: 1.524 m
Weight: complete, with welded steel rotator, M24A1 baseplate and M53 sight, 305 kg
Weight of barrel: M30, 70.89 kg
Weight of bridge assembly: 76.55 kg
Weight of baseplate assembly: M24A1, 87.42 kg; M24, inner baseplate, 53.45 kg; outer baseplate ring assembly, 46.66 kg
Weight of standard assembly: 26.95 kg
Weight of rotator assembly: cast magnesium, 26.05 kg; welded steel, 40.32 kg

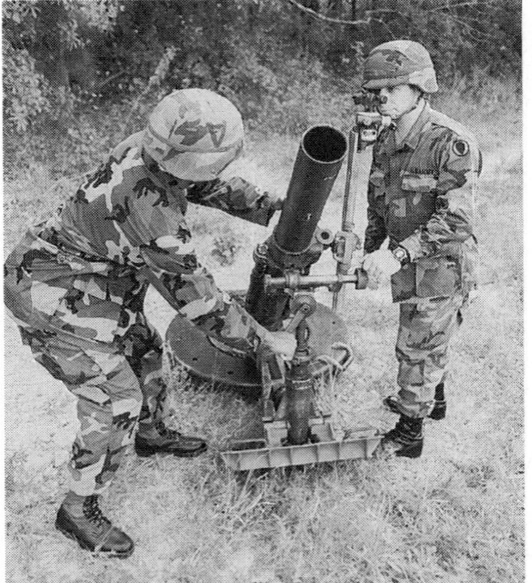

107 mm M30 rifled mortar

Weight of sight equipment: M34A2, 1.81 kg; M53, 2.84 kg
Rifling: 24 grooves; straight to 228.6 mm, thereafter increasing rh twist to 1 turn in 19.98 cal
Rate of fire: max, 18 rds/min for 1 min and 9 rds/min for the next 5 min; this can be followed by a sustained 3 rds/min for prolonged periods.

Status
Production complete.

Service
US National Guard. Also Austria, Belgium, Canada, Greece, Iran, South Korea, Liberia, Netherlands, Norway, Oman, Turkey and Zaïre.

VERIFIED

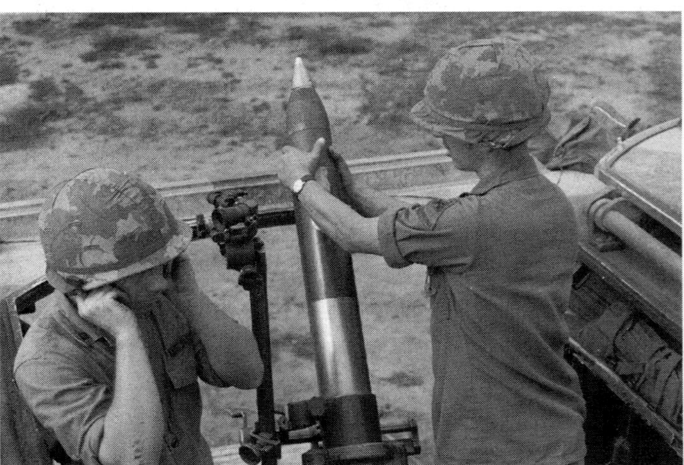

Firing 107 mm mortar from M106A1 carrier

M30 mortar in M106 carrier

120 mm M120/121 Battalion Mortar System

Description

In 1984 the US Army reached the conclusion that the 107 mm M30 mortar was incapable of cost-effective improvement and decided to adopt a 120 mm mortar in its place. After studies of available weapons, in March 1990 the Soltam K6, in towed and carrier-mounted versions, was selected for adoption and type-classified as the Mortar, 120 mm, M285. This designation was later changed to Mortar, 120 mm, M120 for the towed equipment and Mortar, 120 mm, M121 for the carrier-mounted version. Original plans called for the provision of 2,606 mortars, but this was later reduced to 1,725, sufficient to replace existing 107 mm M30 mortars. The larger part of this number involves M121 mortars for use in the M1064 tracked carrier, based on the M113A3 APC chassis.

The M120/121 was co-produced by Soltam and Israel Military Industries (now TAAS - Israel Industries) with ammunition provided by Lockheed Martin.

It has been proposed that a 120 mm mortar could be carried on the rear of a HMMWV, ready to be lowered hydraulically to the ground for firing and raised hydraulically afterwards for transport. This project, known as the HMM20 is still at the demonstration stage, and is intended to be employed by rapid deployment forces. There is also a 120 mm Mortar Weight Reduction (120MWR) programme.

Ammunition provided for the 120 mm mortars includes the M933 HE bomb, M929 Smoke bomb and M930 Illuminating bomb, designs based upon the ammunition originally provided with the Soltam mortars. Future designs are expected to include cargo bombs for bomblets or scatterable mines, a guided anti-tank bomb and a flechette-loaded anti-personnel bomb. Olin Ordnance have developed a 120 mm long-range bomb with a range of 10,000 m.

An 81 mm M303 Insert has been developed for use with 120 mm M120/M121 mortars to reduce costs during training.

Data

M120

Calibre: 120 mm
Length of barrel: including breech piece, 1.758 m
Weight: in firing position, 144.7 kg; travelling, 321 kg
Weight of barrel: 50 kg
Weight of baseplate: 61.87 kg
Weight of bipod: 31.93 kg
Weight of sight: 2.36 kg
Weight of trailer: 181.1 kg
Elevation: +40 to +85°
Traverse: 14°
Range: min, 170 m; max, 7,240 m
Rate of fire: max, 15 rds/min for 1 min; sustained, 4 rds/min

Service

US Army.

UPDATED

120 mm M120 mortar on travelling carriage
(T J Gander)

120 mm M120 battalion mortar with trailer

Lockheed Martin Lightweight 120 mm Mortar

Development/Description

Lockheed Martin has produced two firing prototypes of a lightweight 120 mm mortar and carried out firing demonstrations during April 1994. More test firings have since been conducted at Lockheed Martin's Santa Cruz facility in California. The intention is to reduce the weight of a 120 mm mortar system sufficiently to render it fully man-portable. This has been accomplished by the incorporation of composite materials coupled with a patented firing energy absorber.

As an example of how weight has been saved the lightweight 120 mm mortar barrel weighs 18.1 kg, compared to 50 kg for a conventional steel barrel. The use of composites allows a rate of fire of 12 rds/min for the first minute and 4 rds/min thereafter. It is claimed that the composite barrel has dimensional stability and rigidity and thus improves accuracy over prolonged firing periods. Composite materials are also environmentally inert and corrosion free.

The introduction of the patented energy absorber between the barrel and bipod reduces the normal 249,500 kg peak force produced on firing to 58,970 kg. This enables the lightweight mortar to be fired from all manner of hard and soft terrain and also makes it possible to fire the mortar from the cargo area of a HMMWV light vehicle with all its attendant 'fire and run' mobility.

Lockheed Martin have proposed the extension of composite material techniques to 81 mm mortars and 120 mm mortar munitions.

Data

Calibre: 120 mm
Weight: total, 62.6 kg
Weight of barrel: 18.1 kg
Weight of baseplate: 18.1 kg
Weight of bipod: 13.6 kg
Weight of energy absorber: 12.7 kg
Elevation: +45 to +85°
Traverse: total on bipod, 14°
Range: min, 170 m; max, 7,240 m
Rate of fire: 12 rds/min for first min; 4 rds/min thereafter

Manufacturer

Lockheed Martin Missiles & Space, 1111 Lockheed Way, Sunnyvale, California 94089-3504.

Lockheed Martin Lightweight 120 mm mortar
1996

Status

Prototypes.

NEW ENTRY

YUGOSLAVIA (SERBIA AND MONTENEGRO)

50 mm M-8 mortar

Description

The 50 mm M-8 mortar is similar to the British 2 in (51 mm) mortar but differs in having a supporting and carrying handle at the point of balance. The barrel is a plain steel tube internally screw-threaded at the bottom to take the breech piece which holds the trigger and firing mechanism. The weapon is fired by a lever on the right of the breech piece. The sight is a simple pointer with a bubble to indicate when it is horizontal and a range scale. The baseplate is trough-shaped and lies across the barrel axis.

The M-8 fires an HE bomb weighing about 1 kg, a smoke bomb and an illuminating bomb.

Data

Calibre: 50 mm
Weight: 7.3 kg
Muzzle velocity: 80 m/s
Range: min, 135 m; max, 480 m
Rate of fire: 25-30 rds/min

Holding Company

YUGOIMPORT - SDPR, PO Box 89, Bulevar umetnosti 2, YU-11070 Novi, Belgrade.

Status

Production probably complete.

Service

Yugoslav armed forces.

VERIFIED

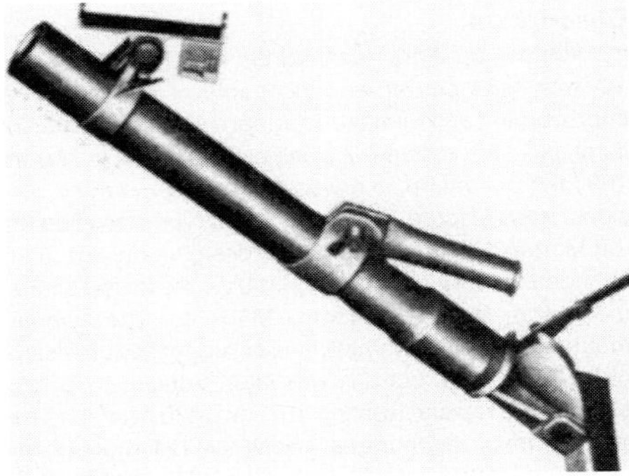

50 mm M-8 mortar

60 mm M-70 mortar

Description

This lightweight 'Commando' mortar consists of a tube and baseplate emplaced for firing by holding the barrel at the desired elevation. The baseplate is a simple spade, there is a canvas grip laced to the tube, and the sight is a bubble adjustable for elevation together with a rifle-type post and notch sight for direction. It fires the same M67 and M73 bombs as the M-57 mortar and is trip-fired by a lever on the breech piece.

Data

Calibre: 60.75 mm
Length of barrel: 650 mm bore; 780 mm with breech piece
Weight: 7.6 kg
Elevation: +5 to +85°
Max pressure: 250 bar
Range: max, 1,630 m
Rate of fire: 20-25 rds/min

Holding Company

YUGOIMPORT - SDPR, PO Box 89, Bulevar umetnosti 2, YU-11070 Novi, Belgrade.

Status

Production probably complete.

Service

Yugoslav armed forces and Croatia.

Licence production

CROATIA
Manufacturer: RH-ALAN doo
Type: Commando 60 mm M70
Remarks: Production not authorised. Offered for export sales.

UPDATED

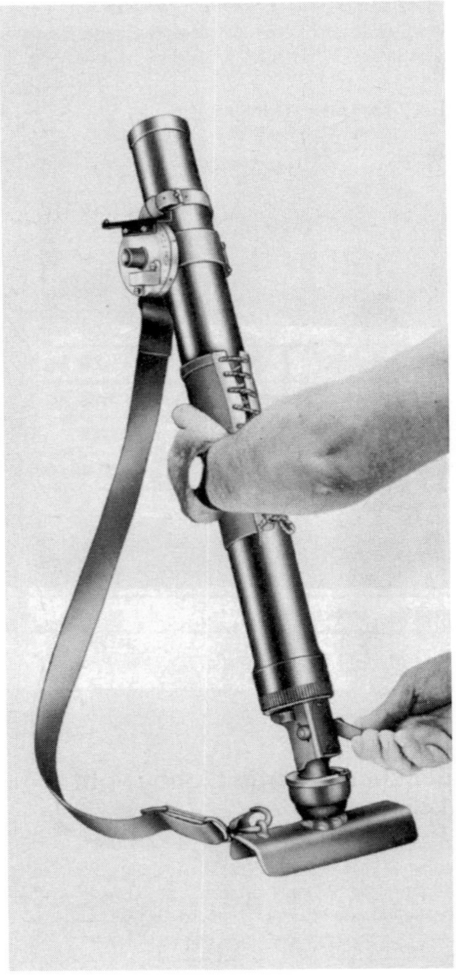

60 mm M-70 mortar

60 mm M-57 mortar

Description

The 60 mm M-57 was developed from the US 60 mm M2 mortar. The breech piece is screw-threaded to fit over the bottom of the barrel, and has a ball-shaped end to fit into the socket of the rectangular baseplate. The bipod has a cross-levelling gear operated by a sleeve on the left leg and connected to a rod attached to the elevating gear. This is attached to the yoke which carries the traversing screw and the shock-absorbing spring cylinders. At the right-hand end of the yoke is the traversing handwheel and on the left is the sight. This has no slipping scale but allows switches from a chosen aiming point.

The mortar fires HE, Smoke and Illuminating bombs.

Data

Calibre: 60.75 mm
Weight of barrel: 5.5 kg
Weight of bipod: 4.5 kg
Weight of baseplate: 8.85 kg
Weight of sight: 1 kg
Muzzle velocity: 159 m/s
Range: min, 75 m; max, 1,700 m
Rate of fire: 25-30 rds/min
Bomb weight: 1.35 kg

Holding Company

YUGOIMPORT - SDPR, PO Box 89, Bulevar umetnosti 2, YU-11070 Novi, Belgrade.

Status

Production probably complete.

Service

Yugoslav armed forces.

VERIFIED

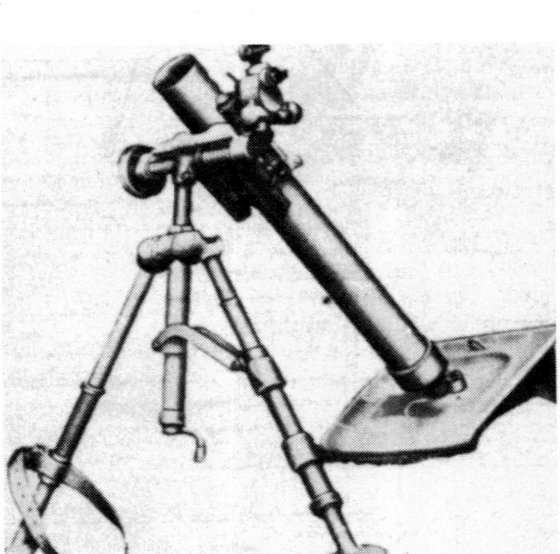

60 mm M-57 mortar

60 mm M-90 long-range mortar system

Description

The 60 mm M-90 long-range mortar was developed to combine the flexibility of the 60 mm light mortar and the firepower of the 81 mm mortar. It differs from the standard version by having a longer and heavier barrel designed to fire bombs to 5,200 m at a rate of fire of over 20 rds/min. For ease of transport it can be broken down into three parts - barrel, baseplate and bipod.

The M-90 will fire the standard range of locally manufactured 60 mm HE, Smoke and Illuminating bombs. In addition, new long-range M-90 HE and Smoke WP bombs were developed for use specifically with this mortar.

Data

Calibre: 60.75 mm
Length of barrel: with breech piece, 1.325 m
Weight: in firing position, 32.6 kg
Elevation: +42 to +85°
Traverse: on bipod, ±4°
Max pressure: 680 bar
Range: min, 150 m; max, 5,208 m
Rate of fire: max, 20 rds/min

Holding Company

YUGOIMPORT - SDPR, PO Box 89, Bulevar umetnosti 2, YU-11070 Novi, Belgrade.

Status

Production probably complete.

Service

Yugoslav armed forces.

Licence production

CROATIA
Manufacturer: RH-ALAN doo
Type: LMB 60 mm M84
Remarks: Production not authorised. Offered for export sales. Mounting incorporates some local modifications.

UPDATED

60 mm M-90 long-range mortar

81 mm M-68 mortar

Description

Although similar in appearance to the Brandt MO-81-61L mortar, the 81 mm M-68 is of local design and construction.

Data

Calibre: 81 mm
Length of barrel: 1.64 m
Weight of barrel: 16 kg
Weight of bipod: 13 kg
Weight of baseplate: 11 kg
Weight of sight: 1.5 kg
Muzzle velocity: 300 m/s
Rate of fire: 20 rds/min
Range: min, 90 m; max, 5,000 m

Holding Company

YUGOIMPORT - SDPR, PO Box 89, Bulevar umetnosti 2, YU-11070 Novi, Belgrade.

Status

Production probably complete.

Service

Yugoslav armed forces.

VERIFIED

81 mm M-68 mortar

82 mm M-69A mortar

Description

Although nominally of different calibre the 82 mm M-69A mortar is a variant of the M-69B (see following entry) and fires the same types of ammunition, though using a different propelling charge. It is transported in three units, barrel, bipod and baseplate, either man-pack or mule-pack. Several of the parts are interchangeable with the M-69B mortar.

The M-69A uses the NSB-3 collimating sight which incorporates azimuth and elevation settings and has tritium illumination for night firing.

Data

Calibre: 82 mm
Length of barrel: 1.2 m
Weight of barrel: 14.5 kg
Weight of bipod: 14 kg
Weight of baseplate: 15 kg
Weight of sight: 1.5 kg
Elevation: +45 to +85°
Max bore pressure: 630 bar
Range: max, 6,050 m
Rate of fire: 20-25 rds/min
Range: max, 6,050 m

Holding Company

YUGOIMPORT - SDPR, PO Box 89, Bulevar umetnosti 2, YU-11070 Novi, Belgrade.

Status

Production probably complete in former Yugoslavia.

Service

Yugoslav armed forces and Croatia.

Licence production

CROATIA
Manufacturer: RH-ALAN doo
Type: LMB 82 mm M93
Remarks: Production not authorised. Offered for export sales.

UPDATED

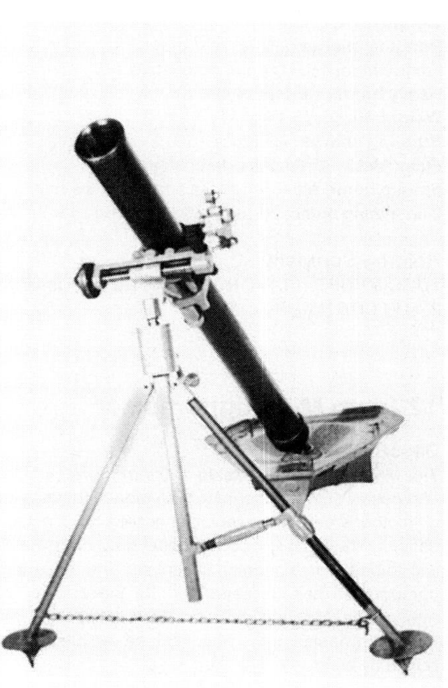

82 mm M-69A mortar

81 mm mortars M-69B(K) and M-69B(D)

Description

These two mortars appear to be slightly improved models of the 81 mm M-68 (see previously). They are of the same general form, tube, baseplate and bipod, with a small spring recuperator between barrel and bipod. They use the same collimating NSB-3 sight as the M-69A and fire the same general range of ammunition as the M-68 and M-69A.

These mortars were designed to be man-carried in three sections, using special carriers, together with a fourth carrier for the sights and accessories; they can also be carried by animal pack, using special carrier saddles.

Two models exist; the M-69B(K) is the short-barrelled version, the M-69B(D) the standard version. Apart from the length of barrel they are otherwise identical.

Data

Model	M-69B(D)	M-69B(K)
Calibre	81 mm	81 mm
Length of barrel	1.45 m	1.15 m
Weight of barrel	16.6 kg	14.5 kg
Weight of bipod	13.5 kg	13.5 kg
Weight of baseplate	16 kg	16 kg
Weight of sight	1.5 kg	1.5 kg
Elevation	+45 to +85°	+45 to +85°
Max range	–	5,400 m
Rate of fire	20-25 rds/min	20-25 rds/min

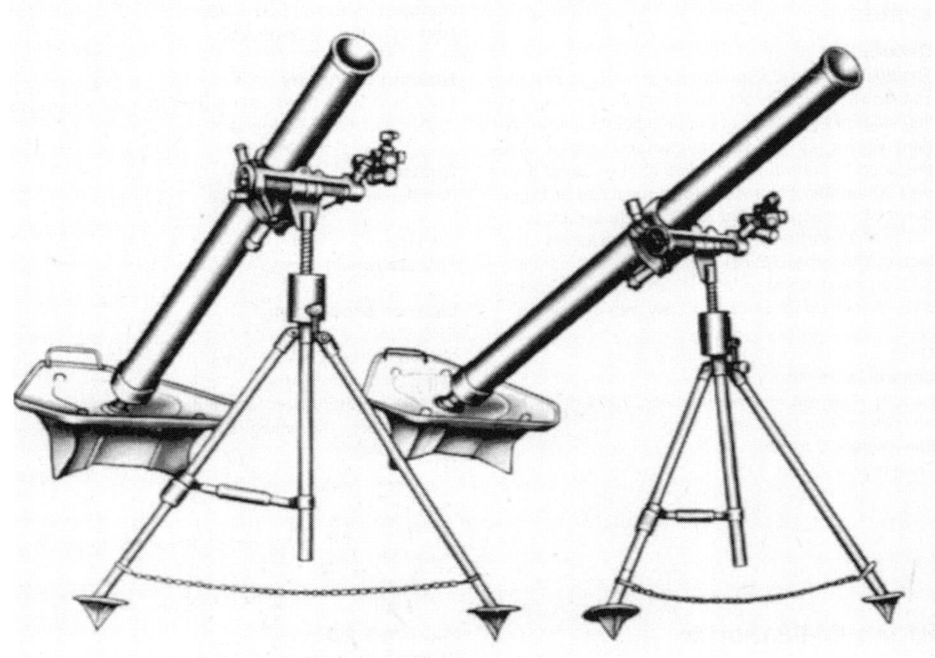

81 mm mortars M-69B(K) (left) and M-69B(D) (right)

Holding Company

YUGOIMPORT - SDPR, PO Box 89, Bulevar umetnosti 2, YU-11070 Novi, Belgrade.

Status

Production probably complete.

Service

Yugoslav armed forces.

VERIFIED

120 mm UBM 52 mortar

Description

The 120 mm UBM 52 is of local design and construction and was been designed for both field and mountain deployment. For field use it is normally towed by its muzzle by a 4 × 4 truck. For transport in rough country it can be quickly broken down into five loads and carried by animals.

When in the firing position the wheels are retained and the hydroelastic recoil system permits immediate commencement of fire once in position. The mortar, which is smoothbore, can also fire CIS 120 mm mortar rounds. It can be fired either by conventional drop-firing or by a trigger. The mortar has a crew of five.

Data

Calibre: 120 mm
Weight: complete equipment, 400 kg
Range: heavy bomb: min, 195 m; max, 4,760 m; light bomb: min, 225 m; max, 6,010 m
Light bomb, HE Model 62
Weight: fuzed, 12.2 kg
Bursting charge: 2.5 kg
Fuze: Model 45UTU point detonating
Heavy bomb, HE Model 49P1
Weight: fuzed, 15.1 kg
Bursting charge: 3.1 kg
Fuze: Model 45TU point detonating
Smoke bomb M64: produces smoke for 2-4 min
Illuminating bomb: burns for 30 s and produces 1.1 Mcd

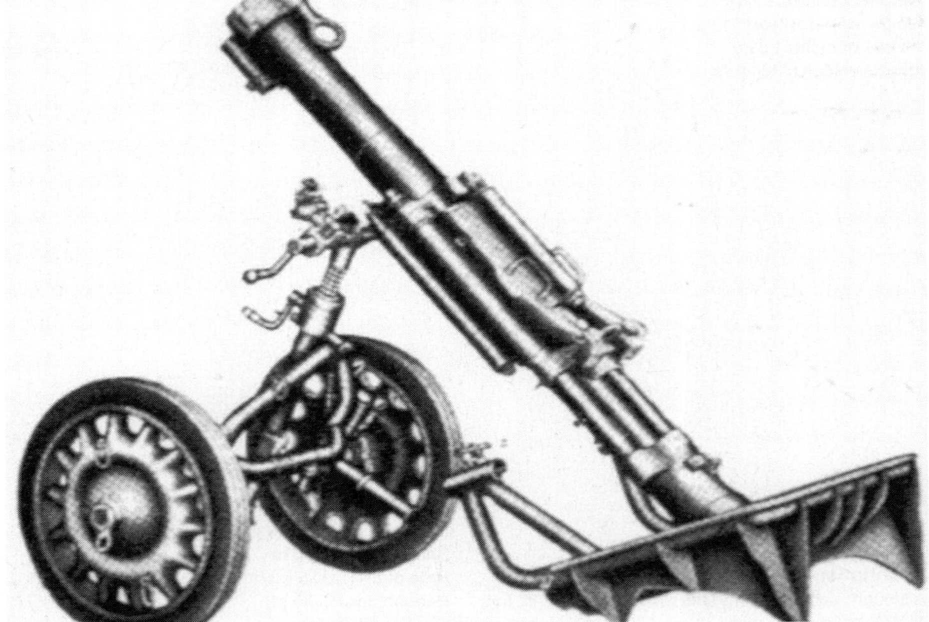

120 mm UBM 52 mortar

Holding Company

YUGOIMPORT - SDPR, PO Box 89, Bulevar umetnosti 2, YU-11070 Novi, Belgrade.

Status

Production probably complete.

Service

Former Yugoslav armed forces.

VERIFIED

120 mm M-74 light mortar

Description

The M-74 is a light, portable 120 mm mortar which is particularly suited to operations in mountainous terrain or in other areas where vehicular access is restricted or difficult. The M-74 is a conventional drop-fired weapon with bipod and triangular baseplate. The three basic components can be separated for carrying, or can remain attached and the mortar be transported on a light two-wheeled trailer which can be vehicle-towed or pulled by hand.

The M-74 fires standard 120 mm ammunition, including light and heavy HE bombs, a rocket-assisted bomb (fired with or without rocket boost), smoke, illuminating and practice bombs.

Data

Calibre: 120 mm
Length of barrel: 1.69 m
Weight: in firing position, 120 kg; travelling, with trailer, 208 kg
Weight of barrel: 45 kg
Weight of bipod: 25 kg
Weight of baseplate: 49 kg
Elevation: +45 to +85°
Traverse: on bipod, 6°
Muzzle velocity: light bomb, 266 m/s
Range: (light bomb) min, 267 m; max, 6,213 m; (heavy bomb) min, 300 m; max, 5,374 m; (rocket-assisted bomb) max, 9,056 m
Rate of fire: max, 12 rds/min; sustained, 35 rds/10 min

Holding Company

YUGOIMPORT - SDPR, PO Box 89, Bulevar umetnosti 2, YU-11070 Novi, Belgrade.

Status
Production probably complete.

Service
Yugoslav armed forces.

UPDATED

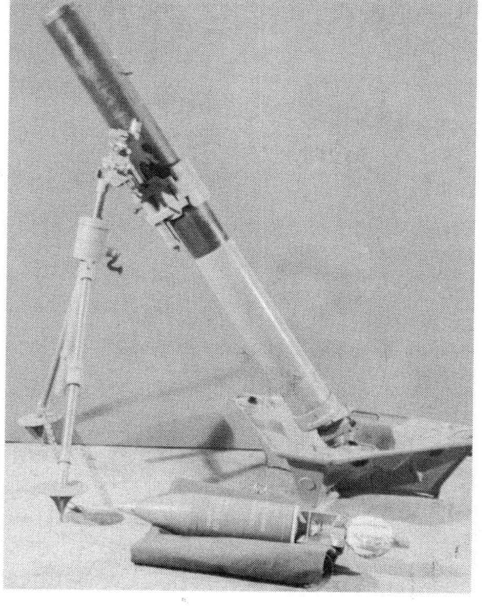
120 mm M-74 light mortar in firing position; rocket-assisted bomb in foreground

120 mm M-74 light mortar in travelling mode

120 mm M-75 light mortar

Description
Although termed as 'light', the 120 mm M-75 is heavier than the M-74 and uses a more substantial circular baseplate in order to withstand the recoil stresses of the heavier bombs. All types of 120 mm ammunition may be fired, as can a rocket-assisted bomb.

The M-75 is of conventional pattern and is transported on the same lightweight trailer as the M-74; a harness rig for towing by a horse is available. It can also be transported by pack animals, helicopter lifted or parachute dropped.

Data
Calibre: 120 mm
Length of barrel: 1.69 m
Weight: in firing position, 177 kg; travelling, with trailer, 263 kg
Weight of barrel: 65 kg
Weight of bipod: 25 kg
Weight of baseplate: 86 kg
Elevation: +45 to +85°
Traverse: on bipod, 6°
Range: light bomb, min, 275 m; max, 6,340 m; heavy bomb, min, 300 m; max, 5,551 m; rocket-assisted bomb, 9,056 m
Rate of fire: max, 15 rds/min

Holding Company
YUGOIMPORT - SDPR, PO Box 89, Bulevar umetnosti 2, YU-11070 Novi, Belgrade.

Status
Production probably complete.

Service
Yugoslav armed forces. In service in Croatia.

Licence production
CROATIA
Manufacturer: RH-ALAN doo
Type: LMB 120 mm M75
Remarks: Production not authorised. Offered for export sales.

UPDATED

120 mm M-75 light mortar in travelling position

1996

120 mm M-75 light mortar in firing order

MORTAR FIRE CONTROL

AUSTRIA

Photonic A-70 mortar sight

Description
The A-70 consists primarily of the sight with elevation and azimuth worm pinions and the sighting telescope (with coarse orientation open sight). The telescope is elbowed and can be axially rotated in its seat. The eyepiece is at 90° to the object lens and allows the gunlayer to view comfortably from above the sight. The telescope can be tilted up or down 25° from the horizontal. The azimuth worm gear can be disengaged for quick adjustments, and the coarse and fine azimuth scales can be adjusted and clamped.

The graticule of the instrument can be illuminated for night firing by means of an attachable night illumination device. In addition, a second light source can be attached in order to illuminate the azimuth and elevation scales and the bubble levels.

Data
Magnification: ×2.5
Field of view: 9°
Exit pupil: 4 mm
Elevation setting range: 700-1,600 mils
Azimuth setting range: 6,400 mils
Dimensions: 104 × 142 × 193 mm
Weight: 1.35 kg

Manufacturer
Photonic Optische Geräte GesmbH, Zeillergasse 20/22, A-1170 Vienna.

Status
In production.

VERIFIED

Photonic A-70 mortar sight

Photonic A-82 mortar sight

Description
The Photonic A-82 mortar sight is to the same general pattern as the A-70 described in the previous entry, but the elbow telescope can be rotated about the axis of the object lens, so that the layer can use the eyepiece from either side of the sight. Instead of an illuminating attachment, all scales, indices and bubble levels are illuminated by tritium for operation at night or in bad visibility.

Data
Magnification: ×2.5
Field of view: 9°
Exit pupil: 4 mm
Elevation setting range: 700-1,600 mils
Azimuth setting range: 6,400 mils
Dimensions: 97 × 142 × 197 mm
Weight: 1.4 kg

Manufacturer
Photonic Optische Geräte GesmbH, Zeillergasse 20/22, A-1170 Vienna.

Status
In production.

VERIFIED

Photonic A-82 mortar sight

Photonic mortar boresight

Description
The Photonic mortar boresight is a device for checking the boresighting adjustment in azimuth and elevation for all types of 81 mm and 120 mm mortars. The calibration is done at an elevation of 1,100 mils and the accuracy is better than 1 mil in both planes.

The objective of the boresighting telescope is focused at 10 m distance on the boresight target. Any combination mortar sight and mortar demands a special target in which two vertical lines are marked at exactly the distance between the axis of the bore and the optical axis of the sight. The mortar is laid at 1,100 mils and the target set up exactly 10 m in front. The boresight is then inserted in the barrel (using the appropriate calibre rod) and its level bubble brought central. The mortar is then moved in azimuth until the boresight vertical graticule is aligned with the right-hand vertical line on the target. The mortar sight should then be aligned with the left-hand line on the target. Failure to so align indicates the need for adjustment of the sight or its mounting.

Manufacturer
Photonic Optische Geräte GesmbH, Zeillergasse 20/22, A-1170 Vienna.

Status
In production.

VERIFIED

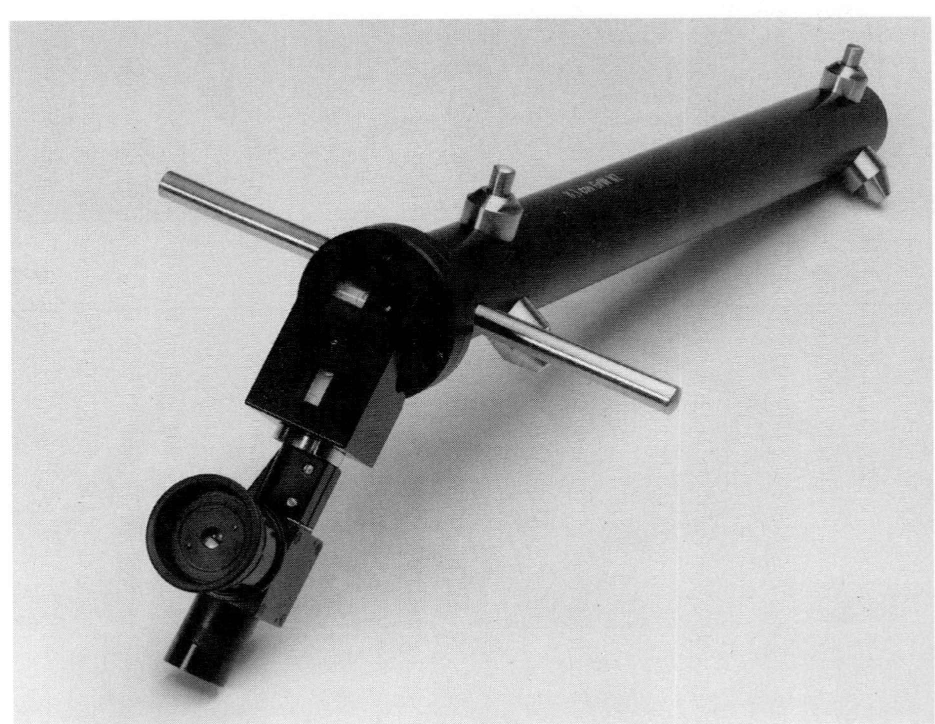

Photonic mortar boresight

CANADA

Mortar sight unit C2

Description

The mortar sight unit C2 was designed for use with a variety of different types of medium mortar and is employed on the Royal Ordnance 81 mm L16A1 mortar. It has also proved to be effective with heavy mortars and can be used to lay a variety of medium and heavy machine guns in the indirect- and sustained fire roles.

The sight unit is a fixed-focus type optical instrument using external light sources to illuminate the elevation and azimuth scales, graticule and level bubbles. It is manufactured in accordance with all the relevant military specifications. The elbow telescope incorporates a dovetail for accepting a periscope extension. This raises the line-of-sight so that the mortar can be fired from a pit or a modified weapons carrier.

Data

Magnification: ×1.8
Field of view: 10°
Illumination: instrument light with batteries
Scales: azimuth, 0-6,400 mils; elevation, mortar, 600-1,600 mils; machine gun, −200 to +600 mils
Accuracy: 1 mil
Weight: with case, 4.9 kg

Manufacturer

Hughes Leitz Optical Technologies Ltd, 328 Ellen Street, Midland, Ontario L4R 2H2.

Status

In production.

Service

Worldwide.

UPDATED

Mortar sight unit C2

Mortar sight unit C2A1

Description

The sight unit C2A1 was designed and developed by Ernst Leitz Canada Limited. It was designed for use with 81 mm mortars and can also be used to lay a variety of machine guns when employed in the indirect-fire role. It is employed on the Royal Ordnance 81 mm L16A1 mortar.

The sight unit is a fixed-focus type optical instrument which uses tritium light sources to illuminate the elevation and azimuth scales, graticule and level vials.

The sight unit elbow telescope incorporates a dovetail to permit a periscope extension (below) to be fitted so as to raise the line-of-sight so that the mortar can be operated from a pit or from a modified weapons carrier.

Data

Magnification: ×1.8
Field of view: 10°
Illumination: tritium (hydrogen 3)
Scales: azimuth, 0-6,400 mils; elevation, mortar, 600-1,600 mils; machine gun, −200 to +600 mils
Accuracy: 1 mil
Weight: with case, 1.4 kg

Manufacturer

Hughes Leitz Optical Technologies Ltd, 328 Ellen Street, Midland, Ontario L4R 2H2.

Status

In production.

Service

Military forces worldwide.

VERIFIED

Leitz sight unit C2A1

Periscope extension C2/C2A1

Description

This periscope extension is used with the C2/C2A1 family of 81 mm mortar sight units in order to raise the line-of-sight by 355 mm so that the mortar can be employed from weapon pits or in modified weapons-carrying vehicles. The periscope extension consists of a periscope optical system in a tubular aluminium body. The extension is designed to be quickly and easily installed and removed from the lamp housing assembly, which is fitted to a dovetailed slot on the elbow telescope of both the C2 and C2A1 sight units.

Data

Length: 379.5 mm
Diameter: 19 mm
Magnification: ×1
Field of view: 8°
Exit pupil: 9 mm
Weight: including case, 270 g

Manufacturer

Hughes Leitz Optical Technologies Ltd, 328 Ellen Street, Midland, Ontario L4R 2H2.

Status

In production.

Service

Military forces worldwide.

VERIFIED

Leitz periscope extension mounted on C2A1 sight unit

Mortar sight unit M53A1

Description

The sight unit M53A1 is the primary medium mortar sight of the US Army and is used to lay the M29 81 mm mortar. It can be used with other weapons, including rocket launchers.

The M53A1 consists of the Mount, Telescope M128A1 and the Telescope, Elbow, M109. The M109 Elbow Telescope is a fixed-focus optical instrument. Illumination of the M128A1 scales and the M109 graticule is provided by the Light Instrument M53E1.

Data

Length: 152 mm
Width: 127 mm
Height: 254 mm
Elevation: 1,600 mils
Azimuth: 0-3,200/6,400 mils
Accuracy: 1 mil
Magnification: ×4
Field of view: 10°

Manufacturer

Hughes Leitz Optical Technologies Ltd, 328 Ellen Street, Midland, Ontario L4R 2H2.

Status

In production.

VERIFIED

Leitz M53A1 mortar sight unit

Mortar sight unit M64A1

Description

The sight unit M64A1 was designed to lay the 60 mm M224 lightweight company mortar of the US Army. It can also be used with the 81 mm M252 mortar and with a variety of other lightweight and medium mortars.

The M64A1 sight unit consists of the Mount, Telescope, M64A1; the Telescope, Elbow, M64A1; and the Ancillary Sighting Equipment Set. The sight unit is a fixed-focus opto-mechanical instrument which uses tritium light sources to illuminate all elevation, azimuth and graticule scales and level vials. Ancillary equipment is also available as part of a complete logistic support package. The M64A1 is built to all applicable military specifications.

Data

Length: 111 mm
Width: 111 mm
Height: 187 mm
Elevation: 700-1,600 mils
Azimuth: 0-6,400 mils
Accuracy: 1 mil
Magnification: unity
Field of view: 8°
Illumination: tritium
Weight: 1.1 kg

Manufacturer

Hughes Leitz Optical Technologies Ltd, 328 Ellen Street, Midland, Ontario L4R 2H2.

Status

In production.

VERIFIED

Leitz M64A1 mortar sight unit

CHINA, PEOPLE'S REPUBLIC

AMC-II mortar fire control computer

Description

The AMC-II computer is a self-contained unit incorporating a computer, keyboard for data entry, display, batteries, illuminator and leather case. The computer is suitable for all types of mortar at all altitudes, can perform routine topographical surveying functions, has high computation speed and short response time, and requires minimum maintenance. It is powered by rechargeable Ni/Cd batteries which can be reused not less than 800 times.

Operation of the computer is simple and similar to that of an ordinary calculator. Without previous knowledge, a user can be instructed to a competent level in no more than two days.

Data

Dimensions: 112 × 243 × 40 mm
Power requirement: 5 V DC
Power supply: 4 × 1.5 V Ni/Cd batteries
Survey accuracy: ±1 m
Range accuracy:<8 m
Azimuth accuracy: 1 mil
Firing data calculation time: 4 s
Weight: with batteries, 1.2 kg

Manufacturer

China North Industries Corporation (NORINCO), 7A Yue Tan Nan Jie, PO Box 2137, Beijing.

Status

In production.

Service

Chinese Army.

VERIFIED

AMC-II mortar fire control computer

FRANCE

Dassault Electronique mortar and gun computer

Description

This is a portable computer for fire control of mortars or guns and can be supplied with ballistic programming for any desired weapon/ammunition combination. It is provided with a shoulder harness and can be operated while on the move if necessary.

The usual topographical and ballistic calculations are provided, and the memory can contain 99 weapon locations, 99 targets, 99 observer locations, 99 protected areas and three weather reports (STANAG 4082, STANAG 4061 or estimated reports). All memory functions are protected against accidental erasure and are retained while the instrument is switched off. There are also programs covering the usual safety routines for range firing and built-in test facilities.

It is possible to link the computer with a suitable radio for data transmission and reception, and links are available for various peripheral equipment such as printers, graphic displays or telecontrol.

Data

Dimensions: 220 × 130 × 100 mm
Power supply: internal 12 V or 24 V Ni/Cd battery or from vehicle battery

Battery life: 4 Ah battery, 8 h
Weight: 3 kg

Manufacturer
Dassault Electronique, 55 quai Marcel Dassault, F-92214 Saint-Cloud.

Status
In production.

VERIFIED

Dassault Electronique mortar and gun computer

Cascem-PC hand-held computer

Description
The Cascem-PC is a hand-held PC-compatible version of the Dassault Electronique mortar and gun computer described above. All the functions available in the portable model are also present in this hand-held version.

Data
Dimensions: 236 × 128 × 43 mm
Power supply: 3 alkaline cells LR6 or 3 rechargeable batteries
Battery life: 30 h operational use (alkaline cells)
Operating system: MS-DOS v3.3
Weight: including batteries, 750 g

Manufacturer
Dassault Electronique, 55 quai Marcel Dassault, F-92214 Saint-Cloud.
Husky Computers Limited, Eden Road, Walsgrave Triangle Business Park, Coventry CV2 2TB, UK.

Status
In production.

VERIFIED

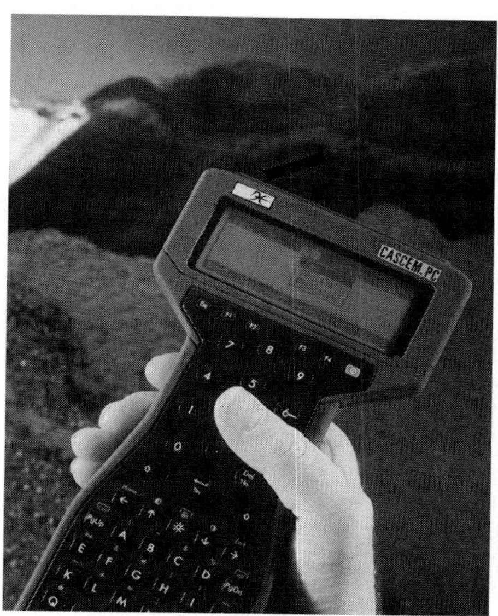

Cascem-PC hand-held mortar computer

TIRAC fire control computer

Description
TIRAC was developed by SECRE in co-operation with the French Direction Technique des Armements Terrestres. For single operator use, the computer can calculate all trajectory data required for gun batteries or mortar platoons.

Firing data is implemented from (1) ballistic data applicable to specific types of ammunition (by means of plug-in ballistic modules); (2) non-standard gun conditions including charge temperature, shell weight and muzzle velocity; (3) survey data covering locations of guns, targets and observers; (4) meteorological data to STANAG 4082, or from datum shooting, or an estimated met message; and (5) possible stored residual corrections following adjustment of fire. Firing data (azimuth, elevation and fuze setting) is determined for each gun of the battery and the charge is selected by the operator on the basis of data displayed by the computer.

The computer can handle up to six guns, 10 targets and 10 registration points. Computation accuracy is to within 1 mil for azimuth and 0.5 PE for range.

SECRE TIRAC gun and mortar computer

In addition TIRAC can calculate observer's adjustments, reduce fired data to map data, determine target co-ordinates from firing data, and perform all routine survey computations.

The equipment is supplied from a DC power source, either a vehicle battery or an internal power pack. Selection of the appropriate ammunition is simply a matter of inserting the correct ballistic module. There are output facilities for data transmission to individual guns or a teleprinter link.

Data
Dimensions: 400 × 330 × 180 mm
Power supply: 19-32 V DC external; 2 × BA58 batteries back-up
Consumption: 25 W
Battery life: 72 h
Weight: 12 kg

Manufacturer
SECRE (subsidiary of Merlin Gerin), 11 rue de Cambrai, F-75945 Paris Cedex 19.

Status
In production.

Service
French Army.

VERIFIED

ISRAEL

Mortano mortar and north-finding computer

Description

The Mortano is a hand-held computer capable of performing all calculations and procedures required in the course of mortar firing. In addition, it has the unique capability of providing the mortar with an accurate north-finding instrument.

The Mortano covers all firing data; it calculates and displays the azimuth and elevation angles and the optimum charge. It can store up to 999 targets in its database; all data on a specific target can be displayed, including name, code, description, range, co-ordinates, height, azimuth and elevation. All standard mortar firing procedures are incorporated.

The NORS 61 is an accurate, easily adaptable, low-cost north-finding module composed of two parts: an attachment to the mortar sight (or battery azimuth instrument) and software built into the Mortano computer. Once north has been determined the sight can then be used as an angle-measuring instrument to permit accurate resection to be carried out so as to determine the position of the mortar. The calculation of position from the observed angles is done by the computer.

The Mortano uses a graphical display, giving 8 lines × 21 characters in alphanumeric mode and 64 × 128 pixels in graphic mode; the graphic mode permits displays in any language. Based on an 8088 CPU the Mortano has calculation speed equivalent to a PC. It has been designed to permit modular extensions, and options will permit a database of 20,000 targets, conversion of data from air photographs to map and vice versa, inclusion of artillery fire procedures, and digital or audio communication links to permit the connection of peripheral equipment such as printers or communications.

Manufacturer

Azimuth Limited, Hakarem Street, Ramot Hashavim 45930.

Status

In production.

VERIFIED

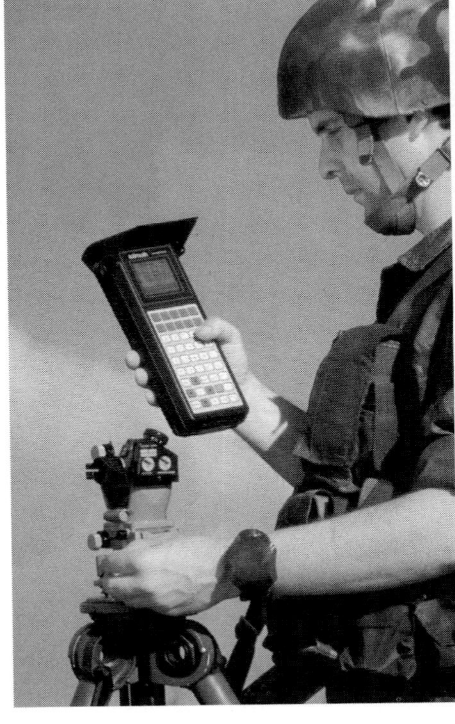

Mortano hand-held computer with NORS 61 north-finding module attached to battery azimuth instrument

ITALY

CMB-8 mortar fire control computer

Description

The CMB-8 is a military, hand-portable computer which automates the procedures needed to compute firing data for a battery deployed as three platoons, each of three or four mortars.

It makes use of all the information stored in its memory to compute firing data for the different mortars of the battery, and can be connected to a Galileo military hand-portable printer for recording of the executed procedures.

An internal rechargeable battery and the possibility of connection to an external power supply give the CMB-8 a long combat autonomy.

Compared with conventional methods of fire control the CMB-8 offers many advantages:

a drastic reduction in time of calculation, together with increased precision

a reduction in the number of members in the fire direction centre and in working time, since a menu sequence directs the operator step by step to completion of the procedures

a possibility of automatic data transfer through a digital link with other units of the battery (forward observers) or with higher echelons of command.

The most important functions performed by the CMB-8 are:

(1) fire preparation, including
 (a) survey preparation, with several programs such as: polar co-ordinates replot, closed traverse replot, change of grid, intersection and resection
 (b) the introduction of positional data for all the mortars in the battery
 (c) ballistic corrections from meteor messages, charge temperature and weight and muzzle velocities
(2) evaluation of engagements taking into account friendly troops, no-fire areas and crest clearance
(3) computation of initial firing data for each mortar
(4) fire adjustment
(5) fire for effect, including fire distribution for smoke screens, barrages and area targets.

CMB-8 mortar computer with optional printer

Manufacturer

Officine Galileo SpA, Via Einstein 35, I-50013 Campo Bisenzio, Firenza.

Status

In production.

Service

Italian Army.

VERIFIED

NEW ZEALAND

MERE mortar fire control computer

Description

MERE (Mortar Elevation and Ranging Equipment) was developed by the New Zealand Department of Scientific and Industrial Research in conjunction with the New Zealand Army and is intended for use with the 81 mm L16 mortar.

MERE consists of a keyboard and display panel, computing circuitry and a clip-on rechargeable battery pack. It will store up to 99 targets, 24 observers or friendly positions, and nine mortar lines. Positions can be specified by grid reference or by azimuth and range from known points. Altitudes can be in feet or metres and NATO format meteor data can be entered and stored. Charge temperature, ammunition variations and similar corrections can also be entered. After storing the mortar location and entering target information, the computer will produce the complete fire mission data of range, charge, azimuth, elevation, fuze length and time of flight. Fall of shot corrections notified by observers can then be inserted to produce corrections to the fire, and at the end of the mission the target can be recorded. Initial firing data is produced in about 12 seconds and corrections in about six seconds. When used in training it is possible to enter range safety information, and the display will indicate any safety hazard; similarly, with crest and friendly force information inserted, the display will produce warnings covering safety of own troops.

Data

Dimensions: 265 × 80 × 125 mm
Power supply: 6 V battery, rechargeable in field
Battery life: continuous, 70 h; intermittent use, 1,000 h
Weight: under 4 kg

Manufacturer

Wormald Vigilant Limited, 211 Maces Road, PO Box 19545, Christchurch.

Status

Production complete.

Service

New Zealand Army.

VERIFIED

MERE mortar fire control computer

NORWAY

Hugin mortar fire control system

Description

The Hugin mortar fire control system was developed by NFT for the Norwegian Army as the core of a total concept for mortar units that will increase survival through dispersion and mobility. The programme includes the KMT 400 Multipurpose Terminal for military use (the hardware) and the mortar fire control application (the software).

The KMT 400 is a compact, ruggedised, modular computer/terminal for field use. The unit has a fully graphic, monochromatic LCD screen with touch data entry based on a menu technique. It is EMP resistant and operates over the full −40 to +55°C temperature range. It can be easily operated by personnel wearing NBC or winter equipment.

Standard interfaces are to handset, field line, radio, RS-232C and digital communications networks. These interfaces are flexible and can be tailored to meet any specification. Ballistic calculations are performed using the Mass Point Model, and necessary software extensions for field artillery applications are available.

Data

Dimensions: 73 × 120 × 232 mm
Processors: 80C88
Memory: FLASH-EPROM 640 kB; RAM 64 kB; EEPROM 8 kB
Modulation: FSK for built-in modem
Speed: 600 bps
Power source: internal, lithium BA-5052N; vehicle, 10-32 V DC
Battery life: 50 h
Weight: <2 kg

Manufacturer

Kongsberg Gruppen AS, PO Box 1003, N-3601 Kongsberg.

Status

Mass production commenced during 1995.

Service

Entering service with Norwegian Army.

UPDATED

Hugin mortar fire control system in use

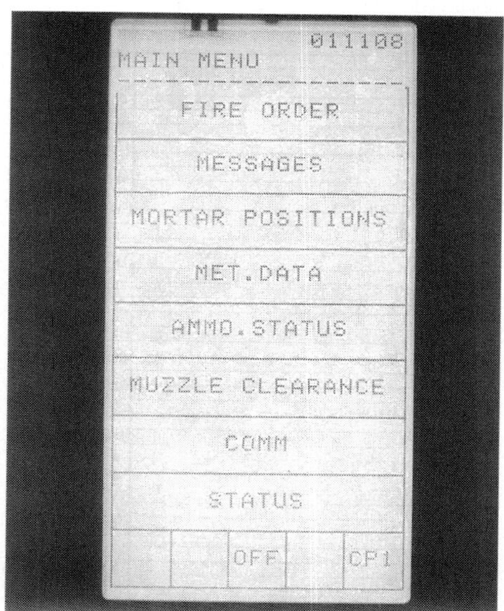

Typical Hugin KMT 400 Multipurpose Terminal display

SOUTH AFRICA

H-019 mortar sight

Description

The H-019 sight and sight mounting were designed for use on various weapons which include the 120 mm M5 mortar, 81 mm M3 mortar, the 60 mm M1 mortar, and the Vektor long-range 60 and 81 mm mortars.

The sight is of modular South African design to ensure interchangeability, adaptability and ease of maintenance. Two variations in sighting methods are available: a lensatic type for use on the 81 mm M3 mortar, 60 mm M1 mortar, and an optical telescope for more accurate sighting for use on the 120 mm M5 mortar.

Mechanically the sight comprises the following:

a dovetail connection with a quick decoupling mechanism

an elevation scale and azimuth scale graduated in 1 mil intervals. The elevation range is between +800 and +1,600 mils for mortar applications and between −400 and +800 mils for machine gun applications

double levelling bubbles for setting in each plane

a decoupling for fast adjustment on the azimuth scales.

The azimuth scales comprise a main and a secondary scale for ease of correction adjustment, both with a 6,400 mil range.

All scales and levelling bubbles are illuminated by Betalights for night use; a cover is also supplied to hide Betalights when not in use at night.

The optical sighting mechanism is adjustable in the vertical plane and is provided with Betalights for night use.

Worm gears are used inside the sight with a built-in self-eliminating play system, to ensure minimum maintenance and accuracy at all times.

Data

Accuracy: within 3 mils
Illumination: 11 Betalights
Weight: 1.5 kg; in transit case complete with accessories, 3.4 kg

Enquiries to

Armscor, Private Bag X337, Pretoria 0001.

Status

In production.

Service

South African National Defence Force and some other armies.

VERIFIED

H-019 mortar sight

SPAIN

Seimor mortar data computer

Description

The Seimor is a compact hand-held computer with high resistance to drop, damage and water. Operation is reliable and simple, the keyboard being colour-coded to facilitate operation.

The display prompts the operator during standard operations, and there is automatic switch-off to conserve battery power. When the computer is switched off, the data is stored and can be recalled. Batteries can be changed without memory loss.

The computer has capacity for 80 targets, 12 observers, 21 mortar locations, up to 21 prohibited areas, seven survey programmes and nine fire for effect procedures. Two simultaneous missions can be computed and there are five initial corrections and five modes to commence firing missions available. Charge selection can be done manually or can be automatic and the display will produce projectile, range, charge, elevation, azimuth, individual corrections and time of flight. There is the facility to attach a printer for hard copy.

Data

Dimensions: 90 × 60 × 180 mm
Operating temperature range: −20 to +60°C
Weight: 500 g

Manufacturer

Seidef SA, Pso de la Castellana 140, E-28046 Madrid.

Status

In production.

Service

Spanish Army.

VERIFIED

Seimor mortar data computer

UNITED KINGDOM

Marconi MORCOS

Description

MORCOS is a self-contained, hand-held unit which incorporates the entry keyboard, visual display, batteries and computing circuitry. The unit is in a polycarbonate case of convenient size and is fully militarised to withstand all battlefield environmental conditions. The keyboard has 24 keys, 10 for the numerical digits and the rest for various functions. After initiating a function, the display then leads the operator through the required sequences for the various operations, either by displaying the parameters for which the operator must insert values or by presenting on the display the names of the various routines which the operator can select. Every entry is displayed and, if incorrect, can be cancelled. The ballistic program can be changed in seconds by removing a plug-in module and substituting another. Modules for most weapon/ammunition combinations are available or can be produced to order. The modules can be changed in the field without impairing the weather-sealing of the instrument. There is a built-in self-test facility.

The memory can accommodate 10 mortar positions, 50 targets, 10 observer locations and 10 locations of own troops. Calculations include normal survey routines, laser rangefinder calculations, firing data, corrections to fall of shot, meteor data, prediction for three parallel missions, and close target routines.

Data

Dimensions: 230 × 110 × 54 mm
Power supply: 9 V DC integral batteries (6 AA cells) or 9 V DC external supply
Weight: 1.35 kg

Manufacturer

Marconi Radar and Control Systems Limited, Chobham Road, Frimley, Camberley, Surrey GU16 5PE.

Status

In production.

Service

Many overseas countries.

VERIFIED

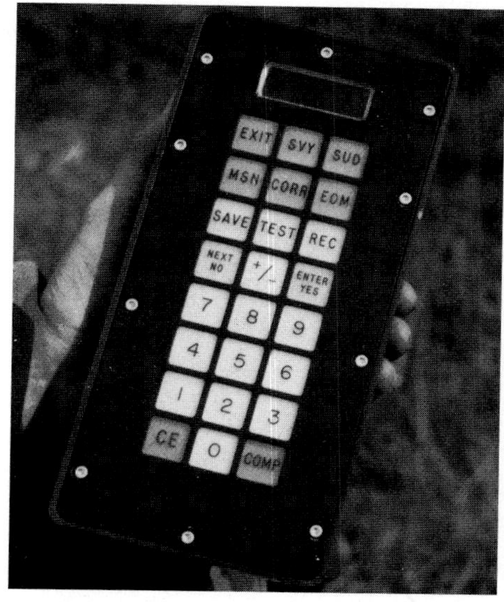

MORCOS fire control computer

Zengrange MORZEN II

Description

MORZEN II is a second-generation instrument which complements the increased sophistication of the latest weapon technology and provides fast, accurate firing data for any mortar, using any ammunition and any sighting system, in the indirect-fire role. It is in service with the British Army for the L16 81 mm mortar, and with an increasing number of other countries.

Lightweight, hand-held, robust and simple to operate, MORZEN II uses standard prompts and procedures familiar to each user. With only a few hours training it can be operated sucessfully in the field to produce faster and more accurate data than any other proven system.

The main features of MORZEN II can be summarised as: all computations to 10-digit accuracy from 6-, 8- or 10-digit reference points; stores up to 314 target locations and nine observation post locations, plus 14 mortar lines for normal or 'shoot and scoot' operations; polar co-ordinates for use with laser rangefinders; easily used datum procedure for accurate location of new mortar lines; without-maps procedures; onboard clock for real time, time relative to H-Hour, splash and co-ordinated illumination tasks; and easily understood error messages generated by miskeying or invalid data input.

MORZEN II computer in use

MORZEN II has a battery life in excess of 80 hours of continuous use, with advance warning of battery failure being given. It was designed for use with NBC or arctic clothing and operates under a wide range of environmental conditions.

MORZEN II passed the British Army testing for high and low temperature (−30 to +55°C), dust implosion, drop testing, vibration, immersion, bounce, nuclear flash burn and electromagnetic compatibility. It is qualified to DEFSTAN 00-1 and DEFSTAN 07-55. Zengrange is qualified to AQAP1 and AQAP13.

Data

Dimensions: 200 × 118 × 58 mm
Target store: 314 max
ML store: 14 max
OP store: 9 max
Power supply: 4 × 1.5 V dry cell
Battery life: >80 h
Accuracy: Range ±1 m; bearing and elevation ±1 mil; fuze setting ±0.5 units; time of flight ±0.5 s; co-ordinates ±1 m
Weight: 625 g

Manufacturer

Zengrange Defence Systems Limited, Greenfield Road, Leeds LS9 8DB.

Status

In production.

Service

British and other armies.

VERIFIED

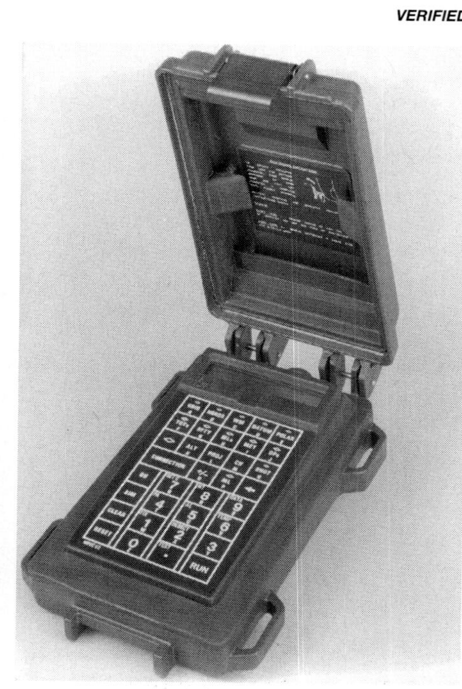

MORZEN II mortar fire control computer

UNITED STATES OF AMERICA

M23 Mortar Ballistic Computer

Description

The M23 is a portable computer which can be adapted by using appropriate software, to the US 60, 81 and 107 mm mortars in current use. It can deal with three concurrent fire missions, three firing locations, 18 individual weapon locations and 12 observer locations. All the usual facilities for ballistic, meteorological and survey computation are available. It is possible to interface with higher level computer systems such as TACFIRE and ACS, and it can receive inputs from most in-service digital message devices.

Data

Dimensions: 267 × 182 × 60 mm
Power supply: external, 20-32 V DC; or battery pack
Weight: 3.2 kg

Manufacturer

Magnavox Electronic Systems Company, 1313 Production Road, Fort Wayne, Indiana 46808.

Status

Production complete.

Service

US Army and US Marine Corps.

VERIFIED

AMMUNITION

Small Arms and Cannon Ammunition
Combat Grenades
Mortar Ammunition
Pyrotechnics

SMALL ARMS AND CANNON AMMUNITION

The following section includes current small arms and cannon ammunition most likely to be encountered in a land-based infantry environment. It cannot claim to be completely comprehensive for the array of ammunition types and natures in production or in service is vast and is covered in more detail in the companion Yearbook *Jane's Ammunition Handbook*, to which the reader is directed should the coverage provided here prove unsuitable for their purposes.

4.73 mm DM 11 Caseless

Synonyms: 4.73 × 33 mm

Armament
Heckler and Koch G11 rifle.

Development
Development of this cartridge was carried out by Dynamit Nobel and Heckler and Koch from about 1970 onwards, specifically for the G11 rifle which attained only very limited service with the Federal German Army. Original versions used nitrocellulose propellant, but this gave rise to cook-off problems because of the absence of a brass case to shield the heat from the chamber walls. The propellant was then changed to a new High Temperature Propellant (HTP) based on a denatured hexogen explosive compound. This raised the cook-off temperature by 100°C and thus rendered the weapon immune to cook-off problems in normal use.

Although development of the G11 rifle, for which this round was intended, is in abeyance, some caseless ammunition development work continues to ensure that the technology involved is not lost. At one time the 4.73 mm DM 11 cartridge was under consideration for the kinetic energy component of the US OICW programme but this proposal was discarded at an early stage of development.

4.73 mm DM 11 Caseless

Description
The 4.73 mm DM 11 has a conventional bullet set into a block of propellant and held rigidly. The block is square in section and longer than the bullet, but the overall length is about half the length of an equivalent conventionally cased cartridge. The primer is a small pellet of initiating mixture loaded into a cavity in the base of the propellant block. Ahead of the primer is a small booster charge of initiatory explosive; this, when ignited by the cap, blows the bullet out of the block and seats it in the rifling before ignition of the propellant takes place. Whilst this seating movement is taking place the block of propellant is shattered and ignited. The object in seating the bullet by means of this booster charge is to ensure that the bullet is firmly lodged in the rifling before the propellant explosion takes place, so avoiding any leak of gas around an unseated bullet. It also ensures that the chamber pressure and gas volume are constant from round to round.

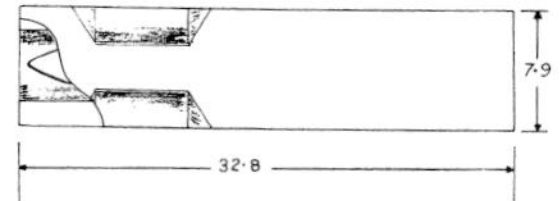

Patr. 4.73 mm DM 11
Vo 930 m/s
Eo 1,380 J

The bullet is streamlined, of conventional form and consists of a lead core, steel jacket and gilding metal envelope. It shows no fragmentation on impact and will pierce 6 mm of mild steel plate at 300 m range and a standard steel helmet at 600 m range.

Data
Round length: 32.8 mm
Round weight: nominal, 5.2 g
Round section: 7.9 mm square
Bullet diameter: 4.73 mm
Bullet weight: 3.2 g
Muzzle velocity: 930 m/s
Muzzle energy: 1,380 J

GERMANY
Manufacturer: Dynamit Nobel
Type: Ball: FMJ; 3.2 g; Vo 930 m/s

UPDATED

5.45 mm Soviet Pistol

Synonyms: 5.45 × 18 mm; PMTs.

Armament
PSM pistol.

Development
Developed in 1979 by Aleksandr Bochin, the 5.45 × 18 mm cartridge was developed orginally for the PSM pistol, although it seems very likely that it was also intended for a sub-machine gun which has either yet to emerge or was not fully developed. The cartridge was developed specifically to penetrate body armour, being capable of penetrating up to 55 layers of Kevlar.

Description
The round uses a bottle-necked rimless case carrying a peculiar jacketed compound bullet with a flat tip. Bullet length is 14.3 mm and weight is 2.6 g. The bullet has a gilding metal plated mild steel jacket surrounding a two-piece core, the front half being mild steel and the rear half lead. This construction has two effects. One is that the steel tip assists penetration while the second is that the main weight of the bullet is to the rear so that the bullet will tend to tumble on striking a target, thus creating a wound potential greater than the small calibre would normally signify. The bullet has a flat tip 2.5 mm in diameter.

The cartridge case is normally brass, containing 0.15 g of an unspecified propellant. The muzzle velocity produced is from 293 to 315 m/s and maximum chamber pressure around 127 MPa.

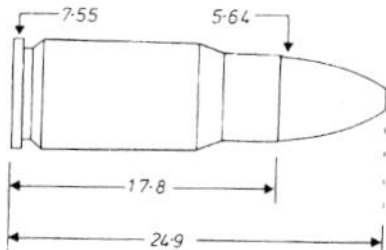

Vo 293-315 m/s
Eo 129J

5.45 × 18 mm Soviet pistol, the PMTs

Data
Round length: 24.9 mm
Round weight: nominal, 4.8 g
Case length: 17.8 mm
Rim diameter: 7.54 mm
Head diameter: 7.55 mm
Bullet diameter: 5.64 mm
Bullet weight: 2.6 g
Muzzle velocity: 293-315 m/s
Muzzle energy: 129 J

COMMONWEALTH OF INDEPENDENT STATES
Manufacturer: State arsenals
Type: Ball: FMJ; 2.6 g; Vo 315 m/s

UPDATED

5.45 mm Soviet Rifle

Synonyms: 5.45 × 39 mm; 5.45 mm M74; 5.45 mm Kalashnikov

Armament
5.45 mm AK-74 rifle, AKSU-74 sub-machine gun, RPK-74 light machine gun.

Development
Soviet ballisticians are known to have been experimenting with reduced calibres prior to 1939, but such work was abandoned during the war years. Following small calibre ammunition trends elsewhere, a group led by Viktor Sabelnikov developed the 5.45 × 39 mm cartridge which, in its basic 7N6 form, has a marked propensity to tumble on striking a target because of the deliberate placing of the centre of gravity towards the base of the bullet.

The 5.45 × 39 mm cartridge was adopted in 1974 for use with the AK-74 assault rifle and became the standard rifle cartridge for most of the front line forces of the old Warsaw Pact, although the former 7.62 × 39 mm cartridge and its associated weapons have yet to be fully supplanted in the manner the 5.56 × 45 mm round has accomplished within NATO and elsewhere.

The 7N10 cartridge was developed to defeat body armour and has been in production since 1993. It has the same ballistic performance as the standard 7N6 cartridge but employs a slightly heavier (3.56 g) bullet carrying a 1.83 g penetrator capable of penetrating a 6Zh85T or 6BZT-M armoured jacket at 100 m. The 7N10 round can also penetrate a 16 mm steel plate at 100 m.

Description
The 5.45 × 39 mm 7N6 round is of conventional rimless bottle-necked form. The case (actual length 39.5 mm)

5.45 × 39 mm 7N6 rifle cartridge

Abridged Ballistic Table: 5.45 × 39.5 mm Ball Type PS

Range	Velocity	Energy	Drop	Elevation	Vertex
0 m	900 m/s	1,391 J	0 mm	0 mils	0 mm
100 m	802 m/s	1,105 J	73 mm	0.82 mils	20 mm
200 m	713 m/s	873 J	296 mm	1.53 mils	82 mm
300 m	628 m/s	677 J	732 mm	2.54 mils	208 mm
400 m	550 m/s	519 J	1,416 mm	3.68 mils	420 mm
500 m	473 m/s	384 J	2,454 mm	5.11 mils	722 mm

is of lacquered steel with a stripe of red lacquer sealing the case/bullet joint. The Berdan primer is of brass and there are two fireholes in the primer pocket. The weight of SF 033FL propellant is 1.45 g.

The streamlined boat-tailed bullet consists of a mild steel core filling about two-thirds of the gilding metal clad steel envelope, and a lead plug about 3 mm long in front of it. The extreme 3 mm of the tip is hollow.

There is a TgS cartridge with a tracer denoted by a green tip on the bullet, and a Blank cartridge with a crimped case mouth. Mention has also been made of an Incendiary round with a red-tipped bullet.

Reference has also been found to a low velocity 5.45 mm round produced specifically for an AKSB-74U silenced sub-machine gun fitted with a PBS-1 suppressor. This cartridge features a 5.15 g 7U1 bullet fired at a muzzle velocity of 303 m/s and capable of piercing a 2 mm steel plate or an SSh-68 steel helmet at a range of 300 m.

Data

7N6

Round length: 56.5 mm
Round weight: 10.5 g
Case length: 39.5 mm
Rim diameter: 9.97 mm
Bullet length: 25.5 mm
Bullet diameter: 5.61 mm
Bullet weight: 3.45 g
Muzzle velocity: 900 m/s
Muzzle energy: 1,383 J

BULGARIA
Manufacturer: Kintex
Type: Ball 7N6: FMJ; steel core, lead sleeve, steel jacket; 3.4 g; Vo 890 m/s. Steel case; 1.45 g charge of SF 033FL powder.

COMMONWEALTH OF INDEPENDENT STATES
Manufacturer: Lugansk Machine Tool Factory and others

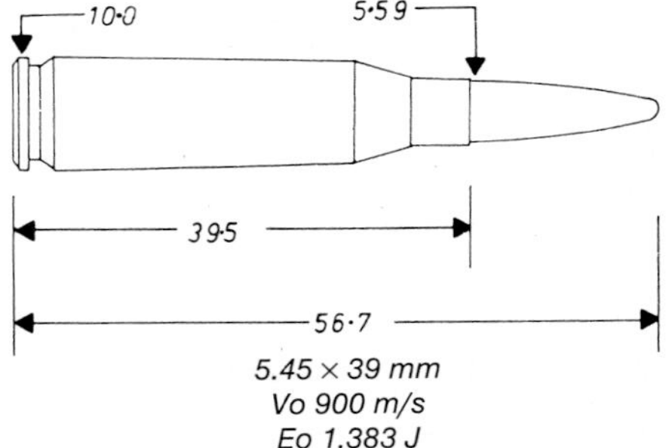

5.45 × 39 mm
Vo 900 m/s
Eo 1,383 J

Type: 7N6: FMJ; 3.45 g; Vo 900 m/s
TgS Tracer: FMJ; red, dark ignition, burns 100 to 400 m
7N10: FMJ; 3.56 g; Vo 900 m/s
Blank: cartridge weight 6.6 g

POLAND
Manufacturer: Mesko Zaklady Metalowe
Type: 7N6: FMJ; steel core; 3.4 g; Vo 880 m/s
Practice: Plastic bullet; no performance details available

YUGOSLAVIA (SERBIA AND MONTENEGRO)
Holding company: YUGOIMPORT - SDPR
Type: Ball: FMJ; steel core; 3.4 g; Vo 969 m/s

UPDATED

0.22 in Long Rifle

Armament
All 0.22 rifles and pistols except those specifically chambered for 0.22 Short cartridges.

Development
The 0.22 Long Rifle rimfire cartridge was developed in 1887 by the J Stevens Arms & Tool Company of the USA, by taking the existing 0.22 Long cartridge and fitting it with a 0.324 g powder charge and a 2.59 g lead bullet instead of the conventional 1.88 g bullet. It was probably first commercially manufactured by the Union Metallic Cartridge Company in 1888. The first high-velocity loadings were developed by Remington in 1930. Over the years it has become the most highly developed and accurate of all rimfire cartridges, and is generally found with either 2.59 g solid lead or 2.4 g hollow point bullets, though innumerable variations can be found. In military hands this cartridge is invariably used for training purposes, though clandestine agencies, resistance forces, Special Forces and similar organisations have, from time to time, used it in combat roles where the low signature and accuracy were of particular value.

Description
The cartridge case is rimmed and usually of brass, copper- or brass-plated steel. The priming composition is distributed around the rim, therefore the ductility of this area is critical. Bullets are of lead, occasionally with a light plating of copper and usually with lubricating cannelures, although lubricant may not always be present. Round-nose or hollow point bullets are the normal types.

Data
Round length: 24.76 mm
Case length: 15.11 mm
Rim diameter: 6.98 mm
Bullet diameter: 5.66 mm
Bullet weight: 2.6 g
Muzzle velocity: 330 m/s
Muzzle energy: 142 J

ARGENTINA
Manufacturer: Cartuchos Orbea
Type: Standard: Lead; 2.6 g; Vo 330 m/s
Hollow Point: Lead, HP; 2.6 g; Vo 380 m/s
Pistol: Lead; 2.6 g; Vo 310 m/s

AUSTRIA
Manufacturer: Hirtenberger AG
Type: Standard: Lead; 2.55 g; Vo 330 m/s

BELGIUM
Manufacturer: Browning SA
Type: Standard Velocity: Lead; 2.59 g; Vo 340 m/s
High Velocity: Lead; 2.59 g; Vo 390 m/s
Match: Lead; 2.59 g; Vo 335 m/s
Zimmer: 2.59 g; Vo 225 m/s

BRAZIL
Manufacturer: Companhia Brasileira de Cartuchos
Type: Standard: Lead; 2.6 g; Vo 351 m/s
High Velocity: Lead, copper-plated; 2.6 g; Vo 383 m/s

BULGARIA
Manufacturer: Kintex
Type: Ball: Lead; 2.5 g; Vo 320 m/s. Brass case, propelling charge 0.083 g pyroxyline

CHINA, PEOPLE'S REPUBLIC
Manufacturer: China North Industries (NORINCO)
Type: Ball: Lead; 2.6 g; Vo 342 m/s
Competition: Lead; 2.6 g; Vo 322 m/s

COMMONWEALTH OF INDEPENDENT STATES
Manufacturer: LVE, Novosibirsk
Type: Ball, Sobol and Junior Steel: Lead; 2.57 g; Vo 320 m/s

Manufacturer: Olimp
Type: Ball R: Lead; 2.6 g; Vo 318 m/s
Ball B: Lead; 2.7 g; Vo 315 m/s

Manufacturer: Sotem
Type: Sniper: Lead; 2.6 g; Vo 320 m/s
Biathlon: Lead; 2.71 g; Vo 315 m/s
Temp Pistol: Lead; 2.6 g; Vo 315 m/s
Junior: Lead; 2.6 g; Vo 320 m/s

Manufacturer: Vostok
Type: Ball: Lead; 2.4 g; Vo 315 m/s
Ball: Lead; 2.4 g; Vo 300 m/s
Ball: Lead; 2.5 g; Vo 300 m/s

CZECH REPUBLIC
Manufacturer: Sellier & Bellot
Type: Rex: Lead; 2.5 g; Vo 345 m/s
Rex Expansive: Lead, HP; 2.4 g; Vo 345 m/s

FINLAND
Manufacturer: Lapua Cartridge Factory
Type: Trainer: Lead; 2.6 g; Vo 335 m/s
Match: Lead; 2.7 g; Vo 345 m/s
Master: Lead; 2.6 g; Vo 325 m/s
Pistol King: Lead; 2.6 g; Vo 315 m/s

0.22 in Long Rifle

Speed Ace: Lead; 2.6 g; Vo 410 m/s
Subsonic HP: Lead; 2.35 g; Vo 315 m/s
Hollow Point: Lead, HP; 2.35 g; Vo 410 m/s
Plinker: Lead; 2.6 g; Vo 328 m/s
Super Club: Lead; 2.6 g; Vo 320 m/s

FRANCE
Manufacturer: Gevched
Type: Grand Precision: Lead; 2.6 g; Vo 340 m/s
Grand Vitesse: Lead; 2.6 g; Vo 400 m/s

Manufacturer: SFM Défense
Type: Standard: Lead; 2.6 g; Vo 340 m/s
Hollow Point: Lead, HP; 2.6 g; Vo 340 m/s
Tracer: Lead, red trace; 2.6 g; Vo 340 m/s

Manufacturer: Survilliers-NCS
Type: Match NCS20: Lead; 2.59 g; Vo 320 m/s
Training NCS30: Lead; 2.59 g; Vo 320 m/s
Pistol Match NCS20: Lead; 2.59 g; Vo 280 m/s
Pistol Training NCS30: Lead; 2.59 g; Vo 280 m/s

GERMANY
Manufacturer: Mauser
Type: KK80: Lead; 2.55 g; Vo 330 m/s

Manufacturer: RWS
Type: R.50: Lead; 2.6 g; Vo 325 m/s
Standard: Lead; 2.55 g; Vo 330 m/s
Tracer: Lead, red trace; 2.55 g; Vo 330 m/s
Sport: Lead; 2.6 g; Vo 325 m/s
High Velocity: Lead, solid or HP; 2.55 g; Vo 400 m/s
Pistol Match: Lead; 2.6 g; Vo 270 m/s
Rifle Match: Lead; 2.6 g; Vo 325 m/s
Biathlon: Lead; 2.6 g; Vo 325 m/s
Subsonic: Lead; 2.6 g; Vo 305 m/s

ITALY
Manufacturer: Fiocchi Munizioni SpA
Type: Maxac: Lead; 2.6 g; Vo 355 m/s
Carabine: Lead; 2.6 g; Vo 330 m/s
Ultrasonic: Lead; 2.6 g; Vo 390 m/s
Expansive: Lead, HP; 2.4 g; Vo 400 m/s

Asonic: Lead; 2.4 g; Vo 310 m/s
Free Pistol: Lead; 2.6 g; Vo 320 m/s
Standard Pistol: Lead; 2.6 g; Vo 290 m/s
Mobile: Lead; 2.6 g; Vo 335 m/s
Biathlon: Lead; 2.6 g; Vo 330 m/s
Competition: Lead; 2.6 g; Vo 325 m/s
Super Speed: Lead; 2.4 g; Vo 490 m/s

KOREA, SOUTH
Manufacturer: Poongsan Metal Corporation
Type: Predator HP: Lead, conoidal, HP; 2.1 g; Vo 457 m/s
Predator HP: Lead, copper-coated; 2.3 g; Vo 430 m/s
Ball HV: Lead, 2.6 g; Vo 310 m/s
Ball Sidewinder: Lead; 2.59 g; Vo 380 m/s
Ball Zapper: Lead; 2.6 g; Vo 390 m/s
Ball Zapper: Lead, HP; 2.5 g; Vo 390 m/s

MEXICO
Manufacturer: Aguila Industrias Tecnos
Type: LRHV: Lead; 2.6 g; Vo 382 m/s
LRHVPH: Lead, HP; 2.3 g; Vo 390 m/s
Standard: Lead; 2.6 g; Vo 350 m/s

PHILIPPINE REPUBLIC
Manufacturer: Armscor
Type: Ball: Lead; 2.6 g; Vo 382 m/s
Ball: Lead, HP; 2.5 g; Vo 390 m/s
Standard Velocity: 2.6 g; Vo 350 m/s

SOUTH AFRICA
Manufacturer: Denel (Pty) Limited
Type: High Velocity: Lead; 2.6 g; Vo 382 m/s
Standard Velocity: Lead; 2.6 g; Vo 350 m/s
Hollow Point: Lead; 2.5 g; Vo 390 m/s

SPAIN
Manufacturer: SANTA BARBARA SA
Type: Standard: Lead; 2.5 g; Vo 330 m/s
Hollow Point: Lead, HP; 2.5 g; Vo 330 m/s

SWEDEN
Manufacturer: Norma AB
Type: Match: Lead; 2.6 g; Vo 330 m/s

UNITED KINGDOM
Manufacturer: Eley Limited
Type: Standard: Lead; 2.59 g; Vo 331 m/s
Tracer: Lead, red trace; 2.27 g; Vo 330 m/s
Moving Target: Lead; 2.59 g; Vo 441 m/s
Silhouex: Lead; 2.59 g; Vo 411 m/s
LRZ: Lead; 2.59 g; Vo 243 m/s
Tenex: Lead; 2.59 g; Vo 331 m/s
Pistol: Lead; 2.59 g; Vo 314 m/s
Pistol Match: Lead; 2.59 g; Vo 305 m/s
HV Solid Point: Lead; 2.59 g; Vo 400 m/s
HV Hollow Point: Lead, HP; 2.27 g; Vo 400 m/s
Subsonic HP: Lead, HP; 2.27 g; Vo 320 m/s

UNITED STATES OF AMERICA
Manufacturer: Cascade Cartridge Company
Type: Standard Velocity: Lead; 2.6 g; Vo 346 m/s
High Velocity: Lead; 2.6 g; Vo 376 m/s
Hollow Point: Lead; 2.4 g; Vo 384 m/s
Super Match: Lead; 2.6 g; Vo 346 m/s

Manufacturer: CCI
Type: Stinger: Lead, HP; 2.08 g; Vo 514 m/s
Mini-Mag: Lead; 2.6 g; Vo 410 m/s
Mini-Mag HP: Lead, HP; 2.4 g; Vo 410 m/s
Mini-Mag +V: Lead; 2.4 g; Vo 435 m/s
Standard Velocity: Lead; 2.6 g; Vo 347 m/s
Competition: Lead; 2.6 g; Vo 347 m/s
Pistol Match: Lead; 2.6 g; Vo 326 m/s
Small Game: 2.5 g; Vo 390 m/s

Manufacturer: Federal Cartridge Company
Type: 710/810: Lead; 2.6 g; Vo 382 m/s
712 Hi-Power: Lead; 2.5 g; Vo 390 m/s
711 Champion: Lead; 2.6 g; Vo 350 m/s
Lightning 510: Lead; 2.6 g; Vo 383 m/s
Spitfire 720: Lead, conoidal; 2.34 g; Vo 430 m/s
Spitfire 722: Lead, conoidal; HP; 2.15 g; Vo 457 m/s
Gold Medal Ultra Match: Lead; 2.6 g; Vo 348 m/s
Gold Medal Target: Lead; 2.6 g; Vo 351 m/s

Manufacturer: Remington Arms Company Inc
Type: High Velocity: Lead; 2.59 g; Vo 407 m/s
High Velocity HP: Lead, HP; 2.33 g; Vo 416 m/s
Target: 2.59 g; Vo 349 m/s
Viper: Lead; 2.34 g; Vo 430 m/s
Yellow Jacket: Lead; 2.14 g; Vo 457 m/s
Thunderbolt: Lead; 2.6 g; Vo 383 m/s
Cyclone: Lead; 2.3 g; Vo 390 m/s
Subsonic: Lead; 2.4 g; Vo 320 m/s

Manufacturer: SK Industries
Type: Rifle Match: Lead; 2.6 g; Vo 320 m/s
Pistol Match: Lead; 2.6 g; Vo 320 m/s
Subsonic: Lead; 2.6 g; Vo 315 m/s
Turbomatch: Lead; 2.6 g; Vo 390 m/s
Hollow Point: Lead, HP; 2.6 g; Vo 385 m/s

Manufacturer: Winchester
Type: Super X: Lead; 2.6 g; Vo 383 m/s
Dynapoint: Lead; 2.6 g; Vo 383 m/s
HP Super X: Lead, HP; 2.4 g; Vo 390 m/s
Target: Lead; 2.6 g; Vo 351 m/s
T22: Lead; 2.6 g; Vo 341 m/s
Laser: Lead, coppered; 2.6 g; Vo 430 m/s
Laser HP: Lead, coppered, HP; 2.4 g; Vo 430 m/s
Xpediter: Lead, coppered; 1.9 g; Vo 505 m/s
Rabbit: Lead, coppered, HP; 2.6 g; Vo 385 m/s
Subsonic: Lead, HP; 2.5 g; Vo 320 m/s
Super Match: Lead; 2.6 g; Vo 323 m/s
Xpert: Lead; 2.6 g; Vo 330 m/s
Pistol: Lead; 2.6 g; Vo 323 m/s
Super Silhouette: Lead; 2.7 g; Vo 372 m/s

YUGOSLAVIA (SERBIA AND MONTENEGRO)
Manufacturer: Pobjeda
Type: Standard: Lead; 2.6 g; Vo 330 m/s
High Velocity: Lead; 2.6 g; Vo 375 m/s

Holding company: YUGOIMPORT - SDPR
Type: Standard: Steel case, lead bullet; 2.56 g; Vo 320 m/s
High Speed: Lead; 2.56 g; Vo 380 m/s
Ball: Lead, HP; 2.56 g; Vo 380 m/s
High Speed: HP; 2.3 g; Vo 380 m/s

NEW ENTRY

5.56 × 45 mm M193

Synonyms: 0.223 Armalite; 0.223 Remington Special

Armament
This cartridge will operate in any suitably chambered rifle or machine gun, but optimum results are obtained when the rifling is one turn in 305 mm. It will retain its accuracy in rifling of one turn in 230 mm, for example the Austrian AUG family of weapons.

Development
The 0.223 cartridge (as it was originally known) was developed as part of the AR-15 rifle programme in 1957. The original design of the AR-15 rifle had been based upon the commercial 0.222 Remington cartridge, but this generated excessive pressures and a new case with slightly greater capacity was designed, this became the 0.223 Remington Magnum. This was slightly longer than desirable, was shortened and became the 0.223 Armalite cartridge. It was finally adopted by the US military as the 'Cartridge, Ball, 5.56 mm M193' in 1964.

Description
The case is rimless and bottle-necked, usually of brass although steel cases have been made outside the USA. Boxer primers are normally used although Berdan primers are common. French service cases have a variant Berdan priming system in which the cap chamber has a small strip of metal punched down from the case, leaving two gaps at each side to act as fireholes.

The M193 ball bullet is streamlined, with a cannelure in the body into which the case mouth is coned on assembly of the round. It is of conventional construction, with a lead antimony core, steel jacket and gilding metal envelope.

The Tracer counterpart for the M193 Ball is the M196 while the Blank Training Cartridge with a rose crimped cartridge case mouth is the M200.

Innumerable experimental bullets have been developed over the years. A 'Duplex' round carrying two bullets was developed in the 1960s by the US Army and used in small quantities in Vietnam, but the idea was later abandoned.

Data
Round length: 57.3 mm
Case length: 44.7 mm
Rim diameter: 9.6 mm
Head diameter: 9.5 mm
Bullet diameter: 5.66 mm
Bullet weight: 3.56 g
Neck diameter: 6.42 mm

ABU DHABI
Manufacturer: ADCOM Manufacturing
Type: Ball: FMJ; 3.52 g; Vo 955 m/s; brass case. Tracer, API and Blank also produced

ARGENTINA
Manufacturer: Direccion General de Fabricaciones Militares
Type: Ball 'C': FMJ; 3.6 g; Vo 952 m/s; brass case

AUSTRIA
Manufacturer: Hirtenberger AG
Type: Ball: FMJ; 3.6 g; Vo 980 m/s
Tracer: FMJ; base tracer, red to 457 m minimum; 3.6 g; Vo 970 m/s

5.56 × 45 mm M193 Ball

Nosler Ball: Fitted with 55 grain Nosler solid base bullet with exceptional stopping power. Intended for police use. Vo 970 m/s
Hollow Point Test: FMJ; 3.6 g; pressure ≥4,400 Bar
Blank: Steel case, crimped
Sporting: FMJ; 3.6 g; Vo 1,010 m/s
Sporting: JSP; 3.6 g; Vo 995 m/s
Sporting: PSP; 3.6 g; Vo 1,000 m/s

Abridged Ballistic Table: 5.45 × 45 mm Ball M193

Range	Velocity	Energy	Drop	Elevation	Vertex
0 m	975 m/s	1,692 J	0 mm	0 mils	0 mm
100 m	852 m/s	1,292 J	63 mm	0.62 mils	17 mm
200 m	735 m/s	962 J	260 mm	1.33 mils	73 mm
300 m	626 m/s	698 J	646 mm	2.25 mils	187 mm
400 m	522 m/s	485 J	1,290 mm	3.38 mils	393 mm
500 m	420 m/s	314 J	2,295 mm	4.84 mils	769 mm

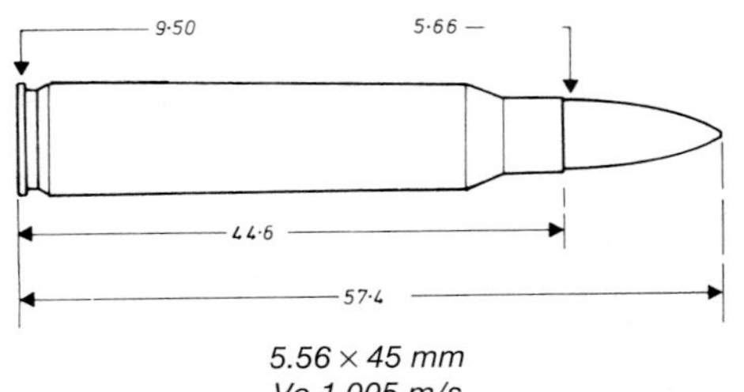

5.56 × 45 mm
Vo 1,005 m/s
Eo 1,692 J

BRAZIL
Manufacturer: Companhia Brasileira de Cartuchos
Type: Ball M193: FMJ; 3.56 g; V_{25} 965 m/s
Tracer M196: FMJ; 3.35 g; V_{25} 950 m/s; dark ignition, traces 70 to 450 m
Ball SS-109(M2811): FMJ; steel core with lead sleeve, SL; 4 g; V_{25} 915 m/s; perforates 3.50 mm SAE 1010 steel (Rockwell B55-70) at 570 m from the muzzle
Blank: Brass case, crimped
Sporting: FMJ; 3.6 g; Vo 988 m/s

BULGARIA
Manufacturer: Kintex
Type: M193 Ball: FMJ; 3.61 g; Vo 1,005 m/s

CHINA, PEOPLE'S REPUBLIC
Manufacturer: China North Industries (NORINCO)
Type: Ball: FMJ; 3.6 g; Vo 991 m/s

CZECH REPUBLIC
Manufacturer: Sellier & Bellot
Type: Ball: FMJ; 3.6 g; Vo 1,006 m/s

EGYPT
Manufacturer: Shoubra Company
Type: Ball M193: FMJ; 3.6 g; Vo 959 m/s

Manufacturer: Maasara Company for Engineering Industries
Type: Ball M193: FMJ; 3.6 g; Vo 959 m/s

FINLAND
Manufacturer: Lapua Cartridge Factory
Type: Ball S350: FMJ; 3.6 g; Vo 955 m/s
Ball E372: JSP; 3.6 g; Vo 955 m/s
Sniper Ball B421: FMJ, SL; 4.7 g.

Manufacturer: Sako Limited
Type: Ball: FMJ; 3.2 g; Vo 985 m/s
Ball: JSP; 3.56 g; Vo 1,015 m/s

FRANCE
Manufacturer: Giat Industries
Type: Ball: FMJ; 3.55 g; Vo 950 m/s
Tracer: FMJ; red trace; 3.55 g; Vo 950 m/s
Blank: Brass case, rose crimp

Manufacturer: SFM Défense
Type: Ball: FMJ; 3.56 g; Vo 950 m/s
Tracer: FMJ, red trace; 3.56 g; Vo 950 m/s
SR-TR Reduced Charge R1: Plastic; 1.17 g; Vo 750 m/s
SR-TR Reduced Charge R2: Plastic; 0.85 g; Vo 915 m/s

GERMANY
Manufacturer: Dynamit Nobel (Geco)
Type: Ball: FMJ; 3.56 g; Vo 1,010 m/s
Tracer: FMJ, red trace; 3.5 g; Vo 1,010 m/s
Practice PT: Brass case, plastic bullet for short-range training. The bullet has a maximum range of 250 m, and a maximum training range of 30 m. Bullet weight 0.2 g; Vo 1,000 m/s

Manufacturer: Metallwerk Elisenhütte GmbH
Type: Special Ball: FMJ; steel and lead core; steel jacket; gilding metal envelope; 4.1 g; Vo 930±15 m/s. Available with single or double base powder, Boxer or Berdan primed
Short-Range KB: A special cartridge for use in training on ranges with restricted safety areas. It can be fired in standard weapons without requiring any change of parts, and at 100 m range the centre of impact corresponds with the standard SS109 round. Fired at point blank, the bullet will strike the ground at approximately 240 m from the weapon. The bullet weighs 1.3 g and is composed of a plastic material with a light metal supporting cup and jacket to give a soft point effect. Vo 1,100 m/s

Manufacturer: RWS
Type: Ball TB: FMJ; 3.56 g; Vo 989 m/s
Ball SPP: JSP; 3.56 g; Vo 989 m/s

INDONESIA
Manufacturer: PT PINDAD (Persero)
Type: Ball MU-4TJ (M193): FMJ; 3.62 g; Vo 989 m/s
Blank MU-5H: Cartridge weight 7.35 g; length 52.8 mm

ISRAEL
Manufacturer: Kalia Israel Cartridge Company Limited
Type: Ball: FMJ; 3.56 g; Vo 882 m/s

Manufacturer: TAAS - Israel Industries Limited
Type: Ball M193: FMJ; 3.62 g; Vo 990 m/s
Tracer M196: 3.5 g; Vo 975 m/s
Hollow Point Test: 6.62 g; Pressure 4,920 kg/cm²
Grenade Launcher: Crimped; Pressure 2,000 kg/cm²
Ball 11AM: FMJ, SL; 3.56 g; Vo 988 m/s
Ball 11BM: JHP, SL; 3.56 g; Vo 988 m/s
Ball 11CM: JSP, SL; 3.56 g; Vo 988 m/s
Ball 13CL: JSP; 4.08 g; Vo 899 m/s

KOREA, SOUTH
Manufacturer: Poongsan Metal Corporation
Type: Ball 223A: FMJ; 3.56 g; Vo 974 m/s
Ball 223B: PSP; 3.56 g; Vo 959 m/s

PORTUGAL
Manufacturer: INDEP
Type: Ball: FMJ; 3.57 g; Vo 990 m/s
Tracer: FMJ, red trace; 3.5 g; Vo 975 m/s
Blank: Brass case, crimped

SINGAPORE
Manufacturer: Chartered Industries of Singapore
Type: Ball M193: Vo 991 m/s; conforms to US MILSPEC MIL-C-9963E
Tracer M196: Vo 975 m/s; conforms to US MILSPEC MIL-C-60111B; red trace to not less than 457 m
Extended Range Ball: FMJ; 4 g; Vo 945 m/s; perforates 3.4 mm NATO plate at 600 m

SOUTH AFRICA
Manufacturer: Denel (Pty) Limited
Type: Ball: FMJ; 3.57 g; Vo 985 m/s
Tracer: FMJ; ? g; Vo 975 m/s; red trace 70 to 450 m
AP: FMJ; steel core; ? g; Vo 960 m/s; penetration 6 mm RHA at 400 m
Hollow Point Test: Generates 483 Mpa chamber pressure

Manufacturer: Musgrave (Pty) Limited
Type: Ball: FMJ, SL; 3.6 g; Vo 980 m/s
Ball: PSP; 3.6 g; Vo 980 m/s

SPAIN
Manufacturer: SANTA BARBARA SA
Type: Ball: FMJ; 3.56 g; Vo 985 m/s
Tracer: FMJ, red trace; 3.56 g; Vo 985 m/s
Survival: Shot charge over a wad; no external bullet; maximum range approximately 50 m
Hollow Point Test: FMJ; pressure 4,350 kg/cm²

UNITED STATES OF AMERICA
Manufacturer: Federal Cartridge Company
Type: FMJ: 3.56 g; Vo 988 m/s

Manufacturer: Glaser Safety Slug Inc
Type: Ball: Glaser Blue; 2.92 g; Vo 1,045 m/s

Manufacturer: Olin Ordnance
Type: Ball M193: FMJ; 3.56 g; V_{25}964 m/s
Tracer M196: FMJ, red trace; 3.5 g; V_{25} 950 m/s
Blank M200: Brass case, rose crimp

Manufacturer: Remington Arms Company Inc
Type: Ball: FMJ; 3.6 g; Vo 987 m/s
Ball: PSP; 3.6 g; Vo 987 m/s
Ball: JHP; 3.6 g; Vo 987 m/s

Manufacturer: US Military Contractors
Type: Ball M193: FMJ; 3.63 g; Vo 990 m/s
Tracer M196: red trace from 75 to 500 m; 3.5 g; Vo 975 m/s
Grenade M195: rose crimp; Vo 43 to 50 m/s with 700 g grenade
Blank M200: rose crimp

Manufacturer: Winchester
Type: Ball: FMJ; 3.6 g; Vo 988 m/s

YUGOSLAVIA (SERBIA AND MONTENEGRO)
Holding company: YUGOIMPORT - SDPR
Type: Ball: FMJ; 3.53 g; Vo 963 m/s
Ball: FMJ; 3.56 g; Vo 1,004 m/s
Ball: FMJ; 4 g; Vo 940 m/s
Ball: JSP; 3.56 g; Vo 999 m/s
Tracer: FMJ; red trace to 450 m; 3.5 g; Vo 956 m/s
HP Test Type 2: FMJ; 3.56 g; pressure 4,810 kg/cm²

UPDATED

5.56 × 45 mm NATO

Synonyms: 5.56 mm SS109

Armament

This round is dimensionally similar to the M193 but carries a heavier and slightly longer bullet. It will therefore chamber in all weapons designed for the M193, and in later weapons using the same chamber dimensions, but is optimised for a faster twist of rifling of one turn in 178 mm. It will, however, retain its accuracy in rifling with one turn in 230 mm.

Development

From 1977 to 1979 the NATO countries held a long series of trials to determine the next generation of small arms ammunition, as a result of which this round was adopted as NATO standard. It is essentially the M193 case with a heavier bullet developed by Fabrique Nationale Herstal of Liege. Trials demonstrated that this bullet had better accuracy and penetrating power, although it demanded a steeper twist of rifling to perform at its best. The resultant cartridge, the FN SS109, was officially adopted by NATO as their standard 5.56 mm cartridge on 24 October 1980.

The standard US Ball round is the M855.

Description

The M855 case is rimless and bottle-necked and made of brass. Boxer primers are normally used. For the M855 the propellent filling is 1.75 g of BALL POWDER although rounds produced elsewhere may use alternative fillings.

The M855 ball bullet is streamlined, with a cannelure in the body into which the case mouth is coned on assembly of the round. The bullet is of two part construction, with a gilding metal envelope containing a hardened steel core behind the tip, followed by a lead base. The bullet is longer (23.2 mm) than the M193 equivalent and is also heavier at 4.02 g.

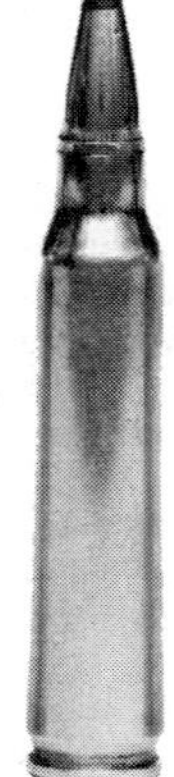

5.56 × 45 mm NATO (M855)

Abridged Ballistic Table: 5.45 × 45 mm NATO Ball

Range	Velocity	Energy	Drop	Elevation	Vertex
0 m	930 m/s	1,708 J	0 mm	0 mils	0 mm
100 m	832 m/s	1,367 J	66 mm	0.65 mils	18 mm
200 m	740 m/s	1,081 J	281 mm	1.41 mils	75 mm
300 m	650 m/s	834 J	675 mm	2.30 mils	187 mm
400 m	574 m/s	650 J	1,296 mm	3.37 mils	382 mm
500 m	500 m/s	494 J	2,268 mm	4.68 mils	681 mm

The M855 Ball can penetrate 3.5 mm of NATO armour plate at 600 m and a steel helmet at 1,100 m.

The Tracer counterpart for the M855 Ball is the M856 while the Blank Training Cartridge with a rose crimped cartridge case mouth is the M862.

Data

M855

Round length: 57.4 mm
Round weight: nominal, 12.3 g
Case length: 44.7 mm
Rim diameter: 9.54 mm
Bullet weight: 4.02 g
Bullet diameter: 5.66 mm
Bullet length: 23.2 mm

AUSTRIA
Manufacturer: Hirtenberger AG
Type: Ball: FMJ; steel core; 4 g; Vo 935 m/s
Frangible Short-Range Ball: 1.35 g bullet using plastic exposed core and CuZn base; Vo 1,030 m/s; Maximum range 750 m; effective range 180 m. Specially designed to function in automatic weapons up to 600 rds/min

BELGIUM
Manufacturer: FN Herstal SA
Type: Ball SS109: Brass or steel case; FMJ; 4 g; Vo 915 m/s
Tracer L110: Brass or steel case; FMJ, red trace; 4.1 g; Vo 890 m/s
AP P-112: Brass case; FMJ, steel core; 4 g; Vo 904 m/s

BRAZIL
Manufacturer: Companhia Brasileira de Cartuchos
Type: Ball M2811 TW7: FMJ; 4 g; Vo 940 m/s

CANADA
Manufacturer: SNC Industrial Technologies Inc
Type: Ball C77: FMJ; 4 g; Vo 910 m/s
Tracer C78: FMJ; red trace; dark ignition, full brightness 140 to 600 m; 4 g; Vo 900 m/s
Blank C79: Brass case, crimped

EGYPT
Manufacturer: Shoubra Company
Type: Ball SS109: FMJ; 4 g; Vo 940 m/s
Tracer L110: FMJ; red trace; 3.6 g; Vo 875 m/s
AP P112: FMJ; steel core; 4 g; Vo 940 m/s

Manufacturer: Maasara Company for Engineering Industries
Type: Ball SS109: FMJ; 4 g; Vo 940 m/s
Tracer L110: FMJ; red trace; 3.6 g; Vo 875 m/s
AP P112: FMJ; steel core; 4 g; Vo 940 m/s

FRANCE
Manufacturer: Giat Industries
Type: Balle O: FMJ; 3.5 g; Vo 957 m/s
Tracer T: FMJ; red trace; 3.5 g; Vo 950 m/s
Balle PPA: FMJ; hard core; 3.6 g; Vo 950 m/s; penetrates 3.5 mm mild steel at 600 m and has the anti-personnel effect of the ordinary Balle O
Balplast F1: Plastic bullet; 0.65 g; Vo 860 m/s
Blank G: Brass case, rose crimp
Grenade launcher: Brass case, rose crimp

Manufacturer: SFM Défense
Type: Balle O: FMJ; 3.5 g; Vo 957 m/s
Training Ball R1: Rilsan-bronze cylindro-conoidal bullet; 1.15 g; Vo 740 m/s. Accuracy to 100 m is comparable to service ball; maximum range 600 m. Case may be brass or aluminium
Training Ball R2: Rilsan-bronze ogival bullet; 0.9 g; Vo 880 m/s. Maximum range 500 m. Case may be brass or aluminium

GERMANY
Manufacturer: Metallwerk Elisenhütte GmbH
Type: Practice KB: This is a short-range practice round using a bullet made of a copper alloy part-jacket and a plastic core, the nose of which is exposed and forms the nose of the bullet. The round feeds in automatic weapons, and the ballistics are such that at 100 m the centre of impact is the same as that of the SS109 ball round. The bullet has a maximum range of approximately 750 m, though at this range the remaining energy is very little. Bullet weight 1.3 g; Vo 1,100 m/s

INDONESIA
Manufacturer: PT PINDAD (Perseo)
Type: Ball: FMJ;

ISRAEL
Manufacturer: TAAS - Israel Industries Limited
Type: Ball MU-5TJ: FMJ; 4.1 g; Vo 915 m/s
{**Tracer MU-5N:** FMJ, lead bullet; 4.17 g; Vo 875 m/s
Blank MU-5H: Cartridge weight 7.35 g; cartridge length 52.8 mm

ITALY
Manufacturer: Europa Metalli
Type: Ball IT-EM/01: FMJ; steel core with lead filler; 4 g; Vo 866 m/s
Tracer IT-EM/04: FMJ; lead core; dark ignition trace 30 to 600+ m. Trajectory crosses ball at 600 m. 4 g; Vo 866 m/s
Short Range: Plastic bullet loaded with bronze to weigh 1.6 g; Vo 1,000 m/s. Shoots within 120 mm of service ball at 100 m range. Maximum range 1,000 m
Plastic Training: Special bullet with Cu/ Zn jacket and exposed plastic core; 1.6 g; Vo 970 m/s; No perforation of 2.5 mm MS plate at 100 m. Maximum range 750 m.

PORTUGAL
Manufacturer: INDEP
Type: Ball: FMJ; 4 g; Vo 932 m/s

SPAIN
Manufacturer: SANTA BARBARA SA
Type: AP: FMJ; round weight 11.8 g; Vo 867 m/s

SWEDEN
Manufacturer: Bofors Carl Gustaf
Type: Ball: FMJ; 4 g; Vo 930 m/s
Tracer: FMJ; 3.9 g; Vo 910 m/s; Red trace; dark ignition; tracer range min 800 m
AP: FMJ with hard core; 3.4 g; Vo 1,000 m/s; penetrates min 12 mm armour plate HB 300 at 200 m; under test for the US Army as the XM995
Blank STAR: brass case, length 49.5 mm

UNITED KINGDOM
Manufacturer: British Aerospace Defence Limited, Royal Ordnance Division
Type: Ball L2A1: FMJ, SL; 4 g; Vo 944 m/s
Tracer L1A2: FMJ; dark ignition; red trace to 600 m; 4.2 g; Vo 865 m/s
Blank L1A1: Brass case, crimped
ROTA Training: Frangible bullet; 1.4 g; Vo 920 m/s. Maximum range 800 m

UNITED STATES OF AMERICA
Manufacturer: Delta Defense Inc
Type: Frangible: 2.14 g; Vo 847 m/s

Manufacturer: Hansen Cartridge Company
Type: Ball: FMJ; steel core; 4 g; Vo 915 m/s

Manufacturer: Olin Defense Systems
Type: Ball M855: FMJ; forward core steel, rear core lead; 4.02 g; V_{25} 914 m/s
Tracer M856: FMJ; red trace; 4.13 g; V_{25} 911 m/s
Short Range Training M862: Plastic bullet; 0.26 g; Vo 1,326 m/s

YUGOSLAVIA (SERBIA AND MONTENEGRO)
Holding company: YUGOIMPORT - SDPR
Type: Ball: FMJ; 4 g; Vo 914 m/s
Blank M200: Brass case, short crimp
Blank PPU18: Brass case, long crimp

UPDATED

5.7 mm FN SS190

Synonyms: 5.7 × 28 mm; SS190

Armament

FN Herstal P90 Personal Weapon and Five SeveN pistol.

Development

The 5.7 mm cartridge was developed in the mid-1980s by FN Herstal for their P90 Personal Defence Weapon. The objective is to produce a new generation close defence weapon system for users who cannot normally carry any weapon other than a pistol but who require something more powerful than the usual 9 mm Parabellum cartridge.

The 5.7 mm SS190 bullet will penetrate a standard M1 steel helmet at 150 m range and 48 layers of Kevlar at over 50 m. The stopping power is claimed to be three times that of the 9 mm Parabellum; the bullet does not fragment but gives up its energy very rapidly on impact. The round is effective up to a range of 150 m.

Description

The 5.7 mm brass case is rimless, bottle-necked, and Berdan primed. The streamlined 2.02 g bullet has a

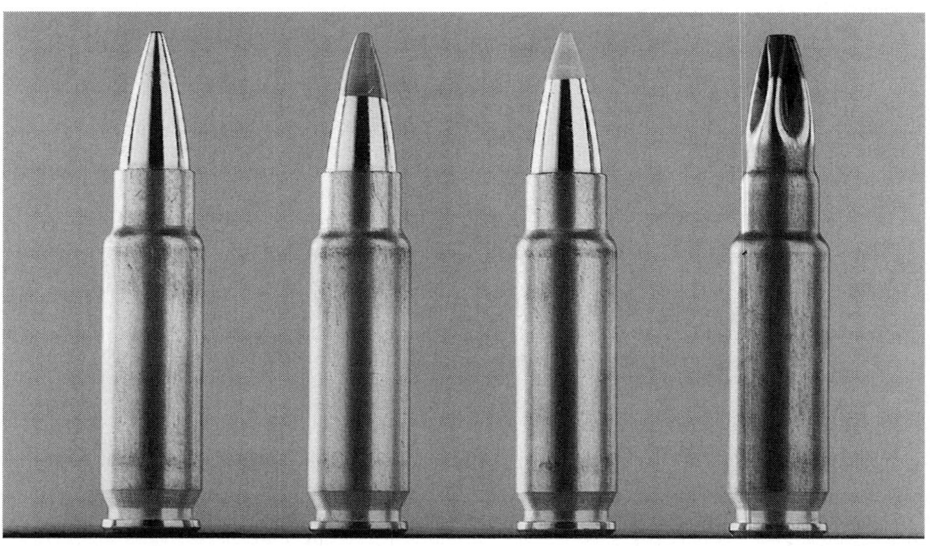

5.7 × 28 mm rounds; from left, BO Ball, BT Tracer, a developmental low-velocity round for suppressors, and BLK Blank

1996

steel jacket containing a steel core behind the tip followed by an aluminium core towards the base.

The SS190 Ball cartridge is known as the BO. The BT, with a 2 g bullet, is a Tracer round, while the BLK is a Blank with a crimped case mouth.

At one time a low-velocity cartridge for possible use with a suppressed P90 was under development but its current status is uncertain.

Data

Round length: 40.5 mm
Cartridge weight: 6 g
Case length: 26.75 mm
Rim diameter: 7.8 mm
Bullet diameter: 5.71 mm
Bullet weight: BO, 2.02 g
Muzzle velocity: 715 m/s
Muzzle energy: 516 J

FRANCE
Manufacturer: Giat Industries
Type: BO Ball: see above
BT Tracer: FMJ; red trace; 2 g; Vo 715 m/s
BLK Blank: Brass case 39 mm long, rose crimp

5.7 × 28 mm
Vo 715 m/s
Eo 516 J

UPDATED

0.243 Winchester

Synonyms: none

Armament

Suitably chambered sniper rifles.

Development

This cartridge was developed commercially by the Winchester company in 1955 as a hunting round. It soon acquired a reputation for accuracy. This resulted in a number of companies producing military sniper rifles chambered for this round, since it produced better accuracy figures than the more usual 7.62 mm NATO cartridge.

Description

The 0.243 Winchester was developed by taking the 7.62 × 51 mm NATO cartridge case and reducing the neck to accommodate a 6 mm bullet. This means that the alterations to a weapon originally designed for the 7.62 mm cartridge are relatively minor except for the different barrel, facilitating the conversion. The case is of brass, bottle-necked and usually Boxer primed. The bullet is of conventional pattern, streamlined, lead-cored, with a steel jacket and gilding metal envelope. Various hunting bullets can be found, although the military standard is a full metal jacket bullet.

Data

Round length: 68.84 mm
Case length: 51.94 mm
Rim diameter: 12.01 mm
Head diameter: 11.94 mm
Bullet diameter: 6.17 mm
Bullet weight: 6.48 g
Neck diameter: 7 mm

AUSTRIA
Manufacturer: Hirtenberger AG
Type: Ball: FMJ; 5.8 g; Vo 945 m/s
Ball: SP; 6.5 g; Vo 900 m/s

CZECH REPUBLIC
Manufacturer: Sellier & Bellot
Type: Ball: SP; 6.5 g; Vo 905 m/s

FINLAND
Manufacturer: Lapua Cartridge Factory
Type: Ball E453: FMJ; 5.8 g; Vo 945 m/s
Ball S342: FMJ; 5.8 g; Vo 885 m/s

Abridged Ballistic Table: 0.243 Winchester 6.48 g ball

Range	Velocity	Energy	Drop	Elevation	Vertex
0 m	836 m/s	2,257 J	0 mm	0 mils	0 mm
100 m	764 m/s	1,885 J	51 mm		
200 m	696 m/s	1,563 J			
300 m	631 m/s	1,285 J	690 mm		
400 m	570 m/s	1,050 J			
500 m	512 m/s	846 J			

Manufacturer: Sako Limited
Type: Ball: FMJ; 5.8 g; Vo 870 m/s
Ball: SP; 5.8 g; Vo 955 m/s

GERMANY
Manufacturer: RWS
Type: Ball TB: FMJ; 4.47 g; Vo 1,000 m/s
Ball CP: FMJ; cone point; 6.22 g; Vo 920 m/s
Ball SPP: PSP; 6.48 g; Vo 936 m/s

ISRAEL
Manufacturer: TAAS - Israel Industries Limited
Type: Ball: SP; 6.5 g; Vo 902 m/s
Ball: HP; 5.2 g; Vo 1,020 m/s

KOREA, SOUTH
Manufacturer: Poongsan Metal Corporation
Type: Ball: PSP; 5.2 g; Vo 896 m/s
Ball: PSP; 6.5 g; Vo 836 m/s

SOUTH AFRICA
Manufacturer: Musgrave (Pty) Limited
Type: Ball: FMJ; 6.5 g; Vo 900 m/s

SWEDEN
Manufacturer: Norma AB
Type: Ball 16002: FMJ; 6.5 g; Vo 945 m/s
Ball 16003: JSP; 6.5 g; Vo 945 m/s

UNITED STATES OF AMERICA
Manufacturer: Federal Cartridge Company
Type: Ball: SP; 5.2 g; Vo 1,042 m/s
Ball: SP; 6.5 g; Vo 902 m/s
Ball: HP; 3.9 g; Vo 1,098 m/s
Ball: Streamlined HP; 5.5 g; Vo 1,012 m/s

Manufacturer: Remington Arms Company Inc
Type: Ball: PSP; 5.2 g; Vo 1,021 m/s
Ball: JHP; 5.2 g; Vo 1,021 m/s
Ball: PSP; 6.05 g; Vo 902 m/s

Manufacturer: Winchester-Olin
Type: Silver Tip: FMJ; 6.5 g; Vo 902 m/s
Ball: SP; 5.2 g; Vo 1,042 m/s
Ball: SP; 6.5 g; Vo 902 m/s

YUGOSLAVIA (SERBIA AND MONTENEGRO)
Holding company: YUGOIMPORT - SDPR
Type: Ball: JSP; 6.5 g; Vo 902 m/s
Hollow Point Test: JSP; 6.5 g

UPDATED

0.243 Winchester

6.35 mm Automatic Pistol

Synonyms: 6.35 × 16 mm; 0.25 Auto

Armament

All suitably chambered automatic pistols; also a limited number of European revolvers of pre-1939 vintage.

Development

Developed by Fabrique Nationale of Liege in association with John Browning, for the Browning Model 1906 pocket automatic pistol. Like many Browning designs it has a semi-rimmed case, allowing it to be used in small revolvers. As a combat round it is relatively useless, but as a personal defence weapon it suffices; the 0.22 Long Rifle cartridge has more power. Automatic pistols in this calibre are, however, carried by senior officers and staff personnel in some armies, and it is used to some extent as a concealed weapon by police forces.

Description

The semi-rimmed case is usually of brass, though steel and alloy have been tried in the past. The standard bullet is a metal-jacketed, lead-cored pattern, round-nosed or ogival, although lead bullets are sometimes seen, these being more suited to revolver use.

Data

Round length: 23.11 mm
Case length: 15.62 mm
Rim diameter: 7.67 mm
Bullet diameter: 6.37 mm
Bullet weight: 3.24 g
Muzzle velocity: 250 m/s
Muzzle energy: 101 J

AUSTRIA
Manufacturer: Hirtenberger AG
Type: Ball: FMJ; 3.2 g; Vo 220 m/s

6.35 mm Automatic Pistol

BRAZIL
Manufacturer: Companhia Brasileira de Cartuchos
Type: Ball: FMJ; 3.2 g; Vo 232 m/s

CZECH REPUBLIC
Manufacturer: Sellier & Bellot
Type: Ball: FMJ; 3.3 g; Vo 238 m/s

FINLAND
Manufacturer: Lapua Cartridge Factory
Type: Ball 4316040: FMJ; 3.25 g; Vo 225 m/s

Manufacturer: Sako Limited
Type: Ball: FMJ; 3.25 g; Vo 245 m/s

FRANCE
Manufacturer: SFM Défense
Type: Ball: FMJ; 3.25 g; Vo 200 m/s

GERMANY
Manufacturer: Geco
Type: Ball: FMJ; 3.2 g; Vo 265 m/s

ITALY
Manufacturer: Fiocchi Munizioni SpA
Type: Ball: FMJ; 3.2 g; Vo 220 m/s

KOREA, SOUTH
Manufacturer: Poongsan Metal Corporation
Type: Ball 25A: FMJ; 3.24 g; Vo 230 m/s

MEXICO
Manufacturer: Aguila Industrias Tecnos
Type: Ball: FMJ; 3.2 g; Vo 239 m/s

SOUTH AFRICA
Manufacturer: Musgrave (Pty) Limited
Type: Ball: FMJ; 3.2 g; Vo 240 m/s

SPAIN
Manufacturer: SANTA BARBARA SA
Type: Ball: FMJ; 3.2 g; Vo 230 m/s
Ball: JSP; 3 g; Vo 230 m/s

SWEDEN
Manufacturer: Norma AB
Type: Ball: FMJ; 3.2 g; Vo 245 m/s

UNITED STATES OF AMERICA
Manufacturer: CCI
Type: Ball: FMJ; 3.2 g; Vo 247 m/s
Ball: JSP; 2.9 g; Vo 260 m/s

Manufacturer: Federal Cartridge Company
Type: Ball: FMJ; 3.2 g; Vo 247 m/s

Manufacturer: Glaser Safety Slug Inc
Type: Ball: Glaser Blue; 2.26 g; Vo 350 m/s

Manufacturer: Remington Arms Company Inc
Type: Ball: FMJ; 3.2 g; Vo 246 m/s

Manufacturer: Winchester-Olin
Type: Ball: FMJ; 3.2 g; Vo 231 m/s
Ball: JSP; 2.92 g; Vo 248 m/s

YUGOSLAVIA (SERBIA AND MONTENEGRO)
Holding company: YUGOIMPORT - SDPR
Type: Ball: FMJ; 3.25 g; Vo 241 m/s

UPDATED

7.62 × 25 mm Soviet Pistol

Synonyms: 7.62 mm Tokarev

Armament
Tokarev TT33 automatic pistol; various obsolete Soviet sub-machine guns (PPD, PPSh, PPS); pistols and sub-machine guns chambered for the 7.63 mm Mauser cartridge will probably operate satisfactorily with this round.

Development
This cartridge began life as the 7.63 mm Mauser automatic pistol cartridge. It was taken into use by the Russian forces in the early 1900s, and the pistol remained popular with the Bolshevik armies, one model being produced specifically for the Soviet government. As a result manufacture of the Mauser cartridge began in Russia, and when in due course the Tokarev automatic pistol was developed, it was designed around the Mauser cartridge. For manufacturing convenience the barrel of the Tokarev was 7.62 mm calibre, and thus the Soviet cartridge lost its Mauser connotation and became known as the 7.62 mm Tokarev. The dimensional differences between the Soviet round and the original Mauser specifications are minute and largely because of the manufacturing processes. As a result it can be expected that any weapon originally using the Mauser cartridge will work with the Soviet pattern.

This cartridge has been manufactured in China and various countries of the former Warsaw Pact, but always to the Soviet specification.

Description
The case is rimless and bottle-necked. The standard ball bullet is round-nosed and lead-cored with a steel jacket.

Data
BALL TYPE P
Round length: 34.55 mm
Round weight: nominal, 10.65 g
Case length: 25.14 mm
Rim diameter: 9.91 mm
Bullet diameter: 7.82 mm
Bullet weight: 5.57 g
Muzzle velocity: 455 m/s
Muzzle energy: 576 J

CHINA, PEOPLE'S REPUBLIC
Manufacturer: China North Industries (NORINCO)
Type: Type 51: FMJ; 5.57 g; Vo 420 m/s

COMMONWEALTH OF INDEPENDENT STATES
Manufacturer: Government arsenals

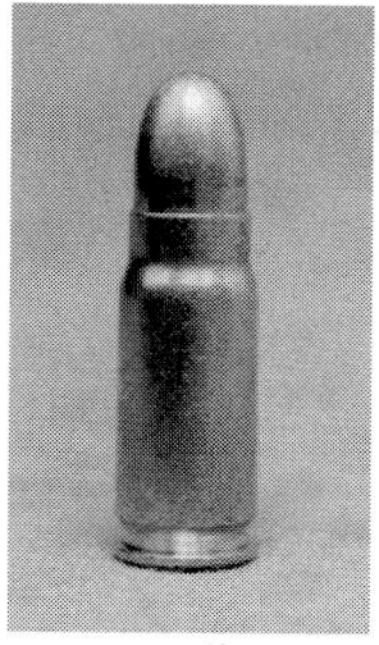

7.62 × 25 mm Soviet Pistol

Type: Ball P: FMJ; 5.57 g; Vo 455 m/s
AP Incendiary P-41: for sub-machine guns; 4.79 g; steel-jacketed bullet with hard steel core and incendiary composition in the nose
Tracer PT: also for sub-machine guns; lead-cored bullet with base tracer

YUGOSLAVIA (SERBIA AND MONTENEGRO)
Holding company: YUGOIMPORT - SDPR
Type: Ball: FMJ; 5.5 g; Vo 520 m/s
Ball: JHP; 5.5 g; Vo 522 m/s

UPDATED

0.30 Carbine M1

Synonyms: 7.62 × 33 mm

Armament
Originally for the US Carbine M1; will operate in Carbines M1, M1A1, M2 and M3 and in commercial copies such as those made by High Standard and other companies. A handful of sub-machine guns, automatic pistols and revolvers have been chambered for this cartridge but none has been very successful, the ballistics not being optimum for short-barrelled weapons.

Development
This cartridge was developed in 1940 by the Winchester Repeating Arms Company to a specification issued by the US Ordnance Department. It was broadly based upon the existing 0.32 Winchester Self-Loading rifle round and was standardised as the 'Cartridge, Carbine, Cal 0.30 M1' on 30 September 1941. Production was entirely by government plants until 1945. After the war numbers of carbines were disposed of commercially, therefore commercial cartridge manufacture began and has continued ever since. Some 6.2 million carbines were made by US Government contracts and several companies have produced commercial copies since the mid-1950s. As a result the weapon is widely distributed and is commonly in use by military and police forces throughout the world.

Description
The case is rimless, straight tapered, and usually of brass, though steel cased ammunition was made in the USA from 1942 to 1945. The bullet is blunt-nosed, having a lead core and gilding metal jacket, and flat-based.

Various tracer and grenade launching rounds were developed but were not widely used.

Data
BALL M1
Round length: 42.5 mm
Case length: 32.7 mm
Rim diameter: 8.99 mm
Bullet diameter: 7.78 mm
Round weight: 12.2 g
Bullet weight: 7 g

Abridged Ballistic Table: 0.30 Carbine Ball M1

Range	Velocity	Energy
0 m	606 m/s	1,303 J
100 m	477 m/s	807 J
200 m	384 m/s	523 J
300 m	315 m/s	352 J
400 m	282 m/s	282 J
500 m	256 m/s	232 J

AUSTRIA
Manufacturer: Hirtenberger AG
Type: Ball: 7.13 g; Vo 610 m/s

BRAZIL
Manufacturer: Companhia Brasileira de Cartuchos
Type: Ball: FMJ; 7.1 g; Vo 607 m/s

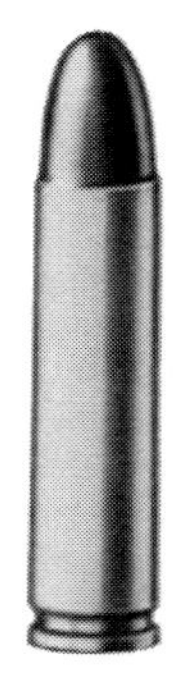

0.30 Carbine M1

GERMANY
Manufacturer: Dynamit Nobel (Geco)
Type: Ball: FMJ; 7.2 g; Vo 600 m/s

ISRAEL
Manufacturer: Kalia Israel Cartridge Company Limited
Type: Ball: FMJ; 8.1 g; Vo 503 m/s

Manufacturer: TAAS - Israel Industries Limited
Type: Ball: FMJ; 7.13 g; Vo 607 m/s

ITALY
Manufacturer: Fiocchi Munizioni SpA
Type: Ball: FMJ; 7.2 g; Vo 580 m/s

KOREA, SOUTH
Manufacturer: Poongsan Metal Corporation
Type: Ball PMC30A: FMJ; 7.12 g; Vo 588 m/s

MEXICO
Manufacturer: Aguila Industrias Tecnos
Type: Ball: FMJ; 7.13 g; Vo 607 m/s

SOUTH AFRICA
Manufacturer: Denel (Pty) Limited
Type: Ball: FMJ; 7.15 g; Vo 607 m/s
Ball: SP; 7.15 g; Vo 607 m/s

SWEDEN
Manufacturer: Norma AB
Type: Ball: JSP; 7.1 g; Vo 600 m/s

UNITED STATES OF AMERICA
Manufacturer: Federal Cartridge Company
Type: Ball: JSP; 7.1 g; Vo 607 m/s

Manufacturer: Government plants
Type: Ball M1: FMJ; see above
Tracer M16: FMJ; lead core, steel jacket; gilding metal envelope; tracer composition; red trace to 500 m; 6.93 g
Tracer M27: FMJ; similar to M16 but with dark ignition 100 to 375 m; 6.54 g

Manufacturer: Hansen Cartridge Company
Type: Ball: FMJ; 7.1 g; Vo 607 m/s

Manufacturer: Remington Arms Company Inc
Type: Ball R30CAR: JSP; 7.13 g; Vo 607 m/s

Manufacturer: Winchester-Olin
Type: Ball X30M2: FMJ; 7.1 g; Vo 607 m/s
Ball: HSP; 7.1 g; Vo 607 m/s

YUGOSLAVIA (SERBIA AND MONTENEGRO)
Holding company: YUGOIMPORT - SDPR
Type: Ball: FMJ; 7.1 g; Vo 610 m/s
Ball: JSP; 7 g; Vo 615 m/s

UPDATED

7.62 mm Soviet M1943

Synonyms: 7.62 × 39 mm; 7.62 mm Kalashnikov; 7.62 mm obr 43g; 57N231.

Armament

Chinese Type 56 carbine, Type 56 and 56-1 rifles, Type 68 rifle, Types 75 and 81 light machine guns; Finnish M60, M62, M76 and M90 rifles, M78 heavy-barrel rifle and M62 machine gun; Hungarian AMD-65 rifle; North Korean Type 68 rifle; Polish PMK, PMK-DGN and PMKM rifles; Soviet AK-47, AKM, SKS rifles, RPD and RPK machine guns; Ruger Mini-Thirty rifle; Yugoslavian M59/66A1, M70B1, M70AB2 rifles and M72, M72AB1 machine guns. In addition, certain early Heckler and Koch rifles and machine guns were offered in this calibre.

Development

Soviet development of an intermediate rifle cartridge began in the late 1930s, parallel with similar work in Finland, Germany and Switzerland, but was dropped in 1939. Probably spurred by the appearance of the German 7.92 × 33 mm 'Kurz' cartridge in late 1942, development was restarted in 1943. A design attributed to N M Elizarov and B V Semin was approved in late 1943 and applied to an experimental carbine by Simonov which later became the SKS. However, the major adoption of the cartridge came with the AK-47 Kalashnikov rifle series, after which it became the standard rifle and light machine gun round for the Warsaw Pact and widely adopted by other countries obtaining arms from the former Soviet Union.

Description

The case is rimless and bottle-necked, of brass, lacquered steel or brass-coated steel, Berdan primed with one or two flash-holes. The standard ball bullet PS is streamlined, with a steel core, steel jacket and gilding metal envelope. Ball bullets manufactured in other countries may be non-streamlined and use a lead core, the weight being adjusted to the PS pattern.

Propellant charges vary in weight and nature. Current CIS production involves 1.6 g of SSNF 50. Bulgarian and Romanian rounds use 1.62 g of SB-43.

Also available are a Tracer cartridge using the T45 bullet, and a Blank with a crimped case mouth. There is also an API-T with the 7.45-7.7 g bullet carrying between 0.4 and 0.43 g of an incendiary composition.

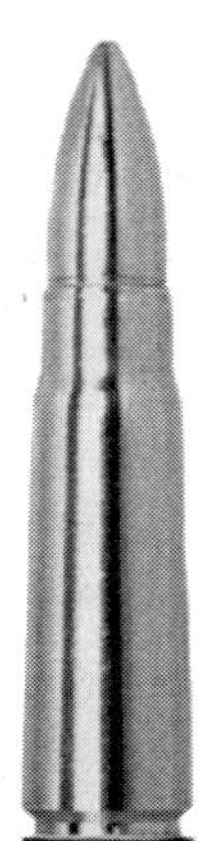

7.62 mm Soviet M1943

Abridged Ballistic Table: 7.62 × 39 mm Ball Type PS

Range	Velocity	Energy	Drop	Elevation	Vertex
0 m	710 m/s	2,010 J	0 mm	0 mils	0 mm
100 m	632 m/s	1,592 J	116 mm	1.03 mils	15 mm
200 m	547 m/s	1,192 J	668 mm	2.68 mils	120 mm
300 m	471 m/s	884 J	1,192 mm	4.63 mils	328 mm
400 m	399 m/s	634 J	2,338 mm	6.27 mils	700 mm
500 m	343 m/s	468 J	4,173 mm	8.85 mils	1,305 mm

The standard 7.62 × 39 mm cartridge is known as the 57N231. There is also a 57N231U with a heavier (12.5 g) bullet and a reduced propellant load to provide a subsonic performance when fired from suppressed weapons at a muzzle velocity of 295 to 310 m/s.

Data

BALL PS
Round length: 55.8 mm
Round weight: nominal, 16.2 g
Case length: 38.65 mm
Rim diameter: 11.35 mm
Bullet diameter: 7.9 mm
Bullet length: 26.8 mm
Bullet weight: 7.97 g
Nominal charge: 1.6 g; SSNF 50 powder
Muzzle velocity: 710-725 m/s
Muzzle energy: 2,010 J

AUSTRIA
Manufacturer: Hirtenberger AG
Type: Ball: FMJ; 7.95 g; Vo 710 m/s
Tracer: FMJ; semi-streamlined; red trace to 800 m minimum; 7.62 g; Vo 700 m/s

BRAZIL
Manufacturer: Companhia Brasileira de Cartuchos
Type: Ball: FMJ; 7.9 g bullet; Vo 690 m/s in 610 mm barrel

BULGARIA
Manufacturer: Arsenal Kazanluk
Type: Heavy Ball: FMJ; steel core in lead jacket and cupro-nickel envelope; steel case; 7.9 g; Vo 715 m/s
Tracer T-45: FMJ; lead core, cupro-nickel envelope, red trace in rear tracing to 800 m; 7.6 g; Vo 715 m/s
Blank: Steel case, rose crimp, 0.77 g; P-125 powder

Manufacturer: Kintex
Type: Ball PSGS: FMJ; steel core in lead jacket and cupro-nickel envelope; steel case; 7.9 g; Vo 715 m/s
Tracer T-45: FMJ; lead core, cupro-nickel envelope, red trace in rear tracing to 800 m; 7.6 g; Vo 715 m/s
Blank: Steel case, rose crimp, 0.77 g; P-125 powder

CHINA PEOPLE'S REPUBLIC
Manufacturer: China North Industries (NORINCO)
Type: Ball Type 56: FMJ; steel core; 7.9 g; Vo 710 m/s
Tracer Type 56: FMJ; partial steel core, red trace; 7.56 g; Vo 710 m/s
Incendiary Type 56: FMJ; steel core with tip exposed; partial gilding metal jacket; incendiary filling in capsule behind core; 6.63 g; Vo 740 m/s
AP-I Type 56: FMJ; steel half-core inside gilding metal jacket; incendiary tracer element in capsule behind core; 7.67 g; Vo 725 m/s

COMMONWEALTH OF INDEPENDENT STATES
Manufacturer: State arsenals
Type: Ball PS: FMJ; steel core; 7.97 g; Vo 710 m/s
Tracer T45: FMJ; 7.45 g non-streamlined bullet; red trace
AP-I BZ: FMJ, SL; steel core with incendiary pellet in the nose; 7.77 g
Incendiary-Ranging ZP: FMJ; with incendiary and tracer elements; 6.61 g

EGYPT
Manufacturer: Aboukir Engineering Industries
Type: Ball: FMJ; lead core; 8 g; Vo 725 m/s

Manufacturer: Shoubra Company
Type: Ball: FMJ; V_{25} 725 m/s
AP-I: FMJ; steel core; V_{25} 735 m/s
Tracer: FMJ; red trace; V_{25} 718 m/s

Manufacturer: Maasara Company for Engineering Industries
Type: Ball: FMJ; V_{25} 725 m/s
AP-I: FMJ; steel core; V_{25} 735 m/s
Tracer: FMJ; red trace; V_{25} 718 m/s

FINLAND
Manufacturer: Lapua Cartridge Factory
Type: Ball S309: FMJ; lead core; 8.04 g
Tracer VJ313: FMJ; red trace; 8.04 g
Ball S405: FMJ; 8 g; Vo 725 m/s
Tracer: FMJ; red trace; 8 g; Vo 725 m/s
AP: FMJ; hard core, steel or gilding metal jacket
Blank: Wooden bullet

Manufacturer: Sako Limited
Type: Ball: FMJ, SL; 8 g; Vo 715 m/s

HUNGARY
Manufacturer: Mátravidéki Fémmüvek
Type: Ball: FMJ; steel core; 8 g; Vo 715 m/s; steel cartridge case
Ball: FMJ; steel core; 8 g; Vo 715 m/s; brass cartridge case

IRAN
Manufacturer: Defence Industries Organisation, Ammunition Group
Type: Ball: FMJ; cartridge weight 18 g; Vo 710 m/s

ISRAEL
Manufacturer: TAAS - Israel Industries Limited
Type: Ball: FMJ; 8 g; Vo 693 m/s; Eo 1,920 J

KOREA, SOUTH
Manufacturer: Poongsan Metal Corporation
Type: Ball: FMJ; 7.9 g; Vo 716 m/s
Ball: PSP; 8.1 g; Vo 707 m/s

POLAND
Manufacturer: Mesko Zaklady Metalowe
Type: Ball: FMJ; 7.9 g; Vo 715 m/s
Tracer: FMJ; red trace; 7.2 g; Vo 715 m/s
Blank: Steel case, star crimp

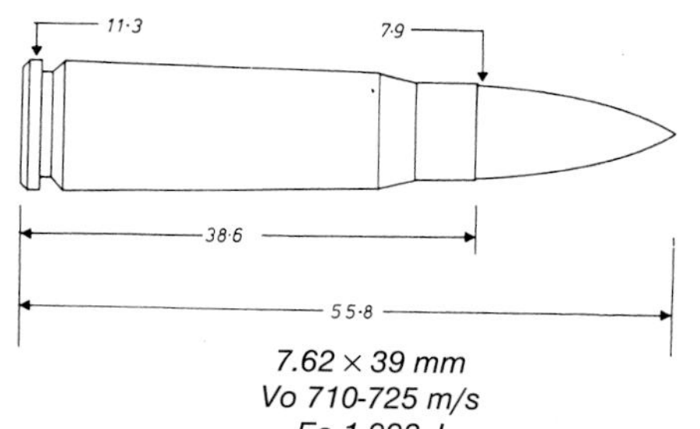

7.62 × 39 mm
Vo 710-725 m/s
Eo 1,993 J

PORTUGAL
Manufacturer: INDEP
Type: Ball: FMJ; 7.95 g; Vo 710 m/s; brass case
Tracer: FMJ; red trace; 7.95 g; Vo 800 m/s; brass case

ROMANIA
Manufacturer: Romtehnica
Type: Ball: FMJ; 7.75-8.05 g; Vo 725 m/s
Tracer: FMJ; 7.47-7.87 g; Vo 740 m/s
API-T: FMJ; 7.47-7.7 g; Vo 725 m/s
Blank: Cartridge weight 7.3-8 g; length 38.7 mm

SOUTH AFRICA
Manufacturer: Denel (Pty) Limited
Type: Ball: FMJ; 8 g; Vo 710 m/s
AP: FMJ; steel core; non-streamlined; Vo 710 m/s; penetration 10 mm RHA at 400 m

SWEDEN
Manufacturer: Norma AB
Type: Ball: SP; 11.7 g; Vo 785 m/s
Ball: SP; 9.7 g; Vo 900 m/s

UNITED STATES OF AMERICA
Manufacturer: Federal Cartridge Company
Type: Ball: SP; 8 g; Vo 701 m/s

Manufacturer: Glaser Safety Slug Inc
Type: Ball: Glaser Blue; 8.42 g; Vo 701 m/s

Manufacturer: Hansen Cartridge Company
Type: Ball: FMJ; 8 g; Vo 708 m/s
Ball: JSP; 8 g; Vo 732 m/s

Manufacturer: Remington Arms Company Inc
Type: Ball: SP; 8 g; Vo 721 m/s

Manufacturer: Winchester-Olin
Type: Ball: SP; 8 g; Vo 721 m/s

YUGOSLAVIA (SERBIA AND MONTENEGRO)
Holding company: YUGOIMPORT - SDPR
Type: Ball M67: FMJ; 8 g; lead-antimony core in gilding metal jacket, non-streamlined; Vo 733 m/s; brass or steel case Boxer or Berdan primed
Tracer M78: 7.7 g; non-streamlined, lead-antimony core in gilding metal jacket with tracer composition at rear end. Dark ignition to 15 m from weapon, then visible from 115 to 800 m. Brass or steel case
API-T M82: 7.55 g; steel core, lead liner, gilding metal jacket, incendiary-tracer composition in capsule at rear end; penetration 7 mm steel at 200 m; Vo 730 m/s; Brass case
Practice M76: The bullet uses an aluminium core with a gilding metal half-jacket which encloses the flat base but leaves the reduced diameter point exposed. The weight is 1.7 g and this, with the odd shape, provides a maximum range of 560 m. Accuracy at 100 m is comparable with the standard ball round. Vo 700 m/s. The standard brass case is used
Grenade Launcher: Crimped
Ball: FMJ; 8 g; Vo 752 m/s
Ball: JSP; 8 g; Vo 747 m/s
Ball: JSP, RN; 8 g; Vo 747 m/s
Subsonic Ball: FMJ; 11.8 g; Vo 296 m/s
HP Test M67 Type 1: FMJ; 8 g; Pressure 3,100 ±100 kg/cm²
HP Test M67 Type 2: FMJ; 9.05 g; Pressure 3,700 ±100 kg/cm²

UPDATED

7.62 mm NATO

Synonyms: 7.62 × 51 mm; .308 Winchester

Armament
All weapons chambered to NATO standard dimensions; notably FN-FAL, G3, M14, BM59 rifles, FN-MAG, L4, MG3, M60 machine guns.

Development
The 7.62 × 51 mm cartridge was devised in the early 1950s as a compromise between the full-sized 0.30-06 and a proposed British 7 mm round; it is little more than the 0.30-06 with a shortened case. As a result it falls between two stools, being less powerful than the rounds it replaced but too powerful to be a practical assault rifle round. Nonetheless, following its adoption by NATO in January 1954, it became widely distributed, having the cachet of NATO approval and remains in widespread use. Production has taken place at one time or another in more than 50 countries and is even manufactured in the CIS nations for competition shooting.

Description
The case is rimless and bottle-necked, brass or lacquered steel, Boxer or Berdan primed; cartridges manufactured to the relevant NATO specification will be marked with a cross-in-circle symbol on the head, forming part of the national pattern of headstamp. The standard ball bullet is lead-antimony cored in a steel jacket with gilding metal or copper envelope.

Data
US BALL M80
Round length: 69.85 mm
Case length: 51.05 mm
Rim diameter: 11.94 mm
Bullet diameter: 7.79 mm
Bullet weight: 9.65 g
Muzzle velocity: 854 m/s
Muzzle energy: 3,519 J

ABU DHABI
Manufacturer: Adcom Manufacturing
Type: Ball: FMJ; Lead core; 9.4 g; Vo 850 m/s

ARGENTINA
Manufacturer: Direccion General de Fabricaciones Militares
Type: Ball FMK 1 Mod 0: FMJ; 9.3 g; Vo 830 m/s
Tracer: FMJ; 8.93 g; Vo 810 m/s; red trace visible to 800 m minimum
AP FMK 2 Mod 0: FMJ; hardened steel core; 9.75 g; Vo 812 m/s
AP-I FMK 3 Mod 0: FMJ; steel core, with incendiary filling in the nose of the jacket; 9 g; Vo 830 m/s
Observing FMK 5 Mod 0: FMJ; steel core, explosive filling so as to produce a smoke ball upon impact. Trajectory matches that of the ball FMK 1 Mod 0; 9 g; Vo 830 m/s
Ball, Special Match, FMK 9 Mod 0: FMJ; 9.3 g; 848 m/s. This is a specially assembled and selected ball cartridge for sniper and competition use in bolt-action rifles
Blank FMK 8 Mod 0: Plastic case, steel extraction rim, case mouth plugged with wax
Blank, Star: Plastic case with extended nose to simulate bullet, steel extraction rim. Alternatively aluminium case with extended crimped nose
Blank FMK 14 Mod 0: Aluminium case, rose crimp
Grenade Launcher FMK 11 Mod 0: Brass case, rose crimp

AUSTRIA
Manufacturer: Hirtenberger AG
Type: Ball: FMJ; 9.45 g; Vo 837 m/s
Tracer: FMJ; base tracer; red trace to 1,000 m; 8.95 g; Vo 834 m/s
AP: FMJ, AP steel core; 9.45 g; Vo 837 m/s
Sniper Ball: FMJ; 9.45 g; Vo 837 m/s
Match Ball: FMJ; 10.9 g; Vo 750 m/s
Short-Range Ball: Alloy core with base cup and part jacket; 1.1 g; Vo 860 m/s
Frangible Short-Range Ball: Plastic core with CuZn base, 4.35 g; Vo 905 m/s. Specially designed to function in automatic weapons up to 600 rds/min
Hollow Point Test: FMJ; 9.45 g; Pressure ≥4,200 Bar
Blank: Steel case, crimped
Bulletted Blank: Brass case, wooden bullet coloured red
Grenade Launcher: Brass case, crimped

BRAZIL
Manufacturer: Companhia Brasileira de Cartuchos
Type: Match Ball: FMJ; 10.5 g; V_{25} 750 m/s
Ball: FMJ, SL; 9.33 g; V_{25}838 m/s
Tracer: FMJ; 8.75 g; Velocity adjusted during manufacture to obtain best match with ball trajectory; V_{25} 838 m/s
AP: FMJ; steel core; 9.55 g; V_{25} 838 m/s
AP-HC: FMJ; tungsten core; 8.3 g; V_{25} 840 m/s
Incendiary M77: FMJ; 9.4 g; V_{25} 808 m/s
Blank: Brass case, crimped

7.62 × 51 mm NATO

CHILE
Manufacturer: FAMAE
Type: Ball: FMJ; 9.5 g; Vo 810 m/s

COMMONWEALTH OF INDEPENDENT STATES
Manufacturer: LVE, Novosobirsk
Type: Competition Ball: FMJ

EGYPT
Manufacturer: Abou Kir Company for Engineering Industries
Type: Ball: FMJ; 8 g; Vo 833 m/s
Tracer: FMJ; rear trace to 770 m; Vo 827 m/s
Manufacturer: Shoubra Company
Type: Ball: FMJ; lead core; 9.45 g; V_{10} 837 m/s
Tracer: FMJ; red trace; 8.95 g; V_{10} 834 m/s
AP: FMJ; steel core; 9.45 g; V_{10} 837 m/s

FINLAND
Manufacturer: Lapua Cartridge Factory
Type: Ball S734: FMJ; 8 g; Vo 895 m/s
Ball FMJ123: FMJ; 8 g; Vo 895 m/s

Abridged Ballistic Table: 7.62 × 51 mm Ball M80

Range	Velocity	Energy	Drop	Elevation	Vertex
0 m	854 m/s	3,519 J	0 mm	0 mils	0 mm
100 m	778 m/s	2,920 J	79 mm	0.75 mils	20 mm
200 m	709 m/s	2,425 J	317 mm	1.60 mils	85 mm
300 m	642 m/s	1,988 J	659 mm	2.63 mils	211 mm
400 m	578 m/s	1,612 J	1,455 mm	3.77 mils	418 mm
500 m	518 m/s	1,295 J	2,456 mm	5.09 mils	730 mm

Ball B448: FMJ; 9.72 g; Vo 905 m/s
Ball GB422: FMJ; 10.85 g; Vo 780 m/s
Ball B436: FMJ; 11 g; Vo 795 m/s
Ball EB423: FMJ, Mega; 12 g; Vo 765 m/s
Ball E375: JSP; 12 g; Vo 765 m/s
Ball E415: FMJ, Mega; 12 g; Vo 765 m/s
Ball GB432: FMJ; 12 g; Vo 755 m/s
Precision Ball D46: FMJ, SL; 12 g; Vo 760 m/s
Precision Ball D47: FMJ, SL; cannelured, for use in automatic rifles; 12 g; Vo 760 m/s
Sniper Ball B406: 12.6 g
Silencer Ball S283: FMJ; 10.8 g
Silencer Ball B416: FMJ; 13 g
AP: FMJ; tungsten carbide core

Manufacturer: Sako Limited
Type: Ball: FMJ; 6 g; Vo 905 m/s
Ball: FMJ; 8 g; Vo 890 m/s
Ball: JSP; 8 g; Vo 925 m/s
Ball: JSP; 10.1 g; Vo 845 m/s
Ball: JHP; 11.7 g; Vo 795 m/s

FRANCE
Manufacturer: Giat Industries
Type: Ball SS77/1: FMJ; 9.3 g; Vo 850 m/s
Ball SS97: FMJ; 9.3 g; Vo 850 m/s
Balle O: FMJ; 9.55 g; Vo 815 m/s
Ball PR (Precision): FMJ; 9.73 g; 820 m/s
Tracer L78: FMJ; base tracer, red trace visible to 777 m; 8.93 g; Vo 850 m/s
Balle T (Tracer): FMJ; red trace; 8.93 g; Vo 820 m/s
AP P80: FMJ; steel core; 9.75 g; Vo 850 m/s
AP-I P86: FMJ; AP steel core, incendiary composition in tip; 9 g; Vo 850 m/s
Balle Plastique: 3.9 g; Vo 810 m/s
Practice Ball: 2.1 g; Vo 1,300 m/s
Short-Range Practice Ball: 2.12 g; Vo 1,100 m/s
Blank: Brass case, crimped

Manufacturer: SFM Défense
Type: Ball: FMJ; ? g; Vo 815 m/s
Tracer: FMJ; red trace; 9 g; Vo 820 m/s
PPI: 9.07 g; Vo 884 m/s. Penetration >10 mm RHA at 300 m
TR Reduced Charge: Plastic; ? g; Vo 675 m/s

GERMANY
Manufacturer: Dynamit Nobel (Geco)
Type: Ball DM111: FMJ; 9.45 g; Vo 850 m/s. Slntox, non-toxic
Tracer: FMJ; red trace; 8.95 g; Vo 850 m/s
AP: FMJ; hard core; 9.55 g; Vo 850 m/s
Match S Ball: FMJ; 12.3 g; Vo 650 m/s
Practice KB: This uses a special short-range bullet composed of a light alloy body with a plastic pointed nose cap. During firing, the nose cap is ejected from the body of the bullet and falls to the ground shortly after leaving the muzzle. The bullet body, now flat-nosed, is strongly retarded because of air resistance and therefore has a maximum range some 50 per cent of a service bullet. Bullet weight 5.55 g; Vo 870 m/s; maximum training range 150 m; maximum flight range 1,150 m
Practice PT: This is a one-piece plastic case/bullet with metal extraction rim. The 'bullet' portion is weakened so as to detach from the 'case' on firing and be ejected from the muzzle. The bullet has a maximum flight range of about 300 m and a maximum training range of 50 m. Bullet weight 0.7 g; Vo 1,100 m/s
Practice PT-T: Similar to the PT round but with a tracer element in the bullet. 1 g; Vo 1,100 m/s

Manufacturer: Metallwerk Elisenhütte GmbH
Type: Ball: FMJ; lead-antimony core, steel jacket, gilding metal envelope; 9.55 g; Vo 837±15 m/s. Single or double base propellant, Berdan primed, non-mercuric
Ball, DM111: FMJ; lead-antimony core, steel jacket, gilding metal envelope. Single or double base propellant. Boxer primed, using a lead- and barium-free composition. 9.55 g; Vo 827±15 m/s
Short-Range KB: This uses a plastic-cored bullet and is generally similar to the 5.56 mm KB round described previously. Bullet weight 3 g; Vo 1,150 m/s; point-blank range 330 m
Blank: Brass case, fluted crimp to represent bullet; 0.75 g single base powder; Berdan primed, non-mercuric. Crimp sealed with black lacquer, primer annulus with green lacquer

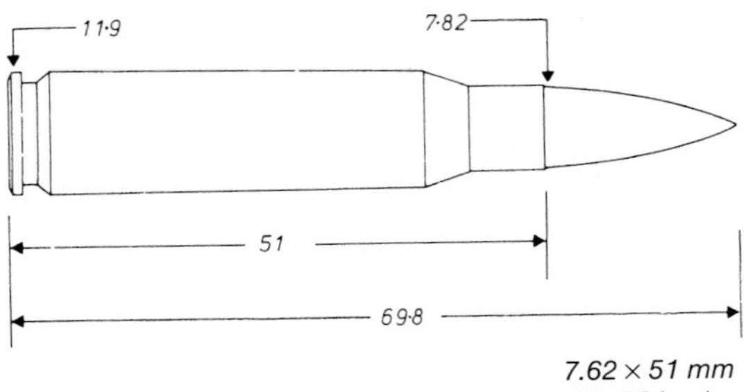

7.62 × 51 mm
Vo 854 m/s
Eo 3,276 J

Manufacturer: RWS
Type: Target: FMJ; 9.52 g; Vo 900 m/s
Cone Point: FMJ; 9.72 g; Vo 870 m/s
Ball: PSP; 9.72 g; Vo 868 m/s
Brenneke Ideal: 9.72 g; Vo 868 m/s
Match: FMJ; 10.87 g; Vo 800 m/s
Cone Point: FMJ; 10.69 g; Vo 820 m/s
H-Mantle: JHP; 11.66 g; Vo 780 m/s
Brenneke Universal: 11.66 g; Vo 777 m/s
Match: FMJ; 12.31 g; Vo 750 m/s

GREECE
Manufacturer: PYRKAL: Greek Powder & Cartridge Company
Type: Ball: FMJ; ? g; Vo 838 m/s
Tracer: FMJ; ? g; Vo 838 m/s
AP-HC: FMJ; steel core; no further details

INDIA
Manufacturer: Indian Ordnance Factory Varangaon
Type: Ball M80: FMJ, SL; 9.65 g; Vo 817±9 m/s
Tracer M62: FMJ; 9.2 g; dark ignition, trace 100 to 800 m; Vo 817±9 m/s

INDONESIA
Manufacturer: PT Pindad (Perseo)
Type: Ball MU-2TJ: FMJ; 9.55 g; Vo 837 m/s
Tracer MU-2N: FMJ; 8.95 g; Vo 837 m/s
Blank MU-2H: brass case; length 63.5 mm

IRAN
Manufacturer: Defence Industries Organisation, Ammunition Group
Type: Ball: cartridge weight 24 g; Vo 855 m/s

ISRAEL
Manufacturer: TAAS - Israel Industries Limited
Type: Ball M80: FMJ; 9.72 g; Vo 852 m/s
Tracer M62: FMJ; 9 g; Vo 837 m/s
AP M61: FMJ; steel core; 9.91 g; Vo 852 m/s
Hollow Point Test: FMJ; 9.72 g; pressure 4,746 kg/cm²
Grenade Launcher: Crimped; pressure 850 kg/cm²
Ball 30AM: FMJ, SL; 9.72 g; Vo 853 m/s
Ball 30CM: JSP, SL; 9.72 g; Vo 853 m/s
Ball 33CM: JSP, SL; 10.69 g; Vo 823 m/s
Ball 36CM: JSP, SL; 11.66 g; Vo 792 m/s

ITALY
Manufacturer: Europa Metalli
Type: Ball: FMJ; lead core; 9.6 g; Vo 780 m/s
Tracer: FMJ; lead core; dark ignition trace visible 40 to 775+ m; trajectory crosses ball at 550 m; 9.2 g; Vo 782 m/s
Short Range: Lead/brass jacket containing a plastic core; 5.7 g; V_{24} 810 m/s; maximum range 1,300 m
Plastic Training: CU/Zn sleeve containing a plastic core, exposed at the tip; 4.6 g; V_{24} 920 m/s. Maximum range 1,100 m
Blank; Brass case elongated and crimped to represent bullet.

KOREA, SOUTH
Manufacturer: Poongsan Metal Corporation
Type: Ball 308A: PSP; 9.72 g; Vo 806 m/s
Ball 308B: FMJ; 9.52 g; Vo 838 m/s
Ball 308C: PSP; 11.66 g; Vo 735 m/s

PAKISTAN
Manufacturer: Pakistan Ordnance Factories
Type: Ball L2A2: FMJ; ? g; Vo 810 m/s
Tracer L5A3: FMJ; red trace; ? g; Vo 795 m/s
Blank L10A2: Star crimp, brass case; 12.75 g
Blank, Rifle Grenade: Taper crimp, brass case; 13.95 g

PORTUGAL
Manufacturer: INDEP
Type: Ball: FMJ, SL; 9.3 g; Vo 837 m/s
Tracer: FMJ; semi-streamlined, red trace; 8.8 g; Vo 834 m/s
AP: FMJ; steel core; 9.22 g; Vo 845 m/s
Hollow Point Test: FMJ; pressure 4,600 kg/cm²
Blank: Brass case, rose or plain crimp
Grenade Launcher: Brass case; chamber pressure ≤750 kg/cm²

SINGAPORE
Manufacturer: Chartered Industries of Singapore
Type: Ball M80: Vo 838 m/s. Conforms to US MIL-SPEC MIL-C-46931D
Tracer M62: Vo 838 m/s. Conforms to US MIL-SPEC MIL-C-46281E
AP M61: Vo 838 m/s. Conforms to US MIL-SPEC MIL-C-60617A. Penetrates 6 mm chrome nickel plate (Brinell 450) at 100 m

SOUTH AFRICA
Manufacturer: Denel (Pty) Limited
Type: Ball: FMJ; 9.27 g; Vo 830 m/s
Tracer: FMJ; 9.1 g; Vo 813 m/s
AP: FMJ; steel core; ? g; Vo 830 m/s; penetration 100 RHA at 500 m

Manufacturer: Musgrave (Pty) Limited
Type: Ball: FMJ, SL; 9.3 g; Vo 840 m/s
Ball: PSP; 9.7 g; Vo 855 m/s
Ball: PSP; 10.89 g; Vo 780 m/s
Ball: PSP; 11.7 g; Vo 775 m/s

SPAIN
Manufacturer: SANTA BARBARA SA
Type: Ball: FMJ; ? g; Vo 815 m/s
Tracer: FMJ; ? g; Vo 820 m/s
AP: FMJ; ? g; Vo 867 m/s
Survival: Shot loading over wad; no external bullet; effective range 50 m
Hollow Point Test: FMJ; pressure 4,200 kg/cm²
Blank: Long or short crimp

SWEDEN
Manufacturer: Bofors Carl Gustaf
Type: Ball: FMJ; 9.5 g; Vo 840 m/s
Tracer: FMJ; 9 g; Vo 820 m/s; red trace; dark ignition; tracer range minimum 1,000 m
FFV AP: FMJ; hard core; 8.3 g; Vo 930 m/s; penetrates min 15 mm armour plate HB 300 at 250 m; to be type classified by US Army as M993
Blank STAR: brass case; length 51 mm

Manufacturer: Norma AB
Type: Ball 17623: JSP; 8.4 g; Vo 884 m/s
Ball 17624: JSP; 9.7 g; Vo 880 m/s
Ball 17679: FMJ; 10.9 g; Vo 777 m/s
Ball 17628: Plastic Point; 11.6 g; Vo 820 m/s
Ball 17635: JSP; 11.6 g; Vo 820 m/s
Ball 17636: JSP; 11.6 g; Vo 820 m/s
Ball 17660: JSP; 11.6 g; Vo 820 m/s
Ball 17681: JSP, SL; 11.6 g; Vo 750 m/s
Ball 17683: JSP; 13 g; Vo 750 m/s

TURKEY
Manufacturer: Makina ve Kimya Endüstrisi Kumuru (MKEK)
Type: Ball: FMJ; 9.3 g; Vo 850 m/s
Tracer: FMJ; red trace; 9 g; Vo 835 m/s
Blank: Brass case, star crimp

UNITED KINGDOM
Manufacturer: British Aerospace Defence Limited, Royal Ordnance Division

Type: Ball L2A2: FMJ; 9.33 g; Vo 838 m/s
Tracer L5A3: FMJ; red trace to 1,000 m; 8.75 g
Blank L13A1: Brass case, crimped nose

UNITED STATES OF AMERICA
Manufacturer: Glaser Safety Slug Inc
Type: Ball: Glaser Blue; 8.42 g; Vo 914 m/s

Manufacturer: Government contractors
Type: Ball M59: FMJ; non-streamlined, lead core; 7 g; Vo 838 m/s
Ball M80: FMJ, SL; lead core; 9.65 g; Vo 838 m/s
Ball, Duplex, M198: Front bullet 5.44 g; rear bullet 5.5 g; Vo 838 m/s (front); 670 m/s (rear)
Ball, frangible, M160: 7.06 g, formed of bakelite and powdered lead. Not to perforate 0.476 mm Dural at 25 m. Vo 402 m/s
Ball, Match, M118: FMJ; 11.37 g; Vo 777 m/s
Tracer M16: FMJ; semi-streamlined, lead tip filler, base tracer; 6.93 g
Tracer M27: FMJ; similar to M16 but dark ignition; 6.45 g
Tracer M62: FMJ; 9.2 g; dark ignition red trace 100 to 800 m; Vo 838 m/s
AP M61: FMJ; steel core; 9.75 g; Vo 838 m/s
Grenade M64: rose crimp; Vo 48 m/s with 700 g grenade

Manufacturer: Hansen Cartridge Company
Type: Ball: FMJ, SL; 9.7 g; Vo 852 m/s
Ball: JSP; 9.7 g; Vo 870 m/s
Ball: JSP; 11.7 g; Vo 780 m/s
Ball: FMJ; 8.1 g; Vo 923 m's

Manufacturer: Olin Ordnance
Type: Ball M80: FMJ; 9.52 g; V_{25}838 m/s
Tracer M62: FMJ; red trace; 9.2 g; V_{25} 838 m/s
Ball, SLAP, M948: 5.56 mm bullet in plastic sabot; 4.02 g; V_{25} 1,220 m/s
Tracer M959: 5.56 mm bullet in plastic sabot; red trace; 3.82 g; V_{25} 1,220 m/s
Dim Tracer XM276: FMJ; trace visible only through night vision devices; 9.07 g; V_{25} 838 m/s
Match Ball M852: FMJ; 10.89 g; V_{25} 777 m/s
Special Match Ball M118: FMJ; 11.15 g; V_{25} 777 m/s
Blank, M82: Brass case extended to simulate bullet; rosette crimp

Manufacturer: Remington Arms Company Inc
Type: Accelerator: APDS Ball; FMJ in plastic sabot; 3.6 g; Vo 1,243 m/s
Ball: PSP; 9.7 g; Vo 860 m/s
Ball: PSP; 11.7 g; Vo 798 m/s
Ball: SP; 11.7 g; Vo 798 m/s

YUGOSLAVIA (SERBIA AND MONTENEGRO)
Holding Cmpany: YUGOIMPORT - SDPR
Type: Ball: FMJ; 9.4 g; Vo 825 m/s
Tracer: FMJ; red trace, dark ignition, visible from 115 m to >830 m; 8.5 g; Vo 800 m/s
Blank: Steel case, crimped
Ball: FMJ; 9.4 g; Vo 865 m/s
Ball: JSP; 9.7 g; Vo 844 m/s
Ball: JSP; 11.7 g; Vo 738 m/s
HP Test Type 1: FMJ; 9.4 g; pressure 4,200±150 kg/cm²
HP Test Type 2: FMJ; 9.4 g; pressure 5,000±100 kg/cm²

UPDATED

7.62 mm Mosin-Nagant

Synonyms: 7.62 × 54R; 7.62 mm Soviet Rimmed; 7.62 mm obr 1891

Armament
All Tsarist Russian and Soviet Mosin-Nagant bolt-action rifles; SVD (Dragunov) sniper rifle; Maxim 1910, DP, DT, RP-46, SG-43, SGM and PK machine guns.

Development
This is the oldest cartridge still in first-line service, having been introduced into Russian service in 1891 with the Mosin-Nagant 'Three-Line' rifle. It was adopted, originally with a round-nose bullet, as the standard infantry cartridge for subsequent rifles and machine guns, and since it has superior long-range performance to the 7.62 × 39 mm cartridge, has been kept in use for machine guns and sniper rifles. It has also been adopted by other countries using Russian weapons, particularly China and Finland, but will be found anywhere where Soviet influence has distributed the appropriate weapons.

Description
The case is rimmed, bottle-necked and may be of brass- or copper-coated steel. The base is part-convex with a Berdan primer. A variety of bullets has been used over the years, but the current standards are the streamlined Ball LPS, steel-cored with a gilding metal jacket and the Heavy Ball D, streamlined and with a lead core in gilding metal jacket.

Recent developments for this calibre have included the 7N13, an armour penetrating round with the 9.41 g bullet having a hardened steel core. The 7N14 is a sniper round intended for use with the SVD rifle. This round uses a brass case and the bullet weight is 8.8 g.

Other developmental rounds in this calibre include the API 7-BZ-1M, the 7BM4 AP which uses a forged tempered steel core to pentrate 10 mm of armour plate at 200 m, the 7-BT-1 AP-T capable of penetrating 5 mm of steel plate at 700 m, an AP cartridge firing the heavy D-90 bullet, and the 7N1M AP intended for sniper applications capable of penetrating 5 mm of steel plate at 650 m.

Data
HEAVY BALL D
Round length: 77.16 mm
Case length: 53.6 mm
Rim diameter: 14.48 mm
Bullet diameter: 7.87 mm
Bullet length: 32.3 mm
Bullet weight: 11.98 g
Muzzle velocity: 818 m/s
Muzzle energy: 4,008 J

BULGARIA
Manufacturer: Kintex
Type: Ball M908/30: FMJ; steel core in lead jacket, cupro-nickel envelope; brass case; 9.6 g; Vo 825 m/s
Heavy Ball M1930: Lead core, steel jacket, gilding metal envelope; 12 g; Vo 810 m/s
Tracer T-46 Mod 1930: FMJ; lead core, steel jacket, gilding metal envelope; red trace to 1,000 m; 9.6 g; Vo 825 m/s
Blank: Steel case, brass-coated, rose crimp

CHINA, PEOPLE'S REPUBLIC
Manufacturer: China North Industries (NORINCO)
Type: Ball Type 53: FMJ; 9.6 g; Vo 840 m/s
Ball Type 53: FMJ; steel core; 9.6 g; Vo 820 m/s
AP-I Type 53: FMJ; AP steel core; 9.95 g; Vo 840 m/s
Tracer Type 53: FMJ; 9.66 g; Vo 800 m/s
Incendiary-Ranging Type 53: FMJ; 10 g; Vo 800 m/s

COMMONWEALTH OF INDEPENDENT STATES
Manufacturer: LVE Novosibirsk
Type: Ball CT: FMJ; hardened steel core; 9.75 g; Vo 835 m/s
Ball Extra: FMJ; 13 g; Vo 740 m/s; for competition shooting
Tracer T-46: FMJ; 9.9 g; Vo 805 m/s
Tracer T-46M: As T-46 but start of tracer delayed until 150 m from muzzle
Sniper Ball: FMJ; lead core; 9.9 g; Vo 835 m/s
Blank: cartridge weight 11.15 g; length 53.72 mm; rose crimp
API 7-BZ-1M: FMJ; in development
AP 7BM4: Can penetrate 10 mm of armour plate at 200 m
AP-T: Uses U-12A subcalibre bullet capable of penetrating 5 mm steel plate at 700 m; entering production

Manufacturer: State arsenals
Type: Ball, Heavy, D: FMJ; 11.98 g; Vo 818 m/s
Ball, Light, LPS: FMJ; 9.65 g; Vo 870 m/s
Tracer T46: FMJ; 9.65 g; non-streamlined; Vo 870 m/s
Tracer T: FMJ; 9.59 g; dark ignition, tracing to 2,000 m
AP B30: FMJ; SL, hard steel core, lead sleeve, steel jacket, gilding metal envelope; 11.92 g; penetration 10 mm at 700 m
AP-I B32: FMJ; SL, steel core with incendiary composition in the nose; 10.04 g; Vo 870 m/s
AP-I BS40: FMJ; non-streamlined, tungsten carbide core with incendiary pellet in the nose; 12.11 g; Vo 805 m/s
AP-I-T BZT: FMJ; non-streamlined, steel core with incendiary composition in nose and tracer in base; 9.2 g
Incendiary-Obs ZP: FMJ; non-streamlined, steel jacket with incendiary pellet in nose, explosive primer beneath and an inertia striker; 10.36 g

EGYPT
Manufacturer: Abou Kir Company for Engineering Industries
Type: Ball: FMJ; 8 g; Vo 780 m/s
Tracer: FMJ, rear tracer to 1.50 seconds. Vo 800 m/s
Blank: Polyethylene dummy bullet

7.62 × 54R Mosin-Nagant

Manufacturer: Maasara Company for Engineering Industries
Type: Ball: FMJ; 8 g; Vo 780 m/s
Tracer: FMJ, rear tracer to 1.50 seconds. Vo 800 m/s
Blank: Polyethylene dummy bullet

FINLAND
Manufacturer: Lapua Cartridge Factory
Ball FMJ123: FMJ; 8 g; Vo 895 m/s
Ball D46: FMJ, SL; 12 g; Vo 780 m/s
Ball E415: FMJ; 12 g; Vo 765 m/s
Ball E375: JSP; 12 g; Vo 765 m/s
Ball D47: FMJ, SL; 12 g; Vo 780 m/s

Manufacturer: Sako Limited
Type: Ball: FMJ; 6 g; Vo 905 m/s
Ball: FMJ; 8 g; Vo 890 m/s
Ball: JSP; 10.1 g; Vo 845 m/s
Ball: JHP; 11.7 g; Vo 795 m/s

IRAN
Manufacturer: Defence Industries Organisation, Ammunition Group
Ball: FMJ; cartridge weight 28 g; Vo 835 m/s

POLAND
Manufacturer: Mesko Zaklady Metalowe
Type: Ball: FMJ; 9.6 g; Vo 827 m/s
Tracer: FMJ; red trace; 9.6 g; Vo 800 m/s
Blank: Brass case, star crimp

ROMANIA
Manufacturer: Romtehnica
Type: Ball: FMJ; steel core bullet; 9.45-9.75 g; Vo 835 m/s

Tracer: FMJ; red trace for 4 s; 9.5-9.9 g; Vo 815 m/s
Blank: Cartridge weight 10.42 g; length 53.72 mm; rose crimp

SWEDEN
Manufacturer: Norma AB
Type: Ball 17634: JSP; 11.6 g; Vo 920 m/s

UNITED STATES OF AMERICA
Manufacturer: Hansen Cartridge Company
Type: Ball: FMJ, SL; 11.8 g; Vo 810 m/s
Ball: JSP; 11.7 g; Vo 805 m/s

YUGOSLAVIA (SERBIA AND MONTENEGRO)
Holding company: YUGOIMPORT - SDPR
Type: Ball CJ M87: FMJ; steel core; 9.6 g; Vo 830 m/s
Light Ball M908: FMJ; 9.7 g; lead-antimony core in steel jacket, gilding metal envelope; Vo 845 m/s
Heavy Ball M30J: FMJ; 11.8 g; lead-antimony core; Vo 785 m/s
Tracer T46: 9.65 g; lead-antimony half-core with tracer element encapsulated behind; red trace to 1,000 m; Vo 815 m/s
AP-I B32: 10.4 g; steel core in gilding metal jacket, with incendiary/tracer capsule in base; Vo 815 m/s

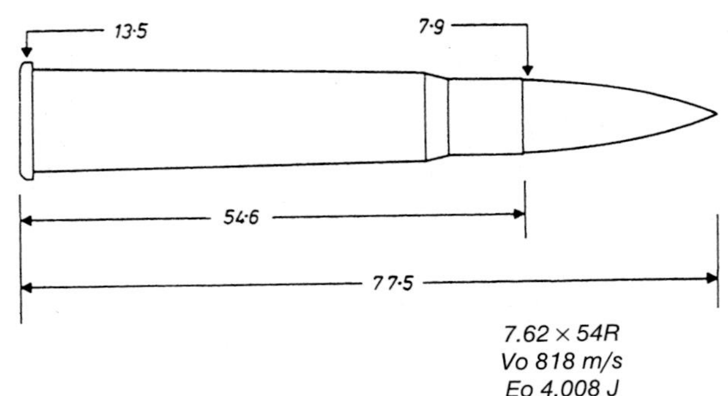

7.62 × 54R
Vo 818 m/s
Eo 4,008 J

Practice M76DT: 2.3 g; bullet having an aluminium core in a short gilding metal envelope so that the point of the core is exposed. The point is sharply tapered, and the combination of light weight and shape gives a maximum range of only 530 m. The standard brass case is used. Restricted to use in machine guns SGMT and PKT only. Vo 770 m/s
Blank M69: Steel case, crimped
Ball: FMJ; 9.6 g; Vo 854 m/s
Ball: FMJ; 9.7 g; Vo 869 m/s
Ball: FMJ; 11.8 g; Vo 804 m/s
Ball: JSP; 11.7 g; Vo 804 m/s
HP Test M30J Type 1: FMJ; 11.8 g; pressure 3,350 kg/cm²
HP Test M30J Type 2: FMJ; 11.8 g; pressure 4,050 kg/cm²

UPDATED

0.30-06 Springfield

Synonyms: 7.62 × 63 mm; 0.30 US Service; 0.30 Browning

Armament

US Springfield M1903 and Enfield M1917 bolt-action rifles; US M1 Garand automatic rifle; Browning M1917, M1919 machine guns; Marlin, Lewis, Savage-Lewis, Chauchat machine guns; Browning Automatic Rifle (BAR).

Development

The 0.30-06 first appeared in 1906, having a pointed bullet to replace the earlier round-nose 0.30-03 cartridge as the service round for the M1903 Springfield rifle. The original bullet was a 9.72 g flat-based type and complaints of lack of machine gun range during the First World War led to the standardisation of the streamlined 11.2 g M1 bullet in 1926. By 1936 complaints had arisen of the excessive danger space required for training with this bullet, and also of malfunctions in the new M1 Garand automatic rifle. This led to the non-streamlined 9.72 g M2 bullet being standardised in 1938. The M2 has remained the standard ball ever since.

Description

The case is rimless, bottle-necked, Berdan-primed and of brass; steel-cased rounds were made from 1942 to 1944 but were never very successful and were never issued for general service. The M2 bullet has a lead core with steel jacket clad in gilding metal, with a cannelure in the parallel section.

Data

Round length: 84.8 mm
Case length: 63.2 mm
Rim diameter: 12 mm
Bullet diameter: 7.82 mm
Bullet weight: 9.72 g

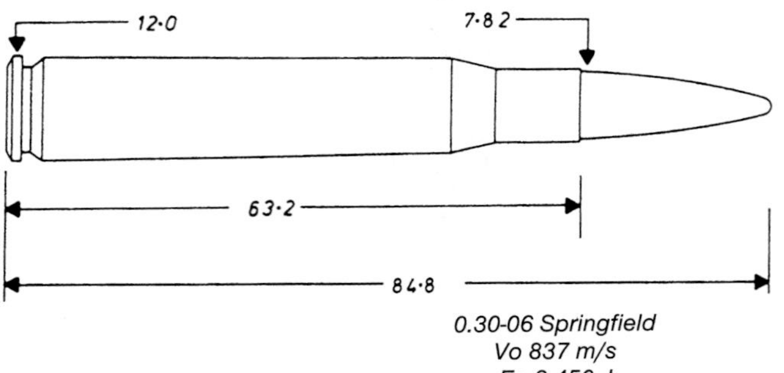

0.30-06 Springfield
Vo 837 m/s
Eo 3,450 J

0.30-06 Remington Accelerator cartridge

0.30-06 Springfield

Abridged Ballistic Table: 0.30-06

Range	Velocity	Energy	Drop
0 m	887 m/s	3,810 J	0 mm
100 m	798 m/s	3,082 J	23 mm
200 m	714 m/s	2,469 J	79 mm
300 m	635 m/s	1,953 J	327 mm
400 m	562 m/s	1,528 J	688 mm
500 m	494 m/s	1,183 J	1,371 mm

AUSTRIA
Manufacturer: Hirtenberger AG
Type: Ball: FMJ; 9.94 g; Vo 830 m/s
Tracer: FMJ; base tracer, red trace to 1,000 m minimum; 9.3 g; Vo 825 m/s

BRAZIL
Manufacturer: Companhia Brasileira de Cartuchos
Type: Ball M2: FMJ; 9.75 g; V_{25} 835 m/s
Tracer M1: FMJ; 9.34 g; V_{25} 812 m/s
AP M2: FMJ; steel core; 10.72 g; V_{25} 730 m/s
Incendiary M1: FMJ; 9.75 g; V_{25} 770 m/s
Blank M1909: Star crimp
Grenade launcher: Star crimp, to deliver Vo 46 m/s with a 560 g grenade

FINLAND
Manufacturer: Lapua Cartridge Factory
Type: Ball S374: FMJ; 8 g; Vo 895 m/s
Ball FMJ123: FMJ; 8 g; Vo 895 m/s
Ball D46: FMJ, SL; 12 g; Vo 810 m/s
Ball E415: FMJ, Mega; 12 g; Vo 800 m/s
Ball EB423: JSP; 12 g; Vo 786 m/s
Ball GB432: FMJ; 12 g; Vo 755 m/s
Ball E401: FMJ, Mega; 13 g; Vo 775 m/s

Manufacturer: Sako Limited
Type: Ball: FMJ; 8 g; Vo 890 m/s
Ball: JSP; 8 g; Vo 950 m/s
Ball: JSP; 10.1 g; Vo 880 m/s
Ball: JHP; 11.7 g; Vo 825 m/s

FRANCE
Manufacturer: SFM Défense
Type: Ball: FMJ; 11.7 g; Vo 820 m/s
Tracer: FMJ; red trace; 11.7 g; Vo 820 m/s

GERMANY
Manufacturer: RWS
Type: Target: FMJ; 9.53 g; Vo 900 m/s
Cone Point: FMJ; 9.72 g; Vo 911 m/s
Ball: PSP; 9.72 g; Vo 911 m/s

Brenneke Ideal: 9.72 g; Vo 911 m/s
Cone Point: FMJ; 10.69 g; Vo 870 m/s
H-Mantle: JHP; 11.66 g; Vo 841 m/s
Standard: JSP; 11.66 g; Vo 840 m/s
Brenneke Universal: 11.66 g; Vo 840 m/s

GREECE
Manufacturer: PYRKAL: Greek Powder & Cartridge Company
Type: Ball M2: FMJ; 9.7 g; Vo 850 m/s
Tracer: FMJ; red trace for 3 seconds minimum; 9.1 g; Vo 850 m/s

HUNGARY
Manufacturer: Mátravidéki Fémmüvek
Type: Ball: FMJ; 9.2 g; Vo 920 m/s
Ball: PSP; 11.7 g; Vo 825 m/s
Ball: SPRN; 9.7 g; Vo 880 m/s
Ball: SPRN; 11.7 g; Vo 825 m/s

ISRAEL
Manufacturer: TAAS - Israel Industries Limited
Type: Ball 30AM: FMJ, SL; 9.72 g; Vo 861 m/s
Ball 30CM: JSP, SL; 9.72 g; Vo 887 m/s
Ball 33CM: JSP, SL; 10.69 g; Vo 853 m/s
Ball 36CM: JSP, SL; 11.66 g; Vo 823 m/s

KOREA, SOUTH
Manufacturer: Poongsan Metal Corporation
Type: Ball 3006A: PSP; 9.72 g; Vo 845 m/s
Ball 3006B: PSP; 11.66 g; Vo 777 m/s
Ball 3006C: FMJ; 9.72 g; Vo 845 m/s

SOUTH AFRICA
Manufacturer: Musgrave (Pty) Limited
Type: Ball: PSP; 9.7 g; Vo 860 m/s
Ball: PSP; 10.89 g; Vo 830 m/s
Ball: PSP; 11.7 g; Vo 800 m/s

SWEDEN
Manufacturer: Norma AB
Type: Ball 17640: JSP; 8.4 g; Vo 977 m/s
Ball 17643: JSP; 9.7 g; Vo 910 m/s
Ball 17648: JSP; 11.6 g; Vo 850 m/s
Ball 17649: JSP, Nosler; 11.6 g; Vo 850 m/s
Ball 17653: Plastic Point; 11.6 g; Vo 850 m/s
Ball 17659: JSP; 11.6 g; Vo 850 m/s
Ball 17682: JSP; 11.6 g; Vo 750 m/s
Ball 17684: JSP; 13 g; Vo 805 m/s

TURKEY
Manufacturer: Makina ve Kimya Endüstrisi Kumuru (MKEK)
Type: Ball M2: FMJ; 9.9 g; Vo 835 m/s
Tracer M25: FMJ; dark ignition, red trace; 9.45 g; 815 m/s
Blank M1909: Star crimp, brass case

UNITED STATES OF AMERICA
Manufacturer: Glaser Safety Slug Inc
Type: Ball: Glaser Blue; 8.42 g; Vo 945 m/s

Manufacturer: Government contractors
Type: Ball M2: FMJ; 9.85 g; Vo 837 m/s
AP M2: FMJ; steel core; 10.69 g; Vo 829 m/s
AP-I M14: FMJ; steel core; 9.72 g; Vo 849 m/s
Incendiary M1: FMJ; 9.07 g; Vo 901 m/s
Tracer M25: FMJ; dark ignition red trace 70 to 850 m; 9.46 g; Vo 814 m/s

Manufacturer: Remington Arms Company Inc
Type: Accelerator: APDS Ball; FMJ in plastic sabot; 3.6 g; Vo 1,243 m/s
Ball: PSP; 8.1 g; Vo 957 m/s
Ball: Bronze Point; 9.7 g; Vo 886 m/s
Ball: Bronze Point; 9.7 g; Vo 886 m/s
Ball: PSP; 10.7 g; Vo 853 m/s
Ball: SP; 11.7 g; Vo 822 m/s
Ball: PSP; 11.7 g; Vo 822 m/s
Ball: Bronze Point; 11.7 g; Vo 822 m/s
Ball: SP; 14.3 g; Vo 735 m/s

YUGOSLAVIA (SERBIA AND MONTENEGRO)
Holding company: YUGOIMPORT - SDPR
Type: Ball: FMJ; 9.7 g; Vo 889 m/s
Tracer: FMJ; red trace, dark ignition, visible from 114 m to >823 m; 9.3 g; Vo 812 m/s
AP-I: FMJ; AP steel core, thermite incendiary composition in nose; 9.6 g; Vo 838 m/s
Ball: JSP; 9.7 g; Vo 910 m/s
Ball: JSP; 11.7 g; Vo 820 m/s

UPDATED

7.62 × 42 mm SP-4

Armament
PSS pistol, NRS scouting knife.

Development
As far as can be discovered the 7.62 × 42 mm SP-4 cartridge is used with only two weapons, the PSS Vul silent pistol developed by Viktor Levchenko and the NRS scouting knife. The round is one of a series of special cartridges, a close relative in development terms being the 7.62 × 62.6 mm SP-3 used with the twin-barrel MSP silent pistol. The SP-3 resembles a conventional (but elongated) cartridge, which the SP-4 does not, but the firing signature is concealed by an internal piston arrangement. By comparison, the SP-4 round has its projectile fully concealed within the cartridge case.

Description
The 7.62 × 42 mm SP-4 cartridge has its bullet fully concealed within a rimless cartridge case. The brass case has thicker walls than a conventional cartridge case while the mouth is necked. The bullet is a blunt-nosed cylindrical alloy steel slug with a copper driving band on the lead edge. It is supported at its base by a dimpled steel disc which acts as a piston. The raised dimple in the disc mates with a recess in the base of the bullet to support and stabilise the bullet as the cartridge is fired. As the cartridge is fired the disc/piston contains the firing noise and flash and retains them within the case as the disc strikes the case neck so the external firing signature is minimal.

The 7.62 × 42 mm SP-4 cartridge has a maximum effective range of 50 m and will penetrate body armour and steel helmets at 20 m and still retain its lethal potential once through. The bullet will penetrate 2 mm of steel plate at 25 m. Muzzle velocity is 200 m/s.

Data
Round length: 42 mm
Bullet weight: 10 g
Cartridge weight: 24 g
Muzzle velocity: 200 m/s

COMMONWEALTH OF INDEPENDENT STATES
Agency: Rosvoorouzhenie
Type: Silent Ball: Steel; 10 g; Vo 200 m/s

NEW ENTRY

0.300 Winchester Magnum

Synonyms: none

Armament
Suitably chambered bolt-action rifles.

Development
This cartridge was introduced in 1963 for a Winchester bolt-action hunting rifle, after which it was used by several American and European rifle makers. It proved to be a very accurate and consistent round with a flat trajectory and because of this has been adopted by a few military and security forces as a sniper round.

Description
The case is belted and bottle-necked, brass, Boxer primed. The belt acts as a positive chambering stop and also reinforces the case against internal pressure. The usual hunting bullet is a pointed soft point, but full metal jacketed bullets have been adopted for military use.

Data
Round length: 83.82 mm
Case length: 66.55 mm
Rim diameter: 13.51 mm
Belt diameter: 13.51 mm
Bullet diameter: 7.82 mm
Bullet weight: 9.72 g
Muzzle velocity: 1,036 m/s
Muzzle energy: 5,216 J

AUSTRIA
Manufacturer: Hirtenberger AG
Type: Ball: FMJ; 10.7 g; Vo 910 m/s

0.300 Winchester Magnum

Ball: PSP; 9.7 g; Vo 975 m/s
Ball: PSP; 11.6 g; Vo 910 m/s

FINLAND
Manufacturer: Lapua Cartridge Factory
Type: Ball: MIRA; 12 g; Vo 895 m/s
Sniper Ball: FMJ, SL; 12.6 g; Vo 870 m/s
Ball: PSP; 11.6 g; Vo 910 m/s
Ball: MEGA; 13 g; Vo 830 m/s

GERMANY
Manufacturer: RWS
Type: Ball: FMJ; 10.7 g; Vo 970 m/s
Ball: FMJ; 10.9 g; Vo 980 m/s
Ball: Brenneke; 11.7 g; Vo 940 m/s

Manufacturer: Metallwerk Elisenhütte GmbH
Type: Ball SF: JHP bullet entirely of gilding metal, the hollow point being plugged and the exterior of the ogive formed in spiral grooves; 10 g; brass case; Boxer non-corrosive primed, single or double base propellant; Vo 950±15 m/s in 600 mm barrel. Accuracy 15 shots in 3.50 cm circle at 100 m

KOREA, SOUTH
Manufacturer: Poongsan Metal Corporation
Type: Ball: PSP; 9.7 g; Vo 961 m/s
Ball: PSP; 11.7 g; Vo 870 m/s

SOUTH AFRICA
Manufacturer: Musgrave (Pty) Limited
Type: Ball: PSP; 11.6 g; Vo 880 m/s
Ball: PSP; 14.3 g; Vo 800 m/s

SWEDEN
Manufacturer: Norma AB
Type: Ball: PSP; 11.6 g; Vo 920 m/s

UNITED STATES OF AMERICA
Manufacturer: A-Square Company Inc
Type: Ball: FMJ; 11.7 g; Vo 933 m/s

Manufacturer: Federal Cartridge Company
Type: Ball: PSP; 11.6 g; Vo 914 m/s
Ball: PSP; 13 g; Vo 863 m/s
Ball: PSP; 11.6 g; Vo 902 m/s

Manufacturer: Remington Arms Company Inc
Type: Ball: PSP; 9.72 g; Vo 1,003 m/s
Ball: PSP; 11.66 g; Vo 914 m/s

Manufacturer: Winchester-Olin
Type: Ball: FMJ; 12.3 g; Vo 879 m/s
Ball: PSP; 9.7 g; Vo 1,003 m/s
Ball: PSP; 11.7 g; Vo 914 m/s
Ball: PSP; 14.3 g; Vo 817 m/s

YUGOSLAVIA (SERBIA AND MONTENEGRO)
Holding company: YUGOIMPORT - SDPR
Type: Ball: PSP; 9.7 g; Vo 1,005 m/s
Ball: PSP; 11.7 g; Vo 915 m/s

NEW ENTRY

7.65 mm Type 64

Synonyms: 7.65 × 17 mm

Armament

Used only in Chinese Type 64 and Type 67 silenced pistols and Type 85 standard and silenced sub-machine guns.

Development

Nothing is known of the development history of this cartridge, which was first known to Western agencies in the 1960s after its use in Vietnam.

Description

The round generally resembles the 7.65 mm Browning but is a true rimless round rather than semi-rimmed and is not therefore, interchangeable. The case is brass, straight tapered and Berdan primed. The bullet is a conventional jacketed round-nose type but the propelling charge is reduced so as to deliver the optimum combination of silence and performance.

Data

Round length: 25 mm
Case length: 17 mm
Rim diameter: 8.42 mm
Bullet diameter: 7.8 mm
Bullet weight: 4.8 g
Muzzle velocity: 274 m/s
Muzzle energy: 180 J

7.65 mm Type 64

CHINA, PEOPLE'S REPUBLIC
Manufacturer: China North Industries (NORINCO)
Type: Ball Type 64: see above

NEW ENTRY

7.65 mm Browning

Synonyms: 7.65 × 17SR; 7.65 mm ACP; 0.32 ACP

Armament

This cartridge will function in any automatic pistol marked '7.65 mm Auto', '0.32 Auto', '0.32 ACP', or '7.65 mm Browning' and also in a number of cheap European revolvers of pre-1939 manufacture which are similarly marked. It is also used in the Czech 'Skorpion' sub-machine gun, for which an armour-piercing bullet was developed.

7.65 mm Browning

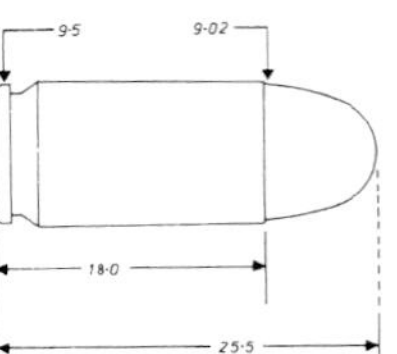

7.65 mm ACP
Vo 300 m/s
Eo 216 J

Development

The 7.65 mm cartridge was developed in the late 1890s by Fabrique National Herstal and John Browning for use in the Browning 1900 automatic pistol, after which it became the most widely used pocket pistol cartridge in the world. It has been estimated that about 65 per cent of the automatic pistols made this century have been chambered for this round.

In military service it has generally been confined to use in staff officers' and second-line troops' pistols, although during both World Wars the German and French armies issued them widely to combat troops in default of sufficient heavier calibres. Its greatest formal application has been as a police and security force weapon in Europe. Being semi-rimmed it has also been used in commercial revolvers.

Description

The case is semi-rimmed, that is, the extraction rim is slightly larger than the body of the case, sufficiently large to provide a positioning stop in the chamber but not large enough to interfere with feeding from a magazine. Cases of brass, steel and light alloy can be found, although brass is the more common. As with any commercial cartridge with such a long history, innumerable different bullets have appeared but the general police/military standard is a round-nosed full jacketed bullet with lead core of about 4.7 g.

Data

Round length: 26.16 mm
Case length: 17.27 mm
Rim diameter: 8.99 mm
Bullet diameter: 7.85 mm
Bullet weight: 4.7 g
Muzzle velocity: 300 m/s
Muzzle energy: 212 J

AUSTRIA
Manufacturer: Hirtenberger AG
Type: Ball: FMJ; 4.6 g; Vo 300 m/s

BRAZIL
Manufacturer: Companhia Brasileira de Cartuchos
Type: Ball: FMJ; 4.6 g; Vo 276 m/s

CHILE
Manufacturer: FAMAE
Type: Ball: 4.75 g; Vo 270 m/s

CZECH REPUBLIC
Manufacturer: Sellier & Bellot
Type: Ball: FMJ; 4.75 g; Vo 318 m/s
MP: FMJ; steel core; 4.75 g; Vo 318 m/s

FINLAND
Manufacturer: Lapua Cartridge Factory
Type: Ball R429: FMJ; 4.8 g; Vo 274 m/s

Manufacturer: Sako Limited
Type: Ball: FMJ; 4.7 g; Vo 305 m/s

FRANCE
Manufacturer: SFM Défense
Type: Ball: FMJ; 4.5 g; Vo 300 m/s
Ball: JSP; 4.5 g; Vo 260 m/s
Ball: THV; 1.65 g; Vo 780 m/s

GERMANY
Manufacturer: Geco
Type: Ball: FMJ; 4.7 g; Vo 330 m/s
Ball: Plastic; 0.2 g; Vo 520 m/s

HUNGARY
Manufacturer: Mátravidéki Fémmüvek
Type: Ball: FMJ; steel jacket; 4.65 g; Vo 300 m/s
Ball: FMJ; Tombak jacket; 4.65 g; Vo 300 m/s

ITALY
Manufacturer: Fiocchi Munizioni SpA
Type: Ball: FMJ; 4.75 g; Vo 290 m/s
Ball: JSP; 4.85 g; Vo 275
Ball: JHP; 3.9 g; Vo 380 m/s

KOREA, SOUTH
Manufacturer: Poongsan Metal Corporation
Type: Ball 32A: FMJ; 4.6 g; Vo 265 m/s

MEXICO
Manufacturer: Aguila Industrias Tecnos
Type: Ball: FMJ; 4.6 g; Vo 274 m/s

SOUTH AFRICA
Manufacturer: Musgrave (Pty) Limited
Type: Ball: FMJ; 4.8 g; Vo 300 m/s

SPAIN
Manufacturer: SANTA BARBARA SA
Type: Ball: FMJ; 4.6 g; Vo 310 m/s
Ball: JSP; 4.4 g; Vo 310 m/s

SWEDEN
Manufacturer: Norma AB
Type: Ball: FMJ; 5 g; Vo 274 m/s

TURKEY
Manufacturer: Makina ve Kimya Endüstrisi Kumuru
Type: Ball: FMJ; 5 g; 275 m/s

UNITED STATES OF AMERICA
Manufacturer: CCI
Type: Ball: FMJ; 4.6 g; Vo 276 m/s

Manufacturer: Federal Cartridge Company
Type: Ball: FMJ; 4.6 g; Vo 293 m/s

Manufacturer: Glaser Safety Slug Inc
Type: Ball: Glaser Blue; 3.56 g; Vo 396 m/s

Manufacturer: Remington Arms Company Inc
Type: Ball: FMJ; 4.6 g; Vo 275 m/s

Manufacturer: Winchester-Olin
Type: Ball: FMJ; 4.6 g; Vo 276 m/s
Ball: JHP; 3.9 g; Vo 296 m/s

YUGOSLAVIA (SERBIA AND MONTENEGRO)
Holding company: YUGOIMPORT - SDPR
Type: Ball: FMJ; 4.6 g; Vo 290 m/s
HP Test: FMJ; 4.6 g. Pressure 2,600 kg/cm^2

UPDATED

7.65 mm Parabellum

Synonyms: 7.65 × 21 mm; 7.65 mm Luger

Armament
Parabellum (Luger) pistols of 7.65 mm calibre; some Beretta, SIG, Walther and Bernardelli pistols; small numbers of early Neuhausen and Solothurn sub-machine guns.

Development
The 7.65 mm Parabellum is a shortened version of the 7.63 mm Mauser cartridge developed by Georg Luger for the original 1900 Parabellum pistol. For many years it was the standard pistol cartridge of Switzerland, Finland, Latvia, Portugal, Brazil and other countries. Although not currently used by any army, it can still be found in use with some police and security forces. Pistols chambered for this round can be made to order.

Description
The brass case is rimless and bottle-necked, Berdan primed. Many commercial bullet types are to be found, and the original military bullet was a cylindro-conoidal flat-tipped jacketed type. The current military standard is a jacketed round-nose bullet of 6.02 g weight.

Data
Round length: 29.21 mm
Case length: 21.6 mm
Rim diameter: 9.93 mm
Bullet diameter: 7.82 mm

7.65 mm Parabellum

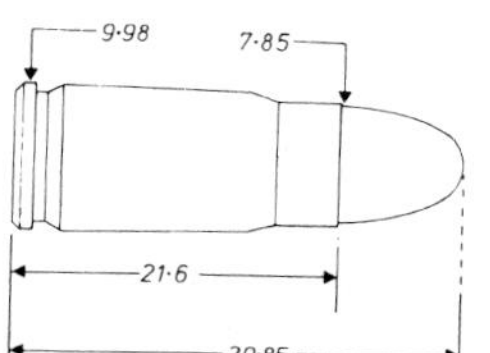

7.65 × 21 mm
Vo 388 m/s
Eo 407 J

Bullet weight: 6.02 g
Muzzle velocity: 368 m/s
Muzzle energy: 408 J

FINLAND
Manufacturer: Lapua Cartridge Factory
Type: Ball 4317530: FMJ; 6 g; Vo 405 m/s

Manufacturer: Sako Limited
Type: Ball: FMJ; 6 g; Vo 390 m/s

GERMANY
Manufacturer: Geco
Type: Ball: FMJ; 6 g; Vo 385 m/s

ITALY
Manufacturer: Fiocchi Munizioni SpA
Type: Ball: FMJ; 6 g; Vo 360 m/s
Ball: JSP; 5.9 g; Vo 350 m/s

PORTUGAL
Manufacturer: INDEP
Type: Ball: FMJ; 6 g; Vo 350 m/s

SOUTH AFRICA
Manufacturer: Musgrave (Pty) Limited
Type: Ball: FMJ; 4.86 g; Vo 300 m/s

SWEDEN
Manufacturer: Norma AB
Type: Ball: FMJ; 6 g; Vo 375 m/s

UNITED STATES OF AMERICA
Manufacturer: Winchester-Olin
Type: Ball: FMJ; 6 g; Vo 372 m/s

NEW ENTRY

0.303 British

Synonyms: 7.7 × 56R; 0.303 Lee-Enfield

Armament
All British service Lee-Enfield rifles, Vickers, Maxim, Lewis, Savage-Lewis, Bren, Vickers-Berthier machine guns, modified Browning M1919 and M2 machine guns, some Madsen and Hotchkiss machine guns.

Development
One of the earliest small calibre rifle rounds, this appeared in 1889 and remained the standard British service round until replaced by the 7.62 mm NATO cartridge in the 1960s. A powerful round, its sole defect was its rim, which complicated the design of automatic weapons and demanded care in loading magazines. Innumerable variations in cartridge and bullet design took place during its life, but the final standard ball rounds were the Mks 7, 7Z, 8 and 8Z.

Description
A rimmed, bottle-necked brass case with Berdan primer. The bullet was a compound design using a core of part lead and part aluminium in order to obtain the correct centre of mass, in a steel jacket clad with copper or nickel. The round was most usually loaded with Cordite propellant, but nitro-cellulose propellants gradually appeared; these all have the Mark number followed by the letter Z. The propellant makes no difference to the shooting, but is indicated for storage and accounting purposes.

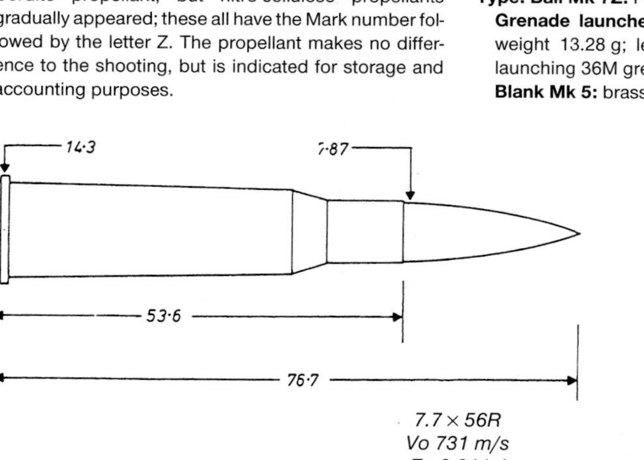

7.7 × 56R
Vo 731 m/s
Eo 3,011 J

Data
BALL MK 7
Round length: 77.47 mm
Case length: 54.61 mm
Rim diameter: 13.46 mm
Bullet diameter: 7.9 mm
Bullet weight: 11.27 g
Muzzle velocity: 731 m/s
Muzzle energy: 3,011 J

CANADA
Manufacturer: SNC Industrial Technologies Inc
Type: Mark 8 CDN Ball: FMJ; 11.4 g; Vo 730 m/s

EGYPT
Manufacturer: Abou Kir Company for Engineering Industries
Type: Ball: FMJ, non-streamlined; Vo 720 m/s

GREECE
Manufacturer: PYRKAL: Greek Powder & Cartridge Company
Type: Ball: FMJ; 11.3 g; Vo 732 m/s

INDIA
Manufacturer: Ordnance Factory Khamaria and Ammunition Factory Khadki Pune
Type: Ball Mk 7Z: FMJ; Vo 730 m/s
Grenade launcher Ballistite M Mk 1Z: Cartridge weight 13.28 g; length 54.81 mm; brass case; for launching 36M grenades
Blank Mk 5: brass case with crimped mouth

PORTUGAL
Manufacturer: INDEP
Type: Ball: FMJ; 11.3 g; Vo 762 m/s
Blank: Brass case, crimped; chamber pressure ≤1,000 kg/cm²

SOUTH AFRICA
Manufacturer: Musgrave (Pty) Limited
Type: Ball: FMJ, SL; 11.3 g; Vo 747 m/s
Ball: PSP; 9.7 g; Vo 825 m/s
Ball: PSP; 11.3 g; Vo 750 m/s

SWEDEN
Manufacturer: Norma AB
Type: Ball 17712: JSP; 9.7 g; Vo 829 m/s

UNITED STATES OF AMERICA
Manufacturer: Remington Arms Company Inc
Type: Ball: SP; 11.7 g; Vo 750 m/s

UPDATED

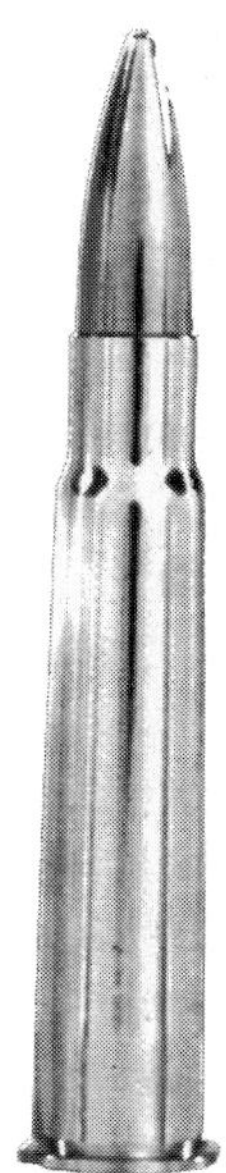

0.303 Ball

7.92 mm Mauser

Synonyms: 7.92 × 57 mm; 8 × 57 mm

Armament
All German service rifles and 7.92 mm machine guns from 1888 to 1945; used also as the standard service calibre by Austria, Bulgaria, Czechoslovakia, Hungary, Poland, Portugal and some South American countries prior to 1945. Used by the UK in the Besa tank machine gun from 1939 to 1950. Now used by the former Yugoslav nations in some sniper rifles and machine guns.

Development
This became the standard German rifle and machine gun cartridge in 1888 and with periodic improvement remained so until 1945. Widely adopted elsewhere it has been made by every major ammunition manufacturer at some time or other.

Description
A rimless, bottle-necked case with Berdan priming, which may be made of brass, brass- or copper-coated, lacquered steel or light alloy. The standard ball bullet varied from country to country but was generally a lead-cored, steel-jacketed, pointed, streamlined type of about 12 g weight.

Data
YUGOSLAV BALL M49
Round length: 80.6 mm
Case length: 57 mm
Rim diameter: 12 mm
Bullet diameter: 8.2 mm
Bullet weight: 12.85 g
Muzzle velocity: 720 m/s
Muzzle energy: 3,330 J

GERMANY
Manufacturer: RWS
Type: Target: FMJ; 12.12 g; Vo 800 m/s
H-Mantle: JHP; 12.12 g; Vo 820 m/s
Ball: JSP, RN; 12.7 g; Vo 798 m/s
Brenneke Ideal: 12.83 g; Vo 798 m/s

PORTUGAL
Manufacturer: INDEP
Type: Blank: Brass case, wooden bullet
Grenade Launcher: Brass case, crimped; chamber pressure $\leq$1,000 kg/cm^2

YUGOSLAVIA (SERBIA AND MONTENEGRO)
Holding company: YUGOIMPORT - SDPR
Type: Ball M49: FMJ; 12.85 g; Vo 720 m/s
Tracer M70: FMJ; red trace, dark ignition; visible 115 m to >900 m; 12.55 g; Vo 705 m/s
AP-I: FMJ; steel core, base tracer, incendiary composition in nose; 11.36 g; Vo 765 m/s
Sniper Ball M75: FMJ; 12.85 g; Vo 720 m/s
Blank: Brass case, crimped
Grenade Launcher: Brass case, crimped
Practice Ball M76: Light alloy bullet with copper base cup and part jacket, conical nose; 1.9 g; Vo 830 m/s
HP Test M49 Type 1: FMJ; 12.85 g; Pressure 3,500 kg/cm^2
HP Test M49 Type 2: FMJ; 12.85 g; Pressure 4,250 kg/cm^2
Ball: FMJ; 12.85 g; Vo 737 m/s
Ball: PSP; 9 g; Vo 807 m/s
Ball: PSP; 12.7 g; Vo 755 m/s
Match Ball: FMJ; 12.85 g; Vo 737 m/s

UPDATED

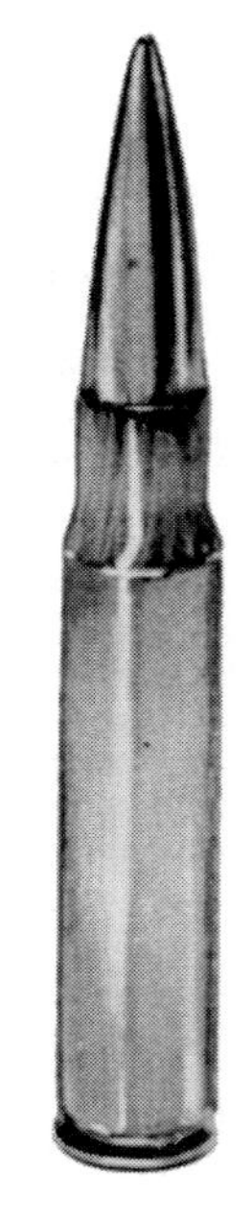

7.92 mm Mauser

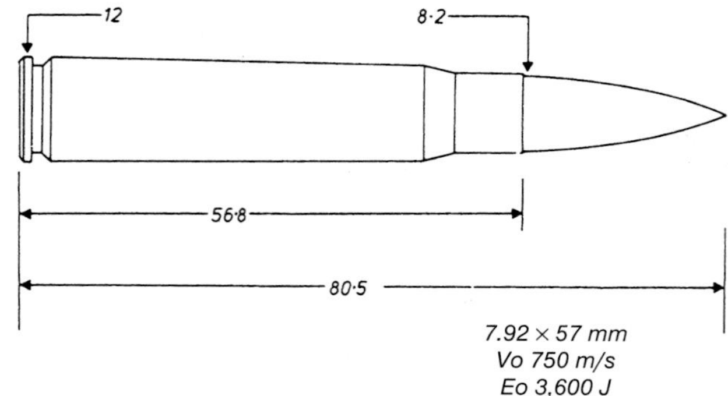

7.92 × 57 mm
Vo 750 m/s
Eo 3,600 J

0.338 Lapua Magnum

Synonyms: 8.6 × 70 mm; 0.338/0.416

Armament
Suitably chambered sniper rifles.

Development
This cartridge was developed in the mid-1980s as a long-range target round, but has since been put forward as a long-range (up to 1,500 m) sniper round. To date relatively few rifles have been produced in this calibre, but the impressive accuracy and consistency at long range makes it highly probable that the cartridge will be increasingly adopted for military and security long-range sniper purposes.

Description
The nomenclature '0.338/.416' indicates that this is the case of the 0.416 Rigby sporting cartridge shortened and necked down to accept the 0.338 bullet. The case is rimless, bottle-necked, brass and Boxer primed. The bullet is a metal jacketed pointed type of conventional construction.

Data
Round length: 91.5 mm
Case length: 69.2 mm
Rim diameter: 14.91 mm
Bullet diameter: 8.61 mm
Bullet weight: 16.2 g
Muzzle velocity: 914 m/s
Muzzle energy: 6,766 J

Abridged Ballistic Table: 0.338 Lapua Magnum

Range	Velocity	Energy	Drop
0 m	914 m/s	6,766 J	0 mm
200 m	810 m/s	5,318 J	250 mm
400 m	713 m/s	4,127 J	1,100 mm
600 m	624 m/s	3,156 J	2,740 mm
800 m	541 m/s	2,375 J	5,380 mm
1,000 m	467 m/s	1,769 J	9,390 mm
1,200 m	403 m/s	1,321 J	15,270 mm
1,400 m	354 m/s	1,015 J	23,640 mm
1,600 m	319 m/s	828 J	35,260 mm
1,800 m	296 m/s	710 J	50,590 mm
2,000 m	278 m/s	626 J	70,110 mm

FINLAND
Manufacturer: Lapua Cartridge Factory
Type: Ball: FMJ; 16.2 g; Vo 914 m/s
Ball: FMJ; 16.2 g; 870 m/s

NEW ENTRY

0.338 Lapua Magnum

---

9 mm Browning Short

Synonyms: 9 × 17 mm; 9 mm Short; 9 mm Corto; 0.380 Auto; 0.380 ACP; 9 mm/0.380 Auto

Armament
All pistols chambered for 9 mm Short or any of the synonyms shown above. The cartridge has also been tried in one or two sub-machine guns, without much success.

Development
Introduced by Colt in the USA in 1908 as the .380 Auto Pistol round. In 1910 it was adopted by Fabrique National Herstal for their Model 1910 pistol and they coined the name 'Browning Short' to distinguish it from the existing 9 mm Browning cartridge, which thereby became the 'Browning Long'. Widely adopted by central European police forces from the outset, it went into use as a military cartridge in the 1920s, notably by Czechoslovakia and Italy. It is still widely used by police

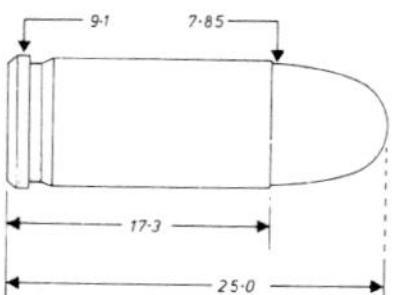

9 × 17 mm
Vo 270 m/s
Eo 224 J

and security forces and is particularly applicable as an airline security round. It combines moderately good stopping power with a velocity low enough to reduce the risk of piercing or ricochet.

Description

A rimless, straight taper, brass case with Berdan or Boxer priming. There are numerous designs of bullet, mostly soft point in order to improve the stopping ability, but the military/police standard is a metal-jacketed, lead-cored, round-nose bullet of 6.15 g weight.

Data

Round length: 24.89 mm
Case length: 17.27 mm
Rim diameter: 9.50 mm
Bullet diameter: 9.04 mm
Bullet weight: 6.15 g
Muzzle velocity: 290 m/s
Muzzle energy: 259 J

ARGENTINA
Manufacturer: Direccion General de Fabricaciones Militares
Type: Ball: FMJ, RN; 6.15 g; Vo 274 m/s

AUSTRIA
Manufacturer: Hirtenberger AG
Type: Ball: FMJ; 6.2 g; Vo 280 m/s

BELGIUM
Manufacturer: Browning SA
Type: Ball: FMJ; 6.05 g; Vo 270 m/s

BRAZIL
Manufacturer: Companhia Brasileira de Cartuchos
Type: Ball: FMJ; 6.2 g; Vo 290 m/s

CHILE
Manufacturer: FAMAE
Type: Ball: FMJ; 6 g; Vo 290 m/s

CZECH REPUBLIC
Manufacturer: Sellier & Bellot
Type: Ball: FMJ; 6 g; Vo 291 m/s
Ball vz.82: 4.5 g; 412 m/s

EGYPT
Manufacturer: Shoubra Company
Type: Ball: FMJ; 6 g; Vo 285 m/s

9 mm Browning Short

FINLAND
Manufacturer: Lapua Cartridge Factory
Type: Ball 4319170: FMJ; 6.1 g; Vo 285 m/s

Manufacturer: Sako Limited
Type: Ball: FMJ; 6.2 g; Vo 280 m/s

GERMANY
Manufacturer: Geco
Type: Ball: FMJ; 6 g; Vo 305 m/s

HUNGARY
Manufacturer: Mátravidéki Fémmüvek
Type: Ball: FMJ; 6 g; Vo 270 m/s

ISRAEL
Manufacturer: Kalia Israel Cartridge Company Limited
Type: Ball: JSP; 5.6 g; Vo 252 m/s

Manufacturer: TAAS - Israel Industries Limited
Type: Ball: FMJ; 6.16 g; Vo 288 m/s
Ball: JHP; 5.83 g; Vo 305 m/s

ITALY
Manufacturer: Fiocchi Munizioni SpA
Type: Ball: FMJ; 6 g; Vo 285 m/s

KOREA, SOUTH
Manufacturer: Poongsan Metal Corporation
Type: Ball: FMJ; 5.8 g; Vo 278 m/s
Ball: JSP; 5.8 g; Vo 280 m/s

MEXICO
Manufacturer: Aguila Industrias Tecnos
Type: Ball: FMJ; 6 g; Vo 288 m/s

PAKISTAN
Manufacturer: Pakistan Ordnance Factories
Type: Ball P11Z: JSP; ? g; Vo 340 m/s

SOUTH AFRICA
Manufacturer: Denel (Pty) Limited
Type: Ball: FMJ; 6 g; Vo 305 m/s

Manufacturer: Musgrave (Pty) Limited
Type: Ball: FMJ; 6.16 g; Vo 288 m/s

SPAIN
Manufacturer: SANTA BARBARA SA
Type: Ball: FMJ; 6.15 g; Vo 310 m/s

SWEDEN
Manufacturer: Norma AB
Type: Ball: FMJ; 6.1 g; Vo 315 m/s

UNITED STATES OF AMERICA
Manufacturer: 3-D
Type: Ball: FMJ; 6.16 g; Vo 274 m/s
Ball: JHP; 6.48 g; Vo 290 m/s
Ball: Lead, RN; 7.45 g; Vo 244 m/s

Manufacturer: CCI
Type: Ball: JSP; 5.72 g; Vo 320 m/s
Ball: FMJ; 6 g; Vo 291 m/s

Manufacturer: Federal Cartridge Company
Type: Ball: FMJ; 6 g; Vo 291 m/s
Ball: JSP; 5.8 g; Vo 305 m/s

Manufacturer: Glaser Safety Slug Inc
Type: Ball: Glaser Blue; 4.54 g; Vo 411 m/s
Ball: Glaser Silver; 4.54 g; Vo 411 m/s

Manufacturer: Remington Arms Company Inc
Type: Ball: FMJ; 6.16 g; Vo 291 m/s
Ball: JHP; 5.72 g; Vo 302 m/s

Manufacturer: Winchester
Type: Ball: FMJ; 6.2 g; Vo 316 m/s
Ball: JSP; 5.5 g; Vo 291 m/s

YUGOSLAVIA (SERBIA AND MONTENEGRO)
Holding company: YUGOIMPORT - SDPR
Type: Ball: FMJ; 6.1 g; Vo 289 m/s
Ball: JHP; 6.1 g; Vo 290 m/s

NEW ENTRY

9 mm Makarov

Synonyms: 9 × 18 mm Soviet; 9 mm Stechkin; 57-N-181S; 9 mm Type 59

Armament

Soviet Makarov and Stechkin pistols; Makarov copies such as the Chinese Type 59, Czech Model 83, East German Pistole M, and other pistols chambered for this round such as the Hungarian PA-63 and Polish P-64.

Development

The development of the 9 mm Makarov cartridge was carried out by Boris Semin. The cartridge was introduced into Soviet service together with the Makarov and Stechkin pistols in 1951 but has always been regarded as a rather low-powered round for most combat purposes, to the extent that a more powerful version has been developed with the higher velocity 57-N-181SM round fired from the Makarov PMM pistol and various sub-machine guns.

Description

A rimless, straight taper cartridge of brass or, more usually, of copper-coated or lacquered steel. The bullet is jacketed, round-nosed and flat-based; early designs used a lead alloy core, more recent manufacture uses a lead core with a small steel insert. Both bullets weigh the same.

The standard 57-N-181S cartridge case contains 0.25 g of P-125 propellant providing a muzzle velocity of 315 m/s. With the 57-N-181SM cartridge, which retains the same overall dimensions as the standard round, a lighter bullet (5.4 to 5.8 g) with a steel penetrator is employed. This, coupled with a revised propellant load, provides a nominal muzzle velocity between 410 and 435 m/s, and thus increased muzzle energy. The 57-N-181SM bullet can penetrate 3.5 mm of steel plate at 10 m and 3 mm at 20 m.

Tracer and hollow point rounds are under development. These include a hollow-point round weighing 8.14 g, of which 7.8 g is the bullet, and a round with a jacketed bullet containing a steel tip and a lead base. The latter, which is being prepared for production, has a round weight of 10.07 g and a bullet weight of 6.15 g. Muzzle velocity is from 290 to 315 m/s. Also being prepared for production is a Tracer round with charcteristics matching those of the steel-tipped round.

Data

57-N-181S
Round length: 24.79 mm
Case length: 18.1 mm
Rim diameter: 9.98 mm
Bullet diameter: 9.23 mm
Bullet length: 12.35 mm
Bullet weight: 5.75-6.15 g
Muzzle velocity: 340 m/s
Muzzle energy: 348 J

BULGARIA
Manufacturer: Kintex
Type: Ball: Steel core, steel jacket, gilding metal envelope; 6.1 g; Vo 315 m/s. Cartridge case steel-, brass- or copper-plated

CHINA, PEOPLE'S REPUBLIC
Manufacturer: China North Industries (NORINCO)
Type: Ball: FMJ; 6.1 g; Vo 315 m/s

COMMONWEALTH OF INDEPENDENT STATES
Manufacturer: LVE Novosibirsk
Type: Ball: FMJ; 6 g; Vo 315 m/s
Ball, Sporting: FMJ; 6.6 g; Vo 300 m/s

9 mm Makarov

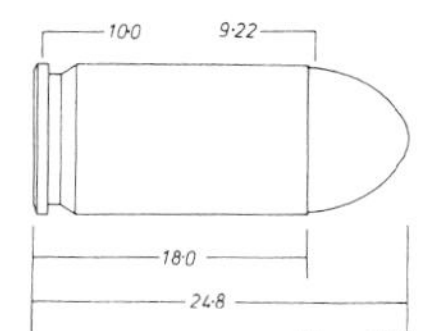

9 × 18 mm Makarov
Vo 340 m/s
Eo 348 J

Manufacturer: State arsenals
Type: Ball: FMJ; 6.02 g; Vo 340 m/s

CZECH REPUBLIC
Manufacturer: Sellier & Bellot
Type: Ball: FMJ; 6.1 g; Vo 310 m/s
Ball: Lead; 4.5 g; Vo 380 m/s

Manufacturer: State arsenals
Type: Ball vz82: FMJ; 4.5 g; Vo 400 m/s. This round was developed specifically for use with the vz82 pistol and a 9 mm version of the Skorpion sub-machine gun. The complete round weighs 8.1 g, instead of the 10 g of the Soviet original, though the length of bullet and round are the same.

HUNGARY
Manufacturer: Technika Foreign Trading Company
Type: Reduced Effect Ball: Polyamid plastic bullet with metal base cup; 1.7 g; Vo 420 m/s. This is a special round in use with Hungarian PA-63 and R-61 pistols for security and police personnel, combining good terminal effect with low ricochet and shoot-through risks
Ball: FMJ; steel core; 6 g; Vo 300 m/s

ITALY
Manufacturer: Fiocchi Munizioni SpA
Type: Ball: FMJ; 6.1 g; Vo 330 m/s

POLAND
Manufacturer: Mesko Zaklady Metalowe
Type: Ball: FMJ, RN; 6 g; Vo 305 m/s
Blank: Star crimp

ROMANIA
Manufacturer: Romtehnica
Type: Ball: FMJ with steel core; Vo 350 m/s

UNITED STATES OF AMERICA
Manufacturer: CCI
Type: Ball: JHP; 5.8 g; Vo 320 m/s

Manufacturer: Hansen Cartridge Company
Type: Ball: FMJ; 6.1 g; Vo 315 m/s

Manufacturer: Hornady Manufacturing Corporation
Type: Ball: JHP; 6.1 g; Vo 305 m/s

YUGOSLAVIA (SERBIA AND MONTENEGRO)
Holding company: YUGOIMPORT - SDPR
Type: Ball: FMJ; 6 g; Vo 325 m/s

UPDATED

9 mm Parabellum

Synonyms: 9 × 19 mm; 9 mm Luger; 9 mm Patrone '08

Armament

Any pistol marked '9 × 19', '9 mm Luger' or any other synonym; the more notable being the Parabellum (Luger), Browning GP35, Beretta 92, SIG P225 and so on, and pistols by Astra, Bernardelli, Star, FEG, Walther, Heckler and Koch, Smith & Wesson, Ruger and others. Also the predominant sub-machine gun cartridge throughout the world. It has been used in revolvers (Ruger, Manurhin and so on) and semi-automatic carbines (Marlin), though in lesser numbers.

Development

Developed in 1902 by Georg Luger in order to improve the stopping power of his pistol; he opened up the mouth of the 7.62 mm Parabellum case and inserted a 9 mm bullet to meet a German Army demand. In original form it used a cylindro-conoidal bullet with a flat tip. This showed a tendency to jam in early sub-machine guns and was replaced during 1917 with an ogival shaped bullet which has remained the military standard; bullets of the original shape are still available commercially.

Since its introduction the 9 mm Parabellum has been universally accepted and manufactured all over the world.

Description

A rimless, straight taper case with Berdan or Boxer priming; because of its broad-based manufacture brass, steel and alloy cases have been made and every sort of bullet has been tried. The military standard is a conventional lead-cored, steel-jacketed with gilding metal envelope, ogival-head pattern weighing 7.45 g.

Data

BRITISH MK 2Z
Round length: 29.28 mm
Case length: 19.35 mm
Rim diameter: 9.94 mm
Bullet diameter: 9 mm
Bullet weight: 7.45 g
Muzzle velocity: 396 m/s
Muzzle energy: 584 J

ABU DHABI
Manufacturer: Adcom Manufacturing
Type: Ball: FMJ; lead core; 7.45 g; Vo 384 m/s

ARGENTINA
Manufacturer: Direccion General de Fabricaciones Militares
Type: Ball FMK 5 Mod 0: FMJ; 8 g; 245 m/s
Tracer FMK 2 Mod 0: FMJ; red trace to over 200 m; 7.18 g; Vo 340 m/s

AUSTRIA
Manufacturer: Hirtenberger AG
Type: Ball: FMJ; 8 g; Vo 380 m/s
Combat Match: FMJ; 7.45 g; Vo 372 m/s
Ball, Police 'Defender': JHP; 8.03 g; Vo 381 m/s
Tracer: FMJ; red trace to 350 m; 7 g; Vo 380 m/s
Frangible, Non-Toxic: 4.8 g bullet made from special non-toxic metal, free from Sb, Ba, Pb; Vo 500±15 m/s
Hollow Point Test: FMJ; 8 g; pressure ≥3,200 Bar
Training Tracer: Special round for use with sub-calibre device in Miniman anti-tank weapon. Truncated conical 5.5 g bullet with red tracer burning 1.5 seconds; Vo 165 m/s

BRAZIL
Manufacturer: Companhia Brasileira de Cartuchos
Type: Ball: FMJ; 8.03 g; V16 384 m/s
Ball: JHP; 7.45 g; V16 400 m/s

CANADA
Manufacturer: SNC Industrial Technologies Inc
Type: Ball CDN Mk 1: FMJ; 7.5 g; Vo 388 m/s
Ball: JHP; 7.45 g; Vo 364 m/s
Ball: JHP; 9.53 g; Vo 297 m/s
Ball: JHP; 7.45 g; Vo 354 m/s
Blank C30: Brass case, star crimped

CHILE
Manufacturer: FAMAE
Type: Ball: FMJ; 7.55 g; Vo 370 m/s

CHINA, PEOPLE'S REPUBLIC
Manufacturer: China North Industries (NORINCO)
Type: Ball: FMJ; 8 g; Vo 345 m/s
Ball: FMJ; 7.5 g; Vo 376V

COMMONWEALTH OF INDEPENDENT STATES
Manufacturer: LVE, Novosibirsk
Type: Ball, Sporting: FMJ; 8 g; Vo 350 m/s

CZECH REPUBLIC
Manufacturer: Sellier & Bellot
Type: Ball: FMJ; steel core; 7.5 g; Vo 398 m/s
Ball: FMJ; steel core; 6.45 g; Vo 452 m/s
Semi-Wadcutter: Lead; 8 g; Vo 372 m/s

EGYPT
Manufacturer: Shoubra Company
Type: Ball: FMJ; 7.45 g; Vo 380 m/s

FINLAND
Manufacturer: Lapua Cartridge Factory
Type: Combat Ball 4319613: FMJ; 8 g; Vo 355 m/s
Ball 4319200: FMJ; 7.5 g; Vo 405 m/s
Ball 4319177: FMJ; 8 g; Vo 320 m/s
Ball 4319230: FMJ; 8 g; Vo 400 m/s
Police Ball L434: FMJ; 3.25 g; Vo 530 m/s
Combat Trainer 4319208: FMJ; 8 g; Vo 355 m/s
Ball: CEPP Super; 7.8 g; Vo 360 m/s
Ball: CEPP Extra; 7.8 g; Vo 360 m/s
Ball: CEPP Extra Subsonic; 9.7 g; Vo 315 m/s

Manufacturer: Sako Limited
Type: Ball: FMJ; 7.5 g; Vo 320 m/s

FRANCE
Manufacturer: Arcane
Type: MP Ball: Conical, pure copper; 4.54 g; Vo 543 m/s

Manufacturer: Giat Industries
Type: Balle O: FMJ; 8 g; Vo 385 m/s
Tracer T: FMJ; red trace; 8 g; Vo 385 m/s
Subsonic ball: 8 g; 285 m/s
Plastic ball: Brass case; 1.2 g plastic bullet; 350 m/s
Blank: Brass case, rose crimp
Blank: Plastic case

Manufacturer: SFM Défense
Type: Ball 'UL': JSP; 3.2 g; Vo 655 m/s
Tracer: 8 g; Vo 418 m/s
THV: 2.9 g; Vo 780 m/s; penetrates 2 mm steel at 7 m
Blank: Plastic case

GERMANY
Manufacturer: Geco
Type: Ball: FMJ; 8 g; Vo 360 m/s, Sintox, non-toxic

9 × 19 mm Parabellum

Tracer: FMJ; red trace; 7.5 g; Vo 375 m/s
Ball: Action; 5.6 g; Vo 430 m/s
Practice PT: Plastic; 0.42 g; Vo 410 m/s

Manufacturer: Metallwerk Elisenhütte GmbH
Type: Ball DM11A1B2: FMJ; lead-antimony core, steel jacket, gilding metal envelope; single or double base propellant, Berdan primed, non-mercuric; 8 g; Vo 395±15 m/s
Schadstoffarm: This is a special lead- and barium-free round designed for practice shooting in indoor ranges. Although the bullet is the standard lead-cored FMJ pattern, the primer cap has a special non-toxic composition which has minimal emission of lead and barium, thus considerably reducing the health hazard common with indoor ranges. 8 g; Vo 385±15 m/s
Subsonic: For use in silenced arms. FMJ; lead-antimony core, steel jacket with gilding metal envelope; 10 g; Vo 300±15 m/s fired in MP5SD sub-machine gun. A 7.5 g bullet is also available
Quick Defence 1 (QD 1): This is a special bullet intended to open out very rapidly on impact, delivering maximum energy as quickly as possible. The bullet is of gilding metal, and is virtually a hollow point type with a ball placed in the mouth of the hollow. The shape thus resembles the normal ogive and feeds through automatic weapons without causing problems. On impact the ball is driven back into the hollow, spreading the walls of the bullet to give a greater impact area. 5.2 g; Vo 440±15 m/s
Quick Defence 2 (QD 2): This is similar to the QD 1 round described previously but is designed so that the deformation of the bullet and hence the energy transfer is somewhat less. 5.7 g; gilding metal bullet; Vo 420±15 m/s

GREECE
Manufacturer: PYRKAL: Greek Powder & Cartridge Company
Type: Ball: FMJ; ogival; 8 g; 330 m/s

HUNGARY
Manufacturer: Mátravidéki Fémmüvek
Type: Ball: FMJ; cylindro-conoidal; 8 g; Vo 360 m/s
Ball: FMJ; ogival; 8 g; Vo 360 m/s

INDIA
Manufacturer: Indian Ordnance Factory Khadki
Type: Ball Mark 2Z: FMJ; ogival, 7.45 g; Vo 396±15 m/s at 18.50 m

INDONESIA
Manufacturer: PT Pinad (Persero)
Type: Ball MU-1TJ: FMJ; 8.1 g; Vo 380 m/s

IRAN
Manufacturer: Defence Industries Organisation, Ammunition Group
Type: Ball: FMJ; Cartridge weight 12.3 g; Vo 385 m/s

ISRAEL
Manufacturer: Kalia Israel Cartridge Company Limited
Type: Ball: FMJ; 7.45 g; Vo 335 m/s
Ball: FMJ; 7.45 g; Vo 351 m/s
Ball: FMJ; 7.45 g; Vo 370 m/s

Manufacturer: TAAS - Israel Industries Limited
Type: Carbine Ball: FMJ; 7.45 g; Vo 420 m/s
Carbine Tracer: FMJ; 7.45 g; Vo 420 m/s
Pistol Ball: FMJ; 7.45 g; Vo 325 m/s
Subsonic Ball: FMJ; 10.24 g; Vo 286 m/s
Subsonic Ball: FMJ; 8.04 g; Vo 314 m/s
Pistol Ball Special: FMJ; 7.45 g; Vo 384 m/s
Pistol Ball +P: FMJ; 7.45 g; Vo ?
Carbine Ball +P: JHP; 7.45 g; Vo 430 m/s
Carbine Ball +P: FMJ; 7.45 g; Vo 420 m/s
Luger Ball: JHP; 7.45 g; Vo 354 m/s
Luger Ball: FMJ; 7.45 g; Vo 343 m/s

ITALY
Manufacturer: Europa Metalli
Type: Ball: FMJ; lead core; 7.5 g; Vo 380 m/s

Manufacturer: Fiocchi Munizioni SpA
Type: Ball: FMJ; 7.45 g; Vo 415 m/s
Ball: FMJ; 8 g; Vo 390 m/s
Ball: Lead; 8 g; Vo 322

Manufacturer: Government arsenals
Type: Ball M38: FMJ; 8 g; Vo 305 m/s

KOREA, SOUTH
Manufacturer: Poongsan Metal Corporation
Type: Ball 9A: FMJ; 7.45 g; Vo 353 m/s
Ball 9B: JHP; 7.45 g; Vo 356 m/s

MEXICO
Manufacturer: Aguila Industrias Tecnos
Type: Ball: FMJ; 8 g; Vo 341 m/s
Ball: FMJ; 7.45 g; Vo 410 m/s

PAKISTAN
Manufacturer: Pakistan Ordnance Factories
Type: Ball 2Z: FMJ; ? g; Vo 395 m/s

POLAND
Manufacturer: Mesko Zaklady Metalowe
Type: Ball: FMJ; ogival; 8 g; Vo 345 m/s
Ball: JSP; 8 g; 340 m/s

PORTUGAL
Manufacturer: INDEP
Type: Ball M947: FMJ; 8 g; Vo 375 m/s
Ball M970: FMJ; 7.45 g; Vo 384 m/s

Manufacturer: Prvi Partizan
Type: Ball: FMJ; 8 g; Vo 390 m/s
Ball: FMJ; 8 g; Vo 365 m/s
Ball: FMJ; 7.5 g; Vo 407 m/s

10·0 9·02 19·45 29·7

9 × 19 mm Parabellum
Vo 396 m/s
Eo 584 J

ROMANIA
Manufacturer: Romtehnica
Type: Ball: FMJ; Vo 385 m/s

SOUTH AFRICA
Manufacturer: Denel (Pty) Limited
Type: Ball: FMJ; 7.45 g; Vo 401 m/s
Tracer: red trace from 14-183 m; FMJ; ? g; Vo 386 m/s

Manufacturer: Musgrave (Pty) Limited
Type: Ball: FMJ; 6.16 g; Vo 405 m/s
Ball: FMJ; 7.45 g; Vo 349 m/s

SPAIN
Manufacturer: SANTA BARBARA SA
Type: Ball: FMJ; ? g; Vo 425 m/s

SWEDEN
Manufacturer: Bofors Carl Gustaf
Type: Ball: FMJ; 8 g; Vo 420 m/s
AP: FMJ; 6.75 g; Vo 450 m/s; penetrates min 3 mm steel plate at more than 50 m

Manufacturer: Norma AB
Type: Ball: Bronze, ogival; 6.7 g; Vo 390 m/s
Ball 19021: JHP; 7.4 g; Vo 355 m/s
Ball 19022: FMJ; 7.5 g; Vo 355 m/s
Ball: Cylindro-coniodal; 7.5 g; Vo 355 m/s
Ball 19026: JSP; 7.5 g; Vo 355 m/s

TURKEY
Manufacturer: Makina ve Kimya Endüstrisi Kumuru (MKEK)
Type: Ball: FMJ; ogival; 8 g; Vo 350 m/s

UNITED KINGDOM
Manufacturer: Cobra Gun Company
Type: Ball: HSA Flechette; Vo 490 m/s

Manufacturer: Conjay Firearms & Ammunition Limited
Type: Pistol Ball: CBX; 5.51 g; Vo ?
Carbine Ball: CBX; 5.51 g; Vo 460 m/s in 254 mm barrel
Carbine Ball: CBXX; 6.8 g; Vo 460 m/s in 254 mm barrel
Pistol AP: CBAP; 5.51 g; Vo 430 m/s; penetrates 3.5 mm RHA (480-530 Brinell, 52 Rockwell)
Carbine AP: CBAP; 5.51 g; Vo 495 m/s; penetration of RHA as for Pistol AP above; it will also drive plugs from the rear of the same material at 5.5 mm thickness

UNITED STATES OF AMERICA
Manufacturer: 3-D
Type: Ball: FMJ: 9.52 g; Vo 290 m/s
Ball: FMJ; 7.45 g; Vo 350 m/s
Ball: JHP; 9.52 g; 294 m/s
Ball: JHP; 7.45 g; Vo 350 m/s
Ball: Lead, RN; 8.1 g; Vo 320 m/s

Manufacturer: CCI
Type: Ball: FMJ; 7.45 g; Vo 350 m/s
Ball: JHP; 7.45 g; Vo 351 m/s
Ball: JSP; 8 g; Vo 352 m/s

Manufacturer: Delta Defense Inc
Type: Frangible: 5.51 g; Vo 434 m/s

Manufacturer: Federal Cartridge Company
Type: Ball: SJHP; 6.2 g; Vo 412 m/s
Ball: FMJ; 8 g; Vo 341 m/s
Ball: JHP; 7.5 g; Vo 354 m/s
Hydra-Shok: 9.5 g; Vo 320 m/s

Manufacturer: Glaser Safety Slug Inc
Type: Ball: Glaser Blue; 5.18 g; Vo 503 m/s
Ball: Glaser Silver; 5.18 g; Vo 503 m/s

Manufacturer: Olin Ordnance
Type: Ball M882: FMJ; 8.03 g; V16 375 m/s

Manufacturer: Remington Arms Company Inc
Type: Ball: JHP; 7.45 g; Vo 354 m/s
Ball: FMJ; 8 g; Vo 341 m/s
Ball: FMJ; 7.45 g; Vo 350 m/s
Subsonic Ball: JHP; 9.5 g; Vo 302 m/s
Ball: JHP; 5.72 g; Vo 457 m/s
Ball: JHP; 7.5 g; Vo 381 m/s

Manufacturer: Winchester
Type: Ball: JSP; 6.17 g; Vo 413 m/s
Ball: FMJ; 7.5 g; Vo 352 m/s
Ball: JHP; 7.5 g; Vo 374 m/s

YUGOSLAVIA (SERBIA AND MONTENEGRO)
Holding company: YUGOIMPORT - SDPR
Type: Ball: FMJ; 7.45 g; Vo 343 m/s
Ball: FMJ; 8 g; Vo 327 m/s
Ball: JHP; 7.45 g; Vo 344 m/s
Ball: Lead, RN; 8 g; Vo 328 m/s
Ball: FMJ; cylindro-conoidal; 8 g; Vo 328 m/s
Subsonic Ball: FMJ; 9.5 g; Vo 296 m/s
Subsonic Ball: JHP; 9.5 g; Vo 297 m/s

UPDATED

9 × 21 mm Gyurza

Armament
9 mm Gyurza pistol.

Development
The 9 × 21 mm Gyurza cartridge was developed specifically for the penetration of body armour and, as far as can be determined, it is employed with only one weapon to date, the 9 mm Gyurza pistol. This pistol together with the cartridge, were developed by the Precision Mechanical Engineering Central Institute (TsNIITochmash).

Description
The 9 × 21 mm Gyurza cartridge has a partially necked brass rimless case firing a 6.7 g semi-enveloped bullet with a blunt-nosed hardened steel or tungsten penetrator protruding from the tip of the bullet. The propellant load provides a muzzle velocity of 415 to 420 m/s and a muzzle energy rating of 608 J which combine to provide an effective range of up to 100 m. At that range it is claimed that the bullet can penetrate body armour made up of one or two 1.4 mm titanium plates and 30 layers of Kevlar. It is also possible to penetrate a 4 mm steel plate at 60 m.

Data
Round length: 33 mm
Case length: 21 mm
Bullet weight: 6.7 g
Cartridge weight: 11 g
Muzzle velocity: 415-420 m/s
Muzzle energy: 608 J

COMMONWEALTH OF INDEPENDENT STATES
Agency: Rosvoorouzhenie
Type: Silent Ball: Steel penetrator; 11 g; Vo 415-420 m/s

NEW ENTRY

9 × 39 mm SP-5 and SP-6

Armament
A-91 sub-machine gun; 9 mm VSS; 9 mm VAL.

Development
Apart from their dimensions the 9 × 39 mm SP-5 and SP-6 rounds are very different but can be mentioned here together. The SP-6 was developed to be utilised in the A-91 sub-machine gun series and the VAL silent assault rifle. The SP-5 is an armour-piercing anti-materiel round intended for use mainly with the VSS silent rifle, part of the BSK-94 sniper rifle complex, although this weapon can also employ the SP-6. The SP-5 cartridge was developed by Nikolai Zabelin and the SP-6 by Yuri Folov.

Description
The 9 × 39 mm SP-5 and SP-6 cartridges resemble conventional rimless cartridges with lacquered steel cases. The cases are 7.62 × 39 mm cases with their necks opened up to accommodate the 9 mm bullets. Both cartridges are intended to be fired at subsonic velocities and have relatively heavy (16.2 g) streamlined bullets. The SP-6 has a conventional lead-based bullet

specially developed for the sniper role while the SP-5 bullet has a hardened steel or dense metal (possibly tungsten) penetrator tip. The latter enables the bullet to penetrate 2 mm of steel plate at 500 m or a 6 mm armour plate at 100 m.

Data
Round length: 56 mm
Case length: 39 mm
Bullet weight: 16.2 g
Cartridge weight: 23 g

COMMONWEALTH OF INDEPENDENT STATES
Agency: Rosvoorouzhenie
Type: Silent Ball: FMJ; 16.2 g

NEW ENTRY

0.357 Magnum

Synonyms: 0.357 Smith & Wesson Magnum

Armament
Suitably chambered revolvers.

Development
Introduced in 1935 by Smith & Wesson, this virtually became the standard law enforcement round in the USA. Whilst the calibre is the same as that of the normal 0.38 cartridge, the nomenclature was changed to 0.357 (the exact, rather than the nominal calibre) to distinguish this more powerful round. In addition the case is about 2.5 mm longer than other 0.38 cases to prevent it being chambered in older revolvers which may not be strong enough to withstand the extra pressure.

Description
A rimmed, straight taper case of brass- or nickel-plated brass, Boxer or Berdan primed. Various types and weights of bullet have been produced, but the standard factory loading is a 10.23 g soft pointed pattern.

Data
Round length: 38.5 mm (depending upon bullet)
Case length: 32.76 mm
Rim diameter: 11.17 mm
Bullet diameter: 9.07 mm
Bullet weight: 10.23 g
Muzzle velocity: 436 m/s (in 4 in barrel)
Muzzle energy: 972 J

AUSTRIA
Manufacturer: Hirtenberger AG
Type: Ball: JSP; 10.2 g; Vo 460 m/s

BRAZIL
Manufacturer: Companhia Brasileira de Cartuchos
Type: Ball: JSP; 10.2 g; Vo 424 m/s
Ball: JHP; 10.2 g; Vo 424 m/s

FINLAND
Manufacturer: Lapua Cartridge Factory
Type: Ball: SWC; 10.2 g; Vo 360 m/s
Ball: JHP; 10.2 g; Vo 470 m/s

Manufacturer: Sako Limited
Type: Ball: Lead, SWC; 10.3 g; Vo 365 m/s
Ball: JSP; 10.2 g; Vo 450 m/s

FRANCE
Manufacturer: SFM Défense
Type: Ball: FMJ; 9.7 g; Vo 475 m/s
Ball: JSP; 10.46 g; Vo 430 m/s
Ball 'UL': JSP; 3.9 g; Vo 600 m/s
THV: 2.9 g; Vo 800 m/s; penetrates 3.50 mm steel at 7 m

GERMANY
Manufacturer: Dynamit Nobel (Geco)
Type: Ball: JSP: 10.25 g; Vo 435 m/s
Ball: FMJ; conical; 10.25 g; Vo 435 m/s

ISRAEL
Manufacturer: Kalia Israel Cartridge Company Limited
Type: Ball: SWC; 9.7 g; Vo 320 m/s
Ball: SWC; 9.7 g; Vo 379 m/s

Manufacturer: TAAS - Israel Industries Limited
Type: Ball: JHP; 8.1 g; Vo 434 m/s
Ball: JHP; 10.24 g; Vo 372 m/s
Ball: JSP; 8.1 g; Vo 434 m/s
Ball: JSP; 10.24 g; Vo 372 m/s

ITALY
Manufacturer: Fiocchi Munizioni SpA
Type: Ball: JSP; 9.6 g; Vo 460 m/s
Ball: SWC; 10.2 g; Vo 375 m/s
Ball: FMJ; 9.2 g; Vo 460 m/s

KOREA, SOUTH
Manufacturer: Poongsan Metal Corporation
Type: Ball: SWC; 10.2 g; Vo 364 m/s
Ball 357A: JSP; 10.24 g; Vo 364 m/s
Ball 357B: JHP; 8.1 g; Vo 364 m/s
Ball 357C: JHP; 9.72 g; Vo 376 m/s
Ball 357D: JHP; 7.13 g; Vo 420 m/s

MEXICO
Manufacturer: Aguila Industrias Tecnos
Type: Ball: JSP; 10.2 g; Vo 473 m/s
Ball: FMJ; 10.2 g; Vo 430 m/s

SOUTH AFRICA
Manufacturer: Denel (Pty) Limited
Type: Ball: JSP; 8.1 g; Vo 377 m/s
Ball: JSP; 8.1 g; Vo 473 m/s

Manufacturer: Musgrave (Pty) Limited
Type: Ball: JSP; 10.24 g; Vo 405 m/s
Ball: JHP; 10.24 g; Vo 405 m/s

SWEDEN
Manufacturer: Norma AB
Type: Ball 19101: JHP; 10.24 g; Vo 442 m/s
Ball 19106: FMJ, SWC; 10.24 g; Vo 442 m/s
Ball 19107: JSP; 10.24 g; Vo 442 m/s

UNITED KINGDOM
Manufacturer: Cobra Gun Company
Type: Ball: HSA Flechette

Manufacturer: Conjay Firearms & Ammunition Limited
Type: Ball: CBX; 5.51 g; Vo 590 m/s in 152 mm barrel
Ball: CBXX; 6.8 g; Vo 495 m/s in 152 mm barrel
Ball: CBAP; 5.51 g; Vo 690 m/s; penetrates 5.5 mm RHA (480-530 Brinell, 52 Rockwell) at angles from normal to 45°

UNITED STATES OF AMERICA
Manufacturer: 3-D
Type: Ball: JHP; 9.07 g; 381 m/s
Ball: Lead, SWC; 10.24 g; Vo 305 m/s

Manufacturer: Black Hills Ammunition Company
Type: Ball: JHP; 8.1 g; Vo 427 m/s
Ball: JHP; 10.2 g; Vo 366 m/s
Ball: SWC; 10.2 g; Vo 320 m/s

Manufacturer: CCI
Type: Ball: JHP; 7.1 g; Vo 421 m/s
Ball: JHP; 8.1 g; Vo 442 m/s
Ball: JSP; 8.1 g; Vo 442 m/s
Ball: JHP; 9 g; Vo 421 m/s
Ball: JHP; 10.2 g; Vo 377 m/s

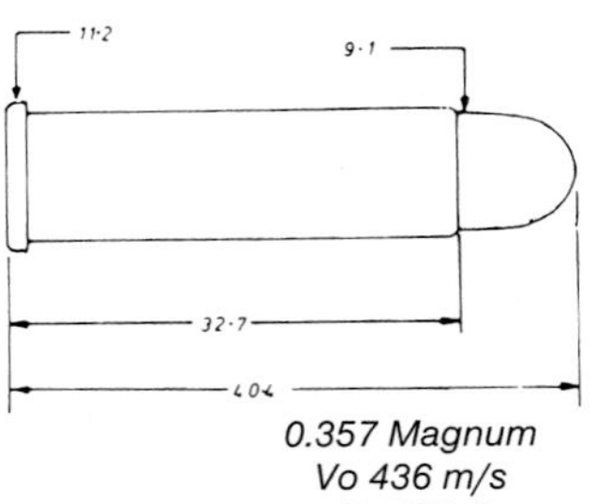

0.357 Magnum
Vo 436 m/s
Eo 972 J

Manufacturer: Cor-Bon
Type: Ball: JHP; 7.5 g; Vo 412 m/s
Ball: JHP; 8.1 g; Vo 442 m/s
Ball: JHP; 11.7 g; Vo 381 m/s
Ball 'Penetrator': FMJ; 13 g; Vo 366 m/s
Ball: FMJ; 13.3 g; Vo 390 m/s
Ball Special: FMJ; 11.7 g; Vo 503 m/s

Manufacturer: Delta Defense Inc
Type: Frangible: 5.51 g; Vo 449 m/s

Manufacturer: Federal Cartridge Company
Type: Ball: JSP; 10.2 g; Vo 376 m/s
Ball: JHP; 8.1 g; Vo 442 m/s
Ball: SWC; 10.2 g; Vo 376 m/s
Ball: JHP; 8.1 g; Vo 442 m/s
Ball: FMJ; 11.7 g; Vo 332 m/s

Manufacturer: Remington Arms Company Inc
Type: Ball: JHP; 8.1 g; Vo 441 m/s
Ball: JHP; 7.1 g; Vo 394 m/s
Ball: JHP; 10.24 g; Vo 376 m/s
Ball: Lead; 10.24 g; Vo 376 m/s
Ball: JSP; 10.24 g; Vo 376 m/s
Ball: FMJ; 10.2 g; Vo 376 m/s

Manufacturer: Ultramax
Type: Ball: JHP; 8.1 g; Vo 436 m/s
Ball: JHP; 10.2 g; Vo 335 m/s
Ball: Lead, SWC; 8.1 g; Vo 250 m/s
Ball: Lead, SWC; 10.2 g; Vo 354 m/s

Manufacturer: Winchester
Type: Ball: JHP; 7.1 g; Vo 394 m/s
Ball: JHP; 8.1 g; Vo 441 m/s
Ball: SWC; 10.2 g; Vo 376 m/s
Ball: Metal-Piercing; 10.2 g; Vo 376 m/s
Ball: JHP; 10.2 g; Vo 376 m/s
Ball: JSP; 10.2 g; Vo 376 m/s
Ball: JSP; 9.42 g; Vo 393 m/s

YUGOSLAVIA (SERBIA AND MONTENEGRO)
Holding company: YUGOIMPORT - SDPR
Type: Ball: FMJ; 10.2 g; Vo 467 m/s
Ball: JHP; 10.2 g; Vo 493 m/s
Hollow Point Test: FMJ; conical or JSP; 10.2 g

UPDATED

0.357 SIG

Synonyms: 0.357 SIG

Armament
Various SIG pistols.

Development
The 0.357 SIG is a recent American development produced by combining the 0.40 Smith & Wesson cartridge case and propellant load with a 0.357 Magnum bullet. The combination provides a powerful high energy cartridge suitable for many police and paramilitary applications with the minimum of production changes on manufacturing lines as a great deal of existing tooling can be utilised with relatively few costly changes. The round can be accommodated in existing 0.40 magazines without modification. Some SIG pistols, the P229 and P239, have already been produced for this cartridge.

Description
The 0.357 SIG cartridge utilises a necked 0.40 S & W rimless cartridge case allied with a coned flat-nosed bullet weighing 8.1 g. The combination produces a muzzle velocity of 411 m/s and a muzzle energy rating of 685 J as well as a flatter trajectory at longer ranges. As an example of the latter the fall below a zero line at 91.2 m for the 0.357 SIG is about 77 mm. For the 0.40 S & W cartridge the value is 127 mm.

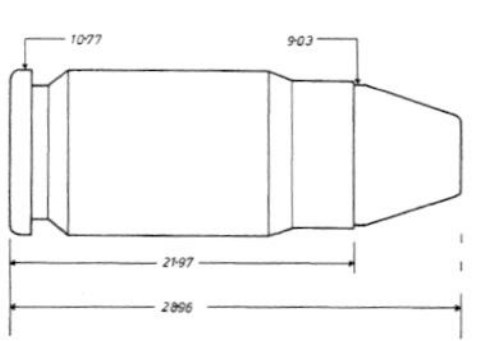
0.357 S & W
Vo 411 m/s
Eo 685 J

Data
Round length: 28.96 mm
Case length: 21.97 mm
Bullet diameter: 9.03 mm
Bullet weight: 8.1 g
Rim diameter: 10.77 mm
Muzzle velocity: 411 m/s
Muzzle energy: 685 J

UNITED STATES OF AMERICA
Manufacturer: Federal Cartridge Company
Type: Ball: JSP; 8.1 g; Vo 411 m/s

NEW ENTRY

0.380 British

Synonyms: 0.38/200; 9.6 × 31R

Armament
British service Webley Mk IV and Enfield revolvers; suitably chambered revolvers of other makes (most revolvers chambered for the 0.38 Smith & Wesson Long or 0.38 Long Colt will fire this round).

Development
Introduced in 1930 when the British Army adopted 0.38 as their standard revolver calibre in place of 0.455. A 12.96 g lead round-nosed bullet was adopted, but in view of its dubious legality in the eyes of the Hague Convention on expanding bullets it was replaced by a jacketed bullet in 1938. Although the 0.38 revolver is no longer in British service it is still widely used in various parts of the past and present British Commonwealth as a police and military revolver.

Description
A rimmed, brass, straight taper case, Berdan primed. As noted, the early standard military bullet was a round-nosed lead type, but this was replaced by a metal jacketed ogival pattern; the envelope may be gilding metal or nickel.

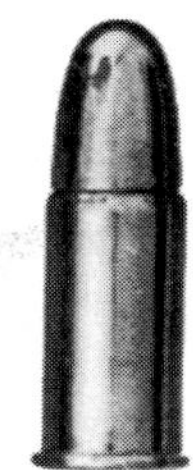

0.38/200 British

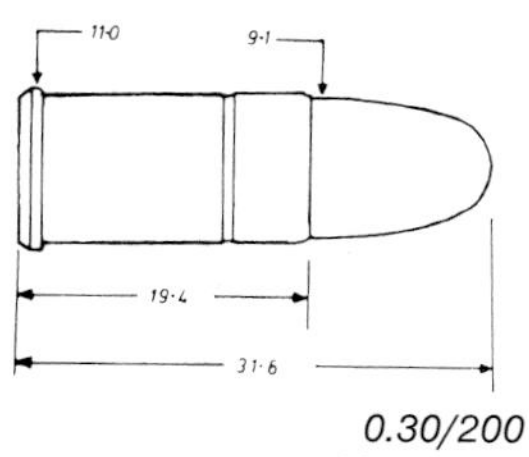

0.30/200
Vo 180 m/s
Eo 186 J

Data
BRITISH SERVICE MK 2
Round length: 31.11 mm
Case length: 19.38 mm
Rim diameter: 11 mm
Bullet diameter: 9.09 mm
Bullet weight: 11.53 g
Muzzle velocity: 180 m/s
Muzzle energy: 186 J

FRANCE
Manufacturer: SFM Défense
Type: Ball: FMJ; 13 g; Vo 185 m/s

SOUTH AFRICA
Manufacturer: Denel (Pty) Limited
Type: Ball: FMJ; 11.53 g; Vo 180 m/s

UPDATED

0.38 Special

Synonyms: 0.38 Smith & Wesson Special

Armament
Suitably chambered revolvers.

Development
This was developed by Smith & Wesson in about 1900 as a possible military round, before the US Army decided that 0.45 was the minimum calibre it would consider for pistols. It was placed on commercial sale in 1902 and attained considerable popularity with police and as a sporting and target round. In more recent years it has been adopted by the US armed forces for military police, aircrew and similar tasks.

Description
A rimmed, straight taper case, brass- or nickel-plated, Boxer or Berdan primed. Various bullets can be found with soft point predominating, but the US military standard Ball M41 uses a lead-cored, jacketed type weighing 8.55 g.

Data
US M41 BALL
Round length: 39.37 mm
Case length: 29.45 mm
Rim diameter: 11 mm
Bullet diameter: 9.04 mm
Bullet weight: 8.55 g
Muzzle velocity: 289 m/s
Muzzle energy: 357 J

ARGENTINA
Manufacturer: Direccion General de Fabricaciones Militares
Type: Ball: Lead-antimony, RN; 10.25 g; Vo 255 m/s (151 mm barrel)

AUSTRIA
Manufacturer: Hirtenberger AG
Type: Ball: JSP; 10.2 g; Vo 385 m/s
Ball: WC; 9.6 g; Vo 225 m/s
Ball: Lead, RN; 10.2 g; Vo 260 m/s
Ball: FMJ; 10.2 g; Vo 260 m/s

BRAZIL
Manufacturer: Companhia Brasileira de Cartuchos
Type: Ball: Lead; 10.2 g; Vo 276 m/s
Ball: JSP; 10.2 g; Vo 276 m/s
Ball: Lead, WC; 9.6 g; Vo 245 m/s

CHILE
Manufacturer: FAMAE
Type: Ball: Lead, RN; 10.15 g; Vo 225 m/s

CZECH REPUBLIC
Manufacturer: Sellier & Bellot
Type: Ball: JHP; 10.25 g; Vo 266 m/s
Ball: Lead, RN; 10.25 g; Vo 266 m/s
Ball: Lead, WC; 9.6 g; Vo 213 m/s

FINLAND
Manufacturer: Lapua Cartridge Factory
Type: Ball: WC; 9.6 g; Vo 240 m/s
Ball: Lead; 10.3 g; Vo 255 m/s

Manufacturer: Sako Limited
Type: Ball: Lead, WC; 9.6 g; Vo 235 m/s
Ball: Lead, SWC; 10.3 g; Vo 235 m/s
Ball: FMJ, RN; 7.5 g; Vo 260 m/s
Ball: JSP; 10.3 g; Vo 275 m/s

FRANCE
Manufacturer: Arcane
Type: MP Ball: Conical, pure copper; 4.54 g; Vo 551 m/s

Manufacturer: SFM Défense
Type: Ball: Lead, RN; 10.4 g; Vo 240 m/s
Ball: Lead, WC; Vo 220 m/s
Ball: JSP; 3.9 g; Vo 470 m/s
Ball: JSP; 10.4 g; Vo 320 m/s
Ball: JHP; 9.7 g; Vo 290 m/s
Ball: JHP; 10.4 g; Vo 280 m/s
THV: 2.9 g; Vo 740 m/s; penetrates 3 mm steel at 7 m

GERMANY
Manufacturer: Dynamit Nobel (Geco)
Type: Ball: WC; 9.6 g; Vo 265 m/s
Ball: Lead, RN; 10.25 g; Vo 265 m/s
Ball: Conical SWC, 10.25 g, Vo335 m/s, Sintox, non-toxic
Ball: Plastic; 0.4 g; Vo 800 m/s
Ball: JSP; 10.25 g; Vo 335 m/s
Ball: FMJ, conical; 10.25 g; Vo 335 m/s

INDONESIA
Manufacturer: PT Pindad (Persero)
Type: Ball MU-6TJ: Lead, RN; 10.31 g; Vo 265 m/s

ISRAEL
Manufacturer: Kalia Israel Cartridge Company Limited
Type: Ball: SWC; 9.72 g; Vo 259 m/s
Ball: WC; 9.4 g; Vo 229 m/s

Manufacturer: TAAS - Israel Industries Limited
Type: Ball: WC; 9.59 g; Vo 219 m/s
Ball: SWC; 10.24 g; Vo 232 m/s
Ball: Lead, HP, SWC; 10.24 g; Vo 232 m/s
Ball +P: JHP; 7.13 g; Vo 402 m/s

0.38 S & W Special

Ball +P: JHP; 8.1 g; Vo 287 m/s
Ball +P: JHP; 10.24 g; Vo 279 m/s
Ball +P: JSP; 10.24 g; Vo 279 m/s
Ball +P: SWC; 10.24 g; Vo 271 m/s
Ball +P: Lead, HP, SWC; 10.24 g; Vo 271 m/s

ITALY
Manufacturer: Fiocchi Munizioni SpA
Type: Ball: FMJ; 10.2 g; Vo 370 m/s
Ball: Lead; 10.2 g; Vo 260 m/s
Ball: WC; 9.55 g; Vo 230 m/s
Ball: JHP; 8.1 g; Vo 350 m/s
Ball: JHP; 9.6 g; Vo 335 m/s

KOREA, SOUTH
Manufacturer: Poongsan Metal Corporation
Type: Ball 38A: Lead, RN; 10.24 g; Vo 250 m/s
Ball 38B: SWC; 10.24 g; Vo 250 m/s
Ball 38C: WC; 9.59 g; Vo 222 m/s
Ball +P 38D: JHP; 8.05 g; Vo 297 m/s
Ball +P 38E: JHP; 7.13 g; Vo 328 m/s
Ball +P 38F: JHP; 9.72 g; Vo 270 m/s
Ball 38G: FMJ; 8.55 g; Vo 256 m/s
Ball 38H: JSP; 8.1 g; Vo 296 m/s
Ball 38AC: Lead, copper-plated, RN; 10.24 g; Vo 242 m/s
Ball 38BC: SWC, copper-plated; 10.24 g; Vo 241 m/s

MEXICO
Manufacturer: Aguila Industrias Tecnos
Type: Ball: FMJ; 10.2 g; Vo 230 m/s
Ball: Lead; 10.2 g; Vo 193 m/s
Ball: Lead, WC; 9.6 g; Vo 216 m/s

POLAND
Manufacturer: Mesko Zaklady Metalowe
Type: Ball: JSP; 10.2 g; Vo 265 m/s
Ball: Lead, RN; 10.3 g; Vo 215 m/s
Target: Wadcutter; 9.6 g; Vo 170 m/s
Target: Plastic, RN; 3 g; Vo 300 m/s

SOUTH AFRICA
Manufacturer: Denel (Pty) Limited
Type: Ball: JHP; 9.1 g; Vo 286 m/s
Ball: WC; 9.6 g; Vo 216 m/s
Ball: SWC; 10.2 g; Vo 230 m/s
Ball: Lead, RN; 10.2 g; Vo 230 m/s
Ball: FMJ; 10.2 g; Vo 230 m/s
Ball: SWC +P; 10.2 g; Vo 279 m/s

Manufacturer: Musgrave (Pty) Limited
Type: Ball: JSP; 10.24 g; Vo 274 m/s
Ball: JHP; 10.24 g; Vo 274 m/s

SPAIN
Manufacturer: SANTA BARBARA SA
Type: Ball: JSP; 10.25 g; Vo 265 m/s
Ball: WC; 9.65 g; Vo 250 m/s

SWEDEN
Manufacturer: Norma AB
Type: Ball 19119: JHP; 7.1 g; Vo 470 m/s
Ball 19110: WC; 9.6 g; Vo 235 m/s
Ball 19112: Lead, RN; 10.2 g; Vo 265 m/s
Ball 19114: FMJ, SWC; 10.2 g; Vo 274 m/s
Ball 19124: JSP; 10.2 g; Vo 274 m/s
Ball 19125: JHP; 10.2 g; Vo 274 m/s

TURKEY
Manufacturer: Makina ve Kimya Endüstrisi Kumuru (MKEK)
Type: Ball: Lead, RN; 10.2 g; Vo 265 m/s
Target: Wadcutter; 9.5 g; Vo 250 m/s

UNITED KINGDOM
Manufacturer: Cobra Gun Company
Type: Ball: HSA Flechette

Manufacturer: Conjay Firearms & Ammunition Limited
Type: Ball +P: CBX; 5.51 g; Vo ?
Ball +P: CBXX; 6.8 g; Vo ?
Ball +P: CBAP; 5.51 g; Vo 690 m/s; penetrates 5.5 mm RHA (480-530 Brinell, 52 Rockwell)

UNITED STATES OF AMERICA
Manufacturer: 3-D
Type: Ball: JHP; 8.1 g; Vo 282 m/s
Ball: Lead, SWC; 10.24 g; Vo 233 m/s
Ball: Lead, SWC; 8.1 g; Vo 251 m/s
Ball: Lead, WC; 6.48 g; Vo244 m/s
Ball: Lead, hollow base WC; 9.59 g; Vo236 m/s

Manufacturer: Black Hills Ammunition Company
Type: Ball +P: JHP; 8.1 g; Vo 320 m/s
Ball: Lead, SWC, 10.2 g; Vo 259 m/s

Manufacturer: CCI
Type: Shot: 7 g; Vo 305 m/s
Ball: JHP; 7.1 g; Vo 305 m/s
Ball: JHP; 8.1 g; Vo 288 m/s
Ball: JHP; 9 g; Vo 283 m/s
Ball: WC; 9.6 g; Vo 217 m/s
Ball: SWC; 10.2 g; Vo 230 m/s
Ball: Lead; 10.2 g; Vo 230 m/s

Manufacturer: Cor-Bon
Type: Ball +P: JHP; 7.5 g; Vo 381 m/s
Ball +P: JHP; 10.2 g; Vo 305 m/s
Ball +P: JHP; 7.5 g; Vo 442 m/s
Ball +P: JHP; 8 g; Vo 412 m/s

Manufacturer: Delta Defense Inc
Type: Frangible: 5.51 g; 386 m/s

Manufacturer: Federal Cartridge Company
Type: Ball: WC; 9.6 g; Vo 216 m/s
Ball: Lead; 10.2 g; Vo 230 m/s
Ball: Lead; 10.2 g; Vo 230 m/s
Ball: Lead, SWC; 10.2 g; Vo 230 m/s
Ball: Lead, SWC; 10.2 g; Vo 279 m/s
Ball: Lead, ogival; 10.2 g; Vo 279 m/s
Ball: JHP; 8.4 g; Vo 288 m/s
Ball: JHP; 7.1 g; Vo 311 m/s
Ball: JSP; 8.1 g; Vo 251 m/s

Manufacturer: Glaser Safety Slug Inc
Type: Ball: Glaser Blue; 5.18 g; Vo 457 m/s
Ball +P: Glaser Blue; 5.18 g; Vo 503 m/s
Ball: Glaser Silver; 5.18 g; Vo 457 m/s
Ball +P: Glaser Silver; 5.18 g; Vo 503 m/s

Manufacturer: Olin Defense Systems
Type: Ball M41: FMJ; 8.42 g; V5 290 m/s

Manufacturer: Remington Arms Company Inc
Type: Ball +P: JHP; 6.2 g; Vo 358 m/s
Ball +P: JHP; 7.1 g; Vo 310 m/s

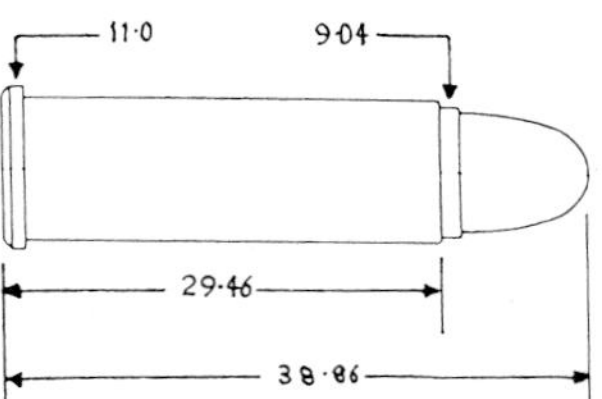

0.38 S & W Special
Vo 260 m/s
Eo 346 J

Ball +P: JHP; 8.1 g; Vo 288 m/s
Ball: Lead, WC; 10.2 g; Vo 230 m/s
Ball: Lead, RN; 10.2 g; Vo 230 m/s
Ball: Lead; 10.2 g; Vo 230 m/s
Ball: SWC; 10.2 g; Vo 230 m/s
Ball +P: Lead; 10.2 g; Vo 287 m/s
Ball +P: Lead, HP; 10.2 g; Vo 278 m/s
Ball: Lead; 13 g; Vo 193 m/s

Manufacturer: Winchester
Type: Ball: JHP; 6.16 g; Vo 288 m/s
Ball: JHP; 7.1 g; Vo 193 m/s
Ball: JHP; 7.1 g; Vo 303 m/s
Ball: JHP; 8.1 g; Vo 303 m/s
Ball: FMJ; 10.2 g; Vo 230 m/s
Ball: MP; 9.7 g; Vo 262 m/s
Ball: MP; 10.2 g; Vo 230 m/s
Ball: JHP; 10.2 g; Vo 230 m/s
Ball: Lead; 9.7 g; Vo 335 m/s
Ball: Lead, SWC; 10.2 g; Vo 230 m/s
Ball: Lead, WC; 9.6 g; Vo 216 m/s
Ball: Lead, 'Match'; 13 g; Vo 190 m/s
Ball: Lead, 'Super Match'; 10.2 g; Vo 230 m/s

YUGOSLAVIA (SERBIA AND MONTENEGRO)
Holding company: YUGOIMPORT - SDPR
Type: Ball: Lead, RN; 10.25 g; Vo 264 m/s
Ball: Lead, WC; 9.6 g; Vo 252 m/s
Ball: SWC; 10.25 g; Vo 264 m/s
Ball: FMJ; 10.2 g; Vo 264 m/s
Ball: JHP; 10.2 g; Vo 254 m/s
Ball +P: Lead, RN; 10.25 g; Vo 312 m/s
Ball +P: SWC; 10.25 g; Vo 312 m/s
Ball +P: JHP; 10.2 g; Vo 312 m/s

UPDATED

0.40 Smith & Wesson

Synonyms: 0.40 Auto

Armament
Suitably chambered Glock, SIG, Smith & Wesson and other pistols and Heckler and Koch sub-machine guns.

Development
This cartridge was developed by Smith & Wesson in the late 1980s as their response to the 10 mm Auto cartridge. The principal difference is the shorter case. This means that a pistol frame originally designed for 9 mm Parabellum can be easily modified to 0.40, whereas modification of a design to take the longer 10 mm Auto round requires a considerable change to the magazine well. The calibre has been adopted by several influential makers and appears to have become more popular than the 10 mm round. It has not yet been formally adopted by any military force, but police and security forces in the USA have taken it into service.

Description
The 0.40 Smith & Wesson is a rebated rim cartridge, the extraction rim being somewhat smaller in diameter than the case. The case is of brass, Boxer primed and straight tapered. Bullets are generally semi-jacketed hollow point, though a full metal jacketed round has been developed for possible military use.

0.40 Smith & Wesson

Data
Round length: 28.7 mm
Case length: 21.6 mm
Rim diameter: 10.77 mm
Bullet diameter: 10.11 mm
Bullet weight: 11.55 g
Muzzle velocity: 285 m/s
Muzzle energy: 469 J

ISRAEL
Manufacturer: TAAS - Israel Industries Limited
Type: Ball: FMJ; 11.66 g; Vo 290 m/s
Ball: JSP; 11.66 g; Vo 290 m/s
Ball: JHP; 11.66 g; Vo 290 m/s

ITALY
Manufacturer: Fiocchi Munizioni Spa
Type: Ball: JHP; 9.4 g; Vo 365 m/s

KOREA, SOUTH
Manufacturer: Poongsan Metal Corporation
Type: Ball: HP; 11.7 g; Vo 300 m/s

UNITED STATES OF AMERICA
Manufacturer: 3-D
Type: Ball: FMJ; 11.66 g; Vo 290 m/s
Ball: JHP; 11.66 g; Vo 290 m/s
Ball: JHP; 10.04 g; Vo 335 m/s
Ball: Lead, SWC; 11.66 g; Vo 274 m/s

Manufacturer: Black Hills Ammunition Company
Type: Ball: JHP; 10.1 g; Vo 351 m/s
Ball: JHP; 11.7 g; Vo 290 m/s

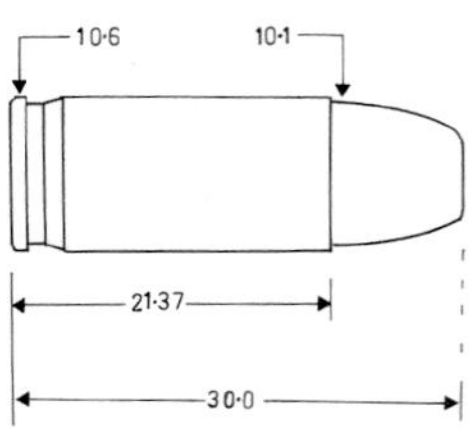

0.40 Smith & Wesson
Vo 260 m/s
Eo 346 J

Manufacturer: CCI
Type: Blazer: JHP; 11.7 g; Vo 300 m/s
Blazer: FMJ, Low Toxicity; 11.7 g; Vo 305 m/s
Blazer: JHP; 10.1 g; Vo 358 m/s
Gold Dot: JHP; 11.7 g; Vo 300 m/s
Gold Dot: JHP; 10.1 g; Vo 358 m/s

Manufacturer: Cor-Bon
Type: Ball +P: JHP; 8.8 g; Vo 396 m/s
Ball +P: JHP; 9.7 g; Vo 366 m/s
Ball +P: JHP; 10.7 g; Vo 343 m/s
Ball +P: JHP; 11.7 g; Vo 320 m/s

Manufacturer: Federal Cartridge Company
Type: Ball: HP; 11.7 g; Vo 300 m/s
Ball: JHP; 10.1 g; Vo 348 m/s

Manufacturer: Glaser Safety Slug Inc
Type: Ball: Glaser Blue; 7.45 g; Vo 472 m/s
Ball: Glaser Silver; 7.45 g; Vo 472 m/s

Manufacturer: Hornady Manufacturing Corporation
Type: Ball: JHP; 10.1 g; Vo 360 m/s
Ball: JHP; 11.7 g; Vo 290 m/s

Manufacturer: Impact Premium Ammunition
Type: Ball: JHP; 11.7 g; Vo 290 m/s
SWC: Lead; 11.7 g; Vo 274 m/s

Manufacturer: Remington Arms Company Inc
Type: Ball: JHP; 10.1 g; Vo 348 m/s
Ball: JHP; 11.7 g; Vo 300 m/s

Manufacturer: Winchester
Type: Ball: JHP; 11.7 g; Vo 300 m/s
Ball: FMJ; 10.1 g; Vo 343 m/s
Subsonic Ball: JHP; 11.7 g; Vo 302 m/s

YUGOSLAVIA (SERBIA AND MONTENEGRO)
Holding company: YUGOIMPORT - SDPR
Type: Ball: JHP; 11.7 g; Vo 286 m/s
Ball: JSP; 12.3 g; Vo 281 m/s

UPDATED

10 mm Auto

Synonyms: 10 mm Bren Ten

Armament

Suitably chambered Glock, SIG, Smith & Wesson, Colt and Heckler and Koch pistols, Heckler and Koch MP5/10 sub-machine gun, Steyr Tactical Machine Pistol.

Development

The 10 mm Auto cartridge was developed with the 10 mm Bren Ten automatic pistol from 1980 to 1983, by Dornaus & Dixon in co-operation with the Norma ammunition company. The object was to provide a pistol round with recoil comparable to the 0.45 ACP but with greater velocity and energy. The Bren Ten pistol failed, but Norma produced ammunition and various companies contemplated pistols. Colt eventually took the lead with their 10 mm Delta, followed by Smith & Wesson, and the US FBI then decided upon the 10 mm as their service round. This gave it a degree of official approval, and other makers began producing pistols and sub-machine guns.

Description

A straight taper rimless cartridge case in brass, Boxer primed. The bullet may be hollow point or full metal jacket with lead core.

10 mm Auto

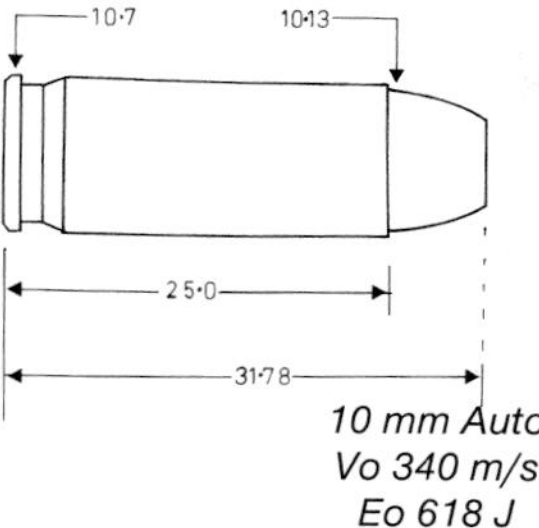

10 mm Auto
Vo 340 m/s
Eo 618 J

Data

NORMA 11002
Round length: 31.78 mm
Case length: 25 mm
Rim diameter: 10.72 mm
Bullet diameter: 10.13 mm
Bullet weight: 10.7 g
Muzzle velocity: 340 m/s
Muzzle energy: 618 J

ITALY
Manufacturer: Fiocchi Munizioni Spa
Type: Ball: FMJ; 13 g; Vo 290 m/s

KOREA, SOUTH
Manufacturer: Poongsan Metal Corporation
Type: Ball: SP; 11.7 g; Vo 290 m/s
Type: Ball: HP; 13 g; Vo 320 m/s
Type: Ball: FMJ; 11 g; Vo 366 m/s
Target: SWC; 13.6 g; Vo 366 m/s

SWEDEN
Manufacturer: Norma AB
Type: Ball 11001: FMJ; 13 g; Vo 340 m/s
Ball 11002: JHP; 13 g; Vo 340 m/s

UNITED STATES OF AMERICA
Manufacturer: Black Hills Ammunition Company
Type: Ball: JHP; 10.1 g; Vo 381 m/s
Ball: JHP; 11.7 g; Vo 290 m/s

Manufacturer: CCI
Type: Ball: FMJ; 13 g; Vo 320 m/s
JHP: 11.7 g; Vo 351 m/s

Manufacturer: Federal Cartridge Company
Type: Ball: JHP; 10.1 g; Vo 404 m/s
Hydra-Shok: 11.7 g; Vo 290 m/s
Ball: JHP; 11.7 g; Vo 314 m/s
Ball: FMJ; 13 g; Vo 290 m/s

Manufacturer: Glaser Safety Slug Inc
Type: Ball: Glaser Blue; 7.45 g; Vo 503 m/s
Ball: Glaser Silver; 7.45 g; Vo 503 m/s

Manufacturer: Hornady Manufacturing Corporation
Type: Ball: HP; 10.1 g; Vo 430 m/s
Ball: HP; 11.7 g; Vo 386 m/s
Ball: HP; 13 g; Vo 351 m/s
Subsonic Ball: HP; 11.7 g; Vo 290 m/s
Ball: FMJ; 13 g; Vo 351 m/s

Manufacturer: Impact Premium Ammunition
Type: Ball: JSP; 11.7 g; Vo 351 m/s
Ball: FMJ; 11.7 g; Vo 328 m/s
Ball: Lead, SWC; 14.3 g; Vo 290 m/s

Manufacturer: Remington Arms Company Inc
Type: High Velocity Ball: FMJ; 11.7 g; Vo 378 m/s
Subsonic Ball: FMJ; 11.7 g; Vo 290 m/s
Ball: SP; 11 g; Vo 410 m/s
Ball: SP; 13 g; Vo 354 m/s

Manufacturer: Winchester
Type: Subsonic Ball: SP; 11.7 g; Vo 302 m/s
Ball: HP; 11.4 g; Vo 393 m/s

YUGOSLAVIA (SERBIA AND MONTENEGRO)
Holding company: YUGOIMPORT - SDPR
Type: Ball: JHP; 11.7 g; Vo 364 m/s
Ball: JSP; 12.3 g; Vo 354 m/s

UPDATED

0.45 Auto Colt

Synonyms: 0.45 ACP; 11.43 × 23 mm Norwegian Colt

Armament

Colt M1911, M1911A1 automatic pistols; Smith & Wesson, Star, Obregon, Hafdasa and pistols based upon the M1911; Smith & Wesson and Colt M1917 revolvers; Thompson, Reising, M2, M3 and M3A1 sub-machine guns.

Development

The 0.45 ACP was the standard US pistol cartridge from 1911 until the adoption of the 9 mm Parabellum round in 1985 and is still in general use. It was adopted by Norway and several Central and South American countries and has been used by the British and Commonwealth forces at various times. A popular commercial round, weapons have been made to suit it in many countries.

Description

A rimless, straight taper case which may be of brass or steel, Boxer or Berdan primed. An enormous variety of bullets has been developed for this round, but the standard military loading has always been a 15.16 g lead-cored, full metal jacket round-nosed type.

Data

US BALL M1911
Round length: 32.19 mm
Case length: 22.79 mm
Rim diameter: 11.86 mm
Bullet diameter: 11.43 mm

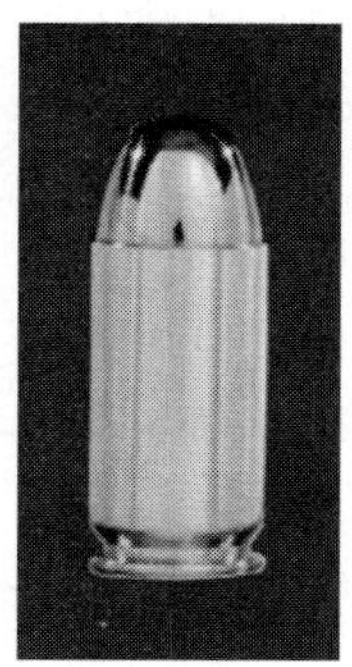

0.45 ACP

Bullet weight: 15.16 g
Muzzle velocity: 250 m/s
Muzzle energy: 474 J

ARGENTINA
Manufacturer: Direccion General de Fabricaciones Militares
Type: Ball FMK 1 Mod C: FMJ, RN; lead core in brass jacket; 14.9 g; Vo 255 m/s

AUSTRIA
Manufacturer: Hirtenberger AG
Type: Ball: FMJ; 14.9 g; Vo 265 m/s

BRAZIL
Manufacturer: Companhia Brasileira de Cartuchos
Type: Ball: FMJ; 14.9 g; Vo 255 m/s
Ball: FMJ, SWC; 12 g; Vo 235 m/s

GERMANY
Manufacturer: Dynamit Nobel (Geco)
Type: Ball: FMJ; 14.95 g; Vo 265 m/s

GREECE
Manufacturer: PYRKAL: Greek Powder & Cartridge Company
Type: Ball: FMJ; 14.9 g; Vo 261 m/s

ISRAEL
Manufacturer: Kalia Israel Cartridge Company Limited
Type: Ball: Lead; 13.6 g; Vo 256 m/s
Ball: Lead, SWC; 12.56 g; Vo 256 m/s

Manufacturer: TAAS - Israel Industries Limited
Type: Ball: FMJ; 14.9 g; Vo 257 m/s
Ball: JHP; 11.99 g; Vo 283 m/s
Ball: Metal-Cased SWC; 11.99 g; Vo 236 m/s
High-Velocity Carbine Ball: Metal-Cased SWC; 11.99 g; Vo 335 m/s

KOREA, SOUTH
Manufacturer: Poongsan Metal Corporation
Type: Ball 45A: FMJ; 14.9 g; Vo 273 m/s
Ball 45B: JHP; 14.9 g; Vo 275 m/s
Ball: FMJ, SWC; 13 g; Vo 259 m/s

SOUTH AFRICA
Manufacturer: Musgrave (Pty) Limited
Type: Ball: FMJ; 14.26 g; Vo 257 m/s
Ball: JHP; 12.96 g; Vo 257 m/s

TURKEY
Manufacturer: Makina ve Kimya Endüstrisi Kumuru (MKEK)

FRANCE
Manufacturer: Arcane
Type: MP Ball: Conical, pure copper; 7.9 g; Vo 493 m/s

Manufacturer: SFM Défense
Type: Ball: FMJ; 14.9 g; Vo 270 m/s
Ball: Lead; 14.9 g; Vo 230 m/s
Ball: THV; 3.9 g; Vo 780 m/s

Type: Ball: FMJ; 14.9 g; Vo 265 m/s

UNITED STATES OF AMERICA
Manufacturer: 3-D
Type: Ball: JHP; 12.96 g; Vo 256 m/s
Ball: JHP; 11.98 g; Vo 312 m/s
Ball: FMJ; 14.9 g; Vo 251 m/s
Ball: Lead, RN; 14.9 g; Vo 236 m/s

Manufacturer: Delta Defense Inc
Type: Frangible: 8.1 g; Vo 307 m/s

Manufacturer: Glaser Safety Slug Inc
Type: Ball: Glaser Blue; 9.4 g; Vo 412 m/s
Ball: Glaser Silver; 9.4 g; Vo 412 m/s

Manufacturer: Government contractors
Ball M1911: FMJ, RN; 15.16 g; 260 m/s
Tracer M26: FMJ; red trace visible 15 to 150 m; 13.15 g; Vo 270 m/s
Blank M9: Mouth tapered and sealed with red lacquered disc

Manufacturer: Olin Defense Systems
Type: Ball M1911: FMJ; 14.9 g; $V_{7.7}$ 260 m/s
Match Ball M1911: FMJ; 14.9 g; $V_{7.7}$ 260 m/s

Manufacturer: Remington Arms Company Inc
Type: Ball: Jacketed WC; 12 g; Vo 235 m/s
Ball: JHP; 12 g; Vo 286 m/s
Ball: FMJ; 15 g; Vo 246 m/s
Targetmaster Ball: FMJ; 15 g; Vo 246 m/s

Manufacturer: Winchester
Type: Ball: FMJ; 14.9 g; Vo 246 m/s
Ball: FMJ, SWC; 12 g; Vo 262 m/s
Ball: JHP; 12 g; Vo 304 m/s

YUGOSLAVIA (SERBIA AND MONTENEGRO)
Holding company: YUGOIMPORT - SDPR
Type: Ball: FMJ; 14.85 g; Vo 257 m/s

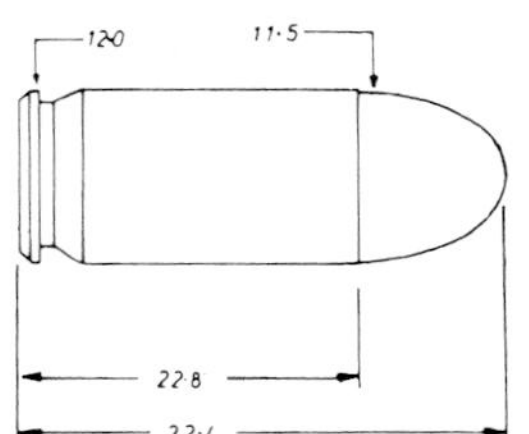

0.45 ACP
Vo 250 m/s
Eo 474 J

Ball: JHP; 12 g; Vo 274 m/s
Ball: FMJ; 13 g; Vo 268 m/s
Ball: Lead, RN; 14.9 g; Vo 257 m/s
Ball: JHP; 12 g; Vo 269 m/s
Ball: FMJ; 13 g; Vo 263 m/s
Ball: FMJ; 14.85 g; Vo 252 m/s
Ball: Lead, RN; 14.9 g; Vo 252 m/s

UPDATED

0.50 Browning Machine Gun

Synonyms: 12.7 × 99 mm

Armament

Browning M2 and M2 HB machine guns; also used in Barrett, Mitchell and similar single-shot sniper rifles. CIS 0.50 machine gun.

Development

The 12.7 × 99 mm cartridge was developed in the 1920s, its design being greatly influenced by the German 13 mm Tank und Flieger (TuF) machine gun round. The 0.50 M2 Browning gun was originally devised as an anti-aircraft weapon, but during 1940 to 1945 it achieved prominence as an aircraft, tank and infantry support gun. Since then it has been widely distributed and adopted by almost every country outside the former Soviet Bloc. Ammunition has been manufactured in many countries but largely to American specifications, although since the 1970s several manufacturers have developed more potent rounds of their own design.

Description

A rimless, brass, bottle-necked case, which may be Berdan or Boxer primed. The standard bullet is ball using a steel core in lead support, steel jacket and gilding metal envelope. In view of the tactical use of this weapon, AP-I and similar bullets with better terminal effect are used more commonly today. One widely used round is the multipurpose NM 140 round which combines both armour piercing and incendiary effects for numerous tactical applications ranging from the attack of light armoured vehicles to explosive ordnance disposal. This round has been taken into US service as the MK211 API.

Standard US rounds include the Ball M33, Tracer M17, Blank M1A1 and API M8. The Saboted Light Armor Penetrating (SLAP) M903 armour-penetrating round has a tungsten penetrator which can penetrate 19 mm of steel at 1,370 m. Its Tracer equivalent is the M962.

Data

BALL M33
Round length: 137.8 mm
Case length: 99.1 mm
Rim diameter: 20.3 mm
Bullet diameter: 12.96 mm
Bullet weight: 42.9 g
Muzzle velocity: 887 m/s
Muzzle energy: 16,876 J

ABU DHABI
Manufacturer: Adcom Manufacturing
Type: Ball: FMJ; lead core; 45 g; Vo 887 m/s
Tracer, AP-I and AP-I-T are also available but no details have been provided.

ARGENTINA
Manufacturer: Direccion General de Fabricaciones Militares
Type: Ball FMK 1 Mod 0: FMJ; steel core in lead sleeve, gilding metal jacket; 45 g; Vo 802 m/s
Tracer FMK 5 Mod 0: FMJ; steel core in lead sleeve, tracer capsule; 43.5 g; Vo 812 m/s
AP FMK 6 Mod 0: FMJ; steel core; 45 g; Vo 800 m/s
AP FMK 9 Mod 0: FMJ; steel core; 45 g; Vo 800 m/s. Fitted with electric primer and used as a subcalibre training round for the 105 mm FMK 4 Mod 0 recoilless gun
Blank FMK 7 Mod 0: Brass case, rose crimped

BELGIUM
Manufacturer: FN Herstal SA
Type: Ball M33: FMJ; 43 g; Vo 925 m/s
Tracer M17: FMJ; base tracer, red trace to 1,500 m; 41.7 g; Vo 925 m/s
AP M8: FMJ; AP steel core; 43 g; Vo 925 m/s; penetrates 15 mm steel plate 450 Brinell at 200 m crosses trajectory with M33 at 500 m
Incendiary: FMJ; 40.33 g; Vo 925 m/s
AP-I M8: FMJ; AP steel core; incendiary composition in tip; 43 g; Vo 925 m/s; penetrates 15 mm steel plate of 450 Brinell at 200 m. Ballistically equivalent to M33
AP-I-T M20: FMJ; base tracer; red trace visible to 900 m; AP steel core; incendiary composition in tip 40.4 g; Vo 925 m/s penetrates 15 mm steel plate of 450 Brinell at 200 m. Crosses trajectory with M33 and M8 at 500 m
AP-E-I FN169: This round uses a bullet with a solid steel body, supporting a heavy alloy penetrative core inside a brass jacket. The space between core and jacket is filled with a non-phosphorus explosive/incendiary mixture. 43 g; Vo 890 m/s. The core will defeat 6 to 11 mm armour steel (400 Brinell) up to 1,000 m range, and will also defeat plastic armour at angles up to 80°. A 1 mm steel or 2 mm dural plate is sufficient to detonate the projectile, which occurs about 20 cm after piercing the target skin.
Blank: Brass case, crimped

BRAZIL
Manufacturer: Companhia Brasileira de Cartuchos
Type: Ball M2: FMJ; 45.2 g; V_{25} 856 m/s
Ball M33: FMJ, SL; steel core; 42.4 g; V_{25} 887 m/s
Tracer M1: FMJ; red trace to 1,600 m; 42.6 g; V_{25}823 m/s
Tracer M17: FMJ; red trace visible to 1,800 m; 40.8 g; V_{25} 872 m/s
Tracer M33: FMJ; red trace; 37.8 g; V_{25} 887 m/s
AP-I-T M20: FMJ, SL; steel core; incendiary tip filler, red trace; 39.7 g; V_{25} 887 m/s
AP M2: FMJ, SL; steel core; 45.5 g; V_{25} 856 m/s
AP-HC: FMJ; alloy steel core; 49.2 g; V_{25} 830 m/s
Incendiary M1: FMJ, SL; steel sleeve; 40.8 g; V_{25}900 m/s
AP-I M8: FMJ, AP; steel core; incendiary tip filler; 42.6 g; V_{25} 887 m/s
AP-I-HC: FMJ; alloy steel core; incendiary composition in nose; 50 g; V_{25} 830 m/s

CANADA
Manufacturer: SNC Industrial Technologies Inc
Type: Ball M2: FMJ; 44.5-44.97 g; Vo 856 m/s

0.50 Browning Ball M33

Ball M33: FMJ; 41.4-42.9 g; Vo 881 m/s
Tracer M17: FMJ; red trace visible to 1,850 m; 43 g; Vo 872 m/s
AP-T C44: FMJ; tungsten carbide core; red trace, dark ignition; visible 150 to 1,200 m; 62 g; Vo 785 m/s; penetration 22 mm/0°/800 m
Blank C48: Brass case, star crimped

EGYPT
Manufacturer: Shoubra Company
Type: Ball: FMJ; S/L; steel core; 43.35 g; Vo 860 m/s
AP-I: FMJ; steel core with incendiary filling in nose; 43.35 g; Vo 860 m/s
Tracer: FMJ; cark ignition tracer, duration 2 seconds. 43.7 g; Vo 845 m/s

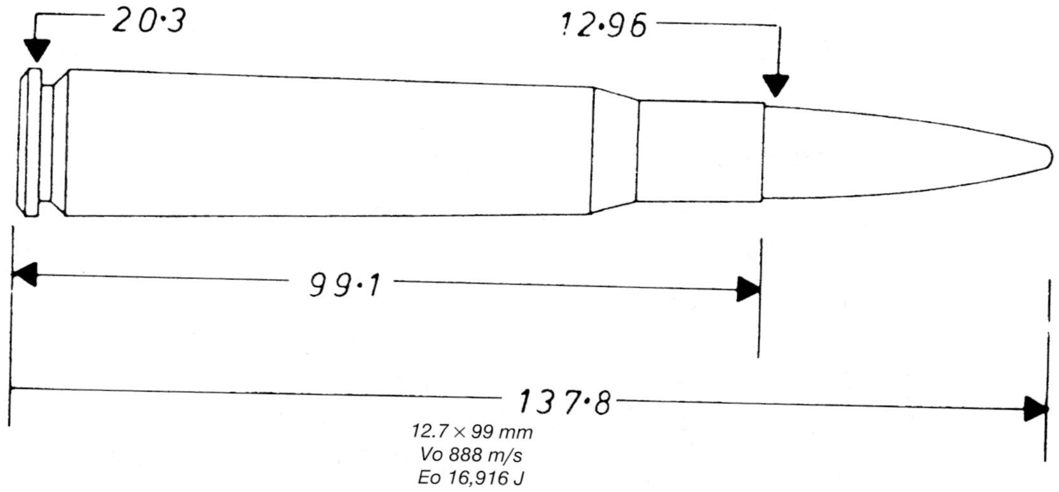

12.7 × 99 mm
Vo 888 m/s
Eo 16,916 J

AP-I-T: FMJ, steel core, dark ignition tracer, duration 2 seconds; 41.6 g; Vo 845 m/s

FRANCE
Manufacturer: Giat Industries (Manurhin)
Type: Ball M33: FMJ, SL; steel core; 41.7 g; Vo 900 m/s
Tracer M33T: FMJ; red trace; 42.7 g; Vo 900 m/s
Anti-Ricochet Balle O: 40.1 g; Vo 910 m/s
Anti-Ricochet Tracer T: 40.1 g; Vo 910 m/s
AP M2: FMJ, SL; steel core; 41.85 g; Vo 900 m/s
AP-I M8: FMJ; AP steel core; incendiary tip filler; 42.2 g; Vo 900 m/s
AP-I-T M20: FMJ, SL; steel core; incendiary tip filler, red trace; 39.66 g; Vo 900 m/s
AP-I-HC: FMJ; hard core; incendiary tip filler; 51 g; Vo 882 m/s

Manufacturer: SFM Défense
Type: Ball: FMJ; ? g; Vo 860 m/s
Tracer: no details
AP: no details
AP-I: no details
AP-I-T: no details
PPI: 47.82 g; Vo 890 m/s; penetration >13 mm RHA at 1,200 m

GERMANY
Manufacturer: Dynamit-Nobel
Type: Plastic Practice Ball M858: This is a one-piece plastic moulded case and bullet attached to a metal extracting rim. The bullet section is solid, but the junction of the bullet section and case mouth section is weak and will shear under the pressure of the propellant gas, so that the bullet breaks free and is fired from the weapon. The round is intended for use on scaled practice ranges and is produced to meet a US military specification. Bullet weight 3.2 g; Vo 1,075 m/s. The bullet has a velocity of 850 m/s at 23.70 m from the gun muzzle and a maximum range of 700 m. The maximum effective training range is 150 m
Practice Plastic Tracer M860: This is similar to the practice ball round described previously but the bullet is counterbored at the base to receive a tracer filling giving a red trace from 20 to 150 m range. 3.2 g; Vo 1,045 m/s

GREECE
Manufacturer: PYRKAL: Greek Powder & Cartridge Company
Type: Ball M33: FMJ, SL; soft steel core; sodium monohydrate point filler; 42.5 g; Vo 880 m/s
AP-I M8: FMJ; manganese molybdenum steel core; point filler of incendiary composition and lead-antimony base seal; 43 g; Vo 880 m/s
AP-I-T M20: As for M8 but tracer in place of base seal; 40.1 g; Vo 880 m/s
Tracer M17: FMJ; lead-antimony core, base tracer; 41.55 g; Vo 880 m/s
Tracer (Pyrkal): Details not known; accuracy better than M17; Vo 880 m/s
Multi-Purpose: Details not known, but presumably licensed version of Raufoss design

INDONESIA
Manufacturer: PT Pindad (Persero)
Type: AP MU-3P: FMJ; 45.9 g; Vo 900 m/s
AP-T MU-3PN: FMJ; 42.9 g; Vo 900 m/s
API-T MU3PBN: FMJ; 42.9 g; 900 m/s

ISRAEL
Manufacturer: TAAS - Israel Industries Limited
Type: Ball M2: FMJ; 45.88 g; Vo 915 m/s
Ball M33: FMJ; steel core; 42.9 g; Vo 915 m/s
Ball M33 Mod: FMJ; hard core; 42.5 g; Vo 915 m/s
Tracer M17: FMJ; 41.66 g; Vo 915 m/s
AP-I-T M20: FMJ; hard core; 40.11 g; Vo 915 m/s
AP-I M8: FMJ; hard core; 42.5 g; Vo 915 m/s
AP M2: FMJ; hard core; 45.88 g; Vo 915 m/s
Hollow Point Test: FMJ; steel core; 42.5 g; pressure 4,700 kg/cm

ITALY
Manufacturer: Europa Metalli
Type: Ball M33: FMJ; steel core; 42 g; V_{24} 890 m/s
Tracer M33T: FMJ; steel core; dark ignition trace visible to 1,500+ m; 40 g; V_{24} 890 m/s
AP-I M8: FMJ; steel core; 43 g; V_{24} 890 m/s. Penetration: 22 mm MS plate at 100 m.
AP-I-T M20: FMJ; steel core; red dark ignition trace 40 to 1,500+ m. 40 g; V_{24} 890 m/s. Penetration: 22 mm MS plate at 100 m.
AP-I-HC (MD-SMI-3 type): FMJ; tungsten carbide core; 46 g; $V_{24?}$ 860 m/s. Penetration: 30 mm MS plate at 0° at 200 m.
Blank: Brass case; rose crimp.

KOREA, SOUTH
Manufacturer: Poongsan Metal Corporation
Type: Ball M33: FMJ; 43 g; Vo 890 m/s
Tracer M17: FMJ; dark ignition red trace to 1,600 m; 43 g; Vo 870 m/s
AP M2: FMJ; steel core; 46 g; Vo 885 m/s
AP-I M8: FMJ; AP steel core; 43 g; Vo 870 m/s
AP-I-T M20: FMJ; steel core; incendiary tip filler, red trace to 1,600 m; 45.5 g; Vo 870 m/s

NETHERLANDS
Manufacturer: Eurometaal NV
Type: AP 2000: FMJ; ballistic conformity to Ball M2, M33, API M8 and so on, but with superior penetrative capability. Will effectively penetrate up to 40 mm RHA or 20 mm HHA plates at various angles and ranges. Post armour effect is significant
API 2000: Similar armour-defeating performance to the AP 2000 but with added incendiary effect

NORWAY
Manufacturer: Raufoss Ammunisjonsfabrikker A/S
Type: Multipurpose NM 140: The exterior of this round conforms to the standard US M2 Ball, but the bullet is a design patented by Raufoss which incorporates penetrative, incendiary and fragmentation effects. NM 140 will defeat 16 mm of armour (Brinell 360) at 30° at 400 m; after penetration it will explode and produce about 20 effective fragments inside the target. There will also be a shower of incendiary particles which are still effective 15 m behind the target plate. No mechanical fuze is used because the need for detonators or sensitive high explosives has been eliminated. Muzzle velocity 900 m/s, bullet weight 43 g
Multipurpose NM 140A1: This is similar to the NM 140 but with some minor changes to meet a US Navy specification

0.50 SLAP round by Chartered Industries of Singapore

Multipurpose Tracer NM 160: This is the NM 140 but with the addition of a base tracer element. The inclusion of this reduces the effective fragmentation by about 30 per cent but does not affect the penetrative or incendiary performance. The tracer is a dark ignition type which is invisible to 50 m from the muzzle, fully visible from 200 to 1,500 m.
AP-S NM 173: This has ballistic conformity to US Ball M2 but is characterised by its penetration effect. Penetration of 20 mm armour plate at 0° NATO at 1,000 m and 11 mm armour plate at 30° NATO at 1,500 m.

PAKISTAN
Manufacturer: Pakistan Ordnance Factories
Type: AP-I: FMJ; hard core; incendiary filling in nose; ? g; Vo 833 m/s
AP-I-T: FMJ; hard core; incendiary and tracer fillings; ? g; Vo 818 m/s

SINGAPORE
Manufacturer: Chartered Industries of Singapore
Type: Ball M33: FMJ; 42.9 g; Vo 887 m/s
Tracer M17: FMJ; red trace to not less than 1,463 m; 41.65 g; Vo 872 m/s
AP-I M8: Vo 887 m/s; conforms to US MIL-SPEC MIL-C-3066B
AP-I-T M20: Vo 887 m/s; conforms to US MIL-SPEC MIL-C-3066B
APHCI: AP hard core incendiary; Vo 870 m/s; penetration 15 mm RHA at 500 m
SLAP: APDS subprojectile in plastic sabot; 27 g; Vo 1,185 m/s; penetration 25 mm RHA (321-375 BH) at 1,000 m. For use in CIS 0.50 machine gun only

SOUTH AFRICA
Manufacturer: Denel (Pty) Limited
Type: Ball: FMJ; ? g; Vo 830 m/s
Tracer: FMJ; ? g; Vo 830 m/s; red trace 15 to >900 m
AP-I: FMJ; steel core; ? g; Vo 876 m/s; penetration 15 mm RHA at 183 m with incendiary effect
API HC: FMJ; hard core; ? g; Vo 876 m/s; penetrates 15 mm RHA at 1,000 m with incendiary effect
Incendiary-Tracer: Brass jacket, tracer and incendiary elements; ? g; Vo 876 m/s; red trace >1,700 m; penetration 10 mm RHA at 1,000 m

SPAIN
Manufacturer: SANTA BARBARA SA
Type: AP-I M8: FMJ; 42.5 g; Vo 895 m/s
AP-I-T M20: FMJ; hard core; 40.1 g; Vo 895 m/s

TURKEY
Manufacturer: Makina ve Kimya Endüstrisi Kumuru (MKEK)
Type: Ball M33: FMJ; 43 g; Vo 890 m/s
Tracer M17: FMJ; red trace to 1,500 m; 43 g; 870 m/s
AP-I M8: FMJ; AP steel core; 43 g; Vo 870 m/s
AP-I-T M20: FMJ; steel core; incendiary tip filler, red trace; 45.3 g; Vo 865 m/s

UNITED STATES OF AMERICA
Manufacturer: Government contractors
Type: Ball M2: FMJ, SL; steel core; 46.01 g; Vo 858 m/s
Ball M33: FMJ, SL; steel core; 42.9 g; Vo 888 m/s
AP M2: FMJ, SL; steel core; 45.88 g; Vo 885 m/s
Tracer M10: FMJ; red trace 205 to 1,450 m; 41.67 g; Vo 873 m/s
Tracer M17: FMJ; red trace 100 to 1,450 m; 41.67 g; Vo 873 m/s
Tracer, Headlight, M21: FMJ; red trace 200 to 500 m; 45.3 g; Vo 867 m/s
AP-I M8: FMJ, SL; steel core; incendiary tip filler; 42.06 g; Vo 888 m/s
AP-I-T M20: FMJ, SL; steel core; incendiary tip filler, red trace 100 to 1,450 m; 39.66 g; Vo 888 m/s
Incendiary M1: FMJ, SL; steel sleeve; 41.02 g; Vo 901 m/s
Incendiary M23: DMJ; steel incendiary container; 33.18 g; Vo 1,036 m/s

Manufacturer: Olin Defense Systems
Type: SLAP M903: APDS, AP subprojectile in amber-coloured plastic sabot; 26.89 g; V_{25} 1,220 m/s
SLAP/T M962: APDS-T, the AP subprojectile carrying a red tracer being fitted into a red coloured sabot. 26.89 g; V_{25}1,173 m/s
AP-I M8: FMJ; steel core with incendiary material in nose; 40.34 g; V_{25}887 m/s
AP-I Mk211 Mod 0 (Multipurpose): Based upon the Norwegian Raufoss NM-140, made under licence. Similar performance. 43.48 g; V_{25}899 m/s
AP-I-T M20: FMJ; steel core with incendiary material in nose, red trace. 40.11 g; V_{25} 887 m/s
Ball M33: FMJ; 42.77 g; V_{25} 887 m/s
Tracer M17: FMJ, red trace 100 to 1,450 m; 40.82 g; V_{25} 872 m/s
Blank M1A1: Brass case, rose crimp.

YUGOSLAVIA (SERBIA AND MONTENEGRO)
Holding company: YUGOIMPORT - SDPR
Type: Ball M33: FMJ; steel core, lead inner tip; 45 g; Vo 887 m/s
Tracer M17: FMJ; red trace, visible to >1,463 m; 40 g; Vo 887 m/s
AP-I M8: FMJ; AP steel core; thermite incendiary composition in tip; 42.8 g; Vo 887 m/s
AP-I-T M20: FMJ; AP steel core; base tracer, thermite composition in tip; red trace, visible from 91 to >1,463 m; 40.2 g; Vo 887 m/s

UPDATED

12.7 mm Soviet machine gun

Synonyms: 12.7 × 107 mm

Armament
Soviet DShK and NSV heavy machine guns; Chinese Type 77 and W-85 machine guns; V-94 sniper rifle; Hungarian 'Gepard' sniper rifle.

Development
The 12.7 × 107 mm was originally developed for the Degtyarev DK heavy machine gun in the late 1920s, and like its US equivalent the 0.50 Browning, leaned heavily upon the German 13 mm TuF round for its inspiration. It was then used for the Degtyarev DShK series of guns and has remained in use to the present, being widely distributed.

Description
The case is rimless and bottle-necked, of brass or lacquered steel and Berdan primed. The standard bullet, as used with the B-32 round, is a non-streamlined armour-piercing/incendiary pattern with a steel core inside a steel jacket with gilding metal envelope. The core is shorter than the jacket and the space at the front filled with incendiary composition. The BZT API-T round uses a slightly lighter bullet.

Data
AP-I B-32
Round length: 146.8 mm
Case length: 105.9 mm
Rim diameter: 21.6 mm
Bullet diameter: 12.95 mm
Bullet weight: 48.28 g
Muzzle velocity: 840 m/s
Muzzle energy: 15,570 J

BULGARIA
Manufacturer: Kintex
Type: AP-I B-32: FMJ; steel core; bimetal jacket, incendiary mixture in nose; 49 g; Vo 850 m/s
Type: AP-I-T BZT: FMJ; steel core; bimetal jacket, incendiary mixture in nose, tracer capsule in base; 44 g; Vo 850 m/s

CHINA, PEOPLE'S REPUBLIC
Manufacturer: China North Industries (NORINCO)
Type: AP-I-T: FMJ, SL; steel core inside a gilding metal jacket, with incendiary composition in the jacket nose, ahead of the core, and a tracer filling at the rear; 43.2 to 44.6 g, 1.15 g incendiary filling; trace to 1,550 m; Vo 810 to 825 m/s
AP-I: Conventional FMJ with solid steel core and a 1.05 g incendiary filling in the jacket nose; 47.4 to 49 g; Vo 810 to 850 m/s; penetrates 20 mm hard steel plate at normal at 100 m range, and ignites fuel behind a 15 mm armour plate at 70 m range
HE-I: The bullet consists of a hollow steel core filled with 2.5 g of explosive and a hardened penetrative sleeve at the front end of the core containing a detonator, all enclosed in a gilding metal jacket. 44 g; Vo 810 to 825 m/s; will function against a 2 mm dural plate and ignite fuel after penetrating
Blank: This cartridge is distinguished by having the mouth coned around a rounded cover, this is ejected from the gun muzzle on firing and can be dangerous up to 15 m from the muzzle

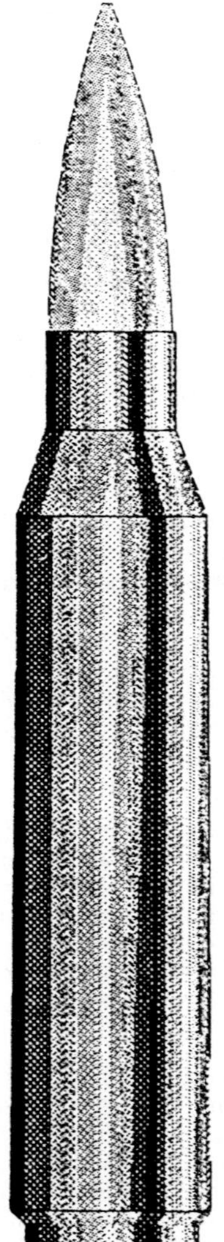

12.7 × 107 mm Soviet

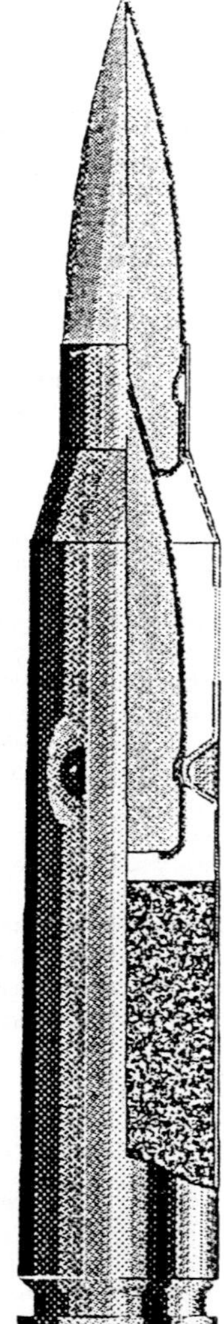

LVE Duplex 12.7 mm round 1SL in part-section. This round was originally developed for use in the Yak aircraft machine gun

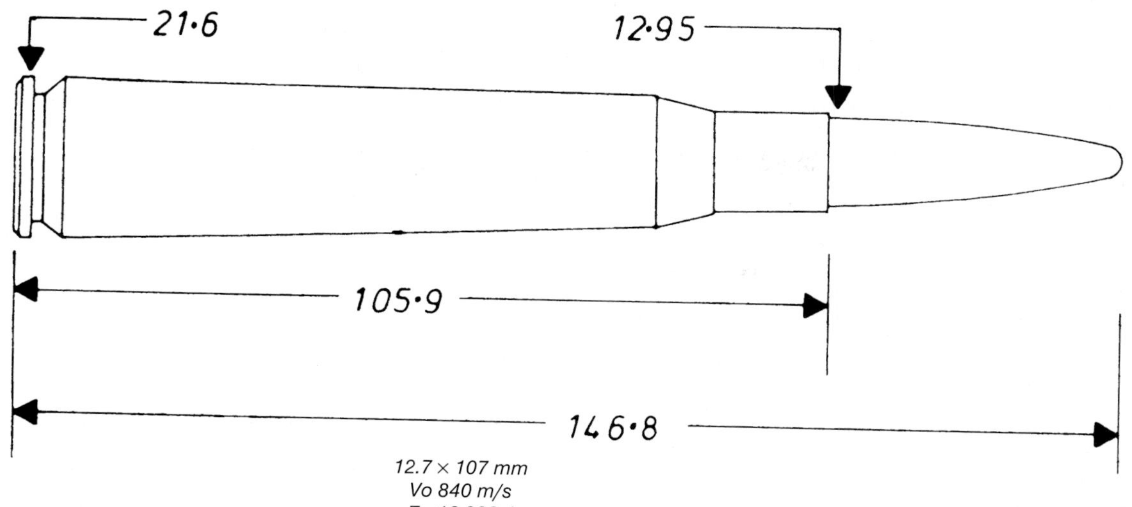

12.7 × 107 mm
Vo 840 m/s
Eo 16,920 J

COMMONWEALTH OF INDEPENDENT STATES
Manufacturer: LVE, Novosibirsk
Type: Duplex AP-I 1SL: This uses the standard cartridge case but carries two bullets; the first bullet is crimped into the neck in the usual way, and has a recessed base into which the tip of the second bullet fits. This second bullet is supported axially by indents in the body of the cartridge case and is flat-based. The bullets conform to the general design of the AP-I B-32, with a hard core and an incendiary charge in the nose. The first bullet weighs about 30 g and has V_{25} of 750 m/s; it can penetrate 5 mm of steel at 1,000 m range. Deviation of the second bullet (which weighs 31 g) is not known
Duplex AP-I-T 1SLT: This is essentially the same as the Duplex 1CL round except that tracer bullets similar to those of the BZT-44 round are used, with an average tracer burning time of 2.9 seconds. The bullets weigh about 27 g each. Velocity is the same as that of the 1CL round.
AP/I B-32: FMJ; steel core; incendiary composition in nose; 49 g; Vo 825 m/s
AP/I/T BZT-44: FMJ; steel core; incendiary composition in nose; base tracer; 44.8 g; Vo 825 m/s
AP/I BS: FMJ; tungsten carbide core with incendiary filling in nose. 56.5 g; Vo 825 m/s

Manufacturer: State arsenals
Type: AP B-30: FMJ, SL; steel core; 51.06 g
AP-I B32: FMJ; steel core, incendiary composition in nose; 48.28 g; Vo 800 m/s
AP-I-T BZT-44: FMJ; steel core, incendiary composition in nose, base tracer; 44.13 g
AP-I: FMJ; steel core, white phosphorus filler; 46.98 g; Vo 800 m/s
AP-I BS-41: FMJ; tungsten carbide core with incendiary composition in nose; 51.9 g; Vo 860 m/s
HE-I: Fuzed bullet filled 1.14 g HE and 1.28 g incendiary composition; 42.76 g; anti-aircraft use

EGYPT
Manufacturer: Shoubra Company
Type: AP-I: FMJ; based on CIS B32; 49.5 g; Vo 815 m/s
AP-I-T: FMJ; based on CIS BZT; 45.5 g; Vo 845 m/s
AP-T B-62: FMJ; tungsten carbide core; red trace to 1,200 m; 62 g; Vo 825 m/s

IRAN
Manufacturer: Defence Industries Organisation
Type: Ball: FMJ; brass case; ? g; Vo 870 m/s

ROMANIA
Manufacturer: Romtehnica
Type: AP-I: FMJ; 43.2-44.6 g; Vo 825 m/s
AP-I-T: FMJ; API-T; 47.4-49 g; Vo 850 m/s
HE-I-T: FMJ; 44 g; Vo 825 m/s
Blank: Cartridge weight 72 g

YUGOSLAVIA (SERBIA AND MONTENEGRO)
Holding company: YUGOIMPORT - SDPR
Type: AP-I B-32: FMJ; AP steel core; thermite composition in tip; 48 g; Vo 825 m/s
AP-I-T BZT-44: FMJ; AP steel core in lead sleeve, base tracer; thermite composition in tip; red trace to 1,000 m; 44 g; Vo 825 m/s
Practice DShK-V: Lead-antimony core in steel jacket; base tracer; 23 g; Vo 1,000 m/s

UPDATED

14.5 mm Soviet machine gun

Synonyms: 14.5 × 114 mm

Armament

KPV heavy machine gun; Chinese Type 75-1 machine gun; Hungarian 'Destroyer' heavy rifle.

Development

The 14.5 × 114 mm cartridge was originally developed in the late 1930s for the contemporary anti-tank rifles. After the rifles became obsolete the round was carried over to the KPV heavy machine gun and has remained in use with this weapon. Gun and ammunition have been widely distributed and ammunition has been made in China, Egypt, North Korea and Romania at various times.

Description

The case is rimless, bottle-necked and of brass, Berdan primed. The standard bullet is an armour-piercing/incendiary pattern with a steel core and incendiary composition in the nose.

Data

AP-I B-32
Round length: 155.8 mm
Case length: 114.3 mm
Rim diameter: 26.9 mm
Bullet diameter: 14.52 mm

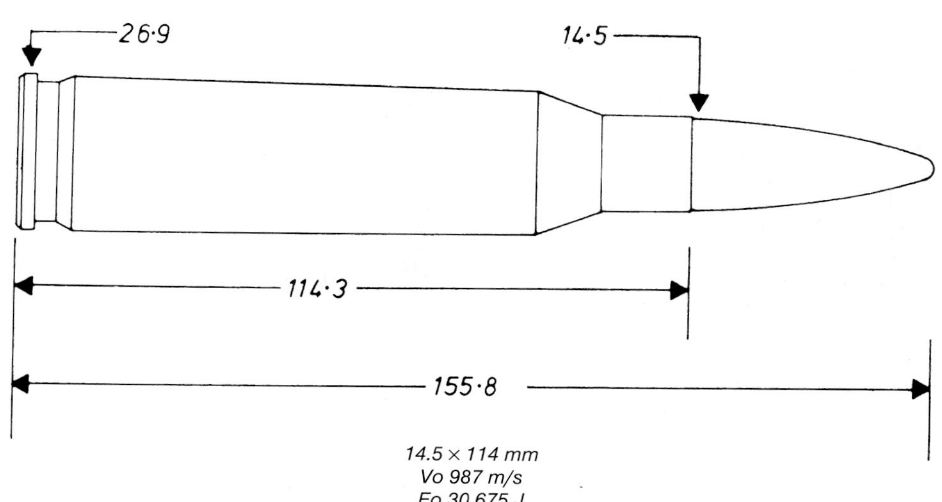

14.5 × 114 mm
Vo 987 m/s
Eo 30,675 J

14.5 × 114 mm AP-I B-32

Bullet length: 68 mm
Bullet weight: 64.1 g
Muzzle velocity: 987 m/s
Muzzle energy: 30,675 J

BULGARIA
Manufacturer: Kintex
Type: AP-I B-32: FMJ; steel core, incendiary composition in the nose; 64.1 g; Vo 987 m/s
AP-I-T BZT: FMJ; 59.6 g; Vo 976 m/s

CHINA, PEOPLE'S REPUBLIC
Manufacturer: China North Industries (NORINCO)
Type: AP-I: FMJ; steel core, incendiary composition in the nose. Appears to be based upon the Soviet B32 pattern. 64 g; Vo 980 to 995 m/s; penetrates 20 mm of armour at 20° at 300 m; ignites fuels behind a 20 mm armour plate at normal at 100 m; cartridge case may be of brass or varnished steel
AP-I-T: FMJ; steel core, tracer capsule at rear, incendiary material in nose; 59 g; Vo 995 to 1,015 m/s; Penetration and incendiary performance as for the AP-I round above. Brass or steel cartridge case
HE-I: Bullet consists of a hollow steel core closed at the rear end and filled with explosive, pressed in two increments with a booster increment on top. The core is closed at the front end by a coned steel plug carrying a detonator and surmounted by a steel striker plug which protrudes from the nose of the bullet. The bullet envelope is of steel with a gilding metal jacket. 59 g; Vo 1,000 to 1,015 m/s; ignites fuels 400 mm beyond a 2 mm dural plate at 200 m and 100 mm behind the plate at 1,500 m range, with considerable fragmentation. Brass or steel cartridge case
Blank: Similar to the 12.7 mm blank, in that the mouth of the steel case is coned around a metal closing cup which is ejected from the muzzle and has a danger area of 25 mm from the muzzle

COMMONWEALTH OF INDEPENDENT STATES
Manufacturer: State arsenals
Type: AP-I B-32: FMJ; steel core; incendiary composition in nose; 63.44 g; Vo 976 m/s; penetration 38 mm at 100 m
AP-I BS-41: FMJ; tungsten carbide core; incendiary composition in nose; 64.41 g; Vo 976 m/s
AP-I-T BZT: FMJ; steel core; incendiary in nose, tracer in base; 59.6 g; Vo 976 m/s
Incdy-T ZP: FMJ; smaller steel core; greater incendiary filling, tracer; 59.6 g; Vo 976 m/s

EGYPT
Manufacturer: Shoubra Company
Type: AP-I: FMJ; based on CIS B-32; 64.2 g; Vo 985 m/s
AP-I-T: FMJ; based on CIS BZT; 59.5 g; Vo 1,000 m/s

POLAND
Manufacturer: Mesko Zaklady Metalowe
Type: AP-I: FMJ; appears to be based on the CIS B32 round but loaded to a higher velocity; 65.1 g; Vo 995 m/s
AP-I-T: FMJ; appears to be based on the CIS BZT round but loaded to a higher velocity; 61.3 g; Vo 1,015 m/s

ROMANIA
Manufacturer: Romtehnica
Type: AP-I: FMJ; based on the CIS B-32 round; 64 g; Vo 995 m/s
AP-I-T: FMJ; based on the CIS BZT round; 59 g; Vo 995 m/s
HEI: FMJ; 59 g; Vo 1,000 m/s
Blank: Domed cover over mouth; cartridge weight 110 g

UPDATED

15.2 mm Steyr AMR

Synonyms: 15.2 mm IWS 2000

Armament
Steyr-Mannlicher 15.2 mm Anti-Materiel Rifle.

Development
Developed from 1988 to 1990 by Steyr-Mannlicher for their Anti-Materiel Rifle (AMR), a heavy rifle designed to attack vulnerable equipment at long range. Development of the weapon and its ammunition is continuing. Originally 15 mm, then changed to 14.5 mm, the calibre has now become 15.2 mm. Development is continuing.

Description
The cartridge is of partially synthetic material, partly metal construction and of conventional bottle-necked form. The projectile is a fin-stabilised 35 g, 5.5 mm tungsten dart with an effective range of 1,000 to 2,000 m depending upon the nature of the target. At 800 m range the dart will penetrate 40 mm of rolled homogeneous armour, with considerable secondary fragmentation behind the plate. The high velocity produces an exceptionally flat trajectory. The vertex at 1,000 m range is only 800 mm above the line of sight. The cartridge uses a long primer to give optimum ignition of the propelling charge, and a pusher plate drives the projectile up the bore.

Data
Round weight: 150 g
Round length: 207 mm
Case diameter: 26 mm
Projectile diameter: 5.5 mm
Projectile weight: 35 g
Muzzle velocity: 1,450 m/s
Muzzle energy: 36,793 J
Max pressure: 4,800 bar

AUSTRIA
Manufacturer: Steyr-Mannlicher GmbH
Type: Fin-Stabilised Discarding Sabot: See above

UPDATED

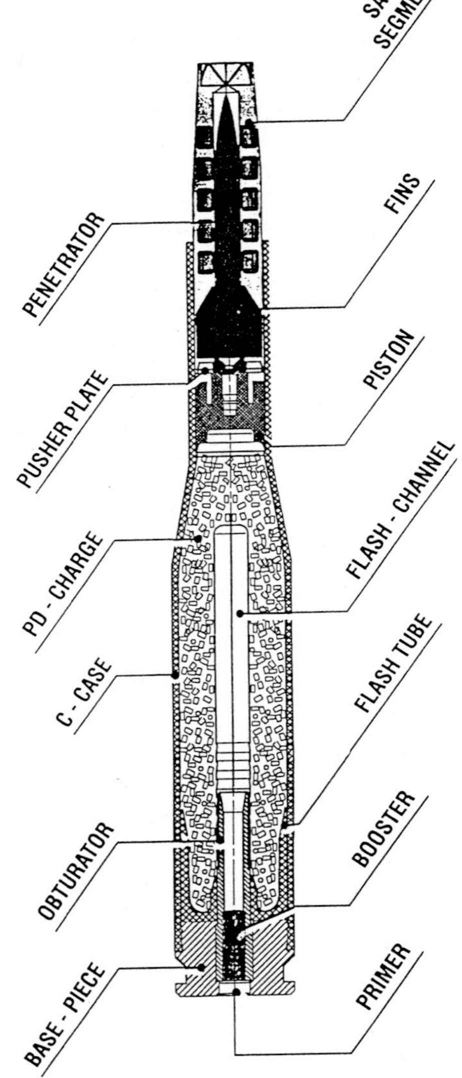

Internal arrangement of the Steyr 15.2 mm FSDS round

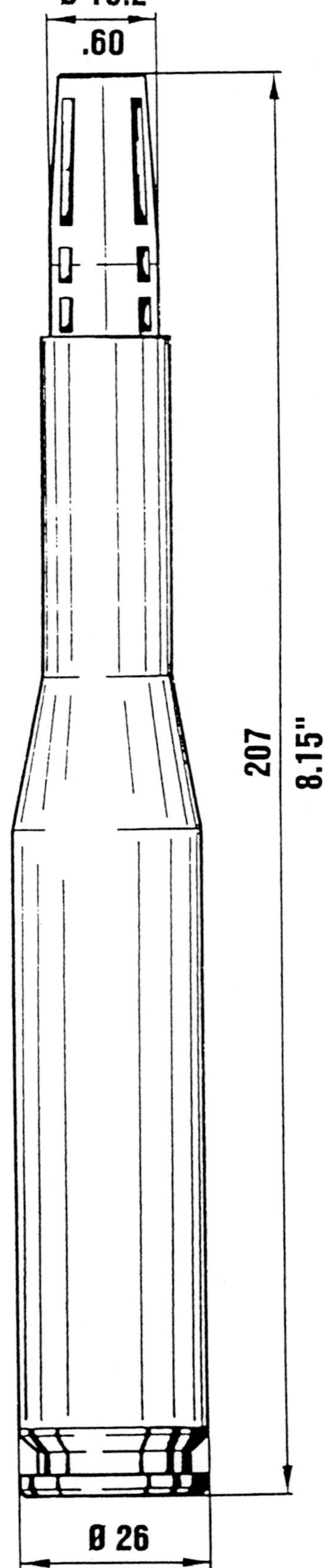

15.2 mm Steyr AMR

20 mm MG151

Synonyms: 20 × 82 mm

Armament

MG 151; ML60; AML Bitube; GA1.

Development

The 20 mm MG151 cartridge was developed by the Mauser company in the 1930s as a possible anti-tank projectile, but was eventually used in aircraft cannon and light anti-aircraft guns used by both Germany and Japan. In post-war years the cannon was manufactured in France and used to arm light armoured vehicles. Manufacture ceased there in about 1970.

Manufacture of the GA1 cannon commenced in South Africa during the late 1980s. The South African cartridge has minor dimensional differences from the original but chambers and functions in the MG 151 and similar weapons.

Description

This cartridge has a short, rimless, bottle-necked brass case, percussion primed. The wartime projectiles were usually those of the 20 × 80RB Oerlikon gun, and post-war production generally followed current Oerlikon designs. South African projectiles use soft iron driving bands to minimise barrel wear.

Data

Round length: 147 mm
Case length: 81.7 mm
Rim diameter: 25.1 mm
Bourrelet diameter: 19.9 mm
Projectile weight: ca 110 g
Muzzle velocity: 720 m/s
Muzzle energy: 28.5 kJ

FRANCE
Manufacturer: Manurhin Defense
Type: HE-I: Steel shell filled Hexal 70/30, nose impact fuze 16/18 Mle 61; 112 g; Vo 720 m/s
HE-I: As above but with nose impact fuze MR21
HEI-T: Steel two-section shell, front filled Hexal 70/30; nose impact fuze 16/18 Mle 61; rear filled red trace to 1,200 m; 112 g; Vo 720 m/s
HE-I-T: As above but with nose impact fuze MR21
AP-I: Pointed steel shell filled incendiary composition from base; 120 g; Vo 720 m/s
AP-T: Pointed steel shell with cavity partly filled inert, remainder filled red trace to 1,500 m; 120 g; Vo 720 m/s; penetration 20 mm at normal engagement ranges
TP: Pointed, flat-nosed shell, empty, 112 g; Vo 720 m/s
TP-T: As above but cavity part-filled inert, remainder with red tracer

SOUTH AFRICA
Manufacturer: Denel (Pty) Limited
Type: HE-I: Steel shell filled Hexal 30, nose impact fuze; soft iron driving band; 110 g; Vo 720 m/s
HE-I-T: Two-section steel shell, front section filled Hexal 30, rear section red tracer; nose impact fuze; 110 g; Vo 720 m/s
SAP-HE-I: Pointed shell filled Hexal P18, with ballistic cap and base fuze; 110 g; Vo 720 m/s; penetration 15 mm armour plate (110 kg/m²) at 100 m range
TP: HE-I shell body filled inert and with dummy fuze; 110 g; Vo 720 m/s
TP-T: As for TP but with tracer; 110 g; Vo 720 m/s

UPDATED

Manurhin 20 mm MG151 complete rounds

South African GA1 HEI and TP rounds

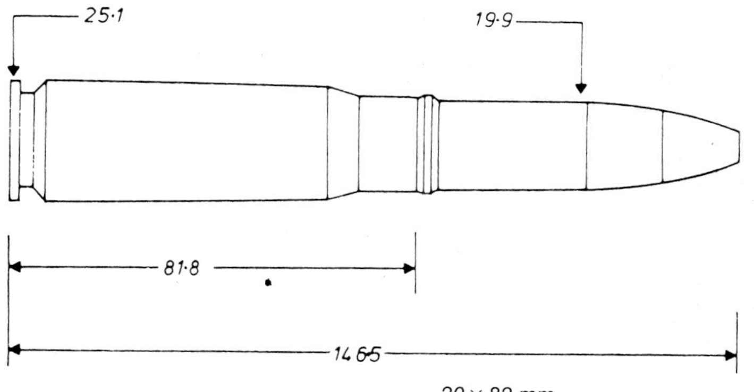

20 × 82 mm

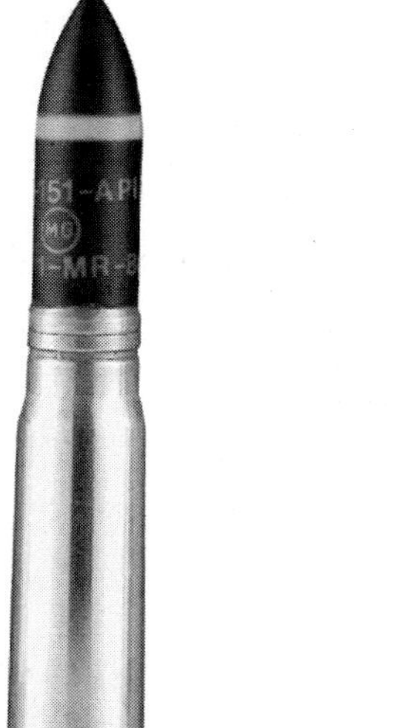
20 mm × 82 mm MG151

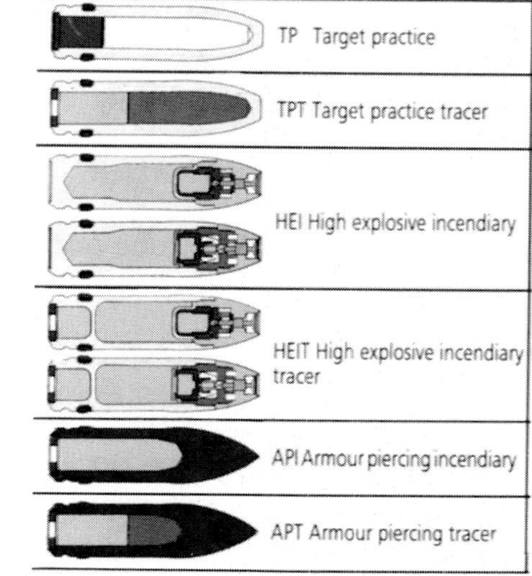

Manurhin 20 mm MG151 projectiles

20 mm Hispano-Suiza HS 404

Synonyms: 20 × 110 mm; 20 mm M24

Armament

Hispano-Suiza HS404, HS804, Mk II and Mk V guns; US M3, M24 and M24A1 guns; UK Mks 2 and 5; Israel TCM-20; Yugoslav M55 guns; Croatian RT-20 sniper rifle.

Development

This cartridge was used extensively by the UK and the USA during the Second World War and large numbers of these guns are still in use. Both countries produced the same range of cartridges in this chambering as their Oerlikon guns; the projectiles were exactly the same, only the cartridge case and propelling charge differed.

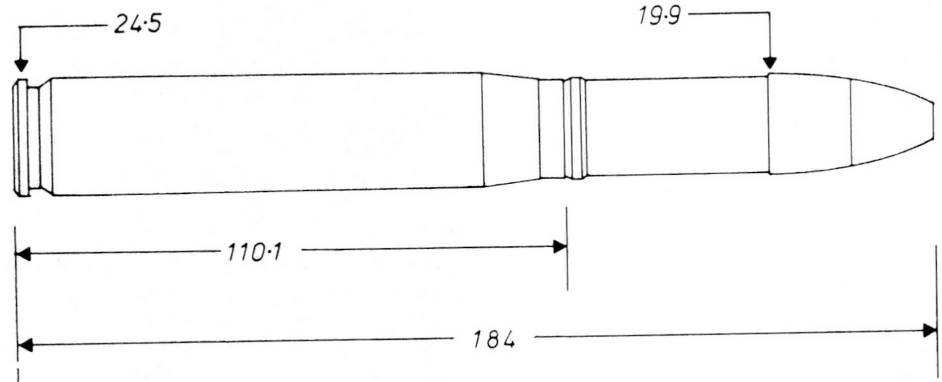

Outline dimensions of 20 mm Hispano-Suiza HS 404 round

Description

The ammunition is almost identical to the 20 mm × 110RB Oerlikon, but the case is a rimless type rather than a rebated rimless. European manufactured ammunition uses Berdan primers, while the American M21, M21A1 and M21A1B1 cases use a Boxer primer. In post-war years the US Air Force developed electrically primed ammunition in this chambering for use in the M24 gun: it was prominently marked 'ELECTRIC' on the side of the cartridge case, and the primer cap exhibits a ring of black insulating material around it. Other nations have adopted electric priming, although the presence of the insulating material is often the only indication.

Projectiles resemble those developed for the Oerlikon 20 × 110RB gun but are notable for having the explosive cavity reduced in diameter at the rear in order to better resist the crushing action set up by the driving band during its passage along the gun bore. For this reason the tracer recess is smaller.

Data

Round length: 184 mm
Case length: 110.1 mm
Rim diameter: 24.5 mm
Bourrelet diameter: 19.9 mm
Projectile weight: 122 g
Muzzle velocity: 844 m/s
Muzzle energy: 43,452 J

ARGENTINA
Manufacturer: Direccion General de Fabricaciones Militares
Type: HE-T FMK 1 Mod 0: Steel shell loaded 5 g Hexogen; tracer capsule in rear; impact fuzed, percussion primed; 118.7 g; Vo 830 m/s. For Oerlikon guns
HE-I FMK 15 Mod 0: Steel shell loaded 4 g of HE-Incendiary material; impact fuzed, electric primed; 110 g; Vo 950 m/s. For EINC guns
HE-I FMK 7 Mod 0: Steel shell loaded 11 g HE-Incendiary material; impact fuzed; percussion primed; 258 g; Vo 845 m/s. For HS 404 guns
TP-T FMK 4 Mod 0: Steel shell loaded 5 g inert ballast; dummy fuze; 124.8 g; Vo 830 m/s. For Oerlikon guns
TP-T FMK 13 Mod 0: Steel shell loaded 4 g inert ballast; dummy fuze; electric primed; 110 g; Vo 950 m/s. For EINC guns
TP-T FMK 6 Mod 0: Steel shell loaded 16 g inert ballast; tracer capsule in rear; percussion primed; 260 g; Vo 845 m/s. For HS 404 guns

A typical Hispano-Suiza HS 404 round

BRAZIL
Manufacturer: Companhia Brasileira de Cartuchos
Type:HE-I M74E1: Steel shell with copper driving band, fuzed impact CBC M74E1 and loaded with 11 g RDX. 100 g; $V_{7.5}$ 810 m/s
HEI-T-SD M72E1: Steel shell with copper driving band, fuzed impact CBC M72E1 and loaded with 7 g RDX. Tracer loaded into the rear of the shell with a heat relay giving self-destruction between 3 and 7 seconds of flight. 101 g; $V_{7.5}$ 875 m/s
TP: Hollow steel shell, flat-nosed, no filling; 123 g; V_{25} 810 m/s
TP-T: As for TP but with tracer capsule inserted in rear

EGYPT
Manufacturer: Maasara Company for Engineering
Type: HE-I: Steel shell, brass case, nose impact fuze. Appears to be based upon the standard Oerlikon HE-I round. No details

FINLAND
Manufacturer: Lapua Cartridge Factory
No information is to hand, beyond the fact that cartridges in this calibre are manufactured by the Lapua company

FRANCE
Manufacturer: Manurhin Defense
Type: HE-I: Steel shell containing 14 g HE/I filling as one prepressed pellet of HE, one of incendiary material and a final pellet of HE, with MR21 or 16/18 Mle 61 impact fuze. 122 g; Vo 840 m/s
HE-I: As above but fuzed MR22 short delay impact fuze
HE-I: As above but fuzed MR2028 impact with self-destruction between 2.5 and 9 seconds
HE-I-T: Steel shell with 8 g filling composed of one pellet of HE and one of incendiary material, with MR212 impact fuze. 128 g; Vo 840 m/s
HE-I-T: As above but fuzed MR22 short delay impact fuze
HE-I-T: As above but fuzed MR2028 impact with self-destruction between 2.5 and 9 seconds
TP: Steel pointed shell, empty; 132 g; Vo 840 m/s
TP-T: Steel pointed shell, part filled inert, remained filled red tracer; 132 g; Vo 840 m/s
AP-T: Hard steel pointed shell, base bored for tracer; 139 g; Vo 830 m/s

ITALY
Manufacturer: SNIA-BPD
Type: HE: Steel shell loaded HE, with self-destruct fuze; 135 g; Vo 810 m/s
HE-I: Steel shell loaded HE-Incendiary mixture, with self-destruct fuze; 135 g; Vo 810 m/s
AP-I: Steel pointed shell with incendiary-filled ballistic cap; 130 g; Vo 815 m/s
AP-T: Steel pointed shot with tracer; 130 g; Vo 815 m/s
SAP-I: Steel shell loaded HE/Incendiary mixture, reinforced nose impact delay fuze; 135 g; Vo 810 m/s
BL: Target practice round. Steel shell shaped to represent HE shell and fuze, hollow, inert; 130 g; Vo 815 m/s
BL-T: As for BL but with tracer capsule in rear; 130 g; Vo 815 m/s
TP: Steel shell body, empty, with dummy fuze; 125 g; Vo 815 m/s

NORWAY
Manufacturer: Raufoss Technology A/S
Type: Multipurpose High Capacity (MPHC-T): Fragmentation and incendiary comparable to standard 20 mm Multipurpose. Optimised penetration capability as a result of the introduction of a hard core penetrator. Tracer optimised for HS 404 guns. 120 g; Vo 904 m/s

SPAIN
Manufacturer: EXPAL SA
Type: HE-I: Steel shell, filled Hexolite, nose impact fuze Mk 26; 122 g; Vo 845 m/s
HE-I-T: Steel shell, filled Hexolite, nose impact fuze Mk 26, red tracer with minimum burning time of 3 seconds; 122 g; Vo 845 m/s
HE-I: Steel shell, filled RDX, nose impact fuze Mk 26; 120 g; Vo 845 m/s
AP-T: Steel pointed shell, filled RDX, unfuzed, red tracer with minimum burning time 3 seconds; 122 g; Vo 845 m/s
TP and TP-T: Use the HE-I and HE-I-T projectile bodies filled inert and with inert fuze; 122 g; Vo 945 m/s

SWITZERLAND
Manufacturer: Oerlikon-Contraves AG
Type: HE-I: Steel shell, filled Hexal, nose impact fuze; 122 g; Vo 830 m/s
HE-I-T: Steel two-section shell, front filled Hexal and with nose impact fuze, rear filled red tracer; 122 g; Vo 830 m/s
AP-I-T: Pointed steel shell with Hexal filling and base delay fuze with red tracer; 113 g; Vo 965 m/s
TP-T: Steel two-section shell body, front empty, rear filled red tracer, dummy fuze; 122 g; Vo 830 m/s

YUGOSLAVIA (SERBIA AND MONTENEGRO)
Holding company: YUGOIMPORT - SDPR
Type: HE-I-T M57: Steel shell filled TNT or RDX/Al; red trace composition in rear, trace for 3 seconds; self-destruct by fuze between 4.5 and 8 seconds; 137 g; Vo 850 m/s
HE-T M57: Steel shell filled TNT or RDX/Al; red trace for 3 seconds; self-destruct between 4.5 and 8 seconds; 137 g; Vo 850 m/s
HE-I M57: Steel shell filled TNT with central filling of incendiary composition; self-destruct 4.5 and 8 seconds; 132 g; Vo 850 m/s
HE M57: Steel shell filled TNT or RDX/Al; self-destruct 4.5 and 8 seconds; 132 g; Vo 840 m/s
AP-I M60: Steel pointed shell filled from the rear with incendiary composition; unfuzed; 142 g; Vo 840 m/s
AP-I-T M60: As for AP-I but with trace burning for 3 seconds; 142 g; Vo 840 m/s
AP-T M60: Steel pointed shot with tracer cavity in rear; red trace for 3 seconds; 142 g; Vo 850 m/s

TP-T M57: This is the same shell as the HE-I M57 but with the forward section filled inert and with dummy fuze. Trace for 3 seconds; 137 g; Vo 850 m/s
TP M57: The same shell as the HE M57 but filled inert and with dummy fuze; 132 g; Vo 850 m/s
TP-T M79: This has the two-section shell of the HE-I M57; the front section is filled with flash composition, and the rear with tracer composition. An impact fuze is fitted. On striking the target the shell explodes with a bright flash. 137 g; Vo 850 m/s. Not for use in cannon: for subcalibre training tubes fitted to larger weapons only
Blank M77: Polyethylene projectile filled with sintered iron powder. For training and test use in M55 AA guns only. The projectile breaks up in the bore, having provided sufficient recoil force to operate the automatic mechanism. Maximum range of fragments and dust 65 m from muzzle. The entire cartridge weighs 195 g

UPDATED

20 mm Oerlikon S

Synonyms: 20 × 110RB

Armament
Oerlikon Mk 4 (US Navy); Oerlikon Mk 2, 3 and 4 (UK); Polsten (UK).

Development
This round was developed by Oerlikon in the 1930s and was subsequently adopted by the UK and the USA during the Second World War. Several other countries took Oerlikon guns of this chambering into service, and ammunition of this pattern is still widely manufactured.

Description
The bottle-necked cartridge case uses a rebated rim, the rim being very obviously smaller than the head of the case. Steel or brass cases may be found, although the former is not common. Ignition is by a conventional percussion cap pressed into the base of the case. Projectiles vary from solid shot to explosive and incendiary projectiles with impact fuzes.

Data
Round length: 181 mm
Case length: 109.8 mm
Rim diameter: 22.2 mm
Bourrelet diameter: 19.8 mm
Projectile weight: 122 g
Muzzle velocity: 830 m/s
Muzzle energy: 42,023 J

BELGIUM
Manufacturer: FN Herstal SA
Type: HE-I FN60: Steel shell, forward end filled TNT, rear of cavity filled incendiary mixture. Compression-ignition strikerless nose impact fuze. 122 g; Vo 830 m/s
HE-I-T FN62: Steel two-part shell, front section filled HE-Incendiary mixture, rear filled tracer. Strikerless nose impact fuze. 120 g; Vo 830 m/s
HE-I-T-SD FN71: As for HE-I-T FN62 but with a heat relay between tracer and explosive for self-destruct; 122 g; Vo 830 m/s
SAP-I-T FN74: Pointed steel shell, filled incendiary composition and red tracer; 122 g; Vo 830 m/s
TP FN65A2: HE-I shell body, filled inert; 122 g; Vo 830 m/s
TP-T FN63A1: HE-I-T shell body, front empty, dummy fuze, rear filled tracer; 120 g; Vo 830 m/s

BRAZIL
Manufacturer: Companhia Brasileira de Cartuchos
Type: HE-I M74E1: Steel shell filled Hexolite, nose impact fuze; 122 g; Vo 845 m/s
HE-I-T M72E1: As for HE-I but two-compartment shell with rear compartment filled red tracer with heat relay for self-destruct. Nose impact fuze. 122 g; Vo 845 m/s
TP: Body of HE-I shell, empty, with dummy fuze; 120 g; Vo 845 m/s
TP-T: Body of HE-I-T shell with dummy fuze and tracer; 123 g; Vo 845 m/s

FRANCE
Manufacturer: Manurhin Defense
Type: HE-I: Steel shell filled 11.5 g Hexal 70/30 and fuzed 16/18 Mle 61 impact fuze; 122 g; Vo 845 m/s
HE-I: As above but with MR21 impact fuze
HE-I: As above but with MR204B impact fuze with self-destruct between 2.5 and 9 seconds
HE-I-T: Steel two-section shell, front filled 4.7 g Hexal 70/30 and with 16/18 Mle 61 impact fuze. Base filled with a red tracer. 122 g; Vo 845 m/s
HE-I-T: As above but with MR21 impact fuze
HE-I-T: As above but with MR204B impact fuze with self-destruct between 2.5 and 9 seconds
AP-T: Steel pointed shell with rear cavity, part filled inert, remainder filled red tracer; 122 g; Vo 845 m/s
TP: Steel shell, filled inert, with plug representing fuze MR21; 122 g; Vo 845 m/s
TP: Steel shell, filled inert, with plug representing fuze MR204B; 122 g; Vo 845 m/s
TP-T: Steel two-section shell; front filled inert, with plug representing fuze MR21; rear section filled red tracer; 122 g; Vo 845 m/s
TP-T: As above but with plug representing fuze MR204B; 122 g; Vo 845 m/s

GREECE
Manufacturer: PYRKAL: Greek Powder & Cartridge Company
Type: HE-I: Steel shell, loaded 11.4 g Hexal, with impact fuze 16/18 Mle 61; 122 g; Vo 845 m/s
HE-I-T: Steel two-section shell, forward section filled 4.7 g Hexal, rear section red tracer; impact fuze 16/18 Mle 61; 123 g; Vo 845 m/s
TP: Body of HE-I shell, empty, with dummy fuze; 120 g; Vo 845 m/s
TP-T: Body of HE-I-T shell with dummy fuze and tracer; 123 g; Vo 845 m/s

ITALY
Manufacturer: SNIA-BPD
Type: HE: Steel shell, loaded Hexal or TNT, nose impact fuze Mk 26; 120 g; Vo 845 m/s
HE-I: As for HE but filled one-third incendiary, two-thirds HE
HE-T: Steel two-section shell, forward section filled HE over incendiary mixture, rear section red tracer; nose impact fuze Mk 26; 122 g; Vo 845 m/s
BL: Target practice round; HE shell body filled inert, with dummy fuze; 120 g; Vo 845 m/s

SPAIN
Manufacturer: EXPAL SA
Type: HE-T: Steel shell, filled RDX, fuzed Mk 26 nose impact fuze; red tracer 3 seconds nominal; 122 g; Vo 845 m/s
HE-I: Steel shell filled Hexolite, nose impact fuze Mk 26; 122 g; Vo 845 m/s
HE-I-T: As for HE-I but two-section shell with rear section filled red tracer burning 3 seconds; nose impact fuze Mk 26; 122 g; Vo 845 m/s
AP-T: Pointed projectile, filled RDX, base fuzed; red trace burning 3 seconds; 122 g; Vo 845 m/s
TP: HE shell body filled inert and with inert fuze; 122 g; Vo 845 m/s
TP-T: HE-I-T shell body with front compartment filled inert, inert fuze; red trace burning 3 seconds; 122 g; Vo 845 m/s

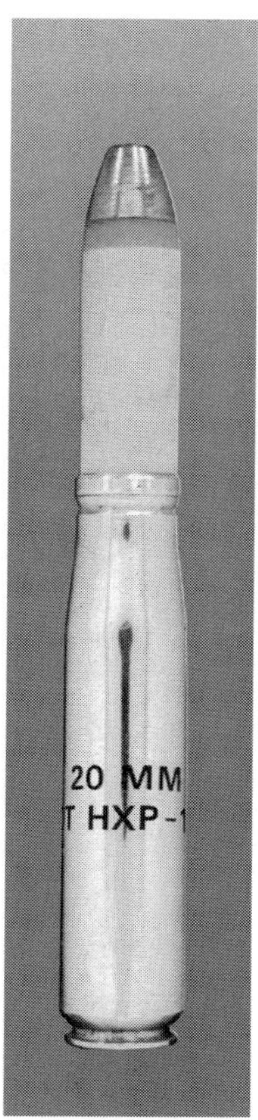

20 mm Oerlikon S

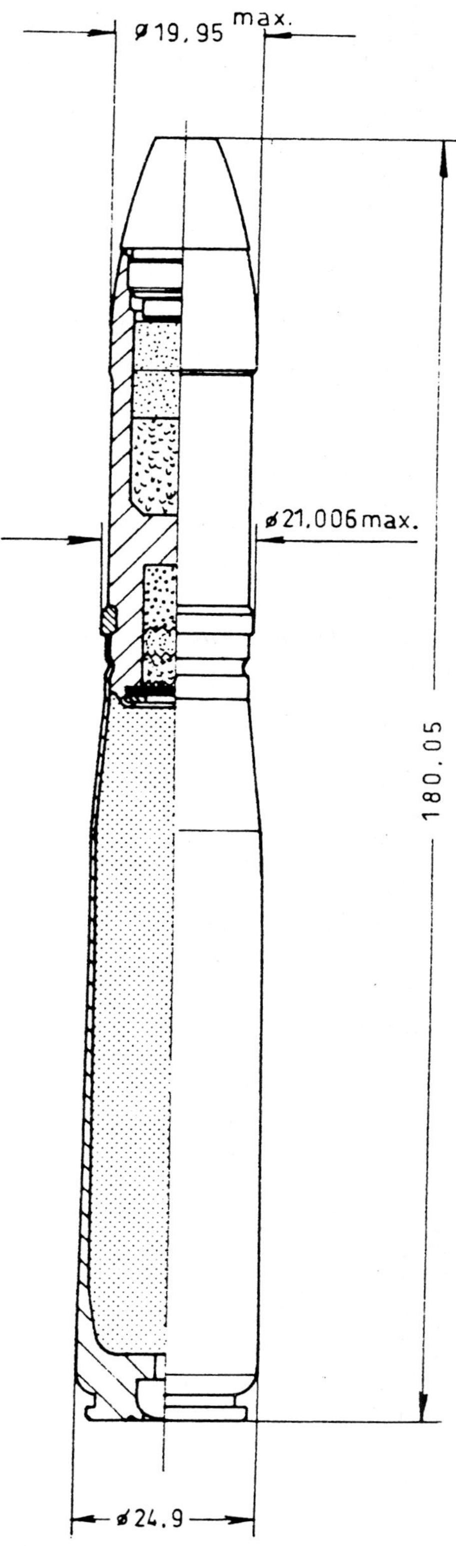

Dimensioned drawing of Expal HE-I-T round showing typical two-section shell construction

UNITED STATES OF AMERICA
Type: HE: Steel shell filled Pentolite or Tetryl, with nose impact fuze Mk 26. 125 g; Vo 830 m/s
HE-I Mk 3: Steel shell filled Pentolite and incendiary composition, nose impact fuze Mk 26. 125 g; Vo 830 m/s
HEI-T Mk 4: As for HE-I but with tracer in shell base and reduced filling of high explosive. 123 g; Vo 830 m/s
AP-T Mk 9: Solid steel shot, bored at rear for tracer. 120 g; Vo 840 m/s

YUGOSLAVIA (SERBIA AND MONTENEGRO)
Holding company: YUGOIMPORT - SDPR
Type: HE: Steel shell filled 12.1 g TNT or RDX/Aluminium, with nose impact fuze; 122 g; Vo 845 m/s
HE-T M62: Steel two-section shell, front filled 6.1 g of TNT or RDX/Aluminium, rear filled red tracer with heat relay for self-destruct after about 3.5 seconds flight; impact nose fuze; 122 g; Vo 845 m/s
HE-I: As for HE but filled 12.1 g RDX/Aluminium and nose impact fuze; 122 g; Vo 845 m/s
HE-I-T: As for HE-T but filled 6.1 g RDX/Aluminium only, with nose impact fuze; 122 g; Vo 845 m/s
AP: Pointed piercing shot with empty balance cavity; 142 g; Vo 844 m/s
AP-T M63: A pointed piercing shot with tracer compartment at the rear; red trace, burning 3.5 seconds; 142 g; Vo 844 m/s; penetration 25 mm armour plate at 100 m
AP-I M63: A pointed piercing shot with a rear compartment filled incendiary composition, unfuzed; 142 g; Vo 844 m/s; penetration 25 mm of armour plate at 100 m
AP-I-T M63: Similar to the AP-I M63 but the rear compartment is partly filled with incendiary composition and the remainder with a 3.5 seconds tracer; 142 g; Vo 844 m/s; penetration 25 mm of armour plate at 100 m
TP: Body of the HE shell, filled inert, with dummy fuze; 122 g; Vo 845 m/s
TP-T M62: Body of the HE-T M62 but with the front compartment filled inert and with an inert fuze; 122 g; Vo 845 m/s

UPDATED

Manurhin 20 × 110RB rounds

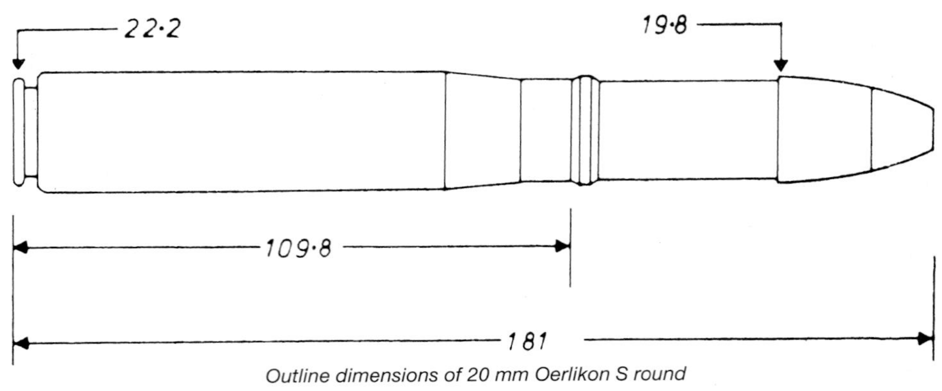

Outline dimensions of 20 mm Oerlikon S round

20 mm Oerlikon KAA

Synonyms: 20 × 128 mm

Armament
Oerlikon KAA, KAB, 5TG, 204GK, GAM-B01, RK206, RK251; Spanish Meroka.

Development
This round was developed by Oerlikon in the 1950s for use in various designs of ground and aircraft guns; rounds for ground guns use percussion primers, while those for aircraft use are electrically primed. In 1972, when Oerlikon and Hispano-Suiza combined, the terminology of various guns was changed, but production continued.

Description
The case is of lacquered steel, with a screwed-in percussion or electric primer. Projectile weights vary between 120 and 130 g, but charges are regulated to give a normal muzzle velocity of 1,050 m/s in an 85 calibre length barrel.

Data
Round length: 203 mm
Case length: 128.7 mm
Rim diameter: 32 mm
Bourrelet diameter: 19.9 mm
Projectile weight: ca 125 g
Muzzle velocity: 1,050 m/s
Muzzle energy: 68.9 kJ

FRANCE
Manufacturer: Manurhin Defense
Type: HE-I: Steel shell, filled Hexal; fuzed MR207 impact fuze with self-destruct between 5 and 9 seconds; 125 g; Vo 1,050 m/s (KAA); 1,100 m/s (KAB)
HE-I-T: Steel two-section shell, front loaded Hexal, rear loaded red tracer; nose impact fuze MR207 giving self-destruct between 5 and 9 seconds; 125 g; Vo 1,050 m/s (KAA); 1,100 m/s (KAB)
AP (HC): Light alloy body with tungsten carbide penetrator; 110 g; Vo 1,150 m/s (KAA); 1,200 m/s (KAB)
AP-T (HC): As for AP (HC) above but with red tracer
SAP-HE-I: Pointed shell with ballistic cap; shell filled HE-Incendiary mixture and fuzed MR208 base fuze with short delay and self-destruct between 5 and 9 seconds; 128 g; Vo 1,050 m/s (KAA); 1,100 m/s (KAB)
SAP-HE-I-T: Characteristics as for SAP-HE-I with the addition of red trace
TP: HE-I shell body, empty, with dummy fuze; 125 g; Vo 1,050 m/s (KAA); 1,100 m/s (KAB)

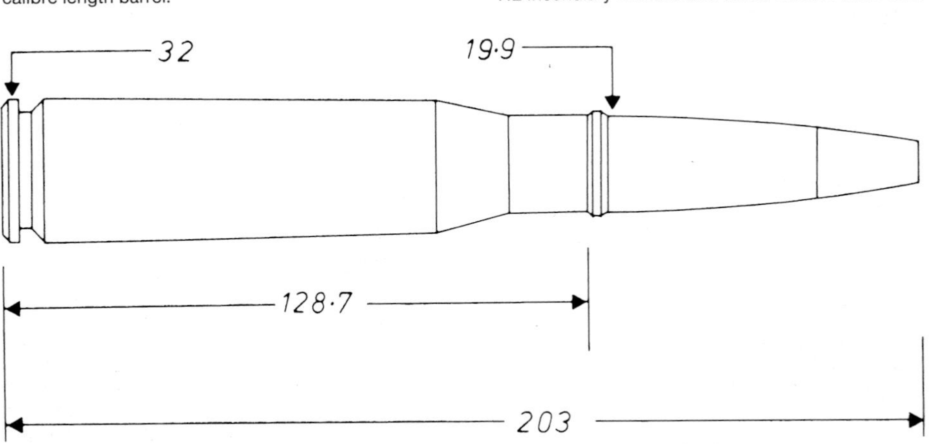

Outline dimensions of 20 × 128 mm Oerlikon KAA round

20 × 128 mm Oerlikon KAA round

TP-T: HE-I-T shell body with front section empty, dummy fuze, rear section filled tracer; 125 g; Vo 1,050 m/s (KAA); 1,100 m/s (KAB)

NETHERLANDS
Manufacturer: NWM de Kruithoorn
Type: Break-up Shot MN13: Plastic shell body containing dust shot. Disintegrates by centrifugal action outside the muzzle. Used for training, practice in confined areas and gun testing. 125 g; Vo 1,050 m/s

SPAIN
Manufacturer: SANTA BARBARA SA
Type: Mine HE-I: Steel thin-walled shell loaded 18 g Hexal, fuzed KZA-199 nose fuze with self-destruct. Electric primed
SAP-HE-I: Steel pointed shell with alloy ballistic cap, loaded 5.4 g Hexal, fuzed BZ-0144 base fuze with self-destruct; electric primed
TP-T: Steel shell with forward section empty and dummy fuze, rear section loaded red tracer; electric primed

SWITZERLAND
Manufacturer: Oerlikon-Contraves AG
Type: HE-I MSA: A steel-bodied shell filled 10 g Hexal and fitted with a nose impact fuze which also provides self-destruct after 4 to 9 seconds; 125 g; Vo (KAA) 1,050 m/s; (KAB) 1,100 m/s
HE-I-T MLA: A steel-bodied shell in two sections, the front section filled with 5.6 g Hexal and fitted with nose fuze providing self-destruct after 4 to 9 seconds, and the rear section filled with tracer composition; 125 g; Vo (KAA) 1,050 m/s; (KAB) 1,100 m/s
APHC-T HLA: Light alloy body carrying a tungsten carbide penetrator and a rear tracer; 110 g; Vo (KAA) 1,150 m/s; (KAB) 1,200 m/s
SAP-HE-I PSA: Pointed steel shell with light alloy ballistic cap, filled 4.7 g Hexal, base delay fuze providing self-destruct; 125 g; Vo (KAA) 1,050 m/s; (KAB) 1,100 m/s
SAP-HE-I-T PLA: As for SAP-HE-I above but with red tracer; 125 g; Vo (KAA) 1,050 m/s; (KAB) 1,100 m/s
TP UGA: Steel shell with empty cavity plugged with dummy fuze; 125 g; Vo (KAA) 1,050 m/s; (KAB)1,100 m/s
TP-T ULA: Similar to the TP UGA but a two-section shell, the rear section being filled with tracer composition. Weight and velocities as for UGA

UPDATED

20 mm Hispano-Suiza HS 820

Synonyms: 20 × 139 mm

Armament
Hispano-Suiza HS 820; Oerlikon KAD, KAD-B; Rheinmetall Rh 202; US M139, French M693/F2; Mauser Model B.

Development
This round was developed by Hispano-Suiza in the late 1940s as a more powerful replacement for the wartime 20 × 110 mm round. The new weapon which accompanied it, the HS 820 gun, became one of the most successful 20 mm designs ever made and was widely purchased throughout the world as an aircraft and ground weapon. As a result of this widespread distribution the cartridge was adopted by other designers who developed suitable guns, notably the German Rheinmetall Rh 202, the French M693 and the American M139.

Description
The rimless, bottle-necked case may be of steel or brass and is percussion primed. A wide variety of projectiles has been made across the years, but those listed below are the current manufacturing standards. Propelling charges are generally regulated to give a muzzle velocity of 1,040 m/s in a 95 calibre barrel.

Data
Round length: 213 mm
Case length: 138.5 mm
Rim diameter: 28.4 mm
Bourrelet diameter: 19.9 mm
Projectile weight: ca 125 g
Muzzle velocity: 1,040 m/s
Muzzle energy: 72.6 kJ

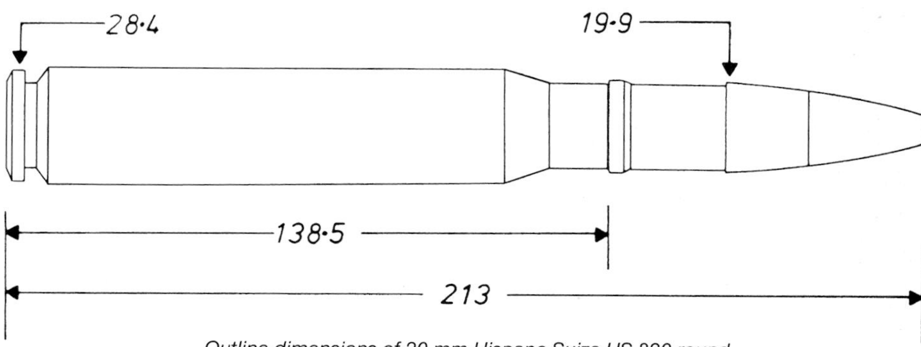

Outline dimensions of 20 mm Hispano-Suiza HS 820 round

ARGENTINA
Manufacturer: Direccion General de Fabricaciones Militares
Type: HE-I: Steel shell with sintered iron driving band; loaded explosive/incendiary mixture; nose impact fuze with self-destruct between 3.5 and 10 seconds; 122 g; 1,041 m/s
AP-T: Hard core, light body, sintered iron driving band; 114 g; 1,090 m/s. Steel cartridge case.
TP: Steel shell, sintered iron driving band, inert aluminium fuze; 122 g; 1,041 m/s
TP-T FMK 8 Mod 0: Steel shell with inert loading and tracer capsule; inert fuze; percussion primed; 317 g; Vo 1,041 m/s

FRANCE
Manufacturer: Manurhin Defense
Type: HE-I: Steel shell filled with 10 g explosive/incendiary mixture. MR26 impact fuze; 120 g; Vo 1,050 m/s
HE-I: As above but fuzed MR201B impact fuze with self-destruct between 3.5 and 9 seconds; 120 g; Vo 1,050 m/s
HE-I: As above but fuzed MR203B delay impact fuze with self-destruct after 3.5 to 9 seconds; 120 g; Vo 1,050 m/s
HE-I-T: Two-section steel shell, front section filled 8.5 g explosive incendiary mixture, rear section filled red tracer. MR26 impact fuze. 120 g; Vo 1,050 m/s
HE-I-T: As above but with MR201B impact fuze with self-destruct between 3.5 and 9 seconds; 120 g; Vo 1,050 m/s
HE-I-T: As above but with MR203B delay impact fuze with self-destruct after 3.5 to 9 seconds; 120 g; Vo 1,050 m/s
AP-I: Hard core in light alloy body with ballistic cap filled with incendiary mixture; 112 g; Vo 1,100 m/s
AP-T: As above but without incendiary filling and body base containing red tracer; 112 g; Vo 1,100 m/s
AP-I-T: As for AP-I but with body base containing red tracer; 112 g; Vo 1,100 m/s
SAP-HE-I: Pointed shell with ballistic cap, filled with explosive/incendiary mixture and with base delay fuze MR205 giving self-destruct between 3.5 and 8 seconds; 120 g; Vo 1,050 m/s
TP: Body of HE-I shell, empty, fitted with dummy fuze; 120 g; 1,050 m/s
TP-T: Body of HE-I-T shell, front section empty, fitted with dummy fuze; 120 g; Vo 1,050 m/s

GERMANY
Manufacturer: Diehl GmbH
Type: Shrapnel DM 111: A hollow shell containing an ejection charge in the base, above which is an ejection tube and an extension tube. Within these tubes are a number of heavy metal balls. The nose of the shell is closed by a ballistic cap. The shell base contains a striker, cap and delay element. On firing, gas pressure drives the striker forward to fire the cap and ignite the delay. After about 40 ms the delay ignites the ejection charge. This forces the ejection tube forward; this presses on the extension tube which shears off the shell nose cap and allows the two tubes and the shot to be ejected. The extension tube is prefragmented and shatters to add to the effect. The ejection tube is retarded, permitting the balls to fly free on a cone of about 3° angle. About 120 fragments are produced, capable of penetrating 2 mm F40 dural at 70 m range. 118 g; Vo 1,055 m/s
HE-I M599: Steel two-section shell, front loaded HE, rear loaded red tracer burning for 3.4 seconds. Fitted nose impact fuze M594 providing self-destruct after 5.5 ±1.5 seconds, or M594A1 providing self-destruct after 5.75 ±2.25 seconds. 122 g; Vo 1,045 m/s
HE-I DM 81: As for M599 but longer shell body and shorter fuze, giving better explosive capacity. Self-destruct by nose fuze after 3.1 or 5.7 seconds nominal time. 120 g; Vo 1,055 m/s
HE-I DM 101: Pointed piercing shell with ballistic cap containing incendiary composition. Shell filled HE, with base fuze giving self-destruct after 3.1 or 5.7 seconds nominal time. 120 g; Vo 1,055 m/s
TP-T DM 48A1: M599 shell body, front filled inert, rear filled red tracer, dummy fuze; 122 g; Vo 1,045 m/s

Manufacturer: Rheinmetall GmbH
Type HE-I: Steel shell filled 6.5 g HE-Incendiary composition, nose impact fuze; 120 g; Vo 1,045 m/s
HE-I-T DM 81: Steel two-section shell, front filled HE-Incendiary composition, nose impact fuze; rear filled red tracer; 120 g; Vo 1,045 m/s
AP-I-T DM 43A1: Hard subprojectile in steel body, light alloy ballistic cap with incendiary filler; red tracer; 111 g; Vo 1,100 m/s. Penetrates 32 mm armour at 1,000 m
APDS-T DM 63: Tungsten carbide subprojectile in light alloy/plastic sabot, red tracer; 108 g; Vo 1,150 m/s
TP DM 48: Steel two-section shell, filled inert, dummy fuze; 120 g; Vo 1,045 m/s
TP-T DM 48A1: Similar to DM 48 but with tracer in rear section; 120 g; Vo 1,045 m/s
Break-up Shot DM 78A1: Plastic body filled with dust shot. When fired it produces sufficient resistance in the bore to actuate automatic and recoil mechanisms but after leaving the bore centrifugal force causes the body to disintegrate within a short distance of the muzzle. For training, practice in confined areas and gun testing. 120 g; Vo 1,045 m/s

GREECE
Manufacturer: PYRKAL: Greek Powder & Cartridge Company
Type: HE-I-T NM75MP: Steel shell with sintered iron driving band, loaded RX51 explosive charge, RS40/41 incendiary charge, and strikerless fuze, with rear tracer burning >4.3 seconds. Self-destruct is provided by means of a heat relay between the tracer and the explosive filling. 120 g; Vo 1,045 m/s
HE-I-T M599: Steel body, sintered iron driving band, 5.8 g of Hexal 70/30 and an impact fuze M594 with self-destruct after 5.5 seconds. A dark ignition tracer is fitted, burning from 300 to about 2,000 m. A tracerless version is also available. 120 g; Vo 1,055 m/s
TP: Steel shell, empty, dummy fuze; 120 g; Vo 1,045 m/s
TP-T: Steel two-section shell, front empty with dummy fuze, rear filled red tracer; 120 g; Vo 1,045 m/s

NETHERLANDS
Manufacturer: NWM de Kruithoorn BV
Type: HE-I-T-SD DM 81: Steel two-section shell, front filled HE-Incendiary composition, rear filled red tracer burning for 3.4 seconds. Nose impact fuze with self-destruct mechanism after 5.5 or 9 seconds. 120 g; Vo 1,045 m/s
AP-I-T DM 43A1: Hard core in steel body, light alloy ballistic cap with incendiary filler; red tracer burning for 1.4 seconds; 111 g; Vo 1,100 m/s

20 mm Hispano-Suiza HS 820 API

APDS-T DM 63: Tungsten carbide subprojectile in plastic discarding sabot, red tracer burning for 1.2 seconds; 108 g; Vo 1,150 m/s
TP DM 48: Steel two-section shell, filled inert, light alloy dummy fuze; 120 g; Vo 1,045 m/s
TP-T DM 48A1: Similar to DM 48 but with tracer in rear section burning for 3.4 seconds; 120 g; Vo 1,045 m/s
TP-T DM 98A2: Steel two-section shell, filled inert; plastic dummy fuze; 120 g; Vo 1,045 m/s
TP-T DM 98A3: Similar to DM 98A2 but with red tracer in rear section burning for 3.4 seconds
Break-up Shot DM 78A2: Plastic body filled with iron powder. Projectile disintegrates within less than 100 m from the muzzle; gun will function as with normal rounds. For training and practice in restricted areas and gun testing. 120 g

NORWAY
Manufacturer: Raufoss Technology A/S
Type: Multipurpose NM75 (MP-T-SD): Steel shell with high-explosive and incendiary filling, the incendiary extending into the nose cap. A red tracer is in the rear, with a heat relay providing self-destruct after 3.7 to 5 seconds flight. The projectile will function against 1.5 mm dural at 1,000 m range, producing a minimum of 10 fragments, and will penetrate 12.7 mm steel plate to 200 m range and an impact angle of 60° NATO. Projectile weight 122 g; Vo 1,045 m/s. Red trace 200 to 2,200 m

SOUTH AFRICA
Manufacturer: Denel (Pty) Limited
Type: APCT: Solid core armour-piercing tracer with ballistic cap; soft iron driving band; steel cartridge case; 110 g; Vo 1,100 m/s; penetration 40 mm/30°/100 m
HE-I: Filled Hexal P30; soft iron driving band, KZA348 nose impact fuze; steel cartridge case; 120 g; Vo 1,050 m/s
HE-I-T: As HE-I but with base tracer
TP: Inert loading; 120 g; Vo 1,050 m/s
TP-T: As for TP but with base tracer

SPAIN
Manufacturer: EXPAL SA
Type: Multipurpose NM75: The shell carries a strikerless compression-ignition fuze in the nose, above a 5.5 g incendiary filling and a 2.8 g explosive filling. On impact the fuze is crushed, igniting the incendiary filling which then ignites the explosive. The explosion gives a forward impulse to the burning incendiary fragments. The rear of the projectile carries a red tracer, which by heat transfer actuates self-destruct between 3.8 and 5.3 seconds. Steel cartridge case DM 1001A1 with DM64 percussion primer. Vo 1,045 m/s
TP-T: Steel two-section shell, front empty with dummy fuze, rear filled red tracer; 120 g; Vo 1,050 m/s

Manufacturer: SANTA BARBARA SA
Type: HE-I: Thin-wall steel shell loaded with 10 g of Hexal and fitted with an impact nose fuze providing self-destruct after 6 to 11 seconds flight; 125 g; Vo 1,040 m/s
SAP-HE-I: Pointed shell, light alloy ballistic cap filled 4.7 g Hexal, base fuze with self-destruct after 6 to 11 seconds; 125 g; Vo 1,040 m/s
TP-T: Steel two-section shell, front empty, rear with tracer; dummy fuze; 125 g; Vo 1,040 m/s

SWITZERLAND
Manufacturer: Oerlikon-Contraves AG
Type: Mine HE-I MSA: A large capacity steel shell loaded with 10 g of Hexal and fitted with an impact nose fuze providing self-destruct after 6 to 11 seconds of flight; 125 g; Vo 1,040 m/s
Mine HE-I-T MLA: Similar to Type MSA but with the shell body divided into two, the front section loaded 5.6 g of Hexal, the rear containing tracer composition; 125 g; Vo 1,040 m/s
SAPHE-I PSA: A pointed shell with light alloy ballistic cap, filled 4.7 g Hexal and fitted with a base fuze providing self-destruct after 6 to 11 seconds of flight; 125 g; Vo 1,040 m/s
SAPHEI-T PLA: Similar to Type PSA but with the addition of a base tracer; 125 g; Vo 1,040 m/s
APHC-T HLA: A light alloy body containing a core of tungsten carbide and a base tracer; 110 g; Vo 1,100 m/s
TP UGA: Uses a similar steel body to the Mine HE-I but of cheaper steel and with a smaller cavity which is left empty and plugged with a dummy fuze; 125 g; Vo 1,040 m/s
TP-T ULA: Uses a similar two-section body to the Mine HE-I-T but with a larger tracer section; plugged with dummy fuze; 125 g; Vo 1,040 m/s

UPDATED

---

23 mm Soviet VYa and ZU-23

Synonyms: 23 × 152B

Armament
ZU-23 and ZSU-23-4 anti-aircraft guns; Volkov-Yarzev VYa-23 aircraft guns; subcalibre devices.

Development
This cartridge was developed for the VYa aircraft cannon in 1940, a weapon widely adopted by the Soviet Air Force, notably in the Ilyushin Il-2 'Stormovik' ground attack fighter. The VYa cannon was replaced by other types, but the cartridge was adopted for the ZU-23 cannon, which first appeared as an aircraft gun but is now used mainly as an air defence gun.

In 1993 it was revealed that the South African National Defence Force had acquired numbers of ZU-23 weapons during the Angolan conflict and that these had been assimilated into South African service as Zumlac self-propelled air defence guns. As a result, manufacture of 23 mm ammunition is carried out in South Africa.

Description
The case is rimless, belted, bottle-necked and of brass or steel. Whilst dimensionally similar, ammunition for the VYa and ZU-23 cannon is not functionally interchangeable. Cartridges for the VYa gun had brass cases and a screwed-in percussion primer. Those for the ZU-23 use steel cases and an extended-tube primer which is pressed in and cannot be removed. Brass-cased VYa cases are still manufactured, being used in a subcalibre training device for tank guns. The details which follow refer to ammunition for the ZSU-23, unless otherwise indicated.

Data
Round length: 234.95 mm
Case length: 151.13 mm
Rim diameter: 33.15 mm
Bourrelet diameter: 22.93 mm
Projectile weight: 185 g
Muzzle velocity: 1,000 m/s
Muzzle energy: 92.5 kJ

COMMONWEALTH OF INDEPENDENT STATES
Manufacturer: State arsenals
Type: HE-I-T OZT: Steel shell filled 18 g HE and fitted with Mk 25 nose impact fuze. Copper driving band. 183 g; Vo 970 to 990 m/s
AP-I BZ: Steel pointed shell with alloy ballistic cap filled with incendiary mixture; 1,895 g; Vo 1,000 m/s

EGYPT
Manufacturer: Maasara Company for Engineering
Type; HE-I-T: Probably based on the Serbian HE-I-T round; no details
AP-I-T: Probably based on the Serbian AP-I-T round; no details
Type: VYa Sub-Calibre: Solid shot, with tracer, with ballistic cap; brass case with screwed-in primer. No details of performance. For use in tank gun subcalibre insert barrels

IRAN
Manufacturer: Defence Industries Organisation
Type: HE-AP-I: In spite of the Iranian designation, this appears to be a conventional HE-I shell with nose impact fuze. It has a red tracer. Vo 970 m/s

POLAND
Manufacturer: Zaklady Metalowe MESKO
Type: HEI-T OFZT: 188.5 g; Vo 970 m/s
APC-I-T BZT: 190 g; 970 m/s

SOUTH AFRICA
Manufacturer: Pretoria Metal Pressings
Type: HE-I-SD: Steel shell loaded Hexal P30 and with nose impact fuze. Soft iron driving band. Steel case, Boxer primed. 187 g; Vo 975 m/s; self-destruct 5 to 12 seconds

23 × 152B BZT

APC-I-T: Steel shot with ballistic cap, incendiary filling inside the cap. Soft iron driving band. Base tracer to 5 seconds minimum. Steel case, Boxer primed. 187 g; 975 m/s; penetration 50 mm

YUGOSLAVIA (SERBIA AND MONTENEGRO)
Holding company: YUGOIMPORT - SDPR
Type: HE: Steel shell loaded 19.6 g RDX/Aluminium. Nose impact fuze. 190 g; Vo 970 m/s
HE-T: Steel shell loaded 16.3 g RDX/Aluminium, with tracer in rear and heat relay giving self-destruct after 3.5 seconds. Nose impact fuze. 190 g; Vo 970 m/s
HE-I: Steel shell loaded 19.6 g RDX/Aluminium. Nose impact fuze. 182 g; Vo 970 m/s
HE-I-T: Steel shell loaded 16.3 g RDX/Aluminium, with tracer in rear and heat relay giving self-destruct after 3.5 seconds. Nose impact fuze. 190 g; Vo 970 m/s
AP-I: Steel pointed shell loaded 5 g TNT/Aluminium with incendiary filling in ballistic cap. Base delay fuze. 190 g; Vo 970 m/s
AP-I-T: As for AP-I but with red tracer; 190 g; Vo 970 m/s
TP: HE shell body, empty, with dummy fuze; 182 g; Vo 970 m/s
TP-T: HE-T shell body, forward section empty, rear section red tracer, dummy fuze; 190 g; Vo 970 m/s

UPDATED

25 mm Oerlikon KBA

Synonyms: 25 × 137 mm; 25 mm GAU-12/U

Armament
Oerlikon KBA; US Bushmaster, GAU-12/U and M242; Mauser Model E; Rheinmetall Rh205; Giat M811; Aden 25.

Development
In the early 1960s the US Army's 'Bushmaster' MICV cannon project resulted in, among other things, the Thompson-Ramo-Wooldridge TRW-6425 cannon, designed by Eugene Stoner of Armalite rifle fame. When Bushmaster was shelved, the European rights to the design were acquired by Oerlikon-Bührle. They made some improvements and produced it as their Model KBA, developing a range of corresponding ammunition. Since then the utility of this calibre has been generally recognised and the Bushmaster project was eventually revived, the Hughes Chain Gun and the GAU-12 Gatling gun were developed and, more recently, the British Aden aircraft gun has been adapted to this calibre.

Description
The lacquered steel case is rimless and bottle-necked, being instantly identifiable by the belt link locating groove just behind the case shoulder. This is expanded by internal pressure when fired, but it leaves an easily discerned mark around the case. The original projectiles were of typical Oerlikon pattern but many new types have been developed in the USA and other countries.

Data
Round length: 223 mm
Case length: 137 mm
Rim diameter: 38 mm
Bourrelet diameter: 24.99 mm
Projectile weight: 180 g
Muzzle velocity: 1,100 m/s
Muzzle energy: 108.9 kJ

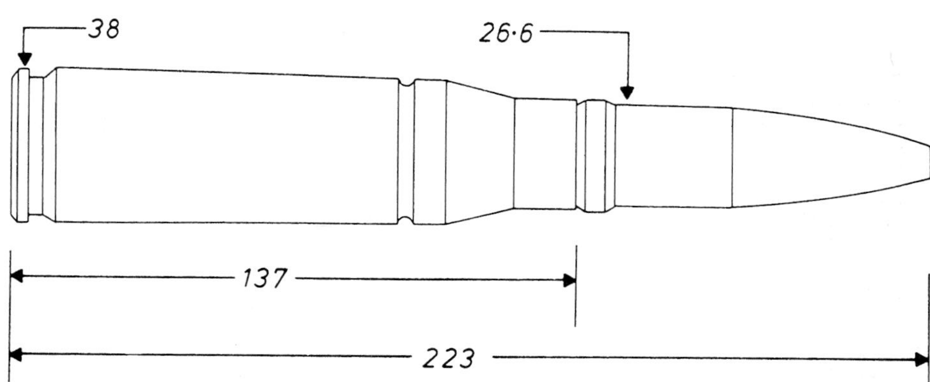

Outline dimensions of 25 × 137 mm KBA round

BELGIUM
Manufacturer: MECAR
Type: APFSDS-T M935A2: Tungsten alloy penetrator; cartridge weight 450 g; Vo 1,440 m/s; can penetrate 42 mm RHA at 60° NATO at 1,000 m

FRANCE
Manufacturer: Manurhin Defense
Type: HE-I: Steel thin-walled shell, filled Hexal, with nose impact fuze MR251 giving self-destruct. 183 g; Vo 1,100 m/s
HE-I-T: Steel thin-walled shell with tracer socket formed on base. Filled Hexal, with nose impact fuze providing self-destruct. Red tracer; 183 g; Vo 1,100 m/s
SAP-HE-I: Pointed steel shell with ballistic cap; shell body filled Hexal with base delay fuze; ballistic cap filled incendiary composition; 185 g; Vo 1,100 m/s
SAP-HE-I-T: As SAP-HE-I but with a tracer socket on the rear of the base fuze; 190 g; Vo 1,100 m/s
AP-HC: Tungsten core in steel body with alloy ballistic cap; 180 g; Vo 1,100 m/s
AP-HC-T: As AP-HC but with red tracer; 178 g; Vo 1,100 m/s
APDS: Tungsten carbide subprojectile in alloy/plastic sabot; 180 g; Vo 1,100 m/s
APDS-T: As APDS but with red tracer; 183 g; Vo 1,100 m/s
TP: Steel shell body, empty, with dummy fuze; 180 g; Vo 1,100 m/s
TP-T: Steel two-section shell body, front empty, rear carrying red tracer; 180 g; Vo 1,100 m/s

GERMANY
Manufacturer: Diehl GmbH
Type: HE-I: 195 g; Vo 1,100 m/s self-destruct 3.7 to 5 seconds
SAP-HE: 195 g; Vo 1,100 m/s self-destruct 3.7 to 5 seconds
TP-T: 195 g; Vo 1,100 m/s

NETHERLANDS
Manufacturer: Eurometaal NV
Type AP-I-T: Steel pointed projectile with alloy ballistic cap containing incendiary mixture. Tail tracer; 185 g; Vo 1,100 m/s
TP-T: Inert projectile representing AP-I-T, with tracer; 180 g; Vo 1,100 m/s
APDS-T: Tungsten penetrator in plastic sabot

NORWAY
Manufacturer: Raufoss Technology A/S
Type: Multipurpose MPT-SD Mark 2: Steel body with high-explosive and incendiary fillings, the incendiary extending into the nose cap. Dark ignition red tracer to 2,500 m in rear, with heat relay to effect self-destruct after 5.3 seconds of flight. 183.5 g; Vo 1,089 m/s
Multipurpose M84A1: Similar to MPT-SD but without tracer; 183.5 g; Vo 1,089 m/s
Multipurpose MPT-SD Low Drag: Similar to MPT-SD Mark 2 but using a new low drag projectile body which improves terminal velocity at 2,000 m by 54 per cent, reduces the time of flight to 2,000 m by 25 per cent and gives increased terminal effects.

SWITZERLAND
Manufacturer: Oerlikon-Contraves AG
Type: HE-I-T PMB 050: Drawn steel shell with soft iron driving band; filled 27 g Hexal P30; base tracer; fuzed KZB335 impact fuze providing self-destruct after 7 seconds; 180 g; Vo 1,100 m/s
SAP-HE-I-T PMB 054: Steel shell body with alloy ballistic cap; filled 11 g Hexal P30 and fitted with base fuze BZB336, providing 200 ms delay and self-destruct after 7 seconds of flight; 180 g; Vo 1,100 m/s
APDS-T PMB 073: Dense metal penetrator core with light alloy ballistic cap, in plastic and alloy sabot, base tracer burning >1.7 seconds; 150 g; Vo 1,335 m/s; penetration 30 mm at 60° at 1,000 m
FAPDS PMB 091: Generally similar to the APDS projectile but with a subprojectile of frangible metal, so that after penetration of the target it disintegrates into a cloud of fragments to cause widespread damage; 150 g; Vo 1,335 m/s
APFSDS-T PMB 090: Fin-stabilised discarding sabot; generally similar to the APDS projectile but with a fin-stabilising tail unit attached to the subprojectile. Performance is maintained with over a 40 per cent range increase over the APDS projectile. 140 g; Vo 1,370 m/s
TP-T PMB 052: The same shell body as the HE-I-T shell SLB 050 but filled inert, with a dummy fuze; tracer in rear, burns >3.4 seconds; 180 g; Vo 1,100 m/s
TPDS-T PMB 072: A target practice discarding sabot shot with tracer; functions in the same manner as APDS but uses simpler materials in its construction. 150 g; Vo 1,335 m/s

UNITED STATES OF AMERICA
Manufacturer: Alliant Techsystems Inc
Type: HE-I-T M792: Steel thin-walled projectile loaded 30.2 g HE and with red tracer. Fuzed PD M758 with self-destruct. 185 g; Vo 1,100 m/s
APDS-T M791: Hard core 102.8 g in alloy/plastic discarding sabot with red tracer; 133 g; Vo 1,345 m/s
TP-T M793: Hollow steel shell with alloy nose cap and rear tracer, ballistically matched to M792. 182 g; Vo 1,100 m/s
TPDS-T M910: Inert subprojectile with tracer, in alloy/plastic discarding sabot. Ballistically matched to M791 to a limited range. 94 g; Vo 1,540 m/s. Tracer visible in excess of 2,000 m; maximum range 8,000 m. Time of flight to 2,000 m is 1.8 seconds; remaining velocity 785 m/s

MECAR APFSDS-T M935A2

1996

TP-T M793: Hollow steel shell with alloy nose cap and rear tracer, empty, ballistically matched to M792; 184 g; Vo 1,100 m/s

TP PGU-23: Similar to M793 but without tracer. For US Navy

TPDS-T M910: Training round for the APFSDS-T M919. Uses similar construction but with a steel subprojectile and tracer; 95 g; Vo 1,540 m/s

TP Frangible PGU-33/U: Solid projectile made of Partially Stabilized Zirconia (PSZ) ceramic material which shatters on impact with the target. For practice purposes; ballistically matched to the PGU-20/U AP projectile

Manufacturer: Olin Ordnance

Type: HE-I-T M792: Steel thin-walled projectile loaded 30.2 g HE and with red tracer. Fuzed PD M758 with self-destruct between 6.2 and 19 seconds. 184 g; Vo 1,100 m/s

HE-I PGU-25: Similar to M792 but without tracer. For US Navy

AP-I PGU-20/U: 150 g depleted uranium (DU) subprojectile in steel body with thin metal ballistic cap; 215 g; Vo 1,000 m/s

APDS-T M791: 102 g tungsten core in alloy/plastic discarding sabot with red tracer; 134 g; Vo 1,345 m/s

APFSDS-T M919: A 96 g depleted uranium long rod penetrator, fin-stabilised, carried in an alloy/plastic sabot. Rear of fin unit contains tracer. 132 g; Vo 1,420 m/s. Time of flight to 2,000 m is 1.56 seconds; remaining velocity 1,149 m/s

UPDATED

25 mm Oerlikon KBB

Synonyms: 25 × 184 mm

Armament

Oerlikon KBB cannon.

Development

The 25 × 184 mm KBB was developed by Oerlikon in the early 1980s, in order to produce a 25 mm cannon with more power than the KBA for air defence or vehicle armament roles.

Description

The case is rimless, bottle-necked, lacquered steel and has a screw-in percussion primer. The projectile is secured to the case by an eight-point crimp. Projectiles are described below.

Data

Round length: 288 mm
Case length: 184 mm
Rim diameter: 38.6 mm
Bourrelet diameter: 24.9 mm
Projectile weight: 230 g
Muzzle velocity: 1,160 m/s
Muzzle energy: 154.7 kJ

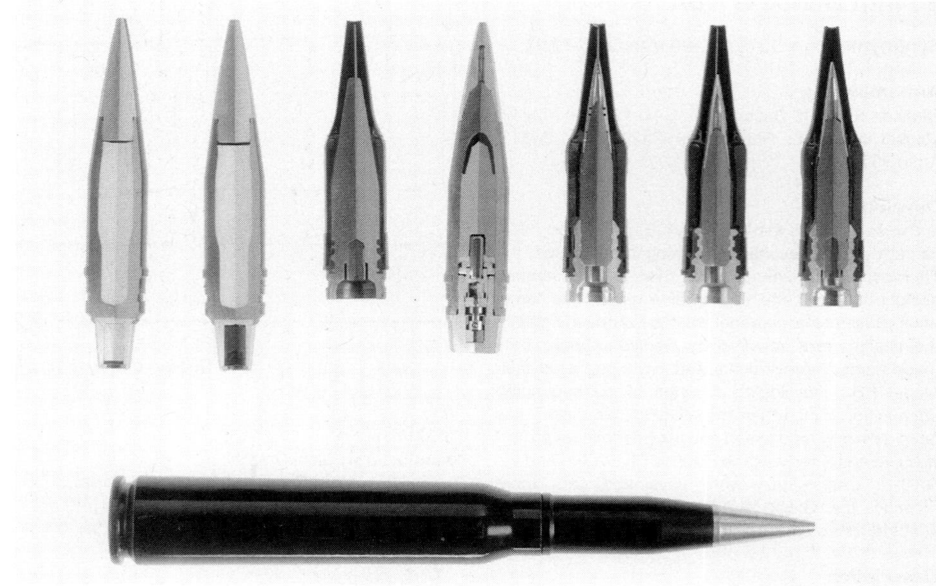

Oerlikon-Contraves ammunition for the KBB gun: (left to right) TP; TP-T; TPDS-T; HE-I; FAPDS; AMDS; APDS-T

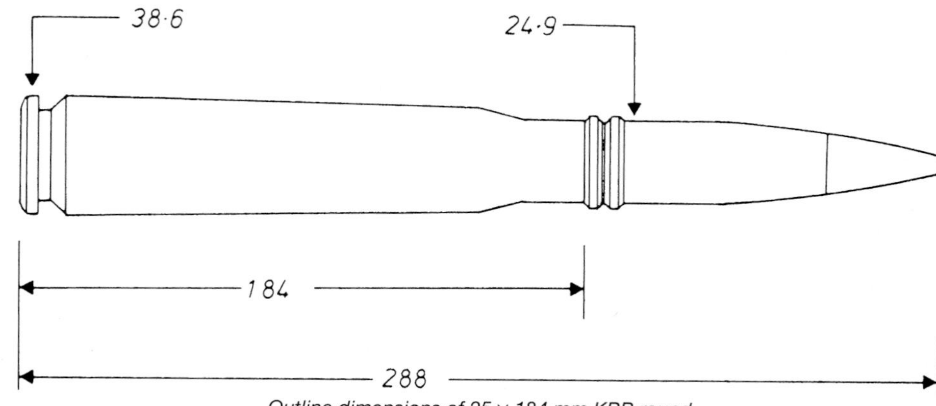

Outline dimensions of 25 × 184 mm KBB round

SWITZERLAND

Manufacturer: Oerlikon-Contraves AG

Type: HE-I SSB 064: Pointed steel shell with ballistic cap, filled 22 g Hexal and fitted with a base fuze providing self-destruct by spin decay after about 9 seconds of flight. 230 g; Vo 1,160 m/s

APDS-T TLB 067: A 156 g heavy metal penetrator core with light alloy ballistic cap, in a plastic/alloy sabot with base tracer; 190 g; Vo 1,285 m/s. Penetration 34 mm at 60° at 1,000 m

AMDS TKB 075: This is virtually the same as the APDS-T projectile TLB 067 but does not contain a tracer. It is intended for the attack of sea-skimming anti-ship missiles. 190 g; Vo 1,270 m/s. This projectile fulfils all the conditions necessary to destroy a missile warhead at ranges exceeding 1,000 m, with the missile closing at speeds from Mach 0.9 to 3.0.

APFSDS-T PMB 0696: This generally resembles the APDS projectile but has a fin-stabilising tail unit attached to the base of the subprojectile. It will defeat 35 mm plate at 60° at 1,500 m range. 140 g; Vo 1,200 m/s

FAPDS: This resembles the APDS in form but uses a frangible subprojectile which, after penetrating the target, disintegrates into fragments to cause widespread damage. 230 g; Vo 1,160 m/s

TP: Steel shell, filled inert, with dummy fuze. 230 g; Vo 1,160 m/s

TP-T: Steel shell, filled inert with red tracer, dummy fuze; 230 g; Vo 1,160 m/s

TPDS-T: Steel subprojectile in alloy/plastic sabot, with tracer; 190 g; Vo 1,285 m/s

UPDATED

Ammunition for 30 mm 2A42 Cannon

Armament

30 mm 2A42 cannon fitted to BMP-2 and Sarath (India) MIFVs and Bulgarian BMP-30 IFV.

30 mm 2A72 cannon fitted to BMP-3 IFV.

30 mm 2A38 and 2A38M cannon mounted on the 2S6 quad 30 mm/SA-19 2K22 Tunguska self-propelled air defence system.

30 mm 2A38M cannon mounted on Strop self-propelled air defence system (Czech Republic).

30 mm M86 and M89 cannon (former Yugoslavia).

Development

The 30 mm 2A42 dual feed cannon first appeared in public during 1982. Intended for use against lightly armoured targets at ranges up to 1,500 m, the 2A42 cannon is also intended for use against low-flying aerial targets such as helicopters. The 30 mm 2A42 has a dual feed system and two rates of fire; the slowest is 200 to 300 rds/min while the faster rate is a minimum of 550 rds/min. By contrast the 30 mm 2A72 cannon is lighter and has a fixed fire rate of around 330 rds/min; the dual feed feature is retained. The 2A38 and 2A38M air defence cannon have a rate of fire of 1,950 to 2,500 rds/min.

The 30 mm ammunition for all these cannon is known to be in production in Russia, the Czech Republic, Slovakia, Ukraine and the former Yugoslavia, although it is probable that production has now ceased in the latter region.

It is possible that some or all of these 30 mm rounds are already being produced in India. A new munitions plant is to be built at Bolangir in Eastern India which, when fully operational, will be able to produce 'substantial quantities' of 30 mm ammunition every year.

Description

The 30 × 165 mm ammunition for the 2A42 cannon is fixed, with the projectile rigidly secured to the tapered cartridge case by two rows of spaced crimps engaging in cannelures in the projectile. The projectiles have a pronounced 'waist' just behind the bourrelet and forward of the drive bands, which can be either copper/gilding metal or sintered iron. Cartridge cases are usually lacquered or corrosion-treated steel.

HEI (OF3) projectiles carry 123 g of A-IX-2 (desensitised RDX/Aluminium) explosive and are fitted with A-670M point detonating fuzes with a self-destruct device functioning between 7.5 and 14.5 seconds. Tracer burn time is not less than 3.5 seconds. A HE-T (OT) round also exists.

The AP-T (BT) projectile weighs 400 g, contains 127 g of explosive and is stated to be able to penetrate up to 18 mm of armour plate set at 60° at a range of

1,000 m (25 mm at 750 m). Penetration at the maximum effective range of 1,500 m is 15 mm.

There is also an APDS-T variant with a muzzle velocity of 1,120 m/s and an armour penetration capability of 25 mm of armour plate set at 60° at a range of 1,500 m. This round is also referred to as a 'Kerner' cartridge.

All cartridge cases are lacquered or corrosion-treated (usually zinc-plated) steel and contain 125 g of pyroxyline powder propellant ignited by a KV-3 primer. The cases are 164 mm long and weigh 328 g.

30 mm 2A42 ammunition produced in the CIS is intended for operation over a temperature range of −50 to +50°C.

TP, TP-T and inert drill rounds are available for both the HEI and AP rounds.

BULGARIA
Manufacturer: Kintex
Type: HE, AP-T
Description: HE complete round weighs 833 g with projectile 389 g. AP-T weighs 853 g with projectile 400 g

COMMONWEALTH OF INDEPENDENT STATES
Marketing agency: VO GED, General Export for Defence
Type: HEI, HE-T, AP-T, APDS-T
Description: Standard specifications as in text and Specifications table

Marketing agency: SSIE 'Pribor'
Type: HEI, HE-T, AP-T, APDS-T
Description: Standard specifications as in text and Specifications table

CZECH REPUBLIC AND SLOVAKIA
Manufacturer: MOEX Vlárské stronjirny Slavičin
Type: HE-T, TP-HE-T, HEI, AP-T
Description: Developed by Prototypa, sp Brno in co-operation with VZU(A) 010 PVVV Slavičin, these rounds differ in some respects from the CIS original specification, especially with the AP-T which can penetrate up to 36 mm of armour set at 0° at 1,000 m. Propellant loads are of the order of 120 to 128 g using an unspecified type of propellant producing a lower than usual pressure curve

UKRAINE
Manufacturer: MINMASHPROM
Type: HE-T, TP-HE-T, HEI

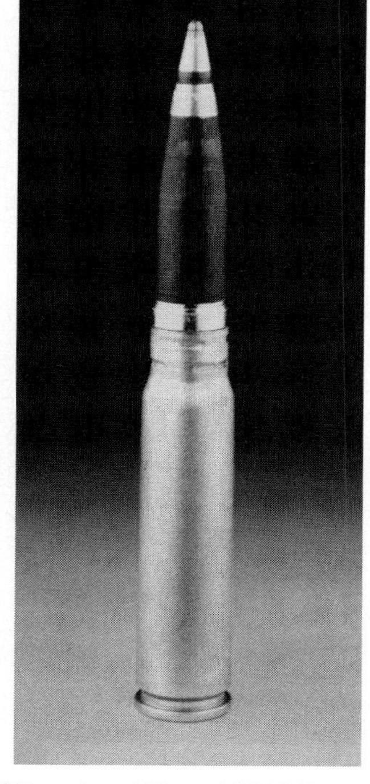

CIS produced 30 mm HEI 2A42 round

CIS produced 30 mm AP-T round

Data

Type	**HEI**	**HE-T**	**AP-T**	**APDS-T**
Weight				
complete round	842 g	835-837 g	853 g	765 g
Lengths				
complete round	291 mm	291 mm	291 mm	291 mm
cartridge case	164 mm	164 mm	164 mm	164 mm
Muzzle velocity	950-970 m/s	950-970 m/s	960-980 m/s	1,120 m/s
Self-destruct time	7.5-14.5 s	7.5-14.5 s	7.5-14.5 s	7.5-14.5 s

Description: Standard specifications as in text and Specifications table. Note that APDS-T is not produced.

YUGOSLAVIA (SERBIA AND MONTENEGRO)
Holding company: YUGOIMPORT - SDPR
Type: HEI-T, HEI, HE, HE-T, AP-T, AP, TP-T, TP
Description: Probably no longer in production but in service with the former Yugoslav armed forces. Can be used with 30 mm M86 and M89 cannon

UPDATED

Ammunition for 30 mm DEFA Guns

Armament

Giat Industries 30 mm DEFA Series 552/553 guns; Giat Industries 30 mm Series 554 gun: Giat Industries 30 mm Mle 781 (30 M 781) cannon.

Development

The French 30 mm DEFA Series 552/553 and Series 554 aircraft guns have the same design origins as the British 30 mm Aden Guns, namely the Second World War German Mauser 213C revolver cannon. The German gun was taken as the base for what became a series of 30 mm aircraft guns, culminating in the Series 552/553 and 554 guns used in the Dassault Mirage series of strike fighters, as well as other aircraft. Following early French cartridge design forays, international design collaboration with the British resulted in the 30 × 113 mm B cartridge, which is nominally interchangeable between the DEFA guns and the 30 mm Aden Gun series. However, DEFA cartridges have significant priming and other differences, plus a higher muzzle velocity which renders them non-compatible with their 30 mm Aden gun counterparts.

Raufoss of Norway produce a 30 mm MultiPurpose (MP) round suitable for DEFA cannon.

DEFA 30 mm ammunition is in production throughout the world, especially with nations receiving French export materiel.

Description

All 30 × 113 mm B DEFA rounds are fixed with the projectiles rigidly crimped to their J-type thermally treated steel cartridge cases, coated with a self-lubricating lacquer. The cartridge case has a raised belt just above the extraction rim. All projectiles have copper or gilding metal drive bands, although sintered iron may be encountered. Primers are electrical, using a special conductive compound. The propellant used is approximately 50 g of a single base multiperforated (19-hole) tubular propellant known as B.19.T and produced by SNPE.

DEFA 30 × 113 mm B ammunition may be put into one of three groups; high explosive, armour-piercing and practice. A further discrimination can be made between ammunition for air-to-air combat and air-to-ground use.

Rounds are fed to the revolver loading mechanism on the Series 553/554 guns in belts using pressed steel links.

The following 30 mm DEFA ammunition types are those produced by Matra Manurhin Defense, later known as Manurhin Defense and now part of Giat Industries.

HE-I - OMEI Known as a 'mine' (Obus Mine Explosif Incendiaire), the round in this air-to-air combat category is the 30 mm OMEI Mle F7572, which utilises a cold drawn steel, thin-walled, flat-based projectile fitted with a nose-mounted MR3081 delay fuze with muzzle and rain safety features. The projectile is filled with 50 g of high temperature tolerant Hexal (RDX/Aluminium). The fuze has a self-destruct mechanism operating between 7 and 15 seconds after firing, using a pyrotechnic chain to time the self-destruct point.

The OMEI projectile weighs 245 g, while a complete round weighs 441 g. Muzzle velocity is 810 m/s.

The training round used to simulate the OMEI is the 30 mm OXL Mle F2570. It is fitted with an inert nose plug, filled with an inert substance and matches the ballistics of the 30 mm OMEI projectiles exactly. Projectile weight is 245 g and the complete round weighs 441 g.

There were two earlier versions of the OMEI round, which are no longer in production, the OMEI Mle F7570 with an ML68 fuze and the OMEI Mle F7571 with an MR381 fuze.

APHEI - OAPEI For this round the AP in APHEI refers to anti-personnel, not armour-piercing. Intended for air-to-ground combat use, the 30 mm OAPEI Mle F5272 has a cold drawn steel, thick-walled, flat-based projectile fitted with a nose-mounted MR3001 point impact fuze, and is filled with 30 g of high temperature tolerant Hexal (RDX/Aluminium). There is no self-destruct mechanism.

The OAPEI projectile weighs 275 g, while the weight of a complete round is 490 g. Muzzle velocity is 775 m/s.

The training round used to simulate the OAPEI is the OXAS Mle F2270. It is fitted with an inert nose plug, filled with an inert substance and matches the ballistics of the OAPEI projectile exactly. Projectile weight is 275 g and the complete round weighs 490 g.

Two earlier versions of the 30 mm OAPEI round, which are no longer in production, were the OAPEI Mle F5270 with an MRX70 fuze and the OAPEI Mle F5271 with an MR31 fuze.

SAPHEI - OSPEI Intended primarily for air-to-air combat, this round is known as the 30 mm Mle F7670. It uses a thick-walled, heat-treated steel projectile with a blunt nose covered by a light alloy windshield to retain a ballistic match with other rounds in the DEFA 30 mm family. The base of the projectile is occupied by a delayed action base fuze known as the MR3005, with an arming distance of 20 m from the muzzle and a self-destruct time of between 7 and 15 seconds. When the fuze functions it detonates the explosive payload, which weighs 18 g, and creates high fragmentation

The complete range of 30 × 113 mm B ammunition produced by Giat Industries for 30 mm DEFA guns: from left, TP; TP; TP-T; HE-I; APHEI; SAPHEI; API-T

effects behind the target armour. The projectile can penetrate up to 20 mm of armour.

The weight of the projectile used with this round is 275 g and the weight of a complete round is 490 g. Muzzle velocity is 775 m/s.

API-T - OPIT This round, known as the 30 mm Mle 5970, is intended for the air-to-ground attack of hard targets such as armoured vehicles. The projectile uses a cold drawn steel body, with thin sidewalls carrying a heat-treated special steel blunt-nose slug which acts as the armour penetrator. It is capable of piercing approximately 20 mm of armour plate. The projectile nose is covered by a light alloy windshield to retain a ballistic match with the other rounds in the DEFA 30 mm family. The projectile body is filled with 20 g of an incendiary composition and a tracer element is mounted in the projectile base.

The projectile used with this round weighs 275 g and the weight of a complete round is 490 g. Muzzle velocity is 765 m/s.

TP-T - OXT This round acts as a general purpose training round for the 30 mm DEFA family. Known as the Mle F3170, it uses a thick-walled steel body with an inert nose plug. The interior is hollow and empty and the base is occupied by the same tracer element assembly as that used on the 30 mm API-T (OPIT) Mle F5970.

The projectile used with this round weighs 245 g and the weight of a complete round is 490 g. Muzzle velocity is 795 m/s.

MP Produced by Raufoss of Norway this 30 mm MP projectile will function reliably against a 2 mm dural plate at impact angles between 0 and 87° NATO but will not function against a 0.5 mm dural plate in front of the gun muzzle. No fuze is involved as the MP projectile relies upon a drop safe pyrotechnic ignition train. When impacting against aircraft type targets the projectiles will detonate approximately 300 mm within the target. The distribution of fragments is approximately 20° on each side of the line of fire; the fragments are heavy and optimised to defeat material type targets. The secondary incendiary effects will ignite JP4 and JP5 or heavy diesel oil in self-sealing tanks. Incendiary and blast effects are contained within the target.

The length of the Raufoss 30 mm MP round is given as 199.88 mm and it weighs 440 g; projectile weight is 245 g. Muzzle velocity is 810 m/s.

Data

FOR SPECIFIC DETAILS OF EACH ROUND SEE TEXT

Weights:
- **complete round OMEI** - 441 g
- **complete round OAPEI, OSPEI, OPIT** - 490 g
- **complete round MP** - 440 g
- **projectile OMEI** - 245 g
- **projectile OAPEI, OSPEI, OPIT** - 275 g
- **projectile MP** - 245 g
- **propellant** - approx 50 g

Lengths:
- **complete round** - 200 mm
- **cartridge case** - 113 mm

Max projectile diameter: 29.99 mm
Max case diameter (over band): 33.8 mm
Max case diameter (over rim): 33.4 mm
Operating temperature range: −54 to +74°C

ARGENTINA
Manufacturer: Direccion General de Fabricaciones Militares
Type: HEI (EINC), HEI-SD (EINC-AD), AP (SPEINC), TP (EJ)
Description: Manufacturers are Fabrica Militar 'Fray Luis Beltran' of Santa Fe. Standard specifications

BELGIUM
Manufacturer: FN Herstal SA
Type: HEI, SAPHEI, API-T, TP
Description: No longer in production but may still be retained for service. Standard specifications

BRAZIL
Manufacturer: Companhia Brasiliera de Cartuchos (CBC)
Type: HEI, API, TP, TP-T
Description: Standard specifications

EGYPT
Manufacturer: Maasara Company for Engineering
Type: HEI (OMEI - GMET F 7570 A/C), TP-T (OXT - F 3170 A/A), TP (OXL - F 2570 A/A)
Description: Standard specifications

FRANCE
Manufacturer: Giat Industries
Type: See text
Description: See text

GREECE
Manufacturer: Hellenic Arms Industry (EBO)
Type: HEI-SD, HEI-SD-T, TP, TP-T
Description: Standard specifications

Manufacturer: Pyrkal Greek Powder & Cartridge Company
Type: HE, HEI, TP
Description: Standard specifications

NORWAY
Manufacturer: Raufoss Ammunisjonsfabrikker A/S
Type: MP
Description: See text

SINGAPORE
Manufacturer: Chartered Industries of Singapore
Type: HEI, TP
Description: Standard specifications

SOUTH AFRICA
Manufacturer: Denel (PMP)
Type: HEI, HEI-T, SAPHEI, APCI M1A1, TP, TP-T
Description: APCI M1A1 has tungsten carbide core which can penetrate 50 mm of armour at 0° NATO at 200 m with strong after-penetration incendiary effects. SAPHEI stated to penetrate 25 mm of armour at 0° NATO at a range of 100 m. HEI and HEI-T are filled with Hexal P 30. Drive bands are sintered iron. Muzzle velocity for all types of round given as 820 m/s. A reloading cartridge is also produced

UNITED STATES OF AMERICA
Manufacturer: Alliant Techsystems
Type: HEI, HEDP, TP
Description: See separate entry

Manufacturer: Olin Ordnance
Type: HEDP, TP
Description: See separate entry

UPDATED

Oerlikon-Contraves 30 mm KCB Ammunition

Armament

HS 661 towed anti-aircraft gun; 30 mm Bushmaster II; AMX-13 and AMX-30 twin self-propelled anti-aircraft guns; Flying Tiger (Biho) twin 30 mm self-propelled anti-aircraft gun system (South Korea); SABRE twin turret fitted in prototype form to AMX-10RC, AMX-30, Chieftain and Steyr chassis; MBT TCM 30 twin naval mounting; Oerlikon-Contraves GCM series of twin naval mountings; DS 30 B naval mounting; Emerlec 30 twin naval mounting.

KCB ammunition can also be fired by 30 mm L21 RARDEN guns.

Development

When originally produced, the series of 30 mm automatic cannon now known as the Oerlikon-Contraves KCB series was designated the Hispano-Suiza HSS 831 L series. The designation was officially changed when Hispano-Suiza became part of Oerlikon-Contraves but is often retained. The 30 mm cannon design involved was virtually a scaled-up version of the 20 mm Type KAD cannon and has been produced in several forms. Modified versions of the 30 × 170 mm ammunition family (with a longer 173 mm case) are used by the Oerlikon-Contraves 30 mm KCA series of aircraft revolver cannon.

The KCB ammunition family was used as the basis for the 30 mm RARDEN gun ammunition family. 30 mm KCB ammunition can be fired from L21 RARDEN guns.

Description

All 30 × 170 mm KCB rounds are fixed rounds, with the steel projectiles rigidly fixed into the lacquered steel cartridge cases (some cases are brass) by 360° crimping rings engaging in cannelures on the projectile body. A single sintered iron drive band is used on the streamlined projectiles.

The same 170.3 mm long-necked cartridge cases are used for all types of round in the KCB family and are each filled with 160 g of a NC single base propellant. A percussion primer is threaded into the base of the cartridge case. All types of projectile are ballistically matched and have a muzzle velocity of 1,080 m/s.

The 30 mm KCB ammunition family includes the following rounds:

HEI MSC The HEI projectile is a thin-walled steel shell pressed from a steel blank by a technique ensuring high fragmentation. The filling is 40 g of Hexal 30 which, when detonated, has been demonstrated to produce an average of 1,133 fragments (of which only 0.05 g is dust) in addition to the blast and incendiary effects. The projectile nose is largely occupied by an impact nose fuze originally known as the Type F-831-L3, now the KZC-L3. The rain safe fuze has a muzzle safety distance of 20 m and self-destructs the projectile after approximately 6 to 11 seconds. Fuze functioning occurs a delayed interval after impact to maximise the blast and fragmentation effects in the target. The explosive filling also produces an incendiary effect.

HEI-T MLC This projectile is much the same as the HEI but the base of the projectile is occupied by a tracer element so the explosive filling is reduced to 25 g of Hexal 30. After firing, the tracer burns for an average of 4 seconds during which time the projectile will have reached a range of approximately 2,700 m.

SAPHEI-T PSC The projectile for this round is manufactured from tempered steel and has relatively thick walls. As the nose of the shell is blunt, for armour penetration efficiency it is covered by a light aluminium windshield to maintain the correct aerodynamic outline. The base of the projectile is occupied by a base impact fuze, known as the BZC-L5, which functions after a short delay following an impact to detonate 20 g of Hexal 30 inside the target armour. In this way blast, fragmentation and incendiary effects are added to the target penetration. The fuze provides projectile self-destruct after 6 to 11 seconds.

TP UGC This training round is a hollow steel shell containing an inert material, with the nose fuze replaced by an aluminium plug with an outline corresponding to operational KCB rounds.

TP-T ULC This uses exactly the same projectile as the TP but the rear of the shell is occupied by a tracer element. After firing, the tracer burns for an average of 4 seconds, during which time the projectile will have reached a range of approximately 2,700 m.

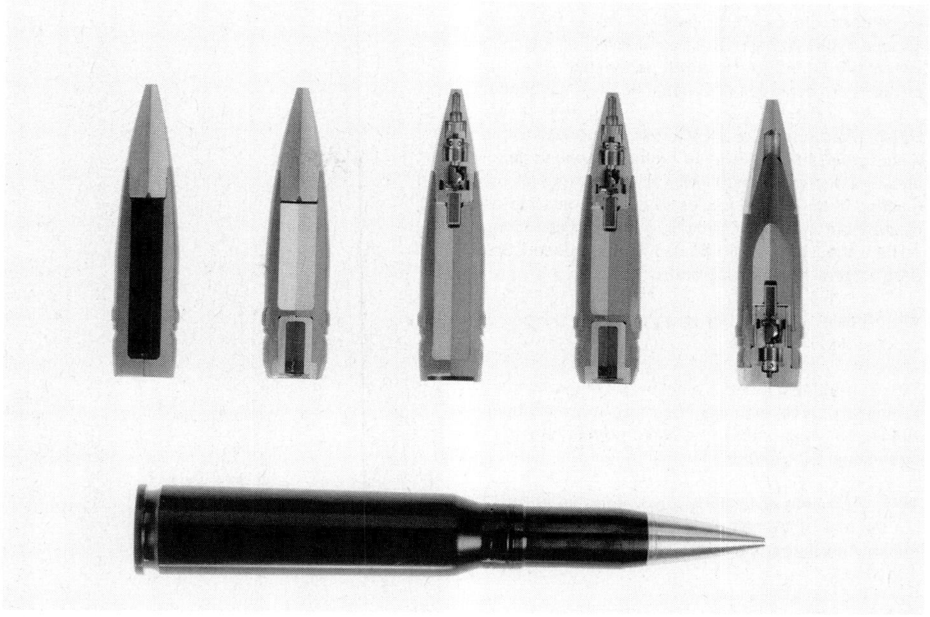

30 mm KCB ammunition projectiles: from left, TP; TP-T; HEI; HEI-T; SAPHEI

Data

Type	HEI	HEI-T	SAPHEI-T	TP	TP-T
Weights					
complete round	870 g	870 g	870 g	870 g	870 g
projectile	360 g	360 g	360 g	360 g	360 g
filling	40 g	25 g	20 g	none	none
Round length	285 mm	285 mm	285 mm	285 mm	285 mm
Cartridge case length	170.3 mm	170.3 mm	170.3 mm	170.3 mm	170.3 mm
Muzzle velocity	1,080 m/s	1,080 m/s	1,080 m/s	1,080 m/s	1,080 m/s
Flight time					
to 1,000 m	1.08 s	1.08 s	1.08 s	1.08 s	1.08 s
to 2,000 m	2.61 s	2.56 s	2.61 s	2.61 s	2.56 s
to 3,000 m	4.93 s	4.83 s	4.93 s	4.93 s	4.83 s
Chamber pressure	4,200 bar	4,200 bar	4,200 bar	4,200 bar	4,200 bar

Break-up NWM de Kruithoorn BV of the Netherlands produce 30 mm break-up ammunition suitable for use in KCB cannon.

ARGENTINA
Manufacturer: Direccion General de Fabricaciones Militares
Type: HEI (EINC-AD), HEI-T-SD (EINC-T-AD), AP (SPEINC), TP-T (EJT)
Description: Manufacturers are Fabrica Militar 'Fray Luis Beltran' of Santa Fe. Standard specifications

FINLAND
Manufacturer: Sako Limited
Type: HEI, HEI-T, TP, TP-T
Description: Standard specifications

FRANCE
Manufacturer: Giat Industries/Manurhin Defense
Type: HEI/SD, HEI-T/SD, TP, TP-T
Description: Brass cartridge cases. Otherwise standard specifications

GREECE
Manufacturer: Hellenic Arms Industry (EBO)
Type: HEI-SD, HEI-SD-T, TP, TP-T
Description: Standard specifications

KOREA, SOUTH
Manufacturer: Daewoo Corporation
Type: HEI, HEI-T, TP-T
Description: Standard specifications

Manufacturer: Poongsan Metal Corporation
Type: HEI, HEI-T, TP-T
Description: Standard specifications

SWITZERLAND
Manufacturer: Oerlikon-Contraves
Type: See text
Description: See text

UNITED KINGDOM
Manufacturer: British Aerospace Defence Limited, Royal Ordnance Division
Type: TP, TP-T, HEI, HEI-T, SAPHEI, API-T
Description: TP RO 277 uses brass cartridge case, RO 673 primer and single base propellant as used in 30 mm RARDEN ammunition family. Matches KCB HEI-T. Weight of complete round 908 g, length 276 mm. Projectile weight 357 g. Muzzle velocity 1,070 m/s. Other rounds, standard specifications

UPDATED

Oerlikon-Contraves 30 × 173 mm KCA Ammunition

Armament

Oerlikon-Contraves 30 mm KCA aircraft cannon.

Development

The Oerlikon-Contraves 30 mm KCA cannon was originally known as the 304Rk and is yet another gas-operated revolver aircraft cannon based on the Second World War German Mauser 213, with a cyclic rate of fire of 1,350 rds/min. An 'Americanized' version of the KCA proposed by Hughes Helicopters for service with the US Air Force was known as the GAU-9. The cartridge fired from the 30 mm KCA cannon is acknowledged to be one of the most powerful of its type and has been used as the basis for other rounds, such as those fired from the 30 × 173 mm Mauser MK 30 cannon and the American 30 mm GAU-8/A ammunition range.

Description

All 30 × 173 mm KCA rounds are fixed rounds with the steel projectiles being rigidly fixed into the lacquered steel cartridge cases (some cases are brass) by 360° crimping rings engaging in cannelures on the projectile body. A single copper or gilding metal drive band is used on the streamlined projectiles. The same 173 mm long-necked cartridge cases are used for all types of round in the KCA family and are filled with 160 g of a graphited nitrocellulose powder propellant. A percussion primer tube is threaded into the base of the cartridge case. All types of projectile are ballistically matched and have a muzzle velocity of 1,030 m/s.

All rounds in this family have storage, handling, transport, barrel and muzzle safeties.

Rounds in the 30 × 170 mm aircraft cannon KCA family include the following:

SAPHEI The projectile for this round is manufactured from tempered steel and has relatively thick walls. As the nose of the shell is blunt, for armour penetration efficiency it is covered by a light aluminium windshield to maintain the correct aerodynamic outline. The base of the projectile is occupied by a base impact fuze which functions after a short delay following an impact to detonate its explosive payload inside the target armour. In this way blast, fragmentation and incendiary effects are added to the target penetration. The fuze provides projectile self-destruct after 6 to 11 seconds.

HEI-T Also known as HEI/T, this round has a thin-walled steel projectile with an explosive payload detonated by a nose-mounted MR301 impact fuze which can function at low-impact angles. A base-mounted tracer element burns for approximately 4 seconds after firing. A HEI without the tracer has also been produced. Self-destruct time is between 5 and 12 seconds.

AP These rounds are apparently no longer in production.

TP This training round is a hollow steel shell not containing explosive, with the nose fuze replaced by an aluminium plug and an outline corresponding to operational KCA rounds.

TP-T This uses exactly the same projectile as the TP but the rear of the hollow steel shell is occupied by a tracer. After firing, the tracer burns for an average of 4 seconds.

Data

Weights:
- **complete round** - 890 g
- **projectile** - 360 g
- **propellant** - 160 g

Lengths:
- **complete round** - 285-290 mm
- **cartridge case** - 173 mm

Muzzle velocity: 1,035 m/s

FRANCE
Manufacturer: Giat Industries/Manurhin Defense
Type: HEI, HEI/T, TP, TP/T
Description: Referred to as 30 × 170 mm Manurhin ammunition for HS 831 A guns. Brass cartridge cases. Otherwise standard specifications

UPDATED

30 × 173 mm KCA ammunition (30 × 170 mm Manurhin) produced by Giat Industries/Manurhin Defense: from left, TP; TP-T; HEI; HEI-T

Ammunition for 30 mm Aden Guns

Armament

All 30 mm Aden Guns; M230 Chain Gun; McDonnell Douglas ASP-30 cannon.

Development

The Mauser MG213C revolver cannon had a profound effect on European aircraft gun designers in the years following 1945. Development of the basic German design eventually resulted in two associated 30 mm gun designs, the British Aden gun series and the French DEFA Series 550 weapons. Following early independent projects, shared development work between the two nations resulted in the 30 × 113 mm B cartridge which is nominally interchangeable between the Aden and DEFA series of guns, although differences between priming, propellant loads and other factors render the two types of ammunition incompatible. In France 30 mm Aden gun ammunition is produced by Giat Industries.

The 30 mm Aden gun has been in British service for well over 30 years. Ammunition production totals reached over 30 million some years ago. Development of 30 mm Aden ammunition continues.

Raufoss Ammunisjonsfabrikker A/S of Norway produces a 30 mm MP for firing from Aden cannon.

Description

All 30 mm Aden gun rounds are fixed with the projectiles rigidly crimped to their J-type brass cartridge cases; lacquered steel cases have also been used. The cartridge case has a prominent raised belt just above the extraction rim. All projectiles have copper drive bands. Primers are electrical, the latest being the RD1658 (operating at 24 or 48 V). The propellant used is 46 g of single base cut tubular NRN 141 AB.

Rounds are fed to the revolver loading mechanism on the Aden gun in belts formed using pressed steel links.

The 30 mm Aden gun ammunition has been, and is still being, produced in many countries. The listing of types provided here consists of the types currently manufactured in the United Kingdom by the Ammunition Division of Royal Ordnance plc.

HE Mk 6*Z This has been the standard air-to-air and air-to-ground projectile fired from Aden guns and used by the Royal Air Force for nearly 30 years. The projectile fired has a cold drawn steel body with a pronounced hemispherical base to maximise the internal volume for explosive filling, in this case 48 g of pour-filled Torpex. On detonation this projectile will produce approximately 950 fragments, with a high blast and incendiary effect to the extent that some manufacturers refer to this round as a HEI. A nose-mounted 944 post impact delay fuze with a safe arming distance of 2.3 m functions after 0.0005 second and will ensure self-destruct 6 to 12 seconds after firing.

The TP Mk 4Z matches the ballistics of the HE Mk 6*Z. It incorporates a unique inert insert which is claimed to make the round very cost competitive.

A further TP-T Mk 1*Z was developed to meet the specific needs of the Royal Navy. It is matched to the ballistics of the HE Mk 6*Z and is fitted with a tracer which burns for approximately 6 seconds.

HE-I, RO 376 The HE-I RO 376 was specifically designed for use in warm climates, the main change being the explosive charge is 26 g of Hexal as opposed to the Torpex of the HE Mk 6*Z. The design of the projectile body is also changed as the HE-I RO 376 has a machined steel body with a flat base. The new body, which weighs 171 g, produces a smaller number of fragments on detonation (approximately 570) but they are larger and more effective on target. The HE-I RO 376 retains the same 944 post impact delay fuze as the HE Mk 6*Z and the ballistics of the two rounds are essentially similar.

MPT, RO 379 Also known as the 30 mm MPT-LD (MultiPurpose Tracer - Low Drag). The streamlined projectile used with this round lacks a fuze as it utilises a pyrotechnic fuze with a post impact delay function for maximum behind-armour effects. It, therefore, combines penetration, high blast and incendiary features with high fragmentation. Fuze operation is achieved down to an 80° (NATO) angle of impact on aluminium. The projectile has a flat base with an insert for a tracer element and has thick walls made from high strength steel. The explosive filling is Hexal. This round meets the requirements of MIL-STD-1466 Safety Criteria and Qualification Requirements for Pyrotechnically Initiated Ammunition (PIA).

MP Produced by Raufoss Ammunisjonsfabrikker A/S of Norway this 30 mm MP projectile will function reliably against a 2 mm dural plate at impact angles between 0 and 87° NATO but will not function against a 0.5 mm dural plate in front of the gun muzzle. No fuze is involved as the MP projectile relies upon a drop safe pyrotechnic ignition train. When impacting against aircraft type targets the projectiles will detonate approximately 300 mm within the aircraft. The distribution of fragments is approximately 20° on each side of the line of fire; the fragments are heavy and optimised to defeat material type targets. The secondary incendiary effects will ignite JP4 and JP5 or heavy diesel oil in self-sealing tanks. Incendiary and blast effects are contained within the target.

The length of the Raufoss 30 mm MP round is given as 197.2 mm and it weighs 500 g; projectile weight is 225 g. Muzzle velocity is 730 m/s.

AP Mk 1*Z This armour-piercing round is no longer produced by Royal Ordnance but may still be encountered. It uses a flat-based projectile with a two-part body assembly protected by a light metal windshield over the nose. The two halves of the body are threaded together and contain a heavy metal penetrator core capable of entering up to 25 mm of armour.

Data

Weights:
- **complete round, HE, HE-I, MPT** - 496 g
- **propellant** - 46 g

Length of complete round: 198 mm
Length of cartridge case: 113 mm
Muzzle velocity: 785 m/s

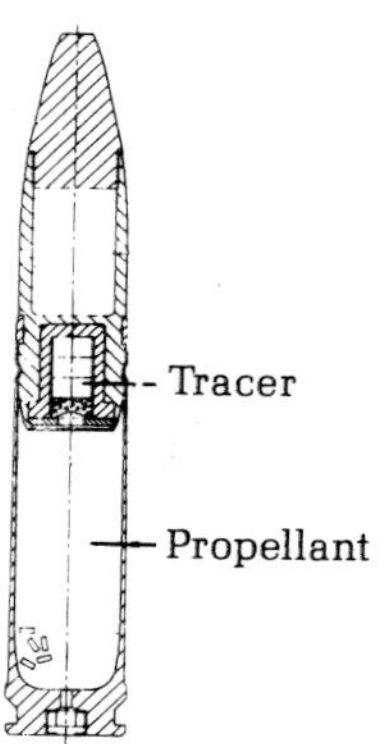

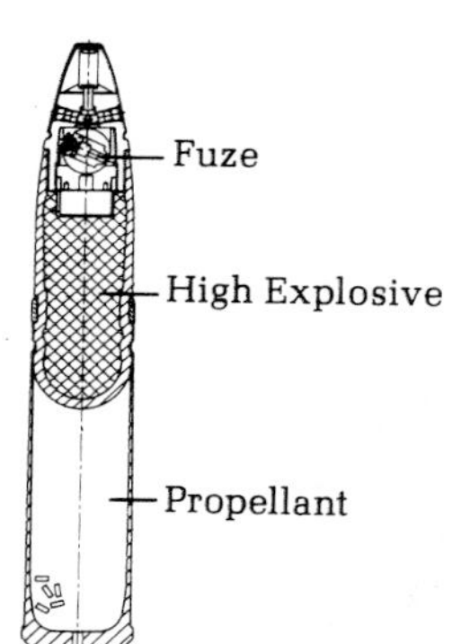

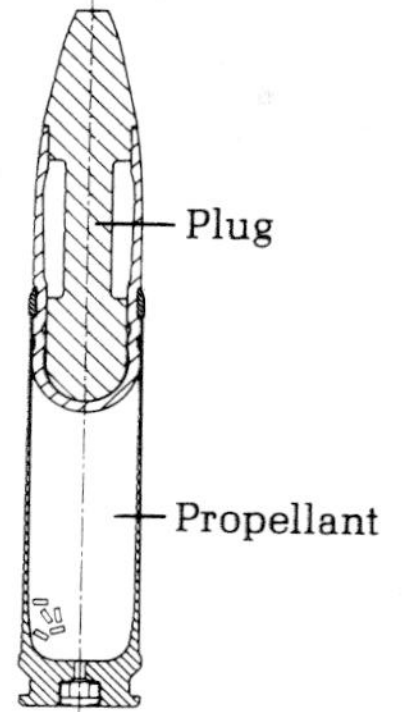

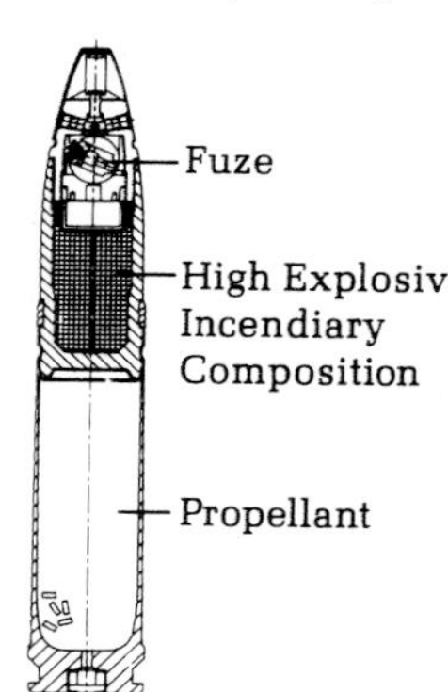

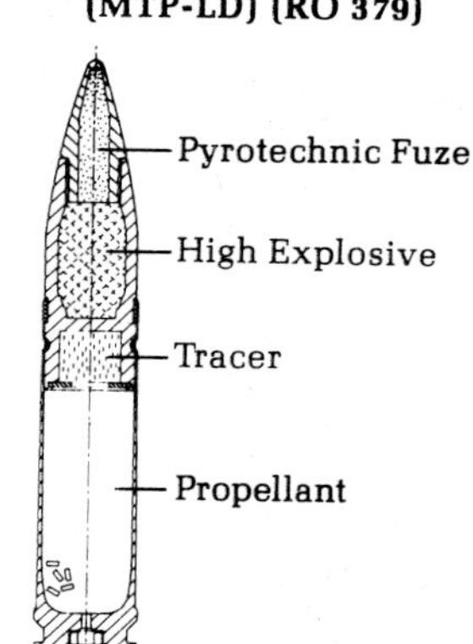

Types of 30 mm Aden gun ammunition originally produced by Royal Ordnance

BELGIUM
Manufacturer: FN Herstal SA
Type: HE-I, HE-I-SD, AP-HE-I-SD, TP
Description: No longer produced but may still be retained in service with some nations. Standard specifications

FINLAND
Manufacturer: Sako Limited
Type: HEI, HEI-SD, HEI-T, SAPHEI, APHC, TP, TP-T
Description: Standard specifications

FRANCE
Manufacturer: Giat Industries
Type: HE-I Mle 5478 and 5479, Mle 7478 and 7479, AP-T Mle 5879, TP Mle 2469
Description: Standard specifications

INDIA
Manufacturer: Indian Ordnance Factories
Type: HE
Description: Produced at Ordnance Factory Khamaria, Jabalpur as 'Upgrade' for Aden Guns. Weight of complete round 432 g ±20 g, and projectile 219 g ±3 g. Muzzle velocity 780.8 m/s

NORWAY
Manufacturer: Raufoss Ammunisjonsfabrikker A/S
Type: MP
Description: See text

SINGAPORE
Manufacturer: Chartered Industries of Singapore
Type: HEI, SAPHEI, EP (TP)
Description: HEI and SAPHEI both have Hexal fillings. The base-fuzed SAPHEI has a muzzle velocity of 765 m/s and can penetrate 15 mm of armour plate at normal impact angles

UNITED KINGDOM
Manufacturer: British Aerospace Defence Limited, Royal Ordnance Division
Type: HE, HE-I, MP-T, TP, TP-T
Description: See text

UPDATED

30 mm RARDEN Gun Ammunition

Armament

RARDEN 30 mm gun L21A1 and L21A2 fitted to FV107 Scorpion and Sabre, Warrior MCV, 4K 7FA MICV 30/1 (Austria, prototype). The 30 mm Bushmaster II cannon can also accommodate RARDEN ammunition.

Development

The first 30 mm RARDEN gun appeared in 1966 and the first service examples, the L21A1, followed during the early 1970s. The design philosophy called for an anti-armour gun with the emphasis on accuracy, as opposed to high rates of fire, combined with the ability to be mounted in light turrets. The ammunition selected for the RARDEN was based on that developed in Switzerland for the 30 mm Hispano-Suiza 831 L (now the Oerlikon-Contraves KCB) cannon family, but subsequent development has resulted in what may be regarded as a separate ammunition family. KCB ammunition can be fired from RARDEN guns.

The UK Design Authority for 30 mm RARDEN ammunition is Royal Ordnance plc.

30 mm RARDEN gun ammunition produced by Giat Industries/Manurhin: from left, TP; TP-T; HEI/SD; HEI-T/SD

Description

All 30 mm RARDEN ammunition involves fixed rounds with the projectiles crimped to the necked drawn brass cartridge cases by a crimping ring engaging in a cannelure just above the boat-tailed base on the projectile. In most cases the drive bands are sintered iron pressed into place. The rounds are issued and loaded in charger clips of three.

The brass cartridge cases are 170.3 mm long and have an L16 (RO 673) percussion primer filled with 0.97 g of gunpowder G20 threaded into the base; the rim diameter is 42.9 mm. The cartridge case is filled with variable amounts of granular NRN 141/RDN propellant.

Ammunition available for the 30 mm RARDEN gun includes the following types:

HE-T L13A1 The original HE-T round was the L8A1 but this has been replaced by the L13A1. The projectile involved is machined from steel bar and contains 25.6 g of Torpex 2. An L86A2 impact fuze is threaded into the projectile nose while the shell base is occupied by an L11 tracer element which burns for a distance of over 2,000 m after leaving the gun muzzle.

APSE-T L5A2 The hardened steel pointed projectile used with this round has no fuze and is fitted with an anodised aluminium ballistic cap. It relies, for its secondary effects, on a payload formed by a pyrotechnic mixture of explosive and CS5390 smoke composition which is ignited by friction once the projectile has penetrated light armour. An L10A1 tracer is fitted in the base and burns for a range of approximately 1,500 m.

APDS-T L14A2 and L14A3 This APDS-T round was jointly developed by Royal Ordnance and PATEC of the USA. Introduced in 1980, it uses a plastic sabot with a lateral cross-section at the nose. Once the projectile leaves the gun muzzle the sabot breaks into four lateral segments, leaving the tungsten penetrator to complete its trajectory, marked for ranges out to at least 2,000 m by a tracer element in the penetrator base. Further result assessment is provided by a pyrophoric cap on the penetrator nose which provides a flash on impact with a hard target. The penetrator can pierce a 40 mm RHA plate angled at 45° out to a range of 1,500 m. Accuracy is stated to be half the dispersion of the HE-T round. Figures of 0.5 mils vertical and horizontal at 1,000 m have been quoted.

The APDS-T L14A3 has a new type of penetrator and was introduced in 1986, the same year that a lower barrel wear propellant was first produced. One further round developed for the armour-piercing role was the APHC L6A1 carrying a tungsten carbide core; it is no longer manufactured.

APFSDS-T An APFSDS-T round for use with the 30 mm RARDEN gun is under development for introduction into service in 1997 to 1998, with Royal Ordnance Radway Green as the lead facility. First demonstrated in 1987 the projectile assembly features a two petal sabot while the tungsten alloy penetrator is understood to have a diameter of 8.6 mm and a velocity at 50 m from the muzzle of 1,165 m/s. The tracer element burns to a range of about 2,000 m. It is anticipated that the APFSDS-T will demonstrate a 60 per cent improvement in armour penetration compared to the APDS-T L14A3, while retaining a similar barrel life capability.

TP-T L12A1 This round, the RO 271, replaces the earlier L7A4 and uses a hollow steel body with an aluminium nose cap. It ballistically matches the HE-T round.

RRTR RO 275 The 30 mm Reduced Range Training Round (RRTR) was developed to provide a low-cost training round for use on ranges with limited areas. It was originally designed to match ballistically the 30 mm L14A2 APDS-T round during the early part of its trajectory. The RRTR is in production by Royal Ordnance to meet a British Army order calling for 660,000 units.

The 30 mm RO 275 RRTR is a fixed round with a tubular projectile carried inside a sabot similar to that used on the 30 mm L14A2 APDS-T. After firing the sabot falls away leaving the tubular projectile to follow closely the trajectory of the L14A2 APDS-T projectile out to a range of approximately 1,500 m. At that range the projectile velocity will have fallen from an initial 1,300 m/s to around 700 m/s, and an aerodynamic effect comes into play to effectively choke the flow of air through the interior of the tube. The projectile then becomes unstable and falls to the ground after travelling approximately 2,000 m. The resultant range safety trace is thus considerably reduced. For example, the safety trace for an L14A2 APDS-T projectile is 11,900 m long and nearly 4,000 m wide. For the RO 275 RRTR the safety trace is 4,500 m long and 2,260 m wide. The RO 275 incorporates a tracer which remains effective out to 1,500 m.

Although the 30 mm RO 275 RRTR is matched to the 30 mm L14A2 APDS-T the same projectile design can be configured to match the 30 mm HE-T and APFSDS (still under development) projectiles. The changes required for such matching involve a change in the launch velocity, minor profile changes to the tubular projectile to alter the coefficient of drag and, in the case of the APFSDS, a change in the pusher material, which will be an aluminium alloy.

Break-up NWM de Kruithoorn BV of the Netherlands produce 30 mm break-up ammunition suitable for use in RARDEN guns.

Data

Type	**HE-T**	**APSE-T**	**APDS-T**	**TP-T**	**RRTR**
Designation	L13A1	L5A2	L14A2	L12A1	RO 275
Weights					
complete round	903.9 g	904.4 g	822 g	903.9 g	760 g
projectile	356.9 g	357.4 g	300 g	356.9 g	270 g
filling	25.6 g	29 g	none	none	none
propellant	160 g	160 g	150 g	160 g	140 g
cartridge case	365 g	365 g	365 g	365 g	365 g
Length					
complete round	285.55 mm	285.55 mm	285.55 mm	285.55 mm	285.55 mm
cartridge case	170.3 mm	170.3 mm	170.3 mm	170.3 mm	170.3 mm
Muzzle velocity	1,070 m/s	1,070 m/s	1,175 m/s	1,070 m/s	1,300 m/s

FRANCE

Manufacturer: Giat Industries/Manurhin
Type: HEI/SD, HEI-T/SD, TP, TP-T
Description: HEI/SD and HEI-T/SD fitted with MR302 PD fuze with self-destruct after 7 to 12 seconds. Otherwise standard specifications

UPDATED

COMBAT GRENADES

ARGENTINA

FM GME-FMK2-MO hand grenade

Description

The GME-FMK2-MO hand grenade was designed by Fabricaciones Militares (FM). The body design is based on a nodular-iron technique which is claimed by the company to be unique. Its final characteristics are the result of its metallic structure (Fe, C, Mg, Mn and so on) and the post-casting heat treatment used.

The fuze body consists of one piece in aluminium which can be produced by drop forging or injection moulding. It incorporates the detonator and firing mechanism and can be removed from the grenade body as a single unit.

On explosion the fuze body is reduced to fragments, in most cases of a weight of 3 to 5 g. These produce a lethal area of about 5 m radius and will not endanger the thrower whose throw in the prone position is limited to about 10 m or to 30 m if kneeling.

Firing is initiated by percussion and the powder train produces a delay of between 3.6 and 4.5 seconds between the throwing of the grenade and the moment of explosion.

The bursting charge consists of a mixture of recrystallised Hexogen and flaked TNT in proportions which can be varied to suit fragmentation requirements. The net weight of the charge varies between 75 and 79 g.

The hand grenade has two safety devices. One is a specially shaped spring-steel wire securing the grenade in transit, and the other consists of two split pins on the safety pin ring. The first can be removed at any time and the grenade can continue to be handled with perfect safety. It is released by firm thumb pressure and must always be removed before issuing the grenades.

The second split pin is the last safety device on the grenade and once it is removed the grenade is ready for throwing. Should the safety lever come away on pulling out the pin, the grenade will explode.

If a launching device is used with the grenade, it can be fired to a distance of between 350 and 400 m, to explode either at ground level or at varying heights, according to the degree of skill acquired by the soldier in practice.

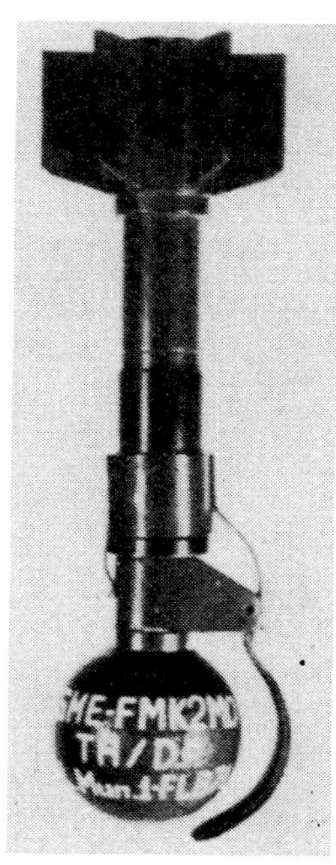

FM grenade and tail for rifle launching

The launcher is provided separately, complete with its propellant cartridge. To fire: after making sure the grenade is firmly in its seating and that the safety lever is in place, the grenade and launcher are put in the muzzle of the automatic light rifle (FAL). There is no need to use a muzzle adaptor. The gas vent plug is rotated to close it, and the rifle is loaded with the propellant cartridge provided with the launcher. The safety split pin is removed, the weapon aimed and fired.

Data

Weight: grenade body, 165 g; fuze body, 40.8 g; charge, 75/79 g

FM GME-FMK2-MO hand grenade

Length overall: ca 120 mm
Body diameter: ca 55 mm
Delay: 3.6-4.5 s

Manufacturer

Direccion General de Fabricaciones Militares, Av Cabildo 65, 1426 Buenos Aires.

Status

In production.

Service

Argentine forces.

UPDATED

FM rifle grenades

Description

There are three types of FM rifle grenade, as follows:

GEAT FRAG 40 TB - AP/AV
GEAT 58 TB - AP
Iluminante Cal 38.1 mm para FAL - Illuminating.

All three grenades are intended to be launched from any 7.62 mm rifle with an appropriate 22 mm diameter muzzle attachment. Propulsion is provided by special ballistite cartridges.

The GEAT FRAG 40 TB is a dual purpose anti-personnel/anti-vehicle (AP/AV) grenade with a steel body containing a RDX/TNT 'shaped charge warhead with a diameter of 40 mm. The shaped charge, initiated by a PIBD fuze, can penetrate up to 125 mm of armour. Maximum range with the launch rifle barrel at an elevation of 45° is 450 m. The lethal radius produced by the fragmentation of the steel body is 15 m.

The GEAT 58 TB is a purely armour penetrating grenade with a 58 mm diameter RDX/TNT shaped charge in an aluminium body. The shaped charge, initiated by a PIBD fuze, can penetrate up to 230 mm of armour. Maximum range with the launch rifle barrel at an elevation of 45° is 400 m.

The Iluminante Cal 38.1 mm para FAL grenade can produce 100,000 candela for 30 seconds, sufficient to illuminate a target radius of 30 m at the maximum range of 200 m. The initial height at which the grenade functions is 100 m, following a delay of 4.5 seconds. This grenade is launched using a special-to-type ballistite cartridge.

Data

Model	AP/AV	AP	III
Warhead calibre	40 mm	58 mm	38.1 mm
Length	381 mm	423 mm	347 mm
Weight	600 g	660 g	380 g

Model	AP/AV	AP	III
Muzzle velocity	68 m/s	63 m/s	70 m/s
Max range	450 m	400 m	200 m

Manufacturer

Direccion General de Fabricaciones Militares, Av Cabildo 65, 1426 Buenos Aires.

Status

In production.

Service

Argentine forces.

NEW ENTRY

AUSTRALIA

ADI grenades

Description

The ADI family of grenades is a totally co-ordinated system which allows individual models to be adapted to specific needs. The grenades share common components and each can be hand thrown or adapted for rifle-firing with ADI's 5.56 mm bullet trap projector. The system eliminates the need to meet the combat and training requirements for hand and rifle grenades with different arrangements. Another benefit is the versatility of the system under combat conditions; the soldier can quickly change from hand throwing to rifle projection with minimum preparation or ancillary equipment. This advanced system is an Australian design to meet the exacting specifications of the Australian Army.

HE fragmentation hand grenade

This is a defensive grenade, to be thrown from cover. Fragmentation is controlled by using several thousand steel balls set in a plastic matrix around the explosive charge and the positioning is designed to ensure even and predictable dispersion. The lethal radius is 6 m and the casualty radius 15 m. Actuation is by the usual 'mousetrap' striker retained by a lever and pin, with the ignitor unit fitting into the centre of the grenade. A gasless primer is used, giving consistent delay times. The delay may be specified by the user; standard delays are 3.5, 4, 4.5 or 5 seconds.

Data

Mass: 370 g
Length: 90 mm
Diameter: 68 mm
Standard delay: 5 s (others optional from 0.1 to 7 s)
Range on projector: 180 m

HE offensive (blast) hand grenade

This is used where fragments are undesirable but flash is required. No prefragmented matrix is present, but extra explosive is added to increase the blast effect. The mass of the bursting charge is double that of the fragmentation grenade. A similar ignition system is used.

Data

Mass: 220 g
Length: 90 mm
Diameter: 68 mm
Standard delay: 5 s (others optional from 0.1 to 7 s)
Range on projector: 200 m

Practice hand grenade

This approximates the fragmentation grenade's weight, throwing and rifle projection characteristics. It can be reused many times by inserting a replaceable low-explosive charge which gives a flash and bang on functioning but has no dangerous effects and does not damage the grenade body.

Data

Mass: 370 g
Length: 90 mm
Diameter: 68 mm
Standard delay: 5 s (others optional from 0.1 to 7 s)
Range on projector: 180 m

Rifle Projector

The Rifle Projector is a tail unit which can be fitted to any of the grenades in the ADI system to allow them to be fired from Steyr AUG or M16 5.56 mm rifles. The design can also be adapted to other 5.56 mm small arms weapons as required. There is a small booster unit inside the projector which ensures accuracy and light recoil for the user. The projector is light and easily carried by the soldier.

To launch a grenade by rifle, the striker housing assembly is removed and replaced by the projector unit. It is then slipped over the muzzle of the rifle and fired to a range of up to 200 m.

Data

Mass: 245 g
Length: including grenade, 265 mm
Diameter: 68 mm

Manufacturer

Australian Defence Industries Limited, Corporate Headquarters, Level 22 Plaza II, Cnr Grosvenor and Grafton Streets, Bondi Junction, NSW 2022.

Service

Australian armed forces.

UPDATED

ADI HE fragmentation grenade F1

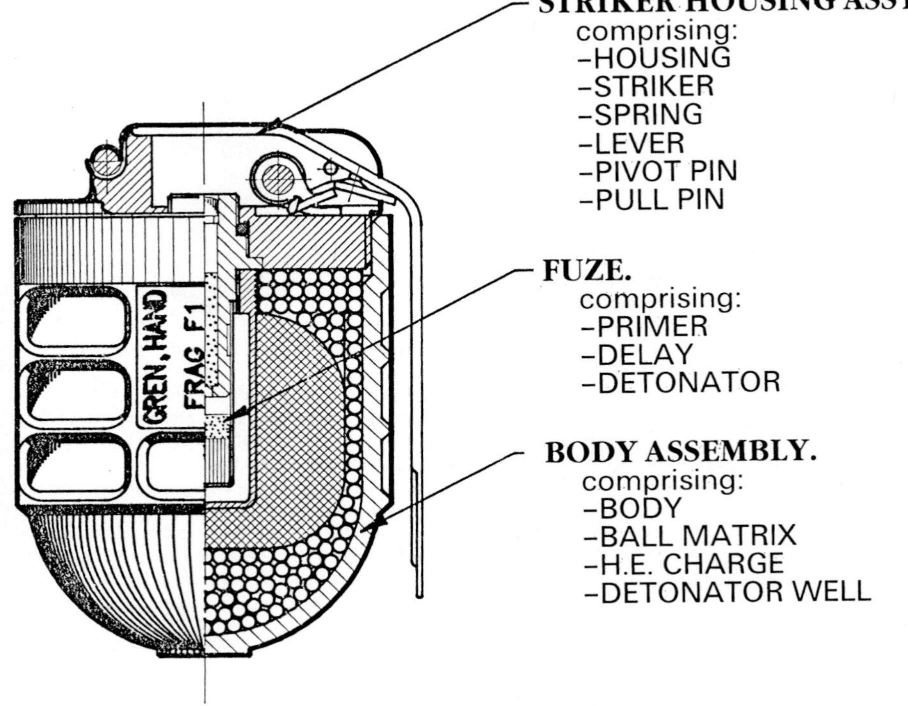

ADI HE fragmentation grenade

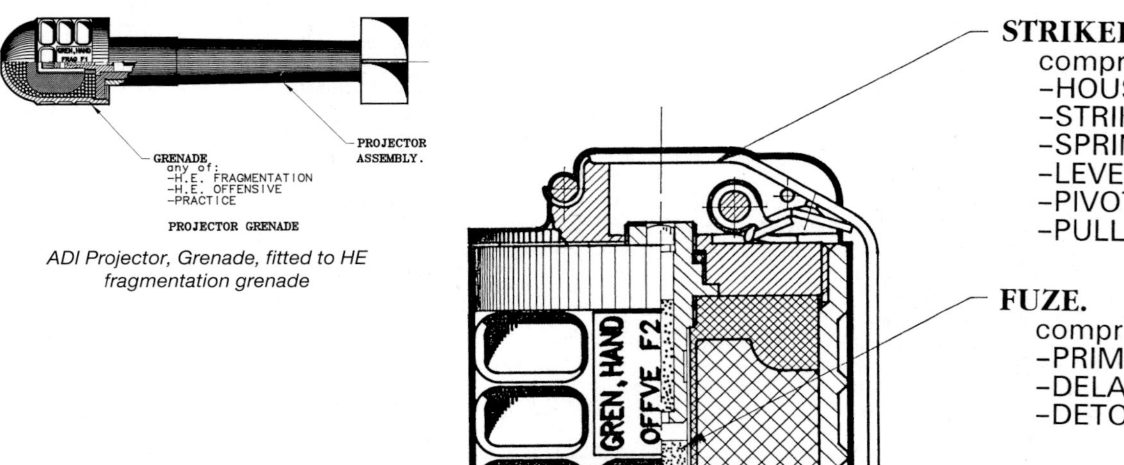

ADI Projector, Grenade, fitted to HE fragmentation grenade

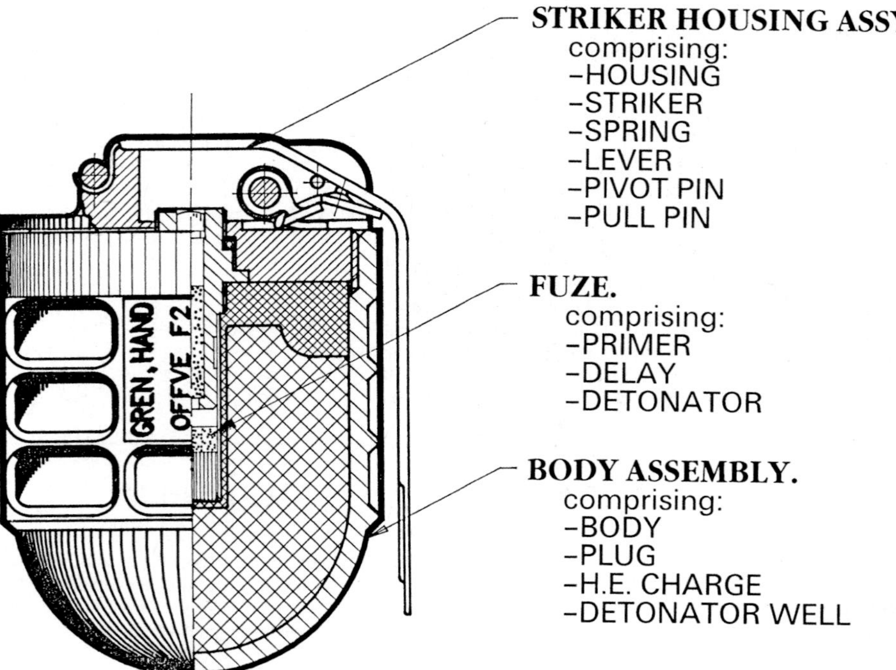

ADI offensive hand grenade

AUSTRIA

Arges Type HG 84 fragmentation grenade

Description

The Arges HG 84 is the largest type in the Arges HG range and is an improved version of the Arges HG 78. The special design of the fragmentation body gives an optimum distribution of fragments and is therefore effective for close combat operations in open terrain. In spite of being an egg-shaped grenade, development has resulted in the same symmetrical fragment distribution as is found in grenades of spherical shape. Detonated in a lying position, 1 m above the ground, at a distance of 5 m from the target, the HG84 gives an average fragment density of 14.3 per m² in 20 mm soft wood.

The Offensive Hand Grenade 84 (OFF HG 84) differs from the HG 84 by having a plastic core without fragments, a smooth 'lemon skin' surface, and being somewhat lighter in weight. In all other respects it resembles the HG 84.

Data

Weight: 480 ±20 g
Length: 115 ±2 mm
Largest diameter: 60 ±1 mm
Charge weight: 95 ±5 g
Fragment diameter: 2-2.3 mm
Number of fragments: ca 5,300
Delay time: 4 s +1.5/−0.5 s at 21°C

Manufacturer

Armaturen GesmbH, A-4690 Schwanenstadt-Rüstorf.

Status

In production.

Service

Austrian Army.

UPDATED

Arges HG 84 fragmentation grenade

Arges Type ÜbHG 91 practice grenade

Description

The Arges Type ÜbHG 91 practice grenade resembles the HG 91 service grenade but is solely for training troops in handling and throwing. The body is of cast iron, with a hole on top into which the practice igniter fits, and a hole at the bottom to release the gas pressure and noise of the igniter when it fires. The grenade body is covered with black, high impact proof plastic material with a 12 mm blue strip moulded in to provide permanent identification.

The practice igniter unit corresponds to a service igniter, but instead of a detonator it carries a pyrotechnic charge giving a loud report. The noise, escaping from the open end of the grenade, gives ample evidence of its functioning and is most realistic. Having once been thrown, the grenade can then be salvaged, cleaned, and a new practice igniter set fitted. With its robust construction, the life of the grenade body is virtually limitless.

Data

Length: 115 ±2 mm
Diameter: 60 ±1 mm
Weight: 520 ±20 g
Delay time: 4 s +1.5/−0.5 s

Manufacturer

Armaturen GesmbH, A-4690 Schwanenstadt-Rüstorf.

Status

In production.

Service

Supplied to Middle and Far East countries and Africa.

VERIFIED

Arges ÜbHG 91 practice grenade

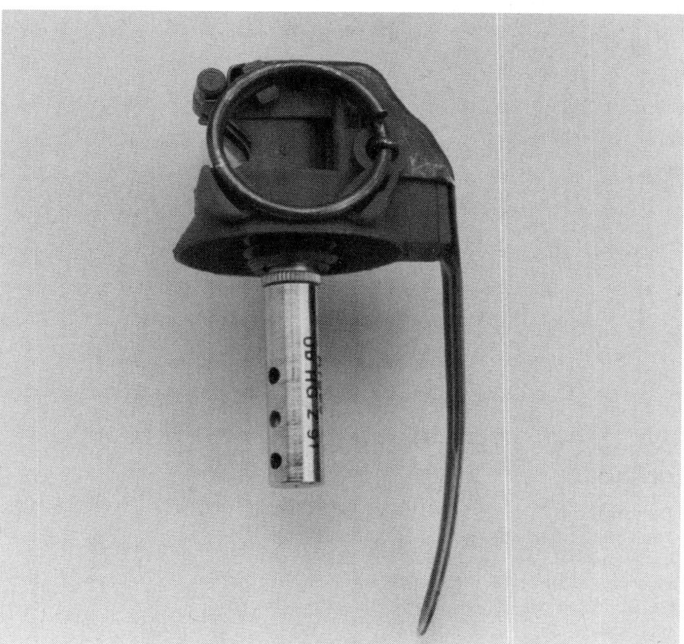

Arges ÜbHG 91 complete practice igniter set ready for insertion in grenade

Arges Type HG 85 fragmentation hand grenade

Description

The Arges HG 85 is the medium size in the Arges HG range and is a further development of the HG 79 design. As with the earlier design it consists of a plastic body, an inner liner carrying the preformed fragments, and a filling of PETN high explosive. Fragment distribution covers the full 360° area around the grenade and an HG 85, detonated 1 m above the ground and 5 m from a 20 mm soft wood target gives an average fragment density of 8.1 per m^2.

An offensive pattern, OFF HG 85, is available and differs from the HG 85 by having a plastic core without fragments, a smooth 'lemon skin' surface, and being slightly less in weight.

Data

Weight: 340 ±20 g
Length: 96 ±2 mm
Diameter: 57 ±1 mm
Weight of explosive charge: 50 ±3 g
Diameter of fragments: 2-2.3 mm
Number of fragments: ca 3,500
Delay time: 4 s +1.5/−0.5 s at 21°C

Manufacturer

Armaturen GesmbH, A-4690 Schwanenstadt-Rüstorf.

Status

In production.

Service

Supplied to Far East countries.

UPDATED

Arges HG 85 fragmentation hand grenade

Arges Type HG 86 Mini fragmentation hand grenade

Description

The Arges HG 86 (Mini 86) is the smallest type in the Arges range and is a further development of the Arges HG 80 Mini. The principal improvement is in the control of fragment distribution. An HG 86 detonated 1 m above the ground, 5 m from a 20 mm soft wood target, averages a fragment density of 3.05 per m^2.

The HG 86 Mini can also be supplied as an offensive grenade (OFF HG 86) in which case it has a plastic core without fragments and a smooth surface.

Data

Weight: 180 ±10 g
Length: 76 ±1 mm
Diameter: 43 ±1 mm
Weight of explosive charge: ca 17 g
Diameter of fragments: 2-2.3 mm
Number of fragments: ca 1,600
Delay time: 4 s +1.5/−0.5 s at 21°C

Manufacturer

Armaturen GesmbH, A-4690 Schwanenstadt-Rüstorf.

Status

In production.

UPDATED

Arges HG 86 Mini fragmentation grenade

Arges 40 × 46 mm Practice 94 grenade

Description

The Practice 94 cartridge fires an inert projectile. It is solely used for training purposes. The ballistic characteristics are identical to those of the Arges HE/Frag 92, HE/DP 92, HE/HL 93, CS 93, CN 93 and Smoke 93 grenades.

Data

Round length: 102 mm
Round weight: 265 g
Projectile weight: 190 g
Payload: inert
Muzzle velocity: 76 m/s
Max range: 400 m

Manufacturer

Armaturen GesmbH, A-4690 Schwanenstadt-Rüstorf.

Status

In production.

VERIFIED

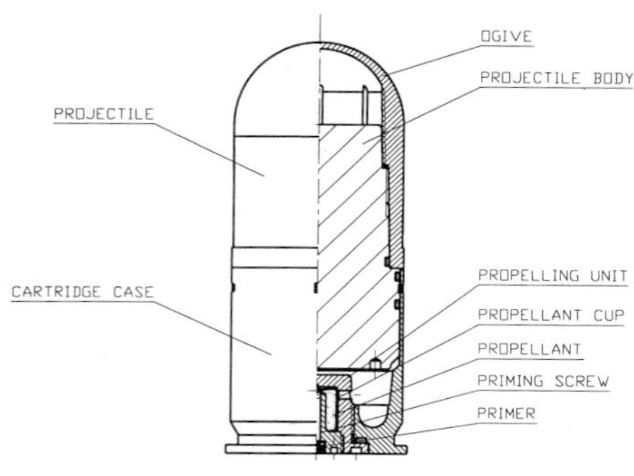

Arges 40 × 46 mm Practice 94 grenade

Arges 40 × 46 mm Practice/ Tracer 94 grenade

Description

The Practice/Tracer 94 grenade fires an inert projectile equipped with a tracer to allow the observation of the projectile's trajectory and its impact on the target. It is solely used for training purposes. The ballistic characteristics are identical to those of the Arges HE/Frag 92, HE/DP 92, HE/HL 93, CS 93, CN 93 and Smoke 93 grenades.

Data

Round length: 102 mm
Round weight: 265 g
Projectile weight: 190 g
Payload: inert
Tracer burning time: ca 4 s
Muzzle velocity: 76 m/s
Max range: 400 m

Manufacturer

Armaturen GesmbH, A-4690 Schwanenstadt-Rüstorf.

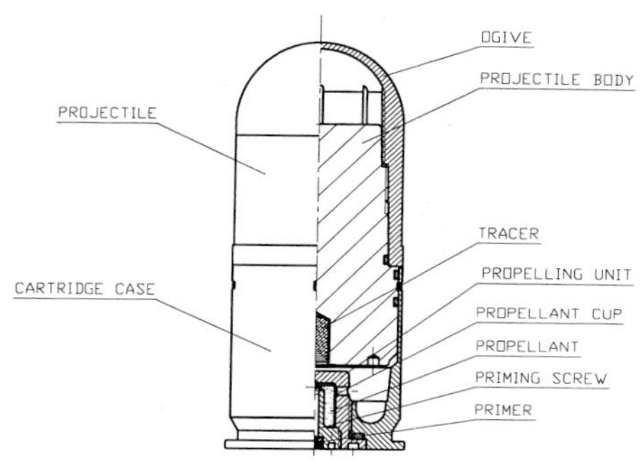

Status

In production.

VERIFIED

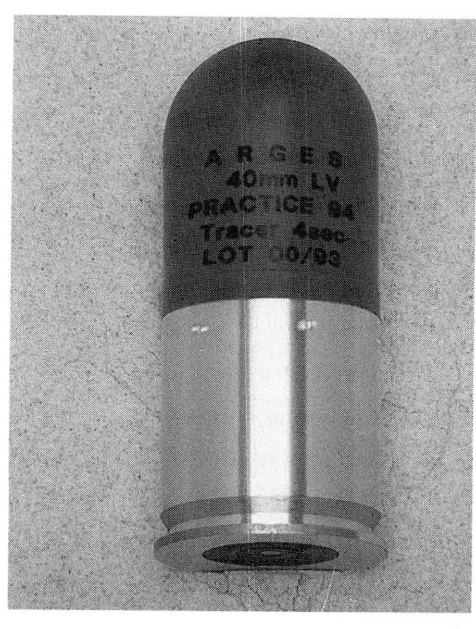

Arges 40 × 46mm Practice/Tracer 94 grenade

Arges 40 × 46 mm Practice 93 grenade

Description

The 40 mm Practice 93 grenade is used for training purposes. It consists of a solid plastic projectile and can therefore be readily disposed of. It is available with propelling charges providing various muzzle velocities from 76 to approximately 100 m/s.

Data

Round length: 102 mm
Round weight: 175 g
Projectile weight: 100 g
Muzzle velocity: 76 m/s
Max range: 400 m

Manufacturer

Armaturen-GesmbH, A-4690 Schwanenstadt-Rüstorf.

Status

In production.

VERIFIED

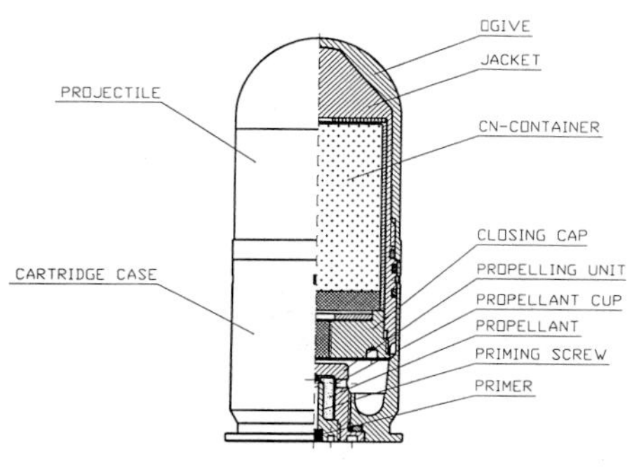

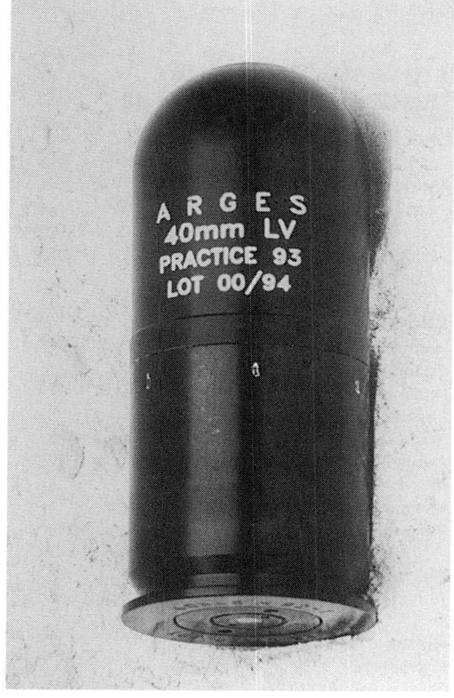

Arges 40 × 46 mm Practice 93 grenade

Arges 40 × 46 mm Drill 93 grenade

Description

The Cartridge 40 mm Drill 93 is used for safe training in the handling of grenade launchers and their ammunition. It is completely inert. To protect the weapon's firing pin from damage, a plastic plug is inserted into the base of the cartridge.

Data

Cartridge length: 102 mm
Cartridge weight: ca 245 g

Manufacturer

Armaturen GesmbH, A-4690 Schwanenstadt/Rüstorf.

Status

In production.

VERIFIED

GRENADE BODY

BOTTOM SCREW

Arges 40 × 46 mm Drill 93 grenade

Arges 40 × 46 mm HE/ Fragmentation 92 grenade

Description

The Arges Fragmentation 92 cartridge is provided with a fragmentation jacket and can be used for combat against soft targets even if under cover. The fragmentation grenade is equipped with an impact fuze and a self-destruct device, detonating either on impact or destroying itself after approximately 15 seconds.

Data

Round length: 102 mm
Round weight: 265 g
Projectile weight: 190 g
Payload: 28 g RDX
Fragments: ca 1,000 steel balls 2-2.3 mm diameter
Arming distance: 11 - 19 m
Muzzle safety: min, 11 m
Muzzle velocity: 76 m/s
Max range: 400 m

Manufacturer

Armaturen GesmbH, A-4690 Schwanenstadt-Rüstorf.

Status

In production.

VERIFIED

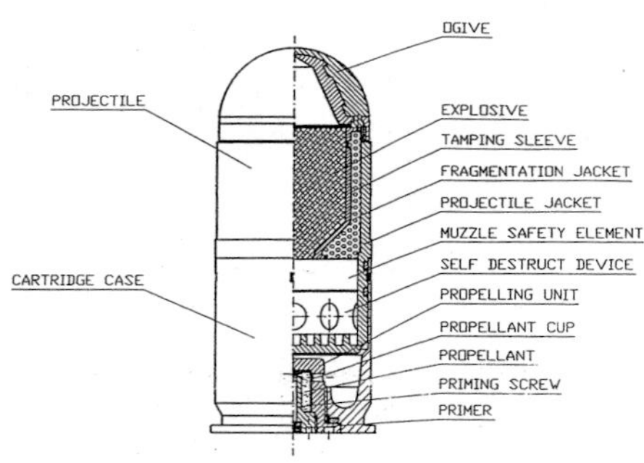

Arges 40 × 46 mm HE/Fragmentation 92 grenade

Arges 40 × 46 mm HE/ Dual-Purpose 92 grenade

Description

The cartridge 40 mm HE/DP 92 is provided with a hollow charge cone and a fragmentation jacket and can thus be used against soft or lightly armoured targets. It is equipped with an impact fuze and a self-destruct device and detonates either on impact or after 15 seconds.

Data

Round length: 102 mm
Round weight: 265 g
Projectile weight: 190 g
Payload: 20 g RDX
Fragments: ca 1,000 steel balls 2-2.3 mm diameter
Arming distance: 11-19 m
Muzzle velocity: 76 m/s
Max range: 400 m
Perforation: >25 mm homogeneous steel armour

Manufacturer

Armaturen GesmbH, A-4690 Schwanenstadt-Rüstorf.

Status

In production.

VERIFIED

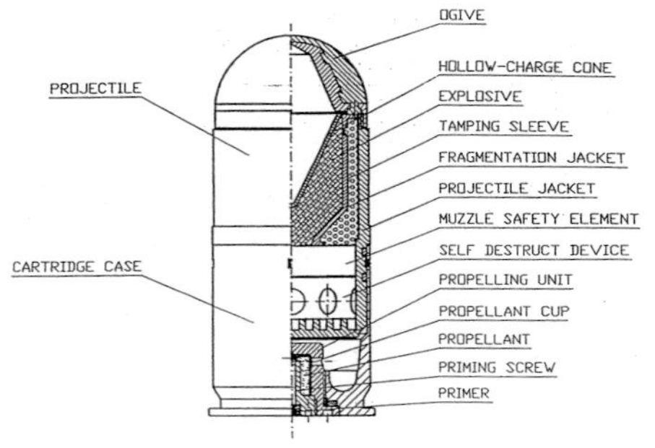

Arges 40 × 46 mm HE/Dual-Purpose 92 grenade

Arges 40 × 46 mm HE/HL 93 hollow charge grenade

Description

The Arges 40 mm HE/HL 93 is a hollow charge grenade designed for employment against armoured targets.

Data

Cartridge length: 102 mm
Cartridge weight: 265 g
Projectile weight: 190 g
Muzzle velocity: 76 m/s
Max range: 400 m
Charge weight: 30 g
Arming distance: 11-19 m
Self-destruct time: ca 15 s
Penetration: min 60 mm armoured steel plate set at normal

Manufacturer

Armaturen GesmbH, A-4690 Schwanenstadt/Rüstorf.

Status

In production

VERIFIED

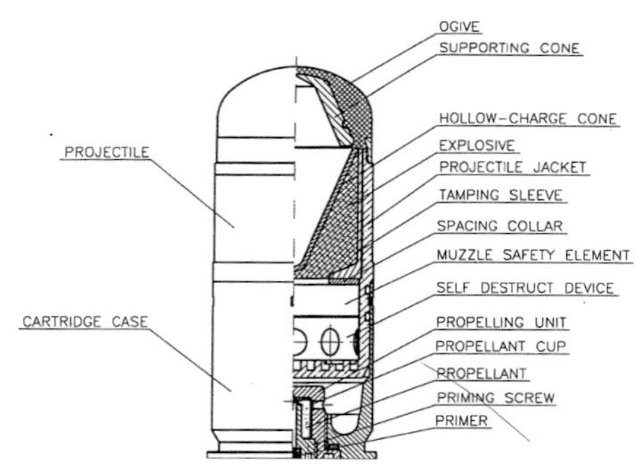

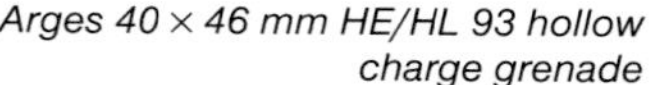

Arges 40 × 46 mm HE/HL 93 hollow charge grenade

Arges 40 × 46 mm Smoke 93 grenade

Description

The cartridge 40 mm Smoke 93 emits smoke for approximately 20 seconds, with a delay of 1 second after firing. It can be used for tactical screening purposes.

Data

Round length: 102 mm
Round weight: 255 g
Projectile weight: ca 180 g
Payload: smoke composition
Muzzle velocity: 76 m/s
Max range: 400 m
Duration of smoke emission: ca 20 s

Manufacturer

Armaturen GesmbH, A-4690 Schwanenstadt-Rüstorf.

Status

In production.

VERIFIED

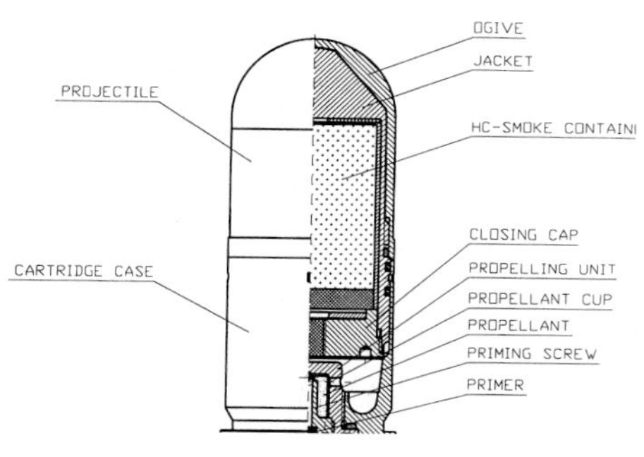

Arges 40 × 46 mm Smoke 93 grenade

Arges 40 × 46 mm RP92 incendiary grenade

Description

The cartridge 40 mm RP92 is an incendiary munition for use at ranges from 100 to 400 m. It will ignite inflammable material at the target and also produces smoke. The projectile will either break on impact, liberating the contents or, on a soft target, is ignited by means of a pyrotechnic charge.

Data

Round length: 112 mm
Round weight: 220 g
Projectile weight: 145 g
Payload: ca 45 g red phosphorus
Muzzle velocity: 76 m/s
Max range: 400 m
Self-destruction: after ca 8 s

Manufacturer

Armaturen GesmbH, A4690 Schwanenstadt/Rüstorf.

Status

In production

VERIFIED

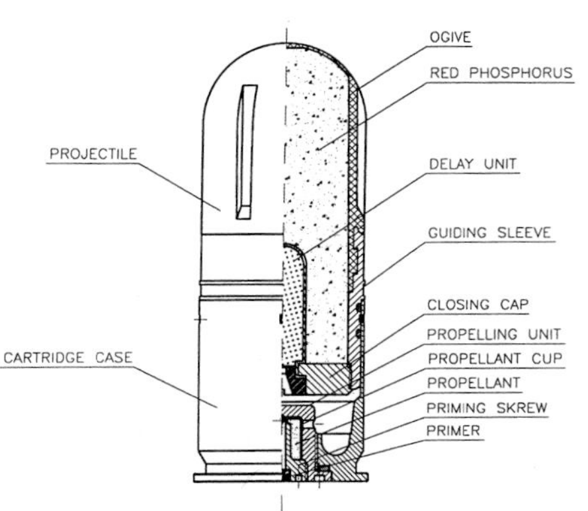

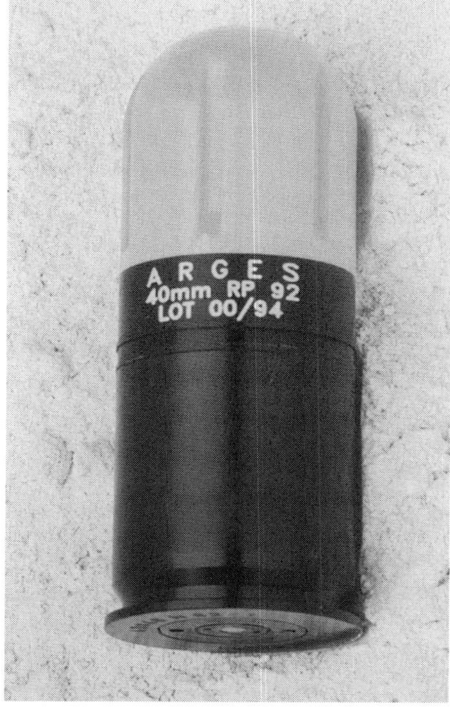

Arges 40 × 46 mm RP92 incendiary grenade

Arges 40 × 46 mm CN 93 Tear Gas grenade

Description

The cartridge 40 mm CN 93 emits CN tear gas for approximately 20 seconds with a delay of one second after firing. It penetrates light obstacles such as insulating glazing, windscreens or plywood doors.

Data

Round length: 102 mm
Round weight: 255 g
Projectile weight: ca 180 g
Payload: CN composition
Muzzle velocity: 76 m/s
Max range: 400 m
Duration of CN emission: ca 20 s

Manufacturer

Armaturen GesmbH, A-4690 Schwanenstadt-Rüstorf.

Status

In production.

VERIFIED

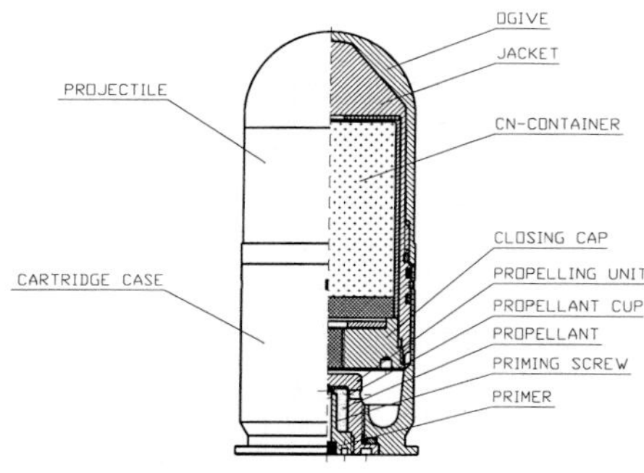

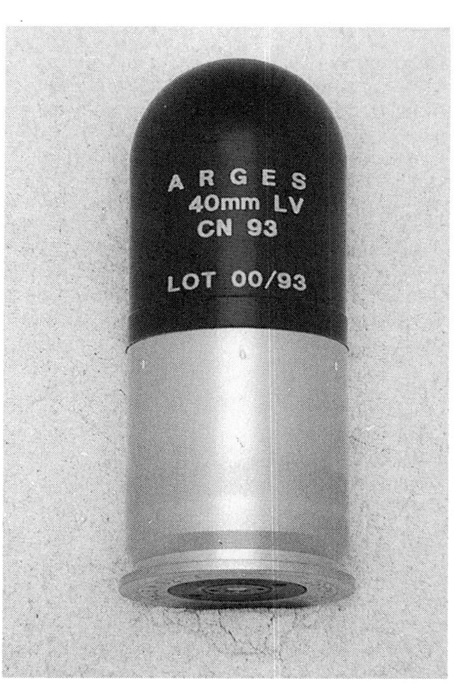

Arges 40 × 46 mm CN 93 Tear Gas grenade

Arges 40 × 46 mm CS95 Tear Gas grenade

Description
The cartridge 40 mm CS95 is a tear gas grenade for use at short ranges. The ogive is weakened, and impact with any obstacle, even a window-pane, will cause it to break off and release the tear gas into the room.

Data
Round length: 102 mm
Round weight: 230 g
Projectile weight: 160 g
Payload: ca 17 g CS composition
Muzzle velocity: 76 m/s
Effective range: 100 m

Manufacturer
Armaturen GesmbH, A-4690 Schwanenstadt/Rüstorf.

Status
In production

VERIFIED

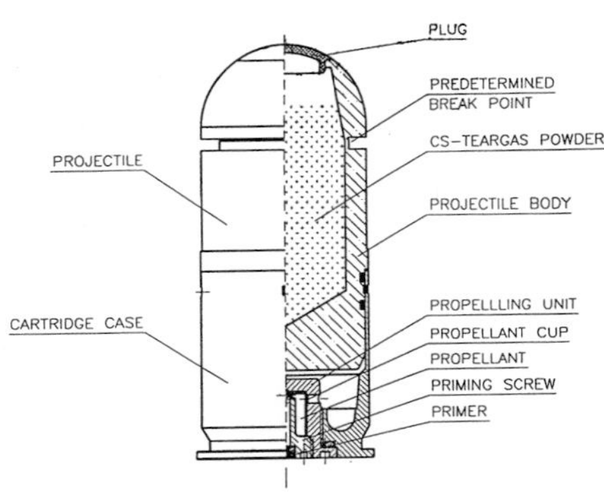

Arges 40 × 46 mm CS95 Tear Gas grenade

Arges 40 × 46 mm Flashbang 93 practice grenade

Description
The Cartridge 40 mm Flashbang 93 is used for training purposes and simulates a detonation by means of noise and flash effects without endangering the users. It is available with various self-destruct times depending upon the desired firing range.

Data
Round length: 102 mm
Round weight: 265 g
Projectile weight: 190 g
Muzzle velocity: 76 m/s
Firing range: 50 m
Self-destruct time: after approx 1.2 s

Manufacturer
Armaturen GesmbH, A-4690 Schwanenstadt/Rüstorf.

Status
In production.

VERIFIED

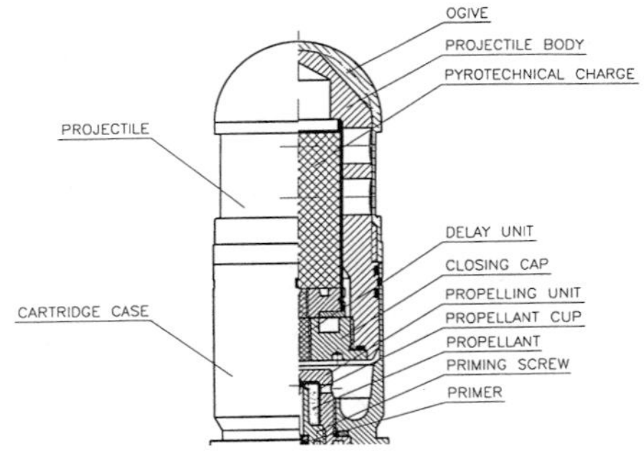

Arges 40 × 46 mm Flashbang 93 practice grenade

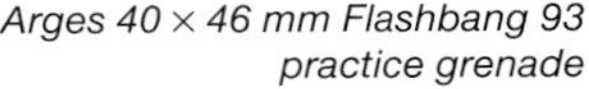

DNW SHG 60 hand grenade

Description
DNW's SHG 60 grenade is manufactured from non-corroding components and is impervious to humidity and temperatures in the range from −40 to +70°C. Even contact with sand will not affect the operation. The body is plastic, and the high explosive filling is surrounded by some 4,000 steel balls to provide the fragmentation effect. A special 'splinter disc' is interposed between the fuze detonator and the high explosive filling which, by carefully controlling the propagation of detonation, ensures an even distribution of fragments and, on test delivers a minimum of 20 hits/m² on a wall circle of 5 m diameter.

Data
Height: 115 mm
Diameter: 60 mm
Weight: 575 g
Fragments: 4,000 steel balls 2.5 mm diameter
Fuze delay: 4.5 ±0.5 s

Manufacturer
Dynamit Nobel Wien GmbH, Tuchlauben 7A, A-1014 Wien.

Status
In production.

VERIFIED

DNW SHG 60 grenade

SplHGr 80/DNW hand grenade

Description

This is a high-explosive grenade consisting of a plastic body into which approximately 4,000 preformed fragments are embedded. Inside this is the plastic explosive bursting charge. A percussion ignition set is installed in the centre of the grenade, controlled by the usual safety pin and fly-off lever. Upon detonation the preformed fragments are effective in all directions, with a danger area of 100 m radius. The SplHGr 80/DNW can be considered as both an offensive and a defensive grenade, since 90 per cent of the fragments are effective within a radius of 20 m.

Data

Length: 115 mm
Diameter: 60 mm
Weight: 585 g
Bursting charge: 75 g
Number of fragments: ca 4,000
Diameter of fragments: 2.5 mm
Delay time: ca 4.5 s
Temperature range: −40 to +70°C

Manufacturer

Dynamit Nobel Wien GmbH, Tuchlauben 7A, A-1014 Wien.

Status

In production.

VERIFIED

SplHGr 80/DNW hand grenade

DNW Smoke grenade HC-75

Description

The DNW HC-75 consists of a green varnished tin plate can, fitted with a fly-off lever type of igniter and loaded with a metal oxide/hexachloroethane smoke compound. The grenade is thrown in the usual manner and a cloud of whitish-grey smoke of high obscuring power is generated in about 10 seconds. It is not recommended that this grenade be used in confined spaces, nor where there is a risk of fire.

Data

Length: 108 mm
Diameter: 73 mm
Weight: 570 g
Delay time: ca 4.5 s
Duration of smoke: 2 min

Manufacturer

Dynamit Nobel Wien GmbH, Tuchlauben 7A, A-1014 Wien.

Status

In production.

Service

Austrian Army.

VERIFIED

DNW Smoke grenade HC-75

DNW illuminating hand grenade LHG-40

Description

The DNW illuminating hand grenade LHG-40 can be used either as a close-area illuminating device or as an incendiary grenade in urban areas. The intensity of illumination is such that most optical and electro-optical devices are dazzled. In order to obtain maximum effectiveness in fog, rain or other poor weather conditions, the emitted light is yellow rather than pure white. The grenade consists of a metal canister containing the flare composition, fitted with the standard type of fly-off lever grenade igniter set.

Data

Length: 145 mm
Diameter: 55 mm
Weight: 500 g
Ignition delay: ca 4.5 s
Duration of flare: ca 35 s
Temperature of flare: 1,800°C
Luminous power: 280,000 candelas over radius of 150 m

Manufacturer

Dynamit Nobel Wien GmbH, Tuchlauben 7A, A-1014 Wien.

Status

In production.

Service

Austrian Army.

UPDATED

DNW illuminating hand grenade LHG-40

DNG incendiary smoke hand grenade

Description

The DNG incendiary smoke hand grenade has an oval plastic casing containing a charge of stabilised red phosphorus composition which gives both incendiary and smoke-producing effects. The grenade is designed as a close-combat weapon for short-range applications against moving or stationary targets. It is particularly effective against fortified positions, houses in street fighting, tanks and other armoured vehicles and against concentrations of soft vehicles.

Instead of the customary spring-actuated percussion ignition system, this grenade uses a friction igniter. Removing the cap reveals a cord attached to the cap and to the central pin of the igniter. Pulling the cord withdraws this pin, which has a roughened surface, through a friction-sensitive substance which is thus ignited. This, in turn, ignites a delay fuze which, after 1.9 seconds, ignites the combustible mass in the body of the grenade. On striking the ground or target, the body of the grenade is shattered, exposing the smouldering composition to the air. This sudden exposure causes the mass to explode with a bright flash, after which it will burn for up to five minutes, giving off dense smoke.

A practice and drill grenade, identical in form, shape and weight but without a fuze and filled with an inert lime dust compound for visual target effect, is also available.

Data
Length: 135 mm
Diameter: 67 mm
Weight: 320 g
Filling: 260 g stabilised red phosphorus
Delay time: ca 1.9 s
Temperature range: −40 to +50°C
Burning temperature: 1,200°C

Manufacturer
Dynamit Nobel Graz GmbH, Annenstrasse 58, A-8020 Graz.

Status
In production.

UPDATED

BELGIUM

FN Bullet-Thru® rifle grenade

Description
The FN Bullet-Thru® (B-T) rifle grenade is a novel concept for which the manufacturers claim four distinct advantages:

The grenade can be launched with any type of ammunition, including armour-piercing and tracer bullets. Instead of the conventional bullet trap, the B-T is made with a clear passage through the axis of the grenade, through which the bullet passes without impact. A specially designed gas trap ensures that the propellant gas is retained and utilised in projecting the grenade.

Having a low mass, it can be comfortably fired from the shoulder in any combat position, thus increasing accuracy and reducing the vulnerable exposure time for the soldier.

Because of the compact method of construction, the rifleman can carry several grenades. Moreover, since the grenades are small, the soldier is more likely to carry them into combat and less likely to discard them as excess weight.

Again, as a result of the compact construction, large numbers of grenades can be stored in smaller areas and transport is utilised to better effect.

The B-T grenade consists of two basic units, the head and the tail. During shipping and transportation the tail is telescoped into the head, giving an overall length of only 189 mm. In this configuration the detonator is out of line with the firing pin and a safety shutter isolates the detonator from the explosive train.

When required for use, the grenade is extended by simply pulling out the tail unit. This gives an overall length of 290 mm and safety is maintained since the firing pin is in the extended tail while the detonator is in the forebody. The extended tail unit also carries the fragmenting sleeve, and this is now withdrawn from proximity with the central explosive charge.

The grenade is placed on the muzzle of the rifle and launched in the usual way, using whatever ammunition is in the rifle chamber. After leaving the muzzle a spring causes the tail unit to retract once again into the body; while doing so it rotates so that the firing pin and detonator are brought into alignment. As the two parts come together, the fragmentation sleeve takes up position around the central explosive charge. During flight the safety devices in the fuze function and the fuze is armed after about 5 m of flight so that the grenade is ready to detonate on impact.

Firing FN Bullet-Thru® rifle grenade; note the carrying case on the soldier's belt

1996

The foregoing description is that of the Anti-Personnel (AP) grenade. Other grenades are included in the FN Bullet-Thru® (B-T) rifle grenade series; these include the AP-X training grenade, which operates in exactly the same manner as the AP grenade but does not contain any explosive filling. It can be reconditioned after firing very easily by means of a repair kit supplied with the grenade and thus reused a number of times, a useful economy measure for training purposes. The Smoke (SMK) grenade is used for the production of screening smoke and has an emission time of 70 seconds. The Illuminating (ILL) grenade carries a conventional type of parachute flare capable of giving 80,000 candela of illumination for 33 seconds. The SMK and ILL grenades are also non-telescoping types.

Data
AP GRENADE
Calibre: 39 mm
Length: retracted, 189 mm; extended, 290 mm
Weight: 320 g
Max indirect fire range: FNC 5.56 mm rifle, ≥300 m; LAR 7.62 mm rifle, ≥400 m

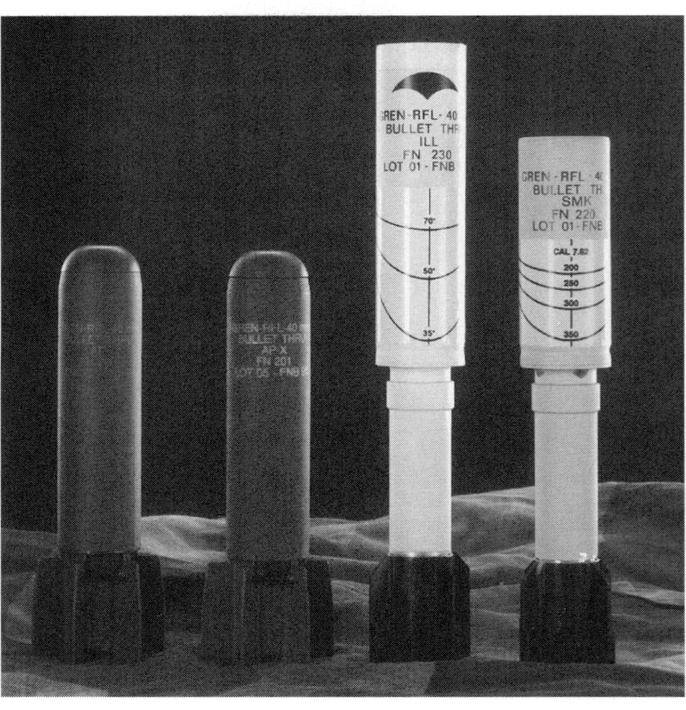

FN Bullet-Thru® rifle grenade range; from left, AP, AP-X, ILL, SMK

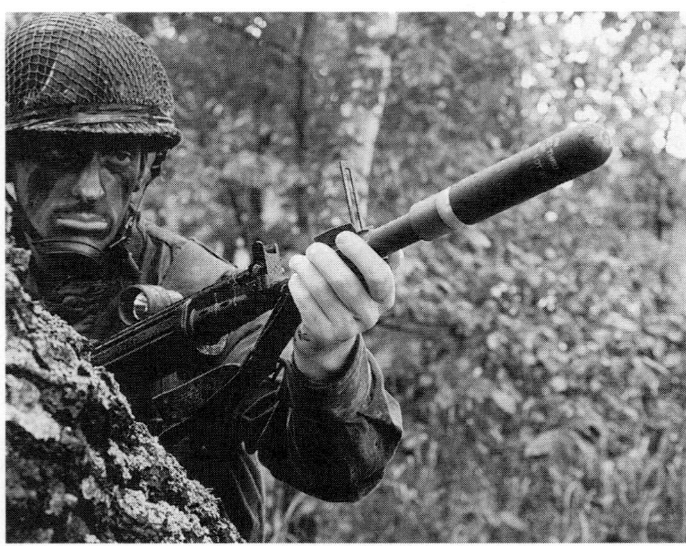

Loading an FN Bullet-Thru® rifle grenade

1996

Recoil energy: FNC 5.56 mm rifle, ≤45 J; LAR 7.62 mm rifle, ≤60 J
Lethal radius: 10 m
Arming distance: 5 m

Manufacturer

FN Herstal SA, Voie de Liège 33, B-4040 Herstal.

Status

In production.

Licence production

INDONESIA
Manufacturer: PT Pindad (Persero)
Type: Granat Senapan Teleskopik
Remarks: Standard specifications as provided in main text.

UPDATED

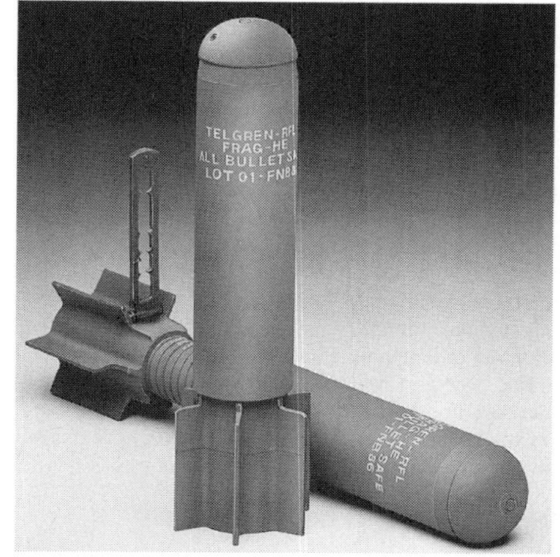

FN Bullet-Thru® rifle grenades
1996

MECAR FRAG-C-M72 controlled fragmentation hand grenade

Description

The MECAR FRAG-C-M72 hand grenade produces a controlled fragmentation because of the number and shape of the splinters and the weight and type of explosive used. The two main characteristics of this grenade are its very high efficiency at close range and the very short safety radius. Personnel 20 m from the point of burst will be unharmed by the explosion of the grenade. The dispersion of the effective splinters is obtained in an homogeneous manner and is uniform in the space around the point of impact, whatever the position of the grenade may be at the moment of detonation. The result is that the same grenade can be used offensively and defensively. This grenade is used with the fuze M72A1 (4 seconds) gasless delay time fuze.

Data

Weight: 230 g
Explosive charge: 60 g Comp B
Number of fragments: ca 900
Lethal radius: 9.5 m
Safety radius: 20 m
Delay: 4 s

Manufacturer

MECAR SA, B-7181 Petit-Roeulx-lez-Nivelles.

Status

Available.

VERIFIED

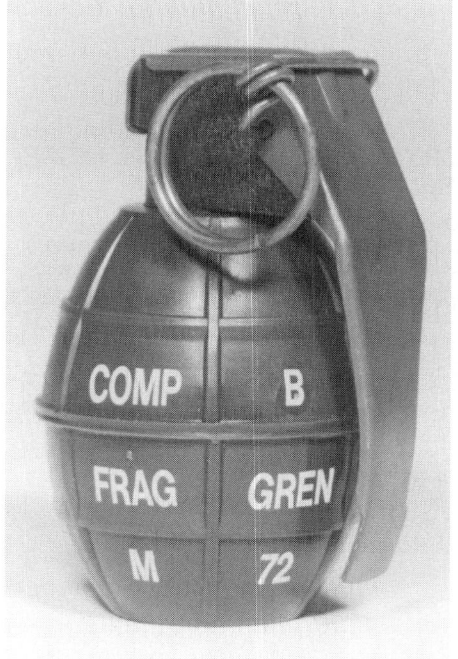

MECAR FRAG-C-M72 controlled fragmentation hand grenade

MECAR M73 practice hand grenade

Description

The practice hand grenade is intended for training soldiers in the use of the live grenade M72. It must be used with the M73A1 fuze. It has the same weight and shape and the same outside aspect, but it looks different from the live grenades because of its colour, marking, materials and elements used.

The effect on functioning is to produce noise, simulating an explosion and a small quantity of smoke. The functioning does not produce any projection of metallic elements, consequently its use does not demand any special safety provisions.

The grenade has two separate components, the body and the time fuze.

The body is of aluminium alloy. This part has a through channel in which is screwed the fuze with its deflagrator and which permits the evacuation of the gases produced when functioning. The body can be reused with no other maintenance than simple cleaning with a rag.

The M73A1 practice time fuze is identical in all respects with the M72A1 gasless delay time fuze used with the live grenade; its use and functioning are the same. The only difference is that the detonator has been replaced by a deflagrator which produces noise and smoke. Like the M72 time fuze the practice time fuze cannot be reused as it is destroyed on functioning.

Manufacturer

MECAR SA, B-7181 Petit-Roeulx-lez-Nivelles.

Status

Available.

VERIFIED

MECAR M73 practice hand grenade

MECAR anti-armour BTU rifle grenade, ARP-RFL-40 BTU M260 series

Description

The ARP-RFL-40 BTU M260 series rifle grenade is used for the defeat of soft and armoured targets by means of a shaped charge HEAT warhead.

The grenade may be launched from any standard military 5.56 or 7.62 mm assault rifle having a 22 mm diameter muzzle, using standard issue ball ammunition. The grenade is launched with the rifle at the shoulder or any other conventional position.

The patented MECAR Bullet Trap Universal (BTU) has been specifically designed to accept soft steel core bullets as well as lead core bullets. Muzzle velocities and recoil energy values vary between rifles.

Data

Length: 330 mm
Weight: 390 g
Fuze: PIBD F55
Sight: weapon-related aiming grid with each grenade
Arming distance: 15 m
Max range: 300-400 m
Operational range: 15-200 m
Probable error at 100 m: H+W 1.25 × 1.25 m
Impact function: 70° of obliquity
Penetration: RHA (BH220), >160 mm; concrete, >300 mm
Operating temperature range: −32 to +52°C

Manufacturer

MECAR SA, B-7181 Petit-Roeulx-lez-Nivelles.

Status

In production.

Service

More than 35 countries including NATO forces.

VERIFIED

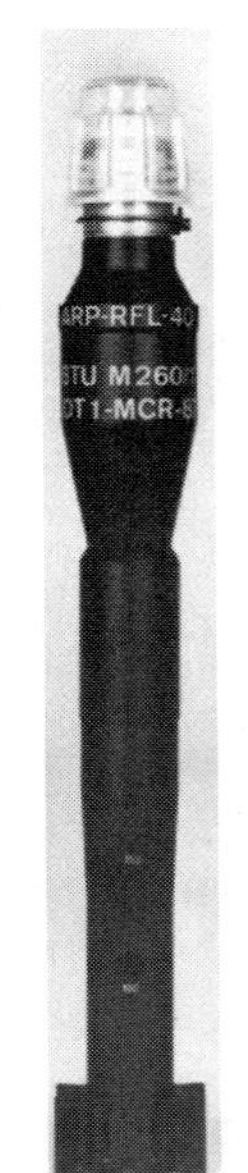

MECAR anti-armour BTU rifle grenade, ARP-RFL-40 BTU M260 series

MECAR BTS rifle grenade, HE-AP-RFL-35, BTS M235 CLAW

Description

The shoulder-launched MECAR BTS HE-AP-RFL-35, BTS M235 CLAW rifle grenade was designed for use with 5.56 mm assault rifles to provide a safe, accurate and effective direct and indirect fire capability. The grenade can be used against troops in the open and light armoured vehicles.

The fuze employs a double safety system in accordance with STANAG 4187 and it can be fired safely using all types of 5.56 mm ammunition.

A practice version, the PRAC-RFL-35 BTS M230 CLAW, has the same ballistic characteristics as the operational grenade. It can be fired once only with 5.56 mm Ball ammunition but is reusable if ballistite cartridges are used.

Data

Length: 320 mm
Head diameter: 35 mm
Weight: 425 g
Fuze: PZ-03
Sight: weapon-related aiming grid with each grenade
Arming distance: 15 m
Max range: direct fire, 150 m; indirect fire 300 m
Penetration: >80 mm RHA
Lethal radius: 10 m

Manufacturer

MECAR SA, B-7181 Petit-Roeulx-lez-Nivelles.

Status

In production.

NEW ENTRY

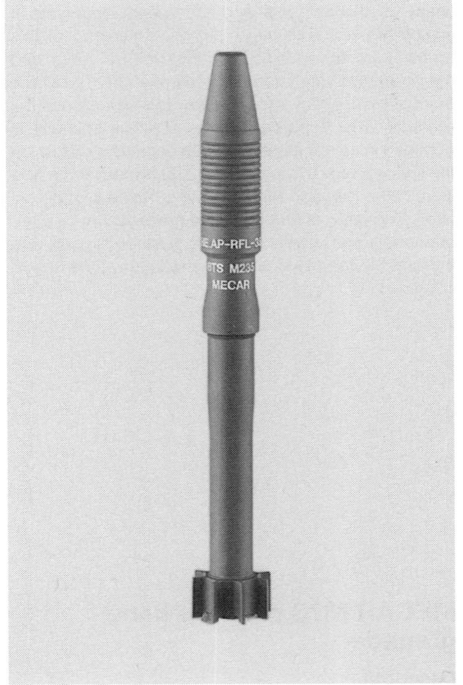

MECAR BTS rifle grenade, HE-AP-RFL-35, BTS M235 CLAW
1996

MECAR blast and fragmentation BTU rifle grenade, HE-RFL-35 BTU M262 series

Description

The shoulder-launched MECAR HE-RFL-35 BTU M262 series blast and fragmentation rifle grenades provide riflemen with the ability to engage and defeat, in both direct- and indirect-fire roles, light structures, material targets and personnel.

The grenade may be launched from any standard military 5.56 or 7.62 mm assault rifle having a 22 mm diameter muzzle, using standard issue ball ammunition.

The patented MECAR Bullet Trap Universal (BTU) has been specifically designed to accept soft steel core bullets as well as lead core bullets. Muzzle velocities and recoil energy values vary between rifles.

Data

Length: 288 mm
Weight: 400 g
Head diameter: 35 mm
Fuze: PD F60/1
Sight: weapon-related aiming grid with each grenade
Arming distance: 15 m
Max range: 300-400 m
Operational range: 15-200 m
Impact function: 70° of obliquity
Lethal radius 10 m
Effective radius: 18 m
Number of fragments: >300
Operating temperature range: −32 to +52°C

Manufacturer

MECAR SA, B-7181 Petit-Roeulx-lez-Nivelles.

Status

In production.

Service

More than 35 countries including NATO forces.

VERIFIED

MECAR blast and fragmentation rifle grenade, HE-RFL-35 BTU M262 series

MECAR parachute flare rifle grenade, PFL-RFL-40 BTU M259 series

Description

The MECAR PFL-RFL-40 BTU M259 series rifle grenade is used when a specific area of operations is required to be illuminated. The flare is ejected from the grenade four seconds after launch and gives an intense yellow illumination for some 30 seconds during its parachute descent. When fired at an angle of 80° the grenade reaches a height of approx 150 m and at 45° approx 100 m to light up an area of 200 m diameter.

The grenade may be launched from any standard military 5.56 or 7.62 mm assault rifle having a 22 mm diameter muzzle, using standard issue ball ammunition. The grenade is launched with the rifle at the shoulder or any other conventional position.

The patented MECAR Bullet Trap Universal (BTU) has been specifically designed to accept soft steel core bullets as well as lead core bullets. Muzzle velocities and recoil energy values vary between rifles.

Data

Length: 359 mm
Weight: 420 g
Fuze: pyrotechnic with 4 s delay
Operational range: 175 m
Function: after 4 s delay
Intensity of illumination: 100,000 candela
Duration of illumination: 30 s
Operating temperature range: −32 to +52°C

Manufacturer

MECAR SA, B-7181 Petit-Roeulx-lez-Nivelles.

Status

In production.

Service

More than 35 countries including NATO forces.

VERIFIED

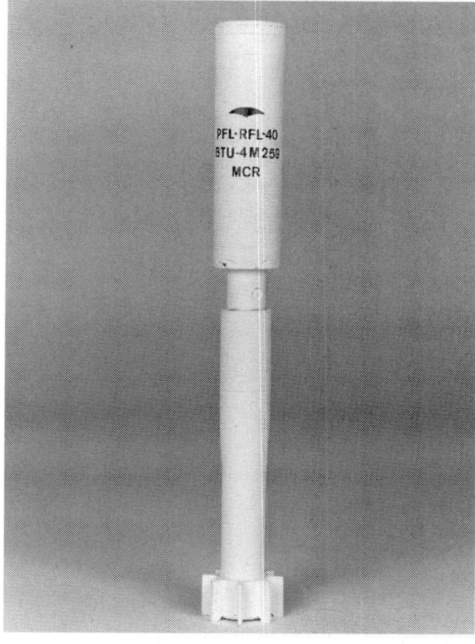

MECAR parachute flare rifle grenade, PFL-RFL-40 BTU M259 series

MECAR Smoke rifle grenade, SMK-RFL-40 BTU-2 M258 series

Description

The MECAR SMK-RFL-40 BTU-2 M258 rifle grenades are used when smoke effects are required. Seven different colours of smoke screen are available with smoke screen duration of 80 seconds.

The grenade may be launched from any standard 5.56 or 7.62 mm rifle having a 22 mm diameter muzzle. The patented MECAR Bullet Trap Universal (BTU) has been specifically designed to accept soft core bullets. Muzzle velocities and recoil energy values vary between rifles.

Data

Length: 289 mm
Weight: 423 g
Filling: white phosphorus
Fuze: pyrotechnic with 4 s delay
Sight: weapon-related aiming grid with each grenade
Max range: to 300 m
Operational range: 200 m
Function: 4 s after launch
Effective radius: >3 m
Operating temperature range: −32 to +52°C

Manufacturer

MECAR SA, B-7181 Petit-Roeulx-lez-Nivelles.

Status

In production.

Service

More than 35 countries including NATO forces.

VERIFIED

MECAR smoke and incendiary rifle grenade, SMK(WP)-RFL-40 BTU-2 M258 (Delay) series

BRAZIL

M3 defensive/offensive hand grenade

Description

The M3 grenade is basically a cylinder filled with HE and detonated by a pyrotechnic delay fuze which is itself ignited by a percussion cap and striker lever. A fragmenting sleeve can be slid over the body of the grenade.

As a plain explosive charge the grenade can be used in the offensive role relying on its blast effect to demoralise the enemy. It can also be a small demolition charge for immediate action in the field. With the fragmentation sleeve in place the M3 becomes an effective defensive grenade which produces over 240 fragments.

Data

Length: 96 mm
Diameter: 40 mm; with fragmentation sleeve, 47 mm
Weight: 415 g; fragmentation sleeve, 200 g
Filling: 90 g Comp B
Delay: 4.5 s
Number of fragments: 240+
Fragment velocity: 2,000 m/s
Lethal radius: 8 m

Manufacturer

Companhia de Explosivos Valparaiba, Praia do Flamengo 200, 20° Andar, 22210 Rio de Janeiro, RJ.

Status

In production.

VERIFIED

M3 grenade with fragmenting sleeve in place

M4 defensive/offensive hand grenade

Description

The M4 is a conventional time-fuzed hand grenade using a plastic body shell lined with notched wire which provides controlled fragmentation. Being lighter in weight than older types of cast iron grenade, it can be thrown further and with greater accuracy.

Data

Length: 87 mm
Diameter: 50.5 mm
Weight: fuzed, 241 g; fragmentation wire, 75 g
Filling: 75 g Comp B
Delay time: 4.5 s
Number of fragments: 800+
Lethal radius: 9.5 m

Manufacturer

Companhia de Explosivos Valparaiba, Praia do Flamengo 200, 20° Andar, 22210 Rio de Janeiro, RJ.

Status

In production.

Service

Brazilian Army.

VERIFIED

M4 hand grenade

M2-CEV rifle grenade

Description

The M2-CEV grenade is an anti-personnel HE type, fired from the muzzle of a 7.62 mm rifle, using a ballistite launching cartridge. It produces fragments on exploding, but also has a useful explosive effect which will penetrate concrete and light armour making it a general purpose munition. Each grenade is packed in a waterproof container with a launching cartridge and an attachable sight for the rifle.

When fitted with Fuze SIG the nomenclature of this grenade changes to M4-CEV.

Data

Length: 323 mm
Diameter: max, 40 mm
Weight: total, 550 g; explosive, 85 g Comp B or Pentolite
Max range: (42°) 380 m
Muzzle velocity: 70 m/s
Fragments: 450+
Penetration: armour, 50 mm

Manufacturer

Companhia de Explosivos Valparaiba, Praia do Flamengo 200, 20° Andar, 22210 Rio de Janeiro, RJ.

Status

In production.

VERIFIED

M2-CEV rifle grenade

M3-CEV anti-tank rifle grenade

Description

The M3-CEV is a larger and heavier version of the M2-CEV, containing a larger HE charge specifically for use against armoured vehicles. It is a muzzle-launched grenade in the same way as the M2-CEV, but because of its greater weight the velocity and range are less. This, however, is not important since it can be expected that engagement ranges will be short. As with the M2-CEV, the grenade is intended for use with the FAL rifle in use with the Brazilian forces.

When fitted with Fuze SIG the nomenclature of this grenade changes to M5-CEV.

Data

Length: 410 mm
Diameter: 65 mm
Weight: total, 770 g
Filling: 265 g Comp B or Pentolite
Max range: 42°, 260 m
Muzzle velocity: 60 m/s
Penetration: armour, 76 mm

Manufacturer

Companhia de Explosivos Valparaiba, Praia do Flamengo 200, 20° Andar, 22210 Rio de Janeiro, RJ.

Status

In production.

VERIFIED

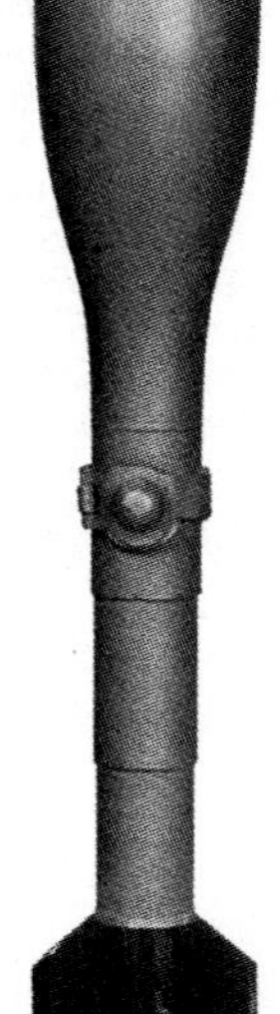

M3-CEV anti-armour grenade

MB-306/T1 vehicular smoke grenade

Description

The MB-306/T1 vehicular smoke grenade is designed to be electrically fired from vehicle smoke dischargers. Loaded with a zinc/hexachloroethane mixture, it emits a dense smoke cloud in order to screen the vehicle. Models are available with ranges of 60 or 100 m.

Data

Length: 181 mm
Diameter: 80 mm
Weight: 1.4 kg
Delay time: nominal, 1 s
Emission time: 2-4 min

Manufacturer

Condor SA Industria Quimica, Rua Armando Dias Pereira 160, Adrianopolis, CEP 26260, Nova Iguacu, Rio de Janiero, RJ.

Status

In production.

VERIFIED

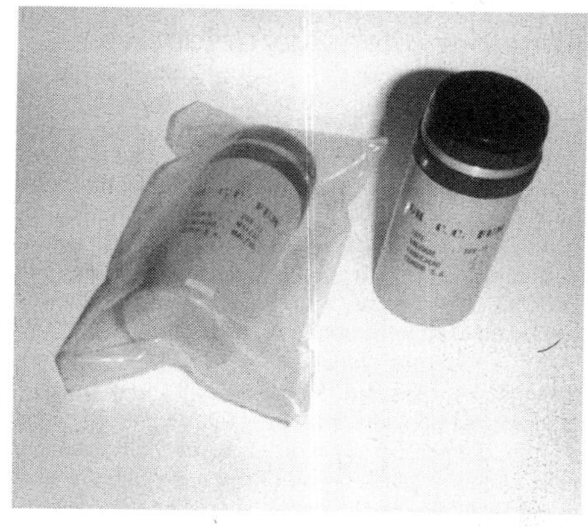

Condor MB-306/T1 vehicular smoke grenade

MB-502 smoke hand grenade

Description

The Condor MB-502 smoke hand grenade has an aluminium cylindrical body with a cap perforated to allow the emission of smoke. It is fitted with a fly-off lever type of igniter and produces a dense grey smoke.

Data

Length: 149 mm
Diameter: 68.5 mm
Weight: 540 g
Delay time: 1.5 - 5.5 s
Emission time: 65 - 115 s

Manufacturer

Condor SA Industria Quimica, Rua Armando Dias Pereira 160, Adrianopolis, CEP 26260, Nova Iguacu, Rio de Janiero, RJ.

Status

In production.

VERIFIED

Condor MB-502 smoke hand grenade

CHILE

FAMAE GM 78-F7 offensive/defensive hand grenade

Description

The FAMAE GM 78-F7 offensive/defensive hand grenade consists of a high impact plastic body containing a TNT-type explosive. The grenade is supplied with a notched steel wire fragmentation coil which may be placed around the body for defensive use or left off for offensive employment. The grenade is armed by a fly-off lever mechanism with the usual safety pin, followed by a pyrotechnic train consisting of an initiator, a chemical delay and a detonator to actuate the percussion fuze after a 3 to 5 second delay. On bursting the wire coil sleeve will scatter about 150 fragments over a 10 m lethal radius.

One feature regarding this grenade is that the fuze arrangement can be replaced by a small pressure plate and fuze to convert the grenade, normally minus the fragmentation coil, into an anti-personnel mine known as the FAMAE MAPT 78-F2.

Data

Length: overall, 100 mm
Diameter: over body, 50 mm; over fragmentation sleeve, 55 mm
Weight: with fragmentation sleeve, 343 g; without fragmentation sleeve, 178 g
Fuze delay: 3-5 s
Lethal radius: 10 m

Manufacturer

Fabricas y Maestranzas del Ejercito (FAMAE), Avenido Pedro Montt 1568/1606, Santiago.

Status

In production.

Service

Chilean Army and other forces.

NEW ENTRY

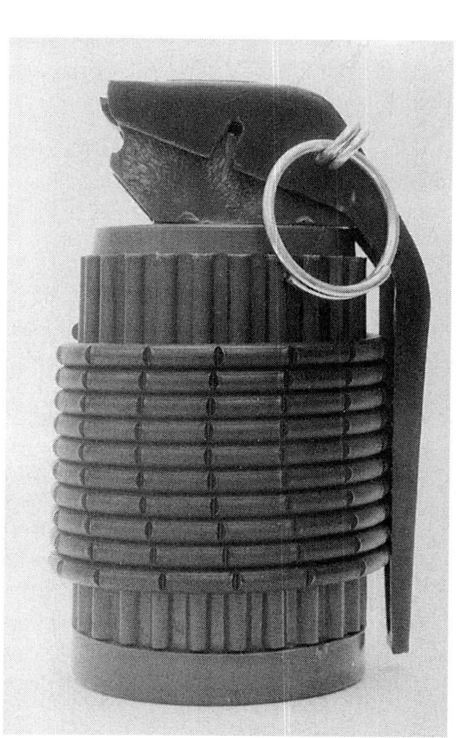

FAMAE GM 78-F7 offensive/defensive hand grenade (T J Gander)
1996

Metalnor offensive/defensive hand grenade

Description

The Metalnor offensive/defensive hand grenade is a cylindrical sheet steel casing containing HE, detonated by a short pyrotechnic delay which is ignited by the usual striker and fly-off safety lever. A steel fragmenting case can be slid over the body for the defensive role, while for offensive use the grenade is thrown without it, relying on blast and shock for its effect.

A feature of this grenade is that the lower surface of the main cylinder is a hollow cone. If the grenade is stood upright on a flat surface it acts as a small hollow charge demolition munition. The manufacturer supplies a special igniter assembly for use when the grenade is used as a booby trap or an anti-personnel mine. This igniter assembly operates on a push-pull principle and is intended to be connected to a trip wire.

Data

Length: 121.5 mm
Diameter: 46.5 mm
Weight: offensive, 210 g; defensive, 500 g
Filling: 100 g TNT
Fuze delay: 5 s

Manufacturer

Metalnor Industria, Metalurgica del Norte Ltda, Division Defensa, Av Providencia 2237, 6° Piso, Santiago.

Status

In production.

Service

Chilean and other armed forces.

UPDATED

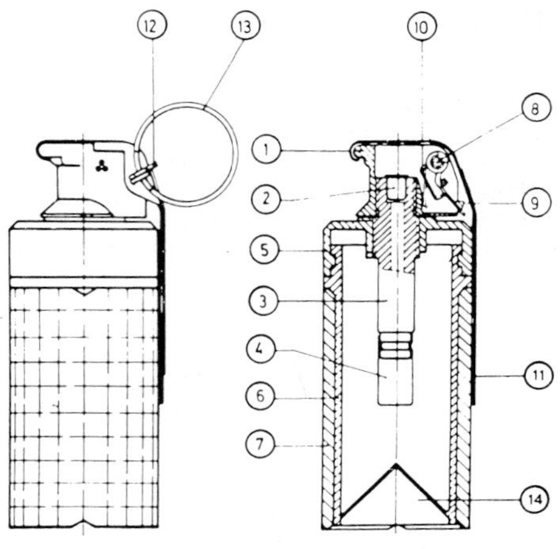

Metalnor offensive/defensive grenades
(1) *fuze head* **(2)** *cap* **(3)** *delay column* **(4)** *detonator* **(5)** *plastic head* **(6)** *interior body* **(7)** *removable fragmentation case* **(8)** *spring axis* **(9)** *striker* **(10)** *spring* **(11)** *safety lever* **(12)** *safety pin* **(13)** *pull ring* **(14)** *shaped charge cone*

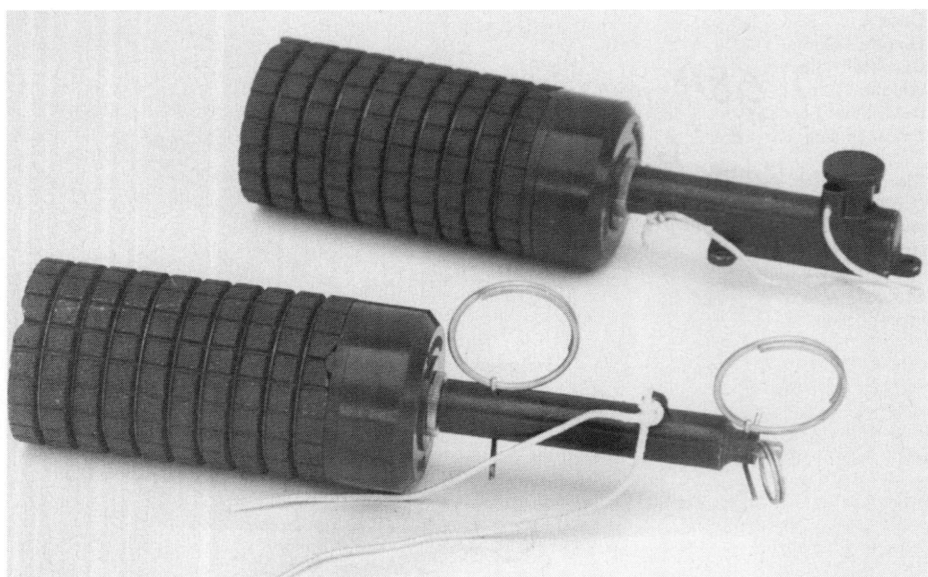

Two Metalnor offensive/defensive grenades set up with pressure and pull igniters for use as booby traps

Metalnor offensive/defensive grenade. Sectioned grenade on right shows the hollow cone in the lower surface

---

Metalnor Type MK-2 hand grenade

Description

The Metalnor Type MK-2 hand grenade bears a close resemblance to the obsolete US Mk 2, having a pear-shaped cast iron body with prominent serrations. The igniter mechanism is the same as that of the offensive/defensive grenade described in the previous entry. This grenade can also be fitted with the Metalnor special igniter for use as a mine or booby trap.

Data

Weight: 662 g
Filling: 115 g HE
Fuze delay: 5 s

Manufacturer

Metalnor Industria, Metalurgica del Norte Ltda, Division Defensa, Av Providencia 2237, 6° Piso, Santiago.

Status

In production.

Service

Chilean and other armed forces.

UPDATED

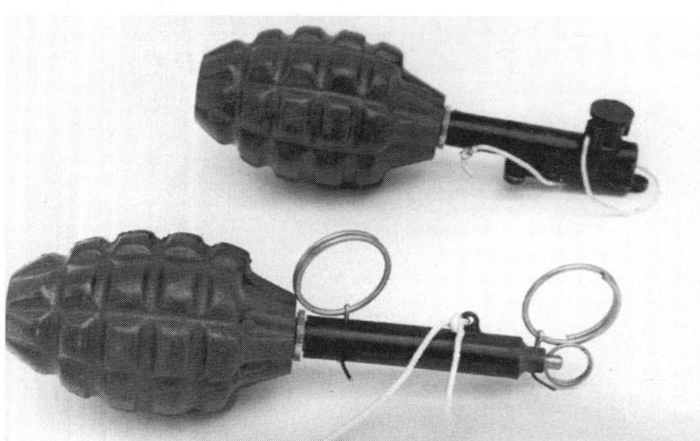

Metalnor Type MK-2 grenades with pressure and pull igniters for use as booby traps

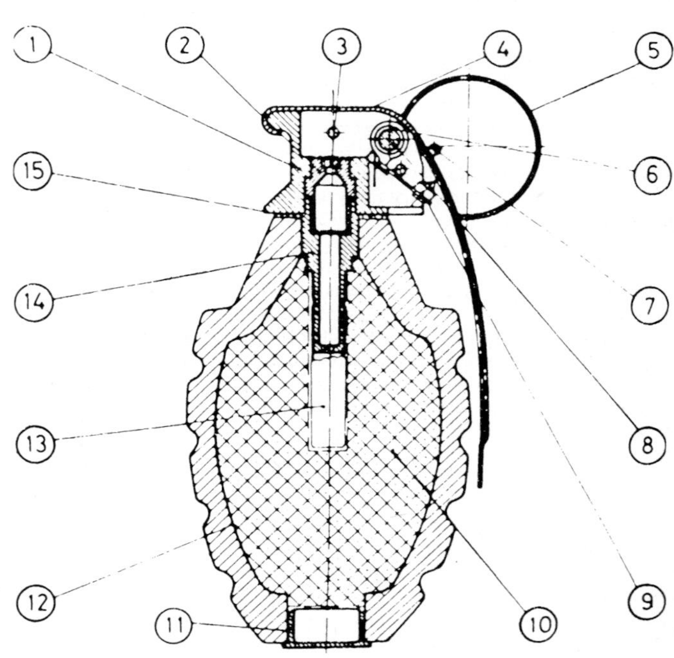

Metalnor Type MK-2 grenade
(1) *fuze head* **(2)** *fulcrum* **(3)** *cap* **(4)** *safety lever* **(5)** *pull ring* **(6)** *striker axis* **(7)** *safety pin* **(8)** *spring* **(9)** *striker* **(10)** *explosive filling* **(11)** *base plug* **(12)** *casing* **(13)** *detonator* **(14)** *delay column* **(15)** *sealing ring*

Metalnor mini hand grenade

Description

This is a small grenade developed by metalnor with a view to obtaining the maximum throw range consistent with good fragmentation effect. The cast iron body is loaded with TNT and provides all-round lethal coverage. The safety ring on the time fuze must be twisted and then pulled to remove; it cannot be withdrawn accidentally. The fuze unit is silent and smokeless in operation; it can be removed and pull or pressure switches inserted so as to convert the grenade into a booby trap or anti-personnel mine.

Data

Weight: 332 g
Height: 79.3 mm
Diameter: 52.8 mm
Filling: 77 g TNT
Delay: 4 s

Manufacturer

Metalnor Industria, Metalurgica del Norte Ltda, Division Defensa, Av Providencia 2237, 6° Piso, Santiago.

Status

In production.

Service

Chilean and other armies.

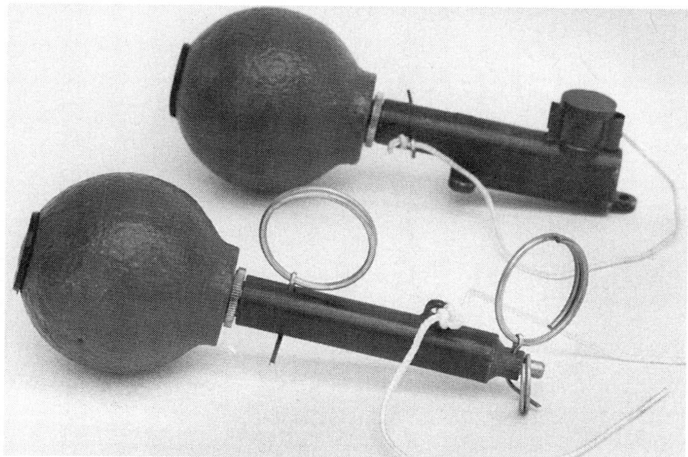

Metalnor mini hand grenades with pressure and pull igniters for use as booby traps

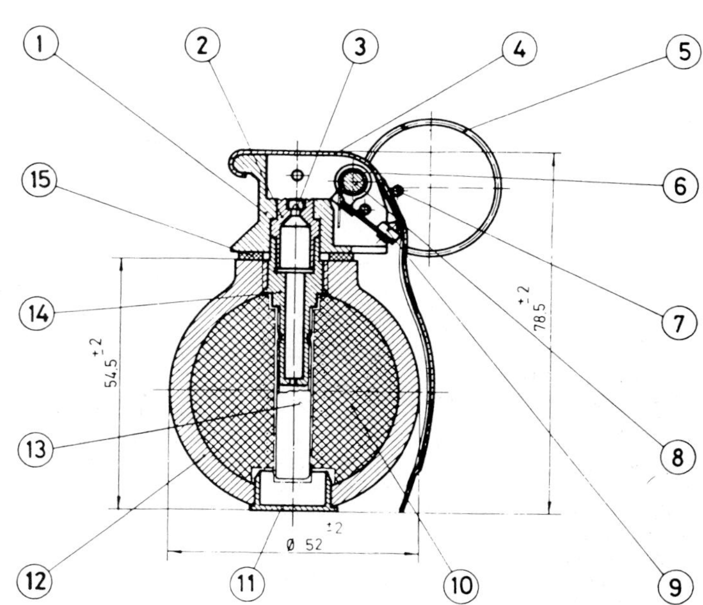

Metalnor mini hand grenade
(1) *fuze head* **(2)** *interior fuze head* **(3)** *primer cap* **(4)** *lever* **(5)** *safety pin pull ring* **(6)** *striker spring* **(7)** *safety device* **(8)** *spring bolt* **(9)** *striker* **(10)** *high explosive* **(11)** *base plug* **(12)** *body* **(13)** *detonator* **(14)** *delay column* **(15)** *rubber*

CHINA, PEOPLE'S REPUBLIC

Chinese grenades

Description

The armed forces of the People's Republic of China use a variety of grenades, some of which are local designs and others, principally older models, are copies of Soviet designs. The Chinese copies are similar to the Soviet models, except that in most cases they have different dimensions and, as the same design is often made by a number of factories, there are minor differences between copies of the same type of grenade. Listed below are some copies or near-copies of Chinese manufacture whose designations are known.

Type 42 offensive/defensive hand grenade
This is a direct copy of the Soviet RG-42 grenade and its method of operation is identical.

Type 1 defensive hand grenade
This is a copy of the Soviet F-1 grenade.

Type 59 defensive hand grenade
This is similar in design to the Soviet RGD-5 grenade and operates in the same manner.

Type 73 prefragmented mini-grenade
This is a smaller version of the Type 59, using an internally prenotched body to which is attached a prominent percussion ignition set similar in operation to that used with the Type 59 and Soviet RGD-5. It weighs 190 g, is 88 mm high and has a diameter of 42 mm. It contains 580 3 mm steel balls.

Manufacturer

China North Industries Corporation (NORINCO), 7A Yue Tan Nan Jie, PO Box 2137, Beijing.

Status

Available.

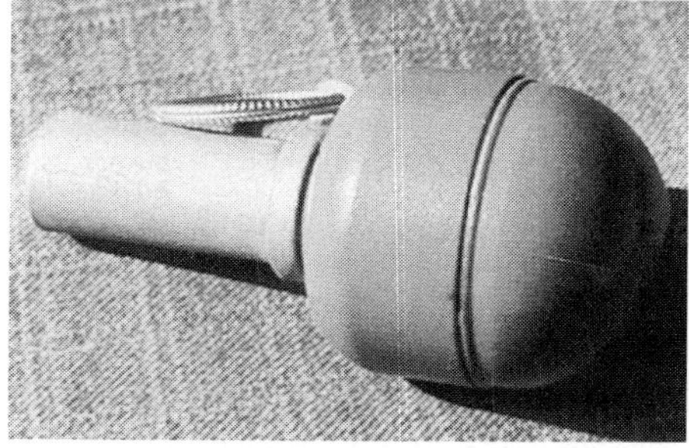

Type 73 prefragmented mini-grenade
1996

Service

Chinese armed forces.

UPDATED

Chinese stick grenades

Description

The Chinese have manufactured a wide variety of stick grenades for defensive operations. Scored, serrated and plain types have been encountered. Their contents have included picric acid, mixtures of TNT or nitroglycerin with potassium nitrate or sawdust and schneiderite. The grenades are operated by pulling the cord of the pull-friction fuze, which is underneath the cap at the end of the throwing handle. This ignites the delay element which lasts between 2.5 and 5 seconds, after which the detonator explodes the main charge.

A typical example of a defensive stick grenade is illustrated here and is known to be still in use. It is a fragmenting type with a serrated head made of grey cast iron. This produces a small number of large fragments and a very large number of fragments so small that they could well be described as 'dust'. The filling is picric acid which was discarded as an explosive filling in the West many years ago, principally because it forms

Chinese smooth fragmentation grenade with cap removed to show pull cord

dangerous and unstable compounds. In the UK it was the explosive Lyddite, used during the First World War. It is essential that the inside of the container is varnished.

Data

Weight: 500 g
Length: 228 mm
Diameter: 50 mm
Weight: 99 g filling
Filling: picric acid
Fuze delay: 2.5-5 s
Effective fragmentation radius: 10 m

Manufacturer

China North Industries Corporation (NORINCO), 7A Yue Tan Nan Jie, PO Box 2137, Beijing.

Status

In production.

Service

Chinese armed forces.

VERIFIED

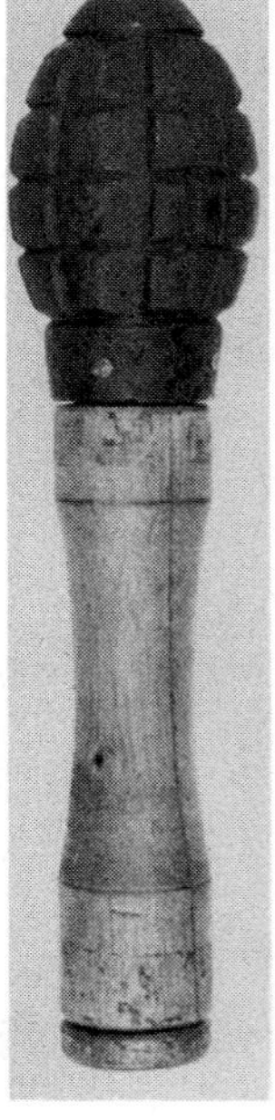

Chinese, serrated, fragmenting, defensive stick grenade

NORINCO Type 77-1 stick grenade

Description

The Type 77-1 stick grenade is commercially available through NORINCO, as well as being on issue to the People's Liberation Army. Like the stick grenades described previously, the 77-1 uses a friction pull igniter, revealed by removing a screwed cap from the plastic handle. The head is of cast metal, smooth and ovoid.

Data

Length: 170-173 mm
Weight: 360-380 g
Diameter: 48.5 mm
Filling: 70 g TNT
Delay time: 2.8-4 s
Lethal radius: 7 m

Manufacturer

China North Industries Corporation (NORINCO), 7A Yue Tan Nan Jie, PO Box 2137, Beijing.

Status

In production.

Service

Chinese armed forces.

VERIFIED

NORINCO Type 82 hand grenades

Description

NORINCO Type 82 grenades are basically of the offensive pattern though the makers claim they can be used in either role. The body is internally serrated to give fragmentation control, and ignition is by the usual percussion 'mousetrap' igniter system.

The Type 82-1 has a smooth, slightly irregular, oval shape and is fitted with a friction pull igniter beneath a removable cap.

The Type 82-2 (illustrated) is built up from two halves and has a prominent joint around the centre. There is a close-fitting protective cap which covers the working parts and retains the safety pin and ring whilst being shipped or carried. After removing this cap the pin is pulled and the grenade thrown in the usual way. Light and compact, it can be thrown for a considerable distance and several can be carried.

The Type 82-3 uses the same body as the Type 82-1 but has a percussion igniter of somewhat different type from that used on the Type 82-2.

Data

TYPE 82-2:
Length: 85 mm
Diameter: 48 mm
Weight: 260 g
Filling: 62 g TNT
Number of fragments: ca 280
Lethal radius: 6 m
Safety radius: 30 m
Delay time: 2.8-3.8 s
Operational temperature range: −40 to +45°C

Manufacturer

China North Industries Corporation (NORINCO), 7A Yue Tan Nan Jie, PO Box 2137, Beijing.

Status

Production for export.

UPDATED

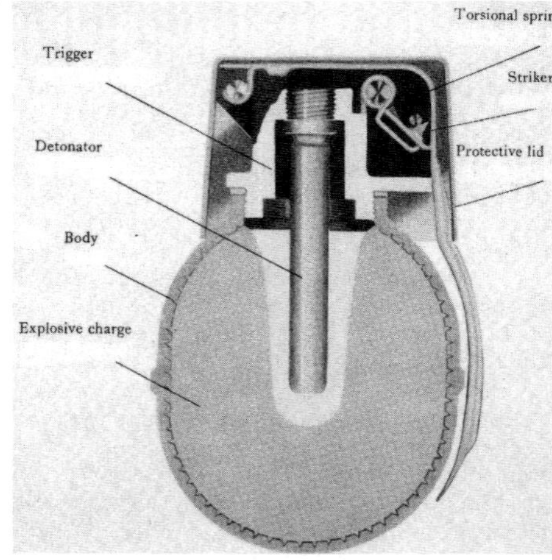

NORINCO Type 82-2 hand grenade

COMMONWEALTH OF INDEPENDENT STATES

RGD-5 offensive hand grenade

Description

The RGD-5 is an ovoid offensive fragmentation grenade with a smooth exterior surface to the two-piece steel body which is prefragmented internally. It is a compact, easily handled grenade which can be thrown slightly further than the earlier Soviet defensive hand grenades. It uses the UZRGM fuze assembly which provides a time delay of 3.2 to 4.2 seconds. (Some UZRGN fuze assemblies may be encountered with delays ranging from 0 to 13 seconds for use in booby traps.) The RGD-5 can also employ the more modern DVM-78 fuze assembly with a time delay of from 3.2 to 4 seconds.

Data

Length: 114 mm
Diameter: 56.8 mm
Weight: 310 g
Filling: 110 g TNT
Fuze delay: 3.2-4.2 s
Range thrown: 40 m
Effective fragment radius: 20-25 m

Status

In production.

Service

Most former Warsaw Pact member nations and their satellite nations.

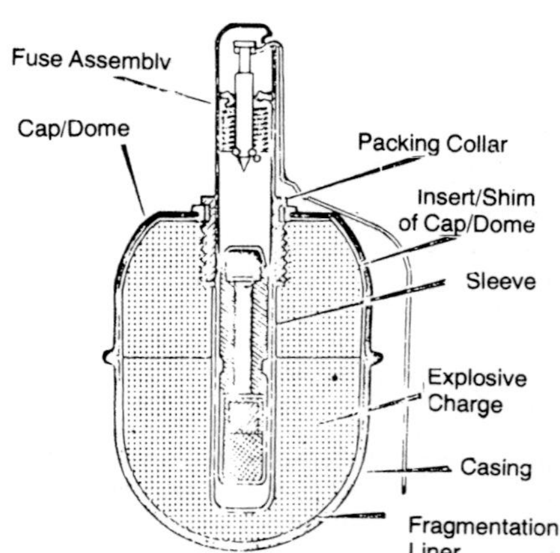

RGD-5 anti-personnel hand grenade

Licence production
BULGARIA
Manufacturer: Kintex
Type: RGD-5
Remarks: Standard specifications
CHINA, PEOPLE'S REPUBLIC
Manufacturer: China North Industries Corporation (NORINCO)
Type: Type 59
Remarks: Essentially similar to RGD-5
POLAND
Manufacturer: Dezamet
Type: RGO
Remarks: Filling is 60 g of A-IX-1

UPDATED

RGD-5 anti-personnel hand grenade
1996

RGO-78 defensive hand grenade

Description

The RGO-78 defensive hand grenade follows the same general body outlines as the RGD-5 in that it is an ovoid offensive fragmentation grenade with a smooth exterior surface to the two-piece steel body. However, it is slightly larger overall with the fragmentation effect created by an internal layer of steel balls set in a resin matrix around the 85 g TNT filling which, on detonation, will spread the balls over a lethal radius of 20 m, creating a lethal area of 266 m².

The RGO-78 may be encountered fitted with the UZRGM fuze but is more likely to be seen with the DVM-78, a more modern design making extensive use of plastics. The latter fuze provides a time delay of 3.2 to 4 seconds.

Data

Length: 138 mm
Diameter: 60 mm
Weight: 450 g
Filling: 85 g TNT
Fuze delay: 3.2-4 s
Range thrown: 30-45 m
Effective fragment radius: 20 m
Operational temperature range: −50 to +50°C

Manufacturer

Various state factories.

Status

In production.

Service

Most former Warsaw Pact member nations and their satellite nations.

Licence production

BULGARIA
Manufacturer: Kintex
Type: RGO-78
Remarks: Standard specifications

NEW ENTRY

RGO-78 defensive hand grenade with DVM-78 fuze assembly
1996

RGN-86 hand grenade

Description

The RGN-86 hand grenade may be regarded as an updated version of the RGD-5. The two grenades are similar in size and appearance but the RGN-86 is fitted with the DVM-78 fuze assembly providing a time delay of 3.2 to 4 seconds. The filling is 57 g of TNT allied with some form of internal fragmentation assembly which is probably of the prefragmented pattern to produce a large number of relatively small fragments with limited mass and lethal range. The lethal area produced by this grenade is about 66 m², making it suitable for close order combat situations.

Data

Length: approx 110 mm
Diameter: approx 57 mm
Weight: 265 g
Filling: 57 g TNT
Fuze delay: 3.2-4 s
Range thrown: 30-35 m
Lethal area: 66 m²
Operational temperature range: −50 to +50°C

Manufacturer

Various state factories.

Status

In production.

Service

Most former Warsaw Pact member nations and their satellite nations.

Licence production

BULGARIA
Manufacturer: Kintex
Type: RGN-86
Remarks: Standard specifications

NEW ENTRY

RGN-86 defensive hand grenade with DVM-78 fuze assembly
1996

F-1 anti-personnel hand grenade

Description

The F-1 anti-personnel hand grenade was introduced during the Great Patriotic War (1941-1945) and became a potent visual symbol of resistance to invasion, a role later carried over to various freedom fighter organisations with whom the F-1 became a favoured weapon. It is no longer regarded as a front-line weapon with any of the former Warsaw Pact nations, other than Poland, but is still likely to be widely encountered. Many underground organisations have produced their own local designs based on the F-1.

The F-1 is a fragmentation grenade with a cast iron body notched into cubes on the external surface in a manner reminiscent of the US Mark 2 or the British No 36M grenade. It suffers from the same defects as the latter and produces a number of fragments from the base plug and filler, which can be lethal out to 200 m, so the thrower would be well advised to throw the grenade from under cover. The immediate lethal area produced by the detonation of 60 g of TNT (other weights and types of explosive have been employed) is 20 m.

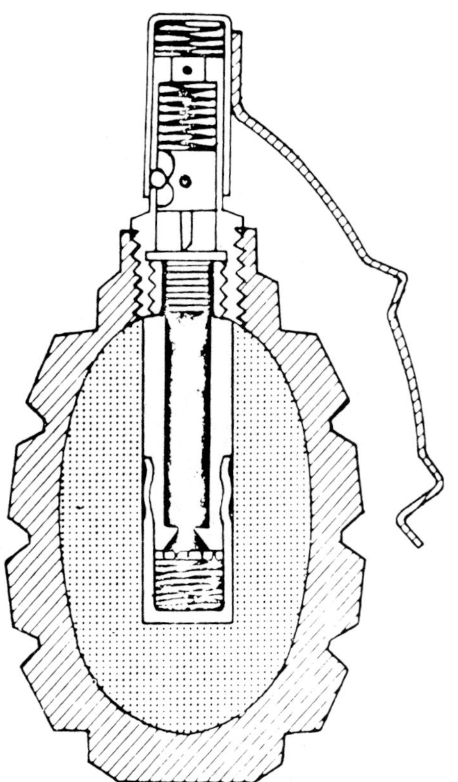

F-1 defensive hand grenade

F-1 fragmentation grenade. This grenade is usually painted olive green

The usual fuze assembly used with the F-1 is the UZRGM but the later DVM-78 is also employed. The UZRGM Bouchon fuze provides a time delay of 3.2 to 4.2 seconds while the DVM-78 mousetrap provides 3.2 to 4 seconds. (Some UZRGN fuze assemblies may be encountered with delays ranging from 0 to 13 seconds for use in booby traps.)

The Polish Army introduced this grenade in a modified form with an impact fuze as the F1/N60 rifle grenade.

Data

Length: 130 mm
Diameter: 55 mm
Weight: 600 g
Filling: 60 g TNT
Fuze delay: 3.2-4.2 s
Range thrown: 30 m
Effective fragment radius: 20 m
Operational temperature range: −50 to +50°C

Manufacturer

Various state factories.

Status

Still in production in some nations.

Service

Still widespread.

Licence production

BULGARIA
Manufacturer: Kintex
Type: F-1
Remarks: Standard specifications
CHINA, PEOPLE'S REPUBLIC
Manufacturer: China North Industries Corporation (NORINCO)
Type: Type 1
Remarks: Standard specifications. May no longer be in production
POLAND
Manufacturer: Zaklady Sprzetu Precyzyjnego
Type: F-1
Remarks: Filling is 50-56 g of TNT

UPDATED

RGO fragmentation hand grenade

Description

The RGO (Ruchnaya Granata Oboronitel'naya) is a defensive hand grenade which consists of a spherical prefragmented steel body formed of four hemispheres; the two inner hemispheres are internally segmented for fragment control. The outer hemispheres are also serrated to give regular fragments; the lower half has external serrations while the upper section is internally serrated and displays a smooth exterior to the top of the grenade body. There is a central detonator well which is threaded to accept the fuze assembly.

The fuze, the UDZS, has a white polythene body and a mousetrap striker which ignites a single pellet. This pellet, in turn, ignites three pyrotechnic delay trains. Two are employed as safety delays for an impact function while the third acts as a conventional delay detonator. As two of the trains burn away detent pins are withdrawn by spring pressure to release a shutter which rotates to free an impact plunger. This occurs 1 to 1.8 seconds after throwing so if the grenade impacts on a target after that time it will detonate under the influence of a spherical impact weight filled with lead shot in a resin matrix. Should the grenade not impact before 3.2 to 4.2 seconds the third pyrotechnic train will ignite the relay detonator and cause the grenade to detonate. Thus it is possible to throw this grenade in such a manner that it will produce an air burst over a target area, should one be required. It is possible to throw this grenade to a distance of 30 to 40 m.

The RGO grenade contains 90 g of A-IX-1 (RDX 96%, wax, 4%) explosive. This, combined with the body fragments, can produce a lethal radius of 20 m, with a danger radius extending to 100 m from the point of detonation.

Data

Length: max, 114.5 mm
Diameter: max, 61 mm
Weight: 520-530 g
Filling: 90 g A-IX-1
Lethal radius: 20 m
Delay time: impact, 1-1.8 s; time, 3.2-4.2 s
Operational temperature range: −50 to +50°C

Manufacturer

GP Zavod imeni J A M Sverdlova, Dzerzhinsk, Nishi Novgorod 606002.

Status

In production.

Service

Former CIS armed forces and some other countries, including Iraq.

UPDATED

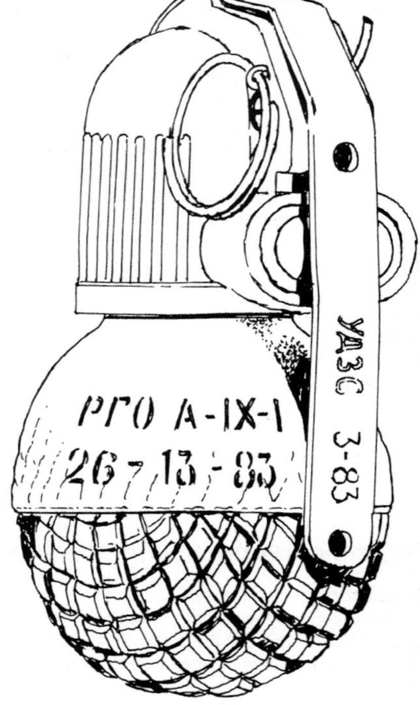

RGO fragmentation hand grenade (L Haywood)
1996

RGN offensive/defensive hand grenade

Description

The RGN (Ruchnaya Granata Nastupatel'naya) can be used as an offensive or defensive hand grenade. It has a spherical prefragmented aluminium alloy casing containing the high explosive bursting charge. The grenade body consists of upper and lower hemispheres, both of which are internally serrated to give the desired fragmentation control. The external body surfaces are smooth.

The RGN uses the same UDZS fuze as the RGO (see previous entry). The UDZS has a white polythene body and a mousetrap striker which ignites a single pellet. This pellet, in turn, ignites three pyrotechnic delay trains. Two are employed as safety delays for an impact function while the third acts as a conventional delay detonator. As two of the trains burn away detent pins are withdrawn by spring pressure to release a shutter which rotates to free an impact plunger. This occurs 1 to 1.8 seconds after throwing so if the grenade impacts on a target after that time it will detonate under the influence of a spherical impact weight filled with lead shot in a resin matrix. Should the grenade not impact before 3.2 to 4.2 seconds the third pyrotechnic train will ignite the relay detonator and cause the grenade to detonate. Thus it is possible to throw this grenade in such a manner that it will produce an airburst over a target area, should one be required. It is possible to throw this grenade to a distance of 30 to 40 m.

The RGO grenade contains 97 g of A-IX-1 (RDX 96%, wax, 4%) explosive. This can produce a lethal radius of 8 to 10 m. The safety radius is 25 m from the point of detonation.

Data

Length: max, 113.5 mm
Diameter: max, 61 mm
Weight: 290 g
Filling: 97 g A-IX-1
Lethal radius: 8-10 m
Safety radius: 25 m
Delay time: impact, 1-1.8 s; time, 3.2-4.2 s
Operational temperature range: −50 to +50°C

Manufacturer

GP Zavod imeni J A M Sverdlova, Dzerzhinsk, Nishi Novgorod 606002.

Status

In production.

Service

Former CIS armed forces and some other countries, including Iraq.

UPDATED

RGN offensive/defensive hand grenade (L Haywood)

1996

RG-42 offensive hand grenade

Description

The RG-42 offensive hand grenade dates back to the Great Patriotic War of 1941-1945 and has largely been regarded as obsolete by many of its former user nations. However, the type is still produced in Poland and it was in production in China, at least until recently. Despite its age the RG-42 is still likely to be encountered almost anywhere, especially within Africa.

The RG-42 is a simple design being little more than a steel sheet cylinder filled with between 110 and 120 g of TNT ignited by a UZRGM Bouchon igniter assembly screwed into a well in the upper surface. The blast effects will be harmful over a radius of at least 10 m.

Data

Length: with fuze, 130 mm; without fuze, 85 mm
Diameter: 55 mm
Weight: with fuze, 420 g; without fuze, 384 g
Filling: 110-120 g TNT
Fuze delay: 3.2-4.2 s
Range thrown: 30 m

Manufacturer

Various state factories.

Status

Still in production in Poland.

Service

Polish armed forces, China People's Liberation Army and others.

Licence production

CHINA, PEOPLE'S REPUBLIC
Manufacturer: China North Industries Corporation (NORINCO)
Type: Type 42
Remarks: Standard specifications
POLAND
Manufacturer: Zaklady Sprzetu Precyzyjnego
Type: RG-42
Remarks: Standard specifications

NEW ENTRY

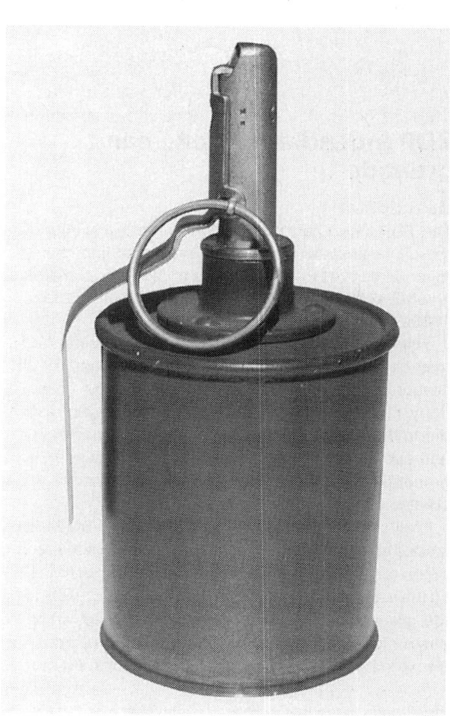
RG-42 offensive hand grenade

1996

RKG-3 anti-tank grenades

Description

The family of RKG-3 grenades replaced the earlier RPG-40, RPG-43 and RPG-6 anti-tank grenades in the Eastern Bloc and China.

The RKG-3 is stabilised in flight by a four-panelled fabric drogue which is pulled out from the handle when the grenade is thrown. The drogue development completes the arming of the grenade. The drogue also ensures that it is possible to drop the grenade on the top of an armoured vehicle.

The earliest member of the family was the RKG-3 and this was capable of penetrating just 125 mm. Two further versions, the 3M and the 3T, have been produced. The RKG-3M has a slightly larger warhead with a copper cone instead of the steel liner of the first version and the 3T also has a different cone liner. The RKG-3M was used extensively in the 1973 Arab/Israeli War and was shown to penetrate 165 mm of armour.

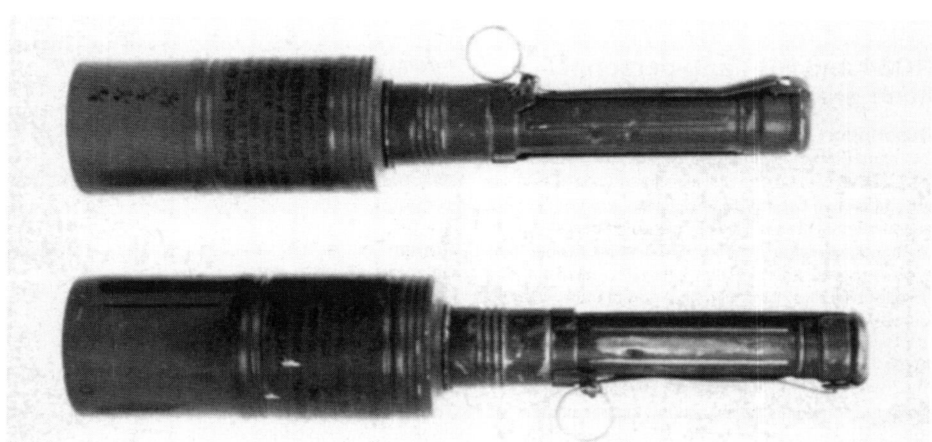
RKG-3 (top) and (lower) RKG-3M anti-tank grenades

The UPG-8 practice grenade is used during training in place of the RKG-3 series of anti-tank grenades.

Data

Length: 362 mm
Diameter: max, 55.6 mm
Weight: with fuze, 1.07 kg
Filling: 567 g TNT/RDX
Penetration: RKG-3, 125 mm; RKG-3M, 165 mm
Type of fuze: instantaneous impact
Effective fragment radius: 20 m

Manufacturer

Various state factories.

Status

Probably no longer in production.

Service

Widespread, especially in the Middle East.

Licence production

YUGOSLAVIA (SERBIA AND MONTENEGRO)
Manufacturer: YUGOIMPORT-SDPR
Type: M79
Remarks: Direct copy of RKG-3M

UPDATED

RDG-2 and RDG-3 smoke hand grenades

Description

The RDG-2 smoke grenade has been in service since the early 1950s and was adopted by all the former Soviet Bloc countries. It is a tactical grenade and is used to conceal the movements of small bodies of infantry or engineers. It is made of cardboard, with a cardboard tube down the centre and is filled with a burning type filler and a friction igniter. The grenade is waxed and is damp-proof but it cannot be used to produce smoke over water. It produces a dense white smoke after about 15 seconds creating a cloud 20 to 25 m long and 8 m wide. Smoke will be produced for about 1 to 1.5 minutes.

The RDG-3 is identical to the RDG-2 but produces orange smoke for marking purposes.

There is also an RDG-2Kh grenade which emits clouds of irritant smoke which used to simulate chemical agent attacks. The RDG-2Ch creates black smoke for various screening purposes. Both grenades are basically similar to the standard RDG-2.

Data

Length: 240 mm
Diameter: 46 mm
Weight: 500 g
Throwing range: 35 m
Smoke area: 160 m^2
Duration of smoke: 1-1.5 min

Manufacturer

Various state factories.

Status

In production.

Licence production

BULGARIA
Manufacturer: Kintex
Type: RDG-2 and RDG-3
Remarks: Standard specifications

UPDATED

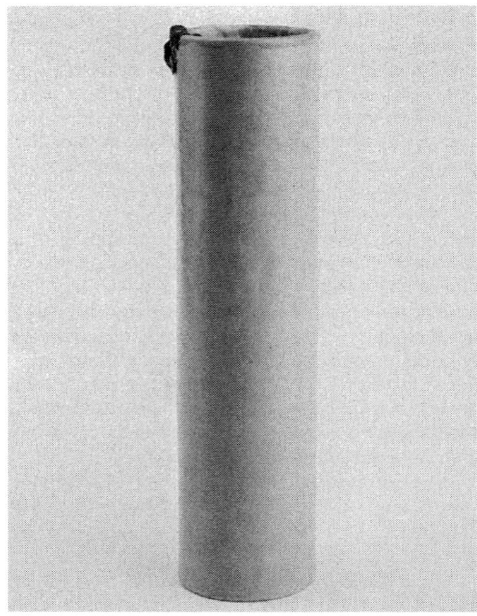

RDG-2 smoke hand grenade
1996

ZDP incendiary smoke hand grenade

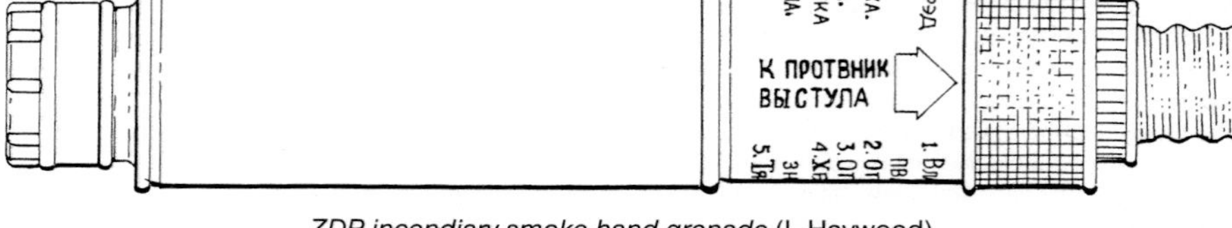

ZDP incendiary smoke hand grenade (L Haywood)
1996

Description

The ZDP incendiary smoke hand grenade is intended for use by assault troops and is small and light enough for an individual to carry several of them for use during an attack. It has been in service since at least the mid 1980s.

The ZDP grenade, which has a tubular metal body, may be used in one of two ways. Instructions for use covering both methods are printed on the grenade body. One method is as a conventional grenade. In this mode the user removes a metallic green cap from one end and throws the grenade in the usual manner. The grenade then ignites an incendiary element which creates clouds of smoke after a short delay.

The second method is to project the incendiary smoke element using a rocket motor. For this mode the green cap is removed from one end as before, followed by the removal of a red plastic cap from the other end of the grenade body. A metal ring is exposed which is eased, together with a short length of cord, from the grenade body and pulled to initiate the rocket motor. It is recommended that the grenade body is steadied against a rifle for the launch. The round has a maximum range of 560 m, although 200 m is understood to be a more practical range.

Data

Length: 290 mm
Diameter: 50 mm
Weight: 750 g
Throwing range: 30 m
Range using rocket motor: max, 560 m; practical, 200 m

Manufacturer

Various state factories.

Status

In production.

Service

Russian armed forces.

NEW ENTRY

CZECH REPUBLIC

RG34 and RG4 anti-personnel hand grenades

Description

CIS hand grenades in current service are also used in the Czech Republic and Slovakia. In addition two native grenades have been produced. These are both impact anti-personnel grenades. They are described as offensive hand grenades but the sheet metal bodies also produce some fragmentation. If required, each of these grenades can be fitted with an external fragmentation sleeve which improves the lethality.

RG34

The RG34 is a steel-bodied cylindrical grenade, distinguished by the fluting around the mid-section.

Data

Length: 76 mm
Diameter: max, 64 mm
Weight: without fragmentation sleeve, 240 g
Body material: steel
Filling: 100 g TNT
Fuze type: impact; all ways
Range thrown: 35 m
Effective fragment radius: 13 m; with fragmentation sleeve, 25 m

Status

Obsolete, but still in stock.

Service

Czech and Slovak forces.

RG4

The RG4 replaced the RG34 in the Czech and Slovak Armies. It has a cylindrical steel body which is completely

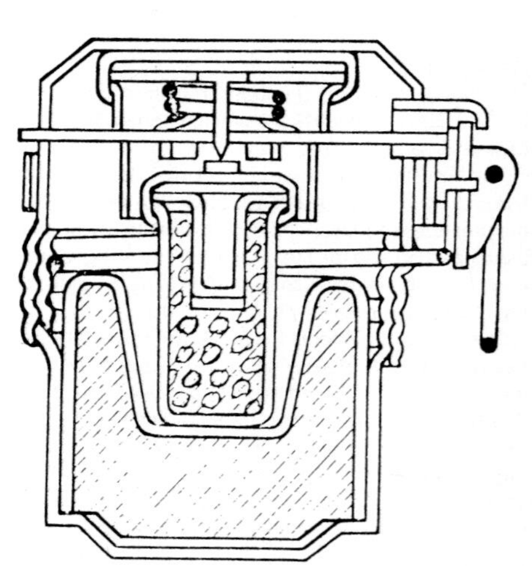

Czech RG4 hand grenade

Czech RG4 hand grenade

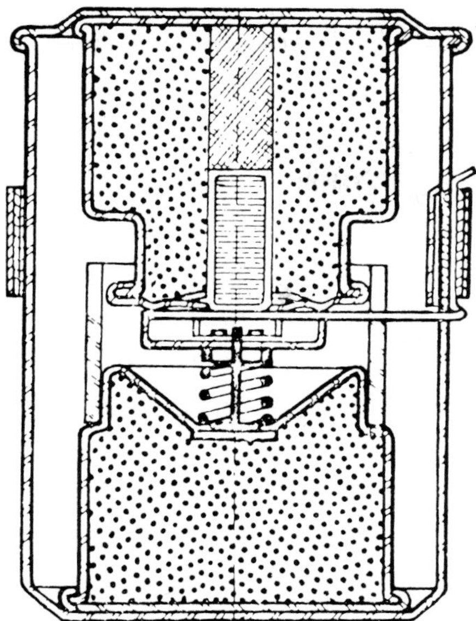

Czech Model RG34 hand grenade

Czech Model RG34 hand grenade

smooth. It can be converted to a defensive grenade by adding a fragmentation sleeve. This grenade is unusual in containing an upper and lower bursting charge.

Data

Type: blast
Length: 84 mm
Max diameter: 53 mm
Weight: without fragmentation sleeve, 320 g
Filling: 105 g TNT
Fuze type: impact
Fragmentation radius: 13 m; with fragmentation sleeve, 25 m

Status

In current use.

Service

Czech and Slovak forces and some guerrilla forces in Africa.

VERIFIED

EGYPT

Kaha No 1 defensive hand grenade

Description

The Kaha No 1 defensive hand grenade has an ovoid plastic body containing 60 g of Composition B around which are packed between 5,000 and 6,000 small steel balls. After the grenade is thrown and the fly-off safety lever has initiated the fuze assembly, there will be a delay of between three and four seconds before the main charge detonates. This scatters the steel balls over a full 360° radius but the size and low mass of the steel balls are such that at a radius of 5 m only about 20 will retain sufficient energy to penetrate 20 mm of wood. At 20 m the number of fragments will be zero.

This grenade can be fired from the muzzle of a shotgun by using a suitable but unspecified attachment. The range with the barrel at an angle of elevation of 45° is 200 m.

Data

Length: 105 mm
Diameter: 59 mm
Weight: 560 g
Filling: 60 g Comp B
Delay time: 4 ±0.5 s
Number of fragments: 5,000-6,000
Lethal radius: 5 m
Safety radius: 20 m

Manufacturer

Kaha Company for Chemical Industries, PO Box 2332, Cairo.

Status

In production.

Service

Egyptian Army.

NEW ENTRY

Kaha No 1 defensive hand grenades
1996

Kaha No 1 offensive hand grenade

Description

The Kaha No 1 (also known as the 270/1) offensive hand grenade has the same ovoid plastic body as the No 1 defensive hand grenade (see previous entry) but the contents are confined to 60 g of Composition B - there is no steel ball content. This produces a powerful blast effect over a radius of 5 m from the point of explosion with only minimal fragmentation past that distance. The grenade retains the same fuze assembly as its defensive counterpart but the delay time is reduced to between two and three seconds.

Data

Length: 105 mm
Diameter: 59 mm
Weight: 240 g
Filling: 60 g Comp B
Delay time: 2.5 ±0.5 s

Manufacturer

Kaha Company for Chemical Industries, PO Box 2332, Cairo.

Status

In production.

Service

Egyptian Army.

NEW ENTRY

Kaha No 1 (270/1) offensive hand grenades

Hossam anti-tank grenade

Description
The Hossam anti-tank grenade was designed and manufactured in Egypt and uses a shaped charge to achieve penetration. The grenade is stabilised in flight by a drag parachute which ensures the impact fuze strikes the target and the shaped charge is correctly oriented.

Data
Length: 192 mm
Diameter: 63 mm
Weight: 575 g
Filling: 149 g Hexogen 90%
Arming distance: 5 m
Penetration: armour, 120 mm; concrete, 400 mm

Manufacturer
Sakr Factory for Developed Industries, PO Box 33, Heliopolis, Cairo.

Status
In production.

Service
Egyptian Army.

VERIFIED

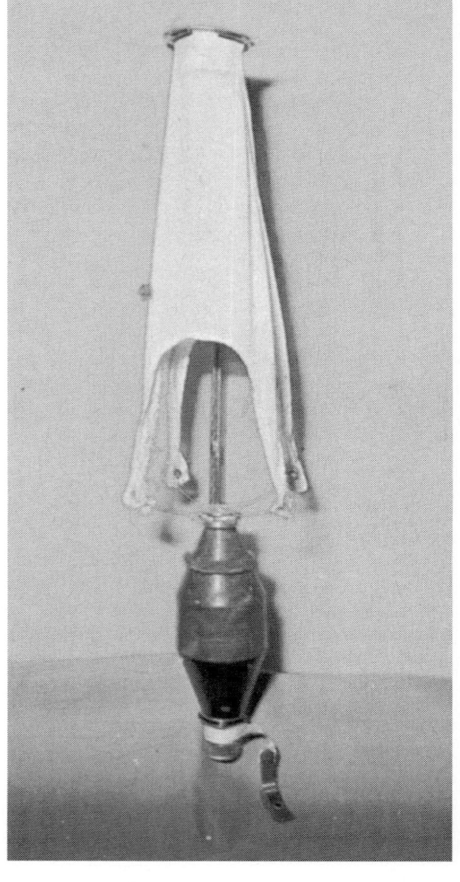

Hossam anti-tank grenade with parachute deployed

FRANCE

Giat 40 mm anti-personnel/anti-vehicle (AP/AV) rifle grenade

Development
The Giat 40 mm anti-personnel/anti-vehicle (AP/AV) rifle grenade has been described as a second generation grenade. It is based on the Luchaire bullet trap design (see following entry) and can be fired from standard 5.56 or 7.62 mm assault rifles provided with a standard 22 mm launching device at the muzzle.

In 1994 the French Army ordered 120,000 of these AP/AV rifle grenades with deliveries commencing at the end of 1994.

Description
The grenade has an aluminium body containing a preformed steel fragmentation sleeve around an anti-personnel high explosive charge behind a shaped charge forming the anti-vehicle component. The warhead has a long ogive with, at its tip, an anti-skid ring which ensures the warhead will operate against any surface. On contact with a target a stab detonator ignites a relay which, in turn, ignites the main booster and the main explosive payload. The shaped charge then forms a jet capable of penetrating 80 mm of RHA steel and the fragmentation sleeve produces some 400 fragments to be scattered over a lethal radius of 12 m.

The AP/AV grenade is launched using standard ball cartridges on the Luchaire bullet trap principle, with different grenade tails to suit the 5.56 and 7.62 mm calibres.

The TP inert training grenade which can be fired using live ball rounds is available. Another inert training grenade, the TPM, is available, this time being launched by crimped cartridge cases without a bullet. TMP grenades, which may be reused up to ten times, provide a puff of smoke to indicate the point of impact. Both types of training grenade have identical ballistic performances to live grenades.

Data
Length: 380 mm
Weight: 5.56 mm, 425 g; 7.62 mm, 440 g
Muzzle velocity: 5.56 mm, 69 m/s; 7.62 mm, 80 m/s
Range: direct, combat, 150 m; max indirect, 5.56 mm, 380 m; max indirect, 7.62 mm, 400 m
Number of fragments: ca 400
Lethal radius: 12 m
Armour penetration: RHA, 80 mm

Manufacturer
Giat Industries, 13 route de la Minière, F-78034 Versailles-Satory Cedex.

Status
In production.

Service
French Army.

NEW ENTRY

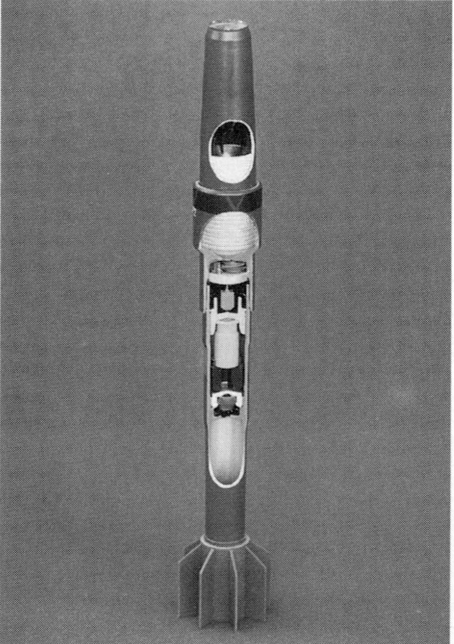

Cutaway example of Giat 40 mm anti-personnel/ anti-vehicle (AP/AV) rifle grenade

1996

Luchaire 40 mm bullet trap rifle grenades

Description
Luchaire 40 mm rifle grenades combine the accuracy of very short-range flat firing and the efficiency of curved fire beyond 400 m range, thus providing an individual weapon system designed for anti-armoured vehicle and anti-personnel applications.

The 40 mm calibre is the result of a technical compromise between ergonomics, carrying power and optimal performance.

Because of its bullet trap the Luchaire 40 mm rifle grenade may be safely and directly launched using any combat cartridge. The infantryman's rifle remains ready for instant use before, during and after launching a grenade.

Being without seal or safety pin the grenade is permanently ready for use. Versatility is assured, since the range comprises six combat warheads and two training rounds. All use the same fuze and tail tube assembly which is compatible with most modern 5.56, 7.5 or 7.62 mm rifles.

Combat grenades
Anti-tank (HEAT)
Anti-personnel/anti-vehicle (HEAT-APERS-FRAG)
Anti-personnel (HE-APERS-FRAG)
Anti-personnel delay (HE-APERS-FRAG-Delay)
Smoke (SMOKE)
Illuminating (ILL)

Training/practice grenades
Training practice (TP)
One-piece training practice with target marker (TPM)

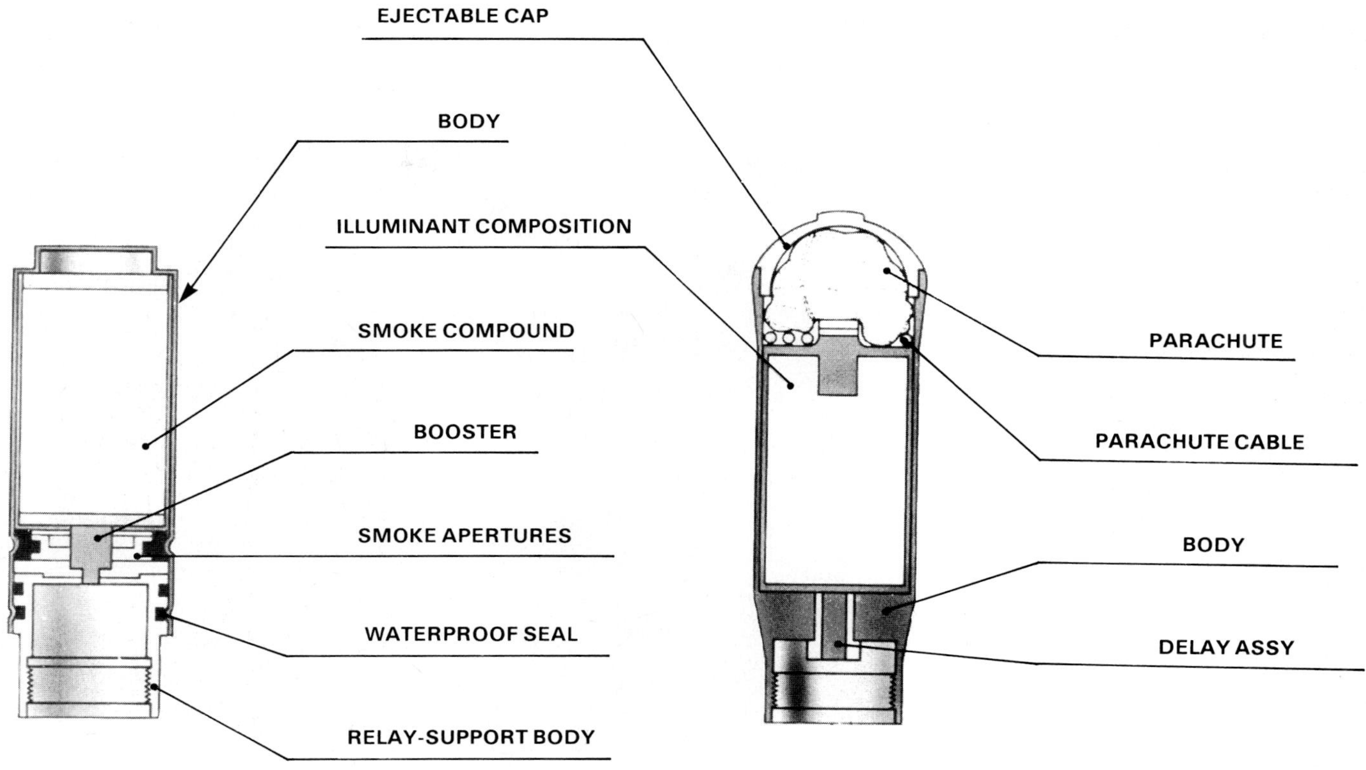

Warheads for the Luchaire smoke and illuminating rifle grenades

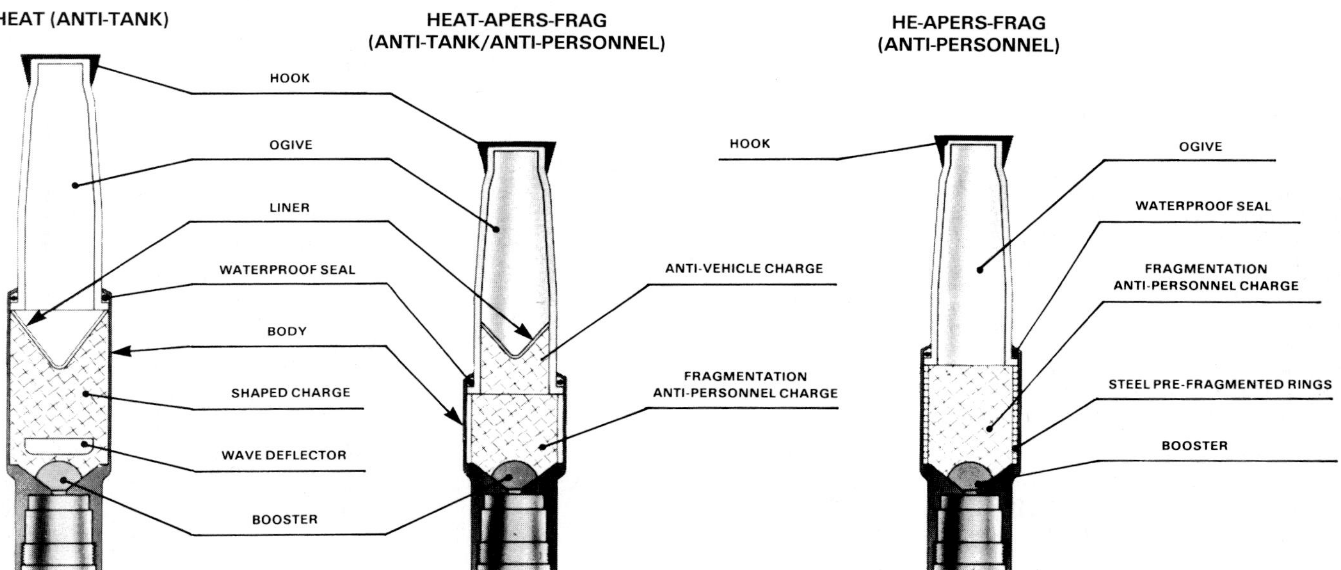

Warheads for the Luchaire AC, AP/AV and AP rifle grenades

Data

	HEAT	HEAT-APERS-FRAG	HE-APERS-FRAG	HE-APERS-FRAG-Delay	SMOKE	ILL	TP	TPM
External calibre	40 mm	40 mm	40 mm	40 mm	40 mm	40 mm	40 mm	40 mm
Length	383 mm	352 mm	357 mm	305 mm	292 mm	306 mm	383 mm	351 mm
Weight	412 g	405 g	412 g	410 g	410 g	390 g	415 g	405 g
Fuze type	SQ	SQ	SQ	Delay	SQ	Delay	–	–
Prefragmented splinters	–	396	468	468	–	–	–	–
Splinter weight	–	0.17 g	0.17 g	0.17 g	–	–	–	–
Muzzle velocity	68 m/s	68 m/s	68 m/s	68 m/s	70 m/s	73 m/s	68 m/s	68 m/s
Muzzle safety				5 or 12 m depending upon timer				
Safe operating range				8 or 15 m depending upon timer				
Ranges:								
Effective combat range								
flat trajectory	100 m	100 m	–	100 m	–		100 m	100 m
Max range, curved trajectory	360 m	360 m	360 m	360 m	(a)		360 m	360 m
Accuracy; standard deviation	<.3	<.3	<.3	<.3			<.3	<.3
Lethal radius		12	12	12				
Penetration, RHA	200 mm	100 mm						
Smoke emission time					25 s			
Smoke screen duration					60 s			
Illumination duration						30-40 s		
Illumination intensity						90,000 cd		

Note (a) - smoke pot opening height and distance for a 45° firing angle is about 200 m

Manufacturer
Giat Industries, 13 route de la Minière, F-78034 Versailles-Satory Cedex.

Status
In production.

Service
HEAT in service with British Army (8,000).

UPDATED

Luchaire LU 213 HE-APERS-FRAG hand grenade

Description

The LU 213 is a controlled fragmentation anti-personnel hand grenade in which particular attention has been paid to the fragmentation process. Perfect reliability and homogeneous fragment distribution is obtained by a coiled sleeve fragment generator and the use of 230 small steel ball pads on the upper and lower parts of the grenade.

At the time of the grenade's detonation, whatever its position in flight, the steel balls and the fragments of the prenotched sleeve are projected at very high speed, providing a dense and uniform pattern all around the point of detonation. The safety radius has been studied so as to allow the thrower to stand unprotected within a radius of 20 m without being endangered.

An inert ballasted hand grenade is available for demonstration and drill purposes. The exterior shape is identical to the LU 213 anti-personnel grenade, but the inert body is coloured orange and equipped with an inert fuze assembly. A safety device is placed at the base of the priming system so as to prevent fitting a live fuze into the inert grenade body.

Data

Length, with fuze: 94 mm
Diameter: 52 mm
Weight: with fuze, 280 g; fuze weight, 55 g; prefragmented sleeve, 100 g; explosive filling, 90 g Comp B or equivalent
Number of fragments: ca 1,100 plus 230 steel balls
Average fragment weight: 0.085 g
Lethal radius: 7 m

Manufacturer

Giat Industries, 13 route de la Minière, F-78034 Versailles-Satory Cedex.

Luchaire LU 213 (left) and LU 216 hand grenade

Status

In production.

UPDATED

Luchaire LU 216 HE blast effect hand grenade

Description

For this version of the Luchaire grenade, design was concentrated on producing maximum blast effect. Of ovoid form, the LU 216 HE differs from the LU 213 by the absence of the prenotched fragmentation sleeve and steel ball pads. It is fitted with the same fuze as that of the LU 213 hand grenade.

For training the LU 219 PRAC is available. The body colour is blue and of identical form to the LU 213 hand grenade. The body is aluminium alloy and may be used up to 100 times. It is equipped with an expendable practice fuze which simulates the functions of the real fuze (handling, firing, functioning and delay) and the terminal effects of the grenade (noise, flash and smoke).

Data

Length: with fuze, 94 mm
Diameter: 52 mm
Weight: with fuze, 140 g; fuze, 55 g
Filling: 60 g Composition B or equivalent

Manufacturer

Giat Industries, 13 route de la Minière, F-78034 Versailles-Satory Cedex.

Status

In production.

UPDATED

ALSETEX defensive and offensive grenades

Description

ALSETEX grenades are made with either a thin casing for offensive use, or a thicker fragmenting case for use as a defensive munition.

These grenades are provided with the ALSETEX Mk F5A fuze. This safety fuze, with fly-off lever and compressed pyrotechnical delay, includes a pyrotechnic chain interrupting device, allowing greater safety in use and storage.

Both models are filled with compressed Tolite (TNT) explosive. There are also practice models containing inert materials.

Data

Grenade	No 1	No 2
Type	offensive	defensive
Case	metal	metal
Diameter	60 mm	55 mm
Height	95 mm	100 mm
Weight	140 g	540 g
Filling weight	90 g	56 g

Manufacturer

SAE ALSETEX, 36 rue Tronchet, F-75009 Paris.

Status

Production complete.

Service

In service with the French Army.

UPDATED

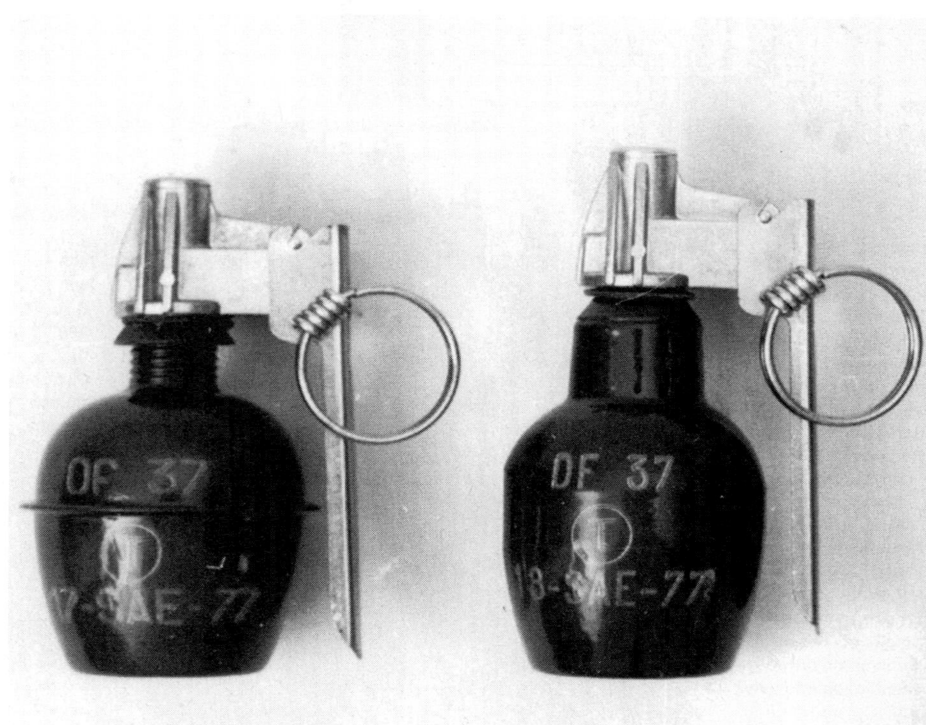

ALSETEX defensive and offensive grenades

ALSETEX SAE 210 offensive grenade

Description

The ALSETEX SAE 210 offensive grenade has a plastic ovoid body with a granular surface not unlike that of an orange skin to provide a good grip. The grenade is delivered with a BALPIC plastic igniter fuze with a fly-off lever, a pyrotechnic chain interruption system, and a 4-second delay. The filling is 100 g of TNT.

The shape and reduced weight of this grenade are claimed to provide good throwing accuracy to a range of about 30 m. The pressure recorded during the detonation of this grenade at a distance of 5 m from the point of explosion is higher than 80 mb. No dangerous fragments are projected beyond a radius of 15 m.

Data

Height: fuzed, 76 mm
Diameter: 55 mm
Weight: 190 g
Filling: 100 g TNT
Delay time: 4 s
Operational temperature range: −40 to +70°C

Manufacturer

SAE ALSETEX, 36 rue Tronchet, F-75009 Paris.

Status

In production.

NEW ENTRY

ALSETEX SAE 210 offensive grenade
1996

ALSETEX SAE 310 controlled fragmentation offensive grenade

Description

The ALSETEX SAE 310 controlled fragmentation offensive grenade has a plastic ovoid body which contains prefragmented splinters and is filled with 90 g of Hexolite (RDX/TNT). The form of the 1,300 splinters and their weight were designed to allow a thrower to remain exposed in safety at a point 20 m from the impact.

The SAE 310 grenade is delivered with a BALPIC plastic igniter fuze with a fly-off lever, a pyrotechnic chain interruption system, and a 4-second delay. With its double action twist and pull pin, the fuze can only be actuated after the total ejection of the fly-off lever, thus preventing any explosion hazard while the grenade is being held.

This grenade can be thrown to a distance of more than 20 m. The body shape ensures that all 1,300 splinters are distributed all around the 5 m lethal radius.

Data

Length: fuzed, 95 mm
Diameter: 57 mm
Weight: 300 g
Filling: 90 g RDX/TNT
Delay time: 4 s
Number of splinters: 1,300
Lethal radius: 5 m
Safety radius: 20 m
Operational temperature range: −40 to +70°C

Manufacturer

SAE ALSETEX, 36 rue Tronchet, F-75009 Paris.

Status

In production.

NEW ENTRY

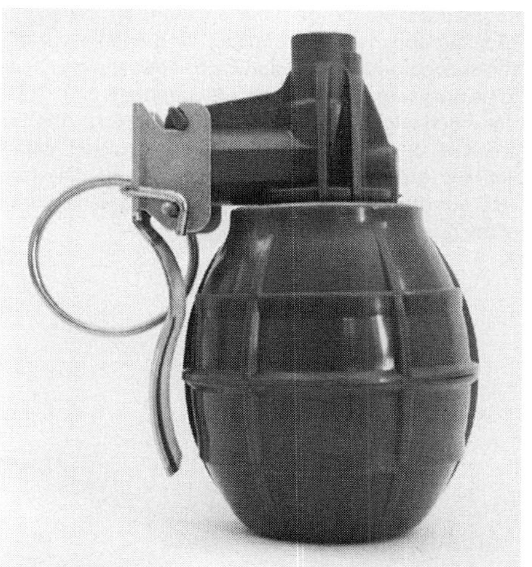

ALSETEX SAE 310 controlled fragmentation offensive grenade
1996

ALSETEX screening smoke grenades

Description

The ALSETEX screening smoke grenades may be used to screen the movement of personnel and vehicles, provide ground-to-ground or ground-to-air signals, indicate wind speed and direction, and provide ground recognition marking for aircraft or helicopters. These grenades are characterised by good operating reliability and simple and rapid use.

These grenades have a waterproof aluminium body with four smoke holes to produce rapidly a very dense smoke cloud during the complete time of emission. The body is filled with a solid block of smoke producing compound and is closed by an obturating plate. The SAE 630 grenade will emit smoke for 30 seconds while the SAE 660 time is 60 seconds. ALSETEX can also produce a screening smoke grenade which can emit smoke for 120 seconds. These grenades are delivered with a plastic igniter fuze with a fly-off lever. With its double action twist and pull pin, the fuze can only be actuated after the total ejection of the fly-off lever; there is a 0.5 second delay. A special fuze without the lever is available.

Data

Length: 134 mm
Diameter: 55 mm
Weight: 300 g
Delay time: 0.5 s
Emission time: SAE 630, 30 s; SAE 660, 60 s

Manufacturer

SAE ALSETEX, 36 rue Tronchet, F-75009 Paris.

Status

In production.

NEW ENTRY

ALSETEX coloured smoke grenades

Description

ALSETEX coloured smoke grenades may be used to screen the movement of personnel and vehicles, provide ground-to-ground or ground-to-air signals, indicate wind speed and direction, and provide ground recognition marking for aircraft or helicopters. These grenades are characterised by good operating reliability and simple and rapid use.

ALSETEX coloured smoke grenades have a waterproof aluminium body with four smoke holes to produce rapidly a very dense smoke cloud during the complete time of emission. The body is filled with a solid block of coloured smoke producing compound and is closed by an obturating plate with the four holes. The SAE 500 series will emit smoke for 30 seconds while the SAE 510 series time is 60 seconds. ALSETEX can also produce coloured smoke grenades which can emit smoke for 120 seconds. These grenades are delivered with a plastic igniter fuze with a fly-off lever. With its double action twist and pull pin, the fuze can

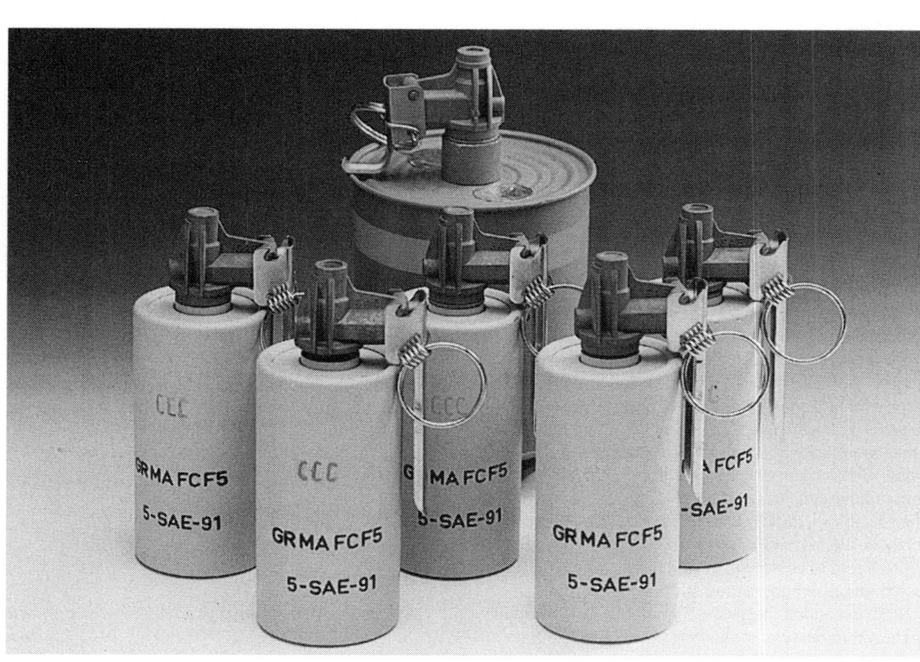

ALSETEX coloured smoke grenades
1996

only be actuated after the total ejection of the fly-off lever; there is a 2.5 seconds delay. A special fuze without the lever is available.

Colours available are as follows:

White: SAE 500 or 510
Yellow: SAE 501 or 511
Red: SAE 502 or 512
Green: SAE 503 or 513
Orange: SAE 504 or 514
Blue: SAE 505 or 515.

Data

Length: 134 mm
Diameter: 55 mm
Weight: 300 g
Filling weight: 100 g
Delay time: 2.5 s
Emission time: SAE 500 series, 30 s; SAE 510 series, 60 s

Manufacturer

SAE ALSETEX, 36 rue Tronchet, F-75009 Paris.

Status

In production.

NEW ENTRY

ALSETEX rifle grenades

Description

ALSETEX rifle grenades are of three types: anti-personnel (AP); anti-personnel and anti-vehicle (AP/AV); and practice. They are capable of being fired from any 5.56 mm assault rifle having a standard muzzle diameter of 22 mm.

The two service grenades utilise a new design of shaped charge, the AP/AV model having a pre-fragmented liner to give the optimum anti-personnel effect around the point of impact. All grenades utilise a shock-absorbent bullet trap which allows the grenades to be fired with ball, tracer or AP cartridges.

A mechanical delayed arming fuze, armed by the gas pressure of the propelling cartridge, provides safety until the grenade is *en route* to the target. The fuze incorporates a safety indicator, allowing a visual check of the arming state prior to firing.

Data

Length: 368 mm
Head diameter: 34 mm
Weight: 420 g
Weight of filling: 55 g
Max range: 400 m
Max launch velocity: 67 m/s
Number of fragments: ca 700
Penetration: 30 mm homogeneous steel armour at 60° incidence
Accuracy: ≤300 mm at 100 m
Arming delay: 5-14 m

Manufacturer

SAE ALSETEX, 36 rue Tronchet, F-75009 Paris.

Status

Ready for production.

Service

In course of formal approval by French Army.

UPDATED

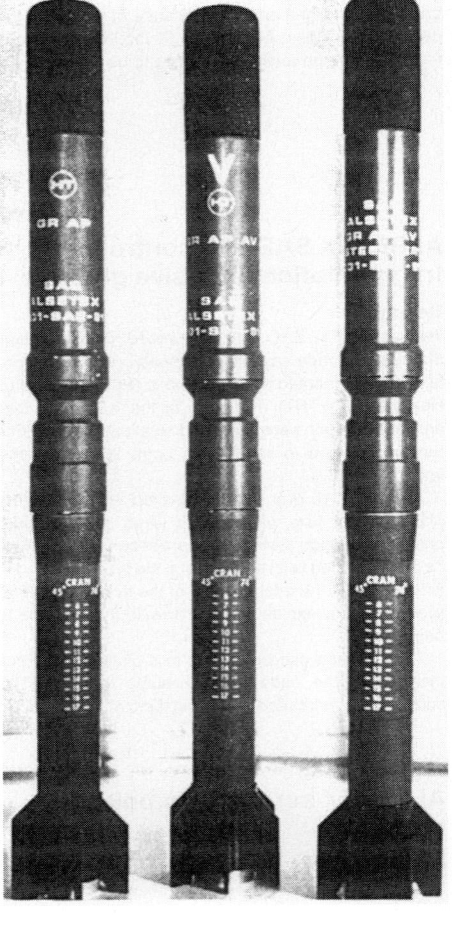

Alsetex rifle grenades: AP; AP/AV; practice

Ruggieri Type 0052 and Type 0052A hand grenades

Description

The Ruggieri Type 0052 and Type 0052A hand grenades are visually identical but vary in their utility. Both have oval high impact plastic two-section bodies with a sealing cap at the base and the same Type 1760 igniter mechanism. However, the Type 0052 is an offensive grenade while the Type 0052A produces preformed fragments to act as both an offensive and defensive grenade (it is described in Ruggieri literature as an anti-personnel grenade).

Both types of grenade have a disabling radius of 5 m, with both being able to be employed in close combat with a considerable degree of safety for the user. The offensive Type 0052 produces disabling blast effects over a radius of 5 m but the thrower will be unaffected if they remain more than 15 m from the point of explosion.

The offensive/defensive Type 0052A contains a 33 g copper-plated fragmentation casing formed into 950 small facets by concave incisions directed towards the centre of the explosive charge. This produces the effects of a series of small shaped charges. As the explosive shock wave strikes each facet, each forms a small independent fragment with a very high initial velocity. Each fragment weighs approximately 0.035 g and rapidly loses its energy with range to the extent that the disabling radius is limited to 5 m. A standing thrower will remain unaffected at a safe distance of 25 m. Tests have demonstrated that a 150 μm sheet of polythene is not perforated at 25 m.

Both grenades utilise the Type 1760 igniter mechanism with a delay of 4 seconds. The Type 1760 igniter permits safe storage and handling of a fuzed grenade in a ready for use state. An alignment interruption system in the igniter/pyrotechnic chain provides an additional safety until after the firing lever is released. The Type 1760 igniter mechanism makes extensive use of plastic materials and thus avoids the production of large unwanted fragments on detonation. Each Type 1760 weighs approximately 50 g. It can be retrofitted to existing grenades.

Data

Diameter: 59 mm
Height: fuzed, 89.5 mm
Weight: Type 0052A, ca 210 g; Type 0052, ca 190 g
Weight of explosive filling: Type 0052A, ca 90 g; Type 0052, ca 100 g
Delay time: 4 s
Lethal radius: 5 m
Safety radius: Type 0052A, 25 m; Type 0052, 15 m

Manufacturer

Ruggieri Technologie, route de Gaudiès, F-09270 Mazères.

Status

In production.

NEW ENTRY

Ruggieri Type 0052 hand grenade
1996

Offensive hand grenade Model 1937

Description

The offensive hand grenade Model 1937 is made from a stamped aluminium body divided into two swaged parts. One of these parts bears a hole into which is fitted a light metal exploder tube, which is then screwed and stuck to the body. The body is filled with nitrate explosive. An Igniter Plug Model F3 Type 1719 screws on to the top of the grenade, the delay and detonator unit being contained in the exploder tube. For transit and storage the igniter plug is not present, the tube being closed by a plastic plug.

Data

Length: fuzed, 100 mm
Diameter: 60 mm
Weight: total, fuzed, 180 g; explosive content, 78 g
Delay time: 4-7 s

Manufacturer

Ruggieri Technologie, route des Gaudiès, F-09270 Mazères.

Status

In current use.

Service

French and other armies.

VERIFIED

Ruggieri offensive hand grenade Model 1937

Defensive hand grenade Model 1937

Description

The defensive hand grenade Model 1937 is made of cast iron. The body has a hole at the upper end into which a light metal exploder tube is screwed and stuck. The Igniter Plug Model F3 Type 1719 then screws into this hole, the detonator and delay assembly fitting into the exploder tube. Upon detonation, the cast iron body shatters to produce a cloud of dangerous fragments.

Data

Length: fuzed, 95 mm
Diameter: 55 mm
Weight: fuzed, 540 g
Delay time: 4-7 s

Manufacturer

Ruggieri Technologie, route des Gaudiès, F-09270 Mazères.

Status

In current use.

Service

French and other armies.

Ruggieri defensive hand grenade Model 1937

GERMANY

Diehl M-DN 11 hand grenade

Description

The M-DN 11 grenade is barrel-shaped and made of plastic. The thick wall contains 3,800 steel balls embedded in the plastic. The outside of the grenade body has longitudinal and transverse ribs which are raised from the surface to improve the user's grip and generally to improve handling under adverse conditions such as subzero temperatures or in mud. The interior is filled with plasticised nitropenta.

At the top a fuze thread is cut into the plastic material for the DM 82A1B1 or M-DN 42 fuze. This has the same external appearance as previous fuzes made by Diehl and consists of a spring-actuated hammer, held back by a long safety lever from hitting the primer cap and a delay pellet above a detonator and a booster in the bursting charge.

The main feature of the fuze, however, is that it incorporates a series of safeties. The first one is that the detonator is physically separated from the booster until the grenade is thrown. Additional safety is provided by a flap valve at the bottom of the delay tube. Should the cap be struck for whatever unlikely reason, or should the delay pellet be missing, this valve ensures that the flash cannot reach the detonator. After 2.5 seconds the soldered joint in the delay tube melts and allows the spring to force the tube into contact with the detonator. The flash material is now close enough to the detonator to ignite it and the flap valve is clear of the throat which held it shut, so that the flash can open it and pass through the fire holes. The entire arrangement is most ingenious and certainly safer than any fuzing that has been in service before. The sequence of events is explained in the following description (see diagram.)

When the safety pin **(1)** is withdrawn the torsion spring, **(2)** in the accompanying drawing, forces the hammer **(3)** round and it hits the DM 1024A1B1 primer cap **(4)**. The spring force throws off the protective cap **(5)** and the safety lever **(6)**. The flash of the percussion primer ignites the delay pellet **(7)** in its delay tube **(8)**. After about 2.5 seconds burning time the heat melts a solder ring **(9)** and so disconnects the delay tube **(8)** from the detonator holder **(10)**. The pressure spring **(11)** then forces the detonator holder and DM 1066B1 detonator **(12)** down on to the DM 1034 booster **(13)**. After four seconds burning time the flash from the delay pellet passes through two holes in the cup **(14)** over the detonator and strikes the flap valve **(15)** bending it up and passing it to reach the detonator **(12)**. The detonator sets off the booster which in turn ignites the main charge.

M-DN 11 hand grenade

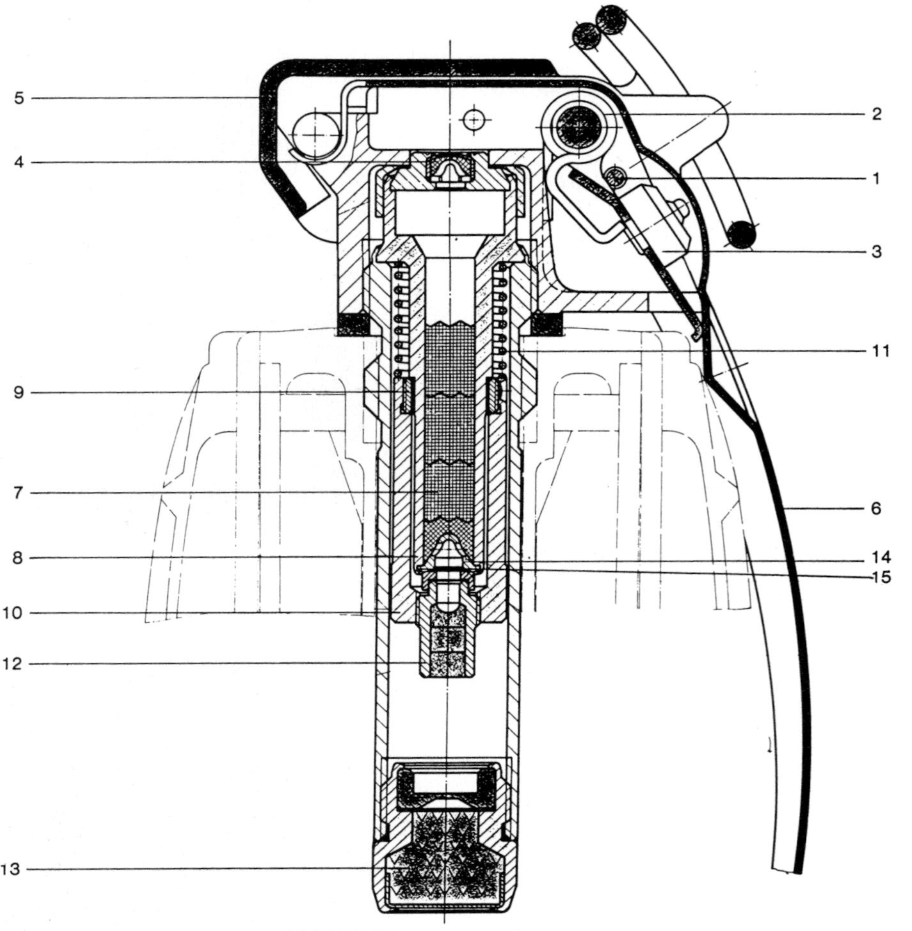

DM 82A1B1 fuze assembly (see text)

Data
Length: with fuze, 97 mm
Diameter: 60 mm
Weight: with fuze, 467 g; grenade body, 364 g; explosive filling, 42.5 g
Ball diameter: 2.5-3 mm
Number of balls: 3,800
Danger area: ca 100 m

Manufacturer
Diehl GmbH & Co, Ammunition Division, Fischbachstrasse 16, D-90552, Röthenbach/Pegnitz.

Status
Available.

VERIFIED

Diehl M-DN 21 hand grenade

Description
The M-DN 21 hand grenade is a smaller edition of the M-DN 11 (see previous entry). It functions in exactly the same way.

Data
Length: fuzed, 85 mm
Diameter: 50 mm
Weight: with fuze, 224 g; grenade body, 118 g; explosive filling, 45 g
Grenade body: empty, 118 g
Filling: 45 g HE
Ball diameter: 2-2.3 mm
Number of balls: 2,200
Danger area: 45 m

Manufacturer
Diehl GmbH & Co, Ammunition Division, Fischbachstrasse 16, D-90552, Röthenbach/Pegnitz.

Status
Available.

VERIFIED

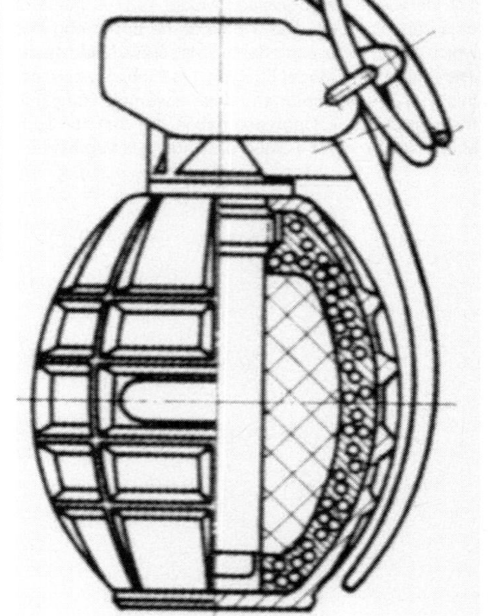

M-DN 21 hand grenade

Diehl M-DN 31 hand grenade

Description
The Diehl M-DN 31 hand grenade is similar in size and function to the M-DN 21 grenade but has more fragments and a greater danger area.

Data
Length: fuzed, 85 mm
Diameter: 50 mm
Weight: with fuze, 247 g; fuze, 61 g; grenade body, 151 g; explosive filling, 35 g
Ball diameter: 2-2.3 mm
Number of balls: 3,000
Danger area: 60 m

Manufacturer
Diehl GmbH & Co, Ammunition Division, Fischbachstrasse 16, D-90552, Röthenbach/Pegnitz.

Status
Available.

UPDATED

M-DN 31 hand grenade

Diehl DM 61 hand grenade

Description

The DM 61 completes the series of Diehl defensive grenades and closely resembles the others except for its size, which is intermediate between the M-DN 21 and the M-DN 11.

Data

Length: fuzed, 109 mm
Diameter: body, 57 mm
Weight: with fuze, 350 g; grenade body, 215 g; explosive filling, 65 g
Ball diameter: 2-2.3 mm
Number of balls: 3,800
Danger area: 75 m

Manufacturer

Diehl GmbH & Co, Ammunition Division, Fischbachstrasse 16, D-90552 Röthenbach/Pegnitz.

Status

Available.

Service

Norwegian armed forces.

UPDATED

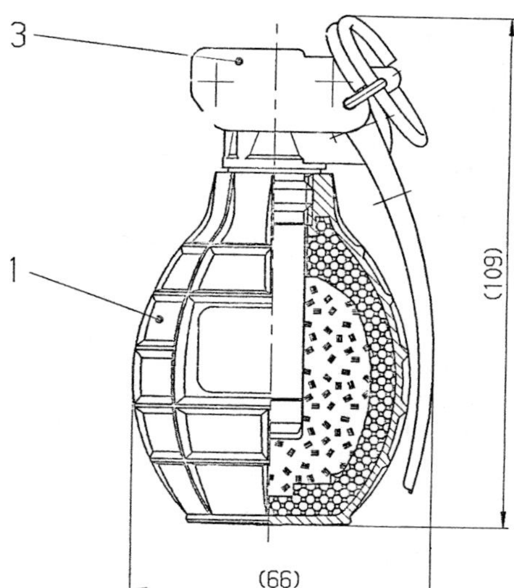

Diehl DM 61 hand grenade
1996

Diehl DM 51 offensive/defensive hand grenade

Description

The Diehl DM 51 dual-purpose grenade consists of two major elements: a high explosive hand grenade body and a fragmentation jacket. The body can be used separately as an offensive grenade and when the jacket is placed around it the two parts together make up a defensive grenade. The parts are connected and held by a bayonet fixing in which the outer jacket is turned through 90° and the base portion of the jacket locks on to the central body. The HE grenade body has the shape of a hexagonal prism. It consists of a watertight plastic container filled with compressed nitropenta. The top end of the container is screw-threaded to take a fuze. The fragmentation surround is cylindrical and is made of plastic, with fragmentation inserts, in the same way as the defensive grenades described above. The two major parts, HE centre and jacket, can be joined and separated as often as required. Several grenade bodies can be joined to make up a cluster charge or connected end to end to make a Bangalore torpedo for assault engineer applications.

The offensive/defensive DM 51 hand grenade is in service with the German Army under the nomenclature 'Handgranate Spreng/Splitter DM 51 mit Handgranaten-Zünder DM 82A1B1'. It is the standard grenade for all units.

Data

Length: fuzed, 107 mm
Diameter: 57 mm
Weight: DM 51 hand grenade, 425 g; DM 82A1B2 fuze, 65 g; fragmentation jacket, 280 g; explosive filling, 60 g
Ball diameter: 2-2.3 mm
Number of balls: 6,500
Danger area: 35 m

Manufacturer

Diehl GmbH & Co, Ammunition Division, Fischbachstrasse 16, D-90552, Röthenbach/Pegnitz.

Status

In production.

Service

German forces.

UPDATED

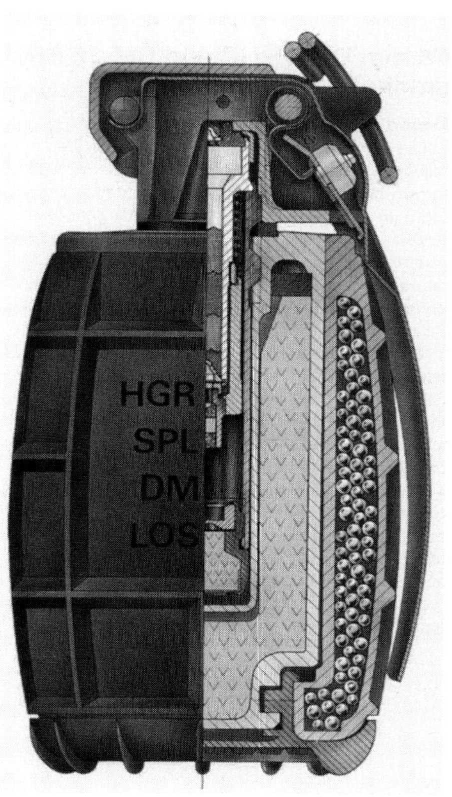

DM 51 offensive/defensive hand grenade

Diehl 76 mm fragmentation grenade

Description

The Diehl fragmentation grenade is designed to be fired from the usual type of 76 mm smoke launcher fitted to most armoured vehicles, though a special launcher unit has been designed and produced by Wegmann and Company for fitting to vehicles requiring it. Since the standard type of launcher is designed to accept the low pressures associated with firing smoke grenades, this fragmentation grenade has a special propelling charge which, whilst maintaining low chamber pressure, ensures regularity of performance and bursts the grenade at the optimum point.

Prior to loading the safety pin **(1)** is removed. On firing, the ignition current flows via contact rings **(2)** and the cable **(3)** to the igniter cap **(4)** which fires the propellant **(5)**. The gas pressure generated by the combustion of the propellant expands through the gas ports **(6)** into the remaining space in the launcher, thus accelerating the grenade to a velocity of about 24 m/s. The initial pressure generated by the propellant shears off a collar **(9)** on the striker **(7)** which then activates the fuze **(8)** and ignites the cap in the pyrotechnic time fuze; this initiates the explosive charge after a flight time of 3.8 seconds. The pyrotechnic fuze gives a delay in arming of 1.3 seconds.

The grenade is intended for use in the close in defence of armoured vehicles, against infantry who are so close that the vehicle weapons cannot be depressed sufficiently to engage them at short range. In addition, it can also be used against personnel in foxholes or behind cover protecting them against direct fire from the vehicle.

Data

Length: 200 mm
Diameter: 76 mm
Weight: 1.7 kg
Filling: RDX/TNT surrounded by 4 mm steel fragments
Arming delay: 1.3 s
Firing delay: 3.8 s
Optimum range: 40 m at 63° elevation
Optimum burst height: 12 m at 63° elevation
Lethal area: ca 40 m² below optimum burst height

Manufacturer

Diehl GmbH & Co, Ammunition Division, Fischbachstrasse 16, D-90552, Röthenbach/Pegnitz.

Status

In production.

VERIFIED

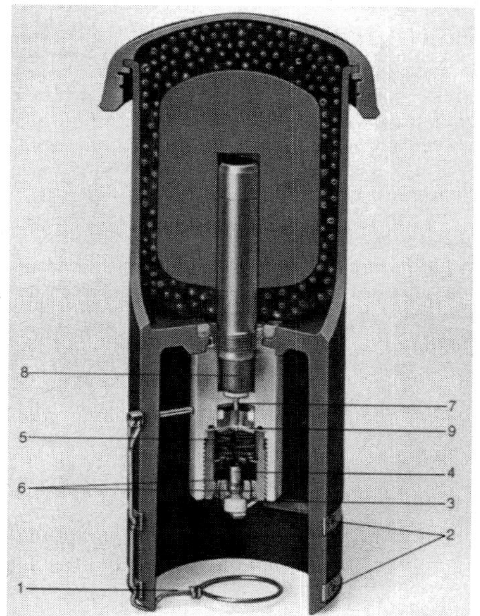

Diehl 76 mm fragmentation grenade for vehicle launching

DM 24 incendiary smoke hand grenade

Description

The DM 24 incendiary smoke hand grenade is the improved successor to the earlier DM 19 and is designed to burst in a brilliant flash on impact and simultaneously produce a cloud of dense smoke. The flash is blinding to the enemy and the smoke induces a severe irritating cough. The incendiary mass burns for about five minutes at a temperature of approximately 1,200°C. This heat ignites any combustible material the burning mass touches.

The grenade is absolutely safe to handle and use because it does not contain an explosive charge. The incendiary mass is only activated by the ignition system. When striking a hard surface or object the plastic body breaks up so that the activated incendiary mass on being exposed to the oxygen of the air instantaneously bursts into a blaze of fire and smoke.

For transport the grenade is packed inside a cylindrical casing.

The training version of the DM 24 has the designation DM 68 and is handled in exactly the same way as the live grenade. The fuze does not have any live elements and, instead of the incendiary charge, it contains an inert lime dust compound for simulation on the target.

Data

Length: 133 mm
Diameter: 67 mm
Weight: 340 g
Filling: 255 g red phosphorus

Manufacturer

Buck Werke GmbH & Company, Postfach 2405, D-83435 Bad Reichenhall.

Status

In production.

Service

Several armies.

VERIFIED

DM 24 and 68 incendiary smoke grenades

76 mm DM 15 HC and DM 35 RP smoke hand grenades

Description

These smoke grenades are designed to be fired from launching tubes on an armoured vehicle as well as thrown by hand, depending on the situation encountered. In either case, some 2.5 seconds after ignition smoke is produced which forms a dense screen preventing the opposing forces from observing any rearrangements of vehicles and personnel or other operation. Smoke emission continues for some 2.5 minutes.

The smoke grenade has two separate ignition systems, the electrical and the mechanical system. Electrical ignition is used with the launching device allowing the grenade to be fired remotely from inside the armoured vehicle. Mechanical ignition is used when throwing the grenade by hand.

The cylindrical, thin-walled, metal body of the DM 15 HC grenade is filled with a smoke charge consisting mainly of hexachloroethane. The DM 35 RP grenade is filled with a red phosphorus composition and has the advantage of producing full cover more rapidly. Both grenades are moisture and shockproof and will function within a temperature range of −40 to +50°C.

Data

Length: DM 15, 175 mm; DM 35, 169.5 mm
Diameter: 76 mm
Weight: DM 15, 1.2 kg; DM 35, 1.1 kg; smoke charge, 880 g
Range of projection under 45°: 40-70 m
Duration of smoke emission: DM 15, 150 + 50 s; DM 35, 120 + 30 s)
Time to full cover: DM 15, 6-8 s; DM 35, 0.5 s)

Manufacturer

Buck Werke GmbH & Company, Postfach 2405, D-83435 Bad Reichenhall.

Status

In production.

Service

German and NATO armies and those of some other countries.

VERIFIED

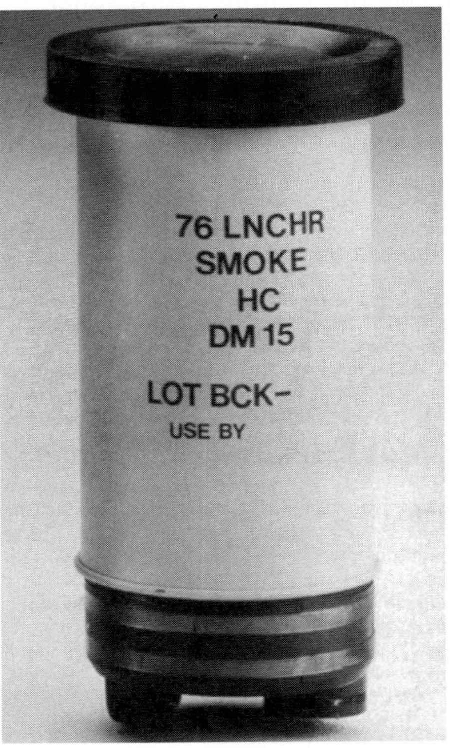

HC smoke grenade

Piepenbrock 76 mm instant tank smoke grenade

Description

The Piepenbrock 76 mm instant tank smoke grenade is suitable for use with 76 mm Wegmann launchers used on many armoured vehicles. It is designed for rapid production of defensive screening smoke, behind which the vehicle can manoeuvre.

The grenade is ignited from inside the vehicle by an electrical remote-control switchboard. After ignition a rocket motor propels the grenade from the launcher; after about 5 m of flight it begins to rotate, which effects a rapid and dispersed distribution of the smoke.

Data

Length: 170 mm
Diameter: 76 mm
Weight: ca 1.95 kg
Emission time: ca 2.5 min
Range: ca 35 m

Manufacturer

Piepenbrock Pyrotechnik GmbH, PO Box 20, D-67306 Göllheim/Pfalz.

Status

Development completed.

VERIFIED

Piepenbrock 76 mm instant tank smoke grenade

Piepenbrock 76 mm infra-red tank screening smoke grenade

Description

The Piepenbrock 76 mm infra-red tank screening tank smoke grenade resembles the instant smoke grenade described previously and is used in the same manner, being fired from the standard 76 mm Wegmann launcher. On firing, however, it is ejected from the launcher by a propelling charge and at about 30 m from the vehicle and a height of approximately 10 m, bursts, releasing a quantity of pellets which fall to the ground and emit smoke. The smoke is effective in screening both optically and across the electro-optical spectrum, so preventing the concealed vehicle being seen by electro-optical surveillance or aiming devices or detected by target-seeking sensors. The smoke itself emits radiation in the thermal band, so confusing thermal imaging sensors.

Data

Length: 170 mm
Diameter: 76 mm
Weight: ca 1.2 kg
Emission time: 30-60 s
Time to full screening: 1-5 s

Manufacturer

Piepenbrock Pyrotechnik GmbH, PO Box 20, D-67306 Göllheim/Pfalz.

Status

Development completed.

VERIFIED

Rheinmetall 40 mm dual-purpose DM 12 grenade

Description

The 40 mm dual-purpose DM 12 grenade has been privately developed by Rheinmetall Industrie GmbH for use in all standard 40 mm grenade launchers. A multipurpose munition, it can be used for the attack of lightly armoured or soft targets. It is described as interoperable with the US M433 HEDP grenade.

The DM 12 cartridge is made up of the grenade with base fuze and main charge, and the cartridge with propelling system. For better fragmentation the main charge comprises a moulded steel sleeve containing a shaped charge, and an aluminium ogive.

A specially shaped copper lining develops the penetrative capacity of the charge. The filling is cyclonite, TNT and wax.

The grenade base is of aluminium alloy with an integrated synthetic driving band, and contains the contact fuze with pyrotechnic self-destruction device. This rapid-acting contact fuze, which allows safe handling and transportation, has an optimum response sensitivity for hard and soft targets and an arming distance of between 9 and 15 m.

The cartridge case is also of aluminium and contains the propelling charge with primer. Rated break points between the case and projectile ensure consistent shot ejection at a specific gas pressure. This provides a consistent muzzle velocity and excellent accuracy, even if different types of launchers are used.

In addition to a fragmentation effect, a shaped charge effect and a reliable self-destruction device, the 40 mm DM 12 offers high functional reliability to a maximum range of about 400 m.

Manufacturer

Rheinmetall Industrie GmbH, Postfach 1663, Pempelfurtstrasse 1, D-40836 Ratingen.

Status

Development completed.

UPDATED

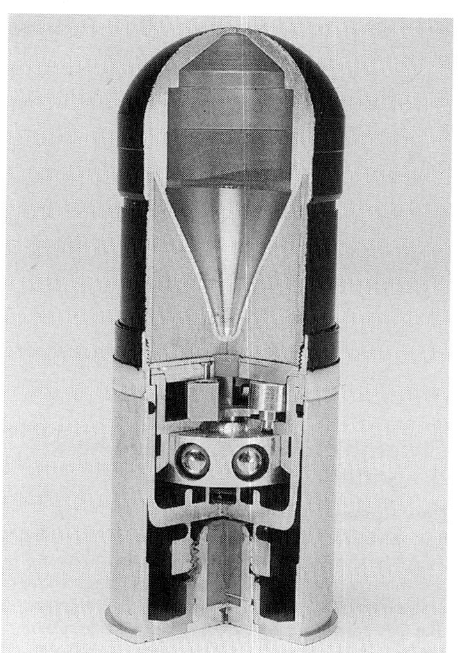

Cutaway example of Rheinmetall 40 mm dual-purpose DM 12 grenade
1996

Nico 40 mm training grenade cartridges

Description

There are two Nico 40 × 46 mm training grenade cartridges, both suitable for use in a wide array of 40 mm grenade launchers. Both employ Nico's patented 40 mm propulsion system which provides predictable accuracy.

One cartridge is a 40 mm TP-T, suitable for many training purposes. The other is a TP with an impact signature provided by the release on impact of a cloud of brightly coloured non toxic powder. This round is suitable for training situations where a tracer would not be suitable or appropriate.

Manufacturer

Nico Pyrotechnik, Hanns-Jürgen Diederichs GmbH & Co KG, Bei der Feuerwerkerei 4, PO Box 1227, D-22946 Trittau.

Status

In production.

Service

Several NATO countries.

NEW ENTRY

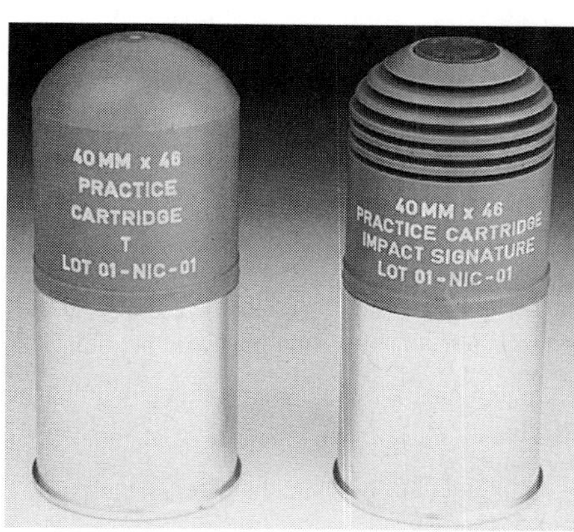

Nico 40 mm training grenade cartridges: (left) *TP-T;* (right) *TP with impact signature*
1996

GREECE

Elviemek EM 01 defensive hand grenade

Description

The defensive model EM 01 has a plastic body containing preformed fragments in a matrix. The size, weight and configuration of the pellets, together with the amount of explosive, give a lethal radius of 15 m. The igniter is largely made of plastic. It has a plastic head containing the metal striker, safety pin and lever and the percussion cap. The delay fuze inside the body is carried in a light metallic tube. The explosive filling is 37 g of 'Pentaplastit' a PETN derivative exclusively produced by Elviemek.

Data

Length: 91 mm
Diameter: 57 mm
Weight: 355 g
Filling: 37 g PETN

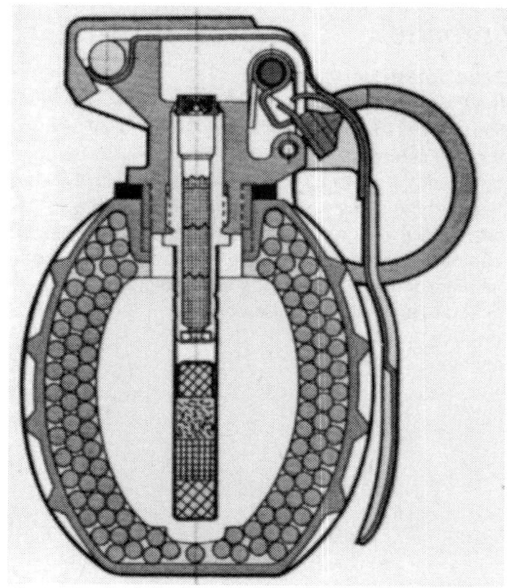

Sectioned drawing of Elviemek EM 01 grenade

Number of pellets: 2,600 ± 50
Diameter of pellets: 2.5 mm
Delay time: nominal, 4 s

Manufacturer
Elviemek SA, Hellenic Explosives and Ammunition Industry, Atrina Centre, 32 Kifissias Avenue, GR-15125 Athens.

Status
In production.

Service
Greek armed forces.

VERIFIED

Elviemek EM 01 defensive hand grenades

Elviemek EM 02 offensive hand grenade

Description
The Elviemek EM 02 offensive hand grenade is designed to produce blast with minimum fragment hazard and is constructed entirely of plastic. It has a danger area of 2 to 3 m. The explosive filling is 'Pentaplastit'. The fuze consists of a plastic housing with metal spring, striker and delay tube, and is held safe by a metal fly-off lever retained by a safety pin. The body of the grenade is smooth so as to differentiate it from the defensive grenade.

Data
Length: 92 mm
Diameter: 57 mm
Weight: 140 g
Filling: 37 g PETN
Delay time: nominal, 4 s

Manufacturer
Elviemek SA, Hellenic Explosives and Ammunition Industry, Atrina Centre, 32 Kifissias Avenue, GR-15125 Athens.

Status
In production.

Service
Greek armed forces.

VERIFIED

Elviemek EM 02 offensive hand grenades

Elviemek EM 03 practice hand grenade

Description
The Elviemek EM 03 practice hand grenade is used to simulate the EM 01 defensive hand grenade. The body is manufactured from cast iron to allow it to be used repeatedly. The EM 03 has a practice fuze with a plastic head, a time delay device, a striker and a metal safety lever. The grenade is handled and thrown in the same manner as the operational grenade but after the nominal time delay of 4 seconds a pyrotechnic charge is ignited to simulate a real detonation.

Data
Length: 91±1 mm
Diameter: 57±1 mm
Weight: 390±20 g
Delay time: 3.5-5 s

Manufacturer
Elviemek SA, Hellenic Explosives and Ammunition Industry, Atrina Centre, 32 Kifissias Avenue, GR-15125 Athens.

Status
In production.

Service
Greek armed forces.

NEW ENTRY

Elviemek EM 03 practice hand grenades
1996

Elviemek EM 04 smoke grenades

Description
The Elviemek EM 04 smoke grenade consists of a metallic cylindrical pot containing the smoke mixture. The igniter system, the same as in the other hand grenade models, is screwed on top of the pot. Elviemek manufactures seven similar models in the following colours: white, red, blue, yellow, green, violet and fog.

Data
Length: 145 mm
Diameter: fog, 66 mm; colours, 60 mm
Weight: fog, 650 g; colours, 450 g
Delay time: nominal, 2.5 s
Emission time: fog, 90-150 s; colours, 100-140 s

Manufacturer
Elviemek SA, Hellenic Explosives and Ammunition Industry, Atrina Centre, 32 Kifissias Avenue, GR-15125 Athens.

Status
In production.

Service
Greek forces.

VERIFIED

Elviemek EM 04 smoke grenades

HUNGARY

M42 hand grenade

Description
The only grenade of Hungarian design in current service is the M42. This is an offensive stick-type hand grenade. It employs a delay fuze. One unusual feature of M42 is the provision of a male thread at the top of the grenade and a female thread at the bottom which permits the junction of several grenades to provide a small demolition charge. There are three 0.5 in (13 mm) red bands around the body.

Data
Length: total, 194 mm; head, 76 mm; stick, 118 mm
Diameter: max, 48 mm
Weight: 310 g
Filling: 134 g TNT
Fuze type: delay
Fuze delay: 3.5-4.5 s
Range: thrown, 30 m

Manufacturer
State arsenals.

Status
In current use.

Service
Believed still in service with the Hungarian armed forces.

VERIFIED

Hungarian M42 grenade with additional head

Hungarian M42 grenade with additional head screwed on

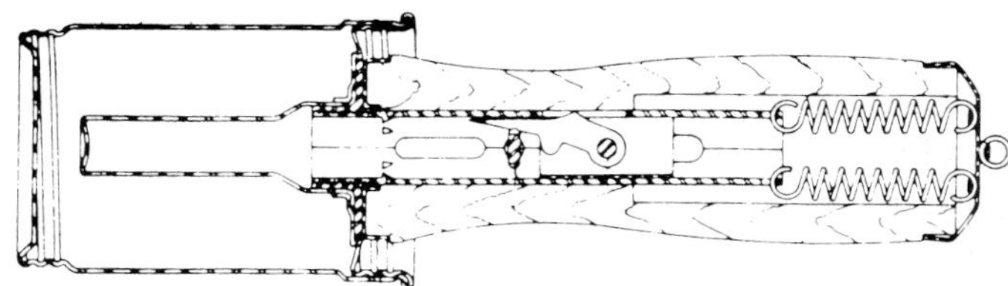

Sectioned view of Hungarian M42 offensive hand grenade, showing percussion fuze mechanism

INDIA

IOF 36M hand and rifle grenades

Description
The 36M grenade is manufactured by the Indian Ordnance Factories (IOF) Ammunition Factory, Khadki, and is offered for export sales. It is the venerable Mills Bomb, the No 36M hand grenade officially introduced into British Army service in 1918, although the earlier No 36 and similar No 23 had been in service for at least a year previously. Production facilities were established in India, probably during the Second World War, and the type remains in production there.

The No 36M grenade can probably be regarded as the prototype of the classic hand grenade. Its Indian counterpart, the 36M, retains the cast iron body with prominent external serrations and utilises a fly-off lever fuze assembly. As the lever flies off a spring-loaded striker pin initiates a pyrotechnic delay train to ignite a detonator assembly which, in turn detonates the main TNT filling. The fuze assembly is screwed into a well on top of the body prior to use while the detonator assembly is inserted into a well in the base before a base plug is screwed into place. The standard fuze, used on the hand grenade, provides a 4 second delay. On detonation the cast iron body breaks up into anything from 40 to 60 major fragments, plus many more of reduced size. The disabling radius is claimed to be 18.3 m. Throwing distance may be as much as 30 m, although this is often reduced to about 23 m for many users.

The Indian Ordnance Factories also produce the 36M as a rifle grenade. Their design owes little to earlier No 36M grenade launching systems as the 36M Mk 1 rifle grenade has a long tube, the 7.62 mm Projector Grenade 1A, fitted to the base. The projector is placed over the muzzle of a 7.62 mm 1A1 assault rifle and launching is accomplished using a Cartridge SA, Rifle Grenade, 7.62 mm HD. Before launch the usual safety pin is removed while the fly-off lever is held in place by a rubber arming ring. The latter is pushed downwards off the lever an instant prior to launch. This is one reason why the 36M rifle grenade is provided with a fuze with a 7-second delay. The range of the 36M rifle grenade is 185 m.

Data

36M HAND GRENADE
Length: ca 100 mm
Max diameter: ca 60 mm
Weight: 773 g
Filling: TNT
Fuze delay: 4 s
Number of fragments: 40-60
Range: thrown, 23-30 m

Manufacturer

Indian Ordnance Factories, Ammunitiôn Factory, Khadki, Pune 411 003.

Status

In production.

Service

Indian armed forces.

NEW ENTRY

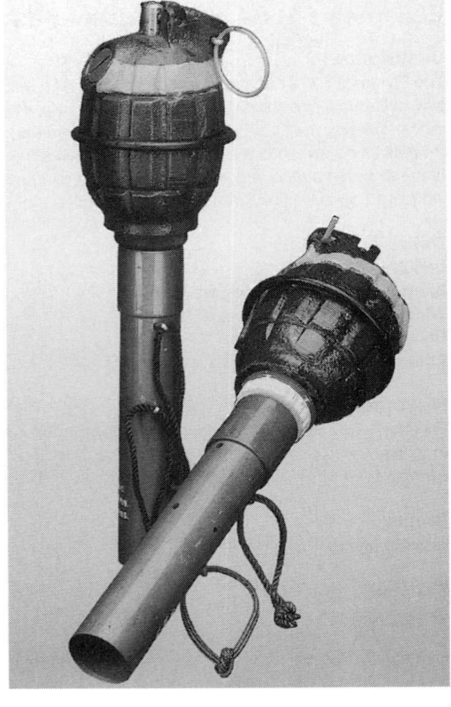

IOF 36M Mk 1 rifle grenades
1996

INDONESIA

GT-5PE A2 hand grenade

Description

The GT-5PE A2 hand grenade is a fragmentation grenade with a serrated cast iron body filled with pressed TNT powder. The grenade employs a No 19 D fly-off lever fuze assembly providing a delay time of 4 to 6 seconds once the fuze lever has been released for throwing. As the grenade detonates it produces approximately 1,050 fragments which are distributed at a density of approximately 7 per m^2 at a radius of 5 m from the point of explosion.

It is claimed that this grenade can be thrown from 30 to 40 m. A projection adaptor allows the grenade to be fired from a rifle.

The GT-5H A2 is a training grenade identical to the operational model but with a smoke or other non-lethal filling to indicate the point of impact. It weighs 380 g and employs a No 19 E fuze assembly.

The GT-D A2 is an inert drill and training grenade with an inert No 19 F fuze assembly.

Data

Length: total, 108 mm
Max diameter: 50 mm
Weight: 450 g
Filling: TNT
Fuze delay: 4-6 s
Number of fragments: ca 1,050
Range: thrown, 30-40 m

Manufacturer

PT Pindad (Persero), Jl Jendral Gatot Subroto, PO Box 807, Bandung 40284.

Status

In production.

Service

Indonesian armed forces.

NEW ENTRY

GT-5PE A2 hand grenades
1996

IRAN

Anti-personnel hand grenade

Description

The exact designation and design origins of the Iranian anti-personnel hand grenade are uncertain but the overall appearance is very similar to that of the Austrian Arges HG 73 defensive grenade and its design derivative the Yugoslav M75. It is a grenade of modern pattern, using an ovoid plastic outer casing and almost certainly contains preformed fragments. If the Arges/M75 design origins are followed the latter are about 3,000 small steel balls packed around the explosive filling which consists of 45 g of Nitropenta. Total weight of the grenade is 430 g. No other data is available.

Manufacturer

Defence Industries Organisation, Ammunition Group, Export Department 14542, Pasdaran Street, PO Box 16765-1835, Teheran 16.

Status

In production.

Service

Iranian armed forces. Offered for export sales.

NEW ENTRY

Iranian anti-personnel hand grenade
1996

ISRAEL

No 5 white smoke hand grenade

Description
The No 5 white smoke hand grenade is used for signalling and also to produce a local smoke screen. The smoke is emitted for slightly less than two minutes and, being of hexachloroethane, has less tendency to pillar than some of the phosphorus smokes. The grenade consists of a tin-plated cylinder and is operated in the conventional fashion with a fly-off lever restrained by a safety pin.

Data
Length: 150 mm
Diameter: 63 mm
Weight: 800 g; mixture, 610 g
Smoke colour: white
Smoke mixture: hexachloroethane
Body material: steel sheet
Functioning: delay fuze, nominal 2 s
Smoke emission: 90-130 s

Manufacturer
TAAS - Israel Industries Limited, PO Box 1044, Ramat Hasharon 47100.

Status
In production.

Service
Israeli forces.

VERIFIED

No 5 white smoke hand grenade

No 14 offensive hand grenade

Description
The No 14 grenade is used by assaulting infantry who need to close with the enemy immediately after the blast and so require the radius of effect of the grenade to be comparatively small without the production of splinters. The grenade operates in the usual way with a fly-off lever and safety pin.

Data
Length: without fuze, 110 mm; with fuze, 135 mm
Diameter: 64 mm
Weight: 325 g
Filling: 200 g TNT flakes
Body material: laminated paper with sheet metal ends
Functioning: delay fuze, 4.5 ± 0.5 s

Manufacturer
TAAS - Israel Industries Limited, PO Box 1044, Ramat Hasharon 47100.

Status
In production.

Service
Israeli forces.

VERIFIED

No 14 offensive hand grenade

M26A2 fragmentation hand grenade

Description
The M26A2 is a copy of the US grenade. It is a fragmentation grenade with a notched coil inside the thin-wall sheet steel body.

Data
Length: 107 mm
Diameter: 61 mm
Weight: 425 g
Filling: 155 g cast Comp B
Body material: sheet steel
Fragmenting material: spirally wound steel coil, prenotched
Number of fragments: 1,000
Functioning: delay fuze, 4.5 ± 0.5 s
Lethality: 50% chance of a hit at 10 m

Manufacturer
TAAS - Israel Industries Limited, PO Box 1044, Ramat Hasharon 47100.

Status
In production.

Service
Israeli forces.

VERIFIED

M26A2 fragmentation hand grenade

No 5 coloured smoke hand grenade

Description

The No 5 coloured smoke hand grenade is employed for ground position indication or ground-to-air indication. It is also used for the production of local red, yellow or green smoke screens.

Data

Length: 150 mm
Diameter: 63 mm
Weight: 480 g; smoke charge, 290 g
Body material: sheet metal
Type of fuze: delayed ignition
Delay time: nominal, 2 s
Smoke emission: 45-85 s

Manufacturer

TAAS - Israel Industries Limited, PO Box 1044, Ramat Hasharon 47100.

Status

In production.

Service

Israeli forces.

VERIFIED

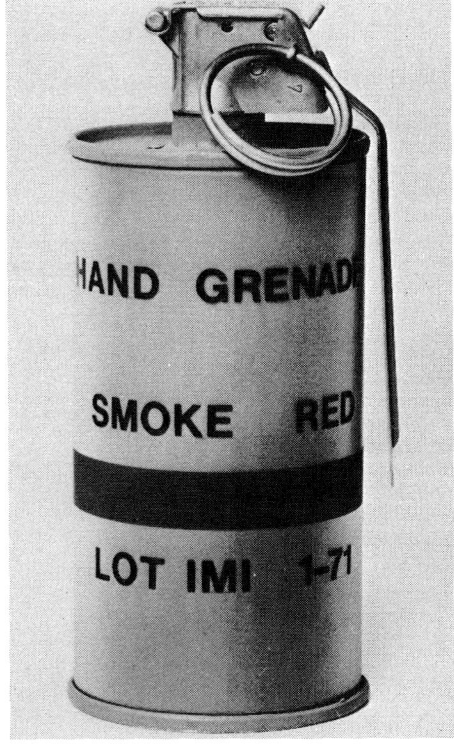

No 5 coloured smoke hand grenade

TAAS bullet trap rifle grenades

Description

The family of TAAS rifle grenades features a bullet trap mechanism allowing the use of any type of 5.56 mm ammunition, including AP, ball and tracer. TAAS bullet trap rifle grenades can be used with all rifles equipped with NATO standard 22 mm flash suppressors.

The grenades are mounted on the rifle barrel, and no adaptors or special mounting devices are required. Aiming is effected by means of a plastic sight which is easily assembled on to the rifle before firing and is discarded after one time use.

The family of bullet-trap grenades includes:
HEAT BT/AT-44
HEAP BT/AP-M1091
ILLUM BT/SGI-50
ILLUM-IR BT/SGIR-50
HC & NT Smoke BT/SGF-40
Instantaneous Smoke BT/RP-80
Incendiary BT/SGIN-50
Incendiary BT/RPIN-80
Ambush ARG/AP50

The above types cover most of the tactical and operational needs of the individual soldier and squad.

These grenades are marketed in the United States by Alliant Techsystems under the general designation of Muzzle Loaded Ordnance (MLO).

In early 1996 it was announcd that the British Ministry of Defence had ordered approximately 50,000 of various types of these grenades.

UPDATED

TAAS BT/AT-44 dual-purpose rifle grenade

Description

The TAAS BT/AT-44 dual-purpose rifle grenade is made in three parts: a fragmentation/penetration warhead, a mechanical fuze and a stabilised tail unit. The warhead contains a shaped charge for armour penetration, with a fragmentation surround for anti-personnel effect. The mechanical fuze gives arming safety and has visual indicator windows permitting the arming state to be checked. The tail unit contains the bullet trap.

Training grenades with the same physical characteristics and ballistic performance are available in two versions.

Data

Length: 403 mm
Diameter: 41 mm
Weight: 490 g
Filling: Comp A5
Range: 280 m
Penetration: steel plate, 160 mm
Lethal radius of fragmentation: 10 m

Manufacturer

TAAS - Israel Industries Limited, PO Box 1044, Ramat Hasharon 47100.

Status

In production.

VERIFIED

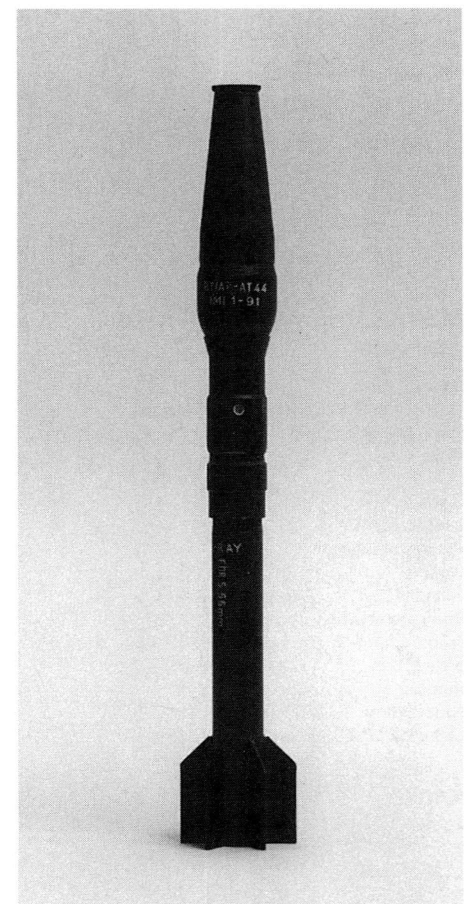

TAAS BT/AT-44 dual-purpose rifle grenade

TAAS BT/AP-M1091 rifle grenade

Description

The TAAS BT/AP-M1091 rifle grenade is for use against troops in the open. It consists of a steel fragmentation warhead with high explosive filling, a tail tube and stabiliser with bullet trap and deflector, and a mechanical fuze with arming safety. It can be fired from 5.56 mm rifles using ball, AP or tracer ammunition, and the tail unit fits the standard 22 mm muzzle launcher. A disposable plastic sight is provided with each grenade.

Data

Length: 320 ±3 mm
Diameter: 41 mm
Weight: 480 g
Filling: pressed Comp A3 Type II
Range: 280 m
Lethal radius: 11 m

Manufacturer

TAAS - Israel Industries Limited, PO Box 1044, Ramat Hasharon 47100.

Status

In production.

VERIFIED

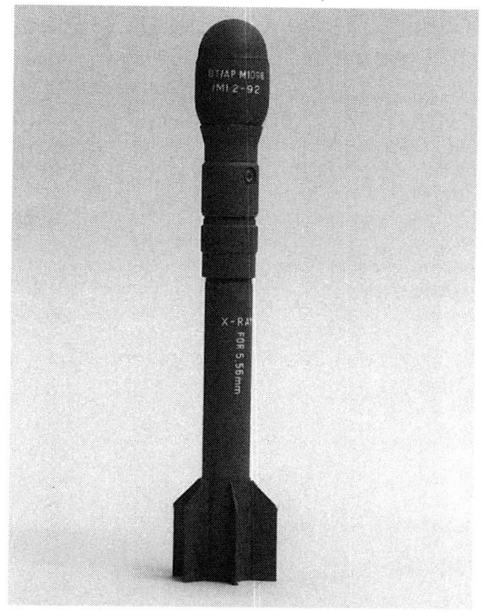

TAAS BT/AP-M1091 rifle grenade

TAAS BT/SGI-50 illuminating rifle grenade

Description

The TAAS BT/SGI-50 grenade consists of a head unit containing a parachute and illumination candle, a pyrotechnic fuze and a stabilised tail unit containing the bullet trap assembly. There is a 3.5 second delay of actuation after firing, after which the fuze ejects the parachute and candle.

The grenade can be launched from 22 mm muzzle launchers on 5.56 mm rifles, using ball, AP or tracer ammunition. A disposable plastic sight is provided with each grenade.

Data

Length: 350 mm
Diameter: 50 mm
Weight: 560 g
Range: 250 m
Illumination: 100,000 candelas for min 30 s

Manufacturer

TAAS - Israel Industries Limited, PO Box 1044, Ramat Hasharon 47100.

Status

In production.

VERIFIED

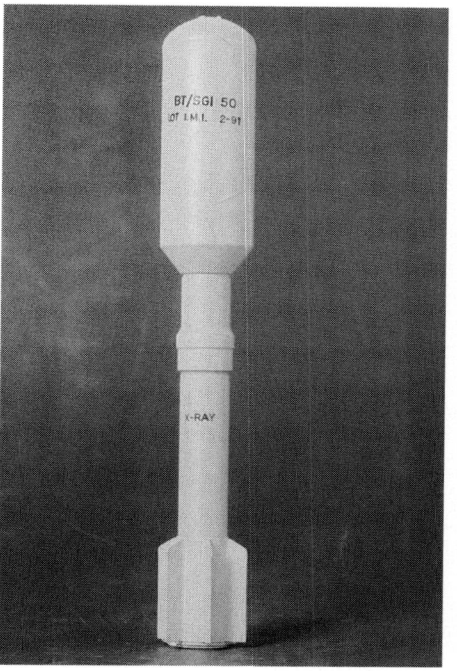

TAAS BT/SGI-50 illuminating rifle grenade

TAAS BT/SGIR-50 illuminating/infra-red rifle grenade

Description

The TAAS BT/SGIR-50 illuminating/infra-red rifle grenade is similar in construction and use to the illuminating grenade BT/SGI-50 rifle grenade described previously but differs in that the candle produces infra-red illumination in the 0.7 to 1.1 μm wavelength.

It can be fired from 5.56 mm assault rifles having the standard 22 mm muzzle dimension, and ball, AP or tracer ammunition can be used. A disposable sight is provided with each grenade.

Data

Length: 350 mm
Diameter: 50 mm
Weight: 560 g
Range: 250 m
Illumination: 100 W/SR, min 30 s
Ratio of visible to infra-red light: 1:50

Manufacturer

TAAS - Israel Industries Limited, PO Box 1044, Ramat Hasharon 47100.

Status

In production.

VERIFIED

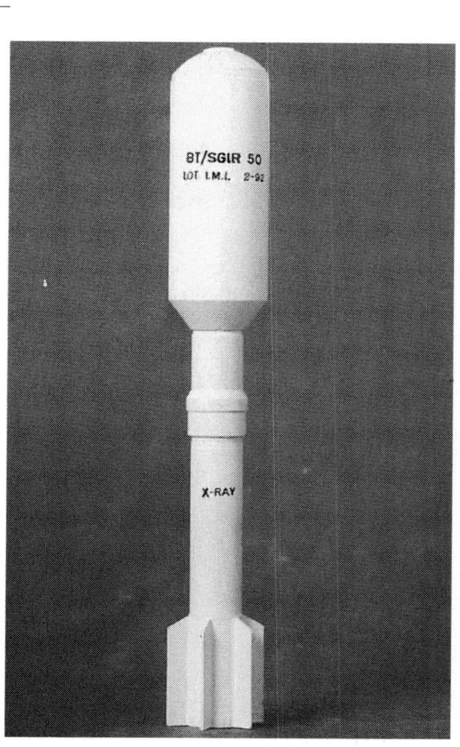

TAAS BT/SGIR-50 illuminating-IR rifle grenade

TAAS BT/SGF-40 HC & NT smoke rifle grenade

Description

The TAAS BT/SGF-40 HC & NT smoke rifle grenade has a smoke generator head, pyrotechnic fuze and bullet trap stabilised tail unit. The absence of mechanical components ensures that no shock or vibration can arm the grenade prior to firing, and the pyrotechnic fuze gives a minimum 3.5 second delay after firing before lighting the smoke generator. The smoke is a non-toxic hexachloroethane type producing a dense white cloud.

Data

Length: 280 mm
Diameter: 50 mm
Weight: 490 g
Range: 300 m
Smoke screen: 10 × 20 m for 70 s

Manufacturer

TAAS - Israel Industries Limited, PO Box 1044, Ramat Hasharon 47100.

Status

In production.

VERIFIED

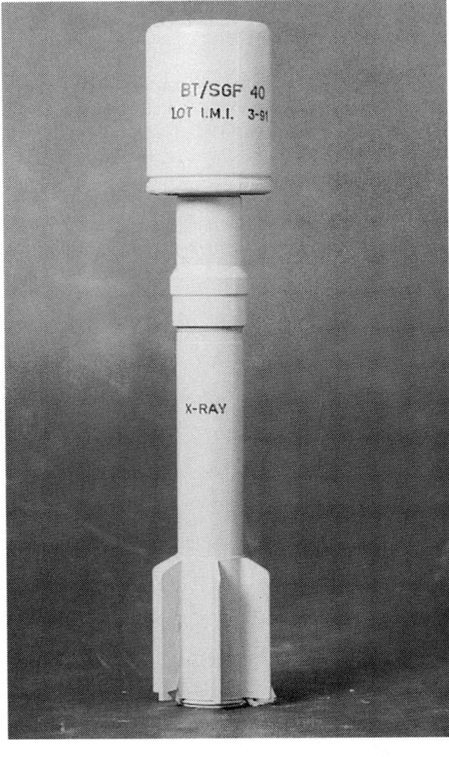

TAAS BT/SGF-40 HC & NT smoke rifle grenade

TAAS BT/RP-80 instantaneous smoke rifle grenade

Description

The TAAS BT/RP-80 instantaneous smoke rifle grenade consists of three components: the smoke/incendiary warhead, a mechanical fuze and the stabilised bullet trap tail unit. The head is filled with a red phosphorus mixture providing both screening smoke and incendiary effects. The fuze is provided with detonation safety and arming safety and also has visible indication windows to permit inspection of the arming state at any time.

Data

Length: 380 mm
Diameter: 56 mm
Weight: 550 g
Range: 250 m
Incendiary effects: 5 m radius
Smoke screen: on impact, 10 × 20 m
Duration of effects: 20 s

Manufacturer

TAAS - Israel Industries Limited, PO Box 1044, Ramat Hasharon 47100.

Status

In production.

VERIFIED

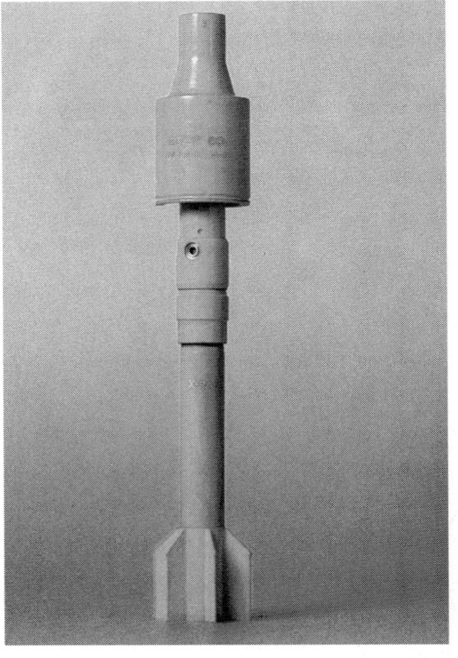

TAAS BT/RP-80 instantaneous smoke rifle grenade

TAAS BT/SGIN-50 incendiary rifle grenade

Description

The TAAS BT/SGIN-50 incendiary rifle grenade is of the usual three-part construction: a warhead which is an incendiary generator, a pyrotechnic fuze and a bullet trap stabilised tail unit. It is intended for special purpose localised incendiary applications on targets, enabling target destruction and incendiary terminal effects on ammunition dumps, fuel depots and command/control installations. Ignition delay is approximately 3.5 seconds after launching.

Data

Length: 350 mm
Diameter: 50 mm
Weight: 560 g
Range: 250 m
Incendiary duration: min, 30 s

Manufacturer

TAAS - Israel Industries Limited, PO Box 1044, Ramat Hasharon 47100.

Status

In production.

VERIFIED

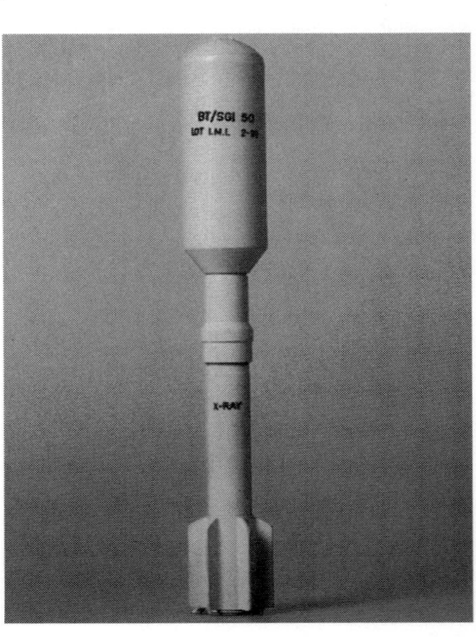

TAAS BT/SGIN-50 incendiary rifle grenade

TAAS BT/RPIN-80 instantaneous incendiary rifle grenade

Description

The TAAS BT/RPIN-80 instantaneous incendiary rifle grenade produces an instantaneous dense cloud of smoke covering an area 10 × 20 m and a significant incendiary terminal effect on impact. It can be used for screening purposes, neutralising bunkers, emplacements and caves, and for marking small targets. The smoke and incendiary effects are generated for a minimum of 20 seconds after the initial explosion.

Data

Length: 380 mm
Diameter: 56 mm
Weight: 550 g
Range: 250 m
Incendiary effects: 7 m radius

Manufacturer

TAAS - Israel Industries Limited, PO Box 1044, Ramat Hasharon 47100.

Status

In production.

VERIFIED

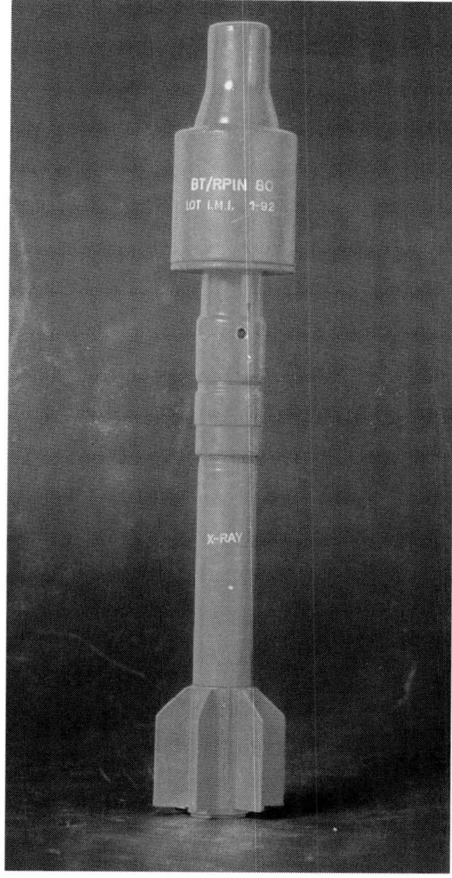

TAAS BT/RPIN-80 instantaneous incendiary grenade

TAAS ARG/AP-50 anti-personnel ambush rifle grenade

Description

The TAAS ARG/AP-50 anti-personnel ambush rifle grenade differs from other rifle grenades in that it does not leave the muzzle of the rifle when discharged; instead, it effectively functions as a rifle-mounted 'Beehive' flechette canister round. Simultaneous fire by three soldiers can cover a 25 m wide lethal area up to a range of 25 m.

The grenade is of conventional cylindrical shape, with a hollow tail which passes over the muzzle of any standard 7.62 or 5.56 mm rifle. It has, however, no fuze or fins. The rifle is then pointed towards the target and fired; the round in the chamber, whether it be ball, tracer or AP, can be fired quite safely. The bullet passes through the grenade body, but the propelling gases cause the grenade's payload of 160 flechettes to be discharged towards the target with considerable velocity. The flechettes disperse through an arc of about 10° from the muzzle of the rifle, have an effective range of 50 m, and can penetrate a 22 mm pine plank or a 1.5 mm mild steel plate.

One advantage of this grenade is that when used by security personnel within a wired compound, it can be fired through the wire fencing; normal grenades would be stopped by the wire. Another advantage is that conventional rounds can be fired immediately after the use of the flechettes.

Data

Length: 275 mm
Diameter: 53 mm
Weight: 640 g
Payload: 160 flechettes
Effective range: up to 50 m
Dispersion: 10° from point of firing
Penetration: pine plank, 22 mm; mild steel plate, 1.5 mm

Manufacturer

TAAS - Israel Industries Limited, PO Box 1044, Ramat Hasharon 47100.

Status

In production.

UPDATED

TAAS ARG/AP-50 anti-personnel ambush rifle grenade

ITALY

MISAR MU-50/G hand grenade

Description

The MISAR MU-50/G is described by its manufacturers as a controlled effects munition. It is a small hand-thrown grenade with a plastic shell containing a matrix of fragments which can be varied in size to obtain the desired effect. Thus, the lethal range can be altered by different sizes of fragment and different quantities and types of explosive.

The grenades are fitted with a silent, flashless and smokeless igniter and delay which leave no trace behind them when thrown. The igniter is activated by the usual spring-loaded striker held down by a safety lever and secured by a pin.

The grenade is quite small and light and this, together with its almost spherical shape makes it easy to throw. The makers claim that it can be thrown with greater precision and to a greater range than any other grenade of the traditional size and shape.

MISAR MU-50/G hand grenade on Franchi SPAS 12 combat shotgun

If required, this grenade can be launched from the muzzle of an assault rifle or Franchi SPAS 12 combat shotgun once an adaptor has been fitted to the grenade.

MISAR also supplies other versions of this grenade with different sizes of fragment so that the lethal range is increased. The MU-50/E is a practice grenade with the same operating characteristics but with no fragments. It produces noise, flash and smoke. The MU-50/I is an inert drill grenade.

Data

Length: 70 mm
Diameter: 46 mm
Weight: 140 g
Filling: 46 g compressed TNT
Lethal range: 5 m
Safety distance: 20 m
Fuze delay: 4 ± 0.5 s
Number of fragments: 1,500 steel balls

Manufacturer

MISAR SpA, SS 236 Goitese, Loc Fascia d'Oro, I-25018 Montichiari, Brescia.

Status

In production, in service.

UPDATED

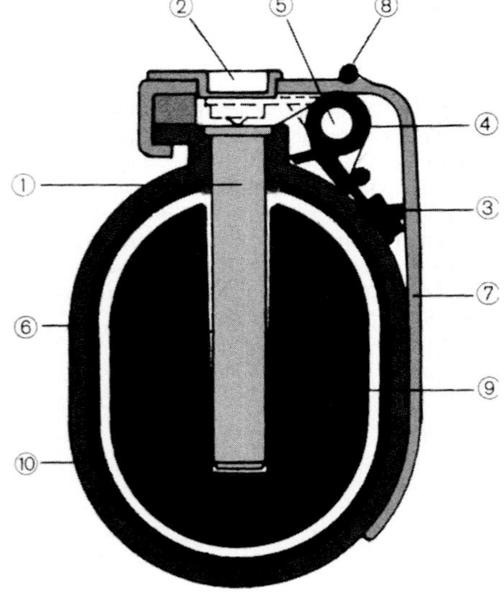

MISAR MU-50/G hand grenade
(1) *detonator* **(2)** *plug* **(3)** *striker* **(4)** *spring* **(5)** *spring rod* **(6)** *grenade body* **(7)** *safety lever* **(8)** *safety* **(9)** *charge* **(10)** *fragmentation*

OD/82 hand grenade

Description

The OD/82 hand grenade may be used as an offensive or defensive grenade without adding to or removing any component parts, since its fragmentation pattern and density is closely controlled by its design.

Data

Length: 83 mm
Diameter: 59 mm
Weight: 286 ± 5 g
Filling: 112 g Comp B
Weight of one fragment: 0.05 g
Safety distance: 20 m

Manufacturer

La Precisa - Stabilimenti di Teano SpA, I-81057 Teano, Caserta.

Status

In production.

VERIFIED

Preparing to throw the OD/82 hand grenade

KOREA, SOUTH

K400 fragmentation hand grenade

Description

The K400 fragmentation hand grenade was developed in South Korea and appears to be based on the American M67 design. The body is steel, internally embossed to give break up for optimum fragmentation and with the usual type of igniter and a K402 fuze screwed into the top. Tests indicate that in excess of 1,300 fragments may be expected from the detonation of one grenade.

Data

Length: 90 mm
Diameter: 60 mm
Weight: 405 g
Filling: 130 g Comp B
Delay: 4-5 s
Lethal radius: 15 m
Range: thrown, 40 m

Manufacturer

Korea Explosives Company Limited, 34 Seosomoon-Dong, Chung-Ku, Seoul.

Status

In production.

Service

South Korean Army.

VERIFIED

K400 fragmentation hand grenade

Smoke hand grenade K-M8

Description

As might be gathered from the nomenclature the K-M8 is based on an old American design. The K-M8 is a screening smoke grenade, filled with HC mixture and fitted with the usual Bouchon igniter set. An aluminium cup under the igniter contains a starter composition and four holes in the top of the grenade, normally tape-sealed, allow the smoke to escape.

Data

Length: 110 mm
Weight: 710 g
Fuze delay: nominal, 1.5 s
Emission time: 105-150 s

Manufacturer

Korea Explosives Company Limited, 34 Seosomoon-Dong, Chung-Ku, Seoul.

Status

In production.

Service

South Korean Army.

VERIFIED

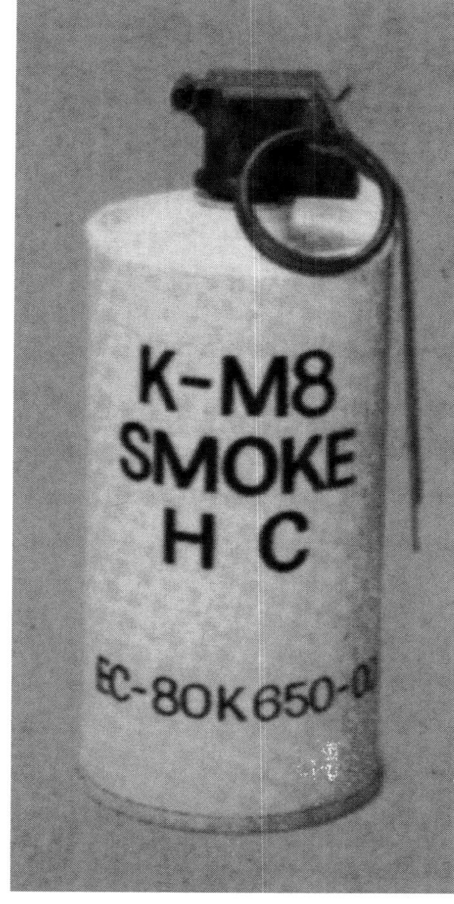

K-M8 smoke grenade

40 mm grenades

Description

Korean 40 mm grenades are based on the American M79/M203 launcher design of low-velocity grenade. They consist of a cartridge case, detonating unit, fuze assembly and sheet-metal ogive. The detonating unit is a spherical grenade body carried in a parallel-walled skirt which, together with the ogive unit, forms a ballistic projectile.

In addition to the high explosive grenade K200 there is also a practice grenade using the same impact fuze K502 and with a smoke pellet to indicate the strike instead of a high explosive bursting charge.

Data

Length: 98.8 mm
Weight: 226 g
Filling: HE, Comp B; TP, plastic ball and smoke pellet
Max range: 400 m
Muzzle velocity: ca 75 m/s

Manufacturer

Korea Explosives Company Limited, 34 Seosomoon-Dong, Chung-Ku, Seoul.

Status

In production.

Service

South Korean Army.

VERIFIED

K200 40 mm fragmentation grenade

40 mm multiple projectile cartridge

Description

The 40 mm multiple projectile cartridge is fired from all standard 40 mm grenade launchers and is intended for close defence and rapid response to ambushes. The cartridge case carries a plastic sabot holding a pellet cup loaded with 20 metal pellets. On firing, the sabot and cup are driven up the bore of the launcher; setback causes the cup to move slightly back in the sabot and thus removes the covering cap. At the muzzle, air resistance causes the sabot and cup to discard, leaving the pellets to fly freely to the target.

Data

Length: 67.2 mm
Weight: round, 115 g; projectile, 24 g
Loading: 20 metal pellets
Effective range: 30 m
Muzzle velocity: 269 m/s

Manufacturer

Korea Explosives Company Limited, 34 Seosomoon-Dong, Chung-Ku, Seoul.

Status

In production.

VERIFIED

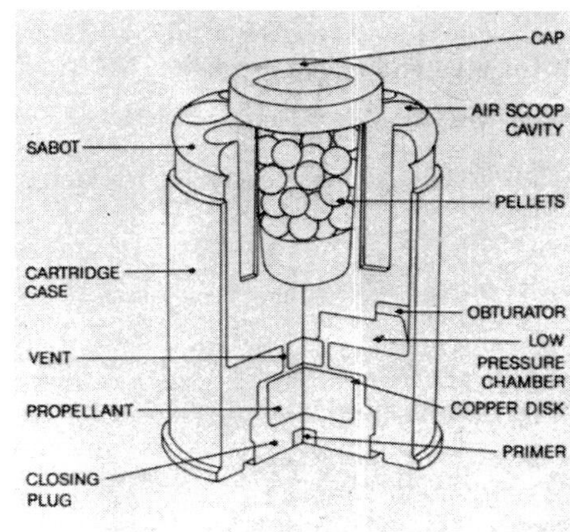

Interior arrangement of 40 mm multiple projectile cartridge

NETHERLANDS

Eurometaal NR20 C1 hand grenade

Description

The Eurometaal NR20 C1 hand grenade, developed for the Royal Netherlands Army, consists of a plastic body with an inner lining of steel balls and an HE charge. It is fitted with the standard 19C3 mechanical fuze with pyrotechnic delay, a member of what has been the standard range of fuzes of the Netherlands Army for some years. The safety pin blocks the lever. This lever in turn blocks the firing pin, which, when released, ignites the percussion primer. The primer then ignites the rest of the explosive train.

During trials it was found that the grenade exploded into about 2,100 fragments at an initial velocity of 1,600 m/s and that it is highly lethal up to a distance of 5 m (almost 100 per cent against personnel in the prone position) and not lethal above 20 m from the point of detonation.

Using targets of mild steel 1.5 mm thick and spruce 25 mm thick the following perforation data were obtained.

Perforations per m² at distance

	3 m	4 m	5 m	20 m
Mild steel 1.5 mm	18	10	6	0
Spruce 25 mm	18	10	6.5	0

Data

Length: 104.5 mm
Diameter: 61 mm
Weight: grenade, with filling, 390 g; fragmentation body, 160 g
Filling: 150 g Comp B

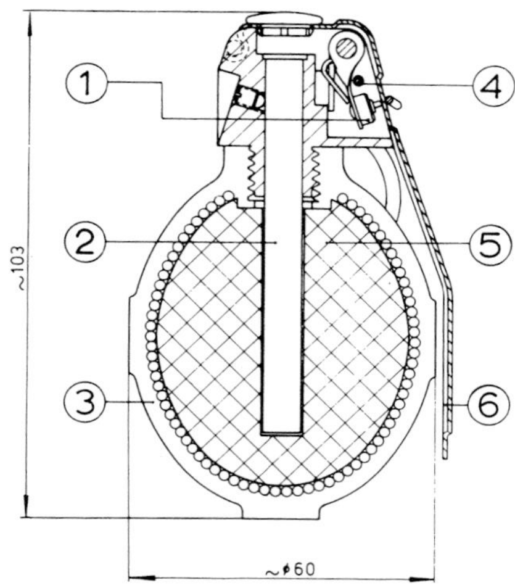

Components of the NR20 C1 Eurometaal hand grenade
(1) *striker* **(2)** *detonator with pyrotechnic delay* **(3)** *fragmentation body* **(4)** *safety pin* **(5)** *explosive filling* **(6)** *safety lever*

Fuze: 19C3
Delay: 3.5 ±0.5 s

Manufacturer

Eurometaal NV, PO Box 419, NL-1500 EK Zaandam.

Status

In production.

Eurometaal NR20 C1 grenade

Service

Royal Netherlands Army.

VERIFIED

PAKISTAN

POF plastic hand grenade

Description

The Pakistan Ordnance Factories (POF) plastic hand grenade is a licence-produced Arges 69 grenade, no longer produced in Austria. The ovoid plastic body contains approximately 3,500 steel balls packed around a central 65 g of plasticised PETN explosive. As the grenade explodes it scatters the steel balls over a lethal radius of 20 m.

The POF plastic hand grenade employs a conventional fly-off fuze assembly providing a time delay of 4 seconds. Each grenade is issued within an individual plastic container.

This grenade is also employed as the warhead for the POF bounding anti-personnel mine.

Data

Length: 115 mm
Diameter: max, 60 mm
Weight: 485 g
Filling: 65 g plasticised PETN
Fuze delay: 4 s
Number of fragments: 3,500 steel balls

Manufacturer

Pakistan Ordnance Factories, Wah Cantt.

Status

In production.

Service

Pakistan armed forces.

NEW ENTRY

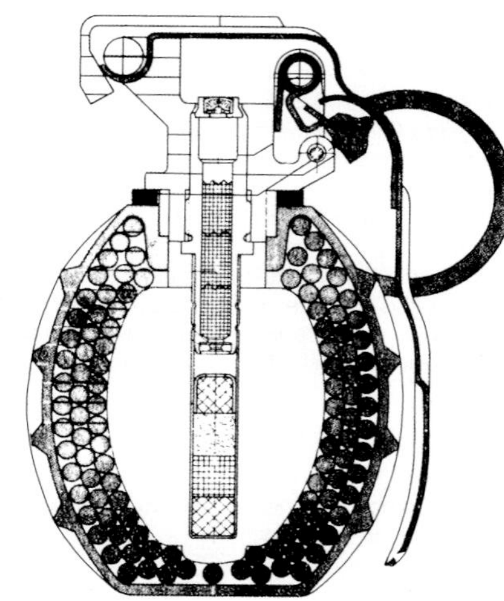

Cross-section drawing of POF plastic hand grenade, the Arges 69
1996

POF metal hand grenade

Description

The Pakistan Ordnance Factories (POF) metal hand grenade is the venerable Mills Bomb, the No 36M Mk 1 hand grenade officially introduced into British Army service in 1918, although the earlier No 36 and similar No 23 had been in service for at least a year before then. Production facilities were established on the Indian subcontinent, probably during the Second World War, and the type remains in production in Pakistan.

The No 36M grenade can probably be regarded as the prototype of the classic hand grenade. The POF version retains the cast iron body with prominent external serrations and utilises a fly-off lever fuze assembly. As the lever flies off a spring-loader striker pin initiates a pyrotechnic delay train to ignite a detonator assembly which, in turn detonates the main TNT filling weighing 66 g.

Prior to use the fuze assembly is screwed into a well on top of the body while the detonator assembly is inserted into a well in the base before a base plug is screwed into place. The fuze provides a 4 second delay. On detonation the cast iron body breaks up into a quoted 25 to 30 major fragments, plus many more of reduced size. The disabling radius is claimed to be 18 m. Throwing distance may be as much as 30 m, although this is often reduced to about 23 m for many users.

Data

Length: 102 mm
Diameter: max, 59 mm
Weight: 765 g
Filling: 66 g TNT
Fuze delay: 4 s
Number of fragments: 25-30
Range: thrown, 23-30 m

Manufacturer

Pakistan Ordnance Factories, Wah Cantt.

Status

In production.

Service

Pakistan armed forces.

NEW ENTRY

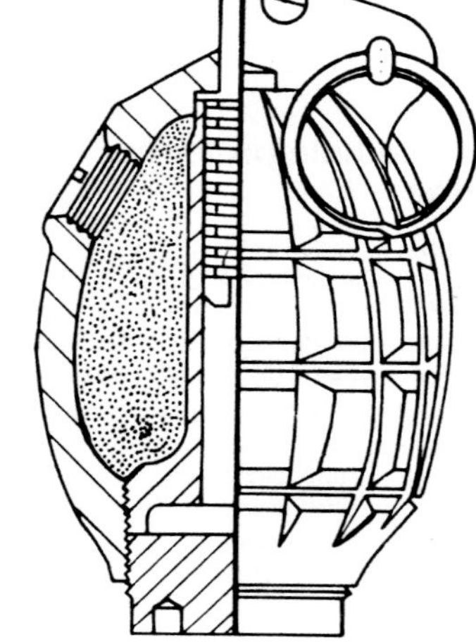

Cutaway drawing of POF metal hand grenade, the No 36M Mk 1
1996

POF 75 mm P-2 Mk 1 HEAT rifle grenade

Description

The Pakistan Ordnance Factories (POF) P-2 Mk 1 HEAT rifle grenade is no longer in production but is still likely to be encountered. It is launched from the muzzle flash hider of a Heckler and Koch G-3 assault rifle (licence manufactured in Pakistan) using a ballistite cartridge. Once launched, a rocket booster motor ignites to maintain the ballistic trajectory. A folding plastic aiming grid on the grenade body is used to aim the grenade to a maximum effective range of 200 m. On target a nose-mounted point detonating fuze sensor functions to ignite the main shaped charge warhead which can penetrate 275 mm of steel plate. The fuze is armed between 2 and 8 m from the muzzle after launch.

Data

Length: 425.4 mm
Diameter: max, 75 mm
Weight: 765 g
Muzzle velocity: 75 m/s
Range: effective, 200 m
Penetration: steel plate, 275 mm

Manufacturer

Pakistan Ordnance Factories, Wah Cantt.

Status

Production complete.

Service

Pakistan armed forces.

NEW ENTRY

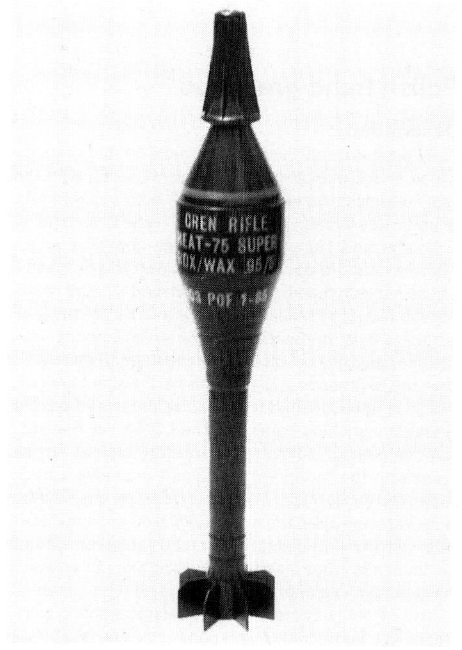

POF 75 mm P-2 Mk 1 HEAT rifle grenade
1996

POF WP P-1 smoke grenade

Description

The Pakistan Ordnance Factories (POF) WP P-1 smoke grenade is used to create smoke screens. Its cylindrical sheet metal body contains white phosphorus (WP) which is ignited using an igniter cap actuated following a 4 second delay initiated by a conventional fly-off lever fuze mechanism. The lever is released as the grenade is thrown. This grenade is of the non-bursting type.

Data

Length: 150 mm
Diameter: max, 62 mm
Weight: 460 g
Filling: WP
Fuze delay: 4 s

Manufacturer

Pakistan Ordnance Factories, Wah Cantt.

Status

In production.

Service

Pakistan armed forces.

NEW ENTRY

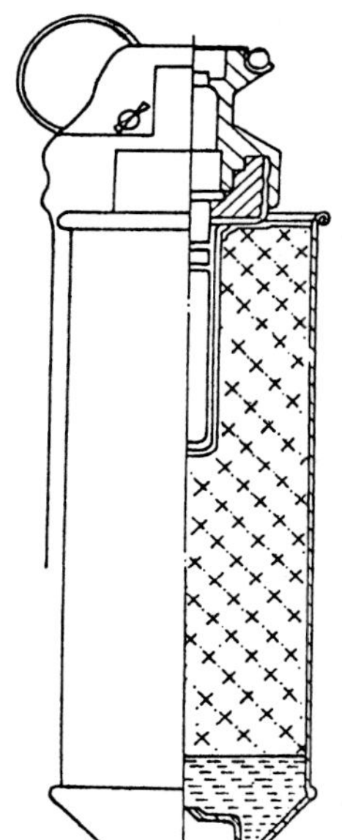

Cutaway drawing of POF WP P-1 smoke grenade
1996

POF target indication grenades

Description

The Pakistan Ordnance Factories (POF) target indication grenades all employ standard cylindrical sheet metal bodies and differ only in the colour of smoke produced. All have a standard fly-off lever fuze assembly producing a 4 second delay. Once the fuze has ignited the coloured smoke composition, smoke is emitted through two holes in the top of the body for between 90 and 120 seconds.

The following POF target indication grenades are produced:

White, P4 Mk 1
Blue, P5 Mk 1
Green, P6 Mk 1
Orange, P7 Mk 1
Red, P8 Mk 1
Yellow, P9 Mk 1.

Data

Length: 127 mm
Diameter: max, 65 mm
Weight: 1 kg
Fuze delay: 4 s
Emission time: 90-120 s

Manufacturer

Pakistan Ordnance Factories, Wah Cantt.

Status

In production.

Service

Pakistan armed forces.

NEW ENTRY

POF target indication grenade
1996

POLAND

Polish hand grenades

Description

The Polish Army still uses the Soviet F-1 hand grenade which is still manufactured in Poland, along with the RGO defensive hand grenade, with both marketed for export sales by the Cenzin Foreign Trade Enterprise.

Poland and Hungary are the only former Warsaw Pact countries to equip their forces with rifle grenades. These are launched from the Polish model of the AK-47 assault rifle, the PMK-DGN-60. The muzzle of the rifle is coned to take the Polish LON-1 grenade launcher and the rifle has a gas cut-off to ensure that the gas energy is devoted to grenade launching and at the same time the piston is not subjected to excessive pressure from the grenade launching cartridge. The LON-1 has an external diameter of 20 mm and therefore should not be used with the more normal 22 mm grenades usually projected from rifles. The two grenades, the F1/N60 anti-personnel grenade and the PGN-60 anti-tank grenade, with their internal diameter tubes of 20 mm cannot be launched from any Western Bloc rifle, all of which have 22 mm launchers.

The F1/N60 rifle grenade is a Polish adaptation of the Soviet F-1 fragmentation grenade (qv). The delay fuze has been replaced with an impact fuze and an unfinned stabilising boom has been added. The sights of the PMK-DGN-60 rifle show ranges of 100-240 m.

The PGN-60 HEAT anti-tank rifle grenade is of Polish design and the shaped charge warhead has a finned stabilising boom, and is employed out to ranges of 100 m.

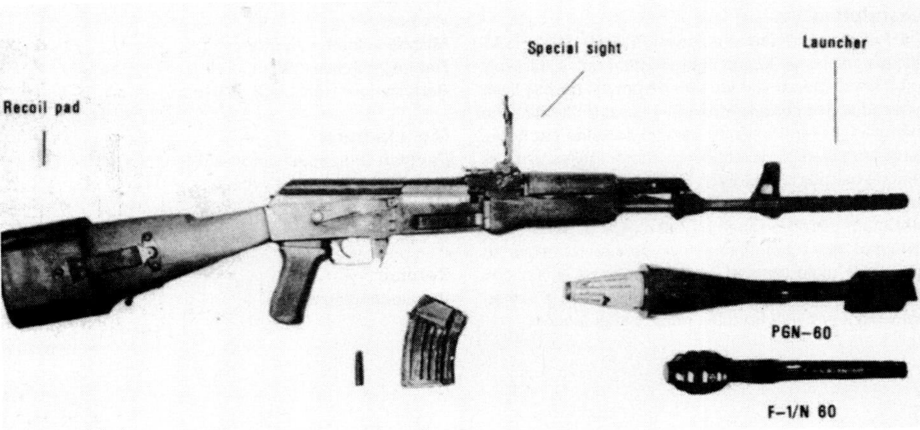

Polish PMK-DGN-60 rifle with associated grenades, PGN-60 and F1/N60

Data

	F1/N60	PGN-60
Length	270 mm	405 mm
Diameter	55 mm	67.5 mm
Weight, fuzed	632 g	580 g
Weight of HE filler	45 g	218 g
Fuze type	impact	impact
Max range	40 m	100 m
Fragment radius	15-20 m	–
Penetration	–	100 mm

Status

In current use.

Service

Polish armed forces.

UPDATED

Dezamet rifle grenades

Description

These grenades, manufactured by Dezamet, are intended to be launched from AKM pattern rifles using ballistite cartridges. The following types are available:

AP/AV GNPO. This grenade has a 40 mm calibre dual purpose warhead with a shaped charge capable of penetrating 100 mm of steel plate. The warhead also fragments for anti-personnel effects.

Incendiary NGZ. This grenade has a 52 mm calibre warhead containing an incendiary composition.

Smoke NGD. This grenade has a 52 mm calibre warhead containing smoke composition.

Illuminating NGOs. This grenade can illuminate an area at a maximum range of 150 m from the point of launch and from a height of 100 m. Body calibre is 43 mm.

Manufacturer

Zaklady Metalowe Dezamet, 39-460 Nowa Dęba, woj: Tarnobrzeg.

Status

In production.

Service

Polish Army and export.

The complete range of Dezamet grenades with the 40 mm grenades in the foreground together with an RGO hand grenade; in the background are the Dezamet rifle grenades

1996

Data

Type	AP/AV	Incendiary	Smoke	Ill
Designation	GNPO	NGZ	NGD	NGOs
Calibre	40 mm	52 mm	52 mm	43 mm
Length	275 mm	220 mm	218 mm	310 mm
Muzzle velocity	58 m/s	58 m/s	58 m/s	58 m/s
Max range	240 m	200 m	200 m	150 m

NEW ENTRY

Dezamet 40 mm grenades

Description

These grenades, manufactured by Dezamet, are the standard low-velocity 40 mm type intended to be fired from the PALLAD launchers; they can also be fired from any other 40 mm grenade launcher. Maximum range of

Data

Type	Frag	Incendiary	PRAC	PRAC
Designation	NGO	NGB	NGZ	NGC
Calibre	40 mm	40 mm	40 mm	40 mm
Weight	250 g	220 g	250 g	250 g
Length	102.7 mm	102.7 mm	102 mm	102 mm
Muzzle velocity	78 m/s	78 m/s	78 m/s	78 m/s
Arming distance	15 m	15 m	n/a	n/a
Max range	400 m	400 m	400 m	400 m

these grenades is 400 m. The following types are available:

- Fragmentation NGO
- Incendiary NGB
- Practice Marker NGZ
- Practice Inert NGC.

Manufacturer
Zaklady Metalowe Dezamet, 39-460 Nowa Dęba, woj: Tarnobrzeg.

Status
In production.

Service
Polish Army and export.

UPDATED

PORTUGAL

M312 fragmentation hand grenade

Description

The M312 fragmentation hand grenade is similar in design to the US M26 grenade, a thin sheet metal body lined with a notched wire fragmentation sleeve. There are certain detail differences in the shape of the detonator pocket and in the base plug being lined with a prefragmented plate. Ignition is by the usual type of fly-off lever percussion unit. Average fragmentation over a 20 m radius is 756 perforations of plywood screens set at varying distances.

Data

Weight: 450 g; fragmentation sleeve, 150 g
Filling: 150 g Comp B
Number of fragments: ca 900
Delay time: 4-5 s
Effective radius: 10.5 m
Safety radius: 30 m

Manufacturer

Sociedade Portugesa de Explosivos SA, av Infante Santo 76, 5°, P-1300 Lisbon.

Status

In production.

VERIFIED

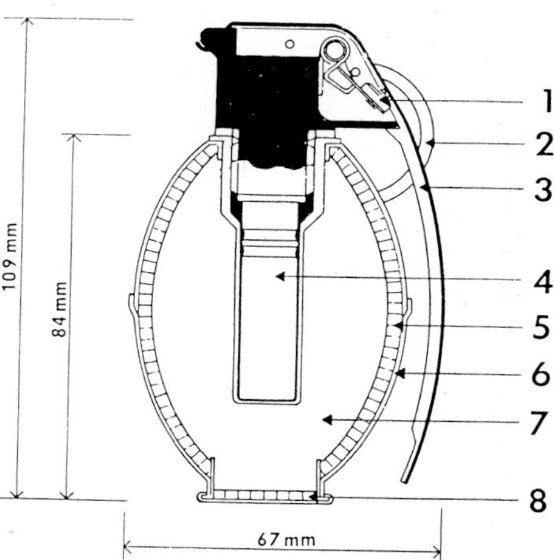

M312 fragmentation grenade
(1) *fuze* **(2)** *safety pin* **(3)** *safety lever* **(4)** *detonator* **(5)** *fragmentation sleeve* **(6)** *metal body* **(7)** *explosive filling* **(8)** *plug*

M313 fragmentation hand grenade

Description

The M313 grenade uses a plastic body inside which is a double layer of steel balls packed into a resin matrix surrounding the plastic explosive filling. There is the usual threaded detonator well, into which a conventional fly-off lever percussion igniter set is screwed. Average fragmentation of the grenade produces over 900 perforations in plywood screens arranged at distances from 2 to 15 m from the point of burst.

Data

Weight: 370 g; fragmentation liner, 230 g
Filling: 51 g plastic HE
Number of fragments: ca 1,800
Delay time: 4-5 s
Lethal radius: 9 m
Safety radius: 20 m

Manufacturer

Sociedade Portugesa de Explosivos SA, av Infante Santo 76, 5°, P-1300 Lisbon.

Status

In production.

VERIFIED

M313 fragmentation grenade
(1) *fuze* **(2)** *safety pin* **(3)** *safety lever* **(4)** *detonator* **(5)** *fragmentation liner* **(6)** *explosive filling* **(7)** *plastic body*

M314 fragmentation hand grenade

Description

The M314 is of similar design to the M313 (previously) but larger and containing a heavier explosive charge and more fragments, a third layer of smaller steel balls being incorporated into the liner. As a ready means of identification this grenade has a smooth exterior finish, whereas the M313 has a roughened finish.

Data

Length: 103 mm
Diameter: 74 mm
Weight: 575 g; fragmentation liner, 400 g; explosive filling, 58 g
Number of fragments: ca 6,500
Delay time: 4-5 s
Lethal radius: 8 m
Safety radius: 20 m

Manufacturer

Sociedade Portugesa de Explosivos SA, av Infante Santo 76, 5°, P-1300 Lisbon.

Status

In production.

VERIFIED

SINGAPORE

SFG 87 fragmentation hand grenade

Description

The SFG 87 is a prefragmented hand grenade, the body of which is of high strength plastic material. The filling is 80 g of Composition B (RDX/TNT) embedded with 2,200 steel balls. The lethal radius is 10 m, the safety radius 25 m. Initiation is by the conventional fly-off lever percussion mechanism.

Data

Length: 92 mm
Diameter: 56 mm
Weight: 300 g
Filling: 80 g Comp B
Fragments: ca 2,200 2 mm diameter steel balls
Delay time: nominal, 4.5 s
Lethal radius: 10 m
Safety radius: 25 m

Manufacturer

Chartered Ammunition Industries (A subsidiary of Chartered Industries of Singapore), 249 Jalan Boon Lay, Singapore 2261.

Status

In production.

UPDATED

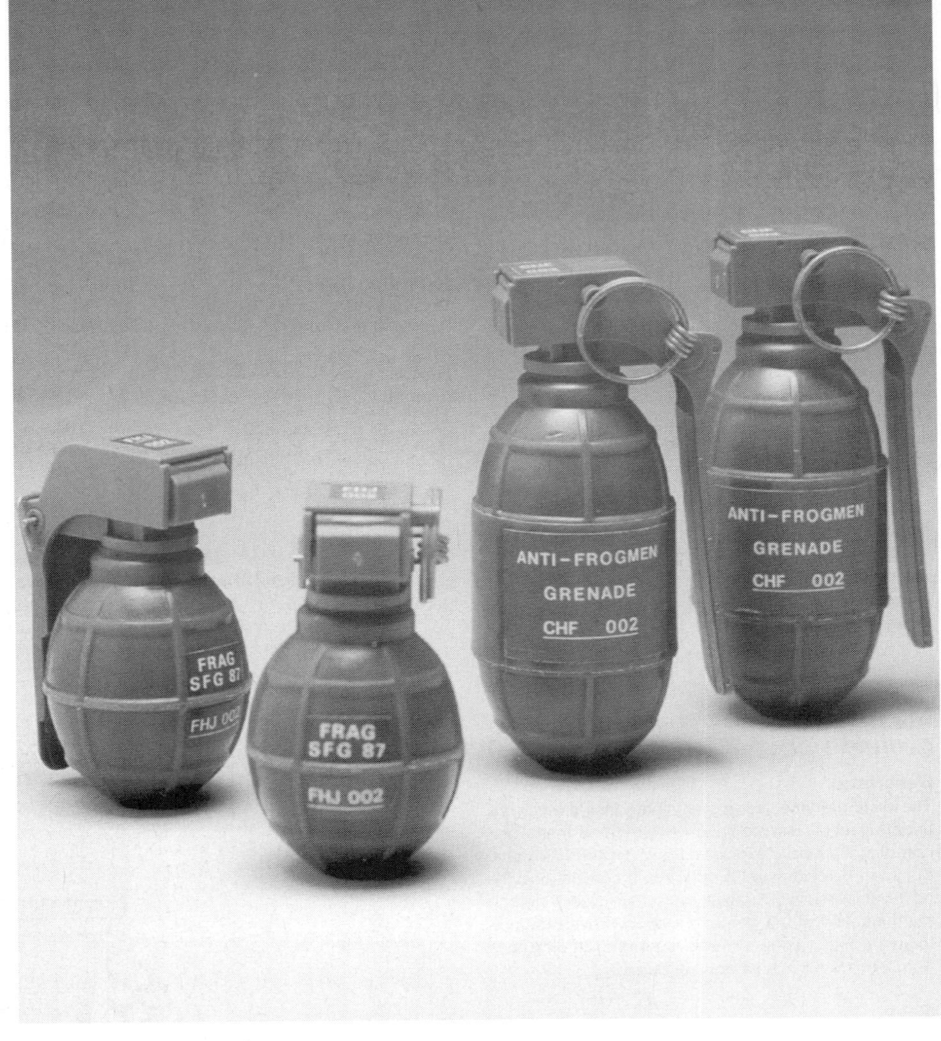

SFG 87 fragmentation (left) *and AFG anti-frogmen grenades*

Anti-frogmen grenade (AFG)

Description

The Anti-Frogmen Grenade (AFG) is a powerful blast grenade made from high strength cross-linked plastic and filled with 180 g of Composition B (RDX/TNT) to develop a powerful blast overpressure in water. The fuze and grenade are entirely watertight and the grenade will detonate at a depth of between 3 and 8 m below the surface. The safety radius is 30 m.

Data

Length: 138 mm
Diameter: 54 mm
Weight: 300 g
Filling: 180 g Comp B
Delay time: nominal, 4.5 s
Lethal radius: 10 m
Safety radius: 30 m

Manufacturer

Chartered Ammunition Industries (A subsidiary of Chartered Industries of Singapore), 249 Jalan Boon Lay, Singapore 2261.

Status

In production.

UPDATED

M8 screening smoke grenade

Description

The M8 screening smoke grenade is a conventional hexachloroethane combustion smoke grenade of cylindrical shape, fitted with a fly-off lever igniter. The design gives excellent waterproofing together with a very rapid development of dense grey smoke.

Data

Length: 141 mm
Diameter: 65 mm
Weight: 630 g
Delay: 5 s or as required
Emission time: 130 s or as required

Manufacturer

Chartered Ammunition Industries (A subsidiary of Chartered Industries of Singapore), 249 Jalan Boon Lay, Singapore 2261.

Status

In production.

VERIFIED

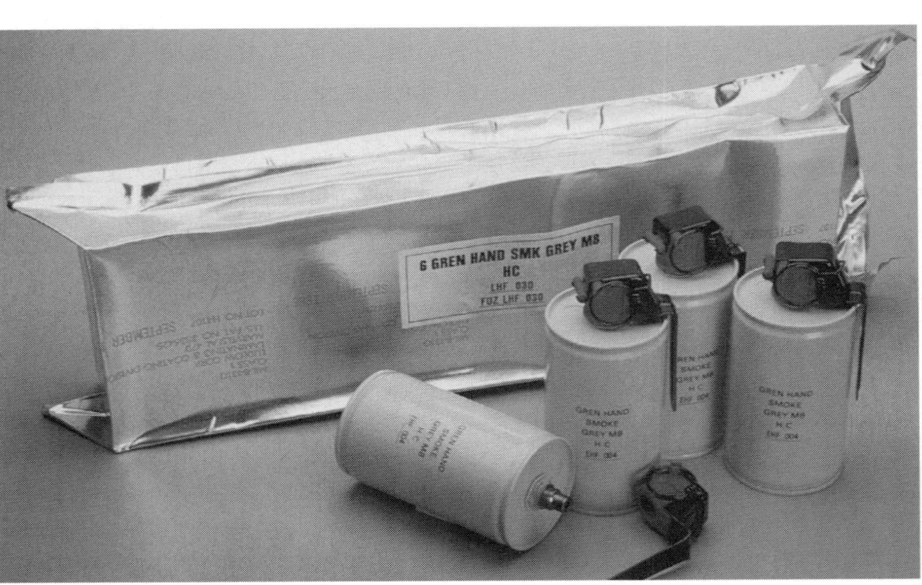

M8 screening smoke grenade

SSG 86 smoke/incendiary grenade

Description

The SSG 86 grenade is designed to produce an instantaneous smoke screen accompanied by burning particles dispersed over a radius of 20 m. These burning particles will ignite any inflammable materials such as clothing, vegetation or wood. The grenade body, fuze and fly-off lever are of high strength plastic material; the body is ribbed and of a size to give a positive grip.

Data

Length: 118 mm
Diameter: 55 mm
Weight: 320 g
Filling: 240 g red phosphorus
Fuze delay: nominal, 4.5 s
Smoke duration: 40 s
Safety radius: 25 m

Manufacturer

Chartered Ammunition Industries (A subsidiary of Chartered Industries of Singapore), 249 Jalan Boon Lay, Singapore 2261.

Status

In production.

UPDATED

SSG 86 smoke/incendiary grenades

HEDP cartridge, low-velocity, 40 mm, S401A

Description

The HEDP S401A is a dual-purpose, shaped charge and fragmentation, explosive grenade for use in low-velocity 40 mm launchers of the CIS 40 GL, M79/M203 and other approved launchers. It incorporates a Point Initiating, Base Detonating (PIBD) fuze capable of penetrating a minimum of 63 mm of mild steel.

As a training aid, the S401 HEDP round is partnered by the Target Practice S406 round which is ballistically similar but disperses an orange dye on impact.

Data

Calibre: 40 mm
Length: 109 mm
Weight: 240 g
Filling: Comp A5
Arming distance: 14 m
Muzzle velocity: 75 m/s
Max range: 400 m
Penetration: 63 mm MS plate at normal impact
Operating temperature range: −20 to +50°C

Manufacturer

Chartered Ammunition Industries (A subsidiary of Chartered Industries of Singapore), 249 Jalan Boon Lay, Singapore 2261.

40 mm HEDP S401A and TP S406 grenades

Status

In production.

UPDATED

HEDP cartridge, high-velocity, 40 mm, S411

Description

The HEDP S411 is a dual-purpose, shaped charge and fragmentation explosive grenade for use in high-velocity automatic grenade launchers such as the CIS 40 AGL, US MK 19 and other approved launchers.

The HEDP S411 projectile, fitted with a Point Initiating, Base Detonating (PIBD) fuze is capable of penetrating a minimum of 50 mm of Rolled Homogeneous Armour (RHA).

Other rounds from the 40 mm high-velocity family include:

HE S412: A high explosive grenade for anti-personnel use.

TP-T S415: A target practice grenade with tracer element, employed for training purposes where the tracer is required for discerning the trajectory.

TP S416A: A target practice grenade for training purposes.

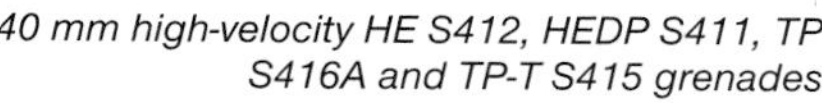

40 mm high-velocity HE S412, HEDP S411, TP S416A and TP-T S415 grenades

Data
Calibre: 40 mm
Length: 112 mm
Weight: 350 g
Filling: Comp A5
Arming distance: 14-40 m
Muzzle velocity: 242 m/s
Max range: 2,200 m

Manufacturer
Chartered Ammunition Industries (A subsidiary of Chartered Industries of Singapore), 249 Jalan Boon Lay, Singapore 2261.

Status
In production.

Licence production
INDONESIA
Manufacturer: PT Pindad (Persero)
Type: Grant 40 mm Kecepatan Tinggi
Remarks: Standard specifications. For licence-produced CIS 40 mm 40-AGL.

UPDATED

SOUTH AFRICA

M26 HE Fragmentation hand grenade

Description
This is a copy of the American M26 grenade, using the same construction of sheet metal body and notched wire fragmentation unit, with an RDX/TNT filling. The fuze unit contains a striker mechanism and 4.5 second pyrotechnic delay, actuated by the usual type of pin and fly-off lever. On detonation the fragmentation unit produces about 1,000 fragments, each with an average weight of 200 mg.

Inert drill grenades and reusable practice grenades are also available.

Data
Length: 113 mm
Diameter: 60 mm
Weight: fuzed, 465 g
Filling: 160 g RDX/TNT
Delay time: 4.5 s
Fragments: 1,000+
Lethal radius: with 50% lethality, 15 m

Manufacturer
Swartklip Products, a Division of Denel (Pty) Limited, PO Box 977, Cape Town 8000.

Status
In production.

Service
South African National Defence Force.

UPDATED

M26 HE Fragmentation hand grenade

M791 anti-personnel rifle grenade

Description
The M791 anti-personnel rifle grenade is the body of the M26 hand grenade attached to a special base fuze unit and tail unit. The fuze unit contains a pyrotechnic delay train which is initiated by setback on firing and ensures that arming only takes place at a safe distance from the rifle (15 m). Until this point the firing pin is securely locked and the grenade cannot explode should it strike any object. Once armed, detonation takes place on impact. The fuze is exceptionally sensitive and detonation will take place even on soft ground or at slight glancing angles of impact. A visual indicator is provided to show whether the grenade is in the safe or armed condition.

The tail unit contains a bullet deflector so that the grenade will fail safe should it be fired with a ball round instead of the correct ballistite cartridge. In such a case the grenade will not explode, but will travel a few metres from the rifle before falling to the ground with the arming device locked in the safe position.

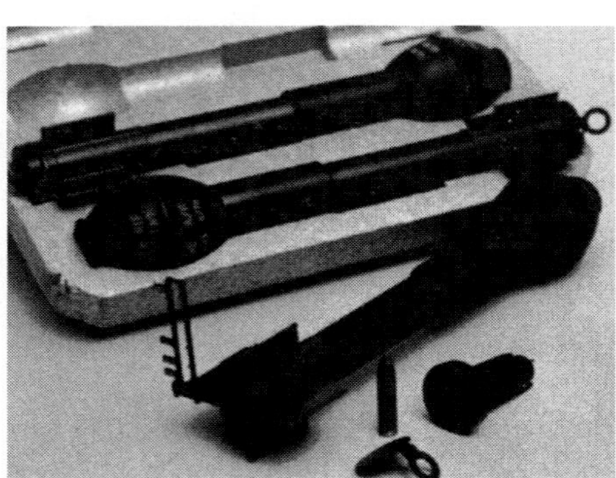
M791 anti-personnel rifle grenade

The grenade can be fired from any suitable 5.56 or 7.62 mm rifle, and is provided with a disposable plastic sight.

Data
Length: 297 mm
Weight: 610 g
Filling: 150 g RDX/TNT
Fragments: 1,000+
Max range: 5.56 mm rifle, 210 m; 7.62 mm rifle, 300 m
Arming distance: 15 m
Lethal radius: with 50% lethality, 15 m

Manufacturer
Swartklip Products, a Division of Denel (Pty) Limited, PO Box 977, Cape Town 8000.

Status
In service.

Service
South African National Defence Force.

UPDATED

Grenade, Rifle, Red Phosphorus

Description
This grenade consists of a warhead, an arming mechanism, and a launcher tube which fits over the rifle barrel. The warhead is a tinplate body containing granular semi-consolidated red phosphorus composition, which is dispersed and ignited when the grenade is detonated. The arming mechanism is located between the warhead and the launcher tube. A pyrotechnic train initiated on setback ensures that arming only occurs at a safe distance from the rifle. This device is very sensitive, and the grenade, when armed, will function even when impacting on soft ground or at glancing angles.

The launcher tube incorporates a bullet deflector so that the grenade will fail safe if a live round is accidentally used to discharge it.

The grenade is intended as an incendiary device in the anti-personnel role. It can be used defensively when screening smoke is required or offensively when the acrid smoke or incendiary effect can be used for bunker or house clearing operations.

Data
Length: 353 mm
Weight: 610 g
Filling: 260 g red phosphorus
Fragments: average, 3,000
Max range: 5.56 mm rifle, 210 m; 7.62 mm rifle, 300 m
Arming distance: min, 15 m
Spreading area: 500 m^2

Manufacturer
Swartklip Products, a Division of Denel (Pty) Limited, PO Box 977, Cape Town 8000.

Status
In production.

Service
South African National Defence Force.

VERIFIED

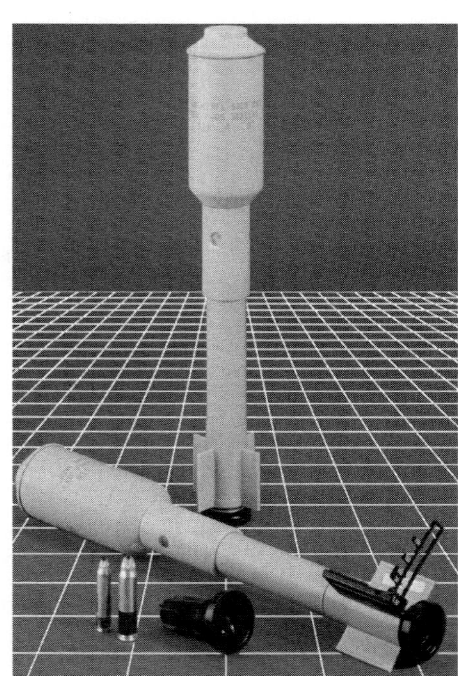
Grenade, Rifle, Red Phosphorus

Grenade, Hand, Red Phosphorus

Description

This grenade consists of a cylindrical tinplate body filled with semi-consolidated red phosphorus composition granules. The spring-loaded striker mechanism is of the fly-off lever type and contains an integral pyrotechnic delay and detonator. The fly-off lever is retained by a safety pin with a safety clip to prevent accidental release. The grenade, when functioned, ignites and disperses the red phosphorus composition after a preset delay interval.

On throwing the grenade there is a 4.5 second delay, after which the detonator ruptures the body and spreads the burning red phosphorus granules over the immediate area. The grenade can be used in a defensive role where screening smoke is required and as an offensive weapon when the acrid smoke and incendiary effect can be used for bunker or room clearance. The burning granules will also ignite various materials.

Data

Length: 136 mm
Diameter: 51 mm
Weight: 420 g
Filling: 220 g red phosphorus
Delay time: 4.5 s
Number of granules: average, 2,600
Spreading area: 500 m^2

Manufacturer

Swartklip Products, a Division of Denel (Pty) Limited, PO Box 977, Cape Town 8000.

Status

In production.

Service

South African National Defence Force.

VERIFIED

Grenade, Hand, Red Phosphorus

Practice rifle grenade

Description

This resembles the anti-personnel rifle grenade (see previously) but is completely inert. The arming mechanism of the live grenade is replaced by a metal spacer, but the bullet deflector is retained so the firer is protected if accidentally using a ball cartridge for launching. The grenade can be fired from most suitable 5.56 and 7.62 mm rifles and the head can be reused up to 10 times, after recovery, by fitting a new tail unit.

Data

Length: 297 mm
Weight: 610 g
Max range: 5.56 mm rifle, 210 m; 7.62 mm rifle, 300 m

Manufacturer

Swartklip Products, a Division of Denel (Pty) Limited, PO Box 977, Cape Town 8000.

Status

In production.

Service

South African National Defence Force.

VERIFIED

Practice rifle grenade

Bullet Trap (BT) rifle grenades

Description

The Bullet Trap (BT) rifle grenades may be considered as the second generation of South African rifle grenades. They are almost identical to the series described above but now have a bullet trap and deflector incorporated in the tail tube. This allows them to be fired without the need to unload the rifle of its combat cartridge, reload with a grenade-launching blank, and then reload with the combat cartridge after firing the grenade. Instead, the grenade is placed over the muzzle and the cartridge in the chamber, ball or tracer, is fired. The bullet is stopped by the bullet trap, and its impact, together with the expanding gas from the propelling charge, provides the impetus to propel the grenade to its target.

In the unlikely event of a failure of the bullet trap, or if the soldier should inadvertently fire an armour-piercing bullet, the added bullet deflector will prevent the bullet arming the grenade.

The range of Bullet Trap rifle grenades includes HE/anti-personnel, red phosphorus, illuminating, smoke and practice grenades, details of which are tabulated following.

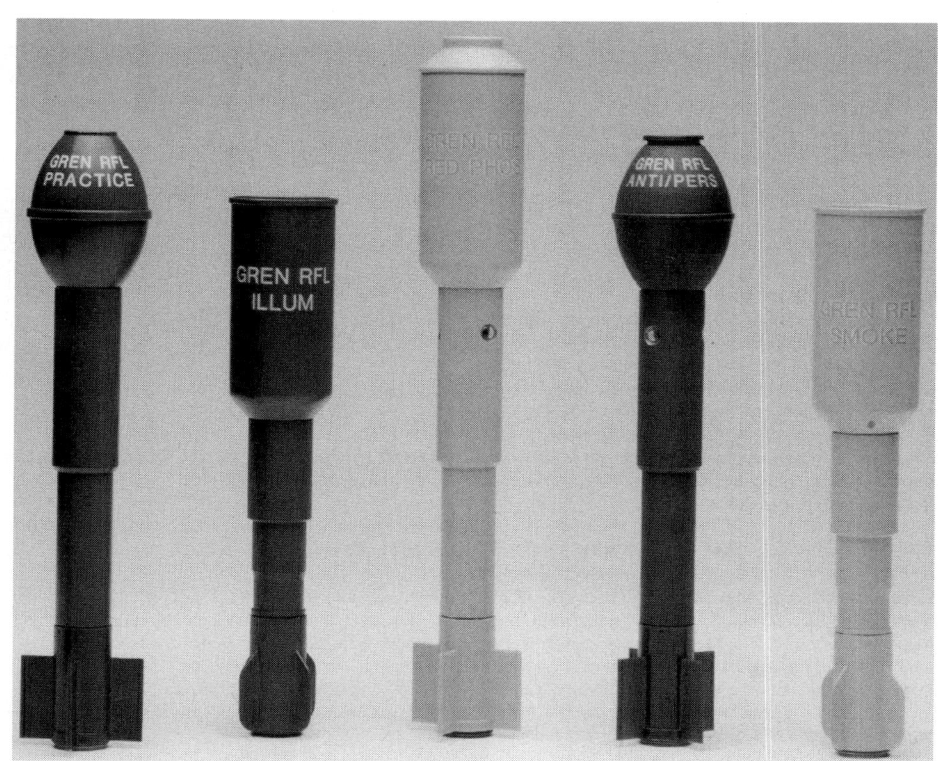

South African Bullet Trap rifle grenades

Data

	HE/AP	RP	Illum	Smoke	Practice
Length	305 mm	360 mm	275 mm	275 mm	305 mm
Warhead diameter	60 mm	–	–	–	–
Weight	660 g	630 g	396 g	545 g	660 g
Weight of filling	160 g	260 g	–	–	–
Nature of filling	Comp B	RP	Flare	HC Smoke	–
Average fragments	1,000	3,000	–	–	–
Effective radius	15 m	500 m²	–	–	–
Illumination	–	–	100,000 cd	–	–
Duration of effect	–	–	30 s	40 s	–
Max range (5.56 mm)	210 m	210 m	100 m	210 m	210 m
Max range (7.62 mm)	300 m	300 m	130 m	300 m	300 m
Arming distance			min 15 m		
Operating temperature			−15 to +65°C		

Manufacturer
Swartklip Products, a Division of Denel (Pty) Limited, PO Box 977, Cape Town 8000.

Service
South African National Defence Force.

VERIFIED

75 mm HEAT rifle grenade

Description

This is a shaped charge grenade, based on the Energa and designed to be fired from most 7.62 mm rifles. The tail unit carries a 1 g incremental charge which augments the propulsive effort derived from the ballistite launching cartridge; the tail unit is pressure-tested to 750 kg/cm² to ensure safety during firing.

The grenade is fitted with a spit-back impact fuze which initiates a detonator at the rear of the shaped charge. This detonator is carried externally in a plastic case fitted to the tail tube and is installed in the grenade before firing by simply unscrewing the tail tube. The fuze is protected by a plastic nose cap which is resistant to dropping or other rough treatment and is removed immediately before firing. There is also an internal safety unit which seals the spit-back from the fuze away from the detonator unless unlocked by setback caused by actual firing.

Penetration is given as a minimum of 275 mm of armour plate with a strike angle of from 0 to 70°.

A Practice grenade was produced.

Data

Weight: fuzed, 720 g
Filling: RDX/WX 93/7 main charge and CH6 booster
Fuze: DA and Graze Type BS02
Max velocity: 62 m/s
Range: max, 375 m; optimum, 75 m
Penetration: 275 mm
Accuracy: H+L 2.5 m at 100 m

Enquiries to

Armscor, Private Bag X337, Pretoria 0001.

Status

No longer in production.

Service

South African National Defence Force.

UPDATED

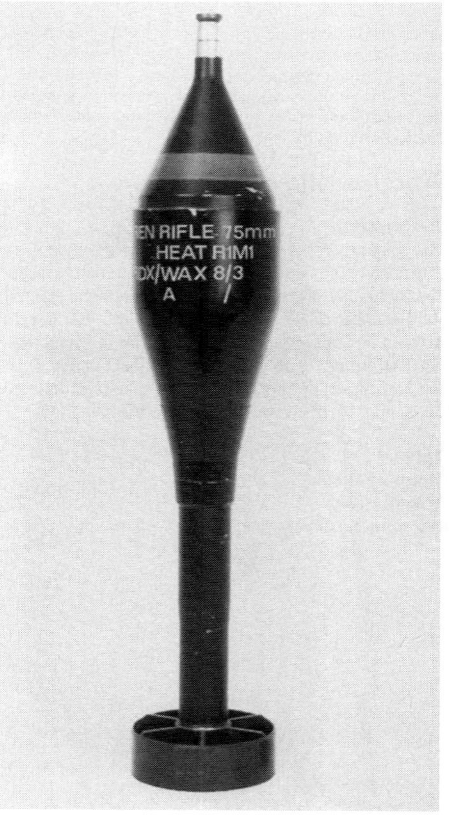

75 mm HEAT rifle grenade

Swartklip 40 mm grenades

Description

Swartklip Products, a Division of Denel (Pty) Limited, produce a range of 40 mm low-velocity grenades suitable for launching from most shoulder-fired 40 mm grenade launchers. The range covers combat grenades as well as riot control munitions and pyrotechnics, and includes the following types.

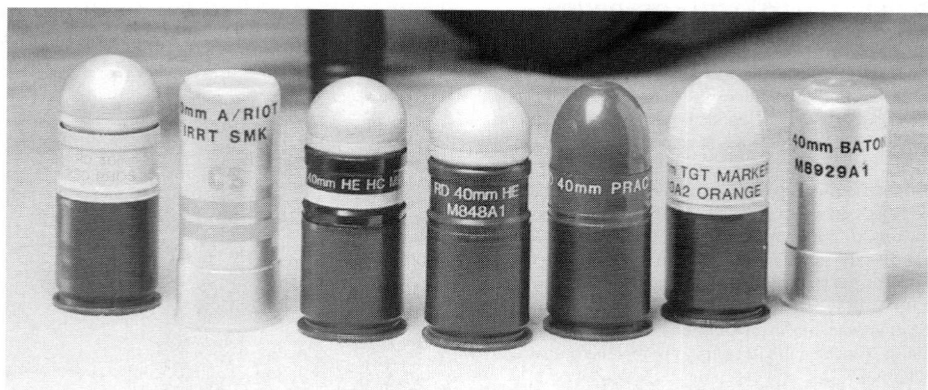

Swartklip 40 mm low-velocity grenades; from left, RP, CS, HE HC, HE, PRAC, TGT and BATON
1996

High Explosive HE M848A1

This is one of the standard combat grenades with a 40 g RDX/TNT filling surrounded by a fragmentation unit of the prepatterned steel cup type producing 425 fragments, each weighing 120 mg. On detonation the fragments are scattered over a casualty radius of at least 5 m. The fuze arms after a distance of from 9 to 28 m from the muzzle so the grenade can be used over an effective range of from 20 to 400 m, the latter being the maximum possible range.

High Explosive Hollow Charge HE HC M9115

For the HE HC M9115 the RDX main charge is surrounded by the same type of prepatterned steel cup fragmentation unit as the HE M848A1 grenade but the charge is shaped to provide an armour penetration capability. The grenade can penetrate a minimum of 50 mm of armour plate or a minimum of 400 mm of concrete as well as scattering 425 anti-personnel fragments over a 5 m casualty radius. This grenade can be used over an effective range of from 28 to 400 m, the latter being the maximum possible range.

Red Phosphorus RP M8931

This grenade is described as a multipurpose munition with tactical applications ranging from producing smoke screens to clearing trenches, bunkers and buildings. The grenade, which contains 32 g of red phosphorus, has an incendiary effect, bursting on impact to scatter burning particles over a radius of 10 m. This grenade can be used over an effective range of from 28 to 400 m, the latter being the maximum possible range.

Target Marker, Orange TGT M8410

This grenade has ballistic characteristics identical to those of the HE M848A1. On impact it emits orange marking smoke for from 20 to 40 seconds. The fuze is armed after a delay of 1.5 seconds so the grenade is effective over a range of from 100 to 400 m, the latter being the maximum possible range.

Practice PRAC

In place of a combat payload this practice grenade carries an inert solid aluminium slug. Maximum range is 400 m.

Irritant Smoke CS

This non-lethal riot control munition produces CS smoke for 20 seconds following a fuze arming delay of 0.6 second. The grenade is effective over a range of from 80 to 170 m, the latter being the maximum possible range for this grenade.

Baton Hard Rubber BATON M8929

The solid hard rubber baton used with this non-lethal riot control munition weighs 75 g and is 45 mm long. Projected with a muzzle velocity of 85 m/s the maximum direct fire range is 100 m.

Manufacturer

Swartklip Products, a Division of Denel (Pty) Limited, PO Box 977, Cape Town 8000.

Status

In production.

Service

South African National Defence Force.

NEW ENTRY

Data

Model	HE	HE HC	RP	TGT
Designation	M848A1	M9115	M8931	M8410
Length overall	103 mm	103 mm	103 mm	101 mm
Weight	230 g	227 g	225 g	227 g
Projectile length	82 mm	82 mm	82 mm	82 mm
Projectile weight	180 g	178 g	175 g	177 g
Fuze arming/delay	9-28 m	9-28 m	9-28 m	1.5 s
Muzzle velocity	76 m/s	76 m/s	76 m/s	76 m/s
Max range	400 m	400 m	400 m	400 m
Effective range	28-400 m	28-400 m	28-400 m	100-400 m

Model	CS	BATON	PRAC
Designation	none	M8929	none
Length overall	99 mm	87 mm	101 mm
Weight	184 g	150 g	205 g
Projectile length	83 mm	45 mm	75 mm
Projectile weight	76 g	75 g	176 g
Fuze arming/delay	0.6 s	n/a	n/a
Muzzle velocity	88 m/s	85 m/s	76 m/s
Max range	170 m	100 m	400 m
Effective range	80-170 m	30-100 m	n/a

Swartklip 40 mm high-velocity grenades

Description

Swartklip Products, a Division of Denel (Pty) Limited, produces a range of 40 mm high-velocity grenades suitable for launching from automatic 40 mm grenade launchers, including the Vektor 40 mm AGL. All grenades are delivered in belts of 20 linked rounds. The range includes the following types.

High Explosive HE

This grenade contains 45 g of RDX surrounded by a fragmentation unit creating a casualty radius of 12 m. At a spin rate of 12,000 rpm the PIBD fuze arms after a distance of from 14 to 61 m from the muzzle. Maximum possible range is 2,200 m.

Hollow Charge HEDP

For the HEDP grenade the 26 g RDX main charge is shaped to provide an armour penetration capability. The grenade can penetrate a minimum of 50 mm of armour plate. Anti-personnel fragments are produced although the casualty radius is reduced to 5 m. The PIBD fuze arms after a distance of from 14 to 61 m from the muzzle. Maximum possible range is 2,200 m.

Practice

In place of a combat payload this practice grenade carries an inert warhead. There is no fuze. Maximum range is 2,200 m.

The operational temperature range for all three grenades is from −40 to +70°C.

Manufacturer

Swartklip Products, a Division of Denel (Pty) Limited, PO Box 977, Cape Town 8000.

Swartklip 40 mm high-velocity grenades

1996

Data

Model	HE	HEDP	PRAC
Weight	340 g	340 g	340 g
Length	112 mm	112 mm	112 mm
Weight of explosive	45 g	26 g	none
Muzzle velocity	242 m/s	242 m/s	242 m/s
Max range	2,200 m	2,200 m	2,200 m
Casualty radius	10 m	5 m	n/a

Status

In production.

NEW ENTRY

SPAIN

EXPAL MB-8 hand grenade

Description

The EXPAL MB-8 is a controlled fragmentation defensive grenade which can easily be converted into an offensive pattern by removing the metal sleeve.

Safety is provided by the delayed action fuze with detonation occurring at 4 ±0.5 seconds after release of the striker. Simplicity is imparted by the ease of conversion from defensive to offensive and vice versa. Effectiveness is achieved by the type and uniform distribution of fragments as well as the ratio of high explosive to fragment material.

A practice version of this grenade is also produced.

EXPAL MB-8 hand grenade dismantled to show ignition set, offensive unit and defensive fragmentation sleeve

The MBS-9 grenade is externally the same as the MB-8 but differs in that it has a trajectory safety device incorporated.

Data
Weight: defensive, 440 g; offensive, 171 g
Length: defensive, 103 mm; offensive, 95 mm
Diameter: defensive, 61 mm; offensive, 50 mm

Manufacturer
EXPAL SA, Avenida del Partenòn 16, 5th Floor, E-28042 Madrid.

Status
In production.

VERIFIED

EXPAL MB-8 hand grenade in defensive form

EXPAL incendiary hand grenades

Description
Explosivos Alaveses SA (EXPAL) manufactures a family of incendiary hand grenades which differ in their filling agent and thus are capable of being matched to different tactical requirements.

The basic GWP grenade is filled with white phosphorus and therefore has applications as a smoke producer, an anti-personnel weapon or as an incendiary grenade. When detonated the WP particles are distributed over an area some 40 m in diameter. Ignition is by the usual type of fly-off lever igniter, which has a nominal 5 second delay. The grenade is 140 mm high, 55 mm in diameter and weighs 750 g.

The GRP grenade is filled with red phosphorus; it thus has a primary role as a smoke producer but it will also act as an incendiary device with easily ignited substances. When detonated, the red phosphorus particles are distributed over an area of 30 m diameter, and the phosphorus continues to burn for about 30 seconds. The same igniter set is used as in the GWP grenade above. The grenade is 142 mm high, 55 mm in diameter and weighs 560 g.

The CTE grenade is filled with thermite and is therefore purely an incendiary device which will ignite anything capable of being burned, since the thermite combustion reaches a temperature of 3,000°C. Burning continues for 40 seconds. The igniter set is of similar pattern to the previous grenades, but the delay time is only 2 seconds. The grenade is 140 mm high, 55 mm in diameter and weighs 580 g.

Manufacturer
EXPAL SA, Avenida del Partenòn 16, 5th Floor, E-28042 Madrid.

Status
In production.

VERIFIED

EXPAL GWP incendiary hand grenade

R-41 Cordo delay hand grenade

Description
The R-41 Cordo delay hand grenade replaced the earlier R-1 in production. It is an offensive pattern grenade, using a plastic casing, a charge of high explosive, and the Oramil pattern of fly-off lever igniter set. The igniter includes the Cordo delayed arming system in which a safety ring, holding the detonator away from the rest of the explosive train, needs to be melted by heat generated by the ignition system, so giving an arming delay of about 1.2 seconds.

When used in the defensive role, the R-41 is enclosed by a separate fragmentation container which carries 3,500 steel balls of 2 mm diameter. It is possible to load the container with balls of different sizes, and accordingly the quantity of fragments will be varied. The container is locked to the body of the grenade by a simple bayonet catch.

In 1987 the R-41 grenade, in competition with several other designs of Spanish and foreign grenades, won a contest organised by the Spanish Ministry of Defence to select a hand grenade for provision to the Spanish Army.

Data
Length: offensive, 110 mm; defensive, 114 mm
Diameter: offensive, 53 mm; defensive, 61.5 mm
Weight: offensive, 242 g; defensive, 400 g
Weight of filling: 125 g
Delay time: 4 s

Manufacturer
Cordo SA, PO Box 202, Villa Urcabe-Baita, E-20180 Oyarzun.

Status
In production.

VERIFIED

R-41 Cordo hand grenade in offensive and defensive forms

Instalaza FTI rifle grenade

Description

The Instalaza FTI rifle grenade is an anti-tank grenade with a shaped charge warhead. It can penetrate up to 250 mm of armour plate and 650 mm of concrete. It can be employed with any rifle with a launcher or flash eliminator of 22 mm.

A practice grenade is produced for training purposes. The tail, made of high strength alloy steel, allows the practice grenade to be used repeatedly.

Data

Length: 395 mm
Diameter: 64 mm
Internal diameter of fin assembly: 22 mm
Weight: ready for firing, 700 g; in carrying case, 870 g
Muzzle velocity: 55.5 m/s
Range: max, 300 m; effective, 100 m

Manufacturer

Instalaza SA, Monreal 27, E-50002, Zaragoza.

Status

Available.

Service

Spanish and other armed forces.

UPDATED

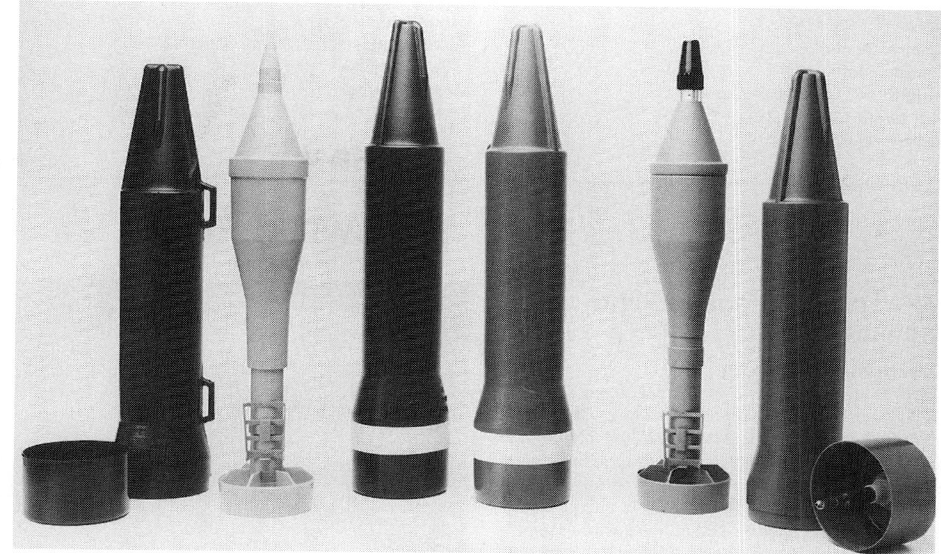

Practice (left) *and operational* (right) *Instalaza FTI rifle grenades and their transit containers*
1996

Instalaza FTII rifle grenade

Description

The Instalaza FTII rifle grenade is a shaped charge, fragmentation grenade for use against tanks and personnel, which will penetrate armour plate of up to 130 mm and up to 350 mm of concrete. It may be used with any rifle with a 22 mm flash eliminator, or a launcher of that dimension.

A reusable practice grenade with a high-strength steel alloy tail is available.

Data

Calibre: 40 mm
Length: 333 mm
Internal diameter of fin assembly: 22 mm
Weight: grenade, 500 g; grenade in case, 600 g
Filling: 88 g HE
Max range: 425 m
Muzzle velocity: 75 m/s
Effective radius against personnel: 30

Manufacturer

Instalaza SA, Monreal 27, E-50002, Zaragoza.

Status

In production.

Service

Spanish and other armed forces.

UPDATED

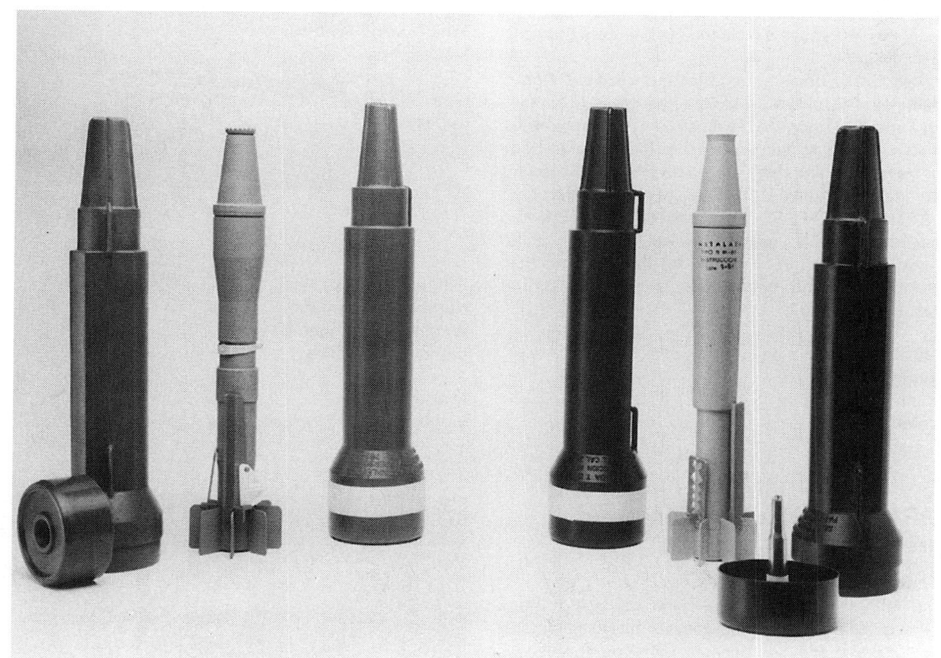

Practice (left) *and operational* (right) *Instalaza FTII rifle grenades and their transit containers*
1996

Instalaza FTV rifle grenade

Description

The Instalaza FTV rifle grenade can be launched from any 5.56 mm rifle with a standard 22 mm flash suppressor and using any kind of live ammunition, including SS109 rounds.

This grenade features a shaped charge warhead inside a fragmentation casing, combining an anti-armour capability with anti-personnel effects. It can penetrate more than 110 mm of armour steel and breaks up into more than 550 lethal fragments. A density of at least one lethal fragment/m^2 is obtained up to a distance of 8 m, and the fragments are still lethal at 20 m distance.

Two different practice grenades are also available, the FTV-INS-1 and the FTV-INS-20. The former is a one-shot grenade, and the latter is reusable, being supplied with the appropriate number of spare launch cartridges and sights.

Instalaza FTV rifle grenade prepared for firing
1996

Data
Calibre: 36 mm
Length: 345 mm
Weight: 440 g; cased, 520 g
Max range: 300 m
Muzzle velocity: 65 m/s
Lethal radius: 8 m
Penetration: steel, 110 mm; concrete, 280 mm

Manufacturer
Instalaza SA, Monreal 27, E-50002, Zaragoza.

Status
Available.

Service
Spanish armed forces.

UPDATED

SWITZERLAND

MFA HG 85 defensive hand grenade

Description
The MFA HG 85 defensive hand grenade, described by MFA as a fragmentation grenade, uses an almost circular cross-section cast steel body with the internal wall area formed into a dimpled surface. This surface is shattered into approximately 2,000 fragments as the main 155 g RDX/TNT 55/45 explosive charge detonates. This occurs after the DM 82 CH igniter assembly operates following a time delay. Each of the fragments weighs about 0.1 g and retains an energy rating of approximately 80 J at a distance of 5 m from the point of explosion.

For training with the HG 85 the EUHG 85 practice grenade is available. This grenade operates in exactly the same manner as the operational grenade but the filling is reduced to 135 g of RDX/TNT 30/70 and the body walls shatter in such a manner that no fragments can travel more than a distance of 10 m from the point of explosion.

Two further training grenades are available. One is the MARK HG 85, a reusable practice grenade containing a sound cartridge to create a sound signature as the igniter assembly functions. The thick metal body remains intact and there are no fragments. The sound cartridge will produce 105 dB at 20 m detonation.

The WURFK HG 85 is a completely inert ballasted practice grenade with a solid moulded igniter assembly. It is employed only for throwing practice, having the same dimensions and weight as an operational grenade.

MFA HG 85 defensive hand grenade

1996

Data
Length: 97 mm
Diameter: 65 mm
Weight: 465 g
Filling: 155 g RDX/TNT 55/45
Number of fragments: ca 2,000
Weight of fragment: 0.1 g
Lethal radius: 5 m

Manufacturer
Eidgenössische Munitionsfabrik, CH-6460 Altdorf.

Status
In production.

Service
Swiss armed forces. Offered for export sales.

NEW ENTRY

MFA OHG 92 offensive hand grenade

Description
The MFA OHG 92 offensive hand grenade has the same circular cross-section body as the HG 85 defensive hand grenade (see previous entry) but the body is plastic, roughened externally to provide a good grip. The grenade contains 170 g of RDX/TNT 55/45 to produce blast effects only. There are no preformed fragments and no fragments travel further than 10 m from the point of explosion. The same DM 82 CH igniter assembly as that used with the HG 85 is employed.

For training, the EUHG 85 practice grenade, MARK HG 85 reusable practice grenade and WURFK HG 85 inert ballasted practice grenade are available. See previous entry for details of these training grenades.

Data
Length: 97 mm
Diameter: 65 mm
Weight: 270 g
Filling: 170 g RDX/TNT 55/45

Manufacturer
Eidgenössische Munitionsfabrik, CH-6460 Altdorf.

Status
In production.

Service
Swiss armed forces. Offered for export sales.

NEW ENTRY

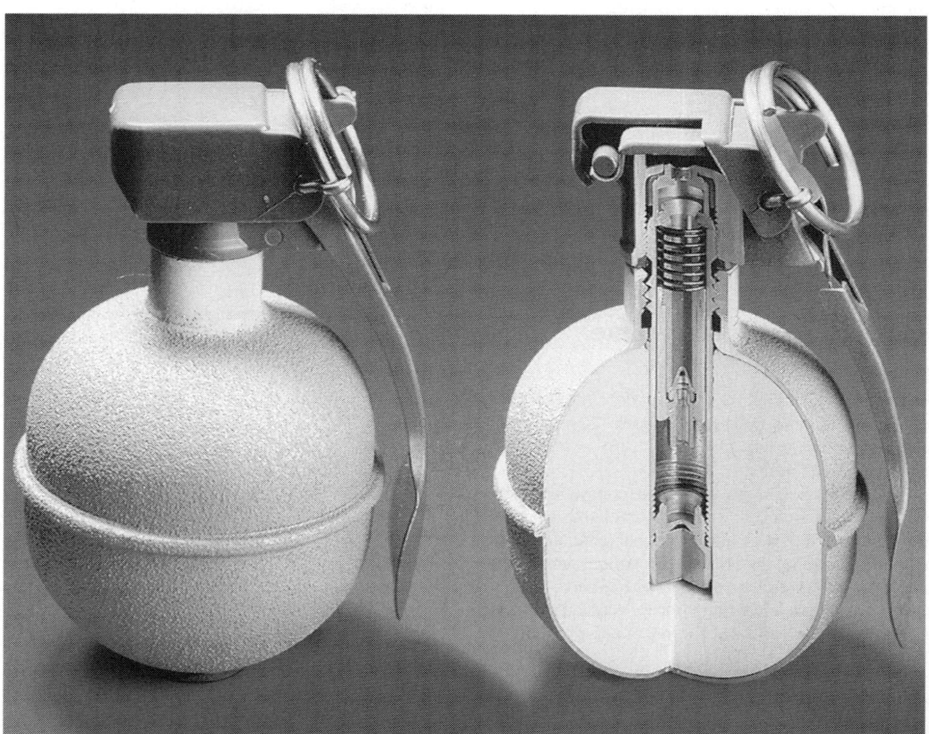

MFA OHG 92 offensive hand grenade

1996

MFA SIM HG 93 training hand grenade

Description
The MFA SIM HG 93 training hand grenade has the same overall dimensions and general appearance as the HG 85 and OHG 92 grenades (see previous entries) but has been developed for non-explosive Tactical Environment Simulation (TES) training based on MILES or similar training systems. The SIM HG 93 is an electronics based training grenade, having the same dimensions as an operational grenade and weighing 385 g.

The SIM HG 93 is handled and thrown in the same manner as an operational grenade. As the electronics-based delay system functions, an optional audible report may be produced. At the same time the grenade will emit infra-red light to actuate MILES or similar TES sensors worn by personnel participating in the training. The infra-red light will provide a disabling signal to TES sensors within a preselected range, thus removing the affected personnel from the exercise.

The SIM HG 93 can be reused many times and is powered by a 5.6 V EPX 27 battery.

Data
Length: 97 mm
Diameter: 65 mm
Weight: 385 g
Battery: 5.6 V EPX 27

Manufacturer
Eidgenössische Munitionsfabrik, CH-6460 Altdorf.

Status
Available.

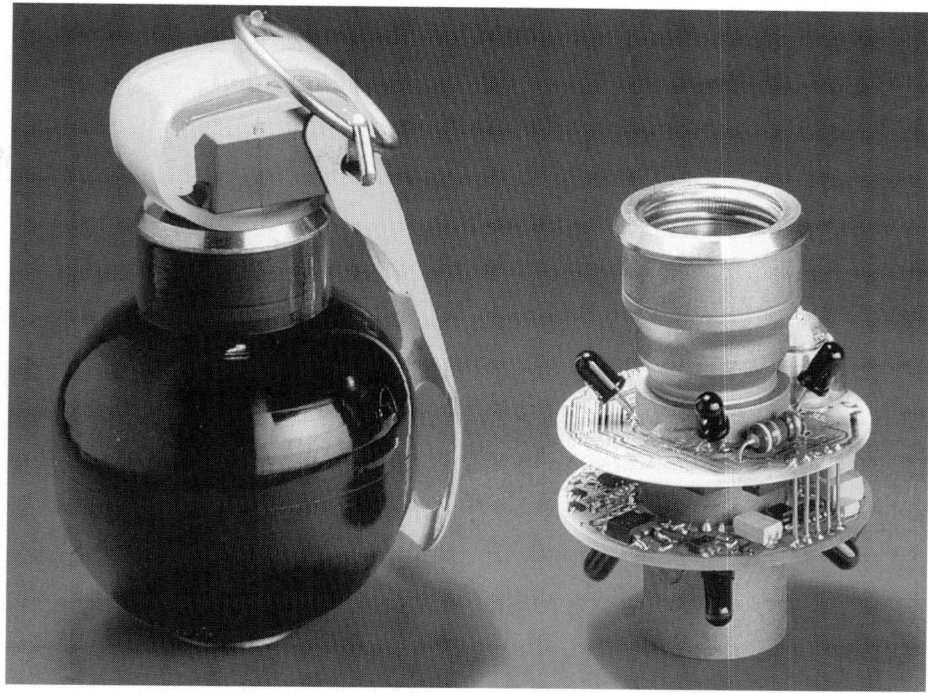

MFA SIM HG 93 training hand grenade

1996

Service
Offered for export sales.

NEW ENTRY

TAIWAN

Taiwanese grenades

Description
The Nationalist Chinese forces use a variety of combat grenades, almost all of US origin. The Hsing Hua Company also manufactures a number of types in substantial quantities for use by the Taiwan army and for export. These include:

Type 67: A small grenade resembling the M67/68 types
Fragmentation grenade: Based on the obsolete American Mk IIA1
Offensive grenade: Cylindrical body
Smoke grenade: Various colours, for screening and marking
Rifle grenade: Resembles the 40 mm MECAR type (see entry under *Belgium*) and has a range of 300 m
Tear gas grenade: Filled with CS, CN or CN/DM mixture

Manufacturer
Hsing Hua Company Limited, PO Box 8746, Taipei.

Status
In production.

Service
Chinese Nationalist Army and others.

VERIFIED

Fragmentation grenades manufactured by Hsing Hua Company

TURKEY

MKEK hand grenades

Description
Makina ve Kimya Endüstri Kurumu (MKEK) produces two types of operational hand grenade, one defensive and the other offensive. Both are based on US designs which have now passed from service elsewhere.

The defensive hand grenade Mk 2 is based on the US Mk 2, a design which originally entered service in 1918 and which was itself based on an earlier French design. The body is cast iron serrated externally to enhance fragmentation and filled with RDX/TNT. The grenade operates in the conventional manner with a fly-off safety lever held in place by a pin for transport. The fuze assembly is the M204A2.

An inert Practice Mk 2 grenade is available with a live M205A2 fuze for training.

The offensive hand grenade is the Mk 3A2 fitted with a M206A2 fuze. The filling is a pellet of RDX/TNT packed within the pressed fibre body.

MKEK hand grenades: from left, Practice Mk 2; Mk 3A2 offensive; Mk 2 defensive

1996

Data

Type	Defensive	Offensive	Practice
Designation	Mk 2	Mk 3A2	Mk 2
Length	120 mm	145 mm	120 mm
Weight	640 g	335 g	565 g
Fuze	M204A2	M206A2	M205A2

Manufacturer
Makina ve Kimya Endüstri Kurumu, 06330 Tandogan, Ankara.

Status
In production.

Service
Turkish armed forces. Offered for export sales.

NEW ENTRY

UNITED KINGDOM

No 80 WP Mark 1 and Mark 1/1 grenade

Description
The No 80 WP is intended to produce screening smoke and may be either thrown or projected from a multi-barrel discharger on an armoured fighting vehicle, using a Fuze, Electric, No F103. It has a tinplate body filled with white phosphorus, Detonator No 75 and Striker Mechanism, Grenade, No 2.

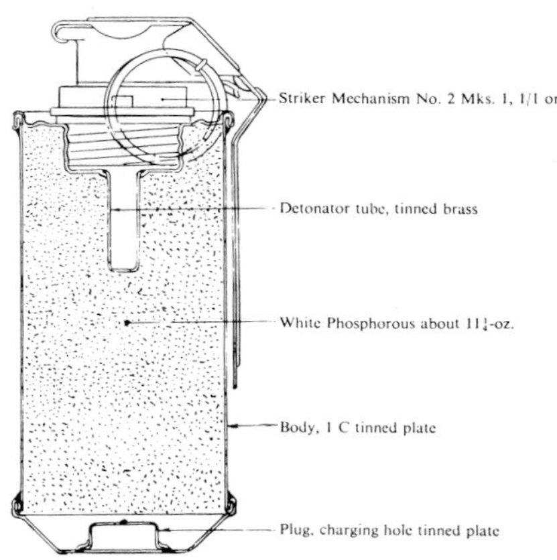

No 80 WP Mark 1 grenade

The Striker Mechanism, Grenade, No 2 consists of an adaptor with a screwed-in housing for a spring-operated striker which is retained in the cocked position by a fly-off lever and safety pin.

The Detonator No 75 consists of a 0.22 in rimfire cap attached to a 1.5 in (38 mm) length of Fuze, Grenade, No 1 and a cap chamber. The fuze gives a delay of 2.5 to 4 seconds and then initiates a Detonator No 63 Mk 2 or No 78 Mk 1.

The Fuze, Electric, No F103 has a brass magazine, containing G 20 gunpowder, closed by a brass cover into which is set a Fuze, Electric, No 53 Mark 2.

The Drill Grenade Hand/Discharger No 80 Mark 1 is the body of a service grenade with an inert filling. The Drill Detonator No 75 Mark 1 is an empty 0.22 rimfire cap in a cap chamber and connected to a dummy fuze and an empty detonator tube. It is painted white with a hole through the tube of the detonator.

Manufacturer
Royal Ordnance plc, Bishopton, Strathclyde.

Status
Obsolete. To be replaced by a grenade producing non-toxic smoke.

Service
British armed forces and others.

UPDATED

No 80 WP Mark 1 hand/discharger grenade

Grenade, Hand, Anti-Personnel, L2A2

Description
The Grenade L2A2 is a high-explosive anti-personnel grenade currently in service with the British Army and other forces worldwide. Recent product development, driven by the requirements of the British Army, has resulted in this grenade and fuze being modified to an L2A3 standard. By 1997 further modifications will arise following further developments involved with another Royal Ordnance design, the RO 01A1 (see following entry).

The L2A2 grenade has a body section and a separate L25A6 grenade fuze. As the body and fuze are both sealed against the ingress of moisture, they may be carried separately by the soldier and assembled just before use. The fuze may be fitted and removed at least six times without adversely affecting performance.

The grenade body consists of a deep bronze-green metal casing, covered in a matt non-slip paint, enclosing a steel fragmentation coil and high explosive filling. The fuze has a conventional fly-off lever striker which ignites a percussion delay detonator. The fragmentation coil produces about 1,200 fragments, each weighing between 0.1 and 0.5 g.

The Grenade, Hand, Drill, L4A2 is an inert grenade suitable for demonstrating the method of operation and for classroom use.

The Grenade, Hand, Practice, L56A1 has a live fuze in an inert body which operates in an identical manner to the fuzed grenade L2A2 but on functioning produces a loud report and a puff of smoke to denote the point of impact.

Data
Length: 106 mm
Diameter: max, 64 mm

L2A2 anti-personnel hand grenade

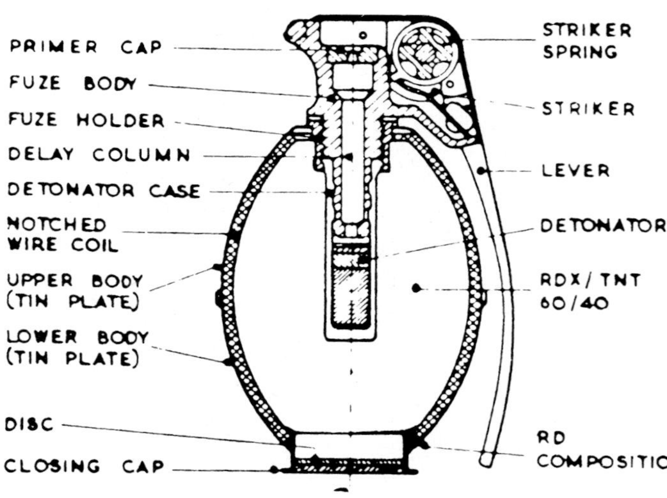

L2A2 anti-personnel hand grenade

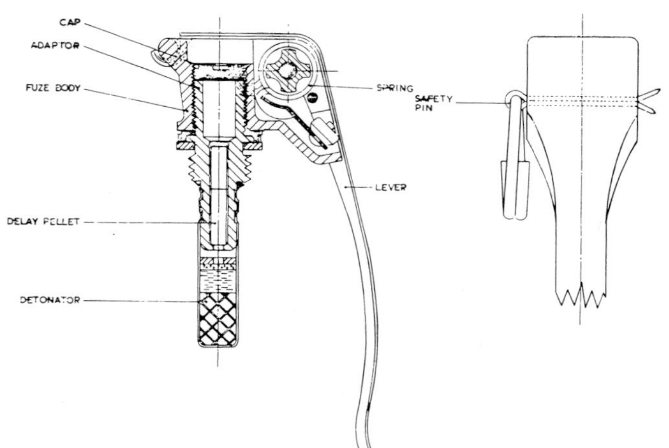

Fuze, grenade, percussion, L25A6

Weight: 395 g
Filling: 170 g RDX/TNT 55/45
Number of fragments: 1,200
Fuze: L25A6
Delay: 4.4 ±0.5 s
Lethal radius: 10 m

Manufacturer
British Aerospace Defence Limited, Royal Ordnance Division, Euxton Lane, Chorley, Lancashire PR7 6AD.

Status
Available.

Service
British armed forces and other military forces worldwide.

UPDATED

Grenade, Hand, Anti-Personnel, RO 01A1

Description
The Grenade RO 01A1 is a high explosive anti-personnel grenade originally designed to replace the L2A2 grenade currently in British service. The grenade is a development of the L2A2 and consists of a metal body and fragmentation coil with an HE filling. In all operational conditions the reliability of the grenade fuze is in excess of 99.99 per cent, providing a high degree of safety for the thrower. The overall reliability of the grenade is in excess of 97.5 per cent.

The Drill Grenade RO 01-01A1 is an inert grenade suitable for demonstrating the method of operation and for classroom training.

The Grenade, Practice, Hand, RO 01-05A1 with Practice Fuze RO 50A1 has an inert reusable body. The separately supplied fuzes are fitted by the user at the time of use. On functioning, the practice grenade produces a sharp report and a puff of smoke to denote the point of impact.

Data
Length: 106 mm
Diameter: max, 64 mm
Weight: total, 395 g; explosive content, 170 g
Delay: 4.3 s
Lethal radius: 10 m

Manufacturer
British Aerospace Defence Limited, Royal Ordnance Division, Euxton Lane, Chorley, Lancashire PR7 6AD.

Status
Development.

UPDATED

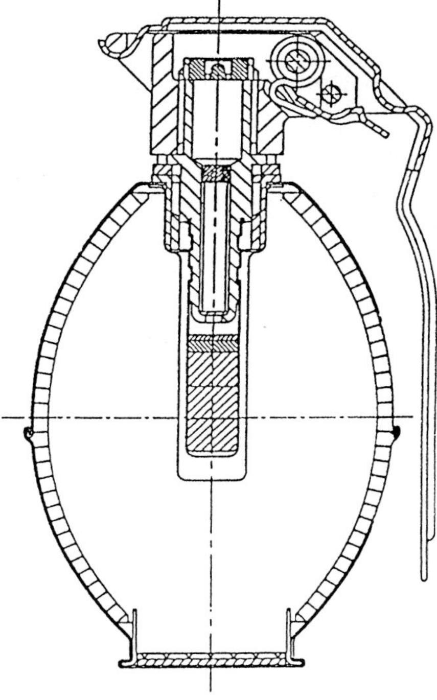

RO 01A1 anti-personnel hand grenade

66 mm screening smoke grenade L8A4

Description
L8A4 grenades are electrically initiated smoke grenades for the screening and protection of armoured vehicles. They are fired from fixed launcher tubes attached to the hull or turret.

The grenade consists of a black rubber body, a closing plug and a metal housing containing the propelling charge and the fuze. The rubber body is filled with a mixture of 95/5 red phosphorus and butyl rubber. A plastic burster tube containing gun powder passes through the body. At its top, in the metal housing, is the delay which is ignited by the propellant. In the metal housing is a small propelling charge of gun powder which throws the complete grenade clear of the vehicle.

When fired, the grenades burst clear of the vehicle and explosively disseminate the ignited RP pellets which then produce a cloud of white smoke round a frontal arc, concealing the vehicle entirely. An effective screen will be provided in approximately two seconds.

Data
Length: 185 mm
Diameter: 66.3 mm
Weight: 680 g
Weight of main filling: 360 g
Range: 25 m
Height of burst: 6 m
Delay time: 0.75 s
Screen width: ca 35 m
Min duration of smoke cloud: wind velocity at 24 km/h, 3 min
Operating voltage: 3 V

Manufacturer
British Aerospace Defence Limited, Royal Ordnance Division, Euxton Lane, Chorley, Lancashire PR7 6AD.

Status
In production.

Service
UK and US armies and forces worldwide.

UPDATED

66 mm screening smoke grenade L8A4

RO 66 mm VIS/IR screening smoke grenade

The Royal Ordnance Visual and Infra-Red screening smoke grenade is designed to give maximum protection to armoured fighting vehicles against modern target acquisition and surveillance techniques. The grenade can be fired interchangeably with the L8A4 screening smoke grenade (previously) and launchers are in service worldwide.

The grenade resembles the L8A4 and consists of a thin metal body filled with a mixture of the red phosphorus/butyl rubber composition similar to that used in the L8A4, and a fine metal powder. When fired the grenades burst in the air, clear of the vehicle, and are therefore unaffected by adverse terrain.

Data
Length: 185 mm
Diameter: 66.3 mm
Screen width: 50 m
Screen height: 10 m
Operating voltage: 3 V DC

Manufacturer
British Aerospace Defence Limited, Royal Ordnance Division, Euxton Lane, Chorley, Lancashire PR7 6AD.

Status
Final development.

VERIFIED

Haley and Weller E105 fragmentation hand grenade

Description

The Haley and Weller E105 is a high-explosive fragmentation grenade for infantry use. It is supplied in two separate parts, one being the body with explosive charge and preformed fragments, the other being the electrical initiation set. The initiation set is of Haley and Weller's patented design which is completely silent in operation. The manipulation of the grenade is the same as for a conventional type: the thrower removing the safety pin and releasing the lever as he throws. However, the lever does not fly off and therefore cannot make a noise when striking the ground. Also there is no percussion element to be struck and cause noise. Initiation and timing is entirely electrical and silent, and the timing is therefore extremely accurate and consistent.

In addition to its normal use for throwing, the grenade can also be fitted with the trip-wire mechanism E190 for use in perimeter defence, and it can be wired into a grenade necklace without requiring alternative detonators. The grenade is completely waterproof and has a shelf life of three years.

Data

Length: 125 mm
Diameter: 51 mm
Weight: 580 g; explosive filling, 51 g
Number of fragments: ca 2,000
Delay time: 4-4.5 s
Lethal radius: 5 m
Operational temperature range: −30 to +75°C

Manufacturer

Haley and Weller Limited, Wilne, Draycott, Derbyshire DE72 3QJ.

Status

In production.

Service

Middle and Far Eastern countries.

UPDATED

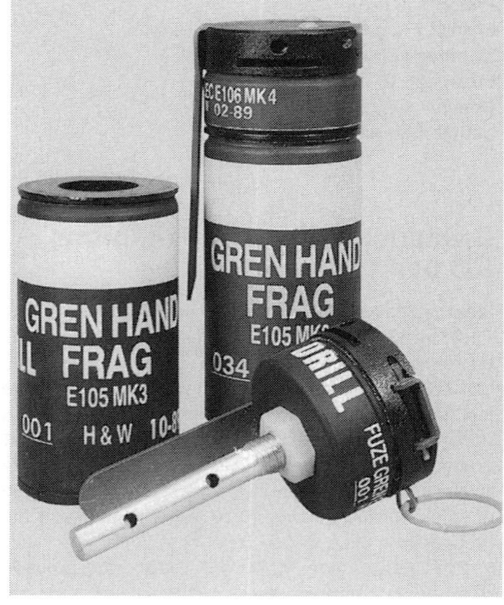

Haley and Weller E105 fragmentation hand grenade with inert drill model
1996

Haley and Weller E108 incendiary grenade

Description

The Haley and Weller E108 incendiary grenade employs the same electrical silent ignition system as the other grenades in this series. It was developed for use as a sabotage and destruction weapon. The grenade functions in the standard manner, but in addition can be fired remotely by means of external terminals. It burns at a temperature in excess at 2,700°C and will melt through 3 mm of steel.

Data

Length: 134 mm
Diameter: 50 mm
Weight: 530 g; incendiary filling, 400 g
Burn time: 60 s
Operational temperature range: −30 to +75°C

Manufacturer

Haley and Weller Limited, Wilne, Draycott, Derbyshire DE72 3QJ.

Status

In production.

Service

Various European countries.

UPDATED

Haley and Weller E108 incendiary grenade

Haley and Weller N110 screening smoke grenade

Description

The Haley and Weller N110 screening smoke grenade may be regarded as an updated version of the venerable No 83 smoke grenade. It is a member of the Haley and Weller N100 range of smoke grenades which can be provided in various colours, with the N110 producing a dense could of grey smoke for 30 seconds. The smoke is produced by using a low-toxicity cinnamic acid composition which has been approved by the UK Ministry of Defence.

The N110 continues to utilise a fly-off lever ignition system on an extruded aluminium casing employing a centre fire percussion cap. A single emission port is located centrally in the base. The grenade is fully waterproof and will function after immersion in 300 mm of water.

Data

Length: 135 mm
Diameter: 63 mm
Weight: 440 g
Filling: 305 g smoke composition
Emission time: nominal, 30 s
Operational temperature range: −30 to +75°C

Manufacturer

Haley and Weller Limited, Wilne, Draycott, Derbyshire DE72 3QJ.

Status

In production.

NEW ENTRY

Haley and Weller N110 screening smoke grenade
1996

Haley and Weller V101 66 mm screening smoke grenade

Description

The Haley and Weller V101 grenade is designed to be fired from standard 66 mm vehicle mountings. The V101 grenade has a metal base and rubber body and is electrically fired. It is filled with smoke composition which has an extremely low incendiary hazard liability.

Data

Length: 185 mm
Diameter: 65.5 mm
Weight: 550 g; explosive filling, 350 g
Range: 20-35 m
Duration of smoke: nominal, 40 s
Operating voltage: min, 1.5 V
Operational temperature range: −30 to +75°C

Manufacturer

Haley and Weller Limited, Wilne, Draycott, Derbyshire DE72 3QJ.

Status

In production.

VERIFIED

Haley and Weller V101 66 mm screening smoke grenade
1996

Haley and Weller V130 66 mm fragmentation grenade

Description

The Haley and Weller V130 66 mm fragmentation grenade is an anti-personnel grenade designed to be launched from standard 66 mm vehicle-mounted launchers, originally intended for discharging smoke grenades. The fragmentation grenade contains several thousand steel ball bearings embedded in a plastic matrix surrounding a central tube containing the bursting charge. The grenade is electrically fired and launched in the same manner as a smoke grenade, but bursts above the ground and distributes the fragments in an all-round pattern. It is an ideal anti-ambush device and is also valuable for defending vehicles against close-in infantry at ranges where the vehicle weapons cannot be brought to bear.

Data

Length: 185 mm
Diameter: 65.5 mm
Weight: 1.4 kg; explosive filling, 51 g
Range: 60-90 m
Lethal area: 5-10 m
Danger area: 50 m
Operating voltage: min, 1.5 V
Operational temperature range: −30 to +75°C

Manufacturer

Haley and Weller Limited, Wilne, Draycott, Derbyshire DE72 3QJ.

Status

In production.

UPDATED

Haley and Weller V130 66 mm fragmentation grenade

Schermuly Mk 4 screening smoke grenade

Description

The Schermuly Mk 4 screening smoke grenade is a hand-thrown grenade consisting of a metal body containing a solid block of smoke-producing composition (HCE) and a fly-off lever ignition system. Smoke emission is through apertures at the top of the body; the smoke is grey-white for maximum obscuration and is formulated to give a rapid build-up. The grenade has a strengthened base to allow it to be fired from a rifle using a discharger cup and launching cartridge.

Ignition of the grenade is instantaneous so as to contribute to the rapid screening effect.

The screening smoke grenade is used for the rapid concealment of troops, vehicles or buildings in any situation where it is necessary to degrade the line of sight.

A new low-toxicity screening smoke grenade is now available from Pains-Wessex. This product, while still maintaining the dense obscuration requirement, is safe for troops to use in training without the effects of conventional screening smokes which are considered to be potentially hazardous to health and the environment.

Data

Length: 138 mm
Diameter: 62 mm
Weight: HCE, 650 g: low toxicity, 463 g
Pyrotechnic weight: HCE, 450 g; low toxicity, 284 g
Burning times: HCE, 120 s or 60 s; low toxicity, 60 s

Manufacturer

Pains-Wessex Schermuly, High Post, Salisbury, Wiltshire SP4 6AS (a member of the Chemring Group plc).

Status

Available.

Service

Worldwide.

VERIFIED

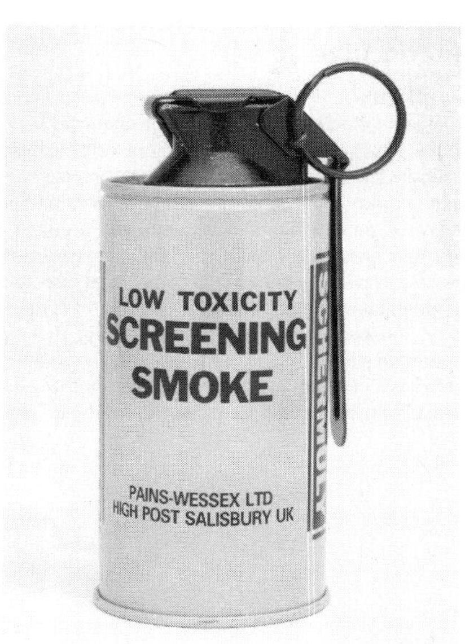

Schermuly Mk 4 screening smoke grenade

Schermuly 66 mm vehicle-discharged screening smoke grenade

The Schermuly 66 mm vehicle-discharged screening smoke grenade was developed to allow a well-proven grenade design to be used from a vehicle. This grenade is intended to be fired from the 66 mm dischargers common to most armoured fighting vehicles.

The grenade consists of a discharger base fitted with the electrical jack plug contact and the grenade payload screwed to the base. Both parts of the grenade are ejected from the discharger. A pull ring is incorporated in the top of the grenade to facilitate withdrawal from the launcher in the event of non-use.

Smoke emission begins on ejection and full volume is achieved by the time the grenade has landed. The screen is produced by a dense greyish-white opaque smoke which develops rapidly. Unlike phosphorus screening smokes, this smoke will not burn exposed parts of the body or start fires and is therefore suitable in both operational and training roles. To enhance training conditions in desert terrain a sand-coloured variant is available.

The grenades are used in situations where an effective screen is required at a distance from the user to prevent accurate enemy fire and conceal one's own withdrawal.

Data

Length: 151 mm
Diameter: 66 mm
Weight: in primary pack, 745 g
Filling: 300 g HCE
Operating voltage: min, 1.5 V
Range: nominal, 30 m
Delay: instantaneous ejection on firing, 5 s from firing to full volume
Burning time: nominal, 30 s

Manufacturer

Pains-Wessex Schermuly, High Post, Salisbury, Wiltshire SP4 6AS (a member of the Chemring Group plc).

Status

Available.

Service

Middle East.

VERIFIED

Schermuly 66 mm vehicle-discharged screening smoke grenade

76 mm and 66 mm vehicle-discharged screening smoke grenade, red phosphorus

Description

The 76 mm screening smoke grenade is based on a red phosphorus composition which gives excellent visual screening with a very rapid build-up of smoke. It is fired from existing 76 mm launcher systems fitted with side electrical connectors and gives an almost instantaneous smoke screen.

In light wind conditions up to 24 km/h, a salvo of eight smoke grenades produces an effective visual screen within two seconds of being fired and over an arc of 80°.

In addition there is a low-toxicity practice version for training.

Both these grenades are also available in 66 mm calibre.

Data

Length: 274 mm
Diameter: 76 mm
Operating voltage: min, 1.5 V
Operating current: min, 0.76 A
No-fire current: max, 0.3 A
Screen duration: nominal, 90 s
Screen height: min, 5 m
Range: from vehicle, 20-30 m

Manufacturer

Pains-Wessex Schermuly, High Post, Salisbury, Wiltshire SP4 6AS (a member of the Chemring Group plc).

Status

Available.

VERIFIED

Schermuly 76 mm vehicle-discharged screening smoke grenade

Brocks screening smoke grenade

Description

This is a hand grenade, using the conventional type of pin-and-lever ignition set, intended to be used for screening purposes. It emits a dense cloud of grey smoke about one second after initiation, and is ideal for screening troop movements for a period of about one minute without the hazards often attendant upon the use of white phosphorus in this role. Robust and waterproof, it can be used in any climatic conditions.

Data

Length: 140 mm
Diameter: 70 mm
Delay: nominal, 1 s
Duration: 40-110 s

Manufacturer

Brocks Pyrotechnics, Sanquhar, Dumfries and Galloway DG4 6JP.

VERIFIED

Ferranti EDNON electronic grenade fuze

Description

EDNON is an electronic delay fuze for chemical hand grenades which provides a safer more reliable fuze than current pyrotechnic fuzes, with more precise control of the time delay within the range 0.55 to 4.05 s. Operation is compatible with conventional grenades, the thrower removes the safety pin, retaining pressure on the safety lever. On release the safety lever operates the magnetic power source and initiates the time delay circuitry. The lever does not fly off on throwing, making the operation silent.

There are numerous safety features in the fuze including: ring pull and split pin lock; safety pin and safety lever; the power source is shorted out until the safety pin is released; electrical energy is generated only after release of the safety lever and safety pin; early impact sterilisation option available to ensure safety in the event of the grenade being accidentally dropped.

The fuze assembly is 55 mm in diameter and 33 mm high; it weighs 80 g.

Manufacturer

Ferranti Technologies Limited, Cairo Mill, Waterhead, Oldham, Lancashire OL4 3JA.

Status

Advanced development.

UPDATED

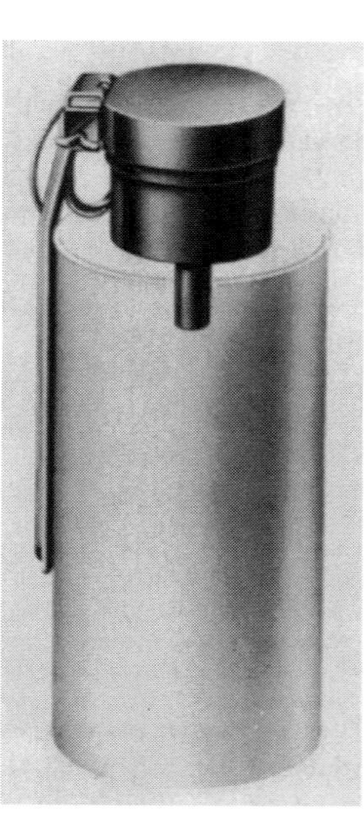

Ferranti EDNON electronic hand grenade fuze

UNITED STATES OF AMERICA

M67 delay fragmentation hand grenade

Description

The M67 grenade is one of the standard grenades in service with the US Army and Marine Corps. It is a small spherical grenade which is time-fuzed and can only be thrown by hand. There are no arrangements for projecting it by any other means.

The steel body is 63.5 mm in diameter and it breaks up on detonation to provide the fragments. The M213 fuze is integral with the body and the grenade is issued with its fuze *in situ*. The fuze is activated by a conventional striker which is held down by the safety lever. The safety lever is retained by two means. The first is the usual split pin which must be pulled out before throwing. The second is a small wire clip, which also holds the lever down. This clip is intended to act as a second safety should the split pin be pulled out unintentionally. In throwing, it is usual to remove the clip first and leave the pin until the last moment.

There is a training version of the M67, known as the M69. The difference is in the colour, the M69 being blue with a brown band, and in the fact that when the fuze burns out it emits a small, sharp crack like a firework and releases a small puff of smoke.

Data

Length: max, 89.7 mm
Diameter: 63.5 mm
Weight: 390 g
Filling: 184.6 g Comp B
Body: steel
Casualty radius: max, 15 m
Throwing range: 40 m
Fuze: model, M213; type, pyrotechnic delay-detonating
Primer: percussion, M42
Detonator: lead azide, lead styphnate, and RDX
Delay time: 4-5 s
Weight: 71 g
Length: 85 mm

Manufacturer

Government arsenals.

Status

In production.

Service

US armed forces.

VERIFIED

M67 fragmentation hand grenade

M61 delay fragmentation hand grenade

Description

The M61 is a standard issue HE fragmentation grenade which is a little larger than the M67. It has been in service for some years and formed the original pattern for the British L2 model. The major improvement of this grenade over the M67 is that it has a notched coil liner and so has regular and predictable fragments in the pattern. It is slightly heavier than the M67 and so cannot be thrown quite as far, but in overall effectiveness is thought to be a good deal better.

The fuzing and safety arrangements are the same as for the M67.

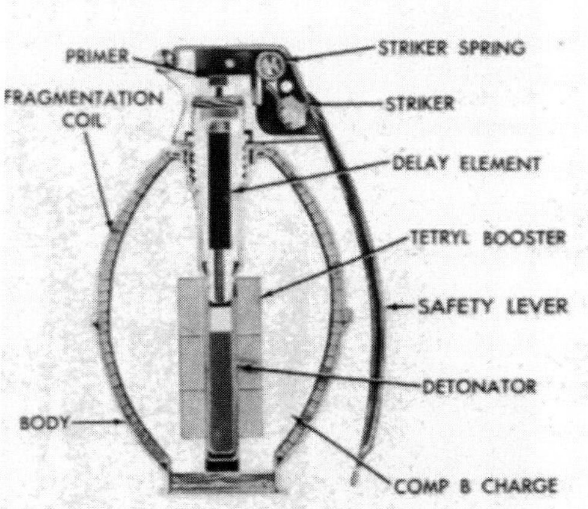

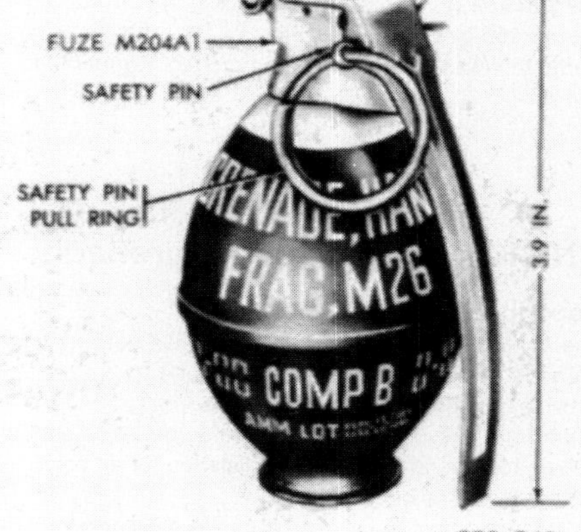

M61 fragmentation hand grenade

Data

Length: max, 99 mm
Diameter: 57 mm
Weight: 450 g
Filling: 156 g Comp B and 8 g tetryl pellets
Body: thin-wall sheet steel with inner fragmentation coil
Effective casualty radius: 15 m
Fuze: models M204A1 and M204A2; type, pyrotechnic delay-detonating
Primer: percussion, M42
Detonator: lead azide, lead styphnate, and RDX
Delay time: 4-5 s
Weight: 73 g
Length: 102 mm

Manufacturer

US Army Materiel Readiness Command.

Status

In production.

Service

US armed forces.

VERIFIED

M68 impact fragmentation hand grenade

Description

The M68 is essentially the M67 fitted with a different fuze. The bodies and fillings are the same, as are the lethal radii and throwing range.

The fuze is the M217 and is activated in exactly the same way as the fuze in the M67: by releasing a safety lever as the grenade is thrown. The M217 fuze, however, is a rarity in that it is an impact fuze with a time delay as a back-up. The impact part of the fuze acts by an electrical detonator and a tiny thermal power supply which is started by a powder train set off by the percussion cap. This thermal power supply requires a second or two to generate sufficient electricity, but after that the detonator will be fired if the grenade strikes a hard surface or is sharply jolted. Should the impact action fail for some reason the powder train continues burning and sets off the detonator by pyrotechnic action within seven seconds.

This complicated fuze is contained within a very small space in the body and is apparently perfectly reliable. The only effect of the size of the fuze is that temperature alters the delay times which are quoted as lying between three seconds at +52°C and seven seconds at −40°C, with a rough mean of 4.5 seconds at normal ambient of 20°C.

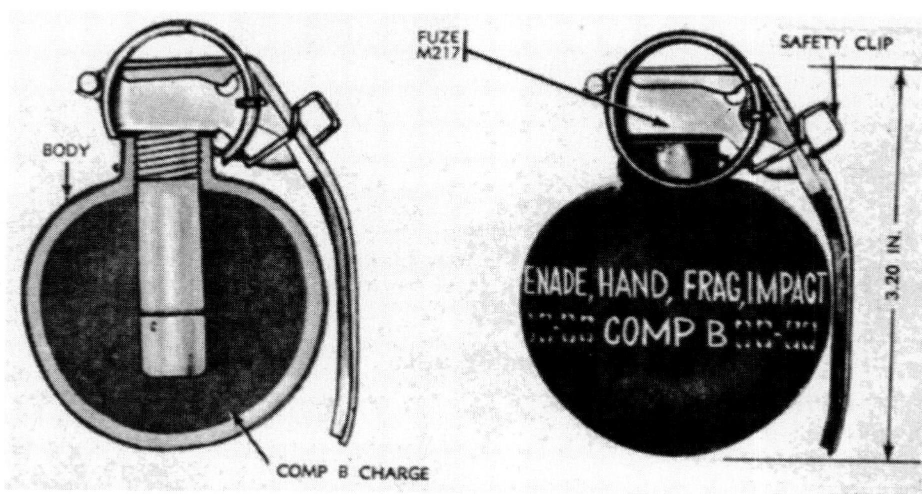

M68 fragmentation hand grenade

Data

Length: max, 126 mm
Diameter: 64 mm
Weight: 0.39 kg
Filling: 184 g Comp B
Body: steel
Fuze: model, M217; type, electrical impact with overriding delay function feature

Primer: M42
Detonator: lead azide, lead styphnate, PETN
Delay time: 3-7 s
Weight: 169 g
Length: 76.2 mm

Manufacturer
US Army Materiel Readiness Command.

Status
In production.

Service
US armed forces.

VERIFIED

M57 and M26A2 impact fragmentation hand grenade

Description
The M57 hand grenade is the M26A2 hand grenade with a safety clip. Each grenade is assembled with an electrical impact fuze.

Grenade body: bodies of the M61, M26A1 and M26 are identical to the M26A2 except the fuze threads are different. The body, constructed of two pieces of thin-wall sheet steel, has a notched fragmentation coil liner. Bodies contain a high explosive filler. Bodies of the M61, M26A1 and M26 contain booster pellets and are longer and narrower than those of the M26A2 and M57 which do not contain booster pellets.

Fuze, hand grenade, M217: see description in the entry for the M68 grenade.

Safety clip: the hand grenade safety clip is designed to keep the safety lever in place, should the safety pin be unintentionally removed from the grenade. It is an additional safety device used in conjunction with the safety pin.

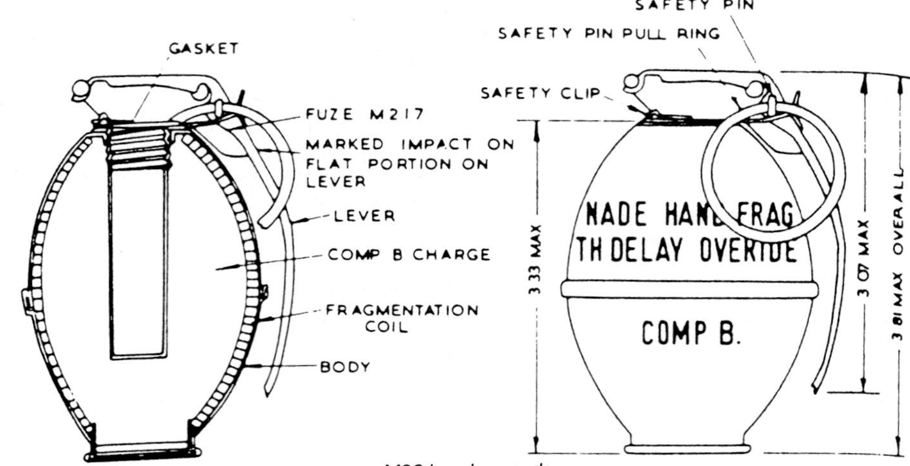

M26 hand grenade

Data
Length: max, 99 mm
Diameter: 57 mm
Weight: 454 g
Filling: 156 g; 8 g tetryl pellets
Body: thin-wall sheet steel with notched fragmentation coil
Filling: Comp B with tetryl pellets
Fuze: model, M217; type, electrical impact with overriding delay function feature
Primer: M42
Detonator: lead azide, lead styphnate, PETN
Delay time: 3-7 s
Weight: 76 g
Length: 76.2 mm

Manufacturer
US Army Materiel Readiness Command.

Status
In production.

Service
US armed forces.

VERIFIED

AN-M8 HC smoke hand grenade

Description
The AN-M8 HC smoke hand grenade is a burning type grenade used to generate white smoke for screening activities of small units. It is also used for ground-to-air signalling. The duration of smoke screen or signal is 105 to 150 seconds. Throwing distance for an average soldier is said to be 30 m.

Grenade body: the grenade body is a cylinder of thin sheet metal. It is filled with HC smoke mixture topped with a starter mixture directly under the fuze opening.

Fuze, hand grenade, M201A1: fuze M201A1 is a pyrotechnic delay-igniting fuze. The body contains a primer, first-fire mixture, pyrotechnic delay column and ignition mixture. Assembled to the body are a striker, striker spring, safety lever and safety pin with pull ring. The split end of the safety pin has an angular spread.

Safety clips: safety clips are not required with these grenades.

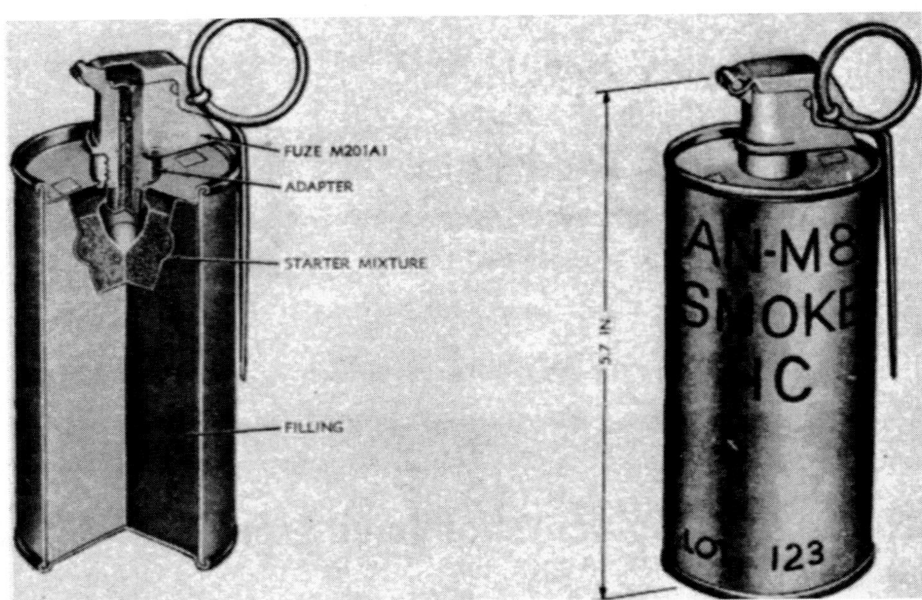

AN-M8 HC smoke hand grenade

Data
Length: 145 mm
Diameter: 63.5 mm
Weight: 681 g
Filling: 539 g HC type C
Model: AN-M8
Body: sheet metal
Burning time: 105-150 s
Fuze: model, M201A1; type, pyrotechnic delay-igniting
Primer: M39A1
Ignition mixture: iron oxide, titanium, zirconium
Delay time: 0.7-2 s
Weight: 43 g
Length: 99 mm

Manufacturer
US Army Materiel Readiness Command.

Status
In production.

Service
US armed forces.

VERIFIED

AN-M14 TH3 incendiary hand grenade

Description
The AN-M14 TH3 incendiary hand grenade is used primarily to provide a source of intense heat to destroy equipment. It generates heat to 2,200°C. The grenade filler will burn from 30 to 45 seconds. This was developed by the Chemical Warfare Service in response to a military requirement for a hand-held incendiary device which would allow the individual soldier to destroy equipment. Over 800,000 M14s were made, with final deliveries by Ordnance Products Incorporated in 1970. The grenade is normally hand thrown, although it may be rifle launched by using a special M2 series projection adaptor. Throwing distance for an average soldier is said to be 25 m.

The grenade body, of thin sheet metal, is cylindrical in shape. It is filled with an incendiary mixture, Thermite TH3 and First Fire Mixture VII.

Data
Length: 145 mm
Diameter: 63.5 mm
Weight: 900 g; filling, 752 g
Body: sheet metal
Filler: type, igniter mixture III, delay mixture V, FF mixture VII; incendiary mixture; Thermite TH3 and Thermite plain
Fuze: model, M201A1; type, pyrotechnic delay-igniting
Primer: M39A1

Ignition mixture: iron oxide, titanium, zirconium
Delay time: 0.7-2 s
Weight: 42 g
Length: 99 mm

Manufacturer
US Army Materiel Readiness Command.

Status
Probably no longer made.

Service
US Army.

VERIFIED

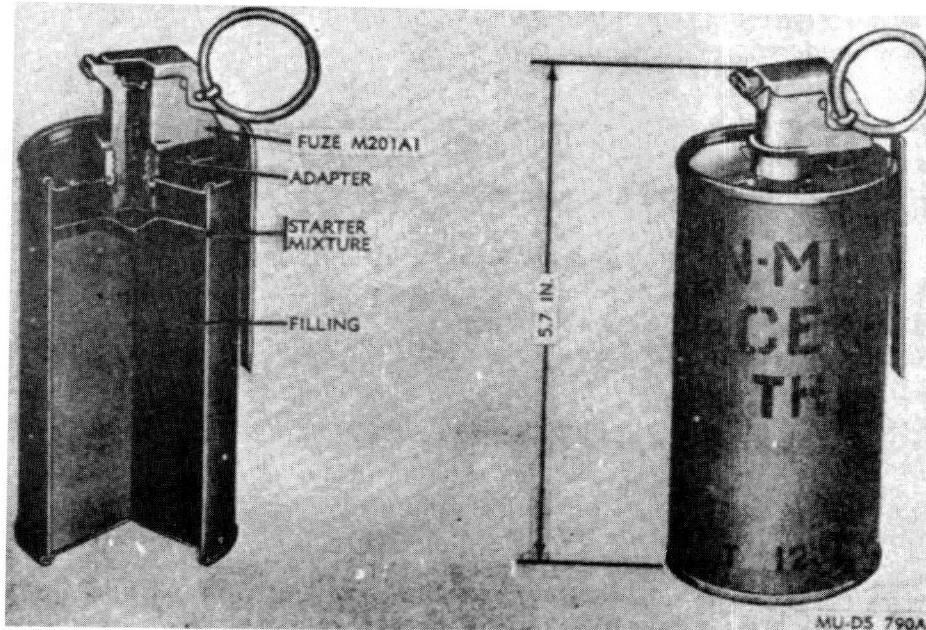

AN-M14 TH3 incendiary hand grenade

M76 vehicle smoke grenade

Description
The M76 grenade was developed by the AAI Corporation under contract to the US Army. It protects armoured vehicles and personnel with an instantaneous smoke screen which is effective against visual and infra-red observation. Though primarily developed for use with the M1 Abrams tank, the M76 can be fired from any system currently using the L8 grenade with either the six-tube M250 or four-tube M243 launcher.

Within two seconds of firing the M76 grenade forms a screen over a 100° arc, 7 m high and 30 m forward of the launcher. The screen remains effective for a minimum of 45 seconds with winds up to 20 km/h.

Data
Length: 238 mm
Diameter: 66.5 mm
Weight: 1.81 kg

Manufacturer
AAI Corporation, PO Box 126, Hunt Valley, Maryland 21030-0126.

Status
In production.

Service
US Army.

VERIFIED

40 mm grenade cartridges

Description
There is a wide range of 40 mm grenade cartridges suitable for use with M79, M203, HK 79 and other types of grenade launcher. The 40 mm cartridge is a fixed munition consisting of a cartridge case and a projectile.

The cartridge case is made of aluminium or plastic and has an integral propellant retainer into which is inserted a thin-walled brass cup, containing the propellant, followed by an aluminium base plug which seals the base of the cartridge case. This arrangement is the basis of the high-low propulsion system required to propel a 40 mm projectile from a shoulder-fired weapon. When the firing pin strikes the cartridge primer the propellant in the brass powder cap is ignited and generates a pressure in the region of 2,500 kg/cm^2. This causes the brass case to rupture at a ring of vent holes in the propellant retainer, allowing gas to flow into the remainder of the cartridge case which thus forms a low pressure chamber (around 200 kg/cm^2). This pressure is adequate to propel the projectile and does not produce excessive recoil. The grenade leaves the launcher

40 MM CARTRIDGES

Model	Type	Fuze	Remarks
M381	HE	M552	Obsolete
M386	HE	M551	Obsolete
M397A1	HE	M536E1	Airburst; obsolete
M406	HE	M551	Current production
M441	HE	M551	Obsolete
M433	HEDP	M550	Fragmentation/shaped charge
M576E1	Multiprojectile	None	Contains 20 balls
M583A1	White star	None	Parachute; obsolete
M585	White star	None	Cluster; obsolete
M661	Green star	None	Parachute
M662	Red star	None	Parachute
M675	Red smoke	None	Current
M676	Yellow smoke	None	Canopy smoke; obsolete
M680	White smoke	None	Canopy smoke; obsolete
M682	Red smoke	None	Canopy smoke; obsolete
M715	Green smoke	None	Current
M717	Yellow smoke	None	Current
M674	CS	None	Current
M382	Practice	M552	Filled yellow dye
M407A1	Practice	M551	Yellow smoke and dye
M781	Practice	None	Yellow or orange dye

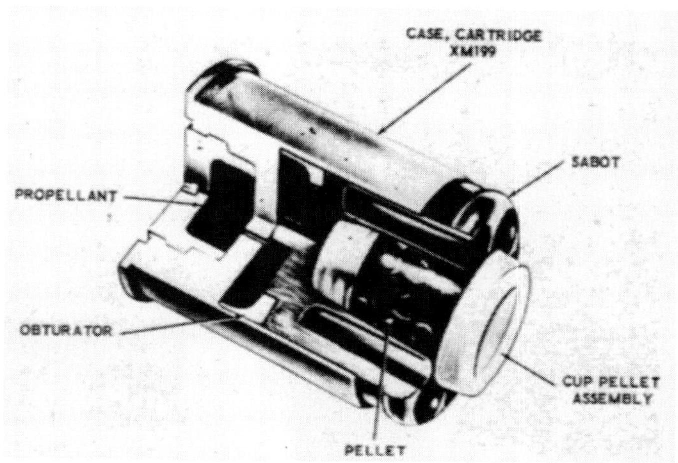

40 mm M576E1 multiple projectile cartridge

40 mm cartridge case and projectile

with a muzzle velocity of 76 m/s and is spin stabilised by the launcher rifling. The spin also provides the rotational force necessary to arm the fuze.

Grenades known to be in the series are listed below and it will be seen that they include high explosive, riot control, practice and a variety of smoke, signalling and illuminating types.

M552 and M551 impact detonating fuzes are used with the HE and the TP rounds. The M552 fuze arms by a spin action and is armed about 3 m from the muzzle. The M551 fuze arms by a spin and setback action and must travel between 14 and 28 m before being armed.

The high explosive grenade has an effective casualty radius of 5 m. The effective casualty radius is defined as the radius of a circle about the point of detonation in which it may be expected that 50 per cent of exposed troops will become casualties.

Manufacturer

Lockheed Martin, Milan Army Ammunition Plant, Tennessee.

40 mm grenade ammunition (and close derivatives) is also manufactured outside the USA in Austria, Germany, Indonesia (under licence from Singapore), South Korea, Poland, Singapore, South Africa and Spain.

Status

In production.

Service

US armed forces and many other nations.

UPDATED

40 mm high-velocity grenade cartridges

Description

The family of 40 mm high-velocity grenade cartridges developed for use in weapons such as the MK 19 Mod 3 automatic grenade launcher are the result of development work carried out at Picatinny Arsenal and elsewhere during the 1960s. The overall design and overall operating principle employed by the high-velocity cartridges resembles that of their lower velocity grenade equivalents (see previous entry) but the cartridge cases are longer and more robust to provide for the larger propellant load which propels the grenades at a muzzle velocity of 244 m/s to a maximum range of 2,200 m.

There are two main operational grenade cartridges, the 40 mm Cartridge, HE, M384, and the 40 mm Cartridge, HEDP, M430. Both have a propellant system which operates along the same lines as the lower velocity cartridges but the grenade warheads and fuzing systems differ. A third cartridge is the 40 mm Cartridge, Practice, M385.

Both operational cartridges are fixed rounds issued in linked belts of 48 or 50 rounds. On the HEDP M430 (the operation of the HE M384 is identical), when the firing pin strikes the FED 215 percussion primer, the propellant charge (4.64 g of M2 propellant) is ignited. Pressure is generated in the M169 cartridge's high pressure chamber and forces gases through vent holes into the low pressure chamber to propel the grenade forward. As the grenade travels down the barrel, rifling in the barrel operates on the grenade's copper drive band to impart stabilising spin at a rate of 12,000 rpm.

The HE M384 warhead has a filling of 54.5 g of Composition A5 ignited by an M533 point detonating fuze and creates blast and fragments. By contrast the HEDP M430 contains 38 g of RDX or Composition A5 ignited by a PIBD M549 fuze. The explosive is a shaped charge behind a copper cone. In addition to the shaped charge on the M430 warhead the steel walls are internally serrated to produce anti-personnel fragments. With both grenades the fuzes are armed some 18 to 30 m from the muzzle.

40 mm high-velocity grenades are also produced in Indonesia (under licence from Singapore), South Korea, Singapore and South Africa.

Data

Model	M384	M430
Type	HE	HEDP
Weight	350 g	340 g
Length	112 mm	113 mm
Filling	54.5 g Comp A5	38 g RDX
Muzzle velocity	244 m/s	244 m/s
Max range	2,200 m	2,200 m
Fuze	PD M533	PIBD M549

Manufacturer

Lockheed Martin, Milan Army Ammunition Plant, Tennessee.

Status

In production.

Service

US armed forces and many other nations.

NEW ENTRY

YUGOSLAVIA (SERBIA AND MONTENEGRO)

Yugoslav grenades

Description

Yugoslavia uses both hand and rifle grenades. The hand grenades are egg-shaped and are basically of the same design. The earliest ones had an impact fuze but the later ones are all fitted with delay fuzes. Typical of the series is the M69 which weighs 600 g and has a 4.5 second delay fuze.

The M59/66 semi-automatic rifle is used to project the new grenades which were developed in the country. The Yugoslav M64 assault rifle, which is equipped with a grenade sight, takes a grenade launcher when the compensator is removed. Erecting the grenade sight automatically cuts off the gas supply to the piston.

There are four Yugoslav rifle grenades, all characterised by their rather small warhead diameters. In addition to the anti-tank and anti-personnel grenades, there are an illuminating and a smoke grenade. The smoke grenade is 330.5 mm long and 40 mm in diameter. It generates smoke for 90 seconds.

Data

	Anti-tank	Anti-personnel
Length	390 mm	307 mm
Diameter	60 mm	30 mm
Weight, fuzed	602 g	520 g

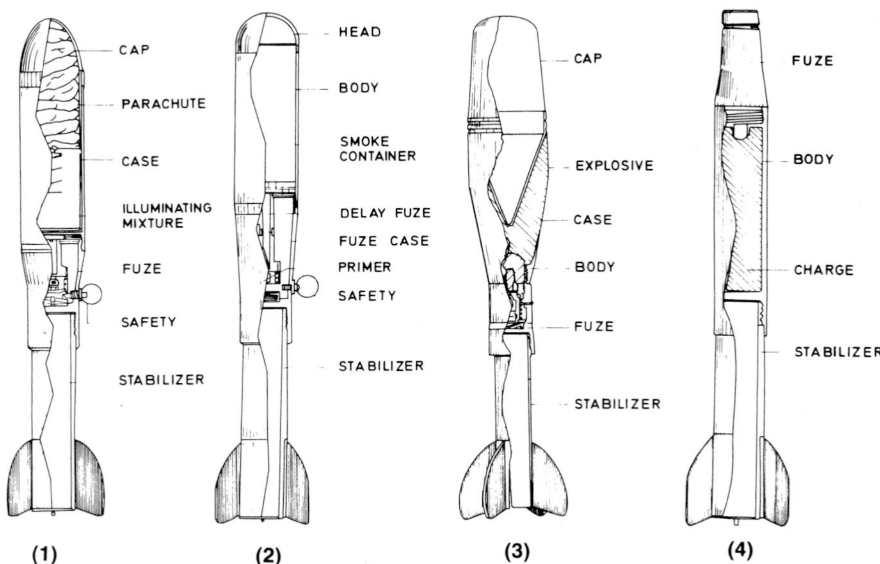

(1) *M62 illuminating rifle grenade* **(2)** *M62 smoke rifle grenade* **(3)** *M60 anti-tank rifle grenade* **(4)** *M60 anti-personnel rifle grenade*

Yugoslav M69 anti-personnel hand grenade

7.62 mm M70B1 rifle with M60 HEAT rifle grenade loaded and M60 HE anti-personnel grenade below

	Anti-tank	Anti-personnel
Weight of HE	235 g	67 g
Fuze type	Impact	Impact
Max range	150 m	400 m
Effect range	–	20-50 m
Penetration	–	–

Holding Company
YUGOIMPORT - SDPR, PO Box 89, Bulevar umetnosti 2, YU-11070 Novi, Belgrade.

Status
Current production status uncertain. The rifle grenades are offered for export sales by Euroinvest of Macedonia and RH-ALAN of Croatia.

Service
Most former Yugoslav states.

UPDATED

M75 hand grenade

Description
The M75 has uncertain design origins but it appears to be a close copy of the Austrian Arges HG 73 grenade. It is a defensive grenade using prefragmented steel balls packed in a plastic body around a 36 g plastic explosive bursting charge. The usual type of fly-off lever and mousetrap fuze is fitted centrally in the grenade. The grenade has a storage life of at least 10 years, will function in temperatures between −30 and +60°C, and is drop safe from a height of 10 m.

Data
Length: 89 mm
Diameter: 57 mm
Weight: 355 g
Filling: 36 g plastic
Steel balls: ca 3,000 of 2.5-2.9 mm diameter
Delay: 3-4 s

Holding Company
YUGOIMPORT - SDPR, PO Box 89, Bulevar umetnosti 2, YU-11070 Novi, Belgrade.

Status
Available.

M75 hand grenade

Service
Has been encountered in Southern Africa.

UPDATED

M79 anti-tank grenade

Description
The M79 is a virtual direct copy of the Soviet RKG-3 series so it is a hand-thrown, drag-stabilised, shaped charge grenade for the attack of armour at close range. The body contains the shaped charge unit and also incorporates the desired standoff distance. The handle carries the ignition system and also a drag parachute which keeps the grenade in a nose-forward attitude to ensure correct operation on impact. There is an impact fuze which is armed by a fly-off lever and by the action of the drag parachute after throwing.

Data
Length: 397 mm
Diameter: 76 mm
Weight: 1.1 kg

M79 shaped charge anti-tank grenade

Throwing range: ca 25 m
Penetration: homogeneous armour, 220 mm

Holding Company
YUGOIMPORT - SDPR, PO Box 89, Bulevar umetnosti 2, YU-11070 Novi, Belgrade.

Status
Available.

UPDATED

BRD M83 smoke grenades

Description
The BRD M83 series of smoke hand grenades are all of similar appearance and include white, red, yellow or green smoke, the contents being indicated by the colour of the base of the grenade body. The coloured smoke grenades are somewhat smaller and lighter than the white smoke model. There are two types of ignition system available, either the conventional fly-off lever percussion system or a pull-type friction igniter. In the latter model the cap is removed, revealing a ring and cord; the ring is pulled so as to drag the friction bar through the igniting composition and the grenade is then thrown.

Data
Length: 130 mm
Diameter: 80 mm
Weight: 1 kg
Filling: 700 g HC
Smoke emission time: 3 min

Holding Company
YUGOIMPORT - SDPR, PO Box 89, Bulevar umetnosti 2, YU-11070 Novi, Belgrade.

Status
Available.

VERIFIED

M83 smoke grenades; (left) *pull-igniter type,* (right) *fly-off lever igniter type*

BRD M83 white smoke hand grenade

Description

The BRD M83 white smoke grenade may be used for screening or signalling purposes. It is filled with a hexachloroethane composition and uses the conventional type of fly-off lever striker mechanism.

Data

Length: 148 mm
Diameter: 60 mm
Weight: 600 g
Filling: 454 g HC
Delay time: 2.5-3.5 s
Emission time: 70-90 s

Manufacturer

Institute of Security, PO Box 574, YU-11005 Belgrade.

Status

In production.

VERIFIED

KD-M79 vehicle-launched smoke grenade

Description

The KD-M79 is a cylindrical grenade designed to be fired from vehicle-mounted smoke dischargers to provide local screening. It may also be electrically initiated as a static smoke generator or, fitted with a friction fuze, thrown by hand.

Data

Length: 152 mm
Diameter: 82 mm
Weight: 1.549 kg
Filling: 920 g HC mixture
Delay: 2.2-3.9 s
Emission time: 3.5 min
Electrical initiation: min, 4.5 V DC 1 A

Manufacturer

Institute of Security, PO Box 574, YU-11005 Belgrade.

Status

In production.

VERIFIED

KD-M79 vehicle-launched smoke grenade

MORTAR AMMUNITION

ARGENTINA

60 mm HE/fragmentation bomb PE 1-60 FMK1 Mod 0

Description

A high explosive projectile with a heat-treated forged steel body, the FMK1 Mod 0 bomb is composed of three basic parts; fuze, body and stabiliser unit. The stabiliser consists of a steel cartridge carrier and eight aluminium fins. The bomb is standard for the 60 mm FMK1 assault and FMK3 standard mortars. The fuze is the Brandt V19P or FMK13 or FMK20 of Argentine origin, but it can be fitted with any other fuze using the same 1.5 in fuze well thread at the user's choice. The explosive filling is 350 g of TNT, although Composition B could be loaded if required. The propelling charge consists of a primary cartridge screwed into the cartridge carrier and six 'horseshoe' secondary charges clipped round the cartridge carrier.

There is also an exercise version of the bomb which is completely inert but loaded to the same weight as the HE bomb.

Data

Length: 384.7 mm
Weight: 1.9 kg
Lethal radius: 15 m

PERFORMANCE:

Charge	Muzzle velocity	Max range
0 (Primary)	65 m/s	415 m
1	100 m/s	903 m
2	127 m/s	1,353 m
3	150 m/s	1,842 m
4	170 m/s	2,220 m
5	189 m/s	2,632 m
6	205 m/s	3,000 m

Manufacturer

Fabrica Militar Rio Tercero, Rio Tercero, Cordoba.

Status

In production.

Service

Argentine forces.

VERIFIED

60 mm HE/fragmentation bomb FMK1 Mod 0

---

60 mm smoke bomb FMK3 Mod 0

Description

This uses the same body as the high explosive bomb FMK1 Mod 0 but the steel is specially heat treated to break into large pieces rather than small fragments, since the requirement is simply to liberate the payload. A central exploder carries a charge of high explosive, initiated by a nose fuze, which shatters the bomb body and distributes the charge of white phosphorus. The tail unit is the same as that of the HE bomb, the same propelling charge system is used and the ballistic performance is the same.

Data

Length: fuzed, 384.7 mm
Weight, fuzed: 1.9 kg
Filling: 270 g white phosphorus
Number of charges: P + 6
Fuze: PD FMK20 or FMK13 of Argentine manufacture, or V19P

Manufacturer

Fabrica Militar Rio Tercero

VERIFIED

60 mm smoke bomb FMK3 Mod 0

---

60 mm HEAT bomb FMK4 Mod 0

Description

This bomb was developed by the Rio Tercero factory to provide an anti-tank projectile for the infantry assault mortar, which is capable of being fired horizontally.

The bomb generally resembles the 60 mm high explosive bomb but is longer, with a distinct parallel-sided section to the body and a conical rather than ogival nose. The nose section is hollow, and the parallel-walled section contains a conventional shaped charge initiated by a PIBD fuze having a piezoelectric element in the nose of the bomb. The tail unit is that of the HE bomb, but only two propellant increments are provided, sufficient to give an effective range of about 200 m. Although intended for short range direct fire, there seems no reason why this bomb could not be used with a conventional charge system for indirect fire and top attack of armoured vehicles.

Data

Length: fuzed, 414 mm
Weight: fuzed, 2 kg
Filling: 250 g TNT or RDX
Number of charges: P + 2
Fuze: PIBD M509 or FMK25
Effective range: 200 m

Manufacturer

Fabrica Militar Rio Tercero, Rio Tercero, Cordoba.

VERIFIED

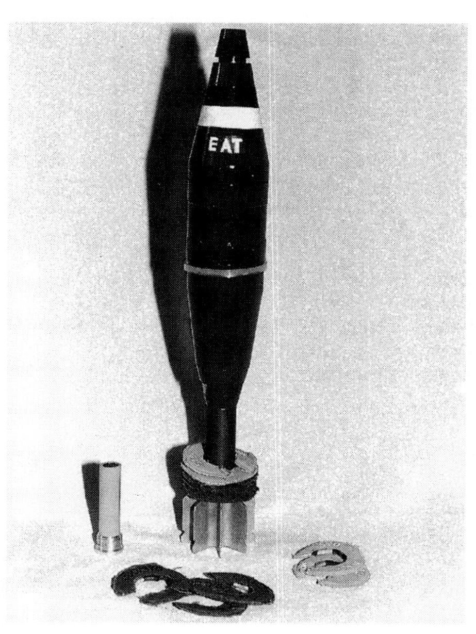

Argentine 60 mm HEAT bomb FMK4 Mod 0

81 mm HE/fragmentation bomb FMK6 Mod 0 High Capacity

Description

This bomb is manufactured under licence from Thomson-Brandt of France and is the Model 81 GC. The body and stabiliser are of steel and it is filled with 2 kg of TNT or Composition B. The fuze may be the Argentine FMK13 or the Thomson-Brandt V19P. The propelling charge consists of a primary cartridge pressed into the cartridge carrier and four secondary charges clipped into the fins. The bomb is used with Thomson-Brandt or Argentine FMRT 81 mm mortars.

An inert version, loaded to the correct weight, is available for practice purposes.

Data

Length: 615 mm
Weight: 6.6 kg
Lethal radius: 30 m

PERFORMANCE:

Charge	Muzzle velocity	Max range
0 (Primary)	45 m/s	200 m
1	75 m/s	535 m
2	98 m/s	900 m
3	120 m/s	1,250 m
4	138 m/s	1,620 m

Manufacturer

Fabrica Militar Rio Tercero, Rio Tercero, Cordoba.

Status

In production.

Service

Argentine forces.

UPDATED

81 mm HE/fragmentation bomb FMK6 Mod 0 High-Capacity and Instructional bomb

81 mm HE/fragmentation bomb FMK13 Mod 0

Description

This bomb is of French origin, originally provided for the Thomson-Brandt mortars in Argentine service and later manufactured under licence. The body is of forged steel and the stabiliser an aluminium casting. The filling is 700 g of TNT, although Composition B can be used if desired. The fuze may be the Argentine FMK13 or the Thomson-Brandt V19P. The propelling charge consists of a primary cartridge screwed into the stabiliser and eight secondary charges clipped around the stabiliser tube.

An inert version, loaded to the correct weight, is available for practice purposes.

Data

Length: 415 mm
Weight: 4.3 kg
Filling: 700 g TNT or Comp B
Lethal radius: 25 m

PERFORMANCE:

Charge	Muzzle velocity	Max range
0 (Primary)	70 m/s	485 m
1	111 m/s	1,160 m
2	143 m/s	1,820 m
3	172 m/s	2,380 m
4	199 m/s	2,940 m
5	224 m/s	3,480 m
6	247 m/s	4,000 m
7	269 m/s	4,470 m
8	290 m/s	4,900 m

Manufacturer

Fabrica Militar Rio Tercero, Rio Tercero, Cordoba.

Status

In production.

Service

Argentine forces.

VERIFIED

81 mm HE/fragmentation bomb FMK13 Mod 0

120 mm HE/fragmentation bomb FMK2 Mod 1 CN

Description

This bomb is of French origin, manufactured under Thomson-Brandt licence, and is the standard bomb for the 120 mm FMK2 Mod 0 and Thomson-Brandt mortars in Argentine service. The body and stabiliser are of steel, the filling is 2.6 kg of TNT or Composition B, and the FMK13 or V19P fuzes may be used. The propelling charge consists of a primary cartridge pressed into the stabiliser tube and seven secondary cartridges clipped around the tube.

An inert bomb is available for practice purposes.

Data

Length: 672 mm
Weight: 13 kg
Filling: 2.6 kg TNT or Comp B
Lethal radius: 30 m

PERFORMANCE:

Charge	Muzzle velocity	Max range
0 (Primary)	119 m/s	1,360 m
1	153 m/s	2,200 m
2	185 m/s	3,050 m
3	217 m/s	3,850 m
4	248 m/s	4,750 m
5	277 m/s	5,550 m
6	305 m/s	6,150 m
7	336 m/s	6,700 m

Manufacturer

Fabrica Militar Rio Tercero, Rio Tercero, Cordoba.

Status

In production.

Service

Argentine forces.

VERIFIED

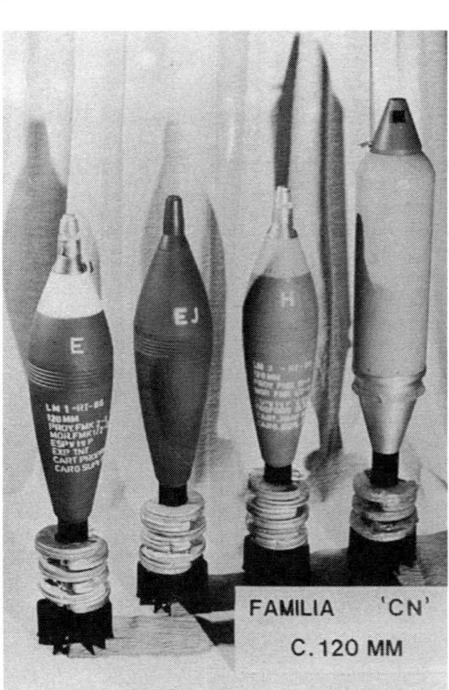

120 mm HE, practice, smoke and illuminating bombs

120 mm HE/fragmentation bomb FMK1 Mod 1 High Capacity

Description

This is another bomb of French origin, manufactured under licence. Body and stabiliser are of steel and the filling is 4.2 kg of TNT or Composition B. The Argentine fuze FMK13 or the Thomson-Brandt fuze V19P may be used. The propelling charge consists of a primary cartridge pressed into the stabiliser tube, together with six secondary charges clipped around the tube.

An inert bomb, weighted to match the service bomb, is available for practice purposes.

Data

Length: 820 mm
Weight: 17.2 kg
Filling: 4.2 kg TNT or Comp B
Lethal radius: 30 m

PERFORMANCE:

Charge	Muzzle velocity	Max range
0 (Primary)	105 m/s	1,050 m
1	137 m/s	1,750 m
2	166 m/s	2,550 m
3	196 m/s	3,300 m
4	222 m/s	4,100 m
5	246 m/s	4,950 m
6	272 m/s	5,650 m

Manufacturer

Fabrica Militar Rio Tercero, Rio Tercero, Cordoba.

Status

In production.

Service

Argentine forces.

VERIFIED

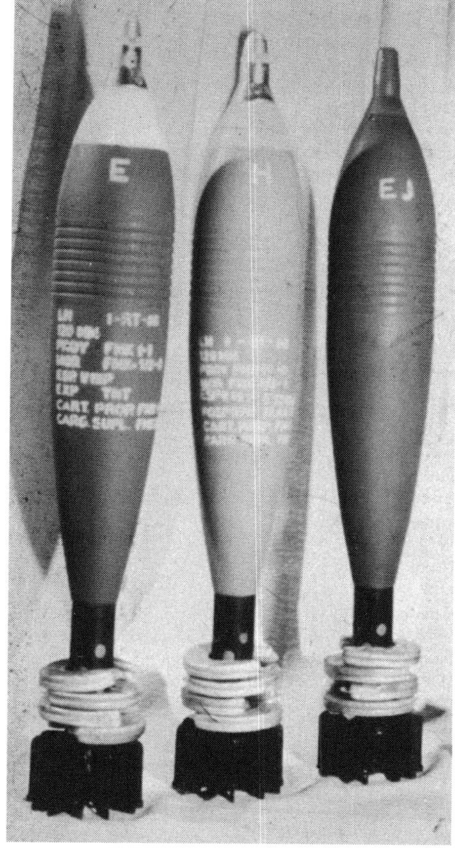

120 mm HE/fragmentation bomb FMK1 Mod 1 High-Capacity

AUSTRIA

DNG 81 mm mortar ammunition

Description

The firm of Dynamit Nobel Graz (DNG) manufactures both 81 mm and 120 mm mortar ammunition.

The 81 mm bombs are a licensed version of the British L15A1 made to close tolerances and high standards. As a result, their fragmentation pattern is very even and their flight regular and predictable. Distribution of the fragments is claimed to be precisely even, so producing the optimal effect from the weight of the bomb.

Data

Length: 487 mm
Weight: 4.15 kg
Weight of casing: 2.9 kg
Filling: 750 g TNT
Number of fragments: ca 1,600 of at least 0.5 g; ca 2,000 of 0.3-0.5 g
Muzzle velocity: 75-300 m/s
Max range: 5,800 m
Max barrel pressure: 750 bars
Fuzes: percussion, M125A1, DM111A2 or equivalent

Manufacturer

Dynamit Nobel Graz, Annenstrasse 58, A-8020 Graz

Status

In production.

Service

Austrian Army.

VERIFIED

81 mm DNG HE bomb without secondary charges

DNG 120 mm HE long-range mortar bomb

Description

The DNG 120 mm HE long-range bomb is a streamlined projectile much resembling the 81 mm in general shape and possessing a good aerodynamic efficiency which gives a long range with minimum dispersion. Unlike the 81 mm bomb the 120 mm does not rely on a plastic sealing ring for obturation, but uses a bourrelet section in the centre of the body with six machined grooves.

The fins are carried on a boom; the primary cartridge is carried inside this boom in the usual manner. Augmenting charges are flat plates covered with coloured silk, and these plates have a slot cut in them to allow them to be slipped over the tail boom as required. These charges run up to number 9, but the normal firing is with Charge 6 since the higher charges can only be used with heavy mortars. The body is made from ferritic cast steel, fully machined over and carefully controlled for weight and all dimensions. The normal fuze is the DM 111 series but any one of several electronic proximity fuzes will also fit the nose adaptor.

Data

Length: overall, 747 mm
Weight: overall, 14.5 kg; shell body, 10.5 kg
Filling: 2.3 kg TNT
Fragments: ca 6,000 effective: 2,800 between 0.3-0.5 g; 3,200 between 0.5-30 g

Charge	Muzzle velocity	Max range	Barrel pressure
1	135 m/s		
6	322 m/s	7,040 m	800 bar
9	403 m/s	9,010 m	1,500 bar

Manufacturer

Dynamit Nobel Graz, Annenstrasse 58, A-8020 Graz

Status

Available.

UPDATED

Hirtenberger mortar ammunition

Description

The Austrian company Hirtenberger manufactures ammunition of many small and large calibres, including four sizes of mortar bomb: 60, 81, 82 and 120 mm.

All bombs are of modern design; their maximum firing ranges and accuracy are due to their aerodynamic shape and favourable ratio of weight to cross-section.

The stabiliser, made from extruded aluminium alloy, is screwed firmly into the body and contains the waterproof propelling cartridge. The horseshoe-shaped augmenting charges are clipped to the tail boom.

The bodies of the high explosive bombs are made of spheroidal graphite cast iron. All high explosive bombs are also available with inert filling and impact fuze or fuze plug as practice versions for training purposes, with identical ballistic performance.

A new design is the Hirtenberger 60 mm LD bomb which, because of its increased length and weight, offers a 25 per cent better fragmentation than the standard version, though the maximum ranges of both types are the same.

For the smoke versions, Hirtenberger uses red phosphorus in the 60 mm calibre, both red and white phosphorus in the 81 mm and 82 mm calibres and white phosphorus or hexachloroethane (HC) in the 120 mm calibre.

The incendiary versions use red phosphorus, which achieves a much higher combustion temperature than conventional white phosphorus and offers an additional long-lasting smoke effect.

All Hirtenberger mortar bombs are suitable for all usual mortar types with corresponding calibres and gas pressure.

Manufacturer

Hirtenberger AG, A-2552 Hirtenberg.

Status

In production.

VERIFIED

Hirtenberger 60 mm mortar bombs

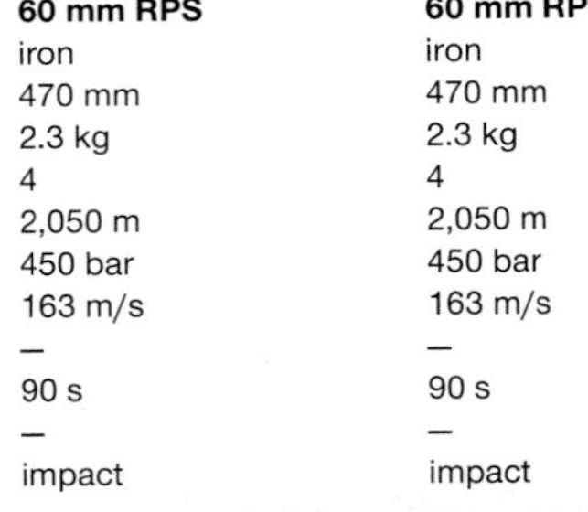

Data	**60 mm HE 80**	**60 mm HE LD**	**60 mm ILL**	**60 mm RPS**	**60 mm RPI**
Body	cast iron	cast iron	iron	iron	iron
Length	300 mm	360 mm	470 mm	470 mm	470 mm
Weight	1.6 kg	1.9 kg	2.3 kg	2.3 kg	2.3 kg
No of propelling charges	4 (5)*	4 (5)*	4	4	4
Max range	2,900 (4,400) m	3,000 (4,400) m	1,900 m	2,050 m	2,050 m
Gas pressure	450 (550) bar	500 (550) bar	450 bar	450 bar	450 bar
Initial velocity	199 (276) m/s	210 (275) m/s	163 m/s	163 m/s	163 m/s
Illuminating power	–	–	400,000 cd	–	–
Burning time	–	–	35 s	90 s	90 s
Rate of descent	–	–	3-4 m/s	–	–
Fuze	impact	impact	time	impact	impact

With one or two propelling charges only, the 60 mm HE 80 mortar bomb is suitable for commando mortars.
*Values in parentheses are for a barrel length of 1,000 mm

VERIFIED

Hirtenberger 60 mm mortar bombs

Hirtenberger 81 mm mortar bombs

Data	**81 mm HE 70**	**81 mm WP**	**81 mm ILL Mk 3**	**81 mm RPS Mk 3**	**81 mm RPI Mk 3**
Body	cast iron	cast iron	aluminium	aluminium	aluminium
Length	487 mm	487 mm	635 mm	635 mm	635 mm
Weight	4.15 kg	4.17 kg	3.8 kg	3.8 kg	3.8 kg
No of propelling charges	6	6	6	6	6
Max range standard mortar	5,800 m	5,800 m	5,200 m	5,500 m	5,500 m
Max range long-range mortar	6,300 m	6,300 m	5,450 m	5,700 m	5,700 m
Gas pressure	850 bar	850 bar	850 bar	850 bar	850 bar
Initial velocity standard mortar	295 m/s	295 m/s	311 m/s	311 m/s	311 m/s
Initial velocity long-range mortar	303 m/s	303 m/s	320 m/s	320 m/s	320 m/s
Illuminating power	–	–	1,200,000 cd	–	–
Burning time	–	–	35 s	90 s	90 s
Rate of descent	–	–	4-5 m	–	–
Fuze	impact or proximity	impact or proximity	time	impact or proximity	impact

The above 81 mm mortar bombs are also available in 82 mm calibre with practically identical ballistic data.

The 81 mm HE 70 can also be equipped with a different set of propelling charges (three big ones, three small ones) and thus offers the same ballistic data as the L15 HE Bomb.

VERIFIED

Hirtenberger 120 mm mortar bombs

Data	120 mm HE 78	120 mm WP	120 mm HC
Body	cast iron	cast iron	cast iron
Length	747 mm	747 mm	747 mm
Max range			
Charge 6 (standard mortar)	7,100 m	7,100 m	7,100 m
Charge 7 (standard mortar)	7,750 m	7,750 m	7,750 m
Charge 7 (long-range mortar)	8,640 m	8,640 m	8,640 m
Charge 9 (long-range mortar)	9,670 m	–	–
Gas pressure			
Charge 6	800 bar	800 bar	800 bar
Charge 7	1,000 bar	1,000 bar	1,000 bar
Charge 9	1,500 bar	–	–
Initial velocity			
Charge 6 (standard mortar)	320 m/s	320 m/s	320 m/s
Charge 7 (standard mortar)	340 m/s	340 m/s	340 m/s
Charge 7 (long-range mortar)	370 m/s	370 m/s	370 m/s
Charge 9 (long-range mortar)	430 m/s	–	–
Duration of smoke effect	–	–	210 s
Height of burst	–	–	100-200 m
Fuze	impact or proximity	impact or proximity	time

VERIFIED

Hirtenberger 120 mm mortar bombs

BELGIUM

81 mm MECAR A1 Series mortar bomb family

Description

The MECAR A1 Series of 81 mm mortar bombs is used in M1 low pressure, M29/29A1 medium pressure and M252 and L16 series high pressure mortar systems and their equivalents. Five models of bomb are available and may be divided into two groups of ballistically matched bombs, that is M511A1 to M513A1 using a common firing table and M514A1 and M515A1 fired to a second firing table. The bombs available are: Smoke (FM) (TTC) M511A1; HE (TNT) M512A1; HE/ICM M514A1; Smoke (WP) M513A1; and Illuminating M515A1.

Bombs are packed in three-round NATO-approved waterproof polymer containers. HE and smoke bombs have a maximum range of some 5,500 m in high pressure mortars, 4,500 m in medium pressure and >2,500 m in low pressure mortar systems.

The A1 propulsion system employs a primary cartridge comprising a screw-threaded primer and shotgun-type ignition cartridge, and up to five horseshoe-type augmenting charges, the number of charges permitted to be used depending upon the pressure capacity of the mortar. The base augmenting charge is coloured blue, and there are three translucent cased charges and a red coloured supercharge.

Charge 0 = Primary cartridge only

Charge 1 = Primary cartridge and base augmenting charge

Charge 2 = Primary cartridge, base augmenting charge and one translucent case charge

(Charge 2 is the maximum for low pressure mortars)

Charge 3 = Primary cartridge, base augmenting charge and two translucent case charges

Charge 4 = Primary cartridge, base augmenting charge and three translucent case charges

(Charge 4 is maximum for medium pressure mortars)

Charge 5 = Primary cartridge, base augmenting charge, three translucent cased charges and the red cased supercharge

(Charge 5 is used only in high pressure mortar systems)

Data

PRIMARY AND AUGMENTING CHARGES

Primary Cartridge: M519; Augmenting Charges (3) M521 (translucent);
Base Augmenting Charge M520 (blue); Supercharge Increment M522 (red)

81 MM HIGH EXPLOSIVE BOMB M512A1

Length: fuzed, 516 mm
Weight: fuzed, 4.1 kg
Filling: 1.03 kg TNT
Body: nodular cast iron
Ranges: 100-4,500 m (Ch 0 to 4); to 5,500 m with Supercharge M522

81 MM SMOKE (TTC) (FM) BOMB M511A1

Length: fuzed, 516 mm
Weight: fuzed, 4.1 kg
Filling: 880 g Titanium Tetrachloride (FM)
Body: nodular cast iron
Ranges: 100-4,500 m (Ch 0 to 4); to 5,500 m with Supercharge M522

SMOKE (WP) BOMB M513A1

Length: fuzed, 516 mm
Weight: fuzed, 4.1 kg
Filling: 880 g white phosphorus (WP)
Body: nodular cast iron
Ranges: 100-4,500 m (Ch 0 to 4); to 5,500 m with Supercharge M522

81 MM HE/ICM BOMB M514A1

Length: fuzed, 630 mm
Weight: fuzed, 4.135 kg
Filling: 12 HEAT/fragmentation submunitions
Body: steel
Ranges: 250-4,100 m (Ch 0 to 4); to 4,600 m with Supercharge M522

81 MM ILLUMINATING BOMB M515A1

Length: fuzed, 625 mm
Weight: fuzed, 4.2 kg
Filling: parachute and flare canister

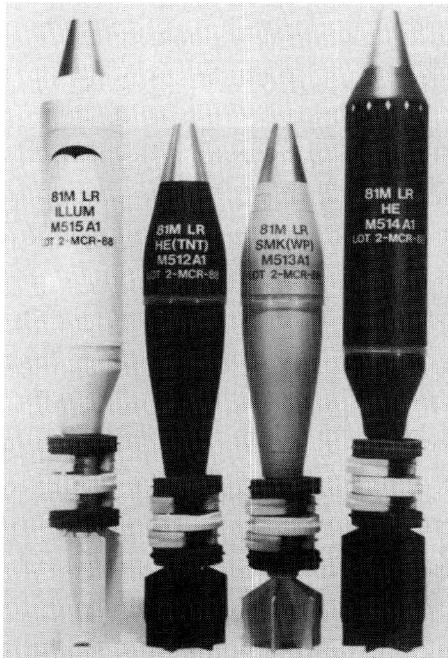

MECAR A1 family of 81 mm mortar bombs

Body: steel
Ranges: 250-4,100 m (Ch 0 to 4); to 4,600 m, with Supercharge M522

Manufacturer

MECAR SA, B-7181 Petit-Roeulx-lez-Nivelles.

Status

In production.

VERIFIED

MECAR 120 mm mortar bombs

Development

This family of 120 mm mortar bombs has been developed by MECAR as an improved performance set of rounds for the Brandt Type LT mortar and similar 120 mm mortars as standard rounds for use in the Royal Ordnance Armoured Mortar System (AMS) breech loading mortar.

Description

The 120 mm family of mortar bombs is currently in production by MECAR and consists of HE (Comp B), Smoke (WP) and Illuminating types.

The aerodynamic profile yields ranges in excess of 7,000 m with the Thomson-Brandt LT mortar and 9,200 m with the Royal Ordnance AMS mortar.

The HE and Smoke bodies are made from high fragmentation cast iron. Obturation is achieved by using a split-type plastic ring which fits into a groove behind the bourrelet. A light alloy tail unit, incorporating the primary cartridge, is screwed on to the body and up to eight augmenting cartridges can be fitted round the tail boom, depending upon which mortar is used.

The bombs are packaged in two-round NATO-approved polymer containers.

Data

HE M530A1

Length: fuzed, 780 mm
Weight: fuzed, 14.3 kg
Filling: 2.6 kg Comp B
Number of charges: Primary M547 + 6 augmenting M546 (Thomson-Brandt LT)
Primary M547 + 8 augmenting M546 (RO AMS)
Fuze: Impact, SQ and delay
Range: >9,200 m
Muzzle velocity: 440 m/s

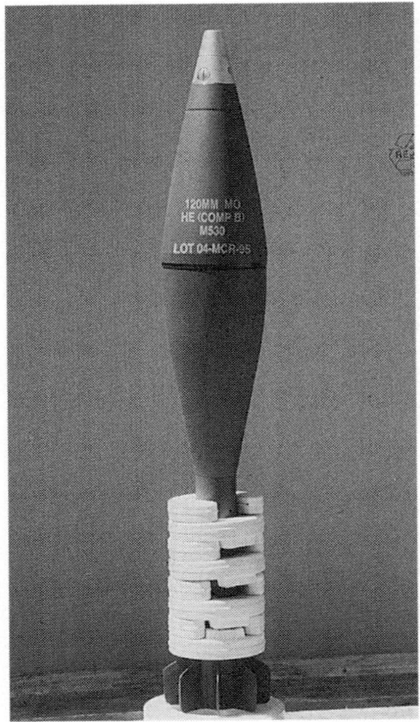

MECAR 120 mm HE bomb M530A1
1996

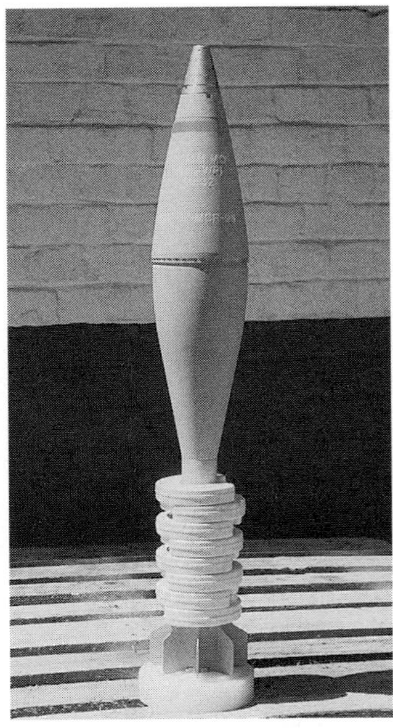

MECAR 120 mm Smoke (WP) bomb M532A1
1996

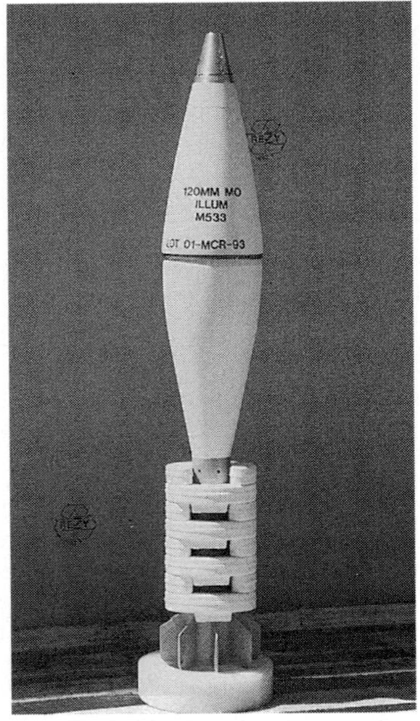

MECAR 120 mm Illuminating bomb M533
1996

SMOKE (WP) M532A1
Length: fuzed, 780 mm
Weight: fuzed, 14.3 kg
Filling: 2.45 kg WP
Number of charges: Primary M547 + 6 augmenting M546 (Thomson-Brandt LT)
Primary M547 + 8 augmenting M546 (RO AMS)
Fuze: Impact, SQ and delay
Max range: >9,200 m
Muzzle velocity: 440 m/s

ILLUMINATING M533A1
Length: filling, 780 mm
Weight: fuzed, 14.3 kg
Filling: parachute and flare
Number of charges: Primary M547 + 6 augmenting M546 (Thomson-Brandt LT)
Primary M547 + 8 augmenting M546 (RO AMS)
Fuze: MTSQ
Illumination intensity: 1 Mcd
Burning time: 50 s
Rate of descent: 6 m/s
Max range: >9,200 m
Muzzle velocity: 440 m/s

Manufacturer
MECAR SA, B-7181 Petit-Roeulx-lez-Nivelles.

Status
In production.

Service
Ordered by Saudi Arabia.

UPDATED

BRAZIL

60 mm HE bomb TIR 60 AE M3

Description
The 60 mm HE bomb TIR 60 AE M3 can be used with all types of smoothbore mortars of corresponding calibre, being designed to explode instantaneously on impact. Its fuze offers total safety during handling and, by means of a safety wire that does not allow the alignment of the explosive train when the round is in the bore, it also ensures complete safety in case of a double feed.

Data
Length: 245 mm
Weight: 1.37 kg
Filling: 150 g TNT or Comp B
Range: 1,800 m
Propellant increments: 1-4
Number of fragments: 600+

Manufacturer
Companhia de Explosivos Valparaiba, Praia do Flamengo, 200-20° Andar, 22210 Rio de Janeiro, RJ.

Status
In production. In service.

VERIFIED

81 mm HE bomb TIR 81 AE M4

Description
This round is suitable for all types of smoothbore 81 mm mortars, incorporating identical safety items to those of the same manufacturer's 60 mm round. Grenades, fuzes, ignition cartridges and propellant increments come in separate packages.

Data
Length: 335 mm
Weight: 3.345 kg
Filling: 500 g TNT or Comp B
Range: 4,050 m
Propellant increments: 1-6
Number of fragments: 1,000+

Manufacturer
Companhia de Explosivos Valparaiba, Praia do Flamengo, 200-20° Andar, 22210 Rio de Janeiro, RJ.

Status
In production. In service.

VERIFIED

81 mm HE bomb TIR 81 AE M7

Description
This is a more modern design than the AE M4, being longer and better streamlined. It uses the same EOP M4-CEV point detonating fuze as other 60 and 81 mm bombs, but the improved weight and shape give it a better ballistic performance and an improved lethal radius of burst.

Data
Length: 400 mm
Weight: 3.875 kg
Filling: 600 g TNT
Range: 5,200 m
Propellant increments: 6

Manufacturer
Companhia de Explosivos Valparaiba, Praia do Flamengo, 200-20° Andar, 22210 Rio de Janeiro, RJ.

Status
In production.

VERIFIED

BULGARIA

82 mm fragmentation bomb O-832DU

Description

The O-832DU fragmentation bomb is a conventional teardrop-shaped bomb with ferro-steel body and a welded steel tail unit with ten fins. The body has five gas-check rings around the bourrelet and is screwed at the nose for a fuze and at the rear end for the tail unit. The primary cartridge fits into the tail boom, and the secondary charges clip around the boom in horseshoe containers.

The complete round is known as the VO-832DU.

Data

Length: fuzed, 330 mm; unfuzed, 280 mm
Weight: 3.18 kg
Filling: 400 g TNT
Muzzle velocity: 225 m/s
Max pressure: 450×10^5 Pa
Min range: 85 m
Max range: 3,040 m

Agency

Kintex, 66 James Boucher str, 1407 Sofia.

Status

In production.

UPDATED

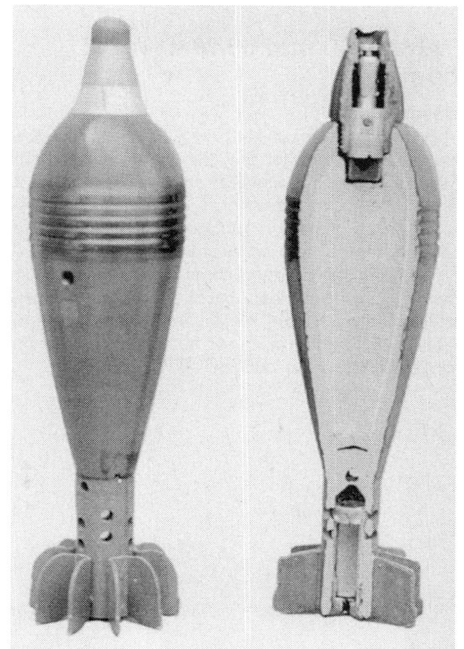

82 mm fragmentation bomb O-832DU

120 mm HE bomb OF-843A

Description

The OF-843A is a well-streamlined conventional bomb for general bombardment purposes. The body is of ferro-steel, with six gas-check grooves around the bourrelet and threaded at the nose for a fuze and at the rear for the tail unit. The tail unit is of welded steel, with eight fins, and the primary cartridge is inserted into the rear. Up to six secondary increments may be clipped around the boom.

Data

Length: fuzed, 665 mm
Weight: 16.5 kg
Filling: 1.6 kg TD-50
Muzzle velocity: 274 m/s
Max pressure: $1,030 \times 10^5$ Pa
Min range: 430 m
Max range: 5,850 m

Agency

Kintex, 66 James Boucher str, 1407 Sofia.

Status

In production.

UPDATED

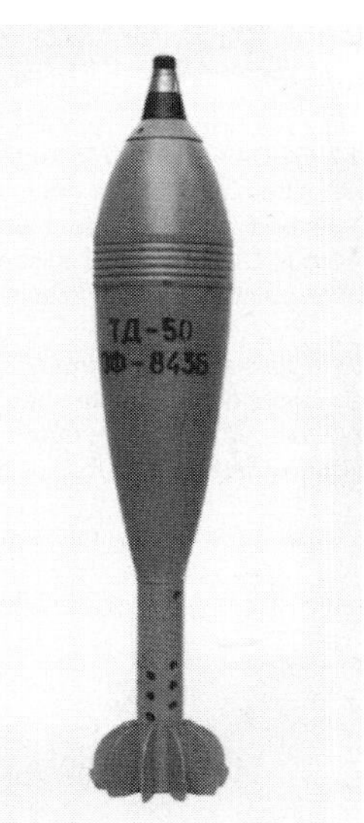

120 mm HE bomb OF-843A

Fuze, percussion, M-6

Description

The M-6 fuze is used with 82 mm HE mortar bombs. It is manufactured with a plastic or aluminium alloy body. Safety is ensured by a safety pin, withdrawn before firing, which locks the striker in the safe condition. A steel ball prevents arming of the fuze until setback occurs on firing. When the fuze arms, the striker is lifted out of a pocket in the bore-safe shutter, which then allows a spring to move the shutter so that a detonator and relay charge are positioned beneath the striker. On impact, the striker is driven into the detonator, and the relay charge and fuze magazine initiate the filling of the bomb.

Data

Length: 83.4 mm
Weight: plastic body, 128 g; aluminium body, 148 g
Diameter: 40 mm
Arming distance: 0.75-10 m

Agency

Kintex, 66 James Boucher str, 1407 Sofia.

Status

In production.

VERIFIED

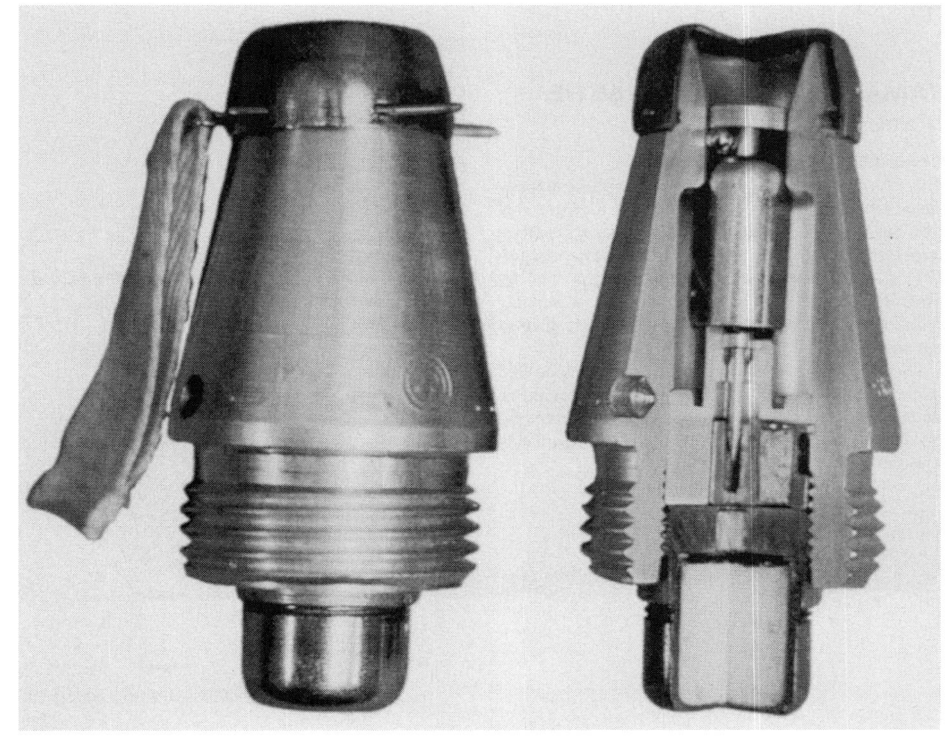

Fuze, percussion, M-6 for 82 mm mortar bombs

CHILE

FAMAE 60 mm 60-M-61-A HE bomb

Description

The FAMAE 60 mm 60-M-61-A HE bomb is a licence-manufactured copy of the French Thomson-Brandt Mark 61 HE bomb and differs from the original in few, if any, respects. The cast iron body contains approximately 260 g of pressed TNT ignited on impact with a target by a nose-mounted V9 point detonating fuze.

The propelling charges consist of a primary type LS cartridge, four secondary ballistite charges and one supplementary charge. When fired from a FAMAE 60 mm Commando or similar mortar the maximum range of this bomb is 1,050 m.

Data

Length: ca 317 mm
Weight: 1.72 kg
Filling: ca 260 g TNT
Max range: 1,050 m

Manufacturer

Fabricas y Maestranas del Ejercito (FAMAE), Avenida Pedro Montt 1568/1606, Santiago.

Status

Available.

Service

Chilean Army.

NEW ENTRY

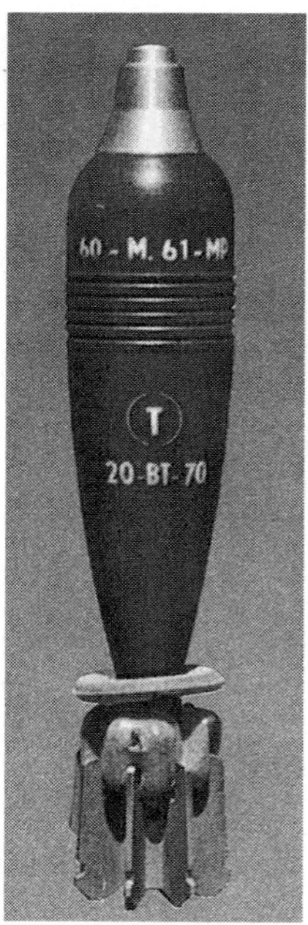

FAMAE 60 mm 60-M-61-A HE bomb
1996

FAMAE 81 mm 81-M-57-DA HE bomb

Description

The FAMAE 60 mm 81-M-57-DA HE bomb follows the classic Thomson-Brandt pattern and uses a one-piece forged steel casting for the body and aluminium for the tail fin assembly. The filling is TNT ignited by a licence-produced V-19P or V-19PA point detonating fuze in the nose.

The propelling system consists of a primary type Z 58 C cartridge plus six secondary ballistite charges. When fired from a FAMAE 81 mm mortar the maximum range of this bomb is approximately 4,200 m.

Data

Length: ca 382 mm
Weight: 3.2 kg
Filling: TNT
Max range: ca 4,200 m

Manufacturer

Fabricas y Maestranas del Ejercito (FAMAE), Avenida Pedro Montt 1568/1606, Santiago.

Status

In production.

Service

Chilean Army.

NEW ENTRY

FAMAE 81 mm 81-M-57-DA HE bomb
1996

FAMAE 120 mm 120-44/66 HE bomb

Description

The FAMAE 120 mm 120-44/6 HE bomb is another licence-manufactured Thomson-Brandt design, this time based on the French Mle 44/66. It has a one-piece forged steel casting for the streamlined body and a perforated laminated steel tube carries the tail fin assembly. The filling is TNT ignited by a licence-produced V-19P or V-19PA point detonating fuze in the nose.

The propelling system consists of a primary type CL3 cartridge plus seven secondary ballistite charges. When fired from a FAMAE 120 mm mortar the maximum range of this bomb is approximately 4,750 m.

Data

Length: ca 679 mm
Weight: 13 kg
Filling: TNT
Max range: ca 4,750 m
Min range: 500 m

Manufacturer

Fabricas y Maestranas del Ejercito (FAMAE), Avenida Pedro Montt 1568/1606, Santiago.

Status

In production.

Service

Chilean Army.

NEW ENTRY

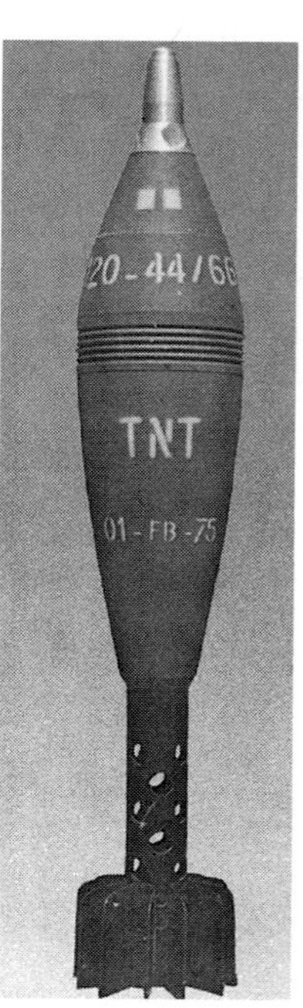

FAMAE 120-44/66 HE bomb
1996

CHINA, PEOPLE'S REPUBLIC

60 mm Type 63-1 fragmentation projectile

Description

There appears to be only one type of projectile used with the NORINCO Type 63-1 60 mm mortar. This is a high explosive bomb with a cast iron body containing 122 g of TNT/Dinal-80. It is fitted with a Pai-1A point detonating fuze and is propelled by a primary ignition cartridge and up to three incremental charges. Maximum range is 1,600 m.

Data

Length: 238.5 mm
Weight: fuzed, 1.35 kg
Filling: 122 g TNT/Dinal-80
Muzzle velocity: 141 m/s
Range: max, 1,600 m; min, 100 m
Fuze: Pai-1A point detonating

Manufacturer

China North Industries Corporation (NORINCO), 7A Yue Tan Nan Jie, PO Box 2137, Beijing.

Status

Available.

Service

People's Liberation Army and many other armed forces. Licence produced in Egypt.

NEW ENTRY

NORINCO 60 mm HE mortar bomb for Type 63-1 mortar
1996

---

82 mm Type 53 HE mortar bomb

Description

The 82 mm Type 53 HE mortar bomb was originally developed for use with the NORINCO Type 53 mortar but as this design has been replaced in service by the later Type 67 it is now more likely to be used with the latter. The Type 53 is a cast iron bomb with an unsophisticated outline apparently based on the Soviet O-832 and containing 392 g of TNT/Dinal-42. It is fitted with a Pai-1A point detonating fuze and is propelled by a primary ignition cartridge and up to three incremental charges. Maximum range is 3,040 m.

Data

Length: 329.3 mm
Weight: fuzed, 3.158 kg
Filling: 392 g TNT/Dinal-42
Muzzle velocity: 211 m/s
Range: max, 3,040 m; min, 85 m
Fuze: Pai-1A point detonating

Manufacturer

China North Industries Corporation (NORINCO), 7A Yue Tan Nan Jie, PO Box 2137, Beijing.

Status

Available.

Service

People's Liberation Army and many other armed forces.

UPDATED

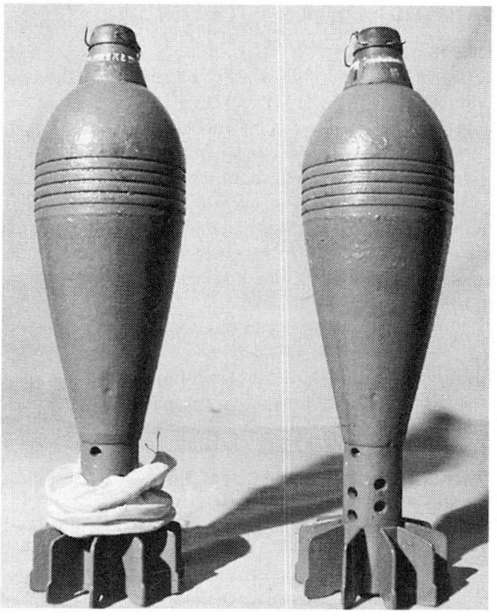

NORINCO 82 mm Type 53 HE mortar bombs
1996

---

82 mm Type 53 smoke bomb

Description

As with its HE equivalent, the 82 mm Type 53 smoke bomb was originally developed for use with the NORINCO Type 53 mortar but as this design has been replaced in service by the later Type 67 it is now more likely to be used with the latter.

The Type 53 smoke bomb is a cast iron bomb fitted with a Pai-1A point detonating fuze. The contents are 386 g of white phosphorus (WP) which is scattered by a small burster charge on impact to create screening smoke. It is propelled by a primary ignition cartridge and up to three incremental charges. Maximum range is 2,900 m.

Data

Length: 312 mm
Weight: fuzed, 3.486 kg
Filling: 386 g WP
Muzzle velocity: 197 m/s
Range: max, 2,900 m
Fuze: Pai-1A point detonating

Manufacturer

China North Industries Corporation (NORINCO), 7A Yue Tan Nan Jie, PO Box 2137, Beijing.

Status

Available.

Service

People's Liberation Army and many other armed forces.

NEW ENTRY

NORINCO 82 mm Type 53 smoke bomb
1996

82 mm Type 53 illuminating mortar bomb

Description

The 82 mm Type 53 illuminating mortar bomb is used primarily with the NORINCO Type 67 mortar. Compared to the other Type 53 mortar bombs it is much longer and heavier as it carries a magnesium-based flare and parachute assembly weighing 750 g. These are ejected from the body at an optimum height of approximately 400 m under the influence of a nose-mounted Shi-3 mechanical time fuze preset at the fire position. As the flare and parachute assembly falls to the ground it emits 250,000 candela for up to 22 seconds. The maximum range at which an effective ejection sequence can be completed is 2,210 m; the minimum is 350 m.

The Type 53 illuminating mortar bomb is propelled by a primary ignition cartridge and up to three incremental charges.

Data

Length: 584 mm
Weight: fuzed, 5.335 kg
Filling: 750 g flare assembly
Muzzle velocity: 180 m/s
Range: max effective, 2,210 m; min, ca 350 m
Intensity: 250,000 cd
Emission time: >22 s
Fuze: Shi-3 MT

Manufacturer

China North Industries Corporation (NORINCO), 7A Yue Tan Nan Jie, PO Box 2137, Beijing.

Status

Available.

Service

People's Liberation Army and many other armed forces.

NEW ENTRY

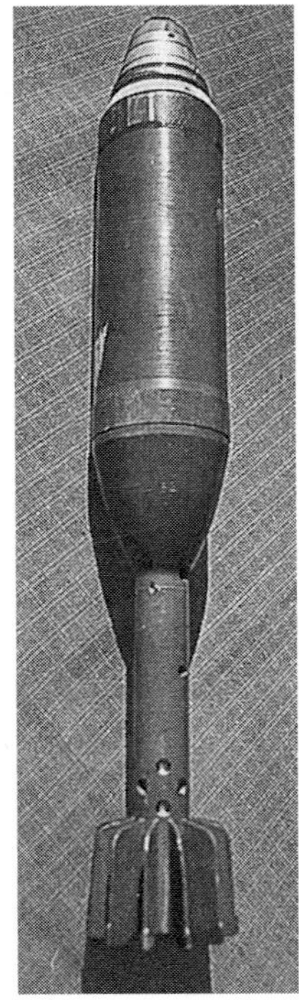
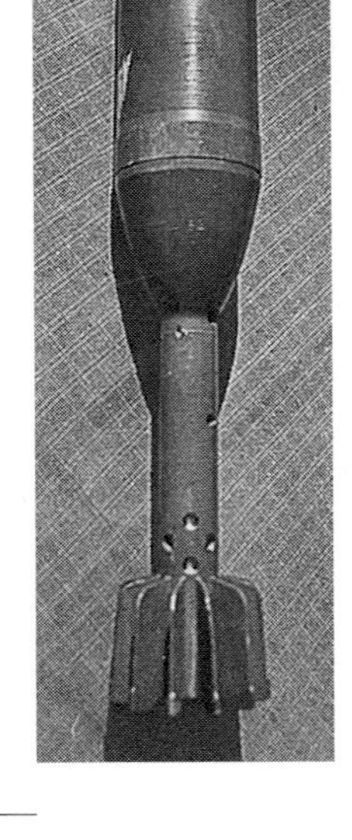

NORINCO 82 mm Type 53 illuminating mortar bomb
1996

NORINCO ammunition for 100 mm Type 71 mortar

Description

There is only one mortar with a calibre of 100 mm, namely the NORINCO Type 71. Exactly why this calibre was selected is unknown as the only users (as far as is understood) are the People's Liberation Army.

There are three main operational bombs in service: high explosive, smoke and illuminating. All three bombs are entirely conventional in design but they are not perfectly ballistically matched. In all three cases the propellant system involves a primary cartridge and up to five incremental charges.

The high explosive bomb has a cast iron body filled with 961 kg of TNT/Dinal. There is a nose-mounted Pai-4 point detonating fuze. Maximum range is 4,750 m

The smoke bomb is similar in appearance to the high explosive bomb and uses the same fuze but it contains 1.078 kg of white phosphorus (WP) plus a small burster charge to scatter the contents and create screening smoke. Maximum range is 4,570 m.

As with other similar mortar projectiles, the illuminating bomb is longer and heavier than the other types as it carries a flare and parachute assembly weighing 850 g. This is ejected at a height of 500 to 700 m over a target area to produce a light intensity of 550,000 candela for up to 40 seconds. The maximum range at which an effective airburst can be produced is about 3,700 m; minimum range is about 800 m. This bomb is normally used with the primary cartridge plus the three most powerful incremental charges.

Manufacturer

China North Industries Corporation (NORINCO), 7A Yue Tan Nan Jie, PO Box 2137, Beijing.

Status

Available.

Service

People's Liberation Army.

NEW ENTRY

Data

Type	HE	Smoke	Illuminating
Length	532.6 mm	538 mm	648.5 mm
Weight	8 kg	8.46 kg	9.05 kg
Filling weight	961 g	1.078 kg	850 g
Muzzle velocity	250 m/s	244 m/s	233 m/s
Range	4,750 m	4,570 m	3,700 m

NORINCO 100 mm illuminating mortar bomb
1996

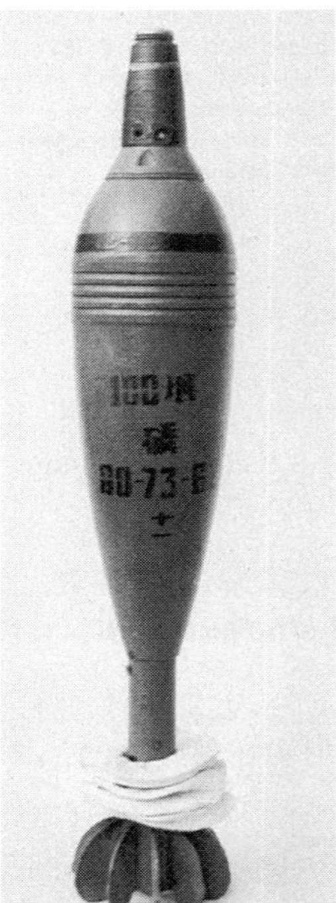

NORINCO 100 mm smoke mortar bomb; the high explosive bomb is almost identical visually
1996

NORINCO ammunition for 120 mm Type 55 mortar

Description

Although these 120 mm mortar bombs were developed primarily for use with the NORINCO Type 55 mortar there seems to be no reason why they could not be used with other 120 mm smoothbore mortar designs. The designs are entirely conventional and based on Soviet technology. These bombs have been licence produced or copied in several countries.

There are three main operational bombs in service: high explosive, smoke and illuminating. All three bombs are entirely conventional in design but they are not perfectly ballistically matched. In all three cases the propellant system involves a primary cartridge and up to six incremental charges.

The high explosive bomb has a cast iron body filled with 1.36 kg of high explosive which may be of several types, usually with a TNT content. There is a nose-mounted Pai-4 point detonating fuze. Maximum range is 5,700 m.

The smoke bomb is similar in appearance to the high explosive bomb and uses the same fuze but it contains 1.547 kg of white phosphorus (WP) plus a small burster charge to scatter the contents and create screening smoke. Maximum range is 5,800 m.

As with other similar mortar projectiles, the illuminating bomb is longer and heavier than the other types as it carries a flare and parachute assembly weighing 2.045 kg. This is ejected over a target area to produce a light intensity of 420,000 candela for up to 33 seconds. The fuze involved is the Shi-3A mechanical time.

Data

Type	HE	Smoke	Illuminating
Length	664.1 mm	669.3 mm	791.9 mm
Weight	16.56 kg	17.467 kg	16.4 kg
Filling weight	1.36 kg	1.547 kg	2.045 kg
Muzzle velocity	272 m/s	265.3	m/s n/avail
Range	5,700 m	5,800 m	n/avail

Manufacturer
China North Industries Corporation (NORINCO), 7A Yue Tan Nan Jie, PO Box 2137, Beijing.

Status
Available.

Service
People's Liberation Army and other armed forces.

NEW ENTRY

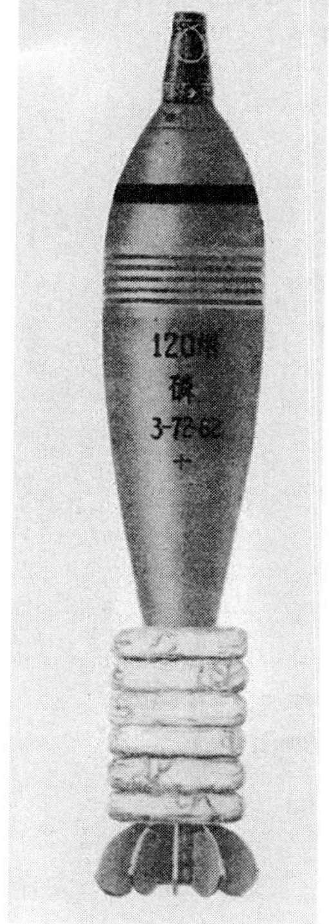

NORINCO 120 mm smoke bomb; the 120 mm high explosive bomb is almost identical visually
1996

COMMONWEALTH OF INDEPENDENT STATES

82 mm Model O-832 DU fragmentation bomb

Data
Calibre: 82 mm
Type: HE/fragmentation
Weight: fuzed, 3.23 kg

Filling: 436 g TNT/dinitronapthalene
Fuze: M-6 point detonating
Known using weapon: mortar M-37 (1942-43 version)
Remarks: also found with M-1, M-2, M-3, M-4, M-5, and MP-82 fuzes

VERIFIED

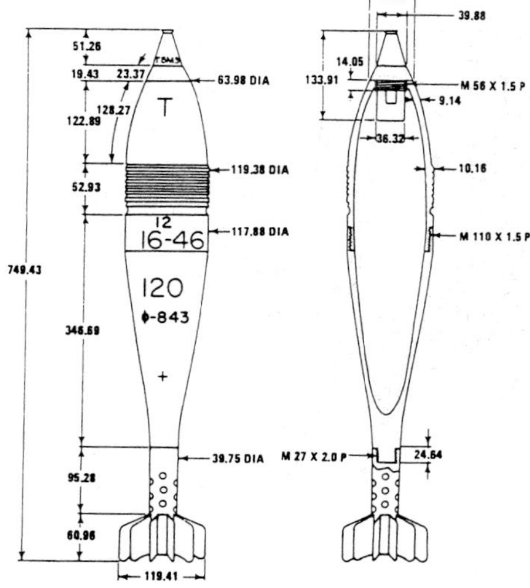

82 mm Model O-832 DU fragmentation bomb

107 mm Model OF-841A HE/ fragmentation bomb

Data
Calibre: 107 mm
Type: HE/fragmentation
Weight: fuzed, 9.1 kg

Filling: 1 kg TNT or Amatol
Fuze: GVMZ-7 point detonating
Known using weapon: mountain-pack regimental mortar M-38
Remarks: also uses Models GVMZ and GVMZ-1 point detonating fuzes. Fuze is shown with safety cap installed

VERIFIED

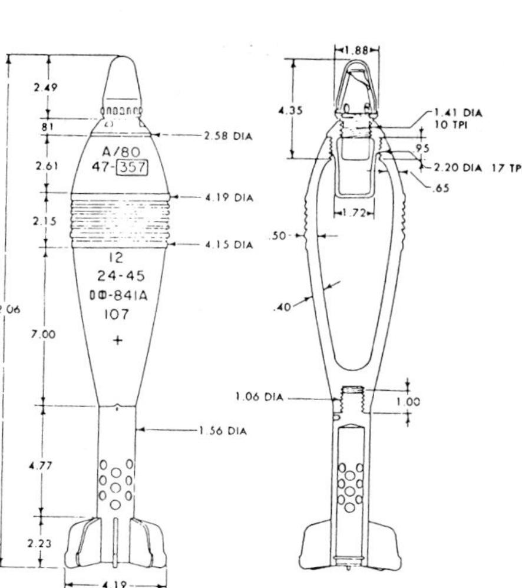

107 mm Model OF-841A HE/fragmentation bomb; dimensions are in inches

120 mm Model OF-843 HE bomb

Data

Calibre: 120 mm
Type: HE/fragmentation
Weight: fuzed, 16.02 kg
Filling: 2.68 kg TNT
Fuze: Model GVMZ point detonating
Known using weapons: regimental mortars M-38 and M-43
Remarks: also found with M-1 or M-4 PD fuzes

VERIFIED

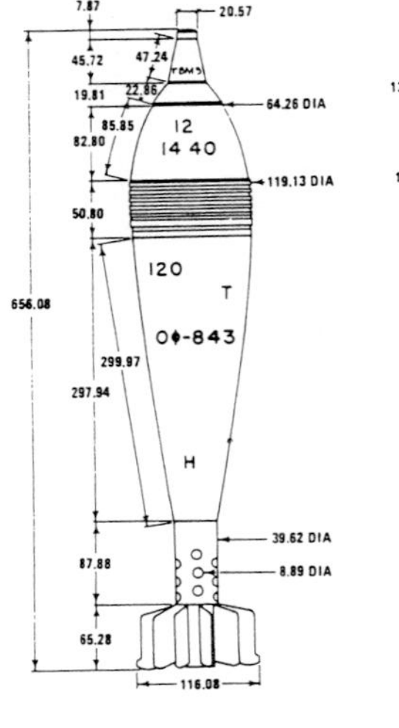

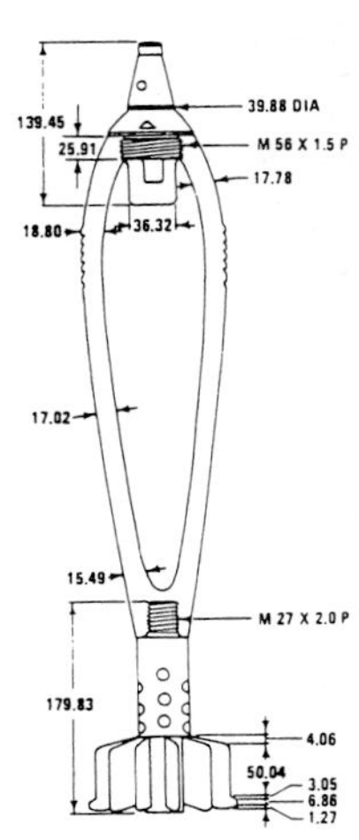

120 mm Model OF-843 HE bomb

---

120 mm Model OF-843A HE/fragmentation bomb

Data

Calibre: 120 mm
Type: HE/fragmentation
Weight: fuzed, 15.98 kg
Filling: 1.58 kg Amatol 80/20
Fuze: Model GVMZ-1 point detonating
Known using weapons: regimental mortars M-38 and M-43
Remarks: also found with M-4 point detonating fuze

VERIFIED

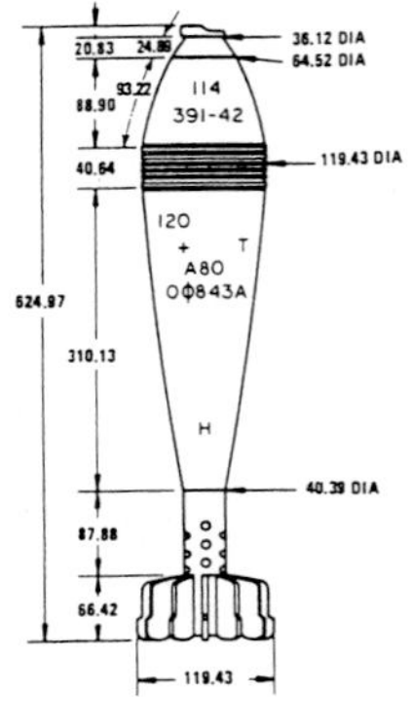

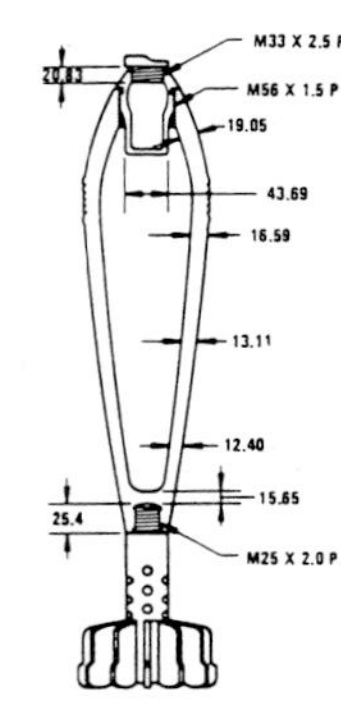

120 mm Model OF-843A HE/fragmentation bomb

---

120 mm Gran laser-guided mortar projectile

Description

The 120 mm Gran (Facet) laser-guided mortar projectile was developed by the KBP at Tula and is intended for the indirect engagement of spot targets such as structures or lightly armoured vehicles by 120 mm mortars when conventional artillery assets are not available. The Gran is apparently usually carried by self-propelled 120 mm mortar vehicles such as the 2S9 Nona-S or 2S23 Nona-SVK, but could also be fired from conventional ground-mounted 120 mm mortars. Maximum range is 7,500 m.

The Gran projectile is 1.225 m long and weighs 25 kg. It resembles an elongated artillery projectile; there are no tail fins. A laser sensor is located in the nose.

For a Gran fire mission a forward observer locates a suitable target and relays target data to a fire control position. At that position a microcomputer is used to produce fire data while at the fire position the fire data is passed to a hand-held microcomputer for further specialised computation relating to preparation of the Gran laser seeker electronics. Once the projectile has been launched and has passed its trajectory apogee the target is illuminated by a laser target designator for the descending Gran projectile to sense and home onto. As far as can be determined trajectory corrections are effected using small thruster rockets close to the projectile's centre of gravity.

On target the 11 kg warhead, of which 5.1 kg is explosive, is of the high explosive fragmentation (HEF) type. It is capable of destroying structures such as bunkers.

Data

Calibre: 120 mm
Length: 1.225 m
Weight: total, 25 kg: warhead, 11 kg
Filling: 5.1 kg HE
Range: 7,500 m

Development Agency

KBP Instrument Design Bureau, Tula.

Status

Development complete. Offered for export sales.

NEW ENTRY

---

160 mm Model F-852 HE bomb

Data

Calibre: 160 mm
Type: HE
Weight: 40 kg
Filling: 7.39 kg TNT
Fuze: Model GVMZ-7 point detonating
Known using weapon: mortar M-43

VERIFIED

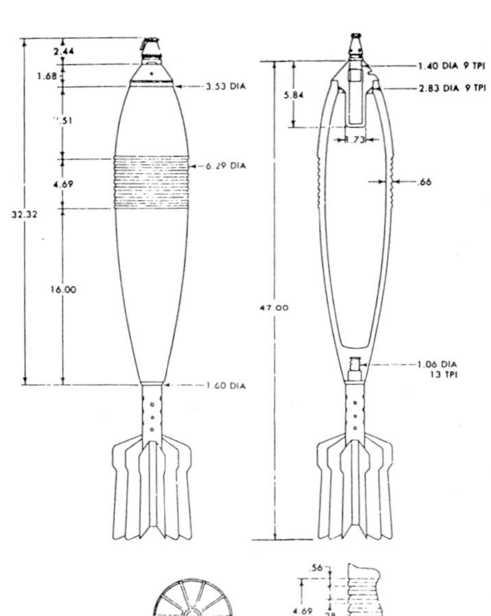

160 mm Model F-852 HE bomb

160 mm Model F-853A HE bomb

Data
Calibre: 160 mm
Type: HE
Weight: 41.18 kg
Filling: 7.73 kg Amatol 80/20
Fuze: Model GVMZ-7 point detonating
Known using weapon: mortar M-160
Remarks: Cast-iron bomb; fuze shown here with safety cap fitted

VERIFIED

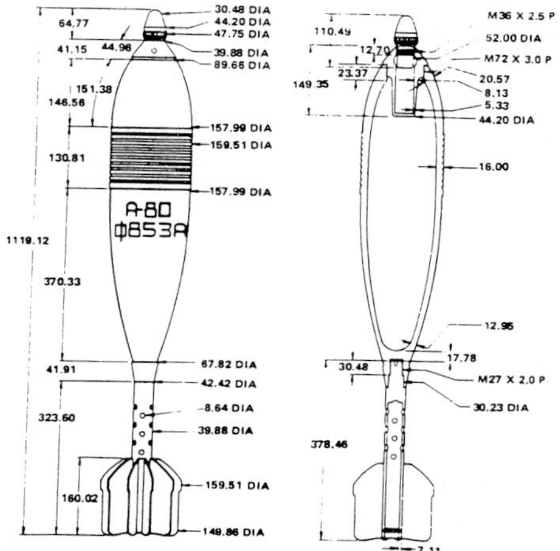

160 mm Model F-853A HE bomb

160 mm Model F-853U HE bomb

Data
Calibre: 160 mm
Type: HE
Weight: 41.18 kg
Filling: 8.989 kg TNT
Fuze: Model GVMZ-7 point detonating
Known using weapon: mortar M-160
Remarks: Steel bomb

VERIFIED

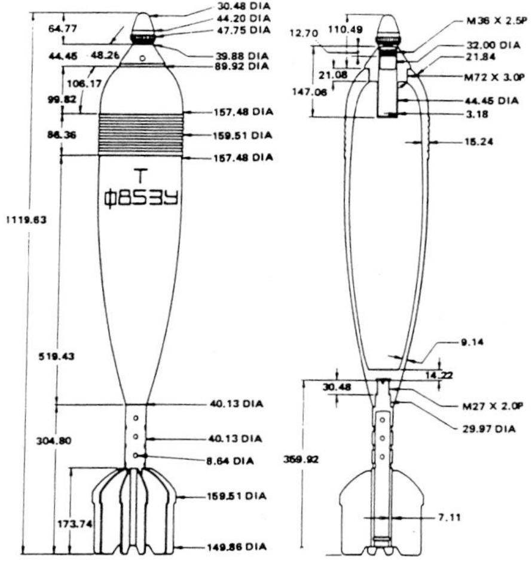

160 mm Model F-853U HE bomb

CZECH REPUBLIC

120 mm HE bomb Model OF-A

Description
Though generally adhering to a CIS pattern, this bomb has signs of Czech variation to the basic design. It is of conventional teardrop shape in cast iron, with three deep gas-check rings at the bourrelet and a steel tail boom with twelve fins. A primary cartridge fits into the tail and six secondary increments, in cloth bags, are tied around the tail boom.

Data
Length: fuzed, 664 mm
Weight: fuzed, 15.33 kg
Filling: 2.043 kg TNT
Fuze: MZ30AV point detonating
Max range: 5,750 m
Muzzle velocity: 275 m/s
Known using weapons: Soviet M1938 and M1943 mortars

VERIFIED

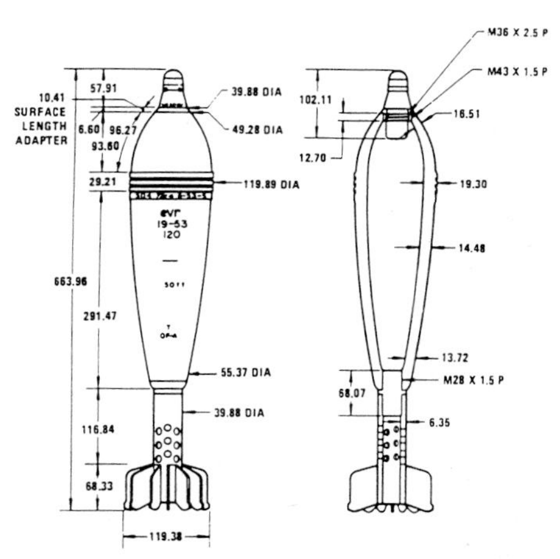

Czech 120 mm HE bomb Model OF-A

FINLAND

Vammas 60 mm HE bombs

Description
These bombs were originally designed by Tampella Defence Division (which was merged with Vammas in 1991) and upgraded by Vammas. They can be used in 60 mm smoothbore Tampella mortars.

The bomb body is made from a steel casting and machined to shape. The tail unit is made from extruded aluminium.

Manufacturer
Vammas Oy, PO Box 18, FIN-38201 Vammala.

Status
Available.

Data	Standard TAM 1.6	Long Range TAM 1.8
Min range	150 m	150 m
Max range	2,600 m	4,000 m
Min velocity	66 m/s	62 m/s
Max velocity	199 m/s	268 m/s
Total length	294 mm	315 mm
Bomb weight	1.6 kg	1.8 kg
Filling (TNT)	220 g	200 g
Charges	5	5

Service
Finnish defence forces.

VERIFIED

Vammas 60 mm TAM 1.8 HE mortar bomb

Vammas 81 mm HE bomb TAM 4.2

Description
This bomb was originally designed by Tampella Defence Division (which was merged with Vammas in 1991) and upgraded by Vammas. It can be used in 81 mm smoothbore Tampella mortars.

The bomb body is made from a steel casting and machined to shape. The tail unit is made from extruded aluminium.

Data
Length: 504 mm
Weight: 4.25 kg
Filling: 520 g TNT
Min range: 150 m
Max range: 6,500 m
Min velocity: 76 m/s
Max velocity: 321 m/s
Charges: 7

Manufacturer
Vammas Oy, PO Box 18, FIN-38201 Vammala.

Status
Available.

Service
Finnish defence forces.

VERIFIED

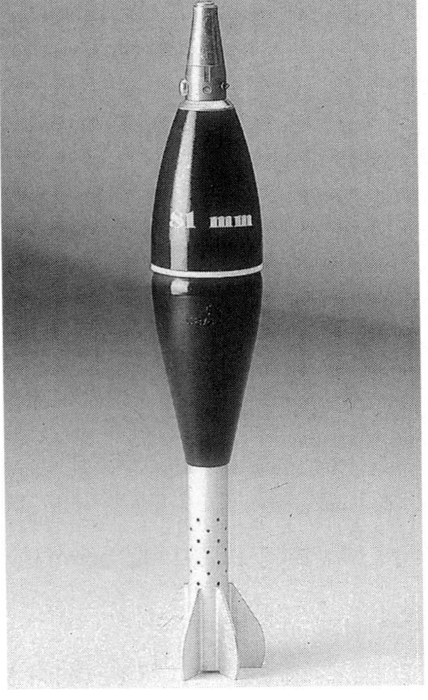

Vammas 81 mm TAM 4.2 HE mortar bomb

Vammas 120 mm HE bombs

Description
The 120 mm mortar bomb TAM 12.8 was originally designed by Tampella Defence Division and has been upgraded by Vammas. The Long-Range Bomb VAM 14.9 was designed by Vammas and is under test. Both can be used in Tampella 120 mm smoothbore mortars.

The body of the TAM 12.8 bomb is made from a steel casting and machined to shape. That of the VAM 14.9 bomb is made from forged steel. The tail unit of both bombs is of extruded aluminium.

Data

	Standard Tam 12.8	Long-Range VAM 14.9
Min range	300 m	300 m
Max range	8,500 m	9,500 m
Min velocity	112 m/s	105 m/s
Max velocity	444 m/s	450 m/s
Length	665 mm	795 mm
Weight	12.8 kg	14.9 kg
Filling (TNT)	2.1 kg	2.6 kg
Charges	8	8

Manufacturer
Vammas Oy, PO Box 18, FIN-38201 Vammala.

Status
Standard bomb in current use; in production.
Long-range bomb under test.

Service
Finnish defence forces.

VERIFIED

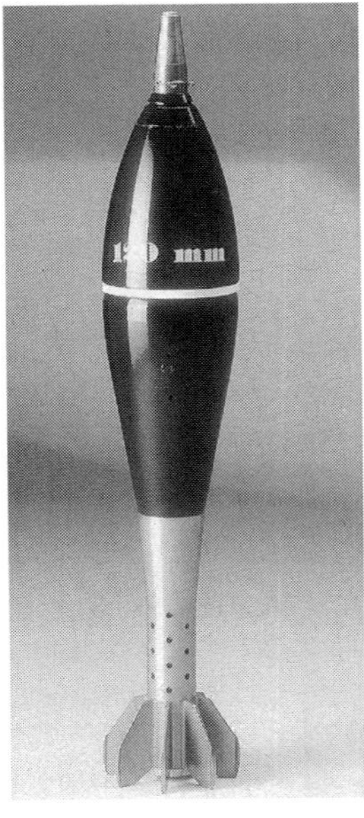
Vammas 120 mm TAM 12.8 HE mortar bomb

FRANCE

Ammunition for the 51 mm FLY-K Individual Weapon System

Description
The FLY-K weapon system (described in the *Mortars* section) uses a new propulsion unit called FLY-K integrated into the projectile stabiliser.

This new concept provides noiseless, smokeless, flashless and heatless firing, giving infra-red undetectability because of the absence of weapon thermal radiation.

The FLY-K propulsion unit is composed of a thin-walled cylinder partially reinforced and of high mechanical strength and a sliding element (piston), hermetically closed at its upper end and containing the propellant charge.

During the expansion of gases from the propellant charge the piston bears on the spigot of the weapon, the effect of which is to propel the ammunition.

During the launching process, and in subsequent flight, the piston ensures a full rearward sealing of the propulsion unit which explains the complete absence of noise, flash and smoke and hence a no-signature weapon system.

Four types of ammunition are currently produced: HE/fragmentation, smoke/incendiary, illuminating and practice. All are generally similar, however, since the bomb does not fully enter the barrel of the launcher, the overall diameter varies.

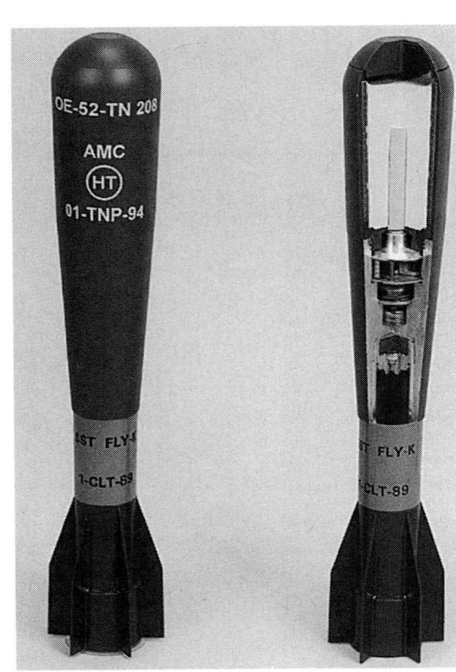

HE/fragmentation bomb TN 208

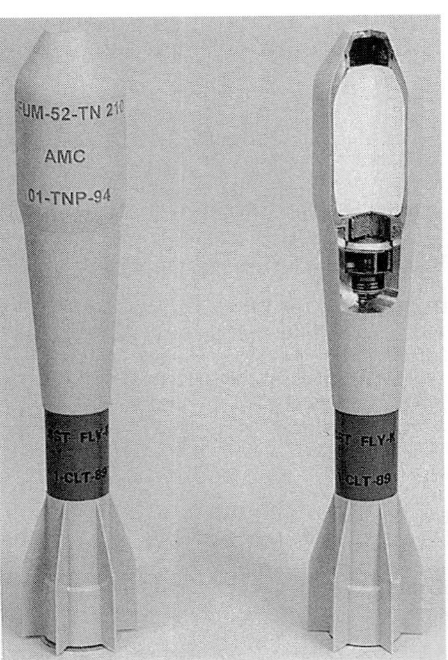

Smoke/incendiary bomb TN 210

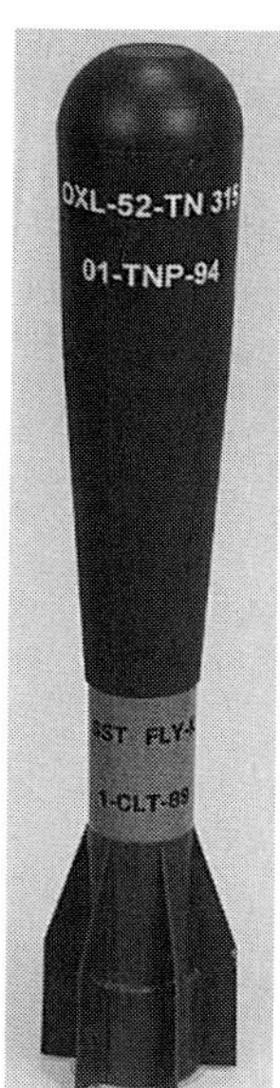

Target practice bomb TN 315

Data

HE/FRAGMENTATION TN 208
Warhead diameter: 51 mm
Length: 311 mm
Weight: 765 g
Filling: 135 g Comp B
Muzzle velocity: 88.5 m/s
Range: nominal, 675 m
Arming distance: 35 m
Lethal radius: 16 m
Number of fragments: ca 580
Propelling charge: 2.25 g
Fuze: base detonating

SMOKE/INCENDIARY TN 210
Warhead diameter: 51 mm
Length: 310 mm
Weight: 708 g
Incendiary charge: 180 g
Muzzle velocity: 91 m/s
Range: nominal, 675 m
Arming distance: 35 m
Propelling charge: 2.25 g

ILLUMINATING BOMB TN 209
Warhead diameter: 47 mm
Length: 380 mm
Weight: 650 g
Muzzle velocity: 91 m/s
Expelling range: horizontal, 500 m
Candle power: min, 350,000 cd
Illuminated range, 2 lux: 115-877 m
Illuminated range, 5 lux: 295-697 m
Burning time: min, 20 s
Propelling charge: 2.4 g

TARGET PRACTICE TN 315
Warhead diameter: 51 mm
Length: 311 mm
Weight: 765 g
Muzzle velocity: 88.5 m/s
Range: nominal, 675 m
Propelling charge: 2.25 g

Manufacturer

Titanite SA, BP15, F-21270 Pontailler-sur-Saone.

Status

In production.

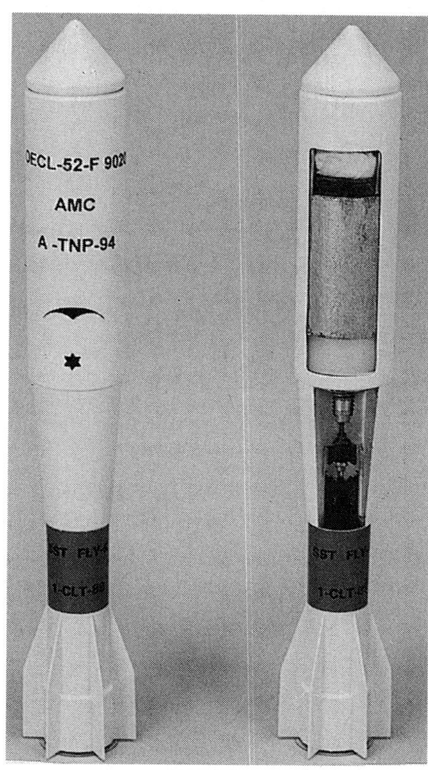

Illuminating bomb TN 209

Service

French armed forces.

VERIFIED

Thomson-Brandt ammunition for 60 mm mortars

Description

Primary cartridges for all Thomson-Brandt 60 mm bombs are 24 mm in diameter and 65 mm long, containing 4.2 g of ballistite. Secondary charges are generally of the horseshoe type, although earlier bombs used secondary charges fitted between the fins. The supercharge is horseshoe-shaped and contains 5 g of ballistite.

Mark 61 HE bomb

The body of this mortar bomb is made of pearlitic, malleable cast-iron. It is filled with 260 g pressed TNT and is fitted with the V9 fuze. The total length of the bomb is 307 mm and, once fired, it weighs 1.73 kg. When fired from the MO-60-63 mortar it is propelled to 2,050 m.

Mark 61 HE colour marker bomb

This bomb is used for ranging and target indication and contains the normal HE filling but around it is colouring agent which may be green, yellow, red or black. The bombs have the same physical characteristics as the normal HE bombs.

Mark 61 ammunition, showing shape and position of secondary charges

Mark 61 smoke bomb

This bomb, filled with liquid titanium chloride or white phosphorus, produces an effective smoke screen. The weight, length and so on are the same as those of the HE bombs. The same fuze is used.

Mark 61 practice bomb

This bomb is ballasted with a dummy head. Again it has the same ballistics as the HE bomb. Either live or inert V9 fuzes may be fitted. Some users fire the Mark 61 colour marker (black) as a practice bomb.

Mark 63 illuminating bomb

This bomb has a magnesium-based filling and is hung on a parachute. The light persists for a minimum of 30 seconds and produces 180,000 candela. It provides illumination over an area with a radius of 150 m. The

60 mm Mark 63 parachute illuminating bomb

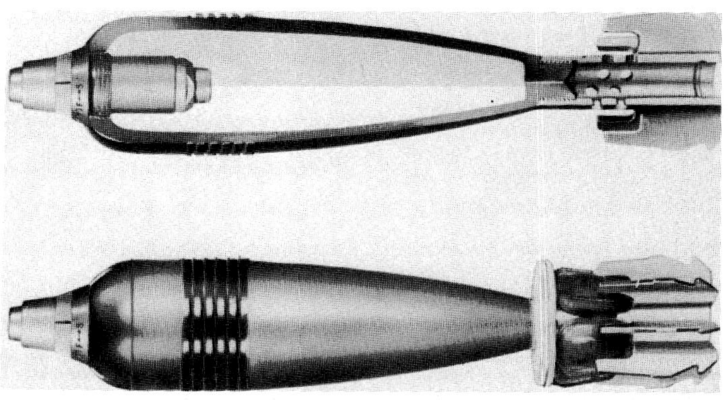

60 mm HE bomb Mark 61 with fuze V9

round is 333 mm long, weighs 1.55 kg and has a clockwork fuze, graduated from 7 to 35 seconds in half-seconds.

60 LP

This is a 60 mm long-range HE projectile with a longer streamlined outline intended to be fired from Thomson-Brandt gun mortars or the MO-60-LP mortar. It can thus have a maximum range of as much as 5,150 m. Filled with TNT it has a fired weight of 2.14 kg.

Manufacturer

Thomson-Brandt Armements, 4 avenue Morane Saulnier, F-78140 Vélizy-Villacoublay.

Status

In production.

Service

The armed forces of at least 20 countries.

UPDATED

Thomson-Brandt 60 mm CC anti-tank mortar bomb

Description

This bomb was developed by Thomson-Brandt to be fired from their 60 mm gun mortars, thus giving these weapons a full range of ammunition; that is HE, illuminating, smoke, practice and anti-tank. The mortar bomb has a piezoelectric fuze which gives it a 10 m muzzle safety; the bomb is not armed until it is 10 m away from the weapon. This fuze is the same type as is fitted to the 68 mm rocket, with the accelerometer modified for mortar loadings.

Although the hollow charge will penetrate almost 200 mm of armour or concrete it retains an anti-personnel capability.

Data

Length: 321 mm
Weight: 1.73 kg
Filling: Hexolite or TNT
Muzzle velocity: 200 m/s
Range: stationary target, 500 m; moving target, 300 m

Manufacturer

Thomson-Brandt Armements, 4 avenue Morane Saulnier, F-78140 Vélizy-Villacoublay.

Status

In production.

UPDATED

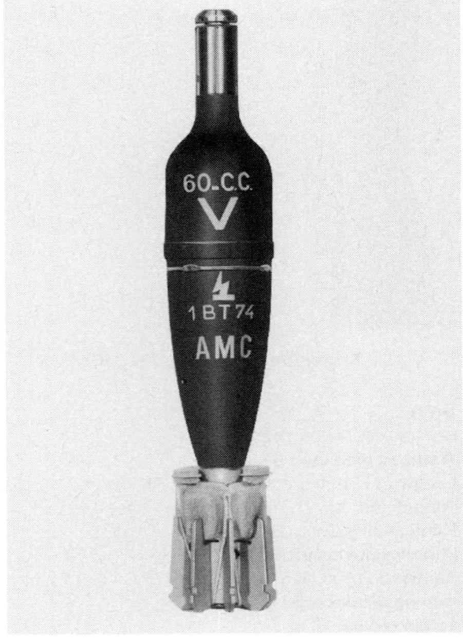

60 mm CC anti-tank mortar bomb

Thomson-Brandt 60 mm canister round

Description

This canister round was designed for the close defence of armoured vehicles equipped with the Thomson-Brandt 60 mm gun mortar; it is also an expedient method of clearing attacking infantry from a nearby armoured vehicle. The round is a one-piece unit which is breech-loaded into the gun mortar and it contains a propelling charge and 132 hardened lead shot, each 8.7 mm in diameter and weighing, in total, 530 g. When fired, the crimped mouth of the case is blown open and the shot ejected. Once the shot has left the gun the empty casing is ejected through the muzzle. At 50 m range the shot charge covers an area of about 25 m^2 and most have sufficient remaining velocity to pierce 27 mm pine boards. The shot will not perforate 5 mm thickness of mild steel at 5 m range, so that it may be fired at a friendly vehicle without risk to the crew.

Data

Length: 217 mm
Weight: 1.175 kg
Max range: 50 m
Filling: 132 hardened lead shot

Manufacturer

Thomson-Brandt Armements, 4 avenue Morane Saulnier, F-78140 Vélizy-Villacoublay.

Status

In production.

UPDATED

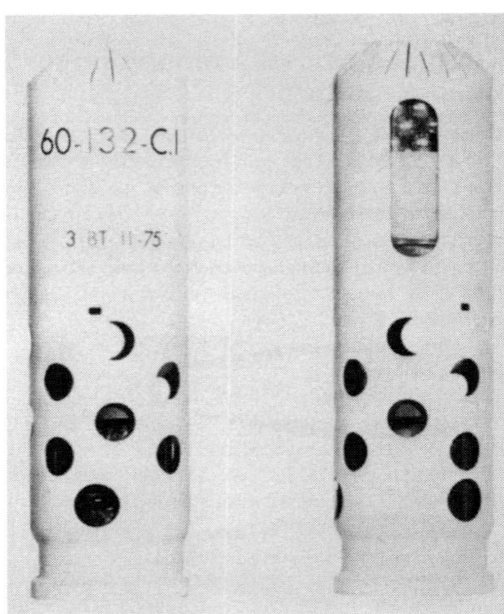

Thomson-Brandt 60 mm canister round

Alsetex 60 mm mortar bombs

Description

Alsetex makes a range of bombs suitable for use with 60 mm mortars of all types, including those that are breech loaded. The bombs are supplied as complete rounds with fuzes and cartridges in position and packed in individual containers which can be coupled for carriage. There are six models shown in the accompanying illustration.

Data

Calibre: 60 mm
Body: French Army type FA Mark 47
Weight: 1.35 kg
Fuze: SNEM Mark F1 with double safety (except ref. **F** - see caption)
Propellant: central cartridge with 4 increments
Range: 100-1,000 m

Manufacturer

SAE Alsetex, 4 rue de Castellane, F-75008 Paris.

Status

In production.

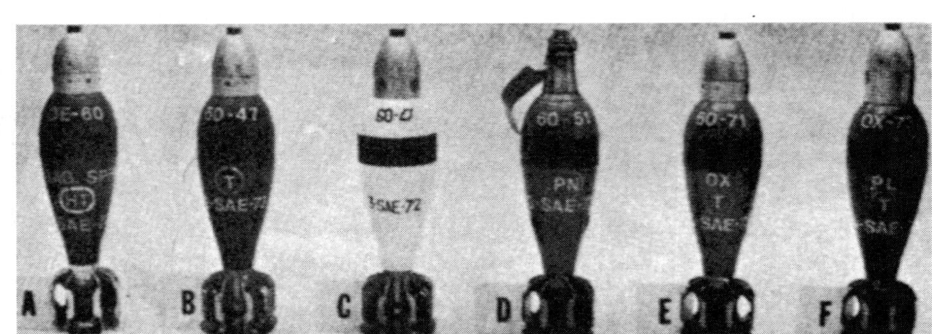

Alsetex 60 mm mortar bombs
(A) *high-efficiency fragmentation, approximately 150 g of RDX producing about 350 fragments;* **(B)** *HE (Mark 47) 150 g of TNT;* **(C)** *inert (Mark 47) 150 g inert ballast;* **(D)** *smoke (Mark 51);* **(E)** *OX 60 PLT Mark F1 practice, inert marking ballast dispersed on impact by small TNT charge, SNEM Mark F1 fuze;* **(F)** *OX 60 PLT Mark F1 (as* **E** *but with 21 × 28 fuze)*

Service

French armed forces.

VERIFIED

Thomson-Brandt ammunition for 81 mm mortars

Description

M57D HE bomb

The Thomson-Brandt M57D is a conventional high explosive bomb with a cast iron body filled with TNT. It is 382 mm long. When fired from the Thomson-Brandt MO-81-LC mortar minimum range is 120 m and the maximum 4,550 m. There is a practice bomb and a similar smoke bomb filled with WP or FM.

Length: 382 mm
Weight: complete, 3.28 kg; after firing, 3.2 kg
Filling: TNT
Range: max, 4,550 m; min, 120 m

M61 HE bomb

This is a newer and heavier bomb than the M57D. Maximum range when fired from an MO-81-LC mortar is 5,000 m and weight, once fired, is 4.2 kg. Versions which produce coloured smoke as they burst are available, as are smoke and practice models.

Length: 414 mm
Weight: complete, 4.33 kg; after firing, 4.2 kg
Filling: TNT
Range: max, 5,000 m; min, ca 85 m

M82 HE bomb

The M82 is a further development in the 81 mm HE series with a revised propellant system. Maximum range when fired from an MO-81-LC mortar is 5,600 m and weight, once fired, is 4.45 kg.

Length: 449 mm
Weight: complete, 4.56 kg; after firing, 4.45 kg
Filling: TNT
Range: max, 5,600 m; min, ca 110 m

81 LP bomb

This 81 mm long-range bomb is still at the preproduction stage and is intended to have a maximum range, when fired from the Thomson-Brandt MO-81-LP long-range mortar, of 7,600 m. Weight, when fired, is 7.1 kg, of which 1 kg is the TNT filling.

Length: 579 mm
Weight: complete, 7.4 kg; after firing, 7.1 kg

Thomson-Brandt 81 mm 81 LP long-range bomb
1996

Filling: 1 kg TNT
Range: max, 7,600 m; min, 150 m

M77A illuminating bomb

The M77A replaced the earlier M77. It releases a magnesium flare and parachute assembly over a target area at an optimum height of 340 m and produces 1 Mcd of light for 45 seconds.

Length: 476 mm
Weight: complete, 4 kg; after firing, 3.87 kg
Filling: illuminating compound, magnesium-based
Power: 1 Mcd

Thomson-Brandt 81 mm M77A illuminating bomb
1996

Burning time: 45 s
Radius of illumination: 580 m

Manufacturer

Thomson-Brandt Armements, 4 avenue Morane Saulnier, F-78140 Vélizy-Villacoublay.

Status

In production.

Service

French Army and several others.

NEW ENTRY

Alsetex 81 mm mortar bombs

Description

SAE Alsetex produces a range of bombs suitable for use with all types of 81 mm mortar. They are supplied as complete rounds, with fuze and cartridge in position, packed in individual containers which may be coupled for carriage. There are five models, shown in the accompanying illustration.

Data

Calibre: 81 mm
Body: French Army type FA Mark 32
Weight: ca 3.5 kg
Fuze: SNEM Mark F1 with double safety (feed and trajectory)
Propellant: central cartridge with 3 or 6 increments
Range: average practical, ca 2,000 m; max, ca 3,000 m

Manufacturer

SAE Alsetex, 4 rue de Castellane, F-75008 Paris.

Status

In production.

Service

French armed forces.

VERIFIED

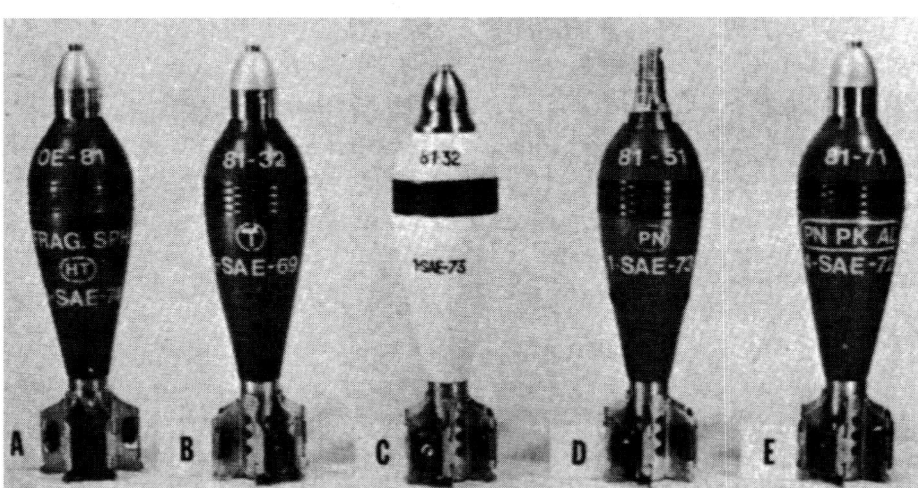

81 mm Alsetex mortar bombs
(A) *high-efficiency fragmentation, RDX filling producing about 300 splinters* **(B)** *HE, containing 550 g cast TNT* **(C)** *inert (Mark 32)* **(D)** *smoke practice (Mark 51)* **(E)** *practice PL PN Mark F1, inert marking, ballast dispersed by small charge*

Alsetex 81 mm Alta 81 mortar fuze

Description

The Alta 81 is a point detonating impact fuze for use with smoothbore 81 mm mortar bombs. Highly sensitive, it will function on any type of ground. Safety during storage and handling is ensured by an interruptor in the firing train and by positive locking of the arming elements. Arming is delayed on firing and the fuze cannot arm within 40 m of the muzzle. A visual indication of the state of arming is provided.

Data

Length: 75 mm
Diameter: 48 mm
Weight: 150 g
Arming impulse: 800-8,500 *g* acceleration
Operating temperature: −40 to +60°C

Manufacturer

SAE Alsetex, 4 rue de Castellane, F-75008 Paris.

Status

Prototype production.

VERIFIED

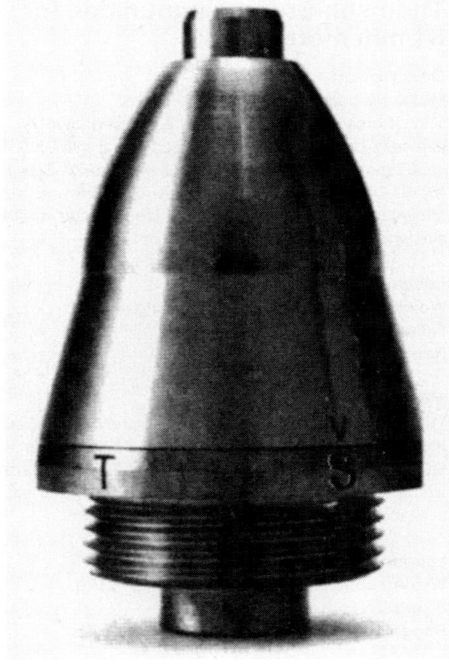

Alsetex Alta 81 mortar fuze

Thomson-Brandt ammunition for 120 mm Brandt mortars

Description

M44 HE bomb range

The Thomson-Brandt M44 is a conventional high explosive bomb, fuzed V19, weighing 13 kg and filled with TNT. It is 679 mm long, minimum range is 500 m and the maximum 7,000 m. There is a practice bomb, a smoke bomb filled with WP or FM and an HE marker bomb with the same weight and performance as the M44 bomb. The M44 bomb was produced up to 1966 by Brandt. Since that time it has been manufactured for the French Army by other contractors and is licence produced in several countries. The current Thomson-Brandt HE bomb is designated M44/66 HE and fuzed V19P.

M852 HE bomb

This is a newer and heavier bomb than the M44 and M44/66 series and involves a revised propellant system. Maximum range when fired from an MO-120-LT mortar is 7,000 m and weight, once fired, is 13.88 kg. Versions which produce coloured smoke as they burst are available, as are smoke and practice models.

Smoke bomb

The standard smoke bomb is the Mark 44/67 which is filled with white phosphorus.

Mark 62-ED illuminating bomb

This has a mechanical time fuze, type FH81-B and spring-loaded tail fins. It has been superseded by the M842.

Data

Length: 800 mm
Weight: 13.65 kg
Filling: illuminating compound, magnesium-based
Power: 700,000 cd
Burning time: 1 min
Radius of illumination: 300 m

M842 illuminating bomb

This bomb can produce 1.4 Mcd of light for 65 seconds. It weighs 14.35 kg once fired and has a maximum range of 7,000 m when fired from the Thomson-Brandt MO-120-LT mortar.

Manufacturer

Thomson-Brandt Armements, 4 avenue Morane Saulnier, F-78140 Vélizy-Villacoublay.

Status

In production.

Service

French Army and several others.

UPDATED

120 mm M44 series mortar bombs

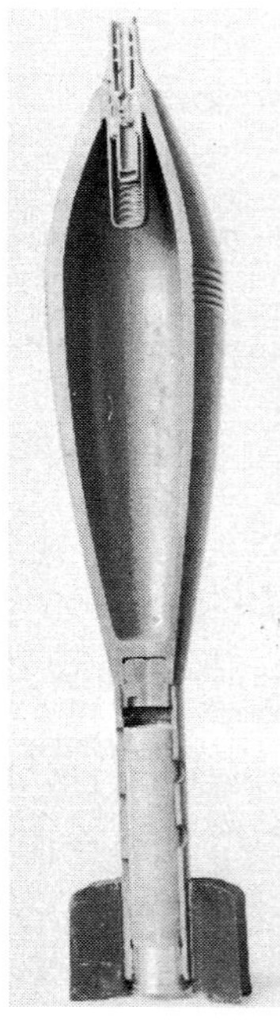

M44 bomb sectioned

Thomson-Brandt 120 mm PEPA rocket-assisted mortar bomb

Description

The 120 mm PEPA bomb consists of a steel body, an HE filling, an internal solid fuel rocket motor and the V19P fuze. The propelling charge, made up of the primary and secondary charges, is at the rear. The body **(3)** is made in two parts, screwed together, with the tail assembly **(10)** screwed on to the rear part of the body.

The HE content **(2)** is 2 kg of hexogen (RDX/TNT) cast inside the cavity and around the solid fuel motor. The solid fuel motor **(5)** of star-shaped section, is housed in a tube **(4)**, externally inhibited, with a venturi at the end **(6)**. The venturi is secured at the tail end by an obturator containing an optional delay **(7)** and a selection lock **(8)** which allows the use of the solid fuel propulsion unit if required. A spring catch **(13)** holds the selection lock in the required position.

The propulsion charge consists of the primary

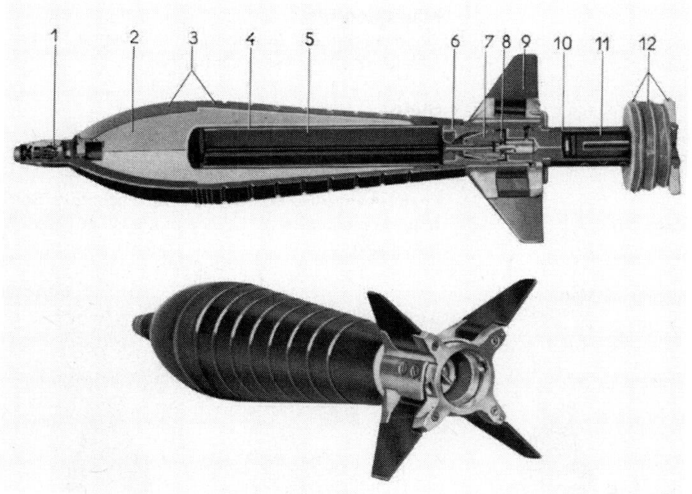

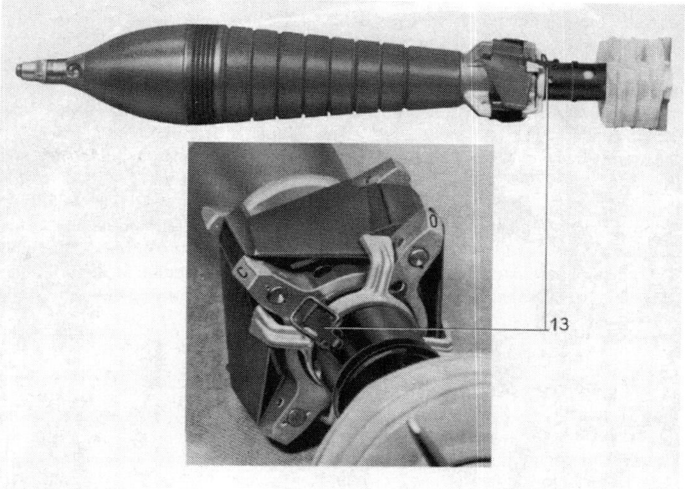

120 mm PEPA bomb
(1) *fuze V19P* **(2)** *HE filling* **(3)** *steel body* **(4)** *tube containing rocket assistance* **(5)** *rocket motor* **(6)** *venturi* **(7)** *optional delay* **(8)** *selection lock* **(9)** *spring-opened tail fin* **(10)** *tail assembly* **(11)** *primary cartridge* **(12)** *secondary charges* **(13)** *spring catch holding selection lock in place*

cartridge **(11)** and up to seven augmenting cartridges. The tube **(10)** holding the primary cartridge contains the flash holes communicating with the secondary charges. This tube is ejected after the bomb is fired.

In use the order of events is as follows. The bomb is removed from the container and the charge system adjusted by the removal of unrequired secondaries. If the rocket assistance is required the tube containing the primary cartridge is rotated clockwise. The fuze V19P is then set for instantaneous action or delay.

The ranges obtainable with the PEPA are shown in the adjacent table. 'R' indicates rocket assistance.

From the table it can be seen that the maximum range without rocket assistance is 4,250 m. The muzzle velocity is 240 m/s. The maximum range with rocket assistance is 6,550 m. The muzzle velocity is increased by 110 m/s. The PEPA/LP bomb was developed for use with the MO-120-M65 and 120-AM-50 mortars. It has a wider range bracket.

Charge	Elevation (max)	Range (min)	Elevation (min)	Range (max)
1	74°	600 m	45°	1,350 m
2	72°	1,000 m	45°	1,900 m
3	69°	1,700 m	45°	2,560 m
4	67°	2,200 m	45°	3,175 m
5	70°	2,500 m	45°	3,560 m
6	69°	2,800 m	45°	4,250 m
3R	66°	4,000 m	55°	4,800 m
4R	66°	4,600 m	55°	5,500 m
5R	63°	5,300 m	50°	6,100 m
6R	62°	5,700 m	45°	6,550 m

Manufacturer
Thomson-Brandt Armements, 4 avenue Morane Saulnier, F-78140 Vélizy-Villacoublay.

Status
In production.

Service
French and other armies.

VERIFIED

Thomson-Brandt ammunition for 120 mm rifled mortars

Description
Although the Thomson-Brandt MO-120-RT-61 is a rifled mortar, it will fire smoothbore bombs except those types having spring-loaded tail fin assemblies with straight fins. Smoothbore bombs are frequently used for bedding in the baseplate (1 round charge 3, 1 charge 5, 1 charge 7) and also for economy in training. Bombs designed for the MO-120-RT-61 are equipped with a tail tube carrying the primary and secondary cartridges. This tube is ejected just after the bomb has left the mortar and falls about 100 m from the muzzle.

Data
PRPA BOMB

The *projectile rayé à propulsion additionelle* (PRPA) has rocket assistance, which comes into action after a delay of 10 seconds, and a pre-engraved driving band.

Length: with tail tube, 918 mm
Weight: before firing, 18.7 kg; after firing, 15.7 kg
Filling: RDX/TNT
Range: max with rocket assistance, 13,000 m; min without rocket assistance, 1,100 m
Fuze: PD, M557 or Brandt VG-29. The latter has the normal Brandt pneumatic delayed arming features but makes use of setback, creep and centrifugal forces in the initiation of this delay

PR 14 BOMB

No rocket assistance. Pre-engraved driving band.

Length: with tail tube, 897 mm
Weight: complete round, 18.6 kg; in flight, 15.7 kg
Filling: RDX/TNT
Range: max, 8,135 m; min, 1,100 m
Fuze: PD, M557 or Brandt VG-29. The latter has the normal Brandt pneumatic delayed arming features but makes use of setback, creep and centrifugal forces in the initiation of this delay

PR AB BOMB

This bomb is described as a high explosive anti-armour projectile as it has a body of prefragmented steel alloy. On impact the body breaks up into high-velocity (1,500 m/s) fragments which can penetrate from 8 to 15 mm of armour plate 10 m from the point of impact. The fragments are scattered over a radius of 25 m.

Length: overall, 897 mm
Weight: complete round, 18.7 kg; in flight, 15.7 kg
Filling: 4,45 kg RDX/TNT
Range: max, 8,135 m; min, 1,100 m

PR CARGO

This projectile, the OGR-120-PR, is still under development. Having been qualified carrying 20 bomblets,

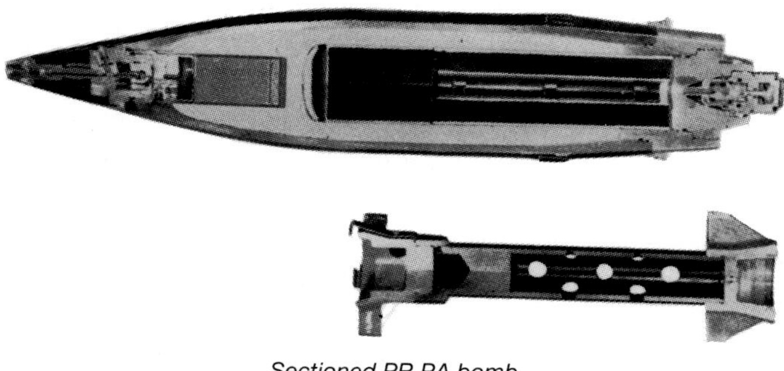

Sectioned PR PA bomb

120 mm PR PA, PR 14, illuminating and anti-armour bombs

current, development work involves a payload of 16 bomblets with self-destruct mechanisms. The projectile delivers its M42 dual-purpose bomblets from an optimum height of 400 to 500 m. Each bomblet can penetrate 75 to 125 mm of armour on impact and scatter approximately 750 anti-personnel fragments over a 25 m radius. Maximum range is 8,135 m.

Length: overall, 890 mm
Weight: complete round, 18.6 kg; in flight, 15.6 kg
Filling: 16 or 20 dual-purpose bomblets
Range: max, 8,135 m; min, 1,100 m

ILLUMINATING BOMB PRECLAIR
This bomb carries a magnesium flare unit which produces 1.5 Mcd for 65 seconds.

Length: overall, 890 mm
Weight: complete round, 18.5 kg; in flight, 15.3 kg
Filling: magnesium-based compound
Fuze: FR 55 A clockwork
Fuze setting: 7-55 s
Optimum height of functioning: 420 m
Rate of fall of illuminant: 5 m/s
Radius of illumination: 700 m
Burning time: min, 65 s producing 1.5 Mcd

Manufacturer
Thomson-Brandt Armements, 4 avenue Morane Saulnier, F-78140 Vélizy-Villacoublay.

Status
In production.

Service
French Army. Also exported to several countries. Some of these projectiles are licence produced in Turkey by MKEK (qv)

UPDATED

120 mm Alsetex mortar bombs

Description
The Alsetex company manufactures a range of 120 mm bombs suitable for use in any conventional mortar. They are of steel, loaded with cast TNT, and supplied complete with fuze, primary cartridge, and seven augmenting cartridges in an airtight container. The basic design is that of the standard French Army Mk 44 bomb, and the fuze is the SNEM Mark F1 with delayed arming and double-loading safety features.

Type A is a high-capacity HE bomb with hexolite filling and a lining of spherical preformed fragments. Type B is a standard HE bomb with a 2.5 kg TNT filling. Type C is an inert practice bomb, and Type D is a practice bomb with a small pyrotechnic burster charge to indicate the point of impact. All types weigh approximately 13 kg and have a maximum range of about 7,000 m.

Manufacturer
SAE Alsetex, 4 rue de Castellane, F-75008 Paris.

Status
In production.

VERIFIED

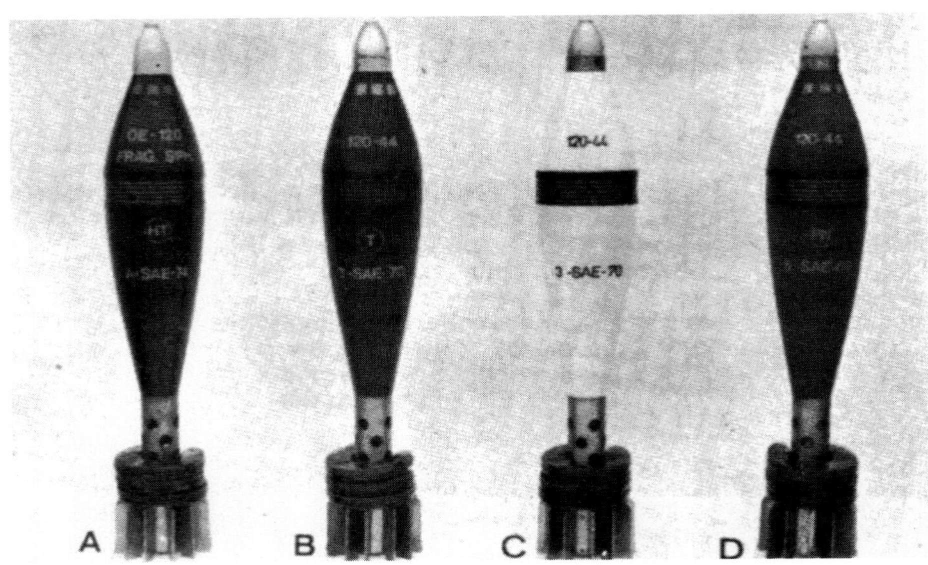

Alsetex 120 mm mortar ammunition
(A) *High explosive, high capacity* **(B)** *High explosive, standard* **(C)** *Inert practice* **(D)** *Practice with burster charge*

GERMANY

120 mm Rheinmetall HE-Improved bomb

Description
The improved performance HE-I 120 mm mortar bomb is effective against both semi-hard targets such as APCs and trucks, and against personnel and field emplacements. The HE-I 120 mm mortar projectile is interoperable with all NATO 120 mm systems and it can be optimised for the US Army's 120 mm Mortar System (enhanced range, blast and firing tables).

Its effect against semi-hard targets is obtained by means of a dish-shaped heavy metal plate in the forward cylindrical section of the projectile. This heavy metal plate consists of prefragmented sphero-hexagonal tungsten fragments. The individual fragments are accelerated by the high-explosive detonation and penetrate the targets, causing high behind-armour effects. The effectiveness against soft targets is achieved by the fragmentation of the bomb body.

When the predetermined optimal burst height in the target area is reached, the ground proximity sensor ignites a separation charge and the ogive is separated and accelerated forward to clear the way for the tungsten fragments. The impulse of the separation decelerates the warhead so that the base fuze (working like an impact fuze) initiates the main explosive charge. The armour of all known semi-hard targets (current and projected) is penetrated with certainty. There is sufficient reserve of energy to ensure penetration of semi-hard targets which may be developed in the future.

As an additional option, the fuze can be set for point detonation on impact.

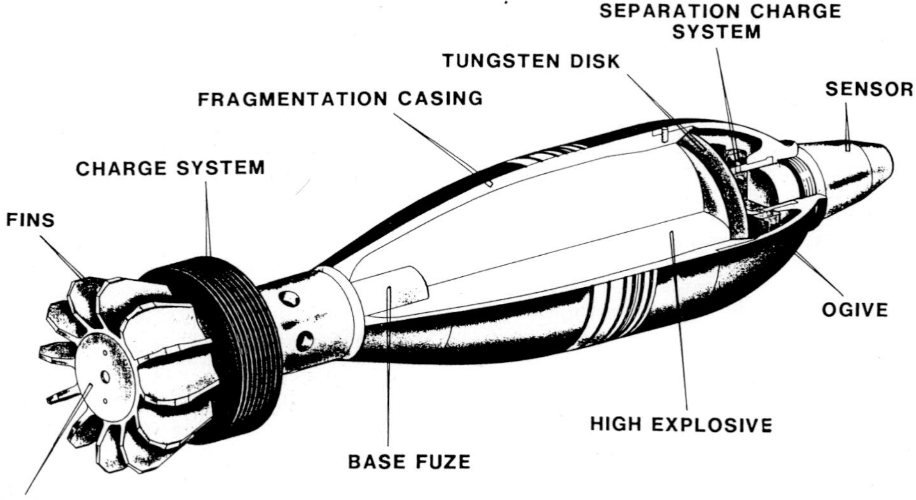

Rheinmetall HE-I Improved HE bomb

Data
Length: fuzed, 645 mm; unfuzed, 595 mm
Weight: 13 kg
Ignition height: ca 17 m
Design gas pressure: 1,200 bar
Max range: ca 6,200 m
Muzzle safety: >2 s

Manufacturer
Rheinmetall Industrie GmbH, Postfach 1663, Pempelfurtstrasse 1, D-40836 Ratingen 1.

Status
Development completed.

UPDATED

Buck 60 mm mortar ammunition

Description

This is a new generation of 60 mm smoke, smoke/incendiary and illuminating mortar bombs. The design has produced a highly efficient bomb which will function with various types of 60 mm mortar including the US M224. Their enhanced payloads permit the achievement of the tactical function with less expenditure of ammunition.

Manufacturer

Buck Werke GmbH & Co, Postfach 2405, D-83435 Bad Reichenall.

Data	60 mm Smoke	60 mm Smoke/ Incendiary	60 mm Illuminating
Weight	2.3 kg	2.3 kg	2.3 kg
Length	467 mm	467 mm	467 mm
Range	1,900 m	1,900 m	1,600 m
Emission time	2 min	–	–
Luminance	–	–	450,000 cd/35 s
Payload	400 g	submunition	430 g

Status

In production.

VERIFIED

Buck 81 mm mortar ammunition

Description

As with the 60 mm rounds, these Buck 81 mm smoke and smoke/incendiary bombs are of modern design and are intended to produce the desired effects for less expenditure of ammunition. They will function in any modern 81 mm mortar including the US M29 and UK L16A1 series.

Manufacturer

Buck Werke GmbH & Co, Postfach 2405, D-83435 Bad Reichenall.

Data	81 mm Smoke	81 mm Smoke/ Incendiary	81 mm Illuminating
Weight	3.9 kg	3.9 kg	4.4 kg
Length	630 mm	630 mm	635 mm
Range	5,600 m	5,600 m	5,400 m
Emission time	2.5 min	–	–
Luminance	–	–	600,000 cd or 1 Mcd/35 s
Payload	1.4 kg	submunition	1.4 kg

Status

In production.

VERIFIED

Buck 120 mm mortar ammunition

Description

These new-generation mortar bombs will function with all types of modern 120 mm mortar including Tampella and Thomson-Brandt types. As with the smaller calibres, the object has been to produce bombs which are capable of delivering the requisite tactical effect with less expenditure of ammunition.

Manufacturer

Buck Werke GmbH & Co, Postfach 2405, D-83435 Bad Reichenall.

Status

In production.

VERIFIED

Data	120 mm Smoke	120 mm Smoke/ Incendiary	120 mm Illuminating	120 mm Practice
Weight	12.9 kg	12.9 kg	12.9 kg	12.9 kg
Length	590 mm	590 mm	583 mm	582 mm
Range	6,000 m	–	–	–
Luminance	3 min	–	1.2 Mcd/50 s	–
Payload	3.65 kg	submunition	1.71 kg	Spotting charge

Buck family of mortar ammunition

120 mm Diehl Bussard terminally guided mortar projectile

See entry under *International*.

DM 111A4 point detonating mortar fuze

Description

The DM 111A4 is a mechanical nose fuze used for smoothbore HE, smoke and TP mortar ammunition of 51 to 160 mm calibre. The fuze is highly sensitive so that it will function even upon impact on fresh snow, marshy ground and water or against shrubs and bushes. Two modes can be set, Super-quick (SQ) or Delay (D). In the safe position the rotor with two detonators is locked out of line by a safety pin and a pulse safety which is only released by firing. The muzzle safety distance is ≥40 m.

The DM 111A4 fuze has been UK qualified as the L127A3 fuze.

Manufacturer

Gebrüder Junghans GmbH, Junghans Feinwerktechnik, Branch Company of Diehl GmbH, PO Box 70, D-78701 Schramberg.

Status

In production.

VERIFIED

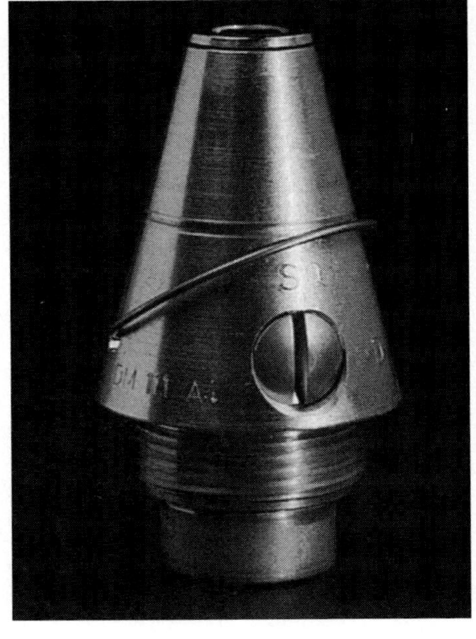

DM 111A4 point detonating mortar fuze

DM 111A5 point detonating mortar fuze

Description

The DM 111A5 is a mechanical nose fuze for unspun HE, smoke and TP mortar ammunition of calibres from 51 to 160 mm. This fuze is not as sensitive as the DM 111A4, so that it can be used in wooded terrain (for example, in jungle areas).

Two modes can be set: Super-quick (SQ) or Delay (D). In the safe position the rotor with two detonators is locked out of line by a safety pin and a pulse safety which is only released by firing. The muzzle safety distance is ≥40 m.

Manufacturer

Gebrüder Junghans GmbH, Junghans Feinwerktechnik, Branch Company of Diehl GmbH, PO Box 70, D-78701 Schramberg.

Status

In production.

VERIFIED

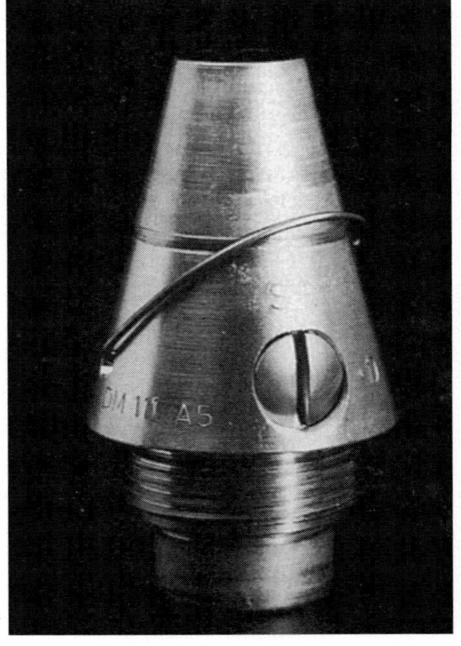

DM 111A5 point detonating mortar fuze

DM 111 AZ-W point detonating mortar fuze

Description

The DM 111 AZ-W point detonating mortar fuze provides a vane type arming mechanism in addition to the usual setback detent system. It thus has two entirely independent safe and arming systems activated by two different physical properties, acceleration and rotation. Because of the vane feature this fuze does not require a safety pin which must be removed before firing, as seen on many conventional PD fuzes. After the bomb has left the muzzle of the mortar, the wind streaming past the fuze rotates the vane and releases the vane safe and arming device.

The DM 111 AZ-W fuze can be used with smoothbore 51 mm to 120 mm mortar HE, smoke and practice bombs. Release criterion is a translational acceleration of 500 *g* or better. The arming distance is at least 40 m using the lowest charge.

All requirements concerning handling, transport, firing and muzzle safety, as well as the provisions of STANAG 3525 and MIL-STD-1316 are fulfilled.

Manufacturer

Gebrüder Junghans GmbH, Junghans Feinwerktechnik, Branch Company of Diehl GmbH, PO Box 70, D-78701 Schramberg.

Status

Licence produced in some countries.

UPDATED

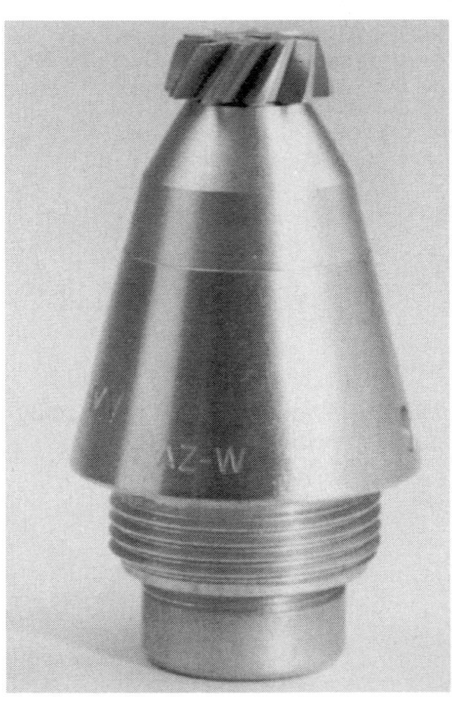

DM 111 AZ-W wind vane arming impact fuze

DM 93/M776 MTSQ mortar fuze

Description

The DM 93 is a mechanical time fuze with an additional impact device, for use with smoke and illuminating mortar projectiles of 51 to 160 mm calibre. It can be set from 6 to 54 seconds with a possible extension up to 67 seconds. At the zero setting the fuze functions upon impact. In the safe position the rotor is out of line and the clockwork is locked by a safety pin and a pulse safety which is released only by firing. The muzzle safety distance is ≥40 m.

Manufacturer

Gebrüder Junghans GmbH, Junghans Feinwerktechnik, Branch Company of Diehl GmbH, PO Box 70, D-78701 Schramberg.

Status

In production.

VERIFIED

DM 93 time and percussion mortar fuze

M772 MTSQ mortar fuze

Description

The M772 is a mechanical time fuze with an additional impact device, for smoothbore smoke and illuminating bombs in 81 mm calibre. It can be set from 4 to 55 seconds. The M772 is a derivation of the DM 93 specially developed for US 81 mm ammunition (M853 and M819) used by the US Army and US Marine Corps. At the zero setting the fuze functions upon impact. In the safe position, the rotor is out of line and the clockwork is locked by a safety pin and a pulse safety which is released only by firing. The muzzle safety distance is ≥40 m.

Manufacturer

Gebrüder Junghans GmbH, Junghans Feinwerktechnik, Branch Company of Diehl GmbH, PO Box 70, D-78701 Schramberg.

Status

In production.

VERIFIED

M772 MTSQ mortar fuze

DM 421 PD mortar fuze

Description

The DM 421 is a mechanical point detonating fuze for smoothbore 51 to 160 mm mortar HE, TP, illuminating and bursting smoke ammunition. This fuze is designed to function in the super-quick mode only. Because of the application of state-of-the-art production technologies and materials (plastics), the fuze is an extremely cost-effective product meeting all current requirements of modern mortar fuzes.

Manufacturer

Gebrüder Junghans GmbH, Junghans Feinwerktechnik, Branch Company of Diehl GmbH, PO Box 70, D-78701 Schramberg.

Status

In production.

UPDATED

DM 421 PD mortar fuze

GREECE

81 mm M374A2 HE bomb

Description

This is the standard US pattern 81 mm HE bomb. It consists of a steel body with plastic obturating ring, filled with Composition B. The fuse is the PD M524A6 with selectable super-quick or delay functioning; any other suitable fuze can be used. The fins are of aluminium alloy and a primary cartridge M285 is fitted into the forward section of the tail tube, with a percussion primer M71A2 screwed into the base of the tail unit. The flash from the primer passes up the centre of the tail unit to ignite the primary, which then vents through holes in the tail tube to ignite the secondary cartridges. There are nine secondaries in the form of long bags filled with flake M9 powder; these are retained by two increment holders. The secondaries are of two types: one Charge A unit of 12 g weight, and eight Charge B units of 10.875 g weight.

Data

Length: 529 mm
Weight: 4.24 kg
Filling: 945 g Comp B
Muzzle velocity: Charge 0, 64 m/s; Charge 9, 261 m/s
Max range: Charge 0, 403 m; Charge 9, 4,500 m

Manufacturer

Greek Powder and Cartridge Company (PYRKAL), 1 Ilioupoleos Avenue, GR-17236 Hymettus, Athens.

Status

In production.

Service

Greek Army.

VERIFIED

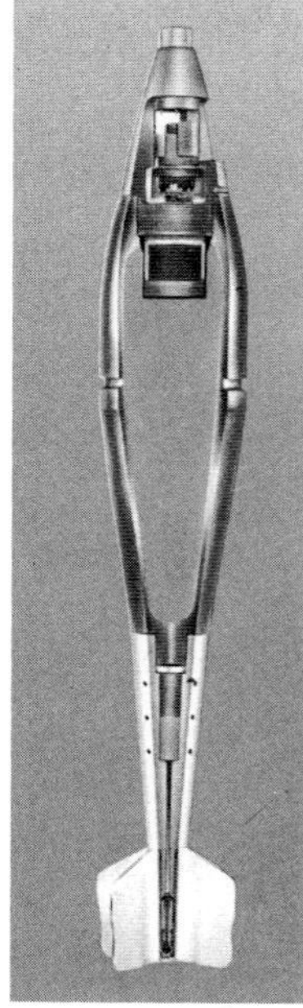

81 mm M374A2 HE bomb

PYRKAL 107 mm HE GRM20 cluster mortar bomb

Description

The Greek Powder and Cartridge Company (PYRKAL) is producing the GRM20 cluster mortar bomb for use with 107 mm (4.2 in) M2 and M30 mortars. The bomb carries 20 M20G prefragmented steel bomblets which contain a 30 g Composition A5 shaped charge for penetration and fragmentation effects. Each 39 mm diameter bomblet is equipped with a GRM3A2 fuze and a stabilising and arming streamer, and functions upon impact with the target.

Targets include personnel and light armoured vehicles, with a penetrating ability of 60 mm of armour plate, a dispersion radius of approximately 70 to 140 m, an effective radius of 15 m around the impact point of each bomblet and a total effective area of some 7,000 m^2.

The projectile fuze (MTSQ M577A1 or M582A1, or Electronic Time Fuze AXA-I) is set to function at 300 m above the target area. At this predetermined point in flight the entire payload is ejected from the rear of the projectile and is radially dispersed over the target area.

Data

Length: fuzed, ca 670 mm
Weight: ca 13 kg
Dispersion radius: 70-140 m around mpi
Max range: 5,500 m

Manufacturer

Greek Powder and Cartridge Company (PYRKAL), 1 Ilioupoleos Avenue, GR-17236 Hymettus, Athens.

Status

In production.

Service

Greek Army and Middle East countries.

UPDATED

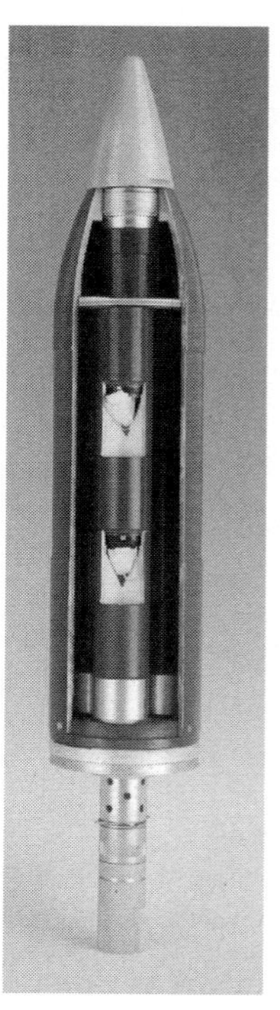

107 mm HE GRM20 cluster mortar bomb

HUNGARY

82 mm dual-purpose mortar bomb

Description

Developed for use with the 82 mm 2B9M automatic mortar (licence produced in Hungary) in both the direct and indirect fire modes, this 82 mm dual-purpose bomb can also be fired from conventional 82 mm mortars. It contains a copper lined two-part shaped charge capable of penetrating 100 mm of armour. As the bomb detonates the cast steel body breaks up to create anti-personnel fragments.

Using a basic primary cartridge this bomb is fired from the 2B9M mortar using a fixed 4D2 propellant charge weighing 75 g around the tail assembly. When fired from conventional mortars up to three ballistite propellant discs can be located around the tail. In both instances the maximum range is 4,300 m.

Data

Length: 330 mm
Weight: 3.1 ±0.2 kg
Muzzle velocity: 2B9M, 270 m/s; conventional mortar, 210 m/s
Max range: 4,300 m
Armour penetration: 100 mm

Manufacturer

Mechanikai Müvek, PO Box 64, H-1518 Budapest.

Status

In production.

Service

Hungarian Army.

NEW ENTRY

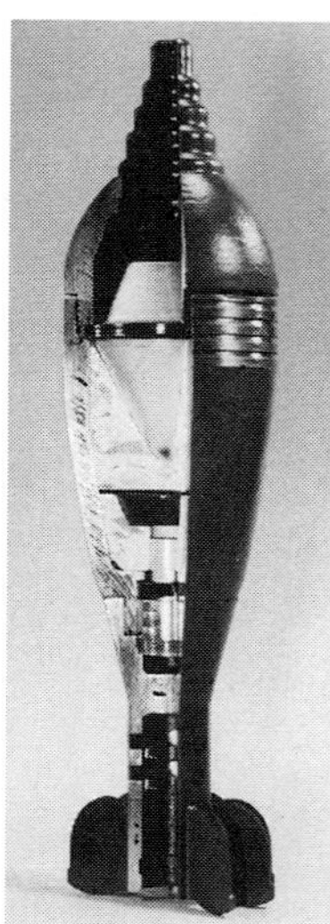

Cross-sectioned example of 82 mm dual-purpose mortar bomb

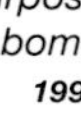

1996

INDIA

IOF 51 mm HE-1A HE bomb

Description

The Indian Ordnance Factories (IOF) 51 mm HE-1A HE bomb is intended for the 51 mm mortar manufactured in India. The light steel body is lined with a notched wire coil to provide controlled fragmentation. The filling is RDX/Wax. A percussion fuze, which appears similar to the UK Fuze, Percussion, 152 is fitted and there is a primary and one secondary cartridge.

Data

Length: 269 mm
Weight: nominal, 950 g
Muzzle velocity: 107 m/s
Min range: 200 m
Max range: 850 m

Manufacturer

Indian Ordnance Factory Board, 10A Auckland Road, Calcutta 700001.

Status

In production.

Service

Indian Army and export.

UPDATED

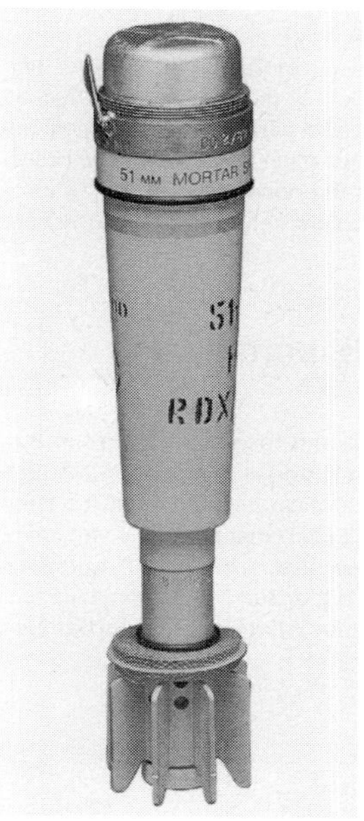

IOF 51 mm HE-1A bomb

IOF 81 mm HE bomb

Description

The IOF 81 mm HE bomb is very similar to the old British 3 in HE bomb but has a raised bourrelet band at the waist and a plastic obturating ring. The tail unit is of machined light alloy and is similar in pattern to that used with bombs for the L16 mortar. The bomb is fitted with a locally manufactured version of the British Fuze, Percussion, No 162, an impact fuze which carries a safety pin beneath the cap and which is armed on firing. The propelling charge consists of a primary cartridge and eight horseshoe secondaries clipped around the tail unit.

Data

Weight: 4.2 kg
Filling: 705 g TNT
Muzzle velocity: 295 m/s
Max range: 5,000 m

Manufacturer

Indian Ordnance Factory Board, 10A Auckland Road, Calcutta 700001.

Status

In production.

Service

Indian Army and export.

VERIFIED

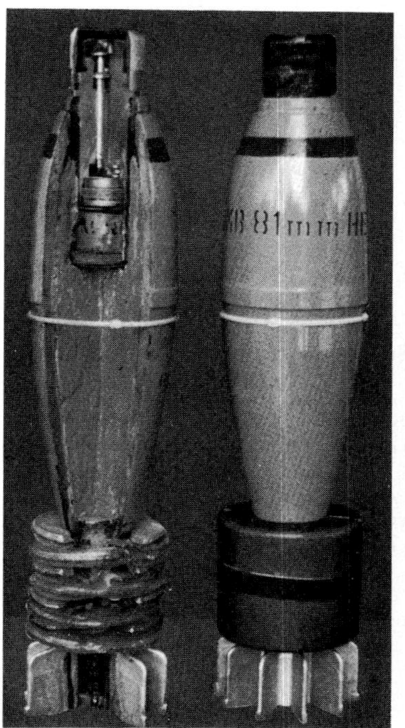

IOF 81 mm HE bomb

IOF 81 mm smoke bomb

Description

This smoke bomb is essentially the same as the IOF HE bomb described previously, except that the filling is Plasticised White Phosphorus (PWP). On impact it takes about 5 seconds to build up a smoke screen which is then emitted for an average of 20 seconds. The same impact fuze is fitted as that on the HE bomb and the eight-unit propelling charge is similar but adjusted to give the bomb the same ballistic performance as the HE bomb.

Data

Weight: 4.4 kg
Muzzle velocity: 295 m/s
Max range: 5,000 m

Manufacturer

Indian Ordnance Factory Board, 10A Auckland Road, Calcutta 700001.

Status

In production.

Service

Indian Army and export.

UPDATED

IOF 81 mm PWP smoke bomb

IOF 81 mm Illuminating mortar bomb

Description

The IOF 81 mm Illuminating mortar bomb is longer than the other IOF mortar bombs (see previous entries) as the body carries a flare and parachute unit. A mechanical time fuze, which appears to be based on the British No 221, is mounted in the nose. When this fuze functions, ideally at an optimum height of 500 m, the flare and parachute are ejected to create a light output of over 10,000 candelas for more than 30 seconds. This is sufficient to illuminate a ground area over 800 m in diameter. Maximum range is 4,300 m.

Data

Weight: 4.2 kg
Filling: flare and parachute assembly
Max range: 4,300 m
Luminosity: 10,000 cd

Manufacturer

Indian Ordnance Factory, Dehu Road, Pune 412113.

Status

In production.

Service

Indian Army and export.

NEW ENTRY

IOF 120 mm HE bomb

Description

The IOF 120 mm HE bomb resembles the Thomson-Brandt M44/66 HE bomb and is of forged steel with a series of obturation grooves around the waist. It is fitted with the Fuze DA4A, a locally manufactured version of the Brandt V19P impact fuze. The propelling charge consists of a primary cartridge in the tail unit, plus four secondaries in horseshoe containers clipped around the tail boom.

Data

Weight: 13.2 kg
Filling: 448 g TNT
Number of charges: 8
Muzzle velocity: Charge 0, 118 m/s; Charge 7, 331 m/s
Max range: 6,650 m

Manufacturer

Indian Ordnance Factory Board, 10A Auckland Road, Calcutta 700001.

Status

In production.

Service

Indian Army and export.

UPDATED

IOF 120 mm HE bomb

IOF 120 mm Illuminating mortar bomb

Description

This 120 mm Illuminating mortar bomb outwardly resembles the IOF 120 mm HE mortar bomb (see previous entry) but the body carries a flare and parachute unit. A mechanical time fuze, which appears to be based on the British No 221, is mounted in the nose. When this fuze functions, ideally at an optimum height of 700 m, the flare and parachute are ejected to create a light output of over 11,000 candelas for over 40 seconds. This is sufficient to illuminate a ground area over 1,200 m in diameter.

The propelling charge consists of a primary cartridge in the tail unit, plus four secondaries in horseshoe containers clipped around the tail boom.

Data

Weight: 13.2 kg
Filling: flare and parachute assembly
Number of charges: 8
Muzzle velocity: Charge 0, 118 m/s; Charge 7, 331 m/s
Max range: 6,700 m
Luminosity: 11,000 cd

Manufacturer

Indian Ordnance Factory, Dehu Road, Pune 412113.

Status

In production.

Service

Indian Army and export.

NEW ENTRY

INDONESIA

60 mm GMO-6 PE A1 mortar bomb

Description

This bomb is intended for use in short Commando type 60 mm mortars and is of conventional pattern with gas-check rings around the waist. It has one propellant charge, composed of the usual ignition cartridge and a single incremental charge fitted around the tail boom. The tail fins are slightly canted to give some degree of roll stabilisation. The design is licensed from Vammas.

Data

Length: 300 mm
Weight: 1.61 kg
Filling: 200 g TNT
Range: max, 800 m; min, 50 m
Muzzle velocity: 94 m/s
Fuze: M111B1 PD

Manufacturer

PT Pindad (Persero), Jl Jendral Gatot Subroto, PO Box 807, Bandung 40284.

Status

In production.

UPDATED

60 mm GMO-6 PE A1 mortar bomb

60 mm GMO-6 PE A2 mortar bomb

Description
This bomb is for use in standard types of 60 mm mortars and is of modern pattern, with streamlined body, canted fins and an obturating ring around the waist. It is in fact the Israeli Soltam M38 produced in Indonesia under licence. The propelling charge consists of a primary cartridge and up to six secondary increments which fit around the tail boom.

Data
Length: 351 mm
Weight: 1.86 kg
Filling: 330 g TNT
Range: max, 4,000 m; min, 200 m
Muzzle velocities: 89-258 m/s
Fuze: M111B1 PD

Manufacturer
PT Pindad (Persero), Jl Jendral Gatot Subroto, PO Box 807, Bandung 40284.

Status
In production.

UPDATED

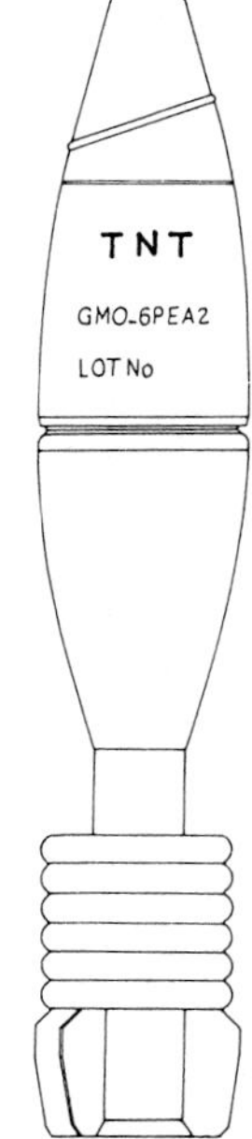

60 mm GMO-6 PE A2 mortar bomb

81 mm GMO-8 PE A1 mortar bomb

Description
The 81 mm GMO-8 PE A1 mortar bomb is a modern design, based on the Soltam M64 and manufactured under licence in Indonesia. The streamlined body has an obturating ring around the waist and has a propelling charge composed of a primary cartridge and up to eight secondary charges fitted around the tail boom.

Data
Length: 490 mm
Weight: 4.63 kg
Filling: 740 g TNT
Range: max, 6,500 m; min, 100 m
Muzzle velocities: 67-353 m/s
Fuze: M111B1 PD

Manufacturer
PT Pindad (Persero), Jl Jendral Gatot Subroto, PO Box 807, Bandung 40284.

Status
In production.

UPDATED

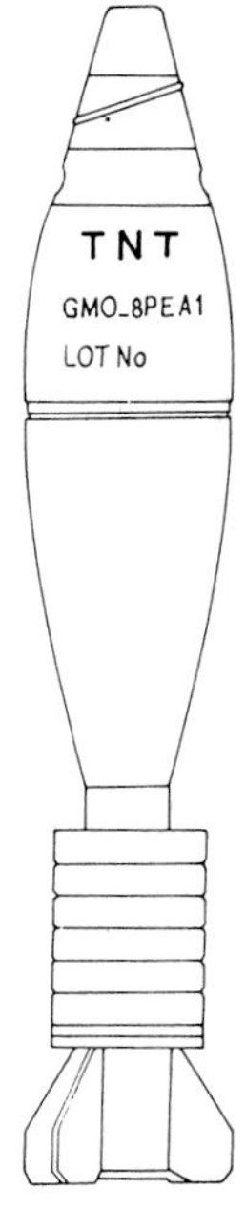

81 mm GMO-8 PE A1 mortar bomb

INTERNATIONAL

120 mm Precision Guided Mortar Munition (PGMM)

Development
In 1975 Diehl GmbH was awarded a contract by the former West German Ministry of Defence to perform a feasibility study into a terminally guided mortar projectile. Bodensee Werk Geratetechnik, AEG and Eltro were also involved in the project. The projectile which resulted from this study was the Diehl Bussard (Buzzard).

The objective of the study was to combine the advantages of traditional gun-launched projectiles with those of guided missiles so that the resulting munition would be used from existing ordnance, would have a high first-round hit probability, even against moving targets, and would permit engaging armoured targets in the top attack mode. Successful firings of the first development models took place in 1983 with, at one point, three out of three direct hits being attained.

By the mid-1980s the Bussard project had been suspended for funding reasons, only to be revived when the US Army issued a requirement for a 120 mm Precision Guided Mortar Munition (PGMM). Lockheed Martin joined with Diehl for the selection contest and in October 1994 was awarded a Phase I critical components demonstration contract to continue development, along with another team formed by British Aerospace, Alliant Techsystems and Rockwell, who proposed an enlarged Merlin 81 mm terminally guided mortar projectile allied with a Hellfire missile warhead. Two other contestants, Saab Missiles and Hercules, were eliminated from the PGMM contest at that stage.

In June 1995 it was announced that the Lockheed Martin/Diehl PGMM submission had been selected as the sole entrant for Phase II of the PGMM programme. A $10.8 million advanced technology demonstration contract was awarded by the US Army Armament Research, Development and Engineering Center (ARDEC) to permit Lockheed Martin/Diehl to manufacture seven prototype rounds for live test firing until FY 1998.

The PGMM can be fired from all existing 120 mm ground-mounted or self-propelled mortars, including the Lockheed Martin 120 mm lightweight composite mortar. No attachments, assembly prior to use or sustainer propellant sections are required.

Description
The 120 mm Precision Guided Mortar Munition (PGMM) is based on the Diehl Bussard but considerably revised to meet the PGMM requirements. It is a passive, gliding, intelligent 120 mm mortar munition with a maximum range of 15,000 m.

The nose of the projectile contains the seeker section, comprising a low-cost cooled mid-wave infra-red sensor, the associated processor, and gyro units. A mid-body section contains thermal batteries, power conditioning and four switch-blade wings. This section is followed by the warhead, comprising a tandem shaped charge, the fuze and the safe and arm unit. The tandem shaped charge (produced by Dynamit Nobel) enables the PGMM to be deployed against targets such as bunkers as well as armoured targets.

The tail section contains the control activation system, the four control fins and the fin deployment mechanism. The propulsion system, including the igniter, is at the rear (base) of the projectile body.

The PGMM can be operated in two modes. One is the Man-in-the-Loop mode where the projectile is guided by a laser designator operated by a forward observer; the projectile uses its imaging infra-red seeker to detect and track the designated target. For the Fire-and-Forget mode the PGMM operates autonomously, using the same imaging infra-red sensor as before but in conjunction with a target image processor. In this mode the projectile can detect and acquire both moving and stationary targets emitting infra-red radiation.

Before deploying the PGMM and following a call for a fire mission, the usual ballistic calculations are made at a fire direction centre and fire orders are given to a 120 mm mortar team. The PGMM is then loaded and fired in the normal manner. The onboard thermal batteries are activated by the launch acceleration and begin to power the system electronics. Immediately after launch the tail fins deploy to provide aerodynamic stability.

After the apogee of the trajectory the wings deploy and the projectile commences a relatively straight line target acquisition glide trajectory toward the target area. For a laser-guided mission the seeker searches for a designated target. If laser radiation is detected the autopilot assumes a Man-in-the-Loop operation and begins homing on the signal, following the laser signal to the target. During an autonomous fire mission the infra-red scene is evaluated for potential targets and the PGMM then homes onto the highest ranked infra-red signature. The seeker can search a 500 × 500 m footprint although this could be upgraded to 1,000 × 1,000 m.

In both modes, close to the end of the homing phase and as the target resolution improves, a target aimpoint is computed prior to the hit.

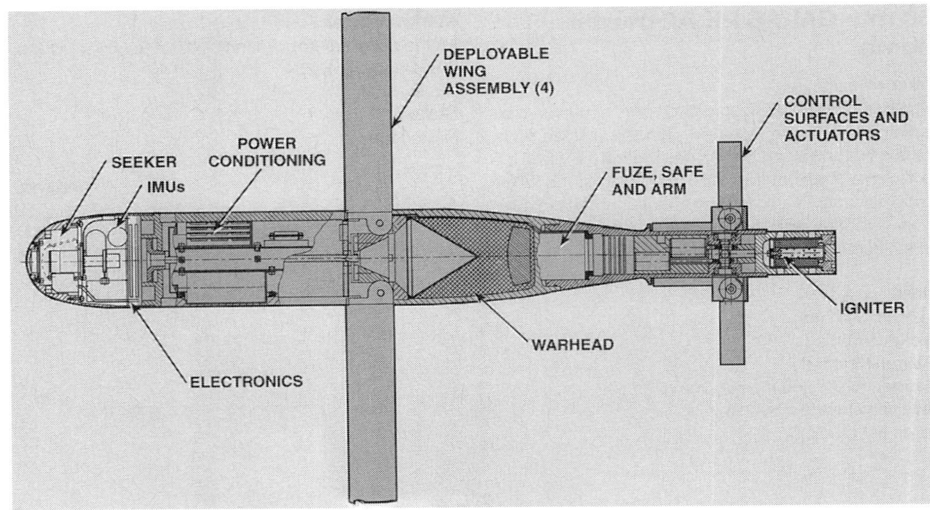

Main components of the 120 mm Precision Guided Mortar Munition (PGMM)

1996

Data

Calibre: 120 mm
Length: 965 mm
Weight: 17.2 kg
Range: max, 15,000 m; min, 500 m
Storage life: 10 years

Manufacturers

Diehl GmbH & Co, Ammunition Division, Fischbachstrasse 16, D-90552, Röthenbach/Pegnitz, Germany.
Lockheed Martin Missiles and Space, 1111 Lockheed Way, Sunnyvale, California 94089-3504, USA.

Status

Advanced development.

NEW ENTRY

ISRAEL

TAAS 52 mm smoke mortar bomb

Data

Length: 240 mm
Weight: 940 g
Filling: 400 g Hexachloroethane, zinc oxide and aluminium powder
Propelling charge: primary cartridge with 4 g of ballistite
Muzzle velocity: 78 m/s
Max range: 450 m
Time delay: 7.5 s
Time of smoke generation: 100 s

Manufacturer

TAAS - Israel Industries Limited, PO Box 1044, Ramat Hasharon 47100.

Status

In production.

Service

Israeli forces.

VERIFIED

TAAS 52 mm smoke mortar bomb

TAAS 52 mm HE mortar fragmentation bomb

Data

Length: 250 mm
Weight: 1.15 kg
Filling: 105 g TNT
Propelling charge: base ignition cartridge containing 4 g of ballistite
Fuze: No 161
Muzzle velocity: 75 m/s
Max range at 45° elevation: 450 m

Manufacturer

TAAS - Israel Industries Limited, PO Box 1044, Ramat Hasharon 47100.

Status

In production.

Service

Israeli forces.

VERIFIED

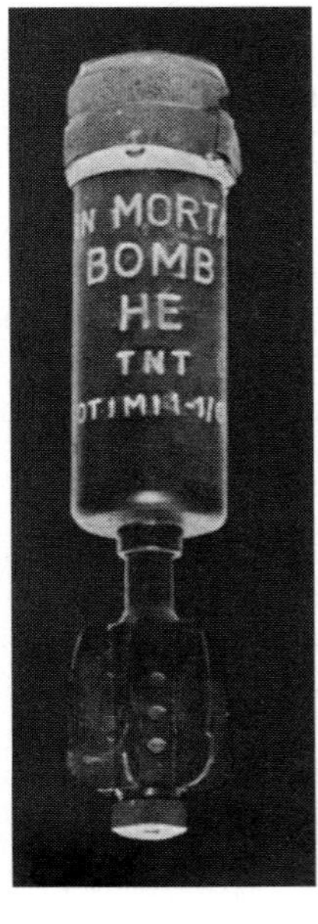

TAAS 52 mm HE mortar fragmentation bomb

TAAS 52 mm illuminating mortar bomb

Data

Length: 271 mm
Weight: 830 g
Filling: 150 g pyrotechnic composition
Propelling charge: primary cartridge with 5.5 g ballistite
Muzzle velocity: 95 m/s
Parachute opening: range, 350 m; height, 100 m
Delay time to burst: ca 6 s
Illuminating power: 100,000 cd
Min burning time: 30 s

Manufacturer

TAAS - Israel Industries Limited, PO Box 1044, Ramat Hasharon 47100.

Status

In production.

Service

Israeli forces.

VERIFIED

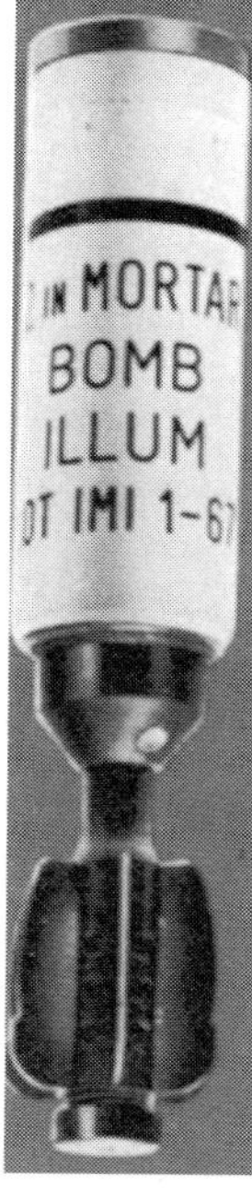

TAAS 52 mm illuminating mortar bomb

Ammunition for 60 mm Soltam mortars

Description

Bombs for 60 mm mortars are of the conventional pattern. The HE bomb is made from a steel forging and machined to shape. The tail unit is made from an extruded aluminium tube. The secondaries are arranged around the tail boom above the fins and when the bomb is dropped into a hot barrel the secondaries do not contact the walls of the tube, and there is no danger of a cook-off.

The long-range bomb has an enhanced aero-ballistic configuration and is equipped with a plastic obturation ring.

Smoke Bombs

Smoke bombs are filled with titanium tetrachloride (FM), White Phosphorus (WP) or Plasticised White Phosphorus (PWP). All other data is as for the long-range HE bomb. They also use the Fuze DM 111.

Manufacturer

Soltam Limited, PO Box 371, Haifa.

Status

Available.

Service

Israeli defence forces.

Ammunition for the 60 mm Mortar: HE bombs

	Long-Range M38	**Standard M61**
Max range	4,000 m	2,550 m
Min range	200 m	150 m
Max velocity	350 m/s	285 m/s
Min velocity	70 m/s	66 m/s
Bomb weight	1.9 kg	1.6 kg
Filling (TNT)	320 g	220 g
Charges	6	4
Fuze	DM 111 PD & SQ	

VERIFIED

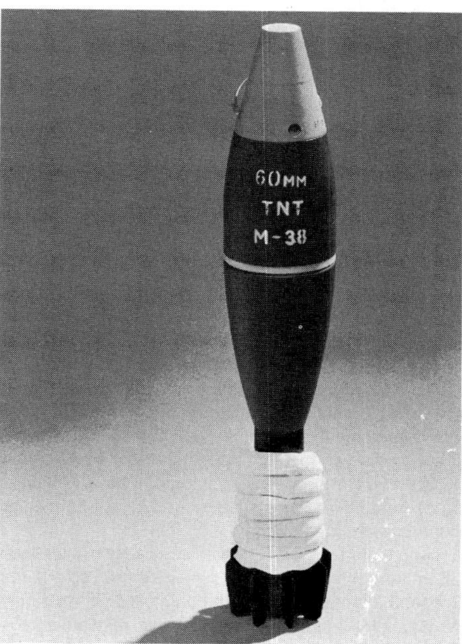

60 mm Soltam HE long-range bomb

TAAS 60 mm long-range illuminating bomb

Description

The TAAS 60 mm long-range illuminating bomb was developed in order to provide effective illumination at double the range of previously available 60 mm bombs. The increased range has been achieved by reduction of weight and by increasing the propellant charge to take advantage of the maximum permissible barrel pressure in the mortar. Illumination power has been increased by 83 per cent.

The long-range bomb incorporates two original design concepts: the parachute assembly is located in the rear section, permitting dedication of the entire interior to the illuminating assembly; and the acceleration/deceleration sequence is designed to permit opening of the main parachute only after the discarded tail section is well clear, making collision impossible.

The functioning of the bomb is initiated by a time fuze which ignites a small burster charge. This separates the two parts of the bomb by shearing the connecting pins, ignites the illuminant and ignites a small rocket motor. This motor gives a forward thrust to the illuminant assembly, ejecting it from the nose of the bomb and causing the small drag parachute to deploy. This drag parachute slows down the illuminant assembly, allowing the empty bomb body to pass it in flight. The drag parachute then extracts the main parachute from its container, which is still attached to the bomb body by a line. The main parachute then deploys, to lower the illuminant to the ground; the drag parachute has by now been burned by the illuminant, and the bomb body, trailing the empty main parachute bag, has continued on its own trajectory and cannot interfere with the descending illuminant unit.

Data

Length: 432 mm
Weight: 1.44 kg
Propelling charge: primary + 2 secondaries
Range: max, 2,270 m
Illuminance: 500,000 cd
Duration: 38 s
Rate of descent: 4 m/s

Manufacturer

TAAS - Israel Industries Limited, PO Box 1044, Ramat Hasharon 47100.

Status

In production.

Service

Israeli forces and overseas sales.

VERIFIED

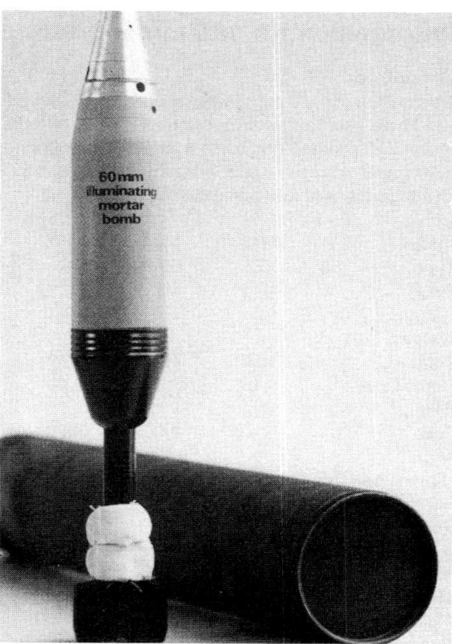

TAAS 60 mm long-range illuminating bomb

Ammunition for 81 mm mortars

HE BOMBS

	Long-Range	Standard
Type	M64	M21
Max range	6,500 m	4,900 m
Min range	200 m	150 m
Max muzzle velocity	350 m/s	285 m/s
Min muzzle velocity	70 m/s	66 m/s
Bomb weight	4.6 kg	3.9 kg
Filling (TNT)	740 g	540 g
No of propelling charges	8	7

Fuze: DM 111 point detonating super-quick and delay; M25 proximity, with PD option
Body: forged steel

SMOKE BOMBS
Smoke bombs are filled with 540 g of titanium tetrachloride, white phosphorus or plastic white phosphorus. All other data is as for the long-range HE bomb.

PRACTICE BOMB
Weight of smoke charge: (to mark point of impact) 100 g
Weight of inert filling: 440 g
All other data as the HE round.

Manufacturer
Soltam Limited, PO Box 371, Haifa.

Status
Available.

Service
Israeli forces. Exported to other countries and can be encountered in East Africa.

VERIFIED

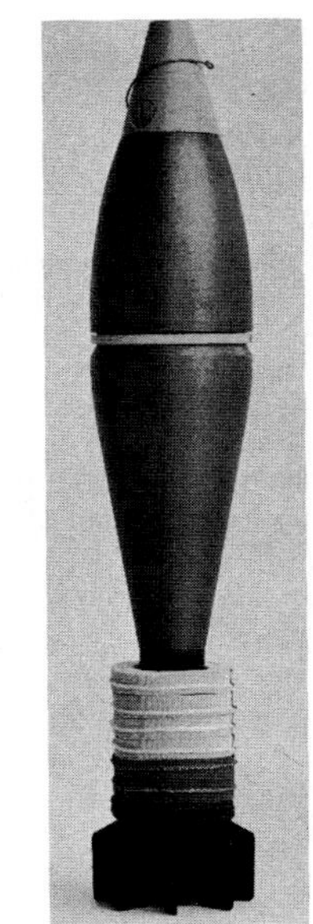

81 mm Soltam HE bombs, (left) *long-range M64*

TAAS 81 mm M2A1 illuminating mortar bomb

Description
The TAAS 81 mm M2A1 illuminating mortar bomb embodies two original concepts in design and construction: the parachute assembly is located in the rear section, permitting dedication of the entire interior to the illuminating assembly; and the acceleration-deceleration sequence is designed to permit opening of the main parachute only after the discarded tail section is well clear, making collision impossible.

The bomb functions with a clockwork fuze which at a predetermined point initiates a small propelling charge. The pressure developed by this charge shears the pins holding the two parts of the bomb body, ignites a small rocket motor and the pyrotechnic charge. The rocket motor drives the illuminant forward and this pulls the auxiliary parachute out of the rear section of the bomb. The main parachute is also pulled out but remains attached to the rear section of the bomb. The auxiliary parachute slows the illuminant assembly and the rear part of the bomb goes past and pulls the main parachute from its bag. The main parachute opens and the illuminant descends gently.

Two versions of this bomb exist; one for use with the US pattern M29A1 mortar and the other for use with the British M252 mortar. Whilst the bomb itself is the same, the propellant increments differ so as to take advantage of the maximum permissible barrel pressure of the two mortars, with increased maximum range as a result.

The illuminating power can be increased by adjusting the composition - with a corresponding decrease in burning time - in accordance with the operational requirements of the user. For example, it can be adapted to provide one million candela for a burning time of up to 40 seconds.

Data
Length: 570 mm
Weight: complete round, 4 kg
Weight of illuminant: 800 g
Propellant weight: M29A1 mortar, 132 g; M242 mortar, 147.5 g
Range, M29A1: from 400 m (Ch 1) to 4,500 m (Ch 4)
Range, M242: from 400 m (Ch 1) to 5,000 m (Ch 4)
Rate of descent: 5 m/s
Illuminating power: 700,000 cd for 55-60 s

Manufacturer
TAAS - Israel Industries Limited, PO Box 1044, Ramat Hasharon 47100.

Status
In production.

Service
Israeli forces and overseas sales.

VERIFIED

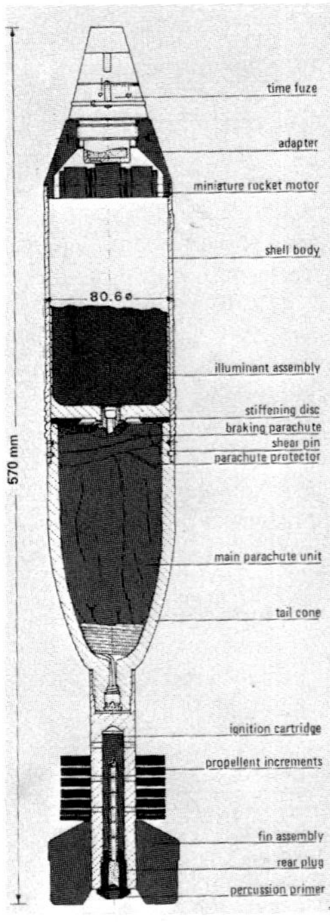

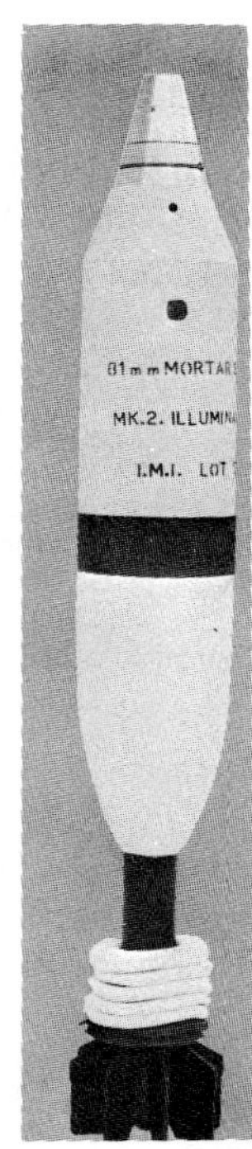

TAAS 81 mm Mark 2 illuminating mortar bomb

Ammunition for 120 mm mortars

Description
Ammunition for 120 mm Soltam mortars is of conventional type, using a streamlined bomb with alloy tail unit. Secondary charges are in the form of cloth-wrapped split rings which can be attached around the tail boom. The usual types of fuze are available.

For all types of ammunition, illuminating and practice rounds are available.

HIGH EXPLOSIVE ROUND M48
Length: 580 mm
Weight: complete round, 12.6 kg; forged steel body, 9.3 kg; aluminium alloy tail unit, 700 g
Fuze: super-quick/delay; arming safety 0.8 s; weight

Model	**M48**	**M84**	**M57**	**M68**	**M98**	**M59**	**M69**	**M100**
Type	HE	Smoke	HE	Smoke	HE	HE	Smoke	HE
Fuze	M111	M111	M111	M111	M111	M111	M111	M111
Weights (kg)								
Total	12.6	12.6	13.2	13.2	13.6	14.75	14.75	14.85
Bomb body	9.18	8.72	9.68	9.68	—	10.86	10.15	—
Filling	2.2	2 (WP)	2.25	2.03 (WP)	3 (Comp B)	2.49	2.46 (WP)	2.5
Fuze	0.205	0.205	0.205	0.205	0.205	0.205	0.205	0.205
Dimensions (mm)								
Length overall	581	581	664	664	709	703	703	787
Body length	333	321	350	350	—	377	377	—
Ballistic Data								
Mortar model	K5, K6			K6		A7		A7
Max prop charges	8			10		15		7
Range	250-6,250			200-7,200		8,500		9,500
Muzzle velocities	113-310			102-320		308		—

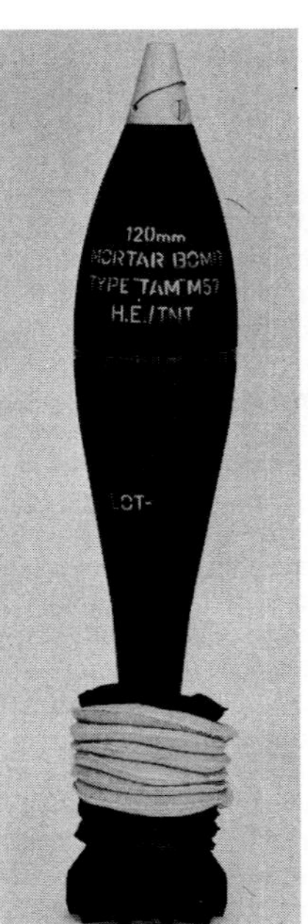

120 mm M57 HE and M58F HE rounds

200 g. The performance of the PD fuze is improved by using the M25 proximity fuze.
Filling: 2.3 kg TNT
Charge system: primary cartridge + additional charge + 4 secondary charges
Muzzle velocity: min 115 m/s; max 310 m/s
Max chamber pressure: 900 kg/cm²
Range: min 400 m; max 6,500 m

SMOKE ROUND M48
FM round. Filling: 2.3 kg titanium tetrachloride.
WP round. Filling: 2.3 kg white phosphorus.
PWP round. Filling: 2.3 kg plastic white phosphorus.
All other data are the same as for the M48 HE round

PRACTICE ROUND M48
Weight of smoke indicator charge: 200 g
Weight of inert filling: 2.1 kg
All other data are the same as for the M48 HE round

M98 Enhanced Ammunition for 120 mm mortars
This type of bomb was designed especially to meet the US Army requirement and was type classified by the US Army in 1991.

The bomb has a 50 per cent better fragmentation performance and is equipped with a waterproofed 4 equal increments propellant system, while the maximum range remains 7,200 m with the light mortar and 7,600 m with the heavy mortar. The smoke and illuminating bombs are ballistically matched to the HE bomb.

The ammunition is packed in an inner cardboard container, two containers in either a wooden box or a sealed metal container at the customer's request.

Data
HIGH EXPLOSIVE ROUND M98
Length: overall, 703 mm
Weight: complete round, 13.6 kg
Fuze: super-quick/delay PD fuze
Filling: 3 kg Comp B
Charge system: primary cartridge plus 4 equal increments
Muzzle velocity: min, 102 m/s; max, 318 m/s
Range: min, 200 m; max, 7,200 m

Manufacturer
Soltam Limited, PO Box 371, Haifa.

Status
Available.

Service
Israeli forces and some other armies.

VERIFIED

TAAS 120 mm HE rocket-assisted mortar bomb

Description
The TAAS 120 mm HE rocket-assisted mortar bomb will reach 10,500 m. It can be fired from any current 120 mm mortar.

Data
Length: overall, 744 mm
Weight: complete round, 16.7 kg; rocket propellant, 1.15 kg
Filling: 2.15 kg Comp B
Propelling charge: primary and 9 increments
Range: max, 10,500 m
Muzzle velocity: max, 280 m/s

Manufacturer
TAAS - Israel Industries Limited, PO Box 1044, Ramat Hasharon 47100.

Status
In production.

Service
Israeli forces and several other armies.

VERIFIED

TAAS 120 mm HE rocket-assisted mortar bomb

TAAS 120 mm CL3144 ICM mortar bomb

Description
The TAAS 120 mm CL3144 ICM mortar bomb carries 24 dual-purpose (anti-personnel/anti-armour) 42 mm diameter Bantam bomblets to be scattered over a target area following an airburst. The Bantam bomblets are carried inside the CL3144 body in six layers of four bomblets.

The CL3144 is handled and loaded in the same way as conventional 120 mm mortar bombs, with the six-increment propellant system being adjusted as required and the nose-mounted DM 93 (or similar) mechanical time fuze preset to ensure an airburst over the target area. Once the bomb has left the mortar muzzle flip-out tail fins deploy to provide aerodynamic stability.

When the fuze functions in flight it ignites an expulsion charge to pressurise the bomb ogive and eject the bomblet payload. The 24 bomblets are scattered to fall to the ground to form an approximate X pattern over a 100 to 110 m radius covering an area of about 4,800 m². A salvo of CL3144 bombs over a target will ensure good target area coverage.

As each Bantam bomblet impacts with the target area its RDX shaped charge detonates to provide a penetration of more than 105 mm of armour. The bomblet body also breaks up into more than 1,200 lethal fragments. Each 42 mm Bantam bomblet weighs 296 g of which approximately 44 g is the RDX shaped charge; each bomblet is 55.65 mm long. Should a Bantam bomblet not detonate on impact there is an integral pyrotechnic self-destruct mechanism initiated as the fuze is armed during its descent from the CL3144 body.

Data
Length: fuzed, 827 mm; unfuzed, 767 mm
Weight: 15 kg
Fin diameter: 270 mm
Filling: 24 Bantam dual-purpose bomblets
Muzzle velocity: 311 m/s
Max range: ca 5,600 m
Dispersion radius: ca 100-110 m
Number of fragments: total per bomb, ca 29,000

Manufacturer
TAAS - Israel Industries Limited, PO Box 1044, Ramat Hasharon 47100.

Status
In production.

Service
Israeli defence forces and offered for export sales.

NEW ENTRY

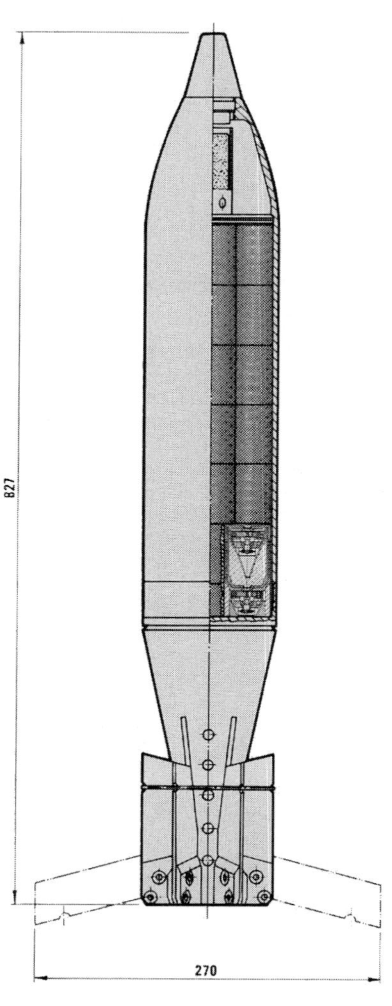

Cross-sectioned drawing of TAAS 120 mm CL3144 ICM mortar bomb
1996

TAAS 120 mm M3 illuminating bomb

The TAAS M3 illuminating bomb is fired from a Tampella, Soltam or Brandt 120 mm mortar. The ballistics of the bomb are identical to those of the comparable HE bomb and the fuze is set in accordance with the standard firing tables. The M3 bomb works in the same manner as the 81 mm M2A1 bomb described previously.

The M3 120 mm bomb embodies the same two original concepts in design and construction: the parachute assembly is located in the rear section, permitting dedication of the entire interior to the illuminating assembly; and the acceleration-deceleration sequence is designed to permit opening of the main parachute only after the discarded tail section is well clear, making collision impossible.

Data

Length: incl fuze, 580 mm
Weight: incl fuze, 12 kg
Filling: illuminant, 1.2 kg
Propelling charge: increments Charges 0-9
Range: from 1,100 m at Charge 1 to 6,100 m
Illuminating power: min, 1.25 Mcd
Illuminating time: min, 45 s
Rate of descent: 5-6 m/s

Manufacturer

TAAS - Israel Industries Limited, PO Box 1044, Ramat Hasharon 47100.

Status

In production.

VERIFIED

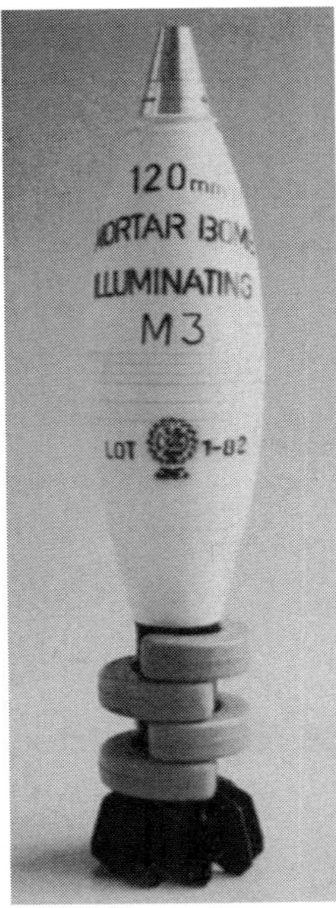

TAAS 120 mm M3 illuminating bomb

Ammunition for the 160 mm Soltam mortar

Description

The HE bomb for the 160 mm Soltam mortar is made from forged steel. The tail unit is made from extruded aluminium. The secondaries, which are flat discs, do not protrude beyond the diameter of the fins and so do not come into contact with the barrel wall. This reduces the chance of a cook-off, even with a barrel temperature as high as 650°C. The high explosive filling is 5 kg of TNT. The propellant system is made up of the primary cartridge and nine secondaries. The Diehl fuze gives a direct action or a slight delay.

There are two kinds of smoke bomb. One has a titanium tetrachloride filling and the other Plasticised White Phosphorus (PWP). The PWP bomb can be used also as a night-ranging bomb.

Data

Weight: 40 kg
Filling: 5 kg TNT
Rate of fire: 5-8 rds/min

Manufacturer

Soltam Limited, PO Box 371, Haifa.

Status

Available.

Service

Israeli Army.

VERIFIED

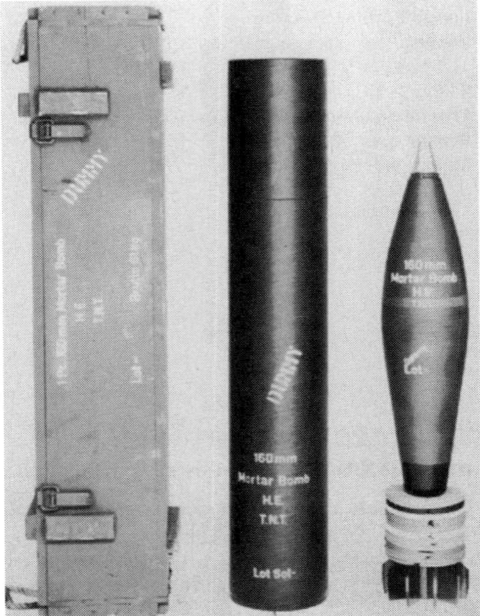

160 mm Soltam HE bomb

Alpha M787 dual-option PPD mortar fuze

Description

The Alpha M787 is a dual-option proximity/point detonating fuze compatible with all 60, 81, 82, 120 and 160 mm mortar bombs, irrespective of their charges, muzzle or terminal velocities.

The M787 fuze is powered by a wind-driven alternator which serves as a second safety in addition to the more usual setback-operated integrator. The safe arming distance is 100 m. The circuitry includes a peak trajectory sensor which activates the fuze only after the vertex has been passed and the bomb is on the downward leg of its trajectory. This feature gives additional own-troops safety and crest clearance safety. A fail-safe feature prevents malfunction: should airspeed precede setback, the fuze will lock in the safe mode.

The fuze was designed and manufactured in accordance with Israeli Defence Forces specifications, as well as all relevant Western safety and performance criteria.

Manufacturer

Reshef Technologies Limited, 7 Haplada Street, PO Box 696, 60256 Or-Yehuda.

Status

In production.

Service

Israeli Defence Force and other countries.

UPDATED

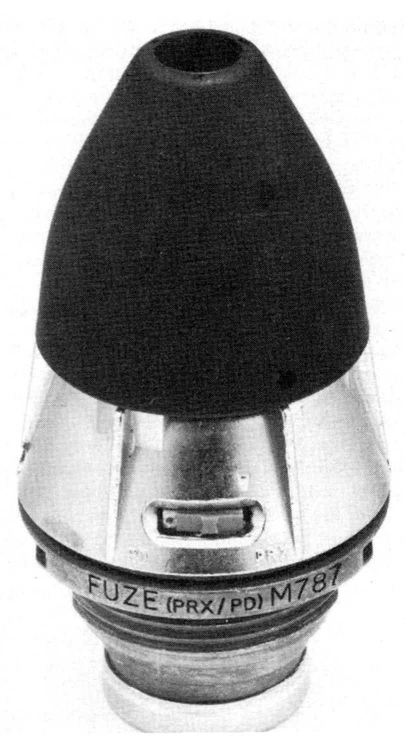

Reshef Alpha M787 dual-option mortar fuze

Lambda M760 electronic time mortar fuze (US-PTF XM 778)

Description

The Lambda M760 is an electronic time fuze compatible with all 60, 81 and 120 mm mortar bombs, irrespective of their charges, muzzle or terminal velocities, for use with smoke, illuminating and carrier bombs (in expelling mode) or for HE and WP bombs (in detonating mode).

The fuze is powered by a wind-driven alternator which serves as a second safety device in addition to the more usual setback-operated integrator. The safe arming distance is 100 m. A fail-safe feature prevents malfunction: should airspeed precede setback, the fuze will lock in the safe condition.

Lambda has an electronic time delay of 3 to 99.8 seconds in 0.1 second increments. The standard deviation of the delay error is less than 0.1 second. Point detonating can be set when 99.9 seconds is selected. The system setting is digital, manually operated (no tools are required) by three setting rings with positive locking.

The fuze was designed in accordance with the relevant Western safety and performance criteria.

Data

Weight: 275 g
Length: 94.8 mm
Cross-section diameter: 49.3 mm
Intrusion: 27.5 mm

Manufacturer

Reshef Technologies Limited, 7 Haplada Street, PO Box 696, 60256 Or-Yehuda.

Status

In production.

UPDATED

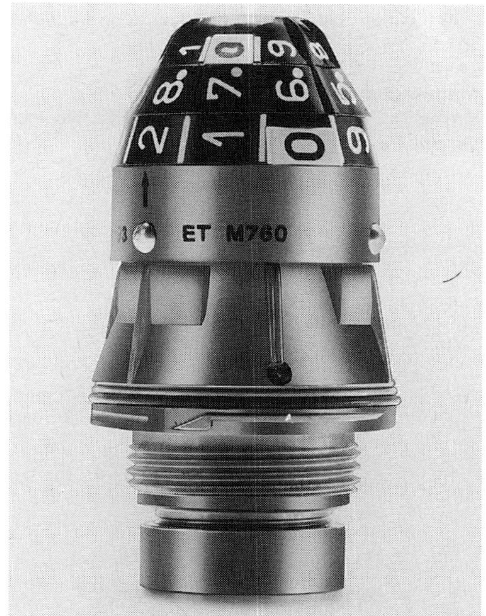

Reshef Lambda M760 electronic time fuze
1996

EPD M797 electronic point detonating mortar fuze

Description

The EPD M797 is an electronic point detonating fuze with a super-quick function and a delay selection mode. It is compatible with 60, 81, 82, 120 and 160 mm mortar bombs, for all charges and terminal velocities and may be used with high explosive and smoke bombs.

The EPD M797 fuze is powered by a wind-driven alternator which serves as a second safety device in addition to the more usual setback-operated integrator. The safe arming distance is 100 m. There is a fail-safe feature that prevents malfunctioning. Should the airspeed mode precede setback the fuze will lock in the safe position. As this fuze has a dual-safety system there is no safety pull wire.

The fuze was designed in accordance with the relevant US and NATO safety and performance criteria.

Data

Weight: 275 g
Length: 95.1 mm
Cross-section diameter: 49 mm
Intrusion: 28.4 mm

Manufacturer

Reshef Technologies Limited, 7 Haplada Street, PO Box 696, 60256 Or-Yehuda.

Status

In production.

NEW ENTRY

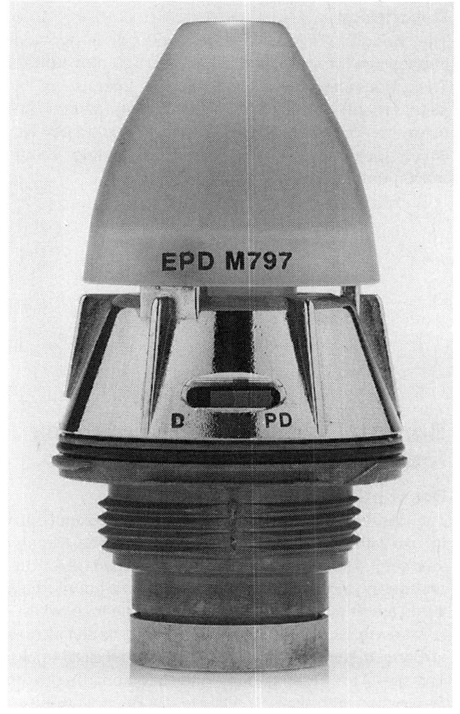

Reshef EPD M797 electronic point detonating mortar fuze
1996

ITALY

Simmel Difesa mortar ammunition

Description

Mortar ammunition in the Simmel Difesa range covers the 81 mm and 120 mm calibres. Each calibre family is identical in ballistic properties and therefore employs a unique firing table for each calibre.

The 81 mm family consists of eight types of bomb as follows:
S1A1 E high-capacity PFF (2,600 splinters); S2A1 WP smoke; RS3A5 Illuminating (1.4 Mcd over 35 seconds); S5A1 HC smoke; S6A2 Cargo SA (9 bomblets, still under development); S7A1 IC; S8A1 Cargo SA TP; and S9A1 FSS (Flash-Sound-Smoke) TP.

The cargo bomb, containing nine bomblets with the ability to penetrate 60 mm of steel armour is still under development.

The RS3A5 has been qualified for British Army service with the L16 mortar family where its designation is L54A1. It is 650 mm long and weighs 4.5 kg.

The Simmel Difesa 120 mm family consists of four types of bomb as follows:
S10B HE (TNT or Composition B); S23B Illuminating (1.2 Mcd over 45 seconds); S23B HC smoke; and S14B FSS TP.

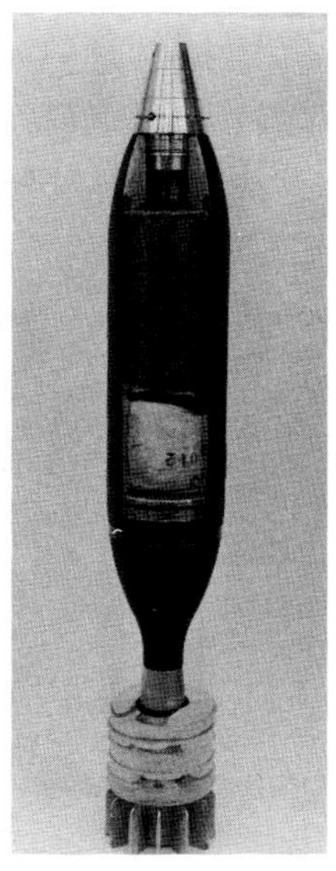

81 mm RS3A5 illuminating bomb

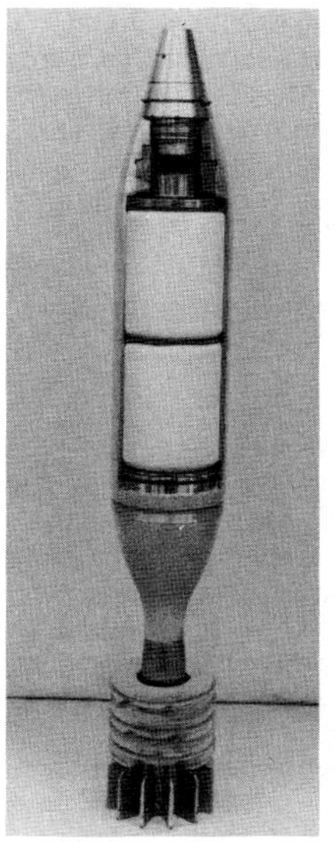

81 mm S5A1 HC smoke bomb

All bombs in this family are 662.8 mm long and weigh 13 kg.

Manufacturers
Simmel Difesa, Borgo Padova 2, I-31033 Castelfranco Veneto, Treviso.

Status
In production.

UPDATED

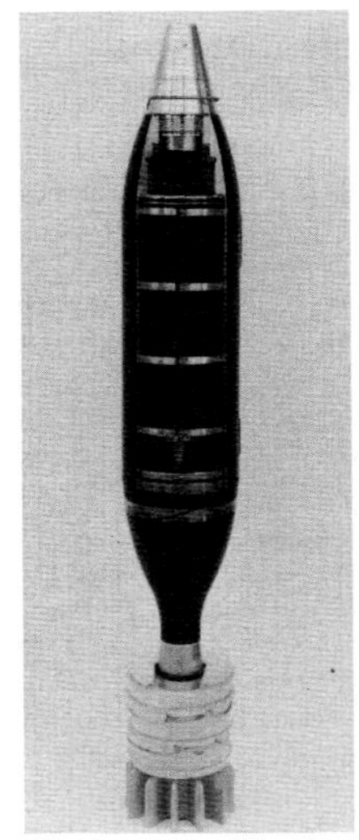

81 mm S6A2 cargo (submunition) bomb

Borletti FB 332 point detonating fuze

Description
The Borletti FB 332 is a conventional super-quick impact fuze for use with all 60, 81 and 120 mm mortars. The mechanism is bore-safe by means of an escapement-controlled delayed arming shutter and safety conforms with MIL-STD-1316. There is a pull-wire which locks the setback detent regulating shutter movement.

Manufacturer
Borletti Division (of Simmel Difesa SpA) Via Verdi 33, I-20010 San Giorgio su Legnano, Milan.

Status
In production.

UPDATED

Borletti FB 332 point detonating fuze

Borletti FB 391 mortar proximity fuze

Description
The Borletti FB 391 is a radio-frequency proximity fuze for use with 60, 81 and 120 mm mortar bombs. Circuitry involving frequency agility and advanced electronic processing techniques ensures electronic countermeasures are relatively ineffective against the fuze, whilst it is immune to mutual interference and battlefield disturbances such as radio emissions, dust and smoke. The signal processing techniques used ensure proximity function with sharper altitude discrimination than is obtainable with an unmodulated RF carrier. The Doppler processor triggers the fuze at a typical height of 4 m, and burst height is virtually independent of soil nature, approach angles and terminal velocities. Proximity action is not affected when the fuze is used with rocket-assisted bombs.

The fuze has a dual-option selector giving the user the choice of either proximity with impact back-up or point detonating effect. Electric power is provided by a wind-driven alternator. The safe and arming system complies with MIL-STD-1316C (the fuze is provided with an inspection window). Before firing, the electric detonator is out of line and short-circuited. A setback integration and several hundred revolutions of the turbine are necessary before arming can take place. An electronic timer, set during manufacture, inhibits fuze action during the initial part of the trajectory, ensuring delayed arming.

Two variant models are made: FB 391A has intrusion and thread to comply with MIL-STD-331A; FB 391B has intrusion and thread interchangeable with the V19P fuze. Other configurations of intrusion and thread are possible on request.

Manufacturer
Borletti Division (of Simmel Difesa SpA) Via Verdi 33, I-20010 San Giorgio su Legnano, Milan.

Status
In production.

Service
Undergoing qualification by Italian Army.

UPDATED

Borletti FB 391 mortar proximity fuze with nose protection

Borletti FB 392 mortar electronic time fuze

Description
The Borletti FB 392 is an Electronic Time (ET) fuze intended for use in smoke, illuminating and cargo rounds for mortar systems of 60, 81 and 120 mm calibre.

Although it is understood that some types of rounds may require some adaptation of the basic configuration (especially at the mechanical interface between fuze and bomb - the fuze hole bush) the basic configuration has been conceived in such a way as to permit the greatest possible compatibility.

The fuze has two operating modes: time, with impact back-up; or PD function only. The fuze is set by hand to whatever mode is desired.

Major subassemblies are: the time setting unit; electronic module; turbine alternator power supply; Safe and Arming; detonating elements; and three plastic rings which allow selection of the required time setting from 3 to 99.9 seconds, in increments of 0.1 second. Setting is performed by aligning the selected number of each ring with a fixed index. When the PD-only mode is required, the fuze is set to zero.

Manufacturer
Borletti Division (of Simmel Difesa SpA) Via Verdi 33, I-20010 San Giorgio su Legnano, Milan.

Status
Advanced development.

UPDATED

Borletti FB 39 electronic time fuze with nose protection

KOREA, SOUTH

60 mm M49A4 HE mortar bomb

Description

The 60 mm M49A4 is a conventional type of HE mortar bomb, based generally on an American original but with an improved tail unit and extension. The primary cartridge is in the tail boom, while the four augmenting charges, sealed in plastic, are clipped between the fins. The bomb body is of pearlitic malleable cast iron, filled with Composition B, and the bomb is fitted with fuze PD M525.

Data

Length: fuzed, 295 mm
Weight: 1.47 kg
Range:

Charge 0	256 m
Charge 1	640 m
Charge 2	1,063 m
Charge 3	1,451 m
Charge 4	1,795 m

Manufacturer

Korea Explosives Company Limited, 34 Seosomoon-Dong, Chung-Ku, Seoul.

Status

In production.

Service

South Korean Army.

VERIFIED

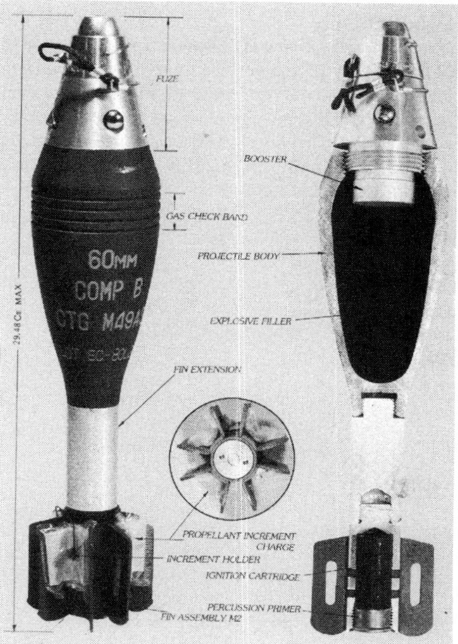

60 mm M49A4 HE bomb

---

60 mm M83A3 illuminating mortar bomb

Description

Designed to be fired from 60 mm M2 or M19 mortars, the 60 mm M83A3 illuminating mortar bomb is based on the American M83 design. The fixed-time fuze M65A1 operates about 15 seconds after firing, igniting the expelling charge which blows off the tail and ignites the illuminant charge. As the contents are ejected from the bomb body, so the parachute deploys and supports the illuminant during its fall.

Data

Length: fuzed, 362.7 mm
Weight: fuzed, 1.88 kg
Burning time: min, 32 s
Intensity: min, 320,000 cd
Max range: 1,006 m

Manufacturers

Korea Explosives Company Limited, 34 Seosomoon-Dong, Chung-Ku, Seoul.
Golden Bell Trading Company Limited, 60-1, 3-Ka, Chungmu-Ro, Chung-Ku, Seoul.

Status

In production.

Service

South Korean Army.

UPDATED

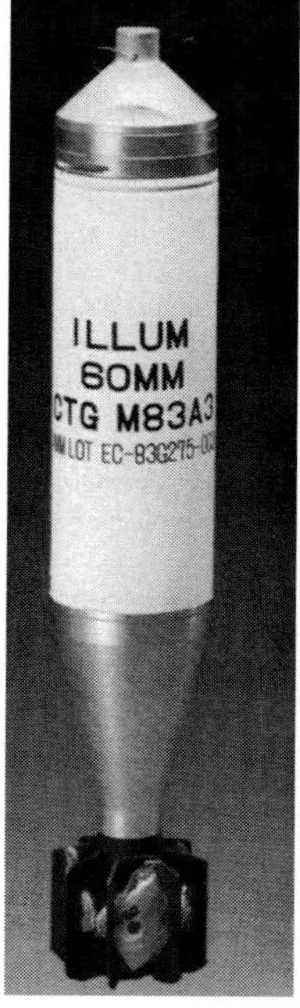

60 mm M83A3 illuminating bomb

---

81 mm M374 HE mortar bomb

Description

The 81 mm M374 HE mortar bomb is based on the US design and may be fired from 81 mm M1, M29 and M29A1 mortars. It consists of a bomb body with obturating ring, a point detonating or proximity fuze, a fin assembly which incorporates the primary cartridge housing and two types of propellant charge enclosed in fabric bags. The body is filled with Composition B, and the fins are canted 5° to produce spin.

Data

Length: 529 mm
Weight: fuzed, 4.35 kg
Max range: Charge 9, 4,500 m

Manufacturer

Korea Explosives Company Limited, 34 Seosomoon-Dong, Chung-Ku, Seoul.

Status

In production.

Service

South Korean Army.

VERIFIED

81 mm M374 HE mortar bomb

81 mm M301A3 illuminating mortar bomb

Description

This illuminating bomb is based on the American M301 pattern and is for use in any 81 mm mortar. The bomb carries an M84 time fuze which is set before loading in accordance with the firing tables. At the set time the tail is blown off and the bomb, the illuminant (weight 620 g) and parachute are expelled. This bomb is not fired at charges less than Charge 3.

Data

Length: 628.3 mm
Weight: 4.58 kg
Burning time: min, 60 s
Intensity: min, 500,000 cd
Max range: 3,150 m

Manufacturers

Korea Explosives Company Limited, 34 Seosomoon-Dong, Chung-Ku, Seoul.
Golden Bell Trading Company Limited, 60-1, 3-Ka, Chungmu-Ro, Chung-Ku, Seoul.

Status

In production.

Service

South Korean Army.

UPDATED

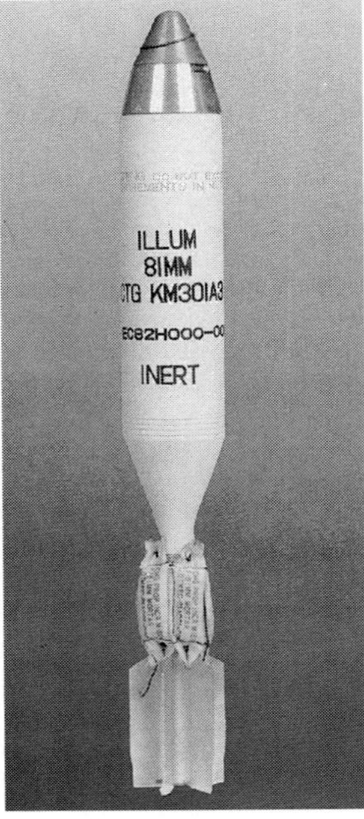

81 mm M301A3 illuminating bomb
1996

NORWAY

81 mm NM 123A1 HE mortar bomb

Description

The 81 mm NM 123A1 HE mortar bomb was developed by Raufoss A/S in co-operation with Norwegian Army Materiel Command specifically for use with the UK L16A1 mortar. The bomb is made from ductile cast iron with an extruded aluminium alloy tail unit. The bomb produces approximately 10,000 effective fragments. The NM 123A1 has a maximum range of 5,863 m.

The bomb can be used with any type of mortar bomb fuze at the option of the user, for example the Norwegian PPD 323 (see following entry). Normally the bomb is fitted with 1.5 in (38 mm) threads, but may, if desired, be produced with 2 in (50 mm) threads, or 2 in threads with an adaptor for 1.5 in threads.

The NM 123A1 has six augmenting charges of equal weight. Identical charges allow simple and safer handling during night operations. The propellant charges are made from ballistite. The bomb has low loading densities and excellent ballistic performances. Probable error at maximum range is 4 m in azimuth and 22 m in range. The bomb is filled with 800 g of Composition B.

The NM 123A1 has a plastic obturating ring which reduces the spread in muzzle velocity. Significantly the ring has been tested at excess pressure of 150 per cent at −40°C.

Data

Length: fuzed, 522 mm
Weight: 4.34 kg; body, 3.04 kg
Filling: 800 g Comp B
Muzzle velocity: max, 297 m/s; min, 74 m/s
Range: max, 5,863 m; min, 183 m

Manufacturer

Raufoss A/S, N-2831 Raufoss.

Status

In production.

Service

Norwegian, Swedish and Canadian armies.

UPDATED

81 mm NM 123A1 HE mortar bomb

PPD-323 and PPD-324 proximity fuzes

Description

The PPD-323 and PPD-324 are multirole fuzes for use in mortar bombs, the PPD-323 for 81 mm bombs and the PPD-324 for 120 mm bombs, providing selectable impact or proximity functions. Power for the proximity function is furnished by a turbo-generator and the action of the turbine also provides mechanical arming which, in connection with an integrated setback detent, ensures full compliance with the safety requirements of STANAG 3525 (MIL-STD-1316B). Mechanical arming does not take place until after 250 m of flight, and electrical emission by the proximity circuits does not take place until the bomb has passed the top of the trajectory.

The electronic circuits are fully protected against countermeasures. This is accomplished by high radiated power, radiation pattern, frequency modulation and protective features in the signal processor. The burst height may be specified by the customer. The PPD-323 and 324 meet all environmental requirements of MIL-STD-331A.

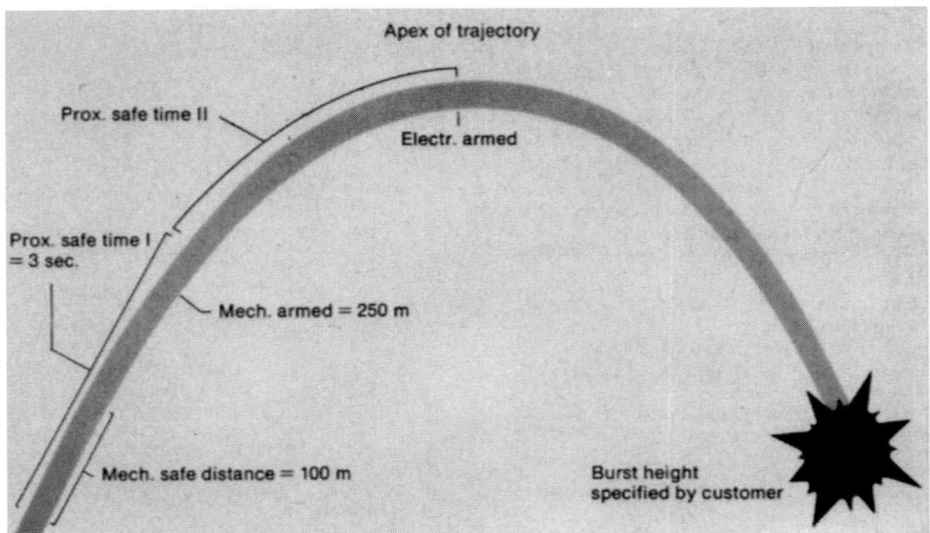

Operating characteristics of PPD-323 and PPD-324 fuzes

Manufacturer
Kongsberg Gruppen AS, PO Box 1003, N-3601 Kongsberg.

Status
In production.

Service
Canadian, German, Norwegian, Swedish, Swiss, and other armies.

UPDATED

PPD-323/324 multirole fuze

PAKISTAN

V19P impact and delay fuze

Description
The V19P is a Thomson-Brandt design manufactured in Pakistan under licence. The fuze is a point detonating impact fuze with selectable delay and has comprehensive safety arrangements which cover parachute-dropping, bore, muzzle and double-loading safeties.

The fuze has a firing pin set in its tip; below this is a pneumatically braked delayed arming system which, until the bomb has travelled some distance from the mortar, does not release the cylindrical bore-safety shutter. Once released, this shutter rotates through 60 or 120° depending upon the setting of the fuze for super-quick action or delay action. Upon impact, the firing pin strikes a detonator and the subsequent flash initiates a further detonator, through a delay if so set, and this in turn initiates the RDX/WX filling of the fuze magazine.

The V19P fuze is for use with 120 mm mortar bombs; a similar fuze, known as V19PA, is for use with 81 mm bombs. The sole difference lies in the size and weight, the V19PA weighing 208 g, the V19P 285 g.

Manufacturer
Alsons Industries (Pvt) Limited, PO Box 4853, Karachi 74000.

Status
In production.

Service
Pakistan Army.

VERIFIED

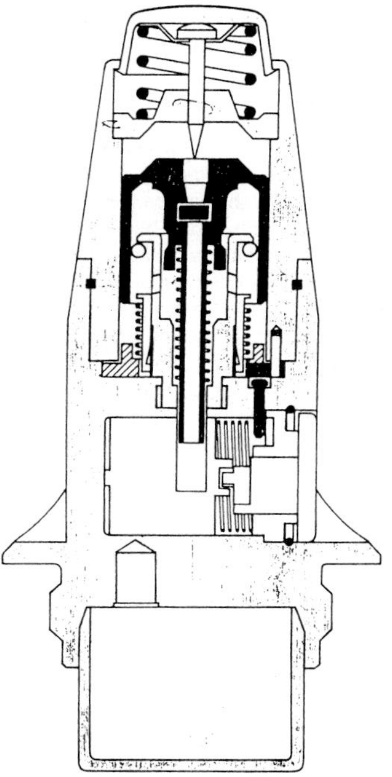

V19P mortar fuze

POF 60 mm HE bomb

Description
The POF 60 mm HE bomb is a standard forged steel or graphite cast iron bomb, apparently based on a Brandt design, filled with 112 g of TNT/DNN 42/58 and fitted with a modified Chinese impact fuze Pai-7, having improved transportation and muzzle safety features. Obturation is performed by four grooves around the waist of the bomb. The tail unit is of steel with welded fins; it carries the primary cartridge and five secondary cartridges in horseshoe form, clipped round the tail boom.

Data
Length: 233-243 mm
Weight: 1.33 kg
Filling: 112 g TNT/DNN 42/58
Muzzle velocity: max, 137 m/s; min, 65 m/s
Max range: 1,410 m

Manufacturer
Pakistan Ordnance Factory, Wah Cantt.

Status
In production.

Service
Pakistan Army and other countries

UPDATED

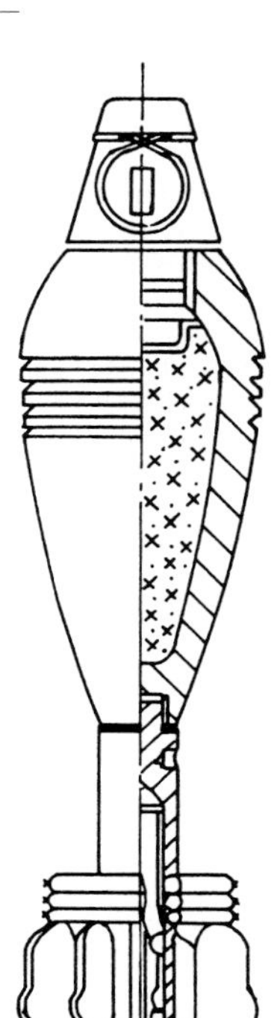

POF 60 mm HE mortar bomb

POF 60 mm smoke bomb

Description

The POF 60 mm smoke bomb matches the HE bomb described previously and is essentially the same bomb but with a filling of white phosphorus and three secondary cartridges in horseshoe form, clipped around the tail boom. It is fuzed with the modified impact fuze Pai-7 Chinese, with improved transporation and muzzle safety features.

Data

Length: ca 280 mm
Weight: 1.5 kg
Muzzle velocity: max, 137 m/s; min, 65 m/s
Max range: 1,410 m

Manufacturer

Pakistan Ordnance Factory, Wah Cantt.

Status

In production.

Service

Pakistan Army and other countries.

UPDATED

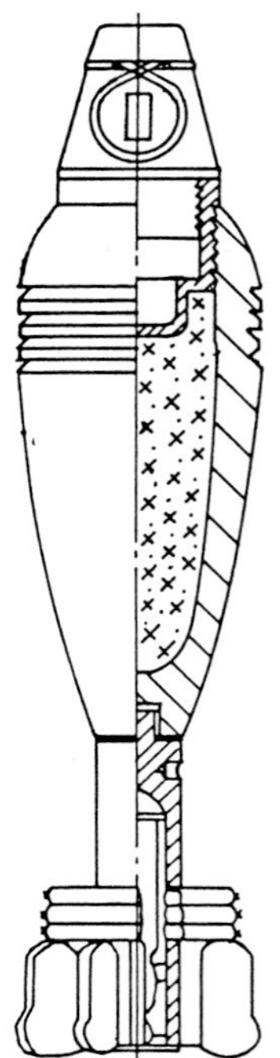

60 mm mortar smoke bomb

60 mm illuminating and signal bombs

Description

This is a cylindrical fuzeless bomb which comes in three variants: a parachute illuminating bomb or a red or green signal bomb. The basic bomb is a canister attached to a tail unit which carries the primary cartridge. At the base of the bomb is a chamber containing a fixed time delay unit which is ignited by the propellant flash. After burning through, the delay ignites an expelling charge in the base of the body; this ignites the pyrotechnic payload and then ejects it, blowing off the front cap of the body to permit exit. In the case of the illuminating bomb, the parachute deploys and lowers the flare unit to the ground. In the case of the signal bombs, the flare unit is thrown clear and falls free, the duration of its fall being sufficient for signalling purposes.

Data

Length: max, 246.9 mm
Weight: 1 kg
Filling: illuminating composition SR 562 and parachute; or green flare composition SR 429AM; or red flare composition SR 406
Delay time: 4 ± 1 s
Duration: illumination 34-45 s; signal 8-12 s

Manufacturer

Pakistan Ordnance Factory, Wah Cantt.

Status

In production.

Service

Pakistan Army and other countries.

UPDATED

POF 60 mm red signal bomb
1996

POF 81 mm HE bomb M-57D Mk 1

Description

This is a copy of the Thomson-Brandt Mle 57D bomb, with a forged steel body and a very short tail boom. The filling is 681 g of Composition B or RDX/TNT 60/40. The fuze is a licence-produced Thomson-Brandt point detonating V19PA.

The Pakistan Ordnance Factories also produce an 82 mm version of this bomb to render it suitable for Eastern Bloc 82 mm mortars. Various Eastern Bloc fuzes can be fitted.

Data

Length: ca 382 mm
Weight: 3.2 kg
Filling: 681 g Comp B or RDX/TNT 60/40
Muzzle velocity: Charge 7, 291 m/s
Max range: 4,250 m
Min range: 100 m

Manufacturer

Pakistan Ordnance Factory, Wah Cantt.

Status

In production.

Service

Pakistan Army and other countries.

UPDATED

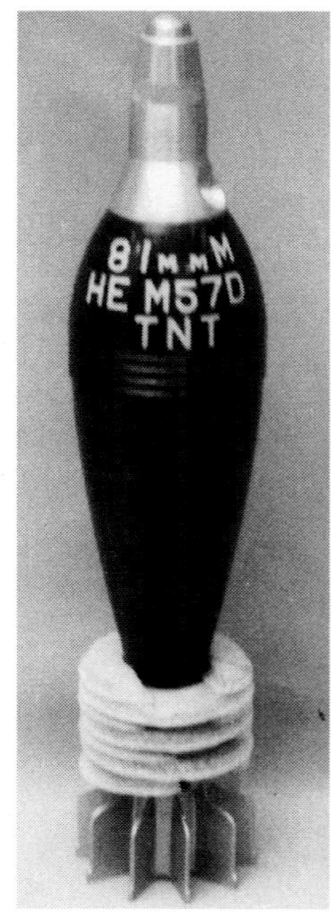

81 mm HE bomb M-57D Mk 1
1996

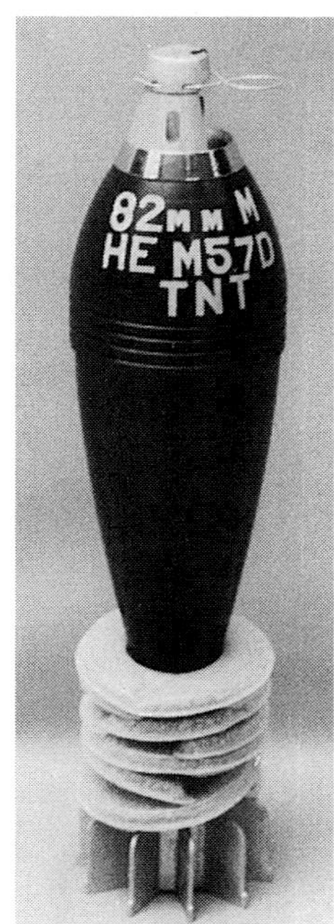

82 mm version of POF HE bomb M-57D Mk 1
1996

POF 81 mm smoke bomb

Description

The POF 81 mm smoke bomb is the partner to the HE bomb described in the previous entry and is of the same general appearance, differing in being somewhat longer and more slender and in the use of white phosphorus as the filling. It uses the same V19PA impact fuze and is ballistically matched to the HE bomb.

The Pakistan Ordnance Factories also produce an 82 mm version of this bomb to render it suitable for Eastern Bloc 82 mm mortars. Various Eastern Bloc fuzes can be fitted.

Data

Length: 382 mm
Weight: 3.2 kg
Muzzle velocity: Charge 7, 291 m/s
Max range: 4,250 m
Min range: 100 m

Manufacturer

Pakistan Ordnance Factory, Wah Cantt.

Status

In production.

Service

Pakistan Army and other countries.

UPDATED

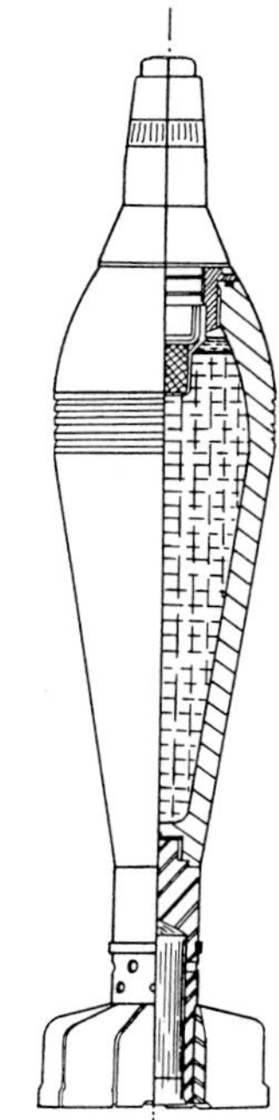

81 mm WP smoke bomb

POF 120 mm HE bomb

Description

The POF 120 mm HE bomb appears to be a locally manufactured copy of a Thomson-Brandt 120 mm bomb. It has a forged steel body containing 2.6 kg of TNT. The bomb is fuzed with a V19P impact fuze and has a rather large exploder system which probably gives the bomb extremely efficient detonation and fragmentation characteristics. The primary cartridge is held inside the tail boom and the seven secondary charges are clipped around the boom in horseshoe units.

Data

Length: fuzed, 676 mm
Weight: 13 kg
Filling: 2.6 kg TNT
Muzzle velocity: Charge 8, 331 m/s
Max range: Charge 8, 6,745 m

Manufacturer

Pakistan Ordnance Factory, Wah Cantt.

Status

In production.

Service

Pakistan Army and other countries.

UPDATED

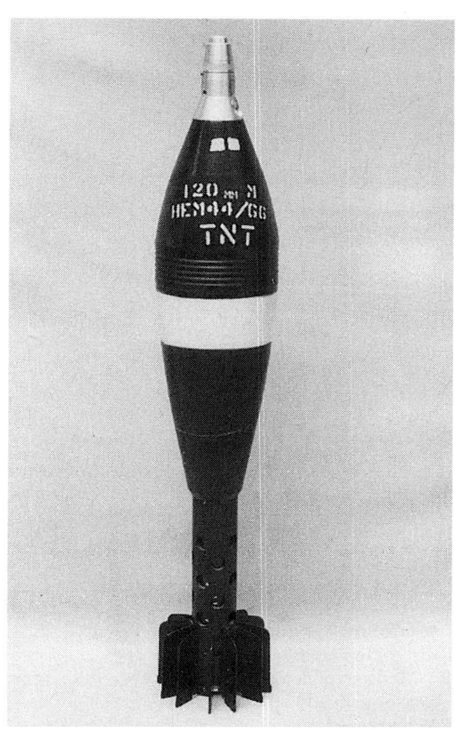

POF 120 mm HE bomb
1996

PORTUGAL

INDEP mortar ammunition

Description

INDEP manufactures three types of mortar ammunition, two in 60 mm and one in 81 mm, all of them high explosive. These are usually supplied with the Fuze PD M525 or the Diehl AZ DM 111A2 fuze manufactured in Portugal under licence. They may, however, be equipped with other fuzes. Both 60 mm rounds may be fired from Brandt breech loading mortars of the type used in armoured fighting vehicles.

Data

81 MM HE
THIS IS SIMILAR TO THE US M43A1B1
Weight: 3.25 kg
Filling: 560 g TNT or Comp B
Min range: 75 m
Max range: with 6 increments 3,500 m; with 8 increments 4,200 m

60 MM HE
THIS IS SIMILAR TO THE US M49A2
Weight: 1.34 kg
Filling: 155 g TNT
Max range: 1,820 m
Min range: 46 m

60 MM HE, APERS, TYPE NR431A1 (UNDER PRB LICENCE)
Weight: 1.36 kg
Filling: 156 g TNT
Max range: 2,100 m
Muzzle velocity: Charge 4, 177 m/s
Min range: 50 m

Manufacturer

Indústrias Nacionais de Defesa EP (INDEP), Rua Fernando Palha, P-1899 Lisbon-Codex.

VERIFIED

SINGAPORE

CIS 60 mm mortar bombs

Description

There are three main types of 60 mm mortar bomb produced by Chartered Industries of Singapore (CIS), two High Explosive (HE) and one smoke.

The two HE bombs are basically similar and differ only in the type of fuze involved. Both have forged steel streamlined bodies of orthodox design filled with 250 g of Grade 1 TNT. One type has an A2 super-quick/delay fuze screwed directly into the nose fuze well; the delay, when selected, is 0.05 second. The other, having an SF1 point detonating fuze, uses an adaptor to allow the smaller diameter and longer fuze body to be accommodated. On both types the tail unit is constructed using extruded aluminium alloy and both have a primary ignition cartridge with up to four incremental charges located around the tail unit. Both bombs are ballistically matched having muzzle velocities from 66 to 199 m/s, the latter providing a maximum range of 2,555 m.

The smoke bomb is filled with 271 g of titanium tetrachloride and utilises the SF1 point detonating fuze. This bomb is otherwise similar in construction and other details to the HE bombs and is ballistically matched to them.

Training and drill rounds are available.

Data

Type	**HE**	**HE**	**Smoke**
Fuze	A2 SQ/D	SF1 PD	SF1 PD
Length	287 mm	318 mm	318 mm
Weight	1.55 kg	1.684 kg	1.684 kg
Filling weight	250 g	250 g	271 g
Muzzle velocity, max	199 m/s	199 m/s	199 m/s
Muzzle velocity, min	66 m/s	66 m/s	66 m/s
Max range	2,555 m	2,555 m	2,555 m
Min range	150 m	150 m	150 m

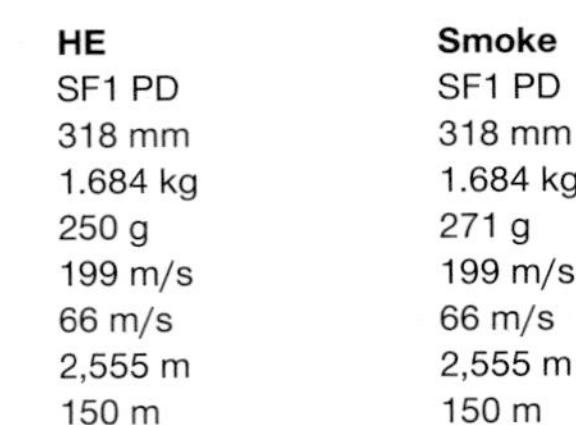

Manufacturer

Chartered Industries of Singapore, 249 Jalan Boon Lay, Singapore 2261.

Status

In production.

NEW ENTRY

CIS 60 mm HE mortar bomb with A2 SQ/D fuze
1996

CIS 60 mm HE mortar bomb with SF1 PD fuze
1996

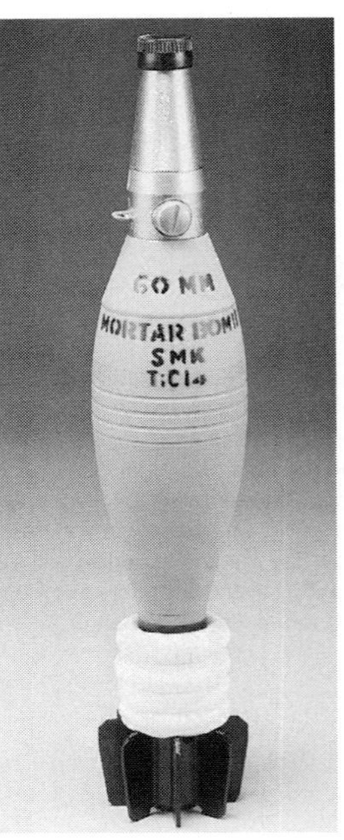

CIS 60 mm smoke bomb with SF1 PD fuze
1996

CIS 81 mm mortar bombs

Description

There are two types of 81 mm mortar bomb produced by Chartered Industries of Singapore (CIS), High Explosive (HE) and smoke. The two bombs are basically similar and differ only in the type of filling involved. Both have forged steel streamlined bodies with plastic obturating rings and an A2 super-quick/delay fuze screwed directly into the nose fuze well. The delay, when selected, is 0.05 second.

Both bombs are ballistically matched having muzzle velocities from 66 to 271 m/s, the latter providing a maximum range of 4,660 m. The propellant system utilises a single primary cartridge and up to seven incremental horseshoe charges placed around the extruded aluminium tail unit.

The HE bomb is filled with 500 g of Grade 1 TNT. The smoke bomb is filled with 396 g of titanium tetrachloride although it is possible to utilise White Phosphorus (WP).

Training and drill rounds are available.

Data

Type	**HE**	**Smoke**
Length	369 mm	369 mm
Weight	3.86 kg	3.86 kg
Filling weight	500 g	396 g
Muzzle velocity, max	271 m/s	271 m/s
Muzzle velocity, min	66 m/s	66 m/s
Max range	4,660 m	4,660 m
Min range	150 m	150 m

Manufacturer

Chartered Industries of Singapore, 249 Jalan Boon Lay, Singapore 2261.

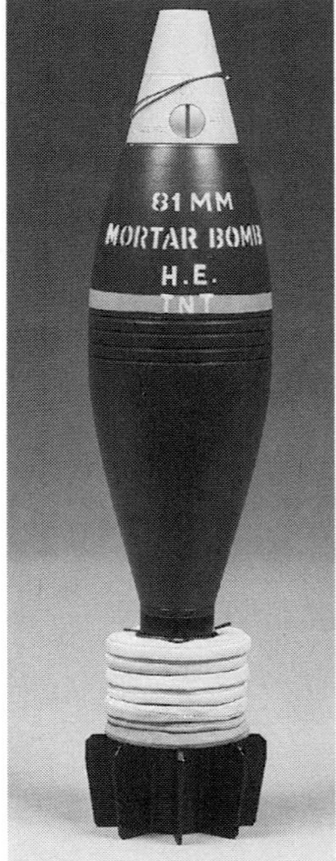

CIS 81 mm HE mortar bomb with A2 SQ/D fuze
1996

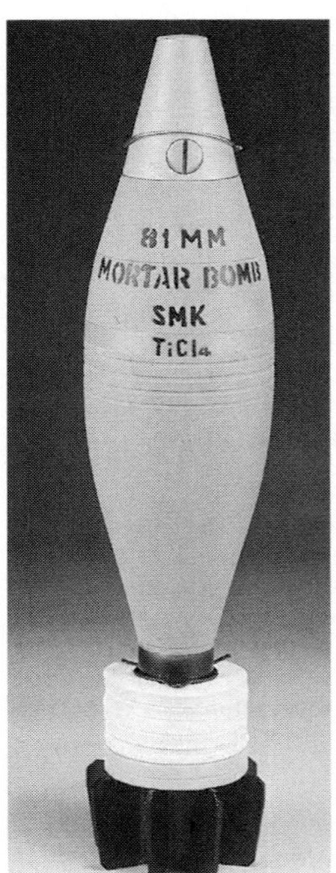

CIS 81 mm smoke bomb
1996

Status

In production.

NEW ENTRY

CIS 81 mm Extended Range mortar bombs

Description

The CIS 81 mm Extended Range (ER) mortar bombs are of advanced aerodynamic design and were produced to increase the maximum range possible from conventional 81 mm mortars. Maximum range using the usual propellant charge system of a primary cartridge and seven incremental charges is 5,850 m.

There are three types of 81 mm extended range mortar bomb, one High Explosive (HE) and two smoke. All three bombs are basically similar and differ only in the type of filling involved. All have elongated forged steel streamlined bodies with plastic obturating rings and an A2 super-quick/delay fuze screwed directly into the nose fuze well. The delay, when selected, is 0.05 second.

All bombs are ballistically matched, having muzzle velocities from 69 to 304 m/s.

The HE bomb is filled with 750 g of Grade 1 TNT. The smoke bombs are filled with 700 g of titanium tetrachloride (FM) or White Phosphorus (WP).

Training and drill rounds are available.

Data

Length: 483 mm
Weight: 4.25 kg
Filling: HE, 750 g TNT; smoke, 700 g WP or FM
Muzzle velocity: max, 304 m/s; min, 69 m/s
Max range: 5,850 m
Min range: 150 m
Max chamber pressure: 670 bar

Manufacturer

Chartered Industries of Singapore, 249 Jalan Boon Lay, Singapore 2261.

Status

In production.

NEW ENTRY

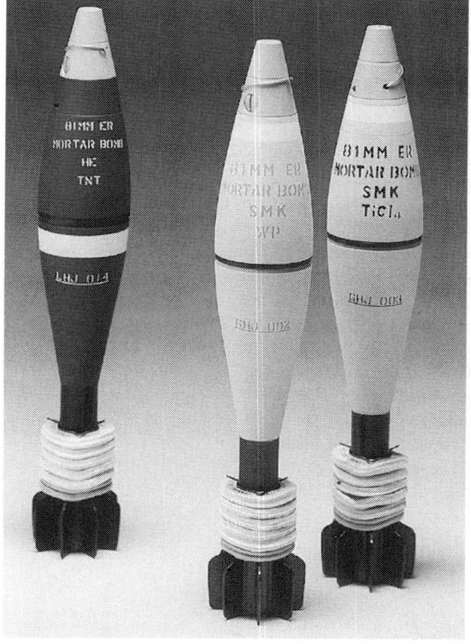

CIS 81 mm Extended Range (ER) mortar bombs, from left, HE, smoke WP, smoke FM
1996

CIS 120 mm mortar bombs

Description

There are two types of 120 mm mortar bomb produced by Chartered Industries of Singapore (CIS), High Explosive (HE) and smoke. The two bombs are basically similar and differ only in the type of filling involved. Both have forged steel streamlined bodies of the classic Thomson-Brandt type with plastic obturating rings and an A2 super-quick/delay fuze screwed directly into the nose fuze well. The delay, when selected, is 0.05 second.

Both bombs are ballistically matched having muzzle velocities from 112 to 307 m/s, the latter providing a maximum range of 6,500 m. The propellant system utilises a single primary cartridge and up to five incremental horseshoe charges placed around the extruded aluminium tail unit.

The HE bomb is filled with 2.2 kg of Grade 1 TNT. The smoke bomb is filled with 2.2 kg of titanium tetrachloride although it is possible to utilise White Phosphorus (WP).

Training and drill rounds are available.

Data

HE AND SMOKE
Length: 581 mm
Weight: 12.6 kg
Filling weight: HE, 2.2 kg TNT; smoke, 2.2 kg FM
Muzzle velocity: max, 307 m/s; min, 112 m/s
Max range: 6,500 m
Min range: 400 m

Manufacturer

Chartered Industries of Singapore, 249 Jalan Boon Lay, Singapore 2261.

Status

In production.

NEW ENTRY

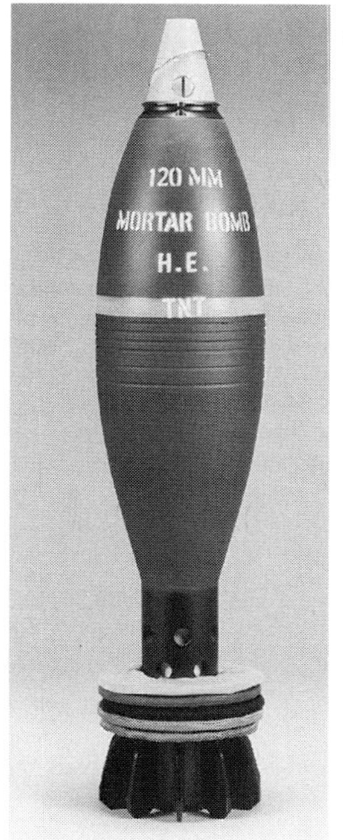

CIS 120 mm HE mortar bomb with A2 SQ/D fuze; the 120 mm smoke bomb is visually identical
1996

CIS 120 mm Extended Range mortar bombs

Description

The CIS 120 mm Extended Range (ER) mortar bombs are of advanced aerodynamic design and were produced to increase the maximum range possible from conventional 120 mm mortars. Maximum range using the usual propellant charge system of a primary cartridge and eight incremental charges is 8,250 m.

There are three types of 120 mm extended range mortar bomb, one High Explosive (HE) and two smoke. All three bombs are basically similar and differ only in the type of filling involved. All have elongated forged steel streamlined bodies with plastic obturating rings and an A2 super-quick/delay fuze screwed directly into the nose fuze well. The delay, when selected, is 0.05 second.

All bombs are ballistically matched, having muzzle velocities from 134 to 369 m/s.

The HE bomb is filled with 2.8 kg of Grade 1 TNT. The smoke bombs are filled with 1.8 kg of titanium tetrachloride (FM) or White Phosphorus (WP).

Training and drill rounds are available.

Data

HE AND SMOKE
Length: 699 mm
Weight: 14.6 kg
Filling: HE, 2.8 kg TNT; smoke, 1.8 kg WP or FM
Muzzle velocity: max, 369 m/s; min, 134 m/s
Max range: 8,250 m
Min range: 400 m
Max chamber pressure: 1,345 bar

Manufacturer

Chartered Industries of Singapore, 249 Jalan Boon Lay, Singapore 2261.

Status

In production.

NEW ENTRY

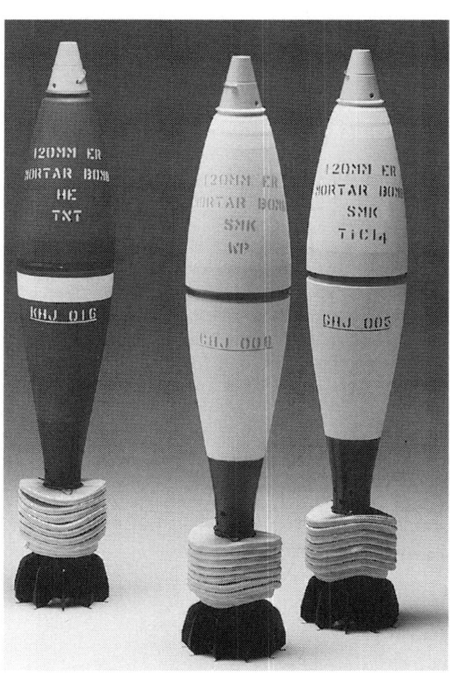

CIS 120 mm Extended Range (ER) mortar bombs, from left, HE, smoke WP, smoke FM
1996

SOUTH AFRICA

60 mm M61 HE bomb

Description
The 60 mm M61 HE bomb can be used in either the standard infantry drop-fired mortars or in the 60 mm gun mortar used with armoured vehicles. It uses a forged steel body and is filled with TNT or RDX/TNT 40/60. A primary cartridge fits into the tail tube and four secondaries fit in between the fins, while a fifth can be clipped around the tail tube ahead of the fins. It can be supplied fuzed with the V9 direct action impact fuze.

Data
Weight: 1.8 kg
Muzzle velocity:
Charge 0 62 m/s
Charge 1 88 m/s
Charge 2 111 m/s
Charge 3 132 m/s
Charge 4 149 m/s
Charge super 171 m/s
Range: Charge super, 2,100 m; Charge primary only, 358 m

Manufacturer
NASCHEM, Private Bag X 1254, Potchefstroom, 2520.

Status
Available.

Service
South African National Defence Force.

UPDATED

60 mm M61 HE bomb

60 mm M61 smoke bomb

Description
The 60 mm M61 smoke bomb is ballistically matched to the M61 HE bomb (see previous entry). It is filled with titanium tetrachloride (FM) which is preferred to White Phosphorus (WP) because of its lower fire risk in bush country. The FM is distributed by a small explosive burster and generates dense white smoke by reaction with water vapour in the air. The direct action impact fuze V9 is fitted as standard.

Data
Weight: 1.8 kg
Muzzle velocity:
Charge 0 62 m/s
Charge 1 88 m/s
Charge 2 111 m/s
Charge 3 132 m/s
Charge 4 149 m/s
Charge super 171 m/s
Range: Charge super, 2,100 m; Charge primary only, 358 m

Manufacturer
NASCHEM, Private Bag X 1254, Potchefstroom, 2520.

Status
Available.

Service
South African National Defence Force.

UPDATED

60 mm M61 smoke bomb

60 mm M61 practice bomb

Description
This is ballistically matched to the 60 mm M61 HE bomb so it may be used for practice when non-explosive bombs are necessary. The bomb is filled with an inert substance but carries a small marker charge of TNT/aluminium which detonates on impact to give an indication of the fall of shot. The direct action impact fuze V9 is fitted as standard.

Data
Weight: 1.8 kg
Muzzle velocity:
Charge 0 62 m/s
Charge 1 88 m/s
Charge 2 111 m/s
Charge 3 132 m/s
Charge 4 149 m/s
Charge super 171 m/s
Range: Charge super, 2,100 m; Charge primary only, 358 m

Manufacturer
NASCHEM, Private Bag X 1254, Potchefstroom, 2520.

Status
Available.

Service
South African National Defence Force.

UPDATED

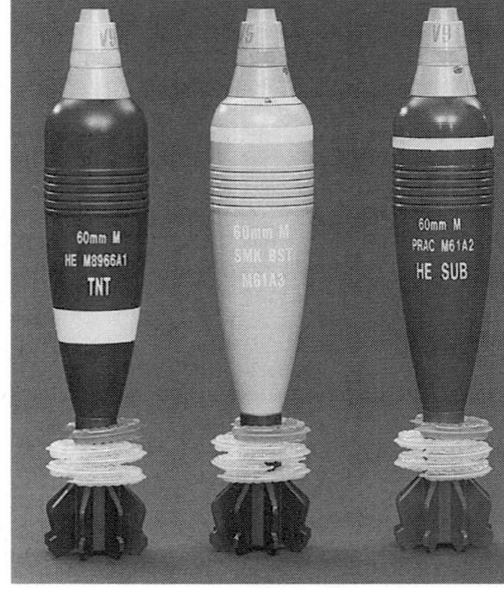

60 mm M61 practice bomb (right) *compared with M61 HE* (left) *and M61 smoke* (centre)
1996

60 mm illuminating bomb M802A2

Description

The 60 mm illuminating bomb M802A2 is of South African design and is of the conventional pattern using a parachute-suspended flare unit to provide illumination. A time fuze is fitted and set before firing; at the end of the set time the fuze ignites an expelling charge which blows off the tail unit and ejects the flare and parachute.

Data

Length: 377 mm
Weight: 1.6 kg
Illuminating power: min, 180,000 cd
Burning time: min, 30 s
Rate of descent: ca 4.5 m/s
Max range: effective, 2,000 m

Manufacturer

NASCHEM, Private Bag X 1254, Potchefstroom, 2520.

Status

Available.

Service

South African National Defence Force.

UPDATED

60 mm illuminating bomb M802A2

60 mm red phosphorus bomb M1A1

Description

The 60 mm red phosphorus bomb M1A1 is fitted with a point detonating fuze which causes the bomb, on impact, to eject a canister containing red phosphorus composition 2 to 6 m into the air. At that height a burster charge ignites the prepelletted red phosphorus granules and distributes them over the immediate area up to 20 m from the point of burst. The red phosphorus composition is formulated to cause a high temperature conflagration of the granules for about 30 seconds.

The bomb is used to provide screening smoke as well as serve as a powerful incendiary device. Deployed in advance of forward troops it can be used as a defensive or offensive munition, or as a means of indicating targets to forward air support.

Data

Length: 377 mm
Weight: total, 1.79 kg
Filling: 320 g red phosphorus in 2,800 granules
Burning time: min, 5 s
Burning temperature: 800°C
Range: 2,100 m
Distribution: min, 700 m^2

Manufacturer

NASCHEM, Private Bag X 1254, Potchefstroom, 2520.

Status

Available.

Service

South African National Defence Force.

UPDATED

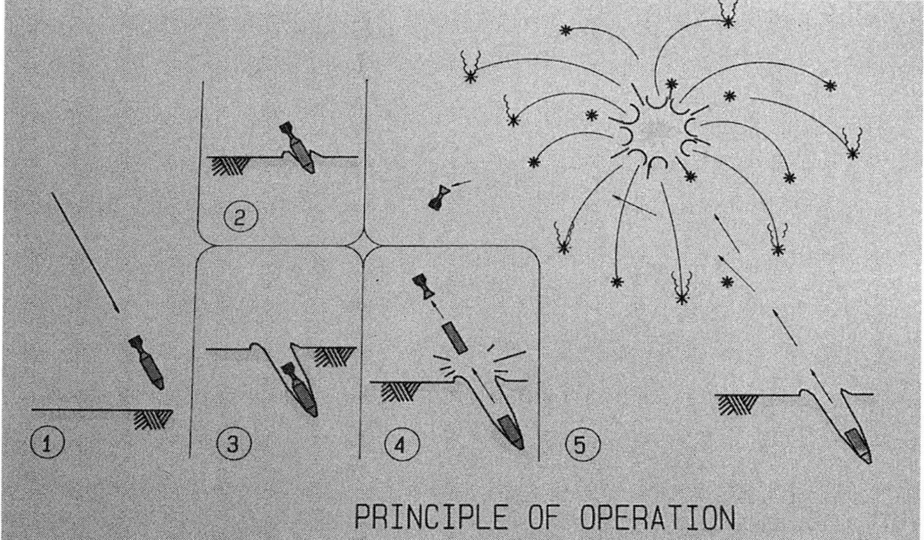

Principle of operation of the 60 mm red phosphorus bomb M1A1, from approach (1) to scattering of the red phosphorus pellets (5)

1996

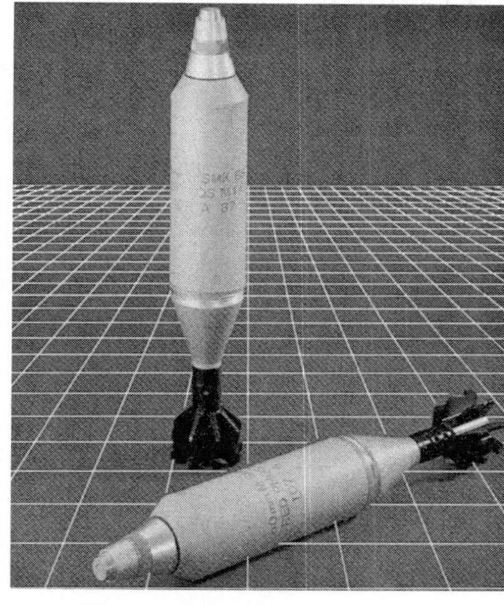

60 mm red phosphorus bomb M1A1

60 mm Long Range mortar bombs

Description

The NASCHEM 60 mm Long Range mortar bomb family is claimed to be the only one of its type in the 60 mm mortar calibre and was developed specifically for the Vektor 60 mm M6 long-range mortar system (qv). There are three bombs in the family, the high explosive HE LR M8917A1, the smoke WP LR M9514, and the practice PRAC LR M8918A2. All three are ballistically matched and can be fired to a maximum range of 6,180 m.

The 60 mm HE LR bomb is claimed to have the same on-target capability as the 81 mm M61 HE bomb (see following entry). It has a streamlined cast iron body filled with TNT and may be provided with a variety of fuzes, including the V9 point detonating as well as the programmable M9148 and various proximity fuzes.

The smoke WP LR is filled with white phosphorus although later bombs may use titanium tetrachloride (FM). This bomb is ballistically matched to the 60 mm HE LR.

The practice bomb has its body filled with a high explosive substitute (HE SUB) but contains a TNT/Aluminium flash pellet for impact marking purposes.

Data

Weight: 2.4 kg
Velocity: 67.5-330 m/s
Max range: 6,180 m
Min range: 170 m

Manufacturer

NASCHEM, Private Bag X 1254, Potchefstroom, 2520.

Status

Available.

Service

South African National Defence Force.

NEW ENTRY

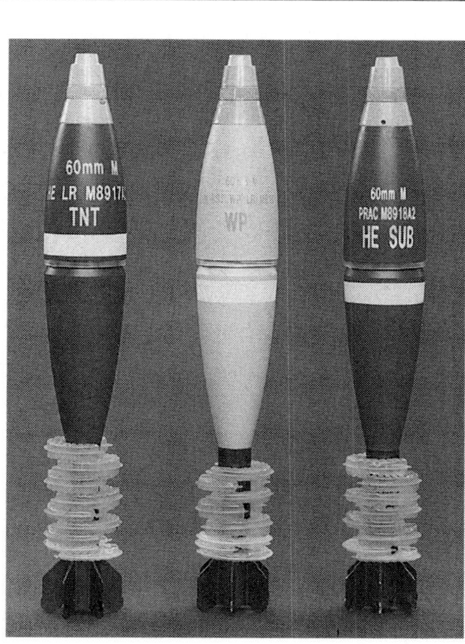

60 mm Long Range mortar bombs, from left, HE, smoke and practice

1996

81 mm M61 HE bomb

Description

The 81 mm M61 HE bomb uses a forged steel body with a Rilsan obturating ring and a light alloy tail tube and fin assembly. It is filled with RDX/TNT 40/60 and fuzed with the direct action V19P fuze which has provision for instantaneous or delay action, selectable before firing. A primary cartridge fits into the tail tube, and secondaries can be clipped around the tube; there are eight charges.

Also available is a group of HE bombs similar to the M61 HE which can produce impact bursts in various colours. Colours available include violet, orange, yellow, green and red.

Data

Weight: 4.43 kg
Filling: RDX/TNT 40/60
Max range: 4,856 m
Min range: 75 m
Velocity: Charge primary, 75 m/s; Charge 8, 279 m/s

Manufacturer

NASCHEM, Private Bag X 1254, Potchefstroom, 2520.

Status

Available.

Service

South African National Defence Force.

UPDATED

81 mm M61 HE bomb

81 mm M61 smoke bomb

Description

The 81 mm M61 smoke bomb is ballistically matched to the M61 HE bomb and is filled with titanium tetrachloride (FM). On impact a small burster charge of explosive ruptures the steel body and distributes the FM, which generates smoke by reaction with atmospheric water vapour. It produces a dense smoke for screening or signalling purposes.

Data

Weight: 4.43 kg
Filling: RDX/TNT 40/60
Max range: 4,856 m
Min range: 75 m
Velocity: Charge primary, 75 m/s; Charge 8, 279 m/s

Manufacturer

NASCHEM, Private Bag X 1254, Potchefstroom, 2520.

Status

Available.

Service

South African National Defence Force.

UPDATED

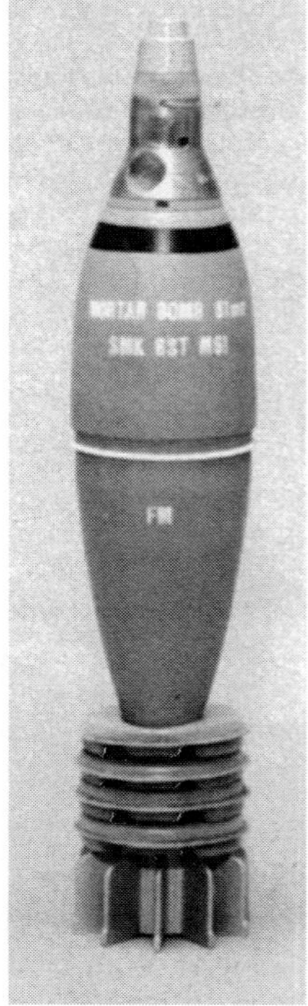

81 mm M61 smoke bomb

81 mm M61 practice bomb

Description

The 81 mm M61 practice bomb is ballistically matched to the 81 mm M61 HE bomb and is filled with an inert substance. There is a small indicating charge of TNT/aluminium which is detonated on impact by the V19P fuze to give indication of fall of shot. It can be used at any charge up to Charge 8.

Data

Weight: 4.43 kg
Max range: Charge 8, 4,856 m
Velocity: Charge primary, 75 m/s; Charge 4, 196 m/s; Charge 8, 279 m/s

Manufacturer

NASCHEM, Private Bag X 1254, Potchefstroom, 2520.

Status

Available.

Service

South African National Defence Force.

UPDATED

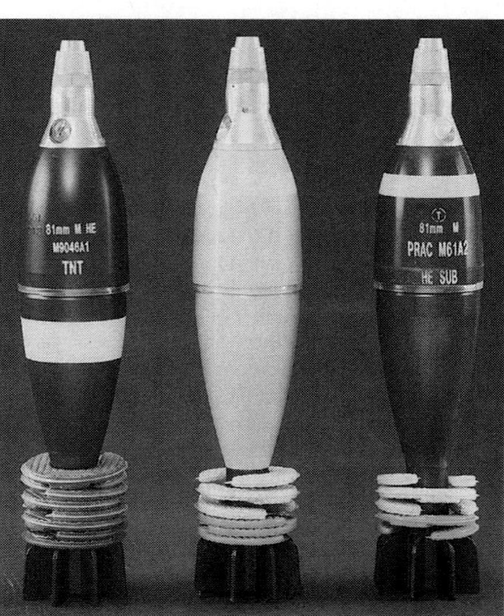

81 mm M61 practice bomb (right) *compared with M61 HE* (left) *and M61 smoke* (centre)
1996

81 mm Long Range mortar bombs

Description

The NASCHEM 81 mm Long Range mortar bomb family was developed specifically for the Vektor 81 mm M3 mortar system (qv) but can also be used with any other 81 mm mortar system. There are three bombs in the family, the high explosive HE LR M9135A2, the smoke SMK LR M9135A1, and the practice PRAC LR M9135A1. All three are ballistically matched and can be fired to a maximum range of 7,263 m.

The 81 mm HE LR bomb has a streamlined cast iron body filled with TNT and may be provided with a variety of fuzes.

The smoke SMK LR is filled with titanium tetrachloride (FM). This bomb is ballistically matched to the 81 mm HE LR.

The 81 mm PRAC LR practice bomb has its body filled with a high explosive substitute (HE SUB) but contains a TNT/aluminium flash pellet for impact marking purposes.

Data

Weight: 4.9 kg
Velocity: 63-363 m/s
Max range: 7,263 m
Min range: 100 m

Manufacturer

NASCHEM, Private Bag X 1254, Potchefstroom, 2520.

Status

Available.

Service

South African National Defence Force.

NEW ENTRY

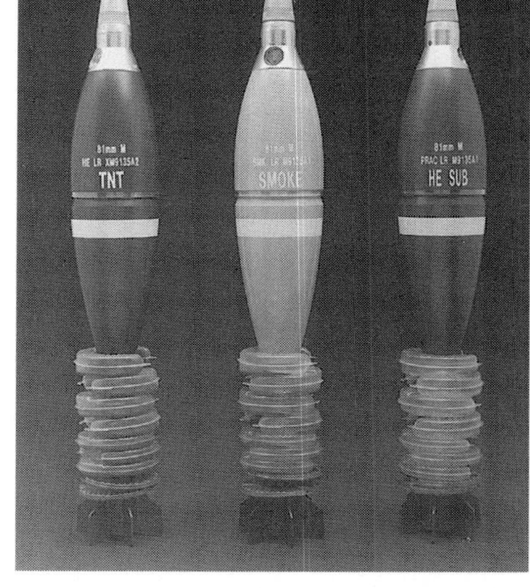

81 mm Long Range mortar bombs, from left, HE, smoke and practice
1996

SPAIN

ECIA ammunition for 60 mm mortars

Description

ECIA manufactures two families of mortar ammunition, a conventional one and the aerodynamic (AE-84) type. Both types include the whole range (HE, smoke HC or WP, illuminating and practice rounds).

The ammunition manufactured by ECIA is obtained from forging, thus giving a knife-like fragment with higher velocity and therefore higher effectiveness. Upon request, ECIA can provide 60 mm bombs made from cast iron.

The Model AE-84 was introduced in 1984 and uses a highly streamlined body to obtain a good weight to calibre ratio and a plastic obturating ring to achieve good gas sealing and ballistic regularity. The tail unit, consisting of stabiliser tube and fins, is of forged aluminium and is attached to the body by a screwed section. The primary cartridge is inserted into the tail tube and the incremental charges are in horse-shoe-shaped containers which clip round the stabiliser tube.

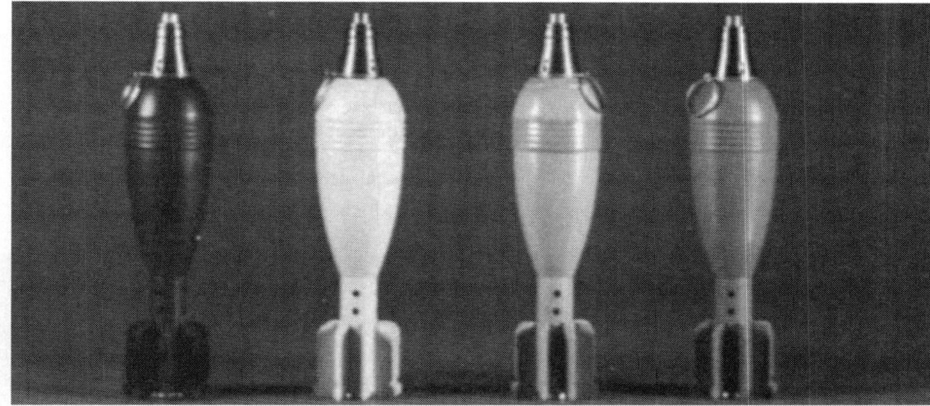

Model N series bombs for 60 mm mortars

Data

HE BOMB MODEL N
Length: 263 mm
Weight: 1.43 kg
Filling: 232 g TNT
Max range: 1,975 m

HE BOMB MODEL AE
Length: 396 mm
Weight: 2.05 kg
Filling: 355 g TNT
Max range: 4,600 m

ILLUMINATING BOMB MODEL N
Length: with fuze, 368 mm
Weight: 1.97 kg
Max range: 1,575 m
Intensity: 250,000 cd
Burning time: 23 s
Rate of descent: 4 m/s

ILLUMINATING BOMB MODEL AE
Weight: without fuze, 1.55 kg
Max range: 4,200 m
Intensity: 250,000 cd
Burning time: 25 s
Rate of descent: 4 m/s

Manufacturer

Esperanza y Cia SA, Marquina, Vizcaya.

Status

Available.

Service

Spanish Army and several others.

VERIFIED

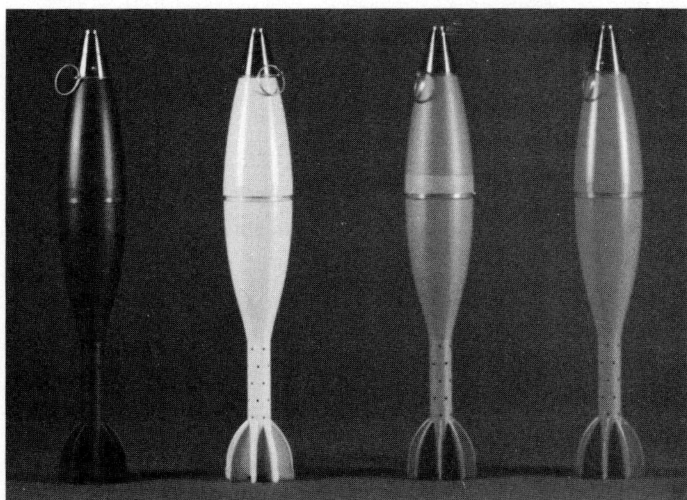

Model AE series bombs for 60 mm mortars

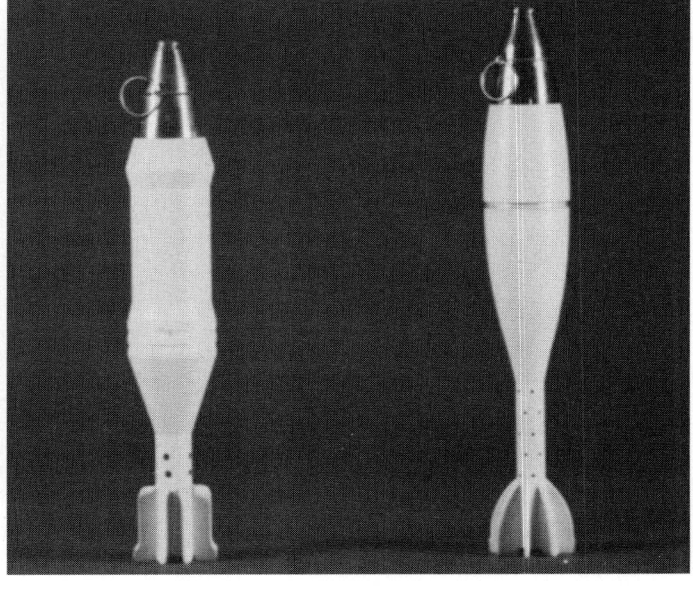

60 mm illuminating bombs Models N and AE

ECIA ammunition for 81 mm mortars

Description

ECIA manufactures two families of mortar ammunition, a conventional one and the aerodynamic (AE-84) type. Both types include the whole range (HE, smoke HC or WP, illuminating and practice rounds).

The ammunition manufactured by ECIA is obtained from forging, thus giving a knife-like fragment with higher velocity and higher effectiveness. Upon request, ECIA can provide 81 mm bombs made from cast iron.

The Model AE-84 was introduced in 1984 and uses a highly streamlined body to obtain a good weight to calibre ratio and a plastic obturating ring to achieve good gas sealing and ballistic regularity. The tail unit, consisting of stabiliser tube and fins, is of forged aluminium and is attached to the body by a screwed section. The primary cartridge is inserted into the tail tube, and the incremental charges are in horseshoe-shaped containers which clip round the stabiliser tube.

82 mm mortar rounds are also available.

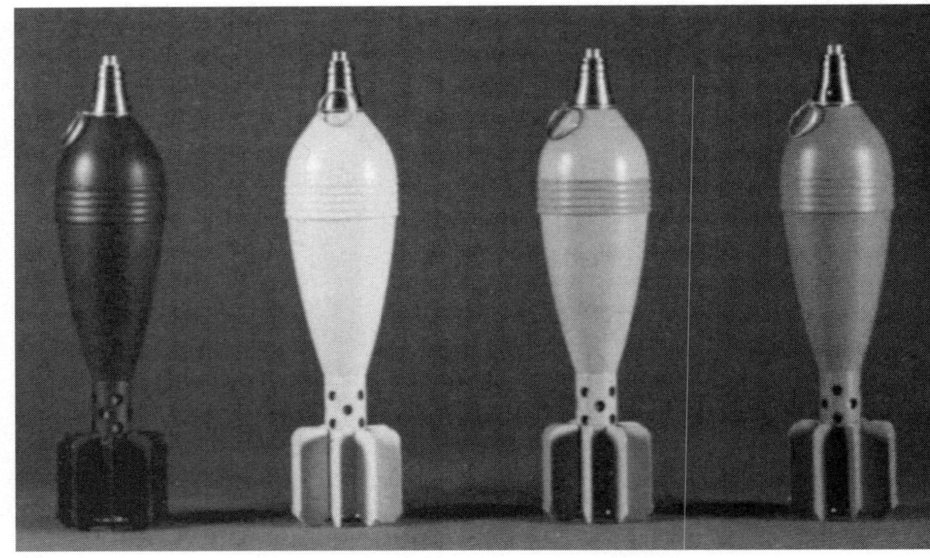

Model NA series bombs for 81 mm mortars

Data

81 MM HE BOMB MODEL NA
Length: overall, 343 mm
Weight: 3.2 kg
Filling: 496 g TNT
Max range: mortar LN, 4,125 m; mortar LL, 4,680 m

81 MM HE BOMB MODEL AE
Weight: 4.5 kg
Filling: 100 g TNT
Max range: mortar LN M-86, 6,200 m; mortar LL M-86, 6,900 m

81 MM ILLUMINATING BOMB MODEL N
Length: overall, 474 mm
Weight: 3.93 kg
Intensity: 550,000 cd
Burning time: 30 s
Rate of descent: 4 m/s
Max range: mortar LN, 3,350 m; mortar LL, 4,000 m

81 MM ILLUMINATING BOMB MODEL AE
Weight: without fuze, 4.25 kg
Intensity: 550,000 or 700,000 cd
Burning time: 35 s
Rate of descent: 4 m/s
Max range: mortar LN M-86, 5,800 m; mortar LL M-86, 6,500 m

Manufacturer

Esperanza y Cia SA, Marquina, Vizcaya.

Status

Available.

Service

Spanish and several other armies.

VERIFIED

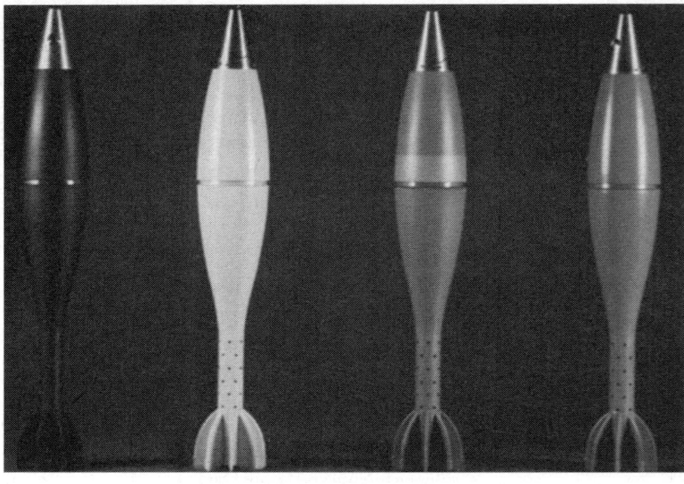

Model AE series bombs for 81 mm mortars

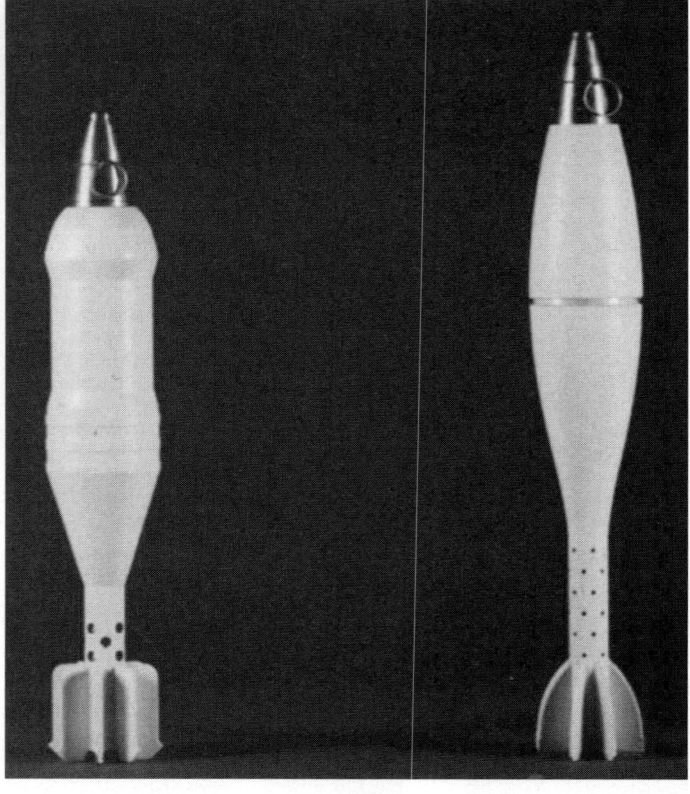

81 mm illuminating bombs Models N and AE

ECIA ammunition for 120 mm mortars

ECIA manufactures two families of 120 mm mortar ammunition, a conventional one and the aerodynamic (AE-84) type. Both types include the whole range (HE, smoke HC or WP, illuminating and practice rounds).

The ammunition manufactured by ECIA is obtained from forging, thus giving a knife-like fragment with higher velocity and therefore higher effectiveness. Upon request, ECIA can provide 120 mm bombs made from cast iron.

The Model AE-84 was introduced in 1984 and uses a highly streamlined body to obtain a good weight to calibre ratio and a plastic obturating ring to achieve good gas sealing and ballistic regularity. The tail unit, consisting of stabiliser tube and fins, is of forged aluminium and is attached to the body by a screwed section. The primary cartridge is inserted into the tail tube and the incremental charges are in horseshoe-shaped containers which clip round the stabiliser tube.

Data

HE BOMB MODEL L
Weight: 13 kg

Filling: 2.391 kg TNT
Max range: with 1.6 m barrel, 6,725 m; with 1.8 m barrel, 7,000 m

HE BOMB MODEL AE
Weight: 14.75 kg
Filling: 3.148 kg TNT
Max range: mortar M-86 (M120-13), 8,000 m; mortar M-86 (M120-15), 8,250 m

ILLUMINATING BOMB MODEL N
Weight: 14.087 kg
Max range: 5,450 m
Intensity: 1 Mcd
Burning time: 60 s
Rate of descent: 4 m/s

ILLUMINATING BOMB MODEL AE
Weight: without fuze, 14 kg
Max range: mortar M-86 (M120-13), 7,300 m; mortar M-86 (M120-15), 7,500 m
Intensity: 1 or 1.5 Mcd
Burning time: 60 s
Rate of descent: 4 m/s

Manufacturer
Esperanza y Cia SA, Marquina, Vizcaya.

Status
In production.

Service
Spanish forces and several others.

VERIFIED

Model L series bombs for 120 mm mortars

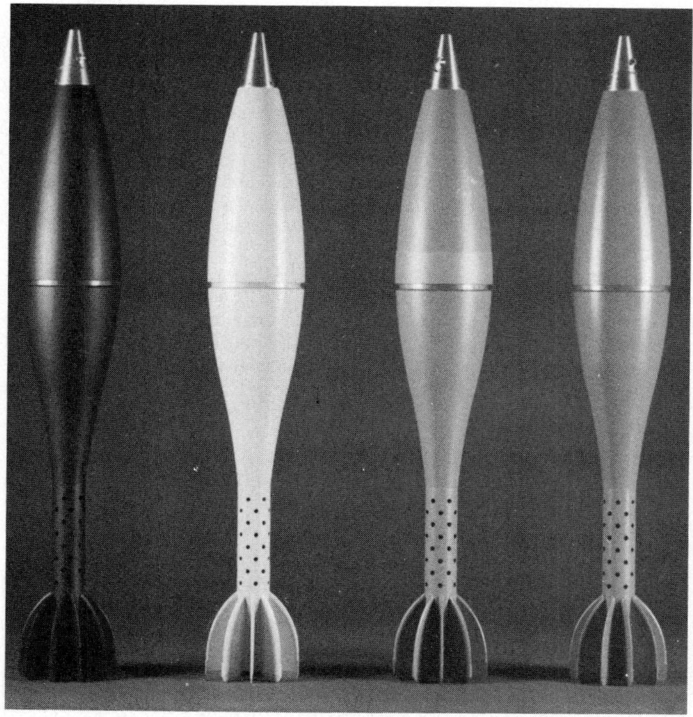

Model AE series bombs for 120 mm mortars

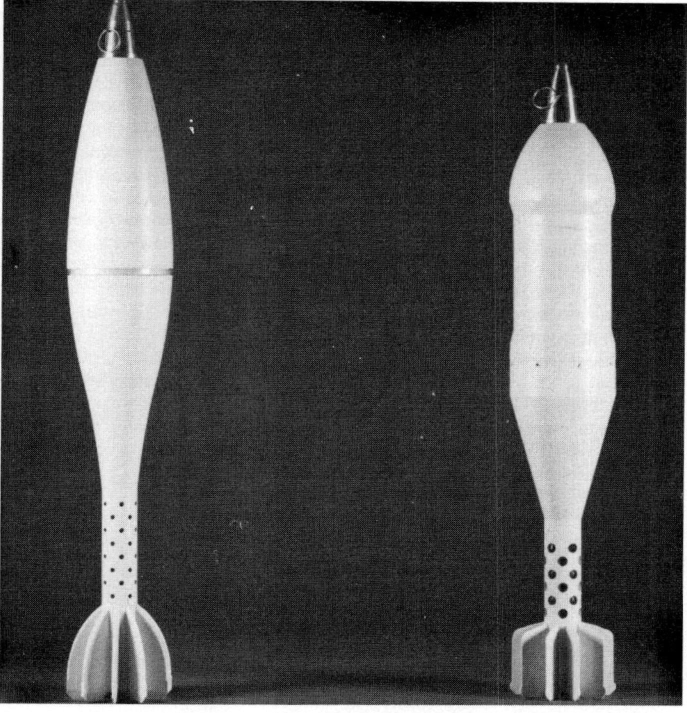

120 mm illuminating bombs Models N and AE

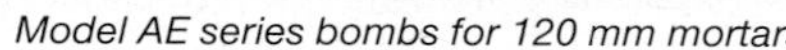

MAT120 cluster mortar projectile

Description
The MAT120 cluster mortar projectile is a further development of the earlier ESPIN cluster projectile, and is intended to enhance the capabilities of 120 mm mortars.

Each round of MAT120 ammunition contains 21 37 mm diameter submunitions which are effective both against armour and personnel. Each submunition is provided with a shaped charge able to perforate 150 mm of steel and also has an anti-personnel capability. The complete projectile is fired in the same manner as any other mortar projectile and bursts by means of a time fuze at a preselected height to release the payload. Maximum range is 5,500 m and the covered area is 2,200 m^2.

Each of the MAT120 bomblets is fitted with a self-destruct mechanism operating after a short delay following impact. Should the self-destruct mechanism not function as intended it subsequently introduces a self-sterilisation state, thereby rendering the submunition safe to handle.

Data
Calibre: 120 mm
Number of submunitions: 21
Total number of fragments produced: ca 13,600

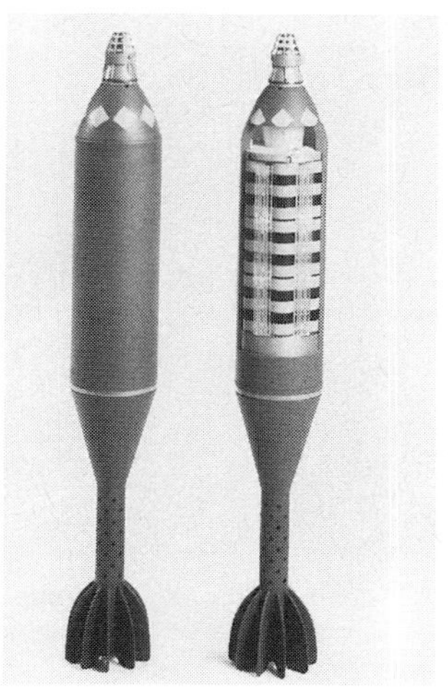

MAT120 cluster mortar projectiles
1996

Max range: 5,500 m
Area covered: 2,200 m²

SUBMUNITIONS
Diameter: 37 mm
Weight: total, 275 g; filling, 50 g
Penetration: steel, 150 mm
Number of fragments: ca 650
Lethal area radius: 6 m
Action radius: 18 m

Manufacturer
Esperanza y Cia SA, Marquina, Vizcaya.

Status
In production.

UPDATED

SWEDEN

Strix 120 mm guided mortar projectile

Development
Development of the Strix 120 mm guided mortar projectile commenced in 1984 as a joint venture between FFV (now part of Bofors AB) and Saab Missiles. The first complete round was delivered in October 1988 and initiated two years of test firings. On June 27 1991 the Swedish Defence Materiel Administration (FMV) signed a SKr 900 million contract for the procurement of Strix for the Swedish Army. Deliveries commenced during 1994.

Strix is the first in-service, mortar-launched, target-seeking projectile to be deployed anywhere in the world. It can be fired from any standard 120 mm mortar and employs autonomous passive infra-red terminal guidance techniques and a hollow charge warhead to attack the top armour of armoured targets. It is intended to be deployed as a mass attack weapon against multiple targets.

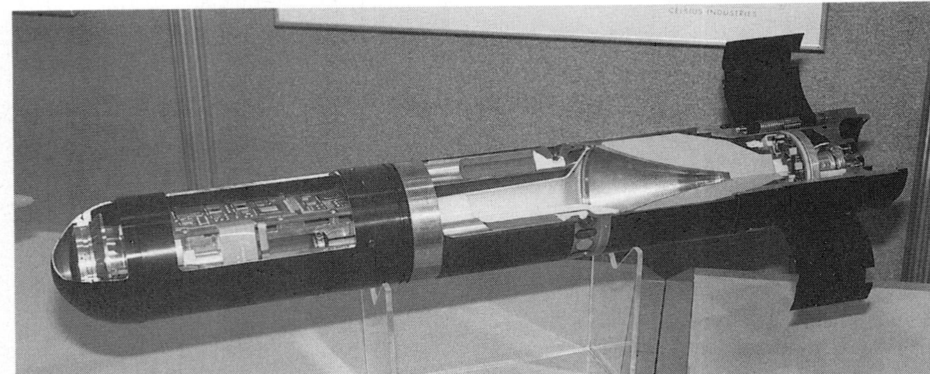

Internal arrangement of the Strix homing projectile (T J Gander)

1996

Description
The main components of the Strix are the projectile, a sustainer and a launch unit. In addition there is a separate hand-held programming unit used to preset some system electronics.

The main parts of the Strix projectile are, from the nose, the target seeker, an electronics and power supply section, a thruster rocket assembly, the warhead, and the fuze system, surrounded by the fin assembly.

The target seeker is of the passive infra-red type with a detector behind a nose-mounted lens. It creates a digital image of a target area which is continuously updated during the terminal phase of the flight. The image is fed to a processor unit where target acquisition and tracking is performed using advanced signal processing algorithms. The system provides effective detection of tank targets against severe backgrounds, even if they are camouflaged. It can also carry out intelligent target selection under all conditions and is highly resistant to active and passive countermeasures. The target seeker can be used both day and night under all climatic conditions.

The electronics and power supply section contains a signal processor unit which performs analogue and digital processing of signals from the target seeker. It also contains a thermal battery unit, activated at the instant of firing by a striker mechanism. There is also a power unit to convert the battery output into a suitable form for powering the various circuits, a safety unit to avoid the accidental firing of any of the thruster rockets, a programming connector, a wiring flexprint cable harness, and a roll gyro.

The electronic elements of the processor unit consist of large scale integrated circuits mounted on ceramic multilayer boards. The unit contains analogue and digital circuits as well as high-performance general purpose microprocessors and memories. Analogue signals are amplified, filtered and converted to digital form before collection in an image memory. The target signature is extracted from the background by image processing techniques and the correct target is selected and tracked, even if it is camouflaged, obscured by smoke or flares and other decoys are present. With the help of the continuously updated target position in the seeker field of view, the correct moment and direction of trajectory correction pulses are calculated, resulting in initiation signals for the thruster rockets. Via power amplifiers in the power unit and via the safety unit, these pulses are routed to the thruster rocket igniters. The gyro unit measures the roll angle which is used to select the correct thruster rocket, bearing in mind that the projectile is spinning.

There are 12 thruster rockets arranged around the periphery of the thruster rocket assembly. Each of the thrusters can be fired individually. As the assembly is close to the centre of gravity of the projectile the course is corrected instantaneously as each rocket is fired. By successively firing several thrusters high-precision steering, and thus guidance, is possible.

The hollow charge warhead is configured to achieve maximum target penetration with behind armour damage such as spalling, over-pressure and heat. It can defeat explosive reactive armour. The fuze is of the point-initiating base-detonating (PIBD) type, with arming taking place from one to two seconds after launch.

The fin assembly around the base has four folding wraparound surfaces which unfold immediately once the projectile has left the mortar barrel to impart a slow rate of spin.

The basic projectile can be used over ranges from 1,000 to 5,000 m. This is achieved using a separate launch unit consisting of a boom containing a primary cartridge plus up to eight secondary charges around the boom. Each secondary charge consists of disc propellant enclosed in a cloth bag. A docking mechanism located in the front of the boom docks the sustainer with the projectile.

If ranges of more than 5,000 m are required a sustainer unit is employed. This is a rocket motor with a burn time of about six seconds. During the launch acceleration phase the sustainer and the projectile are docked. Before the guidance phase commences the

Loading Strix into a 120 mm mortar

1996

Programming Strix at the mortar position

sustainer is separated from the projectile. To avoid error when loading the sustainer, an anti-inversion device is connected to the top of the sustainer unit. As the sustainer is loaded correctly this device falls off.

The Strix electronics contain a time circuit to preset the correct target seeker activation time. Before launch the Strix projectile is programmed using a hand-held unit which enters time of flight until target seeker activation, projectile velocity and angle of descent at the start of the guidance phase, one of two activation levels, and one of three control rocket temperatures. The programming unit is connected to the projectile by a cable.

To initiate a Strix fire mission a forward observer establishes a target position and calls for fire. At the fire position calculations are made to determine the propelling charge, angle of elevation and flight time to target seeker activation. This data is entered into the programming unit. The correct number of secondary charges are fitted and, if required, the sustainer unit is prepared. The programming unit is connected to the projectile and the inserted data is transferred. The mortar is then loaded with the launch unit, then with the sustainer (if needed), and finally with the projectile. A protective cover is maintained in place as the projectile is loaded; it falls off as the projectile passes below the muzzle. The mortar is then fired. The spent launch unit falls within 100 m in front of the mortar. If the sustainer is involved the firing gases initiate the rocket motor via a pyrotechnic delay four seconds after firing.

On firing the thermal battery is activated and the mechanical arming sequence commences. As the projectile leaves the muzzle the four tail fins unfold and impart spin. A few seconds before target seeker activation, the empty sustainer case (if involved) is separated from the projectile by means of a pyrotechnic charge. The warhead initiation circuit is charged at the same instant as target seeker activation.

At a preset height above the target, given by the programmed time when the Strix descends in an almost vertical trajectory, the target seeker is activated and starts to collect infra-red signals emitted from the target area. The target area image is analysed and, according to preset selection criteria, one of the targets in the area is selected for tracking. It is possible for the system to change the selected target when the projectile is at a lower height.

Once a target has been selected the error vector between the centre of the target and the projected impact point is continuously monitored. As soon as this error or error vector time exceeds a preset value, one or several thruster rockets are fired in a direction to bring the value of the error close to zero.

By continuous calculation of the predicted impact point relative to the predicted target position at impact it is possible to use proportional navigation which avoids any influence of target movement, wind effects, and so on. In other words, Strix has a full performance against both stationary and moving targets. The tracking technique makes it possible to make several course corrections in rapid succession. If necessary all 12 thruster rockets can be used with full control during the last few seconds of flight.

No special forward observer fire control procedures are necessary with Strix. For training, a special programming practice projectile is available along with an inert launch unit and sustainer.

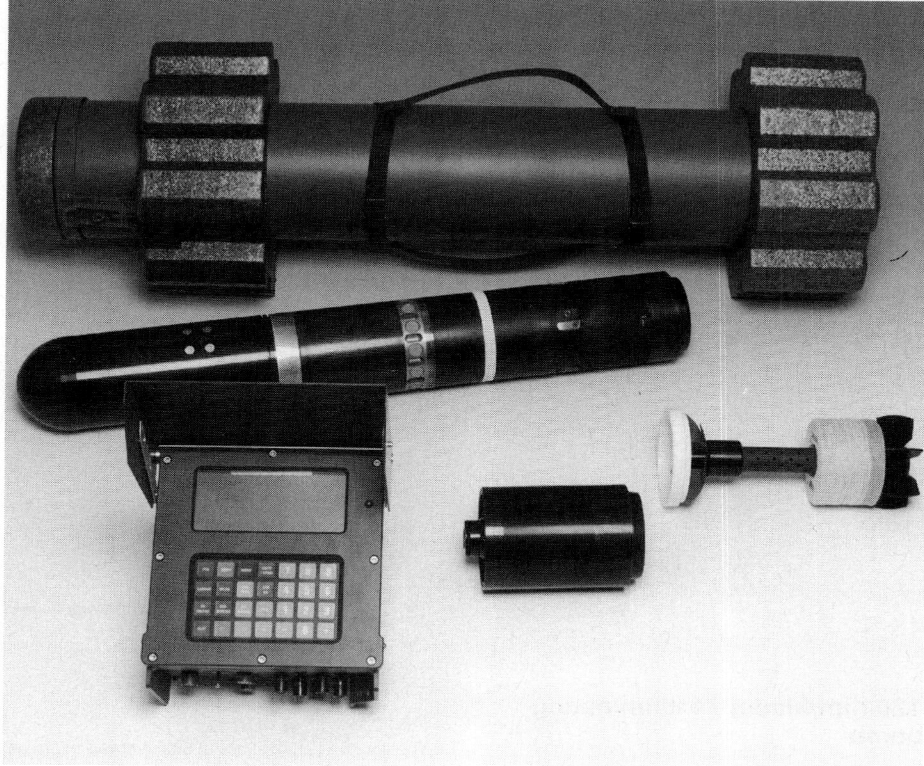

Strix main parts: package, projectile, launch unit (with propellant charges), sustainer and programmer

Data

Calibre: 120 mm
Length: projectile, 842 mm; launch unit, 295 mm; sustainer, 197 mm
Weight: projectile, 18.2 kg; launch unit with secondary charges, 1.8 kg; sustainer, 3.6 kg
Muzzle velocity: 150-320 m/s
Range: without sustainer, 1,000 to 5,000 m; with sustainer, 5,000 to 7,500 m

Manufacturers

Bofors AB, S-691 80 Karlskoga.
Saab Missiles AB, S-581 88 Linköping.

Status

In production.

Service

Swedish Army.

UPDATED

SWITZERLAND

Ammunition for 81 mm mortars

Description

Ammunition for the Model 1972 Swiss mortar is of the conventional streamlined pattern.

Manufacturer

w + f Thun, CH-3602 Thun.

Service

Swiss Army.

VERIFIED

Data

Types	HE	Smoke	High capacity fragmentation bomb
Fuzing	PD or DA	percussion	PD
Weight	3.17 kg	3.67 kg	6.89 kg
Filling	TNT	smoke composition	TNT
Charges	0-6	0-6	1-4
Muzzle velocity	70-260 m/s	70-210 m/s	64-110 m/s
Max range	4,100 m	3,000 m	1,070 m

81 mm Model 73 illuminating bomb

Description

The Model 73 illuminating bomb is of Swedish design and the first batch supplied to the Swiss Army was manufactured in Sweden; however, the bomb is produced at w + f Thun. The mechanical time fuze and propelling cartridges are of Swiss design and manufacture.

The Model 73 is a conventional parachute illuminating bomb in which the flare is ejected at about 300 m above the target area and then illuminates an area of 650 m diameter for up to 30 seconds.

Data

Weight: 3.5 kg
Range: 500-3,250 m
Rate of descent of flare: 4 m/s
Flight duration at max range: 30 s
Time fuze setting: variable between 5 and 60 s
Duration of illumination: 30 s
Charges: 0-6

Manufacturer

w + f Thun, CH-3602 Thun.

Status

In production.

Service

Swiss Army.

VERIFIED

Ammunition for 120 mm mortars

Description
Ammunition for the Models 64 and 74 120 mm mortars is of conventional type.

Data
Weight: 14.33 kg
Type of filling: HE, trotyl; incendiary/smoke, white phosphorus
Charges: 1-8
Muzzle velocity: 128-420 m/s
Max range: 7,500 m
Fuze: HE, point detonation and slight delay; incendiary/smoke, point detonation

Manufacturer
w + f Thun, CH-3602 Thun.

Status
In production.

Service
Swiss Army.

VERIFIED

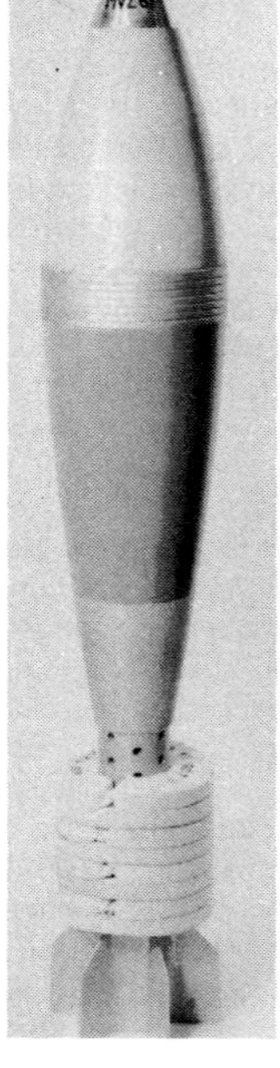

120 mm mortar ammunition (HE)

120 mm Model 74 illuminating bomb

Description
The Model 74 is a conventional parachute illuminating bomb, the functioning of which is controlled by a mechanical time fuze of Swiss manufacture. The body of the bomb, including the flare, is manufactured in Sweden, but the tail assembly, propelling charges and fuze are produced in Switzerland. The Model 74 can be fired from the 120 mm mortars Mw 64 and Mw 74 as well as the statically emplaced Fest Mw 59.

The flare is ejected at about 500 m above the target and illuminates an area of approximately 1,600 m in diameter.

Data
Weight: 15 kg
Rate of descent of flare: 5 m/s
Max range: 7,000 m
Flight duration at max range: 50 s
Fuze time setting: variable from 5 to 60 s

Manufacturer
w + f Thun, CH-3602 Thun.

Status
In production.

Service
Swiss Army.

VERIFIED

Degen K-85 fuze

Description
The Degen K-85 is a single-action point detonating fuze. It can be used with mortar bombs of 60, 81, 82 and 120 mm calibre. It is similar to fuzes M52, M525 and DM 111. Safety is ensured by a shutter which holds the detonator out of alignment prior to firing, and by a safety pull wire which remains in place until removed immediately before loading the bomb. The minimum arming distance is 30 m from the muzzle. A fully supported and guided firing pin ensures detonation at low angles of impact. The fuze meets all requirements of MIL-STD-331.

Manufacturer
Degen and Company AG, CH-4435 Niederdorf.

Status
In production.

VERIFIED

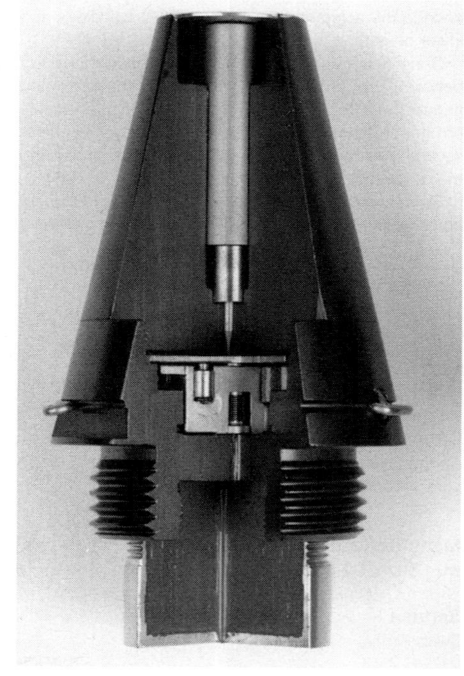

Degen K-85 mortar fuze

TURKEY

MKEK 81 mm mortar bombs

Description
Makina ve Kimya Endüstrisi Kurumu (MKEK) produces 81 mm mortar ammunition of varying types. One HE bomb, intended for use with locally held M1 and M29 mortars, is a licence-produced US M43A1B1 fitted with an AZDM 111A2 point detonating fuze and containing a TNT explosive payload. This bomb has a maximum range of 3,017 m.

By contrast, MKEK also produce a longer range and more modern 81 mm HE mortar bomb, the MKE Mod 214, primarily for use with the MKEK 81 mm UT1 and UT2 mortars. This bomb is much longer than the M43A1B1 and has a more streamlined forged steel body with a plastic obturating ring. The AZDM 111A2 fuze is retained but there is a longer tail assembly housing an MKE Mod 30 primary cartridge and up to six M8 propellant increments. The MKE Mod 214 has a maximum range of 5,850 m. A practice version, the MKE Mod 238, containing an inert payload is also produced. It is ballistically matched to the MKE Mod 214.

MKEK also produces an 81 mm Illuminating round, the M301A2. This is another US licence-produced bomb but it is fitted with a nose-mounted MTSQ DM 93 fuze. Maximum range is approximately 2,150 m and the flare and parachute assembly can produce light for up to 60 seconds.

Data
HE MKE MOD 214
Length: 500.52 mm
Weight: 4.82 kg
Muzzle velocity: 331 m/s
Max range: 5,850 m

Manufacturer
Makina ve Kimya Endüstrisi Kurumu (MKEK), 06330 Tandogan, Ankara.

Status
In production.

Service
Turkish armed forces.

NEW ENTRY

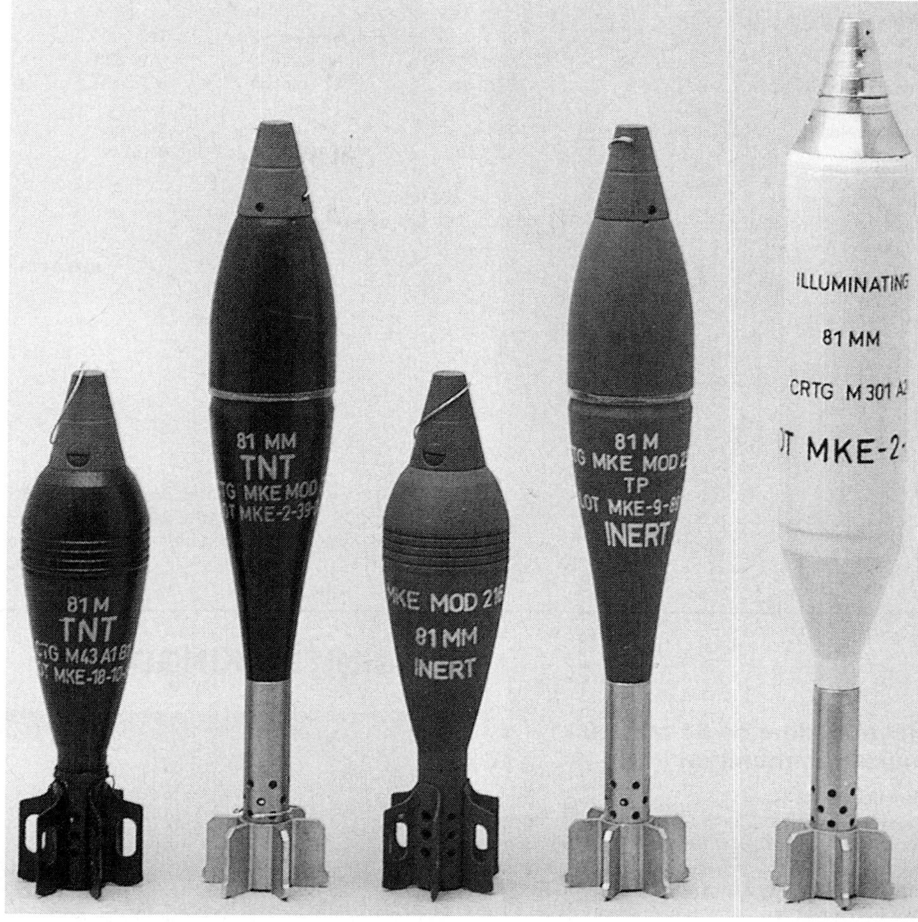

MKEK 81 mm mortar bombs, from left: HE M43A1B1; HE MKE Mod 214; Practice equivalent of HE M43A1B1; Practice MKE Mod 238; Illuminating M301A2
1996

MKEK 107 mm mortar bombs

Description
MKEK licence produces 107 mm mortar bombs for 4.2 in M2 and M30 rifled mortars in service with the Turkish armed forces. The MKEK bombs closely follow the US originals but a few local modifications have been introduced to suit local manufacturing processes and user requirements.

The high explosive round is the M329B1, filled with TNT and fitted with a locally produced M51A5 or M557 point detonating fuze. Maximum range is 5,500 m.

To complement the HE M329B1 for training purposes, MKEK produces the TP MKE Mod 217. This has an inert filling and weighs only 9.9 kg without a dummy fuze. Apart from the usual blue coloration, it is identical to its operational equivalent.

The Illuminating bomb is the M335A2, a direct copy of the US original carrying a flare and parachute assembly capable of producing 850,000 candela for 60 seconds. The fuze is the MTSQ M501.

To complete their 107 mm mortar ammunition suite, MKEK produces the Smoke M328A1. This is another licence-produced US bomb containing White Phosphorus (WP) which is released by a burster charge to create screening smoke when the M51A5 or M557 point detonating fuze impacts on a target area.

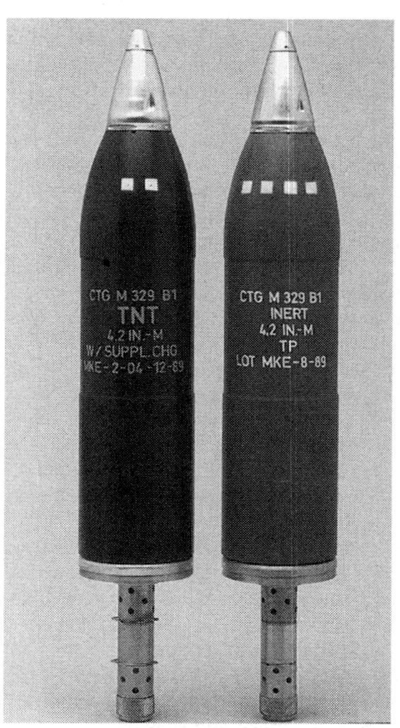

MKEK 107 mm mortar bombs; HE M329B1 (left) *and TP MKE Mod 217*
1996

Data

Type	**HE**	**Ill**	**Smoke**	**TP**
Designation	M329B1	M335A2	M328A1	Mod 217
Length	655 mm	650 mm	655 mm	655 mm
Weight	12.324 kg	12.4 kg	13.6 kg	9.9 kg*
Muzzle velocity	292 m/s	293 m/s	293 m/s	292 m/s
Max range	5,500 m	5,600 m	5,600 m	n/avail

* unfuzed

Manufacturer
Makina ve Kimya Endüstrisi Kurumu (MKEK), 06330 Tandogan, Ankara.

Status
In production.

Service
Turkish armed forces.

NEW ENTRY

MKEK 120 mm mortar bombs

Description
The MKEK 120 mm mortar bombs are licence-produced Thomson-Brandt designs intended for the MKEK 120 mm Tosam HY 12 D1 mortar, itself a licence-produced Thomson-Brandt MO-120-RT rifled mortar. The Turkish rounds thus closely resemble their French equivalents and differ from them only in detail. As such they are spin-stabilised projectiles with a prerifled drive band, a plastic obturating band at the rear and an extension to carry the propellant charges protruding from the base. The extension falls to the ground once the projectile has left the mortar barrel.

The high explosive round is the HE MKE Mod 209, filled with TNT and fitted with a locally produced M51A5 or M557 point detonating fuze. Maximum range is 8,180 m. The closest French equivalent is the Thomson-Brandt PR 14.

To complement the HE MKE Mod 209 for training purposes, MKEK produces the TP MKE Mod 228. This has an inert filling and a dummy fuze. Apart from the usual blue colour, it is identical to the operational HE equivalent.

The Illuminating bomb is the MKE Mod 236, a close derivative of the Thomson-Brandt 120 mm PRECLAIR. It carries a flare and parachute assembly capable of producing 850,000 candela for 60 seconds. The fuze is the MTSQ DM 93 or M501.

To complete the 120 mm rifled mortar ammunition suite, MKEK produces the Smoke WP MKE Mod 226. This contains White Phosphorus (WP) which is released by a burster charge to create screening smoke when the M51A5 or M557 point detonating fuze impacts on a target area.

Data

Type	**HE**	**Ill**	**Smoke**	**TP**
Designation	Mod 209	Mod 236	Mod 226	Mod 228
Length	827 mm	827 mm	827 mm	827 mm
Weight	17 kg	16 kg	17.4 kg	17 kg
Muzzle velocity	365 m/s	365 m/s	365 m/s	365 m/s
Max range	8,180 m	8,132 m	8,180 m	8,132 m

Manufacturer
Makina ve Kimya Endüstrisi Kurumu (MKEK), 06330 Tandogan, Ankara.

Status
In production.

Service
Turkish armed forces.

NEW ENTRY

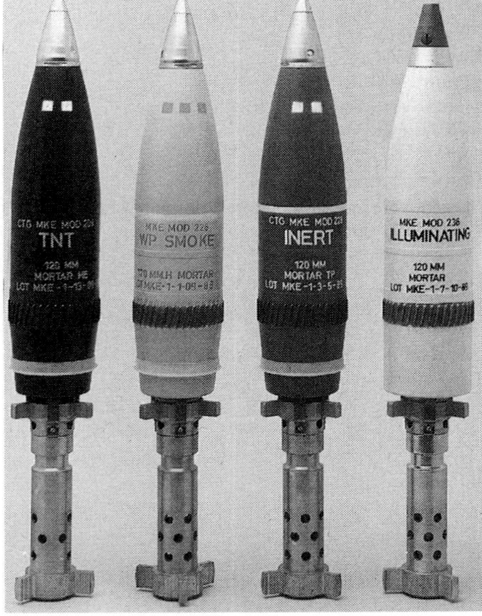

MKEK 120 mm mortar bombs, from left; HE MKE Mod 209, Smoke WP MKE Mod 226, TP MKE Mod 228, Illuminating MKE Mod 236
1996

UNITED KINGDOM

Royal Ordnance 51 mm L9A1 mortar ammunition

Development
Development of the 51 mm L9A1 mortar commenced during the mid-1970s. When the L9A1 entered production in 1982 the ammunition utilised a single propellant charge system with a Short Range Insert (SRI) introduced into the barrel when short ranges were required. In practice the SRI worked but was not deemed a success for several reasons, prompting a change to a two-charge system for the ammunition. In place of the single charge a propellant increment was added around the tail and removed when short ranges were required - the SRI was no longer involved and was withdrawn.

This change of propellant system was also used as a point at which revised mortar bombs could be introduced. The main change involved the HE round. Formerly the HE L1A1 used an aluminium body with an internal fragmentation sleeve. The revised HE bomb, introduced in 1992, has a cast iron body with no fragmentation sleeve. Changes introduced to the Illuminating bomb include an improved flare parachute.

In 1994 a £50 million order was placed with Royal Ordnance for 400,000 of the new bombs for the British Army. Of this order 200,000 are for the Illuminating round, reflecting the role of the L9A1 mortar as a method of illuminating targets for other weapons to engage. The remainder of the order is split 100,000 HE and 100,000 Smoke. First deliveries are due during 1997.

The description provided here relates to the latest types of 51 mm bomb. Details of the original models, which are still in service, will be found in *Jane's Infantry Weapons 1995-96* page 564.

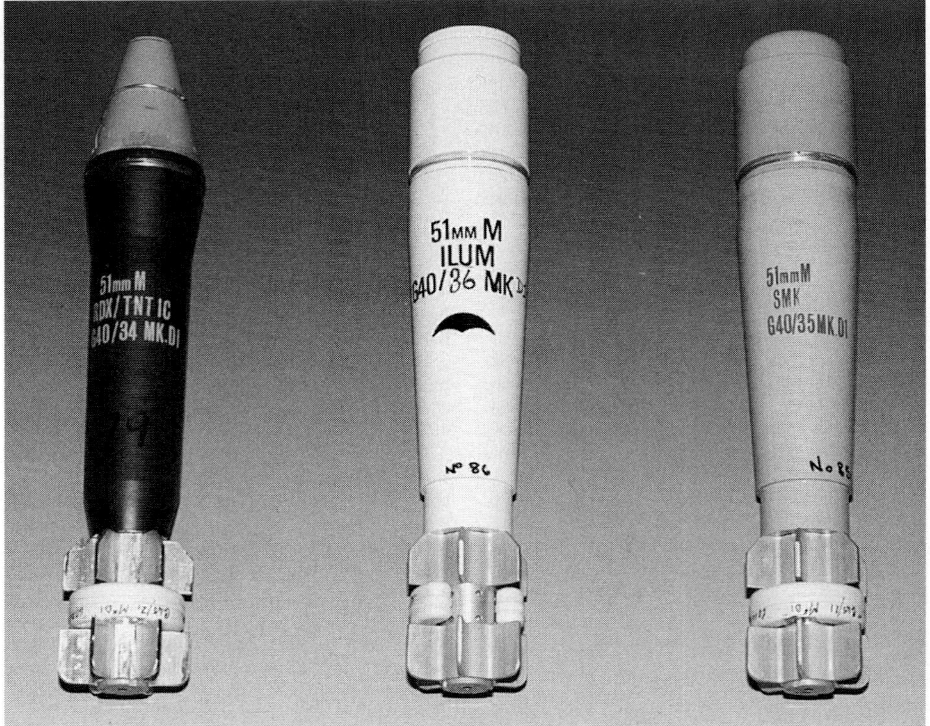

51 mm L9A1 assault mortar bombs, from left, HE, Smoke and Illuminating
1996

Description
The ammunition for the 51 mm L9A1 mortar comprises High Explosive (HE), Smoke and Illuminating bombs. All three bombs have a primary charge in the tail assembly with a secondary propellant increment inserted in a space between the tail fins. The secondary increment is removed when short ranges are required. The HE and Smoke bombs have a maximum range of 800 m with a minimum range of 50 m when firing at high angles with the primary charge only. The Illuminating bomb has a maximum range of 775 m.

The HE bomb weighs 1.1 kg and has a one-piece spheroidal graphite cast iron body filled with RDX/TNT. The fuze is an L127A3 or M935.

The Smoke bomb weighs 950 g and has an aluminium alloy body filled with PN800 smoke composition capable of producing screening smoke for 120 seconds. The fuze involves a fixed time (4 seconds) pyrotechnic delay.

The Illuminating bomb also has an aluminium alloy body. The flare carried produces an average light output of 170,000 candela over the total burning time of 44 seconds. This produces a light on the ground of 2 lux over a 200 m radius throughout the burn. The bomb is intended to burst at a height of 325 m with the height at burnout being 120 m. Rate of descent is 4 to 6 m/s. Bomb weight is 850 g.

Training and drill bombs are available.

Data

HE BOMB
Length: overall, fuzed, 290 mm
Weight: 1.1 kg
Filling: RDX/TNT
Fuze: point detonating, L127A3
Max range: 800 m
Min range: 50 m

SMOKE BOMB
Length: 290 mm
Weight: 950 g
Filling: PN800 smoke composition
Fuze: fixed time delay, 4 s min
Duration of emission: 120 s
Max range: 800 m
Min range: 50 m

ILLUMINATING BOMB
Length: 290 mm
Weight: 850 g
Burst height: 325 m
Rate of descent: 4-6 m/s
Illumination: average, 170,000 cd
Duration of burning: ca 44 s
Max range: 775 m

Manufacturer
British Aerospace Defence Limited, Royal Ordnance Division, Euxton Lane, Chorley, Lancashire PR7 6AD.

Status
In production.

Service
In service with the British Army.

UPDATED

Royal Ordnance 81 mm mortar ammunition

Development

Although the Royal Ordnance 81 mm mortar bombs are optimised for the L16A2 and M242 mortars they can be fired from any standard 81 mm mortar. When used with the L16A2 the maximum range is 5,650 m. Over the years there has been considerable ammunition development which is still continuing. For instance, the original HE bomb was the L15, followed by the L31 and the L36A2, the latter now being the standard HE bomb. The HE L36A2 is ballistically matched by the L40A1 Smoke which contains white phosphorus. In 1991 this was joined by a red phosphorus bomb.

Royal Ordnance developed two 81 mm charge systems, the Marks 4 and 5. The Mark 4 system is in service with the British Army, and the Mark 5 with the US Army. The Mark 4 has six equal secondary charges, the Mark 5 has four equal secondary charges. The maximum charge gives the same range with either system.

Description

The high explosive round L36A2, fuzed L127A3, is streamlined in shape and has been designed to produce the maximum number of fragments of the optimum size. A study of cast iron, forged steel and spheroidal graphite cast iron, showed the third to be the best material. The L36A3 gains in velocity and consistency by the incorporation of a sealing ring. This allows adequate windage as the bomb drops down the tube but the pressure produced by the burning propellant forces the polycarbon ring outwards against the interior wall of the mortar tube and so prevents gas leakage. The ring also centres the bomb to reduce the yaw at the muzzle associated with conventional mortars and which leads to inaccuracy and inconsistency.

The propellant system used by the British Army involves the Primary Cartridge L33A1 plus Augmenting Cartridges L34A1 and L32A1.

HE bomb L36A2
Length: 472.4 mm
Weight: 4.2 kg

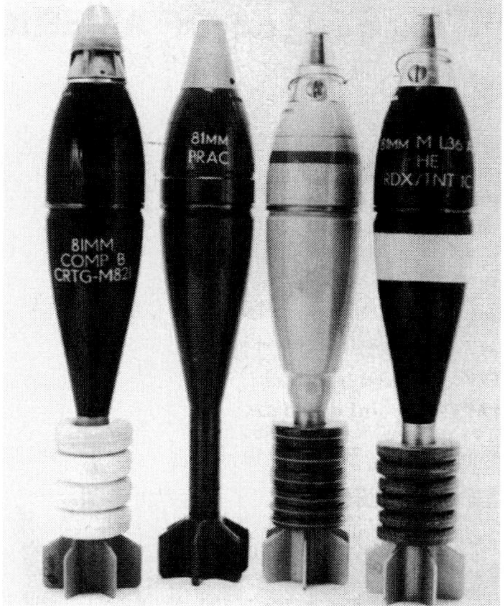

81 mm L16 mortar ammunition and various charge systems. From left to right: HE bomb (US variant) with Mark 5 charge system; recoverable Practice bomb; Smoke bomb; HE bomb (UK variant) with Mark 4 charge system

Filling: 680 g 60/40 RDX/TNT
Muzzle velocity: max, 297 m/s; min, 70 m/s
Fuze: L127A3, US M734 or M935 and other NATO fuzes

Smoke bomb L40A1
The body of the Smoke L40A1 is similar to that of the L36A2 HE bomb and is ballistically matched to it. It is filled with White Phosphorus (WP) to create a smoke screen lasting 60 seconds.

Length: 472.4 mm
Weight: 4.2 kg

Red phosphorus bomb
This bomb is longer than other associated 81 mm bombs, with the body containing 36 pellets of a red phosphorus composition. This bomb is intended to be used to produce airbursts at a height of 100 to 200 m so that the pellets will scatter over a target area and create broad and dense white smoke clouds for at least two minutes, twice as long as conventional WP smoke bombs, plus incendiary effects. The nose-mounted fuze is a DM 93 mechanical time. The bomb is fired using a Primary Cartridge L39A1 plus Augmenting Cartridges L40A1. Nominal maximum range to burst is 5,100 m.

Length: 604 mm
Weight: nominal, 4.3 kg

Illuminating bomb
An illuminating bomb is available which delivers 1 million candela for 35 seconds. It is ballistically matched to the HE and smoke bombs.

Length: 582 mm
Weight: 4.1 kg

Practice bomb L27A1
The practice bomb is a reusable item designed to fire to a maximum range of 80 m when an obturating ring is fitted. It is fired using only the Primary Cartridge L33A1.

Drill bomb
A dummy round, not intended for firing, but inert filled to the correct weight for training purposes.

Manufacturer

British Aerospace Defence Limited, Royal Ordnance Division, Euxton Lane, Chorley, Lancashire PR7 6AD.

Status

In production.

Service

British Army and numerous other armed forces.

UPDATED

Merlin 81 mm terminally guided mortar projectile

Development

The Merlin 81 mm terminally guided mortar projectile has been undergoing development since the mid 1980s, following a number of earlier design projects. Live guided firing commenced during 1987. Development has reached the point where the overall design is undergoing low-rate initial production for industrial qualification prior to full production. It is anticipated that many of the millimetric-wave radar seeker techniques developed for Merlin will be employed with other similar projects including, possibly, a 120 mm terminally guided mortar projectile.

Merlin offers a highly effective fire-and-forget capability to infantry confronted with armoured vehicles and considerably extends the number of anti-armour systems available to infantry forces.

Merlin is marketed in the United States by Alliant Techsystems who will co-produce Merlin for the US market.

Loading a Merlin terminally guided mortar projectile into an 81 mm L16A1 mortar

1996

Description

Merlin is a terminally guided armour-piercing mortar bomb for use with any 81 mm mortar and could be adapted for 82 mm mortars. Rate of fire can be up to 10 rds/min. Merlin is effective over a range from 1,500 to 4,200 m.

Firing is carried out in the normal way, using target information from the Mortar Fire Controller and using standard procedures. After launch, six rear-mounted fins are deployed to provide basic aerodynamic stability, followed by four canard fins to provide directional control. The seeker is switched on as Merlin approaches the vertex and searches first for moving and then for stationary targets.

Once in flight the nose-mounted miniature fully active millimetric seeker provides information to the control system to accomplish de-spin and attitude control. The seeker then carries out a search over a footprint area of 300 × 300 m, giving priority to moving targets. Having acquired a target, the seeker provides the necessary angular error information to the guidance system to ensure impact with the most vulnerable areas on the top of the armoured vehicle. If no moving target is available the seeker will search for stationary targets over a 100 × 100 m area.

The flight angle during search is about 45°, with the dive angle at the target being greater than 60°. Terminal velocity is approximately 140 m/s. On impact the full calibre shaped warhead is at the optimum distance for armour penetration, with the jet provided with a clear path through the centre of the missile body.

The effectiveness of Merlin has been demonstrated in a series of successful firings against armoured vehicles. These have proved the ability of Merlin to carry out the full sequence of acquiring and locking on to moving and stationary targets and then to steer itself out of its ballistic trajectory to attack them.

Specifications

Calibre: 81 mm
Length: 900 mm
Weight: 7 kg
Range: max, 4,200 m; min, 1,500 m
Terminal velocity: 140 m/s

Manufacturer

British Aerospace (Dynamics) Plc, Six Hills Way, Stevenage, Hertfordshire SG1 2DA.

Status

Low rate initial production for industrial qualification prior to full production.

UPDATED

UNITED STATES OF AMERICA

Ammunition for the 60 mm mortar M19

Description

The bombs for the 60 mm mortar are of conventional cast iron pattern, with obturating grooves around the body and steel tail units screwed in. The primary cartridge is of the shotgun type, inserted into the centre of the tail unit; the secondary charges are in the form of leaves of smokeless powder stitched together and are sprung into place between the tail fins as required.

Maximum ranges given below are applicable only to the mortar used with the M5 baseplate. Using the M1 baseplate no more than one incremental charge should be added to the bomb.

M49A4 HE BOMB
Weight: 1.46 kg
Filling: 154 g TNT
Propelling increments: 4
Min range: 45 m
Max range: Charge 4, 1,814 m
Bursting area: 9 × 18 m

M302A2 (WP) SMOKE BOMB
Weight: 2.26 kg
Filling: white phosphorus
Min range: 91 m
Max range: Charge 4, 1,465 m
Bursting area: diameter, 10 m

M83A3 ILLUMINATING BOMB
Weight: 2.27 kg
Min range: 375 m
Max range: Charge 4, 1,000 m
Illuminated area: diameter, 600 m
Illumination duration: 25 s

M50A2 PRACTICE BOMB
Weight: 1.45 kg
Min range: 45 m
Max range: Charge 4, 1,814 m

VERIFIED

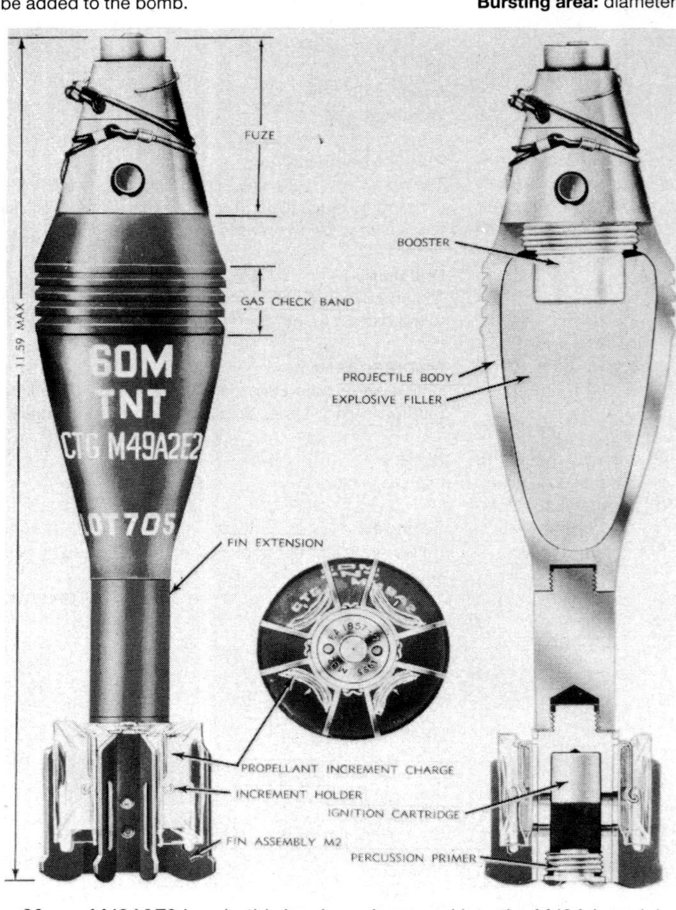

60 mm M49A2E2 bomb; this has been improved into the M49A4 model, though the basic details remain the same

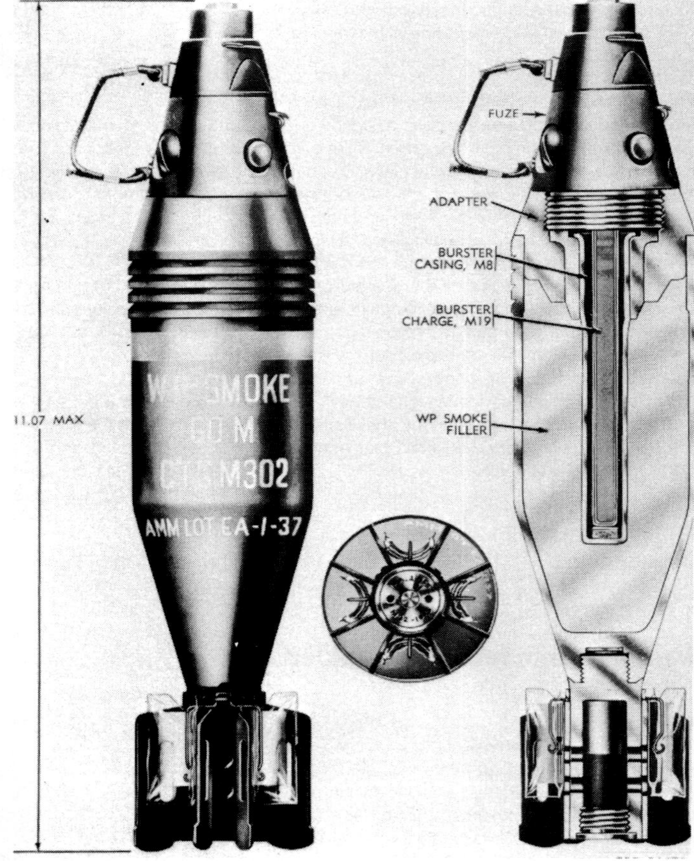

60 mm M302 white phosphorus smoke bomb

Ammunition for the 81 mm mortars M29 and M29A1

Description

The latest ammunition for these mortars is based upon the original British designs with modifications to suit American systems of manufacture and operation. The principal obvious change is the use of cloth bags for the secondary charges, rather than the British plastic horseshoe container, and the canting of the tail fins to produce a slight degree of roll stabilisation. Older designs of ammunition utilise simple forged steel bombs with steel tail units and the bombs have obturating grooves around the body rather than plastic obturating rings.

Authorised rounds:
High explosive, M43A1B1, M362A1, M374A2
White phosphorus, M370, M375A2
Illuminating, M301A1, M301A2, M301A3
Training practice, M43A1
Training (concrete), M798
White phosphorus, M784
Illuminating, M512E1, M652

HIGH EXPLOSIVE
HE M374A2
The pearlitic malleable iron projectile is loaded with approximately 950 g of Composition B. The rear of the bourrelet section of the projectile is fitted with an obturator ring with a circumferential groove. The aluminium fin assembly, M170, consists of an ignition cartridge housing and six extruding fins canted counter-clockwise 5° at the rear to stabilise the round in flight.

Weight: 4.23 kg
Min range: 72 m
Max range: 4,595 m
Bursting area: diameter, 34 m

HE M362A1
The forged steel projectile is loaded with approximately 950 g of Composition B. The aluminium fin assembly is the M141. This bomb is not classified Standard A.

Weight: 4.25 kg
Min range: 46 m
Max range: 3,987 m
Bursting area: 25 × 20 m

HE M43A1B1
The complete round consists of a relatively thin-walled shallow-cavity steel projectile containing a TNT bursting charge and PD fuze.

Weight: 3.22 kg
Min range: 69 m
Max range: 3,890 m
Bursting area: 20 × 15 m

WHITE PHOSPHORUS
WP M375A2
The M375A2 WP round is ballistically and otherwise similar to the HE M374A2 round except that it is loaded with approximately 725 g of white phosphorus, and contains a one-piece aluminium burster casing (M158) prefilled to the forward end of the body. The burster casing houses a central burster tube containing RDX.

Weight: 4.23 kg
Filling: 725 g WP
Min range: 72 m
Max range: 4,737 m
Bursting area: diameter, 20 m

WP M370
The M370 WP round is ballistically and otherwise similar to the HE M362A1 round except for the white phosphorus filler. This round is not classified Standard A.

Weight: 4.23 kg
Min range: 52 m
Max range: 3,987 m
Bursting area: diameter, 20 m

ILLUMINATING AMMUNITION

M301A3

The complete round consists of a time fuze, a thin-walled steel body-tube containing the parachute and illuminant assembly, and a steel tail cone and fin assembly.

This round is designed to be fired with a minimum of two propelling charge increments and not more than eight. It has a burst height of 600 m and will illuminate a 1,200 m² area for a minimum of 60 seconds.

Weight: 4.89 kg
Min range: 100 m
Max range: 3,150 m

M301A2

Illuminating round M301A2 is similar to the M301A3 except it has a tail fin which is 57 mm shorter than the M301A3. The round is designed to be fired with a minimum of two propelling charge increments and not more than four increments. The height of burst is 400 m and it will illuminate a 1,100 m² area for a minimum of 60 seconds. It is not a Standard A round.

Weight: 4.89 kg
Min range: 100 m
Max range: 2,150 m

M301A1

Illuminating round M301A1 is similar to the M301A2 except that it has gas-check bourrelet grooves and some minor dimensional differences in metal parts. It is not a Standard A round.

TARGET PRACTICE AMMUNITION (TP)

M43A1

Target practice round M43A1 is intended for use in training only and is similar to the M43A1 HE except for the projectile filler and its colour. The target practice projectile is loaded with an inert material (plaster of Paris and stearic acid) and a 22.68 g powder pellet. On impact, the black powder pellet and the fuze booster charge provide a spotting charge for observation purposes. The projectile is loaded to simulate the weight of the high explosive projectile and has the same ballistic characteristics.

Fuzes

POINT DETONATING (PD) FUZES

M524 SERIES

Dual-purpose fuze M524A5, super-quick or 0.05 second delay, is used with M362 series HE cartridges and M374 or with WP cartridges M370 and M375.

M525, M525A1

These fuzes are modifications of M52 series fuzes. The modification consists of the substitution of a head assembly containing a delayed arming device in addition to the firing pin mechanism. Those fuzes are used for M43 HE and TP series ammunition.

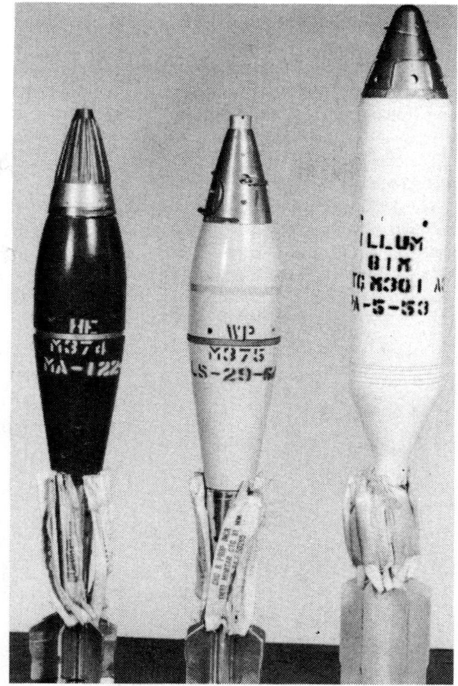

81 mm bombs
(left) *HE bomb M374* (centre) *white phosphorus smoke bomb M375* (right) *illuminating bomb M301A3*

M526 SERIES

These fuzes, which are replacing PD fuze M519, consist of the former M52 series fuzes modified, as in the M525 series, with an arming delay. In addition the fuzes are fitted with an adaptor containing booster pellets for a newer design ammunition. This fuze may be used instead of PD fuze M524A1 in cartridges M362 and M374 HE and M370 and M375 WP.

PD FUZE M519

which is a combination of PD fuze M525A1 and a fuze adaptor, is a single-action type with a direct-action firing device for use with cartridges M362 HE and M370 WP.

PROXIMITY FUZES

M532

The M532 is a radio-Doppler fuze which is standard for the M374 HE round and may also be used on the M362 HE or the M375 WP round. It provides an airburst at or near a height for optimum effectiveness by employment of the radio-Doppler principle of target detection. A clock mechanism provides a nominal nine seconds of safe air travel (610 to 2,340 m travel along trajectory for Charges 0 to 9, respectively). It can also be set to superquick (point detonating) to eliminate the proximity function.

81 mm M43A1 HE bomb

M43A1 bomb

The proximity fuze can be converted to a point detonating fuze action by rotating the top of the fuze more than 120° in either direction.

M517

This proximity fuze is provided for use with the M362 high explosive round. It will not function in the M374 or M375 rounds because of the spin. The M517 fuze's operating principles are similar to the M532, differing primarily in the arming system. The minimum time (after firing) to arm for an impact function is in excess of 1.5 seconds. The minimum time (after firing) to arm for a proximity function is 3.2 seconds. This fuze does not provide a PDSQ option.

MECHANICAL TIME FUZE

M84

This fuze is a single-purpose, powder train, selective time type used with the 81 mm illuminating cartridges M301A1 and M301A2. It has a time setting of up to 25 seconds.

M84A1

This fuze is a single-purpose, tungsten ring, selective time type used with the 81 mm illumination cartridge M301A3. It has a time setting of up to 50 seconds.

UPDATED

HE Ammunition for the 81 mm Mortar M252

Description

There are four models of HE bomb utilised with the 81 mm M252 mortar.

When the M252 was first procured in 1984 the initial ammunition lots were delivered direct from Royal Ordnance of the UK. These bombs are the M821 and M889 which differ only in the type of nose fuze installed. The M821 employs the M734 multi-option fuze while the M889 has the M935 point detonating fuze. Both bombs have their bodies manufactured using machined graphite cast iron. Both are 510 mm long and weigh approximately 4.125 kg in flight. Maximum range is 5,700 m.

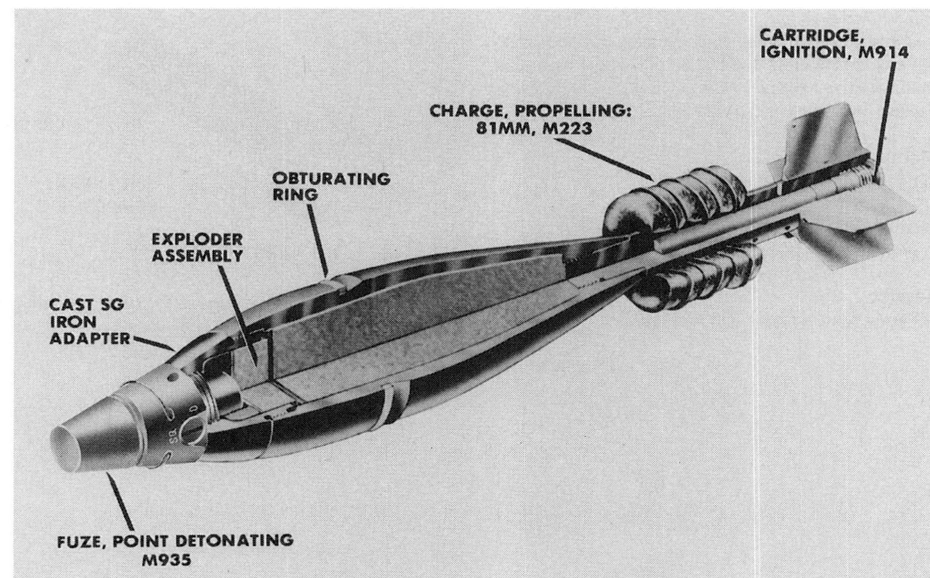

Cartridge, 81 mm: HE, M889 with point detonating M935 fuze

1996

When the M252 licensing agreement was made it included proprietary access to the ammunition. Following development by the US Army Armament Research Development and Engineering Center (ARDEC) slightly revised HE bombs were produced using high fragmentation steel bodies in place of the original cast iron. As with the original UK ammunition there are two US 81 mm HE bombs, the M984 with the M734 multi-option fuze, and the M983 with the M935 point detonating fuze. Both US bombs were type classified in July 1991 and entered production that same year. They will replace the M821 and M889.

Two Practice bombs were developed to match the 81 mm HE bombs.

The M879 is a full range training round ballistically matched to the M983/M984. It produces a flash, bang and smoke signature on impact and has an M751 fuze which simulates the M734 multi-option fuze. Following type classification in 1986, the bomb body is produced at the Milan Army Ammunition Plant.

The M880 is a short range training round with a nominal maximum range of 490 m. It produces a flash, bang and smoke signature on impact and has an M775 practice fuze. This bomb can be retrieved after firing and reused up to ten times by using a refurbishment kit.

Status
M983 and M984 in production.

Service
US Army, Marine Corps and Air Force.

NEW ENTRY

81 mm Smoke RP M819

Description
The full designation of this bomb is Cartridge 81 mm: Red Phosphorus (RP) Smoke M818 w/fuze MTSQ M772. It is intended for use with the 81 mm M252 mortar and was developed in the USA. It was type classified in December 1986 with the first unit equipped in 1991.

The M819 has a thin-walled body filled with 28 pelletised red phosphorus wedges. When the nose-mounted M772 MTSQ fuze functions after a preset time delay the body breaks open and the wedges are scattered to produce a smoke screen rapidly. Three bombs are sufficient to create a fully effective screen. The maximum range at which a screen can be produced is 5,000 m; minimum range is 300 m.

The in-flight weight of the M819 is 4.695 kg and it is 642.5 mm long.

The RP wedges are produced at the Pine Bluff Arsenal and packed into the body tubes. These are then shipped to the Longhorn Army Ammunition Plant for final assembly and packing.

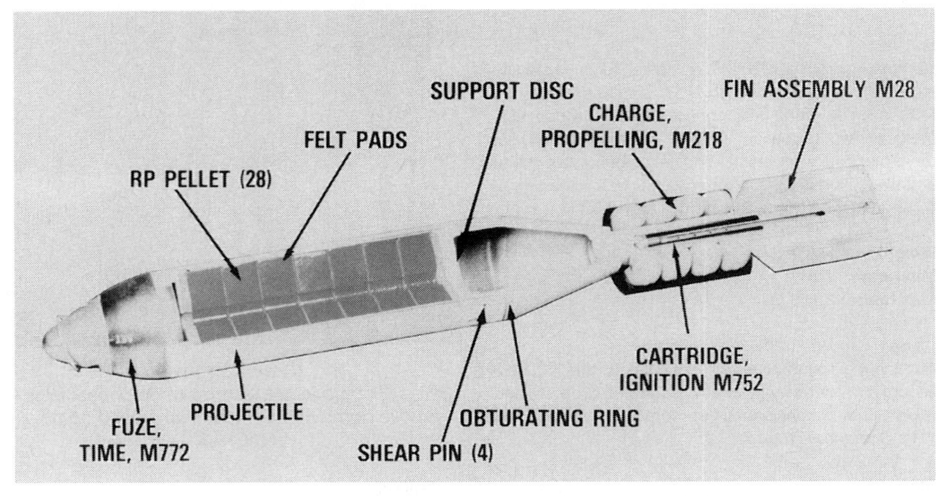

Cartridge, 81 mm: Smoke RP M819 with M772 MTSQ fuze

1996

Manufacturers
Pine Bluff Arsenal.
Longhorn Army Ammunition Plant.

Status
In production.

Service
US Army, Marine Corps and Air Force.

NEW ENTRY

81 mm Illuminating M853A1

Description
The full designation of this bomb is Cartridge 81 mm: Illuminating M853A1 w/fuze MTSQ M772. It is intended for use with the 81 mm M252 mortar and was developed in the USA. It was type classified in December 1986 with the first unit equipped in 1991.

The M853A1 has a thin-walled body containing a parachute and illuminant flare payload. When the nose-mounted M772 MTSQ fuze functions after a preset time delay the payload is base ejected in flight to produce 525,000 candela for a minimum of 50 seconds. The maximum range at which this bomb can be effectively utilised is 5,100 m; minimum range is 300 m. Maximum ground area covered is 320,000 m^2.

The in-flight weight of the M853A1 is 4.03 kg. It is 645 mm long.

Under development is an illuminating payload which will provide infra-red illumination of a target area. The illuminated area will be visible only by using night vision devices and is not visible with the naked eye.

Manufacturer
Longhorn Army Ammunition Plant.

Status
In production.

Service
US Army, Marine Corps and Air Force.

NEW ENTRY

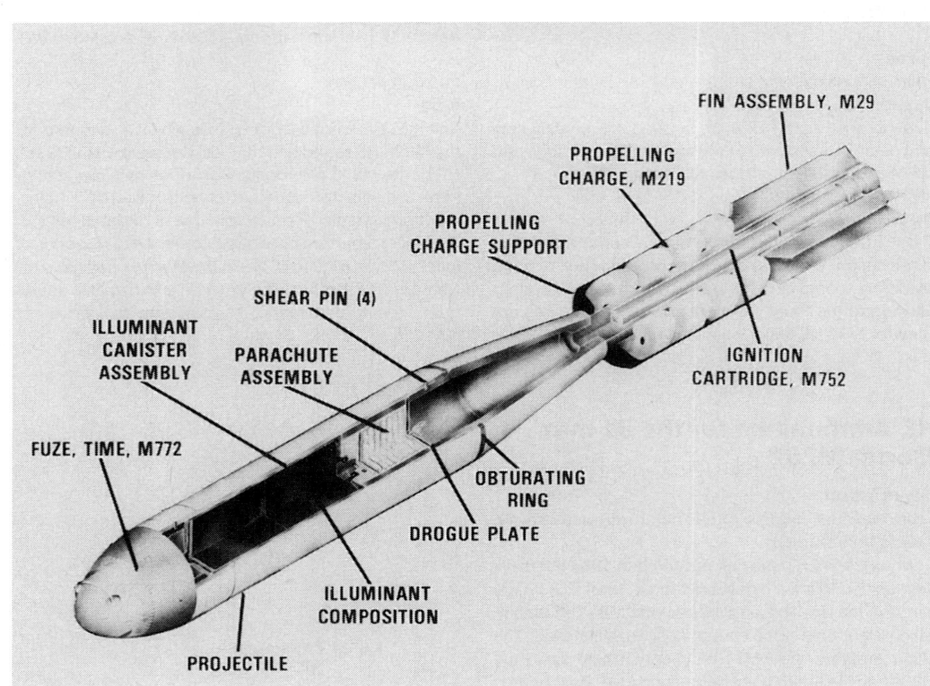

Cartridge, 81 mm: Illuminating M853A1 with M772 MTSQ fuze

1996

Ammunition for the 107 mm mortar M30

Description

The 107 mm mortar M30 is a rifled weapon and the bombs are spin stabilised. The projectiles are fitted with a malleable copper plate, slightly dished, at the base with a flat steel plate behind it. Both plates cover the base of the bomb and surround the cartridge container. The copper plate is of bomb diameter when loaded and therefore the bomb slides down the barrel quite easily. On firing, the pressure of the propellant gases drives the steel plate forward, so flattening the copper plate and forcing its circumference to protrude beyond the body of the bomb so that it is forced to engrave in the rifling of the barrel. The spin so given is transferred to the bomb during its passage up the bore of the mortar.

Ammunition for the 107 mm M30 mortar is issued in the form of complete rounds. The propelling charge consists of an ignition cartridge carried inside the cartridge container and 41 propellant increments assembled in a bag and sheets, which are fitted around the body of the cartridge container. To adjust the charge individual increments are removed. The table below contains information on the standard rounds authorised for the 107 mm mortar M30.

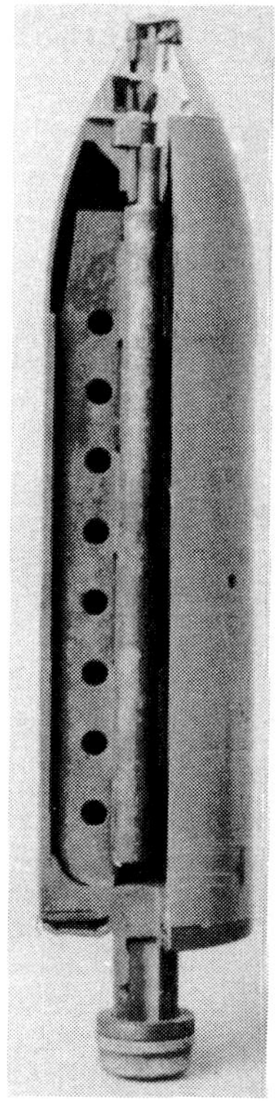

107 mm M328A1 white phosphorus smoke bomb; the construction, apart from the internal baffles, is standard for the 107 mm range of bombs

Type	Range min	Range max	Weight	Effective area	Filler and designation	Notes
HE M329A1	920 m	5,650 m	12.26 kg	40 × 20 m	TNT	–
HE M329A2	770 m	6,800 m	10 kg	–	–	–
HE M34A1	870 m	4,620 m	12.21 kg	40 × 20 m	TNT	–
Smoke M328A1	920 m	5,650 m	12.98 kg	n/a	WP	–
Smoke M2	870 m	4,620 m	11.32 kg	n/a	WP	burning time 90 s
Illumination M335A2	400 m	5,490 m	12.09 kg	1,500 m dia	–	rate of descent 5 m/s
Gas M2A1	870 m	4,540 m	11.17 kg	n/a	HD	withdrawn
Tactical CS XM630	1,540 m	5,650 m	11.64 kg	n/a	CS	–

VERIFIED

Boeing Fiber Optic Mortar Projectile (FOMP)/Infantry Precision Attack Weapon (IPAW)

Description

FOMP is a 120 mm mortar projectile and IPAW is an 81 mm projectile, both of which are fired from a normal mortar in the usual manner. Shortly after launch wings and fins are deployed and as the bomb approaches the vertex of the trajectory, a rocket sustainer motor is ignited. During the flight, the bomb dispenses a fibre optic link which is connected to a control station at the mortar position. The bomb is fitted with a television camera, the view from which is relayed down the fibre optic link to be displayed at the control station. The operator, furnished with the bomb's picture, can then select a target and steer the bomb to impact.

The project has reached the stage where feasibility of the various components has been demonstrated and actual fibre optic links have survived live firing.

Manufacturer

Boeing Defense and Space Group, PO Box 240002, MS JW-49, Huntsville, Alabama 35824-6402.

Status

Development.

VERIFIED

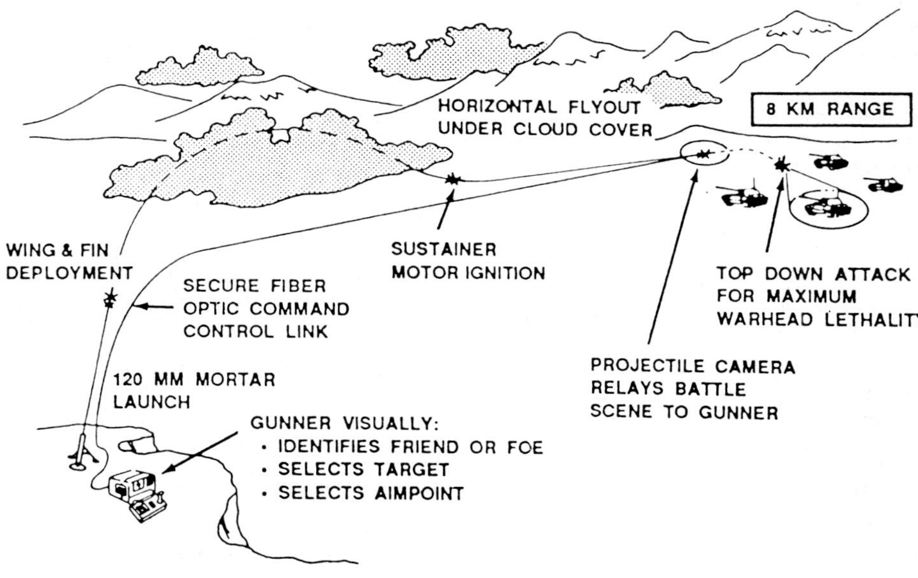

Operational principle of FOMP/IPAW

YUGOSLAVIA (SERBIA AND MONTENEGRO)

Ammunition for 50 mm mortars

Description

The HE bomb M82 is a modern design of streamlined shape. The bursting charge consists of TNT and the bomb is fitted with the direct action impact fuze TK-135. The propelling charge is a single 5 g primary cartridge which is sufficient to send the 950 g bomb to a range of 1,000 m.

Holding Company

YUGOIMPORT - SDPR, PO Box 89, Bulevar umetnosti 2, YU-11070 Novi, Belgrade.

Status

Current production status uncertain.

Service

Yugoslav forces.

VERIFIED

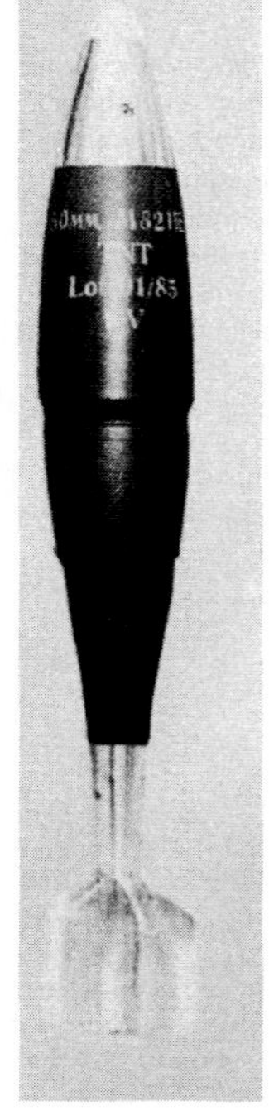

50 mm HE mortar bomb M82

Ammunition for 60 mm mortars

Description

60 mm HE bomb M73

This is a conventional teardrop bomb with a tail unit around which the supplementary charges are fitted in horseshoe form celluloid containers. The TNT-filled bomb is fitted with the direct action impact fuze UT M68P1. A primary cartridge fits into the tail tube to form Charge 0 and four supplementary charges can be fitted, except when using Commando type light mortars, when only two supplementaries are permitted. Each supplementary charge contains about 4.2 g of nitrocellulose powder.

This bomb is produced in Macedonia, by Euroinvest, and in Croatia where the marketing agency is RH-ALAN.

Length: 286 mm
Weight: 1.35 kg
Filling: 220 g TNT
Muzzle safety distance: 8 m

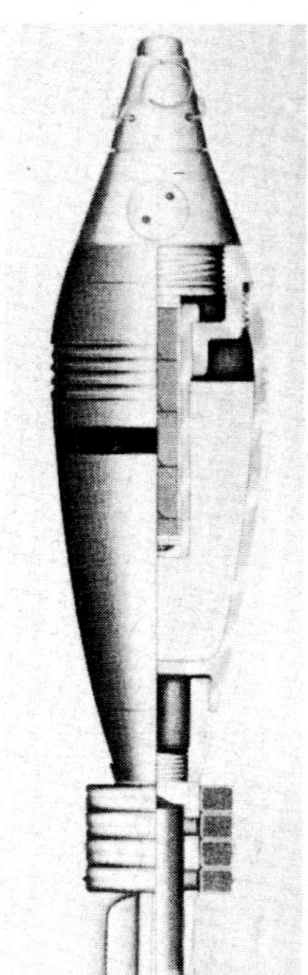

60 mm HE mortar bomb M73

Performance:

Charge	Muzzle velocity	Min range	Max range
0	74 m/s	94 m	523 m
1	111 m/s	184 m	1,072 m
2	143 m/s	283 m	1,632 m
3	170 m/s	365 m	2,136 m
4	193 m/s	440 m	2,538 m

60 mm smoke bomb M73

The M73 smoke bomb is of the white phosphorus bursting type, using a central exploder container and fitted with the direct action impact fuze UT M70P1. It is ballistically matched to the HE bomb M73 and uses the same propelling charges to obtain the same ranges.

This bomb is produced in Macedonia by Euroinvest.

Weight: 1.35 kg
Filling: 190 g white phosphorus
Muzzle safety distance: 8 m

60 mm illuminating bomb M67

The M67 illuminating bomb contains the usual type of magnesium flare unit suspended by a parachute. The bomb is fuzed with the pyrotechnic time fuze M67 which, at the end of the set time, ignites an expelling charge and ejects the parachute and flare unit from the nose of the bomb. The M67 bomb cannot be fired with the Charge 0 (primary cartridge only) and is restricted to Charges 1 and 2 when fired from light Commando type mortars.

This bomb is producted in Macedonia by Euroinvest.

Weight: 1.27 kg
Illumination intensity: 180,000 cd
Illuminated area: diameter, 500 m
Illumination duration: 35 s
Mean height of parachute opening: 180 m
Rate of descent: 2.5 m/s
Performance:

Charge	Muzzle velocity	Min range	Max range
1	117 m/s	200 m	950 m
2	154 m/s	300 m	1,600 m
3	185 m/s		2,100 m
4	210 m/s		2,450 m

60 mm HE bomb M90

This is a streamlined bomb similar to the 50 mm M82 described previously, having a light alloy tail boom and a single obturating ring. The tail boom contains a primer and an ignition cartridge, and up to six secondary increments in horseshoe containers can be attached ahead of the fins.

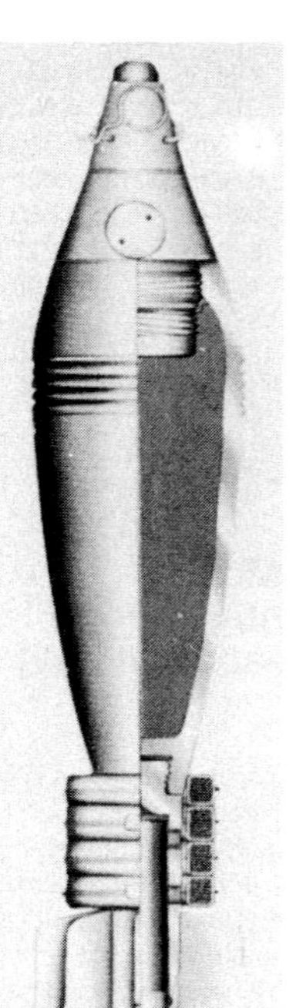

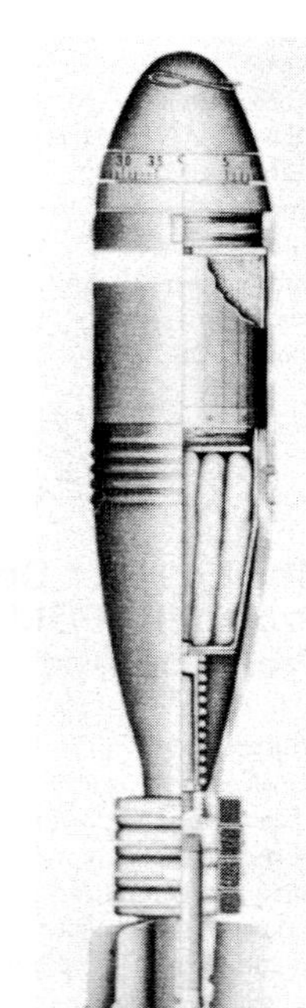

60 mm WP smoke bomb M73 (left)
60 mm illuminating bomb M67

Length: overall, 400 mm
Weight: firing order, 2.1 kg
Filling: 400 g TNT
Fuze: super-quick; muzzle safe to 14 m
Range: 150-5,208 m

60 mm WP smoke bomb M90

The M90 smoke bomb uses the same body as the HE bomb M90 but is filled with white phosphorus and has a central burster initiated by the SQ fuze. Apart from the filling (270 g WP) the dimensions, weight and ballistic performance are exactly the same as for the HE bomb M90.

Holding Company

YUGOIMPORT - SDPR, PO Box 89, Bulevar umetnosti 2, YU-11070 Novi, Belgrade.

Status

Current production status within Yugoslavia uncertain.

Service

Yugoslav forces.

UPDATED

Ammunition for 81 mm mortars

Description

81 mm HE bomb M72

This bomb might well be called a transitional design, since it tends towards the older teardrop shape but with a tapered ogive, giving it a more symmetrical look, and with a single gas-check groove around the waist. It is fitted with the direct action impact fuze UT M68P1, and the propelling charge consists of a primary cartridge and six combinations of 8 and 13 g secondary increments which clip around the tail boom.

Weight: total, 3.05 kg
Filling: 680 g TNT
Muzzle safety distance: 8 m
Performance:

Charge	Muzzle velocity	Min range	Max range
0	72 m/s	88 m	513 m
1	120 m/s	230 m	1,302 m
2	167 m/s	380 m	2,135 m
3	197 m/s	505 m	2,932 m
4	229 m/s	650 m	3,685 m
5	257 m/s	737 m	4,307 m
6	283 m/s	858 m	4,876 m

81 mm HE bomb M72P1

This is similar to the M72 bomb described above but is fitted as standard with the direct action and delay fuze UTU M67. Data and performance are exactly the same as those for the M72.

81 mm HE bomb M71

The M71 bomb is of modern design, being symmetrically streamlined and with a plastic obturating ring fitted into a gas-check groove at the waist. The bomb is fitted with the direct action impact fuze UT M68P1, and the propelling charge consists of a primary cartridge (Charge 0) and eight combinations of 8 or 14 g supplementary increments which fit around the tail boom to give Charges 1 to 8.

Weight: total, 4.1 kg
Filling: 690 g TNT
Muzzle safety distance: 8 m
Performance:

Charge	Muzzle velocity	Min range	Max range
0	73 m/s	90 m	520 m
1	109 m/s	190 m	1,100 m
2	143 m/s	350 m	1,790 m
3	174 m/s	700 m	2,490 m
4	204 m/s	1,200 m	3,210 m
5	230 m/s	1,500 m	3,875 m
6	256 m/s	1,900 m	4,500 m
7	277 m/s	2,300 m	5,000 m
8	297 m/s	2,600 m	5,426 m

These values apply when fired from a barrel of 1.15-1.2 m length.

81 mm HE bomb M84

This is to the same general design as the M72 bomb described above, but is lighter and therefore reaches to a greater range. It is fitted with the direct action impact fuze UT M84P1 and the propelling charge consists of a primary cartridge (Charge 0) and six combinations of 8 and 25 g secondary increments giving Charges 1 to 6. The increased bursting charge is of RDX/TNT (Composition B), giving the bomb a better lethality than previous designs.

Weight: total, 3.8 kg
Filling: 890 g TNT
Muzzle safety distance: 8 m
Performance:

Charge	Muzzle velocity	Max range
1	122 m/s	2,320 m
2	169 m/s	3,210 m
3	212 m/s	4,030 m
4	255 m/s	4,800 m
5	292 m/s	5,500 m
6	318 m/s	6,050 m

These figures apply when fired from a barrel of 1.15-1.2 m length.

81 mm WP smoke bomb M72

The M72 smoke bomb is ballistically matched to the M72 HE bomb and delivers the same range performance with the same charges. The bomb is charged with white phosphorus and has a central HE burster, initiated by a direct action impact fuze UT M70P1. Except for the weight of the smoke charge (600 g), the data and performance figures are exactly as for the HE M72 bomb.

81 mm WP smoke bomb M71

This is a ballistic match for the HE bomb M72 described above, data figures and performance being exactly the same except for the weight of white phosphorus of 630 g. The bomb is fuzed with the direct action impact fuze UT M70P1 which detonates a central high explosive burster to open the bomb and disperse the WP filling.

81 mm illuminating bomb M67

The M67 illuminating bomb is of conventional nose ejection pattern, being fitted with the pyrotechnic time fuze M67. This ignites an expelling charge after the set time, removing the nose of the bomb and causing the parachute and flare unit to be ejected. The propelling charge consists of a primary cartridge and four combinations of 7.6 and 13.8 g secondary increments to give Charges 1 to 4. The bomb cannot be fired with the primary cartridge alone.

Weight: total, 2.95 kg
Illumination intensity: 500,000 cd
Illuminated area: diameter, 800 m
Illumination duration: 40 s
Mean height of parachute opening: 270 m
Rate of descent: 2.4 m/s
Performance:

Charge	Muzzle velocity	Min range	Max range
2	175 m/s	600 m	2,050 m
3	212 m/s	1,400 m	2,770 m
4	244 m/s	2,100 m	3,380 m

These values apply when fired from a 1.15-1.2 m barrel; when fired from a 1.45 m barrel the maximum range with Charge 4 is 3,645 m.

Holding Company

YUGOIMPORT - SDPR, PO Box 89, Bulevar umetnosti 2, YU-11070 Novi, Belgrade.

Status

Current production status uncertain.

Service

Yugoslav forces.

UPDATED

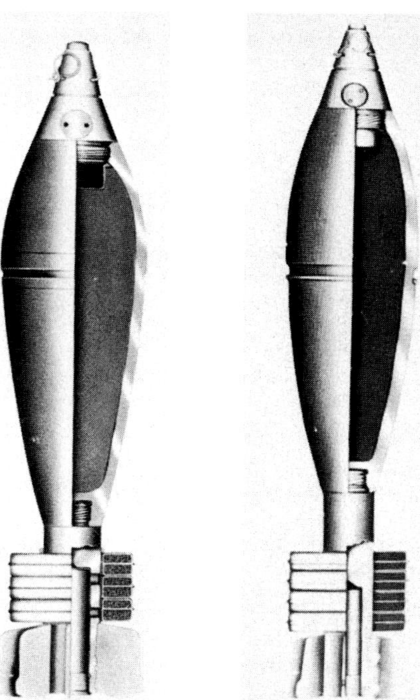

81 mm HE bomb M72 (left)
81 mm HE bomb M71

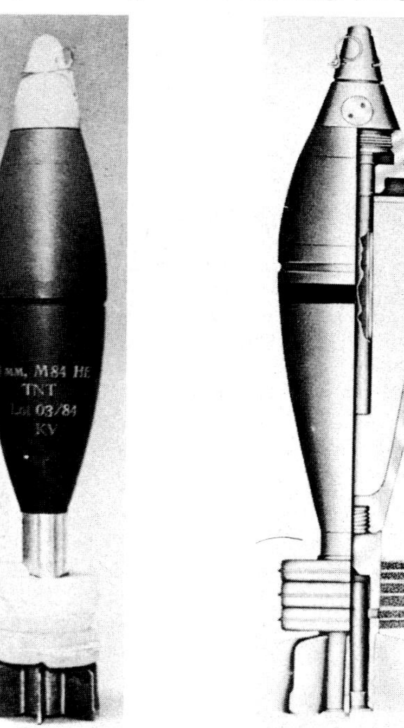

81 mm HE bomb M84 (left)
81 mm WP smoke bomb M72

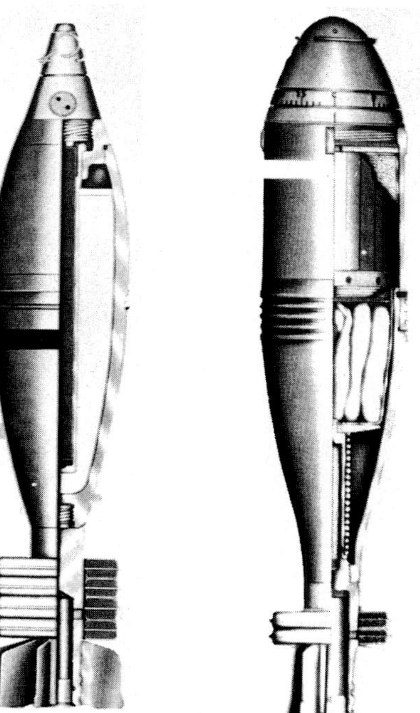

81 mm WP smoke bomb M71 (left)
81 mm illuminating bomb M67

Ammunition for 82 mm mortars

Description

The range of bombs for use in 82 mm mortars is much the same as those described above for the 81 mm mortar, but there are some small differences in nomenclature. Thus, the 82 mm HE bomb M74 is the same as the 81 mm HE M72, but the maximum range is 4,943 m, there being small differences in the ranges achieved with each charge. The 82 mm HE bombs M72P1, M71 and M84 are exactly the same as the similarly numbered 81 mm bombs and have the same data and performance. The 82 mm smoke bomb M74 is similar to the 81 mm smoke bomb M72 but has a maximum range of 4,943 m. The 82 mm smoke bomb M71 and illuminating bomb M67 are the same as the corresponding 81 mm bomb and have the same data and performance.

Some of these bombs are in production in Macedonia, by Euroinvest, and in Croatia where the marketing agency is RH-ALAN.

Holdiing Company
YUGOIMPORT - SDPR, PO Box 89, Bulevar umetnosti 2, YU-11070 Novi, Belgrade.

Status
Current production status within Yugoslavia uncertain.

Service
Yugoslav forces.

UPDATED

Ammunition for 120 mm mortars

Description

120 mm HE bomb, light, M62P1

This is a conventional teardrop bomb with multiple gas-check rings at the waist, and is filled with TNT. It is fitted as standard with a direct action and delay impact fuze and the propelling charge consists of a primary cartridge (not fired alone) and six combinations of 35 and 75 g secondary increments providing Charges 1 to 6.

An essential similar bomb, the 120 mm M62P3 with a 2.35 kg TNT filling, is produced in Croatia and marketed by RH-ALAN.

Weight: total, 12.6 kg
Filling: 2.25 kg TNT
Muzzle safety distance: 10 m
Performance:

Charge	Muzzle velocity	Min range	Max range
1	121 m/s	400 m	1,400 m
2	162 m/s	800 m	2,360 m
3	200 m/s	1,500 m	3,370 m
4	236 m/s	2,000 m	4,400 m
5	267 m/s	2,500 m	5,280 m
6	297 m/s	3,000 m	6,050 m

120 mm rocket-assisted HE bomb M77

This is a cylindrical bomb with an ogival head and a long tail boom which has a set of four folding fins ahead of the charge container. The bomb body is divided into two sections, the forward section containing the HE bursting charge and the rear section containing the rocket motor. A pre-firing adjustment permits the rocket motor to be switched in or out of action as required. The propelling charge consists of the usual primary cartridge (not fired alone) and six 71 g secondary increments giving Charges 1 to 6.

The rocket motor is ignited by means of a pyrotechnic delay train from the primary cartridge and the initial ignition of the rocket motor blows away the cartridge container. This permits the rocket efflux to escape; the forward section of the tail boom remains in place to support the fins which unfold into the slipstream as the bomb leaves the muzzle. These fins are skewed, so as to develop a degree of spin stabilisation during flight.

120 mm illuminating bomb M84

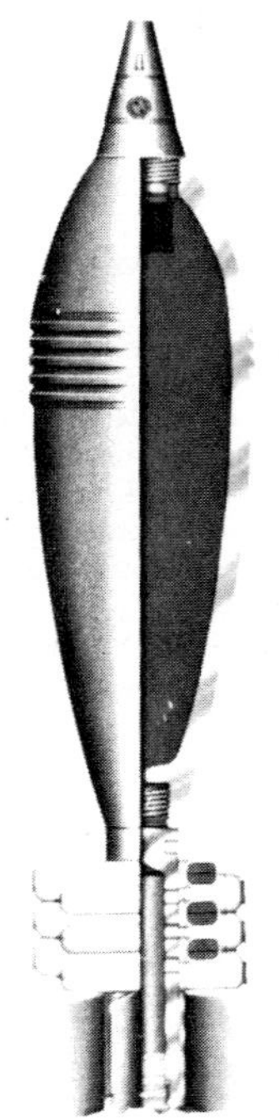
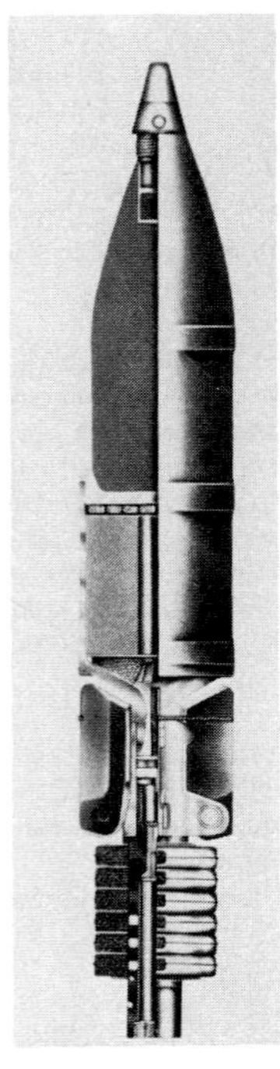
120 mm HE bomb M62P1 (left)
120 mm rocket-assisted HE bomb M77

Weight: total, 13.65 kg
Filling: 2.91 kg HE
Performance:

Charge	Muzzle velocity	Min range	Max range
without rocket assistance:			
1	133 m/s	300 m	1,500 m
2	180 m/s	800 m	2,500 m
3	217 m/s	1,600 m	3,400 m
4	251 m/s	2,600 m	4,100 m
5	280 m/s	3,500 m	4,700 m
6	307 m/s	4,200 m	5,300 m
with rocket assistance:			
3	217 m/s	5,100 m	7,900 m
4	251 m/s	5,100 m	8,500 m
5	280 m/s	5,100 m	8,900 m
6	307 m/s	5,100 m	9,400 m

120 mm smoke bomb M64P1

This is a conventional white phosphorus bursting smoke bomb, fitted with the direct-action impact fuze UT M70P1 and a central HE burster. The propelling charge consists of a primary cartridge and up to six 76 g secondary incremental charges clipped around the tail boom, giving Charges 1 to 6.

This bomb is produced in Macedonia by Euroinvest.

Weight: total, 12.4 kg
Filling: 2.45 kg WP
Muzzle safety distance: 8 m
Performance:

Charge	Muzzle velocity	Min range	Max range
1	123 m/s	255 m	1,410 m
2	165 m/s	435 m	2,375 m
3	212 m/s	625 m	3,400 m
4	240 m/s	810 m	4,400 m
5	271 m/s	970 m	5,250 m
6	302 m/s	1,100 m	6,010 m

These figures apply when fired from the 120 mm mortar M52; when fired from the 120 mm mortar M75 the maximum range is 6,464 m.

120 mm smoke bomb M84

This is a nose ejection bomb using a single container of hexachloroethane-based persistent smoke mixture.

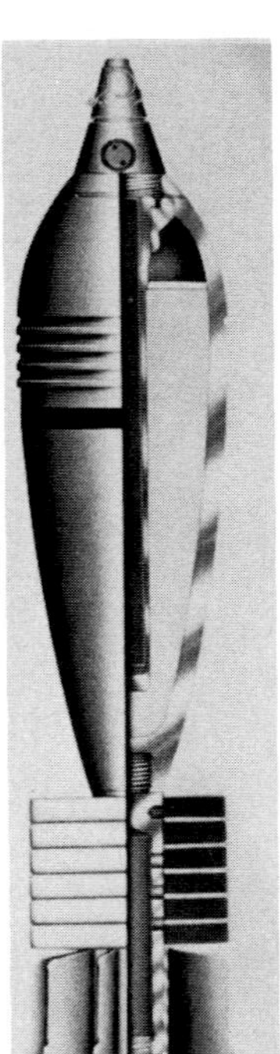

120 mm WP smoke bomb M64P1 (left)
120 mm smoke bomb M84

The bomb is fitted with the pyrotechnic time fuze M84 which, at the set time, ignites an expelling charge which blows off the bomb nose and ejects the ignited smoke canister. The propelling charge consists of a primary cartridge and up to five 76 g secondary increments in silk bags which are tied around the tail boom, giving Charges 1 to 5.

Weight: total, 10.35 kg
Filling: 1.2 kg hexachloroethane-based mix
Duration of smoke emission: >3 min
Performance (in mortar with 1.5 m barrel):

Charge	Muzzle velocity	Min range	Max range
1	137 m/s	230 m	1,260 m
2	185 m/s	900 m	2,560 m
3	229 m/s	1,300 m	3,850 m
4	266 m/s	1,700 m	4,850 m
5	303 m/s	2,000 m	5,850 m

120 mm illuminating bomb M84

As the type number suggests, this is almost identical to the screening smoke bomb, but contains a parachute and flare assembly. The bomb is fitted with the pyrotechnic time fuze M84 which, at the set time, blows off the nose and ejects the burning flare and parachute. The propelling charge consists of a primary and five secondary incremental charges in silk bags.

Weight: total, 10.35 kg
Illumination intensity: >900,000 cd
Area of illumination: diameter, 1,800 m^2
Rate of descent: 3 m/s
Performance: as for smoke bomb M84 above

Holding Company
YUGOIMPORT - SDPR, PO Box 89, Bulevar umetnosti 2, YU-11070 Novi, Belgrade.

Status
Current production status within Yugoslavia uncertain.

Service
Yugoslav forces.

UPDATED

Impact fuze UT M68P1

Description

This is a bore-safe direct-action fuze intended for use in unspun mortar bombs from 60 to 82 mm calibre. Muzzle safety is ensured by means of a spring-driven axial rotor which is locked by the firing pin in the safe condition. On firing, setback releases the firing pin, allowing the rotor to turn and slide until it aligns its detonator with the armed firing pin.

This bomb is produced in Croatia where the marketing agency is RH-ALAN.

Data

Length: 85 mm
Weight: 175 g
Drop safety height: 3 m

Holding Company

YUGOIMPORT - SDPR, PO Box 89, Bulevar umetnosti 2, YU-11070 Novi, Belgrade.

Status

Current production status within Yugoslavia uncertain.

Service

Yugoslav forces.

UPDATED

Impact fuze UT M68P1

Impact fuze UTU M67

Description

The M67 fuze is an impact, super-quick and delay pattern for use with 81 mm and 82 mm unspun mortar bombs. Bore safety is achieved by a rotor, locked in the safe condition by the firing pin. The fuze may be set before firing for delay or instant action; doing so sets or releases a block on the movement of the sliding rotor. In the instant setting, the firing pin is released by setback upon firing, after which the rotor moves across to align a detonator with the firing pin. On striking the target the pin is driven in, ignites the detonator, this in turn fires the fuze magazine and the bursting charge of the bomb. At delay setting, the rotor moves so as to bring a second detonator, with pyrotechnic delay, under the firing pin. When the fuze strikes, the action is as before but the pyrotechnic delay has to burn through before the fuze magazine can be initiated.

Data

Length: 91 mm
Weight: 285 g
Drop safety height: 3 m

Holding Company

YUGOIMPORT - SDPR, PO Box 89, Bulevar umetnosti 2, YU-11070 Novi, Belgrade.

Status

Current production status uncertain.

Service

Yugoslav forces.

VERIFIED

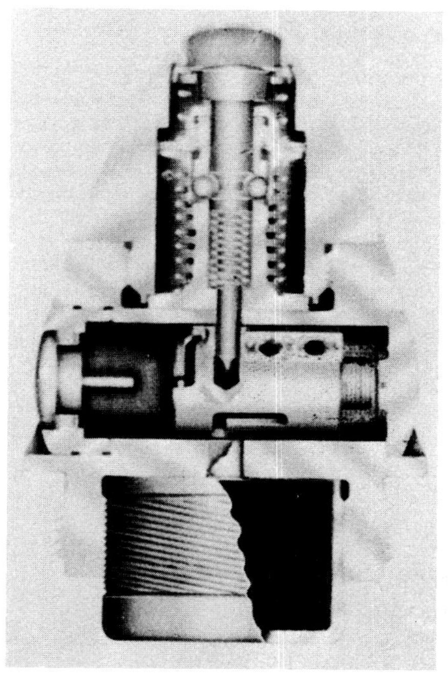

Impact, SQ and delay fuze UTU M67

Impact fuze UTU M78 (AU-29)

Description

This fuze is for use in the 120 mm rocket-assisted bomb, where spin is available to assist in functioning. It uses a similar method of obtaining delay to the US M48 series and has a Semple rotor locked by a centrifugal detent to provide bore safety. The fuze can be set, before loading, for delay or non-delay; this closes or opens a central channel between an impact detonator and the rotor detonator. If this channel is open, the flash from the impact detonator, struck by the impact of the firing pin, passes through to fire the rotor detonator and the fuze magazine. If the channel is closed, then the flash from the impact detonator is channelled via a delay unit to a delay detonator which then fires the rotor detonator.

This bomb is produced in Croatia where the marketing agency is RH-ALAN.

Data

Length: 105 mm
Weight: 430 g
Muzzle safety distance: min, 17 m

Holding Company

YUGOIMPORT - SDPR, PO Box 89, Bulevar umetnosti 2, YU-11070 Novi, Belgrade.

Status

Current production status within Yugoslavia uncertain.

Service

Yugoslav forces.

UPDATED

Impact, SQ and delay fuze UTU M78

Impact fuze UT M70P1

Description

This is precisely the same fuze as the M68P1 described previously except that the magazine and lower portion of the fuze body are of differing dimensions so as to suit the internal arrangements of WP smoke bombs. It is for use with 60, 81 and 82 mm mortars. The fuze weighs 160 g and is 78 mm long.

Holding Company

YUGOIMPORT - SDPR, PO Box 89, Bulevar umetnosti 2, YU-11070 Novi, Belgrade.

Status

Current production status uncertain.

Service

Yugoslav forces.

VERIFIED

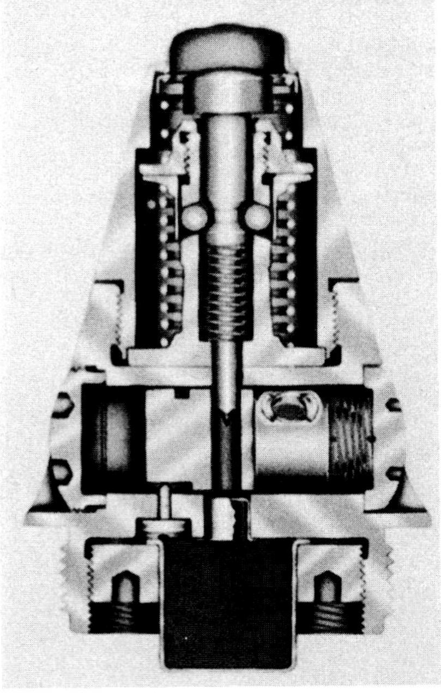

Impact fuze M70P1 for WP smoke bombs

Fuze, Time, TP M67

Description

This is a single-banked, tensioned, igniferous, time fuze for use in nose ejection bombs in 81 and 82 mm mortars. The single time ring offers timing from 5 to 38 seconds and is filled with a zirconium-based powder. A central hammer, suspended on a shear-wire, drops on setback to strike a detonator which lights the time composition. This burns round until it fires a flash channel leading to the fuze magazine, which then explodes and initiates the bomb's functioning.

Data

Length: 85 mm
Weight: 580 g

Holding Company

YUGOIMPORT - SDPR, PO Box 89, Bulevar umetnosti 2, YU-11070 Novi, Belgrade.

Status

Current production status uncertain.

Service

Yugoslav forces.

VERIFIED

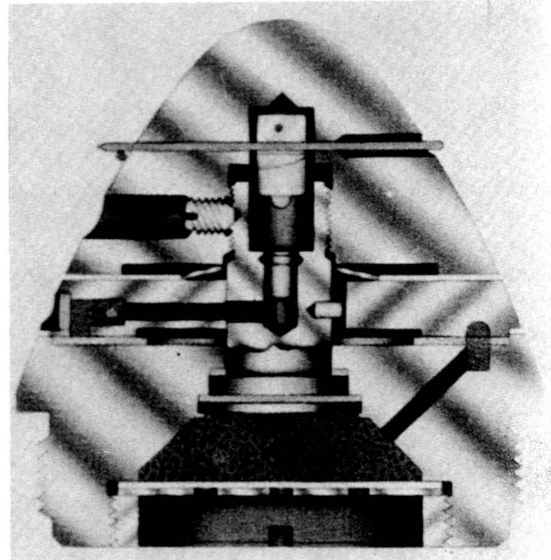

Time fuze TP M67

Fuze, Time, TP M66

Description

This is identical in form to the M67 fuze described previously but is larger and offers timing from 5 to 50 seconds, being intended for use in 120 mm illuminating bombs. The fuze weighs 1.265 kg and is 108 mm long.

Holding Company

YUGOIMPORT - SDPR, PO Box 89, Bulevar umetnosti 2, YU-11070 Novi, Belgrade.

Status

Current production status uncertain.

Service

Yugoslav forces.

VERIFIED

PYROTECHNICS

AUSTRIA

DNW defence smoke cartridge

Description

This is a screening smoke cartridge which can be fired from ground- or vehicle-mounted electrically fired projectors. It is designed so that the smoke rises, concealing the site from aerial or superior observation while leaving the fields of view and fire clear close to the ground. The cartridge is a tinned-plate canister with propulsion charge and a smoke loading of about 1 kg of composition. Electrical ignition fires the propulsion charge to eject the canister to a range of approximately 25 m from the launcher and ignites a pyrotechnic delay which in turn ignites the smoke composition. Smoke emission begins during flight and continues for about 40 seconds.

Data

Diameter: 80 mm
Length: 180 mm
Weight: 2.3 kg
Range: 25 ± 5 m
Duration: 40 ± 10 s

Manufacturer

Dynamit Nobel Wien GesmbH, Postfach 74, Opernring 3-5, A-1015 Wien.

Status

In production.

VERIFIED

DNW defence smoke cartridge

DNW smoke cartridge HC-72

Description

The HC-72 is a vehicle smoke screening device intended to be electrically fired from the Discharger Equipment 69. It consists of a metal canister containing the hexachloroethane/zinc smoke composition in the front end and a propellant charge and electrical igniter in the rear. After loading, a protection cover closes the discharger protecting the cartridge against the weather. To ensure safety there are two parallel firing circuits both of which are safe against stray currents up to 360 mA. On firing, the grenade is projected to a distance of about 50 m and the smoke composition is ignited. In approximately five to seven seconds after firing the smoke reaches full development.

Data

Diameter: 80 mm
Length: 322 mm
Weight: 5 kg
Duration of smoke: ca 70 s
Colour of smoke: grey-white

Manufacturer

Dynamit Nobel Wien GesmbH, Postfach 74, Opernring 3-5, A-1015 Wien.

Status

In production.

VERIFIED

Smoke cartridge HC-72

DNW smoke generator HC-81

Description

The smoke generator HC-81 is designed for the production of screening smoke for periods which can be as long as required. This variable time is achieved by making the generators with mating surfaces so that two or more may be screwed together to make up whatever size is desired. Each individual generator contains sufficient HC composition to burn for about five minutes and where another generator has been connected, transfer of the burning is automatically carried out by a connector which is permanently installed in the base of the generator. It is even possible to add further generators to a group already ignited, provided the last generator of the group has not yet begun to burn.

Ignition is by friction or electric igniter, two of each being provided with each generator.

Data

Diameter: 182 mm
Height: 140 mm
Weight: 4.5 kg
Igniter delay: nominal, 4.5 s
Duration: 5 min

Manufacturer

Dynamit Nobel Wien GesmbH, Postfach 74, Opernring 3-5, A-1015 Wien.

Status

In production.

DNW smoke generators HC-81

Service

Austrian Army.

VERIFIED

DNW smoke generator HC-75

Description

The HC-75 is a large generator developing screening smoke for periods of up to 22 minutes. It is loaded with HC mixture and can be ignited by either a friction or an electric igniter, one of each being supplied with each generator. Emission of smoke is intense and this equipment should not be used in confined spaces; there is also a certain amount of spark emission which should be taken into account when sitting in dry grass areas in training.

Data

Weight: 10.5 kg
Duration: 18-22 min
Ignition delay: friction igniter, nominal, 9 s; electric igniter, nominal, 4.5 s

Manufacturer

Dynamit Nobel Wien GesmbH, Postfach 74, Opernring 3-5, A-1015 Wien.

Status

In production.

Service

Austrian Army.

VERIFIED

Smoke generator HC-75

DNW illuminating mine LK-40

Description

The LK-40 is a trip-operated flare unit. The aluminium body is closed by a cap threaded to accept the Universal Fuze DDZ78 which functions by pull, pressure or cutting of the tripwires. Once the fuze fires, the ignition of the flare composition blows off the top cap so that the burning mass is exposed, illuminating the surrounding area.

The mine is packed with a mounting support, Fuze DDZ78, two rolls of tripwire, six pickets and two nails. A hammer is provided with every ten mines.

Data

Diameter: 55 mm
Height: without fuze, 120 mm
Weight: 450 g; flare composition, 350 g
Duration of flare: nominal, 40 s
Luminous power: 280,000 cd
Colour of flare: yellow or white

Manufacturer

Dynamit Nobel Wien GesmbH, Postfach 74, Opernring 3-5, A-1015 Wien.

Status

In production.

Service

Austrian Army.

VERIFIED

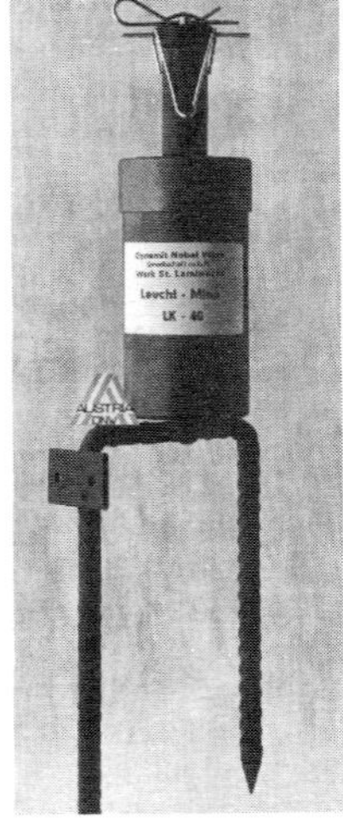

DNW illuminating mine LK-40

BRAZIL

Condor SS-601 coloured smoke hand grenade

Description

The Condor SS-601 grenade consists of an aluminium body with an opening in the cap through which the smoke can escape. A removable plastic cap, which is of the same colour as the emitted smoke, conceals the friction igniter. To operate, the plastic cap is removed, exposing a loop of nylon cord; pulling this cord activates the friction igniter, after which the grenade is thrown.

Data

Diameter: 68.5 mm
Length: 149 mm
Weight: depending upon colour, 260-400 g
Delay time: 2-6 s
Emission time: 60 s
Colours: red, orange, blue, yellow, green, white

Manufacturer

Condor SA - Industria Quimica, Rua Armando Dias Pereira 160, Adrianopolis, CEP 26260, Nova Iguacu, Rio de Janeiro.

Status

In production.

VERIFIED

SS-601 coloured smoke grenades

Condor SS-602 floating smoke generator

Description

The SS-602 is a long-lasting generator for maritime use and releases a dense cloud of orange smoke for about three minutes. A plastic lid covers the pull-ring and fuze assembly. Each smoke generator is protected by a plastic bag.

Data

Diameter: 77 mm
Length: 160 mm
Weight: charge, 560 g
Delay: 1.5-5.5 s
Emission time: 3 min

Manufacturer

Condor SA - Industria Quimica, Rua Armando Dias Pereira 160, Adrianopolis, CEP 26260, Nova Iguacu, Rio de Janeiro.

Status

In production.

VERIFIED

SS-602 floating smoke generator

Condor SS-603 hand-held signal star

Description

The Condor SS-603 consists of a hand-held launching tube containing a propelling charge which fires into the air a white, green or red bright star; the colour of the star is indicated by the colour of the tube. Firing is performed by grasping the lower end, removing the red cap, and pulling the cord which operates a traction-type plastic fuze. The signal device fires immediately.

Data

Diameter: 33 mm
Length: 207 mm
Weight: 140 g
Burning time: 5-7 s
Power: 20,000 cd
Altitude: 80 ± 10 m

Manufacturer

Condor SA - Industria Quimica, Rua Armando Dias Pereira 160, Adrianopolis, CEP 26260, Nova Iguacu, Rio de Janeiro.

Status

In production.

VERIFIED

Condor SS-603 hand-held signal star

Condor SS-604 hand-held signal flare

Description

The Condor SS-604 is a simple non-ejecting flare for signal or distress use. It consists of a plastic body containing a pyrotechnic composition which is ignited by a traction-type plastic fuze in the handle. Once ignited, the device produces a high intensity red flare for about one minute.

Data

Diameter: 44 mm
Length: 242 mm
Weight: 580 g
Power: 15,000 cd

Manufacturer

Condor SA - Industria Quimica, Rua Armando Dias Pereira 160, Adrianopolis, CEP 26260, Nova Iguacu, Rio de Janeiro.

Status

In production.

VERIFIED

Condor SS-604 hand-held signal flare

Condor SS-605 day and night signal

Description

The Condor SS-605 combines day and night signal functions in a single device. It consists of two plastic tubes connected together by a red tape, each having individual signal characteristics. Each unit has a traction-type plastic fuze incorporated. Upon being separated and fired, the day signal emits a dense cloud of orange smoke, while the night signal produces a high intensity red flare. After separation the unused flare can be retained for later use.

Data

Diameter: 44 mm
Length: 252 mm
Weight: 513 g
Burning time: either unit, 25 s
Power of night flare: 15,000 cd

Manufacturer

Condor SA - Industria Quimica, Rua Armando Dias Pereira 160, Adrianopolis, CEP 26260, Nova Iguacu, Rio de Janeiro.

Status

In production.

VERIFIED

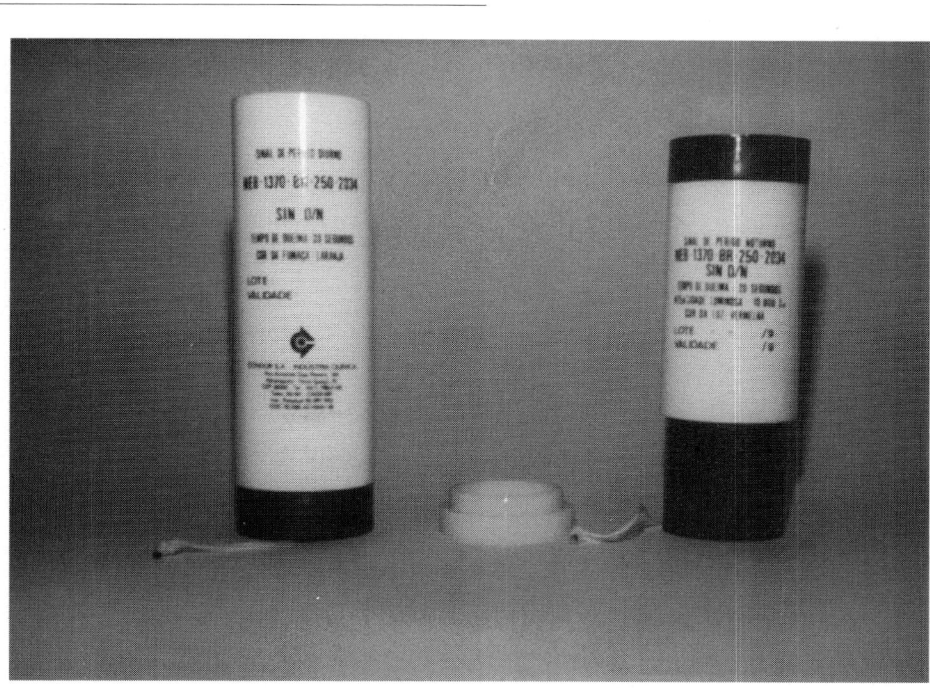

Condor SS-605 day and night signal separated

Condor SS-606 hand-held parachute flare

Description
The Condor SS-606 is a hand-launched parachute flare fired by a traction-type plastic fuze concealed in the base of the plastic launch tube. The igniter has a delay of about two seconds, sufficient to allow the firer to pull the ignition cord and then grasp the launch tube with both hands before the rocket is fired.

Data
Diameter: 44 mm
Length: 325 mm
Weight: 490 g
Burning time: min, 40 s
Power: 30,000 cd
Altitude: min, 300 m

Manufacturer
Condor SA - Industria Quimica, Rua Armando Dias Pereira 160, Adrianopolis, CEP 26260, Nova Iguacu, Rio de Janeiro.

Status
In production.

VERIFIED

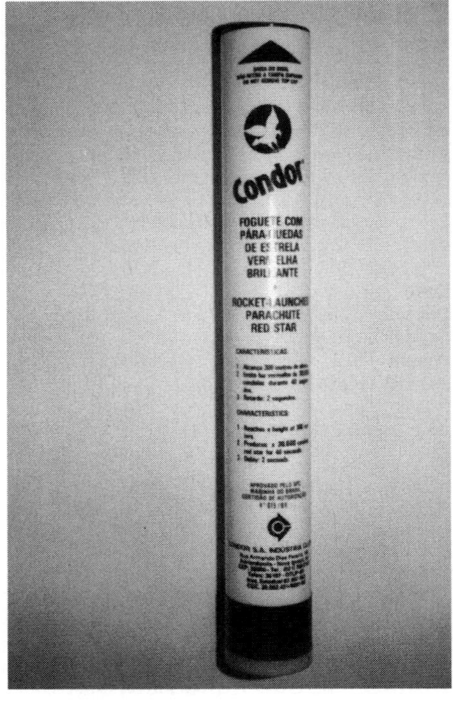

SS-606 hand-held parachute flare

Condor SS-607 hand-held five-star rocket

Description
The Condor SS-607 consists of an aluminium launching tube containing a rocket carrying five stars of a specified colour, which is indicated by the colour of the front closing cap. The rocket is fired by a traction-type plastic fuze concealed under the rear cap. The rocket bursts at the vertex of the trajectory to display the five individual stars.

Data
Diameter: 44 mm
Length: 324 ± 5 mm
Weight: 490 ± 20 g
Firing delay: 1-3 s
Star burning time: 5 s
Power: 15,000 cd
Altitude: min, 300 m
Colours: red, green, white

Manufacturer
Condor SA - Industria Quimica, Rua Armando Dias Pereira 160, Adrianopolis, CEP 26260, Nova Iguacu, Rio de Janeiro.

Status
In production.

VERIFIED

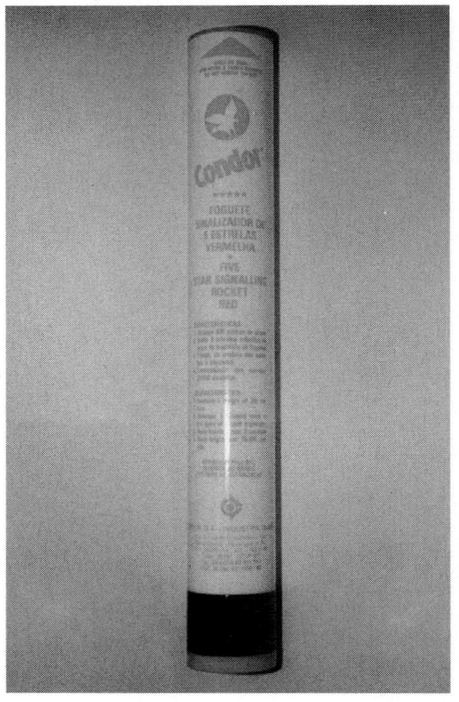

Condor SS-607 hand-held five-star signal rocket

Condor SS-608 40 mm signalling cartridges

Description
The Condor SS-608 is a conventional signal cartridge of the 37/38/40 mm class intended for rescue, survival or military signalling. Available in green, white or red colours.

Data
Diameter: 43 mm
Length: 96 mm
Weight: 135 g
Burning time: 6 s
Power: 15,000 cd
Altitude: 80 m

Manufacturer
Condor SA - Industria Quimica, Rua Armando Dias Pereira 160, Adrianopolis, CEP 26260, Nova Iguacu, Rio de Janeiro.

Status
In production.

VERIFIED

Condor SS-608 40 mm signal cartridge

Condor MB-306/T1 vehicular smoke grenade

Description

The Condor MB-306/T1 is electrically fired from vehicle smoke dischargers. Loaded with a zinc/hexachloroethane mixture, it emits a dense smoke cloud so as to screen the vehicle. Models are available with ranges of 60 or 100 m.

Data

Diameter: 80 mm
Length: 181 mm
Weight: 1.4 kg
Delay time: nominal, 1 s
Emission time: 2-4 min

Manufacturer

Condor SA - Industria Quimica, Rua Armando Dias Pereira 160, Adrianopolis, CEP 26260, Nova Iguacu, Rio de Janeiro.

Status

In production.

VERIFIED

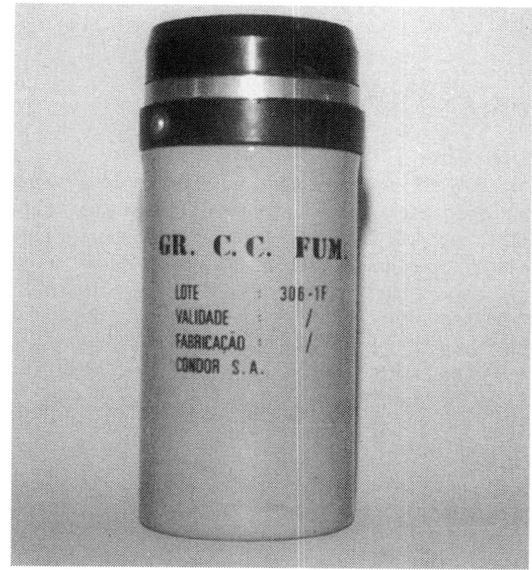

Condor MB-306/T1 vehicular smoke grenade

M5-CEV coloured smoke grenade

Description

The M5-CEV is a general purpose grenade intended mainly for daylight signalling. The coloured smoke is visible for more than 1,000 m in reasonable weather and conditions. Ignition is by a percussion cap and spring striker, held down by a safety lever and pin. It is available in white, red, brick, yellow, blue or green colours.

Data

Diameter: 63 mm
Length: 157 mm
Weight: 600 g
Delay: 4 ± 1.5 s
Emission time: 60 s

Manufacturer

Companhia de Explosivos Valparaiba, Praia do Flamengo, 200, 20° Andar, Rio de Janeiro, 22210 RJ.

Status

In production.

VERIFIED

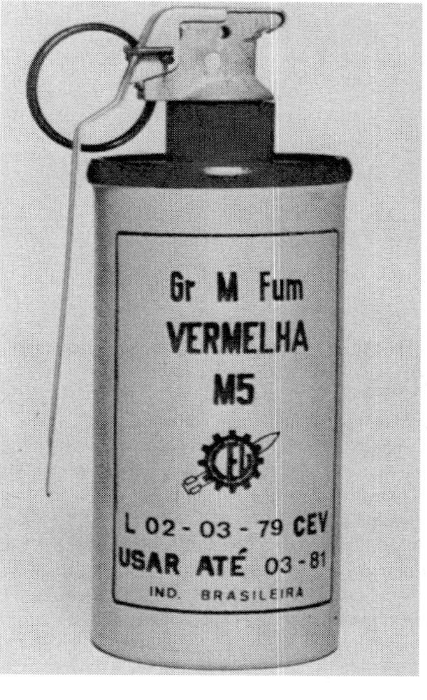

M5-CEV coloured smoke grenade

BULGARIA

26.5 mm signal cartridges

Description

These 26.5 mm signal cartridges are based on the use of cardboard tubes for the cartridge body and metal rimmed bases. They are intended to be fired from any standard 26.5 mm signal pistol and are produced in three basic versions: one-star; two-star; and three-star. All three types can be produced to fire either white, green or red flare stars visible over a distance of 2,000 m during the day and 10,000 m at night. When fired vertically each flare star will be ejected at a height of 90 m and all flare stars will burn for between 6.5 and 9.5 seconds.

Manufacturer

Kintex, 66 James Boucher Street, 1407 Sofia.

Status

In production.

Service

Bulgarian armed forces. Offered for export.

VERIFIED

26.5 mm signal cartridge, three-star

30 mm signal rocket cartridges

Description

These 30 mm signal rocket cartridges are hand held for launching and have cardboard tubes for the cartridge body and metal for the portion held in the hand. They are three basic versions: one-star; two-star; and three-star. All three types can be produced to fire either white, green or red flare stars visible over a distance of 2,000 m during the day and 10,000 m at night. When launched vertically using a striker cap, each flare will be ejected at a height of 110 m. When launched at an angle of 45° each flare star will be ejected at a range of 350 m. The flare stars will burn for between eight and nine seconds.

There is one further 30 mm signal rocket cartridge with an increased rocket-assisted range of 400 m when launched at an angle of 45°; vertical launch height is 150 m. Only white flare stars are available and are visible at a distance of between 2,000 and 2,500 m in daylight and from 10,000 to 15,000 m at night. The hand-held increased range 30 mm signal cartridge is operated using a cord-actuated friction fuze.

Manufacturer

Kintex, 66 James Boucher Street, 1407 Sofia.

Status

In production.

Service

Bulgarian armed forces. Offered for export.

VERIFIED

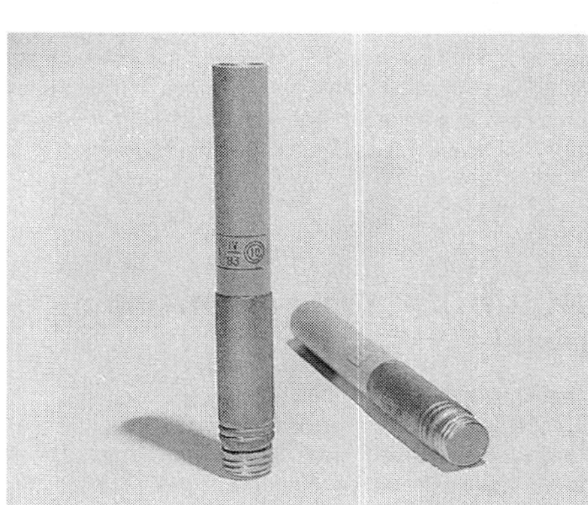

30 mm signal rocket cartridge, one-star

CHILE

FAMAE infiltration alarm

Description

The FAMAE infiltration alarm is a trip-flare which when activated delivers an intense white light for a minimum of 20 seconds, followed by a loud detonation. It consists of an illuminating grenade with its base formed into a socket so that it can be fitted to an upright stake. A trip-wire is then attached to the pull-ring of the igniter safety pin. When the wire is tripped, the safety pin is pulled out and the fly-off lever ignition mechanism acts to fire the flare.

Manufacturer

FAMAE Fabricaciones Militares, Av Pedro Montt 1568/1606, Santiago.

Status

In production.

VERIFIED

FAMAE infiltration alarm

CHINA, PEOPLE'S REPUBLIC

NORINCO 27 mm Type 57 signal pistol

Description

The NORINCO 27 mm Type 57 signal pistol is a single-shot, manually operated pistol for firing the conventional 27 mm pyrotechnic cartridges at optimum vertical angles between 80 and 90° to a height of not less than 100 m. The Type 57 is built of eight assemblies and contains only 30 component parts. The design, which is based on the Soviet SPSh-2 signal pistol, is simple and robust, and the pistol is easy to operate and maintain.

Data

Calibre: 26.65 mm
Length: 220 mm
Weight: 900 g
Rate of fire: possible, 12 rds/min

Manufacturer

China North Industries Corporation (NORINCO), 7A Yue Tan Nan Jie, PO Box 2137, Beijing.

Status

Available.

UPDATED

NORINCO Type 57 signal pistol

NORINCO 27 mm signal cartridges

Description

These signal cartridges are of the conventional 27 mm form, using a brass and plastic case to contain the pyrotechnic loading. They are optimised for use with the NORINCO Type 57 pistol (see previous entry) and are thus fired to a height of not less than 100 m where the signal flares burn for at least 6.5 seconds. At this height they can be detected from a distance of at least 7,000 m.

Data

Calibre: 26.65 mm
Length: 79 mm
Weight: complete round, 55.3 g
Burning time: 6.5 s
Candlepower: ca 50,000 cd
Vertex height: ca 100 m
Colours: red, green, white

Manufacturer

China North Industries Corporation (NORINCO), 7A Yue Tan Nan Jie, PO Box 2137, Beijing.

Status

Available.

UPDATED

NORINCO 11 mm pyrotechnic pistol

Description

The NORINCO 11 mm pyrotechnic pistol is a simple, reusable, single-shot device designed along conventional fountain-pen pyrotechnic pistol lines, being intended for standby use by military and marine personnel. It is cocked by pulling back a button into a detent, thus placing a spring-loaded striker pin under tension. The 11 mm pyrotechnic cartridge is then screwed into the top of the device and the pistol is aimed vertically. The button is pushed from its detent so that the striker pin can hit the base of the cartridge. The pyrotechnic signal or flare is then projected to a maximum height of about 30 m.

Data

Calibre: 11 mm
Length: 116 mm
Weight: 60 g

Range: 30 m
Cartridge height: 22 mm
Cartridge width: 14 mm

Manufacturer
China North Industries Corporation (NORINCO), 7A Yue Tan Nan Jie, PO Box 2137, Beijing.

Status
Available.

NEW ENTRY

NORINCO 11 mm pyrotechnic pistol
1996

COMMONWEALTH OF INDEPENDENT STATES

26.5 mm signal pistols

Description
There are two types of 26.5 mm signal pistols likely to be encountered among former Soviet Union armed forces and others to whom Soviet military aid was extended. Both types employ the same signal cartridges and both require spent cartridge cases to be removed from the barrel manually.

The oldest design of the two is the LP-1, a design influenced by a German pre-1945 Walther signal pistol. The LP-1 has several refinements not always found in signal pistols such as a semi-shrouded hammer and a prominent barrel latching catch in front of the trigger guard.

By contrast the SPSh-2 signal pistol is far simpler and more suited to mass production techniques. The trigger is a semi-recessed stamping while the external hammer profile is the minimum necessary. The trigger will only extend forward once the hammer has been manually cocked. On this pistol the barrel release catch is at the base of the trigger. There is no trigger guard.

On both designs the maximum range is about 120 m. A wide array of single-, twin- and three-flare white and coloured star loadings is available. Cartridges are cardboard tubes with metal rimmed bases.

The Chinese NORINCO Type 57 signal pistol (qv) is based on the SPSh-2 while the Egyptian Abu Redis pistol (qv) appears to be another close copy.

Data
SPSH-2
Calibre: 26.65 mm
Length: 220 mm
Weight: 900 g

Manufacturer
State factories.

Status
Probably no longer in production.

Service
Russian armed forces and many others.

UPDATED

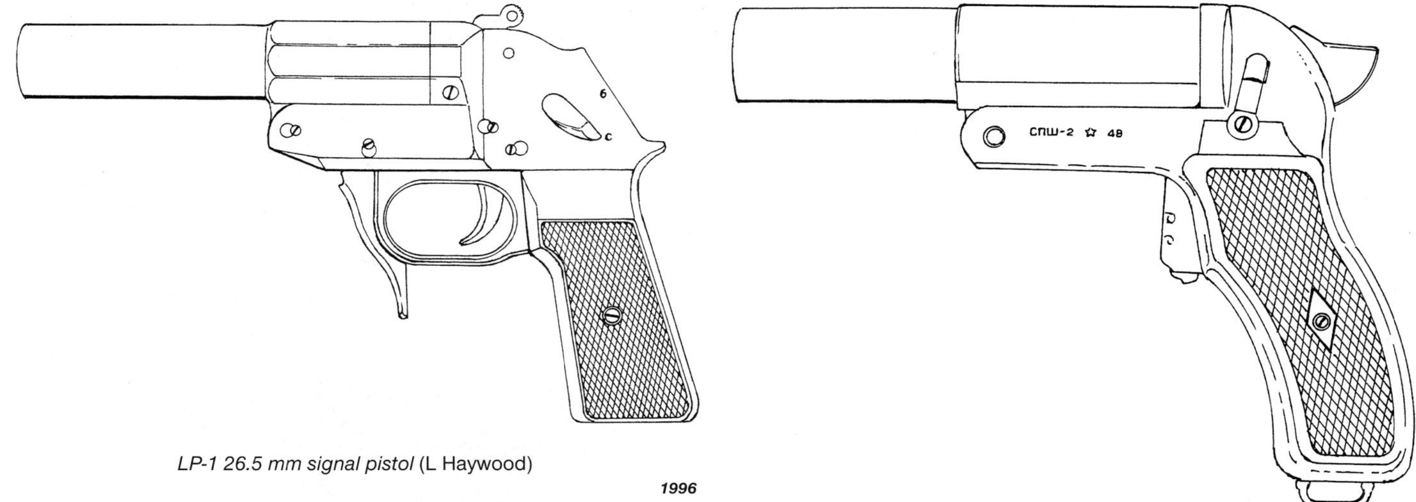

LP-1 26.5 mm signal pistol (L Haywood)
1996

SPSh-2 26.5 mm signal pistol (L Haywood)
1996

Hand-fired illuminating rocket PG 431

Description
The 30 mm calibre PG 431 illuminating rocket is comparable to similar devices used elsewhere. It has a range of approximately 450 m, emits a parachute star unit which burns for about nine seconds, and can illuminate an area of some 600 m^2. There are two known variant models: the illuminating pattern and a special signal for warning of NBC attacks. The latter develops a piercing whistle, audible over long distances, during its upward flight and at the top of its trajectory bursts to display a number of red stars.

To use either of these rockets, the rear cap is removed to expose a pull-ring and lanyard; the device is then grasped and pointed upwards at 45° and the lanyard pulled sharply to ignite the rocket. It can be fired

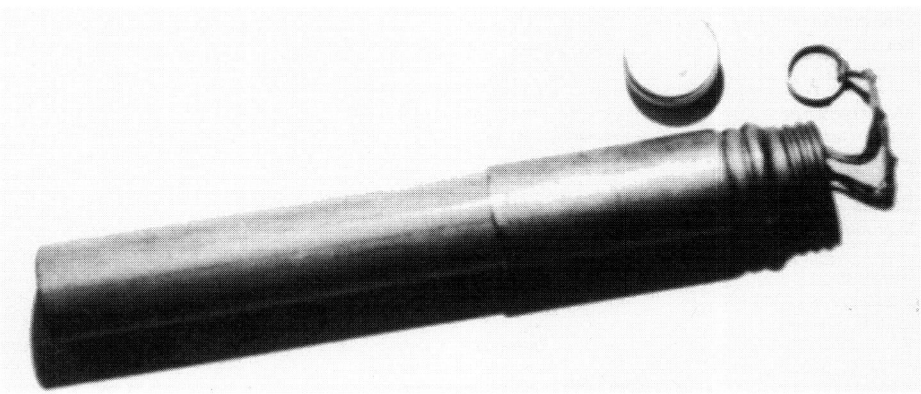

30 mm hand-fired illuminating rocket PG 431

free hand, but the recommended method is to use a rifle as an improvised support. For this the rocket unit is held alongside the rifle barrel, with its rear end braced against the front of the handguard using the left hand and the butt of the rifle resting on the ground. The lanyard is then pulled.

Data

Calibre: 30 mm
Length: 225 mm
Weight: 190 g

Manufacturer

State factories.

Status

Available.

VERIFIED

Rocket illuminators

Description

There are two larger versions of the hand-fired PG 431 (see previous entry). A 40 mm calibre model has a range of 300 m, burns for 20 seconds and will illuminate an area of 400 m^2. It is presumably fired using the rifle as a support, as outlined above.

There is also a 50 mm model provided with a two-stage solid propellant rocket to give greater range. It will reach 1,200 m, burns for 25 to 30 seconds and can cover an area of 600 m^2.

Manufacturer

State factories.

Status

Available.

VERIFIED

CZECH REPUBLIC AND SLOVAKIA

DA25 and DA100 coloured smoke generators

Description

The DA25 coloured smoke generator is the smaller of two similar devices. It consists of a metal canister, the top plate of which has a number of emission holes. This is normally covered by a white plastic cap. On removing the cap a central igniter socket is exposed and any standard igniter can be inserted; the friction igniter 'ČAROZ' is the most common type, but electrical igniters can be used in cases where several generators are to be simultaneously fired. White, black, red, violet, blue, orange, green and yellow smokes are available, the colour of the smoke being indicated by a coloured stripe on the body of the generator. The DA25 emits smoke for about 25 seconds.

The DA100 coloured smoke generator is similar to the DA25 but is larger, emitting smoke for about one minute. The action is the same as the DA25, but the filling is in layered form to improve burning. White, black, red, blue and orange smokes are available.

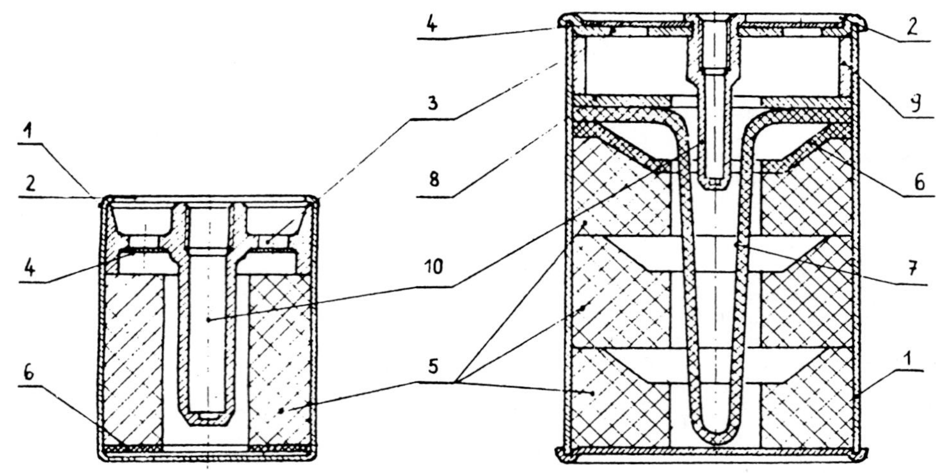

Czech smoke generators DA25 and DA100
(1) *light alloy canister* **(2)** *white plastic cover* **(3)** *emission holes* **(4)** *sealing cover* **(5)** *smoke composition* **(6)** *priming charge*

Status

Probably no longer in production.

Service

In service with the Czech and Slovak armed forces.

VERIFIED

EGYPT

26.5 mm 'Abu Redis' signal pistol

Description

The 26.5 mm 'Abu Redis' signal pistol is a single-shot pistol for firing standard 26.5 mm signal cartridges. The design appears to have been based on the Soviet SPSh-2 pistol.

Loading is accomplished by breaking open the action shotgun-fashion and hand loading a single cartridge directly into the breech. The trigger is normally folded away for carrying and is only released when the hammer is manually actuated. Construction is simple.

Data

Calibre: 26.5 mm
Length: 230 mm
Barrel: 150 mm
Weight: 920 g

Manufacturer

Maadi Company for Engineering Industries, PO Box 414, Maadi, Cairo.

Status

In production.

Service

Egyptian armed forces.

UPDATED

26.5 mm 'Abu Redis' signal pistol

FRANCE

Ruggieri coloured smoke hand grenade Type 188

Description

The Type 188 grenade consists of a light alloy cylindrical body filled with coloured smoke composition. The grenade is fuzed by a Model F8 or F12 igniter set, both being of the fly-off lever type. The grenade is used for daytime signalling, target indication and for indicating wind direction to helicopters and light aircraft.

Data

Diameter: 55 mm
Height: fuzed, 142 mm
Weight: 400 g
Delay time: nominal, 2.5 s
Emission time: ca 2 min
Colours: red, green, yellow, blue, white, orange, violet

Manufacturer

Ruggieri Technologie, route de Gaudies, F-09270 Mazeres.

Status

In production.

Service

French Army.

VERIFIED

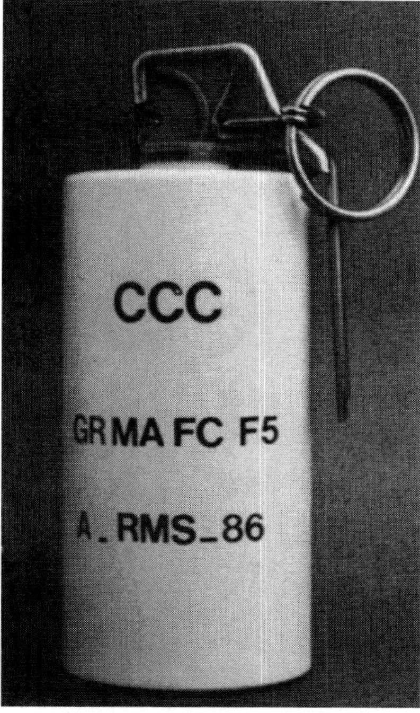

Ruggieri coloured smoke grenade Type 188

Ruggieri flare mine with sound effect Type 424

Description

The Type 424 is a form of trip-flare, actuated by a traction-type ignition device. The mine is buried in the ground up to its igniter and fitted with trip-wires. When the wire is pulled, the igniter initiates the mine, causing a detonating charge to be ejected into the air to a height of about 10 m, where it detonates with a loud report. At the same time the flare pot, in the mine body, is ignited and pushed upwards by a spring, enabling the burning flare to come partially out of the mine body and burn just above the ground surface.

Data

Diameter: 60 mm
Height: without igniter, 373 mm; with igniter, 420 mm
Weight: without igniter, 1.65 kg; igniter 50 g
Luminosity: 40,000 cd

Manufacturer

Ruggieri Technologie, route de Gaudies, F-09270 Mazeres.

Status

In production.

Service

French Air Force.

VERIFIED

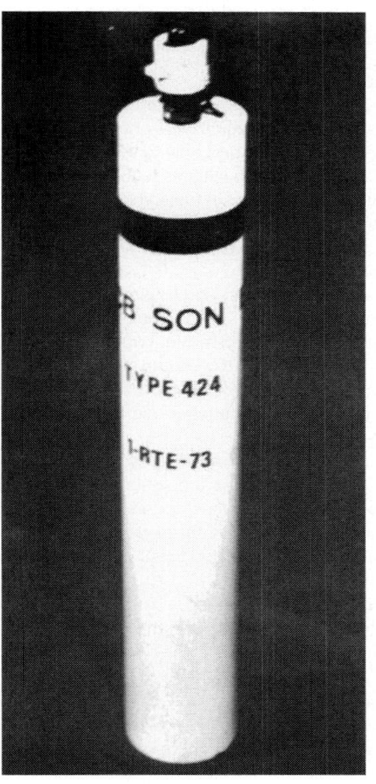

Ruggieri flare mine with sound effect Type 424

Ruggieri day and night distress signal Type 252

Description

The Type 252 was designed to provide a single system which gives an easily seen signal by day or night, on land or sea. It consists of a cylindrical body with a percussion igniter at each end. One end contains an orange smoke composition for use by day, the other a red Bengal Light for use by night. The daylight signal develops a cloud of orange smoke for 15 to 30 seconds, which can be seen up to 10,000 m away in clear weather. The night signal produces a red flame which can be seen from an aircraft 48 km away in clear weather. After using either end of the signal, the other end remains ready for subsequent use.

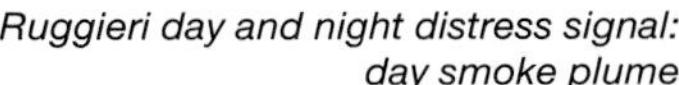

Ruggieri day and night distress signal: day smoke plume

Data
Diameter: 38 mm
Height: 126 mm
Weight: 135 g

Manufacturer
Ruggieri Technologie, route de Gaudies, F-09270 Mazeres.

Status
In production.

Service
French Army.

VERIFIED

Ruggieri day and night distress signal: night flare

27 mm hand signalling device Mle F1

Description
All 27 mm signalling devices Mle F1 are in the form of a plastic tube, containing a signal (with or without a parachute), an ejection charge and a firing system. The latter is a friction igniter which initiates a delay and which is operated by pulling a cord. A fastening cap of plastic material protects the ignition system and, by having embossed markings, permits night identification of the type of device.

The signalling devices are used to allow visual signalling by day or night, by firing single stars, either parachute-suspended or free-falling. With several combinations of colours, a simple code can easily be set up. Since these devices are manually fired, their use is not restricted to personnel carrying signal pistols or dischargers.

The devices will project, approximately three seconds after ignition, a signal to a height of about 90 m. The duration of the signal varies: for devices without a parachute it is seven seconds, for devices with a parachute 15 seconds. The available types of signal are as follows:

Single star: white, green or red
Parachute star: white, green or red

Data
Diameter: 28.5 mm
Length: single star, 146 mm; parachute star, 273 mm
Weight: single star, 85 g; parachute star, 160 g

Manufacturer
Ruggieri Technologie, route de Gaudies, F-09270 Mazeres.

Status
In production.

Service
French Army.

VERIFIED

Ruggieri 27 mm signalling device Mle F1

Type 328 distress signal kit

Description

The Type 328 signal kit consists of four flare signals F428, two flare and radar-detectable signals Type 328, one launcher and one holder. The Type 328 is intended to be used by persons in distress at sea or in hazardous terrain. They can be used to equip rescue buoys or form part of individual or collective packs.

The Type 328 signal ascends to a height of about 150 m, where it releases a red flare signal and a mass of radar detectable dipoles. The flare is visible to 30 km in clear weather, while the dipoles can be detected up to 35 km away.

Data

Weight: kit, 300 g
Dimensions: 145 × 70 × 20 mm

Manufacturer

Ruggieri Technologie, route de Gaudies, F-09270 Mazeres.

Status

In production.

VERIFIED

Type 428 distress signal kit

Description

The Type 428 distress signal kit consists of a launcher and six red flares which, when fired vertically, reach a height of 400 m and can be seen from up to 60 km on a clear night. There is no recoil or spark from the launcher.

Data

Dimensions: kit, 140 × 70 × 15 mm
Weight: kit, 250 g
Duration of illumination: 11 s
Illumination: 5,000 cd
Culminating height: 400 m
Visibility: on a clear night, 60 km

Manufacturer

Ruggieri Technologie, route de Gaudies, F-09270 Mazeres.

Status

Available.

VERIFIED

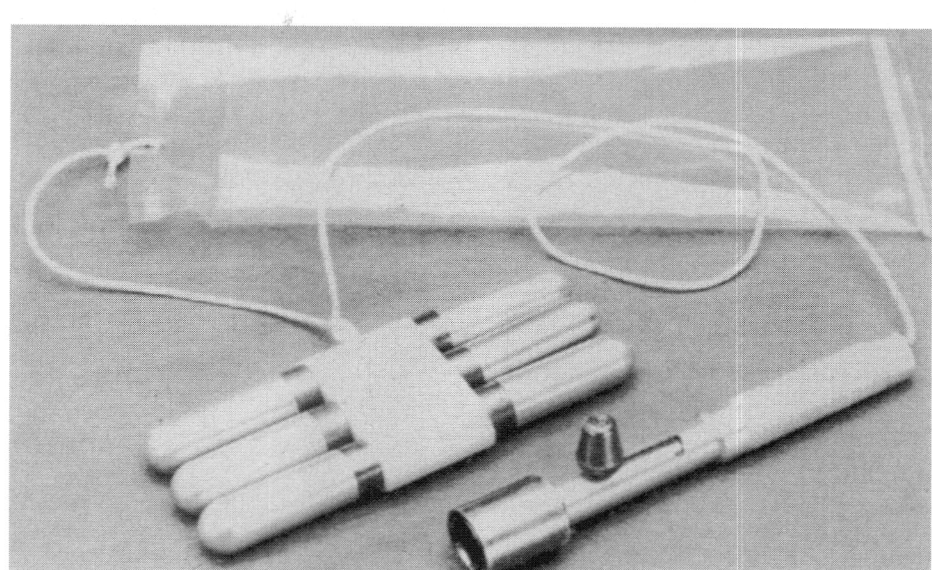

Type 428 distress signal kit

Lacroix signal cartridges

Description

These cartridges can be fired from any 40 mm (1.5 in) signal pistol. The coloured star ignites during the upward part of its trajectory and extinguishes about 15 m above the ground.

Data

Diameter: 40 mm
Length: 100 mm
Weight: 132 g
Illumination time: 6.5 s
Ignition height: >100 m
Candlepower: red, 40,000; green, 25,000; white, 70,000 cd

Manufacturer

Etienne Lacroix, BP 213, 6 boulevard de Joffrery, F-31607 Muret Cedex.

Lacroix signal cartridges

Status

In production.

VERIFIED

Lacroix ASATOX chemical attack warning device

Description

The Lacroix ASATOX chemical attack warning device is a hand-held rocket which emits visual and audible warning of sufficient intensity to alert a platoon or company-sized unit under combat conditions. The device is fired by holding it near vertical, turning the handgrip to the left, and pulling it sharply. After a 1 to 1.5 second delay, so that the firer can ensure the device is pointed correctly, the rocket motor ignites and provides soft ejection of the payload. The signals are fully deployed in about five seconds.

Data

Diameter: 47.5 mm
Length: 347 mm
Weight: 690 g
Radius of warning: 500 m
Audible signal intensity: at 0.5 m from source, 139 dB
Duration of audible signal: 10 s
Duration of visual signal: 20 s
Altitude: ca 200 m

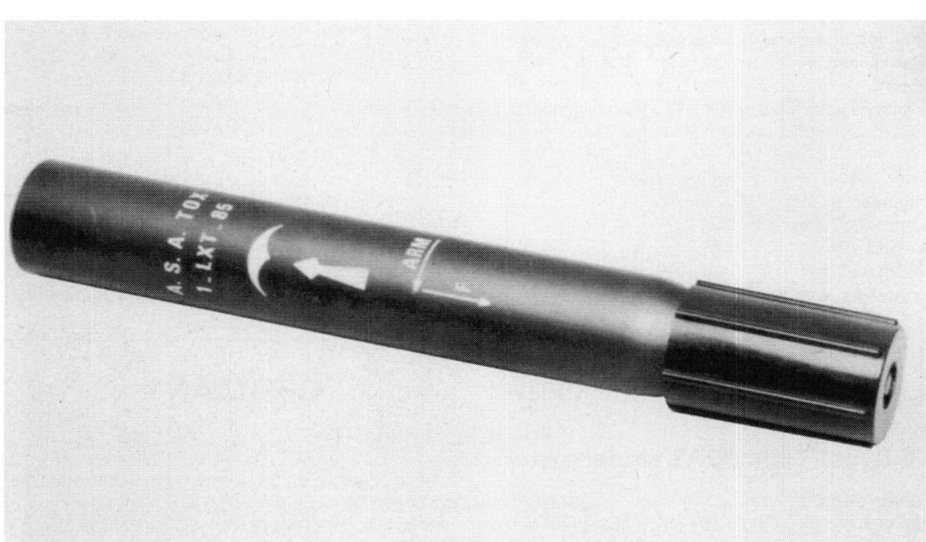

Lacroix ASATOX chemical attack warning device

Manufacturer

Etienne Lacroix, BP 213, 6 boulevard de Joffrery, F-31607 Muret Cedex.

Status

In production.

Service

French Army.

VERIFIED

Lacroix short-range illuminating rocket

Description

This is a hand-held illuminating rocket designed to produce a minimum illumination on the ground of 5 lux within a 300 m diameter area. This is considered to be a sufficient level of illumination to permit identification of a target at night and engagement with any infantry weapon, more particularly anti-tank weapons, without the need for night vision devices.

The device is fired by turning and pulling the handle-grip at the rear end, while holding the tubular body in the desired direction of launch. A rocket is discharged which, at its vertex, deploys a parachute-supported star unit. The recoil stress is short and weak, so the rocket can be launched without danger and the need for special training. There is no visible trail to disclose the firer's position.

Data

Calibre: 40 mm
Length: 323 mm
Weight: complete, 580 g; rocket, 350 g
Range: 300-350 m or 600-1,000 m
Rate of descent: 3.5 s
Duration: 25 s
Power: 200,000 cd
Altitude: ejection, 140 m; burnout, 40 m

Manufacturer

Etienne Lacroix, BP 213, 6 boulevard de Joffrery, F-31607 Muret Cedex.

Status

In production.

Service

French Army, in which it is known as 'Type F2'.

VERIFIED

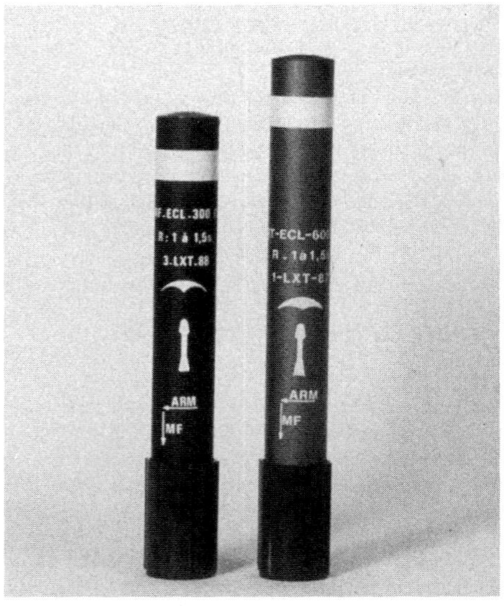

Lacroix short-range illuminating rocket

Lacroix stabilised illuminating device

Description

The Lacroix stabilised illuminating device is a multiple-star discharger which can be set on the ground or partly buried and can then be operated by conventional trip or pull switches as a trip-flare or by remote control at distances up to 2,000 m from the operator. Once triggered, it fires a vertical star every second until the entire 16 stars have been used. As a result it delivers almost constant illumination over a period of about 23 seconds, sufficient to permit the aiming and firing of any infantry weapon. The device is called stabilised because atmospheric conditions (particularly wind) have no effect, as they do in the case of the usual parachute-suspended star devices. The centre of the illuminated zone therefore remains constant during the period of operation. All the stars have identical trajectories and power and they burn out at approximately 40 m from the ground. It would be possible to increase the operational time slightly to suit specific requirements.

Lacroix stabilised illuminating device

Data

Dimensions: 180 × 135 × 110 mm
Weight: unit, 2.2 kg
Illumination level: 5 lux
Radius: 200 m
Duration: 23 s
Number of stars: 16

Manufacturer

Etienne Lacroix, BP 213, 6 boulevard de Joffrery, F-31607 Muret Cedex.

Status

In production.

Service

French Army.

VERIFIED

Long emission time signal mine Type F1B

Description

This is a trip or command type ground flare designed to give almost constant light output for a period of 6 minutes. It will illuminate an area of about 10,000 m². This operation time is achieved by a spring device which keeps moving the illuminant upwards in its container so as to keep the burning area exposed and thus emitting light in all directions. The mine is ignited by a pull-type switch 'Fusée F2'.

Data

Diameter: 65 mm
Length: 397 mm
Weight: 2.36 kg
Duration: 6 min
Light level: average, 4,500 cd; max, 7,500 cd

Manufacturer

Etienne Lacroix, BP 213, 6 boulevard de Joffrery, F-31607 Muret Cedex.

Status

In production.

Service

French Army.

VERIFIED

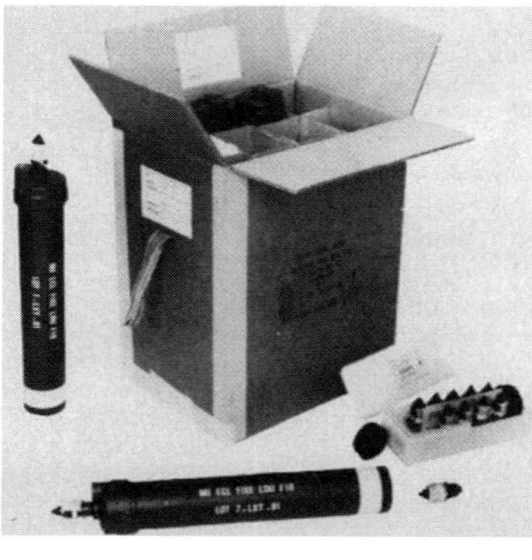

Lacroix F1B signal mine

GERMANY

26.5 mm Type P2A1 signal pistol

Description

The 26.5 mm Type P2A1 is a single-shot break-action signalling pistol which can be adapted also to fire riot control grenades. In its basic form the pistol comprises a seamless drawn steel barrel which is pivoted on a receiver and grip assembly. A locking stud at the rear of the barrel engages with a breech catch on the receiver to lock the barrel in position; to break the weapon for loading or unloading the breech catch at the top of the grip must be depressed. The action of opening the breech automatically partially ejects an empty case by means of the ejector at the bottom of the barrel.

After loading a cartridge in the chamber the barrel is snapped shut and the weapon is loaded and safe. To fire it the hammer must be thumb-cocked. There is no applied safety but the hammer may be lowered under control.

The weapon is easily maintained but the barrel must be cleaned after firing. For protection in store any common gun oil may be used.

Data

Calibre: 26.5 mm
Length: 200 mm
Barrel: 155 mm
Weight: 520 g

Manufacturer
Heckler and Koch GmbH, D-78722 Oberndorf-Neckar.

Status
In production.

Service
German and Swiss armed forces, and those of numerous other countries worldwide.

VERIFIED

P2A1 signal pistol

Heckler and Koch emergency flare kit

Description
This flare launcher was specifically developed for use in emergency. The compact launcher is magazine loaded with up to five 19 mm DM 13 signal cartridges, usually single star red although other colours are available. The flares reach a height of 65 m.

The kit was designed for fast, single-hand operation with the hammer moved to the rear by the thumb, the safety pushed up to fire, again by the thumb, and then the trigger squeezed with the index finger.

Data
Calibre: 19 mm
Length: overall, 80 mm
Height: 146 mm
Width: 37 mm
Weight: loaded, 445 g; launcher, 225 g; filled magazine, 220 g

Manufacturer
Heckler and Koch GmbH, D-78722 Oberndorf-Neckar.

Status
In production.

Service
German armed forces; military and police forces in other countries and wide commercial sales.

VERIFIED

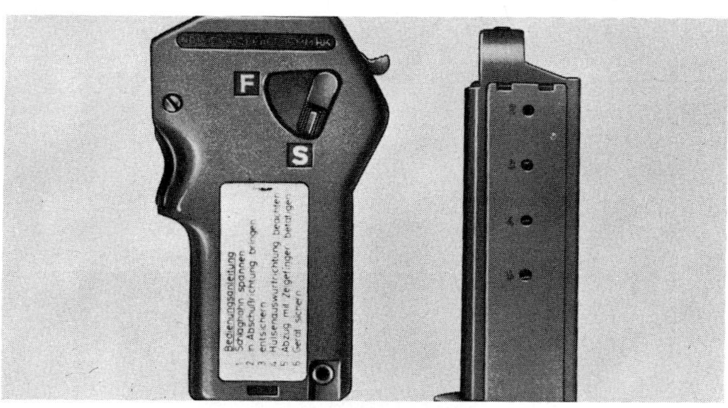

Heckler and Koch emergency flare kit launcher and five-round magazine

19 mm DM 13 cartridge

Nico signalling coloured smokes

Description
For centuries fighting forces have communicated on the battlefield using various forms of smoke signal. This means of communication continues to be employed and the Nico range of signalling smoke products offers flexibility and proven performance.

The brilliant coloured smoke produced by the Nico range is based on organic dyes and is biodegradable. A wide selection of colours, burning times, ignition systems and canister designs are features of the Nico range.

Canister diameter	Canister height (incl igniter)	Ignition type	Burning time	Colours
44 mm	ca 140 mm	Pull cord Fly-off lever Electric	60/90 or 120 s	white, black, grey, green, violet, yellow, blue, orange, red
60 mm	ca 145 mm	Pull cord Fly-off lever Electric	60/90 or 120 s	white, black, grey, green, violet, yellow, blue, orange, red

Manufacturer
Nico Pyrotechnik, Hanns-Jürgen Diederichs GmbH and Co KG, Bei der Feuerwerkerei 4, Postfach 1227, D-22946 Trittau/Hamburg.

Status
In production.

VERIFIED

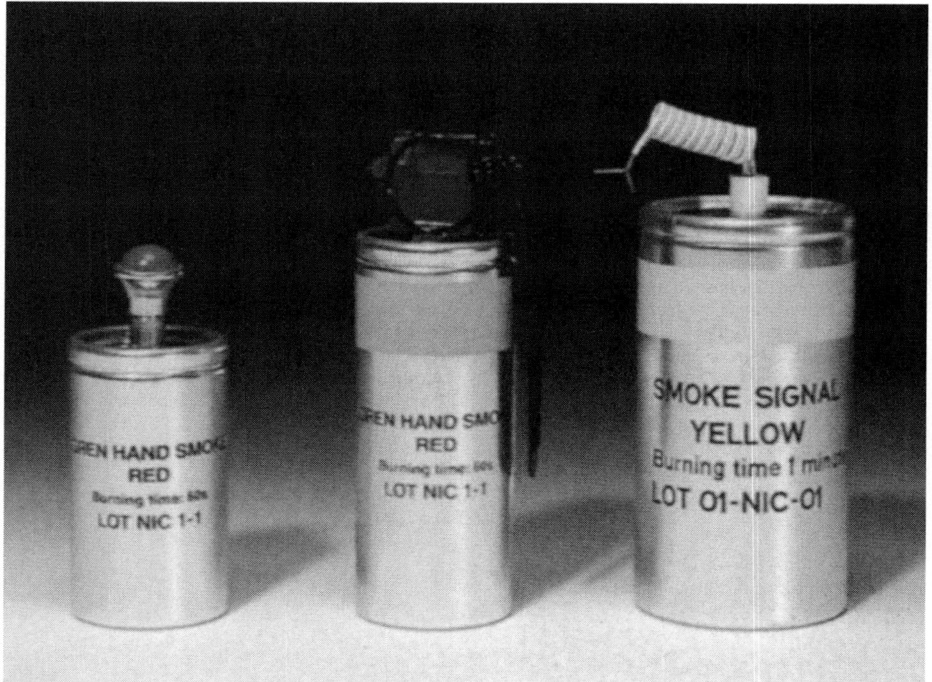

Nico signalling coloured smokes

Nico illumination and signal devices

Description

Nico illuminating and signalling devices are available for signal pistols or as hand-held devices. In both forms the high quality devices ensure maximum performance and visibility under all conditions.

Nico pistol-fired illumination cartridges

A selection of pistol-fired illuminating and signal cartridges (calibre 1 in and 1.5 in) is produced by Nico to meet the needs of customers. The selection includes options on burning time, ascent heights, candela power, colours and number of stars and the cartridges are produced with or without parachutes. Nico's 40 mm illumination cartridge for use with 40 mm grenade launchers is also available.

Nico hand-held illumination devices

Powerful hand-held illumination devices that require no launching pistols are available for a variety of applications. Variants available provide choices of ascent heights, burning times, candela ratings, colours, number of stars and firing systems and they are produced with or without parachute.

Manufacturer

Nico Pyrotechnik, Hanns-Jürgen Diederichs GmbH and Co KG, Bei der Feuerwerkerei 4, Postfach 1227, D-22946 Trittau/Hamburg.

Status

In production.

VERIFIED

Nico illumination and signal devices

Nico 116 mm illumination rocket

Description

The Nico self-contained 116 mm illuminating flare offers users the flexibility to illuminate large areas independently of heavy weapons systems. The self-contained launching tube can be prepared quickly for firing. The range is up to 5,000 m, and the flare emits 1.2 Mcd for a duration of 60 seconds.

Manufacturer

Nico Pyrotechnik, Hanns-Jürgen Diederichs GmbH and Co KG, Bei der Feuerwerkerei 4, Postfach 1227, D-22946 Trittau/Hamburg.

Status

In production.

VERIFIED

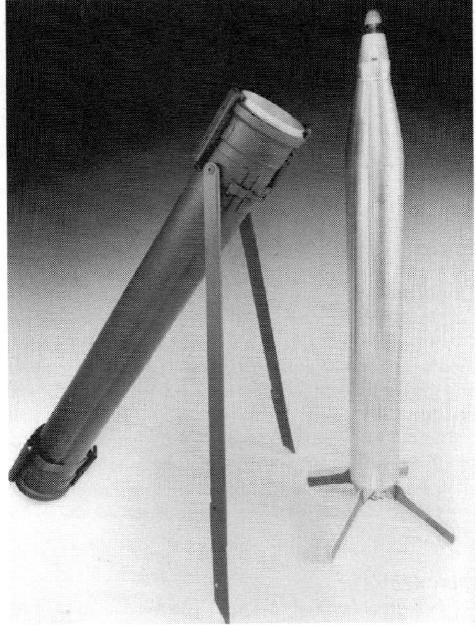

Nico 116 mm illuminating rocket

Nicosignal flare kit

Description

The Nicosignal is a hand-held repeating flare discharger manufactured from stainless steel and plastic materials to ensure resistance to sea water and corrosion. It consists of a handle with security loop, safety catch, trigger and firing mechanism, to which a six-shot revolving flare magazine is clipped. The magazine can be fired rapidly and a fresh magazine can be installed quickly. Cartridges are available in red, green and white, giving various signalling options; the intensity and duration of the ejected flare is in conformity with accepted military standards.

Data

DISCHARGER

Dimensions: 170 × 55 × 50 mm

Weight: with magazine, 230 g; replacement magazine, 90 g

AMMUNITION

Cartridge diameter: 16 mm

Cartridge length: 45 mm

Burning time: ca 6 s

Intensity: red, 10,000 cd

Functioning height: ca 80 m

Colours: red, green, white

Storage life: 3 years

Manufacturer

Nico Pyrotechnik, Hanns-Jürgen Diederichs GmbH and Co KG, Bei der Feuerwerkerei 4, Postfach Box 1227, D-22946 Trittau/Hamburg.

Status

In production.

VERIFIED

Nicosignal flare kit

Nico NT screening smoke

Description

This is a patented smoke composition developed by the Nico Pyrotechnik company which replaces the standard hexachloroethane/zinc (HC) composition. The NT composition has high mechanical strength and resists damage, so that it can be formed into any required shape without requiring any container or canister. The chemical characteristics of NT smoke are such that it has better storage life than HC smokes and the burning rate can be controlled by the method of use, so that it becomes possible to generate a large volume of smoke to start the cloud, then reduce the output to keep the cloud fed and maintain the cover. The absence of a metal container means that less weight needs to be transported to provide a given volume of smoke and there is no corrosion problem in store.

The smoke produced is very white, has a high TOP (Total Obscuring Power) value, is not so weather-dependent as HC or FS smokes and is chemically neutral (pH 5.6 to 6.2) so that it does not irritate soldiers operating in the screen.

The solid material is zinc oxide and ammonium salt based, it is chemically neutral and has a virtually unlimited store life. Combustion is at ca 800/850°C and is flameless. Burning continues even under water spray conditions. The rate of combustion depends upon the free surface area, and the material can be ignited by pyrotechnic inserts, electrical squibs, storm matches, ordinary matches, percussion fuze or even a lighted cigarette.

Standard sizes are 0.5, 1 and 10 kg blocks, but any desired shape or size can be manufactured.

Manufacturer

Nico Pyrotechnik, Hanns-Jürgen Diederichs GmbH and Co KG, Bei der Feuerwerkerei 4, Postfach 1227, D-22946 Trittau/Hamburg.

Status

In production.

VERIFIED

Nico KM screening smoke

Description

KM screening smoke is a new smoke development by Nico and consists of a pyrotechnically generated aerosol, the main ingredients of which are potassium chloride (KCl) and magnesium oxide (MgO). The compound is certified as non-toxic and not harmful to the environment and the smoke produced is of a pure white colour.

KM screening smoke compound can be used in any existing type of smoke canister, including vehicle-launched types, and is available for a variety of burning times. All HC and NT smoke stores can therefore be replaced for training purposes by KM compounds. At the present time Nico has produced 40 mm, 66 mm, 76 mm and 81 mm vehicle-launched grenades, 44 mm, 60 mm and 66 mm grenades and five minute smoke canisters filled with KM composition.

Manufacturer

Nico Pyrotechnik, Hanns-Jürgen Diederichs GmbH and Co KG, Bei der Feuerwerkerei 4, Postfach 1227, D-22946 Trittau/Hamburg.

Status

In production.

VERIFIED

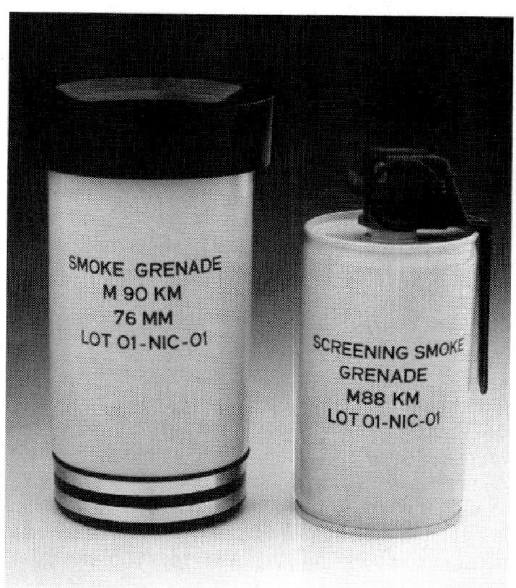

Nico KM screening smoke stores

Nico 76 mm screening smoke grenade M90

Description

The Nico 76 mm screening smoke grenade M90 is for use with vehicle dischargers and is available with various fillings: KM, NT, HC and IR (infra-red screening). The IR screening smoke differs from most others in that it is not designed to build up a heat-developing cloud as a decoy but to provide an optronically impenetrable screen so that the projectile and its sighting system will not be able to penetrate the smoke and find a target. The smoke unit is carried in a cartridge case which remains in the launcher after firing and so protects the launcher against corrosive gases as well as environmental influences.

Manufacturer

Nico Pyrotechnik, Hanns-Jürgen Diederichs GmbH and Co KG, Bei der Feuerwerkerei 4, Postfach 1227, D-22946 Trittau/Hamburg.

Status

In production.

VERIFIED

Nico 76 mm screening smoke grenade M90

Nico 40 mm pyrotechnic grenades

Description

These grenades are for use with all types of 40 mm grenade launcher, but differ slightly in having Nico's patented propulsion system. Among other things this has the projectile fixed to the cartridge case so that it is not released until uniform pressures are achieved in

Nico 40 mm grenade family

both the high and low pressure chambers. When this predetermined pressure is reached, the projectile detaches from the cartridge and is propelled in a single thrust. This provides uniform consistency of ballistics from shot to shot and is instrumental in giving these grenades a high degree of accuracy.

The types of 40 mm grenade currently available are:

40 × 46 mm Practice with Tracer (DM 118)
40 × 46 mm Practice with Impact Signature
40 × 46 mm Illuminating with Parachute
240 × 46 mm Signal (single or multiple star) with or without Parachute
40 × 46 mm NT Smoke
40 × 46 mm Spontaneous Smoke
40 × 46 mm Sound & Flash (Stun) cartridge
40 × 53 mm High velocity

Manufacturer

Nico Pyrotechnik, Hanns-Jürgen Diederichs GmbH and Co KG, Bei der Feuerwerkerei 4, Postfach 1227, D-22946 Trittau/Hamburg.

Status

In production.

VERIFIED

Nico alarm devices

Description

Two specific warning devices have been developed by Nico for perimeter protection, the Alarm Flare and the Alarm Mine.

Alarm Flare

This flare is designed for rapid deployment around areas and perimeters to both alert the defenders and illuminate intruders. The simple igniter device is sensitive to disturbance of the actuating wire, which is itself not under tension but is connected to the laid out and tensioned trip-wire. This arrangement enables the flare to be sensitively but safely linked to undergrowth or tree branches which, when moved, will trigger the flare.

The flare is supplied with a complete kit of wires and fixings. It gives high light intensity and long burning time, and the detonation as the flare ignites serves as an audible warning.

Alarm Mine

The pyrotechnic alarm mine is a simple and reliable warning device and is in use throughout the world. It is installed in position by attaching the holding bracket to a tree stump or other suitable surface; the mine is fixed to the bracket and finally the trip-wire is laid in position and the mine armed by removal of the safety pin.

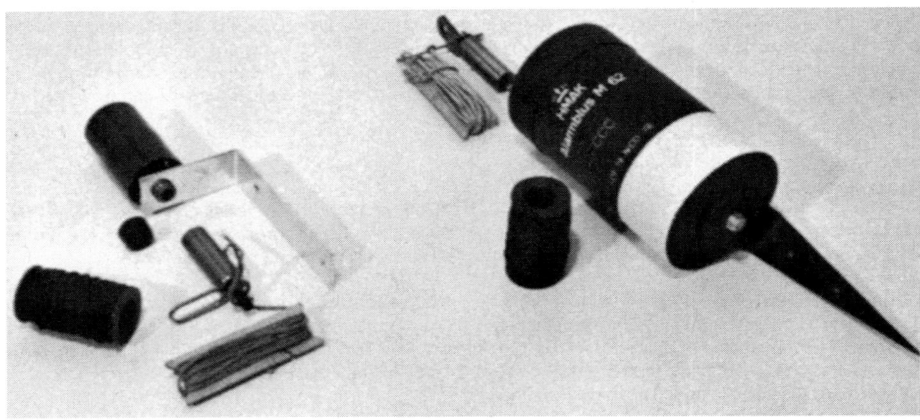

Nico alarm devices

The mine gives a loud alarm signal, a pyrotechnic discharge some 10 m into the air. Then the remaining mine contents burn with an intense continuous light for approximately 10 seconds.

Manufacturer

Nico Pyrotechnik, Hanns-Jürgen Diederichs GmbH and Co KG, Bei der Feuerwerkerei 4, Postfach 1227, D-22946 Trittau/Hamburg.

Status

In production.

VERIFIED

Comet 26.5 mm signal cartridges

Description

Comet 26.5 mm signal cartridges are fired by signal pistols. They are aluminium cased and are provided in four colours (red, white, green and yellow) and as a fulminating cartridge which delivers a flash and smoke signal plus a loud report. Individual types are identified by notches in the cartridge rim, coloured sealed tops and by the model number on the outer shell.

Data

Diameter: 26.5 mm
Length: 80 mm
Weight: 55 g
Burning time: ca 8 s
Brilliance: depending on colour, ca 25,000 cd
Altitude: fired at 90°, ca 120 m

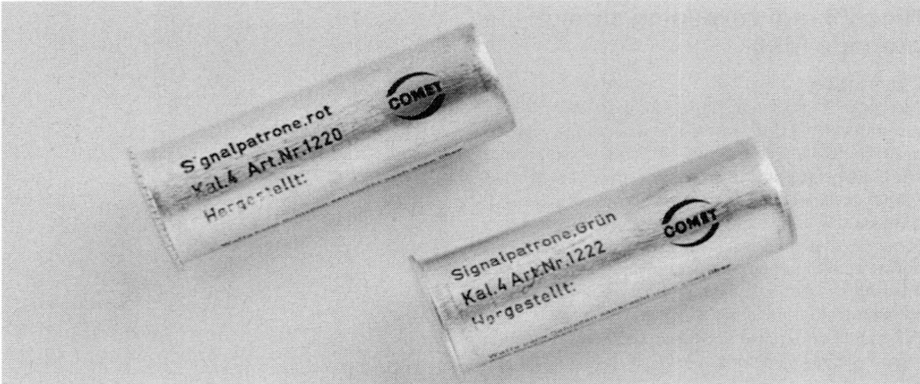

Comet 26.5 mm signal cartridges

Manufacturer

Comet GmbH Pyrotechnik Apparatebau, Postfach 100267, D-27502 Bremerhaven.

Status

In production.

VERIFIED

Comet 26.5 mm parachute signal cartridges

Description

These are elongated cartridges, for use in signal pistols, containing a parachute flare for close-in battlefield illumination or for use in search and rescue operations at sea. White and red flares are available.

These cartridges are also available with a rocket motor using smokeless propellant. This provides a considerable increase in performance while the smokeless propellant means that the trajectory and firing point cannot be detected.

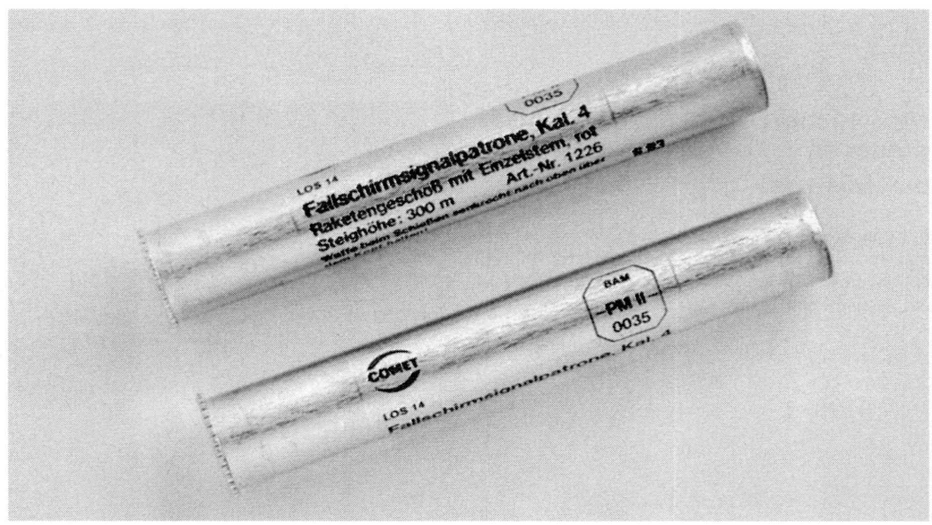

Comet 26.5 mm parachute signal cartridges

Data
Diameter: 26.5 mm
Length: 170 mm
Weight: 120 g
Burning time: 15-30 s
Candlepower: red, 10,000 cd; white, up to 80,000 cd
Illumination: up to 1 lux at 100 m altitude
Altitude: fired at 90°, 300 m

Manufacturer
Comet GmbH Pyrotechnik Apparatebau, Postfach 100267, D-27502 Bremerhaven.

Status
In production.

UPDATED

Comet 26.5 mm smoke signal cartridge

Description
The Comet 26.5 mm smoke signal cartridge ejects a smoke-generating unit which takes effect at the vertex and then continues burning while falling. It is used for indicating wind direction and velocity to aircraft and helicopters, as well as for target indication and marking. Two colours, violet and orange, are available, while other colours may be ordered.

Data
Diameter: 26.5 mm
Length: 148 mm
Weight: 85 g
Smoke duration: 7 s
Altitude: fired at 90°, 120 m

Manufacturer
Comet GmbH Pyrotechnik Apparatebau, Postfach 100267, D-27502 Bremerhaven.

Status
In production.

VERIFIED

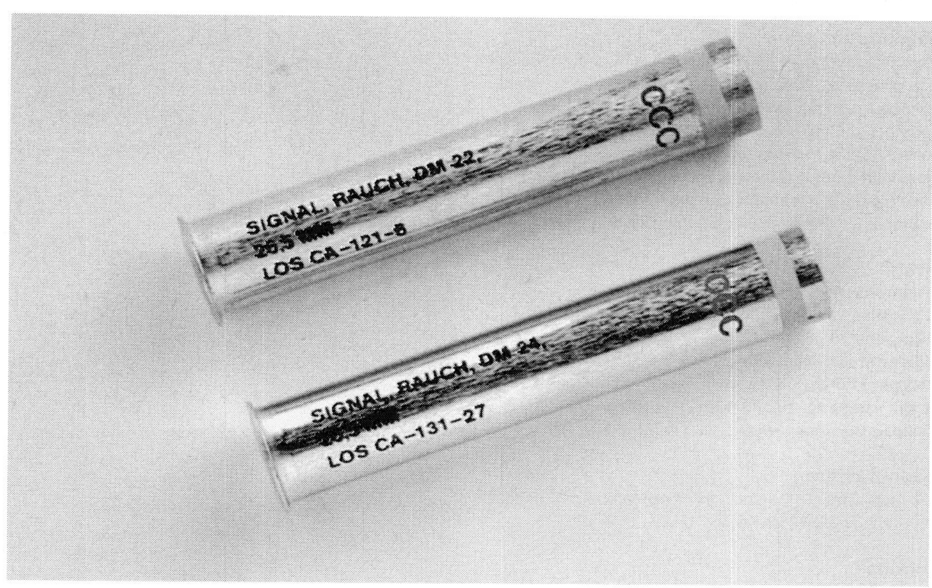

Comet 26.5 mm smoke signal cartridges

Comet colour signal rockets

Description
These are hand-held rockets which, at their vertex, eject a parachute and a single colour flare or a flare cluster. They can be used for signalling, the red version being approved for use as a marine distress signal.

Data
Diameter: 45 mm
Length: 270 mm
Weight: 435 g
Burning times: Type 1232, 40 s; Type 1233, 30 s
Light intensity: Type 1232, 40,000 cd; Type 1233, 80,000 cd
Altitude: fired at 90°, 300 m

Manufacturer
Comet GmbH Pyrotechnik Apparatebau, Postfach 100267, D-27502 Bremerhaven.

Status
In production.

UPDATED

Comet colour signal rockets

Comet white parachute signal rockets

Description
These are hand-held rockets for illumination purposes. Two types exist: The Type 1236 is launched by a pull-wire igniter housed under a removable cap on the base; the Type 1260 is more powerful and is launched by twisting the plastic end cap after removal of a safety pin. This model also incorporates smokeless propellant so that there is no firing signature and the launch point cannot be readily identified.

Data

	Type 1236	Type 1260
Diameter	33 mm	45 mm
Length	295 mm	300 mm
Weight	270 g	550 g
Burning time	30 s	30 s

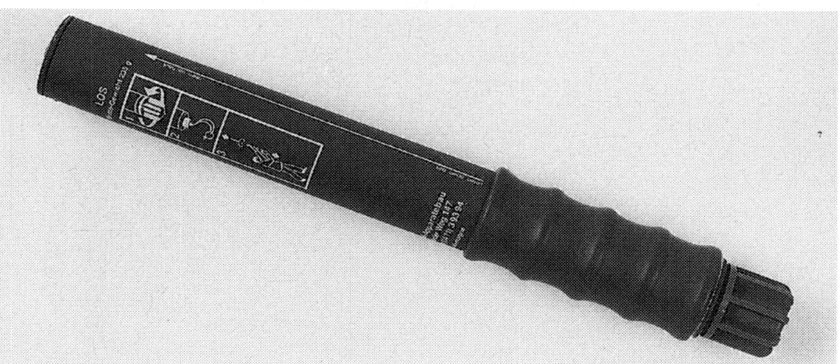

Comet parachute signal rocket Type 1260

1996

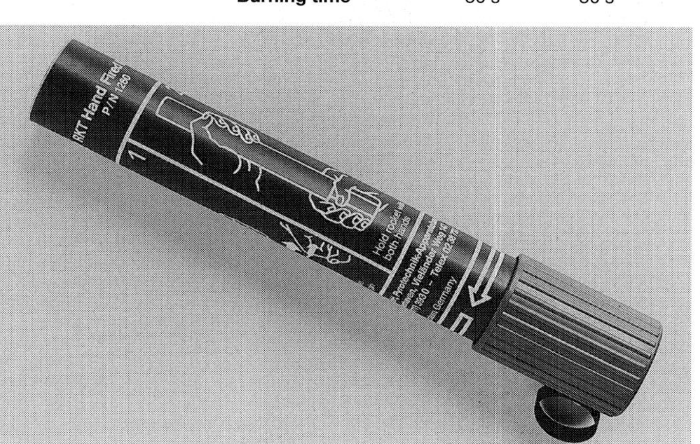

Comet parachute signal rocket Type 1236

1996

	Type 1236	Type 1260
Light intensity	100,000 cd	225,000 cd
Illumination at 100 m	8 lux	20 lux
Altitude fired at 90°	300 m	500 m

Manufacturer
Comet GmbH Pyrotechnik Apparatebau, Postfach 100267, D-27502 Bremerhaven.

Status
In production.

UPDATED

Comet Day and Night signal

Description
The Comet Day and Night signal is a dual-purpose signal or distress device. By removing the brown end cap and pulling the igniter wire so exposed, a cloud of orange smoke is released for signalling by day. By removing the red cap, at the other end, and pulling the exposed wire, a red flare is initiated for use by night. The casing is watertight to a depth of 30 m and is thus suitable for use by divers or waterborne forces.

Data
Diameter: 30 mm
Length: 190 mm
Weight: 170 g
Ignition: pull-wire igniter
Burning time: 20 s
Light intensity: 15,000 cd
Smoke duration: 18 s

Manufacturer
Comet GmbH Pyrotechnik Apparatebau, Postfach 100267, D-27502 Bremerhaven.

Status
In production.

UPDATED

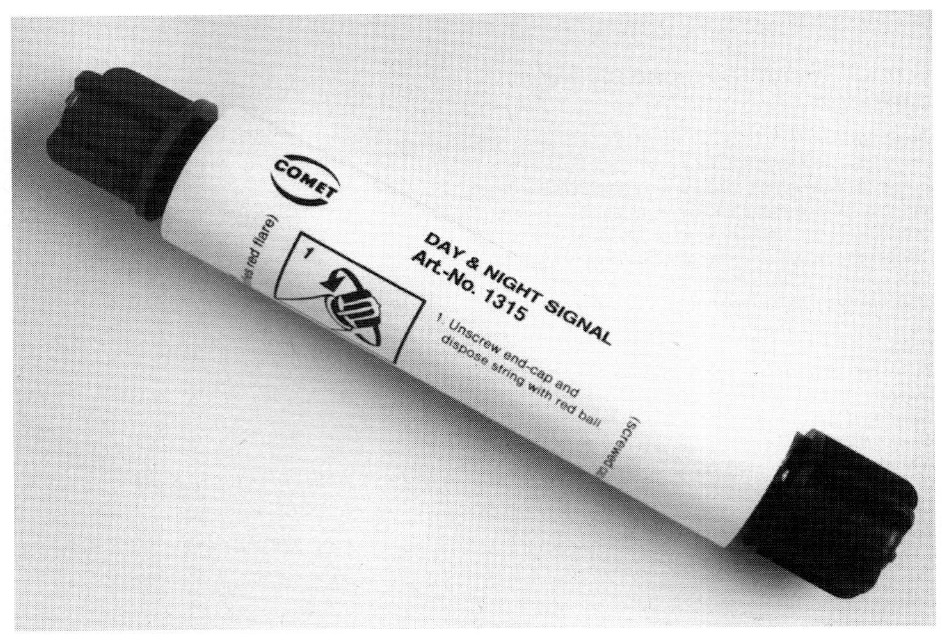

Comet Day and Night signal

Comet ground smoke signal

Description
This is a small smoke generator, ignited by a pull-wire friction igniter concealed under the removable cap. It can be placed on the ground for use as a marker for aircraft or helicopters, as a wind indicator, or it can be thrown or dropped as a target marker. It is normally available in orange although other colours can be provided on request.

Data
Diameter: 50 mm
Length: 95 mm
Weight: 180 g
Ignition delay: 5 s
Smoke duration: 60 s

Manufacturer
Comet GmbH Pyrotechnik Apparatebau, Postfach 100267, D-27502 Bremerhaven.

Status
In production.

UPDATED

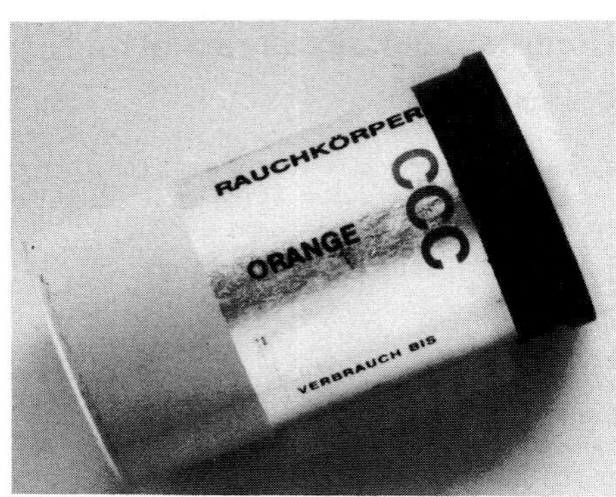

Comet ground smoke signal

Comet signal pistols

Description
Comet GmbH distribute two signal pistols, both are of 26.5 mm calibre single-shot, drop barrel, hammer-fired types but differ principally in their operating method and construction materials.

Data

Type	Comet	Diana
Calibre	26.5 mm	26.5 mm
Length	235 mm	215 mm
Barrel	150 mm	155 mm
Weight	1 kg	650 g

Manufacturer
Comet GmbH Pyrotechnik Apparatebau, Postfach 100267, D-27502 Bremerhaven.

Status
In production.

UPDATED

Diana 26.5 mm signal pistol

Comet 26.5 mm signal pistol

Piepenbrock 26.5 mm signal cartridges

Description

Piepenbrock Pyrotechnik GmbH manufactures a wide range of signal cartridges for the standard type of 26.5 mm signal pistol. The following are the most important types for infantry use:

- Single star, red, white or green
- Multi-star, giving four stars in red, white or green
- Multi-star giving various combinations of stars: 3 red; 3 green; 2 red/1 green; 2 green/1 red; 2 white/1 green; 2 green/1 white; 2 red/1 white; 2 white/1 red. Other combinations of multi-star are possible and can be provided to order.
- Yellow flare, with or without parachute
- Smoke trail, orange or violet
- Flash-report, giving a bright flash and loud explosion
- Parachute signal, giving parachute-suspended flares in red, green or white.

Non-parachute stars operate at 100-120 m altitude and burn for 6-8 seconds. Parachute types operate at about 75 m and burn for up to 12 seconds.

Manufacturer

Piepenbrock Pyrotechnik GmbH, Postfach 20, D-67306 Göllheim/Pfalz.

Status

In production.

VERIFIED

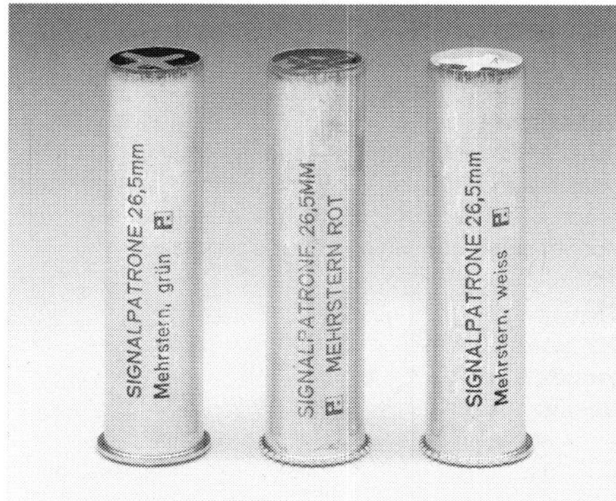

Piepenbrock 26.5 mm multistar signal cartridges

Piepenbrock 37.5 mm signal cartridges

Description

These are single star signals for use with standard 37 mm (1.5 in) signal pistols. All operate at an altitude of about 90 m and burn for seven seconds. The red and white stars are approximately 50,000 candela, the green star 30,000 candela and the yellow star 70,000 candela.

Manufacturer

Piepenbrock Pyrotechnik GmbH, Postfach 20, D-67306 Göllheim/Pfalz.

Status

In production.

VERIFIED

Piepenbrock 40 mm signal cartridges

Description

These 40 mm signal cartridges are fired from 40 mm grenade launchers such as the US M79/M203 or German HK 69A1. The single star cartridge is available in red, white, green or yellow, operates at 100 m altitude and burns for eight seconds. The parachute flare cartridge can be used for signalling or illumination; it ejects a parachute and white flare at 350 m altitude, and burns for approximately 30 seconds, giving an illumination of 100,000 candelas. The single star cartridges are 100 mm long, while the parachute flare cartridge is 250 mm long.

It should be noted that all Piepenbrock cartridge types produced in 1.5 in (37 mm) calibre are also available in 40 mm form.

Manufacturer

Piepenbrock Pyrotechnik GmbH, Postfach 20, D-67306 Göllheim/Pfalz.

Status

In production.

VERIFIED

Piepenbrock 16.5 mm signal projectors

Description

These hand-actuated firing devices are used with 16.5 mm screw-on cartridges. The Model 11010 is a simple spring device operated by pulling back the striker until it is retained by a trigger, after which the trigger is pressed to fire the cartridge. The Model 11009 is equally simple, being fired by pulling back on a cocking piece and releasing it. This model is furnished as part of a kit containing a selection of signal cartridges.

The cartridges available include red, white, yellow and green single stars, and an audible effect cartridge which emits a whistle during its upward flight and then bursts with a loud report.

Manufacturer

Piepenbrock Pyrotechnik GmbH, Postfach 20, D-67306 Göllheim/Pfalz.

Status

In production.

VERIFIED

Piepenbrock hand signal cartridges

Description

Piepenbrock hand signal cartridges produce the same effect as a conventional signal pistol but are hand-held devices which are discarded after firing. They consist of a metal tube with wrist strap and removable end cap. The forward end is closed with a coloured and notched plastic cap indicating the contents. To operate, the strap is looped around the wrist and the tube grasped firmly; then the end cap is removed, whereupon a trigger ring is exposed. Pulling this fires the signal cartridge from the tube.

Types available are: single star in red, green, white or yellow, which reach 100 m altitude and burn for about eight seconds; multistar, with four red, green or white stars, reaching 75 m and burning for six seconds; and parachute single star in red, green, white or yellow, reaching 70 m and burning for about 12 seconds. All cartridges are 28 mm diameter and approximately 200 mm long.

Manufacturer

Piepenbrock Pyrotechnik GmbH, Postfach 20, D-67306 Göllheim/Pfalz.

Status

In production.

VERIFIED

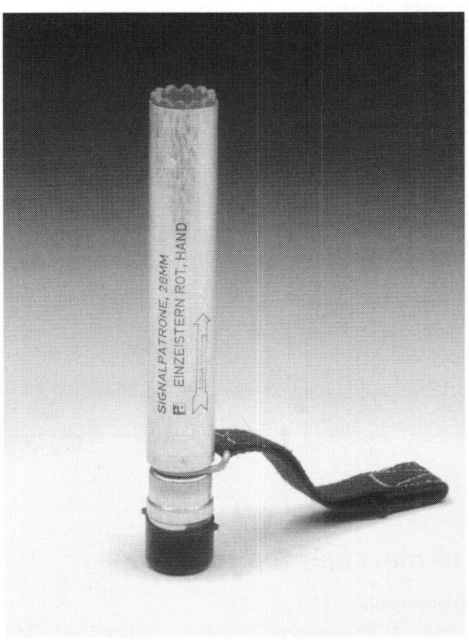

Piepenbrock hand signal cartridge

Piepenbrock 26.5 mm rocket-assisted illuminating cartridges

Description

Piepenbrock 26.5 mm rocket-assisted illuminating cartridges are similar in effect to normal illuminating cartridges but are rocket-boosted during flight so as to give greater range and permit the deployment of the illuminating unit to points beyond the reach of normal signal pistol cartridges. Two types are available, one without parachute and one with parachute. The non-parachute model is 173 mm long, reaches a range of about 400 m when fired at 45° elevation, burns for eight seconds and gives approximately 130,000 candelas. The parachute model is 210 mm long, ranges to about 300 m when fired at 45° elevation, burns for about 15 seconds and provides 160,000 candelas.

Manufacturer

Piepenbrock Pyrotechnik GmbH, Postfach 20, D-67306 Göllheim/Pfalz.

Status

In production.

VERIFIED

Piepenbrock DM 22 smoke generator

Description

The DM 22 generator is used for marking landing areas, signalling, or indicating wind strength and direction. It consists of a light metal canister with friction igniter concealed under a plastic cap.

Data

Diameter: 50 mm
Length: 95 mm
Weight: 185 g
Ignition delay: nominal, 4 s
Duration of smoke: nominal, 60 s
Colour of smoke: orange

Manufacturer

Piepenbrock Pyrotechnik GmbH, Postfach 20, D-67306 Göllheim/Pfalz.

Status

In production.

VERIFIED

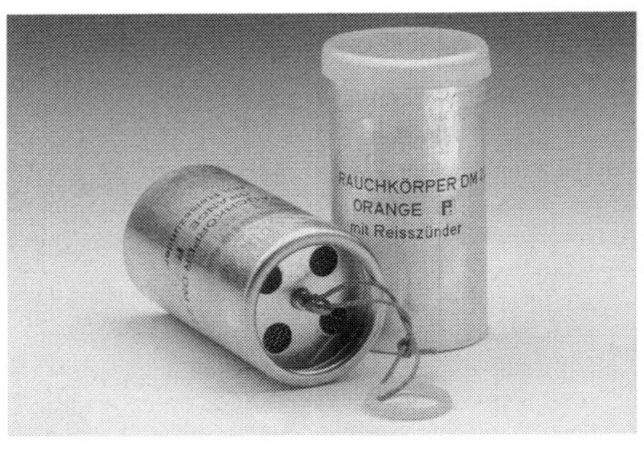

Piepenbrock DM 22 smoke generator

Piepenbrock 40 mm screening smoke cartridge

Description

This 40 mm cartridge is fired from standard 40 mm grenade launchers such as the US M79/M203 or the HK 69A1. On discharge, it launches a rocket-boosted projectile to a range up to 200 m, which then generates light grey smoke for screening purposes.

Data

Calibre: 40 mm
Length: 250 mm
Range: 200 m
Emission time: 30 s

Manufacturer

Piepenbrock Pyrotechnik GmbH, Postfach 20, D-67306 Göllheim/Pfalz.

Status

In production.

VERIFIED

Piepenbrock alarm mine

Description

The Piepenbrock alarm mine is a trip-wire-operated device which, upon being triggered, emits a loud report and a bright flash. It is intended for perimeter alarm and the signal can be heard and seen over a considerable distance. The device is small and easily concealed.

Data

Weight: 50 g
Time of illumination: 10 s
Candlepower: 15,000 cd

Manufacturer

Piepenbrock Pyrotechnik GmbH, Postfach 20, D-67306 Göllheim/Pfalz.

Status

In production.

VERIFIED

Piepenbrock trip-flare

Description

The Piepenbrock trip-flare is a conventional alarm device which can be quickly emplaced and functions when the trip-wire is either pulled or cut. On firing it produces a vivid yellow flare. Two types are available, with different light intensity and duration. In addition to its function as an alarm, it is possible to emplace this flare as an ambush illuminant and ignite it remotely by using an electric igniter.

Data

Weight: 600 g
Candlepower: Type 1, 250,000 cd; Type 2, 100,000 cd
Duration of flare: Type 1, 35 s; Type 2, 60 s

Manufacturer

Piepenbrock Pyrotechnik GmbH, Postfach 20, D-67306 Göllheim/Pfalz.

Status

In production.

VERIFIED

Piepenbrock trip-flare

INDIA

16 mm signal cartridges

Description

These 16 mm signal cartridges are fired from a 13 mm diameter pen-type spring-loaded launcher. The cartridges are screwed onto the top of the launcher, the spring cocked and then released using a button-type trigger. The cartridges fire a red, green or white flare to a height of over 60 m. Each flare burns for more than 3.5 seconds.

Data

Diameter: 16 mm
Height of burning: >60 m
Burn time: over 3.5 s

Manufacturer

Indian Ordnance Factory, Dehu Road, Pune 412 113.

Status

In production.

Service

Indian armed forces. Offered for export.

VERIFIED

16 mm signal cartridges and launcher

38 mm signal cartridges

Description
These 38 mm signal cartridges are fired from 38 mm signal pistols. The cartridges fire a red, green or yellow flare to a height of over 90 m. Each flare burns for more than eight seconds. Weight of each cartridge is 75 g.

Data
Diameter: 38 mm
Height of burning: >90 m
Burn time: >8 s

Manufacturer
Indian Ordnance Factory, Dehu Road, Pune 412 113.

Status
In production.

Service
Indian armed forces. Offered for export.

VERIFIED

38 mm signal cartridges

INDONESIA

25.4 mm illuminating cartridge

Description
Produced by PT Pindad, these illuminating cartridges are fired from standard 25.4 mm (1 in) signal pistols. The cartridges have aluminium bodies with Berdan primers and use black powder to propel aluminium-cased single red, white or green flares which burn for a minimum of four seconds. The flares are visible from a distance of 1,000 m during daylight and 3,000 m at night.

Data
Diameter: cartridge, 27.2 mm
Length: 57 mm
Weight: red, green, 49 g; white, 53 g
Burn time: > 4 s

Manufacturer
PT Pindad (Persero), Jl Jendral Gatot Subroto, PO Box 807, Bandung 40284.

Status
In production.

Service
Indonesian armed forces. Offered for export.

UPDATED

25.4 mm illuminating cartridges

ITALY

Bernardelli 37 mm PS 023 signal pistol

Description
The Bernardelli 37 mm PS 023 signal pistol is a pyrotechnic flare launcher manufactured in conformity to military standards. It can fire any standard 37 mm pyrotechnic cartridge. It can be used in ground operations or, with an adaptor and Mount M1, in aircraft for both signalling and illumination tasks. The PS 023 pistol is a single-shot, manually breech loaded weapon of conventional type.

Data
Cartridge: 37 mm (1.5 in) pyrotechnic
Operation: manual, single shot
Locking: manual top lever
Length: 209 mm
Weight: 1 kg
Height: 215 mm

Manufacturer
Bernardelli SpA, I-25063 Gardone Val Trompia, Brescia.

Status
In production.

VERIFIED

Bernardelli PS 023 signal pistol

PS 023 signal pistol open for loading

Valsella VS-T-86 pull-release illuminating device

Description

The Valsella VS-T-86 illuminating device consists of a cylindrical plastic container, housing a flare, fitted with a trip-wire fuze mounted on the top. The device can be planted in the ground or snapped on to a special plastic stake which can be driven into the ground. Actuation is by pressure on the fuze rods or by traction on trip-wires, which may be up to 15 m long. Actuation by release can be obtained by the fitting of an optional spring.

Data

Diameter: 70 mm
Weight: 470 g; illuminating compound, 350 g
Height: 220 mm
Illumination period: min, 50 s
Flare intensity: at least 15 lux at 57 m distance
Operating force: pressure, 4-10 kg; pull, 3-7 kg
Operating temperature range: −32 to +60°C

Manufacturer

Valsella Meccanotecnica SpA, I-25014 Castenedolo, Brescia.

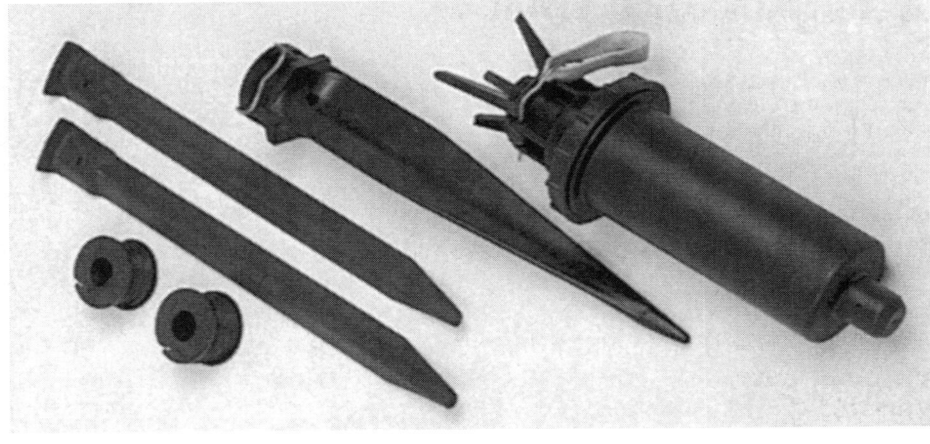

Valsella VS-T-86 illuminating device

Status

In production.

Service

Approved by Italian Army; in service with several countries.

VERIFIED

Valsella VS-TA-90 multifunction illuminating device

Description

The Valsella VS-TA-90 is a combined ground-air, trip-wire-activated, warning illuminating device intended to give acoustic and visual warning of intrusion into a protected area. It is composed of a ground flare, an in-air flare, one plastic rest picket for each flare, two metal pickets for the trip-wires of the ground flare and two spools with appropriate trip-wires. The fuze of the ground flare is triggered by pulling one of its trip-wires or by pressure and/or shock directly applied to the fuze prongs.

The in-air flare is automatically activated, with a proper delay, by the ground flare at the end of its illuminating period.

Several in-air flares could be connected in series so as to repeat the sequence, or in parallel, by means of a special connector, so as to increase the brilliance of the illuminated area.

Optional attachments are available to allow functioning of the device by release, functioning of the ground flare on command by remote control, or conversion of the ground flare from illuminating to smoke discharging.

Valsella VS-TA-90 multifunction illuminating device

Data

DATA FOR GROUND FLARE; THAT FOR AIR FLARE SHOWN IN PARENTHESIS

Diameter: 80 mm; (60 mm)
Weight: 423 g; (270 g)
Height: 180 mm; (140 mm)
Illumination data: 15 lux for >50 s at 57 m; (8 lux for >10 s at 33 m)
Operating force: traction, 5 kg; pressure, 9 kg
Operating temperature limits: −32 to +60°C

Manufacturer

Valsella Meccanotecnica SpA, I-25014 Castenedolo, Brescia.

Status

In production.

VERIFIED

Valsella VS-MK-83 smoke hand grenade

Description

The VS-MK-83 is a signalling device intended for manual activation and positioning. It is fitted with a friction igniter activated by pulling a cord protected by a removable safety cap. Emission of smoke begins approximately three seconds after the igniter is pulled. The grenade can be placed on the ground and ignited, or the igniter can be fired and the grenade then thrown in the normal way.

The VS-MK-83 grenade was developed in accordance with requirements resulting from the environmental and operational conditions foreseen by the Italian Army.

Data

Diameter: 60 mm
Weight: depending upon colour, 250-350 g
Height: 140 mm
Emission time: ca 70 s
Delay time: ca 3 s
Colours available: white, red, green, yellow, orange, violet, grey

Manufacturer

Valsella Meccanotecnica SpA, I-25014 Castenedolo, Brescia.

Status

In production.

VERIFIED

Valsella VS-MK-83 signalling smoke grenade

Stacchini M60F smoke signal grenade

Description

The Stacchini M60F smoke signal grenade was developed to meet Italian Army requirements for several applications including ground-to-ground signalling, ground-to-air signalling, ground recognition marking, wind indication, target identification and tactical screening.

The basic grenade is a hand-thrown canister. A plastic cap is removed before throwing which exposes the pull-type friction igniter cord and also permits the smoke to be emitted from the internal pyrotechnic unit. There is a delay of about 3.5 seconds before smoke is emitted.

If required to be rifle launched, the end cap is removed and a tail tube and vane unit is screwed on in its place. This unit has vents which permit the escape of the smoke. A suitable blank cartridge is packed with the tail unit; this is loaded into the rifle and the grenade is slipped over the muzzle. Any rifle with the NATO standard 22 mm muzzle configuration will accept the grenade.

Data

Diameter: 64 mm
Length: hand, 140 mm; rifle, 340 mm
Burning time: 90 s
Range: rifle-launched, 180 m
Colours: white, grey, red, orange, green, blue, black

Manufacturer

Stacchini Sud SpA, SS Tiburtina Valeria Km 64, I-67063 Oricola (AQ).

Status

In production.

VERIFIED

Stacchini M60F smoke grenade, rifle version

Stacchini 1.57 in signal cartridges

Description

These cartridges are designed for the US Pistol, Signal, Mod 8 and are suitable for all military signalling purposes. The cartridge carries two coloured stars and a propellant charge in an aluminium cartridge case sealed with a weatherproof disc. The colour of the signal is indicated by two coloured stripes on the case and by indentations on the base which can be distinguished in darkness.

Data

Calibre: 1.57 in (39.5 mm)
Length: 97.5 mm
Intensity: ca 25,000 cd
Burning time: 5-7 s
Altitude: 80 m
Colours: white, red, green, yellow

Manufacturer

Stacchini Sud SpA, SS Tiburtina Valeria Km 64, I-67063 Oricola (AQ).

Status

In production.

VERIFIED

Stacchini 1.57 in signal cartridges

Stacchini 1 in signal cartridges

Description

These are designed for the 1 in (nominal 27 mm) Very signal pistol and are suitable for all military signal purposes. The round consists of an aluminium cartridge case which contains a single coloured star and a propellant charge. The case is sealed with a waterproof closing disc which is coloured to indicate the colour of the star; the base rim is also indented to identify the colour in darkness.

In addition to coloured stars, similar cartridges loaded with an illuminating flare, a single smoke pellet, two illuminating flares, or a combination of one flare and one smoke pellet can be provided.

Data

Calibre: nominal, 27 mm
Duration: 5 s
Intensity: according to colour, 10,000-15,000 cd
Altitude: 80 m
Colours: red, white, green, yellow

Manufacturer

Stacchini Sud SpA, SS Tiburtina Valeria Km 64, I-67063 Oricola (AQ).

Status

In production.

VERIFIED

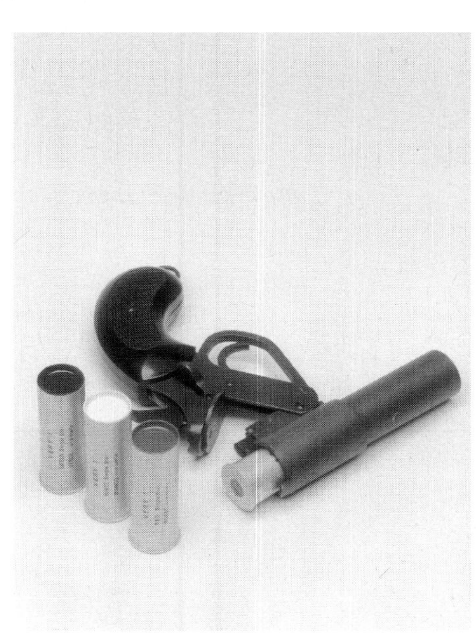

Stacchini 1 in signal cartridges

KOREA, SOUTH

Smoke hand grenade K-M18

Description

As might be gathered from the nomenclature the K-M18 is based on a US design. It is a coloured smoke grenade filled with chlorate/dye composition and is available in red, yellow, green and violet. The top plate of the grenade is coloured appropriately. The filling has a central hole which is lined with starter composition, and a layer of starter composition also covers the top of the smoke mixture. There is a hole in the bottom of the grenade body, normally tape-sealed, through which the smoke can escape.

Data

Weight: 520 g
Height: 118 mm
Fuze delay: nominal, 1.5 s
Emission time: 50-90 s

Manufacturer

Korea Explosives Company Limited, 34 Seosomoon-Dong, Chung-Ku, Seoul.

Status

In production.

Service

South Korean Army.

UPDATED

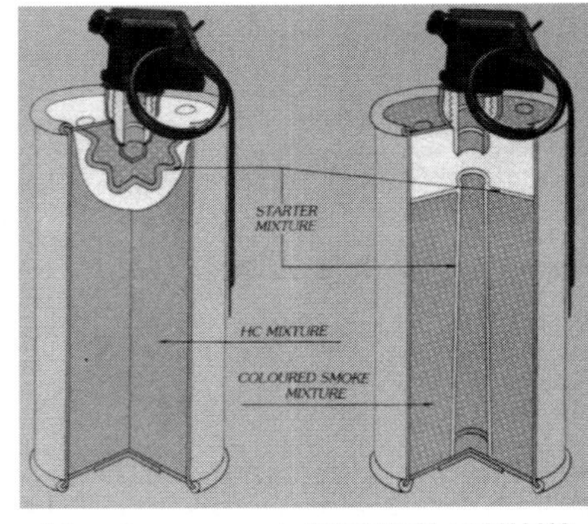

Internal arrangement of K-M8 (left) *and K-M18 grenades*

40 mm illuminating parachute cartridges

Description

These are pyrotechnic cartridges designed to be fired from standard 40 mm grenade launchers such as the US M79/M203 and similar types. They provide a convenient method of signalling and illuminating with less weight and bulk and greater accuracy than comparable hand-held signal devices.

There are three cartridge models: KM583A1 white star, KM661 green star and KM662 red star. Each comprises a flare candle attached to a parachute contained in a projectile unit which has a pyrotechnic delay lit by the propellant explosion. The delay burns through and fires an ejection charge which ignites the flare candle and blows both candle and parachute free from the projectile.

Data

Length: 134 mm
Weight: 220 g
Burst height: at 85° elevation, 183 m
Candlepower: white, 90,000; red, 20,000; green, 8,000 cd
Burning time: ca 40 s
Muzzle velocity: 76 m/s

Manufacturer

Korea Explosives Company Limited, 34 Seosomoon-Dong, Chung-Ku, Seoul.

Status

In production.

VERIFIED

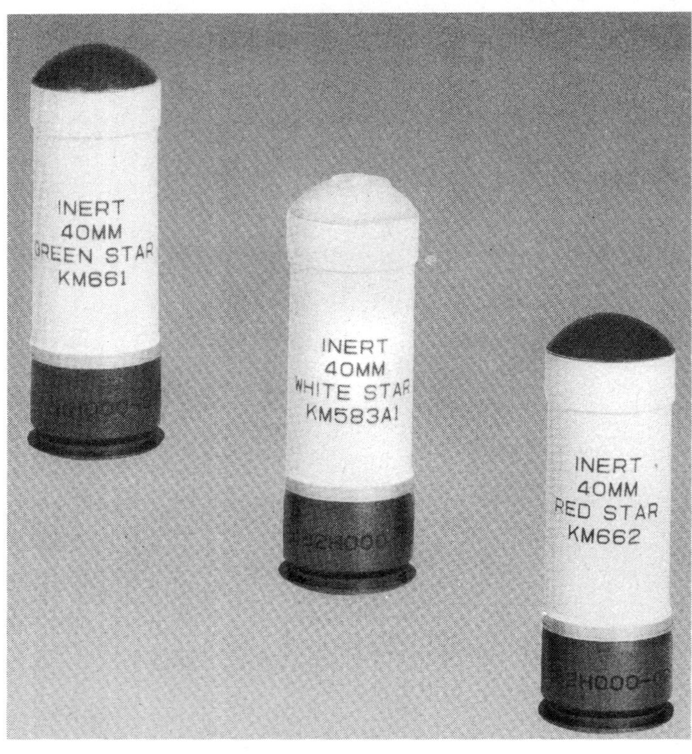

40 mm illuminating parachute cartridges

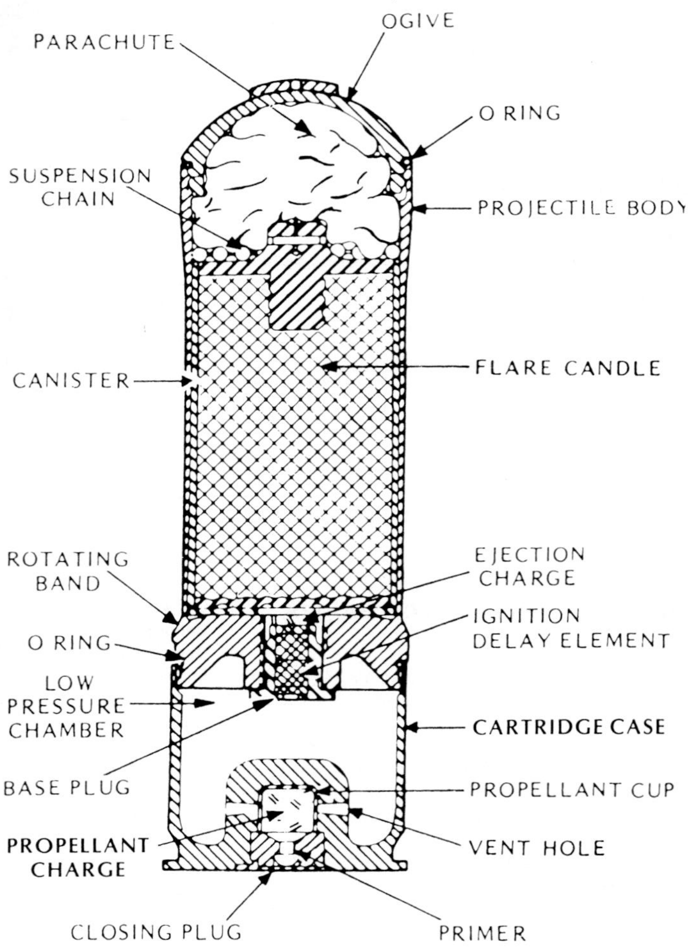

Internal arrangements of 40 mm parachute cartridge

PAKISTAN

POF flare trip-wire Mark 2/2

Description

The Pakistan Ordnance Factories (POF) flare trip-wire Mark 2/2 is a sentry device attached to a stake or upright structure to provide an indication and illumination of an area where intrusions by an enemy may occur. It is a flare body initiated by a trip-wire (provided on a spool with the unit) in such a manner that the flare unit provides a light intensity of 40,000 candela for 60 seconds. The flare composition is SR547 initiated by a Composition ABC percussion type primer.

Data

Length: 92.7 mm
Diameter: 48.6 mm
Weight: 1.84 kg
Burning time: 60 s
Light output: average, 40,000 cd

Manufacturer

Pakistan Ordnance Factories, Wah Cantt.

Status

In production.

Service

Pakistan armed forces.

NEW ENTRY

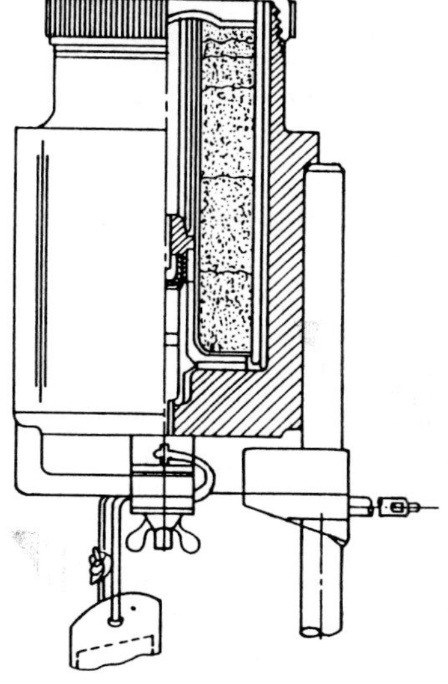

Section drawing of POF flare trip-wire Mark 2/2
1996

PERU

Convertible signal pistol MGP-S2

Description

The MGP-S2 is a rather unusual signal pistol on two counts; firstly because it is designed for 12-gauge signal cartridges rather than the more usual 26/27 mm type, and secondly because it is supplied with a drop-in auxiliary barrel which converts it into a 0.38 Special calibre single-shot survival pistol. The weapon uses a machined steel barrel, hinged to the frame so as to drop down for loading; there is an automatic ejector beneath the chamber. The barrel is locked in the firing position by an external catch which resembles the Webley stirrup lock, after which the hammer is manually cocked and the trigger pressed to fire. With the barrel open, the auxiliary 0.38 calibre barrel can be dropped into the chamber and secured at the muzzle by a screwed collar. The automatic extractor works with this barrel also.

Data

DATA FOR 12-GAUGE; THAT FOR 0.38 SPECIAL IN PARENTHESIS
Cartridge: 12-gauge pyrotechnic; 0.38 Special (see text)
Action: single action, single shot
Weight empty: 585 g; (786 g)
Length: 206.5 mm; (219 mm)
Barrel: 150 mm; (162.5 mm)
Rifling: 0.38 barrel, 12 grooves, rh, 1 turn in 250 mm
Muzzle velocity: 0.38 calibre, 270 m/s

Manufacturer

SIMA-CEFAR, Av Contralmirante Mora 1102, Base Naval, Callao.

SIMA convertible signal pistol MGP-S2

Status

In production.

Service

Peruvian armed forces.

VERIFIED

POLAND

PRONIT 26 mm signal cartridges

Description

The PRONIT Plastics Works at Pionki produces a range of 26 mm signal cartridges for firing from standard Very-type signal pistols. Using standard waxed finish cartridge cases, the range includes a white flare for illumination and red, green or yellow flares for signalling or target marking. All these flares have a burning time of about 6.5 seconds and, by providing up to 50,000 candela, can be seen over a distance of up to 7,000 m. When fired at an angle of 45° the flares can reach a height of 90 m and a range of 120 m. All cartridges of this type are 76 mm long.

Also produced is a 26 mm smoke signal cartridge which can produce red or blue smoke for up to seven seconds. These cartridges have a minimum trajectory

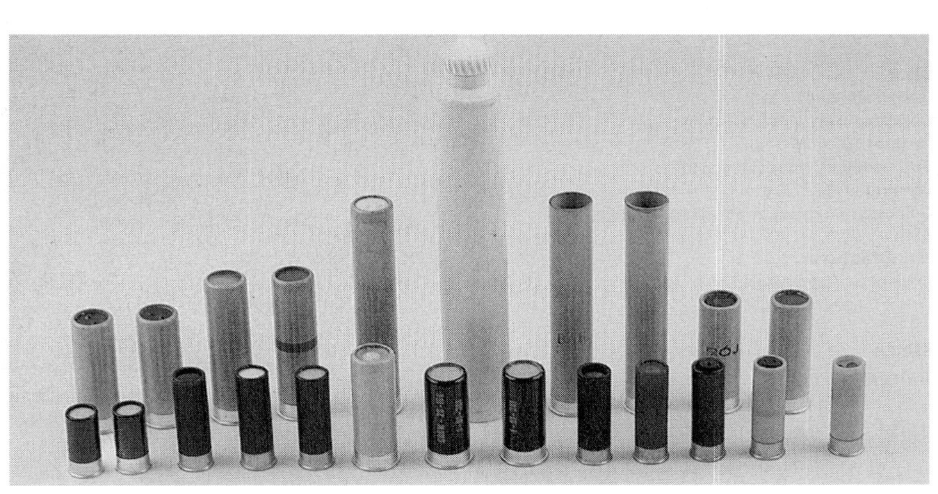

The full range of PRONIT pyrotechnic products

height of 50 m and the smoke can be seen from up to a distance of 2,000 m.

PRONIT also produces a 26 mm thunderflash cartridge for bird-scaring and similar purposes.

There is also a PRONIT 26 mm hand-held flare launcher which can be used for illumination. These launchers are 150 mm long and produce a flare suspended on a small parachute to provide illumination for up to 19 seconds. The flares start to burn at a height of 80 m.

Manufacturer
Zaklady Tworzyw Sztucznych PRONIT, PL-26 940 Pionki, ul Zakladowa 7.

Status
In production.

Service
Polish armed forces. Offered for export.

VERIFIED

40 mm hand-held signal cartridges

Description
PRONIT produces a range of 40 mm signal cartridges which launch four types of pyrotechnic signal. All four signals are launched from an airtight polystyrene tube 259 mm long and 41.6 mm in diameter.

A 40 mm signal cartridge is used to provide warning of chemical attack by firing a flare to a height of 200 m. The flare burns for five seconds while at the same time producing an acoustic effect which is audible for eight seconds and can be heard from a distance of 800 m. This cartridge weighs 425 g.

A 40 mm flare with parachute is launched to a height of 300 m. After a delay of six seconds the flare is ejected to burn under its parachute for 25 seconds, producing a light output of 50,000 candela. A similar cartridge produces a red parachute flare which burns for 40 seconds with an output of 30,000 candela.

A further PRONIT 40 mm signal cartridge produces a flash and bang effect after a delay of nine seconds. This cartridge weighs 345 g.

Manufacturer
Zaklady Tworzyw Sztucznych PRONIT, PL-26 940 Pionki, ul Zakladowa 7.

Status
In production.

Service
Polish armed forces. Offered for export.

VERIFIED

SINGAPORE

Coloured smoke grenades

Description
Singapore Technologies manufactures a variety of coloured smoke grenades suitable for training and signalling purposes. All are of the same pattern, a metal canister filled with various chlorate/dye mixtures, but the size, smoke duration and method of ignition vary as shown in the accompanying table.

Manufacturer
Singapore Technologies, 249 Jalan Boon Lay, Jurong Town, Singapore 2261.

Status
In production.

Coloured smoke grenades

Code	Diameter	Height	Ignition	Burning time	Colours
4412F	44 mm	137 mm	friction	120 s	white, black, grey, yellow, green, violet, blue, red, orange
4412L	44 mm	132 mm	fly-off lever	120 s	as above
446L	44 mm	132 mm	fly-off lever	60 s	white, yellow, green, violet, blue, red
609P	60 mm	164 mm	pull-cord	90 s	as above
606L	60 mm	132 mm	fly-off lever	60 s	yellow, green, violet, red
609L	60 mm	132 mm	fly-off lever	90 s	white, yellow, green, violet, blue, red
9324P	93 mm	175 mm	pull-cord	240 s	orange (other colours available)

VERIFIED

1.5 in signal cartridges

Description
The Singapore Technologies 1.5 in signal cartridge is the standard type of 38 mm metal-cased single star signalling cartridge. It produces a coloured star visible for considerable distances.

Data
Calibre: 38 mm (1.5 in)
Cartridge diameter: 39.7 mm
Length: 100 mm
Vertex height: 100 ±20 m at 90°
Burning time: 7 ±1 s
Colours: red, green, white, illuminating (yellow)

Manufacturer
Singapore Technologies, 249 Jalan Boon Lay, Jurong Town, Singapore 2261.

Status
In production.

VERIFIED

38 mm (1.5 in) signal cartridges

Singapore Technologies Miniflares

Description

Singapore Technologies Miniflares are of the usual screw-threaded canister type, to fit into a pen-type launcher. Two types are provided. The standard pattern is 17 mm in diameter and 32 mm long and ejects a star to a height of 60 to 90 m. The stabilised pattern is 17 mm in diameter and 50 mm long and ejects a star to a maximum height of 110 m. Both types are available in red, white, green or yellow, and both have a burning time of four to six seconds.

Manufacturer

Singapore Technologies, 249 Jalan Boon Lay, Jurong Town, Singapore 2261.

Status

In production.

VERIFIED

Singapore Technologies Miniflares

Singapore Technologies alarm flare

Description

The Singapore Technologies alarm flare is issued as a kit in a sealed case or plastic bag and contains not only the alarm flare unit but also a 25 m reel of plastic coated steel wire to form the trip-wire, a mounting bracket and some other parts. Once the flare unit has been mounted and the igniter inserted into the top of the flare unit it is ready for use as soon as the trip-wire is connected. If the trip-wire is actuated the flare produces a bright light for 35 seconds and also a loud report to draw attention to its functioning.

Data

Height: 110 mm
Diameter: 60 mm
Weight: complete, ca 600 g
Burning time: ca 35 s
Light intensity: 200,000 cd

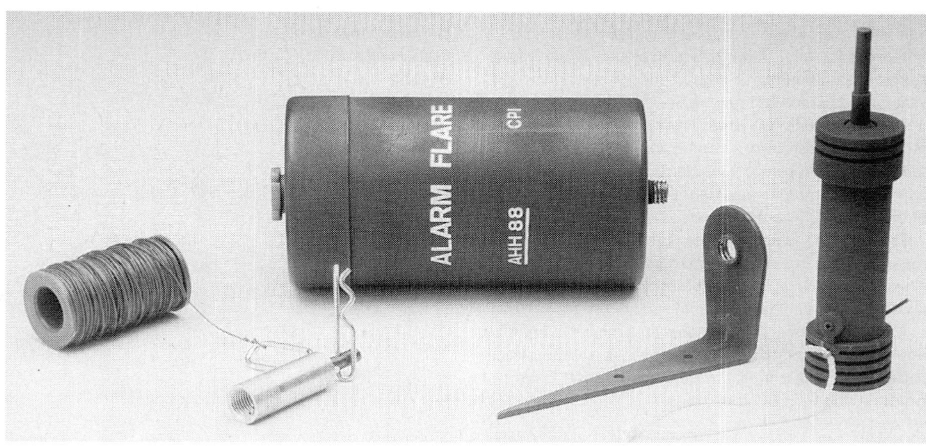

Singapore Technologies alarm flare

1996

Manufacturer

Singapore Technologies, 249 Jalan Boon Lay, Jurong Town, Singapore 2261.

Status

In production.

NEW ENTRY

SOUTH AFRICA

38 mm signal and illuminating cartridges

Description

Red, green or yellow stars or an illuminating flare are available and are identified by a coloured identification band and, at night, by knurling on the case rim. The cartridges are fired from standard 37/38 mm signal pistols or launchers.

Data

Diameter: 38 mm
Length: 100 mm
Weight: pyrotechnic, 70 g
Burning time: 6 s
Vertex: ca 100 m at 90°

Manufacturer

Swartklip Products, PO Box 977, Cape Town 8000.

Status

In production.

Service

South African National Defence Force.

UPDATED

38 mm signal and illuminating cartridges

38 mm coloured smoke cartridge

Description

This 38 mm cartridge consists of an anodised aluminium case carrying a percussion cap and a propellant to eject the aluminium canister containing the coloured smoke composition. The NG/NC propellant minimises muzzle smoke and weapon fouling.

Designed for use in most 38 mm weapons, the cartridge is intended for daylight signalling in a ground-to-air or ground-to-ground role. It can also be employed as a training device in place of CS irritant smoke. The cartridge can also be fired from 40 mm grenade launchers of the M79/M203 pattern, the rim being stepped to suit either weapon.

Data

Case diameter: 39.9 mm
Length: 104 mm
Weight: 192 g; pyrotechnic filling, 87 g
Smoke emission time: min, 20 s
Delay time: 0.6 s
Range: effective, min, 110 m
Colours: red, green, orange, blue

Manufacturer

Swartklip Products, PO Box 977, Cape Town 8000.

Status

In production.

Service

South African National Defence Force.

UPDATED

15 mm signal cartridge

Description

This 15 mm signal cartridge consists of an anodised aluminium case fitted with a 0.22 rimfire cap. The case is threaded and will fit most standard launchers of the 'Pengun' or 'Miniflare' types. There are three colours available, identifiable by coloured closing cups and, at night, by raised ribs on the case: red star, two ribs; green star, one rib; white star, smooth case.

Data

Diameter: 20 mm
Length: 40 mm
Weight: 11.2 g; pyrotechnic filling, 3.8 g
Burning time: 6 s
Max height: 75 m

Manufacturer

Swartklip Products, PO Box 977, Cape Town 8000.

Status

In production.

UPDATED

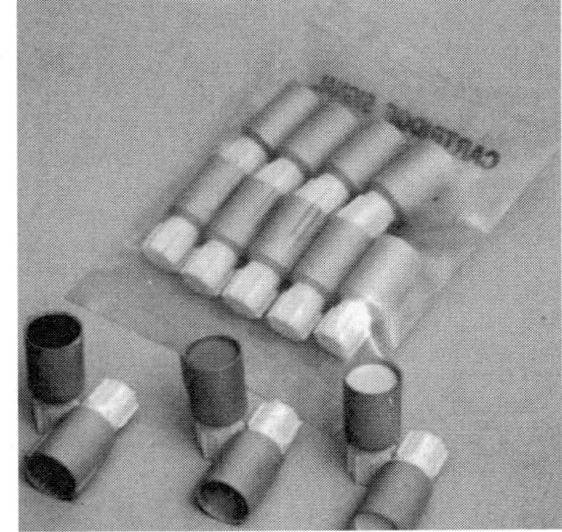

15 mm signal cartridge

Hand rocket flare

Description

This is a hand-held rocket discharger. The rocket has a solid-fuel propellant motor and is spin stabilised for greater accuracy. After firing it ejects a parachute-suspended illuminating flare at the vertex of its flight. When fired vertically it can be used to illuminate a large area or, alternatively, when fired at an angle of 45° a more concentrated light can be achieved over a restricted area. It is therefore suitable for target identification, counter-insurgency operations, search and rescue operations, perimeter defence and signalling.

The flare is designed for ease of operation. On removal of the bottom cap and safety pin, a trigger lever swings free. While holding the rocket firmly this trigger is pressed, resulting in instant ignition.

A training version is available. This model is handled and fired in exactly the same manner as an operational rocket flare but it fires an inert body which does not produce any light or other signal.

Data

Diameter: 48 mm
Length: 267 mm
Weight: 350 g; pyrotechnic filling, 83 g
Burning time: 25-35 s
Luminosity: 80-100,000 cd
Height: 200 m at 45°
Illumination area: 200 m diameter
Range: at 45°, 300 m

Manufacturer

Swartklip Products, PO Box 977, Cape Town 8000.

Status

In production.

UPDATED

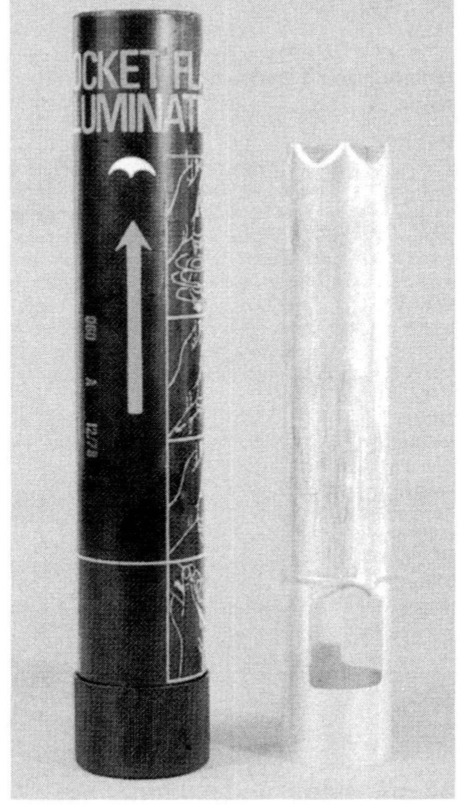

Hand rocket flare

Hand rocket signal

Description

This is a tubular casing containing a solid-fuel rocket which is spin stabilised for greater accuracy. After firing it ejects a free-falling coloured star at the vertex of its flight. It can be fired vertically or at an angle and is suitable or use as a position marker or for various signalling applications.

Data

Diameter: 48 mm
Length: 267 mm
Weight: 300 g; pyrotechnic filling, 83 g
Burning time: 6-10 s
Ejection height: at 45°, 250 m
Colours: red, green, yellow

Manufacturer

Swartklip Products, PO Box 977, Cape Town 8000.

Status

In production.

UPDATED

Hand rocket signal

Trip-wire flare

Description

This trip-wire flare is used for perimeter protection to give warning of intruders. The flare with its integral striker mechanism is mounted on a short steel picket. The trip-wire is attached to the release mechanism and tensioning device, and stretched to another picket some 19 m distant. Pulling or releasing the trip-wire causes the flare to be ignited, thus illuminating the surrounding area. Full light output is almost instantaneous. An adaptor is available for electric ignition. The flare assembly can also be used as a hand-thrown illuminating grenade in an emergency.

Data

Diameter: 43 mm
Height: 173 mm
Weight: 480 g; pyrotechnic, 250 g
Length of trip-wire: 19 m
Length of electric wire: 50 m
Burning time: 80-110 s
Luminosity: min, 80,000 cd

Manufacturer

Swartklip Products, PO Box 977, Cape Town 8000.

Status

In production.

Service

South African National Defence Force.

UPDATED

Swartklip trip-wire flare

Illuminating hand grenade

Description

This illuminating hand grenade consists of an aluminium case containing the illuminating composition, to which is fitted a conventional fly-off lever striker mechanism. The striker mechanism contains the integral delay element which provides the prefunctioning delay. Intended for use by ground forces, the illuminating grenade provides sufficient light for target identification and attack. The grenade can also be used as a light source for emergency conditions when other lights are not available.

Data

Diameter: 43 mm
Length: 173 mm
Weight: 480 g; illuminating composition, 250 g
Delay time: 1.5 s
Burning time: min, 80-110 s
Luminosity: min, 80,000 cd

Manufacturer

Swartklip Products, PO Box 977, Cape Town 8000.

Status

In production.

Service

South African National Defence Force.

UPDATED

Swartklip illuminating hand grenade

Grenade, Hand, Coloured Smoke No 83 Mk 3

Description

The Grenade, Hand, Coloured Smoke No 83 Mk 3 consists of a cylindrical metal body containing the smoke composition, a spring-loaded striker mechanism of the fly-off lever type and a pyrotechnic igniter/delay system. The fly-off lever is retained by a conventional safety pin and pull ring.

The grenade has a variety of signalling applications and may also be used for screening and for training exercises in riot control.

Data

Diameter: 58 mm
Length: 143 mm
Weight: 482 g; pyrotechnic filling, 175 g
Delay time: 1.2-3 s
Emission time: 12 to 30 s
Colours: red, green, yellow, orange, blue, white

Manufacturer

Swartklip Products, PO Box 977, Cape Town 8000.

Status

In production.

Service

South African National Defence Force.

UPDATED

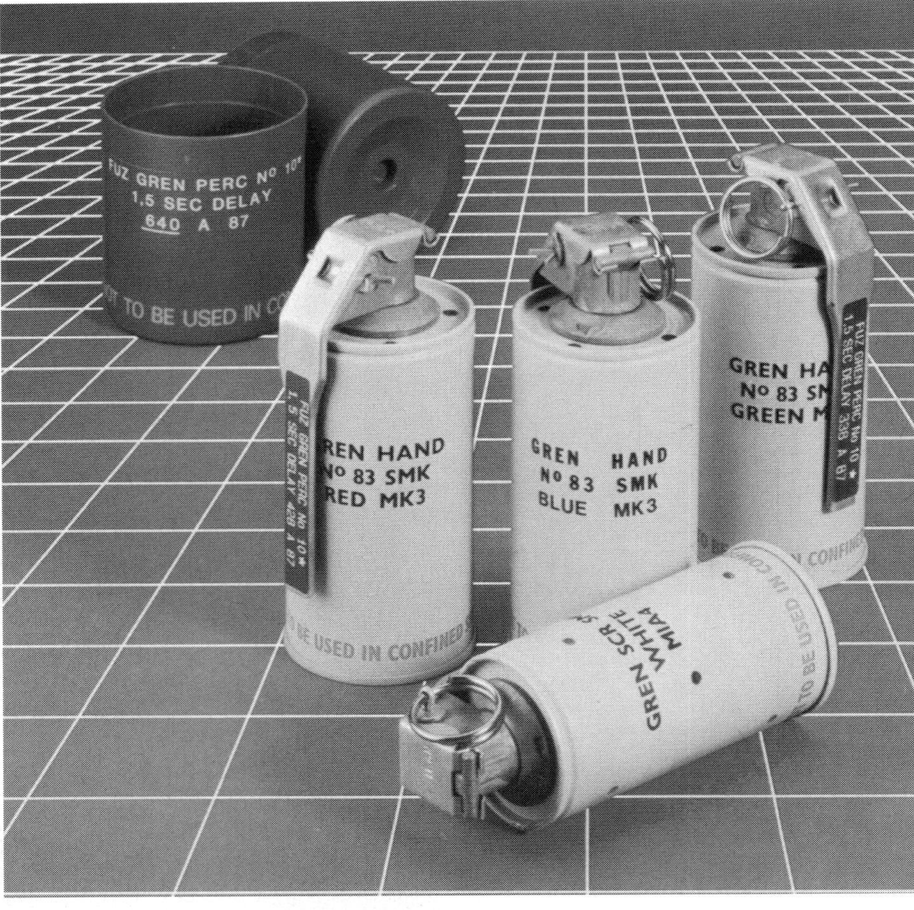

Grenade, Hand, Coloured Smoke No 83 Mk 3

81 mm vehicle-launched smoke generator

Description

This is an electrically initiated munition which is fired from the standard 81 mm launcher fitted to various types of armoured vehicle. It is waterproof and resistant to jolts and vibration, enabling it to remain permanently in the launcher ready for instant use in an emergency.

The device consists of three major parts: the propulsion assembly and the primary and secondary smoke canisters. On firing, the launching charge propels the generator away from the vehicle and simultaneously lights a delay element which controls the time before smoke emission begins. A few metres above the ground a detonator bursts the primary canister, causing an instant cloud of smoke formed by a composition containing Red Phosphorus (RP). Simultaneously, the secondary canister is ignited so that it is already emitting a dense cloud of HC smoke when it reaches the ground.

The generator is intended to provide an instant smoke screen in front of a vehicle subjected to attack. The RP cloud gives instant cover and the HC cloud provides sufficient time for either the vehicle to withdraw or at least for the crew to make their escape.

Data

Diameter: 80.5 mm
Length: 315 mm
Weight: 2.5 kg
Filling: 1.47 kg RP and HC
Delay time: ca 3 s
Range: 40-60 m
Time of smoke emission: 35-50 s

Manufacturer

Swartklip Products, PO Box 977, Cape Town 8000.

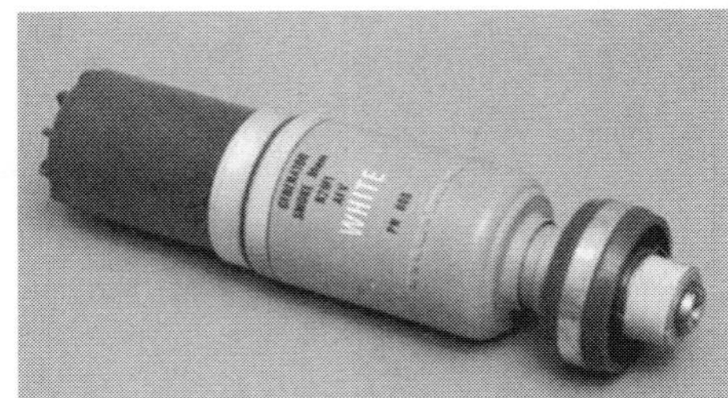

81 mm vehicle-launched smoke generator

Status

In production.

Service

South African National Defence Force.

UPDATED

SPAIN

EXPAL ME-1 penrocket flare

Description

The EXPAL ME-1 consists of a pen-shaped discharger fitted with a spring-driven firing pin and a screwed socket. The flares are screwed into the socket and the firing pin released to fire the flare, which can reach an altitude of 70 m. The flare burns with a coloured light of 6,000 candela for eight seconds and can be used for identification or signalling purposes. The discharger and six flares, two each of green, white and red, are contained in a soft case.

Manufacturer

EXPAL, Avenida del Partenón, 16 5th Floor, E-28042 Madrid.

Status

In production.

VERIFIED

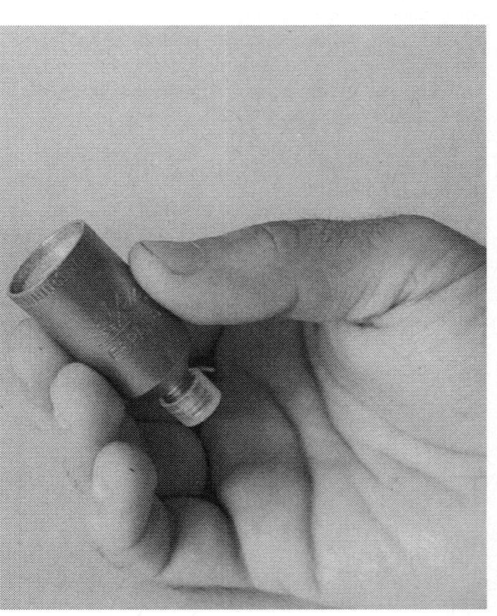

EXPAL ME-1 penrocket flare

EXPAL MRH-1 signal smoke generator

Description

The EXPAL MRH-1 is a hand grenade-type smoke generator which can be used for signalling purposes, for target indication or for indicating wind direction to helicopters. It is available in red, yellow, green or blue colours, burns for 85 seconds and has a delay of 12 seconds after releasing the fly-off lever igniter, to allow the generator to be set carefully and for the operator to move clear.

Manufacturer

EXPAL, Avenida del Partenón, 16 5th Floor, E-28042 Madrid.

Status

In production.

VERIFIED

EXPAL MRH-1 signal smoke generators

EXPAL day and night signal

Description

The EXPAL day and night signal is a double hand-held signal consisting of an aluminium tube with a signal unit at each end with independent ignition. The day unit emits an intense orange smoke cloud for more than 30 seconds; the night unit emits a brilliant red light of over 10,000 candela for more than 30 seconds. Initiation is by percussion, having removed the screwed cap on the appropriate end.

Manufacturer

EXPAL, Avenida del Partenón, 16 5th Floor, E-28042 Madrid.

Status

In production.

VERIFIED

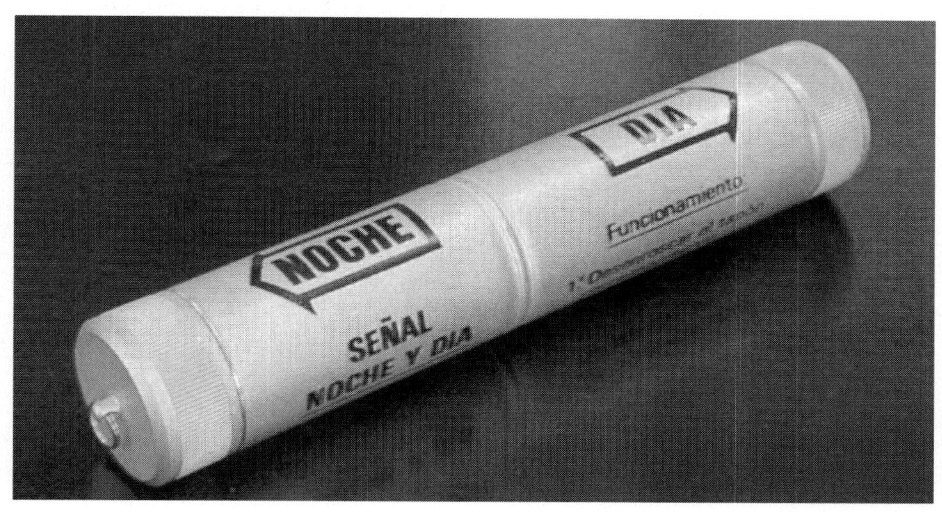

EXPAL day and night signal

EXPAL red parachute flare

Description

This is a hand-launched rocket carrying a red parachute flare for use as a distress or other signal. The rocket is gyroscopically stabilised, and can be initiated and then thrown into water if necessary as it can float and will launch the rocket from the water. The canister is of plastic and removal of the cap exposes the percussion firing device which has a two second delay.

Data

Base diameter: 65 mm
Body diameter: 45 mm
Length: 245 mm
Weight: 420 g
Altitude: 200 m
Luminosity: 25,000 cd
Burning time: 45 s

Manufacturer

EXPAL, Avenida del Partenón, 16 5th Floor, E-28042 Madrid.

Status

In production.

VERIFIED

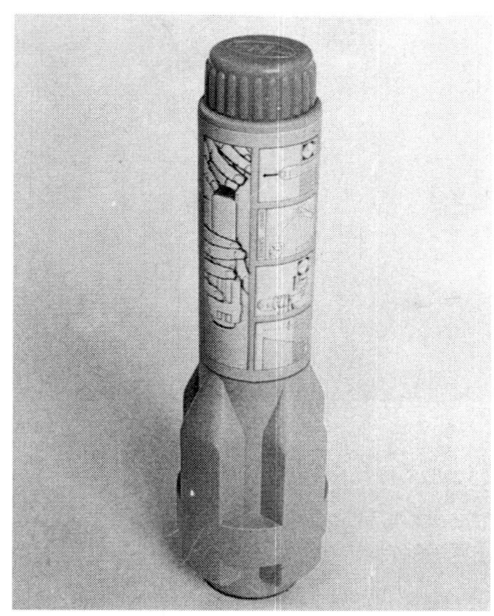

EXPAL red parachute flare

EXPAL red hand flare

Description

This is a very compact red flare for emergency signalling. To use, the plastic tubular cover is unscrewed to expose the flare body and is then reversed and screwed to the base of the body so that it forms a handle. The flare is then ignited by pulling the percussion igniter ring, exposed when the cover was removed.

Data

Diameter: 35 mm
Length: 122 mm
Weight: 110 g
Luminosity: 13,000 cd
Burning time: 45 s

Manufacturer

EXPAL, Avenida del Partenón, 16 5th Floor, E-28042 Madrid.

Status

In production.

VERIFIED

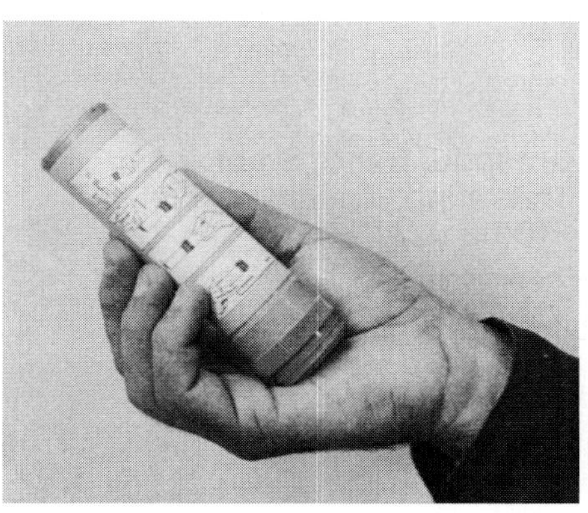

EXPAL red hand flare

EXPAL smoke generators

Description

EXPAL manufactures three standard smoke generators. All are filled with hexachloroethane mixtures for screening or signalling use.

Model HC2 is a hand grenade pattern, weighing 580 g and giving off smoke for two minutes.

Model HC4 is similar to the HC2 but weighs 1.7 kg and emits smoke for four minutes.

Model HC9 is percussion fired, weighs 5.4 kg and emits smoke for nine minutes.

Manufacturer

EXPAL, Avenida del Partenón, 16 5th Floor, E-28042 Madrid.

Status

In production.

VERIFIED

EXPAL smoke generators

SWEDEN

Helios illuminating rocket

Description

The Helios illuminating rocket is a self-contained hand-fired rocket for the illumination of target areas. It will provide 5 lux illumination at ranges up to 400 m when burst at the optimum height of 170 m above the target area. This level of light is adequate for well-aimed effective fire on any type of target.

The Helios rocket is fired by removing the end cap, elevating the tube and pressing the trigger lever against the outer case.

Data

Diameter: 45 mm
Length: 270 mm
Weight: 550 g
Illumination: 380,000 cd
Light duration: 30 s
Rate of descent: 5 m/s

Manufacturer

Norabel AB, PO Box 1133, S-436 00 Askim.

Status

In production.

VERIFIED

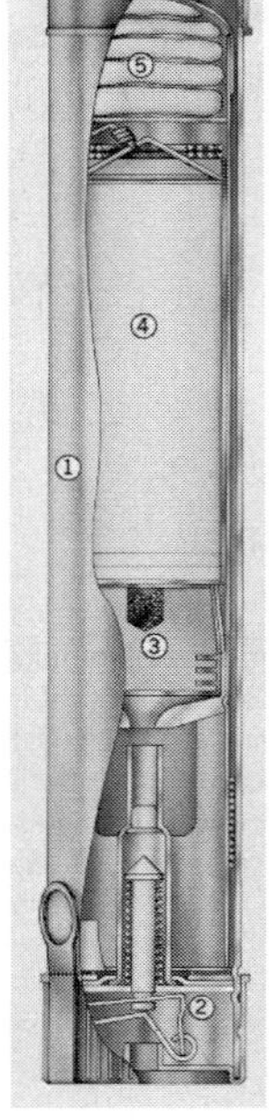

Helios illuminating rocket
(1) *launching tube* **(2)** *firing mechanism* **(3)** *rocket motor* **(4)** *flare* **(5)** *parachute*

Horizon illuminating rocket

Description

The Horizon illuminating rocket is similar to the Helios rocket (see previous entry) but is longer and heavier and differs in internal construction since it is intended for long-range firing.

Data

Diameter: 45 mm
Length: 400 mm
Weight: 750 g
Illumination: 450,000 cd
Duration: 30 s
Range: 1,000 m

Manufacturer

Norabel AB, PO Box 1133, S-436 00 Askim.

Status

In production.

VERIFIED

UNITED KINGDOM

Grenade, Hand, Smoke, Screening, Training, L72A1 (RO 03 02A1)

Description

The RO 03 02A1 grenade is produced for use by forces undergoing training when individual soldiers are exposed to smoke screens and may inhale the smoke.

The grenade comprises a pellet of smoke-producing composition housed in a tinplate container. Initiation of the grenade is by a conventional fly-off lever and striker system. On functioning of the striker mechanism, ignition of the smoke pellet is immediate, but there is a short delay (nominally eight seconds) before the production of effective smoke.

The smoke produced is of low toxicity and the safety of the grenade has been demonstrated during testing by the UK Ordnance Board and acceptance testing by the British Army.

Data

Diameter: 70 mm
Weight: 450 g
Height: 120 mm
Delay time to ignition: nil
Time to effective smoke: 8 s
Duration of smoke: 80 s

Manufacturer
British Aerospace Defence Limited, Royal Ordnance Division, Euxton Lane, Chorley, Lancashire PR7 6AD.

Status
In production.

Service
British Army.

VERIFIED

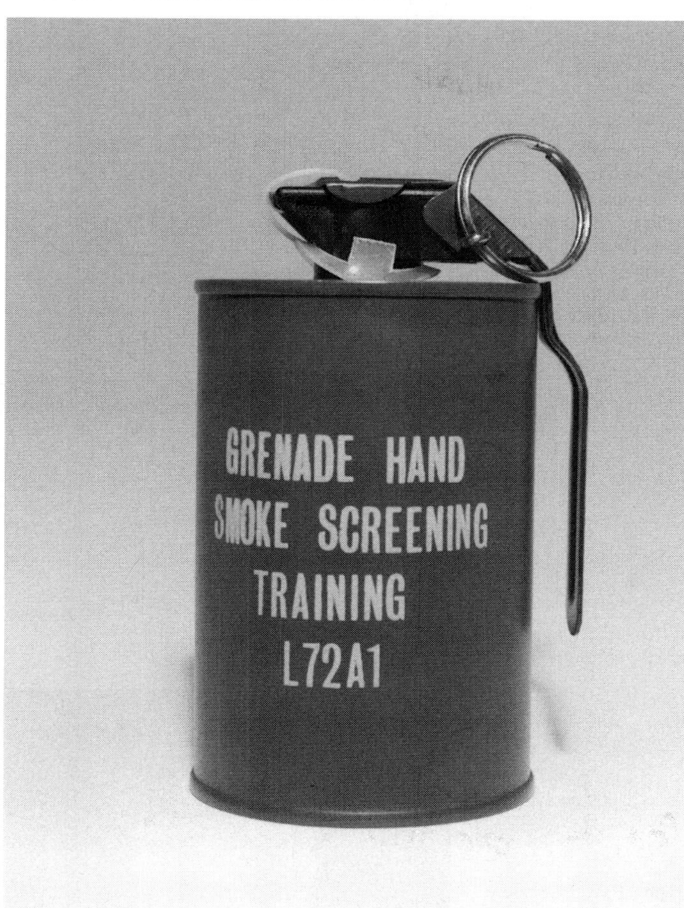

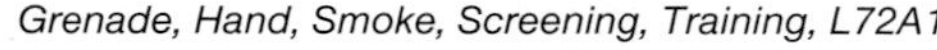

Grenade, Hand, Smoke, Screening, Training, L72A1

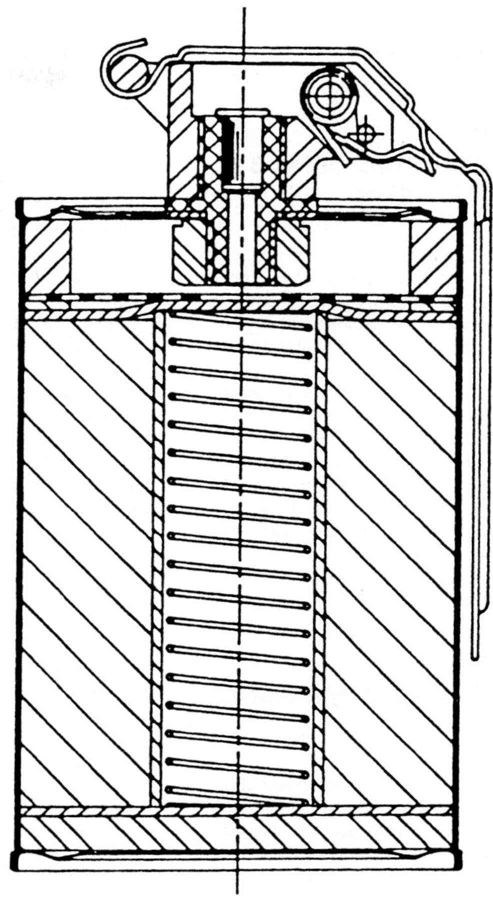

Internal arrangements of L72A1 screening smoke grenade

Generators, Smoke Training, N5 and N6

Description
The Generators, Smoke Training, N5 and N6 are used during breathing apparatus and firefighting training to produce a dense, low toxicity smoke, white without flaming. The generator is operated by removing the top cap and pulling the exposed lanyard igniter. There is a delay of two to four seconds before smoke starts to be generated for between 25 and 30 seconds.

Both generators have the same external dimensions but the N6 is slightly heavier, producing smoke for use in chambers with a volume of 85 m^3 (the volume for the N5 is 43 m^3).

Data
Diameter: 66 mm
Weight: N5, 150 g; N6, 200 g
Height: 105 mm
Delay: 2-4 s
Duration of smoke: 25-30 s

Manufacturer
British Aerospace Defence Limited, Royal Ordnance Division, Euxton Lane, Chorley, Lancashire PR7 6AD.

Status
In production.

Service
British Army.

VERIFIED

Generators, Smoke Training, N5 and N6

Schermuly 38 mm hand-held rockets

Description
Schermuly offers a range of hand-held rockets, all of which are manufactured to the same basic design, have identical firing sequences and differ only in payload. These stores have been well proven in operational service with British and overseas armed forces. The parachute illuminating rocket described below is the one most commonly encountered for military applications. Other versions have payloads of parachute flares, free-falling stars, explosive maroons or radar-reflective dipoles.

Each self-contained rocket consists of an environmentally protected launch tube containing the firing mechanism (uncocked for maximum safety), the rocket assembly and the payload. The rocket assembly itself has a solid fuel motor.

Fired vertically the illuminating rocket can be used to illuminate a large area, or alternatively when fired at an angle of 45°, a more concentrated light can be achieved over a smaller area. It is therefore suitable for a wide range of illuminating applications from battle areas to specific targets, for night reconnaissance and for search and rescue.

The unit was designed for maximum ease of operation. Top and bottom transit caps and the safety pin are removed, thus allowing the trigger lever at the base to swing free. While holding the rocket firmly, the trigger lever is pressed against the outer case. Ignition is instantaneous.

In addition to the parachute illuminating rocket the following payloads are available: multiple free-falling stars - red, green or white, for short duration general or distress signalling purposes; parachute suspended flares - red or green, for longer duration general or distress signalling purposes; maroon, flash and loud report; radar reflective dipoles which enable position location by radar under adverse weather conditions. The signal carries packs of dipoles to provide a radar echo and a free-falling star for visual sighting under more favourable conditions. Parataget provides an airborne orange target for anti-aircraft practice with small arms.

38 mm hand-held parachute illuminating rocket

Data

Diameter: 48 mm
Length: 267 mm
Weight: 345 g
Flare composition: 120 g
Illuminating power: min, 80,000 cd
Ejection altitude: vertical firing, 300 m; 45° firing, 200 m
Burning time: nominal, 30 s

Manufacturer

Pains-Wessex Schermuly, High Post, Salisbury, Wiltshire SP4 6AS (a member of the Chemring Group plc).

Status

Available.

Service

British and other armies in America, Africa, Europe, Middle East and Far East.

VERIFIED

38 mm hand-held rocket Mk 3 Radasound/Radaflare

Description

Each Radasound/Radaflare self-contained rocket consists of an environmentally protective tube containing the uncocked firing mechanism, rocket assembly and radar reflective payload. The rocket assembly has a solid fuel motor, is tail stabilised and rotates on its axis for greater accuracy.

For Radasound the typical payload is five packs of I-band chaff but this may be varied to meet customer requirements. For Radaflare, the payload comprises typically four packs of I-band chaff and one free-falling red star. The red star gives a visual indication of the location of the chaff cloud. Both rockets are fired vertically and are operated in the same way as the 38 mm parachute illuminating L5A4 rocket.

Radaflare and Radasound can be used in a variety of situations including: distress signalling (position location in adverse weather conditions), decoy, radar training and testing, terrain-mapping, radar range calibration, location aid to assist supply dropping.

The maximum theoretical radar cross-section return averages are 1,100 m² for Radasound and 880 m² for Radaflare; tracking is possible for 12 to 25 minutes dependent on wind conditions. Both rockets exclusively use CHEM-CHAFF aluminised glassfibre.

Data

L8A1 RADASOUND
Diameter: 48 mm
Length: 268 mm
Weight: 345 g; explosive content, 46 g
Deployment height: vertical, 300 m; 45°, 200 m
Radar cross-section: 1,100 m² at 9.3 GHz (I-Band)

RADAFLARE
Diameter: 48 mm
Length: 268 mm
Weight: 390 g; explosive content, 100 g
Deployment height: 300 m
Flare burn time: nominal, 11 s
Light intensity: min, 30,000 cd
Radar cross-section: 880 m² at 9.3 GHz (I-Band)

Manufacturers

Pains-Wessex Schermuly, High Post, Salisbury, Wiltshire SP4 6AS.
Chemring Limited, Alchem Works, Fratton Trading Estate, Portsmouth PO4 8SX (members of Chemring Group plc).

Status

Available.

Service

Europe, Far East, UK.

VERIFIED

Schermuly Day and Night distress signal

Description

The Schermuly Day and Night signal is a dual-purpose compact distress signal designed to military specifications. One purpose is an orange smoke to attract attention in daylight, and the other a red flare to summon help during darkness. The signal is visible from at least 5,000 m by day and 8,000 m by night in good conditions.

The signal consists of a yellow protective outer case with an improved tactile marking of two raised ribs around the flare end for easy recognition in darkness. The red protective end caps, which are now interchangeable, are sealed by waterproof 'O' rings.

Either end can be fired independently by means of a pull ring percussion striker mechanism and the unused end stored until it is required.

The Day and Night's versatility and compact shape enables it to be carried easily as a personal distress signal. It is suitable for infantry, for use in aircraft emergency and personal survival packs, especially in maritime situations, and as a diver's personal distress signal.

Data

Length: 139 mm
Width: 42.5 mm
Weight: 228 g
Burn time: flare, nominal, 20 s, >10,000 cd average; smoke, nominal, 18 s

Manufacturer

Pains-Wessex Schermuly, High Post, Salisbury, Wiltshire SP4 6AS (a member of the Chemring Group plc).

Status

In production.

Service

UK Ministry of Defence and other countries worldwide.

VERIFIED

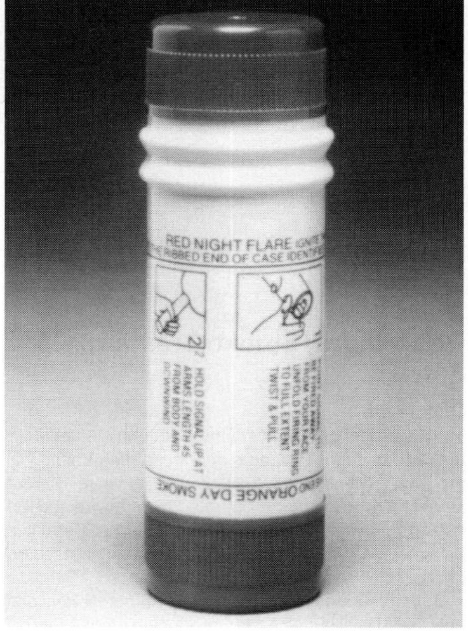

Schermuly Day and Night distress signal

Manroy 25 mm signal pistol

Description

The Manroy 25 mm signal pistol is a standard Very pattern single-shot, drop barrel signal pistol, using a thumb-cocked hammer single-action mechanism. It is of all steel construction with bakelite grips and will accept and fire all types of standard 25 mm (1 in) signal pistol ammunition.

The same pistol is also available in 26.5 mm calibre to suit certain other patterns of signal cartridge.

Data

Calibre: 25 mm (1 in)
Length: 250 mm
Barrel: 145 mm
Weight: 820 g

Manufacturer

Manroy Engineering, Hobbs Lane, Beckley, East Sussex TN31 6TS.

Status

In production.

Service

British Army.

VERIFIED

Manroy 25 mm signal pistol

Haley and Weller screening smoke grenade E101

Description
The E101 grenade ejects six submunitions to give a very rapid build-up of dense smoke for screening purposes. It was developed to supplant the white phosphorus grenade with a non-incendiary device which will provide smoke with equal speed. The ignition is performed by the unique Haley and Weller silent electrical system. This resembles the conventional fly-off lever pattern so far as the user is concerned but the lever does not leave the grenade and there is no noise of a cap being struck. The movement of the lever after throwing permits an electrical circuit to close and send current from a built-in battery to an electrical delay system which then ignites the smoke composition.

Data
Diameter: 50 mm
Length: 134 mm
Weight: 460 g; smoke content, 270 g
Fully effective screen: <3 s
Duration of smoke: 25 s
Operational temperature range: −30 to +75°C

Manufacturer
Haley and Weller Limited, Wilne, Draycott, Derbyshire DE72 3QJ.

Status
In production.

UPDATED

Haley and Weller screening smoke grenade

Signalling smoke grenade E120 range

Description
The grenades in the E120 range resemble the screening smoke grenade described in the previous entry but are filled with coloured smoke composition for signalling and indicating purposes. They use the same electrical ignition system.

Data
Diameter: 50 mm
Length: 134 mm
Weight: 300 g; smoke content, 150 g
Burning time: nominal, 60 s
Colours: red, blue, yellow, green, orange, purple, white, sand
Operational temperature range: −30 to +75°C

Manufacturer
Haley and Weller Limited, Wilne, Draycott, Derbyshire DE72 3QJ.

Status
In production.

UPDATED

Flare, Hand, Illuminating, grenade E140

Description
The E140 grenade provides an instantaneous bright white light either for illuminating purposes or to blind electro-optical night sights by overloading their circuitry. It uses the same silent electrical ignition system as the smoke grenades previously described, making it a particularly effective countermeasure against snipers with night sights.

Data
Diameter: 50 mm
Length: 134 mm
Weight: 297 g; flare content, 105 g
Intensity: 80,000 cd
Duration: 12-15 s
Delay time: 1-1.5 s
Operational temperature range: −30 to +75°C

Manufacturer
Haley and Weller Limited, Wilne, Draycott, Derbyshire DE72 3QJ.

Status
In production.

UPDATED

Haley and Weller illuminating hand grenade

Flare, Ground Indicating, Yellow, Y920

Description
The Y920 is a pyrotechnic flare for indicating landing or dropping zones for parachute troops, or for other indicating or illuminating functions. It is fitted to a spike so that it can be driven into the ground easily, and then ignited by means of a portfire or other flame.

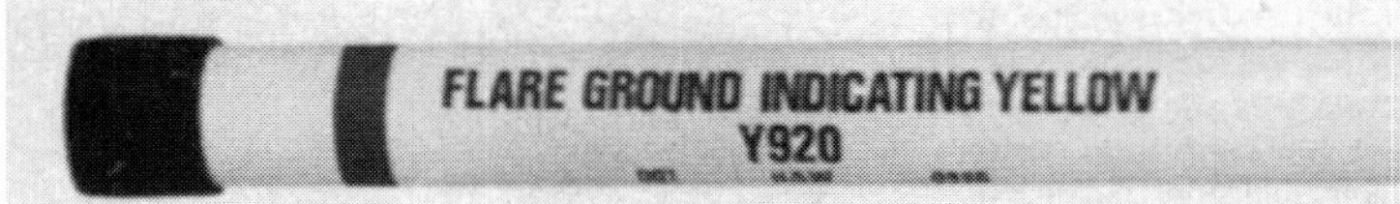

Haley and Weller Type Y920 ground indicating flare

Data
Diameter: 41 mm
Length: 340 mm
Weight: 373 g; flare content, 333 g
Duration: nominal, 3 min
Operational temperature range: −30 to +75°C
Shelf life: 3 yrs, subject to storage conditions

Manufacturer
Haley and Weller Limited, Wilne, Draycott, Derbyshire DE72 3QJ.

Status
In production.

VERIFIED

Haley and Weller 38 mm hand-fired rockets

Description

Haley and Weller 38 mm hand-fired rockets, employing a standard launch tube, rocket assembly and firing mechanism, are manufactured with a variety of payloads to meet any signalling requirements. Continuous product development has produced a reliable and robust range of rockets with excellent performance characteristics in relation to their size and weight.

Data

Diameter: 45 mm; rocket, 38 mm
Length: 275 mm
Operational temperature range: −30 to +75°C
Shelf life: 3 yrs, subject to storage conditions

Manufacturer

Haley and Weller Limited, Wilne, Draycott, Derbyshire DE72 3QJ.

Status

In production.

VERIFIED

Haley and Weller 38 mm hand-fired rockets

Haley and Weller trip-wire mechanism E190

Description

The Haley and Weller trip-wire mechanism E190 (Trip-wire Mechanism Mark 6 E190) is a reusable mechanism designed to meet operational requirements in any theatre worldwide. The design incorporates safety features for user confidence, facilitating safe and simple setting by night. The spring-tensioned trip-wire ensures actuation of the mechanism if tripped or cut.

Data

Length: operational, 220 mm
Weight: 1.25 kg
Operational temperature range: −30 to +75°C
Shelf life: 3 yrs, subject to storage conditions

Manufacturer

Haley and Weller Limited, Wilne, Draycott, Derbyshire DE72 3QJ.

Status

In production.

UPDATED

Haley and Weller trip-wire mechanism E190

Schermuly 38 mm signalling system

Description

The Schermuly 38 mm signalling system consists of the 38 mm (1.5 in) signal pistol and 38 mm signal cartridges.

The Schermuly pistol is chambered to fire 38 mm paper and alloy cased signal cartridges. The pistol is constructed of phosphated steel for light weight and durability, and incorporates a single-action trigger mechanism. It can be held and made ready for loading with one hand, as the breech-opening catch is conveniently located on the side of the pistol body. The pistol handle contains a plastic grip in each side to give a positive firm hold.

To load the weapon it is necessary only to push forward the breech-opening catch and allow the barrel to swing open through 90°, insert the round as far as the extractor will permit and close the pistol until the barrel catch fully engages. The pistol has a tough, smooth, wipe-clean surface, requires minimum maintenance and is simple and safe to use. A lanyard is also provided for extra security.

Schermuly signal cartridges consist of a rimmed aluminium case into which the propellant charge and signalling or illuminating star are inserted and sealed with a weatherproof closing disc. The colour of the star is indicated on the side of the case and by means of the NATO night identification markings embossed on the cap.

Schermuly 38 mm signal pistol and cartridge

Applications for the system include: ignition of gas/oil stack pipes; signalling surface-to-surface, surface-to-air or air-to-surface, and distress and search and rescue tasks. Red cartridges are used for distress signalling; green and white cartridges for warn-off and miscellaneous signalling; illuminating cartridges for location or night reconnaissance. All cartridges are visible from 3,200 to 4,800 m by day or 8,000 m by night, in clear weather conditions.

Data

CARTRIDGES
Calibre: 38 mm
Weight: 86 g
Length: 70 mm
Trajectory height: nominal, 78 m
Burn time: nominal, coloured stars, 6 s
Illuminating time: nominal, 5 s
Candlepower: red, >55,000; green, >33,000; white, >55,000; illuminating, >120,000 cd (average values)

SIGNAL PISTOL
Calibre: 38 mm
Length: 265 mm
Weight: 1.28 kg

Manufacturer

Pains-Wessex Schermuly, High Post, Salisbury, Wiltshire SP4 6AS (a member of the Chemring Group plc).

Status

Available.

Service

Worldwide.

VERIFIED

Schermuly 38 mm 2-star signal cartridges

Description

The Schermuly 38 mm 2-star signal cartridge is designed to be fired from the Schermuly 38 mm signal pistol and other similar pistols. Colours available are red, green and white.

The cartridge is used for both day and night maritime and land signalling purposes. It consists of two coloured stars and a propellant charge inserted in an aluminium case and sealed with a weatherproof closing disc.

The cartridges have been developed to function reliably under high humidity conditions and over a wide temperature range.

Data

Calibre: 38 mm
Length: 85 mm
Trajectory height: nominal, 80 m
Burn time: 6-7 s
Candlepower: red, >25,000; green, >22,000; white, >25,000 cd

Manufacturer

Pains-Wessex Schermuly, High Post, Salisbury, Wiltshire SP4 6AS (a member of the Chemring Group plc).

Status

In production.

Service

Overseas governments.

VERIFIED

Miniflare No 1 Personal Signalling Kit

Description

This kit comprises a pen-sized projector and eight screw-on cartridges in a weatherproof plastic pack. Each cartridge, a self-contained waterproof unit, has a propellant and a star emitting an intense light. For signalling use, red, green and white colour-coded cartridges are available. An important feature of the Miniflare is that when fired vertically as a distress signal, the star burns out well above ground or sea level. With an effective range of 50 to 100 m this system is suitable for small patrol and ambush operations.

The cartridge consists of an aluminium case containing the propellant and star, the case also acting as a projection cup. A waterproofing disc seals the top of the cartridge and the threaded projection at the base contains the percussion cap.

The projector is anodised light alloy with a female thread at the top to take the male-threaded cartridge. A trigger is attached directly to the spring-loaded striker.

The plastic pouch has NATO tactile markings for night identification.

Data

Projector dimensions: 120 × 22 mm
Cartridge dimensions: 32.5 × 16.7 mm
Pack dimensions: 150 × 54 × 19 mm
Weight: complete, 216 g
Altitude: typical, 85 m
Burning time: nominal, 5-7 s
Candlepower: red, 3,000; green, 1,500; white, 3,000 cd

Manufacturer

Pains-Wessex Schermuly, High Post, Salisbury, Wiltshire SP4 6AS (a member of the Chemring Group plc).

Status

In production.

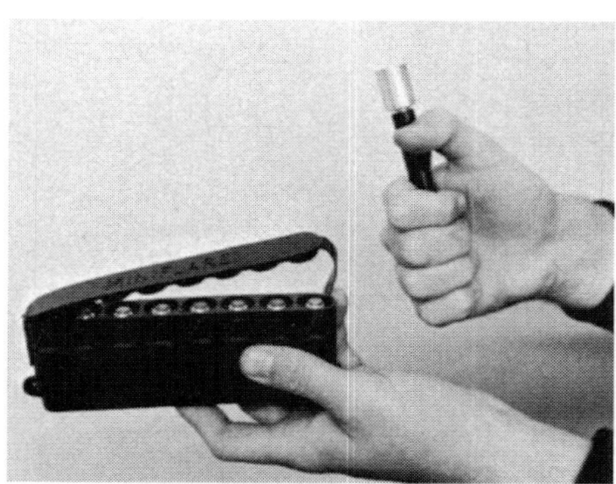

16 mm Miniflare signal kit

Service

UK Ministry of Defence and worldwide.

VERIFIED

Schermuly signal smoke grenade Mk 4

Description

This version of the Schermuly signal smoke grenade incorporates substantial design improvements, including a fly-off lever ignition system which provides immediate generation of smoke though apertures in the top of the grenade body. The body is more compact and has been designed and tested for use under all climatic conditions.

Dense smoke is produced for 45 seconds, providing a visible signal for a minimum of 60 seconds. It is designed for use wherever colour definition and good smoke density are vital.

An FOD (Foreign Object Damage) free version is available for use from helicopters and so on. A plastic loop is fitted to the grenade which attaches the fly-off lever and safety pin to the body of the grenade.

Data

Diameter: 62 mm
Length: 115 mm
Weight: 365 g; smoke composition, 180 g
Burning time: 45-50 s
Colours: red, orange, yellow, green, blue/violet, white

Manufacturer

Pains-Wessex Schermuly, High Post, Salisbury, Wiltshire SP4 6AS (a member of the Chemring Group plc).

Status

Available.

Service

Worldwide.

VERIFIED

Schermuly signal smoke grenade Mk 4

Schermuly signal smoke grenade Mk 5

Description

The coloured signal smoke grenade Mk 5 was designed and developed by Schermuly to meet British Army requirements for a greater and more dense volume of smoke for long distance visual recognition. The grenade consists of a cylindrical metal body with an igniter, a striker and fly-off lever assembly which is held to the body by a plastic strap. The strap enables the grenade to be thrown safely from helicopters without causing Foreign Object Damage (FOD).

When the safety pin is removed and the grenade thrown, the striker fires the percussion cap on release of the fly-off lever. Coloured smoke will emerge from eight apertures in the body lids which are sealed with a plastic disc until after initiation. Dense smoke is emitted for 60 seconds.

Data

Diameter: 66 mm
Length: 138 mm
Weight: 475 g; smoke composition, 215 g
Burn time: 50-60 s
Colours: violet, orange, white, yellow, red, green, blue
UK MoD references: green, L64A1; orange, L65A1; red, L66A1; blue, L67A1

Manufacturer

Pains-Wessex Schermuly, High Post, Salisbury, Wiltshire SP4 6AS (a member of the Chemring Group plc).

Status

In production.

Service

British armed forces and overseas governments.

VERIFIED

Effect of Schermuly signal smoke grenade Mk 5

Schermuly trip-flare kit

Description

The Schermuly trip-flare kit is intended for perimeter security at night. An immediate benefit of the device is the illumination of infiltrating enemy troops by the instantaneous ignition of the flare which produces a high candela for approximately one minute.

The kit basically consists of a flare with a fly-off lever mechanism, two pickets, a spool of wire and a spring. Additional product benefits included in the kit are an anchorage spade to ensure picket stability in light soils and an anti-glare shield which, in its original state, doubles as a packing tin. The flare can also be fitted with an electrical connection for remote firing from a strike detonator. If required, the flare can be used as a hand-thrown incendiary device.

One complete kit consisting of two pickets, one spring and one box containing flare and spool is removed. One picket is forced into the ground in the required position. The second picket is used as an axle to deploy wire from the spool. Anchorage spades are fitted as necessary.

The flare is removed from the packing and one end of the spring assembly is hooked under the leg of the picket and the picket arms are slid into the grooves on the side of the flare. The wing nut is slackened and the wire is bent under and around the bolt. The wire is tensioned until the locking tongue is in the 12 o'clock position (securing the fly-off lever) when the wing nut is tightened. Readjustments are made as necessary. The safety pin is removed ensuring the trigger is still in the set position. The safety pin is kept for disarming. The

Schermuly trip-flare

flare is now armed. The anti-glare shield is fitted as required.

Data

Weight: flare body, total, 360 g; pyrotechnic filling, 115 g
Burn time: nominal, 45 s
Candlepower: 60,000 cd

Manufacturer

Pains-Wessex Schermuly, High Post, Salisbury, Wiltshire SP4 6AS (a member of the Chemring Group plc).

Status

Available.

Service

Worldwide.

VERIFIED

51 mm illuminating rocket and launcher

Description

The 51 mm illuminating rocket and launcher was designed to meet the requirement for an illuminating light source operating at a ground range of 900 m. The 51 mm rocket has a light intensity of 350,000 candela and burns for 30 seconds. Deployed at a height of 250 m the parachute flare will adequately illuminate an area some 300 m in diameter.

The rocket, complete with payload, is supplied in an aluminium container/launch tube incorporating the ignition system and is completely weatherproof. It can be supplied with either a percussion or an electrical ignition device, and for both versions a tripod-mounted launcher weighing only 4 kg is available. The optimum launch angle is 30°.

Data

Diameter: 55 mm
Length: 385 mm
Weight: container/launch tube and rocket, 1.02 kg; infantry-type launcher, 3.9 kg
Light intensity: average, 350,000 cd
Burning time: 30 s
Deployment height: typical, at 32°, 250 m
Range: at ground level with 32° launch angle, 900 m

Manufacturer

Pains-Wessex Schermuly, High Post, Salisbury, Wiltshire SP4 6AS (a member of the Chemring Group plc).

Status

Available.

Service

Armies in Africa, the Americas, Europe and the Middle East.

VERIFIED

Loading infantry launcher with 51 mm parachute illuminating rocket

ML 57 mm rocket parachute illuminator

Description

The 57 mm ML illuminating rocket system is used for target identification in combat areas, where a degree of high mobility is essential. With its lightweight single- or double-barrelled launcher carried in a rugged man-pack, it is immediately deployable in all roles where short-term white light support is required, for example defence perimeters, infantry and armoured support.

The rocket is spin stabilised, highly accurate and, apart from a minimum flash at launch, has the advantage of being signatureless in flight. Rockets can be supplied with fixed ranges, for example 650, 1,000, 1,200 and 1,800 m or with a variable electrical fuze to allow range selection within the 650 to 1,800 m range band. The flare provides an illumination of 250,000 candela when deployed at a height of 250 m.

The rocket is fired from a recoilless, compact launcher weighing only 2 to 3 kg and suited for infantry use, allowing simple one-man operation in both offensive and defensive situations. The launcher systems are mounted on a tripod which folds away for easy carriage. Luminous levelling and sighting devices are provided which enable the firer to set up the launcher very quickly.

Data

Diameter: 57 mm
Length: 300 mm

Loading ML lightweight launcher

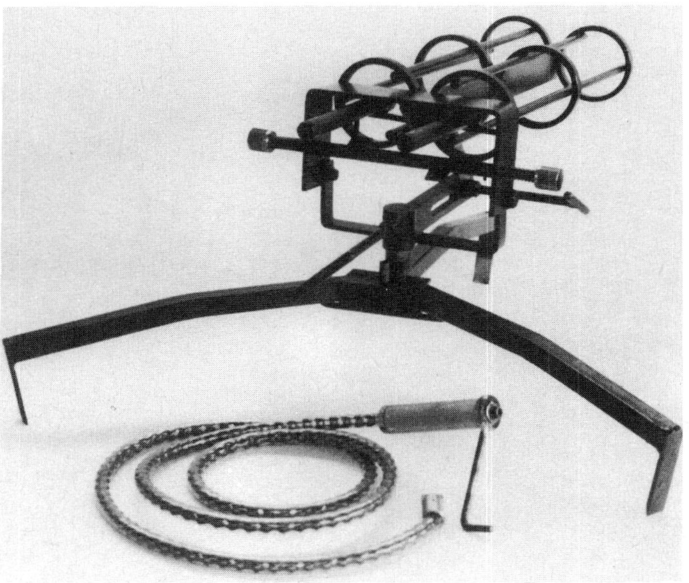

ML 57 mm lightweight launcher, twin-barrelled version

Weight: rocket, 1.8 kg
Burning time: 30 s
Illumination: nominal, 250,000 cd
Range: at 40° QE, 650-1,800 m
Time of flight: ca 7-18 s
Launch ignition: remote mechanical by Bowden cable, electric, electromechanical, or contactless induction

Manufacturer
ML Aviation Limited, Arkay House, Weyhill Road, Andover, Hampshire, SP10 3NR.

Status
In production.

Service
NATO, Middle East, Africa.

VERIFIED

Brocks Ground Indicating Markers

Description
Brocks Explosives Limited, a member of the Explosives Development Group, manufactures a range of signal devices designed to mark an area or light up a temporary landing site in situations where normal signal grenades are ineffective.

The **Arctic Grenade** is designed as a wind reference indicator and ground reference point for helicopter or light aircraft crews in situations where the terrain is covered in snow and in temperatures between −46 and +20°C. By the use of expanding flaps, the grenade lies on the surface of soft snow emitting coloured smoke for a minimum of 25 seconds, whilst the ejected smoke pellets dye the surface of the snow to provide a ground reference mark in white-out conditions. The grenades can be manufactured in alternative colours.

The **Buoyant Grenade** is similar in purpose but is designed to float on water. In addition to delivering a coloured smoke plume (various colours being available) the emitted dyestuff also stains the surface of the adjacent water making a prominent signal mark.

The **Ground Flare, Indicating** is used for indicating landing or dropping areas. It consists of a tin-plate body, holding a flare candle, which has a spike on one end to support the flare when in use. The flare is ignited by a portfire.

Manufacturer
Brocks Explosives Limited, Sanquhar, Dumfries and Galloway DG4 6JP.

Status
In production.

VERIFIED

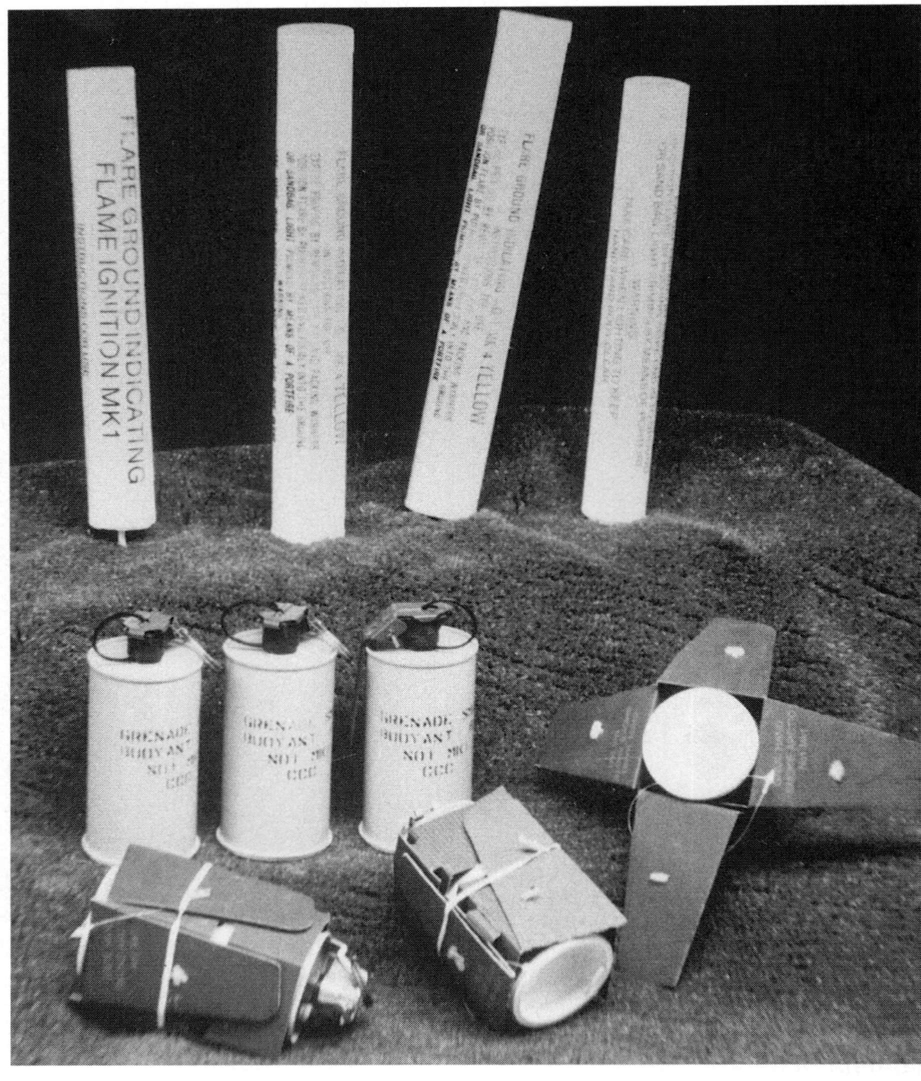

Brocks Ground Indicating Markers

Brocks Smoke Grenades and Generators

Description
Brocks manufactures smoke products for both training and operational purposes. The smoke generator and screening grenades produce a dense cloud of non-toxic grey smoke, screening the movements of infantry and vehicles without the risks associated with white phosphorus.

The signal grenades are manufactured in a range of colours to produce a distinctive plume of coloured smoke, allowing troops to carry out manoeuvres without radio contact.

The **Smoke Generator** produces a dense cloud of grey/white smoke and burns for approximately four to six minutes after ignition. The HCE smoke carries none of the risk sometimes associated with white phosphorus. The unit is electrically ignited and produces smoke almost immediately. It is robust and waterproof and is suitable for use in any climatic conditions.

The **Screening Smoke Grenade** is more fully described in the *Combat Grenades* section.

The **Signal Grenades** are small, lightweight, hand-thrown munitions which produce a dense cloud of coloured smoke shortly after release. They have been approved for British Army service under the following nomenclature:

Grenade, Hand, Signal, Smoke, Blue, L52
Grenade, Hand, Signal, Smoke, Green, L53
Grenade, Hand, Signal, Smoke, Red, L54
Grenade, Hand, Signal, Smoke, Orange, L55.

Brocks Smoke Grenades and Generators

Manufacturer
Brocks Explosives Limited, Sanquhar, Dumfries and Galloway DG4 6JP.

Status
In production.

Service
British Army and export sales.

VERIFIED

UNITED STATES OF AMERICA

M18 coloured smoke hand grenade

Description

The M18 smoke grenade produces dense non-toxic coloured smokes for various signalling, wind direction indication and marker purposes. It is produced by Kilgore Operations and marketed by Alliant Techsystems.

The grenade is a cylindrical metal canister with an M201A1 fuze assembly at the top. A safety lever is held in place with a cotter pin and a pull ring. A smoke emission hole, located at the top of the canister, is sealed with tape to prevent the ingress of moisture.

Each canister contains 196 g of a smoke mixture available in five colours; red, green, white, yellow and violet-blue. All smoke mixture raw materials and dyes involved are non-toxic. The body of the grenades is stencilled to indicate the appropriate smoke colour.

Releasing the safety lever allows the striker to hit the primer. This ignites a 0.7 to 2 second delay element which ignites an ignition mixture to fire the starter mixture and subsequently the smoke mixture. The gases generated blow the tape off the emission hole and allow the smoke to escape. The grenade will produce smoke for a minimum of 50 seconds.

This grenade is produced in South Korea as the K-M18 (qv).

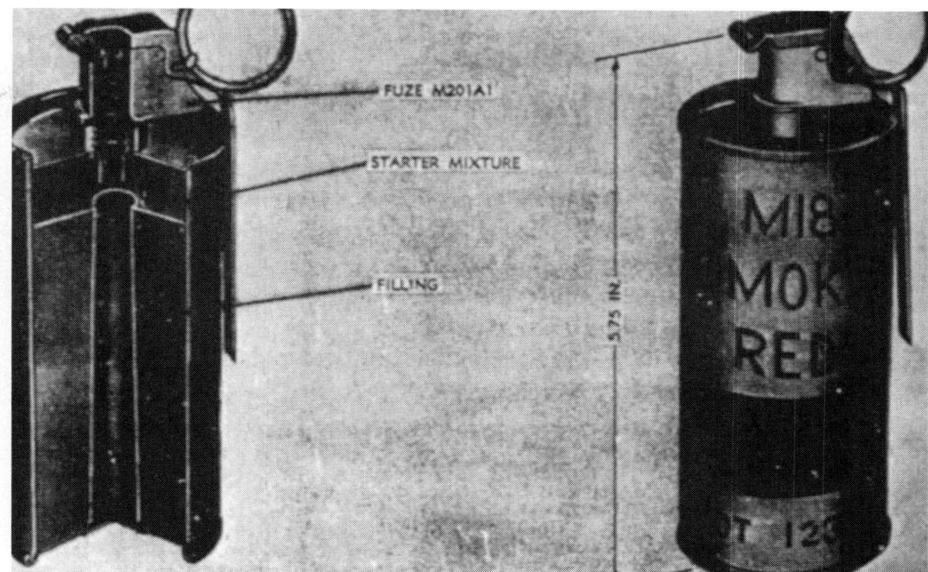

M18 coloured smoke hand grenade

Data

Length: 114 mm
Diameter: 63.5 mm
Weight of filling: 196 g
Time delay: 0.7-2 s
Emission time: at least 50 s

Manufacturer

Kilgore Operations, Kilgore Drive, Toone, Tennessee 38381.

Status

In production.

Service

US Army.

UPDATED

M49A1 surface trip-flare

Description

The M49A1 surface trip-flare is a booby trap flare that illuminates an infiltrating force at night. It is composed of an aluminium case and a cover assembly containing the initiating mechanism. The latter consists of a hinged striker, a lever, a safety clip and a pull ring. It is fitted with a U-shaped mounting bracket and is installed by nailing or clamping it to any upright stake, post or other upright object.

Depending on the installation the flare may be fired by releasing the trigger or by removing a pull pin in the same manner as a grenade pin. When a trip-wire (a 13.7 m length of wire is provided with each unit) is attached to the trigger the flare can be fired by a pressure of as little as 900 g. Cutting or breaking the trip-wire will also fire the flare. The flare is activated instantaneously and produces a 35,000 candela brilliant light for approximately one minute.

The M49A1 surface trip-flare is manufactured by Kilgore Operations and marketed by Alliant Techsystems.

Data

Length: 119 mm
Diameter: 38 mm
Weight: 418 g
Weight of filling: 125 g
Burning time: 55-70 s
Light output: average, 35,000 cd

Manufacturer

Kilgore Operations, Kilgore Drive, Toone, Tennessee 38381.

Status

In production.

Service

US Army and others.

NEW ENTRY

Cyalume Lightsticks

Description

Though not strictly pyrotechnics, since they do not burn, these devices must be considered under this heading since they perform the same function as pyrotechnic flares. The Cyalume Lightstick is a sealed plastic tube containing a liquid chemical composition which is inert until the tube is bent between the fingers and shaken. It then emits light without heat, flames or sparks; depending upon the type of light in use, this emission will last from 30 minutes to 12 hours. The lights are provided in various colours so that they may be used for warning, signalling or illumination purposes. In military use they can be used as minefield, route or perimeter markers, drop-zone markers, river-crossing guides, for map reading or illuminating the interior of command posts or communication centres. The lights work equally well under water and can be used by demolition frogmen. A rugged plastic carrying case (the Combat Light Device) has a clip where it can be attached to clothes or equipment and is fitted with an adjustable shutter which means the light can be regulated or occluded as desired. There is also a Personnel Marker Light which can be attached to flotation devices; this clips to the clothing or life-jacket and carries a permanently fitted Cyalume Lightstick. This can be activated by squeezing it, after which it acts as a marker visible up to 1,600 m and lasts for eight hours.

Types of Lightstick available include:

4 in, General Purpose (green)
6 in, 30 minutes High Intensity (yellow)
6 in, 12 hours General Purpose (green)
6 in, 12 hours Red
6 in, 8 hours Blue
6 in, 12 hours Yellow
6 in, 12 hours Orange
6 in, Non-visible Infra-Red
Personnel Marker Light (green-yellow, 8 hours)
Combat Light Device (holder for 6 in lights).

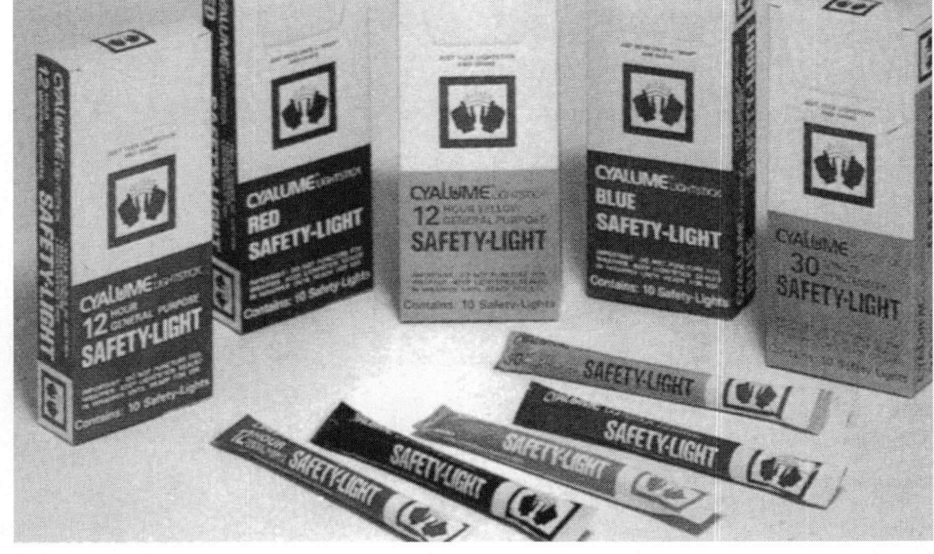

Cyalume Lightsticks

Manufacturer

Omniglow Corporation, 231 West Parkway, Pompton Plains, New Jersey 07444.

Status

In production.

Service

US armed forces and Coast Guard.

VERIFIED

YUGOSLAVIA (SERBIA AND MONTENEGRO)

38 mm coloured signal cartridge

Description

This 38 mm cartridge consists of an aluminium case with primer, ignition charge and a coloured signalling star. It is waterproof and marked for night identification purposes. It can be fired from any standard 38 mm signal pistol or riot gun.

Data

Burning time: 6 s
Light intensity: red, 55,000; green, 35,000; white, 55,000 cd
Vertical range: 100 m

Manufacturer

Institute of Security, PO Box 574, YU-11000 Belgrade.

Status

May no longer be in production.

Service

Yugoslav forces.

VERIFIED

38 mm coloured signal cartridges

38 mm single star illuminating cartridge

Description

This 38 mm cartridge provides instantaneous area illumination. It consists of an aluminium case with primer, ignition charge and an illuminating star. It is waterproof and marked for night identification. It can be fired from any standard 38 mm pistol or riot gun and should be fired at an angle of 45°.

Data

Burning time: 6 s
Light intensity: 120,000 cd
Max vertical range: 80 m
Illuminated area: diameter, 150 m

Manufacturer

Institute of Security, PO Box 574, YU-11000 Belgrade.

Status

May no longer be in production.

Service

Yugoslav forces.

VERIFIED

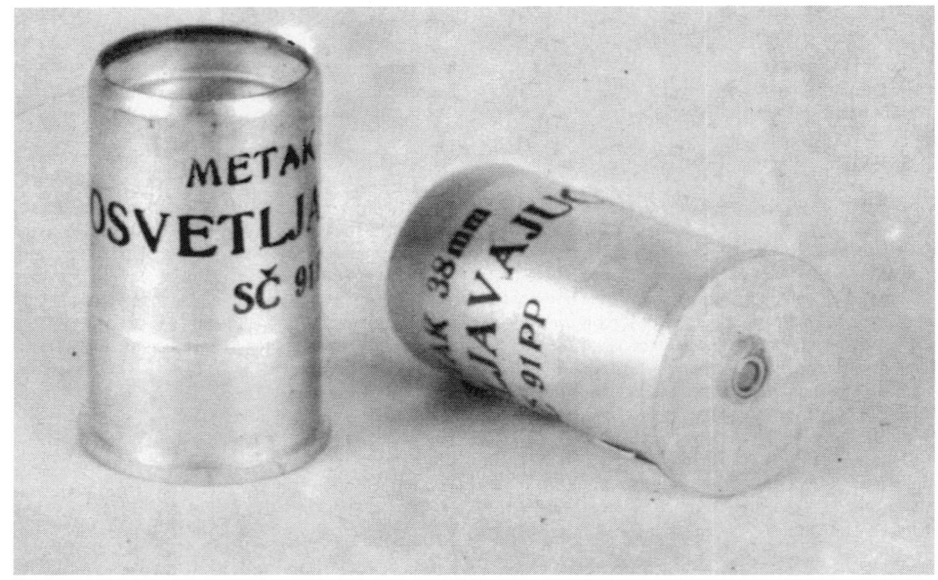

38 mm single star illuminating cartridge

38 mm triple star illuminating cartridge

Description

This 'Tripartite-S' cartridge is intended for the illumination of an area at night and consists of an aluminium case, primer, ignition charge and three illuminating stars. It is waterproof and marked for night identification. It may be fired from any standard 38 mm flare pistol or riot gun at an angle of 45°.

Data

Burning time: 5 s
Light intensity: 120,000 cd
Max vertical range: 80 m

Manufacturer

Institute of Security, PO Box 574, YU-11000 Belgrade.

Status

May no longer be in production.

Service

Yugoslav forces.

VERIFIED

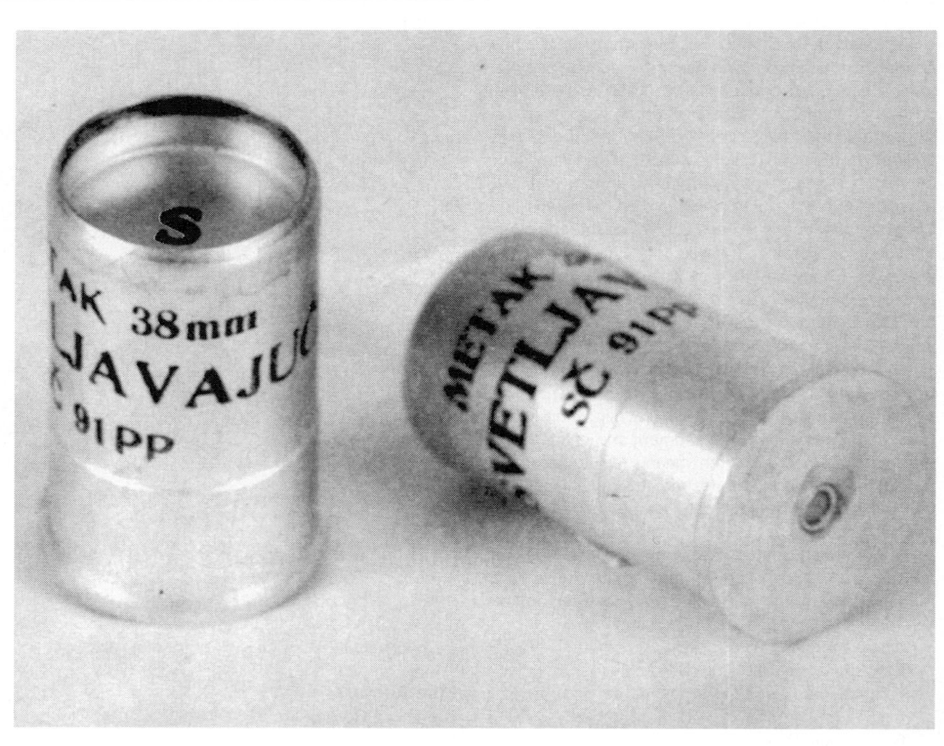

38 mm triple star illuminating cartridge

C-1 red smoke grenade

Description
The C-1 is intended for either daylight signalling and target indication purposes or as a training aid to represent the RBM-1 chemical (CS) grenade. The grenade is the usual metal cylinder with fly-off lever ignition system, and emits a dense red smoke when initiated.

Data
Diameter: 57 mm
Length: 150 mm
Weight: 410 g
Delay time: 2-3 s
Burning time: 30-50 s

Manufacturer
Institute of Security, PO Box 574, YU-11000 Belgrade.

Status
May no longer be in production.

Service
Yugoslav forces.

VERIFIED

C-2 red smoke grenade

Description
The C-2 grenade is a pocket version of the C-1, being smaller in all dimensions and with a correspondingly shorter burning time. It can be used for daylight signalling or as a training grenade representing the RBM-2 or RBM-404 CS grenades. It is of the usual cylindrical type with a fly-off lever ignition system.

Data
Diameter: 35 mm
Length: 105 mm
Weight: 180 g
Delay time: 2-3 s
Burning time: 25-45 s

Manufacturer
Institute of Security, PO Box 574, YU-11000 Belgrade.

Status
May no longer be in production.

Service
Yugoslav forces.

VERIFIED

Ž-1 yellow smoke grenade

Description
The Ž-1 is generally the same as the C-1 red smoke grenade (see separate entry) but emits intense yellow smoke when activated. It may be used for daylight signalling or as a training grenade to represent the RBM-11 CS grenade.

Data
Diameter: 57 mm
Length: 150 mm
Weight: 410 g
Delay time: 2-3 s
Burning time: 30-50 s

Manufacturer
Institute of Security, PO Box 574, YU-11000 Belgrade.

Status
May no longer be in production.

Service
Yugoslav forces.

VERIFIED

Ž-1 yellow smoke grenade

Ž-2 yellow smoke grenade

Description
The Ž-2 yellow smoke grenade is the pocket-sized version of the Ž-1 grenade described previously and is used for similar purposes. The only difference is its smaller size and reduced performance.

Data
Diameter: 35 mm
Length: 105 mm
Weight: 180 g
Delay time: 2-3 s
Burning time: 25-45 s

Manufacturer
Institute of Security, PO Box 574, YU-11000 Belgrade.

Status
May no longer be in production.

Service
Yugoslav forces.

VERIFIED

Ž-2 yellow smoke grenade

SIGHTING EQUIPMENT

SIGHTING EQUIPMENT

AUSTRALIA

Multipurpose Universal Gunner Sight® (MUGS)

Description

The Electro Optic Systems Pty Limited (EOS) Multipurpose Universal Gunner Sight® (MUGS) is a state-of-the-art full solution fire control system which has applications for a wide variety of infantry and armour direct fire weapons. MUGS permits engagement of static or moving targets by day or night with a high hit probability. The sight provides an almost instantaneous new aiming point that takes into account range, target movement and environmental conditions. Tests have shown that the effective range of weapons such as the 106 mm RCL rifle, the MK 19 grenade launcher and the 0.50/12.7 mm M2 machine gun can be doubled, with a significant improvement in accuracy.

MUGS is adaptable to most direct fire infantry weapons and a single unit can be programmed to handle a wide variety of munitions. The Common Module architecture meets modern logistics support requirements and allows a low-cost upgrade path through module replacement.

MUGS 4 Mk II, the culmination of a three-year research and development programme by EOS, is a fully integrated day/night system which incorporates an eye-safe laser rangefinder, a ballistic computer, image intensified night vision, an aimpoint display, a built-in training camera, and externally mounted environmental sensors. The unit can be used with a flat screen display instead of the eyepiece.

Data

Dimensions: 270 × 190 × 140 mm
Weight: 5 kg
Laser: Eye-safe Class IA
Range: 90-3,000 m
Wavelength: 1.54 μm
Pulse repetition rate: 12 ppm
Accuracy: ±5 m
Magnification: ×6, day and night
Night sight: ICCD, 2nd or 3rd generation
Power supply: external battery pack, Ni/Cd, lithium, or external 24 V DC

Manufacturer

Electro Optic Systems Pty Limited, 55A Monaro Street, Queanbeyan, NSW 2620.

Status

Available.

UPDATED

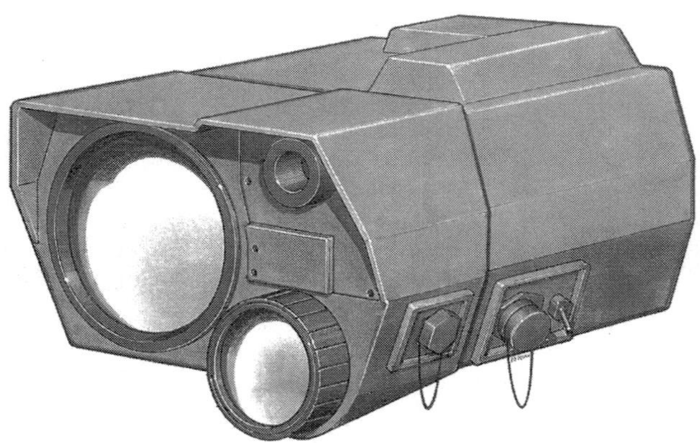

Front view of Multipurpose Universal Gunner Sight® (MUGS)

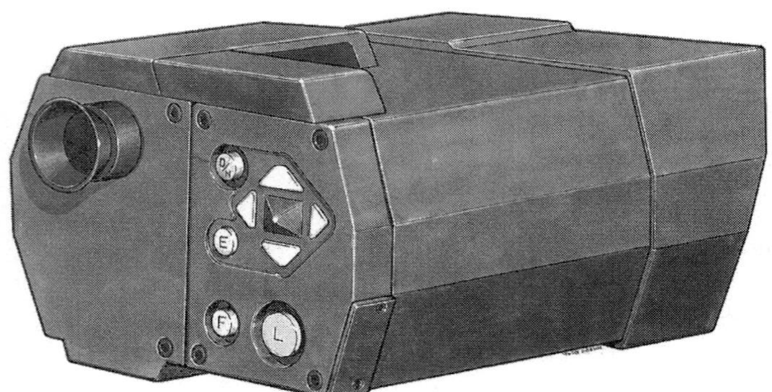

Rear view of Multipurpose Universal Gunner Sight® (MUGS)

AUSTRIA

Photonic telescope sights

Description

These sights were designed to improve the efficiency of infantry weapons; the ×1.5 and ×4 models are intended for use with assault rifles, whilst the ×6 and ×10 models are conceived for sniping rifles. All feature an ample field of view, high optical efficiency with excellent light transmission and accurate colour rendering, optional tritium illumination, aiming graticule or rangefinding reticle as required and a clamp-on or STANAG mount.

Manufacturer

Photonic Optische Geräte GesmbH, Zeillergasse 20/22, A-1170 Vienna.

Status

In production.

UPDATED

Data

Model	1.5 × 14	4 × 25	6 × 42	10 × 42
Magnification	×1.5	×4	×6	×10
Field of view	8.3°	5.7°	4°	2.2°
Length	205 mm	215 mm	320 mm	340 mm
Diameter	33 mm	37 mm	50 mm	50 mm
Weight	300 g	330 g	510 g	520 g

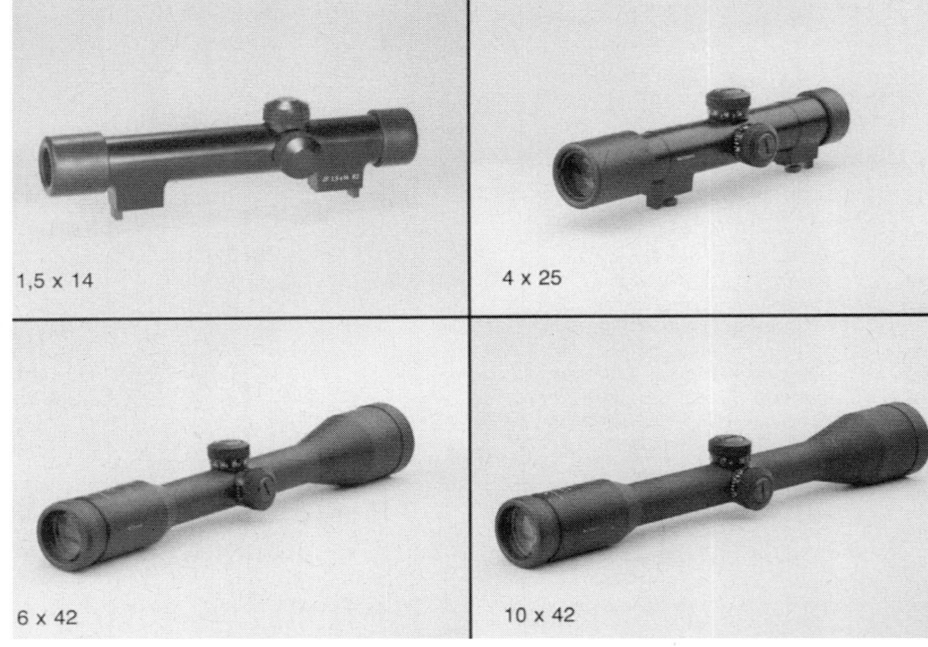

Photonic telescope sights

ATAS Automatic Target Acquisition System

Description

Compared to conventional aiming devices, the ATAS aiming device for 106 mm recoilless rifles significantly increases the first round hit probability against targets up to 1,000 m range by means of opto-electronic measuring techniques.

The ATAS includes the following features: a telescopic day sight; a telescopic night sight; a laser rangefinder; three selectable operating modes for determining lead; a ballistics computer; and automatic sighting. The day and night telescopic sights are equipped with the same object lenses and eyepieces so as to yield the same view in all cases; all numerical values are reflected into the sight. A semi-conductor laser which is eye-safe is used in the laser rangefinder; the device can optionally be equipped with an Nd:YAG laser (not eye-safe) if required.

The ballistic computer enables the ballistic data of various types of ammunition to be stored. The type of ammunition in use can be entered into the computer by means of an external switch.

Three modes of operation to determine lead can be switch-selected. These are as follows:

(1) Automatic measurement of target crossing speed

In this mode the laser is automatically triggered. The computer calculates the lead and elevation based upon the ammunition in use and on the data from the rangefinder and the cross-motion scanner. It then guides the X-Y carriage of the point of aim diode to the proper position. The gunner relays the gun so that the point of aim diode rests on the centre of the target and then fires.

(2) Manual measurement of target crossing speed

The gunner triggers the laser when the sight cross-hair is positioned on the target. The point of aim diode then lights up in the centre of the cross-hair. The gunner follows the target with the point of aim diode by means of an aim adjuster which operates on the X-carriage. The computer calculates the lead and elevation and guides the X-Y carriage of the point of aim diode into the proper position. The gunner relays the gun to place the diode on to the centre of the target and fires.

(3) Operation with time of flight indicator

The gunner triggers the laser when the sight cross-hair is on the target and the point of aim diode lights up in the centre of the cross-hair. The computer calculates the time of flight based on the ammunition used, and extinguishes the laser diode at the end of the calculated time. The gunner must then remember the position of the target relative to the cross-hair and traverse the gun so that the cross-hair lies in front of the target by the same distance. Elevation is adjusted according to the indicated range and the gun is fired. This system corresponds to the conventional method of aiming-off in the absence of any assistance and is, of course, open to all the usual human errors.

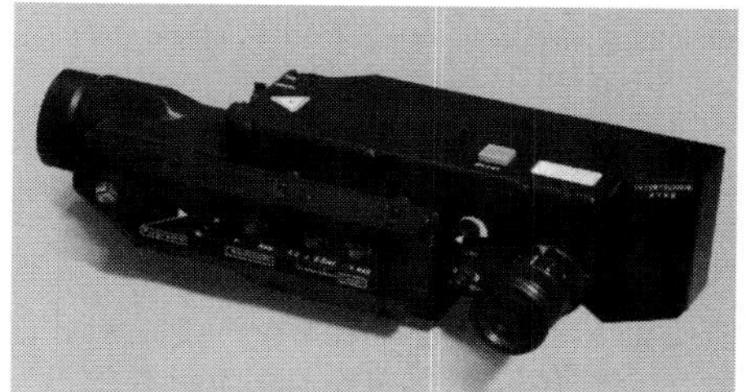

ATAS Automatic Target Acquisition System

Data

Magnification: ×6
Dioptre adjustment: ±5 dioptres
Night vision: 2nd generation; 3rd generation optional
Laser wavelength: 904 nm
Range cover: 100 to 1,000 m
Resolution: 10 m
Power supply: 6 × 1.5 V rechargeable Ni/Cd batteries

Manufacturer

Intertechnik GesmbH, Industriezeile 56, Postfach 100, A-4040 Linz.

Status

In production.

VERIFIED

BELGIUM

Laser Hit Marker

Description

The Aims Optronics Laser Hit Marker is a laser spot projector which, by means of various adaptors, can be attached to any small arm. When switched on it delivers a clearly visible red spot indicating the point of aim of the weapon. The firer need take no conscious aim, but with both eyes open merely lays the spot on the target and fires.

The Marker is a battery-powered helium laser operating at 632.8 nm to deliver a visible spot; a compact model is available for distances up to 100 m and a newly developed laser beam control offers long distance aiming for snipers. For military applications a different laser, generating a spot only visible with night sights or night goggles, to a range of 300 m, can be provided. The Marker can be mounted using any conventional telescope mounting rings to any type of small arm including pistols. Depending on the model of Marker, the laser beam can be adjusted either mechanically or optically in order to boresight the weapon at any selected range. Special X-Y collimators are available for beam control and long distance work.

Different power supplies accommodate the user's requirements in action or for training. Complete marksman training systems are available to improve accuracy, follow-through, or to assess the shooter's overall performance.

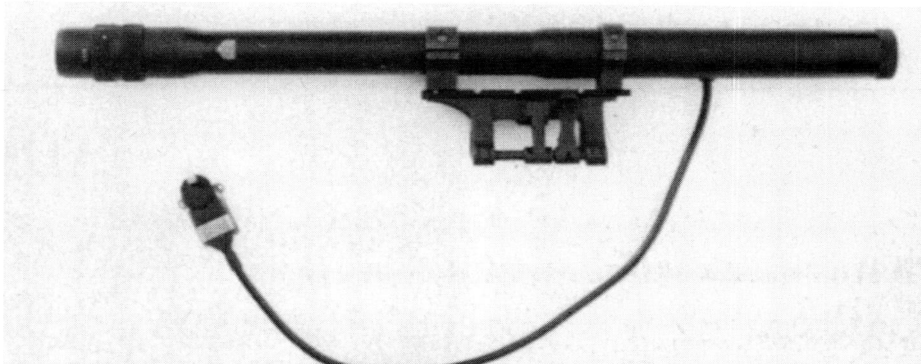

Aims Optronics Laser Hit Marker with mounting bracket

Manufacturer

Aims Optronics NV SA, Rue F Kinnenstraat 30, B-1950 Kraainem.

Status

In production.

VERIFIED

CANADA

CLASS Computerised Laser Sight System

Description

The AN/PSG-501 CLASS Computerised Laser Sight System is a full solution fire control system which provides direct fire weapons with a significantly improved hit capability and a greatly increased effective range. Designed specifically to enhance the performance of direct fire weapons cost-effectively, CLASS is interchangeable between weapon systems and has been successfully used on the 84 mm Carl Gustaf RCL, the 106 mm RCL rifle, the 25 mm M242 cannon, the 30 mm ASP-30 combat support weapon, the 40 mm MK 19 grenade launcher and the 155 mm M198 howitzer.

The three CLASS modules have been integrated into a single system that is rugged, lightweight and easy to operate. The modules consist of a laser rangefinder, an electro-optic module and a ballistic computer. The laser rangefinder is available in either eye-safe or non

CLASS computerised laser sight (AN/PSG-501)
1996

eye-safe versions, and provides a precise, instantaneous target range up to 4,000 m with an accuracy of ±5 m. The electro-optic module provides ×5 magnification by both day and night, and is available with Generation II, II+ or III image intensification tubes. The ballistic computer includes both a cant sensor and an automatic crossing speed rate sensor.

The ballistic computer receives data inputs on the target range and speed, the ballistic characteristics of the ammunition selected, and the meteorological conditions. From these inputs it computes the correct elevation and azimuth of the weapon and, via the beam-steering assembly, drives the Risley counter-rotating prisms until the target image has shifted to the correct position. The gunner then keeps the graticule directly on to the centre of the target, for both moving and stationary targets, and engages.

CLASS has been designed for ease of operation. The key gunner actions are target acquisition, ammunition selection, range determination via the laser rangefinder, repositioning of the sight graticule once the ballistic solution has been computed, and finally the firing of the weapon. All of this is accomplished in less that two seconds for stationary targets and less than six seconds for moving targets. Because CLASS operation is instinctive, training time is reduced to an absolute minimum and gunners master its use very quickly.

Live firing has been conducted in North America, Europe, the Middle East and Asia with various weapon systems for both army and navy requirements. By significantly increasing the probability of a first round hit over extended ranges, CLASS not only enhances weapon system performance, but can also provide a very cost-effective method of upgrading in-service weapon systems to meet changing operational requirements.

Data

Dimensions: 280 × 180 × 90 mm
Weight: 4 kg
Laser type: pulsed Erbium Glass or Nd:YAG
Laser range: to 4,000 m
Range resolution: ±5 m
Electro-optical module display: sight graticule and 5-character 16 segment LED display
Magnification: ×5
Field of view: 8°
Night vision: 2nd, 2nd plus or 3rd generation image intensification
Automatic aimpoint offset: target image shifts to correct aiming point
Automatic target crossing speed calculation: angular rate sensor
Ballistic computer process inputs: ammunition, target range, target speed, weapon cant, meteorological conditions
Fire control solution: full ballistic solution
Aimpoint accuracy: ±0.1 mil
Ammunition memory: data stored for 10 types specified by user

Manufacturer

Computing Devices Canada Limited, PO Box 8508, Ottawa, Ontario K1G 3M9.

Status

In production.

Service

Canadian Forces.

UPDATED

CLASS computerised laser sight on 40 mm MK 19 automatic grenade launcher

1996

CLASS computerised laser sight on 30 mm ASP-30 combat support weapon

1996

CHINA, PEOPLE'S REPUBLIC

Type JWJ machine gun low light level sight

Description

The Type JWJ is a second-generation image intensifying sight intended for use on machine guns and other crew-served weapons and also as an independent surveillance instrument. It is provided with bright light protection circuitry which prevents interference from flashes and bright illumination in the target area. An accessory handle is provided for when the sight is used for hand-held observation.

Data

Weight: 2.1 kg
Magnification: ×4.5
Field of view: 10°
Focusing range: 20 m to infinity
Dioptric adjustment: ±2.5 dioptres
Resolution: 0.5 mrad (E = 10^{-1} lux, C = 0.85); 1.2 mrad (E = 10^{-3} lux, C =0.35)
Night vision range: person, 500 m; vehicle, 1,000 m

Type JWJ night vision sight

Manufacturer

China North Industries Corporation (NORINCO), 7A Yue Tan Nan Jie, PO Box 2137 Beijing.

Status

In production.

VERIFIED

COMMONWEALTH OF INDEPENDENT STATES

CIS Day and Night Vision Devices and Sights

Description

In order to provide an indication of the standard types of day and night vision devices and infantry weapon sights in production in the CIS, the following check list is provided:

Daylight Sights

Model	**1L29**	**10L50**	**MLM44M**	**LAG-17**	**LO 4 × 24**	**ULO-2**
Magnification	×4	×3.6	×2.55	×2.7	×4	×4
Field of view	8°	12.6°	9°	13°	6°	8°
Exit pupil diameter	6.5 mm	8 mm	4 mm	4.5 mm	6 mm	6.5 mm
Eye relief	35 mm	55 mm	20 mm	27 mm	68 mm	35 mm
Dimensions	203 × 80.5 × 178 mm	365 × 96 × 178 mm	97 × 108 × 165 mm	124 × 101 × 145 mm	375 × 136 × 72 mm	275 × 78 × 82 mm
Weight	0.8 kg	1.7 kg	0.9 kg	1 kg	0.6 kg	0.7 kg
Application	AK-74, RPK-74, PKM	NSV	Mortars	AGS-17	AKM series	AKM series

Night Sights

Model	**1LH51**	**1LH52**	**1LH53**	**1LH58**	**1LH84**
Generation	2	2	1	1	2+
Magnification	×3.46	×5.3	×5.9	×3.5	×3.7
Field of view	8.5°	7.6°	5.5°	5°	10°
Identify range					
tank	700 m	700 m	–	600 m	600 m
man	400 m	–	–	400 m	400 m
Voltage	6 V	6 V	6 V	6 V	6 V
Battery life	10 h	10 h	10 h	10 h	10 h
Weight	2.1 kg	3.2 kg	15 kg	2 kg	1.3 kg
Dimensions	300 × 210 × 140 mm	333 × 186 × 183 mm	452 × 305 × 301 mm	458 × 186 × 99 mm	295 × 98 × 90 mm
Applications	AK-74, RPK, SVD	RPG-7	anti-tank gun	AK-74, RPK, SVD	AK-74, RPK

Night Vision Devices

Model	**1LH54***	**TLB-2**	**NHB**	**1N10**	**ML**	**UM8-2**	**T3K**
Magnification	×5.5/5	×15	×15	×4 - 20	×4 - 20	×8	×8/9.9
Field of vision	6°/5°	6°	6°	1.5 - 8°	1.5 - 8°	5°	6/7°
Diameter exit pupil	5 mm	7.33 mm	7.33 mm	7 - 1.4 mm	7 - 1.4 mm	2 mm	3/8 mm
Dioptre setting	±5	−3/+12	−3/+12	−5/+10	−5/+10	±10	±5
Dimensions	544 × 255 × 607 mm	565 × 325 × 545 mm	400 × 415 × 580 mm	765 × 42 × 190 mm	765 × 42 × 190 mm	100 × 45 mm	396 × 423 × 438 mm
Weight	18.5 kg	14.8 kg	30 kg	0.45 kg	0.5 kg	0.12 kg	14.6 kg

*day/ night

Manufacturer

Mainly: Production Amalgamation 'Novosibirsk' Instrument-Making Plant, 630049 Novosibirsk - 49, Russia.

Status

In production.

Service

CIS and former Warsaw Pact armies.

VERIFIED

1LH58 image intensifying sight fitted to an RPG-7 launcher

1LH51 combat night sight mounted on RPK-74 light machine gun
(T J Gander)

FRANCE

Sopelem CLARA night vision goggles

Description

The Sopelem CLARA night vision goggles are claimed to be the only ultra-wide field night observation goggles available. Delivering hands-free night vision, CLARA is claimed to provide optimum performance by providing ×1 magnification, adjustable focus, and NBC equipment compatibility. The device also provides an efficient night aiming capability for infantry when used with an associated laser target designator coupled to the user's weapon. The goggles are watertight to a depth of 1 m.

Data

Weight: 450 g
Binocular: monotube, 2nd or 3rd generation, 18 mm wafer tube
Magnification: ×1
Focus: adjustable from 250 mm to infinity
Resolution: 0.8 pl/mrad
Dioptre adjustment: −3 to +2 dioptres
Power supply: 1 × AA battery
Battery life: up to 20 h
Operating temperature range: −40 to +52°C

Manufacturer

Sopelem-Sofretec, 53 rue Casimir Périer, F-95870 Bezons Cedex.

Status

Available for production.

VERIFIED

Sopelem CLARA night vision goggles

Sopelem TN2-1 night observation and driving sights

Description

The TN2-1 sight is a nocturnal vision instrument provided for observation and movement by night, guiding the driver of an armoured vehicle by the commander (with his head outside the turret), and short-range night firing in combination with the PS 1 spotlight. It is a binocular sight fitted to a mask worn by the observer. It operates on the principle of electronic light intensification.

Data

Weight: 470 g
Binocular: monotube, 2nd generation, 18 mm wafer tube, double focus
Magnification: ×1
Field of view: 40°
Power supply: 2.7 V or 3.5 V military lithium batteries or 2 × 1.5 V standard commercial batteries

Manufacturer

Sopelem-Sofretec, 53 rue Casimir Périer, F-95870 Bezons Cedex.

Status

In production.

Service

French and other special forces.

VERIFIED

TN2-1 night observation and driving sight

Sopelem PS 2000 laser (infra-red) target pointer

Description

This laser pointer is intended to be attached to a weapon and used in conjunction with the TN-2 or similar night vision goggles. The unit is attached to the weapon by means of a mounting bracket and is then boresighted to the weapon. The firer presses a switch to project an infra-red beam from the unit, and observes the projected spot of light by means of his night vision goggles. Once the spot is on target he may fire, with confidence of a hit.

If preferred, the same unit can be supplied projecting a visible red light beam for use in daylight or in naked eye night operations.

Data

Length: 170 mm

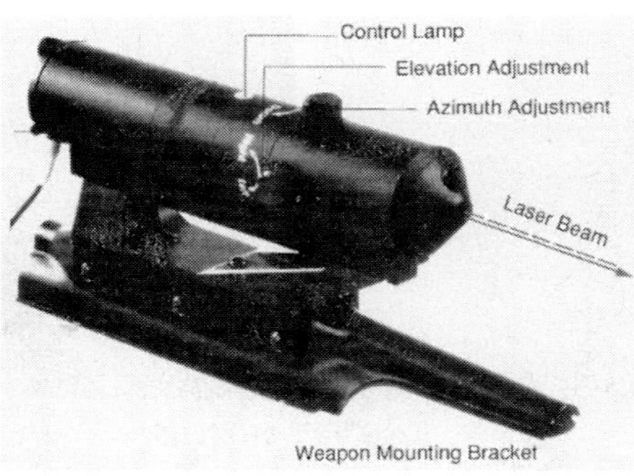

Sopelem PS 2000 Laser (IR) target pointer

Weight: 380 g with batteries
Laser output power: 2.5 mW
Emission wavelength: 820 nm
Beam divergence: 0.5 mrad
Beam diameter: 55 mm at 100 m
Range: >400 m
Power supply: 2 × 1.5 V AA batteries
Power supply: 2 × AA commercial batteries
Battery life: 1.5 h continuous

Manufacturer

Sopelem-Sofretec, 53 rue Casimir Périer, F-95870 Bezons Cedex.

Status

In production.

Service

French Army and other special forces.

VERIFIED

Sopelem OB-50 night firing scope

Description

The Sopelem OB-50 night aiming telescope was designed for observation and aiming, without artificial lighting of the target, by intensification of residual light from the night sky. This technique ensures total discretion and enables instantaneous detection of active infra-red emission. It incorporates a second-, second- Plus, second- Super or third-generation microchannel 18 mm light intensifier tube fitted with double-proximity focus, and with built-in AGC. An illuminated micrometer and an eyepiece shade with shutter are also fitted. Various types of graticule are available, and designs can be adapted to meet special requirements.

The telescope is adaptable to all current types of infantry weapon and is particularly suitable for rifles, machine guns and light anti-tank rocket launchers.

Data

Length: 230 mm
Weight: 900 g with batteries
Magnification: ×3.2
Field of view: 11°
Eyepiece focus: −4 to +2 dioptres
Power supply: 2 × AA 1.5 V; or 1 × BA1567/U 2.7 V; or 1 × BA5765/U 2.7 V; or 1 × PS31 3.6 V
Operating temperature: −45 to +52°C

Manufacturer

Sopelem-Sofretec, 53 rue Casimir Périer, F-95870 Bezons Cedex.

Status

In production.

Service

French Army and other special forces.

VERIFIED

Sopelem OB-50 night firing scope on Panzerfaust 3 firing unit

Sopelem DANTE day/night weapon sight

Description

The DANTE weapon sight was designed for use by infantry and snipers and can be used by both day and night. It is claimed to be accurate and easy to use, with a single switch selecting day or night modes. The day channel is fully protected against the main forms of laser aggression so far identified.

The ×4 magnification feature and an ultra-fine graticule make the DANTE sight suitable for long-range use against small targets. Graticule illumination is adjustable under day and night conditions.

Data

Length: 280 mm
Weight: 1.45 kg
Magnification: ×4
Field of view: 8.5° (15 m to 100 m)
Resolution: day, 0.075 mrad; night, 4.3 pl/mrad
Dioptre adjustment: −4 to +2 dioptres
Image intensifier: 2nd or 3rd generation, 18 mm format
Graticule brightness: adjustable
Weapon alignment arc: 40 mrad
Variation per click: 0.2 mrad
Eyepiece ring extension: 35 mm
Power supply: 2 × AA batteries or military battery
Battery life: 60 h
Operating temperature range: −40 to +52°C

Sopelem DANTE day/night sight

Manufacturer

Sopelem-Sofretec, 53 rue Casimir Périer, F-95870 Bezons Cedex.

Status

Available for production.

VERIFIED

MIRA and MEPHIRA thermal imaging night sights

Description

MIRA (MILAN Infra-Red Attachment) is an add-on sight unit which can be rapidly fitted to the existing MILAN missile firing post sight to convert it to night operation. It is automatically harmonised with the optical path of the standard sight and reflects the infra-red picture into the optical sight so that the normal day sight eyepiece is still used for viewing. Using series-parallel scanning and a CMT detector, the sight picture is produced by LED visualisation. The sensor may be cooled either by an HP compressed gas bottle (Joule-Thomson expansion) or by a built-in cooling unit (Stirling cycle).

A variant model, known as MEPHIRA, has been developed by Thomson TTD Optronique and is currently in production and being introduced into service with French forces. This version is optimised for use with HOT missiles and is installed on VAB-HOT launchers.

Data

Overall dimensions: 530 × 290 × 160 mm
Weight: 8 kg JT version with bottle and power; 9 kg Stirling version

MIRA thermal imaging night sight on VAB-HOT vehicle

Spectral bandwidth: 8-13 μm
Type of detector: Hg Cd Te
Field of view: 6 × 3°
Angular resolution: 0.17 mrad
NETp: 0.1°C
Sight axis harmonising: automatic
Scanning: series parallel
Image frequency: ≤22 Hz
Detection range: 6,000 m typical on vehicle
Reconnaissance range: 3,000 m typical on vehicle
Identification range: 2,000 m typical
Temperature range: −40 to +52°C
Start-up time: <20 s (JT); <4 min (Stirling) from cold, <30 s from standby condition

Manufacturers

Series production is carried out under the general management of: Thomson TTD Optronique, rue Guynemer, F-78283 Guyancourt Cedex, France.
Marconi Radar and Control Systems, PO Box 133, Chobham Road, Frimley, Camberley, Surrey GU16 5PE, UK.
Siemens AG, Hoffmannstrasse 51, D-8000 Munich 70, Germany.

Status

In production.

Service

British, French and German armies and numerous other forces.

UPDATED

GERMANY

AN/PVS-4, AN/TVS-5 weapon sights, EURONOD-2 observation sight

Description

All three of these sights are image intensifying devices using a 25 mm second-generation microchannel plate tube, and are equipped with the same common standardised battery housing and eyepiece/eye guard assembly which guarantees interchangeability. Only the objective lens assemblies are different and serve the required purpose.

The second-generation technology has built-in automatic gain control protecting the tube against muzzle and detonation flash, which results in a uniform image without tendency to flare. Additional manual brightness control allows the operator to adjust screen luminance to his individual requirements and for contrast enhancement.

AN/PVS-4 and AN/TVS-5 are weapon sights with different magnifications, using interchangeable projected graticules for every common weapon fully adjustable for adaptation to the scenery brightness. They can be easily adapted to the weapon by using appropriate brackets.

EURONOD-2 is an observation sight with high magnification, offering a wide field of applications, particularly that of artillery fire control.

AN/PVS-4 rifle sight with G3 rifle mounting

Data

Model	**AN/PVS-4**	**AN/TVS-5**	**EURONOD-2**
Magnification	×3.7	×5.8	×9.7
Field of view	14°	9°	5.6°
Object lens	95 mm f/1.6(T)	155 mm f/1.6(T)	258 mm f/1.7(T)
Focus range	10 m - ∞	25 m - ∞	75 m - ∞
Length	240 mm	310 mm	480 mm
Width	120 mm	160 mm	240 mm
Height	120 mm	170 mm	240 mm
Weight	1.7 kg	3.6 kg	9 kg
Power supply	2 × AA/LR6	2 × AA/LR6	2 × AA/LR6
Battery life	50 h (20°C)	50 h (20°C)	50 h (20°C)
Operating range	to 600 m	to 1,200 m	to 6,000 m

Manufacturer

Euroatlas GmbH, Zum Panrepel 2, D-28307 Bremen.

Status

Available.

VERIFIED

EUROVIS-4 weapon sight

Description

The EUROVIS-4 is a weapon sight of the latest generation, developed to meet military requirements. EUROVIS-4 weighs only 1 kg and is adaptable to any weapon, even very light weapons, by using an appropriate bracket.

It uses an 18 mm image intensifier tube of either second- or third-generation. The illuminated and controllable graticule, together with the focusing device with a positive grip usable with heavy gloves, ensures accurate aiming. Four times magnification gives a wide focusing range from 15 m to infinity, enabling man-sized targets to be detected up to 600 m under normal weather conditions.

The image intensifier tubes use microchannel plates and are provided with automatic gain control against muzzle flash and bright light sources. Excellent contrast and resolution capability, particularly at very low light levels, allow the operator to use the sight over long periods without fatigue. EUROVIS-4 fulfils all military environmental test requirements.

Data

Weight: 1 kg
Magnification: ×4
Field of view: 9°
Objective: 116 mm f/1.8(T)
Focusing range: 15 m to infinity
Power supply: 2 × AA/LR6
Battery life: >50 h at 20°C
Operating range: to 1,500 m

EUROVIS-4 weapon sight

Manufacturer

Euroatlas GmbH, Zum Panrepel 2, D-28307 Bremen.

Status

Available.

VERIFIED

RT 5A laser illuminator

Description

The RT 5A laser illuminator is intended to be used with other Euroatlas night vision devices in the case of absolute darkness or in difficult observing conditions. It is also recommended where light barriers, for example bright light sources, are between the observer and his objective or where a dark objective is between bright areas. This laser torch is light and small and can easily be mounted on any vision device by using the adjusting and mounting bracket provided. The laser diode has a working range up to 1,000 m at low current consumption. There is an adjusting wheel which permits the illumination field to be changed during operation within the ratio of 1:4.

Data
Length: 178 mm
Width: 75 mm
Height: 85 mm without mounting
Weight: 650 g
Wavelength: 840-870 nm
Laser diode: double hetero-structure LB1
Radiant output: 6-10 mW
Pulse duty factor: 5%
Beamwidth: ca 1 mrad
Normal ocular hazard distance: 50 m
Working range: 1,000 m
Power supply: 6 × AA/LR6 batteries

Manufacturer
Euroatlas GmbH, Zum Panrepel 2, D-28307 Bremen.

Status
Available.

VERIFIED

RT 5A laser illuminator

LM-18 target marker

Description
The LM-18 is a covert infra-red target marker (aiming light). It is designed for weapon mountings and features an integral zeroing (boresighting) mechanism which dispenses with the need for costly zeroing mounts. It will enable rapid covert target acquisition by wearers of night vision goggles and other users of night vision systems.

The unit is powered by three readily available 'AA' size alkaline batteries. A quick release bracket interface enables rapid detachment and reattachment to a weapon without the need for rezeroing. Operation is by means of a remote wired 'press for on' switch which can be attached to any convenient point on the weapon. The unit is rugged and reliable.

Data
Length: 120 mm
Diameter: 40 mm
Weight: 350 g
Wavelength: 800-870 nm
Radiant output: 0.6 mW
Working range: up to 300 m
Spot size: 50 mm at 50 m range (others available)
Power supply: 3 × AA/LR6 batteries
Waterproofing: 30 m max depth
Switch: remote, with strap to attach to weapon at any point

Manufacturer
Euroatlas GmbH, Zum Panrepel 2, D-28307 Bremen.

Status
Available.

VERIFIED

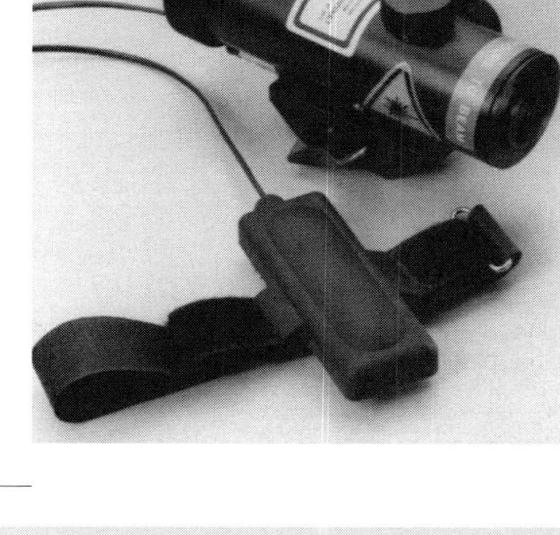

LM-18 target marker, with remote switch

Telescope sight PSG 1

Description
The PSG 1 is a conventional optical telescope sight intended for use with sniper rifles and particularly the Heckler and Koch PSG 1. In addition to serving to aim the weapon it can also be used as a long-range observation instrument. The maximum adjustable shooting distance is 600 m.

Hensoldt PSG 1 telescope sight

Data
Length: 390 mm
Cap diameter: 56 mm
Height above rifle: 130 mm
Weight: 580 g
Magnification: ×6
Object lens: 42 mm
Exit pupil: 7 mm
Eye relief: 70 mm
Field of view: 4°
Elevation adjustment: 100-600 m
Graticule light: adjustable
Batteries: 3 × Varta 60DK

Manufacturer
M Hensoldt and Sohne Optische Werke AG, Postfach 1760, D-35573 Wetzlar.

Status
In production.

VERIFIED

BLITS - Beta Lighted Infantry Telescope System

Description
The BLITS fire control system was developed for use with the M16A2 rifle incorporating a modified upper receiver designed for a quick detachable 'low-scope' mounting. It provides the rifleman with enhanced quick fire control for his individual weapon, thus allowing quick target acquisition and improved target identification. This optic also enables the rifleman to

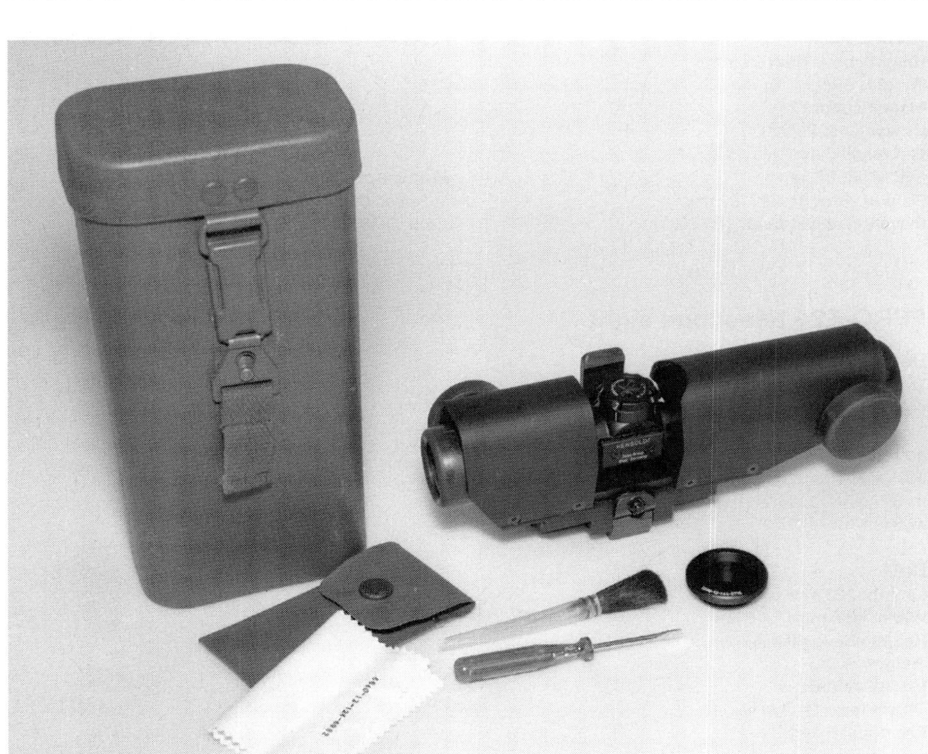

BLITS sight unit with carrying case and accessories

observe the enemy and the effect on the target at extended ranges out to 800 m. In addition, BLITS features an illuminated reticle which enhances target acquisition under low light conditions. It can be adapted to other types of small arms, including the M249 SAW, provided they are fitted with a Weaver type rail mount.

Data

Magnification: ×4
Entrance pupil: 24 mm
Exit pupil: 6 mm
Eye relief: ca 60 mm
Dioptre setting: −0.5 to −0.75 (fixed)
Field of view: 6° - 105 m at 1,000 m

Manufacturer

M Hensoldt and Sohne Optische Werke AG, Postfach 1760, D-35573 Wetzlar.

Status

In production.

VERIFIED

ZF PzF3 telescope sight

Description

The ZF PzF3 is for use with the Panzerfaust 3 anti-tank launcher. It serves the operator for target detection and recognition, distance determination and setting the correct angle of elevation and the lead angle with the appropriate markings of the reticle pattern. It is a monocular instrument with a prism erecting system.

Data

Diameter: 27 mm
Overall length: 150 mm
Weight: 135 g with eye guard
Optical length: 106 mm
Magnification: ×2.5
Entrance pupil: 12 mm
Exit pupil: 4.8 mm
Eye relief: 33 mm
Field of view: 10° - 17.5 m at 100 m
Dioptre setting: −0.6 fixed

Manufacturer

M Hensoldt and Sohne Optische Werke AG, Postfach 1760, D-35573 Wetzlar.

Status

In production.

VERIFIED

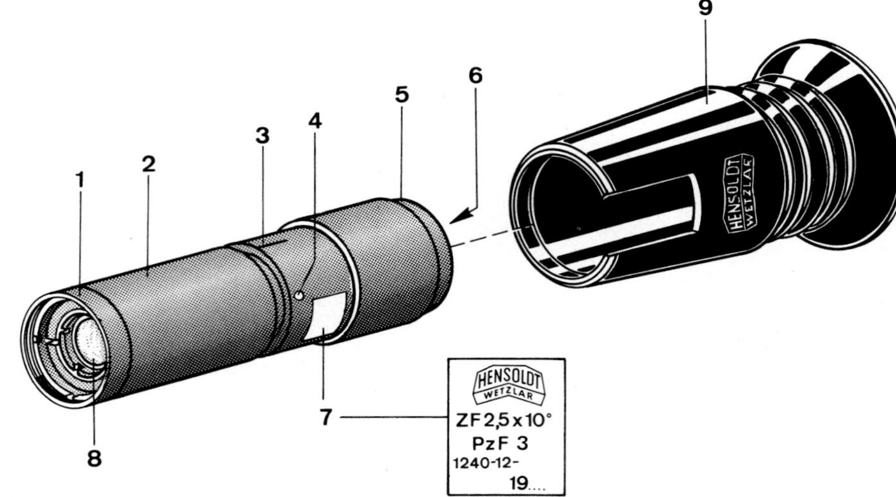

1 end housing
2 telescope/tube
3 fitting position mark
4 threaded coupling for leakage test
5 eyepiece unit
6 eyepiece
7 nameplate/marking
8 objective
9 eyeguard

ZF PzF3 sight with eye guard

ZF 4 × 24 telescope sight

Description

The ZF is a monocular telescope with lens erecting system and ×4 magnification. It is intended for sniping use and also as an observation instrument. All optical surfaces are coated with a wipe-proof reflection reducing coating, and the sight is so well sealed that even under sudden temperature changes and water-spray it remains usable.

Data

Length: 228 mm
Width: 50 mm
Height above rifle: 95 mm
Weight: 350 g
Magnification: ×4
Object lens: 24 mm
Exit pupil: 6 mm
Eye relief: 60 mm
Field of view: 5° 40′ (100 mils)
Elevation adjustment: 100-600 m

Hensoldt ZF telescope sight

Manufacturer

M Hensoldt and Sohne Optische Werke AG, Postfach 1760, D-35573 Wetzlar.

Status

In production.

VERIFIED

FERO-Z24 telescope sight

Description

The FERO-Z24 is a monocular telescope sight with erecting lens system; it is intended for use with sniper rifles and as an observation instrument. It has an illuminated graticule for use under poor light conditions. As with all Hensoldt military telescope sights the mount is to STANAG 2424 and will fit any NATO compatible weapon.

Data

Length: 223 mm
Width: 46.5 mm
Height above rifle: 43 mm
Weight: 300 g
Magnification: ×4
Object lens: 24 mm
Exit pupil: 6 mm

Hensoldt FERO-Z24 telescope sight

Eye relief: 60 mm
Field of view: 6° (107 mils)
Elevation adjustment: 100-600 m at 100 m click-stops

Manufacturer
M Hensoldt and Sohne Optische Werke AG, Postfach 1760, D-35573 Wetzlar.

Status
In production.

VERIFIED

ZF 10 × 42 telescope sight

Description
The ZF 10 × 42 telescope sight is a precision sight intended for shooting to ranges of 1,000 m. It is a monocular/monobjective sight with a lens erecting system and ×10 magnification. The superelevation angle is adjustable for shooting distances from 100 to 1,000 m. All air/glass surfaces of the optical components are coated with a wipe-proof reflection reducing coating. The sight is so well sealed that even under sudden temperature changes and spray-water influences it remains usable. The graticule pattern can be changed to meet user requirements.

Data
Length: 373 mm with caps and eye guard
Diameter: 56 mm
Weight: 429 g
Magnification: ×10
Entrance pupil: 42 mm
Exit pupil: 4.2 mm
Eye relief: 70 mm
Field of view: 2.4° - 42 m at 1,000 m
Dioptre setting: ±2

Manufacturer
M Hensoldt and Sohne Optische Werke AG, Postfach 1760, D-35573 Wetzlar.

Status
In production.

VERIFIED

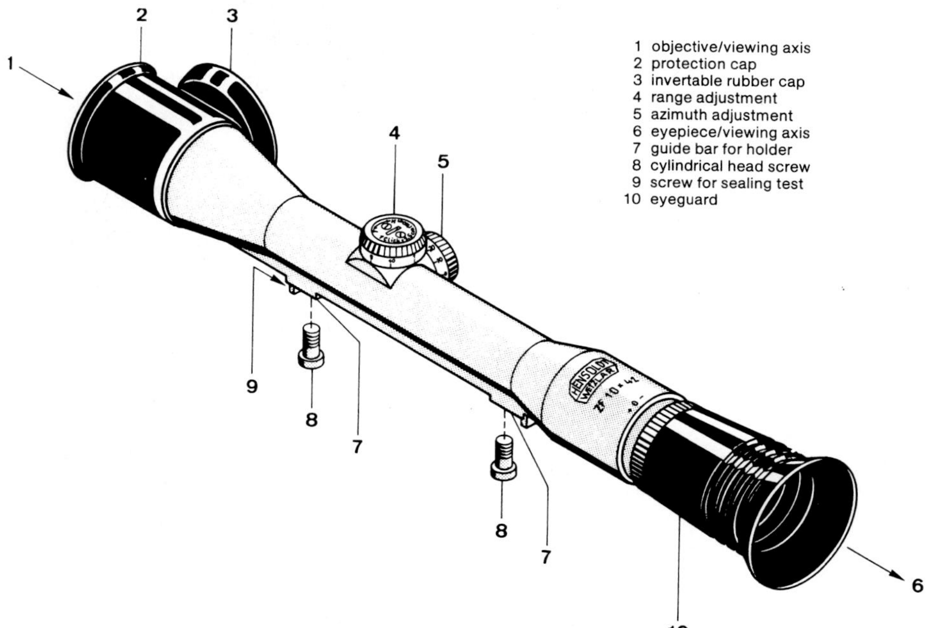

Hensoldt ZF 10 × 42 telescope sight

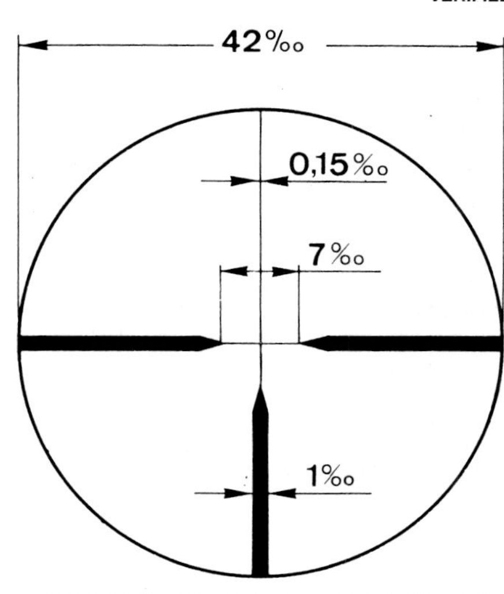

Standard graticule pattern of ZF 10 × 42 sight

ZF 6 × 36 telescope sight

Description
This is a device for precise aiming with a rifle to a range of 1,300 m. It serves the firer for target acquisition and aiming as well as general observation and the determination of target effects. The sight is a monocular/monobjective system and the superelevation angle is adjustable for shooting at ranges from 100 to 600 m. For the 600 to 1,300 m bracket there are sighting marks on the graticule. When shooting at these distances, the range adjustment is set to the 600 m mark and the additional superelevation obtained by selection of the appropriate aiming mark. For use in intense sunlight conditions a grey (neutral density) filter can be fitted to the eyepiece.

Data
Length: with caps, 255 mm
Diameter: 55 mm
Weight: 390 g
Magnification: ×6
Entrance pupil: 36 mm
Exit pupil: 6 mm
Eye relief: 60 mm
Field of view: 4° - 70 m at 1,000 m
Dioptre setting: fixed −0.5

Manufacturer
M Hensoldt and Sohne Optische Werke AG, Postfach 1760, D-35573 Wetzlar.

Status
In production.

VERIFIED

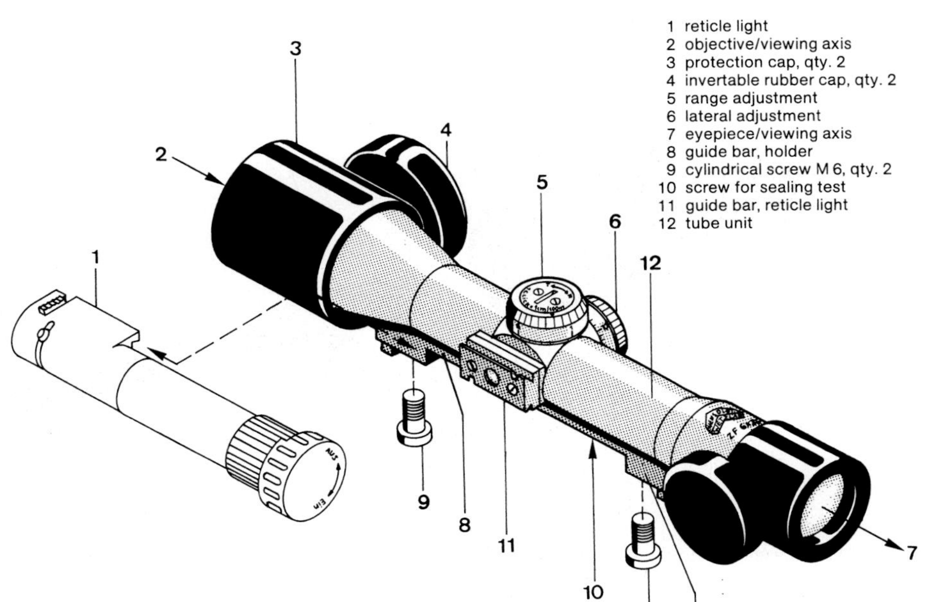

Hensoldt ZF 6 × 36 telescope sight

Standard graticule of ZF 6 × 36 sight, showing long-range aiming marks

ZF 4 × 24 MG3 telescope sight

Description

The ZF 4 × 24 MG3 telescopic sight serves for direct and indirect aiming of a mounted machine gun at ground targets up to a distance of 1,600 m, for observation and for determination of distance when the target size is known. Because of the periscopic design of the sight, the gunner can aim and observe from behind cover.

Data

Length: 210 mm
Height: 195 mm
Weight: 1.65 kg
Magnification: ×4
Eye relief: 45 mm
Field of view: 8° 35′ - 150 m at 1,000 m
Dioptre setting: ±2.5 dioptres

Manufacturer

M Hensoldt and Sohne Optische Werke AG, Postfach 1760, D-35573 Wetzlar.

Status

In production.

VERIFIED

Hensoldt ZF 4 × 24 MG3 telescope sight

ZF 4 × 24 MG1 telescope sight

Description

This is a similar sight to the MG3 (see previous entry) but has a higher periscope tube, allowing for different mountings and heights of cover. It is used for aiming machine guns to ranges up to 1,600 m and for range-finding when the target size is known. Data is similar to that for the MG 3 sight except that the height is 285 mm, length 185 mm and the weight 1.67 kg.

Manufacturer

M Hensoldt and Sohne Optische Werke AG, Postfach 1760, D-35573 Wetzlar.

Status

In production.

VERIFIED

Hensoldt ZF 4 × 24 MG1 telescope sight

Boresight SA

Description

The Boresight SA is a 90° telescope which is used in conjunction with an insert rod for checking and adjusting parallelism of the bore axis of a weapon to the line of sight of the weapon sight. By attaching different insert rods, weapons of various calibres can be accommodated.

The insert rod having been attached to the telescope, the unit is then inserted into the barrel of the weapon and the weapon moved until the telescope is laid on the bore axis marker of the target. The sight should then be aligned with the sight marker of the target.

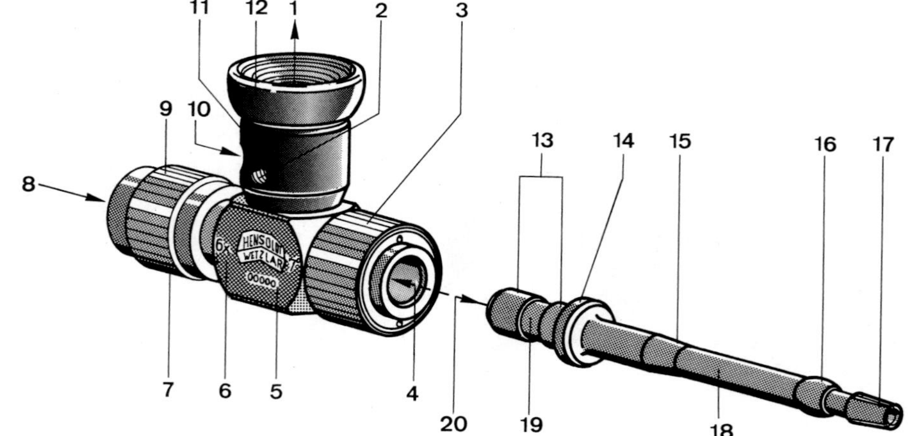

1 viewing axis (eyepiece)
2 azimuth adjustment screw
3 mounting unit
4 sleeve/borehole Ø 12 G 7
5 marking
6 housing unit
7 objective unit
8 line of sight/viewing axis objective
9 range adjustment
10 elevation adjustment screw
11 reticle unit (ring)
12 eyepiece unit
13 fitting cylinder Ø 12 h 6
14 flange
15 cone
16 ball
17 cone, with 3 slots
18 insert rod
19 marking
20 borehole

Hensoldt Boresight SA

Data
Dimensions: without bore insert rod, 96 × 36 × 63 mm
Weight: 200 g
Telescope magnification: ×6
Field of view: 7°
Eye relief: 10 mm
Range adjustment: 0.5 m to infinity
Dioptre setting: fixed at −0.5
Graticule pattern: cross-hairs

Manufacturer
M Hensoldt and Sohne Optische Werke AG, Postfach 1760, D-35573 Wetzlar.

Status
In production.

VERIFIED

Leica SZR/SKS muzzle boresight

Description
The Leica SZR/SKS is a universal muzzle boresight device, designed for efficient and precise calibration of weapons. It consists of an angled telescope and a wide range of interchangeable calibre bars for all standard bores from 7.62 to 155 mm.

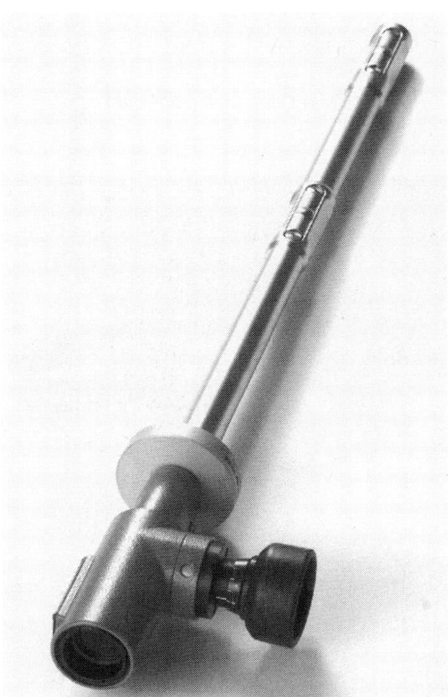

Leica SZR2-1 boresight telescope with SKS20 calibre bar

The SZR2-1 is the MIL-tested version of the boresight telescope. It has a fixed eyepiece and a graticule illumination window for use at night. Supplementary lenses are available for use at 10, 25 and 40 m distance.

The SZR2-3 is a version with a higher magnification.

The SZR2-4 is used for remote viewing with a CCTV camera and monitor. The monitor can be placed at a safe and convenient location so that one man can carry out the calibration.

Leica SZR/SKS muzzle boresight

Data
Dimensions of telescope: 111 × 94 × 42 mm
Weight of telescope: 300 g
Magnification: SZR2-1, ×6; SZR2-3, ×9.2
Field of view: 85 mils
Exit pupil diameter: SZR2-1, 3 mm; SZR2-3, 1.9 mm
Eye relief: SZR2-1, 12 mm; SZR2-3, 8 mm
Dioptric setting: −5 to +5 dioptres
Graticule: cross-hairs with 1 mil graduations
Centring accuracy: ±0.1 mil (telescope to calibre bar)

Manufacturer
Leica Sensortechnik GmbH, D-35578 Wetzlar.

Status
In production.

Service
In service with several NATO armies.

UPDATED

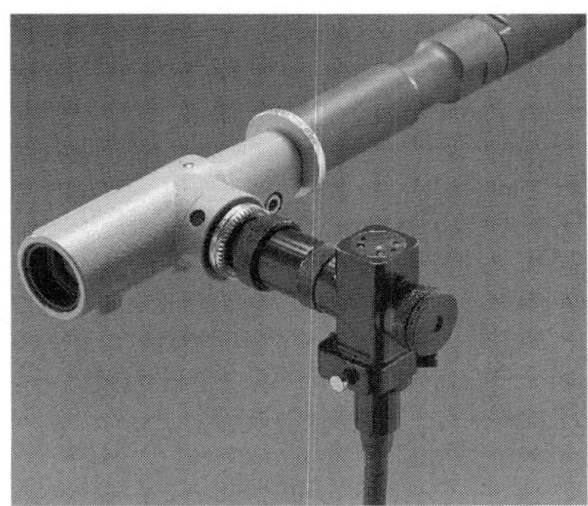

Leica SZR2 boresight telescope with laser eyepiece

Zeiss NSA 80 II night sight attachment

Description
The Zeiss NSA 80 II night sight attachment incorporates the principle of residual light intensification and is an attachment unit suitable for use on various sighting and observation units and the Zeiss HALEM II laser rangefinder. The compact design makes the NSA 80 II a lightweight, reliable unit providing high image quality for numerous applications.

Data
Weight: with batteries, 1.2 kg
Magnification: ×1
Object lens diameter: 85 mm
Exit pupil: 10 mm
Field of view: 8°
Focusing range: 20 m to infinity
Power supply: 2 × 1.2 V rechargeable batteries or 2 × 1.5 V alkaline cells
Operating temperature range: −40 to +50°C

Manufacturer
Carl Zeiss Geschäftsbereich Sondertechnik, PO Box 1380, D-73446 Oberkochen.

Status
In production.

UPDATED

Zeiss NSA 80 II night sight attachment
1996

Zeiss NSV 80 II night sight attachment

Description
The Zeiss NSV 80 II night sight attachment incorporates the principle of residual light intensification and is an attachment unit suitable for use on various sighting and observation units. The compact design makes the NSV 80 II a lightweight, reliable unit providing high image quality for numerous applications.

Data
Weight: with batteries, 1.2 kg
Magnification: ×1
Object lens diameter: 85 mm
Exit pupil: 24 mm
Field of view: 8°
Focusing range: 20 m to infinity
Power supply: 2 × 1.2 V rechargeable batteries or 2 × 1.5 V alkaline cells
Operating temperature range: −40 to +50°C

Manufacturer
Carl Zeiss Geschäftsbereich Sondertechnik, PO Box 1380, D-73446 Oberkochen.

Status
In production.

NEW ENTRY

Zeiss NSV 80 II night sight attachment
1996

Zeiss Orion 80 II image intensifier sight

Description

The Zeiss Orion 80 II is a passive night sight. Standardised dimensions permit the sight to be attached to various weapons using suitable mounts. When fitted with a grip the unit can also be used for hand-held terrain observation.

The passive Orion 80 II night sight uses the principle of residual light intensification. No artificial light sources are required to illuminate the target. All objects which are only weakly illuminated and are invisible or nearly invisible to the human eye are made visible to the observer.

The instrument features a lightweight, compact and sturdy design which makes it very insensitive to shocks, vibrations and climatic fluctuations. The unit is virtually maintenance-free. All glass-to-air surfaces of the optical components have been provided with an abrasion-resistant anti-reflection coating.

Data

Diameter: 112 mm
Length: 212 mm without eye cup
Weight: 1.5 kg including batteries
Magnification: ×5.5
Entrance pupil diameter: 85 mm
Exit pupil diameter: ≥6 mm
Focal length: 125 mm
Focusing range: ca 10 m to infinity
Field of view: 8°
Dioptric adjustment: ±5 dioptres
Resolution: 0.25 mrad
Power supply: 2 Ni/Cd rechargeable batteries or 2 C cells
Battery life: ≥30 h

Manufacturer

Carl Zeiss Geschäftsbereich Sondertechnik, PO Box 1380, D-73446 Oberkochen.

Status

In production.

VERIFIED

Zeiss Orion 80 II night sight

Zeiss HALEM II eye-safe laser rangefinder

Description

The Zeiss HALEM II is a hand-held, lightweight and easily portable laser rangefinder, used against fixed or moving targets at distances between 50 m and 25,000 m. The system is simple to operate and maintain.

The rangefinder is equipped with two eyepieces, one for target detection and the other for the data display. Remote control and data transfer by means of a standard RS-422 serial interface are provided for.

Data

Dimensions: 200 × 190 × 75 mm
Weight: including power pack, 2.5 kg
Type: Raman-shifted Nd:YAG laser
Wavelength: 1,543 nm
Laser class: 3A (IEC 825)
Pulse energy: ≤15 mJ
Beam diameter: ≤15 mm
Beam divergence: ≤1 mrad
Range display: 50-25,000 m
Range accuracy: ±5 m
Target discrimination: ≤20 m
Power supply: rechargeable Ni/Cd, 12 V

Manufacturer

Carl Zeiss Geschäftsbereich Sondertechnik, PO Box 1380, D-73446 Oberkochen.

Status

In production.

UPDATED

Zeiss HALEM II laser rangefinder

1996

Hensoldt aiming point projector

Description

This aiming point projector is primarily used with Heckler and Koch small arms and it is offered by Heckler and Koch as a fitment to its range of sub-machine guns. It is a powerful, narrow beam of white light, projected from a small tubular housing mounted above the weapon. The beam of light illuminates the target and is sufficiently large to allow the operator a limited ability to search with it. In the middle of the beam is a small black area which shows as a black dot. It is this dot which is the actual aiming mark. The operator swings his beam on to his target and refines the aim with the dot. Where the dot rests, the rounds will land.

It is an ideal arrangement for shooters who have to engage fleeting targets in difficult conditions. It enables the firer to shoot from the hip, keeping his head free from the sight pattern and so allowing him an unrestricted view of his target. The beam of light is quickly moved on to a chosen target, and the knowledge that it means that a weapon is pointing directly down the beam acts as a severe deterrent on taking retaliatory or evasive action.

The lamp is a 6 V, 10 W halogen type giving a range of up to 120 m in good weather conditions.

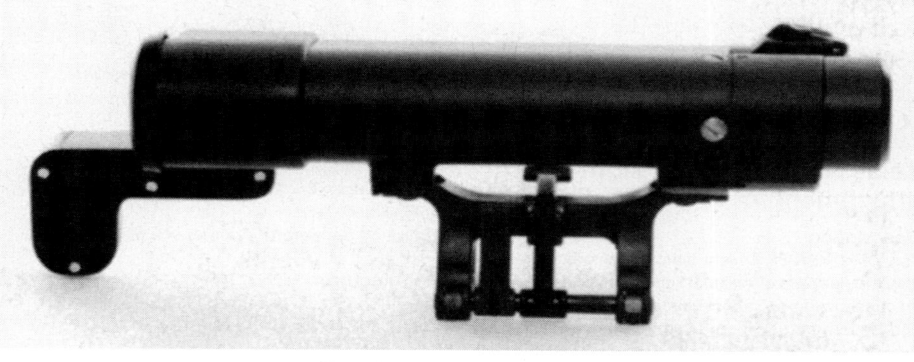

Hensoldt aiming point projector

Data

Dimensions: 290 × 80 × 100 mm
Weight: 1.7 kg complete
Lamp: 6 V, 10 W halogen. Osram No 64225
Lamp life: ca 100 h

Illumination	White light	Black spot
25 m	1 m diameter	150 mm
50 m	2 m	300 mm
75 m	3 m	450 mm
100 m	4 m	600 mm

Range: accuracy, 100 m; search, 120 m
Batteries: rechargeable Ni/Cd accumulators. Varta RSH 1.8, baby size

Manufacturer

Hensoldt-Wetzlar. Marketed by Heckler and Koch GmbH, Postfach 1329, D-78722 Oberndorf/Neckar.

Status

In production.

Service

Several German police forces. Military and police forces in many other countries.

VERIFIED

INDIA

Passive Night Sight

Description

The Passive Night Sight, intended for use on 5.56 and 7.62 mm rifles and light machine guns, was developed by the Defence Research and Development Organisation of the Ministry of Defence, India. It employs an 18 mm proximity focused microchannel 2nd Generation image intensifier tube with power provided by two 1.5 V AA cells or a 3.5 V lithium cell. The performance of the device is such that potential targets can be detected at 300 m and recognised at 200 m.

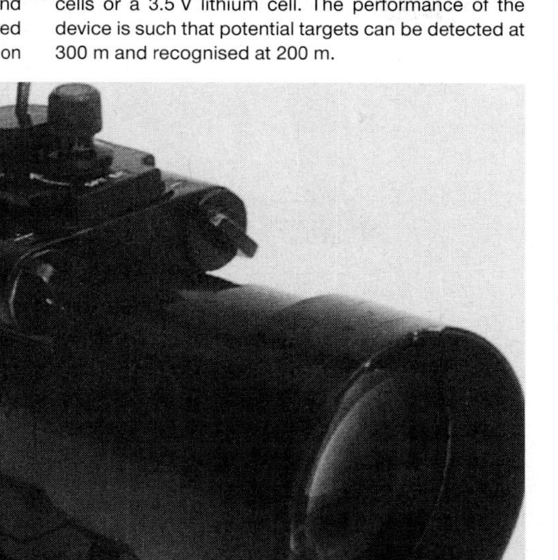

Data

Length: 281 mm
Diameter: 70 mm
Weight: without mount, 1.06 kg
Field of view: 10°
Magnification: ×4
Objective: f/1.5 refractive
Eyepiece dioptre: adjustable from +5 to −5 dioptres
Power supply: 2 × 1.5 V AA batteries or 1 × 3.5 V lithium cell

Development Agency

Defence Research and Development Organisation, 130-E South Block, New Delhi 110011.

Status

Available.

NEW ENTRY

Passive Night Sight for rifles and light machine guns
1996

INTERNATIONAL

UGO day/night goggles

Development

UGO day/night goggles were developed by Thomson-TRT Défense (now Thomson TTD Optronique) using optics developed by Angenieux SA. Using optics produced by Angenieux SA the goggles are manufactured by Pilkington Optronics in the United Kingdom. Marketing is a joint operation between Thomson TTD Optronique and Pilkington Optronics.

Description

The UGO goggle is a self-contained hand-held or head-mounted combined day and night vision system designed to provide a practical solution for various civilian and military tasks. UGO is based on a conventional 8 × 24 day binocular channel utilising high quality optics to provide clear and precise day viewing. The right-hand eyepiece contains a graticule graduated horizontally and vertically in mils for range estimation.

The ×1 magnification night channel is optically and mechanically integrated with the day binocular and is compatible with either second, Super second or third generation image intensifier tubes. Weighing only 750 g the UGO may be hand held or head mounted for either wide field of view night observation or for close-up tasks such as map reading.

For enhanced night viewing a ×4 magnification afocal lens adaptor, which can be fitted by the user in the field, is supplied as standard to provide an improved target recognition capability.

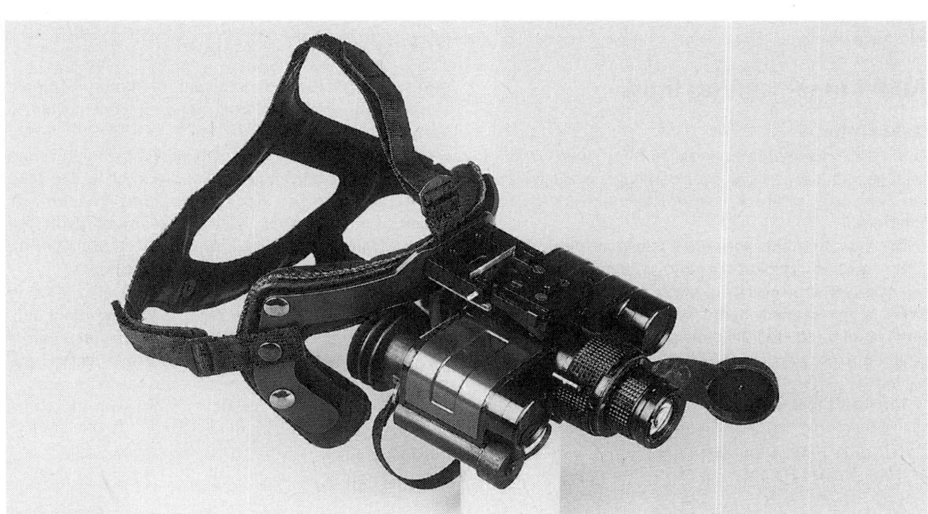

UGO day/night goggles

1996

Data

Length: 153 mm
Height: 62 mm
Width: 140 mm
Weight: 750 g
Field of view: day, 6°; ×1 night, 40°; ×4 night, 10°
Magnification: day, ×8; night, ×1 or ×4
Eye relief: 145 mm min
Eyepiece dioptre: adjustable from +3 to −5 dioptres
Focus range: 250 mm to infinity
Min range: ×1, recognition of a standing man at 300 m in starlight (10^{-3} lux); ×4, recognition of a standing man at 400 m in starlight (10^{-3} lux)
Power supply: 2 × 1.5 V AA batteries
Operating temperature range: −40 to +45°C

Manufacturers

Thomson TTD Optronique, rue Guynemer, BP 55, F-78283 Guyancourt Cedex, France.
Angenieux SA, F-42570 Saint-Heand, France.
Pilkington Optronics, Glascoed Road, St Asaph, Clwyd LL17 OLL, UK.

Status

In production.

UPDATED

ISRAEL

NVL-11 Mk IV fire control night sight

Description

The NVL-11 Mk IV is a computerised fire control night sight for use on small anti-tank rocket launchers and similar short-range direct fire weapons. The NVL-11 Mk IV is highly effective for all night-time rocket-launching applications. A single system can support up to five different weapons or ammunition types, with a firing range of up to 1,000 m.

The sight is based on a high-performance second-generation gated night scope. The fire control system is combined with two low-power laser transmitters, optics and electronic circuits. This compact weapon-mounted system measures the distance to the target and computes ballistic elevation required for the specific weapon and ammunition. The swift and accurate aiming capability yields excellent results which compare favourably with daytime firing, even for long-range targets. A high light detector is installed to avoid damage to the image intensifier from background light intensity over 1 lux.

Data

Dimensions: (L × W × H) 190 × 210 × 130 mm
Weight: excl weapon, adaptor and batteries, 2.1 kg
Field of view: 13.6°
Scope magnification: ×3
Repetition rate: 20/min
Resolution (tube): 32 lp/mm
Ranging laser wavelength: 840 nm
Peak output power: 100 mW
Beam divergence: vertical line, 3.5 mils high, 0.5 mil wide
Performance: range 20 to 990 m; resolution 10 m
Aiming laser wavelength: 840 nm
Peak power output: 10 mW (Hi position); 1 mW (Lo position)
Beam divergence: 0.3 mil
Ballistics: elevation up to 70 mils; accuracy ±0.5 mil
Power source: 4 × 1.5 V AA alkaline batteries
Battery life: 3 h ranging (ca 2,000 rangings of 5 s each)
Construction: anodised aluminium

NVL-11 Mk IV sight mounted on 84 mm Carl Gustaf RCL

Manufacturer

International Technologies (Lasers) Limited, 12 Hachoma Street, PO Box 4099, 75140 Rishon-Letzion.

Status

Available.

VERIFIED

NVL-11 Mk IV sight mounted on 84 mm AT-4 light anti-armour weapon

NVL-11 Mk IV sight mounted on RPG-7 anti-tank grenade launcher

AIM-1 laser aiming light

Description

The AIM-1 family comprises infra-red laser aiming devices for night combat and operations, which may be mounted on a variety of weapons by means of specific adaptors.

The light source of the AIM-1 is a laser diode which emits radiation at near infra-red spectrum (excluding the AIM-1/R which emits a visible red dot). The AIM-1 emits a dot-shaped beam which marks the target with an infra-red dot, visible with suitable night vision equipment, so allowing rapid and accurate night-time aiming.

The AIM-1 family consists of the following options:

AIM-1/D An infra-red aiming light for infantry personal weapons of various calibres, ranging from pistols to assault rifles.

AIM-1/DLR An infra-red aiming light with longer range for infantry standard heavy weapons such as the M60 or MAG machine gun. Adding an optional adaptor with ballistic compensation, it may be mounted on the 40 mm MK 19 grenade launcher, 0.50/12.7 mm machine gun and other crew-served weapons.

AIM-1/MLR A long-range IR aiming light with the same characteristics as AIM-1/DLR but with a different activation switch (on/off/remote) located at the rear side of the unit. Its principal applications are shipboard-mounted and airborne door-mounted machine guns.

AIM-1/EXL A long-range aiming light with the same characteristics as the AIM-1/DLR, designed to be mounted in helicopters. The power supply is obtained directly from the platform power source by cable connected to the rear of the unit. Activation is from the gunner station.

AIM-1/SLR An extra long-range aiming light (10 to 12 km) designed to be installed on extremely long-range observation devices such as long-range FLIRs. The unit is utilised as a pointing device for long-range targets for assault helicopters and special forces units active deep inside enemy territory.

AIM-1/SLX Same as the AIM-1/SLR but with an external power supply option (as AIM-1/EXL) designed to be installed on aerial platforms to increase visibility of target designators.

AIM-1/D mounted on Galil rifle

AIM-1/DLR mounted on Minimi light machine gun

Data

Model	AIM 1/D	AIM 1/DLR	AIM 1/EXL	AIM 1/MLR	AIM 1/SLR	AIM 1/SLX
Range (typical)	500 m	3,000 m	3,000 m	3,000 m	10,000 m	10,000 m
Output power	1.5 mW	15 mW	15 mW	15 mW	50 mW	50 mW
Beam divergance	0.3 mrad	0.3 mrad	0.3 mrad	0.3 mrad	0.3 mrad	0.3 mrad
Emission wavelength	810 nm (IR)	830 nm (IR)	830 nm (IR)	830 nm (IR)	830 nm (IR)	830 nm (IR)
Power source	3 V (2 × AA alkaline cells)	3 V (2 × AA alkaline cells)	24/28 V (ext)	3 V (2 × AA alkaline cells)	3 V (2 × AA alkaline cells)	12/24/28 V (ext)
Boresight + adjustment travel range	min 2.5° end to end in discrete steps (clicks)	same as AIM-1/D	same as AIM-1/D	same as AIM-1/D	same as AIM-1/D	same as AIM-1/D
Step (click) size	0.58 mrad	0.58 mrad	0.58 mrad	0.58 mrad	0.58 mrad	0.58 mrad
Operating time	lo: 50 h hi: 10 h	lo: 50 h hi: 5 h	n/a	lo: 50 h hi: 5 h	lo: 50 h hi: 3 h	n/a
Operational function	5 position switch off, lo, hi lo (remote) hi (remote)	same as AIM-1/D	n/a external control	3 position switch off, on, remote	3 position switch off, lo, hi	n/a external control
Switch position	side	side	n/a	rear	rear	n/a
Remote operation	available	available	available	available	available	available

Manufacturer
International Technologies (Lasers) Limited, 12 Hachoma Street, PO Box 4099, 75140 Rishon-Letzion.

Status
In production.

VERIFIED

AIM-1/R visible laser aiming light

Description
The AIM-1R is a small, lightweight, visible red laser aiming light which allows quick and accurate aiming of weapons by simply marking the target with a red dot. It is effective for indoor and outdoor operations in twilight and night conditions, with an aiming range up to 150 m.

The unit will fit any weapon, from pistols to shoulder arms, and can be operated locally or by remote control cable.

Applications include anti-terrorist units, infantry units, special forces and so on. It can be readily boresighted to any weapon.

AIM-1/R visible laser aiming light

Data
Dimensions: excl weapon adaptor, 92 × 49 × 57 mm
Weight: excl weapon adaptor and batteries, 255 g
Beam divergence: max, 1 mrad
Average output power: 1 mW
Peak wavelength: nominal, 670 nm
Power supply: 2 × 1.5 V AA alkaline
Battery life: ca 5 h

Manufacturer
International Technologies (Lasers) Limited, 12 Hachoma Street, PO Box 4099, 75140 Rishon-Letzion.

Status
In production.

VERIFIED

S-8 target illuminator and designator

Description
The S-8 target illuminator and designator is a high-intensity light projector which mounts beneath the fore-end handgrip of any weapon and is used to illuminate targets up to a range of 100 to 150 m. The light permits ready identification of the target and since the projector is boresighted to the weapon, placement of the light on the target guarantees the aim. Moreover the intensity of the light is such that the target will be disoriented and unable to fire back accurately. The unit, once mounted, becomes part of the weapon and does not materially affect the balance or handling. The operating switch has a safety catch and is spring-loaded for positive pressure only.

S-8 target illuminator and designator

Data
Weight: including power pack, <950 g
Boresight tolerance: ±2°
Beam divergence: 0.7° at 50% peak intensity
Peak power intensity: 900 lux at 5 m
Lamp: 10 W quartz-halogen
Lamp life: 100 h
Dimensions: 165 × 71 mm

Manufacturer
International Technologies (Lasers) Limited, 12 Hachoma Street, PO Box 4099, 75140 Rishon-Letzion.

Status
In production.

VERIFIED

Set Beam target illuminator/searchlight

Description
The Set Beam is a powerful, small-sized, low weight, weapon-mounted Xenon target illuminator and searchlight. It can light areas and/or point targets with visible or (using an infra-red filter) invisible infra-red light.

The main component of the Set Beam is a 150 W Xenon lamp. It is mounted on a shock absorbent system which prevents damage caused by operational activity. It has a built-in flash capability; the projector can flash at a rate of seven times per second, creating a sharp dazzling effect on the target. A change of beam angle can be obtained through an internally installed zoom mechanism.

Control is performed by a built-in control panel or by remote control. The light can be mounted on various machine guns and small calibre cannon as well as on tripods, or it may be hand held.

The Set Beam system consists of the projector, a power supply unit and a rechargeable battery mounted on a backpack or installed in the vehicle.

Data

Weight: excl power supply and battery, 3.2 kg
Beam peak power: 1,000 lux at 50 m
Beam angle: continuously variable from 0.3-6°
Spectral range: visible 0.4-0.7 μm; IR 0.78-2.2 μm; UV 0.3-0.38 μm
Lamp: 150 W short arc Xenon
Operating voltage: adaptors exist for 12, 24-30, 110 and 220 V
Power input: 17 A at 12 V
Flashing rate: 7 ±1 Hz with light pulse width of 400 ± 50 μs

Manufacturer

International Technologies (Lasers) Limited, 12 Hachoma Street, PO Box 4099, 75140 Rishon-Letzion.

Status

In production.

VERIFIED

Set Beam illuminator/searchlight

SL-1 sniper's spotlight

Description

The SL-1 is a laser illuminator for attachment to a rifle which provides the firer with a source of infra-red light capable of illuminating the scene viewed through a night vision sight when ambient light is insufficient to provide the necessary contrast for good viewing. The control gives four light levels to meet varying requirements and can also be remote-controlled so as to allow the scene to be illuminated from an angle or to protect the firer against detection by infra-red sensitive instruments in the target area. In normal use, the spotlight is fitted to the rifle and then aligned with the night sight so as to direct its light accurately into the field of view.

SL-1 spotlight mounted on sniper rifle

Data

Diameter: 51 mm
Length: 155 mm
Weight: with batteries, 550 g
Source: GaAs laser diode
Wavelength: 850 ±20 nm
Average power: 10 mW min
Beam divergence: 2 ± 0.5°
Power source: 4 × 1.5 V AA alkaline batteries
Battery life: 5-15 h, depending upon intensity mode

Manufacturer

International Technologies (Lasers) Limited, 12 Hachoma Street, PO Box 4099, 75140 Rishon-Letzion.

Status

Available.

VERIFIED

S-3 high-intensity target illuminator

Description

S-3 is a compact, lightweight, high-intensity target illuminator which enables rapid and accurate aiming of weapons by spotlighting the target. The projected beam marks the target with a bright and clearly defined spot. Once the target is selected, the S-3, boresighted to the weapon, ensures that the first shot will be fired within a very short time and will hit the target.

Use of the S-3 hinders enemy reaction; the spotlight's blinding glare, together with the rapid target acquisition and aiming process afforded by using the S-3, ensures that the target has neither the time nor the necessary night vision to fire back.

S-3 high-intensity target illuminator

S-3 fits a variety of weapons using different adaptors. Once mounted, it becomes an integral part of the weapon and will not affect the shooting habits of the user or the performance of the weapon. It is powered by a single 3.5 V lithium battery or two 1.5 V alkaline D cells and is operated by a flexible cable switch.

Data

Dimensions: 215 × 52 mm
Weight: 350 g
Light intensity: 200 lux at 5 m
Beam divergence: 7 mrad (140 mm diameter at 20 m distance)

Manufacturer

International Technologies (Lasers) Limited, 12 Hachoma Street, PO Box 4099, 75140 Rishon Letzion

Status

In production.

VERIFIED

BL-2 Borelight

Description

The BL-2 Borelight is a portable electro-optical instrument used for the dry zeroing of weapons equipped with laser aiming lights. The device is equipped with an image intensifier tube which converts the invisible laser infra-red light into a visible picture. It enables rapid and accurate dry zeroing of laser aiming lights under any conditions, day or night, indoors or outdoors, eliminating the need for a firing range, targets and weapon rests.

The BL-2 Borelight can zero a wide range of weapon types with different bore calibres by using different spuds. As a standard, the system is delivered with one spud, selected from 5.56, 7.62 or 9 mm calibres.

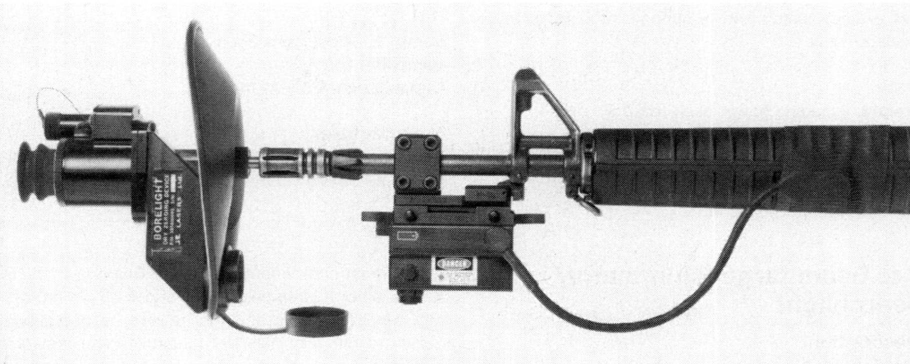

BL-2 Borelight

Data

Dimensions: (L × W × H) 145 × 55 × 145 mm
Weight: less battery and barrel spud, 650 g
Field of view: 4°
Magnification: ×10
Eyepiece adjustment: +2 to −6 dioptres
Calibration accuracy: 0.5 mrad
Resolution (min): 0.2 mrad/lp
Spectral response: 400-920 nm
Operating voltage: 3.5 V DC (one AA lithium battery)

Manufacturer

International Technologies (Lasers) Limited, 12 Hachoma Street, PO Box 4099, 75140 Rishon-Letzion.

Status

Available.

VERIFIED

MINI N/SEAS night single eye acquisition sight

Description

The MINI N/SEAS is a miniature night vision monocular system offering high resolution and a clear bright image in a single eye configuration. The single eye configuration allows the user to achieve a high level of night vision through one eye while maintaining a high level of awareness of immediate surroundings with the other eye. It also eliminates dazzle effects disturbing users after the use of standard dual-eye night vision goggles and cuts the adaptation period to darkness to a minimum.

The MINI N/SEAS can be used as a hand-held pocket scope, mounted on a face mask (under or without a helmet), or attached to a variety of weapons; it is NBC compatible. An optional clip-on ×3 magnifier can be attached by a user during field operations.

The MINI N/SEAS includes an infra-red illumination for close range covert illumination. An indicator in the eyepiece lights when the infra-red illuminator is activated and battery power is low. The system incorporates a high light sensor which automatically shuts off the image intensifier tube for protection against high light environments.

Data

Dimensions: (L × W × H) 115 × 65 × 40 mm
Weight: with battery but not face mask, 330 g; with battery and face mask, 540 g
Field of view: ×1, 40°; ×3, 13.3°
Magnification: ×1 or ×3
Focus range: ×1, 250 mm to infinity; ×3, 10 m to infinity
Dioptre adjustment: −6 to +2
Objective lens: ×1, f/1.2; ×3, f/1.4
Resolution (system): 0.88 lp/mrad, Gen 3
Gain: 2,500, Gen 3
Image intensifier resolution, min: Gen II+, 321 lp/mm; Gen III, 45 lp/mm
Power source: 1 × 1.5 V AA size alkaline or lithium battery; or 1 × 3.5 V AA size lithium battery; or 1 × 1.2 V AA size Ni/Cd rechargeable battery
Battery life: alkaline, 30 h; lithium, 60 h

Manufacturer

International Technologies (Lasers) Limited, 12 Hachoma Street, PO Box 4099, 75140 Rishon-Letzion.

Status

In production.

VERIFIED

MINI N/SEAS night single eye acquisition sight mounted on M16 series assault rifle

MINI N/SEAS night single eye acquisition sight with clip-on ×3 magnifier

N/CROS Mk II night compass rangefinder system

Description

The N/CROS Mk II is a hand-held night binocular combining night vision with laser rangefinding and azimuth measurement. It incorporates a second or third generation image intensifier, an eye-safe laser rangefinder and a digital compass in a compact lightweight system. Described as user-friendly and simple to operate, the N/CROS Mk II has a measurement range of 20 to 2,000 m. Range and azimuth data are displayed in the eyepieces, allowing the user to identify targets and relay fire support data. The system has a built-in data that bank stores data for up to 10 targets (range and azimuth) and an RS-232 communications port offering interface ability to external systems such as GPS and other combat computers. A LPL-30 laser pointer is integrated into the system to allow the user to indicate targets to others at distances of up to 4,000 m.

N/CROS Mk II is used as part of the NIMTAS night multipurpose target acquisition system (see following entry).

Data

Dimensions: (L × W × H) 210 × 200 × 78 mm
Weight: 1.6 kg
Field of view: 13.5°
Magnification: ×1 or ×3
Focus range: ×1, 250 mm to infinity; ×3, 10 m to infinity
Dioptre adjustment: −6 to +2
Objective lens: 75 mm, f/1.3
Resolution (system): 2.3 lp/mrad, Gen 3
Gain: 1,800, Gen 3
Laser rangefinder range: 20 to 2,000 m
Accuracy: up to 200 m, ±1 m; 200 to 500 m, ±2.5 m; over 500 m, ±5 m
Measurement time: 0.6 s
Laser type: GaAs laser diode
Laser output: max, 4 W
Wavelength: 904 nm
Beam divergence: 1.5 mrad
Compass accuracy: ±0.6°
Compass resolution: ±0.2°
Power source: 4 × 3 V 2/3A size lithium batteries Type BA 5123/U; or 6 × 1.5 V AA size alkaline batteries
Battery life: 18 h continuous viewing inc 1,000 range measurements and 1,000 azimuth measurements

Manufacturer

International Technologies (Lasers) Limited, 12 Hachoma Street, PO Box 4099, 75140 Rishon-Letzion.

Status

In production.

VERIFIED

N/CROS Mk II night compass rangefinder

N/CROS Mk II night compass rangefinder

NIMTAS night target acquisition system

Description

The NIMTAS target acquisition system is claimed to be the smallest of its kind to be available off-the-shelf to provide field commanders and reconnaissance teams with real-time target data at night. This data consists of the user's own GPS derived location, the target's exact co-ordinates, and range and azimuth to the target. All this data can be transmitted in real time through the user's transceiver to fire or command posts requiring the data.

The NIMTAS system consists of two main parts: N/CROS Mk II (see previous entry for details); and an Azimuth Limited MP hand-held tactical computer with target acquisition, fire control and GPS navigation functions.

Manufacturers

International Technologies (Lasers) Limited, 12 Hachoma Street, PO Box 4099, 75140 Rishon-Letzion.
Azimuth Limited, Hakarem Street, 45930 Ramot Hashavin.

Status

In production.

VERIFIED

NIMTAS night target acquisition system

ORT-MS4 mini weapon night vision sight

Description

The ORT-MS4 is a lightweight second-generation night sight. It is particularly suited for use on infantry weapons, or as a hand-held night observation device. The ORT-MS4's modular design incorporates an 18 mm second-generation image intensifier.

The instrument is self-contained, battery-powered and is supplied in a storage/carrying case with all necessary accessories.

Data

Dimensions: 266 × 70 mm
Weight: 1.16 kg
Field of view: 10°
Magnification: ×3.75
Eyepiece focal length: 27 mm
Focus adjustment: +2 to −4 dioptres
Objective: 100 mm fixed focus
Viewing range: moonlight, 500 m; starlight, 350 m

Manufacturer

ORTEK Limited, PO Box 388, Sderot 80100.

ORT-MS4 weapon sight

Status

In production.

VERIFIED

ORT-TS 5 crew-served weapon sight

Description

The second-generation ORT-TS 5 night vision sight was primarily designed for heavy machine guns, recoilless rifles and other crew-served weapons. It serves such weapons as the M2 0.50/12.7 mm machine gun, the 106 mm M40 RCL and the B-10 rocket launcher, providing the operator with a military advantage at night. On weapons where access to the conventional eyepiece is hazardous, an optional right angle relay is available.

The sight has an internally mounted adjustable illuminated reticle for boresighting. Different reticle patterns are available. Image intensification is performed by a second-generation 25 mm tube. To compensate for differing levels of ambient light, image tube and reticle brightness can be manually adjusted.

In order to facilitate logistics and maintenance, major assemblies and subassemblies are interchangeable with the AN/PVS-4 individual sight and the AN/TVS-5 crew-served weapon sight.

Data

Diameter: 165 mm
Length: 356 mm
Weight: 3.8 kg
Field of view: 9° (156 mils)
Magnification: ×5.6
Dynamic range: from 10^{-5} to 10^{-2} FC
Viewing range: moonlight, 2,000 m; starlight, 1,200 m
Focal length: 155 mm
Focus adjustment: 25 m to infinity
Reticle adjustment: ±2.5° (in 0.25 mil increments)
Eyepiece focal length: 26.5 mm
Eyepiece adjustment: −6 to +3 dioptres

Manufacturer

ORTEK Limited, PO Box 388, Sderot 80100.

Status

In production.

VERIFIED

ORT-TS 5 crew-served weapon sight on M2 machine gun

ORT-TLS 8 aiming light

Description

The ORT-TLS 8 is a full military standard light source which gives an appropriate solution for special forces activities in darkness. It can be attached to any small arm by using the correct type of adaptor. It is equipped with a pressure-sensitive tape switch, allowing precise on/off control and positioning at any place on the weapon grip or frame that is convenient for the particular user.

Once the light has been mounted on a weapon it is boresighted, allowing it to be used either as an aiming light or simply as a source of illumination. An infra-red filter is optional.

Data
Diameter: 45 mm
Length: 95 mm
Weight: with batteries, up to 300 g
Power supply: 4 × lithium 2/3A

Manufacturer
ORTEK Limited, PO Box 388, Sderot 80100.

Status
In production.

VERIFIED

ORT-TLS 8 aiming light fitted to a Mini-Uzi sub-machine gun

Hit-Eye 1500 optical sight

Description
The Hit-Eye 1500 is an optical sight for assault rifles. It can be incorporated into the design of a new rifle or fitted, by means of mounts, to virtually any type of existing rifle. The graticule pattern is a circle, effective for aiming at targets at a few hundred metres range. Distance can be estimated rapidly with the aid of the graticule pattern, which is always centred in the field of view. Built-in illumination of the graticule permits firing at night and in poor light.

The sight is fitted with external knobs for zeroing adjustment in elevation and windage. External iron sights, with night aiming luminous points, may be fitted as an option.

Hit-Eye 1500 assault rifle sight

Data
Diameter: 21 mm or 26.5 mm, as required
Length: 138 mm
Field of view: 7°
Magnification: ×1.5
Object lens diameter: 12 mm
Eye relief: 72 mm

Manufacturer
ORTEK Limited, PO Box 388, Sderot 80100.

Status
In production.

VERIFIED

Hit-Eye 3000 rifle sight

Description
The Hit-Eye 3000 rifle sight is similar in principle to the Hit-Eye 1500, but of greater magnification and slightly larger. It also has a different graticule pattern, using the traditional cross-wires, and the eyepiece is adjustable. The integrated mount fits into the carrying handle of M16 series rifles, but mounts for other rifles can be provided.

Data
Diameter: 35 mm
Length: 160 mm
Field of view: 6°
Magnification: ×3
Object lens diameter: 21 mm
Eye relief: 90 mm
Dioptric adjustment: −5 to +5 dioptres
Elevation adjustment: (100-200) −500 m

Manufacturer
ORTEK Limited, PO Box 388, Sderot 80100.

Status
In production.

VERIFIED

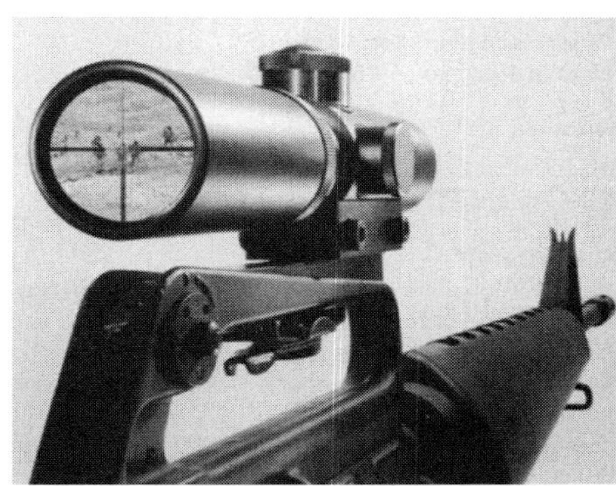

Hit-Eye 3000 rifle sight

Hit-Eye 4000 launcher sight

Description
Hit-Eye 4000 is a telescope sight for fitting to light anti-armour rocket launchers, though it can of course be mounted on virtually any direct fire short-range weapon. Various mounts are available and the sight can be zeroed by means of the mount. The sight is hermetically sealed, preventing internal fogging or fungal growth, is robust, has no moving parts and can withstand the most extreme field conditions.

Hit-Eye 4000 can be provided with patterned graticules according to the ballistics of the weapon in use. The graticule is illuminated by a Betalight source, giving excellent aiming properties in night or poor light conditions.

Data
Dimensions: 155 × 67 × 67 mm
Weight: 470 g
Field of view: 10°
Magnification: ×4
Eye relief: 35 mm

Manufacturer
ORTEK Limited, PO Box 388, Sderot 80100.

Status
In production.

VERIFIED

Hit-Eye 4000 launcher sight

Elbit Falcon optical sight

Description
The Falcon sight uses a compact aircraft-type head-up display mounted on the front end of the rifle. The unique design does not incorporate a tube or telescope array, therefore firing with both eyes open gives the user faster acquisition of a target as well as high precision in harsh battlefield conditions. Aiming is performed by placing a sharply defined luminous red dot, which is projected in the display, on the target. Since parallax has been eliminated, the position of the dot on the screen is of no significance; provided the firer sees the dot anywhere on the screen and superimposes it on the target, he will hit.

The sight was specifically designed for M16 and Galil assault rifles, but mounts are available for other types of rifle to permit fitting without the need for machining or other gunsmithing operations.

Data

Lens diameter: 30 mm
Length: 182 mm
Height: 51 mm
Weight: 300 g
Field of view: unrestricted
Lens magnification: ×1
Boresighting range: ±14 mrad windage, ±14 mrad elevation
Dot diameter: 1.3 mrad
Power supply: 3.6 V lithium AA cell
Battery life: 250 h day, 700 h night
Shelf life: 10 years
Environmental conditions: in accordance with MIL-STD-810C

Manufacturer

Elbit Computers Limited, PO Box 5390, Haifa 31053.

Status

In production.

VERIFIED

Elbit Falcon optical sight

El-Op MALOS miniature laser optical sight

Description

The El-Op MALOS miniature laser optical sight integrates a highly accurate laser rangefinder with a sighting telescope and a fire control computer in one compact instrument. It extends the effective range of sniping rifles and various other types of direct fire weapon. Though slightly larger than a conventional sighting telescope it is nevertheless of convenient size and lightweight.

Range measurement is controlled by a single button which can be mounted anywhere on the weapon. Once the range has been measured the ballistic computer calculates the required elevation for the measured range and displays the required aiming point. Windage and range drums are fitted for boresighting and as manual back-up.

Data

Dimensions: 351 × 85 × 61 mm
Weight: 1.2 kg
Rangefinder range: 50-2,000 m ±5 m
Field of view: 0.6 mrad
Output energy: 3 mJ
Repetition rate: 12 pulses/min
Extinction: 24 dB
Optical sight magnification: ×6
Object lens diameter: 40 mm
Field of view: 3°45′ (66.6 mils)
Exit pupil diameter: 6.7 mm
Power supply: 2 × 3 V AA lithium batteries

MALOS laser optical sight on B-300 anti-tank launcher

Manufacturer

El-Op Electro-Optics Industries, PO Box 1165, Rehovot 76110.

Status

In production.

VERIFIED

El-Op INTIM thermal binoculars

Description

El-Op INTIM thermal binoculars form a lightweight dual field of view night vision device able to detect tank and human targets. INTIM is based on Thermal Imaging Module (TIM) technology originally developed by El-Op for the add-on night observation system fitted to the commander's sight on Leopard 2 MBTs. This involves mercury cadmium telluride photo voltaic techniques. INTIM thermal binoculars are equipped with a miniature Detector Dewar Cooler and focal plane signal preprocessing circuitry in one integral miniature assembly.

Each set of INTIM thermal binoculars weighs 3.5 kg. No further information is yet available.

Manufacturer

El-Op Electro-Optics Industries, PO Box 1165, Rehovot 76110.

Status

Production scheduled to commence during early 1996.

NEW ENTRY

El-Op INTIM thermal binoculars mounted on a tripod together with a laser rangefinder
1996

Meprolight illuminated night sight system

Description

Meprolight's illuminated night sight system was designed as replacement parts for the standard weapon sights, and can be mounted directly with no modifications.

The addition of luminous elements to the standard iron sights greatly improves the low light sighting capability and hit probability. Tritium-illuminated Meprolight sights are clearly visible even in total darkness. Unlike batteries, the tritium power sources function in all temperature conditions and add no weight to the weapon system.

Meprolight has supplied over half a million sets of illuminated sights to military forces on three continents. Sights for virtually all current service rifles, pistols and sub-machine guns can be provided from stock.

Manufacturer

Scopus Light (1990) Limited, Kibbutz Maayan Zvi 30805

Status

In production.

VERIFIED

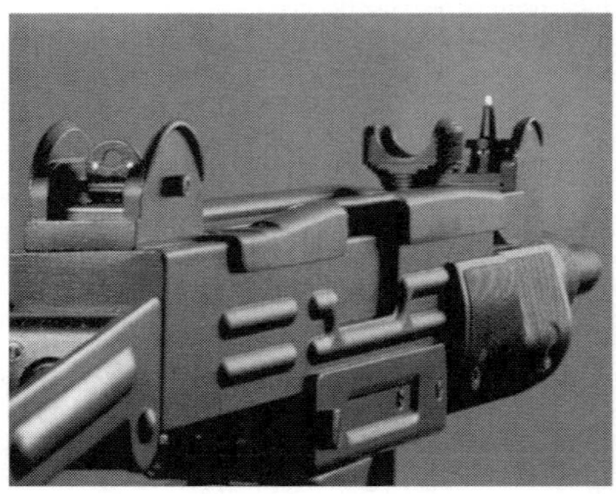

Meprolight night sight on Uzi sub-machine gun

ITALY

M 166 mini weapon sight

Description

The M 166 has been in production since 1977 and is in service with the Italian Army and government agencies. It is a compact, lightweight sight using the cascade tube technique for light intensification and it has an adjustable reticle for sighting.

Data

Diameter: 83 mm
Length: 410 mm
Weight: 2 kg
Field of view: 208 mil (11.7°)
Magnification: ×3
Lens: 81.5 mm
Relative aperture: 1.4
Focus range: 15 m to infinity
Tube: 18 mm
Battery: 3 V lithium cell
Battery life: 60 h continuous operation
Temperature range: −40 to +50°C

Manufacturer

Aeritalia SpA, Avionic Systems and Equipment Group, Viale Europa 1, I-20014 Nerviano, Milan.

Status

In production.

Service

Italian Government and military.

VERIFIED

M 166 and M 193 mini weapon sights

M 193 passive weapon sight

Description

The M 193 is a second-generation version of the M 166 and fits the same range of weapons. It has the usual advantages of a second-generation over a first in that the gain is more uniform, there is less distortion and there is no, or very little, flare in bright lights.

Data

Dimensions: 340 × 83 mm
Weight: 1.9 kg
Field of view: 13°
Magnification: ×4.8
Focus range: 15 m to infinity
Battery: 3 V lithium cell
Battery life: 60 h continuous operation

M 193 Aeritalia second-generation weapon sight

Manufacturer

Aeritalia SpA, Avionic Systems and Equipment Group, Viale Europa 1, I-20014 Nerviano, Milan.

Status

Available.

VERIFIED

M 176 night observation sight

Description

The M 176 was designed for artillery fire control and also as a sight for crew-served weapons such as the M40 106 mm RCL. A version without graticule is also available and is suitable for medium-range surveillance from helicopters, moving vehicles or static positions.

The M 176 has a range of up to 1,500 m in starlight conditions and the weapon sight version has an adjustable illuminated graticule.

Data

Weight: 2.2 kg
Field of view: 10.5°
Magnification: ×5.3
Resolution: 0.4 mil at 1 mlux
Focus range: 15 m to infinity
Objective: 136 mm f/1.2 catadioptric
Power supply: 3 V lithium cell
Battery life: 50 h
Operating temperature: −40 to +50°C

M 176 night observation sight

Manufacturer

Aeritalia SpA, Avionic Systems and Equipment Group, Viale Europa 1, I-20014 Nerviano, Milan.

Status

In production.

VERIFIED

VTG 120 thermal imaging system

Description

VTG 120 is a man-portable apparatus for night surveillance and for the aiming of medium-range weapons. In particular, its application to the TOW missile system has been studied.

The equipment is completely independent of any external source; its battery and air bottle allow more than two hours of continuous operation. Empty bottles and battery can easily be replaced to extend operation.

It can be used to view directly, or can be connected to an external CRT for remote monitoring of the observed scene. An electrically generated graticule can be positioned for aiming purposes.

A complete ancillary support equipment is available for use and maintenance, including the following: battery charger; gas bottle charger; boresight collimator; field test equipment; lens cleaning kit; test and adjustment equipment; cooling system purging kit; tripod mounting interface; and a carrying case.

Standard performances on an operative tank target are detection at 3,000 m and identification at 2,000 m in standard meteorological conditions according to FINABEL 1. R.9.

The same set of modular subsystems has been arranged in different configurations in order to provide appropriate sights for the MILAN and MAF anti-tank missiles.

Data

Field of view: narrow, 20 × 40 mrad; wide, 60 × 120 mrad
Resolution: narrow, 0.16 mrad; wide, 0.5 mrad
IR band: 8-14 μm
Cooling system: Joule-Thomson minicooler
Detector: CMT 60-element array
Number of lines: 120
Power supply: 22-28 V DC Ni/Cd battery

Manufacturer

Officine Galileo SpA, Via Albert Einstein 35, I-50013 Campi Bisenzio.

Status

In production.

VERIFIED

VTG 120 light thermal imager on TOW launcher

OGVN 7 miniaturised night sight

Description

The OGVN 7 is a compact night sight for observation and aiming, suitable for installation on individual weapons. The optical system is equipped with second-generation light intensifier tubes, Automatic Brightness Control (ABC) and automatic switch-off when field of vision brightness is dangerously high. The telescope is equipped with a sighting graticule and boresighting devices.

Data

Dimensions: 227 × 80 × 45 mm
Weight: 1 kg
Field of view: 13°
Magnification: ×2.8
Dioptre adjustment: ±5 dioptres
Focusing: fixed, from 5 m to infinity
Resolution: 1 mrad at 0.001 lux
Power supply: 2 × 1.5 V alkaline-manganese batteries LR6
Battery life: 40 h at +25°C

Manufacturer

Officine Galileo SpA, Via Albert Einstein 35, I-50013 Campi Bisenzio.

Status

In production.

Service

Under evaluation by the Italian Army.

VERIFIED

Officine Galileo OGVN 7 miniaturised night sight

OGVN 3 night vision sight

Description

The OGVN 3 is a compact aiming telescope, suitable for installation on individual precision weapons. The high-performance reflective telescope includes a second-generation microchannel light intensifier tube with automatic brightness control and automatic switch-off when field of vision brightness reaches dangerous levels.

The luminous protected aiming graticule may be adjusted in brightness and position as required for precision aiming.

Data

Dimensions: 330 × 130 × 140 mm
Weight: 2.35 kg
Field of view: 11°
Magnification: ×3.6
Dioptre adjustment: ±4 dioptres
Focusing: fixed, from 50 m to infinity
Resolution: 0.8 mrad at 10^{-3} lux
Power supply: 2 × 1.5 V alkaline-manganese batteries LR14
Battery life: 35 h at +25°C

Manufacturer

Officine Galileo SpA, Via Albert Einstein 35, I-50013 Campi Bisenzio.

Status

In production.

Service

Italian Navy Special Forces.

VERIFIED

Officine Galileo OGVN 3 night vision sight

NETHERLANDS

Type GK4MC passive mini weapon sight

Description

The GK4MC small arms sight is a second-generation lightweight passive night vision device specially designed for accurate weapon aiming and surveillance at night. The sight enables the user to engage the enemy and to observe targets without any artificial illumination, thus presenting full security from detection. The sight can also be used to detect hostile infrared light sources.

GK4MC mini weapon sight

The sight is primarily used as a night sight on basic infantry weapons such as rifles, machine guns and short-range anti-tank weapons. It can also be used as a hand-held or tripod-mounted observation device for direct viewing.

Data

Dimensions: 295 × 144 × 98 mm
Weight: 1.5 kg
Field of view: 10°
Magnification: ×4
Battery power: 2 × 1.5 V DC, size AA, or 2 × Ni/Cd cells, size AA, or one lithium cell size AA or size 2 × AA, with adaptor

Manufacturer

Delft Instruments Electro-Optics BV, Postbus 5083, NL-2600 GB, Delft.

Status

In production.

Service

Several armies.

VERIFIED

Type MS4GT mini weapon sight

Description

The Type MS4GT is a very low-weight second-generation sight particularly suited for use on infantry weapons. It incorporates a catadioptric object lens and an 18 mm microchannel plate amplifier. The graticule may be illuminated, although this cuts the battery life by about 50 per cent. Mounting brackets are available to fit the sight to a variety of rifles, machine guns and platoon weapons. The Type MS4GT can also be used as a surveillance device or as an infra-red detection system.

MS4GT mini weapon sight mounted on Mauser 7.62 mm Model 86 sniper rifle

Data

Dimensions: 260 × 122 × 87 mm
Weight: without mounting bracket, ca 1 kg
Field of view: 11°
Magnification: ×4
Battery: 2 × Ni/Cd cells, size AA; 2 × dry cells, size AA; or, with adaptor, one lithium cell size AA

Manufacturer

Delft Instruments Electro-Optics BV, Postbus 5083, NL-2600 GB, Delft.

Status

In production.

Service

Several armies.

VERIFIED

Type GK7MC aiming device

Description

This passive night vision instrument is a lightweight aiming device suitable for medium-range weapons, such as machine guns, grenade launchers, recoilless guns and other crew-served weapons.

Its passive operation is based on a second-generation image intensifier tube and does not require any infra-red or other artificial light source, thus providing maximum security from detection. Explosion flashes, flares and other bright lights in the field of view will not bloom the image. Tube damage caused by too great a light intensity is prevented by the attenuation characteristics of the microchannel plates incorporated in the image intensifier tube.

It can also be employed as an observation device for long-range detection. The modular construction ensures simple maintenance and service.

GK7MC second-generation passive aiming device

Data

Dimensions: 363 × 174 × 134 mm
Weight: 2.2 kg
Field of view: 6.5°
Magnification: ×6.7
Power supply: 2 × AA Ni/Cd or conventional cells, or one lithium cell with adaptor

Manufacturer

Delft Instruments Electro-Optics BV, Postbus 5083, NL-2600 GB, Delft.

Status

In production.

Service

Several armies.

VERIFIED

Type TM-007 laser target pointer

Description

This small, lightweight, infra-red laser target pointer is specially designed for close range combat with small arms at night, with the use of passive night vision goggles. It enables the user to aim at a target with great accuracy under low light conditions or in complete darkness, without visual recognition.

TM-007 laser target pointer

1996

Data

Dimensions: length, 157 mm; diameter, 34 mm
Weight: with batteries, 350 g
IR laser output power: 2 mW (optional 15 mW)
Wavelength: nominal, 820 nm
Beam diameter: nominal, 5 mm
Beam divergence: typical, 0.35 mrad
Beam adjustment: +10 and −10 mrad in steps of 0.25 mrad
Visibility with goggles: 300 m (spot 100 mm)
Operation temperature: −30 to 60°C
Batteries: 2 alkaline 1.5 V DC size AA
Operating time: ca 4 h continuous (2 mW type)

Manufacturer

Delft Instruments Electro-Optics BV, Postbus 5083, NL-2600 GB, Delft.

Status

In production.

UPDATED

UA 1134 compact night sight

Description

The UA 1134 sight is designed for use on small lightweight assault weapons. The monocular sight incorporates a second-generation, high-performance image intensifier tube, but can also use a second plus- or third-generation tube.

Providing a ×4 magnification and an 8° field of view, the sight has a focus setting of 25 m to infinity. Automatic Brightness Control ensures constant image brightness and Bright Source Protection safeguards the user against dazzle from intense light sources.

For daylight aiming, the daylight protection cap on the object lens has been provided with a small zero-axis hole, allowing the sight to be aimed under daylight conditions.

The sight's components and subassemblies are fully interchangeable, including the fixed-focus object and eyepiece lenses. The aiming mark is automatically illuminated. Once fitted and adjusted to a weapon, the sight remains perfectly accurate in spite of constant firing shocks. The sight controls are operable with gloved hands.

Other than changing the batteries and cleaning the lenses, the UA 1134 is designed to be maintenance-free throughout its whole lifetime.

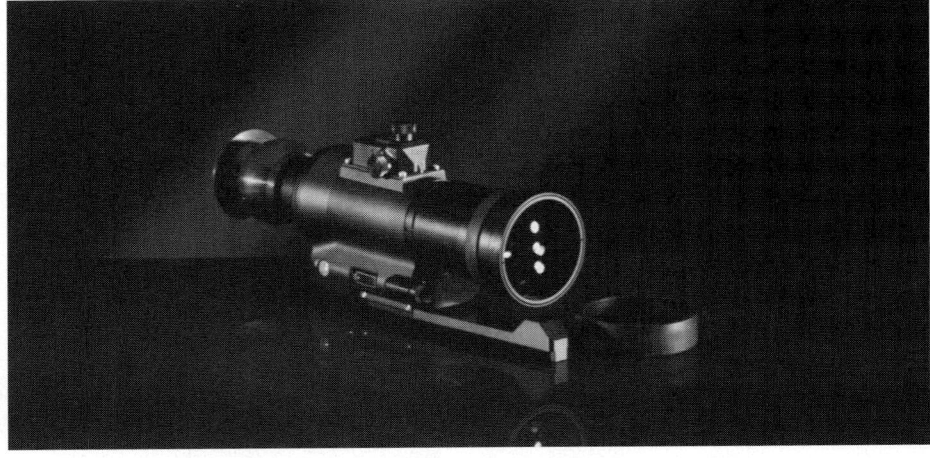

UA 1134 compact night sight

Data

Weight: 1.1 kg
Field of view: 8° nominal
Magnification: ×4
Focus setting: 25 m to infinity
Focusing: fixed at 125 m
Eye relief: 30 mm
Pupil diameter: 10 mm
Controls: on/off switch; brightness aiming mark
Power consumption: 50 mW
Operating temperature range: −45 to +45°C
Battery supply: 2 × AA cells (R6)

Manufacturer

Signaal Usfa, PO Box 6034, NL-5600 HA Eindhoven.

Status

In production.

VERIFIED

UA 1137 self-powered rifle sight

Description

Signaal Usfa's low budget UA 1137 is a versatile self-powered sight for use on hand-held weapons in all ambient light conditions from full daylight to starlight.

A wide horizontal field of view (22°) makes it suitable for both observation and aiming. The vertical field of view is 10° which ensures the user is shielded from sky illumination above and from own-weapon muzzle flash and barrel reflections below. The sight incorporates a shadow-type aiming mark and line of sight adjustment.

Designed to simplify logistics and minimise maintenance, the UA 1137 does not use replaceable batteries. Instead, its image intensifier tube is powered by an internal energy source that can be manually re-energised during use. Thus the user simply presses a noiseless lever 5 to 10 times and the sight will be operational for 30 to 60 minutes.

Resistant to NBC contamination, this watertight rifle sight requires no maintenance whatsoever apart from keeping the lenses clean. The UA 1137 can be stored for at least five years without the necessity to test its performance.

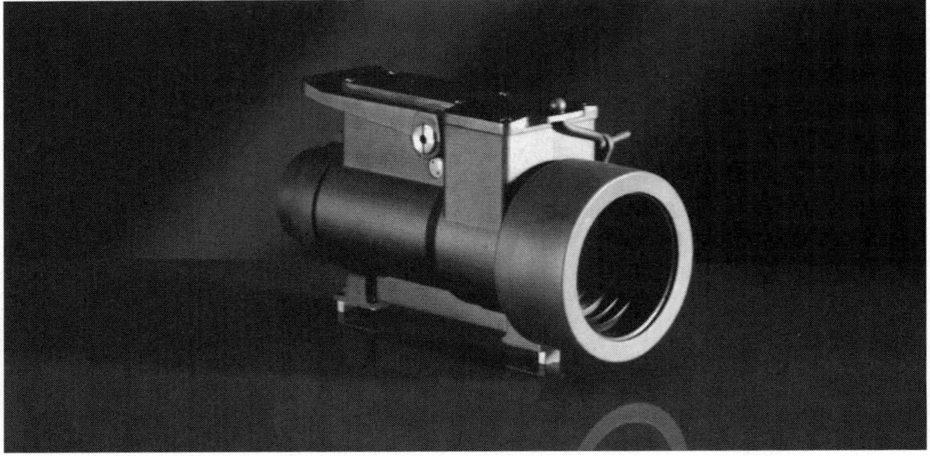

UA 1137 self-powered rifle sight

Data

Weight: <1.5 kg
Field of view: horizontal, 22°; vertical, 10°
Magnification: ×1.5
Focusing: fixed (field 75-400 m)
Aperture: f/1.2
Focal length: 50 mm
Eyepiece setting: fixed at −1 dioptre
Pupil: 6 mm
Power source: hand-activated piezoelectric power supply

Manufacturer

Signaal Usfa, PO Box 6034, NL-5600 HA Eindhoven.

Status

In production.

VERIFIED

NORWAY

Simrad LP101 laser gun sight

Description

The Simrad LP101 is a complete fire control system for mounting on direct fire weapons, providing instant target range and aiming information. The built-in ballistic computer is programmed for the type of weapon and ammunition used. Aiming point and time of flight information are displayed in the eyepiece. The time of flight information enables the gunner to estimate the correct lead angle.

The Simrad LP101 is particularly suited for upgrading the 106 mm M40 RCL gun and similar weapons, dramatically increasing hit probability and effective range. Night capability can be achieved using Simrad KN200 and KN250F image intensifiers (described in the next entries). The add-on principle of these instruments ensures that the eye position of the

Simrad LP101 laser gun sight on 106 mm RCL gun

gunner remains unchanged by day or night, and no boresighting is required.

The LP101 can be easily programmed by the operator for ballistic and environmental data. Parameters displayed as standard are minimum range, boresight marker, four types of ammunition, manual range setting and the dimming of the display. Optional parameters which can be programmed include ammunition stock, compensation for wind, muzzle velocity, air temperature and powder temperature, adjustment of display contrast and switching the time of flight indication on or off.

Data

Dimensions: 320 × 195 × 140 mm
Field of view: 200 mrad (11.3°)
Laser type: Nd:YAG 1.064 μm
Sighting telescope: 4 × 30
Reticle pattern: to customer specification
Range resolution: 5 m
Range: min, 60 m; max, 9,995 m
Power supply: 10–15 V
Battery type: rechargeable Ni/Cd; alkaline optional
Rangings per battery before recharge: up to 400 at 20°C

Manufacturer

Simrad Optronics A/S, PO Box 6114, Etterstad, N-0602 Oslo.

Status

In production.

VERIFIED

Simrad KN200 image intensifier

Description

The Simrad KN200 is a battery-powered image intensifier based on a new design concept. The Simrad KN200 is intended as an add-on unit to optical sights on direct fire weapons, laser rangefinders and other daylight image devices such as television cameras.

The unique design concept has overcome the problems usually connected with night sights and compared to traditional night sights the KN200 offers several advantages. As the eyepiece of the existing day sight is also used at night, the graticule pattern remains the same and the position of the operator's eye remains unchanged day and night. It can be mounted and removed within a few seconds and no boresighting adjustments are necessary.

The use of a 100 mm objective aperture yields a good range performance, enabling the operator to detect and engage targets within the practical range of the weapon during night operations. Sudden illumination does not have any effect on the sighting capabilities.

Simrad KN200 image intensifier mounted on Simrad LP7 laser rangefinder

Change between second- and third-generation tubes requires no modification to the instrument.

Data

Dimensions: 210 × 209 × 120 mm
Weight: including battery, 1.4 kg (1.46 kg with adjustable focus)
Field of view: 10°
Focusing range: fixed or adjustable, 25 m to infinity
Resolution: 0.5 mrad/lp (contrast 30%, illuminance 100 lux)
Objective lens: catadioptric, 100 mm f/1
Power supply: 2 × 1.5 V AA alkaline or 2 × C cells; lithium optional
Battery life: >80 h at 20°C
Operational temperature range: −40 to +52°C

Manufacturer

Simrad Optronics A/S, PO Box 6114, Etterstad, N-0602 Oslo.

Status

In production.

VERIFIED

Simrad KN250F image intensifier

Description

The Simrad KN250F image intensifier is mounted as an add-on unit on telescope sights, laser rangefinders and other optical daylight imaging devices such as video cameras. As no boresighting adjustment is required, the mounting procedure only takes a few seconds.

The KN250 can use both second- and third-generation image intensifier tubes. Change between these tubes requires no modification of the instrument.

With this new concept the gunner can aim through the eyepiece of the day sight both day and night, an advantage not achieved with traditional types of night sight. Sudden illumination of the scene does not have any effect on the sighting capabilities.

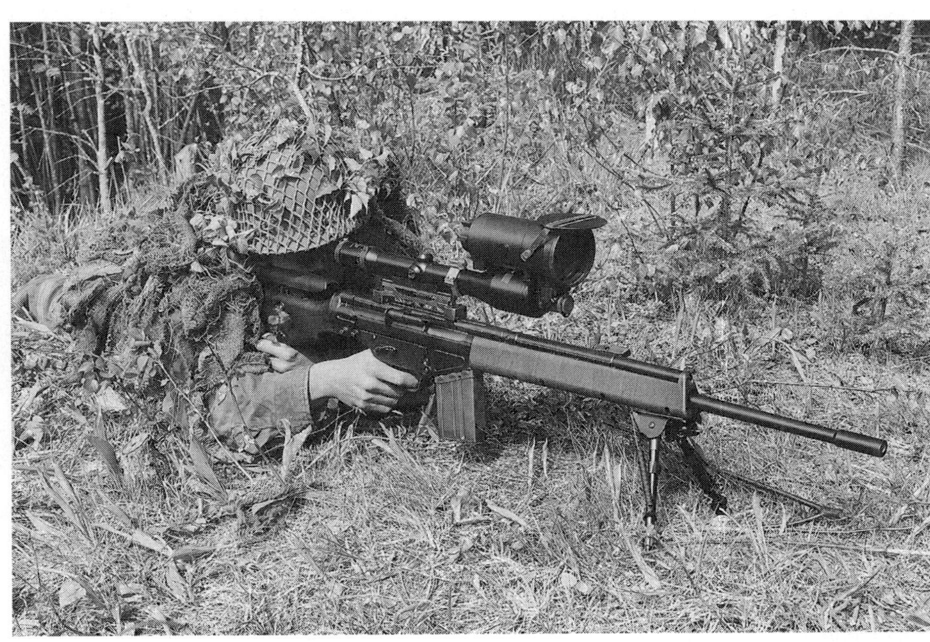

Simrad KN250F image intensifier mounted on an optical day sight

1996

Data

Weight: including batteries, 740 g (790 g with adjustable focus)
Magnification: ×1
Field of view: 12°
Resolution: 0.6 mrad/lp (contrast 30%, illuminance 100 lux)
Objective lens: catadioptric, 80 mm f/1.1 (T/1.4)
Focusing: fixed or variable, between 25 m and infinity
Tube: 2nd or 3rd Gen, 18 mm wafer tube
Power supply: 2 × 1.5 V alkaline AA cells
Battery life: >80 h at 20°C
Operational temperature range: −40 to +52°C

Manufacturer

Simrad Optronics A/S, PO Box 6114, Etterstad, N-0602 Oslo.

Status

In production.

VERIFIED

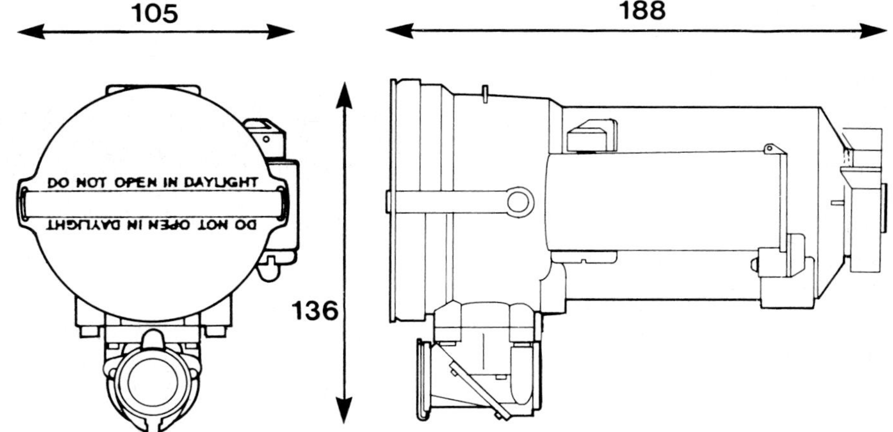

Simrad KN250F image intensifier; dimensions

Simrad IS2000 eye-safe laser gun sight

Description

The Simrad IS2000 eye-safe laser gun sight is a fire control system for direct fire weapons. The unit provides instant target range and aiming information and has a built-in computer programmed for ammunition ballistics. Range and aiming information for moving and stationary targets are displayed. Range is shown in the panel display and aiming point corrected for elevation and lead angle is shown in a head-up display. The IS2000 can be easily reprogrammed by the operator to compensate for changes in battlefield conditions.

The IS2000 is particularly suited for upgrading manportable direct fire weapons such as the 84 mm Carl Gustaf RCL. The unit can be used with Simrad GN1 night vision goggles or can be fitted with a Simrad KN250F image intensifier (see previous entry) for round the clock operations.

Data

Weight: ca 1.5 kg
Dimensions: 210 × 140 × 120 mm
Wavelength: 905 nm
Range: max, 2,000 m; min, 50 m
Resolution: 5 m
Objective diameter: 45 mm
Magnification: ×1
Field of view: 8°
Power supply: 6-9 V Ni/Cd or alkaline AA cells
Ranges before recharging: >600 at 20°C
Remote control: yes
Data communication: RS-485
Eye safety: Class 1 (IEC825)

Simrad IS2000 eye-safe laser gun sight

1996

Manufacturer

Simrad Optronics A/S, PO Box 6114, Etterstad, N-0602 Oslo.

Status

In production.

NEW ENTRY

PAKISTAN

AN/PVS-4A individual weapon sight

Description

The AN/PVS-4A is a licence-produced weapon sight for use on rifles, machine guns and rocket launchers. It uses a 25 mm second-generation image intensifier tube and has an internally adjustable, illuminated graticule projector; the graticule being interchangeable for use with different weapons. The sight has automatic gain control which allows it to cope with very large fluctuations in brightness at the target, and a bright source protection circuit prevents damage caused by inadvertent exposure to daylight. A special cover permits daytime boresighting and training.

Data

Length: 240 mm
Width: 120 mm
Height: 120 mm
Weight: 1.7 kg
Magnification: ×3.6
Field of view: 14.5°
Object lens: 95 mm f/1.6
Dioptre adjustment: +2 to −6
Power source: AA cell
Battery life: 35 h at 25°C

Manufacturer

Institute of Optronics Pakistan, PO Box 1596, Rawalpindi.

AN/PVS-4A weapon sight on MG3 machine gun

Status

In production.

VERIFIED

AN/TVS-5A crew-served weapon sight

Description

This licence-produced sight is intended for use on heavier weapons such as recoilless rifles, heavy machine guns and some types of wire-guided anti-tank missile. It can also be tripod mounted for use as a surveillance device.

The sight uses a 25 mm second-generation image intensifier tube and has automatic gain control so that sudden bright lights in the target area will not affect the system. The internally mounted adjustable illuminated graticule can be interchanged to cater for different weapons or ammunition. There is a bright source protection circuit which prevents damage caused by

AN/TVS-5A weapon sight on 106 mm recoilless rifle

inadvertent exposure to daylight and the sight is provided with a cover which allows daytime boresighting and training.

Data

Length: 310 mm
Width: 160 mm
Height: 170 mm
Weight: 3 kg
Magnification: ×6
Field of view: 9°
Object lens: 155 mm f/1.7
Power supply: AA cells
Battery life: 35 h at 20°C

Manufacturer

Institute of Optronics Pakistan, PO Box 1596, Rawalpindi.

Status

In production.

VERIFIED

SOUTH AFRICA

MS-20 miniature night sight

Description

Equipped with an 18 mm second-generation microchannel intensifier, the MS-20 night sight is optimised for application in counter-insurgency operations.

Evolved from extensive combat experience, this instrument is of compact and rugged design and is easy to use. It features a fixed focus objective and is instantly ready for use as soon as it is switched on. The only adjustments required, prior to action, are boresighting and eyepiece dioptre setting.

Power supply is by means of two readily available penlight batteries. Boresighting adjustment is incorporated in an integral quick-release mounting bracket. Mounts for a variety of rifles are available.

Data

Length: 210 mm
Weight: without mounting bracket, 950 g
Image intensifier: 18 mm microchannel amplifier to MIL-I-49052B
Magnification: ×2.6
Field of view: 16°
Entrance pupil: 40 mm
Exit pupil: 7 mm
Eye relief: 24 mm
Eyepiece adjustment: −4 to +4 dioptres
Boresighting range: 20 mils continuous, azimuth and elevation
Recognition range: man target, 200 m at 10^{-3} lux
Power supply: 2 × 1.5 V AA penlight batteries

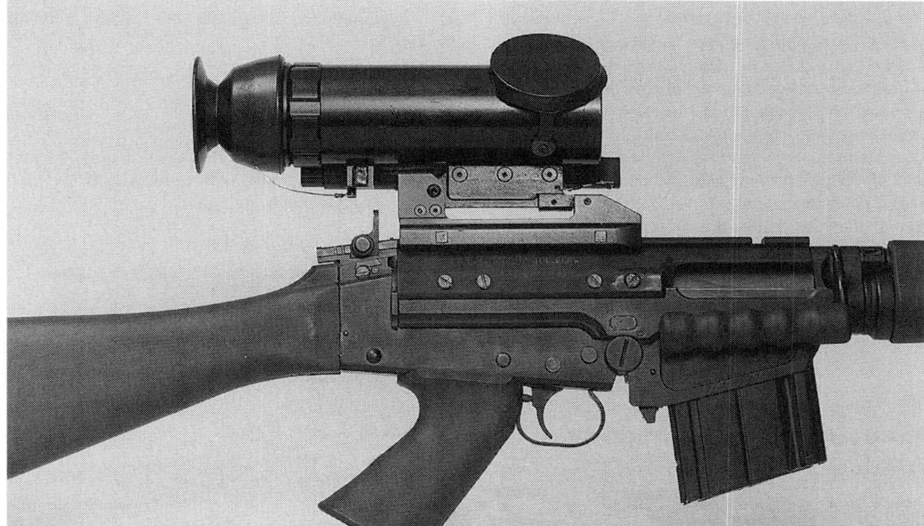

MS-20 sight mounted on receiver cover of FN FAL (R1) rifle

Enquiries to

Armscor, Private Bag X337, Pretoria, 0001.

Status

In production.

Service

South African National Defence Force.

VERIFIED

SPAIN

VNP-004 night sight

Description

The VNP-004 was developed in Spain specifically as a night sight for the MG42 machine gun, although there seems no reason why it should not be readily adaptable to other weapons. The VNP-004 is a second-generation sight using a Philips XX1500 18 mm microchannel amplifier, giving variable gain from ×30,000 to ×70,000 selectable by the operator. The sight is provided with a mounting of standard NATO dimensions.

Data

Dimensions: 340 × 150 × 120 mm
Weight: 3.2 kg
Magnification: ×3.5
Field of view: 9° 30′ (169 mils)

VNP-004 night sight

Resolution: 0.6 mrad at 10^{-3} lux
Dioptre correction: +2 to −5
Focusing: fixed, parallax-free
Zero adjustment: graticule displacement
Power supply: 2 × 1.5 V batteries (R14 or equivalent)
Battery life: over 50 h

Manufacturer

ENOSA, Empresa Nacional de Optica SA, Poligono Industrial La Mina, Parcela 11, E-28770 Colmenar Viejo, Madrid.

Status

Available.

Service

Spanish Army.

VERIFIED

AVN-01 laser aiming light

Description

The AVN-01 is a laser spot projector capable of being fitted to virtually any weapon. When the user is wearing night vision goggles or using a night vision sight, the laser spot allows weapon aiming in conditions of darkness.

Data

Laser: GaAs
Operating mode: pulse
Average power: 50 μW
Beam divergence: 2 mrad
Wavelength: 815 nm
Spectral bandwidth: 8 nm
Power supply: IEC-6F22 battery
Battery life: 100 h

Manufacturer

ENOSA, Empresa Nacional de Optica SA, Poligono Industrial La Mina, Parcela 11, E-28770 Colmenar Viejo, Madrid.

Status

Available.

VERIFIED

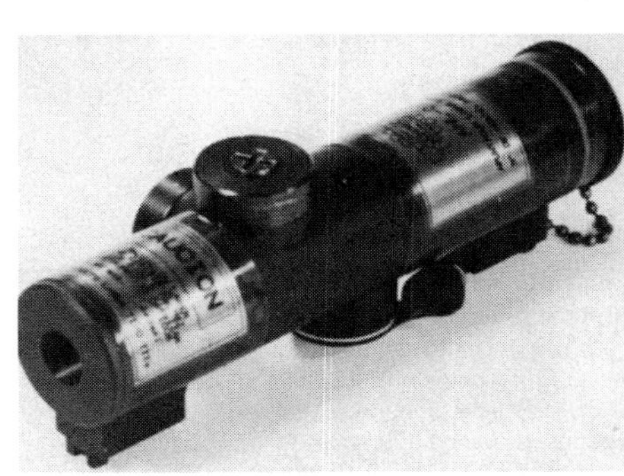

AVN-01 laser aiming light

SWEDEN

Aimpoint 3000 electronic sight unit

Description

The Aimpoint 3000 electronic sight was introduced by Aimpoint AB in Sweden and is claimed to enable the firer to aim more quickly and more accurately than possible with a conventional iron sight.

In the Aimpoint device the iron sights are replaced by a single battery-powered optical and electronic unit. The eyepiece of this unit is designed to be viewed by one eye while the other eye can be either open or closed - but is preferably open because the firer's field of view thus embraces more than can be seen through a telescope eyepiece and the sight cover is less obtrusive than if only one eye is used. The electronics of the system cause a small bright red dot to appear in the field of view coincident with the point of aim of the weapon and all that the firer has to do in order to aim correctly is to bring this dot into coincidence with the desired target.

The position of the dot depends only on the pointing direction of the weapon and is not altered by movement of the firer's head; thus, when it is in coincidence with the target, the system is parallax-free. Other advantages of the system are that it can be used in all light and weather conditions, it gives no signal to a potential enemy, it is ruggedly made and, above all, skill in using it is quickly acquired.

The Aimpoint 3000 is a development of earlier models, to incorporate features which experience has shown to be desirable. Thus the red dot covers only 2.5 in (63.5 mm) at 100 yards (91.4 m) range. Power is from a lithium battery which continues to function in extremes of heat or cold; and the electronics are designed to withstand the most exacting conditions.

Data

Length: 140 mm
Weight: 146 g
Diameter of dot: 63.5 mm at 91.4 m (2.5 in at 100 yds)
Power supply: lithium DL 1/3N or mercury 875
Battery life: 250-500 h average

Manufacturer

Aimpoint AB, Jägershillgatan 15, S-213 75 Malmö.

Status

In production.

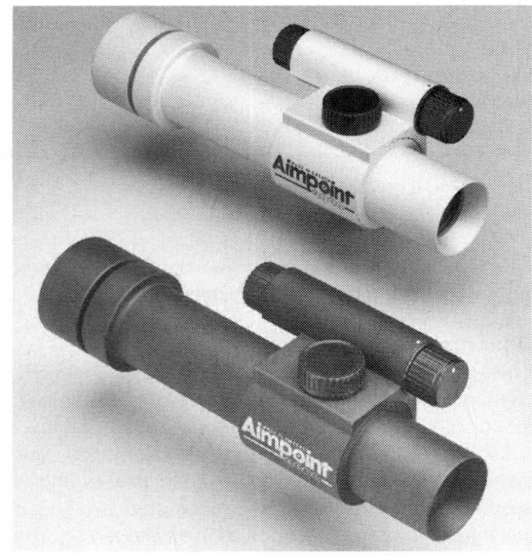

Aimpoint 3000 electronic sight units

Service

Commercial sales worldwide.

VERIFIED

Aimpoint 5000 electronic sight unit

Description

The Aimpoint 5000 is to the same general specification as the Model 3000 described previously, but differs in having a 30 mm field of view and in having the red dot available in two sizes. The sight is consequently rather greater in diameter and is supplied with special 30 mm mounting rings.

Data

Length: 140 mm
Weight: 165 g
Diameter of dot: (A) 76 mm at 91.4 m (3 in at 100 yds); or (B) 254 mm at 91.4 m (10 in at 100 yds)
Mounting: 30 mm rings
Power supply: lithium 2L76 or DL 1/3N or 2 × mercury SP675
Battery life: average, 250-500 h

Manufacturer

Aimpoint AB, Jägershillgatan 15, S-213 75 Malmö.

Status

In production.

Service

Commercial sales worldwide.

VERIFIED

Aimpoint 5000 Two-power electronic sight unit

Description

The Aimpoint 5000 Two-power sight generally resembles the Aimpoint 5000 described previously but has built-in two-power magnification. The large 46 mm object lens and 30 mm tube allow for excellent light transmission.

Data

Length: 178 mm
Weight: 255 g
Diameter of dot: 38 mm at 91.4 m (1.5 in at 100 yds)
Mounting: 30 mm rings
Power supply: lithium SL76 or DL 1/3N or 2 × mercury SP675
Battery life: average, 250-500 h

Manufacturer

Aimpoint AB, Jägershillgatan 15, S-213 75 Malmö.

Status

In production.

Service

Commercial sales worldwide.

VERIFIED

Aimpoint 5000 Two-power electronic sight unit

Aimpoint Laserdot II

Description

Laserdot II allows the shooter to project an intense sharp red dot in to the target quickly and accurately. Designed to withstand recoil and demanding conditions, Laserdot II incorporates the finest optics and strongest power supply available. It can be mounted on almost any firearm currently on the market. Windage and elevation adjustment for boresighting is internal.

Data

Length: 89 mm
Weight: 57 g
Diameter of dot: 38 mm at 91.4 m (1.5 in at 100 yds)
Output beam: Wavelength 670 nm; Class IIIA limit.
Mounting: standard 1 in rings
Power supply: 3 V lithium
Battery life: up to 15 h continuous

Manufacturer

Aimpoint AB, Jägershillgatan 15, S-213 75 Malmö.

Status

In production.

Service

Commercial sales worldwide.

VERIFIED

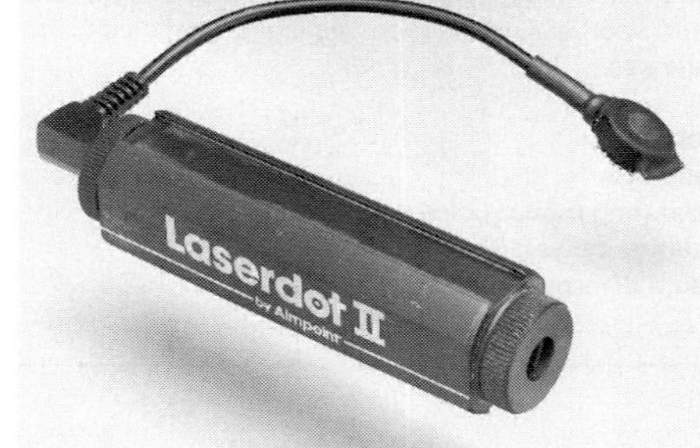

Aimpoint Laserdot II

Weibull direct aiming sights

Description

These sights, which are adaptable to all types of small-calibre direct fire weapons from machine guns to light anti-aircraft guns, have been developed in close co-operation with military experts. The prevailing object has been to develop a unitary sight system which may be used for air, ground or sea targets or any combination of these. The graticule pattern, designed to fit specific weapons, is adapted to the ballistic performance and also to certain patterns of target behaviour which experience has shown can be expected for a large proportion of the time. The philosophy behind the system is that it is better to have a simple and effective system on a large number of weapons than a complicated and expensive system on a smaller number.

The Reflex Sight is equipped with four different graticules mounted on a revolving disc. Two of these are intended for air target combat - one for fast-moving targets and the other for slow. One graticule is designed for seaborne targets, and is also suitable for ground targets, and the last graticule corresponds to a conventional ring sight. The sight embodies an optical collimating system, such that the aimer sees the graticule at an infinite distance, superimposed upon a view of the target area. It is not necessary for the aimer's eye to be in any specific place; provided he can see both the graticule and the target, the aim will be correct. For conditions where ambient light does not permit the graticule to be clearly visible, electric illumination is provided.

Weibull Reflex Sight for 7.62 mm FN MAG machine gun

Custom-made graticules to correspond to any weapon are available, as are mounts and adaptors to fit the sights to all types of weapon.

Manufacturer

Ingeniörsfirman JL Weibull AB, PO Box 43, S-232 20 Åkarp.

Status

In production.

Weibull Reflex Sight for 40 mm anti-aircraft gun

Service

Swedish Army, Coastal Artillery and Navy.

VERIFIED

BILL Thermal Imager (BILL TI)

Description

The BILL Thermal Imager (BILL TI) is an add-on thermal imaging system intended for use in darkness and poor visibility. It is mounted on top of the RBS-56 BILL anti-armour missile day sight so that the thermal picture is mirrored in the front lens of the day sight. It can be mounted very rapidly, with high-sightline accuracy and with no need for boresighting.

BILL TI has two fields of view; a wide field (WFOV) for surveillance and a narrow field (NFOV) for more detailed observation and aiming. The front lens is coated with a diamond-hard anti-reflection coating which has extreme resistance against mechanical abrasion.

The sight is powered by a battery pack and a high pressure air bottle, both rapidly changeable. Alternatively, a mini-compressor can be used to provide the high-pressure air.

The visual reproduction of the thermal picture is red and black; normally black means cold and red means hot, but a polarity switch allows these to be reversed should the target contrast demand it.

Although primarily developed for the BILL missile system, the BILL TI is readily adaptable to other types of day sight by enclosing the mechanical Quick-Lock coupling and the optical mirror bridge in the equipment. Other versions are formed by using only the top part of BILL TI. It can be used as a separate direct view sight by adding a graticule and an eyepiece. In vehicle applications a solid-state TV camera is used for video conversions.

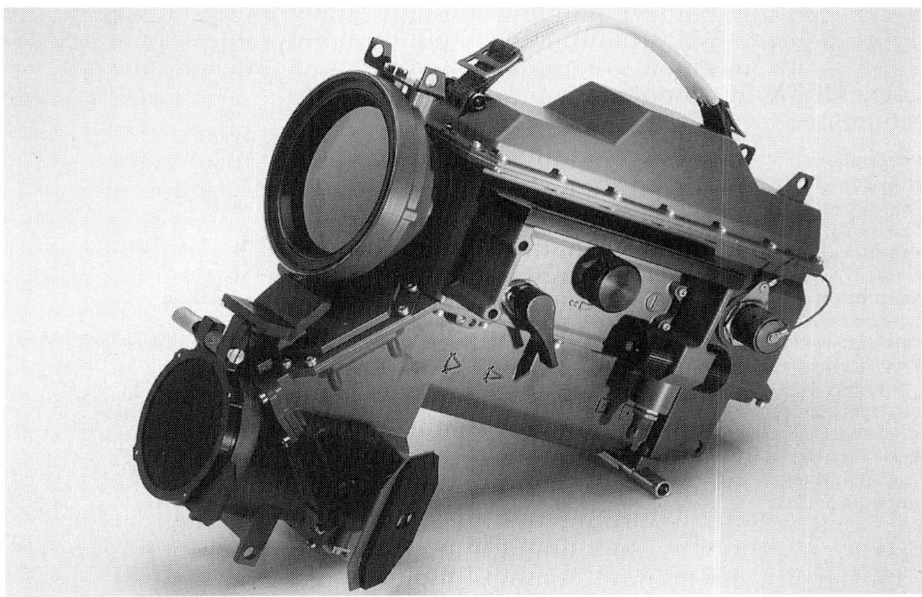

BILL Thermal Imager (BILL TI)

1996

Data

Weight: with batteries and bottle, 9.2 kg
WFOV: 12.5 × 5.5°
NFOV: 6 × 2.5°
Range: (typical) stationary tank: detection (WFOV), 4,000 m; recognition (NFOV), 1,800 m

Manufacturer

CelsiusTech Electronics AB, S-175 88 Järfälla.

Status

In production.

Service

Austria, Sweden.

VERIFIED

SWITZERLAND

FORTIS thermal imaging system

Description

FORTIS (Forward Observation and Reconnaissance Thermal Imaging System) is a portable thermal imager specifically designed for military and paramilitary use. Its large and selectable field of view, high target resolution, image inversion and long range make it ideally suited for observation, reconnaissance and surveillance tasks at night and in daylight.

FORTIS provides long-term observation with minimum operator fatigue thanks to the green image display and biocular viewing lens. It is instantly ready for use and has a high degree of operational autonomy. The complete system can be carried easily by a single person.

Long-range telescope, external video monitor, closed-cycle cooling system and remote-control options are available as additional equipment.

FORTIS is a joint development by Siemens-Albis and Leica.

Data

Dimensions: 460 × 240 × 300 mm
Weight: operational, 13 kg
Wavelength: 8-12 μm
Cooling system: Joule-Thomson, 77° K
Temperature resolution: ca 0.1°K

FORTIS in use on low-set tripod

Field of view: 18.5 × 9.5°/4.5 × 2.3°
Spatial resolution: 0.8 mrad/0.2 mrad
Nominal voltage: 7.2 V
Power consumption: <4 W
Range: NATO target in standard atmosphere, 293°K:
Detect: 6,000 m
Recognise: 2,000 m
Identify: >1,000 m

Manufacturer
Siemens-Albis AG, Defence Electronics, Freilagerstrasse 38, CH-8047 Zurich.

Status
In production.

Service
Swiss armed forces.

UPDATED

FORTIS on tripod showing biocular viewing lens

Self-propelled howitzer seen through FORTIS at night

Leica MRF2000 monocular rangefinder

Description
The Leica MRF2000 monocular rangefinder employs state-of-the-art technology to measure distances from 30 to 3,300 m with 2 m accuracy. The laser diode rangefinder built into the MRF2000 meets the class 1 eye-safety requirements of the Federal Drug Administration and the European Norm regulations for laser radiation; thus assuring that the emitted beam is safe not only to the naked eye but also to observers using binoculars or other magnifying devices.

The MRF 2000 baseplate allows mechanical adaptation to various systems, with a boresighting accuracy of ±0.6 mils in azimuth and ±0.3 mils in elevation. Range data can be transmitted to an external display via an RS-232 interface. Remote triggering and power supply are possible.

Data
Dimensions: 210 × 120 × 80 mm
Weight: with batteries, 1.1 kg
Magnification: ×7
Field of view: 6.4° × 5.1°
Objective aperture: 50 mm
Exit pupil diameter: 7 mm
Eye relief: 18 mm
Dioptre setting: −6 to +6 dioptres
Laser type: 850 nm laser diode
Range: 30 to 3,300 m
Accuracy: ±1 m up to 2,000 m; ±2 m beyond 2,000 m
Measurement time: 0.3 s
Repetition rate: 20/min
Power supply: 2 × 3 V CR 123 lithium cells
Operating temperature range: −35 to +55°C

Manufacturer
Leica AG, Defense & Special Projects, CH-9435 Heerbrugg.

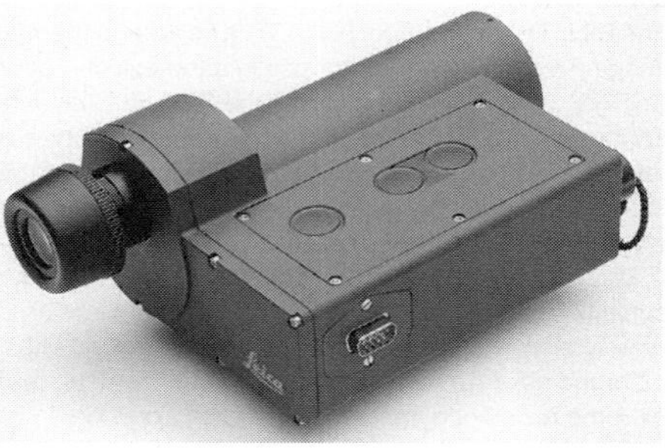

Leica MRF2000 monocular rangefinder

1996

Status
Development and final testing completed. Series production began in second half of 1995.

UPDATED

Leica VECTOR 1500 DAES rangefinding binocular

Description
The Leica VECTOR 1500 DAES rangefinding binocular is the top model in a range of four. It has a wide range of possible applications for, in addition to acting as a high-quality surveillance and general purpose 7 × 42 binocular, it can also act as a rangefinding device for ranges up to 1,500 m, provide azimuth and elevation data, and there is a serial data output. It thus has numerous applications for sniper operations, reconnaissance, and crew-served weapon support.

The rubber armoured body of the Leica VECTOR binocular was ergonomically designed and can be easily handled, even when wearing gloves. It is ruggedised and waterproof. Controls are limited to two push-buttons, the right-hand of which operates the eye-safe Class 1 diode laser rangefinder. Operating the other button provides an azimuth reading from a digital magnetic compass with an inclinometer. Distance, elevation and elevation data is digitally displayed in the field of view via red LEDs for a duration of several seconds. An RS-232 data interface is provided and a method of combining VECTOR data with the Global Positioning System (GPS) is under development.

Power is provided by a 6 V lithium battery. Battery life is about 1,000 measurements.

Data
Dimensions: 205 × 178 × 82 mm
Weight: with battery, 1.71 kg
Magnification: ×7
Field of view: at 1,000 m, 120 m
Objective aperture: 42 mm
Exit pupil diameter: 6 mm
Eye relief: 18.5 mm
Dioptre setting: −6 to +6 dioptres
Range: 25 to 1,500 m
Accuracy: ±2 m
Measurement time: 0.3 s
Repetition rate: 12/min
Power supply: 1 × 6 V 2CR5 lithium cell
Operating temperature range: −35 to +63°C

Manufacturer
Leica AG, Defense & Special Projects, CH-9435 Heerbrugg.

Status
In production.

NEW ENTRY

Leica VECTOR 1500 DAES rangefinding binocular

1996

UNITED KINGDOM

SA-80 optical sight

Description

Heckler and Koch (UK) Limited have developed a lightweight, low-cost optical sight for the SA-80 weapon system. The sight fits the existing SA-80 weapon sight mount rail and is interchangeable with existing sights. A ×3 magnification is standard with a tritium illuminated graticule for low light conditions; an interchangeable ×1.5 magnification sight is also available. The housing is made from a high-impact glass-reinforced polymer and is shaped to act as a carrying handle.

The sight pattern is an annular ring with internal cross. The ring represents a 1.75 m man height at 400 m, with additional points of aim from 200 to 800 m in 200 m steps.

The UK Ministry of Defence has issued a requirement for a sight of this nature.

Heckler and Koch (UK) Limited SA-80 optical sight

1996

Data

Weight: 250 g
Length: 150 mm
Width: 37 mm
Height: 80 mm
Magnification: ×3; ×1.5 optional
Field of view: 70 mils (4°)
Light transmission: 75%
Entrance pupil: 9 mm

Status

Available.

Manufacturer

British Aerospace Defence Limited, Heckler and Koch (UK) Limited, Kings Meadow Road, Nottingham NG2 1EQ.

UPDATED

SUSAT (Sight Unit Small Arms Trilux) L9A1

Description

The SUSAT (Sight Unit Small Arms Trilux) L9A1 was developed by the Royal Armaments Research and Development Establishment (RARDE) to realise the full capabilities of the L85A1 rifle in service with the British Army.

SUSAT is smaller, lighter and more robust than its predecessor SUIT. The one-piece, pressure diecast aluminium body contains enhanced optics to improve twilight surveillance and weapon aiming tasks. Being of fixed focus, with an aiming needle for shooting, the only moving part on SUSAT is the adjuster for the variable intensity Trilux source.

SUSAT can be mounted on to a wide range of rifles, machine guns and recoilless rifles by means of a universal mount. The mount also enables the user to remove SUSAT easily and change to fit an image intensifying or other type of night sight.

Fixed iron sights are located over the sight body for emergency use should the optical sight become unserviceable for any reason.

SUSAT sight on 5.56 mm L86A1 Light Support Weapon

Data

Length: 145 mm
Width: 60 mm
Height: 55 mm
Weight: 417 g
Magnification: ×4
Field of view: 175 mils (10°)
Light transmission: >80%
Entrance pupil diameter: 25.5 mm
Exit pupil diameter: 6 mm
Eye relief: 25 mm
Eyepiece focus: −1 dioptre
Flatness of field: 0.25 dioptre
Veiling glare: 2° max

Manufacturer

United Scientific Instruments, Unit 5, Quinn Close, Manor Park, Whitley, Coventry CV3 4LH.

Status

In production.

Service

British, Cameroon, Oman, Spanish and Swedish armies.

UPDATED

Pilkington Kite individual weapon sight

Description

The Pilkington Kite is a lightweight, high-performance night sight designed to meet a demanding military specification for weapon aiming sights. Made from highly robust advanced materials and with a weight of less than 1.2 kg, the optics afford a magnification of ×4 and a 9° field of view.

The refractive lens system incorporates an injected graticule to assist with accurate weapon aiming. The sight is configured to accept second- or third-generation image intensifier tubes.

Pilkington Optronics Kite weapon sight

Pilkington Kite weapon sight in use on Panzerfaust 3

The Kite is simple to operate with a rotary on/off switch, a graticule brightness adjustment, and a large collar focusing control.

The Kite was selected for the British Army for use with its range of infantry weapons and is also in service in over 40 countries worldwide.

Data

Length: 255 mm
Height: 80 mm
Width: 80 mm
Weight: <1.2 kg
Field of view: 9°
Magnification: ×4
Range: recognition of standing man at 300 m in starlight (10^{-3} lux)
Eye relief: 30 mm
Power supply: 2 × 1.5 V AA batteries
Eyepiece dioptre: fixed at 1 ± 0.35 dioptres; adjustable +2 to −6 dioptres
Focus range: 15 m to infinity

Manufacturer

Pilkington Optronics, Glascoed Road, St Asaph, Clwyd LL17 OLL.

Status

In production.

Service

United Kingdom Ministry of Defence and over 40 other armed forces.

UPDATED

Pilkington Maxi-Kite weapon sight

Description

The Pilkington Maxi-Kite is a compact, lightweight night sight for use as a highly portable long range surveillance device or as a crew-served weapon sight. This fully ruggedised system has a ×6 magnification, 5.5° field of view and provides a high-resolution image for effective observation and accurate weapon aiming at long-range targets. Maxi-Kite is produced by adding an afocal lens assembly to the highly successful Kite individual weapon sight.

Maxi-Kite can be supplied either as a weapon sight or as an upgrade to the standard Kite night sight. It can be fitted with a suitable graticule to interface fully with a weapon's ballistic characteristics. Maxi-Kite is particularly suited for use on sniper rifles, heavy machine guns and shoulder-launched or crew-served weapons. Interface brackets can be provided.

The sight's performance is supported by a simple maintenance philosophy which permits rapid replacement of modular subassemblies.

Data

Length: 360 mm
Height: 90 mm
Width: 95 mm
Weight: 1.5 kg
Magnification: ×6
Field of view: 5.5°
Power supply: 2 × 1.5 V AA batteries
Eyepiece dioptre: fixed at −1 ± 0.35 dioptres; adjustable at +2 to −6 dioptres
Focus range: 25 m to infinity
Eye relief: 30 mm
Min range: recognition of a standing man at 450 m in starlight (10^{-3} lux)

Manufacturer

Pilkington Optronics, Glascoed Road, St Asaph, Clwyd LL17 OLL.

Status

In production.

Service

United Kingdom Ministry of Defence and overseas armed forces.

UPDATED

Pilkington Maxi-Kite weapon sight

Maxi-Kite sight on L96A1 sniper rifle

Pilkington Bino-Kite night vision binocular

Description

The Pilkington Bino-Kite is a compact, lightweight night vision binocular with a magnification of ×4.5, making it suitable for medium- to long-range night surveillance or fire control observation. It may be supplied with either a second- or third-generation image intensifier tube, dependent on the resolution required. The binocular is portable and may be worn around the neck or on a lanyard, or mounted on a tripod for fixed observation purposes.

Bino-Kite was designed as part of a suite of interchangeable night vision equipment. Bino-Kite can be reconfigured as Maxi-Bino-Kite by adding an afocal lens assembly (see following entry) offering greater magnification for long-range surveillance tasks.

Data

Length: 270 mm
Height: 83 mm
Width: 130 mm
Weight: 1.1 kg
Field of view: 8.8°
Magnification: ×4.5
Eye relief: 27 mm
Power supply: 2 × 1.5 V AA batteries
Eyepiece dioptre: adjustable from +2 to −6 dioptres
Focus range: 20 m to infinity
Min range: recognition of a standing man at 400 m in starlight (10^{-3} lux)

Pilkington Bino-Kite night vision binocular mounted on a tripod

Manufacturer

Pilkington Optronics, Glascoed Road, St Asaph, Clwyd LL17 OLL.

Status

In production.

UPDATED

Pilkington Maxi-Bino-Kite night vision binocular

Description

The Pilkington Maxi-Bino-Kite is a compact and lightweight night vision binocular for use as a portable long-range surveillance device. The binocular configuration offers maximum comfort to the end user for surveillance over prolonged periods. Maxi-Bino-Kite is configured to accept either second- or third-generation image intensifier tubes without any need for modification.

Data

Weight: 1.5 kg
Field of view: 5.2°
Magnification: ×6.7
Power supply: 2 × 1.5 V AA batteries
Eyepiece dioptre: adjustable from +2 to −6 dioptres

Manufacturer

Pilkington Optronics, Glascoed Road, St Asaph, Clwyd LL17 OLL.

Status

In production.

NEW ENTRY

Pilkington Maxi-Bino-Kite night vision binocular

1996

Pilkington Jay night vision goggle

Description

The Jay night vision goggle is a self-contained, head-mounted, ×1 magnification night vision system suitable for personal use when walking, driving vehicles, map reading, carrying out first aid, vehicle maintenance and short-range surveillance.

Jay may be supplied with either a second- or third-generation image intensifier tube, dependent on the resolution required. Weighing only 680 g the goggle may be hand held or head mounted. For close-up viewing tasks such as map reading an infra-red emitting light source are incorporated into the goggle to provide illumination.

Maximum user comfort is achieved by using a binocular viewer with adjustable dioptric eyepieces to suit individual user requirements.

The Jay goggle was designed as part of a suite of interchangeable night vision equipment. For medium range surveillance the goggle can be reconfigured as a Pilkington Bino-Kite by replacing the objective lens assembly. An even longer range surveillance equipment can be obtained by adding the Pilkington Maxi-Kite Adaptor.

Data

Weight: 680 g
Field of view: 40°
Magnification: ×1
Eye relief: 15 mm min
Power supply: 2 × 1.5 V AA batteries
Eyepiece dioptre: adjustable from +2 to −6 dioptres
Focus range: 250 mm to infinity
Min range: recognition of a standing man at 300 m in starlight (10^{-3} lux)

Manufacturer

Pilkington Optronics, Glascoed Road, St Asaph, Clwyd LL17 OLL.

Status

In production.

UPDATED

Pilkington Jay night vision goggle in use

RV20 Reflex sight

Description

The RV20 Reflex sight is an optomechanical sight which ensures quick, fatigue-free aiming, insensitive to parallax errors. Both eyes can be used for rapid target acquisition in an unrestricted field of view. In insufficient ambient light conditions, reticle illumination is provided by a self-activated maintenance-free tritium light source.

The RV20 sight meets all police and military requirements and can be fitted to almost all small arms by means of STANAG (NATO), MP5 and Weaver Rail sight mounts.

Data

Dimensions: 120 × 35 × 65 mm
Weight: ca 250 g
Magnification: ×1
Aim correction increments: 0.25 mil
Adjustment range: ±50 increments

Manufacturer

Hall and Watts Defence Optics Limited, 266 Hatfield Road, St Albans, Hertfordshire AL1 4UN.

Watts RV20 Reflex sight on MP5

Status

In production.

Service

German special forces.

VERIFIED

Wildcat optical combat sight

Description

The Wildcat optical combat sight is a third-generation optical sight designed for use in all adverse weather conditions and was developed to supersede earlier sights such as the SUIT and SUSAT (qv). It is in service with the Canadian, Netherlands and other armed forces worldwide.

The Wildcat comprises an optical telescope and mount assembly intended to be an integral component of combat rifles and other weapons. An extra large exit pupil diameter allows fast target acquisition while the optical performance characteristics are claimed to turn dawn and dusk light conditions into day. The sight is effective on machine guns as well as rifles, permitting tracer rounds and fall of shot to be clearly seen at extended ranges.

The Wildcat is fully interchangeable with the Blackcat night vision sight (see following entry).

The standard sight magnification is ×3.4. Available as options are ×6 and ×10. Mounts are available for various weapons, as are various reticle options.

Data

Dimensions: 160 × 72 × 55 mm
Weight: sight, 350 g; mount, 290 g
Field of view: 8°
Magnification: ×3.4
Apparent field of view: 27.2°
Entrance pupil diameter: 28 mm
Exit pupil diameter: 8.5 mm
Eye relief: 70 mm

Wildcat optical combat sight mounted on M4A1 carbine

Manufacturer

Hall and Watts Defence Optics Limited, 266 Hatfield Road, St Albans, Hertfordshire AL1 4UN.

Status

In production.

Service

Canadian and Netherlands armed forces and others worldwide.

VERIFIED

Blackcat night vision sight

Description

The Blackcat night vision sight is fully interchangeable with the Wildcat optical combat sight (see previous entry) as it features a see-in-the-dark performance because of an image intensifier tube installed in the optical housing of the Wildcat sight. The optics were specifically designed to provide optimum performance from such a compact sight; the sight is small enough to be carried in a magazine pouch.

Also available is the HI-LIGHT night vision sight with ×6 magnification and an 80 mm diameter objective lens. Various reticle options also feature.

Data

Dimensions: 160 × 72 × 55 mm
Weight: 900 g
Field of view: 12°
Field of view: apparent, 37°
Magnification: ×3.4
Entrance pupil diameter: 36 mm
Exit pupil diameter: 8.3 mm
Eye relief: 42 mm
Resolution: >32 lp/mm
Fixed focus: 30 m to infinity

Blackcat night vision sight mounted on Minimi machine gun

Power supply: 2 × AA batteries, lithium or alkaline
Battery life: alkaline, nominal, 96 h

Manufacturer

Hall and Watts Defence Optics Limited, 266 Hatfield Road, St Albans, Hertfordshire AL1 4UN.

Status

In production.

Service

Various armed forces worldwide.

VERIFIED

Hall and Watts small arms collimator (DZD)

Description

The Hall and Watts small arms collimator, designated the DZD, was designed to assist in the realignment of optical sights for all types of small arms, without the need to fire the weapon. It is an inexpensive, compact, lightweight unit with a high degree of consistency enabling the user to have assurance that, after zeroing, the alignment of the weapon can be reset quickly and accurately by an operator with very little training. The unit will fit into an ammunition pouch or into a custom-made carrying case which can be attached to a soldier's webbing.

Using mounting blocks and spigots designed for individual weapon types the collimator can be fitted to any weapon with an optical sighting system, with calibres from 5.56 to 12.7 mm. An easy-to-read grid system enables a soldier to record his own setting and reset the optical sight to his individual zero point.

Data

Diameter: 40 mm
Length: configured for FN FAL, 225 mm
Height: configured for FN FAL, 110 mm
Weight: configured for FN FAL, 284 g
Graticule: grid, workshop adjustment
Focus: fixed at infinity
Calibre: 5.56 to 12.7 mm
Temperature range: −40 to +55°C

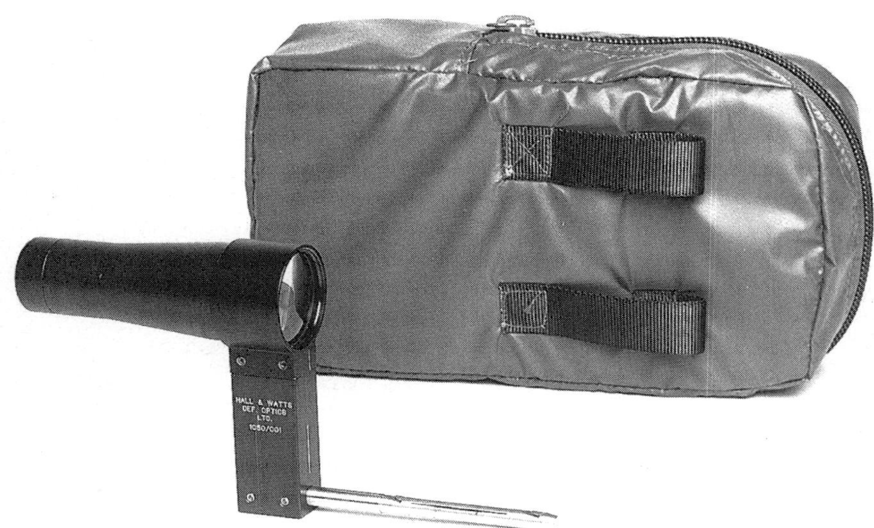

Hall and Watts small arms collimator (DZD)

Manufacturer

Hall and Watts Defence Optics Limited, 266 Hatfield Road, St Albans, Hertfordshire AL1 4UN.

Status

In production.

Service

Various armed forces worldwide.

VERIFIED

Francis Barker Collimators

Description

Specially designed for military, paramilitary, security and police forces, the range of Francis Barker Collimators provides a fast and accurate method for the alignment of weapon sights and laser aiming devices to the correct zero position for a given firing range without the need for live firing or stripping the weapon. They are designed to meet the rigorous standards demanded by the British Armed Forces.

Whenever weapons are deployed under operational conditions, correct zeroing is essential. The need for accuracy has been compounded by the growing use of optical and night sights and laser aiming devices which are more sensitive to rough handling than iron sights and therefore need more frequent checking and setting.

Francis Barker supplies Small Arms Collimators (SAC) to zero optical, night vision and iron sights to the correct zero position or to Personal Zero positions. The SAC is simply fitted into the barrel and the sight aligned with the markings on the graticule within the SAC housing. Sights can be quickly aligned to an accuracy of ±0.25 mil (25 mm in 100 m) by soldiers in the field, day or night. This system has been operationally proven by the British Armed Forces and is the only collimator fully endorsed by them. The SAC is available to suit all weapon, sight and laser designation configurations up to 0.50/12.7 mm calibre.

Francis Barker has also developed a range of Laser Designator Collimators (LDC) which can zero visible and invisible laser light pointers, aimers and designators. The LDC is viewed like a conventional collimator through night vision goggles or a night vision sight. Alignment of a laser pointer is achieved by adjusting the pointer to align with the centre of the collimator graticule. This takes seconds to complete and is accurate. The alignment of both laser pointers and night sights can be undertaken by a single collimator to the same levels of accuracy as the SAC. As with the SAC, the LDC permits rapid zeroing to the correct zero position or to Personal Zero positions.

All these collimator units are passive and require no power source for their operation. They are manufactured to full military specifications and are supplied complete with a carrying case and full instructions for their use.

All collimator measurements and other data are dependent on weapon/sight configurations. The following data should be taken as typical.

Francis Barker Collimators

1996

Data

Length: 200 mm
Height: 150 mm
Width: 39 mm
Weight: 400 g

Manufacturer

Pyser-SGI Limited, Francis Barker Division, Fircroft Way, Edenbridge, Kent TN8 6HA.

Status

In production.

Service

In service with British Army and many other military, paramilitary, security and police forces worldwide.

UPDATED

Armalon AM16ST sight unit

Description

The AM16ST is a high-precision, nitrogen sealed unit with coated glass lenses of the highest optical quality. The adjustments for windage and elevation are engineered to move the image so that the graticule pattern remains centred in the apparent field of view.

The sight has an integral mounting bracket, designed to fit AR-15/M16 pattern rifles. The trajectory compensating elevation drum markings indicate the type of ball ammunition for which the range settings are calibrated (for example M193/SS92) and whether the distances are in hundreds of metres or yards. If the drum is marked M193 or SS92 and not marked to indicate alternative barrel length or twist, the calibration assumes standard barrel length with rifling of one turn in 305 mm. If the drum is marked M855 or SS109, the assumption is a standard length barrel with rifling one turn in 178 mm.

Data

Diameter: 41 mm
Length: 190 mm

Weight: 450 g
Field of view: 96.7 mils
Magnification: ×4
Objective aperture: 21 mm
Eye relief: ca 76 mm
Eyepiece focus: +2 to −3 dioptres, locking
Graticule system: multiple source black, red and green day and night illuminated (tritium)

Manufacturer
Armalon Limited, 44 Harrowby Street, London W1H 5HX.

Status
In production.

VERIFIED

Armalon AM16ST sight on AR-15 rifle

Ring Sight system

Description
All Ring Sights are solid glass optical collimating sights. They are unit power (×1 magnification), allowing rapid target acquisition and an absence of scale effects.

The principle of operation is similar to that of some other collimating sights but the Ring Sight system differs in that the aiming eye can see the target directly as well as the aiming mark which is projected at infinity into the field of view. For daylight firing the aiming mark is a circle which is generated, optically, using light reflected from the direction of the target and so adjusted that it always appears brighter than the target background. This ring marker is supplemented by a tritium source or LED illumination. On some models this is a fixed 24-hour facility, while on others it merely requires the quick flip of a switch.

The fact that the firer, who can use one or both eyes when aiming, can see the target as well as the marker with the aiming eye is claimed by the maker to have the advantage over other types of collimating sight of avoiding the problems of eye wander and retinal rivalry. In addition the unit power enables the firer to improve his natural pointing ability.

Modern armies are now using night vision goggles and need to be able to aim and fire their weapons when using such devices. Ring Sights enable them to do so. With pistols and pintle-mounted machine guns the goggles can be positioned as required. With shoulder-fired weapons the sightline must be in line with the goggles so the sight either has to be placed on an extension or a prism can be added, as with the Ring Sight RC-12 sight on the LAW-80. No special zeroing is needed and the same graticule is used so night vision goggle wearers can fire any weapon using Ring Sights.

The range of sights currently available is as follows:

Pistol and machine pistol sights
The EPC-1, Mach 1 and Hunter sights are subminiature non-magnifying red dot sights. Both eyes can be used allowing unimpaired peripheral vision and fast, accurate target acquisition. The sight is ideal for small arms, particularly pistols, allowing them to be holstered and the snap-action switch allows single-handed operation.

The intensity of the dot is automatically adjusted by a forward-facing sensor, immediately giving optimum dot contrast in quickly changing light conditions.

RC-12 Ring Sight on LAW 80

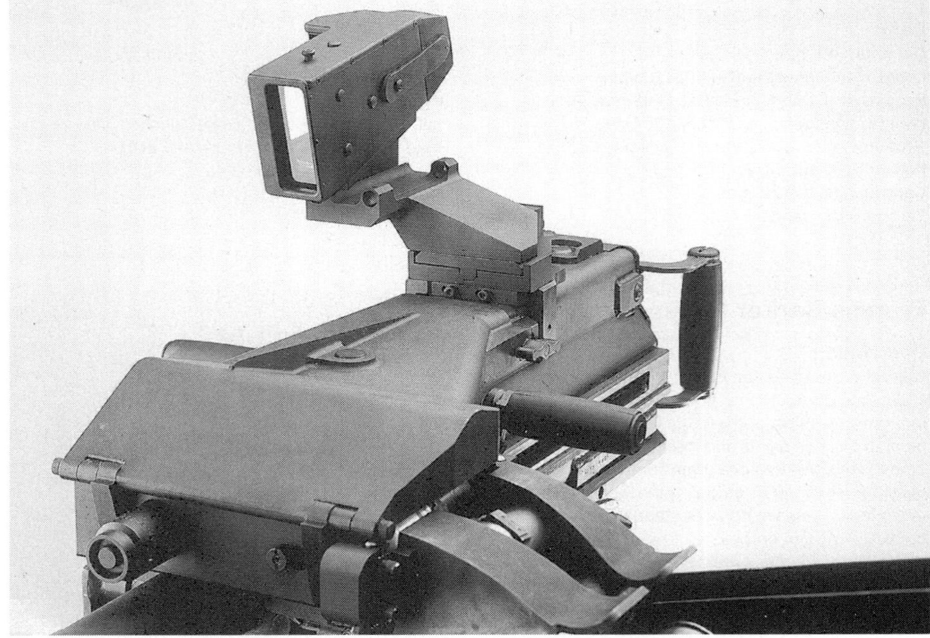

Ring Sight WC-30 mounted on 40 mm MK 19 grenade launcher

1996

The optics are solid in construction with no gas voids, so that fogging is impossible. A further important feature is that the optical faces are flush with the casing so that they can be simply wiped to remove condensation or dust.

The EPC-1 sight is cemented directly to the weapon or a suitable interface. The Mach 1 and Hunter sight versions are zeroable and are fixed mechanically to the weapon. All of these can be used with night vision goggles.

In most applications the iron sights remain visible through the EPC and can still be used as a back-up and as an alignment check.

Ring Sight mounted on M16 rifle

Rifle sights
The HC-10-62 was designed for use on rifles. The base has an integral zeroing mechanism; range setting and windage are optional. The optic is the first of a new range which allows both day and night graticules to be fully lit during the daytime, thereby allowing total flexibility under the most varied and adverse conditions. The graticules are illuminated not only by either the target light or the light from the sky, but by both at the same time. The standard graticule has two daytime rings with four night-time radii. The sight is Night Vision

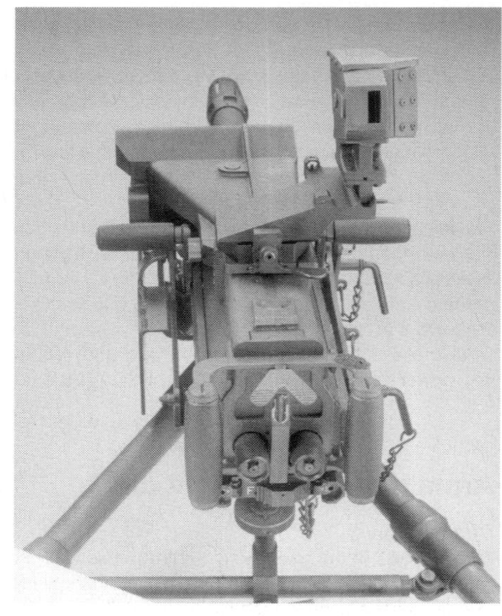

WC-30 sight on 40 mm MK 19 Mod 3 grenade launcher

Data

Model	Application	Aperture	Focal length	Length	Height	Width	Weight	Remarks
LC-7-40	Rifle	7 mm	47 mm	63 mm	32 mm	14 mm	4 g	zeroable
EPC	Pistol	7 mm	47 mm	40 mm	10 mm	17 mm	18 g	no zeroing‡
Mach 1	Pistol	7 mm	47 mm	n/a	n/a	n/a	n/a	zeroable‡
Hunter Sight	Rifle/SMG	7 mm	47 mm	n/a	n/a	n/a	n/a	zeroable‡
LC-14-16	M60 MG	14 mm	56 mm	80 mm	88 mm	62 mm	300 g	zeroable
LC-31-83	MG sight	31 mm	97 mm	94 mm	85 mm	40 mm	485 g	zeroable
LC-40-100	MG & cannon	40 mm	132 mm	190 mm	120 mm	65 mm	1.7 kg	zeroable
LC-40-100-NVG	MG & cannon	40 mm	132 mm	190 mm	120 mm	65 mm	1.7 kg	zeroable‡
HC-10-62	Rifle	10 mm	36 mm	105 mm	65 mm	38 mm	275 g	zeroable
HC-14-62	PDW	14 mm	36 mm	67 mm	30 mm	14 mm	60 g	optic only†
RC-12	LAW	12 mm	100 mm	120 mm	as needed	12 mm	50 g	optic only†
WC-30	40 mm GL	25 mm	n/a	n/a	n/a	n/a	n/a	aligned in production‡
Spot On	Emergency	5 mm	13 mm	25 mm	4 mm	5 mm	1 g	optic only†
MC-12-90	Rifle	10 mm	59 mm	92 mm	20 mm	10 mm	45 g	optic only

† Personal Defence Weapon sights, LAW sights, emergency sights and throwaway barrel sights are best designed in from the start.
‡ LED lit for low light and NVG as standard.
All sights are solid glass or acrylic
All sights can be used with night vision goggles.

Goggle (NVG) compatible and can be supplied with an LED battery module for graticule illumination in difficult lighting conditions such as urban and jungle locations. The sight can alternatively be fitted with a laser pointer module which is zeroed to the optical sight.

Interfaces are available for SA80, M16, FN-FNC and other rifles.

Riot gun sights

The LC-14-46 is zeroable and is ideally suited for riot guns. It is mounted on the weapon using a standard mount and has a flip-down light source for low light and urban conditions. Since the muzzle velocity is low, a range graticule should be fitted; this can be a stadia pattern. The sight is night vision goggle compatible.

Personal defence weapons

The FN P90 has an integral Ring Sight - the HC-14-62 which is similar to the HC-10-62. This sight achieves the precision required to meet the CRISAT hit chance criteria. It can be built into personal defence weapons as an integral sight. It can be used with night vision goggles.

Light and General Purpose machine gun sights

The LC-14-46 was designed for the M60 machine gun and is one of the few, if not the only, optical sights able to fit into the small space available between the interchangeable barrel and the feed cover. It is only 80 mm long and slides straight into the M60 dovetail. It can be supplied with a range of different interfaces for other weapons. It has a flip-down light source for low-light capability and can be supplied with either a cartwheel graticule or an elliptical graticule for anti-helicopter firing. It is inexpensive and easy to install and use.

For other machine guns, such as the M60D or FN MAG, where space is not at a premium, the LC-31-83 (NVG compatible) provides a graticule for surface-to-surface shootings and a cartwheel (or ellipses) for surface-to-air. It is fitted with a zeroing mechanism.

Heavy machine gun and Cannon sights

The LC-40-100 is the largest of the Ring Sight family and was designed as a back-up sight for the BAe Air Defence Gunsight. It has been supplied in zeroable form for use on the 0.50 Browning M2HB machine gun for both ground and air defence roles. The large 40 mm aperture allows the firer great visibility and quicker acquisition of targets and the specially designed graticule gives tangent elevation and lead marks for both surface and air targets. The flip-up light source instantly illuminates the centre two rings for low light, night and flare situations.

The same sight is supplied with a zeroable or non-zeroable interface to fit the 20 mm and 30 mm cannon, both for naval use and land vehicles and in the air defence role. Although the sight gives an upgrade from the standard sight it is inexpensive and easy to install.

The LC-40-100-NVG was specially developed for sideways (and rearwards) firing guns from helicopters both by day and by night using night vision goggles. The chance of a hit has been greatly improved by its use. Interfaces are available for most 7.62 mm machine guns and for other machine guns and cannon. The LC-40-100-NVG, together with the optional ILL/SIM/40 infra-red illuminator, has been ordered by the Royal Air Force for use on helicopters.

40 mm Automatic Grenade Launcher sights

The WC-30 sight is a solid glass, unit power sight with no moving parts and has been specially designed for use with Saco Defense's Mark 19 Mod 3 40 mm grenade launcher. It is fixed directly to the launcher and provides up to 24° Tangent Elevation (TE). The graticule has range markings, one of which the firer selects and lays on the target; this provides the correct TE and drift for that range. Using the graticule he can observe the strike and correct his aim for wind and target movement.

The sight is used with an eye distance of 400 to 450 m but this is not critical. About 4° of the graticule can be seen through the sight at one time; the firer elevates or depresses the launcher to find his selected range, thus putting on the correct TE.

Spot On sight

The Spot On sight is a basic sight, replacing the front sight on conventional weapons. It is especially useful on weapons with a short sight base (such as pistols), providing accuracy as good as other Ring Sights but without an increase in the cross-section of the weapon. It is night vision goggle compatible and should be built as part of the original design (although it can be added later).

Light Anti-armour Weapon sight

The RC-12 and its derivatives provide a sight for light anti-armour weapons; it is the integral sight on the LAW 80. Its basic design can be adapted to other LAWs to suit each one's particular layout. It provides a range and elevation graticule and is night vision goggle compatible.

Manufacturer

Ring Sights Defence Limited, PO Box 22, Bordon, Hampshire GU35 9PD.

Status

In production.

Service

In use by military and police forces of many countries throughout the world.

UPDATED

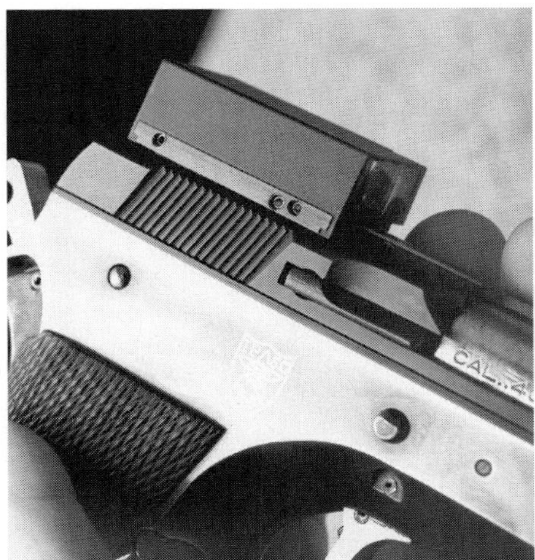

Mach 1 Ring Sight on Tanfoglio pistol

LC-40-100 sight on BAe Air Defence Gunsight

Trilux night sights

Description

Trilux night sights are simple modifications of rifle and machine gun iron sights and are designed to enable troops to fire accurately at dimly distinguishable targets up to the maximum range of visibility. The principle involved is that of artificially enhancing the visibility of the foresight and enlarging the aperture of the rear-sight to allow for the expansion of the eye pupil in darkness.

The way in which the sighting principle is applied to a specific weapon depends on the design of the weapon's iron sights; and an indication of some of the different techniques is given by the accompanying illustrations. In all cases, however, the enhancement of the visibility of the foresight is achieved by incorporating a Betalight self-powered light source in the night foresight. This source, which is visible only to the firer, shows the position of his foresight and enables him to align it with a shadowy target. The light source has a maintenance-free useful life of 10 years.

Manufacturer

Saunders-Roe Developments Limited, Millington Road, Hayes, Middlesex UB3 4NB.

Status

Weapons for which sights are available include the FN-FAL, FN FAL (L1A1), G3 and M16 rifles and the FN MAG. The sights are currently in service with United Kingdom and NATO armies as well as commercial sales to armies throughout the world.

VERIFIED

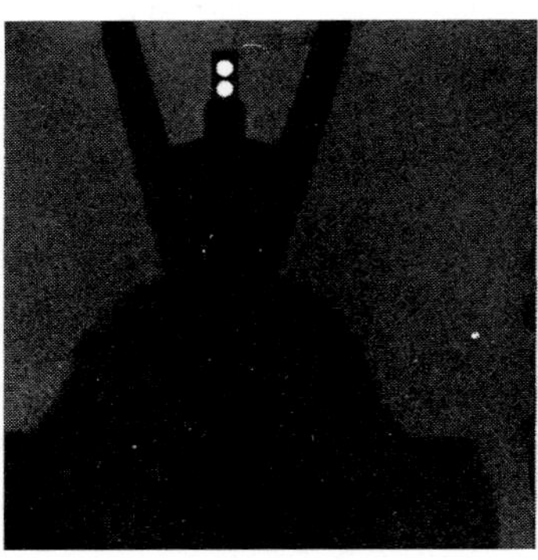

Appearance of Betalight foresight by night and day

Night sights for 7.62 mm FN-FAL rifle

Night sights for 7.62 mm L1A1 rifle

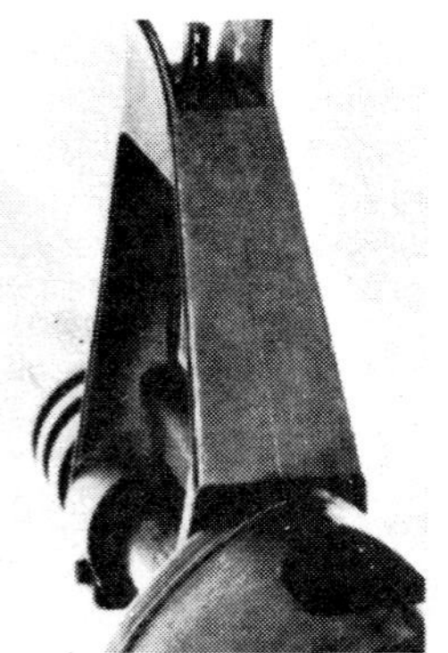

Night sights for 5.56 mm M16 rifle

LS55 laser aiming system

Description

The compact LS55 laser aiming system adds an accurate point-and-shoot capability to almost all standard issue rifles, shotguns, carbines and sub-machine guns in military, police and special forces use worldwide. The target is acquired using an intense red dot of laser light providing pinpoint accuracy, especially useful in security situations. Faster than conventional optical sighting methods and giving the ability to fire with both eyes open from the waist or shoulder level, the LS55 improves first round hit probability.

The LS55 is suitable for any weapon which will take 1 in telescope rings and incorporates elevation and windage to allow zeroing to individual weapons. The LS55 is available in both visible and infra-red versions.

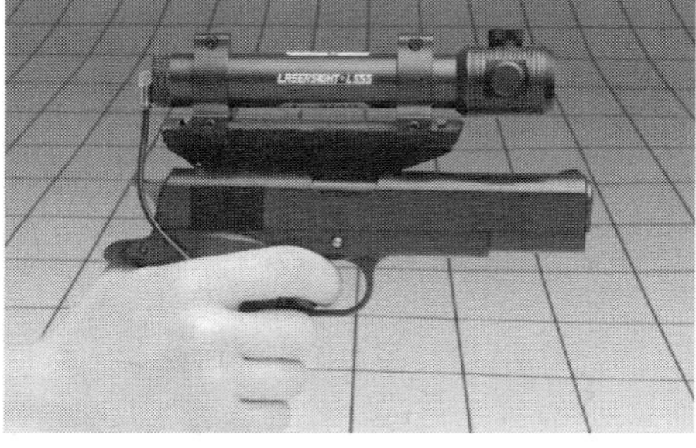

LS55 laser aiming system

Data

Length: 176 mm
Width: 25 mm
Laser output: visible, <5 mW (Class IIIa); IR, <5 mW (Class IIIb)
Wavelength: visible, 670 nm; IR, 800/850 nm
Beam diameter: 25 mm at 50 m; 50 mm at 100 m; 150 mm at 300 m
Divergence: 0.5 mrad
Power supply: 3 × AAA alkaline batteries
Battery life: >8 h continuous use

Manufacturer

Imatronic Limited, Kingfisher Court, Hambridge Road, Newbury, Berkshire RG14 5SJ.

Status

In production.

VERIFIED

Model 1500 night sight

Description

The Model 1500 is a lightweight telescope weapon sight for use with small arms over short and intermediate ranges. The image intensifier used is a second-generation pattern specially designed for this application. It has particularly good resolution, variable gain, good low-light performance and long operational life. It will give good vision in spite of highlights (headlights, street lights, pyrotechnic flares) in the field of view and it is protected against ill effects from short-term high illumination overload. The aiming graticule appears red against the green field of vision, making target acquisition and aiming fast and accurate. Boresighting is easily done by adjusting the position of the graticule by means of two adjuster screws.

Model 1500 night sight

Data

Main body diameter: 59 mm
Max diameter: 76 mm
Length: 265 mm
Weight: 1 kg
Field of view: 10°
Magnification: ×3
Gain: × nominal, 45,000
Object lens: 100 mm f/2
Power supply: 2 × AA cells MN1500 or equivalent
Battery life: ca 80 h
Weapon mount: to requirement

Manufacturer

Alrad Instruments Limited, Alder House, Turnpike Road Industrial Estate, Newbury, Berkshire RG13 2NS.

Status

In production.

VERIFIED

P 840 infra-red pointer

Description

The P 840 is a lightweight infra-red pointer for use in conjunction with night vision devices using image intensification.

The small diameter and light weight mean the pointer may be hand held or fitted in place of a telescope sight. It is designed to suit imperial or metric small-calibre weapon sight mountings. Pitch and azimuth adjustment knobs provide zeroing adjustment.

The P 840 operates from commercially available batteries which can be changed without removing the pointer from the weapon.

Data

Dimensions: 248 × 32 × 46 mm
Weight: 290 g
Light output: near-IR, wavelength 820 nm in the form of a collimated beam 5.4 mm diameter, divergence <0.3 mrad
Beam power: >2 mW max

Manufacturer

Alrad Instruments Limited, Alder House, Turnpike Road Industrial Estate, Newbury, Berkshire RG13 2NS.

Status

Available.

VERIFIED

UNITED STATES OF AMERICA

Pulse Beam laser sight

Description

The Pulse Beam is an advanced laser aiming sight for use with modern firearms. By sealing the component of the laser module in a special epoxy matrix, the Pulse Beam laser is protected from recoil effects, heat and moisture damage. The small size allows the unit to be mounted on most Heckler and Koch firearms and Benelli shotguns without interfering with the standard sights, telescope sights or flashlight-mounted fore-ends. The point of aim is adjustable for windage and elevation by turning two small Allen screws in the laser housing.

Three models are available. The Model 100 mounts on the front of the cocking tube of the MP5, HK33, HK53, G3, HK94, HK91 and SP89 (with modified fore-end). The Model 200 fits on the magazine tube of Benelli M1 and M3 shotguns as well as several other 12-gauge shotguns. The Model 300 fits any Weaver-base telescope mount. All models mount close to the barrel axis for effective alignment between the beam and the firearm's point of impact.

The Class III laser beam pulses at 10 Hz and has an effective range of more than 100 m. The pulsed beam provides superior target acquisition over a constant beam and extends battery life by up to 50 per cent longer than conventional laser sights.

Data

Dimensions: 51 × 51 × 20 mm
Weight: 99 g
Laser type: visible laser diode
Output power: <5 mW
Wavelength: 665-675 nm
Beam colour: deep red
Beam size: 80% of beam intensity in 1 in square at 50 ft
Voltage required: 7 V DC

Manufacturer

Applied Laser Systems, 2160 NW Vine Street, Grants Pass, Oregon 97526.

Pulse Beam laser sight attached to shotgun

Status

In production.

VERIFIED

Contraves Small Arms Common Module Fire Control System II

Description

The Contraves Small Arms Common Module Fire Control System II (SACMFCS II) is a compact, modular, reliable sighting system which increases the first-round hit capability for crew-served weapons. Compared to the original Contraves SACMFCS the SACMFCS II is lighter, smaller, more durable and easier to maintain. Designed and developed to meet the most stringent field and environmental requirements, this fire control system provides precision fire control by combining both day and night sighting capabilities together with a mini-laser rangefinder. Optional modules can be added internally for full solution fire control and thermal imagery.

The laser rangefinder is an eye-safe Erbium laser with ranging capabilities from 90 to 4,000 m. The day/night sighting scope uses Generation III technology and can zoom to adjust the field of view to optimum performance. Rapid battery replacement is possible with either 6 V lithium or C-cell batteries.

The Contraves SACMFCS II was specifically designed to increase the accuracy and effective range of crew-served weapons such as the 40 mm MK 19 grenade launcher, 0.50/12.7 mm M2, 7.62 mm M240 and M60 machine guns, and the 84 mm Carl Gustaf recoilless rifles. It may also be used with 155 mm howitzers in the direct fire mode.

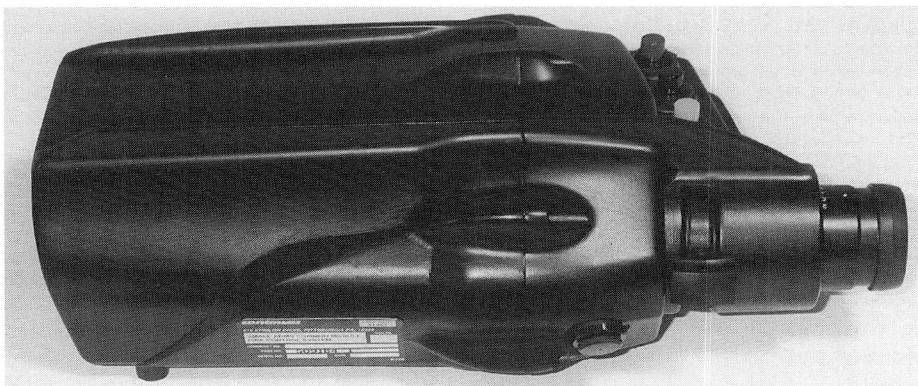

Contraves Small Arms Common Module Fire Control System II (SACMFCS II)

1996

Data

Weight: 2.5 kg
Length: 310 mm
Width: 170 mm
Height: 100 mm
Field of view: 12-4.5°
Zoom: ×1.2-×5
Laser ranging: 90-4,000 m
Tube: Gen III
Battery: 6 V lithium or C-cell

Manufacturer

Contraves USA, 615 Epsilon Drive, Pittsburgh, Pennsylvania 15238.

Status

Advanced development completed. US Army moving to full-scale engineering development.

NEW ENTRY

BUSHNELL HOLOsight

Description

The BUSHNELL HOLOsight is produced by Electro-Optics Technologies (EOTech) and distributed by Bushnell Sports Optics. Also known as the BUSHNELL HOLOsight Model 450LE, this sight involves holographic technology with which a hologram of a reticle pattern is recorded on a head-up display window. When illuminated by coherent (laser) light the holographic image becomes visible at the target plane where it remains in focus with the target. Critical eye alignment is not required and multiplane focusing error is eliminated. Using the sight therefore involves looking through the sight window, placing the image reticle on the target, and firing. The sight uses a 1 in Weaver dovetail mount.

The HOLOsight emits no position-revealing light and is undetectable by night vision systems. All reticle patterns are instantly visible in any light regardless of the shooting angle or position and remain in view between shots or when sweeping a target area. Reticles are designed as large see-through patterns to capture the user's eye instantly without covering the aiming point. As the virtual image of a bright red reticle is on the target, concentration on the target is ensured. The use of holographic technology allows the creation of virtually any image as a reticle pattern in either two or three dimensions. The HOLOsight has interchangeable windows which allow reticle patterns to be changed in less than 30 seconds while still retaining essential zero.

The BUSHNELL HOLOsight involves digital electronics and has recessed push-button controls. An add-on filter element is provided for extreme low light conditions and for use with night vision systems. By using an onboard microprocessor the HOLOsight provides user-selectable features such as a battery check indicator, programmable auto shut down modes, and user-configurable auto brightness start-up settings.

Power is supplied by two Type N alkaline batteries. Battery life is a nominal 50 hours of continuous use. A detachable ruggedised hood is available to protect the HOLOsight.

Data

Weight: 246.65 g
Length: 152 mm
Width: 38 mm
Height: 45.7 mm
Magnification: ×1, unlimited eye relief

Manufacturer

Electro-Optics Technologies (EOTech), PO Box 134010, Ann Arbor, Michigan 48113-4010.

Status

In production.

NEW ENTRY

BUSHNELL HOLOsight mounted on a 9 mm MP5 sub-machine gun

1996

BUSHNELL HOLOsight mounted on a 9 mm MP5 sub-machine gun

1996

Hughes Thermal Weapon Sight (TWS) AN/PAS-13

Development

The Thermal Weapon Sight (TWS) is a versatile man-portable infra-red imaging sensor developed for the US Army Night Vision and Electro-Optics Directorate, Fort Belvoir, Virginia. It is designed for the acquisition and sighting of targets by individual or crew-served weapons during darkness or daylight and under adverse battlefield conditions.

In June 1995 the US Army Communications and Electronics command exercised a $22 million contract option calling for the production of approximately 500 TWS, with second year options for up to 1,500 additional units. The first production sights were scheduld for delivery during 1996. In service with the US Army the TWS will replace the AN/PVS-4 and AN/TVS-5 image intensifier sights.

Description

The TWS operates in the 3-5 μm range and uses a mercury cadmium telluride sensor which is thermoelectrically cooled. Its low operating power consumption enables the TWS to be battery operated. The TWS incorporates binary optics, which require 30 to 50 per cent fewer optical elements, resulting in lower weight and reduced production costs.

In its various configurations the TWS is capable of recognising targets such as groups of men, vehicles, tanks or aircraft at the necessary tactical ranges. In addition, the modular design allows for its application to other electro-optical sensor missions, such as perimeter surveillance, RPV payloads and light armoured vehicle control systems.

Since no single sensor configuration is capable of satisfying all the hardware's intended mission requirements, the TWS design uses modular telescopes which are tailored to a specific application, yet interface into a common sensor body. Thus the basic TWS sensor, weight 1.5 kg, can be used for applications requiring a wide field of view, such as the Stinger air defence missile. The Light Weapon Thermal sight weighs 1.68 kg and can be mounted on M16 rifles and the M4 Carbine, providing a range of 550 m. The Medium Weapon Thermal Sight weighs 1.9 kg and can be used with the M60 machine gun, providing a range of 1,100 m. Then comes the Heavy Weapon Thermal Sight weighing 2.08 kg and intended for weapons such as the 0.50/12.7 mm M2 machine gun, the 40 mm MK 19 automatic grenade launcher and sniper rifles; this version has a range of 2,200 m.

Data

Weight: Light, 1.68 kg: Medium, 1.9 kg; Heavy, 2.08 kg
Detector array: 40 × 16
TE cooler: 175°K

Manufacturer

Hughes Aircraft Company, Electro-Optical Systems, PO Box 902, Building E1/M S E107 El Segundo, California 90245.

Status

In production for the US Army.

UPDATED

The three types of Hughes Thermal Weapon Sight; from left, Heavy Weapon Thermal Sight, Medium Weapon Thermal Sight, and Light Weapon Thermal Sight

1996

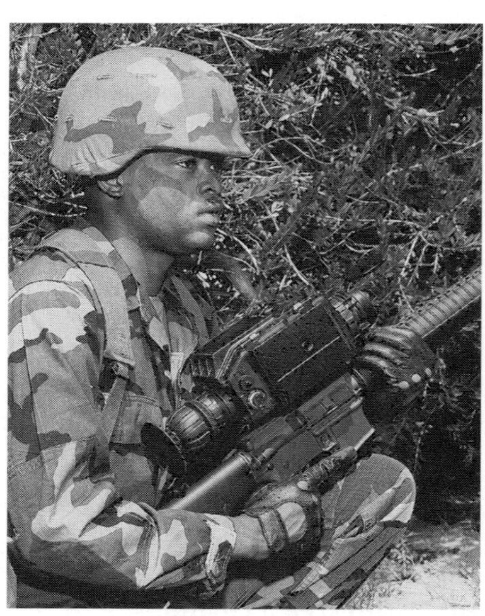

An engineering and manufacturing development model of the Hughes Light Weapon Thermal Sight mounted on an M16 rifle

1996

Magnavox MAG-600 Individual Weapon Thermal Sight

Description

The MAG-600 is a lightweight, multipurpose, thermal imaging sight designed for use on individual weapons, laser designators and as a general purpose surveillance sight. Operating passively in the infra-red spectrum the MAG-600 provides the user with the capability of target acquisition and firing in total darkness and under adverse visibility conditions.

A ×2 electronic zoom is provided. Boresighting the system with an electronic reticle adds to the unit's ease of operation. Push-button controls are incorporated for ease of operation and power is provided by a standard disposable MIL-STD BA 5847/U disposal battery or an external 6 to 24 V DC supply.

The system also features a standard RS-170 video format output (CCIR is also available) for remote viewing and/or recording for training and post engagement analysis.

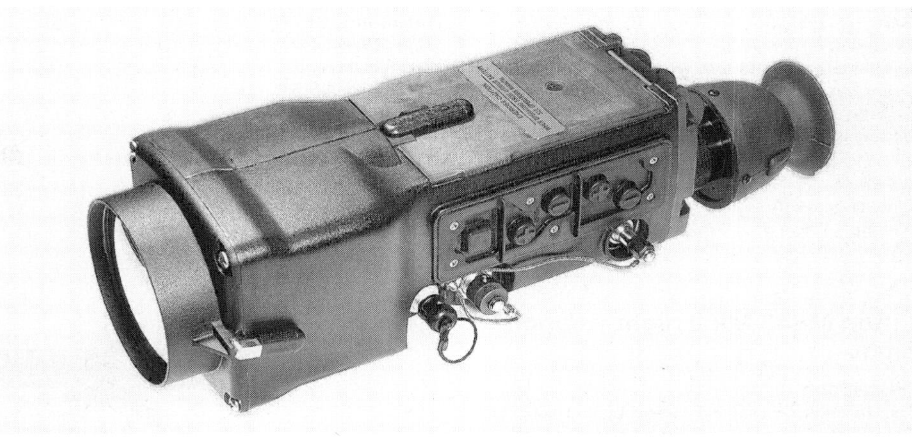

Magnavox MAG-600 Individual Weapon Thermal Sight

1996

Data

Dimensions: 355 × 114 × 114 mm
Weight: <1.8 kg
Field of view: horizontal, 6.2°; vertical, 4.4°
Spectral band: 3-5 μm

Manufacturer

Magnavox Electronic Systems Company, East Coast Division, 1300 MacArthur Boulevard, Mahwah, New Jersey 07430-2052.

Status

In production.

NEW ENTRY

Magnavox AN/PAS-19 Multipurpose Thermal Sight

Description

The AN/PAS-19, also known as the MAG-1200, is a lightweight, multipurpose, thermal imaging sight designed for use on individual and crew-served weapons, laser designators and as a general purpose surveillance sight. Operating passively in the infra-red spectrum the AN/PAS-19 provides the user with the capability of target acquisition and firing in total darkness and under adverse visibility conditions.

A ×2/×4 electronic zoom is provided, as is a memory which can store a number of electronically generated graticules. Boresighting the system with an electronic reticle adds to the unit's ease of operation. Push-button controls are incorporated for ease of operation and power is provided by a standard disposable MIL-STD BA 5847/U disposal battery or an external 6 to 24 V DC supply.

The system also features a standard RS-170 video format output (CCIR is also available) for remote viewing and/or recording for training and post engagement analysis.

Magnavox AN/PAS-19 Multipurpose Thermal Sight

1996

Data

Dimensions: 355 × 114 × 114 mm
Weight: <1.8 kg
Field of view: horizontal, 15.3°; vertical, 7.6°
Spectral band: 3-5 μm

Manufacturer

Magnavox Electronic Systems Company, East Coast Division, 1300 MacArthur Boulevard, Mahwah, New Jersey 07430-2052.

Status

In production.

NEW ENTRY

Litton M992/M993/M994/M995 Ranger™ night weapon sights

Description

The Ranger™ is the latest development from Litton Electronic Devices. It is a high-performance, individual night weapon sight, lightweight, compact and user-friendly. Because of the offset eyepiece design a more conventional head position can be adopted, just as for a day sight. This design eliminates a major disadvantage of older high-profile night sights.

The sight also incorporates a ballistically calibrated elevation knob to allow the user to adjust the estimated range using the knob range scale. In addition, the precision knobs provide repeatable clicks for highly accurate adjustment of both windage and elevation. Only one AA size battery is required to power these rugged, water-resistant, ×4 and ×6 sights, which provide a high resolution intensified image for effective night sighting and aiming. A variable intensity red reticle is optically superimposed on the green intensified image, providing the marksman with an extremely visible, high-contrast, accurate aiming capability.

The Ranger models M992, M993, M994 and M995 differ only in the type of interchangeable image intensifier and magnification. The M992 and M994 incorporate the Litton-developed second plus-generation image intensifier, while the M993 and M995 use the improved third-generation for greater sensitivity and increased resolution under extremely low light levels. All models include splashproof lens covers which incorporate a large neutral density filter for realistic daylight training and accurate daylight weapon zeroing. Magnification is ×4 for the M992 and M993, ×6 for the M994 and M995. All parts except the object lenses are interchangeable.

Litton Ranger™ night sight

1996

Data

M992/M993
Dimensions: 250 × 103 × 88 mm
Weight: 1.2 kg
Object lens: f/1.5 (T/1.6)
Field of view: 10.2°
Magnification: ×4
Resolution: 3.2/3.6

Gain: 1,200/1,500
Eye relief: 30 mm
Dioptre adjustment: +2 to −6
Tube: 2nd plus/3rd generation
Battery: 1 × AA, alkaline
Battery life: 24 h

M994/M995
Dimensions: 310 × 123 × 120 mm
Weight: 1.7 kg
Object lens: f/1.7 (T/1.8)
Field of view: 6.3°
Magnification: ×6
Resolution: 4.5/5.0
Gain: 900/1,100
Eye relief: 30 mm
Dioptre adjustment: +2 to −6
Tube: 2nd plus/3rd generation
Battery: 1 × AA, alkaline
Battery life: 24 h

Manufacturer
Litton Systems Inc, Electron Devices Division, 1215 South 52nd Street, Tempe, Arizona 85281-6987.

Status
Available.

Service
Military and paramilitary use.

VERIFIED

Litton M845 Mk II night weapon sight

Description
The M845 Mk II is a second plus-generation night vision sight, light in weight and powered by two AA alkaline batteries. It has a red dot variable intensity aiming reticle for a simple point of aim and quick reaction. The red dot is variable in intensity as well as adjustable for windage and elevation for zeroing purposes. The system is ideally suited for military use at combat ranges out to 300 m.

This night weapon sight has applications with special forces and Ranger type elements as well as with police SWAT teams. It is simple to boresight and use.

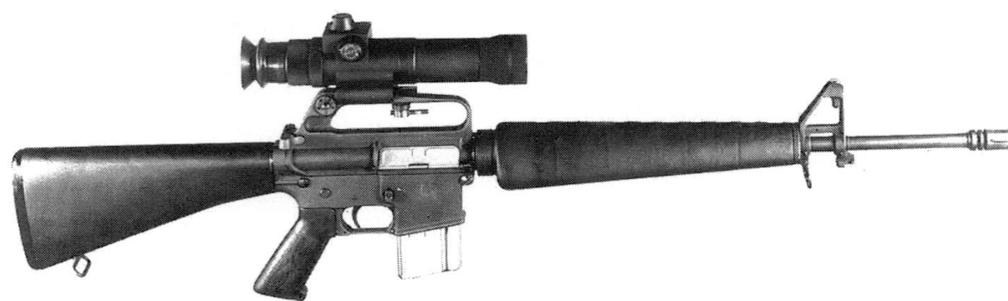

Litton M845 Mk II night weapon sight

Data
Dimensions: 260 × 70 mm
Weight: with batteries, 1.3 kg
Height above mounting surface: 90 mm
Field of view: 13.5°
Magnification: ×1.5
Resolution: 1.5 lp/mm
Gain: nominal, ×1,000
Tube: 2nd plus-generation
Power supply: 2 × 1.5 V alkaline
Battery life: ca 40 h

Manufacturer
Litton Systems Inc, Electron Devices Division, 1215 South 52nd Street, Tempe, Arizona 85281-6987.

Status
Available.

Service
Military and paramilitary use.

VERIFIED

IRAD-600ES eye-safe infra-red aiming light

Description
The IRAD-600ES incorporates lessons learned with aiming lights used in Desert Storm, other combat operations and extensive field training and evaluation. The result is an easy to mount, quick to zero, aiming light with unmatched beam quality and range in an eye-safe device. The IRAD-600ES is manufactured by Insight Technology Inc and is marketed through Litton Systems Inc. It is in service with the US armed forces as the AN/PAQ-4A.

The unit provides a rapid, accurate aim point for personnel engaged in night operations. The IRAD-600ES projects a highly collimated (0.3 mrad) laser beam, invisible to the eye but readily seen with night vision goggles. Once boresighted to a weapon the firer simply puts the laser beam on the target and fires. The unit has a minimum Mil-Spec range of 600 m out to 1,800 m under ideal conditions. Accuracy is equivalent to the finest optical sights, and the ease and rapidity of aiming are unequalled in any conventional sight. The unit incorporates a preset zero setting (neutral position) at which the beam is in precise alignment with the unit's mounting surface. This, combined with the M16 mounting bracket, enables the unit to be nearly zeroed when first mounted on the weapon. Highly accurate elevation and windage controls permit fine zero adjustment.

Advanced optical design and unique optical baffles reduce off-axis detectability to less than, or equal to, 6° and preclude the beam being seen by others using night vision devices.

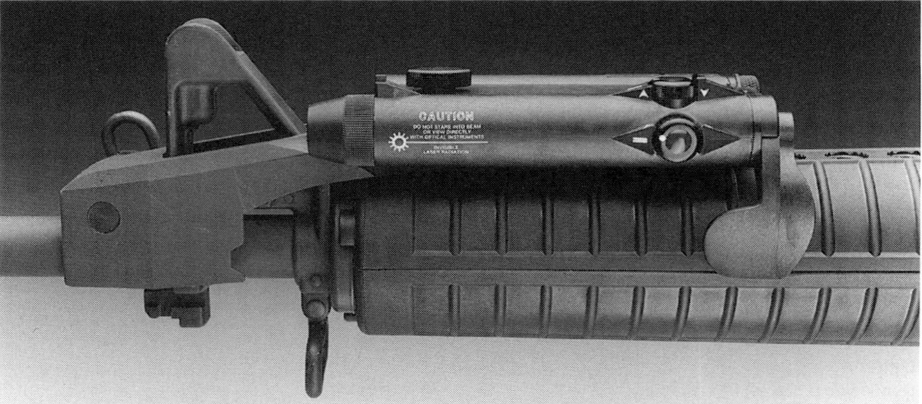

IRAD-600ES eye-safe infra-red aiming light (AN/PAQ-4C) mounted on an M16A2 rifle

1996

Data
Dimensions: 136 × 27 × 56 mm
Weight: 125 g
Divergence: 0.3 mrad
Peak wavelength: 830 nm
Power output: 1 mW, 167 ms 'on', 20% duty cycle
Power supply: 2 × AA

Manufacturer
Insight Technology Inc, 10 Tinker Avenue, Londonderry, New Hampshire 03053.

Marketed by: Litton Systems Inc, Electron Devices Division, 1215 South 52nd Street, Tempe, Arizona 85281-6987.

Status
In production.

Service
US Government and armed forces.

UPDATED

IRAD-2500, 4500, 6500 and 8000 infra-red laser pointers

Description
These Litton laser infra-red aiming devices (for M16 and NATO STANAG mounts) are high-performance versions of the US Military Standard AN/PAQ-4A, intended for users demanding the highest standards of performance and range.

The IRAD projects an intense infra-red laser beam which is invisible to the unaided eye but readily seen with night vision goggles. Once the unit has been boresighted to the weapon, the user simply places the spot on the target and opens fire. The IRADs offer versions with ranges up to 8,000 m, with accuracy comparable to that of the best optical sights and an ease and rapidity of aiming unparalleled in any conventional sight. The extended range of the IRAD makes it suitable for snipers, target designation, secure communications and airborne applications.

The IRAD units incorporate an advanced laser and precision optics in a diffraction-limited optical system to produce a distinct round aiming spot. The collimation of the beam results in a near constant spot size over the

Data

Model	IRAD-2500	IRAD-4500	IRAD-6500	IRAD-8000
Length	170 mm	170 mm	170 mm	170 mm
Diameter	44 mm	44 mm	44 mm	44 mm
Weight	180 g	180 g	180 g	180 g
Divergence	0.3 mrad	0.3 mrad	0.3 mrad	0.3 mrad
Wavelength, peak	830 nm	830 nm	830 nm	830 nm
Power output	3.0 mW	10 mW	20 mW	30 mW
Power supply	2 × AA alkaline	2 × AA alkaline	2 × AA alkaline	2 × AA alkaline
Battery life	50 h	35 h	30 h	20 h
Max range	2,500 m	4,500 m	6,500 m	8,000 m

entire range. Activation is by means of a remote, watertight thumb switch attached to the weapon by Velcro tape.

Precise azimuth and elevation adjustments are incorporated into the IRAD. A compact mandrel and boresight aperture are provided with each unit for fast and accurate boresighting.

Manufacturer
Litton Systems Inc, Electron Devices Division, 1215 South 52nd Street, Tempe, Arizona 85281-6987.

Status
Available.

Service
Military and paramilitary use.

VERIFIED

Litton IRAD infra-red laser pointer

NVS-700 night vision system

Description
The NVS-700 second-generation individual weapon sight is extensively used by military and police authorities in many countries. Both it and the NVS-800 crew-served weapon sight (qv) use the same 25 mm micro-channel plate image intensifier tube and have almost all parts in common except the objective lens and associated fittings. The smaller NVS-700 is suitable for mounting on the 5.56 mm M16 rifle and similar weapons.

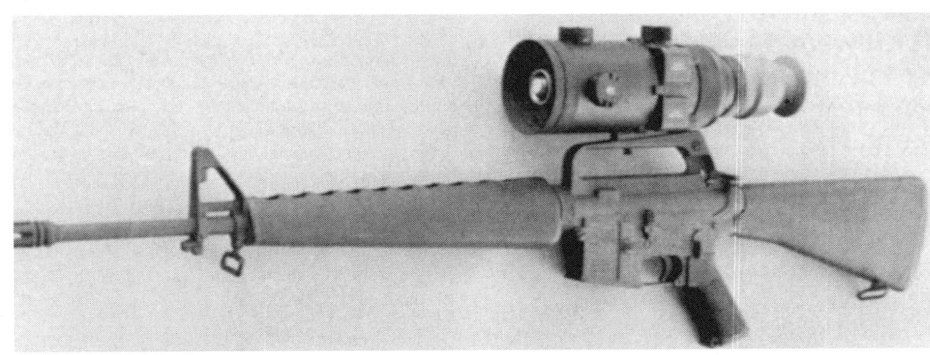

NVS-700 sight on M16 rifle

Data
Diameter: nominal, 101.6 mm
Length: nominal, 292 mm
Weight: 1.814 kg
Field of view: 253 mils
Magnification: ×3.5
Viewing range: moonlight, 25-700 m; starlight, 25-450 m
Objective lens: focal length, 95 mm f/1.7; focus adjustment, 25 m to infinity
Eyepiece: focal length, 26.5 mm; focus adjustment, +3 to −6 dioptres
Operating temperature: with arctic kit, −54 to +52°C
Power supply: 2 × AA mercury cells
Battery life: 60 h

Manufacturer
Optic-Electronic Corporation, 11545 Pagemills Road, Dallas, Texas 75243.

Status
In production.

Service
Widespread military sales to foreign governments.

VERIFIED

NVS-800 night vision system

Description
Designed for use on heavy automatic weapons, recoilless guns and other armament of similar size, the NVS-800 sight closely resembles the NVS-700 (small Starlight scope) individual weapon sight (qv) but has a greater range capability resulting from the use of a larger objective lens. The eyepiece, 25 mm second-generation image intensifier and associated components are common to both sights and the data following relates only to the differences between the two.

NVS-800 night vision system

Data
Diameter: 165 mm
Length: 355.6 mm
Weight: 3.856 kg
Field of view: 156 mils
Magnification: ×6
Viewing range: moonlight, 25-2,000 m; starlight, 25-1,200 m
Objective lens: 155 mm T/1.7
Power supply: 2 × AA mercury cells
Battery life: 60 h

Manufacturer
Optic-Electronic Corporation, 11545 Pagemills Road, Dallas, Texas 75243.

Status
In production.

Service
US and other armies.

VERIFIED

AN/PAQ-4 laser aiming light

Description
The AN/PAQ-4 infra-red laser aiming light provides rapid night-time target acquisition and positive aiming for anti-terrorist squads, police, commando and jungle warfare units.

The light is robust and lightweight, battery-powered and can be mounted and boresighted to virtually all individual weapons including 5.56 and 7.62 mm rifles, light machine guns, recoilless guns and rocket launchers. It utilises hybrid electronics for controlled output and a Gallium Aluminium Arsenide laser diode tuned to an infra-red frequency above human eye level. The system is completely eye-safe, to allow for its use in training without the need for attenuation filters. The non-visible pulsed laser beam can only be seen through night vision goggles or other infra-red sensitive instruments.

Data

Dimensions: 160 × 42.5 × 53 mm
Weight: 300 g
Beam: circular, 2 mrad diameter
Output wavelength: 850 mm
Spectral bandwidth: 8 nm
Power supply: 2 × BA1567/U
Battery life: 100 h

Manufacturer

Optic-Electronic Corporation, 11545 Pagemills Road, Dallas, Texas 75243.

Status

Available.

VERIFIED

AN/PAQ-4 laser aiming light

ITT Integrated Day/Night Scope F7111

Description

The ITT Integrated Day/Night Scope (IDNS) F7111 is intended to provide small arms users with a full 24 hour performance capability. The IDNS uses a Generation III image intensifier to provide advanced night vision capability with the tube having significant performance improvements in the near infra-red spectrum to take advantage of starlight. During the day the same system uses high-performance ×8.3 magnification optics to present a sharp image to the user.

ITT Integrated Day/Night Scope F7111

1996

The IDNS uses an optical arrangement whereby the eye can see both day and night images at the same time. This eliminates concerns over the proper alignment of moving parts within the system and also provides the best image available under changing light conditions without the user having to decide if the day or night capability is better.

Data

PROVISIONAL
Dimensions: 114 × 114 × 381 mm
Weight: <2.27 kg
Field of view: 2°
Magnification: ×8.3
Eye relief: >60 mm
Power supply: AA cell

Manufacturer

ITT Defense, Electro-Optical Products Division, 7635 Plantation Road, Roanoke, Virginia 24019.

Status

Under development. Under test by US Government.

NEW ENTRY

Trijicon Advanced Combat Optical Gunsights (ACOG)

Development

There are two types of Trijicon Advanced Combat Optical Gunsights (ACOG), both with illuminated self-luminous aiming systems designed to improve firing accuracy under low-light conditions. One is the ACOG 4 × 32 and the other the ACOG 3.5 × 35.

The first type to be developed, by Glyn A J Bindon, was the ACOG 4 × 32 which was designed specifically for M16 series rifles but may also be mounted on many other types of 5.56 mm rifle. It was used on the AAI submission for the US Army Advanced Combat Rifle trials held in 1990. In 1995 the ACOG 4 × 32 was selected by US Special Operations Command (SOCOM) as standard issue for all Special Forces units, including the Green Berets, US Navy SEALs and Rangers. In this role the scope is specifically modified for the M4A1 Carbine and is part of the Special Operations Peculiar Modification Kit which allows equipment to be selected for particular missions. This scope is also an option for use with the KAC 5.56 mm Modular Weapon System (qv).

The ACOG 3.5 × 35 has a red dot reticle and longer eye relief suitable for 7.62 mm and similar rifles.

Description

Both models of ACOG are compact sighting scopes with self-luminous aiming reticles. The body uses the same forged 7075-T6 aluminium construction as M16/M4 receivers, with a hard anodised finish. The scopes are waterproofed beyond US Navy SEAL operating depth requirements. All adjustments are effected internally with only the prism housing actually moving. Fine threads and screws give the ACOG an adjustment sensitivity of three or four clicks/in (depending on the model) to provide precise adjustment.

A hole through the ACOG mounting adaptor, which secures the scope on the Picatinny mount, permits use of the normal M16/M4 iron sights without having to remove the scope.

The standard ACOG 4 × 32 has a cross-hair reticle which is black during daytime light conditions but glows red at night by using a tritium lamp light source. (Other reticle patterns are available.) By contrast the ACOG 3.5 × 35 has a glowing red dot reticle that remains luminous under day and night light conditions. In daylight a built-in fibre optic system gathers the exact

Trijicon 3.5 × 35 Advanced Combat Optical Gunsight (ACOG)

1996

Comparison view of the two models of Trijicon Advanced Combat Optical Gunsight (ACOG)

1996

amount of sunlight to illuminate the dot with enough intensity to make high-contrast aiming possible under any daytime conditions.

The ACOG 3.5 × 35 incorporates a feature known as the Bindon Aiming Concept in which the aiming speed of red dot aiming systems using both eyes can be combined with the accuracy and precision of magnified telescopic sights. Using this concept the rifle is raised to the firing position with both eyes remaining open. As the red dot reticle appears and moves in the target area the brain will automatically focus on the ×3.5 magnified view for final aiming. When the red dot is moved the telescopic image blurs more quickly than the unmagnified view but for final aiming the brain selects the non-blurred view automatically. Blur decreases near the target so the brain selects the detailed magnified view. This concept can be rapidly learned to the point of becoming virtually instinctive for 96 per cent of a user population.

Data

Model	**ACOG 4 × 32**	**ACOG 3.5 × 35**
Weight	280 g	397 g
Length	147 mm	203 mm
Field of view	7°	5.5°
Magnification	×4	×3.5

Manufacturer
Trijicon Inc, PO Box 6029, 49385 Shafer Avenue, Wixom, Michigan 49393.

Status
In production.

Service
US Special Forces.

NEW ENTRY

Trijicon ACOG Reflex Sight

Description
The Trijicon ACOG Reflex Sight was developed for military use in close quarters battle situations where both eyes remain open at all times. The 1 × 24 sight features an amber aiming dot that adjusts automatically by glowing more or less brightly according to ambient light conditions. The aiming dot is illuminated both by light from the target area and from a tritium lamp to provide optimum target contrast under all light conditions, from daylight to full night darkness. For daylight shooting a fluorescent fibre optic light gathering system causes the amber dot to glow brightly so it can be clearly seen, and glow less brightly under low light conditions to reduce contrast.

The sight has a 6.5 Minute Of Angle (MOA) dot projected 56 mm above the normal iron sight axis, leaving them available for back-up use. The sight, in a machined aluminium housing, is mounted on an RX10 bracket for M16 series rifles with carrying handles or an RX11 bracket for M16 rifles with Picatinny rail mounts.

To encourage the firer to keep both eyes open at all times when using the ACOG Reflex Sight, a special 24-layer dichroic (two colour) mirror super lens was designed to reflect the amber dot toward the eye and focus it on the target. When aiming with one eye closed is attempted the dichroic mirror takes the red and orange colour out of the target scene, making it look more bluish, with reds turning brown. As the other eye is opened all natural colour is put back into the target scene.

Data
Weight: with RX10 mounting bracket, 232.5 g; sight only, 130 g
Length: 108 mm
Width: 34.3 mm
Height: 42 mm

Manufacturer
Trijicon Inc, PO Box 6029, 49385 Shafer Avenue, Wixom, Michigan 49393.

Status
In production.

NEW ENTRY

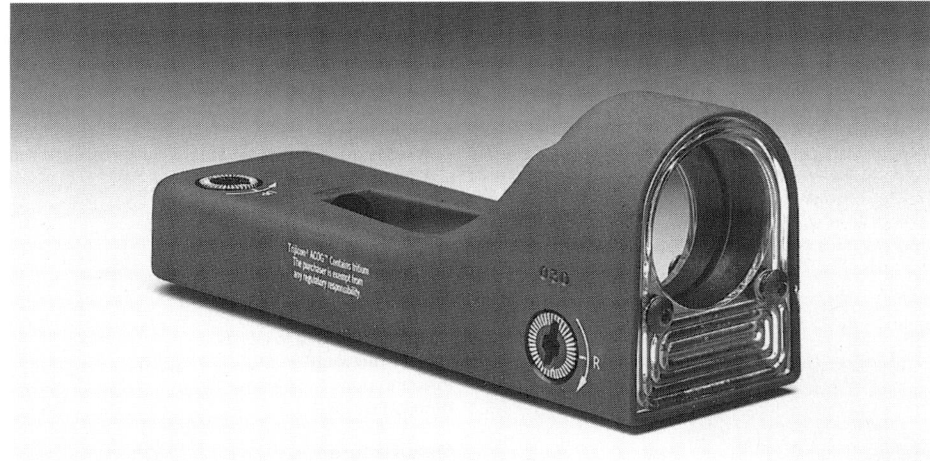

Trijicon ACOG Reflex Sight

1996

Trijicon ACOG Reflex Sight mounted on an M16 rifle

1996

Trijicon Night Sights

Description
Trijicon Night Sights are intended for hand guns but there are rifle, sub-machine gun and shotgun applications. All these sights use a three-dot aiming system in which two dots on the rearsights are aligned with a single dot on the foresight. Using this system aiming is accurate and rapid, even under light conditions so dark that the gun involved cannot be seen by the firer.

The sights contain a miniature glass lamp filled with pressurised tritium gas. Radioactive decay of the tritium gas strikes a layer of phosphor on the inside of the glass to produce a bright green glow that can be seen under low light conditions. The glass lamps are suspended in silicone rubber inside an aluminium cylinder. Each of the three dots involves a polished sapphire to act as a window to the tritium lamp inside and provide a sharply defined circular dot for aiming at night. The sapphire also protects the sights from solvents or puncture.

On most Trijicon Night Sights an inlaid ring of white paint around each sapphire acts as a standard white dot to assist aiming under daylight conditions.

Manufacturer
Trijicon Inc, PO Box 6029, 49385 Shafer Avenue, Wixom, Michigan 49393.

Status
In production.

NEW ENTRY

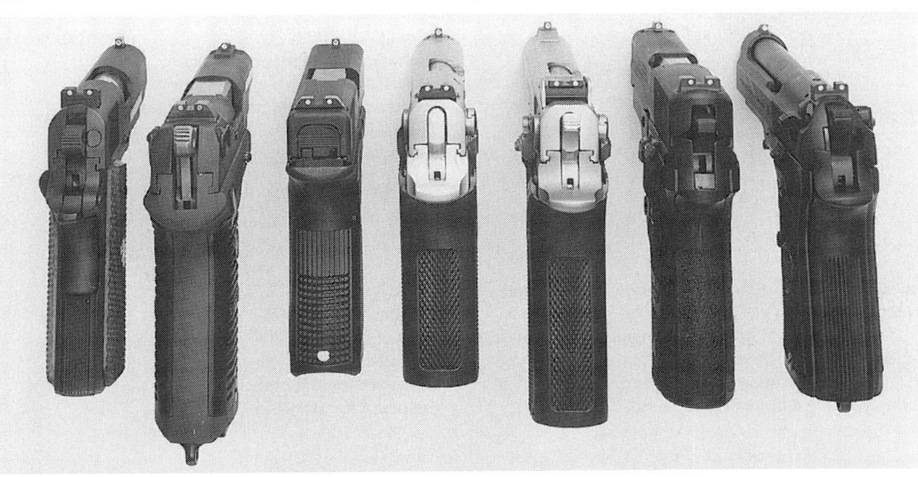

Trijicon Night Sights mounted on a variety of automatic pistols

1996

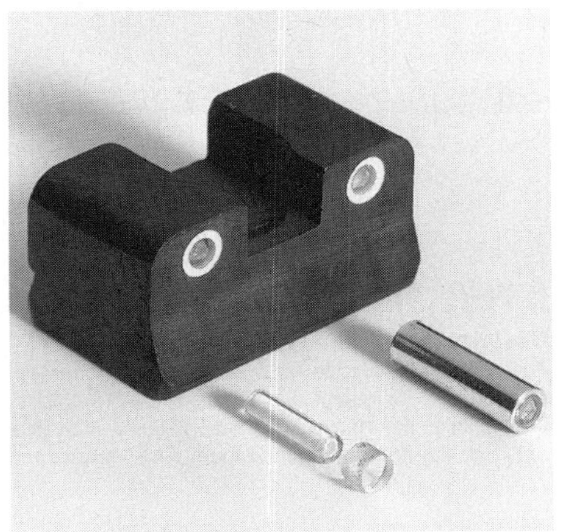

Trijicon Night Sight showing the tritium lamp

1996

AN/PVS-4 second-generation weapon sight

Description

Performing the same range of functions as the earlier first-generation AN/PVS-2 Starlight scope but with superior characteristics, the AN/PVS-4 is a light, passive night vision sight using a 25 mm microchannel plate inverter intensifier tube. The sight is easily attached to a number of different weapons or may be hand held for night reconnaissance.

An adjustable internally projected reticle and interchangeable reticle pattern allows the sight to be boresighted to the various weapons without having to move the sight.

Image tube gain and reticle brightness are manually adjustable to compensate for different levels of ambient lighting.

Automatic gain control circuitry is employed to maintain the viewed scene illumination constant during periods of changing light level conditions, such as the period from sunset to full darkness. This allows the operator of the sight to use the sight without having to readjust the tube gain control every few minutes during this period.

The tube features muzzle flash protection which prevents the tube from being damaged by high-intensity short duration flashes of light. The flash protection circuit is designed to recover in time for the observer to see the round hit the target.

AN/PVS-4 second-generation weapon sight

Data

Dimensions: 240 × 120 × 120 mm
Weight: 1.5 kg
Field of view: 14° 30′
Magnification: ×3.7
Focus range: 25 m to infinity
Objective focal length: 95 mm
Eyepiece focal length: 25 mm
Eye relief: 34 mm
Dioptre range: ±4
Power: 2 × AA batteries
Battery life: 30 h

Manufacturer

Varo Inc, Electron Devices, PO Box 469014, 2203 West Walnut Street, Garland, Texas 75046. (An IMO Industries company.)

Status

In production.

Service

US Army and Marine Corps. Numerous armies around the world.

Licence production

GERMANY
Manufacturer: Euroatlas GmbH
Type: AN/PVS-4
Remarks: See entry under *Germany* for details
PAKISTAN
Manufacturer: Institute of Optronics Pakistan
Type: AN/PVS-4A
Remarks: See entry under *Pakistan* for details

VERIFIED

AN/TVS-5 second-generation crew-served weapon sight

Description

The AN/TVS-5 is a light passive night vision sight which uses a 25 mm microchannel plate inverter intensifier tube. It can be used with a range of different crew-served weapons and can also be tripod mounted for night reconnaissance.

An adjustable internally projected reticle and interchangeable reticle patterns allow the sight to be boresighted to the various weapons without having to move the sight.

Image tube gain and reticle brightness are manually adjustable to compensate for different levels of ambient lighting.

Automatic gain control circuitry is employed to maintain the viewed scene illumination constant during periods of changing light level conditions, such as the period from sunset to full darkness. This allows the operator to use the sight without having to readjust the tube gain control every few minutes during this period.

The tube features muzzle flash protection which prevents the tube from being damaged by high-intensity short duration flashes of light. The flash protection circuit is designed to recover in time for the observer to see the round hit the target.

AN/TVS-5 second-generation crew-served weapon sight mounted on 40 mm MK 19 grenade launcher
1996

Data

Dimensions: 310 × 160 × 170 mm
Weight: 3 kg
Field of view: 9°
Magnification: ×6.2
Focus range: 25 m to infinity
Dioptre range: ±4
Power supply: 2 × AA batteries
Battery life: 30 h

Manufacturer

Varo Inc, Electron Devices, PO Box 469014, 2203 West Walnut Street, Garland, Texas 75046. (An IMO Industries company.)

Status

In production.

Service

US Army and Marine Corps. Also used by 20 other countries.

Licence production

GERMANY
Manufacturer: Euroatlas GmbH
Type: AN/TVS-5
Remarks: See entry under *Germany* for details
KOREA, SOUTH
Manufacturer: Not known
Type: KAN/TVS-5
Remarks: Standard specifications
PAKISTAN
Manufacturer: Institute of Optronics Pakistan
Type: AN/TVS-5A
Remarks: See entry under *Pakistan* for details

UPDATED

Aquila ™ mini weapon sight

Description

The Aquila mini weapon sight is a ×4 passive individual weapon sight for infantry and special forces. The Model 2500 utilises the 18 mm MCP second-generation Plus or Super Gen tube, while the Model 3000 features the third-generation tube. The tubes are optically interchangeable.

This 1.1 kg sight operates on standard AA batteries and incorporates the basic on/off/graticule brightness, azimuth and elevation controls, in addition to the range and eyepiece dioptre adjustments.

Data

Weight: 1.1 kg
Field of view: 8.3° (146 mils)
Magnification: ×4
Objective focal length: 120 mm
Eyepiece focus: +2 to −5 dioptres

Graticule: red LED, 0.2 mil increments
Focus range: 25 m to infinity
Operating temperature: −54 to +52°C

Manufacturer
Varo Inc, Electron Devices, PO Box 469014, 2203 West Walnut Street, Garland, Texas 75046. (An IMO Industries company.)

Status
Available.

Service
US Government agency.

VERIFIED

Varo Aquila mini weapon sight

Varo Model 9886A Infra-red Aiming Light

Description
The Infra-red Aiming Light (IAL) is used in conjunction with night vision goggles to provide an aiming point that will allow the user to deliver accurate fire at night. The IAL consists of an aiming light assembly and a carrying bag. Mounting brackets are available for adapting to most weapons currently in use and can be supplied with the aiming light.

A borelight assembly with either a 5.56 or 7.62 mm mandrel can be supplied as an accessory to the IAL. The borelight assembly is a small lightweight infra-red source which is used to provide a fast and accurate means of boresighting the IAL to the bore of any weapon.

The IAL incorporates a mounting base that is compatible with the standard AN/PVS-4 type adaptor brackets. Additional mounting brackets are available which permit the IAL to be mounted to either the NATO STANAG base or a Weaver type rail.

Controls and adjustments include a remote on/off switch and boresight adjustments. The boresight adjustment includes azimuth and elevation knobs indexed at 0.5 mil increments.

Varo Model 9886A Infra-red Aiming Light

Data
Length: 208 mm
Width: 52 mm
Height: 75 mm
Weight: with batteries, 345 g
Optical output power: 3.2 mW
Output peak wavelength: 820 nm
Output beam size: max, 0.2 mrad
Spectral bandwidth: 12 nm
Range: >400 m
Input DC current: 95 mA
Boresight adjustment: 0.5 mil clicks
Power supply: 4 × AA batteries or 2 × BA5567/U lithium
Operating temperature: −54 to +51°C
Storage temperature: −57 to +65°C

Manufacturer
Varo Inc, Electron Devices, PO Box 469014, 2203 West Walnut Street, Garland, Texas 75046. (An IMO Industries company.)

Status
Available.

VERIFIED

Star-Tron MK-500 Multi-Mission night vision riflescope

Description
The Star-Tron MK-500 riflescope utilises a Mil-Spec 25 mm second-generation inverter tube or a second-generation 20 mm (input) magnifier tube which electro-optically increases system magnification by 150 per cent. When compared with the 18 mm wafer tube used in other night vision sights, both the 25 mm and 20 mm tubes offer substantially higher gain, resolution and field of view. However, when equipped with a comparable focal length object lens the MK-500 is shorter than most wafer tube sights.

The MK-500 can be equipped with an array of field-interchangeable, specially optimised (S20-S25) catadioptric lenses of various focal lengths, permitting its use in any tactical situation. For example, when fitted with a 100 mm f/1.4 lens its compact size and ×3.7 magnification make it ideal for use on assault rifles. When fitted with a 170 mm f/1.5 lens and the 20 mm magnifier tube the system magnification of ×9.5 makes it ideal for anti-terrorist and hostage rescue applications.

The scope employs Star-Tron's unique MTF reticle, which has been designed to couple the visual characteristics of the dark-adapted eye. It is completely passive and does not require any brightness adjustment, yet is distinctly visible under any light conditions. The focusing system is of the parallel type which does not rotate during focusing, therefore obviating any abrasion of the skin around the eye. A dioptre scale is fitted, allowing pre-adjustment to the user's eyesight.

One of the unique features of the MK-500 is the system which permits user-selected reticle adjustment increments that are matched to each focal length lens.

Star-Tron MK-500 Multi-Mission riflescope

1996

The adjustment mechanisms for windage and elevation are housed in removable watertight cartridges, easily field-exchanged using only an Allen wrench. A knob adjustment system is available if preferred.

The scope is provided with a full lens diameter daylight filter, with its own protective metal cover, allowing boresighting in daylight.

Dimension and other details are not given, since these will vary according to the selected lens and configuration.

Manufacturer
Star-Tron Technology Corporation, 526 Alpha Drive, Pittsburgh, Pennsylvania 15238.

Status
In production.

Service
US and foreign armies and security forces.

VERIFIED

Insight M30 Boresighting Equipment

Description

The Insight M30 Boresighting Equipment provides an accurate and easy way to transfer master weapon boresight data to individual crew-served weapons. Once boresighted to a master weapon the operator selects the proper calibre mandrel, installs it in the appropriate sight offset location, inserts the assembled M30 into the barrel of the gun and aligns the weapon sights to the target projected by the M30.

The standard issue M30 enables an operator to dry zero iron sights, optical day sights and night vision sights on M16 and M24 rifles, and M2, M249 and M60 machine guns. In addition, the modular design of the M30 enables it to be adapted to other weapon and sight combinations. The M30 contains an easy to read reticle uniformly illuminated with light specifically optimised for boresighting both second- and third-generation night vision devices as well as iron sights. A rotary switch enables the operator to select the desired illumination level of the reticle for various ambient light conditions.

The M30 body and mounts are manufactured using rugged, chemically resistant composite materials. Weight is 450 g.

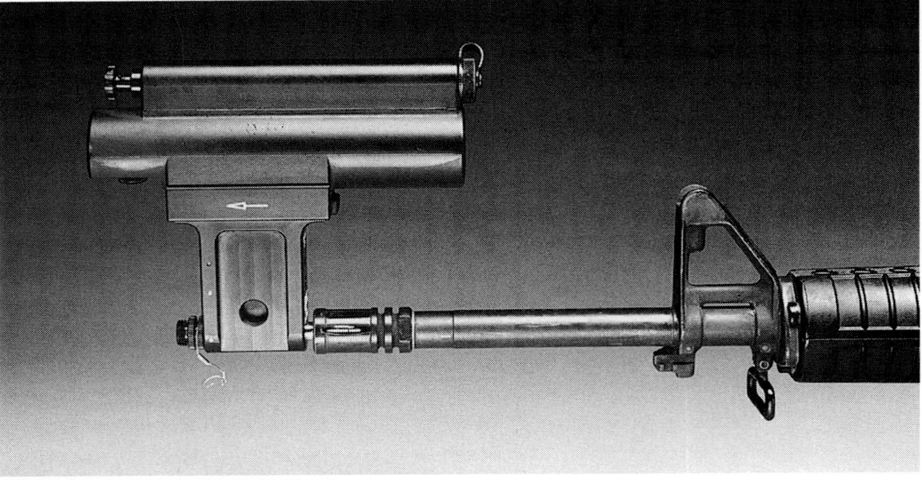

Insight M30 Boresighting Equipment mounted on an M16 rifle

1996

Manufacturer

Insight Technology Inc, 10 Tinker Avenue, Londonderry, New Hampshire 03053.

Status

In production.

NEW ENTRY

DG-1 sight aligner

Description

The DG-1 is an optical device used to implement the military doctrine of dry zero for aligning any number of weapons from a single master weapon. Any weapon whose iron sights range from 15 to 101 mm above the bore axis can be zeroed and the unit can be used with telescope or night vision sights. A variety of barrel adaptors is available, including 5.56 mm, 7.62 mm, 12.7 mm and others up to 40 mm.

After zeroing a master weapon on a firing range in the conventional manner, the unit is inserted into the barrel and the reticle aiming point is aligned with the sight using the DG-1 azimuth and elevation adjustments. The DG-1 is then transferred to a similar weapon to be dry zeroed and the sights are adjusted to match the DG-1 reticle aiming point. Three rounds are normally fired to check the sight setting and make fine adjustment to suit the firer. The DG-1 permits the calibrated reticle settings for each type weapon to be recorded; this allows return calibration recheck after a different calibre or type of weapon has been dry zeroed.

Data

Dimensions: (W × H × L) 63.5 × 127 × 280 mm with adaptor
Weight: with adaptor, 1.58 kg
Reticle: US Army M16 target; other reticles may be provided
Environment: to MIL-STD-810C for vibration, shock transit/drop, temperature, altitude, salt/fog, humidity, chemical exposure, dust and waterproofing

Manufacturer

Metal Dexterity Company Inc, Optronics Division, 624 South F Street, Lake Worth, Florida 33460-4898.

Status

In production.

VERIFIED

DG-1 sight aligner in transit case

DG-1 sight aligner mounted on M16 rifle

SUPPRESSORS

SUPPRESSORS

COMMONWEALTH OF INDEPENDENT STATES

PBS-1 silencer for assault rifles

Description

The PBS-1 silencer was developed for AK-47 and AKM assault rifles and is used in conjunction with special low-power 7.62 × 39 mm cartridges involving heavier than normal bullets. When the PBS-1 is fitted to an AK-47 or AKM rifle the standard muzzle attachment has to be removed and special rearsights are fitted.

The PBS-1 has a two-part cylindrical housing joined by a screw thread. Internally there are up to ten baffle plates preceded by a 20 mm thick rubber plug which has to be pierced by the bullet when fired. The combination of the rubber plug, an expansion chamber and the baffle plates reduces the sound signature considerably and eliminates muzzle flash. However, the rubber plug will require changing after a low number of rounds have been fired if the firing sound level is to be kept to a minimum.

A generally similar silencer has been produced for use with 5.45 mm AK-74 rifles and sub-machine guns, again utilising subsonic ball ammunition.

Manufacturer

State factories.

Status

In service with some Russian special forces.

NEW ENTRY

FINLAND

Valme silencers

Description

Oy Valmenninmetalli AB produces muzzle attachment silencers to a patented design for use with virtually any firearm. Three models have been produced.

The Sniper model is intended for attachment to military and sniper rifles and for pistols with calibres from 4 mm to 11.8 mm. It is claimed that this silencer will reduce the firing signature of any cartridge to that of a 0.22 rimfire cartridge. The device also eliminates flash and reduces recoil.

The Marksman model is smaller and intended for use on 0.22 rimfire weapons.

The Varminter is a fixed silencer and barrel assembly which replaces the standard barrel on 0.22 rimfire firearms. It is claimed that this model is highly efficient, eliminating the sound of a discharge completely.

Data

Model	**Sniper**	**Marksman**	**Varminter**
Weight	500 g	150 g	
Length	350 mm	240 mm	500 mm
Diameter	35 mm	25 mm	25 mm

Manufacturer

Oy Valmenninmetalli AB, Sulantie 3, FIN-04300, Hyrylä.

Status

Available.

NEW ENTRY

SWITZERLAND

BRÜGGER + THOMET IMPULS-II pistol silencer

Description

The BRÜGGER + THOMET IMPULS-II pistol silencer is intended for use with autoloading pistols using the Browning locking system; it can also be used with many other locked pistol systems.

With the base IMPULS-I system the semi-automatic feature on the pistol is retained although it can be disengaged so that the sound of the mechanical locking process is eliminated. Every round thus has to be loaded manually. On the IMPULS-II an active reloading system is located separately at the rear end of the silencer tube. The design permits easy field stripping of the silencer and also allows the use of liquid/grease sound reduction aids.

Data

Weight: 520 g
Length: 280 mm
Diameter: 35 mm
Maintenance: every 1,500 rds

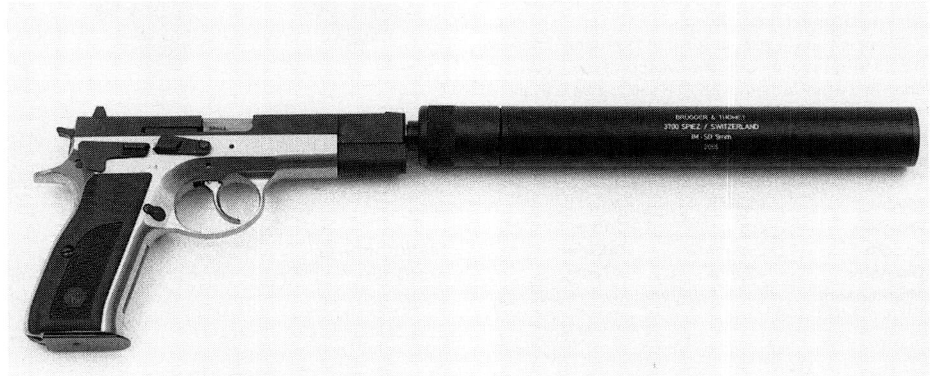

BRÜGGER + THOMET IMPULS-II pistol silencer

1996

Manufacturer

BRÜGGER + THOMET, Thunstrasse 4A, CH-3700 Spiez.

Status

Available.

NEW ENTRY

BRÜGGER + THOMET IMPULS-III pistol silencer

Description

The BRÜGGER + THOMET IMPULS-III pistol silencer was designed for special operation units and is claimed to provide the best sound reduction levels available. It is a combination of the IMPULS-II (see previous entry) with the rubber-plated end cap of the Square Ultra Compact (SUC) silencer (see following entry). Rubber baffles can be added to provide maximum sound reduction.

Data

Weight: 490 g
Length: 250 mm
Diameter: 35 mm
Maintenance: rubber baffles for max reduction, 50 rds; standard, 1,000 rds

Manufacturer

BRÜGGER + THOMET, Thunstrasse 4A, CH-3700 Spiez.

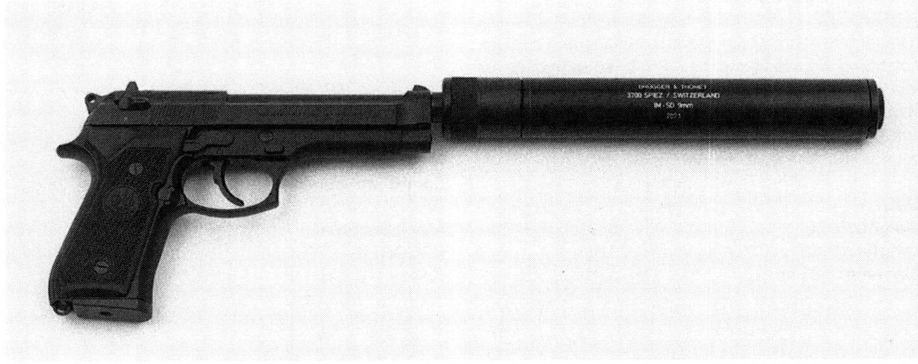

BRÜGGER + THOMET IMPULS-III pistol silencer

1996

Status

Available.

NEW ENTRY

BRÜGGER + THOMET Square Ultra Compact pistol silencers

Description

The BRÜGGER + THOMET Square Ultra Compact (SUC) pistol silencer was designed for missions where short firing ranges are anticipated. The SUC is available for all calibres from 0.22 to 9 mm.

The design of the SUC employs an eccentric silencer body configuration involving high-resistance rubber baffles which limit the accuracy of the pistol involved to a range of 15 m. However the shape of the SUC is such that the standard sights can be employed.

Each SUC is issued with four rubber baffle replacement kits which are used to replace the baffles after every 30 to 50 rounds.

Data

Weight: 210 g
Length: 125 mm
Height: 50 mm
Width: 25 mm
Maintenance: every 30-50 rds

Manufacturer

BRÜGGER + THOMET, Thunstrasse 4A, CH-3700 Spiez.

Status

Available.

NEW ENTRY

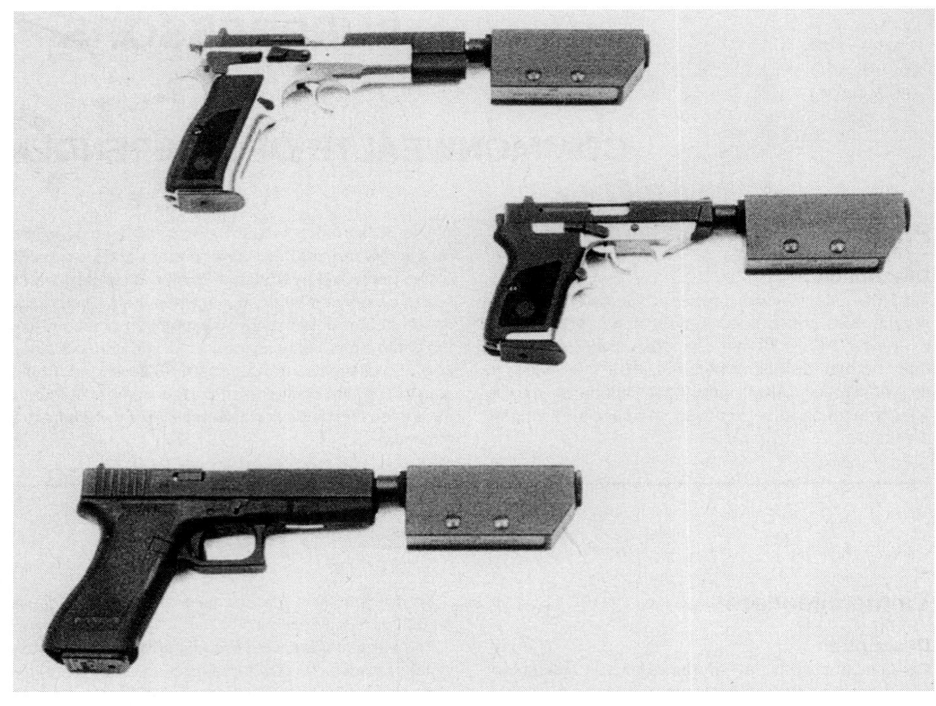

BRÜGGER + THOMET IMPULS-III Square Ultra Compact (SUC) pistol silencers ***1996***

BRÜGGER + THOMET silencers for sub-machine guns

Description

BRÜGGER + THOMET silencers are issued complete with the weapon specific threading. For best nose reduction results it is recommended that subsonic ammunition is used. The silencers can be used for both semi- and full-automatic firing.

Silencers are available for most carbines and sub-machine guns although the weapon type has to be specified when ordering. MOD K silencers are available for calibres from 0.25 ACP to 9 mm, while the MOD L is available for 9 mm to 0.45 ACP.

Data

Model	MOD K	MOD L
Weight	605 g	715 g
Length	252 mm	300 mm
Diameter	42 mm	42 mm
Maintenance	every 5,000 rds	every 5,000 rds

BRÜGGER + THOMET silencer on 9 mm S.A.F. sub-machine gun ***1996***

Manufacturer

BRÜGGER + THOMET, Thunstrasse 4A, CH-3700 Spiez.

Status

Available.

NEW ENTRY

BRÜGGER + THOMET quick detachable silencer for MP5

Description

The BRÜGGER + THOMET quick detachable silencer for the Heckler and Koch MP5 sub-machine gun was developed to meet the requirements of government agencies and armed forces who have a temporary or limited use for a silenced weapon, including for training purposes in urban areas.

BRÜGGER + THOMET have developed a quick detachable mount which allows a rapid, strong and precise fitting of a silencer to the MP5. The mount does not require any modification to the MP5. Standard and short silencer bodies are available. It is recommended that subsonic ammunition is used to achieve the best sound reduction performance.

Data

Weight: 960 g
Length: standard, 325 mm; short, 275 mm
Diameter: 42 mm
Maintenance: every 6,000 rds
Sound reduction: with subsonic ammunition, 28 dB

BRÜGGER + THOMET silencer on Heckler and Koch 9 mm MP5 sub-machine gun ***1996***

Manufacturer

BRÜGGER + THOMET, Thunstrasse 4A, CH-3700 Spiez.

Status

Available.

NEW ENTRY

BRÜGGER + THOMET quick detachable silencers for assault rifles

Description

BRÜGGER + THOMET have developed quick detachable silencers for assault rifles used by both military and law enforcement units. The unique and simple mounting system allows a rapid and secure installation of the stainless steel silencer on all standard types of assault rifle with calibres from 5.56 mm up to 0.308 and provided with Mil-Spec 22 mm diameter flash hiders. Because of the mounting system the silencer is perfectly aligned with the barrel and does not affect the accuracy of the weapon involved. In addition to the sound reduction the silencer eliminates all muzzle flash.

The silencer has successfully passed all durability tests and can be used on light machine guns.

Data
Weight: 860 g
Length: standard, 340 mm
Diameter: 42 mm
Maintenance: every 5,000 rds

Manufacturer
BRÜGGER + THOMET, Thunstrasse 4A, CH-3700 Spiez.

Status
Available.

NEW ENTRY

BRÜGGER + THOMET silencer on Steyr 5.56 mm AUG assault rifle

1996

BRÜGGER + THOMET integrated silencer systems

Description
BRÜGGER + THOMET have developed silencer systems for a variety of firearms. Their Category 3 silencer systems are mounted on military and police sniper rifles. When used in combination with subsonic ammunition involving heavy bullets, targets can be neutralised at ranges up to 200 m. Two standard types are available.

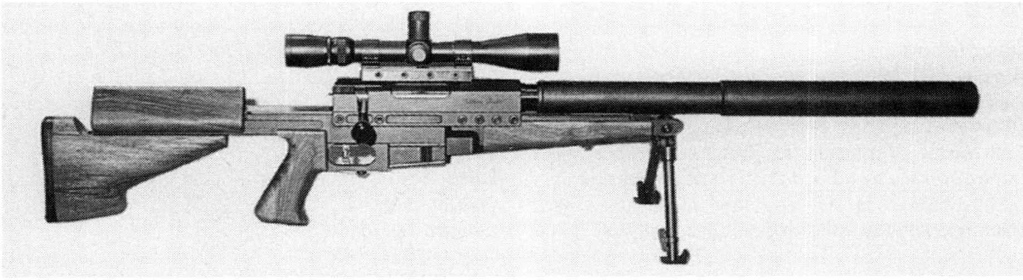

BRÜGGER + THOMET integrated silencer system on PGM UR Intervention sniper rifle

1996

Data
Weight: 1.1 kg or 960 g
Length: 540 mm
Diameter: 50 mm or 42 mm
Maintenance: every 3,000 rds

Manufacturer
BRÜGGER + THOMET, Thunstrasse 4A, CH-3700 Spiez.

Status
Available.

NEW ENTRY

UNITED STATES OF AMERICA

Ciener sound suppressors

Description
Jonathan Arthur Ciener Inc manufactures an array of sound suppressors for specific firearms that are believed to be specially suited to the fitting of such devices. Each unit is specifically designed to suit a particular firearm. The suppressor may be complete with a replacement barrel, or a device to fit an existing barrel, or a combination of suppressor and modified standard barrel as the particular design warrants.

Suppressors are readily available to screw onto weapons such as the AR-15/M-16 series, M14, Heckler and Koch MP5 series, Ruger Mini-14/AC-556, Uzi, M3/M3A1 and the Sten. The company pioneered the fitting of suppressors to long-range sniper firearms up to 0.50/12.7 mm calibre.

Ciener suppressors require no maintenance as their designs do not demand periodical cleaning or rebuilding. Specifications differ according to the parent firearm but suppressors fitted to semi-automatic firearms invariably reduce the noise of discharge to less than the mechanical noise of the firearm.

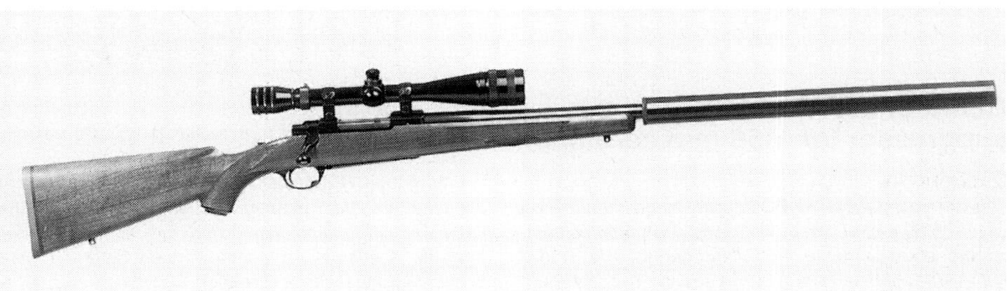

Ciener sound suppressor fitted to Ruger M77V rifle

1996

Manufacturer
Jonathan Arthur Ciener Inc, 8700 Commerce Street, Cape Canaveral, Florida 32920.

Status
In production.

NEW ENTRY

Ciener sound suppressor fitted to Ruger Mk II Government 0.22LR target pistol

1996

Ciener sound suppressor fitted to Heckler and Koch MP5

1996

Gemtech SOS suppressor for pistols

Description
The Gemtech SOS suppressor is stated to be the most compact and lightweight suppressor available for 9 mm semi-automatic pistols and is suitable for use with Browning design pistols without the need for recoil enhancing accessories.

The SOS is constructed of aircraft grade 6061-T6 and 2024-T6 aluminium and does not involve bore obstructing devices. Instead the unit has an advanced vortex generating baffle design which creates an outstanding sound reduction performance while completely suppressing the muzzle flash signature.

The SOS is intended for mounting on a threaded barrel but can also be configured for the Gemtech Quik-Snap rapid detach coupler. All maintenance requirements can be performed without disassembly.

Data
Weight: 100 g
Length: 120 mm
Diameter: 31 mm
Degree of suppression: ca –28 dB

Manufacturer
Gemtech Division of Gemini Technologies Inc, PO Box 3538, Boise, Idaho 83703.

Status
Available.

NEW ENTRY

Gemtech Vortex-9 suppressor for pistols

Description

The Gemtech Vortex-9 suppressor is a compact and lightweight sound suppressor designed to achieve full sound suppression potential absolutely dry. Therefore it does not involve the use of oil, water, grease or other artificial environment technologies.

The Vortex-9 is constructed of aircraft grade 6061-T6 and 2024-T6 aluminium and does not involve bore obstructing devices. Instead the unit has an advanced baffle design.

The SOS is intended for mounting on a threaded barrel but can also be configured for the Gemtech Quik-Snap rapid detach coupler. All maintenance requirements can be performed without disassembly.

Data

Weight: 160 g
Length: 180 mm
Diameter: 31 mm
Degree of suppression: ca –27 dB

Manufacturer

Gemtech Division of Gemini Technologies Inc, PO Box 3538, Boise, Idaho 83703.

Status

Available.

NEW ENTRY

Gemtech MK-9K suppressor for MP5

Description

The Gemtech MK-9K suppressor is intended for use on the Heckler and Koch MP5 sub-machine gun and the MP5K-PDW personal defence weapon, and is a compact version of the Gemtech MK-9 suppressor which carries over the same performance as the full size unit.

The MK-9K has stainless steel baffles in a hard black anodised aluminium body using aircraft grade aluminium alloys. Muzzle velocity and accuracy are not affected by the unit which can accommodate fully automatic fire. No wipes or mesh packings are used and the MK-9K can be disassembled for maintenance.

The MK-9K is designed to mount on the standard Heckler and Koch three-lug barrel utilising a secure three-lug device. Other couplings are available.

Data

Weight: 560 g
Length: 190 mm
Diameter: 51 mm
Degree of suppression: ca –32 dB

Manufacturer

Gemtech Division of Gemini Technologies Inc, PO Box 3538, Boise, Idaho 83703.

Status

Available.

NEW ENTRY

Gemtech MINI-TAC suppressor for MP5

Description

The Gemtech MINI-TAC suppressor is intended for use on the Heckler and Koch MP5 sub-machine gun and is the latest in a line of similar sound suppressors. Gemtech claims that the MINI-TAC can outperform any comparable sized suppressor and many others that are larger and more expensive.

The MINI-TAC has stainless steel baffles in a hard black anodised aluminium body using aircraft grade aluminium alloys. Handling and accuracy are not affected by the unit. No wipes or mesh packings are used. The MINI-TAC can be cleaned by dipping in solvents.

The MINI-TAC is designed to mount on the standard Heckler and Koch three-lug barrel utilising a secure three-lug device. Other couplings are available.

Data

Weight: 300 g
Length: 205 mm
Diameter: 35 mm
Degree of suppression: ca –27 dB

Manufacturer

Gemtech Division of Gemini Technologies Inc, PO Box 3538, Boise, Idaho 83703.

Status

Available.

NEW ENTRY

Gemtech COMMANDO suppressor for 5.56 mm carbines

Description

The Gemtech COMMANDO suppressor is intended for use on 5.56 mm sub-machine guns and carbines such as the short-barrelled Colt Commando, the Ruger Mini-14/AC-556, and similar weapons. The compact unit reduces the muzzle sound signature to a level similar to that produced by a 0.22 rifle and reduces muzzle flash.

The COMMANDO is manufactured from firearms grade alloy steel with a dark phosphate finish although a stainless steel version is available. Muzzle velocity and accuracy are not affected by the unit.

Fitting the COMMANDO unit involves the removal of the standard flash hider on the weapon to allow the COMMANDO unit to screw onto the exposed barrel threads. A special replacement flash hider for M16 series weapons with a Gemtech Quik-Snap coupler is available.

Data

Weight: 680 g
Length: 153 mm
Diameter: 35 mm
Degree of suppression: ca –24 dB

Manufacturer

Gemtech Division of Gemini Technologies Inc, PO Box 3538, Boise, Idaho 83703.

Status

Available.

NEW ENTRY

Gemtech SPEC-OP 2 and 3 suppressors

Description

Gemtech SPEC-OP 2 and 3 suppressors are essentially similar units intended primarily for assault rifles. The SPEC-OP 2 is optimised for weapons with barrel external diameters greater than 0.60 in/15.24 mm. This includes the M16A2 series and heavy-barrel versions of the M16/AR15. The SPEC-OP 3 is intended for use with a variety of rifles firing 7.62 mm NATO ammunition.

Both models utilise a two-point high-accuracy mounting system and do not affect muzzle velocity or accuracy. Both have stainless steel interior components with the external finish being a dark grey optically flat phosphate finish.

Data

Weight: 1.1 kg
Length: 285 mm
Diameter: 41 mm
Degree of suppression: ca –35 dB

Manufacturer

Gemtech Division of Gemini Technologies Inc, PO Box 3538, Boise, Idaho 83703.

Status

Available.

NEW ENTRY

Knight suppressed revolver

Description

This is a modified Ruger GP100 revolver to which a snap-on type suppressor, similar to that used with the Beretta pistol (see following) is fitted. In general, it has been accepted that revolvers cannot be effectively silenced because of the cylinder/barrel gap. However, as can be seen from the photograph, the cylinder and barrel have been modified, and the pistol fires a specially made 5.56 mm telescoped cartridge. This permits sealing of the gap and thus prevents the escape of propellant gas other than through the barrel and suppressor. The cartridge is also extremely accurate over normal hand gun ranges, and the reduction in noise is from 163 dB without suppressor to 121 dB with the suppressor fitted. The complete assembly weighs 1.58 kg and is 330 mm in length. The picture also shows the optional telescope sight and laser spot projector.

Manufacturer

Knight's Armament Company, 7750 9th Street SW, Vero Beach, Florida 32968

Status

In production.

VERIFIED

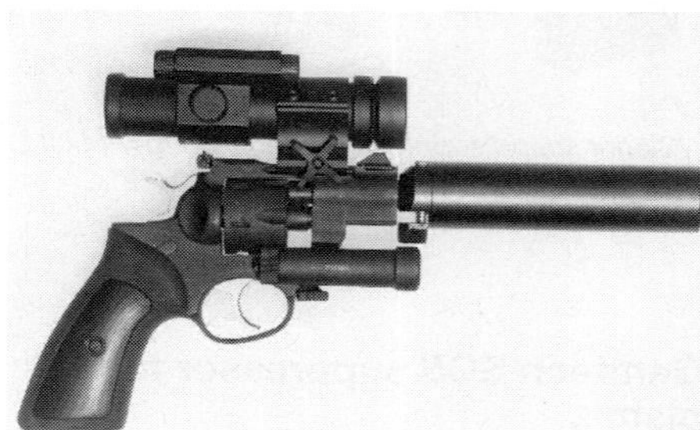

Knight suppressed revolver

Snap-on suppressor for Beretta 92F

This suppressor comes with an optional slide-locking feature and is specifically designed to fit the Beretta 92F pistol. A simple thumb-push on the suppressor's knurled sight base allows it to be easily attached or removed. There is a quick-change wipe container pack that enables field-level replacement of the polyurethane bullet wipes or the whole container pack. This quick and easy change will ensure continued suppressed fire mission capabilities by the individual using the weapon. Knight's Armament Company has also added a reliable slide-locking bar to the Beretta frame as an option, with only minor modifications that do not interfere with the normal handling and operation of the pistol. The slide-lock feature easily and ambidextrously locks the slide when firing in the suppressed mode to eliminate the mechanical noise of the slide opening and closing. Leaving the slide unlocked will allow the normal semi-automatic functioning. Sound readings in the locked slide mode are very quiet by 9 mm standards.

The pistol barrel is modified to accept the 'Snap-on' system, which includes convenient combat-style sights on the suppressor itself.

Data

Weight: 170 g
Length: 152 mm; total length of pistol and suppressor, 350 mm
Diameter: 38 mm
Sound suppression: 32 dB drop
Life: 30 to 40 rds before replacing wipes

Manufacturer

Knight's Armament Company, 7750 9th Street SW, Vero Beach, Florida 32968.

Status

Available.

Service

US Army and Air Force.

VERIFIED

Snap-on suppressor detached from pistol

Snap-on suppressor attached to Beretta 92F pistol

ADDENDA

PISTOLS

COMMONWEALTH OF INDEPENDENT STATES

9 mm PMM self-loading pistol

Development

The 9 mm PMM self-loading pistol is an updated version of the PM Makarov pistol intended to fire the 9 mm 57-N-181SM cartridge which has the same overall dimensions as the standard 9 × 18 mm Makarov cartridge but has greater muzzle velocity, penetration and stopping power. The conversion of the Makarov PM pistol to accommodate the new cartridge was undertaken at the Izhevsk Mechanical Plant, the designers being B Pletsky and R Shogapov. Mass production of the PMM commenced in January 1994.

Description

At first sight the 9 mm PMM self-loading pistol is visually similar to the Makarov PM although the butt is reconfigured to assist the firer's grip as it is 6.5 mm wider to accommodate a 12-round magazine. With the 12-round magazine the pistol is known as the PMM-12 as there is a variant, the PMM-8, which retains the original eight-round magazine.

The overall blowback action of the PM is apparently retained although the chamber now has three spiral grooves in its surface to ensure the pistol will function when firing the standard PM 9 × 18 mm 57-N-181S cartridge.

The new 57-N-181SM bullet has a conical outline with a flat nose. It can penetrate 3.5 mm of steel at 10 m and 3 mm at 20 m. Hollow point and tracer cartridges are under development.

Data

Cartridge: 9 × 18 mm 57-N-181SM
Operation: blowback, self-loading, double action
Feed: PMM-12, 12-round box magazine; PMM-8, 8-round box magazine
Weight: empty, 760 g
Length: 167 mm
Barrel: 93.5 mm
Sights: fore, blade; rear, fixed notch
Muzzle velocity: 410-435 m/s
Muzzle energy: 505 J
Recoil energy: 5.4 J

Manufacturer

IZHMASH, 3 Derjabin Street, 426006 Izhevsk, Russia.

Status

In production.

NEW ENTRY

ROMANIA

Romtehnica 9 mm self-loading pistol

Description

This 9 mm pistol is apparently intended for the export market as it fires the 9 × 19 mm Parabellum NATO cartridge. It appears to be an entirely conventional design based on standard Browning principles, with a box magazine holding 15 cartridges. It has been seen carrying a tactical light or a laser pointer.

Data

Cartridge: 9 × 19 mm Parabellum
Operation: recoil, semi-automatic
Locking: projecting lug
Feed: 15-round box magazine
Weight: 1.08 kg
Length: 208 mm
Barrel: 112 mm
Sights: fore, blade; rear, fixed notch
Muzzle velocity: 384 m/s

Manufacturer

Romtehnica, 9-11 Drumul Taberei Street, 6 Bucharest.

Status

In production.

NEW ENTRY

Romtehnica 9 mm self-loading pistol
1996

RIFLES

SOUTH AFRICA

AEROTEK 20 mm NTW 20 rifle

Description

The AEROTEK 20 mm NTW 20 rifle is an anti-materiel weapon with other possible applications, including Explosive Ordnance Disposal (EOD). It fires the 20 × 82 mm MG151 cartridge, the same round as that fired by the 20 mm GA1 cannon produced in South Africa. Ammunition natures fired by the NTW 20 include HEI, HEI-T and SAPHEI. Muzzle velocity is 720 m/s and the operational range is given as more than 1,200 m. Rounds are fed into the rifle from a three-round box magazine protruding from the left-hand side of the receiver. Spent cases eject to the right as the bolt is withdrawn.

The NTW 20 is a manual bolt-action rifle with the bolt face employing six locking lugs. Recoil is limited by a large double baffle muzzle brake and a combined hydraulic/pneumatic damping and buffer system. Recoiling parts slide within a chassis frame which includes the butt assembly, so the firer can assume a conventional firing position with the butt held into the shoulder by a grip under the frame and a pistol grip for the adjustable pull trigger assembly. The recoiling parts and the barrel can be removed from the chassis frame for maintenance and also for carrying.

AEROTEK 20 mm NTW 20 rifle
1996

The NTW 20 may be carried on two special backpack harnesses. One harness has a collapsible bipod built into the frame and comes with two groundsheets attached to the structure. One of these groundsheets can be folded out under the barrel to suppress the dust signature produced on firing. The other groundsheet has a light sponge mattress sewn into it and unrolls for the user to lie on in the firing position.

An adjustable-height bipod is attached to the rifle for when the backpack frame is not in use. The bipod is mounted on the non-recoiling chassis frame, as is a long eye relief telescopic sight.

Data

Cartridge: 20 × 82 mm MG151
Operation: manual bolt action
Feed: 3-round box magazine
Weight: 26 kg
Length: 1.795 m
Barrel: 1 m
Sights: telescopic, long eye relief
Muzzle velocity: 720 m/s
Operational range: >1,200 m

Manufacturer

AEROTEK, PO Box 395, Pretoria 0001.

Status

Available.

NEW ENTRY

MORTAR FIRE CONTROL

SWITZERLAND

Leica SG12 digital goniometer

Description

The Leica SG12 digital goniometer is a compact, portable, tripod-mounted electro-optical instrument for military survey and alignment tasks. An integral digital magnetic compass with tilt sensors provides autonomous orientation, vertical angle adjustment and eliminates the need to set the tripod horizontal. Menu-guided operation for target determination in UTM grid co-ordinates is included in the basic software. Survey programmes such as 'Azimuth by astronomical observations' or 'Traversing' are available as options.

An RS-232 or -422 interface allows the input of station co-ordinates from a GPS receiver or data transfer to a fire control system.

There are two variants of the SG12. The SG12F is for forward observers and mortar fire controllers, with rapid pointing using a swivel lever. The SG12S is for surveying and gunlaying, with accurate pointing by horizontal and vertical drive knobs.

A recommended adjunct for the SG12S is the Leica MRF2000 rangefinder. The SG12F can be used with one or two accessories such as a laser rangefinder, a thermal imager, a telescope and a camera.

Data

Weight: 5.3 kg
Dimensions: 240 × 220 × 240 mm

Status

Production scheduled to commence during first half of 1996.

Manufacturer

Leica AG, Defence & Special Projects, CH-9435 Heerbrugg.

NEW ENTRY

Leica SG12S digital goniometer with Leica MRF2000 rangefinder

1996

Leica SG12S digital goniometer with laser rangefinder

1996

National Inventories

In this section the infantry weapons believed to be in service with the military and paramilitary forces of various countries are listed under the country headings. Readers having information which will modify or complete any entry are invited to communicate with the Editor.

Pistols	Sub-machine guns	Rifles	Machine guns	Close support weapons	Mortars	Anti-tank
Afghanistan						
7.62 mm Tokarev 82 mm RCL B-10	7.62 mm PPSh41 5.45 mm AK-74	5.45 mm AK-74 7.65 mm CZ vz-61 7.62 mm AK-47, AKM	7.62 mm RPD, RPK 7.62 mm Simonov SKS	30 mm AGS-17 12.7 mm DShK	82 mm M37 120 mm M1943 160 mm M1943	73 mm SPG-9 107 mm M38 Snapper ATGW
Albania						
7.62 mm Tokarev	7.62 mm PPD-40 7.62 mm PPS-41	7.62 mm M44 carbine 7.62 mm Simonov SKS 7.62 mm AK-47, AKM	7.62 mm RP-46 7.62 mm RPD, RPK 12.7 mm DShK 14.5 mm KPV		82 mm M37 120 mm M43 160 mm M43	82 mm RCL T21 14.5 mm PTRS rifle
Algeria						
9 mm MAC Mle 50 9 mm Tokagypt	9 mm Uzi 9 mm Carl Gustaf 45	7.62 mm AK-47, AKM 7.62 mm Simonov SKS 7.62 mm Dragunov SVD 7.62 mm Beretta BM59	7.62 mm RP-46 7.62 mm RPD 12.7 mm DShK		120 mm M43 160 mm M43	Sagger ATGW Snapper ATGW Spigot ATGW Spandrel ATGW Spiral ATGW
Angola						
7.62 mm Tokarev 9 mm Stechkin 9 mm Makarov	7.65 mm Skorpion 9 mm Star Z-45 9 mm Uzi 9 mm FMBP-48	7.62 mm AK-47, AKM 7.62 mm Simonov SKS 7.62 mm FN-FAL 7.63 mm G3	7.62 mm RP-46 7.62 mm RPD 7.62 mm vz/52 12.7 mm DShK	30 mm AGS-17	82 mm M43 120 mm M43	82 mm RCL B-10 107 mm RCL B-11 RPG-2 RPG-7 Sagger ATGW
Argentina						
9 mm FN 35	9 mm FMK3 9 mm Sterling Mk 4 9 mm H&K MP5 0.45 M3A1	5.56 mm FARA-83 7.62 mm FN-FAL 7.62 mm Beretta BM59 7.62 mm Steyr SSG69	7.62 mm FN MAG 0.50 Browning M2HB	40 mm FMAP 40 mm HK69	81 mm Brandt 120 mm Brandt	75 mm RCL M20 89 mm RL M65 90 mm RCL M67 105 mm RCL M1974 SS11, SS12 ATGW CIBEL-2K ATGW
Australia						
9 mm FN 35	9 mm F1	5.56 mm F88 (AUG) 7.62 mm FN-FAL 7.62 mm Parker-Hale 82	5.56 mm F89 (Minimi) 7.62 mm Bren L4 7.62 mm M60 7.62 mm FN MAG 0.50 Browning M2HB	40 mm M79 40 mm M203	81 mm F2 (L16)	66 mm LAW 84 mm Carl-Gustaf 106 mm RCL M40 MILAN ATGW
Austria						
9 mm Glock 17 9 mm Walther P-38	9 mm Steyr MPi69	5.56 mm Steyr AUG 7.62 mm FN-FAL 7.62 mm Steyr SSG69	5.56 mm Steyr AUG/HBAR 7.62 mm MG 42/59 7.62 mm M74 0.50 Browning M2HB	40 mm Steyr M203	60 mm Bohler 81 mm Bohler 81 mm M1, M29 81 mm L16 120 mm Bohler 120 mm M43	66 mm LAW 74 mm Miniman 84 mm Carl-Gustaf 106 mm RCL M40 Panzerfaust 3 RBS 56 BILL ATGW
Bahrain						
9 mm FN 35	9 mm Beretta M12 9 mm H&K MP5 9 mm Sterling Mk 4	7.62 mm Beretta BM59 7.62 mm FN-FAL 7.62 mm G3	7.62 mm FN MAG		81 mm L16	106 mm M40A1 120 mm RCL Mobat TOW ATGW
Bangladesh						
9 mm FN 35	9 mm Sterling L2	7.62 mm FN-FAL 7.62 mm G3A2/3 7.62 mm AK-47, AKM 7.62 mm Simonov SKS	7.62 mm Bren L4 7.62 mm RPD 7.62 mm HK11A1 7.62 mm HK21A1 12.7 mm DShK		82 mm Type 53 120 mm Type 53	106 mm RCL M40
Barbados						
9 mm FN 35 0.38 S&W	9 mm Sterling	7.62 mm FN-FAL	7.62 mm Bren L4 7.62 mm FN MAG		3 in UK	
Belgium						
9 mm FN 35	9 mm Uzi 9 mm H&K MP5	5.56 mm FN-FNC 7.62 mm FN-FAL	5.56 mm Minimi 7.62 mm FN MAG 0.50 Browning M2HB		60 mm NR493 81 mm NR475A1 107 mm M30 120 mm Brandt	66 mm LAW Swingfire ATGW MILAN ATGW
Belize						
9 mm FN 35	9 mm Sterling	5.56 mm M16A1 7.62 mm FN-FAL	7.62 mm Bren L4 7.62 mm FN MAG		81 mm L16	84 mm Carl Gustaf

Pistols	Sub-machine guns	Rifles	Machine guns	Close support weapons	Mortars	Anti-tank
Benin						
7.62 mm Tokarev	9 mm MAT-49	7.5 mm MAS 49/56 7.62 mm AK-47, AKM 7.62 mm Simonov SKS	7.5 mm M24/29 7.5 mm AAT-52 7.62 mm RP-46 7.62 mm RPD 0.50 Browning M2HB 14.5 mm KPV		60 mm Brandt 81 mm MO-81-61	RPG-7
Bolivia						
9 mm FN 35 0.45 M1911A1	9 mm Uzi 9 mm MAT-49 0.45 M3A1	5.56 mm Steyr AUG 5.56 mm Galil 7.62 mm SIG 510-4 7.62 mm FN-FAL 7.62 mm G3	7.62 mm SIG 710-3 7.62 mm FN MAG 7.62 mm M60 0.50 Browning M2HB		60 mm Brandt 81 mm M29 107 mm M30	106 mm RCL M40A1
Botswana						
9 mm FN 35	9 mm Sten 9 mm Sterling	5.56 mm Galil 7.62 mm AK-47, AKM 7.62 mm FN-FAL	7.62 mm Bren L4 7.62 mm FN MAG 7.62 mm SGM		81 mm L16 120 mm M43	84 mm Carl Gustaf RPG-7 TOW ATGW
Brazil						
9 mm Beretta M92 9 mm IMBEL	9 mm Beretta M12S 9 mm Walther MPK 9 mm Mekanika	5.56 mm M16A1 5.56 mm HK33E 7.62 mm FN-FAL	7.62 mm Madsen 7.62 mm FN MAG 0.50 Browning M2HB	40 mm M79	60 mm IMBEL 81 mm IMBEL 81 mm M29 81 mm L16 107 mm M30 120 mm Brandt(?)	3.5 in RL M20 57 mm RCL M18 75 mm RCL M20 106 mm RCL M40
Brunei						
9 mm FN 35	5.56 mm Colt XM177 9 mm Sterling	5.56 mm M16A1 7.62 mm G3	5.56 mm Colt M16A1/HB 7.62 mm SIG 710-3 7.62 mm FN MAG 7.62 mm HK21A1	40 mm M203	81 mm Tampella	SS-11 ATGW
Bulgaria						
7.62 mm Tokarev 9 mm Makarov		7.62 mm AK-47, AKM 7.62 mm Simonov SKS	7.62 mm RP-46 7.62 mm RPD, RPK 12.7 mm DShK 14.5 mm KPV	30 mm AGS-17	82 mm M43 120 mm M43 160 mm M43	82 mm RCL B-10 Snapper ATGW Sagger ATGW Spiral ATGW
Burkina Faso						
7.65 mm Walther PP 9 mm MAB PA-15 0.38 Manurhin MR73	7.65 mm MAS-38 9 mm MAT-49 9 mm Beretta M12	5.56 mm SIG 540 5.56 mm AR70/.223 7.5 mm MAS-49/56 7.62 mm G3	7.5 mm M24/29 7.5 mm AAT-52 7.62 mm FN MAG 0.50 Browning M2HB		60 mm Brandt 81 mm Brandt	75 mm RCL Type 52 106 mm RCL M40 84 mm Carl Gustaf
Burma (Myanmar)						
9 mm FN 35	9 mm Sterling	5.56 mm M16A1 7.62 mm FN-FAL 7.62 mm G3	7.62 mm FN MAG 7.62 mm MG3 0.50 Browning M2HB	40 mm M79 40 mm M203	81 mm M29 82 mm M43 120 mm Tampella	
Burundi						
7.65 mm Browning 1910 9 mm FN 35	9 mm Vigneron 9 mm MAT-49	7.5 mm MAS 49/56 7.62 mm FN-FAL 7.62 mm G3A3, A4	7.62 mm FN MAG 00.30 Browning M1919A4 0.50 Browning M2HB		82 mm M43	75 mm RCL Type 52
Cambodia						
7.62 mm Tokarev 9 mm FN 35	9 mm vz/23, 25	5.56 mm M16A1 7.62 mm FN-FAL 7.62 mm AK-47, AKM	7.62 mm M60 7.62 mm RPD 7.62 mm Type 59 12.7 mm DShK 14.5 mm KPV		82 mm M43 120 mm Brandt 160 mm M43	82 mm RCL B-10 106 mm RCL M40 107 mm RCL B-11
Cameroon						
9 mm FN 35 9 mm MAC Mle 50 0.357 Manurhin MR73	9 mm H&K MP5	5.56 mm SIG 540 5.56 mm Steyr AUG 5.56 mm M16A1 7.62 mm FN-FAL	7.62 mm HK21 0.50 Browning M2HB 14.5 mm KPV	40 mm M203	60 mm Brandt 81 mm Brandt 81 mm L16 120 mm SB Brandt 120 mm rifled Brandt	106 mm RCL M40 MILAN ATGW HOT ATGW TOW ATGW
Canada						
9 mm FN 35	9 mm C1 9 mm H&K MP5A2	5.56 mm C7 5.56 mm C8 7.62 mm FN-FAL 7.62 mm Parker-Hale	5.56 mm C9 7.62 mm FN-FAL/HB 7.62 mm FN MAG 7.62 mm Browning C1 0.50 Browning M2HB		60 mm M19 81 mm C3 107 mm M30 120 mm Brandt	66 mm LAW 84 mm Carl Gustaf 106 mm RCL M40 TOW ATGW Eryx ATGW
Cape Verde						
	9 mm FMBP48	7.62 mm AK-47, AKM 7.62 mm Simonov SKS	7.62 mm RPD 7.62 mm PK, RPK 7.62 mm MG3 7.62 mm SGM 12.7 mm DShK		82 mm M41 120 mm M43	RPG-7
Central African Republic						
9 mm Walther PP 9 mm MAC Mle 50 0.38 Manurhin MR73	9 mm MAT-49 9 mm Uzi	5.56 mm SIG 541 5.56 mm M16A1 7.5 mm MAS 49/56 7.62 mm AK-47, AKM	7.5 mm M24/29 7.5 mm AAT-42 7.62 mm RP-46 7.62 mm RPD, RPK 12.7 mm DShK		60 mm Brandt 81 mm Brandt 120 mm M43	RPG-7 106 mm RCL M40

Pistols	Sub-machine guns	Rifles	Machine guns	Close support weapons	Mortars	Anti-tank
Chad						
7.65 mm Walther PP 9 mm MAC Mle 50 9 mm Tokagypt 9 mm Walther P1 9 mm FN 35 0.38 Manurhin MR73	9 mm MAT-49 9 mm Beretta M12 9 mm Uzi	5.56 mm M16A1 7.62 mm FN-FAL 7.62 mm SIG 542 7.62 mm G3 7.62 mm AK-47, AKM	7.62 mm AAT-52 7.62 mm PK, RPK 7.62 mm RPD 7.62 mm SGM 7.62 mm Yugo M70 12.7 mm DShK 0.50 Browning M2HB	30 mm AGS-17 40 mm M79	60 mm Brandt 81 mm Brandt 82 mm M43 120 mm Brandt	RPG-7 82 mm RCL B-10 106 mm RCL M40 Eryx ATGW TOW ATGW
Chile						
9 mm FN 35 9 mm Beretta M92 9 mm SIG P220 9 mm Walther P1 0.38 FAMAE	9 mm Madsen M53 9 mm Uzi 9 mm SAF 9 mm H&K MP5 9 mm Beretta M12S	5.56 mm Galil 5.56 mm HK33 5.56 mm M16A1 7.62 mm FN-FAL 7.62 mm G3 7.62 mm SIG 542 7.62 mm SIG 510	7.62 mm SIG 710-3 7.62 mm FN-FAL/HB 7.62 mm MG42/59 7.62 mm M60 0.50 Browning M2HB		81 mm M29 120 mm ECIA	3.5 in RL M20 106 mm RCL M40 MILAN ATGW
China, People's Republic						
7.62 mm Types 51, 54 7.62 mm Type 64 9 mm Type 59	7.62 mm Type 43 7.62 mm Type 50 7.62 mm Type 64 7.62 mm Type 68	7.62 mm Carbine Type 53 7.62 mm Type 56 7.62 mm Type 67	7.62 mm Type 53 7.62 mm Type 56 7.62 mm Type 58 12.7 mm Type 57 12.7 mm Type 77	30 mm AGS-17 35 mm W87	60 mm Type 31 75 mm RCL Types 52, 56 82 mm Type 53 120 mm Type 53 160 mm M43	40 mm RL Type 56 40 mm RL Type 69 57 mm RCL Type 36 82 mm RCL Type 65 90 mm RL Type 51 Red Arrow 8 ATGW
Colombia						
9 mm FN 35 0.45 M1911A1	9 mm Madsen M46, 50, 53 9 mm Walther MPK 9 mm Ingram 9 mm Uzi	5.56 mm Galil 7.62 mm FN-FAL 7.62 mm G3 7.62 mm M14	7.62 mm M60 7.62 mm HK21E1 7.62 mm FN MAG 00.30 BAR M1918 00.30 Browning M1919A4 0.50 Browning M2HB	40 mm M79	60 mm M19 81 mm M1 107 mm M30 120 mm Brandt	75 mm RCL M20 106 mm RCL M40 TOW ATGW
Commonwealth of Independent States						
5.45 mm PSM 7.62 mm Tokarev 9 mm PM 9 mm PMM	5.45 mm AKSU-74	5.45 mm AK-74 7.62 mm AKM 7.62 mm Dragunov 7.62 mm SKS 12.7 mm V-94	5.45 mm RPK-74 7.62 mm RPK, RPKS 7.62 mm PK, PKB, PKS 12.7 mm DShK, NSV 14.5 mm KPV	30 mm AGS-17 40 mm GB-25	82 mm 2B14 82 mm 2B9 82 mm M37M 107 mm M107 120 mm M43 120 mm 2B11 160 mm M160	RPG-7V, 7D 73 mm RPG-18 73 mm RCL SPG-9 Swatter ATGW Sagger ATGW Spiral ATGW Spigot ATGW Spandrel ATGW
Comoros						
	9 mm MAT-49	7.5 mm MAS 49/56 7.62 mm AK-47, AKM 7.62 mm SKS	7.5 mm AAT-52 7.62 mm RPD, RPK 7.62 mm Type 58 7.62 mm RP-46 7.62 mm SGM			
Congo						
7.62 mm Tokarev 7.65 mm Walther PP 9 mm MAC Mle 50	9 mm Vigneron 9 mm MAT-49 9 mm Franchi LF-57 9 mm Sola Super	7.5 mm MAS 49/56 7.62 mm FN-FAL 7.62 mm CETME 58 7.62 mm AK-47, AKM	7.5 mm M24/29 7.62 mm RP-46 7.62 mm RPD, RPK 12.7 mm DShK 14.5 mm KPV		82 mm M41, 43 120 mm M43	57 mm RCL M18 RPG-7 Sagger ATGW
Costa Rica						
0.45 M1911A1	9 mm Beretta 38/49	5.56 mm Galil 5.56 mm M16A1 5.56 mm Type 68 7.62 mm M14 7.62 mm FN-FAL	7.62 mm M60 00.30 BAR M1918 00.30 Browning M1919A4	40 mm M79		
Cuba						
9 mm Makarov 9 mm FN 35	9 mm vz/23, 25 9 mm Star Z-45	7.62 mm vz/58 7.62 mm AK-47, AKM	7.62 mm PK, RPK 7.62 mm RP-46 7.62 mm DPM 12.7 mm DShK	30 mm AGS-17	82 mm M41, 43 120 mm M43 160 mm M43	RPG-7 Sagger ATGW Snapper ATGW
Cyprus, Republic of						
9 mm FN 35	9 mm Sterling	7.62 mm G3 7.62 mm FN-FAL 7.62 mm vz/58	7.62 mm FN MAG 7.62 mm SGM 12.7 mm DShK		81 mm M29 82 mm M41, 43 107 mm M2	57 mm RCL M18 106 mm RCL M40 MILAN ATGW HOT ATGW
Czech Republic						
7.62 mm CZ 52 7.65 mm CZ 83	7.62 mm Skorpion	5.45 mm AK74 7.62 mm vz/58	7.62 mm vz/59 12.7 mm DShK 12.7 mm NSV		81 mm M48 81 mm M52 120 mm M43	RPG-7V RPG-75 82 mm RCL M59 Snapper ATGW Swatter ATGW Sagger ATGW Spiral ATGW

Pistols	Sub-machine guns	Rifles	Machine guns	Close support weapons	Mortars	Anti-tank
Denmark						
9 mm FN 35 9 mm SIG P210	9 mm Hovea M49 9 mm H&K MP5A3	5.56 mm C7,C7A1 5.56 mm M16A1 7.62 mm G3	7.62 mm MG42/59 0.50 Browning M2HB		60 mm M51 81 mm M57 120 mm M50	84 mm Carl Gustaf 106 mm RCL M40 TOW ATGW M72 LAW
Djibouti						
9 mm MAC Mle 50 9 mm MAB PA-15	9 mm MAT-49	5.56 mm SIG 540 5.56 mm FAMAS 7.62 mm G3 7.62 mm CETME 58 7.62 mm FN-FAL	7.62 mm AAT-52 7.62 mm RPD, RPK 7.62 mm FN MAG 0.50 Browning M2HB		60 mm Brandt 81 mm Brandt 120 mm Brandt	RPG-7 106 mm RCL M40A1 HOT ATGW
Dominican Republic						
9 mm FB 35 0.45 M1911A1	9 mm Beretta M38/49 9 mm Uzi	7.62 mm G3 7.62 mm FN-FAL 7.62 mm CETME 58 7.62 mm M14	7.62 mm M60 7.62 mm FN MAG 00.30 Browning M1919A4 0.50 Browning M2HB	40 mm M79	81 mm M1 120 mm ECIA SL	106 mm RCL M40A1
Ecuador						
9 mm FN 35 0.45 M1911A1	9 mm Uzi 0.45 M3A1	5.56 mm Steyr AUG 5.56 mm SIG 540 5.56 mm M16A1 5.56 mm HK33 7.62 mm M14 7.62 mm FN-FAL	7.62 mm FN MAG 7.62 mm Yugo MG42 0.50 Browning M2HB	40 mm M203	81 mm M29 107 mm M30 160 mm Soltam	90 mm RCL M67 106 mm RCL M40A1
Egypt						
9 mm Helwan 9 mm Beretta M951	9 mm Beretta 12S 9 mm Star Z-45 9mm Port Said	7.62 mm SKS 7.62 mm AK-47, AKM 7.62 mm SVD 7.92 mm Hakim	7.62 mm RPD 7.62 mm SGM 7.62 mm FN MAG 7.62 mm M60 12.7 mm DShK 0.50 Browning M2HB		60 mm LM 82 mm M37, 43 120 mm M43 120 mm UK2 160 mm M43	RPG-7V 107 mm RCL B-11 Snapper ATGW Sagger ATGW Swingfire ATGW MILAN ATGW HOT ATGW TOW ATGW
El Salvador						
9 mm FN 35	9 mm H&K MP5 9 mm Uzi	5.56 mm HK33 5.56 mm M16A1 5.56 mm Galil 7.62 mm M14 7.62 mm G3	7.62 mm M60 7.62 mm Madsen 00.30 Browning M1919A4 0.50 Browning M2HB	40 mm M79 40 mm M203	60 mm M19 81 mm M29 120 mm UBM-52	90 mm RCL M67 3.5 in RL M20 75 mm RCL M20
Equatorial Guinea						
7.62 mm Tokarev	9 mm MAT-49	7.62 mm CETME 58 7.62 mm AK-47, AKM 7.62 mm SKS	7.62 mm RPD, RPK 7.62 mm SGM 7.62 mm Type 67 7.62 mm RP-46 12.7 mm DShK		82 mm M37	
Ethiopia						
7.62 mm vz/52 9 mm Beretta M34 9 mm Makarov	9 mm Beretta M38/49 9 mm Uzi	7.62 mm Type 56 7.62 mm Beretta BM59 7.62 mm vz/58 7.62 mm M14	7.62 mm RP-46 7.62 mm RPD, RPK 7.62 mm vz/52 00.30 BAR M1918 12.7 mm DShK 0.50 Browning M2HB	40 mm M79	60 mm M19 81 mm M29 82 mm M43 107 mm M30 120 mm M43	Sagger ATGW Spigot ATGW Spandrel ATGW TOW ATGW
Fiji						
0.45 M1911A1		5.56 mm M16A1 7.62 mm FN-FAL	7.62 mm M60	40 mm M79	81 mm M29	
Finland						
7.65 mm Parabellum 9 mm Lahti M35 9 mm FN 35	9 mm Suomi M31 9 mm Suomi M44	7.62 mm M62, M76	7.62 mm M62 7.62 mm RPD 7.62 mm SGM 12.7 mm DShK, NSV		60 mm Tampella 81 mm Tampella 120 mm Tampella 120 mm M43 160 mm M43	M55 RL 66 LAW M72 SS-11 ATGW 95 mm RCL SM58-61 TOW ATGW Apilas
France						
9 mm MAB PA-15 9 mm SIG P220 9 mm MAC Mle 50 9 mm Beretta 92G	9 mm MAT-49 9 mm H&K MP5	5.56 mm FAMAS 5.56 mm SG540 7.62 mm FR-F1, F2 0.50 Barrett	7.62 mm AAT-52 0.50 Browning M2HB		60 mm Brandt 81 mm Brandt 120 mm Brandt	Apilas MILAN ATGW HOT ATGW Eryx ATGW
Gabon						
9 mm MAC Mle 50 9 mm MAB PA-15 0.38 Manurhin MR73	9 mm Beretta M12 9 mm MAT-49 9 mm Sterling 9 mm Uzi	5.56 mm FN-CAL 5.56 mm FAMAS 5.56 mm M16A1 7.62 mm SIG 540 7.62 mm G3 7.62 mm AK-47, AKM	7.62 mm AAT-52 7.62 mm FN MAG 00.30 Browning M1919A4 0.50 Browning M2HB	40 mm M203	81 mm Brandt 120 mm Brandt	106 mm RCL M40 MILAN ATGW

Pistols	Sub-machine guns	Rifles	Machine guns	Close support weapons	Mortars	Anti-tank
Gambia						
0.38 Webley	9 mm Sterling	7.62 mm FN-FAL 7.62 mm FN MAG 0.50 Browning M2HB	7.62 mm Bren L4			
Germany						
9 mm Pistole 1 9 mm Pistole 6 9 mm SIG P226	5.56 mm H&K MP53 9 mm Uzi 9 mm H&K MP5	7.62 mm G3 7.62 mm G3SG1 7.62 mm Mauser SP66	7.62 mm MG3	40 mm Granatpistole 40 mm HK79	81 mm M37M 120 mm M43 120 mm Soltam 120 mm Brandt	RPG-7V, 7D 84 mm Carl Gustaf TOW ATGW MILAN ATGW HOT ATGW Panzerfaust 3
Ghana						
9 mm FN 35	9 mm Sterling 9 mm H&K MP5	5.56 mm M16A1 5.56 mm HK33 7.62 mm FN-FAL 7.62 mm G3	7.62 mm FN MAG 7.62 mm Bren L4 12.7 mm DShK 0.50 Browning M2HB		81 mm Tampella 120 mm Tampella	84 mm Carl Gustaf RPG-2
Greece						
9 mm EP9S 9 mm EP7	9 mm Steyr Mi69 9 mm H&K MP5	5.56 mm M16A2 5.56 mm HK33E 7.62 mm FN-FAL 7.62 mm G3	7.62 mm M60 7.62 mm HK11A1 7.62 mm FN MAG 0.50 Browning M2HB	40 mm M79 40 mm M203	60 mm M2 81 mm M1/M29 107 mm M30	90 mm RCL M67 106 mm RCL M40 TOW ATGW MILAN ATGW
Guatemala						
9 mm Star 9mm FN 35 0.45 M1911A1	9 mm Uzi 9 mm Beretta 12 0.45 M3A1	5.56 mm M16A1 7.62 mm Galil 00.30 Carbine M1	7.62 mm FN MAG 00.30 Browning M1919A4 0.50 Browning M2HB	40 mm M79	60 mm M2 81 mm M1 107 mm M30 120 mm ECIA SL	3.5 in RL M20 106 mm RCL M40
Guinea						
7.62 mm Tokarev	7.62 mm PPSh-41 9 mm MAT-49 9 mm vz/23, 25	7.62 mm vz/58 7.62 mm AK-47, AKM 7.62 mm SKS	7.62 mm PK 7.62 mm SGM 12.7 mm DShK		82 mm M43 120 mm M43	RPG-7 82 mm RCL B-10 Sagger ATGW Swatter ATGW
Guinea-Bissau						
7.62 mm Tokarev 7.62 mm CZ 52	9 mm vz/23, 25 9 mm FMBP M948	7.62 mm AK-47, AKM 7.62 mm vz/52, 57 7.62 mm SKS	7.62 mm PK, RPK 7.62 mm SGM 12.7 mm DShK 14.5 mm KPV		82 mm M43 120 mm M43	3.5 in RL M20 75 mm RCL Type 52 82 mm RCL B-10 Sagger ATGW Spiral ATGW
Guyana						
7.65 mm Walther PPK 9 mm S & W M39	9 mm Sterling 9 mm Beretta M12	7.62 mm FN-FAL 7.62 mm G3 7.62 mm AK-47, AKM 7.62 mm SKS	7.62 mm Bren L4 7.62 mm FN MAG		81 mm L16 82 mm M43 120 mm M43	RPG-7
Haiti						
9 mm Beretta 951 0.45 M1911A1	9 mm Uzi, Mini-Uzi 0.45 Thompson	5.56 mm Galil 5.56 mm M16A1 7.62 mm G3 0.30 Garand M1	7.62 mm M60 00.30 Browning M1919A4		60 mm M2 81 mm M1	57 mm RCL M18 106 mm RCL M40
Honduras						
9 mm FN 35	9 mm Beretta 93R 9 mm Uzi, Mini-Uzi 9 mm H&K MP5	5.56 mm Mini-14 5.56 mm M16A1 7.62 mm FN-FAL 7.62 mm M14	7.62 mm M60 7.62 mm FN MAG 0.50 Browning M2HB	40 mm M79 40 mm M203	81 mm M1 120 mm Soltam 160 mm Soltam	84 mm Carl Gustaf 106 mm RCL M40
Hungary						
7.62 mm M48 7.65 mm M48	7.62 mm 48M	7.62 mm AMD 7.62 mm AK-47, AKM 7.62 mm M48	7.62 mm RPK 7.62 mm PK, PKB, PKS 12.7 mm DShK		82 mm M37M 82 mm 2B9 120 mm M43	RPG-7V, 7D 73 mm SPG-9 107 mm RCL B-11 Snapper ATGW Sagger ATGW Spigot ATGW
India						
9 mm FN 35 9 mm Glock	9 mm Sterling 9 mm H&K MP5K	7.62 mm FN-FAL 7.62 mm AKM	7.62 mm Bren L4 7.62 mm FN MAG 0.50 Browning M2HB		81 mm L16 82 mm M43 120 mm M43 160 mm M43	84 mm Carl Gustaf 106 mm RCL M40 MILAN ATGW Sagger ATGW Spigot ATGW
Indonesia						
9 mm Pindad 9 mm Beretta 92	9 mm Beretta M12	5.56 mm M16A1 5.56 mm SIG 541 5.56 mm FN-FNC 7.62 mm vz/52/57 7.62 mm FN-FAL 7.62 mm Beretta BM59	5.56 mm Minimi 7.62 mm FN MAG 7.62 mm M60 12.7 mm DShK 0.50 Browning M2HB	40 mm M79 40 mm M203 40 mm CIS-AGL	60 mm Pindad 60 mm M2 81 mm Pindad 81 mm M29 120 mm UBM-52	3.5 in RL M20 75 mm RCL BO-10 84 mm Carl Gustaf 105 mm RCL BO-11 106 mm RCL M40 Entac ATGW

Pistols	Sub-machine guns	Rifles	Machine guns	Close support weapons	Mortars	Anti-tank
Iran						
9 mm Beretta 92 9 mm SIG P220 0.45 M1911A1	9 mm Uzi 9 mm H&K MP5 9 mm Beretta M12	7.62 mm G3 7.62 mm AK-47, AKM 7.62 mm Dragunov SVD	7.62 mm MG1A1 7.62 mm PK, RPK 12.7 mm DShK 0.50 Browning M2HB	30 mm AGS-17 40 mm M79	37 mm LM 60 mm M19 81 mm M29 107 mm M30 120 mm Soltam M65	3.5 in RL M20 RPG-7V 57 mm RCL M18 Dragon ATGW Entac ATGW TOW ATGW Sagger ATGW
Iraq						
7.62 mm Tokarev 9 mm FN 35 9 mm CZ 75	9 mm Sterling	7.62 mm AK-47, AKM 7.62 mm SKS 7.62 mm SVD	7.62 mm FN MAG 7.62 mm RPD 7.62 mm SGM 12.7 mm DShK	30 mm AGS-17	81 mm M37 120 mm M43 160 mm M43	RPG-7 Sagger ATGW Spigot ATGW HOT ATGW MILAN ATGW
Ireland						
9 mm FN 35	9 mm FFV M45 9 mm Uzi 9 mm HK53	5.56 mm Steyr AUG 7.62 mm FN-FAL	7.62 mm FN MAG 0.50 Browning M2HB	40 mm M79	60 mm Brandt 81 mm Swedish 120 mm Swedish	84 mm Carl Gustaf 90 mm FFV RCL gun MILAN ATGW
Israel						
9 mm Beretta M951	9 mm Uzi, Mini-Uzi	5.56 mm Galil 5.56 mm M16A1 7.62 mm FN-FAL 7.62 mm M14 7.62 mm AK-47, AKM	5.56 mm Negev 7.62 mm FN-FAL/HB 7.62 mm FN MAG 0.30 Browning M1919A4 12.7 mm DShK 0.50 Browning M2HB	40 mm M79 40 mm M203 40 mm MK19	52 mm Ta'as 81 mm Soltam 120 mm Soltam M65 160 mm M66	84 mm Carl Gustaf 106 mm RCL M40 Dragon ATGW TOW ATGW Sagger ATGW MAPATS ATGW
Italy						
9 mm Beretta M34 9 mm Beretta M951 9 mm Beretta M92	9 mm Beretta M38/49 9 mm Beretta M12S 9 mm Franchi LF-57	5.56 mm AR70/90 7.62 mm BM59	7.62 mm MG42/59 7.62 mm M73 0.50 Browning M2HB		81 mm M62 120 mm Brandt	80 mm Folgore 106 mm RCL M40 TOW ATGW MILAN ATGW Apilas
Ivory Coast						
9 mm MAB PA-15 9 mm MAS Mle 50 0.357 Manurhin MR73	9 mm MAT-49	5.56 mm SIG 540 7.5 mm MAS 49/56 7.62 mm G3 7.62 mm FN 30-11	7.5 mm M24/29 7.5 mm AAT-52 0.50 Browning M2HB		81 mm Brandt 120 mm Brandt	106 mm RCL M40
Jamaica						
9 mm FN 35	9 mm Sterling	5.56 mm M16A1 5.56 mm L85A1 1W 7.62 mm FN-FAL	7.62 mm FN-FAL/HB 7.62 mm FN MAG 0.50 Browning M2HB		81 mm L16A1	
Japan						
9 mm SIG P220 0.38 New Nambu	9 mm H&K MP5 9 mm SCK M66	7.62 mm Type 64	7.62 mm Type 62 0.50 Browning M2HB		60 mm M1 81 mm M1 81 mm Type 64 107 mm M30 120 mm Brandt	84 mm Carl Gustaf 106 mm RCL M40 Type 64 ATGW TOW ATGW KAM-3D ATGW
Jordan						
9 mm FN 35 9 mm Glock	9 mm Sterling 9 mm H&K MP5K	5.45 mm AK-47 5.56 mm AR70/90 5.56 mm M16A1, A2 7.62 mm G3	7.62 mm FN MAG 7.62 mm HK21E 7.62 mm M60 0.50 Browning M2HB	40 mm M79 40 mm M203	81 mm M29 107 mm M30 120 mm Brandt	106 mm RCL M40 TOW ATGW Dragon ATGW HOT ATGW Apilas LAW-80
Kenya						
9 mm FN 35	9 mm Sterling 9 mm Uzi 9 mm H&K MP5	7.62 mm FN-FAL 7.62 mm G3	7.62 mm HK21A1 7.62 mm Bren L4 7.62 mm FN MAG 7.62 mm AAT-52	40 mm M79	81 mm L16 120 mm Brandt	84 mm Carl Gustaf MILAN ATGW Swingfire ATGW
Korea, North						
7.62 mm Type 68 7.65 mm Type 64	7.62 mm Type 49	7.62 mm Type 63 (SKS) 7.62 mm Type 58 (AK-47) 7.62 mm Type 68 (AKM)	7.62 mm Type 64 (RPK) 7.62 mm RPD 7.62 mm PK, PKB, PKS 12.7 mm DShK		82 mm M37 120 mm M43 160 mm M43	RPG-2 RPG-7V 82 mm SPG-82 82 mm RCL B-10 107 mm RCL B-11 Snapper ATGW Sagger ATGW
Korea, South						
9 mm DP51 0.45 M1911A1	0.45 M3A1	5.56 mm K2 5.56 mm M16A1 0.30 Carbine M1 0.30 Garand M1	5.56 mm K3 7.62 mm M60 7.62 mm FN MAG 0.30 Browning M1919A4 0.50 Browning M2HB	40 mm M79 40 mm M203 40 mm MK19	60 mm M2/KM19 81 mm KM29 107 mm M30	3.5 in RL M20 66 mm LAW M72 106 mm RCL M40 TOW ATGW
Kuwait						
9 mm FN 35	9 mm Sterling	5.56 mm M16A1 7.62 mm FN-FAL	7.62 mm FN MAG 0.50 Browning M2HB	40 mm M203	81 mm L16	TOW ATGW HOT ATGW

Pistols	Sub-machine guns	Rifles	Machine guns	Close support weapons	Mortars	Anti-tank
Laos 7.62 mm Tokarev 9 mm CZ 70 9 mm Makarov	7.62 mm PPSh-41 9 mm MAT-49	7.62 mm AK-47, AKM 7.62 mm Type 56 7.62 mm SKS	7.62 mm RP-46, RPD 12.7 mm DShK		81 mm M1 82 mm M43 107 mm M2A1 107 mm M1938 120 mm M43	RPG-7 106 mm RCL M40 107 mm RCL B-11
Lebanon 9 mm FN 35 9 mm Walther P-38 9 mm Colt Commander	9 mm MAT-49 9 mm Sterling	5.56 mm FAMAS 5.56 mm FN-CAL 5.56 mm M16A1 5.56 mm SIG 540 7.62 mm FN-FAL 7.62 mm G3	7.5 mm M24/29 7.62 mm AAT-52 7.62 mm FN MAG 0.50 Browning M2HB	40 mm M79 40 mm M203	60 mm Brandt 81 mm Brandt 81 mm M29 120 mm Brandt	89 mm RL M65 RPG-7 106 mm RCL M40 TOW ATGW MILAN ATGW
Lesotho 0.38 Enfield	9 mm Sterling	5.56 mm Beretta AR70/90 5.56 mm M16A1 5.56 mm Galil 7.62 mm AK-47, AKM	7.62 mm FN MAG 7.62 mm Bren L4 7.62 mm SGM		81 mm L16	
Liberia 9 mm FN 35 0.45 M1911A1	9 mm Uzi	5.56 mm M16A1, A2 7.62 mm FN-FAL 0.30 Carbine M1	5.56 mm M16/HB 7.62 mm FN-FAL/HB 7.62 mm SIG 710-3 7.62 mm M60 0.30 BAR M1918 0.30 Browning M1919A4 0.50 Browning M2HB	40 mm M203	60 mm M31 81 mm M29 107 mm M30	3.5 in RL M20 106 mm RCL M40
Libya 7.62 mm Tokarev 9 mm Beretta M951 9 mm Stechkin 9 mm Makarov	7.65 mm Skorpion 9 mm Beretta M12 9 mm Sterling L34	7.62 mm FN-FAL 9 mm Beretta M12 7.62 mm AK-47, AKM 7.62 mm SKS 7.62 mm G3 7.62 mm vz/58	7.62 mm RP-46 7.62 mm RPD, RPK 7.62 mm FN MAG 12.7 mm DShK		81 mm M1 82 mm M37 107 mm M30 120 mm M43 160 mm M43	RPG-7 84 mm Carl Gustaf 106 mm RCL M40 107 mm RCL B-11 Spigot ATGW Sagger ATGW TOW ATGW
Luxembourg 9 mm FN 35	9 mm Uzi 9 mm H&K MP5	7.62 mm FN-FAL	7.62 mm FN MAG 0.50 Browning M2HB		81 mm M29	66 mm LAW M72 106 mm RCL M40 TOW ATGW
Madagascar 7.62 mm Tokarev 7.65 mm Walther PP 9 mm MAB PA-15	9 mm MAT-49	7.62 mm SKS 7.62 mm AK-47, AKM 7.62 mm Type 68	7.5 mm AAT-52 12.7 mm DShK 0.50 Browning M2HB		82 mm M37 120 mm M43	RPG-7 106 mm RCL M40
Malawi 9 mm FN 35	9 mm Sterling	7.62 mm FN-FAL 7.62 mm G3 7.62 mm CETME L	7.62 mm FN MAG 14.5 mm KPV		81 mm L16	3.5 in RL M20
Malaysia 9 mm FN 35 9 mm H&K P9S	9 mm Sterling	5.56 mm M16A1 5.56 mm AR70/90 5.56 mm HK33E 7.62 mm FN-FAL 7.62 mm G3SG1	5.56 mm AS70/90 7.62 mm Bren L4 7.62 mm FN MAG 7.62 mm HK11A1 7.62 mm HK21E 0.50 Browning M2HB	40 mm M79 40 mm M203	81 mm L16	84 mm Carl Gustaf 3.5 in RL M20 120 mm RCL Mobat 106 mm RCL M40 SS-11 ATGW
Mali 7.65 mm Walther PP	9 mm MAT-49	7.5 mm MAS 49/56 7.62 mm AK-47, AKM	7.5 mm AAT-52 7.62 mm PK, RPK 7.62 mm SGM 12.7 mm DShK		82 mm M43 120 mm M43	RPG-2 RPG-7 Sagger ATGW Swatter ATGW
Malta 7.62 mm Tokarev 9 mm Beretta M34 9 mm Makarov	9 mm H&K MP5K 9 mm Sterling 9 mm Uzi	7.62 mm FN-FAL 7.62 mm Type 56	7.62 mm RPD, RPK 12.7 mm DShK 14.5 mm KPV			RPG-7
Mauritania 7.62 mm Tokarev 9 mm MAC Mle 50 9 mm MAB PA-15	9 mm MAT-49 9 mm Star Z-45	7.5 mm MAS 49/56 7.62 mm FR-F1 0.30 Carbine M1	7.5 mm AAT-52 7.62 mm FN MAG 7.92 mm MG42 0.30 Browning M1919A4 0.50 Browning M2HB		60 mm Brandt 81 mm Brandt 120 mm Brandt 120 mm ECIA SL	RPG-7 57 mm RCL M18 75 mm RCL M20 106 mm RCL M40 MILAN ATGW
Mauritius 7.65 mm Walther PP 0.38 Manurhin MR73	9 mm H&K MP5	5.56 mm SIG 540	7.5 mm AA-52 7.62 mm Bren L4		81 mm Brandt	

Pistols	Sub-machine guns	Rifles	Machine guns	Close support weapons	Mortars	Anti-tank
Mexico						
9 mm H&K P7M13	9 mm Mendoza	5.56 mm M16A1	5.56 mm Ameli		60 mm M2	106 mm RCL M40
0.45 M1911A1	9 mm HK53	5.56 mm HK33E	7.62 mm FN MAG		81 mm M1	MILAN ATGW
	9 mm H&K MP5	7.62 mm FN-FAL	7.62 mm HK21A1		107 mm M30	
		7.62 mm G3	0.50 Browning M2HB		120 mm Brandt	
Morocco						
7.65 mm HK4	9 mm Beretta 38/49	5.56 mm Beretta 70/223	7.62 mm AAT-52		60 mm M2	66 mm M72 LAW
9 mm HK VP70	9 mm MAT-49	5.56 mm M16A1	7.62 mm RPD		81 mm ECIA LN	RPG-7V
9 mm MAC Mle 50	9 mm H&K MP5	5.56 mm Steyr AUG	7.62 mm M60		82 mm M37	3.5 in RL M20
9 mm MAB PA-15		7.62 mm AK-47, AKM	0.50 Browning M2HB		120 mm M43	106 mm RCL M40
		7.62 mm FN-FAL			120 mm Brandt	Dragon ATGW
		7.62 mm G3			120 mm ECIA	MILAN ATGW
		7.62 mm Valmet M76				TOW ATGW
		7.62 mm Beretta BM59				HOT ATGW
Mozambique						
7.62 mm Tokarev	9 mm Star Z-75	7.62 mm AK-47, AKM	7.62 mm PK, RPK	30 mm AGS-17	82 mm M43	82 mm RCL B-10
9 mm Stechkin	9 mm FMBP M48	7.62 mm vz/58	7.62 mm SGM		120 mm M43	107 mm RCL B-12
9 mm Makarov	9 mm FMBP M63	7.62 mm FN-FAL	7.92 mm vz/53			Sagger ATGW
9 mm FN 35	9 mm vz/23, 25	7.62 mm SKS	12.7 mm DShK			Spigot ATGW
9 mm Walther P-38	9 mm Franchi LF-57					
Nepal						
9 mm FN 35	9 mm Sterling	7.62 mm FN-FAL	7.62 mm Bren L4		81 mm M29	
			7.62 mm FN MAG		107 mm M30	
					120 mm M43	
Netherlands						
9 mm FN 35	9 mm Uzi	5.56 mm C7, C7A1	5.56 mm Minimi		81 mm L16A2	66 mm M72 LAW
9 mm Walther P5	9 mm H&K MP5A2/A3	5.56 mm C8	7.62 mm FN MAGH		107 mm M30	84 mm Carl Gustaf
9 mm H&K P9S	9 mm H&K MP5K	5.56 mm HK33	7.62 mm FN-FAL/HB		120 mm Brandt	Dragon ATGW
9 mm Glock 17		7.62 mm FN-FAL	0.50 Browning M2HB			TOW ATGW
		7.62 mm Steyr SSG69				Bofors AT-4
New Zealand						
9 mm FN 35	9 mm Sterling	5.56 mm Steyr AUG	7.62 mm FN-FAL/HB	40 mm M79	81 mm L16	66 mm M72 LAW
	9 mm H&K MP5	5.56 mm M16A1	7.62 mm Bren L4	40 mm M203	81 mm Brandt	84 mm Carl Gustaf
		7.62 mm FN-FAL	7.62 mm FN MAG		120 mm Soltam	106 mm RCL M40
		7.62 mm Parker-Hale 82	0.50 Browning M2HB			
Nicaragua						
9 mm Makarov	9 mm Uzi	5.56 mm M16A1	7.62 mm M60	30 mm AGS-17	82 mm M43	Sagger ATGW
0.45 M1911A1	9 mm vz/23, 25	5.56 mm Galil	7.62 mm RPD, RPK	40 mm M79	120 mm M43	
	9 mm Madsen M50	5.56 mm SIG 540	12.7 mm DShK			
			0.50 Browning M2HB			
Niger						
7.65 mm Walther PP	9 mm MAT-49	7.5 mm AS 49/56	7.62 mm HK11		60 mm Brandt	57 mm RCL M18
9 mm MAB PA-15	9 mm H&K MP5	7.62 mm G3	7.62 mm HK21		81 mm Brandt	75 mm RCL M20
	9 mm Uzi	7.62 mm M14	7.62 mm AAT-52		120 mm SB Brandt	89 mm LRAC
			0.50 Browning M2HB		120 mm rifled Brandt	
Nigeria						
9 mm FN 35	9 mm Sterling	5.56 mm M16A1	5.56 mm AUG/HBAR		60 mm M29	3.5 in RL M20
9 mm SIG P220	9 mm Beretta M12	5.56 mm AR70/223	7.62 mm HK21		81 mm L16	RPG-7
9 mm Beretta M951	9 mm Franchi LF-57	5.56 mm FN-FNC	7.62 mm FN MAG		81 mm M43	84 mm Carl Gustaf
	9 mm vz/23, 25	5.56 mm SIG 540	7.62 mm RP-46			89 mm LRAC
	9 mm H&K MP5	7.62 mm FN-FAL	7.62 mm RPD, RPK			Swingfire ATGW
	9 mm Uzi	7.62 mm G3	12.7 mm DShK			
		7.62 mm Steyr SSG69	0.50 Browning M2HB			
		7.62 mm vz/52				
		7.62 mm Beretta BM59				
Norway						
9 mm Glock	9 mm MP40	7.62 mm G3	7.62 mm MG3	40 mm HK79	81 mm L16	66 mm M72LAW
9 mm Walther P1	9 mm H&K MP5		0.50 Browning M2HB		107 mm M30	84 mm Carl Gustaf
9 mm H&K P7M8						106 mm RCL M40
0.45 M1912						TOW ATGW
						Eryx ATGW
Oman						
9 mm FN 35	9 mm Sterling	5.56 mm Steyr AUG	7.62 mm FN MAG	40 mm M79	60 mm Brandt	TOW ATGW
		5.56 mm M16A1	0.50 Browning M2HB	40 mm M203	81 mm L16	MILAN ATGW
		5.56 mm SIG 540			107 mm M30	
		7.62 mm FN-FAL			120 mm Brandt	
Pakistan						
9 mm Walther P-38	9 mm Sterling	7.62 mm G3	7.62 mm RPD		60 mm PMT	75 mm RCL Type 52
	9 mm H&K MP5	7.62 mm Type 56 (AKM)	7.62 mm MG1A3		81 mm PMT	RPG-7
			12.7 mm Type 54		120 mm PMT	106 mm RCL M40
			0.50 Browning M2HB			Cobra ATGW
						TOW ATGW

Pistols	Sub-machine guns	Rifles	Machine guns	Close support weapons	Mortars	Anti-tank
Panama						
9 mm FN 35	9 mm Uzi	5.56 mm M16A1 5.56 mm Type 65 0.30 Garand M1	5.56 mm M16A1/HB 7.62 mm FN MAG 7.62 mm M60 0.30 Browning M1919A4 0.50 Browning M2HB	40 mm M203	60 mm M19 81 mm M29	
Papua New Guinea						
9 mm FN 35	9 mm Sterling	5.56 mm Steyr AUG (F88) 7.62 mm FN-FAL	7.62 mm FN MAG			
Paraguay						
9 mm FN 35 9 mm H&K P9S 9 mm H&K VP70Z	9 mm Madsen 9 mm Carl Gustaf 9 mm Uzi	5.56 mm AR70/90 5.56 mm SIG 540 7.62 mm FN-FAL	7.62 mm FN-FAL/HB 7.62 mm Madsen 0.50 Browning M2HB	40 mm M79	81 mm Brandt 107 mm M2 107 mm M30	75 mm RCL M20
Peru						
9 mm FN 35 9 mm Star 30M	9 mm Uzi 9 mm SIMA 79 9 mm Star Z-45, Z-62	5.56 mm M16A1 7.62 mm FN-FAL 7.62 mm G3 7.62 mm AK-47, AKM 7.62 mm Steyr SSG69	7.62 mm FN MAG 7.62 mm M60 12.7 mm DShK 0.50 Browning M2HB		81 mm Brandt 120 mm Brandt 120 mm ECIA	106 mm RCL M40 SS-11 ATGW
Philippines						
9 mm FN 35 9 mm Glock 0.45 M1911A1	9 mm Uzi 0.45 M3A1	5.56 mm M16A1 5.56 mm Galil 7.62 mm M14 7.62 mm G3	5.56 mm Ultimax 100 7.62 mm M60 7.62 mm FN MAG 0.50 Browning M2HB	40 mm M79 40 mm M203	81 mm M29 107 mm M30	75 mm RCL M20 90 mm RCL M67 106 mm RCL M40A1
Poland						
9 mm P-64 9 mm Makarov	9 mm PM-63	5.45 mm AK-74 7.62 mm PMK 7.62 mm PMK-DGN 7.62 mm Dragunov SVD	5.45 mm RPK-74 7.62 mm RPD 7.62 mm RPK 7.62 mm PK, PKS	30 mm AGS-17	82 mm M37M 120 mm M43 160 mm M160	RPG-7V 82 mm RCL B-10 Sagger ATGW Spigot ATGW Spandrel ATGW Spiral ATGW Panad
Portugal						
9 mm Walther P1 9 mm FN 35 9 mm H & K VP70M	9 mm FMBP 63, 76 9 mm Uzi 9 mm Sterling 9 mm Star Z-45	5.56 mm HK33E 7.62 mm G3	7.62 mm FN MAG 7.62 mm HK21 7.62 mm MG42/59 0.50 Browning M2HB	40 mm M79	60 mm FBP 81 mm FBP 107 mm M30 120 mm Brandt	3.5 in RL M20 90 mm RCL M67 106 mm RCL M40 TOW ATGW MILAN ATGW
Qatar						
9 mm HK4	9 mm Sterling 9 mm H&K MP5	5.56 mm M16A1 5.56 mm Valmet M76 7.62 mm G3 7.62 mm AK-47, AKM	5.56 mm M16A1/HB 7.62 mm HK21 7.62 mm Valmet 62 7.62 mm FN MAG 0.50 Browning M2HB	40 mm M203	81 mm L16	HOT ATGW MILAN ATGW
Romania						
7.62 mm Tokarev 9 mm Makarov	9 mm vz/24, 26	7.62 mm AK-47, AKM	7.62 mm RPD, RPK 7.62 mm SGM 7.62 mm PK, PKS	40 mm AGL	82 mm M37M 120 mm M43 160 mm M160	RPG-7V Snapper ATGW Sagger ATGW
Rwanda						
7.65 mm Browning M 1910 9 mm FN 35 9 mm MAB PA-15	9 mm Uzi	7.62 mm FN-FAL 9 mm Vigneron	7.5 mm MAS 49/56 7.62 mm FN-FAL/HB 7.62 mm FN MAG 0.50 Browning M2HB		81 mm Brandt	RPG-7
St Vincent and Grenadines						
		5.56 mm M16A1 7.62 mm FN-FAL	7.62 mm M60	40 mm M79		
Sao Tome and Principe						
	7.62 mm PPS-43	7.62 mm AK-47, AKM 7.62 mm SKS 7.62 mm SGM 12.7 mm DShK 14.5 mm KPV	7.62 mm MG3 7.62 mm PK			RPG-7
Saudi Arabia						
9 mm FN 35 9 mm H&K P9S	9 mm MPi69 9 mm Beretta M12 9 mm H&K MP5	5.56 mm Steyr AUG 5.56 mm HK33E 7.62 mm G3 7.62 mm Steyr SSG69	5.56 mm Steyr AUG/ HBAR 7.62 mm FN MAG 7.62 mm MG3 0.50 Browning M2HB	40 mm M79 40 mm HK69	81 mm L16 81 mm M29 107 mm M30	75 mm RCL M20 90 mm RCL M67 TOW ATGW HOT ATGW
Senegal						
7.65 mm Walther PP 9 mm MAB PA-15 0.38 Manurhin MR73	5.56 mmHK53 7.65 mm MAS-38 9 mm MAT-49	5.56 mm SIG 540 5.56 mm KH33E 5.56 mm FAMAS 7.5 mm MAS 49/56 7.62 mm G3	7.5 mm AAT-52 7.62 mm HK21 0.50 Browning M2HB		81 mm Brandt 120 mm Brandt 120 mm Rifled Brandt	RPG-7 MILAN ATGW

Pistols	Sub-machine guns	Rifles	Machine guns	Close support weapons	Mortars	Anti-tank
Seychelles						
7.65 mm Walther PP 0.38 Manurhin MR 73	9 mm MAT-49	5.56 mm SIG 540 7.5 mm mas 49/56 7.62 mm AK-47, AKM 7.62 mm SKS	7.62 mm Bren L4 7.62 mm RPD, RPK 7.62 mm RP-46 7.62 mm FN MAG 7.62 mm AAT-52 12.7 mm DShK		82 mm M43	RPG-7
Sierra Leone						
7.62 mm Tokarev 9 mm FN 35	7.62 mm PPSh-41 7.62 mm PPS-43 9 mm Sterling	7.62 mm AK-47, AKM 7.62 mm SKS 7.62 mm FN-FAL	7.62 mm FN MAG 7.62 mm RPD 12.7 mm DShK		60 mm Brandt 81 mm Brandt	84 mm Carl Gustaf 3.5 in RL M20
Singapore						
9 mm FN 35 9 mm SIG P226 9 mm H&K P7M8	9 mm Sterling 9 mm H&K MP5	5.56 mm M16A1 5.56 mm SAR80 5.56 mm SR88	5.56 mm M16A1/HB 5.56 Ultimax 100 7.62 mm FN MAG 0.50 Browning M2HB	40 mm M203 40 mm CIS-40GL 40 mm CIS-40AGL	60 mm Soltam/ODE 81 mm Soltam 120 mm Soltam 120 mm Tampella	84 mm Carl Gustaf 3.5 in RL M20 106 mm RCL M40 TOW ATGW MILAN ATGW
Somali Republic						
7.62 mm vz/52 7.62 mm Tokarev 9 mm Makarov	9 mm Sterling 9 mm Uzi 9 mm Beretta M12 9 mm vz/23, 25	5.56 mm SAR80 5.56 mm M16A1 7.62 mm AK-47, AKM 7.62 mm vz/58 7.62 mm M14 7.62 mm G3	7.62 mm AAT-52 7.62 mm RPD, RPK 7.62 mm SGM 7.62 mm RP-46 12.7 mm DShK 0.50 Browning M2HB	40 mm M79	81 mm M1941 120 mm M43	RPG-2 RPG-7 89 mm LRAC MILAN ATGW Sagger ATGW
South Africa						
9 mm Z88	9 mm Uzi 9 mm BXP	5.56 mm R4, R5 7.62 mm FN-FAL 7.62 mm G3	7.62 mm SS-77 7.62 mm FN MAG 7.62 mm Browning M1919 0.50 Browning M2HB	40 mm MGL-6	60 mm M1 81 mm M3 120 mm Brandt	RPG-7 XT3 ATGW 92 mm FT-5
Spain						
9 mm Llama 82 9 mm Astra A80 9 mm Star 30M 9 mm H&K P9S	9 mm Star Z-45, 62 9 mm Star Z-70B, 84 9 mm H&K MP5	5.56 mm CETME L, LC 5.56 mm HK33E 7.62 mm CETME C 7.62 mm AW	5.56 mm Ameli 7.62 mm FN MAG 7.62 mm MG1A3 7.62 mm MG 42/59 0.50 Browning M2HB	40 mm SB 40	60 mm ECIA L 81 mm ECIA LL 105 mm ECIA L 120 mm ECIA SL	89 mm RL M69 106 mm RCL M40A1 Dragon ATGW MILAN ATGW TOW ATGW
Sri Lanka						
7.62 mm Type 50 9 mm FN 35	9 mm Sterling 9 mm H&K MP5A3	5.56 mm FN-FNC 5.56 mm SAR80 5.56 mm M16A1 7.62 mm SKS 7.62 mm Type 56	5.56 mm Minimi 7.62 mm Type 58 7.62 mm FN MAG 7.62 mm HK11 7.62 mm HK21	40 mm M203 40 mm HK69	82 mm M43 107 mm M43 120 mm M43	82 mm RCL M60 106 mm RCL M40
Sudan						
9 mm Helwan 9 mm H&K P9S 9 mm FN 35	9 mm Sterling 9 mm H&K MP5 9 mm Beretta M12 9 mm Uzi	7.62 mm SKS 7.62 mm G3 7.62 mm AK-47, AKM	7.62 mm HK21 7.62 mm RPD, RPK 7.62 mm RP-46 7.62 mm M60 7.62 mm SGM 7.62 mm MG3		81 mm M37M 82 mm M43 120 mm SB Brandt 120 mm Rifled Brandt	RPG-7 Swingfire ATGW
Surinam						
9 mm FN 35	9 mm Uzi	7.62 mm FN-FAL 7.62 mm AKM	7.62 mm FN MAG 7.62 mm Bren L4		81 mm M29	106 mm RCL M40
Swaziland						
	9 mm Sterling 9 mm Uzi	5.56 mm Galil 5.56 mm SIG 540 5.56 mm AR-18 7.62 mm FN-FAL	7.62 mm FN MAG 7.62 mm Bren L4			
Sweden						
9 mm Glock 17 9 mm Glock 19	9 mm FFV M45(D)	5.56 mm Ak-5 7.62 mm Ak-4 7.62 mm Psg 90	5.56 Minimi 7.62 mm M36 7.62 mm M39 7.62 mm FN MAG		81 mm M29 120 mm M41D	84 mm AT-4 TOW ATGW RBS 56 BILL ATGW Panzerfaust
Switzerland						
9 mm SIG P210	9 mm SIG 310	5.56 mm Stgw 90 7.5 mm Kar 55 7.5 mm Stgw 57	7.5 mm M25 7.5 mm MG51 0.50 Browning M2 HB		60 mm Brandt 81 mm Mw33 81 mm Mw72	83 mm RL RR50 83 mm RL RR58 106 mm RCL M40 Dragon ATGW TOW ATGW
Syria						
7.62 mm Tokarev 9 mm Makarov	9 mm vz/23, 25	7.62 mm AK-47, AKM 7.62 mm FN-FAL 7.62 mm Steyr SSG69	7.62 mm RPD, RPK 7.62 mm SGM 12.7 mm DShK, NSV		82 mm M41, 43 120 mm M43 160 mm M43	82 mm RCL SPG-9 RPG-7V Sagger ATGW Spigot ATGW MILAN ATGW

Pistols	Sub-machine guns	Rifles	Machine guns	Close support weapons	Mortars	Anti-tank
Taiwan						
9 mm FN 35 0.45 M1911A1	9 mm M3A1 0.45 M3A1	5.56 mm M16A1 5.56 mm Type 65 7.62 mm M14	5.56 mm Type 75 7.62 mm Type 74 7.62 mm M60 0.30 Browning M1919 0.50 Browning M2HB		60 mm M2 81 mm M29 81 mm Type 75	3.5 in RL M20 90 mm RCL M67 106 mm RCL M40 TOW ATGW
Tanzania						
7.62 mm vz/52 9 mm Stechkin 9 mm FN 35	9 mm Sterling 9 mm vz/23, 25	7.62 mm AK-47, AKM 7.62 mm SKS 7.62 mm FN-FAL 7.62 mm G3 7.62 mm vz/58	7.62 mm RP-46 7.62 mm RPD, RPK 7.62 mm FN MAG 7.62 mm SGM 12.7 mm DShK		82 mm M43 120 mm M43	RPG-7 75 mm RCL Type 52 TOW ATGW
Thailand						
9 mm SIG P226 9 mm Glock 0.45 M1911A1	9 mm Uzi 9 mm H&K MP5	5.56 mm HK33 5.56 mm M16A1, A2	5.56 mm Minimi 7.62 mm M60 0.50 Browning M2HB	40 mm M79 40 mm M203	60 mm M19 81 mm M29 107 mm M30 120 mm Brandt	66 mm M72 LAW 106 mm RCL M40 TOW ATGW Dragon ATGW
Togo						
7.65 mm Walther PP 9 mm FN 35 9 mm MAB PA-15 9 mm SIG P240 0.38 Manurhin MR73	9 mm MAT-49 9 mm Uzi	5.56 mm SIG 540 7.62 mm AK-47, AKM 7.62 mm G3 7.62 mm Hakim 7.62 mm Steyr SSG69	7.5 mm AAT-52 7.62 mm MG3 7.62 mm RPD, RP-46 12.7 mm DShK 0.50 Browning M2HB 14.5 mm KPV		81 mm M37M 82 mm M43	57 mm RCL M18 75 mm RCL Type 56 82 mm RCL Type 65 89 mm LRAC
Tonga						
0.38 S&W	9 mm Sten	5.56 mm FN-FNC	0.303 Bren 0.303 Vickers 0.50 Browning M2HB			
Trinidad and Tobago						
9 mm FN 35	9 mm Sterling	5.56 mm M16A1 5.56 mm Galil 7.62 mm FN-FAL	7.62 mm Bren L4 7.62 mm FN MAG 7.62 mm M60		60 mm M2	82 mm B-300
Tunisia						
9 mm Beretta M951 9 mm FN 35 9 mm MAC Mle 50 9 mm MAB PA-15	9 mm MAT-49 9 mm Uzi 9 mm Beretta 12	5.56 mm Steyr AUG 7.62 mm FN-FAL 7.62 mm M14	5.56 mm Steyr AUG/HBAR 7.62 mm M60 7.62 mm AAT-52 0.30 Browning M1919A4 0.50 Browning M2HB		60 mm Brandt 81 mm M29 82 mm M43 107 mm M30 120 mm Brandt	3.5 in RL M20 106 mm RCL M40A1 MILAN ATGW TOW ATGW
Turkey						
9 mm MKE	9 mm Rexim 9 mm H&K MP5 0.45 M3A1	5.56 mm M16A2 7.62 mm G3 7.62 mm FN-FAL	7.62 mm MG3 7.62 mm FN MAG 0.50 Browning M2HB	40 mm M79 40 mm M203	60 mm M2 81 mm UTI 107 mm M2 120 mm HY12D1	66 mm LAW M72 106 mm RCL M40 MILAN ATGW TOW ATGW
Uganda						
7.62 mm Tokarev 9 mm FN 35	7.65 mm Skorpion 9 mm Sterling 9 mm Uzi	5.56 mm M16A1 7.62 mm SKS 7.62 mm FN-FAL 7.62 mm G3	7.62 mm M60 7.62 mm RPD, RPK 7.62 mm PK 7.62 mm Bren L4 7.62 mm FN MAG 12.7 mm DShK		60 mm Brandt 81 mm L16 82 mm M43	Sagger ATGW
United Arab Emirates						
9 mm FN 35 9 mm H&K P7M13	5.56 mm HK53 9 mm H&K MP5 9 mm Sterling	5.56 mm M16A1 5.56 mm FAMAS 7.62 mm G3 7.62 mm FN-FAL 7.62 mm AKM	5.56 mm HK23E 5.56 mm Minimi 7.62 mm FN MAG 0.50 mm Browning M2HB	40 mm M203	81 mm L16 120 mm Brandt	84 mm Carl Gustaf TOW ATGW MILAN ATGW HOT ATGW
United Kingdom						
9 mm FN 35		5.56 mm L85A1 7.62 mm L39A1 7.62 mm L96A1 0.50 Barrett	5.56 mm L86A1 7.62 mm FN MAG (GPMG) L7A2 7.62 mm Bren L4 0.50 Browning M2HB		51 mm L9A1 81 mm L16A1	LAW-80 MILAN ATGW TOW ATGW
United States of America						
9 mm Beretta 92F (M9) 0.45 M1911A1	9 mm Colt 9 mm H&K MP5 0.45 M3A1	5.56 mm M16A1, A2 7.62 mm M14 7.62 mm M21	5.56 mm M249 (Minimi) 7.62 mm M60 7.62 mm M240E4 0.50 Browning M2HB	40 mm M79 40 mm M203 40 mm MK19 Mod 3 SMAW	60 mm M19 81 mm M29A1 81 mm M252 (L16) 107 mm M30 120 mm Tampella	66 mm M72 LAW 106 mm RCL M40A1 Dragon ATGW TOW ATGW FFV AT-4
Uruguay						
9 mm H&K P7M8 9 mm HK4 9 mm FN 35 9 mm SIG P220	9 mm Uzi 9 mm H&K MP5 9 mm Star Z-45	5.56 mm M16A1 7.62 mm FN-FAL 0.30 Garand M1	7.62 mm FN MAG 0.30 Browning M1919A4, A6 0.30 BAR M1918A2 0.50 Browning M2HB		81 mm M1 107 mm M30	106 mm RCL M40A1 MILAN ATGW

Pistols	Sub-machine guns	Rifles	Machine guns	Close support weapons	Mortars	Anti-tank
Venezuela						
9 mm FN 35	9 mm Uzi	5.56 mm FN-FNC	7.62 mm FN MAG		60 mm Brandt	84 mm AT-4
9 mm HK4	9 mm Beretta M12	7.62 mm M14	7.62 mm M60		81 mm Brandt	106 mm RCL M40
	9 mm Walther MPK	7.62 mm FN-FAL	0.50 Browning M2HB		120 mm Brandt	SS-11 ATGW
Vietnam						
7.62 mm Tokarev	7.62 mm K-50M	7.62 mm AK-47, AKM	7.62 mm Type 53, 67		60 mm M19	RPG-2, -7
7.62 mm Type 68	7.62 mm MAT-49	7.62 mm SKS	7.62 mm DPM, RPD		81 mm M29	75 mm RCL type 52, 57
	7.62 mm PPSh-41	7.62 mm Type 56	7.62 mm SGM		82 mm M43	82 mm RCL B-10
			12.7 mm DShK		82 mm RCL B-18	107 mm RCL B-11
					120 mm M43	Sagger ATGW
					160 mm M43	
Yemen						
7.62 mm Beretta M951	9 mm Beretta 38/49	5.56 mm M16A2	7.62 mm RPD, RPK	40 mm M79	81 mm M29	66 mm M72 LAW
9 mm Beretta M951		7.62 mm vz/52	7.62 mm SGM		82 mm M43	RPG-7
		7.62 mm FN-FAL	12.7 mm DShK		120 mm M43	75 mm RCL M20
		7.62 mm SKS	0.50 Browning M2HB		160 mm M43	Dragon ATGW
		7.62 mm AK-47, AKM				TOW ATGW
						Sagger ATGW
Yugoslavia (Serbia and Montenegro)						
7.62 mm M57	7.62 mm M49/57	7.62 mm M59/66	7.62 mm M53	128 mm RL M71	50 mm M8	44 mm RL M57
7.62 mm M70	7.62 mm M56	7.62 mm M70, 70A	7.62 mm M65A, B		60 mm M57	64 mm RL RBR-M80
9 mm M65			7.92 mm M72		81 mm M31	75 mm RCL M20
			0.50 Browning M2HB		81 mm M68	82 mm RCL M60
			12.7 mm NSV		120 mm M74, M75	105 mm RCL M65
						Sagger ATGW
						Snapper ATGW
						Dragon ATGW
						TOW ATGW
Zaire						
9 mm FN 35	9 mm Uzi	5.56 mm FN-CAL	5.56 mm M16A1/HB		60 mm Brandt	RPG-7
9 mm S&W M39	9 mm Franchi LF-57	5.56 mm FN-FNC	7.62 mm Bren L4		81 mm PRB	57 mm RCL M18
9 mm H&K P7M13	9 mm H&K MP5	5.56 mm SIG 540	7.62 mm FN MAG		81 mm M37M	75 mm RCL M20
		5.56 mm M16A1	7.62 mm M60		107 mm M30	89 mm LRAC
		7.62 mm CETME 58	7.62 mm FN-FAL/HB		120 mm SB Brandt	106 mm RCL M40
		7.62 mm AK-47, AKM	12.7 mm DShK		120 mm Rifled Brandt	Entac ATGW
		7.62 mm FN-FAL	0.50 Browning M2HB			
Zambia						
7.62 mm Tokarev	9 mm Sterling	7.62 mm FN-FAL	7.62 mm PK			RPG-7
7.62 mm vz/52	9 mm H&K MP5	7.62 mm G3	7.62 mm SGM			44 mm RM M57
9 mm Stechkin		7.62 mm AK-47, AKM	12.7 mm DShK			57 mm RCL M18
						75 mm RCL M20
						84 mm Carl Gustaf
						Sagger ATGW
Zimbabwe						
7.62 mm Tokarev	9 mm Sterling	5.56 mm AR70/223	5.56 mm Ultimax 100		81 mm L16	RPG-7
7 mm FN 35	9 mm Uzi	5.56 mm R4	7.62 mm FN MAG		82 mm M43	40 mm Type 69
0.45 M1911A1	9 mm Walther MPK	7.62 mm FN-FAL	7.62 mm Bren L4		120 mm M43	106 mm RCL M40
		7.62 mm M14	7.62 mm RPD, RPK			107 mm RCL B-12
		7.62 mm M14	12.7 mm DShK			
		7.62 mm AK-47, AKM	0.50 Browning M2HB			

Manufacturers Index

Note: Telephone, Telex and Facsimile (Teletext) numbers are given in the international convention; they must be prefixed by the appropriate national dialling out code.

Where a discrepancy exists between data on the text pages and in this table, the table should be accepted as the most up-to-date and correct version. This table is composed after the text has been completed, and there are often last minute changes in address and other details which cannot be corrected in the text.

Borletti FB, srl
Ordnance Division, Via Verdi 33/35, I-20010 San Giorgio su Legnano (MI), Italy
Tel: +39 331 402216
Tx: 316812 BORLFB I
Fax: +39 331 404550
Mortar ammunition 566

Breda Meccanica Bresciana SpA
Via Lunga 2, I-25126 Brescia, Italy
Tel: +39 30 37911
Tx: 300056 BREDAR I
Fax: +39 30 322115
Anti-tank weapons 327
Mortars 388

British Aerospace plc, Dynamics Division
Six Hills Way, Stevenage, Hertfordshire SG1 2DA, UK
Tel: +44 1438 312422
Fax: +44 1438 753377
Mortar ammunition 585

British Aerospace Defence Ltd, Royal Ordnance Division
King's Meadow Road, Nottingham NG1 1ER, UK
Tel: +44 115 968 2000
Fax: +44 115 968 2001
Cannon 300, 301
Mortars 402
Sighting Equipment 671

British Aerospace Defence Ltd, Royal Ordnance Ammunition Division
Euxton Lane, Chorley, Lancashire PR7 6AD, UK
Tel: +44 1257 265511
Fax: +44 1257 242609
Combat grenades 522
Mortar ammunition 584
Pyrotechnics 626

Zbrojovka Brno
Narodni Podnik, Czech Republic
Machine guns 259

Brocks Explosives Ltd
Sanquhar, Dumfriesshire DG4 6JP, UK
Tel: +44 1659 50531
Tx: 778965 BROCKS G
Fax: +44 1659 50526
Combat grenades 526
Pyrotechnics 634

Brügger + Thomet Feinmechanik
Thunstrasse 4a, CH-3700 Spiez, Switzerland
Tel: +41 33 549757
Fax: +41 33 549758
Suppressors 691

Brunswick Corporation
3333 Harbor Boulevard, Costa Mesa, California 92628-2009, USA
Tel: +1 714 546 8030
Fax: +1 714 434 7492
Light support weapons 230

Buck Werke GmbH and Co
Postfach 2405, Mozartstrasse 2, D-83435 Bad Reichenall, Germany
Tel: +49 8651 7020
Fax: +49 8651 70270
Combat grenades 496
Light support weapons 214
Mortar ammunition 553

Bushman Ltd
10 Park Industrial Estate, Frogmore, St Albans, Hertfordshire AL2 2DR, UK
Tel: +44 1727 875010
Fax: +44 1727 875011
Sub-machine guns 110

C

CelsiusTech Electronics AB
S-175 88 Järfalla, Sweden
Tel: +46 8 5808 4000
Fax: +46 8 5803 2244
Sighting equipment 669

Cenzin Foreign Trade Enterprise
ul Frascati 2, PL-00489 Warsaw, Poland
Tel: +48 22 296396
Fax: +48 2 628 6356
Anti-tank weapons 330

Ceská Zbrojovka as
688 27 Uherský Brod, Czech Republic
Tel: +42 633 65
Fax: +42 633 3665
Machine guns 259
Pistols 18
Rifles 141
Sub-machine guns 89

Chartered Industries of Singapore
249 Jalan Boon Lay, Jurong Town, Singapore 2261
Tel: +65 265 1066
Fax: +65 264 4606
Anti-tank weapons 331
Light support weapons 221
Machine guns 274
Mortars 392
Mortar ammunition 573
Rifles 172

Chartered Ammunition Industries
(A subsidiary of Chartered Industries of Singapore), 249 Jalan Boon Lay, Jurong Town, Singapore 2261
Tel: +65 265 1066
Fax: +65 268 7579
Combat grenades 512

China North Industries Corporation (NORINCO)
7A Yue Tan Nan Jie, PO Box 2137, Beijing, China
Tel: +86 1 8013716
Fax: +86 1 8033236
Anti-tank weapons 306
Combat grenades 481
Light support weapons 206
Machine guns 248
Mortar ammunition 541
Mortar fire control 414
Mortars 360
Pistols 12
Pyrotechnics 600
Rifles 130
Sighting equipment 642
Sub-machine guns 82

Ciener, Jonathan Arthur, Inc
8700 Commerce Street, Cape Canaveral, Florida 32920, USA
Tel: +1 407 868 2200
Fax: +1 407 868 2201
Light support weapons 237
Pistols 71
Rifles 201
Suppressors 693

CITEFA: (Instituto de Investigaciones Cientificas y Tecnicas de las Fuerzas Armadas), Zufriategui y Varela
1603 Villa Martelli, Buenos Aires, Argentina
Anti-tank weapons 304

CMS
PO Box 896, Shalimar, Florida 32579, USA
Tel: +1 904 651 3948
Fax: +1 904 651 4205
Anti-tank weapons 344
Light support weapons 230

Colt's Manufacturing Co Inc
PO Box 1868, Hartford, Connecticut 06144-1868, USA
Tel: +1 203 236 6311
Fax: +1 203 244 1381
Light Support Weapons 232
Machine guns 291
Rifles 190
Sub-machine guns 111

Combined Service Forces
Hsing-Ho Arsenal, Kaohsuing, Taiwan
Machine guns 278
Rifles 181

Comet GmbH Pyrotechnik-Apparatebau
Postfach 100267, D-27502 Bremerhaven, Germany
Tel: +49 471 3930
Fax: +49 471 39394
Pyrotechnics ... 610

Companhia de Explosivos Valparaiba
Praia do Flamengo 200, 20° Andar, Rio de Janeiro 22210, Brazil
Tel: +55 21 205 6612
Combat grenades ... 477
Mortar ammunition ... 538
Pyrotechnics ... 599
Sub-machine guns ... 77

Computing Devices Canada
PO Box 8508, Ottawa, Ontario K1G 3M9, Canada
Tel: +1 613 596 7000
Fax: +1 613 820 5081
Sighting equipment ... 641

Condor SA - Industria Quimica
Rue Mexico 148, Grupo 801/802, Rio de Janeiro 20031, Brazil
Tel: +55 21 220 6814
Combat grenades ... 479
Pyrotechnics ... 596

Contraves USA
615 Epsilon Drive, Pittsburgh, Pennsylvania 15238, USA
Tel: +1 412 967 7700
Fax: +1 412 967 7273
Sighting equipment ... 679

Cordo SA
PO Box 202, Villa Urcabe-Baita, E-20180 Oyarzun, Spain
Tel: +34 4349 2853
Fax: +34 4349 2862
Combat grenades ... 518

CTA International
7 Route de Guerry, F-18023 Bourges Cedex, France
Tel: +33 48 21 97 60
Fax: +33 48 20 09 27
Machine guns ... 268

D

Daewoo Precision Industries
PO Box 25, Kum-Jeong, Pusan, South Korea
Tel: +82 51 58 2221
Fax: +82 51 512 2229
Light support weapons ... 219
Machine guns ... 272
Pistols ... 45
Rifles ... 168
Sub-machine guns ... 101

Daimler-Benz Aerospace
PO Box 801189, D-81663 Munich, Germany
Tel: +49 89 607 34553
Fax: +49 89 607 34569
Anti-tank weapons ... 325

Dassault Electronique
55 quai Marcel Dassault, F-92214 Saint-Cloud, France
Tel: +33 1 49 11 80 00
Tx: 250787 F ESD SCLOU
Fax: +33 1 46 02 57 58
Mortar fire control ... 414

Defence Industries Organisation
PO Box 17185-617 Tehran, Iran
Tel: +98 21 339000
Fax: +98 21 345021
Combat grenades ... 500
Machine guns ... 268
Mortars ... 379

Degen and Co AG
Grittweg 9, CH-4435 Niederdorf, Switzerland
Tel: +41 61 978282
Tx: 966093 DEGN CH
Fax: +41 61 978490
Mortar ammunition ... 582

Delft Instruments Electro-Optics BV
Postbus 5083, NL-2600 GB, Delft, Netherlands
Tel: +31 15 601 901
Fax: +31 15 145 762
Sighting equipment ... 662

Design Bureau for Instrument Engineering
Shcheglovskaya Zaseka, 300001 Tula, Russia
Sub-machine guns ... 86

Dezamet, Zaklady Metalowe
39-460 Nowa Dęba, woj Tarnobrzeg, Poland
Tel: +48 46 2601
Fax: +48 46 2610
Combat grenades ... 510

Diehl GmbH and Co
Ammunition Division, Fischbachstrasse 16, D-90552, Röthenbach/Pegnitz, Germany
Tel: +49 911 957 0
Fax: +49 911 557 2150
Combat grenades ... 493
Mortar ammunition ... 560

Diemaco
1036 Wilson Avenue, Kitchener, Ontario N2C 1J3, Canada
Tel: +1 519 893 6840
Fax: +1 519 893 3144
Machine guns ... 247
Rifles ... 128

Dynamit Nobel AG
Postfach 1261, Kaiserstrasse 1, D-53839 Troisdorf, Germany
Tel: +49 2241 891294
Tx: 885666 DN D
Fax: +49 2241 891542
Anti-tank weapons ... 323

Dynamit Nobel Graz
Annenstrasse 58, A-8020 Graz, Austria
Tel: +43 316 832631
Fax: +43 316 803 6947
Combat grenades ... 473
Mortar ammunition ... 535

Dynamit Nobel Wien GesmbH
Tuchlauben 7a, A-1014 Vienna, Austria
Tel: +43 222 565646
Combat grenades ... 472
Pyrotechnics ... 595

E

Elbit Computers Ltd
PO Box 5390, IL-31053 Haifa, Israel
Tel: +972 4 517111
Fax: +972 4 520002
Sighting equipment ... 659

Electro Optic Systems Pty Ltd
55A Monaro Street, Queanbeyan, NSW 2620, Australia
Tel: +61 6 299 2470
Fax: +61 6 299 2477
Sighting equipment ... 640

Electro-Optics Technologies
PO Box 134010, Ann Arbor, Michigan 48113 4010, USA
Tel: +1 313 741 8868
Fax: +1 313 741 8221
Sighting Equipment ... 680

El-Op Electro-Optical Industries
PO Box 1165, Rehovot 76110, Israel
Tel: +972 848 6211
Fax: +972 848 6214
Sighting equipment ... 660

Elviemek SA
Atrina Centre, 32 Kifissias Avenue, Athens, Greece
Tel: +30 1 682 8601
Fax: +30 1 684 1524
Combat grenades ... 497

ENOSA Empresa Nacional de Optica SA
Poligono Industrial La Mina, Parcela 11, E-28770 Colmenar Viejo, Madrid, Spain
Tel: +34 1 846 0203
Tx: 42719 OPTI E
Fax: +34 1 846 0213
Sighting equipment ... 667

Enterprise Metallist
127 Frunze Street, Uralsk, Kazakhstan 417820
Tel: +7 31122 4 23 30
Fax: +7 31122 4 21 82
Sub-machine guns ... 85

Esperanza y Cia SA
PO Box 2, Avenida de Xemien 12, Marquina, E-48270 Vizcaya, Spain
Tel: +34 4 686 6025
Fax: +34 4 686 6026
Mortar ammunition ... 577
Mortars ... 395

Euroatlas GmbH
Zum Panrepel 2, Postfach 450241, D-28307 Bremen, Germany
Tel: +49 421 486930
Tx: 244504 EURAT D
Fax: +49 421 486 9341
Sighting equipment ... 646

Eurometaal NV
Postbus 419, 1500 EK, Zaandam, Netherlands
Tel: +31 75 6504 267
Fax: +31 75 6504 261
Combat grenades ... 508

Euromissile
12 rue de la Redoute, F-92260 Fontenay-aux-Roses, France
Tel: +33 1 46 61 73 11
Fax: +33 1 46 61 64 67
Anti-tank weapons ... 325

EXPAL SA
Avenida del Partenòn 16, 5th Floor, E-28042 Madrid, Spain
Tel: +34 1 722 0235
Fax: +34 1 722 0295

Combat grenades 517
Pyrotechnics 624

F

Fabrica Militar de Armas Portatiles 'Domingo Matheu'
Avenida Ovidio Lagos 5250, 2000-Rosario, Argentina
Tel: +54 4183 8383
Tx: 41451 FMDM AR
Fax: +54 4183 7062
Machine guns 241
Pistols 3
Rifles 117
Sub-machine guns 74

Fabrica Militar 'Rio Tercero'
Av Grl Savio s/Nro, (5850) Rio Tercero, Pcia de Cordoba, Argentina
Tel: +54 571 21136
Fax: +54 571 21516
Anti-tank weapons 304
Mortar ammunition 533
Mortars 353

Fabricaciones Militares
Avenida Cabildo 65, 1426 Buenos Aires, Argentina
Combat grenades 465

FAMAE
Avenida Pedro Montt 1606, Santiago, Chile
Tel: +56 2 556 9131
Fax: +56 2 556 9131
Combat grenades 479
Mortars 359
Mortar ammunition 540
Pyrotechnics 600
Rifles 130
Sub-machine guns 81

FEG Arms and Gas Appliances Factory
Soroksari ut 158, H-1095 Budapest, Hungary
Tel: +36 1 477 920
Pistols 29
Sub-machine guns 96

Ferranti Technologies Ltd
Cairo Mill, Waterhead, Oldham, Lancashire OL4 3JA, UK
Tel: +44 161 624 0281
Fax: +44 161 624 5244
Combat grenades 526

FMJ
PO Box 759, Copperhill, Tennessee 37317, USA
Sub-machine guns 113

FN Herstal SA
Voie de Liège 33, B-4040 Herstal, Belgium
Tel: +32 4140 8111
Fax: +32 4140 8679
Combat grenades 474
Machine guns 242
Pistols 6
Rifles 122
Sub-machine guns 76

FN Manufacturing Inc
797 Clemson Road, PO Box 24257, Columbia, South Carolina 29224, USA
Tel: +1 803 736 0522
Fax: +1 803 699 9373
Machine guns 287

Franchi, Luigi, SpA
Via del Serpente, 12 Zona Industriale, I-25131 Fornaci, (Brescia), Italy
Tel: +39 30 3581833
Tx: 300208 FRARM I
Fax: +39 30 3581554
Light support weapons 216

FR Ordnance International
PO Box 64, Waltham Cross, Hertfordshire EN7 5EH, UK
Tel: +44 1707 873566
Fax: +44 1707 872945
Sub-machine guns 111

G

Gemtech Division of Gemini Technologies Inc
PO Box 3538, Boise, Idaho 83703, USA
Tel: +1 208 939 7222
Suppressors 693

Giat Industries
13 route de la Minière, Satory, F-78022 Versailles Cedex, France
Tel: +33 1 39 49 35 30
Fax: +33 1 39 49 82 79
Anti-tank weapons 322
Cannon 295
Combat grenades 488
Light support weapons 213
Machine guns 261
Pistols 21
Rifles 145

Gibbs Rifle Co Inc
Cannon Hill Industrial Park, Hoffman Road, Martinsburg, West Virginia 25401, USA
Tel: +1 304 274 0458
Fax: +1 304 274 0078
Rifles 195

Glock GmbH
PO Box 50, A-2232 Deutsche-Wagram, Austria
Tel: +43 2247 2460
Tx: 133307 GLOCK A
Fax: +43 2247 246012
Pistols 3

GNPP Bazalt
105058 Moscow, Russia
Anti-tank weapons 313

Golden Bell Trading Company Ltd
60-1, 3-Ka, Chungmu-Ro, Chung-Ku, Seoul, South Korea
Mortar ammunition 567

Oy Golden Gun Ltd
Humalistonkatu 9, FIN-20100 Turku, Finland
Tel: +359 21 233 0120
Fax: +359 21 233 0610
Sub-machine guns 92

Greek Powder and Cartridge Co (Pyrkal)
1 Ilioupoleos Avenue, Hymettus, GR-17236 Athens, Greece
Tel: +30 1 975 1857
Fax: +30 1 970 5009
Mortar ammunition 556

Grendel Inc
PO Box 560909, Rockledge, Florida 32956-0909, USA
Tel: +1 407 636 1211
Fax: +1 407 633 6710
Pistols 62
Rifles 197

GUSK
18/1 Ovchinnikovskaya nab, 113324 Moscow, Russia
Tel: +7 095 233 9222
Tx: 412171 GUSK SA
Fax: +7 095 233 1249
Machine guns 257

Gyconsa
Joaquin Rodrigo 11, E-28300 Aranjuez, Madrid, Spain
Tel: +34 1 894 8952
Fax: +34 1 891 7628
Anti-tank weapons 332

H

Haley and Weller Ltd
Wilne, Draycott, Derbyshire DE72 3QJ, UK
Tel: +44 1332 872475
Tx: 378215 HALWEL G
Fax: +44 1332 873046
Combat grenades 524
Pyrotechnics 629

Hall and Watts Defence Optics Ltd
266 Hatfield Road, St Albans, Hertfordshire AL1 4UN, UK
Tel: +44 1727 859288
Fax: +44 1727 835683
Sighting equipment 674

Hawk Engineering Inc
42 Sherwood Terrace, Suite 101, Lake Bluff, Illinois 60044, USA
Tel: +1 312 295 2340
Fax: +1 312 295 3319
Light support weapons 235

Heckler and Koch GmbH
Postfach 1329, D-78722 Oberndorf/Neckar, Germany
Tel: +49 7423 791
Fax: +49 7423 79406
Light support weapons 214
Machine guns 262
Pistols 25
Pyrotechnics 606
Rifles 150
Sub-machine guns 93

Heckler and Koch Inc
21480 Pacific Boulevard, Sterling, Virginia 22170-8903, USA
Tel: +1 703 450 1900
Fax: +1 703 450 8160
Sub-machine guns 113

Heckler and Koch (UK)
Kings Meadow Road, Nottingham, NG2 1EQ, UK
Tel: +44 115 968 2300
Fax: +44 115 968 2001
Light support weapons 226
Machine guns 281
Rifles 182
Sighting equipment 671

Hellenic Arms Industry (EBO) SA
160 Kifissias Avenue, GR-11525 Athens, Greece
Tel: +30 647 2611
Tx: 21 8562 EBO GR
Fax: +30 647 2715
Machine guns 267
Mortars 378
Pistols 29
Rifles 156
Sub-machine guns 96

Helwan Machine Tools
23 Talat Harb Street, PO Box 1582, Cairo, Egypt
Tel: +20 2 3921738
Mortars 369

Hensoldt and Söhne Optische Werke AG
Postfach 1760, D-35573 Wetzlar, Germany
Tel: +49 6441 4041
Tx: 483884
Fax: +49 6441 404203
Sighting equipment 647

Hensoldt-Wetzlar
Germany
Sighting equipment 652

Hirtenberger AG
A-2552 Hirtenberg, Austria
Tel: +43 2256 8184
Fax: +43 222 8114 342
Mortars 355
Mortar ammunition 536

Howa Machinery Ltd
Sukaguchi, Shinkawa-cho, Nishikasugai-Gun, 452 Aichi-ken, Japan
Tel: +81 52 408 1361
Fax: +81 52 400 7107
Rifles 166

Hsing Hua Co
PO Box 8746, Taipei, Taiwan
Tel: +886 2761 5367
Combat grenades 521
Light support weapons 226

Sakr Factory for Developed Industries
PO Box 33, Heliopolis, Cairo, Egypt
Tel: +20 2 963239
Tx: 92175 CERVA UN
Anti-tank weapons ... 320
Combat grenades ... 488
Light support weapons ... 212

SANTA BARBARA
Julián Camarillo 32, Apartado 35087, E-28037 Madrid 17, Spain
Tel: +34 1 585 0273
Tx: 444466 ENSB E
Fax: +34 1 585 0244
Anti-tank weapons ... 334
Light support weapons ... 225
Machine guns ... 276
Rifles ... 175
Sub-machine guns ... 108

Sauer, J P, & Sohn GmbH
Eckernförde, Germany
Pistols ... 55
Rifles ... 178

Saunders-Roe Developments Ltd
Millington Road, Hayes, Middlesex UB3 4BN, UK
Tel: +44 181 573 3800
Fax: +44 181 561 3436
Sighting equipment ... 678

Schermuly see Pains-Wessex (Schermuly)

Scopus Light (1990) Ltd
Kibbutz Maayan Zvi, IL-30805 Israel
Tel: +972 6 390787
Fax: +972 6 396861
Sighting equipment ... 660

SECRE: Société d'Etudes et de Construction Electroniques
11 rue de Cambrai, F-75945 Paris Cedex 19, France
Tel: +33 1 40 89 45 00
Tx: 220169 SECRE F
Fax: +33 1 44 89 45 45
Mortar fire control ... 415

Seidef SA
Pso de la Castellana 140, E-28046 Madrid, Spain
Tel: +34 1404 8661
Tx: 43738 SEDEF E
Fax: +34 1453 6883
Mortar fire control ... 418

Siemens AG
Landshuter Strasse 26, D-85705 Unterschleissheim, Germany
Tel: +49 89 3179 0
Fax: +49 89 3179 2156
Sighting equipment ... 645

Siemens-Albis AG, Defence Electronics
Freilagerstrasse 38, CH-8047 Zurich, Switzerland
Tel: +41 1411 542211
Tx: 52132
Sighting equipment ... 669

SIG Swiss Industrial Company
CH-8212 Neuhausen Rheinfalls, Switzerland
Tel: +41 52 674 6111
Fax: +41 52 674 6601
Pistols ... 54
Rifles ... 177

Signaal-USFA
PO Box 6034, Eindhoven, NL-5600 HA, Netherlands
Tel: +31 4050 3603
Fax: +31 4050 3777
Sighting equipment ... 664

Sima Cifar
Avenida Contralmirante Mora 1102, Base Naval, Callao, Peru
Tel: +51 14 657183
Tx: 26128 SIMA PE
Fax: +51 14 657183
Pyrotechnics ... 619
Sub-machine guns ... 102

Simmel
Borgo Padova 2, I-31033 Castelfranco Veneto, Italy
Tel: +39 423 431241
Mortar ammunition ... 565

Simrad Optronics A/S
PO Box 6114, Etterstad, N-0602 Oslo, Norway
Tel: +47 2 670490
Tx: 76136 SIM N
Fax: +47 2 680704
Sighting equipment ... 664

Singapore Technologies Ordnance
249 Jalan Boon Lay, Jurong Town, Singapore 2261
Tel: +65 265 1066
Fax: +65 261 6932
Pyrotechnics ... 620

SITES SpA
Via Magenta 36, I-10128 Turin, Italy
Tel: +39 11 561 3884
Tx: 212586 SITES I
Fax: +39 11 562 0252
Sub-machine guns ... 100

Smith & Wesson Inc
2100 Roosevelt Avenue, PO Box 2208, Springfield, Massachusetts 01101-2208, USA
Tel: +1 413 781 8300
Pistols ... 67

Sociedade Portuguesa de Explosivos SA
Avenida Infante Santo 76, 5°, 1300 Lisbon, Portugal
Tel: +351 160 3080
Tx: 121398 SPELEX P
Fax: +351 166 1070
Combat grenades ... 511

Soltam
PO Box 1371, Haifa, IL-31013 Israel
Tel: +972 4896 671
Tx: 46277 IL
Fax: +972 4892 046
Mortar ammunition ... 561
Mortars ... 384

Somchem, a division of Denel (Pty) Limited
PO Box 187, Somerset West 7129, South Africa
Tel: +27 24 429 2911
Fax: +27 24 429 2111
Anti-tank weapons ... 331

Sopelem-Sofretec
53 rue Casimir Périer, F-95870 Bezons, France
Tel: +33 1 34 23 30 00
Fax: +33 1 34 23 33 50
Sighting equipment ... 644

Sphinx Engineering SA
Chemin des Grandes-Vies 2, CH-2900 Porrentruy, Switzerland
Tel: +41 6666 7381
Fax: +41 6666 3090
Pistols ... 57

Springfield Armory Inc
420 West Main Street, Geneseo, Illinois 61254, USA
Tel: +1 309 944 5631
TWX: 910 650 2449 SPRINGROK GENS
Fax: +1 309 944 3676
Pistols ... 63
Rifles ... 189, 196

Stacchini Sud SpA
Via Giacomo Puccini 9, I-00198 Rome, Italy
Tel: +39 6854 7135
Tx: 601080 START I
Fax: +39 6854 7136
Pyrotechnics ... 617

Star Bonifacio Echeverria y Cia
PO Box 10, E-20600 Eiber, Spain
Tel: +34 4311 7340
Pistols ... 49
Sub-machine guns ... 106

Star-Tron Technology Corp
526 Alpha Drive, RIDC Industrial Park, Pittsburgh, Pennsylvania 15238, USA
Tel: +1 412 962 7170
Fax: +1 412 963 1552
Sighting equipment ... 687

Steyr-Mannlicher GesmbH
PO Box 1000, Mannlicherstrasse 1, A-4400 Steyr, Austria
Tel: +43 7252 6733 1229
Tx: 61 373 2224 SMGV A
Fax: +43 7252 68621
Machine guns ... 242
Pistols ... 5
Rifles ... 119
Sub-machine guns ... 74

Sturm, Ruger and Co
49 Lacey Place, Southport, Connecticut 06490, USA
Tel: +1 203 259 7843
Fax: +1 203 255 2163
Pistols ... 64
Rifles ... 197
Sub-machine guns ... 114

Sumitomo Heavy Industries
2-1-1 Yato-cho, Tanashi-shi, Tokyo 188, Japan
Tel: +81 424 632121
Tx: 282 2373
Machine guns ... 271

Sun Yat-Sen Scientific Research Institute
PO Box 2, Lung Tan 325, Taiwan
Anti-tank weapons ... 339

Sverdlova, GP Zavod imeni J A M
Dzerzhinsk, Nishi Novogorod 606002, Russia
Combat grenades ... 484

Swartklip Products a division of Denel (Pty) Ltd
PO Box 977, Cape Town 8000, South Africa
Tel: +27 21 376 1500
Fax: +27 21 376 1108
Combat grenades ... 514
Pyrotechnics ... 621

T

TAAS Israel Industries Ltd
PO Box 1044, Ramat Hasharon IL-47100, Israel
Tel: +972 3542 5222
Fax: +972 3540 6908
Anti-tank weapons ... 326
Combat grenades ... 501
Machine guns ... 269
Mortar ammunition ... 560, 563
Mortars ... 382
Pistols ... 31
Rifles ... 162
Sub-machine guns ... 98

Talley Defense Systems
3500 North Greenfield Road, PO Box 849, Mesa, Arizona 85201, USA
Tel: +1 602 898 2200
Fax: +1 602 898 2358
Anti-tank weapons ... 342
Light support weapons ... 231

Tanfoglio, Fratelli, Srl
Via Val Trompia 39/41, I-25063 Gardone Val Trompia (Brescia), Italy
Tel: +39 30 891 0361
Tx: 303429 TARGA I
Fax: +39 30 891 0183
Pistols ... 42

Tarnow, Zaklady Mechaniczne
ul Kochanowskiego 30, PL-33-100 Tarnow, Poland
Tel: +48 14 216001
Fax: +48 14 216496

Alphabetical Index

G

H

I

J

K

N

O

P